# Contents

# 5 Integrals 263

# 6 Applications of Definite Integrals 327

# 7 Transcendental Functions 380

# 8 Techniques of Integration 458

# 9 First-Order Differential Equations 538

# 10 Infinite Sequences and Series 574

# 11 Parametric Equations and Polar Coordinates 655

# 16    Integrals and Vector Fields    952

# 17    Second-Order Differential Equations    online

## Appendices    AP-1

## Answers to Odd-Numbered Exercises    A-1

## Credits    C-1

## Index    I-1

## A Brief Table of Integrals    T-1

## Basic Formulas and Rules    F-1

# Preface

*Thomas' Calculus,* Thirteenth Edition, provides a modern introduction to calculus that focuses on conceptual understanding in developing the essential elements of a traditional course. This material supports a three-semester or four-quarter calculus sequence typically taken by students in mathematics, engineering, and the natural sciences. Precise explanations, thoughtfully chosen examples, superior figures, and time-tested exercise sets are the foundation of this text. We continue to improve this text in keeping with shifts in both the preparation and the ambitions of today's students, and the applications of calculus to a changing world.

Many of today's students have been exposed to the terminology and computational methods of calculus in high school. Despite this familiarity, their acquired algebra and trigonometry skills sometimes limit their ability to master calculus at the college level. In this text, we seek to balance students' prior experience in calculus with the algebraic skill development they may still need, without slowing their progress through calculus itself. We have taken care to provide enough review material (in the text and appendices), detailed solutions, and variety of examples and exercises, to support a complete understanding of calculus for students at varying levels. We present the material in a way to encourage student thinking, going beyond memorizing formulas and routine procedures, and we show students how to generalize key concepts once they are introduced. References are made throughout which tie a new concept to a related one that was studied earlier, or to a generalization they will see later on. After studying calculus from *Thomas*, students will have developed problem solving and reasoning abilities that will serve them well in many important aspects of their lives. Mastering this beautiful and creative subject, with its many practical applications across so many fields of endeavor, is its own reward. But the real gift of studying calculus is acquiring the ability to think logically and factually, and learning how to generalize conceptually. We intend this book to encourage and support those goals.

## New to this Edition

In this new edition we further blend conceptual thinking with the overall logic and structure of single and multivariable calculus. We continue to improve clarity and precision, taking into account helpful suggestions from readers and users of our previous texts. While keeping a careful eye on length, we have created additional examples throughout the text. Numerous new exercises have been added at all levels of difficulty, but the focus in this revision has been on the mid-level exercises. A number of figures have been reworked and new ones added to improve visualization. We have written a new section on probability, which provides an important application of integration to the life sciences.

We have maintained the basic structure of the Table of Contents, and retained improvements from the twelfth edition. In keeping with this process, we have added more improvements throughout, which we detail here:

- **Functions** In discussing the use of software for graphing purposes, we added a brief subsection on least squares curve fitting, which allows students to take advantage of this widely used and available application. Prerequisite material continues to be reviewed in Appendices 1–3.

- **Continuity** We clarified the continuity definitions by confining the term "endpoints" to intervals instead of more general domains, and we moved the subsection on continuous extension of a function to the end of the continuity section.

- **Derivatives** We included a brief geometric insight justifying l'Hôpital's Rule. We also enhanced and clarified the meaning of differentiability for functions of several variables, and added a result on the Chain Rule for functions defined along a path.

- **Integrals** We wrote a new section reviewing basic integration formulas and the Substitution Rule, using them in combination with algebraic and trigonometric identities, before presenting other techniques of integration.

- **Probability** We created a new section applying improper integrals to some commonly used probability distributions, including the exponential and normal distributions. Many examples and exercises apply to the life sciences.

- **Series** We now present the idea of absolute convergence before giving the Ratio and Root Tests, and then state these tests in their stronger form. Conditional convergence is introduced later on with the Alternating Series Test.

- **Multivariable and Vector Calculus** We give more geometric insight into the idea of multiple integrals, and we enhance the meaning of the Jacobian in using substitutions to evaluate them. The idea of surface integrals of vector fields now parallels the notion for line integrals of vector fields. We have improved our discussion of the divergence and curl of a vector field.

- **Exercises and Examples** Strong exercise sets are traditional with *Thomas' Calculus*, and we continue to strengthen them with each new edition. Here, we have updated, changed, and added many new exercises and examples, with particular attention to including more applications to the life science areas and to contemporary problems. For instance, we added new exercises addressing drug concentrations and dosages, estimating the spill rate of a ruptured oil pipeline, and predicting rising costs for college tuition.

- **The Use of SI Units** All the units in this edition have been converted to SI units, except where a non-SI unit is commonly used in scientific, technical, and commercial literature in most regions.

## Continuing Features

RIGOR   The level of rigor is consistent with that of earlier editions. We continue to distinguish between formal and informal discussions and to point out their differences. We think starting with a more intuitive, less formal, approach helps students understand a new or difficult concept so they can then appreciate its full mathematical precision and outcomes. We pay attention to defining ideas carefully and to proving theorems appropriate for calculus students, while mentioning deeper or subtler issues they would study in a more advanced course. Our organization and distinctions between informal and formal discussions give the instructor a degree of flexibility in the amount and depth of coverage of the various topics. For example, while we do not prove the Intermediate Value Theorem or the Extreme Value Theorem for continuous functions on $a \leq x \leq b$, we do state these theorems precisely, illustrate their meanings in numerous examples, and use them to prove other important results. Furthermore, for those instructors who desire greater depth of coverage, in Appendix 6 we discuss the reliance of the validity of these theorems on the completeness of the real numbers.

**WRITING EXERCISES**   Writing exercises placed throughout the text ask students to explore and explain a variety of calculus concepts and applications. In addition, the end of each chapter contains a list of questions for students to review and summarize what they have learned. Many of these exercises make good writing assignments.

**END-OF-CHAPTER REVIEWS AND PROJECTS**   In addition to problems appearing after each section, each chapter culminates with review questions, practice exercises covering the entire chapter, and a series of Additional and Advanced Exercises serving to include more challenging or synthesizing problems. Most chapters also include descriptions of several **Technology Application Projects** that can be worked by individual students or groups of students over a longer period of time. These projects require the use of a computer running *Mathematica* or *Maple* and additional material that is available over the Internet at **www.pearsonglobaleditions/thomas** and in MyMathLab.

**WRITING AND APPLICATIONS**   As always, this text continues to be easy to read, conversational, and mathematically rich. Each new topic is motivated by clear, easy-to-understand examples and is then reinforced by its application to real-world problems of immediate interest to students. A hallmark of this book has been the application of calculus to science and engineering. These applied problems have been updated, improved, and extended continually over the last several editions.

**TECHNOLOGY**   In a course using the text, technology can be incorporated according to the taste of the instructor. Each section contains exercises requiring the use of technology; these are marked with a $\boxed{\text{T}}$ if suitable for calculator or computer use, or they are labeled **Computer Explorations** if a computer algebra system (CAS, such as *Maple* or *Mathematica*) is required.

## Additional Resources

**INSTRUCTOR'S SOLUTIONS MANUAL**
Single Variable Calculus (Chapters 1–11), 1-292-08987-3 | 978-1-292-08987-4
Multivariable Calculus (Chapters 10–16), ISBN 1-292-08988-1 | 978-1-292-08988-1
The *Instructor's Solutions Manual* contains complete worked-out solutions to all of the exercises in *Thomas' Calculus*.

**JUST-IN-TIME ALGEBRA AND TRIGONOMETRY FOR
CALCULUS, Fourth Edition**
ISBN 0-321-67104-X | 978-0-321-67104-2
Sharp algebra and trigonometry skills are critical to mastering calculus, and *Just-in-Time Algebra and Trigonometry for Calculus* by Guntram Mueller and Ronald I. Brent is designed to bolster these skills while students study calculus. As students make their way through calculus, this text is with them every step of the way, showing them the necessary algebra or trigonometry topics and pointing out potential problem spots. The easy-to-use table of contents has algebra and trigonometry topics arranged in the order in which students will need them as they study calculus.

**Technology Resource Manuals**
*Maple Manual* by Marie Vanisko, Carroll College
*Mathematica Manual* by Marie Vanisko, Carroll College
*TI-Graphing Calculator Manual* by Elaine McDonald-Newman, Sonoma State University
These manuals cover *Maple* 17, *Mathematica* 8, and the TI-83 Plus/TI-84 Plus and TI-89, respectively. Each manual provides detailed guidance for integrating a specific software

package or graphing calculator throughout the course, including syntax and commands. These manuals are available to qualified instructors through the *Thomas' Calculus* Web site, **www.pearsonglobaleditions/thomas**, and MyMathLab.

### WEB SITE www.pearsonglobaleditions/thomas

The *Thomas' Calculus* Web site contains the chapter on Second-Order Differential Equations, including odd-numbered answers, and provides the expanded historical biographies and essays referenced in the text. The Technology Resource Manuals and the **Technology Application Projects**, which can be used as projects by individual students or groups of students, are also available.

### MyMathLab® Online Course (access code required)

MyMathLab from Pearson is the world's leading online resource in mathematics, integrating interactive homework, assessment, and media in a flexible, easy-to-use format.

MyMathLab delivers **proven results** in helping individual students succeed.

- MyMathLab has a consistently positive impact on the quality of learning in higher education math instruction. MyMathLab can be successfully implemented in any environment—lab-based, hybrid, fully online, traditional—and demonstrates the quantifiable difference that integrated usage makes in regard to student retention, subsequent success, and overall achievement.

- MyMathLab's comprehensive online gradebook automatically tracks your students' results on tests, quizzes, homework, and in the study plan. You can use the gradebook to quickly intervene if your students have trouble, or to provide positive feedback on a job well done. The data within MyMathLab are easily exported to a variety of spreadsheet programs, such as Microsoft Excel. You can determine which points of data you want to export, and then analyze the results to determine success.

MyMathLab provides **engaging experiences** that personalize, stimulate, and measure learning for each student.

- **"Getting Ready" chapter** includes hundreds of exercises that address prerequisite skills in algebra and trigonometry. Each student can receive remediation for just those skills he or she needs help with.

- **Exercises:** The homework and practice exercises in MyMathLab are correlated to the exercises in the textbook, and they regenerate algorithmically to give students unlimited opportunity for practice and mastery. The software offers immediate, helpful feedback when students enter incorrect answers.

- **Multimedia Learning Aids:** Exercises include guided solutions, sample problems, animations, Java™ applets, and eText access for extra help at point-of-use.

And, MyMathLab comes from an **experienced partner** with educational expertise and an eye on the future.

- Knowing that you are using a Pearson product means knowing that you are using quality content. It means that our eTexts are accurate and our assessment tools work. It also means we are committed to making MyMathLab as accessible as possible.

- Whether you are just getting started with MyMathLab, or have a question along the way, we're here to help you learn about our technologies and how to incorporate them into your course.

To learn more about how MyMathLab combines proven learning applications with powerful assessment, visit **www.mymathlab.com** or contact your Pearson representative.

### TestGen®

TestGen® (**www.pearsoned.com/testgen**) enables instructors to build, edit, print, and administer tests using a computerized bank of questions developed to cover all the objectives of the text. TestGen is algorithmically based, allowing instructors to create multiple but equivalent versions of the same question or test with the click of a button. Instructors can also modify test bank questions or add new questions. The software and test bank are available for download from Pearson Education's online catalog.

### PowerPoint® Lecture Slides

These classroom presentation slides are geared specifically to the sequence and philosophy of the *Thomas' Calculus* series. Key graphics from the book are included to help bring the concepts alive in the classroom. These files are available to qualified instructors through the Pearson Instructor Resource Center, **www.pearsonglobaleditions.com/ thomas**, and MyMathLab.

## Acknowledgments

We would like to express our thanks to the people who made many valuable contributions to this edition as it developed through its various stages:

### Accuracy Checkers

Lisa Collette
Patricia Nelson
Tom Wegleitner

### Reviewers for Recent Editions

Meighan Dillon, *Southern Polytechnic State University*
Anne Dougherty, *University of Colorado*
Said Fariabi, *San Antonio College*
Klaus Fischer, *George Mason University*
Tim Flood, *Pittsburg State University*
Rick Ford, *California State University—Chico*
Robert Gardner, *East Tennessee State University*
Christopher Heil, *Georgia Institute of Technology*
Joshua Brandon Holden, *Rose-Hulman Institute of Technology*
Alexander Hulpke, *Colorado State University*
Jacqueline Jensen, *Sam Houston State University*
Jennifer M. Johnson, *Princeton University*
Hideaki Kaneko, *Old Dominion University*
Przemo Kranz, *University of Mississippi*
Xin Li, *University of Central Florida*
Maura Mast, *University of Massachusetts—Boston*
Val Mohanakumar, *Hillsborough Community College—Dale Mabry Campus*
Aaron Montgomery, *Central Washington University*
Christopher M. Pavone, *California State University at Chico*
Cynthia Piez, *University of Idaho*
Brooke Quinlan, *Hillsborough Community College—Dale Mabry Campus*
Rebecca A. Segal, *Virginia Commonwealth University*
Andrew V. Sills, *Georgia Southern University*
Alex Smith, *University of Wisconsin—Eau Claire*
Mark A. Smith, *Miami University*
Donald Solomon, *University of Wisconsin—Milwaukee*

John Sullivan, *Black Hawk College*
Maria Terrell, *Cornell University*
Blake Thornton, *Washington University in St. Louis*
David Walnut, *George Mason University*
Adrian Wilson, *University of Montevallo*
Bobby Winters, *Pittsburg State University*
Dennis Wortman, *University of Massachusetts—Boston*

## Global Edition Contributors and Reviewers

Pearson would like to thank the following for their contributions to the Global Edition:

### Contributor
Antonio Behn, *Universidad de Chile*

### Additional Contributions
Nalinakshi N., *Atria Institute of Technology*

### Reviewers
B. R. Shankar, *National Institute of Technology Karnataka, India*
Fedor Duzhin, *Nanyang Technological University, Singapore*
José Luis Zuleta Estrugo, *École Polytechnique Fédérale de Lausanne, Switzerland*

# 1

# Functions

**OVERVIEW** Functions are fundamental to the study of calculus. In this chapter we review what functions are and how they are pictured as graphs, how they are combined and transformed, and ways they can be classified. We review the trigonometric functions, and we discuss misrepresentations that can occur when using calculators and computers to obtain a function's graph. The real number system, Cartesian coordinates, straight lines, circles, parabolas, and ellipses are reviewed in the Appendices.

## 1.1 Functions and Their Graphs

Functions are a tool for describing the real world in mathematical terms. A function can be represented by an equation, a graph, a numerical table, or a verbal description; we will use all four representations throughout this book. This section reviews these function ideas.

### Functions; Domain and Range

The temperature at which water boils depends on the elevation above sea level (the boiling point drops as you ascend). The interest paid on a cash investment depends on the length of time the investment is held. The area of a circle depends on the radius of the circle. The distance an object travels at constant speed along a straight-line path depends on the elapsed time.

In each case, the value of one variable quantity, say $y$, depends on the value of another variable quantity, which we might call $x$. We say that "$y$ is a function of $x$" and write this symbolically as

$$y = f(x) \qquad (\text{"$y$ equals $f$ of $x$"}).$$

In this notation, the symbol $f$ represents the function, the letter $x$ is the **independent variable** representing the input value of $f$, and $y$ is the **dependent variable** or output value of $f$ at $x$.

> **DEFINITION** A **function** $f$ from a set $D$ to a set $Y$ is a rule that assigns a *unique* (single) element $f(x) \in Y$ to each element $x \in D$.

The set $D$ of all possible input values is called the **domain** of the function. The set of all output values of $f(x)$ as $x$ varies throughout $D$ is called the **range** of the function. The range may not include every element in the set $Y$. The domain and range of a function can be any sets of objects, but often in calculus they are sets of real numbers interpreted as points of a coordinate line. (In Chapters 13–16, we will encounter functions for which the elements of the sets are points in the coordinate plane or in space.)

Often a function is given by a formula that describes how to calculate the output value from the input variable. For instance, the equation $A = \pi r^2$ is a rule that calculates the area $A$ of a circle from its radius $r$ (so $r$, interpreted as a length, can only be positive in this formula). When we define a function $y = f(x)$ with a formula and the domain is not stated explicitly or restricted by context, the domain is assumed to be the largest set of real $x$-values for which the formula gives real $y$-values, which is called the **natural domain**. If we want to restrict the domain in some way, we must say so. The domain of $y = x^2$ is the entire set of real numbers. To restrict the domain of the function to, say, positive values of $x$, we would write "$y = x^2, x > 0$."

Changing the domain to which we apply a formula usually changes the range as well. The range of $y = x^2$ is $[0, \infty)$. The range of $y = x^2, x \geq 2$, is the set of all numbers obtained by squaring numbers greater than or equal to 2. In set notation (see Appendix 1), the range is $\{x^2 \mid x \geq 2\}$ or $\{y \mid y \geq 4\}$ or $[4, \infty)$.

When the range of a function is a set of real numbers, the function is said to be **real-valued**. The domains and ranges of most real-valued functions of a real variable we consider are intervals or combinations of intervals. The intervals may be open, closed, or half open, and may be finite or infinite. Sometimes the range of a function is not easy to find.

A function $f$ is like a machine that produces an output value $f(x)$ in its range whenever we feed it an input value $x$ from its domain (Figure 1.1). The function keys on a calculator give an example of a function as a machine. For instance, the $\sqrt{x}$ key on a calculator gives an output value (the square root) whenever you enter a nonnegative number $x$ and press the $\sqrt{x}$ key.

A function can also be pictured as an **arrow diagram** (Figure 1.2). Each arrow associates an element of the domain $D$ with a unique or single element in the set $Y$. In Figure 1.2, the arrows indicate that $f(a)$ is associated with $a$, $f(x)$ is associated with $x$, and so on. Notice that a function can have the same *value* at two different input elements in the domain (as occurs with $f(a)$ in Figure 1.2), but each input element $x$ is assigned a *single* output value $f(x)$.

$x \xrightarrow{\quad} \boxed{f} \xrightarrow{\quad} f(x)$

Input (domain)     Output (range)

**FIGURE 1.1** A diagram showing a function as a kind of machine.

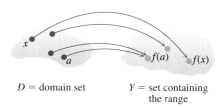

$D$ = domain set     $Y$ = set containing the range

**FIGURE 1.2** A function from a set $D$ to a set $Y$ assigns a unique element of $Y$ to each element in $D$.

**EXAMPLE 1** Let's verify the natural domains and associated ranges of some simple functions. The domains in each case are the values of $x$ for which the formula makes sense.

| Function | Domain ($x$) | Range ($y$) |
|---|---|---|
| $y = x^2$ | $(-\infty, \infty)$ | $[0, \infty)$ |
| $y = 1/x$ | $(-\infty, 0) \cup (0, \infty)$ | $(-\infty, 0) \cup (0, \infty)$ |
| $y = \sqrt{x}$ | $[0, \infty)$ | $[0, \infty)$ |
| $y = \sqrt{4 - x}$ | $(-\infty, 4]$ | $[0, \infty)$ |
| $y = \sqrt{1 - x^2}$ | $[-1, 1]$ | $[0, 1]$ |

**Solution** The formula $y = x^2$ gives a real $y$-value for any real number $x$, so the domain is $(-\infty, \infty)$. The range of $y = x^2$ is $[0, \infty)$ because the square of any real number is nonnegative and every nonnegative number $y$ is the square of its own square root, $y = (\sqrt{y})^2$ for $y \geq 0$.

The formula $y = 1/x$ gives a real $y$-value for every $x$ except $x = 0$. For consistency in the rules of arithmetic, *we cannot divide any number by zero*. The range of $y = 1/x$, the set of reciprocals of all nonzero real numbers, is the set of all nonzero real numbers, since $y = 1/(1/y)$. That is, for $y \neq 0$ the number $x = 1/y$ is the input assigned to the output value $y$.

The formula $y = \sqrt{x}$ gives a real $y$-value only if $x \geq 0$. The range of $y = \sqrt{x}$ is $[0, \infty)$ because every nonnegative number is some number's square root (namely, it is the square root of its own square).

In $y = \sqrt{4 - x}$, the quantity $4 - x$ cannot be negative. That is, $4 - x \geq 0$, or $x \leq 4$. The formula gives real $y$-values for all $x \leq 4$. The range of $\sqrt{4 - x}$ is $[0, \infty)$, the set of all nonnegative numbers.

The formula $y = \sqrt{1 - x^2}$ gives a real $y$-value for every $x$ in the closed interval from $-1$ to $1$. Outside this domain, $1 - x^2$ is negative and its square root is not a real number. The values of $1 - x^2$ vary from 0 to 1 on the given domain, and the square roots of these values do the same. The range of $\sqrt{1 - x^2}$ is $[0, 1]$. ∎

## Graphs of Functions

If $f$ is a function with domain $D$, its **graph** consists of the points in the Cartesian plane whose coordinates are the input-output pairs for $f$. In set notation, the graph is

$$\{(x, f(x)) \mid x \in D\}.$$

The graph of the function $f(x) = x + 2$ is the set of points with coordinates $(x, y)$ for which $y = x + 2$. Its graph is the straight line sketched in Figure 1.3.

The graph of a function $f$ is a useful picture of its behavior. If $(x, y)$ is a point on the graph, then $y = f(x)$ is the height of the graph above (or below) the point $x$. The height may be positive or negative, depending on the sign of $f(x)$ (Figure 1.4).

| $x$ | $y = x^2$ |
|---|---|
| $-2$ | $4$ |
| $-1$ | $1$ |
| $0$ | $0$ |
| $1$ | $1$ |
| $\dfrac{3}{2}$ | $\dfrac{9}{4}$ |
| $2$ | $4$ |

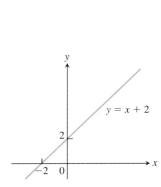

**FIGURE 1.3** The graph of $f(x) = x + 2$ is the set of points $(x, y)$ for which $y$ has the value $x + 2$.

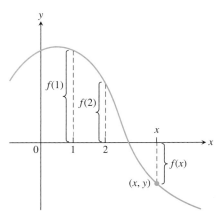

**FIGURE 1.4** If $(x, y)$ lies on the graph of $f$, then the value $y = f(x)$ is the height of the graph above the point $x$ (or below $x$ if $f(x)$ is negative).

**EXAMPLE 2**    Graph the function $y = x^2$ over the interval $[-2, 2]$.

**Solution**    Make a table of $xy$-pairs that satisfy the equation $y = x^2$. Plot the points $(x, y)$ whose coordinates appear in the table, and draw a *smooth* curve (labeled with its equation) through the plotted points (see Figure 1.5). ∎

How do we know that the graph of $y = x^2$ doesn't look like one of these curves?

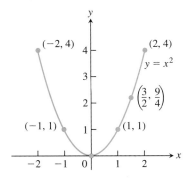

**FIGURE 1.5**    Graph of the function in Example 2.

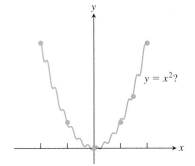

To find out, we could plot more points. But how would we then connect *them*? The basic question still remains: How do we know for sure what the graph looks like between the points we plot? Calculus answers this question, as we will see in Chapter 4. Meanwhile, we will have to settle for plotting points and connecting them as best we can.

### Representing a Function Numerically

We have seen how a function may be represented algebraically by a formula (the area function) and visually by a graph (Example 2). Another way to represent a function is **numerically**, through a table of values. Numerical representations are often used by engineers and experimental scientists. From an appropriate table of values, a graph of the function can be obtained using the method illustrated in Example 2, possibly with the aid of a computer. The graph consisting of only the points in the table is called a **scatterplot**.

**EXAMPLE 3**    Musical notes are pressure waves in the air. The data associated with Figure 1.6 give recorded pressure displacement versus time in seconds of a musical note produced by a tuning fork. The table provides a representation of the pressure function over time. If we first make a scatterplot and then connect approximately the data points $(t, p)$ from the table, we obtain the graph shown in the figure.

| Time | Pressure | Time | Pressure |
|------|----------|------|----------|
| 0.00091 | −0.080 | 0.00362 | 0.217 |
| 0.00108 | 0.200 | 0.00379 | 0.480 |
| 0.00125 | 0.480 | 0.00398 | 0.681 |
| 0.00144 | 0.693 | 0.00416 | 0.810 |
| 0.00162 | 0.816 | 0.00435 | 0.827 |
| 0.00180 | 0.844 | 0.00453 | 0.749 |
| 0.00198 | 0.771 | 0.00471 | 0.581 |
| 0.00216 | 0.603 | 0.00489 | 0.346 |
| 0.00234 | 0.368 | 0.00507 | 0.077 |
| 0.00253 | 0.099 | 0.00525 | −0.164 |
| 0.00271 | −0.141 | 0.00543 | −0.320 |
| 0.00289 | −0.309 | 0.00562 | −0.354 |
| 0.00307 | −0.348 | 0.00579 | −0.248 |
| 0.00325 | −0.248 | 0.00598 | −0.035 |
| 0.00344 | −0.041 | | |

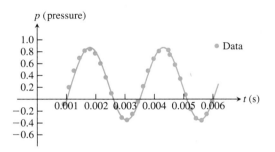

**FIGURE 1.6**   A smooth curve through the plotted points gives a graph of the pressure function represented by the accompanying tabled data (Example 3).

### The Vertical Line Test for a Function

Not every curve in the coordinate plane can be the graph of a function. A function $f$ can have only one value $f(x)$ for each $x$ in its domain, so *no vertical* line can intersect the graph of a function more than once. If $a$ is in the domain of the function $f$, then the vertical line $x = a$ will intersect the graph of $f$ at the single point $(a, f(a))$.

A circle cannot be the graph of a function, since some vertical lines intersect the circle twice. The circle graphed in Figure 1.7a, however, does contain the graphs of functions of $x$, such as the upper semicircle defined by the function $f(x) = \sqrt{1 - x^2}$ and the lower semicircle defined by the function $g(x) = -\sqrt{1 - x^2}$ (Figures 1.7b and 1.7c).

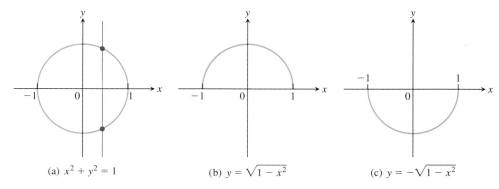

(a) $x^2 + y^2 = 1$      (b) $y = \sqrt{1 - x^2}$      (c) $y = -\sqrt{1 - x^2}$

**FIGURE 1.7** (a) The circle is not the graph of a function; it fails the vertical line test. (b) The upper semicircle is the graph of a function $f(x) = \sqrt{1 - x^2}$. (c) The lower semicircle is the graph of a function $g(x) = -\sqrt{1 - x^2}$.

## Piecewise-Defined Functions

Sometimes a function is described in pieces by using different formulas on different parts of its domain. One example is the **absolute value function**

$$|x| = \begin{cases} x, & x \geq 0 & \text{First formula} \\ -x, & x < 0, & \text{Second formula} \end{cases}$$

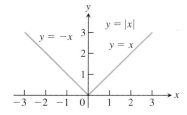

**FIGURE 1.8** The absolute value function has domain $(-\infty, \infty)$ and range $[0, \infty)$.

whose graph is given in Figure 1.8. The right-hand side of the equation means that the function equals $x$ if $x \geq 0$, and equals $-x$ if $x < 0$. Piecewise-defined functions often arise when real-world data are modeled. Here are some other examples.

**EXAMPLE 4**  The function

$$f(x) = \begin{cases} -x, & x < 0 & \text{First formula} \\ x^2, & 0 \leq x \leq 1 & \text{Second formula} \\ 1, & x > 1 & \text{Third formula} \end{cases}$$

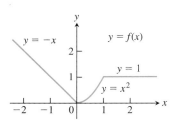

**FIGURE 1.9** To graph the function $y = f(x)$ shown here, we apply different formulas to different parts of its domain (Example 4).

is defined on the entire real line but has values given by different formulas, depending on the position of $x$. The values of $f$ are given by $y = -x$ when $x < 0$, $y = x^2$ when $0 \leq x \leq 1$, and $y = 1$ when $x > 1$. The function, however, is *just one function* whose domain is the entire set of real numbers (Figure 1.9). ◾

**EXAMPLE 5**  The function whose value at any number $x$ is the *greatest integer less than or equal to x* is called the **greatest integer function** or the **integer floor function**. It is denoted $\lfloor x \rfloor$. Figure 1.10 shows the graph. Observe that

$$\lfloor 2.4 \rfloor = 2, \quad \lfloor 1.9 \rfloor = 1, \quad \lfloor 0 \rfloor = 0, \quad \lfloor -1.2 \rfloor = -2,$$
$$\lfloor 2 \rfloor = 2, \quad \lfloor 0.2 \rfloor = 0, \quad \lfloor -0.3 \rfloor = -1, \quad \lfloor -2 \rfloor = -2.$$
◾

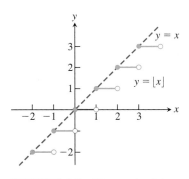

**FIGURE 1.10** The graph of the greatest integer function $y = \lfloor x \rfloor$ lies on or below the line $y = x$, so it provides an integer floor for $x$ (Example 5).

**EXAMPLE 6**  The function whose value at any number $x$ is the *smallest integer greater than or equal to x* is called the **least integer function** or the **integer ceiling function**. It is denoted $\lceil x \rceil$. Figure 1.11 shows the graph. For positive values of $x$, this function might represent, for example, the cost of parking $x$ hours in a parking lot that charges \$1 for each hour or part of an hour. ◾

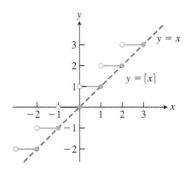

**FIGURE 1.11** The graph of the least integer function $y = \lceil x \rceil$ lies on or above the line $y = x$, so it provides an integer ceiling for $x$ (Example 6).

### Increasing and Decreasing Functions

If the graph of a function *climbs* or *rises* as you move from left to right, we say that the function is *increasing*. If the graph *descends* or *falls* as you move from left to right, the function is *decreasing*.

> **DEFINITIONS** Let $f$ be a function defined on an interval $I$ and let $x_1$ and $x_2$ be any two points in $I$.
>
> 1. If $f(x_2) > f(x_1)$ whenever $x_1 < x_2$, then $f$ is said to be **increasing** on $I$.
> 2. If $f(x_2) < f(x_1)$ whenever $x_1 < x_2$, then $f$ is said to be **decreasing** on $I$.

It is important to realize that the definitions of increasing and decreasing functions must be satisfied for *every* pair of points $x_1$ and $x_2$ in $I$ with $x_1 < x_2$. Because we use the inequality $<$ to compare the function values, instead of $\leq$, it is sometimes said that $f$ is *strictly* increasing or decreasing on $I$. The interval $I$ may be finite (also called bounded) or infinite (unbounded) and by definition never consists of a single point (Appendix 1).

**EXAMPLE 7** The function graphed in Figure 1.9 is decreasing on $(-\infty, 0]$ and increasing on $[0, 1]$. The function is neither increasing nor decreasing on the interval $[1, \infty)$ because of the strict inequalities used to compare the function values in the definitions. ∎

### Even Functions and Odd Functions: Symmetry

The graphs of *even* and *odd* functions have characteristic symmetry properties.

> **DEFINITIONS** A function $y = f(x)$ is an
>
> $$\text{even function of } x \quad \text{if } f(-x) = f(x),$$
> $$\text{odd function of } x \quad \text{if } f(-x) = -f(x),$$
>
> for every $x$ in the function's domain.

(a)

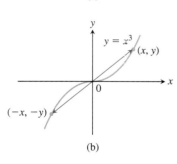

(b)

**FIGURE 1.12** (a) The graph of $y = x^2$ (an even function) is symmetric about the $y$-axis. (b) The graph of $y = x^3$ (an odd function) is symmetric about the origin.

The names *even* and *odd* come from powers of $x$. If $y$ is an even power of $x$, as in $y = x^2$ or $y = x^4$, it is an even function of $x$ because $(-x)^2 = x^2$ and $(-x)^4 = x^4$. If $y$ is an odd power of $x$, as in $y = x$ or $y = x^3$, it is an odd function of $x$ because $(-x)^1 = -x$ and $(-x)^3 = -x^3$.

The graph of an even function is **symmetric about the $y$-axis**. Since $f(-x) = f(x)$, a point $(x, y)$ lies on the graph if and only if the point $(-x, y)$ lies on the graph (Figure 1.12a). A reflection across the $y$-axis leaves the graph unchanged.

The graph of an odd function is **symmetric about the origin**. Since $f(-x) = -f(x)$, a point $(x, y)$ lies on the graph if and only if the point $(-x, -y)$ lies on the graph (Figure 1.12b). Equivalently, a graph is symmetric about the origin if a rotation of 180° about the origin leaves the graph unchanged. Notice that the definitions imply that both $x$ and $-x$ must be in the domain of $f$.

**EXAMPLE 8** Here are several functions illustrating the definition.

$f(x) = x^2$      Even function: $(-x)^2 = x^2$ for all $x$; symmetry about $y$-axis.

$f(x) = x^2 + 1$      Even function: $(-x)^2 + 1 = x^2 + 1$ for all $x$; symmetry about $y$-axis (Figure 1.13a).

$f(x) = x$      Odd function: $(-x) = -x$ for all $x$; symmetry about the origin.

$f(x) = x + 1$      Not odd: $f(-x) = -x + 1$, but $-f(x) = -x - 1$. The two are not equal.

                     Not even: $(-x) + 1 \neq x + 1$ for all $x \neq 0$ (Figure 1.13b). ∎

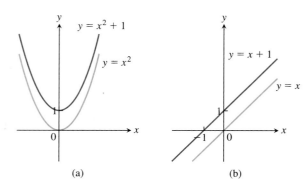

(a)                              (b)

FIGURE 1.13  (a) When we add the constant term 1 to the function $y = x^2$, the resulting function $y = x^2 + 1$ is still even and its graph is still symmetric about the $y$-axis. (b) When we add the constant term 1 to the function $y = x$, the resulting function $y = x + 1$ is no longer odd, since the symmetry about the origin is lost. The function $y = x + 1$ is also not even (Example 8).

## Common Functions

A variety of important types of functions are frequently encountered in calculus. We identify and briefly describe them here.

**Linear Functions**   A function of the form $f(x) = mx + b$, for constants $m$ and $b$, is called a **linear function**. Figure 1.14a shows an array of lines $f(x) = mx$ where $b = 0$, so these lines pass through the origin. The function $f(x) = x$ where $m = 1$ and $b = 0$ is called the **identity function**. Constant functions result when the slope $m = 0$ (Figure 1.14b). A linear function with positive slope whose graph passes through the origin is called a *proportionality* relationship.

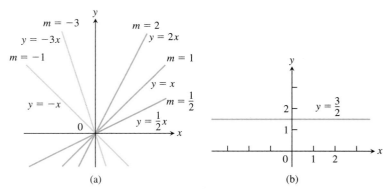

(a)                              (b)

FIGURE 1.14   (a) Lines through the origin with slope $m$. (b) A constant function with slope $m = 0$.

> **DEFINITION**   Two variables $y$ and $x$ are **proportional** (to one another) if one is always a constant multiple of the other; that is, if $y = kx$ for some nonzero constant $k$.

If the variable $y$ is proportional to the reciprocal $1/x$, then sometimes it is said that $y$ is **inversely proportional** to $x$ (because $1/x$ is the multiplicative inverse of $x$).

**Power Functions**   A function $f(x) = x^a$, where $a$ is a constant, is called a **power function**. There are several important cases to consider.

**(a)** $a = n$, a positive integer.

The graphs of $f(x) = x^n$, for $n = 1, 2, 3, 4, 5$, are displayed in Figure 1.15. These functions are defined for all real values of $x$. Notice that as the power $n$ gets larger, the curves tend to flatten toward the $x$-axis on the interval $(-1, 1)$, and to rise more steeply for $|x| > 1$. Each curve passes through the point $(1, 1)$ and through the origin. The graphs of functions with even powers are symmetric about the $y$-axis; those with odd powers are symmetric about the origin. The even-powered functions are decreasing on the interval $(-\infty, 0]$ and increasing on $[0, \infty)$; the odd-powered functions are increasing over the entire real line $(-\infty, \infty)$.

**FIGURE 1.15**   Graphs of $f(x) = x^n$, $n = 1, 2, 3, 4, 5$, defined for $-\infty < x < \infty$.

**(b)** $a = -1$   or   $a = -2$.

The graphs of the functions $f(x) = x^{-1} = 1/x$ and $g(x) = x^{-2} = 1/x^2$ are shown in Figure 1.16. Both functions are defined for all $x \neq 0$ (you can never divide by zero). The graph of $y = 1/x$ is the hyperbola $xy = 1$, which approaches the coordinate axes far from the origin. The graph of $y = 1/x^2$ also approaches the coordinate axes. The graph of the function $f$ is symmetric about the origin; $f$ is decreasing on the intervals $(-\infty, 0)$ and $(0, \infty)$. The graph of the function $g$ is symmetric about the $y$-axis; $g$ is increasing on $(-\infty, 0)$ and decreasing on $(0, \infty)$.

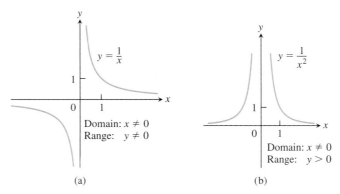

**FIGURE 1.16**   Graphs of the power functions $f(x) = x^a$ for part (a) $a = -1$ and for part (b) $a = -2$.

**(c)** $a = \dfrac{1}{2}, \dfrac{1}{3}, \dfrac{3}{2},$ and $\dfrac{2}{3}$.

The functions $f(x) = x^{1/2} = \sqrt{x}$ and $g(x) = x^{1/3} = \sqrt[3]{x}$ are the **square root** and **cube root** functions, respectively. The domain of the square root function is $[0, \infty)$, but the cube root function is defined for all real $x$. Their graphs are displayed in Figure 1.17, along with the graphs of $y = x^{3/2}$ and $y = x^{2/3}$. (Recall that $x^{3/2} = (x^{1/2})^3$ and $x^{2/3} = (x^{1/3})^2$.)

**Polynomials**   A function $p$ is a **polynomial** if

$$p(x) = a_n x^n + a_{n-1} x^{n-1} + \cdots + a_1 x + a_0$$

where $n$ is a nonnegative integer and the numbers $a_0, a_1, a_2, \ldots, a_n$ are real constants (called the **coefficients** of the polynomial). All polynomials have domain $(-\infty, \infty)$. If the

**FIGURE 1.17** Graphs of the power functions $f(x) = x^a$ for $a = \dfrac{1}{2}, \dfrac{1}{3}, \dfrac{3}{2},$ and $\dfrac{2}{3}$.

leading coefficient $a_n \neq 0$ and $n > 0$, then $n$ is called the **degree** of the polynomial. Linear functions with $m \neq 0$ are polynomials of degree 1. Polynomials of degree 2, usually written as $p(x) = ax^2 + bx + c$, are called **quadratic functions**. Likewise, **cubic functions** are polynomials $p(x) = ax^3 + bx^2 + cx + d$ of degree 3. Figure 1.18 shows the graphs of three polynomials. Techniques to graph polynomials are studied in Chapter 4.

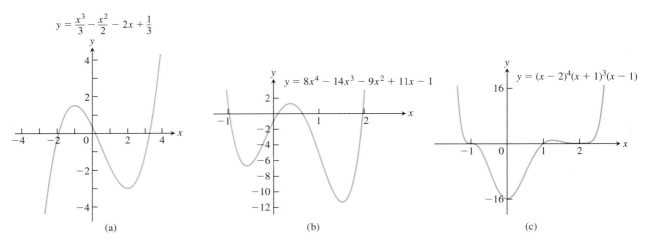

**FIGURE 1.18** Graphs of three polynomial functions.

**Rational Functions** A **rational function** is a quotient or ratio $f(x) = p(x)/q(x)$, where $p$ and $q$ are polynomials. The domain of a rational function is the set of all real $x$ for which $q(x) \neq 0$. The graphs of several rational functions are shown in Figure 1.19.

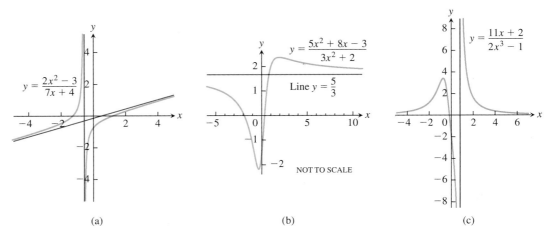

**FIGURE 1.19** Graphs of three rational functions. The straight red lines approached by the graphs are called *asymptotes* and are not part of the graphs. We discuss asymptotes in Section 2.6.

**Algebraic Functions** Any function constructed from polynomials using algebraic operations (addition, subtraction, multiplication, division, and taking roots) lies within the class of **algebraic functions**. All rational functions are algebraic, but also included are more complicated functions (such as those satisfying an equation like $y^3 - 9xy + x^3 = 0$, studied in Section 3.7). Figure 1.20 displays the graphs of three algebraic functions.

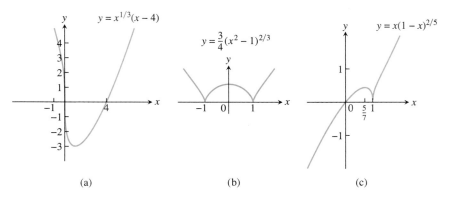

**FIGURE 1.20** Graphs of three algebraic functions.

**Trigonometric Functions** The six basic trigonometric functions are reviewed in Section 1.3. The graphs of the sine and cosine functions are shown in Figure 1.21.

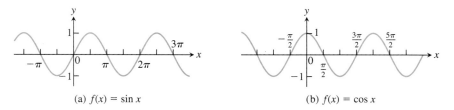

**FIGURE 1.21** Graphs of the sine and cosine functions.

**Exponential Functions** Functions of the form $f(x) = a^x$, where the base $a > 0$ is a positive constant and $a \neq 1$, are called **exponential functions**. All exponential functions have domain $(-\infty, \infty)$ and range $(0, \infty)$, so an exponential function never assumes the value 0. We develop exponential functions in Section 7.3. The graphs of some exponential functions are shown in Figure 1.22.

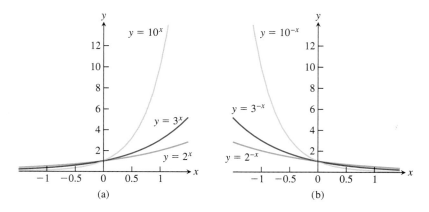

**FIGURE 1.22** Graphs of exponential functions.

**Logarithmic Functions**   These are the functions $f(x) = \log_a x$, where the base $a \neq 1$ is a positive constant. They are the *inverse functions* of the exponential functions, and we define these functions in Section 7.2. Figure 1.23 shows the graphs of four logarithmic functions with various bases. In each case the domain is $(0, \infty)$ and the range is $(-\infty, \infty)$.

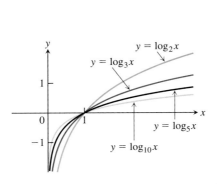

**FIGURE 1.23**   Graphs of four logarithmic functions.

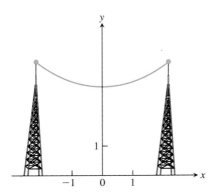

**FIGURE 1.24**   Graph of a catenary or hanging cable. (The Latin word *catena* means "chain.")

**Transcendental Functions**   These are functions that are not algebraic. They include the trigonometric, inverse trigonometric, exponential, and logarithmic functions, and many other functions as well. A particular example of a transcendental function is a **catenary**. Its graph has the shape of a cable, like a telephone line or electric cable, strung from one support to another and hanging freely under its own weight (Figure 1.24). The function defining the graph is discussed in Section 7.7.

# Exercises 1.1

**Functions**

In Exercises 1–6, find the domain and range of each function.

**1.** $f(x) = 1 + x^2$       **2.** $f(x) = 1 - \sqrt{x}$

**3.** $F(x) = \sqrt{5x + 10}$     **4.** $g(x) = \sqrt{x^2 - 3x}$

**5.** $f(t) = \dfrac{4}{3 - t}$        **6.** $G(t) = \dfrac{2}{t^2 - 16}$

In Exercises 7 and 8, which of the graphs are graphs of functions of $x$, and which are not? Give reasons for your answers.

**7. a.**

**b.**

**8. a.**

**b.**

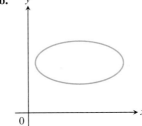

**Finding Formulas for Functions**

**9.** Express the area and perimeter of an equilateral triangle as a function of the triangle's side length $x$.

**10.** Express the side length of a square as a function of the length $d$ of the square's diagonal. Then express the area as a function of the diagonal length.

**11.** Express the edge length of a cube as a function of the cube's diagonal length $d$. Then express the surface area and volume of the cube as a function of the diagonal length.

**12.** A point $P$ in the first quadrant lies on the graph of the function $f(x) = \sqrt{x}$. Express the coordinates of $P$ as functions of the slope of the line joining $P$ to the origin.

**13.** Consider the point $(x, y)$ lying on the graph of the line $2x + 4y = 5$. Let $L$ be the distance from the point $(x, y)$ to the origin $(0, 0)$. Write $L$ as a function of $x$.

**14.** Consider the point $(x, y)$ lying on the graph of $y = \sqrt{x - 3}$. Let $L$ be the distance between the points $(x, y)$ and $(4, 0)$. Write $L$ as a function of $y$.

### Functions and Graphs

Find the natural domain and graph the functions in Exercises 15–20.

**15.** $f(x) = 5 - 2x$      **16.** $f(x) = 1 - 2x - x^2$

**17.** $g(x) = \sqrt{|x|}$      **18.** $g(x) = \sqrt{-x}$

**19.** $F(t) = t/|t|$      **20.** $G(t) = 1/|t|$

**21.** Find the domain of $y = \dfrac{x + 3}{4 - \sqrt{x^2 - 9}}$.

**22.** Find the range of $y = 2 + \dfrac{x^2}{x^2 + 4}$.

**23.** Graph the following equations and explain why they are not graphs of functions of $x$.

   **a.** $|y| = x$      **b.** $y^2 = x^2$

**24.** Graph the following equations and explain why they are not graphs of functions of $x$.

   **a.** $|x| + |y| = 1$      **b.** $|x + y| = 1$

### Piecewise-Defined Functions

Graph the functions in Exercises 25–28.

**25.** $f(x) = \begin{cases} x, & 0 \le x \le 1 \\ 2 - x, & 1 < x \le 2 \end{cases}$

**26.** $g(x) = \begin{cases} 1 - x, & 0 \le x \le 1 \\ 2 - x, & 1 < x \le 2 \end{cases}$

**27.** $F(x) = \begin{cases} 4 - x^2, & x \le 1 \\ x^2 + 2x, & x > 1 \end{cases}$

**28.** $G(x) = \begin{cases} 1/x, & x < 0 \\ x, & 0 \le x \end{cases}$

Find a formula for each function graphed in Exercises 29–32.

**29. a.**     **b.**

**30. a.**     **b.**

**31. a.**     **b.**

**32. a.**     **b.**

### The Greatest and Least Integer Functions

**33.** For what values of $x$ is

   **a.** $\lfloor x \rfloor = 0$?      **b.** $\lceil x \rceil = 0$?

**34.** What real numbers $x$ satisfy the equation $\lfloor x \rfloor = \lceil x \rceil$?

**35.** Does $\lceil -x \rceil = -\lfloor x \rfloor$ for all real $x$? Give reasons for your answer.

**36.** Graph the function

$$f(x) = \begin{cases} \lfloor x \rfloor, & x \ge 0 \\ \lceil x \rceil, & x < 0. \end{cases}$$

Why is $f(x)$ called the *integer part* of $x$?

### Increasing and Decreasing Functions

Graph the functions in Exercises 37–46. What symmetries, if any, do the graphs have? Specify the intervals over which the function is increasing and the intervals where it is decreasing.

**37.** $y = -x^3$      **38.** $y = -\dfrac{1}{x^2}$

**39.** $y = -\dfrac{1}{x}$      **40.** $y = \dfrac{1}{|x|}$

**41.** $y = \sqrt{|x|}$      **42.** $y = \sqrt{-x}$

**43.** $y = x^3/8$      **44.** $y = -4\sqrt{x}$

**45.** $y = -x^{3/2}$      **46.** $y = (-x)^{2/3}$

### Even and Odd Functions

In Exercises 47–58, say whether the function is even, odd, or neither. Give reasons for your answer.

**47.** $f(x) = 3$      **48.** $f(x) = x^{-5}$

**49.** $f(x) = x^2 + 1$      **50.** $f(x) = x^2 + x$

**51.** $g(x) = x^3 + x$      **52.** $g(x) = x^4 + 3x^2 - 1$

**53.** $g(x) = \dfrac{1}{x^2 - 1}$      **54.** $g(x) = \dfrac{x}{x^2 - 1}$

**55.** $h(t) = \dfrac{1}{t - 1}$      **56.** $h(t) = |t^3|$

**57.** $h(t) = 2t + 1$      **58.** $h(t) = 2|t| + 1$

### Theory and Examples

**59.** The variable $s$ is proportional to $t$, and $s = 25$ when $t = 75$. Determine $t$ when $s = 60$.

**60. Kinetic energy** The kinetic energy $K$ of a mass is proportional to the square of its velocity $v$. If $K = 12,960$ joules when $v = 18$ m/s, what is $K$ when $v = 10$ m/s?

**61.** The variables $r$ and $s$ are inversely proportional, and $r = 6$ when $s = 4$. Determine $s$ when $r = 10$.

**62. Boyle's Law** Boyle's Law says that the volume $V$ of a gas at constant temperature increases whenever the pressure $P$ decreases, so that $V$ and $P$ are inversely proportional. If $P = 14.7$ N/cm$^2$ when $V = 1000$ cm$^3$, then what is $V$ when $P = 23.4$ N/cm$^2$?

**63.** A box with an open top is to be constructed from a rectangular piece of cardboard with dimensions 14 cm by 22 cm by cutting out equal squares of side $x$ at each corner and then folding up the sides as in the figure. Express the volume $V$ of the box as a function of $x$.

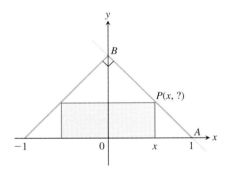

**64.** The accompanying figure shows a rectangle inscribed in an isosceles right triangle whose hypotenuse is 2 units long.

 **a.** Express the $y$-coordinate of $P$ in terms of $x$. (You might start by writing an equation for the line $AB$.)

 **b.** Express the area of the rectangle in terms of $x$.

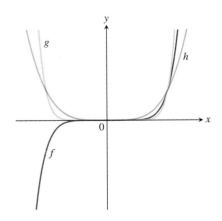

In Exercises 65 and 66, match each equation with its graph. Do not use a graphing device, and give reasons for your answer.

**65. a.** $y = x^4$    **b.** $y = x^7$    **c.** $y = x^{10}$

**66. a.** $y = 5x$    **b.** $y = 5^x$    **c.** $y = x^5$

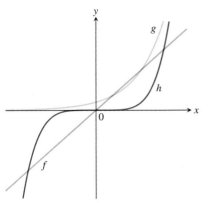

$\boxed{T}$ **67. a.** Graph the functions $f(x) = x/2$ and $g(x) = 1 + (4/x)$ together to identify the values of $x$ for which

$$\frac{x}{2} > 1 + \frac{4}{x}.$$

 **b.** Confirm your findings in part (a) algebraically.

$\boxed{T}$ **68. a.** Graph the functions $f(x) = 3/(x - 1)$ and $g(x) = 2/(x + 1)$ together to identify the values of $x$ for which

$$\frac{3}{x - 1} < \frac{2}{x + 1}.$$

 **b.** Confirm your findings in part (a) algebraically.

**69.** For a curve to be *symmetric about the x-axis*, the point $(x, y)$ must lie on the curve if and only if the point $(x, -y)$ lies on the curve. Explain why a curve that is symmetric about the $x$-axis is not the graph of a function, unless the function is $y = 0$.

**70.** Three hundred books sell for $40 each, resulting in a revenue of $(300)(\$40) = \$12,000$. For each $5 increase in the price, 25 fewer books are sold. Write the revenue $R$ as a function of the number $x$ of $5 increases.

**71.** A pen in the shape of an isosceles right triangle with legs of length $x$ m and hypotenuse of length $h$ m is to be built. If fencing costs $5/m for the legs and $10/m for the hypotenuse, write the total cost $C$ of construction as a function of $h$.

**72. Industrial costs** A power plant sits next to a river where the river is 250 m wide. To lay a new cable from the plant to a location in the city 2 km downstream on the opposite side costs $180 per meter across the river and $100 per meter along the land.

NOT TO SCALE

 **a.** Suppose that the cable goes from the plant to a point $Q$ on the opposite side that is $x$ m from the point $P$ directly opposite the plant. Write a function $C(x)$ that gives the cost of laying the cable in terms of the distance $x$.

 **b.** Generate a table of values to determine if the least expensive location for point $Q$ is less than 300 m or greater than 300 m from point $P$.

# $1.2$ Combining Functions; Shifting and Scaling Graphs

In this section we look at the main ways functions are combined or transformed to form new functions.

## Sums, Differences, Products, and Quotients

Like numbers, functions can be added, subtracted, multiplied, and divided (except where the denominator is zero) to produce new functions. If $f$ and $g$ are functions, then for every $x$ that belongs to the domains of both $f$ and $g$ (that is, for $x \in D(f) \cap D(g)$), we define functions $f + g, f - g,$ and $fg$ by the formulas

$$(f + g)(x) = f(x) + g(x)$$
$$(f - g)(x) = f(x) - g(x)$$
$$(fg)(x) = f(x)g(x).$$

Notice that the $+$ sign on the left-hand side of the first equation represents the operation of addition of *functions*, whereas the $+$ on the right-hand side of the equation means addition of the real numbers $f(x)$ and $g(x)$.

At any point of $D(f) \cap D(g)$ at which $g(x) \neq 0$, we can also define the function $f/g$ by the formula

$$\left(\frac{f}{g}\right)(x) = \frac{f(x)}{g(x)} \qquad \text{(where } g(x) \neq 0\text{)}.$$

Functions can also be multiplied by constants: If $c$ is a real number, then the function $cf$ is defined for all $x$ in the domain of $f$ by

$$(cf)(x) = cf(x).$$

**EXAMPLE 1** The functions defined by the formulas

$$f(x) = \sqrt{x} \qquad \text{and} \qquad g(x) = \sqrt{1 - x}$$

have domains $D(f) = [0, \infty)$ and $D(g) = (-\infty, 1]$. The points common to these domains are the points

$$[0, \infty) \cap (-\infty, 1] = [0, 1].$$

The following table summarizes the formulas and domains for the various algebraic combinations of the two functions. We also write $f \cdot g$ for the product function $fg$.

| Function | Formula | Domain |
|---|---|---|
| $f + g$ | $(f + g)(x) = \sqrt{x} + \sqrt{1 - x}$ | $[0, 1] = D(f) \cap D(g)$ |
| $f - g$ | $(f - g)(x) = \sqrt{x} - \sqrt{1 - x}$ | $[0, 1]$ |
| $g - f$ | $(g - f)(x) = \sqrt{1 - x} - \sqrt{x}$ | $[0, 1]$ |
| $f \cdot g$ | $(f \cdot g)(x) = f(x)g(x) = \sqrt{x(1 - x)}$ | $[0, 1]$ |
| $f/g$ | $\dfrac{f}{g}(x) = \dfrac{f(x)}{g(x)} = \sqrt{\dfrac{x}{1 - x}}$ | $[0, 1)\,(x = 1 \text{ excluded})$ |
| $g/f$ | $\dfrac{g}{f}(x) = \dfrac{g(x)}{f(x)} = \sqrt{\dfrac{1 - x}{x}}$ | $(0, 1]\,(x = 0 \text{ excluded})$ |

■

The graph of the function $f + g$ is obtained from the graphs of $f$ and $g$ by adding the corresponding $y$-coordinates $f(x)$ and $g(x)$ at each point $x \in D(f) \cap D(g)$, as in Figure 1.25. The graphs of $f + g$ and $f \cdot g$ from Example 1 are shown in Figure 1.26.

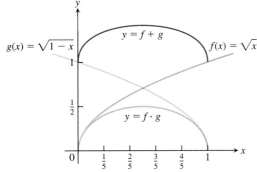

**FIGURE 1.25** Graphical addition of two functions.

**FIGURE 1.26** The domain of the function $f + g$ is the intersection of the domains of $f$ and $g$, the interval $[0, 1]$ on the $x$-axis where these domains overlap. This interval is also the domain of the function $f \cdot g$ (Example 1).

## Composite Functions

Composition is another method for combining functions.

---

**DEFINITION**   If $f$ and $g$ are functions, the **composite** function $f \circ g$ ("$f$ composed with $g$") is defined by

$$(f \circ g)(x) = f(g(x)).$$

The domain of $f \circ g$ consists of the numbers $x$ in the domain of $g$ for which $g(x)$ lies in the domain of $f$.

---

The definition implies that $f \circ g$ can be formed when the range of $g$ lies in the domain of $f$. To find $(f \circ g)(x)$, *first* find $g(x)$ and *second* find $f(g(x))$. Figure 1.27 pictures $f \circ g$ as a machine diagram, and Figure 1.28 shows the composite as an arrow diagram.

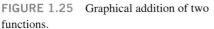

**FIGURE 1.27**   A composite function $f \circ g$ uses the output $g(x)$ of the first function $g$ as the input for the second function $f$.

**FIGURE 1.28**   Arrow diagram for $f \circ g$. If $x$ lies in the domain of $g$ and $g(x)$ lies in the domain of $f$, then the functions $f$ and $g$ can be composed to form $(f \circ g)(x)$.

To evaluate the composite function $g \circ f$ (when defined), we find $f(x)$ first and then $g(f(x))$. The domain of $g \circ f$ is the set of numbers $x$ in the domain of $f$ such that $f(x)$ lies in the domain of $g$.

The functions $f \circ g$ and $g \circ f$ are usually quite different.

**EXAMPLE 2**      If $f(x) = \sqrt{x}$ and $g(x) = x + 1$, find

**(a)** $(f \circ g)(x)$      **(b)** $(g \circ f)(x)$      **(c)** $(f \circ f)(x)$      **(d)** $(g \circ g)(x)$.

**Solution**

| **Composite** | **Domain** |
|---|---|
| **(a)** $(f \circ g)(x) = f(g(x)) = \sqrt{g(x)} = \sqrt{x + 1}$ | $[-1, \infty)$ |
| **(b)** $(g \circ f)(x) = g(f(x)) = f(x) + 1 = \sqrt{x} + 1$ | $[0, \infty)$ |
| **(c)** $(f \circ f)(x) = f(f(x)) = \sqrt{f(x)} = \sqrt{\sqrt{x}} = x^{1/4}$ | $[0, \infty)$ |
| **(d)** $(g \circ g)(x) = g(g(x)) = g(x) + 1 = (x + 1) + 1 = x + 2$ | $(-\infty, \infty)$ |

To see why the domain of $f \circ g$ is $[-1, \infty)$, notice that $g(x) = x + 1$ is defined for all real $x$ but belongs to the domain of $f$ only if $x + 1 \geq 0$, that is to say, when $x \geq -1$.  ∎

Notice that if $f(x) = x^2$ and $g(x) = \sqrt{x}$, then $(f \circ g)(x) = (\sqrt{x})^2 = x$. However, the domain of $f \circ g$ is $[0, \infty)$, not $(-\infty, \infty)$, since $\sqrt{x}$ requires $x \geq 0$.

## Shifting a Graph of a Function

A common way to obtain a new function from an existing one is by adding a constant to each output of the existing function, or to its input variable. The graph of the new function is the graph of the original function shifted vertically or horizontally, as follows.

---

**Shift Formulas**

**Vertical Shifts**

$y = f(x) + k$      Shifts the graph of $f$ *up* $k$ units if $k > 0$
Shifts it *down* $|k|$ units if $k < 0$

**Horizontal Shifts**

$y = f(x + h)$      Shifts the graph of $f$ *left* $h$ units if $h > 0$
Shifts it *right* $|h|$ units if $h < 0$

---

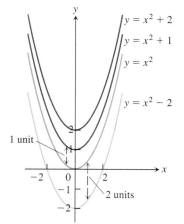

**FIGURE 1.29**   To shift the graph of $f(x) = x^2$ up (or down), we add positive (or negative) constants to the formula for $f$ (Examples 3a and b).

## EXAMPLE 3

**(a)** Adding 1 to the right-hand side of the formula $y = x^2$ to get $y = x^2 + 1$ shifts the graph up 1 unit (Figure 1.29).

**(b)** Adding $-2$ to the right-hand side of the formula $y = x^2$ to get $y = x^2 - 2$ shifts the graph down 2 units (Figure 1.29).

**(c)** Adding 3 to $x$ in $y = x^2$ to get $y = (x + 3)^2$ shifts the graph 3 units to the left, while adding $-2$ shifts the graph 2 units to the right (Figure 1.30).

**(d)** Adding $-2$ to $x$ in $y = |x|$, and then adding $-1$ to the result, gives $y = |x - 2| - 1$ and shifts the graph 2 units to the right and 1 unit down (Figure 1.31).  ∎

## Scaling and Reflecting a Graph of a Function

To scale the graph of a function $y = f(x)$ is to stretch or compress it, vertically or horizontally. This is accomplished by multiplying the function $f$, or the independent variable $x$, by an appropriate constant $c$. Reflections across the coordinate axes are special cases where $c = -1$.

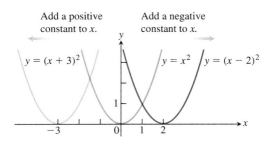

**FIGURE 1.30**  To shift the graph of $y = x^2$ to the left, we add a positive constant to $x$ (Example 3c). To shift the graph to the right, we add a negative constant to $x$.

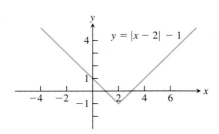

**FIGURE 1.31**  The graph of $y = |x|$ shifted 2 units to the right and 1 unit down (Example 3d).

---

**Vertical and Horizontal Scaling and Reflecting Formulas**

**For $c > 1$, the graph is scaled:**

$y = cf(x)$  $\qquad$ Stretches the graph of $f$ vertically by a factor of $c$.

$y = \dfrac{1}{c}f(x)$  $\qquad$ Compresses the graph of $f$ vertically by a factor of $c$.

$y = f(cx)$  $\qquad$ Compresses the graph of $f$ horizontally by a factor of $c$.

$y = f(x/c)$  $\qquad$ Stretches the graph of $f$ horizontally by a factor of $c$.

**For $c = -1$, the graph is reflected:**

$y = -f(x)$  $\qquad$ Reflects the graph of $f$ across the $x$-axis.

$y = f(-x)$  $\qquad$ Reflects the graph of $f$ across the $y$-axis.

---

**EXAMPLE 4**  Here we scale and reflect the graph of $y = \sqrt{x}$.

(a) **Vertical:** Multiplying the right-hand side of $y = \sqrt{x}$ by 3 to get $y = 3\sqrt{x}$ stretches the graph vertically by a factor of 3, whereas multiplying by $1/3$ compresses the graph by a factor of 3 (Figure 1.32).

(b) **Horizontal:** The graph of $y = \sqrt{3x}$ is a horizontal compression of the graph of $y = \sqrt{x}$ by a factor of 3, and $y = \sqrt{x/3}$ is a horizontal stretching by a factor of 3 (Figure 1.33). Note that $y = \sqrt{3x} = \sqrt{3}\sqrt{x}$ so a horizontal compression *may* correspond to a vertical stretching by a different scaling factor. Likewise, a horizontal stretching may correspond to a vertical compression by a different scaling factor.

(c) **Reflection:** The graph of $y = -\sqrt{x}$ is a reflection of $y = \sqrt{x}$ across the $x$-axis, and $y = \sqrt{-x}$ is a reflection across the $y$-axis (Figure 1.34). ∎

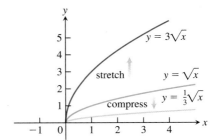

**FIGURE 1.32**  Vertically stretching and compressing the graph $y = \sqrt{x}$ by a factor of 3 (Example 4a).

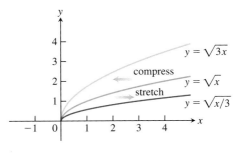

**FIGURE 1.33**  Horizontally stretching and compressing the graph $y = \sqrt{x}$ by a factor of 3 (Example 4b).

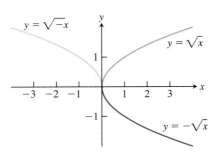

**FIGURE 1.34**  Reflections of the graph $y = \sqrt{x}$ across the coordinate axes (Example 4c).

**EXAMPLE 5**     Given the function $f(x) = x^4 - 4x^3 + 10$ (Figure 1.35a), find formulas to

**(a)** compress the graph horizontally by a factor of 2 followed by a reflection across the $y$-axis (Figure 1.35b).

**(b)** compress the graph vertically by a factor of 2 followed by a reflection across the $x$-axis (Figure 1.35c).

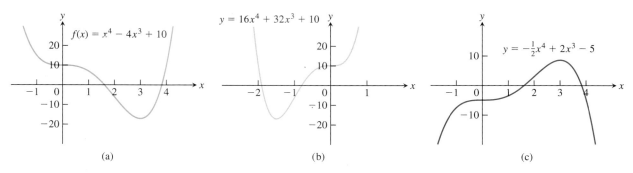

(a)                              (b)                              (c)

**FIGURE 1.35**   (a) The original graph of $f$. (b) The horizontal compression of $y = f(x)$ in part (a) by a factor of 2, followed by a reflection across the $y$-axis. (c) The vertical compression of $y = f(x)$ in part (a) by a factor of 2, followed by a reflection across the $x$-axis (Example 5).

**Solution**

**(a)** We multiply $x$ by 2 to get the horizontal compression, and by $-1$ to give reflection across the $y$-axis. The formula is obtained by substituting $-2x$ for $x$ in the right-hand side of the equation for $f$:

$$y = f(-2x) = (-2x)^4 - 4(-2x)^3 + 10$$
$$= 16x^4 + 32x^3 + 10.$$

**(b)** The formula is

$$y = -\frac{1}{2}f(x) = -\frac{1}{2}x^4 + 2x^3 - 5.$$   ∎

# Exercises **1.2**

**Algebraic Combinations**

In Exercises 1 and 2, find the domains and ranges of $f, g, f + g$, and $f \cdot g$.

**1.** $f(x) = x, \quad g(x) = \sqrt{x - 1}$

**2.** $f(x) = \sqrt{x + 1}, \quad g(x) = \sqrt{x - 1}$

In Exercises 3 and 4, find the domains and ranges of $f, g, f/g$, and $g/f$.

**3.** $f(x) = 2, \quad g(x) = x^2 + 1$

**4.** $f(x) = 1, \quad g(x) = 1 + \sqrt{x}$

**Composites of Functions**

**5.** If $f(x) = x + 5$ and $g(x) = x^2 - 3$, find the following.

  **a.** $f(g(0))$            **b.** $g(f(0))$

  **c.** $f(g(x))$            **d.** $g(f(x))$

  **e.** $f(f(-5))$         **f.** $g(g(2))$

  **g.** $f(f(x))$           **h.** $g(g(x))$

**6.** If $f(x) = x - 1$ and $g(x) = 1/(x + 1)$, find the following.

  **a.** $f(g(1/2))$       **b.** $g(f(1/2))$

  **c.** $f(g(x))$            **d.** $g(f(x))$

  **e.** $f(f(2))$           **f.** $g(g(2))$

  **g.** $f(f(x))$           **h.** $g(g(x))$

In Exercises 7–10, write a formula for $f \circ g \circ h$.

**7.** $f(x) = x + 1, \quad g(x) = 3x, \quad h(x) = 4 - x$

**8.** $f(x) = 3x + 4, \quad g(x) = 2x - 1, \quad h(x) = x^2$

**9.** $f(x) = \sqrt{x + 1}, \quad g(x) = \dfrac{1}{x + 4}, \quad h(x) = \dfrac{1}{x}$

**10.** $f(x) = \dfrac{x + 2}{3 - x}, \quad g(x) = \dfrac{x^2}{x^2 + 1}, \quad h(x) = \sqrt{2 - x}$

Let $f(x) = x - 3$, $g(x) = \sqrt{x}$, $h(x) = x^3$, and $j(x) = 2x$. Express each of the functions in Exercises 11 and 12 as a composite involving one or more of $f$, $g$, $h$, and $j$.

**11. a.** $y = \sqrt{x} - 3$    **b.** $y = 2\sqrt{x}$

    **c.** $y = x^{1/4}$    **d.** $y = 4x$

    **e.** $y = \sqrt{(x - 3)^3}$    **f.** $y = (2x - 6)^3$

**12. a.** $y = 2x - 3$    **b.** $y = x^{3/2}$

    **c.** $y = x^9$    **d.** $y = x - 6$

    **e.** $y = 2\sqrt{x - 3}$    **f.** $y = \sqrt{x^3 - 3}$

**13.** Copy and complete the following table.

| $g(x)$ | $f(x)$ | $(f \circ g)(x)$ |
|---|---|---|
| **a.** $x - 7$ | $\sqrt{x}$ | ? |
| **b.** $x + 2$ | $3x$ | ? |
| **c.** ? | $\sqrt{x - 5}$ | $\sqrt{x^2 - 5}$ |
| **d.** $\dfrac{x}{x - 1}$ | $\dfrac{x}{x - 1}$ | ? |
| **e.** ? | $1 + \dfrac{1}{x}$ | $x$ |
| **f.** $\dfrac{1}{x}$ | ? | $x$ |

**14.** Copy and complete the following table.

| $g(x)$ | $f(x)$ | $(f \circ g)(x)$ |
|---|---|---|
| **a.** $\dfrac{1}{x - 1}$ | $|x|$ | ? |
| **b.** ? | $\dfrac{x - 1}{x}$ | $\dfrac{x}{x + 1}$ |
| **c.** ? | $\sqrt{x}$ | $|x|$ |
| **d.** $\sqrt{x}$ | ? | $|x|$ |

**15.** Evaluate each expression using the given table of values:

| $x$ | $-2$ | $-1$ | $0$ | $1$ | $2$ |
|---|---|---|---|---|---|
| $f(x)$ | 1 | 0 | $-2$ | 1 | 2 |
| $g(x)$ | 2 | 1 | 0 | $-1$ | 0 |

    **a.** $f(g(-1))$    **b.** $g(f(0))$    **c.** $f(f(-1))$

    **d.** $g(g(2))$    **e.** $g(f(-2))$    **f.** $f(g(1))$

**16.** Evaluate each expression using the functions

$$f(x) = 2 - x, \quad g(x) = \begin{cases} -x, & -2 \le x < 0 \\ x - 1, & 0 \le x \le 2. \end{cases}$$

    **a.** $f(g(0))$    **b.** $g(f(3))$    **c.** $g(g(-1))$

    **d.** $f(f(2))$    **e.** $g(f(0))$    **f.** $f(g(1/2))$

In Exercises 17 and 18, **(a)** write formulas for $f \circ g$ and $g \circ f$ and find the **(b)** domain and **(c)** range of each.

**17.** $f(x) = \sqrt{x + 1}$, $g(x) = \dfrac{1}{x}$

**18.** $f(x) = x^2$, $g(x) = 1 - \sqrt{x}$

**19.** Let $f(x) = \dfrac{x}{x - 2}$. Find a function $y = g(x)$ so that $(f \circ g)(x) = x$.

**20.** Let $f(x) = 2x^3 - 4$. Find a function $y = g(x)$ so that $(f \circ g)(x) = x + 2$.

**Shifting Graphs**

**21.** The accompanying figure shows the graph of $y = -x^2$ shifted to two new positions. Write equations for the new graphs.

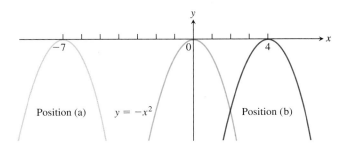

**22.** The accompanying figure shows the graph of $y = x^2$ shifted to two new positions. Write equations for the new graphs.

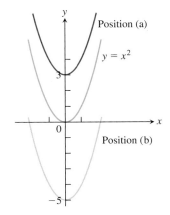

**23.** Match the equations listed in parts (a)–(d) to the graphs in the accompanying figure.

    **a.** $y = (x - 1)^2 - 4$    **b.** $y = (x - 2)^2 + 2$

    **c.** $y = (x + 2)^2 + 2$    **d.** $y = (x + 3)^2 - 2$

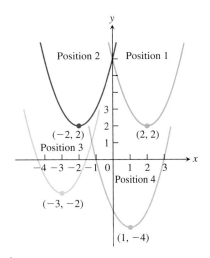

**24.** The accompanying figure shows the graph of $y = -x^2$ shifted to four new positions. Write an equation for each new graph.

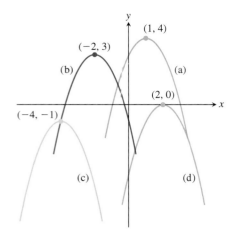

Exercises 25–34 tell how many units and in what directions the graphs of the given equations are to be shifted. Give an equation for the shifted graph. Then sketch the original and shifted graphs together, labeling each graph with its equation.

**25.** $x^2 + y^2 = 49$   Down 3, left 2

**26.** $x^2 + y^2 = 25$   Up 3, left 4

**27.** $y = x^3$   Left 1, down 1

**28.** $y = x^{2/3}$   Right 1, down 1

**29.** $y = \sqrt{x}$   Left 0.81

**30.** $y = -\sqrt{x}$   Right 3

**31.** $y = 2x - 7$   Up 7

**32.** $y = \frac{1}{2}(x + 1) + 5$   Down 5, right 1

**33.** $y = 1/x$   Up 1, right 1

**34.** $y = 1/x^2$   Left 2, down 1

Graph the functions in Exercises 35–54.

**35.** $y = \sqrt{x + 4}$

**36.** $y = \sqrt{9 - x}$

**37.** $y = |x - 2|$

**38.** $y = |1 - x| - 1$

**39.** $y = 1 + \sqrt{x - 1}$

**40.** $y = 1 - \sqrt{x}$

**41.** $y = (x + 1)^{2/3}$

**42.** $y = (x - 8)^{2/3}$

**43.** $y = 1 - x^{2/3}$

**44.** $y + 4 = x^{2/3}$

**45.** $y = \sqrt[3]{x - 1} - 1$

**46.** $y = (x + 2)^{3/2} + 1$

**47.** $y = \frac{1}{x - 2}$

**48.** $y = \frac{1}{x} - 2$

**49.** $y = \frac{1}{x} + 2$

**50.** $y = \frac{1}{x + 2}$

**51.** $y = \frac{1}{(x - 1)^2}$

**52.** $y = \frac{1}{x^2} - 1$

**53.** $y = \frac{1}{x^2} + 1$

**54.** $y = \frac{1}{(x + 1)^2}$

**55.** The accompanying figure shows the graph of a function $f(x)$ with domain $[0, 2]$ and range $[0, 1]$. Find the domains and ranges of the following functions, and sketch their graphs.

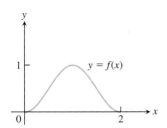

**a.** $f(x) + 2$
**b.** $f(x) - 1$
**c.** $2f(x)$
**d.** $-f(x)$
**e.** $f(x + 2)$
**f.** $f(x - 1)$
**g.** $f(-x)$
**h.** $-f(x + 1) + 1$

**56.** The accompanying figure shows the graph of a function $g(t)$ with domain $[-4, 0]$ and range $[-3, 0]$. Find the domains and ranges of the following functions, and sketch their graphs.

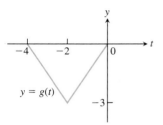

**a.** $g(-t)$
**b.** $-g(t)$
**c.** $g(t) + 3$
**d.** $1 - g(t)$
**e.** $g(-t + 2)$
**f.** $g(t - 2)$
**g.** $g(1 - t)$
**h.** $-g(t - 4)$

**Vertical and Horizontal Scaling**

Exercises 57–66 tell by what factor and direction the graphs of the given functions are to be stretched or compressed. Give an equation for the stretched or compressed graph.

**57.** $y = x^2 - 1$,   stretched vertically by a factor of 3

**58.** $y = x^2 - 1$,   compressed horizontally by a factor of 2

**59.** $y = 1 + \frac{1}{x^2}$,   compressed vertically by a factor of 2

**60.** $y = 1 + \frac{1}{x^2}$,   stretched horizontally by a factor of 3

**61.** $y = \sqrt{x + 1}$,   compressed horizontally by a factor of 4

**62.** $y = \sqrt{x + 1}$,   stretched vertically by a factor of 3

**63.** $y = \sqrt{4 - x^2}$,   stretched horizontally by a factor of 2

**64.** $y = \sqrt{4 - x^2}$,   compressed vertically by a factor of 3

**65.** $y = 1 - x^3$,   compressed horizontally by a factor of 3

**66.** $y = 1 - x^3$,   stretched horizontally by a factor of 2

### Graphing

In Exercises 67–74, graph each function, not by plotting points, but by starting with the graph of one of the standard functions presented in Figures 1.14–1.17 and applying an appropriate transformation.

**67.** $y = -\sqrt{2x + 1}$

**68.** $y = \sqrt{1 - \dfrac{x}{2}}$

**69.** $y = (x - 1)^3 + 2$

**70.** $y = (1 - x)^3 + 2$

**71.** $y = \dfrac{1}{2x} - 1$

**72.** $y = \dfrac{2}{x^2} + 1$

**73.** $y = -\sqrt[3]{x}$

**74.** $y = (-2x)^{2/3}$

**75.** Graph the function $y = |x^2 - 1|$.

**76.** Graph the function $y = \sqrt{|x|}$.

### Combining Functions

**77.** Assume that $f$ is an even function, $g$ is an odd function, and both $f$ and $g$ are defined on the entire real line $(-\infty, \infty)$. Which of the following (where defined) are even? odd?

  **a.** $fg$      **b.** $f/g$      **c.** $g/f$

  **d.** $f^2 = ff$    **e.** $g^2 = gg$    **f.** $f \circ g$

  **g.** $g \circ f$      **h.** $f \circ f$      **i.** $g \circ g$

**78.** Can a function be both even and odd? Give reasons for your answer.

T **79.** (*Continuation of Example 1.*) Graph the functions $f(x) = \sqrt{x}$ and $g(x) = \sqrt{1 - x}$ together with their (a) sum, (b) product, (c) two differences, (d) two quotients.

T **80.** Let $f(x) = x - 7$ and $g(x) = x^2$. Graph $f$ and $g$ together with $f \circ g$ and $g \circ f$.

# 1.3 Trigonometric Functions

This section reviews radian measure and the basic trigonometric functions.

### Angles

**FIGURE 1.36** The radian measure of the central angle $A'CB'$ is the number $\theta = s/r$. For a unit circle of radius $r = 1$, $\theta$ is the length of arc $AB$ that central angle $ACB$ cuts from the unit circle.

Angles are measured in degrees or radians. The number of **radians** in the central angle $A'CB'$ within a circle of radius $r$ is defined as the number of "radius units" contained in the arc $s$ subtended by that central angle. If we denote this central angle by $\theta$ when measured in radians, this means that $\theta = s/r$ (Figure 1.36), or

$$s = r\theta \qquad (\theta \text{ in radians}). \qquad (1)$$

If the circle is a unit circle having radius $r = 1$, then from Figure 1.36 and Equation (1), we see that the central angle $\theta$ measured in radians is just the length of the arc that the angle cuts from the unit circle. Since one complete revolution of the unit circle is $360°$ or $2\pi$ radians, we have

$$\pi \text{ radians} = 180° \qquad (2)$$

and

$$1 \text{ radian} = \frac{180}{\pi} (\approx 57.3) \text{ degrees} \qquad \text{or} \qquad 1 \text{ degree} = \frac{\pi}{180} (\approx 0.017) \text{ radians}.$$

Table 1.1 shows the equivalence between degree and radian measures for some basic angles.

**TABLE 1.1** Angles measured in degrees and radians

| Degrees | $-180$ | $-135$ | $-90$ | $-45$ | 0 | 30 | 45 | 60 | 90 | 120 | 135 | 150 | 180 | 270 | 360 |
|---|---|---|---|---|---|---|---|---|---|---|---|---|---|---|---|
| $\theta$ (radians) | $-\pi$ | $\dfrac{-3\pi}{4}$ | $\dfrac{-\pi}{2}$ | $\dfrac{-\pi}{4}$ | 0 | $\dfrac{\pi}{6}$ | $\dfrac{\pi}{4}$ | $\dfrac{\pi}{3}$ | $\dfrac{\pi}{2}$ | $\dfrac{2\pi}{3}$ | $\dfrac{3\pi}{4}$ | $\dfrac{5\pi}{6}$ | $\pi$ | $\dfrac{3\pi}{2}$ | $2\pi$ |

An angle in the *xy*-plane is said to be in **standard position** if its vertex lies at the origin and its initial ray lies along the positive *x*-axis (Figure 1.37). Angles measured counterclockwise from the positive *x*-axis are assigned positive measures; angles measured clockwise are assigned negative measures.

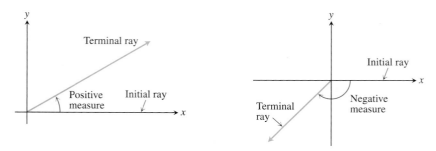

**FIGURE 1.37**    Angles in standard position in the *xy*-plane.

Angles describing counterclockwise rotations can go arbitrarily far beyond $2\pi$ radians or 360°. Similarly, angles describing clockwise rotations can have negative measures of all sizes (Figure 1.38).

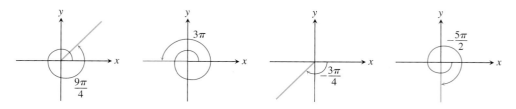

**FIGURE 1.38**    Nonzero radian measures can be positive or negative and can go beyond $2\pi$.

**Angle Convention: Use Radians**    From now on, in this book it is assumed that all angles are measured in radians unless degrees or some other unit is stated explicitly. When we talk about the angle $\pi/3$, we mean $\pi/3$ radians (which is 60°), not $\pi/3$ degrees. We use radians because it simplifies many of the operations in calculus, and some results we will obtain involving the trigonometric functions are not true when angles are measured in degrees.

### The Six Basic Trigonometric Functions

You are probably familiar with defining the trigonometric functions of an acute angle in terms of the sides of a right triangle (Figure 1.39). We extend this definition to obtuse and negative angles by first placing the angle in standard position in a circle of radius *r*. We then define the trigonometric functions in terms of the coordinates of the point $P(x, y)$ where the angle's terminal ray intersects the circle (Figure 1.40).

$$\sin\theta = \frac{\text{opp}}{\text{hyp}} \qquad \csc\theta = \frac{\text{hyp}}{\text{opp}}$$

$$\cos\theta = \frac{\text{adj}}{\text{hyp}} \qquad \sec\theta = \frac{\text{hyp}}{\text{adj}}$$

$$\tan\theta = \frac{\text{opp}}{\text{adj}} \qquad \cot\theta = \frac{\text{adj}}{\text{opp}}$$

**FIGURE 1.39**    Trigonometric ratios of an acute angle.

| | | | |
|---|---|---|---|
| **sine:** | $\sin\theta = \dfrac{y}{r}$ | **cosecant:** | $\csc\theta = \dfrac{r}{y}$ |
| **cosine:** | $\cos\theta = \dfrac{x}{r}$ | **secant:** | $\sec\theta = \dfrac{r}{x}$ |
| **tangent:** | $\tan\theta = \dfrac{y}{x}$ | **cotangent:** | $\cot\theta = \dfrac{x}{y}$ |

These extended definitions agree with the right-triangle definitions when the angle is acute. Notice also that whenever the quotients are defined,

$$\tan\theta = \frac{\sin\theta}{\cos\theta} \qquad \cot\theta = \frac{1}{\tan\theta}$$

$$\sec\theta = \frac{1}{\cos\theta} \qquad \csc\theta = \frac{1}{\sin\theta}$$

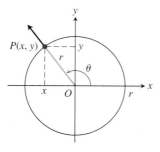

**FIGURE 1.40**    The trigonometric functions of a general angle $\theta$ are defined in terms of *x*, *y*, and *r*.

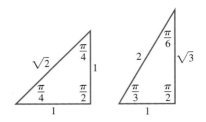

**FIGURE 1.41** Radian angles and side lengths of two common triangles.

As you can see, $\tan \theta$ and $\sec \theta$ are not defined if $x = \cos \theta = 0$. This means they are not defined if $\theta$ is $\pm \pi/2$, $\pm 3\pi/2$, .... Similarly, $\cot \theta$ and $\csc \theta$ are not defined for values of $\theta$ for which $y = 0$, namely $\theta = 0$, $\pm \pi$, $\pm 2\pi$, ....

The exact values of these trigonometric ratios for some angles can be read from the triangles in Figure 1.41. For instance,

$$\sin \frac{\pi}{4} = \frac{1}{\sqrt{2}} \qquad \sin \frac{\pi}{6} = \frac{1}{2} \qquad \sin \frac{\pi}{3} = \frac{\sqrt{3}}{2}$$

$$\cos \frac{\pi}{4} = \frac{1}{\sqrt{2}} \qquad \cos \frac{\pi}{6} = \frac{\sqrt{3}}{2} \qquad \cos \frac{\pi}{3} = \frac{1}{2}$$

$$\tan \frac{\pi}{4} = 1 \qquad \tan \frac{\pi}{6} = \frac{1}{\sqrt{3}} \qquad \tan \frac{\pi}{3} = \sqrt{3}$$

The CAST rule (Figure 1.42) is useful for remembering when the basic trigonometric functions are positive or negative. For instance, from the triangle in Figure 1.43, we see that

$$\sin \frac{2\pi}{3} = \frac{\sqrt{3}}{2}, \qquad \cos \frac{2\pi}{3} = -\frac{1}{2}, \qquad \tan \frac{2\pi}{3} = -\sqrt{3}.$$

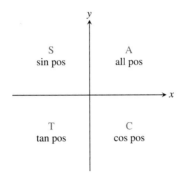

**FIGURE 1.42** The CAST rule, remembered by the statement "Calculus Activates Student Thinking," tells which trigonometric functions are positive in each quadrant.

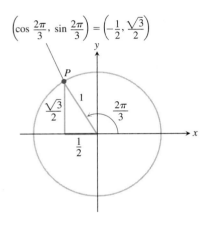

**FIGURE 1.43** The triangle for calculating the sine and cosine of $2\pi/3$ radians. The side lengths come from the geometry of right triangles.

Using a similar method we determined the values of $\sin \theta$, $\cos \theta$, and $\tan \theta$ shown in Table 1.2.

**TABLE 1.2** Values of $\sin \theta$, $\cos \theta$, and $\tan \theta$ for selected values of $\theta$

| Degrees | −180 | −135 | −90 | −45 | 0 | 30 | 45 | 60 | 90 | 120 | 135 | 150 | 180 | 270 | 360 |
|---|---|---|---|---|---|---|---|---|---|---|---|---|---|---|---|
| $\theta$ (radians) | $-\pi$ | $\dfrac{-3\pi}{4}$ | $\dfrac{-\pi}{2}$ | $\dfrac{-\pi}{4}$ | 0 | $\dfrac{\pi}{6}$ | $\dfrac{\pi}{4}$ | $\dfrac{\pi}{3}$ | $\dfrac{\pi}{2}$ | $\dfrac{2\pi}{3}$ | $\dfrac{3\pi}{4}$ | $\dfrac{5\pi}{6}$ | $\pi$ | $\dfrac{3\pi}{2}$ | $2\pi$ |
| $\sin \theta$ | 0 | $\dfrac{-\sqrt{2}}{2}$ | −1 | $\dfrac{-\sqrt{2}}{2}$ | 0 | $\dfrac{1}{2}$ | $\dfrac{\sqrt{2}}{2}$ | $\dfrac{\sqrt{3}}{2}$ | 1 | $\dfrac{\sqrt{3}}{2}$ | $\dfrac{\sqrt{2}}{2}$ | $\dfrac{1}{2}$ | 0 | −1 | 0 |
| $\cos \theta$ | −1 | $\dfrac{-\sqrt{2}}{2}$ | 0 | $\dfrac{\sqrt{2}}{2}$ | 1 | $\dfrac{\sqrt{3}}{2}$ | $\dfrac{\sqrt{2}}{2}$ | $\dfrac{1}{2}$ | 0 | $-\dfrac{1}{2}$ | $\dfrac{-\sqrt{2}}{2}$ | $\dfrac{-\sqrt{3}}{2}$ | −1 | 0 | 1 |
| $\tan \theta$ | 0 | 1 | | −1 | 0 | $\dfrac{\sqrt{3}}{3}$ | 1 | $\sqrt{3}$ | | $-\sqrt{3}$ | −1 | $\dfrac{-\sqrt{3}}{3}$ | 0 | | 0 |

## Periodicity and Graphs of the Trigonometric Functions

When an angle of measure $\theta$ and an angle of measure $\theta + 2\pi$ are in standard position, their terminal rays coincide. The two angles therefore have the same trigonometric function values: $\sin(\theta + 2\pi) = \sin\theta$, $\tan(\theta + 2\pi) = \tan\theta$, and so on. Similarly, $\cos(\theta - 2\pi) = \cos\theta$, $\sin(\theta - 2\pi) = \sin\theta$, and so on. We describe this repeating behavior by saying that the six basic trigonometric functions are *periodic*.

**Periods of Trigonometric Functions**

Period $\pi$: $\quad\tan(x + \pi) = \tan x$
$\qquad\qquad\cot(x + \pi) = \cot x$

Period $2\pi$: $\quad\sin(x + 2\pi) = \sin x$
$\qquad\qquad\cos(x + 2\pi) = \cos x$
$\qquad\qquad\sec(x + 2\pi) = \sec x$
$\qquad\qquad\csc(x + 2\pi) = \csc x$

> **DEFINITION** A function $f(x)$ is **periodic** if there is a positive number $p$ such that $f(x + p) = f(x)$ for every value of $x$. The smallest such value of $p$ is the **period** of $f$.

When we graph trigonometric functions in the coordinate plane, we usually denote the independent variable by $x$ instead of $\theta$. Figure 1.44 shows that the tangent and cotangent functions have period $p = \pi$, and the other four functions have period $2\pi$. Also, the symmetries in these graphs reveal that the cosine and secant functions are even and the other four functions are odd (although this does not prove those results).

**Even**

$\cos(-x) = \cos x$
$\sec(-x) = \sec x$

**Odd**

$\sin(-x) = -\sin x$
$\tan(-x) = -\tan x$
$\csc(-x) = -\csc x$
$\cot(-x) = -\cot x$

Domain: $-\infty < x < \infty$
Range: $-1 \le y \le 1$
Period: $2\pi$
(a)

Domain: $-\infty < x < \infty$
Range: $-1 \le y \le 1$
Period: $2\pi$
(b)

Domain: $x \ne \pm\dfrac{\pi}{2}, \pm\dfrac{3\pi}{2}, \ldots$
Range: $-\infty < y < \infty$
Period: $\pi$
(c)

Domain: $x \ne \pm\dfrac{\pi}{2}, \pm\dfrac{3\pi}{2}, \ldots$
Range: $y \le -1$ or $y \ge 1$
Period: $2\pi$
(d)

Domain: $x \ne 0, \pm\pi, \pm 2\pi, \ldots$
Range: $y \le -1$ or $y \ge 1$
Period: $2\pi$
(e)

Domain: $x \ne 0, \pm\pi, \pm 2\pi, \ldots$
Range: $-\infty < y < \infty$
Period: $\pi$
(f)

**FIGURE 1.44** Graphs of the six basic trigonometric functions using radian measure. The shading for each trigonometric function indicates its periodicity.

## Trigonometric Identities

The coordinates of any point $P(x, y)$ in the plane can be expressed in terms of the point's distance $r$ from the origin and the angle $\theta$ that ray $OP$ makes with the positive $x$-axis (Figure 1.40). Since $x/r = \cos\theta$ and $y/r = \sin\theta$, we have

$$x = r\cos\theta, \qquad y = r\sin\theta.$$

When $r = 1$ we can apply the Pythagorean theorem to the reference right triangle in Figure 1.45 and obtain the equation

$$\cos^2\theta + \sin^2\theta = 1. \tag{3}$$

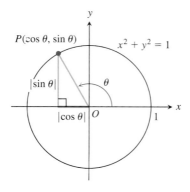

**FIGURE 1.45** The reference triangle for a general angle $\theta$.

This equation, true for all values of $\theta$, is the most frequently used identity in trigonometry. Dividing this identity in turn by $\cos^2 \theta$ and $\sin^2 \theta$ gives

$$1 + \tan^2 \theta = \sec^2 \theta$$
$$1 + \cot^2 \theta = \csc^2 \theta$$

The following formulas hold for all angles $A$ and $B$ (Exercise 58).

**Addition Formulas**

$$\cos (A + B) = \cos A \cos B - \sin A \sin B$$
$$\sin (A + B) = \sin A \cos B + \cos A \sin B$$

(4)

There are similar formulas for $\cos (A - B)$ and $\sin (A - B)$ (Exercises 35 and 36). All the trigonometric identities needed in this book derive from Equations (3) and (4). For example, substituting $\theta$ for both $A$ and $B$ in the addition formulas gives

**Double-Angle Formulas**

$$\cos 2\theta = \cos^2 \theta - \sin^2 \theta$$
$$\sin 2\theta = 2 \sin \theta \cos \theta$$

(5)

Additional formulas come from combining the equations

$$\cos^2 \theta + \sin^2 \theta = 1, \qquad \cos^2 \theta - \sin^2 \theta = \cos 2\theta.$$

We add the two equations to get $2 \cos^2 \theta = 1 + \cos 2\theta$ and subtract the second from the first to get $2 \sin^2 \theta = 1 - \cos 2\theta$. This results in the following identities, which are useful in integral calculus.

**Half-Angle Formulas**

$$\cos^2 \theta = \frac{1 + \cos 2\theta}{2}$$

(6)

$$\sin^2 \theta = \frac{1 - \cos 2\theta}{2}$$

(7)

### The Law of Cosines

If $a$, $b$, and $c$ are sides of a triangle $ABC$ and if $\theta$ is the angle opposite $c$, then

$$c^2 = a^2 + b^2 - 2ab \cos \theta.$$

(8)

This equation is called the **law of cosines**.

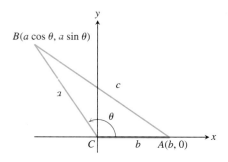

**FIGURE 1.46** The square of the distance between $A$ and $B$ gives the law of cosines.

We can see why the law holds if we introduce coordinate axes with the origin at $C$ and the positive $x$-axis along one side of the triangle, as in Figure 1.46. The coordinates of $A$ are $(b, 0)$; the coordinates of $B$ are $(a\cos\theta, a\sin\theta)$. The square of the distance between $A$ and $B$ is therefore

$$c^2 = (a\cos\theta - b)^2 + (a\sin\theta)^2$$
$$= a^2 \underbrace{(\cos^2\theta + \sin^2\theta)}_{1} + b^2 - 2ab\cos\theta$$
$$= a^2 + b^2 - 2ab\cos\theta.$$

The law of cosines generalizes the Pythagorean theorem. If $\theta = \pi/2$, then $\cos\theta = 0$ and $c^2 = a^2 + b^2$.

## Two Special Inequalities

For any angle $\theta$ measured in radians, the sine and cosine functions satisfy

$$-|\theta| \leq \sin\theta \leq |\theta| \qquad \text{and} \qquad -|\theta| \leq 1 - \cos\theta \leq |\theta|.$$

To establish these inequalities, we picture $\theta$ as a nonzero angle in standard position (Figure 1.47). The circle in the figure is a unit circle, so $|\theta|$ equals the length of the circular arc $AP$. The length of line segment $AP$ is therefore less than $|\theta|$.

Triangle $APQ$ is a right triangle with sides of length

$$QP = |\sin\theta|, \qquad AQ = 1 - \cos\theta.$$

From the Pythagorean theorem and the fact that $AP < |\theta|$, we get

$$\sin^2\theta + (1 - \cos\theta)^2 = (AP)^2 \leq \theta^2. \tag{9}$$

The terms on the left-hand side of Equation (9) are both positive, so each is smaller than their sum and hence is less than or equal to $\theta^2$:

$$\sin^2\theta \leq \theta^2 \qquad \text{and} \qquad (1 - \cos\theta)^2 \leq \theta^2.$$

By taking square roots, this is equivalent to saying that

$$|\sin\theta| \leq |\theta| \qquad \text{and} \qquad |1 - \cos\theta| \leq |\theta|,$$

so

$$-|\theta| \leq \sin\theta \leq |\theta| \qquad \text{and} \qquad -|\theta| \leq 1 - \cos\theta \leq |\theta|.$$

These inequalities will be useful in the next chapter.

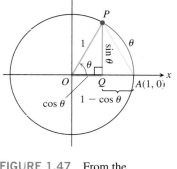

**FIGURE 1.47** From the geometry of this figure, drawn for $\theta > 0$, we get the inequality $\sin^2\theta + (1 - \cos\theta)^2 \leq \theta^2$.

## Transformations of Trigonometric Graphs

The rules for shifting, stretching, compressing, and reflecting the graph of a function summarized in the following diagram apply to the trigonometric functions we have discussed in this section.

Vertical stretch or compression; reflection about $y = d$ if negative

Vertical shift

$$y = af(b(x + c)) + d$$

Horizontal stretch or compression; reflection about $x = -c$ if negative

Horizontal shift

The transformation rules applied to the sine function give the **general sine function** or **sinusoid** formula

$$f(x) = A \sin\left(\frac{2\pi}{B}(x - C)\right) + D,$$

where $|A|$ is the *amplitude*, $|B|$ is the *period*, $C$ is the *horizontal shift*, and $D$ is the *vertical shift*. A graphical interpretation of the various terms is given below.

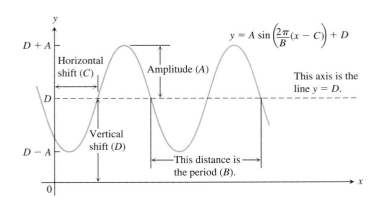

# Exercises 1.3

### Radians and Degrees

1. On a circle of radius 10 m, how long is an arc that subtends a central angle of (**a**) $4\pi/5$ radians? (**b**) $110°$?

2. A central angle in a circle of radius 8 is subtended by an arc of length $10\pi$. Find the angle's radian and degree measures.

3. You want to make an $80°$ angle by marking an arc on the perimeter of a 12-cm-diameter disk and drawing lines from the ends of the arc to the disk's center. To the nearest millimeter, how long should the arc be?

4. If you roll a 1-m-diameter wheel forward 30 cm over level ground, through what angle will the wheel turn? Answer in radians (to the nearest tenth) and degrees (to the nearest degree).

### Evaluating Trigonometric Functions

5. Copy and complete the following table of function values. If the function is undefined at a given angle, enter "UND." Do not use a calculator or tables.

| $\theta$ | $-\pi$ | $-2\pi/3$ | $0$ | $\pi/2$ | $3\pi/4$ |
|---|---|---|---|---|---|
| $\sin \theta$ | | | | | |
| $\cos \theta$ | | | | | |
| $\tan \theta$ | | | | | |
| $\cot \theta$ | | | | | |
| $\sec \theta$ | | | | | |
| $\csc \theta$ | | | | | |

6. Copy and complete the following table of function values. If the function is undefined at a given angle, enter "UND." Do not use a calculator or tables.

| $\theta$ | $-3\pi/2$ | $-\pi/3$ | $-\pi/6$ | $\pi/4$ | $5\pi/6$ |
|---|---|---|---|---|---|
| $\sin \theta$ | | | | | |
| $\cos \theta$ | | | | | |
| $\tan \theta$ | | | | | |
| $\cot \theta$ | | | | | |
| $\sec \theta$ | | | | | |
| $\csc \theta$ | | | | | |

In Exercises 7–12, one of $\sin x$, $\cos x$, and $\tan x$ is given. Find the other two if $x$ lies in the specified interval.

7. $\sin x = \dfrac{3}{5}$, $x \in \left[\dfrac{\pi}{2}, \pi\right]$

8. $\tan x = 2$, $x \in \left[0, \dfrac{\pi}{2}\right]$

9. $\cos x = \dfrac{1}{3}$, $x \in \left[-\dfrac{\pi}{2}, 0\right]$

10. $\cos x = -\dfrac{5}{13}$, $x \in \left[\dfrac{\pi}{2}, \pi\right]$

11. $\tan x = \dfrac{1}{2}$, $x \in \left[\pi, \dfrac{3\pi}{2}\right]$

12. $\sin x = -\dfrac{1}{2}$, $x \in \left[\pi, \dfrac{3\pi}{2}\right]$

### Graphing Trigonometric Functions

Graph the functions in Exercises 13–22. What is the period of each function?

13. $\sin 2x$

14. $\sin (x/2)$

15. $\cos \pi x$

16. $\cos \dfrac{\pi x}{2}$

17. $-\sin \dfrac{\pi x}{3}$

18. $-\cos 2\pi x$

19. $\cos\left(x - \dfrac{\pi}{2}\right)$

20. $\sin\left(x + \dfrac{\pi}{6}\right)$

**21.** $\sin\left(x - \dfrac{\pi}{4}\right) + 1$    **22.** $\cos\left(x + \dfrac{2\pi}{3}\right) - 2$

Graph the functions in Exercises 23–26 in the *ts*-plane (*t*-axis horizontal, *s*-axis vertical). What is the period of each function? What symmetries do the graphs have?

**23.** $s = \cot 2t$    **24.** $s = -\tan \pi t$

**25.** $s = \sec\left(\dfrac{\pi t}{2}\right)$    **26.** $s = \csc\left(\dfrac{t}{2}\right)$

T **27.** **a.** Graph $y = \cos x$ and $y = \sec x$ together for $-3\pi/2 \le x \le 3\pi/2$. Comment on the behavior of $\sec x$ in relation to the signs and values of $\cos x$.

   **b.** Graph $y = \sin x$ and $y = \csc x$ together for $-\pi \le x \le 2\pi$. Comment on the behavior of $\csc x$ in relation to the signs and values of $\sin x$.

T **28.** Graph $y = \tan x$ and $y = \cot x$ together for $-7 \le x \le 7$. Comment on the behavior of $\cot x$ in relation to the signs and values of $\tan x$.

**29.** Graph $y = \sin x$ and $y = \lfloor \sin x \rfloor$ together. What are the domain and range of $\lfloor \sin x \rfloor$?

**30.** Graph $y = \sin x$ and $y = \lceil \sin x \rceil$ together. What are the domain and range of $\lceil \sin x \rceil$?

**Using the Addition Formulas**

Use the addition formulas to derive the identities in Exercises 31–36.

**31.** $\cos\left(x - \dfrac{\pi}{2}\right) = \sin x$    **32.** $\cos\left(x + \dfrac{\pi}{2}\right) = -\sin x$

**33.** $\sin\left(x + \dfrac{\pi}{2}\right) = \cos x$    **34.** $\sin\left(x - \dfrac{\pi}{2}\right) = -\cos x$

**35.** $\cos(A - B) = \cos A \cos B + \sin A \sin B$ (Exercise 57 provides a different derivation.)

**36.** $\sin(A - B) = \sin A \cos B - \cos A \sin B$

**37.** What happens if you take $B = A$ in the trigonometric identity $\cos(A - B) = \cos A \cos B + \sin A \sin B$? Does the result agree with something you already know?

**38.** What happens if you take $B = 2\pi$ in the addition formulas? Do the results agree with something you already know?

In Exercises 39–42, express the given quantity in terms of $\sin x$ and $\cos x$.

**39.** $\cos(\pi + x)$    **40.** $\sin(2\pi - x)$

**41.** $\sin\left(\dfrac{3\pi}{2} - x\right)$    **42.** $\cos\left(\dfrac{3\pi}{2} + x\right)$

**43.** Evaluate $\sin \dfrac{7\pi}{12}$ as $\sin\left(\dfrac{\pi}{4} + \dfrac{\pi}{3}\right)$.

**44.** Evaluate $\cos \dfrac{11\pi}{12}$ as $\cos\left(\dfrac{\pi}{4} + \dfrac{2\pi}{3}\right)$.

**45.** Evaluate $\cos \dfrac{\pi}{12}$.    **46.** Evaluate $\sin \dfrac{5\pi}{12}$.

**Using the Half-Angle Formulas**

Find the function values in Exercises 47–50.

**47.** $\cos^2 \dfrac{\pi}{8}$    **48.** $\cos^2 \dfrac{5\pi}{12}$

**49.** $\sin^2 \dfrac{\pi}{12}$    **50.** $\sin^2 \dfrac{3\pi}{8}$

**Solving Trigonometric Equations**

For Exercises 51–54, solve for the angle $\theta$, where $0 \le \theta \le 2\pi$.

**51.** $\sin^2 \theta = \dfrac{3}{4}$    **52.** $\sin^2 \theta = \cos^2 \theta$

**53.** $\sin 2\theta - \cos \theta = 0$    **54.** $\cos 2\theta + \cos \theta = 0$

**Theory and Examples**

**55.** **The tangent sum formula** The standard formula for the tangent of the sum of two angles is

$$\tan(A + B) = \frac{\tan A + \tan B}{1 - \tan A \tan B}.$$

Derive the formula.

**56.** (*Continuation of Exercise 55.*) Derive a formula for $\tan(A - B)$.

**57.** Apply the law of cosines to the triangle in the accompanying figure to derive the formula for $\cos(A - B)$.

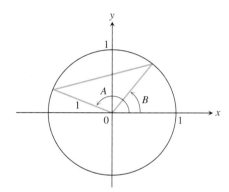

**58.** **a.** Apply the formula for $\cos(A - B)$ to the identity $\sin \theta = \cos\left(\dfrac{\pi}{2} - \theta\right)$ to obtain the addition formula for $\sin(A + B)$.

   **b.** Derive the formula for $\cos(A + B)$ by substituting $-B$ for $B$ in the formula for $\cos(A - B)$ from Exercise 35.

**59.** A triangle has sides $a = 2$ and $b = 3$ and angle $C = 60°$. Find the length of side $c$.

**60.** A triangle has sides $a = 2$ and $b = 3$ and angle $C = 40°$. Find the length of side $c$.

**61.** **The law of sines** *The law of sines* says that if $a$, $b$, and $c$ are the sides opposite the angles $A$, $B$, and $C$ in a triangle, then

$$\frac{\sin A}{a} = \frac{\sin B}{b} = \frac{\sin C}{c}.$$

Use the accompanying figures and the identity $\sin(\pi - \theta) = \sin \theta$, if required, to derive the law.

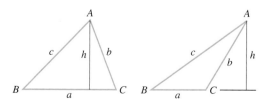

**62.** A triangle has sides $a = 2$ and $b = 3$ and angle $C = 60°$ (as in Exercise 59). Find the sine of angle $B$ using the law of sines.

**63.** A triangle has side $c = 2$ and angles $A = \pi/4$ and $B = \pi/3$. Find the length $a$ of the side opposite $A$.

T **64. The approximation $\sin x \approx x$**   It is often useful to know that, when $x$ is measured in radians, $\sin x \approx x$ for numerically small values of $x$. In Section 3.11, we will see why the approximation holds. The approximation error is less than 1 in 5000 if $|x| < 0.1$.

   **a.** With your grapher in radian mode, graph $y = \sin x$ and $y = x$ together in a viewing window about the origin. What do you see happening as $x$ nears the origin?

   **b.** With your grapher in degree mode, graph $y = \sin x$ and $y = x$ together about the origin again. How is the picture different from the one obtained with radian mode?

### General Sine Curves

For

$$f(x) = A \sin\left(\frac{2\pi}{B}(x - C)\right) + D,$$

identify $A$, $B$, $C$, and $D$ for the sine functions in Exercises 65–68 and sketch their graphs.

**65.** $y = 2\sin(x + \pi) - 1$      **66.** $y = \dfrac{1}{2}\sin(\pi x - \pi) + \dfrac{1}{2}$

**67.** $y = -\dfrac{2}{\pi}\sin\left(\dfrac{\pi}{2}t\right) + \dfrac{1}{\pi}$      **68.** $y = \dfrac{L}{2\pi}\sin\dfrac{2\pi t}{L},\ \ L > 0$

### COMPUTER EXPLORATIONS

In Exercises 69–72, you will explore graphically the general sine function

$$f(x) = A \sin\left(\frac{2\pi}{B}(x - C)\right) + D$$

as you change the values of the constants $A$, $B$, $C$, and $D$. Use a CAS or computer grapher to perform the steps in the exercises.

**69. The period $B$**   Set the constants $A = 3, C = D = 0$.

   **a.** Plot $f(x)$ for the values $B = 1, 3, 2\pi, 5\pi$ over the interval $-4\pi \le x \le 4\pi$. Describe what happens to the graph of the general sine function as the period increases.

   **b.** What happens to the graph for negative values of $B$? Try it with $B = -3$ and $B = -2\pi$.

**70. The horizontal shift $C$**   Set the constants $A = 3, B = 6, D = 0$.

   **a.** Plot $f(x)$ for the values $C = 0, 1$, and 2 over the interval $-4\pi \le x \le 4\pi$. Describe what happens to the graph of the general sine function as $C$ increases through positive values.

   **b.** What happens to the graph for negative values of $C$?

   **c.** What smallest positive value should be assigned to $C$ so the graph exhibits no horizontal shift? Confirm your answer with a plot.

**71. The vertical shift $D$**   Set the constants $A = 3, B = 6, C = 0$.

   **a.** Plot $f(x)$ for the values $D = 0, 1$, and 3 over the interval $-4\pi \le x \le 4\pi$. Describe what happens to the graph of the general sine function as $D$ increases through positive values.

   **b.** What happens to the graph for negative values of $D$?

**72. The amplitude $A$**   Set the constants $B = 6, C = D = 0$.

   **a.** Describe what happens to the graph of the general sine function as $A$ increases through positive values. Confirm your answer by plotting $f(x)$ for the values $A = 1, 5$, and 9.

   **b.** What happens to the graph for negative values of $A$?

# $1.4$ Graphing with Software

Today a number of hardware devices, including computers, calculators, and smartphones, have graphing applications based on software that enables us to graph very complicated functions with high precision. Many of these functions could not otherwise be easily graphed. However, some care must be taken when using such graphing software, and in this section we address some of the issues that may be involved. In Chapter 4 we will see how calculus helps us determine that we are accurately viewing all the important features of a function's graph.

### Graphing Windows

When using software for graphing, a portion of the graph is displayed in a **display** or **viewing window**. Depending on the software, the default window may give an incomplete or misleading picture of the graph. We use the term *square window* when the units or scales used on both axes are the same. This term does not mean that the display window itself is square (usually it is rectangular), but instead it means that the *x*-unit is the same length as the *y*-unit.

When a graph is displayed in the default mode, the *x*-unit may differ from the *y*-unit of scaling in order to capture essential features of the graph. This difference in scaling can cause visual distortions that may lead to erroneous interpretations of the function's behavior.

Some graphing software allows us to set the viewing window by specifying one or both of the intervals, $a \le x \le b$ and $c \le y \le d$, and it may allow for equalizing the scales used for the axes as well. The software selects equally spaced $x$-values in $[a, b]$ and then plots the points $(x, f(x))$. A point is plotted if and only if $x$ lies in the domain of the function and $f(x)$ lies within the interval $[c, d]$. A short line segment is then drawn between each plotted point and its next neighboring point. We now give illustrative examples of some common problems that may occur with this procedure.

**EXAMPLE 1**     Graph the function $f(x) = x^3 - 7x^2 + 28$ in each of the following display or viewing windows:

**(a)** $[-10, 10]$ by $[-10, 10]$     **(b)** $[-4, 4]$ by $[-50, 10]$     **(c)** $[-4, 10]$ by $[-60, 60]$

**Solution**

**(a)** We select $a = -10$, $b = 10$, $c = -10$, and $d = 10$ to specify the interval of $x$-values and the range of $y$-values for the window. The resulting graph is shown in Figure 1.48a. It appears that the window is cutting off the bottom part of the graph and that the interval of $x$-values is too large. Let's try the next window.

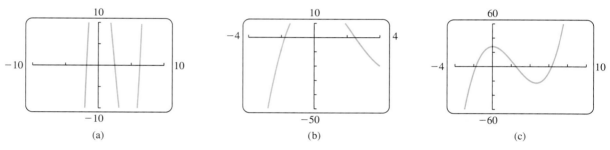

(a)          (b)          (c)

**FIGURE 1.48**    The graph of $f(x) = x^3 - 7x^2 + 28$ in different viewing windows. Selecting a window that gives a clear picture of a graph is often a trial-and-error process (Example 1). The default window used by the software may automatically display the graph in (c).

**(b)** We see some new features of the graph (Figure 1.48b), but the top is missing and we need to view more to the right of $x = 4$ as well. The next window should help.

**(c)** Figure 1.48c shows the graph in this new viewing window. Observe that we get a more complete picture of the graph in this window, and it is a reasonable graph of a third-degree polynomial.    ■

**EXAMPLE 2**     When a graph is displayed, the $x$-unit may differ from the $y$-unit, as in the graphs shown in Figures 1.48b and 1.48c. The result is distortion in the picture, which may be misleading. The display window can be made square by compressing or stretching the units on one axis to match the scale on the other, giving the true graph. Many software systems have built-in options to make the window "square." If yours does not, you may have to bring to your viewing some foreknowledge of the true picture.

Figure 1.49a shows the graphs of the perpendicular lines $y = x$ and $y = -x + 3\sqrt{2}$, together with the semicircle $y = \sqrt{9 - x^2}$, in a nonsquare $[-4, 4]$ by $[-6, 8]$ display window. Notice the distortion. The lines do not appear to be perpendicular, and the semicircle appears to be elliptical in shape.

Figure 1.49b shows the graphs of the same functions in a square window in which the $x$-units are scaled to be the same as the $y$-units. Notice that the scaling on the $x$-axis for Figure 1.49a has been compressed in Figure 1.49b to make the window square. Figure 1.49c gives an enlarged view of Figure 1.49b with a square $[-3, 3]$ by $[0, 4]$ window.    ■

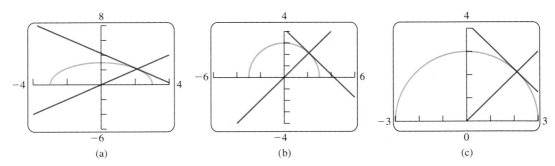

**FIGURE 1.49** Graphs of the perpendicular lines $y = x$ and $y = -x + 3\sqrt{2}$ and of the semicircle $y = \sqrt{9 - x^2}$ appear distorted (a) in a nonsquare window, but clear (b) and (c) in square windows (Example 2). Some software may not provide options for the views in (b) or (c).

If the denominator of a rational function is zero at some $x$-value within the viewing window, graphing software may produce a steep near-vertical line segment from the top to the bottom of the window. Example 3 illustrates steep line segments.

Sometimes the graph of a trigonometric function oscillates very rapidly. When graphing software plots the points of the graph and connects them, many of the maximum and minimum points are actually missed. The resulting graph is then very misleading.

**EXAMPLE 3**   Graph the function $f(x) = \sin 100x$.

**Solution**   Figure 1.50a shows the graph of $f$ in the viewing window $[-12, 12]$ by $[-1, 1]$. We see that the graph looks very strange because the sine curve should oscillate periodically between $-1$ and 1. This behavior is not exhibited in Figure 1.50a. We might experiment with a smaller viewing window, say $[-6, 6]$ by $[-1, 1]$, but the graph is not better (Figure 1.50b). The difficulty is that the period of the trigonometric function $y = \sin 100x$ is very small ($2\pi/100 \approx 0.063$). If we choose the much smaller viewing window $[-0.1, 0.1]$ by $[-1, 1]$ we get the graph shown in Figure 1.50c. This graph reveals the expected oscillations of a sine curve.  ■

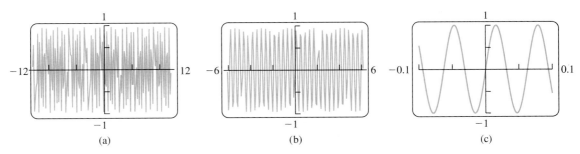

**FIGURE 1.50**   Graphs of the function $y = \sin 100x$ in three viewing windows. Because the period is $2\pi/100 \approx 0.063$, the smaller window in (c) best displays the true aspects of this rapidly oscillating function (Example 3).

**EXAMPLE 4**   Graph the function $y = \cos x + \dfrac{1}{200}\sin 200x$.

**Solution**   In the viewing window $[-6, 6]$ by $[-1, 1]$ the graph appears much like the cosine function with some very small sharp wiggles on it (Figure 1.51a). We get a better look when we significantly reduce the window to $[-0.2, 0.2]$ by $[0.97, 1.01]$, obtaining the graph in Figure 1.51b. We now see the small but rapid oscillations of the second term, $(1/200)\sin 200x$, added to the comparatively larger values of the cosine curve.  ■

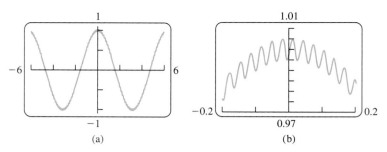

(a)          (b)

**FIGURE 1.51** In (b) we see a close-up view of the function
$y = \cos x + \dfrac{1}{200} \sin 200x$ graphed in (a). The term $\cos x$ clearly dominates
the second term, $\dfrac{1}{200} \sin 200x$, which produces the rapid oscillations along the
cosine curve. Both views are needed for a clear idea of the graph (Example 4).

### Obtaining a Complete Graph

Some graphing software will not display the portion of a graph for $f(x)$ when $x < 0$. Usually that happens because of the algorithm the software is using to calculate the function values. Sometimes we can obtain the complete graph by defining the formula for the function in a different way, as illustrated in the next example.

**EXAMPLE 5** Graph the function $y = x^{1/3}$.

**Solution** Some graphing software displays the graph shown in Figure 1.52a. When we compare it with the graph of $y = x^{1/3} = \sqrt[3]{x}$ in Figure 1.17, we see that the left branch for $x < 0$ is missing. The reason the graphs differ is that the software algorithm calculates $x^{1/3}$ as $e^{(1/3)\ln x}$. Since the logarithmic function is not defined for negative values of $x$, the software can produce only the right branch, where $x > 0$. (Logarithmic and exponential functions are discussed in Chapter 7.)

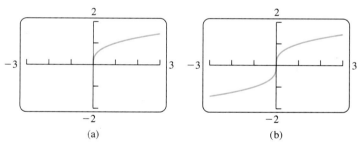

(a)          (b)

**FIGURE 1.52** The graph of $y = x^{1/3}$ is missing the left branch in (a). In (b) we
graph the function $f(x) = \dfrac{x}{|x|} \cdot |x|^{1/3}$, obtaining both branches. (See Example 5.)

To obtain the full picture showing both branches, we can graph the function

$$f(x) = \frac{x}{|x|} \cdot |x|^{1/3}.$$

This function equals $x^{1/3}$ except at $x = 0$ (where $f$ is undefined, although $0^{1/3} = 0$). A graph of $f$ is displayed in Figure 1.52b. ∎

### Capturing the Trend of Collected Data

We have pointed out that applied scientists and analysts often collect data to study a particular issue or phenomenon of interest. If there is no known principle or physical law relating the independent and dependent variables, the data can be plotted in a scatterplot to help find a curve that captures the overall trend of the data points. This process is called **regression analysis**, and the curve is called a **regression curve**.

Many graphing utilities have software that finds the regression curve for a particular type of curve (such as a straight line, a quadratic or other polynomial, or a power curve) and then superimposes the graph of the found curve over the scatterplot. This procedure results in a useful graphical visualization, and often the formula produced for the regression curve can be used to make reasonable estimates or to help explain the issue of interest.

One common method, known as **least squares**, finds the desired regression curve by minimizing the sum of the squares of the vertical distances between the data points and the curve. The least squares method is an *optimization* problem. (In Section 14.7 exercises, we discuss how the regression curve is calculated when fitting a straight line to the data.) Here we present a few examples illustrating the technique by using available software to find the curve. Keep in mind that different software packages may have different ways of entering the data points, and different output features as well.

**EXAMPLE 6**     Table 1.3 shows the annual cost of tuition and fees for a full-time student attending the University of California for the years 1990–2011. The data in the list cite the beginning of the academic year when the corresponding cost was in effect. Use the table to find a regression line capturing the trend of the data points, and use the line to estimate the cost for academic year 2018–19.

**TABLE 1.3** Tuition and fees at the University of California

| Year, $x$ | Cost, $y$ |
|-----------|-----------|
| 1990 | 1,820 |
| 1995 | 4,166 |
| 2000 | 3,964 |
| 2005 | 6,802 |
| 2010 | 11,287 |
| 2011 | 13,218 |

**Solution**   We use regression software that allows for fitting a straight line, and we enter the data from the table to obtain the formula

$$y = 506.25x - 1.0066 \cdot 10^6,$$

where $x$ represents the year and $y$ the cost that took effect that year. Figure 1.53 displays the scatterplot of the data together with the graph of this regression line. From the equation of the line, we find that for $x = 2018$,

$$y = 506.25(2018) - 1.0066 \cdot 10^6 = 15{,}013$$

is the estimated cost (rounded to the nearest dollar) for the academic year 2018–19. The last two data points rise above the trend line in the figure, so this estimate may turn out to be low.      ∎

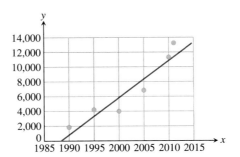

**FIGURE 1.53**   Scatterplot and regression line for the University of California tuition and fees from Table 1.3 (Example 6).

**EXAMPLE 7**     The Centers for Disease Control and Prevention recorded the deaths from tuberculosis in the United States for 1970–2006. We list the data in Table 1.4 for 5-year intervals. Find linear and quadratic regression curves capturing the trend of the data points. Which curve might be the better predictor?

**TABLE 1.4** U.S. deaths from tuberculosis

| Year, $x$ | Deaths, $y$ |
| --- | --- |
| 1970 | 5,217 |
| 1975 | 3,333 |
| 1980 | 1,978 |
| 1985 | 1,752 |
| 1990 | 1,810 |
| 1995 | 1,336 |
| 2000 | 776 |
| 2005 | 648 |

**Solution**   Using regression software that allows us to fit a straight line as well as a quadratic curve, we enter the data to obtain the formulas

$$y = 2.2279 \cdot 10^5 - 111.04x, \qquad \text{line fit}$$

and

$$y = \frac{1451}{350}x^2 - \frac{3{,}483{,}953}{210}x + \frac{464{,}757{,}147}{28}, \qquad \text{quadratic fit}$$

where $x$ represents the year and $y$ represents the number of deaths that occurred. A scatterplot of the data, together with the two trend curves, is displayed in Figure 1.54. In looking at the figure, it would appear that the quadratic curve most closely captures the trend of the data, except for the years 1990 and 1995, and would make the better predictor. However, the quadratic seems to have a minimum value near the year 2000, rising upward thereafter, so it would probably not be a useful tool for making good estimates in the years beyond 2010. This example illustrates the danger of using a regression curve to predict values beyond the range of the data used to construct the curve.  ∎

**FIGURE 1.54**   Scatterplot with the regression line and quadratic curves for tuberculosis deaths in the United States, based on Table 1.4 (Example 7).

# Exercises 1.4

### Choosing a Viewing Window

T  In Exercises 1–4, use graphing software to determine which of the given viewing windows displays the most appropriate graph of the specified function.

**1.** $f(x) = x^4 - 7x^2 + 6x$

   **a.** $[-1, 1]$ by $[-1, 1]$    **b.** $[-2, 2]$ by $[-5, 5]$

   **c.** $[-10, 10]$ by $[-10, 10]$    **d.** $[-5, 5]$ by $[-25, 15]$

**2.** $f(x) = x^3 - 4x^2 - 4x + 16$

   **a.** $[-1, 1]$ by $[-5, 5]$    **b.** $[-3, 3]$ by $[-10, 10]$

   **c.** $[-5, 5]$ by $[-10, 20]$    **d.** $[-20, 20]$ by $[-100, 100]$

**3.** $f(x) = 5 + 12x - x^3$

   **a.** $[-1, 1]$ by $[-1, 1]$    **b.** $[-5, 5]$ by $[-10, 10]$

   **c.** $[-4, 4]$ by $[-20, 20]$    **d.** $[-4, 5]$ by $[-15, 25]$

**4.** $f(x) = \sqrt{5 + 4x - x^2}$

   **a.** $[-2, 2]$ by $[-2, 2]$    **b.** $[-2, 6]$ by $[-1, 4]$

   **c.** $[-3, 7]$ by $[0, 10]$    **d.** $[-10, 10]$ by $[-10, 10]$

### Finding a Viewing Window

T  In Exercises 5–30, find an appropriate graphing software viewing window for the given function and use it to display its graph. The window should give a picture of the overall behavior of the function. There is more than one choice, but incorrect choices can miss important aspects of the function.

**5.** $f(x) = x^4 - 4x^3 + 15$

**6.** $f(x) = \dfrac{x^3}{3} - \dfrac{x^2}{2} - 2x + 1$

**7.** $f(x) = x^5 - 5x^4 + 10$

**8.** $f(x) = 4x^3 - x^4$

**9.** $f(x) = x\sqrt{9 - x^2}$

**10.** $f(x) = x^2(6 - x^3)$

**11.** $y = 2x - 3x^{2/3}$

**12.** $y = x^{1/3}(x^2 - 8)$

**13.** $y = 5x^{2/5} - 2x$

**14.** $y = x^{2/3}(5 - x)$

**15.** $y = |x^2 - 1|$

**16.** $y = |x^2 - x|$

**17.** $y = \dfrac{x + 3}{x + 2}$

**18.** $y = 1 - \dfrac{1}{x + 3}$

**19.** $f(x) = \dfrac{x^2 + 2}{x^2 + 1}$

**20.** $f(x) = \dfrac{x^2 - 1}{x^2 + 1}$

**21.** $f(x) = \dfrac{x - 1}{x^2 - x - 6}$

**22.** $f(x) = \dfrac{8}{x^2 - 9}$

**23.** $f(x) = \dfrac{6x^2 - 15x + 6}{4x^2 - 10x}$

**24.** $f(x) = \dfrac{x^2 - 3}{x - 2}$

**25.** $y = \sin 250x$

**26.** $y = 3 \cos 60x$

**27.** $y = \cos\left(\dfrac{x}{50}\right)$       **28.** $y = \dfrac{1}{10}\sin\left(\dfrac{x}{10}\right)$

**29.** $y = x + \dfrac{1}{10}\sin 30x$       **30.** $y = x^2 + \dfrac{1}{50}\cos 100x$

Use graphing software to graph the functions specified in Exercises 31–36. Select a viewing window that reveals the key features of the function.

**31.** Graph the lower half of the circle defined by the equation $x^2 + 2x = 4 + 4y - y^2$.

**32.** Graph the upper branch of the hyperbola $y^2 - 16x^2 = 1$.

**33.** Graph four periods of the function $f(x) = -\tan 2x$.

**34.** Graph two periods of the function $f(x) = 3\cot\dfrac{x}{2} + 1$.

**35.** Graph the function $f(x) = \sin 2x + \cos 3x$.

**36.** Graph the function $f(x) = \sin^3 x$.

### Regression Lines or Quadratic Curve Fits

T   Use a graphing utility to find the regression curves specified in Exercises 37–42.

**37. Weight of males**   The table shows the average weight for men of medium frame based on height as reported by the Metropolitan Life Insurance Company (1983).

| Height (cm) | Weight (kg) | Height (cm) | Weight (kg) |
|---|---|---|---|
| 157.5 | 61.7 | 177.5 | 71.2 |
| 160 | 62.6 | 180 | 72.6 |
| 162.5 | 64 | 182.5 | 74.2 |
| 165 | 64.2 | 185 | 75.8 |
| 167.5 | 65.8 | 187.5 | 77.6 |
| 170 | 67.1 | 190 | 79.2 |
| 172.5 | 68.5 | 192.5 | 81.2 |
| 175 | 69.9 | | |

    **a.** Make a scatterplot of the data.

    **b.** Find and plot a regression line, and superimpose the line on the scatterplot.

    **c.** Does the regression line reasonably capture the trend of the data? What weight would you predict for a male of height 2 m?

**38. Federal minimum wage**   The federal minimum hourly wage rates have increased over the years. The table shows the rates at the year in which they first took effect, as reported by the U.S. Department of Labor.

| Year | Wage ($) | Year | Wage ($) |
|---|---|---|---|
| 1978 | 2.65 | 1996 | 4.75 |
| 1979 | 2.90 | 1997 | 5.15 |
| 1980 | 3.10 | 2007 | 5.85 |
| 1981 | 3.35 | 2008 | 6.55 |
| 1990 | 3.80 | 2009 | 7.25 |
| 1991 | 4.25 | | |

    **a.** Make a scatterplot of the data.

    **b.** Find and plot a regression line, and superimpose the line on the scatterplot.

    **c.** What do you estimate as the minimum wage for the year 2018?

**39. Median home price**   The median price of single-family homes in the United States increased quite consistently during the years 1976–2000. Then a housing "bubble" occurred for the years 2001–2010, in which prices first rose dramatically for 6 years and then dropped in a steep "crash" over the next 4 years, causing considerable turmoil in the U.S. economy. The table shows some of the data as reported by the National Association of Realtors.

| Year | Price ($) | Year | Price ($) |
|---|---|---|---|
| 1976 | 37400 | 2000 | 122600 |
| 1980 | 56250 | 2002 | 150000 |
| 1984 | 66500 | 2004 | 187500 |
| 1988 | 87500 | 2006 | 247500 |
| 1992 | 95800 | 2008 | 183300 |
| 1996 | 104200 | 2010 | 162500 |

    **a.** Make a scatterplot of the data.

    **b.** Find and plot the regression line for the years 1976–2002, and superimpose the line on the scatterplot in part (a).

    **c.** How would you interpret the meaning of a data point in the housing "bubble"?

**40. Average energy prices**   The table shows the average residential and transportation prices for energy consumption in the United States for the years 2000–2008, as reported by the U.S. Department of Energy. The prices are given as dollars paid for one million BTU (British thermal units) of consumption.

| Year | Residential ($) | Transportation ($) |
|---|---|---|
| 2000 | 15 | 10 |
| 2001 | 16 | 10 |
| 2002 | 15 | 9 |
| 2003 | 16 | 11 |
| 2004 | 18 | 13 |
| 2005 | 19 | 16 |
| 2006 | 21 | 19 |
| 2007 | 21 | 20 |
| 2008 | 23 | 25 |

    **a.** Make a scatterplot of the data sets.

    **b.** Find and plot a regression line for each set of data points, and superimpose the lines on their scatterplots.

    **c.** What do you estimate as the average energy price for residential and transportation use for a million BTU in year 2017?

    **d.** In looking at the trend lines, what do you conclude about the rising costs of energy across the two sectors of usage?

**41. Global annual mean surface air temperature**   A NASA Goddard Institute for Space Studies report gives the annual global mean land-ocean temperature index for the years 1880 to the present. The index number is the difference between the mean temperature over the base years 1951–1980 and the actual temperature for the year recorded. For the recorded year, a positive index is the number of degrees Celsius above the base; a negative index is the number below the base. The table lists the index for the years 1940–2010 in 5-year intervals, reported in the NASA data set.

| Year | Index (°C) | Year | Index (°C) |
|------|-----------|------|-----------|
| 1940 | 0.04 | 1980 | 0.20 |
| 1945 | 0.06 | 1985 | 0.05 |
| 1950 | −0.16 | 1990 | 0.36 |
| 1955 | −0.11 | 1995 | 0.39 |
| 1960 | −0.01 | 2000 | 0.35 |
| 1965 | −0.12 | 2005 | 0.62 |
| 1970 | 0.03 | 2010 | 0.63 |
| 1975 | −0.04 | | |

a. Make a scatterplot of the data.

b. Find and plot a regression line, and superimpose the line on the scatterplot.

c. Find and plot a quadratic curve that captures the trend of the data, and superimpose the curve on the scatterplot.

**42. Growth of yeast cells**  The table shows the amount of yeast cells (measured as *biomass*) growing over a 7-hour period in a nutrient, as recorded by R. Pearl (1927) during a well-known biological experiment.

| Hour | 0 | 1 | 2 | 3 | 4 | 5 | 6 | 7 |
|------|---|---|---|---|---|---|---|---|
| Biomass | 9.6 | 18.3 | 29.0 | 47.2 | 71.1 | 119.1 | 174.6 | 257.3 |

a. Make a scatterplot of the data.

b. Find and plot a regression quadratic, and superimpose the quadratic curve on the scatterplot.

c. What do you estimate as the biomass of yeast in the nutrient after 11 hours?

d. Do you think the quadratic curve would provide a good estimate of the biomass after 18 hours? Give reasons for your answer.

## Chapter 1 Questions to Guide Your Review

1. What is a function? What is its domain? Its range? What is an arrow diagram for a function? Give examples.

2. What is the graph of a real-valued function of a real variable? What is the vertical line test?

3. What is a piecewise-defined function? Give examples.

4. What are the important types of functions frequently encountered in calculus? Give an example of each type.

5. What is meant by an increasing function? A decreasing function? Give an example of each.

6. What is an even function? An odd function? What symmetry properties do the graphs of such functions have? What advantage can we take of this? Give an example of a function that is neither even nor odd.

7. If $f$ and $g$ are real-valued functions, how are the domains of $f + g, f - g, fg$, and $f/g$ related to the domains of $f$ and $g$? Give examples.

8. When is it possible to compose one function with another? Give examples of composites and their values at various points. Does the order in which functions are composed ever matter?

9. How do you change the equation $y = f(x)$ to shift its graph vertically up or down by $|k|$ units? Horizontally to the left or right? Give examples.

10. How do you change the equation $y = f(x)$ to compress or stretch the graph by a factor $c > 1$? Reflect the graph across a coordinate axis? Give examples.

11. What is radian measure? How do you convert from radians to degrees? Degrees to radians?

12. Graph the six basic trigonometric functions. What symmetries do the graphs have?

13. What is a periodic function? Give examples. What are the periods of the six basic trigonometric functions?

14. Starting with the identity $\sin^2 \theta + \cos^2 \theta = 1$ and the formulas for $\cos(A + B)$ and $\sin(A + B)$, show how a variety of other trigonometric identities may be derived.

15. How does the formula for the general sine function $f(x) = A \sin((2\pi/B)(x - C)) + D$ relate to the shifting, stretching, compressing, and reflection of its graph? Give examples. Graph the general sine curve and identify the constants $A$, $B$, $C$, and $D$.

16. Name three issues that arise when functions are graphed using a calculator or computer with graphing software. Give examples.

## Chapter 1 Practice Exercises

### Functions and Graphs

1. Express the area and circumference of a circle as functions of the circle's radius. Then express the area as a function of the circumference.

2. Express the radius of a sphere as a function of the sphere's surface area. Then express the surface area as a function of the volume.

3. A point $P$ in the first quadrant lies on the parabola $y = x^2$. Express the coordinates of $P$ as functions of the angle of inclination of the line joining $P$ to the origin.

4. A hot-air balloon rising straight up from a level field is tracked by a range finder located 500 m from the point of liftoff. Express the balloon's height as a function of the angle the line from the range finder to the balloon makes with the ground.

In Exercises 5–8, determine whether the graph of the function is symmetric about the $y$-axis, the origin, or neither.

**5.** $y = x^{1/5}$

**6.** $y = x^{2/5}$

**7.** $y = x^2 - 2x - 1$

**8.** $y = e^{-x^2}$

In Exercises 9–16, determine whether the function is even, odd, or neither.

**9.** $y = x^2 + 1$

**10.** $y = x^5 - x^3 - x$

**11.** $y = 1 - \cos x$

**12.** $y = \sec x \tan x$

**13.** $y = \dfrac{x^4 + 1}{x^3 - 2x}$

**14.** $y = x - \sin x$

**15.** $y = x + \cos x$

**16.** $y = x \cos x$

**17.** Suppose that $f$ and $g$ are both odd functions defined on the entire real line. Which of the following (where defined) are even? odd?

    **a.** $fg$   **b.** $f^3$   **c.** $f(\sin x)$   **d.** $g(\sec x)$   **e.** $|g|$

**18.** If $f(a - x) = f(a + x)$, show that $g(x) = f(x + a)$ is an even function.

In Exercises 19–28, find the **(a)** domain and **(b)** range.

**19.** $y = |x| - 2$

**20.** $y = -2 + \sqrt{1 - x}$

**21.** $y = \sqrt{16 - x^2}$

**22.** $y = 3^{2-x} + 1$

**23.** $y = 2e^{-x} - 3$

**24.** $y = \tan(2x - \pi)$

**25.** $y = 2\sin(3x + \pi) - 1$

**26.** $y = x^{2/5}$

**27.** $y = \ln(x - 3) + 1$

**28.** $y = -1 + \sqrt[3]{2 - x}$

**29.** State whether each function is increasing, decreasing, or neither.

    **a.** Volume of a sphere as a function of its radius

    **b.** Greatest integer function

    **c.** Height above Earth's sea level as a function of atmospheric pressure (assumed nonzero)

    **d.** Kinetic energy as a function of a particle's velocity

**30.** Find the largest interval on which the given function is increasing.

    **a.** $f(x) = |x - 2| + 1$     **b.** $f(x) = (x + 1)^4$

    **c.** $g(x) = (3x - 1)^{1/3}$     **d.** $R(x) = \sqrt{2x - 1}$

**Piecewise-Defined Functions**

In Exercises 31 and 32, find the **(a)** domain and **(b)** range.

**31.** $y = \begin{cases} \sqrt{-x}, & -4 \le x \le 0 \\ \sqrt{x}, & 0 < x \le 4 \end{cases}$

**32.** $y = \begin{cases} -x - 2, & -2 \le x \le -1 \\ x, & -1 < x \le 1 \\ -x + 2, & 1 < x \le 2 \end{cases}$

In Exercises 33 and 34, write a piecewise formula for the function.

**33.**

**34.**
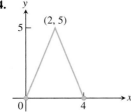

**Composition of Functions**

In Exercises 35 and 36, find

    **a.** $(f \circ g)(-1)$.     **b.** $(g \circ f)(2)$.

    **c.** $(f \circ f)(x)$.     **d.** $(g \circ g)(x)$.

**35.** $f(x) = \dfrac{1}{x}$,    $g(x) = \dfrac{1}{\sqrt{x + 2}}$

**36.** $f(x) = 2 - x$,    $g(x) = \sqrt[3]{x + 1}$

In Exercises 37 and 38, **(a)** write formulas for $f \circ g$ and $g \circ f$ and find the **(b)** domain and **(c)** range of each.

**37.** $f(x) = 2 - x^2$,    $g(x) = \sqrt{x + 2}$

**38.** $f(x) = \sqrt{x}$,    $g(x) = \sqrt{1 - x}$

For Exercises 39 and 40, sketch the graphs of $f$ and $f \circ f$.

**39.** $f(x) = \begin{cases} -x - 2, & -4 \le x \le -1 \\ -1, & -1 < x \le 1 \\ x - 2, & 1 < x \le 2 \end{cases}$

**40.** $f(x) = \begin{cases} x + 1, & -2 \le x < 0 \\ x - 1, & 0 \le x \le 2 \end{cases}$

**Composition with absolute values** In Exercises 41–48, graph $f_1$ and $f_2$ together. Then describe how applying the absolute value function in $f_2$ affects the graph of $f_1$.

| | $f_1(x)$ | $f_2(x)$ |
|---|---|---|
| **41.** | $x$ | $|x|$ |
| **42.** | $x^2$ | $|x|^2$ |
| **43.** | $x^3$ | $|x^3|$ |
| **44.** | $x^2 + x$ | $|x^2 + x|$ |
| **45.** | $4 - x^2$ | $|4 - x^2|$ |
| **46.** | $\dfrac{1}{x}$ | $\dfrac{1}{|x|}$ |
| **47.** | $\sqrt{x}$ | $\sqrt{|x|}$ |
| **48.** | $\sin x$ | $\sin |x|$ |

**Shifting and Scaling Graphs**

**49.** Suppose the graph of $g$ is given. Write equations for the graphs that are obtained from the graph of $g$ by shifting, scaling, or reflecting, as indicated.

    **a.** Up $\dfrac{1}{2}$ unit, right 3

    **b.** Down 2 units, left $\dfrac{2}{3}$

    **c.** Reflect about the $y$-axis

    **d.** Reflect about the $x$-axis

    **e.** Stretch vertically by a factor of 5

    **f.** Compress horizontally by a factor of 5

**50.** Describe how each graph is obtained from the graph of $y = f(x)$.

    **a.** $y = f(x - 5)$     **b.** $y = f(4x)$

    **c.** $y = f(-3x)$     **d.** $y = f(2x + 1)$

    **e.** $y = f\left(\dfrac{x}{3}\right) - 4$     **f.** $y = -3f(x) + \dfrac{1}{4}$

In Exercises 51–54, graph each function, not by plotting points, but by starting with the graph of one of the standard functions presented in Figures 1.15–1.17, and applying an appropriate transformation.

**51.** $y = -\sqrt{1 + \dfrac{x}{2}}$      **52.** $y = 1 - \dfrac{x}{3}$

**53.** $y = \dfrac{1}{2x^2} + 1$      **54.** $y = (-5x)^{1/3}$

**Trigonometry**

In Exercises 55–58, sketch the graph of the given function. What is the period of the function?

**55.** $y = \cos 2x$      **56.** $y = \sin \dfrac{x}{2}$

**57.** $y = \sin \pi x$      **58.** $y = \cos \dfrac{\pi x}{2}$

**59.** Sketch the graph $y = 2 \cos \left( x - \dfrac{\pi}{3} \right)$.

**60.** Sketch the graph $y = 1 + \sin \left( x + \dfrac{\pi}{4} \right)$.

In Exercises 61–64, $ABC$ is a right triangle with the right angle at $C$. The sides opposite angles $A$, $B$, and $C$ are $a$, $b$, and $c$, respectively.

**61. a.** Find $a$ and $b$ if $c = 2$, $B = \pi/3$.
    **b.** Find $a$ and $c$ if $b = 2$, $B = \pi/3$.

**62. a.** Express $a$ in terms of $A$ and $c$.
    **b.** Express $a$ in terms of $A$ and $b$.

**63. a.** Express $a$ in terms of $B$ and $b$.
    **b.** Express $c$ in terms of $A$ and $a$.

**64. a.** Express $\sin A$ in terms of $a$ and $c$.
    **b.** Express $\sin A$ in terms of $b$ and $c$.

**65. Height of a pole** Two wires stretch from the top $T$ of a vertical pole to points $B$ and $C$ on the ground, where $C$ is 10 m closer to the base of the pole than is $B$. If wire $BT$ makes an angle of 35° with the horizontal and wire $CT$ makes an angle of 50° with the horizontal, how high is the pole?

**66. Height of a weather balloon** Observers at positions $A$ and $B$ 2 km apart simultaneously measure the angle of elevation of a weather balloon to be 40° and 70°, respectively. If the balloon is directly above a point on the line segment between $A$ and $B$, find the height of the balloon.

T **67. a.** Graph the function $f(x) = \sin x + \cos(x/2)$.
    **b.** What appears to be the period of this function?
    **c.** Confirm your finding in part (b) algebraically.

T **68. a.** Graph $f(x) = \sin (1/x)$.
    **b.** What are the domain and range of $f$?
    **c.** Is $f$ periodic? Give reasons for your answer.

# Chapter **1**   Additional and Advanced Exercises

**Functions and Graphs**

**1.** Are there two functions $f$ and $g$ such that $f \circ g = g \circ f$? Give reasons for your answer.

**2.** Are there two functions $f$ and $g$ with the following property? The graphs of $f$ and $g$ are not straight lines but the graph of $f \circ g$ is a straight line. Give reasons for your answer.

**3.** If $f(x)$ is odd, can anything be said of $g(x) = f(x) - 2$? What if $f$ is even instead? Give reasons for your answer.

**4.** If $g(x)$ is an odd function defined for all values of $x$, can anything be said about $g(0)$? Give reasons for your answer.

**5.** Graph the equation $|x| + |y| = 1 + x$.

**6.** Graph the equation $y + |y| = x + |x|$.

**Derivations and Proofs**

**7.** Prove the following identities.

    **a.** $\dfrac{1 - \cos x}{\sin x} = \dfrac{\sin x}{1 + \cos x}$    **b.** $\dfrac{1 - \cos x}{1 + \cos x} = \tan^2 \dfrac{x}{2}$

**8.** Explain the following "proof without words" of the law of cosines. (*Source:* Kung, Sidney H., "Proof Without Words: The Law of Cosines," *Mathematics Magazine*, Vol. 63, no. 5, Dec. 1990, p. 342.)

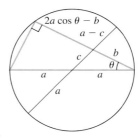

**9.** Show that the area of triangle $ABC$ is given by $(1/2)ab \sin C = (1/2)bc \sin A = (1/2)ca \sin B$.

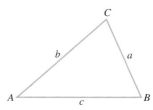

**10.** Show that the area of triangle $ABC$ is given by $\sqrt{s(s - a)(s - b)(s - c)}$ where $s = (a + b + c)/2$ is the semiperimeter of the triangle.

**11.** Show that if $f$ is both even and odd, then $f(x) = 0$ for every $x$ in the domain of $f$.

**12. a. Even-odd decompositions** Let $f$ be a function whose domain is symmetric about the origin, that is, $-x$ belongs to the domain whenever $x$ does. Show that $f$ is the sum of an even function and an odd function:

$$f(x) = E(x) + O(x),$$

where $E$ is an even function and $O$ is an odd function. (*Hint:* Let $E(x) = (f(x) + f(-x))/2$. Show that $E(-x) = E(x)$, so that $E$ is even. Then show that $O(x) = f(x) - E(x)$ is odd.)

**b. Uniqueness** Show that there is only one way to write $f$ as the sum of an even and an odd function. (*Hint:* One way is given in part (a). If also $f(x) = E_1(x) + O_1(x)$ where $E_1$ is even and $O_1$ is odd, show that $E - E_1 = O_1 - O$. Then use Exercise 11 to show that $E = E_1$ and $O = O_1$.)

### Effects of Parameters on Graphs

[T] **13.** What happens to the graph of $y = ax^2 + bx + c$ as

**a.** $a$ changes while $b$ and $c$ remain fixed?

**b.** $b$ changes ($a$ and $c$ fixed, $a \neq 0$)?

**c.** $c$ changes ($a$ and $b$ fixed, $a \neq 0$)?

[T] **14.** What happens to the graph of $y = a(x + b)^3 + c$ as

**a.** $a$ changes while $b$ and $c$ remain fixed?

**b.** $b$ changes ($a$ and $c$ fixed, $a \neq 0$)?

**c.** $c$ changes ($a$ and $b$ fixed, $a \neq 0$)?

### Geometry

**15.** An object's center of mass moves at a constant velocity $v$ along a straight line past the origin. The accompanying figure shows the coordinate system and the line of motion. The dots show positions that are 1 sec apart. Why are the areas $A_1, A_2, \ldots, A_5$ in the figure all equal? As in Kepler's equal area law (see Section 13.6), the line that joins the object's center of mass to the origin sweeps out equal areas in equal times.

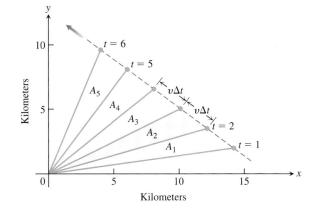

**16. a.** Find the slope of the line from the origin to the midpoint $P$ of side $AB$ in the triangle in the accompanying figure ($a, b > 0$).

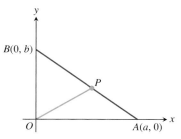

**b.** When is $OP$ perpendicular to $AB$?

**17.** Consider the quarter-circle of radius 1 and right triangles $ABE$ and $ACD$ given in the accompanying figure. Use standard area formulas to conclude that

$$\frac{1}{2}\sin\theta\cos\theta < \frac{\theta}{2} < \frac{1}{2}\frac{\sin\theta}{\cos\theta}.$$

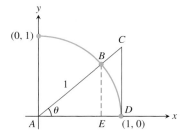

**18.** Let $f(x) = ax + b$ and $g(x) = cx + d$. What condition must be satisfied by the constants $a, b, c, d$ in order that $(f \circ g)(x) = (g \circ f)(x)$ for every value of $x$?

## Chapter 1    Technology Application Projects

### *An Overview of Mathematica*
An overview of *Mathematica* sufficient to complete the *Mathematica* modules appearing on the Web site.

Mathematica/Maple Module:

### *Modeling Change: Springs, Driving Safety, Radioactivity, Trees, Fish, and Mammals*
Construct and interpret mathematical models, analyze and improve them, and make predictions using them.

# 2

# Limits and Continuity

**OVERVIEW** Mathematicians of the seventeenth century were keenly interested in the study of motion for objects on or near the earth and the motion of planets and stars. This study involved both the speed of the object and its direction of motion at any instant, and they knew the direction at a given instant was along a line tangent to the path of motion. The concept of a limit is fundamental to finding the velocity of a moving object and the tangent to a curve. In this chapter we develop the limit, first intuitively and then formally. We use limits to describe the way a function varies. Some functions vary *continuously*; small changes in x produce only small changes in f(x). Other functions can have values that jump, vary erratically, or tend to increase or decrease without bound. The notion of limit gives a precise way to distinguish between these behaviors.

## 2.1 Rates of Change and Tangents to Curves

Calculus is a tool that helps us understand how a change in one quantity is related to a change in another. How does the speed of a falling object change as a function of time? How does the level of water in a barrel change as a function of the amount of liquid poured into it? We see change occurring in nearly everything we observe in the world and universe, and powerful modern instruments help us see more and more. In this section we introduce the ideas of average and instantaneous rates of change, and show that they are closely related to the slope of a curve at a point P on the curve. We give precise developments of these important concepts in the next chapter, but for now we use an informal approach so you will see how they lead naturally to the main idea of this chapter, the *limit*. The idea of a limit plays a foundational role throughout calculus.

### Average and Instantaneous Speed

**HISTORICAL BIOGRAPHY***

Galileo Galilei
(1564–1642)

In the late sixteenth century, Galileo discovered that a solid object dropped from rest (not moving) near the surface of the earth and allowed to fall freely will fall a distance proportional to the square of the time it has been falling. This type of motion is called **free fall**. It assumes negligible air resistance to slow the object down, and that gravity is the only force acting on the falling object. If $y$ denotes the distance fallen in meters after $t$ seconds, then Galileo's law is

$$y = 4.9t^2,$$

where 4.9 is the (approximate) constant of proportionality.

   A moving object's **average speed** during an interval of time is found by dividing the distance covered by the time elapsed. The unit of measure is length per unit time: kilometers per hour, meters per second, or whatever is appropriate to the problem at hand.

---

*To learn more about the historical figures mentioned in the text and the development of many major elements and topics of calculus, visit **www.aw.com/thomas**.

**EXAMPLE 1**     A rock breaks loose from the top of a tall cliff. What is its average speed

**(a)** during the first 2 s of fall?

**(b)** during the 1-s interval between second 1 and second 2?

**Solution**     The average speed of the rock during a given time interval is the change in distance, $\Delta y$, divided by the length of the time interval, $\Delta t$. (Increments like $\Delta y$ and $\Delta t$ are reviewed in Appendix 3, and pronounced "delta y" and "delta t.") Measuring distance in meters and time in seconds, we have the following calculations:

**(a)**   For the first 2 s:      $\dfrac{\Delta y}{\Delta t} = \dfrac{4.9(2)^2 - 4.9(0)^2}{2 - 0} = 9.8\dfrac{m}{s}$

**(b)**   From s 1 to s 2:      $\dfrac{\Delta y}{\Delta t} = \dfrac{4.9(2)^2 - 4.9(1)^2}{2 - 1} = 14.7\dfrac{m}{s}$      ∎

We want a way to determine the speed of a falling object at a single instant $t_0$, instead of using its average speed over an interval of time. To do this, we examine what happens when we calculate the average speed over shorter and shorter time intervals starting at $t_0$. The next example illustrates this process. Our discussion is informal here, but it will be made precise in Chapter 3.

**EXAMPLE 2**     Find the speed of the falling rock in Example 1 at $t = 1$ and $t = 2$ s.

**Solution**     We can calculate the average speed of the rock over a time interval $[t_0, t_0 + h]$, having length $\Delta t = h$, as

$$\frac{\Delta y}{\Delta t} = \frac{4.9(t_0 + h)^2 - 4.9t_0^2}{h}. \tag{1}$$

We cannot use this formula to calculate the "instantaneous" speed at the exact moment $t_0$ by simply substituting $h = 0$, because we cannot divide by zero. But we *can* use it to calculate average speeds over increasingly short time intervals starting at $t_0 = 1$ and $t_0 = 2$. When we do so, by taking smaller and smaller values of $h$, we see a pattern (Table 2.1).

**TABLE 2.1** Average speeds over short time intervals $[t_0, t_0 + h]$

Average speed: $\dfrac{\Delta y}{\Delta t} = \dfrac{4.9(t_0 + h)^2 - 4.9t_0^2}{h}$

| Length of time interval $h$ | Average speed over interval of length $h$ starting at $t_0 = 1$ | Average speed over interval of length $h$ starting at $t_0 = 2$ |
|---|---|---|
| 1 | 14.7 | 24.5 |
| 0.1 | 10.29 | 20.09 |
| 0.01 | 9.849 | 19.649 |
| 0.001 | 9.8049 | 19.6049 |
| 0.0001 | 9.80049 | 19.60049 |

The average speed on intervals starting at $t_0 = 1$ seems to approach a limiting value of 9.8 as the length of the interval decreases. This suggests that the rock is falling at a speed of 9.8 m/s at $t_0 = 1$ s. Let's confirm this algebraically.

If we set $t_0 = 1$ and then expand the numerator in Equation (1) and simplify, we find that

$$\frac{\Delta y}{\Delta t} = \frac{4.9(1 + h)^2 - 4.9(1)^2}{h} = \frac{4.9(1 + 2h + h^2) - 4.9}{h}$$

$$= \frac{9.8h + 4.9h^2}{h} = 9.8 + 4.9h.$$

For values of $h$ different from 0, the expressions on the right and left are equivalent and the average speed is $9.8 + 4.9h$ m/s. We can now see why the average speed has the limiting value $9.8 + 4.9(0) = 9.8$ m/s as $h$ approaches 0.

Similarly, setting $t_0 = 2$ in Equation (1), the procedure yields

$$\frac{\Delta y}{\Delta t} = 19.6 + 4.9h$$

for values of $h$ different from 0. As $h$ gets closer and closer to 0, the average speed has the limiting value 19.6 m/s when $t_0 = 2$ s, as suggested by Table 2.1. ∎

The average speed of a falling object is an example of a more general idea which we discuss next.

## Average Rates of Change and Secant Lines

Given any function $y = f(x)$, we calculate the average rate of change of $y$ with respect to $x$ over the interval $[x_1, x_2]$ by dividing the change in the value of $y$, $\Delta y = f(x_2) - f(x_1)$, by the length $\Delta x = x_2 - x_1 = h$ of the interval over which the change occurs. (We use the symbol $h$ for $\Delta x$ to simplify the notation here and later on.)

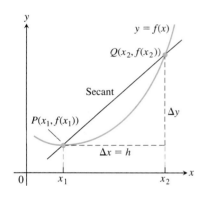

FIGURE 2.1　A secant to the graph $y = f(x)$. Its slope is $\Delta y/\Delta x$, the average rate of change of $f$ over the interval $[x_1, x_2]$.

---

> **DEFINITION**　The **average rate of change** of $y = f(x)$ with respect to $x$ over the interval $[x_1, x_2]$ is
>
> $$\frac{\Delta y}{\Delta x} = \frac{f(x_2) - f(x_1)}{x_2 - x_1} = \frac{f(x_1 + h) - f(x_1)}{h}, \qquad h \neq 0.$$

---

Geometrically, the rate of change of $f$ over $[x_1, x_2]$ is the slope of the line through the points $P(x_1, f(x_1))$ and $Q(x_2, f(x_2))$ (Figure 2.1). In geometry, a line joining two points of a curve is a **secant** to the curve. Thus, the average rate of change of $f$ from $x_1$ to $x_2$ is identical with the slope of secant $PQ$. Let's consider what happens as the point $Q$ approaches the point $P$ along the curve, so the length $h$ of the interval over which the change occurs approaches zero. We will see that this procedure leads to defining the slope of a curve at a point.

## Defining the Slope of a Curve

We know what is meant by the slope of a straight line, which tells us the rate at which it rises or falls—its rate of change as a linear function. But what is meant by the *slope of a curve* at a point $P$ on the curve? If there is a *tangent* line to the curve at $P$—a line that just touches the curve like the tangent to a circle—it would be reasonable to identify *the slope of the tangent* as the slope of the curve at $P$. So we need a precise meaning for the tangent at a point on a curve.

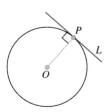

FIGURE 2.2　$L$ is tangent to the circle at $P$ if it passes through $P$ perpendicular to radius $OP$.

For circles, tangency is straightforward. A line $L$ is tangent to a circle at a point $P$ if $L$ passes through $P$ perpendicular to the radius at $P$ (Figure 2.2). Such a line just *touches* the circle. But what does it mean to say that a line $L$ is tangent to some other curve $C$ at a point $P$?

To define tangency for general curves, we need an approach that takes into account the behavior of the secants through $P$ and nearby points $Q$ as $Q$ moves toward $P$ along the curve (Figure 2.3). Here is the idea:

1. Start with what we *can* calculate, namely the slope of the secant $PQ$.

2. Investigate the limiting value of the secant slope as $Q$ approaches $P$ along the curve. (We clarify the *limit* idea in the next section.)

3. If the *limit* exists, take it to be the slope of the curve at $P$ and *define* the tangent to the curve at $P$ to be the line through $P$ with this slope.

**HISTORICAL BIOGRAPHY**

Pierre de Fermat
(1601–1665)

This procedure is what we were doing in the falling-rock problem discussed in Example 2. The next example illustrates the geometric idea for the tangent to a curve.

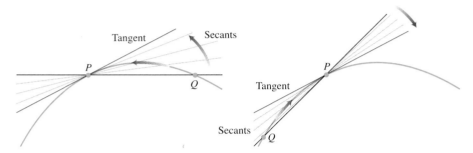

**FIGURE 2.3** The tangent to the curve at $P$ is the line through $P$ whose slope is the limit of the secant slopes as $Q \rightarrow P$ from either side.

**EXAMPLE 3**    Find the slope of the parabola $y = x^2$ at the point $P(2, 4)$. Write an equation for the tangent to the parabola at this point.

**Solution**    We begin with a secant line through $P(2, 4)$ and $Q(2 + h, (2 + h)^2)$ nearby. We then write an expression for the slope of the secant $PQ$ and investigate what happens to the slope as $Q$ approaches $P$ along the curve:

$$\text{Secant slope} = \frac{\Delta y}{\Delta x} = \frac{(2 + h)^2 - 2^2}{h} = \frac{h^2 + 4h + 4 - 4}{h}$$

$$= \frac{h^2 + 4h}{h} = h + 4.$$

If $h > 0$, then $Q$ lies above and to the right of $P$, as in Figure 2.4. If $h < 0$, then $Q$ lies to the left of $P$ (not shown). In either case, as $Q$ approaches $P$ along the curve, $h$ approaches zero and the secant slope $h + 4$ approaches 4. We take 4 to be the parabola's slope at $P$.

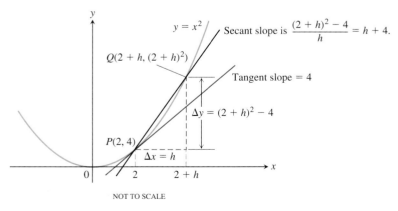

NOT TO SCALE

**FIGURE 2.4**    Finding the slope of the parabola $y = x^2$ at the point $P(2, 4)$ as the limit of secant slopes (Example 3).

The tangent to the parabola at $P$ is the line through $P$ with slope 4:

$$y = 4 + 4(x - 2) \qquad \text{Point-slope equation}$$

$$y = 4x - 4.$$

### Instantaneous Rates of Change and Tangent Lines

The rates at which the rock in Example 2 was falling at the instants $t = 1$ and $t = 2$ are called *instantaneous rates of change*. Instantaneous rates and slopes of tangent lines are closely connected, as we see in the following examples.

**EXAMPLE 4**    Figure 2.5 shows how a population $p$ of fruit flies (*Drosophila*) grew in a 50-day experiment. The number of flies was counted at regular intervals, the counted values plotted with respect to time $t$, and the points joined by a smooth curve (colored blue in Figure 2.5). Find the average growth rate from day 23 to day 45.

**Solution**    There were 150 flies on day 23 and 340 flies on day 45. Thus the number of flies increased by $340 - 150 = 190$ in $45 - 23 = 22$ days. The average rate of change of the population from day 23 to day 45 was

$$\text{Average rate of change: } \frac{\Delta p}{\Delta t} = \frac{340 - 150}{45 - 23} = \frac{190}{22} \approx 8.6 \text{ flies/day.}$$

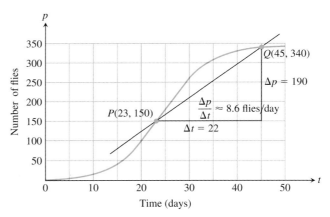

**FIGURE 2.5**    Growth of a fruit fly population in a controlled experiment. The average rate of change over 22 days is the slope $\Delta p/\Delta t$ of the secant line (Example 4).

This average is the slope of the secant through the points $P$ and $Q$ on the graph in Figure 2.5.

The average rate of change from day 23 to day 45 calculated in Example 4 does not tell us how fast the population was changing on day 23 itself. For that we need to examine time intervals closer to the day in question.

**EXAMPLE 5**    How fast was the number of flies in the population of Example 4 growing on day 23?

**Solution**    To answer this question, we examine the average rates of change over increasingly short time intervals starting at day 23. In geometric terms, we find these rates by calculating the slopes of secants from $P$ to $Q$, for a sequence of points $Q$ approaching $P$ along the curve (Figure 2.6).

| $Q$ | Slope of $PQ = \Delta p / \Delta t$ (flies / day) |
|---|---|
| $(45, 340)$ | $\dfrac{340 - 150}{45 - 23} \approx 8.6$ |
| $(40, 330)$ | $\dfrac{330 - 150}{40 - 23} \approx 10.6$ |
| $(35, 310)$ | $\dfrac{310 - 150}{35 - 23} \approx 13.3$ |
| $(30, 265)$ | $\dfrac{265 - 150}{30 - 23} \approx 16.4$ |

**FIGURE 2.6** The positions and slopes of four secants through the point $P$ on the fruit fly graph (Example 5).

The values in the table show that the secant slopes rise from 8.6 to 16.4 as the $t$-coordinate of $Q$ decreases from 45 to 30, and we would expect the slopes to rise slightly higher as $t$ continued on toward 23. Geometrically, the secants rotate counterclockwise about $P$ and seem to approach the red tangent line in the figure. Since the line appears to pass through the points $(14, 0)$ and $(35, 350)$, it has slope

$$\frac{350 - 0}{35 - 14} = 16.7 \text{ flies/day (approximately).}$$

On day 23 the population was increasing at a rate of about 16.7 flies/day. ∎

The instantaneous rates in Example 2 were found to be the values of the average speeds, or average rates of change, as the time interval of length $h$ approached 0. That is, the instantaneous rate is the value the average rate approaches as the length $h$ of the interval over which the change occurs approaches zero. The average rate of change corresponds to the slope of a secant line; the instantaneous rate corresponds to the slope of the tangent line as the independent variable approaches a fixed value. In Example 2, the independent variable $t$ approached the values $t = 1$ and $t = 2$. In Example 3, the independent variable $x$ approached the value $x = 2$. So we see that instantaneous rates and slopes of tangent lines are closely connected. We investigate this connection thoroughly in the next chapter, but to do so we need the concept of a *limit*.

# Exercises 2.1

**Average Rates of Change**

In Exercises 1–6, find the average rate of change of the function over the given interval or intervals.

1. $f(x) = x^3 + 1$
   a. $[2, 3]$   b. $[-1, 1]$
2. $g(x) = x^2 - 2x$
   a. $[1, 3]$   b. $[-2, 4]$
3. $h(t) = \cot t$
   a. $[\pi/4, 3\pi/4]$   b. $[\pi/6, \pi/2]$
4. $g(t) = 2 + \cos t$
   a. $[0, \pi]$   b. $[-\pi, \pi]$

5. $R(\theta) = \sqrt{4\theta + 1}; \quad [0, 2]$
6. $P(\theta) = \theta^3 - 4\theta^2 + 5\theta; \quad [1, 2]$

**Slope of a Curve at a Point**

In Exercises 7–14, use the method in Example 3 to find **(a)** the slope of the curve at the given point $P$, and **(b)** an equation of the tangent line at $P$.

7. $y = x^2 - 5, \quad P(2, -1)$
8. $y = 7 - x^2, \quad P(2, 3)$
9. $y = x^2 - 2x - 3, \quad P(2, -3)$
10. $y = x^2 - 4x, \quad P(1, -3)$
11. $y = x^3, \quad P(2, 8)$

**12.** $y = 2 - x^3$, $P(1, 1)$

**13.** $y = x^3 - 12x$, $P(1, -11)$

**14.** $y = x^3 - 3x^2 + 4$, $P(2, 0)$

Instantaneous Rates of Change

**15. Speed of a car** The accompanying figure shows the time-to-distance graph for a sports car accelerating from a standstill.

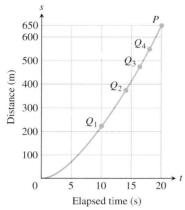

Elapsed time (s)

**a.** Estimate the slopes of secants $PQ_1$, $PQ_2$, $PQ_3$, and $PQ_4$, arranging them in order in a table like the one in Figure 2.6. What are the appropriate units for these slopes?

**b.** Then estimate the car's speed at time $t = 20\,\text{s}$.

**16.** The accompanying figure shows the plot of distance fallen versus time for an object that fell from the lunar landing module a distance 80 m to the surface of the moon.

**a.** Estimate the slopes of the secants $PQ_1$, $PQ_2$, $PQ_3$, and $PQ_4$, arranging them in a table like the one in Figure 2.6.

**b.** About how fast was the object going when it hit the surface?

Elapsed time (s)

**T 17.** The profits of a small company for each of the first five years of its operation are given in the following table:

| Year | Profit in $1000s |
|------|------------------|
| 2010 | 6 |
| 2011 | 27 |
| 2012 | 62 |
| 2013 | 111 |
| 2014 | 174 |

**a.** Plot points representing the profit as a function of year, and join them by as smooth a curve as you can.

**b.** What is the average rate of increase of the profits between 2012 and 2014?

**c.** Use your graph to estimate the rate at which the profits were changing in 2012.

**T 18.** Make a table of values for the function $F(x) = (x + 2)/(x - 2)$ at the points $x = 1.2$, $x = 11/10$, $x = 101/100$, $x = 1001/1000$, $x = 10001/10000$, and $x = 1$.

**a.** Find the average rate of change of $F(x)$ over the intervals $[1, x]$ for each $x \neq 1$ in your table.

**b.** Extending the table if necessary, try to determine the rate of change of $F(x)$ at $x = 1$.

**T 19.** Let $g(x) = \sqrt{x}$ for $x \geq 0$.

**a.** Find the average rate of change of $g(x)$ with respect to $x$ over the intervals $[1, 2]$, $[1, 1.5]$ and $[1, 1 + h]$.

**b.** Make a table of values of the average rate of change of $g$ with respect to $x$ over the interval $[1, 1 + h]$ for some values of $h$ approaching zero, say $h = 0.1, 0.01, 0.001, 0.0001, 0.00001$, and $0.000001$.

**c.** What does your table indicate is the rate of change of $g(x)$ with respect to $x$ at $x = 1$?

**d.** Calculate the limit as $h$ approaches zero of the average rate of change of $g(x)$ with respect to $x$ over the interval $[1, 1 + h]$.

**T 20.** Let $f(t) = 1/t$ for $t \neq 0$.

**a.** Find the average rate of change of $f$ with respect to $t$ over the intervals (i) from $t = 2$ to $t = 3$, and (ii) from $t = 2$ to $t = T$.

**b.** Make a table of values of the average rate of change of $f$ with respect to $t$ over the interval $[2, T]$, for some values of $T$ approaching 2, say $T = 2.1, 2.01, 2.001, 2.0001, 2.00001$, and $2.000001$.

**c.** What does your table indicate is the rate of change of $f$ with respect to $t$ at $t = 2$?

**d.** Calculate the limit as $T$ approaches 2 of the average rate of change of $f$ with respect to $t$ over the interval from 2 to $T$. You will have to do some algebra before you can substitute $T = 2$.

**21.** The accompanying graph shows the total distance $s$ traveled by a bicyclist after $t$ hours.

Elapsed time (h)

**a.** Estimate the bicyclist's average speed over the time intervals $[0, 1]$, $[1, 2.5]$, and $[2.5, 3.5]$.

**b.** Estimate the bicyclist's instantaneous speed at the times $t = \frac{1}{2}$, $t = 2$, and $t = 3$.

**c.** Estimate the bicyclist's maximum speed and the specific time at which it occurs.

**22.** The accompanying graph shows the total amount of gasoline $A$ in the gas tank of a motorcycle after being driven for $t$ days.

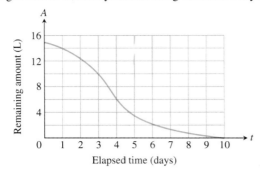

**a.** Estimate the average rate of gasoline consumption over the time intervals $[0, 3]$, $[0, 5]$, and $[7, 10]$.

**b.** Estimate the instantaneous rate of gasoline consumption at the times $t = 1$, $t = 4$, and $t = 8$.

**c.** Estimate the maximum rate of gasoline consumption and the specific time at which it occurs.

# 2.2 Limit of a Function and Limit Laws

In Section 2.1 we saw that limits arise when finding the instantaneous rate of change of a function or the tangent to a curve. Here we begin with an informal definition of *limit* and show how we can calculate the values of limits. A precise definition is presented in the next section.

**HISTORICAL ESSAY**

Limits

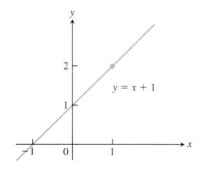

**FIGURE 2.7** The graph of $f$ is identical with the line $y = x + 1$ except at $x = 1$, where $f$ is not defined (Example 1).

## Limits of Function Values

Frequently when studying a function $y = f(x)$, we find ourselves interested in the function's behavior *near* a particular point $c$, but not *at* $c$. This might be the case, for instance, if $c$ is an irrational number, like $\pi$ or $\sqrt{2}$, whose values can only be approximated by "close" rational numbers at which we actually evaluate the function instead. Another situation occurs when trying to evaluate a function at $c$ leads to division by zero, which is undefined. We encountered this last circumstance when seeking the instantaneous rate of change in $y$ by considering the quotient function $\Delta y/h$ for $h$ closer and closer to zero. Here's a specific example in which we explore numerically how a function behaves near a particular point at which we cannot directly evaluate the function.

**EXAMPLE 1**    How does the function

$$f(x) = \frac{x^2 - 1}{x - 1}$$

behave near $x = 1$?

**Solution**    The given formula defines $f$ for all real numbers $x$ except $x = 1$ (we cannot divide by zero). For any $x \neq 1$, we can simplify the formula by factoring the numerator and canceling common factors:

$$f(x) = \frac{(x - 1)(x + 1)}{x - 1} = x + 1 \qquad \text{for} \qquad x \neq 1.$$

The graph of $f$ is the line $y = x + 1$ with the point $(1, 2)$ *removed*. This removed point is shown as a "hole" in Figure 2.7. Even though $f(1)$ is not defined, it is clear that we can make the value of $f(x)$ *as close as we want* to 2 by choosing $x$ *close enough* to 1 (Table 2.2).    ∎

| TABLE 2.2 As $x$ gets closer to 1, $f(x)$ gets closer to 2. | |
| --- | --- |
| $x$ | $f(x) = \dfrac{x^2 - 1}{x - 1}$ |
| 0.9 | 1.9 |
| 1.1 | 2.1 |
| 0.99 | 1.99 |
| 1.01 | 2.01 |
| 0.999 | 1.999 |
| 1.001 | 2.001 |
| 0.999999 | 1.999999 |
| 1.000001 | 2.000001 |

Generalizing the idea illustrated in Example 1, suppose $f(x)$ is defined on an open interval about $c$, *except possibly at c itself*. If $f(x)$ is arbitrarily close to the number $L$ (as close to $L$ as we like) for all $x$ sufficiently close to $c$, we say that $f$ approaches the **limit** $L$ as $x$ approaches $c$, and write

$$\lim_{x \to c} f(x) = L,$$

which is read "the limit of $f(x)$ as $x$ approaches $c$ is $L$." For instance, in Example 1 we would say that $f(x)$ approaches the *limit* 2 as $x$ approaches 1, and write

$$\lim_{x \to 1} f(x) = 2, \qquad \text{or} \qquad \lim_{x \to 1} \frac{x^2 - 1}{x - 1} = 2.$$

Essentially, the definition says that the values of $f(x)$ are close to the number $L$ whenever $x$ is close to $c$ (on either side of $c$).

Our definition here is "informal" because phrases like *arbitrarily close* and *sufficiently close* are imprecise; their meaning depends on the context. (To a machinist manufacturing a piston, *close* may mean *within a few hundredths of a millimeter*. To an astronomer studying distant galaxies, *close* may mean *within a few thousand light-years*.) Nevertheless, the definition is clear enough to enable us to recognize and evaluate limits of many specific functions. We will need the precise definition given in Section 2.3, however, when we set out to prove theorems about limits or study complicated functions. Here are several more examples exploring the idea of limits.

**EXAMPLE 2**     The limit value of a function does not depend on how the function is defined at the point being approached. Consider the three functions in Figure 2.8. The function $f$ has limit 2 as $x \to 1$ even though $f$ is not defined at $x = 1$. The function $g$ has limit 2 as $x \to 1$ even though $2 \neq g(1)$. The function $h$ is the only one of the three functions in Figure 2.8 whose limit as $x \to 1$ equals its value at $x = 1$. For $h$, we have $\lim_{x \to 1} h(x) = h(1)$. This equality of limit and function value is of special importance, and we return to it in Section 2.5. ∎

(a) Identity function

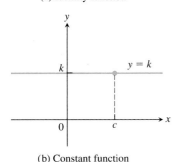

(b) Constant function

**FIGURE 2.9** The functions in Example 3 have limits at all points $c$.

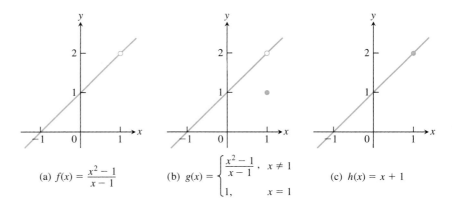

(a) $f(x) = \dfrac{x^2 - 1}{x - 1}$     (b) $g(x) = \begin{cases} \dfrac{x^2 - 1}{x - 1}, & x \neq 1 \\ 1, & x = 1 \end{cases}$     (c) $h(x) = x + 1$

**FIGURE 2.8** The limits of $f(x)$, $g(x)$, and $h(x)$ all equal 2 as $x$ approaches 1. However, only $h(x)$ has the same function value as its limit at $x = 1$ (Example 2).

**EXAMPLE 3**
**(a)** If $f$ is the **identity function** $f(x) = x$, then for any value of $c$ (Figure 2.9a),

$$\lim_{x \to c} f(x) = \lim_{x \to c} x = c.$$

**(b)** If $f$ is the **constant function** $f(x) = k$ (function with the constant value $k$), then for any value of $c$ (Figure 2.9b),

$$\lim_{x \to c} f(x) = \lim_{x \to c} k = k.$$

For instances of each of these rules we have

$$\lim_{x \to 3} x = 3 \quad \text{and} \quad \lim_{x \to -7} (4) = \lim_{x \to 2} (4) = 4.$$

We prove these rules in Example 3 in Section 2.3. ∎

A function may not have a limit at a particular point. Some ways that limits can fail to exist are illustrated in Figure 2.10 and described in the next example.

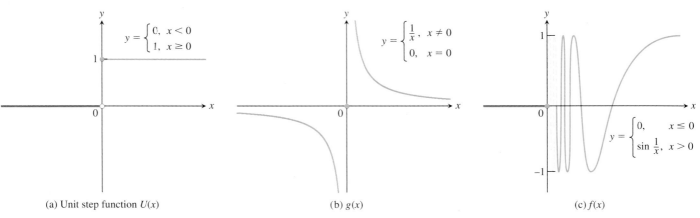

(a) Unit step function $U(x)$        (b) $g(x)$        (c) $f(x)$

**FIGURE 2.10**   None of these functions has a limit as $x$ approaches 0 (Example 4).

**EXAMPLE 4**    Discuss the behavior of the following functions, explaining why they have no limit as $x \to 0$.

**(a)**   $U(x) = \begin{cases} 0, & x < 0 \\ 1, & x \geq 0 \end{cases}$

**(b)**   $g(x) = \begin{cases} \dfrac{1}{x}, & x \neq 0 \\ 0, & x = 0 \end{cases}$

**(c)**   $f(x) = \begin{cases} 0, & x \leq 0 \\ \sin\dfrac{1}{x}, & x > 0 \end{cases}$

**Solution**

**(a)** It *jumps*: The **unit step function** $U(x)$ has no limit as $x \to 0$ because its values jump at $x = 0$. For negative values of $x$ arbitrarily close to zero, $U(x) = 0$. For positive values of $x$ arbitrarily close to zero, $U(x) = 1$. There is no *single* value $L$ approached by $U(x)$ as $x \to 0$ (Figure 2.10a).

**(b)** It *grows too "large" to have a limit*: $g(x)$ has no limit as $x \to 0$ because the values of $g$ grow arbitrarily large in absolute value as $x \to 0$ and do not stay close to *any* fixed real number (Figure 2.10b). We say the function is *not bounded*.

**(c)** It *oscillates too much to have a limit*: $f(x)$ has no limit as $x \to 0$ because the function's values oscillate between $+1$ and $-1$ in every open interval containing 0. The values do not stay close to any one number as $x \to 0$ (Figure 2.10c). ∎

## The Limit Laws

To calculate limits of functions that are arithmetic combinations of functions having known limits, we can use several fundamental rules.

---

**THEOREM 1—Limit Laws** If $L, M, c,$ and $k$ are real numbers and

$$\lim_{x \to c} f(x) = L \quad \text{and} \quad \lim_{x \to c} g(x) = M, \quad \text{then}$$

1. *Sum Rule:* $\quad\lim_{x \to c}(f(x) + g(x)) = L + M$

2. *Difference Rule:* $\quad\lim_{x \to c}(f(x) - g(x)) = L - M$

3. *Constant Multiple Rule:* $\quad\lim_{x \to c}(k \cdot f(x)) = k \cdot L$

4. *Product Rule:* $\quad\lim_{x \to c}(f(x) \cdot g(x)) = L \cdot M$

5. *Quotient Rule:* $\quad\lim_{x \to c} \dfrac{f(x)}{g(x)} = \dfrac{L}{M}, \quad M \neq 0$

6. *Power Rule:* $\quad\lim_{x \to c}[f(x)]^n = L^n, n$ a positive integer

7. *Root Rule:* $\quad\lim_{x \to c} \sqrt[n]{f(x)} = \sqrt[n]{L} = L^{1/n}, n$ a positive integer

(If $n$ is even, we assume that $\lim_{x \to c} f(x) = L > 0$.)

---

In words, the Sum Rule says that the limit of a sum is the sum of the limits. Similarly, the next rules say that the limit of a difference is the difference of the limits; the limit of a constant times a function is the constant times the limit of the function; the limit of a product is the product of the limits; the limit of a quotient is the quotient of the limits (provided that the limit of the denominator is not 0); the limit of a positive integer power (or root) of a function is the integer power (or root) of the limit (provided that the root of the limit is a real number).

It is reasonable that the properties in Theorem 1 are true (although these intuitive arguments do not constitute proofs). If $x$ is sufficiently close to $c$, then $f(x)$ is close to $L$ and $g(x)$ is close to $M$, from our informal definition of a limit. It is then reasonable that $f(x) + g(x)$ is close to $L + M$; $f(x) - g(x)$ is close to $L - M$; $kf(x)$ is close to $kL$; $f(x)g(x)$ is close to $LM$; and $f(x)/g(x)$ is close to $L/M$ if $M$ is not zero. We prove the Sum Rule in Section 2.3, based on a precise definition of limit. Rules 2–5 are proved in Appendix 4. Rule 6 is obtained by applying Rule 4 repeatedly. Rule 7 is proved in more advanced texts. The Sum, Difference, and Product Rules can be extended to any number of functions, not just two.

**EXAMPLE 5** Use the observations $\lim_{x \to c} k = k$ and $\lim_{x \to c} x = c$ (Example 3) and the fundamental rules of limits to find the following limits.

**(a)** $\lim_{x \to c}(x^3 + 4x^2 - 3)$

**(b)** $\lim_{x \to c} \dfrac{x^4 + x^2 - 1}{x^2 + 5}$

**(c)** $\lim_{x \to -2} \sqrt{4x^2 - 3}$

**Solution**

**(a)** $\lim_{x \to c}(x^3 + 4x^2 - 3) = \lim_{x \to c} x^3 + \lim_{x \to c} 4x^2 - \lim_{x \to c} 3$    Sum and Difference Rules

$$= c^3 + 4c^2 - 3$$    Power and Multiple Rules

**(b)** $\lim_{x \to c} \dfrac{x^4 + x^2 - 1}{x^2 + 5} = \dfrac{\lim_{x \to c}(x^4 + x^2 - 1)}{\lim_{x \to c}(x^2 + 5)}$    Quotient Rule

$$= \frac{\lim_{x \to c} x^4 + \lim_{x \to c} x^2 - \lim_{x \to c} 1}{\lim_{x \to c} x^2 + \lim_{x \to c} 5}$$    Sum and Difference Rules

$$= \frac{c^4 + c^2 - 1}{c^2 + 5}$$    Power or Product Rule

**(c)** $\lim_{x \to -2} \sqrt{4x^2 - 3} = \sqrt{\lim_{x \to -2}(4x^2 - 3)}$    Root Rule with $n = 2$

$$= \sqrt{\lim_{x \to -2} 4x^2 - \lim_{x \to -2} 3}$$    Difference Rule

$$= \sqrt{4(-2)^2 - 3}$$    Product and Multiple Rules

$$= \sqrt{16 - 3}$$

$$= \sqrt{13}$$    ∎

Theorem 1 simplifies the task of calculating limits of polynomials and rational functions. To evaluate the limit of a polynomial function as $x$ approaches $c$, merely substitute $c$ for $x$ in the formula for the function. To evaluate the limit of a rational function as $x$ approaches a point $c$ *at which the denominator is not zero*, substitute $c$ for $x$ in the formula for the function. (See Examples 5a and 5b.) We state these results formally as theorems.

---

**THEOREM 2—Limits of Polynomials**

If $P(x) = a_n x^n + a_{n-1} x^{n-1} + \cdots + a_0$, then

$$\lim_{x \to c} P(x) = P(c) = a_n c^n + a_{n-1} c^{n-1} + \cdots + a_0.$$

---

**THEOREM 3—Limits of Rational Functions**

If $P(x)$ and $Q(x)$ are polynomials and $Q(c) \neq 0$, then

$$\lim_{x \to c} \frac{P(x)}{Q(x)} = \frac{P(c)}{Q(c)}.$$

---

**EXAMPLE 6**    The following calculation illustrates Theorems 2 and 3:

$$\lim_{x \to -1} \frac{x^3 + 4x^2 - 3}{x^2 + 5} = \frac{(-1)^3 + 4(-1)^2 - 3}{(-1)^2 + 5} = \frac{0}{6} = 0$$    ∎

**Identifying Common Factors**

It can be shown that if $Q(x)$ is a polynomial and $Q(c) = 0$, then $(x - c)$ is a factor of $Q(x)$. Thus, if the numerator and denominator of a rational function of $x$ are both zero at $x = c$, they have $(x - c)$ as a common factor.

## Eliminating Common Factors from Zero Denominators

Theorem 3 applies only if the denominator of the rational function is not zero at the limit point $c$. If the denominator is zero, canceling common factors in the numerator and denominator may reduce the fraction to one whose denominator is no longer zero at $c$. If this happens, we can find the limit by substitution in the simplified fraction.

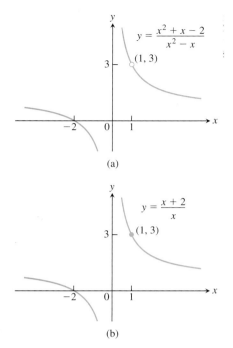

**FIGURE 2.11** The graph of $f(x) = (x^2 + x - 2)/(x^2 - x)$ in part (a) is the same as the graph of $g(x) = (x + 2)/x$ in part (b) except at $x = 1$, where $f$ is undefined. The functions have the same limit as $x \to 1$ (Example 7).

**EXAMPLE 7** Evaluate

$$\lim_{x \to 1} \frac{x^2 + x - 2}{x^2 - x}.$$

**Solution** We cannot substitute $x = 1$ because it makes the denominator zero. We test the numerator to see if it, too, is zero at $x = 1$. It is, so it has a factor of $(x - 1)$ in common with the denominator. Canceling this common factor gives a simpler fraction with the same values as the original for $x \neq 1$:

$$\frac{x^2 + x - 2}{x^2 - x} = \frac{(x - 1)(x + 2)}{x(x - 1)} = \frac{x + 2}{x}, \qquad \text{if } x \neq 1.$$

Using the simpler fraction, we find the limit of these values as $x \to 1$ by Theorem 3:

$$\lim_{x \to 1} \frac{x^2 + x - 2}{x^2 - x} = \lim_{x \to 1} \frac{x + 2}{x} = \frac{1 + 2}{1} = 3.$$

See Figure 2.11. ∎

### Using Calculators and Computers to Estimate Limits

When we cannot use the Quotient Rule in Theorem 1 because the limit of the denominator is zero, we can try using a calculator or computer to guess the limit numerically as $x$ gets closer and closer to $c$. We used this approach in Example 1, but calculators and computers can sometimes give false values and misleading impressions for functions that are undefined at a point or fail to have a limit there. Usually the problem is associated with rounding errors, as we now illustrate.

**EXAMPLE 8** Estimate the value of $\lim\limits_{x \to 0} \dfrac{\sqrt{x^2 + 100} - 10}{x^2}$.

**Solution** Table 2.3 lists values of the function obtained on a calculator for several points approaching $x = 0$. As $x$ approaches 0 through the points $\pm 1$, $\pm 0.5$, $\pm 0.10$, and $\pm 0.01$, the function seems to approach the number 0.05.

As we take even smaller values of $x$, $\pm 0.0005$, $\pm 0.0001$, $\pm 0.00001$, and $\pm 0.000001$, the function appears to approach the number 0.

Is the answer 0.05 or 0, or some other value? We resolve this question in the next example. ∎

**TABLE 2.3** Computed values of $f(x) = \dfrac{\sqrt{x^2 + 100} - 10}{x^2}$ near $x = 0$

| $x$ | $f(x)$ | |
|---|---|---|
| $\pm 1$ | 0.049876 | |
| $\pm 0.5$ | 0.049969 | approaches 0.05? |
| $\pm 0.1$ | 0.049999 | |
| $\pm 0.01$ | 0.050000 | |
| $\pm 0.0005$ | 0.050000 | |
| $\pm 0.0001$ | 0.000000 | approaches 0? |
| $\pm 0.00001$ | 0.000000 | |
| $\pm 0.000001$ | 0.000000 | |

Using a computer or calculator may give ambiguous results, as in the last example. The calculator could not keep track of enough digits to avoid rounding errors in computing the values of $f(x)$ when $x$ is very small. We cannot substitute $x = 0$ in the problem, and the numerator and denominator have no obvious common factors (as they did in Example 7). Sometimes, however, we can create a common factor algebraically.

**EXAMPLE 9**  Evaluate

$$\lim_{x \to 0} \frac{\sqrt{x^2 + 100} - 10}{x^2}.$$

**Solution**  This is the limit we considered in Example 8. We can create a common factor by multiplying both numerator and denominator by the conjugate radical expression $\sqrt{x^2 + 100} + 10$ (obtained by changing the sign after the square root). The preliminary algebra rationalizes the numerator:

$$\frac{\sqrt{x^2 + 100} - 10}{x^2} = \frac{\sqrt{x^2 + 100} - 10}{x^2} \cdot \frac{\sqrt{x^2 + 100} + 10}{\sqrt{x^2 + 100} + 10}$$

$$= \frac{x^2 + 100 - 100}{x^2(\sqrt{x^2 + 100} + 10)}$$

$$= \frac{x^2}{x^2(\sqrt{x^2 + 100} + 10)} \qquad \text{Common factor } x^2$$

$$= \frac{1}{\sqrt{x^2 + 100} + 10}. \qquad \text{Cancel } x^2 \text{ for } x \neq 0.$$

Therefore,

$$\lim_{x \to 0} \frac{\sqrt{x^2 + 100} - 10}{x^2} = \lim_{x \to 0} \frac{1}{\sqrt{x^2 + 100} + 10}$$

$$= \frac{1}{\sqrt{0^2 + 100} + 10} \qquad \begin{array}{l}\text{Denominator not 0 at}\\ x = 0; \text{ substitute.}\end{array}$$

$$= \frac{1}{20} = 0.05.$$

This calculation provides the correct answer, in contrast to the ambiguous computer results in Example 8.  ■

We cannot always algebraically resolve the problem of finding the limit of a quotient where the denominator becomes zero. In some cases the limit might then be found with the aid of some geometry applied to the problem (see the proof of Theorem 7 in Section 2.4), or through methods of calculus (illustrated in Section 7.5). The next theorems give helpful tools by using function comparisons.

## The Sandwich Theorem

The following theorem enables us to calculate a variety of limits. It is called the Sandwich Theorem because it refers to a function $f$ whose values are sandwiched between the values of two other functions $g$ and $h$ that have the same limit $L$ at a point $c$. Being trapped between the values of two functions that approach $L$, the values of $f$ must also approach $L$ (Figure 2.12). You will find a proof in Appendix 4.

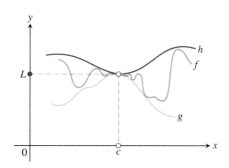

**FIGURE 2.12**  The graph of $f$ is sandwiched between the graphs of $g$ and $h$.

> **THEOREM 4—The Sandwich Theorem** Suppose that $g(x) \leq f(x) \leq h(x)$ for all $x$ in some open interval containing $c$, except possibly at $x = c$ itself. Suppose also that
>
> $$\lim_{x \to c} g(x) = \lim_{x \to c} h(x) = L.$$
>
> Then $\lim_{x \to c} f(x) = L$.

The Sandwich Theorem is also called the Squeeze Theorem or the Pinching Theorem.

**EXAMPLE 10**  Given that

$$1 - \frac{x^2}{4} \leq u(x) \leq 1 + \frac{x^2}{2} \qquad \text{for all } x \neq 0,$$

find $\lim_{x \to 0} u(x)$, no matter how complicated $u$ is.

**Solution**  Since

$$\lim_{x \to 0} \left(1 - (x^2/4)\right) = 1 \qquad \text{and} \qquad \lim_{x \to 0} \left(1 + (x^2/2)\right) = 1,$$

the Sandwich Theorem implies that $\lim_{x \to 0} u(x) = 1$ (Figure 2.13). ∎

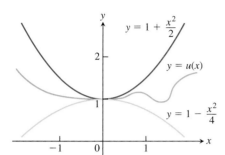

**FIGURE 2.13**  Any function $u(x)$ whose graph lies in the region between $y = 1 + (x^2/2)$ and $y = 1 - (x^2/4)$ has limit 1 as $x \to 0$ (Example 10).

**EXAMPLE 11**  The Sandwich Theorem helps us establish several important limit rules:

(a) $\lim_{\theta \to 0} \sin \theta = 0$  (b) $\lim_{\theta \to 0} \cos \theta = 1$

(c) For any function $f$, $\lim_{x \to c} |f(x)| = 0$ implies $\lim_{x \to c} f(x) = 0$.

**Solution**

(a) In Section 1.3 we established that $-|\theta| \leq \sin \theta \leq |\theta|$ for all $\theta$ (see Figure 2.14a). Since $\lim_{\theta \to 0}(-|\theta|) = \lim_{\theta \to 0} |\theta| = 0$, we have

$$\lim_{\theta \to 0} \sin \theta = 0.$$

(b) From Section 1.3, $0 \leq 1 - \cos \theta \leq |\theta|$ for all $\theta$ (see Figure 2.14b), and we have $\lim_{\theta \to 0}(1 - \cos \theta) = 0$ or

$$\lim_{\theta \to 0} \cos \theta = 1.$$

(c) Since $-|f(x)| \leq f(x) \leq |f(x)|$ and $-|f(x)|$ and $|f(x)|$ have limit 0 as $x \to c$, it follows that $\lim_{x \to c} f(x) = 0$. ∎

Another important property of limits is given by the next theorem. A proof is given in the next section.

(a)

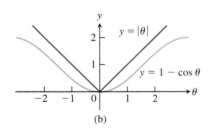

(b)

**FIGURE 2.14**  The Sandwich Theorem confirms the limits in Example 11.

> **THEOREM 5**  If $f(x) \leq g(x)$ for all $x$ in some open interval containing $c$, except possibly at $x = c$ itself, and the limits of $f$ and $g$ both exist as $x$ approaches $c$, then
>
> $$\lim_{x \to c} f(x) \leq \lim_{x \to c} g(x).$$

Caution  The assertion resulting from replacing the less than or equal to ($\leq$) inequality by the strict less than ($<$) inequality in Theorem 5 is false. Figure 2.14a shows that for $\theta \neq 0$, $-|\theta| < \sin \theta < |\theta|$. So $\lim_{\theta \to 0} \sin \theta = 0 = \lim_{\theta \to 0} |\theta|$, not $\lim_{\theta \to 0} \sin \theta < \lim_{\theta \to 0} |\theta|$.

## Exercises 2.2

### Limits from Graphs

**1.** For the function $g(x)$ graphed here, find the following limits or explain why they do not exist.

**a.** $\lim\limits_{x\to 1} g(x)$    **b.** $\lim\limits_{x\to 2} g(x)$    **c.** $\lim\limits_{x\to 3} g(x)$    **d.** $\lim\limits_{x\to 2.5} g(x)$

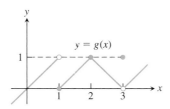

**2.** For the function $f(t)$ graphed here, find the following limits or explain why they do not exist.

**a.** $\lim\limits_{t\to -2} f(t)$    **b.** $\lim\limits_{t\to -1} f(t)$    **c.** $\lim\limits_{t\to 0} f(t)$    **d.** $\lim\limits_{t\to -0.5} f(t)$

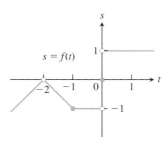

**3.** Which of the following statements about the function $y = f(x)$ graphed here are true, and which are false?

**a.** $\lim\limits_{x\to 0} f(x)$ exists.

**b.** $\lim\limits_{x\to 0} f(x) = 0$

**c.** $\lim\limits_{x\to 0} f(x) = 1$

**d.** $\lim\limits_{x\to 1} f(x) = 1$

**e.** $\lim\limits_{x\to 1} f(x) = 0$

**f.** $\lim\limits_{x\to c} f(x)$ exists at every point $c$ in $(-1, 1)$.

**g.** $\lim\limits_{x\to 1} f(x)$ does not exist.

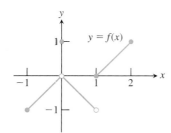

**4.** Which of the following statements about the function $y = f(x)$ graphed here are true, and which are false?

**a.** $\lim\limits_{x\to 2} f(x)$ does not exist.

**b.** $\lim\limits_{x\to 2} f(x) = 2$

**c.** $\lim\limits_{x\to 1} f(x)$ does not exist.

**d.** $\lim\limits_{x\to c} f(x)$ exists at every point $c$ in $(-1, 1)$.

**e.** $\lim\limits_{x\to c} f(x)$ exists at every point $c$ in $(1, 3)$.

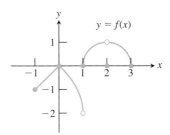

### Existence of Limits

In Exercises 5 and 6, explain why the limits do not exist.

**5.** $\lim\limits_{x\to 0} \dfrac{x}{|x|}$

**6.** $\lim\limits_{x\to 1} \dfrac{1}{x - 1}$

**7.** Suppose that a function $f(x)$ is defined for all real values of $x$ except $x = c$. Can anything be said about the existence of $\lim\limits_{x\to c} f(x)$? Give reasons for your answer.

**8.** Suppose that a function $f(x)$ is defined for all $x$ in $[-1, 1]$. Can anything be said about the existence of $\lim\limits_{x\to 0} f(x)$? Give reasons for your answer.

**9.** If $\lim\limits_{x\to 1} f(x) = 5$, must $f$ be defined at $x = 1$? If it is, must $f(1) = 5$? Can we conclude *anything* about the values of $f$ at $x = 1$? Explain.

**10.** If $f(1) = 5$, must $\lim\limits_{x\to 1} f(x)$ exist? If it does, then must $\lim\limits_{x\to 1} f(x) = 5$? Can we conclude *anything* about $\lim\limits_{x\to 1} f(x)$? Explain.

### Calculating Limits

Find the limits in Exercises 11–22.

**11.** $\lim\limits_{x\to -3} (x^2 - 13)$    **12.** $\lim\limits_{x\to 2}(-x^2 + 5x - 2)$

**13.** $\lim\limits_{t\to 6} 8(t - 5)(t - 7)$    **14.** $\lim\limits_{x\to -2} (x^3 - 2x^2 + 4x + 8)$

**15.** $\lim\limits_{x\to 2} \dfrac{2x + 5}{11 - x^3}$    **16.** $\lim\limits_{s\to 2/3} (8 - 3s)(2s - 1)$

**17.** $\lim\limits_{x\to -1/2} 4x(3x + 4)^2$    **18.** $\lim\limits_{y\to 2} \dfrac{y + 2}{y^2 + 5y + 6}$

**19.** $\lim\limits_{y\to -3} (5 - y)^{4/3}$    **20.** $\lim\limits_{z\to 4} \sqrt{z^2 - 10}$

**21.** $\lim\limits_{h\to 0} \dfrac{3}{\sqrt{3h + 1} + 1}$    **22.** $\lim\limits_{h\to 0} \dfrac{\sqrt{5h + 4} - 2}{h}$

**Limits of quotients**    Find the limits in Exercises 23–42.

**23.** $\lim\limits_{x\to 5} \dfrac{x - 5}{x^2 - 25}$    **24.** $\lim\limits_{x\to -3} \dfrac{x + 3}{x^2 + 4x + 3}$

**25.** $\lim\limits_{x\to -5} \dfrac{x^2 + 3x - 10}{x + 5}$    **26.** $\lim\limits_{x\to 2} \dfrac{x^2 - 7x + 10}{x - 2}$

**27.** $\lim\limits_{t\to 1} \dfrac{t^2 + t - 2}{t^2 - 1}$    **28.** $\lim\limits_{t\to -1} \dfrac{t^2 + 3t + 2}{t^2 - t - 2}$

**29.** $\lim\limits_{x\to -2} \dfrac{-2x - 4}{x^3 + 2x^2}$    **30.** $\lim\limits_{y\to 0} \dfrac{5y^3 + 8y^2}{3y^4 - 16y^2}$

**31.** $\lim\limits_{x \to 1} \dfrac{x^{-1} - 1}{x - 1}$

**32.** $\lim\limits_{x \to 0} \dfrac{\frac{1}{x-1} + \frac{1}{x+1}}{x}$

**33.** $\lim\limits_{u \to 1} \dfrac{u^4 - 1}{u^3 - 1}$

**34.** $\lim\limits_{v \to 2} \dfrac{v^3 - 8}{v^4 - 16}$

**35.** $\lim\limits_{x \to 9} \dfrac{\sqrt{x} - 3}{x - 9}$

**36.** $\lim\limits_{x \to 4} \dfrac{4x - x^2}{2 - \sqrt{x}}$

**37.** $\lim\limits_{x \to 1} \dfrac{x - 1}{\sqrt{x + 3} - 2}$

**38.** $\lim\limits_{x \to -1} \dfrac{\sqrt{x^2 + 8} - 3}{x + 1}$

**39.** $\lim\limits_{x \to 2} \dfrac{\sqrt{x^2 + 12} - 4}{x - 2}$

**40.** $\lim\limits_{x \to -2} \dfrac{x + 2}{\sqrt{x^2 + 5} - 3}$

**41.** $\lim\limits_{x \to -3} \dfrac{2 - \sqrt{x^2 - 5}}{x + 3}$

**42.** $\lim\limits_{x \to 4} \dfrac{4 - x}{5 - \sqrt{x^2 + 9}}$

**Limits with trigonometric functions** Find the limits in Exercises 43–50.

**43.** $\lim\limits_{x \to 0} (2 \sin x - 1)$

**44.** $\lim\limits_{x \to \pi/4} \sin^2 x$

**45.** $\lim\limits_{x \to 0} \sec x$

**46.** $\lim\limits_{x \to \pi/3} \tan x$

**47.** $\lim\limits_{x \to 0} \dfrac{1 + x + \sin x}{3 \cos x}$

**48.** $\lim\limits_{x \to 0} (x^2 - 1)(2 - \cos x)$

**49.** $\lim\limits_{x \to -\pi} \sqrt{x + 4} \cos (x + \pi)$

**50.** $\lim\limits_{x \to 0} \sqrt{7 + \sec^2 x}$

**Using Limit Rules**

**51.** Suppose $\lim\limits_{x \to 0} f(x) = 1$ and $\lim\limits_{x \to 0} g(x) = -5$. Name the rules in Theorem 1 that are used to accomplish steps (a), (b), and (c) of the following calculation.

$$\lim_{x \to 0} \frac{2f(x) - g(x)}{(f(x) + 7)^{2/3}} = \frac{\lim\limits_{x \to 0} (2f(x) - g(x))}{\lim\limits_{x \to 0} (f(x) + 7)^{2/3}} \quad \text{(a)}$$

$$= \frac{\lim\limits_{x \to 0} 2f(x) - \lim\limits_{x \to 0} g(x)}{\left( \lim\limits_{x \to 0} (f(x) + 7) \right)^{2/3}} \quad \text{(b)}$$

$$= \frac{2 \lim\limits_{x \to 0} f(x) - \lim\limits_{x \to 0} g(x)}{\left( \lim\limits_{x \to 0} f(x) + \lim\limits_{x \to 0} 7 \right)^{2/3}} \quad \text{(c)}$$

$$= \frac{(2)(1) - (-5)}{(1 + 7)^{2/3}} = \frac{7}{4}$$

**52.** Let $\lim\limits_{x \to 1} h(x) = 5, \lim\limits_{x \to 1} p(x) = 1,$ and $\lim\limits_{x \to 1} r(x) = 2.$ Name the rules in Theorem 1 that are used to accomplish steps (a), (b), and (c) of the following calculation.

$$\lim_{x \to 1} \frac{\sqrt{5h(x)}}{p(x)(4 - r(x))} = \frac{\lim\limits_{x \to 1} \sqrt{5h(x)}}{\lim\limits_{x \to 1} (p(x)(4 - r(x)))} \quad \text{(a)}$$

$$= \frac{\sqrt{\lim\limits_{x \to 1} 5h(x)}}{\left( \lim\limits_{x \to 1} p(x) \right) \left( \lim\limits_{x \to 1} (4 - r(x)) \right)} \quad \text{(b)}$$

$$= \frac{\sqrt{5 \lim\limits_{x \to 1} h(x)}}{\left( \lim\limits_{x \to 1} p(x) \right) \left( \lim\limits_{x \to 1} 4 - \lim\limits_{x \to 1} r(x) \right)} \quad \text{(c)}$$

$$= \frac{\sqrt{(5)(5)}}{(1)(4 - 2)} = \frac{5}{2}$$

**53.** Suppose $\lim\limits_{x \to c} f(x) = 5$ and $\lim\limits_{x \to c} g(x) = -2$. Find

  **a.** $\lim\limits_{x \to c} f(x)g(x)$

  **b.** $\lim\limits_{x \to c} 2f(x)g(x)$

  **c.** $\lim\limits_{x \to c} (f(x) + 3g(x))$

  **d.** $\lim\limits_{x \to c} \dfrac{f(x)}{f(x) - g(x)}$

**54.** Suppose $\lim\limits_{x \to 4} f(x) = 0$ and $\lim\limits_{x \to 4} g(x) = -3$. Find

  **a.** $\lim\limits_{x \to 4} (g(x) + 3)$

  **b.** $\lim\limits_{x \to 4} xf(x)$

  **c.** $\lim\limits_{x \to 4} (g(x))^2$

  **d.** $\lim\limits_{x \to 4} \dfrac{g(x)}{f(x) - 1}$

**55.** Suppose $\lim\limits_{x \to b} f(x) = 7$ and $\lim\limits_{x \to b} g(x) = -3$. Find

  **a.** $\lim\limits_{x \to b} (f(x) + g(x))$

  **b.** $\lim\limits_{x \to b} f(x) \cdot g(x)$

  **c.** $\lim\limits_{x \to b} 4g(x)$

  **d.** $\lim\limits_{x \to b} f(x)/g(x)$

**56.** Suppose that $\lim\limits_{x \to -2} p(x) = 4, \lim\limits_{x \to -2} r(x) = 0,$ and $\lim\limits_{x \to -2} s(x) = -3$. Find

  **a.** $\lim\limits_{x \to -2} (p(x) + r(x) + s(x))$

  **b.** $\lim\limits_{x \to -2} p(x) \cdot r(x) \cdot s(x)$

  **c.** $\lim\limits_{x \to -2} (-4p(x) + 5r(x))/s(x)$

**Limits of Average Rates of Change**

Because of their connection with secant lines, tangents, and instantaneous rates, limits of the form

$$\lim_{h \to 0} \frac{f(x + h) - f(x)}{h}$$

occur frequently in calculus. In Exercises 57–62, evaluate this limit for the given value of $x$ and function $f$.

**57.** $f(x) = x^2, \quad x = 1$

**58.** $f(x) = x^2, \quad x = -2$

**59.** $f(x) = 3x - 4, \quad x = 2$

**60.** $f(x) = 1/x, \quad x = -2$

**61.** $f(x) = \sqrt{x}, \quad x = 7$

**62.** $f(x) = \sqrt{3x + 1}, \quad x = 0$

**Using the Sandwich Theorem**

**63.** If $\sqrt{5 - 2x^2} \le f(x) \le \sqrt{5 - x^2}$ for $-1 \le x \le 1$, find $\lim\limits_{x \to 0} f(x)$.

**64.** If $2 - x^2 \le g(x) \le 2 \cos x$ for all $x$, find $\lim\limits_{x \to 0} g(x)$.

**65. a.** It can be shown that the inequalities

$$1 - \frac{x^2}{6} < \frac{x \sin x}{2 - 2 \cos x} < 1$$

hold for all values of $x$ close to zero. What, if anything, does this tell you about

$$\lim_{x \to 0} \frac{x \sin x}{2 - 2 \cos x}?$$

Give reasons for your answer.

  T **b.** Graph $y = 1 - (x^2/6), y = (x \sin x)/(2 - 2 \cos x),$ and $y = 1$ together for $-2 \le x \le 2$. Comment on the behavior of the graphs as $x \to 0$.

**66. a.** Suppose that the inequalities

$$\frac{1}{2} - \frac{x^2}{24} < \frac{1 - \cos x}{x^2} < \frac{1}{2}$$

hold for values of $x$ close to zero. (They do, as you will see in Section 9.9.) What, if anything, does this tell you about

$$\lim_{x \to 0} \frac{1 - \cos x}{x^2}?$$

Give reasons for your answer.

[T] **b.** Graph the equations $y = (1/2) - (x^2/24)$, $y = (1 - \cos x)/x^2$, and $y = 1/2$ together for $-2 \le x \le 2$. Comment on the behavior of the graphs as $x \to 0$.

### Estimating Limits

[T] You will find a graphing calculator useful for Exercises 67–74.

**67.** Let $f(x) = (x^2 - 9)/(x + 3)$.

**a.** Make a table of the values of $f$ at the points $x = -3.1$, $-3.01, -3.001$, and so on as far as your calculator can go. Then estimate $\lim_{x \to -3} f(x)$. What estimate do you arrive at if you evaluate $f$ at $x = -2.9, -2.99, -2.999, \ldots$ instead?

**b.** Support your conclusions in part (a) by graphing $f$ near $c = -3$ and using Zoom and Trace to estimate $y$-values on the graph as $x \to -3$.

**c.** Find $\lim_{x \to -3} f(x)$ algebraically, as in Example 7.

**68.** Let $g(x) = (x^2 - 2)/(x - \sqrt{2})$.

**a.** Make a table of the values of $g$ at the points $x = 1.4, 1.41$, $1.414$, and so on through successive decimal approximations of $\sqrt{2}$. Estimate $\lim_{x \to \sqrt{2}} g(x)$.

**b.** Support your conclusion in part (a) by graphing $g$ near $c = \sqrt{2}$ and using Zoom and Trace to estimate $y$-values on the graph as $x \to \sqrt{2}$.

**c.** Find $\lim_{x \to \sqrt{2}} g(x)$ algebraically.

**69.** Let $G(x) = (x + 6)/(x^2 + 4x - 12)$.

**a.** Make a table of the values of $G$ at $x = -5.9, -5.99, -5.999$, and so on. Then estimate $\lim_{x \to -6} G(x)$. What estimate do you arrive at if you evaluate $G$ at $x = -6.1, -6.01$, $-6.001, \ldots$ instead?

**b.** Support your conclusions in part (a) by graphing $G$ and using Zoom and Trace to estimate $y$-values on the graph as $x \to -6$.

**c.** Find $\lim_{x \to -6} G(x)$ algebraically.

**70.** Let $h(x) = (x^2 - 2x - 3)/(x^2 - 4x + 3)$.

**a.** Make a table of the values of $h$ at $x = 2.9, 2.99, 2.999$, and so on. Then estimate $\lim_{x \to 3} h(x)$. What estimate do you arrive at if you evaluate $h$ at $x = 3.1, 3.01, 3.001, \ldots$ instead?

**b.** Support your conclusions in part (a) by graphing $h$ near $c = 3$ and using Zoom and Trace to estimate $y$-values on the graph as $x \to 3$.

**c.** Find $\lim_{x \to 3} h(x)$ algebraically.

**71.** Let $f(x) = (x^2 - 1)/(|x| - 1)$.

**a.** Make tables of the values of $f$ at values of $x$ that approach $c = -1$ from above and below. Then estimate $\lim_{x \to -1} f(x)$.

**b.** Support your conclusion in part (a) by graphing $f$ near $c = -1$ and using Zoom and Trace to estimate $y$-values on the graph as $x \to -1$.

**c.** Find $\lim_{x \to -1} f(x)$ algebraically.

**72.** Let $F(x) = (x^2 + 3x + 2)/(2 - |x|)$.

**a.** Make tables of values of $F$ at values of $x$ that approach $c = -2$ from above and below. Then estimate $\lim_{x \to -2} F(x)$.

**b.** Support your conclusion in part (a) by graphing $F$ near $c = -2$ and using Zoom and Trace to estimate $y$-values on the graph as $x \to -2$.

**c.** Find $\lim_{x \to -2} F(x)$ algebraically.

**73.** Let $g(\theta) = (\sin \theta)/\theta$.

**a.** Make a table of the values of $g$ at values of $\theta$ that approach $\theta_0 = 0$ from above and below. Then estimate $\lim_{\theta \to 0} g(\theta)$.

**b.** Support your conclusion in part (a) by graphing $g$ near $\theta_0 = 0$.

**74.** Let $G(t) = (1 - \cos t)/t^2$.

**a.** Make tables of values of $G$ at values of $t$ that approach $t_0 = 0$ from above and below. Then estimate $\lim_{t \to 0} G(t)$.

**b.** Support your conclusion in part (a) by graphing $G$ near $t_0 = 0$.

### Theory and Examples

**75.** If $x^4 \le f(x) \le x^2$ for $x$ in $[-1, 1]$ and $x^2 \le f(x) \le x^4$ for $x < -1$ and $x > 1$, at what points $c$ do you automatically know $\lim_{x \to c} f(x)$? What can you say about the value of the limit at these points?

**76.** Suppose that $g(x) \le f(x) \le h(x)$ for all $x \ne 2$ and suppose that

$$\lim_{x \to 2} g(x) = \lim_{x \to 2} h(x) = -5.$$

Can we conclude anything about the values of $f$, $g$, and $h$ at $x = 2$? Could $f(2) = 0$? Could $\lim_{x \to 2} f(x) = 0$? Give reasons for your answers.

**77.** If $\lim_{x \to 4} \dfrac{f(x) - 5}{x - 2} = 1$, find $\lim_{x \to 4} f(x)$.

**78.** If $\lim_{x \to -2} \dfrac{f(x)}{x^2} = 1$, find

**a.** $\lim_{x \to -2} f(x)$

**b.** $\lim_{x \to -2} \dfrac{f(x)}{x}$

**79. a.** If $\lim_{x \to 2} \dfrac{f(x) - 5}{x - 2} = 3$, find $\lim_{x \to 2} f(x)$.

**b.** If $\lim_{x \to 2} \dfrac{f(x) - 5}{x - 2} = 4$, find $\lim_{x \to 2} f(x)$.

**80.** If $\lim\limits_{x\to 0}\dfrac{f(x)}{x^2} = 1$, find

   **a.** $\lim\limits_{x\to 0} f(x)$

   **b.** $\lim\limits_{x\to 0}\dfrac{f(x)}{x}$

T **81. a.** Graph $g(x) = x\sin(1/x)$ to estimate $\lim_{x\to 0} g(x)$, zooming in on the origin as necessary.

   **b.** Confirm your estimate in part (a) with a proof.

T **82. a.** Graph $h(x) = x^2\cos(1/x^3)$ to estimate $\lim_{x\to 0} h(x)$, zooming in on the origin as necessary.

   **b.** Confirm your estimate in part (a) with a proof.

**83.** $\lim\limits_{x\to 2}\dfrac{x^4 - 16}{x - 2}$

**84.** $\lim\limits_{x\to -1}\dfrac{x^3 - x^2 - 5x - 3}{(x + 1)^2}$

**85.** $\lim\limits_{x\to 0}\dfrac{\sqrt[3]{1 + x} - 1}{x}$

**86.** $\lim\limits_{x\to 3}\dfrac{x^2 - 9}{\sqrt{x^2 + 7} - 4}$

**87.** $\lim\limits_{x\to 0}\dfrac{1 - \cos x}{x\sin x}$

**88.** $\lim\limits_{x\to 0}\dfrac{2x^2}{3 - 3\cos x}$

## COMPUTER EXPLORATIONS

### Graphical Estimates of Limits

In Exercises 83–88, use a CAS to perform the following steps:

   **a.** Plot the function near the point $c$ being approached.

   **b.** From your plot guess the value of the limit.

# $2.3$ The Precise Definition of a Limit

We now turn our attention to the precise definition of a limit. We replace vague phrases like "gets arbitrarily close to" in the informal definition with specific conditions that can be applied to any particular example. With a precise definition, we can avoid misunderstandings, prove the limit properties given in the preceding section, and establish many important limits.

To show that the limit of $f(x)$ as $x \to c$ equals the number $L$, we need to show that the gap between $f(x)$ and $L$ can be made "as small as we choose" if $x$ is kept "close enough" to $c$. Let us see what this would require if we specified the size of the gap between $f(x)$ and $L$.

**EXAMPLE 1** Consider the function $y = 2x - 1$ near $x = 4$. Intuitively it appears that $y$ is close to 7 when $x$ is close to 4, so $\lim_{x\to 4}(2x - 1) = 7$. However, how close to $x = 4$ does $x$ have to be so that $y = 2x - 1$ differs from 7 by, say, less than 2 units?

**Solution** We are asked: For what values of $x$ is $|y - 7| < 2$? To find the answer we first express $|y - 7|$ in terms of $x$:

$$|y - 7| = |(2x - 1) - 7| = |2x - 8|.$$

The question then becomes: what values of $x$ satisfy the inequality $|2x - 8| < 2$? To find out, we solve the inequality:

$$|2x - 8| < 2$$
$$-2 < 2x - 8 < 2$$
$$6 < 2x < 10$$
$$3 < x < 5 \qquad \text{Solve for } x.$$
$$-1 < x - 4 < 1. \qquad \text{Solve for } x - 4.$$

**FIGURE 2.15** Keeping $x$ within 1 unit of $x = 4$ will keep $y$ within 2 units of $y = 7$ (Example 1).

Keeping $x$ within 1 unit of $x = 4$ will keep $y$ within 2 units of $y = 7$ (Figure 2.15). ∎

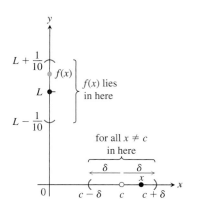

**FIGURE 2.16** How should we define $\delta > 0$ so that keeping $x$ within the interval $(c - \delta, c + \delta)$ will keep $f(x)$ within the interval $\left(L - \dfrac{1}{10}, L + \dfrac{1}{10}\right)$?

In the previous example we determined how close $x$ must be to a particular value $c$ to ensure that the outputs $f(x)$ of some function lie within a prescribed interval about a limit value $L$. To show that the limit of $f(x)$ as $x \to c$ actually equals $L$, we must be able to show that the gap between $f(x)$ and $L$ can be made less than *any prescribed error*, no matter how small, by holding $x$ close enough to $c$.

## Definition of Limit

Suppose we are watching the values of a function $f(x)$ as $x$ approaches $c$ (without taking on the value of $c$ itself). Certainly we want to be able to say that $f(x)$ stays within one-tenth of a unit from $L$ as soon as $x$ stays within some distance $\delta$ of $c$ (Figure 2.16). But that in itself is not enough, because as $x$ continues on its course toward $c$, what is to prevent $f(x)$ from jittering about within the interval from $L - (1/10)$ to $L + (1/10)$ without tending toward $L$?

We can be told that the error can be no more than $1/100$ or $1/1000$ or $1/100,000$. Each time, we find a new $\delta$-interval about $c$ so that keeping $x$ within that interval satisfies the new error tolerance. And each time the possibility exists that $f(x)$ jitters away from $L$ at some stage.

The figures on the next page illustrate the problem. You can think of this as a quarrel between a skeptic and a scholar. The skeptic presents $\epsilon$-challenges to prove that the limit does not exist or, more precisely, that there is room for doubt. The scholar answers every challenge with a $\delta$-interval around $c$ that keeps the function values within $\epsilon$ of $L$.

How do we stop this seemingly endless series of challenges and responses? We can do so by proving that for every error tolerance $\epsilon$ that the challenger can produce, we can present a matching distance $\delta$ that keeps $x$ "close enough" to $c$ to keep $f(x)$ within that $\epsilon$-tolerance of $L$ (Figure 2.17). This leads us to the precise definition of a limit.

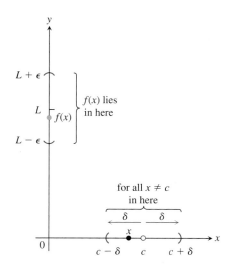

**FIGURE 2.17** The relation of $\delta$ and $\epsilon$ in the definition of limit.

---

**DEFINITION** Let $f(x)$ be defined on an open interval about $c$, except possibly at $c$ itself. We say that the **limit of $f(x)$ as $x$ approaches $c$ is the number $L$**, and write

$$\lim_{x \to c} f(x) = L,$$

if, for every number $\epsilon > 0$, there exists a corresponding number $\delta > 0$ such that for all $x$,

$$0 < |x - c| < \delta \quad \Rightarrow \quad |f(x) - L| < \epsilon.$$

---

One way to think about the definition is to suppose we are machining a generator shaft to a close tolerance. We may try for diameter $L$, but since nothing is perfect, we must be satisfied with a diameter $f(x)$ somewhere between $L - \epsilon$ and $L + \epsilon$. The $\delta$ is the measure of how accurate our control setting for $x$ must be to guarantee this degree of accuracy in the diameter of the shaft. Notice that as the tolerance for error becomes stricter, we may have to adjust $\delta$. That is, the value of $\delta$, how tight our control setting must be, depends on the value of $\epsilon$, the error tolerance.

## Examples: Testing the Definition

The formal definition of limit does not tell how to find the limit of a function, but it enables us to verify that a conjectured limit value is correct. The following examples show how the definition can be used to verify limit statements for specific functions. However, the real purpose of the definition is not to do calculations like this, but rather to prove general theorems so that the calculation of specific limits can be simplified, such as the theorems stated in the previous section.

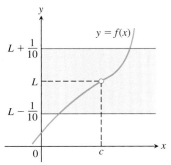

The challenge:

Make $f(x) - L| < \epsilon = \frac{1}{10}$

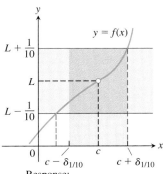

Response:

$|x - c| < \delta_{1/10}$ (a number)

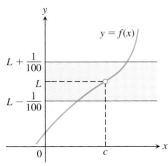

New challenge:

Make $|f(x) - L| < \epsilon = \frac{1}{100}$

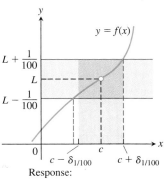

Response:

$|x - c| < \delta_{1/100}$

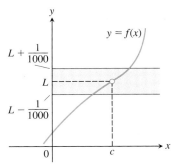

New challenge:

$\epsilon = \frac{1}{1000}$

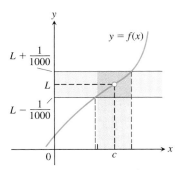

Response:

$|x - c| < \delta_{1/1000}$

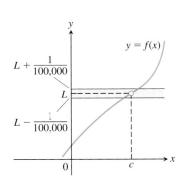

New challenge:

$\epsilon = \frac{1}{100,000}$

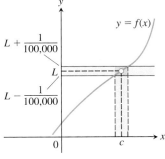

Response:

$|x - c| < \delta_{1/100,000}$

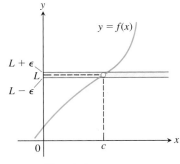

New challenge:

$\epsilon = \cdots$

**EXAMPLE 2**    Show that

$$\lim_{x \to 1} (5x - 3) = 2.$$

**Solution**    Set $c = 1$, $f(x) = 5x - 3$, and $L = 2$ in the definition of limit. For any given $\epsilon > 0$, we have to find a suitable $\delta > 0$ so that if $x \neq 1$ and $x$ is within distance $\delta$ of $c = 1$, that is, whenever

$$0 < |x - 1| < \delta,$$

it is true that $f(x)$ is within distance $\epsilon$ of $L = 2$, so

$$|f(x) - 2| < \epsilon.$$

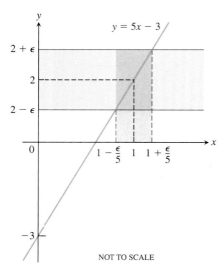

**FIGURE 2.18** If $f(x) = 5x - 3$, then $0 < |x - 1| < \epsilon/5$ guarantees that $|f(x) - 2| < \epsilon$ (Example 2).

We find $\delta$ by working backward from the $\epsilon$-inequality:

$$|(5x - 3) - 2| = |5x - 5| < \epsilon$$
$$5|x - 1| < \epsilon$$
$$|x - 1| < \epsilon/5.$$

Thus, we can take $\delta = \epsilon/5$ (Figure 2.18). If $0 < |x - 1| < \delta = \epsilon/5$, then

$$|(5x - 3) - 2| = |5x - 5| = 5|x - 1| < 5(\epsilon/5) = \epsilon,$$

which proves that $\lim_{x \to 1}(5x - 3) = 2$.

The value of $\delta = \epsilon/5$ is not the only value that will make $0 < |x - 1| < \delta$ imply $|5x - 5| < \epsilon$. Any smaller positive $\delta$ will do as well. The definition does not ask for a "best" positive $\delta$, just one that will work. ∎

**EXAMPLE 3** Prove the following results presented graphically in Section 2.2.

**(a)** $\lim_{x \to c} x = c$

**(b)** $\lim_{x \to c} k = k$    ($k$ constant)

**Solution**

**(a)** Let $\epsilon > 0$ be given. We must find $\delta > 0$ such that for all $x$

$$0 < |x - c| < \delta \qquad \text{implies} \qquad |x - c| < \epsilon.$$

The implication will hold if $\delta$ equals $\epsilon$ or any smaller positive number (Figure 2.19). This proves that $\lim_{x \to c} x = c$.

**(b)** Let $\epsilon > 0$ be given. We must find $\delta > 0$ such that for all $x$

$$0 < |x - c| < \delta \qquad \text{implies} \qquad |k - k| < \epsilon.$$

Since $k - k = 0$, we can use any positive number for $\delta$ and the implication will hold (Figure 2.20). This proves that $\lim_{x \to c} k = k$. ∎

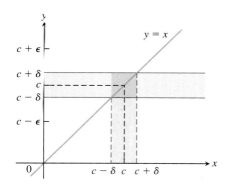

**FIGURE 2.19** For the function $f(x) = x$, we find that $0 < |x - c| < \delta$ will guarantee $|f(x) - c| < \epsilon$ whenever $\delta \leq \epsilon$ (Example 3a).

### Finding Deltas Algebraically for Given Epsilons

In Examples 2 and 3, the interval of values about $c$ for which $|f(x) - L|$ was less than $\epsilon$ was symmetric about $c$ and we could take $\delta$ to be half the length of that interval. When such symmetry is absent, as it usually is, we can take $\delta$ to be the distance from $c$ to the interval's *nearer* endpoint.

**EXAMPLE 4** For the limit $\lim_{x \to 5}\sqrt{x - 1} = 2$, find a $\delta > 0$ that works for $\epsilon = 1$. That is, find a $\delta > 0$ such that for all $x$

$$0 < |x - 5| < \delta \qquad \Rightarrow \qquad |\sqrt{x - 1} - 2| < 1.$$

**Solution** We organize the search into two steps.

**1.** *Solve the inequality* $|\sqrt{x - 1} - 2| < 1$ *to find an interval containing* $x = 5$ *on which the inequality holds for all* $x \neq 5$.

$$|\sqrt{x - 1} - 2| < 1$$
$$-1 < \sqrt{x - 1} - 2 < 1$$
$$1 < \sqrt{x - 1} < 3$$
$$1 < x - 1 < 9$$
$$2 < x < 10$$

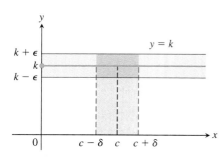

**FIGURE 2.20** For the function $f(x) = k$, we find that $|f(x) - k| < \epsilon$ for any positive $\delta$ (Example 3b).

**FIGURE 2.21** An open interval of radius 3 about $x = 5$ will lie inside the open interval (2, 10).

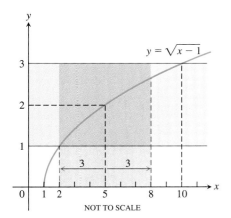

**FIGURE 2.22** The function and intervals in Example 4.

The inequality holds for all $x$ in the open interval (2, 10), so it holds for all $x \neq 5$ in this interval as well.

2. *Find a value of $\delta > 0$ to place the centered interval $5 - \delta < x < 5 + \delta$ (centered at $x = 5$) inside the interval* (2, 10). The distance from 5 to the nearer endpoint of (2, 10) is 3 (Figure 2.21). If we take $\delta = 3$ or any smaller positive number, then the inequality $0 < |x - 5| < \delta$ will automatically place $x$ between 2 and 10 to make $|\sqrt{x - 1} - 2| < 1$ (Figure 2.22):

$$0 < |x - 5| < 3 \quad \Rightarrow \quad |\sqrt{x - 1} - 2| < 1. \quad \blacksquare$$

---

**How to Find Algebraically a $\delta$ for a Given $f$, $L$, $c$, and $\varepsilon > 0$**

The process of finding a $\delta > 0$ such that for all $x$

$$0 < |x - c| < \delta \quad \Rightarrow \quad |f(x) - L| < \epsilon$$

can be accomplished in two steps.

1. *Solve the inequality $|f(x) - L| < \epsilon$ to find an open interval $(a, b)$ containing $c$ on which the inequality holds for all $x \neq c$.*

2. *Find a value of $\delta > 0$ that places the open interval $(c - \delta, c + \delta)$ centered at $c$ inside the interval $(a, b)$. The inequality $|f(x) - L| < \epsilon$ will hold for all $x \neq c$ in this $\delta$-interval.*

---

**EXAMPLE 5**    Prove that $\lim_{x \to 2} f(x) = 4$ if

$$f(x) = \begin{cases} x^2, & x \neq 2 \\ 1, & x = 2. \end{cases}$$

**Solution**    Our task is to show that given $\epsilon > 0$ there exists a $\delta > 0$ such that for all $x$

$$0 < |x - 2| < \delta \quad \Rightarrow \quad |f(x) - 4| < \epsilon.$$

1. *Solve the inequality $|f(x) - 4| < \epsilon$ to find an open interval containing $x = 2$ on which the inequality holds for all $x \neq 2$.*

For $x \neq c = 2$, we have $f(x) = x^2$, and the inequality to solve is $|x^2 - 4| < \epsilon$:

$$|x^2 - 4| < \epsilon$$
$$-\epsilon < x^2 - 4 < \epsilon$$
$$4 - \epsilon < x^2 < 4 + \epsilon$$
$$\sqrt{4 - \epsilon} < |x| < \sqrt{4 + \epsilon} \quad \text{Assumes } \epsilon < 4; \text{ see below.}$$
$$\sqrt{4 - \epsilon} < x < \sqrt{4 + \epsilon}. \quad \text{An open interval about } x = 2$$
$$\text{that solves the inequality}$$

The inequality $|f(x) - 4| < \epsilon$ holds for all $x \neq 2$ in the open interval ($\sqrt{4 - \epsilon}$, $\sqrt{4 + \epsilon}$) (Figure 2.23).

2. *Find a value of $\delta > 0$ that places the centered interval $(2 - \delta, 2 + \delta)$ inside the interval $(\sqrt{4 - \epsilon}, \sqrt{4 + \epsilon})$.*

Take $\delta$ to be the distance from $x = 2$ to the nearer endpoint of $(\sqrt{4 - \epsilon}, \sqrt{4 + \epsilon})$. In other words, take $\delta = \min\{2 - \sqrt{4 - \epsilon}, \sqrt{4 + \epsilon} - 2\}$, the *minimum* (the

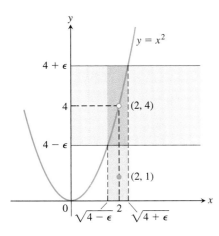

**FIGURE 2.23** An interval containing $x = 2$ so that the function in Example 5 satisfies $|f(x) - 4| < \epsilon$.

smaller) of the two numbers $2 - \sqrt{4 - \epsilon}$ and $\sqrt{4 + \epsilon} - 2$. If $\delta$ has this or any smaller positive value, the inequality $0 < |x - 2| < \delta$ will automatically place $x$ between $\sqrt{4 - \epsilon}$ and $\sqrt{4 + \epsilon}$ to make $|f(x) - 4| < \epsilon$. For all $x$,

$$0 < |x - 2| < \delta \quad \Rightarrow \quad |f(x) - 4| < \epsilon.$$

This completes the proof for $\epsilon < 4$.

If $\epsilon \geq 4$, then we take $\delta$ to be the distance from $x = 2$ to the nearer endpoint of the interval $(0, \sqrt{4 + \epsilon})$. In other words, take $\delta = \min\{2, \sqrt{4 + \epsilon} - 2\}$. (See Figure 2.23.) ∎

## Using the Definition to Prove Theorems

We do not usually rely on the formal definition of limit to verify specific limits such as those in the preceding examples. Rather, we appeal to general theorems about limits, in particular the theorems of Section 2.2. The definition is used to prove these theorems (Appendix 5). As an example, we prove part 1 of Theorem 1, the Sum Rule.

**EXAMPLE 6**    Given that $\lim_{x \to c} f(x) = L$ and $\lim_{x \to c} g(x) = M$, prove that

$$\lim_{x \to c} (f(x) + g(x)) = L + M.$$

**Solution**    Let $\epsilon > 0$ be given. We want to find a positive number $\delta$ such that for all $x$

$$0 < |x - c| < \delta \quad \Rightarrow \quad |f(x) + g(x) - (L + M)| < \epsilon.$$

Regrouping terms, we get

$$|f(x) + g(x) - (L + M)| = |(f(x) - L) + (g(x) - M)| \qquad \text{Triangle Inequality:}$$
$$\leq |f(x) - L| + |g(x) - M|. \qquad |a + b| \leq |a| + |b|$$

Since $\lim_{x \to c} f(x) = L$, there exists a number $\delta_1 > 0$ such that for all $x$

$$0 < |x - c| < \delta_1 \quad \Rightarrow \quad |f(x) - L| < \epsilon/2.$$

Similarly, since $\lim_{x \to c} g(x) = M$, there exists a number $\delta_2 > 0$ such that for all $x$

$$0 < |x - c| < \delta_2 \quad \Rightarrow \quad |g(x) - M| < \epsilon/2.$$

Let $\delta = \min\{\delta_1, \delta_2\}$, the smaller of $\delta_1$ and $\delta_2$. If $0 < |x - c| < \delta$ then $|x - c| < \delta_1$, so $|f(x) - L| < \epsilon/2$, and $|x - c| < \delta_2$, so $|g(x) - M| < \epsilon/2$. Therefore

$$|f(x) + g(x) - (L + M)| < \frac{\epsilon}{2} + \frac{\epsilon}{2} = \epsilon.$$

This shows that $\lim_{x \to c}(f(x) + g(x)) = L + M$. ∎

Next we prove Theorem 5 of Section 2.2.

**EXAMPLE 7**    Given that $\lim_{x \to c} f(x) = L$ and $\lim_{x \to c} g(x) = M$, and that $f(x) \leq g(x)$ for all $x$ in an open interval containing $c$ (except possibly $c$ itself), prove that $L \leq M$.

**Solution**    We use the method of proof by contradiction. Suppose, on the contrary, that $L > M$. Then by the limit of a difference property in Theorem 1,

$$\lim_{x \to c} (g(x) - f(x)) = M - L.$$

Therefore, for any $\epsilon > 0$, there exists $\delta > 0$ such that

$$\left| (g(x) - f(x)) - (M - L) \right| < \epsilon \qquad \text{whenever} \quad 0 < |x - c| < \delta.$$

Since $L - M > 0$ by hypothesis, we take $\epsilon = L - M$ in particular and we have a number $\delta > 0$ such that

$$\left| (g(x) - f(x)) - (M - L) \right| < L - M \qquad \text{whenever} \quad 0 < |x - c| < \delta.$$

Since $a \le |a|$ for any number $a$, we have

$$(g(x) - f(x)) - (M - L) < L - M \qquad \text{whenever} \quad 0 < |x - c| < \delta$$

which simplifies to

$$g(x) < f(x) \qquad \text{whenever} \quad 0 < |x - c| < \delta.$$

But this contradicts $f(x) \le g(x)$. Thus the inequality $L > M$ must be false. Therefore $L \le M$. ∎

---

## Exercises 2.3

### Centering Intervals About a Point

In Exercises 1–6, sketch the interval $(a, b)$ on the x-axis with the point $c$ inside. Then find a value of $\delta > 0$ such that for all $x, 0 < |x - c| < \delta \;\Rightarrow\; a < x < b$.

1. $a = 1, \quad b = 7, \quad c = 5$
2. $a = 1, \quad b = 7, \quad c = 2$
3. $a = -7/2, \quad b = -1/2, \quad c = -3$
4. $a = -7/2, \quad b = -1/2, \quad c = -3/2$
5. $a = 4/9, \quad b = 4/7, \quad c = 1/2$
6. $a = 2.7591, \quad b = 3.2391, \quad c = 3$

### Finding Deltas Graphically

In Exercises 7–14, use the graphs to find a $\delta > 0$ such that for all $x$
$$0 < |x - c| < \delta \;\Rightarrow\; |f(x) - L| < \epsilon.$$

**7.**

NOT TO SCALE

**8.**

NOT TO SCALE

**9.**

**10.**

NOT TO SCALE

**11.**

NOT TO SCALE

**12.**

NOT TO SCALE

**13.**

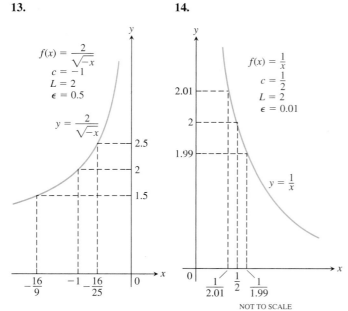

$f(x) = \dfrac{2}{\sqrt{-x}}$
$c = -1$
$L = 2$
$\epsilon = 0.5$

$y = \dfrac{2}{\sqrt{-x}}$

**14.**

$f(x) = \dfrac{1}{x}$
$c = \dfrac{1}{2}$
$L = 2$
$\epsilon = 0.01$

$y = \dfrac{1}{x}$

NOT TO SCALE

## Finding Deltas Algebraically

Each of Exercises 15–30 gives a function $f(x)$ and numbers $L, c$, and $\epsilon > 0$. In each case, find an open interval about $c$ on which the inequality $|f(x) - L| < \epsilon$ holds. Then give a value for $\delta > 0$ such that for all $x$ satisfying $0 < |x - c| < \delta$ the inequality $|f(x) - L| < \epsilon$ holds.

**15.** $f(x) = x + 1, \quad L = 5, \quad c = 4, \quad \epsilon = 0.01$

**16.** $f(x) = 2x - 2, \quad L = -6, \quad c = -2, \quad \epsilon = 0.02$

**17.** $f(x) = \sqrt{x + 1}, \quad L = 1, \quad c = 0, \quad \epsilon = 0.1$

**18.** $f(x) = \sqrt{x}, \quad L = 1/2, \quad c = 1/4, \quad \epsilon = 0.1$

**19.** $f(x) = \sqrt{19 - x}, \quad L = 3, \quad c = 10, \quad \epsilon = 1$

**20.** $f(x) = \sqrt{x - 7}, \quad L = 4, \quad c = 23, \quad \epsilon = 1$

**21.** $f(x) = 1/x, \quad L = 1/4, \quad c = 4, \quad \epsilon = 0.05$

**22.** $f(x) = x^2, \quad L = 3, \quad c = \sqrt{3}, \quad \epsilon = 0.1$

**23.** $f(x) = x^2, \quad L = 4, \quad c = -2, \quad \epsilon = 0.5$

**24.** $f(x) = 1/x, \quad L = -1, \quad c = -1, \quad \epsilon = 0.1$

**25.** $f(x) = x^2 - 5, \quad L = 11, \quad c = 4, \quad \epsilon = 1$

**26.** $f(x) = 120/x, \quad L = 5, \quad c = 24, \quad \epsilon = 1$

**27.** $f(x) = mx, \quad m > 0, \quad L = 2m, \quad c = 2, \quad \epsilon = 0.03$

**28.** $f(x) = mx, \quad m > 0, \quad L = 3m, \quad c = 3, \quad \epsilon = c > 0$

**29.** $f(x) = mx + b, \quad m > 0, \quad L = (m/2) + b,$
$c = 1/2, \quad \epsilon = c > 0$

**30.** $f(x) = mx + b, \quad m > 0, \quad L = m + b, \quad c = 1,$
$\epsilon = 0.05$

## Using the Formal Definition

Each of Exercises 31–36 gives a function $f(x)$, a point $c$, and a positive number $\epsilon$. Find $L = \lim\limits_{x \to c} f(x)$. Then find a number $\delta > 0$ such that for all $x$

$$0 < |x - c| < \delta \quad \Rightarrow \quad |f(x) - L| < \epsilon.$$

**31.** $f(x) = 3 - 2x, \quad c = 3, \quad \epsilon = 0.02$

**32.** $f(x) = -3x - 2, \quad c = -1, \quad \epsilon = 0.03$

**33.** $f(x) = \dfrac{x^2 - 4}{x - 2}, \quad c = 2, \quad \epsilon = 0.05$

**34.** $f(x) = \dfrac{x^2 + 6x + 5}{x + 5}, \quad c = -5, \quad \epsilon = 0.05$

**35.** $f(x) = \sqrt{1 - 5x}, \quad c = -3, \quad \epsilon = 0.5$

**36.** $f(x) = 4/x, \quad c = 2, \quad \epsilon = 0.4$

Prove the limit statements in Exercises 37–50.

**37.** $\lim\limits_{x \to 4} (9 - x) = 5$
**38.** $\lim\limits_{x \to 3} (3x - 7) = 2$

**39.** $\lim\limits_{x \to 9} \sqrt{x - 5} = 2$
**40.** $\lim\limits_{x \to 0} \sqrt{4 - x} = 2$

**41.** $\lim\limits_{x \to 1} f(x) = 1 \quad \text{if} \quad f(x) = \begin{cases} x^2, & x \neq 1 \\ 2, & x = 1 \end{cases}$

**42.** $\lim\limits_{x \to -2} f(x) = 4 \quad \text{if} \quad f(x) = \begin{cases} x^2, & x \neq -2 \\ 1, & x = -2 \end{cases}$

**43.** $\lim\limits_{x \to 1} \dfrac{1}{x} = 1$
**44.** $\lim\limits_{x \to \sqrt{3}} \dfrac{1}{x^2} = \dfrac{1}{3}$

**45.** $\lim\limits_{x \to -3} \dfrac{x^2 - 9}{x + 3} = -6$
**46.** $\lim\limits_{x \to 1} \dfrac{x^2 - 1}{x - 1} = 2$

**47.** $\lim\limits_{x \to 1} f(x) = 2 \quad \text{if} \quad f(x) = \begin{cases} 4 - 2x, & x < 1 \\ 6x - 4, & x \geq 1 \end{cases}$

**48.** $\lim\limits_{x \to 0} f(x) = 0 \quad \text{if} \quad f(x) = \begin{cases} 2x, & x < 0 \\ x/2, & x \geq 0 \end{cases}$

**49.** $\lim\limits_{x \to 0} x \sin \dfrac{1}{x} = 0$

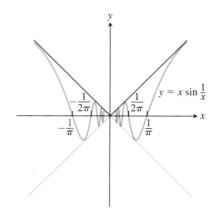

$y = x \sin \dfrac{1}{x}$

**50.** $\lim_{x \to 0} x^2 \sin \frac{1}{x} = 0$

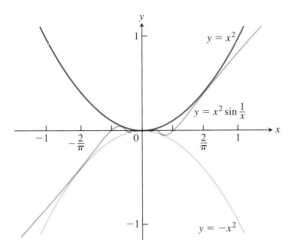

### Theory and Examples

**51.** Define what it means to say that $\lim_{x \to 0} g(x) = k$.

**52.** Prove that $\lim_{x \to c} f(x) = L$ if and only if $\lim_{h \to 0} f(h + c) = L$.

**53.** **A wrong statement about limits** Show by example that the following statement is wrong.

> The number $L$ is the limit of $f(x)$ as $x$ approaches $c$ if $f(x)$ gets closer to $L$ as $x$ approaches $c$.

Explain why the function in your example does not have the given value of $L$ as a limit as $x \to c$.

**54.** **Another wrong statement about limits** Show by example that the following statement is wrong.

> The number $L$ is the limit of $f(x)$ as $x$ approaches $c$ if, given any $\epsilon > 0$, there exists a value of $x$ for which $|f(x) - L| < \epsilon$.

Explain why the function in your example does not have the given value of $L$ as a limit as $x \to c$.

T **55.** **Grinding engine cylinders** Before contracting to grind engine cylinders to a cross-sectional area of $60$ cm², you need to know how much deviation from the ideal cylinder diameter of $c = 8.7404$ cm you can allow and still have the area come within $0.1$ cm² of the required $60$ cm². To find out, you let $A = \pi(x/2)^2$ and look for the interval in which you must hold $x$ to make $|A - 60| \leq 0.1$. What interval do you find?

**56.** **Manufacturing electrical resistors** Ohm's law for electrical circuits like the one shown in the accompanying figure states that $V = RI$. In this equation, $V$ is a constant voltage, $I$ is the current in amperes, and $R$ is the resistance in ohms. Your firm has been asked to supply the resistors for a circuit in which $V$ will be $120$ volts and $I$ is to be $5 \pm 0.1$ amp. In what interval does $R$ have to lie for $I$ to be within $0.1$ amp of the value $I_0 = 5$?

### When Is a Number $L$ Not the Limit of $f(x)$ as $x \to c$?

**Showing $L$ is not a limit** We can prove that $\lim_{x \to c} f(x) \neq L$ by providing an $\epsilon > 0$ such that no possible $\delta > 0$ satisfies the condition

$$\text{for all } x, \quad 0 < |x - c| < \delta \quad \Rightarrow \quad |f(x) - L| < \epsilon.$$

We accomplish this for our candidate $\epsilon$ by showing that for each $\delta > 0$ there exists a value of $x$ such that

$$0 < |x - c| < \delta \quad \text{and} \quad |f(x) - L| \geq \epsilon.$$

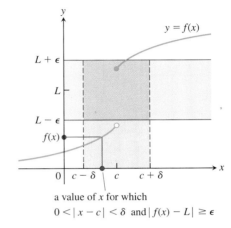

a value of $x$ for which
$$0 < |x - c| < \delta \text{ and } |f(x) - L| \geq \epsilon$$

**57.** Let $f(x) = \begin{cases} x, & x < 1 \\ x + 1, & x > 1. \end{cases}$

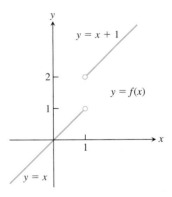

**a.** Let $\epsilon = 1/2$. Show that no possible $\delta > 0$ satisfies the following condition:

> For all $x, \quad 0 < |x - 1| < \delta \quad \Rightarrow \quad |f(x) - 2| < 1/2$.

That is, for each $\delta > 0$ show that there is a value of $x$ such that

$$0 < |x - 1| < \delta \quad \text{and} \quad |f(x) - 2| \geq 1/2.$$

This will show that $\lim_{x \to 1} f(x) \neq 2$.

**b.** Show that $\lim_{x \to 1} f(x) \neq 1$.

**c.** Show that $\lim_{x \to 1} f(x) \neq 1.5$.

**58.** Let $h(x) = \begin{cases} x^2, & x < 2 \\ 3, & x = 2 \\ 2, & x > 2. \end{cases}$

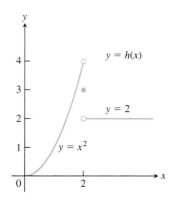

Show that

**a.** $\lim\limits_{x \to 2} h(x) \neq 4$

**b.** $\lim\limits_{x \to 2} h(x) \neq 3$

**c.** $\lim\limits_{x \to 2} h(x) \neq 2$

**59.** For the function graphed here, explain why

**a.** $\lim\limits_{x \to 3} f(x) \neq 4$

**b.** $\lim\limits_{x \to 3} f(x) \neq 4.8$

**c.** $\lim\limits_{x \to 3} f(x) \neq 3$

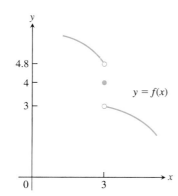

**60. a.** For the function graphed here, show that $\lim_{x \to -1} g(x) \neq 2$.

**b.** Does $\lim_{x \to -1} g(x)$ appear to exist? If so, what is the value of the limit? If not, why not?

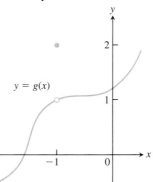

**COMPUTER EXPLORATIONS**

In Exercises 61–66, you will further explore finding deltas graphically. Use a CAS to perform the following steps:

**a.** Plot the function $y = f(x)$ near the point $c$ being approached.

**b.** Guess the value of the limit $L$ and then evaluate the limit symbolically to see if you guessed correctly.

**c.** Using the value $\epsilon = 0.2$, graph the banding lines $y_1 = L - \epsilon$ and $y_2 = L + \epsilon$ together with the function $f$ near $c$.

**d.** From your graph in part (c), estimate a $\delta > 0$ such that for all $x$

$$0 < |x - c| < \delta \quad \Longrightarrow \quad |f(x) - L| < \epsilon.$$

Test your estimate by plotting $f$, $y_1$, and $y_2$ over the interval $0 < |x - c| < \delta$. For your viewing window use $c - 2\delta \leq x \leq c + 2\delta$ and $L - 2\epsilon \leq y \leq L + 2\epsilon$. If any function values lie outside the interval $[L - \epsilon, L + \epsilon]$, your choice of $\delta$ was too large. Try again with a smaller estimate.

**e.** Repeat parts (c) and (d) successively for $\epsilon = 0.1, 0.05,$ and $0.001$.

**61.** $f(x) = \dfrac{x^4 - 81}{x - 3}, \quad c = 3$     **62.** $f(x) = \dfrac{5x^3 + 9x^2}{2x^5 + 3x^2}, \quad c = 0$

**63.** $f(x) = \dfrac{\sin 2x}{3x}, \quad c = 0$     **64.** $f(x) = \dfrac{x(1 - \cos x)}{x - \sin x}, \quad c = 0$

**65.** $f(x) = \dfrac{\sqrt[3]{x} - 1}{x - 1}, \quad c = 1$

**66.** $f(x) = \dfrac{3x^2 - (7x + 1)\sqrt{x} + 5}{x - 1}, \quad c = 1$

# 2.4 One-Sided Limits

In this section we extend the limit concept to *one-sided limits*, which are limits as $x$ approaches the number $c$ from the left-hand side (where $x < c$) or the right-hand side ($x > c$) only.

## Approaching a Limit from One Side

To have a limit $L$ as $x$ approaches $c$, a function $f$ must be defined on *both sides* of $c$ and its values $f(x)$ must approach $L$ as $x$ approaches $c$ from either side. That is, $f$ must be defined in some open interval about $c$, but not necessarily at $c$. Because of this, ordinary limits are called **two-sided**.

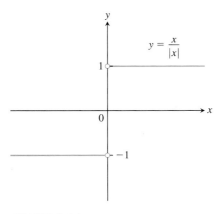

**FIGURE 2.24** Different right-hand and left-hand limits at the origin.

If $f$ fails to have a two-sided limit at $c$, it may still have a one-sided limit, that is, a limit if the approach is only from one side. If the approach is from the right, the limit is a **right-hand limit**. From the left, it is a **left-hand limit**.

The function $f(x) = x/|x|$ (Figure 2.24) has limit 1 as $x$ approaches 0 from the right, and limit $-1$ as $x$ approaches 0 from the left. Since these one-sided limit values are not the same, there is no single number that $f(x)$ approaches as $x$ approaches 0. So $f(x)$ does not have a (two-sided) limit at 0.

Intuitively, if $f(x)$ is defined on an interval $(c, b)$, where $c < b$, and approaches arbitrarily close to $L$ as $x$ approaches $c$ from within that interval, then $f$ has **right-hand limit** $L$ at $c$. We write

$$\lim_{x \to c^+} f(x) = L.$$

The symbol "$x \to c^+$" means that we consider only values of $x$ greater than $c$.

Similarly, if $f(x)$ is defined on an interval $(a, c)$, where $a < c$ and approaches arbitrarily close to $M$ as $x$ approaches $c$ from within that interval, then $f$ has **left-hand limit** $M$ at $c$. We write

$$\lim_{x \to c^-} f(x) = M.$$

The symbol "$x \to c^-$" means that we consider only $x$-values less than $c$.

These informal definitions of one-sided limits are illustrated in Figure 2.25. For the function $f(x) = x/|x|$ in Figure 2.24 we have

$$\lim_{x \to 0^+} f(x) = 1 \qquad \text{and} \qquad \lim_{x \to 0^-} f(x) = -1.$$

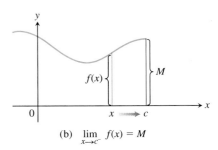

(a) $\lim_{x \to c^+} f(x) = L$  (b) $\lim_{x \to c^-} f(x) = M$

**FIGURE 2.25** (a) Right-hand limit as $x$ approaches $c$. (b) Left-hand limit as $x$ approaches $c$.

**EXAMPLE 1** The domain of $f(x) = \sqrt{4 - x^2}$ is $[-2, 2]$; its graph is the semicircle in Figure 2.26. We have

$$\lim_{x \to -2^+} \sqrt{4 - x^2} = 0 \qquad \text{and} \qquad \lim_{x \to 2^-} \sqrt{4 - x^2} = 0.$$

The function does not have a left-hand limit at $x = -2$ or a right-hand limit at $x = 2$. It does not have a two-sided limit at either $-2$ or 2 because each point does not belong to an open interval over which $f$ is defined. ∎

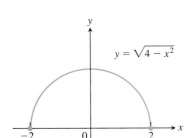

**FIGURE 2.26** The function $f(x) = \sqrt{4 - x^2}$ has right-hand limit 0 at $x = -2$ and left-hand limit 0 at $x = 2$ (Example 1).

One-sided limits have all the properties listed in Theorem 1 in Section 2.2. The right-hand limit of the sum of two functions is the sum of their right-hand limits, and so on. The theorems for limits of polynomials and rational functions hold with one-sided limits, as do the Sandwich Theorem and Theorem 5. One-sided limits are related to limits in the following way.

---

**THEOREM 6** A function $f(x)$ has a limit as $x$ approaches $c$ if and only if it has left-hand and right-hand limits there and these one-sided limits are equal:

$$\lim_{x \to c} f(x) = L \quad \Longleftrightarrow \quad \lim_{x \to c^-} f(x) = L \quad \text{and} \quad \lim_{x \to c^+} f(x) = L.$$

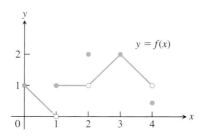

**FIGURE 2.27**   Graph of the function in Example 2.

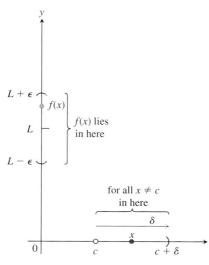

**FIGURE 2.28**   Intervals associated with the definition of right-hand limit.

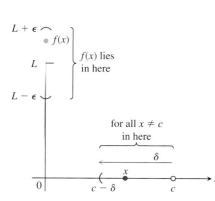

**FIGURE 2.29**   Intervals associated with the definition of left-hand limit.

**EXAMPLE 2**     For the function graphed in Figure 2.27,

At $x = 0$:     $\lim_{x \to 0^+} f(x) = 1$,
$\lim_{x \to 0^-} f(x)$ and $\lim_{x \to 0} f(x)$ do not exist. The function is not defined to the left of $x = 0$.

At $x = 1$:     $\lim_{x \to 1^-} f(x) = 0$ even though $f(1) = 1$,
$\lim_{x \to 1^+} f(x) = 1$,
$\lim_{x \to 1} f(x)$ does not exist. The right- and left-hand limits are not equal.

At $x = 2$:     $\lim_{x \to 2^-} f(x) = 1$,
$\lim_{x \to 2^+} f(x) = 1$,
$\lim_{x \to 2} f(x) = 1$ even though $f(2) = 2$.

At $x = 3$:     $\lim_{x \to 3^-} f(x) = \lim_{x \to 3^+} f(x) = \lim_{x \to 3} f(x) = f(3) = 2$.

At $x = 4$:     $\lim_{x \to 4^-} f(x) = 1$ even though $f(4) \neq 1$,
$\lim_{x \to 4^+} f(x)$ and $\lim_{x \to 4} f(x)$ do not exist. The function is not defined to the right of $x = 4$.

At every other point $c$ in $[0, 4]$, $f(x)$ has limit $f(c)$.     ∎

## Precise Definitions of One-Sided Limits

The formal definition of the limit in Section 2.3 is readily modified for one-sided limits.

---

**DEFINITIONS**   We say that $f(x)$ has **right-hand limit $L$ at $c$**, and write

$$\lim_{x \to c^+} f(x) = L \qquad \text{(see Figure 2.28)}$$

if for every number $\epsilon > 0$ there exists a corresponding number $\delta > 0$ such that for all $x$

$$c < x < c + \delta \quad \Rightarrow \quad |f(x) - L| < \epsilon.$$

We say that $f$ has **left-hand limit $L$ at $c$**, and write

$$\lim_{x \to c^-} f(x) = L \qquad \text{(see Figure 2.29)}$$

if for every number $\epsilon > 0$ there exists a corresponding number $\delta > 0$ such that for all $x$

$$c - \delta < x < c \quad \Rightarrow \quad |f(x) - L| < \epsilon.$$

---

**EXAMPLE 3**     Prove that

$$\lim_{x \to 0^+} \sqrt{x} = 0.$$

**Solution**   Let $\epsilon > 0$ be given. Here $c = 0$ and $L = 0$, so we want to find a $\delta > 0$ such that for all $x$

$$0 < x < \delta \quad \Rightarrow \quad |\sqrt{x} - 0| < \epsilon,$$

or

$$0 < x < \delta \quad \Rightarrow \quad \sqrt{x} < \epsilon.$$

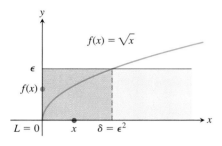

**FIGURE 2.30** $\lim_{x \to 0^+} \sqrt{x} = 0$ in Example 3.

Squaring both sides of this last inequality gives

$$x < \epsilon^2 \quad \text{if} \quad 0 < x < \delta.$$

If we choose $\delta = \epsilon^2$ we have

$$0 < x < \delta = \epsilon^2 \quad \Rightarrow \quad \sqrt{x} < \epsilon,$$

or

$$0 < x < \epsilon^2 \quad \Rightarrow \quad |\sqrt{x} - 0| < \epsilon.$$

According to the definition, this shows that $\lim_{x \to 0^+} \sqrt{x} = 0$ (Figure 2.30). ∎

The functions examined so far have had some kind of limit at each point of interest. In general, that need not be the case.

**EXAMPLE 4**    Show that $y = \sin(1/x)$ has no limit as $x$ approaches zero from either side (Figure 2.31).

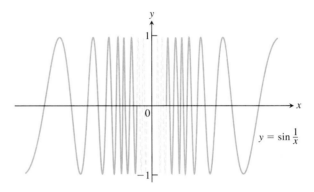

**FIGURE 2.31**    The function $y = \sin(1/x)$ has neither a right-hand nor a left-hand limit as $x$ approaches zero (Example 4). The graph here omits values very near the $y$-axis.

**Solution**    As $x$ approaches zero, its reciprocal, $1/x$, grows without bound and the values of $\sin(1/x)$ cycle repeatedly from $-1$ to $1$. There is no single number $L$ that the function's values stay increasingly close to as $x$ approaches zero. This is true even if we restrict $x$ to positive values or to negative values. The function has neither a right-hand limit nor a left-hand limit at $x = 0$. ∎

## Limits Involving $(\sin \theta)/\theta$

A central fact about $(\sin \theta)/\theta$ is that in radian measure its limit as $\theta \to 0$ is 1. We can see this in Figure 2.32 and confirm it algebraically using the Sandwich Theorem. You will see the importance of this limit in Section 3.5, where instantaneous rates of change of the trigonometric functions are studied.

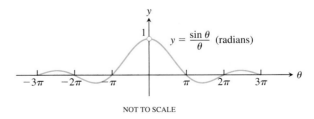

NOT TO SCALE

**FIGURE 2.32**    The graph of $f(\theta) = (\sin \theta)/\theta$ suggests that the right- and left-hand limits as $\theta$ approaches 0 are both 1.

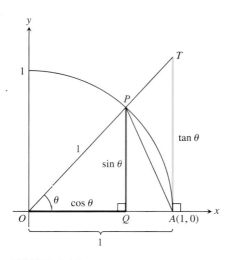

**FIGURE 2.33** The figure for the proof of Theorem 7. By definition, $TA/OA = \tan\theta$, but $OA = 1$, so $TA = \tan\theta$.

Equation (2) is where radian measure comes in: The area of sector $OAP$ is $\theta/2$ only if $\theta$ is measured in radians.

> **THEOREM 7—Limit of the Ratio $\sin\theta/\theta$ as $\theta \to 0$**
>
> $$\lim_{\theta\to 0}\frac{\sin\theta}{\theta} = 1 \qquad (\theta \text{ in radians}) \tag{1}$$

**Proof**    The plan is to show that the right-hand and left-hand limits are both 1. Then we will know that the two-sided limit is 1 as well.

To show that the right-hand limit is 1, we begin with positive values of $\theta$ less than $\pi/2$ (Figure 2.33). Notice that

$$\text{Area } \triangle OAP < \text{ area sector } OAP < \text{ area } \triangle OAT.$$

We can express these areas in terms of $\theta$ as follows:

$$\text{Area } \triangle OAP = \frac{1}{2}\text{base} \times \text{height} = \frac{1}{2}(1)(\sin\theta) = \frac{1}{2}\sin\theta$$

$$\text{Area sector } OAP = \frac{1}{2}r^2\theta = \frac{1}{2}(1)^2\theta = \frac{\theta}{2} \tag{2}$$

$$\text{Area } \triangle OAT = \frac{1}{2}\text{base} \times \text{height} = \frac{1}{2}(1)(\tan\theta) = \frac{1}{2}\tan\theta.$$

Thus,

$$\frac{1}{2}\sin\theta < \frac{1}{2}\theta < \frac{1}{2}\tan\theta.$$

This last inequality goes the same way if we divide all three terms by the number $(1/2)\sin\theta$, which is positive, since $0 < \theta < \pi/2$:

$$1 < \frac{\theta}{\sin\theta} < \frac{1}{\cos\theta}.$$

Taking reciprocals reverses the inequalities:

$$1 > \frac{\sin\theta}{\theta} > \cos\theta.$$

Since $\lim_{\theta\to 0^+}\cos\theta = 1$ (Example 11b, Section 2.2), the Sandwich Theorem gives

$$\lim_{\theta\to 0^+}\frac{\sin\theta}{\theta} = 1.$$

To consider the left-hand limit, we recall that $\sin\theta$ and $\theta$ are both *odd functions* (Section 1.1). Therefore, $f(\theta) = (\sin\theta)/\theta$ is an *even function*, with a graph symmetric about the $y$-axis (see Figure 2.32). This symmetry implies that the left-hand limit at 0 exists and has the same value as the right-hand limit:

$$\lim_{\theta\to 0^-}\frac{\sin\theta}{\theta} = 1 = \lim_{\theta\to 0^+}\frac{\sin\theta}{\theta},$$

so $\lim_{\theta\to 0}(\sin\theta)/\theta = 1$ by Theorem 6.    ■

**EXAMPLE 5**    Show that **(a)** $\lim_{h\to 0}\dfrac{\cos h - 1}{h} = 0$   and   **(b)** $\lim_{x\to 0}\dfrac{\sin 2x}{5x} = \dfrac{2}{5}$.

**Solution**

(a) Using the half-angle formula $\cos h = 1 - 2 \sin^2 (h/2)$, we calculate

$$\lim_{h \to 0} \frac{\cos h - 1}{h} = \lim_{h \to 0} -\frac{2 \sin^2 (h/2)}{h}$$

$$= -\lim_{\theta \to 0} \frac{\sin \theta}{\theta} \sin \theta \qquad \text{Let } \theta = h/2.$$

$$= -(1)(0) = 0. \qquad \text{Eq. (1) and Example 11a in Section 2.2}$$

(b) Equation (1) does not apply to the original fraction. We need a $2x$ in the denominator, not a $5x$. We produce it by multiplying numerator and denominator by $2/5$:

$$\lim_{x \to 0} \frac{\sin 2x}{5x} = \lim_{x \to 0} \frac{(2/5) \cdot \sin 2x}{(2/5) \cdot 5x}$$

$$= \frac{2}{5} \lim_{x \to 0} \frac{\sin 2x}{2x} \qquad \text{Now, Eq. (1) applies with } \theta = 2x.$$

$$= \frac{2}{5}(1) = \frac{2}{5} \qquad \blacksquare$$

**EXAMPLE 6**    Find $\displaystyle \lim_{t \to 0} \frac{\tan t \sec 2t}{3t}$.

**Solution**    From the definition of $\tan t$ and $\sec 2t$, we have

$$\lim_{t \to 0} \frac{\tan t \sec 2t}{3t} = \lim_{t \to 0} \frac{1}{3} \cdot \frac{1}{t} \cdot \frac{\sin t}{\cos t} \cdot \frac{1}{\cos 2t}$$

$$= \frac{1}{3} \lim_{t \to 0} \frac{\sin t}{t} \cdot \frac{1}{\cos t} \cdot \frac{1}{\cos 2t}$$

$$= \frac{1}{3}(1)(1)(1) = \frac{1}{3}. \qquad \text{Eq. (1) and Example 11b in Section 2.2} \qquad \blacksquare$$

# Exercises 2.4

### Finding Limits Graphically

**1.** Which of the following statements about the function $y = f(x)$ graphed here are true, and which are false?

a. $\displaystyle \lim_{x \to -1^+} f(x) = 1$

b. $\displaystyle \lim_{x \to 0^-} f(x) = 0$

c. $\displaystyle \lim_{x \to 0^-} f(x) = 1$

d. $\displaystyle \lim_{x \to 0^-} f(x) = \lim_{x \to 0^+} f(x)$

e. $\displaystyle \lim_{x \to 0} f(x)$ exists.

f. $\displaystyle \lim_{x \to 0} f(x) = 0$

g. $\displaystyle \lim_{x \to 0} f(x) = 1$

h. $\displaystyle \lim_{x \to 1} f(x) = 1$

i. $\displaystyle \lim_{x \to 1} f(x) = 0$

j. $\displaystyle \lim_{x \to 2^-} f(x) = 2$

k. $\displaystyle \lim_{x \to -1^-} f(x)$ does not exist.

l. $\displaystyle \lim_{x \to 2^+} f(x) = 0$

**2.** Which of the following statements about the function $y = f(x)$ graphed here are true, and which are false?

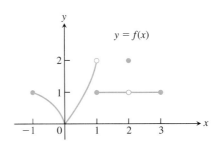

a. $\displaystyle \lim_{x \to -1^+} f(x) = 1$

b. $\displaystyle \lim_{x \to 2} f(x)$ does not exist.

c. $\displaystyle \lim_{x \to 2} f(x) = 2$

d. $\displaystyle \lim_{x \to 1^-} f(x) = 2$

e. $\displaystyle \lim_{x \to 1^+} f(x) = 1$

f. $\displaystyle \lim_{x \to 1} f(x)$ does not exist.

g. $\displaystyle \lim_{x \to 0^+} f(x) = \lim_{x \to 0^-} f(x)$

h. $\displaystyle \lim_{x \to c} f(x)$ exists at every $c$ in the open interval $(-1, 1)$.

i. $\displaystyle \lim_{x \to c} f(x)$ exists at every $c$ in the open interval $(1, 3)$.

j. $\displaystyle \lim_{x \to -1^-} f(x) = 0$

k. $\displaystyle \lim_{x \to 3^+} f(x)$ does not exist.

**3.** Let $f(x) = \begin{cases} 3 - x, & x < 2 \\ \dfrac{x}{2} + 1, & x > 2. \end{cases}$

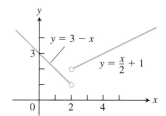

**a.** Find $\lim_{x \to 2^+} f(x)$ and $\lim_{x \to 2^-} f(x)$.

**b.** Does $\lim_{x \to 2} f(x)$ exist? If so, what is it? If not, why not?

**c.** Find $\lim_{x \to 4^-} f(x)$ and $\lim_{x \to 4^+} f(x)$.

**d.** Does $\lim_{x \to 4} f(x)$ exist? If so, what is it? If not, why not?

**4.** Let $f(x) = \begin{cases} 3 - x, & x < 2 \\ 2, & x = 2 \\ \dfrac{x}{2}, & x > 2. \end{cases}$

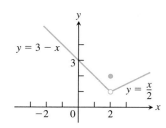

**a.** Find $\lim_{x \to 2^+} f(x)$, $\lim_{x \to 2^-} f(x)$, and $f(2)$.

**b.** Does $\lim_{x \to 2} f(x)$ exist? If so, what is it? If not, why not?

**c.** Find $\lim_{x \to -1^-} f(x)$ and $\lim_{x \to -1^+} f(x)$.

**d.** Does $\lim_{x \to -1} f(x)$ exist? If so, what is it? If not, why not?

**5.** Let $f(x) = \begin{cases} 0, & x \le 0 \\ \sin\dfrac{1}{x}, & x > 0. \end{cases}$

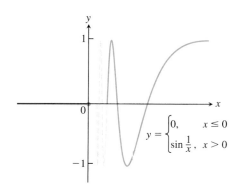

**a.** Does $\lim_{x \to 0^+} f(x)$ exist? If so, what is it? If not, why not?

**b.** Does $\lim_{x \to 0^-} f(x)$ exist? If so, what is it? If not, why not?

**c.** Does $\lim_{x \to 0} f(x)$ exist? If so, what is it? If not, why not?

**6.** Let $g(x) = \sqrt{x} \sin(1/x)$.

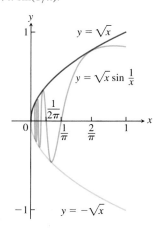

**a.** Does $\lim_{x \to 0^+} g(x)$ exist? If so, what is it? If not, why not?

**b.** Does $\lim_{x \to 0^-} g(x)$ exist? If so, what is it? If not, why not?

**c.** Does $\lim_{x \to 0} g(x)$ exist? If so, what is it? If not, why not?

**7. a.** Graph $f(x) = \begin{cases} x^3, & x \neq 1 \\ 0, & x = 1. \end{cases}$

**b.** Find $\lim_{x \to 1^-} f(x)$ and $\lim_{x \to 1^+} f(x)$.

**c.** Does $\lim_{x \to 1} f(x)$ exist? If so, what is it? If not, why not?

**8. a.** Graph $f(x) = \begin{cases} 1 - x^2, & x \neq 1 \\ 2, & x = 1. \end{cases}$

**b.** Find $\lim_{x \to 1^+} f(x)$ and $\lim_{x \to 1^-} f(x)$.

**c.** Does $\lim_{x \to 1} f(x)$ exist? If so, what is it? If not, why not?

Graph the functions in Exercises 9 and 10. Then answer these questions.

**a.** What are the domain and range of $f$?

**b.** At what points $c$, if any, does $\lim_{x \to c} f(x)$ exist?

**c.** At what points does only the left-hand limit exist?

**d.** At what points does only the right-hand limit exist?

**9.** $f(x) = \begin{cases} \sqrt{1 - x^2}, & 0 \le x < 1 \\ 1, & 1 \le x < 2 \\ 2, & x = 2 \end{cases}$

**10.** $f(x) = \begin{cases} x, & -1 \le x < 0, \text{ or } 0 < x \le 1 \\ 1, & x = 0 \\ 0, & x < -1 \text{ or } x > 1 \end{cases}$

Finding One-Sided Limits Algebraically

Find the limits in Exercises 11–18.

**11.** $\lim_{x \to -0.5^-} \sqrt{\dfrac{x + 2}{x + 1}}$

**12.** $\lim_{x \to 1^+} \sqrt{\dfrac{x - 1}{x + 2}}$

**13.** $\lim_{x \to -2^+} \left(\dfrac{x}{x + 1}\right)\left(\dfrac{2x + 5}{x^2 + x}\right)$

**14.** $\lim_{x \to 1^-} \left(\dfrac{1}{x + 1}\right)\left(\dfrac{x + 6}{x}\right)\left(\dfrac{3 - x}{7}\right)$

**15.** $\lim_{h \to 0^+} \dfrac{\sqrt{h^2 + 4h + 5} - \sqrt{5}}{h}$

**16.** $\lim\limits_{h \to 0^-} \dfrac{\sqrt{6} - \sqrt{5h^2 + 11h + 6}}{h}$

**17. a.** $\lim\limits_{x \to -2^+} (x + 3) \dfrac{|x + 2|}{x + 2}$　　**b.** $\lim\limits_{x \to -2^-} (x + 3) \dfrac{|x + 2|}{x + 2}$

**18. a.** $\lim\limits_{x \to 1^+} \dfrac{\sqrt{2x}(x - 1)}{|x - 1|}$　　**b.** $\lim\limits_{x \to 1^-} \dfrac{\sqrt{2x}(x - 1)}{|x - 1|}$

Use the graph of the greatest integer function $y = \lfloor x \rfloor$, Figure 1.10 in Section 1.1, to help you find the limits in Exercises 19 and 20.

**19. a.** $\lim\limits_{\theta \to 3^-} \dfrac{\lfloor \theta \rfloor}{\theta}$　　**b.** $\lim\limits_{\theta \to 3^-} \dfrac{\lfloor \theta \rfloor}{\theta}$

**20. a.** $\lim\limits_{t \to 4^+} (t - \lfloor t \rfloor)$　　**b.** $\lim\limits_{t \to 4^-} (t - \lfloor t \rfloor)$

Using $\lim\limits_{\theta \to 0} \dfrac{\sin \theta}{\theta} = 1$

Find the limits in Exercises 21–42.

**21.** $\lim\limits_{\theta \to 0} \dfrac{\sin \sqrt{2}\theta}{\sqrt{2}\theta}$　　**22.** $\lim\limits_{t \to 0} \dfrac{\sin kt}{t}$ ($k$ constant)

**23.** $\lim\limits_{y \to 0} \dfrac{\sin 3y}{4y}$　　**24.** $\lim\limits_{h \to 0^-} \dfrac{h}{\sin 3h}$

**25.** $\lim\limits_{x \to 0} \dfrac{\tan 2x}{x}$　　**26.** $\lim\limits_{t \to 0} \dfrac{2t}{\tan t}$

**27.** $\lim\limits_{x \to 0} \dfrac{x \csc 2x}{\cos 5x}$　　**28.** $\lim\limits_{x \to 0} 6x^2(\cot x)(\csc 2x)$

**29.** $\lim\limits_{x \to 0} \dfrac{x + x \cos x}{\sin x \cos x}$　　**30.** $\lim\limits_{x \to 0} \dfrac{x^2 - x + \sin x}{2x}$

**31.** $\lim\limits_{\theta \to 0} \dfrac{1 - \cos \theta}{\sin 2\theta}$　　**32.** $\lim\limits_{x \to 0} \dfrac{x - x \cos x}{\sin^2 3x}$

**33.** $\lim\limits_{t \to 0} \dfrac{\sin(1 - \cos t)}{1 - \cos t}$　　**34.** $\lim\limits_{h \to 0} \dfrac{\sin(\sin h)}{\sin h}$

**35.** $\lim\limits_{\theta \to 0} \dfrac{\sin \theta}{\sin 2\theta}$　　**36.** $\lim\limits_{x \to 0} \dfrac{\sin 5x}{\sin 4x}$

**37.** $\lim\limits_{\theta \to 0} \theta \cos \theta$　　**38.** $\lim\limits_{\theta \to 0} \sin \theta \cot 2\theta$

**39.** $\lim\limits_{x \to 0} \dfrac{\tan 3x}{\sin 8x}$　　**40.** $\lim\limits_{y \to 0} \dfrac{\sin 3y \cot 5y}{y \cot 4y}$

**41.** $\lim\limits_{\theta \to 0} \dfrac{\tan \theta}{\theta^2 \cot 3\theta}$　　**42.** $\lim\limits_{\theta \to 0} \dfrac{\theta \cot 4\theta}{\sin^2 \theta \cot^2 2\theta}$

### Theory and Examples

**43.** Once you know $\lim\limits_{x \to a^+} f(x)$ and $\lim\limits_{x \to a^-} f(x)$ at an interior point of the domain of $f$, do you then know $\lim\limits_{x \to a} f(x)$? Give reasons for your answer.

**44.** If you know that $\lim\limits_{x \to c} f(x)$ exists, can you find its value by calculating $\lim\limits_{x \to c^+} f(x)$? Give reasons for your answer.

**45.** Suppose that $f$ is an odd function of $x$. Does knowing that $\lim\limits_{x \to 0^+} f(x) = 3$ tell you anything about $\lim\limits_{x \to 0^-} f(x)$? Give reasons for your answer.

**46.** Suppose that $f$ is an even function of $x$. Does knowing that $\lim\limits_{x \to -2^-} f(x) = 7$ tell you anything about either $\lim\limits_{x \to -2^-} f(x)$ or $\lim\limits_{x \to -2^+} f(x)$? Give reasons for your answer.

### Formal Definitions of One-Sided Limits

**47.** Given $\epsilon > 0$, find an interval $I = (5, 5 + \delta)$, $\delta > 0$, such that if $x$ lies in $I$, then $\sqrt{x - 5} < \epsilon$. What limit is being verified and what is its value?

**48.** Given $\epsilon > 0$, find an interval $I = (4 - \delta, 4)$, $\delta > 0$, such that if $x$ lies in $I$, then $\sqrt{4 - x} < \epsilon$. What limit is being verified and what is its value?

Use the definitions of right-hand and left-hand limits to prove the limit statements in Exercises 49 and 50.

**49.** $\lim\limits_{x \to 0^-} \dfrac{x}{|x|} = -1$　　**50.** $\lim\limits_{x \to 2^+} \dfrac{x - 2}{|x - 2|} = 1$

**51. Greatest integer function** Find **(a)** $\lim\limits_{x \to 400^+} \lfloor x \rfloor$ and **(b)** $\lim\limits_{x \to 400^-} \lfloor x \rfloor$; then use limit definitions to verify your findings. **(c)** Based on your conclusions in parts (a) and (b), can you say anything about $\lim\limits_{x \to 400} \lfloor x \rfloor$? Give reasons for your answer.

**52. One-sided limits** Let $f(x) = \begin{cases} x^2 \sin(1/x), & x < 0 \\ \sqrt{x}, & x > 0. \end{cases}$

Find **(a)** $\lim\limits_{x \to 0^+} f(x)$ and **(b)** $\lim\limits_{x \to 0^-} f(x)$; then use limit definitions to verify your findings. **(c)** Based on your conclusions in parts (a) and (b), can you say anything about $\lim\limits_{x \to 0} f(x)$? Give reasons for your answer.

# $2.5$ Continuity

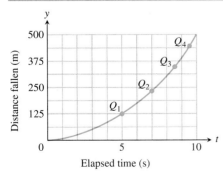

**FIGURE 2.34** Connecting plotted points by an unbroken curve from experimental data $Q_1, Q_2, Q_3, \ldots$ for a falling object.

When we plot function values generated in a laboratory or collected in the field, we often connect the plotted points with an unbroken curve to show what the function's values are likely to have been at the points we did not measure (Figure 2.34). In doing so, we are assuming that we are working with a *continuous function*, so its outputs vary regularly and consistently with the inputs, and do not jump abruptly from one value to another without taking on the values in between. Intuitively, any function $y = f(x)$ whose graph can be sketched over its domain in one unbroken motion is an example of a continuous function. Such functions play an important role in the study of calculus and its applications.

## Continuity at a Point

To understand continuity, it helps to consider a function like that in Figure 2.35, whose limits we investigated in Example 2 in the last section.

**EXAMPLE 1**    At which numbers does the function $f$ in Figure 2.35 appear to be not continuous? Explain why. What occurs at other numbers in the domain?

**Solution**    First we observe that the domain of the function is the closed interval $[0, 4]$, so we will be considering the numbers $x$ within that interval. From the figure, we notice right away that there are breaks in the graph at the numbers $x = 1$, $x = 2$, and $x = 4$. The breaks appear as jumps, which we identify later as "jump discontinuities." These are numbers for which the function is not continuous, and we discuss each in turn.

**Numbers at which the graph of $f$ has breaks:**

At $x = 1$, the function fails to have a limit. It does have both a left-hand limit, $\lim_{x \to 1^-} f(x) = 0$, as well as a right-hand limit, $\lim_{x \to 1^+} f(x) = 1$, but the limit values are different, resulting in a jump in the graph. The function is not continuous at $x = 1$.

At $x = 2$, the function does have a limit, $\lim_{x \to 2} f(x) = 1$, but the value of the function is $f(2) = 2$. The limit and function values are not the same, so there is a break in the graph and $f$ is not continuous at $x = 2$.

At $x = 4$, the function does have a left-hand limit at this right endpoint, $\lim_{x \to 4^-} f(x) = 1$, but again the value of the function $f(4) = \frac{1}{2}$ differs from the value of the limit. We see again a break in the graph of the function at this endpoint and the function is not continuous from the left.

**Numbers at which the graph of $f$ has no breaks:**

At $x = 0$, the function has a right-hand limit at this left endpoint, $\lim_{x \to 0^+} f(x) = 1$, and the value of the function is the same, $f(0) = 1$. So no break occurs in the graph of the function at this endpoint, and the function is continuous from the right at $x = 0$.

At $x = 3$, the function has a limit, $\lim_{x \to 3} f(x) = 2$. Moreover, the limit is the same value as the function there, $f(3) = 2$. No break occurs in the graph and the function is continuous at $x = 3$.

At all other numbers $x = c$ in the domain, which we have not considered, the function has a limit equal to the value of the function at the point, so $\lim_{x \to c} f(x) = f(c)$. For example, $\lim_{x \to 5/2} f(x) = f\left(\frac{5}{2}\right) = \frac{3}{2}$. No breaks appear in the graph of the function at any of these remaining numbers and the function is continuous at each of them.    ∎

The following definitions capture the continuity ideas we observed in Example 1.

> **DEFINITIONS**    Let $c$ be a real number on the $x$-axis.
>
> The function $f$ is **continuous at $c$** if
> $$\lim_{x \to c} f(x) = f(c).$$
> The function $f$ is **right-continuous at $c$ (or continuous from the right)** if
> $$\lim_{x \to c^+} f(x) = f(c).$$
> The function $f$ is **left-continuous at $c$ (or continuous from the left)** if
> $$\lim_{x \to c^-} f(x) = f(c).$$

From Theorem 6, it follows immediately that a function $f$ is continuous at an interior point $c$ of its domain if and only if it is both right-continuous and left-continuous at $c$ (Figure 2.36). We say that a function is **continuous over a closed interval** $[a, b]$ if it is right-continuous at $a$, left-continuous at $b$, and continuous at all interior points of the interval.

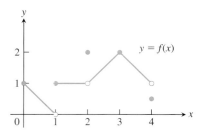

**FIGURE 2.35**    The function is not continuous at $x = 1$, $x = 2$, and $x = 4$ (Example 1).

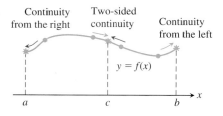

**FIGURE 2.36**    Continuity at points $a$, $b$, and $c$.

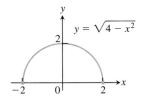

**FIGURE 2.37** A function that is continuous over its domain (Example 2).

**FIGURE 2.38** A function that has a jump discontinuity at the origin (Example 3).

This definition applies to the infinite closed intervals $[a, \infty)$ and $(-\infty, b]$ as well, but only one endpoint is involved. If a function is not continuous at an interior point $c$ of its domain, we say that $f$ is **discontinuous at $c$**, and that $c$ is a point of discontinuity of $f$. Note that a function $f$ can be continuous, right-continuous, or left-continuous only at a point $c$ for which $f(c)$ is defined.

**EXAMPLE 2** The function $f(x) = \sqrt{4 - x^2}$ is continuous over its domain $[-2, 2]$ (Figure 2.37). It is right-continuous at $x = -2$, and left-continuous at $x = 2$. ∎

**EXAMPLE 3** The unit step function $U(x)$, graphed in Figure 2.38, is right-continuous at $x = 0$, but is neither left-continuous nor continuous there. It has a jump discontinuity at $x = 0$. ∎

We summarize continuity at an interior point in the form of a test.

---

**Continuity Test**

A function $f(x)$ is continuous at a point $x = c$ if and only if it meets the following three conditions.

1. $f(c)$ exists          ($c$ lies in the domain of $f$).
2. $\lim_{x \to c} f(x)$ exists          ($f$ has a limit as $x \to c$).
3. $\lim_{x \to c} f(x) = f(c)$          (the limit equals the function value).

---

For one-sided continuity and continuity at an endpoint of an interval, the limits in parts 2 and 3 of the test should be replaced by the appropriate one-sided limits.

**EXAMPLE 4** The function $y = \lfloor x \rfloor$ introduced in Section 1.1 is graphed in Figure 2.39. It is discontinuous at every integer because the left-hand and right-hand limits are not equal as $x \to n$:

$$\lim_{x \to n^-} \lfloor x \rfloor = n - 1 \quad \text{and} \quad \lim_{x \to n^+} \lfloor x \rfloor = n.$$

Since $\lfloor n \rfloor = n$, the greatest integer function is right-continuous at every integer $n$ (but not left-continuous).

The greatest integer function is continuous at every real number other than the integers. For example,

$$\lim_{x \to 1.5} \lfloor x \rfloor = 1 = \lfloor 1.5 \rfloor.$$

In general, if $n - 1 < c < n$, $n$ an integer, then

$$\lim_{x \to c} \lfloor x \rfloor = n - 1 = \lfloor c \rfloor. \qquad ∎$$

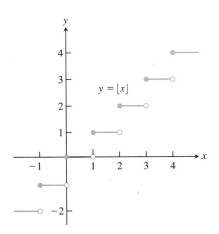

**FIGURE 2.39** The greatest integer function is continuous at every noninteger point. It is right-continuous, but not left-continuous, at every integer point (Example 4).

Figure 2.40 displays several common types of discontinuities. The function in Figure 2.40a is continuous at $x = 0$. The function in Figure 2.40b would be continuous if it had $f(0) = 1$. The function in Figure 2.40c would be continuous if $f(0)$ were 1 instead of 2. The discontinuity in Figure 2.40c is **removable**. The function has a limit as $x \to 0$, and we can remove the discontinuity by setting $f(0)$ equal to this limit.

The discontinuities in Figure 2.40d through f are more serious: $\lim_{x \to 0} f(x)$ does not exist, and there is no way to improve the situation by changing $f$ at 0. The step function in Figure 2.40d has a **jump discontinuity**: The one-sided limits exist but have different values. The function $f(x) = 1/x^2$ in Figure 2.40e has an **infinite discontinuity**. The function in Figure 2.40f has an **oscillating discontinuity**: It oscillates too much to have a limit as $x \to 0$.

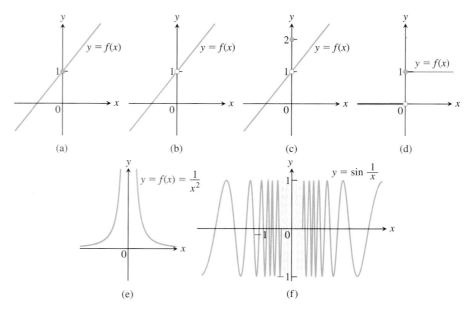

**FIGURE 2.40** The function in (a) is continuous at $x = 0$; the functions in (b) through (f) are not.

## Continuous Functions

Generally, we want to describe the continuity behavior of a function throughout its entire domain, not only at a single point. We know how to do that if the domain is a closed interval. In the same way, we define a **continuous function** as one that is continuous at every point in its domain. This is a property of the *function*. A function always has a specified domain, so if we change the domain, we change the function, and this may change its continuity property as well. If a function is discontinuous at one or more points of its domain, we say it is a **discontinuous** function.

## EXAMPLE 5

(a) The function $y = 1/x$ (Figure 2.41) is a continuous function because it is continuous at every point of its domain. It has a point of discontinuity at $x = 0$, however, because it is not defined there; that is, it is discontinuous on any interval containing $x = 0$.

(b) The identity function $f(x) = x$ and constant functions are continuous everywhere by Example 3, Section 2.3. ∎

Algebraic combinations of continuous functions are continuous wherever they are defined.

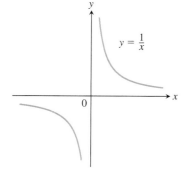

**FIGURE 2.41** The function $y = 1/x$ is continuous over its natural domain. It has a point of discontinuity at the origin, so it is discontinuous on any interval containing $x = 0$ (Example 5).

---

**THEOREM 8—Properties of Continuous Functions** If the functions $f$ and $g$ are continuous at $x = c$, then the following algebraic combinations are continuous at $x = c$.

1. *Sums:* $f + g$
2. *Differences:* $f - g$
3. *Constant multiples:* $k \cdot f,$ for any number $k$
4. *Products:* $f \cdot g$
5. *Quotients:* $f/g,$ provided $g(c) \neq 0$
6. *Powers:* $f^n,$ $n$ a positive integer
7. *Roots:* $\sqrt[n]{f},$ provided it is defined on an open interval containing $c$, where $n$ is a positive integer

---

Most of the results in Theorem 8 follow from the limit rules in Theorem 1, Section 2.2. For instance, to prove the sum property we have

$$\lim_{x \to c}(f + g)(x) = \lim_{x \to c}(f(x) + g(x))$$
$$= \lim_{x \to c} f(x) + \lim_{x \to c} g(x) \qquad \text{Sum Rule, Theorem 1}$$
$$= f(c) + g(c) \qquad \text{Continuity of } f, g \text{ at } c$$
$$= (f + g)(c).$$

This shows that $f + g$ is continuous.

## EXAMPLE 6

**(a)** Every polynomial $P(x) = a_n x^n + a_{n-1}x^{n-1} + \cdots + a_0$ is continuous because $\lim_{x \to c} P(x) = P(c)$ by Theorem 2, Section 2.2.

**(b)** If $P(x)$ and $Q(x)$ are polynomials, then the rational function $P(x)/Q(x)$ is continuous wherever it is defined ($Q(c) \neq 0$) by Theorem 3, Section 2.2. ∎

**EXAMPLE 7** The function $f(x) = |x|$ is continuous. If $x > 0$, we have $f(x) = x$, a polynomial. If $x < 0$, we have $f(x) = -x$, another polynomial. Finally, at the origin, $\lim_{x \to 0} |x| = 0 = |0|$. ∎

The functions $y = \sin x$ and $y = \cos x$ are continuous at $x = 0$ by Example 11 of Section 2.2. Both functions are, in fact, continuous everywhere (see Exercise 70). It follows from Theorem 8 that all six trigonometric functions are then continuous wherever they are defined. For example, $y = \tan x$ is continuous on $\cdots \cup (-\pi/2, \pi/2) \cup (\pi/2, 3\pi/2) \cup \cdots$.

### Composites

All composites of continuous functions are continuous. The idea is that if $f(x)$ is continuous at $x = c$ and $g(x)$ is continuous at $x = f(c)$, then $g \circ f$ is continuous at $x = c$ (Figure 2.42). In this case, the limit as $x \to c$ is $g(f(c))$.

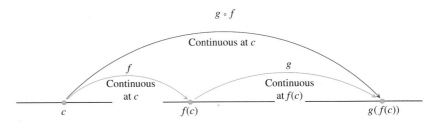

**FIGURE 2.42** Composites of continuous functions are continuous.

**THEOREM 9—Composite of Continuous Functions** If $f$ is continuous at $c$ and $g$ is continuous at $f(c)$, then the composite $g \circ f$ is continuous at $c$.

Intuitively, Theorem 9 is reasonable because if $x$ is close to $c$, then $f(x)$ is close to $f(c)$, and since $g$ is continuous at $f(c)$, it follows that $g(f(x))$ is close to $g(f(c))$.

The continuity of composites holds for any finite number of functions. The only requirement is that each function be continuous where it is applied. For an outline of a proof of Theorem 9, see Exercise 6 in Appendix 4.

**EXAMPLE 8** Show that the following functions are continuous on their natural domains.

(a) $y = \sqrt{x^2 - 2x - 5}$

(b) $y = \dfrac{x^{2/3}}{1 + x^4}$

(c) $y = \left| \dfrac{x - 2}{x^2 - 2} \right|$

(d) $y = \left| \dfrac{x \sin x}{x^2 + 2} \right|$

**Solution**

(a) The square root function is continuous on $[0, \infty)$ because it is a root of the continuous identity function $f(x) = x$ (Part 7, Theorem 8). The given function is then the composite of the polynomial $f(x) = x^2 - 2x - 5$ with the square root function $g(t) = \sqrt{t}$, and is continuous on its natural domain.

(b) The numerator is the cube root of the identity function squared; the denominator is an everywhere-positive polynomial. Therefore, the quotient is continuous.

(c) The quotient $(x - 2)/(x^2 - 2)$ is continuous for all $x \neq \pm\sqrt{2}$, and the function is the composition of this quotient with the continuous absolute value function (Example 7).

(d) Because the sine function is everywhere-continuous (Exercise 70), the numerator term $x \sin x$ is the product of continuous functions, and the denominator term $x^2 + 2$ is an everywhere-positive polynomial. The given function is the composite of a quotient of continuous functions with the continuous absolute value function (Figure 2.43). ∎

Theorem 9 is actually a consequence of a more general result, which we now state and prove.

**FIGURE 2.43** The graph suggests that $y = \left| (x \sin x)/(x^2 + 2) \right|$ is continuous (Example 8d).

---

**THEOREM 10—Limits of Continuous Functions** If $g$ is continuous at the point $b$ and $\lim_{x \to c} f(x) = b$, then

$$\lim_{x \to c} g(f(x)) = g(b) = g(\lim_{x \to c} f(x)).$$

---

**Proof** Let $\epsilon > 0$ be given. Since $g$ is continuous at $b$, there exists a number $\delta_1 > 0$ such that

$$\left| g(y) - g(b) \right| < \epsilon \quad \text{whenever} \quad 0 < \left| y - b \right| < \delta_1.$$

Since $\lim_{x \to c} f(x) = b$, there exists a $\delta > 0$ such that

$$\left| f(x) - b \right| < \delta_1 \quad \text{whenever} \quad 0 < \left| x - c \right| < \delta.$$

If we let $y = f(x)$, we then have that

$$\left| y - b \right| < \delta_1 \quad \text{whenever} \quad 0 < \left| x - c \right| < \delta,$$

which implies from the first statement that $\left| g(y) - g(b) \right| = \left| g(f(x)) - g(b) \right| < \epsilon$ whenever $0 < \left| x - c \right| < \delta$. From the definition of limit, this proves that $\lim_{x \to c} g(f(x)) = g(b)$. ∎

**EXAMPLE 9** As an application of Theorem 10, we have the following calculation:

$$\lim_{x \to \pi/2} \cos\left( 2x + \sin\left( \frac{3\pi}{2} + x \right) \right) = \cos\left( \lim_{x \to \pi/2} 2x + \lim_{x \to \pi/2} \sin\left( \frac{3\pi}{2} + x \right) \right)$$

$$= \cos\left( \pi + \sin 2\pi \right) = \cos \pi = -1. \quad ∎$$

## Intermediate Value Theorem for Continuous Functions

Functions that are continuous on intervals have properties that make them particularly useful in mathematics and its applications. One of these is the *Intermediate Value Property*. A function is said to have the **Intermediate Value Property** if whenever it takes on two values, it also takes on all the values in between.

**THEOREM 11—The Intermediate Value Theorem for Continuous Functions**   If $f$ is a continuous function on a closed interval $[a, b]$, and if $y_0$ is any value between $f(a)$ and $f(b)$, then $y_0 = f(c)$ for some $c$ in $[a, b]$.

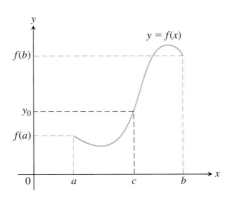

Theorem 11 says that continuous functions over *finite closed* intervals have the Intermediate Value Property. Geometrically, the Intermediate Value Theorem says that any horizontal line $y = y_0$ crossing the $y$-axis between the numbers $f(a)$ and $f(b)$ will cross the curve $y = f(x)$ at least once over the interval $[a, b]$.

The proof of the Intermediate Value Theorem depends on the completeness property of the real number system (Appendix 7) and can be found in more advanced texts.

The continuity of $f$ on the interval is essential to Theorem 11. If $f$ is discontinuous at even one point of the interval, the theorem's conclusion may fail, as it does for the function graphed in Figure 2.44 (choose $y_0$ as any number between 2 and 3).

**A Consequence for Graphing: Connectedness**   Theorem 11 implies that the graph of a function continuous on an interval cannot have any breaks over the interval. It will be **connected**—a single, unbroken curve. It will not have jumps like the graph of the greatest integer function (Figure 2.39), or separate branches like the graph of $1/x$ (Figure 2.41).

**A Consequence for Root Finding**   We call a solution of the equation $f(x) = 0$ a **root** of the equation or **zero** of the function $f$. The Intermediate Value Theorem tells us that if $f$ is continuous, then any interval on which $f$ changes sign contains a zero of the function.

In practical terms, when we see the graph of a continuous function cross the horizontal axis on a computer screen, we know it is not stepping across. There really is a point where the function's value is zero.

**EXAMPLE 10**   Show that there is a root of the equation $x^3 - x - 1 = 0$ between 1 and 2.

**Solution**   Let $f(x) = x^3 - x - 1$. Since $f(1) = 1 - 1 - 1 = -1 < 0$ and $f(2) = 2^3 - 2 - 1 = 5 > 0$, we see that $y_0 = 0$ is a value between $f(1)$ and $f(2)$. Since $f$ is continuous, the Intermediate Value Theorem says there is a zero of $f$ between 1 and 2. Figure 2.45 shows the result of zooming in to locate the root near $x = 1.32$.  ∎

**FIGURE 2.44**   The function
$$f(x) = \begin{cases} 2x - 2, & 1 \le x < 2 \\ 3, & 2 \le x \le 4 \end{cases}$$
does not take on all values between $f(1) = 0$ and $f(4) = 3$; it misses all the values between 2 and 3.

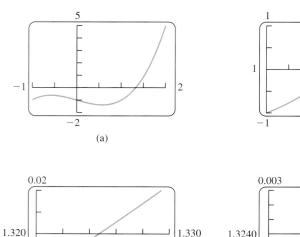

(a)

(b)

(c)

(d)

**FIGURE 2.45** Zooming in on a zero of the function $f(x) = x^3 - x - 1$. The zero is near $x = 1.3247$ (Example 10).

**EXAMPLE 11** Use the Intermediate Value Theorem to prove that the equation

$$\sqrt{2x + 5} = 4 - x^2$$

has a solution (Figure 2.46).

**Solution** We rewrite the equation as

$$\sqrt{2x + 5} + x^2 = 4,$$

and set $f(x) = \sqrt{2x + 5} + x^2$. Now $g(x) = \sqrt{2x + 5}$ is continuous on the interval $[-5/2, \infty)$ since it is the composite of the square root function with the nonnegative linear function $y = 2x + 5$. Then $f$ is the sum of the function $g$ and the quadratic function $y = x^2$, and the quadratic function is continuous for all values of $x$. It follows that $f(x) = \sqrt{2x + 5} + x^2$ is continuous on the interval $[-5/2, \infty)$. By trial and error, we find the function values $f(0) = \sqrt{5} \approx 2.24$ and $f(2) = \sqrt{9} + 4 = 7$, and note that $f$ is also continuous on the finite closed interval $[0, 2] \subset [-5/2, \infty)$. Since the value $y_0 = 4$ is between the numbers 2.24 and 7, by the Intermediate Value Theorem there is a number $c \in [0, 2]$ such that $f(c) = 4$. That is, the number $c$ solves the original equation. ∎

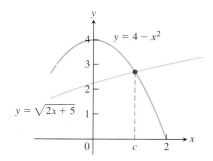

**FIGURE 2.46** The curves $y = \sqrt{2x + 5}$ and $y = 4 - x^2$ have the same value at $x = c$ where $\sqrt{2x + 5} = 4 - x^2$ (Example 11).

## Continuous Extension to a Point

Sometimes the formula that describes a function $f$ does not make sense at a point $x = c$. It might nevertheless be possible to extend the domain of $f$, to include $x = c$, creating a new function that is continuous at $x = c$. For example, the function $y = f(x) = (\sin x)/x$ is continuous at every point except $x = 0$, since the origin is not in its domain. Since $y = (\sin x)/x$ has a finite limit as $x \to 0$ (Theorem 7), we can extend the function's domain to include the point $x = 0$ in such a way that the extended function is continuous at $x = 0$. We define the new function

$$F(x) = \begin{cases} \dfrac{\sin x}{x}, & x \neq 0 \\ 1, & x = 0. \end{cases}$$

The function $F(x)$ is continuous at $x = 0$ because

$$\lim_{x \to 0} \frac{\sin x}{x} = F(0),$$

so it meets the requirements for continuity (Figure 2.47).

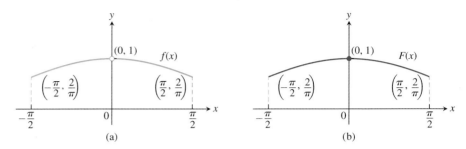

**FIGURE 2.47**   The graph (a) of $f(x) = (\sin x)/x$ for $-\pi/2 \le x \le \pi/2$ does not include the point $(0, 1)$ because the function is not defined at $x = 0$. (b) We can remove the discontinuity from the graph by defining the new function $F(x)$ with $F(0) = 1$ and $F(x) = f(x)$ everywhere else. Note that $F(0) = \lim_{x \to 0} f(x)$.

More generally, a function (such as a rational function) may have a limit at a point where it is not defined. If $f(c)$ is not defined, but $\lim_{x \to c} f(x) = L$ exists, we can define a new function $F(x)$ by the rule

$$F(x) = \begin{cases} f(x), & \text{if } x \text{ is in the domain of } f \\ L, & \text{if } x = c. \end{cases}$$

The function $F$ is continuous at $x = c$. It is called the **continuous extension of $f$** to $x = c$. For rational functions $f$, continuous extensions are often found by canceling common factors in the numerator and denominator.

**EXAMPLE 12**   Show that

$$f(x) = \frac{x^2 + x - 6}{x^2 - 4}, \quad x \ne 2$$

has a continuous extension to $x = 2$, and find that extension.

**Solution**   Although $f(2)$ is not defined, if $x \ne 2$ we have

$$f(x) = \frac{x^2 + x - 6}{x^2 - 4} = \frac{(x - 2)(x + 3)}{(x - 2)(x + 2)} = \frac{x + 3}{x + 2}.$$

The new function

$$F(x) = \frac{x + 3}{x + 2}$$

is equal to $f(x)$ for $x \ne 2$, but is continuous at $x = 2$, having there the value of $5/4$. Thus $F$ is the continuous extension of $f$ to $x = 2$, and

$$\lim_{x \to 2} \frac{x^2 + x - 6}{x^2 - 4} = \lim_{x \to 2} f(x) = \frac{5}{4}.$$

The graph of $f$ is shown in Figure 2.48. The continuous extension $F$ has the same graph except with no hole at $(2, 5/4)$. Effectively, $F$ is the function $f$ with its point of discontinuity at $x = 2$ removed.   ∎

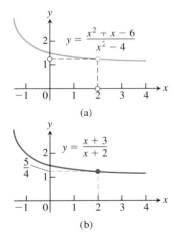

**FIGURE 2.48**   (a) The graph of $f(x)$ and (b) the graph of its continuous extension $F(x)$ (Example 12).

# Exercises 2.5

### Continuity from Graphs

In Exercises 1–4, say whether the function graphed is continuous on $[-1, 3]$. If not, where does it fail to be continuous and why?

**1.**

**2.**

**3.**

**4.**
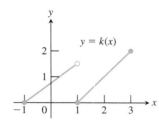

Exercises 5–10 refer to the function

$$f(x) = \begin{cases} x^2 - 1, & -1 \le x < 0 \\ 2x, & 0 < x < 1 \\ 1, & x = 1 \\ -2x + 4, & 1 < x < 2 \\ 0, & 2 < x < 3 \end{cases}$$

graphed in the accompanying figure.

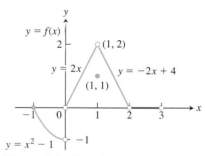

The graph for Exercises 5–10.

**5. a.** Does $f(-1)$ exist?

   **b.** Does $\lim_{x \to -1^+} f(x)$ exist?

   **c.** Does $\lim_{x \to -1^+} f(x) = f(-1)$?

   **d.** Is $f$ continuous at $x = -1$?

**6. a.** Does $f(1)$ exist?

   **b.** Does $\lim_{x \to 1} f(x)$ exist?

   **c.** Does $\lim_{x \to 1} f(x) = f(1)$?

   **d.** Is $f$ continuous at $x = 1$?

**7. a.** Is $f$ defined at $x = 2$? (Look at the definition of $f$.)

   **b.** Is $f$ continuous at $x = 2$?

**8.** At what values of $x$ is $f$ continuous?

**9.** What value should be assigned to $f(2)$ to make the extended function continuous at $x = 2$?

**10.** To what new value should $f(1)$ be changed to remove the discontinuity?

### Applying the Continuity Test

At which points do the functions in Exercises 11 and 12 fail to be continuous? At which points, if any, are the discontinuities removable? Not removable? Give reasons for your answers.

**11.** Exercise 1, Section 2.4    **12.** Exercise 2, Section 2.4

At what points are the functions in Exercises 13–30 continuous?

**13.** $y = \dfrac{1}{x - 2} - 3x$

**14.** $y = \dfrac{1}{(x + 2)^2} + 4$

**15.** $y = \dfrac{x + 1}{x^2 - 4x + 3}$

**16.** $y = \dfrac{x + 3}{x^2 - 3x - 10}$

**17.** $y = |x - 1| + \sin x$

**18.** $y = \dfrac{1}{|x| + 1} - \dfrac{x^2}{2}$

**19.** $y = \dfrac{\cos x}{x}$

**20.** $y = \dfrac{x + 2}{\cos x}$

**21.** $y = \csc 2x$

**22.** $y = \tan \dfrac{\pi x}{2}$

**23.** $y = \dfrac{x \tan x}{x^2 + 1}$

**24.** $y = \dfrac{\sqrt{x^4 + 1}}{1 + \sin^2 x}$

**25.** $y = \sqrt{2x + 3}$

**26.** $y = \sqrt[4]{3x - 1}$

**27.** $y = (2x - 1)^{1/3}$

**28.** $y = (2 - x)^{1/5}$

**29.** $g(x) = \begin{cases} \dfrac{x^2 - x - 6}{x - 3}, & x \ne 3 \\ 5, & x = 3 \end{cases}$

**30.** $f(x) = \begin{cases} \dfrac{x^3 - 8}{x^2 - 4}, & x \ne 2, x \ne -2 \\ 3, & x = 2 \\ 4, & x = -2 \end{cases}$

### Limits Involving Trigonometric Functions

Find the limits in Exercises 31–38. Are the functions continuous at the point being approached?

**31.** $\lim_{x \to \pi} \sin(x - \sin x)$

**32.** $\lim_{t \to 0} \sin\left(\dfrac{\pi}{2} \cos(\tan t)\right)$

**33.** $\lim_{y \to 1} \sec(y \sec^2 y - \tan^2 y - 1)$

**34.** $\lim_{x \to 0} \tan\left(\dfrac{\pi}{4} \cos(\sin x^{1/3})\right)$

**35.** $\lim_{t \to 0} \cos\left(\dfrac{\pi}{\sqrt{19 - 3 \sec 2t}}\right)$

**36.** $\lim_{x \to \pi/6} \sqrt{\csc^2 x + 5\sqrt{3} \tan x}$

**37.** $\lim_{x \to 0} \sin\sqrt{\dfrac{\cos^2 x - \cos x}{x}}$

**38.** $\lim_{x \to 0} \sec\left(\dfrac{\pi(\sin 2x - \sin x)}{3x}\right)$

**Continuous Extensions**

**39.** Define $g(3)$ in a way that extends $g(x) = (x^2 - 9)/(x - 3)$ to be continuous at $x = 3$.

**40.** Define $h(2)$ in a way that extends $h(t) = (t^2 + 3t - 10)/(t - 2)$ to be continuous at $t = 2$.

**41.** Define $f(1)$ in a way that extends $f(s) = (s^3 - 1)/(s^2 - 1)$ to be continuous at $s = 1$.

**42.** Define $g(4)$ in a way that extends

$$g(x) = (x^2 - 16)/(x^2 - 3x - 4)$$

to be continuous at $x = 4$.

**43.** For what value of $a$ is

$$f(x) = \begin{cases} x^2 - 1, & x < 3 \\ 2ax, & x \geq 3 \end{cases}$$

continuous at every $x$?

**44.** For what value of $b$ is

$$g(x) = \begin{cases} x, & x < -2 \\ bx^2, & x \geq -2 \end{cases}$$

continuous at every $x$?

**45.** For what values of $a$ is

$$f(x) = \begin{cases} a^2x - 2a, & x \geq 2 \\ 12, & x < 2 \end{cases}$$

continuous at every $x$?

**46.** For what value of $b$ is

$$g(x) = \begin{cases} \dfrac{x - b}{b + 1}, & x < 0 \\ x^2 + b, & x > 0 \end{cases}$$

continuous at every $x$?

**47.** For what values of $a$ and $b$ is

$$f(x) = \begin{cases} -2, & x \leq -1 \\ ax - b, & -1 < x < 1 \\ 3, & x \geq 1 \end{cases}$$

continuous at every $x$?

**48.** For what values of $a$ and $b$ is

$$g(x) = \begin{cases} ax + 2b, & x \leq 0 \\ x^2 + 3a - b, & 0 < x \leq 2 \\ 3x - 5, & x > 2 \end{cases}$$

continuous at every $x$?

T In Exercises 49–52, graph the function $f$ to see whether it appears to have a continuous extension to the origin. If it does, use Trace and Zoom to find a good candidate for the extended function's value at $x = 0$. If the function does not appear to have a continuous extension, can it be extended to be continuous at the origin from the right or from the left? If so, what do you think the extended function's value(s) should be?

**49.** $f(x) = \dfrac{10^x - 1}{x}$

**50.** $f(x) = \dfrac{10^{|x|} - 1}{x}$

**51.** $f(x) = \dfrac{\sin x}{|x|}$

**52.** $f(x) = (1 + 2x)^{1/x}$

**Theory and Examples**

**53.** A continuous function $y = f(x)$ is known to be negative at $x = 0$ and positive at $x = 1$. Why does the equation $f(x) = 0$ have at least one solution between $x = 0$ and $x = 1$? Illustrate with a sketch.

**54.** Explain why the equation $\cos x = x$ has at least one solution.

**55. Roots of a cubic** Show that the equation $x^3 - 15x + 1 = 0$ has three solutions in the interval $[-4, 4]$.

**56. A function value** Show that the function $F(x) = (x - a)^2 \cdot (x - b)^2 + x$ takes on the value $(a + b)/2$ for some value of $x$.

**57. Solving an equation** If $f(x) = x^3 - 8x + 10$, show that there are values $c$ for which $f(c)$ equals **(a)** $\pi$; **(b)** $-\sqrt{3}$; **(c)** 5,000,000.

**58.** Explain why the following five statements ask for the same information.

   **a.** Find the roots of $f(x) = x^3 - 3x - 1$.

   **b.** Find the $x$-coordinates of the points where the curve $y = x^3$ crosses the line $y = 3x + 1$.

   **c.** Find all the values of $x$ for which $x^3 - 3x = 1$.

   **d.** Find the $x$-coordinates of the points where the cubic curve $y = x^3 - 3x$ crosses the line $y = 1$.

   **e.** Solve the equation $x^3 - 3x - 1 = 0$.

**59. Removable discontinuity** Give an example of a function $f(x)$ that is continuous for all values of $x$ except $x = 2$, where it has a removable discontinuity. Explain how you know that $f$ is discontinuous at $x = 2$, and how you know the discontinuity is removable.

**60. Nonremovable discontinuity** Give an example of a function $g(x)$ that is continuous for all values of $x$ except $x = -1$, where it has a nonremovable discontinuity. Explain how you know that $g$ is discontinuous there and why the discontinuity is not removable.

**61. A function discontinuous at every point**

   **a.** Use the fact that every nonempty interval of real numbers contains both rational and irrational numbers to show that the function

$$f(x) = \begin{cases} 1, & \text{if } x \text{ is rational} \\ 0, & \text{if } x \text{ is irrational} \end{cases}$$

   is discontinuous at every point.

   **b.** Is $f$ right-continuous or left-continuous at any point?

**62.** If functions $f(x)$ and $g(x)$ are continuous for $0 \le x \le 1$, could $f(x)/g(x)$ possibly be discontinuous at a point of $[0, 1]$? Give reasons for your answer.

**63.** If the product function $h(x) = f(x) \cdot g(x)$ is continuous at $x = 0$, must $f(x)$ and $g(x)$ be continuous at $x = 0$? Give reasons for your answer.

**64. Discontinuous composite of continuous functions**  Give an example of functions $f$ and $g$, both continuous at $x = 0$, for which the composite $f \circ g$ is discontinuous at $x = 0$. Does this contradict Theorem 9? Give reasons for your answer.

**65. Never-zero continuous functions**  Is it true that a continuous function that is never zero on an interval never changes sign on that interval? Give reasons for your answer.

**66. Stretching a rubber band**  Is it true that if you stretch a rubber band by moving one end to the right and the other to the left, some point of the band will end up in its original position? Give reasons for your answer.

**67. A fixed point theorem**  Suppose that a function $f$ is continuous on the closed interval $[0, 1]$ and that $0 \le f(x) \le 1$ for every $x$ in $[0, 1]$. Show that there must exist a number $c$ in $[0, 1]$ such that $f(c) = c$ ($c$ is called a **fixed point** of $f$).

**68. The sign-preserving property of continuous functions**  Let $f$ be defined on an interval $(a, b)$ and suppose that $f(c) \ne 0$ at some $c$ where $f$ is continuous. Show that there is an interval $(c - \delta, c + \delta)$ about $c$ where $f$ has the same sign as $f(c)$.

**69.** Prove that $f$ is continuous at $c$ if and only if

$$\lim_{h \to 0} f(c + h) = f(c).$$

**70.** Use Exercise 69 together with the identities

$$\sin(h + c) = \sin h \cos c + \cos h \sin c,$$

$$\cos(h + c) = \cos h \cos c - \sin h \sin c$$

to prove that both $f(x) = \sin x$ and $g(x) = \cos x$ are continuous at every point $x = c$.

### Solving Equations Graphically

T Use the Intermediate Value Theorem in Exercises 71–76 to prove that each equation has a solution. Then use a graphing calculator or computer grapher to solve the equations.

**71.** $x^3 - 3x - 1 = 0$

**72.** $2x^3 - 2x^2 - 2x + 1 = 0$

**73.** $x(x - 1)^2 = 1$  (one root)

**74.** $\sqrt{x} + \sqrt{1 + x} = 4$

**75.** $\cos x = x$  (one root). Make sure you are using radian mode.

**76.** $2 \sin x = x$  (three roots). Make sure you are using radian mode.

# $2.6$ Limits Involving Infinity; Asymptotes of Graphs

In this section we investigate the behavior of a function when the magnitude of the independent variable $x$ becomes increasingly large, or $x \to \pm\infty$. We further extend the concept of limit to *infinite limits*, which are not limits as before, but rather a new use of the term limit. Infinite limits provide useful symbols and language for describing the behavior of functions whose values become arbitrarily large in magnitude. We use these limit ideas to analyze the graphs of functions having *horizontal* or *vertical asymptotes*.

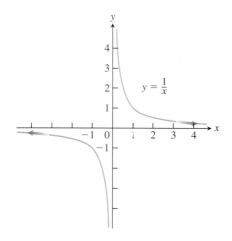

**FIGURE 2.49**  The graph of $y = 1/x$ approaches 0 as $x \to \infty$ or $x \to -\infty$.

### Finite Limits as $x \to \pm\infty$

The symbol for infinity ($\infty$) does not represent a real number. We use $\infty$ to describe the behavior of a function when the values in its domain or range outgrow all finite bounds. For example, the function $f(x) = 1/x$ is defined for all $x \ne 0$ (Figure 2.49). When $x$ is positive and becomes increasingly large, $1/x$ becomes increasingly small. When $x$ is negative and its magnitude becomes increasingly large, $1/x$ again becomes small. We summarize these observations by saying that $f(x) = 1/x$ has limit 0 as $x \to \infty$ or $x \to -\infty$, or that 0 is a *limit of* $f(x) = 1/x$ *at infinity and negative infinity*. Here are precise definitions.

---

**DEFINITIONS**

**1.** We say that $f(x)$ has the **limit $L$ as $x$ approaches infinity** and write

$$\lim_{x \to \infty} f(x) = L$$

if, for every number $\epsilon > 0$, there exists a corresponding number $M$ such that for all $x$

$$x > M \quad \Rightarrow \quad |f(x) - L| < \epsilon.$$

**2.** We say that $f(x)$ has the **limit $L$ as $x$ approaches minus infinity** and write

$$\lim_{x \to -\infty} f(x) = L$$

if, for every number $\epsilon > 0$, there exists a corresponding number $N$ such that for all $x$

$$x < N \quad \Rightarrow \quad |f(x) - L| < \epsilon.$$

---

Intuitively, $\lim_{x \to \infty} f(x) = L$ if, as $x$ moves increasingly far from the origin in the positive direction, $f(x)$ gets arbitrarily close to $L$. Similarly, $\lim_{x \to -\infty} f(x) = L$ if, as $x$ moves increasingly far from the origin in the negative direction, $f(x)$ gets arbitrarily close to $L$.

The strategy for calculating limits of functions as $x \to \pm\infty$ is similar to the one for finite limits in Section 2.2. There we first found the limits of the constant and identity functions $y = k$ and $y = x$. We then extended these results to other functions by applying Theorem 1 on limits of algebraic combinations. Here we do the same thing, except that the starting functions are $y = k$ and $y = 1/x$ instead of $y = k$ and $y = x$.

The basic facts to be verified by applying the formal definition are

$$\lim_{x \to \pm\infty} k = k \qquad \text{and} \qquad \lim_{x \to \pm\infty} \frac{1}{x} = 0. \tag{1}$$

We prove the second result in Example 1, and leave the first to Exercises 87 and 88.

**EXAMPLE 1** Show that

**(a)** $\lim_{x \to \infty} \dfrac{1}{x} = 0$ 
**(b)** $\lim_{x \to -\infty} \dfrac{1}{x} = 0.$

**Solution**

**(a)** Let $\epsilon > 0$ be given. We must find a number $M$ such that for all $x$

$$x > M \quad \Rightarrow \quad \left|\frac{1}{x} - 0\right| = \left|\frac{1}{x}\right| < \epsilon.$$

The implication will hold if $M = 1/\epsilon$ or any larger positive number (Figure 2.50). This proves $\lim_{x \to \infty}(1/x) = 0$.

**(b)** Let $\epsilon > 0$ be given. We must find a number $N$ such that for all $x$

$$x < N \quad \Rightarrow \quad \left|\frac{1}{x} - 0\right| = \left|\frac{1}{x}\right| < \epsilon.$$

The implication will hold if $N = -1/\epsilon$ or any number less than $-1/\epsilon$ (Figure 2.50). This proves $\lim_{x \to -\infty}(1/x) = 0$. ■

Limits at infinity have properties similar to those of finite limits.

---

**THEOREM 12** All the Limit Laws in Theorem 1 are true when we replace $\lim_{x \to c}$ by $\lim_{x \to \infty}$ or $\lim_{x \to -\infty}$. That is, the variable $x$ may approach a finite number $c$ or $\pm\infty$.

---

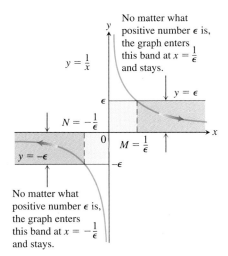

No matter what positive number $\epsilon$ is, the graph enters this band at $x = \frac{1}{\epsilon}$ and stays.

$y = \frac{1}{x}$

$y = \epsilon$

$N = -\frac{1}{\epsilon}$

$y = -\epsilon$

$M = \frac{1}{\epsilon}$

No matter what positive number $\epsilon$ is, the graph enters this band at $x = -\frac{1}{\epsilon}$ and stays.

**FIGURE 2.50** The geometry behind the argument in Example 1.

**EXAMPLE 2** The properties in Theorem 12 are used to calculate limits in the same way as when $x$ approaches a finite number $c$.

(a) $\lim\limits_{x\to\infty}\left(5+\dfrac{1}{x}\right) = \lim\limits_{x\to\infty} 5 + \lim\limits_{x\to\infty}\dfrac{1}{x}$   Sum Rule

$\qquad\qquad = 5 + 0 = 5$   Known limits

(b) $\lim\limits_{x\to-\infty}\dfrac{\pi\sqrt{3}}{x^2} = \lim\limits_{x\to-\infty}\pi\sqrt{3}\cdot\dfrac{1}{x}\cdot\dfrac{1}{x}$

$\qquad\qquad = \lim\limits_{x\to-\infty}\pi\sqrt{3}\cdot\lim\limits_{x\to-\infty}\dfrac{1}{x}\cdot\lim\limits_{x\to-\infty}\dfrac{1}{x}$   Product Rule

$\qquad\qquad = \pi\sqrt{3}\cdot 0\cdot 0 = 0$   Known limits   ∎

### Limits at Infinity of Rational Functions

To determine the limit of a rational function as $x\to\pm\infty$, we first divide the numerator and denominator by the highest power of $x$ in the denominator. The result then depends on the degrees of the polynomials involved.

**EXAMPLE 3** These examples illustrate what happens when the degree of the numerator is less than or equal to the degree of the denominator.

(a) $\lim\limits_{x\to\infty}\dfrac{5x^2+8x-3}{3x^2+2} = \lim\limits_{x\to\infty}\dfrac{5+(8/x)-(3/x^2)}{3+(2/x^2)}$   Divide numerator and denominator by $x^2$.

$\qquad\qquad = \dfrac{5+0-0}{3+0} = \dfrac{5}{3}$   See Fig. 2.51.

(b) $\lim\limits_{x\to-\infty}\dfrac{11x+2}{2x^3-1} = \lim\limits_{x\to-\infty}\dfrac{(11/x^2)+(2/x^3)}{2-(1/x^3)}$   Divide numerator and denominator by $x^3$.

$\qquad\qquad = \dfrac{0+0}{2-0} = 0$   See Fig. 2.52.   ∎

Cases for which the degree of the numerator is greater than the degree of the denominator are illustrated in Examples 9 and 13.

### Horizontal Asymptotes

If the distance between the graph of a function and some fixed line approaches zero as a point on the graph moves increasingly far from the origin, we say that the graph approaches the line asymptotically and that the line is an *asymptote* of the graph.

Looking at $f(x) = 1/x$ (see Figure 2.49), we observe that the $x$-axis is an asymptote of the curve on the right because

$$\lim\limits_{x\to\infty}\frac{1}{x} = 0$$

and on the left because

$$\lim\limits_{x\to-\infty}\frac{1}{x} = 0.$$

We say that the $x$-axis is a *horizontal asymptote* of the graph of $f(x) = 1/x$.

> **DEFINITION** A line $y = b$ is a **horizontal asymptote** of the graph of a function $y = f(x)$ if either
>
> $$\lim\limits_{x\to\infty} f(x) = b \quad\text{or}\quad \lim\limits_{x\to-\infty} f(x) = b.$$

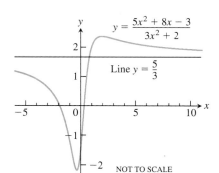

**FIGURE 2.51** The graph of the function in Example 3a. The graph approaches the line $y = 5/3$ as $|x|$ increases.

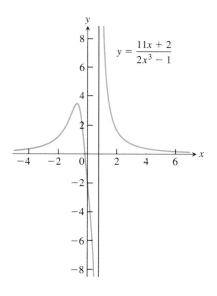

**FIGURE 2.52** The graph of the function in Example 3b. The graph approaches the $x$-axis as $|x|$ increases.

The graph of the function

$$f(x) = \frac{5x^2 + 8x - 3}{3x^2 + 2}$$

sketched in Figure 2.51 (Example 3a) has the line $y = 5/3$ as a horizontal asymptote on both the right and the left because

$$\lim_{x \to \infty} f(x) = \frac{5}{3} \quad \text{and} \quad \lim_{x \to -\infty} f(x) = \frac{5}{3}.$$

**EXAMPLE 4**   Find the horizontal asymptotes of the graph of

$$f(x) = \frac{x^3 - 2}{|x|^3 + 1}.$$

**Solution**   We calculate the limits as $x \to \pm\infty$.

For $x \geq 0$: $\displaystyle\lim_{x \to \infty} \frac{x^3 - 2}{|x|^3 + 1} = \lim_{x \to \infty} \frac{x^3 - 2}{x^3 + 1} = \lim_{x \to \infty} \frac{1 - (2/x^3)}{1 + (1/x^3)} = 1.$

For $x < 0$: $\displaystyle\lim_{x \to -\infty} \frac{x^3 - 2}{|x|^3 + 1} = \lim_{x \to -\infty} \frac{x^3 - 2}{(-x)^3 + 1} = \lim_{x \to -\infty} \frac{1 - (2/x^3)}{-1 + (1/x^3)} = -1.$

The horizontal asymptotes are $y = -1$ and $y = 1$. The graph is displayed in Figure 2.53. Notice that the graph crosses the horizontal asymptote $y = -1$ for a positive value of $x$. ∎

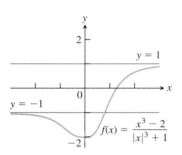

**FIGURE 2.53**   The graph of the function in Example 4 has two horizontal asymptotes.

**EXAMPLE 5**   Find **(a)** $\displaystyle\lim_{x \to \infty} \sin(1/x)$   and   **(b)** $\displaystyle\lim_{x \to \pm\infty} x \sin(1/x)$.

**Solution**
**(a)**   We introduce the new variable $t = 1/x$. From Example 1, we know that $t \to 0^+$ as $x \to \infty$ (see Figure 2.49). Therefore,

$$\lim_{x \to \infty} \sin\frac{1}{x} = \lim_{t \to 0^+} \sin t = 0.$$

**(b)**   We calculate the limits as $x \to \infty$ and $x \to -\infty$:

$$\lim_{x \to \infty} x \sin\frac{1}{x} = \lim_{t \to 0^+} \frac{\sin t}{t} = 1 \quad \text{and} \quad \lim_{x \to -\infty} x \sin\frac{1}{x} = \lim_{t \to 0^-} \frac{\sin t}{t} = 1.$$

The graph is shown in Figure 2.54, and we see that the line $y = 1$ is a horizontal asymptote. ∎

**FIGURE 2.54**   The line $y = 1$ is a horizontal asymptote of the function graphed here (Example 5b).

The Sandwich Theorem also holds for limits as $x \to \pm\infty$. You must be sure, though, that the function whose limit you are trying to find stays between the bounding functions at very large values of $x$ in magnitude consistent with whether $x \to \infty$ or $x \to -\infty$.

**EXAMPLE 6**   Using the Sandwich Theorem, find the horizontal asymptote of the curve

$$y = 2 + \frac{\sin x}{x}.$$

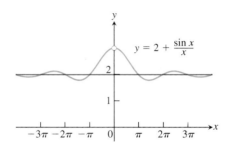

**FIGURE 2.55** A curve may cross one of its asymptotes infinitely often (Example 6).

**Solution** We are interested in the behavior as $x \to \pm\infty$. Since

$$0 \le \left| \frac{\sin x}{x} \right| \le \left| \frac{1}{x} \right|$$

and $\lim_{x \to \pm\infty} |1/x| = 0$, we have $\lim_{x \to \pm\infty} (\sin x)/x = 0$ by the Sandwich Theorem. Hence,

$$\lim_{x \to \pm\infty} \left( 2 + \frac{\sin x}{x} \right) = 2 + 0 = 2,$$

and the line $y = 2$ is a horizontal asymptote of the curve on both left and right (Figure 2.55). This example illustrates that a curve may cross one of its horizontal asymptotes many times.

We can investigate the behavior of $y = f(1/x)$ as $x \to 0$ by investigating $y = f(t)$ as $t \to \pm\infty$, where $t = 1/x$.

**EXAMPLE 7** Find $\lim_{x \to 0^+} x \left\lfloor \frac{1}{x} \right\rfloor$.

**Solution** We let $t = 1/x$ so that

$$\lim_{x \to 0^+} x \left\lfloor \frac{1}{x} \right\rfloor = \lim_{t \to \infty} \frac{1}{t} \lfloor t \rfloor$$

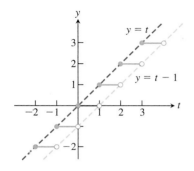

**FIGURE 2.56** The graph of the greatest integer function $y = \lfloor t \rfloor$ is sandwiched between $y = t - 1$ and $y = t$.

From the graph in Figure 2.56, we see that $t - 1 \le \lfloor t \rfloor \le t$, which gives

$$1 - \frac{1}{t} \le \frac{1}{t} \lfloor t \rfloor \le 1 \qquad \text{Multiply inequalities by } \frac{1}{t} > 0.$$

It follows from the Sandwich Theorem that

$$\lim_{t \to \infty} \frac{1}{t} \lfloor t \rfloor = 1,$$

so 1 is the value of the limit we seek.

**EXAMPLE 8** Find $\lim_{x \to \infty} \left( x - \sqrt{x^2 + 16} \right)$.

**Solution** Both of the terms $x$ and $\sqrt{x^2 + 16}$ approach infinity as $x \to \infty$, so what happens to the difference in the limit is unclear (we cannot subtract $\infty$ from $\infty$ because the symbol does not represent a real number). In this situation we can multiply the numerator and the denominator by the conjugate radical expression to obtain an equivalent algebraic result:

$$\lim_{x \to \infty} \left( x - \sqrt{x^2 + 16} \right) = \lim_{x \to \infty} \left( x - \sqrt{x^2 + 16} \right) \frac{x + \sqrt{x^2 + 16}}{x + \sqrt{x^2 + 16}}$$

$$= \lim_{x \to \infty} \frac{x^2 - (x^2 + 16)}{x + \sqrt{x^2 + 16}} = \lim_{x \to \infty} \frac{-16}{x + \sqrt{x^2 + 16}}.$$

As $x \to \infty$, the denominator in this last expression becomes arbitrarily large, so we see that the limit is 0. We can also obtain this result by a direct calculation using the Limit Laws:

$$\lim_{x \to \infty} \frac{-16}{x + \sqrt{x^2 + 16}} = \lim_{x \to \infty} \frac{-\dfrac{16}{x}}{1 + \sqrt{\dfrac{x^2}{x^2} + \dfrac{16}{x^2}}} = \frac{0}{1 + \sqrt{1 + 0}} = 0.$$

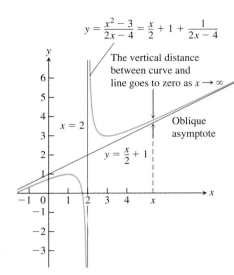

$$y = \frac{x^2 - 3}{2x - 4} = \frac{x}{2} + 1 + \frac{1}{2x - 4}$$

The vertical distance between curve and line goes to zero as $x \to \infty$

Oblique asymptote

$y = \frac{x}{2} + 1$

$x = 2$

**FIGURE 2.57** The graph of the function in Example 9 has an oblique asymptote.

## Oblique Asymptotes

If the degree of the numerator of a rational function is 1 greater than the degree of the denominator, the graph has an **oblique** or **slant line asymptote**. We find an equation for the asymptote by dividing numerator by denominator to express $f$ as a linear function plus a remainder that goes to zero as $x \to \pm\infty$.

**EXAMPLE 9**    Find the oblique asymptote of the graph of

$$f(x) = \frac{x^2 - 3}{2x - 4}$$

in Figure 2.57.

**Solution**    We are interested in the behavior as $x \to \pm\infty$. We divide $(2x - 4)$ into $(x^2 - 3)$:

$$
\begin{array}{r}
\frac{x}{2} + 1 \phantom{xxxx} \\
2x - 4 \overline{)\, x^2 - 3 \phantom{xx}} \\
\underline{x^2 - 2x \phantom{xxx}} \\
2x - 3 \\
\underline{2x - 4} \\
1
\end{array}
$$

This tells us that

$$f(x) = \frac{x^2 - 3}{2x - 4} = \underbrace{\left(\frac{x}{2} + 1\right)}_{\text{linear } g(x)} + \underbrace{\left(\frac{1}{2x - 4}\right)}_{\text{remainder}}.$$

As $x \to \pm\infty$, the remainder, whose magnitude gives the vertical distance between the graphs of $f$ and $g$, goes to zero, making the slanted line

$$g(x) = \frac{x}{2} + 1$$

an asymptote of the graph of $f$ (Figure 2.57). The line $y = g(x)$ is an asymptote both to the right and to the left. The next subsection will confirm that the function $f(x)$ grows arbitrarily large in absolute value as $x \to 2$ (where the denominator is zero), as shown in the graph. ∎

Notice in Example 9 that if the degree of the numerator in a rational function is greater than the degree of the denominator, then the limit as $|x|$ becomes large is $+\infty$ or $-\infty$, depending on the signs assumed by the numerator and denominator.

## Infinite Limits

Let us look again at the function $f(x) = 1/x$. As $x \to 0^+$, the values of $f$ grow without bound, eventually reaching and surpassing every positive real number. That is, given any positive real number $B$, however large, the values of $f$ become larger still (Figure 2.58). Thus, $f$ has no limit as $x \to 0^+$. It is nevertheless convenient to describe the behavior of $f$ by saying that $f(x)$ approaches $\infty$ as $x \to 0^+$. We write

$$\lim_{x \to 0^+} f(x) = \lim_{x \to 0^+} \frac{1}{x} = \infty.$$

In writing this equation, we are *not* saying that the limit exists. Nor are we saying that there is a real number $\infty$, for there is no such number. Rather, we are saying that $\lim_{x \to 0^+}(1/x)$ *does not exist because $1/x$ becomes arbitrarily large and positive as $x \to 0^+$.*

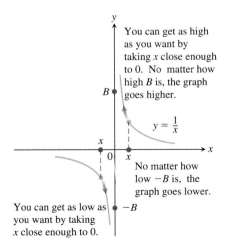

You can get as high as you want by taking $x$ close enough to 0. No matter how high $B$ is, the graph goes higher.

$B$

$y = \frac{1}{x}$

No matter how low $-B$ is, the graph goes lower.

$-B$

You can get as low as you want by taking $x$ close enough to 0.

**FIGURE 2.58** One-sided infinite limits:
$$\lim_{x \to 0^+} \frac{1}{x} = \infty \quad \text{and} \quad \lim_{x \to 0^-} \frac{1}{x} = -\infty.$$

As $x \rightarrow 0^-$, the values of $f(x) = 1/x$ become arbitrarily large and negative. Given any negative real number $-B$, the values of $f$ eventually lie below $-B$. (See Figure 2.58.) We write

$$\lim_{x \rightarrow 0^-} f(x) = \lim_{x \rightarrow 0^-} \frac{1}{x} = -\infty.$$

Again, we are not saying that the limit exists and equals the number $-\infty$. There *is* no real number $-\infty$. We are describing the behavior of a function whose limit as $x \rightarrow 0^-$ *does not exist because its values become arbitrarily large and negative.*

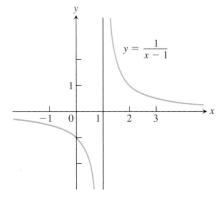

**FIGURE 2.59** Near $x = 1$, the function $y = 1/(x - 1)$ behaves the way the function $y = 1/x$ behaves near $x = 0$. Its graph is the graph of $y = 1/x$ shifted 1 unit to the right (Example 10).

**EXAMPLE 10** Find $\lim_{x \rightarrow 1^+} \dfrac{1}{x - 1}$ and $\lim_{x \rightarrow 1^-} \dfrac{1}{x - 1}$.

**Geometric Solution** The graph of $y = 1/(x - 1)$ is the graph of $y = 1/x$ shifted 1 unit to the right (Figure 2.59). Therefore, $y = 1/(x - 1)$ behaves near 1 exactly the way $y = 1/x$ behaves near 0:

$$\lim_{x \rightarrow 1^+} \frac{1}{x - 1} = \infty \qquad \text{and} \qquad \lim_{x \rightarrow 1^-} \frac{1}{x - 1} = -\infty.$$

**Analytic Solution** Think about the number $x - 1$ and its reciprocal. As $x \rightarrow 1^+$, we have $(x - 1) \rightarrow 0^+$ and $1/(x - 1) \rightarrow \infty$. As $x \rightarrow 1^-$, we have $(x - 1) \rightarrow 0^-$ and $1/(x - 1) \rightarrow -\infty$. ■

**EXAMPLE 11** Discuss the behavior of

$$f(x) = \frac{1}{x^2} \qquad \text{as} \qquad x \rightarrow 0.$$

**Solution** As $x$ approaches zero from either side, the values of $1/x^2$ are positive and become arbitrarily large (Figure 2.60). This means that

$$\lim_{x \rightarrow 0} f(x) = \lim_{x \rightarrow 0} \frac{1}{x^2} = \infty.$$

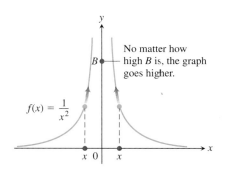

**FIGURE 2.60** The graph of $f(x)$ in Example 11 approaches infinity as $x \rightarrow 0$.

The function $y = 1/x$ shows no consistent behavior as $x \rightarrow 0$. We have $1/x \rightarrow \infty$ if $x \rightarrow 0^+$, but $1/x \rightarrow -\infty$ if $x \rightarrow 0^-$. All we can say about $\lim_{x \rightarrow 0} (1/x)$ is that it does not exist. The function $y = 1/x^2$ is different. Its values approach infinity as $x$ approaches zero from either side, so we can say that $\lim_{x \rightarrow 0} (1/x^2) = \infty$. ■

**EXAMPLE 12** These examples illustrate that rational functions can behave in various ways near zeros of the denominator.

**(a)** $\displaystyle \lim_{x \rightarrow 2} \frac{(x - 2)^2}{x^2 - 4} = \lim_{x \rightarrow 2} \frac{(x - 2)^2}{(x - 2)(x + 2)} = \lim_{x \rightarrow 2} \frac{x - 2}{x + 2} = 0$

**(b)** $\displaystyle \lim_{x \rightarrow 2} \frac{x - 2}{x^2 - 4} = \lim_{x \rightarrow 2} \frac{x - 2}{(x - 2)(x + 2)} = \lim_{x \rightarrow 2} \frac{1}{x + 2} = \frac{1}{4}$

**(c)** $\displaystyle \lim_{x \rightarrow 2^+} \frac{x - 3}{x^2 - 4} = \lim_{x \rightarrow 2^+} \frac{x - 3}{(x - 2)(x + 2)} = -\infty$    The values are negative for $x > 2$, $x$ near 2.

**(d)** $\displaystyle \lim_{x \rightarrow 2^-} \frac{x - 3}{x^2 - 4} = \lim_{x \rightarrow 2^-} \frac{x - 3}{(x - 2)(x + 2)} = \infty$    The values are positive for $x < 2$, $x$ near 2.

**(e)**  $\lim\limits_{x \to 2} \dfrac{x - 3}{x^2 - 4} = \lim\limits_{x \to 2} \dfrac{x - 3}{(x - 2)(x + 2)}$ does not exist.    See parts (c) and (d).

**(f)**  $\lim\limits_{x \to 2} \dfrac{2 - x}{(x - 2)^3} = \lim\limits_{x \to 2} \dfrac{-(x - 2)}{(x - 2)^3} = \lim\limits_{x \to 2} \dfrac{-1}{(x - 2)^2} = -\infty$

In parts (a) and (b) the effect of the zero in the denominator at $x = 2$ is canceled because the numerator is zero there also. Thus a finite limit exists. This is not true in part (f), where cancellation still leaves a zero factor in the denominator.    ∎

**EXAMPLE 13**    Find $\lim\limits_{x \to -\infty} \dfrac{2x^5 - 6x^4 + 1}{3x^2 + x - 7}$.

**Solution**    We are asked to find the limit of a rational function as $x \to -\infty$, so we divide the numerator and denominator by $x^2$, the highest power of $x$ in the denominator:

$$\lim\limits_{x \to -\infty} \frac{2x^5 - 6x^4 + 1}{3x^2 + x - 7} = \lim\limits_{x \to -\infty} \frac{2x^3 - 6x^2 + x^{-2}}{3 + x^{-1} - 7x^{-2}}$$

$$= \lim\limits_{x \to -\infty} \frac{2x^2(x - 3) + x^{-2}}{3 + x^{-1} - 7x^{-2}}$$

$$= -\infty, \qquad x^{-n} \to 0, \; x - 3 \to -\infty$$

because the numerator tends to $-\infty$ while the denominator approaches 3 as $x \to -\infty$.    ∎

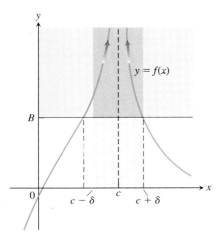

**FIGURE 2.61**    For $c - \delta < x < c + \delta$, the graph of $f(x)$ lies above the line $y = B$.

### Precise Definitions of Infinite Limits

Instead of requiring $f(x)$ to lie arbitrarily close to a finite number $L$ for all $x$ sufficiently close to $c$, the definitions of infinite limits require $f(x)$ to lie arbitrarily far from zero. Except for this change, the language is very similar to what we have seen before. Figures 2.61 and 2.62 accompany these definitions.

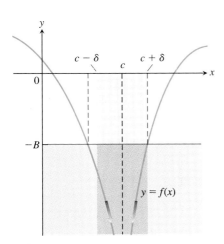

**FIGURE 2.62**    For $c - \delta < x < c + \delta$, the graph of $f(x)$ lies below the line $y = -B$.

---

**DEFINITIONS**

**1.** We say that $f(x)$ **approaches infinity as $x$ approaches $c$**, and write

$$\lim\limits_{x \to c} f(x) = \infty,$$

if for every positive real number $B$ there exists a corresponding $\delta > 0$ such that for all $x$

$$0 < |x - c| < \delta \quad \Rightarrow \quad f(x) > B.$$

**2.** We say that $f(x)$ **approaches minus infinity as $x$ approaches $c$**, and write

$$\lim\limits_{x \to c} f(x) = -\infty,$$

if for every negative real number $-B$ there exists a corresponding $\delta > 0$ such that for all $x$

$$0 < |x - c| < \delta \quad \Rightarrow \quad f(x) < -B.$$

---

The precise definitions of one-sided infinite limits at $c$ are similar and are stated in the exercises.

**EXAMPLE 14**    Prove that $\lim\limits_{x \to 0} \dfrac{1}{x^2} = \infty$.

**Solution**    Given $B > 0$, we want to find $\delta > 0$ such that

$$0 < |x - 0| < \delta \quad \text{implies} \quad \frac{1}{x^2} > B.$$

Now,

$$\frac{1}{x^2} > B \qquad \text{if and only if} \qquad x^2 < \frac{1}{B}$$

or, equivalently,

$$|x| < \frac{1}{\sqrt{B}}.$$

Thus, choosing $\delta = 1/\sqrt{B}$ (or any smaller positive number), we see that

$$|x| < \delta \quad \text{implies} \quad \frac{1}{x^2} > \frac{1}{\delta^2} \geq B.$$

Therefore, by definition,

$$\lim_{x \to 0} \frac{1}{x^2} = \infty. \qquad \blacksquare$$

## Vertical Asymptotes

Notice that the distance between a point on the graph of $f(x) = 1/x$ and the $y$-axis approaches zero as the point moves vertically along the graph and away from the origin (Figure 2.63). The function $f(x) = 1/x$ is unbounded as $x$ approaches 0 because

$$\lim_{x \to 0^+} \frac{1}{x} = \infty \qquad \text{and} \qquad \lim_{x \to 0^-} \frac{1}{x} = -\infty.$$

We say that the line $x = 0$ (the $y$-axis) is a *vertical asymptote* of the graph of $f(x) = 1/x$. Observe that the denominator is zero at $x = 0$ and the function is undefined there.

$y$

Vertical asymptote

$y = \dfrac{1}{x}$

Horizontal asymptote

$1$

$0$   $1$   Horizontal asymptote, $y = 0$

$x$

Vertical asymptote, $x = 0$

**FIGURE 2.63**   The coordinate axes are asymptotes of both branches of the hyperbola $y = 1/x$.

> **DEFINITION**    A line $x = a$ is a **vertical asymptote** of the graph of a function $y = f(x)$ if either
>
> $$\lim_{x \to a^+} f(x) = \pm\infty \qquad \text{or} \qquad \lim_{x \to a^-} f(x) = \pm\infty.$$

**EXAMPLE 15**    Find the horizontal and vertical asymptotes of the curve

$$y = \frac{x + 3}{x + 2}.$$

**Solution**    We are interested in the behavior as $x \to \pm\infty$ and the behavior as $x \to -2$, where the denominator is zero.

The asymptotes are quickly revealed if we recast the rational function as a polynomial with a remainder, by dividing $(x + 2)$ into $(x + 3)$:

$$
\begin{array}{r}
1 \phantom{xxx} \\
x + 2 \overline{)\, x + 3} \\
\underline{x + 2} \phantom{x} \\
1
\end{array}
$$

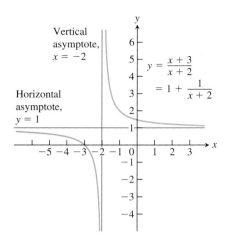

**FIGURE 2.64** The lines $y = 1$ and $x = -2$ are asymptotes of the curve in Example 15.

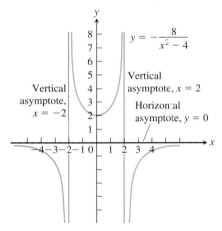

**FIGURE 2.65** Graph of the function in Example 16. Notice that the curve approaches the $x$-axis from only one side. Asymptotes do not have to be two-sided.

This result enables us to rewrite $y$ as:

$$y = 1 + \frac{1}{x + 2}.$$

As $x \to \pm\infty$, the curve approaches the horizontal asymptote $y = 1$; as $x \to -2$, the curve approaches the vertical asymptote $x = -2$. We see that the curve in question is the graph of $f(x) = 1/x$ shifted 1 unit up and 2 units left (Figure 2.64). The asymptotes, instead of being the coordinate axes, are now the lines $y = 1$ and $x = -2$. ∎

**EXAMPLE 16**    Find the horizontal and vertical asymptotes of the graph of

$$f(x) = -\frac{8}{x^2 - 4}.$$

**Solution**    We are interested in the behavior as $x \to \pm\infty$ and as $x \to \pm 2$, where the denominator is zero. Notice that $f$ is an even function of $x$, so its graph is symmetric with respect to the $y$-axis.

(a) *The behavior as $x \to \pm\infty$*. Since $\lim_{x\to\infty} f(x) = 0$, the line $y = 0$ is a horizontal asymptote of the graph to the right. By symmetry it is an asymptote to the left as well (Figure 2.65). Notice that the curve approaches the $x$-axis from only the negative side (or from below). Also, $f(0) = 2$.

(b) *The behavior as $x \to \pm 2$*. Since

$$\lim_{x\to 2^+} f(x) = -\infty \qquad \text{and} \qquad \lim_{x\to 2^-} f(x) = \infty,$$

the line $x = 2$ is a vertical asymptote both from the right and from the left. By symmetry, the line $x = -2$ is also a vertical asymptote.

There are no other asymptotes because $f$ has a finite limit at all other points. ∎

**EXAMPLE 17**    The curves

$$y = \sec x = \frac{1}{\cos x} \qquad \text{and} \qquad y = \tan x = \frac{\sin x}{\cos x}$$

both have vertical asymptotes at odd-integer multiples of $\pi/2$, where $\cos x = 0$ (Figure 2.66).

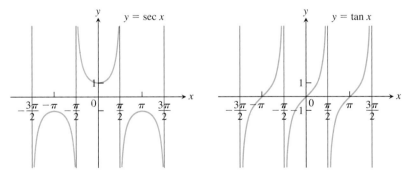

**FIGURE 2.66** The graphs of $\sec x$ and $\tan x$ have infinitely many vertical asymptotes (Example 17). ∎

Dominant Terms

In Example 9 we saw that by long division we could rewrite the function

$$f(x) = \frac{x^2 - 3}{2x - 4}$$

as a linear function plus a remainder term:

$$f(x) = \left(\frac{x}{2} + 1\right) + \left(\frac{1}{2x - 4}\right).$$

This tells us immediately that

$$f(x) \approx \frac{x}{2} + 1 \qquad \text{For } |x| \text{ large, } \frac{1}{2x - 4} \text{ is near 0.}$$

$$f(x) \approx \frac{1}{2x - 4} \qquad \text{For } x \text{ near 2, this term is very large in absolute value.}$$

If we want to know how $f$ behaves, this is the way to find out. It behaves like $y = (x/2) + 1$ when $|x|$ is large and the contribution of $1/(2x - 4)$ to the total value of $f$ is insignificant. It behaves like $1/(2x - 4)$ when $x$ is so close to 2 that $1/(2x - 4)$ makes the dominant contribution.

We say that $(x/2) + 1$ **dominates** when $x$ is numerically large, and we say that $1/(2x - 4)$ dominates when $x$ is near 2. **Dominant terms** like these help us predict a function's behavior.

**EXAMPLE 18**    Let $f(x) = 3x^4 - 2x^3 + 3x^2 - 5x + 6$ and $g(x) = 3x^4$. Show that although $f$ and $g$ are quite different for numerically small values of $x$, they are virtually identical for $|x|$ very large, in the sense that their ratios approach 1 as $x \to \infty$ or $x \to -\infty$.

**Solution**    The graphs of $f$ and $g$ behave quite differently near the origin (Figure 2.67a), but appear as virtually identical on a larger scale (Figure 2.67b).

We can test that the term $3x^4$ in $f$, represented graphically by $g$, dominates the polynomial $f$ for numerically large values of $x$ by examining the ratio of the two functions as $x \to \pm\infty$. We find that

$$\lim_{x \to \pm\infty} \frac{f(x)}{g(x)} = \lim_{x \to \pm\infty} \frac{3x^4 - 2x^3 + 3x^2 - 5x + 6}{3x^4}$$

$$= \lim_{x \to \pm\infty} \left(1 - \frac{2}{3x} + \frac{1}{x^2} - \frac{5}{3x^3} + \frac{2}{x^4}\right)$$

$$= 1,$$

which means that $f$ and $g$ appear nearly identical when $|x|$ is large.    ∎

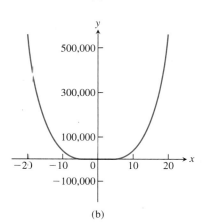

**FIGURE 2.67**    The graphs of $f$ and $g$ are (a) distinct for $|x|$ small, and (b) nearly identical for $|x|$ large (Example 18).

## Summary

In this chapter we presented several important calculus ideas that are made meaningful and precise by the concept of the limit. These include the three ideas of the exact rate of change of a function, the slope of the graph of a function at a point, and the continuity of a function. The primary methods used for calculating limits of many functions are captured in the algebraic Limit Laws of Theorem 1 and in the Sandwich Theorem, all of which are proved from the precise definition of the limit. We saw that these computational rules also apply to one-sided limits and to limits at infinity. However, to calculate complicated limits such as

$$\lim_{x \to 0} \left(1 + \frac{1}{x}\right)^x, \qquad \lim_{x \to 0} \frac{x - \sin x}{x^3}, \qquad \text{and} \qquad \lim_{x \to 0} x^{1/x},$$

techniques other than simple algebra are required. The *derivative* is one of the tools we need to calculate limits such as these, and this notion is the central subject of our next chapter.

# Exercises **2.6**

### Finding Limits

**1.** For the function $f$ whose graph is given, determine the following limits.

　**a.** $\lim_{x \to 2} f(x)$　　**b.** $\lim_{x \to -3^+} f(x)$　　**c.** $\lim_{x \to -3^-} f(x)$

　**d.** $\lim_{x \to -3} f(x)$　　**e.** $\lim_{x \to 0^+} f(x)$　　**f.** $\lim_{x \to 0^-} f(x)$

　**g.** $\lim_{x \to 0} f(x)$　　**h.** $\lim_{x \to \infty} f(x)$　　**i.** $\lim_{x \to -\infty} f(x)$

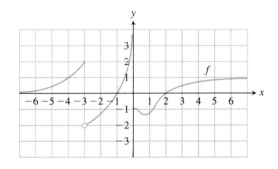

**2.** For the function $f$ whose graph is given, determine the following limits.

　**a.** $\lim_{x \to 4} f(x)$　　**b.** $\lim_{x \to 2^+} f(x)$　　**c.** $\lim_{x \to 2^-} f(x)$

　**d.** $\lim_{x \to 2} f(x)$　　**e.** $\lim_{x \to -3^+} f(x)$　　**f.** $\lim_{x \to -3^-} f(x)$

　**g.** $\lim_{x \to -3} f(x)$　　**h.** $\lim_{x \to 0^+} f(x)$　　**i.** $\lim_{x \to 0^-} f(x)$

　**j.** $\lim_{x \to 0} f(x)$　　**k.** $\lim_{x \to \infty} f(x)$　　**l.** $\lim_{x \to -\infty} f(x)$

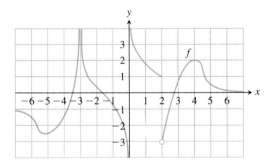

In Exercises 3–8, find the limit of each function **(a)** as $x \to \infty$ and **(b)** as $x \to -\infty$. (You may wish to visualize your answer with a graphing calculator or computer.)

**3.** $f(x) = \dfrac{2}{x} - 3$

**4.** $f(x) = \pi - \dfrac{2}{x^2}$

**5.** $g(x) = \dfrac{1}{2 + (1/x)}$

**6.** $g(x) = \dfrac{1}{8 - (5/x^2)}$

**7.** $h(x) = \dfrac{-5 + (7/x)}{3 - (1/x^2)}$

**8.** $h(x) = \dfrac{3 - (2/x)}{4 + (\sqrt{2}/x^2)}$

Find the limits in Exercises 9–12.

**9.** $\lim_{x \to \infty} \dfrac{\sin 2x}{x}$

**10.** $\lim_{\theta \to -\infty} \dfrac{\cos \theta}{3\theta}$

**11.** $\lim_{t \to -\infty} \dfrac{2 - t + \sin t}{t + \cos t}$

**12.** $\lim_{r \to \infty} \dfrac{r + \sin r}{2r + 7 - 5 \sin r}$

### Limits of Rational Functions

In Exercises 13–22, find the limit of each rational function **(a)** as $x \to \infty$ and **(b)** as $x \to -\infty$.

**13.** $f(x) = \dfrac{2x + 3}{5x + 7}$

**14.** $f(x) = \dfrac{2x^3 + 7}{x^3 - x^2 + x + 7}$

**15.** $f(x) = \dfrac{x + 1}{x^2 + 3}$

**16.** $f(x) = \dfrac{3x + 7}{x^2 - 2}$

**17.** $h(x) = \dfrac{7x^3}{x^3 - 3x^2 + 6x}$

**18.** $h(x) = \dfrac{9x^4 + x}{2x^4 + 5x^2 - x + 6}$

**19.** $g(x) = \dfrac{10x^5 + x^4 + 31}{x^6}$

**20.** $g(x) = \dfrac{x^3 + 7x^2 - 2}{x^2 - x + 1}$

**21.** $f(x) = \dfrac{3x^7 + 5x^2 - 1}{6x^3 - 7x + 3}$

**22.** $h(x) = \dfrac{5x^8 - 2x^3 + 9}{3 + x - 4x^5}$

### Limits as $x \to \infty$ or $x \to -\infty$

The process by which we determine limits of rational functions applies equally well to ratios containing noninteger or negative powers of $x$: Divide numerator and denominator by the highest power of $x$ in the denominator and proceed from there. Find the limits in Exercises 23–36.

**23.** $\lim_{x \to \infty} \sqrt{\dfrac{8x^2 - 3}{2x^2 + x}}$

**24.** $\lim_{x \to -\infty} \left( \dfrac{x^2 + x - 1}{8x^2 - 3} \right)^{1/3}$

**25.** $\lim_{x \to -\infty} \left( \dfrac{1 - x^3}{x^2 + 7x} \right)^5$

**26.** $\lim_{x \to \infty} \sqrt{\dfrac{x^2 - 5x}{x^3 + x - 2}}$

**27.** $\lim_{x \to \infty} \dfrac{2\sqrt{x} + x^{-1}}{3x - 7}$

**28.** $\lim_{x \to \infty} \dfrac{2 + \sqrt{x}}{2 - \sqrt{x}}$

**29.** $\lim_{x \to -\infty} \dfrac{\sqrt[3]{x} - \sqrt[5]{x}}{\sqrt[3]{x} + \sqrt[5]{x}}$

**30.** $\lim_{x \to \infty} \dfrac{x^{-1} + x^{-4}}{x^{-2} - x^{-3}}$

**31.** $\lim_{x \to \infty} \dfrac{2x^{5/3} - x^{1/3} + 7}{x^{8/5} + 3x + \sqrt{x}}$

**32.** $\lim_{x \to -\infty} \dfrac{\sqrt[3]{x} - 5x + 3}{2x + x^{2/3} - 4}$

**33.** $\lim_{x \to \infty} \dfrac{\sqrt{x^2 + 1}}{x + 1}$

**34.** $\lim_{x \to -\infty} \dfrac{\sqrt{x^2 + 1}}{x + 1}$

**35.** $\lim_{x \to \infty} \dfrac{x - 3}{\sqrt{4x^2 + 25}}$

**36.** $\lim_{x \to -\infty} \dfrac{4 - 3x^3}{\sqrt{x^6 + 9}}$

### Infinite Limits

Find the limits in Exercises 37–48.

**37.** $\lim_{x \to 0^+} \dfrac{1}{3x}$

**38.** $\lim_{x \to 0^-} \dfrac{5}{2x}$

**39.** $\lim_{x \to 2^-} \dfrac{3}{x - 2}$

**40.** $\lim_{x \to 3^+} \dfrac{1}{x - 3}$

**41.** $\lim_{x \to -8^+} \dfrac{2x}{x + 8}$

**42.** $\lim_{x \to -5^-} \dfrac{3x}{2x + 10}$

**43.** $\lim_{x \to 7} \dfrac{4}{(x - 7)^2}$

**44.** $\lim_{x \to 0} \dfrac{-1}{x^2(x + 1)}$

**45. a.** $\lim_{x \to 0^+} \dfrac{2}{3x^{1/3}}$　　**b.** $\lim_{x \to 0^-} \dfrac{2}{3x^{1/3}}$

**46. a.** $\lim_{x \to 0^+} \dfrac{2}{x^{1/5}}$    **b.** $\lim_{x \to 0^-} \dfrac{2}{x^{1/5}}$

**47.** $\lim_{x \to 0} \dfrac{4}{x^{2/5}}$    **48.** $\lim_{x \to 0} \dfrac{1}{x^{2/3}}$

Find the limits in Exercises 49–52.

**49.** $\lim_{x \to (\pi/2)^-} \tan x$    **50.** $\lim_{x \to (-\pi/2)^+} \sec x$

**51.** $\lim_{\theta \to 0^-} (1 + \csc \theta)$    **52.** $\lim_{\theta \to 0} (2 - \cot \theta)$

Find the limits in Exercises 53–58.

**53.** $\lim \dfrac{1}{x^2 - 4}$ as

    **a.** $x \to 2^+$    **b.** $x \to 2^-$
    **c.** $x \to -2^+$    **d.** $x \to -2^-$

**54.** $\lim \dfrac{x}{x^2 - 1}$ as

    **a.** $x \to 1^+$    **b.** $x \to 1^-$
    **c.** $x \to -1^+$    **d.** $x \to -1^-$

**55.** $\lim \left( \dfrac{x^2}{2} - \dfrac{1}{x} \right)$ as

    **a.** $x \to 0^+$    **b.** $x \to 0^-$
    **c.** $x \to \sqrt[3]{2}$    **d.** $x \to -1$

**56.** $\lim \dfrac{x^2 - 1}{2x + 4}$ as

    **a.** $x \to -2^+$    **b.** $x \to -2^-$
    **c.** $x \to 1^+$    **d.** $x \to 0^-$

**57.** $\lim \dfrac{x^2 - 3x + 2}{x^3 - 2x^2}$ as

    **a.** $x \to 0^+$    **b.** $x \to 2^+$
    **c.** $x \to 2^-$    **d.** $x \to 2$
    **e.** What, if anything, can be said about the limit as $x \to 0$?

**58.** $\lim \dfrac{x^2 - 3x + 2}{x^3 - 4x}$ as

    **a.** $x \to 2^+$    **b.** $x \to -2^+$
    **c.** $x \to 0^-$    **d.** $x \to 1^+$
    **e.** What, if anything, can be said about the limit as $x \to 0$?

Find the limits in Exercises 59–62.

**59.** $\lim \left( 2 - \dfrac{3}{t^{1/3}} \right)$ as

    **a.** $t \to 0^+$    **b.** $t \to 0^-$

**60.** $\lim \left( \dfrac{1}{t^{3/5}} + 7 \right)$ as

    **a.** $t \to 0^+$    **b.** $t \to 0^-$

**61.** $\lim \left( \dfrac{1}{x^{2/3}} + \dfrac{2}{(x - 1)^{2/3}} \right)$ as

    **a.** $x \to 0^+$    **b.** $x \to 0^-$
    **c.** $x \to 1^+$    **d.** $x \to 1^-$

**62.** $\lim \left( \dfrac{1}{x^{1/3}} - \dfrac{1}{(x - 1)^{4/3}} \right)$ as

    **a.** $x \to 0^+$    **b.** $x \to 0^-$
    **c.** $x \to 1^+$    **d.** $x \to 1^-$

### Graphing Simple Rational Functions

Graph the rational functions in Exercises 63–68. Include the graphs and equations of the asymptotes and dominant terms.

**63.** $y = \dfrac{1}{x - 1}$    **64.** $y = \dfrac{1}{x + 1}$

**65.** $y = \dfrac{1}{2x + 4}$    **66.** $y = \dfrac{-3}{x - 3}$

**67.** $y = \dfrac{x + 3}{x + 2}$    **68.** $y = \dfrac{2x}{x + 1}$

### Inventing Graphs and Functions

In Exercises 69–72, sketch the graph of a function $y = f(x)$ that satisfies the given conditions. No formulas are required—just label the coordinate axes and sketch an appropriate graph. (The answers are not unique, so your graphs may not be exactly like those in the answer section.)

**69.** $f(0) = 0, f(1) = 2, f(-1) = -2, \lim_{x \to -\infty} f(x) = -1$, and
    $\lim_{x \to \infty} f(x) = 1$

**70.** $f(0) = 0, \lim_{x \to \pm\infty} f(x) = 0, \lim_{x \to 0^+} f(x) = 2$, and
    $\lim_{x \to 0^-} f(x) = -2$

**71.** $f(0) = 0, \lim_{x \to \pm\infty} f(x) = 0, \lim_{x \to 1^-} f(x) = \lim_{x \to -1^+} f(x) = \infty$,
    $\lim_{x \to 1^+} f(x) = -\infty$, and $\lim_{x \to -1^-} f(x) = -\infty$

**72.** $f(2) = 1, f(-1) = 0, \lim_{x \to \infty} f(x) = 0, \lim_{x \to 0^+} f(x) = \infty$,
    $\lim_{x \to 0^-} f(x) = -\infty$, and $\lim_{x \to -\infty} f(x) = 1$

In Exercises 73–76, find a function that satisfies the given conditions and sketch its graph. (The answers here are not unique. Any function that satisfies the conditions is acceptable. Feel free to use formulas defined in pieces if that will help.)

**73.** $\lim_{x \to \pm\infty} f(x) = 0, \lim_{x \to 2^-} f(x) = \infty$, and $\lim_{x \to 2^+} f(x) = \infty$

**74.** $\lim_{x \to \pm\infty} g(x) = 0, \lim_{x \to 3^-} g(x) = -\infty$, and $\lim_{x \to 3^+} g(x) = \infty$

**75.** $\lim_{x \to -\infty} h(x) = -1, \lim_{x \to \infty} h(x) = 1, \lim_{x \to 0^-} h(x) = -1$, and
    $\lim_{x \to 0^+} h(x) = 1$

**76.** $\lim_{x \to \pm\infty} k(x) = 1, \lim_{x \to 1^-} k(x) = \infty$, and $\lim_{x \to 1^+} k(x) = -\infty$

**77.** Suppose that $f(x)$ and $g(x)$ are polynomials in $x$ and that $\lim_{x \to \infty} (f(x)/g(x)) = 2$. Can you conclude anything about $\lim_{x \to -\infty} (f(x)/g(x))$? Give reasons for your answer.

**78.** Suppose that $f(x)$ and $g(x)$ are polynomials in $x$. Can the graph of $f(x)/g(x)$ have an asymptote if $g(x)$ is never zero? Give reasons for your answer.

**79.** How many horizontal asymptotes can the graph of a given rational function have? Give reasons for your answer.

### Finding Limits of Differences When $x \to \pm\infty$

Find the limits in Exercises 80–86.

**80.** $\lim_{x \to \infty} \left( \sqrt{x + 9} - \sqrt{x + 4} \right)$

**81.** $\lim_{x \to \infty} \left( \sqrt{x^2 + 25} - \sqrt{x^2 - 1} \right)$

**82.** $\lim_{x \to -\infty} \left( \sqrt{x^2 + 3} + x \right)$

**83.** $\lim_{x \to -\infty} \left( 2x + \sqrt{4x^2 + 3x - 2} \right)$

**84.** $\lim_{x \to \infty} \left( \sqrt{9x^2 - x} - 3x \right)$

**85.** $\lim_{x\to\infty} \left( \sqrt{x^2+3x} - \sqrt{x^2-2x} \right)$

**86.** $\lim_{x\to\infty} \left( \sqrt{x^2+x} - \sqrt{x^2-x} \right)$

### Using the Formal Definitions

Use the formal definitions of limits as $x \to \pm\infty$ to establish the limits in Exercises 87 and 88.

**87.** If $f$ has the constant value $f(x) = k$, then $\lim_{x\to\infty} f(x) = k$.

**88.** If $f$ has the constant value $f(x) = k$, then $\lim_{x\to-\infty} f(x) = k$.

Use formal definitions to prove the limit statements in Exercises 89–92.

**89.** $\lim_{x\to0} \dfrac{-1}{x^2} = -\infty$

**90.** $\lim_{x\to0} \dfrac{1}{|x|} = \infty$

**91.** $\lim_{x\to3} \dfrac{-2}{(x-3)^2} = -\infty$

**92.** $\lim_{x\to-5} \dfrac{1}{(x+5)^2} = \infty$

**93.** Here is the definition of **infinite right-hand limit**.

---

We say that $f(x)$ approaches infinity as $x$ approaches $c$ from the right, and write

$$\lim_{x\to c^+} f(x) = \infty,$$

if, for every positive real number $B$, there exists a corresponding number $\delta > 0$ such that for all $x$

$$c < x < c + \delta \quad \Rightarrow \quad f(x) > B.$$

---

Modify the definition to cover the following cases.

**a.** $\lim_{x\to c^-} f(x) = \infty$

**b.** $\lim_{x\to c^+} f(x) = -\infty$

**c.** $\lim_{x\to c^-} f(x) = -\infty$

Use the formal definitions from Exercise 93 to prove the limit statements in Exercises 94–98.

**94.** $\lim_{x\to0^+} \dfrac{1}{x} = \infty$

**95.** $\lim_{x\to0^-} \dfrac{1}{x} = -\infty$

**96.** $\lim_{x\to2^-} \dfrac{1}{x-2} = -\infty$

**97.** $\lim_{x\to2^+} \dfrac{1}{x-2} = \infty$

**98.** $\lim_{x\to1^-} \dfrac{1}{1-x^2} = \infty$

### Oblique Asymptotes

Graph the rational functions in Exercises 99–104. Include the graphs and equations of the asymptotes.

**99.** $y = \dfrac{x^2}{x-1}$

**100.** $y = \dfrac{x^2+1}{x-1}$

**101.** $y = \dfrac{x^2-4}{x-1}$

**102.** $y = \dfrac{x^2-1}{2x+4}$

**103.** $y = \dfrac{x^2-1}{x}$

**104.** $y = \dfrac{x^3+1}{x^2}$

### Additional Graphing Exercises

T Graph the curves in Exercises 105–108. Explain the relationship between the curve's formula and what you see.

**105.** $y = \dfrac{x}{\sqrt{4-x^2}}$

**106.** $y = \dfrac{-1}{\sqrt{4-x^2}}$

**107.** $y = x^{2/3} + \dfrac{1}{x^{1/3}}$

**108.** $y = \sin\left( \dfrac{\pi}{x^2+1} \right)$

T Graph the functions in Exercises 109 and 110. Then answer the following questions.

**a.** How does the graph behave as $x \to 0^+$?

**b.** How does the graph behave as $x \to \pm\infty$?

**c.** How does the graph behave near $x = 1$ and $x = -1$?

Give reasons for your answers.

**109.** $y = \dfrac{3}{2}\left( x - \dfrac{1}{x} \right)^{2/3}$

**110.** $y = \dfrac{3}{2}\left( \dfrac{x}{x-1} \right)^{2/3}$

---

## Chapter 2 Questions to Guide Your Review

**1.** What is the average rate of change of the function $g(t)$ over the interval from $t = a$ to $t = b$? How is it related to a secant line?

**2.** What limit must be calculated to find the rate of change of a function $g(t)$ at $t = t_0$?

**3.** Give an informal or intuitive definition of the limit

$$\lim_{x\to c} f(x) = L.$$

Why is the definition "informal"? Give examples.

**4.** Does the existence and value of the limit of a function $f(x)$ as $x$ approaches $c$ ever depend on what happens at $x = c$? Explain and give examples.

**5.** What function behaviors might occur for which the limit may fail to exist? Give examples.

**6.** What theorems are available for calculating limits? Give examples of how the theorems are used.

**7.** How are one-sided limits related to limits? How can this relationship sometimes be used to calculate a limit or prove it does not exist? Give examples.

**8.** What is the value of $\lim_{\theta\to0} ((\sin\theta)/\theta)$? Does it matter whether $\theta$ is measured in degrees or radians? Explain.

9. What exactly does $\lim_{x \to c} f(x) = L$ mean? Give an example in which you find a $\delta > 0$ for a given $f$, $L$, $c$, and $\epsilon > 0$ in the precise definition of limit.

10. Give precise definitions of the following statements.

    a. $\lim_{x \to 2^-} f(x) = 5$    b. $\lim_{x \to 2^+} f(x) = 5$

    c. $\lim_{x \to 2} f(x) = \infty$    d. $\lim_{x \to 2} f(x) = -\infty$

11. What conditions must be satisfied by a function if it is to be continuous at an interior point of its domain? At an endpoint?

12. How can looking at the graph of a function help you tell where the function is continuous?

13. What does it mean for a function to be right-continuous at a point? Left-continuous? How are continuity and one-sided continuity related?

14. What does it mean for a function to be continuous on an interval? Give examples to illustrate the fact that a function that is not continuous on its entire domain may still be continuous on selected intervals within the domain.

15. What are the basic types of discontinuity? Give an example of each. What is a removable discontinuity? Give an example.

16. What does it mean for a function to have the Intermediate Value Property? What conditions guarantee that a function has this property over an interval? What are the consequences for graphing and solving the equation $f(x) = 0$?

17. Under what circumstances can you extend a function $f(x)$ to be continuous at a point $x = c$? Give an example.

18. What exactly do $\lim_{x \to \infty} f(x) = L$ and $\lim_{x \to -\infty} f(x) = L$ mean? Give examples.

19. What are $\lim_{x \to \pm\infty} k$ ($k$ a constant) and $\lim_{x \to \pm\infty} (1/x)$? How do you extend these results to other functions? Give examples.

20. How do you find the limit of a rational function as $x \to \pm\infty$? Give examples.

21. What are horizontal and vertical asymptotes? Give examples.

# Chapter **2** Practice Exercises

## Limits and Continuity

1. Graph the function

$$f(x) = \begin{cases} 1, & x \le -1 \\ -x, & -1 < x < 0 \\ 1, & x = 0 \\ -x, & 0 < x < 1 \\ 1, & x \ge 1. \end{cases}$$

Then discuss, in detail, limits, one-sided limits, continuity, and one-sided continuity of $f$ at $x = -1, 0,$ and $1$. Are any of the discontinuities removable? Explain.

2. Repeat the instructions of Exercise 1 for

$$f(x) = \begin{cases} 0, & x \le -1 \\ 1/x, & 0 < |x| < 1 \\ 0, & x = 1 \\ 1, & x > 1. \end{cases}$$

3. Suppose that $f(t)$ and $f(t)$ are defined for all $t$ and that $\lim_{t \to t_0} f(t) = -7$ and $\lim_{t \to t_0} g(t) = 0$. Find the limit as $t \to t_0$ of the following functions.

    a. $3f(t)$    b. $(f(t))^2$

    c. $f(t) \cdot g(t)$    d. $\dfrac{f(t)}{g(t) - 7}$

    e. $\cos(g(t))$    f. $|f(t)|$

    g. $f(t) + g(t)$    h. $1/f(t)$

4. Suppose the functions $f(x)$ and $g(x)$ are defined for all $x$ and that $\lim_{x \to 0} f(x) = 1/2$ and $\lim_{x \to 0} g(x) = \sqrt{2}$. Find the limits as $x \to 0$ of the following functions.

    a. $-g(x)$    b. $g(x) \cdot f(x)$

    c. $f(x) + g(x)$    d. $1/f(x)$

    e. $x + f(x)$    f. $\dfrac{f(x) \cdot \cos x}{x - 1}$

In Exercises 5 and 6, find the value that $\lim_{x \to 0} g(x)$ must have if the given limit statements hold.

5. $\lim_{x \to 0} \left( \dfrac{4 - g(x)}{x} \right) = 1$    6. $\lim_{x \to -4} \left( x \lim_{x \to 0} g(x) \right) = 2$

7. On what intervals are the following functions continuous?

    a. $f(x) = x^{1/3}$    b. $g(x) = x^{3/4}$

    c. $h(x) = x^{-2/3}$    d. $k(x) = x^{-1/6}$

8. On what intervals are the following functions continuous?

    a. $f(x) = \tan x$    b. $g(x) = \csc x$

    c. $h(x) = \dfrac{\cos x}{x - \pi}$    d. $k(x) = \dfrac{\sin x}{x}$

## Finding Limits

In Exercises 9–28, find the limit or explain why it does not exist.

9. $\lim \dfrac{x^2 - 4x + 4}{x^3 + 5x^2 - 14x}$

    a. as $x \to 0$    b. as $x \to 2$

10. $\lim \dfrac{x^2 + x}{x^5 + 2x^4 + x^3}$

    a. as $x \to 0$          b. as $x \to -1$

11. $\lim\limits_{x \to 1} \dfrac{1 - \sqrt{x}}{1 - x}$

12. $\lim\limits_{x \to a} \dfrac{x^2 - a^2}{x^4 - a^4}$

13. $\lim\limits_{h \to 0} \dfrac{(x + h)^2 - x^2}{h}$

14. $\lim\limits_{x \to 0} \dfrac{(x + h)^2 - x^2}{h}$

15. $\lim\limits_{x \to 0} \dfrac{\dfrac{1}{2 + x} - \dfrac{1}{2}}{x}$

16. $\lim\limits_{x \to 0} \dfrac{(2 + x)^3 - 8}{x}$

17. $\lim\limits_{x \to 1} \dfrac{x^{1/3} - 1}{\sqrt{x} - 1}$

18. $\lim\limits_{x \to 64} \dfrac{x^{2/3} - 16}{\sqrt{x} - 8}$

19. $\lim\limits_{x \to 0} \dfrac{\tan(2x)}{\tan(\pi x)}$

20. $\lim\limits_{x \to \pi^-} \csc x$

21. $\lim\limits_{x \to \pi} \sin\left(\dfrac{x}{2} + \sin x\right)$

22. $\lim\limits_{x \to \pi} \cos^2(x - \tan x)$

23. $\lim\limits_{x \to 0} \dfrac{8x}{3 \sin x - x}$

24. $\lim\limits_{x \to 0} \dfrac{\cos 2x - 1}{\sin x}$

In Exercises 25–28, find the limit of $g(x)$ as $x$ approaches the indicated value.

25. $\lim\limits_{x \to 0^+} (4g(x))^{1/3} = 2$

26. $\lim\limits_{x \to \sqrt{5}} \dfrac{1}{x + g(x)} = 2$

27. $\lim\limits_{x \to 1} \dfrac{3x^2 + 1}{g(x)} = \infty$

28. $\lim\limits_{x \to -2} \dfrac{5 - x^2}{\sqrt{g(x)}} = 0$

T **Roots**

29. Let $f(x) = x^3 - x - 1$.

    a. Use the Intermediate Value Theorem to show that $f$ has a zero between $-1$ and $2$.

    b. Solve the equation $f(x) = 0$ graphically with an error of magnitude at most $10^{-8}$.

    c. It can be shown that the exact value of the solution in part (b) is

$$\left(\dfrac{1}{2} + \dfrac{\sqrt{69}}{18}\right)^{1/3} + \left(\dfrac{1}{2} - \dfrac{\sqrt{69}}{18}\right)^{1/3}.$$

    Evaluate this exact answer and compare it with the value you found in part (b).

T 30. Let $f(\theta) = \theta^3 - 2\theta + 2$.

    a. Use the Intermediate Value Theorem to show that $f$ has a zero between $-2$ and $0$.

    b. Solve the equation $f(\theta) = 0$ graphically with an error of magnitude at most $10^{-4}$.

    c. It can be shown that the exact value of the solution in part (b) is

$$\left(\sqrt{\dfrac{19}{27}} - 1\right)^{1/3} - \left(\sqrt{\dfrac{19}{27}} + 1\right)^{1/3}.$$

    Evaluate this exact answer and compare it with the value you found in part (b).

**Continuous Extension**

31. Can $f(x) = x(x^2 - 1)/|x^2 - 1|$ be extended to be continuous at $x = 1$ or $-1$? Give reasons for your answers. (Graph the function—you will find the graph interesting.)

32. Explain why the function $f(x) = \sin(1/x)$ has no continuous extension to $x = 0$.

T In Exercises 33–36, graph the function to see whether it appears to have a continuous extension to the given point $a$. If it does, use Trace and Zoom to find a good candidate for the extended function's value at $a$. If the function does not appear to have a continuous extension, can it be extended to be continuous from the right or left? If so, what do you think the extended function's value should be?

33. $f(x) = \dfrac{x - 1}{x - \sqrt[4]{x}}, \quad a = 1$

34. $g(\theta) = \dfrac{5 \cos \theta}{4\theta - 2\pi}, \quad a = \pi/2$

35. $h(t) = (1 + |t|)^{1/t}, \quad a = 0$

36. $k(x) = \dfrac{x}{1 - 2^{|x|}}, \quad a = 0$

**Limits at Infinity**

Find the limits in Exercises 37–46.

37. $\lim\limits_{x \to \infty} \dfrac{2x + 3}{5x + 7}$

38. $\lim\limits_{x \to -\infty} \dfrac{2x^2 + 3}{5x^2 + 7}$

39. $\lim\limits_{x \to -\infty} \dfrac{x^2 - 4x + 8}{3x^3}$

40. $\lim\limits_{x \to \infty} \dfrac{1}{x^2 - 7x + 1}$

41. $\lim\limits_{x \to -\infty} \dfrac{x^2 - 7x}{x + 1}$

42. $\lim\limits_{x \to \infty} \dfrac{x^4 + x^3}{12x^3 + 128}$

43. $\lim\limits_{x \to \infty} \dfrac{\sin x}{\lfloor x \rfloor}$    (If you have a grapher, try graphing the function for $-5 \le x \le 5$.)

44. $\lim\limits_{\theta \to \infty} \dfrac{\cos \theta - 1}{\theta}$    (If you have a grapher, try graphing $f(x) = x(\cos(1/x) - 1)$ near the origin to "see" the limit at infinity.)

45. $\lim\limits_{x \to \infty} \dfrac{x + \sin x + 2\sqrt{x}}{x + \sin x}$

46. $\lim\limits_{x \to \infty} \dfrac{x^{2/3} + x^{-1}}{x^{2/3} + \cos^2 x}$

**Horizontal and Vertical Asymptotes**

47. Use limits to determine the equations for all vertical asymptotes.

    a. $y = \dfrac{x^2 + 4}{x - 3}$

    b. $f(x) = \dfrac{x^2 - x - 2}{x^2 - 2x + 1}$

    c. $y = \dfrac{x^2 + x - 6}{x^2 + 2x - 8}$

48. Use limits to determine the equations for all horizontal asymptotes.

    a. $y = \dfrac{1 - x^2}{x^2 + 1}$      b. $f(x) = \dfrac{\sqrt{x} + 4}{\sqrt{x} + 4}$

    c. $g(x) = \dfrac{\sqrt{x^2 + 4}}{x}$      d. $y = \sqrt{\dfrac{x^2 + 9}{9x^2 + 1}}$

# Chapter **2**   Additional and Advanced Exercises

[T] **1. Assigning a value to $0^0$**   The rules of exponents tell us that $a^0 = 1$ if $a$ is any number different from zero. They also tell us that $0^n = 0$ if $n$ is any positive number.

If we tried to extend these rules to include the case $0^0$, we would get conflicting results. The first rule would say $0^0 = 1$, whereas the second would say $0^0 = 0$.

We are not dealing with a question of right or wrong here. Neither rule applies as it stands, so there is no contradiction. We could, in fact, define $0^0$ to have any value we wanted as long as we could persuade others to agree.

What value would you like $0^0$ to have? Here is an example that might help you to decide. (See Exercise 2 below for another example.)

**a.** Calculate $x^x$ for $x = 0.1, 0.01, 0.001$, and so on as far as your calculator can go. Record the values you get. What pattern do you see?

**b.** Graph the function $y = x^x$ for $0 < x \le 1$. Even though the function is not defined for $x \le 0$, the graph will approach the $y$-axis from the right. Toward what $y$-value does it seem to be headed? Zoom in to further support your idea.

[T] **2. A reason you might want $0^0$ to be something other than 0 or 1**   As the number $x$ increases through positive values, the numbers $1/x$ and $1/(\ln x)$ both approach zero. What happens to the number

$$f(x) = \left(\frac{1}{x}\right)^{1/(\ln x)}$$

as $x$ increases? Here are two ways to find out.

**a.** Evaluate $f$ for $x = 10, 100, 1000$, and so on as far as your calculator can reasonably go. What pattern do you see?

**b.** Graph $f$ in a variety of graphing windows, including windows that contain the origin. What do you see? Trace the $y$-values along the graph. What do you find?

**3. Lorentz contraction**   In relativity theory, the length of an object, say a rocket, appears to an observer to depend on the speed at which the object is traveling with respect to the observer. If the observer measures the rocket's length as $L_0$ at rest, then at speed $v$ the length will appear to be

$$L = L_0 \sqrt{1 - \frac{v^2}{c^2}}.$$

This equation is the Lorentz contraction formula. Here, $c$ is the speed of light in a vacuum, about $3 \times 10^8$ m/s. What happens to $L$ as $v$ increases? Find $\lim_{v \to c^-} L$. Why was the left-hand limit needed?

**4. Controlling the flow from a draining tank**   Torricelli's law says that if you drain a tank like the one in the figure shown, the rate $y$ at which water runs out is a constant times the square root of the water's depth $x$. The constant depends on the size and shape of the exit valve.

Exit rate $y$ m³/h

Suppose that $y = \sqrt{x}/2$ for a certain tank. You are trying to maintain a fairly constant exit rate by adding water to the tank with a hose from time to time. How deep must you keep the water if you want to maintain the exit rate

**a.** within 0.2 m³/h of the rate $y_0 = 1$ m³/h?

**b.** within 0.1 m³/h of the rate $y_0 = 1$ m³/h?

**5. Thermal expansion in precise equipment**   As you may know, most metals expand when heated and contract when cooled. The dimensions of a piece of laboratory equipment are sometimes so critical that the shop where the equipment is made must be held at the same temperature as the laboratory where the equipment is to be used. A typical aluminum bar that is 10 cm wide at 20°C will be

$$y = 10 + 2(t - 20) \times 10^{-4}$$

centimeters wide at a nearby temperature $t$. Suppose that you are using a bar like this in a gravity wave detector, where its width must stay within 0.0005 cm of the ideal 10 cm. How close to $t_0 = 20°C$ must you maintain the temperature to ensure that this tolerance is not exceeded?

**6. Stripes on a measuring cup**   The interior of a typical 1-L measuring cup is a right circular cylinder of radius 6 cm (see accompanying figure). The volume of water we put in the cup is therefore a function of the level $h$ to which the cup is filled, the formula being

$$V = \pi 6^2 h = 36\pi h.$$

How closely must we measure $h$ to measure out 1 L of water (1000 cm³) with an error of no more than 1% (10 cm³)?

(a)

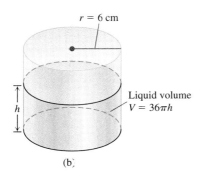

$r = 6$ cm

Liquid volume
$V = 36\pi h$

$h$

(b)

A 1-L measuring cup (a), modeled as a right circular cylinder (b) of radius $r = 6$ cm

## Precise Definition of Limit

In Exercises 7–10, use the formal definition of limit to prove that the function is continuous at c.

**7.** $f(x) = x^2 - 7, \quad c = 1$

**8.** $g(x) = 1/(2x), \quad c = 1/4$

**9.** $h(x) = \sqrt{2x - 3}, \quad c = 2$

**10.** $F(x) = \sqrt{9 - x}, \quad c = 5$

**11. Uniqueness of limits** Show that a function cannot have two different limits at the same point. That is, if $\lim_{x \to c} f(x) = L_1$ and $\lim_{x \to c} f(x) = L_2$, then $L_1 = L_2$.

**12.** Prove the limit Constant Multiple Rule:

$$\lim_{x \to c} kf(x) = k \lim_{x \to c} f(x) \quad \text{for any constant } k.$$

**13. One-sided limits** If $\lim_{x \to 0^+} f(x) = A$ and $\lim_{x \to 0^-} f(x) = B$, find

**a.** $\lim_{x \to 0^+} f(x^3 - x)$

**b.** $\lim_{x \to 0^-} f(x^3 - x)$

**c.** $\lim_{x \to 0^+} f(x^2 - x^4)$

**d.** $\lim_{x \to 0^-} f(x^2 - x^4)$

**14. Limits and continuity** Which of the following statements are true, and which are false? If true, say why; if false, give a counterexample (that is, an example confirming the falsehood).

**a.** If $\lim_{x \to c} f(x)$ exists but $\lim_{x \to c} g(x)$ does not exist, then $\lim_{x \to c}(f(x) + g(x))$ does not exist.

**b.** If neither $\lim_{x \to c} f(x)$ nor $\lim_{x \to c} g(x)$ exists, then $\lim_{x \to c} (f(x) + g(x))$ does not exist.

**c.** If $f$ is continuous at $x$, then so is $|f|$.

**d.** If $|f|$ is continuous at $c$, then so is $f$.

In Exercises 15 and 16, use the formal definition of limit to prove that the function has a continuous extension to the given value of $x$.

**15.** $f(x) = \dfrac{x^2 - 1}{x + 1}, \quad x = -1$

**16.** $g(x) = \dfrac{x^2 - 2x - 3}{2x - 6}, \quad x = 3$

**17. A function continuous at only one point** Let

$$f(x) = \begin{cases} x, & \text{if } x \text{ is rational} \\ 0, & \text{if } x \text{ is irrational.} \end{cases}$$

**a.** Show that $f$ is continuous at $x = 0$.

**b.** Use the fact that every nonempty open interval of real numbers contains both rational and irrational numbers to show that $f$ is not continuous at any nonzero value of $x$.

**18. The Dirichlet ruler function** If $x$ is a rational number, then $x$ can be written in a unique way as a quotient of integers $m/n$ where $n > 0$ and $m$ and $n$ have no common factors greater than 1. (We say that such a fraction is in *lowest terms*. For example, $6/4$ written in lowest terms is $3/2$.) Let $f(x)$ be defined for all $x$ in the interval $[0, 1]$ by

$$f(x) = \begin{cases} 1/n, & \text{if } x = m/n \text{ is a rational number in lowest terms} \\ 0, & \text{if } x \text{ is irrational.} \end{cases}$$

For instance, $f(0) = f(1) = 1, f(1/2) = 1/2, f(1/3) = f(2/3) = 1/3, f(1/4) = f(3/4) = 1/4$, and so on.

**a.** Show that $f$ is discontinuous at every rational number in $[0, 1]$.

**b.** Show that $f$ is continuous at every irrational number in $[0, 1]$. (*Hint:* If $\epsilon$ is a given positive number, show that there are only finitely many rational numbers $r$ in $[0, 1]$ such that $f(r) \geq \epsilon$.)

**c.** Sketch the graph of $f$. Why do you think $f$ is called the "ruler function"?

**19. Antipodal points** Is there any reason to believe that there is always a pair of antipodal (diametrically opposite) points on Earth's equator where the temperatures are the same? Explain.

**20.** If $\lim_{x \to c} (f(x) + g(x)) = 3$ and $\lim_{x \to c} (f(x) - g(x)) = -1$, find $\lim_{x \to c} f(x)g(x)$.

**21. Roots of a quadratic equation that is almost linear** The equation $ax^2 + 2x - 1 = 0$, where $a$ is a constant, has two roots if $a > -1$ and $a \neq 0$, one positive and one negative:

$$r_+(a) = \frac{-1 + \sqrt{1 + a}}{a}, \qquad r_-(a) = \frac{-1 - \sqrt{1 + a}}{a},$$

**a.** What happens to $r_+(a)$ as $a \to 0$? As $a \to -1^+$?

**b.** What happens to $r_-(a)$ as $a \to 0$? As $a \to -1^+$?

**c.** Support your conclusions by graphing $r_+(a)$ and $r_-(a)$ as functions of $a$. Describe what you see.

**d.** For added support, graph $f(x) = ax^2 + 2x - 1$ simultaneously for $a = 1, 0.5, 0.2, 0.1$, and $0.05$.

**22. Root of an equation** Show that the equation $x + 2 \cos x = 0$ has at least one solution.

**23. Bounded functions** A real-valued function $f$ is **bounded from above** on a set $D$ if there exists a number $N$ such that $f(x) \leq N$ for all $x$ in $D$. We call $N$, when it exists, an **upper bound** for $f$ on $D$ and say that $f$ is bounded from above by $N$. In a similar manner, we say that $f$ is **bounded from below** on $D$ if there exists a number $M$ such that $f(x) \geq M$ for all $x$ in $D$. We call $M$, when it exists, a **lower bound** for $f$ on $D$ and say that $f$ is bounded from below by $M$. We say that $f$ is **bounded** on $D$ if it is bounded from both above and below.

**a.** Show that $f$ is bounded on $D$ if and only if there exists a number $B$ such that $|f(x)| \leq B$ for all $x$ in $D$.

Stripes about 1 mm wide

**b.** Suppose that $f$ is bounded from above by $N$. Show that if $\lim_{x \to c} f(x) = L$, then $L \leq N$.

**c.** Suppose that $f$ is bounded from below by $M$. Show that if $\lim_{x \to c} f(x) = L$, then $L \geq M$.

**24. Max $\{a, b\}$ and min $\{a, b\}$**

**a.** Show that the expression

$$\max \{a, b\} = \frac{a + b}{2} + \frac{|a - b|}{2}$$

equals $a$ if $a \geq b$ and equals $b$ if $b \geq a$. In other words, max $\{a, b\}$ gives the larger of the two numbers $a$ and $b$.

**b.** Find a similar expression for min $\{a, b\}$, the smaller of $a$ and $b$.

**Generalized Limits Involving** $\dfrac{\sin \theta}{\theta}$

The formula $\lim_{\theta \to 0} (\sin \theta)/\theta = 1$ can be generalized. If $\lim_{x \to c} f(x) = 0$ and $f(x)$ is never zero in an open interval containing the point $x = c$, except possibly $c$ itself, then

$$\lim_{x \to c} \frac{\sin f(x)}{f(x)} = 1.$$

Here are several examples.

**a.** $\lim_{x \to 0} \dfrac{\sin x^2}{x^2} = 1$

**b.** $\lim_{x \to 0} \dfrac{\sin x^2}{x} = \lim_{x \to 0} \dfrac{\sin x^2}{x^2} \lim_{x \to 0} \dfrac{x^2}{x} = 1 \cdot 0 = 0$

**c.** $\lim_{x \to -1} \dfrac{\sin (x^2 - x - 2)}{x + 1} = \lim_{x \to -1} \dfrac{\sin (x^2 - x - 2)}{(x^2 - x - 2)} \cdot$

$\lim_{x \to -1} \dfrac{(x^2 - x - 2)}{x + 1} = 1 \cdot \lim_{x \to -1} \dfrac{(x + 1)(x - 2)}{x + 1} = -3$

**d.** $\lim_{x \to 1} \dfrac{\sin (1 - \sqrt{x})}{x - 1} = \lim_{x \to 1} \dfrac{\sin (1 - \sqrt{x})}{1 - \sqrt{x}} \dfrac{1 - \sqrt{x}}{x - 1} =$

$1 \cdot \lim_{x \to 1} \dfrac{(1 - \sqrt{x})(1 + \sqrt{x})}{(x - 1)(1 + \sqrt{x})} = \lim_{x \to 1} \dfrac{1 - x}{(x - 1)(1 + \sqrt{x})} = -\dfrac{1}{2}$

Find the limits in Exercises 25–30.

**25.** $\lim_{x \to 0} \dfrac{\sin (1 - \cos x)}{x}$

**26.** $\lim_{x \to 0^+} \dfrac{\sin x}{\sin \sqrt{x}}$

**27.** $\lim_{x \to 0} \dfrac{\sin (\sin x)}{x}$

**28.** $\lim_{x \to 0} \dfrac{\sin (x^2 + x)}{x}$

**29.** $\lim_{x \to 2} \dfrac{\sin (x^2 - 4)}{x - 2}$

**30.** $\lim_{x \to 9} \dfrac{\sin (\sqrt{x} - 3)}{x - 9}$

**Oblique Asymptotes**

Find all possible oblique asymptotes in Exercises 31–34.

**31.** $y = \dfrac{2x^{3/2} + 2x - 3}{\sqrt{x} + 1}$

**32.** $y = x + x \sin \dfrac{1}{x}$

**33.** $y = \sqrt{x^2 + 1}$

**34.** $y = \sqrt{x^2 + 2x}$

---

# Chapter **2** Technology Application Projects

# 3

# Derivatives

**OVERVIEW** In the beginning of Chapter 2, we discussed how to determine the slope of a curve at a point and how to measure the rate at which a function changes. Now that we have studied limits, we can define these ideas precisely and see that both are interpretations of the *derivative* of a function at a point. We then extend this concept from a single point to the *derivative function*, and we develop rules for finding this derivative function easily, without having to calculate any limits directly. These rules are used to find derivatives of many of the common functions reviewed in Chapter 1, as well as various combinations of them.

The derivative is one of the key ideas in calculus, and is used to study a wide range of problems in mathematics, science, economics, and medicine. These problems include finding points where a continuous function is zero, calculating the velocity and acceleration of a moving object, determining how the rate of flow of a liquid into a container changes the level of the liquid within it, describing the path followed by a light ray going from a point in air to a point in water, finding the number of items a manufacturing company should produce in order to maximize its profits, studying the spread of an infectious disease within a given population, or calculating the amount of blood the heart pumps in a minute based on how well the lungs are functioning.

## 3.1 Tangents and the Derivative at a Point

In this section we define the slope and tangent to a curve at a point, and the derivative of a function at a point. The derivative gives a way to find both the slope of a graph and the instantaneous rate of change of a function.

### Finding a Tangent to the Graph of a Function

To find a tangent to an arbitrary curve $y = f(x)$ at a point $P(x_0, f(x_0))$, we use the procedure introduced in Section 2.1. We calculate the slope of the secant through $P$ and a nearby point $Q(x_0 + h, f(x_0 + h))$. We then investigate the limit of the slope as $h \to 0$ (Figure 3.1). If the limit exists, we call it the slope of the curve at $P$ and define the tangent at $P$ to be the line through $P$ having this slope.

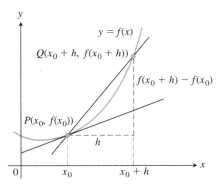

**FIGURE 3.1** The slope of the tangent line at $P$ is $\lim\limits_{h \to 0} \dfrac{f(x_0 + h) - f(x_0)}{h}$.

---

**DEFINITIONS** The **slope of the curve** $y = f(x)$ at the point $P(x_0, f(x_0))$ is the number

$$m = \lim_{h \to 0} \frac{f(x_0 + h) - f(x_0)}{h} \qquad \text{(provided the limit exists).}$$

The **tangent line** to the curve at $P$ is the line through $P$ with this slope.

---

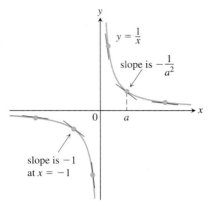

**FIGURE 3.2**   The tangent slopes, steep near the origin, become more gradual as the point of tangency moves away (Example 1).

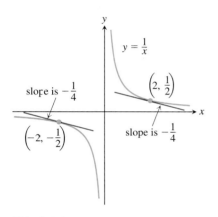

**FIGURE 3.3**   The two tangent lines to $y = 1/x$ having slope $-1/4$ (Example 1).

In Section 2.1, Example 3, we applied these definitions to find the slope of the parabola $f(x) = x^2$ at the point $P(2, 4)$ and the tangent line to the parabola at $P$. Let's look at another example.

### EXAMPLE 1

**(a)** Find the slope of the curve $y = 1/x$ at any point $x = a \neq 0$. What is the slope at the point $x = -1$?

**(b)** Where does the slope equal $-1/4$?

**(c)** What happens to the tangent to the curve at the point $(a, 1/a)$ as $a$ changes?

**Solution**

**(a)** Here $f(x) = 1/x$. The slope at $(a, 1/a)$ is

$$\lim_{h \to 0} \frac{f(a + h) - f(a)}{h} = \lim_{h \to 0} \frac{\dfrac{1}{a + h} - \dfrac{1}{a}}{h} = \lim_{h \to 0} \frac{1}{h} \frac{a - (a + h)}{a(a + h)}$$

$$= \lim_{h \to 0} \frac{-h}{ha(a + h)} = \lim_{h \to 0} \frac{-1}{a(a + h)} = -\frac{1}{a^2}.$$

Notice how we had to keep writing "$\lim_{h \to 0}$" before each fraction until the stage at which we could evaluate the limit by substituting $h = 0$. The number $a$ may be positive or negative, but not 0. When $a = -1$, the slope is $-1/(-1)^2 = -1$ (Figure 3.2).

**(b)** The slope of $y = 1/x$ at the point where $x = a$ is $-1/a^2$. It will be $-1/4$ provided that

$$-\frac{1}{a^2} = -\frac{1}{4}.$$

This equation is equivalent to $a^2 = 4$, so $a = 2$ or $a = -2$. The curve has slope $-1/4$ at the two points $(2, 1/2)$ and $(-2, -1/2)$ (Figure 3.3).

**(c)** The slope $-1/a^2$ is always negative if $a \neq 0$. As $a \to 0^+$, the slope approaches $-\infty$ and the tangent becomes increasingly steep (Figure 3.2). We see this situation again as $a \to 0^-$. As $a$ moves away from the origin in either direction, the slope approaches 0 and the tangent levels off becoming more and more horizontal. ∎

### Rates of Change: Derivative at a Point

The expression

$$\frac{f(x_0 + h) - f(x_0)}{h}, \quad h \neq 0$$

is called the **difference quotient of $f$ at $x_0$ with increment $h$**. If the difference quotient has a limit as $h$ approaches zero, that limit is given a special name and notation.

---

The notation $f'(x_0)$ is read "$f$ prime of $x_0$."

> **DEFINITION**   The **derivative of a function $f$ at a point $x_0$**, denoted $f'(x_0)$, is
>
> $$f'(x_0) = \lim_{h \to 0} \frac{f(x_0 + h) - f(x_0)}{h}$$
>
> provided this limit exists.

If we interpret the difference quotient as the slope of a secant line, then the derivative gives the slope of the curve $y = f(x)$ at the point $P(x_0, f(x_0))$. Exercise 33 shows that the derivative of the linear function $f(x) = mx + b$ at any point $x_0$ is simply the slope of the line, so

$$f'(x_0) = m,$$

which is consistent with our definition of slope.

If we interpret the difference quotient as an average rate of change (Section 2.1), the derivative gives the function's instantaneous rate of change with respect to $x$ at the point $x = x_0$. We study this interpretation in Section 3.4.

**EXAMPLE 2**    In Examples 1 and 2 in Section 2.1, we studied the speed of a rock falling freely from rest near the surface of the earth. We knew that the rock fell $y = 4.9t^2$ meters during the first $t$ seconds, and we used a sequence of average rates over increasingly short intervals to estimate the rock's speed at the instant $t = 1$. What was the rock's *exact* speed at this time?

**Solution**    We let $f(t) = 4.9t^2$. The average speed of the rock over the interval between $t = 1$ and $t = 1 + h$ seconds, for $h > 0$, was found to be

$$\frac{f(1 + h) - f(1)}{h} = \frac{4.9(1 + h)^2 - 4.9(1)^2}{h} = \frac{4.9(h^2 + 2h)}{h} = 4.9(h + 2).$$

The rock's speed at the instant $t = 1$ is then

$$f'(1) = \lim_{h \to 0} 4.9(h + 2) = 4.9(0 + 2) = 9.8 \text{ m/s}.$$

Our original estimate of 9.8 m/s in Section 2.1 was right.    ■

## Summary

We have been discussing slopes of curves, lines tangent to a curve, the rate of change of a function, and the derivative of a function at a point. All of these ideas refer to the same limit.

---

The following are all interpretations for the limit of the difference quotient,

$$\lim_{h \to 0} \frac{f(x_0 + h) - f(x_0)}{h}.$$

1. The slope of the graph of $y = f(x)$ at $x = x_0$
2. The slope of the tangent to the curve $y = f(x)$ at $x = x_0$
3. The rate of change of $f(x)$ with respect to $x$ at $x = x_0$
4. The derivative $f'(x_0)$ at a point

---

In the next sections, we allow the point $x_0$ to vary across the domain of the function $f$.

# Exercises 3.1

## Slopes and Tangent Lines

In Exercises 1–4, use the grid and a straight edge to make a rough estimate of the slope of the curve (in $y$-units per $x$-unit) at the points $P_1$ and $P_2$.

**1.**

**2.**

**3.**

**4.**

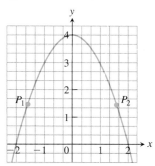

In Exercises 5–10, find an equation for the tangent to the curve at the given point. Then sketch the curve and tangent together.

**5.** $y = 4 - x^2$, $(-1, 3)$

**6.** $y = (x - 1)^2 + 1$, $(1, 1)$

**7.** $y = 2\sqrt{x}$, $(1, 2)$

**8.** $y = \dfrac{1}{x^2}$, $(-1, 1)$

**9.** $y = x^3$, $(-2, -8)$

**10.** $y = \dfrac{1}{x^3}$, $\left(-2, -\dfrac{1}{8}\right)$

In Exercises 11–18, find the slope of the function's graph at the given point. Then find an equation for the line tangent to the graph there.

**11.** $f(x) = x^2 + 1$, $(2, 5)$

**12.** $f(x) = x - 2x^2$, $(1, -1)$

**13.** $g(x) = \dfrac{x}{x - 2}$, $(3, 3)$

**14.** $g(x) = \dfrac{8}{x^2}$, $(2, 2)$

**15.** $h(t) = t^3$, $(2, 8)$

**16.** $h(t) = t^3 + 3t$, $(1, 4)$

**17.** $f(x) = \sqrt{x}$, $(4, 2)$

**18.** $f(x) = \sqrt{x + 1}$, $(8, 3)$

In Exercises 19–22, find the slope of the curve at the point indicated.

**19.** $y = 5x - 3x^2$, $x = 1$

**20.** $y = x^3 - 2x + 7$, $x = -2$

**21.** $y = \dfrac{1}{x - 1}$, $x = 3$

**22.** $y = \dfrac{x - 1}{x + 1}$, $x = 0$

## Interpreting Derivative Values

**23. Growth of yeast cells** In a controlled laboratory experiment, yeast cells are grown in an automated cell culture system that counts the number $P$ of cells present at hourly intervals. The number after $t$ hours is shown in the accompanying figure.

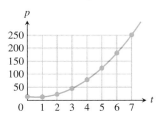

**a.** Explain what is meant by the derivative $P'(5)$. What are its units?

**b.** Which is larger, $P'(2)$ or $P'(3)$? Give a reason for your answer.

**c.** The quadratic curve capturing the trend of the data points (see Section 1.4) is given by $P(t) = 6.10t^2 - 9.28t + 16.43$. Find the instantaneous rate of growth when $t = 5$ hours.

**24. Effectiveness of a drug** On a scale from 0 to 1, the effectiveness $E$ of a pain-killing drug $t$ hours after entering the bloodstream is displayed in the accompanying figure.

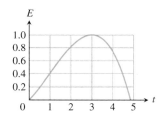

**a.** At what times does the effectiveness appear to be increasing? What is true about the derivative at those times?

**b.** At what time would you estimate that the drug reaches its maximum effectiveness? What is true about the derivative at that time? What is true about the derivative as time increases in the 1 hour *before* your estimated time?

At what points do the graphs of the functions in Exercises 25 and 26 have horizontal tangents?

**25.** $f(x) = x^2 + 4x - 1$

**26.** $g(x) = x^3 - 3x$

**27.** Find equations of all lines having slope $-1$ that are tangent to the curve $y = 1/(x - 1)$.

**28.** Find an equation of the straight line having slope $1/4$ that is tangent to the curve $y = \sqrt{x}$.

## Rates of Change

**29. Object dropped from a tower** An object is dropped from the top of a 100-m-high tower. Its height above ground after $t$ s is $100 - 4.9t^2$ m. How fast is it falling 2 s after it is dropped?

**30. Speed of a rocket**  At $t$ seconds after liftoff, the height of a rocket is $3t^2$ m. How fast is the rocket climbing 10 s after liftoff?

**31. Circle's changing area**  What is the rate of change of the area of a circle $(A = \pi r^2)$ with respect to the radius when the radius is $r = 3$?

**32. Ball's changing volume**  What is the rate of change of the volume of a ball $(V = (4/3)\pi r^3)$ with respect to the radius when the radius is $r = 2$?

**33.** Show that the line $y = mx + b$ is its own tangent line at any point $(x_0, mx_0 + b)$.

**34.** Find the slope of the tangent to the curve $y = 1/\sqrt{x}$ at the point where $x = 4$.

## Testing for Tangents

**35.** Does the graph of

$$f(x) = \begin{cases} x^2 \sin(1/x), & x \neq 0 \\ 0, & x = 0 \end{cases}$$

have a tangent at the origin? Give reasons for your answer.

**36.** Does the graph of

$$g(x) = \begin{cases} x \sin(1/x), & x \neq 0 \\ 0, & x = 0 \end{cases}$$

have a tangent at the origin? Give reasons for your answer.

## Vertical Tangents

We say that a continuous curve $y = f(x)$ has a **vertical tangent** at the point where $x = x_0$ if the limit of the difference quotient is $\infty$ or $-\infty$. For example, $y = x^{1/3}$ has a vertical tangent at $x = 0$ (see accompanying figure):

$$\lim_{h \to 0} \frac{f(0 + h) - f(0)}{h} = \lim_{h \to 0} \frac{h^{1/3} - 0}{h}$$

$$= \lim_{h \to 0} \frac{1}{h^{2/3}} = \infty.$$

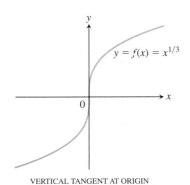

VERTICAL TANGENT AT ORIGIN

However, $y = x^{2/3}$ has *no* vertical tangent at $x = 0$ (see next figure):

$$\lim_{h \to 0} \frac{g(0 + h) - g(0)}{h} = \lim_{h \to 0} \frac{h^{2/3} - 0}{h}$$

$$= \lim_{h \to 0} \frac{1}{h^{1/3}}$$

does not exist, because the limit is $\infty$ from the right and $-\infty$ from the left.

NO VERTICAL TANGENT AT ORIGIN

**37.** Does the graph of

$$f(x) = \begin{cases} -1, & x < 0 \\ 0, & x = 0 \\ 1, & x > 0 \end{cases}$$

have a vertical tangent at the origin? Give reasons for your answer.

**38.** Does the graph of

$$U(x) = \begin{cases} 0, & x < 0 \\ 1, & x \geq 0 \end{cases}$$

have a vertical tangent at the point $(0, 1)$? Give reasons for your answer.

T  Graph the curves in Exercises 39–48.

    **a.** Where do the graphs appear to have vertical tangents?

    **b.** Confirm your findings in part (a) with limit calculations. But before you do, read the introduction to Exercises 37 and 38.

**39.** $y = x^{2/5}$             **40.** $y = x^{4/5}$

**41.** $y = x^{1/5}$             **42.** $y = x^{3/5}$

**43.** $y = 4x^{2/5} - 2x$     **44.** $y = x^{5/3} - 5x^{2/3}$

**45.** $y = x^{2/3} - (x - 1)^{1/3}$    **46.** $y = x^{1/3} + (x - 1)^{1/3}$

**47.** $y = \begin{cases} -\sqrt{|x|}, & x \leq 0 \\ \sqrt{x}, & x > 0 \end{cases}$   **48.** $y = \sqrt{|4 - x|}$

## COMPUTER EXPLORATIONS

Use a CAS to perform the following steps for the functions in Exercises 49–52:

    **a.** Plot $y = f(x)$ over the interval $(x_0 - 1/2) \leq x \leq (x_0 + 3)$.

    **b.** Holding $x_0$ fixed, the difference quotient

$$q(h) = \frac{f(x_0 + h) - f(x_0)}{h}$$

    at $x_0$ becomes a function of the step size $h$. Enter this function into your CAS workspace.

    **c.** Find the limit of $q$ as $h \to 0$.

    **d.** Define the secant lines $y = f(x_0) + q \cdot (x - x_0)$ for $h = 3, 2,$ and 1. Graph them together with $f$ and the tangent line over the interval in part (a).

**49.** $f(x) = x^3 + 2x, \quad x_0 = 0$    **50.** $f(x) = x + \dfrac{5}{x}, \quad x_0 = 1$

**51.** $f(x) = x + \sin(2x), \quad x_0 = \pi/2$

**52.** $f(x) = \cos x + 4 \sin(2x), \quad x_0 = \pi$

# 3.2 The Derivative as a Function

**HISTORICAL ESSAY**
The Derivative

In the last section we defined the derivative of $y = f(x)$ at the point $x = x_0$ to be the limit

$$f'(x_0) = \lim_{h \to 0} \frac{f(x_0 + h) - f(x_0)}{h}.$$

We now investigate the derivative as a *function* derived from $f$ by considering the limit at each point $x$ in the domain of $f$.

---

**DEFINITION** The **derivative** of the function $f(x)$ with respect to the variable $x$ is the function $f'$ whose value at $x$ is

$$f'(x) = \lim_{h \to 0} \frac{f(x + h) - f(x)}{h},$$

provided the limit exists.

---

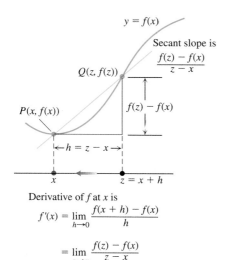

Derivative of $f$ at $x$ is

$$f'(x) = \lim_{h \to 0} \frac{f(x + h) - f(x)}{h}$$

$$= \lim_{z \to x} \frac{f(z) - f(x)}{z - x}$$

**FIGURE 3.4** Two forms for the difference quotient.

We use the notation $f(x)$ in the definition to emphasize the independent variable $x$ with respect to which the derivative function $f'(x)$ is being defined. The domain of $f'$ is the set of points in the domain of $f$ for which the limit exists, which means that the domain may be the same as or smaller than the domain of $f$. If $f'$ exists at a particular $x$, we say that $f$ is **differentiable (has a derivative) at $x$**. If $f'$ exists at every point in the domain of $f$, we call $f$ **differentiable**.

If we write $z = x + h$, then $h = z - x$ and $h$ approaches 0 if and only if $z$ approaches $x$. Therefore, an equivalent definition of the derivative is as follows (see Figure 3.4). This formula is sometimes more convenient to use when finding a derivative function, and focuses on the point $z$ that approaches $x$.

---

**Alternative Formula for the Derivative**

$$f'(x) = \lim_{z \to x} \frac{f(z) - f(x)}{z - x}$$

---

## Calculating Derivatives from the Definition

The process of calculating a derivative is called **differentiation**. To emphasize the idea that differentiation is an operation performed on a function $y = f(x)$, we use the notation

$$\frac{d}{dx} f(x)$$

as another way to denote the derivative $f'(x)$. Example 1 of Section 3.1 illustrated the differentiation process for the function $y = 1/x$ when $x = a$. For $x$ representing any point in the domain, we get the formula

$$\frac{d}{dx}\left(\frac{1}{x}\right) = -\frac{1}{x^2}.$$

Here are two more examples in which we allow $x$ to be any point in the domain of $f$.

**Derivative of the Reciprocal Function**

$$\frac{d}{dx}\left(\frac{1}{x}\right) = -\frac{1}{x^2}, \quad x \neq 0$$

**EXAMPLE 1**    Differentiate $f(x) = \dfrac{x}{x-1}$.

**Solution**   We use the definition of derivative, which requires us to calculate $f(x+h)$ and then subtract $f(x)$ to obtain the numerator in the difference quotient. We have

$$f(x) = \frac{x}{x-1} \quad \text{and} \quad f(x+h) = \frac{(x+h)}{(x+h)-1}, \text{ so}$$

$$f'(x) = \lim_{h \to 0} \frac{f(x+h) - f(x)}{h} \qquad \text{Definition}$$

$$= \lim_{h \to 0} \frac{\dfrac{x+h}{x+h-1} - \dfrac{x}{x-1}}{h}$$

$$= \lim_{h \to 0} \frac{1}{h} \cdot \frac{(x+h)(x-1) - x(x+h-1)}{(x+h-1)(x-1)} \qquad \frac{a}{b} - \frac{c}{d} = \frac{ad - cb}{bd}$$

$$= \lim_{h \to 0} \frac{1}{h} \cdot \frac{-h}{(x+h-1)(x-1)} \qquad \text{Simplify.}$$

$$= \lim_{h \to 0} \frac{-1}{(x+h-1)(x-1)} = \frac{-1}{(x-1)^2}. \qquad \text{Cancel } h \neq 0. \qquad \blacksquare$$

**EXAMPLE 2**
**(a)** Find the derivative of $f(x) = \sqrt{x}$ for $x > 0$.
**(b)** Find the tangent line to the curve $y = \sqrt{x}$ at $x = 4$.

**Solution**
**(a)** We use the alternative formula to calculate $f'$:

$$f'(x) = \lim_{z \to x} \frac{f(z) - f(x)}{z - x}$$

$$= \lim_{z \to x} \frac{\sqrt{z} - \sqrt{x}}{z - x}$$

$$= \lim_{z \to x} \frac{\sqrt{z} - \sqrt{x}}{(\sqrt{z} - \sqrt{x})(\sqrt{z} + \sqrt{x})}$$

$$= \lim_{z \to x} \frac{1}{\sqrt{z} + \sqrt{x}} = \frac{1}{2\sqrt{x}}.$$

**(b)** The slope of the curve at $x = 4$ is

$$f'(4) = \frac{1}{2\sqrt{4}} = \frac{1}{4}.$$

The tangent is the line through the point $(4, 2)$ with slope $1/4$ (Figure 3.5):

$$y = 2 + \frac{1}{4}(x - 4)$$

$$y = \frac{1}{4}x + 1. \qquad \blacksquare$$

**Derivative of the Square Root Function**

$$\frac{d}{dx}\sqrt{x} = \frac{1}{2\sqrt{x}}, \quad x > 0$$

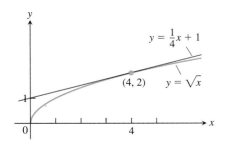

**FIGURE 3.5**   The curve $y = \sqrt{x}$ and its tangent at $(4, 2)$. The tangent's slope is found by evaluating the derivative at $x = 4$ (Example 2).

**Notations**

There are many ways to denote the derivative of a function $y = f(x)$, where the independent variable is $x$ and the dependent variable is $y$. Some common alternative notations for the derivative are

$$f'(x) = y' = \frac{dy}{dx} = \frac{df}{dx} = \frac{d}{dx}f(x) = D(f)(x) = D_x f(x).$$

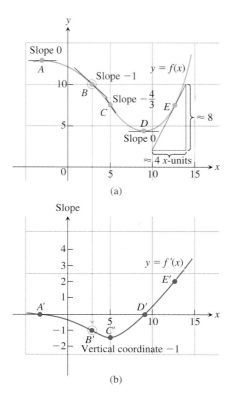

**FIGURE 3.6** We made the graph of $y = f'(x)$ in (b) by plotting slopes from the graph of $y = f(x)$ in (a). The vertical coordinate of $B'$ is the slope at $B$ and so on. The slope at $E$ is approximately $8/4 = 2$. In (b) we see that the rate of change of $f$ is negative for $x$ between $A'$ and $D'$; the rate of change is positive for $x$ to the right of $D'$.

The symbols $d/dx$ and $D$ indicate the operation of differentiation. We read $dy/dx$ as "the derivative of $y$ with respect to $x$," and $df/dx$ and $(d/dx)f(x)$ as "the derivative of $f$ with respect to $x$." The "prime" notations $y'$ and $f'$ come from notations that Newton used for derivatives. The $d/dx$ notations are similar to those used by Leibniz. The symbol $dy/dx$ should not be regarded as a ratio (until we introduce the idea of "differentials" in Section 3.9).

To indicate the value of a derivative at a specified number $x = a$, we use the notation

$$f'(a) = \frac{dy}{dx}\bigg|_{x=a} = \frac{df}{dx}\bigg|_{x=a} = \frac{d}{dx}f(x)\bigg|_{x=a}.$$

For instance, in Example 2

$$f'(4) = \frac{d}{dx}\sqrt{x}\bigg|_{x=4} = \frac{1}{2\sqrt{x}}\bigg|_{x=4} = \frac{1}{2\sqrt{4}} = \frac{1}{4}.$$

## Graphing the Derivative

We can often make a reasonable plot of the derivative of $y = f(x)$ by estimating the slopes on the graph of $f$. That is, we plot the points $(x, f'(x))$ in the $xy$-plane and connect them with a smooth curve, which represents $y = f'(x)$.

**EXAMPLE 3** Graph the derivative of the function $y = f(x)$ in Figure 3.6a.

**Solution** We sketch the tangents to the graph of $f$ at frequent intervals and use their slopes to estimate the values of $f'(x)$ at these points. We plot the corresponding $(x, f'(x))$ pairs and connect them with a smooth curve as sketched in Figure 3.6b. ∎

What can we learn from the graph of $y = f'(x)$? At a glance we can see

1. where the rate of change of $f$ is positive, negative, or zero;
2. the rough size of the growth rate at any $x$ and its size in relation to the size of $f(x)$;
3. where the rate of change itself is increasing or decreasing.

## Differentiable on an Interval; One-Sided Derivatives

A function $y = f(x)$ is **differentiable on an open interval** (finite or infinite) if it has a derivative at each point of the interval. It is **differentiable on a closed interval** $[a, b]$ if it is differentiable on the interior $(a, b)$ and if the limits

$$\lim_{h \to 0^+} \frac{f(a + h) - f(a)}{h} \qquad \textbf{Right-hand derivative at } a$$

$$\lim_{h \to 0^-} \frac{f(b + h) - f(b)}{h} \qquad \textbf{Left-hand derivative at } b$$

exist at the endpoints (Figure 3.7).

Right-hand and left-hand derivatives may be defined at any point of a function's domain. Because of Theorem 6, Section 2.4, a function has a derivative at a point if and only if it has left-hand and right-hand derivatives there, and these one-sided derivatives are equal.

**EXAMPLE 4** Show that the function $y = |x|$ is differentiable on $(-\infty, 0)$ and $(0, \infty)$ but has no derivative at $x = 0$.

**Solution** From Section 3.1, the derivative of $y = mx + b$ is the slope $m$. Thus, to the right of the origin,

$$\frac{d}{dx}(|x|) = \frac{d}{dx}(x) = \frac{d}{dx}(1 \cdot x) = 1. \qquad \frac{d}{dx}(mx + b) = m, |x| = x$$

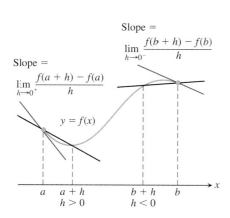

**FIGURE 3.7** Derivatives at endpoints of a closed interval are one-sided limits.

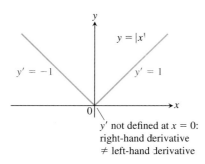

$y' = -1$    $y' = 1$

$y'$ not defined at $x = 0$:
right-hand derivative
$\neq$ left-hand derivative

**FIGURE 3.8** The function $y = |x|$ is not differentiable at the origin where the graph has a "corner" (Example 4).

To the left,

$$\frac{d}{dx}(|x|) = \frac{d}{dx}(-x) = \frac{d}{dx}(-1 \cdot x) = -1 \qquad |x| = -x$$

(Figure 3.8). There is no derivative at the origin because the one-sided derivatives differ there:

$$\text{Right-hand derivative of } |x| \text{ at zero} = \lim_{h \to 0^+} \frac{|0 + h| - |0|}{h} = \lim_{h \to 0^+} \frac{|h|}{h}$$

$$= \lim_{h \to 0^+} \frac{h}{h} \qquad |h| = h \text{ when } h > 0$$

$$= \lim_{h \to 0^+} 1 = 1$$

$$\text{Left-hand derivative of } |x| \text{ at zero} = \lim_{h \to 0^-} \frac{|0 + h| - |0|}{h} = \lim_{h \to 0^-} \frac{|h|}{h}$$

$$= \lim_{h \to 0^-} \frac{-h}{h} \qquad |h| = -h \text{ when } h < 0$$

$$= \lim_{h \to 0^-} -1 = -1. \qquad \blacksquare$$

**EXAMPLE 5**    In Example 2 we found that for $x > 0$,

$$\frac{d}{dx}\sqrt{x} = \frac{1}{2\sqrt{x}}.$$

We apply the definition to examine if the derivative exists at $x = 0$:

$$\lim_{h \to 0^+} \frac{\sqrt{0 + h} - \sqrt{0}}{h} = \lim_{h \to 0^+} \frac{1}{\sqrt{h}} = \infty.$$

Since the (right-hand) limit is not finite, there is no derivative at $x = 0$. Since the slopes of the secant lines joining the origin to the points $(h, \sqrt{h})$ on a graph of $y = \sqrt{x}$ approach $\infty$, the graph has a *vertical tangent* at the origin. (See Figure 1.17 on page 9.)    $\blacksquare$

## When Does a Function *Not* Have a Derivative at a Point?

A function has a derivative at a point $x_0$ if the slopes of the secant lines through $P(x_0, f(x_0))$ and a nearby point $Q$ on the graph approach a finite limit as $Q$ approaches $P$. Whenever the secants fail to take up a limiting position or become vertical as $Q$ approaches $P$, the derivative does not exist. Thus differentiability is a "smoothness" condition on the graph of $f$. A function can fail to have a derivative at a point for many reasons, including the existence of points where the graph has

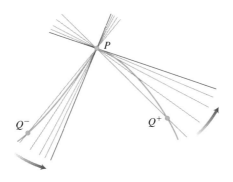

**1.** a *corner*, where the one-sided derivatives differ.

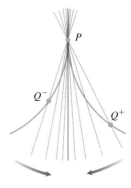

**2.** a *cusp*, where the slope of $PQ$ approaches $\infty$ from one side and $-\infty$ from the other.

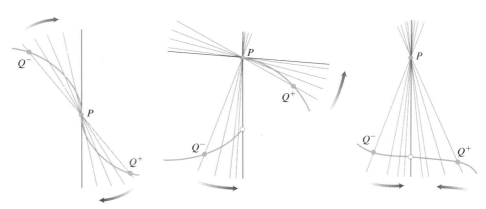

**3.** a *vertical tangent*, where the slope of $PQ$ approaches $\infty$ from both sides or approaches $-\infty$ from both sides (here, $-\infty$).

**4.** a *discontinuity* (two examples shown).

Another case in which the derivative may fail to exist occurs when the function's slope is oscillating rapidly near $P$, as with $f(x) = \sin(1/x)$ near the origin, where it is discontinuous (see Figure 2.31).

## Differentiable Functions Are Continuous

A function is continuous at every point where it has a derivative.

---

**THEOREM 1—Differentiability Implies Continuity** If $f$ has a derivative at $x = c$, then $f$ is continuous at $x = c$.

---

**Proof** Given that $f'(c)$ exists, we must show that $\lim_{x \to c} f(x) = f(c)$, or equivalently, that $\lim_{h \to 0} f(c + h) = f(c)$. If $h \neq 0$, then

$$f(c + h) = f(c) + (f(c + h) - f(c))$$
$$= f(c) + \frac{f(c + h) - f(c)}{h} \cdot h.$$

Now take limits as $h \to 0$. By Theorem 1 of Section 2.2,

$$\lim_{h \to 0} f(c + h) = \lim_{h \to 0} f(c) + \lim_{h \to 0} \frac{f(c + h) - f(c)}{h} \cdot \lim_{h \to 0} h$$
$$= f(c) + f'(c) \cdot 0$$
$$= f(c) + 0$$
$$= f(c). \qquad \blacksquare$$

Similar arguments with one-sided limits show that if $f$ has a derivative from one side (right or left) at $x = c$, then $f$ is continuous from that side at $x = c$.

Theorem 1 says that if a function has a discontinuity at a point (for instance, a jump discontinuity), then it cannot be differentiable there. The greatest integer function $y = \lfloor x \rfloor$ fails to be differentiable at every integer $x = n$ (Example 4, Section 2.5).

Caution The converse of Theorem 1 is false. A function need not have a derivative at a point where it is continuous, as we saw with the absolute value function in Example 4.

# Exercises 3.2

### Finding Derivative Functions and Values

Using the definition, calculate the derivatives of the functions in Exercises 1–6. Then find the values of the derivatives as specified.

**1.** $f(x) = 4 - x^2$;  $f'(-3), f'(0), f'(1)$

**2.** $F(x) = (x - 1)^2 + 1$;  $F'(-1), F'(0), F'(2)$

**3.** $g(t) = \dfrac{1}{t^2}$;  $g'(-1), g'(2), g'(\sqrt{3})$

**4.** $k(z) = \dfrac{1 - z}{2z}$;  $k'(-1), k'(1), k'(\sqrt{2})$

**5.** $p(\theta) = \sqrt{3\theta}$;  $p'(1), p'(3), p'(2/3)$

**6.** $r(s) = \sqrt{2s + 1}$;  $r'(0), r'(1), r'(1/2)$

In Exercises 7–12, find the indicated derivatives.

**7.** $\dfrac{dy}{dx}$ if $y = 2x^3$      **8.** $\dfrac{dr}{ds}$ if $r = s^3 - 2s^2 + 3$

**9.** $\dfrac{ds}{dt}$ if $s = \dfrac{t}{2t + 1}$    **10.** $\dfrac{dv}{dt}$ if $v = t - \dfrac{1}{t}$

**11.** $\dfrac{dp}{dq}$ if $p = q^{3/2}$    **12.** $\dfrac{dz}{dw}$ if $z = \dfrac{1}{\sqrt{w^2 - 1}}$

### Slopes and Tangent Lines

In Exercises 13–16, differentiate the functions and find the slope of the tangent line at the given value of the independent variable.

**13.** $f(x) = x + \dfrac{9}{x}$,  $x = -3$    **14.** $k(x) = \dfrac{1}{2 + x}$,  $x = 2$

**15.** $s = t^3 - t^2$,  $t = -1$    **16.** $y = \dfrac{x + 3}{1 - x}$,  $x = -2$

In Exercises 17–18, differentiate the functions. Then find an equation of the tangent line at the indicated point on the graph of the function.

**17.** $y = f(x) = \dfrac{8}{\sqrt{x - 2}}$,  $(x, y) = (6, 4)$

**18.** $w = g(z) = 1 + \sqrt{4 - z}$,  $(z, w) = (3, 2)$

In Exercises 19–22, find the values of the derivatives.

**19.** $\dfrac{ds}{dt}\Big|_{t=-1}$ if $s = 1 - 3t^2$    **20.** $\dfrac{dy}{dx}\Big|_{x=\sqrt{3}}$ if $y = 1 - \dfrac{1}{x}$

**21.** $\dfrac{dr}{d\theta}\Big|_{\theta=0}$ if $r = \dfrac{2}{\sqrt{4 - \theta}}$    **22.** $\dfrac{dw}{dz}\Big|_{z=4}$ if $w = z + \sqrt{z}$

### Using the Alternative Formula for Derivatives

Use the formula

$$f'(x) = \lim_{z \to x} \frac{f(z) - f(x)}{z - x}$$

to find the derivative of the functions in Exercises 23–26.

**23.** $f(x) = \dfrac{1}{x + 2}$      **24.** $f(x) = x^2 - 3x + 4$

**25.** $g(x) = \dfrac{x}{x - 1}$      **26.** $g(x) = 1 + \sqrt{x}$

### Graphs

Match the functions graphed in Exercises 27–30 with the derivatives graphed in the accompanying figures (a)–(d).

(a)

(b)

(c)

(d)

**27.**

**28.**

**29.**

**30.**

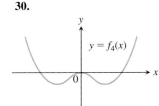

**31. a.** The graph in the accompanying figure is made of line segments joined end to end. At which points of the interval $[-4, 6]$ is $f'$ not defined? Give reasons for your answer.

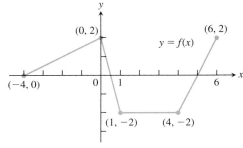

**b.** Graph the derivative of $f$.
The graph should show a step function.

**32. Recovering a function from its derivative**

a. Use the following information to graph the function $f$ over the closed interval $[-2, 5]$.

   i) The graph of $f$ is made of closed line segments joined end to end.

   ii) The graph starts at the point $(-2, 3)$.

   iii) The derivative of $f$ is the step function in the figure shown here.

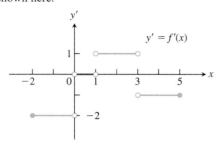

b. Repeat part (a), assuming that the graph starts at $(-2, 0)$ instead of $(-2, 3)$.

**33. Growth in the economy**    The graph in the accompanying figure shows the average annual percentage change $y = f(t)$ in the U.S. gross national product (GNP) for the years 2005–2011. Graph $dy/dt$ (where defined).

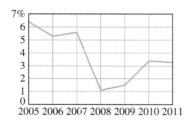

**34. Fruit flies**    (*Continuation of Example 4, Section 2.1.*) Populations starting out in closed environments grow slowly at first, when there are relatively few members, then more rapidly as the number of reproducing individuals increases and resources are still abundant, then slowly again as the population reaches the carrying capacity of the environment.

a. Use the graphical technique of Example 3 to graph the derivative of the fruit fly population. The graph of the population is reproduced here.

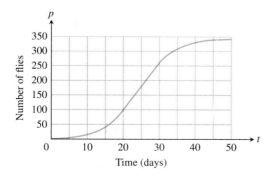

b. During what days does the population seem to be increasing fastest? Slowest?

**35. Temperature**    The given graph shows the temperature $T$ in °C at Davis, CA, on April 18, 2008, between 6 A.M. and 6 P.M.

a. Estimate the rate of temperature change at the times

   **i)** 7 A.M.    **ii)** 9 A.M.    **iii)** 2 P.M.    **iv)** 4 P.M.

b. At what time does the temperature increase most rapidly? Decrease most rapidly? What is the rate for each of those times?

c. Use the graphical technique of Example 3 to graph the derivative of temperature $T$ versus time $t$.

**36. Weight loss**    Jared Fogle, the former spokesman for the Subway restaurants (http://en.wikipedia.org/wiki/Jared_Fogle), weighed 193 kg in 1997 before losing more than 108 kg in 12 months. A chart showing his possible dramatic weight loss is given in the accompanying figure.

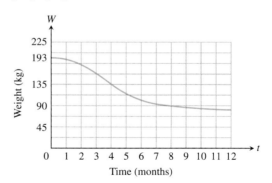

a. Estimate Jared's rate of weight loss when

   **i)** $t = 1$    **ii)** $t = 4$    **iii)** $t = 11$

b. When does Jared lose weight most rapidly and what is this rate of weight loss?

c. Use the graphical technique of Example 3 to graph the derivative of weight $W$.

**One-Sided Derivatives**

Compute the right-hand and left-hand derivatives as limits to show that the functions in Exercises 37–40 are not differentiable at the point $P$.

**37.**                                    **38.**

**39.**

**40.**

In Exercises 41 and 42, determine if the piecewise-defined function is differentiable at the origin.

**41.** $f(x) = \begin{cases} 2x - 1, & x \geq 0 \\ x^2 + 2x + 7, & x < 0 \end{cases}$

**42.** $g(x) = \begin{cases} x^{2/3}, & x \geq 0 \\ x^{1/3}, & x < 0 \end{cases}$

### Differentiability and Continuity on an Interval

Each figure in Exercises 43–48 shows the graph of a function over a closed interval $D$. At what domain points does the function appear to be

   **a.** differentiable?

   **b.** continuous but not differentiable?

   **c.** neither continuous nor differentiable?

Give reasons for your answers.

**43.**

**44.**

**45.**

**46.**

**47.**

**48.**

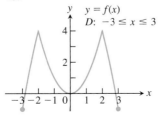

### Theory and Examples

In Exercises 49–52,

   **a.** Find the derivative $f'(x)$ of the given function $y = f(x)$.

   **b.** Graph $y = f(x)$ and $y = f'(x)$ side by side using separate sets of coordinate axes, and answer the following questions.

   **c.** For what values of $x$, if any, is $f'$ positive? Zero? Negative?

   **d.** Over what intervals of $x$-values, if any, does the function $y = f(x)$ increase as $x$ increases? Decrease as $x$ increases? How is this related to what you found in part (c)? (We will say more about this relationship in Section 4.3.)

**49.** $y = -x^2$           **50.** $y = -1/x$

**51.** $y = x^3/3$         **52.** $y = x^4/4$

**53. Tangent to a parabola** Does the parabola $y = 2x^2 - 13x + 5$ have a tangent whose slope is $-1$? If so, find an equation for the line and the point of tangency. If not, why not?

**54. Tangent to $y = \sqrt{x}$** Does any tangent to the curve $y = \sqrt{x}$ cross the $x$-axis at $x = -1$? If so, find an equation for the line and the point of tangency. If not, why not?

**55. Derivative of $-f$** Does knowing that a function $f(x)$ is differentiable at $x = x_0$ tell you anything about the differentiability of the function $-f$ at $x = x_0$? Give reasons for your answer.

**56. Derivative of multiples** Does knowing that a function $g(t)$ is differentiable at $t = 7$ tell you anything about the differentiability of the function $3g$ at $t = 7$? Give reasons for your answer.

**57. Limit of a quotient** Suppose that functions $g(t)$ and $h(t)$ are defined for all values of $t$ and $g(0) = h(0) = 0$. Can $\lim_{t\to 0} (g(t))/(h(t))$ exist? If it does exist, must it equal zero? Give reasons for your answers.

**58. a.** Let $f(x)$ be a function satisfying $|f(x)| \leq x^2$ for $-1 \leq x \leq 1$. Show that $f$ is differentiable at $x = 0$ and find $f'(0)$.

   **b.** Show that

$$f(x) = \begin{cases} x^2 \sin\frac{1}{x}, & x \neq 0 \\ 0, & x = 0 \end{cases}$$

is differentiable at $x = 0$ and find $f'(0)$.

**T** **59.** Graph $y = 1/(2\sqrt{x})$ in a window that has $0 \leq x \leq 2$. Then, on the same screen, graph

$$y = \frac{\sqrt{x + h} - \sqrt{x}}{h}$$

for $h = 1, 0.5, 0.1$. Then try $h = -1, -0.5, -0.1$. Explain what is going on.

**T** **60.** Graph $y = 3x^2$ in a window that has $-2 \leq x \leq 2, 0 \leq y \leq 3$. Then, on the same screen, graph

$$y = \frac{(x + h)^3 - x^3}{h}$$

for $h = 2, 1, 0.2$. Then try $h = -2, -1, -0.2$. Explain what is going on.

**61. Derivative of $y = |x|$** Graph the derivative of $f(x) = |x|$. Then graph $y = (|x| - 0)/(x - 0) = |x|/x$. What can you conclude?

**62. Weierstrass's nowhere differentiable continuous function**
The sum of the first eight terms of the Weierstrass function $f(x) = \sum_{n=0}^{\infty} (2/3)^n \cos(9^n \pi x)$ is

$$g(x) = \cos(\pi x) + (2/3)^1 \cos(9\pi x) + (2/3)^2 \cos(9^2 \pi x)$$
$$+ (2/3)^3 \cos(9^3 \pi x) + \cdots + (2/3)^7 \cos(9^7 \pi x).$$

Graph this sum. Zoom in several times. How wiggly and bumpy is this graph? Specify a viewing window in which the displayed portion of the graph is smooth.

**COMPUTER EXPLORATIONS**
Use a CAS to perform the following steps for the functions in Exercises 63–68.

**a.** Plot $y = f(x)$ to see that function's global behavior.

**b.** Define the difference quotient $q$ at a general point $x$, with general step size $h$.

**c.** Take the limit as $h \to 0$. What formula does this give?

**d.** Substitute the value $x = x_0$ and plot the function $y = f(x)$ together with its tangent line at that point.

**e.** Substitute various values for $x$ larger and smaller than $x_0$ into the formula obtained in part (c). Do the numbers make sense with your picture?

**f.** Graph the formula obtained in part (c). What does it mean when its values are negative? Zero? Positive? Does this make sense with your plot from part (a)? Give reasons for your answer.

**63.** $f(x) = x^3 + x^2 - x, \quad x_0 = 1$

**64.** $f(x) = x^{1/3} + x^{2/3}, \quad x_0 = 1$

**65.** $f(x) = \dfrac{4x}{x^2 + 1}, \quad x_0 = 2$

**66.** $f(x) = \dfrac{x - 1}{3x^2 + 1}, \quad x_0 = -1$

**67.** $f(x) = \sin 2x, \quad x_0 = \pi/2$

**68.** $f(x) = x^2 \cos x, \quad x_0 = \pi/4$

# 3.3 Differentiation Rules

This section introduces several rules that allow us to differentiate constant functions, power functions, polynomials, rational functions, and certain combinations of them, simply and directly, without having to take limits each time.

### Powers, Multiples, Sums, and Differences

A simple rule of differentiation is that the derivative of every constant function is zero.

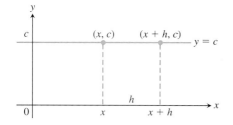

**FIGURE 3.9** The rule $(d/dx)(c) = 0$ is another way to say that the values of constant functions never change and that the slope of a horizontal line is zero at every point.

---

**Derivative of a Constant Function**

If $f$ has the constant value $f(x) = c$, then

$$\frac{df}{dx} = \frac{d}{dx}(c) = 0.$$

---

**Proof** We apply the definition of the derivative to $f(x) = c$, the function whose outputs have the constant value $c$ (Figure 3.9). At every value of $x$, we find that

$$f'(x) = \lim_{h \to 0} \frac{f(x + h) - f(x)}{h} = \lim_{h \to 0} \frac{c - c}{h} = \lim_{h \to 0} 0 = 0. \quad \blacksquare$$

From Section 3.1, we know that

$$\frac{d}{dx}\left(\frac{1}{x}\right) = -\frac{1}{x^2}, \quad \text{or} \quad \frac{d}{dx}(x^{-1}) = -x^{-2}.$$

From Example 2 of the last section we also know that

$$\frac{d}{dx}\left(\sqrt{x}\right) = \frac{1}{2\sqrt{x}}, \quad \text{or} \quad \frac{d}{dx}(x^{1/2}) = \frac{1}{2}x^{-1/2}.$$

These two examples illustrate a general rule for differentiating a power $x^n$. We first prove the rule when $n$ is a positive integer.

---

**Derivative of a Positive Integer Power**

If $n$ is a positive integer, then

$$\frac{d}{dx}x^n = nx^{n-1}.$$

---

**HISTORICAL BIOGRAPHY**

Richard Courant
(1888–1972)

**Proof of the Positive Integer Power Rule**     The formula

$$z^n - x^n = (z - x)(z^{n-1} + z^{n-2}x + \cdots + zx^{n-2} + x^{n-1})$$

can be verified by multiplying out the right-hand side. Then from the alternative formula for the definition of the derivative,

$$\begin{aligned}
f'(x) &= \lim_{z \to x} \frac{f(z) - f(x)}{z - x} = \lim_{z \to x} \frac{z^n - x^n}{z - x} \\
&= \lim_{z \to x}(z^{n-1} + z^{n-2}x + \cdots + zx^{n-2} + x^{n-1}) \qquad \text{$n$ terms} \\
&= nx^{n-1}.
\end{aligned}$$

The Power Rule is actually valid for all real numbers $n$. We have seen examples for a negative integer and fractional power, but $n$ could be an irrational number as well. To apply the Power Rule, we subtract 1 from the original exponent $n$ and multiply the result by $n$. Here we state the general version of the rule, but postpone its proof until Chapter 7.

---

**Power Rule (General Version)**

If $n$ is any real number, then

$$\frac{d}{dx}x^n = nx^{n-1},$$

for all $x$ where the powers $x^n$ and $x^{n-1}$ are defined.

---

**EXAMPLE 1**     Differentiate the following powers of $x$.

**(a)** $x^3$     **(b)** $x^{2/3}$     **(c)** $x^{\sqrt{2}}$     **(d)** $\dfrac{1}{x^4}$     **(e)** $x^{-4/3}$     **(f)** $\sqrt{x^{2+\pi}}$

**Solution**

**Applying the Power Rule**

Subtract 1 from the exponent and multiply the result by the original exponent.

**(a)** $\dfrac{d}{dx}(x^3) = 3x^{3-1} = 3x^2$

**(b)** $\dfrac{d}{dx}(x^{2/3}) = \dfrac{2}{3}x^{(2/3)-1} = \dfrac{2}{3}x^{-1/3}$

**(c)** $\dfrac{d}{dx}(x^{\sqrt{2}}) = \sqrt{2}x^{\sqrt{2}-1}$

**(d)** $\dfrac{d}{dx}\left(\dfrac{1}{x^4}\right) = \dfrac{d}{dx}(x^{-4}) = -4x^{-4-1} = -4x^{-5} = -\dfrac{4}{x^5}$

**(e)** $\dfrac{d}{dx}(x^{-4/3}) = -\dfrac{4}{3}x^{-(4/3)-1} = -\dfrac{4}{3}x^{-7/3}$

**(f)** $\dfrac{d}{dx}\left(\sqrt{x^{2+\pi}}\right) = \dfrac{d}{dx}\left(x^{1+(\pi/2)}\right) = \left(1 + \dfrac{\pi}{2}\right)x^{1+(\pi/2)-1} = \dfrac{1}{2}(2 + \pi)\sqrt{x^{\pi}}$

The next rule says that when a differentiable function is multiplied by a constant, its derivative is multiplied by the same constant.

---

**Derivative Constant Multiple Rule**

If $u$ is a differentiable function of $x$, and $c$ is a constant, then

$$\frac{d}{dx}(cu) = c\frac{du}{dx}.$$

---

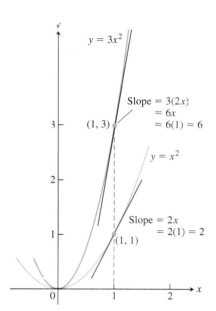

**FIGURE 3.10** The graphs of $y = x^2$ and $y = 3x^2$. Tripling the $y$-coordinate triples the slope (Example 2).

In particular, if $n$ is any real number, then

$$\frac{d}{dx}(cx^n) = cnx^{n-1}.$$

**Proof**

$$\frac{d}{dx}cu = \lim_{h \to 0}\frac{cu(x + h) - cu(x)}{h} \qquad \text{Derivative definition with } f(x) = cu(x)$$

$$= c\lim_{h \to 0}\frac{u(x + h) - u(x)}{h} \qquad \text{Constant Multiple Limit Property}$$

$$= c\frac{du}{dx} \qquad u \text{ is differentiable.} \qquad \blacksquare$$

**EXAMPLE 2**

**(a)** The derivative formula

$$\frac{d}{dx}(3x^2) = 3 \cdot 2x = 6x$$

says that if we rescale the graph of $y = x^2$ by multiplying each $y$-coordinate by 3, then we multiply the slope at each point by 3 (Figure 3.10).

**(b)** **Negative of a function**

The derivative of the negative of a differentiable function $u$ is the negative of the function's derivative. The Constant Multiple Rule with $c = -1$ gives

$$\frac{d}{dx}(-u) = \frac{d}{dx}(-1 \cdot u) = -1 \cdot \frac{d}{dx}(u) = -\frac{du}{dx}. \qquad \blacksquare$$

The next rule says that the derivative of the sum of two differentiable functions is the sum of their derivatives.

**Denoting Functions by $u$ and $v$**
The functions we are working with when we need a differentiation formula are likely to be denoted by letters like $f$ and $g$. We do not want to use these same letters when stating general differentiation rules, so instead we use letters like $u$ and $v$ that are not likely to be already in use.

---

**Derivative Sum Rule**

If $u$ and $v$ are differentiable functions of $x$, then their sum $u + v$ is differentiable at every point where $u$ and $v$ are both differentiable. At such points,

$$\frac{d}{dx}(u + v) = \frac{du}{dx} + \frac{dv}{dx}.$$

---

For example, if $y = x^4 + 12x$, then $y$ is the sum of $u(x) = x^4$ and $v(x) = 12x$. We then have

$$\frac{dy}{dx} = \frac{d}{dx}(x^4) + \frac{d}{dx}(12x) = 4x^3 + 12.$$

**Proof**   We apply the definition of the derivative to $f(x) = u(x) + v(x)$:

$$\frac{d}{dx}[u(x) + v(x)] = \lim_{h \to 0} \frac{[u(x + h) + v(x + h)] - [u(x) + v(x)]}{h}$$

$$= \lim_{h \to 0}\left[\frac{u(x + h) - u(x)}{h} + \frac{v(x + h) - v(x)}{h}\right]$$

$$= \lim_{h \to 0} \frac{u(x + h) - u(x)}{h} + \lim_{h \to 0} \frac{v(x + h) - v(x)}{h} = \frac{du}{dx} + \frac{dv}{dx}. \qquad \blacksquare$$

Combining the Sum Rule with the Constant Multiple Rule gives the **Difference Rule**, which says that the derivative of a *difference* of differentiable functions is the difference of their derivatives:

$$\frac{d}{dx}(u - v) = \frac{d}{dx}[u + (-1)v] = \frac{du}{dx} + (-1)\frac{dv}{dx} = \frac{du}{dx} - \frac{dv}{dx}.$$

The Sum Rule also extends to finite sums of more than two functions. If $u_1, u_2, \ldots, u_n$ are differentiable at $x$, then so is $u_1 + u_2 + \cdots + u_n$, and

$$\frac{d}{dx}(u_1 + u_2 + \cdots + u_n) = \frac{du_1}{dx} + \frac{du_2}{dx} + \cdots + \frac{du_n}{dx}.$$

For instance, to see that the rule holds for three functions we compute

$$\frac{d}{dx}(u_1 + u_2 + u_3) = \frac{d}{dx}((u_1 + u_2) + u_3) = \frac{d}{dx}(u_1 + u_2) + \frac{du_3}{dx} = \frac{du_1}{dx} + \frac{du_2}{dx} + \frac{du_3}{dx}.$$

A proof by mathematical induction for any finite number of terms is given in Appendix 2.

**EXAMPLE 3**   Find the derivative of the polynomial $y = x^3 + \frac{4}{3}x^2 - 5x + 1$.

**Solution**   $\dfrac{dy}{dx} = \dfrac{d}{dx}x^3 + \dfrac{d}{dx}\left(\dfrac{4}{3}x^2\right) - \dfrac{d}{dx}(5x) + \dfrac{d}{dx}(1)$      Sum and Difference Rules

$$= 3x^2 + \frac{4}{3} \cdot 2x - 5 + 0 = 3x^2 + \frac{8}{3}x - 5 \qquad \blacksquare$$

We can differentiate any polynomial term by term, the way we differentiated the polynomial in Example 3. All polynomials are differentiable at all values of $x$.

**EXAMPLE 4**   Does the curve $y = x^4 - 2x^2 + 2$ have any horizontal tangents? If so, where?

**Solution**   The horizontal tangents, if any, occur where the slope $dy/dx$ is zero. We have

$$\frac{dy}{dx} = \frac{d}{dx}(x^4 - 2x^2 + 2) = 4x^3 - 4x.$$

Now solve the equation $\dfrac{dy}{dx} = 0$ for $x$:

$$4x^3 - 4x = 0$$
$$4x(x^2 - 1) = 0$$
$$x = 0, 1, -1.$$

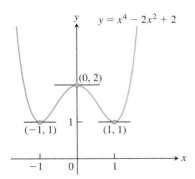

**FIGURE 3.11**   The curve in Example 4 and its horizontal tangents.

The curve $y = x^4 - 2x^2 + 2$ has horizontal tangents at $x = 0, 1,$ and $-1$. The corresponding points on the curve are $(0, 2), (1, 1),$ and $(-1, 1)$. See Figure 3.11. We will see in Chapter 4 that finding the values of $x$ where the derivative of a function is equal to zero is an important and useful procedure. $\qquad \blacksquare$

## Products and Quotients

While the derivative of the sum of two functions is the sum of their derivatives, the derivative of the product of two functions is *not* the product of their derivatives. For instance,

$$\frac{d}{dx}(x \cdot x) = \frac{d}{dx}(x^2) = 2x, \qquad \text{while} \qquad \frac{d}{dx}(x) \cdot \frac{d}{dx}(x) = 1 \cdot 1 = 1.$$

The derivative of a product of two functions is the sum of *two* products, as we now explain.

---

**Derivative Product Rule**

If $u$ and $v$ are differentiable at $x$, then so is their product $uv$, and

$$\frac{d}{dx}(uv) = u\frac{dv}{dx} + v\frac{du}{dx}.$$

---

Equation (3) is equivalent to saying that

$$(fg)' = f'g + fg'.$$

This form of the Product Rule is useful and applies to *dot products* and *cross products* of *vector-valued functions,* studied in Chapter 13.

The derivative of the product $uv$ is $u$ times the derivative of $v$ plus $v$ times the derivative of $u$. In *prime notation,* $(uv)' = uv' + vu'$. In function notation,

$$\frac{d}{dx}[f(x)g(x)] = f(x)g'(x) + g(x)f'(x). \tag{3}$$

**EXAMPLE 5**   Find the derivative of $y = (x^2 + 1)(x^3 + 3)$.

**Solution**

**(a)** From the Product Rule with $u = x^2 + 1$ and $v = x^3 + 3$, we find

$$\frac{d}{dx}\left[(x^2 + 1)(x^3 + 3)\right] = (x^2 + 1)(3x^2) + (x^3 + 3)(2x) \qquad \scriptstyle \frac{d}{dx}(uv) = u\frac{dv}{dx} + v\frac{du}{dx}$$

$$= 3x^4 + 3x^2 + 2x^4 + 6x$$
$$= 5x^4 + 3x^2 + 6x.$$

**Picturing the Product Rule**

Suppose $u(x)$ and $v(x)$ are positive and increase when $x$ increases, and $h > 0$.

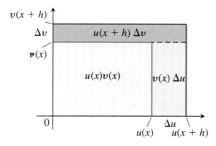

Then the change in the product $uv$ is the difference in areas of the larger and smaller "squares," which is the sum of the upper and right-hand reddish-shaded rectangles. That is,

$$\Delta(uv) = u(x + h)v(x + h) - u(x)v(x)$$
$$= u(x + h)\Delta v + v(x)\Delta u.$$

Division by $h$ gives

$$\frac{\Delta(uv)}{h} = u(x + h)\frac{\Delta v}{h} + v(x)\frac{\Delta u}{h}.$$

The limit as $h \to 0^+$ gives the Product Rule.

**(b)** This particular product can be differentiated as well (perhaps better) by multiplying out the original expression for $y$ and differentiating the resulting polynomial:

$$y = (x^2 + 1)(x^3 + 3) = x^5 + x^3 + 3x^2 + 3$$
$$\frac{dy}{dx} = 5x^4 + 3x^2 + 6x.$$

This is in agreement with our first calculation.   ∎

**Proof of the Derivative Product Rule**

$$\frac{d}{dx}(uv) = \lim_{h \to 0}\frac{u(x + h)v(x + h) - u(x)v(x)}{h}$$

To change this fraction into an equivalent one that contains difference quotients for the derivatives of $u$ and $v$, we subtract and add $u(x + h)v(x)$ in the numerator:

$$\frac{d}{dx}(uv) = \lim_{h \to 0}\frac{u(x + h)v(x + h) - u(x + h)v(x) + u(x + h)v(x) - u(x)v(x)}{h}$$

$$= \lim_{h \to 0}\left[u(x + h)\frac{v(x + h) - v(x)}{h} + v(x)\frac{u(x + h) - u(x)}{h}\right]$$

$$= \lim_{h \to 0}u(x + h) \cdot \lim_{h \to 0}\frac{v(x + h) - v(x)}{h} + v(x) \cdot \lim_{h \to 0}\frac{u(x + h) - u(x)}{h}.$$

As $h$ approaches zero, $u(x + h)$ approaches $u(x)$ because $u$, being differentiable at $x$, is continuous at $x$. The two fractions approach the values of $dv/dx$ at $x$ and $du/dx$ at $x$. Therefore,

$$\frac{d}{dx}(uv) = u\frac{dv}{dx} + v\frac{du}{dx}. \qquad \blacksquare$$

The derivative of the quotient of two functions is given by the Quotient Rule.

---

**Derivative Quotient Rule**

If $u$ and $v$ are differentiable at $x$ and if $v(x) \neq 0$, then the quotient $u/v$ is differentiable at $x$, and

$$\frac{d}{dx}\left(\frac{u}{v}\right) = \frac{v\dfrac{du}{dx} - u\dfrac{dv}{dx}}{v^2}.$$

---

In function notation,

$$\frac{d}{dx}\left[\frac{f(x)}{g(x)}\right] = \frac{g(x)f'(x) - f(x)g'(x)}{g^2(x)}.$$

**EXAMPLE 6**    Find the derivative of $y = \dfrac{t^2 - 1}{t^3 + 1}$.

**Solution**    We apply the Quotient Rule with $u = t^2 - 1$ and $v = t^3 + 1$:

$$\frac{dy}{dt} = \frac{(t^3 + 1) \cdot 2t - (t^2 - 1) \cdot 3t^2}{(t^3 + 1)^2} \qquad \frac{d}{dt}\left(\frac{u}{v}\right) = \frac{v(du/dt) - u(dv/dt)}{v^2}$$

$$= \frac{2t^4 + 2t - 3t^4 + 3t^2}{(t^3 + 1)^2}$$

$$= \frac{-t^4 + 3t^2 + 2t}{(t^3 + 1)^2}. \qquad \blacksquare$$

**Proof of the Derivative Quotient Rule**

$$\frac{d}{dx}\left(\frac{u}{v}\right) = \lim_{h \to 0} \frac{\dfrac{u(x + h)}{v(x + h)} - \dfrac{u(x)}{v(x)}}{h}$$

$$= \lim_{h \to 0} \frac{v(x)u(x + h) - u(x)v(x + h)}{hv(x + h)v(x)}$$

To change the last fraction into an equivalent one that contains the difference quotients for the derivatives of $u$ and $v$, we subtract and add $v(x)u(x)$ in the numerator. We then get

$$\frac{d}{dx}\left(\frac{u}{v}\right) = \lim_{h \to 0} \frac{v(x)u(x + h) - v(x)u(x) + v(x)u(x) - u(x)v(x + h)}{hv(x + h)v(x)}$$

$$= \lim_{h \to 0} \frac{v(x)\dfrac{u(x + h) - u(x)}{h} - u(x)\dfrac{v(x + h) - v(x)}{h}}{v(x + h)v(x)}.$$

Taking the limits in the numerator and denominator now gives the Quotient Rule. Exercise 74 outlines another proof. $\qquad \blacksquare$

The choice of which rules to use in solving a differentiation problem can make a difference in how much work you have to do. Here is an example.

**EXAMPLE 7**    Find the derivative of

$$y = \frac{(x-1)(x^2 - 2x)}{x^4}.$$

**Solution**    Using the Quotient Rule here will result in a complicated expression with many terms. Instead, use some algebra to simplify the expression. First expand the numerator and divide by $x^4$:

$$y = \frac{(x-1)(x^2 - 2x)}{x^4} = \frac{x^3 - 3x^2 + 2x}{x^4} = x^{-1} - 3x^{-2} + 2x^{-3}.$$

Then use the Sum and Power Rules:

$$\frac{dy}{dx} = -x^{-2} - 3(-2)x^{-3} + 2(-3)x^{-4}$$

$$= -\frac{1}{x^2} + \frac{6}{x^3} - \frac{6}{x^4}. \qquad\blacksquare$$

## Second- and Higher-Order Derivatives

If $y = f(x)$ is a differentiable function, then its derivative $f'(x)$ is also a function. If $f'$ is also differentiable, then we can differentiate $f'$ to get a new function of $x$ denoted by $f''$. So $f'' = (f')'$. The function $f''$ is called the **second derivative** of $f$ because it is the derivative of the first derivative. It is written in several ways:

$$f''(x) = \frac{d^2y}{dx^2} = \frac{d}{dx}\left(\frac{dy}{dx}\right) = \frac{dy'}{dx} = y'' = D^2(f)(x) = D_x^2 f(x).$$

The symbol $D^2$ means the operation of differentiation is performed twice.

If $y = x^6$, then $y' = 6x^5$ and we have

$$y'' = \frac{dy'}{dx} = \frac{d}{dx}(6x^5) = 30x^4.$$

Thus $D^2(x^6) = 30x^4$.

If $y''$ is differentiable, its derivative, $y''' = dy''/dx = d^3y/dx^3$, is the **third derivative** of $y$ with respect to $x$. The names continue as you imagine, with

$$y^{(n)} = \frac{d}{dx}y^{(n-1)} = \frac{d^ny}{dx^n} = D^n y$$

denoting the ***n*th derivative** of $y$ with respect to $x$ for any positive integer $n$.

We can interpret the second derivative as the rate of change of the slope of the tangent to the graph of $y = f(x)$ at each point. You will see in the next chapter that the second derivative reveals whether the graph bends upward or downward from the tangent line as we move off the point of tangency. In the next section, we interpret both the second and third derivatives in terms of motion along a straight line.

> **How to Read the Symbols for Derivatives**
>
> $y'$    "y prime"
>
> $y''$    "y double prime"
>
> $\dfrac{d^2y}{dx^2}$    "d squared y dx squared"
>
> $y'''$    "y triple prime"
>
> $y^{(n)}$    "y super n"
>
> $\dfrac{d^ny}{dx^n}$    "d to the n of y by dx to the n"
>
> $D^n$    "D to the n"

**EXAMPLE 8**    The first four derivatives of $y = x^3 - 3x^2 + 2$ are

| | |
|---|---|
| First derivative: | $y' = 3x^2 - 6x$ |
| Second derivative: | $y'' = 6x - 6$ |
| Third derivative: | $y''' = 6$ |
| Fourth derivative: | $y^{(4)} = 0.$ |

All polynomial functions have derivatives of all orders. In this example, the fifth and later derivatives are all zero.    $\blacksquare$

## Exercises 3.3

**Derivative Calculations**

In Exercises 1–12, find the first and second derivatives.

**1.** $y = -x^2 + 3$          **2.** $y = x^2 + x + 8$

**3.** $s = 5t^3 - 3t^5$          **4.** $w = 3z^7 - 7z^3 + 21z^2$

**5.** $y = \dfrac{4x^3}{3} - x$          **6.** $y = \dfrac{x^3}{3} + \dfrac{x^2}{2} + \dfrac{x}{4}$

**7.** $w = 3z^{-2} - \dfrac{1}{z}$          **8.** $s = -2t^{-1} + \dfrac{4}{t^2}$

**9.** $y = 6x^2 - 10x - 5x^{-2}$          **10.** $y = 4 - 2x - x^{-3}$

**11.** $r = \dfrac{1}{3s^2} - \dfrac{5}{2s}$          **12.** $r = \dfrac{12}{\theta} - \dfrac{4}{\theta^3} + \dfrac{1}{\theta^4}$

In Exercises 13–16, find $y'$ **(a)** by applying the Product Rule and **(b)** by multiplying the factors to produce a sum of simpler terms to differentiate.

**13.** $y = (3 - x^2)(x^3 - x + 1)$  **14.** $y = (2x + 3)(5x^2 - 4x)$

**15.** $y = (x^2 + 1)\left(x + 5 + \dfrac{1}{x}\right)$ **16.** $y = (1 + x^2)(x^{3/4} - x^{-3})$

Find the derivatives of the functions in Exercises 17–28.

**17.** $y = \dfrac{2x + 5}{3x - 2}$          **18.** $z = \dfrac{4 - 3x}{3x^2 + x}$

**19.** $g(x) = \dfrac{x^2 - 4}{x + 0.5}$          **20.** $f(t) = \dfrac{t^2 - 1}{t^2 + t - 2}$

**21.** $v = (1 - t)(1 + t^2)^{-1}$  **22.** $w = (2x - 7)^{-1}(x + 5)$

**23.** $f(s) = \dfrac{\sqrt{s} - 1}{\sqrt{s} + 1}$          **24.** $u = \dfrac{5x + 1}{2\sqrt{x}}$

**25.** $v = \dfrac{1 + x - 4\sqrt{x}}{x}$          **26.** $r = 2\left(\dfrac{1}{\sqrt{\theta}} + \sqrt{\theta}\right)$

**27.** $y = \dfrac{1}{(x^2 - 1)(x^2 + x + 1)}$ **28.** $y = \dfrac{(x + 1)(x + 2)}{(x - 1)(x - 2)}$

Find the derivatives of all orders of the functions in Exercises 29–32.

**29.** $y = \dfrac{x^4}{2} - \dfrac{3}{2}x^2 - x$          **30.** $y = \dfrac{x^5}{120}$

**31.** $y = (x - 1)(x + 2)(x + 3)$  **32.** $y = (4x^2 + 3)(2 - x)x$

Find the first and second derivatives of the functions in Exercises 33–40.

**33.** $y = \dfrac{x^3 + 7}{x}$          **34.** $s = \dfrac{t^2 + 5t - 1}{t^2}$

**35.** $r = \dfrac{(\theta - 1)(\theta^2 + \theta + 1)}{\theta^3}$  **36.** $u = \dfrac{(x^2 + x)(x^2 - x + 1)}{x^4}$

**37.** $w = \left(\dfrac{1 + 3z}{3z}\right)(3 - z)$  **38.** $w = (z + 1)(z - 1)(z^2 + 1)$

**39.** $p = \left(\dfrac{q^2 + 3}{12q}\right)\left(\dfrac{q^4 - 1}{q^3}\right)$  **40.** $p = \dfrac{q^2 + 3}{(q - 1)^3 + (q + 1)^3}$

**41.** Suppose $u$ and $v$ are functions of $x$ that are differentiable at $x = 0$ and that

$$u(0) = 5, \quad u'(0) = -3, \quad v(0) = -1, \quad v'(0) = 2.$$

Find the values of the following derivatives at $x = 0$.

**a.** $\dfrac{d}{dx}(uv)$   **b.** $\dfrac{d}{dx}\left(\dfrac{u}{v}\right)$   **c.** $\dfrac{d}{dx}\left(\dfrac{v}{u}\right)$   **d.** $\dfrac{d}{dx}(7v - 2u)$

**42.** Suppose $u$ and $v$ are differentiable functions of $x$ and that

$$u(1) = 2, \quad u'(1) = 0, \quad v(1) = 5, \quad v'(1) = -1.$$

Find the values of the following derivatives at $x = 1$.

**a.** $\dfrac{d}{dx}(uv)$   **b.** $\dfrac{d}{dx}\left(\dfrac{u}{v}\right)$   **c.** $\dfrac{d}{dx}\left(\dfrac{v}{u}\right)$   **d.** $\dfrac{d}{dx}(7v - 2u)$

**Slopes and Tangents**

**43. a. Normal to a curve**  Find an equation for the line perpendicular to the tangent to the curve $y = x^3 - 4x + 1$ at the point $(2, 1)$.

  **b. Smallest slope**  What is the smallest slope on the curve? At what point on the curve does the curve have this slope?

  **c. Tangents having specified slope**  Find equations for the tangents to the curve at the points where the slope of the curve is 8.

**44. a. Horizontal tangents**  Find equations for the horizontal tangents to the curve $y = x^3 - 3x - 2$. Also find equations for the lines that are perpendicular to these tangents at the points of tangency.

  **b. Smallest slope**  What is the smallest slope on the curve? At what point on the curve does the curve have this slope? Find an equation for the line that is perpendicular to the curve's tangent at this point.

**45.** Find the tangents to *Newton's serpentine* (graphed here) at the origin and the point $(1, 2)$.

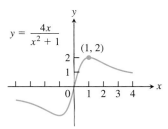

**46.** Find the tangent to the *Witch of Agnesi* (graphed here) at the point $(2, 1)$.

**47. Quadratic tangent to identity function**  The curve $y = ax^2 + bx + c$ passes through the point $(1, 2)$ and is tangent to the line $y = x$ at the origin. Find $a$, $b$, and $c$.

**48. Quadratics having a common tangent**  The curves $y = x^2 + ax + b$ and $y = cx - x^2$ have a common tangent line at the point $(1, 0)$. Find $a$, $b$, and $c$.

**49.** Find all points $(x, y)$ on the graph of $f(x) = 3x^2 - 4x$ with tangent lines parallel to the line $y = 8x + 5$.

**50.** Find all points $(x, y)$ on the graph of $g(x) = \frac{1}{3}x^3 - \frac{3}{2}x^2 + 1$ with tangent lines parallel to the line $8x - 2y = 1$.

**51.** Find all points $(x, y)$ on the graph of $y = x/(x - 2)$ with tangent lines perpendicular to the line $y = 2x + 3$.

**52.** Find all points $(x, y)$ on the graph of $f(x) = x^2$ with tangent lines passing through the point $(3, 8)$.

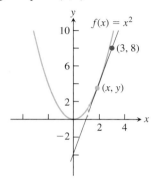

**53. a.** Find an equation for the line that is tangent to the curve $y = x^3 - x$ at the point $(-1, 0)$.

   [T] **b.** Graph the curve and tangent line together. The tangent intersects the curve at another point. Use Zoom and Trace to estimate the point's coordinates.

   [T] **c.** Confirm your estimates of the coordinates of the second intersection point by solving the equations for the curve and tangent simultaneously (Solver key).

**54. a.** Find an equation for the line that is tangent to the curve $y = x^3 - 6x^2 + 5x$ at the origin.

   [T] **b.** Graph the curve and tangent together. The tangent intersects the curve at another point. Use Zoom and Trace to estimate the point's coordinates.

   [T] **c.** Confirm your estimates of the coordinates of the second intersection point by solving the equations for the curve and tangent simultaneously (Solver key).

**Theory and Examples**

For Exercises 55 and 56 evaluate each limit by first converting each to a derivative at a particular $x$-value.

**55.** $\lim\limits_{x \to 1} \dfrac{x^{50} - 1}{x - 1}$

**56.** $\lim\limits_{x \to -1} \dfrac{x^{2/9} - 1}{x + 1}$

**57.** Find the value of $a$ that makes the following function differentiable for all $x$-values.

$$g(x) = \begin{cases} ax, & \text{if } x < 0 \\ x^2 - 3x, & \text{if } x \ge 0 \end{cases}$$

**58.** Find the values of $a$ and $b$ that make the following function differentiable for all $x$-values.

$$f(x) = \begin{cases} ax + b, & x > -1 \\ bx^2 - 3, & x \le -1 \end{cases}$$

**59.** The general polynomial of degree $n$ has the form

$$P(x) = a_n x^n + a_{n-1} x^{n-1} + \cdots + a_2 x^2 + a_1 x + a_0$$

where $a_n \ne 0$. Find $P'(x)$.

**60. The body's reaction to medicine**  The reaction of the body to a dose of medicine can sometimes be represented by an equation of the form

$$R = M^2 \left( \frac{C}{2} - \frac{M}{3} \right),$$

where $C$ is a positive constant and $M$ is the amount of medicine absorbed in the blood. If the reaction is a change in blood pressure, $R$ is measured in millimeters of mercury. If the reaction is a change in temperature, $R$ is measured in degrees, and so on.

Find $dR/dM$. This derivative, as a function of $M$, is called the sensitivity of the body to the medicine. In Section 4.5, we will see how to find the amount of medicine to which the body is most sensitive.

**61.** Suppose that the function $v$ in the Derivative Product Rule has a constant value $c$. What does the Derivative Product Rule then say? What does this say about the Derivative Constant Multiple Rule?

**62. The Reciprocal Rule**

   **a.** The *Reciprocal Rule* says that at any point where the function $v(x)$ is differentiable and different from zero,

   $$\frac{d}{dx} \left( \frac{1}{v} \right) = -\frac{1}{v^2} \frac{dv}{dx}.$$

   Show that the Reciprocal Rule is a special case of the Derivative Quotient Rule.

   **b.** Show that the Reciprocal Rule and the Derivative Product Rule together imply the Derivative Quotient Rule.

**63. Generalizing the Product Rule**  The Derivative Product Rule gives the formula

$$\frac{d}{dx} (uv) = u \frac{dv}{dx} + v \frac{du}{dx}$$

for the derivative of the product $uv$ of two differentiable functions of $x$.

   **a.** What is the analogous formula for the derivative of the product $uvw$ of *three* differentiable functions of $x$?

   **b.** What is the formula for the derivative of the product $u_1 u_2 u_3 u_4$ of *four* differentiable functions of $x$?

   **c.** What is the formula for the derivative of a product $u_1 u_2 u_3 \cdots u_n$ of a finite number $n$ of differentiable functions of $x$?

**64. Power Rule for negative integers**  Use the Derivative Quotient Rule to prove the Power Rule for negative integers, that is,

$$\frac{d}{dx} (x^{-m}) = -mx^{-m-1}$$

where $m$ is a positive integer.

**65. Cylinder pressure**   If gas in a cylinder is maintained at a constant temperature $T$, the pressure $P$ is related to the volume $V$ by a formula of the form

$$P = \frac{nRT}{V - nb} - \frac{an^2}{V^2},$$

in which $a$, $b$, $n$, and $R$ are constants. Find $dP/dV$. (See accompanying figure.)

**66. The best quantity to order**   One of the formulas for inventory management says that the average weekly cost of ordering, paying for, and holding merchandise is

$$A(q) = \frac{km}{q} + cm + \frac{hq}{2},$$

where $q$ is the quantity you order when things run low (shoes, TVs, brooms, or whatever the item might be); $k$ is the cost of placing an order (the same, no matter how often you order); $c$ is the cost of one item (a constant); $m$ is the number of items sold each week (a constant); and $h$ is the weekly holding cost per item (a constant that takes into account things such as space, utilities, insurance, and security). Find $dA/dq$ and $d^2A/dq^2$.

# $3.4$ The Derivative as a Rate of Change

In Section 2.1 we introduced average and instantaneous rates of change. In this section we study further applications in which derivatives model the rates at which things change. It is natural to think of a quantity changing with respect to time, but other variables can be treated in the same way. For example, an economist may want to study how the cost of producing steel varies with the number of tonnes produced, or an engineer may want to know how the power output of a generator varies with its temperature.

## Instantaneous Rates of Change

If we interpret the difference quotient $(f(x + h) - f(x))/h$ as the average rate of change in $f$ over the interval from $x$ to $x + h$, we can interpret its limit as $h \to 0$ as the rate at which $f$ is changing at the point $x$.

---

**DEFINITION**   The **instantaneous rate of change** of $f$ with respect to $x$ at $x_0$ is the derivative

$$f'(x_0) = \lim_{h \to 0} \frac{f(x_0 + h) - f(x_0)}{h},$$

provided the limit exists.

---

Thus, instantaneous rates are limits of average rates.

It is conventional to use the word *instantaneous* even when $x$ does not represent time. The word is, however, frequently omitted. When we say *rate of change*, we mean *instantaneous rate of change*.

**EXAMPLE 1**   The area $A$ of a circle is related to its diameter by the equation

$$A = \frac{\pi}{4}D^2.$$

How fast does the area change with respect to the diameter when the diameter is 10 m?

**Solution** The rate of change of the area with respect to the diameter is

$$\frac{dA}{dD} = \frac{\pi}{4} \cdot 2D = \frac{\pi D}{2}.$$

When $D = 10$ m, the area is changing with respect to the diameter at the rate of $(\pi/2)10 = 5\pi$ m$^2$/m $\approx 15.71$ m$^2$/m. ■

### Motion Along a Line: Displacement, Velocity, Speed, Acceleration, and Jerk

Suppose that an object (or body, considered as a whole mass) is moving along a coordinate line (an $s$-axis), usually horizontal or vertical, so that we know its position $s$ on that line as a function of time $t$:

$$s = f(t).$$

The **displacement** of the object over the time interval from $t$ to $t + \Delta t$ (Figure 3.12) is

$$\Delta s = f(t + \Delta t) - f(t),$$

Position at time $t$ ...    and at time $t + \Delta t$

$s = f(t)$      $s + \Delta s = f(t + \Delta t)$

**FIGURE 3.12** The positions of a body moving along a coordinate line at time $t$ and shortly later at time $t + \Delta t$. Here the coordinate line is horizontal.

and the **average velocity** of the object over that time interval is

$$v_{av} = \frac{\text{displacement}}{\text{travel time}} = \frac{\Delta s}{\Delta t} = \frac{f(t + \Delta t) - f(t)}{\Delta t}.$$

To find the body's velocity at the exact instant $t$, we take the limit of the average velocity over the interval from $t$ to $t + \Delta t$ as $\Delta t$ shrinks to zero. This limit is the derivative of $f$ with respect to $t$.

---

**DEFINITION**    **Velocity (instantaneous velocity)** is the derivative of position with respect to time. If a body's position at time $t$ is $s = f(t)$, then the body's velocity at time $t$ is

$$v(t) = \frac{ds}{dt} = \lim_{\Delta t \to 0} \frac{f(t + \Delta t) - f(t)}{\Delta t}.$$

---

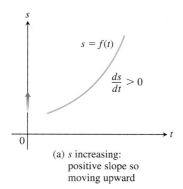

(a) $s$ increasing: positive slope so moving upward

Besides telling how fast an object is moving along the horizontal line in Figure 3.12, its velocity tells the direction of motion. When the object is moving forward ($s$ increasing), the velocity is positive; when the object is moving backward ($s$ decreasing), the velocity is negative. If the coordinate line is vertical, the object moves upward for positive velocity and downward for negative velocity. The blue curves in Figure 3.13 represent position along the line over time; they do not portray the path of motion, which lies along the vertical $s$-axis.

If we drive to a friend's house and back at 50 km/h, say, the speedometer will show 50 on the way over but it will not show $-50$ on the way back, even though our distance from home is decreasing. The speedometer always shows *speed*, which is the absolute value of velocity. Speed measures the rate of progress regardless of direction.

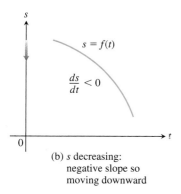

(b) $s$ decreasing: negative slope so moving downward

**FIGURE 3.13** For motion $s = f(t)$ along a straight line (the vertical axis), $v = ds/dt$ is (a) positive when $s$ increases and (b) negative when $s$ decreases.

---

**DEFINITION**    **Speed** is the absolute value of velocity.

$$\text{Speed} = |v(t)| = \left| \frac{ds}{dt} \right|$$

---

**EXAMPLE 2**    Figure 3.14 shows the graph of the velocity $v = f'(t)$ of a particle moving along a horizontal line (as opposed to showing a position function $s = f(t)$ such as in Figure 3.13). In the graph of the velocity function, it's not the slope of the curve that tells us if the particle is moving forward or backward along the line (which is not shown in the figure), but rather

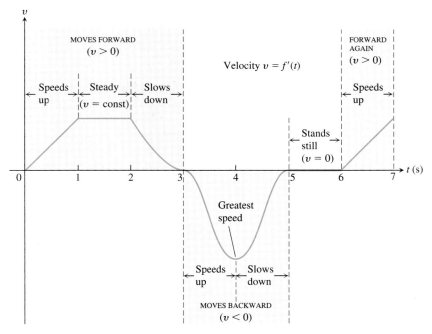

**FIGURE 3.14**   The velocity graph of a particle moving along a horizontal line, discussed in Example 2.

the sign of the velocity. Looking at Figure 3.14, we see that the particle moves forward for the first 3 s (when the velocity is positive), moves backward for the next 2 s (the velocity is negative), stands motionless for a full second, and then moves forward again. The particle is speeding up when its positive velocity increases during the first second, moves at a steady speed during the next second, and then slows down as the velocity decreases to zero during the third second. It stops for an instant at $t = 3$ s (when the velocity is zero) and reverses direction as the velocity starts to become negative. The particle is now moving backward and gaining in speed until $t = 4$ s, at which time it achieves its greatest speed during its backward motion. Continuing its backward motion at time $t = 4$, the particle starts to slow down again until it finally stops at time $t = 5$ (when the velocity is once again zero). The particle now remains motionless for one full second, and then moves forward again at $t = 6$ s, speeding up during the final second of the forward motion indicated in the velocity graph.     ■

**HISTORICAL BIOGRAPHY**
Bernard Bolzano
(1781–1848)

The rate at which a body's velocity changes is the body's *acceleration*. The acceleration measures how quickly the body picks up or loses speed. In Chapter 13 we will study motion in the plane and in space, where acceleration of an object may also lead to a change in direction.

A sudden change in acceleration is called a *jerk*. When a ride in a car or a bus is jerky, it is not that the accelerations involved are necessarily large but that the changes in acceleration are abrupt.

---

**DEFINITIONS**   **Acceleration** is the derivative of velocity with respect to time. If a body's position at time $t$ is $s = f(t)$, then the body's acceleration at time $t$ is

$$a(t) = \frac{dv}{dt} = \frac{d^2s}{dt^2}.$$

**Jerk** is the derivative of acceleration with respect to time:

$$j(t) = \frac{da}{dt} = \frac{d^3s}{dt^3}.$$

---

Near the surface of Earth all bodies fall with the same constant acceleration. Galileo's experiments with free fall (see Section 2.1) lead to the equation

$$s = \frac{1}{2}gt^2,$$

where $s$ is the distance fallen and $g$ is the acceleration due to Earth's gravity. This equation holds in a vacuum, where there is no air resistance, and closely models the fall of dense, heavy objects, such as rocks or steel tools, for the first few seconds of their fall, before the effects of air resistance are significant.

The value of $g$ in the equation $s = (1/2)gt^2$ depends on the units used to measure $t$ and $s$. With $t$ in seconds (the usual unit), the value of $g$ determined by measurement at sea level is approximately 9.8 m/s$^2$ (meters per second squared). (This gravitational constant depends on the distance from Earth's center of mass, and is slightly lower on top of Mt. Everest, for example.)

The jerk associated with the constant acceleration of gravity ($g = 9.8$ m/s$^2$) is zero:

$$j = \frac{d}{dt}(g) = 0.$$

An object does not exhibit jerkiness during free fall.

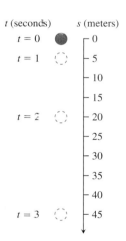

$t$ (seconds)     $s$ (meters)

**FIGURE 3.15** A ball bearing falling from rest (Example 3).

**EXAMPLE 3**   Figure 3.15 shows the free fall of a heavy ball bearing released from rest at time $t = 0$ s.

**(a)** How many meters does the ball fall in the first 3 s?

**(b)** What is its velocity, speed, and acceleration when $t = 3$?

**Solution**

**(a)** The metric free-fall equation is $s = 4.9t^2$. During the first 3 s, the ball falls

$$s(3) = 4.9(3)^2 = 44.1 \text{ m}.$$

**(b)** At any time $t$, *velocity* is the derivative of position:

$$v(t) = s'(t) = \frac{d}{dt}(4.9t^2) = 9.8t.$$

At $t = 3$, the velocity is

$$v(3) = 29.4 \text{ m/s}$$

in the downward (increasing $s$) direction. The *speed* at $t = 3$ is

$$\text{speed} = |v(3)| = 29.4 \text{ m/s}.$$

The *acceleration* at any time $t$ is

$$a(t) = v'(t) = s''(t) = 9.8 \text{ m/s}^2.$$

At $t = 3$, the acceleration is 9.8 m/s$^2$.  ■

**EXAMPLE 4**   A dynamite blast blows a heavy rock straight up with a launch velocity of 49 m/s (176.4 km/h) (Figure 3.16a). It reaches a height of $s = 49t - 4.9t^2$ m after $t$ seconds.

**(a)** How high does the rock go?

**(b)** What are the velocity and speed of the rock when it is 78.4 m above the ground on the way up? On the way down?

**(c)** What is the acceleration of the rock at any time $t$ during its flight (after the blast)?

**(d)** When does the rock hit the ground again?

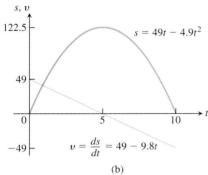

**FIGURE 3.16** (a) The rock in Example 4. (b) The graphs of $s$ and $v$ as functions of time; $s$ is largest when $v = ds/dt = 0$. The graph of $s$ is *not* the path of the rock: It is a plot of height versus time. The slope of the plot is the rock's velocity, graphed here as a straight line.

**Solution**

(a) In the coordinate system we have chosen, $s$ measures height from the ground up, so the velocity is positive on the way up and negative on the way down. The instant the rock is at its highest point is the one instant during the flight when the velocity is 0. To find the maximum height, all we need to do is to find when $v = 0$ and evaluate $s$ at this time.

At any time $t$ during the rock's motion, its velocity is

$$v = \frac{ds}{dt} = \frac{d}{dt}(49t - 4.9t^2) = 49 - 9.8t \text{ m/s}.$$

The velocity is zero when

$$49 - 9.8t = 0 \quad \text{or} \quad t = 5 \text{ s}.$$

The rock's height at $t = 5$ s is

$$s_{\max} = s(5) = 49(5) - 4.9(5)^2 = 245 - 122.5 = 122.5 \text{ m}.$$

See Figure 3.16b.

(b) To find the rock's velocity at 78.4 m on the way up and again on the way down, we first find the two values of $t$ for which

$$s(t) = 49t - 4.9t^2 = 78.4.$$

To solve this equation, we write

$$4.9t^2 - 49t + 78.4 = 0$$
$$4.9(t^2 - 10t + 16) = 0$$
$$(t - 2)(t - 8) = 0$$
$$t = 2 \text{ s}, t = 8 \text{ s}.$$

The rock is 78.4 m above the ground 2 s after the explosion and again 8 s after the explosion. The rock's velocities at these times are

$$v(2) = 49 - 9.8(2) = 49 - 19.6 = 29.4 \text{ m/s}.$$
$$v(8) = 49 - 9.8(8) = 49 - 78.4 = -29.4 \text{ m/s}.$$

At both instants, the rock's speed is 29.4 m/s. Since $v(2) > 0$, the rock is moving upward ($s$ is increasing) at $t = 2$ s; it is moving downward ($s$ is decreasing) at $t = 8$ because $v(8) < 0$.

(c) At any time during its flight following the explosion, the rock's acceleration is a constant

$$a = \frac{dv}{dt} = \frac{d}{dt}(49 - 9.8t) = -9.8 \text{ m/s}^2.$$

The acceleration is always downward and is the effect of gravity on the rock. As the rock rises, it slows down; as it falls, it speeds up.

(d) The rock hits the ground at the positive time $t$ for which $s = 0$. The equation $49t - 4.9t^2 = 0$ factors to give $4.9t(10 - t) = 0$, so it has solutions $t = 0$ and $t = 10$. At $t = 0$, the blast occurred and the rock was thrown upward. It returned to the ground 10 s later. ∎

### Derivatives in Economics

Engineers use the terms *velocity* and *acceleration* to refer to the derivatives of functions describing motion. Economists, too, have a specialized vocabulary for rates of change and derivatives. They call them *marginals*.

In a manufacturing operation, the *cost of production* $c(x)$ is a function of $x$, the number of units produced. The **marginal cost of production** is the rate of change of cost with respect to level of production, so it is $dc/dx$.

Cost $y$ (dollars)

**FIGURE 3.17** Weekly steel production: $c(x)$ is the cost of producing $x$ tonnes per week. The cost of producing an additional $h$ tonnes is $c(x + h) - c(x)$.

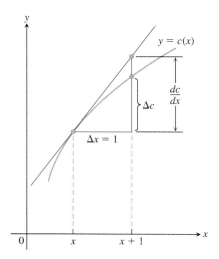

**FIGURE 3.18** The marginal cost $dc/dx$ is approximately the extra cost $\Delta c$ of producing $\Delta x = 1$ more unit.

Suppose that $c(x)$ represents the dollars needed to produce $x$ tonnes of steel in one week. It costs more to produce $x + h$ tonnes per week, and the cost difference, divided by $h$, is the average cost of producing each additional tonne:

$$\frac{c(x + h) - c(x)}{h} = \begin{array}{l} \text{average cost of each of the additional} \\ h \text{ tonnes of steel produced.} \end{array}$$

The limit of this ratio as $h \to 0$ is the *marginal cost* of producing more steel per week when the current weekly production is $x$ tonnes (Figure 3.17):

$$\frac{dc}{dx} = \lim_{h \to 0} \frac{c(x + h) - c(x)}{h} = \text{marginal cost of production.}$$

Sometimes the marginal cost of production is loosely defined to be the extra cost of producing one additional unit:

$$\frac{\Delta c}{\Delta x} = \frac{c(x + 1) - c(x)}{1},$$

which is approximated by the value of $dc/dx$ at $x$. This approximation is acceptable if the slope of the graph of $c$ does not change quickly near $x$. Then the difference quotient will be close to its limit $dc/dx$, which is the rise in the tangent line if $\Delta x = 1$ (Figure 3.18). The approximation works best for large values of $x$.

Economists often represent a total cost function by a cubic polynomial

$$c(x) = \alpha x^3 + \beta x^2 + \gamma x + \delta$$

where $\delta$ represents *fixed costs*, such as rent, heat, equipment capitalization, and management costs. The other terms represent *variable costs*, such as the costs of raw materials, taxes, and labor. Fixed costs are independent of the number of units produced, whereas variable costs depend on the quantity produced. A cubic polynomial is usually adequate to capture the cost behavior on a realistic quantity interval.

**EXAMPLE 5**    Suppose that it costs

$$c(x) = x^3 - 6x^2 + 15x$$

dollars to produce $x$ radiators when 8 to 30 radiators are produced and that

$$r(x) = x^3 - 3x^2 + 12x$$

gives the dollar revenue from selling $x$ radiators. Your shop currently produces 10 radiators a day. About how much extra will it cost to produce one more radiator a day, and what is your estimated increase in revenue for selling 11 radiators a day?

**Solution**    The cost of producing one more radiator a day when 10 are produced is about $c'(10)$:

$$c'(x) = \frac{d}{dx}\left(x^3 - 6x^2 + 15x\right) = 3x^2 - 12x + 15$$

$$c'(10) = 3(100) - 12(10) + 15 = 195.$$

The additional cost will be about \$195. The marginal revenue is

$$r'(x) = \frac{d}{dx}(x^3 - 3x^2 + 12x) = 3x^2 - 6x + 12.$$

The marginal revenue function estimates the increase in revenue that will result from selling one additional unit. If you currently sell 10 radiators a day, you can expect your revenue to increase by about

$$r'(10) = 3(100) - 6(10) + 12 = \$252$$

if you increase sales to 11 radiators a day.    ∎

**EXAMPLE 6**     To get some feel for the language of marginal rates, consider marginal tax rates. If your marginal income tax rate is 28% and your income increases by $1000, you can expect to pay an extra $280 in taxes. This does not mean that you pay 28% of your entire income in taxes. It just means that at your current income level $I$, the rate of increase of taxes $T$ with respect to income is $dT/dI = 0.28$. You will pay $0.28 in taxes out of every extra dollar you earn. Of course, if you earn a lot more, you may land in a higher tax bracket and your marginal rate will increase.  ∎

## Sensitivity to Change

When a small change in $x$ produces a large change in the value of a function $f(x)$, we say that the function is relatively **sensitive** to changes in $x$. The derivative $f'(x)$ is a measure of this sensitivity.

## EXAMPLE 7     Genetic Data and Sensitivity to Change

The Austrian monk Gregor Johann Mendel (1822–1884), working with garden peas and other plants, provided the first scientific explanation of hybridization.

His careful records showed that if $p$ (a number between 0 and 1) is the frequency of the gene for smooth skin in peas (dominant) and $(1 - p)$ is the frequency of the gene for wrinkled skin in peas, then the proportion of smooth-skinned peas in the next generation will be

$$y = 2p(1 - p) + p^2 = 2p - p^2.$$

The graph of $y$ versus $p$ in Figure 3.19a suggests that the value of $y$ is more sensitive to a change in $p$ when $p$ is small than when $p$ is large. Indeed, this fact is borne out by the derivative graph in Figure 3.19b, which shows that $dy/dp$ is close to 2 when $p$ is near 0 and close to 0 when $p$ is near 1.

The implication for genetics is that introducing a few more smooth skin genes into a population where the frequency of wrinkled skin peas is large will have a more dramatic effect on later generations than will a similar increase when the population has a large proportion of smooth skin peas.  ∎

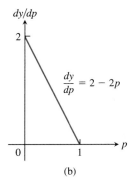

**FIGURE 3.19**   (a) The graph of $y = 2p - p^2$, describing the proportion of smooth-skinned peas in the next generation. (b) The graph of $dy/dp$ (Example 7).

# Exercises 3.4

**Motion Along a Coordinate Line**
Exercises 1–6 give the positions $s = f(t)$ of a body moving on a coordinate line, with $s$ in meters and $t$ in seconds.

  **a.** Find the body's displacement and average velocity for the given time interval.

  **b.** Find the body's speed and acceleration at the endpoints of the interval.

  **c.** When, if ever, during the interval does the body change direction?

1. $s = t^2 - 3t + 2,\quad 0 \le t \le 2$
2. $s = 6t - t^2,\quad 0 \le t \le 6$
3. $s = -t^3 + 3t^2 - 3t,\quad 0 \le t \le 3$
4. $s = (t^4/4) - t^3 + t^2,\quad 0 \le t \le 3$
5. $s = \dfrac{25}{t^2} - \dfrac{5}{t},\quad 1 \le t \le 5$
6. $s = \dfrac{25}{t + 5},\quad -4 \le t \le 0$

7. **Particle motion**   At time $t$, the position of a body moving along the $s$-axis is $s = t^3 - 6t^2 + 9t$ m.
  **a.** Find the body's acceleration each time the velocity is zero.
  **b.** Find the body's speed each time the acceleration is zero.
  **c.** Find the total distance traveled by the body from $t = 0$ to $t = 2$.
8. **Particle motion**   At time $t \ge 0$, the velocity of a body moving along the horizontal $s$-axis is $v = t^2 - 4t + 3$.
  **a.** Find the body's acceleration each time the velocity is zero.
  **b.** When is the body moving forward? Backward?
  **c.** When is the body's velocity increasing? Decreasing?

**Free-Fall Applications**
9. **Free fall on Mars and Jupiter**   The equations for free fall at the surfaces of Mars and Jupiter ($s$ in meters, $t$ in seconds) are $s = 1.86t^2$ on Mars and $s = 11.44t^2$ on Jupiter. How long does it take a rock falling from rest to reach a velocity of 27.8 m/s (about 100 km/h) on each planet?

**10. Lunar projectile motion**  A rock thrown vertically upward from the surface of the moon at a velocity of 24 m/s (about 86 km/h) reaches a height of $s = 24t - 0.8t^2$ m in $t$ s.

 **a.** Find the rock's velocity and acceleration at time $t$. (The acceleration in this case is the acceleration of gravity on the moon.)

 **b.** How long does it take the rock to reach its highest point?

 **c.** How high does the rock go?

 **d.** How long does it take the rock to reach half its maximum height?

 **e.** How long is the rock aloft?

**11. Finding $g$ on a small airless planet**  Explorers on a small airless planet used a spring gun to launch a ball bearing vertically upward from the surface at a launch velocity of 15 m/s. Because the acceleration of gravity at the planet's surface was $g_s$ m/s², the explorers expected the ball bearing to reach a height of $s = 15t - (1/2)g_s t^2$ m $t$ seconds later. The ball bearing reached its maximum height 20 s after being launched. What was the value of $g_s$?

**12. Speeding bullet**  A 45-caliber bullet shot straight up from the surface of the moon would reach a height of $s = 250t - 0.8t^2$ m after $t$ seconds. On Earth, in the absence of air, its height would be $s = 250t - 4.9t^2$ m after $t$ seconds. How long will the bullet be aloft in each case? How high will the bullet go?

**13. Free fall from the Tower of Pisa**  Had Galileo dropped a cannonball from the Tower of Pisa, 56 m above the ground, the ball's height above the ground $t$ seconds into the fall would have been $s = 56 - 4.9t^2$.

 **a.** What would have been the ball's velocity, speed, and acceleration at time $t$?

 **b.** About how long would it have taken the ball to hit the ground?

 **c.** What would have been the ball's velocity at the moment of impact?

**14. Galileo's free-fall formula**  Galileo developed a formula for a body's velocity during free fall by rolling balls from rest down increasingly steep inclined planks and looking for a limiting formula that would predict a ball's behavior when the plank was vertical and the ball fell freely; see part (a) of the accompanying figure. He found that, for any given angle of the plank, the ball's velocity $t$ seconds into motion was a constant multiple of $t$. That is, the velocity was given by a formula of the form $v = kt$. The value of the constant $k$ depended on the inclination of the plank.

 In modern notation—part (b) of the figure—with distance in meters and time in seconds, what Galileo determined by experiment was that, for any given angle $\theta$, the ball's velocity $t$ s into the roll was

$$v = 9.8(\sin\theta)t \text{ m/s}.$$

Free-fall position

(a)                              (b)

 **a.** What is the equation for the ball's velocity during free fall?

 **b.** Building on your work in part (a), what constant acceleration does a freely falling body experience near the surface of Earth?

**Understanding Motion from Graphs**

**15.** The accompanying figure shows the velocity $v = ds/dt = f(t)$ (m/s) of a body moving along a coordinate line.

 **a.** When does the body reverse direction?

 **b.** When (approximately) is the body moving at a constant speed?

 **c.** Graph the body's speed for $0 \le t \le 10$.

 **d.** Graph the acceleration, where defined.

**16.** A particle $P$ moves on the number line shown in part (a) of the accompanying figure. Part (b) shows the position of $P$ as a function of time $t$.

(a)

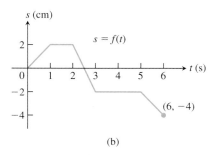

(b)

 **a.** When is $P$ moving to the left? Moving to the right? Standing still?

 **b.** Graph the particle's velocity and speed (where defined).

**17. Launching a rocket**  When a model rocket is launched, the propellant burns for a few seconds, accelerating the rocket upward. After burnout, the rocket coasts upward for a while and then begins to fall. A small explosive charge pops out a parachute shortly after the rocket starts down. The parachute slows the rocket to keep it from breaking when it lands.

 The figure here shows velocity data from the flight of the model rocket. Use the data to answer the following.

 **a.** How fast was the rocket climbing when the engine stopped?

 **b.** For how many seconds did the engine burn?

c. When did the rocket reach its highest point? What was its velocity then?

d. When did the parachute pop out? How fast was the rocket falling then?

e. How long did the rocket fall before the parachute opened?

f. When was the rocket's acceleration greatest?

g. When was the acceleration constant? What was its value then (to the nearest integer)?

18. The accompanying figure shows the velocity $v = f(t)$ of a particle moving on a horizontal coordinate line.

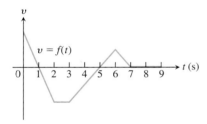

a. When does the particle move forward? Move backward? Speed up? Slow down?

b. When is the particle's acceleration positive? Negative? Zero?

c. When does the particle move at its greatest speed?

d. When does the particle stand still for more than an instant?

19. **Two falling balls** The multiflash photograph in the accompanying figure shows two balls falling from rest. The vertical rulers are marked in centimeters. Use the equation $s = 490t^2$ (the free-fall equation for $s$ in centimeters and $t$ in seconds) to answer the following questions. (*Source: PSSC Physics*, 2nd ed., Reprinted by permission of Education Development Center, Inc.)

a. How long did it take the balls to fall the first 160 cm? What was their average velocity for the period?

b. How fast were the balls falling when they reached the 160-cm mark? What was their acceleration then?

c. About how fast was the light flashing (flashes per second)?

20. **A cruising motorcycle** The accompanying graph shows the position $s$ of a motorcycle cruising along a highway. The motorcycle starts at $t = 0$ and returns 15 h later at $t = 15$.

a. Use the technique described in Section 3.2, Example 3, to graph the motorcycle's velocity $v = ds/dt$ for $0 \le t \le 15$. Then repeat the process, with the velocity curve, to graph the motorcycle's acceleration $dv/dt$.

b. Suppose that $s = 15t^2 - t^3$. Graph $ds/dt$ and $d^2s/dt^2$ and compare your graphs with those in part (a).

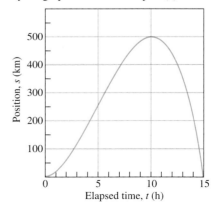

21. The graphs in the accompanying figure show the position $s$, velocity $v = ds/dt$, and acceleration $a = d^2s/dt^2$ of a body moving along a coordinate line as functions of time $t$. Which graph is which? Give reasons for your answers.

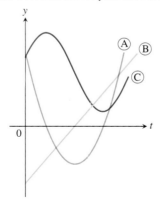

22. The graphs in the accompanying figure show the position $s$, the velocity $v = ds/dt$, and the acceleration $a = d^2s/dt^2$ of a body moving along a coordinate line as functions of time $t$. Which graph is which? Give reasons for your answers.

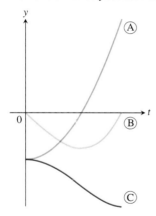

**Economics**

23. **Marginal cost** Suppose that the dollar cost of producing $x$ washing machines is $c(x) = 2000 + 100x - 0.1x^2$.

    a. Find the average cost per machine of producing the first 100 washing machines.

    b. Find the marginal cost when 100 washing machines are produced.

    c. Show that the marginal cost when 100 washing machines are produced is approximately the cost of producing one more washing machine after the first 100 have been made, by calculating the latter cost directly.

24. **Marginal revenue** Suppose that the revenue from selling $x$ washing machines is

    $$r(x) = 20{,}000\left(1 - \frac{1}{x}\right)$$

    dollars.

    a. Find the marginal revenue when 100 machines are produced.

    b. Use the function $r'(x)$ to estimate the increase in revenue that will result from increasing production from 100 machines a week to 101 machines a week.

    c. Find the limit of $r'(x)$ as $x \to \infty$. How would you interpret this number?

Additional Applications

25. **Bacterium population** When a bactericide was added to a nutrient broth in which bacteria were growing, the bacterium population continued to grow for a while, but then stopped growing and began to decline. The size of the population at time $t$ (hours) was $b = 10^6 + 10^4t - 10^3t^2$. Find the growth rates at

    a. $t = 0$ hours.

    b. $t = 5$ hours.

    c. $t = 10$ hours.

26. **Body surface area** A typical male's body surface area $S$ in square meters is often modeled by the formula $S = \frac{1}{60}\sqrt{wh}$, where $h$ is the height in cm, and $w$ the weight in kg, of the person. Find the rate of change of body surface area with respect to weight for males of constant height $h = 180$ cm. Does $S$ increase more rapidly with respect to weight at lower or higher body weights? Explain.

T 27. **Draining a tank** It takes 12 hours to drain a storage tank by opening the valve at the bottom. The depth $y$ of fluid in the tank $t$ hours after the valve is opened is given by the formula

    $$y = 6\left(1 - \frac{t}{12}\right)^2 \text{ m.}$$

    a. Find the rate $dy/dt$ (m/h) at which the tank is draining at time $t$.

    b. When is the fluid level in the tank falling fastest? Slowest? What are the values of $dy/dt$ at these times?

    c. Graph $y$ and $dy/dt$ together and discuss the behavior of $y$ in relation to the signs and values of $dy/dt$.

28. **Draining a tank** The number of liters of water in a tank $t$ minutes after the tank has started to drain is $Q(t) = 200(30 - t)^2$. How fast is the water running out at the end of 10 min? What is the average rate at which the water flows out during the first 10 min?

29. **Vehicular stopping distance** Based on data from the U.S. Bureau of Public Roads, a model for the total stopping distance of a moving car in terms of its speed is

    $$s = 0.21v + 0.00636v^2,$$

    where $s$ is measured in meters and $v$ in km/h. The linear term $0.21v$ models the distance the car travels during the time the driver perceives a need to stop until the brakes are applied, and the quadratic term $0.00636v^2$ models the additional braking distance once they are applied. Find $ds/dv$ at $v = 50$ and $v = 100$ km/h, and interpret the meaning of the derivative.

30. **Inflating a balloon** The volume $V = (4/3)\pi r^3$ of a spherical balloon changes with the radius.

    a. At what rate (m³/m) does the volume change with respect to the radius when $r = 2$ m?

    b. By approximately how much does the volume increase when the radius changes from 2 to 2.2 m?

**31. Airplane takeoff**   Suppose that the distance an aircraft travels along a runway before takeoff is given by $D = (10/9)t^2$, where $D$ is measured in meters from the starting point and $t$ is measured in seconds from the time the brakes are released. The aircraft will become airborne when its speed reaches 200 km/h. How long will it take to become airborne, and what distance will it travel in that time?

**32. Volcanic lava fountains**   Although the November 1959 Kilauea Iki eruption on the island of Hawaii began with a line of fountains along the wall of the crater, activity was later confined to a single vent in the crater's floor, which at one point shot lava 580 m straight into the air (a Hawaiian record). What was the lava's exit velocity in meters per second? In kilometers per hour? (*Hint:* If $v_0$ is the exit velocity of a particle of lava, its height $t$ seconds later will be $s = v_0 t - 4.9t^2$ m. Begin by finding the time at which $ds/dt = 0$. Neglect air resistance.)

**Analyzing Motion Using Graphs**

T  Exercises 33–36 give the position function $s = f(t)$ of an object moving along the $s$-axis as a function of time $t$. Graph $f$ together with the velocity function $v(t) = ds/dt = f'(t)$ and the acceleration function $a(t) = d^2s/dt^2 = f''(t)$. Comment on the object's behavior in relation to the signs and values of $v$ and $a$. Include in your commentary such topics as the following:

    **a.** When is the object momentarily at rest?

    **b.** When does it move to the left (down) or to the right (up)?

    **c.** When does it change direction?

    **d.** When does it speed up and slow down?

    **e.** When is it moving fastest (highest speed)? Slowest?

    **f.** When is it farthest from the axis origin?

**33.** $s = 60t - 4.9t^2$,   $0 \le t \le 12.5$ (a heavy object fired straight up from Earth's surface at 60 m/s)

**34.** $s = t^2 - 3t + 2$,   $0 \le t \le 5$

**35.** $s = t^3 - 6t^2 + 7t$,   $0 \le t \le 4$

**36.** $s = 4 - 7t + 6t^2 - t^3$,   $0 \le t \le 4$

# 3.5 Derivatives of Trigonometric Functions

Many phenomena of nature are approximately periodic (electromagnetic fields, heart rhythms, tides, weather). The derivatives of sines and cosines play a key role in describing periodic changes. This section shows how to differentiate the six basic trigonometric functions.

## Derivative of the Sine Function

To calculate the derivative of $f(x) = \sin x$, for $x$ measured in radians, we combine the limits in Example 5a and Theorem 7 in Section 2.4 with the angle sum identity for the sine function:

$$\sin(x + h) = \sin x \cos h + \cos x \sin h.$$

If $f(x) = \sin x$, then

$$f'(x) = \lim_{h \to 0} \frac{f(x + h) - f(x)}{h} = \lim_{h \to 0} \frac{\sin(x + h) - \sin x}{h} \qquad \text{Derivative definition}$$

$$= \lim_{h \to 0} \frac{(\sin x \cos h + \cos x \sin h) - \sin x}{h}$$

$$= \lim_{h \to 0} \frac{\sin x (\cos h - 1) + \cos x \sin h}{h}$$

$$= \lim_{h \to 0} \left( \sin x \cdot \frac{\cos h - 1}{h} \right) + \lim_{h \to 0} \left( \cos x \cdot \frac{\sin h}{h} \right)$$

$$= \sin x \cdot \underbrace{\lim_{h \to 0} \frac{\cos h - 1}{h}}_{\text{limit } 0} + \cos x \cdot \underbrace{\lim_{h \to 0} \frac{\sin h}{h}}_{\text{limit } 1} = \sin x \cdot 0 + \cos x \cdot 1 = \cos x.$$

Example 5a and Theorem 7, Section 2.4

---

**The derivative of the sine function is the cosine function:**

$$\frac{d}{dx}(\sin x) = \cos x.$$

**EXAMPLE 1** We find derivatives of the sine function involving differences, products, and quotients.

**(a)** $y = x^2 - \sin x$:

$$\frac{dy}{dx} = 2x - \frac{d}{dx}(\sin x) \qquad \text{Difference Rule}$$

$$= 2x - \cos x$$

**(b)** $y = x^2 \sin x$:

$$\frac{dy}{dx} = x^2 \frac{d}{dx}(\sin x) + 2x \sin x \qquad \text{Product Rule}$$

$$= x^2 \cos x + 2x \sin x.$$

**(c)** $y = \dfrac{\sin x}{x}$:

$$\frac{dy}{dx} = \frac{x \cdot \dfrac{d}{dx}(\sin x) - \sin x \cdot 1}{x^2} \qquad \text{Quotient Rule}$$

$$= \frac{x \cos x - \sin x}{x^2}$$

### Derivative of the Cosine Function

With the help of the angle sum formula for the cosine function,

$$\cos(x + h) = \cos x \cos h - \sin x \sin h,$$

we can compute the limit of the difference quotient:

$$\frac{d}{dx}(\cos x) = \lim_{h \to 0} \frac{\cos(x + h) - \cos x}{h} \qquad \text{Derivative definition}$$

$$= \lim_{h \to 0} \frac{(\cos x \cos h - \sin x \sin h) - \cos x}{h} \qquad \begin{array}{l}\text{Cosine angle sum}\\\text{identity}\end{array}$$

$$= \lim_{h \to 0} \frac{\cos x(\cos h - 1) - \sin x \sin h}{h}$$

$$= \lim_{h \to 0} \cos x \cdot \frac{\cos h - 1}{h} - \lim_{h \to 0} \sin x \cdot \frac{\sin h}{h}$$

$$= \cos x \cdot \lim_{h \to 0} \frac{\cos h - 1}{h} - \sin x \cdot \lim_{h \to 0} \frac{\sin h}{h}$$

$$= \cos x \cdot 0 - \sin x \cdot 1 \qquad \begin{array}{l}\text{Example 5a and}\\\text{Theorem 7, Section 2.4}\end{array}$$

$$= -\sin x.$$

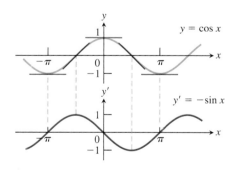

**FIGURE 3.20** The curve $y' = -\sin x$ as the graph of the slopes of the tangents to the curve $y = \cos x$.

---

> **The derivative of the cosine function is the negative of the sine function:**
>
> $$\frac{d}{dx}(\cos x) = -\sin x.$$

Figure 3.20 shows a way to visualize this result in the same way we did for graphing derivatives in Section 3.2, Figure 3.6.

**EXAMPLE 2** We find derivatives of the cosine function in combinations with other functions.

**(a)** $y = 5x + \cos x$:

$$\frac{dy}{dx} = \frac{d}{dx}(5x) + \frac{d}{dx}(\cos x) \qquad \text{Sum Rule}$$

$$= 5 - \sin x.$$

**(b)** $y = \sin x \cos x$:

$$\frac{dy}{dx} = \sin x \frac{d}{dx}(\cos x) + \cos x \frac{d}{dx}(\sin x) \qquad \text{Product Rule}$$

$$= \sin x(-\sin x) + \cos x(\cos x)$$

$$= \cos^2 x - \sin^2 x$$

**(c)** $y = \dfrac{\cos x}{1 - \sin x}$:

$$\frac{dy}{dx} = \frac{(1 - \sin x)\frac{d}{dx}(\cos x) - \cos x \frac{d}{dx}(1 - \sin x)}{(1 - \sin x)^2} \qquad \text{Quotient Rule}$$

$$= \frac{(1 - \sin x)(-\sin x) - \cos x(0 - \cos x)}{(1 - \sin x)^2}$$

$$= \frac{1 - \sin x}{(1 - \sin x)^2} \qquad\qquad \sin^2 x + \cos^2 x = 1$$

$$= \frac{1}{1 - \sin x} \qquad\qquad\qquad \blacksquare$$

## Simple Harmonic Motion

The motion of an object or weight bobbing freely up and down with no resistance on the end of a spring is an example of *simple harmonic motion*. The motion is periodic and repeats indefinitely, so we represent it using trigonometric functions. The next example describes a case in which there are no opposing forces such as friction to slow the motion.

**EXAMPLE 3** A weight hanging from a spring (Figure 3.21) is stretched down 5 units beyond its rest position and released at time $t = 0$ to bob up and down. Its position at any later time $t$ is

$$s = 5 \cos t.$$

What are its velocity and acceleration at time $t$?

**FIGURE 3.21** A weight hanging from a vertical spring and then displaced oscillates above and below its rest position (Example 3).

**Solution** We have

Position: $\quad s = 5 \cos t$

Velocity: $\quad v = \dfrac{ds}{dt} = \dfrac{d}{dt}(5 \cos t) = -5 \sin t$

Acceleration: $\quad a = \dfrac{dv}{dt} = \dfrac{d}{dt}(-5 \sin t) = -5 \cos t.$

Notice how much we can learn from these equations:

1. As time passes, the weight moves down and up between $s = -5$ and $s = 5$ on the $s$-axis. The amplitude of the motion is 5. The period of the motion is $2\pi$, the period of the cosine function.

2. The velocity $v = -5 \sin t$ attains its greatest magnitude, 5, when $\cos t = 0$, as the graphs show in Figure 3.22. Hence, the speed of the weight, $|v| = 5|\sin t|$, is greatest when $\cos t = 0$, that is, when $s = 0$ (the rest position). The speed of the weight is zero when $\sin t = 0$. This occurs when $s = 5 \cos t = \pm 5$, at the endpoints of the interval of motion.

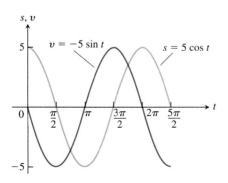

3. The weight is acted on by the spring and by gravity. When the weight is below the rest position, the combined forces pull it up, and when it is above the rest position, they pull it down. The weight's acceleration is always proportional to the negative of its displacement. This property of springs is called *Hooke's Law*, and is studied further in Section 6.5.

**FIGURE 3.22** The graphs of the position and velocity of the weight in Example 3.

4.  The acceleration, $a = -5 \cos t$, is zero only at the rest position, where $\cos t = 0$ and the force of gravity and the force from the spring balance each other. When the weight is anywhere else, the two forces are unequal and acceleration is nonzero. The acceleration is greatest in magnitude at the points farthest from the rest position, where $\cos t = \pm 1$.    ∎

**EXAMPLE 4**   The jerk associated with the simple harmonic motion in Example 3 is

$$ j = \frac{da}{dt} = \frac{d}{dt}(-5 \cos t) = 5 \sin t. $$

It has its greatest magnitude when $\sin t = \pm 1$, not at the extremes of the displacement but at the rest position, where the acceleration changes direction and sign.    ∎

### Derivatives of the Other Basic Trigonometric Functions

Because $\sin x$ and $\cos x$ are differentiable functions of $x$, the related functions

$$ \tan x = \frac{\sin x}{\cos x}, \qquad \cot x = \frac{\cos x}{\sin x}, \qquad \sec x = \frac{1}{\cos x}, \qquad \text{and} \qquad \csc x = \frac{1}{\sin x} $$

are differentiable at every value of $x$ at which they are defined. Their derivatives, calculated from the Quotient Rule, are given by the following formulas. Notice the negative signs in the derivative formulas for the cofunctions.

---

**The derivatives of the other trigonometric functions:**

$$ \frac{d}{dx}(\tan x) = \sec^2 x \qquad\qquad \frac{d}{dx}(\cot x) = -\csc^2 x $$

$$ \frac{d}{dx}(\sec x) = \sec x \tan x \qquad\qquad \frac{d}{dx}(\csc x) = -\csc x \cot x $$

---

To show a typical calculation, we find the derivative of the tangent function. The other derivations are left to Exercise 60.

**EXAMPLE 5**   Find $d(\tan x)/dx$.

**Solution**   We use the Derivative Quotient Rule to calculate the derivative:

$$ \frac{d}{dx}(\tan x) = \frac{d}{dx}\left(\frac{\sin x}{\cos x}\right) = \frac{\cos x \dfrac{d}{dx}(\sin x) - \sin x \dfrac{d}{dx}(\cos x)}{\cos^2 x} \qquad \text{Quotient Rule} $$

$$ = \frac{\cos x \cos x - \sin x (-\sin x)}{\cos^2 x} $$

$$ = \frac{\cos^2 x + \sin^2 x}{\cos^2 x} $$

$$ = \frac{1}{\cos^2 x} = \sec^2 x. \qquad ∎ $$

**EXAMPLE 6**   Find $y''$ if $y = \sec x$.

**Solution**   Finding the second derivative involves a combination of trigonometric derivatives.

$$ y = \sec x $$

$$ y' = \sec x \tan x \qquad\qquad \text{Derivative rule for secant function} $$

$$y'' = \frac{d}{dx}(\sec x \tan x)$$

$$= \sec x \frac{d}{dx}(\tan x) + \tan x \frac{d}{dx}(\sec x) \qquad \text{Derivative Product Rule}$$

$$= \sec x (\sec^2 x) + \tan x (\sec x \tan x) \qquad \text{Derivative rules}$$

$$= \sec^3 x + \sec x \tan^2 x \qquad\qquad\qquad ■$$

The differentiability of the trigonometric functions throughout their domains gives another proof of their continuity at every point in their domains (Theorem 1, Section 3.2). So we can calculate limits of algebraic combinations and composites of trigonometric functions by direct substitution.

**EXAMPLE 7**    We can use direct substitution in computing limits provided there is no division by zero, which is algebraically undefined.

$$\lim_{x \to 0} \frac{\sqrt{2 + \sec x}}{\cos(\pi - \tan x)} = \frac{\sqrt{2 + \sec 0}}{\cos(\pi - \tan 0)} = \frac{\sqrt{2 + 1}}{\cos(\pi - 0)} = \frac{\sqrt{3}}{-1} = -\sqrt{3} \qquad ■$$

# Exercises 3.5

## Derivatives

In Exercises 1–18, find $dy/dx$.

**1.** $y = -10x + 3\cos x$

**2.** $y = \frac{3}{x} + 5\sin x$

**3.** $y = x^2 \cos x$

**4.** $y = \sqrt{x}\sec x + 3$

**5.** $y = \csc x - 4\sqrt{x} + 7$

**6.** $y = x^2 \cot x - \frac{1}{x^2}$

**7.** $f(x) = \sin x \tan x$

**8.** $g(x) = \frac{\cos x}{\sin^2 x}$

**9.** $y = x \sec x + \frac{1}{x}$

**10.** $y = (\sin x + \cos x)\sec x$

**11.** $y = \frac{\cot x}{1 + \cot x}$

**12.** $y = \frac{\cos x}{1 + \sin x}$

**13.** $y = \frac{4}{\cos x} + \frac{1}{\tan x}$

**14.** $y = \frac{\cos x}{x} + \frac{x}{\cos x}$

**15.** $y = (\sec x + \tan x)(\sec x - \tan x)$

**16.** $y = x^2 \cos x - 2x \sin x - 2\cos x$

**17.** $f(x) = x^3 \sin x \cos x$

**18.** $g(x) = (2 - x)\tan^2 x$

In Exercises 19–22, find $ds/dt$.

**19.** $s = \tan t - t$

**20.** $s = t^2 - \sec t + 1$

**21.** $s = \frac{1 + \csc t}{1 - \csc t}$

**22.** $s = \frac{\sin t}{1 - \cos t}$

In Exercises 23–26, find $dr/d\theta$.

**23.** $r = 4 - \theta^2 \sin \theta$

**24.** $r = \theta \sin \theta + \cos \theta$

**25.** $r = \sec \theta \csc \theta$

**26.** $r = (1 + \sec \theta)\sin \theta$

In Exercises 27–32, find $dp/dq$.

**27.** $p = 5 + \frac{1}{\cot q}$

**28.** $p = (1 + \csc q)\cos q$

**29.** $p = \frac{\sin q + \cos q}{\cos q}$

**30.** $p = \frac{\tan q}{1 + \tan q}$

**31.** $p = \frac{q \sin q}{q^2 - 1}$

**32.** $p = \frac{3q + \tan q}{q \sec q}$

**33.** Find $y''$ if

   **a.** $y = \csc x$.

   **b.** $y = \sec x$.

**34.** Find $y^{(4)} = d^4 y/dx^4$ if

   **a.** $y = -2\sin x$.

   **b.** $y = 9\cos x$.

## Tangent Lines

In Exercises 35–38, graph the curves over the given intervals, together with their tangents at the given values of $x$. Label each curve and tangent with its equation.

**35.** $y = \sin x, \quad -3\pi/2 \le x \le 2\pi$

   $x = -\pi, 0, 3\pi/2$

**36.** $y = \tan x, \quad -\pi/2 < x < \pi/2$

   $x = -\pi/3, 0, \pi/3$

**37.** $y = \sec x, \quad -\pi/2 < x < \pi/2$

   $x = -\pi/3, \pi/4$

**38.** $y = 1 + \cos x, \quad -3\pi/2 \le x \le 2\pi$

   $x = -\pi/3, 3\pi/2$

T Do the graphs of the functions in Exercises 39–42 have any horizontal tangents in the interval $0 \le x \le 2\pi$? If so, where? If not, why not? Visualize your findings by graphing the functions with a grapher.

**39.** $y = x + \sin x$

**40.** $y = 2x + \sin x$

**41.** $y = x - \cot x$

**42.** $y = x + 2\cos x$

**43.** Find all points on the curve $y = \tan x$, $-\pi/2 < x < \pi/2$, where the tangent line is parallel to the line $y = 2x$. Sketch the curve and tangent(s) together, labeling each with its equation.

**44.** Find all points on the curve $y = \cot x$, $0 < x < \pi$, where the tangent line is parallel to the line $y = -x$. Sketch the curve and tangent(s) together, labeling each with its equation.

In Exercises 45 and 46, find an equation for **(a)** the tangent to the curve at $P$ and **(b)** the horizontal tangent to the curve at $Q$.

**45.**

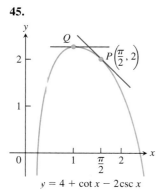

$$y = 4 + \cot x - 2\csc x$$

**46.**

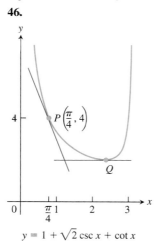

$$y = 1 + \sqrt{2} \csc x + \cot x$$

**Trigonometric Limits**

Find the limits in Exercises 47–54.

**47.** $\displaystyle\lim_{x \to 2} \sin\left(\frac{1}{x} - \frac{1}{2}\right)$

**48.** $\displaystyle\lim_{z \to -\pi/6} \sqrt{1 + \cos(\pi \csc x)}$

**49.** $\displaystyle\lim_{\theta \to \pi/6} \frac{\sin\theta - \frac{1}{2}}{\theta - \frac{\pi}{6}}$

**50.** $\displaystyle\lim_{\theta \to \pi/4} \frac{\tan\theta - 1}{\theta - \frac{\pi}{4}}$

**51.** $\displaystyle\lim_{x \to 0} \sec\left[\cos x + \pi \tan\left(\frac{\pi}{4 \sec x}\right) - 1\right]$

**52.** $\displaystyle\lim_{x \to 0} \sin\left(\frac{\pi + \tan x}{\tan x - 2 \sec x}\right)$

**53.** $\displaystyle\lim_{t \to 0} \tan\left(1 - \frac{\sin t}{t}\right)$

**54.** $\displaystyle\lim_{\theta \to 0} \cos\left(\frac{\pi\theta}{\sin\theta}\right)$

**Theory and Examples**

The equations in Exercises 55 and 56 give the position $s = f(t)$ of a body moving on a coordinate line ($s$ in meters, $t$ in seconds). Find the body's velocity, speed, acceleration, and jerk at time $t = \pi/4$ s.

**55.** $s = 2 - 2 \sin t$      **56.** $s = \sin t + \cos t$

**57.** Is there a value of $c$ that will make

$$f(x) = \begin{cases} \dfrac{\sin^2 3x}{x^2}, & x \neq 0 \\ c, & x = 0 \end{cases}$$

continuous at $x = 0$? Give reasons for your answer.

**58.** Is there a value of $b$ that will make

$$g(x) = \begin{cases} x + b, & x < 0 \\ \cos x, & x \geq 0 \end{cases}$$

continuous at $x = 0$? Differentiable at $x = 0$? Give reasons for your answers.

**59.** By computing the first few derivatives and looking for a pattern, find $d^{999}/dx^{999}(\cos x)$.

**60.** Derive the formula for the derivative with respect to $x$ of

    **a.** $\sec x$.     **b.** $\csc x$.     **c.** $\cot x$.

**61.** A weight is attached to a spring and reaches its equilibrium position ($x = 0$). It is then set in motion resulting in a displacement of

$$x = 10 \cos t,$$

where $x$ is measured in centimeters and $t$ is measured in seconds. See the accompanying figure.

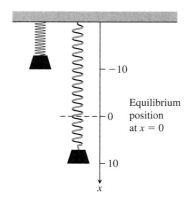

    **a.** Find the spring's displacement when $t = 0$, $t = \pi/3$, and $t = 3\pi/4$.

    **b.** Find the spring's velocity when $t = 0$, $t = \pi/3$, and $t = 3\pi/4$.

**62.** Assume that a particle's position on the $x$-axis is given by

$$x = 3 \cos t + 4 \sin t,$$

where $x$ is measured in meters and $t$ is measured in seconds.

    **a.** Find the particle's position when $t = 0$, $t = \pi/2$, and $t = \pi$.

    **b.** Find the particle's velocity when $t = 0$, $t = \pi/2$, and $t = \pi$.

**T 63.** Graph $y = \cos x$ for $-\pi \leq x \leq 2\pi$. On the same screen, graph

$$y = \frac{\sin(x + h) - \sin x}{h}$$

for $h = 1, 0.5, 0.3$, and $0.1$. Then, in a new window, try $h = -1, -0.5$, and $-0.3$. What happens as $h \to 0^+$? As $h \to 0^-$? What phenomenon is being illustrated here?

T **64.** Graph $y = -\sin x$ for $-\pi \le x \le 2\pi$. On the same screen, graph

$$y = \frac{\cos(x + h) - \cos x}{h}$$

for $h = 1, 0.5, 0.3$, and $0.1$. Then, in a new window, try $h = -1, -0.5$, and $-0.3$. What happens as $h \to 0^+$? As $h \to 0^-$? What phenomenon is being illustrated here?

T **65. Centered difference quotients**   The *centered difference quotient*

$$\frac{f(x + h) - f(x - h)}{2h}$$

is used to approximate $f'(x)$ in numerical work because (1) its limit as $h \to 0$ equals $f'(x)$ when $f'(x)$ exists, and (2) it usually gives a better approximation of $f'(x)$ for a given value of $h$ than the difference quotient

$$\frac{f(x + h) - f(x)}{h}.$$

See the accompanying figure.

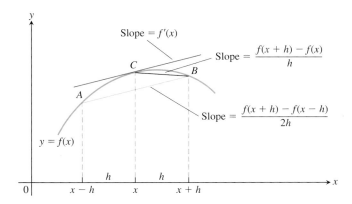

**a.** To see how rapidly the centered difference quotient for $f(x) = \sin x$ converges to $f'(x) = \cos x$, graph $y = \cos x$ together with

$$y = \frac{\sin(x + h) - \sin(x - h)}{2h}$$

over the interval $[-\pi, 2\pi]$ for $h = 1, 0.5$, and $0.3$. Compare the results with those obtained in Exercise 63 for the same values of $h$.

**b.** To see how rapidly the centered difference quotient for $f(x) = \cos x$ converges to $f'(x) = -\sin x$, graph $y = -\sin x$ together with

$$y = \frac{\cos(x + h) - \cos(x - h)}{2h}$$

over the interval $[-\pi, 2\pi]$ for $h = 1, 0.5$, and $0.3$. Compare the results with those obtained in Exercise 64 for the same values of $h$.

**66. A caution about centered difference quotients**   (*Continuation of Exercise 65.*) The quotient

$$\frac{f(x + h) - f(x - h)}{2h}$$

may have a limit as $h \to 0$ when $f$ has no derivative at $x$. As a case in point, take $f(x) = |x|$ and calculate

$$\lim_{h \to 0} \frac{|0 + h| - |0 - h|}{2h}.$$

As you will see, the limit exists even though $f(x) = |x|$ has no derivative at $x = 0$. *Moral:* Before using a centered difference quotient, be sure the derivative exists.

T **67. Slopes on the graph of the tangent function**   Graph $y = \tan x$ and its derivative together on $(-\pi/2, \pi/2)$. Does the graph of the tangent function appear to have a smallest slope? A largest slope? Is the slope ever negative? Give reasons for your answers.

T **68. Slopes on the graph of the cotangent function**   Graph $y = \cot x$ and its derivative together for $0 < x < \pi$. Does the graph of the cotangent function appear to have a smallest slope? A largest slope? Is the slope ever positive? Give reasons for your answers.

T **69. Exploring $(\sin kx)/x$**   Graph $y = (\sin x)/x$, $y = (\sin 2x)/x$, and $y = (\sin 4x)/x$ together over the interval $-2 \le x \le 2$. Where does each graph appear to cross the $y$-axis? Do the graphs really intersect the axis? What would you expect the graphs of $y = (\sin 5x)/x$ and $y = (\sin(-3x))/x$ to do as $x \to 0$? Why? What about the graph of $y = (\sin kx)/x$ for other values of $k$? Give reasons for your answers.

T **70. Radians versus degrees: degree mode derivatives**   What happens to the derivatives of $\sin x$ and $\cos x$ if $x$ is measured in degrees instead of radians? To find out, take the following steps.

**a.** With your graphing calculator or computer grapher in *degree mode*, graph

$$f(h) = \frac{\sin h}{h}$$

and estimate $\lim_{h \to 0} f(h)$. Compare your estimate with $\pi/180$. Is there any reason to believe the limit *should* be $\pi/180$?

**b.** With your grapher still in degree mode, estimate

$$\lim_{h \to 0} \frac{\cos h - 1}{h}.$$

**c.** Now go back to the derivation of the formula for the derivative of $\sin x$ in the text and carry out the steps of the derivation using degree-mode limits. What formula do you obtain for the derivative?

**d.** Work through the derivation of the formula for the derivative of $\cos x$ using degree-mode limits. What formula do you obtain for the derivative?

**e.** The disadvantages of the degree-mode formulas become apparent as you start taking derivatives of higher order. Try it. What are the second and third degree-mode derivatives of $\sin x$ and $\cos x$?

# 3.6 The Chain Rule

C: $y$ turns   B: $u$ turns   A: $x$ turns

**FIGURE 3.23**   When gear A makes $x$ turns, gear B makes $u$ turns and gear C makes $y$ turns. By comparing circumferences or counting teeth, we see that $y = u/2$ (C turns one-half turn for each B turn) and $u = 3x$ (B turns three times for A's one), so $y = 3x/2$. Thus, $dy/dx = 3/2 = (1/2)(3) = (dy/du)(du/dx)$.

How do we differentiate $F(x) = \sin(x^2 - 4)$? This function is the composite $f \circ g$ of two functions $y = f(u) = \sin u$ and $u = g(x) = x^2 - 4$ that we know how to differentiate. The answer, given by the *Chain Rule*, says that the derivative is the product of the derivatives of $f$ and $g$. We develop the rule in this section.

## Derivative of a Composite Function

The function $y = \dfrac{3}{2}x = \dfrac{1}{2}(3x)$ is the composite of the functions $y = \dfrac{1}{2}u$ and $u = 3x$. We have

$$\frac{dy}{dx} = \frac{3}{2}, \qquad \frac{dy}{du} = \frac{1}{2}, \qquad \text{and} \qquad \frac{du}{dx} = 3.$$

Since $\dfrac{3}{2} = \dfrac{1}{2} \cdot 3$, we see in this case that

$$\frac{dy}{dx} = \frac{dy}{du} \cdot \frac{du}{dx}.$$

If we think of the derivative as a rate of change, our intuition allows us to see that this relationship is reasonable. If $y = f(u)$ changes half as fast as $u$ and $u = g(x)$ changes three times as fast as $x$, then we expect $y$ to change $3/2$ times as fast as $x$. This effect is much like that of a multiple gear train (Figure 3.23). Let's look at another example.

**EXAMPLE 1**    The function

$$y = (3x^2 + 1)^2$$

is the composite of $y = f(u) = u^2$ and $u = g(x) = 3x^2 + 1$. Calculating derivatives, we see that

$$\frac{dy}{du} \cdot \frac{du}{dx} = 2u \cdot 6x$$

$$= 2(3x^2 + 1) \cdot 6x \qquad \text{Substitute for } u$$

$$= 36x^3 + 12x.$$

Calculating the derivative from the expanded formula $(3x^2 + 1)^2 = 9x^4 + 6x^2 + 1$ gives the same result:

$$\frac{dy}{dx} = \frac{d}{dx}(9x^4 + 6x^2 + 1)$$

$$= 36x^3 + 12x. \qquad \blacksquare$$

The derivative of the composite function $f(g(x))$ at $x$ is the derivative of $f$ at $g(x)$ times the derivative of $g$ at $x$. This is known as the Chain Rule (Figure 3.24).

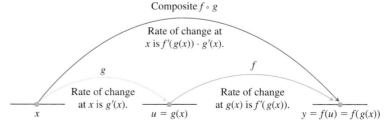

**FIGURE 3.24**   Rates of change multiply: The derivative of $f \circ g$ at $x$ is the derivative of $f$ at $g(x)$ times the derivative of $g$ at $x$.

> **THEOREM 2—The Chain Rule**   If $f(u)$ is differentiable at the point $u = g(x)$ and $g(x)$ is differentiable at $x$, then the composite function $(f \circ g)(x) = f(g(x))$ is differentiable at $x$, and
>
> $$(f \circ g)'(x) = f'(g(x)) \cdot g'(x).$$
>
> In Leibniz's notation, if $y = f(u)$ and $u = g(x)$, then
>
> $$\frac{dy}{dx} = \frac{dy}{du} \cdot \frac{du}{dx},$$
>
> where $dy/du$ is evaluated at $u = g(x)$.

### A Proof of One Case of the Chain Rule:

Let $\Delta u$ be the change in $u$ when x changes by $\Delta x$, so that

$$\Delta u = g(x + \Delta x) - g(x).$$

Then the corresponding change in $y$ is

$$\Delta y = f(u + \Delta u) - f(u).$$

If $\Delta u \neq 0$, we can write the fraction $\Delta y / \Delta x$ as the product

$$\frac{\Delta y}{\Delta x} = \frac{\Delta y}{\Delta u} \cdot \frac{\Delta u}{\Delta x} \tag{1}$$

and take the limit as $\Delta x \to 0$:

$$\begin{aligned}
\frac{dy}{dx} &= \lim_{\Delta x \to 0} \frac{\Delta y}{\Delta x} \\
&= \lim_{\Delta x \to 0} \frac{\Delta y}{\Delta u} \cdot \frac{\Delta u}{\Delta x} \\
&= \lim_{\Delta x \to 0} \frac{\Delta y}{\Delta u} \cdot \lim_{\Delta x \to 0} \frac{\Delta u}{\Delta x} \\
&= \lim_{\Delta u \to 0} \frac{\Delta y}{\Delta u} \cdot \lim_{\Delta x \to 0} \frac{\Delta u}{\Delta x} \qquad \text{(Note that } \Delta u \to 0 \text{ as } \Delta x \to 0 \\
& \qquad\qquad\qquad\qquad\qquad\qquad \text{ since } g \text{ is continuous.)} \\
&= \frac{dy}{du} \cdot \frac{du}{dx}.
\end{aligned}$$

The problem with this argument is that if the function $g(x)$ oscillates rapidly near $x$, then $\Delta u$ can be zero even when $\Delta x \neq 0$, so the cancelation of $\Delta u$ in Equation (1) would be invalid. A complete proof requires a different approach that avoids this problem, and we give one such proof in Section 3.9. ∎

**EXAMPLE 2**    An object moves along the $x$-axis so that its position at any time $t \geq 0$ is given by $x(t) = \cos(t^2 + 1)$. Find the velocity of the object as a function of $t$.

**Solution**   We know that the velocity is $dx/dt$. In this instance, $x$ is a composite function: $x = \cos(u)$ and $u = t^2 + 1$. We have

$$\frac{dx}{du} = -\sin(u) \qquad x = \cos(u)$$

$$\frac{du}{dt} = 2t. \qquad u = t^2 + 1$$

By the Chain Rule,

$$\frac{dx}{dt} = \frac{dx}{du} \cdot \frac{du}{dt}$$

$$= -\sin(u) \cdot 2t \qquad \frac{dx}{du} \text{ evaluated at } u$$

$$= -\sin(t^2 + 1) \cdot 2t$$

$$= -2t \sin(t^2 + 1). \qquad \blacksquare$$

---

**Ways to Write the Chain Rule**

$$(f \circ g)'(x) = f'(g(x)) \cdot g'(x)$$

$$\frac{dy}{dx} = \frac{dy}{du} \cdot \frac{du}{dx}$$

$$\frac{dy}{dx} = f'(g(x)) \cdot g'(x)$$

$$\frac{d}{dx} f(u) = f'(u) \frac{du}{dx}$$

---

### "Outside-Inside" Rule

A difficulty with the Leibniz notation is that it doesn't state specifically where the derivatives in the Chain Rule are supposed to be evaluated. So it sometimes helps to think about the Chain Rule using functional notation. If $y = f(g(x))$, then

$$\frac{dy}{dx} = f'(g(x)) \cdot g'(x).$$

In words, differentiate the "outside" function $f$ and evaluate it at the "inside" function $g(x)$ left alone; then multiply by the derivative of the "inside function."

**EXAMPLE 3** Differentiate $\sin(x^2 + x)$ with respect to $x$.

**Solution** We apply the Chain Rule directly and find

$$\frac{d}{dx} \sin \underbrace{(x^2 + x)}_{\text{inside}} = \cos \underbrace{(x^2 + x)}_{\substack{\text{inside} \\ \text{left alone}}} \cdot \underbrace{(2x + 1)}_{\substack{\text{derivative of} \\ \text{the inside}}}. \qquad \blacksquare$$

### Repeated Use of the Chain Rule

We sometimes have to use the Chain Rule two or more times to find a derivative.

**EXAMPLE 4** Find the derivative of $g(t) = \tan(5 - \sin 2t)$.

**Solution** Notice here that the tangent is a function of $5 - \sin 2t$, whereas the sine is a function of $2t$, which is itself a function of $t$. Therefore, by the Chain Rule,

$$g'(t) = \frac{d}{dt}(\tan(5 - \sin 2t))$$

$$= \sec^2(5 - \sin 2t) \cdot \frac{d}{dt}(5 - \sin 2t) \qquad \begin{array}{l} \text{Derivative of } \tan u \text{ with} \\ u = 5 - \sin 2t \end{array}$$

$$= \sec^2(5 - \sin 2t) \cdot \left( 0 - \cos 2t \cdot \frac{d}{dt}(2t) \right) \qquad \begin{array}{l} \text{Derivative of } 5 - \sin u \\ \text{with } u = 2t \end{array}$$

$$= \sec^2(5 - \sin 2t) \cdot (-\cos 2t) \cdot 2$$

$$= -2(\cos 2t) \sec^2(5 - \sin 2t). \qquad \blacksquare$$

**HISTORICAL BIOGRAPHY**

Johann Bernoulli
(1667–1748)

### The Chain Rule with Powers of a Function

If $f$ is a differentiable function of $u$ and if $u$ is a differentiable function of $x$, then substituting $y = f(u)$ into the Chain Rule formula

$$\frac{dy}{dx} = \frac{dy}{du} \cdot \frac{du}{dx}$$

leads to the formula

$$\frac{d}{dx} f(u) = f'(u) \frac{du}{dx}.$$

If $n$ is any real number and $f$ is a power function, $f(u) = u^n$, the Power Rule tells us that $f'(u) = nu^{n-1}$. If $u$ is a differentiable function of $x$, then we can use the Chain Rule to extend this to the **Power Chain Rule**:

$$\frac{d}{dx}(u^n) = nu^{n-1}\frac{du}{dx}. \qquad \frac{d}{du}(u^n) = nu^{n-1}$$

**EXAMPLE 5**     The Power Chain Rule simplifies computing the derivative of a power of an expression.

**(a)** $\frac{d}{dx}(5x^3 - x^4)^7 = 7(5x^3 - x^4)^6\frac{d}{dx}(5x^3 - x^4)$     Power Chain Rule with $u = 5x^3 - x^4, n = 7$

$$= 7(5x^3 - x^4)^6(5 \cdot 3x^2 - 4x^3)$$
$$= 7(5x^3 - x^4)^6(15x^2 - 4x^3)$$

**(b)** $\frac{d}{dx}\left(\frac{1}{3x - 2}\right) = \frac{d}{dx}(3x - 2)^{-1}$

$$= -1(3x - 2)^{-2}\frac{d}{dx}(3x - 2) \qquad \text{Power Chain Rule with } u = 3x - 2, n = -1$$
$$= -1(3x - 2)^{-2}(3)$$
$$= -\frac{3}{(3x - 2)^2}$$

In part (b) we could also find the derivative with the Derivative Quotient Rule.

**(c)** $\frac{d}{dx}(\sin^5 x) = 5\sin^4 x \cdot \frac{d}{dx}\sin x$     Power Chain Rule with $u = \sin x, n = 5$, because $\sin^n x$ means $(\sin x)^n, n \neq -1$.

$$= 5\sin^4 x\cos x \qquad\qquad\blacksquare$$

**EXAMPLE 6**     In Section 3.2, we saw that the absolute value function $y = |x|$ is not differentiable at $x = 0$. However, the function is differentiable at all other real numbers, as we now show. Since $|x| = \sqrt{x^2}$, we can derive the following formula:

$$\frac{d}{dx}(|x|) = \frac{d}{dx}\sqrt{x^2}$$
$$= \frac{1}{2\sqrt{x^2}} \cdot \frac{d}{dx}(x^2) \qquad \text{Power Chain Rule with } u = x^2, n = 1/2, x \neq 0$$
$$= \frac{1}{2|x|} \cdot 2x \qquad\qquad \sqrt{x^2} = |x|$$
$$= \frac{x}{|x|}, \quad x \neq 0. \qquad\qquad\blacksquare$$

> **Derivative of the Absolute Value Function**
>
> $$\frac{d}{dx}(|x|) = \frac{x}{|x|}, \quad x \neq 0$$
> $$= \begin{cases} 1, & x > 0 \\ -1, & x < 0 \end{cases}$$

**EXAMPLE 7**     Show that the slope of every line tangent to the curve $y = 1/(1 - 2x)^3$ is positive.

**Solution**     We find the derivative:

$$\frac{dy}{dx} = \frac{d}{dx}(1 - 2x)^{-3}$$
$$= -3(1 - 2x)^{-4} \cdot \frac{d}{dx}(1 - 2x) \qquad \text{Power Chain Rule with } u = (1 - 2x), n = -3$$
$$= -3(1 - 2x)^{-4} \cdot (-2)$$
$$= \frac{6}{(1 - 2x)^4}.$$

At any point $(x, y)$ on the curve, the coordinate $x$ is not $1/2$ and the slope of the tangent line is

$$\frac{dy}{dx} = \frac{6}{(1 - 2x)^4},$$

which is the quotient of two positive numbers. ∎

**EXAMPLE 8** The formulas for the derivatives of both $\sin x$ and $\cos x$ were obtained under the assumption that $x$ is measured in radians, *not* degrees. The Chain Rule gives us new insight into the difference between the two. Since $180° = \pi$ radians, $x° = \pi x/180$ radians where $x°$ is the size of the angle measured in degrees.

By the Chain Rule,

$$\frac{d}{dx}\sin(x°) = \frac{d}{dx}\sin\left(\frac{\pi x}{180}\right) = \frac{\pi}{180}\cos\left(\frac{\pi x}{180}\right) = \frac{\pi}{180}\cos(x°).$$

See Figure 3.25. Similarly, the derivative of $\cos(x°)$ is $-(\pi/180)\sin(x°)$.

The factor $\pi/180$ would compound with repeated differentiation, showing an advantage for the use of radian measure in computations. ∎

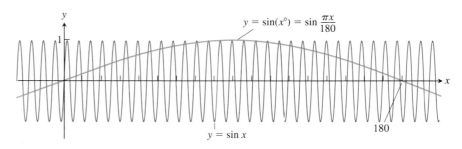

**FIGURE 3.25** The function $\sin(x°)$ oscillates only $\pi/180$ times as often as $\sin x$ oscillates. Its maximum slope is $\pi/180$ at $x = 0$ (Example 8).

# Exercises 3.6

## Derivative Calculations

In Exercises 1–8, given $y = f(u)$ and $u = g(x)$, find $dy/dx = f'(g(x))g'(x)$.

**1.** $y = 6u - 9, \quad u = (1/2)x^4$    **2.** $y = 2u^3, \quad u = 8x - 1$

**3.** $y = \sin u, \quad u = 3x + 1$    **4.** $y = \cos u, \quad u = -x/3$

**5.** $y = \sqrt{u}, \quad u = \sin x$    **6.** $y = \sin u, \quad u = x - \cos x$

**7.** $y = \tan u, \quad u = \pi x^2$    **8.** $y = -\sec u, \quad u = \dfrac{1}{x} + 7x$

In Exercises 9–18, write the function in the form $y = f(u)$ and $u = g(x)$. Then find $dy/dx$ as a function of $x$.

**9.** $y = (2x + 1)^5$    **10.** $y = (4 - 3x)^9$

**11.** $y = \left(1 - \dfrac{x}{7}\right)^{-7}$    **12.** $y = \left(\dfrac{\sqrt{x}}{2} - 1\right)^{-10}$

**13.** $y = \left(\dfrac{x^2}{8} + x - \dfrac{1}{x}\right)^4$    **14.** $y = \sqrt{3x^2 - 4x + 6}$

**15.** $y = \sec(\tan x)$    **16.** $y = \cot\left(\pi - \dfrac{1}{x}\right)$

**17.** $y = \tan^3 x$    **18.** $y = 5\cos^{-4} x$

Find the derivatives of the functions in Exercises 19–40.

**19.** $p = \sqrt{3 - t}$    **20.** $q = \sqrt[3]{2r - r^2}$

**21.** $s = \dfrac{4}{3\pi}\sin 3t + \dfrac{4}{5\pi}\cos 5t$    **22.** $s = \sin\left(\dfrac{3\pi t}{2}\right) + \cos\left(\dfrac{3\pi t}{2}\right)$

**23.** $r = (\csc \theta + \cot \theta)^{-1}$    **24.** $r = 6(\sec \theta - \tan \theta)^{3/2}$

**25.** $y = x^2 \sin^4 x + x \cos^{-2} x$    **26.** $y = \dfrac{1}{x}\sin^{-5} x - \dfrac{x}{3}\cos^3 x$

**27.** $y = \dfrac{1}{18}(3x - 2)^6 + \left(4 - \dfrac{1}{2x^2}\right)^{-1}$

**28.** $y = (5 - 2x)^{-3} + \dfrac{1}{8}\left(\dfrac{2}{x} + 1\right)^4$

**29.** $y = (4x + 3)^4(x + 1)^{-3}$    **30.** $y = (2x - 5)^{-1}(x^2 - 5x)^6$

**31.** $h(x) = x \tan\left(2\sqrt{x}\right) + 7$    **32.** $k(x) = x^2 \sec\left(\dfrac{1}{x}\right)$

**33.** $f(x) = \sqrt{7 + x \sec x}$    **34.** $g(x) = \dfrac{\tan 3x}{(x + 7)^4}$

**35.** $f(\theta) = \left(\dfrac{\sin \theta}{1 + \cos \theta}\right)^2$    **36.** $g(t) = \left(\dfrac{1 + \sin 3t}{3 - 2t}\right)^{-1}$

**37.** $r = \sin(\theta^2)\cos(2\theta)$

**38.** $r = \sec\sqrt{\theta}\tan\left(\dfrac{1}{\theta}\right)$

**39.** $q = \sin\left(\dfrac{t}{\sqrt{t+1}}\right)$

**40.** $q = \cot\left(\dfrac{\sin t}{t}\right)$

In Exercises 41–58, find $dy/dt$.

**41.** $y = \sin^2(\pi t - 2)$

**42.** $y = \sec^2 \pi t$

**43.** $y = (1 + \cos 2t)^{-4}$

**44.** $y = (1 + \cot(t/2))^{-2}$

**45.** $y = (t\tan t)^{10}$

**46.** $y = (t^{-3/4}\sin t)^{4/3}$

**47.** $y = \left(\dfrac{t^2}{t^3 - 4t}\right)^3$

**48.** $y = \left(\dfrac{3t - 4}{5t + 2}\right)^{-5}$

**49.** $y = \sin(\cos(2t - 5))$

**50.** $y = \cos\left(5\sin\left(\dfrac{t}{3}\right)\right)$

**51.** $y = \left(1 + \tan^4\left(\dfrac{t}{12}\right)\right)^3$

**52.** $y = \dfrac{1}{6}(1 + \cos^2(7t))^3$

**53.** $y = \sqrt{1 + \cos(t^2)}$

**54.** $y = 4\sin\left(\sqrt{1 + \sqrt{t}}\right)$

**55.** $y = \tan^2(\sin^3 t)$

**56.** $y = \cos^4(\sec^2 3t)$

**57.** $y = 3t(2t^2 - 5)^4$

**58.** $y = \sqrt{3t + \sqrt{2 + \sqrt{1 - t}}}$

### Second Derivatives
Find $y''$ in Exercises 59–64.

**59.** $y = \left(1 + \dfrac{1}{x}\right)^3$

**60.** $y = (1 - \sqrt{x})^{-1}$

**61.** $y = \dfrac{1}{9}\cot(3x - 1)$

**62.** $y = 9\tan\left(\dfrac{x}{3}\right)$

**63.** $y = x(2x + 1)^4$

**64.** $y = x^2(x^3 - 1)^5$

### Finding Derivative Values
In Exercises 65–70, find the value of $(f \circ g)'$ at the given value of $x$.

**65.** $f(u) = u^5 + 1, \quad u = g(x) = \sqrt{x}, \quad x = 1$

**66.** $f(u) = 1 - \dfrac{1}{u}, \quad u = g(x) = \dfrac{1}{1 - x}, \quad x = -1$

**67.** $f(u) = \cot\dfrac{\pi u}{10}, \quad u = g(x) = 5\sqrt{x}, \quad x = 1$

**68.** $f(u) = u + \dfrac{1}{\cos^2 u}, \quad u = g(x) = \pi x, \quad x = 1/4$

**69.** $f(u) = \dfrac{2u}{u^2 + 1}, \quad u = g(x) = 10x^2 + x + 1, \quad x = 0$

**70.** $f(u) = \left(\dfrac{u - 1}{u + 1}\right)^2, \quad u = g(x) = \dfrac{1}{x^2} - 1, \quad x = -1$

**71.** Assume that $f'(3) = -1, g'(2) = 5, g(2) = 3$, and $y = f(g(x))$. What is $y'$ at $x = 2$?

**72.** If $r = \sin(f(t)), f(0) = \pi/3$, and $f'(0) = 4$, then what is $dr/dt$ at $t = 0$?

**73.** Suppose that functions $f$ and $g$ and their derivatives with respect to $x$ have the following values at $x = 2$ and $x = 3$.

| $x$ | $f(x)$ | $g(x)$ | $f'(x)$ | $g'(x)$ |
|---|---|---|---|---|
| 2 | 8 | 2 | 1/3 | −3 |
| 3 | 3 | −4 | $2\pi$ | 5 |

Find the derivatives with respect to $x$ of the following combinations at the given value of $x$.

**a.** $2f(x), \quad x = 2$

**b.** $f(x) + g(x), \quad x = 3$

**c.** $f(x) \cdot g(x), \quad x = 3$

**d.** $f(x)/g(x), \quad x = 2$

**e.** $f(g(x)), \quad x = 2$

**f.** $\sqrt{f(x)}, \quad x = 2$

**g.** $1/g^2(x), \quad x = 3$

**h.** $\sqrt{f^2(x) + g^2(x)}, \quad x = 2$

**74.** Suppose that the functions $f$ and $g$ and their derivatives with respect to $x$ have the following values at $x = 0$ and $x = 1$.

| $x$ | $f(x)$ | $g(x)$ | $f'(x)$ | $g'(x)$ |
|---|---|---|---|---|
| 0 | 1 | 1 | 5 | 1/3 |
| 1 | 3 | −4 | −1/3 | −8/3 |

Find the derivatives with respect to $x$ of the following combinations at the given value of $x$.

**a.** $5f(x) - g(x), \quad x = 1$

**b.** $f(x)g^3(x), \quad x = 0$

**c.** $\dfrac{f(x)}{g(x) + 1}, \quad x = 1$

**d.** $f(g(x)), \quad x = 0$

**e.** $g(f(x)), \quad x = 0$

**f.** $(x^{11} + f(x))^{-2}, \quad x = 1$

**g.** $f(x + g(x)), \quad x = 0$

**75.** Find $ds/dt$ when $\theta = 3\pi/2$ if $s = \cos\theta$ and $d\theta/dt = 5$.

**76.** Find $dy/dt$ when $x = 1$ if $y = x^2 + 7x - 5$ and $dx/dt = 1/3$.

### Theory and Examples
What happens if you can write a function as a composite in different ways? Do you get the same derivative each time? The Chain Rule says you should. Try it with the functions in Exercises 77 and 78.

**77.** Find $dy/dx$ if $y = x$ by using the Chain Rule with $y$ as a composite of

**a.** $y = (u/5) + 7$ and $u = 5x - 35$

**b.** $y = 1 + (1/u)$ and $u = 1/(x - 1)$.

**78.** Find $dy/dx$ if $y = x^{3/2}$ by using the Chain Rule with $y$ as a composite of

**a.** $y = u^3$ and $u = \sqrt{x}$

**b.** $y = \sqrt{u}$ and $u = x^3$.

**79.** Find the tangent to $y = ((x - 1)/(x + 1))^2$ at $x = 0$.

**80.** Find the tangent to $y = \sqrt{x^2 - x + 7}$ at $x = 2$.

**81. a.** Find the tangent to the curve $y = 2\tan(\pi x/4)$ at $x = 1$.

**b. Slopes on a tangent curve** What is the smallest value the slope of the curve can ever have on the interval $-2 < x < 2$? Give reasons for your answer.

**82. Slopes on sine curves**

**a.** Find equations for the tangents to the curves $y = \sin 2x$ and $y = -\sin(x/2)$ at the origin. Is there anything special about how the tangents are related? Give reasons for your answer.

**b.** Can anything be said about the tangents to the curves $y = \sin mx$ and $y = -\sin(x/m)$ at the origin ($m$ a constant $\neq 0$)? Give reasons for your answer.

**c.** For a given $m$, what are the largest values the slopes of the curves $y = \sin mx$ and $y = -\sin(x/m)$ can ever have? Give reasons for your answer.

**d.** The function $y = \sin x$ completes one period on the interval $[0, 2\pi]$, the function $y = \sin 2x$ completes two periods, the function $y = \sin(x/2)$ completes half a period, and so on. Is there any relation between the number of periods $y = \sin mx$ completes on $[0, 2\pi]$ and the slope of the curve $y = \sin mx$ at the origin? Give reasons for your answer.

**83. Running machinery too fast** Suppose that a piston is moving straight up and down and that its position at time $t$ s is

$$s = A\cos(2\pi bt),$$

with $A$ and $b$ positive. The value of $A$ is the amplitude of the motion, and $b$ is the frequency (number of times the piston moves up and down each second). What effect does doubling the frequency have on the piston's velocity, acceleration, and jerk? (Once you find out, you will know why some machinery breaks when you run it too fast.)

**84. Temperatures in Fairbanks, Alaska** The graph in the accompanying figure shows the average Celsius temperature in Fairbanks, Alaska, during a typical 365-day year. The equation that approximates the temperature on day $x$ is

$$y = 20\sin\left[\frac{2\pi}{365}(x - 101)\right] - 4$$

and is graphed in the accompanying figure.

**a.** On what day is the temperature increasing the fastest?

**b.** About how many degrees per day is the temperature increasing when it is increasing at its fastest?

**85. Particle motion** The position of a particle moving along a coordinate line is $s = \sqrt{1 + 4t}$, with $s$ in meters and $t$ in seconds. Find the particle's velocity and acceleration at $t = 6$ s.

**86. Constant acceleration** Suppose that the velocity of a falling body is $v = k\sqrt{s}$ m/s ($k$ a constant) at the instant the body has fallen $s$ m from its starting point. Show that the body's acceleration is constant.

**87. Falling meteorite** The velocity of a heavy meteorite entering Earth's atmosphere is inversely proportional to $\sqrt{s}$ when it is $s$ km from Earth's center. Show that the meteorite's acceleration is inversely proportional to $s^2$.

**88. Particle acceleration** A particle moves along the $x$-axis with velocity $dx/dt = f(x)$. Show that the particle's acceleration is $f(x)f'(x)$.

**89. Temperature and the period of a pendulum** For oscillations of small amplitude (short swings), we may safely model the relationship between the period $T$ and the length $L$ of a simple pendulum with the equation

$$T = 2\pi\sqrt{\frac{L}{g}},$$

where $g$ is the constant acceleration of gravity at the pendulum's location. If we measure $g$ in centimeters per second squared, we measure $L$ in centimeters and $T$ in seconds. If the pendulum is made of metal, its length will vary with temperature, either increasing or decreasing at a rate that is roughly proportional to $L$. In symbols, with $u$ being temperature and $k$ the proportionality constant,

$$\frac{dL}{du} = kL.$$

Assuming this to be the case, show that the rate at which the period changes with respect to temperature is $kT/2$.

**90. Chain Rule** Suppose that $f(x) = x^2$ and $g(x) = |x|$. Then the composites

$$(f \circ g)(x) = |x|^2 = x^2 \quad \text{and} \quad (g \circ f)(x) = |x^2| = x^2$$

are both differentiable at $x = 0$ even though $g$ itself is not differentiable at $x = 0$. Does this contradict the Chain Rule? Explain.

**T** **91. The derivative of sin 2x** Graph the function $y = 2\cos 2x$ for $-2 \le x \le 3.5$. Then, on the same screen, graph

$$y = \frac{\sin 2(x + h) - \sin 2x}{h}$$

for $h = 1.0, 0.5$, and $0.2$. Experiment with other values of $h$, including negative values. What do you see happening as $h \to 0$? Explain this behavior.

**92. The derivative of cos($x^2$)** Graph $y = -2x\sin(x^2)$ for $-2 \le x \le 3$. Then, on the same screen, graph

$$y = \frac{\cos((x + h)^2) - \cos(x^2)}{h}$$

for $h = 1.0, 0.7$, and $0.3$. Experiment with other values of $h$. What do you see happening as $h \to 0$? Explain this behavior.

Using the Chain Rule, show that the Power Rule $(d/dx)x^n = nx^{n-1}$ holds for the functions $x^n$ in Exercises 93 and 94.

**93.** $x^{1/4} = \sqrt{\sqrt{x}}$ **94.** $x^{3/4} = \sqrt{x\sqrt{x}}$

### COMPUTER EXPLORATIONS
Trigonometric Polynomials

**95.** As the accompanying figure shows, the trigonometric "polynomial"

$$s = f(t) = 0.78540 - 0.63662\cos 2t - 0.07074\cos 6t$$
$$- 0.02546\cos 10t - 0.01299\cos 14t$$

gives a good approximation of the sawtooth function $s = g(t)$ on the interval $[-\pi, \pi]$. How well does the derivative of $f$ approximate the derivative of $g$ at the points where $dg/dt$ is defined? To find out, carry out the following steps.

**a.** Graph $dg/dt$ (where defined) over $[-\pi, \pi]$.

**b.** Find $df/dt$.

**c.** Graph $df/dt$. Where does the approximation of $dg/dt$ by $df/dt$ seem to be best? Least good? Approximations by trigonometric polynomials are important in the theories of heat and oscillation, but we must not expect too much of them, as we see in the next exercise.

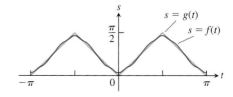

**96.** (*Continuation of Exercise 95.*) In Exercise 95, the trigonometric polynomial $f(t)$ that approximated the sawtooth function $g(t)$ on $[-\pi, \pi]$ had a derivative that approximated the derivative of the sawtooth function. It is possible, however, for a trigonometric polynomial to approximate a function in a reasonable way without its derivative approximating the function's derivative at all well. As a case in point, the trigonometric "polynomial"

$$s = h(t) = 1.2732 \sin 2t + 0.4244 \sin 6t + 0.25465 \sin 10t$$
$$+ 0.18189 \sin 14t + 0.14147 \sin 18t$$

graphed in the accompanying figure approximates the step function $s = k(t)$ shown there. Yet the derivative of $h$ is nothing like the derivative of $k$.

**a.** Graph $dk/dt$ (where defined) over $[-\pi, \pi]$.

**b.** Find $dh/dt$.

**c.** Graph $dh/dt$ to see how badly the graph fits the graph of $dk/dt$. Comment on what you see.

# 3.7 Implicit Differentiation

Most of the functions we have dealt with so far have been described by an equation of the form $y = f(x)$ that expresses $y$ explicitly in terms of the variable $x$. We have learned rules for differentiating functions defined in this way. Another situation occurs when we encounter equations like

$$x^3 + y^3 - 9xy = 0, \qquad y^2 - x = 0, \qquad \text{or} \qquad x^2 + y^2 - 25 = 0.$$

(See Figures 3.26, 3.27, and 3.28.) These equations define an *implicit* relation between the variables $x$ and $y$. In some cases we may be able to solve such an equation for $y$ as an explicit function (or even several functions) of $x$. When we cannot put an equation $F(x, y) = 0$ in the form $y = f(x)$ to differentiate it in the usual way, we may still be able to find $dy/dx$ by *implicit differentiation*. This section describes the technique.

## Implicitly Defined Functions

We begin with examples involving familiar equations that we can solve for $y$ as a function of $x$ to calculate $dy/dx$ in the usual way. Then we differentiate the equations implicitly, and find the derivative to compare the two methods. Following the examples, we summarize the steps involved in the new method. In the examples and exercises, it is always assumed that the given equation determines $y$ implicitly as a differentiable function of $x$ so that $dy/dx$ exists.

**EXAMPLE 1** Find $dy/dx$ if $y^2 = x$.

**Solution** The equation $y^2 = x$ defines two differentiable functions of $x$ that we can actually find, namely $y_1 = \sqrt{x}$ and $y_2 = -\sqrt{x}$ (Figure 3.27). We know how to calculate the derivative of each of these for $x > 0$:

$$\frac{dy_1}{dx} = \frac{1}{2\sqrt{x}} \qquad \text{and} \qquad \frac{dy_2}{dx} = -\frac{1}{2\sqrt{x}}.$$

But suppose that we knew only that the equation $y^2 = x$ defined $y$ as one or more differentiable functions of $x$ for $x > 0$ without knowing exactly what these functions were. Could we still find $dy/dx$?

The answer is yes. To find $dy/dx$, we simply differentiate both sides of the equation $y^2 = x$ with respect to $x$, treating $y = f(x)$ as a differentiable function of $x$:

$$y^2 = x$$

$$2y\frac{dy}{dx} = 1$$

$$\frac{dy}{dx} = \frac{1}{2y}.$$

The Chain Rule gives $\dfrac{d}{dx}(y^2) =$ $\dfrac{d}{dx}[f(x)]^2 = 2f(x)f'(x) = 2y\dfrac{dy}{dx}.$

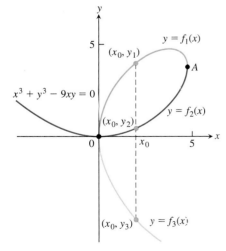

**FIGURE 3.26** The curve $x^3 + y^3 - 9xy = 0$ is not the graph of any one function of $x$. The curve can, however, be divided into separate arcs that *are* the graphs of functions of $x$. This particular curve, called a *folium*, dates to Descartes in 1638.

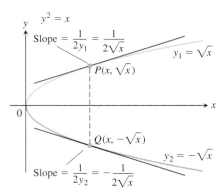

**FIGURE 3.27** The equation $y^2 - x = 0$, or $y^2 = x$ as it is usually written, defines two differentiable functions of $x$ on the interval $x > 0$. Example 1 shows how to find the derivatives of these functions without solving the equation $y^2 = x$ for $y$.

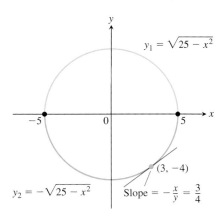

**FIGURE 3.28** The circle combines the graphs of two functions. The graph of $y_2$ is the lower semicircle and passes through $(3, -4)$.

This one formula gives the derivatives we calculated for *both* explicit solutions $y_1 = \sqrt{x}$ and $y_2 = -\sqrt{x}$:

$$\frac{dy_1}{dx} = \frac{1}{2y_1} = \frac{1}{2\sqrt{x}} \quad \text{and} \quad \frac{dy_2}{dx} = \frac{1}{2y_2} = \frac{1}{2\left(-\sqrt{x}\right)} = -\frac{1}{2\sqrt{x}}. \qquad \blacksquare$$

**EXAMPLE 2** Find the slope of the circle $x^2 + y^2 = 25$ at the point $(3, -4)$.

**Solution** The circle is not the graph of a single function of $x$. Rather, it is the combined graphs of two differentiable functions, $y_1 = \sqrt{25 - x^2}$ and $y_2 = -\sqrt{25 - x^2}$ (Figure 3.28). The point $(3, -4)$ lies on the graph of $y_2$, so we can find the slope by calculating the derivative directly, using the Power Chain Rule:

$$\left.\frac{dy_2}{dx}\right|_{x=3} = -\left.\frac{-2x}{2\sqrt{25 - x^2}}\right|_{x=3} = -\frac{-6}{2\sqrt{25 - 9}} = \frac{3}{4}. \qquad \begin{aligned}&\frac{d}{dx}\left(-(25 - x^2)^{1/2}\right) = \\[4pt] &-\frac{1}{2}(25 - x^2)^{-1/2}(-2x)\end{aligned}$$

We can solve this problem more easily by differentiating the given equation of the circle implicitly with respect to $x$:

$$\frac{d}{dx}(x^2) + \frac{d}{dx}(y^2) = \frac{d}{dx}(25)$$

$$2x + 2y\frac{dy}{dx} = 0 \qquad \text{See Example 1.}$$

$$\frac{dy}{dx} = -\frac{x}{y}.$$

The slope at $(3, -4)$ is $\left.-\frac{x}{y}\right|_{(3,-4)} = -\frac{3}{-4} = \frac{3}{4}.$

Notice that unlike the slope formula for $dy_2/dx$, which applies only to points below the $x$-axis, the formula $dy/dx = -x/y$ applies everywhere the circle has a slope; that is, at all circle points $(x, y)$ where $y \neq 0$. Notice also that the derivative involves *both* variables $x$ and $y$, not just the independent variable $x$. $\qquad \blacksquare$

To calculate the derivatives of other implicitly defined functions, we proceed as in Examples 1 and 2: We treat $y$ as a differentiable implicit function of $x$ and apply the usual rules to differentiate both sides of the defining equation.

---

**Implicit Differentiation**

1. Differentiate both sides of the equation with respect to $x$, treating $y$ as a differentiable function of $x$.

2. Collect the terms with $dy/dx$ on one side of the equation and solve for $dy/dx$.

---

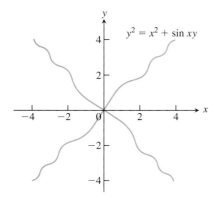

$y^2 = x^2 + \sin xy$

**FIGURE 3.29** The graph of the equation in Example 3.

**EXAMPLE 3**     Find $dy/dx$ if $y^2 = x^2 + \sin xy$ (Figure 3.29).

**Solution**     We differentiate the equation implicitly.

$$y^2 = x^2 + \sin xy$$

$$\frac{d}{dx}\left(y^2\right) = \frac{d}{dx}\left(x^2\right) + \frac{d}{dx}\left(\sin xy\right) \qquad \text{Differentiate both sides with respect to } x \ldots$$

$$2y\frac{dy}{dx} = 2x + (\cos xy)\frac{d}{dx}\left(xy\right) \qquad \ldots \text{treating } y \text{ as a function of } x \text{ and using the Chain Rule.}$$

$$2y\frac{dy}{dx} = 2x + (\cos xy)\left(y + x\frac{dy}{dx}\right) \qquad \text{Treat } xy \text{ as a product.}$$

$$2y\frac{dy}{dx} - (\cos xy)\left(x\frac{dy}{dx}\right) = 2x + (\cos xy)y \qquad \text{Collect terms with } dy/dx.$$

$$(2y - x\cos xy)\frac{dy}{dx} = 2x + y\cos xy$$

$$\frac{dy}{dx} = \frac{2x + y\cos xy}{2y - x\cos xy} \qquad \text{Solve for } dy/dx.$$

Notice that the formula for $dy/dx$ applies everywhere that the implicitly defined curve has a slope. Notice again that the derivative involves *both* variables $x$ and $y$, not just the independent variable $x$. ∎

## Derivatives of Higher Order

Implicit differentiation can also be used to find higher derivatives.

**EXAMPLE 4**     Find $d^2y/dx^2$ if $2x^3 - 3y^2 = 8$.

**Solution**     To start, we differentiate both sides of the equation with respect to $x$ in order to find $y' = dy/dx$.

$$\frac{d}{dx}\left(2x^3 - 3y^2\right) = \frac{d}{dx}(8)$$

$$6x^2 - 6yy' = 0 \qquad \text{Treat } y \text{ as a function of } x.$$

$$y' = \frac{x^2}{y}, \qquad \text{when } y \neq 0 \qquad \text{Solve for } y'.$$

We now apply the Quotient Rule to find $y''$.

$$y'' = \frac{d}{dx}\left(\frac{x^2}{y}\right) = \frac{2xy - x^2y'}{y^2} = \frac{2x}{y} - \frac{x^2}{y^2}\cdot y'$$

Finally, we substitute $y' = x^2/y$ to express $y''$ in terms of $x$ and $y$.

$$y'' = \frac{2x}{y} - \frac{x^2}{y^2}\left(\frac{x^2}{y}\right) = \frac{2x}{y} - \frac{x^4}{y^3}, \qquad \text{when } y \neq 0 \qquad ∎$$

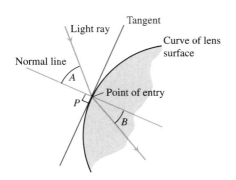

**FIGURE 3.30** The profile of a lens, showing the bending (refraction) of a ray of light as it passes through the lens surface.

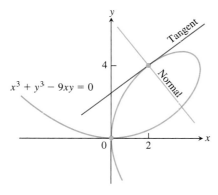

**FIGURE 3.31** Example 5 shows how to find equations for the tangent and normal to the folium of Descartes at (2, 4).

## Lenses, Tangents, and Normal Lines

In the law that describes how light changes direction as it enters a lens, the important angles are the angles the light makes with the line perpendicular to the surface of the lens at the point of entry (angles $A$ and $B$ in Figure 3.30). This line is called the *normal* to the surface at the point of entry. In a profile view of a lens like the one in Figure 3.30, the **normal** is the line perpendicular (also said to be *orthogonal*) to the tangent of the profile curve at the point of entry.

**EXAMPLE 5**    Show that the point $(2, 4)$ lies on the curve $x^3 + y^3 - 9xy = 0$. Then find the tangent and normal to the curve there (Figure 3.31).

**Solution**    The point $(2, 4)$ lies on the curve because its coordinates satisfy the equation given for the curve: $2^3 + 4^3 - 9(2)(4) = 8 + 64 - 72 = 0$.

To find the slope of the curve at $(2, 4)$, we first use implicit differentiation to find a formula for $dy/dx$:

$$x^3 + y^3 - 9xy = 0$$

$$\frac{d}{dx}(x^3) + \frac{d}{dx}(y^3) - \frac{d}{dx}(9xy) = \frac{d}{dx}(0)$$

Differentiate both sides with respect to $x$.

$$3x^2 + 3y^2\frac{dy}{dx} - 9\left(x\frac{dy}{dx} + y\frac{dx}{dx}\right) = 0$$

Treat $xy$ as a product and $y$ as a function of $x$.

$$(3y^2 - 9x)\frac{dy}{dx} + 3x^2 - 9y = 0$$

$$3(y^2 - 3x)\frac{dy}{dx} = 9y - 3x^2$$

$$\frac{dy}{dx} = \frac{3y - x^2}{y^2 - 3x}.$$

Solve for $dy/dx$.

We then evaluate the derivative at $(x, y) = (2, 4)$:

$$\left.\frac{dy}{dx}\right|_{(2,4)} = \left.\frac{3y - x^2}{y^2 - 3x}\right|_{(2,4)} = \frac{3(4) - 2^2}{4^2 - 3(2)} = \frac{8}{10} = \frac{4}{5}.$$

The tangent at $(2, 4)$ is the line through $(2, 4)$ with slope $4/5$:

$$y = 4 + \frac{4}{5}(x - 2)$$

$$y = \frac{4}{5}x + \frac{12}{5}.$$

The normal to the curve at $(2, 4)$ is the line perpendicular to the tangent there, the line through $(2, 4)$ with slope $-5/4$:

$$y = 4 - \frac{5}{4}(x - 2)$$

$$y = -\frac{5}{4}x + \frac{13}{2}.$$    ∎

## Exercises 3.7

### Differentiating Implicitly

Use implicit differentiation to find $dy/dx$ in Exercises 1–14.

**1.** $x^2y + xy^2 = 6$  
**2.** $x^3 + y^3 = 18xy$

**3.** $2xy + y^2 = x + y$  
**4.** $x^3 - xy + y^3 = 1$

**5.** $x^2(x - y)^2 = x^2 - y^2$  
**6.** $(3xy + 7)^2 = 6y$

**7.** $y^2 = \dfrac{x - 1}{x + 1}$  
**8.** $x^3 = \dfrac{2x - y}{x + 3y}$

**9.** $x = \sec y$  
**10.** $xy = \cot(xy)$

**11.** $x + \tan(xy) = 0$  
**12.** $x^4 + \sin y = x^3y^2$

**13.** $y \sin\left(\dfrac{1}{y}\right) = 1 - xy$  
**14.** $x \cos(2x + 3y) = y \sin x$

Find $dr/d\theta$ in Exercises 15–18.

**15.** $\theta^{1/2} + r^{1/2} = 1$  
**16.** $r - 2\sqrt{\theta} = \dfrac{3}{2}\theta^{2/3} + \dfrac{4}{3}\theta^{3/4}$

**17.** $\sin(r\theta) = \dfrac{1}{2}$  
**18.** $\cos r + \cot \theta = r\theta$

### Second Derivatives

In Exercises 19–24, use implicit differentiation to find $dy/dx$ and then $d^2y/dx^2$.

**19.** $x^2 + y^2 = 1$  
**20.** $x^{2/3} + y^{2/3} = 1$

**21.** $y^2 = x^2 + 2x$  
**22.** $y^2 - 2x = 1 - 2y$

**23.** $2\sqrt{y} = x - y$  
**24.** $xy + y^2 = 1$

**25.** If $x^3 + y^3 = 16$, find the value of $d^2y/dx^2$ at the point $(2, 2)$.

**26.** If $xy + y^2 = 1$, find the value of $d^2y/dx^2$ at the point $(0, -1)$.

In Exercises 27 and 28, find the slope of the curve at the given points.

**27.** $y^2 + x^2 = y^4 - 2x$ at $(-2, 1)$ and $(-2, -1)$

**28.** $(x^2 + y^2)^2 = (x - y)^2$ at $(1, 0)$ and $(1, -1)$

### Slopes, Tangents, and Normals

In Exercises 29–38, verify that the given point is on the curve and find the lines that are **(a)** tangent and **(b)** normal to the curve at the given point.

**29.** $x^2 + xy - y^2 = 1$,  $(2, 3)$

**30.** $x^2 + y^2 = 25$,  $(3, -4)$

**31.** $x^2y^2 = 9$,  $(-1, 3)$

**32.** $y^2 - 2x - 4y - 1 = 0$,  $(-2, 1)$

**33.** $6x^2 + 3xy + 2y^2 + 17y - 6 = 0$,  $(-1, 0)$

**34.** $x^2 - \sqrt{3}xy + 2y^2 = 5$,  $\left(\sqrt{3}, 2\right)$

**35.** $2xy + \pi \sin y = 2\pi$,  $(1, \pi/2)$

**36.** $x \sin 2y = y \cos 2x$,  $(\pi/4, \pi/2)$

**37.** $y = 2 \sin(\pi x - y)$,  $(1, 0)$

**38.** $x^2 \cos^2 y - \sin y = 0$,  $(0, \pi)$

**39. Parallel tangents** Find the two points where the curve $x^2 + xy + y^2 = 7$ crosses the $x$-axis, and show that the tangents to the curve at these points are parallel. What is the common slope of these tangents?

**40. Normals parallel to a line** Find the normals to the curve $xy + 2x - y = 0$ that are parallel to the line $2x + y = 0$.

**41. The eight curve** Find the slopes of the curve $y^4 = y^2 - x^2$ at the two points shown here.

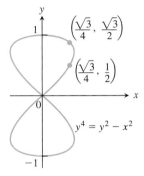

**42. The cissoid of Diocles (from about 200 B.C.)** Find equations for the tangent and normal to the cissoid of Diocles $y^2(2 - x) = x^3$ at $(1, 1)$.

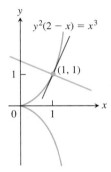

**43. The devil's curve (Gabriel Cramer, 1750)** Find the slopes of the devil's curve $y^4 - 4y^2 = x^4 - 9x^2$ at the four indicated points.

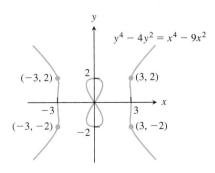

**44. The folium of Descartes**   (See Figure 3.26.)

   **a.** Find the slope of the folium of Descartes $x^3 + y^3 - 9xy = 0$ at the points $(4, 2)$ and $(2, 4)$.

   **b.** At what point other than the origin does the folium have a horizontal tangent?

   **c.** Find the coordinates of the point $A$ in Figure 3.26 where the folium has a vertical tangent.

### Theory and Examples

**45. Intersecting normal**   The line that is normal to the curve $x^2 + 2xy - 3y^2 = 0$ at $(1, 1)$ intersects the curve at what other point?

**46. Power rule for rational exponents**   Let $p$ and $q$ be integers with $q > 0$. If $y = x^{p/q}$, differentiate the equivalent equation $y^q = x^p$ implicitly and show that, for $y \neq 0$,

$$\frac{d}{dx} x^{p/q} = \frac{p}{q} x^{(p/q)-1}.$$

**47. Normals to a parabola**   Show that if it is possible to draw three normals from the point $(a, 0)$ to the parabola $x = y^2$ shown in the accompanying diagram, then $a$ must be greater than $1/2$. One of the normals is the $x$-axis. For what value of $a$ are the other two normals perpendicular?

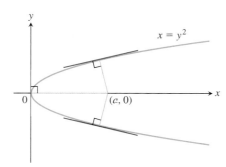

**48.** Is there anything special about the tangents to the curves $y^2 = x^3$ and $2x^2 + 3y^2 = 5$ at the points $(1, \pm 1)$? Give reasons for your answer.

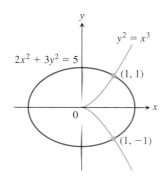

**49.** Verify that the following pairs of curves meet orthogonally.

   **a.** $x^2 + y^2 = 4, \quad x^2 = 3y^2$

   **b.** $x = 1 - y^2, \quad x = \frac{1}{3}y^2$

**50.** The graph of $y^2 = x^3$ is called a **semicubical parabola** and is shown in the accompanying figure. Determine the constant $b$ so that the line $y = -\frac{1}{3}x + b$ meets this graph orthogonally.

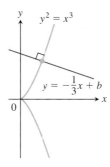

[T] In Exercises 51 and 52, find both $dy/dx$ (treating $y$ as a differentiable function of $x$) and $dx/dy$ (treating $x$ as a differentiable function of $y$). How do $dy/dx$ and $dx/dy$ seem to be related? Explain the relationship geometrically in terms of the graphs.

**51.** $xy^3 + x^2y = 6$       **52.** $x^3 + y^2 = \sin^2 y$

### COMPUTER EXPLORATIONS

Use a CAS to perform the following steps in Exercises 53–60.

   **a.** Plot the equation with the implicit plotter of a CAS. Check to see that the given point $P$ satisfies the equation.

   **b.** Using implicit differentiation, find a formula for the derivative $dy/dx$ and evaluate it at the given point $P$.

   **c.** Use the slope found in part (b) to find an equation for the tangent line to the curve at $P$. Then plot the implicit curve and tangent line together on a single graph.

**53.** $x^3 - xy + y^3 = 7, \quad P(2, 1)$

**54.** $x^5 + y^3x + yx^2 + y^4 = 4, \quad P(1, 1)$

**55.** $y^2 + y = \dfrac{2 + x}{1 - x}, \quad P(0, 1)$

**56.** $y^3 + \cos xy = x^2, \quad P(1, 0)$

**57.** $x + \tan\left(\dfrac{y}{x}\right) = 2, \quad P\left(1, \dfrac{\pi}{4}\right)$

**58.** $xy^3 + \tan(x + y) = 1, \quad P\left(\dfrac{\pi}{4}, 0\right)$

**59.** $2y^2 + (xy)^{1/3} = x^2 + 2, \quad P(1, 1)$

**60.** $x\sqrt{1 + 2y} + y = x^2, \quad P(1, 0)$

# 3.8 Related Rates

In this section we look at problems that ask for the rate at which some variable changes when it is known how the rate of some other related variable (or perhaps several variables) changes. The problem of finding a rate of change from other known rates of change is called a *related rates problem*.

## Related Rates Equations

Suppose we are pumping air into a spherical balloon. Both the volume and radius of the balloon are increasing over time. If $V$ is the volume and $r$ is the radius of the balloon at an instant of time, then

$$V = \frac{4}{3}\pi r^3.$$

Using the Chain Rule, we differentiate both sides with respect to $t$ to find an equation relating the rates of change of $V$ and $r$,

$$\frac{dV}{dt} = \frac{dV}{dr}\frac{dr}{dt} = 4\pi r^2 \frac{dr}{dt}.$$

So if we know the radius $r$ of the balloon and the rate $dV/dt$ at which the volume is increasing at a given instant of time, then we can solve this last equation for $dr/dt$ to find how fast the radius is increasing at that instant. Note that it is easier to directly measure the rate of increase of the volume (the rate at which air is being pumped into the balloon) than it is to measure the increase in the radius. The related rates equation allows us to calculate $dr/dt$ from $dV/dt$.

Very often the key to relating the variables in a related rates problem is drawing a picture that shows the geometric relations between them, as illustrated in the following example.

**EXAMPLE 1**    Water runs into a conical tank at the rate of 0.25 m³/min. The tank stands point down and has a height of 3 m and a base radius of 1.5 m. How fast is the water level rising when the water is 1.8 m deep?

**Solution**    Figure 3.32 shows a partially filled conical tank. The variables in the problem are

$$V = \text{volume (m}^3\text{) of the water in the tank at time } t \text{ (min)}$$
$$x = \text{radius (m) of the surface of the water at time } t$$
$$y = \text{depth (m) of the water in the tank at time } t.$$

We assume that $V$, $x$, and $y$ are differentiable functions of $t$. The constants are the dimensions of the tank. We are asked for $dy/dt$ when

$$y = 1.8 \text{ m} \qquad \text{and} \qquad \frac{dV}{dt} = 0.25 \text{ m}^3/\text{min}.$$

The water forms a cone with volume

$$V = \frac{1}{3}\pi x^2 y.$$

This equation involves $x$ as well as $V$ and $y$. Because no information is given about $x$ and $dx/dt$ at the time in question, we need to eliminate $x$. The similar triangles in Figure 3.32 give us a way to express $x$ in terms of $y$:

$$\frac{x}{y} = \frac{1.5}{3} \qquad \text{or} \qquad x = \frac{y}{2}.$$

Therefore, we find

$$V = \frac{1}{3}\pi \left(\frac{y}{2}\right)^2 y = \frac{\pi}{12}y^3$$

to give the derivative

$$\frac{dV}{dt} = \frac{\pi}{12} \cdot 3y^2 \frac{dy}{dt} = \frac{\pi}{4}y^2 \frac{dy}{dt}.$$

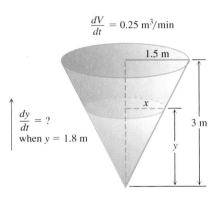

$\dfrac{dV}{dt} = 0.25 \text{ m}^3/\text{min}$

1.5 m

$x$

$\dfrac{dy}{dt} = ?$
when $y = 1.8$ m

3 m

$y$

**FIGURE 3.32**    The geometry of the conical tank and the rate at which water fills the tank determine how fast the water level rises (Example 1).

Finally, use $y = 1.8$ and $dV/dt = 0.25$ to solve for $dy/dt$.

$$0.25 = \frac{\pi}{4}(1.8)^2 \frac{dy}{dt}$$

$$\frac{dy}{dt} = \frac{1}{3.24\pi} \approx 0.098$$

At the moment in question, the water level is rising at about 0.098 m/min. ∎

---

### Related Rates Problem Strategy

1. *Draw a picture and name the variables and constants*. Use $t$ for time. Assume that all variables are differentiable functions of $t$.

2. *Write down the numerical information* (in terms of the symbols you have chosen).

3. *Write down what you are asked to find* (usually a rate, expressed as a derivative).

4. *Write an equation that relates the variables*. You may have to combine two or more equations to get a single equation that relates the variable whose rate you want to the variables whose rates you know.

5. *Differentiate with respect to t*. Then express the rate you want in terms of the rates and variables whose values you know.

6. *Evaluate*. Use known values to find the unknown rate.

---

$\frac{d\theta}{dt} = 0\ 14$ rad/min when $\theta = \pi/4$

$\frac{dy}{dt} = ?$ when $\theta = \pi/4$

Range finder

150 m

**FIGURE 3.33** The rate of change of the balloon's height is related to the rate of change of the angle the range finder makes with the ground (Example 2).

**EXAMPLE 2** A hot air balloon rising straight up from a level field is tracked by a range finder 150 m from the liftoff point. At the moment the range finder's elevation angle is $\pi/4$, the angle is increasing at the rate of 0.14 rad/min. How fast is the balloon rising at that moment?

**Solution** We answer the question in the six strategy steps.

1. *Draw a picture and name the variables and constants* (Figure 3.33). The variables in the picture are

   $\theta =$ the angle in radians the range finder makes with the ground.

   $y =$ the height in meters of the balloon above the ground.

   We let $t$ represent time in minutes and assume that $\theta$ and $y$ are differentiable functions of $t$.

   The one constant in the picture is the distance from the range finder to the liftoff point (150 m). There is no need to give it a special symbol.

2. *Write down the additional numerical information*.

   $$\frac{d\theta}{dt} = 0.14 \text{ rad/min} \qquad \text{when} \qquad \theta = \frac{\pi}{4}$$

3. *Write down what we are to find*. We want $dy/dt$ when $\theta = \pi/4$.

4. *Write an equation that relates the variables $y$ and $\theta$*.

   $$\frac{y}{150} = \tan \theta \qquad \text{or} \qquad y = 150 \tan \theta$$

5. *Differentiate with respect to t using the Chain Rule*. The result tells how $dy/dt$ (which we want) is related to $d\theta/dt$ (which we know).

   $$\frac{dy}{dt} = 150 \, (\sec^2 \theta) \frac{d\theta}{dt}$$

6. *Evaluate with $\theta = \pi/4$ and $d\theta/dt = 0.14$ to find $dy/dt$*.

   $$\frac{dy}{dt} = 150 \left( \sqrt{2} \right)^2 (0.14) = 42 \qquad \sec \frac{\pi}{4} = \sqrt{2}$$

At the moment in question, the balloon is rising at the rate of 42 m/min. ∎

**FIGURE 3.34** The speed of the car is related to the speed of the police cruiser and the rate of change of the distance $s$ between them (Example 3).

**EXAMPLE 3**     A police cruiser, approaching a right-angled intersection from the north, is chasing a speeding car that has turned the corner and is now moving straight east. When the cruiser is 0.6 km north of the intersection and the car is 0.8 km to the east, the police determine with radar that the distance between them and the car is increasing at 30 km/h. If the cruiser is moving at 100 km/h at the instant of measurement, what is the speed of the car?

**Solution**     We picture the car and cruiser in the coordinate plane, using the positive $x$-axis as the eastbound highway and the positive $y$-axis as the southbound highway (Figure 3.34). We let $t$ represent time and set

$$x = \text{position of car at time } t$$
$$y = \text{position of cruiser at time } t$$
$$s = \text{distance between car and cruiser at time } t.$$

We assume that $x$, $y$, and $s$ are differentiable functions of $t$.
    We want to find $dx/dt$ when

$$x = 0.8 \text{ km}, \qquad y = 0.6 \text{ km}, \qquad \frac{dy}{dt} = -100 \text{ km/h}, \qquad \frac{ds}{dt} = 30 \text{ km/h}.$$

Note that $dy/dt$ is negative because $y$ is decreasing.
    We differentiate the distance equation between the car and the cruiser,

$$s^2 = x^2 + y^2$$

(we could also use $s = \sqrt{x^2 + y^2}$), and obtain

$$2s\frac{ds}{dt} = 2x\frac{dx}{dt} + 2y\frac{dy}{dt}$$

$$\frac{ds}{dt} = \frac{1}{s}\left(x\frac{dx}{dt} + y\frac{dy}{dt}\right)$$

$$= \frac{1}{\sqrt{x^2 + y^2}}\left(x\frac{dx}{dt} + y\frac{dy}{dt}\right).$$

Finally, we use $x = 0.8$, $y = 0.6$, $dy/dt = -100$, $ds/dt = 30$, and solve for $dx/dt$.

$$30 = \frac{1}{\sqrt{(0.8)^2 + (0.6)^2}}\left(0.8\frac{dx}{dt} + (0.6)(-100)\right)$$

$$\frac{dx}{dt} = \frac{30\sqrt{(0.8)^2 + (0.6)^2} + (0.6)(100)}{0.8} = 112.5$$

At the moment in question, the car's speed is 112.5 km/h.

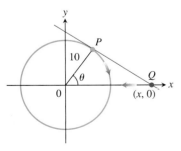

**FIGURE 3.35** The particle $P$ travels clockwise along the circle (Example 4).

**EXAMPLE 4**     A particle $P$ moves clockwise at a constant rate along a circle of radius 10 m centered at the origin. The particle's initial position is (0, 10) on the $y$-axis, and its final destination is the point (10, 0) on the $x$-axis. Once the particle is in motion, the tangent line at $P$ intersects the $x$-axis at a point $Q$ (which moves over time). If it takes the particle 30 s to travel from start to finish, how fast is the point $Q$ moving along the $x$-axis when it is 20 m from the center of the circle?

**Solution**     We picture the situation in the coordinate plane with the circle centered at the origin (see Figure 3.35). We let $t$ represent time and let $\theta$ denote the angle from the $x$-axis to the radial line joining the origin to $P$. Since the particle travels from start to finish in 30 s, it is traveling along the circle at a constant rate of $\pi/2$ radians in $1/2$ min, or $\pi$ rad/min. In other words, $d\theta/dt = -\pi$, with $t$ being measured in minutes. The negative sign appears because $\theta$ is decreasing over time.

Setting $x(t)$ to be the distance at time $t$ from the point $Q$ to the origin, we want to find $dx/dt$ when

$$x = 20 \text{ m} \qquad \text{and} \qquad \frac{d\theta}{dt} = -\pi \text{ rad/min.}$$

To relate the variables $x$ and $\theta$, we see from Figure 3.35 that $x \cos \theta = 10$, or $x = 10 \sec \theta$. Differentiation of this last equation gives

$$\frac{dx}{dt} = 10 \sec \theta \tan \theta \frac{d\theta}{dt} = -10\pi \sec \theta \tan \theta.$$

Note that $dx/dt$ is negative because $x$ is decreasing ($Q$ is moving toward the origin).

When $x = 20$, $\cos \theta = 1/2$ and $\sec \theta = 2$. Also, $\tan \theta = \sqrt{\sec^2 \theta - 1} = \sqrt{3}$. It follows that

$$\frac{dx}{dt} = (-10\pi)(2)\left(\sqrt{3}\right) = -20\sqrt{3}\pi.$$

At the moment in question, the point $Q$ is moving toward the origin at the speed of $20\sqrt{3}\pi \approx 109$ m/min. ∎

**EXAMPLE 5**  A jet airliner is flying at a constant altitude of 10,000 m above sea level as it approaches a Pacific island. The aircraft comes within the direct line of sight of a radar station located on the island, and the radar indicates the initial angle between sea level and its line of sight to the aircraft is 30°. How fast (in kilometers per hour) is the aircraft approaching the island when first detected by the radar instrument if it is turning upward (counterclockwise) at the rate of 1/3 deg/s in order to keep the aircraft within its direct line of sight?

**Solution**  The aircraft $A$ and radar station $R$ are pictured in the coordinate plane, using the positive $x$-axis as the horizontal distance at sea level from $R$ to $A$, and the positive $y$-axis as the vertical altitude above sea level. We let $t$ represent time and observe that $y = 10,000$ is a constant. The general situation and line-of-sight angle $\theta$ are depicted in Figure 3.36. We want to find $dx/dt$ when $\theta = \pi/6$ rad and $d\theta/dt = 1/3$ deg/s.

From Figure 3.36, we see that

$$\frac{10,000}{x} = \tan \theta \qquad \text{or} \qquad x = 10,000 \cot \theta.$$

Using kilometers instead of meters for our distance units, the last equation translates to

$$x = \frac{10,000}{1000} \cot \theta.$$

Differentiation with respect to $t$ gives

$$\frac{dx}{dt} = -10 \csc^2 \theta \frac{d\theta}{dt}.$$

When $\theta = \pi/6$, $\sin^2 \theta = 1/4$, so $\csc^2 \theta = 4$. Converting $d\theta/dt = 1/3$ deg/s to radians per hour, we find

$$\frac{d\theta}{dt} = \frac{1}{3}\left(\frac{\pi}{180}\right)(3600) \text{ rad/h.} \qquad \text{1 h = 3600 s, 1 deg = } \pi/180 \text{ rad}$$

Substitution into the equation for $dx/dt$ then gives

$$\frac{dx}{dt} = (-10)(4)\left(\frac{1}{3}\right)\left(\frac{\pi}{180}\right)(3600) \approx -838.$$

The negative sign appears because the distance $x$ is decreasing, so the aircraft is approaching the island at a speed of approximately 838 km/h when first detected by the radar. ∎

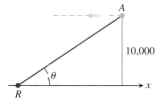

**FIGURE 3.36**  Jet airliner $A$ traveling at constant altitude toward radar station $R$ (Example 5).

(a)

(b)

**FIGURE 3.37** A worker at $M$ walks to the right, pulling the weight $W$ upward as the rope moves through the pulley $P$ (Example 6).

**EXAMPLE 6**    Figure 3.37a shows a rope running through a pulley at $P$ and bearing a weight $W$ at one end. The other end is held 1.5 m above the ground in the hand $M$ of a worker. Suppose the pulley is 7.5 m above ground, the rope is 13.5 m long, and the worker is walking rapidly away from the vertical line $PW$ at the rate of 1.2 m/s. How fast is the weight being raised when the worker's hand is 6.3 m away from $PW$?

**Solution**    We let $OM$ be the horizontal line of length $x$ m from a point $O$ directly below the pulley to the worker's hand $M$ at any instant of time (Figure 3.37). Let $h$ be the height of the weight $W$ above $O$, and let $z$ denote the length of rope from the pulley $P$ to the worker's hand. We want to know $dh/dt$ when $x = 6.3$ given that $dx/dt = 1.2$. Note that the height of $P$ above $O$ is 6 m because $O$ is 1.5 m above the ground. We assume the angle at $O$ is a right angle.

At any instant of time $t$ we have the following relationships (see Figure 3.37b):

$$6 - h + z = 13.5 \qquad \text{Total length of rope is 13.5 m.}$$
$$6^2 + x^2 = z^2. \qquad \text{Angle at } O \text{ is a right angle.}$$

If we solve for $z = 7.5 + h$ in the first equation, and substitute into the second equation, we have

$$6^2 + x^2 = (7.5 + h)^2. \tag{1}$$

Differentiating both sides with respect to $t$ gives

$$2x\frac{dx}{dt} = 2(7.5 + h)\frac{dh}{dt},$$

and solving this last equation for $dh/dt$ we find

$$\frac{dh}{dt} = \frac{x}{7.5 + h}\frac{dx}{dt}. \tag{2}$$

Since we know $dx/dt$, it remains only to find $7.5 + h$ at the instant when $x = 6.3$. From Equation (1),

$$6^2 + 6.3^2 = (7.5 + h)^2$$

so that

$$(7.5 + h)^2 = 75.69, \qquad \text{or} \qquad 7.5 + h = 8.7.$$

Equation (2) now gives

$$\frac{dh}{dt} = \frac{6.3}{8.7} \cdot 1.2 = \frac{756}{870} \approx 0.87 \text{ m/s}$$

as the rate at which the weight is being raised when $x = 6.3$ m.    ∎

## Exercises **3.8**

1. **Area**    Suppose that the radius $r$ and area $A = \pi r^2$ of a circle are differentiable functions of $t$. Write an equation that relates $dA/dt$ to $dr/dt$.

2. **Surface area**    Suppose that the radius $r$ and surface area $S = 4\pi r^2$ of a sphere are differentiable functions of $t$. Write an equation that relates $dS/dt$ to $dr/dt$.

3. Assume that $y = 5x$ and $dx/dt = 2$. Find $dy/dt$.

4. Assume that $2x + 3y = 12$ and $dy/dt = -2$. Find $dx/dt$.

5. If $y = x^2$ and $dx/dt = 3$, then what is $dy/dt$ when $x = -1$?

6. If $x = y^3 - y$ and $dy/dt = 5$, then what is $dx/dt$ when $y = 2$?

7. If $x^2 + y^2 = 25$ and $dx/dt = -2$, then what is $dy/dt$ when $x = 3$ and $y = -4$?

8. If $x^2y^3 = 4/27$ and $dy/dt = 1/2$, then what is $dx/dt$ when $x = 2$?

9. If $L = \sqrt{x^2 + y^2}$, $dx/dt = -1$, and $dy/dt = 3$, find $dL/dt$ when $x = 5$ and $y = 12$.

10. If $r + s^2 + v^3 = 12$, $dr/dt = 4$, and $ds/dt = -3$, find $dv/dt$ when $r = 3$ and $s = 1$.

11. If the original 24 m edge length $x$ of a cube decreases at the rate of 5 m/min, when $x = 3$ m at what rate does the cube's

    a. surface area change?

    b. volume change?

12. A cube's surface area increases at the rate of 72 cm²/s. At what rate is the cube's volume changing when the edge length is $x = 3$ cm?

13. **Volume** The radius $r$ and height $h$ of a right circular cylinder are related to the cylinder's volume $V$ by the formula $V = \pi r^2 h$.

    a. How is $dV/dt$ related to $dh/dt$ if $r$ is constant?

    b. How is $dV/dt$ related to $dr/dt$ if $h$ is constant?

    c. How is $dV/dt$ related to $dr/dt$ and $dh/dt$ if neither $r$ nor $h$ is constant?

14. **Volume** The radius $r$ and height $h$ of a right circular cone are related to the cone's volume $V$ by the equation $V = (1/3)\pi r^2 h$.

    a. How is $dV/dt$ related to $dh/dt$ if $r$ is constant?

    b. How is $dV/dt$ related to $dr/dt$ if $h$ is constant?

    c. How is $dV/dt$ related to $dr/dt$ and $dh/dt$ if neither $r$ nor $h$ is constant?

15. **Changing voltage** The voltage $V$ (volts), current $I$ (amperes), and resistance $R$ (ohms) of an electric circuit like the one shown here are related by the equation $V = IR$. Suppose that $V$ is increasing at the rate of 1 volt/s while $I$ is decreasing at the rate of 1/3 amp/s. Let $t$ denote time in seconds.

    a. What is the value of $dV/dt$?

    b. What is the value of $dI/dt$?

    c. What equation relates $dR/dt$ to $dV/dt$ and $dI/dt$?

    d. Find the rate at which $R$ is changing when $V = 12$ volts and $I = 2$ amps. Is $R$ increasing, or decreasing?

16. **Electrical power** The power $P$ (watts) of an electric circuit is related to the circuit's resistance $R$ (ohms) and current $I$ (amperes) by the equation $P = RI^2$.

    a. How are $dP/dt$, $dR/dt$, and $dI/dt$ related if none of $P$, $R$, and $I$ are constant?

    b. How is $dR/dt$ related to $dI/dt$ if $P$ is constant?

17. **Distance** Let $x$ and $y$ be differentiable functions of $t$ and let $s = \sqrt{x^2 + y^2}$ be the distance between the points $(x, 0)$ and $(0, y)$ in the $xy$-plane.

    a. How is $ds/dt$ related to $dx/dt$ if $y$ is constant?

    b. How is $ds/dt$ related to $dx/dt$ and $dy/dt$ if neither $x$ nor $y$ is constant?

    c. How is $dx/dt$ related to $dy/dt$ if $s$ is constant?

18. **Diagonals** If $x$, $y$, and $z$ are lengths of the edges of a rectangular box, the common length of the box's diagonals is $s = \sqrt{x^2 + y^2 + z^2}$.

    a. Assuming that $x$, $y$, and $z$ are differentiable functions of $t$, how is $ds/dt$ related to $dx/dt$, $dy/dt$, and $dz/dt$?

    b. How is $ds/dt$ related to $dy/dt$ and $dz/dt$ if $x$ is constant?

    c. How are $dx/dt$, $dy/dt$, and $dz/dt$ related if $s$ is constant?

19. **Area** The area $A$ of a triangle with sides of lengths $a$ and $b$ enclosing an angle of measure $\theta$ is

$$A = \frac{1}{2}ab \sin \theta.$$

    a. How is $dA/dt$ related to $d\theta/dt$ if $a$ and $b$ are constant?

    b. How is $dA/dt$ related to $d\theta/dt$ and $da/dt$ if only $b$ is constant?

    c. How is $dA/dt$ related to $d\theta/dt$, $da/dt$, and $db/dt$ if none of $a$, $b$, and $\theta$ are constant?

20. **Heating a plate** When a circular plate of metal is heated in an oven, its radius increases at the rate of 0.01 cm/min. At what rate is the plate's area increasing when the radius is 50 cm?

21. **Changing dimensions in a rectangle** The length $l$ of a rectangle is decreasing at the rate of 2 cm/s while the width $w$ is increasing at the rate of 2 cm/s. When $l = 12$ cm and $w = 5$ cm, find the rates of change of **(a)** the area, **(b)** the perimeter, and **(c)** the lengths of the diagonals of the rectangle. Which of these quantities are decreasing, and which are increasing?

22. **Changing dimensions in a rectangular box** Suppose that the edge lengths $x$, $y$, and $z$ of a closed rectangular box are changing at the following rates:

$$\frac{dx}{dt} = 1 \text{ m/s}, \quad \frac{dy}{dt} = -2 \text{ m/s}, \quad \frac{dz}{dt} = 1 \text{ m/s}.$$

Find the rates at which the box's **(a)** volume, **(b)** surface area, and **(c)** diagonal length $s = \sqrt{x^2 + y^2 + z^2}$ are changing at the instant when $x = 4$, $y = 3$, and $z = 2$.

23. **A sliding ladder** A 3.9-m ladder is leaning against a house when its base starts to slide away (see accompanying figure). By the time the base is 3.6 m from the house, the base is moving at the rate of 1.5 m/s.

    a. How fast is the top of the ladder sliding down the wall then?

    b. At what rate is the area of the triangle formed by the ladder, wall, and ground changing then?

    c. At what rate is the angle $\theta$ between the ladder and the ground changing then?

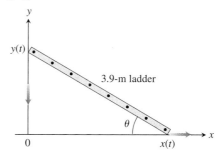

24. **Commercial air traffic** Two commercial airplanes are flying at an altitude of 12,000 m along straight-line courses that intersect at right angles. Plane $A$ is approaching the intersection point at a speed of 442 knots (nautical miles per hour; a nautical mile is 1852 m). Plane $B$ is approaching the intersection at 481 knots. At what rate is the distance between the planes changing when $A$ is 5

nautical miles from the intersection point and $B$ is 12 nautical miles from the intersection point?

**25. Flying a kite**    A girl flies a kite at a height of 90 m, the wind carrying the kite horizontally away from her at a rate of 7.5 m/s. How fast must she let out the string when the kite is 150 m away from her?

**26. Boring a cylinder**    The mechanics at Lincoln Automotive are reboring a 15-cm-deep cylinder to fit a new piston. The machine they are using increases the cylinder's radius one-thousandth of a centimeter every 3 min. How rapidly is the cylinder volume increasing when the bore (diameter) is 10 cm?

**27. A growing sand pile**    Sand falls from a conveyor belt at the rate of 10 m³/min onto the top of a conical pile. The height of the pile is always three-eighths of the base diameter. How fast are the **(a)** height and **(b)** radius changing when the pile is 4 m high? Answer in centimeters per minute.

**28. A draining conical reservoir**    Water is flowing at the rate of 50 m³/min from a shallow concrete conical reservoir (vertex down) of base radius 45 m and height 6 m.

    **a.** How fast (centimeters per minute) is the water level falling when the water is 5 m deep?

    **b.** How fast is the radius of the water's surface changing then? Answer in centimeters per minute.

**29. A draining hemispherical reservoir**    Water is flowing at the rate of 6 m³/min from a reservoir shaped like a hemispherical bowl of radius 13 m, shown here in profile. Answer the following questions, given that the volume of water in a hemispherical bowl of radius $R$ is $V = (\pi/3)y^2(3R - y)$ when the water is $y$ meters deep.

Center of sphere

    **a.** At what rate is the water level changing when the water is 8 m deep?

    **b.** What is the radius $r$ of the water's surface when the water is $y$ m deep?

    **c.** At what rate is the radius $r$ changing when the water is 8 m deep?

**30. A growing raindrop**    Suppose that a drop of mist is a perfect sphere and that, through condensation, the drop picks up moisture at a rate proportional to its surface area. Show that under these circumstances the drop's radius increases at a constant rate.

**31. The radius of an inflating balloon**    A spherical balloon is inflated with helium at the rate of $100\pi$ m³/min. How fast is the balloon's radius increasing at the instant the radius is 5 m? How fast is the surface area increasing?

**32. Hauling in a dinghy**    A dinghy is pulled toward a dock by a rope from the bow through a ring on the dock 2 m above the bow. The rope is hauled in at the rate of 0.5 m/s.

    **a.** How fast is the boat approaching the dock when 3 m of rope are out?

    **b.** At what rate is the angle $\theta$ changing at this instant (see the figure)?

Ring at edge of dock

**33. A balloon and a bicycle**    A balloon is rising vertically above a level, straight road at a constant rate of 0.3 m/s. Just when the balloon is 20 m above the ground, a bicycle moving at a constant rate of 5 m/s passes under it. How fast is the distance $s(t)$ between the bicycle and balloon increasing 3 s later?

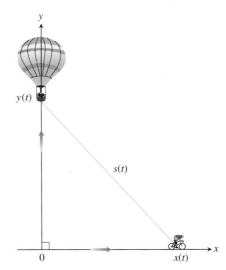

**34. Making coffee**    Coffee is draining from a conical filter into a cylindrical coffeepot at the rate of 160 cm³/min.

    **a.** How fast is the level in the pot rising when the coffee in the cone is 12 cm deep?

    **b.** How fast is the level in the cone falling then?

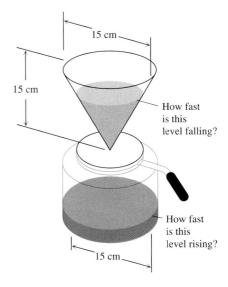

**35. Cardiac output** In the late 1860s, Adolf Fick, a professor of physiology in the Faculty of Medicine in Würzberg, Germany, developed one of the methods we use today for measuring how much blood your heart pumps in a minute. Your cardiac output as you read this sentence is probably about 7 L/min. At rest it is likely to be a bit under 6 L/min. If you are a trained marathon runner running a marathon, your cardiac output can be as high as 30 L/min.

Your cardiac output can be calculated with the formula

$$y = \frac{Q}{D},$$

where $Q$ is the number of milliliters of $CO_2$ you exhale in a minute and $D$ is the difference between the $CO_2$ concentration (mL/L) in the blood pumped to the lungs and the $CO_2$ concentration in the blood returning from the lungs. With $Q = 233$ mL/min and $D = 97 - 56 = 41$ mL/L,

$$y = \frac{233 \text{ mL/min}}{41 \text{ mL/L}} \approx 5.68 \text{ L/min},$$

fairly close to the 6 L/min that most people have at basal (resting) conditions. (Data courtesy of J. Kenneth Herd, M.D., Quillan College of Medicine, East Tennessee State University.)

Suppose that when $Q = 233$ and $D = 41$, we also know that $D$ is decreasing at the rate of 2 units a minute but that $Q$ remains unchanged. What is happening to the cardiac output?

**36. Moving along a parabola** A particle moves along the parabola $y = x^2$ in the first quadrant in such a way that its $x$-coordinate (measured in meters) increases at a steady 10 m/s. How fast is the angle of inclination $\theta$ of the line joining the particle to the origin changing when $x = 3$ m?

**37. Motion in the plane** The coordinates of a particle in the metric $xy$-plane are differentiable functions of time $t$ with $dx/dt = -1$ m/s and $dy/dt = -5$ m/s. How fast is the particle's distance from the origin changing as it passes through the point $(5, 12)$?

**38. Videotaping a moving car** You are videotaping a race from a stand 40 m from the track, following a car that is moving at 288 km/h (80 m/s), as shown in the accompanying figure. How fast will your camera angle $\theta$ be changing when the car is right in front of you? A half second later?

**39. A moving shadow** A light shines from the top of a pole 15 m high. A ball is dropped from the same height from a point 9 m away from the light. (See accompanying figure.) How fast is the shadow of the ball moving along the ground 1/2 s later? (Assume the ball falls a distance $s = 4.9t^2$ m in $t$ seconds.)

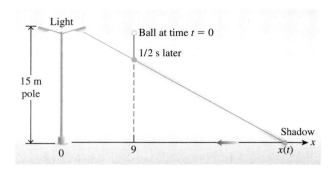

**40. A building's shadow** On a morning of a day when the sun will pass directly overhead, the shadow of an 24 m building on level ground is 18 m long. At the moment in question, the angle $\theta$ the sun makes with the ground is increasing at the rate of $0.27°$/min. At what rate is the shadow decreasing? (Remember to use radians. Express your answer in centimeters per minute, to the nearest tenth.)

**41. A melting ice layer** A spherical iron ball 8 cm in diameter is coated with a layer of ice of uniform thickness. If the ice melts at the rate of 10 cm³/min, how fast is the thickness of the ice decreasing when it is 2 cm thick? How fast is the outer surface area of ice decreasing?

**42. Highway patrol** A highway patrol plane flies 3 km above a level, straight road at a steady 120 km/h. The pilot sees an oncoming car and with radar determines that at the instant the line-of-sight distance from plane to car is 5 km, the line-of-sight distance is decreasing at the rate of 160 km/h. Find the car's speed along the highway.

**43. Baseball players** A baseball diamond is a square 27 m on a side. A player runs from first base to second at a rate of 5 m/s.

    **a.** At what rate is the player's distance from third base changing when the player is 9 m from first base?

    **b.** At what rates are angles $\theta_1$ and $\theta_2$ (see the figure) changing at that time?

**c.** The player slides into second base at the rate of 4.5 m/s. At what rates are angles $\theta_1$ and $\theta_2$ changing as the player touches base?

**44. Ships**  Two ships are steaming straight away from a point $O$ along routes that make a 120° angle. Ship $A$ moves at 14 knots (nautical miles per hour; a nautical mile is 1852 m). Ship $B$ moves at 21 knots. How fast are the ships moving apart when $OA = 5$ and $OB = 3$ nautical miles?

**45. Clock's moving hands**  At what rate is the angle between a clock's minute and hour hands changing at 4 o'clock in the afternoon?

**46. Oil spill**  An explosion at an oil rig located in gulf waters causes an elliptical oil slick to spread on the surface from the rig. The slick is a constant 20 cm thick. After several days, when the major axis of the slick is 2 km long and the minor axis is 3/4 km wide, it is determined that its length is increasing at the rate of 9 m/h, and its width is increasing at the rate of 3 m/h. At what rate (in cubic meters per hour) is oil flowing from the site of the rig at that time?

# 3.9 Linearization and Differentials

Sometimes we can approximate complicated functions with simpler ones that give the accuracy we want for specific applications and are easier to work with. The approximating functions discussed in this section are called *linearizations*, and they are based on tangent lines. Other approximating functions, such as polynomials, are discussed in Chapter 10.

We introduce new variables $dx$ and $dy$, called *differentials*, and define them in a way that makes Leibniz's notation for the derivative $dy/dx$ a true ratio. We use $dy$ to estimate error in measurement, which then provides for a precise proof of the Chain Rule (Section 3.6).

## Linearization

As you can see in Figure 3.38, the tangent to the curve $y = x^2$ lies close to the curve near the point of tangency. For a brief interval to either side, the $y$-values along the tangent line

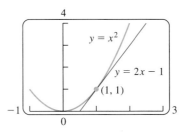

$y = x^2$ and its tangent $y = 2x - 1$ at $(1, 1)$.

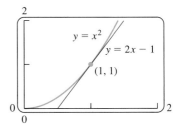

Tangent and curve very close near $(1, 1)$.

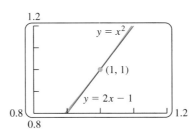

Tangent and curve very close throughout entire $x$-interval shown.

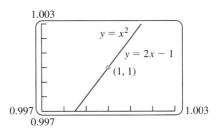

Tangent and curve closer still. Computer screen cannot distinguish tangent from curve on this $x$-interval.

**FIGURE 3.38**  The more we magnify the graph of a function near a point where the function is differentiable, the flatter the graph becomes and the more it resembles its tangent.

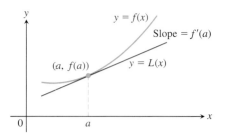

**FIGURE 3.39** The tangent to the curve $y = f(x)$ at $x = a$ is the line $L(x) = f(a) + f'(a)(x - a)$.

give good approximations to the $y$-values on the curve. We observe this phenomenon by zooming in on the two graphs at the point of tangency or by looking at tables of values for the difference between $f(x)$ and its tangent line near the $x$-coordinate of the point of tangency. The phenomenon is true not just for parabolas; every differentiable curve behaves locally like its tangent line.

In general, the tangent to $y = f(x)$ at a point $x = a$, where $f$ is differentiable (Figure 3.39), passes through the point $(a, f(a))$, so its point-slope equation is

$$y = f(a) + f'(a)(x - a).$$

Thus, this tangent line is the graph of the linear function

$$L(x) = f(a) + f'(a)(x - a).$$

For as long as this line remains close to the graph of $f$ as we move off the point of tangency, $L(x)$ gives a good approximation to $f(x)$.

---

**DEFINITIONS** If $f$ is differentiable at $x = a$, then the approximating function

$$L(x) = f(a) + f'(a)(x - a)$$

is the **linearization** of $f$ at $a$. The approximation

$$f(x) \approx L(x)$$

of $f$ by $L$ is the **standard linear approximation** of $f$ at $a$. The point $x = a$ is the **center** of the approximation.

---

**EXAMPLE 1** Find the linearization of $f(x) = \sqrt{1 + x}$ at $x = 0$ (Figure 3.40).

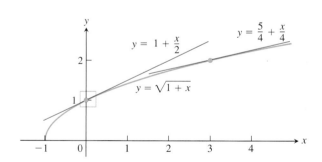

**FIGURE 3.40** The graph of $y = \sqrt{1 + x}$ and its linearizations at $x = 0$ and $x = 3$. Figure 3.41 shows a magnified view of the small window about 1 on the $y$-axis.

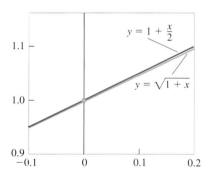

**FIGURE 3.41** Magnified view of the window in Figure 3.40.

**Solution** Since

$$f'(x) = \frac{1}{2}(1 + x)^{-1/2},$$

we have $f(0) = 1$ and $f'(0) = 1/2$, giving the linearization

$$L(x) = f(a) + f'(a)(x - a) = 1 + \frac{1}{2}(x - 0) = 1 + \frac{x}{2}.$$

See Figure 3.41. ∎

The following table shows how accurate the approximation $\sqrt{1 + x} \approx 1 + (x/2)$ from Example 1 is for some values of $x$ near 0. As we move away from zero, we lose accuracy. For example, for $x = 2$, the linearization gives 2 as the approximation for $\sqrt{3}$, which is not even accurate to one decimal place.

| Approximation | True value | \|True value $-$ approximation\| |
|---|---|---|
| $\sqrt{1.2} \approx 1 + \dfrac{0.2}{2} = 1.10$ | 1.095445 | $0.004555 < 10^{-2}$ |
| $\sqrt{1.05} \approx 1 + \dfrac{0.05}{2} = 1.025$ | 1.024695 | $0.000305 < 10^{-3}$ |
| $\sqrt{1.005} \approx 1 + \dfrac{0.005}{2} = 1.00250$ | 1.002497 | $0.000003 < 10^{-5}$ |

Do not be misled by the preceding calculations into thinking that whatever we do with a linearization is better done with a calculator. In practice, we would never use a linearization to find a particular square root. The utility of a linearization is its ability to replace a complicated formula by a simpler one over an entire interval of values. If we have to work with $\sqrt{1 + x}$ for $x$ close to 0 and can tolerate the small amount of error involved, we can work with $1 + (x/2)$ instead. Of course, we then need to know how much error there is. We further examine the estimation of error in Chapter 10.

A linear approximation normally loses accuracy away from its center. As Figure 3.40 suggests, the approximation $\sqrt{1 + x} \approx 1 + (x/2)$ will probably be too crude to be useful near $x = 3$. There, we need the linearization at $x = 3$.

**EXAMPLE 2**    Find the linearization of $f(x) = \sqrt{1 + x}$ at $x = 3$.

**Solution**    We evaluate the equation defining $L(x)$ at $a = 3$. With

$$f(3) = 2, \qquad f'(3) = \frac{1}{2}(1 + x)^{-1/2}\Big|_{x=3} = \frac{1}{4},$$

we have

$$L(x) = 2 + \frac{1}{4}(x - 3) = \frac{5}{4} + \frac{x}{4}. \qquad ∎$$

At $x = 3.2$, the linearization in Example 2 gives

$$\sqrt{1 + x} = \sqrt{1 + 3.2} \approx \frac{5}{4} + \frac{3.2}{4} = 1.250 + 0.800 = 2.050,$$

which differs from the true value $\sqrt{4.2} \approx 2.04939$ by less than one one-thousandth. The linearization in Example 1 gives

$$\sqrt{1 + x} = \sqrt{1 + 3.2} \approx 1 + \frac{3.2}{2} = 1 + 1.6 = 2.6,$$

a result that is off by more than 25%.

**EXAMPLE 3**    Find the linearization of $f(x) = \cos x$ at $x = \pi/2$ (Figure 3.42).

**Solution**    Since $f(\pi/2) = \cos(\pi/2) = 0$, $f'(x) = -\sin x$, and $f'(\pi/2) = -\sin(\pi/2) = -1$, we find the linearization at $a = \pi/2$ to be

$$L(x) = f(a) + f'(a)(x - a)$$

$$= 0 + (-1)\left(x - \frac{\pi}{2}\right)$$

$$= -x + \frac{\pi}{2}. \qquad ∎$$

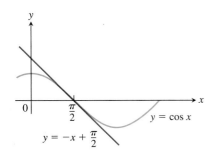

**FIGURE 3.42**    The graph of $f(x) = \cos x$ and its linearization at $x = \pi/2$. Near $x = \pi/2$, $\cos x \approx -x + (\pi/2)$ (Example 3).

An important linear approximation for roots and powers is

$$(1 + x)^k \approx 1 + kx \qquad (x \text{ near } 0; \text{ any number } k)$$

(Exercise 15). This approximation, good for values of $x$ sufficiently close to zero, has broad application. For example, when $x$ is small,

$$\sqrt{1 + x} \approx 1 + \frac{1}{2}x \qquad\qquad\qquad k = 1/2$$

$$\frac{1}{1 - x} = (1 - x)^{-1} \approx 1 + (-1)(-x) = 1 + x \qquad k = -1; \text{ replace } x \text{ by } -x.$$

$$\sqrt[3]{1 + 5x^4} = (1 + 5x^4)^{1/3} \approx 1 + \frac{1}{3}(5x^4) = 1 + \frac{5}{3}x^4 \qquad k = 1/3; \text{ replace } x \text{ by } 5x^4.$$

$$\frac{1}{\sqrt{1 - x^2}} = (1 - x^2)^{-1/2} \approx 1 + \left(-\frac{1}{2}\right)(-x^2) = 1 + \frac{1}{2}x^2 \qquad \begin{array}{l} k = -1/2; \\ \text{replace } x \text{ by } -x^2. \end{array}$$

---

**Approximations Near $x = 0$**

$\sqrt{1 + x} \approx 1 + \dfrac{x}{2}$

$\dfrac{1}{1 - x} \approx 1 + x$

$\dfrac{1}{\sqrt{1 - x^2}} \approx 1 + \dfrac{x^2}{2}$

---

### Differentials

We sometimes use the Leibniz notation $dy/dx$ to represent the derivative of $y$ with respect to $x$. Contrary to its appearance, it is not a ratio. We now introduce two new variables $dx$ and $dy$ with the property that when their ratio exists, it is equal to the derivative.

---

> **DEFINITION** Let $y = f(x)$ be a differentiable function. The **differential $dx$** is an independent variable. The **differential $dy$** is
>
> $$dy = f'(x)\, dx.$$

---

Unlike the independent variable $dx$, the variable $dy$ is always a dependent variable. It depends on both $x$ and $dx$. If $dx$ is given a specific value and $x$ is a particular number in the domain of the function $f$, then these values determine the numerical value of $dy$. Often the variable $dx$ is chosen to be $\Delta x$, the change in $x$.

### EXAMPLE 4
(a) Find $dy$ if $y = x^5 + 37x$.

(b) Find the value of $dy$ when $x = 1$ and $dx = 0.2$.

**Solution**

(a) $dy = (5x^4 + 37)\, dx$

(b) Substituting $x = 1$ and $dx = 0.2$ in the expression for $dy$, we have

$$dy = (5 \cdot 1^4 + 37)0.2 = 8.4. \qquad\qquad\blacksquare$$

The geometric meaning of differentials is shown in Figure 3.43. Let $x = a$ and set $dx = \Delta x$. The corresponding change in $y = f(x)$ is

$$\Delta y = f(a + dx) - f(a).$$

The corresponding change in the tangent line $L$ is

$$\begin{aligned} \Delta L &= L(a + dx) - L(a) \\ &= \underbrace{f(a) + f'(a)[(a + dx) - a]}_{L(a + dx)} - \underbrace{f(a)}_{L(a)} \\ &= f'(a)\, dx. \end{aligned}$$

**FIGURE 3.43** Geometrically, the differential $dy$ is the change $\Delta L$ in the linearization of $f$ when $x = a$ changes by an amount $dx = \Delta x$.

That is, the change in the linearization of $f$ is precisely the value of the differential $dy$ when $x = a$ and $dx = \Delta x$. Therefore, $dy$ represents the amount the tangent line rises or falls when $x$ changes by an amount $dx = \Delta x$.

If $dx \neq 0$, then the quotient of the differential $dy$ by the differential $dx$ is equal to the derivative $f'(x)$ because

$$dy \div dx = \frac{f'(x)\,dx}{dx} = f'(x) = \frac{dy}{dx}.$$

We sometimes write

$$df = f'(x)\,dx$$

in place of $dy = f'(x)\,dx$, calling $df$ the **differential of $f$**. For instance, if $f(x) = 3x^2 - 6$, then

$$df = d(3x^2 - 6) = 6x\,dx.$$

Every differentiation formula like

$$\frac{d(u + v)}{dx} = \frac{du}{dx} + \frac{dv}{dx} \qquad \text{or} \qquad \frac{d(\sin u)}{dx} = \cos u\,\frac{du}{dx}$$

has a corresponding differential form like

$$d(u + v) = du + dv \qquad \text{or} \qquad d(\sin u) = \cos u\,du.$$

**EXAMPLE 5**    We can use the Chain Rule and other differentiation rules to find differentials of functions.

**(a)**  $d(\tan 2x) = \sec^2(2x)\,d(2x) = 2\sec^2 2x\,dx$

**(b)**  $d\left(\dfrac{x}{x + 1}\right) = \dfrac{(x + 1)\,dx - x\,d(x + 1)}{(x + 1)^2} = \dfrac{x\,dx + dx - x\,dx}{(x + 1)^2} = \dfrac{dx}{(x + 1)^2}$    ∎

### Estimating with Differentials

Suppose we know the value of a differentiable function $f(x)$ at a point $a$ and want to estimate how much this value will change if we move to a nearby point $a + dx$. If $dx = \Delta x$ is small, then we can see from Figure 3.43 that $\Delta y$ is approximately equal to the differential $dy$. Since

$$f(a + dx) = f(a) + \Delta y, \qquad \Delta x = dx$$

the differential approximation gives

$$f(a + dx) \approx f(a) + dy$$

when $dx = \Delta x$. Thus the approximation $\Delta y \approx dy$ can be used to estimate $f(a + dx)$ when $f(a)$ is known, $dx$ is small, and $dy = f'(a)\,dx$.

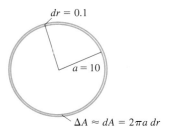

dr = 0.1

a = 10

$\Delta A \approx dA = 2\pi a\,dr$

**FIGURE 3.44** When $dr$ is small compared with $a$, the differential $dA$ gives the estimate $A(a + dr) = \pi a^2 + dA$ (Example 6).

**EXAMPLE 6** The radius $r$ of a circle increases from $a = 10$ m to 10.1 m (Figure 3.44). Use $dA$ to estimate the increase in the circle's area $A$. Estimate the area of the enlarged circle and compare your estimate to the true area found by direct calculation.

**Solution** Since $A = \pi r^2$, the estimated increase is

$$dA = A'(a)\,dr = 2\pi a\,dr = 2\pi(10)(0.1) = 2\pi \text{ m}^2.$$

Thus, since $A(r + \Delta r) \approx A(r) + dA$, we have

$$A(10 + 0.1) \approx A(10) + 2\pi$$
$$= \pi(10)^2 + 2\pi = 102\pi.$$

The area of a circle of radius 10.1 m is approximately $102\pi \text{ m}^2$.

The true area is

$$A(10.1) = \pi(10.1)^2$$
$$= 102.01\pi \text{ m}^2.$$

The error in our estimate is $0.01\pi \text{ m}^2$, which is the difference $\Delta A - dA$. ∎

**EXAMPLE 7** Use differentials to estimate

**(a)** $7.97^{1/3}$

**(b)** $\sin(\pi/6 + 0.01)$.

**Solution**
**(a)** The differential associated with the cube root function $y = x^{1/3}$ is

$$dy = \frac{1}{3x^{2/3}}dx.$$

We set $a = 8$, the closest number near 7.97 where we can easily compute $f(a)$ and $f'(a)$. To arrange that $a + dx = 7.97$, we choose $dx = -0.03$. Approximating with the differential gives

$$f(7.97) = f(a + dx) \approx f(a) + dy$$
$$= 8^{1/3} + \frac{1}{3(8)^{2/3}}(-0.03)$$
$$= 2 + \frac{1}{12}(-0.03) = 1.9975$$

This gives an approximation to the true value of $7.97^{1/3}$, which is 1.997497 to 6 decimals.

**(b)** The differential associated with $y = \sin x$ is

$$dy = \cos x\,dx.$$

To estimate $\sin(\pi/6 + 0.01)$, we set $a = \pi/6$ and $dx = 0.01$. Then

$$f(\pi/6 + 0.01) = f(a + dx) \approx f(a) + dy$$

$$= \sin\frac{\pi}{6} + \left(\cos\frac{\pi}{6}\right)(0.01)$$

$$= \frac{1}{2} + \frac{\sqrt{3}}{2}(0.01) \approx 0.5087$$

$$\sin(a + dx) \approx \sin a + (\cos a)\,dx$$

For comparison, the true value of $\sin(\pi/6 + 0.01)$ to 6 decimals is 0.508635.  ◼

The method in part (b) of Example 7 is used by some calculator and computer algorithms to give values of trigonometric functions. The algorithms store a large table of sine and cosine values between 0 and $\pi/4$. Values between these stored values are computed using differentials as in Example 7b. Values outside of $[0, \pi/4]$ are computed from values in this interval using trigonometric identities.

### Error in Differential Approximation

Let $f(x)$ be differentiable at $x = a$ and suppose that $dx = \Delta x$ is an increment of $x$. We have two ways to describe the change in $f$ as $x$ changes from $a$ to $a + \Delta x$:

|  |  |
|---|---|
| The true change: | $\Delta f = f(a + \Delta x) - f(a)$ |
| The differential estimate: | $df = f'(a)\,\Delta x.$ |

How well does $df$ approximate $\Delta f$?

We measure the approximation error by subtracting $df$ from $\Delta f$:

$$\text{Approximation error} = \Delta f - df$$

$$= \Delta f - f'(a)\Delta x$$

$$= \underbrace{f(a + \Delta x) - f(a)}_{\Delta f} - f'(a)\Delta x$$

$$= \underbrace{\left(\frac{f(a + \Delta x) - f(a)}{\Delta x} - f'(a)\right)}_{\text{Call this part } \epsilon.} \cdot \Delta x$$

$$= \epsilon \cdot \Delta x.$$

As $\Delta x \to 0$, the difference quotient

$$\frac{f(a + \Delta x) - f(a)}{\Delta x}$$

approaches $f'(a)$ (remember the definition of $f'(a)$), so the quantity in parentheses becomes a very small number (which is why we called it $\epsilon$). In fact, $\epsilon \to 0$ as $\Delta x \to 0$. When $\Delta x$ is small, the approximation error $\epsilon\,\Delta x$ is smaller still.

$$\underbrace{\Delta f}_{\substack{\text{true} \\ \text{change}}} = \underbrace{f'(a)\Delta x}_{\substack{\text{estimated} \\ \text{change}}} + \underbrace{\epsilon\,\Delta x}_{\text{error}}$$

Although we do not know the exact size of the error, it is the product $\epsilon \cdot \Delta x$ of two small quantities that both approach zero as $\Delta x \to 0$. For many common functions, whenever $\Delta x$ is small, the error is still smaller.

> **Change in $y = f(x)$ near $x = a$**
>
> If $y = f(x)$ is differentiable at $x = a$ and $x$ changes from $a$ to $a + \Delta x$, the change $\Delta y$ in $f$ is given by
>
> $$\Delta y = f'(a)\,\Delta x + \epsilon\,\Delta x \qquad (1)$$
>
> in which $\epsilon \to 0$ as $\Delta x \to 0$.

In Example 6 we found that

$$\Delta A = \pi(10.1)^2 - \pi(10)^2 = (102.01 - 100)\pi = (\underbrace{2\pi}_{dA} + \underbrace{0.01\pi}_{\text{error}})\ \text{m}^2$$

so the approximation error is $\Delta A - dA = \epsilon\Delta r = 0.01\pi$ and $\epsilon = 0.01\pi/\Delta r = 0.01\pi/0.1 = 0.1\pi$ m.

### Proof of the Chain Rule

Equation (1) enables us to prove the Chain Rule correctly. Our goal is to show that if $f(u)$ is a differentiable function of $u$ and $u = g(x)$ is a differentiable function of $x$, then the composite $y = f(g(x))$ is a differentiable function of $x$. Since a function is differentiable if and only if it has a derivative at each point in its domain, we must show that whenever $g$ is differentiable at $x_0$ and $f$ is differentiable at $g(x_0)$, then the composite is differentiable at $x_0$ and the derivative of the composite satisfies the equation

$$\left.\frac{dy}{dx}\right|_{x=x_0} = f'(g(x_0)) \cdot g'(x_0).$$

Let $\Delta x$ be an increment in $x$ and let $\Delta u$ and $\Delta y$ be the corresponding increments in $u$ and $y$. Applying Equation (1) we have

$$\Delta u = g'(x_0)\Delta x + \epsilon_1\,\Delta x = (g'(x_0) + \epsilon_1)\Delta x,$$

where $\epsilon_1 \to 0$ as $\Delta x \to 0$. Similarly,

$$\Delta y = f'(u_0)\Delta u + \epsilon_2\,\Delta u = (f'(u_0) + \epsilon_2)\Delta u,$$

where $\epsilon_2 \to 0$ as $\Delta u \to 0$. Notice also that $\Delta u \to 0$ as $\Delta x \to 0$. Combining the equations for $\Delta u$ and $\Delta y$ gives

$$\Delta y = (f'(u_0) + \epsilon_2)(g'(x_0) + \epsilon_1)\Delta x,$$

so

$$\frac{\Delta y}{\Delta x} = f'(u_0)g'(x_0) + \epsilon_2 g'(x_0) + f'(u_0)\epsilon_1 + \epsilon_2\epsilon_1.$$

Since $\epsilon_1$ and $\epsilon_2$ go to zero as $\Delta x$ goes to zero, the last three terms on the right vanish in the limit, leaving

$$\left.\frac{dy}{dx}\right|_{x=x_0} = \lim_{\Delta x \to 0}\frac{\Delta y}{\Delta x} = f'(u_0)g'(x_0) = f'(g(x_0)) \cdot g'(x_0). \qquad \blacksquare$$

Sensitivity to Change

The equation $df = f'(x)\,dx$ tells how *sensitive* the output of $f$ is to a change in input at different values of $x$. The larger the value of $f'$ at $x$, the greater the effect of a given change $dx$. As we move from $a$ to a nearby point $a + dx$, we can describe the change in $f$ in three ways:

| | True | Estimated |
|---|---|---|
| Absolute change | $\Delta f = f(a + dx) - f(a)$ | $df = f'(a)\,dx$ |
| Relative change | $\dfrac{\Delta f}{f(a)}$ | $\dfrac{df}{f(a)}$ |
| Percentage change | $\dfrac{\Delta f}{f(a)} \times 100$ | $\dfrac{df}{f(a)} \times 100$ |

**EXAMPLE 8**  You want to calculate the depth of a well from the equation $s = 4.9t^2$ by timing how long it takes a heavy stone you drop to splash into the water below. How sensitive will your calculations be to a 0.1-s error in measuring the time?

**Solution**  The size of $ds$ in the equation

$$ds = 9.8t\,dt$$

depends on how big $t$ is. If $t = 2$ s, the change caused by $dt = 0.1$ is about

$$ds = 9.8(2)(0.1) = 1.96 \text{ m.}$$

Three seconds later at $t = 5$ s, the change caused by the same $dt$ is

$$ds = 9.8(5)(0.1) = 4.9 \text{ m.}$$

For a fixed error in the time measurement, the error in using $ds$ to estimate the depth is larger when it takes a longer time before the stone splashes into the water. That is, the estimate is more sensitive to the effect of the error for larger values of $t$.  ∎

**EXAMPLE 9**  Newton's second law,

$$F = \frac{d}{dt}(mv) = m\frac{dv}{dt} = ma,$$

is stated with the assumption that mass is constant, but we know this is not strictly true because the mass of an object increases with velocity. In Einstein's corrected formula, mass has the value

$$m = \frac{m_0}{\sqrt{1 - v^2/c^2}},$$

where the "rest mass" $m_0$ represents the mass of an object that is not moving and $c$ is the speed of light, which is about 300,000 km/s. Use the approximation

$$\frac{1}{\sqrt{1 - x^2}} \approx 1 + \frac{1}{2}x^2 \tag{2}$$

to estimate the increase $\Delta m$ in mass resulting from the added velocity $v$.

**Solution**  When $v$ is very small compared with $c$, $v^2/c^2$ is close to zero and it is safe to use the approximation

$$\frac{1}{\sqrt{1 - v^2/c^2}} \approx 1 + \frac{1}{2}\left(\frac{v^2}{c^2}\right) \qquad \text{Eq. (2) with } x = \frac{v}{c}$$

to obtain

$$m = \frac{m_0}{\sqrt{1 - v^2/c^2}} \approx m_0 \left[ 1 + \frac{1}{2} \left( \frac{v^2}{c^2} \right) \right] = m_0 + \frac{1}{2} m_0 v^2 \left( \frac{1}{c^2} \right),$$

or

$$m \approx m_0 + \frac{1}{2} m_0 v^2 \left( \frac{1}{c^2} \right). \qquad (3)$$

Equation (3) expresses the increase in mass that results from the added velocity $v$. ∎

### Converting Mass to Energy

Equation (3) derived in Example 9 has an important interpretation. In Newtonian physics, $(1/2)m_0 v^2$ is the kinetic energy (KE) of the object, and if we rewrite Equation (3) in the form

$$(m - m_0) c^2 \approx \frac{1}{2} m_0 v^2,$$

we see that

$$(m - m_0) c^2 \approx \frac{1}{2} m_0 v^2 = \frac{1}{2} m_0 v^2 - \frac{1}{2} m_0 (0)^2 = \Delta(\text{KE}),$$

or

$$(\Delta m) c^2 \approx \Delta(\text{KE}).$$

So the change in kinetic energy $\Delta(\text{KE})$ in going from velocity 0 to velocity $v$ is approximately equal to $(\Delta m) c^2$, the change in mass times the square of the speed of light. Using $c \approx 3 \times 10^8$ m/sec, we see that a small change in mass can create a large change in energy.

## Exercises 3.9

### Finding Linearizations

In Exercises 1–5, find the linearization $L(x)$ of $f(x)$ at $x = a$.

1. $f(x) = x^3 - 2x + 3$, $a = 2$
2. $f(x) = \sqrt{x^2 + 9}$, $a = -4$
3. $f(x) = x + \frac{1}{x}$, $a = 1$
4. $f(x) = \sqrt[3]{x}$, $a = -8$
5. $f(x) = \tan x$, $a = \pi$
6. **Common linear approximations at $x = 0$** Find the linearizations of the following functions at $x = 0$.
   a. $\sin x$   b. $\cos x$   c. $\tan x$

### Linearization for Approximation

In Exercises 7–12, find a linearization at a suitably chosen integer near $a$ at which the given function and its derivative are easy to evaluate.

7. $f(x) = x^2 + 2x$, $a = 0.1$
8. $f(x) = x^{-1}$, $a = 0.9$
9. $f(x) = 2x^2 + 3x - 3$, $a = -0.9$
10. $f(x) = 1 + x$, $a = 8.1$
11. $f(x) = \sqrt[3]{x}$, $a = 8.5$
12. $f(x) = \frac{x}{x + 1}$, $a = 1.3$

13. Show that the linearization of $f(x) = (1 + x)^k$ at $x = 0$ is $L(x) = 1 + kx$.

14. Use the linear approximation $(1 + x)^k \approx 1 + kx$ to find an approximation for the function $f(x)$ for values of $x$ near zero.

   a. $f(x) = (1 - x)^6$
   b. $f(x) = \frac{2}{1 - x}$
   c. $f(x) = \frac{1}{\sqrt{1 + x}}$
   d. $f(x) = \sqrt{2 + x^2}$
   e. $f(x) = (4 + 3x)^{1/3}$
   f. $f(x) = \sqrt[3]{\left( 1 - \frac{x}{2 + x} \right)^2}$

15. **Faster than a calculator** Use the approximation $(1 + x)^k \approx 1 + kx$ to estimate the following.

   a. $(1.0002)^{50}$
   b. $\sqrt[3]{1.009}$

16. Find the linearization of $f(x) = \sqrt{x + 1} + \sin x$ at $x = 0$. How is it related to the individual linearizations of $\sqrt{x + 1}$ and $\sin x$ at $x = 0$?

**Derivatives in Differential Form**

In Exercises 17–28, find $dy$.

**17.** $y = x^3 - 3\sqrt{x}$

**18.** $y = x\sqrt{1 - x^2}$

**19.** $y = \dfrac{2x}{1 + x^2}$

**20.** $y = \dfrac{2\sqrt{x}}{3(1 + \sqrt{x})}$

**21.** $2y^{3/2} + xy - x = 0$

**22.** $xy^2 - 4x^{3/2} - y = 0$

**23.** $y = \sin(5\sqrt{x})$

**24.** $y = \cos(x^2)$

**25.** $y = 4\tan(x^3/3)$

**26.** $y = \sec(x^2 - 1)$

**27.** $y = 3\csc\left(1 - 2\sqrt{x}\right)$

**28.** $y = 2\cot\left(\dfrac{1}{\sqrt{x}}\right)$

**Approximation Error**

In Exercises 29–34, each function $f(x)$ changes value when $x$ changes from $x_0$ to $x_0 + dx$. Find

  **a.** the change $\Delta f = f(x_0 + dx) - f(x_0)$;

  **b.** the value of the estimate $df = f'(x_0)\,dx$; and

  **c.** the approximation error $|\Delta f - df|$.

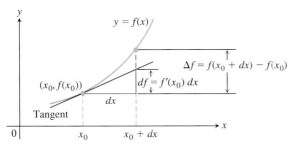

**29.** $f(x) = x^2 + 2x, \quad x_0 = 1, \quad dx = 0.1$

**30.** $f(x) = 2x^2 + 4x - 3, \quad x_0 = -1, \quad dx = 0.1$

**31.** $f(x) = x^3 - x, \quad x_0 = 1, \quad dx = 0.1$

**32.** $f(x) = x^4, \quad x_0 = 1, \quad dx = 0.1$

**33.** $f(x) = x^{-1}, \quad x_0 = 0.5, \quad dx = 0.1$

**34.** $f(x) = x^3 - 2x + 3, \quad x_0 = 2, \quad dx = 0.1$

**Differential Estimates of Change**

In Exercises 35–40, write a differential formula that estimates the given change in volume or surface area.

**35.** The change in the volume $V = (4/3)\pi r^3$ of a sphere when the radius changes from $r_0$ to $r_0 + dr$

**36.** The change in the volume $V = x^3$ of a cube when the edge lengths change from $x_0$ to $x_0 + dx$

**37.** The change in the surface area $S = 6x^2$ of a cube when the edge lengths change from $x_0$ to $x_0 + dx$

**38.** The change in the lateral surface area $S = \pi r\sqrt{r^2 + h^2}$ of a right circular cone when the radius changes from $r_0$ to $r_0 + dr$ and the height does not change

**39.** The change in the volume $V = \pi r^2 h$ of a right circular cylinder when the radius changes from $r_0$ to $r_0 + dr$ and the height does not change

**40.** The change in the lateral surface area $S = 2\pi rh$ of a right circular cylinder when the height changes from $h_0$ to $h_0 + dh$ and the radius does not change

**Applications**

**41.** The radius of a circle is increased from 2.00 to 2.02 m.

  **a.** Estimate the resulting change in area.

  **b.** Express the estimate as a percentage of the circle's original area.

**42.** The diameter of a tree was 25 cm. During the following year, the circumference increased 5 cm. About how much did the tree's diameter increase? The tree's cross-sectional area?

**43. Estimating volume** Estimate the volume of material in a cylindrical shell with length 30 cm, radius 6 cm, and shell thickness 0.5 cm.

**44. Estimating height of a building** A surveyor, standing 9 m from the base of a building, measures the angle of elevation to the top of the building to be 75°. How accurately must the angle be measured for the percentage error in estimating the height of the building to be less than 4%?

**45.** The radius $r$ of a circle is measured with an error of at most 2%. What is the maximum corresponding percentage error in computing the circle's

  **a.** circumference?        **b.** area?

**46.** The edge $x$ of a cube is measured with an error of at most 0.5%. What is the maximum corresponding percentage error in computing the cube's

  **a.** surface area?        **b.** volume?

**47. Tolerance** The height and radius of a right circular cylinder are equal, so the cylinder's volume is $V = \pi h^3$. The volume is to be calculated with an error of no more than 1% of the true value. Find approximately the greatest error that can be tolerated in the measurement of $h$, expressed as a percentage of $h$.

**48. Tolerance**

  **a.** About how accurately must the interior diameter of a 10-m-high cylindrical storage tank be measured to calculate the tank's volume to within 1% of its true value?

  **b.** About how accurately must the tank's exterior diameter be measured to calculate the amount of paint it will take to paint the side of the tank to within 5% of the true amount?

**49.** The diameter of a sphere is measured as $100 \pm 1$ cm and the volume is calculated from this measurement. Estimate the percentage error in the volume calculation.

**50.** Estimate the allowable percentage error in measuring the diameter $D$ of a sphere if the volume is to be calculated correctly to within 3%.

**51. The effect of flight maneuvers on the heart** The amount of work done by the heart's main pumping chamber, the left ventricle, is given by the equation

$$W = PV + \frac{V\delta v^2}{2g},$$

where $W$ is the work per unit time, $P$ is the average blood pressure, $V$ is the volume of blood pumped out during the unit of time, $\delta$ ("delta") is the weight density of the blood, $v$ is the average velocity of the exiting blood, and $g$ is the acceleration of gravity.

When $P$, $V$, $\delta$, and $v$ remain constant, $W$ becomes a function of $g$, and the equation takes the simplified form

$$W = a + \frac{b}{g} \quad (a, b \text{ constant}).$$

As a member of NASA's medical team, you want to know how sensitive $W$ is to apparent changes in $g$ caused by flight maneuvers, and this depends on the initial value of $g$. As part of your investigation, you decide to compare the effect on $W$ of a given change $dg$ on the moon, where $g = 1.6 \text{ m/s}^2$, with the effect the same change $dg$ would have on Earth, where $g = 9.8 \text{ m/s}^2$. Use the simplified equation above to find the ratio of $dW_{\text{moon}}$ to $dW_{\text{Earth}}$.

**52. Drug concentration** The concentration $C$ in milligrams per milliliter (mg/mL) of a certain drug in a person's bloodstream $t$ hours after a pill is swallowed is modeled by the approximation

$$C(t) = \frac{4t}{1 + t^3} + 0.06t.$$

Estimate the change in concentration when $t$ changes from 20 to 30 min.

**53. Unclogging arteries** The formula $V = kr^4$, discovered by the physiologist Jean Poiseuille (1797–1869), allows us to predict how much the radius of a partially clogged artery has to be expanded in order to restore normal blood flow. The formula says that the volume $V$ of blood flowing through the artery in a unit of time at a fixed pressure is a constant $k$ times the radius of the artery to the fourth power. How will a 10% increase in $r$ affect $V$?

**54. Measuring acceleration of gravity** When the length $L$ of a clock pendulum is held constant by controlling its temperature, the pendulum's period $T$ depends on the acceleration of gravity $g$. The period will therefore vary slightly as the clock is moved from place to place on the earth's surface, depending on the change in $g$. By keeping track of $\Delta T$, we can estimate the variation in $g$ from the equation $T = 2\pi(L/g)^{1/2}$ that relates $T$, $g$, and $L$.

a. With $L$ held constant and $g$ as the independent variable, calculate $dT$ and use it to answer parts (b) and (c).

b. If $g$ increases, will $T$ increase or decrease? Will a pendulum clock speed up or slow down? Explain.

c. A clock with a 100-cm pendulum is moved from a location where $g = 980 \text{ cm/s}^2$ to a new location. This increases the period by $dT = 0.001$ s. Find $dg$ and estimate the value of $g$ at the new location.

**55. Quadratic approximations**

a. Let $Q(x) = b_0 + b_1(x - a) + b_2(x - a)^2$ be a quadratic approximation to $f(x)$ at $x = a$ with the properties:

   i) $Q(a) = f(a)$

   ii) $Q'(a) = f'(a)$

   iii) $Q''(a) = f''(a)$.

Determine the coefficients $b_0$, $b_1$, and $b_2$.

b. Find the quadratic approximation to $f(x) = 1/(1 - x)$ at $x = 0$.

T c. Graph $f(x) = 1/(1 - x)$ and its quadratic approximation at $x = 0$. Then zoom in on the two graphs at the point $(0, 1)$. Comment on what you see.

T d. Find the quadratic approximation to $g(x) = 1/x$ at $x = 1$. Graph $g$ and its quadratic approximation together. Comment on what you see.

T e. Find the quadratic approximation to $h(x) = \sqrt{1 + x}$ at $x = 0$. Graph $h$ and its quadratic approximation together. Comment on what you see.

f. What are the linearizations of $f$, $g$, and $h$ at the respective points in parts (b), (d), and (e)?

**56. The linearization is the best linear approximation** Suppose that $y = f(x)$ is differentiable at $x = a$ and that $g(x) = m(x - a) + c$ is a linear function in which $m$ and $c$ are constants. If the error $E(x) = f(x) - g(x)$ were small enough near $x = a$, we might think of using $g$ as a linear approximation of $f$ instead of the linearization $L(x) = f(a) + f'(a)(x - a)$. Show that if we impose on $g$ the conditions

1. $E(a) = 0$      The approximation error is zero at $x = a$.

2. $\displaystyle \lim_{x \to a} \frac{E(x)}{x - a} = 0$    The error is negligible when compared with $x - a$.

then $g(x) = f(a) + f'(a)(x - a)$. Thus, the linearization $L(x)$ gives the only linear approximation whose error is both zero at $x = a$ and negligible in comparison with $x - a$.

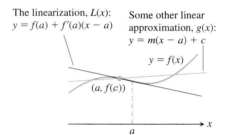

The linearization, $L(x)$:
$y = f(a) + f'(a)(x - a)$

Some other linear approximation, $g(x)$:
$y = m(x - a) + c$

$y = f(x)$

$(a, f(c))$

$a$

$x$

**COMPUTER EXPLORATIONS**

In Exercises 57–60, use a CAS to estimate the magnitude of the error in using the linearization in place of the function over a specified interval $I$. Perform the following steps:

a. Plot the function $f$ over $I$.

b. Find the linearization $L$ of the function at the point $a$.

c. Plot $f$ and $L$ together on a single graph.

d. Plot the absolute error $|f(x) - L(x)|$ over $I$ and find its maximum value.

e. From your graph in part (d), estimate as large a $\delta > 0$ as you can, satisfying

$$|x - a| < \delta \quad \Rightarrow \quad |f(x) - L(x)| < \epsilon$$

for $\epsilon = 0.5, 0.1$, and $0.01$. Then check graphically to see if your $\delta$-estimate holds true.

**57.** $f(x) = x^3 + x^2 - 2x$,   $[-1, 2]$,   $a = 1$

**58.** $f(x) = \dfrac{x - 1}{4x^2 + 1}$,   $\left[-\dfrac{3}{4}, 1\right]$,   $a = \dfrac{1}{2}$

**59.** $f(x) = x^{2/3}(x - 2)$,   $[-2, 3]$,   $a = 2$

**60.** $f(x) = \sqrt{x} - \sin x$,   $[0, 2\pi]$,   $a = 2$

# Chapter **3** Questions to Guide Your Review

1. What is the derivative of a function $f$? How is its domain related to the domain of $f$? Give examples.

2. What role does the derivative play in defining slopes, tangents, and rates of change?

3. How can you sometimes graph the derivative of a function when all you have is a table of the function's values?

4. What does it mean for a function to be differentiable on an open interval? On a closed interval?

5. How are derivatives and one-sided derivatives related?

6. Describe geometrically when a function typically does *not* have a derivative at a point.

7. How is a function's differentiability at a point related to its continuity there, if at all?

8. What rules do you know for calculating derivatives? Give some examples.

9. Explain how the three formulas

   **a.** $\dfrac{d}{dx}(x^n) = nx^{n-1}$ **b.** $\dfrac{d}{dx}(cu) = c\dfrac{du}{dx}$

   **c.** $\dfrac{d}{dx}(u_1 + u_2 + \cdots + u_n) = \dfrac{du_1}{dx} + \dfrac{du_2}{dx} + \cdots + \dfrac{du_n}{dx}$

   enable us to differentiate any polynomial.

10. What formula do we need, in addition to the three listed in Question 9, to differentiate rational functions?

11. What is a second derivative? A third derivative? How many derivatives do the functions you know have? Give examples.

12. What is the relationship between a function's average and instantaneous rates of change? Give an example.

13. How do derivatives arise in the study of motion? What can you learn about an object's motion along a line by examining the derivatives of the object's position function? Give examples.

14. How can derivatives arise in economics?

15. Give examples of still other applications of derivatives.

16. What do the limits $\lim_{h\to 0}((\sin h)/h)$ and $\lim_{h\to 0}((\cos h - 1)/h)$ have to do with the derivatives of the sine and cosine functions? What *are* the derivatives of these functions?

17. Once you know the derivatives of $\sin x$ and $\cos x$, how can you find the derivatives of $\tan x$, $\cot x$, $\sec x$, and $\csc x$? What *are* the derivatives of these functions?

18. At what points are the six basic trigonometric functions continuous? How do you know?

19. What is the rule for calculating the derivative of a composite of two differentiable functions? How is such a derivative evaluated? Give examples.

20. If $u$ is a differentiable function of $x$, how do you find $(d/dx)(u^n)$ if $n$ is an integer? If $n$ is a real number? Give examples.

21. What is implicit differentiation? When do you need it? Give examples.

22. How do related rates problems arise? Give examples.

23. Outline a strategy for solving related rates problems. Illustrate with an example.

24. What is the linearization $L(x)$ of a function $f(x)$ at a point $x = a$? What is required of $f$ at $a$ for the linearization to exist? How are linearizations used? Give examples.

25. If $x$ moves from $a$ to a nearby value $a + dx$, how do you estimate the corresponding change in the value of a differentiable function $f(x)$? How do you estimate the relative change? The percentage change? Give an example.

# Chapter **3** Practice Exercises

## Derivatives of Functions

Find the derivatives of the functions in Exercises 1–40.

1. $y = x^5 - 0.125x^2 + 0.25x$   2. $y = 3 - 0.7x^3 + 0.3x^7$

3. $y = x^3 - 3(x^2 + \pi^2)$   4. $y = x^7 + \sqrt{7}x - \dfrac{1}{\pi + 1}$

5. $y = (x + 1)^2(x^2 + 2x)$   6. $y = (2x - 5)(4 - x)^{-1}$

7. $y = (\theta^2 + \sec\theta + 1)^3$   8. $y = \left(-1 - \dfrac{\csc\theta}{2} - \dfrac{\theta^2}{4}\right)^2$

9. $s = \dfrac{\sqrt{t}}{1 + \sqrt{t}}$   10. $s = \dfrac{1}{\sqrt{t} - 1}$

11. $y = 2\tan^2 x - \sec^2 x$   12. $y = \dfrac{1}{\sin^2 x} - \dfrac{2}{\sin x}$

13. $s = \cos^4(1 - 2t)$   14. $s = \cot^3\left(\dfrac{2}{t}\right)$

15. $s = (\sec t + \tan t)^5$   16. $s = \csc^5(1 - t + 3t^2)$

17. $r = \sqrt{2\theta \sin\theta}$   18. $r = 2\theta\sqrt{\cos\theta}$

19. $r = \sin\sqrt{2\theta}$   20. $r = \sin\left(\theta + \sqrt{\theta + 1}\right)$

21. $y = \dfrac{1}{2}x^2 \csc\dfrac{2}{x}$   22. $y = 2\sqrt{x}\sin\sqrt{x}$

23. $y = x^{-1/2}\sec(2x)^2$   24. $y = \sqrt{x}\csc(x + 1)^3$

25. $y = 5\cot x^2$   26. $y = x^2\cot 5x$

27. $y = x^2\sin^2(2x^2)$   28. $y = x^{-2}\sin^2(x^3)$

29. $s = \left(\dfrac{4t}{t + 1}\right)^{-2}$   30. $s = \dfrac{-1}{15(15t - 1)^3}$

31. $y = \left(\dfrac{\sqrt{x}}{1 + x}\right)^2$   32. $y = \left(\dfrac{2\sqrt{x}}{2\sqrt{x} + 1}\right)^2$

**33.** $y = \sqrt{\dfrac{x^2 + x}{x^2}}$

**34.** $y = 4x\sqrt{x} + \sqrt{x}$

**35.** $r = \left(\dfrac{\sin\theta}{\cos\theta - 1}\right)^2$

**36.** $r = \left(\dfrac{1 + \sin\theta}{1 - \cos\theta}\right)^2$

**37.** $y = (2x + 1)\sqrt{2x + 1}$

**38.** $y = 20(3x - 4)^{1/4}(3x - 4)^{-1/5}$

**39.** $y = \dfrac{3}{(5x^2 + \sin 2x)^{3/2}}$

**40.** $y = (3 + \cos^3 3x)^{-1/3}$

### Implicit Differentiation

In Exercises 41–48, find $dy/dx$ by implicit differentiation.

**41.** $xy + 2x + 3y = 1$

**42.** $x^2 + xy + y^2 - 5x = 2$

**43.** $x^3 + 4xy - 3y^{4/3} = 2x$

**44.** $5x^{4/5} + 10y^{6/5} = 15$

**45.** $\sqrt{xy} = 1$

**46.** $x^2y^2 = 1$

**47.** $y^2 = \dfrac{x}{x + 1}$

**48.** $y^2 = \sqrt{\dfrac{1 + x}{1 - x}}$

In Exercises 49 and 50, find $dp/dq$.

**49.** $p^3 + 4pq - 3q^2 = 2$

**50.** $q = (5p^2 + 2p)^{-3/2}$

In Exercises 51 and 52, find $dr/ds$.

**51.** $r\cos 2s + \sin^2 s = \pi$

**52.** $2rs - r - s + s^2 = -3$

**53.** Find $d^2y/dx^2$ by implicit differentiation:

   **a.** $x^3 + y^3 = 1$

   **b.** $y^2 = 1 - \dfrac{2}{x}$

**54. a.** By differentiating $x^2 - y^2 = 1$ implicitly, show that $dy/dx = x/y$.

   **b.** Then show that $d^2y/dx^2 = -1/y^3$.

### Numerical Values of Derivatives

**55.** Suppose that functions $f(x)$ and $g(x)$ and their first derivatives have the following values at $x = 0$ and $x = 1$.

| $x$ | $f(x)$ | $g(x)$ | $f'(x)$ | $g'(x)$ |
|---|---|---|---|---|
| 0 | 1 | 1 | −3 | 1/2 |
| 1 | 3 | 5 | 1/2 | −4 |

Find the first derivatives of the following combinations at the given value of $x$.

   **a.** $6f(x) - g(x), \quad x = 1$

   **b.** $f(x)g^2(x), \quad x = 0$

   **c.** $\dfrac{f(x)}{g(x) + 1}, \quad x = 1$

   **d.** $f(g(x)), \quad x = 0$

   **e.** $g(f(x)), \quad x = 0$

   **f.** $(x + f(x))^{3/2}, \quad x = 1$

   **g.** $f(x + g(x)), \quad x = 0$

**56.** Suppose that the function $f(x)$ and its first derivative have the following values at $x = 0$ and $x = 1$.

| $x$ | $f(x)$ | $f'(x)$ |
|---|---|---|
| 0 | 9 | −2 |
| 1 | −3 | 1/5 |

Find the first derivatives of the following combinations at the given value of $x$.

   **a.** $\sqrt{x}\,f(x), \quad x = 1$

   **b.** $\sqrt{f(x)}, \quad x = 0$

   **c.** $f(\sqrt{x}), \quad x = 1$

   **d.** $f(1 - 5\tan x), \quad x = 0$

   **e.** $\dfrac{f(x)}{2 + \cos x}, \quad x = 0$

   **f.** $10\sin\left(\dfrac{\pi x}{2}\right)f^2(x), \quad x = 1$

**57.** Find the value of $dy/dt$ at $t = 0$ if $y = 3\sin 2x$ and $x = t^2 + \pi$.

**58.** Find the value of $ds/du$ at $u = 2$ if $s = t^2 + 5t$ and $t = (u^2 + 2u)^{1/3}$.

**59.** Find the value of $dw/ds$ at $s = 0$ if $w = \sin\left(\sqrt{r} - 2\right)$ and $r = 8\sin(s + \pi/6)$.

**60.** Find the value of $dr/dt$ at $t = 0$ if $r = (\theta^2 + 7)^{1/3}$ and $\theta^2 t + \theta = 1$.

**61.** If $y^3 + y = 2\cos x$, find the value of $d^2y/dx^2$ at the point $(0, 1)$.

**62.** If $x^{1/3} + y^{1/3} = 4$, find $d^2y/dx^2$ at the point $(8, 8)$.

### Applying the Derivative Definition

In Exercises 63 and 64, find the derivative using the definition.

**63.** $f(t) = \dfrac{1}{2t + 1}$

**64.** $g(x) = 2x^2 + 1$

**65. a.** Graph the function

$$f(x) = \begin{cases} x^2, & -1 \le x < 0 \\ -x^2, & 0 \le x \le 1. \end{cases}$$

   **b.** Is $f$ continuous at $x = 0$?

   **c.** Is $f$ differentiable at $x = 0$?

Give reasons for your answers.

**66. a.** Graph the function

$$f(x) = \begin{cases} x, & -1 \le x < 0 \\ \tan x, & 0 \le x \le \pi/4. \end{cases}$$

   **b.** Is $f$ continuous at $x = 0$?

   **c.** Is $f$ differentiable at $x = 0$?

Give reasons for your answers.

**67. a.** Graph the function

$$f(x) = \begin{cases} x, & 0 \le x \le 1 \\ 2 - x, & 1 < x \le 2. \end{cases}$$

   **b.** Is $f$ continuous at $x = 1$?

   **c.** Is $f$ differentiable at $x = 1$?

Give reasons for your answers.

**68.** For what value or values of the constant $m$, if any, is

$$f(x) = \begin{cases} \sin 2x, & x \le 0 \\ mx, & x > 0 \end{cases}$$

   **a.** continuous at $x = 0$?

   **b.** differentiable at $x = 0$?

Give reasons for your answers.

### Slopes, Tangents, and Normals

**69. Tangents with specified slope** Are there any points on the curve $y = (x/2) + 1/(2x - 4)$ where the slope is $-3/2$? If so, find them.

**70. Tangents with specified slope** Are there any points on the curve $y = x - 1/2x$ where the slope is 2? If so, find them.

**71. Horizontal tangents** Find the points on the curve $y = 2x^3 - 3x^2 - 12x + 20$ where the tangent is parallel to the $x$-axis.

**72. Tangent intercepts** Find the $x$- and $y$-intercepts of the line that is tangent to the curve $y = x^3$ at the point $(-2, -8)$.

**73. Tangents perpendicular or parallel to lines** Find the points on the curve $y = 2x^3 - 3x^2 - 12x + 20$ where the tangent is

  **a.** perpendicular to the line $y = 1 - (x/24)$.

  **b.** parallel to the line $y = \sqrt{2} - 12x$.

**74. Intersecting tangents** Show that the tangents to the curve $y = (\pi \sin x)/x$ at $x = \pi$ and $x = -\pi$ intersect at right angles.

**75. Normals parallel to a line** Find the points on the curve $y = \tan x, -\pi/2 < x < \pi/2$, where the normal is parallel to the line $y = -x/2$. Sketch the curve and normals together, labeling each with its equation.

**76. Tangent and normal lines** Find equations for the tangent and normal to the curve $y = 1 + \cos x$ at the point $(\pi/2, 1)$. Sketch the curve, tangent, and normal together, labeling each with its equation.

**77. Tangent parabola** The parabola $y = x^2 + C$ is to be tangent to the line $y = x$. Find $C$.

**78. Slope of tangent** Show that the tangent to the curve $y = x^3$ at any point $(a, a^3)$ meets the curve again at a point where the slope is four times the slope at $(a, a^3)$.

**79. Tangent curve** For what value of $c$ is the curve $y = c/(x + 1)$ tangent to the line through the points $(0, 3)$ and $(5, -2)$?

**80. Normal to a circle** Show that the normal line at any point of the circle $x^2 + y^2 = a^2$ passes through the origin.

In Exercises 81–86, find equations for the lines that are tangent and normal to the curve at the given point.

**81.** $x^2 + 2y^2 = 9$, $(1, 2)$

**82.** $x^3 + y^2 = 2$, $(1, 1)$

**83.** $xy + 2x - 5y = 2$, $(3, 2)$

**84.** $(y - x)^2 = 2x + 4$, $(6, 2)$

**85.** $x + \sqrt{xy} = 6$, $(4, 1)$

**86.** $x^{3/2} + 2y^{3/2} = 17$, $(1, 4)$

**87.** Find the slope of the curve $x^3y^3 + y^2 = x + y$ at the points $(1, 1)$ and $(1, -1)$.

**88.** The graph shown suggests that the curve $y = \sin(x - \sin x)$ might have horizontal tangents at the $x$-axis. Does it? Give reasons for your answer.

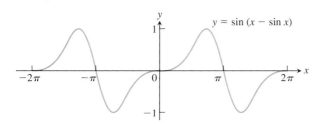

**Analyzing Graphs**

Each of the figures in Exercises 89 and 90 shows two graphs, the graph of a function $y = f(x)$ together with the graph of its derivative $f'(x)$. Which graph is which? How do you know?

**89.**

**90.**

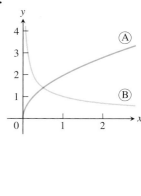

**91.** Use the following information to graph the function $y = f(x)$ for $-1 \le x \le 6$.

  **i)** The graph of $f$ is made of line segments joined end to end.

  **ii)** The graph starts at the point $(-1, 2)$.

  **iii)** The derivative of $f$, where defined, agrees with the step function shown here.

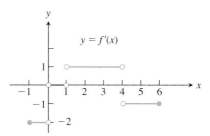

**92.** Repeat Exercise 91, supposing that the graph starts at $(-1, 0)$ instead of $(-1, 2)$.

Exercises 93 and 94 are about the accompanying graphs. The graphs in part (a) show the numbers of rabbits and foxes in a small arctic population. They are plotted as functions of time for 200 days. The number of rabbits increases at first, as the rabbits reproduce. But the foxes prey on rabbits and, as the number of foxes increases, the rabbit population levels off and then drops. Part (b) shows the graph of the derivative of the rabbit population, made by plotting slopes.

**93. a.** What is the value of the derivative of the rabbit population when the number of rabbits is largest? Smallest?

  **b.** What is the size of the rabbit population when its derivative is largest? Smallest (negative value)?

**94.** In what units should the slopes of the rabbit and fox population curves be measured?

(a)

Derivative of the rabbit population

(b)

*Source:* NCPMF "Differentiation" by W.U. Walton et al., Project CALC. Reprinted by permission of Educational Development Center, Inc.

Trigonometric Limits

Find the limits in Exercises 95–102.

**95.** $\lim\limits_{x \to 0} \dfrac{\sin x}{2x^2 - x}$

**96.** $\lim\limits_{x \to 0} \dfrac{3x - \tan 7x}{2x}$

**97.** $\lim\limits_{r \to 0} \dfrac{\sin r}{\tan 2r}$

**98.** $\lim\limits_{\theta \to 0} \dfrac{\sin (\sin \theta)}{\theta}$

**99.** $\lim\limits_{\theta \to (\pi/2)^-} \dfrac{4\tan^2 \theta + \tan \theta + 1}{\tan^2 \theta + 5}$

**100.** $\lim\limits_{\theta \to 0^+} \dfrac{1 - 2\cot^2 \theta}{5\cot^2 \theta - 7\cot \theta - 8}$

**101.** $\lim\limits_{x \to 0} \dfrac{x \sin x}{2 - 2\cos x}$

**102.** $\lim\limits_{\theta \to 0} \dfrac{1 - \cos \theta}{\theta^2}$

Show how to extend the functions in Exercises 103 and 104 to be continuous at the origin.

**103.** $g(x) = \dfrac{\tan (\tan x)}{\tan x}$

**104.** $f(x) = \dfrac{\tan (\tan x)}{\sin (\sin x)}$

Related Rates

**105. Right circular cylinder** The total surface area $S$ of a right circular cylinder is related to the base radius $r$ and height $h$ by the equation $S = 2\pi r^2 + 2\pi rh$.

 **a.** How is $dS/dt$ related to $dr/dt$ if $h$ is constant?

 **b.** How is $dS/dt$ related to $dh/dt$ if $r$ is constant?

 **c.** How is $dS/dt$ related to $dr/dt$ and $dh/dt$ if neither $r$ nor $h$ is constant?

 **d.** How is $dr/dt$ related to $dh/dt$ if $S$ is constant?

**106. Right circular cone** The lateral surface area $S$ of a right circular cone is related to the base radius $r$ and height $h$ by the equation $S = \pi r \sqrt{r^2 + h^2}$.

 **a.** How is $dS/dt$ related to $dr/dt$ if $h$ is constant?

 **b.** How is $dS/dt$ related to $dh/dt$ if $r$ is constant?

 **c.** How is $dS/dt$ related to $dr/dt$ and $dh/dt$ if neither $r$ nor $h$ is constant?

**107. Circle's changing area** The radius of a circle is changing at the rate of $-2/\pi$ m/s. At what rate is the circle's area changing when $r = 10$ m?

**108. Cube's changing edges** The volume of a cube is increasing at the rate of 1200 cm³/min at the instant its edges are 20 cm long. At what rate are the lengths of the edges changing at that instant?

**109. Resistors connected in parallel** If two resistors of $R_1$ and $R_2$ ohms are connected in parallel in an electric circuit to make an $R$-ohm resistor, the value of $R$ can be found from the equation

$$\frac{1}{R} = \frac{1}{R_1} + \frac{1}{R_2}.$$

If $R_1$ is decreasing at the rate of 1 ohm/s and $R_2$ is increasing at the rate of 0.5 ohm/s, at what rate is $R$ changing when $R_1 = 75$ ohms and $R_2 = 50$ ohms?

**110. Impedance in a series circuit** The impedance $Z$ (ohms) in a series circuit is related to the resistance $R$ (ohms) and reactance $X$ (ohms) by the equation $Z = \sqrt{R^2 + X^2}$. If $R$ is increasing at 3 ohms/s and $X$ is decreasing at 2 ohms/s, at what rate is $Z$ changing when $R = 10$ ohms and $X = 20$ ohms?

**111. Speed of moving particle** The coordinates of a particle moving in the metric $xy$-plane are differentiable functions of time $t$ with $dx/dt = 10$ m/s and $dy/dt = 5$ m/s. How fast is the particle moving away from the origin as it passes through the point $(3, -4)$?

**112. Motion of a particle** A particle moves along the curve $y = x^{3/2}$ in the first quadrant in such a way that its distance from the origin increases at the rate of 11 units per second. Find $dx/dt$ when $x = 3$.

**113. Draining a tank** Water drains from the conical tank shown in the accompanying figure at the rate of $0.2 \text{ m}^3/\text{min}$.

   **a.** What is the relation between the variables $h$ and $r$ in the figure?

   **b.** How fast is the water level dropping when $h = 2$ m?

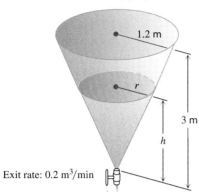

Exit rate: $0.2 \text{ m}^3/\text{min}$

**114. Rotating spool** As television cable is pulled from a large spool to be strung from the telephone poles along a street, it unwinds from the spool in layers of constant radius (see accompanying figure). If the truck pulling the cable moves at a steady 2 m/s (a touch over 7 km/h), use the equation $s = r\theta$ to find how fast (radians per second) the spool is turning when the layer of radius 0.4 m is being unwound.

0.4 m

**115. Moving searchlight beam** The figure shows a boat 1 km offshore, sweeping the shore with a searchlight. The light turns at a constant rate, $d\theta/dt = -0.6$ rad/s.

   **a.** How fast is the light moving along the shore when it reaches point $A$?

   **b.** How many revolutions per minute is 0.6 rad/s?

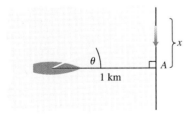

**116. Points moving on coordinate axes** Points $A$ and $B$ move along the $x$- and $y$-axes, respectively, in such a way that the distance $r$ (meters) along the perpendicular from the origin to the line $AB$ remains constant. How fast is $OA$ changing, and is it increasing, or decreasing, when $OB = 2r$ and $B$ is moving toward $O$ at the rate of $0.3r$ m/s?

Linearization

**117.** Find the linearizations of

   **a.** $\tan x$ at $x = -\pi/4$      **b.** $\sec x$ at $x = -\pi/4$.

   Graph the curves and linearizations together.

**118.** We can obtain a useful linear approximation of the function $f(x) = 1/(1 + \tan x)$ at $x = 0$ by combining the approximations

$$\frac{1}{1 + x} \approx 1 - x \quad \text{and} \quad \tan x \approx x$$

to get

$$\frac{1}{1 + \tan x} \approx 1 - x.$$

Show that this result is the standard linear approximation of $1/(1 + \tan x)$ at $x = 0$.

**119.** Find the linearization of $f(x) = \sqrt{1 + x} + \sin x - 0.5$ at $x = 0$.

**120.** Find the linearization of $f(x) = 2/(1 - x) + \sqrt{1 + x} - 3.1$ at $x = 0$.

Differential Estimates of Change

**121. Surface area of a cone** Write a formula that estimates the change that occurs in the lateral surface area of a right circular cone when the height changes from $h_0$ to $h_0 + dh$ and the radius does not change.

$$V = \frac{1}{3}\pi r^2 h$$
$$S = \pi r\sqrt{r^2 + h^2}$$
(Lateral surface area)

**122. Controlling error**

   **a.** How accurately should you measure the edge of a cube to be reasonably sure of calculating the cube's surface area with an error of no more than 2%?

   **b.** Suppose that the edge is measured with the accuracy required in part (a). About how accurately can the cube's volume be calculated from the edge measurement? To find out, estimate the percentage error in the volume calculation that might result from using the edge measurement.

**123. Compounding error**   The circumference of the equator of a sphere is measured as 10 cm with a possible error of 0.4 cm. This measurement is used to calculate the radius. The radius is then used to calculate the surface area and volume of the sphere. Estimate the percentage errors in the calculated values of

   **a.** the radius.

   **b.** the surface area.

   **c.** the volume.

**124. Finding height**   To find the height of a lamppost (see accompanying figure), you stand a 1.8 m pole 10 m from the lamp and measure the length $a$ of its shadow, finding it to be 4.5 m, give or take a centimeter. Calculate the height of the lamppost using the value $a = 4.5$ and estimate the possible error in the result.

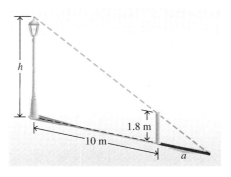

# Chapter 3   Additional and Advanced Exercises

**1.** An equation like $\sin^2\theta + \cos^2\theta = 1$ is called an **identity** because it holds for all values of $\theta$. An equation like $\sin\theta = 0.5$ is not an identity because it holds only for selected values of $\theta$, not all. If you differentiate both sides of a trigonometric identity in $\theta$ with respect to $\theta$, the resulting new equation will also be an identity.

   Differentiate the following to show that the resulting equations hold for all $\theta$.

   **a.** $\sin 2\theta = 2\sin\theta\cos\theta$

   **b.** $\cos 2\theta = \cos^2\theta - \sin^2\theta$

**2.** If the identity $\sin(x + a) = \sin x \cos a + \cos x \sin a$ is differentiated with respect to $x$, is the resulting equation also an identity? Does this principle apply to the equation $x^2 - 2x - 8 = 0$? Explain.

**3.** **a.** Find values for the constants $a$, $b$, and $c$ that will make

$$f(x) = \cos x \quad \text{and} \quad g(x) = a + bx + cx^2$$

   satisfy the conditions

$$f(0) = g(0), \quad f'(0) = g'(0), \quad \text{and} \quad f''(0) = g''(0).$$

   **b.** Find values for $b$ and $c$ that will make

$$f(x) = \sin(x + a) \quad \text{and} \quad g(x) = b\sin x + c\cos x$$

   satisfy the conditions

$$f(0) = g(0) \quad \text{and} \quad f'(0) = g'(0).$$

   **c.** For the determined values of $a$, $b$, and $c$, what happens for the third and fourth derivatives of $f$ and $g$ in each of parts (a) and (b)?

**4. Solutions to differential equations**

   **a.** Show that $y = \sin x$, $y = \cos x$, and $y = a\cos x + b\sin x$ ($a$ and $b$ constants) all satisfy the equation

$$y'' + y = 0.$$

   **b.** How would you modify the functions in part (a) to satisfy the equation

$$y'' + 4y = 0?$$

   Generalize this result.

**5. An osculating circle**   Find the values of $h$, $k$, and $a$ that make the circle $(x - h)^2 + (y - k)^2 = a^2$ tangent to the parabola $y = x^2 + 1$ at the point $(1, 2)$ and that also make the second derivatives $d^2y/dx^2$ have the same value on both curves there. Circles like this one that are tangent to a curve and have the same second derivative as the curve at the point of tangency are called *osculating circles* (from the Latin *osculari*, meaning "to kiss"). We encounter them again in Chapter 13.

**6. Marginal revenue**   A bus will hold 60 people. The number $x$ of people per trip who use the bus is related to the fare charged ($p$ dollars) by the law $p = [3 - (x/40)]^2$. Write an expression for the total revenue $r(x)$ per trip received by the bus company. What number of people per trip will make the marginal revenue $dr/dx$ equal to zero? What is the corresponding fare? (This fare is the one that maximizes the revenue.)

**7. Industrial production**

   **a.** Economists often use the expression "rate of growth" in relative rather than absolute terms. For example, let $u = f(t)$ be the number of people in the labor force at time $t$ in a given industry. (We treat this function as though it were differentiable even though it is an integer-valued step function.)

     Let $v = g(t)$ be the average production per person in the labor force at time $t$. The total production is then $y = uv$. If the labor force is growing at the rate of 4% per year ($du/dt = 0.04u$) and the production per worker is growing at the rate of 5% per year ($dv/dt = 0.05v$), find the rate of growth of the total production, $y$.

   **b.** Suppose that the labor force in part (a) is decreasing at the rate of 2% per year while the production per person is increasing at the rate of 3% per year. Is the total production increasing, or is it decreasing, and at what rate?

8. **Designing a gondola** The designer of a 10-m-diameter spherical hot air balloon wants to suspend the gondola 2.5 m below the bottom of the balloon with cables tangent to the surface of the balloon, as shown. Two of the cables are shown running from the top edges of the gondola to their points of tangency, $(-4, -3)$ and $(4, -3)$. How wide should the gondola be?

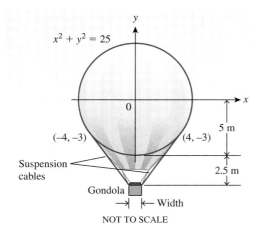

NOT TO SCALE

9. **Pisa by parachute** On August 5, 1988, Mike McCarthy of London jumped from the top of the Tower of Pisa. He then opened his parachute in what he said was a world record low-level parachute jump of 54.6 m. Make a rough sketch to show the shape of the graph of his speed during the jump. (*Source: Boston Globe*, Aug. 6, 1988.)

10. **Motion of a particle** The position at time $t \geq 0$ of a particle moving along a coordinate line is

$$s = 10\cos(t + \pi/4).$$

a. What is the particle's starting position ($t = 0$)?

b. What are the points farthest to the left and right of the origin reached by the particle?

c. Find the particle's velocity and acceleration at the points in part (b).

d. When does the particle first reach the origin? What are its velocity, speed, and acceleration then?

11. **Shooting a paper clip** On Earth, you can easily shoot a paper clip 19.6 m straight up into the air with a rubber band. In $t$ s after firing, the paper clip is $s = 19.6t - 4.9t^2$ m above your hand.

a. How long does it take the paper clip to reach its maximum height? With what velocity does it leave your hand?

b. On the moon, the same acceleration will send the paper clip to a height of $s = 19.6t - 0.8t^2$ m in $t$ s. About how long will it take the paper clip to reach its maximum height, and how high will it go?

12. **Velocities of two particles** At time $t$ s, the positions of two particles on a coordinate line are $s_1 = 3t^3 - 12t^2 + 18t + 5$ m and $s_2 = -t^3 + 9t^2 - 12t$ m. When do the particles have the same velocities?

13. **Velocity of a particle** A particle of constant mass $m$ moves along the $x$-axis. Its velocity $v$ and position $x$ satisfy the equation

$$\frac{1}{2}m(v^2 - v_0{}^2) = \frac{1}{2}k(x_0{}^2 - x^2),$$

where $k$, $v_0$, and $x_0$ are constants. Show that whenever $v \neq 0$,

$$m\frac{dv}{dt} = -kx.$$

14. **Average and instantaneous velocity**

a. Show that if the position $x$ of a moving point is given by a quadratic function of $t$, $x = At^2 + Bt + C$, then the average velocity over any time interval $[t_1, t_2]$ is equal to the instantaneous velocity at the midpoint of the time interval.

b. What is the geometric significance of the result in part (a)?

15. Find all values of the constants $m$ and $b$ for which the function

$$y = \begin{cases} \sin x, & x < \pi \\ mx + b, & x \geq \pi \end{cases}$$

is

a. continuous at $x = \pi$.

b. differentiable at $x = \pi$.

16. Does the function

$$f(x) = \begin{cases} \dfrac{1 - \cos x}{x}, & x \neq 0 \\ 0, & x = 0 \end{cases}$$

have a derivative at $x = 0$? Explain.

17. a. For what values of $a$ and $b$ will

$$f(x) = \begin{cases} ax, & x < 2 \\ ax^2 - bx + 3, & x \geq 2 \end{cases}$$

be differentiable for all values of $x$?

b. Discuss the geometry of the resulting graph of $f$.

18. a. For what values of $a$ and $b$ will

$$g(x) = \begin{cases} ax + b, & x \leq -1 \\ ax^3 + x + 2b, & x > -1 \end{cases}$$

be differentiable for all values of $x$?

b. Discuss the geometry of the resulting graph of $g$.

19. **Odd differentiable functions** Is there anything special about the derivative of an odd differentiable function of $x$? Give reasons for your answer.

20. **Even differentiable functions** Is there anything special about the derivative of an even differentiable function of $x$? Give reasons for your answer.

21. Suppose that the functions $f$ and $g$ are defined throughout an open interval containing the point $x_0$, that $f$ is differentiable at $x_0$, that $f(x_0) = 0$, and that $g$ is continuous at $x_0$. Show that the product $fg$ is differentiable at $x_0$. This process shows, for example, that although $|x|$ is not differentiable at $x = 0$, the product $x|x|$ is differentiable at $x = 0$.

22. (*Continuation of Exercise 21.*) Use the result of Exercise 21 to show that the following functions are differentiable at $x = 0$.

a. $|x| \sin x$    b. $x^{2/3} \sin x$    c. $\sqrt[3]{x}(1 - \cos x)$

d. $h(x) = \begin{cases} x^2 \sin(1/x), & x \neq 0 \\ 0, & x = 0 \end{cases}$

**23.** Is the derivative of

$$h(x) = \begin{cases} x^2 \sin (1/x), & x \neq 0 \\ 0, & x = 0 \end{cases}$$

continuous at $x = 0$? How about the derivative of $k(x) = xh(x)$? Give reasons for your answers.

**24.** Suppose that a function $f$ satisfies the following conditions for all real values of $x$ and $y$:

   **i)** $f(x + y) = f(x) \cdot f(y)$.

   **ii)** $f(x) = 1 + xg(x)$, where $\lim_{x \to 0} g(x) = 1$.

Show that the derivative $f'(x)$ exists at every value of $x$ and that $f'(x) = f(x)$.

**25. The generalized product rule**    Use mathematical induction to prove that if $y = u_1 u_2 \cdots u_n$ is a finite product of differentiable functions, then $y$ is differentiable on their common domain and

$$\frac{dy}{dx} = \frac{du_1}{dx} u_2 \cdots u_n + u_1 \frac{du_2}{dx} \cdots u_n + \cdots + u_1 u_2 \cdots u_{n-1} \frac{du_n}{dx}.$$

**26. Leibniz's rule for higher-order derivatives of products**    Leibniz's rule for higher-order derivatives of products of differentiable functions says that

   **a.** $\dfrac{d^2(uv)}{dx^2} = \dfrac{d^2u}{dx^2} v + 2\dfrac{du}{dx}\dfrac{dv}{dx} + u\dfrac{d^2v}{dx^2}.$

   **b.** $\dfrac{d^3(uv)}{dx^3} = \dfrac{d^3u}{dx^3} v + 3\dfrac{d^2u}{dx^2}\dfrac{dv}{dx} + 3\dfrac{du}{dx}\dfrac{d^2v}{dx^2} + u\dfrac{d^3v}{dx^3}.$

   **c.** $\dfrac{d^n(uv)}{dx^n} = \dfrac{d^nu}{dx^n} v + n\dfrac{d^{n-1}u}{dx^{n-1}}\dfrac{dv}{dx} + \cdots$

   $$+ \frac{n(n-1)\cdots(n-k+1)}{k!} \frac{d^{n-k}u}{dx^{n-k}} \frac{d^k v}{dx^k}$$

   $$+ \cdots + u\frac{d^n v}{dx^n}.$$

The equations in parts (a) and (b) are special cases of the equation in part (c). Derive the equation in part (c) by mathematical induction, using

$$\binom{m}{k} + \binom{m}{k+1} = \frac{m!}{k!(m-k)!} + \frac{m!}{(k+1)!(m-k-1)!}.$$

**27. The period of a clock pendulum**    The period $T$ of a clock pendulum (time for one full swing and back) is given by the formula $T^2 = 4\pi^2 L/g$, where $T$ is measured in seconds, $g = 9.8 \text{ m/s}^2$, and $L$, the length of the pendulum, is measured in meters. Find approximately

   **a.** the length of a clock pendulum whose period is $T = 1 \text{ s}$.

   **b.** the change $dT$ in $T$ if the pendulum in part (a) is lengthened 0.01 m.

   **c.** the amount the clock gains or loses in a day as a result of the period's changing by the amount $dT$ found in part (b).

**28. The melting ice cube**    Assume that an ice cube retains its cubical shape as it melts. If we call its edge length $s$, its volume is $V = s^3$ and its surface area is $6s^2$. We assume that $V$ and $s$ are differentiable functions of time $t$. We assume also that the cube's volume decreases at a rate that is proportional to its surface area. (This latter assumption seems reasonable enough when we think that the melting takes place at the surface: Changing the amount of surface changes the amount of ice exposed to melt.) In mathematical terms,

$$\frac{dV}{dt} = -k(6s^2), \qquad k > 0.$$

The minus sign indicates that the volume is decreasing. We assume that the proportionality factor $k$ is constant. (It probably depends on many things, such as the relative humidity of the surrounding air, the air temperature, and the incidence or absence of sunlight, to name only a few.) Assume a particular set of conditions in which the cube lost $1/4$ of its volume during the first hour, and that the volume is $V_0$ when $t = 0$. How long will it take the ice cube to melt?

---

## Chapter **3**    Technology Application Projects

**Mathematica/Maple Modules:**

### Convergence of Secant Slopes to the Derivative Function
You will visualize the secant line between successive points on a curve and observe what happens as the distance between them becomes small. The function, sample points, and secant lines are plotted on a single graph, while a second graph compares the slopes of the secant lines with the derivative function.

### Derivatives, Slopes, Tangent Lines, and Making Movies
**Parts I–III.** You will visualize the derivative at a point, the linearization of a function, and the derivative of a function. You learn how to plot the function and selected tangents on the same graph.
**Part IV (Plotting Many Tangents)**
**Part V (Making Movies).** Parts IV and V of the module can be used to animate tangent lines as one moves along the graph of a function.

### Convergence of Secant Slopes to the Derivative Function
You will visualize right-hand and left-hand derivatives.

### Motion Along a Straight Line: Position → Velocity → Acceleration
Observe dramatic animated visualizations of the derivative relations among the position, velocity, and acceleration functions. Figures in the text can be animated.

# 4

# Applications of Derivatives

**OVERVIEW** One of the most important applications of the derivative is its use as a tool for finding the optimal (best) solutions to problems. Optimization problems abound in mathematics, physical science and engineering, business and economics, and biology and medicine. For example, what are the height and diameter of the cylinder of largest volume that can be inscribed in a given sphere? What are the dimensions of the strongest rectangular wooden beam that can be cut from a cylindrical log of given diameter? Based on production costs and sales revenue, how many items should a manufacturer produce to maximize profit? How much does the trachea (windpipe) contract to expel air at the maximum speed during a cough? What is the branching angle at which blood vessels minimize the energy loss due to friction as blood flows through the branches?

In this chapter we use derivatives to find extreme values of functions, to determine and analyze the shapes of graphs, and to solve equations numerically. We also introduce the idea of recovering a function from its derivative. The key to many of these applications is the Mean Value Theorem, which paves the way to integral calculus.

## 4.1 Extreme Values of Functions

This section shows how to locate and identify extreme (maximum or minimum) values of a function from its derivative. Once we can do this, we can solve a variety of optimization problems (see Section 4.5). The domains of the functions we consider are intervals or unions of separate intervals.

> **DEFINITIONS** Let $f$ be a function with domain $D$. Then $f$ has an **absolute maximum** value on $D$ at a point $c$ if
> $$f(x) \leq f(c) \qquad \text{for all } x \text{ in } D$$
> and an **absolute minimum** value on $D$ at $c$ if
> $$f(x) \geq f(c) \qquad \text{for all } x \text{ in } D.$$

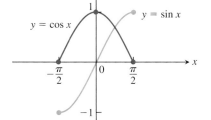

**FIGURE 4.1** Absolute extrema for the sine and cosine functions on $[-\pi/2, \pi/2]$. These values can depend on the domain of a function.

Maximum and minimum values are called **extreme values** of the function $f$. Absolute maxima or minima are also referred to as **global** maxima or minima.

For example, on the closed interval $[-\pi/2, \pi/2]$ the function $f(x) = \cos x$ takes on an absolute maximum value of 1 (once) and an absolute minimum value of 0 (twice). On the same interval, the function $g(x) = \sin x$ takes on a maximum value of 1 and a minimum value of $-1$ (Figure 4.1).

Functions with the same defining rule or formula can have different extrema (maximum or minimum values), depending on the domain. We see this in the following example.

**EXAMPLE 1**    The absolute extrema of the following functions on their domains can be seen in Figure 4.2. Each function has the same defining equation, $y = x^2$, but the domains vary. Notice that a function might not have a maximum or minimum if the domain is unbounded or fails to contain an endpoint.

| Function rule | Domain $D$ | Absolute extrema on $D$ |
|---|---|---|
| **(a)** $y = x^2$ | $(-\infty, \infty)$ | No absolute maximum |
| | | Absolute minimum of 0 at $x = 0$ |
| **(b)** $y = x^2$ | $[0, 2]$ | Absolute maximum of 4 at $x = 2$ |
| | | Absolute minimum of 0 at $x = 0$ |
| **(c)** $y = x^2$ | $(0, 2]$ | Absolute maximum of 4 at $x = 2$ |
| | | No absolute minimum |
| **(d)** $y = x^2$ | $(0, 2)$ | No absolute extrema |

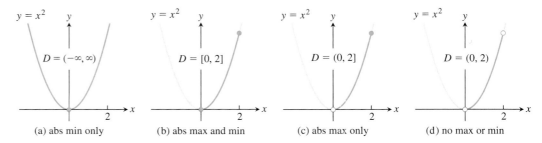

**(a)** abs min only    **(b)** abs max and min    **(c)** abs max only    **(d)** no max or min

**FIGURE 4.2**    Graphs for Example 1.

**HISTORICAL BIOGRAPHY**

Daniel Bernoulli
(1700–1789)

Some of the functions in Example 1 did not have a maximum or a minimum value. The following theorem asserts that a function which is *continuous* over (or on) a finite *closed* interval $[a, b]$ has an absolute maximum and an absolute minimum value on the interval. We look for these extreme values when we graph a function.

> **THEOREM 1—The Extreme Value Theorem**    If $f$ is continuous on a closed interval $[a, b]$, then $f$ attains both an absolute maximum value $M$ and an absolute minimum value $m$ in $[a, b]$. That is, there are numbers $x_1$ and $x_2$ in $[a, b]$ with $f(x_1) = m$, $f(x_2) = M$, and $m \leq f(x) \leq M$ for every other $x$ in $[a, b]$.

The proof of the Extreme Value Theorem requires a detailed knowledge of the real number system (see Appendix 6) and we will not give it here. Figure 4.3 illustrates possible locations for the absolute extrema of a continuous function on a closed interval $[a, b]$. As we observed for the function $y = \cos x$, it is possible that an absolute minimum (or absolute maximum) may occur at two or more different points of the interval.

The requirements in Theorem 1 that the interval be closed and finite, and that the function be continuous, are key ingredients. Without them, the conclusion of the theorem

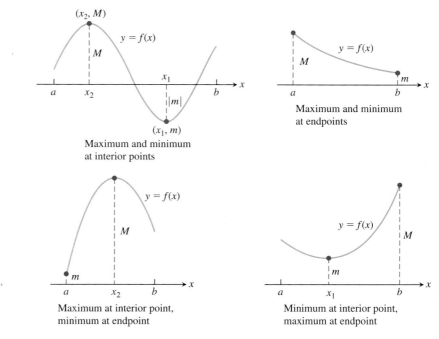

**FIGURE 4.3** Some possibilities for a continuous function's maximum and minimum on a closed interval $[a, b]$.

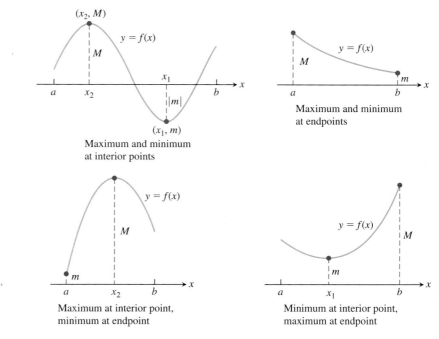

**FIGURE 4.4** Even a single point of discontinuity can keep a function from having either a maximum or minimum value on a closed interval. The function

$$y = \begin{cases} x, & 0 \le x < 1 \\ 0, & x = 1 \end{cases}$$

is continuous at every point of $[0, 1]$ except $x = 1$, yet its graph over $[0, 1]$ does not have a highest point.

need not hold. Example 1 shows that an absolute extreme value may not exist if the interval fails to be both closed and finite. The function $y = x$ over $(-\infty, \infty)$ shows that neither extreme value need exist on an infinite interval. Figure 4.4 shows that the continuity requirement cannot be omitted.

## Local (Relative) Extreme Values

Figure 4.5 shows a graph with five points where a function has extreme values on its domain $[a, b]$. The function's absolute minimum occurs at $a$ even though at $e$ the function's value is smaller than at any other point *nearby*. The curve rises to the left and falls to the right around $c$, making $f(c)$ a maximum locally. The function attains its absolute maximum at $d$. We now define what we mean by local extrema.

---

**DEFINITIONS** A function $f$ has a **local maximum** value at a point $c$ within its domain $D$ if $f(x) \le f(c)$ for all $x \in D$ lying in some open interval containing $c$.

A function $f$ has a **local minimum** value at a point $c$ within its domain $D$ if $f(x) \ge f(c)$ for all $x \in D$ lying in some open interval containing $c$.

---

If the domain of $f$ is the closed interval $[a, b]$, then $f$ has a local maximum at the endpoint $x = a$, if $f(x) \le f(a)$ for all $x$ in some half-open interval $[a, a + \delta), \delta > 0$. Likewise, $f$ has a local maximum at an interior point $x = c$ if $f(x) \le f(c)$ for all $x$ in some open interval $(c - \delta, c + \delta), \delta > 0$, and a local maximum at the endpoint $x = b$ if $f(x) \le f(b)$ for all $x$ in some half-open interval $(b - \delta, b], \delta > 0$. The inequalities are reversed for local minimum values. In Figure 4.5, the function $f$ has local maxima at $c$ and $d$ and local minima at $a$, $e$, and $b$. Local extrema are also called **relative extrema**. Some functions can have infinitely many local extrema, even over a finite interval. One example is the function $f(x) = \sin(1/x)$ on the interval $(0, 1]$. (We graphed this function in Figure 2.40.)

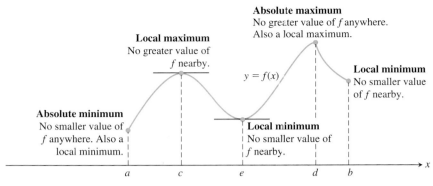

**FIGURE 4.5** How to identify types of maxima and minima for a function with domain $a \leq x \leq b$.

An absolute maximum is also a local maximum. Being the largest value overall, it is also the largest value in its immediate neighborhood. Hence, *a list of all local maxima will automatically include the absolute maximum if there is one.* Similarly, *a list of all local minima will include the absolute minimum if there is one.*

### Finding Extrema

The next theorem explains why we usually need to investigate only a few values to find a function's extrema.

> **THEOREM 2—The First Derivative Theorem for Local Extreme Values** If $f$ has a local maximum or minimum value at an interior point $c$ of its domain, and if $f'$ is defined at $c$, then
>
> $$f'(c) = 0.$$

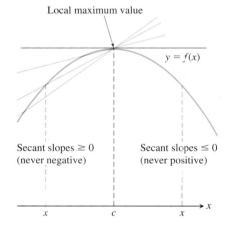

**FIGURE 4.6** A curve with a local maximum value. The slope at $c$, simultaneously the limit of nonpositive numbers and nonnegative numbers, is zero.

**Proof** To prove that $f'(c)$ is zero at a local extremum, we show first that $f'(c)$ cannot be positive and second that $f'(c)$ cannot be negative. The only number that is neither positive nor negative is zero, so that is what $f'(c)$ must be.

To begin, suppose that $f$ has a local maximum value at $x = c$ (Figure 4.6) so that $f(x) - f(c) \leq 0$ for all values of $x$ near enough to $c$. Since $c$ is an interior point of $f$'s domain, $f'(c)$ is defined by the two-sided limit

$$\lim_{x \to c} \frac{f(x) - f(c)}{x - c}.$$

This means that the right-hand and left-hand limits both exist at $x = c$ and equal $f'(c)$. When we examine these limits separately, we find that

$$f'(c) = \lim_{x \to c^+} \frac{f(x) - f(c)}{x - c} \leq 0. \qquad \text{Because } (x - c) > 0 \text{ and } f(x) \leq f(c) \qquad (1)$$

Similarly,

$$f'(c) = \lim_{x \to c^-} \frac{f(x) - f(c)}{x - c} \geq 0. \qquad \text{Because } (x - c) < 0 \text{ and } f(x) \leq f(c) \qquad (2)$$

Together, Equations (1) and (2) imply $f'(c) = 0$.

This proves the theorem for local maximum values. To prove it for local minimum values, we simply use $f(x) \geq f(c)$, which reverses the inequalities in Equations (1) and (2). ∎

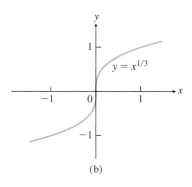

**FIGURE 4.7** Critical points without extreme values. (a) $y' = 3x^2$ is 0 at $x = 0$, but $y = x^3$ has no extremum there. (b) $y' = (1/3)x^{-2/3}$ is undefined at $x = 0$, but $y = x^{1/3}$ has no extremum there.

Theorem 2 says that a function's first derivative is always zero at an interior point where the function has a local extreme value and the derivative is defined. If we recall that all the domains we consider are intervals or unions of separate intervals, the only places where a function $f$ can possibly have an extreme value (local or global) are

1. interior points where $f' = 0$,  At $x = c$ and $x = e$ in Fig. 4.5
2. interior points where $f'$ is undefined,  At $x = d$ in Fig. 4.5
3. endpoints of the domain of $f$.  At $x = a$ and $x = b$ in Fig. 4.5

The following definition helps us to summarize these results.

> **DEFINITION** An interior point of the domain of a function $f$ where $f'$ is zero or undefined is a **critical point** of $f$.

Thus the only domain points where a function can assume extreme values are critical points and endpoints. However, be careful not to misinterpret what is being said here. A function may have a critical point at $x = c$ without having a local extreme value there. For instance, both of the functions $y = x^3$ and $y = x^{1/3}$ have critical points at the origin, but neither function has a local extreme value at the origin. Instead, each function has a *point of inflection* there (see Figure 4.7). We define and explore inflection points in Section 4.4.

Most problems that ask for extreme values call for finding the absolute extrema of a continuous function on a closed and finite interval. Theorem 1 assures us that such values exist; Theorem 2 tells us that they are taken on only at critical points and endpoints. Often we can simply list these points and calculate the corresponding function values to find what the largest and smallest values are, and where they are located. Of course, if the interval is not closed or not finite (such as $a < x < b$ or $a < x < \infty$), we have seen that absolute extrema need not exist. If an absolute maximum or minimum value does exist, it must occur at a critical point or at an included right- or left-hand endpoint of the interval.

> **How to Find the Absolute Extrema of a Continuous Function $f$ on a Finite Closed Interval**
> 1. Evaluate $f$ at all critical points and endpoints.
> 2. Take the largest and smallest of these values.

**EXAMPLE 2** Find the absolute maximum and minimum values of $f(x) = x^2$ on $[-2, 1]$.

**Solution** The function is differentiable over its entire domain, so the only critical point is where $f'(x) = 2x = 0$, namely $x = 0$. We need to check the function's values at $x = 0$ and at the endpoints $x = -2$ and $x = 1$:

Critical point value:  $f(0) = 0$

Endpoint values:  $f(-2) = 4$
  $f(1) = 1.$

The function has an absolute maximum value of 4 at $x = -2$ and an absolute minimum value of 0 at $x = 0$.

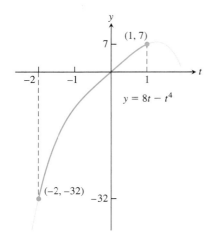

**FIGURE 4.8** The extreme values of $g(t) = 8t - t^4$ on $[-2, 1]$ (Example 3).

**EXAMPLE 3** Find the absolute maximum and minimum values of $g(t) = 8t - t^4$ on $[-2, 1]$.

**Solution** The function is differentiable on its entire domain, so the only critical points occur where $g'(t) = 0$. Solving this equation gives

$$8 - 4t^3 = 0 \quad \text{or} \quad t = \sqrt[3]{2} > 1,$$

a point not in the given domain. The function's absolute extrema therefore occur at the endpoints, $g(-2) = -32$ (absolute minimum), and $g(1) = 7$ (absolute maximum). See Figure 4.8. ■

**EXAMPLE 4** Find the absolute maximum and minimum values of $f(x) = x^{2/3}$ on the interval $[-2, 3]$.

**Solution** We evaluate the function at the critical points and endpoints and take the largest and smallest of the resulting values.

The first derivative

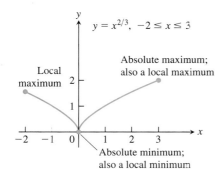

$$f'(x) = \frac{2}{3}x^{-1/3} = \frac{2}{3\sqrt[3]{x}}$$

has no zeros but is undefined at the interior point $x = 0$. The values of $f$ at this one critical point and at the endpoints are

Critical point value:    $f(0) = 0$

Endpoint values:    $f(-2) = (-2)^{2/3} = \sqrt[3]{4}$
$f(3) = (3)^{2/3} = \sqrt[3]{9}.$

**FIGURE 4.9** The extreme values of $f(x) = x^{2/3}$ on $[-2, 3]$ occur at $x = 0$ and $x = 3$ (Example 4).

We can see from this list that the function's absolute maximum value is $\sqrt[3]{9} \approx 2.08$, and it occurs at the right endpoint $x = 3$. The absolute minimum value is 0, and it occurs at the interior point $x = 0$ where the graph has a cusp (Figure 4.9). ■

## Exercises 4.1

**Finding Extrema from Graphs**
In Exercises 1–6, determine from the graph whether the function has any absolute extreme values on $[a, b]$. Then explain how your answer is consistent with .Theorem 1

**1.**

**2.**

**3.** **4.**

**5.**

**6.**

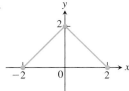

In Exercises 7–10, find the absolute extreme values and where they occur.

**7.**

**8.**

**9.**

**10.**

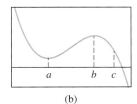

In Exercises 11–14, match the table with a graph.

**11.**

| $x$ | $f'(x)$ |
|---|---|
| $a$ | 0 |
| $b$ | 0 |
| $c$ | 5 |

**12.**

| $x$ | $f'(x)$ |
|---|---|
| $a$ | 0 |
| $b$ | 0 |
| $c$ | $-5$ |

**13.**

| $x$ | $f'(x)$ |
|---|---|
| $a$ | does not exist |
| $b$ | 0 |
| $c$ | $-2$ |

**14.**

| $x$ | $f'(x)$ |
|---|---|
| $a$ | does not exist |
| $b$ | does not exist |
| $c$ | $-1.7$ |

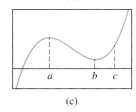

In Exercises 15–20, sketch the graph of each function and determine whether the function has any absolute extreme values on its domain. Explain how your answer is consistent with Theorem 1.

**15.** $f(x) = |x|, \quad -1 < x < 2$

**16.** $y = \dfrac{6}{x^2 + 2}, \quad -1 < x < 1$

**17.** $g(x) = \begin{cases} -x, & 0 \le x < 1 \\ x - 1, & 1 \le x \le 2 \end{cases}$

**18.** $h(x) = \begin{cases} \dfrac{1}{x}, & -1 \le x < 0 \\ \sqrt{x}, & 0 \le x \le 4 \end{cases}$

**19.** $y = 3 \sin x, \quad 0 < x < 2\pi$

**20.** $f(x) = \begin{cases} x + 1, & -1 \le x < 0 \\ \cos x, & 0 < x \le \dfrac{\pi}{2} \end{cases}$

**Absolute Extrema on Finite Closed Intervals**
In Exercises 21–40, find the absolute maximum and minimum values of each function on the given interval. Then graph the function. Identify the points on the graph where the absolute extrema occur, and include their coordinates.

**21.** $f(x) = \dfrac{2}{3}x - 5, \quad -2 \le x \le 3$

**22.** $f(x) = -x - 4, \quad -4 \le x \le 1$

**23.** $f(x) = x^2 - 1, \quad -1 \le x \le 2$

**24.** $f(x) = 4 - x^3, \quad -2 \le x \le 1$

**25.** $F(x) = -\dfrac{1}{x^2}, \quad 0.5 \le x \le 2$

**26.** $F(x) = -\dfrac{1}{x}, \quad -2 \le x \le -1$

**27.** $h(x) = \sqrt[3]{x}, \quad -1 \le x \le 8$

**28.** $h(x) = -3x^{2/3}, \quad -1 \le x \le 1$

**29.** $g(x) = \sqrt{4 - x^2}, \quad -2 \le x \le 1$

**30.** $g(x) = -\sqrt{5 - x^2}, \quad -\sqrt{5} \le x \le 0$

**31.** $f(\theta) = \sin \theta, \quad -\dfrac{\pi}{2} \le \theta \le \dfrac{5\pi}{6}$

**32.** $f(\theta) = \tan \theta, \quad -\dfrac{\pi}{3} \le \theta \le \dfrac{\pi}{4}$

**33.** $g(x) = \csc x, \quad \dfrac{\pi}{3} \le x \le \dfrac{2\pi}{3}$

**34.** $g(x) = \sec x, \quad -\dfrac{\pi}{3} \le x \le \dfrac{\pi}{6}$

**35.** $f(t) = 2 - |t|, \quad -1 \le t \le 3$

**36.** $f(t) = |t - 5|, \quad 4 \le t \le 7$

In Exercises 37–40, find the function's absolute maximum and minimum values and say where they are assumed.

**37.** $f(x) = x^{4/3}, \quad -1 \le x \le 8$

**38.** $f(x) = x^{5/3}, \quad -1 \le x \le 8$

**39.** $g(\theta) = \theta^{3/5}, \quad -32 \le \theta \le 1$

**40.** $h(\theta) = 3\theta^{2/3}, \quad -27 \le \theta \le 8$

## Finding Critical Points
In Exercises 41–48, determine all critical points for each function.

**41.** $y = x^2 - 6x + 7$

**42.** $f(x) = 6x^2 - x^3$

**43.** $f(x) = x(4 - x)^3$

**44.** $g(x) = (x - 1)^2(x - 3)^2$

**45.** $y = x^2 + \dfrac{2}{x}$

**46.** $f(x) = \dfrac{x^2}{x - 2}$

**47.** $y = x^2 - 32\sqrt{x}$

**48.** $g(x) = \sqrt{2x - x^2}$

## Finding Extreme Values
In Exercises 49–58, find the extreme values (absolute and local) of the function over its natural domain, and where they occur.

**49.** $y = 2x^2 - 8x + 9$

**50.** $y = x^3 - 2x + 4$

**51.** $y = x^3 + x^2 - 8x + 5$

**52.** $y = x^3(x - 5)^2$

**53.** $y = \sqrt{x^2 - 1}$

**54.** $y = x - 4\sqrt{x}$

**55.** $y = \dfrac{1}{\sqrt[3]{1 - x^2}}$

**56.** $y = \sqrt{3 + 2x - x^2}$

**57.** $y = \dfrac{x}{x^2 + 1}$

**58.** $y = \dfrac{x + 1}{x^2 + 2x + 2}$

## Local Extrema and Critical Points
In Exercises 59–66, find the critical points, domain endpoints, and extreme values (absolute and local) for each function.

**59.** $y = x^{2/3}(x + 2)$

**60.** $y = x^{2/3}(x^2 - 4)$

**61.** $y = x\sqrt{4 - x^2}$

**62.** $y = x^2\sqrt{3 - x}$

**63.** $y = \begin{cases} 4 - 2x, & x \le 1 \\ x + 1, & x > 1 \end{cases}$

**64.** $y = \begin{cases} 3 - x, & x < 0 \\ 3 + 2x - x^2, & x \ge 0 \end{cases}$

**65.** $y = \begin{cases} -x^2 - 2x + 4, & x \le 1 \\ -x^2 + 6x - 4, & x > 1 \end{cases}$

**66.** $y = \begin{cases} -\dfrac{1}{4}x^2 - \dfrac{1}{2}x + \dfrac{15}{4}, & x \le 1 \\ x^3 - 6x^2 + 8x, & x > 1 \end{cases}$

In Exercises 67 and 68, give reasons for your answers.

**67.** Let $f(x) = (x - 2)^{2/3}$.

**a.** Does $f'(2)$ exist?

**b.** Show that the only local extreme value of $f$ occurs at $x = 2$.

**c.** Does the result in part (b) contradict the Extreme Value Theorem?

**d.** Repeat parts (a) and (b) for $f(x) = (x - a)^{2/3}$, replacing 2 by $a$.

**68.** Let $f(x) = |x^3 - 9x|$.

**a.** Does $f'(0)$ exist?

**b.** Does $f'(3)$ exist?

**c.** Does $f'(-3)$ exist?

**d.** Determine all extrema of $f$.

## Theory and Examples
**69. A minimum with no derivative**   The function $f(x) = |x|$ has an absolute minimum value at $x = 0$ even though $f$ is not differentiable at $x = 0$. Is this consistent with Theorem 2? Give reasons for your answer.

**70. Even functions**   If an even function $f(x)$ has a local maximum value at $x = c$, can anything be said about the value of $f$ at $x = -c$? Give reasons for your answer.

**71. Odd functions**   If an odd function $g(x)$ has a local minimum value at $x = c$, can anything be said about the value of $g$ at $x = -c$? Give reasons for your answer.

**72. No critical points or endpoints exist**   We know how to find the extreme values of a continuous function $f(x)$ by investigating its values at critical points and endpoints. But what if there *are* no critical points or endpoints? What happens then? Do such functions really exist? Give reasons for your answers.

**73.** The function
$$V(x) = x(10 - 2x)(16 - 2x), \qquad 0 < x < 5,$$
models the volume of a box.

**a.** Find the extreme values of $V$.

**b.** Interpret any values found in part (a) in terms of the volume of the box.

**74. Cubic functions**   Consider the cubic function
$$f(x) = ax^3 + bx^2 + cx + d.$$

**a.** Show that $f$ can have 0, 1, or 2 critical points. Give examples and graphs to support your argument.

**b.** How many local extreme values can $f$ have?

**75. Maximum height of a vertically moving body**   The height of a body moving vertically is given by
$$s = -\frac{1}{2}gt^2 + v_0 t + s_0, \qquad g > 0,$$
with $s$ in meters and $t$ in seconds. Find the body's maximum height.

**76. Peak alternating current**   Suppose that at any given time $t$ (in seconds) the current $i$ (in amperes) in an alternating current circuit is $i = 2\cos t + 2\sin t$. What is the peak current for this circuit (largest magnitude)?

**T**  Graph the functions in Exercises 77–80. Then find the extreme values of the function on the interval and say where they occur.

**77.** $f(x) = |x - 2| + |x + 3|, \quad -5 \le x \le 5$

**78.** $g(x) = |x - 1| - |x - 5|, \quad -2 \le x \le 7$

**79.** $h(x) = |x + 2| - |x - 3|, \quad -\infty < x < \infty$

**80.** $k(x) = |x + 1| + |x - 3|, \quad -\infty < x < \infty$

## COMPUTER EXPLORATIONS
In Exercises 81–86, you will use a CAS to help find the absolute extrema of the given function over the specified closed interval. Perform the following steps.

**a.** Plot the function over the interval to see its general behavior there.

**b.** Find the interior points where $f' = 0$. (In some exercises, you may have to use the numerical equation solver to approximate a solution.) You may want to plot $f'$ as well.

**c.** Find the interior points where $f'$ does not exist.

**d.** Evaluate the function at all points found in parts (b) and (c) and at the endpoints of the interval.

**e.** Find the function's absolute extreme values on the interval and identify where they occur.

**81.** $f(x) = x^4 - 8x^2 + 4x + 2, \quad [-20/25, 64/25]$

**82.** $f(x) = -x^4 + 4x^3 - 4x + 1, \quad [-3/4, 3]$

**83.** $f(x) = x^{2/3}(3 - x), \quad [-2, 2]$

**84.** $f(x) = 2 + 2x - 3x^{2/3}, \quad [-1, 10/3]$

**85.** $f(x) = \sqrt{x} + \cos x, \quad [0, 2\pi]$

**86.** $f(x) = x^{3/4} - \sin x + \dfrac{1}{2}, \quad [0, 2\pi]$

# $4.2$ The Mean Value Theorem

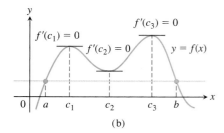

**FIGURE 4.10** Rolle's Theorem says that a differentiable curve has at least one horizontal tangent between any two points where it crosses a horizontal line. It may have just one (a), or it may have more (b).

We know that constant functions have zero derivatives, but could there be a more complicated function whose derivative is always zero? If two functions have identical derivatives over an interval, how are the functions related? We answer these and other questions in this chapter by applying the Mean Value Theorem. First we introduce a special case, known as Rolle's Theorem, which is used to prove the Mean Value Theorem.

## Rolle's Theorem

As suggested by its graph, if a differentiable function crosses a horizontal line at two different points, there is at least one point between them where the tangent to the graph is horizontal and the derivative is zero (Figure 4.10). We now state and prove this result.

---

**THEOREM 3—Rolle's Theorem**   Suppose that $y = f(x)$ is continuous over the closed interval $[a, b]$ and differentiable at every point of its interior $(a, b)$. If $f(a) = f(b)$, then there is at least one number $c$ in $(a, b)$ at which $f'(c) = 0$.

---

**Proof**   Being continuous, $f$ assumes absolute maximum and minimum values on $[a, b]$ by Theorem 1. These can occur only

**1.** at interior points where $f'$ is zero,

**2.** at interior points where $f'$ does not exist,

**3.** at endpoints of the function's domain, in this case $a$ and $b$.

By hypothesis, $f$ has a derivative at every interior point. That rules out possibility (2), leaving us with interior points where $f' = 0$ and with the two endpoints $a$ and $b$.

If either the maximum or the minimum occurs at a point $c$ between $a$ and $b$, then $f'(c) = 0$ by Theorem 2 in Section 4.1, and we have found a point for Rolle's Theorem.

If both the absolute maximum and the absolute minimum occur at the endpoints, then because $f(a) = f(b)$ it must be the case that $f$ is a constant function with $f(x) = f(a) = f(b)$ for every $x \in [a, b]$. Therefore $f'(x) = 0$ and the point $c$ can be taken anywhere in the interior $(a, b)$. ∎

The hypotheses of Theorem 3 are essential. If they fail at even one point, the graph may not have a horizontal tangent (Figure 4.11).

Rolle's Theorem may be combined with the Intermediate Value Theorem to show when there is only one real solution of an equation $f(x) = 0$, as we illustrate in the next example.

**HISTORICAL BIOGRAPHY**

Michel Rolle
(1652–1719)

**EXAMPLE 1**   Show that the equation

$$x^3 + 3x + 1 = 0$$

has exactly one real solution.

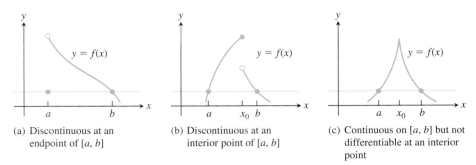

(a) Discontinuous at an endpoint of $[a, b]$

(b) Discontinuous at an interior point of $[a, b]$

(c) Continuous on $[a, b]$ but not differentiable at an interior point

**FIGURE 4.11** There may be no horizontal tangent if the hypotheses of Rolle's Theorem do not hold.

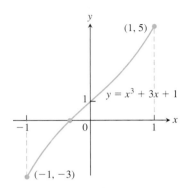

**FIGURE 4.12** The only real zero of the polynomial $y = x^3 + 3x + 1$ is the one shown here where the curve crosses the $x$-axis between $-1$ and $0$ (Example 1).

**Solution** We define the continuous function

$$f(x) = x^3 + 3x + 1.$$

Since $f(-1) = -3$ and $f(0) = 1$, the Intermediate Value Theorem tells us that the graph of $f$ crosses the $x$-axis somewhere in the open interval $(-1, 0)$. (See Figure 4.12.) Now, if there were even two points $x = a$ and $x = b$ where $f(x)$ was zero, Rolle's Theorem would guarantee the existence of a point $x = c$ in between them where $f'$ was zero. However, the derivative

$$f'(x) = 3x^2 + 3$$

is never zero (because it is always positive). Therefore, $f$ has no more than one zero. ∎

Our main use of Rolle's Theorem is in proving the Mean Value Theorem.

### The Mean Value Theorem

The Mean Value Theorem, which was first stated by Joseph-Louis Lagrange, is a slanted version of Rolle's Theorem (Figure 4.13). The Mean Value Theorem guarantees that there is a point where the tangent line is parallel to the secant joining $A$ and $B$.

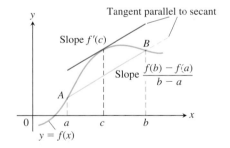

**FIGURE 4.13** Geometrically, the Mean Value Theorem says that somewhere between $a$ and $b$ the curve has at least one tangent parallel to the secant joining $A$ and $B$.

**THEOREM 4—The Mean Value Theorem** Suppose $y = f(x)$ is continuous over a closed interval $[a, b]$ and differentiable on the interval's interior $(a, b)$. Then there is at least one point $c$ in $(a, b)$ at which

$$\frac{f(b) - f(a)}{b - a} = f'(c). \tag{1}$$

**Proof** We picture the graph of $f$ and draw a line through the points $A(a, f(a))$ and $B(b, f(b))$. (See Figure 4.14.) The secant line is the graph of the function

$$g(x) = f(a) + \frac{f(b) - f(a)}{b - a}(x - a) \tag{2}$$

(point-slope equation). The vertical difference between the graphs of $f$ and $g$ at $x$ is

$$h(x) = f(x) - g(x)$$

$$= f(x) - f(a) - \frac{f(b) - f(a)}{b - a}(x - a). \tag{3}$$

Figure 4.15 shows the graphs of $f$, $g$, and $h$ together.

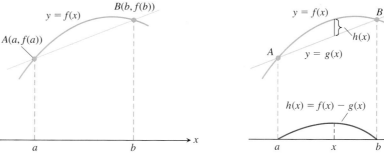

**FIGURE 4.14** The graph of $f$ and the secant $AB$ over the interval $[a, b]$.

**FIGURE 4.15** The secant $AB$ is the graph of the function $g(x)$. The function $h(x) = f(x) - g(x)$ gives the vertical distance between the graphs of $f$ and $g$ at $x$.

**FIGURE 4.16** The function $f(x) = \sqrt{1 - x^2}$ satisfies the hypotheses (and conclusion) of the Mean Value Theorem on $[-1, 1]$ even though $f$ is not differentiable at $-1$ and 1.

The function $h$ satisfies the hypotheses of Rolle's Theorem on $[a, b]$. It is continuous on $[a, b]$ and differentiable on $(a, b)$ because both $f$ and $g$ are. Also, $h(a) = h(b) = 0$ because the graphs of $f$ and $g$ both pass through $A$ and $B$. Therefore $h'(c) = 0$ at some point $c \in (a, b)$. This is the point we want for Equation (1) in the theorem.

To verify Equation (1), we differentiate both sides of Equation (3) with respect to $x$ and then set $x = c$:

$$h'(x) = f'(x) - \frac{f(b) - f(a)}{b - a} \qquad \text{Derivative of Eq. (3)} \ldots$$

$$h'(c) = f'(c) - \frac{f(b) - f(a)}{b - a} \qquad \ldots \text{with } x = c$$

$$0 = f'(c) - \frac{f(b) - f(a)}{b - a} \qquad h'(c) = 0$$

$$f'(c) = \frac{f(b) - f(a)}{b - a}, \qquad \text{Rearranged}$$

which is what we set out to prove. ∎

The hypotheses of the Mean Value Theorem do not require $f$ to be differentiable at either $a$ or $b$. One-sided continuity at $a$ and $b$ is enough (Figure 4.16).

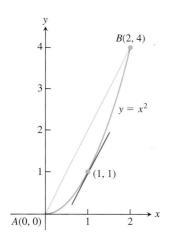

**FIGURE 4.17** As we find in Example 2, $c = 1$ is where the tangent is parallel to the secant line.

**EXAMPLE 2** The function $f(x) = x^2$ (Figure 4.17) is continuous for $0 \le x \le 2$ and differentiable for $0 < x < 2$. Since $f(0) = 0$ and $f(2) = 4$, the Mean Value Theorem says that at some point $c$ in the interval, the derivative $f'(x) = 2x$ must have the value $(4 - 0)/(2 - 0) = 2$. In this case we can identify $c$ by solving the equation $2c = 2$ to get $c = 1$. However, it is not always easy to find $c$ algebraically, even though we know it always exists. ∎

## A Physical Interpretation

We can think of the number $(f(b) - f(a))/(b - a)$ as the average change in $f$ over $[a, b]$ and $f'(c)$ as an instantaneous change. Then the Mean Value Theorem says that at some interior point the instantaneous change must equal the average change over the entire interval.

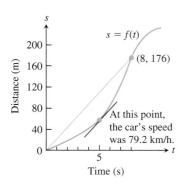

**FIGURE 4.18** Distance versus elapsed time for the car in Example 3.

**EXAMPLE 3** If a car accelerating from zero takes 8 s to go 176 m, its average velocity for the 8-s interval is $176/8 = 22$ m/s. The Mean Value Theorem says that at some point during the acceleration the speedometer must read exactly 79.2 km/h (22 m/s) (Figure 4.18). ∎

### Mathematical Consequences

At the beginning of the section, we asked what kind of function has a zero derivative over an interval. The first corollary of the Mean Value Theorem provides the answer that only constant functions have zero derivatives.

---

**COROLLARY 1** If $f'(x) = 0$ at each point $x$ of an open interval $(a, b)$, then $f(x) = C$ for all $x \in (a, b)$, where $C$ is a constant.

---

**Proof** We want to show that $f$ has a constant value on the interval $(a, b)$. We do so by showing that if $x_1$ and $x_2$ are any two points in $(a, b)$ with $x_1 < x_2$, then $f(x_1) = f(x_2)$. Now $f$ satisfies the hypotheses of the Mean Value Theorem on $[x_1, x_2]$: It is differentiable at every point of $[x_1, x_2]$ and hence continuous at every point as well. Therefore,

$$\frac{f(x_2) - f(x_1)}{x_2 - x_1} = f'(c)$$

at some point $c$ between $x_1$ and $x_2$. Since $f' = 0$ throughout $(a, b)$, this equation implies successively that

$$\frac{f(x_2) - f(x_1)}{x_2 - x_1} = 0, \qquad f(x_2) - f(x_1) = 0, \qquad \text{and} \qquad f(x_1) = f(x_2). \qquad \blacksquare$$

At the beginning of this section, we also asked about the relationship between two functions that have identical derivatives over an interval. The next corollary tells us that their values on the interval have a constant difference.

---

**COROLLARY 2** If $f'(x) = g'(x)$ at each point $x$ in an open interval $(a, b)$, then there exists a constant $C$ such that $f(x) = g(x) + C$ for all $x \in (a, b)$. That is, $f - g$ is a constant function on $(a, b)$.

---

**Proof** At each point $x \in (a, b)$ the derivative of the difference function $h = f - g$ is

$$h'(x) = f'(x) - g'(x) = 0.$$

Thus, $h(x) = C$ on $(a, b)$ by Corollary 1. That is, $f(x) - g(x) = C$ on $(a, b)$, so $f(x) = g(x) + C$. $\blacksquare$

Corollaries 1 and 2 are also true if the open interval $(a, b)$ fails to be finite. That is, they remain true if the interval is $(a, \infty)$, $(-\infty, b)$, or $(-\infty, \infty)$.

Corollary 2 plays an important role when we discuss antiderivatives in Section 4.7. It tells us, for instance, that since the derivative of $f(x) = x^2$ on $(-\infty, \infty)$ is $2x$, any other function with derivative $2x$ on $(-\infty, \infty)$ must have the formula $x^2 + C$ for some value of $C$ (Figure 4.19).

**EXAMPLE 4** Find the function $f(x)$ whose derivative is $\sin x$ and whose graph passes through the point $(0, 2)$.

**Solution** Since the derivative of $g(x) = -\cos x$ is $g'(x) = \sin x$, we see that $f$ and $g$ have the same derivative. Corollary 2 then says that $f(x) = -\cos x + C$ for some

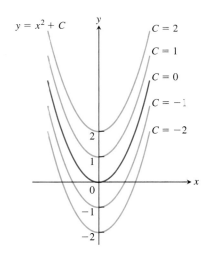

$y = x^2 + C$

**FIGURE 4.19** From a geometric point of view, Corollary 2 of the Mean Value Theorem says that the graphs of functions with identical derivatives on an interval can differ only by a vertical shift there. The graphs of the functions with derivative $2x$ are the parabolas $y = x^2 + C$, shown here for selected values of $C$.

constant $C$. Since the graph of $f$ passes through the point $(0, 2)$, the value of $C$ is determined from the condition that $f(0) = 2$:

$$f(0) = -\cos(0) + C = 2, \quad \text{so} \quad C = 3.$$

The function is $f(x) = -\cos x + 3$. ∎

### Finding Velocity and Position from Acceleration

We can use Corollary 2 to find the velocity and position functions of an object moving along a vertical line. Assume the object or body is falling freely from rest with acceleration $9.8 \text{ m/s}^2$. We assume the position $s(t)$ of the body is measured positive downward from the rest position (so the vertical coordinate line points *downward*, in the direction of the motion, with the rest position at 0).

We know that the velocity $v(t)$ is some function whose derivative is 9.8. We also know that the derivative of $g(t) = 9.8t$ is 9.8. By Corollary 2,

$$v(t) = 9.8t + C$$

for some constant $C$. Since the body falls from rest, $v(0) = 0$. Thus

$$9.8(0) + C = 0, \quad \text{and} \quad C = 0.$$

The velocity function must be $v(t) = 9.8t$. What about the position function $s(t)$?

We know that $s(t)$ is some function whose derivative is $9.8t$. We also know that the derivative of $f(t) = 4.9t^2$ is $9.8t$. By Corollary 2,

$$s(t) = 4.9t^2 + C$$

for some constant $C$. Since $s(0) = 0$,

$$4.9(0)^2 + C = 0, \quad \text{and} \quad C = 0.$$

The position function is $s(t) = 4.9t^2$ until the body hits the ground.

The ability to find functions from their rates of change is one of the very powerful tools of calculus. As we will see, it lies at the heart of the mathematical developments in Chapter 5.

## Exercises 4.2

### Checking the Mean Value Theorem
Find the value or values of $c$ that satisfy the equation

$$\frac{f(b) - f(a)}{b - a} = f'(c)$$

in the conclusion of the Mean Value Theorem for the functions and intervals in Exercises 1–6.

1. $f(x) = x^2 + 2x - 1, \quad [0, 1]$
2. $f(x) = x^{2/3}, \quad [0, 1]$
3. $f(x) = x + \dfrac{1}{x}, \quad \left[\dfrac{1}{2}, 2\right]$
4. $f(x) = \sqrt{x - 1}, \quad [1, 3]$
5. $f(x) = x^3 - x^2, \quad [-1, 2]$
6. $g(x) = \begin{cases} x^3, & -2 \le x \le 0 \\ x^2, & 0 < x \le 2 \end{cases}$

Which of the functions in Exercises 7–12 satisfy the hypotheses of the Mean Value Theorem on the given interval, and which do not? Give reasons for your answers.

7. $f(x) = x^{2/3}, \quad [-1, 8]$
8. $f(x) = x^{4/5}, \quad [0, 1]$
9. $f(x) = \sqrt{x(1 - x)}, \quad [0, 1]$
10. $f(x) = \begin{cases} \dfrac{\sin x}{x}, & -\pi \le x < 0 \\ 0, & x = 0 \end{cases}$
11. $f(x) = \begin{cases} x^2 - x, & -2 \le x \le -1 \\ 2x^2 - 3x - 3, & -1 < x \le 0 \end{cases}$
12. $f(x) = \begin{cases} 2x - 3, & 0 \le x \le 2 \\ 6x - x^2 - 7, & 2 < x \le 3 \end{cases}$
13. The function

$$f(x) = \begin{cases} x, & 0 \le x < 1 \\ 0, & x = 1 \end{cases}$$

is zero at $x = 0$ and $x = 1$ and differentiable on $(0, 1)$, but its derivative on $(0, 1)$ is never zero. How can this be? Doesn't Rolle's Theorem say the derivative has to be zero somewhere in $(0, 1)$? Give reasons for your answer.

**14.** For what values of $a$, $m$, and $b$ does the function

$$f(x) = \begin{cases} 3, & x = 0 \\ -x^2 + 3x + a, & 0 < x < 1 \\ mx + b, & 1 \le x \le 2 \end{cases}$$

satisfy the hypotheses of the Mean Value Theorem on the interval $[0, 2]$?

**Roots (Zeros)**

**15. a.** Plot the zeros of each polynomial on a line together with the zeros of its first derivative.

   **i)** $y = x^2 - 4$

   **ii)** $y = x^2 + 8x + 15$

   **iii)** $y = x^3 - 3x^2 + 4 = (x + 1)(x - 2)^2$

   **iv)** $y = x^3 - 33x^2 + 216x = x(x - 9)(x - 24)$

   **b.** Use Rolle's Theorem to prove that between every two zeros of $x^n + a_{n-1}x^{n-1} + \cdots + a_1x + a_0$ there lies a zero of

   $$nx^{n-1} + (n - 1)a_{n-1}x^{n-2} + \cdots + a_1.$$

**16.** Suppose that $f''$ is continuous on $[a, b]$ and that $f$ has three zeros in the interval. Show that $f''$ has at least one zero in $(a, b)$. Generalize this result.

**17.** Show that if $f'' > 0$ throughout an interval $[a, b]$, then $f'$ has at most one zero in $[a, b]$. What if $f'' < 0$ throughout $[a, b]$ instead?

**18.** Show that a cubic polynomial can have at most three real zeros.

Show that the functions in Exercises 19–26 have exactly one zero in the given interval.

**19.** $f(x) = x^4 + 3x + 1, \quad [-2, -1]$

**20.** $f(x) = x^3 + \dfrac{4}{x^2} + 7, \quad (-\infty, 0)$

**21.** $g(t) = \sqrt{t} + \sqrt{1 + t} - 4, \quad (0, \infty)$

**22.** $g(t) = \dfrac{1}{1 - t} + \sqrt{1 + t} - 3.1, \quad (-1, 1)$

**23.** $r(\theta) = \theta + \sin^2\left(\dfrac{\theta}{3}\right) - 8, \quad (-\infty, \infty)$

**24.** $r(\theta) = 2\theta - \cos^2 \theta + \sqrt{2}, \quad (-\infty, \infty)$

**25.** $r(\theta) = \sec \theta - \dfrac{1}{\theta^3} + 5, \quad (0, \pi/2)$

**26.** $r(\theta) = \tan \theta - \cot \theta - \theta, \quad (0, \pi/2)$

**Finding Functions from Derivatives**

**27.** Suppose that $f(-1) = 3$ and that $f'(x) = 0$ for all $x$. Must $f(x) = 3$ for all $x$? Give reasons for your answer.

**28.** Suppose that $f(0) = 5$ and that $f'(x) = 2$ for all $x$. Must $f(x) = 2x + 5$ for all $x$? Give reasons for your answer.

**29.** Suppose that $f'(x) = 2x$ for all $x$. Find $f(2)$ if

   **a.** $f(0) = 0$   **b.** $f(1) = 0$   **c.** $f(-2) = 3$.

**30.** What can be said about functions whose derivatives are constant? Give reasons for your answer.

In Exercises 31–36, find all possible functions with the given derivative.

**31. a.** $y' = x$   **b.** $y' = x^2$   **c.** $y' = x^3$

**32. a.** $y' = 2x$   **b.** $y' = 2x - 1$   **c.** $y' = 3x^2 + 2x - 1$

**33. a.** $y' = -\dfrac{1}{x^2}$   **b.** $y' = 1 - \dfrac{1}{x^2}$   **c.** $y' = 5 + \dfrac{1}{x^2}$

**34. a.** $y' = \dfrac{1}{2\sqrt{x}}$   **b.** $y' = \dfrac{1}{\sqrt{x}}$   **c.** $y' = 4x - \dfrac{1}{\sqrt{x}}$

**35. a.** $y' = \sin 2t$   **b.** $y' = \cos\dfrac{t}{2}$   **c.** $y' = \sin 2t + \cos\dfrac{t}{2}$

**36. a.** $y' = \sec^2 \theta$   **b.** $y' = \sqrt{\theta}$   **c.** $y' = \sqrt{\theta} - \sec^2 \theta$

In Exercises 37–40, find the function with the given derivative whose graph passes through the point $P$.

**37.** $f'(x) = 2x - 1, \quad P(0, 0)$

**38.** $g'(x) = \dfrac{1}{x^2} + 2x, \quad P(-1, 1)$

**39.** $r'(\theta) = 8 - \csc^2 \theta, \quad P\left(\dfrac{\pi}{4}, 0\right)$

**40.** $r'(t) = \sec t \tan t - 1, \quad P(0, 0)$

**Finding Position from Velocity or Acceleration**

Exercises 41–44 give the velocity $v = ds/dt$ and initial position of an object moving along a coordinate line. Find the object's position at time $t$.

**41.** $v = 9.8t + 5, \quad s(0) = 10$

**42.** $v = 32t - 2, \quad s(0.5) = 4$

**43.** $v = \sin \pi t, \quad s(0) = 0$

**44.** $v = \dfrac{2}{\pi} \cos \dfrac{2t}{\pi}, \quad s(\pi^2) = 1$

Exercises 45–48 give the acceleration $a = d^2s/dt^2$, initial velocity, and initial position of an object moving on a coordinate line. Find the object's position at time $t$.

**45.** $a = 32, \quad v(0) = 20, \quad s(0) = 5$

**46.** $a = 9.8, \quad v(0) = -3, \quad s(0) = 0$

**47.** $a = -4 \sin 2t, \quad v(0) = 2, \quad s(0) = -3$

**48.** $a = \dfrac{9}{\pi^2} \cos \dfrac{3t}{\pi}, \quad v(0) = 0, \quad s(0) = -1$

**Applications**

**49. Temperature change** It took 14 s for a mercury thermometer to rise from $-19°C$ to $100°C$ when it was taken from a freezer and placed in boiling water. Show that somewhere along the way the mercury was rising at the rate of $8.5°C/s$.

**50.** A trucker handed in a ticket at a toll booth showing that in 2 hours she had covered 230 km on a toll road with speed limit 100 km/h. The trucker was cited for speeding. Why?

**51.** Classical accounts tell us that a 170-oar trireme (ancient Greek or Roman warship) once covered 184 nautical miles (a nautical mile is 1852 m) in 24 hours. Explain why at some point during this feat the trireme's speed exceeded 7.5 knots (nautical miles per hour).

**52.** A marathoner ran the 42 km New York City Marathon in 2.2 hours. Show that at least twice the marathoner was running at exactly 18 km/h, assuming the initial and final speeds are zero.

**53.** Show that at some instant during a 2-hour automobile trip the car's speedometer reading will equal the average speed for the trip.

**54. Free fall on the moon** On our moon, the acceleration of gravity is $1.6$ m/s². If a rock is dropped into a crevasse, how fast will it be going just before it hits bottom 30 s later?

Theory and Examples

**55. The geometric mean of $a$ and $b$** The *geometric mean* of two positive numbers $a$ and $b$ is the number $\sqrt{ab}$. Show that the value of $c$ in the conclusion of the Mean Value Theorem for $f(x) = 1/x$ on an interval of positive numbers $[a, b]$ is $c = \sqrt{ab}$.

**56. The arithmetic mean of $a$ and $b$** The *arithmetic mean* of two numbers $a$ and $b$ is the number $(a + b)/2$. Show that the value of $c$ in the conclusion of the Mean Value Theorem for $f(x) = x^2$ on any interval $[a, b]$ is $c = (a + b)/2$.

**T 57.** Graph the function

$$f(x) = \sin x \sin (x + 2) - \sin^2 (x + 1).$$

What does the graph do? Why does the function behave this way? Give reasons for your answers.

**58. Rolle's Theorem**

**a.** Construct a polynomial $f(x)$ that has zeros at $x = -2, -1, 0, 1,$ and $2$.

**b.** Graph $f$ and its derivative $f'$ together. How is what you see related to Rolle's Theorem?

**c.** Do $g(x) = \sin x$ and its derivative $g'$ illustrate the same phenomenon as $f$ and $f'$?

**59. Unique solution** Assume that $f$ is continuous on $[a, b]$ and differentiable on $(a, b)$. Also assume that $f(a)$ and $f(b)$ have opposite signs and that $f' \neq 0$ between $a$ and $b$. Show that $f(x) = 0$ exactly once between $a$ and $b$.

**60. Parallel tangents** Assume that $f$ and $g$ are differentiable on $[a, b]$ and that $f(a) = g(a)$ and $f(b) = g(b)$. Show that there is at least one point between $a$ and $b$ where the tangents to the graphs of $f$ and $g$ are parallel or the same line. Illustrate with a sketch.

**61.** Suppose that $f'(x) \leq 1$ for $1 \leq x \leq 4$. Show that $f(4) - f(1) \leq 3$.

**62.** Suppose that $0 < f'(x) < 1/2$ for all $x$-values. Show that $f(-1) < f(1) < 2 + f(-1)$.

**63.** Show that $|\cos x - 1| \leq |x|$ for all $x$-values. (*Hint:* Consider $f(t) = \cos t$ on $[0, x]$.)

**64.** Show that for any numbers $a$ and $b$, the sine inequality $|\sin b - \sin a| \leq |b - a|$ is true.

**65.** If the graphs of two differentiable functions $f(x)$ and $g(x)$ start at the same point in the plane and the functions have the same rate of change at every point, do the graphs have to be identical? Give reasons for your answer.

**66.** If $|f(w) - f(x)| \leq |w - x|$ for all values $w$ and $x$ and $f$ is a differentiable function, show that $-1 \leq f'(x) \leq 1$ for all $x$-values.

**67.** Assume that $f$ is differentiable on $a \leq x \leq b$ and that $f(b) < f(a)$. Show that $f'$ is negative at some point between $a$ and $b$.

**68.** Let $f$ be a function defined on an interval $[a, b]$. What conditions could you place on $f$ to guarantee that

$$\min f' \leq \frac{f(b) - f(a)}{b - a} \leq \max f',$$

where $\min f'$ and $\max f'$ refer to the minimum and maximum values of $f'$ on $[a, b]$? Give reasons for your answers.

**T 69.** Use the inequalities in Exercise 68 to estimate $f(0.1)$ if $f'(x) = 1/(1 + x^4 \cos x)$ for $0 \leq x \leq 0.1$ and $f(0) = 1$.

**T 70.** Use the inequalities in Exercise 68 to estimate $f(0.1)$ if $f'(x) = 1/(1 - x^4)$ for $0 \leq x \leq 0.1$ and $f(0) = 2$.

**71.** Let $f$ be differentiable at every value of $x$ and suppose that $f(1) = 1$, that $f' < 0$ on $(-\infty, 1)$, and that $f' > 0$ on $(1, \infty)$.

**a.** Show that $f(x) \geq 1$ for all $x$.

**b.** Must $f'(1) = 0$? Explain.

**72.** Let $f(x) = px^2 + qx + r$ be a quadratic function defined on a closed interval $[a, b]$. Show that there is exactly one point $c$ in $(a, b)$ at which $f$ satisfies the conclusion of the Mean Value Theorem.

# $4.3$ Monotonic Functions and the First Derivative Test

In sketching the graph of a differentiable function, it is useful to know where it increases (rises from left to right) and where it decreases (falls from left to right) over an interval. This section gives a test to determine where it increases and where it decreases. We also show how to test the critical points of a function to identify whether local extreme values are present.

## Increasing Functions and Decreasing Functions

As another corollary to the Mean Value Theorem, we show that functions with positive derivatives are increasing functions and functions with negative derivatives are decreasing functions. A function that is increasing or decreasing on an interval is said to be **monotonic** on the interval.

---

**COROLLARY 3** Suppose that $f$ is continuous on $[a, b]$ and differentiable on $(a, b)$.

If $f'(x) > 0$ at each point $x \in (a, b)$, then $f$ is increasing on $[a, b]$.

If $f'(x) < 0$ at each point $x \in (a, b)$, then $f$ is decreasing on $[a, b]$.

---

**Proof** Let $x_1$ and $x_2$ be any two points in $[a, b]$ with $x_1 < x_2$. The Mean Value Theorem applied to $f$ on $[x_1, x_2]$ says that

$$f(x_2) - f(x_1) = f'(c)(x_2 - x_1)$$

for some $c$ between $x_1$ and $x_2$. The sign of the right-hand side of this equation is the same as the sign of $f'(c)$ because $x_2 - x_1$ is positive. Therefore, $f(x_2) > f(x_1)$ if $f'$ is positive on $(a, b)$ and $f(x_2) < f(x_1)$ if $f'$ is negative on $(a, b)$. ∎

Corollary 3 tells us that $f(x) = \sqrt{x}$ is increasing on the interval $[0, b]$ for any $b > 0$ because $f'(x) = 1/\sqrt{x}$ is positive on $(0, b)$. The derivative does not exist at $x = 0$, but Corollary 3 still applies. The corollary is valid for infinite as well as finite intervals, so $f(x) = \sqrt{x}$ is increasing on $[0, \infty)$.

To find the intervals where a function $f$ is increasing or decreasing, we first find all of the critical points of $f$. If $a < b$ are two critical points for $f$, and if the derivative $f'$ is continuous but never zero on the interval $(a, b)$, then by the Intermediate Value Theorem applied to $f'$, the derivative must be everywhere positive on $(a, b)$, or everywhere negative there. One way we can determine the sign of $f'$ on $(a, b)$ is simply by evaluating the derivative at a single point $c$ in $(a, b)$. If $f'(c) > 0$, then $f'(x) > 0$ for all $x$ in $(a, b)$ so $f$ is increasing on $[a, b]$ by Corollary 3; if $f'(c) < 0$, then $f$ is decreasing on $[a, b]$. The next example illustrates how we use this procedure.

**EXAMPLE 1** Find the critical points of $f(x) = x^3 - 12x - 5$ and identify the open intervals on which $f$ is increasing and on which $f$ is decreasing.

**Solution** The function $f$ is everywhere continuous and differentiable. The first derivative

$$f'(x) = 3x^2 - 12 = 3(x^2 - 4)$$
$$= 3(x + 2)(x - 2)$$

is zero at $x = -2$ and $x = 2$. These critical points subdivide the domain of $f$ to create non-overlapping open intervals $(-\infty, -2)$, $(-2, 2)$, and $(2, \infty)$ on which $f'$ is either positive or negative. We determine the sign of $f'$ by evaluating $f'$ at a convenient point in each subinterval. The behavior of $f$ is determined by then applying Corollary 3 to each subinterval. The results are summarized in the following table, and the graph of $f$ is given in Figure 4.20.

| Interval | $-\infty < x < -2$ | $-2 < x < 2$ | $2 < x < \infty$ |
|---|---|---|---|
| $f'$ evaluated | $f'(-3) = 15$ | $f'(0) = -12$ | $f'(3) = 15$ |
| Sign of $f'$ | $+$ | $-$ | $+$ |
| Behavior of $f$ | increasing | decreasing | increasing |

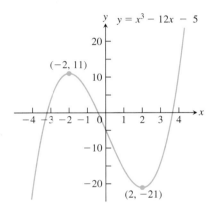

**FIGURE 4.20** The function $f(x) = x^3 - 12x - 5$ is monotonic on three separate intervals (Example 1).

We used "strict" less-than inequalities to identify the intervals in the summary table for Example 1, since open intervals were specified. Corollary 3 says that we could use $\leq$ inequalities as well. That is, the function $f$ in the example is increasing on $-\infty < x \leq -2$, decreasing on $-2 \leq x \leq 2$, and increasing on $2 \leq x < \infty$. We do not talk about whether a function is increasing or decreasing at a single point.

### First Derivative Test for Local Extrema

In Figure 4.21, at the points where $f$ has a minimum value, $f' < 0$ immediately to the left and $f' > 0$ immediately to the right. (If the point is an endpoint, there is only one side to consider.) Thus, the function is decreasing on the left of the minimum value and it is

**HISTORICAL BIOGRAPHY**

Edmund Halley
(1656–1742)

increasing on its right. Similarly, at the points where $f$ has a maximum value, $f' > 0$ immediately to the left and $f' < 0$ immediately to the right. Thus, the function is increasing on the left of the maximum value and decreasing on its right. In summary, at a local extreme point, the sign of $f'(x)$ changes.

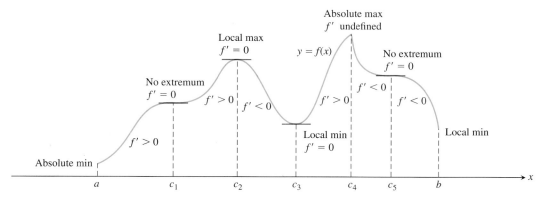

**FIGURE 4.21** The critical points of a function locate where it is increasing and where it is decreasing. The first derivative changes sign at a critical point where a local extremum occurs.

These observations lead to a test for the presence and nature of local extreme values of differentiable functions.

---

**First Derivative Test for Local Extrema**

Suppose that $c$ is a critical point of a continuous function $f$, and that $f$ is differentiable at every point in some interval containing $c$ except possibly at $c$ itself. Moving across this interval from left to right,

1. if $f'$ changes from negative to positive at $c$, then $f$ has a local minimum at $c$;
2. if $f'$ changes from positive to negative at $c$, then $f$ has a local maximum at $c$;
3. if $f'$ does not change sign at $c$ (that is, $f'$ is positive on both sides of $c$ or negative on both sides), then $f$ has no local extremum at $c$.

---

The test for local extrema at endpoints is similar, but there is only one side to consider in determining whether $f$ is increasing or decreasing, based on the sign of $f'$.

**Proof of the First Derivative Test** Part (1). Since the sign of $f'$ changes from negative to positive at $c$, there are numbers $a$ and $b$ such that $a < c < b$, $f' < 0$ on $(a, c)$, and $f' > 0$ on $(c, b)$. If $x \in (a, c)$, then $f(c) < f(x)$ because $f' < 0$ implies that $f$ is decreasing on $[a, c]$. If $x \in (c, b)$, then $f(c) < f(x)$ because $f' > 0$ implies that $f$ is increasing on $[c, b]$. Therefore, $f(x) \geq f(c)$ for every $x \in (a, b)$. By definition, $f$ has a local minimum at $c$.

Parts (2) and (3) are proved similarly. ∎

**EXAMPLE 2** Find the critical points of

$$f(x) = x^{1/3}(x - 4) = x^{4/3} - 4x^{1/3}.$$

Identify the open intervals on which $f$ is increasing and decreasing. Find the function's local and absolute extreme values.

**Solution** The function $f$ is continuous at all $x$ since it is the product of two continuous functions, $x^{1/3}$ and $(x - 4)$. The first derivative

$$f'(x) = \frac{d}{dx}(x^{4/3} - 4x^{1/3}) = \frac{4}{3}x^{1/3} - \frac{4}{3}x^{-2/3}$$

$$= \frac{4}{3}x^{-2/3}(x - 1) = \frac{4(x - 1)}{3x^{2/3}}$$

is zero at $x = 1$ and undefined at $x = 0$. There are no endpoints in the domain, so the critical points $x = 0$ and $x = 1$ are the only places where $f$ might have an extreme value.

The critical points partition the $x$-axis into open intervals on which $f'$ is either positive or negative. The sign pattern of $f'$ reveals the behavior of $f$ between and at the critical points, as summarized in the following table.

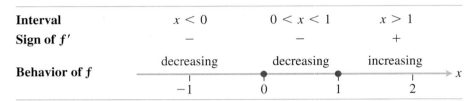

| Interval | $x < 0$ | $0 < x < 1$ | $x > 1$ |
|---|---|---|---|
| **Sign of $f'$** | $-$ | $-$ | $+$ |
| **Behavior of $f$** | decreasing | decreasing | increasing |

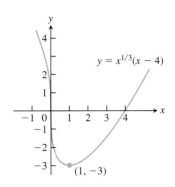

**FIGURE 4.22** The function $f(x) = x^{1/3}(x - 4)$ decreases when $x < 1$ and increases when $x > 1$ (Example 2).

Corollary 3 to the Mean Value Theorem implies that $f$ decreases on $(-\infty, 0)$, decreases on $(0, 1)$, and increases on $(1, \infty)$. The First Derivative Test for Local Extrema tells us that $f$ does not have an extreme value at $x = 0$ ($f'$ does not change sign) and that $f$ has a local minimum at $x = 1$ ($f'$ changes from negative to positive).

The value of the local minimum is $f(1) = 1^{1/3}(1 - 4) = -3$. This is also an absolute minimum since $f$ is decreasing on $(-\infty, 1)$ and increasing on $(1, \infty)$. Figure 4.22 shows this value in relation to the function's graph.

Note that $\lim_{x \to 0} f'(x) = -\infty$, so the graph of $f$ has a vertical tangent at the origin. ∎

**EXAMPLE 3** Within the interval $0 \le x \le 2\pi$, find the critical points of

$$f(x) = \sin^2 x - \sin x - 1.$$

Identify the open intervals on which $f$ is increasing and decreasing. Find the function's local and absolute extreme values.

**Solution** The function $f$ is continuous over $[0, 2\pi]$ and differentiable over $(0, 2\pi)$, so the critical points occur at the zeros of $f'$ in $(0, 2\pi)$. We find

$$f'(x) = 2 \sin x \cos x - \cos x = (2 \sin x - 1)(\cos x).$$

The first derivative is zero if and only if $\sin x = \frac{1}{2}$ or $\cos x = 0$. So the critical points of $f$ in $(0, 2\pi)$ are $x = \pi/6$, $x = 5\pi/6$, $x = \pi/2$, and $x = 3\pi/2$. They partition $[0, 2\pi]$ into open intervals as follows.

| Interval | $\left(0, \frac{\pi}{6}\right)$ | $\left(\frac{\pi}{6}, \frac{\pi}{2}\right)$ | $\left(\frac{\pi}{2}, \frac{5\pi}{6}\right)$ | $\left(\frac{5\pi}{6}, \frac{3\pi}{2}\right)$ | $\left(\frac{3\pi}{2}, 2\pi\right)$ |
|---|---|---|---|---|---|
| **Sign of $f'$** | $-$ | $+$ | $-$ | $+$ | $-$ |
| **Behavior of $f$** | dec | inc | dec | increasing | decreasing |

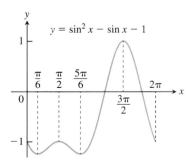

**FIGURE 4.23** The graph of the function in Example 3.

The table displays the open intervals on which $f$ is increasing and decreasing. We can deduce from the table that there is a local minimum value of $f(\pi/6) = \frac{1}{4} - \frac{1}{2} - 1 = -\frac{5}{4}$, a local maximum value of $f(\pi/2) = 1 - 1 - 1 = -1$, another local minimum value of $f(5\pi/6) = -\frac{5}{4}$, and another local maximum value of $f(3\pi/2) = 1 - (-1) - 1 = 1$. The endpoint values are $f(0) = f(2\pi) = -1$. The absolute minimum in $[0, 2\pi]$ is $-\frac{5}{4}$ occurring at $x = \pi/6$ and $x = 5\pi/6$; the absolute maximum is 1 occurring at $x = 3\pi/2$. Figure 4.23 shows the graph. ∎

## Exercises 4.3

### Analyzing Functions from Derivatives

Answer the following questions about the functions whose derivatives are given in Exercises 1–14:

**a.** What are the critical points of $f$?

**b.** On what open intervals is $f$ increasing or decreasing?

**c.** At what points, if any, does $f$ assume local maximum and minimum values?

**1.** $f'(x) = x(x - 1)$

**2.** $f'(x) = (x - 1)(x + 2)$

**3.** $f'(x) = (x - 1)^2(x + 2)$

**4.** $f'(x) = (x - 1)^2(x + 2)^2$

**5.** $f'(x) = (x - 1)(x + 2)(x - 3)$

**6.** $f'(x) = (x - 7)(x + 1)(x + 5)$

**7.** $f'(x) = \dfrac{x^2(x - 1)}{x + 2}, \quad x \neq -2$

**8.** $f'(x) = \dfrac{(x - 2)(x + 4)}{(x + 1)(x - 3)}, \quad x \neq -1, 3$

**9.** $f'(x) = 1 - \dfrac{4}{x^2}, \quad x \neq 0$

**10.** $f'(x) = 3 - \dfrac{6}{\sqrt{x}}, \quad x \neq 0$

**11.** $f'(x) = x^{-1/3}(x + 2)$

**12.** $f'(x) = x^{-1/2}(x - 3)$

**13.** $f'(x) = (\sin x - 1)(2 \cos x + 1), 0 \leq x \leq 2\pi$

**14.** $f'(x) = (\sin x + \cos x)(\sin x - \cos x), 0 \leq x \leq 2\pi$

### Identifying Extrema

In Exercises 15–40:

**a.** Find the open intervals on which the function is increasing and decreasing.

**b.** Identify the function's local and absolute extreme values, if any, saying where they occur.

**15.**

**16.**

**17.**

**18.**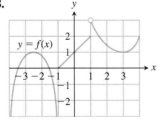

**19.** $g(t) = -t^2 - 3t + 3$

**20.** $g(t) = -3t^2 + 9t + 5$

**21.** $h(x) = -x^3 + 2x^2$

**22.** $h(x) = 2x^3 - 18x$

**23.** $f(\theta) = 3\theta^2 - 4\theta^3$

**24.** $f(\theta) = 6\theta - \theta^3$

**25.** $f(r) = 3r^3 + 16r$

**26.** $h(r) = (r + 7)^3$

**27.** $f(x) = x^4 - 8x^2 + 16$

**28.** $g(x) = x^4 - 4x^3 + 4x^2$

**29.** $H(t) = \dfrac{3}{2}t^4 - t^6$

**30.** $K(t) = 15t^3 - t^5$

**31.** $f(x) = x - 6\sqrt{x - 1}$

**32.** $g(x) = 4\sqrt{x} - x^2 + 3$

**33.** $g(x) = x\sqrt{8 - x^2}$

**34.** $g(x) = x^2\sqrt{5 - x}$

**35.** $f(x) = \dfrac{x^2 - 3}{x - 2}, \quad x \neq 2$

**36.** $f(x) = \dfrac{x^3}{3x^2 + 1}$

**37.** $f(x) = x^{1/3}(x + 8)$

**38.** $g(x) = x^{2/3}(x + 5)$

**39.** $h(x) = x^{1/3}(x^2 - 4)$

**40.** $k(x) = x^{2/3}(x^2 - 4)$

In Exercises 41–52:

**a.** Identify the function's local extreme values in the given domain, and say where they occur.

**b.** Which of the extreme values, if any, are absolute?

**T c.** Support your findings with a graphing calculator or computer grapher.

**41.** $f(x) = 2x - x^2, \quad -\infty < x \leq 2$

**42.** $f(x) = (x + 1)^2, \quad -\infty < x \leq 0$

**43.** $g(x) = x^2 - 4x + 4, \quad 1 \leq x < \infty$

**44.** $g(x) = -x^2 - 6x - 9, \quad -4 \leq x < \infty$

**45.** $f(t) = 12t - t^3, \quad -3 \leq t < \infty$

**46.** $f(t) = t^3 - 3t^2, \quad -\infty < t \leq 3$

**47.** $h(x) = \dfrac{x^3}{3} - 2x^2 + 4x, \quad 0 \leq x < \infty$

**48.** $k(x) = x^3 + 3x^2 + 3x + 1, \quad -\infty < x \leq 0$

**49.** $f(x) = \sqrt{25 - x^2}, \quad -5 \leq x \leq 5$

**50.** $f(x) = \sqrt{x^2 - 2x - 3}, \quad 3 \leq x < \infty$

**51.** $g(x) = \dfrac{x - 2}{x^2 - 1}, \quad 0 \leq x < 1$

**52.** $g(x) = \dfrac{x^2}{4 - x^2}, \quad -2 < x \leq 1$

In Exercises 53–60:

**a.** Find the local extrema of each function on the given interval, and say where they occur.

**T b.** Graph the function and its derivative together. Comment on the behavior of $f$ in relation to the signs and values of $f'$.

**53.** $f(x) = \sin 2x, \quad 0 \leq x \leq \pi$

**54.** $f(x) = \sin x - \cos x, \quad 0 \leq x \leq 2\pi$

**55.** $f(x) = \sqrt{3} \cos x + \sin x, \quad 0 \leq x \leq 2\pi$

**56.** $f(x) = -2x + \tan x, \quad \dfrac{-\pi}{2} < x < \dfrac{\pi}{2}$

**57.** $f(x) = \dfrac{x}{2} - 2 \sin \dfrac{x}{2}, \quad 0 \leq x \leq 2\pi$

**58.** $f(x) = -2 \cos x - \cos^2 x, \quad -\pi \leq x \leq \pi$

**59.** $f(x) = \csc^2 x - 2 \cot x, \quad 0 < x < \pi$

**60.** $f(x) = \sec^2 x - 2 \tan x, \quad \dfrac{-\pi}{2} < x < \dfrac{\pi}{2}$

**Theory and Examples**

Show that the functions in Exercises 61 and 62 have local extreme values at the given values of $\theta$, and say which kind of local extreme the function has.

**61.** $h(\theta) = 3 \cos \dfrac{\theta}{2}, \quad 0 \le \theta \le 2\pi, \quad$ at $\theta = 0$ and $\theta = 2\pi$

**62.** $h(\theta) = 5 \sin \dfrac{\theta}{2}, \quad 0 \le \theta \le \pi, \quad$ at $\theta = 0$ and $\theta = \pi$

**63.** Sketch the graph of a differentiable function $y = f(x)$ through the point $(1, 1)$ if $f'(1) = 0$ and
 **a.** $f'(x) > 0$ for $x < 1$ and $f'(x) < 0$ for $x > 1$;
 **b.** $f'(x) < 0$ for $x < 1$ and $f'(x) > 0$ for $x > 1$;
 **c.** $f'(x) > 0$ for $x \ne 1$;
 **d.** $f'(x) < 0$ for $x \ne 1$.

**64.** Sketch the graph of a differentiable function $y = f(x)$ that has
 **a.** a local minimum at $(1, 1)$ and a local maximum at $(3, 3)$;
 **b.** a local maximum at $(1, 1)$ and a local minimum at $(3, 3)$;
 **c.** local maxima at $(1, 1)$ and $(3, 3)$;
 **d.** local minima at $(1, 1)$ and $(3, 3)$.

**65.** Sketch the graph of a continuous function $y = g(x)$ such that
 **a.** $g(2) = 2, 0 < g' < 1$ for $x < 2, g'(x) \to 1^-$ as $x \to 2^-$, $-1 < g' < 0$ for $x > 2$, and $g'(x) \to -1^+$ as $x \to 2^+$;

**b.** $g(2) = 2, g' < 0$ for $x < 2, g'(x) \to -\infty$ as $x \to 2^-$, $g' > 0$ for $x > 2$, and $g'(x) \to \infty$ as $x \to 2^+$.

**66.** Sketch the graph of a continuous function $y = h(x)$ such that
 **a.** $h(0) = 0, -2 \le h(x) \le 2$ for all $x, h'(x) \to \infty$ as $x \to 0^-$, and $h'(x) \to \infty$ as $x \to 0^+$;
 **b.** $h(0) = 0, -2 \le h(x) \le 0$ for all $x, h'(x) \to \infty$ as $x \to 0^-$, and $h'(x) \to -\infty$ as $x \to 0^+$.

**67.** Discuss the extreme-value behavior of the function $f(x) = x \sin(1/x), x \ne 0$. How many critical points does this function have? Where are they located on the $x$-axis? Does $f$ have an absolute minimum? An absolute maximum? (See Exercise 49 in Section 2.3.)

**68.** Find the open intervals on which the function $f(x) = ax^2 + bx + c, a \ne 0$, is increasing and decreasing. Describe the reasoning behind your answer.

**69.** Determine the values of constants $a$ and $b$ so that $f(x) = ax^2 + bx$ has an absolute maximum at the point $(1, 2)$.

**70.** Determine the values of constants $a$, $b$, $c$, and $d$ so that $f(x) = ax^3 + bx^2 + cx + d$ has a local maximum at the point $(0, 0)$ and a local minimum at the point $(1, -1)$.

# $4.4$ Concavity and Curve Sketching

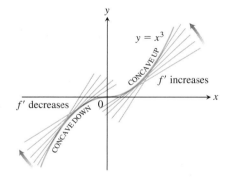

**FIGURE 4.24** The graph of $f(x) = x^3$ is concave down on $(-\infty, 0)$ and concave up on $(0, \infty)$ (Example 1a).

We have seen how the first derivative tells us where a function is increasing, where it is decreasing, and whether a local maximum or local minimum occurs at a critical point. In this section we see that the second derivative gives us information about how the graph of a differentiable function bends or turns. With this knowledge about the first and second derivatives, coupled with our previous understanding of symmetry and asymptotic behavior studied in Sections 1.1 and 2.6, we can now draw an accurate graph of a function. By organizing all of these ideas into a coherent procedure, we give a method for sketching graphs and revealing visually the key features of functions. Identifying and knowing the locations of these features is of major importance in mathematics and its applications to science and engineering, especially in the graphical analysis and interpretation of data.

## Concavity

As you can see in Figure 4.24, the curve $y = x^3$ rises as $x$ increases, but the portions defined on the intervals $(-\infty, 0)$ and $(0, \infty)$ turn in different ways. As we approach the origin from the left along the curve, the curve turns to our right and falls below its tangents. The slopes of the tangents are decreasing on the interval $(-\infty, 0)$. As we move away from the origin along the curve to the right, the curve turns to our left and rises above its tangents. The slopes of the tangents are increasing on the interval $(0, \infty)$. This turning or bending behavior defines the *concavity* of the curve.

---

**DEFINITION** The graph of a differentiable function $y = f(x)$ is

**(a)** **concave up** on an open interval $I$ if $f'$ is increasing on $I$;

**(b)** **concave down** on an open interval $I$ if $f'$ is decreasing on $I$.

---

If $y = f(x)$ has a second derivative, we can apply Corollary 3 of the Mean Value Theorem to the first derivative function. We conclude that $f'$ increases if $f'' > 0$ on $I$, and decreases if $f'' < 0$.

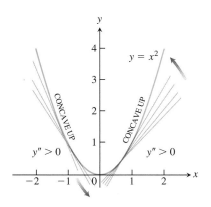

**FIGURE 4.25** The graph of $f(x) = x^2$ is concave up on every interval (Example 1b).

If $y = f(x)$ is twice-differentiable, we will use the notations $f''$ and $y''$ interchangeably when denoting the second derivative.

**EXAMPLE 1**

(a) The curve $y = x^3$ (Figure 4.24) is concave down on $(-\infty, 0)$ where $y'' = 6x < 0$ and concave up on $(0, \infty)$ where $y'' = 6x > 0$.

(b) The curve $y = x^2$ (Figure 4.25) is concave up on $(-\infty, \infty)$ because its second derivative $y'' = 2$ is always positive. ∎

**EXAMPLE 2**    Determine the concavity of $y = 3 + \sin x$ on $[0, 2\pi]$.

**Solution**    The first derivative of $y = 3 + \sin x$ is $y' = \cos x$, and the second derivative is $y'' = -\sin x$. The graph of $y = 3 + \sin x$ is concave down on $(0, \pi)$, where $y'' = -\sin x$ is negative. It is concave up on $(\pi, 2\pi)$, where $y'' = -\sin x$ is positive (Figure 4.26). ∎

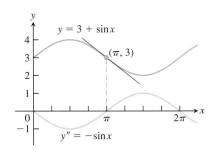

**FIGURE 4.26** Using the sign of $y''$ to determine the concavity of $y$ (Example 2).

### Points of Inflection

The curve $y = 3 + \sin x$ in Example 2 changes concavity at the point $(\pi, 3)$. Since the first derivative $y' = \cos x$ exists for all $x$, we see that the curve has a tangent line of slope $-1$ at the point $(\pi, 3)$. This point is called a *point of inflection* of the curve. Notice from Figure 4.26 that the graph crosses its tangent line at this point and that the second derivative $y'' = -\sin x$ has value 0 when $x = \pi$. In general, we have the following definition.

**DEFINITION**    A point $(c, f(c))$ where the graph of a function has a tangent line and where the concavity changes is a **point of inflection**.

We observed that the second derivative of $f(x) = 3 + \sin x$ is equal to zero at the inflection point $(\pi, 3)$. Generally, if the second derivative exists at a point of inflection $(c, f(c))$, then $f''(c) = 0$. This follows immediately from the Intermediate Value Theorem whenever $f''$ is continuous over an interval containing $x = c$ because the second derivative changes sign moving across this interval. Even if the continuity assumption is dropped, it is still true that $f''(c) = 0$, provided the second derivative exists (although a more advanced argument is required in this noncontinuous case). Since a tangent line must exist at the point of inflection, either the first derivative $f'(c)$ exists (is finite) or the graph has a vertical tangent at the point. At a vertical tangent neither the first nor second derivative exists. In summary, we conclude the following result.

At a point of inflection $(c, f(c))$, either $f''(c) = 0$ or $f''(c)$ fails to exist.

The next example illustrates a function having a point of inflection where the first derivative exists, but the second derivative fails to exist.

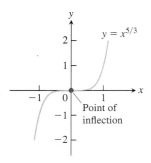

**FIGURE 4.27** The graph of $f(x) = x^{5/3}$ has a horizontal tangent at the origin where the concavity changes, although $f''$ does not exist at $x = 0$ (Example 3).

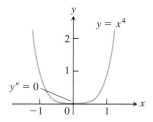

**FIGURE 4.28** The graph of $y = x^4$ has no inflection point at the origin, even though $y'' = 0$ there (Example 4).

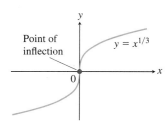

**FIGURE 4.29** A point of inflection where $y'$ and $y''$ fail to exist (Example 5).

**EXAMPLE 3** The graph of $f(x) = x^{5/3}$ has a horizontal tangent at the origin because $f'(x) = (5/3)x^{2/3} = 0$ when $x = 0$. However, the second derivative

$$f''(x) = \frac{d}{dx}\left(\frac{5}{3}x^{2/3}\right) = \frac{10}{9}x^{-1/3}$$

fails to exist at $x = 0$. Nevertheless, $f''(x) < 0$ for $x < 0$ and $f''(x) > 0$ for $x > 0$, so the second derivative changes sign at $x = 0$ and there is a point of inflection at the origin. The graph is shown in Figure 4.27. ∎

Here is an example showing that an inflection point need not occur even though both derivatives exist and $f'' = 0$.

**EXAMPLE 4** The curve $y = x^4$ has no inflection point at $x = 0$ (Figure 4.28). Even though the second derivative $y'' = 12x^2$ is zero there, it does not change sign. ∎

As our final illustration, we show a situation in which a point of inflection occurs at a vertical tangent to the curve where neither the first nor the second derivative exists.

**EXAMPLE 5** The graph of $y = x^{1/3}$ has a point of inflection at the origin because the second derivative is positive for $x < 0$ and negative for $x > 0$:

$$y'' = \frac{d^2}{dx^2}\left(x^{1/3}\right) = \frac{d}{dx}\left(\frac{1}{3}x^{-2/3}\right) = -\frac{2}{9}x^{-5/3}.$$

However, both $y' = x^{-2/3}/3$ and $y''$ fail to exist at $x = 0$, and there is a vertical tangent there. See Figure 4.29. ∎

**Caution** Example 4 in Section 4.1 (Figure 4.9) shows that the function $f(x) = x^{2/3}$ does not have a second derivative at $x = 0$ and does not have a point of inflection there (there is no change in concavity at $x = 0$). Combined with the behavior of the function in Example 5 above, we see that when the second derivative does not exist at $x = c$, an inflection point may or may not occur there. So we need to be careful about interpreting functional behavior whenever first or second derivatives fail to exist at a point. At such points the graph can have vertical tangents, corners, cusps, or various discontinuities.

To study the motion of an object moving along a line as a function of time, we often are interested in knowing when the object's acceleration, given by the second derivative, is positive or negative. The points of inflection on the graph of the object's position function reveal where the acceleration changes sign.

**EXAMPLE 6** A particle is moving along a horizontal coordinate line (positive to the right) with position function

$$s(t) = 2t^3 - 14t^2 + 22t - 5, \qquad t \geq 0.$$

Find the velocity and acceleration, and describe the motion of the particle.

**Solution** The velocity is

$$v(t) = s'(t) = 6t^2 - 28t + 22 = 2(t - 1)(3t - 11),$$

and the acceleration is

$$a(t) = v'(t) = s''(t) = 12t - 28 = 4(3t - 7).$$

When the function $s(t)$ is increasing, the particle is moving to the right; when $s(t)$ is decreasing, the particle is moving to the left.

Notice that the first derivative ($v = s'$) is zero at the critical points $t = 1$ and $t = 11/3$.

| Interval | $0 < t < 1$ | $1 < t < 11/3$ | $11/3 < t$ |
|---|---|---|---|
| **Sign of** $v = s'$ | $+$ | $-$ | $+$ |
| **Behavior of** $s$ | increasing | decreasing | increasing |
| **Particle motion** | right | left | right |

The particle is moving to the right in the time intervals $[0, 1)$ and $(11/3, \infty)$, and moving to the left in $(1, 11/3)$. It is momentarily stationary (at rest) at $t = 1$ and $t = 11/3$.

The acceleration $a(t) = s''(t) = 4(3t - 7)$ is zero when $t = 7/3$.

| Interval | $0 < t < 7/3$ | $7/3 < t$ |
|---|---|---|
| **Sign of** $a = s''$ | $-$ | $+$ |
| **Graph of** $s$ | concave down | concave up |

The particle starts out moving to the right while slowing down, and then reverses and begins moving to the left at $t = 1$ under the influence of the leftward acceleration over the time interval $[0, 7/3)$. The acceleration then changes direction at $t = 7/3$ but the particle continues moving leftward, while slowing down under the rightward acceleration. At $t = 11/3$ the particle reverses direction again: moving to the right in the same direction as the acceleration, so it is speeding up. ∎

### Second Derivative Test for Local Extrema

Instead of looking for sign changes in $f'$ at critical points, we can sometimes use the following test to determine the presence and nature of local extrema.

---

**THEOREM 5—Second Derivative Test for Local Extrema**  Suppose $f''$ is continuous on an open interval that contains $x = c$.

1. If $f'(c) = 0$ and $f''(c) < 0$, then $f$ has a local maximum at $x = c$.

2. If $f'(c) = 0$ and $f''(c) > 0$, then $f$ has a local minimum at $x = c$.

3. If $f'(c) = 0$ and $f''(c) = 0$, then the test fails. The function $f$ may have a local maximum, a local minimum, or neither.

---

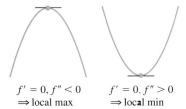

$f' = 0, f'' < 0$
$\Rightarrow$ local max

$f' = 0. f'' > 0$
$\Rightarrow$ local min

**Proof**  Part (1). If $f''(c) < 0$, then $f''(x) < 0$ on some open interval $I$ containing the point $c$, since $f''$ is continuous. Therefore, $f'$ is decreasing on $I$. Since $f'(c) = 0$, the sign of $f'$ changes from positive to negative at $c$ so $f$ has a local maximum at $c$ by the First Derivative Test.

The proof of Part (2) is similar.

For Part (3), consider the three functions $y = x^4$, $y = -x^4$, and $y = x^3$. For each function, the first and second derivatives are zero at $x = 0$. Yet the function $y = x^4$ has a local minimum there, $y = -x^4$ has a local maximum, and $y = x^3$ is increasing in any open interval containing $x = 0$ (having neither a maximum nor a minimum there). Thus the test fails. ∎

This test requires us to know $f''$ *only at c itself* and not in an interval about $c$. This makes the test easy to apply. That's the good news. The bad news is that the test is inconclusive if $f'' = 0$ or if $f''$ does not exist at $x = c$. When this happens, use the First Derivative Test for local extreme values.

Together $f'$ and $f''$ tell us the shape of the function's graph—that is, where the critical points are located and what happens at a critical point, where the function is increasing and where it is decreasing, and how the curve is turning or bending as defined by its concavity. We use this information to sketch a graph of the function that captures its key features.

**EXAMPLE 7**    Sketch a graph of the function

$$f(x) = x^4 - 4x^3 + 10$$

using the following steps.

(a) Identify where the extrema of $f$ occur.

(b) Find the intervals on which $f$ is increasing and the intervals on which $f$ is decreasing.

(c) Find where the graph of $f$ is concave up and where it is concave down.

(d) Sketch the general shape of the graph for $f$.

(e) Plot some specific points, such as local maximum and minimum points, points of inflection, and intercepts. Then sketch the curve.

**Solution**    The function $f$ is continuous since $f'(x) = 4x^3 - 12x^2$ exists. The domain of $f$ is $(-\infty, \infty)$, and the domain of $f'$ is also $(-\infty, \infty)$. Thus, the critical points of $f$ occur only at the zeros of $f'$. Since

$$f'(x) = 4x^3 - 12x^2 = 4x^2(x - 3),$$

the first derivative is zero at $x = 0$ and $x = 3$. We use these critical points to define intervals where $f$ is increasing or decreasing.

| Interval | $x < 0$ | $0 < x < 3$ | $3 < x$ |
|---|---|---|---|
| **Sign of $f'$** | $-$ | $-$ | $+$ |
| **Behavior of $f$** | decreasing | decreasing | increasing |

(a) Using the First Derivative Test for local extrema and the table above, we see that there is no extremum at $x = 0$ and a local minimum at $x = 3$.

(b) Using the table above, we see that $f$ is decreasing on $(-\infty, 0]$ and $[0, 3]$, and increasing on $[3, \infty)$.

(c) $f''(x) = 12x^2 - 24x = 12x(x - 2)$ is zero at $x = 0$ and $x = 2$. We use these points to define intervals where $f$ is concave up or concave down.

| Interval | $x < 0$ | $0 < x < 2$ | $2 < x$ |
|---|---|---|---|
| **Sign of $f''$** | $+$ | $-$ | $+$ |
| **Behavior of $f$** | concave up | concave down | concave up |

We see that $f$ is concave up on the intervals $(-\infty, 0)$ and $(2, \infty)$, and concave down on $(0, 2)$.

(d) Summarizing the information in the last two tables, we obtain the following.

| $x < 0$ | $0 < x < 2$ | $2 < x < 3$ | $3 < x$ |
|---|---|---|---|
| decreasing | decreasing | decreasing | increasing |
| concave up | concave down | concave up | concave up |

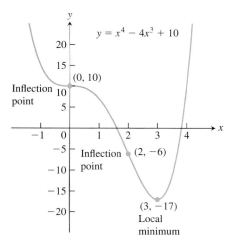

$y = x^4 - 4x^3 + 10$

(0, 10)

Inflection point

Inflection point

(2, −6)

(3, −17)
Local
minimum

**FIGURE 4.30** The graph of $f(x) = x^4 - 4x^3 + 10$ (Example 7).

The general shape of the curve is shown in the accompanying figure.

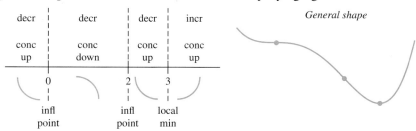

| decr | decr | decr | incr |
| conc up | conc down | conc up | conc up |

0

2   3

infl point

infl point

local min

*General shape*

(e) Plot the curve's intercepts (if possible) and the points where $y'$ and $y''$ are zero. Indicate any local extreme values and inflection points. Use the general shape as a guide to sketch the curve. (Plot additional points as needed.) Figure 4.30 shows the graph of $f$. ∎

The steps in Example 7 give a procedure for graphing the key features of a function. Asymptotes were defined and discussed in Section 2.6. We can find them for rational functions, and the methods in the next section give tools to help find them for more general functions.

---

**Procedure for Graphing** $y = f(x)$

1. Identify the domain of $f$ and any symmetries the curve may have.
2. Find the derivatives $y'$ and $y''$.
3. Find the critical points of $f$, if any, and identify the function's behavior at each one.
4. Find where the curve is increasing and where it is decreasing.
5. Find the points of inflection, if any occur, and determine the concavity of the curve.
6. Identify any asymptotes that may exist.
7. Plot key points, such as the intercepts and the points found in Steps 3–5, and sketch the curve together with any asymptotes that exist.

---

**EXAMPLE 8**    Sketch the graph of $f(x) = \dfrac{(x+1)^2}{1 + x^2}$.

**Solution**

1. The domain of $f$ is $(-\infty, \infty)$ and there are no symmetries about either axis or the origin (Section 1.1).

2. *Find $f'$ and $f''$*.

$$f(x) = \frac{(x+1)^2}{1 + x^2}$$    x-intercept at $x = -1$, y-intercept ($y = 1$) at $x = 0$

$$f'(x) = \frac{(1 + x^2) \cdot 2(x + 1) - (x + 1)^2 \cdot 2x}{(1 + x^2)^2}$$

$$= \frac{2(1 - x^2)}{(1 + x^2)^2}$$    Critical points: $x = -1, x = 1$

$$f''(x) = \frac{(1 + x^2)^2 \cdot 2(-2x) - 2(1 - x^2)[2(1 + x^2) \cdot 2x]}{(1 + x^2)^4}$$

$$= \frac{4x(x^2 - 3)}{(1 + x^2)^3}$$    After some algebra

3. *Behavior at critical points.* The critical points occur only at $x = \pm 1$ where $f'(x) = 0$ (Step 2) since $f'$ exists everywhere over the domain of $f$. At $x = -1$, $f''(-1) = 1 > 0$, yielding a relative minimum by the Second Derivative Test. At $x = 1$, $f''(1) = -1 < 0$, yielding a relative maximum by the Second Derivative test.

4. *Increasing and decreasing.* We see that on the interval $(-\infty, -1)$ the derivative $f'(x) < 0$, and the curve is decreasing. On the interval $(-1, 1)$, $f'(x) > 0$ and the curve is increasing; it is decreasing on $(1, \infty)$ where $f'(x) < 0$ again.

5. *Inflection points.* Notice that the denominator of the second derivative (Step 2) is always positive. The second derivative $f''$ is zero when $x = -\sqrt{3}, 0$, and $\sqrt{3}$. The second derivative changes sign at each of these points: negative on $(-\infty, -\sqrt{3})$, positive on $(-\sqrt{3}, 0)$, negative on $(0, \sqrt{3})$, and positive again on $(\sqrt{3}, \infty)$. Thus each point is a point of inflection. The curve is concave down on the interval $(-\infty, -\sqrt{3})$, concave up on $(-\sqrt{3}, 0)$, concave down on $(0, \sqrt{3})$, and concave up again on $(\sqrt{3}, \infty)$.

6. *Asymptotes.* Expanding the numerator of $f(x)$ and then dividing both numerator and denominator by $x^2$ gives

$$f(x) = \frac{(x + 1)^2}{1 + x^2} = \frac{x^2 + 2x + 1}{1 + x^2} \qquad \text{Expanding numerator}$$

$$= \frac{1 + (2/x) + (1/x^2)}{(1/x^2) + 1}. \qquad \text{Dividing by } x^2$$

We see that $f(x) \to 1^+$ as $x \to \infty$ and that $f(x) \to 1^-$ as $x \to -\infty$. Thus, the line $y = 1$ is a horizontal asymptote.

Since $f$ decreases on $(-\infty, -1)$ and then increases on $(-1, 1)$, we know that $f(-1) = 0$ is a local minimum. Although $f$ decreases on $(1, \infty)$, it never crosses the horizontal asymptote $y = 1$ on that interval (it approaches the asymptote from above). So the graph never becomes negative, and $f(-1) = 0$ is an absolute minimum as well. Likewise, $f(1) = 2$ is an absolute maximum because the graph never crosses the asymptote $y = 1$ on the interval $(-\infty, -1)$, approaching it from below. Therefore, there are no vertical asymptotes (the range of $f$ is $0 \le y \le 2$).

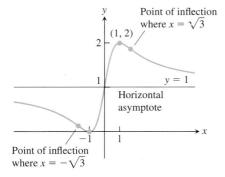

**FIGURE 4.31** The graph of $y = \dfrac{(x + 1)^2}{1 + x^2}$ (Example 8).

7. The graph of $f$ is sketched in Figure 4.31. Notice how the graph is concave down as it approaches the horizontal asymptote $y = 1$ as $x \to -\infty$, and concave up in its approach to $y = 1$ as $x \to \infty$. ∎

**EXAMPLE 9** Sketch the graph of $f(x) = \dfrac{x^2 + 4}{2x}$.

**Solution**

1. The domain of $f$ is all nonzero real numbers. There are no intercepts because neither $x$ nor $f(x)$ can be zero. Since $f(-x) = -f(x)$, we note that $f$ is an odd function, so the graph of $f$ is symmetric about the origin.

2. We calculate the derivatives of the function, but first rewrite it in order to simplify our computations:

$$f(x) = \frac{x^2 + 4}{2x} = \frac{x}{2} + \frac{2}{x} \qquad \text{Function simplified for differentiation}$$

$$f'(x) = \frac{1}{2} - \frac{2}{x^2} = \frac{x^2 - 4}{2x^2} \qquad \text{Combine fractions to solve easily } f'(x) = 0.$$

$$f''(x) = \frac{4}{x^3} \qquad \text{Exists throughout the entire domain of } f$$

3. The critical points occur at $x = \pm 2$ where $f'(x) = 0$. Since $f''(-2) < 0$ and $f''(2) > 0$, we see from the Second Derivative Test that a relative maximum occurs at $x = -2$ with $f(-2) = -2$, and a relative minimum occurs at $x = 2$ with $f(2) = 2$.

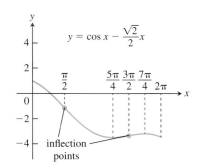

**FIGURE 4.32** The graph of $y = \dfrac{x^2 + 4}{2x}$ (Example 9).

**4.** On the interval $(-\infty, -2)$ the derivative $f'$ is positive because $x^2 - 4 > 0$ so the graph is increasing; on the interval $(-2, 0)$ the derivative is negative and the graph is decreasing. Similarly, the graph is decreasing on the interval $(0, 2)$ and increasing on $(2, \infty)$.

**5.** There are no points of inflection because $f''(x) < 0$ whenever $x < 0$, $f''(x) > 0$ whenever $x > 0$, and $f''$ exists everywhere and is never zero throughout the domain of $f$. The graph is concave down on the interval $(-\infty, 0)$ and concave up on the interval $(0, \infty)$.

**6.** From the rewritten formula for $f(x)$, we see that

$$\lim_{x \to 0^+} \left( \frac{x}{2} + \frac{2}{x} \right) = +\infty \quad \text{and} \quad \lim_{x \to 0^-} \left( \frac{x}{2} + \frac{2}{x} \right) = -\infty,$$

so the $y$-axis is a vertical asymptote. Also, as $x \to \infty$ or as $x \to -\infty$, the graph of $f(x)$ approaches the line $y = x/2$. Thus $y = x/2$ is an oblique asymptote.

**7.** The graph of $f$ is sketched in Figure 4.32. ∎

**EXAMPLE 10** Sketch the graph of $f(x) = \cos x - \dfrac{\sqrt{2}}{2}x$ over $0 \le x \le 2\pi$.

**Solution** The derivatives of $f$ are

$$f'(x) = -\sin x - \frac{\sqrt{2}}{2} \quad \text{and} \quad f''(x) = -\cos x.$$

Both derivatives exist everywhere over the interval $(0, 2\pi)$. Within that open interval, the first derivative is zero when $\sin x = -\sqrt{2}/2$, so the critical points are $x = 5\pi/4$ and $x = 7\pi/4$. Since $f''(5\pi/4) = -\cos(5\pi/4) = \sqrt{2}/2 > 0$, the function has a local minimum value of $f(5\pi/4) \approx -3.48$ (evaluated with a calculator) by The Second Derivative Test. Also, $f''(7\pi/4) = -\cos(7\pi/4) = -\sqrt{2}/2 < 0$, so the function has a local maximum value of $f(7\pi/4) \approx -3.18$.

Examining the second derivative, we find that $f'' = 0$ when $x = \pi/2$ or $x = 3\pi/2$. We conclude that $(\pi/2, f(\pi/2)) \approx (\pi/2, -1.11)$ and $(3\pi/2, f(3\pi/2)) \approx (3\pi/2, -3.33)$ are points of inflection.

Finally, we evaluate $f$ at the endpoints of the interval to find $f(0) = 1$ and $f(2\pi) \approx -3.44$. Therefore, the values $f(0) = 1$ and $f(5\pi/4) \approx -3.48$ are the absolute maximum and absolute minimum values of $f$ over the closed interval $[0, 2\pi]$. The graph of $f$ is sketched in Figure 4.33. ∎

**FIGURE 4.33** The graph of the function in Example 10.

### Graphical Behavior of Functions from Derivatives

As we saw in Examples 7–10, we can learn much about a twice-differentiable function $y = f(x)$ by examining its first derivative. We can find where the function's graph rises and falls and where any local extrema are located. We can differentiate $y'$ to learn how the graph bends as it passes over the intervals of rise and fall. We can determine the shape of the function's graph. Information we cannot get from the derivative is how to place the graph in the $xy$-plane. But, as we discovered in Section 4.2, the only additional information we need to position the graph is the value of $f$ at one point. Information about the asymptotes is found using limits (Section 2.6). The following figure summarizes how the first derivative and second derivative affect the shape of a graph.

| | | |
|---|---|---|
| $y = f(x)$ | $y = f(x)$ | $y = f(x)$ |
| Differentiable $\Rightarrow$ smooth, connected; graph may rise and fall | $y' > 0 \Rightarrow$ rises from left to right; may be wavy | $y' < 0 \Rightarrow$ falls from left to right; may be wavy |
| or | or | $+$ $-$ |
| $y'' > 0 \Rightarrow$ concave up throughout; no waves; graph may rise or fall | $y'' < 0 \Rightarrow$ concave down throughout; no waves; graph may rise or fall | $y''$ changes sign at an inflection point |
| $+$ $-$ or $-$ $+$ | | |
| $y'$ changes sign $\Rightarrow$ graph has local maximum or local minimum | $y' = 0$ and $y'' < 0$ at a point; graph has local maximum | $y' = 0$ and $y'' > 0$ at a point; graph has local minimum |

## Exercises 4.4

### Analyzing Functions from Graphs

Identify the inflection points and local maxima and minima of the functions graphed in Exercises 1–8. Identify the intervals on which the functions are concave up and concave down.

**1.** $y = \dfrac{x^3}{3} - \dfrac{x^2}{2} - 2x + \dfrac{1}{3}$

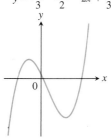

**2.** $y = \dfrac{x^4}{4} - 2x^2 + 4$

**3.** $y = \dfrac{3}{4}(x^2 - 1)^{2/3}$

**4.** $y = \dfrac{9}{14}x^{1/3}(x^2 - 7)$

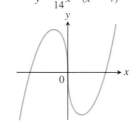

**5.** $y = x + \sin 2x, \quad -\dfrac{2\pi}{3} \le x \le \dfrac{2\pi}{3}$

**6.** $y = \tan x - 4x, \quad -\dfrac{\pi}{2} < x < \dfrac{\pi}{2}$

**7.** $y = \sin|x|, \quad -2\pi \le x \le 2\pi$

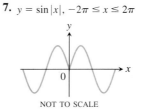

NOT TO SCALE

**8.** $y = 2\cos x - \sqrt{2}x, \quad -\pi \le x \le \dfrac{3\pi}{2}$

### Graphing Functions

In Exercises 9–48, identify the coordinates of any local and absolute extreme points and inflection points. Graph the function.

**9.** $y = x^2 - 4x + 3$

**10.** $y = 6 - 2x - x^2$

**11.** $y = x^3 - 3x + 3$

**12.** $y = x(6 - 2x)^2$

**13.** $y = -2x^3 + 6x^2 - 3$

**14.** $y = 1 - 9x - 6x^2 - x^3$

**15.** $y = (x - 2)^3 + 1$

**16.** $y = 1 - (x + 1)^3$

**17.** $y = x^4 - 2x^2 = x^2(x^2 - 2)$

**18.** $y = -x^4 + 6x^2 - 4 = x^2(6 - x^2) - 4$

**19.** $y = 4x^3 - x^4 = x^3(4 - x)$

**20.** $y = x^4 + 2x^3 = x^3(x + 2)$

**21.** $y = x^5 - 5x^4 = x^4(x - 5)$

**22.** $y = x\left(\dfrac{x}{2} - 5\right)^4$

**23.** $y = x + \sin x, \quad 0 \le x \le 2\pi$

**24.** $y = x - \sin x, \quad 0 \le x \le 2\pi$

**25.** $y = \sqrt{3}x - 2\cos x, \quad 0 \le x \le 2\pi$

**26.** $y = \dfrac{4}{3}x - \tan x, \quad \dfrac{-\pi}{2} < x < \dfrac{\pi}{2}$

**27.** $y = \sin x \cos x, \quad 0 \le x \le \pi$

**28.** $y = \cos x + \sqrt{3} \sin x, \quad 0 \le x \le 2\pi$

**29.** $y = x^{1/5}$

**30.** $y = x^{2/5}$

**31.** $y = \dfrac{x}{\sqrt{x^2 + 1}}$

**32.** $y = \dfrac{\sqrt{1 - x^2}}{2x + 1}$

**33.** $y = 2x - 3x^{2/3}$

**34.** $y = 5x^{2/5} - 2x$

**35.** $y = x^{2/3}\left(\dfrac{5}{2} - x\right)$

**36.** $y = x^{2/3}(x - 5)$

**37.** $y = x\sqrt{8 - x^2}$

**38.** $y = (2 - x^2)^{3/2}$

**39.** $y = \sqrt{16 - x^2}$

**40.** $y = x^2 + \dfrac{2}{x}$

**41.** $y = \dfrac{x^2 - 3}{x - 2}$

**42.** $y = \sqrt[3]{x^3 + 1}$

**43.** $y = \dfrac{8x}{x^2 + 4}$

**44.** $y = \dfrac{5}{x^4 + 5}$

**45.** $y = |x^2 - 1|$

**46.** $y = |x^2 - 2x|$

**47.** $y = \sqrt{|x|} = \begin{cases} \sqrt{-x}, & x < 0 \\ \sqrt{x}, & x \ge 0 \end{cases}$

**48.** $y = \sqrt{|x - 4|}$

### Sketching the General Shape, Knowing $y'$

Each of Exercises 49–70 gives the first derivative of a continuous function $y = f(x)$. Find $y''$ and then use Steps 2–4 of the graphing procedure on page 209 to sketch the general shape of the graph of $f$.

**49.** $y' = 2 + x - x^2$

**50.** $y' = x^2 - x - 6$

**51.** $y' = x(x - 3)^2$

**52.** $y' = x^2(2 - x)$

**53.** $y' = x(x^2 - 12)$

**54.** $y' = (x - 1)^2(2x + 3)$

**55.** $y' = (8x - 5x^2)(4 - x)^2$

**56.** $y' = (x^2 - 2x)(x - 5)^2$

**57.** $y' = \sec^2 x, \quad -\dfrac{\pi}{2} < x < \dfrac{\pi}{2}$

**58.** $y' = \tan x, \quad -\dfrac{\pi}{2} < x < \dfrac{\pi}{2}$

**59.** $y' = \cot \dfrac{\theta}{2}, \quad 0 < \theta < 2\pi$ **60.** $y' = \csc^2 \dfrac{\theta}{2}, \quad 0 < \theta < 2\pi$

**61.** $y' = \tan^2 \theta - 1, \quad -\dfrac{\pi}{2} < \theta < \dfrac{\pi}{2}$

**62.** $y' = 1 - \cot^2 \theta, \quad 0 < \theta < \pi$

**63.** $y' = \cos t, \quad 0 \le t \le 2\pi$

**64.** $y' = \sin t, \quad 0 \le t \le 2\pi$

**65.** $y' = (x + 1)^{-2/3}$ **66.** $y' = (x - 2)^{-1/3}$

**67.** $y' = x^{-2/3}(x - 1)$ **68.** $y' = x^{-4/5}(x + 1)$

**69.** $y' = 2|x| = \begin{cases} -2x, & x \le 0 \\ 2x, & x > 0 \end{cases}$

**70.** $y' = \begin{cases} -x^2, & x \le 0 \\ x^2, & x > 0 \end{cases}$

### Sketching $y$ from Graphs of $y'$ and $y''$

Each of Exercises 71–74 shows the graphs of the first and second derivatives of a function $y = f(x)$. Copy the picture and add to it a sketch of the approximate graph of $f$, given that the graph passes through the point $P$.

**71.**

**72.**

**73.**

**74.**
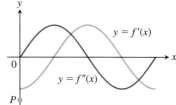

### Graphing Rational Functions

Graph the rational functions in Exercises 75–92 using all the steps in the graphing procedure on page 209.

**75.** $y = \dfrac{2x^2 + x - 1}{x^2 - 1}$

**76.** $y = \dfrac{x^2 - 49}{x^2 + 5x - 14}$

**77.** $y = \dfrac{x^4 + 1}{x^2}$

**78.** $y = \dfrac{x^2 - 4}{2x}$

**79.** $y = \dfrac{1}{x^2 - 1}$

**80.** $y = \dfrac{x^2}{x^2 - 1}$

**81.** $y = -\dfrac{x^2 - 2}{x^2 - 1}$

**82.** $y = \dfrac{x^2 - 4}{x^2 - 2}$

**83.** $y = \dfrac{x^2}{x + 1}$

**84.** $y = -\dfrac{x^2 - 4}{x + 1}$

**85.** $y = \dfrac{x^2 - x + 1}{x - 1}$

**86.** $y = -\dfrac{x^2 - x + 1}{x - 1}$

**87.** $y = \dfrac{x^3 - 3x^2 + 3x - 1}{x^2 + x - 2}$

**88.** $y = \dfrac{x^3 + x - 2}{x - x^2}$

**89.** $y = \dfrac{x}{x^2 - 1}$

**90.** $y = \dfrac{x - 1}{x^2(x - 2)}$

**91.** $y = \dfrac{8}{x^2 + 4}$    (Agnesi's witch)

**92.** $y = \dfrac{4x}{x^2 + 4}$    (Newton's serpentine)

**Theory and Examples**

**93.** The accompanying figure shows a portion of the graph of a twice-differentiable function $y = f(x)$. At each of the five labeled points, classify $y'$ and $y''$ as positive, negative, or zero.

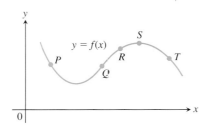

**94.** Sketch a smooth connected curve $y = f(x)$ with

$f(-2) = 8,$ $\qquad$ $f'(2) = f'(-2) = 0,$

$f(0) = 4,$ $\qquad$ $f'(x) < 0$ for $|x| < 2,$

$f(2) = 0,$ $\qquad$ $f''(x) < 0$ for $x < 0,$

$f'(x) > 0$ for $|x| > 2,$ $\qquad$ $f''(x) > 0$ for $x > 0.$

**95.** Sketch the graph of a twice-differentiable function $y = f(x)$ with the following properties. Label coordinates where possible.

| $x$ | $y$ | Derivatives |
|---|---|---|
| $x < 2$ | | $y' < 0, \quad y'' > 0$ |
| $2$ | $1$ | $y' = 0, \quad y'' > 0$ |
| $2 < x < 4$ | | $y' > 0, \quad y'' > 0$ |
| $4$ | $4$ | $y' > 0, \quad y'' = 0$ |
| $4 < x < 6$ | | $y' > 0, \quad y'' < 0$ |
| $6$ | $7$ | $y' = 0, \quad y'' < 0$ |
| $x > 6$ | | $y' < 0, \quad y'' < 0$ |

**96.** Sketch the graph of a twice-differentiable function $y = f(x)$ that passes through the points $(-2, 2)$, $(-1, 1)$, $(0, 0)$, $(1, 1)$, and $(2, 2)$ and whose first two derivatives have the following sign patterns.

$$y': \quad \overset{+}{\underset{-2}{\quad}} \overset{-}{\underset{0}{\quad}} \overset{+}{\underset{2}{\quad}} \overset{-}{\quad}$$

$$y'': \quad \overset{-}{\underset{-1}{\quad}} \overset{+}{\underset{1}{\quad}} \overset{-}{\quad}$$

**Motion Along a Line**    The graphs in Exercises 97 and 98 show the position $s = f(t)$ of an object moving up and down on a coordinate line. **(a)** When is the object moving away from the origin? Toward the origin? At approximately what times is the **(b)** velocity equal to zero? **(c)** Acceleration equal to zero? **(d)** When is the acceleration positive? Negative?

**97.**

**98.**

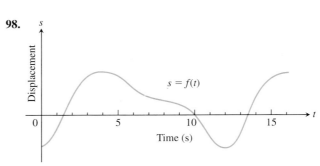

**99. Marginal cost**    The accompanying graph shows the hypothetical cost $c = f(x)$ of manufacturing $x$ items. At approximately what production level does the marginal cost change from decreasing to increasing?

**100.** The accompanying graph shows the monthly revenue of the Widget Corporation for the past 12 years. During approximately what time intervals was the marginal revenue increasing? Decreasing?

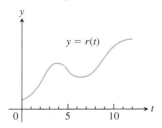

**101.** Suppose the derivative of the function $y = f(x)$ is

$$y' = (x - 1)^2(x - 2).$$

At what points, if any, does the graph of $f$ have a local minimum, local maximum, or point of inflection? (*Hint:* Draw the sign pattern for $y'$.)

**102.** Suppose the derivative of the function $y = f(x)$ is

$$y' = (x - 1)^2(x - 2)(x - 4).$$

At what points, if any, does the graph of $f$ have a local minimum, local maximum, or point of inflection?

**103.** For $x > 0$, sketch a curve $y = f(x)$ that has $f(1) = 0$ and $f'(x) = 1/x$. Can anything be said about the concavity of such a curve? Give reasons for your answer.

**104.** Can anything be said about the graph of a function $y = f(x)$ that has a continuous second derivative that is never zero? Give reasons for your answer.

**105.** If $b$, $c$, and $d$ are constants, for what value of $b$ will the curve $y = x^3 + bx^2 + cx + d$ have a point of inflection at $x = 1$? Give reasons for your answer.

**106. Parabolas**

    **a.** Find the coordinates of the vertex of the parabola $y = ax^2 + bx + c, a \neq 0$.

    **b.** When is the parabola concave up? Concave down? Give reasons for your answers.

**107. Quadratic curves** What can you say about the inflection points of a quadratic curve $y = ax^2 + bx + c, a \neq 0$? Give reasons for your answer.

**108. Cubic curves** What can you say about the inflection points of a cubic curve $y = ax^3 + bx^2 + cx + d, a \neq 0$? Give reasons for your answer.

**109.** Suppose that the second derivative of the function $y = f(x)$ is

$$y'' = (x + 1)(x - 2).$$

For what $x$-values does the graph of $f$ have an inflection point?

**110.** Suppose that the second derivative of the function $y = f(x)$ is

$$y'' = x^2(x - 2)^3(x + 3).$$

For what $x$-values does the graph of $f$ have an inflection point?

**111.** Find the values of constants $a$, $b$, and $c$ so that the graph of $y = ax^3 + bx^2 + cx$ has a local maximum at $x = 3$, local minimum at $x = -1$, and inflection point at $(1, 11)$.

**112.** Find the values of constants $a$, $b$, and $c$ so that the graph of $y = (x^2 + a)/(bx + c)$ has a local minimum at $x = 3$ and a local maximum at $(-1, -2)$.

**COMPUTER EXPLORATIONS**

In Exercises 113–116, find the inflection points (if any) on the graph of the function and the coordinates of the points on the graph where the function has a local maximum or local minimum value. Then graph the function in a region large enough to show all these points simultaneously. Add to your picture the graphs of the function's first and second derivatives. How are the values at which these graphs intersect the $x$-axis related to the graph of the function? In what other ways are the graphs of the derivatives related to the graph of the function?

**113.** $y = x^5 - 5x^4 - 240$      **114.** $y = x^3 - 12x^2$

**115.** $y = \frac{4}{5}x^5 + 16x^2 - 25$

**116.** $y = \frac{x^4}{4} - \frac{x^3}{3} - 4x^2 + 12x + 20$

**117.** Graph $f(x) = 2x^4 - 4x^2 + 1$ and its first two derivatives together. Comment on the behavior of $f$ in relation to the signs and values of $f'$ and $f''$.

**118.** Graph $f(x) = x \cos x$ and its second derivative together for $0 \leq x \leq 2\pi$. Comment on the behavior of the graph of $f$ in relation to the signs and values of $f''$.

# $4.5$ Applied Optimization

What are the dimensions of a rectangle with fixed perimeter having *maximum area*? What are the dimensions for the *least expensive* cylindrical can of a given volume? How many items should be produced for the *most profitable* production run? Each of these questions asks for the best, or optimal, value of a given function. In this section we use derivatives to solve a variety of optimization problems in mathematics, physics, economics, and business.

---

**Solving Applied Optimization Problems**

**1.** *Read the problem.* Read the problem until you understand it. What is given? What is the unknown quantity to be optimized?

**2.** *Draw a picture.* Label any part that may be important to the problem.

**3.** *Introduce variables.* List every relation in the picture and in the problem as an equation or algebraic expression, and identify the unknown variable.

**4.** *Write an equation for the unknown quantity.* If you can, express the unknown as a function of a single variable or in two equations in two unknowns. This may require considerable manipulation.

**5.** *Test the critical points and endpoints in the domain of the unknown.* Use what you know about the shape of the function's graph. Use the first and second derivatives to identify and classify the function's critical points.

(a)

(b)

**FIGURE 4.34** An open box made by cutting the corners from a square sheet of tin. What size corners maximize the box's volume (Example 1)?

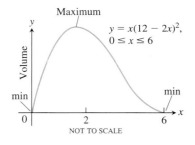

**FIGURE 4.35** The volume of the box in Figure 4.34 graphed as a function of $x$.

**FIGURE 4.36** This one-liter can uses the least material when $h = 2r$ (Example 2).

**EXAMPLE 1** An open-top box is to be made by cutting small congruent squares from the corners of a 12-cm-by-12-cm sheet of tin and bending up the sides. How large should the squares cut from the corners be to make the box hold as much as possible?

**Solution** We start with a picture (Figure 4.34). In the figure, the corner squares are $x$ in. on a side. The volume of the box is a function of this variable:

$$V(x) = x(12 - 2x)^2 = 144x - 48x^2 + 4x^3. \qquad \text{\scriptsize } V = hlw$$

Since the sides of the sheet of tin are only 12 cm long, $x \le 6$ and the domain of $V$ is the interval $0 \le x \le 6$.

A graph of $V$ (Figure 4.35) suggests a minimum value of 0 at $x = 0$ and $x = 6$ and a maximum near $x = 2$. To learn more, we examine the first derivative of $V$ with respect to $x$:

$$\frac{dV}{dx} = 144 - 96x + 12x^2 = 12(12 - 8x + x^2) = 12(2 - x)(6 - x).$$

Of the two zeros, $x = 2$ and $x = 6$, only $x = 2$ lies in the interior of the function's domain and makes the critical-point list. The values of $V$ at this one critical point and two endpoints are

$$\text{Critical point value:} \quad V(2) = 128$$

$$\text{Endpoint values:} \quad V(0) = 0, \quad V(6) = 0.$$

The maximum volume is 128 cm³. The cutout squares should be 2 cm on a side. ∎

**EXAMPLE 2** You have been asked to design a one-liter can shaped like a right circular cylinder (Figure 4.36). What dimensions will use the least material?

**Solution** *Volume of can:* If $r$ and $h$ are measured in centimeters, then the volume of the can in cubic centimeters is

$$\pi r^2 h = 1000. \qquad \text{\scriptsize 1 liter = 1000 cm}^3$$

*Surface area of can:* $A = \underbrace{2\pi r^2}_{\substack{\text{circular} \\ \text{ends}}} + \underbrace{2\pi rh}_{\substack{\text{cylindrical} \\ \text{wall}}}$

How can we interpret the phrase "least material"? For a first approximation we can ignore the thickness of the material and the waste in manufacturing. Then we ask for dimensions $r$ and $h$ that make the total surface area as small as possible while satisfying the constraint $\pi r^2 h = 1000 \text{ cm}^3$.

To express the surface area as a function of one variable, we solve for one of the variables in $\pi r^2 h = 1000$ and substitute that expression into the surface area formula. Solving for $h$ is easier:

$$h = \frac{1000}{\pi r^2}.$$

Thus,

$$A = 2\pi r^2 + 2\pi rh$$

$$= 2\pi r^2 + 2\pi r\left(\frac{1000}{\pi r^2}\right)$$

$$= 2\pi r^2 + \frac{2000}{r}.$$

Our goal is to find a value of $r > 0$ that minimizes the value of $A$. Figure 4.37 suggests that such a value exists.

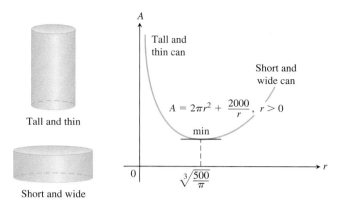

**FIGURE 4.37**    The graph of $A = 2\pi r^2 + 2000/r$ is concave up.

Notice from the graph that for small $r$ (a tall, thin cylindrical container), the term $2000/r$ dominates (see Section 2.6) and $A$ is large. For large $r$ (a short, wide cylindrical container), the term $2\pi r^2$ dominates and $A$ again is large.

Since $A$ is differentiable on $r > 0$, an interval with no endpoints, it can have a minimum value only where its first derivative is zero.

$$\frac{dA}{dr} = 4\pi r - \frac{2000}{r^2}$$

$$0 = 4\pi r - \frac{2000}{r^2} \qquad \text{Set } dA/dr = 0.$$

$$4\pi r^3 = 2000 \qquad \text{Multiply by } r^2.$$

$$r = \sqrt[3]{\frac{500}{\pi}} \approx 5.42 \qquad \text{Solve for } r.$$

What happens at $r = \sqrt[3]{500/\pi}$?
The second derivative

$$\frac{d^2A}{dr^2} = 4\pi + \frac{4000}{r^3}$$

is positive throughout the domain of $A$. The graph is therefore everywhere concave up and the value of $A$ at $r = \sqrt[3]{500/\pi}$ is an absolute minimum.
The corresponding value of $h$ (after a little algebra) is

$$h = \frac{1000}{\pi r^2} = 2\sqrt[3]{\frac{500}{\pi}} = 2r.$$

The one-liter can that uses the least material has height equal to twice the radius, here with $r \approx 5.42$ cm and $h \approx 10.84$ cm.

**Examples from Mathematics and Physics**

**EXAMPLE 3**    A rectangle is to be inscribed in a semicircle of radius 2. What is the largest area the rectangle can have, and what are its dimensions?

**Solution**    Let $\left(x, \sqrt{4 - x^2}\right)$ be the coordinates of the corner of the rectangle obtained by placing the circle and rectangle in the coordinate plane (Figure 4.38). The length, height, and area of the rectangle can then be expressed in terms of the position $x$ of the lower right-hand corner:

$$\text{Length: } 2x, \qquad \text{Height: } \sqrt{4 - x^2}, \qquad \text{Area: } 2x\sqrt{4 - x^2}.$$

Notice that the values of $x$ are to be found in the interval $0 \le x \le 2$, where the selected corner of the rectangle lies.

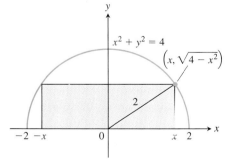

**FIGURE 4.38**    The rectangle inscribed in the semicircle in Example 3.

Our goal is to find the absolute maximum value of the function

$$A(x) = 2x\sqrt{4 - x^2}$$

on the domain $[0, 2]$.

The derivative

$$\frac{dA}{dx} = \frac{-2x^2}{\sqrt{4 - x^2}} + 2\sqrt{4 - x^2}$$

is not defined when $x = 2$ and is equal to zero when

$$\frac{-2x^2}{\sqrt{4 - x^2}} + 2\sqrt{4 - x^2} = 0$$
$$-2x^2 + 2(4 - x^2) = 0$$
$$8 - 4x^2 = 0$$
$$x^2 = 2$$
$$x = \pm\sqrt{2}.$$

Of the two zeros, $x = \sqrt{2}$ and $x = -\sqrt{2}$, only $x = \sqrt{2}$ lies in the interior of $A$'s domain and makes the critical-point list. The values of $A$ at the endpoints and at this one critical point are

Critical point value:  $A(\sqrt{2}) = 2\sqrt{2}\sqrt{4 - 2} = 4$

Endpoint values:  $A(0) = 0, \quad A(2) = 0.$

The area has a maximum value of 4 when the rectangle is $\sqrt{4 - x^2} = \sqrt{2}$ units high and $2x = 2\sqrt{2}$ units long. ∎

**HISTORICAL BIOGRAPHY**

Willebrord Snell van Royen
(1580–1626)

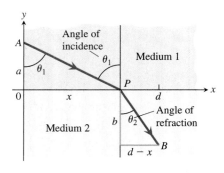

**FIGURE 4.39** A light ray refracted (deflected from its path) as it passes from one medium to a denser medium (Example 4).

**EXAMPLE 4**  The speed of light depends on the medium through which it travels, and is generally slower in denser media.

**Fermat's principle in optics** states that light travels from one point to another along a path for which the time of travel is a minimum. Describe the path that a ray of light will follow in going from a point $A$ in a medium where the speed of light is $c_1$ to a point $B$ in a second medium where its speed is $c_2$.

**Solution**  Since light traveling from $A$ to $B$ follows the quickest route, we look for a path that will minimize the travel time. We assume that $A$ and $B$ lie in the $xy$-plane and that the line separating the two media is the $x$-axis (Figure 4.39).

In a uniform medium, where the speed of light remains constant, "shortest time" means "shortest path," and the ray of light will follow a straight line. Thus the path from $A$ to $B$ will consist of a line segment from $A$ to a boundary point $P$, followed by another line segment from $P$ to $B$. Distance traveled equals rate times time, so

$$\text{Time} = \frac{\text{distance}}{\text{rate}}.$$

From Figure 4.39, the time required for light to travel from $A$ to $P$ is

$$t_1 = \frac{AP}{c_1} = \frac{\sqrt{a^2 + x^2}}{c_1}.$$

From $P$ to $B$, the time is

$$t_2 = \frac{PB}{c_2} = \frac{\sqrt{b^2 + (d - x)^2}}{c_2}.$$

The time from $A$ to $B$ is the sum of these:

$$t = t_1 + t_2 = \frac{\sqrt{a^2 + x^2}}{c_1} + \frac{\sqrt{b^2 + (d - x)^2}}{c_2}.$$

This equation expresses $t$ as a differentiable function of $x$ whose domain is $[0, d]$. We want to find the absolute minimum value of $t$ on this closed interval. We find the derivative

$$\frac{dt}{dx} = \frac{x}{c_1\sqrt{a^2 + x^2}} - \frac{d - x}{c_2\sqrt{b^2 + (d - x)^2}}$$

and observe that it is continuous. In terms of the angles $\theta_1$ and $\theta_2$ in Figure 4.39,

$$\frac{dt}{dx} = \frac{\sin \theta_1}{c_1} - \frac{\sin \theta_2}{c_2}.$$

FIGURE 4.40  The sign pattern of $dt/dx$ in Example 4.

The function $t$ has a negative derivative at $x = 0$ and a positive derivative at $x = d$. Since $dt/dx$ is continuous over the interval $[0, d]$, by the Intermediate Value Theorem for continuous functions (Section 2.5), there is a point $x_0 \in [0, d]$ where $dt/dx = 0$ (Figure 4.40). There is only one such point because $dt/dx$ is an increasing function of $x$ (Exercise 62). At this unique point we then have

$$\frac{\sin \theta_1}{c_1} = \frac{\sin \theta_2}{c_2}.$$

This equation is **Snell's Law** or the **Law of Refraction**, and is an important principle in the theory of optics. It describes the path the ray of light follows. ◼

### Examples from Economics

Suppose that

$r(x) = $ the revenue from selling $x$ items

$c(x) = $ the cost of producing the $x$ items

$p(x) = r(x) - c(x) = $ the profit from producing and selling $x$ items.

Although $x$ is usually an integer in many applications, we can learn about the behavior of these functions by defining them for all nonzero real numbers and by assuming they are differentiable functions. Economists use the terms **marginal revenue, marginal cost**, and **marginal profit** to name the derivatives $r'(x)$, $c'(x)$, and $p'(x)$ of the revenue, cost, and profit functions. Let's consider the relationship of the profit $p$ to these derivatives.

If $r(x)$ and $c(x)$ are differentiable for $x$ in some interval of production possibilities, and if $p(x) = r(x) - c(x)$ has a maximum value there, it occurs at a critical point of $p(x)$ or at an endpoint of the interval. If it occurs at a critical point, then $p'(x) = r'(x) - c'(x) = 0$ and we see that $r'(x) = c'(x)$. In economic terms, this last equation means that

At a production level yielding maximum profit, marginal revenue equals marginal cost (Figure 4.41).

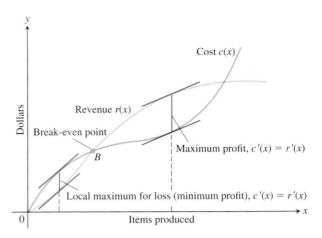

**FIGURE 4.41** The graph of a typical cost function starts concave down and later turns concave up. It crosses the revenue curve at the break-even point $B$. To the left of $B$, the company operates at a loss. To the right, the company operates at a profit, with the maximum profit occurring where $c'(x) = r'(x)$. Farther to the right, cost exceeds revenue (perhaps because of a combination of rising labor and material costs and market saturation) and production levels become unprofitable again.

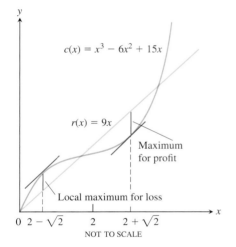

**FIGURE 4.42** The cost and revenue curves for Example 5.

**EXAMPLE 5** Suppose that $r(x) = 9x$ and $c(x) = x^3 - 6x^2 + 15x$, where $x$ represents millions of MP3 players produced. Is there a production level that maximizes profit? If so, what is it?

**Solution** Notice that $r'(x) = 9$ and $c'(x) = 3x^2 - 12x + 15$.

$$3x^2 - 12x + 15 = 9 \qquad \text{Set } c'(x) = r'(x).$$
$$3x^2 - 12x + 6 = 0$$

The two solutions of the quadratic equation are

$$x_1 = \frac{12 - \sqrt{72}}{6} = 2 - \sqrt{2} \approx 0.586 \qquad \text{and}$$

$$x_2 = \frac{12 + \sqrt{72}}{6} = 2 + \sqrt{2} \approx 3.414.$$

The possible production levels for maximum profit are $x \approx 0.586$ million MP3 players or $x \approx 3.414$ million. The second derivative of $p(x) = r(x) - c(x)$ is $p''(x) = -c''(x)$ since $r''(x)$ is everywhere zero. Thus, $p''(x) = 6(2 - x)$, which is negative at $x = 2 + \sqrt{2}$ and positive at $x = 2 - \sqrt{2}$. By the Second Derivative Test, a maximum profit occurs at about $x = 3.414$ (where revenue exceeds costs) and maximum loss occurs at about $x = 0.586$. The graphs of $r(x)$ and $c(x)$ are shown in Figure 4.42. ∎

**EXAMPLE 6** A cabinetmaker uses mahogany wood to produce 5 desks each day. Each delivery of one container of wood is $5000, whereas the storage of that material is $10 per day per unit stored, where a unit is the amount of material needed by her to produce 1 desk. How much material should be ordered each time, and how often should the material be delivered, to minimize her average daily cost in the production cycle between deliveries?

**Solution** If she asks for a delivery every $x$ days, then she must order $5x$ units to have enough material for that delivery cycle. The *average* amount in storage is approximately

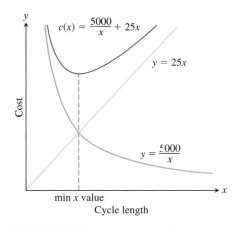

**FIGURE 4.43** The average daily cost $c(x)$ is the sum of a hyperbola and a linear function (Example 6).

one-half of the delivery amount, or $5x/2$. Thus, the cost of delivery and storage for each cycle is approximately

$$\text{Cost per cycle} = \text{delivery costs} + \text{storage costs}$$

$$\text{Cost per cycle} = \underbrace{5000}_{\substack{\text{delivery} \\ \text{cost}}} + \underbrace{\left(\frac{5x}{2}\right)}_{\substack{\text{average} \\ \text{amount stored}}} \cdot \underbrace{x}_{\substack{\text{number of} \\ \text{days stored}}} \cdot \underbrace{10}_{\substack{\text{storage cost} \\ \text{per day}}}$$

We compute the *average daily cost* $c(x)$ by dividing the cost per cycle by the number of days $x$ in the cycle (see Figure 4.43).

$$c(x) = \frac{5000}{x} + 25x, \qquad x > 0.$$

As $x \to 0$ and as $x \to \infty$, the average daily cost becomes large. So we expect a minimum to exist, but where? Our goal is to determine the number of days $x$ between deliveries that provides the absolute minimum cost.

We find the critical points by determining where the derivative is equal to zero:

$$c'(x) = -\frac{500}{x^2} + 25 = 0$$

$$x = \pm \sqrt{200} \approx \pm 14.14.$$

Of the two critical points, only $\sqrt{200}$ lies in the domain of $c(x)$. The critical point value of the average daily cost is

$$c\left(\sqrt{200}\right) = \frac{5000}{\sqrt{200}} + 25\sqrt{200} = 500\sqrt{2} \approx \$707.11.$$

We note that $c(x)$ is defined over the open interval $(0, \infty)$ with $c''(x) = 10000/x^3 > 0$. Thus, an absolute minimum exists at $x = \sqrt{200} \approx 14.14$ days.

The cabinetmaker should schedule a delivery of $5(14) = 70$ units of the mahogany wood every 14 days. ∎

# Exercises 4.5

## Mathematical Applications

Whenever you are maximizing or minimizing a function of a single variable, we urge you to graph it over the domain that is appropriate to the problem you are solving. The graph will provide insight before you calculate and will furnish a visual context for understanding your answer.

1. **Minimizing perimeter** What is the smallest perimeter possible for a rectangle whose area is $16 \text{ cm}^2$, and what are its dimensions?

2. Show that among all rectangles with an 8-m perimeter, the one with largest area is a square.

3. The figure shows a rectangle inscribed in an isosceles right triangle whose hypotenuse is 2 units long.

    a. Express the $y$-coordinate of $P$ in terms of $x$. (*Hint:* Write an equation for the line $AB$.)

    b. Express the area of the rectangle in terms of $x$.

    c. What is the largest area the rectangle can have, and what are its dimensions?

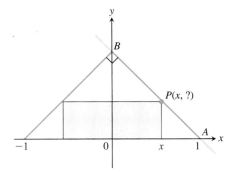

4. A rectangle has its base on the *x*-axis and its upper two vertices on the parabola $y = 12 - x^2$. What is the largest area the rectangle can have, and what are its dimensions?

5. You are planning to make an open rectangular box from an 24-cm-by-45-cm piece of cardboard by cutting congruent squares from the corners and folding up the sides. What are the dimensions of the box of the largest volume you can make this way, and what is its volume?

6. You are planning to close off a corner of the first quadrant with a line segment 20 units long running from $(a, 0)$ to $(0, b)$. Show that the area of the triangle enclosed by the segment is largest when $a = b$.

7. **The best fencing plan** A rectangular plot of farmland will be bounded on one side by a river and on the other three sides by a single-strand electric fence. With 800 m of wire at your disposal, what is the largest area you can enclose, and what are its dimensions?

8. **The shortest fence** A 216 m² rectangular pea patch is to be enclosed by a fence and divided into two equal parts by another fence parallel to one of the sides. What dimensions for the outer rectangle will require the smallest total length of fence? How much fence will be needed?

9. **Designing a tank** Your iron works has contracted to design and build a 4 m³, square-based, open-top, rectangular steel holding tank for a paper company. The tank is to be made by welding thin stainless steel plates together along their edges. As the production engineer, your job is to find dimensions for the base and height that will make the tank weigh as little as possible.

   **a.** What dimensions do you tell the shop to use?

   **b.** Briefly describe how you took weight into account.

10. **Catching rainwater** A 20 m³ open-top rectangular tank with a square base *x* m on a side and *y* m deep is to be built with its top flush with the ground to catch runoff water. The costs associated with the tank involve not only the material from which the tank is made but also an excavation charge proportional to the product *xy*.

    **a.** If the total cost is

    $$c = 5(x^2 + 4xy) + 10xy,$$

    what values of *x* and *y* will minimize it?

    **b.** Give a possible scenario for the cost function in part (a).

11. **Designing a poster** You are designing a rectangular poster to contain 312.5 cm² of printing with a 10 cm margin at the top and bottom and a 5 cm margin at each side. What overall dimensions will minimize the amount of paper used?

12. Find the volume of the largest right circular cone that can be inscribed in a sphere of radius 3.

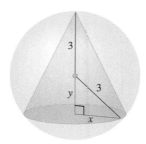

13. Two sides of a triangle have lengths *a* and *b*, and the angle between them is $\theta$. What value of $\theta$ will maximize the triangle's area? (*Hint:* $A = (1/2)ab \sin\theta$.)

14. **Designing a can** What are the dimensions of the lightest open-top right circular cylindrical can that will hold a volume of 1000 cm³? Compare the result here with the result in Example 2.

15. **Designing a can** You are designing a 1000 cm³ right circular cylindrical can whose manufacture will take waste into account. There is no waste in cutting the aluminum for the side, but the top and bottom of radius *r* will be cut from squares that measure 2*r* units on a side. The total amount of aluminum used up by the can will therefore be

    $$A = 8r^2 + 2\pi rh$$

    rather than the $A = 2\pi r^2 + 2\pi rh$ in Example 2. In Example 2, the ratio of *h* to *r* for the most economical can was 2 to 1. What is the ratio now?

T 16. **Designing a box with a lid** A piece of cardboard measures 30 cm by 45 cm. Two equal squares are removed from the corners of a 30 cm side as shown in the figure. Two equal rectangles are removed from the other corners so that the tabs can be folded to form a rectangular box with lid.

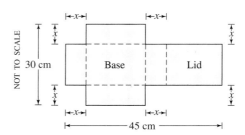

   **a.** Write a formula $V(x)$ for the volume of the box.

   **b.** Find the domain of *V* for the problem situation and graph *V* over this domain.

   **c.** Use a graphical method to find the maximum volume and the value of *x* that gives it.

   **d.** Confirm your result in part (c) analytically.

T 17. **Designing a suitcase** A 60-cm-by-90-cm sheet of cardboard is folded in half to form a 60-cm-by-45-cm rectangle as shown in the accompanying figure. Then four congruent squares of side length *x* are cut from the corners of the folded rectangle. The sheet is unfolded, and the six tabs are folded up to form a box with sides and a lid.

   **a.** Write a formula $V(x)$ for the volume of the box.

   **b.** Find the domain of *V* for the problem situation and graph *V* over this domain.

   **c.** Use a graphical method to find the maximum volume and the value of *x* that gives it.

   **d.** Confirm your result in part (c) analytically.

**e.** Find a value of $x$ that yields a volume of 17,500 cm$^3$

**f.** Write a paragraph describing the issues that arise in part (b).

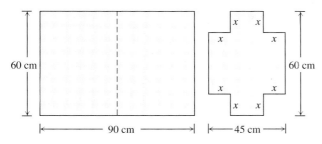

The sheet is then unfolded.

**18.** A rectangle is to be inscribed under the arch of the curve $y = 4 \cos(0.5x)$ from $x = -\pi$ to $x = \pi$. What are the dimensions of the rectangle with largest area, and what is the largest area?

**19.** Find the dimensions of a right circular cylinder of maximum volume that can be inscribed in a sphere of radius 10 cm. What is the maximum volume?

**20. a.** A certain postal service will accept a box for domestic shipment only if the sum of its length and girth (distance around) does not exceed 276 cm. What dimensions will give a box with a square end the largest possible volume?

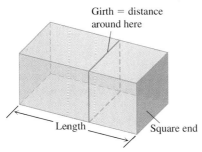

T **b.** Graph the volume of a 276 cm box (length plus girth equals 276 cm) as a function of its length and compare what you see with your answer in part (a).

**21.** (*Continuation of Exercise 20.*)

**a.** Suppose that instead of having a box with square ends you have a box with square sides so that its dimensions are $h$ by $h$ by $w$ and the girth is $2h + 2w$. What dimensions will give the box its largest volume now?

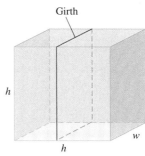

T **b.** Graph the volume as a function of $h$ and compare what you see with your answer in part (a).

**22.** A window is in the form of a rectangle surmounted by a semicircle. The rectangle is of clear glass, whereas the semicircle is of tinted glass that transmits only half as much light per unit area as clear glass does. The total perimeter is fixed. Find the proportions of the window that will admit the most light. Neglect the thickness of the frame.

**23.** A silo (base not included) is to be constructed in the form of a cylinder surmounted by a hemisphere. The cost of construction per square unit of surface area is twice as great for the hemisphere as it is for the cylindrical sidewall. Determine the dimensions to be used if the volume is fixed and the cost of construction is to be kept to a minimum. Neglect the thickness of the silo and waste in construction.

**24.** The trough in the figure is to be made to the dimensions shown. Only the angle $\theta$ can be varied. What value of $\theta$ will maximize the trough's volume?

**25. Paper folding** A rectangular sheet of 21.6-cm-by-28-cm paper is placed on a flat surface. One of the corners is placed on the opposite longer edge, as shown in the figure, and held there as the paper is smoothed flat. The problem is to make the length of the crease as small as possible. Call the length $L$. Try it with paper.

**a.** Show that $L^2 = 2x^3/(2x - 21.6)$.

**b.** What value of $x$ minimizes $L^2$?

**c.** What is the minimum value of $L$?

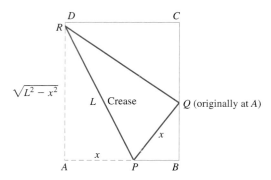

26. **Constructing cylinders** Compare the answers to the following two construction problems.

   **a.** A rectangular sheet of perimeter 36 cm and dimensions $x$ cm by $y$ cm is to be rolled into a cylinder as shown in part (a) of the figure. What values of $x$ and $y$ give the largest volume?

   **b.** The same sheet is to be revolved about one of the sides of length $y$ to sweep out the cylinder as shown in part (b) of the figure. What values of $x$ and $y$ give the largest volume?

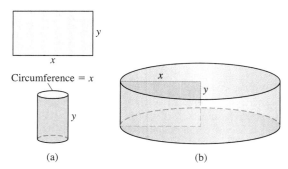

27. **Constructing cones** A right triangle whose hypotenuse is $\sqrt{3}$ m long is revolved about one of its legs to generate a right circular cone. Find the radius, height, and volume of the cone of greatest volume that can be made this way.

28. Find the point on the line $\dfrac{x}{a} + \dfrac{y}{b} = 1$ that is closest to the origin.

29. Find a positive number for which the sum of it and its reciprocal is the smallest (least) possible.

30. Find a positive number for which the sum of its reciprocal and four times its square is the smallest possible.

31. A wire $b$ m long is cut into two pieces. One piece is bent into an equilateral triangle and the other is bent into a circle. If the sum of the areas enclosed by each part is a minimum, what is the length of each part?

32. Answer Exercise 31 if one piece is bent into a square and the other into a circle.

33. Determine the dimensions of the rectangle of largest area that can be inscribed in the right triangle shown in the accompanying figure.

34. Determine the dimensions of the rectangle of largest area that can be inscribed in a semicircle of radius 3. (See accompanying figure.)

35. What value of $a$ makes $f(x) = x^2 + (a/x)$ have

   **a.** a local minimum at $x = 2$?

   **b.** a point of inflection at $x = 1$?

36. What values of $a$ and $b$ make $f(x) = x^3 + ax^2 + bx$ have

   **a.** a local maximum at $x = -1$ and a local minimum at $x = 3$?

   **b.** a local minimum at $x = 4$ and a point of inflection at $x = 1$?

**Physical Applications**

37. **Vertical motion** The height above ground of an object moving vertically is given by

$$s = -4.9t^2 + 29.4t + 34.3,$$

with $s$ in meters and $t$ in seconds. Find

   **a.** the object's velocity when $t = 0$;

   **b.** its maximum height and when it occurs;

   **c.** its velocity when $s = 0$.

38. **Quickest route** Jane is 2 km offshore in a boat and wishes to reach a coastal village 6 km down a straight shoreline from the point nearest the boat. She can row 2 km/h and can walk 5 km/h. Where should she land her boat to reach the village in the least amount of time?

39. **Shortest beam** The 2-m wall shown here stands 5 m from the building. Find the length of the shortest straight beam that will reach to the side of the building from the ground outside the wall.

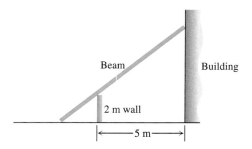

40. **Motion on a line** The positions of two particles on the $s$-axis are $s_1 = \sin t$ and $s_2 = \sin (t + \pi/3)$, with $s_1$ and $s_2$ in meters and $t$ in seconds.

   **a.** At what time(s) in the interval $0 \le t \le 2\pi$ do the particles meet?

   **b.** What is the farthest apart that the particles ever get?

   **c.** When in the interval $0 \le t \le 2\pi$ is the distance between the particles changing the fastest?

41. The intensity of illumination at any point from a light source is proportional to the square of the reciprocal of the distance between the point and the light source. Two lights, one having an intensity eight times that of the other, are 6 m apart. How far from the stronger light is the total illumination least?

42. **Projectile motion** The *range* $R$ of a projectile fired from the origin over horizontal ground is the distance from the origin to the point of impact. If the projectile is fired with an initial velocity $v_0$ at an angle $\alpha$ with the horizontal, then in Chapter 13 we find that

$$R = \frac{v_0^2}{g} \sin 2\alpha,$$

where $g$ is the downward acceleration due to gravity. Find the angle $\alpha$ for which the range $R$ is the largest possible.

T **43. Strength of a beam** The strength $S$ of a rectangular wooden beam is proportional to its width times the square of its depth. (See the accompanying figure.)

a. Find the dimensions of the strongest beam that can be cut from a 30-cm-diameter cylindrical log.

b. Graph $S$ as a function of the beam's width $w$, assuming the proportionality constant to be $k = 1$. Reconcile what you see with your answer in part (a).

c. On the same screen, graph $S$ as a function of the beam's depth $d$, again taking $k = 1$. Compare the graphs with one another and with your answer in part (a). What would be the effect of changing to some other value of $k$? Try it.

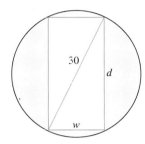

T **44. Stiffness of a beam** The stiffness $S$ of a rectangular beam is proportional to its width times the cube of its depth.

a. Find the dimensions of the stiffest beam that can be cut from a 30-cm-diameter cylindrical log.

b. Graph $S$ as a function of the beam's width $w$, assuming the proportionality constant to be $k = 1$. Reconcile what you see with your answer in part (a).

c. On the same screen, graph $S$ as a function of the beam's depth $d$, again taking $k = 1$. Compare the graphs with one another and with your answer in part (a). What would be the effect of changing to some other value of $k$? Try it.

**45. Frictionless cart** A small frictionless cart, attached to the wall by a spring, is pulled 10 cm from its rest position and released at time $t = 0$ to roll back and forth for 4 s. Its position at time $t$ is $s = 10 \cos \pi t$.

a. What is the cart's maximum speed? When is the cart moving that fast? Where is it then? What is the magnitude of the acceleration then?

b. Where is the cart when the magnitude of the acceleration is greatest? What is the cart's speed then?

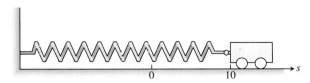

**46.** Two masses hanging side by side from springs have positions $s_1 = 2 \sin t$ and $s_2 = \sin 2t$, respectively.

a. At what times in the interval $0 < t$ do the masses pass each other? (*Hint:* $\sin 2t = 2 \sin t \cos t$.)

b. When in the interval $0 \le t \le 2\pi$ is the vertical distance between the masses the greatest? What is this distance? (*Hint:* $\cos 2t = 2 \cos^2 t - 1$.)

**47. Distance between two ships** At noon, ship $A$ was 12 nautical miles due north of ship $B$. Ship $A$ was sailing south at 12 knots (nautical miles per hour; a nautical mile is 1852 m) and continued to do so all day. Ship $B$ was sailing east at 8 knots and continued to do so all day.

a. Start counting time with $t = 0$ at noon and express the distance $s$ between the ships as a function of $t$.

b. How rapidly was the distance between the ships changing at noon? One hour later?

c. The visibility that day was 5 nautical miles. Did the ships ever sight each other?

T d. Graph $s$ and $ds/dt$ together as functions of $t$ for $-1 \le t \le 3$, using different colors if possible. Compare the graphs and reconcile what you see with your answers in parts (b) and (c).

e. The graph of $ds/dt$ looks as if it might have a horizontal asymptote in the first quadrant. This in turn suggests that $ds/dt$ approaches a limiting value as $t \to \infty$. What is this value? What is its relation to the ships' individual speeds?

**48. Fermat's principle in optics** Light from a source $A$ is reflected by a plane mirror to a receiver at point $B$, as shown in the accompanying figure. Show that for the light to obey Fermat's principle, the angle of incidence must equal the angle of reflection, both measured from the line normal to the reflecting surface. (This result can also be derived without calculus. There is a purely geometric argument, which you may prefer.)

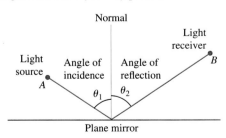

**49. Tin pest** When metallic tin is kept below 13.2°C, it slowly becomes brittle and crumbles to a gray powder. Tin objects eventually crumble to this gray powder spontaneously if kept in a cold climate for years. The Europeans who saw tin organ pipes in their churches crumble away years ago called the change *tin pest* because it seemed to be contagious, and indeed it was, for the gray powder is a catalyst for its own formation.

A *catalyst* for a chemical reaction is a substance that controls the rate of reaction without undergoing any permanent change in itself. An *autocatalytic reaction* is one whose product is a catalyst for its own formation. Such a reaction may proceed slowly at first if the amount of catalyst present is small and slowly again at the end, when most of the original substance is used up. But in between, when both the substance and its catalyst product are abundant, the reaction proceeds at a faster pace.

In some cases, it is reasonable to assume that the rate $v = dx/dt$ of the reaction is proportional both to the amount of the original substance present and to the amount of product. That is, $v$ may be considered to be a function of $x$ alone, and

$$v = kx(a - x) = kax - kx^2,$$

where

$x =$ the amount of product
$a =$ the amount of substance at the beginning
$k =$ a positive constant.

At what value of $x$ does the rate $v$ have a maximum? What is the maximum value of $v$?

**50. Airplane landing path** An airplane is flying at altitude $H$ when it begins its descent to an airport runway that is at horizontal ground distance $L$ from the airplane, as shown in the figure. Assume that the landing path of the airplane is the graph of a cubic polynomial function $y = ax^3 + bx^2 + cx + d$, where $y(-L) = H$ and $y(0) = 0$.

**a.** What is $dy/dx$ at $x = 0$?

**b.** What is $dy/dx$ at $x = -L$?

**c.** Use the values for $dy/dx$ at $x = 0$ and $x = -L$ together with $y(0) = 0$ and $y(-L) = H$ to show that

$$y(x) = H\left[2\left(\frac{x}{L}\right)^3 + 3\left(\frac{x}{L}\right)^2\right].$$

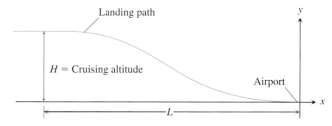

Landing path

$H =$ Cruising altitude

Airport

$L$

**Business and Economics**

**51.** It costs you $c$ dollars each to manufacture and distribute backpacks. If the backpacks sell at $x$ dollars each, the number sold is given by

$$n = \frac{a}{x - c} + b(100 - x),$$

where $a$ and $b$ are positive constants. What selling price will bring a maximum profit?

**52.** You operate a tour service that offers the following rates:

$200 per person if 50 people (the minimum number to book the tour) go on the tour.

For each additional person, up to a maximum of 80 people total, the rate per person is reduced by $2.

It costs $6000 (a fixed cost) plus $32 per person to conduct the tour. How many people does it take to maximize your profit?

**53. Wilson lot size formula** One of the formulas for inventory management says that the average weekly cost of ordering, paying for, and holding merchandise is

$$A(q) = \frac{km}{q} + cm + \frac{hq}{2},$$

where $q$ is the quantity you order when things run low (shoes, radios, brooms, or whatever the item might be), $k$ is the cost of placing an order (the same, no matter how often you order), $c$ is the cost of one item (a constant), $m$ is the number of items sold each week (a constant), and $h$ is the weekly holding cost per item (a constant that takes into account things such as space, utilities, insurance, and security).

**a.** Your job, as the inventory manager for your store, is to find the quantity that will minimize $A(q)$. What is it? (The formula you get for the answer is called the *Wilson lot size formula*.)

**b.** Shipping costs sometimes depend on order size. When they do, it is more realistic to replace $k$ by $k + bq$, the sum of $k$ and a constant multiple of $q$. What is the most economical quantity to order now?

**54. Production level** Prove that the production level (if any) at which average cost is smallest is a level at which the average cost equals marginal cost.

**55.** Show that if $r(x) = 6x$ and $c(x) = x^3 - 6x^2 + 15x$ are your revenue and cost functions, then the best you can do is break even (have revenue equal cost).

**56. Production level** Suppose that $c(x) = x^3 - 20x^2 + 20,000x$ is the cost of manufacturing $x$ items. Find a production level that will minimize the average cost of making $x$ items.

**57.** You are to construct an open rectangular box with a square base and a volume of 6 m³. If material for the bottom costs $60/m² and material for the sides costs $40/m², what dimensions will result in the least expensive box? What is the minimum cost?

**58.** The 800-room Mega Motel chain is filled to capacity when the room charge is $50 per night. For each $10 increase in room charge, 40 fewer rooms are filled each night. What charge per room will result in the maximum revenue per night?

**Biology**

**59. Sensitivity to medicine** (*Continuation of Exercise 72, Section 3.3.*) Find the amount of medicine to which the body is most sensitive by finding the value of $M$ that maximizes the derivative $dR/dM$, where

$$R = M^2\left(\frac{C}{2} - \frac{M}{3}\right)$$

and $C$ is a constant.

**60. How we cough**

**a.** When we cough, the trachea (windpipe) contracts to increase the velocity of the air going out. This raises the questions of how much it should contract to maximize the velocity and whether it really contracts that much when we cough.

Under reasonable assumptions about the elasticity of the tracheal wall and about how the air near the wall is slowed by friction, the average flow velocity $v$ can be modeled by the equation

$$v = c(r_0 - r)r^2 \text{ cm/s}, \qquad \frac{r_0}{2} \le r \le r_0,$$

where $r_0$ is the rest radius of the trachea in centimeters and $c$ is a positive constant whose value depends in part on the length of the trachea.

Show that $v$ is greatest when $r = (2/3)r_0$; that is, when the trachea is about 33% contracted. The remarkable fact is that X-ray photographs confirm that the trachea contracts about this much during a cough.

T **b.** Take $r_0$ to be 0.5 and $c$ to be 1 and graph $v$ over the interval $0 \le r \le 0.5$. Compare what you see with the claim that $v$ is at a maximum when $r = (2/3)r_0$.

### Theory and Examples

**61. An inequality for positive integers** Show that if $a$, $b$, $c$, and $d$ are positive integers, then

$$\frac{(a^2 + 1)(b^2 + 1)(c^2 + 1)(d^2 + 1)}{abcd} \ge 16.$$

**62. The derivative $dt/dx$ in Example 4**

**a.** Show that

$$f(x) = \frac{x}{\sqrt{a^2 + x^2}}$$

is an increasing function of $x$.

**b.** Show that

$$g(x) = \frac{d - x}{\sqrt{b^2 + (d - x)^2}}$$

is a decreasing function of $x$.

**c.** Show that

$$\frac{dt}{dx} = \frac{x}{c_1 \sqrt{a^2 + x^2}} - \frac{d - x}{c_2 \sqrt{b^2 + (d - x)^2}}$$

is an increasing function of $x$.

**63.** Let $f(x)$ and $g(x)$ be the differentiable functions graphed here. Point $c$ is the point where the vertical distance between the curves is the greatest. Is there anything special about the tangents to the two curves at $c$? Give reasons for your answer.

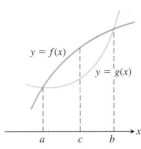

**64.** You have been asked to determine whether the function $f(x) = 3 + 4 \cos x + \cos 2x$ is ever negative.

**a.** Explain why you need to consider values of $x$ only in the interval $[0, 2\pi]$.

**b.** Is $f$ ever negative? Explain.

**65. a.** The function $y = \cot x - \sqrt{2} \csc x$ has an absolute maximum value on the interval $0 < x < \pi$. Find it.

T **b.** Graph the function and compare what you see with your answer in part (a).

**66. a.** The function $y = \tan x + 3 \cot x$ has an absolute minimum value on the interval $0 < x < \pi/2$. Find it.

T **b.** Graph the function and compare what you see with your answer in part (a).

**67. a.** How close does the curve $y = \sqrt{x}$ come to the point $(3/2, 0)$? (*Hint:* If you minimize the *square* of the distance, you can avoid square roots.)

T **b.** Graph the distance function $D(x)$ and $y = \sqrt{x}$ together and reconcile what you see with your answer in part (a).

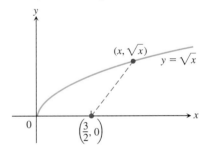

**68. a.** How close does the semicircle $y = \sqrt{16 - x^2}$ come to the point $\left(1, \sqrt{3}\right)$?

T **b.** Graph the distance function and $y = \sqrt{16 - x^2}$ together and reconcile what you see with your answer in part (a).

# $4.6$ Newton's Method

In this section we study a numerical method, called *Newton's method* or the *Newton–Raphson method*, which is a technique to approximate the solution to an equation $f(x) = 0$. Essentially it uses tangent lines of the graph of $y = f(x)$ near the points where $f$ is zero to estimate the solution. (A value of $x$ where $f$ is zero is a *root* of the function $f$ and a *solution* of the equation $f(x) = 0$.)

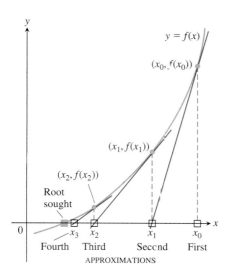

**FIGURE 4.44** Newton's method starts with an initial guess $x_0$ and (under favorable circumstances) improves the guess one step at a time.

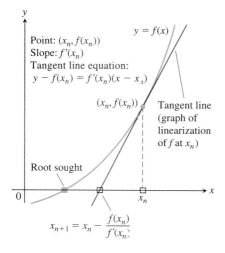

**FIGURE 4.45** The geometry of the successive steps of Newton's method. From $x_n$ we go up to the curve and follow the tangent line down to find $x_{n+1}$.

## Procedure for Newton's Method

The goal of Newton's method for estimating a solution of an equation $f(x) = 0$ is to produce a sequence of approximations that approach the solution. We pick the first number $x_0$ of the sequence. Then, under favorable circumstances, the method does the rest by moving step by step toward a point where the graph of $f$ crosses the $x$-axis (Figure 4.44). At each step the method approximates a zero of $f$ with a zero of one of its linearizations. Here is how it works.

The initial estimate, $x_0$, may be found by graphing or just plain guessing. The method then uses the tangent to the curve $y = f(x)$ at $(x_0, f(x_0))$ to approximate the curve, calling the point $x_1$ where the tangent meets the $x$-axis (Figure 4.44). The number $x_1$ is usually a better approximation to the solution than is $x_0$. The point $x_2$ where the tangent to the curve at $(x_1, f(x_1))$ crosses the $x$-axis is the next approximation in the sequence. We continue on, using each approximation to generate the next, until we are close enough to the root to stop.

We can derive a formula for generating the successive approximations in the following way. Given the approximation $x_n$, the point-slope equation for the tangent to the curve at $(x_n, f(x_n))$ is

$$y = f(x_n) + f'(x_n)(x - x_n).$$

We can find where it crosses the $x$-axis by setting $y = 0$ (Figure 4.45):

$$0 = f(x_n) + f'(x_n)(x - x_n)$$

$$-\frac{f(x_n)}{f'(x_n)} = x - x_n$$

$$x = x_n - \frac{f(x_n)}{f'(x_n)} \qquad \text{If } f'(x_n) \neq 0$$

This value of $x$ is the next approximation $x_{n+1}$. Here is a summary of Newton's method.

---

### Newton's Method

1. Guess a first approximation to a solution of the equation $f(x) = 0$. A graph of $y = f(x)$ may help.

2. Use the first approximation to get a second, the second to get a third, and so on, using the formula

$$x_{n+1} = x_n - \frac{f(x_n)}{f'(x_n)}, \qquad \text{if } f'(x_n) \neq 0. \tag{1}$$

---

## Applying Newton's Method

Applications of Newton's method generally involve many numerical computations, making them well suited for computers or calculators. Nevertheless, even when the calculations are done by hand (which may be very tedious), they give a powerful way to find solutions of equations.

In our first example, we find decimal approximations to $\sqrt{2}$ by estimating the positive root of the equation $f(x) = x^2 - 2 = 0$.

**EXAMPLE 1** Find the positive root of the equation

$$f(x) = x^2 - 2 = 0.$$

**Solution** With $f(x) = x^2 - 2$ and $f'(x) = 2x$, Equation (1) becomes

$$x_{n+1} = x_n - \frac{x_n{}^2 - 2}{2x_n}$$

$$= x_n - \frac{x_n}{2} + \frac{1}{x_n}$$

$$= \frac{x_n}{2} + \frac{1}{x_n}.$$

The equation

$$x_{n+1} = \frac{x_n}{2} + \frac{1}{x_n}$$

enables us to go from each approximation to the next with just a few keystrokes. With the starting value $x_0 = 1$, we get the results in the first column of the following table. (To five decimal places, $\sqrt{2} = 1.41421$.)

| | | Error | Number of correct digits |
|---|---|---|---|
| $x_0 = 1$ | | $-0.41421$ | 1 |
| $x_1 = 1.5$ | | $0.08579$ | 1 |
| $x_2 = 1.41667$ | | $0.00246$ | 3 |
| $x_3 = 1.41422$ | | $0.00001$ | 5 |

Newton's method is the method used by most software applications to calculate roots because it converges so fast (more about this later). If the arithmetic in the table in Example 1 had been carried to 13 decimal places instead of 5, then going one step further would have given $\sqrt{2}$ correctly to more than 10 decimal places.

**EXAMPLE 2**   Find the $x$-coordinate of the point where the curve $y = x^3 - x$ crosses the horizontal line $y = 1$.

**Solution**   The curve crosses the line when $x^3 - x = 1$ or $x^3 - x - 1 = 0$. When does $f(x) = x^3 - x - 1$ equal zero? Since $f(1) = -1$ and $f(2) = 5$, we know by the Intermediate Value Theorem there is a root in the interval $(1, 2)$ (Figure 4.46).

We apply Newton's method to $f$ with the starting value $x_0 = 1$. The results are displayed in Table 4.1 and Figure 4.47.

At $n = 5$, we come to the result $x_6 = x_5 = 1.3247\ 17957$. When $x_{n+1} = x_n$, Equation (1) shows that $f(x_n) = 0$. We have found a solution of $f(x) = 0$ to nine decimals. ∎

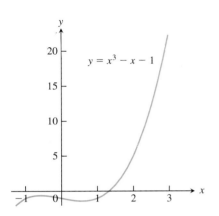

**FIGURE 4.46**   The graph of $f(x) = x^3 - x - 1$ crosses the $x$-axis once; this is the root we want to find (Example 2).

**FIGURE 4.47**   The first three $x$-values in Table 4.1 (four decimal places).

**TABLE 4.1** The result of applying Newton's method to $f(x) = x^3 - x - 1$ with $x_0 = 1$

| $n$ | $x_n$ | $f(x_n)$ | $f'(x_n)$ | $x_{n+1} = x_n - \dfrac{f(x_n)}{f'(x_n)}$ |
|---|---|---|---|---|
| 0 | 1 | $-1$ | 2 | 1.5 |
| 1 | 1.5 | 0.875 | 5.75 | 1.3478 26087 |
| 2 | 1.3478 26087 | 0.1006 82173 | 4.4499 05482 | 1.3252 00399 |
| 3 | 1.3252 00399 | 0.0020 58362 | 4.2684 68292 | 1.3247 18174 |
| 4 | 1.3247 18174 | 0.0000 00924 | 4.2646 34722 | 1.3247 17957 |
| 5 | 1.3247 17957 | $-1.8672\text{E-}13$ | 4.2646 32999 | 1.3247 17957 |

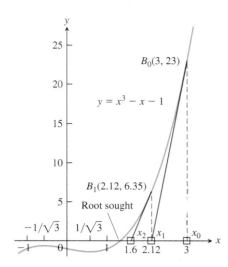

**FIGURE 4.48** Any starting value $x_0$ to the right of $x = 1/\sqrt{3}$ will lead to the root in Example 2.

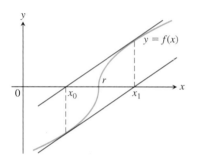

**FIGURE 4.49** Newton's method fails to converge. You go from $x_0$ to $x_1$ and back to $x_0$, never getting any closer to $r$.

In Figure 4.48 we have indicated that the process in Example 2 might have started at the point $B_0(3, 23)$ on the curve, with $x_0 = 3$. Point $B_0$ is quite far from the $x$-axis, but the tangent at $B_0$ crosses the $x$-axis at about $(2.12, 0)$, so $x_1$ is still an improvement over $x_0$. If we use Equation (1) repeatedly as before, with $f(x) = x^3 - x - 1$ and $f'(x) = 3x^2 - 1$, we obtain the nine-place solution $x_7 = x_6 = 1.3247\ 17957$ in seven steps.

## Convergence of the Approximations

In Chapter 10 we define precisely the idea of *convergence* for the approximations $x_n$ in Newton's method. Intuitively, we mean that as the number $n$ of approximations increases without bound, the values $x_n$ get arbitrarily close to the desired root $r$. (This notion is similar to the idea of the limit of a function $g(t)$ as $t$ approaches infinity, as defined in Section 2.6.)

In practice, Newton's method usually gives convergence with impressive speed, but this is not guaranteed. One way to test convergence is to begin by graphing the function to estimate a good starting value for $x_0$. You can test that you are getting closer to a zero of the function by evaluating $|f(x_n)|$, and check that the approximations are converging by evaluating $|x_n - x_{n+1}|$.

*Newton's method does not always converge.* For instance, if

$$f(x) = \begin{cases} -\sqrt{r - x}, & x < r \\ \sqrt{x - r}, & x \ge r, \end{cases}$$

the graph will be like the one in Figure 4.49. If we begin with $x_0 = r - h$, we get $x_1 = r + h$, and successive approximations go back and forth between these two values. No amount of iteration brings us closer to the root than our first guess.

*If Newton's method does converge, it converges to a root.* Be careful, however. There are situations in which the method appears to converge but no root is there. Fortunately, such situations are rare.

*When Newton's method converges to a root, it may not be the root you have in mind.* Figure 4.50 shows two ways this can happen.

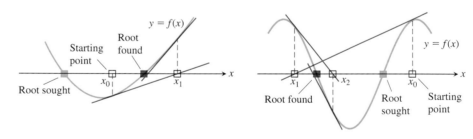

**FIGURE 4.50** If you start too far away, Newton's method may miss the root you want.

# Exercises 4.6

## Root Finding

1. Use Newton's method to estimate the solutions of the equation $x^2 + x - 1 = 0$. Start with $x_0 = -1$ for the left-hand solution and with $x_0 = 1$ for the solution on the right. Then, in each case, find $x_2$.

2. Use Newton's method to estimate the one real solution of $x^3 + 3x + 1 = 0$. Start with $x_0 = 0$ and then find $x_2$.

3. Use Newton's method to estimate the two zeros of the function $f(x) = x^4 + x - 3$. Start with $x_0 = -1$ for the left-hand zero and with $x_0 = 1$ for the zero on the right. Then, in each case, find $x_2$.

4. Use Newton's method to estimate the two zeros of the function $f(x) = 2x - x^2 + 1$. Start with $x_0 = 0$ for the left-hand zero and with $x_0 = 2$ for the zero on the right. Then, in each case, find $x_2$.

5. Use Newton's method to find the positive fourth root of 2 by solving the equation $x^4 - 2 = 0$. Start with $x_0 = 1$ and find $x_2$.

6. Use Newton's method to find the negative fourth root of 2 by solving the equation $x^4 - 2 = 0$. Start with $x_0 = -1$ and find $x_2$.

7. **Guessing a root** Suppose that your first guess is lucky, in the sense that $x_0$ is a root of $f(x) = 0$. Assuming that $f'(x_0)$ is defined and not 0, what happens to $x_1$ and later approximations?

**8. Estimating pi**   You plan to estimate $\pi/2$ to five decimal places by using Newton's method to solve the equation $\cos x = 0$. Does it matter what your starting value is? Give reasons for your answer.

**Theory and Examples**

**9. Oscillation**   Show that if $h > 0$, applying Newton's method to

$$f(x) = \begin{cases} \sqrt{x}, & x \geq 0 \\ \sqrt{-x}, & x < 0 \end{cases}$$

leads to $x_1 = -h$ if $x_0 = h$ and to $x_1 = h$ if $x_0 = -h$. Draw a picture that shows what is going on.

**10. Approximations that get worse and worse**   Apply Newton's method to $f(x) = x^{1/3}$ with $x_0 = 1$ and calculate $x_1, x_2, x_3,$ and $x_4$. Find a formula for $|x_n|$. What happens to $|x_n|$ as $n \to \infty$? Draw a picture that shows what is going on.

**11.** Explain why the following four statements ask for the same information:

**i)** Find the roots of $f(x) = x^3 - 3x - 1$.

**ii)** Find the $x$-coordinates of the intersections of the curve $y = x^3$ with the line $y = 3x + 1$.

**iii)** Find the $x$-coordinates of the points where the curve $y = x^3 - 3x$ crosses the horizontal line $y = 1$.

**iv)** Find the values of $x$ where the derivative of $g(x) = (1/4)x^4 - (3/2)x^2 - x + 5$ equals zero.

**12. Locating a planet**   To calculate a planet's space coordinates, we have to solve equations like $x = 1 + 0.5 \sin x$. Graphing the function $f(x) = x - 1 - 0.5 \sin x$ suggests that the function has a root near $x = 1.5$. Use one application of Newton's method to improve this estimate. That is, start with $x_0 = 1.5$ and find $x_1$. (The value of the root is 1.49870 to five decimal places.) Remember to use radians.

**T 13. Intersecting curves**   The curve $y = \tan x$ crosses the line $y = 2x$ between $x = 0$ and $x = \pi/2$. Use Newton's method to find where.

**T 14. Real solutions of a quartic**   Use Newton's method to find the two real solutions of the equation $x^4 - 2x^3 - x^2 - 2x + 2 = 0$.

**T 15. a.** How many solutions does the equation $\sin 3x = 0.99 - x^2$ have?

**b.** Use Newton's method to find them.

**16. Intersection of curves**

**a.** Does $\cos 3x$ ever equal $x$? Give reasons for your answer.

**b.** Use Newton's method to find where.

**17.** Find the four real zeros of the function $f(x) = 2x^4 - 4x^2 + 1$.

**T 18. Estimating pi**   Estimate $\pi$ to as many decimal places as your calculator will display by using Newton's method to solve the equation $\tan x = 0$ with $x_0 = 3$.

**19. Intersection of curves**   At what value(s) of $x$ does $\cos x = 2x$?

**20. Intersection of curves**   At what value(s) of $x$ does $\cos x = -x$?

**21.** The graphs of $y = x^2(x + 1)$ and $y = 1/x$ $(x > 0)$ intersect at one point $x = r$. Use Newton's method to estimate the value of $r$ to four decimal places.

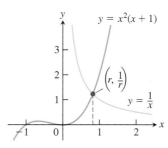

**22.** The graphs of $y = \sqrt{x}$ and $y = 3 - x^2$ intersect at one point $x = r$. Use Newton's method to estimate the value of $r$ to four decimal places.

**23.** Use the Intermediate Value Theorem from Section 2.5 to show that $f(x) = x^3 + 2x - 4$ has a root between $x = 1$ and $x = 2$. Then find the root to five decimal places.

**24. Factoring a quartic**   Find the approximate values of $r_1$ through $r_4$ in the factorization

$$8x^4 - 14x^3 - 9x^2 + 11x - 1 = 8(x - r_1)(x - r_2)(x - r_3)(x - r_4).$$

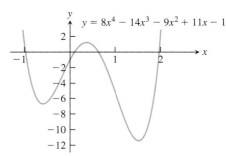

**T 25. Converging to different zeros**   Use Newton's method to find the zeros of $f(x) = 4x^4 - 4x^2$ using the given starting values.

**a.** $x_0 = -2$ and $x_0 = -0.8$, lying in $\left(-\infty, -\sqrt{2}/2\right)$

**b.** $x_0 = -0.5$ and $x_0 = 0.25$, lying in $\left(-\sqrt{21}/7, \sqrt{21}/7\right)$

**c.** $x_0 = 0.8$ and $x_0 = 2$, lying in $\left(\sqrt{2}/2, \infty\right)$

**d.** $x_0 = -\sqrt{21}/7$ and $x_0 = \sqrt{21}/7$

**26. The sonobuoy problem**   In submarine location problems, it is often necessary to find a submarine's closest point of approach (CPA) to a sonobuoy (sound detector) in the water. Suppose that the submarine travels on the parabolic path $y = x^2$ and that the buoy is located at the point $(2, -1/2)$.

**a.** Show that the value of $x$ that minimizes the distance between the submarine and the buoy is a solution of the equation $x = 1/(x^2 + 1)$.

**b.** Solve the equation $x = 1/(x^2 + 1)$ with Newton's method.

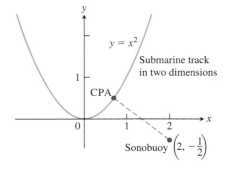

T **27. Curves that are nearly flat at the root** Some curves are so flat that, in practice, Newton's method stops too far from the root to give a useful estimate. Try Newton's method on $f(x) = (x - 1)^{40}$ with a starting value of $x_0 = 2$ to see how close your machine comes to the root $x = 1$. See the accompanying graph.

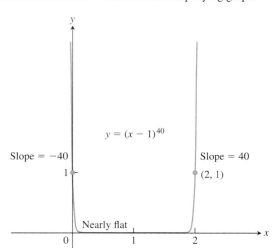

**28.** The accompanying figure shows a circle of radius r with a chord of length 2 and an arc $s$ of length 3. Use Newton's method to solve for $r$ and $\theta$ (radians) to four decimal places. Assume $0 < \theta < \pi$.

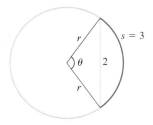

# 4.7 Antiderivatives

We have studied how to find the derivative of a function and how to use it to solve a wide range of problems. However, many other problems require that we recover a function from its known derivative (from its known rate of change). For instance, the laws of physics tell us the acceleration of an object falling from an initial height, and we can use this to compute its velocity and its height at any time. More generally, starting with a function $f$, we want to find a function $F$ whose derivative is $f$. If such a function $F$ exists, it is called an *antiderivative* of $f$. We will see in the next chapter that antiderivatives are the link connecting the two major elements of calculus: derivatives and definite integrals.

## Finding Antiderivatives

> **DEFINITION** A function $F$ is an **antiderivative** of $f$ on an interval $I$ if $F'(x) = f(x)$ for all $x$ in $I$.

The process of recovering a function $F(x)$ from its derivative $f(x)$ is called *antidifferentiation*. We use capital letters such as $F$ to represent an antiderivative of a function $f$, $G$ to represent an antiderivative of $g$, and so forth.

**EXAMPLE 1** Find an antiderivative for each of the following functions.

(a) $f(x) = 2x$ (b) $g(x) = \cos x$ (c) $h(x) = \sec^2 x + \dfrac{1}{2\sqrt{x}}$

**Solution** We need to think backward here: What function do we know has a derivative equal to the given function?

(a) $F(x) = x^2$ (b) $G(x) = \sin x$ (c) $H(x) = \tan x + \sqrt{x}$

Each answer can be checked by differentiating. The derivative of $F(x) = x^2$ is $2x$. The derivative of $G(x) = \sin x$ is $\cos x$, and the derivative of $H(x) = \tan x + \sqrt{x}$ is $\sec^2 x + \left(1/2\sqrt{x}\right)$. ∎

The function $F(x) = x^2$ is not the only function whose derivative is $2x$. The function $x^2 + 1$ has the same derivative. So does $x^2 + C$ for any constant $C$. Are there others?

Corollary 2 of the Mean Value Theorem in Section 4.2 gives the answer: Any two antiderivatives of a function differ by a constant. So the functions $x^2 + C$, where $C$ is an **arbitrary constant**, form *all* the antiderivatives of $f(x) = 2x$. More generally, we have the following result.

---

**THEOREM 8**  If $F$ is an antiderivative of $f$ on an interval $I$, then the most general antiderivative of $f$ on $I$ is

$$F(x) + C$$

where $C$ is an arbitrary constant.

---

Thus the most general antiderivative of $f$ on $I$ is a *family* of functions $F(x) + C$ whose graphs are vertical translations of one another. We can select a particular antiderivative from this family by assigning a specific value to $C$. Here is an example showing how such an assignment might be made.

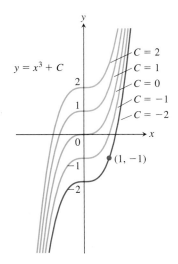

**EXAMPLE 2**  Find an antiderivative of $f(x) = 3x^2$ that satisfies $F(1) = -1$.

**Solution**  Since the derivative of $x^3$ is $3x^2$, the general antiderivative

$$F(x) = x^3 + C$$

gives all the antiderivatives of $f(x)$. The condition $F(1) = -1$ determines a specific value for $C$. Substituting $x = 1$ into $F(x) = x^3 + C$ gives

$$F(1) = (1)^3 + C = 1 + C.$$

Since $F(1) = -1$, solving $1 + C = -1$ for $C$ gives $C = -2$. So

$$F(x) = x^3 - 2$$

is the antiderivative satisfying $F(1) = -1$. Notice that this assignment for $C$ selects the particular curve from the family of curves $y = x^3 + C$ that passes through the point $(1, -1)$ in the plane (Figure 4.51). ∎

**FIGURE 4.51**  The curves $y = x^3 + C$ fill the coordinate plane without overlapping. In Example 2, we identify the curve $y = x^3 - 2$ as the one that passes through the given point $(1, -1)$.

By working backward from assorted differentiation rules, we can derive formulas and rules for antiderivatives. In each case there is an arbitrary constant $C$ in the general expression representing all antiderivatives of a given function. Table 4.2 gives antiderivative formulas for a number of important functions.

The rules in Table 4.2 are easily verified by differentiating the general antiderivative formula to obtain the function to its left. For example, the derivative of $(\tan kx)/k + C$ is $\sec^2 kx$, whatever the value of the constants $C$ or $k \neq 0$, and this establishes Formula 4 for the most general antiderivative of $\sec^2 kx$.

**TABLE 4.2** Antiderivative formulas, $k$ a nonzero constant

| Function | General antiderivative |
|---|---|
| 1. $x^n$ | $\dfrac{1}{n+1}x^{n+1} + C, \quad n \neq -1$ |
| 2. $\sin kx$ | $-\dfrac{1}{k}\cos kx + C$ |
| 3. $\cos kx$ | $\dfrac{1}{k}\sin kx + C$ |
| 4. $\sec^2 kx$ | $\dfrac{1}{k}\tan kx + C$ |
| 5. $\csc^2 kx$ | $-\dfrac{1}{k}\cot kx + C$ |
| 6. $\sec kx \tan kx$ | $\dfrac{1}{k}\sec kx + C$ |
| 7. $\csc kx \cot kx$ | $-\dfrac{1}{k}\csc kx + C$ |

**EXAMPLE 3**    Find the general antiderivative of each of the following functions.

(a) $f(x) = x^5$        (b) $g(x) = \sqrt{x}$

(c) $h(x) = \sin 2x$        (d) $i(x) = \cos \dfrac{x}{2}$

**Solution**    In each case, we can use one of the formulas listed in Table 4.2.

(a) $F(x) = \dfrac{x^6}{6} + C$ 
    *Formula 1 with $n = 5$*

(b) $g(x) = x^{1/2}$, so

$$G(x) = \dfrac{x^{3/2}}{3/2} + C = \dfrac{2}{3}x^{3/2} + C$$
    *Formula 1 with $n = 1/2$*

(c) $H(x) = \dfrac{-\cos 2x}{2} + C$ 
    *Formula 2 with $k = 2$*

(d) $I(x) = \dfrac{\sin(x/2)}{1/2} + C = 2\sin \dfrac{x}{2} + C$ 
    *Formula 3 with $k = 1/2$* ∎

Other derivative rules also lead to corresponding antiderivative rules. We can add and subtract antiderivatives and multiply them by constants.

**TABLE 4.3** Antiderivative linearity rules

| | Function | General antiderivative |
|---|---|---|
| 1. *Constant Multiple Rule:* | $kf(x)$ | $kF(x) + C, \quad k$ a constant |
| 2. *Negative Rule:* | $-f(x)$ | $-F(x) + C$ |
| 3. *Sum or Difference Rule:* | $f(x) \pm g(x)$ | $F(x) \pm G(x) + C$ |

The formulas in Table 4.3 are easily proved by differentiating the antiderivatives and verifying that the result agrees with the original function. Formula 2 is the special case $k = -1$ in Formula 1.

**EXAMPLE 4**    Find the general antiderivative of

$$f(x) = 3\sqrt{x} + \sin 2x.$$

**Solution**    We have that $f(x) = 3g(x) + h(x)$ for the functions $g$ and $h$ in Example 3. Since $G(x) = 2x^{3/2}/3$ is an antiderivative of $g(x)$ from Example 3b, it follows from the Constant Multiple Rule for antiderivatives that $3G(x) = 3(2x^{3/2}/3) = 2x^{3/2}$ is an antiderivative of $3g(x) = 3\sqrt{x}$. Likewise, from Example 3c we know that $H(x) = (-1/2)\cos 2x$ is an antiderivative of $h(x) = \sin 2x$. From the Sum Rule for antiderivatives, we then get that

$$F(x) = 3G(x) + H(x) + C$$
$$= 2x^{3/2} - \frac{1}{2}\cos 2x + C$$

is the general antiderivative formula for $f(x)$, where $C$ is an arbitrary constant.    ■

### Initial Value Problems and Differential Equations

Antiderivatives play several important roles in mathematics and its applications. Methods and techniques for finding them are a major part of calculus, and we take up that study in Chapter 8. Finding an antiderivative for a function $f(x)$ is the same problem as finding a function $y(x)$ that satisfies the equation

$$\frac{dy}{dx} = f(x).$$

This is called a **differential equation**, since it is an equation involving an unknown function $y$ that is being differentiated. To solve it, we need a function $y(x)$ that satisfies the equation. This function is found by taking the antiderivative of $f(x)$. We can fix the arbitrary constant arising in the antidifferentiation process by specifying an initial condition

$$y(x_0) = y_0.$$

This condition means the function $y(x)$ has the value $y_0$ when $x = x_0$. The combination of a differential equation and an initial condition is called an **initial value problem**. Such problems play important roles in all branches of science.

The most general antiderivative $F(x) + C$ (such as $x^3 + C$ in Example 2) of the function $f(x)$ gives the **general solution** $y = F(x) + C$ of the differential equation $dy/dx = f(x)$. The general solution gives *all* the solutions of the equation (there are infinitely many, one for each value of $C$). We **solve** the differential equation by finding its general solution. We then solve the initial value problem by finding the **particular solution** that satisfies the initial condition $y(x_0) = y_0$. In Example 2, the function $y = x^3 - 2$ is the particular solution of the differential equation $dy/dx = 3x^2$ satisfying the initial condition $y(1) = -1$.

### Antiderivatives and Motion

We have seen that the derivative of the position function of an object gives its velocity, and the derivative of its velocity function gives its acceleration. If we know an object's acceleration, then by finding an antiderivative we can recover the velocity, and from an

**FIGURE 4.52** A package dropped from a rising hot-air balloon (Example 5).

antiderivative of the velocity we can recover its position function. This procedure was used as an application of Corollary 2 in Section 4.2. Now that we have a terminology and conceptual framework in terms of antiderivatives, we revisit the problem from the point of view of differential equations.

**EXAMPLE 5**    A hot-air balloon ascending at the rate of 3.6 m/s is at a height 24.5 m above the ground when a package is dropped. How long does it take the package to reach the ground?

**Solution**    Let $v(t)$ denote the velocity of the package at time $t$, and let $s(t)$ denote its height above the ground. The acceleration of gravity near the surface of the earth is 9.8 m/s². Assuming no other forces act on the dropped package, we have

$$\frac{dv}{dt} = -9.8. \qquad \text{Negative because gravity acts in the direction of decreasing } s$$

This leads to the following initial value problem (Figure 4.52):

$$\text{Differential equation:} \qquad \frac{dv}{dt} = -9.8$$

$$\text{Initial condition:} \qquad v(0) = 3.6. \qquad \text{Balloon initially rising}$$

This is our mathematical model for the package's motion. We solve the initial value problem to obtain the velocity of the package.

1. *Solve the differential equation:* The general formula for an antiderivative of $-9.8$ is

$$v = -9.8t + C.$$

Having found the general solution of the differential equation, we use the initial condition to find the particular solution that solves our problem.

2. *Evaluate C:*

$$3.6 = -9.8(0) + C \qquad \text{Initial condition } v(0) = 3.6$$
$$C = 3.6.$$

The solution of the initial value problem is

$$v = -9.8t + 3.6.$$

Since velocity is the derivative of height, and the height of the package is 24.5 m at time $t = 0$ when it is dropped, we now have a second initial value problem:

$$\text{Differential equation:} \qquad \frac{ds}{dt} = -9.8t + 3.6 \qquad \text{Set } v = ds/dt \text{ in the previous equation.}$$

$$\text{Initial condition:} \qquad s(0) = 24.5.$$

We solve this initial value problem to find the height as a function of $t$.

1. *Solve the differential equation:* Finding the general antiderivative of $-9.8t + 3.6$ gives

$$s = -4.9t^2 + 3.6t + C.$$

2. *Evaluate C:*

$$24.5 = -4.9(0)^2 + 3.6(0) + C \qquad \text{Initial condition } s(0) = 24.5$$
$$C = 24.5.$$

The package's height above ground at time $t$ is

$$s = -4.9t^2 + 3.6t + 24.5.$$

*Use the solution*: To find how long it takes the package to reach the ground, we set $s$ equal to 0 and solve for $t$:

$$-4.9t^2 + 3.6t + 24.5 = 0$$

$$t = \frac{-3.6 \pm \sqrt{493.16}}{-9.8} \qquad \text{Quadratic formula}$$

$$t \approx -1.90, \qquad t \approx 2.63.$$

The package hits the ground about 2.63 s after it is dropped from the balloon. (The negative root has no physical meaning.)    ■

### Indefinite Integrals

A special symbol is used to denote the collection of all antiderivatives of a function $f$.

---

**DEFINITION**    The collection of all antiderivatives of $f$ is called the **indefinite integral** of $f$ with respect to $x$, and is denoted by

$$\int f(x)\, dx.$$

The symbol $\int$ is an **integral sign**. The function $f$ is the **integrand** of the integral, and $x$ is the **variable of integration**.

---

After the integral sign in the notation we just defined, the integrand function is always followed by a differential to indicate the variable of integration. We will have more to say about why this is important in Chapter 5. Using this notation, we restate the solutions of Example 1, as follows:

$$\int 2x\, dx = x^2 + C,$$

$$\int \cos x\, dx = \sin x + C,$$

$$\int \left( \sec^2 x + \frac{1}{2\sqrt{x}} \right) dx = \tan x + \sqrt{x} + C$$

This notation is related to the main application of antiderivatives, which will be explored in Chapter 5. Antiderivatives play a key role in computing limits of certain infinite sums, an unexpected and wonderfully useful role that is described in a central result of Chapter 5, called the Fundamental Theorem of Calculus.

**EXAMPLE 6**    Evaluate

$$\int (x^2 - 2x + 5)\, dx.$$

**Solution**    If we recognize that $(x^3/3) - x^2 + 5x$ is an antiderivative of $x^2 - 2x + 5$, we can evaluate the integral as

$$\int (x^2 - 2x + 5)\, dx = \underbrace{\frac{x^3}{3} - x^2 + 5x}_{\text{antiderivative}} + \underset{\text{arbitrary constant}}{\underline{C}}.$$

If we do not recognize the antiderivative right away, we can generate it term-by-term with the Sum, Difference, and Constant Multiple Rules:

$$\int (x^2 - 2x + 5)\, dx = \int x^2\, dx - \int 2x\, dx + \int 5\, dx$$

$$= \int x^2\, dx - 2\int x\, dx + 5\int 1\, dx$$

$$= \left(\frac{x^3}{3} + C_1\right) - 2\left(\frac{x^2}{2} + C_2\right) + 5(x + C_3)$$

$$= \frac{x^3}{3} + C_1 - x^2 - 2C_2 + 5x + 5C_3.$$

This formula is more complicated than it needs to be. If we combine $C_1$, $-2C_2$, and $5C_3$ into a single arbitrary constant $C = C_1 - 2C_2 + 5C_3$, the formula simplifies to

$$\frac{x^3}{3} - x^2 + 5x + C$$

and *still* gives all the possible antiderivatives there are. For this reason, we recommend that you go right to the final form even if you elect to integrate term-by-term. Write

$$\int (x^2 - 2x + 5)\, dx = \int x^2\, dx - \int 2x\, dx + \int 5\, dx$$

$$= \frac{x^3}{3} - x^2 + 5x + C.$$

Find the simplest antiderivative you can for each part and add the arbitrary constant of integration at the end. ∎

# Exercises 4.7

## Finding Antiderivatives

In Exercises 1–16, find an antiderivative for each function. Do as many as you can mentally. Check your answers by differentiation.

**1. a.** $2x$    **b.** $x^2$    **c.** $x^2 - 2x + 1$

**2. a.** $6x$    **b.** $x^7$    **c.** $x^7 - 6x + 8$

**3. a.** $-3x^{-4}$    **b.** $x^{-4}$    **c.** $x^{-4} + 2x + 3$

**4. a.** $2x^{-3}$    **b.** $\dfrac{x^{-3}}{2} + x^2$    **c.** $-x^{-3} + x - 1$

**5. a.** $\dfrac{1}{x^2}$    **b.** $\dfrac{5}{x^2}$    **c.** $2 - \dfrac{5}{x^2}$

**6. a.** $-\dfrac{2}{x^3}$    **b.** $\dfrac{1}{2x^3}$    **c.** $x^3 - \dfrac{1}{x^3}$

**7. a.** $\dfrac{3}{2}\sqrt{x}$    **b.** $\dfrac{1}{2\sqrt{x}}$    **c.** $\sqrt{x} + \dfrac{1}{\sqrt{x}}$

**8. a.** $\dfrac{4}{3}\sqrt[3]{x}$    **b.** $\dfrac{1}{3\sqrt[3]{x}}$    **c.** $\sqrt[3]{x} + \dfrac{1}{\sqrt[3]{x}}$

**9. a.** $\dfrac{2}{3}x^{-1/3}$    **b.** $\dfrac{1}{3}x^{-2/2}$    **c.** $-\dfrac{1}{3}x^{-4/3}$

**10. a.** $x^{\sqrt{3}}$    **b.** $x^{\pi}$    **c.** $x^{\sqrt{2}-1}$

**11. a.** $-\pi \sin \pi x$    **b.** $3 \sin x$    **c.** $\sin \pi x - 3 \sin 3x$

**12. a.** $\pi \cos \pi x$    **b.** $\dfrac{\pi}{2}\cos\dfrac{\pi x}{2}$    **c.** $\cos\dfrac{\pi x}{2} + \pi \cos x$

**13. a.** $\dfrac{1}{2}\sec^2 x$    **b.** $\dfrac{2}{3}\sec^2\dfrac{x}{3}$    **c.** $-\sec^2\dfrac{3x}{2}$

**14. a.** $\csc^2 x$    **b.** $-\dfrac{3}{2}\csc^2\dfrac{3x}{2}$    **c.** $1 - 8\csc^2 2x$

**15. a.** $\csc x \cot x$    **b.** $-\csc 5x \cot 5x$    **c.** $-\pi \csc\dfrac{\pi x}{2}\cot\dfrac{\pi x}{2}$

**16. a.** $\sec x \tan x$    **b.** $4\sec 3x \tan 3x$    **c.** $\sec\dfrac{\pi x}{2}\tan\dfrac{\pi x}{2}$

## Finding Indefinite Integrals

In Exercises 17–56, find the most general antiderivative or indefinite integral. You may need to try a solution and then adjust your guess. Check your answers by differentiation.

**17.** $\displaystyle\int (x + 1)\, dx$      **18.** $\displaystyle\int (5 - 6x)\, dx$

**19.** $\displaystyle\int \left(3t^2 + \frac{t}{2}\right) dt$      **20.** $\displaystyle\int \left(\frac{t^2}{2} + 4t^3\right) dt$

**21.** $\displaystyle\int (2x^3 - 5x + 7)\, dx$      **22.** $\displaystyle\int (1 - x^2 - 3x^5)\, dx$

**23.** $\displaystyle\int \left(\frac{1}{x^2} - x^2 - \frac{1}{3}\right) dx$      **24.** $\displaystyle\int \left(\frac{1}{5} - \frac{2}{x^3} + 2x\right) dx$

**25.** $\displaystyle\int x^{-1/3}\, dx$      **26.** $\displaystyle\int x^{-5/4}\, dx$

**27.** $\int \left( \sqrt{x} + \sqrt[3]{x} \right) dx$

**28.** $\int \left( \frac{\sqrt{x}}{2} + \frac{2}{\sqrt{x}} \right) dx$

**29.** $\int \left( 8y - \frac{2}{y^{1/4}} \right) dy$

**30.** $\int \left( \frac{1}{7} - \frac{1}{y^{5/4}} \right) dy$

**31.** $\int 2x(1 - x^{-3}) \, dx$

**32.** $\int x^{-3}(x + 1) \, dx$

**33.** $\int \frac{t\sqrt{t} + \sqrt{t}}{t^2} dt$

**34.** $\int \frac{4 + \sqrt{t}}{t^3} dt$

**35.** $\int (-2 \cos t) \, dt$

**36.** $\int (-5 \sin t) \, dt$

**37.** $\int 7 \sin \frac{\theta}{3} \, d\theta$

**38.** $\int 3 \cos 5\theta \, d\theta$

**39.** $\int (-3 \csc^2 x) \, dx$

**40.** $\int \left( -\frac{\sec^2 x}{3} \right) dx$

**41.** $\int \frac{\csc \theta \cot \theta}{2} d\theta$

**42.** $\int \frac{2}{5} \sec \theta \tan \theta \, d\theta$

**43.** $\int (4 \sec x \tan x - 2 \sec^2 x) \, dx$

**44.** $\int \frac{1}{2}(\csc^2 x - \csc x \cot x) \, dx$

**45.** $\int (\sin 2x - \csc^2 x) \, dx$

**46.** $\int (2 \cos 2x - 3 \sin 3x) \, dx$

**47.** $\int \frac{1 + \cos 4t}{2} dt$

**48.** $\int \frac{1 - \cos 6t}{2} dt$

**49.** $\int 3x^{\sqrt{3}} \, dx$

**50.** $\int x^{\sqrt{2}-1} \, dx$

**51.** $\int (1 + \tan^2 \theta) \, d\theta$

**52.** $\int (2 + \tan^2 \theta) \, d\theta$

(*Hint:* $1 + \tan^2 \theta = \sec^2 \theta$)

**53.** $\int \cot^2 x \, dx$

**54.** $\int (1 - \cot^2 x) \, dx$

(*Hint:* $1 + \cot^2 x = \csc^2 x$)

**55.** $\int \cos \theta (\tan \theta + \sec \theta) \, d\theta$  **56.** $\int \frac{\csc \theta}{\csc \theta - \sin \theta} d\theta$

**Checking Antiderivative Formulas**
Verify the formulas in Exercises 57–62 by differentiation.

**57.** $\int (7x - 2)^3 \, dx = \frac{(7x - 2)^4}{28} + C$

**58.** $\int (3x + 5)^{-2} \, dx = -\frac{(3x + 5)^{-1}}{3} + C$

**59.** $\int \sec^2 (5x - 1) \, dx = \frac{1}{5} \tan (5x - 1) + C$

**60.** $\int \csc^2 \left( \frac{x - 1}{3} \right) dx = -3 \cot \left( \frac{x - 1}{3} \right) + C$

**61.** $\int \frac{1}{(x + 1)^2} dx = -\frac{1}{x + 1} + C$

**62.** $\int \frac{1}{(x + 1)^2} dx = \frac{x}{x + 1} + C$

**63.** Right, or wrong? Say which for each formula and give a brief reason for each answer.

**a.** $\int x \sin x \, dx = \frac{x^2}{2} \sin x + C$

**b.** $\int x \sin x \, dx = -x \cos x + C$

**c.** $\int x \sin x \, dx = -x \cos x + \sin x + C$

**64.** Right, or wrong? Say which for each formula and give a brief reason for each answer.

**a.** $\int \tan \theta \sec^2 \theta \, d\theta = \frac{\sec^3 \theta}{3} + C$

**b.** $\int \tan \theta \sec^2 \theta \, d\theta = \frac{1}{2} \tan^2 \theta + C$

**c.** $\int \tan \theta \sec^2 \theta \, d\theta = \frac{1}{2} \sec^2 \theta + C$

**65.** Right, or wrong? Say which for each formula and give a brief reason for each answer.

**a.** $\int (2x + 1)^2 \, dx = \frac{(2x + 1)^3}{3} + C$

**b.** $\int 3(2x + 1)^2 \, dx = (2x + 1)^3 + C$

**c.** $\int 6(2x + 1)^2 \, dx = (2x + 1)^3 + C$

**66.** Right, or wrong? Say which for each formula and give a brief reason for each answer.

**a.** $\int \sqrt{2x + 1} \, dx = \sqrt{x^2 + x} + C$

**b.** $\int \sqrt{2x + 1} \, dx = \sqrt{x^2 + x} + C$

**c.** $\int \sqrt{2x + 1} \, dx = \frac{1}{3}\left( \sqrt{2x + 1} \right)^3 + C$

**67.** Right, or wrong? Give a brief reason why.

$$\int \frac{-15(x + 3)^2}{(x - 2)^4} dx = \left( \frac{x + 3}{x - 2} \right)^3 + C$$

**68.** Right, or wrong? Give a brief reason why.

$$\int \frac{x \cos (x^2) - \sin (x^2)}{x^2} dx = \frac{\sin (x^2)}{x} + C$$

Initial Value Problems

**69.** Which of the following graphs shows the solution of the initial value problem

$$\frac{dy}{dx} = 2x, \quad y = 4 \text{ when } x = 1?$$

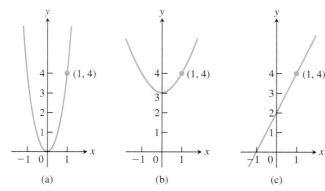

(a)  (b)  (c)

Give reasons for your answer.

**70.** Which of the following graphs shows the solution of the initial value problem

$$\frac{dy}{dx} = -x, \quad y = 1 \text{ when } x = -1?$$

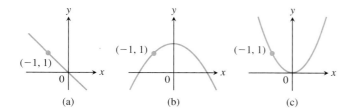

(a)  (b)  (c)

Give reasons for your answer.

Solve the initial value problems in Exercises 71–90.

**71.** $\dfrac{dy}{dx} = 2x - 7, \quad y(2) = 0$

**72.** $\dfrac{dy}{dx} = 10 - x, \quad y(0) = -1$

**73.** $\dfrac{dy}{dx} = \dfrac{1}{x^2} + x, \quad x > 0; \quad y(2) = 1$

**74.** $\dfrac{dy}{dx} = 9x^2 - 4x + 5, \quad y(-1) = 0$

**75.** $\dfrac{dy}{dx} = 3x^{-2/3}, \quad y(-1) = -5$

**76.** $\dfrac{dy}{dx} = \dfrac{1}{2\sqrt{x}}, \quad y(4) = 0$

**77.** $\dfrac{ds}{dt} = 1 + \cos t, \quad s(0) = 4$

**78.** $\dfrac{ds}{dt} = \cos t + \sin t, \quad s(\pi) = 1$

**79.** $\dfrac{dr}{d\theta} = -\pi \sin \pi\theta, \quad r(0) = 0$

**80.** $\dfrac{dr}{d\theta} = \cos \pi\theta, \quad r(0) = 1$

**81.** $\dfrac{dv}{dt} = \dfrac{1}{2} \sec t \tan t, \quad v(0) = 1$

**82.** $\dfrac{dv}{dt} = 8t + \csc^2 t, \quad v\left(\dfrac{\pi}{2}\right) = -7$

**83.** $\dfrac{d^2y}{dx^2} = 2 - 6x; \quad y'(0) = 4, \quad y(0) = 1$

**84.** $\dfrac{d^2y}{dx^2} = 0; \quad y'(0) = 2, \quad y(0) = 0$

**85.** $\dfrac{d^2r}{dt^2} = \dfrac{2}{t^3}; \quad \left.\dfrac{dr}{dt}\right|_{t=1} = 1, \quad r(1) = 1$

**86.** $\dfrac{d^2s}{dt^2} = \dfrac{3t}{8}; \quad \left.\dfrac{ds}{dt}\right|_{t=4} = 3, \quad s(4) = 4$

**87.** $\dfrac{d^3y}{dx^3} = 6; \quad y''(0) = -8, \quad y'(0) = 0, \quad y(0) = 5$

**88.** $\dfrac{d^3\theta}{dt^3} = 0; \quad \theta''(0) = -2, \quad \theta'(0) = -\dfrac{1}{2}, \quad \theta(0) = \sqrt{2}$

**89.** $y^{(4)} = -\sin t + \cos t;$
$y'''(0) = 7, \quad y''(0) = y'(0) = -1, \quad y(0) = 0$

**90.** $y^{(4)} = -\cos x + 8 \sin 2x;$
$y'''(0) = 0, \quad y''(0) = y'(0) = 1, \quad y(0) = 3$

**91.** Find the curve $y = f(x)$ in the $xy$-plane that passes through the point $(9, 4)$ and whose slope at each point is $3\sqrt{x}$.

**92. a.** Find a curve $y = f(x)$ with the following properties:

   **i)** $\dfrac{d^2y}{dx^2} = 6x$

   **ii)** Its graph passes through the point $(0, 1)$ and has a horizontal tangent there.

**b.** How many curves like this are there? How do you know?

Solution (Integral) Curves

Exercises 93–96 show solution curves of differential equations. In each exercise, find an equation for the curve through the labeled point.

**93.**    **94.**

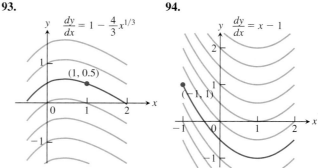

**95.**

$$\frac{dy}{dx} = \sin x - \cos x$$

$(-\pi, -1)$

**96.**

$$\frac{dy}{dx} = \frac{1}{2\sqrt{x}} + \pi \sin \pi x$$

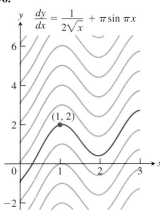

$(1, 2)$

### Applications

**97. Finding displacement from an antiderivative of velocity**

a. Suppose that the velocity of a body moving along the $s$-axis is

$$\frac{ds}{dt} = v = 9.8t - 3.$$

   i) Find the body's displacement over the time interval from $t = 1$ to $t = 3$ given that $s = 5$ when $t = 0$.

   ii) Find the body's displacement from $t = 1$ to $t = 3$ given that $s = -2$ when $t = 0$.

   iii) Now find the body's displacement from $t = 1$ to $t = 3$ given that $s = s_0$ when $t = 0$.

b. Suppose that the position $s$ of a body moving along a coordinate line is a differentiable function of time $t$. Is it true that once you know an antiderivative of the velocity function $ds/dt$ you can find the body's displacement from $t = a$ to $t = b$ even if you do not know the body's exact position at either of those times? Give reasons for your answer.

**98. Liftoff from Earth** A rocket lifts off the surface of Earth with a constant acceleration of $20 \text{ m/s}^2$. How fast will the rocket be going 1 min later?

**99. Stopping a car in time** You are driving along a highway at a steady $108 \text{ km/h}$ ($30 \text{ m/s}$) when you see an accident ahead and slam on the brakes. What constant deceleration is required to stop your car in 75 m? To find out, carry out the following steps.

   **1.** Solve the initial value problem

   Differential equation: $\dfrac{d^2 s}{dt^2} = -k$    ($k$ constant)

   Initial conditions: $\dfrac{ds}{dt} = 30$ and $s = 0$ when $t = 0$.
   
   Measuring time and distance from when the brakes are applied

   **2.** Find the value of $t$ that makes $ds/dt = 0$. (The answer will involve $k$.)

   **3.** Find the value of $k$ that makes $s = 75$ for the value of $t$ you found in Step 2.

**100. Stopping a motorcycle** The State of Illinois Cycle Rider Safety Program requires motorcycle riders to be able to brake from $48 \text{ km/h}$ ($13.3 \text{ m/s}$) to 0 in 13.7 m. What constant deceleration does it take to do that?

**101. Motion along a coordinate line** A particle moves on a coordinate line with acceleration $a = d^2 s/dt^2 = 15\sqrt{t} - \left(3/\sqrt{t}\right)$, subject to the conditions that $ds/dt = 4$ and $s = 0$ when $t = 1$. Find

   a. the velocity $v = ds/dt$ in terms of $t$

   b. the position $s$ in terms of $t$.

T **102. The hammer and the feather** When *Apollo 15* astronaut David Scott dropped a hammer and a feather on the moon to demonstrate that in a vacuum all bodies fall with the same (constant) acceleration, he dropped them from about 1.2 m above the ground. The television footage of the event shows the hammer and the feather falling more slowly than on Earth, where, in a vacuum, they would have taken only half a second to fall the 1.2 m. How long did it take the hammer and feather to fall 1.2 m on the moon? To find out, solve the following initial value problem for $s$ as a function of $t$. Then find the value of $t$ that makes $s$ equal to 0.

   Differential equation:    $\dfrac{d^2 s}{dt^2} = -1.6 \text{ m/s}^2$

   Initial conditions:    $\dfrac{ds}{dt} = 0$ and $s = 1.2$ when $t = 0$

**103. Motion with constant acceleration** The standard equation for the position $s$ of a body moving with a constant acceleration $a$ along a coordinate line is

$$s = \frac{a}{2} t^2 + v_0 t + s_0, \tag{1}$$

where $v_0$ and $s_0$ are the body's velocity and position at time $t = 0$. Derive this equation by solving the initial value problem

   Differential equation:    $\dfrac{d^2 s}{dt^2} = a$

   Initial conditions:    $\dfrac{ds}{dt} = v_0$ and $s = s_0$ when $t = 0$.

**104. Free fall near the surface of a planet** For free fall near the surface of a planet where the acceleration due to gravity has a constant magnitude of $g$ length-units/s$^2$, Equation (1) in Exercise 103 takes the form

$$s = -\frac{1}{2} g t^2 + v_0 t + s_0, \tag{2}$$

where $s$ is the body's height above the surface. The equation has a minus sign because the acceleration acts downward, in the direction of decreasing $s$. The velocity $v_0$ is positive if the object is rising at time $t = 0$ and negative if the object is falling.

Instead of using the result of Exercise 103, you can derive Equation (2) directly by solving an appropriate initial value problem. What initial value problem? Solve it to be sure you have the right one, explaining the solution steps as you go along.

**105.** Suppose that

$$f(x) = \frac{d}{dx}\left(1 - \sqrt{x}\right) \quad \text{and} \quad g(x) = \frac{d}{dx}(x + 2).$$

Find:

**a.** $\displaystyle\int f(x)\,dx$        **b.** $\displaystyle\int g(x)\,dx$

**c.** $\displaystyle\int [-f(x)]\,dx$      **d.** $\displaystyle\int [-g(x)]\,dx$

**e.** $\displaystyle\int [f(x) + g(x)]\,dx$     **f.** $\displaystyle\int [f(x) - g(x)]\,dx$

**106. Uniqueness of solutions** If differentiable functions $y = F(x)$ and $y = g(x)$ both solve the initial value problem

$$\frac{dy}{dx} = f(x), \qquad y(x_0) = y_0,$$

on an interval $I$, must $F(x) = G(x)$ for every $x$ in $I$? Give reasons for your answer.

## COMPUTER EXPLORATIONS

Use a CAS to solve the initial value problems in Exercises 107–110. Plot the solution curves.

**107.** $y' = \cos^2 x + \sin x, \quad y(\pi) = 1$

**108.** $y' = \frac{1}{x} + x, \quad y(1) = -1$

**109.** $y' = \frac{1}{\sqrt{4 - x^2}}, \quad y(0) = 2$

**110.** $y'' = \frac{2}{x} + \sqrt{x}, \quad y(1) = 0, \quad y'(1) = 0$

---

# Chapter 4   Questions to Guide Your Review

**1.** What can be said about the extreme values of a function that is continuous on a closed interval?

**2.** What does it mean for a function to have a local extreme value on its domain? An absolute extreme value? How are local and absolute extreme values related, if at all? Give examples.

**3.** How do you find the absolute extrema of a continuous function on a closed interval? Give examples.

**4.** What are the hypotheses and conclusion of Rolle's Theorem? Are the hypotheses really necessary? Explain.

**5.** What are the hypotheses and conclusion of the Mean Value Theorem? What physical interpretations might the theorem have?

**6.** State the Mean Value Theorem's three corollaries.

**7.** How can you sometimes identify a function $f(x)$ by knowing $f'$ and knowing the value of $f$ at a point $x = x_0$? Give an example.

**8.** What is the First Derivative Test for Local Extreme Values? Give examples of how it is applied.

**9.** How do you test a twice-differentiable function to determine where its graph is concave up or concave down? Give examples.

**10.** What is an inflection point? Give an example. What physical significance do inflection points sometimes have?

**11.** What is the Second Derivative Test for Local Extreme Values? Give examples of how it is applied.

**12.** What do the derivatives of a function tell you about the shape of its graph?

**13.** List the steps you would take to graph a polynomial function. Illustrate with an example.

**14.** What is a cusp? Give examples.

**15.** List the steps you would take to graph a rational function. Illustrate with an example.

**16.** Outline a general strategy for solving max-min problems. Give examples.

**17.** Describe Newton's method for solving equations. Give an example. What is the theory behind the method? What are some of the things to watch out for when you use the method?

**18.** Can a function have more than one antiderivative? If so, how are the antiderivatives related? Explain.

**19.** What is an indefinite integral? How do you evaluate one? What general formulas do you know for finding indefinite integrals?

**20.** How can you sometimes solve a differential equation of the form $dy/dx = f(x)$?

**21.** What is an initial value problem? How do you solve one? Give an example.

**22.** If you know the acceleration of a body moving along a coordinate line as a function of time, what more do you need to know to find the body's position function? Give an example.

# Chapter **4** Practice Exercises

## Extreme Values

**1.** Does $f(x) = x^3 + 2x + \tan x$ have any local maximum or minimum values? Give reasons for your answer.

**2.** Does $g(x) = \csc x + 2 \cot x$ have any local maximum values? Give reasons for your answer.

**3.** Does $f(x) = (7 + x)(11 - 3x)^{1/3}$ have an absolute minimum value? An absolute maximum? If so, find them or give reasons why they fail to exist. List all critical points of $f$.

**4.** Find values of $a$ and $b$ such that the function

$$f(x) = \frac{ax + b}{x^2 - 1}$$

has a local extreme value of 1 at $x = 3$. Is this extreme value a local maximum, or a local minimum? Give reasons for your answer.

**5.** The greatest integer function $f(x) = \lfloor x \rfloor$, defined for all values of $x$, assumes a local maximum value of 0 at each point of $[0, 1)$. Could any of these local maximum values also be local minimum values of $f$? Give reasons for your answer.

**6. a.** Give an example of a differentiable function $f$ whose first derivative is zero at some point $c$ even though $f$ has neither a local maximum nor a local minimum at $c$.

    **b.** How is this consistent with Theorem 2 in Section 4.1? Give reasons for your answer.

**7.** The function $y = 1/x$ does not take on either a maximum or a minimum on the interval $0 < x < 1$ even though the function is continuous on this interval. Does this contradict the Extreme Value Theorem for continuous functions? Why?

**8.** What are the maximum and minimum values of the function $y = |x|$ on the interval $-1 \le x < 1$? Notice that the interval is not closed. Is this consistent with the Extreme Value Theorem for continuous functions? Why?

T **9.** A graph that is large enough to show a function's global behavior may fail to reveal important local features. The graph of $f(x) = (x^8/8) - (x^6/2) - x^5 + 5x^3$ is a case in point.

    **a.** Graph $f$ over the interval $-2.5 \le x \le 2.5$. Where does the graph appear to have local extreme values or points of inflection?

    **b.** Now factor $f'(x)$ and show that $f$ has a local maximum at $x = \sqrt[5]{5} \approx 1.70998$ and local minima at $x = \pm\sqrt{3} \approx \pm 1.73205$.

    **c.** Zoom in on the graph to find a viewing window that shows the presence of the extreme values at $x = \sqrt[5]{5}$ and $x = \sqrt{3}$.

       The moral here is that without calculus the existence of two of the three extreme values would probably have gone unnoticed. On any normal graph of the function, the values would lie close enough together to fall within the dimensions of a single pixel on the screen.

       (*Source*: *Uses of Technology in the Mathematics Curriculum*, by Benny Evans and Jerry Johnson, Oklahoma State University, published in 1990 under a grant from the National Science Foundation, USE-8950044.)

T **10.** (*Continuation of Exercise 9.*)

    **a.** Graph $f(x) = (x^8/8) - (2/5)x^5 - 5x - (5/x^2) + 11$ over the interval $-2 \le x \le 2$. Where does the graph appear to have local extreme values or points of inflection?

    **b.** Show that $f$ has a local maximum value at $x = \sqrt[7]{5} \approx 1.2585$ and a local minimum value at $x = \sqrt[3]{2} \approx 1.2599$.

    **c.** Zoom in to find a viewing window that shows the presence of the extreme values at $x = \sqrt[7]{5}$ and $x = \sqrt[3]{2}$.

## The Mean Value Theorem

**11. a.** Show that $g(t) = \sin^2 t - 3t$ decreases on every interval in its domain.

    **b.** How many solutions does the equation $\sin^2 t - 3t = 5$ have? Give reasons for your answer.

**12. a.** Show that $y = \tan \theta$ increases on every open interval in its domain.

    **b.** If the conclusion in part (a) is really correct, how do you explain the fact that $\tan \pi = 0$ is less than $\tan(\pi/4) = 1$?

**13. a.** Show that the equation $x^4 + 2x^2 - 2 = 0$ has exactly one solution on $[0, 1]$.

T    **b.** Find the solution to as many decimal places as you can.

**14. a.** Show that $f(x) = x/(x + 1)$ increases on every open interval in its domain.

    **b.** Show that $f(x) = x^3 + 2x$ has no local maximum or minimum values.

**15. Water in a reservoir** As a result of a heavy rain, the volume of water in a reservoir increased by a million cubic meters in 24 hours. Show that at some instant during that period the reservoir's volume was increasing at a rate in excess of 500,000 L/min.

**16.** The formula $F(x) = 3x + C$ gives a different function for each value of $C$. All of these functions, however, have the same derivative with respect to $x$, namely $F'(x) = 3$. Are these the only differentiable functions whose derivative is 3? Could there be any others? Give reasons for your answers.

**17.** Show that

$$\frac{d}{dx}\left(\frac{x}{x + 1}\right) = \frac{d}{dx}\left(-\frac{1}{x + 1}\right)$$

even though

$$\frac{x}{x + 1} \ne -\frac{1}{x + 1}.$$

Doesn't this contradict Corollary 2 of the Mean Value Theorem? Give reasons for your answer.

**18.** Calculate the first derivatives of $f(x) = x^2/(x^2 + 1)$ and $g(x) = -1/(x^2 + 1)$. What can you conclude about the graphs of these functions?

**Analyzing Graphs**

In Exercises 19 and 20, use the graph to answer the questions.

**19.** Identify any global extreme values of $f$ and the values of $x$ at which they occur.

**20.** Estimate the open intervals on which the function $y = f(x)$ is

  **a.** increasing.

  **b.** decreasing.

  **c.** Use the given graph of $f'$ to indicate where any local extreme values of the function occur, and whether each extreme is a relative maximum or minimum.

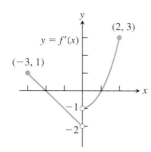

Each of the graphs in Exercises 21 and 22 is the graph of the position function $s = f(t)$ of an object moving on a coordinate line ($t$ represents time). At approximately what times (if any) is each object's **(a)** velocity equal to zero? **(b)** Acceleration equal to zero? During approximately what time intervals does the object move **(c)** forward? **(d)** Backward?

**21.**

**22.**

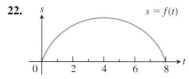

**Graphs and Graphing**

Graph the curves in Exercises 23–32.

**23.** $y = x^2 - (x^3/6)$
**24.** $y = x^3 - 3x^2 + 3$
**25.** $y = -x^3 + 6x^2 - 9x + 3$
**26.** $y = (1/8)(x^3 + 3x^2 - 9x - 27)$
**27.** $y = x^3(8 - x)$
**28.** $y = x^2(2x^2 - 9)$
**39.** $y = x - 3x^{2/3}$
**30.** $y = x^{1/3}(x - 4)$
**31.** $y = x\sqrt{3 - x}$
**32.** $y = x\sqrt{4 - x^2}$

Each of Exercises 33–38 gives the first derivative of a function $y = f(x)$. **(a)** At what points, if any, does the graph of $f$ have a local maximum, local minimum, or inflection point? **(b)** Sketch the general shape of the graph.

**33.** $y' = 16 - x^2$
**34.** $y' = x^2 - x - 6$
**35.** $y' = 6x(x + 1)(x - 2)$
**36.** $y' = x^2(6 - 4x)$
**37.** $y' = x^4 - 2x^2$
**38.** $y' = 4x^2 - x^4$

In Exercises 39–42, graph each function. Then use the function's first derivative to explain what you see.

**39.** $y = x^{2/3} + (x - 1)^{1/3}$
**40.** $y = x^{2/3} + (x - 1)^{2/3}$
**41.** $y = x^{1/3} + (x - 1)^{1/3}$
**42.** $y = x^{2/3} - (x - 1)^{1/3}$

Sketch the graphs of the rational functions in Exercises 43–50.

**43.** $y = \dfrac{x + 1}{x - 3}$
**44.** $y = \dfrac{2x}{x + 5}$
**45.** $y = \dfrac{x^2 + 1}{x}$
**46.** $y = \dfrac{x^2 - x + 1}{x}$
**47.** $y = \dfrac{x^3 + 2}{2x}$
**48.** $y = \dfrac{x^4 - 1}{x^2}$
**49.** $y = \dfrac{x^2 - 4}{x^2 - 3}$
**50.** $y = \dfrac{x^2}{x^2 - 4}$

**Optimization**

**51.** The sum of two nonnegative numbers is 36. Find the numbers if

  **a.** the difference of their square roots is to be as large as possible.

  **b.** the sum of their square roots is to be as large as possible.

**52.** The sum of two nonnegative numbers is 20. Find the numbers

  **a.** if the product of one number and the square root of the other is to be as large as possible.

  **b.** if one number plus the square root of the other is to be as large as possible.

**53.** An isosceles triangle has its vertex at the origin and its base parallel to the $x$-axis with the vertices above the axis on the curve $y = 27 - x^2$. Find the largest area the triangle can have.

**54.** A customer has asked you to design an open-top rectangular stainless steel vat. It is to have a square base and a volume of $1 \text{ m}^3$, to be welded from 6-mm-thick plate, and to weigh no more than necessary. What dimensions do you recommend?

**55.** Find the height and radius of the largest right circular cylinder that can be put in a sphere of radius $\sqrt{3}$.

**56.** The figure here shows two right circular cones, one upside down inside the other. The two bases are parallel, and the vertex of the smaller cone lies at the center of the larger cone's base. What values of $r$ and $h$ will give the smaller cone the largest possible volume?

**57. Manufacturing tires** Your company can manufacture $x$ hundred grade A tires and $y$ hundred grade B tires a day, where $0 \le x \le 4$ and

$$y = \frac{40 - 10x}{5 - x}.$$

Your profit on a grade A tire is twice your profit on a grade B tire. What is the most profitable number of each kind to make?

**58. Particle motion** The positions of two particles on the $s$-axis are $s_1 = \cos t$ and $s_2 = \cos(t + \pi/4)$.

   **a.** What is the farthest apart the particles ever get?

   **b.** When do the particles collide?

**T** **59. Open-top box** An open-top rectangular box is constructed from a 25-cm-by-40-cm piece of cardboard by cutting squares of equal side length from the corners and folding up the sides. Find analytically the dimensions of the box of largest volume and the maximum volume. Support your answers graphically.

**60. The ladder problem** What is the approximate length (in meters) of the longest ladder you can carry horizontally around the corner of the corridor shown here? Round your answer down to the nearest meter.

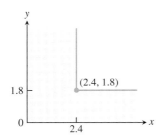

**Newton's Method**

**61.** Let $f(x) = 3x - x^3$. Show that the equation $f(x) = -4$ has a solution in the interval $[2, 3]$ and use Newton's method to find it.

**62.** Let $f(x) = x^4 - x^3$. Show that the equation $f(x) = 75$ has a solution in the interval $[3, 4]$ and use Newton's method to find it.

Finding Indefinite Integrals
Find the indefinite integrals (most general antiderivatives) in Exercises 63–78. You may need to try a solution and then adjust your guess. Check your answers by differentiation.

**63.** $\displaystyle\int (x^5 + 4x^2 + 9)\, dx$

**64.** $\displaystyle\int \left( 6t^5 - \frac{t^3}{4} + t \right) dt$

**65.** $\displaystyle\int \left( t\sqrt{t} + \frac{5}{t^3} \right) dt$

**66.** $\displaystyle\int \left( \frac{5}{t\sqrt{t}} - \frac{4}{t^3} \right) dt$

**67.** $\displaystyle\int \frac{dr}{(r - 3)^3}$

**68.** $\displaystyle\int \frac{5\, dr}{(r + \sqrt{3})^2}$

**69.** $\displaystyle\int 9\theta^2 \sqrt{\theta^3 - 1}\, d\theta$

**70.** $\displaystyle\int \frac{x}{\sqrt{5 - x^2}}\, dx$

**71.** $\displaystyle\int x^4 (1 - x^5)^{-1/5}\, dx$

**72.** $\displaystyle\int (5 - x)^{5/7}\, dx$

**73.** $\displaystyle\int \sec^2 \frac{s}{10}\, ds$

**74.** $\displaystyle\int \csc^2 es\, ds$

**75.** $\displaystyle\int \csc \sqrt{3}\theta \cot \sqrt{3}\theta\, d\theta$

**76.** $\displaystyle\int \sec \frac{t}{5} \tan \frac{t}{5}\, dt$

**77.** $\displaystyle\int \cos^2 \frac{x}{6}\, dx$ $\left( \text{Hint: } \cos^2 x = \dfrac{1 + \cos 2x}{2} \right)$

**78.** $\displaystyle\int \sec^2 \frac{x}{2}\, dx$

Initial Value Problems
Solve the initial value problems in Exercises 79–82.

**79.** $\dfrac{dy}{dx} = \dfrac{x^2 + 1}{x^2}, \quad y(1) = -1$

**80.** $\dfrac{dy}{dx} = \left( x + \dfrac{1}{x} \right)^2, \quad y(1) = 1$

**81.** $\dfrac{d^2 r}{dt^2} = 15\sqrt{t} + \dfrac{3}{\sqrt{t}}; \quad r'(1) = 8, \quad r(1) = 0$

**82.** $\dfrac{d^3 r}{dt^3} = -\cos t; \quad r''(0) = r'(0) = 0, \quad r(0) = -1$

---

# Chapter **4** Additional and Advanced Exercises

**Functions and Derivatives**

**1.** What can you say about a function whose maximum and minimum values on an interval are equal? Give reasons for your answer.

**2.** Is it true that a discontinuous function cannot have both an absolute maximum and an absolute minimum value on a closed interval? Give reasons for your answer.

**3.** Can you conclude anything about the extreme values of a continuous function on an open interval? On a half-open interval? Give reasons for your answer.

**4. Local extrema** Use the sign pattern for the derivative

$$\frac{df}{dx} = 6(x - 1)(x - 2)^2(x - 3)^3(x - 4)^4$$

to identify the points where $f$ has local maximum and minimum values.

**5. Local extrema**

   **a.** Suppose that the first derivative of $y = f(x)$ is

   $$y' = 6(x + 1)(x - 2)^2.$$

   At what points, if any, does the graph of $f$ have a local maximum, local minimum, or point of inflection?

**b.** Suppose that the first derivative of $y = f(x)$ is

$$y' = 6x(x + 1)(x - 2).$$

At what points, if any, does the graph of $f$ have a local maximum, local minimum, or point of inflection?

**6.** If $f'(x) \leq 2$ for all $x$, what is the most the values of $f$ can increase on $[0, 6]$? Give reasons for your answer.

**7. Bounding a function** Suppose that $f$ is continuous on $[a, b]$ and that $c$ is an interior point of the interval. Show that if $f'(x) \leq 0$ on $[a, c)$ and $f'(x) \geq 0$ on $(c, b]$, then $f(x)$ is never less than $f(c)$ on $[a, b]$.

**8. An inequality**

**a.** Show that $-1/2 \leq x/(1 + x^2) \leq 1/2$ for every value of $x$.

**b.** Suppose that $f$ is a function whose derivative is $f'(x) = x/(1 + x^2)$. Use the result in part (a) to show that

$$|f(b) - f(a)| \leq \frac{1}{2}|b - a|$$

for any $a$ and $b$.

**9.** The derivative of $f(x) = x^2$ is zero at $x = 0$, but $f$ is not a constant function. Doesn't this contradict the corollary of the Mean Value Theorem that says that functions with zero derivatives are constant? Give reasons for your answer.

**10. Extrema and inflection points** Let $h = fg$ be the product of two differentiable functions of $x$.

**a.** If $f$ and $g$ are positive, with local maxima at $x = a$, and if $f'$ and $g'$ change sign at $a$, does $h$ have a local maximum at $a$?

**b.** If the graphs of $f$ and $g$ have inflection points at $x = a$, does the graph of $h$ have an inflection point at $a$?

In either case, if the answer is yes, give a proof. If the answer is no, give a counterexample.

**11. Finding a function** Use the following information to find the values of $a$, $b$, and $c$ in the formula $f(x) = (x + a)/(bx^2 + cx + 2)$.

  **i)** The values of $a$, $b$, and $c$ are either 0 or 1.

  **ii)** The graph of $f$ passes through the point $(-1, 0)$.

  **iii)** The line $y = 1$ is an asymptote of the graph of $f$.

**12. Horizontal tangent** For what value or values of the constant $k$ will the curve $y = x^3 + kx^2 + 3x - 4$ have exactly one horizontal tangent?

**Optimization**

**13. Largest inscribed triangle** Points $A$ and $B$ lie at the ends of a diameter of a unit circle and point $C$ lies on the circumference. Is it true that the area of triangle $ABC$ is largest when the triangle is isosceles? How do you know?

**14. Proving the second derivative test** The Second Derivative Test for Local Maxima and Minima (Section 4.4) says:

**a.** $f$ has a local maximum value at $x = c$ if $f'(c) = 0$ and $f''(c) < 0$

**b.** $f$ has a local minimum value at $x = c$ if $f'(c) = 0$ and $f''(c) > 0$.

To prove statement (a), let $\epsilon = (1/2)|f''(c)|$. Then use the fact that

$$f''(c) = \lim_{h \to 0} \frac{f'(c + h) - f'(c)}{h} = \lim_{h \to 0} \frac{f'(c + h)}{h}$$

to conclude that for some $\delta > 0$,

$$0 < |h| < \delta \quad \Rightarrow \quad \frac{f'(c + h)}{h} < f''(c) + \epsilon < 0.$$

Thus, $f'(c + h)$ is positive for $-\delta < h < 0$ and negative for $0 < h < \delta$. Prove statement (b) in a similar way.

**15. Hole in a water tank** You want to bore a hole in the side of the tank shown here at a height that will make the stream of water coming out hit the ground as far from the tank as possible. If you drill the hole near the top, where the pressure is low, the water will exit slowly but spend a relatively long time in the air. If you drill the hole near the bottom, the water will exit at a higher velocity but have only a short time to fall. Where is the best place, if any, for the hole? (*Hint:* How long will it take an exiting droplet of water to fall from height $y$ to the ground?)

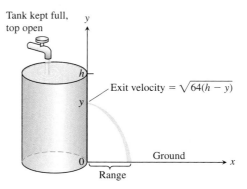

**16. Kicking a field goal** An American football player wants to kick a field goal with the ball being on a right hash mark. Assume that the goal posts are $b$ meters apart and that the hash mark line is a distance $a > 0$ meters from the right goal post. (See the accompanying figure.) Find the distance $h$ from the goal post line that gives the kicker his largest angle $\beta$. Assume that the football field is flat.

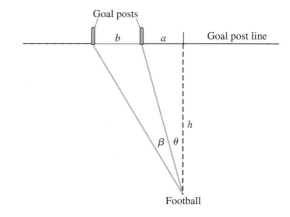

**17. A max-min problem with a variable answer** Sometimes the solution of a max-min problem depends on the proportions of the shapes involved. As a case in point, suppose that a right circular cylinder of radius $r$ and height $h$ is inscribed in a right circular cone of radius $R$ and height $H$, as shown here. Find the value of $r$ (in terms of $R$ and $H$) that maximizes the total surface area of the cylinder (including top and bottom). As you will see, the solution depends on whether $H \leq 2R$ or $H > 2R$.

**18. Minimizing a parameter** Find the smallest value of the positive constant $m$ that will make $mx - 1 + (1/x)$ greater than or equal to zero for all positive values of $x$

**Theory and Examples**

**19.** Suppose that it costs a company $y = a + bx$ dollars to produce $x$ units per week. It can sell $x$ units per week at a price of $P = c - ex$ dollars per unit. Each of $a$, $b$, $c$, and $e$ represents a positive constant. **(a)** What production level maximizes the profit? **(b)** What is the corresponding price? **(c)** What is the weekly profit at this level of production? **(d)** At what price should each item be sold to maximize profits if the government imposes a tax of $t$ dollars per item sold? Comment on the difference between this price and the price before the tax.

**20. Estimating reciprocals without division** You can estimate the value of the reciprocal of a number $a$ without ever dividing by $a$ if you apply Newton's method to the function $f(x) = (1/x) - a$. For example, if $a = 3$, the function involved is $f(x) = (1/x) - 3$.

  **a.** Graph $y = (1/x) - 3$. Where does the graph cross the $x$-axis?

  **b.** Show that the recursion formula in this case is

$$x_{n+1} = x_n(2 - 3x_n),$$

  so there is no need for division.

**21.** To find $x = \sqrt[q]{a}$, we apply Newton's method to $f(x) = x^q - a$. Here we assume that $a$ is a positive real number and $q$ is a positive integer. Show that $x_1$ is a "weighted average" of $x_0$ and $a/x_0^{q-1}$, and find the coefficients $m_0$, $m_1$ such that

$$x_1 = m_0 x_0 + m_1 \left( \frac{a}{x_0^{q-1}} \right), \quad \begin{matrix} m_0 > 0, m_1 > 0, \\ m_0 + m_1 = 1. \end{matrix}$$

What conclusion would you reach if $x_0$ and $a/x_0^{q-1}$ were equal? What would be the value of $x_1$ in that case?

**22.** The family of straight lines $y = ax + b$ ($a$, $b$ arbitrary constants) can be characterized by the relation $y'' = 0$. Find a similar relation satisfied by the family of all circles

$$(x - h)^2 + (y - h)^2 = r^2,$$

where $h$ and $r$ are arbitrary constants. (*Hint:* Eliminate $h$ and $r$ from the set of three equations including the given one and two obtained by successive differentiation.)

**23.** Assume that the brakes of an automobile produce a constant deceleration of $k$ m/s$^2$. **(a)** Determine what $k$ must be to bring an automobile traveling 108 km/h (30 m/s) to rest in a distance of 30 m from the point where the brakes are applied. **(b)** With the same $k$, how far would a car traveling 54 km/h go before being brought to a stop?

**24.** Let $f(x)$, $g(x)$ be two continuously differentiable functions satisfying the relationships $f'(x) = g(x)$ and $f''(x) = -f(x)$. Let $h(x) = f^2(x) + g^2(x)$. If $h(0) = 5$, find $h(10)$.

**25.** Can there be a curve satisfying the following conditions? $d^2y/dx^2$ is everywhere equal to zero and, when $x = 0, y = 0$ and $dy/dx = 1$. Give a reason for your answer.

**26.** Find the equation for the curve in the $xy$-plane that passes through the point $(1, -1)$ if its slope at $x$ is always $3x^2 + 2$.

**27.** A particle moves along the $x$-axis. Its acceleration is $a = -t^2$. At $t = 0$, the particle is at the origin. In the course of its motion, it reaches the point $x = b$, where $b > 0$, but no point beyond $b$. Determine its velocity at $t = 0$.

**28.** A particle moves with acceleration $a = \sqrt{t} - (1/\sqrt{t})$. Assuming that the velocity $v = 4/3$ and the position $s = -4/15$ when $t = 0$, find

  **a.** the velocity $v$ in terms of $t$.

  **b.** the position $s$ in terms of $t$.

**29.** Given $f(x) = ax^2 + 2bx + c$ with $a > 0$. By considering the minimum, prove that $f(x) \geq 0$ for all real $x$ if and only if $b^2 - ac \leq 0$.

**30. Schwarz's inequality**

  **a.** In Exercise 29, let

$$f(x) = (a_1 x + b_1)^2 + (a_2 x + b_2)^2 + \cdots + (a_n x + b_n)^2,$$

  and deduce Schwarz's inequality:

$$(a_1 b_1 + a_2 b_2 + \cdots + a_n b_n)^2$$
$$\leq \left( a_1^2 + a_2^2 + \cdots + a_n^2 \right)\left( b_1^2 + b_2^2 + \cdots + b_n^2 \right).$$

  **b.** Show that equality holds in Schwarz's inequality only if there exists a real number $x$ that makes $a_i x$ equal $-b_i$ for every value of $i$ from 1 to $n$.

# Chapter 4 Technology Application Projects

**Mathematica/Maple Modules:**

*Motion Along a Straight Line: Position → Velocity → Acceleration*
You will observe the shape of a graph through dramatic animated visualizations of the derivative relations among the position, velocity, and acceleration. Figures in the text can be animated.

*Newton's Method: Estimate $\pi$ to How Many Places?*
Plot a function, observe a root, pick a starting point near the root, and use Newton's Iteration Procedure to approximate the root to a desired accuracy. The numbers $\pi$, $e$, and $\sqrt{2}$ are approximated.

# 5

# Integrals

**OVERVIEW** A great achievement of classical geometry was obtaining formulas for the areas and volumes of triangles, spheres, and cones. In this chapter we develop a method to calculate the areas and volumes of very general shapes. This method, called *integration*, is a way to calculate much more than areas and volumes. The *definite integral* is the key tool in calculus for defining and calculating many important quantities, such as areas, volumes, lengths of curved paths, probabilities, averages, energy consumption, the weights of various objects, and the forces against a dam's floodgates, just to mention a few. Many of these applications are studied in subsequent chapters.

As with the derivative, the definite integral also arises as a limit, this time of increasingly fine approximations to the quantity of interest. The idea behind the integral is that we can effectively compute such quantities by breaking them into small pieces, and then summing the contributions from each piece. We then consider what happens when more and more, smaller and smaller pieces are taken in the summation process. As the number of terms contributing to the sum approaches infinity and we take the limit of these sums in a way described in Section 5.3, the result is a definite integral. By considering the rate of change of the area under a graph, we prove that definite integrals are connected to antiderivatives, a connection that gives one of the most important relationships in calculus.

## 5.1 Area and Estimating with Finite Sums

The basis for formulating definite integrals is the construction of appropriate approximations by finite sums. In this section we consider three examples of this construction process: finding the area under a graph, the distance traveled by a moving object, and the average value of a function. Although we need to define precisely what we mean by the area of a general region in the plane, or the average value of a function over a closed interval, we do have intuitive ideas of what these notions mean. So in this section we begin our approach to integration by *approximating* these quantities with finite sums. We also consider what happens when we take more and more terms in the summation process. In subsequent sections we look at taking the limit of these sums as the number of terms goes to infinity, which then leads to precise definitions of the quantities being approximated here.

### Area

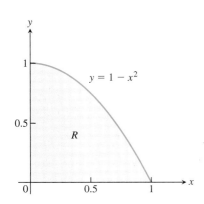

**FIGURE 5.1** The area of the region $R$ cannot be found by a simple formula.

Suppose we want to find the area of the shaded region $R$ that lies above the $x$-axis, below the graph of $y = 1 - x^2$, and between the vertical lines $x = 0$ and $x = 1$ (Figure 5.1). Unfortunately, there is no simple geometric formula for calculating the areas of general shapes having curved boundaries like the region $R$. How, then, can we find the area of $R$?

While we do not yet have a method for determining the exact area of $R$, we can approximate it in a simple way. Figure 5.2a shows two rectangles that together contain the

**263**

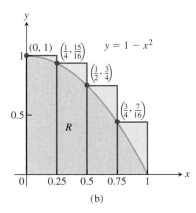

**FIGURE 5.2** (a) We get an upper estimate of the area of $R$ by using two rectangles containing $R$. (b) Four rectangles give a better upper estimate. Both estimates overshoot the true value for the area by the amount shaded in light red.

region $R$. Each rectangle has width $1/2$ and they have heights 1 and $3/4$, moving from left to right. The height of each rectangle is the maximum value of the function $f$ in each subinterval. Because the function $f$ is decreasing, the height is its value at the left endpoint of the subinterval of $[0, 1]$ forming the base of the rectangle. The total area of the two rectangles approximates the area $A$ of the region $R$,

$$A \approx 1 \cdot \frac{1}{2} + \frac{3}{4} \cdot \frac{1}{2} = \frac{7}{8} = 0.875.$$

This estimate is larger than the true area $A$ since the two rectangles contain $R$. We say that 0.875 is an **upper sum** because it is obtained by taking the height of each rectangle as the maximum (uppermost) value of $f(x)$ for a point $x$ in the base interval of the rectangle. In Figure 5.2b, we improve our estimate by using four thinner rectangles, each of width $1/4$, which taken together contain the region $R$. These four rectangles give the approximation

$$A \approx 1 \cdot \frac{1}{4} + \frac{15}{16} \cdot \frac{1}{4} + \frac{3}{4} \cdot \frac{1}{4} + \frac{7}{16} \cdot \frac{1}{4} = \frac{25}{32} = 0.78125,$$

which is still greater than $A$ since the four rectangles contain $R$.

Suppose instead we use four rectangles contained *inside* the region $R$ to estimate the area, as in Figure 5.3a. Each rectangle has width $1/4$ as before, but the rectangles are shorter and

 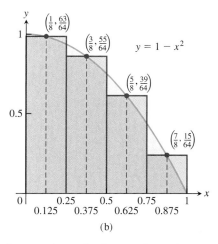

**FIGURE 5.3** (a) Rectangles contained in $R$ give an estimate for the area that undershoots the true value by the amount shaded in light blue. (b) The midpoint rule uses rectangles whose height is the value of $y = f(x)$ at the midpoints of their bases. The estimate appears closer to the true value of the area because the light red overshoot areas roughly balance the light blue undershoot areas.

lie entirely beneath the graph of $f$. The function $f(x) = 1 - x^2$ is decreasing on $[0, 1]$, so the height of each of these rectangles is given by the value of $f$ at the right endpoint of the subinterval forming its base. The fourth rectangle has zero height and therefore contributes no area. Summing these rectangles with heights equal to the minimum value of $f(x)$ for a point $x$ in each base subinterval gives a **lower sum** approximation to the area,

$$A \approx \frac{15}{16} \cdot \frac{1}{4} + \frac{3}{4} \cdot \frac{1}{4} + \frac{7}{16} \cdot \frac{1}{4} + 0 \cdot \frac{1}{4} = \frac{17}{32} = 0.53125.$$

This estimate is smaller than the area $A$ since the rectangles all lie inside of the region $R$. The true value of $A$ lies somewhere between these lower and upper sums:

$$0.53125 < A < 0.78125.$$

By considering both lower and upper sum approximations, we get not only estimates for the area, but also a bound on the size of the possible error in these estimates, since the true value of the area lies somewhere between them. Here the error cannot be greater than the difference $0.78125 - 0.53125 = 0.25$.

Yet another estimate can be obtained by using rectangles whose heights are the values of $f$ at the midpoints of their bases (Figure 5.3b). This method of estimation is called the **midpoint rule** for approximating the area. The midpoint rule gives an estimate that is between a lower sum and an upper sum, but it is not quite so clear whether it overestimates or underestimates the true area. With four rectangles of width $1/4$ as before, the midpoint rule estimates the area of $R$ to be

$$A \approx \frac{63}{64} \cdot \frac{1}{4} + \frac{55}{64} \cdot \frac{1}{4} + \frac{39}{64} \cdot \frac{1}{4} + \frac{15}{64} \cdot \frac{1}{4} = \frac{172}{64} \cdot \frac{1}{4} = 0.671875.$$

In each of our computed sums, the interval $[a, b]$ over which the function $f$ is defined was subdivided into $n$ subintervals of equal width (also called length) $\Delta x = (b - a)/n$, and $f$ was evaluated at a point in each subinterval: $c_1$ in the first subinterval, $c_2$ in the second subinterval, and so on. The finite sums then all take the form

$$f(c_1)\,\Delta x + f(c_2)\,\Delta x + f(c_3)\,\Delta x + \cdots + f(c_n)\,\Delta x.$$

By taking more and more rectangles, with each rectangle thinner than before, it appears that these finite sums give better and better approximations to the true area of the region $R$.

Figure 5.4a shows a lower sum approximation for the area of $R$ using 16 rectangles of equal width. The sum of their areas is 0.634765625, which appears close to the true area, but is still smaller since the rectangles lie inside $R$.

Figure 5.4b shows an upper sum approximation using 16 rectangles of equal width. The sum of their areas is 0.697265625, which is somewhat larger than the true area because the rectangles taken together contain $R$. The midpoint rule for 16 rectangles gives a total area approximation of 0.6669921875, but it is not immediately clear whether this estimate is larger or smaller than the true area.

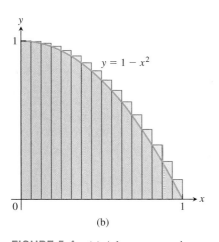

**FIGURE 5.4** (a) A lower sum using 16 rectangles of equal width $\Delta x = 1/16$. (b) An upper sum using 16 rectangles.

**EXAMPLE 1** Table 5.1 shows the values of upper and lower sum approximations to the area of $R$, using up to 1000 rectangles. In Section 5.2 we will see how to get an exact value of the areas of regions such as $R$ by taking a limit as the base width of each rectangle goes to zero and the number of rectangles goes to infinity. With the techniques developed there, we will be able to show that the area of $R$ is exactly $2/3$. ∎

### Distance Traveled

Suppose we know the velocity function $v(t)$ of a car moving down a highway, without changing direction, and want to know how far it traveled between times $t = a$ and $t = b$. The position function $s(t)$ of the car has derivative $v(t)$. If we can find an antiderivative $F(t)$

**TABLE 5.1** Finite approximations for the area of $R$

| Number of subintervals | Lower sum | Midpoint sum | Upper sum |
|---|---|---|---|
| 2 | 0.375 | 0.6875 | 0.875 |
| 4 | 0.53125 | 0.671875 | 0.78125 |
| 16 | 0.634765625 | 0.6669921875 | 0.697265625 |
| 50 | 0.6566 | 0.6667 | 0.6766 |
| 100 | 0.66165 | 0.666675 | 0.67165 |
| 1000 | 0.6661665 | 0.66666675 | 0.6671665 |

of $v(t)$ then we can find the car's position function $s(t)$ by setting $s(t) = F(t) + C$. The distance traveled can then be found by calculating the change in position, $s(b) - s(a) = F(b) - F(a)$. If the velocity function is known only by the readings at various times of a speedometer on the car, then we have no formula from which to obtain an antiderivative function for velocity. So what do we do in this situation?

When we don't know an antiderivative for the velocity function $v(t)$, we can approximate the distance traveled with finite sums in a way similar to our estimates for area discussed before. We subdivide the interval $[a, b]$ into short time intervals on each of which the velocity is considered to be fairly constant. Then we approximate the distance traveled on each time subinterval with the usual distance formula

$$\text{distance} = \text{velocity} \times \text{time}$$

and add the results across $[a, b]$.

Suppose the subdivided interval looks like

with the subintervals all of equal length $\Delta t$. Pick a number $t_1$ in the first interval. If $\Delta t$ is so small that the velocity barely changes over a short time interval of duration $\Delta t$, then the distance traveled in the first time interval is about $v(t_1) \Delta t$. If $t_2$ is a number in the second interval, the distance traveled in the second time interval is about $v(t_2) \Delta t$. The sum of the distances traveled over all the time intervals is

$$D \approx v(t_1) \Delta t + v(t_2) \Delta t + \cdots + v(t_n) \Delta t,$$

where $n$ is the total number of subintervals.

**EXAMPLE 2** The velocity function of a projectile fired straight into the air is $f(t) = 160 - 9.8t$ m/s. Use the summation technique just described to estimate how far the projectile rises during the first 3 seconds. How close do the sums come to the exact value of 435.9 m? (You will learn how to compute the exact value easily in Section 5.4.)

**Solution** We explore the results for different numbers of intervals and different choices of evaluation points. Notice that $f(t)$ is decreasing, so choosing left endpoints gives an upper sum estimate; choosing right endpoints gives a lower sum estimate.

(**a**) *Three subintervals of length 1, with f evaluated at left endpoints giving an upper sum:*

With $f$ evaluated at $t = 0, 1$, and $2$, we have

$$D \approx f(t_1)\,\Delta t + f(t_2)\,\Delta t + f(t_3)\,\Delta t$$
$$= [\,160 - 9.8(0)\,](1) + [\,160 - 9.8(1)\,](1) + [\,160 - 9.8(2)\,](1)$$
$$= 450.6.$$

**(b)** *Three subintervals of length 1, with $f$ evaluated at right endpoints giving a lower sum:*

With $f$ evaluated at $t = 1, 2$, and $3$, we have

$$D \approx f(t_1)\,\Delta t + f(t_2)\,\Delta t + f(t_3)\,\Delta t$$
$$= [\,160 - 9.8(1)\,](1) + [\,160 - 9.8(2)\,](1) + [\,160 - 9.8(3)\,](1)$$
$$= 421.2.$$

**(c)** *With six subintervals of length $1/2$, we get*

These estimates give an upper sum using left endpoints: $D \approx 443.25$; and a lower sum using right endpoints: $D \approx 428.55$. These six-interval estimates are somewhat closer than the three-interval estimates. The results improve as the subintervals get shorter.

As we can see in Table 5.2, the left-endpoint upper sums approach the true value 435.9 from above, whereas the right-endpoint lower sums approach it from below. The true value lies between these upper and lower sums. The magnitude of the error in the closest entries is 0.23, a small percentage of the true value.

$$\text{Error magnitude} = \big|\text{true value} - \text{calculated value}\big|$$
$$= \big|435.9 - 435.67\big| = 0.23.$$

$$\text{Error percentage} = \frac{0.23}{435.9} \approx 0.05\%.$$

It would be reasonable to conclude from the table's last entries that the projectile rose about 436 m during its first 3 s of flight. ∎

TABLE 5.2 Travel-distance estimates

| Number of subintervals | Length of each subinterval | Upper sum | Lower sum |
|---|---|---|---|
| 3 | 1 | 450.6 | 421.2 |
| 6 | 1/2 | 443.25 | 428.55 |
| 12 | 1/4 | 439.58 | 432.23 |
| 24 | 1/8 | 437.74 | 434.06 |
| 48 | 1/16 | 436.82 | 434.98 |
| 96 | 1/32 | 436.36 | 435.44 |
| 192 | 1/64 | 436.13 | 435.67 |

## Displacement Versus Distance Traveled

If an object with position function $s(t)$ moves along a coordinate line without changing direction, we can calculate the total distance it travels from $t = a$ to $t = b$ by summing the distance traveled over small intervals, as in Example 2. If the object reverses direction one or more times during the trip, then we need to use the object's *speed* $|v(t)|$, which is the absolute value of its velocity function, $v(t)$, to find the total distance traveled. Using the velocity itself, as in Example 2, gives instead an estimate to the object's **displacement**, $s(b) - s(a)$, the difference between its initial and final positions.

To see why using the velocity function in the summation process gives an estimate to the displacement, partition the time interval $[a, b]$ into small enough equal subintervals $\Delta t$ so that the object's velocity does not change very much from time $t_{k-1}$ to $t_k$. Then $v(t_k)$ gives a good approximation of the velocity throughout the interval. Accordingly, the change in the object's position coordinate, which is its displacement during the time interval, is about

$$v(t_k)\,\Delta t.$$

The change is positive if $v(t_k)$ is positive and negative if $v(t_k)$ is negative.

In either case, the distance traveled by the object during the subinterval is about

$$|v(t_k)|\,\Delta t.$$

The **total distance traveled** over the time interval is approximately the sum

$$|v(t_1)|\,\Delta t + |v(t_2)|\,\Delta t + \cdots + |v(t_n)|\,\Delta t.$$

We revisit these ideas in Section 5.4.

**EXAMPLE 3** In Example 4 in Section 3.4, we analyzed the motion of a heavy rock blown straight up by a dynamite blast. In that example, we found the velocity of the rock at any time during its motion to be $v(t) = 49 - 9.8t$ m/s. The rock was 78.4 m above the ground 2 s after the explosion, continued upward to reach a maximum height of 122.5 m at 5 s after the explosion, and then fell back down to reach the height of 78.4 m again at $t = 8$ s after the explosion. (See Figure 5.5.)

If we follow a procedure like that presented in Example 2, and use the velocity function $v(t)$ in the summation process over the time interval $[0, 8]$, we will obtain an estimate to the rock's 78.4 m *height* above the ground at $t = 8$. The positive upward motion (which yields a positive distance change of 44.1 m from the height of 78.4 m to the maximum height) is canceled by the negative downward motion (giving a negative change of 44.1 m from the maximum height down to 78.4 m again), so the displacement or height above the ground is estimated from the velocity function.

On the other hand, if the absolute value $|v(t)|$ is used in the summation process, we will obtain an estimate to the *total distance* the rock has traveled: the maximum height reached of 122.5 m plus the additional distance of 44.1 m it has fallen back down from that maximum when it again reaches the height of 78.4 m at $t = 8$ s. That is, using the absolute value of the velocity function in the summation process over the time interval $[0, 8]$, we obtain an estimate to 166.6 m, the total distance up and down that the rock has traveled in 8 s. There is no cancelation of distance changes due to sign changes in the velocity function, so we estimate distance traveled rather than displacement when we use the absolute value of the velocity function (that is, the speed of the rock).

As an illustration of our discussion, we subdivide the interval $[0, 8]$ into sixteen subintervals of length $\Delta t = 1/2$ and take the right endpoint of each subinterval in our calculations. Table 5.3 shows the values of the velocity function at these endpoints.

Using $v(t)$ in the summation process, we estimate the displacement at $t = 8$:

$$(44.1 + 39.2 + 34.3 + 29.4 + 24.5 + 19.6 + 14.7 + 9.8 + 4.9$$

$$+ 0 - 4.9 - 9.8 - 14.7 - 19.6 - 24.5 - 29.4) \cdot \frac{1}{2} = 58.8$$

$$\text{Error magnitude} = 78.4 - 58.8 = 19.6$$

**FIGURE 5.5** The rock in Example 3. The height $s = 78.4$ m is reached at $t = 2$ and $t = 8$ s. The rock falls 44.1 m from its maximum height when $t = 8$.

**TABLE 5.3** Velocity function

| $t$ | $v(t)$ | $t$ | $v(t)$ |
|-----|--------|-----|--------|
| 0 | 49 | 4.5 | 4.9 |
| 0.5 | 44.1 | 5.0 | 0 |
| 1.0 | 39.2 | 5.5 | −4.9 |
| 1.5 | 34.3 | 6.0 | −9.8 |
| 2.0 | 29.4 | 6.5 | −14.7 |
| 2.5 | 24.5 | 7.0 | −19.6 |
| 3.0 | 19.6 | 7.5 | −24.5 |
| 3.5 | 14.7 | 8.0 | −29.4 |
| 4.0 | 9.8 | | |

Using $|v(t)|$ in the summation process, we estimate the total distance traveled over the time interval $[0, 8]$:

$$(44.1 + 39.2 + 34.3 + 29.4 + 24.5 + 19.6 + 14.7 + 9.8 + 4.9$$

$$+ 0 + 4.9 + 9.8 + 14.7 + 19.6 + 24.5 + 29.4) \cdot \frac{1}{2} = 161.7$$

$$\text{Error magnitude} = 166.6 - 161.7 = 4.9$$

If we take more and more subintervals of $[0, 8]$ in our calculations, the estimates to these values 78.4 m and 166.6 m improve, approaching them as shown in Table 5.4. ∎

| TABLE 5.4 | Travel estimates for a rock blown straight up during the time interval $[0, 8]$ | | |
|---|---|---|---|
| Number of subintervals | Length of each subinterval | Displacement | Total distance |
| 16 | 1/2 | 58.8 | 161.7 |
| 32 | 1/4 | 68.6 | 164.15 |
| 64 | 1/8 | 73.5 | 165.375 |
| 128 | 1/16 | 75.95 | 165.9875 |
| 256 | 1/32 | 77.175 | 166.29375 |
| 512 | 1/64 | 77.7875 | 166.446875 |

## Average Value of a Nonnegative Continuous Function

The average value of a collection of $n$ numbers $x_1, x_2, \ldots, x_n$ is obtained by adding them together and dividing by $n$. But what is the average value of a continuous function $f$ on an interval $[a, b]$? Such a function can assume infinitely many values. For example, the temperature at a certain location in a town is a continuous function that goes up and down each day. What does it mean to say that the average temperature in the town over the course of a day is 23°C?

When a function is constant, this question is easy to answer. A function with constant value $c$ on an interval $[a, b]$ has average value $c$. When $c$ is positive, its graph over $[a, b]$ gives a rectangle of height $c$. The average value of the function can then be interpreted geometrically as the area of this rectangle divided by its width $b - a$ (Figure 5.6a).

What if we want to find the average value of a nonconstant function, such as the function $g$ in Figure 5.6b? We can think of this graph as a snapshot of the height of some water

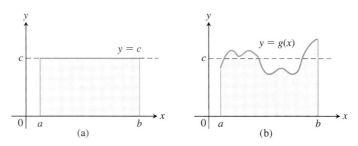

FIGURE 5.6    (a) The average value of $f(x) = c$ on $[a, b]$ is the area of the rectangle divided by $b - a$. (b) The average value of $g(x)$ on $[a, b]$ is the area beneath its graph divided by $b - a$.

that is sloshing around in a tank between enclosing walls at $x = a$ and $x = b$. As the water moves, its height over each point changes, but its average height remains the same. To get the average height of the water, we let it settle down until it is level and its height is constant. The resulting height $c$ equals the area under the graph of $g$ divided by $b - a$. We are led to *define* the average value of a nonnegative function on an interval $[a, b]$ to be the area under its graph divided by $b - a$. For this definition to be valid, we need a precise understanding of what is meant by the area under a graph. This will be obtained in Section 5.3, but for now we look at an example.

**FIGURE 5.7** Approximating the area under $f(x) = \sin x$ between 0 and $\pi$ to compute the average value of $\sin x$ over $[0, \pi]$, using eight rectangles (Example 4).

**EXAMPLE 4**    Estimate the average value of the function $f(x) = \sin x$ on the interval $[0, \pi]$.

**Solution**    Looking at the graph of $\sin x$ between 0 and $\pi$ in Figure 5.7, we can see that its average height is somewhere between 0 and 1. To find the average, we need to calculate the area $A$ under the graph and then divide this area by the length of the interval, $\pi - 0 = \pi$.

We do not have a simple way to determine the area, so we approximate it with finite sums. To get an upper sum approximation, we add the areas of eight rectangles of equal width $\pi/8$ that together contain the region beneath the graph of $y = \sin x$ and above the $x$-axis on $[0, \pi]$. We choose the heights of the rectangles to be the largest value of $\sin x$ on each subinterval. Over a particular subinterval, this largest value may occur at the left endpoint, the right endpoint, or somewhere between them. We evaluate $\sin x$ at this point to get the height of the rectangle for an upper sum. The sum of the rectangular areas then estimates the total area (Figure 5.7):

$$A \approx \left( \sin\frac{\pi}{8} + \sin\frac{\pi}{4} + \sin\frac{3\pi}{8} + \sin\frac{\pi}{2} + \sin\frac{\pi}{2} + \sin\frac{5\pi}{8} + \sin\frac{3\pi}{4} + \sin\frac{7\pi}{8} \right) \cdot \frac{\pi}{8}$$

$$\approx (.38 + .71 + .92 + 1 + 1 + .92 + .71 + .38) \cdot \frac{\pi}{8} = (6.02) \cdot \frac{\pi}{8} \approx 2.364.$$

To estimate the average value of $\sin x$ on $[0, \pi]$ we divide the estimated area by the length $\pi$ of the interval and obtain the approximation $2.364/\pi \approx 0.753$.

Since we used an upper sum to approximate the area, this estimate is greater than the actual average value of $\sin x$ over $[0, \pi]$. If we use more and more rectangles, with each rectangle getting thinner and thinner, we get closer and closer to the true average value as shown in Table 5.5. Using the techniques covered in Section 5.3, we will show that the true average value is $2/\pi \approx 0.63662$.

As before, we could just as well have used rectangles lying under the graph of $y = \sin x$ and calculated a lower sum approximation, or we could have used the midpoint rule. In Section 5.3 we will see that in each case, the approximations are close to the true area if all the rectangles are sufficiently thin. ∎

**TABLE 5.5** Average value of $\sin x$ on $0 \le x \le \pi$

| Number of subintervals | Upper sum estimate |
|---|---|
| 8 | 0.75342 |
| 16 | 0.69707 |
| 32 | 0.65212 |
| 50 | 0.64657 |
| 100 | 0.64161 |
| 1000 | 0.63712 |

## Summary

The area under the graph of a positive function, the distance traveled by a moving object that doesn't change direction, and the average value of a nonnegative function over an interval can all be approximated by finite sums constructed in a certain way. First we subdivide the interval into subintervals, treating some function $f$ as if it were constant over each particular subinterval. Then we multiply the width of each subinterval by the value of $f$ at some point within it, and add these products together. If the interval $[a, b]$ is subdivided into $n$ subintervals of equal widths $\Delta x = (b - a)/n$, and if $f(c_k)$ is the value of $f$ at the chosen point $c_k$ in the $k$th subinterval, this process gives a finite sum of the form

$$f(c_1) \Delta x + f(c_2) \Delta x + f(c_3) \Delta x + \cdots + f(c_n) \Delta x.$$

The choices for the $c_k$ could maximize or minimize the value of $f$ in the $k$th subinterval, or give some value in between. The true value lies somewhere between the approximations given by upper sums and lower sums. The finite sum approximations we looked at improved as we took more subintervals of thinner width.

## Exercises 5.1

### Area

In Exercises 1–4, use finite approximations to estimate the area under the graph of the function using

    **a.** a lower sum with two rectangles of equal width.

    **b.** a lower sum with four rectangles of equal width.

    **c.** an upper sum with two rectangles of equal width.

    **d.** an upper sum with four rectangles of equal width.

**1.** $f(x) = x^2$ between $x = 0$ and $x = 1$.

**2.** $f(x) = x^3$ between $x = 0$ and $x = 1$.

**3.** $f(x) = 1/x$ between $x = 1$ and $x = 5$.

**4.** $f(x) = 4 - x^2$ between $x = -2$ and $x = 2$.

Using rectangles each of whose height is given by the value of the function at the midpoint of the rectangle's base (*the midpoint rule*), estimate the area under the graphs of the following functions, using first two and then four rectangles.

**5.** $f(x) = x^2$ between $x = 0$ and $x = 1$.

**6.** $f(x) = x^3$ between $x = 0$ and $x = 1$.

**7.** $f(x) = 1/x$ between $x = 1$ and $x = 5$.

**8.** $f(x) = 4 - x^2$ between $x = -2$ and $x = 2$.

### Distance

**9. Distance traveled** The accompanying table shows the velocity of a model train engine moving along a track for 10 s. Estimate the distance traveled by the engine using 10 subintervals of length 1 with

    **a.** left-endpoint values.

    **b.** right-endpoint values.

| Time (s) | Velocity (cm/s) | Time (s) | Velocity (cm/s) |
|---|---|---|---|
| 0 | 0 | 6 | 11 |
| 1 | 12 | 7 | 6 |
| 2 | 22 | 8 | 2 |
| 3 | 10 | 9 | 6 |
| 4 | 5 | 10 | 0 |
| 5 | 13 | | |

**10. Distance traveled upstream** You are sitting on the bank of a tidal river watching the incoming tide carry a bottle upstream. You record the velocity of the flow every 5 minutes for an hour, with the results shown in the accompanying table. About how far upstream did the bottle travel during that hour? Find an estimate using 12 subintervals of length 5 with

    **a.** left-endpoint values.

    **b.** right-endpoint values.

| Time (min) | Velocity (m/s) | Time (min) | Velocity (m/s) |
|---|---|---|---|
| 0 | 1 | 35 | 1.2 |
| 5 | 1.2 | 40 | 1.0 |
| 10 | 1.7 | 45 | 1.8 |
| 15 | 2.0 | 50 | 1.5 |
| 20 | 1.8 | 55 | 1.2 |
| 25 | 1.6 | 60 | 0 |
| 30 | 1.4 | | |

**11. Length of a road** You and a companion are about to drive a twisty stretch of dirt road in a car whose speedometer works but whose odometer (kilometer counter) is broken. To find out how long this particular stretch of road is, you record the car's velocity at 10-s intervals, with the results shown in the accompanying table. Estimate the length of the road using

    **a.** left-endpoint values.

    **b.** right-endpoint values.

| Time (s) | Velocity (converted to m/s) (36 km/h = 10 m/s) | Time (s) | Velocity (converted to m/s) (36 km/h = 10 m/s) |
|---|---|---|---|
| 0 | 0 | 70 | 5 |
| 10 | 15 | 80 | 7 |
| 20 | 5 | 90 | 12 |
| 30 | 12 | 100 | 15 |
| 40 | 10 | 110 | 10 |
| 50 | 15 | 120 | 12 |
| 60 | 12 | | |

**12. Distance from velocity data** The accompanying table gives data for the velocity of a vintage sports car accelerating from 0 to 228 km/h in 36 s (10 thousandths of an hour).

| Time (h) | Velocity (km/h) | Time (h) | Velocity (km/h) |
|---|---|---|---|
| 0.0 | 0 | 0.006 | 187 |
| 0.001 | 64 | 0.007 | 201 |
| 0.002 | 100 | 0.008 | 212 |
| 0.003 | 132 | 0.009 | 220 |
| 0.004 | 154 | 0.010 | 228 |
| 0.005 | 174 | | |

a. Use rectangles to estimate how far the car traveled during the 36 s it took to reach 228 km/h.

b. Roughly how many seconds did it take the car to reach the halfway point? About how fast was the car going then?

**13. Free fall with air resistance** An object is dropped straight down from a helicopter. The object falls faster and faster but its acceleration (rate of change of its velocity) decreases over time because of air resistance. The acceleration is measured in m/s² and recorded every second after the drop for 5 s, as shown:

| $t$ | 0 | 1 | 2 | 3 | 4 | 5 |
|---|---|---|---|---|---|---|
| $a$ | 9.8 | 5.944 | 3.605 | 2.187 | 1.326 | 0.805 |

a. Find an upper estimate for the speed when $t = 5$.

b. Find a lower estimate for the speed when $t = 5$.

c. Find an upper estimate for the distance fallen when $t = 3$.

**14. Distance traveled by a projectile** An object is shot straight upward from sea level with an initial velocity of 122.5 m/s.

a. Assuming that gravity is the only force acting on the object, give an upper estimate for its velocity after 5 s have elapsed. Use $g = 9.8$ m/s² for the gravitational acceleration.

b. Find a lower estimate for the height attained after 5 s.

## Average Value of a Function

In Exercises 15–18, use a finite sum to estimate the average value of $f$ on the given interval by partitioning the interval into four subintervals of equal length and evaluating $f$ at the subinterval midpoints.

**15.** $f(x) = x^3$ on $[0, 2]$

**16.** $f(x) = 1/x$ on $[1, 9]$

**17.** $f(t) = (1/2) + \sin^2 \pi t$ on $[0, 2]$

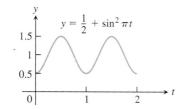

**18.** $f(t) = 1 - \left(\cos \dfrac{\pi t}{4}\right)^4$ on $[0, 4]$

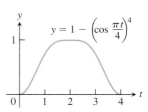

## Examples of Estimations

**19. Water pollution** Oil is leaking out of a tanker damaged at sea. The damage to the tanker is worsening as evidenced by the increased leakage each hour, recorded in the following table.

| Time (h) | 0 | 1 | 2 | 3 | 4 |
|---|---|---|---|---|---|
| Leakage (L/h) | 50 | 70 | 97 | 136 | 190 |

| Time (h) | 5 | 6 | 7 | 8 |
|---|---|---|---|---|
| Leakage (L/h) | 265 | 369 | 516 | 720 |

a. Give an upper and a lower estimate of the total quantity of oil that has escaped after 5 hours.

b. Repeat part (a) for the quantity of oil that has escaped after 8 hours.

c. The tanker continues to leak 720 L/h after the first 8 hours. If the tanker originally contained 25,000 L of oil, approximately how many more hours will elapse in the worst case before all the oil has spilled? In the best case?

**20. Air pollution** A power plant generates electricity by burning oil. Pollutants produced as a result of the burning process are removed by scrubbers in the smokestacks. Over time, the scrubbers become less efficient and eventually they must be replaced when the amount of pollution released exceeds government standards. Measurements are taken at the end of each month determining the rate at which pollutants are released into the atmosphere, recorded as follows.

| Month | Jan | Feb | Mar | Apr | May | Jun |
|---|---|---|---|---|---|---|
| Pollutant release rate (tonnes/day) | 0.20 | 0.25 | 0.27 | 0.34 | 0.45 | 0.52 |

| Month | Jul | Aug | Sep | Oct | Nov | Dec |
|---|---|---|---|---|---|---|
| Pollutant release rate (tonnes/day) | 0.63 | 0.70 | 0.81 | 0.85 | 0.89 | 0.95 |

a. Assuming a 30-day month and that new scrubbers allow only 0.05 tonne/day to be released, give an upper estimate of the total tonnage of pollutants released by the end of June. What is a lower estimate?

b. In the best case, approximately when will a total of 125 tonnes of pollutants have been released into the atmosphere?

**21.** Inscribe a regular $n$-sided polygon inside a circle of radius 1 and compute the area of the polygon for the following values of $n$:

**a.** 4 (square)    **b.** 8 (octagon)    **c.** 16

**d.** Compare the areas in parts (a), (b), and (c) with the area of the circle.

**22.** (*Continuation of Exercise 21.*)

**a.** Inscribe a regular $n$-sided polygon inside a circle of radius 1 and compute the area of one of the $n$ congruent triangles formed by drawing radii to the vertices of the polygon.

**b.** Compute the limit of the area of the inscribed polygon as $n \to \infty$.

**c.** Repeat the computations in parts (a) and (b) for a circle of radius $r$.

**COMPUTER EXPLORATIONS**

In Exercises 23–26, use a CAS to perform the following steps.

**a.** Plot the functions over the given interval.

**b.** Subdivide the interval into $n = 100, 200$, and 1000 subintervals of equal length and evaluate the function at the midpoint of each subinterval.

**c.** Compute the average value of the function values generated in part (b).

**d.** Solve the equation $f(x) =$ (average value) for $x$ using the average value calculated in part (c) for the $n = 1000$ partitioning.

**23.** $f(x) = \sin x$ on $[0, \pi]$    **24.** $f(x) = \sin^2 x$ on $[0, \pi]$

**25.** $f(x) = x \sin \dfrac{1}{x}$ on $\left[\dfrac{\pi}{4}, \pi\right]$    **26.** $f(x) = x \sin^2 \dfrac{1}{x}$ on $\left[\dfrac{\pi}{4}, \pi\right]$

# $5.2$ Sigma Notation and Limits of Finite Sums

In estimating with finite sums in Section 5.1, we encountered sums with many terms (up to 1000 in Table 5.1, for instance). In this section we introduce a more convenient notation for sums with a large number of terms. After describing the notation and stating several of its properties, we look at what happens to a finite sum approximation as the number of terms approaches infinity.

## Finite Sums and Sigma Notation

**Sigma notation** enables us to write a sum with many terms in the compact form

$$\sum_{k=1}^{n} a_k = a_1 + a_2 + a_3 + \cdots + a_{n-1} + a_n.$$

The Greek letter $\Sigma$ (capital sigma, corresponding to our letter S), stands for "sum." The **index of summation** $k$ tells us where the sum begins (at the number below the $\Sigma$ symbol) and where it ends (at the number above $\Sigma$). Any letter can be used to denote the index, but the letters $i, j$, and $k$ are customary.

The index $k$ ends at $k = n$.

The summation symbol (Greek letter sigma) $\displaystyle\sum_{k=1}^{n} a_k$    $a_k$ is a formula for the $k$th term.

The index $k$ starts at $k = 1$.

Thus we can write

$$1^2 + 2^2 + 3^2 + 4^2 + 5^2 + 6^2 + 7^2 + 8^2 + 9^2 + 10^2 + 11^2 = \sum_{k=1}^{11} k^2,$$

and

$$f(1) + f(2) + f(3) + \cdots + f(100) = \sum_{i=1}^{100} f(i).$$

The lower limit of summation does not have to be 1; it can be any integer.

**EXAMPLE 1**

| A sum in sigma notation | The sum written out, one term for each value of $k$ | The value of the sum |
|---|---|---|
| $\displaystyle\sum_{k=1}^{5} k$ | $1 + 2 + 3 + 4 + 5$ | $15$ |
| $\displaystyle\sum_{k=1}^{3} (-1)^k k$ | $(-1)^1(1) + (-1)^2(2) + (-1)^3(3)$ | $-1 + 2 - 3 = -2$ |
| $\displaystyle\sum_{k=1}^{2} \frac{k}{k+1}$ | $\dfrac{1}{1+1} + \dfrac{2}{2+1}$ | $\dfrac{1}{2} + \dfrac{2}{3} = \dfrac{7}{6}$ |
| $\displaystyle\sum_{k=4}^{5} \frac{k^2}{k-1}$ | $\dfrac{4^2}{4-1} + \dfrac{5^2}{5-1}$ | $\dfrac{16}{3} + \dfrac{25}{4} = \dfrac{139}{12}$ |

■

**EXAMPLE 2**  Express the sum $1 + 3 + 5 + 7 + 9$ in sigma notation.

**Solution**  The formula generating the terms changes with the lower limit of summation, but the terms generated remain the same. It is often simplest to start with $k = 0$ or $k = 1$, but we can start with any integer.

$$\text{Starting with } k = 0: \qquad 1 + 3 + 5 + 7 + 9 = \sum_{k=0}^{4} (2k + 1)$$

$$\text{Starting with } k = 1: \qquad 1 + 3 + 5 + 7 + 9 = \sum_{k=1}^{5} (2k - 1)$$

$$\text{Starting with } k = 2: \qquad 1 + 3 + 5 + 7 + 9 = \sum_{k=2}^{6} (2k - 3)$$

$$\text{Starting with } k = -3: \qquad 1 + 3 + 5 + 7 + 9 = \sum_{k=-3}^{1} (2k + 7)$$

■

When we have a sum such as

$$\sum_{k=1}^{3} (k + k^2)$$

we can rearrange its terms,

$$\sum_{k=1}^{3} (k + k^2) = (1 + 1^2) + (2 + 2^2) + (3 + 3^2)$$

$$= (1 + 2 + 3) + (1^2 + 2^2 + 3^2) \qquad \text{Regroup terms.}$$

$$= \sum_{k=1}^{3} k + \sum_{k=1}^{3} k^2.$$

This illustrates a general rule for finite sums:

$$\sum_{k=1}^{n} (a_k + b_k) = \sum_{k=1}^{n} a_k + \sum_{k=1}^{n} b_k.$$

Four such rules are given below. A proof that they are valid can be obtained using mathematical induction (see Appendix 2).

---

**Algebra Rules for Finite Sums**

**1.** *Sum Rule:*

$$\sum_{k=1}^{n}(a_k + b_k) = \sum_{k=1}^{n}a_k + \sum_{k=1}^{n}b_k$$

**2.** *Difference Rule:*

$$\sum_{k=1}^{n}(a_k - b_k) = \sum_{k=1}^{n}a_k - \sum_{k=1}^{n}b_k$$

**3.** *Constant Multiple Rule:*

$$\sum_{k=1}^{n}ca_k = c \cdot \sum_{k=1}^{n}a_k \qquad \text{(Any number } c\text{)}$$

**4.** *Constant Value Rule:*

$$\sum_{k=1}^{n}c = n \cdot c \qquad \text{(}c \text{ is any constant value.)}$$

---

**EXAMPLE 3**    We demonstrate the use of the algebra rules.

**(a)** $\displaystyle\sum_{k=1}^{n}(3k - k^2) = 3\sum_{k=1}^{n}k - \sum_{k=1}^{n}k^2$     Difference Rule and Constant Multiple Rule

**(b)** $\displaystyle\sum_{k=1}^{n}(-a_k) = \sum_{k=1}^{n}(-1) \cdot a_k = -1 \cdot \sum_{k=1}^{n}a_k = -\sum_{k=1}^{n}a_k$     Constant Multiple Rule

**(c)** $\displaystyle\sum_{k=1}^{3}(k + 4) = \sum_{k=1}^{3}k + \sum_{k=1}^{3}4$     Sum Rule

$\qquad\qquad = (1 + 2 + 3) + (3 \cdot 4)$     Constant Value Rule

$\qquad\qquad = 6 + 12 = 18$

**(d)** $\displaystyle\sum_{k=1}^{n}\frac{1}{n} = n \cdot \frac{1}{n} = 1$     Constant Value Rule ($1/n$ is constant) ■

Over the years people have discovered a variety of formulas for the values of finite sums. The most famous of these are the formula for the sum of the first $n$ integers (Gauss is said to have discovered it at age 8) and the formulas for the sums of the squares and cubes of the first $n$ integers.

**HISTORICAL BIOGRAPHY**

Carl Friedrich Gauss
(1777–1855)

**EXAMPLE 4**    Show that the sum of the first $n$ integers is

$$\sum_{k=1}^{n}k = \frac{n(n + 1)}{2}.$$

**Solution**    The formula tells us that the sum of the first 4 integers is

$$\frac{(4)(5)}{2} = 10.$$

Addition verifies this prediction:

$$1 + 2 + 3 + 4 = 10.$$

To prove the formula in general, we write out the terms in the sum twice, once forward and once backward.

$$\begin{array}{ccccccccc} 1 & + & 2 & + & 3 & + & \cdots & + & n \\ n & + & (n-1) & + & (n-2) & + & \cdots & + & 1 \end{array}$$

If we add the two terms in the first column we get $1 + n = n + 1$. Similarly, if we add the two terms in the second column we get $2 + (n - 1) = n + 1$. The two terms in any column sum to $n + 1$. When we add the $n$ columns together we get $n$ terms, each equal to $n + 1$, for a total of $n(n + 1)$. Since this is twice the desired quantity, the sum of the first $n$ integers is $(n)(n + 1)/2$. ■

Formulas for the sums of the squares and cubes of the first $n$ integers are proved using mathematical induction (see Appendix 2). We state them here.

---

The first $n$ squares:
$$\sum_{k=1}^{n} k^2 = \frac{n(n + 1)(2n + 1)}{6}$$

The first $n$ cubes:
$$\sum_{k=1}^{n} k^3 = \left(\frac{n(n + 1)}{2}\right)^2$$

---

### Limits of Finite Sums

The finite sum approximations we considered in Section 5.1 became more accurate as the number of terms increased and the subinterval widths (lengths) narrowed. The next example shows how to calculate a limiting value as the widths of the subintervals go to zero and their number grows to infinity.

**EXAMPLE 5**    Find the limiting value of lower sum approximations to the area of the region $R$ below the graph of $y = 1 - x^2$ and above the interval $[0, 1]$ on the $x$-axis using equal-width rectangles whose widths approach zero and whose number approaches infinity. (See Figure 5.4a.)

**Solution**    We compute a lower sum approximation using $n$ rectangles of equal width $\Delta x = (1 - 0)/n$, and then we see what happens as $n \to \infty$. We start by subdividing $[0, 1]$ into $n$ equal width subintervals

$$\left[0, \frac{1}{n}\right], \left[\frac{1}{n}, \frac{2}{n}\right], \ldots, \left[\frac{n - 1}{n}, \frac{n}{n}\right].$$

Each subinterval has width $1/n$. The function $1 - x^2$ is decreasing on $[0, 1]$, and its smallest value in a subinterval occurs at the subinterval's right endpoint. So a lower sum is constructed with rectangles whose height over the subinterval $[(k - 1)/n, k/n]$ is $f(k/n) = 1 - (k/n)^2$, giving the sum

$$\left[f\left(\frac{1}{n}\right)\right]\left(\frac{1}{n}\right) + \left[f\left(\frac{2}{n}\right)\right]\left(\frac{1}{n}\right) + \cdots + \left[f\left(\frac{k}{n}\right)\right]\left(\frac{1}{n}\right) + \cdots + \left[f\left(\frac{n}{n}\right)\right]\left(\frac{1}{n}\right).$$

We write this in sigma notation and simplify,

$$\sum_{k=1}^{n} f\left(\frac{k}{n}\right)\left(\frac{1}{n}\right) = \sum_{k=1}^{n}\left(1 - \left(\frac{k}{n}\right)^2\right)\left(\frac{1}{n}\right)$$

$$= \sum_{k=1}^{n}\left(\frac{1}{n} - \frac{k^2}{n^3}\right)$$

$$= \sum_{k=1}^{n}\frac{1}{n} - \sum_{k=1}^{n}\frac{k^2}{n^3} \qquad \text{Difference Rule}$$

$$= n \cdot \frac{1}{n} - \frac{1}{n^3}\sum_{k=1}^{n}k^2 \qquad \text{Constant Value and Constant Multiple Rules}$$

$$= 1 - \left(\frac{1}{n^3}\right)\frac{(n)(n + 1)(2n + 1)}{6} \qquad \text{Sum of the First } n \text{ Squares}$$

$$= 1 - \frac{2n^3 + 3n^2 + n}{6n^3}. \qquad \text{Numerator expanded}$$

We have obtained an expression for the lower sum that holds for any $n$. Taking the limit of this expression as $n \to \infty$, we see that the lower sums converge as the number of subintervals increases and the subinterval widths approach zero:

$$\lim_{n \to \infty} \left( 1 - \frac{2n^3 + 3n^2 + n}{6n^3} \right) = 1 - \frac{2}{6} = \frac{2}{3}.$$

The lower sum approximations converge to $2/3$. A similar calculation shows that the upper sum approximations also converge to $2/3$. Any finite sum approximation $\sum_{k=1}^{n} f(c_k)(1/n)$ also converges to the same value, $2/3$. This is because it is possible to show that any finite sum approximation is trapped between the lower and upper sum approximations. For this reason we are led to *define* the area of the region $R$ as this limiting value. In Section 5.3 we study the limits of such finite approximations in a general setting. ∎

## Riemann Sums

The theory of limits of finite approximations was made precise by the German mathematician Bernhard Riemann. We now introduce the notion of a *Riemann sum*, which underlies the theory of the definite integral studied in the next section.

We begin with an arbitrary bounded function $f$ defined on a closed interval $[a, b]$. Like the function pictured in Figure 5.8, $f$ may have negative as well as positive values. We subdivide the interval $[a, b]$ into subintervals, not necessarily of equal widths (or lengths), and form sums in the same way as for the finite approximations in Section 5.1. To do so, we choose $n - 1$ points $\{x_1, x_2, x_3, \dots, x_{n-1}\}$ between $a$ and $b$ satisfying

$$a < x_1 < x_2 < \cdots < x_{n-1} < b.$$

To make the notation consistent, we denote $a$ by $x_0$ and $b$ by $x_n$, so that

$$a = x_0 < x_1 < x_2 < \cdots < x_{n-1} < x_n = b.$$

The set

$$P = \{x_0, x_1, x_2, \dots, x_{n-1}, x_n\}$$

is called a **partition** of $[a, b]$.

The partition $P$ divides $[a, b]$ into $n$ closed subintervals

$$[x_0, x_1], [x_1, x_2], \dots, [x_{n-1}, x_n].$$

The first of these subintervals is $[x_0, x_1]$, the second is $[x_1, x_2]$, and the **$k$th subinterval** of $P$ is $[x_{k-1}, x_k]$, for $k$ an integer between 1 and $n$.

The width of the first subinterval $[x_0, x_1]$ is denoted $\Delta x_1$, the width of the second $[x_1, x_2]$ is denoted $\Delta x_2$, and the width of the $k$th subinterval is $\Delta x_k = x_k - x_{k-1}$. If all $n$ subintervals have equal width, then the common width $\Delta x$ is equal to $(b - a)/n$.

In each subinterval we select some point. The point chosen in the $k$th subinterval $[x_{k-1}, x_k]$ is called $c_k$. Then on each subinterval we stand a vertical rectangle that stretches from the $x$-axis to touch the curve at $(c_k, f(c_k))$. These rectangles can be above or below the $x$-axis, depending on whether $f(c_k)$ is positive or negative, or on the $x$-axis if $f(c_k) = 0$ (Figure 5.9).

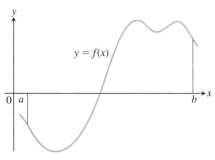

**FIGURE 5.8** A typical continuous function $y = f(x)$ over a closed interval $[a, b]$.

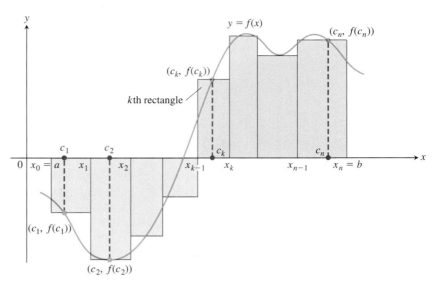

**FIGURE 5.9** The rectangles approximate the region between the graph of the function $y = f(x)$ and the $x$-axis. Figure 5.8 has been enlarged to enhance the partition of $[a, b]$ and selection of points $c_k$ that produce the rectangles.

On each subinterval we form the product $f(c_k) \cdot \Delta x_k$. This product is positive, negative, or zero, depending on the sign of $f(c_k)$. When $f(c_k) > 0$, the product $f(c_k) \cdot \Delta x_k$ is the area of a rectangle with height $f(c_k)$ and width $\Delta x_k$. When $f(c_k) < 0$, the product $f(c_k) \cdot \Delta x_k$ is a negative number, the negative of the area of a rectangle of width $\Delta x_k$ that drops from the $x$-axis to the negative number $f(c_k)$.

Finally we sum all these products to get

$$S_P = \sum_{k=1}^{n} f(c_k) \, \Delta x_k.$$

The sum $S_P$ is called a **Riemann sum for $f$ on the interval $[a, b]$**. There are many such sums, depending on the partition $P$ we choose, and the choices of the points $c_k$ in the subintervals. For instance, we could choose $n$ subintervals all having equal width $\Delta x = (b - a)/n$ to partition $[a, b]$, and then choose the point $c_k$ to be the right-hand endpoint of each subinterval when forming the Riemann sum (as we did in Example 5). This choice leads to the Riemann sum formula

$$S_n = \sum_{k=1}^{n} f\left(a + k\frac{(b - a)}{n}\right) \cdot \left(\frac{b - a}{n}\right).$$

Similar formulas can be obtained if instead we choose $c_k$ to be the left-hand endpoint, or the midpoint, of each subinterval.

In the cases in which the subintervals all have equal width $\Delta x = (b - a)/n$, we can make them thinner by simply increasing their number $n$. When a partition has subintervals of varying widths, we can ensure they are all thin by controlling the width of a widest (longest) subinterval. We define the **norm** of a partition $P$, written $\|P\|$, to be the largest of all the subinterval widths. If $\|P\|$ is a small number, then all of the subintervals in the partition $P$ have a small width. Let's look at an example of these ideas.

(a)

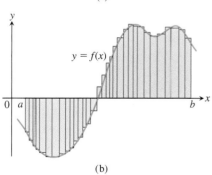

(b)

**FIGURE 5.10** The curve of Figure 5.9 with rectangles from finer partitions of $[a, b]$. Finer partitions create collections of rectangles with thinner bases that approximate the region between the graph of $f$ and the $x$-axis with increasing accuracy.

**EXAMPLE 6** The set $P = \{0, 0.2, 0.6, 1, 1.5, 2\}$ is a partition of $[0, 2]$. There are five subintervals of $P$: $[0, 0.2]$, $[0.2, 0.6]$, $[0.6, 1]$, $[1, 1.5]$, and $[1.5, 2]$:

The lengths of the subintervals are $\Delta x_1 = 0.2$, $\Delta x_2 = 0.4$, $\Delta x_3 = 0.4$, $\Delta x_4 = 0.5$, and $\Delta x_5 = 0.5$. The longest subinterval length is 0.5, so the norm of the partition is $\|P\| = 0.5$. In this example, there are two subintervals of this length.    ■

Any Riemann sum associated with a partition of a closed interval $[a, b]$ defines rectangles that approximate the region between the graph of a continuous function $f$ and the $x$-axis. Partitions with norm approaching zero lead to collections of rectangles that approximate this region with increasing accuracy, as suggested by Figure 5.10. We will see in the next section that if the function $f$ is continuous over the closed interval $[a, b]$, then no matter how we choose the partition $P$ and the points $c_k$ in its subintervals to construct a Riemann sum, a single limiting value is approached as the subinterval widths, controlled by the norm of the partition, approach zero.

## Exercises 5.2

### Sigma Notation

Write the sums in Exercises 1–6 without sigma notation. Then evaluate them.

**1.** $\sum_{k=1}^{2} \frac{6k}{k + 1}$
**2.** $\sum_{k=1}^{3} \frac{k - 1}{k}$

**3.** $\sum_{k=1}^{4} \cos k\pi$
**4.** $\sum_{k=1}^{5} \sin k\pi$

**5.** $\sum_{k=1}^{3} (-1)^{k+1} \sin \frac{\pi}{k}$
**6.** $\sum_{k=1}^{4} (-1)^{k} \cos k\pi$

**7.** Which of the following express $1 + 2 + 4 + 8 + 16 + 32$ in sigma notation?

**a.** $\sum_{k=1}^{6} 2^{k-1}$
**b.** $\sum_{k=0}^{5} 2^{k}$
**c.** $\sum_{k=-1}^{4} 2^{k+1}$

**8.** Which of the following express $1 - 2 + 4 - 8 + 16 - 32$ in sigma notation?

**a.** $\sum_{k=1}^{6} (-2)^{k-1}$
**b.** $\sum_{k=0}^{5} (-1)^{k} 2^{k}$
**c.** $\sum_{k=-2}^{3} (-1)^{k+1} 2^{k+2}$

**9.** Which formula is not equivalent to the other two?

**a.** $\sum_{k=2}^{4} \frac{(-1)^{k-1}}{k - 1}$
**b.** $\sum_{k=0}^{2} \frac{(-1)^{k}}{k + 1}$
**c.** $\sum_{k=-1}^{1} \frac{(-1)^{k}}{k + 2}$

**10.** Which formula is not equivalent to the other two?

**a.** $\sum_{k=1}^{4} (k - 1)^2$
**b.** $\sum_{k=-1}^{3} (k + 1)^2$
**c.** $\sum_{k=-3}^{-1} k^2$

Express the sums in Exercises 11–16 in sigma notation. The form of your answer will depend on your choice of the lower limit of summation.

**11.** $1 + 2 + 3 + 4 + 5 + 6$
**12.** $1 + 4 + 9 + 16$

**13.** $\frac{1}{2} + \frac{1}{4} + \frac{1}{8} + \frac{1}{16}$
**14.** $2 + 4 + 6 + 8 + 10$

**15.** $1 - \frac{1}{2} + \frac{1}{3} - \frac{1}{4} + \frac{1}{5}$
**16.** $-\frac{1}{5} + \frac{2}{5} - \frac{3}{5} + \frac{4}{5} - \frac{5}{5}$

### Values of Finite Sums

**17.** Suppose that $\sum_{k=1}^{n} a_k = -5$ and $\sum_{k=1}^{n} b_k = 6$. Find the values of

**a.** $\sum_{k=1}^{n} 3a_k$
**b.** $\sum_{k=1}^{n} \frac{b_k}{6}$
**c.** $\sum_{k=1}^{n} (a_k + b_k)$

**d.** $\sum_{k=1}^{n} (a_k - b_k)$
**e.** $\sum_{k=1}^{n} (b_k - 2a_k)$

**18.** Suppose that $\sum_{k=1}^{n} a_k = 0$ and $\sum_{k=1}^{n} b_k = 1$. Find the values of

**a.** $\sum_{k=1}^{n} 8a_k$
**b.** $\sum_{k=1}^{n} 250b_k$

**c.** $\sum_{k=1}^{n} (a_k + 1)$
**d.** $\sum_{k=1}^{n} (b_k - 1)$

Evaluate the sums in Exercises 19–32.

**19. a.** $\sum_{k=1}^{10} k$
**b.** $\sum_{k=1}^{10} k^2$
**c.** $\sum_{k=1}^{10} k^3$

**20. a.** $\sum_{k=1}^{13} k$
**b.** $\sum_{k=1}^{13} k^2$
**c.** $\sum_{k=1}^{13} k^3$

**21.** $\sum_{k=1}^{7} (-2k)$
**22.** $\sum_{k=1}^{5} \frac{\pi k}{15}$

**23.** $\sum_{k=1}^{6} (3 - k^2)$
**24.** $\sum_{k=1}^{6} (k^2 - 5)$

**25.** $\sum_{k=1}^{5} k(3k + 5)$
**26.** $\sum_{k=1}^{7} k(2k + 1)$

**27.** $\sum_{k=1}^{5} \frac{k^3}{225} + \left(\sum_{k=1}^{5} k\right)^3$
**28.** $\left(\sum_{k=1}^{7} k\right)^2 - \sum_{k=1}^{7} \frac{k^3}{4}$

**29. a.** $\sum_{k=1}^{7} 3$
**b.** $\sum_{k=1}^{500} 7$
**c.** $\sum_{k=3}^{264} 10$

**30. a.** $\sum_{k=9}^{36} k$
**b.** $\sum_{k=3}^{17} k^2$
**c.** $\sum_{k=18}^{71} k(k - 1)$

**31. a.** $\sum_{k=1}^{n} 4$
**b.** $\sum_{k=1}^{n} c$
**c.** $\sum_{k=1}^{n} (k - 1)$

**32. a.** $\sum_{k=1}^{n} \left(\frac{1}{n} + 2n\right)$
**b.** $\sum_{k=1}^{n} \frac{c}{n}$
**c.** $\sum_{k=1}^{n} \frac{k}{n^2}$

## Riemann Sums

In Exercises 33–36, graph each function $f(x)$ over the given interval. Partition the interval into four subintervals of equal length. Then add to your sketch the rectangles associated with the Riemann sum $\sum_{k=1}^{4} f(c_k) \, \Delta x_k$, given that $c_k$ is the **(a)** left-hand endpoint, **(b)** right-hand endpoint, **(c)** midpoint of the $k$th subinterval. (Make a separate sketch for each set of rectangles.)

**33.** $f(x) = x^2 - 1,$ $[0, 2]$

**34.** $f(x) = -x^2,$ $[0, 1]$

**35.** $f(x) = \sin x,$ $[-\pi, \pi]$

**36.** $f(x) = \sin x + 1,$ $[-\pi, \pi]$

**37.** Find the norm of the partition $P = \{0, 1.2, 1.5, 2.3, 2.6, 3\}$.

**38.** Find the norm of the partition $P = \{-2, -1.6, -0.5, 0, 0.8, 1\}$.

## Limits of Riemann Sums

For the functions in Exercises 39–46, find a formula for the Riemann sum obtained by dividing the interval $[a, b]$ into $n$ equal subintervals and using the right-hand endpoint for each $c_k$. Then take a limit of these sums as $n \to \infty$ to calculate the area under the curve over $[a, b]$.

**39.** $f(x) = 1 - x^2$ over the interval $[0, 1]$.

**40.** $f(x) = 2x$ over the interval $[0, 3]$.

**41.** $f(x) = x^2 + 1$ over the interval $[0, 3]$.

**42.** $f(x) = 3x^2$ over the interval $[0, 1]$.

**43.** $f(x) = x + x^2$ over the interval $[0, 1]$.

**44.** $f(x) = 3x + 2x^2$ over the interval $[0, 1]$.

**45.** $f(x) = 2x^3$ over the interval $[0, 1]$.

**46.** $f(x) = x^2 - x^3$ over the interval $[-1, 0]$.

# $5.3$ The Definite Integral

In Section 5.2 we investigated the limit of a finite sum for a function defined over a closed interval $[a, b]$ using $n$ subintervals of equal width (or length), $(b - a)/n$. In this section we consider the limit of more general Riemann sums as the norm of the partitions of $[a, b]$ approaches zero. For general Riemann sums, the subintervals of the partitions need not have equal widths. The limiting process then leads to the definition of the *definite integral* of a function over a closed interval $[a, b]$.

## Definition of the Definite Integral

The definition of the definite integral is based on the idea that for certain functions, as the norm of the partitions of $[a, b]$ approaches zero, the values of the corresponding Riemann sums approach a limiting value $J$. What we mean by this limit is that a Riemann sum will be close to the number $J$ provided that the norm of its partition is sufficiently small (so that all of its subintervals have thin enough widths). We introduce the symbol $\epsilon$ as a small positive number that specifies how close to $J$ the Riemann sum must be, and the symbol $\delta$ as a second small positive number that specifies how small the norm of a partition must be in order for convergence to happen. We now define this limit precisely.

> **DEFINITION** Let $f(x)$ be a function defined on a closed interval $[a, b]$. We say that a number $J$ is the **definite integral of $f$ over $[a, b]$** and that $J$ is the limit of the Riemann sums $\sum_{k=1}^{n} f(c_k) \, \Delta x_k$ if the following condition is satisfied:
> Given any number $\epsilon > 0$ there is a corresponding number $\delta > 0$ such that for every partition $P = \{x_0, x_1, \ldots, x_n\}$ of $[a, b]$ with $\|P\| < \delta$ and any choice of $c_k$ in $[x_{k-1}, x_k]$, we have
> $$\left| \sum_{k=1}^{n} f(c_k) \, \Delta x_k - J \right| < \epsilon.$$

The definition involves a limiting process in which the norm of the partition goes to zero.

We have many choices for a partition $P$ with norm going to zero, and many choices of points $c_k$ for each partition. The definite integral exists when we always get the same limit $J$, no matter what choices are made. When the limit exists we write it as the definite integral

$$J = \lim_{\|P\| \to 0} \sum_{k=1}^{n} f(c_k) \, \Delta x_k.$$

The limit of any Riemann sum is always taken as the norm of the partitions approaches zero and the number of subintervals goes to infinity.

Leibniz introduced a notation for the definite integral that captures its construction as a limit of Riemann sums. He envisioned the finite sums $\sum_{k=1}^{n} f(c_k)\,\Delta x_k$ becoming an infinite sum of function values $f(x)$ multiplied by "infinitesimal" subinterval widths $dx$. The sum symbol $\sum$ is replaced in the limit by the integral symbol $\int$, whose origin is in the letter "S." The function values $f(c_k)$ are replaced by a continuous selection of function values $f(x)$. The subinterval widths $\Delta x_k$ become the differential $dx$. It is as if we are summing all products of the form $f(x) \cdot dx$ as $x$ goes from $a$ to $b$. While this notation captures the process of constructing an integral, it is Riemann's definition that gives a precise meaning to the definite integral.

The symbol for the number $J$ in the definition of the **definite integral** is

$$\int_a^b f(x)\,dx,$$

which is read as "the integral from $a$ to $b$ of $f$ of $x$ dee $x$" or sometimes as "the integral from $a$ to $b$ of $f$ of $x$ with respect to $x$." The component parts in the integral symbol also have names:

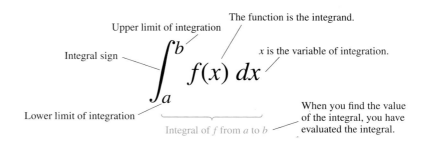

When the condition in the definition is satisfied, we say that the Riemann sums of $f$ on $[a, b]$ **converge** to the definite integral $J = \int_a^b f(x)\,dx$ and that $f$ is **integrable** over $[a, b]$.

In the cases where the subintervals all have equal width $\Delta x = (b - a)/n$, we can form each Riemann sum as

$$S_n = \sum_{k=1}^{n} f(c_k)\,\Delta x_k = \sum_{k=1}^{n} f(c_k)\left(\frac{b - a}{n}\right), \qquad \Delta x_k = \Delta x = (b - a)/n \text{ for all } k$$

where $c_k$ is chosen in the $k$th subinterval. When the limit of these Riemann sums as $n \to \infty$ exists and is equal to $J$, then $J$ is the definite integral of $f$ over $[a, b]$, so

$$J = \int_a^b f(x)\,dx = \lim_{n \to \infty} \sum_{k=1}^{n} f(c_k)\left(\frac{b - a}{n}\right) \qquad \|P\| \to 0 \text{ means } n \to \infty.$$

If we pick the point $c_k$ at the right endpoint of the $k$th subinterval, so $c_k = a + k\,\Delta x = a + k(b - a)/n$, then the formula for the definite integral becomes

$$\int_a^b f(x)\,dx = \lim_{n \to \infty} \sum_{k=1}^{n} f\left(a + k\frac{(b - a)}{n}\right)\left(\frac{b - a}{n}\right) \qquad (1)$$

Equation (1) gives one explicit formula that can be used to compute definite integrals. Other choices of partitions and locations of points $c_k$ result in the same value for the definite integral when we take the limit as $n \to \infty$ provided that the norm of the partition approaches zero.

The value of the definite integral of a function over any particular interval depends on the function, not on the letter we choose to represent its independent variable. If we decide to use $t$ or $u$ instead of $x$, we simply write the integral as

$$\int_a^b f(t)\, dt \qquad \text{or} \qquad \int_a^b f(u)\, du \qquad \text{instead of} \qquad \int_a^b f(x)\, dx.$$

No matter how we write the integral, it is still the same number that is defined as a limit of Riemann sums. Since it does not matter what letter we use, the variable of integration is called a **dummy variable** representing the real numbers in the closed interval $[a, b]$.

## Integrable and Nonintegrable Functions

Not every function defined over the closed interval $[a, b]$ is integrable there, even if the function is bounded. That is, the Riemann sums for some functions may not converge to the same limiting value, or to any value at all. A full development of exactly which functions defined over $[a, b]$ are integrable requires advanced mathematical analysis, but fortunately most functions that commonly occur in applications are integrable. In particular, every *continuous* function over $[a, b]$ is integrable over this interval, and so is every function having no more than a finite number of jump discontinuities on $[a, b]$. (See Figures 1.9 and 1.10. The latter functions are called *piecewise-continuous functions*, and they are defined in Additional Exercises 11–18 at the end of this chapter.) The following theorem, which is proved in more advanced courses, establishes these results.

> **THEOREM 1—Integrability of Continuous Functions**    If a function $f$ is continuous over the interval $[a, b]$, or if $f$ has at most finitely many jump discontinuities there, then the definite integral $\int_a^b f(x)\, dx$ exists and $f$ is integrable over $[a, b]$.

The idea behind Theorem 1 for continuous functions is given in Exercises 86 and 87. Briefly, when $f$ is continuous we can choose each $c_k$ so that $f(c_k)$ gives the maximum value of $f$ on the subinterval $[x_{k-1}, x_k]$, resulting in an upper sum. Likewise, we can choose $c_k$ to give the minimum value of $f$ on $[x_{k-1}, x_k]$ to obtain a lower sum. The upper and lower sums can be shown to converge to the same limiting value as the norm of the partition $P$ tends to zero. Moreover, every Riemann sum is trapped between the values of the upper and lower sums, so every Riemann sum converges to the same limit as well. Therefore, the number $J$ in the definition of the definite integral exists, and the continuous function $f$ is integrable over $[a, b]$.

For integrability to fail, a function needs to be sufficiently discontinuous that the region between its graph and the $x$-axis cannot be approximated well by increasingly thin rectangles. Our first example shows a function that is not integrable over a closed interval.

**EXAMPLE 1**    The function

$$f(x) = \begin{cases} 1, & \text{if } x \text{ is rational} \\ 0, & \text{if } x \text{ is irrational} \end{cases}$$

has no Riemann integral over $[0, 1]$. Underlying this is the fact that between any two numbers there is both a rational number and an irrational number. Thus the function jumps up and down too erratically over $[0, 1]$ to allow the region beneath its graph and above the $x$-axis to be approximated by rectangles, no matter how thin they are. We show, in fact, that upper sum approximations and lower sum approximations converge to different limiting values.

If we pick a partition $P$ of $[0, 1]$ and choose $c_k$ to be the point giving the maximum value for $f$ on $[x_{k-1}, x_k]$ then the corresponding Riemann sum is

$$U = \sum_{k=1}^{n} f(c_k) \, \Delta x_k = \sum_{k=1}^{n} (1) \, \Delta x_k = 1,$$

since each subinterval $[x_{k-1}, x_k]$ contains a rational number where $f(c_k) = 1$. Note that the lengths of the intervals in the partition sum to 1, $\sum_{k=1}^{n} \Delta x_k = 1$. So each such Riemann sum equals 1, and a limit of Riemann sums using these choices equals 1.

On the other hand, if we pick $c_k$ to be the point giving the minimum value for $f$ on $[x_{k-1}, x_k]$, then the Riemann sum is

$$L = \sum_{k=1}^{n} f(c_k) \, \Delta x_k = \sum_{k=1}^{n} (0) \, \Delta x_k = 0,$$

since each subinterval $[x_{k-1}, x_k]$ contains an irrational number $c_k$ where $f(c_k) = 0$. The limit of Riemann sums using these choices equals zero. Since the limit depends on the choices of $c_k$, the function $f$ is not integrable.    ■

Theorem 1 says nothing about how to *calculate* definite integrals. A method of calculation will be developed in Section 5.4, through a connection to knowing an antiderivative of the integrand function $f$.

## Properties of Definite Integrals

In defining $\int_a^b f(x) \, dx$ as a limit of sums $\sum_{k=1}^{n} f(c_k) \, \Delta x_k$, we moved from left to right across the interval $[a, b]$. What would happen if we instead move right to left, starting with $x_0 = b$ and ending at $x_n = a$? Each $\Delta x_k$ in the Riemann sum would change its sign, with $x_k - x_{k-1}$ now negative instead of positive. With the same choices of $c_k$ in each subinterval, the sign of any Riemann sum would change, as would the sign of the limit, the integral $\int_b^a f(x) \, dx$. Since we have not previously given a meaning to integrating backward, we are led to define

$$\int_b^a f(x) \, dx = -\int_a^b f(x) \, dx.$$

Although we have only defined the integral over an interval $[a, b]$ when $a < b$, it is convenient to have a definition for the integral over $[a, b]$ when $a = b$, that is, for the integral over an interval of zero width. Since $a = b$ gives $\Delta x = 0$, whenever $f(a)$ exists we define

$$\int_a^a f(x) \, dx = 0.$$

Theorem 2 states basic properties of integrals, given as rules that they satisfy, including the two just discussed. These rules, listed in Table 5.6, become very useful in the process of computing integrals. We will refer to them repeatedly to simplify our calculations. Rules 2 through 7 have geometric interpretations, shown in Figure 5.11. The graphs in these figures are of positive functions, but the rules apply to general integrable functions.

---

**THEOREM 2**    When $f$ and $g$ are integrable over the interval $[a, b]$, the definite integral satisfies the rules in Table 5.6.

---

While Rules 1 and 2 are definitions, Rules 3 to 7 of Table 5.6 must be proved. The following is a proof of Rule 6. Similar proofs can be given to verify the other properties in Table 5.6.

**TABLE 5.6** Rules satisfied by definite integrals

**1.** *Order of Integration:* $\displaystyle\int_b^a f(x)\,dx = -\int_a^b f(x)\,dx$     A definition

**2.** *Zero Width Interval:* $\displaystyle\int_a^a f(x)\,dx = 0$     A definition when $f(a)$ exists

**3.** *Constant Multiple:* $\displaystyle\int_a^b kf(x)\,dx = k\int_a^b f(x)\,dx$     Any constant $k$

**4.** *Sum and Difference:* $\displaystyle\int_a^b (f(x) \pm g(x))\,dx = \int_a^b f(x)\,dx \pm \int_a^b g(x)\,dx$

**5.** *Additivity:* $\displaystyle\int_a^b f(x)\,dx + \int_b^c f(x)\,dx = \int_a^c f(x)\,dx$

**6.** *Max-Min Inequality:* If $f$ has maximum value max $f$ and minimum value min $f$ on $[a, b]$, then

$$\text{min } f \cdot (b - a) \le \int_a^b f(x)\,dx \le \text{ max } f \cdot (b - a).$$

**7.** *Domination:* $\displaystyle f(x) \ge g(x) \text{ on } [a, b] \Rightarrow \int_a^b f(x)\,dx \ge \int_a^b g(x)\,dx$

$$f(x) \ge 0 \text{ on } [a, b] \Rightarrow \int_a^b f(x)\,dx \ge 0 \quad \text{(Special case)}$$

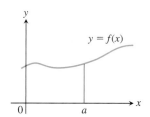

(a) *Zero Width Interval:*

$$\int_a^a f(x)\,dx = 0$$

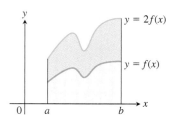

(b) *Constant Multiple: (k = 2)*

$$\int_a^b kf(x)\,dx = k\int_a^b f(x)\,dx$$

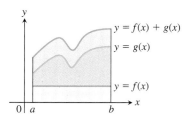

(c) *Sum: (areas add)*

$$\int_a^b (f(x) + g(x))\,dx = \int_a^b f(x)\,dx + \int_a^b g(x)\,dx$$

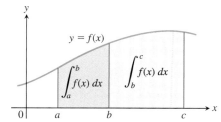

(d) *Additivity for Definite Integrals:*

$$\int_a^b f(x)\,dx + \int_b^c f(x)\,dx = \int_a^c f(x)\,dx$$

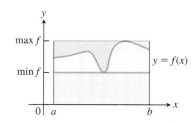

(e) *Max-Min Inequality:*

$$\text{min } f \cdot (b - a) \le \int_a^b f(x)\,dx$$
$$\le \text{ max } f \cdot (b - a)$$

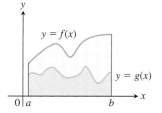

(f) *Domination:*

$$f(x) \ge g(x) \text{ on } [a, b]$$
$$\Rightarrow \int_a^b f(x)\,dx \ge \int_a^b g(x)\,dx$$

**FIGURE 5.11** Geometric interpretations of Rules 2–7 in Table 5.6.

**Proof of Rule 6**    Rule 6 says that the integral of $f$ over $[a, b]$ is never smaller than the minimum value of $f$ times the length of the interval and never larger than the maximum value of $f$ times the length of the interval. The reason is that for every partition of $[a, b]$ and for every choice of the points $c_k$,

$$\min f \cdot (b - a) = \min f \cdot \sum_{k=1}^{n} \Delta x_k \qquad \sum_{k=1}^{n} \Delta x_k = b - a$$

$$= \sum_{k=1}^{n} \min f \cdot \Delta x_k \qquad \text{Constant Multiple Rule}$$

$$\leq \sum_{k=1}^{n} f(c_k) \, \Delta x_k \qquad \min f \leq f(c_k)$$

$$\leq \sum_{k=1}^{n} \max f \cdot \Delta x_k \qquad f(c_k) \leq \max f$$

$$= \max f \cdot \sum_{k=1}^{n} \Delta x_k \qquad \text{Constant Multiple Rule}$$

$$= \max f \cdot (b - a).$$

In short, all Riemann sums for $f$ on $[a, b]$ satisfy the inequality

$$\min f \cdot (b - a) \leq \sum_{k=1}^{n} f(c_k) \, \Delta x_k \leq \max f \cdot (b - a).$$

Hence their limit, the integral, does too.    ∎

**EXAMPLE 2**    To illustrate some of the rules, we suppose that

$$\int_{-1}^{1} f(x) \, dx = 5, \qquad \int_{1}^{4} f(x) \, dx = -2, \quad \text{and} \quad \int_{-1}^{1} h(x) \, dx = 7.$$

Then

**1.** $\displaystyle \int_{4}^{1} f(x) \, dx = -\int_{1}^{4} f(x) \, dx = -(-2) = 2$    Rule 1

**2.** $\displaystyle \int_{-1}^{1} [\, 2f(x) + 3h(x) \,] \, dx = 2\int_{-1}^{1} f(x) \, dx + 3\int_{-1}^{1} h(x) \, dx$    Rules 3 and 4

$$= 2(5) + 3(7) = 31$$

**3.** $\displaystyle \int_{-1}^{4} f(x) \, dx = \int_{-1}^{1} f(x) \, dx + \int_{1}^{4} f(x) \, dx = 5 + (-2) = 3$    Rule 5    ∎

**EXAMPLE 3**    Show that the value of $\int_{0}^{1} \sqrt{1 + \cos x} \, dx$ is less than or equal to $\sqrt{2}$.

**Solution**    The Max-Min Inequality for definite integrals (Rule 6) says that $\min f \cdot (b - a)$ is a *lower bound* for the value of $\int_{a}^{b} f(x) \, dx$ and that $\max f \cdot (b - a)$ is an *upper bound*. The maximum value of $\sqrt{1 + \cos x}$ on $[0, 1]$ is $\sqrt{1 + 1} = \sqrt{2}$, so

$$\int_{0}^{1} \sqrt{1 + \cos x} \, dx \leq \sqrt{2} \cdot (1 - 0) = \sqrt{2}.$$    ∎

### Area Under the Graph of a Nonnegative Function

We now return to the problem that started this chapter, that of defining what we mean by the *area* of a region having a curved boundary. In Section 5.1 we approximated the area under the graph of a nonnegative continuous function using several types of finite sums of areas of rectangles capturing the region—upper sums, lower sums, and sums using the midpoints of each subinterval—all being cases of Riemann sums constructed in special ways. Theorem 1 guarantees that all of these Riemann sums converge to a single definite integral as the norm of the partitions approaches zero and the number of subintervals goes to infinity. As a result, we can now *define* the area under the graph of a nonnegative integrable function to be the value of that definite integral.

---

**DEFINITION** If $y = f(x)$ is nonnegative and integrable over a closed interval $[a, b]$, then the **area under the curve $y = f(x)$ over $[a, b]$** is the integral of $f$ from $a$ to $b$,

$$A = \int_a^b f(x)\, dx.$$

---

For the first time we have a rigorous definition for the area of a region whose boundary is the graph of any continuous function. We now apply this to a simple example, the area under a straight line, where we can verify that our new definition agrees with our previous notion of area.

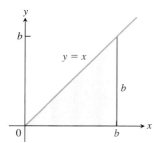

**FIGURE 5.12** The region in Example 4 is a triangle.

**EXAMPLE 4** Compute $\int_0^b x\, dx$ and find the area $A$ under $y = x$ over the interval $[0, b]$, $b > 0$.

**Solution** The region of interest is a triangle (Figure 5.12). We compute the area in two ways.

**(a)** To compute the definite integral as the limit of Riemann sums, we calculate $\lim_{\|P\| \to 0} \sum_{k=1}^n f(c_k)\, \Delta x_k$ for partitions whose norms go to zero. Theorem 1 tells us that it does not matter how we choose the partitions or the points $c_k$ as long as the norms approach zero. All choices give the exact same limit. So we consider the partition $P$ that subdivides the interval $[0, b]$ into $n$ subintervals of equal width $\Delta x = (b - 0)/n = b/n$, and we choose $c_k$ to be the right endpoint in each subinterval. The partition is

$$P = \left\{ 0, \frac{b}{n}, \frac{2b}{n}, \frac{3b}{n}, \cdots, \frac{nb}{n} \right\} \quad \text{and} \quad c_k = \frac{kb}{n}. \text{ So}$$

$$\sum_{k=1}^n f(c_k)\, \Delta x = \sum_{k=1}^n \frac{kb}{n} \cdot \frac{b}{n} \qquad f(c_k) = c_k$$

$$= \sum_{k=1}^n \frac{kb^2}{n^2}$$

$$= \frac{b^2}{n^2} \sum_{k=1}^n k \qquad \text{Constant Multiple Rule}$$

$$= \frac{b^2}{n^2} \cdot \frac{n(n + 1)}{2} \qquad \text{Sum of First } n \text{ Integers}$$

$$= \frac{b^2}{2}\left(1 + \frac{1}{n}\right).$$

As $n \to \infty$ and $\|P\| \to 0$, this last expression on the right has the limit $b^2/2$. Therefore,

$$\int_0^b x \, dx = \frac{b^2}{2}.$$

**(b)** Since the area equals the definite integral for a nonnegative function, we can quickly derive the definite integral by using the formula for the area of a triangle having base length $b$ and height $y = b$. The area is $A = (1/2) b \cdot b = b^2/2$. Again we conclude that $\int_0^b x \, dx = b^2/2$. ∎

Example 4 can be generalized to integrate $f(x) = x$ over any closed interval $[a, b], 0 < a < b$.

$$\int_a^b x \, dx = \int_a^0 x \, dx + \int_0^b x \, dx \qquad \text{Rule 5}$$

$$= -\int_0^a x \, dx + \int_0^b x \, dx \qquad \text{Rule 1}$$

$$= -\frac{a^2}{2} + \frac{b^2}{2}. \qquad \text{Example 4}$$

In conclusion, we have the following rule for integrating $f(x) = x$:

$$\int_a^b x \, dx = \frac{b^2}{2} - \frac{a^2}{2}, \qquad a < b \qquad (2)$$

This computation gives the area of the trapezoid in Figure 5.13a. Equation (2) remains valid when $a$ and $b$ are negative, but the interpretation of the definite integral changes. When $a < b < 0$, the definite integral value $(b^2 - a^2)/2$ is a negative number, the negative of the area of a trapezoid dropping down to the line $y = x$ below the $x$-axis (Figure 5.13b). When $a < 0$ and $b > 0$, Equation (2) is still valid and the definite integral gives the difference between two areas, the area under the graph and above $[0, b]$ minus the area below $[a, 0]$ and over the graph (Figure 5.13c).

The following results can also be established using a Riemann sum calculation similar to that in Example 4 (Exercises 63 and 65).

$$\int_a^b c \, dx = c(b - a), \qquad c \text{ any constant} \qquad (3)$$

$$\int_a^b x^2 \, dx = \frac{b^3}{3} - \frac{a^3}{3}, \qquad a < b \qquad (4)$$

### Average Value of a Continuous Function Revisited

In Section 5.1 we introduced informally the average value of a nonnegative continuous function $f$ over an interval $[a, b]$, leading us to define this average as the area under the graph of $y = f(x)$ divided by $b - a$. In integral notation we write this as

$$\text{Average} = \frac{1}{b - a} \int_a^b f(x) \, dx.$$

We can use this formula to give a precise definition of the average value of any continuous (or integrable) function, whether positive, negative, or both.

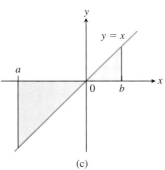

**FIGURE 5.13** (a) The area of this trapezoidal region is $A = (b^2 - a^2)/2$. (b) The definite integral in Equation (2) gives the negative of the area of this trapezoidal region. (c) The definite integral in Equation (2) gives the area of the blue triangular region added to the negative of the area of the tan triangular region.

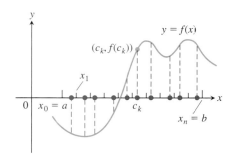

**FIGURE 5.14** A sample of values of a function on an interval $[a, b]$.

Alternatively, we can use the following reasoning. We start with the idea from arithmetic that the average of $n$ numbers is their sum divided by $n$. A continuous function $f$ on $[a, b]$ may have infinitely many values, but we can still sample them in an orderly way. We divide $[a, b]$ into $n$ subintervals of equal width $\Delta x = (b - a)/n$ and evaluate $f$ at a point $c_k$ in each (Figure 5.14). The average of the $n$ sampled values is

$$\frac{f(c_1) + f(c_2) + \cdots + f(c_n)}{n} = \frac{1}{n} \sum_{k=1}^{n} f(c_k)$$

$$= \frac{\Delta x}{b - a} \sum_{k=1}^{n} f(c_k) \qquad \Delta x = \frac{b-a}{n}, \text{ so } \frac{1}{n} = \frac{\Delta x}{b-a}$$

$$= \frac{1}{b - a} \sum_{k=1}^{n} f(c_k) \, \Delta x. \qquad \text{Constant Multiple Rule}$$

The average is obtained by dividing a Riemann sum for $f$ on $[a, b]$ by $(b - a)$. As we increase the size of the sample and let the norm of the partition approach zero, the average approaches $(1/(b - a)) \int_a^b f(x) \, dx$. Both points of view lead us to the following definition.

---

**DEFINITION** If $f$ is integrable on $[a, b]$, then its **average value on $[a, b]$**, also called its **mean**, is

$$\text{av}(f) = \frac{1}{b - a} \int_a^b f(x) \, dx.$$

---

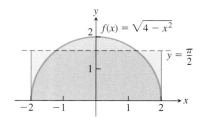

**FIGURE 5.15** The average value of $f(x) = \sqrt{4 - x^2}$ on $[-2, 2]$ is $\pi/2$ (Example 5). The area of the rectangle shown here is $4 \cdot (\pi/2) = 2\pi$, which is also the area of the semicircle.

**EXAMPLE 5** Find the average value of $f(x) = \sqrt{4 - x^2}$ on $[-2, 2]$.

**Solution** We recognize $f(x) = \sqrt{4 - x^2}$ as a function whose graph is the upper semicircle of radius 2 centered at the origin (Figure 5.15).

Since we know the area inside a circle, we do not need to take the limit of Riemann sums. The area between the semicircle and the $x$-axis from $-2$ to 2 can be computed using the geometry formula

$$\text{Area} = \frac{1}{2} \cdot \pi r^2 = \frac{1}{2} \cdot \pi (2)^2 = 2\pi.$$

Because $f$ is nonnegative, the area is also the value of the integral of $f$ from $-2$ to 2,

$$\int_{-2}^{2} \sqrt{4 - x^2} \, dx = 2\pi.$$

Therefore, the average value of $f$ is

$$\text{av}(f) = \frac{1}{2 - (-2)} \int_{-2}^{2} \sqrt{4 - x^2} \, dx = \frac{1}{4} (2\pi) = \frac{\pi}{2}.$$

Notice that the average value of $f$ over $[-2, 2]$ is the same as the height of a rectangle over $[-2, 2]$ whose area equals the area of the upper semicircle (see Figure 5.15). ∎

# Exercises 5.3

## Interpreting Limits of Sums as Integrals
Express the limits in Exercises 1–8 as definite integrals.

**1.** $\lim_{\|P\| \to 0} \sum_{k=1}^{n} c_k^2 \, \Delta x_k$, where $P$ is a partition of $[0, 2]$

**2.** $\lim_{\|P\| \to 0} \sum_{k=1}^{n} 2c_k^3 \, \Delta x_k$, where $P$ is a partition of $[-1, 0]$

**3.** $\lim_{\|P\| \to 0} \sum_{k=1}^{n} (c_k^2 - 3c_k) \, \Delta x_k$, where $P$ is a partition of $[-7, 5]$

**4.** $\lim_{\|P\| \to 0} \sum_{k=1}^{n} \left( \frac{1}{c_k} \right) \Delta x_k$, where $P$ is a partition of $[1, 4]$

**5.** $\lim_{\|P\| \to 0} \sum_{k=1}^{n} \frac{1}{1 - c_k} \, \Delta x_k$, where $P$ is a partition of $[2, 3]$

**6.** $\lim\limits_{\|P\|\to 0} \sum\limits_{k=1}^{n} \sqrt{4 - c_k{}^2}\,\Delta x_k$, where $P$ is a partition of $[0, 1]$

**7.** $\lim\limits_{\|P\|\to 0} \sum\limits_{k=1}^{n} (\sec c_k)\,\Delta x_k$, where $P$ is a partition of $[-\pi/4, 0]$

**8.** $\lim\limits_{\|P\|\to 0} \sum\limits_{k=1}^{n} (\tan c_k)\,\Delta x_k$, where $P$ is a partition of $[0, \pi/4]$

**Using the Definite Integral Rules**

**9.** Suppose that $f$ and $g$ are integrable and that

$$\int_1^2 f(x)\,dx = -4, \quad \int_1^5 f(x)\,dx = 6, \quad \int_1^5 g(x)\,dx = 8.$$

Use the rules in Table 5.6 to find

**a.** $\int_2^2 g(x)\,dx$  **b.** $\int_5^1 g(x)\,dx$

**c.** $\int_1^2 3f(x)\,dx$  **d.** $\int_2^5 f(x)\,dx$

**e.** $\int_1^5 [f(x) - g(x)]\,dx$  **f.** $\int_1^5 [4f(x) - g(x)]\,dx$

**10.** Suppose that $f$ and $h$ are integrable and that

$$\int_1^9 f(x)\,dx = -1, \quad \int_7^9 f(x)\,dx = 5, \quad \int_7^9 h(x)\,dx = 4.$$

Use the rules in Table 5.6 to find

**a.** $\int_1^9 -2f(x)\,dx$  **b.** $\int_7^9 [f(x) + h(x)]\,dx$

**c.** $\int_7^9 [2f(x) - 3h(x)]\,dx$  **d.** $\int_9^1 f(x)\,dx$

**e.** $\int_1^7 f(x)\,dx$  **f.** $\int_9^7 [h(x) - f(x)]\,dx$

**11.** Suppose that $\int_1^2 f(x)\,dx = 5$. Find

**a.** $\int_1^2 f(u)\,du$  **b.** $\int_1^2 \sqrt{3}f(z)\,dz$

**c.** $\int_2^1 f(t)\,dt$  **d.** $\int_1^2 [-f(x)]\,dx$

**12.** Suppose that $\int_{-3}^0 g(t)\,dt = \sqrt{2}$. Find

**a.** $\int_0^{-3} g(t)\,dt$  **b.** $\int_{-3}^0 g(u)\,du$

**c.** $\int_{-3}^0 [-g(x)]\,dx$  **d.** $\int_{-3}^0 \dfrac{g(r)}{\sqrt{2}}\,dr$

**13.** Suppose that $f$ is integrable and that $\int_0^3 f(z)\,dz = 3$ and $\int_0^4 f(z)\,dz = 7$. Find

**a.** $\int_3^4 f(z)\,dz$  **b.** $\int_4^3 f(t)\,dt$

**14.** Suppose that $h$ is integrable and that $\int_{-1}^1 h(r)\,dr = 0$ and $\int_{-1}^3 h(r)\,dr = 6$. Find

**a.** $\int_1^3 h(r)\,dr$  **b.** $-\int_3^1 h(u)\,du$

**Using Known Areas to Find Integrals**

In Exercises 15–22, graph the integrands and use known area formulas to evaluate the integrals.

**15.** $\int_{-2}^4 \left(\dfrac{x}{2} + 3\right) dx$  **16.** $\int_{1/2}^{3/2} (-2x + 4)\,dx$

**17.** $\int_{-3}^3 \sqrt{9 - x^2}\,dx$  **18.** $\int_{-4}^0 \sqrt{16 - x^2}\,dx$

**19.** $\int_{-2}^1 |x|\,dx$  **20.** $\int_{-1}^1 (1 - |x|)\,dx$

**21.** $\int_{-1}^1 (2 - |x|)\,dx$  **22.** $\int_{-1}^1 \left(1 + \sqrt{1 - x^2}\right) dx$

Use known area formulas to evaluate the integrals in Exercises 23–28.

**23.** $\int_0^b \dfrac{x}{2}\,dx, \quad b > 0$  **24.** $\int_0^b 4x\,dx, \quad b > 0$

**25.** $\int_a^b 2s\,ds, \quad 0 < a < b$  **26.** $\int_a^b 3t\,dt, \quad 0 < a < b$

**27.** $f(x) = \sqrt{4 - x^2}$ on **a.** $[-2, 2]$, **b.** $[0, 2]$
**28.** $f(x) = 3x + \sqrt{1 - x^2}$ on **a.** $[-1, 0]$, **b.** $[-1, 1]$

**Evaluating Definite Integrals**

Use the results of Equations (2) and (4) to evaluate the integrals in Exercises 29–40.

**29.** $\int_1^{\sqrt{2}} x\,dx$  **30.** $\int_{0.5}^{2.5} x\,dx$  **31.** $\int_\pi^{2\pi} \theta\,d\theta$

**32.** $\int_{\sqrt{2}}^{5\sqrt{2}} r\,dr$  **33.** $\int_0^{\sqrt[3]{7}} x^2\,dx$  **34.** $\int_0^{0.3} s^2\,ds$

**35.** $\int_0^{1/2} t^2\,dt$  **36.** $\int_0^{\pi/2} \theta^2\,d\theta$  **37.** $\int_a^{2a} x\,dx$

**38.** $\int_a^{\sqrt{3}a} x\,dx$  **39.** $\int_0^{\sqrt[3]{b}} x^2\,dx$  **40.** $\int_0^{3b} x^2\,dx$

Use the rules in Table 5.6 and Equations (2)–(4) to evaluate the integrals in Exercises 41–50.

**41.** $\int_3^1 7\,dx$  **42.** $\int_0^2 5x\,dx$

**43.** $\int_0^2 (2t - 3)\,dt$  **44.** $\int_0^{\sqrt{2}} \left(t - \sqrt{2}\right) dt$

**45.** $\int_2^1 \left(1 + \dfrac{z}{2}\right) dz$  **46.** $\int_3^0 (2z - 3)\,dz$

**47.** $\int_1^2 3u^2\,du$  **48.** $\int_{1/2}^1 24u^2\,du$

**49.** $\int_0^2 (3x^2 + x - 5)\,dx$  **50.** $\int_1^0 (3x^2 + x - 5)\,dx$

**Finding Area by Definite Integrals**

In Exercises 51–54, use a definite integral to find the area of the region between the given curve and the $x$-axis on the interval $[0, b]$.

**51.** $y = 3x^2$  **52.** $y = \pi x^2$

**53.** $y = 2x$  **54.** $y = \dfrac{x}{2} + 1$

In Exercises 55–62, graph the function and find its average value over the given interval.

**55.** $f(x) = x^2 - 1$ on $[0, \sqrt{3}]$

**56.** $f(x) = -\dfrac{x^2}{2}$ on $[0, 3]$

**57.** $f(x) = -3x^2 - 1$ on $[0, 1]$

**58.** $f(x) = 3x^2 - 3$ on $[0, 1]$

**59.** $f(t) = (t - 1)^2$ on $[0, 3]$

**60.** $f(t) = t^2 - t$ on $[-2, 1]$

**61.** $g(x) = |x| - 1$ on **a.** $[-1, 1]$, **b.** $[1, 3]$, and **c.** $[-1, 3]$

**62.** $h(x) = -|x|$ on **a.** $[-1, 0]$, **b.** $[0, 1]$, and **c.** $[-1, 1]$

Definite Integrals as Limits of Sums

Use the method of Example 4a or Equation (1) to evaluate the definite integrals in Exercises 63–70.

**63.** $\displaystyle\int_a^b c\, dx$

**64.** $\displaystyle\int_0^2 (2x + 1)\, dx$

**65.** $\displaystyle\int_a^b x^2\, dx, \quad a < b$

**66.** $\displaystyle\int_{-1}^0 (x - x^2)\, dx$

**67.** $\displaystyle\int_{-1}^2 (3x^2 - 2x + 1)\, dx$

**68.** $\displaystyle\int_{-1}^1 x^3\, dx$

**69.** $\displaystyle\int_a^b x^3\, dx, \quad a < b$

**70.** $\displaystyle\int_0^1 (3x - x^3)\, dx$

Theory and Examples

**71.** What values of $a$ and $b$ maximize the value of

$$\int_a^b (x - x^2)\, dx?$$

(*Hint:* Where is the integrand positive?)

**72.** What values of $a$ and $b$ minimize the value of

$$\int_a^b (x^4 - 2x^2)\, dx?$$

**73.** Use the Max-Min Inequality to find upper and lower bounds for the value of

$$\int_0^1 \frac{1}{1 + x^2}\, dx.$$

**74.** (*Continuation of Exercise 73.*) Use the Max-Min Inequality to find upper and lower bounds for

$$\int_0^{0.5} \frac{1}{1 + x^2}\, dx \quad \text{and} \quad \int_{0.5}^1 \frac{1}{1 + x^2}\, dx.$$

Add these to arrive at an improved estimate of

$$\int_0^1 \frac{1}{1 + x^2}\, dx.$$

**75.** Show that the value of $\displaystyle\int_0^1 \sin(x^2)\, dx$ cannot possibly be 2.

**76.** Show that the value of $\displaystyle\int_0^1 \sqrt{x + 8}\, dx$ lies between $2\sqrt{2} \approx 2.8$ and 3.

**77. Integrals of nonnegative functions** Use the Max-Min Inequality to show that if $f$ is integrable then

$$f(x) \geq 0 \quad \text{on} \quad [a, b] \quad \Rightarrow \quad \int_a^b f(x)\, dx \geq 0.$$

**78. Integrals of nonpositive functions** Show that if $f$ is integrable then

$$f(x) \leq 0 \quad \text{on} \quad [a, b] \quad \Rightarrow \quad \int_a^b f(x)\, dx \leq 0.$$

**79.** Use the inequality $\sin x \leq x$, which holds for $x \geq 0$, to find an upper bound for the value of $\displaystyle\int_0^1 \sin x\, dx$.

**80.** The inequality $\sec x \geq 1 + (x^2/2)$ holds on $(-\pi/2, \pi/2)$. Use it to find a lower bound for the value of $\displaystyle\int_0^1 \sec x\, dx$.

**81.** If $\operatorname{av}(f)$ really is a typical value of the integrable function $f(x)$ on $[a, b]$, then the constant function $\operatorname{av}(f)$ should have the same integral over $[a, b]$ as $f$. Does it? That is, does

$$\int_a^b \operatorname{av}(f)\, dx = \int_a^b f(x)\, dx?$$

Give reasons for your answer.

**82.** It would be nice if average values of integrable functions obeyed the following rules on an interval $[a, b]$.

**a.** $\operatorname{av}(f + g) = \operatorname{av}(f) + \operatorname{av}(g)$

**b.** $\operatorname{av}(kf) = k\operatorname{av}(f)$  (any number $k$)

**c.** $\operatorname{av}(f) \leq \operatorname{av}(g)$ if $f(x) \leq g(x)$ on $[a, b]$.

Do these rules ever hold? Give reasons for your answers.

**83. Upper and lower sums for increasing functions**

**a.** Suppose the graph of a continuous function $f(x)$ rises steadily as $x$ moves from left to right across an interval $[a, b]$. Let $P$ be a partition of $[a, b]$ into $n$ subintervals of equal length $\Delta x = (b - a)/n$. Show by referring to the accompanying figure that the difference between the upper and lower sums for $f$ on this partition can be represented graphically as the area of a rectangle $R$ whose dimensions are $[f(b) - f(a)]$ by $\Delta x$. (*Hint:* The difference $U - L$ is the sum of areas of rectangles whose diagonals $Q_0Q_1, Q_1Q_2, \ldots, Q_{n-1}Q_n$ lie approximately along the curve. There is no overlapping when these rectangles are shifted horizontally onto $R$.)

**b.** Suppose that instead of being equal, the lengths $\Delta x_k$ of the subintervals of the partition of $[a, b]$ vary in size. Show that

$$U - L \leq |f(b) - f(a)|\, \Delta x_{\max},$$

where $\Delta x_{\max}$ is the norm of $P$, and hence that $\lim_{\|P\| \to 0} (U - L) = 0$.

**84. Upper and lower sums for decreasing functions** (*Continuation of Exercise 83.*)

**a.** Draw a figure like the one in Exercise 83 for a continuous function $f(x)$ whose values decrease steadily as $x$ moves from left to right across the interval $[a, b]$. Let $P$ be a partition of $[a, b]$ into subintervals of equal length. Find an expression for $U - L$ that is analogous to the one you found for $U - L$ in Exercise 83a.

**b.** Suppose that instead of being equal, the lengths $\Delta x_k$ of the subintervals of $P$ vary in size. Show that the inequality

$$U - L \leq |f(b) - f(a)| \, \Delta x_{\max}$$

of Exercise 83b still holds and hence that $\lim_{\|P\| \to 0} (U - L) = 0$.

**85.** Use the formula

$$\sin h + \sin 2h + \sin 3h + \cdots + \sin mh$$
$$= \frac{\cos(h/2) - \cos((m + (1/2))h)}{2 \sin (h/2)}$$

to find the area under the curve $y = \sin x$ from $x = 0$ to $x = \pi/2$ in two steps:

**a.** Partition the interval $[0, \pi/2]$ into $n$ subintervals of equal length and calculate the corresponding upper sum $U$; then

**b.** Find the limit of $U$ as $n \to \infty$ and $\Delta x = (b - a)/n \to 0$.

**86.** Suppose that $f$ is continuous and nonnegative over $[a, b]$, as in the accompanying figure. By inserting points

$$x_1, x_2, \ldots, x_{k-1}, x_k, \ldots, x_{n-1}$$

as shown, divide $[a, b]$ into $n$ subintervals of lengths $\Delta x_1 = x_1 - a$, $\Delta x_2 = x_2 - x_1, \ldots, \Delta x_n = b - x_{n-1}$, which need not be equal.

**a.** If $m_k = \min \{f(x) \text{ for } x \text{ in the } k\text{th subinterval}\}$, explain the connection between the **lower sum**

$$L = m_1 \, \Delta x_1 + m_2 \, \Delta x_2 + \cdots + m_n \, \Delta x_n$$

and the shaded regions in the first part of the figure.

**b.** If $M_k = \max \{f(x) \text{ for } x \text{ in the } k\text{th subinterval}\}$, explain the connection between the **upper sum**

$$U = M_1 \, \Delta x_1 + M_2 \, \Delta x_2 + \cdots + M_n \, \Delta x_n$$

and the shaded regions in the second part of the figure.

**c.** Explain the connection between $U - L$ and the shaded regions along the curve in the third part of the figure.

**87.** We say $f$ is **uniformly continuous** on $[a, b]$ if given any $\epsilon > 0$, there is a $\delta > 0$ such that if $x_1, x_2$ are in $[a, b]$ and $|x_1 - x_2| < \delta$, then $|f(x_1) - f(x_2)| < \epsilon$. It can be shown that a continuous function on $[a, b]$ is uniformly continuous. Use this and the figure for Exercise 86 to show that if $f$ is continuous and $\epsilon > 0$ is given, it is possible to make $U - L \leq \epsilon \cdot (b - a)$ by making the largest of the $\Delta x_k$'s sufficiently small.

**88.** If you average 48 km/h on a 240-km trip and then return over the same 240 km at the rate of 80 km/h, what is your average speed for the trip? Give reasons for your answer.

**COMPUTER EXPLORATIONS**

If your CAS can draw rectangles associated with Riemann sums, use it to draw rectangles associated with Riemann sums that converge to the integrals in Exercises 89–94. Use $n = 4$, 10, 20, and 50 subintervals of equal length in each case.

**89.** $\displaystyle\int_0^1 (1 - x) \, dx = \frac{1}{2}$

**90.** $\displaystyle\int_0^1 (x^2 + 1)\, dx = \frac{4}{3}$    **91.** $\displaystyle\int_{-\pi}^{\pi} \cos x\, dx = 0$

**92.** $\displaystyle\int_0^{\pi/4} \sec^2 x\, dx = 1$    **93.** $\displaystyle\int_{-1}^1 |x|\, dx = 1$

**94.** $\displaystyle\int_1^2 \frac{1}{x}\, dx$   (The integral's value is about 0.693.)

In Exercises 95–98, use a CAS to perform the following steps:

**a.** Plot the functions over the given interval.

**b.** Partition the interval into $n = 100, 200$, and $1000$ subintervals of equal length, and evaluate the function at the midpoint of each subinterval.

**c.** Compute the average value of the function values generated in part (b).

**d.** Solve the equation $f(x) =$ (average value) for $x$ using the average value calculated in part (c) for the $n = 1000$ partitioning.

**95.** $f(x) = \sin x$   on   $[0, \pi]$

**96.** $f(x) = \sin^2 x$   on   $[0, \pi]$

**97.** $f(x) = x \sin \frac{1}{x}$   on   $\left[\frac{\pi}{4}, \pi\right]$

**98.** $f(x) = x \sin^2 \frac{1}{x}$   on   $\left[\frac{\pi}{4}, \pi\right]$

# 5.4 The Fundamental Theorem of Calculus

**HISTORICAL BIOGRAPHY**

Sir Isaac Newton
(1642–1727)

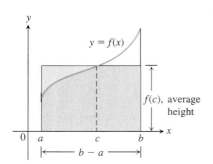

**FIGURE 5.16** The value $f(c)$ in the Mean Value Theorem is, in a sense, the average (or *mean*) height of $f$ on $[a, b]$. When $f \geq 0$, the area of the rectangle is the area under the graph of $f$ from $a$ to $b$,

$$f(c)(b - a) = \int_a^b f(x)\, dx.$$

In this section we present the Fundamental Theorem of Calculus, which is the central theorem of integral calculus. It connects integration and differentiation, enabling us to compute integrals using an antiderivative of the integrand function rather than by taking limits of Riemann sums as we did in Section 5.3. Leibniz and Newton exploited this relationship and started mathematical developments that fueled the scientific revolution for the next 200 years.

Along the way, we present an integral version of the Mean Value Theorem, which is another important theorem of integral calculus and is used to prove the Fundamental Theorem. We also find that the net change of a function over an interval is the integral of its rate of change, as suggested by Example 3 in Section 5.1.

## Mean Value Theorem for Definite Integrals

In the previous section we defined the average value of a continuous function over a closed interval $[a, b]$ as the definite integral $\int_a^b f(x)\, dx$ divided by the length or width $b - a$ of the interval. The Mean Value Theorem for Definite Integrals asserts that this average value is *always* taken on at least once by the function $f$ in the interval.

The graph in Figure 5.16 shows a *positive* continuous function $y = f(x)$ defined over the interval $[a, b]$. Geometrically, the Mean Value Theorem says that there is a number $c$ in $[a, b]$ such that the rectangle with height equal to the average value $f(c)$ of the function and base width $b - a$ has exactly the same area as the region beneath the graph of $f$ from $a$ to $b$.

> **THEOREM 3—The Mean Value Theorem for Definite Integrals**   If $f$ is continuous on $[a, b]$, then at some point $c$ in $[a, b]$,
>
> $$f(c) = \frac{1}{b - a}\int_a^b f(x)\, dx.$$

**Proof**    If we divide both sides of the Max-Min Inequality (Table 5.6, Rule 6) by $(b - a)$, we obtain

$$\min f \leq \frac{1}{b - a}\int_a^b f(x)\, dx \leq \max f.$$

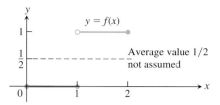

**FIGURE 5.17** A discontinuous function need not assume its average value.

Since $f$ is continuous, the Intermediate Value Theorem for Continuous Functions (Section 2.5) says that $f$ must assume every value between min $f$ and max $f$. It must therefore assume the value $(1/(b - a)) \int_a^b f(x)\, dx$ at some point $c$ in $[a, b]$.    ∎

The continuity of $f$ is important here. It is possible that a discontinuous function never equals its average value (Figure 5.17).

**EXAMPLE 1**    Show that if $f$ is continuous on $[a, b]$, $a \neq b$, and if

$$\int_a^b f(x)\, dx = 0,$$

then $f(x) = 0$ at least once in $[a, b]$.

**Solution**    The average value of $f$ on $[a, b]$ is

$$\mathrm{av}(f) = \frac{1}{b - a} \int_a^b f(x)\, dx = \frac{1}{b - a} \cdot 0 = 0.$$

By the Mean Value Theorem, $f$ assumes this value at some point $c \in [a, b]$.    ∎

## Fundamental Theorem, Part 1

It can be very difficult to compute definite integrals by taking the limit of Riemann sums. We now develop a powerful new method for evaluating definite integrals, based on using antiderivatives. This method combines the two strands of calculus. One strand involves the idea of taking the limits of finite sums to obtain a definite integral, and the other strand contains derivatives and antiderivatives. They come together in the Fundamental Theorem of Calculus. We begin by considering how to differentiate a certain type of function that is described as an integral.

If $f(t)$ is an integrable function over a finite interval $I$, then the integral from any fixed number $a \in I$ to another number $x \in I$ defines a new function $F$ whose value at $x$ is

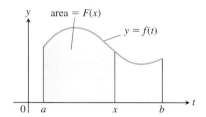

**FIGURE 5.18** The function $F(x)$ defined by Equation (1) gives the area under the graph of $f$ from $a$ to $x$ when $f$ is nonnegative and $x > a$.

$$F(x) = \int_a^x f(t)\, dt. \tag{1}$$

For example, if $f$ is nonnegative and $x$ lies to the right of $a$, then $F(x)$ is the area under the graph from $a$ to $x$ (Figure 5.18). The variable $x$ is the upper limit of integration of an integral, but $F$ is just like any other real-valued function of a real variable. For each value of the input $x$, there is a well-defined numerical output, in this case the definite integral of $f$ from $a$ to $x$.

Equation (1) gives a way to define new functions (as we will see in Section 7.2), but its importance now is the connection it makes between integrals and derivatives. If $f$ is any continuous function, then the Fundamental Theorem asserts that $F$ is a differentiable function of $x$ whose derivative is $f$ itself. At every value of $x$, it asserts that

$$\frac{d}{dx} F(x) = f(x).$$

To gain some insight into why this result holds, we look at the geometry behind it.

If $f \geq 0$ on $[a, b]$, then the computation of $F'(x)$ from the definition of the derivative means taking the limit as $h \to 0$ of the difference quotient

$$\frac{F(x + h) - F(x)}{h}.$$

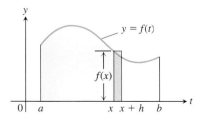

**FIGURE 5.19** In Equation (1), $F(x)$ is the area to the left of $x$. Also, $F(x + h)$ is the area to the left of $x + h$. The difference quotient $[F(x + h) - F(x)]/h$ is then approximately equal to $f(x)$, the height of the rectangle shown here.

For $h > 0$, the numerator is obtained by subtracting two areas, so it is the area under the graph of $f$ from $x$ to $x + h$ (Figure 5.19). If $h$ is small, this area is approximately equal to the area of the rectangle of height $f(x)$ and width $h$, which can be seen from Figure 5.19. That is,

$$F(x + h) - F(x) \approx h f(x).$$

Dividing both sides of this approximation by $h$ and letting $h \to 0$, it is reasonable to expect that

$$F'(x) = \lim_{h \to 0} \frac{F(x + h) - F(x)}{h} = f(x).$$

This result is true even if the function $f$ is not positive, and it forms the first part of the Fundamental Theorem of Calculus.

> **THEOREM 4—The Fundamental Theorem of Calculus, Part 1** If $f$ is continuous on $[a, b]$, then $F(x) = \int_a^x f(t)\, dt$ is continuous on $[a, b]$ and differentiable on $(a, b)$ and its derivative is $f(x)$:
>
> $$F'(x) = \frac{d}{dx} \int_a^x f(t)\, dt = f(x). \tag{2}$$

Before proving Theorem 4, we look at several examples to gain a good understanding of what it says. In each example, notice that the independent variable appears in a limit of integration, possibly in a formula.

**EXAMPLE 2**  Use the Fundamental Theorem to find $dy/dx$ if

**(a)** $y = \displaystyle\int_a^x (t^3 + 1)\, dt$  **(b)** $y = \displaystyle\int_x^5 3t \sin t\, dt$

**(c)** $y = \displaystyle\int_1^{x^2} \cos t\, dt$  **(d)** $y = \displaystyle\int_{1+3x^2}^4 \frac{1}{2+t}\, dt$

**Solution** We calculate the derivatives with respect to the independent variable $x$.

**(a)** $\dfrac{dy}{dx} = \dfrac{d}{dx} \displaystyle\int_a^x (t^3 + 1)\, dt = x^3 + 1$    Eq. (2) with $f(t) = t^3 + 1$

**(b)** $\dfrac{dy}{dx} = \dfrac{d}{dx} \displaystyle\int_x^5 3t \sin t\, dt = \dfrac{d}{dx}\left( -\displaystyle\int_5^x 3t \sin t\, dt \right)$    Table 5.6, Rule 1

$\qquad = -\dfrac{d}{dx} \displaystyle\int_5^x 3t \sin t\, dt$

$\qquad = -3x \sin x$    Eq. (2) with $f(t) = 3t \sin t$

**(c)** The upper limit of integration is not $x$ but $x^2$. This makes $y$ a composite of the two functions,

$$y = \int_1^u \cos t\, dt \qquad \text{and} \qquad u = x^2.$$

We must therefore apply the Chain Rule when finding $dy/dx$.

$$\frac{dy}{dx} = \frac{dy}{du} \cdot \frac{du}{dx}$$

$$= \left( \frac{d}{du} \int_1^u \cos t\, dt \right) \cdot \frac{du}{dx}$$

$$= \cos u \cdot \frac{du}{dx}$$

$$= \cos(x^2) \cdot 2x$$

$$= 2x \cos x^2$$

**(d)** $\dfrac{d}{dx} \displaystyle\int_{1+3x^2}^4 \frac{1}{2+t}\, dt = \dfrac{d}{dx}\left( -\displaystyle\int_4^{1+3x^2} \frac{1}{2+t}\, dt \right)$    Rule 1

$\qquad = -\dfrac{d}{dx} \displaystyle\int_4^{1+3x^2} \frac{1}{2+t}\, dt$

$\qquad = -\dfrac{1}{2 + (1 + 3x^2)} \cdot \dfrac{d}{dx}(1 + 3x^2)$    Eq. (2) and the Chain Rule

$\qquad = -\dfrac{2x}{1 + x^2}$

**Proof of Theorem 4**    We prove the Fundamental Theorem, Part 1, by applying the definition of the derivative directly to the function $F(x)$, when $x$ and $x + h$ are in $(a, b)$. This means writing out the difference quotient

$$\frac{F(x + h) - F(x)}{h} \tag{3}$$

and showing that its limit as $h \to 0$ is the number $f(x)$ for each $x$ in $(a, b)$. Doing so, we find

$$F'(x) = \lim_{h \to 0} \frac{F(x + h) - F(x)}{h}$$

$$= \lim_{h \to 0} \frac{1}{h} \left[ \int_a^{x+h} f(t)\, dt - \int_a^x f(t)\, dt \right]$$

$$= \lim_{h \to 0} \frac{1}{h} \int_x^{x+h} f(t)\, dt. \qquad \text{Table 5.6, Rule 5}$$

According to the Mean Value Theorem for Definite Integrals, the value before taking the limit in the last expression is one of the values taken on by $f$ in the interval between $x$ and $x + h$. That is, for some number $c$ in this interval,

$$\frac{1}{h} \int_x^{x+h} f(t)\, dt = f(c). \tag{4}$$

As $h \to 0$, $x + h$ approaches $x$, forcing $c$ to approach $x$ also (because $c$ is trapped between $x$ and $x + h$). Since $f$ is continuous at $x$, $f(c)$ approaches $f(x)$:

$$\lim_{h \to 0} f(c) = f(x). \tag{5}$$

In conclusion, we have

$$F'(x) = \lim_{h \to 0} \frac{1}{h} \int_x^{x+h} f(t)\, dt$$

$$= \lim_{h \to 0} f(c) \qquad \text{Eq. (4)}$$

$$= f(x). \qquad \text{Eq. (5)}$$

If $x = a$ or $b$, then the limit of Equation (3) is interpreted as a one-sided limit with $h \to 0^+$ or $h \to 0^-$, respectively. Then Theorem 1 in Section 3.2 shows that $F$ is continuous over $[a, b]$. This concludes the proof.    ∎

## Fundamental Theorem, Part 2 (The Evaluation Theorem)

We now come to the second part of the Fundamental Theorem of Calculus. This part describes how to evaluate definite integrals without having to calculate limits of Riemann sums. Instead we find and evaluate an antiderivative at the upper and lower limits of integration.

---

**THEOREM 4 (Continued)—The Fundamental Theorem of Calculus, Part 2**

If $f$ is continuous over $[a, b]$ and $F$ is any antiderivative of $f$ on $[a, b]$, then

$$\int_a^b f(x)\, dx = F(b) - F(a).$$

---

**Proof**    Part 1 of the Fundamental Theorem tells us that an antiderivative of $f$ exists, namely

$$G(x) = \int_a^x f(t)\, dt.$$

Thus, if $F$ is *any* antiderivative of $f$, then $F(x) = G(x) + C$ for some constant $C$ for $a < x < b$ (by Corollary 2 of the Mean Value Theorem for Derivatives, Section 4.2).

Since both $F$ and $G$ are continuous on $[a, b]$, we see that $F(x) = G(x) + C$ also holds when $x = a$ and $x = b$ by taking one-sided limits (as $x \to a^+$ and $x \to b^-$).

Evaluating $F(b) - F(a)$, we have

$$
\begin{aligned}
F(b) - F(a) &= [G(b) + C] - [G(a) + C] \\
&= G(b) - G(a) \\
&= \int_a^b f(t)\, dt - \int_a^a f(t)\, dt \\
&= \int_a^b f(t)\, dt - 0 \\
&= \int_a^b f(t)\, dt. \qquad \blacksquare
\end{aligned}
$$

The Evaluation Theorem is important because it says that to calculate the definite integral of $f$ over an interval $[a, b]$ we need do only two things:

1. Find an antiderivative $F$ of $f$, and
2. Calculate the number $F(b) - F(a)$, which is equal to $\int_a^b f(x)\, dx$.

This process is much easier than using a Riemann sum computation. The power of the theorem follows from the realization that the definite integral, which is defined by a complicated process involving all of the values of the function $f$ over $[a, b]$, can be found by knowing the values of *any* antiderivative $F$ at only the two endpoints $a$ and $b$. The usual notation for the difference $F(b) - F(a)$ is

$$
F(x) \Big]_a^b \qquad \text{or} \qquad \left[ F(x) \right]_a^b,
$$

depending on whether $F$ has one or more terms.

**EXAMPLE 3**    We calculate several definite integrals using the Evaluation Theorem, rather than by taking limits of Riemann sums.

**(a)** $\displaystyle \int_0^\pi \cos x\, dx = \sin x \Big]_0^\pi$ $\qquad\qquad \dfrac{d}{dx}\sin x = \cos x$

$\qquad\qquad = \sin \pi - \sin 0 = 0 - 0 = 0$

**(b)** $\displaystyle \int_{-\pi/4}^0 \sec x \tan x\, dx = \sec x \Big]_{-\pi/4}^0$ $\qquad \dfrac{d}{dx}\sec x = \sec x \tan x$

$\qquad\qquad = \sec 0 - \sec \left(-\dfrac{\pi}{4}\right) = 1 - \sqrt{2}$

**(c)** $\displaystyle \int_1^4 \left(\dfrac{3}{2}\sqrt{x} - \dfrac{4}{x^2}\right) dx = \left[ x^{3/2} + \dfrac{4}{x}\right]_1^4$ $\qquad \dfrac{d}{dx}\left( x^{3/2} + \dfrac{4}{x}\right) = \dfrac{3}{2}x^{1/2} - \dfrac{4}{x^2}$

$\qquad\qquad = \left[ (4)^{3/2} + \dfrac{4}{4}\right] - \left[ (1)^{3/2} + \dfrac{4}{1}\right]$

$\qquad\qquad = [8 + 1] - [5] = 4 \qquad \blacksquare$

Exercise 72 offers another proof of the Evaluation Theorem, bringing together the ideas of Riemann sums, the Mean Value Theorem, and the definition of the definite integral.

The Integral of a Rate

We can interpret Part 2 of the Fundamental Theorem in another way. If $F$ is any antiderivative of $f$, then $F' = f$. The equation in the theorem can then be rewritten as

$$\int_a^b F'(x)\, dx = F(b) - F(a).$$

Now $F'(x)$ represents the rate of change of the function $F(x)$ with respect to $x$, so the last equation asserts that the integral of $F'$ is just the *net change* in $F$ as $x$ changes from $a$ to $b$. Formally, we have the following result.

> **THEOREM 5—The Net Change Theorem**  The net change in a differentiable function $F(x)$ over an interval $a \le x \le b$ is the integral of its rate of change:
>
> $$F(b) - F(a) = \int_a^b F'(x)\, dx. \tag{6}$$

**EXAMPLE 4**    Here are several interpretations of the Net Change Theorem.

(a) If $c(x)$ is the cost of producing $x$ units of a certain commodity, then $c'(x)$ is the marginal cost (Section 3.4). From Theorem 5,

$$\int_{x_1}^{x_2} c'(x)\, dx = c(x_2) - c(x_1),$$

which is the cost of increasing production from $x_1$ units to $x_2$ units.

(b) If an object with position function $s(t)$ moves along a coordinate line, its velocity is $v(t) = s'(t)$. Theorem 5 says that

$$\int_{t_1}^{t_2} v(t)\, dt = s(t_2) - s(t_1),$$

so the integral of velocity is the **displacement** over the time interval $t_1 \le t \le t_2$. On the other hand, the integral of the speed $|v(t)|$ is the **total distance traveled** over the time interval. This is consistent with our discussion in Section 5.1.    ∎

If we rearrange Equation (6) as

$$F(b) = F(a) + \int_a^b F'(x)\, dx,$$

we see that the Net Change Theorem also says that the final value of a function $F(x)$ over an interval $[a, b]$ equals its initial value $F(a)$ plus its net change over the interval. So if $v(t)$ represents the velocity function of an object moving along a coordinate line, this means that the object's final position $s(t_2)$ over a time interval $t_1 \le t \le t_2$ is its initial position $s(t_1)$ plus its net change in position along the line (see Example 4b).

**EXAMPLE 5**    Consider again our analysis of a heavy rock blown straight up from the ground by a dynamite blast (Example 3, Section 5.1). The velocity of the rock at any time $t$ during its motion was given as $v(t) = 49 - 9.8t$ m/s.

(a) Find the displacement of the rock during the time period $0 \le t \le 8$.

(b) Find the total distance traveled during this time period.

**Solution**

**(a)** From Example 4b, the displacement is the integral

$$\int_0^8 v(t)\, dt = \int_0^8 (49 - 9.8t)\, dt = \left[ 49t - 4.9t^2 \right]_0^8$$

$$= (49)(8) - (4.9)(64) = 78.4.$$

This means that the height of the rock is 78.4 m above the ground 8 s after the explosion, which agrees with our conclusion in Example 3, Section 5.1.

**(b)** As we noted in Table 5.3, the velocity function $v(t)$ is positive over the time interval $[0, 5]$ and negative over the interval $[5, 8]$. Therefore, from Example 4b, the total distance traveled is the integral

$$\int_0^8 |v(t)|\, dt = \int_0^5 |v(t)|\, dt + \int_5^8 |v(t)|\, dt$$

$$= \int_0^5 (49 - 9.8t)\, dt - \int_5^8 (49 - 9.8t)\, dt$$

$$= \left[ 49t - 4.9t^2 \right]_0^5 - \left[ 49t - 4.9t^2 \right]_5^8$$

$$= \left[ (49)(5) - (4.9)(25) \right] - \left[ (49)(8) - (4.9)(64) - ((49)(5) - (4.9)(25)) \right]$$

$$= 122.5 - (-44.1) = 166.6.$$

Again, this calculation agrees with our conclusion in Example 3, Section 5.1. That is, the total distance of 166.6 m traveled by the rock during the time period $0 \le t \le 8$ is (i) the maximum height of 122.5 m it reached over the time interval $[0, 5]$ plus (ii) the additional distance of 44.1 m the rock fell over the time interval $[5, 8]$. ∎

### The Relationship Between Integration and Differentiation

The conclusions of the Fundamental Theorem tell us several things. Equation (2) can be rewritten as

$$\frac{d}{dx} \int_a^x f(t)\, dt = f(x),$$

which says that if you first integrate the function $f$ and then differentiate the result, you get the function $f$ back again. Likewise, replacing $b$ by $x$ and $x$ by $t$ in Equation (6) gives

$$\int_a^x F'(t)\, dt = F(x) - F(a),$$

so that if you first differentiate the function $F$ and then integrate the result, you get the function $F$ back (adjusted by an integration constant). In a sense, the processes of integration and differentiation are "inverses" of each other. The Fundamental Theorem also says that every continuous function $f$ has an antiderivative $F$. It shows the importance of finding antiderivatives in order to evaluate definite integrals easily. Furthermore, it says that the differential equation $dy/dx = f(x)$ has a solution (namely, any of the functions $y = F(x) + C$) for every continuous function $f$.

### Total Area

Area is always a nonnegative quantity. The Riemann sum contains terms such as $f(c_k)\, \Delta x_k$ that give the area of a rectangle when $f(c_k)$ is positive. When $f(c_k)$ is negative, then the product $f(c_k)\, \Delta x_k$ is the negative of the rectangle's area. When we add up such terms for a negative function, we get the negative of the area between the curve and the x-axis. If we then take the absolute value, we obtain the correct positive area.

**EXAMPLE 6**     Figure 5.20 shows the graph of $f(x) = x^2 - 4$ and its mirror image $g(x) = 4 - x^2$ reflected across the x-axis. For each function, compute

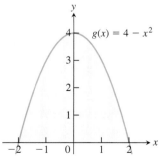

**FIGURE 5.20** These graphs enclose the same amount of area with the x-axis, but the definite integrals of the two functions over $[-2, 2]$ differ in sign (Example 6).

(a) the definite integral over the interval $[-2, 2]$, and

(b) the area between the graph and the $x$-axis over $[-2, 2]$.

**Solution**

(a) $\int_{-2}^{2} f(x)\, dx = \left[ \frac{x^3}{3} - 4x \right]_{-2}^{2} = \left( \frac{8}{3} - 8 \right) - \left( -\frac{8}{3} + 8 \right) = -\frac{32}{3},$

and

$$\int_{-2}^{2} g(x)\, dx = \left[ 4x - \frac{x^3}{3} \right]_{-2}^{2} = \frac{32}{3}.$$

(b) In both cases, the area between the curve and the $x$-axis over $[-2, 2]$ is 32/3 square units. Although the definite integral of $f(x)$ is negative, the area is still positive. ■

To compute the area of the region bounded by the graph of a function $y = f(x)$ and the $x$-axis when the function takes on both positive and negative values, we must be careful to break up the interval $[a, b]$ into subintervals on which the function doesn't change sign. Otherwise we might get cancelation between positive and negative signed areas, leading to an incorrect total. The correct total area is obtained by adding the absolute value of the definite integral over each subinterval where $f(x)$ does not change sign. The term "area" will be taken to mean this *total area*.

**EXAMPLE 7**    Figure 5.21 shows the graph of the function $f(x) = \sin x$ between $x = 0$ and $x = 2\pi$. Compute

(a) the definite integral of $f(x)$ over $[0, 2\pi]$.

(b) the area between the graph of $f(x)$ and the $x$-axis over $[0, 2\pi]$.

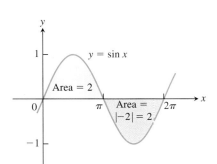

**FIGURE 5.21** The total area between $y = \sin x$ and the $x$-axis for $0 \leq x \leq 2\pi$ is the sum of the absolute values of two integrals (Example 7).

**Solution**

(a) The definite integral for $f(x) = \sin x$ is given by

$$\int_{0}^{2\pi} \sin x\, dx = -\cos x \Big]_{0}^{2\pi} = -[\cos 2\pi - \cos 0] = -[1 - 1] = 0.$$

The definite integral is zero because the portions of the graph above and below the $x$-axis make canceling contributions.

(b) The area between the graph of $f(x)$ and the $x$-axis over $[0, 2\pi]$ is calculated by breaking up the domain of $\sin x$ into two pieces: the interval $[0, \pi]$ over which it is nonnegative and the interval $[\pi, 2\pi]$ over which it is nonpositive.

$$\int_{0}^{\pi} \sin x\, dx = -\cos x \Big]_{0}^{\pi} = -[\cos \pi - \cos 0] = -[-1 - 1] = 2$$

$$\int_{\pi}^{2\pi} \sin x\, dx = -\cos x \Big]_{\pi}^{2\pi} = -[\cos 2\pi - \cos \pi] = -[1 - (-1)] = -2$$

The second integral gives a negative value. The area between the graph and the axis is obtained by adding the absolute values,

$$\text{Area} = |2| + |-2| = 4. \qquad ■$$

---

**Summary:**

To find the area between the graph of $y = f(x)$ and the $x$-axis over the interval $[a, b]$:

1. Subdivide $[a, b]$ at the zeros of $f$.

2. Integrate $f$ over each subinterval.

3. Add the absolute values of the integrals.

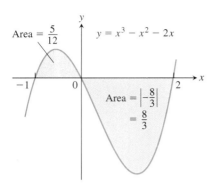

**FIGURE 5.22** The region between the curve $y = x^3 - x^2 - 2x$ and the $x$-axis (Example 8).

**EXAMPLE 8**    Find the area of the region between the $x$-axis and the graph of $f(x) = x^3 - x^2 - 2x$, $-1 \le x \le 2$.

**Solution**    First find the zeros of $f$. Since

$$f(x) = x^3 - x^2 - 2x = x(x^2 - x - 2) = x(x + 1)(x - 2),$$

the zeros are $x = 0, -1$, and 2 (Figure 5.22). The zeros subdivide $[-1, 2]$ into two subintervals: $[-1, 0]$, on which $f \ge 0$, and $[0, 2]$, on which $f \le 0$. We integrate $f$ over each subinterval and add the absolute values of the calculated integrals.

$$\int_{-1}^{0} (x^3 - x^2 - 2x)\,dx = \left[ \frac{x^4}{4} - \frac{x^3}{3} - x^2 \right]_{-1}^{0} = 0 - \left[ \frac{1}{4} + \frac{1}{3} - 1 \right] = \frac{5}{12}$$

$$\int_{0}^{2} (x^3 - x^2 - 2x)\,dx = \left[ \frac{x^4}{4} - \frac{x^3}{3} - x^2 \right]_{0}^{2} = \left[ 4 - \frac{8}{3} - 4 \right] - 0 = -\frac{8}{3}$$

The total enclosed area is obtained by adding the absolute values of the calculated integrals.

$$\text{Total enclosed area} = \frac{5}{12} + \left| -\frac{8}{3} \right| = \frac{37}{12} \qquad \blacksquare$$

## Exercises 5.4

**Evaluating Integrals**

Evaluate the integrals in Exercises 1–28.

**1.** $\displaystyle\int_{0}^{2} x(x - 3)\,dx$

**2.** $\displaystyle\int_{-1}^{1} (x^2 - 2x + 3)\,dx$

**3.** $\displaystyle\int_{-2}^{2} \frac{3}{(x + 3)^4}\,dx$

**4.** $\displaystyle\int_{-1}^{1} x^{299}\,dx$

**5.** $\displaystyle\int_{1}^{4} \left( 3x^2 - \frac{x^3}{4} \right)\,dx$

**6.** $\displaystyle\int_{-2}^{3} (x^3 - 2x + 3)\,dx$

**7.** $\displaystyle\int_{0}^{1} \left( x^2 + \sqrt{x} \right)\,dx$

**8.** $\displaystyle\int_{1}^{32} x^{-6/5}\,dx$

**9.** $\displaystyle\int_{0}^{\pi/3} 2 \sec^2 x\,dx$

**10.** $\displaystyle\int_{0}^{\pi} (1 + \cos x)\,dx$

**11.** $\displaystyle\int_{\pi/4}^{3\pi/4} \csc \theta \cot \theta\,d\theta$

**12.** $\displaystyle\int_{0}^{\pi/3} 4 \frac{\sin u}{\cos^2 u}\,du$

**13.** $\displaystyle\int_{\pi/2}^{0} \frac{1 + \cos 2t}{2}\,dt$

**14.** $\displaystyle\int_{-\pi/3}^{\pi/3} \sin^2 t\,dt$

**15.** $\displaystyle\int_{0}^{\pi/4} \tan^2 x\,dx$

**16.** $\displaystyle\int_{0}^{\pi/6} (\sec x + \tan x)^2\,dx$

**17.** $\displaystyle\int_{0}^{\pi/8} \sin 2x\,dx$

**18.** $\displaystyle\int_{-\pi/3}^{-\pi/4} \left( 4 \sec^2 t + \frac{\pi}{t^2} \right)\,dt$

**19.** $\displaystyle\int_{1}^{-1} (r + 1)^2\,dr$

**20.** $\displaystyle\int_{-\sqrt{3}}^{\sqrt{3}} (t + 1)(t^2 + 4)\,dt$

**21.** $\displaystyle\int_{\sqrt{2}}^{1} \left( \frac{u^7}{2} - \frac{1}{u^5} \right)\,du$

**22.** $\displaystyle\int_{-3}^{-1} \frac{y^5 - 2y}{y^3}\,dy$

**23.** $\displaystyle\int_{1}^{\sqrt{2}} \frac{s^2 + \sqrt{s}}{s^2}\,ds$

**24.** $\displaystyle\int_{1}^{8} \frac{(x^{1/3} + 1)(2 - x^{2/3})}{x^{1/3}}\,dx$

**25.** $\displaystyle\int_{\pi/2}^{\pi} \frac{\sin 2x}{2 \sin x}\,dx$

**26.** $\displaystyle\int_{0}^{\pi/3} (\cos x + \sec x)^2\,dx$

**27.** $\displaystyle\int_{-4}^{4} |x|\,dx$

**28.** $\displaystyle\int_{0}^{\pi} \frac{1}{2}(\cos x + |\cos x|)\,dx$

In Exercises 29–32, guess an antiderivative for the integrand function. Validate your guess by differentiation and then evaluate the given definite integral. (*Hint:* Keep in mind the Chain Rule in guessing an antiderivative. You will learn how to find such antiderivatives in the next section.)

**29.** $\displaystyle\int_{0}^{\sqrt{\pi/2}} x \cos x^2\,dx$

**30.** $\displaystyle\int_{1}^{\pi^2} \frac{\sin \sqrt{x}}{\sqrt{x}}\,dx$

**31.** $\displaystyle\int_{2}^{5} \frac{x\,dx}{\sqrt{1 + x^2}}$

**32.** $\displaystyle\int_{0}^{\pi/3} \sin^2 x \cos x\,dx$

**Derivatives of Integrals**

Find the derivatives in Exercises 33–38.

**a.** by evaluating the integral and differentiating the result.

**b.** by differentiating the integral directly.

**33.** $\dfrac{d}{dx}\displaystyle\int_{0}^{\sqrt{x}} \cos t\, dt$

**34.** $\dfrac{d}{dx}\displaystyle\int_{1}^{\sin x} 3t^2\, dt$

**35.** $\dfrac{d}{dt}\displaystyle\int_{0}^{t^4} \sqrt{u}\, du$

**36.** $\dfrac{d}{d\theta}\displaystyle\int_{0}^{\tan\theta} \sec^2 y\, dy$

**37.** $\dfrac{d}{dx}\displaystyle\int_{0}^{x^3} t^{-2/3}\, dt$

**38.** $\dfrac{d}{dt}\displaystyle\int_{1}^{\sqrt{t}} \left(x^4 + \dfrac{3}{x^3}\right) dx$

Find $dy/dx$ in Exercises 39–46.

**39.** $y = \displaystyle\int_{0}^{x} \sqrt{1 + t^2}\, dt$

**40.** $y = \displaystyle\int_{1}^{x} \dfrac{1}{t}\, dt, \quad x > 0$

**41.** $y = \displaystyle\int_{\sqrt{x}}^{0} \sin(t^2)\, dt$

**42.** $y = x \displaystyle\int_{2}^{x^2} \sin(t^3)\, dt$

**43.** $y = \displaystyle\int_{-1}^{x} \dfrac{t^2}{t^2 + 4}\, dt - \displaystyle\int_{3}^{x} \dfrac{t^2}{t^2 + 4}\, dt$

**44.** $y = \left(\displaystyle\int_{0}^{x} (t^3 + 1)^{10}\, dt\right)^3$

**45.** $y = \displaystyle\int_{0}^{\sin x} \dfrac{dt}{\sqrt{1 - t^2}}, \quad |x| < \dfrac{\pi}{2}$

**46.** $y = \displaystyle\int_{\tan x}^{0} \dfrac{dt}{1 + t^2}$

**Area**

In Exercises 47–50, find the total area between the region and the $x$-axis.

**47.** $y = -x^2 - 2x, \quad -3 \le x \le 2$

**48.** $y = 3x^2 - 3, \quad -2 \le x \le 2$

**49.** $y = x^3 - 3x^2 + 2x, \quad 0 \le x \le 2$

**50.** $y = x^{1/3} - x, \quad -1 \le x \le 8$

Find the areas of the shaded regions in Exercises 51–54.

**51.**

**52.**

**53.**

**54.**

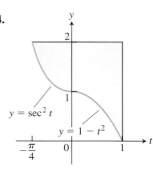

**Initial Value Problems**

Each of the following functions solves one of the initial value problems in Exercises 55–58. Which function solves which problem? Give brief reasons for your answers.

**a.** $y = \displaystyle\int_{1}^{x} \dfrac{1}{t}\, dt - 3$

**b.** $y = \displaystyle\int_{0}^{x} \sec t\, dt + 4$

**c.** $y = \displaystyle\int_{-1}^{x} \sec t\, dt + 4$

**d.** $y = \displaystyle\int_{\pi}^{x} \dfrac{1}{t}\, dt - 3$

**55.** $\dfrac{dy}{dx} = \dfrac{1}{x}, \quad y(\pi) = -3$

**56.** $y' = \sec x, \quad y(-1) = 4$

**57.** $y' = \sec x, \quad y(0) = 4$

**58.** $y' = \dfrac{1}{x}, \quad y(1) = -3$

Express the solutions of the initial value problems in Exercises 59 and 60 in terms of integrals.

**59.** $\dfrac{dy}{dx} = \sec x, \quad y(2) = 3$

**60.** $\dfrac{dy}{dx} = \sqrt{1 + x^2}, \quad y(1) = -2$

**Theory and Examples**

**61. Archimedes' area formula for parabolic arches** Archimedes (287–212 B.C.), inventor, military engineer, physicist, and the greatest mathematician of classical times in the Western world, discovered that the area under a parabolic arch is two-thirds the base times the height. Sketch the parabolic arch $y = h - (4h/b^2)x^2$, $-b/2 \le x \le b/2$, assuming that $h$ and $b$ are positive. Then use calculus to find the area of the region enclosed between the arch and the $x$-axis.

**62.** Show that if $k$ is a positive constant, then the area between the $x$-axis and one arch of the curve $y = \sin kx$ is $2/k$.

**63. Cost from marginal cost** The marginal cost of printing a poster when $x$ posters have been printed is

$$\dfrac{dc}{dx} = \dfrac{1}{2\sqrt{x}}$$

dollars. Find $c(100) - c(1)$, the cost of printing posters 2–100.

**64. Revenue from marginal revenue** Suppose that a company's marginal revenue from the manufacture and sale of eggbeaters is

$$\dfrac{dr}{dx} = 2 - 2/(x + 1)^2,$$

where $r$ is measured in thousands of dollars and $x$ in thousands of units. How much money should the company expect from a production run of $x = 3$ thousand eggbeaters? To find out, integrate the marginal revenue from $x = 0$ to $x = 3$.

**65.** The temperature $T(°C)$ of a room at time $t$ minutes is given by

$$T = 30 - 2\sqrt{25 - t} \quad \text{for} \quad 0 \le t \le 25.$$

**a.** Find the room's temperature when $t = 0$, $t = 16$, and $t = 25$.

**b.** Find the room's average temperature for $0 \le t \le 25$.

**66.** The height $H$(m) of a palm tree after growing for $t$ years is given by

$$H = \sqrt{t + 1} + 5t^{1/3} \quad \text{for} \quad 0 \le t \le 8.$$

**a.** Find the tree's height when $t = 0$, $t = 4$, and $t = 8$.

**b.** Find the tree's average height for $0 \le t \le 8$.

**67.** Suppose that $\int_1^x f(t)\, dt = x^2 - 2x + 1$. Find $f(x)$.

**68.** Find $f(4)$ if $\int_0^x f(t)\, dt = x \cos \pi x$.

**69.** Find the linearization of

$$f(x) = 2 - \int_2^{x+1} \frac{9}{1 + t}\, dt$$

at $x = 1$.

**70.** Find the linearization of

$$g(x) = 3 + \int_1^{x^2} \sec(t - 1)\, dt$$

at $x = -1$.

**71.** Suppose that $f$ has a positive derivative for all values of $x$ and that $f(1) = 0$. Which of the following statements must be true of the function

$$g(x) = \int_0^x f(t)\, dt?$$

Give reasons for your answers.

**a.** $g$ is a differentiable function of $x$.

**b.** $g$ is a continuous function of $x$.

**c.** The graph of $g$ has a horizontal tangent at $x = 1$.

**d.** $g$ has a local maximum at $x = 1$.

**e.** $g$ has a local minimum at $x = 1$.

**f.** The graph of $g$ has an inflection point at $x = 1$.

**g.** The graph of $dg/dx$ crosses the $x$-axis at $x = 1$.

**72. Another proof of the Evaluation Theorem**

**a.** Let $a = x_0 < x_1 < x_2 \cdots < x_n = b$ be any partition of $[a, b]$, and let $F$ be any antiderivative of $f$. Show that

$$F(b) - F(a) = \sum_{i=1}^{n} [F(x_i) - F(x_{i-1})].$$

**b.** Apply the Mean Value Theorem to each term to show that $F(x_i) - F(x_{i-1}) = f(c_i)(x_i - x_{i-1})$ for some $c_i$ in the interval $(x_{i-1}, x_i)$. Then show that $F(b) - F(a)$ is a Riemann sum for $f$ on $[a, b]$.

**c.** From part (b) and the definition of the definite integral, show that

$$F(b) - F(a) = \int_a^b f(x)\, dx.$$

**73.** Suppose that $f$ is the differentiable function shown in the accompanying graph and that the position at time $t$ (s) of a particle moving along a coordinate axis is

$$s = \int_0^t f(x)\, dx$$

meters. Use the graph to answer the following questions. Give reasons for your answers.

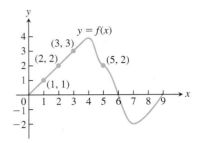

**a.** What is the particle's velocity at time $t = 5$?

**b.** Is the acceleration of the particle at time $t = 5$ positive, or negative?

**c.** What is the particle's position at time $t = 3$?

**d.** At what time during the first 9 s does $s$ have its largest value?

**e.** Approximately when is the acceleration zero?

**f.** When is the particle moving toward the origin? Away from the origin?

**g.** On which side of the origin does the particle lie at time $t = 9$?

**74.** The marginal cost of manufacturing $x$ units of an electronic device is $0.001x^2 - 0.5x + 115$. If 600 units are produced, what is the production cost per unit?

**COMPUTER EXPLORATIONS**

In Exercises 75–78, let $F(x) = \int_a^x f(t)\, dt$ for the specified function $f$ and interval $[a, b]$. Use a CAS to perform the following steps and answer the questions posed.

**a.** Plot the functions $f$ and $F$ together over $[a, b]$.

**b.** Solve the equation $F'(x) = 0$. What can you see to be true about the graphs of $f$ and $F$ at points where $F'(x) = 0$? Is your observation borne out by Part 1 of the Fundamental Theorem coupled with information provided by the first derivative? Explain your answer.

**c.** Over what intervals (approximately) is the function $F$ increasing and decreasing? What is true about $f$ over those intervals?

**d.** Calculate the derivative $f'$ and plot it together with $F$. What can you see to be true about the graph of $F$ at points where $f'(x) = 0$? Is your observation borne out by Part 1 of the Fundamental Theorem? Explain your answer.

**75.** $f(x) = x^3 - 4x^2 + 3x, \quad [0, 4]$

**76.** $f(x) = 2x^4 - 17x^3 + 46x^2 - 43x + 12, \quad \left[0, \frac{9}{2}\right]$

**77.** $f(x) = \sin 2x \cos \frac{x}{3}, \quad [0, 2\pi]$

**78.** $f(x) = x \cos \pi x, \quad [0, 2\pi]$

In Exercises 79–82, let $F(x) = \int_a^{u(x)} f(t)\, dt$ for the specified $a$, $u$, and $f$. Use a CAS to perform the following steps and answer the questions posed.

**a.** Find the domain of $F$.

**b.** Calculate $F'(x)$ and determine its zeros. For what points in its domain is $F$ increasing? Decreasing?

**c.** Calculate $F''(x)$ and determine its zero. Identify the local extrema and the points of inflection of $F$.

**d.** Using the information from parts (a)–(c), draw a rough hand-sketch of $y = F(x)$ over its domain. Then graph $F(x)$ on your CAS to support your sketch.

**79.** $a = 1$, $u(x) = x^2$, $f(x) = \sqrt{1 - x^2}$

**80.** $a = 0$, $u(x) = x^2$, $f(x) = \sqrt{1 - x^2}$

**81.** $a = 0$, $u(x) = 1 - x$, $f(x) = x^2 - 2x - 3$

**82.** $a = 0$, $u(x) = 1 - x^2$, $f(x) = x^2 - 2x - 3$

In Exercises 83 and 84, assume that $f$ is continuous and $u(x)$ is twice-differentiable.

**83.** Calculate $\dfrac{d}{dx}\displaystyle\int_a^{u(x)} f(t)\, dt$ and check your answer using a CAS.

**84.** Calculate $\dfrac{d^2}{dx^2}\displaystyle\int_a^{u(x)} f(t)\, dt$ and check your answer using a CAS.

# 5.5 Indefinite Integrals and the Substitution Method

The Fundamental Theorem of Calculus says that a definite integral of a continuous function can be computed directly if we can find an antiderivative of the function. In Section 4.8 we defined the **indefinite integral** of the function $f$ with respect to $x$ as the set of *all* antiderivatives of $f$, symbolized by $\int f(x)\, dx$. Since any two antiderivatives of $f$ differ by a constant, the indefinite integral $\int$ notation means that for any antiderivative $F$ of $f$,

$$\int f(x)\, dx = F(x) + C,$$

where $C$ is any arbitrary constant. The connection between antiderivatives and the definite integral stated in the Fundamental Theorem now explains this notation:

$$\int_a^b f(x)\, dx = F(b) - F(a) = \left[F(x) + C\right]_a^b = \left[\int f(x)\, dx\right]_a^b.$$

When finding the indefinite integral of a function $f$, remember that it always includes an arbitrary constant $C$.

We must distinguish carefully between definite and indefinite integrals. A definite integral $\int_a^b f(x)\, dx$ is a *number*. An indefinite integral $\int f(x)\, dx$ is a *function* plus an arbitrary constant $C$.

So far, we have only been able to find antiderivatives of functions that are clearly recognizable as derivatives. In this section we begin to develop more general techniques for finding antiderivatives of functions we can't easily recognize as a derivative.

## Substitution: Running the Chain Rule Backwards

If $u$ is a differentiable function of $x$ and $n$ is any number different from $-1$, the Chain Rule tells us that

$$\frac{d}{dx}\left(\frac{u^{n+1}}{n+1}\right) = u^n \frac{du}{dx}.$$

From another point of view, this same equation says that $u^{n+1}/(n+1)$ is one of the antiderivatives of the function $u^n(du/dx)$. Therefore,

$$\int u^n \frac{du}{dx}\, dx = \frac{u^{n+1}}{n+1} + C. \tag{1}$$

The integral in Equation (1) is equal to the simpler integral

$$\int u^n \, du = \frac{u^{n+1}}{n+1} + C,$$

which suggests that the simpler expression $du$ can be substituted for $(du/dx)\, dx$ when computing an integral. Leibniz, one of the founders of calculus, had the insight that indeed this substitution could be done, leading to the *substitution method* for computing integrals. As with differentials, when computing integrals we have

$$du = \frac{du}{dx} \, dx.$$

**EXAMPLE 1** Find the integral $\int (x^3 + x)^5 (3x^2 + 1) \, dx.$

**Solution** We set $u = x^3 + x$. Then

$$du = \frac{du}{dx} \, dx = (3x^2 + 1) \, dx,$$

so that by substitution we have

$$\int (x^3 + x)^5 (3x^2 + 1) \, dx = \int u^5 \, du \qquad \text{Let } u = x^3 + x, \, du = (3x^2 + 1) \, dx.$$

$$= \frac{u^6}{6} + C \qquad \text{Integrate with respect to } u.$$

$$= \frac{(x^3 + x)^6}{6} + C \qquad \text{Substitute } x^3 + x \text{ for } u. \qquad ■$$

**EXAMPLE 2** Find $\int \sqrt{2x + 1} \, dx.$

**Solution** The integral does not fit the formula

$$\int u^n \, du,$$

with $u = 2x + 1$ and $n = 1/2$, because

$$du = \frac{du}{dx} \, dx = 2 \, dx$$

is not precisely $dx$. The constant factor 2 is missing from the integral. However, we can introduce this factor after the integral sign if we compensate for it by a factor of $1/2$ in front of the integral sign. So we write

$$\int \sqrt{2x + 1} \, dx = \frac{1}{2} \int \underbrace{\sqrt{2x + 1}}_{u} \cdot \underbrace{2 \, dx}_{du}$$

$$= \frac{1}{2} \int u^{1/2} \, du \qquad \text{Let } u = 2x + 1, \, du = 2 \, dx.$$

$$= \frac{1}{2} \frac{u^{3/2}}{3/2} + C \qquad \text{Integrate with respect to } u.$$

$$= \frac{1}{3} (2x + 1)^{3/2} + C \qquad \text{Substitute } 2x + 1 \text{ for } u. \qquad ■$$

The substitutions in Examples 1 and 2 are instances of the following general rule.

> **THEOREM 6—The Substitution Rule**   If $u = g(x)$ is a differentiable function whose range is an interval $I$, and $f$ is continuous on $I$, then
>
> $$\int f(g(x))g'(x)\,dx = \int f(u)\,du.$$

**Proof**    By the Chain Rule, $F(g(x))$ is an antiderivative of $f(g(x)) \cdot g'(x)$ whenever $F$ is an antiderivative of $f$:

$$\frac{d}{dx}F(g(x)) = F'(g(x)) \cdot g'(x) \qquad \text{Chain Rule}$$

$$= f(g(x)) \cdot g'(x). \qquad F' = f$$

If we make the substitution $u = g(x)$, then

$$\int f(g(x))g'(x)\,dx = \int \frac{d}{dx}F(g(x))\,dx$$

$$= F(g(x)) + C \qquad \text{Theorem 8 in Chapter 4}$$

$$= F(u) + C \qquad u = g(x)$$

$$= \int F'(u)\,du \qquad \text{Theorem 8 in Chapter 4}$$

$$= \int f(u)\,du. \qquad F' = f \qquad \blacksquare$$

The use of the variable $u$ in the Substitution Rule is traditional (sometimes it is referred to as $u$-substitution), but any letter can be used, such as $v$, $t$, $\theta$ and so forth. The rule provides a method for evaluating an integral of the form $\int f(g(x))g'(x)\,dx$ given that the conditions of Theorem 6 are satisfied. The primary challenge is deciding what expression involving $x$ you want to substitute for in the integrand. Our examples to follow give helpful ideas.

> **The Substitution Method to evaluate** $\int f(g(x))g'(x)\,dx$
>
> **1.** Substitute $u = g(x)$ and $du = (du/dx)\,dx = g'(x)\,dx$ to obtain $\int f(u)\,du$.
> **2.** Integrate with respect to $u$.
> **3.** Replace $u$ by $g(x)$.

**EXAMPLE 3**    Find $\displaystyle\int \sec^2(5x + 1) \cdot 5\,dx$

**Solution**    We substitute $u = 5x + 1$ and $du = 5\,dx$. Then,

$$\int \sec^2(5x + 1) \cdot 5\,dx = \int \sec^2 u\,du \qquad \text{Let } u = 5x + 1,\, du = 5\,dx.$$

$$= \tan u + C \qquad \frac{d}{du}\tan u = \sec^2 u$$

$$= \tan(5x + 1) + C. \qquad \text{Substitute } 5x + 1 \text{ for } u. \qquad \blacksquare$$

**EXAMPLE 4**    Find $\int \cos(7\theta + 3)\, d\theta$.

**Solution**    We let $u = 7\theta + 3$ so that $du = 7\, d\theta$. The constant factor 7 is missing from the $d\theta$ term in the integral. We can compensate for it by multiplying and dividing by 7, using the same procedure as in Example 2. Then,

$$\int \cos(7\theta + 3)\, d\theta = \frac{1}{7} \int \cos(7\theta + 3) \cdot 7\, d\theta \qquad \text{Place factor } 1/7 \text{ in front of integral.}$$

$$= \frac{1}{7} \int \cos u\, du \qquad \text{Let } u = 7\theta + 3, du = 7\, d\theta.$$

$$= \frac{1}{7} \sin u + C \qquad \text{Integrate.}$$

$$= \frac{1}{7} \sin(7\theta + 3) + C. \qquad \text{Substitute } 7\theta + 3 \text{ for } u.$$

There is another approach to this problem. With $u = 7\theta + 3$ and $du = 7\, d\theta$ as before, we solve for $d\theta$ to obtain $d\theta = (1/7)\, du$. Then the integral becomes

$$\int \cos(7\theta + 3)\, d\theta = \int \cos u \cdot \frac{1}{7}\, du \qquad \text{Let } u = 7\theta + 3, du = 7\, d\theta, \text{ and } d\theta = (1/7)\, du.$$

$$= \frac{1}{7} \sin u + C \qquad \text{Integrate.}$$

$$= \frac{1}{7} \sin(7\theta + 3) + C. \qquad \text{Substitute } 7\theta + 3 \text{ for } u.$$

We can verify this solution by differentiating and checking that we obtain the original function $\cos(7\theta + 3)$.                                                           ∎

**EXAMPLE 5**    Sometimes we observe that a power of $x$ appears in the integrand that is one less than the power of $x$ appearing in the argument of a function we want to integrate. This observation immediately suggests we try a substitution for the higher power of $x$. This situation occurs in the following integration.

$$\int x^2 \cos x^3\, dx = \int \cos x^3 \cdot x^2\, dx$$

$$= \int \cos u \cdot \frac{1}{3}\, du \qquad \begin{array}{l} \text{Let } u = x^3, du = 3x^2\, dx, \\ (1/3)\, du = x^2\, dx. \end{array}$$

$$= \frac{1}{3} \int \cos u\, du$$

$$= \frac{1}{3} \sin u + C \qquad \text{Integrate with respect to } u.$$

$$= \frac{1}{3} \sin x^3 + C \qquad \text{Replace } u \text{ by } x^3. \qquad ∎$$

It may happen that an extra factor of $x$ appears in the integrand when we try a substitution $u = g(x)$. In that case, it may be possible to solve the equation $u = g(x)$ for $x$ in terms of $u$. Replacing the extra factor of $x$ with that expression may then allow for an integral we can evaluate. Here's an example of this situation.

**EXAMPLE 6**    Evaluate $\int x\sqrt{2x+1}\,dx$.

**Solution**    Our previous integration in Example 2 suggests the substitution $u = 2x + 1$ with $du = 2\,dx$. Then,

$$\sqrt{2x+1}\,dx = \frac{1}{2}\sqrt{u}\,du.$$

However, in this case the integrand contains an extra factor of $x$ multiplying the term $\sqrt{2x+1}$. To adjust for this, we solve the substitution equation $u = 2x + 1$ to obtain $x = (u-1)/2$, and find that

$$x\sqrt{2x+1}\,dx = \frac{1}{2}(u-1)\cdot\frac{1}{2}\sqrt{u}\,du.$$

The integration now becomes

$$\int x\sqrt{2x+1}\,dx = \frac{1}{4}\int(u-1)\sqrt{u}\,du = \frac{1}{4}\int(u-1)u^{1/2}\,du \qquad \text{Substitute.}$$

$$= \frac{1}{4}\int(u^{3/2} - u^{1/2})\,du \qquad \text{Multiply terms.}$$

$$= \frac{1}{4}\left(\frac{2}{5}u^{5/2} - \frac{2}{3}u^{3/2}\right) + C \qquad \text{Integrate.}$$

$$= \frac{1}{10}(2x+1)^{5/2} - \frac{1}{6}(2x+1)^{3/2} + C. \qquad \text{Replace } u \text{ by } 2x + 1. \ \blacksquare$$

**EXAMPLE 7**    Sometimes we can use trigonometric identities to transform integrals we do not know how to evaluate into ones we can evaluate using the Substitution Rule.

**(a)**  $\displaystyle\int \sin^2 x\,dx = \int \frac{1 - \cos 2x}{2}\,dx$ $\qquad\qquad\qquad \sin^2 x = \frac{1 - \cos 2x}{2}$

$$= \frac{1}{2}\int(1 - \cos 2x)\,dx$$

$$= \frac{1}{2}x - \frac{1}{2}\frac{\sin 2x}{2} + C = \frac{x}{2} - \frac{\sin 2x}{4} + C$$

**(b)**  $\displaystyle\int \cos^2 x\,dx = \int \frac{1 + \cos 2x}{2}\,dx = \frac{x}{2} + \frac{\sin 2x}{4} + C$ $\qquad \cos^2 x = \frac{1 + \cos 2x}{2}$

**(c)**  $\displaystyle\int (1 - 2\sin^2 x)\sin 2x\,dx = \int (\cos^2 x - \sin^2 x)\sin 2x\,dx$

$$= \int \cos 2x \sin 2x\,dx \qquad \cos 2x = \cos^2 x - \sin^2 x$$

$$= \int \frac{1}{2}\sin 4x\,dx = \int \frac{1}{8}\sin u\,du \qquad u = 4x,\ du = 4\,dx$$

$$= -\cos 4x + C. \qquad\qquad\qquad\qquad \blacksquare$$

### Trying Different Substitutions

The success of the substitution method depends on finding a substitution that changes an integral we cannot evaluate directly into one that we can. Finding the right substitution gets easier with practice and experience. If the first substitution fails, try another substitution, possibly coupled with other algebraic or trigonometric simplifications to the integrand. Several of these more complicated substitutions are studied in Chapter 8.

**EXAMPLE 8**    Evaluate $\displaystyle\int \frac{2z\,dz}{\sqrt[3]{z^2 + 1}}$.

**Solution**    We can use the substitution method of integration as an exploratory tool: Substitute for the most troublesome part of the integrand and see how things work out. For the integral here, we might try $u = z^2 + 1$ or we might even press our luck and take $u$ to be the entire cube root. Here is what happens in each case, and both substitutions are successful.

**Method 1:** Substitute $u = z^2 + 1$.

$$\int \frac{2z\,dz}{\sqrt[3]{z^2 + 1}} = \int \frac{du}{u^{1/3}} \qquad \text{Let } u = z^2 + 1,\ du = 2z\,dz.$$

$$= \int u^{-1/3}\,du \qquad \text{In the form } \int u^n\,du$$

$$= \frac{u^{2/3}}{2/3} + C \qquad \text{Integrate.}$$

$$= \frac{3}{2}u^{2/3} + C$$

$$= \frac{3}{2}(z^2 + 1)^{2/3} + C \qquad \text{Replace } u \text{ by } z^2 + 1.$$

**Method 2:** Substitute $u = \sqrt[3]{z^2 + 1}$ instead.

$$\int \frac{2z\,dz}{\sqrt[3]{z^2 + 1}} = \int \frac{3u^2\,du}{u} \qquad \text{Let } u = \sqrt[3]{z^2 + 1},\ u^3 = z^2 + 1,\ 3u^2\,du = 2z\,dz.$$

$$= 3\int u\,du$$

$$= 3 \cdot \frac{u^2}{2} + C \qquad \text{Integrate.}$$

$$= \frac{3}{2}(z^2 + 1)^{2/3} + C \qquad \text{Replace } u \text{ by } (z^2 + 1)^{1/3}. \qquad \blacksquare$$

## Exercises 5.5

### Evaluating Indefinite Integrals

Evaluate the indefinite integrals in Exercises 1–16 by using the given substitutions to reduce the integrals to standard form.

**1.** $\displaystyle\int 2(2x + 4)^5\,dx,\quad u = 2x + 4$

**2.** $\displaystyle\int 7\sqrt{7x - 1}\,dx,\quad u = 7x - 1$

**3.** $\displaystyle\int 2x(x^2 + 5)^{-4}\,dx,\quad u = x^2 + 5$

**4.** $\displaystyle\int \frac{4x^3}{(x^4 + 1)^2}\, dx, \quad u = x^4 + 1$

**5.** $\displaystyle\int (3x + 2)(3x^2 + 4x)^4\, dx, \quad u = 3x^2 + 4x$

**6.** $\displaystyle\int \frac{\left(1 + \sqrt{x}\right)^{1/3}}{\sqrt{x}}\, dx, \quad u = 1 + \sqrt{x}$

**7.** $\displaystyle\int \sin 3x\, dx, \quad u = 3x$

**8.** $\displaystyle\int x \sin (2x^2)\, dx, \quad u = 2x^2$

**9.** $\displaystyle\int \sec 2t \tan 2t\, dt, \quad u = 2t$

**10.** $\displaystyle\int \left(1 - \cos \frac{t}{2}\right)^2 \sin \frac{t}{2}\, dt, \quad u = 1 - \cos \frac{t}{2}$

**11.** $\displaystyle\int \frac{9r^2\, dr}{\sqrt{1 - r^3}}, \quad u = 1 - r^3$

**12.** $\displaystyle\int 12(y^4 + 4y^2 + 1)^2(y^3 + 2y)\, dy, \quad u = y^4 + 4y^2 + 1$

**13.** $\displaystyle\int \sqrt{x} \sin^2 (x^{3/2} - 1)\, dx, \quad u = x^{3/2} - 1$

**14.** $\displaystyle\int \frac{1}{x^2} \cos^2 \left(\frac{1}{x}\right) dx, \quad u = -\frac{1}{x}$

**15.** $\displaystyle\int \csc^2 2\theta \cot 2\theta\, d\theta$

    **a.** Using $u = \cot 2\theta$     **b.** Using $u = \csc 2\theta$

**16.** $\displaystyle\int \frac{dx}{\sqrt{5x + 8}}$

    **a.** Using $u = 5x + 8$     **b.** Using $u = \sqrt{5x + 8}$

Evaluate the integrals in Exercises 17–50.

**17.** $\displaystyle\int \sqrt{3 - 2s}\, ds$

**18.** $\displaystyle\int \frac{1}{\sqrt{5s + 4}}\, ds$

**19.** $\displaystyle\int \theta \sqrt[4]{1 - \theta^2}\, d\theta$

**20.** $\displaystyle\int 3y\sqrt{7 - 3y^2}\, dy$

**21.** $\displaystyle\int \frac{1}{\sqrt{x}\left(1 + \sqrt{x}\right)^2}\, dx$

**22.** $\displaystyle\int \sqrt{\sin x}\, \cos^3 x\, dx$

**23.** $\displaystyle\int \sec^2 (3x + 2)\, dx$

**24.** $\displaystyle\int \tan^2 x \sec^2 x\, dx$

**25.** $\displaystyle\int \sin^5 \frac{x}{3} \cos \frac{x}{3}\, dx$

**26.** $\displaystyle\int \tan^7 \frac{x}{2} \sec^2 \frac{x}{2}\, dx$

**27.** $\displaystyle\int r^2 \left(\frac{r^3}{18} - 1\right)^5 dr$

**28.** $\displaystyle\int r^4 \left(7 - \frac{r^5}{10}\right)^3 dr$

**29.** $\displaystyle\int x^{1/2} \sin (x^{3/2} + 1)\, dx$

**30.** $\displaystyle\int \csc \left(\frac{v - \pi}{2}\right) \cot \left(\frac{v - \pi}{2}\right) dv$

**31.** $\displaystyle\int \frac{\sin (2t + 1)}{\cos^2 (2t + 1)}\, dt$

**32.** $\displaystyle\int \frac{\sec z \tan z}{\sqrt{\sec z}}\, dz$

**33.** $\displaystyle\int \frac{1}{t^2} \cos \left(\frac{1}{t} - 1\right) dt$

**34.** $\displaystyle\int \frac{1}{\sqrt{t}} \cos \left(\sqrt{t} + 3\right) dt$

**35.** $\displaystyle\int \frac{1}{\theta^2} \sin \frac{1}{\theta} \cos \frac{1}{\theta}\, d\theta$

**36.** $\displaystyle\int \frac{\cos \sqrt{\theta}}{\sqrt{\theta} \sin^2 \sqrt{\theta}}\, d\theta$

**37.** $\displaystyle\int \frac{x}{\sqrt{1 + x}}\, dx$

**38.** $\displaystyle\int \sqrt{\frac{x - 1}{x^5}}\, dx$

**39.** $\displaystyle\int \frac{1}{x^2} \sqrt{2 - \frac{1}{x}}\, dx$

**40.** $\displaystyle\int \frac{1}{x^3} \sqrt{\frac{x^2 - 1}{x^2}}\, dx$

**41.** $\displaystyle\int \sqrt{\frac{x^3 - 3}{x^{11}}}\, dx$

**42.** $\displaystyle\int \sqrt{\frac{x^4}{x^3 - 1}}\, dx$

**43.** $\displaystyle\int x(x - 1)^{10}\, dx$

**44.** $\displaystyle\int x\sqrt{4 - x}\, dx$

**45.** $\displaystyle\int (x + 1)^2(1 - x)^5\, dx$

**46.** $\displaystyle\int (x + 5)(x - 5)^{1/3}\, dx$

**47.** $\displaystyle\int x^3\sqrt{x^2 + 1}\, dx$

**48.** $\displaystyle\int 3x^5\sqrt{x^3 + 1}\, dx$

**49.** $\displaystyle\int \frac{x}{(x^2 - 4)^3}\, dx$

**50.** $\displaystyle\int \frac{x}{(2x - 1)^{2/3}}\, dx$

If you do not know what substitution to make, try reducing the integral step by step, using a trial substitution to simplify the integral a bit and then another to simplify it some more. You will see what we mean if you try the sequences of substitutions in Exercises 51 and 52.

**51.** $\displaystyle\int \frac{18 \tan^2 x \sec^2 x}{(2 + \tan^3 x)^2}\, dx$

    **a.** $u = \tan x$, followed by $v = u^3$, then by $w = 2 + v$

    **b.** $u = \tan^3 x$, followed by $v = 2 + u$

    **c.** $u = 2 + \tan^3 x$

**52.** $\displaystyle\int \sqrt{1 + \sin^2 (x - 1)}\, \sin (x - 1) \cos (x - 1)\, dx$

    **a.** $u = x - 1$, followed by $v = \sin u$, then by $w = 1 + v^2$

    **b.** $u = \sin (x - 1)$, followed by $v = 1 + u^2$

    **c.** $u = 1 + \sin^2 (x - 1)$

Evaluate the integrals in Exercises 53 and 54.

**53.** $\displaystyle\int \frac{(2r - 1) \cos \sqrt{3(2r - 1)^2 + 6}}{\sqrt{3(2r - 1)^2 + 6}}\, dr$

**54.** $\displaystyle\int \frac{\sin \sqrt{\theta}}{\sqrt{\theta} \cos^3 \sqrt{\theta}}\, d\theta$

**Initial Value Problems**

Solve the initial value problems in Exercises 55–60.

**55.** $\displaystyle\frac{ds}{dt} = 12t\, (3t^2 - 1)^3, \quad s(1) = 3$

**56.** $\displaystyle\frac{dy}{dx} = 4x\, (x^2 + 8)^{-1/3}, \quad y(0) = 0$

**57.** $\displaystyle\frac{ds}{dt} = 8 \sin^2 \left(t + \frac{\pi}{12}\right), \quad s(0) = 8$

**58.** $\dfrac{dr}{d\theta} = 3\cos^2\left(\dfrac{\pi}{4} - \theta\right), \quad r(0) = \dfrac{\pi}{8}$

**59.** $\dfrac{d^2s}{dt^2} = -4\sin\left(2t - \dfrac{\pi}{2}\right), \quad s'(0) = 100, \quad s(0) = 0$

**60.** $\dfrac{d^2y}{dx^2} = 4\sec^2 2x \tan 2x, \quad y'(0) = 4, \quad y(0) = -1$

**61.** The velocity of a particle moving back and forth on a line is $v = ds/dt = 6\sin 2t$ m/s for all $t$. If $s = 0$ when $t = 0$, find the value of $s$ when $t = \pi/2$ s.

**62.** The acceleration of a particle moving back and forth on a line is $a = d^2s/dt^2 = \pi^2 \cos \pi t$ m/s$^2$ for all $t$. If $s = 0$ and $v = 8$ m/s when $t = 0$, find $s$ when $t = 1$ s.

# $5.6$ Definite Integral Substitutions and the Area Between Curves

There are two methods for evaluating a definite integral by substitution. One method is to find an antiderivative using substitution and then to evaluate the definite integral by applying the Evaluation Theorem. The other method extends the process of substitution directly to *definite* integrals by changing the limits of integration. We apply the new formula introduced here to the problem of computing the area between two curves.

## The Substitution Formula

The following formula shows how the limits of integration change when the variable of integration is changed by substitution.

---

**THEOREM 7—Substitution in Definite Integrals**   If $g'$ is continuous on the interval $[a, b]$ and $f$ is continuous on the range of $g(x) = u$, then

$$\int_a^b f(g(x)) \cdot g'(x)\, dx = \int_{g(a)}^{g(b)} f(u)\, du.$$

---

**Proof**   Let $F$ denote any antiderivative of $f$. Then,

$$\int_a^b f(g(x)) \cdot g'(x)\, dx = F(g(x)) \Big]_{x=a}^{x=b} \qquad \begin{aligned} &\frac{d}{dx} F(g(x)) \\ &= F'(g(x))g'(x) \\ &= f(g(x))g'(x) \end{aligned}$$

$$= F(g(b)) - F(g(a))$$

$$= F(u) \Big]_{u=g(a)}^{u=g(b)}$$

$$= \int_{g(a)}^{g(b)} f(u)\, du. \qquad \begin{aligned} &\text{Fundamental} \\ &\text{Theorem, Part 2} \end{aligned} \qquad \blacksquare$$

To use the formula, make the same $u$-substitution $u = g(x)$ and $du = g'(x)\, dx$ you would use to evaluate the corresponding indefinite integral. Then integrate the transformed integral with respect to $u$ from the value $g(a)$ (the value of $u$ at $x = a$) to the value $g(b)$ (the value of $u$ at $x = b$).

**EXAMPLE 1**   Evaluate $\displaystyle\int_{-1}^{1} 3x^2\sqrt{x^3 + 1}\, dx.$

**Solution**    We have two choices.

**Method 1:** Transform the integral and evaluate the transformed integral with the transformed limits given in Theorem 7.

$$\int_{-1}^{1} 3x^2 \sqrt{x^3 + 1}\, dx$$

Let $u = x^3 + 1$, $du = 3x^2\, dx$.
When $x = -1$, $u = (-1)^3 + 1 = 0$.
When $x = 1$, $u = (1)^3 + 1 = 2$.

$$= \int_{0}^{2} \sqrt{u}\, du$$

$$= \frac{2}{3} u^{3/2} \bigg]_{0}^{2}$$   Evaluate the new definite integral.

$$= \frac{2}{3}\left[ 2^{3/2} - 0^{3/2} \right] = \frac{2}{3}\left[ 2\sqrt{2} \right] = \frac{4\sqrt{2}}{3}$$

**Method 2:** Transform the integral as an indefinite integral, integrate, change back to $x$, and use the original $x$-limits.

$$\int 3x^2 \sqrt{x^3 + 1}\, dx = \int \sqrt{u}\, du$$   Let $u = x^3 + 1$, $du = 3x^2\, dx$.

$$= \frac{2}{3} u^{3/2} + C$$   Integrate with respect to $u$.

$$= \frac{2}{3}(x^3 + 1)^{3/2} + C$$   Replace $u$ by $x^3 + 1$.

$$\int_{-1}^{1} 3x^2 \sqrt{x^3 + 1}\, dx = \frac{2}{3}(x^3 + 1)^{3/2} \bigg]_{-1}^{1}$$   Use the integral just found, with limits of integration for $x$.

$$= \frac{2}{3}\left[ ((1)^3 + 1)^{3/2} - ((-1)^3 + 1)^{3/2} \right]$$

$$= \frac{2}{3}\left[ 2^{3/2} - 0^{3/2} \right] = \frac{2}{3}\left[ 2\sqrt{2} \right] = \frac{4\sqrt{2}}{3}$$   ∎

Which method is better—evaluating the transformed definite integral with transformed limits using Theorem 7, or transforming the integral, integrating, and transforming back to use the original limits of integration? In Example 1, the first method seems easier, but that is not always the case. Generally, it is best to know both methods and to use whichever one seems better at the time.

**EXAMPLE 2**    We use the method of transforming the limits of integration.

**(a)** $\displaystyle \int_{\pi/4}^{\pi/2} \cot\theta \csc^2\theta\, d\theta = \int_{1}^{0} u \cdot (-du)$

Let $u = \cot\theta$, $du = -\csc^2\theta\, d\theta$,
    $-du = \csc^2\theta\, d\theta$.
When $\theta = \pi/4$, $u = \cot(\pi/4) = 1$.
When $\theta = \pi/2$, $u = \cot(\pi/2) = 0$.

$$= -\int_{1}^{0} u\, du$$

$$= -\left[ \frac{u^2}{2} \right]_{1}^{0}$$

$$= -\left[ \frac{(0)^2}{2} - \frac{(1)^2}{2} \right] = \frac{1}{2}$$

**(b)** $\displaystyle\int_{-\pi/4}^{\pi/4} \tan x \, dx = \int_{-\pi/4}^{\pi/4} \frac{\sin x}{\cos x} \, dx$

$$= -\int_{\sqrt{2}/2}^{\sqrt{2}/2} \frac{du}{u}$$

Let $u = \cos x$, $du = -\sin x \, dx$.
When $x = -\pi/4$, $u = \sqrt{2}/2$.
When $x = \pi/4$, $u = \sqrt{2}/2$.

$$= 0$$

Zero width interval

## Definite Integrals of Symmetric Functions

The Substitution Formula in Theorem 7 simplifies the calculation of definite integrals of even and odd functions (Section 1.1) over a symmetric interval $[-a, a]$ (Figure 5.23).

---

**THEOREM 8** Let $f$ be continuous on the symmetric interval $[-a, a]$.

**(a)** If $f$ is even, then $\displaystyle\int_{-a}^{a} f(x) \, dx = 2 \int_{0}^{a} f(x) \, dx.$

**(b)** If $f$ is odd, then $\displaystyle\int_{-a}^{a} f(x) \, dx = 0.$

---

(a)

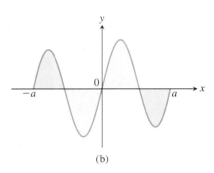

(b)

**FIGURE 5.23** (a) For $f$ an even function, the integral from $-a$ to $a$ is twice the integral from 0 to a. (b) For $f$ an odd function, the integral from $-a$ to $a$ equals 0.

### Proof of Part (a)

$$\int_{-a}^{a} f(x) \, dx = \int_{-a}^{0} f(x) \, dx + \int_{0}^{a} f(x) \, dx$$

Additivity Rule for Definite Integrals

$$= -\int_{0}^{-a} f(x) \, dx + \int_{0}^{a} f(x) \, dx$$

Order of Integration Rule

$$= -\int_{0}^{a} f(-u)(-du) + \int_{0}^{a} f(x) \, dx$$

Let $u = -x$, $du = -dx$.
When $x = 0$, $u = 0$.
When $x = -a$, $u = a$.

$$= \int_{0}^{a} f(-u) \, du + \int_{0}^{a} f(x) \, dx$$

$$= \int_{0}^{a} f(u) \, du + \int_{0}^{a} f(x) \, dx$$

$f$ is even, so
$f(-u) = f(u)$.

$$= 2\int_{0}^{a} f(x) \, dx$$

The proof of part (b) is entirely similar and you are asked to give it in Exercise 114. ∎

The assertions of Theorem 8 remain true when $f$ is an integrable function (rather than having the stronger property of being continuous).

**EXAMPLE 3** Evaluate $\displaystyle\int_{-2}^{2} (x^4 - 4x^2 + 6) \, dx.$

**Solution** Since $f(x) = x^4 - 4x^2 + 6$ satisfies $f(-x) = f(x)$, it is even on the symmetric interval $[-2, 2]$, so

$$\int_{-2}^{2}(x^4 - 4x^2 + 6)\,dx = 2\int_{0}^{2}(x^4 - 4x^2 + 6)\,dx$$

$$= 2\left[\frac{x^5}{5} - \frac{4}{3}x^3 + 6x\right]_{0}^{2}$$

$$= 2\left(\frac{32}{5} - \frac{32}{3} + 12\right) = \frac{232}{15}.\qquad\blacksquare$$

## Areas Between Curves

Suppose we want to find the area of a region that is bounded above by the curve $y = f(x)$, below by the curve $y = g(x)$, and on the left and right by the lines $x = a$ and $x = b$ (Figure 5.24). The region might accidentally have a shape whose area we could find with geometry, but if $f$ and $g$ are arbitrary continuous functions, we usually have to find the area with an integral.

To see what the integral should be, we first approximate the region with $n$ vertical rectangles based on a partition $P = \{x_0, x_1, \ldots, x_n\}$ of $[a, b]$ (Figure 5.25). The area of the $k$th rectangle (Figure 5.26) is

$$\Delta A_k = \text{height} \times \text{width} = [f(c_k) - g(c_k)]\,\Delta x_k.$$

We then approximate the area of the region by adding the areas of the $n$ rectangles:

$$A \approx \sum_{k=1}^{n}\Delta A_k = \sum_{k=1}^{n}[f(c_k) - g(c_k)]\,\Delta x_k. \qquad \text{Riemann sum}$$

As $\|P\| \to 0$, the sums on the right approach the limit $\int_{a}^{b}[f(x) - g(x)]\,dx$ because $f$ and $g$ are continuous. We take the area of the region to be the value of this integral. That is,

$$A = \lim_{\|P\|\to 0}\sum_{k=1}^{n}[f(c_k) - g(c_k)]\,\Delta x_k = \int_{a}^{b}[f(x) - g(x)]\,dx.$$

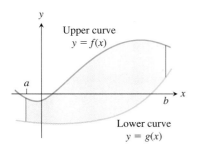

**FIGURE 5.24** The region between the curves $y = f(x)$ and $y = g(x)$ and the lines $x = a$ and $x = b$.

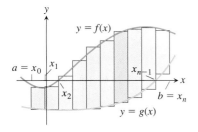

**FIGURE 5.25** We approximate the region with rectangles perpendicular to the $x$-axis.

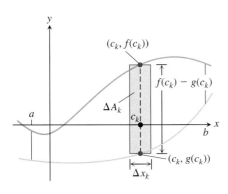

**FIGURE 5.26** The area $\Delta A_k$ of the $k$th rectangle is the product of its height, $f(c_k) - g(c_k)$, and its width, $\Delta x_k$.

---

**DEFINITION** If $f$ and $g$ are continuous with $f(x) \geq g(x)$ throughout $[a, b]$, then the **area of the region between the curves $y = f(x)$ and $y = g(x)$ from $a$ to $b$** is the integral of $(f - g)$ from $a$ to $b$:

$$A = \int_{a}^{b}[f(x) - g(x)]\,dx.$$

---

When applying this definition it is helpful to graph the curves. The graph reveals which curve is the upper curve $f$ and which is the lower curve $g$. It also helps you find the limits of integration if they are not given. You may need to find where the curves intersect to determine the limits of integration, and this may involve solving the equation $f(x) = g(x)$ for values of $x$. Then you can integrate the function $f - g$ for the area between the intersections.

**EXAMPLE 4** Find the area of the region enclosed by the parabola $y = 2 - x^2$ and the line $y = -x$.

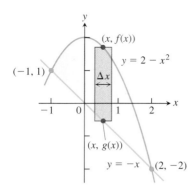

**FIGURE 5.27** The region in Example 4 with a typical approximating rectangle.

**Solution** First we sketch the two curves (Figure 5.27). The limits of integration are found by solving $y = 2 - x^2$ and $y = -x$ simultaneously for $x$.

$$2 - x^2 = -x \qquad \text{Equate } f(x) \text{ and } g(x).$$

$$x^2 - x - 2 = 0 \qquad \text{Rewrite.}$$

$$(x + 1)(x - 2) = 0 \qquad \text{Factor.}$$

$$x = -1, \qquad x = 2. \qquad \text{Solve.}$$

The region runs from $x = -1$ to $x = 2$. The limits of integration are $a = -1$, $b = 2$. The area between the curves is

$$A = \int_a^b \left[ f(x) - g(x) \right] dx = \int_{-1}^2 \left[ (2 - x^2) - (-x) \right] dx$$

$$= \int_{-1}^2 (2 + x - x^2) \, dx = \left[ 2x + \frac{x^2}{2} - \frac{x^3}{3} \right]_{-1}^2$$

$$= \left( 4 + \frac{4}{2} - \frac{8}{3} \right) - \left( -2 + \frac{1}{2} + \frac{1}{3} \right) = \frac{9}{2}. \qquad \blacksquare$$

If the formula for a bounding curve changes at one or more points, we subdivide the region into subregions that correspond to the formula changes and apply the formula for the area between curves to each subregion.

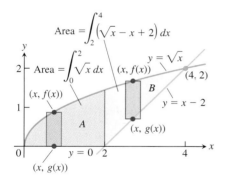

**FIGURE 5.28** When the formula for a bounding curve changes, the area integral changes to become the sum of integrals to match, one integral for each of the shaded regions shown here for Example 6.

**EXAMPLE 5** Find the area of the region in the first quadrant that is bounded above by $y = \sqrt{x}$ and below by the $x$-axis and the line $y = x - 2$.

**Solution** The sketch (Figure 5.28) shows that the region's upper boundary is the graph of $f(x) = \sqrt{x}$. The lower boundary changes from $g(x) = 0$ for $0 \le x \le 2$ to $g(x) = x - 2$ for $2 \le x \le 4$ (both formulas agree at $x = 2$). We subdivide the region at $x = 2$ into subregions $A$ and $B$, shown in Figure 5.28.

The limits of integration for region $A$ are $a = 0$ and $b = 2$. The left-hand limit for region $B$ is $a = 2$. To find the right-hand limit, we solve the equations $y = \sqrt{x}$ and $y = x - 2$ simultaneously for $x$:

$$\sqrt{x} = x - 2 \qquad \text{Equate } f(x) \text{ and } g(x).$$

$$x = (x - 2)^2 = x^2 - 4x + 4 \qquad \text{Square both sides.}$$

$$x^2 - 5x + 4 = 0 \qquad \text{Rewrite.}$$

$$(x - 1)(x - 4) = 0 \qquad \text{Factor.}$$

$$x = 1, \qquad x = 4. \qquad \text{Solve.}$$

Only the value $x = 4$ satisfies the equation $\sqrt{x} = x - 2$. The value $x = 1$ is an extraneous root introduced by squaring. The right-hand limit is $b = 4$.

For $0 \le x \le 2$: $\qquad f(x) - g(x) = \sqrt{x} - 0 = \sqrt{x}$

For $2 \le x \le 4$: $\qquad f(x) - g(x) = \sqrt{x} - (x - 2) = \sqrt{x} - x + 2$

We add the areas of subregions $A$ and $B$ to find the total area:

$$\text{Total area} = \underbrace{\int_0^2 \sqrt{x} \, dx}_{\text{area of } A} + \underbrace{\int_2^4 \left( \sqrt{x} - x + 2 \right) dx}_{\text{area of } B}$$

$$= \left[ \frac{2}{3} x^{3/2} \right]_0^2 + \left[ \frac{2}{3} x^{3/2} - \frac{x^2}{2} + 2x \right]_2^4$$

$$= \frac{2}{3}(2)^{3/2} - 0 + \left(\frac{2}{3}(4)^{3/2} - 8 + 8\right) - \left(\frac{2}{3}(2)^{3/2} - 2 + 4\right)$$

$$= \frac{2}{3}(8) - 2 = \frac{10}{3}.$$ ■

### Integration with Respect to y

If a region's bounding curves are described by functions of $y$, the approximating rectangles are horizontal instead of vertical and the basic formula has $y$ in place of $x$.
For regions like these:

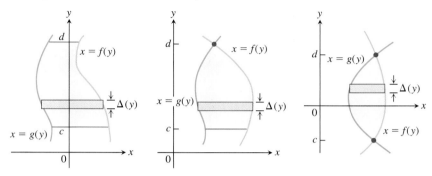

use the formula

$$A = \int_c^d [f(y) - g(y)] \, dy.$$

In this equation $f$ always denotes the right-hand curve and $g$ the left-hand curve, so $f(y) - g(y)$ is nonnegative.

**EXAMPLE 6**    Find the area of the region in Example 5 by integrating with respect to $y$.

**Solution**    We first sketch the region and a typical *horizontal* rectangle based on a partition of an interval of $y$-values (Figure 5.29). The region's right-hand boundary is the line $x = y + 2$, so $f(y) = y + 2$. The left-hand boundary is the curve $x = y^2$, so $g(y) = y^2$. The lower limit of integration is $y = 0$. We find the upper limit by solving $x = y + 2$ and $x = y^2$ simultaneously for $y$:

$$y + 2 = y^2 \qquad \text{Equate } f(y) = y + 2 \text{ and } g(y) = y^2.$$
$$y^2 - y - 2 = 0 \qquad \text{Rewrite.}$$
$$(y + 1)(y - 2) = 0 \qquad \text{Factor.}$$
$$y = -1, \qquad y = 2 \qquad \text{Solve.}$$

The upper limit of integration is $b = 2$. (The value $y = -1$ gives a point of intersection *below* the $x$-axis.)

The area of the region is

$$A = \int_c^d [f(y) - g(y)] \, dy = \int_0^2 [y + 2 - y^2] \, dy$$

$$= \int_0^2 [2 + y - y^2] \, dy$$

$$= \left[2y + \frac{y^2}{2} - \frac{y^3}{3}\right]_0^2$$

$$= 4 + \frac{4}{2} - \frac{8}{3} = \frac{10}{3}.$$

This is the result of Example 6, found with less work. ■

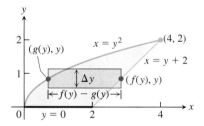

**FIGURE 5.29**    It takes two integrations to find the area of this region if we integrate with respect to $x$. It takes only one if we integrate with respect to $y$ (Example 6).

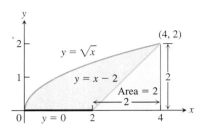

**FIGURE 5.30** The area of the blue region is the area under the parabola $y = \sqrt{x}$ minus the area of the triangle.

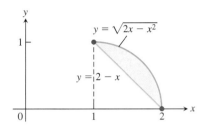

**FIGURE 5.31** The region described by the curves in Example 7.

Although it was easier to find the area in Example 5 by integrating with respect to $y$ rather than $x$ (just as we did in Example 6), there is an easier way yet. Looking at Figure 5.30, we see that the area we want is the area between the curve $y = \sqrt{x}$ and the $x$-axis for $0 \le x \le 4$, *minus* the area of an isosceles triangle of base and height equal to 2. So by combining calculus with some geometry, we find

$$\text{Area} = \int_0^4 \sqrt{x}\, dx - \frac{1}{2}(2)(2)$$

$$= \frac{2}{3}x^{3/2}\Big]_0^4 - 2$$

$$= \frac{2}{3}(8) - 0 - 2 = \frac{10}{3}.$$

**EXAMPLE 7** Find the area of the region bounded below by the line $y = 2 - x$ and above by the curve $y = \sqrt{2x - x^2}$.

**Solution** A sketch of the region is displayed in Figure 5.31, and we see that the line and curve intersect at the points $(1, 1)$ and $(2, 0)$. Using vertical rectangles, the area of the region is given by

$$A = \int_1^2 \left(\sqrt{2x - x^2} + x - 2\right) dx.$$

However, we don't know how to find an antiderivative for the term involving the radical, and no simple substitution is apparent.

To use horizontal rectangles, we first need to express each bounding curve as a function of the variable $y$. The line on the left is easily found to be $x = 2 - y$. For the curve $y = \sqrt{2x - x^2}$ on the right-hand side in Figure 5.31, we have

$$y^2 = 2x - x^2$$
$$= -(x^2 - 2x + 1) + 1 \qquad \text{Complete the square.}$$
$$= -(x - 1)^2 + 1.$$

Solving for $x$,

$$(x - 1)^2 = 1 - y^2,$$
$$x = 1 + \sqrt{1 - y^2}. \qquad x \ge 1,\ 0 \le y \le 1$$

The area of the region is then given by

$$A = \int_0^1 \left[\left(1 + \sqrt{1 - y^2}\right) - (2 - y)\right] dy$$

$$= \int_0^1 \left(\sqrt{1 - y^2} + y - 1\right) dy.$$

Again, we don't know yet how to integrate the radical term (although we will see how to do that in Section 8.4). We conclude that neither vertical nor horizontal rectangles lead to an integral we currently can evaluate.

Nevertheless, as we found with Example 6, sometimes a little observation proves to be helpful. If we look again at the algebra for expressing the right-hand side curve $y = \sqrt{2x - x^2}$ as a function of $y$, we see that $(x - 1)^2 + y^2 = 1$, which is the equation of the unit circle with center shifted to the point $(1, 0)$. From Figure 5.31, we can then see that the area of the region we want is the area of the upper right quarter of the unit circle minus the area of the triangle with vertices $(1, 1)$, $(1, 0)$, and $(2, 0)$. That is, the area is given by

$$A = \frac{\pi}{4} - \frac{1}{2} = \frac{\pi - 2}{4} \approx 0.285.$$

## Exercises 5.6

**Evaluating Definite Integrals**
Use the Substitution Formula in Theorem 7 to evaluate the integrals in Exercises 1–24.

**1. a.** $\int_0^3 \sqrt{y+1}\, dy$  **b.** $\int_{-1}^0 \sqrt{y+1}\, dy$

**2. a.** $\int_0^1 r\sqrt{1-r^2}\, dr$  **b.** $\int_{-1}^1 r\sqrt{1-r^2}\, dr$

**3. a.** $\int_0^{\pi/4} \tan x \sec^2 x\, dx$  **b.** $\int_{-\pi/4}^0 \tan x \sec^2 x\, dx$

**4. a.** $\int_0^{\pi} 3\cos^2 x \sin x\, dx$  **b.** $\int_{2\pi}^{3\pi} 3\cos^2 x \sin x\, dx$

**5. a.** $\int_0^1 t^3(1+t^4)^3\, dt$  **b.** $\int_{-1}^1 t^3(1+t^4)^3\, dt$

**6. a.** $\int_0^{\sqrt{7}} t(t^2+1)^{1/3}\, dt$  **b.** $\int_{-\sqrt{7}}^0 t(t^2+1)^{1/3}\, dt$

**7. a.** $\int_{-1}^1 \frac{5r}{(4+r^2)^2}\, dr$  **b.** $\int_0^1 \frac{5r}{(4+r^2)^2}\, dr$

**8. a.** $\int_0^1 \frac{10\sqrt{v}}{(1+v^{3/2})^2}\, dv$  **b.** $\int_1^4 \frac{10\sqrt{v}}{(1+v^{3/2})^2}\, dv$

**9. a.** $\int_0^{\sqrt{3}} \frac{4x}{\sqrt{x^2+1}}\, dx$  **b.** $\int_{-\sqrt{3}}^{\sqrt{3}} \frac{4x}{\sqrt{x^2+1}}\, dx$

**10. a.** $\int_0^1 \frac{x^3}{\sqrt{x^4+9}}\, dx$  **b.** $\int_{-1}^0 \frac{x^3}{\sqrt{x^4+9}}\, dx$

**11. a.** $\int_0^1 t\sqrt{4+5t}\, dt$  **b.** $\int_1^9 t\sqrt{4+5t}\, dt$

**12. a.** $\int_0^{\pi/6} (1-\cos 3t)\sin 3t\, dt$  **b.** $\int_{\pi/6}^{\pi/3} (1-\cos 3t)\sin 3t\, dt$

**13. a.** $\int_0^{2\pi} \frac{\cos z}{\sqrt{4+3\sin z}}\, dz$  **b.** $\int_{-\pi}^{\pi} \frac{\cos z}{\sqrt{4+3\sin z}}\, dz$

**14. a.** $\int_{-\pi/2}^0 \left(2+\tan\frac{t}{2}\right)\sec^2\frac{t}{2}\, dt$  **b.** $\int_{-\pi/2}^{\pi/2} \left(2+\tan\frac{t}{2}\right)\sec^2\frac{t}{2}\, dt$

**15.** $\int_0^1 \sqrt{t^5+2t}\,(5t^4+2)\, dt$  **16.** $\int_1^4 \frac{dy}{2\sqrt{y}(1+\sqrt{y})^2}$

**17.** $\int_0^{\pi/6} \cos^{-3} 2\theta \sin 2\theta\, d\theta$  **18.** $\int_{\pi}^{3\pi/2} \cot^5\left(\frac{\theta}{6}\right)\sec^2\left(\frac{\theta}{6}\right)\, d\theta$

**19.** $\int_0^{\pi} 5(5-4\cos t)^{1/4}\sin t\, dt$  **20.** $\int_0^{\pi/4} (1-\sin 2t)^{3/2}\cos 2t\, dt$

**21.** $\int_0^1 (4y-y^2+4y^3+1)^{-2/3}(12y^2-2y+4)\, dy$

**22.** $\int_0^1 (y^3+6y^2-12y+9)^{-1/2}(y^2+4y-4)\, dy$

**23.** $\int_0^{\sqrt[3]{\pi^2}} \sqrt{\theta}\cos^2(\theta^{3/2})\, d\theta$  **24.** $\int_{-1}^{-1/2} t^{-2}\sin^2\left(1+\frac{1}{t}\right)\, dt$

**Area**
Find the total areas of the shaded regions in Exercises 25–40.

**25.**

**26.**

**27.**

**28.** $y=\frac{\pi}{2}(\cos x)(\sin(\pi+\pi\sin x))$

**29.**

**30.**

**31.**

NOT TO SCALE

**32.**

**33.**

**34.**  **35.**

**36.**

**37.**  **38.**

**39.**  **40.**

Find the areas of the regions enclosed by the lines and curves in Exercises 41–50.

**41.** $y = x^2 - 2$ and $y = 2$  **42.** $y = 2x - x^2$ and $y = -3$

**43.** $y = x^4$ and $y = 8x$  **44.** $y = x^2 - 2x$ and $y = x$

**45.** $y = x^2$ and $y = -x^2 + 4x$

**46.** $y = 7 - 2x^2$ and $y = x^2 + 4$

**47.** $y = x^4 - 4x^2 + 4$ and $y = x^2$

**48.** $y = x\sqrt{a^2 - x^2}$, $a > 0$, and $y = 0$

**49.** $y = \sqrt{|x|}$ and $5y = x + 6$ (How many intersection points are there?)

**50.** $y = |x^2 - 4|$ and $y = (x^2/2) + 4$

Find the areas of the regions enclosed by the lines and curves in Exercises 51–58.

**51.** $x = 2y^2$, $x = 0$, and $y = 3$

**52.** $x = y^2$ and $x = y + 2$

**53.** $y^2 - 4x = 4$ and $4x - y = 16$

**54.** $x - y^2 = 0$ and $x + 2y^2 = 3$

**55.** $x + y^2 = 0$ and $x + 3y^2 = 2$

**56.** $x - y^{2/3} = 0$ and $x + y^4 = 2$

**57.** $x = y^2 - 1$ and $x = |y|\sqrt{1 - y^2}$

**58.** $x = y^3 - y^2$ and $x = 2y$

Find the areas of the regions enclosed by the curves in Exercises 59–62.

**59.** $4x^2 + y = 4$ and $x^4 - y = 1$

**60.** $x^3 - y = 0$ and $3x^2 - y = 4$

**61.** $x + 4y^2 = 4$ and $x + y^4 = 1$, for $x \geq 0$

**62.** $x + y^2 = 3$ and $4x + y^2 = 0$

Find the areas of the regions enclosed by the lines and curves in Exercises 63–70.

**63.** $y = 2 \sin x$ and $y = \sin 2x$, $0 \leq x \leq \pi$

**64.** $y = 8 \cos x$ and $y = \sec^2 x$, $-\pi/3 \leq x \leq \pi/3$

**65.** $y = \cos (\pi x/2)$ and $y = 1 - x^2$

**66.** $y = \sin (\pi x/2)$ and $y = x$

**67.** $y = \sec^2 x$, $y = \tan^2 x$, $x = -\pi/4$, and $x = \pi/4$

**68.** $x = \tan^2 y$ and $x = -\tan^2 y$, $-\pi/4 \leq y \leq \pi/4$

**69.** $x = 3 \sin y \sqrt{\cos y}$ and $x = 0$, $0 \leq y \leq \pi/2$

**70.** $y = \sec^2 (\pi x/3)$ and $y = x^{1/3}$, $-1 \leq x \leq 1$

**Area Between Curves**

**71.** Find the area of the propeller-shaped region enclosed by the curve $x - y^3 = 0$ and the line $x - y = 0$.

**72.** Find the area of the propeller-shaped region enclosed by the curves $x - y^{1/3} = 0$ and $x - y^{1/5} = 0$.

**73.** Find the area of the region in the first quadrant bounded by the line $y = x$, the line $x = 2$, the curve $y = 1/x^2$, and the $x$-axis.

**74.** Find the area of the "triangular" region in the first quadrant bounded on the left by the $y$-axis and on the right by the curves $y = \sin x$ and $y = \cos x$.

**75.** The region bounded below by the parabola $y = x^2$ and above by the line $y = 4$ is to be partitioned into two subsections of equal area by cutting across it with the horizontal line $y = c$.

**a.** Sketch the region and draw a line $y = c$ across it that looks about right. In terms of $c$, what are the coordinates of the points where the line and parabola intersect? Add them to your figure.

**b.** Find $c$ by integrating with respect to $y$. (This puts $c$ in the limits of integration.)

**c.** Find $c$ by integrating with respect to $x$. (This puts $c$ into the integrand as well.)

**76.** Find the area of the region between the curve $y = 3 - x^2$ and the line $y = -1$ by integrating with respect to **a.** $x$, **b.** $y$.

**77.** Find the area of the region in the first quadrant bounded on the left by the $y$-axis, below by the line $y = x/4$, above left by the curve $y = 1 + \sqrt{x}$, and above right by the curve $y = 2/\sqrt{x}$.

**78.** Find the area of the region in the first quadrant bounded on the left by the $y$-axis, below by the curve $x = 2\sqrt{y}$, above left by the curve $x = (y - 1)^2$, and above right by the line $x = 3 - y$.

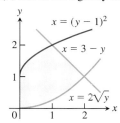

**79.** The figure here shows triangle $AOC$ inscribed in the region cut from the parabola $y = x^2$ by the line $y = a^2$. Find the limit of the ratio of the area of the triangle to the area of the parabolic region as $a$ approaches zero.

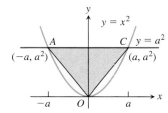

**80.** Suppose the area of the region between the graph of a positive continuous function $f$ and the $x$-axis from $x = a$ to $x = b$ is 4 square units. Find the area between the curves $y = f(x)$ and $y = 2f(x)$ from $x = a$ to $x = b$.

**81.** Which of the following integrals, if either, calculates the area of the shaded region shown here? Give reasons for your answer.

**a.** $\displaystyle\int_{-1}^{1} (x - (-x))\, dx = \int_{-1}^{1} 2x\, dx$

**b.** $\displaystyle\int_{-1}^{1} (-x - (x))\, dx = \int_{-1}^{1} -2x\, dx$

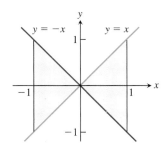

**82.** True, sometimes true, or never true? The area of the region between the graphs of the continuous functions $y = f(x)$ and $y = g(x)$ and the vertical lines $x = a$ and $x = b$ ($a < b$) is

$$\int_a^b [f(x) - g(x)]\, dx.$$

Give reasons for your answer.

**Theory and Examples**

**83.** Suppose that $F(x)$ is an antiderivative of $f(x) = (\sin x)/x$, $x > 0$. Express

$$\int_1^3 \frac{\sin 2x}{x}\, dx$$

in terms of $F$.

**84.** Show that if $f$ is continuous, then

$$\int_0^1 f(x)\, dx = \int_0^1 f(1 - x)\, dx.$$

**85.** Suppose that

$$\int_0^1 f(x)\, dx = 3.$$

Find

$$\int_{-1}^0 f(x)\, dx$$

if **a.** $f$ is odd, **b.** $f$ is even.

**86. a.** Show that if $f$ is odd on $[-a, a]$, then

$$\int_{-a}^a f(x)\, dx = 0.$$

**b.** Test the result in part (a) with $f(x) = \sin x$ and $a = \pi/2$.

**87.** If $f$ is a continuous function, find the value of the integral

$$I = \int_0^a \frac{f(x)\, dx}{f(x) + f(a - x)}$$

by making the substitution $u = a - x$ and adding the resulting integral to $I$.

**88.** By using a substitution, prove that for all positive numbers $x$ and $y$,

$$\int_x^{xy} \frac{1}{t}\, dt = \int_1^y \frac{1}{t}\, dt.$$

**The Shift Property for Definite Integrals**  A basic property of definite integrals is their invariance under translation, as expressed by the equation

$$\int_a^b f(x)\, dx = \int_{a-c}^{b-c} f(x + c)\, dx. \tag{1}$$

The equation holds whenever $f$ is integrable and defined for the necessary values of $x$. For example in the accompanying figure, show that

$$\int_{-2}^{-1} (x + 2)^3\, dx = \int_0^1 x^3\, dx$$

because the areas of the shaded regions are congruent.

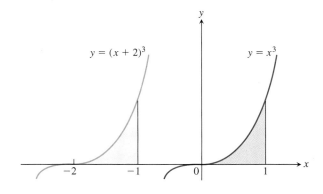

**89.** Use a substitution to verify Equation (1).

**90.** For each of the following functions, graph $f(x)$ over $[a, b]$ and $f(x + c)$ over $[a - c, b - c]$ to convince yourself that Equation (1) is reasonable.

  **a.** $f(x) = x^2$,  $a = 0$,  $b = 1$,  $c = 1$

  **b.** $f(x) = \sin x$,  $a = 0$,  $b = \pi$,  $c = \pi/2$

  **c.** $f(x) = \sqrt{x - 4}$,  $a = 4$,  $b = 8$,  $c = 5$

**COMPUTER EXPLORATIONS**

In Exercises 91–94, you will find the area between curves in the plane when you cannot find their points of intersection using simple algebra. Use a CAS to perform the following steps:

  **a.** Plot the curves together to see what they look like and how many points of intersection they have.

  **b.** Use the numerical equation solver in your CAS to find all the points of intersection.

  **c.** Integrate $|f(x) - g(x)|$ over consecutive pairs of intersection values.

  **d.** Sum together the integrals found in part (c).

**91.** $f(x) = \dfrac{x^3}{3} - \dfrac{x^2}{2} - 2x + \dfrac{1}{3}$,  $g(x) = x - 1$

**92.** $f(x) = \dfrac{x^4}{2} - 3x^3 + 10$,  $g(x) = 8 - 12x$

**93.** $f(x) = x + \sin (2x)$,  $g(x) = x^3$

**94.** $f(x) = x^2 \cos x$,  $g(x) = x^3 - x$

---

# Chapter 5 Questions to Guide Your Review

**1.** How can you sometimes estimate quantities like distance traveled, area, and average value with finite sums? Why might you want to do so?

**2.** What is sigma notation? What advantage does it offer? Give examples.

**3.** What is a Riemann sum? Why might you want to consider such a sum?

**4.** What is the norm of a partition of a closed interval?

**5.** What is the definite integral of a function $f$ over a closed interval $[a, b]$? When can you be sure it exists?

**6.** What is the relation between definite integrals and area? Describe some other interpretations of definite integrals.

**7.** What is the average value of an integrable function over a closed interval? Must the function assume its average value? Explain.

**8.** Describe the rules for working with definite integrals (Table 5.6). Give examples.

**9.** What is the Fundamental Theorem of Calculus? Why is it so important? Illustrate each part of the theorem with an example.

**10.** What is the Net Change Theorem? What does it say about the integral of velocity? The integral of marginal cost?

**11.** Discuss how the processes of integration and differentiation can be considered as "inverses" of each other.

**12.** How does the Fundamental Theorem provide a solution to the initial value problem $dy/dx = f(x)$, $y(x_0) = y_0$, when $f$ is continuous?

**13.** How is integration by substitution related to the Chain Rule?

**14.** How can you sometimes evaluate indefinite integrals by substitution? Give examples.

**15.** How does the method of substitution work for definite integrals? Give examples.

**16.** How do you define and calculate the area of the region between the graphs of two continuous functions? Give an example.

---

# Chapter 5 Practice Exercises

**Finite Sums and Estimates**

**1.** The accompanying figure shows the graph of the velocity (m/s) of a model rocket for the first 8 s after launch. The rocket accelerated straight up for the first 2 s and then coasted to reach its maximum height at $t = 8$ s.

Time after launch (s)

**a.** Assuming that the rocket was launched from ground level, about how high did it go? (This is the rocket in Section 3.4, Exercise 17, but you do not need to do Exercise 17 to do the exercise here.)

**b.** Sketch a graph of the rocket's height above ground as a function of time for $0 \le t \le 8$.

**2. a.** The accompanying figure shows the velocity (m/s) of a body moving along the $s$-axis during the time interval from $t = 0$ to $t = 10$ s. About how far did the body travel during those 10 s?

**b.** Sketch a graph of $s$ as a function of $t$ for $0 \le t \le 10$, assuming $s(0) = 0$.

**3.** Suppose that $\sum_{k=1}^{10} a_k = -2$ and $\sum_{k=1}^{10} b_k = 25$. Find the value of

**a.** $\sum_{k=1}^{10} \dfrac{a_k}{4}$

**b.** $\sum_{k=1}^{10} (b_k - 3a_k)$

**c.** $\sum_{k=1}^{10} (a_k + b_k - 1)$

**d.** $\sum_{k=1}^{10} \left( \dfrac{5}{2} - b_k \right)$

**4.** Suppose that $\sum_{k=1}^{20} a_k = 0$ and $\sum_{k=1}^{20} b_k = 7$. Find the values of

**a.** $\sum_{k=1}^{20} 3a_k$

**b.** $\sum_{k=1}^{20} (a_k + b_k)$

**c.** $\sum_{k=1}^{20} \left( \dfrac{1}{2} - \dfrac{2b_k}{7} \right)$

**d.** $\sum_{k=1}^{20} (a_k - 2)$

### Definite Integrals

In Exercises 5–8, express each limit as a definite integral. Then evaluate the integral to find the value of the limit. In each case, $P$ is a partition of the given interval and the numbers $c_k$ are chosen from the subintervals of $P$.

**5.** $\lim_{\|P\| \to 0} \sum_{k=1}^{n} (2c_k - 1)^{-1/2} \, \Delta x_k$, where $P$ is a partition of $[1, 5]$

**6.** $\lim_{\|P\| \to 0} \sum_{k=1}^{n} c_k(c_k^2 - 1)^{1/3} \, \Delta x_k$, where $P$ is a partition of $[1, 3]$

**7.** $\lim_{\|P\| \to 0} \sum_{k=1}^{n} \left( \cos\left( \dfrac{c_k}{2} \right) \right) \Delta x_k$, where $P$ is a partition of $[-\pi, 0]$

**8.** $\lim_{\|P\| \to 0} \sum_{k=1}^{n} (\sin c_k)(\cos c_k) \, \Delta x_k$, where $P$ is a partition of $[0, \pi/2]$

**9.** If $\int_{-2}^{2} 3f(x) \, dx = 12$, $\int_{-2}^{5} f(x) \, dx = 6$, and $\int_{-2}^{5} g(x) \, dx = 2$, find the values of the following.

**a.** $\int_{-2}^{2} f(x) \, dx$

**b.** $\int_{2}^{5} f(x) \, dx$

**c.** $\int_{5}^{-2} g(x) \, dx$

**d.** $\int_{-2}^{5} (-\pi g(x)) \, dx$

**e.** $\int_{-2}^{5} \left( \dfrac{f(x) + g(x)}{5} \right) dx$

**10.** If $\int_{0}^{2} f(x) \, dx = \pi$, $\int_{0}^{2} 7g(x) \, dx = 7$, and $\int_{0}^{1} g(x) \, dx = 2$, find the values of the following.

**a.** $\int_{0}^{2} g(x) \, dx$

**b.** $\int_{1}^{2} g(x) \, dx$

**c.** $\int_{2}^{0} f(x) \, dx$

**d.** $\int_{0}^{2} \sqrt{2} f(x) \, dx$

**e.** $\int_{0}^{2} (g(x) - 3f(x)) \, dx$

### Area

In Exercises 11–14, find the total area of the region between the graph of $f$ and the $x$-axis.

**11.** $f(x) = x^2 - 4x + 3$, $0 \le x \le 3$

**12.** $f(x) = 1 - (x^2/4)$, $-2 \le x \le 3$

**13.** $f(x) = 5 - 5x^{2/3}$, $-1 \le x \le 8$

**14.** $f(x) = 1 - \sqrt{x}$, $0 \le x \le 4$

Find the areas of the regions enclosed by the curves and lines in Exercises 15–26.

**15.** $y = x$, $y = 1/x^2$, $x = 2$

**16.** $y = x$, $y = 1/\sqrt{x}$, $x = 2$

**17.** $\sqrt{x} + \sqrt{y} = 1$, $x = 0$, $y = 0$

**18.** $x^3 + \sqrt{y} = 1$, $x = 0$, $y = 0$, for $0 \le x \le 1$

**19.** $x = 2y^2$, $x = 0$, $y = 3$    **20.** $x = 4 - y^2$, $x = 0$

**21.** $y^2 = 4x$, $y = 4x - 2$

**22.** $y^2 = 4x + 4$, $y = 4x - 16$

**23.** $y = \sin x$, $y = x$, $0 \le x \le \pi/4$

**24.** $y = |\sin x|$, $y = 1$, $-\pi/2 \le x \le \pi/2$

**25.** $y = 2 \sin x$, $y = \sin 2x$, $0 \le x \le \pi$

**26.** $y = 8 \cos x$, $y = \sec^2 x$, $-\pi/3 \le x \le \pi/3$

**27.** Find the area of the "triangular" region bounded on the left by $x + y = 2$, on the right by $y = x^2$, and above by $y = 2$.

**28.** Find the area of the "triangular" region bounded on the left by $y = \sqrt{x}$, on the right by $y = 6 - x$, and below by $y = 1$.

**29.** Find the extreme values of $f(x) = x^3 - 3x^2$ and find the area of the region enclosed by the graph of $f$ and the $x$-axis.

**30.** Find the area of the region cut from the first quadrant by the curve $x^{1/2} + y^{1/2} = a^{1/2}$.

**31.** Find the total area of the region enclosed by the curve $x = y^{2/3}$ and the lines $x = y$ and $y = -1$.

**32.** Find the total area of the region between the curves $y = \sin x$ and $y = \cos x$ for $0 \le x \le 3\pi/2$.

### Initial Value Problems

**33.** Show that $y = x^2 + \int_1^x \frac{1}{t} dt$ solves the initial value problem

$$\frac{d^2 y}{dx^2} = 2 - \frac{1}{x^2}; \quad y'(1) = 3, \quad y(1) = 1.$$

**34.** Show that $y = \int_0^x (1 + 2\sqrt{\sec t}) \, dt$ solves the initial value problem

$$\frac{d^2 y}{dx^2} = \sqrt{\sec x} \tan x; \quad y'(0) = 3, \quad y(0) = 0.$$

Express the solutions of the initial value problems in Exercises 35 and 36 in terms of integrals.

**35.** $\dfrac{dy}{dx} = \dfrac{\sin x}{x}, \quad y(5) = -3$

**36.** $\dfrac{dy}{dx} = \sqrt{2 - \sin^2 x}, \quad y(-1) = 2$

### Evaluating Indefinite Integrals
Evaluate the integrals in Exercises 37–44.

**37.** $\displaystyle \int 8(\sin x)^{-3/2} \cos x \, dx$

**38.** $\displaystyle \int (\tan x)^{-5/2} \sec^2 x \, dx$

**39.** $\displaystyle \int \left[ 6\theta - 1 - 4\sin(4\theta + 1) \right] d\theta$

**40.** $\displaystyle \int \left[ \frac{2}{\sqrt{3\theta + \pi}} - 3\csc^2(3\theta + \pi) \right] d\theta$

**41.** $\displaystyle \int \left( x - \frac{5}{x} \right)\left( x + \frac{5}{x} \right) dx$

**42.** $\displaystyle \int \frac{(t - 2)^2 + 2}{t^5} dt$

**43.** $\displaystyle \int \sqrt{t} \cos(6t^{3/2}) \, dt$

**44.** $\displaystyle \int \csc \theta \cot \theta \sqrt{1 - \csc \theta} \, d\theta$

### Evaluating Definite Integrals
Evaluate the integrals in Exercises 45–66.

**45.** $\displaystyle \int_{-2}^{2} (4x^3 - 2x + 9) \, dx$

**46.** $\displaystyle \int_0^3 (6s^5 - 9s^2 + 7) \, ds$

**47.** $\displaystyle \int_1^3 \frac{9}{x^3} dx$

**48.** $\displaystyle \int_1^8 x^{-4/3} \, dx$

**49.** $\displaystyle \int_4^9 \frac{dx}{x\sqrt{x}}$

**50.** $\displaystyle \int_1^9 \frac{(5 - \sqrt{x})^{1/3}}{\sqrt{x}} dx$

**51.** $\displaystyle \int_0^1 \frac{72 \, dx}{(4x + 3)^3}$

**52.** $\displaystyle \int_0^2 \frac{dr}{\sqrt[3]{(9 + 7r)^2}}$

**53.** $\displaystyle \int_{1/2}^1 x^{-1/4}(1 - x^{3/4})^{1/3} \, dx$

**54.** $\displaystyle \int_0^{3/2} x^4(1 + 5x^5)^{-5/2} \, dx$

**55.** $\displaystyle \int_0^{\pi} \cos^2 8r \, dr$

**56.** $\displaystyle \int_0^{\pi/4} \sin^2\left( 2t + \frac{\pi}{4} \right) dt$

**57.** $\displaystyle \int_0^{\pi/2} \cos^2 \theta \, d\theta$

**58.** $\displaystyle \int_{\pi/3}^{5\pi/6} \sin^2 \theta \, d\theta$

**59.** $\displaystyle \int_{\pi}^{2\pi} \tan^2 \frac{x}{6} \, dx$

**60.** $\displaystyle \int_{-\pi}^{\pi} \cot^2 \frac{\theta}{3} \, d\theta$

**61.** $\displaystyle \int_{-\pi/3}^{0} \sin x \cos x \, dx$

**62.** $\displaystyle \int_{\pi/3}^{5\pi/6} \sec z \tan z \, dz$

**63.** $\displaystyle \int_0^{\pi/2} 7(\cos x)^{5/2} \sin x \, dx$

**64.** $\displaystyle \int_{-\pi/3}^{\pi/3} 12\cos^2 4x \sin 4x \, dx$

**65.** $\displaystyle \int_0^{\pi/2} \frac{5\cos x \sin x}{\sqrt{4 + 5\cos^2 x}} dx$

**66.** $\displaystyle \int_0^{\pi/4} \frac{\sec^2 x}{(8 + 19\tan x)^{4/3}} dx$

### Average Values
**67.** Find the average value of $f(x) = mx + b$
   **a.** over $[-1, 1]$   **b.** over $[-k, k]$

**68.** Find the average value of
   **a.** $y = \sqrt{3x}$ over $[0, 3]$   **b.** $y = \sqrt{ax}$ over $[0, a]$

**69.** Let $f$ be a function that is differentiable on $[a, b]$. In Chapter 2 we defined the average rate of change of $f$ over $[a, b]$ to be

$$\frac{f(b) - f(a)}{b - a}$$

and the instantaneous rate of change of $f$ at $x$ to be $f'(x)$. In this chapter we defined the average value of a function. For the new definition of average to be consistent with the old one, we should have

$$\frac{f(b) - f(a)}{b - a} = \text{average value of } f' \text{ on } [a, b].$$

Is this the case? Give reasons for your answer.

**70.** Is it true that the average value of an integrable function over an interval of length 2 is half the function's integral over the interval? Give reasons for your answer.

**T** **71.** Compute the average value of the temperature function

$$f(x) = 20 \sin\left( \frac{2\pi}{365}(x - 101) \right) - 4$$

for a 365-day year. (See Exercise 84, Section 3.6.) This is one way to estimate the annual mean air temperature in Fairbanks, Alaska. The National Weather Service's official figure, a numerical average of the daily normal mean air temperatures for the year, is $-3.5°C$, which is slightly higher than the average value of $f(x)$.

**T** **72.** **Specific heat of a gas** Specific heat $C_v$ is the amount of heat required to raise the temperature of one mole (gram molecule) of a gas with constant volume by $1°C$. The specific heat of oxygen depends on its temperature $T$ and satisfies the formula

$$C_v = 8.27 + 10^{-5} (26T - 1.87T^2).$$

Find the average value of $C_v$ for $20° \le T \le 675°C$ and the temperature at which it is attained.

## Differentiating Integrals

In Exercises 73–76, find $dy/dx$.

**73.** $y = \int_2^x \sqrt{2 + \cos^3 t}\, dt$   **74.** $y = \int_2^{7x^2} \sqrt{2 + \cos^3 t}\, dt$

**75.** $y = \int_x^1 \dfrac{6}{3 + t^4}\, dt$   **76.** $y = \int_{\sec x}^2 \dfrac{1}{t^2 + 1}\, dt$

## Theory and Examples

**77.** Is it true that every function $y = f(x)$ that is differentiable on $[a, b]$ is itself the derivative of some function on $[a, b]$? Give reasons for your answer

**78.** Suppose that $f(x)$ is an antiderivative of $f(x) = \sqrt{1 + x^4}$. Express $\int_0^1 \sqrt{1 + x^4}\, dx$ in terms of $F$ and give a reason for your answer.

**79.** Find $dy/dx$ if $y = \int_x^1 \sqrt{1 + t^2}\, dt$. Explain the main steps in your calculation.

**80.** Find $dy/dx$ if $y = \int_{\cos x}^0 (1/(1 - t^2))\, dt$. Explain the main steps in your calculation.

**81. A new parking lot**   To meet the demand for parking, your town has allocated the area shown here. As the town engineer, you have been asked by the town council to find out if the lot can be built for $10,000. The cost to clear the land will be $1.00 a square meter, and the lot will cost $20.00 a square meter to pave. Can the job be done for $10,000? Use a lower sum estimate to see. (Answers may vary slightly, depending on the estimate used.)

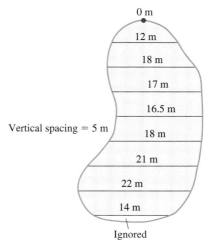

Vertical spacing = 5 m

0 m
12 m
18 m
17 m
16.5 m
18 m
21 m
22 m
14 m

Ignored

**82.** Skydivers A and B are in a helicopter hovering at 2000 m. Skydiver A jumps and descends for 4 s before opening her parachute. The helicopter then climbs to 2200 m and hovers there. Forty-five seconds after A leaves the aircraft, B jumps and descends for 13 s before opening his parachute. Both skydivers descend at 4.9 m/s with parachutes open. Assume that the skydivers fall freely (no effective air resistance) before their parachutes open.

**a.** At what altitude does A's parachute open?

**b.** At what altitude does B's parachute open?

**c.** Which skydiver lands first?

---

## Chapter 5   Additional and Advanced Exercises

## Theory and Examples

**1. a.** If $\int_0^1 7f(x)\, dx = 7$, does $\int_0^1 f(x)\, dx = 1$?

**b.** If $\int_0^1 f(x)\, dx = 4$ and $f(x) \geq 0$, does

$$\int_0^1 \sqrt{f(x)}\, dx = \sqrt{4} = 2?$$

Give reasons for your answers.

**2.** Suppose $\int_{-2}^2 f(x)\, dx = 4,\ \int_2^5 f(x)\, dx = 3,\ \int_{-2}^5 g(x)\, dx = 2.$

Which, if any, of the following statements are true?

**a.** $\int_5^2 f(x)\, dx = -3$   **b.** $\int_{-2}^5 (f(x) + g(x)) = 9$

**c.** $f(x) \leq g(x)$ on the interval $-2 \leq x \leq 5$

**3. Initial value problem**   Show that

$$y = \frac{1}{a} \int_0^x f(t) \sin a(x - t)\, dt$$

solves the initial value problem

$$\frac{d^2 y}{dx^2} + a^2 y = f(x), \qquad \frac{dy}{dx} = 0 \text{ and } y = 0 \text{ when } x = 0.$$

(*Hint:* $\sin(ax - at) = \sin ax \cos at - \cos ax \sin at.$)

**4. Proportionality**   Suppose that $x$ and $y$ are related by the equation

$$x = \int_0^y \frac{1}{\sqrt{1 + 4t^2}}\, dt.$$

Show that $d^2 y/dx^2$ is proportional to $y$ and find the constant of proportionality.

**5.** Find $f(4)$ if

**a.** $\int_0^{x^2} f(t)\, dt = x \cos \pi x$   **b.** $\int_0^{f(x)} t^2\, dt = x \cos \pi x.$

**6.** Find $f(\pi/2)$ from the following information.

**i)** $f$ is positive and continuous.

**ii)** The area under the curve $y = f(x)$ from $x = 0$ to $x = a$ is

$$\frac{a^2}{2} + \frac{a}{2} \sin a + \frac{\pi}{2} \cos a.$$

**7.** The area of the region in the $xy$-plane enclosed by the $x$-axis, the curve $y = f(x)$, $f(x) \geq 0$, and the lines $x = 1$ and $x = b$ is equal to $\sqrt{b^2 + 1} - \sqrt{2}$ for all $b > 1$. Find $f(x)$.

**8.** Prove that

$$\int_0^x \left( \int_0^u f(t) \, dt \right) du = \int_0^x f(u)(x - u) \, du.$$

(*Hint:* Express the integral on the right-hand side as the difference of two integrals. Then show that both sides of the equation have the same derivative with respect to $x$.)

**9. Finding a curve** Find the equation for the curve in the $xy$-plane that passes through the point $(1, -1)$ if its slope at $x$ is always $3x^2 + 2$.

**10. Shoveling dirt** You sling a shovelful of dirt up from the bottom of a hole with an initial velocity of 9.8 m/s. The dirt must rise 5.2 m above the release point to clear the edge of the hole. Is that enough speed to get the dirt out, or had you better duck?

### Piecewise Continuous Functions

Although we are mainly interested in continuous functions, many functions in applications are piecewise continuous. A function $f(x)$ is **piecewise continuous on a closed interval $I$** if $f$ has only finitely many discontinuities in $I$, the limits

$$\lim_{x \to c^-} f(x) \qquad \text{and} \qquad \lim_{x \to c^+} f(x)$$

exist and are finite at every interior point of $I$, and the appropriate one-sided limits exist and are finite at the endpoints of $I$. All piecewise continuous functions are integrable. The points of discontinuity subdivide $I$ into open and half-open subintervals on which $f$ is continuous, and the limit criteria above guarantee that $f$ has a continuous extension to the closure of each subinterval. To integrate a piecewise continuous function, we integrate the individual extensions and add the results. The integral of

$$f(x) = \begin{cases} 1 - x, & -1 \leq x < 0 \\ x^2, & 0 \leq x < 2 \\ -1, & 2 \leq x \leq 3 \end{cases}$$

(Figure 5.32) over $[-1, 3]$ is

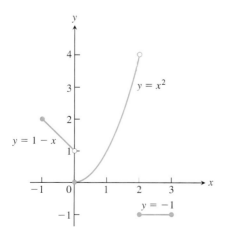

**FIGURE 5.32** Piecewise continuous functions like this are integrated piece by piece.

$$\int_{-1}^3 f(x) \, dx = \int_{-1}^0 (1 - x) \, dx + \int_0^2 x^2 \, dx + \int_2^3 (-1) \, dx$$

$$= \left[ x - \frac{x^2}{2} \right]_{-1}^0 + \left[ \frac{x^3}{3} \right]_0^2 + \left[ -x \right]_2^3$$

$$= \frac{3}{2} + \frac{8}{3} - 1 = \frac{19}{6}.$$

The Fundamental Theorem applies to piecewise continuous functions with the restriction that $(d/dx) \int_a^x f(t) \, dt$ is expected to equal $f(x)$ only at values of $x$ at which $f$ is continuous. There is a similar restriction on Leibniz's Rule (see Exercises 27–29).

Graph the functions in Exercises 11–16 and integrate them over their domains.

**11.** $f(x) = \begin{cases} x^{2/3}, & -8 \leq x < 0 \\ -4, & 0 \leq x \leq 3 \end{cases}$

**12.** $f(x) = \begin{cases} \sqrt{-x}, & -4 \leq x < 0 \\ x^2 - 4, & 0 \leq x \leq 3 \end{cases}$

**13.** $g(t) = \begin{cases} t, & 0 \leq t < 1 \\ \sin \pi t, & 1 \leq t \leq 2 \end{cases}$

**14.** $h(z) = \begin{cases} \sqrt{1 - z}, & 0 \leq z < 1 \\ (7z - 6)^{-1/3}, & 1 \leq z \leq 2 \end{cases}$

**15.** $f(x) = \begin{cases} 1, & -2 \leq x < -1 \\ 1 - x^2, & -1 \leq x < 1 \\ 2, & 1 \leq x \leq 2 \end{cases}$

**16.** $h(r) = \begin{cases} r, & -1 \leq r < 0 \\ 1 - r^2, & 0 \leq r < 1 \\ 1, & 1 \leq r \leq 2 \end{cases}$

**17.** Find the average value of the function graphed in the accompanying figure.

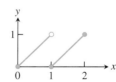

**18.** Find the average value of the function graphed in the accompanying figure.

### Approximating Finite Sums with Integrals

In many applications of calculus, integrals are used to approximate finite sums—the reverse of the usual procedure of using finite sums to approximate integrals.

For example, let's estimate the sum of the square roots of the first $n$ positive integers, $\sqrt{1} + \sqrt{2} + \cdots + \sqrt{n}$. The integral

$$\int_0^1 \sqrt{x} \, dx = \frac{2}{3} x^{3/2} \Big]_0^1 = \frac{2}{3}$$

is the limit of the upper sums

$$S_n = \sqrt{\frac{1}{n}} \cdot \frac{1}{n} + \sqrt{\frac{2}{n}} \cdot \frac{1}{n} + \cdots + \sqrt{\frac{n}{n}} \cdot \frac{1}{n}$$

$$= \frac{\sqrt{1} + \sqrt{2} + \cdots + \sqrt{n}}{n^{3/2}}.$$

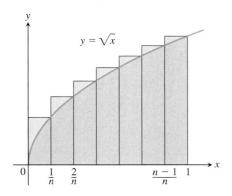

Therefore, when $n$ is large, $S_n$ will be close to $2/3$ and we will have

$$\text{Root sum} = \sqrt{1} + \sqrt{2} + \cdots + \sqrt{n} = S_n \cdot n^{3/2} \approx \frac{2}{3} n^{3/2}.$$

The following table shows how good the approximation can be.

| $n$ | Root sum | $(2/3)n^{3/2}$ | Relative error |
|-----|----------|----------------|----------------|
| 10 | 22.468 | 21.082 | $1.386/22.468 \approx 6\%$ |
| 50 | 239.04 | 235.70 | 1.4% |
| 100 | 671.46 | 666.67 | 0.7% |
| 1000 | 21,097 | 21,082 | 0.07% |

**19.** Evaluate

$$\lim_{n\to\infty} \frac{1^5 + 2^5 + 3^5 + \cdots + n^5}{n^6}$$

by showing that the limit is

$$\int_0^1 x^5 \, dx$$

and evaluating the integral.

**20.** See Exercise 19. Evaluate

$$\lim_{n\to\infty} \frac{1}{n^4} (1^3 + 2^3 + 3^3 + \cdots + n^3).$$

**21.** Let $f(x)$ be a continuous function. Express

$$\lim_{n\to\infty} \frac{1}{n}\left[ f\left(\frac{1}{n}\right) + f\left(\frac{2}{n}\right) + \cdots + f\left(\frac{n}{n}\right) \right]$$

as a definite integral.

**22.** Use the result of Exercise 21 to evaluate

**a.** $\displaystyle\lim_{n\to\infty} \frac{1}{n^2} (2 + 4 + 6 + \cdots + 2n),$

**b.** $\displaystyle\lim_{n\to\infty} \frac{1}{n^{16}} (1^{15} + 2^{15} + 3^{15} + \cdots + n^{15}),$

**c.** $\displaystyle\lim_{n\to\infty} \frac{1}{n} \left( \sin\frac{\pi}{n} + \sin\frac{2\pi}{n} + \sin\frac{3\pi}{n} + \cdots + \sin\frac{n\pi}{n} \right).$

What can be said about the following limits?

**d.** $\displaystyle\lim_{n\to\infty} \frac{1}{n^{17}} (1^{15} + 2^{15} + 3^{15} + \cdots + n^{15})$

**e.** $\displaystyle\lim_{n\to\infty} \frac{1}{n^{15}} (1^{15} + 2^{15} + 3^{15} + \cdots + n^{15})$

**23. a.** Show that the area $A_n$ of an $n$-sided regular polygon in a circle of radius $r$ is

$$A_n = \frac{nr^2}{2} \sin\frac{2\pi}{n}.$$

**b.** Find the limit of $A_n$ as $n \to \infty$. Is this answer consistent with what you know about the area of a circle?

**24.** Let

$$S_n = \frac{1^2}{n^3} + \frac{2^2}{n^3} + \cdots + \frac{(n-1)^2}{n^3}.$$

To calculate $\lim_{n\to\infty} S_n$, show that

$$S_n = \frac{1}{n}\left[ \left(\frac{1}{n}\right)^2 + \left(\frac{2}{n}\right)^2 + \cdots + \left(\frac{n-1}{n}\right)^2 \right]$$

and interpret $S_n$ as an approximating sum of the integral

$$\int_0^1 x^2 \, dx.$$

(*Hint:* Partition $[0, 1]$ into $n$ intervals of equal length and write out the approximating sum for inscribed rectangles.)

Defining Functions Using the Fundamental Theorem

**25. A function defined by an integral** The graph of a function $f$ consists of a semicircle and two line segments as shown. Let $g(x) = \int_1^x f(t) \, dt.$

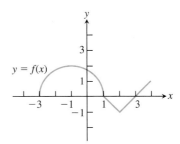

**a.** Find $g(1)$. **b.** Find $g(3)$. **c.** Find $g(-1)$.

**d.** Find all values of $x$ on the open interval $(-3, 4)$ at which $g$ has a relative maximum.

**e.** Write an equation for the line tangent to the graph of $g$ at $x = -1$.

**f.** Find the $x$-coordinate of each point of inflection of the graph of $g$ on the open interval $(-3, 4)$.

**g.** Find the range of $g$.

**26. A differential equation** Show that both of the following conditions are satisfied by $y = \sin x + \int_x^\pi \cos 2t \, dt + 1$:

**i)** $y'' = -\sin x + 2 \sin 2x$

**ii)** $y = 1$ and $y' = -2$ when $x = \pi$.

**Leibniz's Rule**   In applications, we sometimes encounter functions defined by integrals that have variable upper limits of integration and variable lower limits of integration at the same time. We can find the derivative of such an integral by a formula called **Leibniz's Rule**.

---

**Leibniz's Rule**

If $f$ is continuous on $[a, b]$ and if $u(x)$ and $v(x)$ are differentiable functions of $x$ whose values lie in $[a, b]$, then

$$\frac{d}{dx}\int_{u(x)}^{v(x)} f(t)\, dt = f(v(x))\frac{dv}{dx} - f(u(x))\frac{du}{dx}.$$

---

To prove the rule, let $F$ be an antiderivative of $f$ on $[a, b]$. Then

$$\int_{u(x)}^{v(x)} f(t)\, dt = F(v(x)) - F(u(x)).$$

Differentiating both sides of this equation with respect to $x$ gives the equation we want:

$$\frac{d}{dx}\int_{u(x)}^{v(x)} f(t)\, dt = \frac{d}{dx}\big[F(v(x)) - F(u(x))\big]$$

$$= F'(v(x))\frac{dv}{dx} - F'(u(x))\frac{du}{dx} \quad \text{Chain Rule}$$

$$= f(v(x))\frac{dv}{dx} - f(u(x))\frac{du}{dx}.$$

Use Leibniz's Rule to find the derivatives of the functions in Exercises 27–29.

**27.** $f(x) = \displaystyle\int_{1/x}^{x} \frac{1}{t}\, dt$

**28.** $f(x) = \displaystyle\int_{\cos x}^{\sin x} \frac{1}{1 - t^2}\, dt$

**29.** $g(y) = \displaystyle\int_{\sqrt{y}}^{2\sqrt{y}} \sin t^2\, dt$

**30.** Use Leibniz's Rule to find the value of $x$ that maximizes the value of the integral

$$\int_{x}^{x+3} t(5 - t)\, dt.$$

---

## Chapter **5**    Technology Application Projects

**Mathematica/Maple Modules:**

***Using Riemann Sums to Estimate Areas, Volumes, and Lengths of Curves***
Visualize and approximate areas and volumes in Part I.

***Riemann Sums, Definite Integrals, and the Fundamental Theorem of Calculus***
Parts I, II, and III develop Riemann sums and definite integrals. Part IV continues the development of the Riemann sum and definite integral using the Fundamental Theorem to solve problems previously investigated.

***Rain Catchers, Elevators, and Rockets***
Part I illustrates that the area under a curve is the same as the area of an appropriate rectangle for examples taken from the chapter. You will compute the amount of water accumulating in basins of different shapes as the basin is filled and drained.

***Motion Along a Straight Line, Part II***
You will observe the shape of a graph through dramatic animated visualizations of the derivative relations among position, velocity, and acceleration. Figures in the text can be animated using this software.

***Bending of Beams***
Study bent shapes of beams, determine their maximum deflections, concavity, and inflection points, and interpret the results in terms of a beam's compression and tension.

# 6

# Applications of Definite Integrals

**OVERVIEW** In Chapter 5 we saw that a continuous function over a closed interval has a definite integral, which is the limit of any Riemann sum for the function. We proved that we could evaluate definite integrals using the Fundamental Theorem of Calculus. We also found that the area under a curve and the area between two curves could be defined and computed as definite integrals.

In this chapter we extend the applications of definite integrals to defining and finding volumes, lengths of plane curves, and areas of surfaces of revolution. We also use integrals to solve physical problems involving the work done by a force, and to find the location of an object's center of mass. Each application comes from a process leading to an approximation by a Riemann sum, and then taking a limit to obtain an appropriate definite integral. These applications are important to mathematics, science, and engineering. We also use integrals to compute probabilities and their applications to the life sciences in Chapter 8.

## 6.1 Volumes Using Cross-Sections

In this section we define volumes of solids using the areas of their cross-sections. A **cross-section** of a solid $S$ is the plane region formed by intersecting $S$ with a plane (Figure 6.1). We present three different methods for obtaining the cross-sections appropriate to finding the volume of a particular solid: the method of slicing, the disk method, and the washer method.

Suppose we want to find the volume of a solid $S$ like the one in Figure 6.1. We begin by extending the definition of a cylinder from classical geometry to cylindrical solids with arbitrary bases (Figure 6.2). If the cylindrical solid has a known base area $A$ and height $h$, then the volume of the cylindrical solid is

$$\text{Volume} = \text{area} \times \text{height} = A \cdot h.$$

This equation forms the basis for defining the volumes of many solids that are not cylinders, like the one in Figure 6.1. If the cross-section of the solid $S$ at each point $x$ in the interval $[a, b]$ is a region $S(x)$ of area $A(x)$, and $A$ is a continuous function of $x$, we can

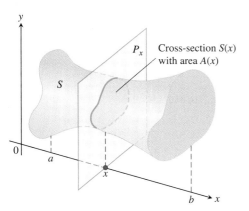

**FIGURE 6.1** A cross-section $S(x)$ of the solid $S$ formed by intersecting $S$ with a plane $P_x$ perpendicular to the $x$-axis through the point $x$ in the interval $[a, b]$.

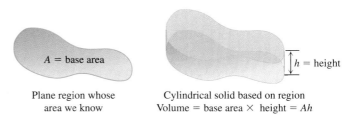

Plane region whose area we know

Cylindrical solid based on region
Volume = base area × height = $Ah$

**FIGURE 6.2** The volume of a cylindrical solid is always defined to be its base area times its height.

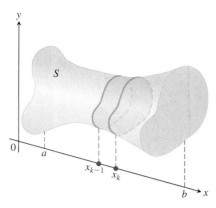

**FIGURE 6.3** A typical thin slab in the solid $S$.

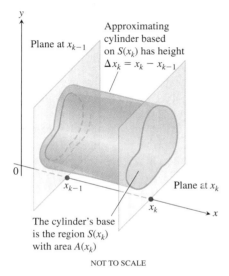

**FIGURE 6.4** The solid thin slab in Figure 6.3 is shown enlarged here. It is approximated by the cylindrical solid with base $S(x_k)$ having area $A(x_k)$ and height $\Delta x_k = x_k - x_{k-1}$.

define and calculate the volume of the solid $S$ as the definite integral of $A(x)$. We now show how this integral is obtained by the **method of slicing**.

### Slicing by Parallel Planes

We partition $[a, b]$ into subintervals of width (length) $\Delta x_k$ and slice the solid, as we would a loaf of bread, by planes perpendicular to the $x$-axis at the partition points $a = x_0 < x_1 < \cdots < x_n = b$. The planes $P_{x_k}$, perpendicular to the $x$-axis at the partition points, slice $S$ into thin "slabs" (like thin slices of a loaf of bread). A typical slab is shown in Figure 6.3. We approximate the slab between the plane at $x_{k-1}$ and the plane at $x_k$ by a cylindrical solid with base area $A(x_k)$ and height $\Delta x_k = x_k - x_{k-1}$ (Figure 6.4). The volume $V_k$ of this cylindrical solid is $A(x_k) \cdot \Delta x_k$, which is approximately the same volume as that of the slab:

$$\text{Volume of the } k\text{th slab} \approx V_k = A(x_k) \, \Delta x_k.$$

The volume $V$ of the entire solid $S$ is therefore approximated by the sum of these cylindrical volumes,

$$V \approx \sum_{k=1}^{n} V_k = \sum_{k=1}^{n} A(x_k) \, \Delta x_k.$$

This is a Riemann sum for the function $A(x)$ on $[a, b]$. We expect the approximations from these sums to improve as the norm of the partition of $[a, b]$ goes to zero. Taking a partition of $[a, b]$ into $n$ subintervals with $\|P\| \to 0$ gives

$$\lim_{n \to \infty} \sum_{k=1}^{n} A(x_k) \, \Delta x_k = \int_a^b A(x) \, dx.$$

So we define the limiting definite integral of the Riemann sum to be the volume of the solid $S$.

> **DEFINITION** The **volume** of a solid of integrable cross-sectional area $A(x)$ from $x = a$ to $x = b$ is the integral of $A$ from $a$ to $b$,
>
> $$V = \int_a^b A(x) \, dx.$$

This definition applies whenever $A(x)$ is integrable, and in particular when it is continuous. To apply the definition to calculate the volume of a solid using cross-sections perpendicular to the $x$-axis, take the following steps:

> **Calculating the Volume of a Solid**
> 1. *Sketch the solid and a typical cross-section.*
> 2. *Find a formula for $A(x)$, the area of a typical cross-section.*
> 3. *Find the limits of integration.*
> 4. *Integrate $A(x)$ to find the volume.*

**EXAMPLE 1** A pyramid 3 m high has a square base that is 3 m on a side. The cross-section of the pyramid perpendicular to the altitude $x$ m down from the vertex is a square $x$ m on a side. Find the volume of the pyramid.

**Solution**
1. *A sketch.* We draw the pyramid with its altitude along the $x$-axis and its vertex at the origin and include a typical cross-section (Figure 6.5).

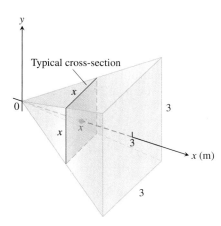

**FIGURE 6.5** The cross-sections of the pyramid in Example 1 are squares.

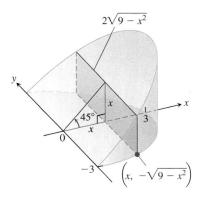

**FIGURE 6.6** The wedge of Example 2, sliced perpendicular to the *x*-axis. The cross-sections are rectangles.

**2.** *A formula for A(x).* The cross-section at *x* is a square *x* meters on a side, so its area is

$$A(x) = x^2.$$

**3.** *The limits of integration.* The squares lie on the planes from $x = 0$ to $x = 3$.

**4.** *Integrate to find the volume:*

$$V = \int_0^3 A(x)\,dx = \int_0^3 x^2\,dx = \frac{x^3}{3}\Bigg]_0^3 = 9\ \text{m}^3.$$

**EXAMPLE 2** A curved wedge is cut from a circular cylinder of radius 3 by two planes. One plane is perpendicular to the axis of the cylinder. The second plane crosses the first plane at a 45° angle at the center of the cylinder. Find the volume of the wedge.

**Solution** We draw the wedge and sketch a typical cross-section perpendicular to the *x*-axis (Figure 6.6). The base of the wedge in the figure is the semicircle with $x \geq 0$ that is cut from the circle $x^2 + y^2 = 9$ by the 45° plane when it intersects the *y*-axis. For any *x* in the interval $[0, 3]$, the *y*-values in this semicircular base vary from $y = -\sqrt{9 - x^2}$ to $y = \sqrt{9 - x^2}$. When we slice through the wedge by a plane perpendicular to the *x*-axis, we obtain a cross-section at *x* which is a rectangle of height *x* whose width extends across the semicircular base. The area of this cross-section is

$$A(x) = (\text{height})(\text{width}) = (x)\left(2\sqrt{9 - x^2}\right)$$
$$= 2x\sqrt{9 - x^2}.$$

The rectangles run from $x = 0$ to $x = 3$, so we have

$$V = \int_a^b A(x)\,dx = \int_0^3 2x\sqrt{9 - x^2}\,dx$$

$$= -\frac{2}{3}(9 - x^2)^{3/2}\Bigg]_0^3 \qquad \text{Let } u = 9 - x^2,$$
$$\qquad\qquad\qquad\qquad\qquad du = -2x\,dx, \text{ integrate,}$$
$$\qquad\qquad\qquad\qquad\qquad \text{and substitute back.}$$

$$= 0 + \frac{2}{3}(9)^{3/2}$$

$$= 18.$$

**EXAMPLE 3** Cavalieri's principle says that solids with equal altitudes and identical cross-sectional areas at each height have the same volume (Figure 6.7). This follows immediately from the definition of volume, because the cross-sectional area function $A(x)$ and the interval $[a, b]$ are the same for both solids.

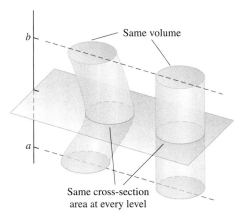

**FIGURE 6.7** *Cavalieri's principle:* These solids have the same volume, which can be illustrated with stacks of coins.

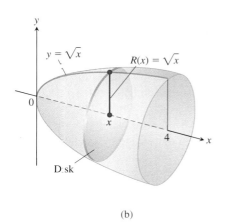

**FIGURE 6.8**   The region (a) and solid of revolution (b) in Example 4.

## Solids of Revolution: The Disk Method

The solid generated by rotating (or revolving) a plane region about an axis in its plane is called a **solid of revolution**. To find the volume of a solid like the one shown in Figure 6.8, we need only observe that the cross-sectional area $A(x)$ is the area of a disk of radius $R(x)$, the distance of the planar region's boundary from the axis of revolution. The area is then

$$A(x) = \pi(\text{radius})^2 = \pi [\, R(x) \,]^2.$$

So the definition of volume in this case gives

---

**Volume by Disks for Rotation About the $x$-axis**

$$V = \int_a^b A(x)\, dx = \int_a^b \pi [\, R(x) \,]^2\, dx.$$

---

This method for calculating the volume of a solid of revolution is often called the **disk method** because a cross-section is a circular disk of radius $R(x)$.

**EXAMPLE 4**    The region between the curve $y = \sqrt{x},\, 0 \le x \le 4$, and the $x$-axis is revolved about the $x$-axis to generate a solid. Find its volume.

**Solution**   We draw figures showing the region, a typical radius, and the generated solid (Figure 6.8). The volume is

$$V = \int_a^b \pi [\, R(x) \,]^2\, dx$$

$$= \int_0^4 \pi [\, \sqrt{x} \,]^2\, dx \qquad \text{Radius } R(x) = \sqrt{x} \text{ for rotation around } x\text{-axis.}$$

$$= \pi \int_0^4 x\, dx = \pi \frac{x^2}{2} \Big]_0^4 = \pi \frac{(4)^2}{2} = 8\pi. \qquad ■$$

**EXAMPLE 5**    The circle

$$x^2 + y^2 = a^2$$

is rotated about the $x$-axis to generate a sphere. Find its volume.

**Solution**   We imagine the sphere cut into thin slices by planes perpendicular to the $x$-axis (Figure 6.9). The cross-sectional area at a typical point $x$ between $-a$ and $a$ is

$$A(x) = \pi y^2 = \pi(a^2 - x^2). \qquad R(x) = \sqrt{a^2 - x^2} \text{ for rotation around } x\text{-axis.}$$

Therefore, the volume is

$$V = \int_{-a}^a A(x)\, dx = \int_{-a}^a \pi(a^2 - x^2)\, dx = \pi \left[ a^2 x - \frac{x^3}{3} \right]_{-a}^a = \frac{4}{3}\pi a^3. \qquad ■$$

The axis of revolution in the next example is not the $x$-axis, but the rule for calculating the volume is the same: Integrate $\pi(\text{radius})^2$ between appropriate limits.

**EXAMPLE 6**    Find the volume of the solid generated by revolving the region bounded by $y = \sqrt{x}$ and the lines $y = 1,\, x = 4$ about the line $y = 1$.

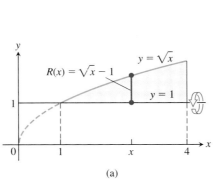

**FIGURE 6.9** The sphere generated by rotating the circle $x^2 + y^2 = a^2$ about the $x$-axis. The radius is $R(x) = y = \sqrt{a^2 - x^2}$ (Example 5).

**Solution** We draw figures showing the region, a typical radius, and the generated solid (Figure 6.10). The volume is

$$V = \int_1^4 \pi \left[ R(x) \right]^2 dx$$

$$= \int_1^4 \pi \left[ \sqrt{x} - 1 \right]^2 dx \qquad \text{Radius } R(x) = \sqrt{x} - 1 \text{ for rotation around } y = 1.$$

$$= \pi \int_1^4 \left[ x - 2\sqrt{x} + 1 \right] dx \qquad \text{Expand integrand.}$$

$$= \pi \left[ \frac{x^2}{2} - 2 \cdot \frac{2}{3} x^{3/2} + x \right]_1^4 = \frac{7\pi}{6}. \qquad \text{Integrate.}$$

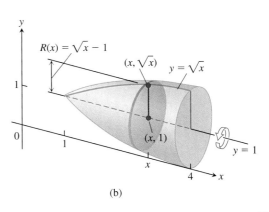

(a)                                                   (b)

**FIGURE 6.10** The region (a) and solid of revolution (b) in Example 6.

To find the volume of a solid generated by revolving a region between the $y$-axis and a curve $x = R(y)$, $c \le y \le d$, about the $y$-axis, we use the same method with $x$ replaced by $y$. In this case, the area of the circular cross-section is

$$A(y) = \pi \left[ \text{radius} \right]^2 = \pi \left[ R(y) \right]^2,$$

and the definition of volume gives

**Volume by Disks for Rotation About the y-axis**

$$V = \int_c^d A(y)\, dy = \int_c^d \pi\,[\,R(y)\,]^2\, dy.$$

**EXAMPLE 7**    Find the volume of the solid generated by revolving the region between the y-axis and the curve $x = 2/y$, $1 \le y \le 4$, about the y-axis.

**Solution**    We draw figures showing the region, a typical radius, and the generated solid (Figure 6.11). The volume is

$$V = \int_1^4 \pi\,[\,R(y)\,]^2\, dy$$

$$= \int_1^4 \pi \left(\frac{2}{y}\right)^2 dy \qquad \text{Radius } R(y) = \frac{2}{y} \text{ for rotation around } y\text{-axis.}$$

$$= \pi \int_1^4 \frac{4}{y^2}\, dy = 4\pi \left[-\frac{1}{y}\right]_1^4 = 4\pi \left[\frac{3}{4}\right] = 3\pi. \qquad ■$$

**EXAMPLE 8**    Find the volume of the solid generated by revolving the region between the parabola $x = y^2 + 1$ and the line $x = 3$ about the line $x = 3$.

**Solution**    We draw figures showing the region, a typical radius, and the generated solid (Figure 6.12). Note that the cross-sections are perpendicular to the line $x = 3$ and have y-coordinates from $y = -\sqrt{2}$ to $y = \sqrt{2}$. The volume is

$$V = \int_{-\sqrt{2}}^{\sqrt{2}} \pi\,[\,R(y)\,]^2\, dy \qquad y = \pm\sqrt{2} \text{ when } x = 3$$

$$= \int_{-\sqrt{2}}^{\sqrt{2}} \pi\,[\,2 - y^2\,]^2\, dy \qquad \begin{array}{l}\text{Radius } R(y) = 3 - (y^2 + 1) \\ \text{for rotation around axis } x = 3.\end{array}$$

$$= \pi \int_{-\sqrt{2}}^{\sqrt{2}} [\,4 - 4y^2 + y^4\,]\, dy \qquad \text{Expand integrand.}$$

$$= \pi \left[4y - \frac{4}{3}y^3 + \frac{y^5}{5}\right]_{-\sqrt{2}}^{\sqrt{2}} \qquad \text{Integrate.}$$

$$= \frac{64\pi\sqrt{2}}{15}.$$

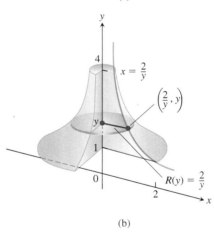

**FIGURE 6.11**    The region (a) and part of the solid of revolution (b) in Example 7.

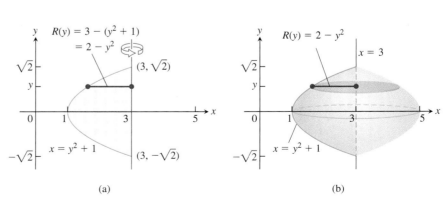

(a)                    (b)

**FIGURE 6.12**    The region (a) and solid of revolution (b) in Example 8.    ■

**FIGURE 6.13** The cross-sections of the solid of revolution generated here are washers, not disks, so the integral $\int_a^b A(x)\,dx$ leads to a slightly different formula.

## Solids of Revolution: The Washer Method

If the region we revolve to generate a solid does not border on or cross the axis of revolution, the solid has a hole in it (Figure 6.13). The cross-sections perpendicular to the axis of revolution are *washers* (the purplish circular surface in Figure 6.13) instead of disks. The dimensions of a typical washer are

$$\text{Outer radius:} \quad R(x)$$
$$\text{Inner radius:} \quad r(x)$$

The washer's area is

$$A(x) = \pi\,[\,R(x)\,]^2 - \pi\,[\,r(x)\,]^2 = \pi([\,R(x)\,]^2 - [\,r(x)\,]^2).$$

Consequently, the definition of volume in this case gives

> **Volume by Washers for Rotation About the $x$-axis**
>
> $$V = \int_a^b A(x)\,dx = \int_a^b \pi([\,R(x)\,]^2 - [\,r(x)\,]^2)\,dx.$$

This method for calculating the volume of a solid of revolution is called the **washer method** because a thin slab of the solid resembles a circular washer of outer radius $R(x)$ and inner radius $r(x)$.

**EXAMPLE 9** The region bounded by the curve $y = x^2 + 1$ and the line $y = -x + 3$ is revolved about the $x$-axis to generate a solid. Find the volume of the solid.

**Solution** We use the four steps for calculating the volume of a solid as discussed early in this section.

1. Draw the region and sketch a line segment across it perpendicular to the axis of revolution (the red segment in Figure 6.14a).

2. Find the outer and inner radii of the washer that would be swept out by the line segment if it were revolved about the $x$-axis along with the region.

   These radii are the distances of the ends of the line segment from the axis of revolution (Figure 6.14).

$$\text{Outer radius:} \quad R(x) = -x + 3$$
$$\text{Inner radius:} \quad r(x) = x^2 + 1$$

(a)

Washer cross-section
Outer radius: $R(x) = -x + 3$
Inner radius: $r(x) = x^2 + 1$

(b)

**FIGURE 6.14** (a) The region in Example 9 spanned by a line segment perpendicular to the axis of revolution. (b) When the region is revolved about the $x$-axis, the line segment generates a washer.

3. Find the limits of integration by finding the $x$-coordinates of the intersection points of the curve and line in Figure 6.14a.

$$x^2 + 1 = -x + 3$$
$$x^2 + x - 2 = 0$$
$$(x + 2)(x - 1) = 0$$
$$x = -2, \quad x = 1 \qquad \text{Limits of integration}$$

4. Evaluate the volume integral.

$$V = \int_a^b \pi ( [R(x)]^2 - [r(x)]^2 ) \, dx \qquad \text{Rotation around } x\text{-axis}$$

$$= \int_{-2}^1 \pi((-x + 3)^2 - (x^2 + 1)^2) \, dx \qquad \begin{array}{l}\text{Values from Steps 2}\\ \text{and 3}\end{array}$$

$$= \pi \int_{-2}^1 (8 - 6x - x^2 - x^4) \, dx \qquad \text{Simplify algebraically.}$$

$$= \pi \left[ 8x - 3x^2 - \frac{x^3}{3} - \frac{x^5}{5} \right]_{-2}^1 = \frac{117\pi}{5} \qquad \text{Integrate.} \qquad ∎$$

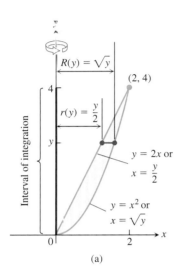

To find the volume of a solid formed by revolving a region about the $y$-axis, we use the same procedure as in Example 9, but integrate with respect to $y$ instead of $x$. In this situation the line segment sweeping out a typical washer is perpendicular to the $y$-axis (the axis of revolution), and the outer and inner radii of the washer are functions of $y$.

**EXAMPLE 10**    The region bounded by the parabola $y = x^2$ and the line $y = 2x$ in the first quadrant is revolved about the $y$-axis to generate a solid. Find the volume of the solid.

**Solution**    First we sketch the region and draw a line segment across it perpendicular to the axis of revolution (the $y$-axis). See Figure 6.15a.

The radii of the washer swept out by the line segment are $R(y) = \sqrt{y}, \, r(y) = y/2$ (Figure 6.15).

The line and parabola intersect at $y = 0$ and $y = 4$, so the limits of integration are $c = 0$ and $d = 4$. We integrate to find the volume:

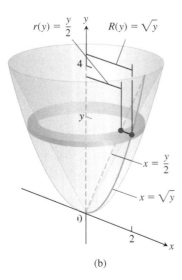

$$V = \int_c^d \pi ( [R(y)]^2 - [r(y)]^2 ) \, dy \qquad \text{Rotation around } y\text{-axis}$$

$$= \int_0^4 \pi \left( [\sqrt{y}]^2 - \left[\frac{y}{2}\right]^2 \right) dy \qquad \begin{array}{l}\text{Substitute for radii and}\\ \text{limits of integration.}\end{array}$$

$$= \pi \int_0^4 \left( y - \frac{y^2}{4} \right) dy = \pi \left[ \frac{y^2}{2} - \frac{y^3}{12} \right]_0^4 = \frac{8}{3}\pi. \qquad ∎$$

**FIGURE 6.15**    (a) The region being rotated about the $y$-axis, the washer radii, and limits of integration in Example 10. (b) The washer swept out by the line segment in part (a).

# Exercises 6.1

### Volumes by Slicing

Find the volumes of the solids in Exercises 1–10.

1. The solid lies between planes perpendicular to the $x$-axis at $x = 0$ and $x = 4$. The cross-sections perpendicular to the axis on the interval $0 \leq x \leq 4$ are squares whose diagonals run from the parabola $y = -\sqrt{x}$ to the parabola $y = \sqrt{x}$.

2. The solid lies between planes perpendicular to the $x$-axis at $x = -1$ and $x = 1$. The cross-sections perpendicular to the $x$-axis are circular disks whose diameters run from the parabola $y = x^2$ to the parabola $y = 2 - x^2$.

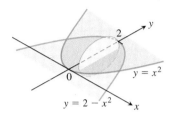

3. The solid lies between planes perpendicular to the $x$-axis at $x = -1$ and $x = 1$. The cross-sections perpendicular to the $x$-axis between these planes are squares whose bases run from the semicircle $y = -\sqrt{1 - x^2}$ to the semicircle $y = \sqrt{1 - x^2}$.

4. The solid lies between planes perpendicular to the $x$-axis at $x = -1$ and $x = 1$. The cross-sections perpendicular to the $x$-axis between these planes are squares whose diagonals run from the semicircle $y = -\sqrt{1 - x^2}$ to the semicircle $y = \sqrt{1 - x^2}$.

5. The base of a solid is the region between the curve $y = 2\sqrt{\sin x}$ and the interval $[0, \pi]$ on the $x$-axis. The cross-sections perpendicular to the $x$-axis are

   a. equilateral triangles with bases running from the $x$-axis to the curve as shown in the accompanying figure.

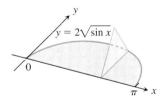

   b. squares with bases running from the $x$-axis to the curve.

6. The solid lies between planes perpendicular to the $x$-axis at $x = -\pi/3$ and $x = \pi/3$. The cross-sections perpendicular to the $x$-axis are

   a. circular disks with diameters running from the curve $y = \tan x$ to the curve $y = \sec x$.

   b. squares whose bases run from the curve $y = \tan x$ to the curve $y = \sec x$.

7. The base of a solid is the region bounded by the graphs of $y = 3x$, $y = 6$, and $x = 0$. The cross-sections perpendicular to the $x$-axis are

   a. rectangles of height 10.

   b. rectangles of perimeter 20.

8. The base of a solid is the region bounded by the graphs of $y = \sqrt{x}$ and $y = x/2$. The cross-sections perpendicular to the $x$-axis are

   a. isosceles triangles of height 6.

   b. semicircles with diameters running across the base of the solid.

9. The solid lies between planes perpendicular to the $y$-axis at $y = 0$ and $y = 2$. The cross-sections perpendicular to the $y$-axis are circular disks with diameters running from the $y$-axis to the parabola $x = \sqrt{5}y^2$.

10. The base of the solid is the disk $x^2 + y^2 \leq 1$. The cross-sections by planes perpendicular to the $y$-axis between $y = -1$ and $y = 1$ are isosceles right triangles with one leg in the disk.

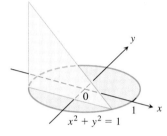

11. Find the volume of the given right tetrahedron. (*Hint:* Consider slices perpendicular to one of the labeled edges.)

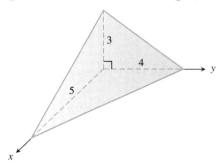

12. Find the volume of the given pyramid, which has a square base of area 9 and height 5.

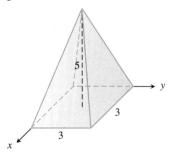

13. **A twisted solid** A square of side length $s$ lies in a plane perpendicular to a line $L$. One vertex of the square lies on $L$. As this square moves a distance $h$ along $L$, the square turns one revolution about $L$ to generate a corkscrew-like column with square cross-sections.

    a. Find the volume of the column.

    b. What will the volume be if the square turns twice instead of once? Give reasons for your answer.

14. **Cavalieri's principle** A solid lies between planes perpendicular to the x-axis at $x = 0$ and $x = 12$. The cross-sections by planes perpendicular to the x-axis are circular disks whose diameters run from the line $y = x/2$ to the line $y = x$ as shown in the accompanying figure. Explain why the solid has the same volume as a right circular cone with base radius 3 and height 12.

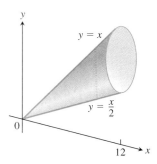

## Volumes by the Disk Method

In Exercises 15–18, find the volume of the solid generated by revolving the shaded region about the given axis.

15. About the x-axis

16. About the y-axis

17. About the y-axis

18. About the x-axis

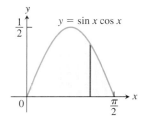

Find the volumes of the solids generated by revolving the regions bounded by the lines and curves in Exercises 19–28 about the x-axis.

19. $y = x^2$, $y = 0$, $x = 2$
20. $y = x^3$, $y = 0$, $x = 2$
21. $y = \sqrt{9 - x^2}$, $y = 0$
22. $y = x - x^2$, $y = 0$
23. $y = \sqrt{\cos x}$, $0 \le x \le \pi/2$, $y = 0$, $x = 0$
24. $y = \sec x$, $y = 0$, $x = -\pi/4$, $x = \pi/4$

In Exercises 25 and 26, find the volume of the solid generated by revolving the region about the given line.

25. The region in the first quadrant bounded above by the line $y = \sqrt{2}$, below by the curve $y = \sec x \tan x$, and on the left by the y-axis, about the line $y = \sqrt{2}$

26. The region in the first quadrant bounded above by the line $y = 2$, below by the curve $y = 2 \sin x, 0 \le x \le \pi/2$, and on the left by the y-axis, about the line $y = 2$

Find the volumes of the solids generated by revolving the regions bounded by the lines and curves in Exercises 27–32 about the y-axis.

27. The region enclosed by $x = \sqrt{5}y^2$, $x = 0$, $y = -1$, $y = 1$
28. The region enclosed by $x = y^{3/2}$, $x = 0$, $y = 2$
29. The region enclosed by $x = \sqrt{2 \sin 2y}$, $0 \le y \le \pi/2$, $x = 0$
30. The region enclosed by $x = \sqrt{\cos(\pi y/4)}$, $-2 \le y \le 0$, $x = 0$
31. $x = y^{1/3}$, $x = y^3$, $0 \le y \le 1$
32. $x = \sqrt{2y}/(y^2 + 1)$, $x = 0$, $y = 1$

## Volumes by the Washer Method

Find the volumes of the solids generated by revolving the shaded regions in Exercises 33 and 34 about the indicated axes.

33. The x-axis

34. The y-axis

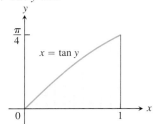

Find the volumes of the solids generated by revolving the regions bounded by the lines and curves in Exercises 35–40 about the x-axis.

35. $y = x$, $y = 1$, $x = 0$
36. $y = 2\sqrt{x}$, $y = 2$, $x = 0$
37. $y = x^2 + 1$, $y = x + 3$
38. $y = 4 - x^2$, $y = 2 - x$
39. $y = \sec x$, $y = \sqrt{2}$, $-\pi/4 \le x \le \pi/4$
40. $y = \sec x$, $y = \tan x$, $x = 0$, $x = 1$

In Exercises 41–44, find the volume of the solid generated by revolving each region about the y-axis.

41. The region enclosed by the triangle with vertices $(1, 0)$, $(2, 1)$, and $(1, 1)$

42. The region enclosed by the triangle with vertices $(0, 1)$, $(1, 0)$, and $(1, 1)$

43. The region in the first quadrant bounded above by the parabola $y = x^2$, below by the x-axis, and on the right by the line $x = 2$

44. The region in the first quadrant bounded on the left by the circle $x^2 + y^2 = 3$, on the right by the line $x = \sqrt{3}$, and above by the line $y = \sqrt{3}$

In Exercises 45 and 46, find the volume of the solid generated by revolving each region about the given axis.

45. The region in the first quadrant bounded above by the curve $y = x^2$, below by the x-axis, and on the right by the line $x = 1$, about the line $x = -1$

**46.** The region in the second quadrant bounded above by the curve $y = -x^3$, below by the $x$-axis, and on the left by the line $x = -1$, about the line $x = -2$

**Volumes of Solids of Revolution**

**47.** Find the volume of the solid generated by revolving the region bounded by $y = \sqrt{x}$ and the lines $y = 2$ and $x = 0$ about

   **a.** the $x$-axis.     **b.** the $y$-axis.

   **c.** the line $y = 2$.     **d.** the line $x = 4$.

**48.** Find the volume of the solid generated by revolving the triangular region bounded by the lines $y = 2x$, $y = 0$, and $x = 1$ about

   **a.** the line $x = 1$.     **b.** the line $x = 2$.

**49.** Find the volume of the solid generated by revolving the region bounded by the parabola $y = x^2$ and the line $y = 1$ about

   **a.** the line $y = 1$.     **b.** the line $y = 2$.

   **c.** the line $y = -1$.

**50.** By integration, find the volume of the solid generated by revolving the triangular region with vertices $(0, 0)$, $(b, 0)$, $(0, h)$ about

   **a.** the $x$-axis.     **b.** the $y$-axis.

**Theory and Applications**

**51. The volume of a torus** The disk $x^2 + y^2 \le a^2$ is revolved about the line $x = b$ $(b > a)$ to generate a solid shaped like a doughnut and called a *torus*. Find its volume. (*Hint:* $\int_{-a}^{a} \sqrt{a^2 - y^2}\, dy = \pi a^2/2$, since it is the area of a semicircle of radius $a$.)

**52. Volume of a bowl** A bowl has a shape that can be generated by revolving the graph of $y = x^2/2$ between $y = 0$ and $y = 5$ about the $y$-axis.

   **a.** Find the volume of the bowl.

   **b. Related rates** If we fill the bowl with water at a constant rate of 3 cubic units per second, how fast will the water level in the bowl be rising when the water is 4 units deep?

**53. Volume of a bowl**

   **a.** A hemispherical bowl of radius $a$ contains water to a depth $h$. Find the volume of water in the bowl.

   **b. Related rates** Water runs into a sunken concrete hemispherical bowl of radius 5 m at the rate of 0.2 m³/s. How fast is the water level in the bowl rising when the water is 4 m deep?

**54.** Explain how you could estimate the volume of a solid of revolution by measuring the shadow cast on a table parallel to its axis of revolution by a light shining directly above it.

**55. Volume of a hemisphere** Derive the formula $V = (2/3)\pi R^3$ for the volume of a hemisphere of radius $R$ by comparing its cross-sections with the cross-sections of a solid right circular cylinder of radius $R$ and height $R$ from which a solid right circular cone of base radius $R$ and height $R$ has been removed, as suggested by the accompanying figure.

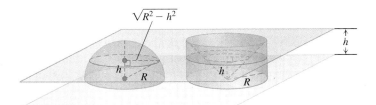

**56. Designing a plumb bob** Having been asked to design a brass plumb bob that will weigh in the neighborhood of 190 g, you decide to shape it like the solid of revolution shown here. Find the plumb bob's volume. If you specify a brass that weighs 8.5 g/cm³, how much will the plumb bob weigh (to the nearest gram)?

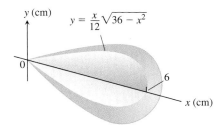

**57. Designing a wok** You are designing a wok frying pan that will be shaped like a spherical bowl with handles. A bit of experimentation at home persuades you that you can get one that holds about 3 L if you make it 9 cm deep and give the sphere a radius of 16 cm. To be sure, you picture the wok as a solid of revolution, as shown here, and calculate its volume with an integral. To the nearest cubic centimeter, what volume do you really get? (1 L = 1000 cm³)

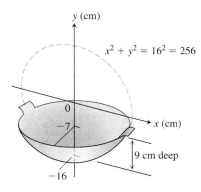

**58. Max-min** The arch $y = \sin x$, $0 \le x \le \pi$, is revolved about the line $y = c$, $0 \le c \le 1$, to generate the solid in the accompanying figure.

   **a.** Find the value of $c$ that minimizes the volume of the solid. What is the minimum volume?

   **b.** What value of $c$ in $[0, 1]$ maximizes the volume of the solid?

   **T c.** Graph the solid's volume as a function of $c$, first for $0 \le c \le 1$ and then on a larger domain. What happens to the volume of the solid as $c$ moves away from $[0, 1]$? Does this make sense physically? Give reasons for your answers.

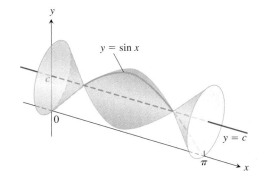

**59.** Consider the region $R$ bounded by the graphs of $y = f(x) > 0$, $x = a > 0$, $x = b > a$, and $y = 0$ (see accompanying figure). If the volume of the solid formed by revolving $R$ about the $x$-axis is $4\pi$, and the volume of the solid formed by revolving $R$ about the line $y = -1$ is $8\pi$, find the area of $R$.

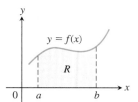

**60.** Consider the region $R$ given in Exercise 63. If the volume of the solid formed by revolving $R$ around the $x$-axis is $6\pi$, and the volume of the solid formed by revolving $R$ around the line $y = -2$ is $10\pi$, find the area of $R$.

# 6.2 Volumes Using Cylindrical Shells

In Section 6.1 we defined the volume of a solid as the definite integral $V = \int_a^b A(x)\,dx$, where $A(x)$ is an integrable cross-sectional area of the solid from $x = a$ to $x = b$. The area $A(x)$ was obtained by slicing through the solid with a plane perpendicular to the $x$-axis. However, this method of slicing is sometimes awkward to apply, as we will illustrate in our first example. To overcome this difficulty, we use the same integral definition for volume, but obtain the area by slicing through the solid in a different way.

### Slicing with Cylinders

Suppose we slice through the solid using circular cylinders of increasing radii, like cookie cutters. We slice straight down through the solid so that the axis of each cylinder is parallel to the $y$-axis. The vertical axis of each cylinder is the same line, but the radii of the cylinders increase with each slice. In this way the solid is sliced up into thin cylindrical shells of constant thickness that grow outward from their common axis, like circular tree rings. Unrolling a cylindrical shell shows that its volume is approximately that of a rectangular slab with area $A(x)$ and thickness $\Delta x$. This slab interpretation allows us to apply the same integral definition for volume as before. The following example provides some insight before we derive the general method.

**EXAMPLE 1** The region enclosed by the $x$-axis and the parabola $y = f(x) = 3x - x^2$ is revolved about the vertical line $x = -1$ to generate a solid (Figure 6.16). Find the volume of the solid.

**Solution** Using the washer method from Section 6.1 would be awkward here because we would need to express the $x$-values of the left and right sides of the parabola in Figure 6.16a in terms of $y$. (These $x$-values are the inner and outer radii for a typical washer, requiring us to solve $y = 3x - x^2$ for $x$, which leads to complicated formulas.) Instead of rotating a horizontal strip of thickness $\Delta y$, we rotate a *vertical strip* of thickness $\Delta x$. This rotation produces a *cylindrical shell* of height $y_k$ above a point $x_k$ within the base of the vertical strip and of thickness $\Delta x$. An example of a cylindrical shell is shown as the orange-shaded region in Figure 6.17. We can think of the cylindrical shell shown in the figure as approximating a slice of the solid obtained by cutting straight down through it, parallel to the axis of revolution, all the way around close to the inside hole. We then cut another cylindrical slice around the enlarged hole, then another, and so on, obtaining $n$ cylinders. The radii of the cylinders gradually increase, and the heights of the cylinders follow the contour of the parabola: shorter to taller, then back to shorter (Figure 6.16a).

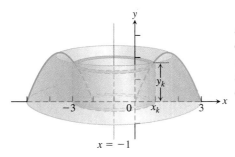

**FIGURE 6.16** (a) The graph of the region in Example 1, before revolution. (b) The solid formed when the region in part (a) is revolved about the axis of revolution $x = -1$.

**FIGURE 6.17** A cylindrical shell of height $y_k$ obtained by rotating a vertical strip of thickness $\Delta x_k$ about the line $x = -1$. The outer radius of the cylinder occurs at $x_k$, where the height of the parabola is $y_k = 3x_k - x_k{}^2$ (Example 1).

Each slice is sitting over a subinterval of the $x$-axis of length (width) $\Delta x_k$. Its radius is approximately $(1 + x_k)$, and its height is approximately $3x_k - x_k{}^2$. If we unroll the cylinder at $x_k$ and flatten it out, it becomes (approximately) a rectangular slab with thickness $\Delta x_k$ (Figure 6.18). The outer circumference of the $k$th cylinder is $2\pi \cdot \text{radius} = 2\pi(1 + x_k)$, and this is the length of the rolled-out rectangular slab. Its volume is approximated by that of a rectangular solid, the area of the rectangle times its thickness,

$$\Delta V_k = \text{circumference} \times \text{height} \times \text{thickness}$$
$$= 2\pi(1 + x_k) \cdot \left(3x_k - x_k{}^2\right) \cdot \Delta x_k.$$

Summing together the volumes $\Delta V_k$ of the individual cylindrical shells over the interval $[0, 3]$ gives the Riemann sum

$$\sum_{k=1}^{n} \Delta V_k = \sum_{k=1}^{n} 2\pi(x_k + 1)\left(3x_k - x_k{}^2\right)\Delta x_k.$$

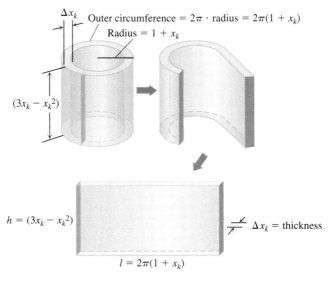

**FIGURE 6.18** Cutting and unrolling a cylindrical shell gives a nearly rectangular solid (Example 1).

Taking the limit as the thickness $\Delta x_k \to 0$ and $n \to \infty$ gives the volume integral

$$V = \lim_{n \to \infty} \sum_{k=1}^{n} 2\pi(x_k + 1)\left(3x_k - x_k^2\right) \Delta x_k$$

$$= \int_0^3 2\pi(x + 1)(3x - x^2) \, dx$$

$$= \int_0^3 2\pi(3x^2 + 3x - x^3 - x^2) \, dx$$

$$= 2\pi \int_0^3 (2x^2 + 3x - x^3) \, dx$$

$$= 2\pi\left[\frac{2}{3}x^3 + \frac{3}{2}x^2 - \frac{1}{4}x^4\right]_0^3 = \frac{45\pi}{2}. \qquad \blacksquare$$

We now generalize the procedure used in Example 1.

### The Shell Method

Suppose the region bounded by the graph of a nonnegative continuous function $y = f(x)$ and the $x$-axis over the finite closed interval $[a, b]$ lies to the right of the vertical line $x = L$ (Figure 6.19a). We assume $a \geq L$, so the vertical line may touch the region, but not pass through it. We generate a solid $S$ by rotating this region about the vertical line $L$.

Let $P$ be a partition of the interval $[a, b]$ by the points $a = x_0 < x_1 < \cdots < x_n = b$, and let $c_k$ be the midpoint of the $k$th subinterval $[x_{k-1}, x_k]$. We approximate the region in Figure 6.19a with rectangles based on this partition of $[a, b]$. A typical approximating rectangle has height $f(c_k)$ and width $\Delta x_k = x_k - x_{k-1}$. If this rectangle is rotated about the vertical line $x = L$, then a shell is swept out, as in Figure 6.19b. A formula from geometry tells us that the volume of the shell swept out by the rectangle is

> The volume of a cylindrical shell of height $h$ with inner radius $r$ and outer radius $R$ is
>
> $$\pi R^2 h - \pi r^2 h = 2\pi\left(\frac{R + r}{2}\right)(h)(R - r).$$

$$\Delta V_k = 2\pi \times \text{average shell radius} \times \text{shell height} \times \text{thickness}$$

$$= 2\pi \cdot (c_k - L) \cdot f(c_k) \cdot \Delta x_k. \qquad R = x_k - L \text{ and } r = x_{k-1} - L$$

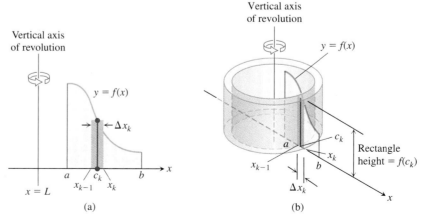

**FIGURE 6.19** When the region shown in (a) is revolved about the vertical line $x = L$, a solid is produced which can be sliced into cylindrical shells. A typical shell is shown in (b).

We approximate the volume of the solid $S$ by summing the volumes of the shells swept out by the $n$ rectangles based on $P$:

$$V \approx \sum_{k=1}^{n} \Delta V_k.$$

The limit of this Riemann sum as each $\Delta x_k \to 0$ and $n \to \infty$ gives the volume of the solid as a definite integral:

$$V = \lim_{n \to \infty} \sum_{k=1}^{n} \Delta V_k = \int_a^b 2\pi(\text{shell radius})(\text{shell height})\, dx$$

$$= \int_a^b 2\pi(x - L)f(x)\, dx.$$

We refer to the variable of integration, here $x$, as the **thickness variable**. We use the first integral, rather than the second containing a formula for the integrand, to emphasize the *process* of the shell method. This will allow for rotations about a horizontal line $L$ as well.

---

**Shell Formula for Revolution About a Vertical Line**

The volume of the solid generated by revolving the region between the $x$-axis and the graph of a continuous function $y = f(x) \geq 0, L \leq a \leq x \leq b$, about a vertical line $x = L$ is

$$V = \int_a^b 2\pi \begin{pmatrix} \text{shell} \\ \text{radius} \end{pmatrix} \begin{pmatrix} \text{shell} \\ \text{height} \end{pmatrix} dx.$$

---

**EXAMPLE 2**    The region bounded by the curve $y = \sqrt{x}$, the $x$-axis, and the line $x = 4$ is revolved about the $y$-axis to generate a solid. Find the volume of the solid.

**Solution**    Sketch the region and draw a line segment across it *parallel* to the axis of revolution (Figure 6.20a). Label the segment's height (shell height) and distance from the axis of revolution (shell radius). (We drew the shell in Figure 6.20b, but you need not do that.)

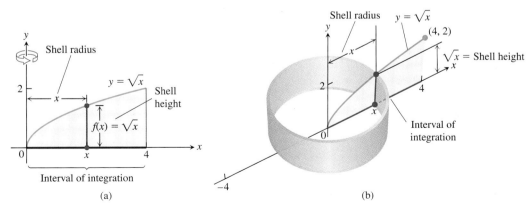

**FIGURE 6.20**    (a) The region, shell dimensions, and interval of integration in Example 2. (b) The shell swept out by the vertical segment in part (a) with a width $\Delta x$.

The shell thickness variable is $x$, so the limits of integration for the shell formula are $a = 0$ and $b = 4$ (Figure 6.20). The volume is then

$$V = \int_a^b 2\pi \begin{pmatrix} \text{shell} \\ \text{radius} \end{pmatrix} \begin{pmatrix} \text{shell} \\ \text{height} \end{pmatrix} dx$$

$$= \int_0^4 2\pi (x)\left( \sqrt{x} \right) dx$$

$$= 2\pi \int_0^4 x^{3/2} \, dx = 2\pi \left[ \frac{2}{5} x^{5/2} \right]_0^4 = \frac{128\pi}{5}.$$

So far, we have used vertical axes of revolution. For horizontal axes, we replace the $x$'s with $y$'s.

**EXAMPLE 3**    The region bounded by the curve $y = \sqrt{x}$, the $x$-axis, and the line $x = 4$ is revolved about the $x$-axis to generate a solid. Find the volume of the solid by the shell method.

**Solution**    This is the solid whose volume was found by the disk method in Example 4 of Section 6.1. Now we find its volume by the shell method. First, sketch the region and draw a line segment across it *parallel* to the axis of revolution (Figure 6.21a). Label the segment's length (shell height) and distance from the axis of revolution (shell radius). (We drew the shell in Figure 6.21b, but you need not do that.)

In this case, the shell thickness variable is $y$, so the limits of integration for the shell formula method are $a = 0$ and $b = 2$ (along the $y$-axis in Figure 6.21). The volume of the solid is

$$V = \int_a^b 2\pi \begin{pmatrix} \text{shell} \\ \text{radius} \end{pmatrix} \begin{pmatrix} \text{shell} \\ \text{height} \end{pmatrix} dy$$

$$= \int_0^2 2\pi (y)(4 - y^2) \, dy$$

$$= 2\pi \int_0^2 (4y - y^3) \, dy$$

$$= 2\pi \left[ 2y^2 - \frac{y^4}{4} \right]_0^2 = 8\pi.$$

(a)

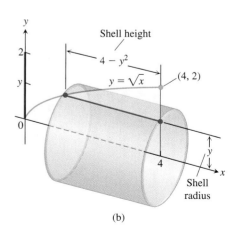
(b)

**FIGURE 6.21**    (a) The region, shell dimensions, and interval of integration in Example 3. (b) The shell swept out by the horizontal segment in part (a) with a width $\Delta y$.

---

**Summary of the Shell Method**

Regardless of the position of the axis of revolution (horizontal or vertical), the steps for implementing the shell method are these.

1. *Draw the region and sketch a line segment* across it *parallel* to the axis of revolution. *Label* the segment's height or length (shell height) and distance from the axis of revolution (shell radius).
2. *Find the limits of integration* for the thickness variable.
3. *Integrate* the product $2\pi$ (shell radius) (shell height) with respect to the thickness variable ($x$ or $y$) to find the volume.

---

The shell method gives the same answer as the washer method when both are used to calculate the volume of a region. We do not prove that result here, but it is illustrated in Exercises 37 and 38. (Exercise 45 outlines a proof.) Both volume formulas are actually special cases of a general volume formula we will look at when studying double and triple integrals in Chapter 15. That general formula also allows for computing volumes of solids other than those swept out by regions of revolution.

## Exercises 6.2

### Revolution About the Axes

In Exercises 1–6, use the shell method to find the volumes of the solids generated by revolving the shaded region about the indicated axis.

**1.**

**2.**

**3.**

**4.**

**5.** The $y$-axis

**6.** The $y$-axis

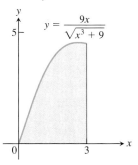

### Revolution About the $y$-Axis

Use the shell method to find the volumes of the solids generated by revolving the regions bounded by the curves and lines in Exercises 7–12 about the $y$-axis.

**7.** $y = x$,   $y = -x/2$,   $x = 2$

**8.** $y = 2x$,   $y = x/2$,   $x = 1$

**9.** $y = x^2$,   $y = 2 - x$,   $x = 0$,   for $x \geq 0$

**10.** $y = 2 - x^2$,   $y = x^2$,   $x = 0$

**11.** $y = 2x - 1$,   $y = \sqrt{x}$,   $x = 0$

**12.** $y = 3/(2\sqrt{x})$,   $y = 0$,   $x = 1$,   $x = 4$

**13.** Let $f(x) = \begin{cases} (\sin x)/x, & 0 < x \le \pi \\ 1, & x = 0 \end{cases}$

  **a.** Show that $x f(x) = \sin x, 0 \le x \le \pi$.

  **b.** Find the volume of the solid generated by revolving the shaded region about the $y$-axis in the accompanying figure.

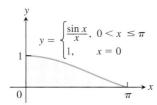

**14.** Let $g(x) = \begin{cases} (\tan x)^2/x, & 0 < x \le \pi/4 \\ 0, & x = 0 \end{cases}$

  **a.** Show that $x\, g(x) = (\tan x)^2, 0 \le x \le \pi/4$.

  **b.** Find the volume of the solid generated by revolving the shaded region about the $y$-axis in the accompanying figure.

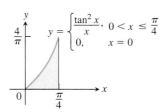

### Revolution About the $x$-Axis

Use the shell method to find the volumes of the solids generated by revolving the regions bounded by the curves and lines in Exercises 15–22 about the $x$-axis.

**15.** $x = \sqrt{y}, \quad x = -y, \quad y = 2$

**16.** $x = y^2, \quad x = -y, \quad y = 2, \quad y \ge 0$

**17.** $x = 2y - y^2, \quad x = 0$     **18.** $x = 2y - y^2, \quad x = y$

**19.** $y = |x|, \quad y = 1$       **20.** $y = x, \quad y = 2x, \quad y = 2$

**21.** $y = \sqrt{x}, \quad y = 0, \quad y = x - 2$

**22.** $y = \sqrt{x}, \quad y = 0, \quad y = 2 - x$

### Revolution About Horizontal and Vertical Lines

In Exercises 23–26, use the shell method to find the volumes of the solids generated by revolving the regions bounded by the given curves about the given lines.

**23.** $y = 3x, \quad y = 0, \quad x = 2$

  **a.** The $y$-axis           **b.** The line $x = 4$

  **c.** The line $x = -1$       **d.** The $x$-axis

  **e.** The line $y = 7$        **f.** The line $y = -2$

**24.** $y = x^3, \quad y = 8, \quad x = 0$

  **a.** The $y$-axis           **b.** The line $x = 3$

  **c.** The line $x = -2$       **d.** The $x$-axis

  **e.** The line $y = 8$        **f.** The line $y = -1$

**25.** $y = x + 2, \quad y = x^2$

  **a.** The line $x = 2$        **b.** The line $x = -1$

  **c.** The $x$-axis           **d.** The line $y = 4$

**26.** $y = x^4, \quad y = 4 - 3x^2$

  **a.** The line $x = 1$         **b.** The $x$-axis

In Exercises 27 and 28, use the shell method to find the volumes of the solids generated by revolving the shaded regions about the indicated axes.

**27.** **a.** The $x$-axis           **b.** The line $y = 1$

    **c.** The line $y = 8/5$      **d.** The line $y = -2/5$

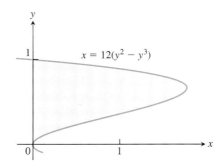

**28.** **a.** The $x$-axis           **b.** The line $y = 2$

    **c.** The line $y = 5$        **d.** The line $y = -5/8$

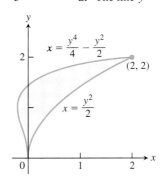

### Choosing the Washer Method or Shell Method

For some regions, both the washer and shell methods work well for the solid generated by revolving the region about the coordinate axes, but this is not always the case. When a region is revolved about the $y$-axis, for example, and washers are used, we must integrate with respect to $y$. It may not be possible, however, to express the integrand in terms of $y$. In such a case, the shell method allows us to integrate with respect to $x$ instead. Exercises 29 and 30 provide some insight.

**29.** Compute the volume of the solid generated by revolving the region bounded by $y = x$ and $y = x^2$ about each coordinate axis using

  **a.** the shell method.        **b.** the washer method.

**30.** Compute the volume of the solid generated by revolving the triangular region bounded by the lines $2y = x + 4, y = x$, and $x = 0$ about

  **a.** the $x$-axis using the washer method.

  **b.** the $y$-axis using the shell method.

  **c.** the line $x = 4$ using the shell method.

  **d.** the line $y = 8$ using the washer method.

In Exercises 31–36, find the volumes of the solids generated by revolving the regions about the given axes. If you think it would be better to use washers in any given instance, feel free to do so.

**31.** The triangle with vertices $(1, 1)$, $(1, 2)$, and $(2, 2)$ about

    **a.** the $x$-axis             **b.** the $y$-axis

    **c.** the line $x = 10/3$     **d.** the line $y = 1$

**32.** The region bounded by $y = \sqrt{x}$, $y = 2$, $x = 0$ about

    **a.** the $x$-axis             **b.** the $y$-axis

    **c.** the line $x = 4$         **d.** the line $y = 2$

**33.** The region in the first quadrant bounded by the curve $x = y - y^3$ and the $y$-axis about

    **a.** the $x$-axis             **b.** the line $y = 1$

**34.** The region in the first quadrant bounded by $x = y - y^3$, $x = 1$, and $y = 1$ about

    **a.** the $x$-axis             **b.** the $y$-axis

    **c.** the line $x = 1$         **d.** the line $y = 1$

**35.** The region bounded by $y = \sqrt{x}$ and $y = x^2/8$ about

    **a.** the $x$-axis             **b.** the $y$-axis

**36.** The region bounded by $y = 2x - x^2$ and $y = x$ about

    **a.** the $y$-axis             **b.** the line $x = 1$

**37.** The region in the first quadrant that is bounded above by the curve $y = 1/x^{1/4}$, on the left by the line $x = 1/16$, and below by the line $y = 1$ is revolved about the $x$-axis to generate a solid. Find the volume of the solid by

    **a.** the washer method.     **b.** the shell method.

**38.** The region in the first quadrant that is bounded above by the curve $y = 1/\sqrt{x}$, on the left by the line $x = 1/4$, and below by the line $y = 1$ is revolved about the $y$-axis to generate a solid. Find the volume of the solid by

    **a.** the washer method.     **b.** the shell method.

**Theory and Examples**

**39.** The region shown here is to be revolved about the $x$-axis to generate a solid. Which of the methods (disk, washer, shell) could you use to find the volume of the solid? How many integrals would be required in each case? Explain.

**40.** The region shown here is to be revolved about the $y$-axis to generate a solid. Which of the methods (disk, washer, shell) could you use to find the volume of the solid? How many integrals would be required in each case? Give reasons for your answers.

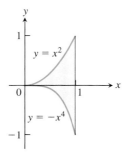

**41.** A bead is formed from a sphere of radius 5 by drilling through a diameter of the sphere with a drill bit of radius 3.

    **a.** Find the volume of the bead.

    **b.** Find the volume of the removed portion of the sphere.

**42.** A Bundt cake, well known for having a ringed shape, is formed by revolving around the $y$-axis the region bounded by the graph of $y = \sin(x^2 - 1)$ and the $x$-axis over the interval $1 \le x \le \sqrt{1 + \pi}$. Find the volume of the cake.

**43.** Derive the formula for the volume of a right circular cone of height $h$ and radius $r$ using an appropriate solid of revolution.

**44.** Derive the equation for the volume of a sphere of radius $r$ using the shell method.

# 6.3 Arc Length

We know what is meant by the length of a straight-line segment, but without calculus, we have no precise definition of the length of a general winding curve. If the curve is the graph of a continuous function defined over an interval, then we can find the length of the curve using a procedure similar to that we used for defining the area between the curve and the $x$-axis. This procedure results in a division of the curve from point $A$ to point $B$ into many pieces and joining successive points of division by straight-line segments. We then sum the lengths of all these line segments and define the length of the curve to be the limiting value of this sum as the number of segments goes to infinity.

## Length of a Curve $y = f(x)$

Suppose the curve whose length we want to find is the graph of the function $y = f(x)$ from $x = a$ to $x = b$. In order to derive an integral formula for the length of the curve, we assume that $f$ has a continuous derivative at every point of $[a, b]$. Such a function is called **smooth**, and its graph is a **smooth curve** because it does not have any breaks, corners, or cusps.

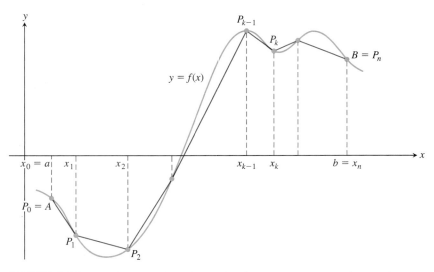

**FIGURE 6.22** The length of the polygonal path $P_0P_1P_2 \cdots P_n$ approximates the length of the curve $y = f(x)$ from point $A$ to point $B$.

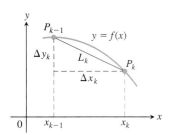

**FIGURE 6.23** The arc $P_{k-1}P_k$ of the curve $y = f(x)$ is approximated by the straight-line segment shown here, which has length $L_k = \sqrt{(\Delta x_k)^2 + (\Delta y_k)^2}$.

We partition the interval $[a, b]$ into $n$ subintervals with $a = x_0 < x_1 < x_2 < \cdots < x_n = b$. If $y_k = f(x_k)$, then the corresponding point $P_k(x_k, y_k)$ lies on the curve. Next we connect successive points $P_{k-1}$ and $P_k$ with straight-line segments that, taken together, form a polygonal path whose length approximates the length of the curve (Figure 6.22). If $\Delta x_k = x_k - x_{k-1}$ and $\Delta y_k = y_k - y_{k-1}$, then a representative line segment in the path has length (see Figure 6.23)

$$L_k = \sqrt{(\Delta x_k)^2 + (\Delta y_k)^2},$$

so the length of the curve is approximated by the sum

$$\sum_{k=1}^{n} L_k = \sum_{k=1}^{n} \sqrt{(\Delta x_k)^2 + (\Delta y_k)^2}. \qquad (1)$$

We expect the approximation to improve as the partition of $[a, b]$ becomes finer. Now, by the Mean Value Theorem, there is a point $c_k$, with $x_{k-1} < c_k < x_k$, such that

$$\Delta y_k = f'(c_k)\, \Delta x_k.$$

With this substitution for $\Delta y_k$, the sums in Equation (1) take the form

$$\sum_{k=1}^{n} L_k = \sum_{k=1}^{n} \sqrt{(\Delta x_k)^2 + (f'(c_k)\Delta x_k)^2} = \sum_{k=1}^{n} \sqrt{1 + [f'(c_k)]^2}\, \Delta x_k. \qquad (2)$$

Because $\sqrt{1 + [f'(x)]^2}$ is continuous on $[a, b]$, the limit of the Riemann sum on the right-hand side of Equation (2) exists as the norm of the partition goes to zero, giving

$$\lim_{n \to \infty} \sum_{k=1}^{n} L_k = \lim_{n \to \infty} \sum_{k=1}^{n} \sqrt{1 + [f'(c_k)]^2}\, \Delta x_k = \int_a^b \sqrt{1 + [f'(x)]^2}\, dx.$$

We define the value of this limiting integral to be the length of the curve.

---

**DEFINITION** If $f'$ is continuous on $[a, b]$, then the **length (arc length)** of the curve $y = f(x)$ from the point $A = (a, f(a))$ to the point $B = (b, f(b))$ is the value of the integral

$$L = \int_a^b \sqrt{1 + [f'(x)]^2}\, dx = \int_a^b \sqrt{1 + \left(\frac{dy}{dx}\right)^2}\, dx. \qquad (3)$$

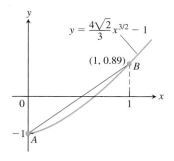

**FIGURE 6.24** The length of the curve is slightly larger than the length of the line segment joining points $A$ and $B$ (Example 1).

**EXAMPLE 1**    Find the length of the curve (Figure 6.24)

$$y = \frac{4\sqrt{2}}{3}x^{3/2} - 1, \qquad 0 \le x \le 1.$$

**Solution**   We use Equation (3) with $a = 0$, $b = 1$, and

$$y = \frac{4\sqrt{2}}{3}x^{3/2} - 1 \qquad \text{\small{$x = 1, y \approx 0.89$}}$$

$$\frac{dy}{dx} = \frac{4\sqrt{2}}{3} \cdot \frac{3}{2}x^{1/2} = 2\sqrt{2}x^{1/2}$$

$$\left(\frac{dy}{dx}\right)^2 = \left(2\sqrt{2}x^{1/2}\right)^2 = 8x.$$

The length of the curve over $x = 0$ to $x = 1$ is

$$L = \int_0^1 \sqrt{1 + \left(\frac{dy}{dx}\right)^2}\, dx = \int_0^1 \sqrt{1 + 8x}\, dx \qquad \text{\small Eq. (3) with}$$
$$\text{\small $a = 0, b = 1$.}$$
$$\text{\small Let $u = 1 + 8x$,}$$
$$= \frac{2}{3} \cdot \frac{1}{8}(1 + 8x)^{3/2} \Big]_0^1 = \frac{13}{6} \approx 2.17. \qquad \text{\small integrate, and}$$
$$\text{\small replace $u$ by}$$
$$\text{\small $1 + 8x$.}$$

Notice that the length of the curve is slightly larger than the length of the straight-line segment joining the points $A = (0, -1)$ and $B = \left(1, 4\sqrt{2}/3 - 1\right)$ on the curve (see Figure 6.24):

$$2.17 > \sqrt{1^2 + (1.89)^2} \approx 2.14. \qquad \text{\small Decimal approximations} \qquad \blacksquare$$

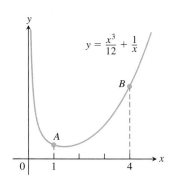

**FIGURE 6.25** The curve in Example 2, where $A = (1, 13/12)$ and $B = (4, 67/12)$.

**EXAMPLE 2**    Find the length of the graph of

$$f(x) = \frac{x^3}{12} + \frac{1}{x}, \qquad 1 \le x \le 4.$$

**Solution**   A graph of the function is shown in Figure 6.25. To use Equation (3), we find

$$f'(x) = \frac{x^2}{4} - \frac{1}{x^2}$$

so

$$1 + [f'(x)]^2 = 1 + \left(\frac{x^2}{4} - \frac{1}{x^2}\right)^2 = 1 + \left(\frac{x^4}{16} - \frac{1}{2} + \frac{1}{x^4}\right)$$

$$= \frac{x^4}{16} + \frac{1}{2} + \frac{1}{x^4} = \left(\frac{x^2}{4} + \frac{1}{x^2}\right)^2.$$

The length of the graph over $[1, 4]$ is

$$L = \int_1^4 \sqrt{1 + [f'(x)]^2}\, dx = \int_1^4 \left(\frac{x^2}{4} + \frac{1}{x^2}\right) dx$$

$$= \left[\frac{x^3}{12} - \frac{1}{x}\right]_1^4 = \left(\frac{64}{12} - \frac{1}{4}\right) - \left(\frac{1}{12} - 1\right) = \frac{72}{12} = 6. \qquad \blacksquare$$

**Dealing with Discontinuities in $dy/dx$**

At a point on a curve where $dy/dx$ fails to exist, $dx/dy$ may exist. In this case, we may be able to find the curve's length by expressing $x$ as a function of $y$ and applying the following analogue of Equation (3):

> **Formula for the Length of** $x = g(y), c \leq y \leq d$
>
> If $g'$ is continuous on $[c, d]$, the length of the curve $x = g(y)$ from $A = (g(c), c)$ to $B = (g(d), d)$ is
>
> $$L = \int_c^d \sqrt{1 + \left(\frac{dx}{dy}\right)^2} \, dy = \int_c^d \sqrt{1 + [g'(y)]^2} \, dy. \qquad (4)$$

**EXAMPLE 3**  Find the length of the curve $y = (x/2)^{2/3}$ from $x = 0$ to $x = 2$.

**Solution**  The derivative

$$\frac{dy}{dx} = \frac{2}{3}\left(\frac{x}{2}\right)^{-1/3}\left(\frac{1}{2}\right) = \frac{1}{3}\left(\frac{2}{x}\right)^{1/3}$$

is not defined at $x = 0$, so we cannot find the curve's length with Equation (3).

We therefore rewrite the equation to express $x$ in terms of $y$:

$$y = \left(\frac{x}{2}\right)^{2/3}$$

$$y^{3/2} = \frac{x}{2} \qquad \text{Raise both sides to the power } 3/2.$$

$$x = 2y^{3/2}. \qquad \text{Solve for } x.$$

From this we see that the curve whose length we want is also the graph of $x = 2y^{3/2}$ from $y = 0$ to $y = 1$ (Figure 6.26).

The derivative

$$\frac{dx}{dy} = 2\left(\frac{3}{2}\right)y^{1/2} = 3y^{1/2}$$

is continuous on $[0, 1]$. We may therefore use Equation (4) to find the curve's length:

$$L = \int_c^d \sqrt{1 + \left(\frac{dx}{dy}\right)^2} \, dy = \int_0^1 \sqrt{1 + 9y} \, dy \qquad \begin{array}{l} \text{Eq. (4) with} \\ c = 0, d = 1. \\ \text{Let } u = 1 + 9y, \\ du/9 = dy, \\ \text{integrate, and} \\ \text{substitute back.} \end{array}$$

$$= \frac{1}{9}\cdot\frac{2}{3}(1 + 9y)^{3/2}\bigg]_0^1$$

$$= \frac{2}{27}\left(10\sqrt{10} - 1\right) \approx 2.27. \qquad \blacksquare$$

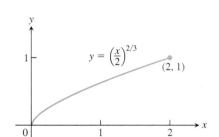

**FIGURE 6.26** The graph of $y = (x/2)^{2/3}$ from $x = 0$ to $x = 2$ is also the graph of $x = 2y^{3/2}$ from $y = 0$ to $y = 1$ (Example 3).

## The Differential Formula for Arc Length

If $y = f(x)$ and if $f'$ is continuous on $[a, b]$, then by the Fundamental Theorem of Calculus we can define a new function

$$s(x) = \int_a^x \sqrt{1 + [f'(t)]^2} \, dt. \qquad (5)$$

From Equation (3) and Figure 6.22, we see that this function $s(x)$ is continuous and measures the length along the curve $y = f(x)$ from the initial point $P_0(a, f(a))$ to the point $Q(x, f(x))$ for each $x \in [a, b]$. The function $s$ is called the **arc length function** for $y = f(x)$. From the Fundamental Theorem, the function $s$ is differentiable on $(a, b)$ and

$$\frac{ds}{dx} = \sqrt{1 + [f'(x)]^2} = \sqrt{1 + \left(\frac{dy}{dx}\right)^2}.$$

Then the differential of arc length is

$$ds = \sqrt{1 + \left(\frac{dy}{dx}\right)^2}\, dx. \tag{6}$$

A useful way to remember Equation (6) is to write

$$ds = \sqrt{dx^2 + dy^2}, \tag{7}$$

which can be integrated between appropriate limits to give the total length of a curve. From this point of view, all the arc length formulas are simply different expressions for the equation $L = \int ds$. Figure 6.27a gives the exact interpretation of $ds$ corresponding to Equation (7). Figure 6.27b is not strictly accurate, but is to be thought of as a simplified approximation of Figure 6.27a. That is, $ds \approx \Delta s$.

**EXAMPLE 4**    Find the arc length function for the curve in Example 2, taking $A = (1, 13/12)$ as the starting point (see Figure 6.25).

**Solution**    In the solution to Example 2, we found that

$$1 + [f'(x)]^2 = \left(\frac{x^2}{4} + \frac{1}{x^2}\right)^2.$$

Therefore the arc length function is given by

$$s(x) = \int_1^x \sqrt{1 + [f'(t)]^2}\, dt = \int_1^x \left(\frac{t^2}{4} + \frac{1}{t^2}\right) dt$$

$$= \left[\frac{t^3}{12} - \frac{1}{t}\right]_1^x = \frac{x^3}{12} - \frac{1}{x} + \frac{11}{12}.$$

To compute the arc length along the curve from $A = (1, 13/12)$ to $B = (4, 67/12)$, for instance, we simply calculate

$$s(4) = \frac{4^3}{12} - \frac{1}{4} + \frac{11}{12} = 6.$$

This is the same result we obtained in Example 2.    ∎

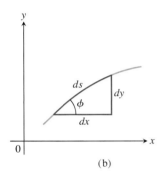

**FIGURE 6.27** Diagrams for remembering the equation $ds = \sqrt{dx^2 + dy^2}$.

# Exercises 6.3

**Finding Lengths of Curves**
Find the lengths of the curves in Exercises 1–12. If you have a grapher, you may want to graph these curves to see what they look like.

1. $y = (1/3)(x^2 + 2)^{3/2}$  from  $x = 0$ to $x = 3$

2. $y = x^{3/2}$  from  $x = 0$ to $x = 4$

3. $x = (y^3/3) + 1/(4y)$  from  $y = 1$ to $y = 3$

4. $x = (y^{3/2}/3) - y^{1/2}$  from  $y = 1$ to $y = 9$

5. $x = (y^4/4) + 1/(8y^2)$  from  $y = 1$ to $y = 2$

6. $x = (y^3/6) + 1/(2y)$  from  $y = 2$ to $y = 3$

7. $y = (3/4)x^{4/3} - (3/8)x^{2/3} + 5,\quad 1 \le x \le 8$

8. $y = (x^3/3) + x^2 + x + 1/(4x + 4),\quad 0 \le x \le 2$

9. $y = \frac{x^3}{3} + \frac{1}{4x},\quad 1 \le x \le 3$

10. $y = \frac{x^5}{5} + \frac{1}{12x^3},\quad \frac{1}{2} \le x \le 1$

11. $x = \int_0^y \sqrt{\sec^4 t - 1}\, dt,\quad -\pi/4 \le y \le \pi/4$

12. $y = \int_{-2}^x \sqrt{3t^4 - 1}\, dt,\quad -2 \le x \le -1$

T **Finding Integrals for Lengths of Curves**
In Exercises 13–20, do the following.

    **a.** Set up an integral for the length of the curve.

    **b.** Graph the curve to see what it looks like.

    **c.** Use your grapher's or computer's integral evaluator to find the curve's length numerically.

13. $y = x^2,\quad -1 \le x \le 2$

14. $y = \tan x,\quad -\pi/3 \le x \le 0$

15. $x = \sin y,\quad 0 \le y \le \pi$

16. $x = \sqrt{1 - y^2},\quad -1/2 \le y \le 1/2$

**17.** $y^2 + 2y = 2x + 1$   from   $(-1, -1)$ to $(7, 3)$

**18.** $y = \sin x - x \cos x, \quad 0 \le x \le \pi$

**19.** $y = \displaystyle\int_0^x \tan t \, dt, \quad 0 \le x \le \pi/6$

**20.** $x = \displaystyle\int_0^y \sqrt{\sec^2 t - 1} \, dt, \quad -\pi/3 \le y \le \pi/4$

**Theory and Examples**

**21. a.** Find a curve with a positive derivative through the point $(1, 1)$ whose length integral (Equation 3) is

$$L = \int_1^4 \sqrt{1 + \frac{1}{4x}} \, dx.$$

  **b.** How many such curves are there? Give reasons for your answer.

**22. a.** Find a curve with a positive derivative through the point $(0, 1)$ whose length integral (Equation 4) is

$$L = \int_1^2 \sqrt{1 + \frac{1}{y^4}} \, dy.$$

  **b.** How many such curves are there? Give reasons for your answer.

**23.** Find the length of the curve

$$y = \int_0^x \sqrt{\cos 2t} \, dt$$

from $x = 0$ to $x = \pi/4$.

**24. The length of an astroid** The graph of the equation $x^{2/3} + y^{2/3} = 1$ is one of a family of curves called *astroids* (not "asteroids") because of their starlike appearance (see the accompanying figure). Find the length of this particular astroid by finding the length of half the first-quadrant portion, $y = (1 - x^{2/3})^{3/2}$, $\sqrt{2}/4 \le x \le 1$, and multiplying by 8.

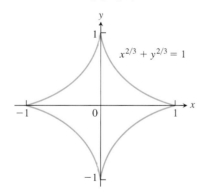

**25. Length of a line segment** Use the arc length formula (Equation 3) to find the length of the line segment $y = 3 - 2x, 0 \le x \le 2$. Check your answer by finding the length of the segment as the hypotenuse of a right triangle.

**26. Circumference of a circle** Set up an integral to find the circumference of a circle of radius $r$ centered at the origin. You will learn how to evaluate the integral in Section 8.4.

**27.** If $9x^2 = y(y - 3)^2$, show that

$$ds^2 = \frac{(y + 1)^2}{4y} \, dy^2.$$

**28.** If $4x^2 - y^2 = 64$, show that

$$ds^2 = \frac{4}{y^2} (5x^2 - 16) \, dx^2.$$

**29.** Is there a smooth (continuously differentiable) curve $y = f(x)$ whose length over the interval $0 \le x \le a$ is always $\sqrt{2a}$? Give reasons for your answer.

**30. Using tangent fins to derive the length formula for curves** Assume that $f$ is smooth on $[a, b]$ and partition the interval $[a, b]$ in the usual way. In each subinterval $[x_{k-1}, x_k]$, construct the *tangent fin* at the point $(x_{k-1}, f(x_{k-1}))$, as shown in the accompanying figure.

  **a.** Show that the length of the $k$th tangent fin over the interval $[x_{k-1}, x_k]$ equals $\sqrt{(\Delta x_k)^2 + (f'(x_{k-1}) \Delta x_k)^2}$.

  **b.** Show that

$$\lim_{n \to \infty} \sum_{k=1}^n (\text{length of } k\text{th tangent fin}) = \int_a^b \sqrt{1 + (f'(x))^2} \, dx,$$

which is the length $L$ of the curve $y = f(x)$ from $a$ to $b$.

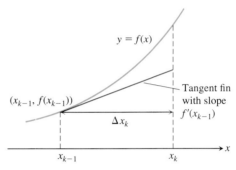

**31.** Approximate the arc length of one-quarter of the unit circle (which is $\pi/2$) by computing the length of the polygonal approximation with $n = 4$ segments (see accompanying figure).

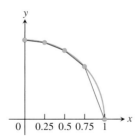

**32. Distance between two points** Assume that the two points $(x_1, y_1)$ and $(x_2, y_2)$ lie on the graph of the straight line $y = mx + b$. Use the arc length formula (Equation 3) to find the distance between the two points.

**33.** Find the arc length function for the graph of $f(x) = 2x^{3/2}$ using $(0, 0)$ as the starting point. What is the length of the curve from $(0, 0)$ to $(1, 2)$?

**34.** Find the arc length function for the curve in Exercise 8, using $(0, 1/4)$ as the starting point. What is the length of the curve from $(0, 1/4)$ to $(1, 59/24)$?

## COMPUTER EXPLORATIONS

In Exercises 35–40, use a CAS to perform the following steps for the given graph of the function over the closed interval.

**a.** Plot the curve together with the polygonal path approximations for $n = 2, 4, 8$ partition points over the interval. (See Figure 6.22.)

**b.** Find the corresponding approximation to the length of the curve by summing the lengths of the line segments.

**c.** Evaluate the length of the curve using an integral. Compare your approximations for $n = 2, 4, 8$ with the actual length given by the integral. How does the actual length compare with the approximations as $n$ increases? Explain your answer.

**35.** $f(x) = \sqrt{1 - x^2}, \quad -1 \le x \le 1$

**36.** $f(x) = x^{1/3} + x^{2/3}, \quad 0 \le x \le 2$

**37.** $f(x) = \sin(\pi x^2), \quad 0 \le x \le \sqrt{2}$

**38.** $f(x) = x^2 \cos x, \quad 0 \le x \le \pi$

**39.** $f(x) = \dfrac{x - 1}{4x^2 + 1}, \quad -\dfrac{1}{2} \le x \le 1$

**40.** $f(x) = x^3 - x^2, \quad -1 \le x \le 1$

# $6.4$ Areas of Surfaces of Revolution

When you jump rope, the rope sweeps out a surface in the space around you similar to what is called a *surface of revolution*. The surface surrounds a volume of revolution, and many applications require that we know the area of the surface rather than the volume it encloses. In this section we define areas of surfaces of revolution. More general surfaces are treated in Chapter 16.

### Defining Surface Area

If you revolve a region in the plane that is bounded by the graph of a function over an interval, it sweeps out a solid of revolution, as we saw earlier in the chapter. However, if you revolve only the bounding curve itself, it does not sweep out any interior volume but rather a surface that surrounds the solid and forms part of its boundary. Just as we were interested in defining and finding the length of a curve in the last section, we are now interested in defining and finding the area of a surface generated by revolving a curve about an axis.

Before considering general curves, we begin by rotating horizontal and slanted line segments about the *x*-axis. If we rotate the horizontal line segment $AB$ having length $\Delta x$ about the *x*-axis (Figure 6.28a), we generate a cylinder with surface area $2\pi y \Delta x$. This area is the same as that of a rectangle with side lengths $\Delta x$ and $2\pi y$ (Figure 6.28b). The length $2\pi y$ is the circumference of the circle of radius $y$ generated by rotating the point $(x, y)$ on the line $AB$ about the *x*-axis.

Suppose the line segment $AB$ has length $L$ and is slanted rather than horizontal. Now when $AB$ is rotated about the *x*-axis, it generates a frustum of a cone (Figure 6.29a). From classical geometry, the surface area of this frustum is $2\pi y^* L$, where $y^* = (y_1 + y_2)/2$ is the average height of the slanted segment $AB$ above the *x*-axis. This surface area is the same as that of a rectangle with side lengths $L$ and $2\pi y^*$ (Figure 6.29b).

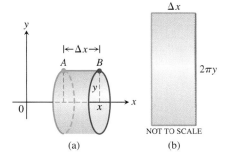

**FIGURE 6.28** (a) A cylindrical surface generated by rotating the horizontal line segment $AB$ of length $\Delta x$ about the *x*-axis has area $2\pi y \Delta x$. (b) The cut and rolled-out cylindrical surface as a rectangle.

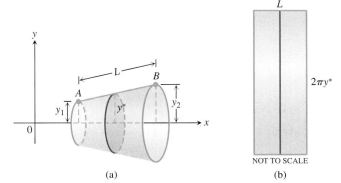

**FIGURE 6.29** (a) The frustum of a cone generated by rotating the slanted line segment $AB$ of length $L$ about the *x*-axis has area $2\pi y^* L$. (b) The area of the rectangle for $y^* = \dfrac{y_1 + y_2}{2}$, the average height of $AB$ above the *x*-axis.

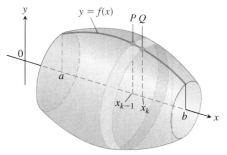

**FIGURE 6.30** The surface generated by revolving the graph of a nonnegative function $y = f(x)$, $a \le x \le b$, about the $x$-axis. The surface is a union of bands like the one swept out by the arc $PQ$.

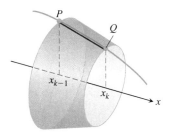

**FIGURE 6.31** The line segment joining $P$ and $Q$ sweeps out a frustum of a cone.

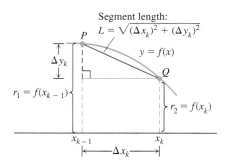

**FIGURE 6.32** Dimensions associated with the arc and line segment $PQ$.

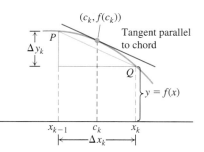

**FIGURE 6.33** If $f$ is smooth, the Mean Value Theorem guarantees the existence of a point $c_k$ where the tangent is parallel to segment $PQ$.

Let's build on these geometric principles to define the area of a surface swept out by revolving more general curves about the $x$-axis. Suppose we want to find the area of the surface swept out by revolving the graph of a nonnegative continuous function $y = f(x)$, $a \le x \le b$, about the $x$-axis. We partition the closed interval $[a, b]$ in the usual way and use the points in the partition to subdivide the graph into short arcs. Figure 6.30 shows a typical arc $PQ$ and the band it sweeps out as part of the graph of $f$.

As the arc $PQ$ revolves about the $x$-axis, the line segment joining $P$ and $Q$ sweeps out a frustum of a cone whose axis lies along the $x$-axis (Figure 6.31). The surface area of this frustum approximates the surface area of the band swept out by the arc $PQ$. The surface area of the frustum of the cone shown in Figure 6.31 is $2\pi y^* L$, where $y^*$ is the average height of the line segment joining $P$ and $Q$, and $L$ is its length (just as before). Since $f \ge 0$, from Figure 6.32 we see that the average height of the line segment is $y^* = (f(x_{k-1}) + f(x_k))/2$, and the slant length is $L = \sqrt{(\Delta x_k)^2 + (\Delta y_k)^2}$. Therefore,

$$\text{Frustum surface area} = 2\pi \cdot \frac{f(x_{k-1}) + f(x_k)}{2} \cdot \sqrt{(\Delta x_k)^2 + (\Delta y_k)^2}$$

$$= \pi(f(x_{k-1}) + f(x_k))\sqrt{(\Delta x_k)^2 + (\Delta y_k)^2}.$$

The area of the original surface, being the sum of the areas of the bands swept out by arcs like arc $PQ$, is approximated by the frustum area sum

$$\sum_{k=1}^{n} \pi(f(x_{k-1}) + f(x_k))\sqrt{(\Delta x_k)^2 + (\Delta y_k)^2}. \tag{1}$$

We expect the approximation to improve as the partition of $[a, b]$ becomes finer. Moreover, if the function $f$ is differentiable, then by the Mean Value Theorem, there is a point $(c_k, f(c_k))$ on the curve between $P$ and $Q$ where the tangent is parallel to the segment $PQ$ (Figure 6.33). At this point,

$$f'(c_k) = \frac{\Delta y_k}{\Delta x_k},$$

$$\Delta y_k = f'(c_k)\, \Delta x_k.$$

With this substitution for $\Delta y_k$, the sums in Equation (1) take the form

$$\sum_{k=1}^{n} \pi(f(x_{k-1}) + f(x_k))\sqrt{(\Delta x_k)^2 + (f'(c_k)\,\Delta x_k)^2}$$

$$= \sum_{k=1}^{n} \pi(f(x_{k-1}) + f(x_k))\sqrt{1 + (f'(c_k))^2}\, \Delta x_k. \tag{2}$$

These sums are not the Riemann sums of any function because the points $x_{k-1}$, $x_k$, and $c_k$ are not the same. However, it can be proved that as the norm of the partition of $[a, b]$ goes to zero, the sums in Equation (2) converge to the integral

$$\int_a^b 2\pi f(x)\sqrt{1 + (f'(x))^2}\, dx.$$

We therefore define this integral to be the area of the surface swept out by the graph of $f$ from $a$ to $b$.

---

**DEFINITION** If the function $f(x) \ge 0$ is continuously differentiable on $[a, b]$, the **area of the surface** generated by revolving the graph of $y = f(x)$ about the $x$-axis is

$$S = \int_a^b 2\pi y \sqrt{1 + \left(\frac{dy}{dx}\right)^2}\, dx = \int_a^b 2\pi f(x)\sqrt{1 + (f'(x))^2}\, dx. \tag{3}$$

The square root in Equation (3) is the same one that appears in the formula for the arc length differential of the generating curve in Equation (6) of Section 6.3.

**EXAMPLE 1** Find the area of the surface generated by revolving the curve $y = 2\sqrt{x}$, $1 \le x \le 2$, about the $x$-axis (Figure 6.34).

**Solution** We evaluate the formula

$$S = \int_a^b 2\pi y \sqrt{1 + \left(\frac{dy}{dx}\right)^2}\, dx \qquad \text{Eq. (3)}$$

with

$$a = 1, \qquad b = 2, \qquad y = 2\sqrt{x}, \qquad \frac{dy}{dx} = \frac{1}{\sqrt{x}}.$$

First, we perform some algebraic manipulation on the radical in the integrand to transform it into an expression that is easier to integrate.

$$\sqrt{1 + \left(\frac{dy}{dx}\right)^2} = \sqrt{1 + \left(\frac{1}{\sqrt{x}}\right)^2}$$

$$= \sqrt{1 + \frac{1}{x}} = \sqrt{\frac{x+1}{x}} = \frac{\sqrt{x+1}}{\sqrt{x}}$$

With these substitutions, we have

$$S = \int_1^2 2\pi \cdot 2\sqrt{x}\, \frac{\sqrt{x+1}}{\sqrt{x}}\, dx = 4\pi \int_1^2 \sqrt{x+1}\, dx$$

$$= 4\pi \cdot \frac{2}{3}(x+1)^{3/2}\Big]_1^2 = \frac{8\pi}{3}\left(3\sqrt{3} - 2\sqrt{2}\right). \qquad \blacksquare$$

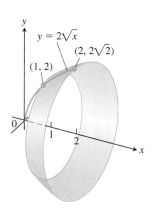

**FIGURE 6.34** In Example 1 we calculate the area of this surface.

### Revolution About the *y*-Axis

For revolution about the $y$-axis, we interchange $x$ and $y$ in Equation (3).

---

**Surface Area for Revolution About the *y*-Axis**

If $x = g(y) \ge 0$ is continuously differentiable on $[c, d]$, the area of the surface generated by revolving the graph of $x = g(y)$ about the $y$-axis is

$$S = \int_c^d 2\pi x \sqrt{1 + \left(\frac{dx}{dy}\right)^2}\, dy = \int_c^d 2\pi g(y) \sqrt{1 + (g'(y))^2}\, dy. \qquad (4)$$

---

**EXAMPLE 2** The line segment $x = 1 - y$, $0 \le y \le 1$, is revolved about the $y$-axis to generate the cone in Figure 6.35. Find its lateral surface area (which excludes the base area).

**Solution** Here we have a calculation we can check with a formula from geometry:

$$\text{Lateral surface area} = \frac{\text{base circumference}}{2} \times \text{slant height} = \pi\sqrt{2}.$$

To see how Equation (4) gives the same result, we take

$$c = 0, \qquad d = 1, \qquad x = 1 - y, \qquad \frac{dx}{dy} = -1,$$

$$\sqrt{1 + \left(\frac{dx}{dy}\right)^2} = \sqrt{1 + (-1)^2} = \sqrt{2}$$

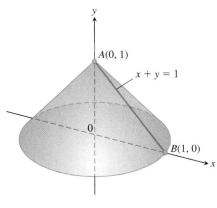

**FIGURE 6.35** Revolving line segment *AB* about the *y*-axis generates a cone whose lateral surface area we can now calculate in two different ways (Example 2).

and calculate

$$S = \int_c^d 2\pi x \sqrt{1 + \left(\frac{dx}{dy}\right)^2} \, dy = \int_0^1 2\pi(1 - y)\sqrt{2} \, dy$$

$$= 2\pi\sqrt{2}\left[ y - \frac{y^2}{2} \right]_0^1 = 2\pi\sqrt{2}\left(1 - \frac{1}{2}\right)$$

$$= \pi\sqrt{2}.$$

The results agree, as they should. ∎

## Exercises 6.4

### Finding Integrals for Surface Area

In Exercises 1–8:

  **a.** Set up an integral for the area of the surface generated by revolving the given curve about the indicated axis.

  T **b.** Graph the curve to see what it looks like. If you can, graph the surface too.

  T **c.** Use your utility's integral evaluator to find the surface's area numerically.

1. $y = \tan x$, $0 \le x \le \pi/4$; x-axis

2. $y = x^2$, $0 \le x \le 2$; x-axis

3. $xy = 1$, $1 \le y \le 2$; y-axis

4. $x = \sin y$, $0 \le y \le \pi$; y-axis

5. $x^{1/2} + y^{1/2} = 3$ from (4, 1) to (1, 4); x-axis

6. $y + 2\sqrt{y} = x$, $1 \le y \le 2$; y-axis

7. $x = \int_0^y \tan t \, dt$, $0 \le y \le \pi/3$; y-axis

8. $y = \int_1^x \sqrt{t^2 - 1} \, dt$, $1 \le x \le \sqrt{5}$; x-axis

### Finding Surface Area

9. Find the lateral (side) surface area of the cone generated by revolving the line segment $y = x/2, 0 \le x \le 4$, about the x-axis. Check your answer with the geometry formula

  Lateral surface area $= \frac{1}{2} \times$ base circumference $\times$ slant height.

10. Find the lateral surface area of the cone generated by revolving the line segment $y = x/2, 0 \le x \le 4$, about the y-axis. Check your answer with the geometry formula

  Lateral surface area $= \frac{1}{2} \times$ base circumference $\times$ slant height.

11. Find the surface area of the cone frustum generated by revolving the line segment $y = (x/2) + (1/2), 1 \le x \le 3$, about the x-axis. Check your result with the geometry formula

  Frustum surface area $= \pi(r_1 + r_2) \times$ slant height.

12. Find the surface area of the cone frustum generated by revolving the line segment $y = (x/2) + (1/2), 1 \le x \le 3$, about the y-axis. Check your result with the geometry formula

  Frustum surface area $= \pi(r_1 + r_2) \times$ slant height.

Find the areas of the surfaces generated by revolving the curves in Exercises 13–23 about the indicated axes. If you have a grapher, you may want to graph these curves to see what they look like.

13. $y = x^3/9$, $0 \le x \le 2$; x-axis

14. $y = \sqrt{x}$, $3/4 \le x \le 15/4$; x-axis

15. $y = \sqrt{2x - x^2}$, $0.5 \le x \le 1.5$; x-axis

16. $y = \sqrt{x + 1}$, $1 \le x \le 5$; x-axis

17. $x = y^3/3$, $0 \le y \le 1$; y-axis

18. $x = (1/3)y^{3/2} - y^{1/2}$, $1 \le y \le 3$; y-axis

19. $x = 2\sqrt{4 - y}$, $0 \le y \le 15/4$; y-axis

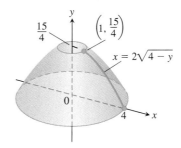

20. $x = \sqrt{2y - 1}$, $5/8 \le y \le 1$; y-axis

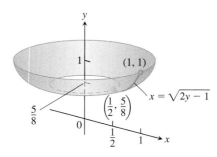

**21.** $y = (1/2)(x^2 + 1), \quad 0 \le x \le 1; \quad y\text{-axis}$

**22.** $y = (1/3)(x^2 + 2)^{3/2}, \quad 0 \le x \le \sqrt{2}; \quad y\text{-axis}$ (*Hint:* Express $ds = \sqrt{dx^2 + dy^2}$ in terms of $dx$, and evaluate the integral $S = \int 2\pi x \, ds$ with appropriate limits.)

**23.** $x = (y^4/4) + 1/(8y^2), \quad 1 \le y \le 2; \quad x\text{-axis}$ (*Hint:* Express $ds = \sqrt{dx^2 + dy^2}$ in terms of $dy$, and evaluate the integral $S = \int 2\pi y \, ds$ with appropriate limits.)

**24.** Write an integral for the area of the surface generated by revolving the curve $y = \cos x, -\pi/2 \le x \le \pi/2$, about the $x$-axis. In Section 8.4 we will see how to evaluate such integrals.

**25. Testing the new definition** Show that the surface area of a sphere of radius $a$ is still $4\pi a^2$ by using Equation (3) to find the area of the surface generated by revolving the curve $y = \sqrt{a^2 - x^2}, -a \le x \le a$, about the $x$-axis.

**26. Testing the new definition** The lateral (side) surface area of a cone of height $h$ and base radius $r$ should be $\pi r \sqrt{r^2 + h^2}$, the semiperimeter of the base times the slant height. Show that this is still the case by finding the area of the surface generated by revolving the line segment $y = (r/h)x, 0 \le x \le h$, about the $x$-axis.

**T** **27. Enameling woks** Your company decided to put out a deluxe version of a wok you designed. The plan is to coat it inside with white enamel and outside with blue enamel. Each enamel will be sprayed on 0.5 mm thick before baking. (See accompanying figure.) Your manufacturing department wants to know how much enamel to have on hand for a production run of 5000 woks. What do you tell them? (Neglect waste and unused material and give your answer in liters. Remember that $1 \, \text{cm}^3 = 1 \, \text{mL}$, so $1 \, \text{L} = 1000 \, \text{cm}^3$.)

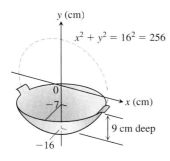

**28. Slicing bread** Did you know that if you cut a spherical loaf of bread into slices of equal width, each slice will have the same amount of crust? To see why, suppose the semicircle $y = \sqrt{r^2 - x^2}$ shown here is revolved about the $x$-axis to generate a sphere. Let $AB$ be an arc of the semicircle that lies above an interval of length $h$ on the $x$-axis. Show that the area swept out by $AB$ does not depend on the location of the interval. (It does depend on the length of the interval.)

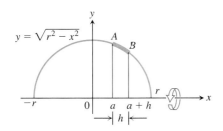

**29.** The shaded band shown here is cut from a sphere of radius $R$ by parallel planes $h$ units apart. Show that the surface area of the band is $2\pi Rh$.

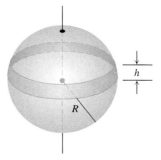

**30.** Here is a schematic drawing of the 30 m dome used by the U.S. National Weather Service to house radar in Bozeman, Montana.

  **a.** How much outside surface is there to paint (not counting the bottom)?

**T** **b.** Express the answer to the nearest square meter.

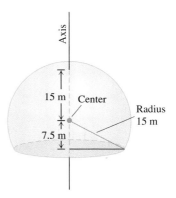

**31. An alternative derivation of the surface area formula** Assume $f$ is smooth on $[a, b]$ and partition $[a, b]$ in the usual way. In the $k$th subinterval $[x_{k-1}, x_k]$, construct the tangent line to the curve at the midpoint $m_k = (x_{k-1} + x_k)/2$, as in the accompanying figure.

  **a.** Show that

$$r_1 = f(m_k) - f'(m_k)\frac{\Delta x_k}{2} \quad \text{and} \quad r_2 = f(m_k) + f'(m_k)\frac{\Delta x_k}{2}.$$

  **b.** Show that the length $L_k$ of the tangent line segment in the $k$th subinterval is $L_k = \sqrt{(\Delta x_k)^2 + (f'(m_k) \, \Delta x_k)^2}$.

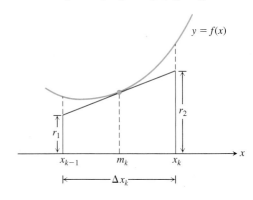

c. Show that the lateral surface area of the frustum of the cone swept out by the tangent line segment as it revolves about the $x$-axis is $2\pi f(m_k)\sqrt{1 + (f'(m_k))^2}\,\Delta x_k$.

d. Show that the area of the surface generated by revolving $y = f(x)$ about the $x$-axis over $[a, b]$ is

$$\lim_{n\to\infty} \sum_{k=1}^{n} \left(\begin{array}{c}\text{lateral surface area}\\\text{of }k\text{th frustum}\end{array}\right) = \int_{a}^{b} 2\pi f(x)\sqrt{1 + (f'(x))^2}\,dx.$$

**32. The surface of an astroid** Find the area of the surface generated by revolving about the $x$-axis the portion of the astroid $x^{2/3} + y^{2/3} = 1$ shown in the accompanying figure.

(*Hint:* Revolve the first-quadrant portion $y = (1 - x^{2/3})^{3/2}$, $0 \le x \le 1$, about the $x$-axis and double your result.)

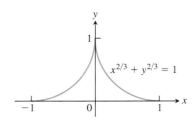

# 6.5 Work and Fluid Forces

In everyday life, *work* means an activity that requires muscular or mental effort. In science, the term refers specifically to a force acting on an object and the object's subsequent displacement. This section shows how to calculate work. The applications run from compressing railroad car springs and emptying subterranean tanks to forcing subatomic particles to collide and lifting satellites into orbit.

## Work Done by a Constant Force

When an object moves a distance $d$ along a straight line as a result of being acted on by a force of constant magnitude $F$ in the direction of motion, we define the **work** $W$ done by the force on the object with the formula

$$W = Fd \qquad \text{(Constant-force formula for work).} \qquad (1)$$

From Equation (1) we see that the unit of work in any system is the unit of force multiplied by the unit of distance. In SI units (SI stands for *Système International*, or International System), the unit of force is a newton, the unit of distance is a meter, and the unit of work is a newton-meter (N · m). The last combination appears so often that it has a special name—the **joule**.

**Joules**

The joule, abbreviated J, is named after the English physicist James Prescott Joule (1818–1889). The defining equation is

1 joule = (1 newton)(1 meter).

In symbols, 1 J = 1 N · m.

**EXAMPLE 1** Suppose you jack up the side of a 1000 kg car 35 cm to change a tire. The jack applies a constant vertical force of about 5000 N in lifting the side of the car (but because of the mechanical advantage of the jack, the force you apply to the jack itself is only about 150 N). The total work performed by the jack on the car is 5000 × 0.35 = 1750 J. ■

## Work Done by a Variable Force Along a Line

If the force you apply varies along the way, as it will if you are stretching or compressing a spring, the formula $W = Fd$ has to be replaced by an integral formula that takes the variation in $F$ into account.

Suppose that the force performing the work acts on an object moving along a straight line, which we take to be the $x$-axis. We assume that the magnitude of the force is a continuous function $F$ of the object's position $x$. We want to find the work done over the interval from $x = a$ to $x = b$. We partition $[a, b]$ in the usual way and choose an arbitrary point $c_k$ in each subinterval $[x_{k-1}, x_k]$. If the subinterval is short enough, the continuous function $F$ will not vary much from $x_{k-1}$ to $x_k$. The amount of work done across the interval will be about $F(c_k)$ times the distance $\Delta x_k$, the same as it would be if $F$ were constant and we could apply Equation (1). The total work done from $a$ to $b$ is therefore approximated by the Riemann sum

$$\text{Work} \approx \sum_{k=1}^{n} F(c_k)\, \Delta x_k.$$

We expect the approximation to improve as the norm of the partition goes to zero, so we define the work done by the force from $a$ to $b$ to be the integral of $F$ from $a$ to $b$:

$$\lim_{n\to\infty} \sum_{k=1}^{n} F(c_k)\, \Delta x_k = \int_{a}^{b} F(x)\, dx.$$

---

**DEFINITION**  The **work** done by a variable force $F(x)$ in moving an object along the $x$-axis from $x = a$ to $x = b$ is

$$W = \int_{a}^{b} F(x)\, dx. \tag{2}$$

---

The units of the integral are joules if $F$ is in newtons and $x$ is in meters, and foot-pounds if $F$ is in pounds and $x$ is in feet. So the work done by a force of $F(x) = 1/x^2$ newtons in moving an object along the $x$-axis from $x = 1$ m to $x = 10$ m is

$$W = \int_{1}^{10} \frac{1}{x^2}\, dx = -\frac{1}{x}\Big]_{1}^{10} = -\frac{1}{10} + 1 = 0.9 \text{ J}.$$

### Hooke's Law for Springs: $F = kx$

One calculation for work arises in finding the work required to stretch or compress a spring. **Hooke's Law** says that the force required to hold a stretched or compressed spring $x$ units from its natural (unstressed) length is proportional to $x$. In symbols,

$$F = kx. \tag{3}$$

The constant $k$, measured in force units per unit length, is a characteristic of the spring, called the **force constant** (or **spring constant**) of the spring. Hooke's Law, Equation (3), gives good results as long as the force doesn't distort the metal in the spring. We assume that the forces in this section are too small to do that.

**EXAMPLE 2**  Find the work required to compress a spring from its natural length of 30 cm to a length of 20 cm if the force constant is $k = 240$ N/m.

**Solution**  We picture the uncompressed spring laid out along the $x$-axis with its movable end at the origin and its fixed end at $x = 0.3$ m (Figure 6.36). This enables us to describe the force required to compress the spring from 0 to $x$ with the formula $F = 240x$. To compress the spring from 0 to 0.1 m, the force must increase from

$$F(0) = 240 \cdot 0 = 0 \text{ N} \qquad \text{to} \qquad F(0.1) = 240 \cdot 0.1 = 24 \text{ N}.$$

The work done by $F$ over this interval is

$$W = \int_{0}^{0.1} 240x\, dx = 120x^2\Big]_{0}^{0.1} = 1.2 \text{ J}. \qquad \begin{array}{l}\text{Eq. (2) with}\\ a = 0, b = 0.1,\\ F(x) = 240x\end{array}$$

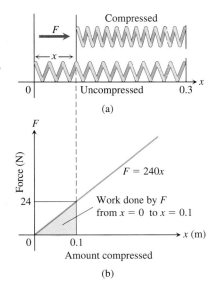

**FIGURE 6.36**  The force $F$ needed to hold a spring under compression increases linearly as the spring is compressed (Example 2).

**FIGURE 6.37** A 24-N weight stretches this spring 0.8 m beyond its unstressed length (Example 3).

**EXAMPLE 3** A spring has a natural length of 1 m. A force of 24 N holds the spring stretched to a total length of 1.8 m.

(a) Find the force constant $k$.

(b) How much work will it take to stretch the spring 2 m beyond its natural length?

(c) How far will a 45-N force stretch the spring?

**Solution**

(a) *The force constant.* We find the force constant from Equation (3). A force of 24 N maintains the spring at a position where it is stretched 0.8 m from its natural length, so

$$24 = k(0.8)$$
$$k = 24/0.8 = 30 \text{ N/m.}$$

Eq. (3) with $F = 24, x = 0.8$

(b) *The work to stretch the spring* 2 m. We imagine the unstressed spring hanging along the $x$-axis with its free end at $x = 0$ (Figure 6.37). The force required to stretch the spring $x$ m beyond its natural length is the force required to hold the free end of the spring $x$ units from the origin. Hooke's Law with $k = 30$ says that this force is

$$F(x) = 30x.$$

The work done by $F$ on the spring from $x = 0$ m to $x = 2$ m is

$$W = \int_0^2 30x \, dx = 15x^2 \Big]_0^2 = 60 \text{ J.}$$

(c) *How far will a 45-N force stretch the spring?* We substitute $F = 45$ in the equation $F = 30x$ to find

$$45 = 30x, \quad \text{or} \quad x = 1.5 \text{ m.}$$

A 45-N force will keep the spring stretched 1.5 m beyond its natural length. ∎

## Lifting Objects and Pumping Liquids from Containers

The work integral is useful to calculate the work done in lifting objects whose weights vary with their elevation.

**EXAMPLE 4** A 2-kg bucket is lifted from the ground into the air by pulling in 6 m of rope at a constant speed (Figure 6.38). The rope weighs 0.1 kg/m. How much work was spent lifting the bucket and rope?

**Solution** The bucket has constant weight, so the work done lifting it alone is weight × distance $= 2 \cdot 9.8 \cdot 6 = 117.6$ J.

The weight of the rope varies with the bucket's elevation, because less of it is freely hanging. When the bucket is $x$ m off the ground, the remaining proportion of the rope still being lifted weighs $(0.1 \cdot 9.8) \cdot (6 - x)$ N. So the work in lifting the rope is

$$\text{Work on rope} = \int_0^6 (0.98)(6 - x) \, dx = \int_0^6 (5.88 - 0.98x) \, dx$$

$$= \big[ 5.88x - 0.49x^2 \big]_0^6 = 35.28 - 17.64 = 17.64 \text{ J.}$$

The total work for the bucket and rope combined is

$$117.6 + 17.64 = 135.24 \text{ J.} \quad ∎$$

**FIGURE 6.38** Lifting the bucket in Example 4.

How much work does it take to pump all or part of the liquid from a container? Engineers often need to know the answer in order to design or choose the right pump, or to compute the cost, to transport water or some other liquid from one place to another. To find out how much work is required to pump the liquid, we imagine lifting the liquid out one thin horizontal slab at a time and applying the equation $W = Fd$ to each slab. We then evaluate

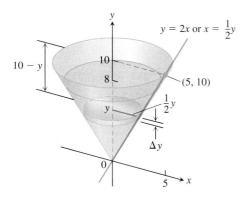

**FIGURE 6.39** The olive oil and tank in Example 5.

the integral this leads to as the slabs become thinner and more numerous. The integral we get each time depends on the weight of the liquid and the dimensions of the container, but the way we find the integral is always the same. The next example shows what to do.

**EXAMPLE 5**    The conical tank in Figure 6.39 is filled to within 2 m of the top with olive oil weighing 0.9 g/cm³ or 8820 N/m³. How much work does it take to pump the oil to the rim of the tank?

**Solution**    We imagine the oil divided into thin slabs by planes perpendicular to the y-axis at the points of a partition of the interval $[0, 8]$.

The typical slab between the planes at $y$ and $y + \Delta y$ has a volume of about

$$\Delta V = \pi(\text{radius})^2(\text{thickness}) = \pi\left(\frac{1}{2}y\right)^2 \Delta y = \frac{\pi}{4}y^2 \, \Delta y \text{ m}^3.$$

The force $F(y)$ required to lift this slab is equal to its weight,

$$F(y) = 8820 \, \Delta V = \frac{8820\pi}{4}y^2 \, \Delta y \text{ N.} \qquad \begin{array}{l}\text{Weight} = (\text{weight per unit} \\ \text{volume}) \times \text{volume}\end{array}$$

The distance through which $F(y)$ must act to lift this slab to the level of the rim of the cone is about $(10 - y)$ m, so the work done lifting the slab is about

$$\Delta W = \frac{8820\pi}{4}(10 - y)y^2 \, \Delta y \text{ J.}$$

Assuming there are $n$ slabs associated with the partition of $[0, 8]$, and that $y = y_k$ denotes the plane associated with the $k$th slab of thickness $\Delta y_k$, we can approximate the work done lifting all of the slabs with the Riemann sum

$$W \approx \sum_{k=1}^{n} \frac{8820\pi}{4}(10 - y_k)y_k^2 \, \Delta y_k \text{ J.}$$

The work of pumping the oil to the rim is the limit of these sums as the norm of the partition goes to zero and the number of slabs tends to infinity:

$$W = \lim_{n\to\infty} \sum_{k=1}^{n} \frac{8820\pi}{4}(10 - y_k)y_k^2 \, \Delta y_k = \int_0^8 \frac{8820\pi}{4}(10 - y)y^2 \, dy$$

$$= \frac{8820\pi}{4} \int_0^8 (10y^2 - y^3) \, dy$$

$$= \frac{8820\pi}{4}\left[\frac{10y^3}{3} - \frac{y^4}{4}\right]_0^8 \approx 4{,}728{,}977 \text{ J.} \quad \blacksquare$$

**FIGURE 6.40** To withstand the increasing pressure, dams are built thicker as they go down.

## Fluid Pressure and Forces

Dams are built thicker at the bottom than at the top (Figure 6.40) because the pressure against them increases with depth. The pressure at any point on a dam depends only on how far below the surface the point is and not on how much the surface of the dam happens to be tilted at that point. The pressure, in newtons per square meter at a point $h$ meters below the surface, is always $9800h$. The number 9800 is the weight-density of freshwater in newtons per cubic meter. The pressure $h$ meters below the surface of any fluid is the fluid's *weight-density* times $h$.

**Weight-density**

A fluid's weight-density $w$ is its weight per unit volume. Typical values (N/m³) are listed below.

| | |
|---|---|
| Gasoline | 6600 |
| Mercury | 133,000 |
| Milk | 10,100 |
| Molasses | 15,700 |
| Olive oil | 8820 |
| Seawater | 10,050 |
| Freshwater | 9800 |

**The Pressure-Depth Equation**

In a fluid that is standing still, the pressure $p$ at depth $h$ is the fluid's weight-density $w$ times $h$:

$$p = wh. \tag{4}$$

**FIGURE 6.41** These containers are filled with water to the same depth and have the same base area. The total force is therefore the same on the bottom of each container. The containers' shapes do not matter here.

In a container of fluid with a flat horizontal base, the total force exerted by the fluid against the base can be calculated by multiplying the area of the base by the pressure at the base. We can do this because total force equals force per unit area (pressure) times area. (See Figure 6.41.) If $F$, $p$, and $A$ are the total force, pressure, and area, then

$$F = \text{total force} = \text{force per unit area} \times \text{area}$$
$$= \text{pressure} \times \text{area} = pA$$
$$= whA. \qquad\qquad p = wh \text{ from Eq. (4)}$$

---

**Fluid Force on a Constant-Depth Surface**

$$F = pA = whA \qquad\qquad (5)$$

---

For example, the weight-density of freshwater is 9800 N/m$^3$, so the fluid force at the bottom of a 3 m $\times$ 6 m rectangular swimming pool 1 m deep is

$$F = whA = (9800 \text{ N/m}^3)(1 \text{ m})(3 \cdot 6 \text{ m}^2)$$
$$= 176{,}400 \text{ N}.$$

For a flat plate submerged *horizontally*, like the bottom of the swimming pool just discussed, the downward force acting on its upper face due to liquid pressure is given by Equation (5). If the plate is submerged *vertically*, however, then the pressure against it will be different at different depths and Equation (5) no longer is usable in that form (because $h$ varies).

Suppose we want to know the force exerted by a fluid against one side of a vertical plate submerged in a fluid of weight-density $w$. To find it, we model the plate as a region extending from $y = a$ to $y = b$ in the $xy$-plane (Figure 6.42). We partition $[a, b]$ in the usual way and imagine the region to be cut into thin horizontal strips by planes perpendicular to the $y$-axis at the partition points. The typical strip from $y$ to $y + \Delta y$ is $\Delta y$ units wide by $L(y)$ units long. We assume $L(y)$ to be a continuous function of $y$.

The pressure varies across the strip from top to bottom. If the strip is narrow enough, however, the pressure will remain close to its bottom-edge value of $w \times$ (strip depth). The force exerted by the fluid against one side of the strip will be about

$$\Delta F = (\text{pressure along bottom edge}) \times (\text{area})$$
$$= w \cdot (\text{strip depth}) \cdot L(y)\,\Delta y.$$

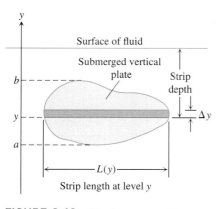

**FIGURE 6.42** The force exerted by a fluid against one side of a thin, flat horizontal strip is about $\Delta F = \text{pressure} \times \text{area} = w \times (\text{strip depth}) \times L(y)\,\Delta y$.

Assume there are $n$ strips associated with the partition of $a \le y \le b$ and that $y_k$ is the bottom edge of the $k$th strip having length $L(y_k)$ and width $\Delta y_k$. The force against the entire plate is approximated by summing the forces against each strip, giving the Riemann sum

$$F \approx \sum_{k=1}^{n} (w \cdot (\text{strip depth})_k \cdot L(y_k))\,\Delta y_k. \qquad (6)$$

The sum in Equation (6) is a Riemann sum for a continuous function on $[a, b]$, and we expect the approximations to improve as the norm of the partition goes to zero. The force against the plate is the limit of these sums:

$$\lim_{n \to \infty} \sum_{k=1}^{n} (w \cdot (\text{strip depth})_k \cdot L(y_k))\,\Delta y_k = \int_{a}^{b} w \cdot (\text{strip depth}) \cdot L(y)\,dy.$$

---

**The Integral for Fluid Force Against a Vertical Flat Plate**

Suppose that a plate submerged vertically in fluid of weight-density $w$ runs from $y = a$ to $y = b$ on the $y$-axis. Let $L(y)$ be the length of the horizontal strip measured from left to right along the surface of the plate at level $y$. Then the force exerted by the fluid against one side of the plate is

$$F = \int_a^b w \cdot (\text{strip depth}) \cdot L(y) \, dy. \qquad (7)$$

---

**EXAMPLE 6**    A flat isosceles right-triangular plate with base 2 m and height 1 m is submerged vertically, base up, 0.6 m below the surface of a swimming pool. Find the force exerted by the water against one side of the plate.

**Solution**    We establish a coordinate system to work in by placing the origin at the plate's bottom vertex and running the $y$-axis upward along the plate's axis of symmetry (Figure 6.43). The surface of the pool lies along the line $y = 1.6$ and the plate's top edge along the line $y = 1$. The plate's right-hand edge lies along the line $y = x$, with the upper-right vertex at $(1, 1)$. The length of a thin strip at level $y$ is

$$L(y) = 2x = 2y.$$

The depth of the strip beneath the surface is $(1.6 - y)$. The force exerted by the water against one side of the plate is therefore

$$\begin{aligned}
F &= \int_a^b w \cdot \left(\begin{array}{c}\text{strip}\\\text{depth}\end{array}\right) \cdot L(y) \, dy \qquad \text{Eq. (7)}\\
&= \int_0^1 9800(1.6 - y)2y \, dy\\
&= 19{,}600 \int_0^1 (1.6y - y^2) \, dy\\
&= 19{,}600 \left[0.8y^2 - \frac{y^3}{3}\right]_0^1 \approx 9147 \text{ N.} \qquad \blacksquare
\end{aligned}$$

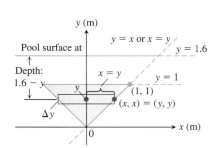

**FIGURE 6.43**    To find the force on one side of the submerged plate in Example 6, we can use a coordinate system like the one here.

---

# Exercises 6.5

**Springs**

**1. Spring constant**    It took 1800 J of work to stretch a spring from its natural length of 2 m to a length of 5 m. Find the spring's force constant.

**2. Stretching a spring**    A spring has a natural length of 10 cm. An 800-N force stretches the spring to 14 cm.

   **a.** Find the force constant.

   **b.** How much work is done in stretching the spring from 10 cm to 12 cm?

   **c.** How far beyond its natural length will a 1600-N force stretch the spring?

**3. Stretching a rubber band**    A force of 2 N will stretch a rubber band 2 cm (0.02 m). Assuming that Hooke's Law applies, how far will a 4-N force stretch the rubber band? How much work does it take to stretch the rubber band this far?

**4. Stretching a spring**    If a force of 90 N stretches a spring 1 m beyond its natural length, how much work does it take to stretch the spring 5 m beyond its natural length?

**5. Subway car springs**    It takes a force of 96,000 N to compress a coil spring assembly on a New York City Transit Authority subway car from its free height of 20 cm to its fully compressed height of 12 cm.

   **a.** What is the assembly's force constant?

   **b.** How much work does it take to compress the assembly the first centimeter? the second centimeter? Answer to the nearest joule.

**6. Bathroom scale**    A bathroom scale is compressed 1.5 mm when a 70 kg person stands on it. Assuming that the scale behaves like a spring that obeys Hooke's Law, how much does someone who compresses the scale 3 mm weigh? How much work is done compressing the scale 3 mm?

## Work Done by a Variable Force

**7. Lifting a rope** A mountain climber is about to haul up a 50-m length of hanging rope. How much work will it take if the rope weighs 0.624 N/m?

**8. Leaky sandbag** A bag of sand originally weighing 600 N was lifted at a constant rate. As it rose, sand also leaked out at a constant rate. The sand was half gone by the time the bag had been lifted to 6 m. How much work was done lifting the sand this far? (Neglect the weight of the bag and lifting equipment.)

**9. Lifting an elevator cable** An electric elevator with a motor at the top has a multistrand cable weighing 60 N/m. When the car is at the first floor, 60 m of cable are paid out, and effectively 0 m are out when the car is at the top floor. How much work does the motor do just lifting the cable when it takes the car from the first floor to the top?

**10. Force of attraction** When a particle of mass $m$ is at $(x, 0)$, it is attracted toward the origin with a force whose magnitude is $k/x^2$. If the particle starts from rest at $x = b$ and is acted on by no other forces, find the work done on it by the time it reaches $x = a$, $0 < a < b$.

**11. Leaky bucket** Assume the bucket in Example 4 is leaking. It starts with 8 L of water (78 N) and leaks at a constant rate. It finishes draining just as it reaches the top. How much work was spent lifting the water alone? (*Hint:* Do not include the rope and bucket, and find the proportion of water left at elevation $x$ m.)

**12. (*Continuation of Exercise 11.*)** The workers in Example 4 and Exercise 11 changed to a larger bucket that held 20 L (195 N) of water, but the new bucket had an even larger leak so that it, too, was empty by the time it reached the top. Assuming that the water leaked out at a steady rate, how much work was done lifting the water alone? (Do not include the rope and bucket.)

## Pumping Liquids from Containers

**13. Pumping water** The rectangular tank shown here, with its top at ground level, is used to catch runoff water. Assume that the water weighs 9800 N/m³.

a. How much work does it take to empty the tank by pumping the water back to ground level once the tank is full?

b. If the water is pumped to ground level with a 5-horsepower (hp) motor (work output 3678 W), how long will it take to empty the full tank (to the nearest minute)?

c. Show that the pump in part (b) will lower the water level 10 m (halfway) during the first 266 min of pumping.

d. **The weight of water** What are the answers to parts (a) and (b) in a location where water weighs 9780 N/m³? 9820 N/m³?

**14. Emptying a cistern** The rectangular cistern (storage tank for rainwater) shown has its top 3 m below ground level. The cistern, currently full, is to be emptied for inspection by pumping its contents to ground level.

a. How much work will it take to empty the cistern?

b. How long will it take a 1/2-hp pump, rated at 370 W, to pump the tank dry?

c. How long will it take the pump in part (b) to empty the tank halfway? (It will be less than half the time required to empty the tank completely.)

d. **The weight of water** What are the answers to parts (a) through (c) in a location where water weighs 9780 N/m³? 9820 N/m³?

**15. Pumping oil** How much work would it take to pump oil from the tank in Example 5 to the level of the top of the tank if the tank were completely full?

**16. Pumping a half-full tank** Suppose that, instead of being full, the tank in Example 5 is only half full. How much work does it take to pump the remaining oil to a level 1 m above the top of the tank?

**17. Emptying a tank** A vertical right-circular cylindrical tank measures 9 m high and 6 m in diameter. It is full of kerosene weighing 7840 N/m³. How much work does it take to pump the kerosene to the level of the top of the tank?

**18. a. Pumping milk** Suppose that the conical container in Example 5 contains milk (weighing 10,100 N/m³) instead of olive oil. How much work will it take to pump the contents to the rim?

  b. **Pumping oil** How much work will it take to pump the oil in Example 5 to a level 1 m above the cone's rim?

**19.** The graph of $y = x^2$ on $0 \le x \le 2$ is revolved about the $y$-axis to form a tank that is then filled with salt water from the Dead Sea (weighing approximately 11,500 N/m³). How much work does it take to pump all of the water to the top of the tank?

**20.** A right-circular cylindrical tank of height 3 m and radius 1.5 m is lying horizontally and is full of diesel fuel weighing 8300 N/m³. How much work is required to pump all of the fuel to a point 4.5 m above the top of the tank?

**21. Emptying a water reservoir** We model pumping from spherical containers the way we do from other containers, with the axis of integration along the vertical axis of the sphere. Use the figure here to find how much work it takes to empty a full hemispherical water reservoir of radius 5 m by pumping the water to a height of 4 m above the top of the reservoir. Water weighs 9800 N/m³.

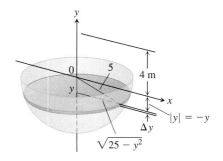

**22.** You are in charge of the evacuation and repair of the storage tank shown here. The tank is a hemisphere of radius 3 m and is full of benzene weighing 8800 N/m³. A firm you contacted says it can empty the tank for 0.4¢ per joule of work. Find the work required to empty the tank by pumping the benzene to an outlet 0.6 m above the top of the tank. If you have $5000 budgeted for the job, can you afford to hire the firm?

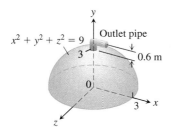

### Work and Kinetic Energy

**23. Kinetic energy** If a variable force of magnitude $F(x)$ moves an object of mass $m$ along the $x$-axis from $x_1$ to $x_2$, the object's velocity $v$ can be written as $dx/dt$ (where $t$ represents time). Use Newton's second law of motion $F = m(dv/dt)$ and the Chain Rule

$$\frac{dv}{dt} = \frac{dv}{dx}\frac{dx}{dt} = v\frac{dv}{dx}$$

to show that the net work done by the force in moving the object from $x_1$ to $x_2$ is

$$W = \int_{x_1}^{x_2} F(x)\,dx = \frac{1}{2}mv_2^2 - \frac{1}{2}mv_1^2,$$

where $v_1$ and $v_2$ are the object's velocities at $x_1$ and $x_2$. In physics, the expression $(1/2)mv^2$ is called the *kinetic energy* of an object of mass $m$ moving with velocity $v$. Therefore, *the work done by the force equals the change in the object's kinetic energy*, and we can find the work by calculating this change.

In Exercises 24–28, use the result of Exercise 23.

**24. Tennis** A 60 g tennis ball was served at 50 m/s (180 km/h). How much work was done on the ball to make it go this fast?

**25. Baseball** How many joules of work does it take to throw a baseball at 144 km/h? A baseball's mass is 150 g.

**26. Golf** A 50 g golf ball is driven off the tee at a speed of 84 m/s (302.4 km/h). How many joules of work are done on the ball getting it into the air?

**27.** On June 11, 2004, in a tennis match between Andy Roddick and Paradorn Srichaphan at the Stella Artois tournament in London, England, Roddick hit a serve measured at 244.8 km/h. How much work was required by Andy to serve a 60 g tennis ball at that speed?

**28. Softball** How much work has to be performed on a 200 g softball to pitch it 40 m/s (144 km/h)?

**29. Drinking a milkshake** The truncated conical container shown here is full of strawberry milkshake which has a density of 0.8 g/cm³. As you can see, the container is 18 cm deep, 6 cm across at the base, and 9 cm across at the top (a standard size at Brigham's in Boston). The straw sticks up 3 cm above the top.

About how much work does it take to suck up the milkshake through the straw (neglecting friction)?

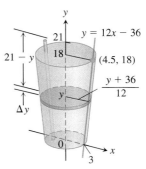

Dimensions in centimeters

**30. Water tower** Your town has decided to drill a well to increase its water supply. As the town engineer, you have determined that a water tower will be necessary to provide the pressure needed for distribution, and you have designed the system shown here. The water is to be pumped from a 90-m well through a vertical 10 cm pipe into the base of a cylindrical tank 6 m in diameter and 7.5 m high. The base of the tank will be 18 m above ground. The pump is a 3-hp pump, rated at 2000 W (J/s). To the nearest hour, how long will it take to fill the tank the first time? (Include the time it takes to fill the pipe.) Assume that water weighs 9800 N/m³.

NOT TO SCALE

**31. Putting a satellite in orbit** The strength of Earth's gravitational field varies with the distance $r$ from Earth's center, and the magnitude of the gravitational force experienced by a satellite of mass $m$ during and after launch is

$$F(r) = \frac{mMG}{r^2}.$$

Here, $M = 5.975 \times 10^{24}$ kg is Earth's mass, $G = 6.6720 \times 10^{-11}$ N·m² kg⁻² is the universal gravitational constant, and $r$ is measured in meters. The work it takes to lift a 1000-kg satellite from Earth's surface to a circular orbit 35,780 km above Earth's center is therefore given by the integral

$$\text{Work} = \int_{6,370,000}^{35,780,000} \frac{1000MG}{r^2}\,dr \text{ joules.}$$

Evaluate the integral. The lower limit of integration is Earth's radius in meters at the launch site. (This calculation does not take into account energy spent lifting the launch vehicle or energy spent bringing the satellite to orbit velocity.)

**32. Forcing electrons together**  Two electrons $r$ meters apart repel each other with a force of

$$F = \frac{23 \times 10^{-29}}{r^2} \text{ newtons.}$$

a. Suppose one electron is held fixed at the point $(1, 0)$ on the $x$-axis (units in meters). How much work does it take to move a second electron along the $x$-axis from the point $(-1, 0)$ to the origin?

b. Suppose an electron is held fixed at each of the points $(-1, 0)$ and $(1, 0)$. How much work does it take to move a third electron along the $x$-axis from $(5, 0)$ to $(3, 0)$?

### Finding Fluid Forces

**33. Triangular plate**  Calculate the fluid force on one side of the plate in Example 6 using the coordinate system shown here.

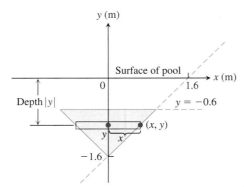

**34. Triangular plate**  Calculate the fluid force on one side of the plate in Example 6 using the coordinate system shown here.

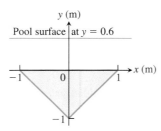

**35. Rectangular plate**  In a pool filled with water to a depth of 3 m, calculate the fluid force on one side of a 0.9 m by 1.2 m rectangular plate if the plate rests vertically at the bottom of the pool

a. on its 1.2-m edge.    b. on its 0.9-m edge.

**36. Semicircular plate**  Calculate the fluid force on one side of a semicircular plate of radius 5 m that rests vertically on its diameter at the bottom of a pool filled with water to a depth of 6 m.

**37. Triangular plate**  The isosceles triangular plate shown here is submerged vertically 1 m below the surface of a freshwater lake.

a. Find the fluid force against one face of the plate.

b. What would be the fluid force on one side of the plate if the water were seawater instead of freshwater?

**38. Rotated triangular plate**  The plate in Exercise 37 is revolved $180°$ about line $AB$ so that part of the plate sticks out of the lake, as shown here. What force does the water exert on one face of the plate now?

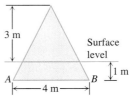

**39. New England Aquarium**  The viewing portion of the rectangular glass window in a typical fish tank at the New England Aquarium in Boston is 1.6 m wide and runs from 0.01 m below the water's surface to 0.85 m below the surface. Find the fluid force against this portion of the window. The weight-density of seawater is $10,050 \text{ N/m}^3$. (In case you were wondering, the glass is 2 cm thick and the tank walls extend 10 cm above the water to keep the fish from jumping out.)

**40. Semicircular plate**  A semicircular plate 2 m in diameter sticks straight down into freshwater with the diameter along the surface. Find the force exerted by the water on one side of the plate.

**41. Tilted plate**  Calculate the fluid force on one side of a 1 m by 1 m square plate if the plate is at the bottom of a pool filled with water to a depth of 2 m and

a. lying flat on its 1 m by 1 m face.

b. resting vertically on a 1 m edge.

c. resting on a 1 m edge and tilted at $45°$ to the bottom of the pool.

**42. Tilted plate**  Calculate the fluid force on one side of a right-triangular plate with edges 3 m, 4 m, and 5 m if the plate sits at the bottom of a pool filled with water to a depth of 6 m on its 3-m edge and tilted at $60°$ to the bottom of the pool.

**43.**  The cubical metal tank shown here has a parabolic gate held in place by bolts and designed to withstand a fluid force of 25,000 N without rupturing. The liquid you plan to store has a weight-density of $8000 \text{ N/m}^3$.

a. What is the fluid force on the gate when the liquid is 2 m deep?

b. What is the maximum height to which the container can be filled without exceeding the gate's design limitation?

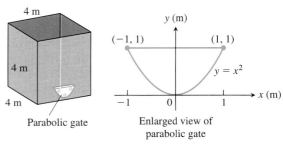

Parabolic gate

Enlarged view of parabolic gate

**44.** The end plates of the trough shown here were designed to withstand a fluid force of 25,000 N. How many cubic meters of water can the tank hold without exceeding this limitation? Round down to the nearest cubic meter. What is the value of $h$?

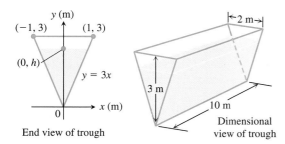

End view of trough

Dimensional view of trough

**45.** A vertical rectangular plate $a$ units long by $b$ units wide is submerged in a fluid of weight-density $w$ with its long edges parallel to the fluid's surface. Find the average value of the pressure along the vertical dimension of the plate. Explain your answer.

**46.** (*Continuation of Exercise 45.*) Show that the force exerted by the fluid on one side of the plate is the average value of the pressure (found in Exercise 45) times the area of the plate.

**47.** Water pours into the tank shown here at the rate of 0.5 m³/min. The tank's cross-sections are 2-m-diameter semicircles. One end of the tank is movable, but moving it to increase the volume

compresses a spring. The spring constant is $k = 3000 \, \text{N/m}$. If the end of the tank moves 2.5 m against the spring, the water will drain out of a safety hole in the bottom at the rate of 0.6 m³/min. Will the movable end reach the hole before the tank overflows?

Side view

**48. Watering trough** The vertical ends of a watering trough are squares 1 m on a side.

**a.** Find the fluid force against the ends when the trough is full.

**b.** How many centimeters do you have to lower the water level in the trough to reduce the fluid force by 25%?

# 6.6 Moments and Centers of Mass

Many structures and mechanical systems behave as if their masses were concentrated at a single point, called the *center of mass* (Figure 6.44). It is important to know how to locate this point, and doing so is basically a mathematical enterprise. Here we consider masses distributed along a line or region in the plane. Masses distributed across a region or curve in three-dimensional space are treated in Chapters 15 and 16.

## Masses Along a Line

We develop our mathematical model in stages. The first stage is to imagine masses $m_1$, $m_2$, and $m_3$ on a rigid $x$-axis supported by a fulcrum at the origin.

The resulting system might balance, or it might not, depending on how large the masses are and how they are arranged along the $x$-axis.

Each mass $m_k$ exerts a downward force $m_k g$ (the weight of $m_k$) equal to the magnitude of the mass times the acceleration due to gravity. Note that gravitational acceleration is downward, hence negative. Each of these forces has a tendency to turn the $x$-axis about the origin, the way a child turns a seesaw. This turning effect, called a **torque**, is measured by multiplying the force $m_k g$ by the signed distance $x_k$ from the point of application to the origin. By convention, a positive torque induces a counterclockwise turn. Masses to the left of the origin exert positive (counterclockwise) torque. Masses to the right of the origin exert negative (clockwise) torque.

The sum of the torques measures the tendency of a system to rotate about the origin. This sum is called the **system torque**.

$$\text{System torque} = m_1 g x_1 + m_2 g x_2 + m_3 g x_3 \tag{1}$$

The system will balance if and only if its torque is zero.

If we factor out the $g$ in Equation (1), we see that the system torque is

$$\underbrace{g}_{\substack{\text{a feature of the}\\\text{environment}}} \cdot \underbrace{(m_1x_1 + m_2x_2 + m_3x_3).}_{\substack{\text{a feature of}\\\text{the system}}}$$

Thus, the torque is the product of the gravitational acceleration $g$, which is a feature of the environment in which the system happens to reside, and the number $(m_1x_1 + m_2x_2 + m_3x_3)$, which is a feature of the system itself, a constant that stays the same no matter where the system is placed.

The number $(m_1x_1 + m_2x_2 + m_3x_3)$ is called the **moment of the system about the origin**. It is the sum of the **moments** $m_1x_1, m_2x_2, m_3x_3$ of the individual masses.

$$M_0 = \text{Moment of system about origin} = \sum m_k x_k$$

(We shift to sigma notation here to allow for sums with more terms.)

We usually want to know where to place the fulcrum to make the system balance, that is, at what point $\bar{x}$ to place it to make the torques add to zero.

FIGURE 6.44 A wrench gliding on ice turning about its center of mass as the center glides in a vertical line. (*Source: PSSC Physics*, 2nd ed., Reprinted by permission of Education Development Center, Inc.)

The torque of each mass about the fulcrum in this special location is

$$\text{Torque of } m_k \text{ about } \bar{x} = \begin{pmatrix} \text{signed distance} \\ \text{of } m_k \text{ from } \bar{x} \end{pmatrix}\begin{pmatrix} \text{downward} \\ \text{force} \end{pmatrix}$$

$$= (x_k - \bar{x})m_k g.$$

When we write the equation that says that the sum of these torques is zero, we get an equation we can solve for $\bar{x}$:

$$\sum (x_k - \bar{x})m_k g = 0 \qquad \text{Sum of the torques equals zero.}$$

$$\bar{x} = \frac{\sum m_k x_k}{\sum m_k}. \qquad \text{Solved for } \bar{x}$$

This last equation tells us to find $\bar{x}$ by dividing the system's moment about the origin by the system's total mass:

$$\bar{x} = \frac{\sum m_k x_k}{\sum m_k} = \frac{\text{system moment about origin}}{\text{system mass}}. \qquad (2)$$

The point $\bar{x}$ is called the system's **center of mass**.

## Masses Distributed over a Plane Region

Suppose that we have a finite collection of masses located in the plane, with mass $m_k$ at the point $(x_k, y_k)$ (see Figure 6.45). The mass of the system is

$$\text{System mass:} \qquad M = \sum m_k.$$

FIGURE 6.45 Each mass $m_k$ has a moment about each axis.

Each mass $m_k$ has a moment about each axis. Its moment about the $x$-axis is $m_k y_k$, and its moment about the $y$-axis is $m_k x_k$. The moments of the entire system about the two axes are

$$\text{Moment about } x\text{-axis:} \qquad M_x = \sum m_k y_k,$$

$$\text{Moment about } y\text{-axis:} \qquad M_y = \sum m_k x_k.$$

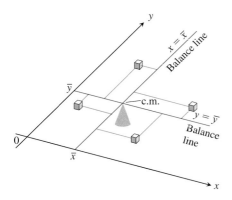

**FIGURE 6.46** A two-dimensional array of masses balances on its center of mass.

The $x$-coordinate of the system's center of mass is defined to be

$$\bar{x} = \frac{M_y}{M} = \frac{\sum m_k x_k}{\sum m_k}. \tag{3}$$

With this choice of $\bar{x}$, as in the one-dimensional case, the system balances about the line $x = \bar{x}$ (Figure 6.46).

The $y$-coordinate of the system's center of mass is defined to be

$$\bar{y} = \frac{M_x}{M} = \frac{\sum m_k y_k}{\sum m_k}. \tag{4}$$

With this choice of $\bar{y}$, the system balances about the line $y = \bar{y}$ as well. The torques exerted by the masses about the line $y = \bar{y}$ cancel out. Thus, as far as balance is concerned, the system behaves as if all its mass were at the single point $(\bar{x}, \bar{y})$. We call this point the system's **center of mass**.

### Thin, Flat Plates

In many applications, we need to find the center of mass of a thin, flat plate: a disk of aluminum, say, or a triangular sheet of steel. In such cases, we assume the distribution of mass to be continuous, and the formulas we use to calculate $\bar{x}$ and $\bar{y}$ contain integrals instead of finite sums. The integrals arise in the following way.

Imagine that the plate occupying a region in the $xy$-plane is cut into thin strips parallel to one of the axes (in Figure 6.47, the $y$-axis). The center of mass of a typical strip is $(\tilde{x}, \tilde{y})$. We treat the strip's mass $\Delta m$ as if it were concentrated at $(\tilde{x}, \tilde{y})$. The moment of the strip about the $y$-axis is then $\tilde{x}\,\Delta m$. The moment of the strip about the $x$-axis is $\tilde{y}\,\Delta m$. Equations (3) and (4) then become

$$\bar{x} = \frac{M_y}{M} = \frac{\sum \tilde{x}\,\Delta m}{\sum \Delta m}, \qquad \bar{y} = \frac{M_x}{M} = \frac{\sum \tilde{y}\,\Delta m}{\sum \Delta m}.$$

The sums are Riemann sums for integrals and approach these integrals as limiting values as the strips into which the plate is cut become narrower and narrower. We write these integrals symbolically as

$$\bar{x} = \frac{\int \tilde{x}\,dm}{\int dm} \qquad \text{and} \qquad \bar{y} = \frac{\int \tilde{y}\,dm}{\int dm}.$$

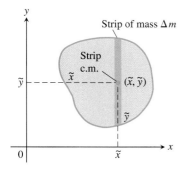

**FIGURE 6.47** A plate cut into thin strips parallel to the $y$-axis. The moment exerted by a typical strip about each axis is the moment its mass $\Delta m$ would exert if concentrated at the strip's center of mass $(\tilde{x}, \tilde{y})$.

---

**Moments, Mass, and Center of Mass of a Thin Plate Covering a Region in the $xy$-Plane**

Moment about the $x$-axis: $\quad M_x = \int \tilde{y}\,dm$

Moment about the $y$-axis: $\quad M_y = \int \tilde{x}\,dm$

Mass: $\quad M = \int dm$

Center of mass: $\quad \bar{x} = \dfrac{M_y}{M}, \quad \bar{y} = \dfrac{M_x}{M}$

(5)

**Density of a plate**

A material's density is its mass per unit area. For wires, rods, and narrow strips, we use mass per unit length.

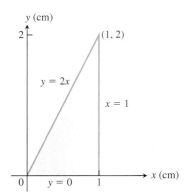

**FIGURE 6.48** The plate in Example 1.

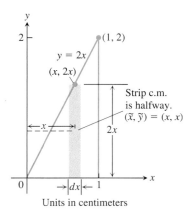

**FIGURE 6.49** Modeling the plate in Example 1 with vertical strips.

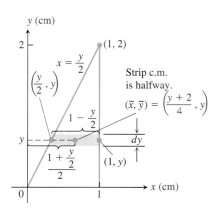

**FIGURE 6.50** Modeling the plate in Example 1 with horizontal strips.

The differential $dm$ is the mass of the strip. For this section, we assume the density $\delta$ of the plate is a constant or a continuous function of $x$. Then $dm = \delta\, dA$, which is the mass per unit area $\delta$ times the area $dA$ of the strip.

To evaluate the integrals in Equations (5), we picture the plate in the coordinate plane and sketch a strip of mass parallel to one of the coordinate axes. We then express the strip's mass $dm$ and the coordinates $(\tilde{x}, \tilde{y})$ of the strip's center of mass in terms of $x$ or $y$. Finally, we integrate $\tilde{y}\, dm$, $\tilde{x}\, dm$, and $dm$ between limits of integration determined by the plate's location in the plane.

**EXAMPLE 1** The triangular plate shown in Figure 6.48 has a constant density of $\delta = 3$ g/cm$^2$. Find

**(a)** the plate's moment $M_y$ about the $y$-axis.

**(b)** the plate's mass $M$.

**(c)** the $x$-coordinate of the plate's center of mass (c.m.).

**Solution    Method 1: Vertical Strips**    (Figure 6.49)

**(a)** The moment $M_y$: The typical vertical strip has the following relevant data.

$$
\begin{aligned}
&\text{center of mass (c.m.):} && (\tilde{x}, \tilde{y}) = (x, x) \\
&\text{length:} && 2x \\
&\text{width:} && dx \\
&\text{area:} && dA = 2x\, dx \\
&\text{mass:} && dm = \delta\, dA = 3 \cdot 2x\, dx = 6x\, dx \\
&\text{distance of c.m. from } y\text{-axis:} && \tilde{x} = x
\end{aligned}
$$

The moment of the strip about the $y$-axis is

$$\tilde{x}\, dm = x \cdot 6x\, dx = 6x^2\, dx.$$

The moment of the plate about the $y$-axis is therefore

$$M_y = \int \tilde{x}\, dm = \int_0^1 6x^2\, dx = 2x^3 \Big]_0^1 = 2 \text{ g} \cdot \text{cm}.$$

**(b)** The plate's mass:

$$M = \int dm = \int_0^1 6x\, dx = 3x^2 \Big]_0^1 = 3 \text{ g}.$$

**(c)** The $x$-coordinate of the plate's center of mass:

$$\bar{x} = \frac{M_y}{M} = \frac{2 \text{ g} \cdot \text{cm}}{3 \text{ g}} = \frac{2}{3} \text{ cm}.$$

By a similar computation, we could find $M_x$ and $\bar{y} = M_x/M$.

**Method 2: Horizontal Strips**    (Figure 6.50)

**(a)** The moment $M_y$: The $y$-coordinate of the center of mass of a typical horizontal strip is $y$ (see the figure), so

$$\tilde{y} = y.$$

The $x$-coordinate is the $x$-coordinate of the point halfway across the triangle. This makes it the average of $y/2$ (the strip's left-hand $x$-value) and 1 (the strip's right-hand $x$-value):

$$\tilde{x} = \frac{(y/2) + 1}{2} = \frac{y}{4} + \frac{1}{2} = \frac{y + 2}{4}.$$

We also have

$$
\text{length:} \quad 1 - \frac{y}{2} = \frac{2-y}{2}
$$

$$
\text{width:} \quad dy
$$

$$
\text{area:} \quad dA = \frac{2-y}{2}\, dy
$$

$$
\text{mass:} \quad dm = \delta\, dA = 3 \cdot \frac{2-y}{2}\, dy
$$

$$
\text{distance of c.m. to } y\text{-axis:} \quad \widetilde{x} = \frac{y+2}{4}.
$$

The moment of the strip about the $y$-axis is

$$
\widetilde{x}\, dm = \frac{y+2}{4} \cdot 3 \cdot \frac{2-y}{2}\, dy = \frac{3}{8}(4 - y^2)\, dy.
$$

The moment of the plate about the $y$-axis is

$$
M_y = \int \widetilde{x}\, dm = \int_0^2 \frac{3}{8}(4 - y^2)\, dy = \frac{3}{8}\left[ 4y - \frac{y^3}{3} \right]_0^2 = \frac{3}{8}\left( \frac{16}{3} \right) = 2 \text{ g} \cdot \text{cm}.
$$

**(b)** The plate's mass:

$$
M = \int dm = \int_0^2 \frac{3}{2}(2 - y)\, dy = \frac{3}{2}\left[ 2y - \frac{y^2}{2} \right]_0^2 = \frac{3}{2}(4 - 2) = 3 \text{ g}.
$$

**(c)** The $x$-coordinate of the plate's center of mass:

$$
\bar{x} = \frac{M_y}{M} = \frac{2 \text{ g} \cdot \text{cm}}{3 \text{ g}} = \frac{2}{3} \text{ cm}.
$$

By a similar computation, we could find $M_x$ and $\bar{y}$. ◼

If the distribution of mass in a thin, flat plate has an axis of symmetry, the center of mass will lie on this axis. If there are two axes of symmetry, the center of mass will lie at their intersection. These facts often help to simplify our work.

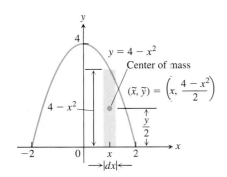

**FIGURE 6.51** Modeling the plate in Example 2 with vertical strips.

**EXAMPLE 2**    Find the center of mass of a thin plate covering the region bounded above by the parabola $y = 4 - x^2$ and below by the $x$-axis (Figure 6.51). Assume the density of the plate at the point $(x, y)$ is $\delta = 2x^2$, which is twice the square of the distance from the point to the $y$-axis.

**Solution**    The mass distribution is symmetric about the $y$-axis, so $\bar{x} = 0$. We model the distribution of mass with vertical strips, since the density is given as a function of the variable $x$. The typical vertical strip (see Figure 6.51) has the following relevant data.

$$
\text{center of mass (c.m.):} \quad (\widetilde{x}, \widetilde{y}) = \left( x, \frac{4 - x^2}{2} \right)
$$

$$
\text{length:} \quad 4 - x^2
$$

$$
\text{width:} \quad dx
$$

$$
\text{area:} \quad dA = (4 - x^2)\, dx
$$

$$
\text{mass:} \quad dm = \delta\, dA = \delta(4 - x^2)\, dx
$$

$$
\text{distance from c.m. to } x\text{-axis:} \quad \widetilde{y} = \frac{4 - x^2}{2}
$$

The moment of the strip about the $x$-axis is

$$
\widetilde{y}\, dm = \frac{4 - x^2}{2} \cdot \delta(4 - x^2)\, dx = \frac{\delta}{2}(4 - x^2)^2\, dx.
$$

The moment of the plate about the $x$-axis is

$$M_x = \int \tilde{y} \, dm = \int_{-2}^{2} \frac{\delta}{2}(4 - x^2)^2 \, dx = \int_{-2}^{2} x^2(4 - x^2)^2 \, dx$$

$$= \int_{-2}^{2} (16x^2 - 8x^4 + x^6) \, dx = \frac{2048}{105}$$

$$M = \int dm = \int_{-2}^{2} \delta(4 - x^2) \, dx = \int_{-2}^{2} 2x^2(4 - x^2) \, dx$$

$$= \int_{-2}^{2} (8x^2 - 2x^4) \, dx = \frac{256}{15}.$$

Therefore,

$$\bar{y} = \frac{M_x}{M} = \frac{2048}{105} \cdot \frac{15}{256} = \frac{8}{7}.$$

The plate's center of mass is

$$(\bar{x}, \bar{y}) = \left(0, \frac{8}{7}\right). \qquad\blacksquare$$

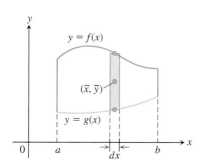

**FIGURE 6.52** Modeling the plate bounded by two curves with vertical strips. The strip c.m. is halfway, so $\tilde{y} = \frac{1}{2}[f(x) + g(x)]$.

## Plates Bounded by Two Curves

Suppose a plate covers a region that lies between two curves $y = g(x)$ and $y = f(x)$, where $f(x) \geq g(x)$ and $a \leq x \leq b$. The typical vertical strip (see Figure 6.52) has

$$\begin{aligned}
\text{center of mass (c.m.):} \quad & (\tilde{x}, \tilde{y}) = \left(x, \tfrac{1}{2}[f(x) + g(x)]\right) \\
\text{length:} \quad & f(x) - g(x) \\
\text{width:} \quad & dx \\
\text{area:} \quad & dA = [f(x) - g(x)] \, dx \\
\text{mass:} \quad & dm = \delta \, dA = \delta[f(x) - g(x)] \, dx.
\end{aligned}$$

The moment of the plate about the $y$-axis is

$$M_y = \int x \, dm = \int_{a}^{b} x\delta[f(x) - g(x)] \, dx,$$

and the moment about the $x$-axis is

$$M_x = \int y \, dm = \int_{a}^{b} \frac{1}{2}[f(x) + g(x)] \cdot \delta[f(x) - g(x)] \, dx$$

$$= \int_{a}^{b} \frac{\delta}{2}[f^2(x) - g^2(x)] \, dx.$$

These moments give the formulas

$$\bar{x} = \frac{1}{M} \int_{a}^{b} \delta x [f(x) - g(x)] \, dx \qquad (6)$$

$$\bar{y} = \frac{1}{M} \int_{a}^{b} \frac{\delta}{2} [f^2(x) - g^2(x)] \, dx \qquad (7)$$

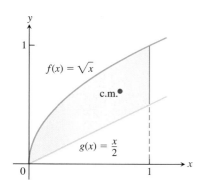

**FIGURE 6.53** The region in Example 3.

**EXAMPLE 3**    Find the center of mass for the thin plate bounded by the curves $g(x) = x/2$ and $f(x) = \sqrt{x}, 0 \le x \le 1$ (Figure 6.53), using Equations (6) and (7) with the density function $\delta(x) = x^2$.

**Solution**    We first compute the mass of the plate, where $dm = \delta[f(x) - g(x)]\, dx$:

$$M = \int_0^1 x^2\left(\sqrt{x} - \frac{x}{2}\right) dx = \int_0^1 \left(x^{5/2} - \frac{x^3}{2}\right) dx = \left[\frac{2}{7}x^{7/2} - \frac{1}{8}x^4\right]_0^1 = \frac{9}{56}.$$

Then from Equations (6) and (7) we get

$$\bar{x} = \frac{56}{9}\int_0^1 x^2 \cdot x\left(\sqrt{x} - \frac{x}{2}\right) dx$$

$$= \frac{56}{9}\int_0^1 \left(x^{7/2} - \frac{x^4}{2}\right) dx$$

$$= \frac{56}{9}\left[\frac{2}{9}x^{9/2} - \frac{1}{10}x^5\right]_0^1 = \frac{308}{405},$$

and

$$\bar{y} = \frac{56}{9}\int_0^1 \frac{x^2}{2}\left(x - \frac{x^2}{4}\right) dx$$

$$= \frac{28}{9}\int_0^1 \left(x^3 - \frac{x^4}{4}\right) dx$$

$$= \frac{28}{9}\left[\frac{1}{4}x^4 - \frac{1}{20}x^5\right]_0^1 = \frac{252}{405}.$$

The center of mass is shown in Figure 6.53.    ∎

## Centroids

The center of mass in Example 3 is not located at the geometric center of the region. This is due to the region's nonuniform density. When the density function is constant, it cancels out of the numerator and denominator of the formulas for $\bar{x}$ and $\bar{y}$. Thus, when the density is constant, the location of the center of mass is a feature of the geometry of the object and not of the material from which it is made. In such cases, engineers may call the center of mass the **centroid** of the shape, as in "Find the centroid of a triangle or a solid cone." To do so, just set $\delta$ equal to 1 and proceed to find $\bar{x}$ and $\bar{y}$ as before, by dividing moments by masses.

**EXAMPLE 4**    Find the center of mass (centroid) of a thin wire of constant density $\delta$ shaped like a semicircle of radius $a$.

**Solution**    We model the wire with the semicircle $y = \sqrt{a^2 - x^2}$ (Figure 6.54). The distribution of mass is symmetric about the $y$-axis, so $\bar{x} = 0$. To find $\bar{y}$, we imagine the wire divided into short subarc segments. If $(\tilde{x}, \tilde{y})$ is the center of mass of a subarc and $\theta$ is the angle between the $x$-axis and the radial line joining the origin to $(\tilde{x}, \tilde{y})$, then $\tilde{y} = a \sin \theta$ is a function of the angle $\theta$ measured in radians (see Figure 6.54a). The length $ds$ of the subarc containing $(\tilde{x}, \tilde{y})$ subtends an angle of $d\theta$ radians, so $ds = a\, d\theta$. Thus a typical subarc segment has these relevant data for calculating $\bar{y}$:

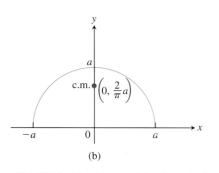

**FIGURE 6.54** The semicircular wire in Example 4. (a) The dimensions and variables used in finding the center of mass. (b) The center of mass does not lie on the wire.

| length: | $ds = a\, d\theta$ | |
|---|---|---|
| mass: | $dm = \delta\, ds = \delta a\, d\theta$ | Mass per unit length times length |
| distance of c.m. to $x$-axis: | $\tilde{y} = a \sin\theta.$ | |

Hence,

$$\bar{y} = \frac{\int \tilde{y}\, dm}{\int dm} = \frac{\int_0^\pi a\sin\theta \cdot \delta a\, d\theta}{\int_0^\pi \delta a\, d\theta} = \frac{\delta a^2\left[-\cos\theta\right]_0^\pi}{\delta a\pi} = \frac{2}{\pi}a.$$

The center of mass lies on the axis of symmetry at the point $(0, 2a/\pi)$, about two-thirds of the way up from the origin (Figure 6.54b). Notice how $\delta$ cancels in the equation for $\bar{y}$, so we could have set $\delta = 1$ everywhere and obtained the same value for $\bar{y}$.  ■

In Example 4 we found the center of mass of a thin wire lying along the graph of a differentiable function in the $xy$-plane. In Chapter 16 we will learn how to find the center of mass of a wire lying along a more general smooth curve in the plane or in space.

## Fluid Forces and Centroids

Surface level of fluid

$\bar{h}$ = centroid depth

Plate centroid

**FIGURE 6.55**  The force against one side of the plate is $w \cdot \bar{h} \cdot$ plate area.

If we know the location of the centroid of a submerged flat vertical plate (Figure 6.55), we can take a shortcut to find the force against one side of the plate. From Equation (7) in Section 6.5, and the definition of the moment about the $x$-axis, we have

$$F = \int_a^b w \times (\text{strip depth}) \times L(y)\, dy$$

$$= w\int_a^b (\text{strip depth}) \times L(y)\, dy$$

$$= w \times (\text{moment about surface level line of region occupied by plate})$$

$$= w \times (\text{depth of plate's centroid}) \times (\text{area of plate}).$$

> **Fluid Forces and Centroids**
>
> The force of a fluid of weight-density $w$ against one side of a submerged flat vertical plate is the product of $w$, the distance $\bar{h}$ from the plate's centroid to the fluid surface, and the plate's area:
>
> $$F = w\bar{h}A. \qquad (8)$$

**EXAMPLE 5**    A flat isosceles triangular plate with base 2 m and height 1 m is submerged vertically, base up with its vertex at the origin, so that the base is 0.6 m below the surface of a swimming pool. (This is Example 6, Section 6.5.) Use Equation (8) to find the force exerted by the water against one side of the plate.

**Solution**    The centroid of the triangle (Figure 6.43) lies on the $y$-axis, one-third of the way from the base to the vertex, so $\bar{h} = 2.8/3$ (where $y = 2/3$), since the pool's surface is $y = 1.6$. The triangle's area is

$$A = \frac{1}{2}(\text{base})(\text{height}) = \frac{1}{2}(2)(1) = 1.$$

Hence,

$$F = w\bar{h}A = (9800)\left(\frac{2.8}{3}\right)(1) \approx 9147\text{ N}.$$   ■

## The Theorems of Pappus

In the fourth century, an Alexandrian Greek named Pappus discovered two formulas that relate centroids to surfaces and solids of revolution. The formulas provide shortcuts to a number of otherwise lengthy calculations.

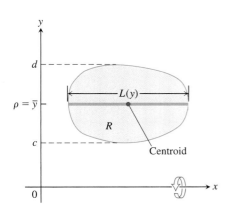

**FIGURE 6.56** The region $R$ is to be revolved (once) about the $x$-axis to generate a solid. A 1700-year-old theorem says that the solid's volume can be calculated by multiplying the region's area by the distance traveled by its centroid during the revolution.

---

> **THEOREM 1 Pappus's Theorem for Volumes** If a plane region is revolved once about a line in the plane that does not cut through the region's interior, then the volume of the solid it generates is equal to the region's area times the distance traveled by the region's centroid during the revolution. If $\rho$ is the distance from the axis of revolution to the centroid, then
>
> $$V = 2\pi\rho A. \tag{9}$$

**Proof** We draw the axis of revolution as the $x$-axis with the region $R$ in the first quadrant (Figure 6.56). We let $L(y)$ denote the length of the cross-section of $R$ perpendicular to the $y$-axis at $y$. We assume $L(y)$ to be continuous.

By the method of cylindrical shells, the volume of the solid generated by revolving the region about the $x$-axis is

$$V = \int_c^d 2\pi(\text{shell radius})(\text{shell height})\,dy = 2\pi\int_c^d y\,L(y)\,dy. \tag{10}$$

The $y$-coordinate of $R$'s centroid is

$$\bar{y} = \frac{\int_c^d \tilde{y}\,dA}{A} = \frac{\int_c^d y\,L(y)\,dy}{A}, \qquad \tilde{y} = y,\, dA = L(y)\,dy$$

so that

$$\int_c^d y\,L(y)\,dy = A\bar{y}.$$

Substituting $A\bar{y}$ for the last integral in Equation (10) gives $V = 2\pi\bar{y}A$. With $\rho$ equal to $\bar{y}$, we have $V = 2\pi\rho A$. ∎

**EXAMPLE 6** Find the volume of the torus (doughnut) generated by revolving a circular disk of radius $a$ about an axis in its plane at a distance $b \geq a$ from its center (Figure 6.57).

**Solution** We apply Pappus's Theorem for volumes. The centroid of a disk is located at its center, the area is $A = \pi a^2$, and $\rho = b$ is the distance from the centroid to the axis of revolution (see Figure 6.57). Substituting these values into Equation (9), we find the volume of the torus to be

$$V = 2\pi(b)(\pi a^2) = 2\pi^2 b a^2.$$

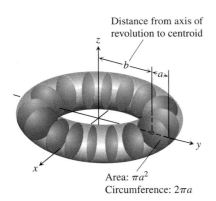

**FIGURE 6.57** With Pappus's first theorem, we can find the volume of a torus without having to integrate (Example 6).

The next example shows how we can use Equation (9) in Pappus's Theorem to find one of the coordinates of the centroid of a plane region of known area $A$ when we also know the volume $V$ of the solid generated by revolving the region about the other coordinate axis. That is, if $\bar{y}$ is the coordinate we want to find, we revolve the region around the $x$-axis so that $\bar{y} = \rho$ is the distance from the centroid to the axis of revolution. The idea is that the rotation generates a solid of revolution whose volume $V$ is an already known quantity. Then we can solve Equation (9) for $\rho$, which is the value of the centroid's coordinate $\bar{y}$.

**EXAMPLE 7** Locate the centroid of a semicircular region of radius $a$.

**Solution** We consider the region between the semicircle $y = \sqrt{a^2 - x^2}$ (Figure 6.58) and the $x$-axis and imagine revolving the region about the $x$-axis to generate a solid sphere. By symmetry, the $x$-coordinate of the centroid is $\bar{x} = 0$. With $\bar{y} = \rho$ in Equation (9), we have

$$\bar{y} = \frac{V}{2\pi A} = \frac{(4/3)\pi a^3}{2\pi(1/2)\pi a^2} = \frac{4}{3\pi}a.$$

**FIGURE 6.58** With Pappus's first theorem, we can locate the centroid of a semicircular region without having to integrate (Example 7).

> **THEOREM 2—Pappus's Theorem for Surface Areas** If an arc of a smooth plane curve is revolved once about a line in the plane that does not cut through the arc's interior, then the area of the surface generated by the arc equals the length $L$ of the arc times the distance traveled by the arc's centroid during the revolution. If $\rho$ is the distance from the axis of revolution to the centroid, then
>
> $$S = 2\pi\rho L. \tag{11}$$

The proof we give assumes that we can model the axis of revolution as the $x$-axis and the arc as the graph of a continuously differentiable function of $x$.

**Proof**     We draw the axis of revolution as the $x$-axis with the arc extending from $x = a$ to $x = b$ in the first quadrant (Figure 6.59). The area of the surface generated by the arc is

$$S = \int_{x=a}^{x=b} 2\pi y \, ds = 2\pi \int_{x=a}^{x=b} y \, ds. \tag{12}$$

The $y$-coordinate of the arc's centroid is

$$\bar{y} = \frac{\int_{x=a}^{x=b} \tilde{y} \, ds}{\int_{x=a}^{x=b} ds} = \frac{\int_{x=a}^{x=b} y \, ds}{L}. \qquad \begin{array}{l}\text{$L = \int ds$ is the arc's}\\ \text{length and $\tilde{y} = y$.}\end{array}$$

Hence

$$\int_{x=a}^{x=b} y \, ds = \bar{y} L.$$

Substituting $\bar{y}L$ for the last integral in Equation (12) gives $S = 2\pi\bar{y}L$. With $\rho$ equal to $\bar{y}$, we have $S = 2\pi\rho L$. ∎

**EXAMPLE 8**     Use Pappus's area theorem to find the surface area of the torus in Example 6.

**Solution**     From Figure 6.57, the surface of the torus is generated by revolving a circle of radius $a$ about the $z$-axis, and $b \geq a$ is the distance from the centroid to the axis of revolution. The arc length of the smooth curve generating this surface of revolution is the circumference of the circle, so $L = 2\pi a$. Substituting these values into Equation (11), we find the surface area of the torus to be

$$S = 2\pi(b)(2\pi a) = 4\pi^2 ba. \qquad \blacksquare$$

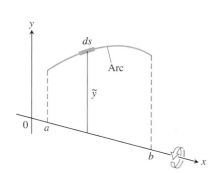

**FIGURE 6.59** Figure for proving Pappus's Theorem for surface area. The arc length differential $ds$ is given by Equation (6) in Section 6.3.

# Exercises **6.6**

**Thin Plates with Constant Density**

In Exercises 1–12, find the center of mass of a thin plate of constant density $\delta$ covering the given region.

**1.** The region bounded by the parabola $y = x^2$ and the line $y = 4$

**2.** The region bounded by the parabola $y = 25 - x^2$ and the $x$-axis

**3.** The region bounded by the parabola $y = x - x^2$ and the line $y = -x$

**4.** The region enclosed by the parabolas $y = x^2 - 3$ and $y = -2x^2$

**5.** The region bounded by the $y$-axis and the curve $x = y - y^3$, $0 \leq y \leq 1$

**6.** The region bounded by the parabola $x = y^2 - y$ and the line $y = x$

**7.** The region bounded by the $x$-axis and the curve $y = \cos x$, $-\pi/2 \leq x \leq \pi/2$

**8.** The region between the curve $y = \sec^2 x$, $-\pi/4 \leq x \leq \pi/4$ and the $x$-axis

**9.** The region bounded by the parabolas $y = 2x^2 - 4x$ and $y = 2x - x^2$.

**10. a.** The region cut from the first quadrant by the circle $x^2 + y^2 = 9$

   **b.** The region bounded by the $x$-axis and the semicircle $y = \sqrt{9 - x^2}$

   Compare your answer in part (b) with the answer in part (a).

11. The "triangular" region in the first quadrant between the circle $x^2 + y^2 = 9$ and the lines $x = 3$ and $y = 3$. (*Hint:* Use geometry to find the area.)

12. The region bounded above by the curve $y = 1/x^3$, below by the curve $y = -1/x^3$, and on the left and right by the lines $x = 1$ and $x = a > 1$. Also, find $\lim_{a \to \infty} \bar{x}$.

### Thin Plates with Varying Density

13. Find the center of mass of a thin plate covering the region between the $x$-axis and the curve $y = 2/x^2$, $1 \le x \le 2$, if the plate's density at the point $(x, y)$ is $\delta(x) = x^2$.

14. Find the center of mass of a thin plate covering the region bounded below by the parabola $y = x^2$ and above by the line $y = x$ if the plate's density at the point $(x, y)$ is $\delta(x) = 12x$.

15. The region bounded by the curves $y = \pm 4/\sqrt{x}$ and the lines $x = 1$ and $x = 4$ is revolved about the $y$-axis to generate a solid.

    a. Find the volume of the solid.

    b. Find the center of mass of a thin plate covering the region if the plate's density at the point $(x, y)$ is $\delta(x) = 1/x$.

    c. Sketch the plate and show the center of mass in your sketch.

16. The region between the curve $y = 2/x$ and the $x$-axis from $x = 1$ to $x = 4$ is revolved about the $x$-axis to generate a solid.

    a. Find the volume of the solid.

    b. Find the center of mass of a thin plate covering the region if the plate's density at the point $(x, y)$ is $\delta(x) = \sqrt{x}$.

    c. Sketch the plate and show the center of mass in your sketch.

### Centroids of Triangles

17. **The centroid of a triangle lies at the intersection of the triangle's medians**   You may recall that the point inside a triangle that lies one-third of the way from each side toward the opposite vertex is the point where the triangle's three medians intersect. Show that the centroid lies at the intersection of the medians by showing that it too lies one-third of the way from each side toward the opposite vertex. To do so, take the following steps.

    i) Stand one side of the triangle on the $x$-axis as in part (b) of the accompanying figure. Express $dm$ in terms of $L$ and $dy$.

    ii) Use similar triangles to show that $L = (b/h)(h - y)$. Substitute this expression for $L$ in your formula for $dm$.

    iii) Show that $\bar{y} = h/3$.

    iv) Extend the argument to the other sides.

(a)

(b)

Use the result in Exercise 17 to find the centroids of the triangles whose vertices appear in Exercises 18–22. Assume $a, b > 0$.

18. $(-1, 0)$, $(1, 0)$, $(0, 3)$

19. $(0, 0)$, $(1, 0)$, $(0, 1)$

20. $(0, 0)$, $(a, 0)$, $(0, a)$

21. $(0, 0)$, $(a, 0)$, $(0, b)$

22. $(0, 0)$, $(a, 0)$, $(a/2, b)$

### Thin Wires

23. **Constant density**   Find the moment about the $x$-axis of a wire of constant density that lies along the curve $y = \sqrt{x}$ from $x = 0$ to $x = 2$.

24. **Constant density**   Find the moment about the $x$-axis of a wire of constant density that lies along the curve $y = x^3$ from $x = 0$ to $x = 1$.

25. **Variable density**   Suppose that the density of the wire in Example 4 is $\delta = k \sin \theta$ ($k$ constant). Find the center of mass.

26. **Variable density**   Suppose that the density of the wire in Example 4 is $\delta = 1 + k|\cos \theta|$ ($k$ constant). Find the center of mass.

### Plates Bounded by Two Curves

In Exercises 27–30, find the centroid of the thin plate bounded by the graphs of the given functions. Use Equations (6) and (7) with $\delta = 1$ and $M$ = area of the region covered by the plate.

27. $g(x) = x^2$   and   $f(x) = x + 6$

28. $g(x) = x^2(x + 1)$,   $f(x) = 2$,   and   $x = 0$

29. $g(x) = x^2(x - 1)$   and   $f(x) = x^2$

30. $g(x) = 0$,   $f(x) = 2 + \sin x$,   $x = 0$,   and   $x = 2\pi$

(*Hint:* $\displaystyle\int x \sin x \, dx = \sin x - x \cos x + C$.)

### Theory and Examples

Verify the statements and formulas in Exercises 31 and 32.

31. The coordinates of the centroid of a differentiable plane curve are
$$\bar{x} = \frac{\int x \, ds}{\text{length}}, \qquad \bar{y} = \frac{\int y \, ds}{\text{length}}.$$

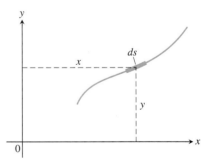

32. Whatever the value of $p > 0$ in the equation $y = x^2/(4p)$, the $y$-coordinate of the centroid of the parabolic segment shown here is $\bar{y} = (3/5)a$.

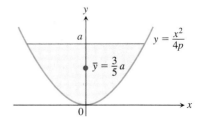

**The Theorems of Pappus**

**33.** The square region with vertices $(0, 2)$, $(2, 0)$, $(4, 2)$, and $(2, 4)$ is revolved about the $x$-axis to generate a solid. Find the volume and surface area of the solid.

**34.** Use a theorem of Pappus to find the volume generated by revolving about the line $x = 5$ the triangular region bounded by the coordinate axes and the line $2x + y = 6$ (see Exercise 17).

**35.** Find the volume of the torus generated by revolving the circle $(x - 2)^2 + y^2 = 1$ about the $y$-axis.

**36.** Use the theorems of Pappus to find the lateral surface area and the volume of a right-circular cone.

**37.** Use Pappus's Theorem for surface area and the fact that the surface area of a sphere of radius $a$ is $4\pi a^2$ to find the centroid of the semicircle $y = \sqrt{a^2 - x^2}$.

**38.** As found in Exercise 37, the centroid of the semicircle $y = \sqrt{a^2 - x^2}$ lies at the point $(0, 2a/\pi)$. Find the area of the surface swept out by revolving the semicircle about the line $y = a$.

**39.** The area of the region $R$ enclosed by the semiellipse $y = (b/a)\sqrt{a^2 - x^2}$ and the $x$-axis is $(1/2)\pi ab$, and the volume of the ellipsoid generated by revolving $R$ about the $x$-axis is $(4/3)\pi ab^2$. Find the centroid of $R$. Notice that the location is independent of $a$.

**40.** As found in Example 7, the centroid of the region enclosed by the $x$-axis and the semicircle $y = \sqrt{a^2 - x^2}$ lies at the point $(0, 4a/3\pi)$. Find the volume of the solid generated by revolving this region about the line $y = -a$.

**41.** The region of Exercise 40 is revolved about the line $y = x - a$ to generate a solid. Find the volume of the solid.

**42.** As found in Exercise 37, the centroid of the semicircle $y = \sqrt{a^2 - x^2}$ lies at the point $(0, 2a/\pi)$. Find the area of the surface generated by revolving the semicircle about the line $y = x - a$.

In Exercises 43 and 44, use a theorem of Pappus to find the centroid of the given triangle. Use the fact that the volume of a cone of radius $r$ and height $h$ is $V = \frac{1}{3}\pi r^2 h$.

**43.**      **44.**

---

**1.** How do you define and calculate the volumes of solids by the method of slicing? Give an example.

**2.** How are the disk and washer methods for calculating volumes derived from the method of slicing? Give examples of volume calculations by these methods.

**3.** Describe the method of cylindrical shells. Give an example.

**4.** How do you find the length of the graph of a smooth function over a closed interval? Give an example. What about functions that do not have continuous first derivatives?

**5.** How do you define and calculate the area of the surface swept out by revolving the graph of a smooth function $y = f(x)$, $a \le x \le b$, about the $x$-axis? Give an example.

**6.** How do you define and calculate the work done by a variable force directed along a portion of the $x$-axis? How do you calculate the work it takes to pump a liquid from a tank? Give examples.

**7.** How do you calculate the force exerted by a liquid against a portion of a flat vertical wall? Give an example.

**8.** What is a center of mass? a centroid?

**9.** How do you locate the center of mass of a thin flat plate of material? Give an example.

**10.** How do you locate the center of mass of a thin plate bounded by two curves $y = f(x)$ and $y = g(x)$ over $a \le x \le b$?

---

**Volumes**

Find the volumes of the solids in Exercises 1–16.

**1.** The solid lies between planes perpendicular to the $x$-axis at $x = 0$ and $x = 1$. The cross-sections perpendicular to the $x$-axis between these planes are circular disks whose diameters run from the parabola $y = x^2$ to the parabola $y = \sqrt{x}$.

**2.** The base of the solid is the region in the first quadrant between the line $y = x$ and the parabola $y = 2\sqrt{x}$. The cross-sections of the solid perpendicular to the $x$-axis are equilateral triangles whose bases stretch from the line to the curve.

**3.** The solid lies between planes perpendicular to the $x$-axis at $x = \pi/4$ and $x = 5\pi/4$. The cross-sections between these planes are circular disks whose diameters run from the curve $y = 2\cos x$ to the curve $y = 2\sin x$.

**4.** The solid lies between planes perpendicular to the $x$-axis at $x = 0$ and $x = 6$. The cross-sections between these planes are squares whose bases run from the $x$-axis up to the curve $x^{1/2} + y^{1/2} = \sqrt{6}$.

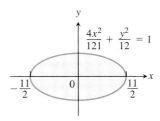

$$x^{1/2} + y^{1/2} = \sqrt{6}$$

**5.** The solid lies between planes perpendicular to the $x$-axis at $x = 0$ and $x = 4$. The cross-sections of the solid perpendicular to the $x$-axis between these planes are circular disks whose diameters run from the curve $x^2 = 4y$ to the curve $y^2 = 4x$.

**6.** The base of the solid is the region bounded by the parabola $y^2 = 4x$ and the line $x = 1$ in the $xy$-plane. Each cross-section perpendicular to the $x$-axis is an equilateral triangle with one edge in the plane. (The triangles all lie on the same side of the plane.)

**7.** Find the volume of the solid generated by revolving the region bounded by the $x$-axis, the curve $y = 3x^4$, and the lines $x = 1$ and $x = -1$ about **(a)** the $x$-axis; **(b)** the $y$-axis; **(c)** the line $x = 1$; **(d)** the line $y = 3$.

**8.** Find the volume of the solid generated by revolving the "triangular" region bounded by the curve $y = 4/x^3$ and the lines $x = 1$ and $y = 1/2$ about **(a)** the $x$-axis; **(b)** the $y$-axis; **(c)** the line $x = 2$; **(d)** the line $y = 4$.

**9.** Find the volume of the solid generated by revolving the region bounded on the left by the parabola $x = y^2 + 1$ and on the right by the line $x = 5$ about **(a)** the $x$-axis; **(b)** the $y$-axis; **(c)** the line $x = 5$.

**10.** Find the volume of the solid generated by revolving the region bounded by the parabola $y^2 = 4x$ and the line $y = x$ about **(a)** the $x$-axis; **(b)** the $y$-axis; **(c)** the line $x = 4$; **(d)** the line $y = 4$.

**11.** Find the volume of the solid generated by revolving the "triangular" region bounded by the $x$-axis, the line $x = \pi/3$, and the curve $y = \tan x$ in the first quadrant about the $x$-axis.

**12.** Find the volume of the solid generated by revolving the region bounded by the curve $y = \sin x$ and the lines $x = 0$, $x = \pi$, and $y = 2$ about the line $y = 2$.

**13.** Find the volume of the solid generated by revolving the region between the $x$-axis and the curve $y = x^2 - 2x$ about **(a)** the $x$-axis; **(b)** the line $y = -1$; **(c)** the line $x = 2$; **(d)** the line $y = 2$.

**14.** Find the volume of the solid generated by revolving about the $x$-axis the region bounded by $y = 2\tan x$, $y = 0$, $x = -\pi/4$, and $x = \pi/4$. (The region lies in the first and third quadrants and resembles a skewed bowtie.)

**15. Volume of a solid sphere hole** A round hole of radius $\sqrt{3}$ m is bored through the center of a solid sphere of a radius 2 m. Find the volume of material removed from the sphere.

**16. Volume of a balloon** The profile of a balloon resembles the ellipse shown here. Find the balloon's volume to the nearest cubic centimeter.

$$\frac{4x^2}{121} + \frac{y^2}{12} = 1$$

$-\dfrac{11}{2}$ $\quad 0 \quad$ $\dfrac{11}{2}$

**Lengths of Curves**

Find the lengths of the curves in Exercises 17–20.

**17.** $y = x^{1/2} - (1/3)x^{3/2}, \quad 1 \le x \le 4$

**18.** $x = y^{2/3}, \quad 1 \le y \le 8$

**19.** $y = (5/12)x^{6/5} - (5/8)x^{4/5}, \quad 1 \le x \le 32$

**20.** $x = (y^3/12) + (1/y), \quad 1 \le y \le 2$

**Areas of Surfaces of Revolution**

In Exercises 21–24, find the areas of the surfaces generated by revolving the curves about the given axes.

**21.** $y = \sqrt{2x + 1}, \quad 0 \le x \le 3; \quad x$-axis

**22.** $y = x^3/3, \quad 0 \le x \le 1; \quad x$-axis

**23.** $x = \sqrt{4y - y^2}, \quad 1 \le y \le 2; \quad y$-axis

**24.** $x = \sqrt{y}, \quad 2 \le y \le 6; \quad y$-axis

**Work**

**25. Lifting equipment** A rock climber is about to haul up 100 N of equipment that has been hanging beneath her on 40 m of rope that weighs 0.8 newton per meter. How much work will it take? (*Hint:* Solve for the rope and equipment separately, then add.)

**26. Leaky tank truck** You drove a 4000-L tank truck of water from the base of Mt. Washington to the summit and discovered on arrival that the tank was only half full. You started with a full tank, climbed at a steady rate, and accomplished the 1500-m elevation change in 50 min. Assuming that the water leaked out at a steady rate, how much work was spent in carrying water to the top? Do not count the work done in getting yourself and the truck there. Water weighs 9.8 N/L.

**27. Earth's attraction** The force of attraction on an object below Earth's surface is directly proportional to its distance from Earth's center. Find the work done in moving a weight of $w$ N located $a$ km below Earth's surface up to the surface itself. Assume Earth's radius is a constant $r$ km.

**28. Garage door spring** A force of 200 N will stretch a garage door spring 0.8 m beyond its unstressed length. How far will a 300-N force stretch the spring? How much work does it take to stretch the spring this far from its unstressed length?

**29. Pumping a reservoir** A reservoir shaped like a right-circular cone, point down, 20 m across the top and 8 m deep, is full of water. How much work does it take to pump the water to a level 6 m above the top?

**30. Pumping a reservoir** (*Continuation of Exercise 29.*) The reservoir is filled to a depth of 5 m, and the water is to be pumped to the same level as the top. How much work does it take?

**31. Pumping a conical tank** A right-circular conical tank, point down, with top radius 5 m and height 10 m is filled with a liquid whose weight-density is 9000 N/m³. How much work does it take to pump the liquid to a point 2 m above the tank? If the pump is driven by a motor rated at 41,250 J/s, how long will it take to empty the tank?

**32. Pumping a cylindrical tank** A storage tank is a right-circular cylinder 6 m long and 2.5 m in diameter with its axis horizontal. If the tank is half full of olive oil weighing 8950 N/m³, find the work done in emptying it through a pipe that runs from the bottom of the tank to an outlet that is 2 m above the top of the tank.

### Centers of Mass and Centroids

**33.** Find the centroid of a thin, flat plate covering the region enclosed by the parabolas $y = 2x^2$ and $y = 3 - x^2$.

**34.** Find the centroid of a thin, flat plate covering the region enclosed by the $x$-axis, the lines $x = 2$ and $x = -2$, and the parabola $y = x^2$.

**35.** Find the centroid of a thin, flat plate covering the "triangular" region in the first quadrant bounded by the $y$-axis, the parabola $y = x^2/4$, and the line $y = 4$.

**36.** Find the centroid of a thin, flat plate covering the region enclosed by the parabola $y^2 = x$ and the line $x = 2y$.

**37.** Find the center of mass of a thin, flat plate covering the region enclosed by the parabola $y^2 = x$ and the line $x = 2y$ if the density function is $\delta(y) = 1 + y$. (Use horizontal strips.)

**38. a.** Find the center of mass of a thin plate of constant density covering the region between the curve $y = 3/x^{3/2}$ and the $x$-axis from $x = 1$ to $x = 9$.

**b.** Find the plate's center of mass if, instead of being constant, the density is $\delta(x) = x$. (Use vertical strips.)

### Fluid Force

**39. Trough of water** The vertical triangular plate shown here is the end plate of a trough full of water ($w = 9800$). What is the fluid force against the plate?

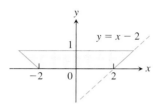

UNITS IN METERS

**40. Trough of maple syrup** The vertical trapezoidal plate shown here is the end plate of a trough full of maple syrup weighing 11,000 N/m³. What is the force exerted by the syrup against the end plate of the trough when the syrup is 0.5 m deep?

**41. Force on a parabolic gate** A flat vertical gate in the face of a dam is shaped like the parabolic region between the curve $y = 4x^2$ and the line $y = 4$, with measurements in meters. The top of the gate lies 5 m below the surface of the water. Find the force exerted by the water against the gate ($w = 9800$).

**T** **42.** You plan to store mercury ($w = 133,350$ N/m³) in a vertical rectangular tank with a 0.3 m square base side whose interior side wall can withstand a total fluid force of 150,000 N. About how many cubic meters of mercury can you store in the tank at any one time?

---

## Chapter **6** Additional and Advanced Exercises

### Volume and Length

**1.** A solid is generated by revolving about the $x$-axis the region bounded by the graph of the positive continuous function $y = f(x)$, the $x$-axis, the fixed line $x = a$, and the variable line $x = b$, $b > a$. Its volume, for all $b$, is $b^2 - ab$. Find $f(x)$.

**2.** A solid is generated by revolving about the $x$-axis the region bounded by the graph of the positive continuous function $y = f(x)$, the $x$-axis, and the lines $x = 0$ and $x = a$. Its volume, for all $a > 0$, is $a^2 + a$. Find $f(x)$.

**3.** Suppose that the increasing function $f(x)$ is smooth for $x \geq 0$ and that $f(0) = a$. Let $s(x)$ denote the length of the graph of $f$ from $(0, a)$ to $(x, f(x))$, $x > 0$. Find $f(x)$ if $s(x) = Cx$ for some constant $C$. What are the allowable values for $C$?

**4. a.** Show that for $0 < \alpha \leq \pi/2$,

$$\int_0^\alpha \sqrt{1 + \cos^2\theta}\, d\theta > \sqrt{\alpha^2 + \sin^2\alpha}.$$

**b.** Generalize the result in part (a).

**5.** Find the volume of the solid formed by revolving the region bounded by the graphs of $y = x$ and $y = x^2$ about the line $y = x$.

**6.** Consider a right-circular cylinder of diameter 1. Form a wedge by making one slice parallel to the base of the cylinder completely through the cylinder, and another slice at an angle of 45° to the first slice and intersecting the first slice at the opposite edge of the cylinder (see accompanying diagram). Find the volume of the wedge.

45° wedge

$r = \frac{1}{2}$

## Surface Area

**7.** At points on the curve $y = 2\sqrt{x}$, line segments of length $h = y$ are drawn perpendicular to the $xy$-plane. (See accompanying figure.) Find the area of the surface formed by these perpendiculars from $(0, 0)$ to $\left(3, 2\sqrt{3}\right)$.

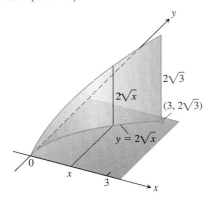

**8.** At points on a circle of radius $a$, line segments are drawn perpendicular to the plane of the circle, the perpendicular at each point $P$ being of length $ks$, where $s$ is the length of the arc of the circle measured counterclockwise from $(a, 0)$ to $P$ and $k$ is a positive constant, as shown here. Find the area of the surface formed by the perpendiculars along the arc beginning at $(a, 0)$ and extending once around the circle.

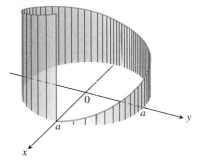

## Work

**9.** A particle of mass $m$ starts from rest at time $t = 0$ and is moved along the $x$-axis with constant acceleration $a$ from $x = 0$ to $x = h$ against a variable force of magnitude $F(t) = t^2$. Find the work done.

**10. Work and kinetic energy** Suppose a 50 g golf ball is placed on a vertical spring with force constant $k = 2\,\text{N/cm}$. The spring is compressed 15 cm and released. About how high does the ball go (measured from the spring's rest position)?

## Centers of Mass

**11.** Find the centroid of the region bounded below by the $x$-axis and above by the curve $y = 1 - x^n$, $n$ an even positive integer. What is the limiting position of the centroid as $n \to \infty$?

**12.** If you haul a telephone pole on a two-wheeled carriage behind a truck, you want the wheels to be 1 m or so behind the pole's center of mass to provide an adequate "tongue" weight. The 12 m wooden telephone poles used by Verizon have a 66 cm circumference at the top and a 104 cm circumference at the base. About how far from the top is the center of mass?

**13.** Suppose that a thin metal plate of area $A$ and constant density $\delta$ occupies a region $R$ in the $xy$-plane, and let $M_y$ be the plate's moment about the $y$-axis. Show that the plate's moment about the line $x = b$ is

   **a.** $M_y - b\delta A$ if the plate lies to the right of the line, and

   **b.** $b\delta A - M_y$ if the plate lies to the left of the line.

**14.** Find the center of mass of a thin plate covering the region bounded by the curve $y^2 = 4ax$ and the line $x = a$, $a =$ positive constant, if the density at $(x, y)$ is directly proportional to **(a)** $x$, **(b)** $|y|$.

**15. a.** Find the centroid of the region in the first quadrant bounded by two concentric circles and the coordinate axes, if the circles have radii $a$ and $b$, $0 < a < b$, and their centers are at the origin.

   **b.** Find the limits of the coordinates of the centroid as $a$ approaches $b$ and discuss the meaning of the result.

**16.** A triangular corner is cut from a square 40 cm on a side. The area of the triangle removed is 400 cm². If the centroid of the remaining region is 22 cm from one side of the original square, how far is it from the remaining sides?

## Fluid Force

**17.** A triangular plate $ABC$ is submerged in water with its plane vertical. The side $AB$, 4 m long, is 6 m below the surface of the water, while the vertex $C$ is 2 m below the surface. Find the force exerted by the water on one side of the plate.

**18.** A vertical rectangular plate is submerged in a fluid with its top edge parallel to the fluid's surface. Show that the force exerted by the fluid on one side of the plate equals the average value of the pressure up and down the plate times the area of the plate.

---

## Chapter **6** Technology Application Projects

### Mathematica/Maple Modules:

***Using Riemann Sums to Estimate Areas, Volumes, and Lengths of Curves***
Visualize and approximate areas and volumes in **Part I** and **Part II**: Volumes of Revolution; and **Part III**: Lengths of Curves.

***Modeling a Bungee Cord Jump***
Collect data (or use data previously collected) to build and refine a model for the force exerted by a jumper's bungee cord. Use the work-energy theorem to compute the distance fallen for a given jumper and a given length of bungee cord.

# 7 Transcendental Functions

**OVERVIEW** There are a wide variety of phenomena of interest to science, engineering, mathematics, and economics which are best modeled by classes of functions we have not yet investigated. Chief among these are the *exponential* functions. They are useful in studying the growth of a biological population, the spread of a disease or of information throughout a human community, drug dosages, radioactive elements and their role in dating fossils, the earth's atmospheric pressure, temperature changes, waves, electrical circuits, the vibrations in bridges, interest rates, and probabilities.

In this chapter, we use calculus to obtain rigorous and precise definitions and properties of the exponential functions. We first define the natural logarithm function $y = \ln x$ as a certain integral, and then the natural exponential function $y = e^x$ as its *inverse* function. These two functions are foundational, and they form the basis for the full array of logarithmic and exponential functions. We also introduce the inverse trigonometric functions, as well as the hyperbolic functions and their inverses, with their applications to integration and the description of hanging cables. Like the trigonometric functions, all of these functions belong to the category called the transcendental functions.

## 7.1 Inverse Functions and Their Derivatives

A function that undoes, or inverts, the effect of a function $f$ is called the *inverse* of $f$. Many common functions, though not all, are paired with an inverse. Important inverse functions often show up in applications. Inverse functions also play a key role in the development and properties of the exponential functions. To have an inverse, a function must possess a special property over its domain.

### One-to-One Functions

A function is a rule that assigns a value from its range to each element in its domain. Some functions assign the same range value to more than one element in the domain. The function $f(x) = x^2$ assigns the same value, 1, to both of the numbers $-1$ and $+1$; the sines of $\pi/3$ and $2\pi/3$ are both $\sqrt{3}/2$. Other functions assume each value in their range no more than once. The square roots and cubes of different numbers are always different. A function that has distinct values at distinct elements in its domain is called one-to-one. These functions take on any one value in their range exactly once.

> **DEFINITION** A function $f(x)$ is **one-to-one** on a domain $D$ if $f(x_1) \neq f(x_2)$ whenever $x_1 \neq x_2$ in $D$.

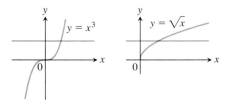

(a) One-to-one: Graph meets each horizontal line at most once.

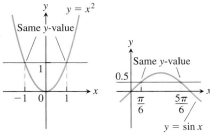

(b) Not one-to-one: Graph meets one or more horizontal lines more than once.

**FIGURE 7.1** (a) $y = x^3$ and $y = \sqrt{x}$ are one-to-one on their domains $(-\infty, \infty)$ and $[0, \infty)$. (b) $y = x^2$ and $y = \sin x$ are not one-to-one on their domains $(-\infty, \infty)$.

**EXAMPLE 1** Some functions are one-to-one on their entire natural domain. Other functions are not one-to-one on their entire domain, but by restricting the function to a smaller domain we can create a function that is one-to-one. The original and restricted functions are not the same functions, because they have different domains. However, the two functions have the same values on the smaller domain, so the original function is an extension of the restricted function from its smaller domain to the larger domain.

**(a)** $f(x) = \sqrt{x}$ is one-to-one on any domain of nonnegative numbers because $\sqrt{x_1} \neq \sqrt{x_2}$ whenever $x_1 \neq x_2$.

**(b)** $g(x) = \sin x$ is *not* one-to-one on the interval $[0, \pi]$ because $\sin(\pi/6) = \sin(5\pi/6)$. In fact, for each element $x_1$ in the subinterval $[0, \pi/2)$ there is a corresponding element $x_2$ in the subinterval $(\pi/2, \pi]$ satisfying $\sin x_1 = \sin x_2$, so distinct elements in the domain are assigned to the same value in the range. The sine function *is* one-to-one on $[0, \pi/2]$, however, because it is an increasing function on $[0, \pi/2]$ giving distinct outputs for distinct inputs. ∎

The graph of a one-to-one function $y = f(x)$ can intersect a given horizontal line at most once. If the function intersects the line more than once, it assumes the same $y$-value for at least two different $x$-values and is therefore not one-to-one (Figure 7.1).

> **The Horizontal Line Test for One-to-One Functions**
>
> A function $y = f(x)$ is one-to-one if and only if its graph intersects each horizontal line at most once.

## Inverse Functions

Since each output of a one-to-one function comes from just one input, the effect of the function can be inverted to send an output back to the input from which it came.

> **DEFINITION** Suppose that $f$ is a one-to-one function on a domain $D$ with range $R$. The **inverse function** $f^{-1}$ is defined by
> $$f^{-1}(b) = a \quad \text{if} \quad f(a) = b.$$
> The domain of $f^{-1}$ is $R$ and the range of $f^{-1}$ is $D$.

**Caution** Do not confuse the inverse function $f^{-1}$ with the reciprocal function $1/f$.

The symbol $f^{-1}$ for the inverse of $f$ is read "$f$ inverse." The "$-1$" in $f^{-1}$ is *not* an exponent; $f^{-1}(x)$ does not mean $1/f(x)$. Notice that the domains and ranges of $f$ and $f^{-1}$ are interchanged.

**EXAMPLE 2** Suppose a one-to-one function $y = f(x)$ is given by a table of values

| $x$ | 1 | 2 | 3 | 4 | 5 | 6 | 7 | 8 |
|------|---|-----|---|------|----|------|----|------|
| $f(x)$ | 3 | 4.5 | 7 | 10.5 | 15 | 20.5 | 27 | 34.5 |

A table for the values of $x = f^{-1}(y)$ can then be obtained by simply interchanging the values in the columns (or rows) of the table for $f$:

| $y$ | 3 | 4.5 | 7 | 10.5 | 15 | 20.5 | 27 | 34.5 |
|--------|---|-----|---|---|---|---|---|---|
| $f^{-1}(y)$ | 1 | 2 | 3 | 4 | 5 | 6 | 7 | 8 |

∎

If we apply $f$ to send an input $x$ to the output $f(x)$ and follow by applying $f^{-1}$ to $f(x)$, we get right back to $x$, just where we started. Similarly, if we take some number $y$ in the range of $f$, apply $f^{-1}$ to it, and then apply $f$ to the resulting value $f^{-1}(y)$, we get back the value $y$ with which we began. Composing a function and its inverse has the same effect as doing nothing.

$$(f^{-1} \circ f)(x) = x, \qquad \text{for all } x \text{ in the domain of } f$$
$$(f \circ f^{-1})(y) = y, \qquad \text{for all } y \text{ in the domain of } f^{-1} \text{ (or range of } f)$$

Only a one-to-one function can have an inverse. The reason is that if $f(x_1) = y$ and $f(x_2) = y$ for two distinct inputs $x_1$ and $x_2$, then there is no way to assign a value to $f^{-1}(y)$ that satisfies both $f^{-1}(f(x_1)) = x_1$ and $f^{-1}(f(x_2)) = x_2$.

A function that is increasing on an interval satisfies the inequality $f(x_2) > f(x_1)$ when $x_2 > x_1$, so it is one-to-one and has an inverse. Decreasing functions also have an inverse. Functions that are neither increasing nor decreasing may still be one-to-one and have an inverse, as with the function $f(x) = 1/x$ for $x \neq 0$ and $f(0) = 0$, defined on $(-\infty, \infty)$ and passing the horizontal line test.

## Finding Inverses

The graphs of a function and its inverse are closely related. To read the value of a function from its graph, we start at a point $x$ on the $x$-axis, go vertically to the graph, and then move horizontally to the $y$-axis to read the value of $y$. The inverse function can be read from the graph by reversing this process. Start with a point $y$ on the $y$-axis, go horizontally to the graph of $y = f(x)$, and then move vertically to the $x$-axis to read the value of $x = f^{-1}(y)$ (Figure 7.2).

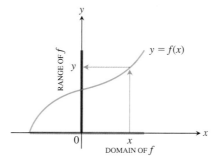

(a) To find the value of $f$ at $x$, we start at $x$, go up to the curve, and then over to the $y$-axis.

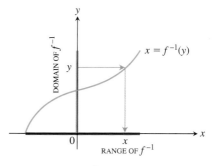

(b) The graph of $f^{-1}$ is the graph of $f$, but with $x$ and $y$ interchanged. To find the $x$ that gave $y$, we start at $y$ and go over to the curve and down to the $x$-axis. The domain of $f^{-1}$ is the range of $f$. The range of $f^{-1}$ is the domain of $f$.

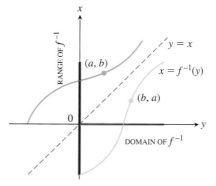

(c) To draw the graph of $f^{-1}$ in the more usual way, we reflect the system across the line $y = x$.

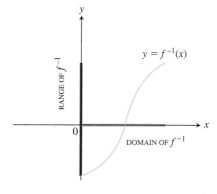

(d) Then we interchange the letters $x$ and $y$. We now have a normal-looking graph of $f^{-1}$ as a function of $x$.

**FIGURE 7.2** The graph of $y = f^{-1}(x)$ is obtained by reflecting the graph of $y = f(x)$ about the line $y = x$.

We want to set up the graph of $f^{-1}$ so that its input values lie along the $x$-axis, as is usually done for functions, rather than on the $y$-axis. To achieve this we interchange the $x$- and $y$-axes by reflecting across the 45° line $y = x$. After this reflection we have a new graph that represents $f^{-1}$. The value of $f^{-1}(x)$ can now be read from the graph in the usual way, by starting with a point $x$ on the $x$-axis, going vertically to the graph, and then horizontally to the $y$-axis to get the value of $f^{-1}(x)$. Figure 7.2 indicates the relationship between the graphs of $f$ and $f^{-1}$. The graphs are interchanged by reflection through the line $y = x$.

The process of passing from $f$ to $f^{-1}$ can be summarized as a two-step procedure.

1. Solve the equation $y = f(x)$ for $x$. This gives a formula $x = f^{-1}(y)$ where $x$ is expressed as a function of $y$.

2. Interchange $x$ and $y$, obtaining a formula $y = f^{-1}(x)$ where $f^{-1}$ is expressed in the conventional format with $x$ as the independent variable and $y$ as the dependent variable.

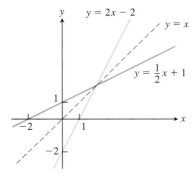

**FIGURE 7.3** Graphing the functions $f(x) = (1/2)x + 1$ and $f^{-1}(x) = 2x - 2$ together shows the graphs' symmetry with respect to the line $y = x$ (Example 3).

**EXAMPLE 3**   Find the inverse of $y = \dfrac{1}{2}x + 1$, expressed as a function of $x$.

**Solution**

1. *Solve for $x$ in terms of $y$:*   $y = \dfrac{1}{2}x + 1$

   The graph is a straight line satisfying the horizontal line test (Fig. 7.3).

$$2y = x + 2$$
$$x = 2y - 2.$$

2. *Interchange $x$ and $y$:*   $y = 2x - 2.$

The inverse of the function $f(x) = (1/2)x + 1$ is the function $f^{-1}(x) = 2x - 2$. (See Figure 7.3.) To check, we verify that both composites give the identity function:

$$f^{-1}(f(x)) = 2\left(\dfrac{1}{2}x + 1\right) - 2 = x + 2 - 2 = x$$

$$f(f^{-1}(x)) = \dfrac{1}{2}(2x - 2) + 1 = x - 1 + 1 = x. \qquad ■$$

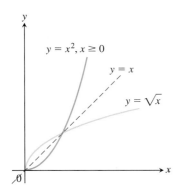

**FIGURE 7.4** The functions $y = \sqrt{x}$ and $y = x^2, x \ge 0$, are inverses of one another (Example 4).

**EXAMPLE 4**   Find the inverse of the function $y = x^2, x \ge 0$, expressed as a function of $x$.

**Solution**   For $x \ge 0$, the graph satisfies the horizontal line test, so the function is one-to-one and has an inverse. To find the inverse, we first solve for $x$ in terms of $y$:

$$y = x^2$$
$$\sqrt{y} = \sqrt{x^2} = |x| = x \qquad |x| = x \text{ because } x \ge 0$$

We then interchange $x$ and $y$, obtaining

$$y = \sqrt{x}.$$

The inverse of the function $y = x^2, x \ge 0$, is the function $y = \sqrt{x}$ (Figure 7.4).

Notice that the function $y = x^2, x \ge 0$, with domain *restricted* to the nonnegative real numbers, *is* one-to-one (Figure 7.4) and has an inverse. On the other hand, the function $y = x^2$, with no domain restrictions, *is not* one-to-one (Figure 7.1b) and therefore has no inverse. ■

### Derivatives of Inverses of Differentiable Functions

We calculated the inverse of the function $f(x) = (1/2)x + 1$ as $f^{-1}(x) = 2x - 2$ in Example 3. If we calculate their derivatives, we see that

$$\frac{d}{dx} f(x) = \frac{d}{dx}\left(\frac{1}{2}x + 1\right) = \frac{1}{2}$$

$$\frac{d}{dx} f^{-1}(x) = \frac{d}{dx}(2x - 2) = 2.$$

The derivatives are reciprocals of one another, so the slope of one line is the reciprocal of the slope of its inverse line. (See Figure 7.3.)

This is not a special case. Reflecting any nonhorizontal or nonvertical line across the line $y = x$ always inverts the line's slope. If the original line has slope $m \neq 0$, the reflected line has slope $1/m$.

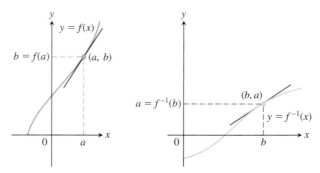

The slopes are reciprocal: $(f^{-1})'(b) = \dfrac{1}{f'(a)}$ or $(f^{-1})'(b) = \dfrac{1}{f'(f^{-1}(b))}$

**FIGURE 7.5** The graphs of inverse functions have reciprocal slopes at corresponding points.

The reciprocal relationship between the slopes of $f$ and $f^{-1}$ holds for other functions as well, but we must be careful to compare slopes at corresponding points. If the slope of $y = f(x)$ at the point $(a, f(a))$ is $f'(a)$ and $f'(a) \neq 0$, then the slope of $y = f^{-1}(x)$ at the point $(f(a), a)$ is the reciprocal $1/f'(a)$ (Figure 7.5). If we set $b = f(a)$, then

$$(f^{-1})'(b) = \frac{1}{f'(a)} = \frac{1}{f'(f^{-1}(b))}.$$

If $y = f(x)$ has a horizontal tangent line at $(a, f(a))$, then the inverse function $f^{-1}$ has a vertical tangent line at $(f(a), a)$, and this infinite slope implies that $f^{-1}$ is not differentiable at $f(a)$. Theorem 1 gives the conditions under which $f^{-1}$ is differentiable in its domain (which is the same as the range of $f$).

---

**THEOREM 1—The Derivative Rule for Inverses** If $f$ has an interval $I$ as domain and $f'(x)$ exists and is never zero on $I$, then $f^{-1}$ is differentiable at every point in its domain (the range of $f$). The value of $(f^{-1})'$ at a point $b$ in the domain of $f^{-1}$ is the reciprocal of the value of $f'$ at the point $a = f^{-1}(b)$:

$$(f^{-1})'(b) = \frac{1}{f'(f^{-1}(b))} \tag{1}$$

or

$$\left.\frac{df^{-1}}{dx}\right|_{x=b} = \frac{1}{\left.\dfrac{df}{dx}\right|_{x=f^{-1}(b)}}.$$

---

Theorem 1 makes two assertions. The first of these has to do with the conditions under which $f^{-1}$ is differentiable; the second assertion is a formula for the derivative of $f^{-1}$ when it exists. While we omit the proof of the first assertion, the second one is proved in the following way:

$$f(f^{-1}(x)) = x \qquad \text{Inverse function relationship}$$

$$\frac{d}{dx} f(f^{-1}(x)) = 1 \qquad \text{Differentiating both sides}$$

$$f'(f^{-1}(x)) \cdot \frac{d}{dx} f^{-1}(x) = 1 \qquad \text{Chain Rule}$$

$$\frac{d}{dx} f^{-1}(x) = \frac{1}{f'(f^{-1}(x))}. \qquad \text{Solving for the derivative}$$

**EXAMPLE 5** The function $f(x) = x^2, x > 0$ and its inverse $f^{-1}(x) = \sqrt{x}$ have derivatives $f'(x) = 2x$ and $(f^{-1})'(x) = 1/(2\sqrt{x})$.

Let's verify that Theorem 1 gives the same formula for the derivative of $f^{-1}(x)$:

$$(f^{-1})'(x) = \frac{1}{f'(f^{-1}(x))}$$

$$= \frac{1}{2(f^{-1}(x))} \qquad \begin{array}{l} f'(x) = 2x \text{ with } x \text{ replaced} \\ \text{by } f^{-1}(x) \end{array}$$

$$= \frac{1}{2(\sqrt{x})}.$$

Theorem 1 gives a derivative that agrees with the known derivative of the square root function.

Let's examine Theorem 1 at a specific point. We pick $x = 2$ (the number $a$) and $f(2) = 4$ (the value $b$). Theorem 1 says that the derivative of $f$ at 2, which is $f'(2) = 4$, and the derivative of $f^{-1}$ at $f(2)$, which is $(f^{-1})'(4)$, are reciprocals. It states that

$$(f^{-1})'(4) = \frac{1}{f'(f^{-1}(4))} = \frac{1}{f'(2)} = \frac{1}{2x}\bigg|_{x=2} = \frac{1}{4}.$$

See Figure 7.6. ∎

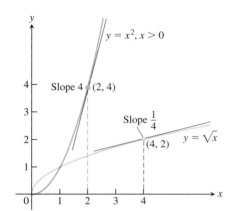

**FIGURE 7.6** The derivative of $f^{-1}(x) = \sqrt{x}$ at the point (4, 2) is the reciprocal of the derivative of $f(x) = x^2$ at (2, 4) (Example 5).

We will use the procedure illustrated in Example 5 to calculate formulas for the derivatives of many inverse functions throughout this chapter. Equation (1) sometimes enables us to find specific values of $df^{-1}/dx$ without knowing a formula for $f^{-1}$.

**EXAMPLE 6** Let $f(x) = x^3 - 2, x > 0$. Find the value of $df^{-1}/dx$ at $x = 6 = f(2)$ without finding a formula for $f^{-1}(x)$.

**Solution** We apply Theorem 1 to obtain the value of the derivative of $f^{-1}$ at $x = 6$:

$$\frac{df}{dx}\bigg|_{x=2} = 3x^2\bigg|_{x=2} = 12$$

$$\frac{df^{-1}}{dx}\bigg|_{x=f(2)} = \frac{1}{\dfrac{df}{dx}\bigg|_{x=2}} = \frac{1}{12}. \qquad \text{Eq. (1)}$$

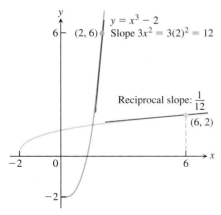

**FIGURE 7.7** The derivative of $f(x) = x^3 - 2$ at $x = 2$ tells us the derivative of $f^{-1}$ at $x = 6$ (Example 6).

See Figure 7.7. ∎

# Exercises **7.1**

## Identifying One-to-One Functions Graphically

Which of the functions graphed in Exercises 1–6 are one-to-one, and which are not?

**1.**

$y = -3x^3$

**2.**

$y = x^4 - x^2$

**3.**

$y = 2|x|$

**4.**
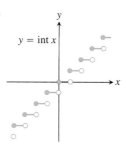
$y = \text{int } x$

**5.**

$y = \frac{1}{x}$

**6.**
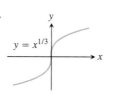
$y = x^{1/3}$

In Exercises 7–10, determine from its graph if the function is one-to-one.

**7.** $f(x) = \begin{cases} 3 - x, & x < 0 \\ 3, & x \geq 0 \end{cases}$

**8.** $f(x) = \begin{cases} 2x + 6, & x \leq -3 \\ x + 4, & x > -3 \end{cases}$

**9.** $f(x) = \begin{cases} 1 - \frac{x}{2}, & x \leq 0 \\ \dfrac{x}{x + 2}, & x > 0 \end{cases}$

**10.** $f(x) = \begin{cases} 2 - x^2, & x \leq 1 \\ x^2, & x > 1 \end{cases}$

## Graphing Inverse Functions

Each of Exercises 11–16 shows the graph of a function $y = f(x)$. Copy the graph and draw in the line $y = x$. Then use symmetry with respect to the line $y = x$ to add the graph of $f^{-1}$ to your sketch. (It is not necessary to find a formula for $f^{-1}$.) Identify the domain and range of $f^{-1}$.

**11.**
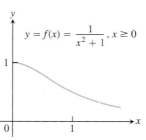
$y = f(x) = \dfrac{1}{x^2 + 1}, \; x \geq 0$

**12.**
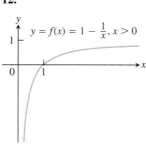
$y = f(x) = 1 - \dfrac{1}{x}, \; x > 0$

**13.**
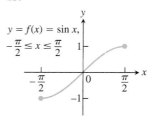
$y = f(x) = \sin x, \; -\dfrac{\pi}{2} \leq x \leq \dfrac{\pi}{2}$

**14.**
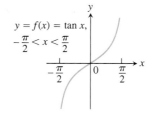
$y = f(x) = \tan x, \; -\dfrac{\pi}{2} < x < \dfrac{\pi}{2}$

**15.**
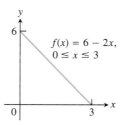
$f(x) = 6 - 2x, \; 0 \leq x \leq 3$

**16.**
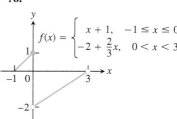
$f(x) = \begin{cases} x + 1, & -1 \leq x \leq 0 \\ -2 + \frac{2}{3}x, & 0 < x < 3 \end{cases}$

**17. a.** Graph the function $f(x) = \sqrt{1 - x^2}, 0 \leq x \leq 1$. What symmetry does the graph have?

    **b.** Show that $f$ is its own inverse. (Remember that $\sqrt{x^2} = x$ if $x \geq 0$.)

**18. a.** Graph the function $f(x) = 1/x$. What symmetry does the graph have?

    **b.** Show that $f$ is its own inverse.

## Formulas for Inverse Functions

Each of Exercises 19–24 gives a formula for a function $y = f(x)$ and shows the graphs of $f$ and $f^{-1}$. Find a formula for $f^{-1}$ in each case.

**19.** $f(x) = x^2 + 1, \; x \geq 0$

**20.** $f(x) = x^2, \; x \leq 0$

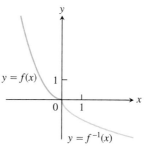

**21.** $f(x) = x^3 - 1$

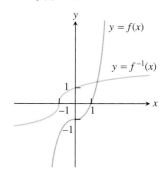

**22.** $f(x) = x^2 - 2x + 1, \quad x \geq 1$

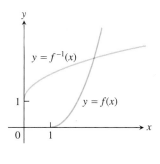

**40. a.** Show that $h(x) = x^3/4$ and $k(x) = (4x)^{1/3}$ are inverses of one another.

    **b.** Graph $h$ and $k$ over an $x$-interval large enough to show the graphs intersecting at $(2, 2)$ and $(-2, -2)$. Be sure the picture shows the required symmetry about the line $y = x$.

    **c.** Find the slopes of the tangents to the graphs of $h$ and $k$ at $(2, 2)$ and $(-2, -2)$.

    **d.** What lines are tangent to the curves at the origin?

**41.** Let $f(x) = x^3 - 3x^2 - 1, x \geq 2$. Find the value of $df^{-1}/dx$ at the point $x = -1 = f(3)$.

**42.** Let $f(x) = x^2 - 4x - 5, x > 2$. Find the value of $df^{-1}/dx$ at the point $x = 0 = f(5)$.

**43.** Suppose that the differentiable function $y = f(x)$ has an inverse and that the graph of $f$ passes through the point $(2, 4)$ and has a slope of $1/3$ there. Find the value of $df^{-1}/dx$ at $x = 4$.

**44.** Suppose that the differentiable function $y = g(x)$ has an inverse and that the graph of $g$ passes through the origin with slope 2. Find the slope of the graph of $g^{-1}$ at the origin.

**23.** $f(x) = (x + 1)^2, \quad x \geq -1$    **24.** $f(x) = x^{2/3}, \quad x \geq 0$

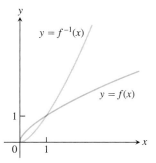

### Inverses of Lines

**45. a.** Find the inverse of the function $f(x) = mx$, where $m$ is a constant different from zero.

    **b.** What can you conclude about the inverse of a function $y = f(x)$ whose graph is a line through the origin with a non-zero slope $m$?

**46.** Show that the graph of the inverse of $f(x) = mx + b$, where $m$ and $b$ are constants and $m \neq 0$, is a line with slope $1/m$ and $y$-intercept $-b/m$.

**47. a.** Find the inverse of $f(x) = x + 1$. Graph $f$ and its inverse together. Add the line $y = x$ to your sketch, drawing it with dashes or dots for contrast.

    **b.** Find the inverse of $f(x) = x + b$ ($b$ constant). How is the graph of $f^{-1}$ related to the graph of $f$?

    **c.** What can you conclude about the inverses of functions whose graphs are lines parallel to the line $y = x$?

**48. a.** Find the inverse of $f(x) = -x + 1$. Graph the line $y = -x + 1$ together with the line $y = x$. At what angle do the lines intersect?

    **b.** Find the inverse of $f(x) = -x + b$ ($b$ constant). What angle does the line $y = -x + b$ make with the line $y = x$?

    **c.** What can you conclude about the inverses of functions whose graphs are lines perpendicular to the line $y = x$?

### Derivatives of Inverse Functions

Each of Exercises 25–34 gives a formula for a function $y = f(x)$. In each case, find $f^{-1}(x)$ and identify the domain and range of $f^{-1}$. As a check, show that $f(f^{-1}(x)) = f^{-1}(f(x)) = x$.

**25.** $f(x) = x^5$      **26.** $f(x) = x^4, \quad x \geq 0$

**27.** $f(x) = x^3 + 1$      **28.** $f(x) = (1/2)x - 7/2$

**29.** $f(x) = 1/x^2, \quad x > 0$    **30.** $f(x) = 1/x^3, \quad x \neq 0$

**31.** $f(x) = \dfrac{x + 3}{x - 2}$      **32.** $f(x) = \dfrac{\sqrt{x}}{\sqrt{x} - 3}$

**33.** $f(x) = x^2 - 2x, \quad x \leq 1$ (*Hint:* Complete the square.)

**34.** $f(x) = (2x^3 + 1)^{1/5}$

In Exercises 35–38:

**a.** Find $f^{-1}(x)$.

**b.** Graph $f$ and $f^{-1}$ together.

**c.** Evaluate $df/dx$ at $x = a$ and $df^{-1}/dx$ at $x = f(a)$ to show that at these points $df^{-1}/dx = 1/(df/dx)$.

**35.** $f(x) = 2x + 3, \quad a = -1$    **36.** $f(x) = (1/5)x + 7, \quad a = -1$

**37.** $f(x) = 5 - 4x, \quad a = 1/2$    **38.** $f(x) = 2x^2, \quad x \geq 0, \quad a = 5$

**39. a.** Show that $f(x) = x^3$ and $g(x) = \sqrt[3]{x}$ are inverses of one another.

    **b.** Graph $f$ and $g$ over an $x$-interval large enough to show the graphs intersecting at $(1, 1)$ and $(-1, -1)$. Be sure the picture shows the required symmetry about the line $y = x$.

    **c.** Find the slopes of the tangents to the graphs of $f$ and $g$ at $(1, 1)$ and $(-1, -1)$ (four tangents in all).

    **d.** What lines are tangent to the curves at the origin?

### Increasing and Decreasing Functions

**49.** Show that increasing functions and decreasing functions are one-to-one. That is, show that for any $x_1$ and $x_2$ in $I$, $x_2 \neq x_1$ implies $f(x_2) \neq f(x_1)$.

Use the results of Exercise 49 to show that the functions in Exercises 50–54 have inverses over their domains. Find a formula for $df^{-1}/dx$ using Theorem 1.

**50.** $f(x) = (1/3)x + (5/6)$      **51.** $f(x) = 27x^3$

**52.** $f(x) = 1 - 8x^3$      **53.** $f(x) = (1 - x)^3$

**54.** $f(x) = x^{5/3}$

## Theory and Applications

**55.** If $f(x)$ is one-to-one, can anything be said about $g(x) = -f(x)$? Is it also one-to-one? Give reasons for your answer.

**56.** If $f(x)$ is one-to-one and $f(x)$ is never zero, can anything be said about $h(x) = 1/f(x)$? Is it also one-to-one? Give reasons for your answer.

**57.** Suppose that the range of $g$ lies in the domain of $f$ so that the composite $f \circ g$ is defined. If $f$ and $g$ are one-to-one, can anything be said about $f \circ g$? Give reasons for your answer.

**58.** If a composite $f \circ g$ is one-to-one, must $g$ be one-to-one? Give reasons for your answer.

**59.** Assume that $f$ and $g$ are differentiable functions that are inverses of one another so that $(g \circ f)(x) = x$. Differentiate both sides of this equation with respect to $x$ using the Chain Rule to express $(g \circ f)'(x)$ as a product of derivatives of $g$ and $f$. What do you find? (This is not a proof of Theorem 1 because we assume here the theorem's conclusion that $g = f^{-1}$ is differentiable.)

**60. Equivalence of the washer and shell methods for finding volume** Let $f$ be differentiable and increasing on the interval $a \leq x \leq b$, with $a > 0$, and suppose that $f$ has a differentiable inverse, $f^{-1}$. Revolve about the $y$-axis the region bounded by the graph of $f$ and the lines $x = a$ and $y = f(b)$ to generate a solid. Then the values of the integrals given by the washer and shell methods for the volume have identical values:

$$\int_{f(a)}^{f(b)} \pi((f^{-1}(y))^2 - a^2)\, dy = \int_{a}^{b} 2\pi x(f(b) - f(x))\, dx.$$

To prove this equality, define

$$W(t) = \int_{f(a)}^{f(t)} \pi((f^{-1}(y))^2 - a^2)\, dy$$

$$S(t) = \int_{a}^{t} 2\pi x(f(t) - f(x))\, dx.$$

Then show that the functions $W$ and $S$ agree at a point of $[a, b]$ and have identical derivatives on $[a, b]$. As you saw in Section 4.7, Exercise 90, this will guarantee $W(t) = S(t)$ for all $t$ in $[a, b]$. In particular, $W(b) = S(b)$. (*Source:* "Disks and Shells Revisited," by Walter Carlip, *American Mathematical Monthly*, Vol. 98, No. 2, Feb. 1991, pp. 154–156.)

## COMPUTER EXPLORATIONS

In Exercises 61–68, you will explore some functions and their inverses together with their derivatives and linear approximating functions at specified points. Perform the following steps using your CAS:

**a.** Plot the function $y = f(x)$ together with its derivative over the given interval. Explain why you know that $f$ is one-to-one over the interval.

**b.** Solve the equation $y = f(x)$ for $x$ as a function of $y$, and name the resulting inverse function $g$.

**c.** Find the equation for the tangent line to $f$ at the specified point $(x_0, f(x_0))$.

**d.** Find the equation for the tangent line to $g$ at the point $(f(x_0), x_0)$ located symmetrically across the 45° line $y = x$ (which is the graph of the identity function). Use Theorem 1 to find the slope of this tangent line.

**e.** Plot the functions $f$ and $g$, the identity, the two tangent lines, and the line segment joining the points $(x_0, f(x_0))$ and $(f(x_0), x_0)$. Discuss the symmetries you see across the main diagonal.

**61.** $y = \sqrt{3x - 2}$, $\dfrac{2}{3} \leq x \leq 4$, $x_0 = 3$

**62.** $y = \dfrac{3x + 2}{2x - 11}$, $-2 \leq x \leq 2$, $x_0 = 1/2$

**63.** $y = \dfrac{4x}{x^2 + 1}$, $-1 \leq x \leq 1$, $x_0 = 1/2$

**64.** $y = \dfrac{x^3}{x^2 + 1}$, $-1 \leq x \leq 1$, $x_0 = 1/2$

**65.** $y = x^3 - 3x^2 - 1$, $2 \leq x \leq 5$, $x_0 = \dfrac{27}{10}$

**66.** $y = 2 - x - x^3$, $-2 \leq x \leq 2$, $x_0 = \dfrac{3}{2}$

**67.** $y = e^x$, $-3 \leq x \leq 5$, $x_0 = 1$

**68.** $y = \sin x$, $-\dfrac{\pi}{2} \leq x \leq \dfrac{\pi}{2}$, $x_0 = 1$

In Exercises 69 and 70, repeat the steps above to solve for the functions $y = f(x)$ and $x = f^{-1}(y)$ defined implicitly by the given equations over the interval.

**69.** $y^{1/3} - 1 = (x + 2)^3$, $-5 \leq x \leq 5$, $x_0 = -3/2$

**70.** $\cos y = x^{1/5}$, $0 \leq x \leq 1$, $x_0 = 1/2$

# $7.2$ Natural Logarithms

Historically, logarithms played important roles in arithmetic computations, making possible the great seventeenth-century advances in offshore navigation and celestial mechanics. In this section we define the natural logarithm as an integral using the Fundamental Theorem of Calculus. While this indirect approach may at first seem strange, it provides an elegant and rigorous way to obtain the key characteristics of logarithmic and exponential functions.

## Definition of the Natural Logarithm Function

The natural logarithm of any positive number $x$, written as $\ln x$, is defined as an integral.

> **DEFINITION**   The **natural logarithm** is the function given by
>
> $$\ln x = \int_1^x \frac{1}{t}\,dt, \qquad x > 0.$$

From the Fundamental Theorem of Calculus, $\ln x$ is a continuous function. Geometrically, if $x > 1$, then $\ln x$ is the area under the curve $y = 1/t$ from $t = 1$ to $t = x$ (Figure 7.8). For $0 < x < 1$, $\ln x$ gives the negative of the area under the curve from $x$ to 1, and the function is not defined for $x \le 0$. From the Zero Width Interval Rule for definite integrals, we also have

$$\ln 1 = \int_1^1 \frac{1}{t}\,dt = 0.$$

**FIGURE 7.8**   The graph of $y = \ln x$ and its relation to the function $y = 1/x, x > 0$. The graph of the logarithm rises above the $x$-axis as $x$ moves from 1 to the right, and it falls below the axis as $x$ moves from 1 to the left.

Notice that we show the graph of $y = 1/x$ in Figure 7.8 but use $y = 1/t$ in the integral. Using $x$ for everything would have us writing

$$\ln x = \int_1^x \frac{1}{x}\,dx,$$

with $x$ meaning two different things. So we change the variable of integration to $t$.

By using rectangles to obtain finite approximations of the area under the graph of $y = 1/t$ and over the interval between $t = 1$ and $t = x$, as in Section 5.1, we can approximate the values of the function $\ln x$. Several values are given in Table 7.1. There is an important number between $x = 2$ and $x = 3$ whose natural logarithm equals 1. This number, which we now define, exists because $\ln x$ is a continuous function and therefore satisfies the Intermediate Value Theorem on $[2, 3]$.

**TABLE 7.1** Typical 2-place values of ln $x$

| $x$ | $\ln x$ |
| --- | --- |
| 0 | undefined |
| 0.05 | $-3.00$ |
| 0.5 | $-0.69$ |
| 1 | 0 |
| 2 | 0.69 |
| 3 | 1.10 |
| 4 | 1.39 |
| 10 | 2.30 |

> **DEFINITION**   The **number $e$** is that number in the domain of the natural logarithm satisfying
>
> $$\ln (e) = \int_1^e \frac{1}{t}\,dt = 1.$$

Interpreted geometrically, the number $e$ corresponds to the point on the $x$-axis for which the area under the graph of $y = 1/t$ and above the interval $[1, e]$ equals the area of the unit square. That is, the area of the region shaded blue in Figure 7.1 is 1 sq unit when $x = e$. In the next section, we will see that the number $e$ can be calculated as a limit and has the numerical value $e \approx 2.718281828459045$ to 15 decimal places.

### The Derivative of $y = \ln x$

By the first part of the Fundamental Theorem of Calculus (Section 5.4),

$$\frac{d}{dx} \ln x = \frac{d}{dx} \int_1^x \frac{1}{t} dt = \frac{1}{x}.$$

For every positive value of $x$, we have

$$\frac{d}{dx} \ln x = \frac{1}{x}. \tag{1}$$

Therefore, the function $y = \ln x$ is a solution to the initial value problem $dy/dx = 1/x$, $x > 0$, with $y(1) = 0$. Notice that the derivative is always positive.

If $u$ is a differentiable function of $x$ whose values are positive, so that $\ln u$ is defined, then applying the Chain Rule we obtain

$$\frac{d}{dx} \ln u = \frac{1}{u} \frac{du}{dx}, \qquad u > 0. \tag{2}$$

**EXAMPLE 1** We use Equation (2) to find derivatives.

(a) $\dfrac{d}{dx} \ln 2x = \dfrac{1}{2x} \dfrac{d}{dx} (2x) = \dfrac{1}{2x} (2) = \dfrac{1}{x}, \quad x > 0$

(b) Equation (2) with $u = x^2 + 3$ gives

$$\frac{d}{dx} \ln (x^2 + 3) = \frac{1}{x^2 + 3} \cdot \frac{d}{dx} (x^2 + 3) = \frac{1}{x^2 + 3} \cdot 2x = \frac{2x}{x^2 + 3}.$$

(c) Equation (2) with $u = |x|$ gives an important derivative:

$$\begin{aligned}
\frac{d}{dx} \ln |x| &= \frac{d}{du} \ln u \cdot \frac{du}{dx} && u = |x|, x \neq 0 \\
&= \frac{1}{u} \cdot \frac{x}{|x|} && \frac{d}{dx}(|x|) = \frac{x}{|x|} \\
&= \frac{1}{|x|} \cdot \frac{x}{|x|} && \text{Substitute for } u. \\
&= \frac{x}{x^2} \\
&= \frac{1}{x}.
\end{aligned}$$

**Derivative of $\ln |x|$**

$$\frac{d}{dx} \ln |x| = \frac{1}{x}, x \neq 0$$

So $1/x$ is the derivative of $\ln x$ on the domain $x > 0$, and the derivative of $\ln(-x)$ on the domain $x < 0$. ∎

Notice from Example 1a that the function $y = \ln 2x$ has the same derivative as the function $y = \ln x$. This is true of $y = \ln bx$ for any constant $b$, provided that $bx > 0$:

$$\frac{d}{dx} \ln bx = \frac{1}{x}, \quad bx > 0$$

$$\frac{d}{dx} \ln bx = \frac{1}{bx} \cdot \frac{d}{dx} (bx) = \frac{1}{bx} (b) = \frac{1}{x}.$$

## Properties of Logarithms

Logarithms, invented by John Napier, were the single most important improvement in arithmetic calculation before the modern electronic computer. What made them so useful is that the properties of logarithms reduce multiplication of positive numbers to addition of their logarithms, division of positive numbers to subtraction of their logarithms, and exponentiation of a number to multiplying its logarithm by the exponent.

---

**THEOREM 2—Algebraic Properties of the Natural Logarithm**    For any numbers $b > 0$ and $x > 0$, the natural logarithm satisfies the following rules:

1. *Product Rule*:        $\ln bx = \ln b + \ln x$

2. *Quotient Rule*:       $\ln \dfrac{b}{x} = \ln b - \ln x$

3. *Reciprocal Rule*:     $\ln \dfrac{1}{x} = -\ln x$        Rule 2 with $b = 1$

4. *Power Rule*:          $\ln x^r = r \ln x$        For $r$ rational

---

For now we consider only rational exponents in Rule 4. In Section 7.3 we will see that the rule holds for all real exponents as well.

**EXAMPLE 2**    Here we demonstrate each of the rules in Theorem 2.

**(a)** $\ln 4 + \ln \sin x = \ln (4 \sin x)$        Product Rule

**(b)** $\ln \dfrac{x + 1}{2x - 3} = \ln (x + 1) - \ln (2x - 3)$        Quotient Rule

**(c)** $\ln \dfrac{1}{8} = -\ln 8$        Reciprocal Rule

$\qquad = -\ln 2^3 = -3 \ln 2$        Power Rule        ■

We now give the proof of Theorem 2. The properties are proved by applying Corollary 2 of the Mean Value Theorem to each of them.

**Proof that $\ln bx = \ln b + \ln x$**    The argument starts by observing that $\ln bx$ and $\ln x$ have the same derivative:

$$\frac{d}{dx}\ln (bx) = \frac{b}{bx} = \frac{1}{x} = \frac{d}{dx}\ln x.$$

According to Corollary 2 of the Mean Value Theorem, the functions must differ by a constant, which means that

$$\ln bx = \ln x + C$$

for some constant $C$.

Since this last equation holds for all positive values of $x$, it must hold for $x = 1$. Hence,

$$\ln (b \cdot 1) = \ln 1 + C$$
$$\ln b = 0 + C \qquad \ln 1 = 0$$
$$C = \ln b.$$

By substituting we conclude that

$$\ln bx = \ln b + \ln x.$$        ■

**Proof that ln $x^r$ = r ln $x$ (assuming $r$ rational)** We use the same-derivative argument again. For all positive values of $x$,

$$\frac{d}{dx}\ln x^r = \frac{1}{x^r}\frac{d}{dx}(x^r) \qquad \text{Eq. (2) with } u = x^r$$

$$= \frac{1}{x^r}rx^{r-1} \qquad \text{General Power Rule for derivatives, } r \text{ rational}$$

$$= r\cdot\frac{1}{x} = \frac{d}{dx}(r\ln x).$$

Since $\ln x^r$ and $r\ln x$ have the same derivative,

$$\ln x^r = r\ln x + C$$

for some constant $C$. Taking $x$ to be 1 identifies $C$ as zero, and we're done. (Exercise 46 in Section 3.7 indicates a proof of the General Power Rule for derivatives when $r$ is rational.)

You are asked to prove Rule 2 in Exercise 86. Rule 3 is a special case of Rule 2, obtained by setting $b = 1$ and noting that $\ln 1 = 0$. This covers all cases of Theorem 2. ∎

We have not yet proved Rule 4 for $r$ irrational; however, the rule does hold for all $r$, rational or irrational. We will show this in the next section after we define exponential functions and irrational exponents.

### The Graph and Range of ln $x$

The derivative $d(\ln x)/dx = 1/x$ is positive for $x > 0$, so $\ln x$ is an increasing function of $x$. The second derivative, $-1/x^2$, is negative, so the graph of $\ln x$ is concave down. (See Figure 7.9a.)

We can estimate the value of $\ln 2$ by considering the area under the graph of $y = 1/x$ and above the interval $[1, 2]$. In Figure 7.9(b) a rectangle of height $1/2$ over the interval $[1, 2]$ fits under the graph. Therefore the area under the graph, which is $\ln 2$, is greater than the area, $1/2$, of the rectangle. So $\ln 2 > 1/2$. Knowing this we have

$$\ln 2^n = n\ln 2 > n\left(\frac{1}{2}\right) = \frac{n}{2}.$$

This result shows that $\ln(2^n) \to \infty$ as $n \to \infty$. Since $\ln x$ is an increasing function, we get that

$$\lim_{x\to\infty}\ln x = \infty.$$

We also have

$$\lim_{x\to0^+}\ln x = \lim_{t\to\infty}\ln t^{-1} = \lim_{t\to\infty}(-\ln t) = -\infty. \qquad x = 1/t = t^{-1}$$

We defined $\ln x$ for $x > 0$, so the domain of $\ln x$ is the set of positive real numbers. The above discussion and the Intermediate Value Theorem show that its range is the entire real line, giving the graph of $y = \ln x$ shown in Figure 7.9(a).

(a)

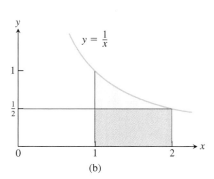

(b)

**FIGURE 7.9** (a) The graph of the natural logarithm. (b) The rectangle of height $y = 1/2$ fits beneath the graph of $y = 1/x$ for the interval $1 \le x \le 2$.

### The Integral $\int(1/u)\,du$

Example 2c leads to the following integral formula.

If $u$ is a differentiable function that is never zero,

$$\int\frac{1}{u}du = \ln|u| + C. \qquad (3)$$

Equation (3) applies anywhere on the domain of $1/u$, the points where $u \neq 0$. It says that integrals of a certain *form* lead to logarithms. If $u = f(x)$, then $du = f'(x)\,dx$ and

$$\int \frac{f'(x)}{f(x)}\,dx = \ln|f(x)| + C$$

whenever $f(x)$ is a differentiable function that is never zero.

**EXAMPLE 3** Here we recognize an integral of the form $\int \dfrac{du}{u}$.

$$\int_{-\pi/2}^{\pi/2} \frac{4\cos\theta}{3 + 2\sin\theta}\,d\theta = \int_{1}^{5} \frac{2}{u}\,du \qquad \begin{array}{l} u = 3 + 2\sin\theta, \quad du = 2\cos\theta\,d\theta, \\ u(-\pi/2) = 1, \quad u(\pi/2) = 5 \end{array}$$

$$= 2\ln|u|\,\Big]_{1}^{5}$$

$$= 2\ln|5| - 2\ln|1| = 2\ln 5$$

Note that $u = 3 + 2\sin\theta$ is always positive on $[-\pi/2, \pi/2]$, so Equation (3) applies. ∎

### The Integrals of tan $x$, cot $x$, sec $x$, and csc $x$

Equation (3) tells us how to integrate these trigonometric functions.

$$\int \tan x\,dx = \int \frac{\sin x}{\cos x}\,dx = \int \frac{-du}{u} \qquad \begin{array}{l} u = \cos x > 0 \text{ on } (-\pi/2\ \pi/2), \\ du = -\sin x\,dx \end{array}$$

$$= -\ln|u| + C = -\ln|\cos x| + C$$

$$= \ln\frac{1}{|\cos x|} + C = \ln|\sec x| + C. \qquad \text{Reciprocal Rule}$$

For the cotangent,

$$\int \cot x\,dx = \int \frac{\cos x\,dx}{\sin x} = \int \frac{du}{u} \qquad \begin{array}{l} u = \sin x, \\ du = \cos x\,dx \end{array}$$

$$= \ln|u| + C = \ln|\sin x| + C = -\ln|\csc x| + C.$$

To integrate sec $x$, we multiply and divide by $(\sec x + \tan x)$ as an algebraic form of 1.

$$\int \sec x\,dx = \int \sec x \frac{(\sec x + \tan x)}{(\sec x + \tan x)}\,dx = \int \frac{\sec^2 x + \sec x \tan x}{\sec x + \tan x}\,dx$$

$$= \int \frac{du}{u} = \ln|u| + C = \ln|\sec x + \tan x| + C \qquad \begin{array}{l} u = \sec x + \tan x, \\ du = (\sec x \tan x + \sec^2 x)dx \end{array}$$

For csc $x$, we multiply and divide by $(\csc x + \cot x)$ as an algebraic form of 1.

$$\int \csc x\,dx = \int \csc x \frac{(\csc x + \cot x)}{(\csc x + \cot x)}\,dx = \int \frac{\csc^2 x + \csc x \cot x}{\csc x + \cot x}\,dx$$

$$= \int \frac{-du}{u} = -\ln|u| + C = -\ln|\csc x + \cot x| + C \qquad \begin{array}{l} u = \csc x + \cot x, \\ du = (-\csc x \cot x - \csc^2 x)\,dx \end{array}$$

In summary, we have the following results.

---

**Integrals of the tangent, cotangent, secant, and cosecant functions**

$$\int \tan u \, du = \ln |\sec u| + C \qquad \int \sec u \, du = \ln |\sec u + \tan u| + C$$

$$\int \cot u \, du = \ln |\sin u| + C \qquad \int \csc u \, du = -\ln |\csc u + \cot u| + C$$

---

### EXAMPLE 4

$$\int_0^{\pi/6} \tan 2x \, dx = \int_0^{\pi/3} \tan u \cdot \frac{du}{2} = \frac{1}{2} \int_0^{\pi/3} \tan u \, du$$

Substitute $u = 2x$,
$dx = du/2$,
$u(0) = 0$,
$u(\pi/6) = \pi/3$

$$= \frac{1}{2} \ln |\sec u| \Big]_0^{\pi/3} = \frac{1}{2}(\ln 2 - \ln 1) = \frac{1}{2}\ln 2 \qquad ■$$

### Logarithmic Differentiation

The derivatives of positive functions given by formulas that involve products, quotients, and powers can often be found more quickly if we take the natural logarithm of both sides before differentiating. This enables us to use the laws of logarithms to simplify the formulas before differentiating. The process, called **logarithmic differentiation**, is illustrated in the next example.

### EXAMPLE 5 Find $dy/dx$ if

$$y = \frac{(x^2 + 1)(x + 3)^{1/2}}{x - 1}, \qquad x > 1.$$

**Solution** We take the natural logarithm of both sides and simplify the result with the properties of logarithms:

$$\ln y = \ln \frac{(x^2 + 1)(x + 3)^{1/2}}{x - 1}$$

$$= \ln \left((x^2 + 1)(x + 3)^{1/2}\right) - \ln (x - 1) \qquad \text{Quotient Rule}$$

$$= \ln (x^2 + 1) + \ln (x + 3)^{1/2} - \ln (x - 1) \qquad \text{Product Rule}$$

$$= \ln (x^2 + 1) + \frac{1}{2}\ln (x + 3) - \ln (x - 1). \qquad \text{Power Rule}$$

We then take derivatives of both sides with respect to $x$, using Equation (2) on the left:

$$\frac{1}{y}\frac{dy}{dx} = \frac{1}{x^2 + 1} \cdot 2x + \frac{1}{2} \cdot \frac{1}{x + 3} - \frac{1}{x - 1}.$$

Next we solve for $dy/dx$:

$$\frac{dy}{dx} = y\left(\frac{2x}{x^2 + 1} + \frac{1}{2x + 6} - \frac{1}{x - 1}\right).$$

Finally, we substitute for $y$ from the original equation:

$$\frac{dy}{dx} = \frac{(x^2 + 1)(x + 3)^{1/2}}{x - 1}\left(\frac{2x}{x^2 + 1} + \frac{1}{2x + 6} - \frac{1}{x - 1}\right). \qquad ■$$

A direct computation in Example 5, using the Derivative Quotient and Product Rules, would be much longer.

# Exercises 7.2

## Using the Algebraic Properties—Theorem 2

1. Express the following logarithms in terms of ln 2 and ln 3.

   **a.** ln 0.75     **b.** ln (4/9)     **c.** ln (1/2)

   **d.** ln $\sqrt[3]{9}$     **e.** ln 3$\sqrt{2}$     **f.** ln $\sqrt{13.5}$

2. Express the following logarithms in terms of ln 5 and ln 7.

   **a.** ln (1/125)     **b.** ln 9.8     **c.** ln 7$\sqrt{7}$

   **d.** ln 1225     **e.** ln 0.056

   **f.** (ln 35 + ln (1/7))/(ln 25)

Use the properties of logarithms to simplify the expressions in Exercises 3 and 4.

3. **a.** $\ln \sin \theta - \ln \left(\dfrac{\sin \theta}{5}\right)$     **b.** $\ln (3x^2 - 9x) + \ln \left(\dfrac{1}{3x}\right)$

   **c.** $\dfrac{1}{2}\ln (4t^4) - \ln 2$

4. **a.** $\ln \sec \theta + \ln \cos \theta$     **b.** $\ln (8x + 4) - 2 \ln 2$

   **c.** $3\ln \sqrt[3]{t^2 - 1} - \ln (t + 1)$

## Finding Derivatives

In Exercises 5–36, find the derivative of $y$ with respect to $x$, $t$, or $\theta$, as appropriate.

5. $y = \ln 3x$

6. $y = \ln kx$, $k$ constant

7. $y = \ln (t^2)$

8. $y = \ln (t^{3/2})$

9. $y = \ln \dfrac{3}{x}$

10. $y = \ln \dfrac{10}{x}$

11. $y = \ln (\theta + 1)$

12. $y = \ln (2\theta + 2)$

13. $y = \ln x^3$

14. $y = (\ln x)^3$

15. $y = t(\ln t)^2$

16. $y = t\sqrt{\ln t}$

17. $y = \dfrac{x^4}{4}\ln x - \dfrac{x^4}{16}$

18. $y = (x^2 \ln x)^4$

19. $y = \dfrac{\ln t}{t}$

20. $y = \dfrac{1 + \ln t}{t}$

21. $y = \dfrac{\ln x}{1 + \ln x}$

22. $y = \dfrac{x \ln x}{1 + \ln x}$

23. $y = \ln (\ln x)$

24. $y = \ln (\ln (\ln x))$

25. $y = \theta (\sin (\ln \theta) + \cos (\ln \theta))$

26. $y = \ln (\sec \theta + \tan \theta)$

27. $y = \ln \dfrac{1}{x\sqrt{x + 1}}$

28. $y = \dfrac{1}{2}\ln \dfrac{1 + x}{1 - x}$

29. $y = \dfrac{1 + \ln t}{1 - \ln t}$

30. $y = \sqrt{\ln \sqrt{t}}$

31. $y = \ln (\sec (\ln \theta))$

32. $y = \ln \left(\dfrac{\sqrt{\sin \theta \cos \theta}}{1 + 2 \ln \theta}\right)$

33. $y = \ln \left(\dfrac{(x^2 + 1)^5}{\sqrt{1 - x}}\right)$

34. $y = \ln \sqrt{\dfrac{(x + 1)^5}{(x + 2)^{20}}}$

35. $y = \displaystyle\int_{x^2/2}^{x^2} \ln \sqrt{t}\, dt$

36. $y = \displaystyle\int_{\sqrt{x}}^{\sqrt[3]{x}} \ln t\, dt$

## Evaluating Integrals

Evaluate the integrals in Exercises 37–54.

37. $\displaystyle\int_{-3}^{-2} \dfrac{dx}{x}$

38. $\displaystyle\int_{-1}^{0} \dfrac{3\, dx}{3x - 2}$

39. $\displaystyle\int \dfrac{2y\, dy}{y^2 - 25}$

40. $\displaystyle\int \dfrac{8r\, dr}{4r^2 - 5}$

41. $\displaystyle\int_{0}^{\pi} \dfrac{\sin t}{2 - \cos t}\, dt$

42. $\displaystyle\int_{0}^{\pi/3} \dfrac{4 \sin \theta}{1 - 4 \cos \theta}\, d\theta$

43. $\displaystyle\int_{1}^{2} \dfrac{2 \ln x}{x}\, dx$

44. $\displaystyle\int_{2}^{4} \dfrac{dx}{x \ln x}$

45. $\displaystyle\int_{2}^{4} \dfrac{dx}{x(\ln x)^2}$

46. $\displaystyle\int_{2}^{16} \dfrac{dx}{2x\sqrt{\ln x}}$

47. $\displaystyle\int \dfrac{3 \sec^2 t}{6 + 3 \tan t}\, dt$

48. $\displaystyle\int \dfrac{\sec y \tan y}{2 + \sec y}\, dy$

49. $\displaystyle\int_{0}^{\pi/2} \tan \dfrac{x}{2}\, dx$

50. $\displaystyle\int_{\pi/4}^{\pi/2} \cot t\, dt$

51. $\displaystyle\int_{\pi/2}^{\pi} 2 \cot \dfrac{\theta}{3}\, d\theta$

52. $\displaystyle\int_{0}^{\pi/12} 6 \tan 3x\, dx$

53. $\displaystyle\int \dfrac{dx}{2\sqrt{x} + 2x}$

54. $\displaystyle\int \dfrac{\sec x\, dx}{\sqrt{\ln (\sec x + \tan x)}}$

## Logarithmic Differentiation

In Exercises 55–68, use logarithmic differentiation to find the derivative of $y$ with respect to the given independent variable.

55. $y = \sqrt{x(x + 1)}$

56. $y = \sqrt{(x^2 + 1)(x - 1)^2}$

57. $y = \sqrt{\dfrac{t}{t + 1}}$

58. $y = \sqrt{\dfrac{1}{t(t + 1)}}$

59. $y = \sqrt{\theta + 3} \sin \theta$

60. $y = (\tan \theta)\sqrt{2\theta + 1}$

61. $y = t(t + 1)(t + 2)$

62. $y = \dfrac{1}{t(t + 1)(t + 2)}$

63. $y = \dfrac{\theta + 5}{\theta \cos \theta}$

64. $y = \dfrac{\theta \sin \theta}{\sqrt{\sec \theta}}$

65. $y = \dfrac{x\sqrt{x^2 + 1}}{(x + 1)^{2/3}}$

66. $y = \sqrt{\dfrac{(x + 1)^{10}}{(2x + 1)^5}}$

67. $y = \sqrt[3]{\dfrac{x(x - 2)}{x^2 + 1}}$

68. $y = \sqrt[3]{\dfrac{x(x + 1)(x - 2)}{(x^2 + 1)(2x + 3)}}$

## Theory and Applications

69. Locate and identify the absolute extreme values of

    **a.** ln (cos x) on $[-\pi/4, \pi/3]$,

    **b.** cos (ln x) on $[1/2, 2]$.

70. **a.** Prove that $f(x) = x - \ln x$ is increasing for $x > 1$.

    **b.** Using part (a), show that ln x < x if x > 1.

71. Find the area between the curves $y = \ln x$ and $y = \ln 2x$ from $x = 1$ to $x = 5$.

**72.** Find the area between the curve $y = \tan x$ and the $x$-axis from $x = -\pi/4$ to $x = \pi/3$.

**73.** The region in the first quadrant bounded by the coordinate axes, the line $y = 3$, and the curve $x = 2/\sqrt{y + 1}$ is revolved about the $y$-axis to generate a solid. Find the volume of the solid.

**74.** The region between the curve $y = \sqrt{\cot x}$ and the $x$-axis from $x = \pi/6$ to $x = \pi/2$ is revolved about the $x$-axis to generate a solid. Find the volume of the solid.

**75.** The region between the curve $y = 1/x^2$ and the $x$-axis from $x = 1/2$ to $x = 2$ is revolved about the $y$-axis to generate a solid. Find the volume of the solid.

**76.** In Section 6.2, Exercise 6, we revolved about the $y$-axis the region between the curve $y = 9x/\sqrt{x^3 + 9}$ and the $x$-axis from $x = 0$ to $x = 3$ to generate a solid of volume $36\pi$. What volume do you get if you revolve the region about the $x$-axis instead? (See Section 6.2, Exercise 6, for a graph.)

**77.** Find the lengths of the following curves.

 **a.** $y = (x^2/8) - \ln x$, $\quad 4 \le x \le 8$

 **b.** $x = (y/4)^2 - 2 \ln (y/4)$, $\quad 4 \le y \le 12$

**78.** Find a curve through the point $(1, 0)$ whose length from $x = 1$ to $x = 2$ is

$$L = \int_1^2 \sqrt{1 + \frac{1}{x^2}}\, dx.$$

**T** **79. a.** Find the centroid of the region between the curve $y = 1/x$ and the $x$-axis from $x = 1$ to $x = 2$. Give the coordinates to two decimal places.

 **b.** Sketch the region and show the centroid in your sketch.

**80. a.** Find the center of mass of a thin plate of constant density covering the region between the curve $y = 1/\sqrt{x}$ and the $x$-axis from $x = 1$ to $x = 16$.

 **b.** Find the center of mass if, instead of being constant, the density function is $\delta(x) = 4/\sqrt{x}$.

**81.** Use a derivative to show that $f(x) = \ln (x^3 - 1)$ is one-to-one.

**82.** Use a derivative to show that $g(x) = \sqrt{x^2 + \ln x}$ is one-to-one.

Solve the initial value problems in Exercises 83 and 84.

**83.** $\dfrac{dy}{dx} = 1 + \dfrac{1}{x}$, $\quad y(1) = 3$

**84.** $\dfrac{d^2y}{dx^2} = \sec^2 x$, $\quad y(0) = 0$ $\quad$ and $\quad y'(0) = 1$

**T** **85.** **The linearization of $\ln (1 + x)$ at $x = 0$** Instead of approximating $\ln x$ near $x = 1$, we approximate $\ln (1 + x)$ near $x = 0$. We get a simpler formula this way.

 **a.** Derive the linearization $\ln (1 + x) \approx x$ at $x = 0$.

 **b.** Estimate to five decimal places the error involved in replacing $\ln (1 + x)$ by $x$ on the interval $[0, 0.1]$.

 **c.** Graph $\ln (1 + x)$ and $x$ together for $0 \le x \le 0.5$. Use different colors, if available. At what points does the approximation of $\ln (1 + x)$ seem best? Least good? By reading coordinates from the graphs, find as good an upper bound for the error as your grapher will allow.

**86.** Use the same-derivative argument, as was done to prove Rules 1 and 4 of Theorem 2, to prove the Quotient Rule property of logarithms.

**T** **87. a.** Graph $y = \sin x$ and the curves $y = \ln (a + \sin x)$ for $a = 2, 4, 8, 20$, and $50$ together for $0 \le x \le 23$.

 **b.** Why do the curves flatten as $a$ increases? (*Hint:* Find an $a$-dependent upper bound for $|y'|$.)

**T** **88.** Does the graph of $y = \sqrt{x} - \ln x$, $x > 0$, have an inflection point? Try to answer this question **(a)** by graphing, **(b)** by using calculus.

# $7.3$ Exponential Functions

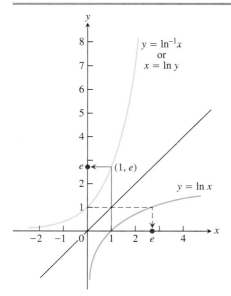

**FIGURE 7.10** The graphs of $y = \ln x$ and $y = \ln^{-1} x = \exp x$. The number $e$ is $\ln^{-1} 1 = \exp (1)$.

Having developed the theory of the function $\ln x$, we introduce its inverse, the exponential function $\exp x = e^x$. We study its properties and compute its derivative and integral. We prove the power rule for derivatives involving general real exponents. Finally, we introduce general exponential functions, $a^x$, and general logarithmic functions, $\log_a x$.

## The Inverse of ln x and the Number e

The function $\ln x$, being an increasing function of $x$ with domain $(0, \infty)$ and range $(-\infty, \infty)$, has an inverse $\ln^{-1} x$ with domain $(-\infty, \infty)$ and range $(0, \infty)$. The graph of $\ln^{-1} x$ is the graph of $\ln x$ reflected across the line $y = x$. As you can see in Figure 7.10,

$$\lim_{x \to \infty} \ln^{-1} x = \infty \qquad \text{and} \qquad \lim_{x \to -\infty} \ln^{-1} x = 0.$$

The function $\ln^{-1} x$ is usually denoted as $\exp x$. We now show that $\exp x$ is an exponential function with base $e$.

The number $e$ was defined to satisfy the equation $\ln (e) = 1$, so $e = \exp (1)$. We can raise the number $e$ to a rational power $r$ in the usual algebraic way:

$$e^2 = e \cdot e, \qquad e^{-2} = \frac{1}{e^2}, \qquad e^{1/2} = \sqrt{e},$$

Typical values of $e^x$

| $x$ | $e^x$ (rounded) |
|---|---|
| $-1$ | 0.37 |
| 0 | 1 |
| 1 | 2.72 |
| 2 | 7.39 |
| 10 | 22026 |
| 100 | $2.6881 \times 10^{43}$ |

and so on. Since $e$ is positive, $e^r$ is positive too, so we can take the logarithm of $e^r$. When we do, we find that for $r$ rational

$$\ln e^r = r \ln e = r \cdot 1 = r. \qquad \text{\small Theorem 2, Rule 4}$$

Then applying the function $\ln^{-1}$ to both sides of the equation $\ln e^r = r$, we find that

$$e^r = \exp r \qquad \text{for } r \text{ rational.} \qquad \text{\small exp is } \ln^{-1}. \qquad (1)$$

We have not yet found a way to give an obvious meaning to $e^x$ for $x$ irrational. But $\ln^{-1} x$ has meaning for any $x$, rational or irrational. So Equation (1) provides a way to extend the definition of $e^x$ to irrational values of $x$. The function $\exp x$ is defined for all $x$, so we use it to assign a value to $e^x$ at every point.

> **DEFINITION** For every real number $x$, we define the **natural exponential function** to be $e^x = \exp x$.

The notations $\ln^{-1} x$, $\exp x$, and $e^x$ all refer to the natural exponential function.

For the first time we have a precise meaning for an irrational exponent—we are raising a specific number $e$ to any real power $x$, rational or irrational. Since the functions $\ln x$ and $e^x$ are inverses of one another, we have the following relationships.

> **Inverse Equations for $e^x$ and $\ln x$**
>
> $$e^{\ln x} = x \qquad (\text{all } x > 0)$$
> $$\ln (e^x) = x \qquad (\text{all } x)$$

**EXAMPLE 1** Solve the equation $e^{2x-6} = 4$ for $x$.

**Solution** We take the natural logarithm of both sides of the equation and use the second inverse equation:

$$\ln (e^{2x-6}) = \ln 4$$
$$2x - 6 = \ln 4 \qquad\qquad \text{\small Inverse relationship}$$
$$2x = 6 + \ln 4$$
$$x = 3 + \frac{1}{2} \ln 4 = 3 + \ln 4^{1/2}$$
$$x = 3 + \ln 2 \qquad\qquad\qquad\qquad\qquad \blacksquare$$

**EXAMPLE 2** A line with slope $m$ passes through the origin and is tangent to the graph of $y = \ln x$. What is the value of $m$?

**Solution** Suppose the point of tangency occurs at the unknown point $x = a > 0$. Then we know that the point $(a, \ln a)$ lies on the graph and that the tangent line at that point has slope $m = 1/a$ (Figure 7.11). Since the tangent line passes through the origin, its slope is

$$m = \frac{\ln a - 0}{a - 0} = \frac{\ln a}{a}.$$

Setting these two formulas for $m$ equal to each other, we have

$$\frac{\ln a}{a} = \frac{1}{a}$$
$$\ln a = 1$$
$$e^{\ln a} = e^1$$
$$a = e$$
$$m = \frac{1}{e}. \qquad\qquad\qquad \blacksquare$$

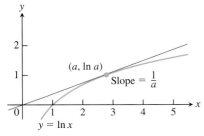

**FIGURE 7.11** The tangent line intersects the curve at some point $(a, \ln a)$, where the slope of the curve is $1/a$ (Example 2).

### The Derivative and Integral of $e^x$

According to Theorem 1, the natural exponential function is differentiable because it is the inverse of a differentiable function whose derivative is never zero. We calculate its derivative using the inverse relationship and Chain Rule:

$$\ln(e^x) = x \qquad \text{Inverse relationship}$$

$$\frac{d}{dx}\ln(e^x) = 1 \qquad \text{Differentiate both sides.}$$

$$\frac{1}{e^x} \cdot \frac{d}{dx}(e^x) = 1 \qquad \text{Eq. (2), Section 7.2, with } u = e^x$$

$$\frac{d}{dx}e^x = e^x. \qquad \text{Solve for the derivative.}$$

That is, for $y = e^x$, we find that $dy/dx = e^x$ so the natural exponential function $e^x$ is its own derivative. Moreover, if $f(x) = e^x$, then $f'(0) = e^0 = 1$. This means that the natural exponential function $e^x$ has slope 1 as it crosses the $y$-axis at $x = 0$.

The Chain Rule extends the derivative result for the natural exponential function to a more general form involving a function $u(x)$:

---

If $u$ is any differentiable function of $x$, then

$$\frac{d}{dx}e^u = e^u\frac{du}{dx}. \qquad\qquad (2)$$

---

**EXAMPLE 3**    We find derivatives of the exponential using Equation (2).

**(a)** $\dfrac{d}{dx}(5e^x) = 5\dfrac{d}{dx}e^x = 5e^x$

**(b)** $\dfrac{d}{dx}e^{-x} = e^{-x}\dfrac{d}{dx}(-x) = e^{-x}(-1) = -e^{-x} \qquad \text{Eq. (2) with } u = -x$

**(c)** $\dfrac{d}{dx}e^{\sin x} = e^{\sin x}\dfrac{d}{dx}(\sin x) = e^{\sin x} \cdot \cos x \qquad \text{Eq. (2) with } u = \sin x$

**(d)** $\dfrac{d}{dx}\left(e^{\sqrt{3x+1}}\right) = e^{\sqrt{3x+1}} \cdot \dfrac{d}{dx}\left(\sqrt{3x+1}\right) \qquad \text{Eq. (2) with } u = \sqrt{3x+1}$

$$= e^{\sqrt{3x+1}} \cdot \frac{1}{2}(3x+1)^{-1/2} \cdot 3 = \frac{3}{2\sqrt{3x+1}}e^{\sqrt{3x+1}} \qquad \blacksquare$$

Since $e^x$ is its own derivative, it is also its own antiderivative. So the integral equivalent of Equation (2) is the following.

---

**The general antiderivative of the exponential function**

$$\int e^u \, du = e^u + C$$

---

**EXAMPLE 4**

**(a)** $\displaystyle\int_0^{\ln 2} e^{3x}\, dx = \int_0^{\ln 8} e^u \cdot \frac{1}{3}\, du \qquad \begin{array}{l} u = 3x, \ \frac{1}{3}du = dx, \ u(0) = 0, \\ u(\ln 2) = 3\ln 2 = \ln 2^3 = \ln 8 \end{array}$

$$= \frac{1}{3}\int_0^{\ln 8} e^u\, du$$

$$= \frac{1}{3} e^u \bigg]_0^{\ln 8}$$

$$= \frac{1}{3}(8 - 1) = \frac{7}{3}$$

**(b)** $\displaystyle\int_0^{\pi/2} e^{\sin x} \cos x \, dx = e^{\sin x} \bigg]_0^{\pi/2}$    Antiderivative from Example 2c

$$= e^1 - e^0 = e - 1$$ ■

The derivative of $e^x$ exists and is everywhere positive, confirming that it is a continuous and increasing function as shown in Figure 7.10. Since the second derivative of $e^x$ is also $e^x$ and everywhere positive, the graph is concave up. Moreover, Figure 7.10 shows that the exponential function has the limits

$$\lim_{x \to -\infty} e^x = 0 \qquad \text{and} \qquad \lim_{x \to \infty} e^x = \infty.$$

From the first of these limits we see that the $x$-axis is a horizontal asymptote of the graph $y = e^x$.

## Laws of Exponents

Even though $e^x$ is defined in a seemingly roundabout way as $\ln^{-1} x$, it obeys the familiar laws of exponents from algebra. The following Theorem 3 shows us that these laws are consequences of the definitions of $\ln x$ and $e^x$.

> **THEOREM 3**    For all numbers $x$, $x_1$, and $x_2$, the natural exponential $e^x$ obeys the following laws:
>
> **1.** $e^{x_1} \cdot e^{x_2} = e^{x_1 + x_2}$    **2.** $e^{-x} = \dfrac{1}{e^x}$
>
> **3.** $\dfrac{e^{x_1}}{e^{x_2}} = e^{x_1 - x_2}$    **4.** $(e^{x_1})^r = e^{rx_1}$,    if $r$ is rational

**Transcendental Numbers and Transcendental Functions**

Numbers that are solutions of polynomial equations with rational coefficients are called **algebraic**: $-2$ is algebraic because it satisfies the equation $x + 2 = 0$, and $\sqrt{3}$ is algebraic because it satisfies the equation $x^2 - 3 = 0$. Numbers such as $e$ and $\pi$ that are not algebraic are called **transcendental**.

We call a function $y = f(x)$ algebraic if it satisfies an equation of the form

$$P_n y^n + \cdots + P_1 y + P_0 = 0$$

in which the $P$'s are polynomials in $x$ with rational coefficients. The function $y = 1/\sqrt{x + 1}$ is algebraic because it satisfies the equation $(x + 1)y^2 - 1 = 0$. Here the polynomials are $P_2 = x + 1$, $P_1 = 0$, and $P_0 = -1$. Functions that are not algebraic are called transcendental.

**Proof of Law 1**    Let $y_1 = e^{x_1}$ and $y_2 = e^{x_2}$. Then

$$x_1 = \ln y_1 \quad \text{and} \quad x_2 = \ln y_2 \qquad \text{Inverse equations}$$

$$x_1 + x_2 = \ln y_1 + \ln y_2$$

$$= \ln y_1 y_2 \qquad \text{Product Rule for logarithms}$$

$$e^{x_1 + x_2} = e^{\ln y_1 y_2} \qquad \text{Exponentiate.}$$

$$= y_1 y_2 \qquad \text{$e^{\ln u} = u$}$$

$$= e^{x_1} e^{x_2}. \qquad ■$$

**Proof of Law 4**    Let $y = (e^{x_1})^r$. Then

$$\ln y = \ln (e^{x_1})^r$$

$$= r \ln (e^{x_1}) \qquad \text{Power Rule for logarithms, rational $r$}$$

$$= rx_1 \qquad \text{$\ln e^u = u$ with $u = x_1$}$$

Thus, exponentiating each side,

$$y = e^{rx_1}. \qquad \text{$e^{\ln y} = y$} \qquad ■$$

Laws 2 and 3 follow from Law 1. Like the Power Rule for logarithms, Law 4 holds for all real numbers $r$.

## The General Exponential Function $a^x$

Since $a = e^{\ln a}$ for any positive number $a$, we can think of $a^x$ as $(e^{\ln a})^x = e^{x \ln a}$. We therefore use the function $e^x$ to define the other exponential functions, which allow us to raise *any* positive number to an irrational exponent.

> **DEFINITION** For any numbers $a > 0$ and $x$, the **exponential function with base $a$** is
> $$a^x = e^{x \ln a}.$$

When $a = e$, the definition gives $a^x = e^{x \ln a} = e^{x \ln e} = e^{x \cdot 1} = e^x$.

Theorem 3 is also valid for $a^x$, the exponential function with base $a$. For example,

$$
\begin{aligned}
a^{x_1} \cdot a^{x_2} &= e^{x_1 \ln a} \cdot e^{x_2 \ln a} && \text{Definition of } a^x \\
&= e^{x_1 \ln a + x_2 \ln a} && \text{Law 1} \\
&= e^{(x_1 + x_2) \ln a} && \text{Factor } \ln a \\
&= a^{x_1 + x_2}. && \text{Definition of } a^x
\end{aligned}
$$

In particular, $a^n \cdot a^{-1} = a^{n-1}$ for any real number $n$.

## Proof of the Power Rule (General Version)

The definition of the general exponential function enables us to make sense of raising any positive number to a real power $n$, rational or irrational. That is, we can define the power function $y = x^n$ for any exponent $n$.

> **DEFINITION** For any $x > 0$ and for any real number $n$,
> $$x^n = e^{n \ln x}.$$

Because the logarithm and exponential functions are inverses of each other, the definition gives

$$\ln x^n = n \ln x, \quad \text{for all real numbers } n.$$

That is, the Power Rule for the natural logarithm holds for *all* real exponents $n$, not just for rational exponents as previously stated in Theorem 2.

The definition of the power function also enables us to establish the derivative Power Rule for any real power $n$, as stated in Section 3.3.

> **General Power Rule for Derivatives**
>
> For $x > 0$ and any real number $n$,
> $$\frac{d}{dx} x^n = nx^{n-1}.$$
>
> If $x \le 0$, then the formula holds whenever the derivative, $x^n$, and $x^{n-1}$ all exist.

**Proof**   Differentiating $x^n$ with respect to $x$ gives

$$\frac{d}{dx} x^n = \frac{d}{dx} e^{n \ln x} \qquad \text{Definition of } x^n, x > 0$$

$$= e^{n \ln x} \cdot \frac{d}{dx} (n \ln x) \qquad \text{Chain Rule for } e^u. \text{ Eq. (2)}$$

$$= x^n \cdot \frac{n}{x} \qquad \text{Definition and derivative of } \ln x$$

$$= nx^{n-1}. \qquad x^n \cdot x^{-1} = x^{n-1}$$

In short, whenever $x > 0$,

$$\frac{d}{dx} x^n = nx^{n-1}.$$

For $x < 0$, if $y = x^n$, $y'$, and $x^{n-1}$ all exist, then

$$\ln |y| = \ln |x|^n = n \ln |x|.$$

Using implicit differentiation (which *assumes* the existence of the derivative $y'$) and Equation (3) in Section 7.2, we have

$$\frac{y'}{y} = \frac{n}{x}.$$

Solving for the derivative,

$$y' = n \frac{y}{x} = n \frac{x^n}{x} = nx^{n-1}.$$

It can be shown directly from the definition of the derivative that the derivative equals 0 when $x = 0$ and $n \geq 1$. This completes the proof of the general version of the Power Rule for all values of $x$. ∎

**EXAMPLE 5**   Differentiate $f(x) = x^x$, $x > 0$.

**Solution**   We cannot apply the power rule here because the exponent is the *variable x* rather than being a constant value $n$ (rational or irrational). However, from the definition of the general exponential function we note that $f(x) = x^x = e^{x \ln x}$, and differentiation gives

$$f'(x) = \frac{d}{dx} (e^{x \ln x})$$

$$= e^{x \ln x} \frac{d}{dx} (x \ln x) \qquad \text{Eq. (2) with } u = x \ln x$$

$$= e^{x \ln x} \left( \ln x + x \cdot \frac{1}{x} \right)$$

$$= x^x (\ln x + 1). \qquad x > 0 \qquad ∎$$

### The Number *e* Expressed as a Limit

We have defined the number $e$ as the number for which $\ln e = 1$, or equivalently, the value exp (1). We see that $e$ is an important constant for the logarithmic and exponential functions, but what is its numerical value? The next theorem shows one way to calculate $e$ as a limit.

---

**THEOREM 4—The Number *e* as a Limit**   The number $e$ can be calculated as the limit

$$e = \lim_{x \to 0} (1 + x)^{1/x}.$$

---

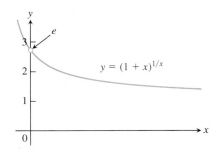

**FIGURE 7.12** The number $e$ is the limit of the function graphed here as $x \to 0$.

**Proof** If $f(x) = \ln x$, then $f'(x) = 1/x$, so $f'(1) = 1$. But, by the definition of derivative,

$$f'(1) = \lim_{h \to 0} \frac{f(1 + h) - f(1)}{h} = \lim_{x \to 0} \frac{f(1 + x) - f(1)}{x}$$

$$= \lim_{x \to 0} \frac{\ln (1 + x) - \ln 1}{x} = \lim_{x \to 0} \frac{1}{x} \ln (1 + x) \qquad \ln 1 = 0$$

$$= \lim_{x \to 0} \ln (1 + x)^{1/x} = \ln \left[ \lim_{x \to 0} (1 + x)^{1/x} \right]. \qquad \begin{array}{l} \text{ln is continuous;} \\ \text{use Theorem 10} \\ \text{in Chapter 2.} \end{array}$$

Because $f'(1) = 1$, we have

$$\ln \left[ \lim_{x \to 0} (1 + x)^{1/x} \right] = 1$$

Therefore, exponentiating both sides we get (see Figure 7.12)

$$\lim_{x \to 0} (1 + x)^{1/x} = e. \qquad \blacksquare$$

Approximating the limit in Theorem 4 by taking $x$ very small gives approximations to $e$. Its value is $e \approx 2.718281828459045$ to 15 decimal places as noted before.

### The Derivative of $a^u$

To find this derivative, we start with the defining equation $a^x = e^{x \ln a}$. Then we have

$$\frac{d}{dx} a^x = \frac{d}{dx} e^{x \ln a} = e^{x \ln a} \cdot \frac{d}{dx} (x \ln a) \qquad \frac{d}{dx} e^u = e^u \frac{du}{dx}$$

$$= a^x \ln a.$$

We now see why $e^x$ is the exponential function preferred in calculus. If $a = e$, then $\ln a = 1$ and the derivative of $a^x$ simplifies to

$$\frac{d}{dx} e^x = e^x \ln e = e^x.$$

With the Chain Rule, we get the following form for the derivative of the general exponential function.

---

If $a > 0$ and $u$ is a differentiable function of $x$, then $a^u$ is a differentiable function of $x$ and

$$\frac{d}{dx} a^u = a^u \ln a \frac{du}{dx}. \qquad (3)$$

---

The integral equivalent of this last result gives the general antiderivative

---

$$\int a^u \, du = \frac{a^u}{\ln a} + C. \qquad (4)$$

---

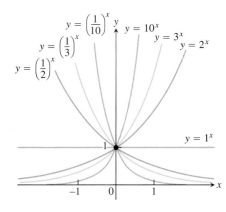

**FIGURE 7.13** Exponential functions decrease if $0 < a < 1$ and increase if $a > 1$. As $x \to \infty$, we have $a^x \to 0$ if $0 < a < 1$ and $a^x \to \infty$ if $a > 1$. As $x \to -\infty$, we have $a^x \to \infty$ if $0 < a < 1$ and $a^x \to 0$ if $a > 1$.

From Equation (3) with $u = x$, we see that the derivative of $a^x$ is positive if $\ln a > 0$, or $a > 1$, and negative if $\ln a < 0$, or $0 < a < 1$. Thus, $a^x$ is an increasing function of $x$ if $a > 1$ and a decreasing function of $x$ if $0 < a < 1$. In each case, $a^x$ is one-to-one. The second derivative

$$\frac{d^2}{dx^2} (a^x) = \frac{d}{dx} (a^x \ln a) = (\ln a)^2 \, a^x$$

is positive for all $x$, so the graph of $a^x$ is concave up on every interval of the real line. Figure 7.13 displays the graphs of several exponential functions.

**EXAMPLE 6**    We find derivatives and integrals using Equations (3) and (4).

**(a)** $\dfrac{d}{dx}3^x = 3^x \ln 3$                                      Eq. (3) with $a = 3$, $u = x$

**(b)** $\dfrac{d}{dx}3^{-x} = 3^{-x}(\ln 3)\dfrac{d}{dx}(-x) = -3^{-x}\ln 3$           Eq. (3) with $a = 3$, $u = -x$

**(c)** $\dfrac{d}{dx}3^{\sin x} = 3^{\sin x}(\ln 3)\dfrac{d}{dx}(\sin x) = 3^{\sin x}(\ln 3)\cos x$        $\ldots$, $u = \sin x$

**(d)** $\displaystyle\int 2^x \, dx = \dfrac{2^x}{\ln 2} + C$                            Eq. (4) with $a = 2$, $u = x$

**(e)** $\displaystyle\int 2^{\sin x} \cos x \, dx = \int 2^u \, du = \dfrac{2^u}{\ln 2} + C$        $u = \sin x$, $du = \cos x \, dx$, and Eq. (4)

$\qquad\qquad = \dfrac{2^{\sin x}}{\ln 2} + C$                            $u$ replaced by $\sin x$    ∎

## Logarithms with Base $a$

If $a$ is any positive number other than 1, the function $a^x$ is one-to-one and has a nonzero derivative at every point. It therefore has a differentiable inverse. We call the inverse the **logarithm of $x$ with base $a$** and denote it by **$\log_a x$**.

---

**DEFINITION**    For any positive number $a \neq 1$,

$$\log_a x \text{ is the inverse function of } a^x.$$

---

The graph of $y = \log_a x$ can be obtained by reflecting the graph of $y = a^x$ across the 45° line $y = x$ (Figure 7.14). When $a = e$, we have $\log_e x = $ inverse of $e^x = \ln x$. (The function $\log_{10} x$ is sometimes written simply as $\log x$ and is called the **common logarithm** of $x$.) Since $\log_a x$ and $a^x$ are inverses of one another, composing them in either order gives the identity function.

**FIGURE 7.14**    The graph of $2^x$ and its inverse, $\log_2 x$.

---

**Inverse Equations for $a^x$ and $\log_a x$**

$$a^{\log_a x} = x \qquad (x > 0)$$

$$\log_a(a^x) = x \qquad (\text{all } x)$$

---

The function $\log_a x$ is actually just a numerical multiple of $\ln x$. To see this, we let $y = \log_a x$ and then take the natural logarithm of both sides of the equivalent equation $a^y = x$ to obtain $y \ln a = \ln x$. Solving for $y$ gives

---

$$\log_a x = \frac{\ln x}{\ln a}. \tag{5}$$

---

The algebraic rules satisfied by $\log_a x$ are the same as the ones for $\ln x$. These rules, given in Table 7.2, can be proved using Equation (5) by dividing the corresponding rules for the natural logarithm function by $\ln a$. For example,

$$\ln xy = \ln x + \ln y \qquad \text{Rule 1 for natural logarithms} \ldots$$

$$\frac{\ln xy}{\ln a} = \frac{\ln x}{\ln a} + \frac{\ln y}{\ln a} \qquad \ldots \text{divided by } \ln a \ldots$$

$$\log_a xy = \log_a x + \log_a y. \qquad \ldots \text{gives Rule 1 for base } a \text{ logarithms.}$$

---

**TABLE 7.2** Rules for base $a$ logarithms

For any numbers $x > 0$ and $y > 0$,

1. *Product Rule:*

   $\log_a xy = \log_a x + \log_a y$

2. *Quotient Rule:*

   $\log_a \dfrac{x}{y} = \log_a x - \log_a y$

3. *Reciprocal Rule:*

   $\log_a \dfrac{1}{y} = -\log_a y$

4. *Power Rule:*

   $\log_a x^y = y \log_a x$

## Derivatives and Integrals Involving $\log_a x$

To find derivatives or integrals involving base $a$ logarithms, we convert them to natural logarithms. If $u$ is a positive differentiable function of $x$, then

$$\frac{d}{dx}(\log_a u) = \frac{d}{dx}\left(\frac{\ln u}{\ln a}\right) = \frac{1}{\ln a}\frac{d}{dx}(\ln u) = \frac{1}{\ln a}\cdot\frac{1}{u}\frac{du}{dx}.$$

$$\frac{d}{dx}(\log_a u) = \frac{1}{\ln a}\cdot\frac{1}{u}\frac{du}{dx}$$

### EXAMPLE 7

**(a)** $\dfrac{d}{dx}\log_{10}(3x + 1) = \dfrac{1}{\ln 10}\cdot\dfrac{1}{3x+1}\dfrac{d}{dx}(3x+1) = \dfrac{3}{(\ln 10)(3x+1)}$

**(b)** $\displaystyle\int\frac{\log_2 x}{x}dx = \frac{1}{\ln 2}\int\frac{\ln x}{x}dx$  $\qquad \log_2 x = \frac{\ln x}{\ln 2}$

$\qquad\qquad = \dfrac{1}{\ln 2}\displaystyle\int u\, du$  $\qquad u = \ln x, \quad du = \frac{1}{x}dx$

$\qquad\qquad = \dfrac{1}{\ln 2}\dfrac{u^2}{2} + C = \dfrac{1}{\ln 2}\dfrac{(\ln x)^2}{2} + C = \dfrac{(\ln x)^2}{2\ln 2} + C$ ∎

# Exercises 7.3

### Solving Exponential Equations
In Exercises 1–4, solve for $t$.

**1. a.** $e^{-0.3t} = 27$  **b.** $e^{kt} = \dfrac{1}{2}$  **c.** $e^{(\ln 0.2)t} = 0.4$

**2. a.** $e^{-0.01t} = 1000$  **b.** $e^{kt} = \dfrac{1}{10}$  **c.** $e^{(\ln 2)t} = \dfrac{1}{2}$

**3.** $e^{\sqrt{t}} = x^2$  **4.** $e^{(x^2)}e^{(2x+1)} = e^t$

### Finding Derivatives
In Exercises 5–24, find the derivative of $y$ with respect to $x$, $t$, or $\theta$, as appropriate.

**5.** $y = e^{-5x}$  **6.** $y = e^{2x/3}$

**7.** $y = e^{5-7x}$  **8.** $y = e^{(4\sqrt{x}+x^2)}$

**9.** $y = xe^x - e^x$  **10.** $y = (1 + 2x)e^{-2x}$

**11.** $y = (x^2 - 2x + 2)e^x$  **12.** $y = (9x^2 - 6x + 2)e^{3x}$

**13.** $y = e^\theta(\sin\theta + \cos\theta)$  **14.** $y = \ln(3\theta e^{-\theta})$

**15.** $y = \cos(e^{-\theta^2})$  **16.** $y = \theta^3 e^{-2\theta}\cos 5\theta$

**17.** $y = \ln(3te^{-t})$  **18.** $y = \ln(2e^{-t}\sin t)$

**19.** $y = \ln\left(\dfrac{e^\theta}{1+e^\theta}\right)$  **20.** $y = \ln\left(\dfrac{\sqrt{\theta}}{1+\sqrt{\theta}}\right)$

**21.** $y = e^{(\cos t + \ln t)}$  **22.** $y = e^{\sin t}(\ln t^2 + 1)$

**23.** $y = \displaystyle\int_0^{\ln x}\sin e^t\, dt$  **24.** $y = \displaystyle\int_{e^{4\sqrt{x}}}^{e^{2x}}\ln t\, dt$

In Exercises 25–28, find $dy/dx$.

**25.** $\ln y = e^y\sin x$  **26.** $\ln xy = e^{x+y}$

**27.** $e^{2x} = \sin(x + 3y)$  **28.** $\tan y = e^x + \ln x$

### Finding Integrals
Evaluate the integrals in Exercises 29–50.

**29.** $\displaystyle\int(e^{3x} + 5e^{-x})\, dx$  **30.** $\displaystyle\int(2e^x - 3e^{-2x})\, dx$

**31.** $\displaystyle\int_{\ln 2}^{\ln 3}e^x\, dx$  **32.** $\displaystyle\int_{-\ln 2}^0 e^{-x}\, dx$

**33.** $\displaystyle\int 8e^{(x+1)}\, dx$  **34.** $\displaystyle\int 2e^{(2x-1)}\, dx$

**35.** $\displaystyle\int_{\ln 4}^{\ln 9}e^{x/2}\, dx$  **36.** $\displaystyle\int_0^{\ln 16}e^{x/4}\, dx$

**37.** $\displaystyle\int\frac{e^{\sqrt{r}}}{\sqrt{r}}dr$  **38.** $\displaystyle\int\frac{e^{-\sqrt{r}}}{\sqrt{r}}dr$

**39.** $\displaystyle\int 2t\, e^{-t^2}\, dt$  **40.** $\displaystyle\int t^3 e^{(t^4)}\, dt$

**41.** $\displaystyle\int\frac{e^{1/x}}{x^2}dx$  **42.** $\displaystyle\int\frac{e^{-1/x^2}}{x^3}dx$

**43.** $\displaystyle\int_0^{\pi/4}(1 + e^{\tan\theta})\sec^2\theta\, d\theta$  **44.** $\displaystyle\int_{\pi/4}^{\pi/2}(1 + e^{\cot\theta})\csc^2\theta\, d\theta$

**45.** $\displaystyle\int e^{\sec \pi t} \sec \pi t \tan \pi t \, dt$

**46.** $\displaystyle\int e^{\csc(\pi + t)} \csc(\pi + t) \cot(\pi + t) \, dt$

**47.** $\displaystyle\int_{\ln(\pi/6)}^{\ln(\pi/2)} 2e^v \cos e^v \, dv$    **48.** $\displaystyle\int_0^{\sqrt{\ln \pi}} 2x \, e^{x^2} \cos(e^{x^2}) \, dx$

**49.** $\displaystyle\int \frac{e^r}{1 + e^r} \, dr$    **50.** $\displaystyle\int \frac{dx}{1 + e^x}$

### Initial Value Problems
Solve the initial value problems in Exercises 51–54.

**51.** $\dfrac{dy}{dt} = e^t \sin(e^t - 2), \quad y(\ln 2) = 0$

**52.** $\dfrac{dy}{dt} = e^{-t} \sec^2(\pi e^{-t}), \quad y(\ln 4) = 2/\pi$

**53.** $\dfrac{d^2y}{dx^2} = 2e^{-x}, \quad y(0) = 1 \quad \text{and} \quad y'(0) = 0$

**54.** $\dfrac{d^2y}{dt^2} = 1 - e^{2t}, \quad y(1) = -1 \quad \text{and} \quad y'(1) = 0$

### Differentiation
In Exercises 55–82, find the derivative of $y$ with respect to the given independent variable.

**55.** $y = 2^x$    **56.** $y = 3^{-x}$

**57.** $y = 5^{\sqrt{s}}$    **58.** $y = 2^{(s^2)}$

**59.** $y = x^\pi$    **60.** $y = t^{1-e}$

**61.** $y = (\cos \theta)^{\sqrt{2}}$    **62.** $y = (\ln \theta)^\pi$

**63.** $y = 7^{\sec \theta} \ln 7$    **64.** $y = 3^{\tan \theta} \ln 3$

**65.** $y = 2^{\sin 3t}$    **66.** $y = 5^{-\cos 2t}$

**67.** $y = \log_2 5\theta$    **68.** $y = \log_3(1 + \theta \ln 3)$

**69.** $y = \log_4 x + \log_4 x^2$    **70.** $y = \log_{25} e^x - \log_5 \sqrt{x}$

**71.** $y = x^3 \log_{10} x$    **72.** $y = \log_3 r \cdot \log_9 r$

**73.** $y = \log_3\left(\left(\dfrac{x+1}{x-1}\right)^{\ln 3}\right)$    **74.** $y = \log_5 \sqrt{\left(\dfrac{7x}{3x+2}\right)^{\ln 5}}$

**75.** $y = \theta \sin(\log_7 \theta)$    **76.** $y = \log_7\left(\dfrac{\sin \theta \cos \theta}{e^\theta \, 2^\theta}\right)$

**77.** $y = \log_{10} e^x$    **78.** $y = \dfrac{\theta 5^\theta}{2 - \log_5 \theta}$

**79.** $y = 3^{\log_2 t}$    **80.** $y = 3 \log_8(\log_2 t)$

**81.** $y = \log_2(8t^{\ln 2})$    **82.** $y = t \log_3\left(e^{(\sin t)(\ln 3)}\right)$

### Integration
Evaluate the integrals in Exercises 83–92.

**83.** $\displaystyle\int 5^x \, dx$    **84.** $\displaystyle\int \frac{3^x}{3 - 3^x} \, dx$

**85.** $\displaystyle\int_0^1 2^{-\theta} \, d\theta$    **86.** $\displaystyle\int_{-2}^0 5^{-\theta} \, d\theta$

**87.** $\displaystyle\int_1^{\sqrt{2}} x 2^{(x^2)} \, dx$    **88.** $\displaystyle\int_1^4 \frac{2^{\sqrt{x}}}{\sqrt{x}} \, dx$

**89.** $\displaystyle\int_0^{\pi/2} 7^{\cos t} \sin t \, dt$    **90.** $\displaystyle\int_0^{\pi/4} \left(\frac{1}{3}\right)^{\tan t} \sec^2 t \, dt$

**91.** $\displaystyle\int_2^4 x^{2x}(1 + \ln x) \, dx$    **92.** $\displaystyle\int \frac{x 2^{x^2}}{1 + 2^{x^2}} \, dx$

Evaluate the integrals in Exercises 93–106.

**93.** $\displaystyle\int 3x^{\sqrt{3}} \, dx$    **94.** $\displaystyle\int x^{\sqrt{2}-1} \, dx$

**95.** $\displaystyle\int_0^3 (\sqrt{2} + 1)x^{\sqrt{2}} \, dx$    **96.** $\displaystyle\int_1^e x^{(\ln 2)-1} \, dx$

**97.** $\displaystyle\int \frac{\log_{10} x}{x} \, dx$    **98.** $\displaystyle\int_1^4 \frac{\log_2 x}{x} \, dx$

**99.** $\displaystyle\int_1^4 \frac{\ln 2 \log_2 x}{x} \, dx$    **100.** $\displaystyle\int_1^e \frac{2 \ln 10 \log_{10} x}{x} \, dx$

**101.** $\displaystyle\int_0^2 \frac{\log_2(x+2)}{x+2} \, dx$    **102.** $\displaystyle\int_{1/10}^{10} \frac{\log_{10}(10x)}{x} \, dx$

**103.** $\displaystyle\int_0^9 \frac{2 \log_{10}(x+1)}{x+1} \, dx$    **104.** $\displaystyle\int_2^3 \frac{2 \log_2(x-1)}{x-1} \, dx$

**105.** $\displaystyle\int \frac{dx}{x \log_{10} x}$    **106.** $\displaystyle\int \frac{dx}{x(\log_8 x)^2}$

Evaluate the integrals in Exercises 107–110.

**107.** $\displaystyle\int_1^{\ln x} \frac{1}{t} \, dt, \quad x > 1$    **108.** $\displaystyle\int_1^{e^x} \frac{1}{t} \, dt$

**109.** $\displaystyle\int_1^{1/x} \frac{1}{t} \, dt, \quad x > 0$    **110.** $\dfrac{1}{\ln a} \displaystyle\int_1^x \frac{1}{t} \, dt, \quad x > 0$

### Logarithmic Differentiation
In Exercises 111–118, use logarithmic differentiation to find the derivative of $y$ with respect to the given independent variable.

**111.** $y = (x + 1)^x$    **112.** $y = x^2 + x^{2x}$

**113.** $y = (\sqrt{t})^t$    **114.** $y = t^{\sqrt{t}}$

**115.** $y = (\sin x)^x$    **116.** $y = x^{\sin x}$

**117.** $y = \sin x^x$    **118.** $y = (\ln x)^{\ln x}$

### Theory and Applications
**119.** Find the absolute maximum and minimum values of $f(x) = e^x - 2x$ on $[0, 1]$.

**120.** Where does the periodic function $f(x) = 2e^{\sin(x/2)}$ take on its extreme values and what are these values?

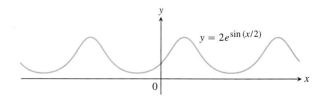
$y = 2e^{\sin(x/2)}$

**121.** Let $f(x) = xe^{-x}$.

    **a.** Find all absolute extreme values for $f$.

    **b.** Find all inflection points for $f$.

**122.** Let $f(x) = \dfrac{e^x}{1 + e^{2x}}$.

    **a.** Find all absolute extreme values for $f$.

    **b.** Find all inflection points for $f$.

**123.** Find the absolute maximum value of $f(x) = x^2 \ln (1/x)$ and say where it is assumed.

**T** **124.** Graph $f(x) = (x - 3)^2 e^x$ and its first derivative together. Comment on the behavior of $f$ in relation to the signs and values of $f'$. Identify significant points on the graphs with calculus, as necessary.

**125.** Find the area of the "triangular" region in the first quadrant that is bounded above by the curve $y = e^{2x}$, below by the curve $y = e^x$, and on the right by the line $x = \ln 3$.

**126.** Find the area of the "triangular" region in the first quadrant that is bounded above by the curve $y = e^{x/2}$, below by the curve $y = e^{-x/2}$, and on the right by the line $x = 2 \ln 2$.

**127.** Find a curve through the origin in the $xy$-plane whose length from $x = 0$ to $x = 1$ is

$$L = \int_0^1 \sqrt{1 + \frac{1}{4}e^x}\, dx.$$

**128.** Find the area of the surface generated by revolving the curve $x = (e^y + e^{-y})/2, 0 \le y \le \ln 2$, about the $y$-axis.

In Exercises 129–132, find the length of each curve.

**129.** $y = \dfrac{1}{2}(e^x + e^{-x})$ from $x = 0$ to $x = 1$

**130.** $y = \ln (e^x - 1) - \ln (e^x + 1)$ from $x = \ln 2$ to $x = \ln 3$

**131.** $y = \ln (\cos x)$ from $x = 0$ to $x = \pi/4$

**132.** $y = \ln (\csc x)$ from $x = \pi/6$ to $x = \pi/4$

**133. a.** Show that $\displaystyle\int \ln x\, dx = x \ln x - x + C$.

    **b.** Find the average value of $\ln x$ over $[1, e]$.

**134.** Find the average value of $f(x) = 1/x$ on $[1, 2]$.

**135. The linearization of $e^x$ at $x = 0$**

    **a.** Derive the linear approximation $e^x \approx 1 + x$ at $x = 0$.

    **T** **b.** Estimate to five decimal places the magnitude of the error involved in replacing $e^x$ by $1 + x$ on the interval $[0, 0.2]$.

    **T** **c.** Graph $e^x$ and $1 + x$ together for $-2 \le x \le 2$. Use different colors, if available. On what intervals does the approximation appear to overestimate $e^x$? Underestimate $e^x$?

**136. The geometric, logarithmic, and arithmetic mean inequality**

    **a.** Show that the graph of $e^x$ is concave up over every interval of $x$-values.

    **b.** Show, by reference to the accompanying figure, that if $0 < a < b$, then

$$e^{(\ln a + \ln b)/2} \cdot (\ln b - \ln a) < \int_{\ln a}^{\ln b} e^x\, dx < \frac{e^{\ln a} + e^{\ln b}}{2} \cdot (\ln b - \ln a).$$

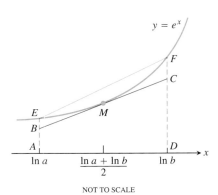

NOT TO SCALE

    **c.** Use the inequality in part (b) to conclude that

$$\sqrt{ab} < \frac{b - a}{\ln b - \ln a} < \frac{a + b}{2}.$$

    This inequality says that the geometric mean of two positive numbers is less than their logarithmic mean, which in turn is less than their arithmetic mean.

**137.** Find the area of the region between the curve $y = 2x/(1 + x^2)$ and the interval $-2 \le x \le 2$ of the $x$-axis.

**138.** Find the area of the region between the curve $y = 2^{1-x}$ and the interval $-1 \le x \le 1$ of the $x$-axis.

**T** **139.** The equation $x^2 = 2^x$ has three solutions: $x = 2, x = 4$, and one other. Estimate the third solution as accurately as you can by graphing.

**T** **140.** Could $x^{\ln 2}$ possibly be the same as $2^{\ln x}$ for $x > 0$? Graph the two functions and explain what you see.

**141. The linearization of $2^x$**

    **a.** Find the linearization of $f(x) = 2^x$ at $x = 0$. Then round its coefficients to two decimal places.

    **T** **b.** Graph the linearization and function together for $-3 \le x \le 3$ and $-1 \le x \le 1$.

**142. The linearization of $\log_3 x$**

    **a.** Find the linearization of $f(x) = \log_3 x$ at $x = 3$. Then round its coefficients to two decimal places.

    **T** **b.** Graph the linearization and function together in the windows $0 \le x \le 8$ and $2 \le x \le 4$.

**T** **143. Which is bigger, $\pi^e$ or $e^\pi$?** Calculators have taken some of the mystery out of this once-challenging question. (Go ahead and check; you will see that it is a very close call.) You can answer the question without a calculator, though.

**a.** Find an equation for the line through the origin tangent to the graph of $y = \ln x$.

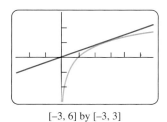

[–3, 6] by [–3, 3]

**b.** Give an argument based on the graphs of $y = \ln x$ and the tangent line to explain why $\ln x < x/e$ for all positive $x \neq e$.

**c.** Show that $\ln(x^e) < x$ for all positive $x \neq e$.

**d.** Conclude that $x^e < e^x$ for all positive $x \neq e$.

**e.** So which is bigger, $\pi^e$ or $e^\pi$?

T **144. A decimal representation of $e$**  Find $e$ to as many decimal places as your calculator allows by solving the equation $\ln x = 1$ using Newton's method in Section 4.6.

# 7.4 Exponential Change and Separable Differential Equations

Exponential functions increase or decrease very rapidly with changes in the independent variable. They describe growth or decay in many natural and human-made situations. The variety of models based on these functions partly accounts for their importance. We now investigate the basic proportionality assumption that leads to such *exponential change.*

## Exponential Change

In modeling many real-world situations, a quantity $y$ increases or decreases at a rate proportional to its size at a given time $t$. Examples of such quantities include the size of a population, the amount of a decaying radioactive material, and the temperature difference between a hot object and its surrounding medium. Such quantities are said to undergo **exponential change**.

If the amount present at time $t = 0$ is called $y_0$, then we can find $y$ as a function of $t$ by solving the following initial value problem:

$$\text{Differential equation:} \quad \frac{dy}{dt} = ky \tag{1a}$$

$$\text{Initial condition:} \quad y = y_0 \quad \text{when} \quad t = 0. \tag{1b}$$

If $y$ is positive and increasing, then $k$ is positive, and we use Equation (1a) to say that the rate of growth is proportional to what has already been accumulated. If $y$ is positive and decreasing, then $k$ is negative, and we use Equation (1a) to say that the rate of decay is proportional to the amount still left.

We see right away that the constant function $y = 0$ is a solution of Equation (1a) if $y_0 = 0$. To find the nonzero solutions, we divide Equation (1a) by $y$:

$$\frac{1}{y} \cdot \frac{dy}{dt} = k \qquad y \neq 0$$

$$\int \frac{1}{y} \frac{dy}{dt} \, dt = \int k \, dt \qquad \text{Integrate with respect to } t; \\ \int (1/u) \, du = \ln |u| + C.$$

$$\ln |y| = kt + C$$

$$|y| = e^{kt+C} \qquad \text{Exponentiate.}$$

$$|y| = e^C \cdot e^{kt} \qquad e^{a+b} = e^a \cdot e^b$$

$$y = \pm e^C e^{kt} \qquad \text{If } |y| = r, \text{ then } y = \pm r.$$

$$y = Ae^{kt}. \qquad A \text{ is a shorter name for } \pm e^C.$$

By allowing $A$ to take on the value 0 in addition to all possible values $\pm e^C$, we can include the solution $y = 0$ in the formula.

We find the value of $A$ for the initial value problem by solving for $A$ when $y = y_0$ and $t = 0$:

$$y_0 = Ae^{k \cdot 0} = A.$$

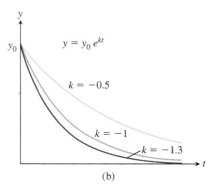

**FIGURE 7.15** Graphs of (a) exponential growth and (b) exponential decay. As $|k|$ increases, the growth ($k > 0$) or decay ($k < 0$) intensifies.

The solution of the initial value problem

$$\frac{dy}{dt} = ky, \qquad y(0) = y_0$$

is

$$y = y_0 e^{kt}. \qquad (2)$$

Quantities changing in this way are said to undergo **exponential growth** if $k > 0$ and **exponential decay** if $k < 0$. The number $k$ is called the **rate constant** of the change. (See Figure 7.15.)

The derivation of Equation (2) shows also that the only functions that are their own derivatives (so $k = 1$) are constant multiples of the exponential function.

Before presenting several examples of exponential change, let's consider the process we used to derive it.

### Separable Differential Equations

Exponential change is modeled by a differential equation of the form $dy/dx = ky$ for some nonzero constant $k$. More generally, suppose we have a differential equation of the form

$$\frac{dy}{dx} = f(x, y), \qquad (3)$$

where $f$ is a function of *both* the independent and dependent variables. A **solution** of the equation is a differentiable function $y = y(x)$ defined on an interval of $x$-values (perhaps infinite) such that

$$\frac{d}{dx} y(x) = f(x, y(x))$$

on that interval. That is, when $y(x)$ and its derivative $y'(x)$ are substituted into the differential equation, the resulting equation is true for all $x$ in the solution interval. The **general solution** is a solution $y(x)$ that contains all possible solutions and it always contains an arbitrary constant.

Equation (3) is **separable** if $f$ can be expressed as a product of a function of $x$ and a function of $y$. The differential equation then has the form

$$\frac{dy}{dx} = g(x)H(y). \qquad \begin{array}{l} g \text{ is a function of } x; \\ H \text{ is a function of } y. \end{array}$$

When we rewrite this equation in the form

$$\frac{dy}{dx} = \frac{g(x)}{h(y)}, \qquad H(y) = \frac{1}{h(y)}$$

its differential form allows us to collect all $y$ terms with $dy$ and all $x$ terms with $dx$:

$$h(y)\, dy = g(x)\, dx.$$

Now we simply integrate both sides of this equation:

$$\int h(y)\, dy = \int g(x)\, dx. \qquad (4)$$

After completing the integrations, we obtain the solution $y$ defined implicitly as a function of $x$.

The justification that we can simply integrate both sides in Equation (4) is based on the Substitution Rule (Section 5.5):

$$\int h(y) \, dy = \int h(y(x)) \frac{dy}{dx} dx$$

$$= \int h(y(x)) \frac{g(x)}{h(y(x))} dx \qquad \frac{dy}{dx} = \frac{g(x)}{h(y)}$$

$$= \int g(x) \, dx.$$

**EXAMPLE 1**    Solve the differential equation

$$\frac{dy}{dx} = (1 + y)e^x, \quad y > -1.$$

**Solution**    Since $1 + y$ is never zero for $y > -1$, we can solve the equation by separating the variables.

$$\frac{dy}{dx} = (1 + y)e^x$$

$$dy = (1 + y)e^x \, dx \qquad \text{Treat } dy/dx \text{ as a quotient of differentials and multiply both sides by } dx.$$

$$\frac{dy}{1 + y} = e^x \, dx \qquad \text{Divide by } (1 + y).$$

$$\int \frac{dy}{1 + y} = \int e^x \, dx \qquad \text{Integrate both sides.}$$

$$\ln(1 + y) = e^x + C \qquad \begin{array}{l} C \text{ represents the combined} \\ \text{constants of integration.} \end{array}$$

The last equation gives $y$ as an implicit function of $x$.    ■

**EXAMPLE 2**    Solve the equation $y(x + 1) \dfrac{dy}{dx} = x(y^2 + 1)$.

**Solution**    We change to differential form, separate the variables, and integrate:

$$y(x + 1) \, dy = x(y^2 + 1) \, dx$$

$$\frac{y \, dy}{y^2 + 1} = \frac{x \, dx}{x + 1} \qquad x \neq -1$$

$$\int \frac{y \, dy}{1 + y^2} = \int \left(1 - \frac{1}{x + 1}\right) dx \qquad \text{Divide } x \text{ by } x + 1.$$

$$\frac{1}{2}\ln(1 + y^2) = x - \ln|x + 1| + C.$$

The last equation gives the solution y as an implicit function of x.    ■

The initial value problem

$$\frac{dy}{dt} = ky, \qquad y(0) = y_0$$

involves a separable differential equation, and the solution $y = y_0 e^{kt}$ expresses exponential change. We now present several examples of such change.

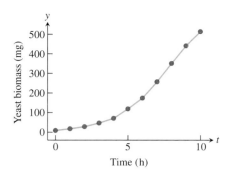

**FIGURE 7.16** Graph of the growth of a yeast population over a 10-hour period, based on the data in Example 3.

| Time (h) | Yeast biomass (mg) |
|---|---|
| 0 | 9.6 |
| 1 | 18.3 |
| 2 | 29.0 |
| 3 | 47.2 |
| 4 | 71.1 |
| 5 | 119.1 |
| 6 | 174.6 |
| 7 | 257.3 |
| 8 | 350.7 |
| 9 | 441.0 |
| 10 | 513.3 |

## Unlimited Population Growth

Strictly speaking, the number of individuals in a population (of people, plants, animals, or bacteria, for example) is a discontinuous function of time because it takes on discrete values. However, when the number of individuals becomes large enough, the population can be approximated by a continuous function. Differentiability of the approximating function is another reasonable hypothesis in many settings, allowing for the use of calculus to model and predict population sizes.

If we assume that the proportion of reproducing individuals remains constant and assume a constant fertility, then at any instant $t$ the birth rate is proportional to the number $y(t)$ of individuals present. Let's assume, too, that the death rate of the population is stable and proportional to $y(t)$. If, further, we neglect departures and arrivals, the growth rate $dy/dt$ is the birth rate minus the death rate, which is the difference of the two proportionalities under our assumptions. In other words, $dy/dt = ky$ so that $y = y_0 e^{kt}$, where $y_0$ is the size of the population at time $t = 0$. As with all kinds of growth, there may be limitations imposed by the surrounding environment, but we will not go into these here. (We treat one model imposing such limitations in Section 9.4.) When $k$ is positive, the proportionality $dy/dt = ky$ models *unlimited population growth*. (See Figure 7.16.)

**EXAMPLE 3** The biomass of a yeast culture in an experiment is initially 29 grams. After 30 minutes the mass is 37 grams. Assuming that the equation for unlimited population growth gives a good model for the growth of the yeast when the mass is below 100 grams, how long will its take for the mass to double from its initial value?

**Solution** Let $y(t)$ be the yeast biomass after $t$ minutes. We use the exponential growth model $dy/dt = ky$ for unlimited population growth, with solution $y = y_0 e^{kt}$.

We have $y_0 = y(0) = 29$. We are also told that

$$y(30) = 29e^{k(30)} = 37.$$

Solving this equation for $k$, we find

$$e^{k(30)} = \frac{37}{29}$$

$$30k = \ln\left(\frac{37}{29}\right)$$

$$k = \frac{1}{30}\ln\left(\frac{37}{29}\right) \approx 0.008118.$$

Then the mass of the yeast in grams after $t$ minutes is given by the equation

$$y = 29e^{(0.008118)t}.$$

To solve the problem we find the time $t$ for which $y(t) = 58$, which is twice the initial amount.

$$29e^{(0.008118)t} = 58$$

$$(0.008118)t = \ln\left(\frac{58}{29}\right)$$

$$t = \frac{\ln 2}{0.008118} \approx 85.38$$

It takes about 85 minutes for the yeast population to double. ∎

In the next example we model the number of people within a given population who are infected by a disease which is being eradicated from the population. Here the constant of proportionality $k$ is negative, and the model describes an exponentially decaying number of infected individuals.

**EXAMPLE 4**    One model for the way diseases die out when properly treated assumes that the rate $dy/dt$ at which the number of infected people changes is proportional to the number $y$. The number of people cured is proportional to the number $y$ that are infected with the disease. Suppose that in the course of any given year the number of cases of a disease is reduced by 20%. If there are 10,000 cases today, how many years will it take to reduce the number to 1000?

**Solution**    We use the equation $y = y_0 e^{kt}$. There are three things to find: the value of $y_0$, the value of $k$, and the time $t$ when $y = 1000$.

*The value of $y_0$.* We are free to count time beginning anywhere we want. If we count from today, then $y = 10{,}000$ when $t = 0$, so $y_0 = 10{,}000$. Our equation is now

$$y = 10{,}000 e^{kt}. \tag{5}$$

*The value of $k$.* When $t = 1$ year, the number of cases will be 80% of its present value, or 8000. Hence,

$$8000 = 10{,}000 e^{k(1)} \qquad \text{Eq. (5) with } t = 1 \text{ and } y = 8000$$
$$e^k = 0.8$$
$$\ln(e^k) = \ln 0.8 \qquad \text{Logs of both sides}$$
$$k = \ln 0.8 < 0. \qquad \ln 0.8 \approx -0.223$$

At any given time $t$,

$$y = 10{,}000 e^{(\ln 0.8)t}. \tag{6}$$

*The value of $t$ that makes $y = 1000$.* We set $y$ equal to 1000 in Equation (6) and solve for $t$:

$$1000 = 10{,}000 e^{(\ln 0.8)t}$$
$$e^{(\ln 0.8)t} = 0.1$$
$$(\ln 0.8)t = \ln 0.1 \qquad \text{Logs of both sides}$$
$$t = \frac{\ln 0.1}{\ln 0.8} \approx 10.32 \text{ years}.$$

It will take a little more than 10 years to reduce the number of cases to 1000. (See Figure 7.17.)    ■

**FIGURE 7.17**    A graph of the number of people infected by a disease exhibits exponential decay (Example 4).

## Radioactivity

Some atoms are unstable and can spontaneously emit mass or radiation. This process is called **radioactive decay**, and an element whose atoms go spontaneously through this process is called **radioactive**. Sometimes when an atom emits some of its mass through this process of radioactivity, the remainder of the atom re-forms to make an atom of some new element. For example, radioactive carbon-14 decays into nitrogen; radium, through a number of intermediate radioactive steps, decays into lead.

Experiments have shown that at any given time the rate at which a radioactive element decays (as measured by the number of nuclei that change per unit time) is approximately proportional to the number of radioactive nuclei present. Thus, the decay of a radioactive element is described by the equation $dy/dt = -ky, k > 0$. It is conventional to use $-k$, with $k > 0$, to emphasize that $y$ is decreasing. If $y_0$ is the number of radioactive nuclei present at time zero, the number still present at any later time $t$ will be

$$y = y_0 e^{-kt}, \qquad k > 0.$$

For radon-222 gas, $t$ is measured in days and $k = 0.18$. For radium-226, which used to be painted on watch dials to make them glow at night (a dangerous practice), $t$ is measured in years and $k = 4.3 \times 10^{-4}$.

The **half-life** of a radioactive element is the time required for half of the radioactive nuclei present in a sample to decay. It is an interesting fact that the half-life is a constant that does not depend on the number of radioactive nuclei initially present in the sample, but only on the radioactive substance.

To see why, let $y_0$ be the number of radioactive nuclei initially present in the sample. Then the number $y$ present at any later time $t$ will be $y = y_0 e^{-kt}$. We seek the value of $t$ at which the number of radioactive nuclei present equals half the original number:

$$y_0 e^{-kt} = \frac{1}{2} y_0$$

$$e^{-kt} = \frac{1}{2}$$

$$-kt = \ln \frac{1}{2} = -\ln 2 \qquad \text{Reciprocal Rule for logarithms}$$

$$t = \frac{\ln 2}{k}.$$

This value of $t$ is the half-life of the element. It depends only on the value of $k$; the number $y_0$ does not have any effect.

$$\boxed{\text{Half-life} = \frac{\ln 2}{k} \qquad\qquad (7)}$$

The effective radioactive lifetime of polonium-210 is so short that we measure it in days rather than years. The number of radioactive atoms remaining after $t$ days in a sample that starts with $y_0$ radioactive atoms is

$$y = y_0 e^{-5 \times 10^{-3} t}.$$

The element's half-life is

$$\text{Half-life} = \frac{\ln 2}{k} \qquad \text{Eq. (7)}$$

$$= \frac{\ln 2}{5 \times 10^{-3}} \qquad \text{The } k \text{ from polonium's decay equation}$$

$$\approx 139 \text{ days.}$$

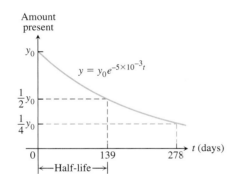

**FIGURE 7.18** Amount of polonium-210 present at time $t$, where $y_0$ represents the number of radioactive atoms initially present.

This means that after 139 days, $1/2$ of $y_0$ radioactive atoms remain; after another 139 days (or 278 days altogether) half of those remain, or $1/4$ of $y_0$ radioactive atoms remain, and so on (see Figure 7.18).

Carbon-14 dating uses the half-life of 5730 years.

**EXAMPLE 5**    The decay of radioactive elements can sometimes be used to date events from Earth's past. In a living organism, the ratio of radioactive carbon, carbon-14, to ordinary carbon stays fairly constant during the lifetime of the organism, being approximately equal to the ratio in the organism's atmosphere at the time. After the organism's death, however, no new carbon is ingested, and the proportion of carbon-14 in the organism's remains decreases as the carbon-14 decays.

Scientists who do carbon-14 dating often use a figure of 5730 years for its half-life. Find the age of a sample in which 10% of the radioactive nuclei originally present have decayed.

**Solution**    We use the decay equation $y = y_0 e^{-kt}$. There are two things to find: the value of $k$ and the value of $t$ when $y$ is $0.9 y_0$ (90% of the radioactive nuclei are still present). That is, find $t$ when $y_0 e^{-kt} = 0.9 y_0$, or $e^{-kt} = 0.9$.

*The value of k.* We use the half-life Equation (7):

$$k = \frac{\ln 2}{\text{half-life}} = \frac{\ln 2}{5730} \qquad \text{(about } 1.2 \times 10^{-4}\text{)}.$$

*The value of t that makes* $e^{-kt} = 0.9.$

$$e^{-kt} = 0.9$$

$$e^{-(\ln 2/5730)t} = 0.9$$

$$-\frac{\ln 2}{5730}t = \ln 0.9 \qquad\qquad \text{Logs of both sides}$$

$$t = -\frac{5730 \ln 0.9}{\ln 2} \approx 871 \text{ years}$$

The sample is about 871 years old.    ■

## Heat Transfer: Newton's Law of Cooling

Hot soup left in a tin cup cools to the temperature of the surrounding air. A hot silver bar immersed in a large tub of water cools to the temperature of the surrounding water. In situations like these, the rate at which an object's temperature is changing at any given time is roughly proportional to the difference between its temperature and the temperature of the surrounding medium. This observation is called *Newton's Law of Cooling*, although it applies to warming as well.

If $H$ is the temperature of the object at time $t$ and $H_S$ is the constant surrounding temperature, then the differential equation is

$$\frac{dH}{dt} = -k(H - H_S). \tag{8}$$

If we substitute $y$ for $(H - H_S)$, then

$$\frac{dy}{dt} = \frac{d}{dt}(H - H_S) = \frac{dH}{dt} - \frac{d}{dt}(H_S)$$

$$= \frac{dH}{dt} - 0 \qquad\qquad H_S \text{ is a constant.}$$

$$= \frac{dH}{dt}$$

$$= -k(H - H_S) \qquad\qquad \text{Eq. (8)}$$

$$= -ky. \qquad\qquad H - H_S = y$$

Now we know that the solution of $dy/dt = -ky$ is $y = y_0 e^{-kt}$, where $y(0) = y_0$. Substituting $(H - H_S)$ for $y$, this says that

$$H - H_S = (H_0 - H_S)e^{-kt}, \tag{9}$$

where $H_0$ is the temperature at $t = 0$. This equation is the solution to Newton's Law of Cooling.

**EXAMPLE 6**    A hard-boiled egg at 98°C is put in a sink of 18°C water. After 5 min, the egg's temperature is 38°C. Assuming that the water has not warmed appreciably, how much longer will it take the egg to reach 20°C?

**Solution**    We find how long it would take the egg to cool from 98°C to 20°C and subtract the 5 min that have already elapsed. Using Equation (9) with $H_S = 18$ and $H_0 = 98$, the egg's temperature $t$ min after it is put in the sink is

$$H = 18 + (98 - 18)e^{-kt} = 18 + 80e^{-kt}.$$

To find $k$, we use the information that $H = 38$ when $t = 5$:

$$38 = 18 + 80e^{-5k}$$

$$e^{-5k} = \frac{1}{4}$$

$$-5k = \ln\frac{1}{4} = -\ln 4$$

$$k = \frac{1}{5}\ln 4 = 0.2 \ln 4 \qquad \text{(about 0.28)}.$$

The egg's temperature at time $t$ is $H = 18 + 80e^{-(0.2 \ln 4)t}$. Now find the time $t$ when $H = 20$:

$$20 = 18 + 80e^{-(0.2\ln 4)t}$$

$$80e^{-(0.2\ln 4)t} = 2$$

$$e^{-(0.2\ln 4)t} = \frac{1}{40}$$

$$-(0.2\ln 4)t = \ln\frac{1}{40} = -\ln 40$$

$$t = \frac{\ln 40}{0.2\ln 4} \approx 13 \text{ min}.$$

The egg's temperature will reach 20°C about 13 min after it is put in the water to cool. Since it took 5 min to reach 38°C, it will take about 8 min more to reach 20°C. ■

## Exercises 7.4

### Verifying Solutions

In Exercises 1–4, show that each function $y = f(x)$ is a solution of the accompanying differential equation.

**1.** $2y' + 3y = e^{-x}$

   **a.** $y = e^{-x}$          **b.** $y = e^{-x} + e^{-(3/2)x}$

   **c.** $y = e^{-x} + Ce^{-(3/2)x}$

**2.** $y' = y^2$

   **a.** $y = -\frac{1}{x}$   **b.** $y = -\frac{1}{x+3}$   **c.** $y = -\frac{1}{x+C}$

**3.** $y = \frac{1}{x}\int_1^x \frac{e^t}{t}\,dt, \quad x^2 y' + xy = e^x$

**4.** $y = \frac{1}{\sqrt{1+x^4}}\int_1^x \sqrt{1+t^4}\,dt, \quad y' + \frac{2x^3}{1+x^4}y = 1$

### Initial Value Problems

In Exercises 5–8, show that each function is a solution of the given initial value problem.

| Differential equation | Initial equation | Solution candidate |
|---|---|---|
| **5.** $y' + y = \dfrac{2}{1+4e^{2x}}$ | $y(-\ln 2) = \dfrac{\pi}{2}$ | $y = e^{-x}\tan^{-1}(2e^x)$ |
| **6.** $y' = e^{-x^2} - 2xy$ | $y(2) = 0$ | $y = (x-2)e^{-x^2}$ |
| **7.** $xy' + y = -\sin x,$ $x > 0$ | $y\left(\dfrac{\pi}{2}\right) = 0$ | $y = \dfrac{\cos x}{x}$ |
| **8.** $x^2 y' = xy - y^2,$ $x > 1$ | $y(e) = e$ | $y = \dfrac{x}{\ln x}$ |

### Separable Differential Equations

Solve the differential equations in Exercises 9–22.

**9.** $2\sqrt{xy}\dfrac{dy}{dx} = 1, \quad x, y > 0$   **10.** $\dfrac{dy}{dx} = x^2\sqrt{y}, \quad y > 0$

**11.** $\dfrac{dy}{dx} = e^{x-y}$          **12.** $\dfrac{dy}{dx} = 3x^2 e^{-y}$

**13.** $\dfrac{dy}{dx} = \sqrt{y}\cos^2\sqrt{y}$    **14.** $\sqrt{2xy}\dfrac{dy}{dx} = 1$

**15.** $\sqrt{x}\dfrac{dy}{dx} = e^{y+\sqrt{x}}, \quad x > 0$   **16.** $(\sec x)\dfrac{dy}{dx} = e^{y+\sin x}$

**17.** $\dfrac{dy}{dx} = 2x\sqrt{1-y^2}, \quad -1 < y < 1$

**18.** $\dfrac{dy}{dx} = \dfrac{e^{2x-y}}{e^{x+y}}$

**19.** $y^2\dfrac{dy}{dx} = 3x^2 y^3 - 6x^2$   **20.** $\dfrac{dy}{dx} = xy + 3x - 2y - 6$

**21.** $\dfrac{1}{x}\dfrac{dy}{dx} = ye^{x^2} + 2\sqrt{y}\,e^{x^2}$   **22.** $\dfrac{dy}{dx} = e^{x-y} + e^x + e^{-y} + 1$

### Applications and Examples

The answers to most of the following exercises are in terms of logarithms and exponentials. A calculator can be helpful, enabling you to express the answers in decimal form.

**23. Human evolution continues** The analysis of tooth shrinkage by C. Loring Brace and colleagues at the University of Michigan's Museum of Anthropology indicates that human tooth size is continuing to decrease and that the evolutionary process did not come to a halt some 30,000 years ago, as many scientists contend. In northern Europeans, for example, tooth size reduction now has a rate of 1% per 1000 years.

**a.** If $t$ represents time in years and $y$ represents tooth size, use the condition that $y = 0.99y_0$ when $t = 1000$ to find the value of $k$ in the equation $y = y_0 e^{kt}$. Then use this value of $k$ to answer the following questions.

**b.** In about how many years will human teeth be 90% of their present size?

**c.** What will be our descendants' tooth size 20,000 years from now (as a percentage of our present tooth size)?

**24. Atmospheric pressure** The earth's atmospheric pressure $p$ is often modeled by assuming that the rate $dp/dh$ at which $p$ changes with the altitude $h$ above sea level is proportional to $p$. Suppose that the pressure at sea level is 1013 hectopascals and that the pressure at an altitude of 20 km is 90 hectopascals.

**a.** Solve the initial value problem

Differential equation: $dp/dh = kp$  ($k$ a constant)

Initial condition: $p = p_0$ when $h = 0$

to express $p$ in terms of $h$. Determine the values of $p_0$ and $k$ from the given altitude-pressure data.

**b.** What is the atmospheric pressure at $h = 50$ km?

**c.** At what altitude does the pressure equal 900 hectopascals?

**25. First-order chemical reactions** In some chemical reactions, the rate at which the amount of a substance changes with time is proportional to the amount present. For the change of δ-glucono lactone into gluconic acid, for example,

$$\frac{dy}{dt} = -0.6y$$

when $t$ is measured in hours. If there are 100 grams of δ-glucono lactone present when $t = 0$, how many grams will be left after the first hour?

**26. The inversion of sugar** The processing of raw sugar has a step called "inversion" that changes the sugar's molecular structure. Once the process has begun, the rate of change of the amount of raw sugar is proportional to the amount of raw sugar remaining. If 1000 kg of raw sugar reduces to 800 kg of raw sugar during the first 10 hours, how much raw sugar will remain after another 14 hours?

**27. Working underwater** The intensity $L(x)$ of light $x$ meters beneath the surface of the ocean satisfies the differential equation

$$\frac{dL}{dx} = -kL.$$

As a diver, you know from experience that diving to 6 meters in the Caribbean Sea cuts the intensity in half. You cannot work without artificial light when the intensity falls below one-tenth of the surface value. About how deep can you expect to work without artificial light?

**28. Voltage in a discharging capacitor** Suppose that electricity is draining from a capacitor at a rate that is proportional to the voltage $V$ across its terminals and that, if $t$ is measured in seconds,

$$\frac{dV}{dt} = -\frac{1}{40}V.$$

Solve this equation for $V$, using $V_0$ to denote the value of $V$ when $t = 0$. How long will it take the voltage to drop to 10% of its original value?

**29. Cholera bacteria** Suppose that the bacteria in a colony can grow unchecked, by the law of exponential change. The colony starts with 1 bacterium and doubles every half-hour. How many bacteria will the colony contain at the end of 24 hours? (Under favorable laboratory conditions, the number of cholera bacteria can double every 30 min. In an infected person, many bacteria are destroyed, but this example helps explain why a person who feels well in the morning may be dangerously ill by evening.)

**30. Growth of bacteria** A colony of bacteria is grown under ideal conditions in a laboratory so that the population increases exponentially with time. At the end of 3 hours there are 10,000 bacteria. At the end of 5 hours there are 40,000. How many bacteria were present initially?

**31. The incidence of a disease** (*Continuation of Example 4.*) Suppose that in any given year the number of cases can be reduced by 25% instead of 20%.

**a.** How long will it take to reduce the number of cases to 1000?

**b.** How long will it take to eradicate the disease, that is, reduce the number of cases to less than 1?

**32. Drug concentration** An antibiotic is administered intravenously into the bloodstream at a constant rate $r$. As the drug flows through the patient's system and acts on the infection that is present, it is removed from the bloodstream at a rate proportional to the amount in the bloodstream at that time. Since the amount of blood in the patient is constant, this means that the concentration $y = y(t)$ of the antibiotic in the bloodstream can be modeled by the differential equation

$$\frac{dy}{dt} = r - ky, \quad k > 0 \text{ and constant.}$$

**a.** If $y(0) = y_0$, find the concentration $y(t)$ at any time $t$.

**b.** Assume that $y_0 < (r/k)$ and find $\lim_{y \to \infty} y(t)$. Sketch the solution curve for the concentration.

**33. Endangered species** Biologists consider a species of animal or plant to be endangered if it is expected to become extinct within 20 years. If a certain species of wildlife is counted to have 1147 members at the present time, and the population has been steadily declining exponentially at an annual rate averaging 39% over the past 7 years, do you think the species is endangered? Explain your answer.

**34. The U.S. population** The U.S. Census Bureau keeps a running clock totaling the U.S. population. On September 20, 2012, the total was increasing at the rate of 1 person every 12 s. The population figure for 8:11 P.M. EST on that day was 314,419,198.

**a.** Assuming exponential growth at a constant rate, find the rate constant for the population's growth (people per 365-day year).

**b.** At this rate, what will the U.S. population be at 8:11 P.M. EST on September 20, 2019?

**35. Oil depletion** Suppose the amount of oil pumped from one of the canyon wells in southern California decreases at the continuous rate of 10% per year. When will the well's output fall to one-fifth of its present value?

**36. Continuous price discounting** To encourage buyers to place 100-unit orders, your firm's sales department applies a continuous discount that makes the unit price a function $p(x)$ of the number of units $x$ ordered. The discount decreases the price at the rate of $0.01 per unit ordered. The price per unit for a 100-unit order is $p(100) = \$20.09$.

**a.** Find $p(x)$ by solving the following initial value problem:

Differential equation: $\dfrac{dp}{dx} = -\dfrac{1}{100}p$

Initial condition: $p(100) = 20.09$.

**b.** Find the unit price $p(10)$ for a 10-unit order and the unit price $p(90)$ for a 90-unit order.

**c.** The sales department has asked you to find out if it is discounting so much that the firm's revenue, $r(x) = x \cdot p(x)$, will actually be less for a 100-unit order than, say, for a 90-unit order. Reassure them by showing that $r$ has its maximum value at $x = 100$.

**d.** Graph the revenue function $r(x) = xp(x)$ for $0 \le x \le 200$.

**37. Plutonium-239** The half-life of the plutonium isotope is 24,360 years. If 10 g of plutonium is released into the atmosphere by a nuclear accident, how many years will it take for 80% of the isotope to decay?

**38. Polonium-210** The half-life of polonium is 139 days, but your sample will not be useful to you after 95% of the radioactive nuclei present on the day the sample arrives has disintegrated. For about how many days after the sample arrives will you be able to use the polonium?

**39. The mean life of a radioactive nucleus** Physicists using the radioactivity equation $y = y_0 e^{-kt}$ call the number $1/k$ the *mean life* of a radioactive nucleus. The mean life of a radon nucleus is about $1/0.18 = 5.6$ days. The mean life of a carbon-14 nucleus is more than 8000 years. Show that 95% of the radioactive nuclei originally present in a sample will disintegrate within three mean lifetimes, i.e., by time $t = 3/k$. Thus, the mean life of a nucleus gives a quick way to estimate how long the radioactivity of a sample will last.

**40. Californium-252** What costs $27 million per gram and can be used to treat brain cancer, analyze coal for its sulfur content, and detect explosives in luggage? The answer is californium-252, a radioactive isotope so rare that only 8 g of it have been made in the Western world since its discovery by Glenn Seaborg in 1950. The half-life of the isotope is 2.645 years—long enough for a useful service life and short enough to have a high radioactivity per unit mass. One microgram of the isotope releases 170 million neutrons per minute.

**a.** What is the value of $k$ in the decay equation for this isotope?

**b.** What is the isotope's mean life? (See Exercise 39.)

**c.** How long will it take 95% of a sample's radioactive nuclei to disintegrate?

**41. Cooling soup** Suppose that a cup of soup cooled from 90°C to 60°C after 10 min in a room whose temperature was 20°C. Use Newton's Law of Cooling to answer the following questions.

**a.** How much longer would it take the soup to cool to 35°C?

**b.** Instead of being left to stand in the room, the cup of 90°C soup is put in a freezer whose temperature is $-15$°C. How long will it take the soup to cool from 90°C to 35°C?

**42. A beam of unknown temperature** An aluminum beam was brought from the outside cold into a machine shop where the temperature was held at 18°C. After 10 min, the beam warmed to 2°C and after another 10 min it was 10°C. Use Newton's Law of Cooling to estimate the beam's initial temperature.

**43. Surrounding medium of unknown temperature** A pan of warm water (46°C) was put in a refrigerator. Ten minutes later, the water's temperature was 39°C; 10 min after that, it was 33°C. Use Newton's Law of Cooling to estimate how cold the refrigerator was.

**44. Silver cooling in air** The temperature of an ingot of silver is 60°C above room temperature right now. Twenty minutes ago, it was 70°C above room temperature. How far above room temperature will the silver be

**a.** 15 min from now?

**b.** 2 hours from now?

**c.** When will the silver be 10°C above room temperature?

**45. The age of Crater Lake** The charcoal from a tree killed in the volcanic eruption that formed Crater Lake in Oregon contained 44.5% of the carbon-14 found in living matter. About how old is Crater Lake?

**46. The sensitivity of carbon-14 dating to measurement** To see the effect of a relatively small error in the estimate of the amount of carbon-14 in a sample being dated, consider this hypothetical situation:

**a.** A bone fragment found in central Illinois in the year 2000 contains 17% of its original carbon-14 content. Estimate the year the animal died.

**b.** Repeat part (a), assuming 18% instead of 17%.

**c.** Repeat part (a), assuming 16% instead of 17%.

**47. Carbon-14** The oldest known frozen human mummy, discovered in the Schnalstal glacier of the Italian Alps in 1991 and called *Otzi*, was found wearing straw shoes and a leather coat with goat fur, and holding a copper ax and stone dagger. It was estimated that Otzi died 5000 years before he was discovered in the melting glacier. How much of the original carbon-14 remained in Otzi at the time of his discovery?

**48. Art forgery** A painting attributed to Vermeer (1632–1675), which should contain no more than 96.2% of its original carbon-14, contains 99.5% instead. About how old is the forgery?

**49. Lascaux Cave paintings** Prehistoric cave paintings of animals were found in the Lascaux Cave in France in 1940. Scientific analysis revealed that only 15% of the original carbon-14 in the paintings remained. What is an estimate of the age of the paintings?

**50. Incan mummy** The frozen remains of a young Incan woman were discovered by archeologist Johan Reinhard on Mt. Ampato in Peru during an expedition in 1995.

**a.** How much of the original carbon-14 was present if the estimated age of the "Ice Maiden" was 500 years?

**b.** If a 1% error can occur in the carbon-14 measurement, what is the oldest possible age for the Ice Maiden?

# 7.5 Indeterminate Forms and L'Hôpital's Rule

**HISTORICAL BIOGRAPHY**

Guillaume François Antoine de l'Hôpital
(1661–1704)
Johann Bernoulli
(1667–1748)

John (Johann) Bernoulli discovered a rule using derivatives to calculate limits of fractions whose numerators and denominators both approach zero or $+\infty$. The rule is known today as **l'Hôpital's Rule**, after Guillaume de l'Hôpital. He was a French nobleman who wrote the first introductory differential calculus text, where the rule first appeared in print. Limits involving transcendental functions often require some use of the rule for their calculation.

## Indeterminate Form 0/0

If we want to know how the function

$$F(x) = \frac{x - \sin x}{x^3}$$

behaves *near* $x = 0$ (where it is undefined), we can examine the limit of $F(x)$ as $x \to 0$. We cannot apply the Quotient Rule for limits (Theorem 1 of Chapter 2) because the limit of the denominator is 0. Moreover, in this case, *both* the numerator and denominator approach 0, and $0/0$ is undefined. Such limits may or may not exist in general, but the limit does exist for the function $F(x)$ under discussion by applying l'Hôpital's Rule, as we will see in Example 1d.

If the continuous functions $f(x)$ and $g(x)$ are both zero at $x = a$, then

$$\lim_{x \to a} \frac{f(x)}{g(x)}$$

cannot be found by substituting $x = a$. The substitution produces $0/0$, a meaningless expression, which we cannot evaluate. We use $0/0$ as a notation for an expression known as an **indeterminate form**. Other meaningless expressions often occur, such as $\infty/\infty$, $\infty \cdot 0$, $\infty - \infty$, $0^0$, and $1^\infty$, which cannot be evaluated in a consistent way; these are called indeterminate forms as well. Sometimes, but not always, limits that lead to indeterminate forms may be found by cancelation, rearrangement of terms, or other algebraic manipulations. This was our experience in Chapter 2. It took considerable analysis in Section 2.4 to find $\lim_{x \to 0}(\sin x)/x$. But we have had success with the limit

$$f'(a) = \lim_{x \to a} \frac{f(x) - f(a)}{x - a},$$

from which we calculate derivatives and which produces the indeterminant form $0/0$ when we attempt to substitute $x = a$. L'Hôpital's Rule enables us to draw on our success with derivatives to evaluate limits that otherwise lead to indeterminate forms.

---

**THEOREM 5—L'Hôpital's Rule**    Suppose that $f(a) = g(a) = 0$, that $f$ and $g$ are differentiable on an open interval $I$ containing $a$, and that $g'(x) \neq 0$ on $I$ if $x \neq a$. Then

$$\lim_{x \to a} \frac{f(x)}{g(x)} = \lim_{x \to a} \frac{f'(x)}{g'(x)},$$

assuming that the limit on the right side of this equation exists.

---

We give a proof of Theorem 5 at the end of this section.

**Caution**

To apply l'Hôpital's Rule to $f/g$, divide the derivative of $f$ by the derivative of $g$. Do not fall into the trap of taking the derivative of $f/g$. The quotient to use is $f'/g'$, not $(f/g)'$.

**EXAMPLE 1**    The following limits involve $0/0$ indeterminate forms, so we apply l'Hôpital's Rule. In some cases, it must be applied repeatedly.

(a) $\displaystyle \lim_{x \to 0} \frac{3x - \sin x}{x} = \lim_{x \to 0} \frac{3 - \cos x}{1} = \frac{3 - \cos x}{1}\bigg|_{x=0} = 2$

(b) $\displaystyle \lim_{x \to 0} \frac{\sqrt{1 + x} - 1}{x} = \lim_{x \to 0} \frac{\dfrac{1}{2\sqrt{1 + x}}}{1} = \frac{1}{2}$

(c) $\displaystyle \lim_{x \to 0} \frac{\sqrt{1 + x} - 1 - x/2}{x^2}$    $\frac{0}{0}$; apply l'Hôpital's Rule.

$\displaystyle = \lim_{x \to 0} \frac{(1/2)(1 + x)^{-1/2} - 1/2}{2x}$    Still $\frac{0}{0}$; apply l'Hôpital's Rule again.

$\displaystyle = \lim_{x \to 0} \frac{-(1/4)(1 + x)^{-3/2}}{2} = -\frac{1}{8}$    Not $\frac{0}{0}$; limit is found.

(d) $\displaystyle \lim_{x \to 0} \frac{x - \sin x}{x^3}$    $\frac{0}{0}$; apply l'Hôpital's Rule.

$\displaystyle = \lim_{x \to 0} \frac{1 - \cos x}{3x^2}$    Still $\frac{0}{0}$; apply l'Hôpital's Rule again.

$\displaystyle = \lim_{x \to 0} \frac{\sin x}{6x}$    Still $\frac{0}{0}$; apply l'Hôpital's Rule again.

$\displaystyle = \lim_{x \to 0} \frac{\cos x}{6} = \frac{1}{6}$    Not $\frac{0}{0}$; limit is found. ∎

Here is a summary of the procedure we followed in Example 1.

---

**Using L'Hôpital's Rule**

To find

$$\lim_{x \to a} \frac{f(x)}{g(x)}$$

by l'Hôpital's Rule, we continue to differentiate $f$ and $g$, so long as we still get the form $0/0$ at $x = a$. But as soon as one or the other of these derivatives is different from zero at $x = a$ we stop differentiating. L'Hôpital's Rule does not apply when either the numerator or denominator has a finite nonzero limit.

---

**EXAMPLE 2**    Be careful to apply l'Hôpital's Rule correctly:

$$\lim_{x \to 0} \frac{1 - \cos x}{x + x^2}$$    $\frac{0}{0}$

$$= \lim_{x \to 0} \frac{\sin x}{1 + 2x}$$    Not $\frac{0}{0}$

It is tempting to try to apply l'Hôpital's Rule again, which would result in

$$\lim_{x \to 0} \frac{\cos x}{2} = \frac{1}{2},$$

but this is not the correct limit. L'Hôpital's Rule can be applied only to limits that give indeterminate forms, and $\lim_{x \to 0} (\sin x)/(1 + 2x)$ does not give an indeterminate form. Instead, this limit is $0/1 = 0$, and the correct answer for the original limit is 0.

L'Hôpital's Rule applies to one-sided limits as well.

**EXAMPLE 3**    In this example the one-sided limits are different.

**(a)**  $\displaystyle\lim_{x\to 0^+}\frac{\sin x}{x^2}$    $\dfrac{0}{0}$

$\qquad\quad = \displaystyle\lim_{x\to 0^+}\frac{\cos x}{2x} = \infty$    Positive for $x > 0$

**(b)**  $\displaystyle\lim_{x\to 0^-}\frac{\sin x}{x^2}$    $\dfrac{0}{0}$

$\qquad\quad = \displaystyle\lim_{x\to 0^-}\frac{\cos x}{2x} = -\infty$    Negative for $x < 0$    ■

Recall that $\infty$ and $+\infty$ mean the same thing.

## Indeterminate Forms $\infty/\infty$, $\infty \cdot 0$, $\infty - \infty$

Sometimes when we try to evaluate a limit as $x \to a$ by substituting $x = a$ we get an indeterminant form like $\infty/\infty$, $\infty \cdot 0$, or $\infty - \infty$, instead of $0/0$. We first consider the form $\infty/\infty$.

More advanced treatments of calculus prove that l'Hôpital's Rule applies to the indeterminate form $\infty/\infty$, as well as to $0/0$. If $f(x) \to \pm\infty$ and $g(x) \to \pm\infty$ as $x \to a$, then

$$\lim_{x\to a}\frac{f(x)}{g(x)} = \lim_{x\to a}\frac{f'(x)}{g'(x)}$$

provided the limit on the right exists. In the notation $x \to a$, $a$ may be either finite or infinite. Moreover, $x \to a$ may be replaced by the one-sided limits $x \to a^+$ or $x \to a^-$.

**EXAMPLE 4**    Find the limits of these $\infty/\infty$ forms:

**(a)**  $\displaystyle\lim_{x\to \pi/2}\frac{\sec x}{1 + \tan x}$    **(b)**  $\displaystyle\lim_{x\to \infty}\frac{\ln x}{2\sqrt{x}}$    **(c)**  $\displaystyle\lim_{x\to \infty}\frac{e^x}{x^2}.$

**Solution**

**(a)** The numerator and denominator are discontinuous at $x = \pi/2$, so we investigate the one-sided limits there. To apply l'Hôpital's Rule, we can choose $I$ to be any open interval with $x = \pi/2$ as an endpoint.

$$\lim_{x\to (\pi/2)^-}\frac{\sec x}{1 + \tan x} \qquad \frac{\infty}{\infty}\text{ from the left so we apply l'Hôpital's Rule.}$$

$$= \lim_{x\to (\pi/2)^-}\frac{\sec x \tan x}{\sec^2 x} = \lim_{x\to (\pi/2)^-}\sin x = 1$$

The right-hand limit is 1 also, with $(-\infty)/(-\infty)$ as the indeterminate form. Therefore, the two-sided limit is equal to 1.

**(b)**  $\displaystyle\lim_{x\to \infty}\frac{\ln x}{2\sqrt{x}} = \lim_{x\to \infty}\frac{1/x}{1/\sqrt{x}} = \lim_{x\to \infty}\frac{1}{\sqrt{x}} = 0$    $\dfrac{1/x}{1/\sqrt{x}} = \dfrac{\sqrt{x}}{x} = \dfrac{1}{\sqrt{x}}$

**(c)**  $\displaystyle\lim_{x\to \infty}\frac{e^x}{x^2} = \lim_{x\to \infty}\frac{e^x}{2x} = \lim_{x\to \infty}\frac{e^x}{2} = \infty$    ■

Next we turn our attention to the indeterminate forms $\infty \cdot 0$ and $\infty - \infty$. Sometimes these forms can be handled by using algebra to convert them to a $0/0$ or $\infty/\infty$ form. Here again we do not mean to suggest that $\infty \cdot 0$ or $\infty - \infty$ is a number. They are only notations for functional behaviors when considering limits. Here are examples of how we might work with these indeterminate forms.

**EXAMPLE 5** Find the limits of these $\infty \cdot 0$ forms:

**(a)** $\lim\limits_{x\to\infty}\left(x \sin \dfrac{1}{x}\right)$ **(b)** $\lim\limits_{x\to 0^+} \sqrt{x}\, \ln x$

**Solution**

**(a)** $\lim\limits_{x\to\infty}\left(x \sin \dfrac{1}{x}\right) = \lim\limits_{h\to 0^+}\left(\dfrac{1}{h}\sin h\right) = \lim\limits_{h\to 0^+}\dfrac{\sin h}{h} = 1$   $\infty \cdot 0$; let $h = 1/x.$

**(b)** $\lim\limits_{x\to 0^+} \sqrt{x}\,\ln x = \lim\limits_{x\to 0^+}\dfrac{\ln x}{1/\sqrt{x}}$   $\infty \cdot 0$ converted to $\infty/\infty$

$\qquad = \lim\limits_{x\to 0^+}\dfrac{1/x}{-1/2x^{3/2}}$   l'Hôpital's Rule applied

$\qquad = \lim\limits_{x\to 0^+}\left(-2\sqrt{x}\right) = 0$ ∎

**EXAMPLE 6** Find the limit of this $\infty - \infty$ form:

$$\lim\limits_{x\to 0}\left(\dfrac{1}{\sin x} - \dfrac{1}{x}\right).$$

**Solution** If $x \to 0^+$, then $\sin x \to 0^+$ and

$$\dfrac{1}{\sin x} - \dfrac{1}{x} \to \infty - \infty.$$

Similarly, if $x \to 0^-$, then $\sin x \to 0^-$ and

$$\dfrac{1}{\sin x} - \dfrac{1}{x} \to -\infty - (-\infty) = -\infty + \infty.$$

Neither form reveals what happens in the limit. To find out, we first combine the fractions:

$$\dfrac{1}{\sin x} - \dfrac{1}{x} = \dfrac{x - \sin x}{x \sin x}.$$   Common denominator is $x \sin x.$

Then we apply l'Hôpital's Rule to the result:

$$\lim\limits_{x\to 0}\left(\dfrac{1}{\sin x} - \dfrac{1}{x}\right) = \lim\limits_{x\to 0}\dfrac{x - \sin x}{x \sin x}$$   $\dfrac{0}{0}$

$$= \lim\limits_{x\to 0}\dfrac{1 - \cos x}{\sin x + x \cos x}$$   Still $\dfrac{0}{0}$

$$= \lim\limits_{x\to 0}\dfrac{\sin x}{2\cos x - x \sin x} = \dfrac{0}{2} = 0.$$ ∎

### Indeterminate Powers

Limits that lead to the indeterminate forms $1^{\infty}$, $0^0$, and $\infty^0$ can sometimes be handled by first taking the logarithm of the function. We use l'Hôpital's Rule to find the limit of the logarithm expression and then exponentiate the result to find the original function limit. This procedure is justified by the continuity of the exponential function and Theorem 10 in Section 2.5, and it is formulated as follows. (The formula is also valid for one-sided limits.)

---

If $\lim\limits_{x\to a} \ln f(x) = L$, then

$$\lim\limits_{x\to a} f(x) = \lim\limits_{x\to a} e^{\ln f(x)} = e^{L}.$$

Here $a$ may be either finite or infinite.

---

**EXAMPLE 7**  Apply l'Hôpital's Rule to show that $\lim_{x \to 0^+} (1 + x)^{1/x} = e$.

**Solution**  The limit leads to the indeterminate form $1^\infty$. We let $f(x) = (1 + x)^{1/x}$ and find $\lim_{x \to 0^+} \ln f(x)$. Since

$$\ln f(x) = \ln (1 + x)^{1/x} = \frac{1}{x} \ln (1 + x),$$

l'Hôpital's Rule now applies to give

$$\lim_{x \to 0^+} \ln f(x) = \lim_{x \to 0^+} \frac{\ln (1 + x)}{x} \qquad \frac{0}{0}$$

$$= \lim_{x \to 0^+} \frac{\dfrac{1}{1 + x}}{1} \qquad \text{l'Hôpital's Rule applied}$$

$$= \frac{1}{1} = 1.$$

Therefore, $\displaystyle\lim_{x \to 0^+} (1 + x)^{1/x} = \lim_{x \to 0^+} f(x) = \lim_{x \to 0^+} e^{\ln f(x)} = e^1 = e$. ∎

**EXAMPLE 8**  Find $\lim_{x \to \infty} x^{1/x}$.

**Solution**  The limit leads to the indeterminate form $\infty^0$. We let $f(x) = x^{1/x}$ and find $\lim_{x \to \infty} \ln f(x)$. Since

$$\ln f(x) = \ln x^{1/x} = \frac{\ln x}{x},$$

l'Hôpital's Rule gives

$$\lim_{x \to \infty} \ln f(x) = \lim_{x \to \infty} \frac{\ln x}{x} \qquad \frac{\infty}{\infty}$$

$$= \lim_{x \to \infty} \frac{1/x}{1} \qquad \text{l'Hôpital's Rule applied}$$

$$= \frac{0}{1} = 0.$$

Therefore $\displaystyle\lim_{x \to \infty} x^{1/x} = \lim_{x \to \infty} f(x) = \lim_{x \to \infty} e^{\ln f(x)} = e^0 = 1$. ∎

### Proof of L'Hôpital's Rule

Before we prove l'Hôpital's Rule, we consider a special case to provide some geometric insight for its reasonableness. Consider the two functions $f(x)$ and $g(x)$ having *continuous* derivatives and satisfying $f(a) = g(a) = 0$, $g'(a) \neq 0$. The graphs of $f(x)$ and $g(x)$, together with their linearizations $y = f'(a)(x - a)$ and $y = g'(a)(x - a)$, are shown in Figure 7.19. We know that near $x = a$, the linearizations provide good approximations to the functions. In fact,

$$f(x) = f'(a)(x - a) + \epsilon_1(x - a) \quad \text{and} \quad g(x) = g'(a)(x - a) + \epsilon_2(x - a)$$

where $\epsilon_1 \to 0$ and $\epsilon_2 \to 0$ as $x \to a$. So, as Figure 7.19 suggests,

$$\lim_{x \to a} \frac{f(x)}{g(x)} = \lim_{x \to a} \frac{f'(a)(x - a) + \epsilon_1(x - a)}{g'(a)(x - a) + \epsilon_2(x - a)}$$

$$= \lim_{x \to a} \frac{f'(a) + \epsilon_1}{g'(a) + \epsilon_2} = \frac{f'(a)}{g'(a)} \qquad g'(a) \neq 0$$

$$= \lim_{x \to a} \frac{f'(x)}{g'(x)}, \qquad \text{Continuous derivatives}$$

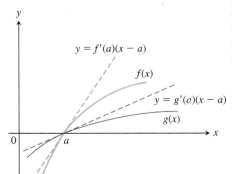

**FIGURE 7.19**  The two functions in l'Hôpital's Rule, graphed with their linear approximations at $x = a$.

When $g(x) = x$, Theorem 6 is the Mean Value Theorem in Chapter 4.

as asserted by l'Hôpital's Rule. We now proceed to a proof of the rule based on the more general assumptions stated in Theorem 5, which do not require that $g'(a) \neq 0$ and that the two functions have *continuous* derivatives.

The proof of l'Hôpital's Rule is based on Cauchy's Mean Value Theorem, an extension of the Mean Value Theorem that involves two functions instead of one. We prove Cauchy's Theorem first and then show how it leads to l'Hôpital's Rule.

---

**THEOREM 6—Cauchy's Mean Value Theorem** Suppose functions $f$ and $g$ are continuous on $[a, b]$ and differentiable throughout $(a, b)$ and also suppose $g'(x) \neq 0$ throughout $(a, b)$. Then there exists a number $c$ in $(a, b)$ at which

$$\frac{f'(c)}{g'(c)} = \frac{f(b) - f(a)}{g(b) - g(a)}.$$

---

**Proof** We apply the Mean Value Theorem of Section 4.2 twice. First we use it to show that $g(a) \neq g(b)$. For if $g(b)$ did equal $g(a)$, then the Mean Value Theorem would give

$$g'(c) = \frac{g(b) - g(a)}{b - a} = 0$$

for some $c$ between $a$ and $b$, which cannot happen because $g'(x) \neq 0$ in $(a, b)$.

We next apply the Mean Value Theorem to the function

$$F(x) = f(x) - f(a) - \frac{f(b) - f(a)}{g(b) - g(a)}\left[ g(x) - g(a) \right].$$

This function is continuous and differentiable where $f$ and $g$ are, and $F(b) = F(a) = 0$. Therefore, there is a number $c$ between $a$ and $b$ for which $F'(c) = 0$. When expressed in terms of f and $g$, this equation becomes

$$F'(c) = f'(c) - \frac{f(b) - f(a)}{g(b) - g(a)}\left[ g'(c) \right] = 0$$

so that

$$\frac{f'(c)}{g'(c)} = \frac{f(b) - f(a)}{g(b) - g(a)}. \qquad \blacksquare$$

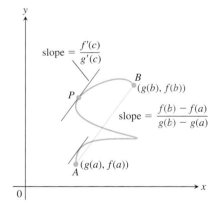

**FIGURE 7.20** There is at least one point $P$ on the curve $C$ for which the slope of the tangent to the curve at $P$ is the same as the slope of the secant line joining the points $A(g(a), f(a))$ and $B(g(b), f(b))$.

Cauchy's Mean Value Theorem has a geometric interpretation for a general winding curve $C$ in the plane joining the two points $A = (g(a), f(a))$ and $B = (g(b), f(b))$. In Chapter 11 you will learn how the curve $C$ can be formulated to show that there is at least one point $P$ on the curve for which the tangent to the curve at $P$ is parallel to the secant line joining the points $A$ and $B$. The slope of that tangent line turns out to be the quotient $f'/g'$ evaluated at the number $c$ in the interval $(a, b)$, which is the left-hand side of the equation in Theorem 6. Because the slope of the secant line joining $A$ and $B$ is

$$\frac{f(b) - f(a)}{g(b) - g(a)},$$

the equation in Cauchy's Mean Value Theorem says that the slope of the tangent line equals the slope of the secant line. This geometric interpretation is shown in Figure 7.20. Notice from the figure that it is possible for more than one point on the curve $C$ to have a tangent line that is parallel to the secant line joining $A$ and $B$.

**Proof of l'Hôpital's Rule** We first establish the limit equation for the case $x \to a^+$. The method needs almost no change to apply to $x \to a^-$, and the combination of these two cases establishes the result.

Suppose that $x$ lies to the right of $a$. Then $g'(x) \neq 0$, and we can apply Cauchy's Mean Value Theorem to the closed interval from $a$ to $x$. This step produces a number $c$ between $a$ and $x$ such that

$$\frac{f'(c)}{g'(c)} = \frac{f(x) - f(a)}{g(x) - g(a)}.$$

But $f(a) = g(a) = 0$, so

$$\frac{f'(c)}{g'(c)} = \frac{f(x)}{g(x)}.$$

As $x$ approaches $a$, $c$ approaches $a$ because it always lies between $a$ and $x$. Therefore,

$$\lim_{x \to a^+} \frac{f(x)}{g(x)} = \lim_{c \to a^+} \frac{f'(c)}{g'(c)} = \lim_{x \to a^+} \frac{f'(x)}{g'(x)},$$

which establishes l'Hôpital's Rule for the case where $x$ approaches $a$ from above. The case where $x$ approaches $a$ from below is proved by applying Cauchy's Mean Value Theorem to the closed interval $[x, a]$, $x < a$. ∎

# Exercises 7.5

### Finding Limits in Two Ways
In Exercises 1–6, use l'Hôpital's Rule to evaluate the limit. Then evaluate the limit using a method studied in Chapter 2.

**1.** $\lim\limits_{x \to -2} \dfrac{x + 2}{x^2 - 4}$

**2.** $\lim\limits_{x \to 0} \dfrac{\sin 5x}{x}$

**3.** $\lim\limits_{x \to \infty} \dfrac{5x^2 - 3x}{7x^2 + 1}$

**4.** $\lim\limits_{x \to 1} \dfrac{x^3 - 1}{4x^3 - x - 3}$

**5.** $\lim\limits_{x \to 0} \dfrac{1 - \cos x}{x^2}$

**6.** $\lim\limits_{x \to \infty} \dfrac{2x^2 + 3x}{x^3 + x + 1}$

### Applying l'Hôpital's Rule
Use l'Hôpital's rule to find the limits in Exercises 7–50.

**7.** $\lim\limits_{x \to 2} \dfrac{x - 2}{x^2 - 4}$

**8.** $\lim\limits_{x \to -5} \dfrac{x^2 - 25}{x + 5}$

**9.** $\lim\limits_{t \to -3} \dfrac{t^3 - 4t + 15}{t^2 - t - 12}$

**10.** $\lim\limits_{t \to -1} \dfrac{3t^3 + 3}{4t^3 - t + 3}$

**11.** $\lim\limits_{x \to \infty} \dfrac{5x^3 - 2x}{7x^3 + 3}$

**12.** $\lim\limits_{x \to \infty} \dfrac{x - 8x^2}{12x^2 + 5x}$

**13.** $\lim\limits_{t \to 0} \dfrac{\sin t^2}{t}$

**14.** $\lim\limits_{t \to 0} \dfrac{\sin 5t}{2t}$

**15.** $\lim\limits_{x \to 0} \dfrac{8x^2}{\cos x - 1}$

**16.** $\lim\limits_{x \to 0} \dfrac{\sin x - x}{x^3}$

**17.** $\lim\limits_{\theta \to \pi/2} \dfrac{2\theta - \pi}{\cos(2\pi - \theta)}$

**18.** $\lim\limits_{\theta \to -\pi/3} \dfrac{3\theta + \pi}{\sin(\theta + (\pi/3))}$

**19.** $\lim\limits_{\theta \to \pi/2} \dfrac{1 - \sin \theta}{1 + \cos 2\theta}$

**20.** $\lim\limits_{x \to 1} \dfrac{x - 1}{\ln x - \sin \pi x}$

**21.** $\lim\limits_{x \to 0} \dfrac{x^2}{\ln(\sec x)}$

**22.** $\lim\limits_{x \to \pi/2} \dfrac{\ln(\csc x)}{(x - (\pi/2))^2}$

**23.** $\lim\limits_{t \to 0} \dfrac{t(1 - \cos t)}{t - \sin t}$

**24.** $\lim\limits_{t \to 0} \dfrac{t \sin t}{1 - \cos t}$

**25.** $\lim\limits_{x \to (\pi/2)^-} \left(x - \dfrac{\pi}{2}\right) \sec x$

**26.** $\lim\limits_{x \to (\pi/2)^-} \left(\dfrac{\pi}{2} - x\right) \tan x$

**27.** $\lim\limits_{\theta \to 0} \dfrac{3^{\sin \theta} - 1}{\theta}$

**28.** $\lim\limits_{\theta \to 0} \dfrac{(1/2)^\theta - 1}{\theta}$

**29.** $\lim\limits_{x \to 0} \dfrac{x2^x}{2^x - 1}$

**30.** $\lim\limits_{x \to 0} \dfrac{3^x - 1}{2^x - 1}$

**31.** $\lim\limits_{x \to \infty} \dfrac{\ln(x + 1)}{\log_2 x}$

**32.** $\lim\limits_{x \to \infty} \dfrac{\log_2 x}{\log_3(x + 3)}$

**33.** $\lim\limits_{x \to 0^+} \dfrac{\ln(x^2 + 2x)}{\ln x}$

**34.** $\lim\limits_{x \to 0^+} \dfrac{\ln(e^x - 1)}{\ln x}$

**35.** $\lim\limits_{y \to 0} \dfrac{\sqrt{5y + 25} - 5}{y}$

**36.** $\lim\limits_{y \to 0} \dfrac{\sqrt{ay + a^2} - a}{y}, \quad a > 0$

**37.** $\lim\limits_{x \to \infty} (\ln 2x - \ln(x + 1))$

**38.** $\lim\limits_{x \to 0^+} (\ln x - \ln \sin x)$

**39.** $\lim\limits_{x \to 0^+} \dfrac{(\ln x)^2}{\ln(\sin x)}$

**40.** $\lim\limits_{x \to 0^+} \left(\dfrac{3x + 1}{x} - \dfrac{1}{\sin x}\right)$

**41.** $\lim\limits_{x \to 1^+} \left(\dfrac{1}{x - 1} - \dfrac{1}{\ln x}\right)$

**42.** $\lim\limits_{x \to 0^+} (\csc x - \cot x + \cos x)$

**43.** $\lim\limits_{\theta \to 0} \dfrac{\cos \theta - 1}{e^\theta - \theta - 1}$

**44.** $\lim\limits_{h \to 0} \dfrac{e^h - (1 + h)}{h^2}$

**45.** $\lim\limits_{t \to \infty} \dfrac{e^t + t^2}{e^t - t}$

**46.** $\lim\limits_{x \to \infty} x^2 e^{-x}$

**47.** $\lim\limits_{x \to 0} \dfrac{x - \sin x}{x \tan x}$

**48.** $\lim\limits_{x \to 0} \dfrac{(e^x - 1)^2}{x \sin x}$

**49.** $\lim\limits_{\theta \to 0} \dfrac{\theta - \sin \theta \cos \theta}{\tan \theta - \theta}$

**50.** $\lim\limits_{x \to 0} \dfrac{\sin 3x - 3x + x^2}{\sin x \sin 2x}$

### Indeterminate Powers and Products
Find the limits in Exercises 51–66.

**51.** $\lim\limits_{x \to 1^+} x^{1/(1-x)}$

**52.** $\lim\limits_{x \to 1^+} x^{1/(x-1)}$

**53.** $\lim\limits_{x \to \infty} (\ln x)^{1/x}$

**54.** $\lim\limits_{x \to e^+} (\ln x)^{1/(x-e)}$

**55.** $\lim\limits_{x \to 0^+} x^{-1/\ln x}$

**56.** $\lim\limits_{x \to \infty} x^{1/\ln x}$

**57.** $\lim\limits_{x\to\infty} (1 + 2x)^{1/(2 \ln x)}$  **58.** $\lim\limits_{x\to 0} (e^x + x)^{1/x}$  **59.** $\lim\limits_{x\to 0^+} x^x$

**60.** $\lim\limits_{x\to 0^+} \left(1 + \dfrac{1}{x}\right)^x$  **61.** $\lim\limits_{x\to\infty} \left(\dfrac{x + 2}{x - 1}\right)^x$  **62.** $\lim\limits_{x\to\infty} \left(\dfrac{x^2 + 1}{x + 2}\right)^{1/x}$

**63.** $\lim\limits_{x\to 0^+} x^2 \ln x$  **64.** $\lim\limits_{x\to 0^+} x (\ln x)^2$

**65.** $\lim\limits_{x\to 0^+} x \tan\left(\dfrac{\pi}{2} - x\right)$  **66.** $\lim\limits_{x\to 0^+} \sin x \cdot \ln x$

### Theory and Applications

L'Hôpital's Rule does not help with the limits in Exercises 67–74. Try it—you just keep on cycling. Find the limits some other way.

**67.** $\lim\limits_{x\to\infty} \dfrac{\sqrt{9x + 1}}{\sqrt{x + 1}}$  **68.** $\lim\limits_{x\to 0^+} \dfrac{\sqrt{x}}{\sqrt{\sin x}}$  **69.** $\lim\limits_{x\to (\pi/2)^-} \dfrac{\sec x}{\tan x}$

**70.** $\lim\limits_{x\to 0^+} \dfrac{\cot x}{\csc x}$  **71.** $\lim\limits_{x\to\infty} \dfrac{2^x - 3^x}{3^x + 4^x}$  **72.** $\lim\limits_{x\to -\infty} \dfrac{2^x + 4^x}{5^x - 2^x}$

**73.** $\lim\limits_{x\to\infty} \dfrac{e^{x^2}}{xe^x}$  **74.** $\lim\limits_{x\to 0^+} \dfrac{x}{e^{-1/x}}$

**75.** Which one is correct, and which one is wrong? Give reasons for your answers.

  **a.** $\lim\limits_{x\to 3} \dfrac{x - 3}{x^2 - 3} = \lim\limits_{x\to 3} \dfrac{1}{2x} = \dfrac{1}{6}$  **b.** $\lim\limits_{x\to 3} \dfrac{x - 3}{x^2 - 3} = \dfrac{0}{6} = 0$

**76.** Which one is correct, and which one is wrong? Give reasons for your answers.

  **a.** $\lim\limits_{x\to 0} \dfrac{x^2 - 2x}{x^2 - \sin x} = \lim\limits_{x\to 0} \dfrac{2x - 2}{2x - \cos x}$

  $= \lim\limits_{x\to 0} \dfrac{2}{2 + \sin x} = \dfrac{2}{2 + 0} = 1$

  **b.** $\lim\limits_{x\to 0} \dfrac{x^2 - 2x}{x^2 - \sin x} = \lim\limits_{x\to 0} \dfrac{2x - 2}{2x - \cos x} = \dfrac{-2}{0 - 1} = 2$

**77.** Only one of these calculations is correct. Which one? Why are the others wrong? Give reasons for your answers.

  **a.** $\lim\limits_{x\to 0^+} x \ln x = 0 \cdot (-\infty) = 0$

  **b.** $\lim\limits_{x\to 0^+} x \ln x = 0 \cdot (-\infty) = -\infty$

  **c.** $\lim\limits_{x\to 0^+} x \ln x = \lim\limits_{x\to 0^+} \dfrac{\ln x}{(1/x)} = \dfrac{-\infty}{\infty} = -1$

  **d.** $\lim\limits_{x\to 0^+} x \ln x = \lim\limits_{x\to 0^+} \dfrac{\ln x}{(1/x)}$

  $= \lim\limits_{x\to 0^+} \dfrac{(1/x)}{(-1/x^2)} = \lim\limits_{x\to 0^+} (-x) = 0$

**78.** Find all values of $c$ that satisfy the conclusion of Cauchy's Mean Value Theorem for the given functions and interval.

  **a.** $f(x) = x$,    $g(x) = x^2$,    $(a, b) = (-2, 0)$

  **b.** $f(x) = x$,    $g(x) = x^2$,    $(a, b)$ arbitrary

  **c.** $f(x) = x^3/3 - 4x$,    $g(x) = x^2$,    $(a, b) = (0, 3)$

**79. Continuous extension**   Find a value of $c$ that makes the function

$$f(x) = \begin{cases} \dfrac{9x - 3 \sin 3x}{5x^3}, & x \neq 0 \\ c, & x = 0 \end{cases}$$

continuous at $x = 0$. Explain why your value of $c$ works.

**80.** For what values of $a$ and $b$ is

$$\lim\limits_{x\to 0} \left(\dfrac{\tan 2x}{x^3} + \dfrac{a}{x^2} + \dfrac{\sin bx}{x}\right) = 0?$$

T **81.** $\infty - \infty$ **Form**

  **a.** Estimate the value of

  $$\lim\limits_{x\to\infty} \left(x - \sqrt{x^2 + x}\right)$$

  by graphing $f(x) = x - \sqrt{x^2 + x}$ over a suitably large interval of $x$-values.

  **b.** Now confirm your estimate by finding the limit with l'Hôpital's Rule. As the first step, multiply $f(x)$ by the fraction $\left(x + \sqrt{x^2 + x}\right)/\left(x + \sqrt{x^2 + x}\right)$ and simplify the new numerator.

**82.** Find $\lim\limits_{x\to\infty} \left(\sqrt{x^2 + 1} - \sqrt{x}\right)$.

T **83.** **0/0 Form**   Estimate the value of

$$\lim\limits_{x\to 1} \dfrac{2x^2 - (3x + 1)\sqrt{x} + 2}{x - 1}$$

by graphing. Then confirm your estimate with l'Hôpital's Rule.

**84.** This exercise explores the difference between the limit

$$\lim\limits_{x\to\infty} \left(1 + \dfrac{1}{x^2}\right)^x$$

and the limit

$$\lim\limits_{x\to\infty} \left(1 + \dfrac{1}{x}\right)^x = e.$$

  **a.** Use l'Hôpital's Rule to show that

  $$\lim\limits_{x\to\infty} \left(1 + \dfrac{1}{x}\right)^x = e.$$

T **b.** Graph

  $$f(x) = \left(1 + \dfrac{1}{x^2}\right)^x \quad \text{and} \quad g(x) = \left(1 + \dfrac{1}{x}\right)^x$$

  together for $x \geq 0$. How does the behavior of $f$ compare with that of $g$? Estimate the value of $\lim\limits_{x\to\infty} f(x)$.

  **c.** Confirm your estimate of $\lim\limits_{x\to\infty} f(x)$ by calculating it with l'Hôpital's Rule.

**85.** Show that

$$\lim\limits_{k\to\infty} \left(1 + \dfrac{r}{k}\right)^k = e^r.$$

**86.** Given that $x > 0$, find the maximum value, if any, of

  **a.** $x^{1/x}$    **b.** $x^{1/x^2}$    **c.** $x^{1/x^n}$ ($n$ a positive integer)

  **d.** Show that $\lim\limits_{x\to\infty} x^{1/x^n} = 1$ for every positive integer $n$.

**87.** Use limits to find horizontal asymptotes for each function.

  **a.** $y = x \tan\left(\dfrac{1}{x}\right)$    **b.** $y = \dfrac{3x + e^{2x}}{2x + e^{3x}}$

**88.** Find $f'(0)$ for $f(x) = \begin{cases} e^{-1/x^2}, & x \neq 0 \\ 0, & x = 0. \end{cases}$

T **89. The continuous extension of $(\sin x)^x$ to $[0, \pi]$**

    **a.** Graph $f(x) = (\sin x)^x$ on the interval $0 \le x \le \pi$. What value would you assign to $f$ to make it continuous at $x = 0$?

    **b.** Verify your conclusion in part (a) by finding $\lim_{x \to 0^+} f(x)$ with l'Hôpital's Rule.

    **c.** Returning to the graph, estimate the maximum value of $f$ on $[0, \pi]$. About where is max $f$ taken on?

    **d.** Sharpen your estimate in part (c) by graphing $f'$ in the same window to see where its graph crosses the $x$-axis. To simplify your work, you might want to delete the exponential factor from the expression for $f'$ and graph just the factor that has a zero.

T **90. The function $(\sin x)^{\tan x}$** *(Continuation of Exercise 89.)*

    **a.** Graph $f(x) = (\sin x)^{\tan x}$ on the interval $-7 \le x \le 7$. How do you account for the gaps in the graph? How wide are the gaps?

    **b.** Now graph $f$ on the interval $0 \le x \le \pi$. The function is not defined at $x = \pi/2$, but the graph has no break at this point. What is going on? What value does the graph appear to give for $f$ at $x = \pi/2$? (*Hint:* Use l'Hôpital's Rule to find lim $f$ as $x \to (\pi/2)^-$ and $x \to (\pi/2)^+$.)

    **c.** Continuing with the graphs in part (b), find max $f$ and min $f$ as accurately as you can and estimate the values of $x$ at which they are taken on.

# 7.6 Inverse Trigonometric Functions

Inverse trigonometric functions arise when we want to calculate angles from side measurements in triangles. They also provide useful antiderivatives and appear frequently in the solutions of differential equations. This section shows how these functions are defined, graphed, and evaluated, how their derivatives are computed, and why they appear as important antiderivatives.

## Defining the Inverses

The six basic trigonometric functions are not one-to-one (their values repeat periodically). However, we can restrict their domains to intervals on which they are one-to-one. The sine function increases from $-1$ at $x = -\pi/2$ to $+1$ at $x = \pi/2$. By restricting its domain to the interval $[-\pi/2, \pi/2]$ we make it one-to-one, so that it has an inverse $\sin^{-1} x$ (Figure 7.21). Similar domain restrictions can be applied to all six trigonometric functions.

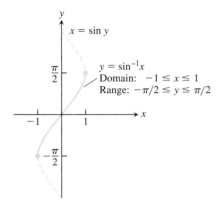

**FIGURE 7.21** The graph of $y = \sin^{-1} x$.

Domain restrictions that make the trigonometric functions one-to-one

$y = \sin x$
Domain: $[-\pi/2, \pi/2]$
Range: $[-1, 1]$

$y = \cos x$
Domain: $[0, \pi]$
Range: $[-1, 1]$

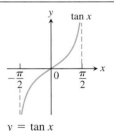

$y = \tan x$
Domain: $(-\pi/2, \pi/2)$
Range: $(-\infty, \infty)$

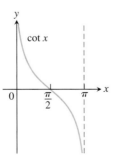

$y = \cot x$
Domain: $(0, \pi)$
Range: $(-\infty, \infty)$

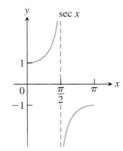

$y = \sec x$
Domain: $[0, \pi/2) \cup (\pi/2, \pi]$
Range: $(-\infty, -1] \cup [1, \infty)$

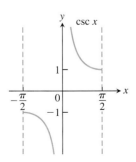

$y = \csc x$
Domain: $[-\pi/2, 0) \cup (0, \pi/2]$
Range: $(-\infty, -1] \cup [1, \infty)$

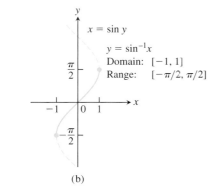

**FIGURE 7.22** The graphs of
(a) $y = \sin x, -\pi/2 \le x \le \pi/2$, and
(b) its inverse, $y = \sin^{-1} x$. The graph
of $\sin^{-1} x$, obtained by reflection across
the line $y = x$, is a portion of the curve
$x = \sin y$.

Since these restricted functions are now one-to-one, they have inverses, which we denote by

$$y = \sin^{-1} x \quad \text{or} \quad y = \arcsin x$$
$$y = \cos^{-1} x \quad \text{or} \quad y = \arccos x$$
$$y = \tan^{-1} x \quad \text{or} \quad y = \arctan x$$
$$y = \cot^{-1} x \quad \text{or} \quad y = \text{arccot } x$$
$$y = \sec^{-1} x \quad \text{or} \quad y = \text{arcsec } x$$
$$y = \csc^{-1} x \quad \text{or} \quad y = \text{arccsc } x$$

These equations are read "$y$ equals the arcsine of $x$" or "$y$ equals arcsin $x$" and so on.

**Caution** The $-1$ in the expressions for the inverse means "inverse." It does *not* mean reciprocal. For example, the *reciprocal* of sin $x$ is $(\sin x)^{-1} = 1/\sin x = \csc x$.

The graphs of the six inverse trigonometric functions are obtained by reflecting the graphs of the restricted trigonometric functions through the line $y = x$. Figure 7.22b shows the graph of $y = \sin^{-1} x$ and Figure 7.23 shows the graphs of all six functions. We now take a closer look at these functions.

Domain: $-1 \le x \le 1$
Range: $-\dfrac{\pi}{2} \le y \le \dfrac{\pi}{2}$

(a)

Domain: $-1 \le x \le 1$
Range: $0 \le y \le \pi$

(b)

Domain: $-\infty < x < \infty$
Range: $-\dfrac{\pi}{2} < y < \dfrac{\pi}{2}$

(c)

Domain: $x \le -1$ or $x \ge 1$
Range: $0 \le y \le \pi, y \ne \dfrac{\pi}{2}$

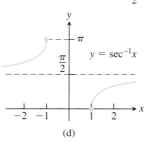

(d)

Domain: $x \le -1$ or $x \ge 1$
Range: $-\dfrac{\pi}{2} \le y \le \dfrac{\pi}{2}, y \ne 0$

(e)

Domain: $-\infty < x < \infty$
Range: $0 < y < \pi$

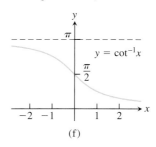

(f)

**FIGURE 7.23** Graphs of the six basic inverse trigonometric functions.

**The "Arc" in Arcsine and Arccosine**

For a unit circle and radian angles, the arc length equation $s = r\theta$ becomes $s = \theta$, so central angles and the arcs they subtend have the same measure. If $x = \sin y$, then, in addition to being the angle whose sine is $x$, $y$ is also the length of arc on the unit circle that subtends an angle whose sine is $x$. So we call $y$ "the arc whose sine is $x$."

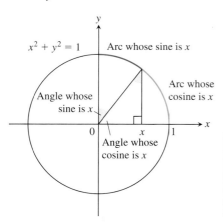

## The Arcsine and Arccosine Functions

We define the arcsine and arccosine as functions whose values are angles (measured in radians) that belong to restricted domains of the sine and cosine functions.

---

**DEFINITION**

$y = \sin^{-1} x$ is the number in $[-\pi/2, \pi/2]$ for which $\sin y = x$.

$y = \cos^{-1} x$ is the number in $[0, \pi]$ for which $\cos y = x$.

---

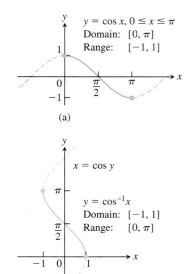

(a)

(b)

**FIGURE 7.24** The graphs of
(a) $y = \cos x$, $0 \leq x \leq \pi$, and (b) its
inverse, $y = \cos^{-1} x$. The graph of $\cos^{-1} x$,
obtained by reflection across the line
$y = x$, is a portion of the curve $x = \cos y$.

The graph of $y = \sin^{-1} x$ (Figure 7.22b) is symmetric about the origin (it lies along the graph of $x = \sin y$). The arcsine is therefore an odd function:

$$\sin^{-1}(-x) = -\sin^{-1} x. \tag{1}$$

The graph of $y = \cos^{-1} x$ (Figure 7.24b) has no such symmetry.

**EXAMPLE 1**  Evaluate  **(a)** $\sin^{-1}\left(\dfrac{\sqrt{3}}{2}\right)$  and  **(b)** $\cos^{-1}\left(-\dfrac{1}{2}\right)$.

**Solution**

**(a)** We see that

$$\sin^{-1}\left(\frac{\sqrt{3}}{2}\right) = \frac{\pi}{3}$$

because $\sin(\pi/3) = \sqrt{3}/2$ and $\pi/3$ belongs to the range $[-\pi/2, \pi/2]$ of the arcsine function. See Figure 7.25a.

**(b)** We have

$$\cos^{-1}\left(-\frac{1}{2}\right) = \frac{2\pi}{3}$$

because $\cos(2\pi/3) = -1/2$ and $2\pi/3$ belongs to the range $[0, \pi]$ of the arccosine function. See Figure 7.25b. ∎

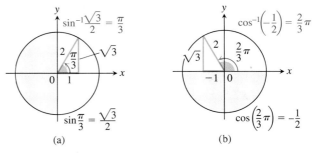

(a)                    (b)

**FIGURE 7.25**  Values of the arcsine and arccosine functions
(Example 1).

Using the same procedure illustrated in Example 1, we can create the following table of common values for the arcsine and arccosine functions.

| $x$ | $\sin^{-1} x$ | $\cos^{-1} x$ |
|---|---|---|
| $\sqrt{3}/2$ | $\pi/3$ | $\pi/6$ |
| $\sqrt{2}/2$ | $\pi/4$ | $\pi/4$ |
| $1/2$ | $\pi/6$ | $\pi/3$ |
| $-1/2$ | $-\pi/6$ | $2\pi/3$ |
| $-\sqrt{2}/2$ | $-\pi/4$ | $3\pi/4$ |
| $-\sqrt{3}/2$ | $-\pi/3$ | $5\pi/6$ |

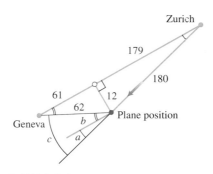

**FIGURE 7.26** Diagram for drift correction (Example 2), with distances rounded to the nearest mile (drawing not to scale).

**EXAMPLE 2** During a 240 km airplane flight from Zurich to Geneva, after flying 180 km the navigator determines that the plane is 12 km off course, as shown in Figure 7.26. Find the angle $a$ for a course parallel to the original correct course, the angle $b$, and the drift correction angle $c = a + b$.

**Solution** From the Pythagorean theorem and given information, we compute an approximate hypothetical flight distance of 179 km, had the plane been flying along the original correct course (see Figure 7.26). Knowing the flight distance from Zurich to Geneva, we next calculate the remaining leg of the original course to be 61 km. Applying the Pythagorean theorem again then gives an approximate distance of 62 km from the position of the plane to Geneva. Finally, from Figure 7.26, we see that $180 \sin a = 12$ and $62 \sin b = 12$, so

$$a = \sin^{-1}\frac{12}{180} \approx 0.067 \text{ radian } \approx 3.8°$$

$$b = \sin^{-1}\frac{12}{62} \approx 0.195 \text{ radian } \approx 11.2°$$

$$c = a + b \approx 15°.$$

### Identities Involving Arcsine and Arccosine

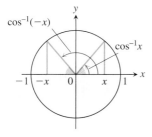

**FIGURE 7.27** $\cos^{-1}x$ and $\cos^{-1}(-x)$ are supplementary angles (so their sum is $\pi$).

As we can see from Figure 7.27, the arccosine of $x$ satisfies the identity

$$\cos^{-1}x + \cos^{-1}(-x) = \pi, \qquad (2)$$

or

$$\cos^{-1}(-x) = \pi - \cos^{-1}x. \qquad (3)$$

Also, we can see from the triangle in Figure 7.28 that for $x > 0$,

$$\sin^{-1}x + \cos^{-1}x = \pi/2. \qquad (4)$$

Equation (4) holds for the other values of $x$ in $[-1, 1]$ as well, but we cannot conclude this from the triangle in Figure 7.28. It is, however, a consequence of Equations (1) and (3) (Exercise 113).

### Inverses of tan x, cot x, sec x, and csc x

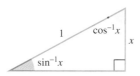

**FIGURE 7.28** $\sin^{-1}x$ and $\cos^{-1}x$ are complementary angles (so their sum is $\pi/2$).

The arctangent of $x$ is a radian angle whose tangent is $x$. The arccotangent of $x$ is an angle whose cotangent is $x$, and so forth. The angles belong to the restricted domains of the tangent, cotangent, secant, and cosecant functions.

**DEFINITIONS**

$y = \tan^{-1}x$ is the number in $(-\pi/2, \pi/2)$ for which $\tan y = x$.

$y = \cot^{-1}x$ is the number in $(0, \pi)$ for which $\cot y = x$.

$y = \sec^{-1}x$ is the number in $[0, \pi/2) \cup (\pi/2, \pi]$ for which $\sec y = x$.

$y = \csc^{-1}x$ is the number in $[-\pi/2, 0) \cup (0, \pi/2]$ for which $\csc y = x$.

We use open or half-open intervals to avoid values for which the tangent, cotangent, secant, and cosecant functions are undefined. (See Figure 7.23.)

The graph of $y = \tan^{-1}x$ is symmetric about the origin because it is a branch of the graph $x = \tan y$ that is symmetric about the origin (Figure 7.23c). Algebraically this means that

$$\tan^{-1}(-x) = -\tan^{-1}x;$$

the arctangent is an odd function. The graph of $y = \cot^{-1}x$ has no such symmetry (Figure 7.23f). Notice from Figure 7.23c that the graph of the arctangent function has two horizontal asymptotes: one at $y = \pi/2$ and the other at $y = -\pi/2$.

Domain: $|x| \geq 1$
Range: $0 \leq y \leq \pi, y \neq \dfrac{\pi}{2}$

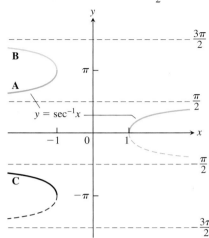

**FIGURE 7.29** There are several logical choices for the left-hand branch of $y = \sec^{-1} x$. With choice **A**, $\sec^{-1} x = \cos^{-1}(1/x)$, a useful identity employed by many calculators.

The inverses of the restricted forms of $\sec x$ and $\csc x$ are chosen to be the functions graphed in Figures 7.23d and 7.23e.

**Caution**   There is no general agreement about how to define $\sec^{-1} x$ for negative values of $x$. We chose angles in the second quadrant between $\pi/2$ and $\pi$. This choice makes $\sec^{-1} x = \cos^{-1}(1/x)$. It also makes $\sec^{-1} x$ an increasing function on each interval of its domain. Some texts choose $\sec^{-1} x$ to lie in $[-\pi, -\pi/2)$ for $x < 0$ and some texts choose it to lie in $[\pi, 3\pi/2)$ (Figure 7.29). These choices simplify the formula for the derivative (our formula needs absolute value signs) but fail to satisfy the computational equation $\sec^{-1} x = \cos^{-1}(1/x)$. From this, we can derive the identity

$$\sec^{-1} x = \cos^{-1}\left(\frac{1}{x}\right) = \frac{\pi}{2} - \sin^{-1}\left(\frac{1}{x}\right) \tag{5}$$

by applying Equation (4).

**EXAMPLE 3**   The accompanying figures show two values of $\tan^{-1} x$.

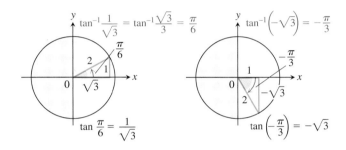

| $x$ | $\tan^{-1} x$ |
|:---:|:---:|
| $\sqrt{3}$ | $\pi/3$ |
| $1$ | $\pi/4$ |
| $\sqrt{3}/3$ | $\pi/6$ |
| $-\sqrt{3}/3$ | $-\pi/6$ |
| $-1$ | $-\pi/4$ |
| $-\sqrt{3}$ | $-\pi/3$ |

The angles come from the first and fourth quadrants because the range of $\tan^{-1} x$ is $(-\pi/2, \pi/2)$. ∎

## The Derivative of $y = \sin^{-1} u$

We know that the function $x = \sin y$ is differentiable in the interval $-\pi/2 < y < \pi/2$ and that its derivative, the cosine, is positive there. Theorem 1 in Section 7.1 therefore assures us that the inverse function $y = \sin^{-1} x$ is differentiable throughout the interval $-1 < x < 1$. We cannot expect it to be differentiable at $x = 1$ or $x = -1$ because the tangents to the graph are vertical at these points (see Figure 7.30).

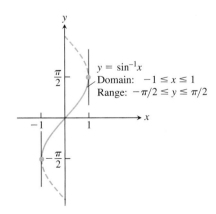

**FIGURE 7.30**   The graph of $y = \sin^{-1} x$ has vertical tangents at $x = -1$ and $x = 1$.

We find the derivative of $y = \sin^{-1}x$ by applying Theorem 1 with $f(x) = \sin x$ and $f^{-1}(x) = \sin^{-1}x$:

$$(f^{-1})'(x) = \frac{1}{f'(f^{-1}(x))} \qquad \text{Theorem 1}$$

$$= \frac{1}{\cos\left(\sin^{-1}x\right)} \qquad f'(u) = \cos u$$

$$= \frac{1}{\sqrt{1 - \sin^2\left(\sin^{-1}x\right)}} \qquad \cos u = \sqrt{1 - \sin^2 u}$$

$$= \frac{1}{\sqrt{1 - x^2}}. \qquad \sin\left(\sin^{-1}x\right) = x$$

If $u$ is a differentiable function of $x$ with $|u| < 1$, we apply the Chain Rule to get the general formula

$$\frac{d}{dx}(\sin^{-1}u) = \frac{1}{\sqrt{1 - u^2}}\frac{du}{dx}, \qquad |u| < 1.$$

**EXAMPLE 4**     Using the Chain Rule, we calculate the derivative

$$\frac{d}{dx}(\sin^{-1}x^2) = \frac{1}{\sqrt{1 - (x^2)^2}} \cdot \frac{d}{dx}(x^2) = \frac{2x}{\sqrt{1 - x^4}}. \qquad \blacksquare$$

### The Derivative of $y = \tan^{-1}u$

We find the derivative of $y = \tan^{-1}x$ by applying Theorem 1 with $f(x) = \tan x$ and $f^{-1}(x) = \tan^{-1}x$. Theorem 1 can be applied because the derivative of $\tan x$ is positive for $-\pi/2 < x < \pi/2$:

$$(f^{-1})'(x) = \frac{1}{f'(f^{-1}(x))} \qquad \text{Theorem 1}$$

$$= \frac{1}{\sec^2\left(\tan^{-1}x\right)} \qquad f'(u) = \sec^2 u$$

$$= \frac{1}{1 + \tan^2\left(\tan^{-1}x\right)} \qquad \sec^2 u = 1 + \tan^2 u$$

$$= \frac{1}{1 + x^2}. \qquad \tan\left(\tan^{-1}x\right) = x$$

The derivative is defined for all real numbers. If $u$ is a differentiable function of $x$, we get the Chain Rule form:

$$\frac{d}{dx}(\tan^{-1}u) = \frac{1}{1 + u^2}\frac{du}{dx}.$$

### The Derivative of $y = \sec^{-1}u$

Since the derivative of $\sec x$ is positive for $0 < x < \pi/2$ and $\pi/2 < x < \pi$, Theorem 1 says that the inverse function $y = \sec^{-1}x$ is differentiable. Instead of applying the formula in Theorem 1 directly, we find the derivative of $y = \sec^{-1} x$, $|x| > 1$, using implicit differentiation and the Chain Rule as follows:

$$y = \sec^{-1} x$$
$$\sec y = x \qquad \text{Inverse function relationship}$$
$$\frac{d}{dx}(\sec y) = \frac{d}{dx} x \qquad \text{Differentiate both sides.}$$
$$\sec y \tan y \frac{dy}{dx} = 1 \qquad \text{Chain Rule}$$
$$\frac{dy}{dx} = \frac{1}{\sec y \tan y}. \qquad \begin{array}{l} \text{Since } |x| > 1, y \text{ lies in} \\ (0, \pi/2) \cup (\pi/2, \pi) \text{ and} \\ \sec y \tan y \neq 0. \end{array}$$

To express the result in terms of $x$, we use the relationships

$$\sec y = x \qquad \text{and} \qquad \tan y = \pm\sqrt{\sec^2 y - 1} = \pm\sqrt{x^2 - 1}$$

to get

$$\frac{dy}{dx} = \pm \frac{1}{x\sqrt{x^2 - 1}}.$$

Can we do anything about the $\pm$ sign? A glance at Figure 7.31 shows that the slope of the graph $y = \sec^{-1} x$ is always positive. Thus,

$$\frac{d}{dx} \sec^{-1} x = \begin{cases} +\dfrac{1}{x\sqrt{x^2 - 1}} & \text{if } x > 1 \\[2ex] -\dfrac{1}{x\sqrt{x^2 - 1}} & \text{if } x < -1. \end{cases}$$

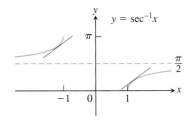

**FIGURE 7.31** The slope of the curve $y = \sec^{-1} x$ is positive for both $x < -1$ and $x > 1$.

With the absolute value symbol, we can write a single expression that eliminates the "$\pm$" ambiguity:

$$\frac{d}{dx} \sec^{-1} x = \frac{1}{|x|\sqrt{x^2 - 1}}.$$

If $u$ is a differentiable function of $x$ with $|u| > 1$, we have the formula

$$\frac{d}{dx}(\sec^{-1} u) = \frac{1}{|u|\sqrt{u^2 - 1}} \frac{du}{dx}, \qquad |u| > 1.$$

**EXAMPLE 5** Using the Chain Rule and derivative of the arcsecant function, we find

$$\frac{d}{dx} \sec^{-1}(5x^4) = \frac{1}{|5x^4|\sqrt{(5x^4)^2 - 1}} \frac{d}{dx}(5x^4)$$

$$= \frac{1}{5x^4\sqrt{25x^8 - 1}}(20x^3) \qquad 5x^4 > 1 > 0$$

$$= \frac{4}{x\sqrt{25x^8 - 1}}. \qquad \blacksquare$$

### Derivatives of the Other Three Inverse Trigonometric Functions

We could use the same techniques to find the derivatives of the other three inverse trigonometric functions—arccosine, arccotangent, and arccosecant—but there is an easier way, thanks to the following identities.

---

**Inverse Function–Inverse Cofunction Identities**

$$\cos^{-1}x = \pi/2 - \sin^{-1}x$$
$$\cot^{-1}x = \pi/2 - \tan^{-1}x$$
$$\csc^{-1}x = \pi/2 - \sec^{-1}x$$

---

We saw the first of these identities in Equation (4). The others are derived in a similar way. It follows easily that the derivatives of the inverse cofunctions are the negatives of the derivatives of the corresponding inverse functions. For example, the derivative of $\cos^{-1}x$ is calculated as follows:

$$\frac{d}{dx}(\cos^{-1}x) = \frac{d}{dx}\left(\frac{\pi}{2} - \sin^{-1}x\right) \qquad \text{Identity}$$

$$= -\frac{d}{dx}(\sin^{-1}x)$$

$$= -\frac{1}{\sqrt{1 - x^2}}. \qquad \text{Derivative of arcsine}$$

The derivatives of the inverse trigonometric functions are summarized in Table 7.3.

---

**TABLE 7.3** Derivatives of the inverse trigonometric functions

**1.** $\dfrac{d(\sin^{-1}u)}{dx} = \dfrac{1}{\sqrt{1 - u^2}}\dfrac{du}{dx}, \quad |u| < 1$

**2.** $\dfrac{d(\cos^{-1}u)}{dx} = -\dfrac{1}{\sqrt{1 - u^2}}\dfrac{du}{dx}, \quad |u| < 1$

**3.** $\dfrac{d(\tan^{-1}u)}{dx} = \dfrac{1}{1 + u^2}\dfrac{du}{dx}$

**4.** $\dfrac{d(\cot^{-1}u)}{dx} = -\dfrac{1}{1 + u^2}\dfrac{du}{dx}$

**5.** $\dfrac{d(\sec^{-1}u)}{dx} = \dfrac{1}{|u|\sqrt{u^2 - 1}}\dfrac{du}{dx}, \quad |u| > 1$

**6.** $\dfrac{d(\csc^{-1}u)}{dx} = -\dfrac{1}{|u|\sqrt{u^2 - 1}}\dfrac{du}{dx}, \quad |u| > 1$

---

### Integration Formulas

The derivative formulas in Table 7.3 yield three useful integration formulas in Table 7.4. The formulas are readily verified by differentiating the functions on the right-hand sides.

---

**TABLE 7.4** Integrals evaluated with inverse trigonometric functions

The following formulas hold for any constant $a > 0$.

**1.** $\displaystyle\int \frac{du}{\sqrt{a^2 - u^2}} = \sin^{-1}\left(\frac{u}{a}\right) + C$ (Valid for $u^2 < a^2$)

**2.** $\displaystyle\int \frac{du}{a^2 + u^2} = \frac{1}{a}\tan^{-1}\left(\frac{u}{a}\right) + C$ (Valid for all $u$)

**3.** $\displaystyle\int \frac{du}{u\sqrt{u^2 - a^2}} = \frac{1}{a}\sec^{-1}\left|\frac{u}{a}\right| + C$ (Valid for $|u| > a > 0$)

---

The derivative formulas in Table 7.3 have $a = 1$, but in most integrations $a \neq 1$, and the formulas in Table 7.4 are more useful.

**EXAMPLE 6**  These examples illustrate how we use Table 7.4.

**(a)** $\displaystyle\int_{\sqrt{2}/2}^{\sqrt{3}/2} \frac{dx}{\sqrt{1 - x^2}} = \sin^{-1}x \Big]_{\sqrt{2}/2}^{\sqrt{3}/2}$   $a = 1, u = x$ in Table 7.4, Formula 1

$$= \sin^{-1}\left(\frac{\sqrt{3}}{2}\right) - \sin^{-1}\left(\frac{\sqrt{2}}{2}\right) = \frac{\pi}{3} - \frac{\pi}{4} = \frac{\pi}{12}$$

**(b)** $\displaystyle\int \frac{dx}{\sqrt{3 - 4x^2}} = \frac{1}{2}\int \frac{du}{\sqrt{a^2 - u^2}}$   $a = \sqrt{3}, u = 2x,$ and $du/2 = dx$

$$= \frac{1}{2}\sin^{-1}\left(\frac{u}{a}\right) + C$$   Table 7.4, Formula 1

$$= \frac{1}{2}\sin^{-1}\left(\frac{2x}{\sqrt{3}}\right) + C$$

**(c)** $\displaystyle\int \frac{dx}{\sqrt{e^{2x} - 6}} = \int \frac{du/u}{\sqrt{u^2 - a^2}}$   $u = e^x, du = e^x\,dx,$ $dx = du/e^x = du/u,$ $a = \sqrt{6}$

$$= \int \frac{du}{u\sqrt{u^2 - a^2}}$$

$$= \frac{1}{a}\sec^{-1}\left|\frac{u}{a}\right| + C$$   Table 7.4, Formula 3

$$= \frac{1}{\sqrt{6}}\sec^{-1}\left(\frac{e^x}{\sqrt{6}}\right) + C$$   ∎

**EXAMPLE 7**  Evaluate

**(a)** $\displaystyle\int \frac{dx}{\sqrt{4x - x^2}}$   **(b)** $\displaystyle\int \frac{dx}{4x^2 + 4x + 2}$

**Solution**

**(a)** The expression $\sqrt{4x - x^2}$ does not match any of the formulas in Table 7.4, so we first rewrite $4x - x^2$ by completing the square:

$$4x - x^2 = -(x^2 - 4x) = -(x^2 - 4x + 4) + 4 = 4 - (x - 2)^2.$$

Then we substitute $a = 2$, $u = x - 2$, and $du = dx$ to get

$$\int \frac{dx}{\sqrt{4x - x^2}} = \int \frac{dx}{\sqrt{4 - (x - 2)^2}}$$

$$= \int \frac{du}{\sqrt{a^2 - u^2}} \qquad a = 2, u = x - 2, \text{ and } du = dx$$

$$= \sin^{-1}\left(\frac{u}{a}\right) + C \qquad \text{Table 7.4, Formula 1}$$

$$= \sin^{-1}\left(\frac{x - 2}{2}\right) + C$$

**(b)** We complete the square on the binomial $4x^2 + 4x$:

$$4x^2 + 4x + 2 = 4(x^2 + x) + 2 = 4\left(x^2 + x + \frac{1}{4}\right) + 2 - \frac{4}{4}$$

$$= 4\left(x + \frac{1}{2}\right)^2 + 1 = (2x + 1)^2 + 1.$$

Then,

$$\int \frac{dx}{4x^2 + 4x + 2} = \int \frac{dx}{(2x + 1)^2 + 1} = \frac{1}{2}\int \frac{du}{u^2 + a^2} \qquad \begin{array}{l} a = 1, u = 2x + 1, \\ \text{and } du/2 = dx \end{array}$$

$$= \frac{1}{2} \cdot \frac{1}{a}\tan^{-1}\left(\frac{u}{a}\right) + C \qquad \text{Table 7.4, Formula 2}$$

$$= \frac{1}{2}\tan^{-1}(2x + 1) + C \qquad a = 1, u = 2x + 1 \qquad \blacksquare$$

## Exercises 7.6

### Common Values

Use reference triangles like those in Examples 1 and 3 to find the angles in Exercises 1–8.

**1. a.** $\tan^{-1} 1$      **b.** $\tan^{-1}\left(-\sqrt{3}\right)$      **c.** $\tan^{-1}\left(\frac{1}{\sqrt{3}}\right)$

**2. a.** $\tan^{-1}(-1)$      **b.** $\tan^{-1}\sqrt{3}$      **c.** $\tan^{-1}\left(\frac{-1}{\sqrt{3}}\right)$

**3. a.** $\sin^{-1}\left(\frac{-1}{2}\right)$      **b.** $\sin^{-1}\left(\frac{1}{\sqrt{2}}\right)$      **c.** $\sin^{-1}\left(\frac{-\sqrt{3}}{2}\right)$

**4. a.** $\sin^{-1}\left(\frac{1}{2}\right)$      **b.** $\sin^{-1}\left(\frac{-1}{\sqrt{2}}\right)$      **c.** $\sin^{-1}\left(\frac{\sqrt{3}}{2}\right)$

**5. a.** $\cos^{-1}\left(\frac{1}{2}\right)$      **b.** $\cos^{-1}\left(\frac{-1}{\sqrt{2}}\right)$      **c.** $\cos^{-1}\left(\frac{\sqrt{3}}{2}\right)$

**6. a.** $\csc^{-1}\sqrt{2}$      **b.** $\csc^{-1}\left(\frac{-2}{\sqrt{3}}\right)$      **c.** $\csc^{-1} 2$

**7. a.** $\sec^{-1}\left(-\sqrt{2}\right)$      **b.** $\sec^{-1}\left(\frac{2}{\sqrt{3}}\right)$      **c.** $\sec^{-1}(-2)$

**8. a.** $\cot^{-1}(-1)$      **b.** $\cot^{-1}\left(\sqrt{3}\right)$      **c.** $\cot^{-1}\left(\frac{-1}{\sqrt{3}}\right)$

### Evaluations

Find the values in Exercises 9–12.

**9.** $\sin\left(\cos^{-1}\left(\frac{\sqrt{2}}{2}\right)\right)$      **10.** $\sec\left(\cos^{-1}\frac{1}{2}\right)$

**11.** $\tan\left(\sin^{-1}\left(-\frac{1}{2}\right)\right)$      **12.** $\cot\left(\sin^{-1}\left(-\frac{\sqrt{3}}{2}\right)\right)$

### Limits

Find the limits in Exercises 13–20. (If in doubt, look at the function's graph.)

**13.** $\lim\limits_{x \to 1^-} \sin^{-1} x$      **14.** $\lim\limits_{x \to -1^+} \cos^{-1} x$

**15.** $\lim\limits_{x \to \infty} \tan^{-1} x$      **16.** $\lim\limits_{x \to -\infty} \tan^{-1} x$

**17.** $\lim\limits_{x \to \infty} \sec^{-1} x$        **18.** $\lim\limits_{x \to -\infty} \sec^{-1} x$

**19.** $\lim\limits_{x \to \infty} \csc^{-1} x$        **20.** $\lim\limits_{x \to -\infty} \csc^{-1} x$

### Finding Derivatives

In Exercises 21–42, find the derivative of $y$ with respect to the appropriate variable.

**21.** $y = \cos^{-1}(x^2)$      **22.** $y = \cos^{-1}(1/x)$

**23.** $y = \sin^{-1}\sqrt{2}\,t$      **24.** $y = \sin^{-1}(1 - t)$

**25.** $y = \sec^{-1}(2s + 1)$      **26.** $y = \sec^{-1} 5s$

**27.** $y = \csc^{-1}(x^2 + 1), \; x > 0$      **28.** $y = \csc^{-1}\dfrac{x}{2}$

**29.** $y = \sec^{-1}\dfrac{1}{t}, \; 0 < t < 1$      **30.** $y = \sin^{-1}\dfrac{3}{t^2}$

**31.** $y = \cot^{-1}\sqrt{t}$      **32.** $y = \cot^{-1}\sqrt{t - 1}$

**33.** $y = \ln(\tan^{-1} x)$      **34.** $y = \tan^{-1}(\ln x)$

**35.** $y = \csc^{-1}(e^t)$      **36.** $y = \cos^{-1}(e^{-t})$

**37.** $y = s\sqrt{1 - s^2} + \cos^{-1} s$      **38.** $y = \sqrt{s^2 - 1} - \sec^{-1} s$

**39.** $y = \tan^{-1}\sqrt{x^2 - 1} + \csc^{-1} x, \quad x > 1$

**40.** $y = \cot^{-1}\dfrac{1}{x} - \tan^{-1} x$      **41.** $y = x\sin^{-1} x + \sqrt{1 - x^2}$

**42.** $y = \ln(x^2 + 4) - x\tan^{-1}\left(\dfrac{x}{2}\right)$

### Evaluating Integrals

Evaluate the integrals in Exercises 43–66.

**43.** $\displaystyle\int \dfrac{dx}{\sqrt{9 - x^2}}$      **44.** $\displaystyle\int \dfrac{dx}{\sqrt{1 - 4x^2}}$

**45.** $\displaystyle\int \dfrac{dx}{17 + x^2}$      **46.** $\displaystyle\int \dfrac{dx}{9 + 3x^2}$

**47.** $\displaystyle\int \dfrac{dx}{x\sqrt{25x^2 - 2}}$      **48.** $\displaystyle\int \dfrac{dx}{x\sqrt{5x^2 - 4}}$

**49.** $\displaystyle\int_0^1 \dfrac{4\,ds}{\sqrt{4 - s^2}}$      **50.** $\displaystyle\int_0^{3\sqrt{2}/4} \dfrac{ds}{\sqrt{9 - 4s^2}}$

**51.** $\displaystyle\int_0^2 \dfrac{dt}{8 + 2t^2}$      **52.** $\displaystyle\int_{-2}^2 \dfrac{dt}{4 + 3t^2}$

**53.** $\displaystyle\int_{-1}^{-\sqrt{2}/2} \dfrac{dy}{y\sqrt{4y^2 - 1}}$      **54.** $\displaystyle\int_{-2/3}^{-\sqrt{2}/3} \dfrac{dy}{y\sqrt{9y^2 - 1}}$

**55.** $\displaystyle\int \dfrac{3\,dr}{\sqrt{1 - 4(r - 1)^2}}$      **56.** $\displaystyle\int \dfrac{6\,dr}{\sqrt{4 - (r + 1)^2}}$

**57.** $\displaystyle\int \dfrac{dx}{2 + (x - 1)^2}$      **58.** $\displaystyle\int \dfrac{dx}{1 + (3x + 1)^2}$

**59.** $\displaystyle\int \dfrac{dx}{(2x - 1)\sqrt{(2x - 1)^2 - 4}}$

**60.** $\displaystyle\int \dfrac{dx}{(x + 3)\sqrt{(x + 3)^2 - 25}}$

**61.** $\displaystyle\int_{-\pi/2}^{\pi/2} \dfrac{2\cos\theta\,d\theta}{1 + (\sin\theta)^2}$      **62.** $\displaystyle\int_{\pi/6}^{\pi/4} \dfrac{\csc^2 x\,dx}{1 + (\cot x)^2}$

**63.** $\displaystyle\int_0^{\ln\sqrt{3}} \dfrac{e^x\,dx}{1 + e^{2x}}$      **64.** $\displaystyle\int_1^{e^{\pi/4}} \dfrac{4\,dt}{t(1 + \ln^2 t)}$

**65.** $\displaystyle\int \dfrac{y\,dy}{\sqrt{1 - y^4}}$      **66.** $\displaystyle\int \dfrac{\sec^2 y\,dy}{\sqrt{1 - \tan^2 y}}$

Evaluate the integrals in Exercises 67–80.

**67.** $\displaystyle\int \dfrac{dx}{\sqrt{-x^2 + 4x - 3}}$      **68.** $\displaystyle\int \dfrac{dx}{\sqrt{2x - x^2}}$

**69.** $\displaystyle\int_{-1}^0 \dfrac{6\,dt}{\sqrt{3 - 2t - t^2}}$      **70.** $\displaystyle\int_{1/2}^1 \dfrac{6\,dt}{\sqrt{3 + 4t - 4t^2}}$

**71.** $\displaystyle\int \dfrac{dy}{y^2 - 2y + 5}$      **72.** $\displaystyle\int \dfrac{dy}{y^2 + 6y + 10}$

**73.** $\displaystyle\int_1^2 \dfrac{8\,dx}{x^2 - 2x + 2}$      **74.** $\displaystyle\int_2^4 \dfrac{2\,dx}{x^2 - 6x + 10}$

**75.** $\displaystyle\int \dfrac{x + 4}{x^2 + 4}\,dx$      **76.** $\displaystyle\int \dfrac{t - 2}{t^2 - 6t + 10}\,dt$

**77.** $\displaystyle\int \dfrac{x^2 + 2x - 1}{x^2 + 9}\,dx$      **78.** $\displaystyle\int \dfrac{t^3 - 2t^2 + 3t - 4}{t^2 + 1}\,dt$

**79.** $\displaystyle\int \dfrac{dx}{(x + 1)\sqrt{x^2 + 2x}}$      **80.** $\displaystyle\int \dfrac{dx}{(x - 2)\sqrt{x^2 - 4x + 3}}$

Evaluate the integrals in Exercises 81–90.

**81.** $\displaystyle\int \dfrac{e^{\sin^{-1} x}\,dx}{\sqrt{1 - x^2}}$      **82.** $\displaystyle\int \dfrac{e^{\cos^{-1} x}\,dx}{\sqrt{1 - x^2}}$

**83.** $\displaystyle\int \dfrac{(\sin^{-1} x)^2\,dx}{\sqrt{1 - x^2}}$      **84.** $\displaystyle\int \dfrac{\sqrt{\tan^{-1} x}\,dx}{1 + x^2}$

**85.** $\displaystyle\int \dfrac{dy}{(\tan^{-1} y)(1 + y^2)}$      **86.** $\displaystyle\int \dfrac{dy}{(\sin^{-1} y)\sqrt{1 - y^2}}$

**87.** $\displaystyle\int_{\sqrt{2}}^2 \dfrac{\sec^2(\sec^{-1} x)\,dx}{x\sqrt{x^2 - 1}}$      **88.** $\displaystyle\int_{2/\sqrt{3}}^2 \dfrac{\cos(\sec^{-1} x)\,dx}{x\sqrt{x^2 - 1}}$

**89.** $\displaystyle\int \dfrac{1}{\sqrt{x}(x + 1)\left((\tan^{-1}\sqrt{x})^2 + 9\right)}\,dx$

**90.** $\displaystyle\int \dfrac{e^x\sin^{-1} e^x}{\sqrt{1 - e^{2x}}}\,dx$

### L'Hôpital's Rule

Find the limits in Exercises 91–98.

**91.** $\lim\limits_{x \to 0} \dfrac{\sin^{-1} 5x}{x}$      **92.** $\lim\limits_{x \to 1^+} \dfrac{\sqrt{x^2 - 1}}{\sec^{-1} x}$

**93.** $\lim\limits_{x \to \infty} x\tan^{-1}\dfrac{2}{x}$      **94.** $\lim\limits_{x \to 0} \dfrac{2\tan^{-1} 3x^2}{7x^2}$

**95.** $\lim\limits_{x \to 0} \dfrac{\tan^{-1} x^2}{x\sin^{-1} x}$      **96.** $\lim\limits_{x \to \infty} \dfrac{e^x\tan^{-1} e^x}{e^{2x} + x}$

**97.** $\lim\limits_{x \to 0^+} \dfrac{(\tan^{-1}\sqrt{x})^2}{x\sqrt{x + 1}}$      **98.** $\lim\limits_{x \to 0^+} \dfrac{\sin^{-1} x^2}{(\sin^{-1} x)^2}$

## Integration Formulas

Verify the integration formulas in Exercises 99–102.

**99.** $\displaystyle\int \frac{\tan^{-1} x}{x^2}\,dx = \ln x - \frac{1}{2}\ln(1 + x^2) - \frac{\tan^{-1} x}{x} + C$

**100.** $\displaystyle\int x^3 \cos^{-1} 5x\,dx = \frac{x^4}{4}\cos^{-1} 5x + \frac{5}{4}\int \frac{x^4\,dx}{\sqrt{1 - 25x^2}}$

**101.** $\displaystyle\int (\sin^{-1} x)^2\,dx = x(\sin^{-1} x)^2 - 2x + 2\sqrt{1 - x^2}\,\sin^{-1} x + C$

**102.** $\displaystyle\int \ln(a^2 + x^2)\,dx = x\ln(a^2 + x^2) - 2x + 2a\tan^{-1}\frac{x}{a} + C$

## Initial Value Problems

Solve the initial value problems in Exercises 103–106.

**103.** $\dfrac{dy}{dx} = \dfrac{1}{\sqrt{1 - x^2}}, \quad y(0) = 0$

**104.** $\dfrac{dy}{dx} = \dfrac{1}{x^2 + 1} - 1, \quad y(0) = 1$

**105.** $\dfrac{dy}{dx} = \dfrac{1}{x\sqrt{x^2 - 1}}, \quad x > 1; \quad y(2) = \pi$

**106.** $\dfrac{dy}{dx} = \dfrac{1}{1 + x^2} - \dfrac{2}{\sqrt{1 - x^2}}, \quad y(0) = 2$

## Applications and Theory

**107.** You are sitting in a classroom next to the wall looking at the blackboard at the front of the room. The blackboard is 4 m long and starts 1 m from the wall you are sitting next to.

**a.** Show that your viewing angle is

$$\alpha = \cot^{-1}\frac{x}{5} - \cot^{-1} x$$

if you are $x$ m from the front wall.

**b.** Find $x$ so that $\alpha$ is as large as possible.

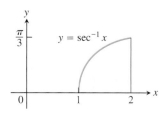

**108.** The region between the curve $y = \sec^{-1} x$ and the $x$-axis from $x = 1$ to $x = 2$ (shown here) is revolved about the $y$-axis to generate a solid. Find the volume of the solid.

**109.** The slant height of the cone shown here is 3 m. How large should the indicated angle be to maximize the cone's volume?

What angle here gives the largest volume?

**110.** Find the angle $\alpha$.

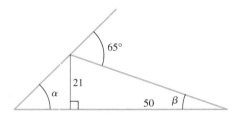

**111.** Here is an informal proof that $\tan^{-1} 1 + \tan^{-1} 2 + \tan^{-1} 3 = \pi$. Explain what is going on.

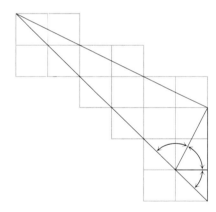

**112. Two derivations of the identity $\sec^{-1}(-x) = \pi - \sec^{-1} x$**

**a.** (*Geometric*) Here is a pictorial proof that $\sec^{-1}(-x) = \pi - \sec^{-1} x$. See if you can tell what is going on.

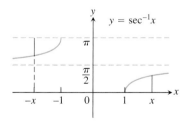

**b.** (*Algebraic*) Derive the identity $\sec^{-1}(-x) = \pi - \sec^{-1} x$ by combining the following two equations from the text:

$$\cos^{-1}(-x) = \pi - \cos^{-1} x \qquad \text{Eq. (3)}$$
$$\sec^{-1} x = \cos^{-1}(1/x) \qquad \text{Eq. (5)}$$

**113. The identity $\sin^{-1}x + \cos^{-1}x = \pi/2$** Figure 7.28 establishes the identity for $0 < x < 1$. To establish it for the rest of $[-1, 1]$, verify by direct calculation that it holds for $x = 1, 0,$ and $-1$. Then, for values of $x$ in $(-1, 0)$, let $x = -a, a > 0,$ and apply Eqs. (1) and (3) to the sum $\sin^{-1}(-a) + \cos^{-1}(-a)$.

**114.** Show that the sum $\tan^{-1}x + \tan^{-1}(1/x)$ is constant.

**115.** Use the identity

$$\csc^{-1}u = \frac{\pi}{2} - \sec^{-1}u$$

to derive the formula for the derivative of $\csc^{-1}u$ in Table 7.3 from the formula for the derivative of $\sec^{-1}u$.

**116.** Derive the formula

$$\frac{dy}{dx} = \frac{1}{1 + x^2}$$

for the derivative of $y = \tan^{-1}x$ by differentiating both sides of the equivalent equation $\tan y = x$.

**117.** Use the Derivative Rule, Theorem 1, to derive

$$\frac{d}{dx}\sec^{-1}x = \frac{1}{|x|\sqrt{x^2 - 1}}, \quad |x| > 1.$$

**118.** Use the identity

$$\cot^{-1}u = \frac{\pi}{2} - \tan^{-1}u$$

to derive the formula for the derivative of $\cot^{-1}u$ in Table 7.3 from the formula for the derivative of $\tan^{-1}u$.

**119.** What is special about the functions

$$f(x) = \sin^{-1}\frac{x - 1}{x + 1}, \quad x \geq 0, \quad \text{and} \quad g(x) = 2\tan^{-1}\sqrt{x}?$$

Explain.

**120.** What is special about the functions

$$f(x) = \sin^{-1}\frac{1}{\sqrt{x^2 + 1}} \quad \text{and} \quad g(x) = \tan^{-1}\frac{1}{x}?$$

Explain.

**121.** Find the volume of the solid of revolution shown here.

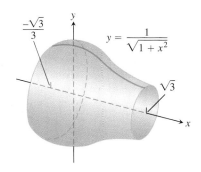

$\dfrac{-\sqrt{3}}{3}$

$y = \dfrac{1}{\sqrt{1 + x^2}}$

$\sqrt{3}$

**122. Arc length** Find the circumference of a circle of radius $r$ using Eq. (3) in Section 6.3.

Find the volumes of the solids in Exercises 123 and 124.

**123.** The solid lies between planes perpendicular to the $x$-axis at $x = -1$ and $x = 1$. The cross-sections perpendicular to the $x$-axis are

    **a.** circles whose diameters stretch from the curve $y = -1/\sqrt{1 + x^2}$ to the curve $y = 1/\sqrt{1 + x^2}$.

    **b.** vertical squares whose base edges run from the curve $y = -1/\sqrt{1 + x^2}$ to the curve $y = 1/\sqrt{1 + x^2}$.

**124.** The solid lies between planes perpendicular to the $x$-axis at $x = -\sqrt{2}/2$ and $x = \sqrt{2}/2$. The cross-sections perpendicular to the $x$-axis are

    **a.** circles whose diameters stretch from the $x$-axis to the curve $y = 2/\sqrt[4]{1 - x^2}$.

    **b.** squares whose diagonals stretch from the $x$-axis to the curve $y = 2/\sqrt[4]{1 - x^2}$.

[T] **125.** Find the values of the following.

    **a.** $\sec^{-1}1.5$     **b.** $\csc^{-1}(-1.5)$     **c.** $\cot^{-1}2$

[T] **126.** Find the values of the following.

    **a.** $\sec^{-1}(-3)$     **b.** $\csc^{-1}1.7$     **c.** $\cot^{-1}(-2)$

[T] In Exercises 127–129, find the domain and range of each composite function. Then graph the composites on separate screens. Do the graphs make sense in each case? Give reasons for your answers. Comment on any differences you see.

**127. a.** $y = \tan^{-1}(\tan x)$     **b.** $y = \tan(\tan^{-1}x)$

**128. a.** $y = \sin^{-1}(\sin x)$     **b.** $y = \sin(\sin^{-1}x)$

**129. a.** $y = \cos^{-1}(\cos x)$     **b.** $y = \cos(\cos^{-1}x)$

[T] Use your graphing utility for Exercises 130–134.

**130.** Graph $y = \sec(\sec^{-1}x) = \sec(\cos^{-1}(1/x))$. Explain what you see.

**131. Newton's serpentine** Graph $y = 4x/(x^2 + 1)$, known as Newton's serpentine. Then graph $y = 2\sin(2\tan^{-1}x)$ in the same graphing window. What do you see? Explain.

**132.** Graph the rational function $y = (2 - x^2)/x^2$. Then graph $y = \cos(2\sec^{-1}x)$ in the same graphing window. What do you see? Explain.

**133.** Graph $f(x) = \sin^{-1}x$ together with its first two derivatives. Comment on the behavior of $f$ and the shape of its graph in relation to the signs and values of $f'$ and $f''$.

**134.** Graph $f(x) = \tan^{-1}x$ together with its first two derivatives. Comment on the behavior of $f$ and the shape of its graph in relation to the signs and values of $f'$ and $f''$.

# 7.7 Hyperbolic Functions

The hyperbolic functions are formed by taking combinations of the two exponential functions $e^x$ and $e^{-x}$. The hyperbolic functions simplify many mathematical expressions and occur frequently in mathematical and engineering applications. In this section we give a brief introduction to these functions, their graphs, their derivatives, their integrals, and their inverse functions.

## Definitions and Identities

The hyperbolic sine and hyperbolic cosine functions are defined by the equations

$$\sinh x = \frac{e^x - e^{-x}}{2} \qquad \text{and} \qquad \cosh x = \frac{e^x + e^{-x}}{2}.$$

We pronounce $\sinh x$ as "cinch $x$," rhyming with "pinch $x$," and $\cosh x$ as "kosh $x$," rhyming with "gosh $x$." From this basic pair, we define the hyperbolic tangent, cotangent, secant, and cosecant functions. The defining equations and graphs of these functions are shown in Table 7.5. We will see that the hyperbolic functions bear many similarities to the trigonometric functions after which they are named.

Hyperbolic functions satisfy the identities in Table 7.6. Except for differences in sign, these resemble identities we know for the trigonometric functions. The identities are proved directly from the definitions, as we show here for the second one:

$$2 \sinh x \cosh x = 2\left(\frac{e^x - e^{-x}}{2}\right)\left(\frac{e^x + e^{-x}}{2}\right)$$

$$= \frac{e^{2x} - e^{-2x}}{2}$$

$$= \sinh 2x.$$

**TABLE 7.5** The six basic hyperbolic functions

(a)

**Hyperbolic sine:**

$$\sinh x = \frac{e^x - e^{-x}}{2}$$

(b)

**Hyperbolic cosine:**

$$\cosh x = \frac{e^x + e^{-x}}{2}$$

(c)

**Hyperbolic tangent:**

$$\tanh x = \frac{\sinh x}{\cosh x} = \frac{e^x - e^{-x}}{e^x + e^{-x}}$$

**Hyperbolic cotangent:**

$$\coth x = \frac{\cosh x}{\sinh x} = \frac{e^x + e^{-x}}{e^x - e^{-x}}$$

(d)

**Hyperbolic secant:**

$$\text{sech } x = \frac{1}{\cosh x} = \frac{2}{e^x + e^{-x}}$$

(e)

**Hyperbolic cosecant:**

$$\text{csch } x = \frac{1}{\sinh x} = \frac{2}{e^x - e^{-x}}$$

**TABLE 7.6** Identities for hyperbolic functions

$$\cosh^2 x - \sinh^2 x = 1$$

$$\sinh 2x = 2 \sinh x \cosh x$$

$$\cosh 2x = \cosh^2 x + \sinh^2 x$$

$$\cosh^2 x = \frac{\cosh 2x + 1}{2}$$

$$\sinh^2 x = \frac{\cosh 2x - 1}{2}$$

$$\tanh^2 x = 1 - \text{sech}^2 x$$

$$\coth^2 x = 1 + \text{csch}^2 x$$

The other identities are obtained similarly, by substituting in the definitions of the hyperbolic functions and using algebra. Like many standard functions, hyperbolic functions and their inverses are easily evaluated with calculators, which often have special keys for that purpose.

For any real number $u$, we know the point with coordinates $(\cos u, \sin u)$ lies on the unit circle $x^2 + y^2 = 1$. So the trigonometric functions are sometimes called the *circular* functions. Because of the first identity

$$\cosh^2 u - \sinh^2 u = 1,$$

with $u$ substituted for $x$ in Table 7.6, the point having coordinates $(\cosh u, \sinh u)$ lies on the right-hand branch of the hyperbola $x^2 - y^2 = 1$. This is where the *hyperbolic* functions get their names (see Exercise 86).

Hyperbolic functions are useful in finding integrals, which we will see in Chapter 8. They play an important role in science and engineering as well. The hyperbolic cosine describes the shape of a hanging cable or wire that is strung between two points at the same height and hanging freely (see Exercise 83). The shape of the St. Louis Arch is an inverted hyperbolic cosine. The hyperbolic tangent occurs in the formula for the velocity of an ocean wave moving over water having a constant depth, and the inverse hyperbolic tangent describes how relative velocities sum according to Einstein's Law in the Special Theory of Relativity.

### Derivatives and Integrals of Hyperbolic Functions

The six hyperbolic functions, being rational combinations of the differentiable functions $e^x$ and $e^{-x}$, have derivatives at every point at which they are defined (Table 7.7). Again, there are similarities with trigonometric functions.

The derivative formulas are derived from the derivative of $e^u$:

$$\frac{d}{dx}(\sinh u) = \frac{d}{dx}\left(\frac{e^u - e^{-u}}{2}\right) \qquad \text{Definition of } \sinh u$$

$$= \frac{e^u \, du/dx + e^{-u} \, du/dx}{2} \qquad \text{Derivative of } e^u$$

$$= \cosh u \frac{du}{dx}. \qquad \text{Definition of } \cosh u$$

This gives the first derivative formula. From the definition, we can calculate the derivative of the hyperbolic cosecant function, as follows:

$$\frac{d}{dx}(\operatorname{csch} u) = \frac{d}{dx}\left(\frac{1}{\sinh u}\right) \qquad \text{Definition of csch } u$$

$$= -\frac{\cosh u \, du}{\sinh^2 u \, dx} \qquad \text{Quotient Rule for derivatives}$$

$$= -\frac{1}{\sinh u}\frac{\cosh u \, du}{\sinh u \, dx} \qquad \text{Rearrange terms.}$$

$$= -\operatorname{csch} u \coth u \frac{du}{dx} \qquad \text{Definitions of csch } u \text{ and coth } u$$

The other formulas in Table 7.7 are obtained similarly.

The derivative formulas lead to the integral formulas in Table 7.8.

**TABLE 7.7** Derivatives of hyperbolic functions

$$\frac{d}{dx}(\sinh u) = \cosh u \frac{du}{dx}$$

$$\frac{d}{dx}(\cosh u) = \sinh u \frac{du}{dx}$$

$$\frac{d}{dx}(\tanh u) = \operatorname{sech}^2 u \frac{du}{dx}$$

$$\frac{d}{dx}(\coth u) = -\operatorname{csch}^2 u \frac{du}{dx}$$

$$\frac{d}{dx}(\operatorname{sech} u) = -\operatorname{sech} u \tanh u \frac{du}{dx}$$

$$\frac{d}{dx}(\operatorname{csch} u) = -\operatorname{csch} u \coth u \frac{du}{dx}$$

**TABLE 7.8** Integral formulas for hyperbolic functions

$$\int \sinh u \, du = \cosh u + C$$

$$\int \cosh u \, du = \sinh u + C$$

$$\int \operatorname{sech}^2 u \, du = \tanh u + C$$

$$\int \operatorname{csch}^2 u \, du = -\coth u + C$$

$$\int \operatorname{sech} u \tanh u \, du = -\operatorname{sech} u + C$$

$$\int \operatorname{csch} u \coth u \, du = -\operatorname{csch} u + C$$

**EXAMPLE 1** We illustrate the derivative and integral formulas.

**(a)** $\dfrac{d}{dt}\left(\tanh \sqrt{1+t^2}\right) = \text{sech}^2 \sqrt{1+t^2} \cdot \dfrac{d}{dt}\left(\sqrt{1+t^2}\right)$

$$= \frac{t}{\sqrt{1+t^2}}\text{sech}^2 \sqrt{1+t^2}$$

**(b)** $\displaystyle\int \coth 5x \, dx = \int \dfrac{\cosh 5x}{\sinh 5x}dx = \dfrac{1}{5}\int \dfrac{du}{u}$     $u = \sinh 5x,$
$du = 5 \cosh 5x \, dx$

$$= \frac{1}{5}\ln |u| + C = \frac{1}{5}\ln |\sinh 5x| + C$$

**(c)** $\displaystyle\int_0^1 \sinh^2 x \, dx = \int_0^1 \dfrac{\cosh 2x - 1}{2}dx$   Table 7.6

$$= \frac{1}{2}\int_0^1 (\cosh 2x - 1)\, dx = \frac{1}{2}\left[\frac{\sinh 2x}{2} - x\right]_0^1$$

$$= \frac{\sinh 2}{4} - \frac{1}{2} \approx 0.40672$$   Evaluate with a calculator.

**(d)** $\displaystyle\int_0^{\ln 2} 4e^x \sinh x \, dx = \int_0^{\ln 2} 4e^x \dfrac{e^x - e^{-x}}{2}dx = \int_0^{\ln 2} (2e^{2x} - 2)\, dx$

$$= \left[e^{2x} - 2x\right]_0^{\ln 2} = (e^{2\ln 2} - 2\ln 2) - (1 - 0)$$

$$= 4 - 2\ln 2 - 1 \approx 1.6137$$   ∎

## Inverse Hyperbolic Functions

The inverses of the six basic hyperbolic functions are very useful in integration (see Chapter 8). Since $d(\sinh x)/dx = \cosh x > 0$, the hyperbolic sine is an increasing function of $x$. We denote its inverse by

$$y = \sinh^{-1} x.$$

For every value of $x$ in the interval $-\infty < x < \infty$, the value of $y = \sinh^{-1} x$ is the number whose hyperbolic sine is $x$. The graphs of $y = \sinh x$ and $y = \sinh^{-1} x$ are shown in Figure 7.32a.

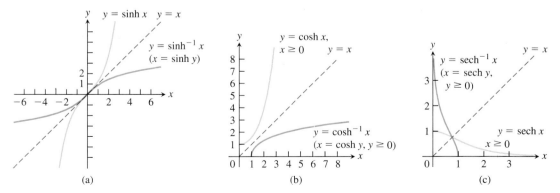

**FIGURE 7.32** The graphs of the inverse hyperbolic sine, cosine, and secant of $x$. Notice the symmetries about the line $y = x$.

The function $y = \cosh x$ is not one-to-one because its graph in Table 7.5 does not pass the horizontal line test. The restricted function $y = \cosh x, x \geq 0$, however, is one-to-one and therefore has an inverse, denoted by

$$y = \cosh^{-1} x.$$

For every value of $x \geq 1$, $y = \cosh^{-1} x$ is the number in the interval $0 \leq y < \infty$ whose hyperbolic cosine is $x$. The graphs of $y = \cosh x, x \geq 0$, and $y = \cosh^{-1} x$ are shown in Figure 7.32b.

Like $y = \cosh x$, the function $y = \operatorname{sech} x = 1/\cosh x$ fails to be one-to-one, but its restriction to nonnegative values of $x$ does have an inverse, denoted by

$$y = \operatorname{sech}^{-1} x.$$

For every value of $x$ in the interval $(0, 1]$, $y = \operatorname{sech}^{-1} x$ is the nonnegative number whose hyperbolic secant is $x$. The graphs of $y = \operatorname{sech} x, x \geq 0$, and $y = \operatorname{sech}^{-1} x$ are shown in Figure 7.32c.

The hyperbolic tangent, cotangent, and cosecant are one-to-one on their domains and therefore have inverses, denoted by

$$y = \tanh^{-1} x, \qquad y = \coth^{-1} x, \qquad y = \operatorname{csch}^{-1} x.$$

These functions are graphed in Figure 7.33.

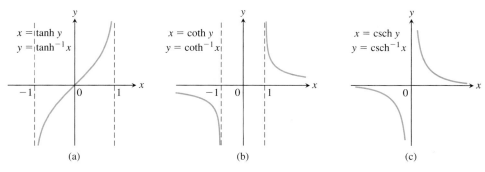

**FIGURE 7.33**   The graphs of the inverse hyperbolic tangent, cotangent, and cosecant of $x$.

## Useful Identities

**TABLE 7.9** Identities for inverse hyperbolic functions

$$\operatorname{sech}^{-1} x = \cosh^{-1} \frac{1}{x}$$

$$\operatorname{csch}^{-1} x = \sinh^{-1} \frac{1}{x}$$

$$\coth^{-1} x = \tanh^{-1} \frac{1}{x}$$

We use the identities in Table 7.9 to calculate the values of $\operatorname{sech}^{-1} x$, $\operatorname{csch}^{-1} x$, and $\coth^{-1} x$ on calculators that give only $\cosh^{-1} x$, $\sinh^{-1} x$, and $\tanh^{-1} x$. These identities are direct consequences of the definitions. For example, if $0 < x \leq 1$, then

$$\operatorname{sech}\left(\cosh^{-1}\left(\frac{1}{x}\right)\right) = \frac{1}{\cosh\left(\cosh^{-1}\left(\frac{1}{x}\right)\right)} = \frac{1}{\left(\frac{1}{x}\right)} = x.$$

We also know that $\operatorname{sech}(\operatorname{sech}^{-1} x) = x$, so because the hyperbolic secant is one-to-one on $(0, 1]$, we have

$$\cosh^{-1}\left(\frac{1}{x}\right) = \operatorname{sech}^{-1} x.$$

### Derivatives of Inverse Hyperbolic Functions

An important use of inverse hyperbolic functions lies in antiderivatives that reverse the derivative formulas in Table 7.10.

**TABLE 7.10** Derivatives of inverse hyperbolic functions

$$\frac{d(\sinh^{-1} u)}{dx} = \frac{1}{\sqrt{1 + u^2}} \frac{du}{dx}$$

$$\frac{d(\cosh^{-1} u)}{dx} = \frac{1}{\sqrt{u^2 - 1}} \frac{du}{dx}, \qquad u > 1$$

$$\frac{d(\tanh^{-1} u)}{dx} = \frac{1}{1 - u^2} \frac{du}{dx}, \qquad |u| < 1$$

$$\frac{d(\coth^{-1} u)}{dx} = \frac{1}{1 - u^2} \frac{du}{dx}, \qquad |u| > 1$$

$$\frac{d(\operatorname{sech}^{-1} u)}{dx} = -\frac{1}{u\sqrt{1 - u^2}} \frac{du}{dx}, \qquad 0 < u < 1$$

$$\frac{d(\operatorname{csch}^{-1} u)}{dx} = -\frac{1}{|u|\sqrt{1 + u^2}} \frac{du}{dx}, \qquad u \neq 0$$

The restrictions $|u| < 1$ and $|u| > 1$ on the derivative formulas for $\tanh^{-1} u$ and $\coth^{-1} u$ come from the natural restrictions on the values of these functions. (See Figure 7.33a and b.) The distinction between $|u| < 1$ and $|u| > 1$ becomes important when we convert the derivative formulas into integral formulas.

We illustrate how the derivatives of the inverse hyperbolic functions are found in Example 2, where we calculate $d(\cosh^{-1} u)/dx$. The other derivatives are obtained by similar calculations.

**EXAMPLE 2**     Show that if $u$ is a differentiable function of $x$ whose values are greater than 1, then

$$\frac{d}{dx}(\cosh^{-1} u) = \frac{1}{\sqrt{u^2 - 1}} \frac{du}{dx}.$$

**Solution**   First we find the derivative of $y = \cosh^{-1} x$ for $x > 1$ by applying Theorem 1 of Section 7.1 with $f(x) = \cosh x$ and $f^{-1}(x) = \cosh^{-1} x$. Theorem 1 can be applied because the derivative of $\cosh x$ is positive for $0 < x$.

$$(f^{-1})'(x) = \frac{1}{f'(f^{-1}(x))} \qquad \text{Theorem 1}$$

$$= \frac{1}{\sinh(\cosh^{-1} x)} \qquad f'(u) = \sinh u$$

$$= \frac{1}{\sqrt{\cosh^2(\cosh^{-1} x) - 1}} \qquad \begin{array}{l} \cosh^2 u - \sinh^2 u = 1, \\ \sinh u = \sqrt{\cosh^2 u - 1} \end{array}$$

$$= \frac{1}{\sqrt{x^2 - 1}} \qquad \cosh(\cosh^{-1} x) = x$$

The Chain Rule gives the final result:

$$\frac{d}{dx}(\cosh^{-1} u) = \frac{1}{\sqrt{u^2 - 1}}\frac{du}{dx}.$$

With appropriate substitutions, the derivative formulas in Table 7.10 lead to the integration formulas in Table 7.11. Each of the formulas in Table 7.11 can be verified by differentiating the expression on the right-hand side.

---

**TABLE 7.11** Integrals leading to inverse hyperbolic functions

1. $\displaystyle\int \frac{du}{\sqrt{a^2 + u^2}} = \sinh^{-1}\left(\frac{u}{a}\right) + C,$    $a > 0$

2. $\displaystyle\int \frac{du}{\sqrt{u^2 - a^2}} = \cosh^{-1}\left(\frac{u}{a}\right) + C,$    $u > a > 0$

3. $\displaystyle\int \frac{du}{a^2 - u^2} = \begin{cases} \dfrac{1}{a}\tanh^{-1}\left(\dfrac{u}{a}\right) + C, & u^2 < a^2 \\[2mm] \dfrac{1}{a}\coth^{-1}\left(\dfrac{u}{a}\right) + C, & u^2 > a^2 \end{cases}$

4. $\displaystyle\int \frac{du}{u\sqrt{a^2 - u^2}} = -\frac{1}{a}\operatorname{sech}^{-1}\left(\frac{u}{a}\right) + C,$    $0 < u < a$

5. $\displaystyle\int \frac{du}{u\sqrt{a^2 + u^2}} = -\frac{1}{a}\operatorname{csch}^{-1}\left|\frac{u}{a}\right| + C,$    $u \neq 0$ and $a > 0$

---

**EXAMPLE 3**    Evaluate

$$\int_0^1 \frac{2\,dx}{\sqrt{3 + 4x^2}}.$$

**Solution**    The indefinite integral is

$$\int \frac{2\,dx}{\sqrt{3 + 4x^2}} = \int \frac{du}{\sqrt{a^2 + u^2}} \qquad u = 2x, \quad du = 2\,dx, \quad a = \sqrt{3}$$

$$= \sinh^{-1}\left(\frac{u}{a}\right) + C \qquad \text{Formula from Table 7.11}$$

$$= \sinh^{-1}\left(\frac{2x}{\sqrt{3}}\right) + C.$$

Therefore,

$$\int_0^1 \frac{2\,dx}{\sqrt{3 + 4x^2}} = \sinh^{-1}\left(\frac{2x}{\sqrt{3}}\right)\Big]_0^1 = \sinh^{-1}\left(\frac{2}{\sqrt{3}}\right) - \sinh^{-1}(0)$$

$$= \sinh^{-1}\left(\frac{2}{\sqrt{3}}\right) - 0 \approx 0.98665.$$

## Exercises 7.7

### Values and Identities

Each of Exercises 1–4 gives a value of $\sinh x$ or $\cosh x$. Use the definitions and the identity $\cosh^2 x - \sinh^2 x = 1$ to find the values of the remaining five hyperbolic functions.

1. $\sinh x = -\dfrac{3}{4}$

2. $\sinh x = \dfrac{4}{3}$

3. $\cosh x = \dfrac{17}{15}, \quad x > 0$

4. $\cosh x = \dfrac{13}{5}, \quad x > 0$

Rewrite the expressions in Exercises 5–10 in terms of exponentials and simplify the results as much as you can.

5. $2 \cosh (\ln x)$

6. $\sinh (2 \ln x)$

7. $\cosh 5x + \sinh 5x$

8. $\cosh 3x - \sinh 3x$

9. $(\sinh x + \cosh x)^4$

10. $\ln (\cosh x + \sinh x) + \ln (\cosh x - \sinh x)$

11. Prove the identities

$$\sinh (x + y) = \sinh x \cosh y + \cosh x \sinh y,$$
$$\cosh (x + y) = \cosh x \cosh y + \sinh x \sinh y.$$

Then use them to show that

a. $\sinh 2x = 2 \sinh x \cosh x.$

b. $\cosh 2x = \cosh^2 x + \sinh^2 x.$

12. Use the definitions of $\cosh x$ and $\sinh x$ to show that

$$\cosh^2 x - \sinh^2 x = 1.$$

### Finding Derivatives

In Exercises 13–24, find the derivative of $y$ with respect to the appropriate variable.

13. $y = 6 \sinh \dfrac{x}{3}$

14. $y = \dfrac{1}{2} \sinh (2x + 1)$

15. $y = 2\sqrt{t} \tanh \sqrt{t}$

16. $y = t^2 \tanh \dfrac{1}{t}$

17. $y = \ln (\sinh z)$

18. $y = \ln (\cosh z)$

19. $y = \operatorname{sech} \theta (1 - \ln \operatorname{sech} \theta)$

20. $y = \operatorname{csch} \theta (1 - \ln \operatorname{csch} \theta)$

21. $y = \ln \cosh v - \dfrac{1}{2} \tanh^2 v$

22. $y = \ln \sinh v - \dfrac{1}{2} \coth^2 v$

23. $y = (x^2 + 1) \operatorname{sech} (\ln x)$

(*Hint:* Before differentiating, express in terms of exponentials and simplify.)

24. $y = (4x^2 - 1) \operatorname{csch} (\ln 2x)$

In Exercises 25–36, find the derivative of $y$ with respect to the appropriate variable.

25. $y = \sinh^{-1} \sqrt{x}$

26. $y = \cosh^{-1} 2\sqrt{x + 1}$

27. $y = (1 - \theta) \tanh^{-1} \theta$

28. $y = (\theta^2 + 2\theta) \tanh^{-1} (\theta + 1)$

29. $y = (1 - t) \coth^{-1} \sqrt{t}$

30. $y = (1 - t^2) \coth^{-1} t$

31. $y = \cos^{-1} x - x \operatorname{sech}^{-1} x$

32. $y = \ln x + \sqrt{1 - x^2} \operatorname{sech}^{-1} x$

33. $y = \operatorname{csch}^{-1} \left( \dfrac{1}{2} \right)^{\theta}$

34. $y = \operatorname{csch}^{-1} 2^{\theta}$

35. $y = \sinh^{-1} (\tan x)$

36. $y = \cosh^{-1} (\sec x), \quad 0 < x < \pi/2$

### Integration Formulas

Verify the integration formulas in Exercises 37–40.

37. a. $\displaystyle \int \operatorname{sech} x \, dx = \tan^{-1} (\sinh x) + C$

   b. $\displaystyle \int \operatorname{sech} x \, dx = \sin^{-1} (\tanh x) + C$

38. $\displaystyle \int x \operatorname{sech}^{-1} x \, dx = \dfrac{x^2}{2} \operatorname{sech}^{-1} x - \dfrac{1}{2} \sqrt{1 - x^2} + C$

39. $\displaystyle \int x \coth^{-1} x \, dx = \dfrac{x^2 - 1}{2} \coth^{-1} x + \dfrac{x}{2} + C$

40. $\displaystyle \int \tanh^{-1} x \, dx = x \tanh^{-1} x + \dfrac{1}{2} \ln (1 - x^2) + C$

### Evaluating Integrals

Evaluate the integrals in Exercises 41–60.

41. $\displaystyle \int \sinh 2x \, dx$

42. $\displaystyle \int \sinh \dfrac{x}{5} \, dx$

43. $\displaystyle \int 6 \cosh \left( \dfrac{x}{2} - \ln 3 \right) dx$

44. $\displaystyle \int 4 \cosh (3x - \ln 2) \, dx$

45. $\displaystyle \int \tanh \dfrac{x}{7} \, dx$

46. $\displaystyle \int \coth \dfrac{\theta}{\sqrt{3}} \, d\theta$

47. $\displaystyle \int \operatorname{sech}^2 \left( x - \dfrac{1}{2} \right) dx$

48. $\displaystyle \int \operatorname{csch}^2 (5 - x) \, dx$

49. $\displaystyle \int \dfrac{\operatorname{sech} \sqrt{t} \tanh \sqrt{t} \, dt}{\sqrt{t}}$

50. $\displaystyle \int \dfrac{\operatorname{csch} (\ln t) \coth (\ln t) \, dt}{t}$

51. $\displaystyle \int_{\ln 2}^{\ln 4} \coth x \, dx$

52. $\displaystyle \int_{0}^{\ln 2} \tanh 2x \, dx$

53. $\displaystyle \int_{-\ln 4}^{-\ln 2} 2e^{\theta} \cosh \theta \, d\theta$

54. $\displaystyle \int_{0}^{\ln 2} 4e^{-\theta} \sinh \theta \, d\theta$

55. $\displaystyle \int_{-\pi/4}^{\pi/4} \cosh (\tan \theta) \sec^2 \theta \, d\theta$

56. $\displaystyle \int_{0}^{\pi/2} 2 \sinh (\sin \theta) \cos \theta \, d\theta$

57. $\displaystyle \int_{1}^{2} \dfrac{\cosh (\ln t)}{t} \, dt$

58. $\displaystyle \int_{1}^{4} \dfrac{8 \cosh \sqrt{x}}{\sqrt{x}} \, dx$

59. $\displaystyle \int_{-\ln 2}^{0} \cosh^2 \left( \dfrac{x}{2} \right) dx$

60. $\displaystyle \int_{0}^{\ln 10} 4 \sinh^2 \left( \dfrac{x}{2} \right) dx$

## Inverse Hyperbolic Functions and Integrals

When hyperbolic function keys are not available on a calculator, it is still possible to evaluate the inverse hyperbolic functions by expressing them as logarithms, as shown here. (We give one derivation in Section 8.4.)

$$\sinh^{-1}x = \ln\left(x + \sqrt{x^2 + 1}\right), \quad -\infty < x < \infty$$

$$\cosh^{-1}x = \ln\left(x + \sqrt{x^2 - 1}\right), \quad x \geq 1$$

$$\tanh^{-1}x = \frac{1}{2}\ln\frac{1 + x}{1 - x}, \quad |x| < 1$$

$$\text{sech}^{-1}x = \ln\left(\frac{1 + \sqrt{1 - x^2}}{x}\right), \quad 0 < x \leq 1$$

$$\text{csch}^{-1}x = \ln\left(\frac{1}{x} + \frac{\sqrt{1 + x^2}}{|x|}\right), \quad x \neq 0$$

$$\coth^{-1}x = \frac{1}{2}\ln\frac{x + 1}{x - 1}, \quad |x| > 1$$

Use the formulas in the box here to express the numbers in Exercises 61–66 in terms of natural logarithms.

**61.** $\sinh^{-1}(-5/12)$    **62.** $\cosh^{-1}(5/3)$

**63.** $\tanh^{-1}(-1/2)$    **64.** $\coth^{-1}(5/4)$

**65.** $\text{sech}^{-1}(3/5)$    **66.** $\text{csch}^{-1}\left(-1/\sqrt{3}\right)$

Evaluate the integrals in Exercises 67–74 in terms of

    **a.** inverse hyperbolic functions.

    **b.** natural logarithms.

**67.** $\displaystyle\int_0^{2\sqrt{3}} \frac{dx}{\sqrt{4 + x^2}}$    **68.** $\displaystyle\int_0^{1/3} \frac{6\,dx}{\sqrt{1 + 9x^2}}$

**69.** $\displaystyle\int_{5/4}^2 \frac{dx}{1 - x^2}$    **70.** $\displaystyle\int_0^{1/2} \frac{dx}{1 - x^2}$

**71.** $\displaystyle\int_{1/5}^{3/13} \frac{dx}{x\sqrt{1 - 16x^2}}$    **72.** $\displaystyle\int_1^2 \frac{dx}{x\sqrt{4 + x^2}}$

**73.** $\displaystyle\int_0^{\pi} \frac{\cos x\,dx}{\sqrt{1 + \sin^2 x}}$    **74.** $\displaystyle\int_1^e \frac{dx}{x\sqrt{1 + (\ln x)^2}}$

## Applications and Examples

**75.** Show that if a function $f$ is defined on an interval symmetric about the origin (so that $f$ is defined at $-x$ whenever it is defined at $x$), then

$$f(x) = \frac{f(x) + f(-x)}{2} + \frac{f(x) - f(-x)}{2}. \tag{1}$$

Then show that $(f(x) + f(-x))/2$ is even and that $(f(x) - f(-x))/2$ is odd.

**76.** Derive the formula $\sinh^{-1}x = \ln\left(x + \sqrt{x^2 + 1}\right)$ for all real $x$. Explain in your derivation why the plus sign is used with the square root instead of the minus sign.

**77. Skydiving** If a body of mass $m$ falling from rest under the action of gravity encounters an air resistance proportional to the square of the velocity, then the body's velocity $t$ s into the fall satisfies the differential equation

$$m\frac{dv}{dt} = mg - kv^2,$$

where $k$ is a constant that depends on the body's aerodynamic properties and the density of the air. (We assume that the fall is short enough so that the variation in the air's density will not affect the outcome significantly.)

    **a.** Show that

$$v = \sqrt{\frac{mg}{k}}\tanh\left(\sqrt{\frac{gk}{m}}\,t\right)$$

    satisfies the differential equation and the initial condition that $v = 0$ when $t = 0$.

    **b.** Find the body's *limiting velocity*, $\lim_{t\to\infty} v$.

    **c.** For a 75 kg skydiver ($mg = 735$ N), with time in seconds and distance in meters, a typical value for $k$ is 0.235. What is the diver's limiting velocity?

**78. Accelerations whose magnitudes are proportional to displacement** Suppose that the position of a body moving along a coordinate line at time $t$ is

    **a.** $s = a\cos kt + b\sin kt$.

    **b.** $s = a\cosh kt + b\sinh kt$.

Show in both cases that the acceleration $d^2s/dt^2$ is proportional to $s$ but that in the first case it is directed toward the origin, whereas in the second case it is directed away from the origin.

**79. Volume** A region in the first quadrant is bounded above by the curve $y = \cosh x$, below by the curve $y = \sinh x$, and on the left and right by the $y$-axis and the line $x = 2$, respectively. Find the volume of the solid generated by revolving the region about the $x$-axis.

**80. Volume** The region enclosed by the curve $y = \text{sech}\,x$, the $x$-axis, and the lines $x = \pm\ln\sqrt{3}$ is revolved about the $x$-axis to generate a solid. Find the volume of the solid.

**81. Arc length** Find the length of the graph of $y = (1/2)\cosh 2x$ from $x = 0$ to $x = \ln\sqrt{5}$.

**82.** Use the definitions of the hyperbolic functions to find each of the following limits.

    **a.** $\lim_{x\to\infty} \tanh x$    **b.** $\lim_{x\to-\infty} \tanh x$

    **c.** $\lim_{x\to\infty} \sinh x$    **d.** $\lim_{x\to-\infty} \sinh x$

    **e.** $\lim_{x\to\infty} \text{sech}\,x$    **f.** $\lim_{x\to\infty} \coth x$

    **g.** $\lim_{x\to0^+} \coth x$    **h.** $\lim_{x\to0^-} \coth x$

    **i.** $\lim_{x\to-\infty} \text{csch}\,x$

**83. Hanging cables** Imagine a cable, like a telephone line or TV cable, strung from one support to another and hanging freely. The cable's weight per unit length is a constant $w$ and the horizontal tension at its lowest point is a *vector* of length $H$. If we choose a coordinate system for the plane of the cable in which the $x$-axis is horizontal, the force of gravity is straight down, the positive $y$-axis points straight up, and the lowest point of the cable lies at the point $y = H/w$ on the $y$-axis (see accompanying figure), then it can be shown that the cable lies along the graph of the hyperbolic cosine

$$y = \frac{H}{w}\cosh\frac{w}{H}x.$$

Such a curve is sometimes called a **chain curve** or a **catenary**, the latter deriving from the Latin *catena*, meaning "chain."

**a.** Let $P(x, y)$ denote an arbitrary point on the cable. The next accompanying figure displays the tension at $P$ as a vector of length (magnitude) $T$, as well as the tension $H$ at the lowest point $A$. Show that the cable's slope at $P$ is

$$\tan\phi = \frac{dy}{dx} = \sinh\frac{w}{H}x.$$

**b.** Using the result from part (a) and the fact that the horizontal tension at $P$ must equal $H$ (the cable is not moving), show that $T = wy$. Hence, the magnitude of the tension at $P(x, y)$ is exactly equal to the weight of $y$ units of cable.

**84.** (*Continuation of Exercise 83.*) The length of arc $AP$ in the Exercise 83 figure is $s = (1/a)\sinh ax$, where $a = w/H$. Show that the coordinates of $P$ may be expressed in terms of $s$ as

$$x = \frac{1}{a}\sinh^{-1}as, \qquad y = \sqrt{s^2 + \frac{1}{a^2}}.$$

**85. Area** Show that the area of the region in the first quadrant enclosed by the curve $y = (1/a)\cosh ax$, the coordinate axes, and the line $x = b$ is the same as the area of a rectangle of height $1/a$ and length $s$, where $s$ is the length of the curve from $x = 0$ to $x = b$. Draw a figure illustrating this result.

**86. The hyperbolic in hyperbolic functions** Just as $x = \cos u$ and $y = \sin u$ are identified with points $(x, y)$ on the unit circle, the functions $x = \cosh u$ and $y = \sinh u$ are identified with points $(x, y)$ on the right-hand branch of the unit hyperbola, $x^2 - y^2 = 1$.

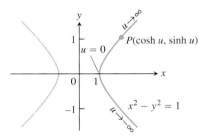

Since $\cosh^2 u - \sinh^2 u = 1$, the point $(\cosh u, \sinh u)$ lies on the right-hand branch of the hyperbola $x^2 - y^2 = 1$ for every value of $u$.

Another analogy between hyperbolic and circular functions is that the variable $u$ in the coordinates $(\cosh u, \sinh u)$ for the points of the right-hand branch of the hyperbola $x^2 - y^2 = 1$ is twice the area of the sector $AOP$ pictured in the accompanying figure. To see why this is so, carry out the following steps.

**a.** Show that the area $A(u)$ of sector $AOP$ is

$$A(u) = \frac{1}{2}\cosh u \sinh u - \int_{1}^{\cosh u}\sqrt{x^2 - 1}\,dx.$$

**b.** Differentiate both sides of the equation in part (a) with respect to $u$ to show that

$$A'(u) = \frac{1}{2}.$$

**c.** Solve this last equation for $A(u)$. What is the value of $A(0)$? What is the value of the constant of integration $C$ in your solution? With $C$ determined, what does your solution say about the relationship of $u$ to $A(u)$?

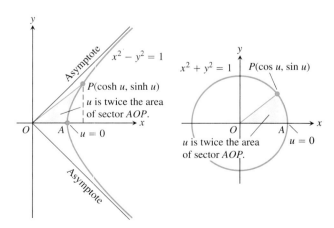

One of the analogies between hyperbolic and circular functions is revealed by these two diagrams (Exercise 86).

# 7.8 Relative Rates of Growth

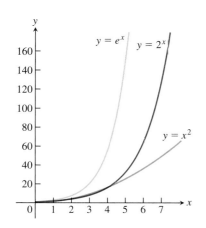

**FIGURE 7.34**    The graphs of $e^x$, $2^x$, and $x^2$.

It is often important in mathematics, computer science, and engineering to compare the rates at which functions of $x$ grow as $x$ becomes large. Exponential functions are important in these comparisons because of their very fast growth, and logarithmic functions because of their very slow growth. In this section we introduce the *little-oh* and *big-oh* notation used to describe the results of these comparisons. We restrict our attention to functions whose values eventually become and remain positive as $x \to \infty$.

## Growth Rates of Functions

You may have noticed that exponential functions like $2^x$ and $e^x$ seem to grow more rapidly as $x$ gets large than do polynomials and rational functions. These exponentials certainly grow more rapidly than $x$ itself, and you can see $2^x$ outgrowing $x^2$ as $x$ increases in Figure 7.34. In fact, as $x \to \infty$, the functions $2^x$ and $e^x$ grow faster than any power of $x$, even $x^{1,000,000}$ (Exercise 19). In contrast, logarithmic functions like $y = \log_2 x$ and $y = \ln x$ grow more slowly as $x \to \infty$ than any positive power of $x$ (Exercise 21).

To get a feeling for how rapidly the values of $y = e^x$ grow with increasing $x$, think of graphing the function on a large blackboard, with the axes scaled in centimeters. At $x = 1$ cm, the graph is $e^1 \approx 3$ cm above the $x$-axis. At $x = 6$ cm, the graph is $e^6 \approx 403$ cm $\approx 4$ m high (it is about to go through the ceiling if it hasn't done so already). At $x = 10$ cm, the graph is $e^{10} \approx 22{,}026$ cm $\approx 220$ m high, higher than most buildings. At $x = 24$ cm, the graph is more than halfway to the moon, and at $x = 43$ cm from the origin, the graph is high enough to reach past the sun's closest stellar neighbor, the red dwarf star Proxima Centauri. By contrast, with axes scaled in centimeters, you have to go nearly 5 light-years out on the $x$-axis to find a point where the graph of $y = \ln x$ is even $y = 43$ cm high. See Figure 7.35.

These important comparisons of exponential, polynomial, and logarithmic functions can be made precise by defining what it means for a function $f(x)$ to grow faster than another function $g(x)$ as $x \to \infty$.

**FIGURE 7.35**    Scale drawings of the graphs of $e^x$ and $\ln x$.

---

**DEFINITION**    Let $f(x)$ and $g(x)$ be positive for $x$ sufficiently large.

1. $f$ **grows faster than** $g$ as $x \to \infty$ if

$$\lim_{x \to \infty} \frac{f(x)}{g(x)} = \infty$$

or, equivalently, if

$$\lim_{x \to \infty} \frac{g(x)}{f(x)} = 0.$$

We also say that $g$ **grows slower than** $f$ as $x \to \infty$.

2. $f$ and $g$ **grow at the same rate** as $x \to \infty$ if

$$\lim_{x \to \infty} \frac{f(x)}{g(x)} = L$$

where $L$ is finite and positive.

According to these definitions, $y = 2x$ does not grow faster than $y = x$. The two functions grow at the same rate because

$$\lim_{x \to \infty} \frac{2x}{x} = \lim_{x \to \infty} 2 = 2,$$

which is a finite, positive limit. The reason for this departure from more common usage is that we want "$f$ grows faster than $g$" to mean that for large $x$-values $g$ is negligible when compared with $f$.

**EXAMPLE 1** Let's compare the growth rates of several common functions.

**(a)** $e^x$ grows faster than $x^2$ as $x \to \infty$ because

$$\underbrace{\lim_{x \to \infty} \frac{e^x}{x^2}}_{\infty/\infty} = \underbrace{\lim_{x \to \infty} \frac{e^x}{2x}}_{\infty/\infty} = \lim_{x \to \infty} \frac{e^x}{2} = \infty. \qquad \text{Using l'Hôpital's Rule twice}$$

**(b)** $3^x$ grows faster than $2^x$ as $x \to \infty$ because

$$\lim_{x \to \infty} \frac{3^x}{2^x} = \lim_{x \to \infty} \left(\frac{3}{2}\right)^x = \infty.$$

**(c)** $x^2$ grows faster than $\ln x$ as $x \to \infty$ because

$$\lim_{x \to \infty} \frac{x^2}{\ln x} = \lim_{x \to \infty} \frac{2x}{1/x} = \lim_{x \to \infty} 2x^2 = \infty. \qquad \text{l'Hôpital's Rule}$$

**(d)** $\ln x$ grows slower than $x^{1/n}$ as $x \to \infty$ for any positive integer $n$ because

$$\lim_{x \to \infty} \frac{\ln x}{x^{1/n}} = \lim_{x \to \infty} \frac{1/x}{(1/n)x^{(1/n)-1}} \qquad \text{l'Hôpital's Rule}$$

$$= \lim_{x \to \infty} \frac{n}{x^{1/n}} = 0. \qquad \text{$n$ is constant.}$$

**(e)** As part (b) suggests, exponential functions with different bases never grow at the same rate as $x \to \infty$. If $a > b > 0$, then $a^x$ grows faster than $b^x$. Since $(a/b) > 1$,

$$\lim_{x \to \infty} \frac{a^x}{b^x} = \lim_{x \to \infty} \left(\frac{a}{b}\right)^x = \infty.$$

**(f)** In contrast to exponential functions, logarithmic functions with different bases $a > 1$ and $b > 1$ always grow at the same rate as $x \to \infty$:

$$\lim_{x \to \infty} \frac{\log_a x}{\log_b x} = \lim_{x \to \infty} \frac{\ln x / \ln a}{\ln x / \ln b} = \frac{\ln b}{\ln a}.$$

The limiting ratio is always finite and never zero. ■

If $f$ grows at the same rate as $g$ as $x \to \infty$, and $g$ grows at the same rate as $h$ as $x \to \infty$, then $f$ grows at the same rate as $h$ as $x \to \infty$. The reason is that

$$\lim_{x \to \infty} \frac{f}{g} = L_1 \qquad \text{and} \qquad \lim_{x \to \infty} \frac{g}{h} = L_2$$

together imply

$$\lim_{x \to \infty} \frac{f}{h} = \lim_{x \to \infty} \frac{f}{g} \cdot \frac{g}{h} = L_1 L_2.$$

If $L_1$ and $L_2$ are finite and nonzero, then so is $L_1 L_2$.

**EXAMPLE 2**    Show that $\sqrt{x^2 + 5}$ and $(2\sqrt{x} - 1)^2$ grow at the same rate as $x \to \infty$.

**Solution**    We show that the functions grow at the same rate by showing that they both grow at the same rate as the function $g(x) = x$:

$$\lim_{x \to \infty} \frac{\sqrt{x^2 + 5}}{x} = \lim_{x \to \infty} \sqrt{1 + \frac{5}{x^2}} = 1,$$

$$\lim_{x \to \infty} \frac{(2\sqrt{x} - 1)^2}{x} = \lim_{x \to \infty} \left(\frac{2\sqrt{x} - 1}{\sqrt{x}}\right)^2 = \lim_{x \to \infty} \left(2 - \frac{1}{\sqrt{x}}\right)^2 = 4. \qquad \blacksquare$$

### Order and Oh-Notation

The "little-oh" and "big-oh" notation was invented by number theorists a hundred years ago and is now commonplace in mathematical analysis and computer science. Notice that saying $f = o(g)$ as $x \to \infty$ is another way to say that $f$ grows slower than $g$ as $x \to \infty$.

---

**DEFINITION**    A function $f$ is **of smaller order than** $g$ as $x \to \infty$ if $\lim_{x \to \infty} \frac{f(x)}{g(x)} = 0$. We indicate this by writing $\boldsymbol{f = o(g)}$ ("$f$ is little-oh of $g$").

---

**EXAMPLE 3**    Here we use little-oh notation.

**(a)** $\ln x = o(x)$ as $x \to \infty$   because   $\lim_{x \to \infty} \frac{\ln x}{x} = 0$

**(b)** $x^2 = o(x^3 + 1)$ as $x \to \infty$   because   $\lim_{x \to \infty} \frac{x^2}{x^3 + 1} = 0$ $\qquad \blacksquare$

---

**DEFINITION**    Let $f(x)$ and $g(x)$ be positive for $x$ sufficiently large. Then $f$ is **of at most the order of** $g$ as $x \to \infty$ if there is a positive integer $M$ for which

$$\frac{f(x)}{g(x)} \le M,$$

for $x$ sufficiently large. We indicate this by writing $\boldsymbol{f = O(g)}$ ("$f$ is big-oh of $g$").

---

**EXAMPLE 4**    Here we use big-oh notation.

**(a)** $x + \sin x = O(x)$ as $x \to \infty$   because   $\frac{x + \sin x}{x} \le 2$ for $x$ sufficiently large.

**(b)** $e^x + x^2 = O(e^x)$ as $x \to \infty$   because   $\frac{e^x + x^2}{e^x} \to 1$ as $x \to \infty$.

**(c)** $x = O(e^x)$ as $x \to \infty$   because   $\frac{x}{e^x} \to 0$ as $x \to \infty$. $\qquad \blacksquare$

If you look at the definitions again, you will see that $f = o(g)$ implies $f = O(g)$ for functions that are positive for $x$ sufficiently large. Also, if $f$ and $g$ grow at the same rate, then $f = O(g)$ and $g = O(f)$ (Exercise 11).

### Sequential vs. Binary Search

Computer scientists often measure the efficiency of an algorithm by counting the number of steps a computer must take to execute the algorithm. There can be significant differences in how efficiently algorithms perform, even if they are designed to accomplish the same task. These differences are often described in big-oh notation. Here is an example.

*Webster's International Dictionary* lists about 26,000 words that begin with the letter *a*. One way to look up a word, or to learn if it is not there, is to read through the list one word at a time until you either find the word or determine that it is not there. This method, called **sequential search**, makes no particular use of the words' alphabetical arrangement in the list. You are sure to get an answer, but it might take 26,000 steps.

Another way to find the word or to learn it is not there is to go straight to the middle of the list (give or take a few words). If you do not find the word, then go to the middle of the half that contains it and forget about the half that does not. (You know which half contains it because you know the list is ordered alphabetically.) This method, called a **binary search**, eliminates roughly 13,000 words in a single step. If you do not find the word on the second try, then jump to the middle of the half that contains it. Continue this way until you have either found the word or divided the list in half so many times there are no words left. How many times do you have to divide the list to find the word or learn that it is not there? At most 15, because

$$(26{,}000/2^{15}) < 1.$$

That certainly beats a possible 26,000 steps.

For a list of length *n*, a sequential search algorithm takes on the order of *n* steps to find a word or determine that it is not in the list. A binary search, as the second algorithm is called, takes on the order of $\log_2 n$ steps. The reason is that if $2^{m-1} < n \le 2^m$, then $m - 1 < \log_2 n \le m$, and the number of bisections required to narrow the list to one word will be at most $m = \lceil \log_2 n \rceil$, the integer ceiling for $\log_2 n$.

Big-oh notation provides a compact way to say all this. The number of steps in a sequential search of an ordered list is $O(n)$; the number of steps in a binary search is $O(\log_2 n)$. In our example, there is a big difference between the two (26,000 vs. 15), and the difference can only increase with *n* because *n* grows faster than $\log_2 n$ as $n \to \infty$.

### Summary

The integral definition of the natural logarithm function $\ln x$ in Section 7.2 is the key to obtaining precisely the exponential and logarithmic functions $a^x$ and $\log_a x$ for any base $a > 0$. The differentiability and increasing behavior of $\ln x$ allow us to define its differentiable inverse, the natural exponential function $e^x$, through Theorem 1. Then $e^x$ provides for the definition of the differentiable function $a^x = e^{x \ln a}$, giving a simple and precise meaning of irrational exponents, and from which we see that *every* exponential function is just $e^x$ raised to an appropriate power, $\ln a$. The increasing (or decreasing) behavior of $a^x$ gives its differentiable inverse $\log_a x$, using Theorem 1 again. Moreover, we saw that $\log_a x = (\ln x)/(\ln a)$ is just a multiple of the natural logarithm function. So $e^x$ and $\ln x$ give the entire array of exponential and logarithmic functions using the algebraic operations of taking constant powers and constant multiples. Furthermore, the differentiability of $e^x$ and $a^x$ establishes the existence of the limits

$$\lim_{h \to 0} \frac{e^h - 1}{h} = 1 \quad \text{and} \quad \lim_{h \to 0} \frac{a^h - 1}{h} = \ln a$$

as the slopes of those functions where they cross the *y*-axis.

In this chapter we have seen the important roles the exponential and logarithmic functions play in analyzing problems associated with growth and decay, in comparing the growth rates of various functions, and in measuring the efficiency of a computer algorithm. In Chapters 9 and 17 we will see that exponential functions play a major role in the solutions to differential equations.

# Exercises 7.8

## Comparisons with the Exponential $e^x$

**1.** Which of the following functions grow faster than $e^x$ as $x \to \infty$? Which grow at the same rate as $e^x$? Which grow slower?

  **a.** $x - 3$              **b.** $x^3 + \sin^2 x$

  **c.** $\sqrt{x}$                  **d.** $4^x$

  **e.** $(3/2)^x$           **f.** $e^{x/2}$

  **g.** $e^x/2$              **h.** $\log_{10} x$

**2.** Which of the following functions grow faster than $e^x$ as $x \to \infty$? Which grow at the same rate as $e^x$? Which grow slower?

  **a.** $10x^4 + 30x + 1$     **b.** $x \ln x - x$

  **c.** $\sqrt{1 + x^4}$         **d.** $(5/2)^x$

  **e.** $e^{-x}$               **f.** $xe^x$

  **g.** $e^{\cos x}$            **h.** $e^{x-1}$

## Comparisons with the Power $x^2$

**3.** Which of the following functions grow faster than $x^2$ as $x \to \infty$? Which grow at the same rate as $x^2$? Which grow slower?

  **a.** $x^2 + 4x$          **b.** $x^5 - x^2$

  **c.** $\sqrt{x^4 + x^3}$      **d.** $(x + 3)^2$

  **e.** $x \ln x$           **f.** $2^x$

  **g.** $x^3 e^{-x}$          **h.** $8x^2$

**4.** Which of the following functions grow faster than $x^2$ as $x \to \infty$? Which grow at the same rate as $x^2$? Which grow slower?

  **a.** $x^2 + \sqrt{x}$        **b.** $10x^2$

  **c.** $x^2 e^{-x}$          **d.** $\log_{10}(x^2)$

  **e.** $x^3 - x^2$         **f.** $(1/10)^x$

  **g.** $(1.1)^x$           **h.** $x^2 + 100x$

## Comparisons with the Logarithm $\ln x$

**5.** Which of the following functions grow faster than $\ln x$ as $x \to \infty$? Which grow at the same rate as $\ln x$? Which grow slower?

  **a.** $\log_3 x$           **b.** $\ln 2x$

  **c.** $\ln \sqrt{x}$          **d.** $\sqrt{x}$

  **e.** $x$                 **f.** $5 \ln x$

  **g.** $1/x$              **h.** $e^x$

**6.** Which of the following functions grow faster than $\ln x$ as $x \to \infty$? Which grow at the same rate as $\ln x$? Which grow slower?

  **a.** $\log_2(x^2)$         **b.** $\log_{10} 10x$

  **c.** $1/\sqrt{x}$          **d.** $1/x^2$

  **e.** $x - 2 \ln x$      **f.** $e^{-x}$

  **g.** $\ln(\ln x)$        **h.** $\ln(2x + 5)$

## Ordering Functions by Growth Rates

**7.** Order the following functions from slowest growing to fastest growing as $x \to \infty$.

  **a.** $e^x$                **b.** $x^x$

  **c.** $(\ln x)^x$          **d.** $e^{x/2}$

**8.** Order the following functions from slowest growing to fastest growing as $x \to \infty$.

  **a.** $2^x$                **b.** $x^2$

  **c.** $(\ln 2)^x$          **d.** $e^x$

## Big-oh and Little-oh; Order

**9.** True, or false? As $x \to \infty$,

  **a.** $x = o(x)$          **b.** $x = o(x + 5)$

  **c.** $x = O(x + 5)$    **d.** $x = O(2x)$

  **e.** $e^x = o(e^{2x})$      **f.** $x + \ln x = O(x)$

  **g.** $\ln x = o(\ln 2x)$   **h.** $\sqrt{x^2 + 5} = O(x)$

**10.** True, or false? As $x \to \infty$,

  **a.** $\dfrac{1}{x + 3} = O\left(\dfrac{1}{x}\right)$     **b.** $\dfrac{1}{x} + \dfrac{1}{x^2} = O\left(\dfrac{1}{x}\right)$

  **c.** $\dfrac{1}{x} - \dfrac{1}{x^2} = o\left(\dfrac{1}{x}\right)$     **d.** $2 + \cos x = O(2)$

  **e.** $e^x + x = O(e^x)$     **f.** $x \ln x = o(x^2)$

  **g.** $\ln(\ln x) = O(\ln x)$   **h.** $\ln(x) = o(\ln(x^2 + 1))$

**11.** Show that if positive functions $f(x)$ and $g(x)$ grow at the same rate as $x \to \infty$, then $f = O(g)$ and $g = O(f)$.

**12.** When is a polynomial $f(x)$ of smaller order than a polynomial $g(x)$ as $x \to \infty$? Give reasons for your answer.

**13.** When is a polynomial $f(x)$ of at most the order of a polynomial $g(x)$ as $x \to \infty$? Give reasons for your answer.

**14.** What do the conclusions we drew in Section 2.8 about the limits of rational functions tell us about the relative growth of polynomials as $x \to \infty$?

## Other Comparisons

**T 15.** Investigate

$$\lim_{x \to \infty} \frac{\ln(x + 1)}{\ln x} \quad \text{and} \quad \lim_{x \to \infty} \frac{\ln(x + 999)}{\ln x}.$$

Then use l'Hôpital's Rule to explain what you find.

**16.** (*Continuation of Exercise 15.*) Show that the value of

$$\lim_{x \to \infty} \frac{\ln(x + a)}{\ln x}$$

is the same no matter what value you assign to the constant $a$. What does this say about the relative rates at which the functions $f(x) = \ln(x + a)$ and $g(x) = \ln x$ grow?

**17.** Show that $\sqrt{10x + 1}$ and $\sqrt{x + 1}$ grow at the same rate as $x \to \infty$ by showing that they both grow at the same rate as $\sqrt{x}$ as $x \to \infty$.

**18.** Show that $\sqrt{x^4 + x}$ and $\sqrt{x^4 - x^3}$ grow at the same rate as $x \to \infty$ by showing that they both grow at the same rate as $x^2$ as $x \to \infty$.

**19.** Show that $e^x$ grows faster as $x \to \infty$ than $x^n$ for any positive integer $n$, even $x^{1,000,000}$. (*Hint:* What is the $n$th derivative of $x^n$?)

**20. The function $e^x$ outgrows any polynomial** Show that $e^x$ grows faster as $x \to \infty$ than any polynomial

$$a_n x^n + a_{n-1} x^{n-1} + \cdots + a_1 x + a_0.$$

**21. a.** Show that $\ln x$ grows slower as $x \to \infty$ than $x^{1/n}$ for any positive integer $n$, even $x^{1/1,000,000}$.

   T **b.** Although the values of $x^{1/1,000,000}$ eventually overtake the values of $\ln x$, you have to go way out on the $x$-axis before this happens. Find a value of $x$ greater than 1 for which $x^{1/1,000,000} > \ln x$. You might start by observing that when $x > 1$ the equation $\ln x = x^{1/1,000,000}$ is equivalent to the equation $\ln (\ln x) = (\ln x)/1,000,000$.

   T **c.** Even $x^{1/10}$ takes a long time to overtake $\ln x$. Experiment with a calculator to find the value of $x$ at which the graphs of $x^{1/10}$ and $\ln x$ cross, or, equivalently, at which $\ln x = 10 \ln (\ln x)$. Bracket the crossing point between powers of 10 and then close in by successive halving.

   T **d.** (*Continuation of part (c).*) The value of $x$ at which $\ln x = 10 \ln (\ln x)$ is too far out for some graphers and root finders to identify. Try it on the equipment available to you and see what happens.

**22. The function $\ln x$ grows slower than any polynomial** Show that $\ln x$ grows slower as $x \to \infty$ than any nonconstant polynomial.

**Algorithms and Searches**

**23. a.** Suppose you have three different algorithms for solving the same problem and each algorithm takes a number of steps that is of the order of one of the functions listed here:

$$n \log_2 n, \quad n^{3/2}, \quad n(\log_2 n)^2.$$

Which of the algorithms is the most efficient in the long run? Give reasons for your answer.

   T **b.** Graph the functions in part (a) together to get a sense of how rapidly each one grows.

**24.** Repeat Exercise 23 for the functions

$$n, \quad \sqrt{n} \log_2 n, \quad (\log_2 n)^2.$$

T **25.** Suppose you are looking for an item in an ordered list one million items long. How many steps might it take to find that item with a sequential search? A binary search?

T **26.** You are looking for an item in an ordered list 450,000 items long (the length of *Webster's Third New International Dictionary*). How many steps might it take to find the item with a sequential search? A binary search?

---

# Chapter **7**  Questions to Guide Your Review

1. What functions have inverses? How do you know if two functions $f$ and $g$ are inverses of one another? Give examples of functions that are (are not) inverses of one another.

2. How are the domains, ranges, and graphs of functions and their inverses related? Give an example.

3. How can you sometimes express the inverse of a function of $x$ as a function of $x$?

4. Under what circumstances can you be sure that the inverse of a function $f$ is differentiable? How are the derivatives of $f$ and $f^{-1}$ related?

5. What is the natural logarithm function? What are its domain, range, and derivative? What arithmetic properties does it have? Comment on its graph.

6. What is logarithmic differentiation? Give an example.

7. What integrals lead to logarithms? Give examples. What are the integrals of $\tan x$, $\cot x$, $\sec x$, and $\csc x$?

8. How is the exponential function $e^x$ defined? What are its domain, range, and derivative? What laws of exponents does it obey? Comment on its graph.

9. How are the functions $a^x$ and $\log_a x$ defined? Are there any restrictions on $a$? How is the graph of $\log_a x$ related to the graph of $\ln x$? What truth is there in the statement that there is really only one exponential function and one logarithmic function?

10. How do you solve separable first-order differential equations?

11. What is the law of exponential change? How can it be derived from an initial value problem? What are some of the applications of the law?

12. Describe l'Hôpital's Rule. How do you know when to use the rule and when to stop? Give an example.

13. How can you sometimes handle limits that lead to indeterminate forms $\infty/\infty$, $\infty \cdot 0$, and $\infty - \infty$? Give examples.

14. How can you sometimes handle limits that lead to indeterminate forms $1^\infty$, $0^0$, and $\infty^\infty$? Give examples.

15. How are the inverse trigonometric functions defined? How can you sometimes use right triangles to find values of these functions? Give examples.

16. What are the derivatives of the inverse trigonometric functions? How do the domains of the derivatives compare with the domains of the functions?

17. What integrals lead to inverse trigonometric functions? How do substitution and completing the square broaden the application of these integrals?

18. What are the six basic hyperbolic functions? Comment on their domains, ranges, and graphs. What are some of the identities relating them?

19. What are the derivatives of the six basic hyperbolic functions? What are the corresponding integral formulas? What similarities do you see here with the six basic trigonometric functions?

**20.** How are the inverse hyperbolic functions defined? Comment on their domains, ranges, and graphs. How can you find values of $\operatorname{sech}^{-1} x$, $\operatorname{csch}^{-1} x$, and $\operatorname{coth}^{-1} x$ using a calculator's keys for $\cosh^{-1} x$, $\sinh^{-1} x$, and $\tanh^{-1} x$?

**21.** What integrals lead naturally to inverse hyperbolic functions?

**22.** How do you compare the growth rates of positive functions as $x \to \infty$?

**23.** What roles do the functions $e^x$ and $\ln x$ play in growth comparisons?

**24.** Describe big-oh and little-oh notation. Give examples.

**25.** Which is more efficient—a sequential search or a binary search? Explain.

## Chapter **7** Practice Exercises

### Finding Derivatives

In Exercises 1–24, find the derivative of $y$ with respect to the appropriate variable.

**1.** $y = 10e^{-x/5}$

**2.** $y = \sqrt{2}e^{\sqrt{2}x}$

**3.** $y = \frac{1}{4}xe^{4x} - \frac{1}{16}e^{4x}$

**4.** $y = x^2 e^{-2/x}$

**5.** $y = \ln(\sin^2 \theta)$

**6.** $y = \ln(\sec^2 \theta)$

**7.** $y = \log_2(x^2/2)$

**8.** $y = \log_5(3x - 7)$

**9.** $y = 8^{-t}$

**10.** $y = 9^{2t}$

**11.** $y = 5x^{3.6}$

**12.** $y = \sqrt{2}x^{-\sqrt{2}}$

**13.** $y = (x + 2)^{x+2}$

**14.** $y = 2(\ln x)^{x/2}$

**15.** $y = \sin^{-1}\sqrt{1 - u^2}$, $\quad 0 < u < 1$

**16.** $y = \sin^{-1}\left(\frac{1}{\sqrt{v}}\right)$, $\quad v > 1$

**17.** $y = \ln \cos^{-1} x$

**18.** $y = z \cos^{-1} z - \sqrt{1 - z^2}$

**19.** $y = t \tan^{-1} t - \frac{1}{2}\ln t$

**20.** $y = (1 + t^2)\cot^{-1} 2t$

**21.** $y = z \sec^{-1} z - \sqrt{z^2 - 1}$, $\quad z > 1$

**22.** $y = 2\sqrt{x - 1}\sec^{-1}\sqrt{x}$

**23.** $y = \csc^{-1}(\sec \theta)$, $\quad 0 < \theta < \pi/2$

**24.** $y = (1 + x^2)e^{\tan^{-1} x}$

### Logarithmic Differentiation

In Exercises 25–30, use logarithmic differentiation to find the derivative of $y$ with respect to the appropriate variable.

**25.** $y = \dfrac{2(x^2 + 1)}{\sqrt{\cos 2x}}$

**26.** $y = \sqrt[10]{\dfrac{3x + 4}{2x - 4}}$

**27.** $y = \left(\dfrac{(t + 1)(t - 1)}{(t - 2)(t + 3)}\right)^5$, $\quad t > 2$

**28.** $y = \dfrac{2u2^u}{\sqrt{u^2 + 1}}$

**29.** $y = (\sin \theta)^{\sqrt{\theta}}$

**30.** $y = (\ln x)^{1/(\ln x)}$

### Evaluating Integrals

Evaluate the integrals in Exercises 31–78.

**31.** $\displaystyle\int e^x \cos(e^x)\, dx$

**32.** $\displaystyle\int e^x \sin(5e^x - 7)\, dx$

**33.** $\displaystyle\int e^x \sec^2(e^x - 7)\, dx$

**34.** $\displaystyle\int e^y \csc(e^y + 1)\cot(e^y + 1)\, dy$

**35.** $\displaystyle\int \sec^2(x)e^{\tan x}\, dx$

**36.** $\displaystyle\int \csc^2 x\, e^{\cot x}\, dx$

**37.** $\displaystyle\int_{-1}^{1} \frac{dx}{3x - 4}$

**38.** $\displaystyle\int_{1}^{e} \frac{\sqrt{\ln x}}{x}\, dx$

**39.** $\displaystyle\int_{0}^{\pi} \tan \frac{x}{3}\, dx$

**40.** $\displaystyle\int_{1/6}^{1/4} 2\cot \pi x\, dx$

**41.** $\displaystyle\int_{0}^{4} \frac{2t}{t^2 - 25}\, dt$

**42.** $\displaystyle\int_{-\pi/2}^{\pi/6} \frac{\cos t}{1 - \sin t}\, dt$

**43.** $\displaystyle\int \frac{\tan(\ln v)}{v}\, dv$

**44.** $\displaystyle\int \frac{dv}{v \ln v}$

**45.** $\displaystyle\int \frac{(\ln x)^{-3}}{x}\, dx$

**46.** $\displaystyle\int \frac{\ln(x + 9)}{x + 9}\, dx$

**47.** $\displaystyle\int \frac{1}{r}\csc^2(1 + \ln r)\, dr$

**48.** $\displaystyle\int \frac{\sin(2 + \ln x)}{x}\, dx$

**49.** $\displaystyle\int x3^{x^2}\, dx$

**50.** $\displaystyle\int 2^{\tan x}\sec^2 x\, dx$

**51.** $\displaystyle\int_{1}^{9} \frac{5}{x}\, dx$

**52.** $\displaystyle\int_{1}^{81} \frac{1}{4x}\, dx$

**53.** $\displaystyle\int_{1}^{4} \left(\frac{x}{8} + \frac{1}{2x}\right) dx$

**54.** $\displaystyle\int_{1}^{8} \left(\frac{2}{3x} - \frac{8}{x^2}\right) dx$

**55.** $\displaystyle\int_{-2}^{-1} e^{-(x+1)}\, dx$

**56.** $\displaystyle\int_{-\ln 2}^{0} e^{2w}\, dw$

**57.** $\displaystyle\int_{0}^{\ln 5} e^r(3e^r + 1)^{-3/2}\, dr$

**58.** $\displaystyle\int_{0}^{\ln 9} e^\theta(e^\theta - 1)^{1/2}\, d\theta$

**59.** $\displaystyle\int_{1}^{e} \frac{1}{x}(1 + 7\ln x)^{-1/3}\, dx$

**60.** $\displaystyle\int_{e^2}^{e^3} \frac{5}{x\sqrt{\ln x}}\, dx$

**61.** $\displaystyle\int_{1}^{3} \frac{(\ln(v+1))^2}{v+1}\,dv$

**62.** $\displaystyle\int_{3}^{9} (1+\ln t)t\ln t\,dt$

**63.** $\displaystyle\int_{1}^{8} \frac{\log_4 \theta}{\theta}\,d\theta$

**64.** $\displaystyle\int_{1}^{e} \frac{8\ln 3\log_3 \theta}{\theta}\,d\theta$

**65.** $\displaystyle\int_{-3/4}^{3/4} \frac{6\,dx}{\sqrt{9-4x^2}}$

**66.** $\displaystyle\int_{-1/5}^{1/5} \frac{6\,dx}{\sqrt{4-25x^2}}$

**67.** $\displaystyle\int_{-2}^{2} \frac{3\,dt}{4+3t^2}$

**68.** $\displaystyle\int_{\sqrt{3}}^{3} \frac{dt}{3+t^2}$

**69.** $\displaystyle\int \frac{dy}{y\sqrt{4y^2-1}}$

**70.** $\displaystyle\int \frac{24\,dy}{y\sqrt{y^2-16}}$

**71.** $\displaystyle\int_{\sqrt{2}/3}^{2/3} \frac{dy}{|y|\sqrt{9y^2-1}}$

**72.** $\displaystyle\int_{-2/\sqrt{5}}^{-\sqrt{6}/\sqrt{5}} \frac{dy}{|y|\sqrt{5y^2-3}}$

**73.** $\displaystyle\int \frac{dx}{\sqrt{-2x-x^2}}$

**74.** $\displaystyle\int \frac{dx}{\sqrt{-x^2+4x-1}}$

**75.** $\displaystyle\int_{-2}^{-1} \frac{2\,dv}{v^2+4v+5}$

**76.** $\displaystyle\int_{-1}^{1} \frac{3\,dv}{4v^2+4v+4}$

**77.** $\displaystyle\int \frac{dt}{(t+1)\sqrt{t^2+2t-8}}$

**78.** $\displaystyle\int \frac{dt}{(3t+1)\sqrt{9t^2+6t}}$

### Solving Equations
In Exercises 79–84, solve for y.

**79.** $3^y = 2^{y+1}$

**80.** $4^{-y} = 3^{y+2}$

**81.** $9e^{2y} = x^2$

**82.** $3^y = 3\ln x$

**83.** $\ln(y-1) = x + \ln y$

**84.** $\ln(10\ln y) = \ln 5x$

### Applying L'Hôpital's Rule
Use l'Hôpital's Rule to find the limits in Exercises 85–108.

**85.** $\displaystyle\lim_{x\to 1} \frac{x^2+3x-4}{x-1}$

**86.** $\displaystyle\lim_{x\to 1} \frac{x^a-1}{x^b-1}$

**87.** $\displaystyle\lim_{x\to \pi} \frac{\tan x}{x}$

**88.** $\displaystyle\lim_{x\to 0} \frac{\tan x}{x+\sin x}$

**89.** $\displaystyle\lim_{x\to 0} \frac{\sin^2 x}{\tan(x^2)}$

**90.** $\displaystyle\lim_{x\to 0} \frac{\sin mx}{\sin nx}$

**91.** $\displaystyle\lim_{x\to \pi/2^-} \sec 7x\cos 3x$

**92.** $\displaystyle\lim_{x\to 0^+} \sqrt{x}\sec x$

**93.** $\displaystyle\lim_{x\to 0} (\csc x - \cot x)$

**94.** $\displaystyle\lim_{x\to 0} \left(\frac{1}{x^4} - \frac{1}{x^2}\right)$

**95.** $\displaystyle\lim_{x\to \infty} \left(\sqrt{x^2+x+1} - \sqrt{x^2-x}\right)$

**96.** $\displaystyle\lim_{x\to \infty} \left(\frac{x^3}{x^2-1} - \frac{x^3}{x^2+1}\right)$

**97.** $\displaystyle\lim_{x\to 0} \frac{10^x-1}{x}$

**98.** $\displaystyle\lim_{\theta\to 0} \frac{3^\theta-1}{\theta}$

**99.** $\displaystyle\lim_{x\to 0} \frac{2^{\sin x}-1}{e^x-1}$

**100.** $\displaystyle\lim_{x\to 0} \frac{2^{-\sin x}-1}{e^x-1}$

**101.** $\displaystyle\lim_{x\to 0} \frac{5-5\cos x}{e^x-x-1}$

**102.** $\displaystyle\lim_{x\to 0} \frac{x\sin x^2}{\tan^3 x}$

**103.** $\displaystyle\lim_{t\to 0^+} \frac{t-\ln(1+2t)}{t^2}$

**104.** $\displaystyle\lim_{x\to 4} \frac{\sin^2(\pi x)}{e^{x-4}+3-x}$

**105.** $\displaystyle\lim_{t\to 0^+} \left(\frac{e^t}{t} - \frac{1}{t}\right)$

**106.** $\displaystyle\lim_{y\to 0^+} e^{-1/y}\ln y$

**107.** $\displaystyle\lim_{x\to \infty} \left(\frac{e^x+1}{e^x-1}\right)^{\ln x}$

**108.** $\displaystyle\lim_{x\to 0^+} \left(1+\frac{3}{x}\right)^x$

### Comparing Growth Rates of Functions

**109.** Does $f$ grow faster, slower, or at the same rate as $g$ as $x\to\infty$? Give reasons for your answers.

   **a.** $f(x) = \log_2 x,$       $g(x) = \log_3 x$

   **b.** $f(x) = x,$            $g(x) = x + \frac{1}{x}$

   **c.** $f(x) = x/100,$    $g(x) = xe^{-x}$

   **d.** $f(x) = x,$            $g(x) = \tan^{-1} x$

   **e.** $f(x) = \csc^{-1} x,$    $g(x) = 1/x$

   **f.** $f(x) = \sinh x,$      $g(x) = e^x$

**110.** Does $f$ grow faster, slower, or at the same rate as $g$ as $x\to\infty$? Give reasons for your answers.

   **a.** $f(x) = 3^{-x},$        $g(x) = 2^{-x}$

   **b.** $f(x) = \ln 2x,$      $g(x) = \ln x^2$

   **c.** $f(x) = 10x^3+2x^2,$   $g(x) = e^x$

   **d.** $f(x) = \tan^{-1}(1/x),$   $g(x) = 1/x$

   **e.** $f(x) = \sin^{-1}(1/x),$   $g(x) = 1/x^2$

   **f.** $f(x) = \operatorname{sech} x,$      $g(x) = e^{-x}$

**111.** True, or false? Give reasons for your answers.

   **a.** $\dfrac{1}{x^2}+\dfrac{1}{x^4} = O\left(\dfrac{1}{x^2}\right)$     **b.** $\dfrac{1}{x^2}+\dfrac{1}{x^4} = O\left(\dfrac{1}{x^4}\right)$

   **c.** $x = o(x+\ln x)$           **d.** $\ln(\ln x) = o(\ln x)$

   **e.** $\tan^{-1} x = O(1)$          **f.** $\cosh x = O(e^x)$

**112.** True, or false? Give reasons for your answers.

   **a.** $\dfrac{1}{x^4} = O\left(\dfrac{1}{x^2}+\dfrac{1}{x^4}\right)$    **b.** $\dfrac{1}{x^4} = o\left(\dfrac{1}{x^2}+\dfrac{1}{x^4}\right)$

   **c.** $\ln x = o(x+1)$           **d.** $\ln 2x = O(\ln x)$

   **e.** $\sec^{-1} x = O(1)$         **f.** $\sinh x = O(e^x)$

### Theory and Applications

**113.** The function $f(x) = e^x + x$, being differentiable and one-to-one, has a differentiable inverse $f^{-1}(x)$. Find the value of $df^{-1}/dx$ at the point $f(\ln 2)$.

**114.** Find the inverse of the function $f(x) = 1 + (1/x)$, $x \neq 0$. Then show that $f^{-1}(f(x)) = f(f^{-1}(x)) = x$ and that

$$\left.\frac{df^{-1}}{dx}\right|_{f(x)} = \frac{1}{f'(x)}.$$

In Exercises 115 and 116, find the absolute maximum and minimum values of each function on the given interval.

**115.** $y = x\ln 2x - x,$    $\left[\dfrac{1}{2e}, \dfrac{e}{2}\right]$

**116.** $y = 10x(2-\ln x),$    $(0, e^2]$

**117. Area**  Find the area between the curve $y = 2(\ln x)/x$ and the $x$-axis from $x = 1$ to $x = e$.

**118. a.** Show that the area between the curve $y = 1/x$ and the $x$-axis from $x = 10$ to $x = 20$ is the same as the area between the curve and the $x$-axis from $x = 1$ to $x = 2$.

   **b.** Show that the area between the curve $y = 1/x$ and the $x$-axis from $ka$ to $kb$ is the same as the area between the curve and the $x$-axis from $x = a$ to $x = b$ $(0 < a < b, k > 0)$.

**119.** A particle is traveling upward and to the right along the curve $y = \ln x$. Its $x$-coordinate is increasing at the rate $(dx/dt) = \sqrt{x}$ m/s. At what rate is the $y$-coordinate changing at the point $(e^2, 2)$?

**120.** A girl is sliding down a slide shaped like the curve $y = 3e^{-x/3}$. Her $y$-coordinate is changing at the rate $dy/dt = (-1/4)\sqrt{3 - y}$ m/s. At approximately what rate is her $x$-coordinate changing when she reaches the bottom of the slide at $x = 3$ m?

**121.** The rectangle shown here has one side on the positive $y$-axis, one side on the positive $x$-axis, and its upper right-hand vertex on the curve $y = e^{-x^2}$. What dimensions give the rectangle its largest area, and what is that area?

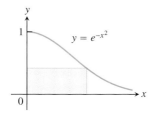

**122.** The rectangle shown here has one side on the positive $y$-axis, one side on the positive $x$-axis, and its upper right-hand vertex on the curve $y = (\ln x)/x^2$. What dimensions give the rectangle its largest area, and what is that area?

 T **123.** Graph the following functions and use what you see to locate and estimate the extreme values, identify the coordinates of the inflection points, and identify the intervals on which the graphs are concave up and concave down. Then confirm your estimates by working with the functions' derivatives.

   **a.** $y = (\ln x)/\sqrt{x}$

   **b.** $y = e^{-x^2}$

   **c.** $y = (1 + x)e^{-x}$

 T **124.** Graph $f(x) = x \ln x$. Does the function appear to have an absolute minimum value? Confirm your answer with calculus.

In Exercises 125–128 solve the differential equation.

**125.** $\dfrac{dy}{dx} = \sqrt{y} \cos^2 \sqrt{y}$

**126.** $y' = \dfrac{3y(x + 1)^2}{y - 1}$

**127.** $yy' = \sec y^2 \sec^2 x$

**128.** $y \cos^2 x \, dy + \sin x \, dx = 0$

In Exercises 129–132 solve the initial value problem.

**129.** $\dfrac{dy}{dx} = e^{-x-y-2}$,  $y(0) = -2$

**130.** $\dfrac{dy}{dx} = \dfrac{y \ln y}{1 + x^2}$,  $y(0) = e^2$

**131.** $x \, dy - \left(y + \sqrt{y}\right) dx = 0$,  $y(1) = 1$

**132.** $y^{-2} \dfrac{dx}{dy} = \dfrac{e^x}{e^{2x} + 1}$,  $y(0) = 1$

**133.** What is the age of a sample of charcoal in which 90% of the carbon-14 originally present has decayed?

**134. Cooling a pie**  A deep-dish apple pie, whose internal temperature was 104°C when removed from the oven, was set out on a breezy 5°C porch to cool. Fifteen minutes later, the pie's internal temperature was 82°C. How long did it take the pie to cool from there to 21°C?

**135. Locating a solar station**  You are under contract to build a solar station at ground level on the east–west line between the two buildings shown here. How far from the taller building should you place the station to maximize the number of hours it will be in the sun on a day when the sun passes directly overhead? Begin by observing that

$$\theta = \pi - \cot^{-1}\frac{x}{60} - \cot^{-1}\frac{50 - x}{30}.$$

Then find the value of $x$ that maximizes $\theta$.

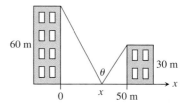

**136.** A round underwater transmission cable consists of a core of copper wires surrounded by nonconducting insulation. If $x$ denotes the ratio of the radius of the core to the thickness of the insulation, it is known that the speed of the transmission signal is given by the equation $v = x^2 \ln (1/x)$. If the radius of the core is 1 cm, what insulation thickness $h$ will allow the greatest transmission speed?

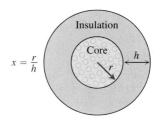

# Chapter **7** Additional and Advanced Exercises

## Limits

Find the limits in Exercises 1–6.

**1.** $\lim\limits_{b \to 1^-} \int_0^b \dfrac{dx}{\sqrt{1 - x^2}}$

**2.** $\lim\limits_{x \to \infty} \dfrac{1}{x} \int_0^x \tan^{-1} t \, dt$

**3.** $\lim\limits_{x \to 0^+} (\cos \sqrt{x})^{1/x}$

**4.** $\lim\limits_{x \to \infty} (x + e^x)^{2/x}$

**5.** $\lim\limits_{n \to \infty} \left( \dfrac{1}{n + 1} + \dfrac{1}{n + 2} + \cdots + \dfrac{1}{2n} \right)$

**6.** $\lim\limits_{n \to \infty} \dfrac{1}{n} \left( e^{1/n} + e^{2/n} + \cdots + e^{(n-1)/n} + e^{n/n} \right)$

**7.** Let $A(t)$ be the area of the region in the first quadrant enclosed by the coordinate axes, the curve $y = e^{-x}$, and the vertical line $x = t, t > 0$. Let $V(t)$ be the volume of the solid generated by revolving the region about the $x$-axis. Find the following limits.

    **a.** $\lim\limits_{t \to \infty} A(t)$    **b.** $\lim\limits_{t \to \infty} V(t)/A(t)$    **c.** $\lim\limits_{t \to 0^+} V(t)/A(t)$

**8. Varying a logarithm's base**

    **a.** Find $\lim \log_a 2$ as $a \to 0^+, 1^-, 1^+,$ and $\infty$.

    [T] **b.** Graph $y = \log_a 2$ as a function of $a$ over the interval $0 < a \le 4$.

## Theory and Examples

**9.** Find the areas between the curves $y = 2(\log_2 x)/x$ and $y = 2(\log_4 x)/x$ and the $x$-axis from $x = 1$ to $x = e$. What is the ratio of the larger area to the smaller?

[T] **10.** Graph $f(x) = \tan^{-1} x + \tan^{-1}(1/x)$ for $-5 \le x \le 5$. Then use calculus to explain what you see. How would you expect $f$ to behave beyond the interval $[-5, 5]$? Give reasons for your answer.

**11.** For what $x > 0$ does $x^{(x^x)} = (x^x)^x$? Give reasons for your answer.

[T] **12.** Graph $f(x) = (\sin x)^{\sin x}$ over $[0, 3\pi]$. Explain what you see.

**13.** Find $f'(2)$ if $f(x) = e^{g(x)}$ and $g(x) = \displaystyle\int_2^x \dfrac{t}{1 + t^4} \, dt.$

**14. a.** Find $df/dx$ if

$$f(x) = \int_1^{e^x} \frac{2 \ln t}{t} \, dt.$$

    **b.** Find $f(0)$.

    **c.** What can you conclude about the graph of $f$? Give reasons for your answer.

**15. Even-odd decompositions**

    **a.** Suppose that $g$ is an even function of $x$ and $h$ is an odd function of $x$. Show that if $g(x) + h(x) = 0$ for all $x$ then $g(x) = 0$ for all $x$ and $h(x) = 0$ for all $x$.

    **b.** If $f(x) = f_E(x) + f_O(x)$ is the sum of an even function $f_E(x)$ and an odd function $f_O(x)$, then show that

$$f_E(x) = \frac{f(x) + f(-x)}{2} \quad \text{and} \quad f_O(x) = \frac{f(x) - f(-x)}{2}.$$

    **c.** What is the significance of the result in part (b)?

**16.** Let $g$ be a function that is differentiable throughout an open interval containing the origin. Suppose $g$ has the following properties:

    **i)** $g(x + y) = \dfrac{g(x) + g(y)}{1 - g(x)g(y)}$ for all real numbers $x, y,$ and $x + y$ in the domain of $g$.

    **ii)** $\lim\limits_{h \to 0} g(h) = 0$

    **iii)** $\lim\limits_{h \to 0} \dfrac{g(h)}{h} = 1$

    **a.** Show that $g(0) = 0.$

    **b.** Show that $g'(x) = 1 + [g(x)]^2.$

    **c.** Find $g(x)$ by solving the differential equation in part (b).

**17. Center of mass** Find the center of mass of a thin plate of constant density covering the region in the first and fourth quadrants enclosed by the curves $y = 1/(1 + x^2)$ and $y = -1/(1 + x^2)$ and by the lines $x = 0$ and $x = 1.$

**18. Solid of revolution** The region between the curve $y = 1/(2\sqrt{x})$ and the $x$-axis from $x = 1/4$ to $x = 4$ is revolved about the $x$-axis to generate a solid.

    **a.** Find the volume of the solid.

    **b.** Find the centroid of the region.

**19. The best branching angles for blood vessels and pipes** When a smaller pipe branches off from a larger one in a flow system, we may want it to run off at an angle that is best from some energy-saving point of view. We might require, for instance, that energy loss due to friction be minimized along the section $AOB$ shown in the accompanying figure. In this diagram, $B$ is a given point to be reached by the smaller pipe, $A$ is a point in the larger pipe upstream from $B$, and $O$ is the point where the branching occurs. A law due to Poiseuille states that the loss of energy due to friction in nonturbulent flow is proportional to the length of the path and inversely proportional to the fourth power of the radius. Thus, the loss along $AO$ is $(kd_1)/R^4$ and along $OB$ is $(kd_2)/r^4$, where $k$ is a constant, $d_1$ is the length of $AO$, $d_2$ is the length of $OB$, $R$ is the radius of the larger pipe, and $r$ is the radius of the smaller pipe. The angle $\theta$ is to be chosen to minimize the sum of these two losses:

$$L = k\frac{d_1}{R^4} + k\frac{d_2}{r^4}.$$

In our model, we assume that $AC = a$ and $BC = b$ are fixed. Thus we have the relations

$$d_1 + d_2 \cos \theta = a \qquad d_2 \sin \theta = b$$

so that

$$d_2 = b \csc \theta,$$
$$d_1 = a - d_2 \cos \theta = a - b \cot \theta.$$

We can express the total loss $L$ as a function of $\theta$:

$$L = k\left(\frac{a - b \cot \theta}{R^4} + \frac{b \csc \theta}{r^4}\right).$$

**a.** Show that the critical value of $\theta$ for which $dL/d\theta$ equals zero is

$$\theta_c = \cos^{-1}\frac{r^4}{R^4}.$$

**b.** If the ratio of the pipe radii is $r/R = 5/6$, estimate to the nearest degree the optimal branching angle given in part (a).

The mathematical analysis described here is also used to explain the angles at which arteries branch in an animal's body.

**20. Urban gardening** A vegetable garden 15 m wide is to be grown between two buildings, which are 150 m apart along an east-west line. If the buildings are 60 m and 105 m tall, where should the garden be placed in order to receive the maximum number of hours of sunlight exposure? (*Hint:* Determine the value of $x$ in the accompanying figure that maximizes sunlight exposure for the garden.)

# 8

# Techniques of Integration

**OVERVIEW** The Fundamental Theorem tells us how to evaluate a definite integral once we have an antiderivative for the integrand function. However, finding antiderivatives (or indefinite integrals) is not as straightforward as finding derivatives. We need to develop some techniques to help us. Nevertheless, we note that it is not always possible to find an antiderivative expressed in terms of elementary functions.

In this chapter we study a number of important techniques which apply to finding integrals for specialized classes of functions such as trigonometric functions, products of certain functions, and rational functions. Since we cannot always find an antiderivative, we also develop some numerical methods for calculating definite integrals. Finally, we extend the idea of the definite integral to *improper integrals*, and we apply them to finding probabilities.

## 8.1 Using Basic Integration Formulas

Table 8.1 summarizes the forms of indefinite integrals for many of the functions we have studied so far, and the substitution method helps us use the table to evaluate more complicated functions involving these basic ones. In this section we combine the Substitution Rules (studied in Chapter 5) with algebraic methods and trigonometric identities to help us use Table 8.1. A more extensive Table of Integrals is given at the back of the book, and we discuss its use in Section 8.6.

Sometimes we have to rewrite an integral to match it to a standard form in Table 8.1. We have used this procedure before, but here is another example.

**EXAMPLE 1** Evaluate the integral

$$\int_3^5 \frac{2x - 3}{\sqrt{x^2 - 3x + 1}}\, dx.$$

**Solution** We rewrite the integral and apply the Substitution Rule for Definite Integrals presented in Section 5.6, to find

$$\int_3^5 \frac{2x - 3}{\sqrt{x^2 - 3x + 1}}\, dx = \int_1^{11} \frac{du}{\sqrt{u}} \qquad u = x^2 - 3x + 1,\ du = (2x - 3)\, dx;$$
$$u = 1 \text{ when } x = 3,\ u = 11 \text{ when } x = 5$$

$$= \int_1^{11} u^{-1/2}\, du$$

$$= 2\sqrt{u}\,\Big]_1^{11} = 2\left(\sqrt{11} - 1\right) \approx 4.63. \qquad \text{Table 8.1, Formula 2} \quad ■$$

---

**TABLE 8.1** Basic integration formulas

**1.** $\int k\,dx = kx + C$ (any number $k$)

**2.** $\int x^n\,dx = \dfrac{x^{n+1}}{n+1} + C$ $(n \neq -1)$

**3.** $\int \dfrac{dx}{x} = \ln|x| + C$

**4.** $\int e^x\,dx = e^x + C$

**5.** $\int a^x\,dx = \dfrac{a^x}{\ln a} + C$ $(a > 0,\ a \neq 1)$

**6.** $\int \sin x\,dx = -\cos x + C$

**7.** $\int \cos x\,dx = \sin x + C$

**8.** $\int \sec^2 x\,dx = \tan x + C$

**9.** $\int \csc^2 x\,dx = -\cot x + C$

**10.** $\int \sec x \tan x\,dx = \sec x + C$

**11.** $\int \csc x \cot x\,dx = -\csc x + C$

**12.** $\int \tan x\,dx = \ln|\sec x| + C$

**13.** $\int \cot x\,dx = \ln|\sin x| + C$

**14.** $\int \sec x\,dx = \ln|\sec x + \tan x| + C$

**15.** $\int \csc x\,dx = -\ln|\csc x + \cot x| + C$

**16.** $\int \sinh x\,dx = \cosh x + C$

**17.** $\int \cosh x\,dx = \sinh x + C$

**18.** $\int \dfrac{dx}{\sqrt{a^2 - x^2}} = \sin^{-1}\left(\dfrac{x}{a}\right) + C$

**19.** $\int \dfrac{dx}{a^2 + x^2} = \dfrac{1}{a}\tan^{-1}\left(\dfrac{x}{a}\right) + C$

**20.** $\int \dfrac{dx}{x\sqrt{x^2 - a^2}} = \dfrac{1}{a}\sec^{-1}\left|\dfrac{x}{a}\right| + C$

**21.** $\int \dfrac{dx}{\sqrt{a^2 + x^2}} = \sinh^{-1}\left(\dfrac{x}{a}\right) + C$ $(a > 0)$

**22.** $\int \dfrac{dx}{\sqrt{x^2 - a^2}} = \cosh^{-1}\left(\dfrac{x}{a}\right) + C$ $(x > a > 0)$

---

**EXAMPLE 2** Complete the square to evaluate

$$\int \frac{dx}{\sqrt{8x - x^2}}.$$

**Solution** We complete the square to simplify the denominator:

$$8x - x^2 = -(x^2 - 8x) = -(x^2 - 8x + 16 - 16)$$
$$= -(x^2 - 8x + 16) + 16 = 16 - (x - 4)^2.$$

Then

$$\int \frac{dx}{\sqrt{8x - x^2}} = \int \frac{dx}{\sqrt{16 - (x - 4)^2}}$$

$$= \int \frac{du}{\sqrt{a^2 - u^2}} \qquad a = 4,\ u = (x - 4),$$
$$\qquad\qquad\qquad\qquad du = dx$$

$$= \sin^{-1}\left(\frac{u}{a}\right) + C \qquad \text{Table 8.1, Formula 18}$$

$$= \sin^{-1}\left(\frac{x - 4}{4}\right) + C.$$ ∎

**EXAMPLE 3**    Evaluate the integral

$$\int (\cos x \sin 2x + \sin x \cos 2x) \, dx.$$

**Solution**    Here we can replace the integrand with an equivalent trigonometric expression using the Sine Addition Formula to obtain a simple substitution:

$$\int (\cos x \sin 2x + \sin x \cos 2x) \, dx = \int (\sin (x + 2x)) \, dx$$

$$= \int \sin 3x \, dx$$

$$= \int \frac{1}{3} \sin u \, du \qquad\qquad u = 3x, \ du = 3 \, dx$$

$$= -\frac{1}{3} \cos 3x + C. \qquad \text{Table 8.1, Formula 6} \quad\blacksquare$$

In Section 7.2 we found the indefinite integral of the secant function by multiplying it by a fractional form identically equal to one, and then integrating the equivalent result. We can use that same procedure in other instances as well, which we illustrate next.

**EXAMPLE 4**    Find $\displaystyle\int_0^{\pi/4} \frac{dx}{1 - \sin x}$.

**Solution**    We multiply the numerator and denominator of the integrand by $1 + \sin x$, which is simply a multiplication by a form of the number one. This procedure transforms the integral into one we can evaluate:

$$\int_0^{\pi/4} \frac{dx}{1 - \sin x} = \int_0^{\pi/4} \frac{1}{1 - \sin x} \cdot \frac{1 + \sin x}{1 + \sin x} \, dx$$

$$= \int_0^{\pi/4} \frac{1 + \sin x}{1 - \sin^2 x} \, dx$$

$$= \int_0^{\pi/4} \frac{1 + \sin x}{\cos^2 x} \, dx$$

$$= \int_0^{\pi/4} (\sec^2 x + \sec x \tan x) \, dx \qquad \begin{array}{l}\text{Use Table 8.1,}\\ \text{Formulas 8 and 10}\end{array}$$

$$= \Big[ \tan x + \sec x \Big]_0^{\pi/4} = \left(1 + \sqrt{2} - (0 + 1)\right) = \sqrt{2}. \qquad\blacksquare$$

**EXAMPLE 5**    Evaluate

$$\int \frac{3x^2 - 7x}{3x + 2} \, dx.$$

$$\begin{array}{r} x - 3 \\ 3x + 2 \overline{\smash{)}\, 3x^2 - 7x} \\ \underline{3x^2 + 2x} \\ -9x \\ \underline{-9x - 6} \\ + 6 \end{array}$$

**Solution**    The integrand is an improper fraction since the degree of the numerator is greater than the degree of the denominator. To integrate it, we perform long division to obtain a quotient plus a remainder that is a proper fraction:

$$\frac{3x^2 - 7x}{3x + 2} = x - 3 + \frac{6}{3x + 2}.$$

Therefore,

$$\int \frac{3x^2 - 7x}{3x + 2}\, dx = \int \left( x - 3 + \frac{6}{3x + 2} \right) dx = \frac{x^2}{2} - 3x + 2 \ln |3x + 2| + C. \quad \blacksquare$$

Reducing an improper fraction by long division (Example 5) does not always lead to an expression we can integrate directly. We see what to do about that in Section 8.5.

**EXAMPLE 6**    Evaluate

$$\int \frac{3x + 2}{\sqrt{1 - x^2}}\, dx.$$

**Solution**    We first separate the integrand to get

$$\int \frac{3x + 2}{\sqrt{1 - x^2}}\, dx = 3 \int \frac{x\, dx}{\sqrt{1 - x^2}} + 2 \int \frac{dx}{\sqrt{1 - x^2}}.$$

In the first of these new integrals, we substitute

$$u = 1 - x^2, \qquad du = -2x\, dx, \qquad \text{so} \qquad x\, dx = -\frac{1}{2}\, du.$$

Then we obtain

$$3 \int \frac{x\, dx}{\sqrt{1 - x^2}} = 3 \int \frac{(-1/2)\, du}{\sqrt{u}} = -\frac{3}{2} \int u^{-1/2}\, du$$

$$= -\frac{3}{2} \cdot \frac{u^{1/2}}{1/2} + C_1 = -3\sqrt{1 - x^2} + C_1.$$

The second of the new integrals is a standard form,

$$2 \int \frac{dx}{\sqrt{1 - x^2}} = 2 \sin^{-1} x + C_2. \qquad \text{Table 8.1, Formula 18}$$

Combining these results and renaming $C_1 + C_2$ as $C$ gives

$$\int \frac{3x + 2}{\sqrt{1 - x^2}}\, dx = -3\sqrt{1 - x^2} + 2 \sin^{-1} x + C. \quad \blacksquare$$

The question of what to substitute for in an integrand is not always quite so clear. Sometimes we simply proceed by trial-and-error, and if nothing works out, we then try another method altogether. The next several sections of the text present some of these new methods, but substitution works in the next example.

**EXAMPLE 7**    Evaluate

$$\int \frac{dx}{\left( 1 + \sqrt{x} \right)^3}.$$

**Solution**    We might try substituting for the term $\sqrt{x}$, but we quickly realize the derivative factor $1/\sqrt{x}$ is missing from the integrand, so this substitution will not help. The other possibility is to substitute for $\left( 1 + \sqrt{x} \right)$, and it turns out this works:

$$\int \frac{dx}{\left( 1 + \sqrt{x} \right)^3} = \int \frac{2(u - 1)\, du}{u^3} \qquad \begin{array}{l} u = 1 + \sqrt{x},\ du = \dfrac{1}{2\sqrt{x}}\, dx; \\[2mm] dx = 2\sqrt{x}\, du = 2(u - 1)\, du \end{array}$$

$$= \int \left( \frac{2}{u^2} - \frac{2}{u^3} \right) du$$

$$= -\frac{2}{u} + \frac{1}{u^2} + C$$

$$= \frac{1 - 2u}{u^2} + C$$

$$= \frac{1 - 2(1 + \sqrt{x})}{(1 + \sqrt{x})^2} + C$$

$$= C - \frac{1 + 2\sqrt{x}}{(1 + \sqrt{x})^2}.$$    ■

When evaluating definite integrals, a property of the integrand may help us in calculating the result.

**EXAMPLE 8**    Evaluate $\displaystyle\int_{-\pi/2}^{\pi/2} x^3 \cos x \, dx.$

**Solution**    No substitution or algebraic manipulation is clearly helpful here. But we observe that the interval of integration is the symmetric interval $[-\pi/2, \pi/2]$. Moreover, the factor $x^3$ is an odd function, and $\cos x$ is an even function, so their product is odd. Therefore,

$$\int_{-\pi/2}^{\pi/2} x^3 \cos x \, dx = 0.$$    Theorem 8, Section 5.6    ■

## Exercises 8.1

### Assorted Integrations

The integrals in Exercises 1–40 are in no particular order. Evaluate each integral using any algebraic method or trigonometric identity you think is appropriate, and then use a substitution to reduce it to a standard form.

**1.** $\displaystyle\int_0^1 \frac{16x}{8x^2 + 2} \, dx$

**2.** $\displaystyle\int \frac{x^2}{x^2 + 1} \, dx$

**3.** $\displaystyle\int (\sec x - \tan x)^2 \, dx$

**4.** $\displaystyle\int_{\pi/4}^{\pi/3} \frac{dx}{\cos^2 x \tan x}$

**5.** $\displaystyle\int \frac{1 - x}{\sqrt{1 - x^2}} \, dx$

**6.** $\displaystyle\int \frac{dx}{x - \sqrt{x}}$

**7.** $\displaystyle\int \frac{e^{-\cot z}}{\sin^2 z} \, dz$

**8.** $\displaystyle\int \frac{2^{\ln z^3}}{16z} \, dz$

**9.** $\displaystyle\int \frac{dz}{e^z + e^{-z}}$

**10.** $\displaystyle\int_1^2 \frac{8 \, dx}{x^2 - 2x + 2}$

**11.** $\displaystyle\int_{-1}^0 \frac{4 \, dx}{1 + (2x + 1)^2}$

**12.** $\displaystyle\int_{-1}^3 \frac{4x^2 - 7}{2x + 3} \, dx$

**13.** $\displaystyle\int \frac{dt}{1 - \sec t}$

**14.** $\displaystyle\int \csc t \sin 3t \, dt$

**15.** $\displaystyle\int_0^{\pi/4} \frac{1 + \sin \theta}{\cos^2 \theta} \, d\theta$

**16.** $\displaystyle\int \frac{d\theta}{\sqrt{2\theta - \theta^2}}$

**17.** $\displaystyle\int \frac{\ln y}{y + 4y \ln^2 y} \, dy$

**18.** $\displaystyle\int \frac{2^{\sqrt{y}} \, dy}{2\sqrt{y}}$

**19.** $\displaystyle\int \frac{d\theta}{\sec \theta + \tan \theta}$

**20.** $\displaystyle\int \frac{dt}{t\sqrt{3 + t^2}}$

**21.** $\displaystyle\int \frac{4t^3 - t^2 + 16t}{t^2 + 4} \, dt$

**22.** $\displaystyle\int \frac{x + 2\sqrt{x} - 1}{2x\sqrt{x} - 1} \, dx$

**23.** $\displaystyle\int_0^{\pi/2} \sqrt{1 - \cos \theta} \, d\theta$

**24.** $\displaystyle\int (\sec t + \cot t)^2 \, dt$

**25.** $\displaystyle\int \frac{dy}{\sqrt{e^{2y} - 1}}$

**26.** $\displaystyle\int \frac{6 \, dy}{\sqrt{y}(1 + y)}$

**27.** $\displaystyle\int \frac{2 \, dx}{x\sqrt{1 - 4 \ln^2 x}}$

**28.** $\displaystyle\int \frac{dx}{(x - 2)\sqrt{x^2 - 4x + 3}}$

**29.** $\displaystyle\int (\csc x - \sec x)(\sin x + \cos x) \, dx$

**30.** $\displaystyle\int 3 \sinh \left(\frac{x}{2} + \ln 5\right) dx$

**31.** $\displaystyle\int_{\sqrt{2}}^3 \frac{2x^3}{x^2 - 1} \, dx$

**32.** $\displaystyle\int_{-1}^1 \sqrt{1 + x^2} \sin x \, dx$

**33.** $\displaystyle\int_{-1}^0 \sqrt{\frac{1 + y}{1 - y}} \, dy$

**34.** $\displaystyle\int e^{z + e^z} \, dz$

**35.** $\displaystyle\int \frac{7\,dx}{(x-1)\sqrt{x^2-2x-48}}$    **36.** $\displaystyle\int \frac{dx}{(2x+1)\sqrt{4x+4x^2}}$

**37.** $\displaystyle\int \frac{2\theta^3-7\theta^2+7\theta}{2\theta-5}\,d\theta$    **38.** $\displaystyle\int \frac{d\theta}{\cos\theta-1}$

**39.** $\displaystyle\int \frac{dx}{1+e^x}$    **40.** $\displaystyle\int \frac{\sqrt{x}}{1+x^3}\,dx$

   *Hint:* Use long division.    *Hint:* Let $u=x^{3/2}$.

**Theory and Examples**

**41. Area**   Find the area of the region bounded above by $y=2\cos x$ and below by $y=\sec x$, $-\pi/4 \le x \le \pi/4$.

**42. Volume**   Find the volume of the solid generated by revolving the region in Exercise 41 about the $x$-axis.

**43. Arc length**   Find the length of the curve $y=\ln(\cos x)$, $0 \le x \le \pi/3$.

**44. Arc length**   Find the length of the curve $y=\ln(\sec x)$, $0 \le x \le \pi/4$.

**45. Centroid**   Find the centroid of the region bounded by the $x$-axis, the curve $y=\sec x$, and the lines $x=-\pi/4$, $x=\pi/4$.

**46. Centroid**   Find the centroid of the region bounded by the $x$-axis, the curve $y=\csc x$, and the lines $x=\pi/6$, $x=5\pi/6$.

**47.** The functions $y=e^{x^3}$ and $y=x^3e^{x^3}$ do not have elementary antiderivatives, but $y=(1+3x^3)e^{x^3}$ does.

Evaluate

$$\int (1+3x^3)e^{x^3}dx.$$

**48.** Use the substitution $u=\tan x$ to evaluate the integral

$$\int \frac{dx}{1+\sin^2 x}.$$

**49.** Use the substitution $u=x^4+1$ to evaluate the integral

$$\int x^7\sqrt{x^4+1}\,dx.$$

**50. Using different substitutions**   Show that the integral

$$\int ((x^2-1)(x+1))^{-2/3}dx$$

can be evaluated with any of the following substitutions.

  **a.** $u=1/(x+1)$

  **b.** $u=((x-1)/(x+1))^k$ for $k=1,\ 1/2,\ 1/3,\ -1/3,\ -2/3$, and $-1$

  **c.** $u=\tan^{-1}x$    **d.** $u=\tan^{-1}\sqrt{x}$

  **e.** $u=\tan^{-1}((x-1)/2)$    **f.** $u=\cos^{-1}x$

  **g.** $u=\cosh^{-1}x$

What is the value of the integral?

# 8.2 Integration by Parts

Integration by parts is a technique for simplifying integrals of the form

$$\int f(x)g(x)\,dx.$$

It is useful when $f$ can be differentiated repeatedly and $g$ can be integrated repeatedly without difficulty. The integrals

$$\int x\cos x\,dx \qquad \text{and} \qquad \int x^2e^x\,dx$$

are such integrals because $f(x)=x$ or $f(x)=x^2$ can be differentiated repeatedly to become zero, and $g(x)=\cos x$ or $g(x)=e^x$ can be integrated repeatedly without difficulty. Integration by parts also applies to integrals like

$$\int \ln x\,dx \qquad \text{and} \qquad \int e^x\cos x\,dx.$$

In the first case, $f(x)=\ln x$ is easy to differentiate and $g(x)=1$ easily integrates to $x$. In the second case, each part of the integrand appears again after repeated differentiation or integration.

## Product Rule in Integral Form

If $f$ and $g$ are differentiable functions of $x$, the Product Rule says that

$$\frac{d}{dx}[f(x)g(x)] = f'(x)g(x) + f(x)g'(x).$$

In terms of indefinite integrals, this equation becomes

$$\int \frac{d}{dx}[f(x)g(x)]\,dx = \int [f'(x)g(x) + f(x)g'(x)]\,dx$$

or

$$\int \frac{d}{dx}[f(x)g(x)]\,dx = \int f'(x)g(x)\,dx + \int f(x)g'(x)\,dx.$$

Rearranging the terms of this last equation, we get

$$\int f(x)g'(x)\,dx = \int \frac{d}{dx}[f(x)g(x)]\,dx - \int f'(x)g(x)\,dx,$$

leading to the **integration by parts** formula

$$\int f(x)g'(x)\,dx = f(x)g(x) - \int f'(x)g(x)\,dx \qquad (1)$$

Sometimes it is easier to remember the formula if we write it in differential form. Let $u = f(x)$ and $v = g(x)$. Then $du = f'(x)\,dx$ and $dv = g'(x)\,dx$. Using the Substitution Rule, the integration by parts formula becomes

**Integration by Parts Formula**

$$\int u\,dv = uv - \int v\,du \qquad (2)$$

This formula expresses one integral, $\int u\,dv$, in terms of a second integral, $\int v\,du$. With a proper choice of $u$ and $v$, the second integral may be easier to evaluate than the first. In using the formula, various choices may be available for $u$ and $dv$. The next examples illustrate the technique. To avoid mistakes, we always list our choices for $u$ and $dv$, then we add to the list our calculated new terms $du$ and $v$, and finally we apply the formula in Equation (2).

**EXAMPLE 1**  Find

$$\int x \cos x\,dx.$$

**Solution**  We use the formula $\int u\,dv = uv - \int v\,du$ with

$$u = x, \qquad dv = \cos x\,dx,$$
$$du = dx, \qquad v = \sin x. \qquad \text{Simplest antiderivative of } \cos x$$

Then

$$\int x \cos x\,dx = x \sin x - \int \sin x\,dx = x \sin x + \cos x + C. \qquad \blacksquare$$

There are four apparent choices available for $u$ and $dv$ in Example 1:

**1.** Let $u = 1$ and $dv = x \cos x\,dx$.  **2.** Let $u = x$ and $dv = \cos x\,dx$.

**3.** Let $u = x \cos x$ and $dv = dx$.  **4.** Let $u = \cos x$ and $dv = x\,dx$.

Choice 2 was used in Example 1. The other three choices lead to integrals we don't know how to integrate. For instance, Choice 3, with $du = (\cos x - x \sin x)\, dx$, leads to the integral

$$\int (x \cos x - x^2 \sin x)\, dx.$$

The goal of integration by parts is to go from an integral $\int u\, dv$ that we don't see how to evaluate to an integral $\int v\, du$ that we can evaluate. Generally, you choose $dv$ first to be as much of the integrand, including $dx$, as you can readily integrate; $u$ is the leftover part. When finding $v$ from $dv$, any antiderivative will work and we usually pick the simplest one; no arbitrary constant of integration is needed in $v$ because it would simply cancel out of the right-hand side of Equation (2).

**EXAMPLE 2**    Find

$$\int \ln x\, dx.$$

**Solution**    Since $\int \ln x\, dx$ can be written as $\int \ln x \cdot 1\, dx$, we use the formula $\int u\, dv = uv - \int v\, du$ with

$u = \ln x$ Simplifies when differentiated          $dv = dx$ Easy to integrate

$du = \dfrac{1}{x}\, dx,$          $v = x.$ Simplest antiderivative

Then from Equation (2),

$$\int \ln x\, dx = x \ln x - \int x \cdot \frac{1}{x}\, dx = x \ln x - \int dx = x \ln x - x + C. \quad \blacksquare$$

Sometimes we have to use integration by parts more than once.

**EXAMPLE 3**    Evaluate

$$\int x^2 e^x\, dx.$$

**Solution**    With $u = x^2$, $dv = e^x\, dx$, $du = 2x\, dx$, and $v = e^x$, we have

$$\int x^2 e^x\, dx = x^2 e^x - 2 \int x e^x\, dx.$$

The new integral is less complicated than the original because the exponent on $x$ is reduced by one. To evaluate the integral on the right, we integrate by parts again with $u = x$, $dv = e^x\, dx$. Then $du = dx$, $v = e^x$, and

$$\int x e^x\, dx = x e^x - \int e^x\, dx = x e^x - e^x + C.$$

Using this last evaluation, we then obtain

$$\int x^2 e^x\, dx = x^2 e^x - 2 \int x e^x\, dx$$

$$= x^2 e^x - 2x e^x + 2e^x + C,$$

where the constant of integration is renamed after substituting for the integral on the right. $\blacksquare$

The technique of Example 3 works for any integral $\int x^n e^x\, dx$ in which $n$ is a positive integer, because differentiating $x^n$ will eventually lead to zero and integrating $e^x$ is easy.

Integrals like the one in the next example occur in electrical engineering. Their evaluation requires two integrations by parts, followed by solving for the unknown integral.

**EXAMPLE 4**   Evaluate

$$\int e^x \cos x \, dx.$$

**Solution**   Let $u = e^x$ and $dv = \cos x \, dx$. Then $du = e^x \, dx$, $v = \sin x$, and

$$\int e^x \cos x \, dx = e^x \sin x - \int e^x \sin x \, dx.$$

The second integral is like the first except that it has $\sin x$ in place of $\cos x$. To evaluate it, we use integration by parts with

$$u = e^x, \qquad dv = \sin x \, dx, \qquad v = -\cos x, \qquad du = e^x \, dx.$$

Then

$$\int e^x \cos x \, dx = e^x \sin x - \left( -e^x \cos x - \int (-\cos x)(e^x \, dx) \right)$$

$$= e^x \sin x + e^x \cos x - \int e^x \cos x \, dx.$$

The unknown integral now appears on both sides of the equation. Adding the integral to both sides and adding the constant of integration give

$$2 \int e^x \cos x \, dx = e^x \sin x + e^x \cos x + C_1.$$

Dividing by 2 and renaming the constant of integration give

$$\int e^x \cos x \, dx = \frac{e^x \sin x + e^x \cos x}{2} + C. \qquad \blacksquare$$

**EXAMPLE 5**   Obtain a formula that expresses the integral

$$\int \cos^n x \, dx$$

in terms of an integral of a lower power of $\cos x$.

**Solution**   We may think of $\cos^n x$ as $\cos^{n-1} x \cdot \cos x$. Then we let

$$u = \cos^{n-1} x \qquad \text{and} \qquad dv = \cos x \, dx,$$

so that

$$du = (n - 1) \cos^{n-2} x \, (-\sin x \, dx) \qquad \text{and} \qquad v = \sin x.$$

Integration by parts then gives

$$\int \cos^n x \, dx = \cos^{n-1} x \sin x + (n - 1) \int \sin^2 x \cos^{n-2} x \, dx$$

$$= \cos^{n-1} x \sin x + (n - 1) \int (1 - \cos^2 x) \cos^{n-2} x \, dx$$

$$= \cos^{n-1} x \sin x + (n - 1) \int \cos^{n-2} x \, dx - (n - 1) \int \cos^n x \, dx.$$

If we add

$$(n - 1) \int \cos^n x \, dx$$

to both sides of this equation, we obtain

$$n \int \cos^n x \, dx = \cos^{n-1} x \sin x + (n-1) \int \cos^{n-2} x \, dx.$$

We then divide through by $n$, and the final result is

$$\int \cos^n x \, dx = \frac{\cos^{n-1} x \sin x}{n} + \frac{n-1}{n} \int \cos^{n-2} x \, dx. \qquad ■$$

The formula found in Example 5 is called a **reduction formula** because it replaces an integral containing some power of a function with an integral of the same form having the power reduced. When $n$ is a positive integer, we may apply the formula repeatedly until the remaining integral is easy to evaluate. For example, the result in Example 5 tells us that

$$\int \cos^3 x \, dx = \frac{\cos^2 x \sin x}{3} + \frac{2}{3} \int \cos x \, dx$$

$$= \frac{1}{3} \cos^2 x \sin x + \frac{2}{3} \sin x + C.$$

### Evaluating Definite Integrals by Parts

The integration by parts formula in Equation (1) can be combined with Part 2 of the Fundamental Theorem in order to evaluate definite integrals by parts. Assuming that both $f'$ and $g'$ are continuous over the interval $[a, b]$, Part 2 of the Fundamental Theorem gives

---

**Integration by Parts Formula for Definite Integrals**

$$\int_a^b f(x)g'(x) \, dx = f(x)g(x) \Big]_a^b - \int_a^b f'(x)g(x) \, dx \qquad (3)$$

---

**EXAMPLE 6**   Find the area of the region bounded by the curve $y = xe^{-x}$ and the $x$-axis from $x = 0$ to $x = 4$.

**Solution**   The region is shaded in Figure 8.1. Its area is

$$\int_0^4 xe^{-x} \, dx.$$

Let $u = x$, $dv = e^{-x} \, dx$, $v = -e^{-x}$, and $du = dx$. Then,

$$\int_0^4 xe^{-x} \, dx = -xe^{-x} \Big]_0^4 - \int_0^4 (-e^{-x}) \, dx$$

$$= [-4e^{-4} - (-0e^{-0})] + \int_0^4 e^{-x} \, dx$$

$$= -4e^{-4} - e^{-x} \Big]_0^4$$

$$= -4e^{-4} - (e^{-4} - e^{-0}) = 1 - 5e^{-4} \approx 0.91. \qquad ■$$

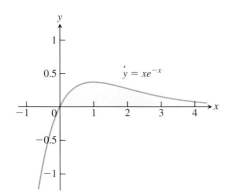

**FIGURE 8.1**   The region in Example 6.

### Tabular Integration Can Simplify Repeated Integrations

We have seen that integrals of the form $\int f(x)g(x) \, dx$, in which $f$ can be differentiated repeatedly to become zero and $g$ can be integrated repeatedly without difficulty, are natural candidates for integration by parts. However, if many repetitions are required, the notation and calculations can be cumbersome; or, you choose substitutions for a repeated integration by parts that just ends up giving back the original integral you were trying to find. In situations like these, there is a nice way to organize the calculations that prevents these pitfalls and simplifies the work. It is called **tabular integration** and is illustrated in the next examples.

**EXAMPLE 7** Evaluate

$$\int x^2 e^x \, dx.$$

**Solution** With $f(x) = x^2$ and $g(x) = e^x$, we list:

| $f(x)$ **and its derivatives** | | $g(x)$ **and its integrals** |
|---|---|---|
| $x^2$ | $(+)$ | $e^x$ |
| $2x$ | $(-)$ | $e^x$ |
| $2$ | $(+)$ | $e^x$ |
| $0$ | | $e^x$ |

We combine the products of the functions connected by the arrows according to the operation signs above the arrows to obtain

$$\int x^2 e^x \, dx = x^2 e^x - 2xe^x + 2e^x + C.$$

Compare this with the result in Example 3.

**EXAMPLE 8** Find the integral

$$\frac{1}{\pi} \int_{-\pi}^{\pi} f(x) \cos nx \, dx$$

for $f(x) = 1$ on $[-\pi, 0)$ and $f(x) = x^3$ on $[0, \pi]$, where $n$ is a positive integer.

**Solution** The integral is

$$\frac{1}{\pi} \int_{-\pi}^{\pi} f(x) \cos nx \, dx = \frac{1}{\pi} \int_{-\pi}^{0} \cos nx \, dx + \frac{1}{\pi} \int_{0}^{\pi} x^3 \cos nx \, dx$$

$$= \frac{1}{n\pi} \sin nx \bigg]_{-\pi}^{0} + \frac{1}{\pi} \int_{0}^{\pi} x^3 \cos nx \, dx$$

$$= \frac{1}{\pi} \int_{0}^{\pi} x^3 \cos nx \, dx.$$

Using tabular integration to find an antiderivative, we have

| $f(x)$ **and its derivatives** | | $g(x)$ **and its integrals** |
|---|---|---|
| $x^3$ | $(+)$ | $\cos nx$ |
| $3x^2$ | $(-)$ | $\dfrac{1}{n} \sin nx$ |
| $6x$ | $(+)$ | $-\dfrac{1}{n^2} \cos nx$ |
| $6$ | $(-)$ | $-\dfrac{1}{n^3} \sin nx$ |
| $0$ | | $\dfrac{1}{n^4} \cos nx$ |

$$\frac{1}{\pi} \int_0^\pi x^3 \cos nx \, dx$$

$$= \frac{1}{\pi} \left[ \frac{x^3}{n} \sin nx + \frac{3x^3}{n^2} \cos nx - \frac{6x}{n^3} \sin nx - \frac{6}{n^4} \cos nx \right]_0^\pi$$

$$= \frac{1}{\pi} \left( \frac{3\pi^2 \cos n\pi}{n^2} - \frac{6 \cos n\pi}{n^4} + \frac{6}{n^4} \right)$$

$$= \frac{3}{\pi} \left( \frac{\pi^2 n^2(-1)^n + 2(-1)^{n+1} + 2}{n^4} \right). \qquad \cos n\pi = (-1)^n$$

Integrals like those in Example 8 occur frequently in electrical engineering.

## Exercises 8.2

### Integration by Parts
Evaluate the integrals in Exercises 1–24 using integration by parts.

1. $\int x \sin \frac{x}{2} \, dx$
2. $\int \theta \cos \pi\theta \, d\theta$
3. $\int t^2 \cos t \, dt$
4. $\int x^2 \sin x \, dx$
5. $\int_1^2 x \ln x \, dx$
6. $\int_1^e x^3 \ln x \, dx$
7. $\int xe^x \, dx$
8. $\int xe^{3x} \, dx$
9. $\int x^2 e^{-x} \, dx$
10. $\int (x^2 - 2x + 1)e^{2x} \, dx$
11. $\int \tan^{-1} y \, dy$
12. $\int \sin^{-1} y \, dy$
13. $\int x \sec^2 x \, dx$
14. $\int 4x \sec^2 2x \, dx$
15. $\int x^3 e^x \, dx$
16. $\int p^4 e^{-p} \, dp$
17. $\int (x^2 - 5x)e^x \, dx$
18. $\int (r^2 + r + 1)e^r \, dr$
19. $\int x^5 e^x \, dx$
20. $\int t^2 e^{4t} \, dt$
21. $\int e^\theta \sin \theta \, d\theta$
22. $\int e^{-y} \cos y \, dy$
23. $\int e^{2x} \cos 3x \, dx$
24. $\int e^{-2x} \sin 2x \, dx$

### Using Substitution
Evaluate the integrals in Exercise 25–30 by using a substitution prior to integration by parts.

25. $\int e^{\sqrt{3s+9}} \, ds$
26. $\int_0^1 x\sqrt{1-x} \, dx$

27. $\int_0^{\pi/3} x \tan^2 x \, dx$
28. $\int \ln (x + x^2) \, dx$
29. $\int \sin (\ln x) \, dx$
30. $\int z(\ln z)^2 \, dz$

### Evaluating Integrals
Evaluate the integrals in Exercises 31–52. Some integrals do not require integration by parts.

31. $\int x \sec x^2 \, dx$
32. $\int \frac{\cos \sqrt{x}}{\sqrt{x}} \, dx$
33. $\int x (\ln x)^2 \, dx$
34. $\int \frac{1}{x (\ln x)^2} \, dx$
35. $\int \frac{\ln x}{x^2} \, dx$
36. $\int \frac{(\ln x)^3}{x} \, dx$
37. $\int x^3 e^{x^4} \, dx$
38. $\int x^5 e^{x^3} \, dx$
39. $\int x^3 \sqrt{x^2 + 1} \, dx$
40. $\int x^2 \sin x^3 \, dx$
41. $\int \sin 3x \cos 2x \, dx$
42. $\int \sin 2x \cos 4x \, dx$
43. $\int \sqrt{x} \ln x \, dx$
44. $\int \frac{e^{\sqrt{x}}}{\sqrt{x}} \, dx$
45. $\int \cos \sqrt{x} \, dx$
46. $\int \sqrt{x} e^{\sqrt{x}} \, dx$
47. $\int_0^{\pi/2} \theta^2 \sin 2\theta \, d\theta$
48. $\int_0^{\pi/2} x^3 \cos 2x \, dx$
49. $\int_{2/\sqrt{3}}^2 t \sec^{-1} t \, dt$
50. $\int_0^{1/\sqrt{2}} 2x \sin^{-1} (x^2) \, dx$
51. $\int x \tan^{-1} x \, dx$
52. $\int x^2 \tan^{-1} \frac{x}{2} \, dx$

**Theory and Examples**

**53. Finding area** Find the area of the region enclosed by the curve $y = x \sin x$ and the $x$-axis (see the accompanying figure) for

**a.** $0 \le x \le \pi$.

**b.** $\pi \le x \le 2\pi$.

**c.** $2\pi \le x \le 3\pi$.

**d.** What pattern do you see here? What is the area between the curve and the $x$-axis for $n\pi \le x \le (n + 1)\pi$, $n$ an arbitrary nonnegative integer? Give reasons for your answer.

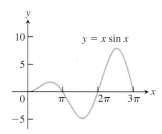

**54. Finding area** Find the area of the region enclosed by the curve $y = x \cos x$ and the $x$-axis (see the accompanying figure) for

**a.** $\pi/2 \le x \le 3\pi/2$.

**b.** $3\pi/2 \le x \le 5\pi/2$.

**c.** $5\pi/2 \le x \le 7\pi/2$.

**d.** What pattern do you see? What is the area between the curve and the $x$-axis for

$$\left(\frac{2n - 1}{2}\right)\pi \le x \le \left(\frac{2n + 1}{2}\right)\pi,$$

$n$ an arbitrary positive integer? Give reasons for your answer.

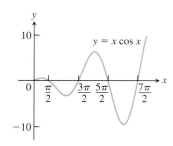

**55. Finding volume** Find the volume of the solid generated by revolving the region in the first quadrant bounded by the coordinate axes, the curve $y = e^x$, and the line $x = \ln 2$ about the line $x = \ln 2$.

**56. Finding volume** Find the volume of the solid generated by revolving the region in the first quadrant bounded by the coordinate axes, the curve $y = e^{-x}$, and the line $x = 1$

**a.** about the $y$-axis.

**b.** about the line $x = 1$.

**57. Finding volume** Find the volume of the solid generated by revolving the region in the first quadrant bounded by the coordinate axes and the curve $y = \cos x$, $0 \le x \le \pi/2$, about

**a.** the $y$-axis.

**b.** the line $x = \pi/2$.

**58. Finding volume** Find the volume of the solid generated by revolving the region bounded by the $x$-axis and the curve $y = x \sin x$, $0 \le x \le \pi$, about

**a.** the $y$-axis.

**b.** the line $x = \pi$.

(See Exercise 53 for a graph.)

**59.** Consider the region bounded by the graphs of $y = \ln x$, $y = 0$, and $x = e$.

**a.** Find the area of the region.

**b.** Find the volume of the solid formed by revolving this region about the $x$-axis.

**c.** Find the volume of the solid formed by revolving this region about the line $x = -2$.

**d.** Find the centroid of the region.

**60.** Consider the region bounded by the graphs of $y = \tan^{-1} x$, $y = 0$, and $x = 1$.

**a.** Find the area of the region.

**b.** Find the volume of the solid formed by revolving this region about the $y$-axis.

**61. Average value** A retarding force, symbolized by the dashpot in the accompanying figure, slows the motion of the weighted spring so that the mass's position at time $t$ is

$$y = 2e^{-t} \cos t, \qquad t \ge 0.$$

Find the average value of $y$ over the interval $0 \le t \le 2\pi$.

**62. Average value** In a mass-spring-dashpot system like the one in Exercise 61, the mass's position at time $t$ is

$$y = 4e^{-t}(\sin t - \cos t), \qquad t \ge 0.$$

Find the average value of $y$ over the interval $0 \le t \le 2\pi$.

**Reduction Formulas**

In Exercises 63–67, use integration by parts to establish the reduction formula.

**63.** $\displaystyle \int x^n \cos x \, dx = x^n \sin x - n \int x^{n-1} \sin x \, dx$

**64.** $\displaystyle \int x^n \sin x \, dx = -x^n \cos x + n \int x^{n-1} \cos x \, dx$

**65.** $\int x^n e^{ax} \, dx = \dfrac{x^n e^{ax}}{a} - \dfrac{n}{a} \int x^{n-1} e^{ax} \, dx, \quad a \neq 0$

**66.** $\int (\ln x)^n \, dx = x(\ln x)^n - n \int (\ln x)^{n-1} \, dx$

**67.** $\int x^m (\ln x)^n \, dx = \dfrac{x^{m+1}}{m+1} (\ln x)^n - \dfrac{n}{m+1} \cdot$

$\int x^m (\ln x)^{n-1} \, dx, \quad m \neq -1$

**68.** Use Example 5 to show that

$$\int_0^{\pi/2} \sin^n x \, dx = \int_0^{\pi/2} \cos^n x \, dx$$

$$= \begin{cases} \left(\dfrac{\pi}{2}\right) \dfrac{1 \cdot 3 \cdot 5 \cdots (n-1)}{2 \cdot 4 \cdot 6 \cdots n}, & n \text{ even} \\[2ex] \dfrac{2 \cdot 4 \cdot 6 \cdots (n-1)}{1 \cdot 3 \cdot 5 \cdots n}, & n \text{ odd} \end{cases}$$

**69.** Show that

$$\int_a^b \left( \int_x^b f(t) \, dt \right) dx = \int_a^b (x - a) f(x) \, dx.$$

**70.** Use integration by parts to obtain the formula

$$\int \sqrt{1 - x^2} \, dx = \frac{1}{2} x \sqrt{1 - x^2} + \frac{1}{2} \int \frac{1}{\sqrt{1 - x^2}} \, dx.$$

**Integrating Inverses of Functions**
Integration by parts leads to a rule for integrating inverses that usually gives good results:

$$\int f^{-1}(x) \, dx = \int y f'(y) \, dy \qquad \begin{matrix} y = f^{-1}(x), \quad x = f(y) \\ dx = f'(y) \, dy \end{matrix}$$

$$= y f(y) - \int f(y) \, dy \qquad \begin{matrix} \text{Integration by parts with} \\ u = y, \, dv = f'(y) \, dy \end{matrix}$$

$$= x f^{-1}(x) - \int f(y) \, dy$$

The idea is to take the most complicated part of the integral, in this case $f^{-1}(x)$, and simplify it first. For the integral of $\ln x$, we get

$$\int \ln x \, dx = \int y e^y \, dy \qquad \begin{matrix} y = \ln x, \quad x = e^y \\ dx = e^y \, dy \end{matrix}$$

$$= y e^y - e^y + C$$

$$= x \ln x - x + C.$$

For the integral of $\cos^{-1} x$ we get

$$\int \cos^{-1} x \, dx = x \cos^{-1} x - \int \cos y \, dy \qquad y = \cos^{-1} x$$

$$= x \cos^{-1} x - \sin y + C$$

$$= x \cos^{-1} x - \sin (\cos^{-1} x) + C.$$

Use the formula

$$\int f^{-1}(x) \, dx = x f^{-1}(x) - \int f(y) \, dy \qquad y = f^{-1}(x) \quad (4)$$

to evaluate the integrals in Exercises 71–74. Express your answers in terms of $x$.

**71.** $\int \sin^{-1} x \, dx$          **72.** $\int \tan^{-1} x \, dx$

**73.** $\int \sec^{-1} x \, dx$          **74.** $\int \log_2 x \, dx$

Another way to integrate $f^{-1}(x)$ (when $f^{-1}$ is integrable, of course) is to use integration by parts with $u = f^{-1}(x)$ and $dv = dx$ to rewrite the integral of $f^{-1}$ as

$$\int f^{-1}(x) \, dx = x f^{-1}(x) - \int x \left( \frac{d}{dx} f^{-1}(x) \right) dx. \quad (5)$$

Exercises 75 and 76 compare the results of using Equations (4) and (5).

**75.** Equations (4) and (5) give different formulas for the integral of $\cos^{-1} x$:

**a.** $\int \cos^{-1} x \, dx = x \cos^{-1} x - \sin (\cos^{-1} x) + C$    Eq. (4)

**b.** $\int \cos^{-1} x \, dx = x \cos^{-1} x - \sqrt{1 - x^2} + C$    Eq. (5)

Can both integrations be correct? Explain.

**76.** Equations (4) and (5) lead to different formulas for the integral of $\tan^{-1} x$:

**a.** $\int \tan^{-1} x \, dx = x \tan^{-1} x - \ln \sec (\tan^{-1} x) + C$    Eq. (4)

**b.** $\int \tan^{-1} x \, dx = x \tan^{-1} x - \ln \sqrt{1 + x^2} + C$    Eq. (5)

Can both integrations be correct? Explain.

Evaluate the integrals in Exercises 77 and 78 with (**a**) Eq. (4) and (**b**) Eq. (5). In each case, check your work by differentiating your answer with respect to $x$.

**77.** $\int \sinh^{-1} x \, dx$          **78.** $\int \tanh^{-1} x \, dx$

# 8.3 Trigonometric Integrals

Trigonometric integrals involve algebraic combinations of the six basic trigonometric functions. In principle, we can always express such integrals in terms of sines and cosines, but it is often simpler to work with other functions, as in the integral

$$\int \sec^2 x \, dx = \tan x + C.$$

The general idea is to use identities to transform the integrals we have to find into integrals that are easier to work with.

## Products of Powers of Sines and Cosines

We begin with integrals of the form

$$\int \sin^m x \cos^n x \, dx,$$

where $m$ and $n$ are nonnegative integers (positive or zero). We can divide the appropriate substitution into three cases according to $m$ and $n$ being odd or even.

---

**Case 1** If $m$ **is odd**, we write $m$ as $2k + 1$ and use the identity $\sin^2 x = 1 - \cos^2 x$ to obtain

$$\sin^m x = \sin^{2k+1} x = (\sin^2 x)^k \sin x = (1 - \cos^2 x)^k \sin x. \qquad (1)$$

Then we combine the single $\sin x$ with $dx$ in the integral and set $\sin x \, dx$ equal to $-d(\cos x)$.

**Case 2** If $m$ **is even and** $n$ **is odd** in $\int \sin^m x \cos^n x \, dx$, we write $n$ as $2k + 1$ and use the identity $\cos^2 x = 1 - \sin^2 x$ to obtain

$$\cos^n x = \cos^{2k+1} x = (\cos^2 x)^k \cos x = (1 - \sin^2 x)^k \cos x.$$

We then combine the single $\cos x$ with $dx$ and set $\cos x \, dx$ equal to $d(\sin x)$.

**Case 3** If **both** $m$ **and** $n$ **are even** in $\int \sin^m x \cos^n x \, dx$, we substitute

$$\sin^2 x = \frac{1 - \cos 2x}{2}, \qquad \cos^2 x = \frac{1 + \cos 2x}{2} \qquad (2)$$

to reduce the integrand to one in lower powers of $\cos 2x$.

---

Here are some examples illustrating each case.

**EXAMPLE 1**    Evaluate

$$\int \sin^3 x \cos^2 x \, dx.$$

**Solution**    This is an example of Case 1.

$$\int \sin^3 x \cos^2 x \, dx = \int \sin^2 x \cos^2 x \sin x \, dx \qquad \text{\small $m$ is odd.}$$

$$= \int (1 - \cos^2 x)(\cos^2 x)(-d(\cos x)) \qquad \text{\small $\sin x \, dx = -d(\cos x)$}$$

$$= \int (1 - u^2)(u^2)(-du) \qquad \text{\small $u = \cos x$}$$

$$= \int (u^4 - u^2) \, du \qquad \text{\small Multiply terms.}$$

$$= \frac{u^5}{5} - \frac{u^3}{3} + C = \frac{\cos^5 x}{5} - \frac{\cos^3 x}{3} + C \qquad \blacksquare$$

**EXAMPLE 2**    Evaluate

$$\int \cos^5 x \, dx.$$

**Solution**    This is an example of Case 2, where $m = 0$ is even and $n = 5$ is odd.

$$\int \cos^5 x \, dx = \int \cos^4 x \cos x \, dx = \int (1 - \sin^2 x)^2 \, d(\sin x) \qquad \cos x \, dx = d(\sin x)$$

$$= \int (1 - u^2)^2 \, du \qquad\qquad\qquad u = \sin x$$

$$= \int (1 - 2u^2 + u^4) \, du \qquad\qquad \text{Square } 1 - u^2.$$

$$= u - \frac{2}{3} u^3 + \frac{1}{5} u^5 + C = \sin x - \frac{2}{3} \sin^3 x + \frac{1}{5} \sin^5 x + C \qquad \blacksquare$$

**EXAMPLE 3**    Evaluate

$$\int \sin^2 x \cos^4 x \, dx.$$

**Solution**    This is an example of Case 3.

$$\int \sin^2 x \cos^4 x \, dx = \int \left(\frac{1 - \cos 2x}{2}\right)\left(\frac{1 + \cos 2x}{2}\right)^2 dx \qquad m \text{ and } n \text{ both even}$$

$$= \frac{1}{8} \int (1 - \cos 2x)(1 + 2\cos 2x + \cos^2 2x) \, dx$$

$$= \frac{1}{8} \int (1 + \cos 2x - \cos^2 2x - \cos^3 2x) \, dx$$

$$= \frac{1}{8} \left[ x + \frac{1}{2} \sin 2x - \int (\cos^2 2x + \cos^3 2x) \, dx \right]$$

For the term involving $\cos^2 2x$, we use

$$\int \cos^2 2x \, dx = \frac{1}{2} \int (1 + \cos 4x) \, dx$$

$$= \frac{1}{2} \left( x + \frac{1}{4} \sin 4x \right). \qquad \begin{array}{l}\text{Omitting the constant of}\\\text{integration until the final result}\end{array}$$

For the $\cos^3 2x$ term, we have

$$\int \cos^3 2x \, dx = \int (1 - \sin^2 2x) \cos 2x \, dx \qquad \begin{array}{l} u = \sin 2x, \\ du = 2\cos 2x \, dx \end{array}$$

$$= \frac{1}{2} \int (1 - u^2) \, du = \frac{1}{2} \left( \sin 2x - \frac{1}{3} \sin^3 2x \right). \qquad \text{Again omitting } C$$

Combining everything and simplifying, we get

$$\int \sin^2 x \cos^4 x \, dx = \frac{1}{16} \left( x - \frac{1}{4} \sin 4x + \frac{1}{3} \sin^3 2x \right) + C. \qquad \blacksquare$$

### Eliminating Square Roots

In the next example, we use the identity $\cos^2 \theta = (1 + \cos 2\theta)/2$ to eliminate a square root.

**EXAMPLE 4**   Evaluate

$$\int_0^{\pi/4} \sqrt{1 + \cos 4x} \, dx.$$

**Solution**   To eliminate the square root, we use the identity

$$\cos^2 \theta = \frac{1 + \cos 2\theta}{2} \quad \text{or} \quad 1 + \cos 2\theta = 2 \cos^2 \theta.$$

With $\theta = 2x$, this becomes

$$1 + \cos 4x = 2 \cos^2 2x.$$

Therefore,

$$\int_0^{\pi/4} \sqrt{1 + \cos 4x} \, dx = \int_0^{\pi/4} \sqrt{2 \cos^2 2x} \, dx = \int_0^{\pi/4} \sqrt{2} \sqrt{\cos^2 2x} \, dx$$

$$= \sqrt{2} \int_0^{\pi/4} |\cos 2x| \, dx = \sqrt{2} \int_0^{\pi/4} \cos 2x \, dx \qquad \begin{matrix} \cos 2x \geq 0 \text{ on} \\ [0, \pi/4] \end{matrix}$$

$$= \sqrt{2} \left[ \frac{\sin 2x}{2} \right]_0^{\pi/4} = \frac{\sqrt{2}}{2} [1 - 0] = \frac{\sqrt{2}}{2}. \qquad \blacksquare$$

### Integrals of Powers of tan $x$ and sec $x$

We know how to integrate the tangent and secant and their squares. To integrate higher powers, we use the identities $\tan^2 x = \sec^2 x - 1$ and $\sec^2 x = \tan^2 x + 1$, and integrate by parts when necessary to reduce the higher powers to lower powers.

**EXAMPLE 5**   Evaluate

$$\int \tan^4 x \, dx.$$

**Solution**

$$\int \tan^4 x \, dx = \int \tan^2 x \cdot \tan^2 x \, dx = \int \tan^2 x \cdot (\sec^2 x - 1) \, dx$$

$$= \int \tan^2 x \sec^2 x \, dx - \int \tan^2 x \, dx$$

$$= \int \tan^2 x \sec^2 x \, dx - \int (\sec^2 x - 1) \, dx$$

$$= \int \tan^2 x \sec^2 x \, dx - \int \sec^2 x \, dx + \int dx$$

In the first integral, we let

$$u = \tan x, \qquad du = \sec^2 x \, dx$$

and have

$$\int u^2 \, du = \frac{1}{3} u^3 + C_1.$$

The remaining integrals are standard forms, so

$$\int \tan^4 x \, dx = \frac{1}{3} \tan^3 x - \tan x + x + C.$$ ∎

**EXAMPLE 6**     Evaluate

$$\int \sec^3 x \, dx.$$

**Solution**     We integrate by parts using

$$u = \sec x, \qquad dv = \sec^2 x \, dx, \qquad v = \tan x, \qquad du = \sec x \tan x \, dx.$$

Then

$$\int \sec^3 x \, dx = \sec x \tan x - \int (\tan x)(\sec x \tan x \, dx)$$

$$= \sec x \tan x - \int (\sec^2 x - 1) \sec x \, dx \qquad \tan^2 x = \sec^2 x - 1$$

$$= \sec x \tan x + \int \sec x \, dx - \int \sec^3 x \, dx.$$

Combining the two secant-cubed integrals gives

$$2 \int \sec^3 x \, dx = \sec x \tan x + \int \sec x \, dx$$

and

$$\int \sec^3 x \, dx = \frac{1}{2} \sec x \tan x + \frac{1}{2} \ln |\sec x + \tan x| + C.$$ ∎

**EXAMPLE 7**     Evaluate

$$\int \tan^4 x \sec^4 x \, dx.$$

**Solution**

$$\int (\tan^4 x)(\sec^4 x) \, dx = \int (\tan^4 x)(1 + \tan^2 x)(\sec^2 x) \, dx \qquad \sec^2 x = 1 + \tan^2 x$$

$$= \int (\tan^4 x + \tan^6 x)(\sec^2 x) \, dx$$

$$= \int (\tan^4 x)(\sec^2 x) \, dx + \int (\tan^6 x)(\sec^2 x) \, dx$$

$$= \int u^4 \, du + \int u^6 \, du = \frac{u^5}{5} + \frac{u^7}{7} + C \qquad \begin{array}{l} u = \tan x, \\ du = \sec^2 x \, dx \end{array}$$

$$= \frac{\tan^5 x}{5} + \frac{\tan^7 x}{7} + C$$ ∎

## Products of Sines and Cosines

The integrals

$$\int \sin mx \sin nx \, dx, \qquad \int \sin mx \cos nx \, dx, \qquad \text{and} \qquad \int \cos mx \cos nx \, dx$$

arise in many applications involving periodic functions. We can evaluate these integrals through integration by parts, but two such integrations are required in each case. It is simpler to use the identities

$$\sin mx \sin nx = \frac{1}{2} \left[ \cos (m - n)x - \cos (m + n)x \right], \qquad (3)$$

$$\sin mx \cos nx = \frac{1}{2} \left[ \sin (m - n)x + \sin (m + n)x \right], \qquad (4)$$

$$\cos mx \cos nx = \frac{1}{2} \left[ \cos (m - n)x + \cos (m + n)x \right]. \qquad (5)$$

These identities come from the angle sum formulas for the sine and cosine functions (Section 1.3). They give functions whose antiderivatives are easily found.

**EXAMPLE 8** Evaluate

$$\int \sin 3x \cos 5x \, dx.$$

**Solution** From Equation (4) with $m = 3$ and $n = 5$, we get

$$\int \sin 3x \cos 5x \, dx = \frac{1}{2} \int [\sin(-2x) + \sin 8x] \, dx$$

$$= \frac{1}{2} \int (\sin 8x - \sin 2x) \, dx$$

$$= -\frac{\cos 8x}{16} + \frac{\cos 2x}{4} + C.$$ ∎

## Exercises 8.3

**Powers of Sines and Cosines**
Evaluate the integrals in Exercises 1–22.

**1.** $\int \cos 2x \, dx$

**2.** $\int_0^\pi 3 \sin \frac{x}{3} \, dx$

**3.** $\int \cos^3 x \sin x \, dx$

**4.** $\int \sin^4 2x \cos 2x \, dx$

**5.** $\int \sin^3 x \, dx$

**6.** $\int \cos^3 4x \, dx$

**7.** $\int \sin^5 x \, dx$

**8.** $\int_0^\pi \sin^5 \frac{x}{2} \, dx$

**9.** $\int \cos^3 x \, dx$

**10.** $\int_0^{\pi/6} 3 \cos^5 3x \, dx$

**11.** $\int \sin^3 x \cos^3 x \, dx$

**12.** $\int \cos^3 2x \sin^5 2x \, dx$

**13.** $\int \cos^2 x \, dx$

**14.** $\int_0^{\pi/2} \sin^2 x \, dx$

**15.** $\int_0^{\pi/2} \sin^7 y \, dy$

**16.** $\int 7 \cos^7 t \, dt$

**17.** $\int_0^\pi 8 \sin^4 x \, dx$

**18.** $\int 8 \cos^4 2\pi x \, dx$

**19.** $\int 16 \sin^2 x \cos^2 x \, dx$

**20.** $\int_0^\pi 8 \sin^4 y \cos^2 y \, dy$

**21.** $\int 8 \cos^3 2\theta \sin 2\theta \, d\theta$

**22.** $\int_0^{\pi/2} \sin^2 2\theta \cos^3 2\theta \, d\theta$

**Integrating Square Roots**
Evaluate the integrals in Exercises 23–32.

**23.** $\int_0^{2\pi} \sqrt{\frac{1 - \cos x}{2}} \, dx$

**24.** $\int_0^\pi \sqrt{1 - \cos 2x} \, dx$

**25.** $\int_0^\pi \sqrt{1 - \sin^2 t} \, dt$

**26.** $\int_0^\pi \sqrt{1 - \cos^2 \theta} \, d\theta$

**27.** $\int_{\pi/3}^{\pi/2} \frac{\sin^2 x}{\sqrt{1 - \cos x}} \, dx$

**28.** $\int_0^{\pi/6} \sqrt{1 + \sin x} \, dx$

$\left( \textit{Hint: } \text{Multiply by } \sqrt{\frac{1 - \sin x}{1 - \sin x}} \right)$

**29.** $\int_{5\pi/6}^\pi \frac{\cos^4 x}{\sqrt{1 - \sin x}} \, dx$

**30.** $\int_{\pi/2}^{3\pi/4} \sqrt{1 - \sin 2x} \, dx$

**31.** $\int_0^{\pi/2} \theta \sqrt{1 - \cos 2\theta} \, d\theta$

**32.** $\int_{-\pi}^\pi (1 - \cos^2 t)^{3/2} \, dt$

**Powers of Tangents and Secants**
Evaluate the integrals in Exercises 33–50.

**33.** $\int \sec^2 x \tan x \, dx$

**34.** $\int \sec x \tan^2 x \, dx$

**35.** $\int \sec^3 x \tan x \, dx$

**36.** $\int \sec^3 x \tan^3 x \, dx$

**37.** $\int \sec^2 x \tan^2 x \, dx$

**38.** $\int \sec^4 x \tan^2 x \, dx$

**39.** $\int_{-\pi/3}^0 2 \sec^3 x \, dx$

**40.** $\int e^x \sec^3 e^x \, dx$

**41.** $\int \sec^4 \theta \, d\theta$

**42.** $\int 3 \sec^4 3x \, dx$

**43.** $\int_{\pi/4}^{\pi/2} \csc^4 \theta \, d\theta$

**44.** $\int \sec^6 x \, dx$

**45.** $\int 4 \tan^3 x \, dx$

**46.** $\int_{-\pi/4}^{\pi/4} 6 \tan^4 x \, dx$

**47.** $\int \tan^5 x \, dx$

**48.** $\int \cot^6 2x \, dx$

**49.** $\int_{\pi/6}^{\pi/3} \cot^3 x \, dx$

**50.** $\int 8 \cot^4 t \, dt$

## Products of Sines and Cosines

Evaluate the integrals in Exercises 51–56.

**51.** $\int \sin 3x \cos 2x \, dx$

**52.** $\int \sin 2x \cos 3x \, dx$

**53.** $\int_{-\pi}^{\pi} \sin 3x \sin 3x \, dx$

**54.** $\int_{0}^{\pi/2} \sin x \cos x \, dx$

**55.** $\int \cos 3x \cos 4x \, dx$

**56.** $\int_{-\pi/2}^{\pi/2} \cos x \cos 7x \, dx$

Exercises 57–62 require the use of various trigonometric identities before you evaluate the integrals.

**57.** $\int \sin^2 \theta \cos 3\theta \, d\theta$

**58.** $\int \cos^2 2\theta \sin \theta \, d\theta$

**59.** $\int \cos^3 \theta \sin 2\theta \, d\theta$

**60.** $\int \sin^3 \theta \cos 2\theta \, d\theta$

**61.** $\int \sin \theta \cos \theta \cos 3\theta \, d\theta$

**62.** $\int \sin \theta \sin 2\theta \sin 3\theta \, d\theta$

## Assorted Integrations

Use any method to evaluate the integrals in Exercises 63–68.

**63.** $\int \frac{\sec^3 x}{\tan x} \, dx$

**64.** $\int \frac{\sin^3 x}{\cos^4 x} \, dx$

**65.** $\int \frac{\tan^2 x}{\csc x} \, dx$

**66.** $\int \frac{\cot x}{\cos^2 x} \, dx$

**67.** $\int x \sin^2 x \, dx$

**68.** $\int x \cos^3 x \, dx$

## Applications

**69. Arc length**  Find the length of the curve

$$y = \ln (\sec x), \quad 0 \le x \le \pi/4.$$

**70. Center of gravity**  Find the center of gravity of the region bounded by the $x$-axis, the curve $y = \sec x$, and the lines $x = -\pi/4$, $x = \pi/4$.

**71. Volume**  Find the volume generated by revolving one arch of the curve $y = \sin x$ about the $x$-axis.

**72. Area**  Find the area between the $x$-axis and the curve $y = \sqrt{1 + \cos 4x}$, $0 \le x \le \pi$.

**73. Centroid**  Find the centroid of the region bounded by the graphs of $y = x + \cos x$ and $y = 0$ for $0 \le x \le 2\pi$.

**74. Volume**  Find the volume of the solid formed by revolving the region bounded by the graphs of $y = \sin x + \sec x$, $y = 0$, $x = 0$, and $x = \pi/3$ about the $x$-axis.

# $8.4$ Trigonometric Substitutions

Trigonometric substitutions occur when we replace the variable of integration by a trigonometric function. The most common substitutions are $x = a \tan \theta$, $x = a \sin \theta$, and $x = a \sec \theta$. These substitutions are effective in transforming integrals involving $\sqrt{a^2 + x^2}$, $\sqrt{a^2 - x^2}$, and $\sqrt{x^2 - a^2}$ into integrals we can evaluate directly since they come from the reference right triangles in Figure 8.2.

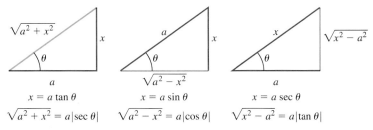

$$x = a \tan \theta \qquad x = a \sin \theta \qquad x = a \sec \theta$$
$$\sqrt{a^2 + x^2} = a|\sec \theta| \qquad \sqrt{a^2 - x^2} = a|\cos \theta| \qquad \sqrt{x^2 - a^2} = a|\tan \theta|$$

**FIGURE 8.2**  Reference triangles for the three basic substitutions identifying the sides labeled $x$ and $a$ for each substitution.

With $x = a \tan \theta$,

$$a^2 + x^2 = a^2 + a^2 \tan^2 \theta = a^2(1 + \tan^2 \theta) = a^2 \sec^2 \theta.$$

With $x = a \sin \theta$,

$$a^2 - x^2 = a^2 - a^2 \sin^2 \theta = a^2(1 - \sin^2 \theta) = a^2 \cos^2 \theta.$$

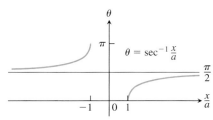

**FIGURE 8.3** The arctangent, arcsine, and arcsecant of $x/a$, graphed as functions of $x/a$.

With $x = a \sec \theta$,

$$x^2 - a^2 = a^2 \sec^2 \theta - a^2 = a^2(\sec^2 \theta - 1) = a^2 \tan^2 \theta.$$

We want any substitution we use in an integration to be reversible so that we can change back to the original variable afterward. For example, if $x = a \tan \theta$, we want to be able to set $\theta = \tan^{-1}(x/a)$ after the integration takes place. If $x = a \sin \theta$, we want to be able to set $\theta = \sin^{-1}(x/a)$ when we're done, and similarly for $x = a \sec \theta$.

As we know from Section 7.6, the functions in these substitutions have inverses only for selected values of $\theta$ (Figure 8.3). For reversibility,

$$x = a \tan \theta \quad \text{requires} \quad \theta = \tan^{-1}\left(\frac{x}{a}\right) \quad \text{with} \quad -\frac{\pi}{2} < \theta < \frac{\pi}{2},$$

$$x = a \sin \theta \quad \text{requires} \quad \theta = \sin^{-1}\left(\frac{x}{a}\right) \quad \text{with} \quad -\frac{\pi}{2} \le \theta \le \frac{\pi}{2},$$

$$x = a \sec \theta \quad \text{requires} \quad \theta = \sec^{-1}\left(\frac{x}{a}\right) \quad \text{with} \quad \begin{cases} 0 \le \theta < \dfrac{\pi}{2} & \text{if} \quad \dfrac{x}{a} \ge 1, \\[2mm] \dfrac{\pi}{2} < \theta \le \pi & \text{if} \quad \dfrac{x}{a} \le -1. \end{cases}$$

To simplify calculations with the substitution $x = a \sec \theta$, we will restrict its use to integrals in which $x/a \ge 1$. This will place $\theta$ in $[0, \pi/2)$ and make $\tan \theta \ge 0$. We will then have $\sqrt{x^2 - a^2} = \sqrt{a^2 \tan^2 \theta} = |a \tan \theta| = a \tan \theta$, free of absolute values, provided $a > 0$.

---

**Procedure for a Trigonometric Substitution**

1. Write down the substitution for $x$, calculate the differential $dx$, and specify the selected values of $\theta$ for the substitution.

2. Substitute the trigonometric expression and the calculated differential into the integrand, and then simplify the results algebraically.

3. Integrate the trigonometric integral, keeping in mind the restrictions on the angle $\theta$ for reversibility.

4. Draw an appropriate reference triangle to reverse the substitution in the integration result and convert it back to the original variable $x$.

---

**EXAMPLE 1**    Evaluate

$$\int \frac{dx}{\sqrt{4 + x^2}}.$$

**Solution**    We set

$$x = 2 \tan \theta, \qquad dx = 2 \sec^2 \theta \, d\theta, \qquad -\frac{\pi}{2} < \theta < \frac{\pi}{2},$$

$$4 + x^2 = 4 + 4 \tan^2 \theta = 4(1 + \tan^2 \theta) = 4 \sec^2 \theta.$$

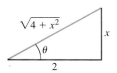

**FIGURE 8.4** Reference triangle for
$x = 2 \tan \theta$ (Example 1):

$$\tan \theta = \frac{x}{2}$$

and

$$\sec \theta = \frac{\sqrt{4 + x^2}}{2}.$$

Then

$$\int \frac{dx}{\sqrt{4 + x^2}} = \int \frac{2 \sec^2 \theta \, d\theta}{\sqrt{4 \sec^2 \theta}} = \int \frac{\sec^2 \theta \, d\theta}{|\sec \theta|} \qquad \sqrt{\sec^2 \theta} = |\sec \theta|$$

$$= \int \sec \theta \, d\theta \qquad\qquad \sec \theta > 0 \text{ for } -\frac{\pi}{2} < \theta < \frac{\pi}{2}$$

$$= \ln |\sec \theta + \tan \theta| + C$$

$$= \ln \left| \frac{\sqrt{4 + x^2}}{2} + \frac{x}{2} \right| + C. \qquad \text{From Fig. 8.4}$$

Notice how we expressed $\ln|\sec \theta + \tan \theta|$ in terms of $x$: We drew a reference triangle for the original substitution $x = 2 \tan \theta$ (Figure 8.4) and read the ratios from the triangle. ∎

**EXAMPLE 2** Here we find an expression for the inverse hyperbolic sine function in terms of the natural logarithm. Following the same procedure as in Example 1, we find that

$$\int \frac{dx}{\sqrt{a^2 + x^2}} = \int \sec \theta \, d\theta \qquad x = a \tan \theta, \; dx = a \sec^2 \theta \, d\theta$$

$$= \ln |\sec \theta + \tan \theta| + C$$

$$= \ln \left| \frac{\sqrt{a^2 + x^2}}{a} + \frac{x}{a} \right| + C \qquad \text{Fig. 8.2}$$

From Table 7.11, $\sinh^{-1}(x/a)$ is also an antiderivative of $1/\sqrt{a^2 + x^2}$, so the two antiderivatives differ by a constant, giving

$$\sinh^{-1} \frac{x}{a} = \ln \left| \frac{\sqrt{a^2 + x^2}}{a} + \frac{x}{a} \right| + C.$$

Setting $x = 0$ in this last equation, we find $0 = \ln |1| + C$, so $C = 0$. Since $\sqrt{a^2 + x^2} > |x|$, we conclude that

$$\sinh^{-1} \frac{x}{a} = \ln \left( \frac{\sqrt{a^2 + x^2}}{a} + \frac{x}{a} \right)$$

(See also Exercise 76 in Section 7.7.) ∎

**EXAMPLE 3** Evaluate

$$\int \frac{x^2 \, dx}{\sqrt{9 - x^2}}.$$

**Solution** We set

$$x = 3 \sin \theta, \qquad dx = 3 \cos \theta \, d\theta, \qquad -\frac{\pi}{2} < \theta < \frac{\pi}{2}$$

$$9 - x^2 = 9 - 9 \sin^2 \theta = 9(1 - \sin^2 \theta) = 9 \cos^2 \theta.$$

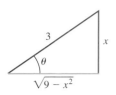

**FIGURE 8.5** Reference triangle for $x = 3 \sin \theta$ (Example 3):

$$\sin \theta = \frac{x}{3}$$

and

$$\cos \theta = \frac{\sqrt{9 - x^2}}{3}.$$

Then

$$\int \frac{x^2 \, dx}{\sqrt{9 - x^2}} = \int \frac{9 \sin^2 \theta \cdot 3 \cos \theta \, d\theta}{|3 \cos \theta|}$$

$$= 9 \int \sin^2 \theta \, d\theta \qquad \cos \theta > 0 \text{ for } -\frac{\pi}{2} < \theta < \frac{\pi}{2}$$

$$= 9 \int \frac{1 - \cos 2\theta}{2} \, d\theta$$

$$= \frac{9}{2} \left( \theta - \frac{\sin 2\theta}{2} \right) + C$$

$$= \frac{9}{2} (\theta - \sin \theta \cos \theta) + C \qquad \sin 2\theta = 2 \sin \theta \cos \theta$$

$$= \frac{9}{2} \left( \sin^{-1} \frac{x}{3} - \frac{x}{3} \cdot \frac{\sqrt{9 - x^2}}{3} \right) + C \qquad \text{From Fig. 8.5}$$

$$= \frac{9}{2} \sin^{-1} \frac{x}{3} - \frac{x}{2} \sqrt{9 - x^2} + C. \qquad ■$$

**EXAMPLE 4** Evaluate

$$\int \frac{dx}{\sqrt{25x^2 - 4}}, \qquad x > \frac{2}{5}.$$

**Solution** We first rewrite the radical as

$$\sqrt{25x^2 - 4} = \sqrt{25 \left( x^2 - \frac{4}{25} \right)}$$

$$= 5 \sqrt{x^2 - \left( \frac{2}{5} \right)^2}$$

to put the radicand in the form $x^2 - a^2$. We then substitute

$$x = \frac{2}{5} \sec \theta, \qquad dx = \frac{2}{5} \sec \theta \tan \theta \, d\theta, \qquad 0 < \theta < \frac{\pi}{2}$$

$$x^2 - \left( \frac{2}{5} \right)^2 = \frac{4}{25} \sec^2 \theta - \frac{4}{25}$$

$$= \frac{4}{25} (\sec^2 \theta - 1) = \frac{4}{25} \tan^2 \theta$$

$$\sqrt{x^2 - \left( \frac{2}{5} \right)^2} = \frac{2}{5} |\tan \theta| = \frac{2}{5} \tan \theta. \qquad \begin{array}{l} \tan \theta > 0 \text{ for} \\ 0 < \theta < \pi/2 \end{array}$$

With these substitutions, we have

$$\int \frac{dx}{\sqrt{25x^2 - 4}} = \int \frac{dx}{5 \sqrt{x^2 - (4/25)}} = \int \frac{(2/5) \sec \theta \tan \theta \, d\theta}{5 \cdot (2/5) \tan \theta}$$

$$= \frac{1}{5} \int \sec \theta \, d\theta = \frac{1}{5} \ln |\sec \theta + \tan \theta| + C$$

$$= \frac{1}{5} \ln \left| \frac{5x}{2} + \frac{\sqrt{25x^2 - 4}}{2} \right| + C. \qquad \text{From Fig. 8.6} \qquad ■$$

**FIGURE 8.6** If $x = (2/5)\sec \theta$, $0 < \theta < \pi/2$, then $\theta = \sec^{-1}(5x/2)$, and we can read the values of the other trigonometric functions of $\theta$ from this right triangle (Example 4).

# Exercises 8.4

## Using Trigonometric Substitutions
Evaluate the integrals in Exercises 1–14.

**1.** $\displaystyle\int \frac{dx}{\sqrt{9+x^2}}$

**2.** $\displaystyle\int \frac{3\,dx}{\sqrt{1+9x^2}}$

**3.** $\displaystyle\int_{-2}^{2} \frac{dx}{4+x^2}$

**4.** $\displaystyle\int_{0}^{2} \frac{dx}{8+2x^2}$

**5.** $\displaystyle\int_{0}^{3/2} \frac{dx}{\sqrt{9-x^2}}$

**6.** $\displaystyle\int_{0}^{1/2\sqrt{2}} \frac{2\,dx}{\sqrt{1-4x^2}}$

**7.** $\displaystyle\int \sqrt{25-t^2}\,dt$

**8.** $\displaystyle\int \sqrt{1-9t^2}\,dt$

**9.** $\displaystyle\int \frac{dx}{\sqrt{4x^2-49}},\quad x>\frac{7}{2}$

**10.** $\displaystyle\int \frac{5\,dx}{\sqrt{25x^2-9}},\quad x>\frac{3}{5}$

**11.** $\displaystyle\int \frac{\sqrt{y^2-49}}{y}\,dy,\quad y>7$

**12.** $\displaystyle\int \frac{\sqrt{y^2-25}}{y^3}\,dy,\quad y>5$

**13.** $\displaystyle\int \frac{dx}{x^2\sqrt{x^2-1}},\quad x>1$

**14.** $\displaystyle\int \frac{2\,dx}{x^3\sqrt{x^2-1}},\quad x>1$

## Assorted Integrations
Use any method to evaluate the integrals in Exercises 15–34. Most will require trigonometric substitutions, but some can be evaluated by other methods.

**15.** $\displaystyle\int \frac{x}{\sqrt{9-x^2}}\,dx$

**16.** $\displaystyle\int \frac{x^2}{4+x^2}\,dx$

**17.** $\displaystyle\int \frac{x^3\,dx}{\sqrt{x^2+4}}$

**18.** $\displaystyle\int \frac{dx}{x^2\sqrt{x^2+1}}$

**19.** $\displaystyle\int \frac{8\,dw}{w^2\sqrt{4-w^2}}$

**20.** $\displaystyle\int \frac{\sqrt{9-w^2}}{w^2}\,dw$

**21.** $\displaystyle\int \sqrt{\frac{x+1}{1-x}}\,dx$

**22.** $\displaystyle\int x\sqrt{x^2-4}\,dx$

**23.** $\displaystyle\int_{0}^{\sqrt{3}/2} \frac{4x^2\,dx}{(1-x^2)^{3/2}}$

**24.** $\displaystyle\int_{0}^{1} \frac{dx}{(4-x^2)^{3/2}}$

**25.** $\displaystyle\int \frac{dx}{(x^2-1)^{3/2}},\quad x>1$

**26.** $\displaystyle\int \frac{x^2\,dx}{(x^2-1)^{5/2}},\quad x>1$

**27.** $\displaystyle\int \frac{(1-x^2)^{3/2}}{x^6}\,dx$

**28.** $\displaystyle\int \frac{(1-x^2)^{1/2}}{x^4}\,dx$

**29.** $\displaystyle\int \frac{8\,dx}{(4x^2+1)^2}$

**30.** $\displaystyle\int \frac{6\,dt}{(9t^2+1)^2}$

**31.** $\displaystyle\int \frac{x^3\,dx}{x^2-1}$

**32.** $\displaystyle\int \frac{x\,dx}{25+4x^2}$

**33.** $\displaystyle\int \frac{v^2\,dv}{(1-v^2)^{5/2}}$

**34.** $\displaystyle\int \frac{(1-r^2)^{5/2}}{r^8}\,dr$

In Exercises 35–48, use an appropriate substitution and then a trigonometric substitution to evaluate the integrals.

**35.** $\displaystyle\int_{0}^{\ln 4} \frac{e^t\,dt}{\sqrt{e^{2t}+9}}$

**36.** $\displaystyle\int_{\ln (3/4)}^{\ln (4/3)} \frac{e^t\,dt}{(1+e^{2t})^{3/2}}$

**37.** $\displaystyle\int_{1/12}^{1/4} \frac{2\,dt}{\sqrt{t}+4t\sqrt{t}}$

**38.** $\displaystyle\int_{1}^{e} \frac{dy}{y\sqrt{1+(\ln y)^2}}$

**39.** $\displaystyle\int \frac{dx}{x\sqrt{x^2-1}}$

**40.** $\displaystyle\int \frac{dx}{1+x^2}$

**41.** $\displaystyle\int \frac{x\,dx}{\sqrt{x^2-1}}$

**42.** $\displaystyle\int \frac{dx}{\sqrt{1-x^2}}$

**43.** $\displaystyle\int \frac{x\,dx}{\sqrt{1+x^4}}$

**44.** $\displaystyle\int \frac{\sqrt{1-(\ln x)^2}}{x\ln x}\,dx$

**45.** $\displaystyle\int \sqrt{\frac{4-x}{x}}\,dx$

**46.** $\displaystyle\int \sqrt{\frac{x}{1-x^3}}\,dx$

　　(*Hint:* Let $x=u^2$.)　　　　(*Hint:* Let $u=x^{3/2}$.)

**47.** $\displaystyle\int \sqrt{x}\,\sqrt{1-x}\,dx$

**48.** $\displaystyle\int \frac{\sqrt{x-2}}{\sqrt{x-1}}\,dx$

## Initial Value Problems
Solve the initial value problems in Exercises 49–52 for $y$ as a function of $x$.

**49.** $x\dfrac{dy}{dx}=\sqrt{x^2-4},\quad x\ge 2,\quad y(2)=0$

**50.** $\sqrt{x^2-9}\,\dfrac{dy}{dx}=1,\quad x>3,\quad y(5)=\ln 3$

**51.** $(x^2+4)\dfrac{dy}{dx}=3,\quad y(2)=0$

**52.** $(x^2+1)^2\dfrac{dy}{dx}=\sqrt{x^2+1},\quad y(0)=1$

## Applications and Examples
**53. Area** Find the area of the region in the first quadrant that is enclosed by the coordinate axes and the curve $y=\sqrt{9-x^2}/3$.

**54. Area** Find the area enclosed by the ellipse
$$\frac{x^2}{a^2}+\frac{y^2}{b^2}=1.$$

**55.** Consider the region bounded by the graphs of $y=\sin^{-1}x$, $y=0$, and $x=1/2$.
　　**a.** Find the area of the region.
　　**b.** Find the centroid of the region.

**56.** Consider the region bounded by the graphs of $y=\sqrt{x}\,\tan^{-1}x$ and $y=0$ for $0\le x\le 1$. Find the volume of the solid formed by revolving this region about the $x$-axis (see accompanying figure).

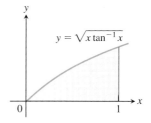

**57.** Evaluate $\int x^3 \sqrt{1 - x^2}\, dx$ using

   **a.** integration by parts.

   **b.** a *u*-substitution.

   **c.** a trigonometric substitution.

**58. Path of a water skier** Suppose that a boat is positioned at the origin with a water skier tethered to the boat at the point (10, 0) on a rope 10 m long. As the boat travels along the positive *y*-axis, the skier is pulled behind the boat along an unknown path $y = f(x)$, as shown in the accompanying figure.

   **a.** Show that $f'(x) = \dfrac{-\sqrt{100 - x^2}}{x}$.

(*Hint:* Assume that the skier is always pointed directly at the boat and the rope is on a line tangent to the path $y = f(x)$.)

**b.** Solve the equation in part (a) for $f(x)$, using $f(10) = 0$.

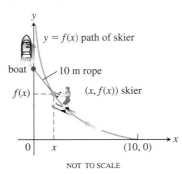

NOT TO SCALE

# 8.5 Integration of Rational Functions by Partial Fractions

This section shows how to express a rational function (a quotient of polynomials) as a sum of simpler fractions, called *partial fractions*, which are easily integrated. For instance, the rational function $(5x - 3)/(x^2 - 2x - 3)$ can be rewritten as

$$\frac{5x - 3}{x^2 - 2x - 3} = \frac{2}{x + 1} + \frac{3}{x - 3}.$$

You can verify this equation algebraically by placing the fractions on the right side over a common denominator $(x + 1)(x - 3)$. The skill acquired in writing rational functions as such a sum is useful in other settings as well (for instance, when using certain transform methods to solve differential equations). To integrate the rational function $(5x - 3)/(x^2 - 2x - 3)$ on the left side of our previous expression, we simply sum the integrals of the fractions on the right side:

$$\int \frac{5x - 3}{(x + 1)(x - 3)}\, dx = \int \frac{2}{x + 1}\, dx + \int \frac{3}{x - 3}\, dx$$

$$= 2 \ln|x + 1| + 3 \ln|x - 3| + C.$$

The method for rewriting rational functions as a sum of simpler fractions is called **the method of partial fractions**. In the case of the preceding example, it consists of finding constants *A* and *B* such that

$$\frac{5x - 3}{x^2 - 2x - 3} = \frac{A}{x + 1} + \frac{B}{x - 3}. \tag{1}$$

(Pretend for a moment that we do not know that $A = 2$ and $B = 3$ will work.) We call the fractions $A/(x + 1)$ and $B/(x - 3)$ **partial fractions** because their denominators are only part of the original denominator $x^2 - 2x - 3$. We call *A* and *B* **undetermined coefficients** until suitable values for them have been found.

To find *A* and *B*, we first clear Equation (1) of fractions and regroup in powers of *x*, obtaining

$$5x - 3 = A(x - 3) + B(x + 1) = (A + B)x - 3A + B.$$

This will be an identity in *x* if and only if the coefficients of like powers of *x* on the two sides are equal:

$$A + B = 5, \qquad -3A + B = -3.$$

Solving these equations simultaneously gives $A = 2$ and $B = 3$.

### General Description of the Method

Success in writing a rational function $f(x)/g(x)$ as a sum of partial fractions depends on two things:

- *The degree of $f(x)$ must be less than the degree of $g(x)$.* That is, the fraction must be proper. If it isn't, divide $f(x)$ by $g(x)$ and work with the remainder term. Example 3 of this section illustrates such a case.
- *We must know the factors of $g(x)$.* In theory, any polynomial with real coefficients can be written as a product of real linear factors and real quadratic factors. In practice, the factors may be hard to find.

Here is how we find the partial fractions of a proper fraction $f(x)/g(x)$ when the factors of $g$ are known. A quadratic polynomial (or factor) is **irreducible** if it cannot be written as the product of two linear factors with real coefficients. That is, the polynomial has no real roots.

---

**Method of Partial Fractions when $f(x)/g(x)$ is Proper**

1. Let $x - r$ be a linear factor of $g(x)$. Suppose that $(x - r)^m$ is the highest power of $x - r$ that divides $g(x)$. Then, to this factor, assign the sum of the $m$ partial fractions:

$$\frac{A_1}{(x - r)} + \frac{A_2}{(x - r)^2} + \cdots + \frac{A_m}{(x - r)^m}.$$

Do this for each distinct linear factor of $g(x)$.

2. Let $x^2 + px + q$ be an irreducible quadratic factor of $g(x)$ so that $x^2 + px + q$ has no real roots. Suppose that $(x^2 + px + q)^n$ is the highest power of this factor that divides $g(x)$. Then, to this factor, assign the sum of the $n$ partial fractions:

$$\frac{B_1 x + C_1}{(x^2 + px + q)} + \frac{B_2 x + C_2}{(x^2 + px + q)^2} + \cdots + \frac{B_n x + C_n}{(x^2 + px + q)^n}.$$

Do this for each distinct quadratic factor of $g(x)$.

3. Set the original fraction $f(x)/g(x)$ equal to the sum of all these partial fractions. Clear the resulting equation of fractions and arrange the terms in decreasing powers of $x$.

4. Equate the coefficients of corresponding powers of $x$ and solve the resulting equations for the undetermined coefficients.

---

**EXAMPLE 1**    Use partial fractions to evaluate

$$\int \frac{x^2 + 4x + 1}{(x - 1)(x + 1)(x + 3)}\, dx.$$

**Solution**    The partial fraction decomposition has the form

$$\frac{x^2 + 4x + 1}{(x - 1)(x + 1)(x + 3)} = \frac{A}{x - 1} + \frac{B}{x + 1} + \frac{C}{x + 3}.$$

To find the values of the undetermined coefficients $A$, $B$, and $C$, we clear fractions and get

$$\begin{aligned}
x^2 + 4x + 1 &= A(x + 1)(x + 3) + B(x - 1)(x + 3) + C(x - 1)(x + 1) \\
&= A(x^2 + 4x + 3) + B(x^2 + 2x - 3) + C(x^2 - 1) \\
&= (A + B + C)x^2 + (4A + 2B)x + (3A - 3B - C).
\end{aligned}$$

The polynomials on both sides of the above equation are identical, so we equate coefficients of like powers of $x$, obtaining

$$
\begin{array}{lll}
\text{Coefficient of } x^2: & A + B + C = 1 \\
\text{Coefficient of } x^1: & 4A + 2B = 4 \\
\text{Coefficient of } x^0: & 3A - 3B - C = 1
\end{array}
$$

There are several ways of solving such a system of linear equations for the unknowns $A$, $B$, and $C$, including elimination of variables or the use of a calculator or computer. Whatever method is used, the solution is $A = 3/4$, $B = 1/2$, and $C = -1/4$. Hence we have

$$
\int \frac{x^2 + 4x + 1}{(x - 1)(x + 1)(x + 3)} \, dx = \int \left[ \frac{3}{4} \frac{1}{x - 1} + \frac{1}{2} \frac{1}{x + 1} - \frac{1}{4} \frac{1}{x + 3} \right] dx
$$

$$
= \frac{3}{4} \ln |x - 1| + \frac{1}{2} \ln |x + 1| - \frac{1}{4} \ln |x + 3| + K,
$$

where $K$ is the arbitrary constant of integration (to avoid confusion with the undetermined coefficient we labeled as $C$). ∎

**EXAMPLE 2**     Use partial fractions to evaluate

$$
\int \frac{6x + 7}{(x + 2)^2} \, dx.
$$

**Solution**     First we express the integrand as a sum of partial fractions with undetermined coefficients.

$$
\frac{6x + 7}{(x + 2)^2} = \frac{A}{x + 2} + \frac{B}{(x + 2)^2}
$$

$$
6x + 7 = A(x + 2) + B \qquad \text{Multiply both sides by } (x + 2)^2.
$$

$$
= Ax + (2A + B)
$$

Equating coefficients of corresponding powers of $x$ gives

$$
A = 6 \quad \text{and} \quad 2A + B = 12 + B = 7, \quad \text{or} \quad A = 6 \quad \text{and} \quad B = -5.
$$

Therefore,

$$
\int \frac{6x + 7}{(x + 2)^2} \, dx = \int \left( \frac{6}{x + 2} - \frac{5}{(x + 2)^2} \right) dx
$$

$$
= 6 \int \frac{dx}{x + 2} - 5 \int (x + 2)^{-2} \, dx
$$

$$
= 6 \ln |x + 2| + 5(x + 2)^{-1} + C. \qquad ∎
$$

The next example shows how to handle the case when $f(x)/g(x)$ is an improper fraction. It is a case where the degree of $f$ is larger than the degree of $g$.

**EXAMPLE 3**     Use partial fractions to evaluate

$$
\int \frac{2x^3 - 4x^2 - x - 3}{x^2 - 2x - 3} \, dx.
$$

**Solution**     First we divide the denominator into the numerator to get a polynomial plus a proper fraction.

$$
\begin{array}{r}
2x \phantom{XXXXXXXX} \\
x^2 - 2x - 3 \overline{)2x^3 - 4x^2 - x - 3} \\
\underline{2x^3 - 4x^2 - 6x - 3} \\
5x - 3
\end{array}
$$

Then we write the improper fraction as a polynomial plus a proper fraction.

$$\frac{2x^3 - 4x^2 - x - 3}{x^2 - 2x - 3} = 2x + \frac{5x - 3}{x^2 - 2x - 3}$$

We found the partial fraction decomposition of the fraction on the right in the opening example, so

$$\int \frac{2x^3 - 4x^2 - x - 3}{x^2 - 2x - 3} \, dx = \int 2x \, dx + \int \frac{5x - 3}{x^2 - 2x - 3} \, dx$$

$$= \int 2x \, dx + \int \frac{2}{x + 1} \, dx + \int \frac{3}{x - 3} \, dx$$

$$= x^2 + 2 \ln |x + 1| + 3 \ln |x - 3| + C. \qquad \blacksquare$$

**EXAMPLE 4**     Use partial fractions to evaluate

$$\int \frac{-2x + 4}{(x^2 + 1)(x - 1)^2} \, dx.$$

**Solution**   The denominator has an irreducible quadratic factor as well as a repeated linear factor, so we write

$$\frac{-2x + 4}{(x^2 + 1)(x - 1)^2} = \frac{Ax + B}{x^2 + 1} + \frac{C}{x - 1} + \frac{D}{(x - 1)^2}. \qquad (2)$$

Clearing the equation of fractions gives

$$-2x + 4 = (Ax + B)(x - 1)^2 + C(x - 1)(x^2 + 1) + D(x^2 + 1)$$
$$= (A + C)x^3 + (-2A + B - C + D)x^2$$
$$+ (A - 2B + C)x + (B - C + D).$$

Equating coefficients of like terms gives

Coefficients of $x^3$:     $0 = A + C$
Coefficients of $x^2$:     $0 = -2A + B - C + D$
Coefficients of $x^1$:     $-2 = A - 2B + C$
Coefficients of $x^0$:     $4 = B - C + D$

We solve these equations simultaneously to find the values of $A$, $B$, $C$, and $D$:

$-4 = -2A$,     $A = 2$     Subtract fourth equation from second.
$C = -A = -2$     From the first equation
$B = (A + C + 2)/2 = 1$     From the third equation and $C = -A$
$D = 4 - B + C = 1.$     From the fourth equation.

We substitute these values into Equation (2), obtaining

$$\frac{-2x + 4}{(x^2 + 1)(x - 1)^2} = \frac{2x + 1}{x^2 + 1} - \frac{2}{x - 1} + \frac{1}{(x - 1)^2}.$$

Finally, using the expansion above we can integrate:

$$\int \frac{-2x + 4}{(x^2 + 1)(x - 1)^2} \, dx = \int \left( \frac{2x + 1}{x^2 + 1} - \frac{2}{x - 1} + \frac{1}{(x - 1)^2} \right) dx$$

$$= \int \left( \frac{2x}{x^2 + 1} + \frac{1}{x^2 + 1} - \frac{2}{x - 1} + \frac{1}{(x - 1)^2} \right) dx$$

$$= \ln (x^2 + 1) + \tan^{-1} x - 2 \ln |x - 1| - \frac{1}{x - 1} + C. \qquad \blacksquare$$

**EXAMPLE 5** Use partial fractions to evaluate

$$\int \frac{dx}{x(x^2 + 1)^2}.$$

**Solution** The form of the partial fraction decomposition is

$$\frac{1}{x(x^2 + 1)^2} = \frac{A}{x} + \frac{Bx + C}{x^2 + 1} + \frac{Dx + E}{(x^2 + 1)^2}.$$

Multiplying by $x(x^2 + 1)^2$, we have

$$\begin{aligned}
1 &= A(x^2 + 1)^2 + (Bx + C)x(x^2 + 1) + (Dx + E)x \\
&= A(x^4 + 2x^2 + 1) + B(x^4 + x^2) + C(x^3 + x) + Dx^2 + Ex \\
&= (A + B)x^4 + Cx^3 + (2A + B + D)x^2 + (C + E)x + A.
\end{aligned}$$

If we equate coefficients, we get the system

$$A + B = 0, \quad C = 0, \quad 2A + B + D = 0, \quad C + E = 0, \quad A = 1.$$

Solving this system gives $A = 1$, $B = -1$, $C = 0$, $D = -1$, and $E = 0$. Thus,

$$\begin{aligned}
\int \frac{dx}{x(x^2 + 1)^2} &= \int \left[ \frac{1}{x} + \frac{-x}{x^2 + 1} + \frac{-x}{(x^2 + 1)^2} \right] dx \\
&= \int \frac{dx}{x} - \int \frac{x\,dx}{x^2 + 1} - \int \frac{x\,dx}{(x^2 + 1)^2} \\
&= \int \frac{dx}{x} - \frac{1}{2}\int \frac{du}{u} - \frac{1}{2}\int \frac{du}{u^2} \qquad \begin{aligned} u &= x^2 + 1, \\ du &= 2x\,dx \end{aligned} \\
&= \ln|x| - \frac{1}{2}\ln|u| + \frac{1}{2u} + K \\
&= \ln|x| - \frac{1}{2}\ln(x^2 + 1) + \frac{1}{2(x^2 + 1)} + K \\
&= \ln \frac{|x|}{\sqrt{x^2 + 1}} + \frac{1}{2(x^2 + 1)} + K.
\end{aligned}$$

**HISTORICAL BIOGRAPHY**

Oliver Heaviside
(1850–1925)

### The Heaviside "Cover-up" Method for Linear Factors

When the degree of the polynomial $f(x)$ is less than the degree of $g(x)$ and

$$g(x) = (x - r_1)(x - r_2) \cdots (x - r_n)$$

is a product of $n$ distinct linear factors, each raised to the first power, there is a quick way to expand $f(x)/g(x)$ by partial fractions.

**EXAMPLE 6** Find $A$, $B$, and $C$ in the partial fraction expansion

$$\frac{x^2 + 1}{(x - 1)(x - 2)(x - 3)} = \frac{A}{x - 1} + \frac{B}{x - 2} + \frac{C}{x - 3}. \tag{3}$$

**Solution** If we multiply both sides of Equation (3) by $(x - 1)$ to get

$$\frac{x^2 + 1}{(x - 2)(x - 3)} = A + \frac{B(x - 1)}{x - 2} + \frac{C(x - 1)}{x - 3}$$

and set $x = 1$, the resulting equation gives the value of $A$:

$$\frac{(1)^2 + 1}{(1 - 2)(1 - 3)} = A + 0 + 0,$$

$$A = 1.$$

Thus, the value of $A$ is the number we would have obtained if we had covered the factor $(x - 1)$ in the denominator of the original fraction

$$\frac{x^2 + 1}{(x - 1)(x - 2)(x - 3)} \tag{4}$$

and evaluated the rest at $x = 1$:

$$A = \frac{(1)^2 + 1}{\boxed{(x - 1)}\,(1 - 2)(1 - 3)} = \frac{2}{(-1)(-2)} = 1.$$

$$\underset{\text{Cover}}{\Uparrow}$$

Similarly, we find the value of $B$ in Equation (3) by covering the factor $(x - 2)$ in Expression (4) and evaluating the rest at $x = 2$:

$$B = \frac{(2)^2 + 1}{(2 - 1)\,\boxed{(x - 2)}\,(2 - 3)} = \frac{5}{(1)(-1)} = -5.$$

$$\underset{\text{Cover}}{\Uparrow}$$

Finally, $C$ is found by covering the $(x - 3)$ in Expression (4) and evaluating the rest at $x = 3$:

$$C = \frac{(3)^2 + 1}{(3 - 1)(3 - 2)\,\boxed{(x - 3)}} = \frac{10}{(2)(1)} = 5. \qquad ∎$$

$$\underset{\text{Cover}}{\Uparrow}$$

---

**Heaviside Method**

**1.** *Write the quotient with* $g(x)$ *factored:*

$$\frac{f(x)}{g(x)} = \frac{f(x)}{(x - r_1)(x - r_2) \cdots (x - r_n)}.$$

**2.** *Cover the factors* $(x - r_i)$ *of* $g(x)$ *one at a time*, *each time replacing all the uncovered* $x$'s *by the number* $r_i$. *This gives a number* $A_i$ *for each root* $r_i$:

$$A_1 = \frac{f(r_1)}{(r_1 - r_2) \cdots (r_1 - r_n)}$$

$$A_2 = \frac{f(r_2)}{(r_2 - r_1)(r_2 - r_3) \cdots (r_2 - r_n)}$$

$$\vdots$$

$$A_n = \frac{f(r_n)}{(r_n - r_1)(r_n - r_2) \cdots (r_n - r_{n-1})}.$$

**3.** *Write the partial fraction expansion of* $f(x)/g(x)$ *as*

$$\frac{f(x)}{g(x)} = \frac{A_1}{(x - r_1)} + \frac{A_2}{(x - r_2)} + \cdots + \frac{A_n}{(x - r_n)}.$$

---

**EXAMPLE 7**      Use the Heaviside Method to evaluate

$$\int \frac{x + 4}{x^3 + 3x^2 - 10x}\, dx.$$

**Solution**  The degree of $f(x) = x + 4$ is less than the degree of the cubic polynomial $g(x) = x^3 + 3x^2 - 10x$, and, with $g(x)$ factored,

$$\frac{x + 4}{x^3 + 3x^2 - 10x} = \frac{x + 4}{x(x - 2)(x + 5)}.$$

The roots of $g(x)$ are $r_1 = 0$, $r_2 = 2$, and $r_3 = -5$. We find

$$A_1 = \frac{0 + 4}{\boxed{x}\,(0 - 2)(0 + 5)} = \frac{4}{(-2)(5)} = -\frac{2}{5}$$

$\Uparrow$
Cover

$$A_2 = \frac{2 + 4}{2\,\boxed{(x - 2)}\,(2 + 5)} = \frac{6}{(2)(7)} = \frac{3}{7}$$

$\Uparrow$
Cover

$$A_3 = \frac{-5 + 4}{(-5)(-5 - 2)\,\boxed{(x + 5)}} = \frac{-1}{(-5)(-7)} = -\frac{1}{35}.$$

$\Uparrow$
Cover

Therefore,

$$\frac{x + 4}{x(x - 2)(x + 5)} = -\frac{2}{5x} + \frac{3}{7(x - 2)} - \frac{1}{35(x + 5)},$$

and

$$\int \frac{x + 4}{x(x - 2)(x + 5)}\, dx = -\frac{2}{5}\ln|x| + \frac{3}{7}\ln|x - 2| - \frac{1}{35}\ln|x + 5| + C. \qquad \blacksquare$$

### Other Ways to Determine the Coefficients

Another way to determine the constants that appear in partial fractions is to differentiate, as in the next example. Still another is to assign selected numerical values to $x$.

**EXAMPLE 8**  Find $A$, $B$, and $C$ in the equation

$$\frac{x - 1}{(x + 1)^3} = \frac{A}{x + 1} + \frac{B}{(x + 1)^2} + \frac{C}{(x + 1)^3}$$

by clearing fractions, differentiating the result, and substituting $x = -1$.

**Solution**  We first clear fractions:

$$x - 1 = A(x + 1)^2 + B(x + 1) + C.$$

Substituting $x = -1$ shows $C = -2$. We then differentiate both sides with respect to $x$, obtaining

$$1 = 2A(x + 1) + B.$$

Substituting $x = -1$ shows $B = 1$. We differentiate again to get $0 = 2A$, which shows $A = 0$. Hence,

$$\frac{x - 1}{(x + 1)^3} = \frac{1}{(x + 1)^2} - \frac{2}{(x + 1)^3}. \qquad \blacksquare$$

In some problems, assigning small values to $x$, such as $x = 0, \pm 1, \pm 2$, to get equations in $A$, $B$, and $C$ provides a fast alternative to other methods.

**EXAMPLE 9** Find $A$, $B$, and $C$ in the expression

$$\frac{x^2 + 1}{(x - 1)(x - 2)(x - 3)} = \frac{A}{x - 1} + \frac{B}{x - 2} + \frac{C}{x - 3}$$

by assigning numerical values to $x$.

**Solution** Clear fractions to get

$$x^2 + 1 = A(x - 2)(x - 3) + B(x - 1)(x - 3) + C(x - 1)(x - 2).$$

Then let $x = 1, 2, 3$ successively to find $A$, $B$, and $C$:

$$x = 1: \qquad (1)^2 + 1 = A(-1)(-2) + B(0) + C(0)$$
$$2 = 2A$$
$$A = 1$$
$$x = 2: \qquad (2)^2 + 1 = A(0) + B(1)(-1) + C(0)$$
$$5 = -B$$
$$B = -5$$
$$x = 3: \qquad (3)^2 + 1 = A(0) + B(0) + C(2)(1)$$
$$10 = 2C$$
$$C = 5.$$

Conclusion:

$$\frac{x^2 + 1}{(x - 1)(x - 2)(x - 3)} = \frac{1}{x - 1} - \frac{5}{x - 2} + \frac{5}{x - 3}.$$

■

# Exercises 8.5

**Expanding Quotients into Partial Fractions**
Expand the quotients in Exercises 1–8 by partial fractions.

1. $\dfrac{5x - 13}{(x - 3)(x - 2)}$

2. $\dfrac{5x - 7}{x^2 - 3x + 2}$

3. $\dfrac{x + 4}{(x + 1)^2}$

4. $\dfrac{2x + 2}{x^2 - 2x + 1}$

5. $\dfrac{z + 1}{z^2(z - 1)}$

6. $\dfrac{z}{z^3 - z^2 - 6z}$

7. $\dfrac{t^2 + 8}{t^2 - 5t + 6}$

8. $\dfrac{t^4 + 9}{t^4 + 9t^2}$

**Nonrepeated Linear Factors**
In Exercises 9–16, express the integrand as a sum of partial fractions and evaluate the integrals.

9. $\displaystyle\int \frac{dx}{1 - x^2}$

10. $\displaystyle\int \frac{dx}{x^2 + 2x}$

11. $\displaystyle\int \frac{x + 4}{x^2 + 5x - 6}\, dx$

12. $\displaystyle\int \frac{2x + 1}{x^2 - 7x + 12}\, dx$

13. $\displaystyle\int_4^8 \frac{y\, dy}{y^2 - 2y - 3}$

14. $\displaystyle\int_{1/2}^1 \frac{y + 4}{y^2 + y}\, dy$

15. $\displaystyle\int \frac{dt}{t^3 + t^2 - 2t}$

16. $\displaystyle\int \frac{x + 3}{2x^3 - 8x}\, dx$

**Repeated Linear Factors**
In Exercises 17–20, express the integrand as a sum of partial fractions and evaluate the integrals.

17. $\displaystyle\int_0^1 \frac{x^3\, dx}{x^2 + 2x + 1}$

18. $\displaystyle\int_{-1}^0 \frac{x^3\, dx}{x^2 - 2x + 1}$

19. $\displaystyle\int \frac{dx}{(x^2 - 1)^2}$

20. $\displaystyle\int \frac{x^2\, dx}{(x - 1)(x^2 + 2x + 1)}$

**Irreducible Quadratic Factors**
In Exercises 21–32, express the integrand as a sum of partial fractions and evaluate the integrals.

21. $\displaystyle\int_0^1 \frac{dx}{(x + 1)(x^2 + 1)}$

22. $\displaystyle\int_1^{\sqrt{3}} \frac{3t^2 + t + 4}{t^3 + t}\, dt$

23. $\displaystyle\int \frac{y^2 + 2y + 1}{(y^2 + 1)^2}\, dy$

24. $\displaystyle\int \frac{8x^2 + 8x + 2}{(4x^2 + 1)^2}\, dx$

25. $\displaystyle\int \frac{2s + 2}{(s^2 + 1)(s - 1)^3}\, ds$

26. $\displaystyle\int \frac{s^4 + 81}{s(s^2 + 9)^2}\, ds$

27. $\displaystyle\int \frac{x^2 - x + 2}{x^3 - 1}\, dx$

28. $\displaystyle\int \frac{1}{x^4 + x}\, dx$

29. $\displaystyle\int \frac{x^2}{x^4 - 1}\, dx$

30. $\displaystyle\int \frac{x^2 + x}{x^4 - 3x^2 - 4}\, dx$

31. $\displaystyle\int \frac{2\theta^3 + 5\theta^2 + 8\theta + 4}{(\theta^2 + 2\theta + 2)^2}\, d\theta$

32. $\displaystyle\int \frac{\theta^4 - 4\theta^3 + 2\theta^2 - 3\theta + 1}{(\theta^2 + 1)^3}\, d\theta$

## Improper Fractions

In Exercises 33–38, perform long division on the integrand, write the proper fraction as a sum of partial fractions, and then evaluate the integral.

**33.** $\displaystyle\int \frac{2x^3 - 2x^2 + 1}{x^2 - x} \, dx$

**34.** $\displaystyle\int \frac{x^4}{x^2 - 1} \, dx$

**35.** $\displaystyle\int \frac{9x^3 - 3x + 1}{x^3 - x^2} \, dx$

**36.** $\displaystyle\int \frac{16x^3}{4x^2 - 4x + 1} \, dx$

**37.** $\displaystyle\int \frac{y^4 + y^2 - 1}{y^3 + y} \, dy$

**38.** $\displaystyle\int \frac{2y^4}{y^3 - y^2 + y - 1} \, dy$

## Evaluating Integrals

Evaluate the integrals in Exercises 39–50.

**39.** $\displaystyle\int \frac{e^t \, dt}{e^{2t} + 3e^t + 2}$

**40.** $\displaystyle\int \frac{e^{4t} + 2e^{2t} - e^t}{e^{2t} + 1} \, dt$

**41.** $\displaystyle\int \frac{\cos y \, dy}{\sin^2 y + \sin y - 6}$

**42.** $\displaystyle\int \frac{\sin \theta \, d\theta}{\cos^2 \theta + \cos \theta - 2}$

**43.** $\displaystyle\int \frac{(x - 2)^2 \tan^{-1}(2x) - 12x^3 - 3x}{(4x^2 + 1)(x - 2)^2} \, dx$

**44.** $\displaystyle\int \frac{(x + 1)^2 \tan^{-1}(3x) + 9x^3 + x}{(9x^2 + 1)(x + 1)^2} \, dx$

**45.** $\displaystyle\int \frac{1}{x^{3/2} - \sqrt{x}} \, dx$

**46.** $\displaystyle\int \frac{1}{(x^{1/3} - 1) \sqrt{x}} \, dx$
(*Hint:* Let $x = u^6$.)

**47.** $\displaystyle\int \frac{\sqrt{x + 1}}{x} \, dx$
(*Hint:* Let $x + 1 = u^2$.)

**48.** $\displaystyle\int \frac{1}{x\sqrt{x + 9}} \, dx$

**49.** $\displaystyle\int \frac{1}{x(x^4 + 1)} \, dx$

**50.** $\displaystyle\int \frac{1}{x^6(x^5 + 4)} \, dx$

$\left(\text{\textit{Hint:} Multiply by } \dfrac{x^3}{x^3}.\right)$

## Initial Value Problems

Solve the initial value problems in Exercises 51–54 for $x$ as a function of $t$.

**51.** $(t^2 - 3t + 2) \dfrac{dx}{dt} = 1 \quad (t > 2), \quad x(3) = 0$

**52.** $(3t^4 + 4t^2 + 1) \dfrac{dx}{dt} = 2\sqrt{3}, \quad x(1) = -\pi\sqrt{3}/4$

**53.** $(t^2 + 2t) \dfrac{dx}{dt} = 2x + 2 \quad (t, x > 0), \quad x(1) = 1$

**54.** $(t + 1) \dfrac{dx}{dt} = x^2 + 1 \quad (t > -1), \quad x(0) = 0$

## Applications and Examples

In Exercises 55 and 56, find the volume of the solid generated by revolving the shaded region about the indicated axis.

**55.** The $x$-axis

**56.** The $y$-axis

**T 57.** Find, to two decimal places, the $x$-coordinate of the centroid of the region in the first quadrant bounded by the $x$-axis, the curve $y = \tan^{-1} x$, and the line $x = \sqrt{3}$.

**T 58.** Find the $x$-coordinate of the centroid of this region to two decimal places.

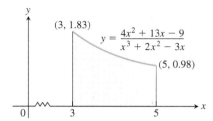

**T 59. Social diffusion** Sociologists sometimes use the phrase "social diffusion" to describe the way information spreads through a population. The information might be a rumor, a cultural fad, or news about a technical innovation. In a sufficiently large population, the number of people $x$ who have the information is treated as a differentiable function of time $t$, and the rate of diffusion, $dx/dt$, is assumed to be proportional to the number of people who have the information times the number of people who do not. This leads to the equation

$$\frac{dx}{dt} = kx(N - x),$$

where $N$ is the number of people in the population.

Suppose $t$ is in days, $k = 1/250$, and two people start a rumor at time $t = 0$ in a population of $N = 1000$ people.

**a.** Find $x$ as a function of $t$.

**b.** When will half the population have heard the rumor? (This is when the rumor will be spreading the fastest.)

**T 60. Second-order chemical reactions** Many chemical reactions are the result of the interaction of two molecules that undergo a change to produce a new product. The rate of the reaction typically depends on the concentrations of the two kinds of molecules. If $a$ is the amount of substance $A$ and $b$ is the amount of substance $B$ at time $t = 0$, and if $x$ is the amount of product at time $t$, then the rate of formation of $x$ may be given by the differential equation

$$\frac{dx}{dt} = k(a - x)(b - x),$$

or

$$\frac{1}{(a - x)(b - x)} \frac{dx}{dt} = k,$$

where $k$ is a constant for the reaction. Integrate both sides of this equation to obtain a relation between $x$ and $t$ **(a)** if $a = b$, and **(b)** if $a \neq b$. Assume in each case that $x = 0$ when $t = 0$.

# $8.6$ Integral Tables and Computer Algebra Systems

In this section we discuss how to use tables and computer algebra systems (CAS) to evaluate integrals.

## Integral Tables

A Brief Table of Integrals is provided at the back of the book, after the index. (More extensive tables appear in compilations such as *CRC Mathematical Tables*, which contain thousands of integrals.) The integration formulas are stated in terms of constants $a, b, c, m, n$, and so on. These constants can usually assume any real value and need not be integers. Occasional limitations on their values are stated with the formulas. Formula 21 requires $n \neq -1$, for example, and Formula 27 requires $n \neq -2$.

The formulas also assume that the constants do not take on values that require dividing by zero or taking even roots of negative numbers. For example, Formula 24 assumes that $a \neq 0$, and Formulas 29a and 29b cannot be used unless $b$ is positive.

**EXAMPLE 1**    Find

$$\int x(2x + 5)^{-1} \, dx.$$

**Solution**    We use Formula 24 at the back of the book (not 22, which requires $n \neq -1$):

$$\int x(ax + b)^{-1} \, dx = \frac{x}{a} - \frac{b}{a^2} \ln |ax + b| + C.$$

With $a = 2$ and $b = 5$, we have

$$\int x(2x + 5)^{-1} \, dx = \frac{x}{2} - \frac{5}{4} \ln |2x + 5| + C.$$

**EXAMPLE 2**    Find

$$\int \frac{dx}{x\sqrt{2x - 4}}.$$

**Solution**    We use Formula 29b:

$$\int \frac{dx}{x\sqrt{ax - b}} = \frac{2}{\sqrt{b}} \tan^{-1} \sqrt{\frac{ax - b}{b}} + C.$$

With $a = 2$ and $b = 4$, we have

$$\int \frac{dx}{x\sqrt{2x - 4}} = \frac{2}{\sqrt{4}} \tan^{-1} \sqrt{\frac{2x - 4}{4}} + C = \tan^{-1} \sqrt{\frac{x - 2}{2}} + C.$$

**EXAMPLE 3**    Find

$$\int x \sin^{-1} x \, dx.$$

**Solution**    We begin by using Formula 106:

$$\int x^n \sin^{-1} ax \, dx = \frac{x^{n+1}}{n + 1} \sin^{-1} ax - \frac{a}{n + 1} \int \frac{x^{n+1} \, dx}{\sqrt{1 - a^2 x^2}}, \qquad n \neq -1.$$

With $n = 1$ and $a = 1$, we have

$$\int x \sin^{-1} x \, dx = \frac{x^2}{2} \sin^{-1} x - \frac{1}{2} \int \frac{x^2 \, dx}{\sqrt{1 - x^2}}.$$

Next we use Formula 49 to find the integral on the right:

$$\int \frac{x^2}{\sqrt{a^2 - x^2}} \, dx = \frac{a^2}{2} \sin^{-1} \left( \frac{x}{a} \right) - \frac{1}{2} x \sqrt{a^2 - x^2} + C.$$

With $a = 1$,

$$\int \frac{x^2 \, dx}{\sqrt{1 - x^2}} = \frac{1}{2} \sin^{-1} x - \frac{1}{2} x \sqrt{1 - x^2} + C.$$

The combined result is

$$\int x \sin^{-1} x \, dx = \frac{x^2}{2} \sin^{-1} x - \frac{1}{2} \left( \frac{1}{2} \sin^{-1} x - \frac{1}{2} x \sqrt{1 - x^2} + C \right)$$

$$= \left( \frac{x^2}{2} - \frac{1}{4} \right) \sin^{-1} x + \frac{1}{4} x \sqrt{1 - x^2} + C'. \qquad \blacksquare$$

### Reduction Formulas

The time required for repeated integrations by parts can sometimes be shortened by applying reduction formulas like

$$\int \tan^n x \, dx = \frac{1}{n - 1} \tan^{n-1} x - \int \tan^{n-2} x \, dx \qquad (1)$$

$$\int (\ln x)^n \, dx = x(\ln x)^n - n \int (\ln x)^{n-1} \, dx \qquad (2)$$

$$\int \sin^n x \cos^m x \, dx = -\frac{\sin^{n-1} x \cos^{m+1} x}{m + n} + \frac{n - 1}{m + n} \int \sin^{n-2} x \cos^m x \, dx \qquad (n \neq -m).$$

$$\qquad (3)$$

By applying such a formula repeatedly, we can eventually express the original integral in terms of a power low enough to be evaluated directly. The next example illustrates this procedure.

**EXAMPLE 4**    Find

$$\int \tan^5 x \, dx.$$

**Solution**    We apply Equation (1) with $n = 5$ to get

$$\int \tan^5 x \, dx = \frac{1}{4} \tan^4 x - \int \tan^3 x \, dx.$$

We then apply Equation (1) again, with $n = 3$, to evaluate the remaining integral:

$$\int \tan^3 x \, dx = \frac{1}{2} \tan^2 x - \int \tan x \, dx = \frac{1}{2} \tan^2 x + \ln |\cos x| + C.$$

The combined result is

$$\int \tan^5 x \, dx = \frac{1}{4} \tan^4 x - \frac{1}{2} \tan^2 x - \ln |\cos x| + C'. \qquad \blacksquare$$

As their form suggests, reduction formulas are derived using integration by parts. (See Example 5 in Section 8.2.)

## Integration with a CAS

A powerful capability of computer algebra systems is their ability to integrate symbolically. This is performed with the **integrate command** specified by the particular system (for example, **int** in Maple, **Integrate** in Mathematica).

**EXAMPLE 5**   Suppose that you want to evaluate the indefinite integral of the function

$$f(x) = x^2\sqrt{a^2 + x^2}.$$

Using Maple, you first define or name the function:

$$> f := x^\wedge 2 * \text{sqrt}\,(a^\wedge 2 + x^\wedge 2);$$

Then you use the integrate command on $f$, identifying the variable of integration:

$$> \text{int}(f, x);$$

Maple returns the answer

$$\frac{1}{4}x(a^2 + x^2)^{3/2} - \frac{1}{8}a^2x\sqrt{a^2 + x^2} - \frac{1}{8}a^4\ln\left(x + \sqrt{a^2 + x^2}\right).$$

If you want to see if the answer can be simplified, enter

$$> \text{simplify}(\%);$$

Maple returns

$$\frac{1}{8}a^2x\sqrt{a^2 + x^2} + \frac{1}{4}x^3\sqrt{a^2 + x^2} - \frac{1}{8}a^4\ln\left(x + \sqrt{a^2 + x^2}\right).$$

If you want the definite integral for $0 \le x \le \pi/2$, you can use the format

$$> \text{int}(f, x = 0..\text{Pi}/2);$$

Maple will return the expression

$$\frac{1}{64}\pi(4a^2 + \pi^2)^{(3/2)} - \frac{1}{32}a^2\pi\sqrt{4a^2 + \pi^2} + \frac{1}{8}a^4\ln(2)$$
$$- \frac{1}{8}a^4\ln\left(\pi + \sqrt{4a^2 + \pi^2}\right) + \frac{1}{16}a^4\ln(a^2).$$

You can also find the definite integral for a particular value of the constant $a$:

$$> a := 1;$$
$$> \text{int}(f, x = 0..1);$$

Maple returns the numerical answer

$$\frac{3}{8}\sqrt{2} + \frac{1}{8}\ln\left(\sqrt{2} - 1\right).$$   ∎

**EXAMPLE 6**   Use a CAS to find

$$\int \sin^2 x \cos^3 x \, dx.$$

**Solution**   With Maple, we have the entry

$$> \text{int}\,((\sin^\wedge 2)(x) * (\cos^\wedge 3)(x), x);$$

with the immediate return

$$-\frac{1}{5}\sin(x)\cos(x)^4 + \frac{1}{15}\cos(x)^2\sin(x) + \frac{2}{15}\sin(x).$$   ∎

Computer algebra systems vary in how they process integrations. We used Maple in Examples 5 and 6. Mathematica would have returned somewhat different results:

1. In Example 5, given

$$In \ [1] := \text{Integrate } \left[ x^{\wedge}2 * \text{Sqrt } \left[ a^{\wedge}2 + x^{\wedge}2 \right], x \right]$$

Mathematica returns

$$Out \ [1] = \sqrt{a^2 + x^2} \left( \frac{a^2 x}{8} + \frac{x^3}{4} \right) - \frac{1}{8} a^4 \text{Log} \left[ x + \sqrt{a^2 + x^2} \right]$$

without having to simplify an intermediate result. The answer is close to Formula 22 in the integral tables.

2. The Mathematica answer to the integral

$$In \ [2] := \text{Integrate } \left[ \text{Sin } \left[ x \right]^{\wedge}2 * \text{Cos } \left[ x \right]^{\wedge}3, x \right]$$

in Example 6 is

$$Out \ [2] = \frac{\text{Sin } \left[ x \right]}{8} - \frac{1}{48} \text{Sin } \left[ 3 x \right] - \frac{1}{80} \text{Sin } \left[ 5 x \right]$$

differing from the Maple answer. Both answers are correct.

Although a CAS is very powerful and can aid us in solving difficult problems, each CAS has its own limitations. There are even situations where a CAS may further complicate a problem (in the sense of producing an answer that is extremely difficult to use or interpret). Note, too, that neither Maple nor Mathematica returns an arbitrary constant $+C$. On the other hand, a little mathematical thinking on your part may reduce the problem to one that is quite easy to handle. We provide an example in Exercise 67.

Many hardware devices have an availability to integration applications, based on software (like Maple or Mathematica), that provide for symbolic input of the integrand to return symbolic output of the indefinite integral. Many of these software applications calculate definite integrals as well. These applications give another tool for finding integrals, aside from using integral tables. However, in some instances, the integration software may not provide an output answer at all.

## Nonelementary Integrals

The development of computers and calculators that find antiderivatives by symbolic manipulation has led to a renewed interest in determining which antiderivatives can be expressed as finite combinations of elementary functions (the functions we have been studying) and which cannot. Integrals of functions that do not have elementary antiderivatives are called **nonelementary** integrals. These integrals can sometimes be expressed with infinite series (Chapter 10) or approximated using numerical methods for their evaluation (Section 8.7). Examples of nonelementary integrals include the error function (which measures the probability of random errors)

$$\text{erf} (x) = \frac{2}{\sqrt{\pi}} \int_0^x e^{-t^2} \, dt$$

and integrals such as

$$\int \sin x^2 \, dx \qquad \text{and} \qquad \int \sqrt{1 + x^4} \, dx$$

that arise in engineering and physics. These and a number of others, such as

$$\int \frac{e^x}{x} \, dx, \qquad \int e^{(e^x)} \, dx, \qquad \int \frac{1}{\ln x} \, dx, \qquad \int \ln (\ln x) \, dx, \qquad \int \frac{\sin x}{x} \, dx,$$

$$\int \sqrt{1 - k^2 \sin^2 x} \, dx, \qquad 0 < k < 1,$$

look so easy they tempt us to try them just to see how they turn out. It can be proved, however, that there is no way to express these integrals as finite combinations of elementary functions. The same applies to integrals that can be changed into these by substitution. The integrands all have antiderivatives, as a consequence of the Fundamental Theorem of Calculus, Part 1, because they are continuous. However, none of the antiderivatives are elementary. The integrals you are asked to evaluate in the present chapter have elementary antiderivatives, but you may encounter nonelementary integrals in your other work.

# Exercises 8.6

## Using Integral Tables
Use the table of integrals at the back of the book to evaluate the integrals in Exercises 1–26.

**1.** $\displaystyle \int \frac{dx}{x\sqrt{x-3}}$

**2.** $\displaystyle \int \frac{dx}{x\sqrt{x+4}}$

**3.** $\displaystyle \int \frac{x\,dx}{\sqrt{x-2}}$

**4.** $\displaystyle \int \frac{x\,dx}{(2x+3)^{3/2}}$

**5.** $\displaystyle \int x\sqrt{2x-3}\,dx$

**6.** $\displaystyle \int x(7x+5)^{3/2}\,dx$

**7.** $\displaystyle \int \frac{\sqrt{9-4x}}{x^2}\,dx$

**8.** $\displaystyle \int \frac{dx}{x^2\sqrt{4x-9}}$

**9.** $\displaystyle \int x\sqrt{4x-x^2}\,dx$

**10.** $\displaystyle \int \frac{\sqrt{x-x^2}}{x}\,dx$

**11.** $\displaystyle \int \frac{dx}{x\sqrt{7+x^2}}$

**12.** $\displaystyle \int \frac{dx}{x\sqrt{7-x^2}}$

**13.** $\displaystyle \int \frac{\sqrt{4-x^2}}{x}\,dx$

**14.** $\displaystyle \int \frac{\sqrt{x^2-4}}{x}\,dx$

**15.** $\displaystyle \int e^{2t}\cos 3t\,dt$

**16.** $\displaystyle \int e^{-3t}\sin 4t\,dt$

**17.** $\displaystyle \int x\cos^{-1}x\,dx$

**18.** $\displaystyle \int x\tan^{-1}x\,dx$

**19.** $\displaystyle \int x^2\tan^{-1}x\,dx$

**20.** $\displaystyle \int \frac{\tan^{-1}x}{x^2}\,dx$

**21.** $\displaystyle \int \sin 3x\cos 2x\,dx$

**22.** $\displaystyle \int \sin 2x\cos 3x\,dx$

**23.** $\displaystyle \int 8\sin 4t\sin \frac{t}{2}\,dt$

**24.** $\displaystyle \int \sin \frac{t}{3}\sin \frac{t}{6}\,dt$

**25.** $\displaystyle \int \cos \frac{\theta}{3}\cos \frac{\theta}{4}\,d\theta$

**26.** $\displaystyle \int \cos \frac{\theta}{2}\cos 7\theta\,d\theta$

## Substitution and Integral Tables
In Exercises 27–40, use a substitution to change the integral into one you can find in the table. Then evaluate the integral.

**27.** $\displaystyle \int \frac{x^3+x+1}{(x^2+1)^2}\,dx$

**28.** $\displaystyle \int \frac{x^2+6x}{(x^2+3)^2}\,dx$

**29.** $\displaystyle \int \sin^{-1}\sqrt{x}\,dx$

**30.** $\displaystyle \int \frac{\cos^{-1}\sqrt{x}}{\sqrt{x}}\,dx$

**31.** $\displaystyle \int \frac{\sqrt{x}}{\sqrt{1-x}}\,dx$

**32.** $\displaystyle \int \frac{\sqrt{2-x}}{\sqrt{x}}\,dx$

**33.** $\displaystyle \int \cot t\sqrt{1-\sin^2 t}\,dt, \quad 0 < t < \pi/2$

**34.** $\displaystyle \int \frac{dt}{\tan t\sqrt{4-\sin^2 t}}$

**35.** $\displaystyle \int \frac{dy}{y\sqrt{3+(\ln y)^2}}$

**36.** $\displaystyle \int \tan^{-1}\sqrt{y}\,dy$

**37.** $\displaystyle \int \frac{1}{\sqrt{x^2+2x+5}}\,dx$
(*Hint:* Complete the square.)

**38.** $\displaystyle \int \frac{x^2}{\sqrt{x^2-4x+5}}\,dx$

**39.** $\displaystyle \int \sqrt{5-4x-x^2}\,dx$

**40.** $\displaystyle \int x^2\sqrt{2x-x^2}\,dx.$

## Using Reduction Formulas
Use reduction formulas to evaluate the integrals in Exercises 41–50.

**41.** $\displaystyle \int \sin^5 2x\,dx$

**42.** $\displaystyle \int 8\cos^4 2\pi t\,dt$

**43.** $\displaystyle \int \sin^2 2\theta\cos^3 2\theta\,d\theta$

**44.** $\displaystyle \int 2\sin^2 t\sec^4 t\,dt$

**45.** $\displaystyle \int 4\tan^3 2x\,dx$

**46.** $\displaystyle \int 8\cot^4 t\,dt$

**47.** $\displaystyle \int 2\sec^3 \pi x\,dx$

**48.** $\displaystyle \int 3\sec^4 3x\,dx$

**49.** $\displaystyle \int \csc^5 x\,dx$

**50.** $\displaystyle \int 16x^3(\ln x)^2\,dx$

Evaluate the integrals in Exercises 51–56 by making a substitution (possibly trigonometric) and then applying a reduction formula.

**51.** $\displaystyle \int e^t\sec^3 (e^t-1)\,dt$

**52.** $\displaystyle \int \frac{\csc^3 \sqrt{\theta}}{\sqrt{\theta}}\,d\theta$

**53.** $\displaystyle \int_0^1 2\sqrt{x^2+1}\,dx$

**54.** $\displaystyle \int_0^{\sqrt{3}/2} \frac{dy}{(1-y^2)^{5/2}}$

**55.** $\displaystyle \int_1^2 \frac{(r^2-1)^{3/2}}{r}\,dr$

**56.** $\displaystyle \int_0^{1/\sqrt{3}} \frac{dt}{(t^2+1)^{7/2}}$

**Applications**

**57. Surface area** Find the area of the surface generated by revolving the curve $y = \sqrt{x^2 + 2}$, $0 \le x \le \sqrt{2}$, about the $x$-axis.

**58. Arc length** Find the length of the curve $y = x^2$, $0 \le x \le \sqrt{3}/2$.

**59. Centroid** Find the centroid of the region cut from the first quadrant by the curve $y = 1/\sqrt{x + 1}$ and the line $x = 3$.

**60. Moment about y-axis** A thin plate of constant density $\delta = 1$ occupies the region enclosed by the curve $y = 36/(2x + 3)$ and the line $x = 3$ in the first quadrant. Find the moment of the plate about the $y$-axis.

T **61.** Use the integral table and a calculator to find to two decimal places the area of the surface generated by revolving the curve $y = x^2$, $-1 \le x \le 1$, about the $x$-axis.

**62. Volume** The head of your firm's accounting department has asked you to find a formula she can use in a computer program to calculate the year-end inventory of gasoline in the company's tanks. A typical tank is shaped like a right circular cylinder of radius $r$ and length $L$, mounted horizontally, as shown in the accompanying figure. The data come to the accounting office as depth measurements taken with a vertical measuring stick marked in centimeters.

   **a.** Show, in the notation of the figure, that the volume of gasoline that fills the tank to a depth $d$ is

$$V = 2L \int_{-r}^{-r+d} \sqrt{r^2 - y^2} \, dy.$$

   **b.** Evaluate the integral.

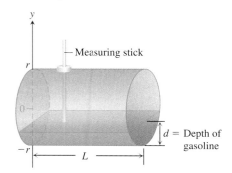

**63.** What is the largest value

$$\int_a^b \sqrt{x - x^2} \, dx$$

can have for any $a$ and $b$? Give reasons for your answer.

**64.** What is the largest value

$$\int_a^b x\sqrt{2x - x^2} \, dx$$

can have for any $a$ and $b$? Give reasons for your answer.

**COMPUTER EXPLORATIONS**

In Exercises 65 and 66, use a CAS to perform the integrations.

**65.** Evaluate the integrals

   **a.** $\displaystyle\int x \ln x \, dx$  **b.** $\displaystyle\int x^2 \ln x \, dx$  **c.** $\displaystyle\int x^3 \ln x \, dx$.

   **d.** What pattern do you see? Predict the formula for $\int x^4 \ln x \, dx$ and then see if you are correct by evaluating it with a CAS.

   **e.** What is the formula for $\int x^n \ln x \, dx$, $n \ge 1$? Check your answer using a CAS.

**66.** Evaluate the integrals

   **a.** $\displaystyle\int \frac{\ln x}{x^2} \, dx$  **b.** $\displaystyle\int \frac{\ln x}{x^3} \, dx$  **c.** $\displaystyle\int \frac{\ln x}{x^4} \, dx$.

   **d.** What pattern do you see? Predict the formula for

$$\int \frac{\ln x}{x^5} \, dx$$

and then see if you are correct by evaluating it with a CAS.

   **e.** What is the formula for

$$\int \frac{\ln x}{x^n} \, dx, \quad n \ge 2?$$

Check your answer using a CAS.

**67. a.** Use a CAS to evaluate

$$\int_0^{\pi/2} \frac{\sin^n x}{\sin^n x + \cos^n x} \, dx$$

where $n$ is an arbitrary positive integer. Does your CAS find the result?

   **b.** In succession, find the integral when $n = 1, 2, 3, 5$, and 7. Comment on the complexity of the results.

   **c.** Now substitute $x = (\pi/2) - u$ and add the new and old integrals. What is the value of

$$\int_0^{\pi/2} \frac{\sin^n x}{\sin^n x + \cos^n x} \, dx?$$

This exercise illustrates how a little mathematical ingenuity solves a problem not immediately amenable to solution by a CAS.

# 8.7 Numerical Integration

The antiderivatives of some functions, like $\sin(x^2)$, $1/\ln x$, and $\sqrt{1 + x^4}$, have no elementary formulas. When we cannot find a workable antiderivative for a function $f$ that we have to integrate, we can partition the interval of integration, replace $f$ by a closely fitting polynomial on each subinterval, integrate the polynomials, and add the results to approximate the definite integral of $f$. This procedure is an example of numerical integration. In this section we study two such methods, the *Trapezoidal Rule* and *Simpson's Rule*. In our presentation we assume that $f$ is positive, but the only requirement is for it to be continuous over the interval of integration $[a, b]$.

### Trapezoidal Approximations

The Trapezoidal Rule for the value of a definite integral is based on approximating the region between a curve and the $x$-axis with trapezoids instead of rectangles, as in Figure 8.7. It is not necessary for the subdivision points $x_0, x_1, x_2, \ldots, x_n$ in the figure to be evenly spaced, but the resulting formula is simpler if they are. We therefore assume that the length of each subinterval is

$$\Delta x = \frac{b - a}{n}.$$

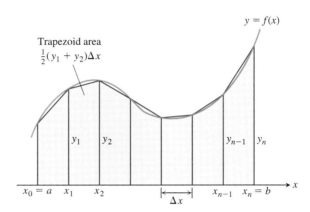

**FIGURE 8.7** The Trapezoidal Rule approximates short stretches of the curve $y = f(x)$ with line segments. To approximate the integral of $f$ from $a$ to $b$, we add the areas of the trapezoids made by joining the ends of the segments to the $x$-axis.

The length $\Delta x = (b - a)/n$ is called the **step size** or **mesh size**. The area of the trapezoid that lies above the $i$th subinterval is

$$\Delta x \left( \frac{y_{i-1} + y_i}{2} \right) = \frac{\Delta x}{2} (y_{i-1} + y_i),$$

where $y_{i-1} = f(x_{i-1})$ and $y_i = f(x_i)$. (See Figure 8.7.) The area below the curve $y = f(x)$ and above the $x$-axis is then approximated by adding the areas of all the trapezoids:

$$T = \frac{1}{2} (y_0 + y_1) \Delta x + \frac{1}{2} (y_1 + y_2) \Delta x + \cdots$$

$$+ \frac{1}{2} (y_{n-2} + y_{n-1}) \Delta x + \frac{1}{2} (y_{n-1} + y_n) \Delta x$$

$$= \Delta x \left( \frac{1}{2} y_0 + y_1 + y_2 + \cdots + y_{n-1} + \frac{1}{2} y_n \right)$$

$$= \frac{\Delta x}{2} (y_0 + 2y_1 + 2y_2 + \cdots + 2y_{n-1} + y_n),$$

where

$$y_0 = f(a), \qquad y_1 = f(x_1), \qquad \ldots, \qquad y_{n-1} = f(x_{n-1}), \qquad y_n = f(b).$$

The Trapezoidal Rule says: Use $T$ to estimate the integral of $f$ from $a$ to $b$.

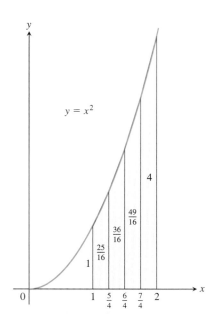

**FIGURE 8.8** The trapezoidal approximation of the area under the graph of $y = x^2$ from $x = 1$ to $x = 2$ is a slight overestimate (Example 1).

| TABLE 8.2 | |
|---|---|
| $x$ | $y = x^2$ |
| 1 | 1 |
| $\dfrac{5}{4}$ | $\dfrac{25}{16}$ |
| $\dfrac{6}{4}$ | $\dfrac{36}{16}$ |
| $\dfrac{7}{4}$ | $\dfrac{49}{16}$ |
| 2 | 4 |

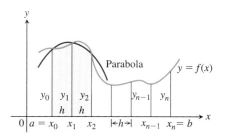

**FIGURE 8.9** Simpson's Rule approximates short stretches of the curve with parabolas.

---

> **The Trapezoidal Rule**
>
> To approximate $\int_a^b f(x)\,dx$, use
>
> $$T = \frac{\Delta x}{2}\left(y_0 + 2y_1 + 2y_2 + \cdots + 2y_{n-1} + y_n\right).$$
>
> The $y$'s are the values of $f$ at the partition points
>
> $$x_0 = a,\; x_1 = a + \Delta x,\; x_2 = a + 2\Delta x,\; \ldots,\; x_{n-1} = a + (n-1)\Delta x,\; x_n = b,$$
>
> where $\Delta x = (b - a)/n$.

**EXAMPLE 1** Use the Trapezoidal Rule with $n = 4$ to estimate $\int_1^2 x^2\,dx$. Compare the estimate with the exact value.

**Solution** Partition $[1, 2]$ into four subintervals of equal length (Figure 8.8). Then evaluate $y = x^2$ at each partition point (Table 8.2).

Using these $y$-values, $n = 4$, and $\Delta x = (2 - 1)/4 = 1/4$ in the Trapezoidal Rule, we have

$$T = \frac{\Delta x}{2}\left(y_0 + 2y_1 + 2y_2 + 2y_3 + y_4\right)$$

$$= \frac{1}{8}\left(1 + 2\left(\frac{25}{16}\right) + 2\left(\frac{36}{16}\right) + 2\left(\frac{49}{16}\right) + 4\right)$$

$$= \frac{75}{32} = 2.34375.$$

Since the parabola is concave *up*, the approximating segments lie above the curve, giving each trapezoid slightly more area than the corresponding strip under the curve. The exact value of the integral is

$$\int_1^2 x^2\,dx = \left.\frac{x^3}{3}\right]_1^2 = \frac{8}{3} - \frac{1}{3} = \frac{7}{3}.$$

The $T$ approximation overestimates the integral by about half a percent of its true value of $7/3$. The percentage error is $(2.34375 - 7/3)/(7/3) \approx 0.00446$, or $0.446\%$. ∎

## Simpson's Rule: Approximations Using Parabolas

Another rule for approximating the definite integral of a continuous function results from using parabolas instead of the straight-line segments that produced trapezoids. As before, we partition the interval $[a, b]$ into $n$ subintervals of equal length $h = \Delta x = (b - a)/n$, but this time we require that $n$ be an *even* number. On each consecutive pair of intervals we approximate the curve $y = f(x) \geq 0$ by a parabola, as shown in Figure 8.9. A typical parabola passes through three consecutive points $(x_{i-1}, y_{i-1})$, $(x_i, y_i)$, and $(x_{i+1}, y_{i+1})$ on the curve.

Let's calculate the shaded area beneath a parabola passing through three consecutive points. To simplify our calculations, we first take the case where $x_0 = -h$, $x_1 = 0$, and

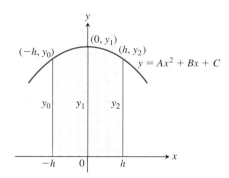

**FIGURE 8.10** By integrating from $-h$ to $h$, we find the shaded area to be

$$\frac{h}{3}(y_0 + 4y_1 + y_2).$$

$x_2 = h$ (Figure 8.10), where $h = \Delta x = (b - a)/n$. The area under the parabola will be the same if we shift the $y$-axis to the left or right. The parabola has an equation of the form

$$y = Ax^2 + Bx + C,$$

so the area under it from $x = -h$ to $x = h$ is

$$A_p = \int_{-h}^{h} (Ax^2 + Bx + C)\, dx$$

$$= \left[ \frac{Ax^3}{3} + \frac{Bx^2}{2} + Cx \right]_{-h}^{h}$$

$$= \frac{2Ah^3}{3} + 2Ch = \frac{h}{3}(2Ah^2 + 6C).$$

Since the curve passes through the three points $(-h, y_0)$, $(0, y_1)$, and $(h, y_2)$, we also have

$$y_0 = Ah^2 - Bh + C, \qquad y_1 = C, \qquad y_2 = Ah^2 + Bh + C,$$

from which we obtain

$$C = y_1,$$
$$Ah^2 - Bh = y_0 - y_1,$$
$$Ah^2 + Bh = y_2 - y_1,$$
$$2Ah^2 = y_0 + y_2 - 2y_1.$$

Hence, expressing the area $A_p$ in terms of the ordinates $y_0$, $y_1$, and $y_2$, we have

$$A_p = \frac{h}{3}(2Ah^2 + 6C) = \frac{h}{3}((y_0 + y_2 - 2y_1) + 6y_1) = \frac{h}{3}(y_0 + 4y_1 + y_2).$$

Now shifting the parabola horizontally to its shaded position in Figure 8.9 does not change the area under it. Thus the area under the parabola through $(x_0, y_0)$, $(x_1, y_1)$, and $(x_2, y_2)$ in Figure 8.9 is still

$$\frac{h}{3}(y_0 + 4y_1 + y_2).$$

Similarly, the area under the parabola through the points $(x_2, y_2)$, $(x_3, y_3)$, and $(x_4, y_4)$ is

$$\frac{h}{3}(y_2 + 4y_3 + y_4).$$

Computing the areas under all the parabolas and adding the results gives the approximation

$$\int_a^b f(x)\, dx \approx \frac{h}{3}(y_0 + 4y_1 + y_2) + \frac{h}{3}(y_2 + 4y_3 + y_4) + \cdots$$

$$+ \frac{h}{3}(y_{n-2} + 4y_{n-1} + y_n)$$

$$= \frac{h}{3}(y_0 + 4y_1 + 2y_2 + 4y_3 + 2y_4 + \cdots + 2y_{n-2} + 4y_{n-1} + y_n).$$

**HISTORICAL BIOGRAPHY**

Thomas Simpson
(1720–1761)

The result is known as Simpson's Rule. The function need not be positive, as in our derivation, but the number $n$ of subintervals must be even to apply the rule because each parabolic arc uses two subintervals.

> **Simpson's Rule**
>
> To approximate $\int_a^b f(x)\,dx$, use
>
> $$S = \frac{\Delta x}{3}(y_0 + 4y_1 + 2y_2 + 4y_3 + \cdots + 2y_{n-2} + 4y_{n-1} + y_n).$$
>
> The $y$'s are the values of $f$ at the partition points
>
> $$x_0 = a, x_1 = a + \Delta x, x_2 = a + 2\Delta x, \ldots, x_{n-1} = a + (n-1)\Delta x, x_n = b.$$
>
> The number $n$ is even, and $\Delta x = (b-a)/n$.

Note the pattern of the coefficients in the above rule: 1, 4, 2, 4, 2, 4, 2, ..., 4, 1.

**EXAMPLE 2** Use Simpson's Rule with $n = 4$ to approximate $\int_0^2 5x^4\,dx$.

**Solution** Partition $[0, 2]$ into four subintervals and evaluate $y = 5x^4$ at the partition points (Table 8.3). Then apply Simpson's Rule with $n = 4$ and $\Delta x = 1/2$:

$$S = \frac{\Delta x}{3}\left(y_0 + 4y_1 + 2y_2 + 4y_3 + y_4\right)$$

$$= \frac{1}{6}\left(0 + 4\left(\frac{5}{16}\right) + 2(5) + 4\left(\frac{405}{16}\right) + 80\right)$$

$$= 32\frac{1}{12}.$$

This estimate differs from the exact value (32) by only $1/12$, a percentage error of less than three-tenths of one percent, and this was with just four subintervals. ∎

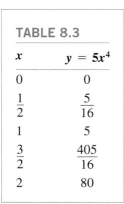

TABLE 8.3

| $x$ | $y = 5x^4$ |
|-----|------------|
| 0 | 0 |
| $\frac{1}{2}$ | $\frac{5}{16}$ |
| 1 | 5 |
| $\frac{3}{2}$ | $\frac{405}{16}$ |
| 2 | 80 |

## Error Analysis

Whenever we use an approximation technique, the issue arises as to how accurate the approximation might be. The following theorem gives formulas for estimating the errors when using the Trapezoidal Rule and Simpson's Rule. The **error** is the difference between the approximation obtained by the rule and the actual value of the definite integral $\int_a^b f(x)\,dx$.

> **THEOREM 1—Error Estimates in the Trapezoidal and Simpson's Rules**
> If $f''$ is continuous and $M$ is any upper bound for the values of $|f''|$ on $[a, b]$, then the error $E_T$ in the trapezoidal approximation of the integral of $f$ from $a$ to $b$ for $n$ steps satisfies the inequality
>
> $$|E_T| \le \frac{M(b-a)^3}{12n^2}. \qquad \text{Trapezoidal Rule}$$
>
> If $f^{(4)}$ is continuous and $M$ is any upper bound for the values of $|f^{(4)}|$ on $[a, b]$, then the error $E_S$ in the Simpson's Rule approximation of the integral of $f$ from $a$ to $b$ for $n$ steps satisfies the inequality
>
> $$|E_S| \le \frac{M(b-a)^5}{180n^4}. \qquad \text{Simpson's Rule}$$

To see why Theorem 1 is true in the case of the Trapezoidal Rule, we begin with a result from advanced calculus, which says that if $f''$ is continuous on the interval $[a, b]$, then

$$\int_a^b f(x)\,dx = T - \frac{b-a}{12}\cdot f''(c)(\Delta x)^2.$$

for some number $c$ between $a$ and $b$. Thus, as $\Delta x$ approaches zero, the error defined by

$$E_T = -\frac{b-a}{12} \cdot f''(c)(\Delta x)^2$$

approaches *zero* as the *square* of $\Delta x$.

The inequality

$$|E_T| \leq \frac{b-a}{12} \max|f''(x)|(\Delta x)^2,$$

where max refers to the interval $[a, b]$, gives an upper bound for the magnitude of the error. In practice, we usually cannot find the exact value of $\max|f''(x)|$ and have to estimate an upper bound or "worst case" value for it instead. If $M$ is any upper bound for the values of $|f''(x)|$ on $[a, b]$, so that $|f''(x)| \leq M$ on $[a, b]$, then

$$|E_T| \leq \frac{b-a}{12} M(\Delta x)^2.$$

If we substitute $(b - a)/n$ for $\Delta x$, we get

$$|E_T| \leq \frac{M(b-a)^3}{12n^2}.$$

To estimate the error in Simpson's Rule, we start with a result from advanced calculus that says that if the fourth derivative $f^{(4)}$ is continuous, then

$$\int_a^b f(x)\,dx = S - \frac{b-a}{180} \cdot f^{(4)}(c)(\Delta x)^4$$

for some point $c$ between $a$ and $b$. Thus, as $\Delta x$ approaches zero, the error,

$$E_S = -\frac{b-a}{180} \cdot f^{(4)}(c)(\Delta x)^4,$$

approaches zero as the *fourth power* of $\Delta x$. (This helps to explain why Simpson's Rule is likely to give better results than the Trapezoidal Rule.)

The inequality

$$|E_S| \leq \frac{b-a}{180} \max|f^{(4)}(x)| \, (\Delta x)^4,$$

where max refers to the interval $[a, b]$, gives an upper bound for the magnitude of the error. As with $\max|f''|$ in the error formula for the Trapezoidal Rule, we usually cannot find the exact value of $\max|f^{(4)}(x)|$ and have to replace it with an upper bound. If $M$ is any upper bound for the values of $|f^{(4)}|$ on $[a, b]$, then

$$|E_S| \leq \frac{b-a}{180} M(\Delta x)^4.$$

Substituting $(b - a)/n$ for $\Delta x$ in this last expression gives

$$|E_S| \leq \frac{M(b-a)^5}{180n^4}.$$

**EXAMPLE 3**    Find an upper bound for the error in estimating $\int_0^2 5x^4\,dx$ using Simpson's Rule with $n = 4$ (Example 2).

**Solution**    To estimate the error, we first find an upper bound $M$ for the magnitude of the fourth derivative of $f(x) = 5x^4$ on the interval $0 \leq x \leq 2$. Since the fourth derivative has the constant value $f^{(4)}(x) = 120$, we take $M = 120$. With $b - a = 2$ and $n = 4$, the error estimate for Simpson's Rule gives

$$|E_S| \le \frac{M(b - a)^5}{180n^4} = \frac{120(2)^5}{180 \cdot 4^4} = \frac{1}{12}.$$

This estimate is consistent with the result of Example 2. ■

Theorem 1 can also be used to estimate the number of subintervals required when using the Trapezoidal or Simpson's Rule if we specify a certain tolerance for the error.

**EXAMPLE 4**    Estimate the minimum number of subintervals needed to approximate the integral in Example 3 using Simpson's Rule with an error of magnitude less than $10^{-4}$.

**Solution**    Using the inequality in Theorem 1, if we choose the number of subintervals $n$ to satisfy

$$\frac{M(b - a)^5}{180n^4} < 10^{-4},$$

then the error $E_S$ in Simpson's Rule satisfies $|E_S| < 10^{-4}$ as required.

From the solution in Example 3, we have $M = 120$ and $b - a = 2$, so we want $n$ to satisfy

$$\frac{120(2)^5}{180n^4} < \frac{1}{10^4}$$

or, equivalently,

$$n^4 > \frac{64 \cdot 10^4}{3}.$$

It follows that

$$n > 10 \left( \frac{64}{3} \right)^{1/4} \approx 21.5.$$

Since $n$ must be even in Simpson's Rule, we estimate the minimum number of subintervals required for the error tolerance to be $n = 22$. ■

**EXAMPLE 5**    As we saw in Chapter 7, the value of ln 2 can be calculated from the integral

$$\ln 2 = \int_1^2 \frac{1}{x} \, dx.$$

Table 8.4 shows $T$ and $S$ values for approximations of $\int_1^2 (1/x) \, dx$ using various values of $n$. Notice how Simpson's Rule dramatically improves over the Trapezoidal Rule.

**TABLE 8.4** Trapezoidal Rule approximations ($T_n$) and Simpson's Rule approximations ($S_n$) of $\ln 2 = \int_1^2 (1/x) \, dx$

| $n$ | $T_n$ | **|Error|** less than ... | $S_n$ | **|Error|** less than ... |
|---|---|---|---|---|
| 10 | 0.6937714032 | 0.0006242227 | 0.6931502307 | 0.0000030502 |
| 20 | 0.6933033818 | 0.0001562013 | 0.6931473747 | 0.0000001942 |
| 30 | 0.6932166154 | 0.0000694349 | 0.6931472190 | 0.0000000385 |
| 40 | 0.6931862400 | 0.0000390595 | 0.6931471927 | 0.0000000122 |
| 50 | 0.6931721793 | 0.0000249988 | 0.6931471856 | 0.0000000050 |
| 100 | 0.6931534305 | 0.0000062500 | 0.6931471809 | 0.0000000004 |

In particular, notice that when we double the value of $n$ (thereby halving the value of $h = \Delta x$), the $T$ error is divided by 2 *squared*, whereas the $S$ error is divided by 2 *to the fourth*.

This has a dramatic effect as $\Delta x = (2 - 1)/n$ gets very small. The Simpson approximation for $n = 50$ rounds accurately to seven places and for $n = 100$ agrees to nine decimal places (billionths)! ∎

If $f(x)$ is a polynomial of degree less than four, then its fourth derivative is zero, and

$$E_S = -\frac{b - a}{180} f^{(4)}(c)(\Delta x)^4 = -\frac{b - a}{180} (0)(\Delta x)^4 = 0.$$

Thus, there will be no error in the Simpson approximation of any integral of $f$. In other words, if $f$ is a constant, a linear function, or a quadratic or cubic polynomial, Simpson's Rule will give the value of any integral of $f$ exactly, whatever the number of subdivisions. Similarly, if $f$ is a constant or a linear function, then its second derivative is zero, and

$$E_T = -\frac{b - a}{12} f''(c)(\Delta x)^2 = -\frac{b - a}{12} (0)(\Delta x)^2 = 0.$$

The Trapezoidal Rule will therefore give the exact value of any integral of $f$. This is no surprise, for the trapezoids fit the graph perfectly.

Although decreasing the step size $\Delta x$ reduces the error in the Simpson and Trapezoidal approximations in theory, it may fail to do so in practice. When $\Delta x$ is very small, say $\Delta x = 10^{-8}$, computer or calculator round-off errors in the arithmetic required to evaluate $S$ and $T$ may accumulate to such an extent that the error formulas no longer describe what is going on. Shrinking $\Delta x$ below a certain size can actually make things worse. You should consult a text on numerical analysis for more sophisticated methods if you are having problems with round-off error using the rules discussed in this section.

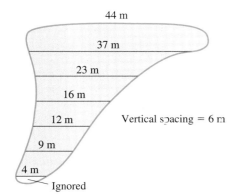

44 m

37 m

23 m

16 m

12 m    Vertical spacing = 6 m

9 m

4 m

Ignored

**FIGURE 8.11** The dimensions of the swamp in Example 6.

**EXAMPLE 6**    A town wants to drain and fill a polluted swamp (Figure 8.11). The swamp averages 1.5 m deep. About how many cubic meters of dirt will it take to fill the area after the swamp is drained?

**Solution**    To calculate the volume of the swamp, we estimate the surface area and multiply by 1.5. To estimate the area, we use Simpson's Rule with $\Delta x = 6$ m and the $y$'s equal to the distances measured across the swamp, as shown in Figure 8.11.

$$S = \frac{\Delta x}{3} (y_0 + 4y_1 + 2y_2 + 4y_3 + 2y_4 + 4y_5 + y_6)$$

$$= \frac{6}{3} (44 + 148 + 46 + 64 + 24 + 36 + 4) = 732$$

The volume is about $(732)(1.5) = 1098$ m³. ∎

# Exercises 8.7

**Estimating Definite Integrals**

The instructions for the integrals in Exercises 1–10 have two parts, one for the Trapezoidal Rule and one for Simpson's Rule.

**I. Using the Trapezoidal Rule**

    **a.** Estimate the integral with $n = 4$ steps and find an upper bound for $|E_T|$.

    **b.** Evaluate the integral directly and find $|E_T|$.

    **c.** Use the formula $(|E_T|/(\text{true value})) \times 100$ to express $|E_T|$ as a percentage of the integral's true value.

**II. Using Simpson's Rule**

    **a.** Estimate the integral with $n = 4$ steps and find an upper bound for $|E_S|$.

    **b.** Evaluate the integral directly and find $|E_S|$.

    **c.** Use the formula $(|E_S|/(\text{true value})) \times 100$ to express $|E_S|$ as a percentage of the integral's true value.

1. $\displaystyle\int_1^2 x \, dx$             2. $\displaystyle\int_1^3 (2x - 1) \, dx$

**3.** $\int_{-1}^{1} (x^2 + 1)\,dx$

**4.** $\int_{-2}^{0} (x^2 - 1)\,dx$

**5.** $\int_{0}^{2} (t^3 + t)\,dt$

**6.** $\int_{-1}^{1} (t^3 + 1)\,dt$

**7.** $\int_{1}^{2} \frac{1}{s^2}\,ds$

**8.** $\int_{2}^{4} \frac{1}{(s-1)^2}\,ds$

**9.** $\int_{0}^{\pi} \sin t\,dt$

**10.** $\int_{0}^{1} \sin \pi t\,dt$

### Estimating the Number of Subintervals

In Exercises 11–22, estimate the minimum number of subintervals needed to approximate the integrals with an error of magnitude less than $10^{-4}$ by **(a)** the Trapezoidal Rule and **(b)** Simpson's Rule. (The integrals in Exercises 11–18 are the integrals from Exercises 1–8.)

**11.** $\int_{1}^{2} x\,dx$

**12.** $\int_{1}^{3} (2x - 1)\,dx$

**13.** $\int_{-1}^{1} (x^2 + 1)\,dx$

**14.** $\int_{-2}^{0} (x^2 - 1)\,dx$

**15.** $\int_{0}^{2} (t^3 + t)\,dt$

**16.** $\int_{-1}^{1} (t^3 + 1)\,dt$

**17.** $\int_{1}^{2} \frac{1}{s^2}\,ds$

**18.** $\int_{2}^{4} \frac{1}{(s-1)^2}\,ds$

**19.** $\int_{0}^{3} \sqrt{x + 1}\,dx$

**20.** $\int_{0}^{3} \frac{1}{\sqrt{x + 1}}\,dx$

**21.** $\int_{0}^{2} \sin (x + 1)\,dx$

**22.** $\int_{-1}^{1} \cos (x + \pi)\,dx$

### Estimates with Numerical Data

**23. Volume of water in a swimming pool** A rectangular swimming pool is 5 m wide and 10 m long. The accompanying table shows the depth $h(x)$ of the water at 1-m intervals from one end of the pool to the other. Estimate the volume of water in the pool using the Trapezoidal Rule with $n = 10$ applied to the integral

$$V = \int_{0}^{10} 5 \cdot h(x)\,dx.$$

| Position (m) $x$ | Depth (m) $h(x)$ | Position (m) $x$ | Depth (m) $h(x)$ |
|---|---|---|---|
| 0 | 1.20 | 6 | 2.30 |
| 1 | 1.64 | 7 | 2.38 |
| 2 | 1.82 | 8 | 2.46 |
| 3 | 1.98 | 9 | 2.54 |
| 4 | 2.10 | 10 | 2.60 |
| 5 | 2.20 | | |

**24. Distance traveled** The accompanying table shows time-to-speed data for a car accelerating from rest to 130 km/h. How far had the car traveled by the time it reached this speed? (Use trapezoids to estimate the area under the velocity curve, but be careful: The time intervals vary in length.)

| Speed change | Time (sec) |
|---|---|
| Zero to 30 km/h | 2.2 |
| 40 km/h | 3.2 |
| 50 km/h | 4.5 |
| 60 km/h | 5.9 |
| 70 km/h | 7.8 |
| 80 km/h | 10.2 |
| 90 km/h | 12.7 |
| 100 km/h | 16.0 |
| 110 km/h | 20.6 |
| 120 km/h | 26.2 |
| 130 km/h | 37.1 |

**25. Wing design** The design of a new airplane requires a gasoline tank of constant cross-sectional area in each wing. A scale drawing of a cross-section is shown here. The tank must hold 2000 kg of gasoline, which has a density of 673 kg/m³. Estimate the length of the tank by Simpson's Rule.

$y_0 = 0.5$ m, $y_1 = 0.55$ m, $y_2 = 0.6$ m, $y_3 = 0.65$ m, $y_4 = 0.7$ m, $y_5 = y_6 = 0.75$ m Horizontal spacing $= 0.3$ m

**26. Oil consumption on Pathfinder Island** A diesel generator runs continuously, consuming oil at a gradually increasing rate until it must be temporarily shut down to have the filters replaced. Use the Trapezoidal Rule to estimate the amount of oil consumed by the generator during that week.

| Day | Oil consumption rate (L/h) |
|---|---|
| Sun | 0.019 |
| Mon | 0.020 |
| Tue | 0.021 |
| Wed | 0.023 |
| Thu | 0.025 |
| Fri | 0.028 |
| Sat | 0.031 |
| Sun | 0.035 |

### Theory and Examples

**27. Usable values of the sine-integral function** *The sine-integral function,*

$$\text{Si}(x) = \int_{0}^{x} \frac{\sin t}{t}\,dt, \qquad \text{"Sine integral of } x\text{"}$$

is one of the many functions in engineering whose formulas cannot be simplified. There is no elementary formula for the antiderivative of $(\sin t)/t$. The values of Si($x$), however, are readily estimated by numerical integration.

Although the notation does not show it explicitly, the function being integrated is

$$f(t) = \begin{cases} \dfrac{\sin t}{t}, & t \neq 0 \\ 1, & t = 0, \end{cases}$$

the continuous extension of $(\sin t)/t$ to the interval $[0, x]$. The function has derivatives of all orders at every point of its domain. Its graph is smooth, and you can expect good results from Simpson's Rule.

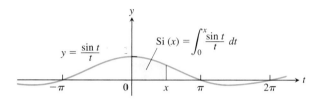

**a.** Use the fact that $|f^{(4)}| \leq 1$ on $[0, \pi/2]$ to give an upper bound for the error that will occur if

$$\text{Si}\left(\frac{\pi}{2}\right) = \int_0^{\pi/2} \frac{\sin t}{t}\, dt$$

is estimated by Simpson's Rule with $n = 4$.

**b.** Estimate Si($\pi/2$) by Simpson's Rule with $n = 4$.

**c.** Express the error bound you found in part (a) as a percentage of the value you found in part (b).

**28. The error function**   *The error function,*

$$\text{erf}(x) = \frac{2}{\sqrt{\pi}} \int_0^x e^{-t^2}\, dt,$$

important in probability and in the theories of heat flow and signal transmission, must be evaluated numerically because there is no elementary expression for the antiderivative of $e^{-t^2}$.

**a.** Use Simpson's Rule with $n = 10$ to estimate erf($1$).

**b.** In $[0, 1]$,

$$\left| \frac{d^4}{dt^4}\left(e^{-t^2}\right) \right| \leq 12.$$

Give an upper bound for the magnitude of the error of the estimate in part (a).

**29.** Prove that the sum $T$ in the Trapezoidal Rule for $\int_a^b f(x)\, dx$ is a Riemann sum for $f$ continuous on $[a, b]$. (*Hint:* Use the Intermediate Value Theorem to show the existence of $c_k$ in the subinterval $[x_{k-1}, x_k]$ satisfying $f(c_k) = (f(x_{k-1}) + f(x_k))/2$.)

**30.** Prove that the sum $S$ in Simpson's Rule for $\int_a^b f(x)\, dx$ is a Riemann sum for $f$ continuous on $[a, b]$. (See Exercise 29.)

**T  31. Elliptic integrals**   The length of the ellipse

$$\frac{x^2}{a^2} + \frac{y^2}{b^2} = 1$$

turns out to be

$$\text{Length} = 4a \int_0^{\pi/2} \sqrt{1 - e^2 \cos^2 t}\, dt,$$

where $e = \sqrt{a^2 - b^2}/a$ is the ellipse's eccentricity. The integral in this formula, called an *elliptic integral*, is nonelementary except when $e = 0$ or 1.

**a.** Use the Trapezoidal Rule with $n = 10$ to estimate the length of the ellipse when $a = 1$ and $e = 1/2$.

**b.** Use the fact that the absolute value of the second derivative of $f(t) = \sqrt{1 - e^2 \cos^2 t}$ is less than 1 to find an upper bound for the error in the estimate you obtained in part (a).

Applications

**T  32.** The length of one arch of the curve $y = \sin x$ is given by

$$L = \int_0^{\pi} \sqrt{1 + \cos^2 x}\, dx.$$

Estimate $L$ by Simpson's Rule with $n = 8$.

**T  33.** Your metal fabrication company is bidding for a contract to make sheets of corrugated iron roofing like the one shown here. The cross-sections of the corrugated sheets are to conform to the curve

$$y = \sin\frac{3\pi}{20}x, \quad 0 \leq x \leq 20 \text{ cm}.$$

If the roofing is to be stamped from flat sheets by a process that does not stretch the material, how wide should the original material be? To find out, use numerical integration to approximate the length of the sine curve to two decimal places.

**T  34.** Your engineering firm is bidding for the contract to construct the tunnel shown here. The tunnel is 90 m long and 15 m wide at the base. The cross-section is shaped like one arch of the curve $y = 7.5 \cos(\pi x/15)$. Upon completion, the tunnel's inside surface (excluding the roadway) will be treated with a waterproof sealer that costs $26.11 per square meter to apply. How much will it cost to apply the sealer? (*Hint:* Use numerical integration to find the length of the cosine curve.)

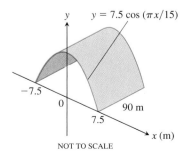

NOT TO SCALE

Find, to two decimal places, the areas of the surfaces generated by revolving the curves in Exercises 35 and 36 about the x-axis.

**35.** $y = \sin x$, $\quad 0 \leq x \leq \pi$

**36.** $y = x^2/4$, $\quad 0 \leq x \leq 2$

**37.** Use numerical integration to estimate the value of

$$\sin^{-1} 0.6 = \int_0^{0.6} \frac{dx}{\sqrt{1-x^2}}.$$

For reference, $\sin^{-1} 0.6 = 0.64350$ to five decimal places.

**38.** Use numerical integration to estimate the value of

$$\pi = 4 \int_0^1 \frac{1}{1+x^2} dx.$$

**39. Drug assimilation** An average adult under age 60 years assimilates a 12-h cold medicine into his or her system at a rate modeled by

$$\frac{dy}{dt} = 6 - \ln(2t^2 - 3t + 3),$$

where $y$ is measured in milligrams and $t$ is the time in hours since the medication was taken. What amount of medicine is absorbed into a person's system over a 12-h period?

**40. Effects of an antihistamine** The concentration of an antihistamine in the bloodstream of a healthy adult is modeled by

$$C = 12.5 - 4 \ln(t^2 - 3t + 4),$$

where $C$ is measured in grams per liter and $t$ is the time in hours since the medication was taken. What is the average level of concentration in the bloodstream over a 6-h period?

# 8.8 Improper Integrals

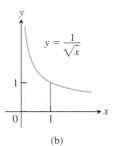

**FIGURE 8.12** Are the areas under these infinite curves finite? We will see that the answer is yes for both curves.

Up to now, we have required definite integrals to have two properties. First, the domain of integration $[a, b]$ must be finite. Second, the range of the integrand must be finite on this domain. In practice, we may encounter problems that fail to meet one or both of these conditions. The integral for the area under the curve $y = (\ln x)/x^2$ from $x = 1$ to $x = \infty$ is an example for which the domain is infinite (Figure 8.12a). The integral for the area under the curve of $y = 1/\sqrt{x}$ between $x = 0$ and $x = 1$ is an example for which the range of the integrand is infinite (Figure 8.12b). In either case, the integrals are said to be *improper* and are calculated as limits. We will see in Section 8.9 that improper integrals play an important role in probability. They are also useful when investigating the convergence of certain infinite series in Chapter 10.

## Infinite Limits of Integration

Consider the infinite region (unbounded on the right) that lies under the curve $y = e^{-x/2}$ in the first quadrant (Figure 8.13a). You might think this region has infinite area, but we will see that the value is finite. We assign a value to the area in the following way. First find the area $A(b)$ of the portion of the region that is bounded on the right by $x = b$ (Figure 8.13b).

$$A(b) = \int_0^b e^{-x/2} dx = -2e^{-x/2} \Big]_0^b = -2e^{-b/2} + 2$$

Then find the limit of $A(b)$ as $b \to \infty$

$$\lim_{b \to \infty} A(b) = \lim_{b \to \infty} \left(-2e^{-b/2} + 2\right) = 2.$$

The value we assign to the area under the curve from 0 to $\infty$ is

$$\int_0^{\infty} e^{-x/2} dx = \lim_{b \to \infty} \int_0^b e^{-x/2} dx = 2.$$

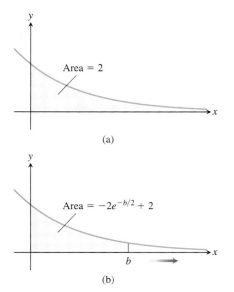

**FIGURE 8.13**  (a) The area in the first quadrant under the curve $y = e^{-x/2}$. (b) The area is an improper integral of the first type.

**DEFINITION**  Integrals with infinite limits of integration are **improper integrals of Type I**.

1.  If $f(x)$ is continuous on $[a, \infty)$, then

$$\int_a^\infty f(x)\, dx = \lim_{b \to \infty} \int_a^b f(x)\, dx.$$

2.  If $f(x)$ is continuous on $(-\infty, b\,]$, then

$$\int_{-\infty}^b f(x)\, dx = \lim_{a \to -\infty} \int_a^b f(x)\, dx.$$

3.  If $f(x)$ is continuous on $(-\infty, \infty)$, then

$$\int_{-\infty}^\infty f(x)\, dx = \int_{-\infty}^c f(x)\, dx + \int_c^\infty f(x)\, dx,$$

where $c$ is any real number.

In each case, if the limit is finite we say that the improper integral **converges** and that the limit is the **value** of the improper integral. If the limit fails to exist, the improper integral **diverges**.

It can be shown that the choice of $c$ in Part 3 of the definition is unimportant. We can evaluate or determine the convergence or divergence of $\int_{-\infty}^\infty f(x)\, dx$ with any convenient choice.

Any of the integrals in the above definition can be interpreted as an area if $f \geq 0$ on the interval of integration. For instance, we interpreted the improper integral in Figure 8.13 as an area. In that case, the area has the finite value 2. If $f \geq 0$ and the improper integral diverges, we say the area under the curve is **infinite**.

**EXAMPLE 1**  Is the area under the curve $y = (\ln x)/x^2$ from $x = 1$ to $x = \infty$ finite? If so, what is its value?

**Solution**  We find the area under the curve from $x = 1$ to $x = b$ and examine the limit as $b \to \infty$. If the limit is finite, we take it to be the area under the curve (Figure 8.14). The area from 1 to $b$ is

$$\int_1^b \frac{\ln x}{x^2}\, dx = \left[ (\ln x)\left(-\frac{1}{x}\right) \right]_1^b - \int_1^b \left(-\frac{1}{x}\right)\left(\frac{1}{x}\right) dx$$

Integration by parts with
$u = \ln x,\ dv = dx/x^2,$
$du = dx/x,\ v = -1/x$

$$= -\frac{\ln b}{b} - \left[ \frac{1}{x} \right]_1^b$$

$$= -\frac{\ln b}{b} - \frac{1}{b} + 1.$$

**FIGURE 8.14**  The area under this curve is an improper integral (Example 1).

The limit of the area as $b \to \infty$ is

$$\int_1^\infty \frac{\ln x}{x^2} \, dx = \lim_{b \to \infty} \int_1^b \frac{\ln x}{x^2} \, dx$$

$$= \lim_{b \to \infty} \left[ -\frac{\ln b}{b} - \frac{1}{b} + 1 \right]$$

$$= -\left[ \lim_{b \to \infty} \frac{\ln b}{b} \right] - 0 + 1$$

$$= -\left[ \lim_{b \to \infty} \frac{1/b}{1} \right] + 1 = 0 + 1 = 1. \qquad \text{l'Hôpital's Rule}$$

Thus, the improper integral converges and the area has finite value 1. ∎

**EXAMPLE 2**   Evaluate

$$\int_{-\infty}^\infty \frac{dx}{1 + x^2}.$$

**HISTORICAL BIOGRAPHY**

Lejeune Dirichlet
(1805–1859)

**Solution**   According to the definition (Part 3), we can choose $c = 0$ and write

$$\int_{-\infty}^\infty \frac{dx}{1 + x^2} = \int_{-\infty}^0 \frac{dx}{1 + x^2} + \int_0^\infty \frac{dx}{1 + x^2}.$$

Next we evaluate each improper integral on the right side of the equation above.

$$\int_{-\infty}^0 \frac{dx}{1 + x^2} = \lim_{a \to -\infty} \int_a^0 \frac{dx}{1 + x^2}$$

$$= \lim_{a \to -\infty} \tan^{-1} x \Big]_a^0$$

$$= \lim_{a \to -\infty} (\tan^{-1} 0 - \tan^{-1} a) = 0 - \left( -\frac{\pi}{2} \right) = \frac{\pi}{2}$$

$$\int_0^\infty \frac{dx}{1 + x^2} = \lim_{b \to \infty} \int_0^b \frac{dx}{1 + x^2}$$

$$= \lim_{b \to \infty} \tan^{-1} x \Big]_0^b$$

$$= \lim_{b \to \infty} (\tan^{-1} b - \tan^{-1} 0) = \frac{\pi}{2} - 0 = \frac{\pi}{2}$$

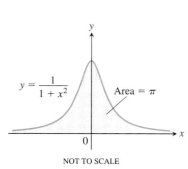

$$y = \frac{1}{1 + x^2} \qquad \text{Area} = \pi$$

NOT TO SCALE

**FIGURE 8.15**   The area under this curve is finite (Example 2).

Thus,

$$\int_{-\infty}^\infty \frac{dx}{1 + x^2} = \frac{\pi}{2} + \frac{\pi}{2} = \pi.$$

Since $1/(1 + x^2) > 0$, the improper integral can be interpreted as the (finite) area beneath the curve and above the $x$-axis (Figure 8.15). ∎

### The Integral $\int_1^\infty \dfrac{dx}{x^p}$

The function $y = 1/x$ is the boundary between the convergent and divergent improper integrals with integrands of the form $y = 1/x^p$. As the next example shows, the improper integral converges if $p > 1$ and diverges if $p \leq 1$.

**EXAMPLE 3**    For what values of $p$ does the integral $\int_1^\infty dx/x^p$ converge? When the integral does converge, what is its value?

**Solution**   If $p \neq 1$,

$$\int_1^b \frac{dx}{x^p} = \frac{x^{-p+1}}{-p+1}\Bigg]_1^b = \frac{1}{1-p}(b^{-p+1} - 1) = \frac{1}{1-p}\left(\frac{1}{b^{p-1}} - 1\right).$$

Thus,

$$\int_1^\infty \frac{dx}{x^p} = \lim_{b\to\infty} \int_1^b \frac{dx}{x^p}$$

$$= \lim_{b\to\infty}\left[\frac{1}{1-p}\left(\frac{1}{b^{p-1}} - 1\right)\right] = \begin{cases} \dfrac{1}{p-1}, & p > 1 \\ \infty, & p < 1 \end{cases}$$

because

$$\lim_{b\to\infty} \frac{1}{b^{p-1}} = \begin{cases} 0, & p > 1 \\ \infty, & p < 1. \end{cases}$$

Therefore, the integral converges to the value $1/(p-1)$ if $p > 1$ and it diverges if $p < 1$.

If $p = 1$, the integral also diverges:

$$\int_1^\infty \frac{dx}{x^p} = \int_1^\infty \frac{dx}{x}$$

$$= \lim_{b\to\infty} \int_1^b \frac{dx}{x}$$

$$= \lim_{b\to\infty} \ln x\Bigg]_1^b$$

$$= \lim_{b\to\infty}(\ln b - \ln 1) = \infty. \qquad \blacksquare$$

### Integrands with Vertical Asymptotes

Another type of improper integral arises when the integrand has a vertical asymptote—an infinite discontinuity—at a limit of integration or at some point between the limits of integration. If the integrand $f$ is positive over the interval of integration, we can again interpret the improper integral as the area under the graph of $f$ and above the $x$-axis between the limits of integration.

Consider the region in the first quadrant that lies under the curve $y = 1/\sqrt{x}$ from $x = 0$ to $x = 1$ (Figure 8.12b). First we find the area of the portion from $a$ to 1 (Figure 8.16):

$$\int_a^1 \frac{dx}{\sqrt{x}} = 2\sqrt{x}\Bigg]_a^1 = 2 - 2\sqrt{a}.$$

**FIGURE 8.16**   The area under this curve is an example of an improper integral of the second kind.

Then we find the limit of this area as $a \to 0^+$:

$$\lim_{a \to 0^+} \int_a^1 \frac{dx}{\sqrt{x}} = \lim_{a \to 0^+} \left( 2 - 2\sqrt{a} \right) = 2.$$

Therefore the area under the curve from 0 to 1 is finite and is defined to be

$$\int_0^1 \frac{dx}{\sqrt{x}} = \lim_{a \to 0^+} \int_a^1 \frac{dx}{\sqrt{x}} = 2.$$

---

**DEFINITION** Integrals of functions that become infinite at a point within the interval of integration are **improper integrals of Type II**.

1. If $f(x)$ is continuous on $(a, b]$ and discontinuous at $a$, then

$$\int_a^b f(x)\, dx = \lim_{c \to a^+} \int_c^b f(x)\, dx.$$

2. If $f(x)$ is continuous on $[a, b)$ and discontinuous at $b$, then

$$\int_a^b f(x)\, dx = \lim_{c \to b^-} \int_a^c f(x)\, dx.$$

3. If $f(x)$ is discontinuous at $c$, where $a < c < b$, and continuous on $[a, c) \cup (c, b]$, then

$$\int_a^b f(x)\, dx = \int_a^c f(x)\, dx + \int_c^b f(x)\, dx.$$

In each case, if the limit is finite we say the improper integral **converges** and that the limit is the **value** of the improper integral. If the limit does not exist, the integral **diverges**.

---

In Part 3 of the definition, the integral on the left side of the equation converges if *both* integrals on the right side converge; otherwise it diverges.

**EXAMPLE 4** Investigate the convergence of

$$\int_0^1 \frac{1}{1 - x}\, dx.$$

**Solution** The integrand $f(x) = 1/(1 - x)$ is continuous on $[0, 1)$ but is discontinuous at $x = 1$ and becomes infinite as $x \to 1^-$ (Figure 8.17). We evaluate the integral as

$$\lim_{b \to 1^-} \int_0^b \frac{1}{1 - x}\, dx = \lim_{b \to 1^-} \left[ -\ln |1 - x| \right]_0^b$$

$$= \lim_{b \to 1^-} \left[ -\ln (1 - b) + 0 \right] = \infty.$$

The limit is infinite, so the integral diverges.

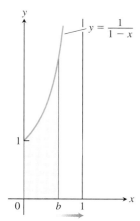

**FIGURE 8.17** The area beneath the curve and above the $x$-axis for $[0, 1)$ is not a real number (Example 4).

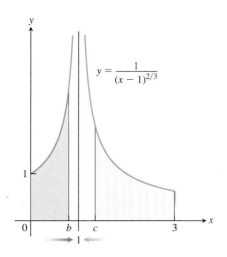

$y = \dfrac{1}{(x-1)^{2/3}}$

**FIGURE 8.18** Example 5 shows that the area under the curve exists (so it is a real number).

**EXAMPLE 5**  Evaluate

$$\int_0^3 \frac{dx}{(x-1)^{2/3}}.$$

**Solution**  The integrand has a vertical asymptote at $x = 1$ and is continuous on $[0, 1)$ and $(1, 3]$ (Figure 8.18). Thus, by Part 3 of the definition above,

$$\int_0^3 \frac{dx}{(x-1)^{2/3}} = \int_0^1 \frac{dx}{(x-1)^{2/3}} + \int_1^3 \frac{dx}{(x-1)^{2/3}}.$$

Next, we evaluate each improper integral on the right-hand side of this equation.

$$\int_0^1 \frac{dx}{(x-1)^{2/3}} = \lim_{b \to 1^-} \int_0^b \frac{dx}{(x-1)^{2/3}}$$

$$= \lim_{b \to 1^-} 3(x-1)^{1/3} \Big]_0^b$$

$$= \lim_{b \to 1^-} \left[ 3(b-1)^{1/3} + 3 \right] = 3$$

$$\int_1^3 \frac{dx}{(x-1)^{2/3}} = \lim_{c \to 1^+} \int_c^3 \frac{dx}{(x-1)^{2/3}}$$

$$= \lim_{c \to 1^+} 3(x-1)^{1/3} \Big]_c^3$$

$$= \lim_{c \to 1^+} \left[ 3(3-1)^{1/3} - 3(c-1)^{1/3} \right] = 3\sqrt[3]{2}$$

We conclude that

$$\int_0^3 \frac{dx}{(x-1)^{2/3}} = 3 + 3\sqrt[3]{2}. \qquad \blacksquare$$

## Improper Integrals with a CAS

Computer algebra systems can evaluate many convergent improper integrals. To evaluate the integral

$$\int_2^\infty \frac{x+3}{(x-1)(x^2+1)} \, dx$$

(which converges) using Maple, enter

> $f := (x + 3)/((x - 1) * (x\wedge 2 + 1));$

Then use the integration command

> $int(f, x = 2..\text{infinity});$

Maple returns the answer

$$-\frac{1}{2}\pi + \ln(5) + \arctan(2).$$

To obtain a numerical result, use the evaluation command **evalf** and specify the number of digits as follows:

> $evalf(\%, 6);$

The symbol % instructs the computer to evaluate the last expression on the screen, in this case $(-1/2)\pi + \ln(5) + \arctan(2)$. Maple returns 1.14579.

Using Mathematica, entering

$$In\ [1]:= \text{Integrate}\ [(x + 3)/((x - 1)(x\hat{}2 + 1)), \{x, 2, \text{Infinity}\}]$$

returns

$$Out\ [1]= \frac{-\pi}{2} + \text{ArcTan}\ [2] + \text{Log}\ [5].$$

To obtain a numerical result with six digits, use the command "N[ %, 6 ]"; it also yields 1.14579.

## Tests for Convergence and Divergence

When we cannot evaluate an improper integral directly, we try to determine whether it converges or diverges. If the integral diverges, that's the end of the story. If it converges, we can use numerical methods to approximate its value. The principal tests for convergence or divergence are the Direct Comparison Test and the Limit Comparison Test.

**EXAMPLE 6**    Does the integral $\int_1^\infty e^{-x^2}\,dx$ converge?

**Solution**    By definition,

$$\int_1^\infty e^{-x^2}\,dx = \lim_{b\to\infty}\int_1^b e^{-x^2}\,dx.$$

We cannot evaluate this integral directly because it is nonelementary. But we *can* show that its limit as $b \to \infty$ is finite. We know that $\int_1^b e^{-x^2}\,dx$ is an increasing function of $b$. Therefore either it becomes infinite as $b \to \infty$ or it has a finite limit as $b \to \infty$. It does not become infinite: For every value of $x \geq 1$, we have $e^{-x^2} \leq e^{-x}$ (Figure 8.19) so that

$$\int_1^b e^{-x^2}\,dx \leq \int_1^b e^{-x}\,dx = -e^{-b} + e^{-1} < e^{-1} \approx 0.36788.$$

Hence,

$$\int_1^\infty e^{-x^2}\,dx = \lim_{b\to\infty}\int_1^b e^{-x^2}\,dx$$

converges to some definite finite value. We do not know exactly what the value is except that it is something positive and less than 0.37. Here we are relying on the completeness property of the real numbers, discussed in Appendix 6.    ■

The comparison of $e^{-x^2}$ and $e^{-x}$ in Example 6 is a special case of the following test.

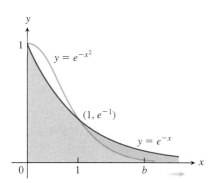

**FIGURE 8.19**    The graph of $e^{-x^2}$ lies below the graph of $e^{-x}$ for $x > 1$ (Example 6).

---

**THEOREM 2—Direct Comparison Test**    Let $f$ and $g$ be continuous on $[a, \infty)$ with $0 \leq f(x) \leq g(x)$ for all $x \geq a$. Then

1. $\displaystyle\int_a^\infty f(x)\,dx$    converges if    $\displaystyle\int_a^\infty g(x)\,dx$    converges.

2. $\displaystyle\int_a^\infty g(x)\,dx$    diverges if    $\displaystyle\int_a^\infty f(x)\,dx$    diverges.

---

**Proof**      The reasoning behind the argument establishing Theorem 2 is similar to that in Example 6. If $0 \leq f(x) \leq g(x)$ for $x \geq a$, then from Rule 7 in Theorem 2 of Section 5.3 we have

$$\int_a^b f(x)\,dx \leq \int_a^b g(x)\,dx, \qquad b > a.$$

From this it can be argued, as in Example 6, that

$$\int_a^{\infty} f(x)\,dx \qquad \text{converges if} \qquad \int_a^{\infty} g(x)\,dx \qquad \text{converges.}$$

Turning this around says that

$$\int_a^{\infty} g(x)\,dx \qquad \text{diverges if} \qquad \int_a^{\infty} f(x)\,dx \qquad \text{diverges.} \qquad \blacksquare$$

Although the theorem is stated for Type I improper integrals, a similar result is true for integrals of Type II as well.

**HISTORICAL BIOGRAPHY**
Karl Weierstrass
(1815–1897)

**EXAMPLE 7**      These examples illustrate how we use Theorem 2.

**(a)** $\displaystyle\int_1^{\infty} \frac{\sin^2 x}{x^2}\,dx$ converges because

$$0 \leq \frac{\sin^2 x}{x^2} \leq \frac{1}{x^2} \quad \text{on} \quad [1, \infty) \quad \text{and} \quad \int_1^{\infty} \frac{1}{x^2}\,dx \quad \text{converges.} \qquad \text{\small Example 3}$$

**(b)** $\displaystyle\int_1^{\infty} \frac{1}{\sqrt{x^2 - 0.1}}\,dx$ diverges because

$$\frac{1}{\sqrt{x^2 - 0.1}} \geq \frac{1}{x} \quad \text{on} \quad [1, \infty) \quad \text{and} \quad \int_1^{\infty} \frac{1}{x}\,dx \quad \text{diverges.} \qquad \text{\small Example 3}$$

**(c)** $\displaystyle\int_0^{\pi/2} \frac{\cos x}{\sqrt{x}}\,dx$ converges because

$$0 \leq \frac{\cos x}{\sqrt{x}} \leq \frac{1}{\sqrt{x}} \quad \text{on} \quad \left[0, \frac{\pi}{2}\right],$$

and

$$\int_0^{\pi/2} \frac{dx}{\sqrt{x}} = \lim_{a \to 0^+} \int_a^{\pi/2} \frac{dx}{\sqrt{x}}$$

$$= \lim_{a \to 0^+} \sqrt{4x}\,\Big]_a^{\pi/2} \qquad \text{\small $2\sqrt{x} = \sqrt{4x}$}$$

$$= \lim_{a \to 0^+} \left(\sqrt{2\pi} - \sqrt{4a}\right) = \sqrt{2\pi} \qquad \text{converges.} \qquad \blacksquare$$

---

**THEOREM 3—Limit Comparison Test**   If the positive functions $f$ and $g$ are continuous on $[a, \infty)$, and if

$$\lim_{x \to \infty} \frac{f(x)}{g(x)} = L, \qquad 0 < L < \infty,$$

then

$$\int_a^{\infty} f(x)\,dx \qquad \text{and} \qquad \int_a^{\infty} g(x)\,dx$$

both converge or both diverge.

We omit the proof of Theorem 3.

Although the improper integrals of two functions from $a$ to $\infty$ may both converge, this does not mean that their integrals necessarily have the same value, as the next example shows.

**EXAMPLE 8**  Show that

$$\int_1^\infty \frac{dx}{1 + x^2}$$

converges by comparison with $\int_1^\infty (1/x^2)\,dx$. Find and compare the two integral values.

**Solution**  The functions $f(x) = 1/x^2$ and $g(x) = 1/(1 + x^2)$ are positive and continuous on $[1, \infty)$. Also,

$$\lim_{x\to\infty} \frac{f(x)}{g(x)} = \lim_{x\to\infty} \frac{1/x^2}{1/(1 + x^2)} = \lim_{x\to\infty} \frac{1 + x^2}{x^2}$$

$$= \lim_{x\to\infty} \left(\frac{1}{x^2} + 1\right) = 0 + 1 = 1,$$

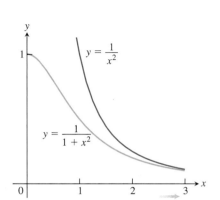

$y = \dfrac{1}{x^2}$

$y = \dfrac{1}{1 + x^2}$

**FIGURE 8.20** The functions in Example 8.

a positive finite limit (Figure 8.20). Therefore, $\displaystyle\int_1^\infty \frac{dx}{1 + x^2}$ converges because $\displaystyle\int_1^\infty \frac{dx}{x^2}$ converges.

The integrals converge to different values, however:

$$\int_1^\infty \frac{dx}{x^2} = \frac{1}{2 - 1} = 1 \qquad \text{Example 3}$$

and

$$\int_1^\infty \frac{dx}{1 + x^2} = \lim_{b\to\infty} \int_1^b \frac{dx}{1 + x^2}$$

$$= \lim_{b\to\infty} \left[\tan^{-1} b - \tan^{-1} 1\right] = \frac{\pi}{2} - \frac{\pi}{4} = \frac{\pi}{4}. \qquad \blacksquare$$

**EXAMPLE 9**  Investigate the convergence of $\displaystyle\int_1^\infty \frac{1 - e^{-x}}{x}\,dx$.

**Solution**  The integrand suggests a comparison of $f(x) = (1 - e^{-x})/x$ with $g(x) = 1/x$. However, we cannot use the Direct Comparison Test because $f(x) \le g(x)$ and the integral of $g(x)$ *diverges*. On the other hand, using the Limit Comparison Test we find that

$$\lim_{x\to\infty} \frac{f(x)}{g(x)} = \lim_{x\to\infty} \left(\frac{1 - e^{-x}}{x}\right)\left(\frac{x}{1}\right) = \lim_{x\to\infty} (1 - e^{-x}) = 1,$$

which is a positive finite limit. Therefore, $\displaystyle\int_1^\infty \frac{1 - e^{-x}}{x}\,dx$ diverges because $\displaystyle\int_1^\infty \frac{dx}{x}$ diverges. Approximations to the improper integral are given in Table 8.5. Note that the values do not appear to approach any fixed limiting value as $b \to \infty$.  $\blacksquare$

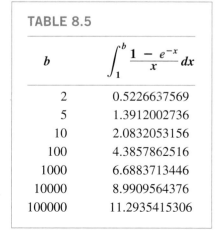

**TABLE 8.5**

| $b$ | $\displaystyle\int_1^b \frac{1 - e^{-x}}{x}\,dx$ |
|---|---|
| 2 | 0.5226637569 |
| 5 | 1.3912002736 |
| 10 | 2.0832053156 |
| 100 | 4.3857862516 |
| 1000 | 6.6883713446 |
| 10000 | 8.9909564376 |
| 100000 | 11.2935415306 |

# Exercises 8.8

### Evaluating Improper Integrals

The integrals in Exercises 1–34 converge. Evaluate the integrals without using tables.

1. $\displaystyle\int_0^\infty \frac{dx}{x^2 + 1}$

2. $\displaystyle\int_1^\infty \frac{dx}{x^{1.001}}$

3. $\displaystyle\int_0^1 \frac{dx}{\sqrt{x}}$

4. $\displaystyle\int_0^4 \frac{dx}{\sqrt{4 - x}}$

5. $\displaystyle\int_{-1}^1 \frac{dx}{x^{2/3}}$

6. $\displaystyle\int_{-8}^1 \frac{dx}{x^{1/3}}$

7. $\displaystyle\int_0^1 \frac{dx}{\sqrt{1 - x^2}}$

8. $\displaystyle\int_0^1 \frac{dr}{r^{0.999}}$

9. $\displaystyle\int_{-\infty}^{-2} \frac{2\,dx}{x^2 - 1}$

10. $\displaystyle\int_{-\infty}^2 \frac{2\,dx}{x^2 + 4}$

11. $\displaystyle\int_2^\infty \frac{2}{v^2 - v}\,dv$

12. $\displaystyle\int_2^\infty \frac{2\,dt}{t^2 - 1}$

13. $\displaystyle\int_{-\infty}^\infty \frac{2x\,dx}{(x^2 + 1)^2}$

14. $\displaystyle\int_{-\infty}^\infty \frac{x\,dx}{(x^2 + 4)^{3/2}}$

15. $\displaystyle\int_0^1 \frac{\theta + 1}{\sqrt{\theta^2 + 2\theta}}\,d\theta$

16. $\displaystyle\int_0^2 \frac{s + 1}{\sqrt{4 - s^2}}\,ds$

17. $\displaystyle\int_0^\infty \frac{dx}{(1 + x)\sqrt{x}}$

18. $\displaystyle\int_1^\infty \frac{1}{x\sqrt{x^2 - 1}}\,dx$

19. $\displaystyle\int_0^\infty \frac{dv}{(1 + v^2)(1 + \tan^{-1} v)}$

20. $\displaystyle\int_0^\infty \frac{16 \tan^{-1} x}{1 + x^2}\,dx$

21. $\displaystyle\int_{-\infty}^0 \theta e^\theta\,d\theta$

22. $\displaystyle\int_0^\infty 2e^{-\theta} \sin\theta\,d\theta$

23. $\displaystyle\int_{-\infty}^0 e^{-|x|}\,dx$

24. $\displaystyle\int_{-\infty}^\infty 2xe^{-x^2}\,dx$

25. $\displaystyle\int_0^1 x \ln x\,dx$

26. $\displaystyle\int_0^1 (-\ln x)\,dx$

27. $\displaystyle\int_0^2 \frac{ds}{\sqrt{4 - s^2}}$

28. $\displaystyle\int_0^1 \frac{4r\,dr}{\sqrt{1 - r^4}}$

29. $\displaystyle\int_1^2 \frac{ds}{s\sqrt{s^2 - 1}}$

30. $\displaystyle\int_2^4 \frac{dt}{t\sqrt{t^2 - 4}}$

31. $\displaystyle\int_{-1}^4 \frac{dx}{\sqrt{|x|}}$

32. $\displaystyle\int_0^2 \frac{dx}{\sqrt{|x - 1|}}$

33. $\displaystyle\int_{-1}^\infty \frac{d\theta}{\theta^2 + 5\theta + 6}$

34. $\displaystyle\int_0^\infty \frac{dx}{(x + 1)(x^2 + 1)}$

### Testing for Convergence

In Exercises 35–64, use integration, the Direct Comparison Test, or the Limit Comparison Test to test the integrals for convergence. If more than one method applies, use whatever method you prefer.

35. $\displaystyle\int_0^{\pi/2} \tan\theta\,d\theta$

36. $\displaystyle\int_0^{\pi/2} \cot\theta\,d\theta$

37. $\displaystyle\int_0^1 \frac{\ln x}{x^2}\,dx$

38. $\displaystyle\int_1^2 \frac{dx}{x \ln x}$

39. $\displaystyle\int_0^{\ln 2} x^{-2} e^{-1/x}\,dx$

40. $\displaystyle\int_0^1 \frac{e^{-\sqrt{x}}}{\sqrt{x}}\,dx$

41. $\displaystyle\int_0^\pi \frac{dt}{\sqrt{t + \sin t}}$

42. $\displaystyle\int_0^1 \frac{dt}{t - \sin t}$ (Hint: $t \geq \sin t$ for $t \geq 0$)

43. $\displaystyle\int_0^2 \frac{dx}{1 - x^2}$

44. $\displaystyle\int_0^2 \frac{dx}{1 - x}$

45. $\displaystyle\int_{-1}^1 \ln |x|\,dx$

46. $\displaystyle\int_{-1}^1 -x \ln |x|\,dx$

47. $\displaystyle\int_1^\infty \frac{dx}{x^3 + 1}$

48. $\displaystyle\int_4^\infty \frac{dx}{\sqrt{x - 1}}$

49. $\displaystyle\int_2^\infty \frac{dv}{\sqrt{v - 1}}$

50. $\displaystyle\int_0^\infty \frac{d\theta}{1 + e^\theta}$

51. $\displaystyle\int_0^\infty \frac{dx}{\sqrt{x^6 + 1}}$

52. $\displaystyle\int_2^\infty \frac{dx}{\sqrt{x^2 - 1}}$

53. $\displaystyle\int_1^\infty \frac{\sqrt{x + 1}}{x^2}\,dx$

54. $\displaystyle\int_2^\infty \frac{x\,dx}{\sqrt{x^4 - 1}}$

55. $\displaystyle\int_\pi^\infty \frac{2 + \cos x}{x}\,dx$

56. $\displaystyle\int_\pi^\infty \frac{1 + \sin x}{x^2}\,dx$

57. $\displaystyle\int_4^\infty \frac{2\,dt}{t^{3/2} - 1}$

58. $\displaystyle\int_2^\infty \frac{1}{\ln x}\,dx$

59. $\displaystyle\int_1^\infty \frac{e^x}{x}\,dx$

60. $\displaystyle\int_{e^x}^\infty \ln(\ln x)\,dx$

61. $\displaystyle\int_1^\infty \frac{1}{\sqrt{e^x - x}}\,dx$

62. $\displaystyle\int_1^\infty \frac{1}{e^x - 2^x}\,dx$

63. $\displaystyle\int_{-\infty}^\infty \frac{dx}{\sqrt{x^4 + 1}}$

64. $\displaystyle\int_{-\infty}^\infty \frac{dx}{e^x + e^{-x}}$

### Theory and Examples

65. Find the values of $p$ for which each integral converges.

a. $\displaystyle\int_1^2 \frac{dx}{x(\ln x)^p}$  b. $\displaystyle\int_2^\infty \frac{dx}{x(\ln x)^p}$

66. $\int_{-\infty}^\infty f(x)\,dx$ **may not equal** $\lim_{b\to\infty} \int_{-b}^b f(x)\,dx$  Show that

$$\int_0^\infty \frac{2x\,dx}{x^2 + 1}$$

diverges and hence that

$$\int_{-\infty}^\infty \frac{2x\,dx}{x^2 + 1}$$

diverges. Then show that

$$\lim_{b\to\infty} \int_{-b}^b \frac{2x\,dx}{x^2 + 1} = 0.$$

Exercises 67–70 are about the infinite region in the first quadrant between the curve $y = e^{-x}$ and the $x$-axis.

67. Find the area of the region.

68. Find the centroid of the region.

69. Find the volume of the solid generated by revolving the region about the $y$-axis.

**70.** Find the volume of the solid generated by revolving the region about the *x*-axis.

**71.** Find the area of the region that lies between the curves $y = \sec x$ and $y = \tan x$ from $x = 0$ to $x = \pi/2$.

**72.** The region in Exercise 71 is revolved about the *x*-axis to generate a solid.

   **a.** Find the volume of the solid.

   **b.** Show that the inner and outer surfaces of the solid have infinite area.

**73.** Evaluate the integrals.

   **a.** $\displaystyle\int_0^1 \frac{dt}{\sqrt{t}(1 + t)}$       **b.** $\displaystyle\int_0^\infty \frac{dt}{\sqrt{t}(1 + t)}$

**74.** Evaluate $\displaystyle\int_3^\infty \frac{dx}{x\sqrt{x^2 - 9}}$.

**75. Estimating the value of a convergent improper integral whose domain is infinite**

   **a.** Show that

$$\int_3^\infty e^{-3x}\, dx = \frac{1}{3}e^{-9} < 0.000042,$$

   and hence that $\int_3^\infty e^{-x^2}\, dx < 0.000042$. Explain why this means that $\int_0^\infty e^{-x^2}\, dx$ can be replaced by $\int_0^3 e^{-x^2}\, dx$ without introducing an error of magnitude greater than 0.000042.

   ⊤ **b.** Evaluate $\int_0^3 e^{-x^2}\, dx$ numerically.

**76. The infinite paint can or Gabriel's horn** As Example 3 shows, the integral $\int_1^\infty (dx/x)$ diverges. This means that the integral

$$\int_1^\infty 2\pi \frac{1}{x}\sqrt{1 + \frac{1}{x^4}}\, dx,$$

which measures the *surface area* of the solid of revolution traced out by revolving the curve $y = 1/x$, $1 \le x$, about the *x*-axis, diverges also. By comparing the two integrals, we see that, for every finite value $b > 1$,

$$\int_1^b 2\pi \frac{1}{x}\sqrt{1 + \frac{1}{x^4}}\, dx > 2\pi \int_1^b \frac{1}{x}\, dx.$$

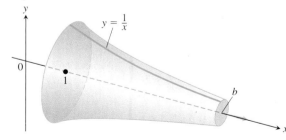

However, the integral

$$\int_1^\infty \pi\left(\frac{1}{x}\right)^2 dx$$

for the *volume* of the solid converges.

   **a.** Calculate it.

   **b.** This solid of revolution is sometimes described as a can that does not hold enough paint to cover its own interior. Think

about that for a moment. It is common sense that a finite amount of paint cannot cover an infinite surface. But if we fill the horn with paint (a finite amount), then we *will* have covered an infinite surface. Explain the apparent contradiction.

**77. Sine-integral function** The integral

$$\text{Si}(x) = \int_0^x \frac{\sin t}{t}\, dt,$$

called the *sine-integral function*, has important applications in optics.

   ⊤ **a.** Plot the integrand $(\sin t)/t$ for $t > 0$. Is the sine-integral function everywhere increasing or decreasing? Do you think $\text{Si}(x) = 0$ for $x > 0$? Check your answers by graphing the function $\text{Si}(x)$ for $0 \le x \le 25$.

   **b.** Explore the convergence of

$$\int_0^\infty \frac{\sin t}{t}\, dt.$$

   If it converges, what is its value?

**78. Error function** The function

$$\text{erf}(x) = \int_0^x \frac{2e^{-t^2}}{\sqrt{\pi}}\, dt,$$

called the *error function*, has important applications in probability and statistics.

   ⊤ **a.** Plot the error function for $0 \le x \le 25$.

   **b.** Explore the convergence of

$$\int_0^\infty \frac{2e^{-t^2}}{\sqrt{\pi}}\, dt.$$

   If it converges, what appears to be its value? You will see how to confirm your estimate in Section 15.4, Exercise 41.

**79. Normal probability distribution** The function

$$f(x) = \frac{1}{\sigma\sqrt{2\pi}} e^{-\frac{1}{2}\left(\frac{x-\mu}{\sigma}\right)^2}$$

is called the *normal probability density function* with mean $\mu$ and standard deviation $\sigma$. The number $\mu$ tells where the distribution is centered, and $\sigma$ measures the "scatter" around the mean. (See Section 8.9.)

From the theory of probability, it is known that

$$\int_{-\infty}^\infty f(x)\, dx = 1.$$

In what follows, let $\mu = 0$ and $\sigma = 1$.

   ⊤ **a.** Draw the graph of $f$. Find the intervals on which $f$ is increasing, the intervals on which $f$ is decreasing, and any local extreme values and where they occur.

   **b.** Evaluate

$$\int_{-n}^n f(x)\, dx$$

   for $n = 1, 2,$ and 3.

**c.** Give a convincing argument that

$$\int_{-\infty}^{\infty} f(x)\, dx = 1.$$

(*Hint:* Show that $0 < f(x) < e^{-x/2}$ for $x > 1$, and for $b > 1$,

$$\int_{b}^{\infty} e^{-x/2}\, dx \to 0 \quad \text{as} \quad b \to \infty.)$$

**80.** Show that if $f(x)$ is integrable on every interval of real numbers and $a$ and $b$ are real numbers with $a < b$, then

**a.** $\int_{-\infty}^{a} f(x)\, dx$ and $\int_{a}^{\infty} f(x)\, dx$ both converge if and only if

$\int_{-\infty}^{b} f(x)\, dx$ and $\int_{b}^{\infty} f(x)\, dx$ both converge.

**b.** $\int_{-\infty}^{a} f(x)\, dx + \int_{a}^{\infty} f(x)\, dx = \int_{-\infty}^{b} f(x)\, dx + \int_{b}^{\infty} f(x)\, dx$
when the integrals involved converge.

**COMPUTER EXPLORATIONS**

In Exercises 81–84, use a CAS to explore the integrals for various values of $p$ (include noninteger values). For what values of $p$ does the integral converge? What is the value of the integral when it does converge? Plot the integrand for various values of $p$.

**81.** $\int_{0}^{e} x^{p} \ln x\, dx$    **82.** $\int_{e}^{\infty} x^{p} \ln x\, dx$

**83.** $\int_{0}^{\infty} x^{p} \ln x\, dx$    **84.** $\int_{-\infty}^{\infty} x^{p} \ln |x|\, dx$

Use a CAS to evaluate the integrals.

**85.** $\int_{0}^{2/\pi} \sin \frac{1}{x}\, dx$    **86.** $\int_{0}^{2/\pi} x \sin \frac{1}{x}\, dx$

# 8.9 Probability

The outcome of some events, such as a heavy rock falling from a great height, can be modeled so that we can predict with high accuracy what will happen. On the other hand, many events have more than one possible outcome and which one of them will occur is uncertain. If we toss a coin, a head or a tail will result with each outcome being equally likely, but we do not know in advance which one it will be. If we randomly select and then weigh a person from a large population, there are many possible weights the person might have, and it is not certain whether the weight will be between 80 and 85 kg. We are told it is highly likely, but not known for sure, that an earthquake of magnitude 6.0 or greater on the Richter scale will occur near a major population area in California within the next one hundred years. Events having more than one possible outcome are *probabilistic* in nature, and when modeling them we assign a *probability* to the likelihood that a particular outcome may occur. In this section we show how calculus plays a central role in making predictions with probabilistic models.

## Random Variables

We begin our discussion with some familiar examples of uncertain events for which the collection of all possible outcomes is finite.

## EXAMPLE 1

**(a)** If we toss a coin once, there are two possible outcomes $\{H, T\}$, where H represents the coin landing head face up and T a tail landing face up. If we toss a coin three times, there are eight possible outcomes, taking into account the order in which a head or tail occurs. The set of outcomes is $\{HHH, HHT, HTH, THH, HTT, THT, TTH, TTT\}$.

**(b)** If we roll a six-sided die once, the set of possible outcomes is $\{1, 2, 3, 4, 5, 6\}$ representing the six faces of the die.

**(c)** If we select at random two cards from a 52-card deck, there are 52 possible outcomes for the first card drawn and then 51 possibilities for the second card. Since the order of the cards does not matter, there are $(52 \cdot 51)/2 = 1{,}326$ possible outcomes altogether. ∎

It is customary to refer to the set of all possible outcomes as the *sample space* for an event. With an uncertain event we are usually interested in which outcomes, if any, are more likely to occur than others, and to how large an extent. In tossing a coin three times, is it more likely that two heads or that one head will result? To answer such questions, we need a way to quantify the outcomes.

> **DEFINITION** A **random variable** is a function $X$ that assigns a numerical value to each outcome in a sample space.

Random variables that have only finitely many values are called **discrete** random variables. A **continuous random variable** can take on values in an entire interval, and it is associated with a *distribution function*, which we explain later.

## EXAMPLE 2

(a) Suppose we toss a coin three times giving the possible outcomes $\{$HHH, HHT, HTH, THH, HTT, THT, TTH, TTT$\}$. Define the random variable $X$ to be the number of heads that appear. So $X(\text{HHT}) = 2$, $X(\text{THT}) = 1$, and so forth. Since $X$ can only assume the values 0, 1, 2, or 3, it is a discrete random variable.

(b) We spin an arrow anchored by a pin located at the origin. The arrow can wind up pointing in any possible direction and we define the random variable $X$ as the radian angle the arrow makes with the positive $x$-axis, measured counterclockwise. In this case, $X$ is a continuous random variable that can take on any value in the interval $[0, 2\pi)$.

(c) The weight of a randomly selected person in a given population is a continuous random variable $W$. The cholesterol level of a randomly chosen person, and the waiting time for service of a person in a queue at a bank, are also continuous random variables.

(d) The scores on the national ACT Examination for college admissions in a particular year are described by a discrete random variable $S$ taking on integer values between 1 and 36. If the number of outcomes is large, or for reasons involving statistical analysis, discrete random variables such as test scores are often modeled as continuous random variables (Example 13).

(e) We roll a pair of dice and define the random variable $X$ to be the sum of the numbers on the top faces. This sum can only assume the integer values from 2 through 12, so $X$ is a discrete random variable.

(f) A tire company produces tires for mid-sized sedans. The tires are guaranteed to last for 50,000 kilometers, but some will fail sooner and some will last many more kilometers beyond 50,000 kilometers. The lifetime in kilometers of a tire is described by a continuous random variable $L$. ∎

## Probability Distributions

A *probability distribution* describes the probabilistic behavior of a random variable. Our chief interest is in probability distributions associated with continuous random variables, but to gain some perspective we first consider a distribution for a discrete random variable.

Suppose we toss a coin three times, with each side H or T equally likely to occur on a given toss. We define the discrete random variable $X$ that assigns the number of heads appearing in each outcome, giving

$$\{\text{HHH, HHT, HTH, HTT, THH, THT, TTH, TTT}\}$$

$$X \quad \downarrow \quad \downarrow \quad \downarrow \quad \downarrow \quad \downarrow \quad \downarrow \quad \downarrow \quad \downarrow$$

$$3 \quad 2 \quad 2 \quad 1 \quad 2 \quad 1 \quad 1 \quad 0$$

Next we count the *frequency* or number of times a specific value of $X$ occurs. Because each of the eight outcomes is equally likely to occur, we can calculate the probability of the random variable $X$ by dividing the frequency of each value by the total number of outcomes. We summarize our results as follows:

| Value of X | 0 | 1 | 2 | 3 |
|---|---|---|---|---|
| **Frequency** | 1 | 3 | 3 | 1 |
| *P(X)* | 1/8 | 3/8 | 3/8 | 1/8 |

**FIGURE 8.21** Probability bar graph for the random variable $X$ when tossing a fair coin three times.

We display this information in a probability bar graph of the discrete random variable $X$, as shown in Figure 8.21. The values of $X$ are portrayed by intervals of length 1 on the $x$-axis so the area of each bar in the graph is the probability of the corresponding outcome. For instance, the probability that exactly two heads occurs in the three tosses of the coin is the area of the bar associated with the value $X = 2$, which is $3/8$. Similarly, the probability that two or more heads occurs is the sum of areas of the bars associated with the values $X = 2$ and $X = 3$, or $4/8$. The probability that either zero or three heads occurs is $\frac{1}{8} + \frac{1}{8} = \frac{1}{4}$, and so forth. Note that the total area of all the bars in the graph is 1, which is the sum of all the probabilities for $X$.

With a continuous random variable, even when the outcomes are equally likely, we cannot simply count the number of outcomes in the sample space or the frequencies of outcomes that lead to a specific value of $X$. In fact, the probability that $X$ takes on any particular one of its values is zero. What *is* meaningful to ask is how probable it is that the random variable takes on a value within some specified *interval* of values.

We capture the information we need about the probabilities of $X$ in a function whose graph behaves much like the bar graph in Figure 8.21. That is, we take a nonnegative function $f$ defined over the range of the random variable with the property that the total area beneath the graph of $f$ is 1. The probability that a value of the random variable $X$ lies within some specified interval $[c, d]$ is then the area under the graph of $f$ over that interval. The following definition assumes the range of the continuous random variable $X$ is any real value, but the definition is general enough to account for random variables having a range of finite length.

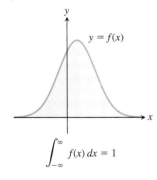

$$\int_{-\infty}^{\infty} f(x)\,dx = 1$$

---

**DEFINITIONS**    A **probability density function** for a continuous random variable is a function $f$ defined over $(-\infty, \infty)$ and having the following properties:

1. $f$ is continuous, except possibly at a finite number of points.

2. $f$ is nonnegative, so $f \geq 0$.

3. $\displaystyle\int_{-\infty}^{\infty} f(x)\,dx = 1.$

If $X$ is a continuous random variable with probability density function $f$, the **probability** that $X$ assumes a value in the interval between $X = c$ and $X = d$ is the area integral

$$P(c \leq X \leq d) = \int_{c}^{d} f(X)\,dX.$$

---

We note that the probability a continuous random variable $X$ assumes a particular real value $c$ is $P(X = c) = \int_{c}^{c} f(X)\,dX = 0$, consistent with our previous assertion. Since the area under the graph of $f$ over the interval $[c, d]$ is only a portion of the total area beneath the graph, the probability $P(c \leq X \leq d)$ is always a number between zero and one. Figure 8.22 illustrates a probability density function.

A probability density function for a random variable $X$ resembles the density function for a wire of varying density. To obtain the mass of a segment of the wire, we integrate the

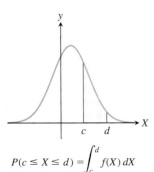

$$P(c \leq X \leq d) = \int_{c}^{d} f(X)\,dX$$

**FIGURE 8.22** A probability density function for the continuous random variable $X$.

density of the wire over an interval. To obtain the probability that a random variable has values in a particular interval, we integrate the probability density function over that interval.

**EXAMPLE 3** Let $f(x) = 2e^{-2x}$ if $0 \le x < \infty$ and $f(x) = 0$ for all negative values of $x$.

(a) Verify that $f$ is a probability density function.

(b) The time $T$ in hours until a car passes a spot on a remote road is described by the probability density function $f$. Find the probability $P(T \le 1)$ that a hitchhiker at that spot will see a car within one hour.

(c) Find the probability $P(T = 1)$ that a car passes by the spot after precisely one hour.

**Solution**

(a) The function $f$ is continuous except at $x = 0$, and is everywhere nonnegative. Moreover,

$$\int_{-\infty}^{\infty} f(x)\,dx = \int_{0}^{\infty} 2e^{-2x}\,dx = \lim_{b \to \infty} \int_{0}^{b} 2e^{-2x}\,dx = \lim_{b \to \infty} \left(1 - e^{-2b}\right) = 1.$$

So all of the conditions are satisfied and we have shown that $f$ is a probability density function.

(b) The probability that a car comes after a time lapse between zero and one hour is given by integrating the probability density function over the interval $[0, 1]$. So

$$P(T \le 1) = \int_{0}^{1} 2e^{-2T}\,dT = -e^{-2T}\Big]_{0}^{1} = 1 - e^{-2} \approx 0.865.$$

This result can be interpreted to mean that if 100 people were to hitchhike at that spot, about 87 of them can expect to see a car within one hour.

(c) This probability is the integral $\int_{1}^{1} f(T)\,dT$, which equals zero. We interpret this to mean that a sufficiently accurate measurement of the time until a car comes by the spot would have no possibility of being precisely equal to one hour. It might be very close, perhaps, but it would not be exactly one hour. ∎

We can extend the definition to finite intervals. If $f$ is a nonnegative function with at most finitely many discontinuities over the interval $[a, b]$, and its extension $F$ to $(-\infty, \infty)$, obtained by defining $F$ to be 0 outside of $[a, b]$, satisfies the definition for a probability density function, then $f$ is a **probability density function for $[a, b]$**. This means that $\int_{a}^{b} f(x)\,dx = 1$. Similar definitions can be made for the intervals $(a, b)$, $(a, b]$, and $[a, b)$.

**EXAMPLE 4** Show that $f(x) = \dfrac{4}{27}x^2(3 - x)$ is a probability density function over the interval $[0, 3]$.

**Solution** The function $f$ is continuous and nonnegative over $[0, 3]$. Also,

$$\int_{0}^{3} \frac{4}{27}x^2(3 - x)\,dx = \frac{4}{27}\left[x^3 - \frac{1}{4}x^4\right]_{0}^{3} = \frac{4}{27}\left(27 - \frac{81}{4}\right) = 1.$$

We conclude that $f$ is a probability density function over $[0, 3]$. ∎

## Exponentially Decreasing Distributions

The distribution in Example 3 is called an *exponentially decreasing probability density function*. These probability density functions always take on the form

$$f(x) = \begin{cases} 0 & \text{if } x < 0 \\ ce^{-cx} & \text{if } x \ge 0 \end{cases}$$

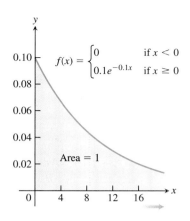

$$f(x) = \begin{cases} 0 & \text{if } x < 0 \\ 0.1e^{-0.1x} & \text{if } x \geq 0 \end{cases}$$

Area = 1

**FIGURE 8.23** An exponentially decreasing probability density function.

(see Exercise 23). Exponential density functions can provide models for describing random variables such as the lifetimes of light bulbs, radioactive particles, tooth crowns, and many kinds of electronic components. They also model the amount of time until some specific event occurs, such as the time until a pollinator arrives at a flower, the arrival times of a bus at a stop, the time between individuals joining a queue, the waiting time between phone calls at a help desk, and even the lengths of the phone calls themselves. A graph of an exponential density function is shown in Figure 8.23.

Random variables with exponential distributions are *memoryless*. If we think of $X$ as describing the lifetime of some object, then the probability that the object survives for at least $s + t$ hours, given that it has survived $t$ hours, is the same as the initial probability that it survives for at least $s$ hours. For instance, the current age $t$ of a radioactive particle does not change the probability that it will survive for at least another time period of length $s$. Sometimes the exponential distribution is used as a model when the memoryless principle is violated, because it provides reasonable approximations that are good enough for their intended use. For instance, this might be the case when predicting the lifetime of an artificial hip replacement or heart valve for a particular patient. Here is an application illustrating the exponential distribution.

**EXAMPLE 5** An electronics company models the lifetime $T$ in years of a chip they manufacture with the exponential density function

$$f(T) = \begin{cases} 0 & \text{if } T < 0 \\ 0.1e^{-0.1T} & \text{if } T \geq 0. \end{cases}$$

Using this model,

**(a)** Find the probability $P(T > 2)$ that a chip will last for more than two years.

**(b)** Find the probability $P(4 \leq T \leq 5)$ that a chip will fail in the fifth year.

**(c)** If 1000 chips are shipped to a customer, how many can be expected to fail within three years?

**Solution**

**(a)** The probability that a chip lasts at least two years is

$$P(T > 2) = \int_2^{\infty} 0.1e^{-0.1T}\, dT = \lim_{b \to \infty} \int_2^b 0.1e^{-0.1T}\, dT$$

$$= \lim_{b \to \infty} \left[ e^{-0.2} - e^{-0.1b} \right] = e^{-0.2} \approx 0.819.$$

That is, about 82% of the chips last more than two years.

**(b)** The probability is

$$P(4 \leq T \leq 5) = \int_4^5 0.1e^{-0.1T}\, dT = -e^{-0.1T}\Big|_4^5 = e^{-0.4} - e^{-0.5} \approx 0.064$$

which means that about 6% of the chips fail during the fifth year.

**(c)** We want the probability

$$P(0 \leq T \leq 3) = \int_0^3 0.1e^{-0.1T}\, dT = -e^{-0.1T}\Big|_0^3 = 1 - e^{-0.3} \approx 0.259.$$

We can expect that about 259 of the 1000 chips will fail within three years. ∎

### Expected Values, Means, and Medians

Suppose the weight in kg of a steer raised on a cattle ranch is a continuous random variable $W$ with probability density function $f(W)$ and that the rancher can sell a steer for $g(W)$ dollars. How much can the rancher expect to earn for a randomly chosen steer on the ranch?

To answer this question, we consider a small interval $[W_i, W_{i+1}]$ of width $\Delta W_i$ and note that the probability a steer has weight in this interval is

$$\int_{W_i}^{W_{i+1}} f(W)\, dW \approx f(W_i)\, \Delta W_i.$$

The earning on a steer in this interval is approximately $g(W_i)$. The Riemann sum

$$\sum g(W_i)\, f(W_i)\, \Delta W_i$$

then approximates the amount the rancher would receive for a steer. We assume that steers have a maximum weight, so $f$ is zero outside some finite interval $[0, b]$. Then taking the limit of the Riemann sum as the width of each interval approaches zero gives the integral

$$\int_{-\infty}^{\infty} g(W)\, f(W)\, dW.$$

This integral estimates how much the rancher can expect to earn for a typical steer on the ranch and is the *expected value of the function g*.

The expected values of certain functions of a random variable $X$ have particular importance in probability and statistics. One of the most important of these functions is the expected value of the function $g(X) = X$.

---

**DEFINITION** The **expected value** or **mean** of a continuous random variable $X$ with probability density function $f$ is the number

$$\mu = \mathrm{E}(X) = \int_{-\infty}^{\infty} X f(X)\, dx.$$

---

The expected value $\mathrm{E}(X)$ can be thought of as a weighted average of the random variable $X$, where each value of $X$ is weighted by $f(X)$. The mean can also be interpreted as the long-run average value of the random variable $X$, and it is one measure of the centrality of the random variable $X$.

**EXAMPLE 6** Find the mean of the random variable $X$ with exponential probability density function

$$f(x) = \begin{cases} 0 & \text{if } X < 0 \\ ce^{-cX} & \text{if } X \geq 0. \end{cases}$$

**Solution** From the definition we have

$$\mu = \int_{-\infty}^{\infty} X f(X)\, dX = \int_{0}^{\infty} X c e^{-cX}\, dX$$

$$= \lim_{b\to\infty} \int_0^b X c e^{-cX}\, dX = \lim_{b\to\infty}\left(-X e^{-cX}\Big]_0^b + \int_0^b e^{-cX}\, dX\right)$$

$$= \lim_{b\to\infty}\left(-b e^{-cb} - \frac{1}{c}e^{-cb} + \frac{1}{c}\right) = \frac{1}{c}. \qquad \text{l'Hôpital's Rule on first term}$$

Therefore, the mean is $\mu = 1/c$. ∎

From the result in Example 6, knowing the mean or expected value $\mu$ of a random variable $X$ having an exponential density function allows us to write its entire formula.

Exponential Density Function for a Random Variable $X$ with Mean $\mu$

$$f(X) = \begin{cases} 0 & \text{if } X < 0 \\ \mu^{-1}e^{-X/\mu} & \text{if } X \geq 0 \end{cases}$$

**EXAMPLE 7**    Suppose the time $T$ before a chip fails in Example 5 is modeled instead by the exponential density function with a mean of eight years. Find the probability that a chip will fail within five years.

**Solution**    The exponential density function with mean $\mu = 8$ is

$$f(T) = \begin{cases} 0 & \text{if } T < 0 \\ \dfrac{1}{8}e^{-T/8} & \text{if } T \geq 0 \end{cases}$$

Then the probability a chip will fail within five years is the definite integral

$$P(0 \leq T \leq 5) = \int_0^5 0.125e^{-0.125T}\, dT = -e^{-0.125T}\Big]_0^5 = 1 - e^{-0.625} \approx 0.465$$

so about 47% of the chips can be expected to fail within five years.    ■

**EXAMPLE 8**    Find the expected value for the random variable $X$ with probability density function given by Example 4.

**Solution**    The expected value is

$$\mu = E(X) = \int_0^3 \frac{4}{27}X^3(3-X)\, dX = \frac{4}{27}\left[\frac{3}{4}X^4 - \frac{1}{5}X^5\right]_0^3$$

$$= \frac{4}{27}\left(\frac{243}{4} - \frac{243}{5}\right) = 1.8$$

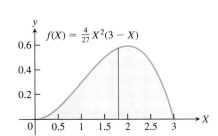

**FIGURE 8.24**    The expected value of a random variable with this probability density function is $\mu = 1.8$ (Example 8).

From Figure 8.24, you can see that this expected value is reasonable because the region beneath the probability density function appears to be balanced about the vertical line $X = 1.8$. That is, the horizontal coordinate of the centroid of a plate described by the region is $\overline{X} = 1.8$.    ■

There are other ways to measure the centrality of a random variable with a given probability density function.

**DEFINITION**    The **median** of a continuous random variable $X$ with probability density function $f$ is the number $m$ for which

$$\int_{-\infty}^{m} f(X)\, dX = \frac{1}{2} \qquad \text{and} \qquad \int_{m}^{\infty} f(X)\, dX = \frac{1}{2}.$$

The definition of the median means that there is an equal likelihood that the random variable $X$ will be smaller than $m$ or larger than $m$.

**EXAMPLE 9**    Find the median of a random variable $X$ with exponential probability density function

$$f(X) = \begin{cases} 0 & \text{if } X < 0 \\ ce^{-cX} & \text{if } X \geq 0. \end{cases}$$

**Solution**    The median $m$ must satisfy

$$\frac{1}{2} = \int_0^m ce^{-cX}\,dX = -e^{-cX}\Big]_0^m = 1 - e^{-cm}.$$

It follows that

$$e^{-cm} = \frac{1}{2} \quad \text{or} \quad m = \frac{1}{c}\ln 2.$$

Also,

$$\frac{1}{2} = \lim_{b \to \infty} \int_m^b ce^{-cX}\,dX = \lim_{b \to \infty}\left(-e^{-cX}\right]_m^b = \lim_{b \to \infty}\left(e^{-cm} - e^{-cb}\right) = e^{-cm}$$

giving the same value for $m$. Since $1/c$ is the mean $\mu$ of $X$ with an exponential distribution, we conclude that the median is $m = \mu \ln 2$. The mean and median differ because the probability density function is skewed and spreads toward the right.    ■

### Variance and Standard Deviation

Random variables with exactly the same mean $\mu$ but different distributions can behave very differently (see Figure 8.25). The *variance* of a random variable $X$ measures how spread out the values of $X$ are in relation to the mean, and we measure this dispersion by the expected value of $(X - \mu)^2$. Since the variance measures the expected square of the difference from the mean, we often work instead with its square root.

**FIGURE 8.25** Probability density functions with the same mean can have different spreads in relation to the mean. The blue and red regions under the curves have equal area.

> **DEFINITIONS**    The **variance** of a random variable $X$ with probability density function $f$ is the expected value of $(X - \mu)^2$:
>
> $$\text{Var}(X) = \int_{-\infty}^{\infty} (X - \mu)^2 f(X)\,dX$$
>
> The **standard deviation** of $X$ is
>
> $$\sigma_X = \sqrt{\text{Var}(X)} = \sqrt{\int_{-\infty}^{\infty} (X - \mu)^2 f(X)\,dX}.$$

**EXAMPLE 10**    Find the standard deviation of the random variable $T$ in Example 5, and find the probability that $T$ lies within one standard deviation of the mean.

**Solution**    The probability density function is the exponential density function with mean $\mu = 10$ by Example 6. To find the standard deviation we first calculate the variance integral:

$$\int_{-\infty}^{\infty} (T - \mu)^2 f(T)\,dT = \int_0^{\infty} (T - 10)^2(0.1e^{-0.1T})\,dT$$

$$= \lim_{b \to \infty} \int_0^b (T - 10)^2(0.1e^{-0.1T})\,dT$$

$$= \lim_{b \to \infty} \left[ \left( -(T - 10)^2 - 20(T - 10) \right) e^{-0.1T} \right]_0^b$$

$$+ \lim_{b \to \infty} \int_0^b 20 e^{-0.1T} \, dT \qquad \text{\small Integrating by parts}$$

$$= \left[ 0 + (-10)^2 + 20(-10) \right] - 20 \lim_{b \to \infty} \left( 10 e^{-0.1T} \right)_0^b$$

$$= -100 - 200 \lim_{b \to \infty} \left( e^{-0.1b} - 1 \right) = 100.$$

The standard deviation is the square root of the variance, so $\sigma = 10.0$.

To find the probability that $T$ lies within one standard deviation of the mean, we find the probability $P(\mu - \sigma \le T \le \mu + \sigma)$. For this example, we have

$$P(10 - 10 \le T \le 10 + 10) = \int_0^{20} 0.1 e^{-0.1T} \, dT = -e^{-0.1T} \Big]_0^{20} = 1 - e^{-2} \approx 0.865$$

This means that about 87% of the chips will fail within twenty years. ∎

### Uniform Distributions

The **uniform distribution** is very simple, but it occurs commonly in applications. The probability density function for this distribution on the interval $[a, b]$ is

$$f(x) = \frac{1}{b - a}, \quad a \le x \le b.$$

If each outcome in the sample space is equally likely to occur, then the random variable $X$ has a uniform distribution. Since $f$ is constant on $[a, b]$, a random variable with a uniform distribution is just as likely to be in one subinterval of a fixed length as in any other of the same length. The probability that $X$ assumes a value in a subinterval of $[a, b]$ is the length of that subinterval divided by $(b - a)$.

**EXAMPLE 11**    An anchored arrow is spun around the origin, and the random variable $X$ is the radian angle the arrow makes with the positive $x$-axis, measured within the interval $[0, 2\pi)$. Assuming there is equal probability for the arrow pointing in any direction, find the probability density function and the probability that the arrow ends up pointing between North and East.

**Solution**    We model the probability density function with the uniform distribution $f(x) = 1/2\pi$, $0 \le x < 2\pi$, and $f(x) = 0$ elsewhere.

The probability that the arrow ends up pointing between North and East is given by

$$P\left( 0 \le X \le \frac{\pi}{2} \right) = \int_0^{\pi/2} \frac{1}{2\pi} \, dx = \frac{1}{4}. \qquad ∎$$

### Normal Distributions

Numerous applications use the **normal distribution,** which is defined by the probability density function

$$f(x) = \frac{1}{\sigma \sqrt{2\pi}} e^{-(x - \mu)^2 / 2\sigma^2}.$$

It can be shown that the mean of a random variable $X$ with this probability density function is $\mu$ and its standard deviation is $\sigma$. In applications the values of $\mu$ and $\sigma$ are often estimated using large sets of data. The function is graphed in Figure 8.26, and the graph is

$$f(x) = \frac{1}{\sigma \sqrt{2\pi}} e^{-(x - \mu)^2 / 2\sigma^2}$$

**FIGURE 8.26**  The normal probability density function with mean $\mu$ and standard deviation $\sigma$.

sometimes called a *bell curve* because of its shape. Since the curve is symmetric about the mean, the median for $X$ is the same as its mean. It is often observed in practice that many random variables have approximately a normal distribution. Some examples illustrating this phenomenon are the height of a man, the annual rainfall in a certain region, an individual's blood pressure, the serum cholesterol level in the blood, the brain weights in a certain population of adults, and the amount of growth in a given period for a population of sunflower seeds.

The normal probability density function does not have an antiderivative expressible in terms of familiar functions. Once $\mu$ and $\sigma$ are fixed, however, an integral involving the normal probability density function can be computed using numerical integration methods. Usually we use the numerical integration capability of a computer or calculator to estimate the values of these integrals. Such computations show that for any normal distribution, we get the following values for the probability that the random variable $X$ lies within $k = 1, 2, 3$, or 4 standard deviations of the mean:

$$P(\mu - \sigma < X < \mu + \sigma) \approx 0.68269$$
$$P(\mu - 2\sigma < X < \mu + 2\sigma) \approx 0.95450$$
$$P(\mu - 3\sigma < X < \mu + 3\sigma) \approx 0.99730$$
$$P(\mu - 4\sigma < X < \mu + 4\sigma) \approx 0.99994$$

This means, for instance, that the random variable $X$ will take on a value within two standard deviations of the mean about 95% of the time. About 68% of the time, $X$ will lie within one standard deviation of the mean (see Figure 8.27).

**FIGURE 8.27** Probabilities of the normal distribution within its standard deviation bands.

**EXAMPLE 12**    An individual's blood pressure is an important indicator of overall health. A medical study of healthy individuals between 14 and 70 years of age modeled their systolic blood pressure using a normal distribution with mean 119.7 mm Hg and standard deviation 10.9 mm Hg.

**(a)** Using this model, what percentage of the population has a systolic blood pressure between 140 and 160 mm Hg, the levels set by the American Heart Association for Stage 1 hypertension?

**(b)** What percentage has a blood pressure between 160 and 180 mm Hg, the levels set by the American Heart Association for Stage 2 hypertension?

**(c)** What percentage has a blood pressure in the normal range of $90 - 120$, as set by the American Heart Association?

**Solution**

**(a)** Since we cannot find an antiderivative, we use a computer to evaluate the probability integral of the normal probability density function with $\mu = 119.7$ and $\sigma = 10.9$:

$$P(140 \leq X \leq 160) = \int_{140}^{160} \frac{1}{10.9\sqrt{2\pi}} e^{-(X-119.7)^2/2(10.9)^2} \, dX \approx 0.03117.$$

This means that about 3% of the population in the studied age range have Stage 1 hypertension.

**(b)** Again we use a computer to calculate the probability that the blood pressure is between 160 and 180 mm Hg:

$$P(160 \leq X \leq 180) = \int_{160}^{180} \frac{1}{10.9\sqrt{2\pi}} e^{-(X-119.7)^2/2(10.9)^2} \, dX \approx 0.00011.$$

We conclude that about 0.011% of the population has Stage 2 hypertension.

**(c)** The probability that the blood pressure falls in the normal range is

$$P(90 \leq X \leq 120) = \int_{90}^{120} \frac{1}{10.9\sqrt{2\pi}} e^{-(X-119.7)^2/2(10.9)^2} \, dX \approx 0.50776.$$

That is, about 51% of the population has a normal systolic blood pressure.    ■

Many national tests are standardized using the normal distribution. The following example illustrates modeling the discrete random variable for scores on a test using the normal distribution function for a continuous random variable.

**EXAMPLE 13**    The ACT is a standardized test taken by high school students seeking admission to many colleges and universities. The test measures knowledge skills and proficiency in the areas of English, math, and science, with scores ranging over the interval $[1, 36]$. Nearly 1.5 million high school students took the test in 2009, and the composite mean score across the academic areas was $\mu = 21.1$ with standard deviation $\sigma = 5.1$.

(a) What percentage of the population had an ACT score between 18 and 24?

(b) What is the ranking of a student who scored 27 on the test?

(c) What is the minimal integer score a student needed to get in order to be in the top 8% of the scoring population?

**Solution**

(a) We use a computer to evaluate the probability integral of the normal probability density function with $\mu = 21.1$ and $\sigma = 5.1$:

$$P(18 \leq X \leq 24) = \int_{18}^{24} \frac{1}{5.1\sqrt{2\pi}} e^{-(X-21.1)^2/2(5.1)^2} \, dX \approx 0.44355.$$

This means that about 44% of the students had an ACT score between 18 and 24.

(b) Again we use a computer to calculate the probability of a student getting a score lower than 27 on the test:

$$P(1 \leq X < 27) = \int_{1}^{27} \frac{1}{5.1\sqrt{2\pi}} e^{-(X-21.1)^2/2(5.1)^2} \, dX \approx 0.87630.$$

We conclude that about 88% of the students scored below a score of 27, so the student ranked in the top 12% of the population.

(c) We look at how many students had a mark higher than 28:

$$P(28 < X \leq 36) = \int_{28}^{36} \frac{1}{5.1\sqrt{2\pi}} e^{-(X-21.1)^2/2(5.1)^2} \, dX \approx 0.0863.$$

Since this number gives more than 8% of the students, we look at the next higher integer score:

$$P(29 < X \leq 36) = \int_{29}^{36} \frac{1}{5.1\sqrt{2\pi}} e^{-(X-21.1)^2/2(5.1)^2} \, dX \approx 0.0595.$$

Therefore, 29 is the lowest integer score a student could get in order to score in the top 8% of the population (and actually scoring here in the top 6%). ◼

The simplest form for a normal distribution of $X$ occurs when its mean is zero and its standard deviation is one. The *standard normal probability density function f* giving mean $\mu = 0$ and standard deviation $\sigma = 1$ is

$$f(X) = \frac{1}{\sqrt{2\pi}} e^{-X^2/2}.$$

Note that the substitution $z = (X - \mu)/\sigma$ gives the equivalent integrals

$$\int_{a}^{b} \frac{1}{\sigma\sqrt{2\pi}} e^{-((X-\mu)/\sigma)^2/2} \, dX = \int_{\alpha}^{\beta} \frac{1}{\sqrt{2\pi}} e^{-z^2/2} \, dz,$$

where $\alpha = (a - \mu)/\sigma$ and $\beta = (b - \mu)/\sigma$. So we can convert random variable values to the "z-values" to standardize a normal distribution, and then use the integral on the right-hand side of the last equation to calculate probabilities for the original random variable normal distribution with mean $\mu$ and standard deviation $\sigma$. In a normal distribution, we know that 95.5% of the population lies within two standard deviations of the mean, so a random variable $X$ converted to a $z$-value has more than a 95% chance of occurring in the interval $[-2, 2]$.

## Exercises 8.9

### Probability Density Functions

In Exercises 1–8, determine which are probability density functions and justify your answer.

**1.** $f(x) = \dfrac{1}{18}x$ over $[4, 8]$

**2.** $f(x) = \dfrac{1}{2}(2 - x)$ over $[0, 2]$

**3.** $f(x) = 2^x$ over $\left[0, \dfrac{\ln(1 + \ln 2)}{\ln 2}\right]$

**4.** $f(x) = x - 1$ over $[0, 1 + \sqrt{3}]$

**5.** $f(x) = \begin{cases} \dfrac{1}{x^2} & x \geq 1 \\ 0 & x < 1 \end{cases}$

**6.** $f(x) = \begin{cases} \dfrac{8}{\pi(4 + x^2)} & x \geq 0 \\ 0 & x < 0 \end{cases}$

**7.** $f(x) = 2\cos 2x$ over $\left[0, \dfrac{\pi}{4}\right]$

**8.** $f(x) = \dfrac{1}{x}$ over $(0, e]$

**9.** Let $f$ be the probability density function for the random variable $L$ in Example 2f. Explain the meaning of each integral.

**a.** $\displaystyle\int_{25,000}^{32,000} f(L)\, dL$    **b.** $\displaystyle\int_{30,000}^{\infty} f(L)\, dL$

**c.** $\displaystyle\int_{0}^{20,000} f(L)\, dL$    **d.** $\displaystyle\int_{-\infty}^{15,000} f(L)\, dL$

**10.** Let $f(X)$ be the uniform distribution for the random variable $X$ in Example 11. Express the following probabilities as integrals.

**a.** The probability that the arrow points either between South and West or between North and West.

**b.** The probability that the arrow makes an angle of at least 2 radians.

Verify that the functions in Exercises 11–16 are probability density functions for a continuous random variable $X$ over the given interval. Determine the specified probability.

**11.** $f(x) = xe^{-x}$ over $[0, \infty)$, $P(1 \leq X \leq 3)$

**12.** $f(x) = \dfrac{\ln x}{x^2}$ over $[1, \infty)$, $P(2 < X < 15)$

**13.** $f(x) = \dfrac{3}{2}x(2 - x)$ over $[0, 1]$, $P(0.5 > X)$

**14.** $f(x) = \dfrac{\sin^2 \pi x}{\pi x^2}$ over $\left[\dfrac{200}{1059}, \infty\right)$, $P(X < \pi/6)$

**15.** $f(x) = \begin{cases} \dfrac{2}{x^3} & x > 1 \\ 0 & x \leq 1 \end{cases}$ over $(-\infty, \infty)$, $P(4 \leq X < 9)$

**16.** $f(x) = \sin x$ over $[0, \pi/2]$, $P\left(\dfrac{\pi}{6} < X \leq \dfrac{\pi}{4}\right)$

In Exercises 17–20, find the value of the constant $c$ so that the given function is a probability density function for a random variable over the specified interval.

**17.** $f(x) = \dfrac{1}{6}x$ over $[3, c]$    **18.** $f(x) = \dfrac{1}{x}$ over $[c, c + 1]$

**19.** $f(x) = 4e^{-2x}$ over $[0, c]$

**20.** $f(x) = cx\sqrt{25 - x^2}$ over $[0, 5]$

**21.** Let $f(x) = \dfrac{c}{1 + x^2}$. Find the value of $c$ so that $f$ is a probability density function. If $f$ is a probability density function for the random variable $X$, find the probability $P(1 \leq X < 2)$.

**22.** Find the value of $c$ so that $f(x) = c\sqrt{x}(1 - x)$ is a probability density function for the random variable $X$ over $[0, 1]$, and find the probability $P(0.25 \leq X \leq 0.5)$.

**23.** Show that if the exponentially decreasing function
$$f(x) = \begin{cases} 0 & \text{if } x < 0 \\ Ae^{-cx} & \text{if } x \geq 0 \end{cases}$$
is a probability density function, then $A = c$.

**24.** Suppose $f$ is a probability density function for the random variable $X$ with mean $\mu$. Show that its variance satisfies
$$\text{Var}(X) = \int_{-\infty}^{\infty} X^2 f(X)\, dX - \mu^2.$$

Compute the mean and median for a random variable with the probability density functions in Exercises 25–28.

**25.** $f(x) = \dfrac{1}{8}x$ over $[0, 4]$    **26.** $f(x) = \dfrac{1}{9}x^2$ over $[0, 3]$

**27.** $f(x) = \begin{cases} \dfrac{2}{x^3} & x \geq 1 \\ 0 & x < 1 \end{cases}$    **28.** $f(x) = \begin{cases} \dfrac{1}{x} & 1 \leq x \leq e \\ 0 & \text{Otherwise} \end{cases}$

## Exponential Distributions

**29. Digestion time**   The digestion time in hours of a fixed amount of food is exponentially distributed with a mean of 1 hour. What is the probability that the food is digested in less than 30 minutes?

**30. Pollinating flowers**   A biologist models the time in minutes until a bee arrives at a flowering plant with an exponential distribution having a mean of 4 minutes. If 1000 flowers are in a field, how many can be expected to be pollinated within 5 minutes?

**31. Lifetime of light bulbs**   A manufacturer of light bulbs finds that the mean lifetime of a bulb is 1200 hours. Assume the life of a bulb is exponentially distributed.

**a.** Find the probability that a bulb will last less than its guaranteed lifetime of 1000 hours.

**b.** In a batch of light bulbs, what is the expected time until half the light bulbs in the batch fail?

**32. Lifetime of an electronic component**   The life expectancy in years of a component in a microcomputer is exponentially distributed, and $1/3$ of the components fail in the first 3 years. The company that manufactures the component offers a 1 year warranty. What is the probability that a component will fail during the warranty period?

**33. Lifetime of an organism**   A *hydra* is a small fresh-water animal, and studies have shown that its probability of dying does not increase with the passage of time. The lack of influence of age on mortality rates for this species indicates that an exponential distribution is an appropriate model for the mortality of hydra. A biologist studies a population of 500 hydra and observes that 200 of them die within the first 2 years. How many of the hydra would you expect to die within the first six months?

**34. Car accidents**   The number of days that elapse between the beginning of a calendar year and the moment a high-risk driver is involved in an accident is exponentially distributed. Based on historical data, an insurance company expects that 30% of high-risk drivers will be involved in an accident during the first 50 days of the calendar year. In a group of 100 high-risk drivers, how many do you expect to be involved in an accident during the first 80 days of the calendar year?

**35. Customer service time**   The mean waiting time to get served after walking into a bakery is 30 seconds. Assume that an exponential density function describes the waiting times.

**a.** What is the probability a customer waits 15 seconds or less?

**b.** What is the probability a customer waits longer than one minute?

**c.** What is the probability a customer waits exactly 5 minutes?

**d.** If 200 customers come to the bakery in a day, how many are likely to be served within three minutes?

**36. Airport waiting time**   According to the U.S. Customs and Border Protection Agency, the average airport wait time at Chicago's O'Hare International airport is 16 minutes for a traveler arriving during the hours 7–8 A.M., and 32 minutes for arrival during the hours 4–5 P.M. The wait time is defined as the total processing time from arrival at the airport until the completion of a passenger's security screening. Assume the wait time is exponentially distributed.

**a.** What is the probability of waiting between 10 and 30 minutes for a traveler arriving during the 7–8 A.M. hour?

**b.** What is the probability of waiting more than 25 minutes for a traveler arriving during the 7–8 A.M. hour?

**c.** What is the probability of waiting between 35 and 50 minutes for a traveler arriving during the 4–5 P.M. hour?

**d.** What is the probability of waiting less than 20 minutes for a traveler arriving during the 4–5 P.M. hour?

**37. Printer lifetime**   The lifetime of a $200 printer is exponentially distributed with a mean of 2 years. The manufacturer agrees to pay a full refund to a buyer if the printer fails during the first year following its purchase, and a one-half refund if it fails during the second year. If the manufacturer sells 100 printers, how much should it expect to pay in refunds?

**38. Failure time**   The time between failures of a photocopier is exponentially distributed. Half of the copiers at a university require service during the first 2 years of operations. If the university purchased 150 copiers, how many do you expect to require service during the first year of their operation?

## Normal Distributions

**39. Cholesterol levels**   The serum cholesterol levels of children aged 12 to 14 years follows a normal distribution with mean $\mu = 4.19$ mmol/L and standard deviation $\sigma = 0.72$ mmol/L. In a population of 1000 of these children, how many would you expect to have serum cholesterol levels between 4.27 and 4.99? between 3.83 and 4.32?

**40. Annual rainfall**   The annual rainfall in mm for San Francisco, California, is approximately a normal random variable with mean 510 mm and standard deviation 120 mm. What is the probability that next year's rainfall will exceed 444 mm?

**41. Manufacturing time**   The assembly time in minutes for a component at an electronic manufacturing plant is normally distributed with a mean of $\mu = 55$ and standard deviation $\sigma = 4$. What is the probability that a component will be made in less than one hour?

**42. Lifetime of a tire**   Assume the random variable $L$ in Example 2f is normally distributed with mean $\mu = 35,000$ kilometers and $\sigma = 6,000$ kilometers.

**a.** In a batch of 4000 tires, how many can be expected to last for at least 29,000 kilometers?

**b.** What is the minimum number of kilometers you would expect to find as the lifetime for 90% of the tires?

**43. Height**   The average height of American females aged 18–24 is normally distributed with mean $\mu = 166$ cm and $\sigma = 6$ cm.

**a.** What percentage of females are taller than 172 cm?

**b.** What is the probability a female is between 155 cm and 163 cm tall?

**44. Life expectancy**   At birth, a French citizen has an average life expectancy of 81 years with a standard deviation of 7 years. If 100 newly born French babies are selected at random, how many would you expect to live between 75 and 85 years? Assume life expectancy is normally distributed.

**45. Length of pregnancy**   A team of medical practitioners determines that in a population of 1000 females with ages ranging from 20 to 35 years, the length of pregnancy from conception to birth is approximately normally distributed with a mean of

266 days and a standard deviation of 16 days. How many of these females would you expect to have a pregnancy lasting from 36 weeks to 40 weeks?

46. **Brain weights**   In a population of 500 adult Swedish males, medical researchers find their brain weights to be approximately normally distributed with mean $\mu = 1400$ g and standard deviation $\sigma = 100$ g.

   a. What percentage of brain weights are between 1325 and 1450 g?

   b. How many males in the population would you expect to have a brain weight exceeding 1480 g?

47. **Blood pressure**   Diastolic blood pressure in adults is normally distributed with $\mu = 80$ mm Hg and $\sigma = 12$ mm Hg. In a random sample of 300 adults, how many would be expected to have a diastolic blood pressure below 70 mm Hg?

48. **Albumin levels**   Serum albumin in healthy 20-year-old males is normally distributed with $\mu = 4.4$ and $\sigma = 0.2$. How likely is it for a healthy 20-year-old male to have a level in the range 4.3 to 4.45?

49. **Quality control**   A manufacturer of generator shafts finds that it needs to add additional weight to its shafts in order to achieve proper static and dynamic balance. Based on experimental tests, the average weight it needs to add is $\mu = 35$ g with $\sigma = 9$ g. Assuming a normal distribution, from 1000 randomly selected shafts, how many would be expected to need an added weight in excess of 40 g?

50. **Kilometers driven**   A taxicab company in New York City analyzed the daily number of kilometers driven by each of its drivers. It found the average distance was 300 km with a standard deviation of 50 km. Assuming a normal distribution, what prediction can we make about the percentage of drivers who will log in either more than 400 km or less than 250 km?

51. **Germination of sunflower seeds**   The germination rate of a particular seed is the percentage of seeds in the batch which successfully emerge as plants. Assume that the germination rate for a batch of sunflower seeds is 80%, and that among a large population of $n$ seeds the number of successful germinations is normally distributed with mean $\mu = 0.8n$ and $\sigma = 0.4\sqrt{n}$.

   a. In a batch of $n = 2500$ seeds, what is the probability that at least 1960 will successfully germinate?

   b. In a batch of $n = 2500$ seeds, what is the probability that at most 1980 will successfully germinate?

   c. In a batch of $n = 2500$ seeds, what is the probability that between 1940 and 2020 will successfully germinate?

52. Suppose you toss a fair coin $n$ times and record the number of heads that land. Assume that $n$ is large and approximate the discrete random variable $X$ with a continuous random variable that is normally distributed with $\mu = n/2$ and $\sigma = \sqrt{n}/2$. If $n = 400$, find the given probabilities.

   a. $P(190 \le X < 210)$         b. $P(X < 170)$

   c. $P(X > 220)$                  d. $P(X = 300)$

### Discrete Random Variables

53. A fair coin is tossed four times and the random variable $X$ assigns the number of tails that appear in each outcome.

   a. Determine the set of possible outcomes.

   b. Find the value of $X$ for each outcome.

   c. Create a probability bar graph for $X$, as in Figure 8.21. What is the probability that at least two heads appear in the four tosses of the coin?

54. You roll a pair of six-sided dice, and the random variable $X$ assigns to each outcome the sum of the number of dots showing on each face, as in Example 2e.

   a. Find the set of possible outcomes.

   b. Create a probability bar graph for $X$.

   c. What is the probability that $X = 8$?

   d. What is the probability that $X \le 5$? $X > 9$?

55. Three people are asked their opinion in a poll about a particular brand of a common product found in grocery stores. They can answer in one of three ways: "Like the product brand" (L), "Dislike the product brand" (D), or "Undecided" (U). For each outcome, the random variable $X$ assigns the number of L's that appear.

   a. Find the set of possible outcomes and the range of $X$.

   b. Create a probability bar graph for $X$.

   c. What is the probability that at least two people like the product brand?

   d. What is the probability that no more than one person dislikes the product brand?

56. **Spacecraft components**   A component of a spacecraft has both a main system and a backup system operating throughout a flight. The probability that both systems fail sometime during the flight is 0.0148. Assuming that each system separately has the same failure rate, what is the probability that the main system fails during the flight?

---

## Chapter **8**     Questions to Guide Your Review

1. What is the formula for integration by parts? Where does it come from? Why might you want to use it?

2. When applying the formula for integration by parts, how do you choose the $u$ and $dv$? How can you apply integration by parts to an integral of the form $\int f(x)\, dx$?

3. If an integrand is a product of the form $\sin^n x \cos^m x$, where $m$ and $n$ are nonnegative integers, how do you evaluate the integral? Give a specific example of each case.

4. What substitutions are made to evaluate integrals of $\sin mx \sin nx$, $\sin mx \cos nx$, and $\cos mx \cos nx$? Give an example of each case.

**5.** What substitutions are sometimes used to transform integrals involving $\sqrt{a^2 - x^2}$, $\sqrt{a^2 + x^2}$, and $\sqrt{x^2 - a^2}$ into integrals that can be evaluated directly? Give an example of each case.

**6.** What restrictions can you place on the variables involved in the three basic trigonometric substitutions to make sure the substitutions are reversible (have inverses)?

**7.** What is the goal of the method of partial fractions?

**8.** When the degree of a polynomial $f(x)$ is less than the degree of a polynomial $g(x)$, how do you write $f(x)/g(x)$ as a sum of partial fractions if $g(x)$

    **a.** is a product of distinct linear factors?

    **b.** consists of a repeated linear factor?

    **c.** contains an irreducible quadratic factor?

    What do you do if the degree of $f$ is *not* less than the degree of $g$?

**9.** How are integral tables typically used? What do you do if a particular integral you want to evaluate is not listed in the table?

**10.** What is a reduction formula? How are reduction formulas used? Give an example.

**11.** How would you compare the relative merits of Simpson's Rule and the Trapezoidal Rule?

**12.** What is an improper integral of Type I? Type II? How are the values of various types of improper integrals defined? Give examples.

**13.** What tests are available for determining the convergence and divergence of improper integrals that cannot be evaluated directly? Give examples of their use.

**14.** What is a random variable? What is a continuous random variable? Give some specific examples.

**15.** What is a probability density function? What is the probability that a continuous random variable has a value in the interval $[c, d]$?

**16.** What is an exponentially decreasing probability density function? What are some typical events that might be modeled by this distribution? What do we mean when we say such distributions are *memoryless*?

**17.** What is the expected value of a continuous random variable? What is the expected value of an exponentially distributed random variable?

**18.** What is the median of a continuous random variable? What is the median of an exponential distribution?

**19.** What does the variance of a random variable measure? What is the standard deviation of a continuous random variable $X$?

**20.** What probability density function describes the normal distribution? What are some examples typically modeled by a normal distribution? How do we usually calculate probabilities for a normal distribution?

**21.** In a normal distribution, what percentage of the population lies within 1 standard deviation of the mean? Within 2 standard deviations?

# Chapter **8** Practice Exercises

### Integration by Parts
Evaluate the integrals in Exercises 1–8 using integration by parts.

**1.** $\displaystyle\int \ln (x + 1)\, dx$

**2.** $\displaystyle\int x^2 \ln x\, dx$

**3.** $\displaystyle\int \sin^{-1} 2x\, dx$

**4.** $\displaystyle\int \tan^{-1}\left(\frac{x}{3}\right) dx$

**5.** $\displaystyle\int (1 + x^2)e^x\, dx$

**6.** $\displaystyle\int x^2 \cos (2 - x)\, dx$

**7.** $\displaystyle\int e^x \cos 2x\, dx$

**8.** $\displaystyle\int x \sin x \cos x\, dx$

### Partial Fractions
Evaluate the integrals in Exercises 9–28. It may be necessary to use a substitution first.

**9.** $\displaystyle\int \frac{x}{x^2 - 5x + 6}\, dx$

**10.** $\displaystyle\int \frac{x}{x^2 + 4x - 5}\, dx$

**11.** $\displaystyle\int \frac{dx}{x(x - 2)^2}$

**12.** $\displaystyle\int \frac{x + 3}{x^2(x - 2)}\, dx$

**13.** $\displaystyle\int \frac{\cos x}{(1 - \sin x)(2 - \sin x)}\, dx$

**14.** $\displaystyle\int \frac{\sin x}{\cos^2 x + 3 \cos x + 2}\, dx$

**15.** $\displaystyle\int \frac{5x^2 + 7x + 9}{x^3 - x}\, dx$

**16.** $\displaystyle\int \frac{9x\, dx}{x^3 + 9x}$

**17.** $\displaystyle\int \frac{x - 5}{3x^3 - 12x}\, dx$

**18.** $\displaystyle\int \frac{5x + 9}{(x - 2)(x + 3)(x + 4)}\, dx$

**19.** $\displaystyle\int \frac{d\theta}{\theta^4 + 5\theta^2 + 4}$

**20.** $\displaystyle\int \frac{t\, dt}{t^4 + t^2 - 2}$

**21.** $\displaystyle\int \frac{x^3 + x^2}{x^2 + x - 2}\, dx$

**22.** $\displaystyle\int \frac{x^3 + 1}{x^3 - x}\, dx$

**23.** $\displaystyle\int \frac{x^3 + 4x^2}{x^2 + 4x + 3}\, dx$

**24.** $\displaystyle\int \frac{2x^3 + x^2 - 21x + 24}{x^2 + 2x - 8}\, dx$

**25.** $\displaystyle\int \frac{dx}{x(3\sqrt{x} + 1)}$

**26.** $\displaystyle\int \frac{dx}{x\left(1 + \sqrt[3]{x}\right)}$

**27.** $\displaystyle\int \frac{ds}{e^s - 1}$

**28.** $\displaystyle\int \frac{ds}{\sqrt{e^s + 1}}$

### Trigonometric Substitutions
Evaluate the integrals in Exercises 29–32 **(a)** without using a trigonometric substitution, **(b)** using a trigonometric substitution.

**29.** $\displaystyle\int \frac{y\, dy}{\sqrt{16 - y^2}}$

**30.** $\displaystyle\int \frac{x\, dx}{\sqrt{4 + x^2}}$

**31.** $\displaystyle\int \frac{x\, dx}{4 - x^2}$

**32.** $\displaystyle\int \frac{t\, dt}{\sqrt{4t^2 - 1}}$

Evaluate the integrals in Exercises 33–36.

**33.** $\displaystyle\int \frac{x\,dx}{9 - x^2}$

**34.** $\displaystyle\int \frac{dx}{x(9 - x^2)}$

**35.** $\displaystyle\int \frac{dx}{9 - x^2}$

**36.** $\displaystyle\int \frac{dx}{\sqrt{9 - x^2}}$

**Trigonometric Integrals**

Evaluate the integrals in Exercises 37–44.

**37.** $\displaystyle\int \sin^5 x \cos^6 x \, dx$

**38.** $\displaystyle\int \cos^3 x \sin^3 x \, dx$

**39.** $\displaystyle\int \tan^6 x \sec^2 x \, dx$

**40.** $\displaystyle\int \tan^3 x \sec^3 x \, dx$

**41.** $\displaystyle\int \sin 7\theta \cos 5\theta \, d\theta$

**42.** $\displaystyle\int \csc^2 \theta \cos^3 \theta \, d\theta$

**43.** $\displaystyle\int \sqrt{1 - \cos (t/2)} \, dt$

**44.** $\displaystyle\int e^\theta \sqrt{1 - \cos^2 e^\theta} \, d\theta$

**Numerical Integration**

**45.** According to the error-bound formula for Simpson's Rule, how many subintervals should you use to be sure of estimating the value of

$$\ln 3 = \int_1^3 \frac{1}{x} \, dx$$

by Simpson's Rule with an error of no more than $10^{-4}$ in absolute value? (Remember that for Simpson's Rule, the number of subintervals has to be even.)

**46.** A brief calculation shows that if $0 \le x \le 1$, then the second derivative of $f(x) = \sqrt{1 + x^4}$ lies between 0 and 8. Based on this, about how many subdivisions would you need to estimate the integral of $f$ from 0 to 1 with an error no greater than $10^{-3}$ in absolute value using the Trapezoidal Rule?

**47.** A direct calculation shows that

$$\int_0^\pi 2 \sin^2 x \, dx = \pi.$$

How close do you come to this value by using the Trapezoidal Rule with $n = 6$? Simpson's Rule with $n = 6$? Try them and find out.

**48.** You are planning to use Simpson's Rule to estimate the value of the integral

$$\int_1^2 f(x) \, dx$$

with an error magnitude less than $10^{-5}$. You have determined that $|f^{(4)}(x)| \le 3$ throughout the interval of integration. How many subintervals should you use to ensure the required accuracy? (Remember that for Simpson's Rule the number has to be even.)

T **49. Mean temperature** Compute the average value of the temperature function

$$f(x) = 20 \sin \left( \frac{2\pi}{365} (x - 101) \right) - 4$$

for a 365-day year. This is one way to estimate the annual mean air temperature in Fairbanks, Alaska. The National Weather Service's

official figure, a numerical average of the daily normal mean air temperatures for the year, is $-3.5°C$, which is slightly higher than the average value of $f(x)$.

**50. Heat capacity of a gas** Heat capacity $C_v$ is the amount of heat required to raise the temperature of a given mass of gas with constant volume by 1 K, measured in units of J/mol·K (joules per mole per degree). The heat capacity of oxygen depends on its temperature $T$ and satisfies the formula

$$C_v = 31.32 - 20.74 \cdot 10^{-3}\,T + 39.87 \cdot 10^{-6}\,T^2.$$

Find the average value of $C_v$ for $300 \le T \le 650$ K and the temperature at which it is attained.

**51. Fuel efficiency** An automobile computer gives a digital readout of fuel consumption in liters per hour. During a trip, a passenger recorded the fuel consumption every 5 min for a full hour of travel.

| Time | L/h | Time | L/h |
|------|-----|------|-----|
| 0 | 2.5 | 35 | 2.5 |
| 5 | 2.4 | 40 | 2.4 |
| 10 | 2.3 | 45 | 2.3 |
| 15 | 2.4 | 50 | 2.4 |
| 20 | 2.4 | 55 | 2.4 |
| 25 | 2.5 | 60 | 2.3 |
| 30 | 2.6 | | |

**a.** Use the Trapezoidal Rule to approximate the total fuel consumption during the hour.

**b.** If the automobile covered 60 km in the hour, what was its fuel efficiency (in kilometers per liter) for that portion of the trip?

**52. A new parking lot** To meet the demand for parking, your town has allocated the area shown here. As the town engineer, you have been asked by the town council to find out if the lot can be built for $11,000. The cost to clear the land will be $1.00 a square meter, and the lot will cost $20.00 a square meter to pave. Use Simpson's Rule to find out if the job can be done for $11,000.

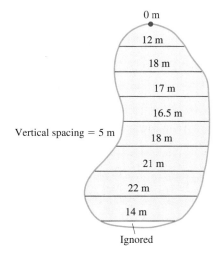

0 m

12 m

18 m

17 m

16.5 m

Vertical spacing = 5 m

18 m

21 m

22 m

14 m

Ignored

## Improper Integrals

Evaluate the improper integrals in Exercises 53–62.

**53.** $\displaystyle\int_0^3 \frac{dx}{\sqrt{9-x^2}}$

**54.** $\displaystyle\int_0^1 \ln x \, dx$

**55.** $\displaystyle\int_0^2 \frac{dy}{(y-1)^{2/3}}$

**56.** $\displaystyle\int_{-2}^0 \frac{d\theta}{(\theta+1)^{3/5}}$

**57.** $\displaystyle\int_3^\infty \frac{2\,du}{u^2-2u}$

**58.** $\displaystyle\int_1^\infty \frac{3v-1}{4v^3-v^2}\,dv$

**59.** $\displaystyle\int_0^\infty x^2 e^{-x}\,dx$

**60.** $\displaystyle\int_{-\infty}^0 xe^{3x}\,dx$

**61.** $\displaystyle\int_{-\infty}^\infty \frac{dx}{4x^2+9}$

**62.** $\displaystyle\int_{-\infty}^\infty \frac{4\,dx}{x^2+16}$

Which of the improper integrals in Exercises 63–68 converge and which diverge?

**63.** $\displaystyle\int_6^\infty \frac{d\theta}{\sqrt{\theta^2+1}}$

**64.** $\displaystyle\int_0^\infty e^{-u}\cos u\,du$

**65.** $\displaystyle\int_1^\infty \frac{\ln z}{z}\,dz$

**66.** $\displaystyle\int_1^\infty \frac{e^{-t}}{\sqrt{t}}\,dt$

**67.** $\displaystyle\int_{-\infty}^\infty \frac{2\,dx}{e^x+e^{-x}}$

**68.** $\displaystyle\int_{-\infty}^\infty \frac{dx}{x^2(1+e^x)}$

## Assorted Integrations

Evaluate the integrals in Exercises 69–116. The integrals are listed in random order so you need to decide which integration technique to use.

**69.** $\displaystyle\int \frac{x\,dx}{1+\sqrt{x}}$

**70.** $\displaystyle\int \frac{x^3+2}{4-x^2}\,dx$

**71.** $\displaystyle\int \sqrt{2x-x^2}\,dx$

**72.** $\displaystyle\int \frac{dx}{\sqrt{-2x-x^2}}$

**73.** $\displaystyle\int \frac{2-\cos x+\sin x}{\sin^2 x}\,dx$

**74.** $\displaystyle\int \sin^2\theta \cos^5\theta\,d\theta$

**75.** $\displaystyle\int \frac{9\,dv}{81-v^4}$

**76.** $\displaystyle\int_2^\infty \frac{dx}{(x-1)^2}$

**77.** $\displaystyle\int \theta\cos(2\theta+1)\,d\theta$

**78.** $\displaystyle\int \frac{x^3\,dx}{x^2-2x+1}$

**79.** $\displaystyle\int \frac{\sin 2\theta\,d\theta}{(1+\cos 2\theta)^2}$

**80.** $\displaystyle\int_{\pi/4}^{\pi/2} \sqrt{1+\cos 4x}\,dx$

**81.** $\displaystyle\int \frac{x\,dx}{\sqrt{2-x}}$

**82.** $\displaystyle\int \frac{\sqrt{1-v^2}}{v^2}\,dv$

**83.** $\displaystyle\int \frac{dy}{y^2-2y+2}$

**84.** $\displaystyle\int \frac{x\,dx}{\sqrt{8-2x^2-x^4}}$

**85.** $\displaystyle\int \frac{z+1}{z^2(z^2+4)}\,dz$

**86.** $\displaystyle\int x^2(x-1)^{1/3}\,dx$

**87.** $\displaystyle\int \frac{t\,dt}{\sqrt{9-4t^2}}$

**88.** $\displaystyle\int \frac{\tan^{-1}x}{x^2}\,dx$

**89.** $\displaystyle\int \frac{e^t\,dt}{e^{2t}+3e^t+2}$

**90.** $\displaystyle\int \tan^3 t\,dt$

**91.** $\displaystyle\int_1^\infty \frac{\ln y}{y^3}\,dy$

**92.** $\displaystyle\int y^{3/2}(\ln y)^2\,dy$

**93.** $\displaystyle\int e^{\ln\sqrt{x}}\,dx$

**94.** $\displaystyle\int e^\theta\sqrt{3+4e^\theta}\,d\theta$

**95.** $\displaystyle\int \frac{\sin 5t\,dt}{1+(\cos 5t)^2}$

**96.** $\displaystyle\int \frac{dv}{\sqrt{e^{2v}-1}}$

**97.** $\displaystyle\int \frac{dr}{1+\sqrt{r}}$

**98.** $\displaystyle\int \frac{4x^3-20x}{x^4-10x^2+9}\,dx$

**99.** $\displaystyle\int \frac{x^3}{1+x^2}\,dx$

**100.** $\displaystyle\int \frac{x^2}{1+x^3}\,dx$

**101.** $\displaystyle\int \frac{1+x^2}{1+x^3}\,dx$

**102.** $\displaystyle\int \frac{1+x^2}{(1+x)^3}\,dx$

**103.** $\displaystyle\int \sqrt{x}\cdot\sqrt{1+\sqrt{x}}\,dx$

**104.** $\displaystyle\int \sqrt{1+\sqrt{1+x}}\,dx$

**105.** $\displaystyle\int \frac{1}{\sqrt{x}\cdot\sqrt{1+x}}\,dx$

**106.** $\displaystyle\int_0^{1/2} \sqrt{1+\sqrt{1-x^2}}\,dx$

**107.** $\displaystyle\int \frac{\ln x}{x+x\ln x}\,dx$

**108.** $\displaystyle\int \frac{1}{x\cdot\ln x\cdot\ln(\ln x)}\,dx$

**109.** $\displaystyle\int \frac{x^{\ln x}\ln x}{x}\,dx$

**110.** $\displaystyle\int (\ln x)^{\ln x}\left[\frac{1}{x}+\frac{\ln(\ln x)}{x}\right]\,dx$

**111.** $\displaystyle\int \frac{1}{x\sqrt{1-x^4}}\,dx$

**112.** $\displaystyle\int \frac{\sqrt{1-x}}{x}\,dx$

**113. a.** Show that $\displaystyle\int_0^a f(x)\,dx = \int_0^a f(a-x)\,dx$.

**b.** Use part (a) to evaluate

$$\int_0^{\pi/2} \frac{\sin x}{\sin x+\cos x}\,dx.$$

**114.** $\displaystyle\int \frac{\sin x}{\sin x+\cos x}\,dx$

**115.** $\displaystyle\int \frac{\sin^2 x}{1+\sin^2 x}\,dx$

**116.** $\displaystyle\int \frac{1-\cos x}{1+\cos x}\,dx$

---

## Chapter 8    Additional and Advanced Exercises

### Evaluating Integrals

Evaluate the integrals in Exercises 1–6.

**1.** $\displaystyle\int (\sin^{-1}x)^2\,dx$

**2.** $\displaystyle\int \frac{dx}{x(x+1)(x+2)\cdots(x+m)}$

**3.** $\displaystyle\int x\sin^{-1}x\,dx$

**4.** $\displaystyle\int \sin^{-1}\sqrt{y}\,dy$

**5.** $\int \dfrac{dt}{t - \sqrt{1 - t^2}}$   **6.** $\int \dfrac{dx}{x^4 + 4}$

Evaluate the limits in Exercises 7 and 8.

**7.** $\displaystyle\lim_{x \to \infty} \int_{-x}^{x} \sin t \, dt$   **8.** $\displaystyle\lim_{x \to 0^+} x \int_{x}^{1} \dfrac{\cos t}{t^2} \, dt$

Evaluate the limits in Exercises 9 and 10 by identifying them with definite integrals and evaluating the integrals.

**9.** $\displaystyle\lim_{n \to \infty} \sum_{k=1}^{n} \ln \sqrt[n]{1 + \dfrac{k}{n}}$   **10.** $\displaystyle\lim_{n \to \infty} \sum_{k=0}^{n-1} \dfrac{1}{\sqrt{n^2 - k^2}}$

## Applications

**11. Finding arc length**   Find the length of the curve

$$y = \int_0^x \sqrt{\cos 2t} \, dt, \quad 0 \le x \le \pi/4.$$

**12. Finding arc length**   Find the length of the graph of the function $y = \ln(1 - x^2)$, $0 \le x \le 1/2$.

**13. Finding volume**   The region in the first quadrant that is enclosed by the x-axis and the curve $y = 3x\sqrt{1 - x}$ is revolved about the y-axis to generate a solid. Find the volume of the solid.

**14. Finding volume**   The region in the first quadrant that is enclosed by the x-axis, the curve $y = 5/\big(x\sqrt{5 - x}\big)$, and the lines $x = 1$ and $x = 4$ is revolved about the x-axis to generate a solid. Find the volume of the solid.

**15. Finding volume**   The region in the first quadrant enclosed by the coordinate axes, the curve $y = e^x$, and the line $x = 1$ is revolved about the y-axis to generate a solid. Find the volume of the solid.

**16. Finding volume**   The region in the first quadrant that is bounded above by the curve $y = e^x - 1$, below by the x-axis, and on the right by the line $x = \ln 2$ is revolved about the line $x = \ln 2$ to generate a solid. Find the volume of the solid.

**17. Finding volume**   Let R be the "triangular" region in the first quadrant that is bounded above by the line $y = 1$, below by the curve $y = \ln x$, and on the left by the line $x = 1$. Find the volume of the solid generated by revolving R about

   **a.** the x-axis.   **b.** the line $y = 1$.

**18. Finding volume**   (*Continuation of Exercise 17.*) Find the volume of the solid generated by revolving the region R about

   **a.** the y-axis.   **b.** the line $x = 1$.

**19. Finding volume**   The region between the x-axis and the curve

$$y = f(x) = \begin{cases} 0, & x = 0 \\ x \ln x, & 0 < x \le 2 \end{cases}$$

is revolved about the x-axis to generate the solid shown here.

   **a.** Show that $f$ is continuous at $x = 0$.

   **b.** Find the volume of the solid.

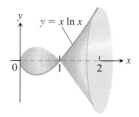

**20. Finding volume**   The infinite region bounded by the coordinate axes and the curve $y = -\ln x$ in the first quadrant is revolved about the x-axis to generate a solid. Find the volume of the solid.

**21. Centroid of a region**   Find the centroid of the region in the first quadrant that is bounded below by the x-axis, above by the curve $y = \ln x$, and on the right by the line $x = e$.

**22. Centroid of a region**   Find the centroid of the region in the plane enclosed by the curves $y = \pm(1 - x^2)^{-1/2}$ and the lines $x = 0$ and $x = 1$.

**23. Length of a curve**   Find the length of the curve $y = \ln x$ from $x = 1$ to $x = e$.

**24. Finding surface area**   Find the area of the surface generated by revolving the curve in Exercise 23 about the y-axis.

**25. The surface generated by an astroid**   The graph of the equation $x^{2/3} + y^{2/3} = 1$ is an *astroid* (see accompanying figure). Find the area of the surface generated by revolving the curve about the x-axis.

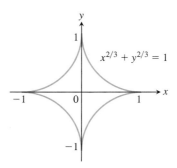

**26. Length of a curve**   Find the length of the curve

$$y = \int_1^x \sqrt{\sqrt{t} - 1} \, dt, \quad 1 \le x \le 16.$$

**27.** For what value or values of $a$ does

$$\int_1^{\infty} \left( \dfrac{ax}{x^2 + 1} - \dfrac{1}{2x} \right) dx$$

converge? Evaluate the corresponding integral(s).

**28.** For each $x > 0$, let $G(x) = \int_0^{\infty} e^{-xt} \, dt$. Prove that $xG(x) = 1$ for each $x > 0$.

**29. Infinite area and finite volume**   What values of $p$ have the following property: The area of the region between the curve $y = x^{-p}$, $1 \le x < \infty$, and the x-axis is infinite but the volume of the solid generated by revolving the region about the x-axis is finite.

**30. Infinite area and finite volume**   What values of $p$ have the following property: The area of the region in the first quadrant enclosed by the curve $y = x^{-p}$, the y-axis, the line $x = 1$, and the interval $[0, 1]$ on the x-axis is infinite but the volume of the solid generated by revolving the region about one of the coordinate axes is finite.

**31. Integrating the square of the derivative** If $f$ is continuously differentiable on $[0, 1]$ and $f(1) = f(0) = -1/6$, prove that

$$\int_0^1 (f'(x))^2 \, dx \geq 2 \int_0^1 f(x) \, dx + \frac{1}{4}.$$

*Hint:* Consider the inequality $0 \leq \int_0^1 \left( f'(x) + x - \frac{1}{2} \right)^2 dx.$

*Source: Mathematics Magazine,* vol. 84, no. 4, Oct. 2011.

**32.** (*Continuation of Exercise 31.*) If $f$ is continuously differentiable on $[0, a]$ for $a > 0$, and $f'(a) = f(0) = b$, prove that

$$\int_0^a (f'(x))^2 \, dx \geq 2 \int_0^a f(x) \, dx - \left( 2ab + \frac{a^3}{12} \right).$$

*Hint:* Consider the inequality $0 \leq \int_0^a \left( f'(x) + x - \frac{a}{2} \right)^2 dx.$

*Source: Mathematics Magazine,* vol. 84, no. 4, Oct. 2011.

### Tabular Integration

The technique of tabular integration also applies to integrals of the form $\int f(x)g(x) \, dx$ when neither function can be differentiated repeatedly to become zero. For example, to evaluate

$$\int e^{2x} \cos x \, dx$$

we begin as before with a table listing successive derivatives of $e^{2x}$ and integrals of $\cos x$:

| $e^{2x}$ and its derivatives | | $\cos x$ and its integrals |
|---|---|---|
| $e^{2x}$ | (+) | $\cos x$ |
| $2e^{2x}$ | (−) | $\sin x$ |
| $4e^{2x}$ | (+) | $-\cos x$ |

← *Stop here:* Row is same as first row except for multiplicative constants (4 on the left, −1 on the right).

We stop differentiating and integrating as soon as we reach a row that is the same as the first row except for multiplicative constants. We interpret the table as saying

$$\int e^{2x} \cos x \, dx = +(e^{2x} \sin x) - (2e^{2x}(-\cos x))$$

$$+ \int (4e^{2x})(-\cos x) \, dx.$$

We take signed products from the diagonal arrows and a signed integral for the last horizontal arrow. Transposing the integral on the right-hand side over to the left-hand side now gives

$$5 \int e^{2x} \cos x \, dx = e^{2x} \sin x + 2e^{2x} \cos x$$

or

$$\int e^{2x} \cos x \, dx = \frac{e^{2x} \sin x + 2e^{2x} \cos x}{5} + C,$$

after dividing by 5 and adding the constant of integration.

Use tabular integration to evaluate the integrals in Exercises 33–40.

**33.** $\int e^{2x} \cos 3x \, dx$    **34.** $\int e^{3x} \sin 4x \, dx$

**35.** $\int \sin 3x \sin x \, dx$    **36.** $\int \cos 5x \sin 4x \, dx$

**37.** $\int e^{ax} \sin bx \, dx$    **38.** $\int e^{ax} \cos bx \, dx$

**39.** $\int \ln (ax) \, dx$    **40.** $\int x^2 \ln (ax) \, dx$

### The Substitution $z = \tan (x/2)$

The substitution

$$z = \tan \frac{x}{2} \tag{1}$$

reduces the problem of integrating a rational expression in $\sin x$ and $\cos x$ to a problem of integrating a rational function of $z$. This in turn can be integrated by partial fractions.

From the accompanying figure

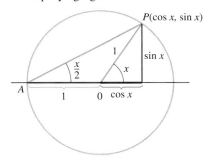

we can read the relation

$$\tan \frac{x}{2} = \frac{\sin x}{1 + \cos x}.$$

To see the effect of the substitution, we calculate

$$\cos x = 2 \cos^2 \left( \frac{x}{2} \right) - 1 = \frac{2}{\sec^2 (x/2)} - 1$$

$$= \frac{2}{1 + \tan^2 (x/2)} - 1 = \frac{2}{1 + z^2} - 1$$

$$\cos x = \frac{1 - z^2}{1 + z^2}, \tag{2}$$

and

$$\sin x = 2 \sin \frac{x}{2} \cos \frac{x}{2} = 2 \frac{\sin (x/2)}{\cos (x/2)} \cdot \cos^2 \left( \frac{x}{2} \right)$$

$$= 2 \tan \frac{x}{2} \cdot \frac{1}{\sec^2 (x/2)} = \frac{2 \tan (x/2)}{1 + \tan^2 (x/2)}$$

$$\sin x = \frac{2z}{1 + z^2}. \tag{3}$$

Finally, $x = 2 \tan^{-1} z$, so

$$dx = \frac{2\,dz}{1 + z^2}. \qquad (4)$$

Examples

**a.** $\displaystyle\int \frac{1}{1 + \cos x}\,dx = \int \frac{1 + z^2}{2}\,\frac{2\,dz}{1 + z^2}$

$$= \int dz = z + C$$

$$= \tan\left(\frac{x}{2}\right) - C$$

**b.** $\displaystyle\int \frac{1}{2 + \sin x}\,dx = \int \frac{1 + z^2}{2 + 2z + 2z^2}\,\frac{2\,dz}{1 + z^2}$

$$= \int \frac{dz}{z^2 + z + 1} = \int \frac{dz}{(z + (1/2))^2 + 3/4}$$

$$= \int \frac{du}{u^2 + a^2}$$

$$= \frac{1}{a}\tan^{-1}\left(\frac{u}{a}\right) + C$$

$$= \frac{2}{\sqrt{3}}\tan^{-1}\frac{2z + 1}{\sqrt{3}} + C$$

$$= \frac{2}{\sqrt{3}}\tan^{-1}\frac{1 + 2\tan(x/2)}{\sqrt{3}} + C$$

Use the substitutions in Equations (1)–(4) to evaluate the integrals in Exercises 41–48. Integrals like these arise in calculating the average angular velocity of the output shaft of a universal joint when the input and output shafts are not aligned.

**41.** $\displaystyle\int \frac{dx}{1 - \sin x}$

**42.** $\displaystyle\int \frac{dx}{1 + \sin x + \cos x}$

**43.** $\displaystyle\int_0^{\pi/2} \frac{dx}{1 + \sin x}$

**44.** $\displaystyle\int_{\pi/3}^{\pi/2} \frac{dx}{1 - \cos x}$

**45.** $\displaystyle\int_0^{\pi/2} \frac{d\theta}{2 + \cos \theta}$

**46.** $\displaystyle\int_{\pi/2}^{2\pi/3} \frac{\cos\theta\,d\theta}{\sin\theta\cos\theta + \sin\theta}$

**47.** $\displaystyle\int \frac{dt}{\sin t - \cos t}$

**48.** $\displaystyle\int \frac{\cos t\,dt}{1 - \cos t}$

Use the substitution $z = \tan(\theta/2)$ to evaluate the integrals in Exercises 49 and 50.

**49.** $\displaystyle\int \sec\theta\,d\theta$

**50.** $\displaystyle\int \csc\theta\,d\theta$

### The Gamma Function and Stirling's Formula

Euler's gamma function $\Gamma(x)$ ("gamma of $x$"; $\Gamma$ is a Greek capital g) uses an integral to extend the factorial function from the nonnegative integers to other real values. The formula is

$$\Gamma(x) = \int_0^\infty t^{x-1}e^{-t}\,dt, \quad x > 0.$$

For each positive $x$, the number $\Gamma(x)$ is the integral of $t^{x-1}e^{-t}$ with respect to $t$ from 0 to $\infty$. Figure 8.28 shows the graph of $\Gamma$ near the origin. You will see how to calculate $\Gamma(1/2)$ if you do Additional Exercise 23 in Chapter 15.

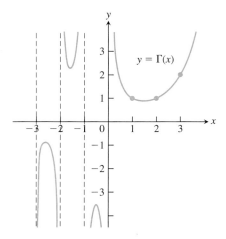

**FIGURE 8.28** Euler's gamma function $\Gamma(x)$ is a continuous function of $x$ whose value at each positive integer $n + 1$ is $n!$. The defining integral formula for $\Gamma$ is valid only for $x > 0$, but we can extend $\Gamma$ to negative noninteger values of $x$ with the formula $\Gamma(x) = (\Gamma(x + 1))/x$, which is the subject of Exercise 51.

**51.** **If $n$ is a nonnegative integer, $\Gamma(n + 1) = n!$**

**a.** Show that $\Gamma(1) = 1$.

**b.** Then apply integration by parts to the integral for $\Gamma(x + 1)$ to show that $\Gamma(x + 1) = x\Gamma(x)$. This gives

$$\Gamma(2) = 1\Gamma(1) = 1$$
$$\Gamma(3) = 2\Gamma(2) = 2$$
$$\Gamma(4) = 3\Gamma(3) = 6$$
$$\vdots$$
$$\Gamma(n + 1) = n\,\Gamma(n) = n! \qquad (5)$$

**c.** Use mathematical induction to verify Equation (5) for every nonnegative integer $n$.

**52.** **Stirling's formula** Scottish mathematician James Stirling (1692–1770) showed that

$$\lim_{x\to\infty}\left(\frac{e}{x}\right)^x\sqrt{\frac{x}{2\pi}}\,\Gamma(x) = 1,$$

so, for large $x$,

$$\Gamma(x) = \left(\frac{x}{e}\right)^x\sqrt{\frac{2\pi}{x}}\,(1 + \epsilon(x)), \qquad \epsilon(x) \to 0 \text{ as } x \to \infty. \qquad (6)$$

Dropping $\epsilon(x)$ leads to the approximation

$$\Gamma(x) \approx \left(\frac{x}{e}\right)^x\sqrt{\frac{2\pi}{x}} \qquad \textbf{(Stirling's formula).} \qquad (7)$$

**a. Stirling's approximation for $n!$**  Use Equation (7) and the fact that $n! = n\Gamma(n)$ to show that

$$n! \approx \left(\frac{n}{e}\right)^n \sqrt{2n\pi} \qquad \textbf{(Stirling's approximation)}. \qquad (8)$$

As you will see if you do Exercise 104 in Section 10.1, Equation (8) leads to the approximation

$$\sqrt[n]{n!} \approx \frac{n}{e}. \qquad (9)$$

T **b.** Compare your calculator's value for $n!$ with the value given by Stirling's approximation for $n = 10, 20, 30, \ldots$, as far as your calculator can go.

T **c.** A refinement of Equation (6) gives

$$\Gamma(x) = \left(\frac{x}{e}\right)^x \sqrt{\frac{2\pi}{x}}\, e^{1/(12x)}(1 + \epsilon(x))$$

or

$$\Gamma(x) \approx \left(\frac{x}{e}\right)^x \sqrt{\frac{2\pi}{x}}\, e^{1/(12x)},$$

which tells us that

$$n! \approx \left(\frac{n}{e}\right)^n \sqrt{2n\pi}\, e^{1/(12n)}. \qquad (10)$$

Compare the values given for 10! by your calculator, Stirling's approximation, and Equation (10).

# Chapter 8  Technology Application Projects

## Mathematica/Maple Modules:

### Riemann, Trapezoidal, and Simpson Approximations
**Part I**: Visualize the error involved in using Riemann sums to approximate the area under a curve.
**Part II**: Build a table of values and compute the relative magnitude of the error as a function of the step size $\Delta x$.
**Part III**: Investigate the effect of the derivative function on the error.
**Parts IV and V**: Trapezoidal Rule approximations.
**Part VI**: Simpson's Rule approximations.

### Games of Chance: Exploring the Monte Carlo Probabilistic Technique for Numerical Integration
Graphically explore the Monte Carlo method for approximating definite integrals.

### Computing Probabilities with Improper Integrals
More explorations of the Monte Carlo method for approximating definite integrals.

# 9

# First-Order Differential Equations

**OVERVIEW** Many real-world problems, when formulated mathematically, lead to differential equations. We encountered a number of these equations in previous chapters when studying phenomena such as the motion of an object moving along a straight line, the decay of a radioactive material, the growth of a population, and the cooling of a heated object placed within a medium of lower temperature.

In Section 4.7 we introduced differential equations of the form $dy/dx = f(x)$, where $f$ is given and $y$ is an unknown function of $x$. When $f$ is continuous over some interval, we learned that the general solution $y(x)$ was found directly by integration, $y = \int f(x)\,dx$. Next, in Section 7.4, we investigated differential equations of the form $dy/dx = f(x, y)$, where $f$ is a function of both the independent variable $x$ and the dependent variable $y$. There we learned how to find the general solution when the differential equation is separable. In this chapter we further extend our study to include other commonly occurring *first-order* differential equations. They involve only first derivatives of the unknown function $y(x)$, and model phenomena such as simple electrical circuits, or the resulting concentration of a chemical being added and mixed with some other fluid in a container. Differential equations involving second derivatives are studied in Chapter 17.

## 9.1 Solutions, Slope Fields, and Euler's Method

We begin this section by defining general differential equations involving first derivatives. We then look at slope fields, which give a geometric picture of the solutions to such equations. Many differential equations cannot be solved by obtaining an explicit formula for the solution. However, we can often find numerical approximations to solutions. We present one such method here, called *Euler's method*, which is the basis for many other numerical methods as well.

### General First-Order Differential Equations and Solutions

A **first-order differential equation** is an equation

$$\frac{dy}{dx} = f(x, y) \tag{1}$$

in which $f(x, y)$ is a function of two variables defined on a region in the $xy$-plane. The equation is of *first order* because it involves only the first derivative $dy/dx$ (and not higher-order derivatives). We point out that the equations

$$y' = f(x, y) \qquad \text{and} \qquad \frac{d}{dx}y = f(x, y)$$

are equivalent to Equation (1) and all three forms will be used interchangeably in the text.

A **solution** of Equation (1) is a differentiable function $y = y(x)$ defined on an interval $I$ of $x$-values (perhaps infinite) such that

$$\frac{d}{dx} y(x) = f(x, y(x))$$

on that interval. That is, when $y(x)$ and its derivative $y'(x)$ are substituted into Equation (1), the resulting equation is true for all $x$ over the interval $I$. The **general solution** to a first-order differential equation is a solution that contains all possible solutions. The general solution always contains an arbitrary constant, but having this property doesn't mean a solution is the general solution. That is, a solution may contain an arbitrary constant without being the general solution. Establishing that a solution *is* the general solution may require deeper results from the theory of differential equations and is best studied in a more advanced course.

**EXAMPLE 1**    Show that every member of the family of functions

$$y = \frac{C}{x} + 2$$

is a solution of the first-order differential equation

$$\frac{dy}{dx} = \frac{1}{x}(2 - y)$$

on the interval $(0, \infty)$, where $C$ is any constant.

**Solution**    Differentiating $y = C/x + 2$ gives

$$\frac{dy}{dx} = C \frac{d}{dx}\left(\frac{1}{x}\right) + 0 = -\frac{C}{x^2}.$$

We need to show that the differential equation is satisfied when we substitute into it the expressions $(C/x) + 2$ for $y$, and $-C/x^2$ for $dy/dx$. That is, we need to verify that for all $x \in (0, \infty)$,

$$-\frac{C}{x^2} = \frac{1}{x}\left[2 - \left(\frac{C}{x} + 2\right)\right].$$

This last equation follows immediately by expanding the expression on the right-hand side:

$$\frac{1}{x}\left[2 - \left(\frac{C}{x} + 2\right)\right] = \frac{1}{x}\left(-\frac{C}{x}\right) = -\frac{C}{x^2}.$$

Therefore, for every value of $C$, the function $y = C/x + 2$ is a solution of the differential equation. ∎

As was the case in finding antiderivatives, we often need a *particular* rather than the general solution to a first-order differential equation $y' = f(x, y)$. The **particular solution** satisfying the initial condition $y(x_0) = y_0$ is the solution $y = y(x)$ whose value is $y_0$ when $x = x_0$. Thus the graph of the particular solution passes through the point $(x_0, y_0)$ in the $xy$-plane. A **first-order initial value problem** is a differential equation $y' = f(x, y)$ whose solution must satisfy an initial condition $y(x_0) = y_0$.

**EXAMPLE 2**    Show that the function

$$y = (x + 1) - \frac{1}{3}e^x$$

is a solution to the first-order initial value problem

$$\frac{dy}{dx} = y - x, \qquad y(0) = \frac{2}{3}.$$

**Solution**    The equation

$$\frac{dy}{dx} = y - x$$

is a first-order differential equation with $f(x, y) = y - x$.

*On the left side of the equation:*

$$\frac{dy}{dx} = \frac{d}{dx}\left(x + 1 - \frac{1}{3}e^x\right) = 1 - \frac{1}{3}e^x.$$

*On the right side of the equation:*

$$y - x = (x + 1) - \frac{1}{3}e^x - x = 1 - \frac{1}{3}e^x.$$

The function satisfies the initial condition because

$$y(0) = \left[(x + 1) - \frac{1}{3}e^x\right]_{x=0} = 1 - \frac{1}{3} = \frac{2}{3}.$$

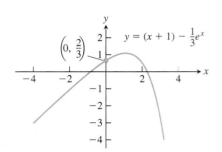

**FIGURE 9.1**    Graph of the solution to the initial value problem in Example 2.

The graph of the function is shown in Figure 9.1.    ∎

### Slope Fields: Viewing Solution Curves

Each time we specify an initial condition $y(x_0) = y_0$ for the solution of a differential equation $y' = f(x, y)$, the **solution curve** (graph of the solution) is required to pass through the point $(x_0, y_0)$ and to have slope $f(x_0, y_0)$ there. We can picture these slopes graphically by drawing short line segments of slope $f(x, y)$ at selected points $(x, y)$ in the region of the $xy$-plane that constitutes the domain of $f$. Each segment has the same slope as the solution curve through $(x, y)$ and so is tangent to the curve there. The resulting picture is called a **slope field** (or **direction field**) and gives a visualization of the general shape of the solution curves. Figure 9.2a shows a slope field, with a particular solution sketched into it in Figure 9.2b. We see how these line segments indicate the direction the solution curve takes at each point it passes through.

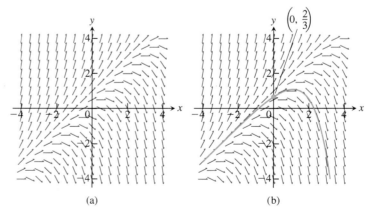

**FIGURE 9.2**    (a) Slope field for $\frac{dy}{dx} = y - x$. (b) The particular solution curve through the point $\left(0, \frac{2}{3}\right)$ (Example 2).

Figure 9.3 shows three slope fields and we see how the solution curves behave by following the tangent line segments in these fields. Slope fields are useful because they display the overall behavior of the family of solution curves for a given differential equation.

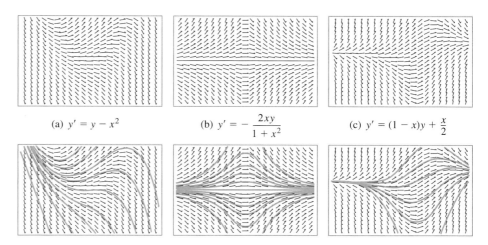

(a) $y' = y - x^2$    (b) $y' = -\dfrac{2xy}{1 + x^2}$    (c) $y' = (1 - x)y + \dfrac{x}{2}$

**FIGURE 9.3** Slope fields (top row) and selected solution curves (bottom row). In computer renditions, slope segments are sometimes portrayed with arrows, as they are here, but they should be considered as just tangent line segments.

For instance, the slope field in Figure 9.3b reveals that every solution $y(x)$ to the differential equation specified in the figure satisfies $\lim_{x \to \pm\infty} y(x) = 0$. We will see that knowing the overall behavior of the solution curves is often critical to understanding and predicting outcomes in a real-world system modeled by a differential equation.

Constructing a slope field with pencil and paper can be quite tedious. All our examples were generated by computer software.

### Euler's Method

If we do not require or cannot immediately find an *exact* solution giving an explicit formula for an initial value problem $y' = f(x, y)$, $y(x_0) = y_0$, we can often use a computer to generate a table of approximate numerical values of $y$ for values of $x$ in an appropriate interval. Such a table is called a **numerical solution** of the problem, and the method by which we generate the table is called a **numerical method**.

Given a differential equation $dy/dx = f(x, y)$ and an initial condition $y(x_0) = y_0$, we can approximate the solution $y = y(x)$ by its linearization

$$L(x) = y(x_0) + y'(x_0)(x - x_0) \qquad \text{or} \qquad L(x) = y_0 + f(x_0, y_0)(x - x_0).$$

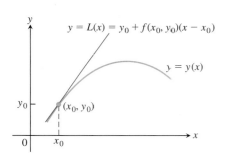

**FIGURE 9.4** The linearization $L(x)$ of $y = y(x)$ at $x = x_0$.

The function $L(x)$ gives a good approximation to the solution $y(x)$ in a short interval about $x_0$ (Figure 9.4). The basis of Euler's method is to patch together a string of linearizations to approximate the curve over a longer stretch. Here is how the method works.

We know the point $(x_0, y_0)$ lies on the solution curve. Suppose that we specify a new value for the independent variable to be $x_1 = x_0 + dx$. (Recall that $dx = \Delta x$ in the definition of differentials.) If the increment $dx$ is small, then

$$y_1 = L(x_1) = y_0 + f(x_0, y_0)\, dx$$

is a good approximation to the exact solution value $y = y(x_1)$. So from the point $(x_0, y_0)$, which lies *exactly* on the solution curve, we have obtained the point $(x_1, y_1)$, which lies very close to the point $(x_1, y(x_1))$ on the solution curve (Figure 9.5).

Using the point $(x_1, y_1)$ and the slope $f(x_1, y_1)$ of the solution curve through $(x_1, y_1)$, we take a second step. Setting $x_2 = x_1 + dx$, we use the linearization of the solution curve through $(x_1, y_1)$ to calculate

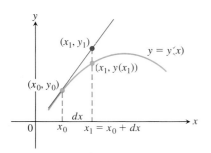

**FIGURE 9.5** The first Euler step approximates $y(x_1)$ with $y_1 = L(x_1)$.

$$y_2 = y_1 + f(x_1, y_1)\, dx.$$

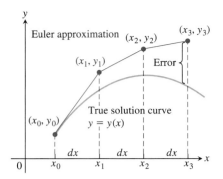

**FIGURE 9.6** Three steps in the Euler approximation to the solution of the initial value problem $y' = f(x, y)$, $y(x_0) = y_0$. As we take more steps, the errors involved usually accumulate, but not in the exaggerated way shown here.

**HISTORICAL BIOGRAPHY**

Leonhard Euler
(1703–1783)

This gives the next approximation $(x_2, y_2)$ to values along the solution curve $y = y(x)$ (Figure 9.6). Continuing in this fashion, we take a third step from the point $(x_2, y_2)$ with slope $f(x_2, y_2)$ to obtain the third approximation

$$y_3 = y_2 + f(x_2, y_2) \, dx,$$

and so on. We are literally building an approximation to one of the solutions by following the direction of the slope field of the differential equation.

The steps in Figure 9.6 are drawn large to illustrate the construction process, so the approximation looks crude. In practice, $dx$ would be small enough to make the red curve hug the blue one and give a good approximation throughout.

**EXAMPLE 3** Find the first three approximations $y_1$, $y_2$, $y_3$ using Euler's method for the initial value problem

$$y' = 1 + y, \qquad y(0) = 1,$$

starting at $x_0 = 0$ with $dx = 0.1$.

**Solution** We have the starting values $x_0 = 0$ and $y_0 = 1$. Next we determine the values of $x$ at which the Euler approximations will take place: $x_1 = x_0 + dx = 0.1$, $x_2 = x_0 + 2\,dx = 0.2$, and $x_3 = x_0 + 3\,dx = 0.3$. Then we find

$$
\begin{aligned}
\textit{First:} \quad y_1 &= y_0 + f(x_0, y_0) \, dx \\
&= y_0 + (1 + y_0) \, dx \\
&= 1 + (1 + 1)(0.1) = 1.2 \\[6pt]
\textit{Second:} \quad y_2 &= y_1 + f(x_1, y_1) \, dx \\
&= y_1 + (1 + y_1) \, dx \\
&= 1.2 + (1 + 1.2)(0.1) = 1.42 \\[6pt]
\textit{Third:} \quad y_3 &= y_2 + f(x_2, y_2) \, dx \\
&= y_2 + (1 + y_2) \, dx \\
&= 1.42 + (1 + 1.42)(0.1) = 1.662
\end{aligned}
$$

The step-by-step process used in Example 3 can be continued easily. Using equally spaced values for the independent variable in the table for the numerical solution, and generating $n$ of them, set

$$
\begin{aligned}
x_1 &= x_0 + dx \\
x_2 &= x_1 + dx \\
&\;\;\vdots \\
x_n &= x_{n-1} + dx.
\end{aligned}
$$

Then calculate the approximations to the solution,

$$
\begin{aligned}
y_1 &= y_0 + f(x_0, y_0) \, dx \\
y_2 &= y_1 + f(x_1, y_1) \, dx \\
&\;\;\vdots \\
y_n &= y_{n-1} + f(x_{n-1}, y_{n-1}) \, dx.
\end{aligned}
$$

The number of steps $n$ can be as large as we like, but errors can accumulate if $n$ is too large.

Euler's method is easy to implement on a computer or calculator. The software program generates a table of numerical solutions to an initial value problem, allowing us to input $x_0$ and $y_0$, the number of steps $n$, and the step size $dx$. It then calculates the approximate solution values $y_1$, $y_2$, . . . , $y_n$ in iterative fashion, as just described.

Solving the separable equation in Example 3, we find that the exact solution to the initial value problem is $y = 2e^x - 1$. We use this information in Example 4.

**EXAMPLE 4**    Use Euler's method to solve

$$y' = 1 + y, \qquad y(0) = 1,$$

on the interval $0 \le x \le 1$, starting at $x_0 = 0$ and taking **(a)** $dx = 0.1$ and **(b)** $dx = 0.05$. Compare the approximations with the values of the exact solution $y = 2e^x - 1$.

**Solution**
**(a)** We used a computer to generate the approximate values in Table 9.1. The "error" column is obtained by subtracting the unrounded Euler values from the unrounded values found using the exact solution. All entries are then rounded to four decimal places.

| TABLE 9.1 Euler solution of $y' = 1 + y$, $y(0) = 1$, step size $dx = 0.1$ | | | |
|---|---|---|---|
| $x$ | $y$ (Euler) | $y$ (exact) | Error |
| 0 | 1 | 1 | 0 |
| 0.1 | 1.2 | 1.2103 | 0.0103 |
| 0.2 | 1.42 | 1.4428 | 0.0228 |
| 0.3 | 1.662 | 1.6997 | 0.0377 |
| 0.4 | 1.9282 | 1.9836 | 0.0554 |
| 0.5 | 2.2210 | 2.2974 | 0.0764 |
| 0.6 | 2.5431 | 2.6442 | 0.1011 |
| 0.7 | 2.8974 | 3.0275 | 0.1301 |
| 0.8 | 3.2872 | 3.4511 | 0.1639 |
| 0.9 | 3.7159 | 3.9192 | 0.2033 |
| 1.0 | 4.1875 | 4.4366 | 0.2491 |

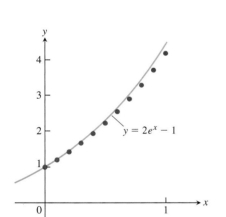

**FIGURE 9.7**    The graph of $y = 2e^x - 1$ superimposed on a scatterplot of the Euler approximations shown in Table 9.1 (Example 4).

By the time we reach $x = 1$ (after 10 steps), the error is about 5.6% of the exact solution. A plot of the exact solution curve with the scatterplot of Euler solution points from Table 9.1 is shown in Figure 9.7.

**(b)** One way to try to reduce the error is to decrease the step size. Table 9.2 shows the results and their comparisons with the exact solutions when we decrease the step size to 0.05, doubling the number of steps to 20. As in Table 9.1, all computations are performed before rounding. This time when we reach $x = 1$, the relative error is only about 2.9%.

It might be tempting to reduce the step size even further in Example 4 to obtain greater accuracy. Each additional calculation, however, not only requires additional computer time but more importantly adds to the buildup of round-off errors due to the approximate representations of numbers inside the computer.

The analysis of error and the investigation of methods to reduce it when making numerical calculations are important but are appropriate for a more advanced course. There are numerical methods more accurate than Euler's method, usually presented in a further study of differential equations or in a numerical analysis course.

**TABLE 9.2** Euler solution of $y' = 1 + y$, $y(0) = 1$, step size $dx = 0.05$

| $x$ | $y$ (Euler) | $y$ (exact) | Error |
|-----|-------------|-------------|-------|
| 0 | 1 | 1 | 0 |
| 0.05 | 1.1 | 1.1025 | 0.0025 |
| 0.10 | 1.205 | 1.2103 | 0.0053 |
| 0.15 | 1.3153 | 1.3237 | 0.0084 |
| 0.20 | 1.4310 | 1.4428 | 0.0118 |
| 0.25 | 1.5526 | 1.5681 | 0.0155 |
| 0.30 | 1.6802 | 1.6997 | 0.0195 |
| 0.35 | 1.8142 | 1.8381 | 0.0239 |
| 0.40 | 1.9549 | 1.9836 | 0.0287 |
| 0.45 | 2.1027 | 2.1366 | 0.0340 |
| 0.50 | 2.2578 | 2.2974 | 0.0397 |
| 0.55 | 2.4207 | 2.4665 | 0.0458 |
| 0.60 | 2.5917 | 2.6442 | 0.0525 |
| 0.65 | 2.7713 | 2.8311 | 0.0598 |
| 0.70 | 2.9599 | 3.0275 | 0.0676 |
| 0.75 | 3.1579 | 3.2340 | 0.0761 |
| 0.80 | 3.3657 | 3.4511 | 0.0853 |
| 0.85 | 3.5840 | 3.6793 | 0.0953 |
| 0.90 | 3.8132 | 3.9192 | 0.1060 |
| 0.95 | 4.0539 | 4.1714 | 0.1175 |
| 1.00 | 4.3066 | 4.4366 | 0.1300 |

# Exercises 9.1

### Slope Fields

In Exercises 1–4, match the differential equations with their slope fields, graphed here.

(a)

(b)

(c)

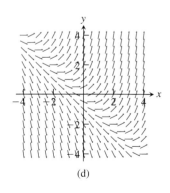

(d)

**1.** $y' = x + y$

**2.** $y' = y + 1$

**3.** $y' = -\dfrac{x}{y}$

**4.** $y' = y^2 - x^2$

In Exercises 5 and 6, copy the slope fields and sketch in some of the solution curves.

**5.** $y' = (y + 2)(y - 2)$

**6.** $y' = y(y + 1)(y - 1)$

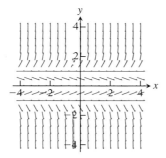

### Integral Equations

In Exercises 7–10, write an equivalent first-order differential equation and initial condition for $y$.

**7.** $y = -1 + \displaystyle\int_{1}^{x} (t - y(t)) \, dt$

**8.** $y = \displaystyle\int_{1}^{x} \frac{1}{t} \, dt$

**9.** $y = 2 - \displaystyle\int_{0}^{x} (1 + y(t)) \sin t \, dt$

**10.** $y = 1 + \displaystyle\int_{0}^{x} y(t) \, dt$

### Using Euler's Method

In Exercises 11–16, use Euler's method to calculate the first three approximations to the given initial value problem for the specified increment size. Calculate the exact solution and investigate the accuracy of your approximations. Round your results to four decimal places.

**11.** $y' = 1 - \dfrac{y}{x}, \quad y(2) = -1, \quad dx = 0.5$

**12.** $y' = x(1 - y), \quad y(1) = 0, \quad dx = 0.2$

**13.** $y' = 2xy + 2y, \quad y(0) = 3, \quad dx = 0.2$

**14.** $y' = y^2(1 + 2x), \quad y(-1) = 1, \quad dx = 0.5$

T **15.** $y' = 2xe^{x^2}, \quad y(0) = 2, \quad dx = 0.1$

T **16.** $y' = ye^x, \quad y(0) = 2, \quad dx = 0.5$

**17.** Use the Euler method with $dx = 0.2$ to estimate $y(1)$ if $y' = y$ and $y(0) = 1$. What is the exact value of $y(1)$?

**18.** Use the Euler method with $dx = 0.2$ to estimate $y(2)$ if $y' = y/x$ and $y(1) = 2$. What is the exact value of $y(2)$?

**19.** Use the Euler method with $dx = 0.5$ to estimate $y(5)$ if $y' = y^2/\sqrt{x}$ and $y(1) = -1$. What is the exact value of $y(5)$?

**20.** Use the Euler method with $dx = 1/3$ to estimate $y(2)$ if $y' = x \sin y$ and $y(0) = 1$. What is the exact value of $y(2)$?

**21.** Show that the solution of the initial value problem

$$y' = x + y, \qquad y(x_0) = y_0$$

is

$$y = -1 - x + (1 + x_0 + y_0) \, e^{x - x_0}.$$

**22.** What integral equation is equivalent to the initial value problem $y' = f(x), y(x_0) = y_0$?

### COMPUTER EXPLORATIONS

In Exercises 23–28, obtain a slope field and add to it graphs of the solution curves passing through the given points.

**23.** $y' = y$ with

    **a.** $(0, 1)$    **b.** $(0, 2)$    **c.** $(0, -1)$

**24.** $y' = 2(y - 4)$ with

    **a.** $(0, 1)$    **b.** $(0, 4)$    **c.** $(0, 5)$

**25.** $y' = y(x + y)$ with

    **a.** $(0, 1)$    **b.** $(0, -2)$    **c.** $(0, 1/4)$    **d.** $(-1, -1)$

**26.** $y' = y^2$ with

    **a.** $(0, 1)$    **b.** $(0, 2)$    **c.** $(0, -1)$    **d.** $(0, 0)$

**27.** $y' = (y - 1)(x + 2)$ with

    **a.** $(0, -1)$    **b.** $(0, 1)$    **c.** $(0, 3)$    **d.** $(1, -1)$

**28.** $y' = \dfrac{xy}{x^2 + 4}$ with

    **a.** $(0, 2)$    **b.** $(0, -6)$    **c.** $\left(-2\sqrt{3}, -4\right)$

In Exercises 29 and 30, obtain a slope field and graph the particular solution over the specified interval. Use your CAS DE solver to find the general solution of the differential equation.

**29. A logistic equation**   $y' = y(2 - y), \; y(0) = 1/2; \; 0 \le x \le 4, \; 0 \le y \le 3$

**30.** $y' = (\sin x)(\sin y), \quad y(0) = 2; \; -6 \le x \le 6, \; -6 \le y \le 6$

Exercises 31 and 32 have no explicit solution in terms of elementary functions. Use a CAS to explore graphically each of the differential equations.

**31.** $y' = \cos(2x - y), \quad y(0) = 2; \; 0 \le x \le 5, \; 0 \le y \le 5$

**32. A Gompertz equation**   $y' = y(1/2 - \ln y), \quad y(0) = 1/3; \; 0 \le x \le 4, \; 0 \le y \le 3$

**33.** Use a CAS to find the solutions of $y' + y = f(x)$ subject to the initial condition $y(0) = 0$, if $f(x)$ is

    **a.** $2x$    **b.** $\sin 2x$    **c.** $3e^{x/2}$    **d.** $2e^{-x/2} \cos 2x$.

Graph all four solutions over the interval $-2 \le x \le 6$ to compare the results.

**34. a.** Use a CAS to plot the slope field of the differential equation

$$y' = \frac{3x^2 + 4x + 2}{2(y - 1)}$$

over the region $-3 \le x \le 3$ and $-3 \le y \le 3$.

    **b.** Separate the variables and use a CAS integrator to find the general solution in implicit form.

c. Using a CAS implicit function grapher, plot solution curves for the arbitrary constant values $C = -6, -4, -2, 0, 2, 4, 6$.

d. Find and graph the solution that satisfies the initial condition $y(0) = -1$.

In Exercises 35–38, use Euler's method with the specified step size to estimate the value of the solution at the given point $x^*$. Find the value of the exact solution at $x^*$.

**35.** $y' = 2xe^{x^2}$, $y(0) = 2$, $dx = 0.1$, $x^* = 1$

**36.** $y' = 2y^2(x - 1)$, $y(2) = -1/2$, $dx = 0.1$, $x^* = 3$

**37.** $y' = \sqrt{x}/y$, $y > 0$, $y(0) = 1$, $dx = 0.1$, $x^* = 1$

**38.** $y' = 1 + y^2$, $y(0) = 0$, $dx = 0.1$, $x^* = 1$

Use a CAS to explore graphically each of the differential equations in Exercises 39–42. Perform the following steps to help with your explorations.

a. Plot a slope field for the differential equation in the given $xy$-window.

b. Find the general solution of the differential equation using your CAS DE solver.

c. Graph the solutions for the values of the arbitrary constant $C = -2, -1, 0, 1, 2$ superimposed on your slope field plot.

d. Find and graph the solution that satisfies the specified initial condition over the interval $[0, b]$.

e. Find the Euler numerical approximation to the solution of the initial value problem with 4 subintervals of the $x$-interval and plot the Euler approximation superimposed on the graph produced in part (d).

f. Repeat part (e) for 8, 16, and 32 subintervals. Plot these three Euler approximations superimposed on the graph from part (e).

g. Find the error ($y$(exact) $-$ $y$(Euler)) at the specified point $x = b$ for each of your four Euler approximations. Discuss the improvement in the percentage error.

**39.** $y' = x + y$, $y(0) = -7/10$; $-4 \le x \le 4$, $-4 \le y \le 4$; $b = 1$

**40.** $y' = -x/y$, $y(0) = 2$; $-3 \le x \le 3$, $-3 \le y \le 3$; $b = 2$

**41.** $y' = y(2 - y)$, $y(0) = 1/2$; $0 \le x \le 4, 0 \le y \le 3$; $b = 3$

**42.** $y' = (\sin x)(\sin y)$, $y(0) = 2$; $-6 \le x \le 6$, $-6 \le y \le 6$; $b = 3\pi/2$

# $9.2$ First-Order Linear Equations

A first-order **linear** differential equation is one that can be written in the form

$$\frac{dy}{dx} + P(x)y = Q(x), \tag{1}$$

where $P$ and $Q$ are continuous functions of $x$. Equation (1) is the linear equation's **standard form**. Since the exponential growth/decay equation $dy/dx = ky$ (Section 7.2) can be put in the standard form

$$\frac{dy}{dx} - ky = 0,$$

we see it is a linear equation with $P(x) = -k$ and $Q(x) = 0$. Equation (1) is *linear* (in $y$) because $y$ and its derivative $dy/dx$ occur only to the first power, they are not multiplied together, nor do they appear as the argument of a function $\left( \text{such as } \sin y, e^y, \text{ or } \sqrt{dy/dx} \right)$.

**EXAMPLE 1**  Put the following equation in standard form:

$$x\frac{dy}{dx} = x^2 + 3y, \qquad x > 0.$$

**Solution**

$$x\frac{dy}{dx} = x^2 + 3y$$

$$\frac{dy}{dx} = x + \frac{3}{x}y \qquad \text{Divide by } x.$$

$$\frac{dy}{dx} - \frac{3}{x}y = x \qquad \text{Standard form with } P(x) = -3/x \text{ and } Q(x) = x$$

Notice that $P(x)$ is $-3/x$, not $+3/x$. The standard form is $y' + P(x)y = Q(x)$, so the minus sign is part of the formula for $P(x)$. ∎

### Solving Linear Equations

We solve the equation

$$\frac{dy}{dx} + P(x)y = Q(x)$$

by multiplying both sides by a *positive* function $v(x)$ that transforms the left-hand side into the derivative of the product $v(x) \cdot y$. We will show how to find $v$ in a moment, but first we want to show how, once found, it provides the solution we seek.

Here is why multiplying by $v(x)$ works:

$$\frac{dy}{dx} + P(x)y = Q(x) \qquad \text{Original equation is in standard form.}$$

$$v(x)\frac{dy}{dx} + P(x)v(x)y = v(x)Q(x) \qquad \text{Multiply by positive } v(x).$$

$$\frac{d}{dx}(v(x) \cdot y) = v(x)Q(x) \qquad \begin{array}{l} v(x) \text{ is chosen to make} \\ v\dfrac{dy}{dx} + Pvy = \dfrac{d}{dx}(v \cdot y). \end{array}$$

$$v(x) \cdot y = \int v(x)Q(x)\, dx \qquad \begin{array}{l} \text{Integrate with respect} \\ \text{to } x. \end{array}$$

$$y = \frac{1}{v(x)}\int v(x)Q(x)\, dx \qquad (2)$$

Equation (2) expresses the solution of Equation (1) in terms of the functions $v(x)$ and $Q(x)$. We call $v(x)$ an **integrating factor** for Equation (1) because its presence makes the equation integrable.

Why doesn't the formula for $P(x)$ appear in the solution as well? It does, but indirectly, in the construction of the positive function $v(x)$. We have

$$\frac{d}{dx}(vy) = v\frac{dy}{dx} + Pvy \qquad \text{Condition imposed on } v$$

$$v\frac{dy}{dx} + y\frac{dv}{dx} = v\frac{dy}{dx} + Pvy \qquad \text{Derivative Product Rule}$$

$$y\frac{dv}{dx} = Pvy \qquad \text{The terms } v\dfrac{dy}{dx} \text{ cancel.}$$

This last equation will hold if

$$\frac{dv}{dx} = Pv$$

$$\frac{dv}{v} = P\, dx \qquad \text{Variables separated, } v > 0$$

$$\int \frac{dv}{v} = \int P\, dx \qquad \text{Integrate both sides.}$$

$$\ln v = \int P\, dx \qquad \begin{array}{l} \text{Since } v > 0, \text{ we do not need absolute} \\ \text{value signs in } \ln v. \end{array}$$

$$e^{\ln v} = e^{\int P\, dx} \qquad \text{Exponentiate both sides to solve for } v.$$

$$v = e^{\int P\, dx} \qquad (3)$$

Thus a formula for the general solution to Equation (1) is given by Equation (2), where $v(x)$ is given by Equation (3). However, rather than memorizing the formula, just remember how to find the integrating factor once you have the standard form so $P(x)$ is correctly identified. Any antiderivative of $P$ works for Equation (3).

> To solve the linear equation $y' + P(x)y = Q(x)$, multiply both sides by the integrating factor $v(x) = e^{\int P(x)\,dx}$ and integrate both sides.

When you integrate the product on the left-hand side in this procedure, you always obtain the product $v(x)y$ of the integrating factor and solution function $y$ because of the way $v$ is defined.

**EXAMPLE 2** Solve the equation

$$x\frac{dy}{dx} = x^2 + 3y, \qquad x > 0.$$

**HISTORICAL BIOGRAPHY**
Adrien Marie Legendre
(1752–1833)

**Solution** First we put the equation in standard form (Example 1):

$$\frac{dy}{dx} - \frac{3}{x}y = x,$$

so $P(x) = -3/x$ is identified.

The integrating factor is

$$
\begin{aligned}
v(x) = e^{\int P(x)\,dx} &= e^{\int (-3/x)\,dx} && \text{Constant of integration is 0,} \\
&= e^{-3\ln|x|} && \text{so } v \text{ is as simple as possible.} \\
&= e^{-3\ln x} && x > 0 \\
&= e^{\ln x^{-3}} = \frac{1}{x^3}.
\end{aligned}
$$

Next we multiply both sides of the standard form by $v(x)$ and integrate:

$$\frac{1}{x^3} \cdot \left( \frac{dy}{dx} - \frac{3}{x}y \right) = \frac{1}{x^3} \cdot x$$

$$\frac{1}{x^3}\frac{dy}{dx} - \frac{3}{x^4}y = \frac{1}{x^2}$$

$$\frac{d}{dx}\left( \frac{1}{x^3}y \right) = \frac{1}{x^2} \qquad \text{Left-hand side is } \frac{d}{dx}(v \cdot y).$$

$$\frac{1}{x^3}y = \int \frac{1}{x^2}\,dx \qquad \text{Integrate both sides.}$$

$$\frac{1}{x^3}y = -\frac{1}{x} + C.$$

Solving this last equation for $y$ gives the general solution:

$$y = x^3\left( -\frac{1}{x} + C \right) = -x^2 + Cx^3, \qquad x > 0. \qquad \blacksquare$$

**EXAMPLE 3** Find the particular solution of

$$3xy' - y = \ln x + 1, \qquad x > 0,$$

satisfying $y(1) = -2$.

**Solution** With $x > 0$, we write the equation in standard form:

$$y' - \frac{1}{3x}y = \frac{\ln x + 1}{3x}.$$

Then the integrating factor is given by

$$v = e^{\int -dx/3x} = e^{(-1/3)\ln x} = x^{-1/3}. \qquad x > 0$$

Thus

$$x^{-1/3}y = \frac{1}{3}\int (\ln x + 1)x^{-4/3}\, dx. \qquad \text{Left-hand side is } vy.$$

Integration by parts of the right-hand side gives

$$x^{-1/3}y = -x^{-1/3}(\ln x + 1) + \int x^{-4/3}\, dx + C.$$

Therefore

$$x^{-1/3}y = -x^{-1/3}(\ln x + 1) - 3x^{-1/3} + C$$

or, solving for $y$,

$$y = -(\ln x + 4) + Cx^{1/3}.$$

When $x = 1$ and $y = -2$ this last equation becomes

$$-2 = -(0 + 4) + C,$$

so

$$C = 2.$$

Substitution into the equation for $y$ gives the particular solution

$$y = 2x^{1/3} - \ln x - 4. \qquad ■$$

In solving the linear equation in Example 2, we integrated both sides of the equation after multiplying each side by the integrating factor. However, we can shorten the amount of work, as in Example 3, by remembering that the left-hand side *always* integrates into the product $v(x) \cdot y$ of the integrating factor times the solution function. From Equation (2) this means that

$$v(x)y = \int v(x)Q(x)\, dx. \qquad (4)$$

We need only integrate the product of the integrating factor $v(x)$ with $Q(x)$ on the right-hand side of Equation (1) and then equate the result with $v(x)y$ to obtain the general solution. Nevertheless, to emphasize the role of $v(x)$ in the solution process, we sometimes follow the complete procedure as illustrated in Example 2.

Observe that if the function $Q(x)$ is identically zero in the standard form given by Equation (1), the linear equation is separable and can be solved by the method of Section 7.4:

$$\frac{dy}{dx} + P(x)y = Q(x)$$

$$\frac{dy}{dx} + P(x)y = 0 \qquad Q(x) = 0$$

$$\frac{dy}{y} = -P(x)\, dx \qquad \text{Separating the variables}$$

**FIGURE 9.8** The *RL* circuit in Example 4.

## *RL* Circuits

The diagram in Figure 9.8 represents an electrical circuit whose total resistance is a constant *R* ohms and whose self-inductance, shown as a coil, is *L* henries, also a constant. There is a switch whose terminals at *a* and *b* can be closed to connect a constant electrical source of *V* volts.

Ohm's Law, $V = RI$, has to be augmented for such a circuit. The correct equation accounting for both resistance and inductance is

$$L\frac{di}{dt} + Ri = V, \qquad (5)$$

where *i* is the current in amperes and *t* is the time in seconds. By solving this equation, we can predict how the current will flow after the switch is closed.

**EXAMPLE 4** The switch in the *RL* circuit in Figure 9.8 is closed at time $t = 0$. How will the current flow as a function of time?

**Solution** Equation (5) is a first-order linear differential equation for *i* as a function of *t*. Its standard form is

$$\frac{di}{dt} + \frac{R}{L}i = \frac{V}{L}, \qquad (6)$$

and the corresponding solution, given that $i = 0$ when $t = 0$, is

$$i = \frac{V}{R} - \frac{V}{R}e^{-(R/L)t}. \qquad (7)$$

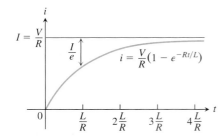

**FIGURE 9.9** The growth of the current in the *RL* circuit in Example 4. *I* is the current's steady-state value. The number $t = L/R$ is the time constant of the circuit. The current gets to within 5% of its steady-state value in 3 time constants (Exercise 27).

(We leave the calculation of the solution for you to do in Exercise 28.) Since *R* and *L* are positive, $-(R/L)$ is negative and $e^{-(R/L)t} \to 0$ as $t \to \infty$. Thus,

$$\lim_{t\to\infty} i = \lim_{t\to\infty}\left(\frac{V}{R} - \frac{V}{R}e^{-(R/L)t}\right) = \frac{V}{R} - \frac{V}{R}\cdot 0 = \frac{V}{R}.$$

At any given time, the current is theoretically less than $V/R$, but as time passes, the current approaches the **steady-state value** $V/R$. According to the equation

$$L\frac{di}{dt} + Ri = V,$$

$I = V/R$ is the current that will flow in the circuit if either $L = 0$ (no inductance) or $di/dt = 0$ (steady current, $i = $ constant) (Figure 9.9).

Equation (7) expresses the solution of Equation (6) as the sum of two terms: a steady-state solution $V/R$ and a transient solution $-(V/R)e^{-(R/L)t}$ that tends to zero as $t \to \infty$. ∎

---

## Exercises 9.2

### First-Order Linear Equations

Solve the differential equations in Exercises 1–14.

**1.** $x\dfrac{dy}{dx} + y = e^x, \quad x > 0$

**2.** $e^x\dfrac{dy}{dx} + 2e^x y = 1$

**3.** $xy' + 3y = \dfrac{\sin x}{x^2}, \quad x > 0$

**4.** $y' + (\tan x)y = \cos^2 x, \quad -\pi/2 < x < \pi/2$

**5.** $x\dfrac{dy}{dx} + 2y = 1 - \dfrac{1}{x}, \quad x > 0$

**6.** $(1 + x)\,y' + y = \sqrt{x}$

**7.** $2y' = e^{x/2} + y$

**8.** $e^{2x}\,y' + 2e^{2x}\,y = 2x$

**9.** $xy' - y = 2x \ln x$

**10.** $x\dfrac{dy}{dx} = \dfrac{\cos x}{x} - 2y, \quad x > 0$

**11.** $(t - 1)^3 \dfrac{ds}{dt} + 4(t - 1)^2 s = t + 1, \quad t > 1$

**12.** $(t + 1) \dfrac{ds}{dt} + 2s = 3(t + 1) + \dfrac{1}{(t + 1)^2}, \quad t > -1$

**13.** $\sin \theta \dfrac{dr}{d\theta} + (\cos \theta) r = \tan \theta, \quad 0 < \theta < \pi/2$

**14.** $\tan \theta \dfrac{dr}{d\theta} + r = \sin^2 \theta, \quad 0 < \theta < \pi/2$

Solving Initial Value Problems
Solve the initial value problems in Exercises 15–20.

**15.** $\dfrac{dy}{dt} + 2y = 3, \quad y(0) = 1$

**16.** $t \dfrac{dy}{dt} + 2y = t^3, \quad t > 0, \quad y(2) = 1$

**17.** $\theta \dfrac{dy}{d\theta} + y = \sin \theta, \quad \theta > 0, \quad y(\pi/2) = 1$

**18.** $\theta \dfrac{dy}{d\theta} - 2y = \theta^3 \sec \theta \tan \theta, \quad \theta > 0, \quad y(\pi/3) = 2$

**19.** $(x + 1) \dfrac{dy}{dx} - 2(x^2 + x)y = \dfrac{e^{x^2}}{x + 1}, \quad x > -1, \quad y(0) = 5$

**20.** $\dfrac{dy}{dx} + xy = x, \quad y(0) = -6$

**21.** Solve the exponential growth/decay initial value problem for $y$ as a function of $t$ by thinking of the differential equation as a first-order linear equation with $P(x) = -k$ and $Q(x) = 0$:

$$\dfrac{dy}{dt} = ky \quad (k \text{ constant}), \quad y(0) = y_0$$

**22.** Solve the following initial value problem for $u$ as a function of $t$:

$$\dfrac{du}{dt} + \dfrac{k}{m}u = 0 \quad (k \text{ and } m \text{ positive constants}), \quad u(0) = u_0$$

**a.** as a first-order linear equation.

**b.** as a separable equation.

Theory and Examples
**23.** Is either of the following equations correct? Give reasons for your answers.

**a.** $x \displaystyle\int \dfrac{1}{x} dx = x \ln |x| + C$ **b.** $x \displaystyle\int \dfrac{1}{x} dx = x \ln |x| + Cx$

**24.** Is either of the following equations correct? Give reasons for your answers.

**a.** $\dfrac{1}{\cos x} \displaystyle\int \cos x \, dx = \tan x + C$

**b.** $\dfrac{1}{\cos x} \displaystyle\int \cos x \, dx = \tan x + \dfrac{C}{\cos x}$

**25. Current in a closed RL circuit** How many seconds after the switch in an $RL$ circuit is closed will it take the current $i$ to reach half of its steady-state value? Notice that the time depends on $R$ and $L$ and not on how much voltage is applied.

**26. Current in an open RL circuit** If the switch is thrown open after the current in an $RL$ circuit has built up to its steady-state value $I = V/R$, the decaying current (see accompanying figure) obeys the equation

$$L \dfrac{di}{dt} + Ri = 0,$$

which is Equation (5) with $V = 0$.

**a.** Solve the equation to express $i$ as a function of $t$.

**b.** How long after the switch is thrown will it take the current to fall to half its original value?

**c.** Show that the value of the current when $t = L/R$ is $I/e$. (The significance of this time is explained in the next exercise.)

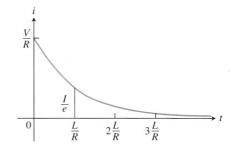

**27. Time constants** Engineers call the number $L/R$ the *time constant* of the $RL$ circuit in Figure 9.9. The significance of the time constant is that the current will reach 95% of its final value within 3 time constants of the time the switch is closed (Figure 9.9). Thus, the time constant gives a built-in measure of how rapidly an individual circuit will reach equilibrium.

**a.** Find the value of $i$ in Equation (7) that corresponds to $t = 3L/R$ and show that it is about 95% of the steady-state value $I = V/R$.

**b.** Approximately what percentage of the steady-state current will be flowing in the circuit 2 time constants after the switch is closed (i.e., when $t = 2L/R$)?

**28. Derivation of Equation (7) in Example 4**

**a.** Show that the solution of the equation

$$\dfrac{di}{dt} + \dfrac{R}{L}i = \dfrac{V}{L}$$

is

$$i = \dfrac{V}{R} + Ce^{-(R/L)t}.$$

**b.** Then use the initial condition $i(0) = 0$ to determine the value of $C$. This will complete the derivation of Equation (7).

**c.** Show that $i = V/R$ is a solution of Equation (6) and that $i = Ce^{-(R/L)t}$ satisfies the equation

$$\dfrac{di}{dt} + \dfrac{R}{L}i = 0.$$

**HISTORICAL BIOGRAPHY**

James Bernoulli
(1654–1705)

A **Bernoulli differential equation** is of the form

$$\frac{dy}{dx} + P(x)y = Q(x)y^n.$$

Observe that, if $n = 0$ or 1, the Bernoulli equation is linear. For other values of $n$, the substitution $u = y^{1-n}$ transforms the Bernoulli equation into the linear equation

$$\frac{du}{dx} + (1 - n)P(x)u = (1 - n)Q(x).$$

For example, in the equation

$$\frac{dy}{dx} - y = e^{-x}y^2$$

we have $n = 2$, so that $u = y^{1-2} = y^{-1}$ and $du/dx = -y^{-2}\,dy/dx$. Then $dy/dx = -y^2\,du/dx = -u^{-2}\,du/dx$. Substitution into the original equation gives

$$-u^{-2}\frac{du}{dx} - u^{-1} = e^{-x}u^{-2}$$

or, equivalently,

$$\frac{du}{dx} + u = -e^{-x}.$$

This last equation is linear in the (unknown) dependent variable $u$.

Solve the Bernoulli equations in Exercises 29–32.

**29.** $y' - y = -y^2$      **30.** $y' - y = xy^2$

**31.** $xy' + y = y^{-2}$      **32.** $x^2y' + 2xy = y^3$

# $9.3$ Applications

We now look at four applications of first-order differential equations. The first application analyzes an object moving along a straight line while subject to a force opposing its motion. The second is a model of population growth. The third application considers a curve or curves intersecting each curve in a second family of curves *orthogonally* (that is, at right angles). The final application analyzes chemical concentrations entering and leaving a container. The various models involve separable or linear first-order equations.

## Motion with Resistance Proportional to Velocity

In some cases it is reasonable to assume that the resistance encountered by a moving object, such as a car coasting to a stop, is proportional to the object's velocity. The faster the object moves, the more its forward progress is resisted by the air through which it passes. Picture the object as a mass $m$ moving along a coordinate line with position function $s$ and velocity $v$ at time $t$. From Newton's second law of motion, the resisting force opposing the motion is

$$\text{Force} = \text{mass} \times \text{acceleration} = m\frac{dv}{dt}.$$

If the resisting force is proportional to velocity, we have

$$m\frac{dv}{dt} = -kv \qquad \text{or} \qquad \frac{dv}{dt} = -\frac{k}{m}v \qquad (k > 0).$$

This is a separable differential equation representing exponential change. The solution to the equation with initial condition $v = v_0$ at $t = 0$ is (Section 7.4)

$$v = v_0 e^{-(k/m)t}. \tag{1}$$

What can we learn from Equation (1)? For one thing, we can see that if $m$ is something large, like the mass of a 20,000-tonne ore boat in Lake Michigan, it will take a long time for the velocity to approach zero (because $t$ must be large in the exponent of the equation in order to make $kt/m$ large enough for $v$ to be small). We can learn even more if we integrate Equation (1) to find the position $s$ as a function of time $t$.

Suppose that an object is coasting to a stop and the only force acting on it is a resistance proportional to its speed. How far will it coast? To find out, we start with Equation (1) and solve the initial value problem

$$\frac{ds}{dt} = v_0 e^{-(k/m)t}, \qquad s(0) = 0.$$

Integrating with respect to $t$ gives

$$s = -\frac{v_0 m}{k}e^{-(k/m)t} + C.$$

Substituting $s = 0$ when $t = 0$ gives

$$0 = -\frac{v_0 m}{k} + C \qquad \text{and} \qquad C = \frac{v_0 m}{k}.$$

The body's position at time $t$ is therefore

$$s(t) = -\frac{v_0 m}{k}e^{-(k/m)t} + \frac{v_0 m}{k} = \frac{v_0 m}{k}(1 - e^{-(k/m)t}). \qquad (2)$$

To find how far the body will coast, we find the limit of $s(t)$ as $t \to \infty$. Since $-(k/m) < 0$, we know that $e^{-(k/m)t} \to 0$ as $t \to \infty$, so that

$$\lim_{t\to\infty} s(t) = \lim_{t\to\infty}\frac{v_0 m}{k}(1 - e^{-(k/m)t})$$

$$= \frac{v_0 m}{k}(1 - 0) = \frac{v_0 m}{k}.$$

Thus,

$$\text{Distance coasted} = \frac{v_0 m}{k}. \qquad (3)$$

The number $v_0 m/k$ is only an upper bound (albeit a useful one). It is true to life in one respect, at least: If $m$ is large, the body will coast a long way.

**EXAMPLE 1**    For a 90-kg ice skater, the $k$ in Equation (1) is about 5 kg/s. How long will it take the skater to coast from 3.3 m/s (11.88 km/h) to 0.3 m/s? How far will the skater coast before coming to a complete stop?

**Solution**    We answer the first question by solving Equation (1) for $t$:

$$3.3e^{-t/18} = 0.3 \qquad\qquad \text{Eq. (1) with } k = 5,$$
$$e^{-t/18} = 1/11 \qquad\qquad m = 90, v_0 = 3.3, v = 0.3$$
$$-t/18 = \ln(1/11) = -\ln 11$$
$$t = 18\ln 11 \approx 43\,\text{s}.$$

We answer the second question with Equation (3):

$$\text{Distance coasted} = \frac{v_0 m}{k} = \frac{3.3 \cdot 90}{5}$$

$$= 59.4 \text{ m.} \qquad\blacksquare$$

## Inaccuracy of the Exponential Population Growth Model

In Section 7.4 we modeled population growth with the Law of Exponential Change:

$$\frac{dP}{dt} = kP, \qquad P(0) = P_0$$

where $P$ is the population at time $t$, $k > 0$ is a constant growth rate, and $P_0$ is the size of the population at time $t = 0$. In Section 7.4 we found the solution $P = P_0 e^{kt}$ to this model.

To assess the model, notice that the exponential growth differential equation says that

$$\frac{dP/dt}{P} = k \qquad (4)$$

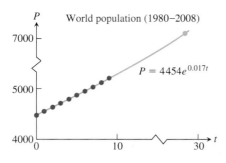

**FIGURE 9.10** Notice that the value of the solution $P = 4454e^{0.017t}$ is 7169 when $t = 28$, which is nearly 7% more than the actual population in 2008.

**TABLE 9.3** World population (midyear)

| Year | Population (millions) | $\Delta P/P$ |
|------|-----------------------|--------------|
| 1980 | 4454 | $76/4454 \approx 0.0171$ |
| 1981 | 4530 | $80/4530 \approx 0.0177$ |
| 1982 | 4610 | $80/4610 \approx 0.0174$ |
| 1983 | 4690 | $80/4690 \approx 0.0171$ |
| 1984 | 4770 | $81/4770 \approx 0.0170$ |
| 1985 | 4851 | $82/4851 \approx 0.0169$ |
| 1986 | 4933 | $85/4933 \approx 0.0172$ |
| 1987 | 5018 | $87/5018 \approx 0.0173$ |
| 1988 | 5105 | $85/5105 \approx 0.0167$ |
| 1989 | 5190 | |

*Source*: U.S. Bureau of the Census (Sept., 2007): www.census .gov/ipc/www/idb.

is constant. This rate is called the **relative growth rate**. Now, Table 9.3 gives the world population at midyear for the years 1980 to 1989. Taking $dt = 1$ and $dP \approx \Delta P$, we see from the table that the relative growth rate in Equation (4) is approximately the constant 0.017. Thus, based on the tabled data with $t = 0$ representing 1980, $t = 1$ representing 1981, and so forth, the world population could be modeled by the initial value problem,

$$\frac{dP}{dt} = 0.017P, \qquad P(0) = 4454.$$

The solution to this initial value problem gives the population function $P = 4454e^{0.017t}$. In year 2008 (so $t = 28$), the solution predicts the world population in midyear to be about 7169 million, or 7.2 billion (Figure 9.10), which is more than the actual population of 6707 million from the U.S. Bureau of the Census. A more realistic model would consider environmental and other factors affecting the growth rate, which has been steadily declining to about 0.012 since 1987. We consider one such model in Section 9.4.

**FIGURE 9.11** An orthogonal trajectory intersects the family of curves at right angles, or orthogonally.

## Orthogonal Trajectories

An **orthogonal trajectory** of a family of curves is a curve that intersects each curve of the family at right angles, or *orthogonally* (Figure 9.11). For instance, each straight line through the origin is an orthogonal trajectory of the family of circles $x^2 + y^2 = a^2$, centered at the origin (Figure 9.12). Such mutually orthogonal systems of curves are of particular importance in physical problems related to electrical potential, where the curves in one family correspond to strength of an electric field and those in the other family correspond to constant electric potential. They also occur in hydrodynamics and heat-flow problems.

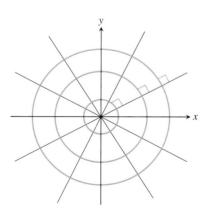

**FIGURE 9.12** Every straight line through the origin is orthogonal to the family of circles centered at the origin.

**EXAMPLE 2** Find the orthogonal trajectories of the family of curves $xy = a$, where $a \neq 0$ is an arbitrary constant.

**Solution** The curves $xy = a$ form a family of hyperbolas having the coordinate axes as asymptotes. First we find the slopes of each curve in this family, or their $dy/dx$ values. Differentiating $xy = a$ implicitly gives

$$x\frac{dy}{dx} + y = 0 \qquad \text{or} \qquad \frac{dy}{dx} = -\frac{y}{x}.$$

Thus the slope of the tangent line at any point $(x, y)$ on one of the hyperbolas $xy = a$ is $y' = -y/x$. On an orthogonal trajectory the slope of the tangent line at this same point must be the negative reciprocal, or $x/y$. Therefore, the orthogonal trajectories must satisfy the differential equation

$$\frac{dy}{dx} = \frac{x}{y}.$$

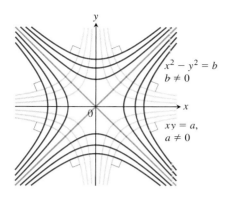

**FIGURE 9.13** Each curve is orthogonal to every curve it meets in the other family (Example 2).

This differential equation is separable and we solve it as in Section 7.4:

$$y \, dy = x \, dx \qquad \text{Separate variables.}$$

$$\int y \, dy = \int x \, dx \qquad \text{Integrate both sides.}$$

$$\frac{1}{2}y^2 = \frac{1}{2}x^2 + C$$

$$y^2 - x^2 = b, \tag{5}$$

where $b = 2C$ is an arbitrary constant. The orthogonal trajectories are the family of hyperbolas given by Equation (5) and sketched in Figure 9.13. ∎

## Mixture Problems

Suppose a chemical in a liquid solution (or dispersed in a gas) runs into a container holding the liquid (or the gas) with, possibly, a specified amount of the chemical dissolved as well. The mixture is kept uniform by stirring and flows out of the container at a known rate. In this process, it is often important to know the concentration of the chemical in the container at any given time. The differential equation describing the process is based on the formula

$$\begin{pmatrix} \text{Rate of change} \\ \text{of amount} \\ \text{in container} \end{pmatrix} = \begin{pmatrix} \text{rate at which} \\ \text{chemical} \\ \text{arrives} \end{pmatrix} - \begin{pmatrix} \text{rate at which} \\ \text{chemical} \\ \text{departs.} \end{pmatrix}. \tag{6}$$

If $y(t)$ is the amount of chemical in the container at time $t$ and $V(t)$ is the total volume of liquid in the container at time $t$, then the departure rate of the chemical at time $t$ is

$$\text{Departure rate} = \frac{y(t)}{V(t)} \cdot (\text{outflow rate})$$

$$= \begin{pmatrix} \text{concentration in} \\ \text{container at time } t \end{pmatrix} \cdot (\text{outflow rate}). \tag{7}$$

Accordingly, Equation (6) becomes

$$\frac{dy}{dt} = (\text{chemical's arrival rate}) - \frac{y(t)}{V(t)} \cdot (\text{outflow rate}). \tag{8}$$

If, say, $y$ is measured in kilograms, $V$ in liters, and $t$ in minutes, the units in Equation (8) are

$$\frac{\text{kilograms}}{\text{minutes}} = \frac{\text{kilograms}}{\text{minutes}} - \frac{\text{kilograms}}{\text{liters}} \cdot \frac{\text{liters}}{\text{minutes}}.$$

**EXAMPLE 3**    In an oil refinery, a storage tank contains 10,000 L of gasoline that initially has 50 kg of an additive dissolved in it. In preparation for winter weather, gasoline containing 0.2 kg of additive per liter is pumped into the tank at a rate of 200 L/min.

The well-mixed solution is pumped out at a rate of 220 L/min. How much of the additive is in the tank 20 min after the pumping process begins (Figure 9.14)?

200 L/min containing 0.2 kg/L

220 L/min containing $\frac{y}{V}$ kg/L

**FIGURE 9.14**  The storage tank in Example 3 mixes input liquid with stored liquid to produce an output liquid.

**Solution**  Let $y$ be the amount (in kilograms) of additive in the tank at time $t$. We know that $y = 50$ when $t = 0$. The number of liters of gasoline and additive in solution in the tank at any time $t$ is

$$V(t) = 10{,}000 \text{ L} + \left(200\frac{\text{L}}{\text{min}} - 220\frac{\text{L}}{\text{min}}\right)(t \text{ min})$$

$$= (10{,}000 - 20t) \text{ L}.$$

Therefore,

$$\text{Rate out} = \frac{y(t)}{V(t)} \cdot \text{outflow rate} \qquad \text{Eq. (7)}$$

$$= \left(\frac{y}{10{,}000 - 20t}\right)220 \qquad \begin{array}{l}\text{Outflow rate is 220 L/min}\\ \text{and } V = 10{,}000 - 20t.\end{array}$$

$$= \frac{220y}{10{,}000 - 20t}\frac{\text{kg}}{\text{min}}.$$

Also,

$$\text{Rate in} = \left(0.2\frac{\text{kg}}{\text{L}}\right)\left(200\frac{\text{L}}{\text{min}}\right)$$

$$= 40\frac{\text{kg}}{\text{min}}.$$

The differential equation modeling the mixture process is

$$\frac{dy}{dt} = 40 - \frac{220y}{10{,}000 - 20t} \qquad \text{Eq. (8)}$$

in kilograms per minute.

To solve this differential equation, we first write it in standard linear form:

$$\frac{dy}{dt} + \frac{220}{10{,}000 - 20t}y = 40.$$

Thus, $P(t) = 220/(10{,}000 - 20t)$ and $Q(t) = 40$. The integrating factor is

$$v(t) = e^{\int P\,dt} = e^{\int \frac{220}{10{,}000 - 20t}dt}$$

$$= e^{-11 \ln (10{,}000 - 20t)} \qquad 10{,}000 - 20t > 0$$

$$= (10{,}000 - 20t)^{-11}.$$

Multiplying both sides of the standard equation by $v(t)$ and integrating both sides gives

$$(10{,}000 - 20t)^{-11} \cdot \left( \frac{dy}{dt} + \frac{220}{10{,}000 - 20t} y \right) = 40(10{,}000 - 20t)^{-11}$$

$$(10{,}000 - 20t)^{-11} \frac{dy}{dt} + 220(10{,}000 - 20t)^{-12} y = 40(10{,}000 - 20t)^{-11}$$

$$\frac{d}{dt} \left[ (10{,}000 - 20t)^{-11} y \right] = 40(10{,}000 - 20t)^{-11}$$

$$(10{,}000 - 20t)^{-11} y = \int 40(10{,}000 - 20t)^{-11} \, dt$$

$$(10{,}000 - 20t)^{-11} y = 40 \cdot \frac{(10{,}000 - 20t)^{-10}}{(-10)(-20)} + C.$$

The general solution is

$$y = 0.2(10{,}000 - 20t) + C(10{,}000 - 20t)^{11}.$$

Because $y = 50$ when $t = 0$, we can determine the value of $C$:

$$50 = 0.2(10{,}000 - 0) + C(10{,}000 - 0)^{11}$$

$$C = -\frac{1950}{(10{,}000)^{11}}.$$

The particular solution of the initial value problem is

$$y = 0.2(10{,}000 - 20t) - \frac{1950}{(10{,}000)^{11}} (10{,}000 - 20t)^{11}.$$

The amount of additive in the tank 20 min after the pumping begins is

$$y(20) = 0.2[\,10{,}000 - 20(20)\,] - \frac{1950}{(10{,}000)^{11}} [\,10{,}000 - 20(20)\,]^{11} \approx 675 \text{ kg}. \qquad \blacksquare$$

# Exercises 9.3

## Motion Along a Line

1. **Coasting bicycle** A 66-kg cyclist on a 7-kg bicycle starts coasting on level ground at 9 m/s. The $k$ in Equation (1) is about 3.9 kg/s.

   a. About how far will the cyclist coast before reaching a complete stop?

   b. How long will it take the cyclist's speed to drop to 1 m/s?

2. **Coasting battleship** Suppose that a battleship has mass around 51,000,000 kg and a $k$ value in Equation (1) of about 59,000 kg/s.

Assume that the ship loses power when it is moving at a speed of 9 m/s.

   a. About how far will the ship coast before it is dead in the water?

   b. About how long will it take the ship's speed to drop to 1 m/s?

3. The data in Table 9.4 were collected with a motion detector and a CBL™ by Valerie Sharritts, then a mathematics teacher at St. Francis DeSales High School in Columbus, Ohio. The table shows the distance $s$ (meters) coasted on inline skates in $t$ (seconds) by her daughter Ashley when she was 10 years old. Find a model

for Ashley's position given by the data in Table 9.4 in the form of Equation (2). Her initial velocity was $v_0 = 2.75$ m/s, her mass $m = 39.92$ kg, and her total coasting distance was 4.91 m.

**TABLE 9.4** Ashley Sharritts skating data

| $t$ (s) | $s$ (m) | $t$ (s) | $s$ (m) | $t$ (s) | $s$ (m) |
|---|---|---|---|---|---|
| 0 | 0 | 2.24 | 3.05 | 4.48 | 4.77 |
| 0.16 | 0.31 | 2.40 | 3.22 | 4.64 | 4.82 |
| 0.32 | 0.57 | 2.56 | 3.38 | 4.80 | 4.84 |
| 0.48 | 0.80 | 2.72 | 3.52 | 4.96 | 4.86 |
| 0.64 | 1.05 | 2.88 | 3.67 | 5.12 | 4.88 |
| 0.80 | 1.28 | 3.04 | 3.82 | 5.28 | 4.89 |
| 0.96 | 1.50 | 3.20 | 3.96 | 5.44 | 4.90 |
| 1.12 | 1.72 | 3.36 | 4.08 | 5.60 | 4.90 |
| 1.28 | 1.93 | 3.52 | 4.18 | 5.76 | 4.91 |
| 1.44 | 2.09 | 3.68 | 4.31 | 5.92 | 4.90 |
| 1.60 | 2.30 | 3.84 | 4.41 | 6.08 | 4.91 |
| 1.76 | 2.53 | 4.00 | 4.52 | 6.24 | 4.90 |
| 1.92 | 2.73 | 4.16 | 4.63 | 6.40 | 4.91 |
| 2.08 | 2.89 | 4.32 | 4.69 | 6.56 | 4.91 |

4. **Coasting to a stop** Table 9.5 shows the distance $s$ (meters) coasted on inline skates in terms of time $t$ (seconds) by Kelly Schmitzer. Find a model for her position in the form of Equation (2). Her initial velocity was $v_0 = 0.80$ m/s, her mass $m = 49.90$ kg, and her total coasting distance was 1.32 m.

**TABLE 9.5** Kelly Schmitzer skating data

| $t$ (s) | $s$ (m) | $t$ (s) | $s$ (m) | $t$ (s) | $s$ (m) |
|---|---|---|---|---|---|
| 0 | 0 | 1.5 | 0.89 | 3.1 | 1.30 |
| 0.1 | 0.07 | 1.7 | 0.97 | 3.3 | 1.31 |
| 0.3 | 0.22 | 1.9 | 1.05 | 3.5 | 1.32 |
| 0.5 | 0.36 | 2.1 | 1.11 | 3.7 | 1.32 |
| 0.7 | 0.49 | 2.3 | 1.17 | 3.9 | 1.32 |
| 0.9 | 0.60 | 2.5 | 1.22 | 4.1 | 1.32 |
| 1.1 | 0.71 | 2.7 | 1.25 | 4.3 | 1.32 |
| 1.3 | 0.81 | 2.9 | 1.28 | 4.5 | 1.32 |

**Orthogonal Trajectories**

In Exercises 5–10, find the orthogonal trajectories of the family of curves. Sketch several members of each family.

5. $y = mx$

6. $y = cx^2$

7. $kx^2 + y^2 = 1$

8. $2x^2 + y^2 = c^2$

9. $y = ce^{-x}$

10. $y = e^{kx}$

11. Show that the curves $2x^2 + 3y^2 = 5$ and $y^2 = x^3$ are orthogonal.

12. Find the family of solutions of the given differential equation and the family of orthogonal trajectories. Sketch both families.

   a. $x\,dx + y\,dy = 0$   b. $x\,dy - 2y\,dx = 0$

**Mixture Problems**

13. **Salt mixture** A tank initially contains 400 L of brine in which 20 kg/L of salt are dissolved. A brine containing 0.2 kg/L of salt runs into the tank at the rate of 20 L/min. The mixture is kept uniform by stirring and flows out of the tank at the rate of 16 L/min.

   a. At what rate (kilograms per minute) does salt enter the tank at time $t$?

   b. What is the volume of brine in the tank at time $t$?

   c. At what rate (kilograms per minute) does salt leave the tank at time $t$?

   d. Write down and solve the initial value problem describing the mixing process.

   e. Find the concentration of salt in the tank 25 min after the process starts.

14. **Mixture problem** An 800-L tank is half full of distilled water. At time $t = 0$, a solution containing 50 grams/L of concentrate enters the tank at the rate of 20 L/min, and the well-stirred mixture is withdrawn at the rate of 12 L/min.

   a. At what time will the tank be full?

   b. At the time the tank is full, how many kilograms of concentrate will it contain?

15. **Fertilizer mixture** A tank contains 400 L of fresh water. A solution containing 0.1 kg/L of soluble lawn fertilizer runs into the tank at the rate of 4 L/min, and the mixture is pumped out of the tank at the rate of 12 L/min. Find the maximum amount of fertilizer in the tank and the time required to reach the maximum.

16. **Carbon monoxide pollution** An executive conference room of a corporation contains 120 m$^3$ of air initially free of carbon monoxide. Starting at time $t = 0$, cigarette smoke containing 4% carbon monoxide is blown into the room at the rate of 0.008 m$^3$/min. A ceiling fan keeps the air in the room well circulated and the air leaves the room at the same rate of 0.008 m$^3$/min. Find the time when the concentration of carbon monoxide in the room reaches 0.01%.

# $9.4$ Graphical Solutions of Autonomous Equations

In Chapter 4 we learned that the sign of the first derivative tells where the graph of a function is increasing and where it is decreasing. The sign of the second derivative tells the concavity of the graph. We can build on our knowledge of how derivatives determine the shape of a graph to solve differential equations graphically. We will see that the ability to discern physical behavior from graphs is a powerful tool in understanding real-world systems. The starting ideas for a graphical solution are the notions of *phase line* and *equilibrium value*. We arrive at these notions by investigating, from a point of view quite different from that studied in Chapter 4, what happens when the derivative of a differentiable function is zero.

## Equilibrium Values and Phase Lines

When we differentiate implicitly the equation

$$\frac{1}{5}\ln(5y - 15) = x + 1,$$

we obtain

$$\frac{1}{5}\left(\frac{5}{5y - 15}\right)\frac{dy}{dx} = 1.$$

Solving for $y' = dy/dx$ we find $y' = 5y - 15 = 5(y - 3)$. In this case the derivative $y'$ is a function of $y$ only (the dependent variable) and is zero when $y = 3$.

A differential equation for which $dy/dx$ is a function of $y$ only is called an **autonomous** differential equation. Let's investigate what happens when the derivative in an autonomous equation equals zero. We assume any derivatives are continuous.

---

**DEFINITION**  If $dy/dx = g(y)$ is an autonomous differential equation, then the values of $y$ for which $dy/dx = 0$ are called **equilibrium values** or **rest points**.

---

Thus, equilibrium values are those at which no change occurs in the dependent variable, so $y$ is at *rest*. The emphasis is on the value of $y$ where $dy/dx = 0$, not the value of $x$, as we studied in Chapter 4. For example, the equilibrium values for the autonomous differential equation

$$\frac{dy}{dx} = (y + 1)(y - 2)$$

are $y = -1$ and $y = 2$.

To construct a graphical solution to an autonomous differential equation, we first make a **phase line** for the equation, a plot on the $y$-axis that shows the equation's equilibrium values along with the intervals where $dy/dx$ and $d^2y/dx^2$ are positive and negative. Then we know where the solutions are increasing and decreasing, and the concavity of the solution curves. These are the essential features we found in Section 4.4, so we can determine the shapes of the solution curves without having to find formulas for them.

**EXAMPLE 1**    Draw a phase line for the equation

$$\frac{dy}{dx} = (y + 1)(y - 2)$$

and use it to sketch solutions to the equation.

**Solution**

1.  *Draw a number line for $y$ and mark the equilibrium values $y = -1$ and $y = 2$, where $dy/dx = 0$.*

2. *Identify and label the intervals where $y' > 0$ and $y' < 0$.* This step resembles what we did in Section 4.3, only now we are marking the $y$-axis instead of the $x$-axis.

We can encapsulate the information about the sign of $y'$ on the phase line itself. Since $y' > 0$ on the interval to the left of $y = -1$, a solution of the differential equation with a $y$-value less than $-1$ will increase from there toward $y = -1$. We display this information by drawing an arrow on the interval pointing to $-1$.

Similarly, $y' < 0$ between $y = -1$ and $y = 2$, so any solution with a value in this interval will decrease toward $y = -1$.

For $y > 2$, we have $y' > 0$, so a solution with a $y$-value greater than 2 will increase from there without bound.

In short, solution curves below the horizontal line $y = -1$ in the $xy$-plane rise toward $y = -1$. Solution curves between the lines $y = -1$ and $y = 2$ fall away from $y = 2$ toward $y = -1$. Solution curves above $y = 2$ rise away from $y = 2$ and keep going up.

3. *Calculate $y''$ and mark the intervals where $y'' > 0$ and $y'' < 0$.* To find $y''$, we differentiate $y'$ with respect to $x$, using implicit differentiation.

$$y' = (y + 1)(y - 2) = y^2 - y - 2 \qquad \text{Formula for } y' \ldots$$

$$y'' = \frac{d}{dx}(y') = \frac{d}{dx}(y^2 - y - 2)$$

$$= 2yy' - y' \qquad \text{differentiated implicitly with respect to } x$$

$$= (2y - 1)y'$$

$$= (2y - 1)(y + 1)(y - 2).$$

From this formula, we see that $y''$ changes sign at $y = -1$, $y = 1/2$, and $y = 2$. We add the sign information to the phase line.

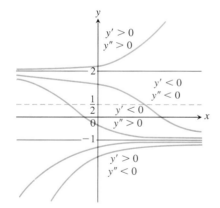

**FIGURE 9.15** Graphical solutions from Example 1 include the horizontal lines $y = -1$ and $y = 2$ through the equilibrium values. No two solution curves can ever cross or touch each other.

4. *Sketch an assortment of solution curves in the $xy$-plane.* The horizontal lines $y = -1$, $y = 1/2$, and $y = 2$ partition the plane into horizontal bands in which we know the signs of $y'$ and $y''$. In each band, this information tells us whether the solution curves rise or fall and how they bend as $x$ increases (Figure 9.15).

The "equilibrium lines" $y = -1$ and $y = 2$ are also solution curves. (The constant functions $y = -1$ and $y = 2$ satisfy the differential equation.) Solution curves that cross the line $y = 1/2$ have an inflection point there. The concavity changes from concave down (above the line) to concave up (below the line).

As predicted in Step 2, solutions in the middle and lower bands approach the equilibrium value $y = -1$ as $x$ increases. Solutions in the upper band rise steadily away from the value $y = 2$. ∎

## Stable and Unstable Equilibria

Look at Figure 9.15 once more, in particular at the behavior of the solution curves near the equilibrium values. Once a solution curve has a value near $y = -1$, it tends steadily toward that value; $y = -1$ is a **stable equilibrium**. The behavior near $y = 2$ is just the opposite: All solutions except the equilibrium solution $y = 2$ itself move *away* from it as $x$ increases. We call $y = 2$ an **unstable equilibrium**. If the solution is *at* that value, it stays, but if it is off by any amount, no matter how small, it moves away. (Sometimes an equilibrium value is unstable because a solution moves away from it only on one side of the point.)

Now that we know what to look for, we can already see this behavior on the initial phase line (the second diagram in Step 2 of Example 1). The arrows lead away from $y = 2$ and, once to the left of $y = 2$, toward $y = -1$.

We now present several applied examples for which we can sketch a family of solution curves to the differential equation models using the method in Example 1.

### Newton's Law of Cooling

**FIGURE 9.16**  First step in constructing the phase line for Newton's Law of Cooling. The temperature tends toward the equilibrium (surrounding-medium) value in the long run.

In Section 7.4 we solved analytically the differential equation

$$\frac{dH}{dt} = -k(H - H_S), \qquad k > 0$$

modeling Newton's Law of Cooling. Here $H$ is the temperature of an object at time $t$ and $H_S$ is the constant temperature of the surrounding medium.

Suppose that the surrounding medium (say, a room in a house) has a constant Celsius temperature of 15°C. We can then express the difference in temperature as $H(t) - 15$. Assuming $H$ is a differentiable function of time $t$, by Newton's Law of Cooling, there is a constant of proportionality $k > 0$ such that

$$\frac{dH}{dt} = -k(H - 15) \tag{1}$$

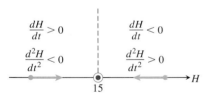

**FIGURE 9.17**  The complete phase line for Newton's Law of Cooling.

(*minus* $k$ to give a negative derivative when $H > 15$).

Since $dH/dt = 0$ at $H = 15$, the temperature 15°C is an equilibrium value. If $H > 15$, Equation (1) tells us that $(H - 15) > 0$ and $dH/dt < 0$. If the object is hotter than the room, it will get cooler. Similarly, if $H < 15$, then $(H - 15) < 0$ and $dH/dt > 0$. An object cooler than the room will warm up. Thus, the behavior described by Equation (1) agrees with our intuition of how temperature should behave. These observations are captured in the initial phase line diagram in Figure 9.16. The value $H = 15$ is a stable equilibrium.

We determine the concavity of the solution curves by differentiating both sides of Equation (1) with respect to $t$:

$$\frac{d}{dt}\left(\frac{dH}{dt}\right) = \frac{d}{dt}(-k(H - 15))$$

$$\frac{d^2H}{dt^2} = -k\frac{dH}{dt}.$$

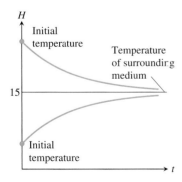

**FIGURE 9.18**  Temperature versus time. Regardless of initial temperature, the object's temperature $H(t)$ tends toward 15°C, the temperature of the surrounding medium.

Since $-k$ is negative, we see that $d^2H/dt^2$ is positive when $dH/dt < 0$ and negative when $dH/dt > 0$. Figure 9.17 adds this information to the phase line.

The completed phase line shows that if the temperature of the object is above the equilibrium value of 15°C, the graph of $H(t)$ will be decreasing and concave upward. If the temperature is below 15°C (the temperature of the surrounding medium), the graph of $H(t)$ will be increasing and concave downward. We use this information to sketch typical solution curves (Figure 9.18).

From the upper solution curve in Figure 9.18, we see that as the object cools down, the rate at which it cools slows down because $dH/dt$ approaches zero. This observation is implicit in Newton's Law of Cooling and contained in the differential equation, but the flattening of the graph as time advances gives an immediate visual representation of the phenomenon.

### A Falling Body Encountering Resistance

Newton observed that the rate of change in momentum encountered by a moving object is equal to the net force applied to it. In mathematical terms,

$$F = \frac{d}{dt}(mv), \tag{2}$$

where $F$ is the net force acting on the object, and $m$ and $v$ are the object's mass and velocity. If $m$ varies with time, as it will if the object is a rocket burning fuel, the right-hand side of Equation (2) expands to

$$m\frac{dv}{dt} + v\frac{dm}{dt}$$

using the Derivative Product Rule. In many situations, however, $m$ is constant, $dm/dt = 0$, and Equation (2) takes the simpler form

$$F = m\frac{dv}{dt} \quad \text{or} \quad F = ma, \tag{3}$$

known as *Newton's second law of motion* (see Section 9.3).

In free fall, the constant acceleration due to gravity is denoted by $g$ and the one force propelling the body downward is

$$F_p = mg,$$

the force due to gravity. If, however, we think of a real body falling through the air—say, a penny from a great height or a parachutist from an even greater height—we know that at some point air resistance is a factor in the speed of the fall. A more realistic model of free fall would include air resistance, shown as a force $F_r$ in the schematic diagram in Figure 9.19.

For low speeds well below the speed of sound, physical experiments have shown that $F_r$ is approximately proportional to the body's velocity. The net force on the falling body is therefore

$$F = F_p - F_r,$$

giving

$$m\frac{dv}{dt} = mg - kv$$

$$\frac{dv}{dt} = g - \frac{k}{m}v. \tag{4}$$

We can use a phase line to analyze the velocity functions that solve this differential equation.

The equilibrium point, obtained by setting the right-hand side of Equation (4) equal to zero, is

$$v = \frac{mg}{k}.$$

If the body is initially moving faster than this, $dv/dt$ is negative and the body slows down. If the body is moving at a velocity below $mg/k$, then $dv/dt > 0$ and the body speeds up. These observations are captured in the initial phase line diagram in Figure 9.20.

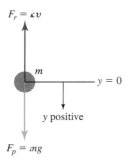

**FIGURE 9.19** An object falling under the propulsion due to gravity, with a resistive force assumed to be proportional to the velocity.

**FIGURE 9.20** Initial phase line for the falling body encountering resistance.

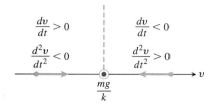

**FIGURE 9.21** The completed phase line for the falling body.

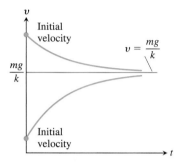

**FIGURE 9.22** Typical velocity curves for a falling body encountering resistance. The value $v = mg/k$ is the terminal velocity.

We determine the concavity of the solution curves by differentiating both sides of Equation (4) with respect to $t$:

$$\frac{d^2v}{dt^2} = \frac{d}{dt}\left(g - \frac{k}{m}v\right) = -\frac{k}{m}\frac{dv}{dt}.$$

We see that $d^2v/dt^2 < 0$ when $v < mg/k$ and $d^2v/dt^2 > 0$ when $v > mg/k$. Figure 9.21 adds this information to the phase line. Notice the similarity to the phase line for Newton's Law of Cooling (Figure 9.17). The solution curves are similar as well (Figure 9.22).

Figure 9.22 shows two typical solution curves. Regardless of the initial velocity, we see the body's velocity tending toward the limiting value $v = mg/k$. This value, a stable equilibrium point, is called the body's **terminal velocity**. Skydivers can vary their terminal velocity from 153 km/h to 290 km/h by changing the amount of body area opposing the fall, which affects the value of $k$.

### Logistic Population Growth

In Section 9.3 we examined population growth using the model of exponential change. That is, if $P$ represents the number of individuals and we neglect departures and arrivals, then

$$\frac{dP}{dt} = kP, \tag{5}$$

where $k > 0$ is the birth rate minus the death rate per individual per unit time.

Because the natural environment has only a limited number of resources to sustain life, it is reasonable to assume that only a maximum population $M$ can be accommodated. As the population approaches this **limiting population** or **carrying capacity**, resources become less abundant and the growth rate $k$ decreases. A simple relationship exhibiting this behavior is

$$k = r(M - P),$$

where $r > 0$ is a constant. Notice that $k$ decreases as $P$ increases toward $M$ and that $k$ is negative if $P$ is greater than $M$. Substituting $r(M - P)$ for $k$ in Equation (5) gives the differential equation

$$\frac{dP}{dt} = r(M - P)P = rMP - rP^2. \tag{6}$$

The model given by Equation (6) is referred to as **logistic growth**.

We can forecast the behavior of the population over time by analyzing the phase line for Equation (6). The equilibrium values are $P = M$ and $P = 0$, and we can see that $dP/dt > 0$ if $0 < P < M$ and $dP/dt < 0$ if $P > M$. These observations are recorded on the phase line in Figure 9.23.

We determine the concavity of the population curves by differentiating both sides of Equation (6) with respect to $t$:

$$\begin{aligned}\frac{d^2P}{dt^2} &= \frac{d}{dt}\left(rMP - rP^2\right) \\ &= rM\frac{dP}{dt} - 2rP\frac{dP}{dt} \\ &= r(M - 2P)\frac{dP}{dt}. \end{aligned} \tag{7}$$

**FIGURE 9.23** The initial phase line for logistic growth (Equation 6).

If $P = M/2$, then $d^2P/dt^2 = 0$. If $P < M/2$, then $(M - 2P)$ and $dP/dt$ are positive and $d^2P/dt^2 > 0$. If $M/2 < P < M$, then $(M - 2P) < 0$, $dP/dt > 0$, and $d^2P/dt^2 < 0$.

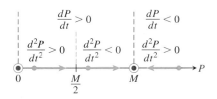

**FIGURE 9.24** The completed phase line for logistic growth (Equation 6).

If $P > M$, then $(M - 2P)$ and $dP/dt$ are both negative and $d^2P/dt^2 > 0$. We add this information to the phase line (Figure 9.24).

The lines $P = M/2$ and $P = M$ divide the first quadrant of the $tP$-plane into horizontal bands in which we know the signs of both $dP/dt$ and $d^2P/dt^2$. In each band, we know how the solution curves rise and fall, and how they bend as time passes. The equilibrium lines $P = 0$ and $P = M$ are both population curves. Population curves crossing the line $P = M/2$ have an inflection point there, giving them a **sigmoid** shape (curved in two directions like a letter S). Figure 9.25 displays typical population curves. Notice that each population curve approaches the limiting population $M$ as $t \to \infty$.

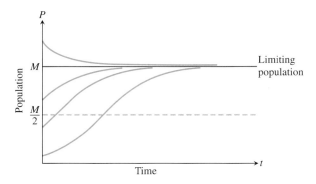

**FIGURE 9.25** Population curves for logistic growth.

# Exercises 9.4

**Phase Lines and Solution Curves**

In Exercises 1–8,

  **a.** Identify the equilibrium values. Which are stable and which are unstable?

  **b.** Construct a phase line. Identify the signs of $y'$ and $y''$.

  **c.** Sketch several solution curves.

**1.** $\dfrac{dy}{dx} = (y + 2)(y - 3)$      **2.** $\dfrac{dy}{dx} = y^2 - 4$

**3.** $\dfrac{dy}{dx} = y^3 - y$      **4.** $\dfrac{dy}{dx} = y^2 - 2y$

**5.** $y' = \sqrt{y}, \quad y > 0$      **6.** $y' = y - \sqrt{y}, \quad y > 0$

**7.** $y' = (y - 1)(y - 2)(y - 3)$    **8.** $y' = y^3 - y^2$

**Models of Population Growth**

The autonomous differential equations in Exercises 9–12 represent models for population growth. For each exercise, use a phase line analysis to sketch solution curves for $P(t)$, selecting different starting values $P(0)$. Which equilibria are stable, and which are unstable?

**9.** $\dfrac{dP}{dt} = 1 - 2P$      **10.** $\dfrac{dP}{dt} = P(1 - 2P)$

**11.** $\dfrac{dP}{dt} = 2P(P - 3)$      **12.** $\dfrac{dP}{dt} = 3P(1 - P)\left(P - \dfrac{1}{2}\right)$

**13. Catastrophic change in logistic growth** Suppose that a healthy population of some species is growing in a limited environment

and that the current population $P_0$ is fairly close to the carrying capacity $M_0$. You might imagine a population of fish living in a freshwater lake in a wilderness area. Suddenly a catastrophe such as the Mount St. Helens volcanic eruption contaminates the lake and destroys a significant part of the food and oxygen on which the fish depend. The result is a new environment with a carrying capacity $M_1$ considerably less than $M_0$ and, in fact, less than the current population $P_0$. Starting at some time before the catastrophe, sketch a "before-and-after" curve that shows how the fish population responds to the change in environment.

**14. Controlling a population** The fish and game department in a certain state is planning to issue hunting permits to control the deer population (one deer per permit). It is known that if the deer population falls below a certain level $m$, the deer will become extinct. It is also known that if the deer population rises above the carrying capacity $M$, the population will decrease back to $M$ through disease and malnutrition.

  **a.** Discuss the reasonableness of the following model for the growth rate of the deer population as a function of time:

$$\frac{dP}{dt} = rP(M - P)(P - m),$$

  where $P$ is the population of the deer and $r$ is a positive constant of proportionality. Include a phase line.

  **b.** Explain how this model differs from the logistic model $dP/dt = rP(M - P)$. Is it better or worse than the logistic model?

**c.** Show that if $P > M$ for all $t$, then $\lim_{t \to \infty} P(t) = M$.

**d.** What happens if $P < m$ for all $t$?

**e.** Discuss the solutions to the differential equation. What are the equilibrium points of the model? Explain the dependence of the steady-state value of $P$ on the initial values of $P$. About how many permits should be issued?

### Applications and Examples

**15. Skydiving** If a body of mass $m$ falling from rest under the action of gravity encounters an air resistance proportional to the square of velocity, then the body's velocity $t$ seconds into the fall satisfies the equation

$$m\frac{dv}{dt} = mg - kv^2, \qquad k > 0$$

where $k$ is a constant that depends on the body's aerodynamic properties and the density of the air. (We assume that the fall is too short to be affected by changes in the air's density.)

**a.** Draw a phase line for the equation.

**b.** Sketch a typical velocity curve.

**c.** For a 45-kg skydiver ($mg = 441$) and with time in seconds and distance in meters, a typical value of $k$ is 0.15. What is the diver's terminal velocity? Repeat for an 80-kg skydiver.

**16. Resistance proportional to $\sqrt{v}$** A body of mass $m$ is projected vertically downward with initial velocity $v_0$. Assume that the resisting force is proportional to the square root of the velocity and find the terminal velocity from a graphical analysis.

**17. Sailing** A sailboat is running along a straight course with the wind providing a constant forward force of 200 N. The only other force acting on the boat is resistance as the boat moves through the water. The resisting force is numerically equal to fifty times the boat's speed, and the initial velocity is 1 m/s. What is the maximum velocity in meters per second of the boat under this wind?

**18. The spread of information** Sociologists recognize a phenomenon called *social diffusion*, which is the spreading of a piece of information, technological innovation, or cultural fad among a population. The members of the population can be divided into two classes: those who have the information and those who do not. In a fixed population whose size is known, it is reasonable to assume that the rate of diffusion is proportional to the number who have the information times the number yet to receive it. If $X$ denotes the number of individuals who have the information in a population of $N$ people, then a mathematical model for social diffusion is given by

$$\frac{dX}{dt} = kX(N - X),$$

where $t$ represents time in days and $k$ is a positive constant.

**a.** Discuss the reasonableness of the model.

**b.** Construct a phase line identifying the signs of $X'$ and $X''$.

**c.** Sketch representative solution curves.

**d.** Predict the value of $X$ for which the information is spreading most rapidly. How many people eventually receive the information?

**19. Current in an *RL* circuit** The accompanying diagram represents an electrical circuit whose total resistance is a constant $R$ ohms and whose self-inductance, shown as a coil, is $L$ henries, also a constant. There is a switch whose terminals at $a$ and $b$ can be closed to connect a constant electrical source of $V$ volts. From Section 9.2, we have

$$L\frac{di}{dt} + Ri = V,$$

where $i$ is the current in amperes and $t$ is the time in seconds.

Use a phase line analysis to sketch the solution curve assuming that the switch in the *RL* circuit is closed at time $t = 0$. What happens to the current as $t \to \infty$? This value is called the *steady-state solution*.

**20. A pearl in shampoo** Suppose that a pearl is sinking in a thick fluid, like shampoo, subject to a frictional force opposing its fall and proportional to its velocity. Suppose that there is also a resistive buoyant force exerted by the shampoo. According to *Archimedes' principle*, the buoyant force equals the weight of the fluid displaced by the pearl. Using $m$ for the mass of the pearl and $P$ for the mass of the shampoo displaced by the pearl as it descends, complete the following steps.

**a.** Draw a schematic diagram showing the forces acting on the pearl as it sinks, as in Figure 9.19.

**b.** Using $v(t)$ for the pearl's velocity as a function of time $t$, write a differential equation modeling the velocity of the pearl as a falling body.

**c.** Construct a phase line displaying the signs of $v'$ and $v''$.

**d.** Sketch typical solution curves.

**e.** What is the terminal velocity of the pearl?

# 9.5 Systems of Equations and Phase Planes

In some situations we are led to consider not one, but several, first-order differential equations. Such a collection is called a **system** of differential equations. In this section we present an approach to understanding systems through a graphical procedure known as a *phase-plane analysis*. We present this analysis in the context of modeling the populations of trout and bass living in a common pond.

## Phase Planes

A general system of two first-order differential equations may take the form

$$\frac{dx}{dt} = F(x, y),$$

$$\frac{dy}{dt} = G(x, y).$$

Such a system of equations is called **autonomous** because $dx/dt$ and $dy/dt$ do not depend on the independent variable time $t$, but only on the dependent variables $x$ and $y$. A **solution** of such a system consists of a pair of functions $x(t)$ and $y(t)$ that satisfies both of the differential equations simultaneously for every $t$ over some time interval (finite or infinite).

We cannot look at just one of these equations in isolation to find solutions $x(t)$ or $y(t)$ since each derivative depends on both $x$ and $y$. To gain insight into the solutions, we look at both dependent variables together by plotting the points $(x(t), y(t))$ in the $xy$-plane starting at some specified point. Therefore the solution functions define a solution curve through the specified point, called a **trajectory** of the system. The $xy$-plane itself, in which these trajectories reside, is referred to as the **phase plane**. Thus we consider both solutions together and study the behavior of all the solution trajectories in the phase plane. It can be proved that two trajectories can never cross or touch each other. (Solution trajectories are examples of *parametric curves*, which are studied in detail in Chapter 11.)

## A Competitive-Hunter Model

Imagine two species of fish, say trout and bass, competing for the same limited resources (such as food and oxygen) in a certain pond. We let $x(t)$ represent the number of trout and $y(t)$ the number of bass living in the pond at time $t$. In reality $x(t)$ and $y(t)$ are always integer valued, but we will approximate them with real-valued differentiable functions. This allows us to apply the methods of differential equations.

Several factors affect the rates of change of these populations. As time passes, each species breeds, so we assume its population increases proportionally to its size. Taken by itself, this would lead to exponential growth in each of the two populations. However, there is a countervailing effect from the fact that the two species are in competition. A large number of bass tends to cause a decrease in the number of trout, and vice versa. Our model takes the size of this effect to be proportional to the frequency with which the two species interact, which in turn is proportional to $xy$, the product of the two populations. These considerations lead to the following model for the growth of the trout and bass in the pond:

$$\frac{dx}{dt} = (a - by)x, \tag{1a}$$

$$\frac{dy}{dt} = (m - nx)y. \tag{1b}$$

Here $x(t)$ represents the trout population, $y(t)$ the bass population, and $a, b, m, n$ are positive constants. A solution of this system then consists of a pair of functions $x(t)$ and $y(t)$ that gives the population of each fish species at time $t$. Each equation in (1) contains both of the unknown functions $x$ and $y$, so we are unable to solve them individually. Instead, we will use a graphical analysis to study the solution trajectories of this **competitive-hunter model**.

We now examine the nature of the phase plane in the trout-bass population model. We will be interested in the 1st quadrant of the $xy$-plane, where $x \geq 0$ and $y \geq 0$, since populations cannot be negative. First, we determine where the bass and trout populations are both constant. Noting that the $(x(t), y(t))$ values remain unchanged when $dx/dt = 0$ and $dy/dt = 0$, Equations (1a and 1b) then become

$$(a - by)x = 0,$$

$$(m - nx)y = 0.$$

This pair of simultaneous equations has two solutions: $(x, y) = (0, 0)$ and $(x, y) = (m/n, a/b)$. At these $(x, y)$ values, called **equilibrium** or **rest points**, the two populations remain at constant values over all time. The point $(0, 0)$ represents a pond containing no members of either fish species; the point $(m/n, a/b)$ corresponds to a pond with an unchanging number of each fish species.

Next, we note that if $y = a/b$, then Equation (1a) implies $dx/dt = 0$, so the trout population $x(t)$ is constant. Similarly, if $x = m/n$, then Equation (1b) implies $dy/dt = 0$, and the bass population $y(t)$ is constant. This information is recorded in Figure 9.26.

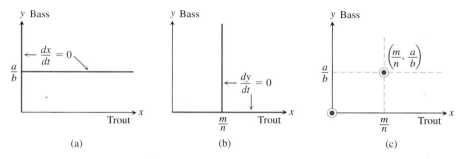

**FIGURE 9.26** Rest points in the competitive-hunter model given by Equations (1a) and (1b).

In setting up our competitive-hunter model, precise values of the constants $a$, $b$, $m$, $n$ will not generally be known. Nonetheless, we can analyze the system of Equations (1) to learn the nature of its solution trajectories. We begin by determining the signs of $dx/dt$ and $dy/dt$ throughout the phase plane. Although $x(t)$ represents the number of trout and $y(t)$ the number of bass at time $t$, we are thinking of the pair of values $(x(t), y(t))$ as a point tracing out a trajectory curve in the phase plane. When $dx/dt$ is positive, $x(t)$ is increasing and the point is moving to the right in the phase plane. If $dx/dt$ is negative, the point is moving to the left. Likewise, the point is moving upward where $dy/dt$ is positive and downward where $dy/dt$ is negative.

We saw that $dy/dt = 0$ along the vertical line $x = m/n$. To the left of this line, $dy/dt$ is positive since $dy/dt = (m - nx)y$ and $x < m/n$. So the trajectories on this side of the line are directed upward. To the right of this line, $dy/dt$ is negative and the trajectories point downward. The directions of the associated trajectories are indicated in Figure 9.27. Similarly, above the horizontal line $y = a/b$, we have $dx/dt < 0$ and the trajectories head leftward; below this line they head rightward, as shown in Figure 9.28. Combining this information gives four distinct regions in the plane $A$, $B$, $C$, $D$, with their respective trajectory directions shown in Figure 9.29.

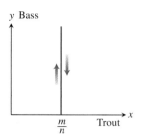

**FIGURE 9.27** To the left of the line $x = m/n$ the trajectories move upward, and to the right they move downward.

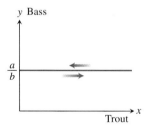

**FIGURE 9.28** Above the line $y = a/b$ the trajectories move to the left, and below it they move to the right.

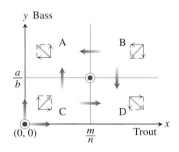

**FIGURE 9.29** Composite graphical analysis of the trajectory directions in the four regions determined by $x = m/n$ and $y = a/b$.

Next, we examine what happens near the two equilibrium points. The trajectories near $(0, 0)$ point away from it, upward and to the right. The behavior near the equilibrium point $(m/n, a/b)$ depends on the region in which a trajectory begins. If it starts in region $B$, for instance, then it will move downward and leftward toward the equilibrium point. Depending on where the trajectory begins, it may move downward into region $D$, leftward into region $A$, or perhaps straight into the equilibrium point. If it enters into regions $A$ or $D$,

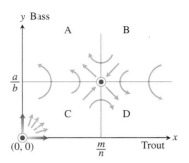

**FIGURE 9.30** Motion along the trajectories near the rest points $(0, 0)$ and $(m/n, a/b)$.

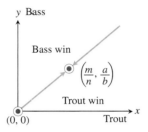

**FIGURE 9.31** Qualitative results of analyzing the competitive-hunter model. There are exactly two trajectories approaching the point $(m/n, a/b)$.

**FIGURE 9.32** Trajectory direction near the rest point $(0, 0)$.

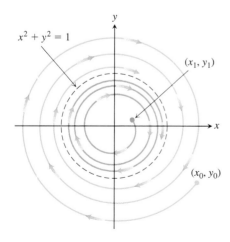

**FIGURE 9.34** The solution $x^2 + y^2 = 1$ is a limit cycle.

then it will continue to move away from the rest point. We say that both rest points are **unstable**, meaning (in this setting) there are trajectories near each point that head away from them. These features are indicated in Figure 9.30.

It turns out that in each of the half-planes above and below the line $y = a/b$, there is exactly one trajectory approaching the equilibrium point $(m/n, a/b)$ (see Exercise 7). Above these two trajectories the bass population increases and below them it decreases. The two trajectories approaching the equilibrium point are suggested in Figure 9.31.

Our graphical analysis leads us to conclude that, under the assumptions of the competitive-hunter model, it is unlikely that both species will reach equilibrium levels. This is because it would be almost impossible for the fish populations to move exactly along one of the two approaching trajectories for all time. Furthermore, the initial populations point $(x_0, y_0)$ determines which of the two species is likely to survive over time, and mutual coexistence of the species is highly improbable.

## Limitations of the Phase-Plane Analysis Method

Unlike the situation for the competitive-hunter model, it is not always possible to determine the behavior of trajectories near a rest point. For example, suppose we know that the trajectories near a rest point, chosen here to be the origin $(0, 0)$, behave as in Figure 9.32. The information provided by Figure 9.32 is not sufficient to distinguish between the three possible trajectories shown in Figure 9.33. Even if we could determine that a trajectory near an equilibrium point resembles that of Figure 9.33c, we would still not know how the other trajectories behave. It could happen that a trajectory closer to the origin behaves like the motions displayed in Figure 9.33a or 9.33b. The spiraling trajectory in Figure 9.33b can never actually reach the rest point in a finite time period.

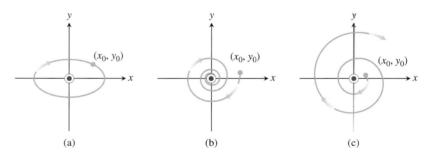

**FIGURE 9.33** Three possible trajectory motions: (a) periodic motion, (b) motion toward an asymptotically stable rest point, and (c) motion near an unstable rest point.

## Another Type of Behavior

The system

$$\frac{dx}{dt} = y + x - x(x^2 + y^2), \tag{2a}$$

$$\frac{dy}{dt} = -x + y - y(x^2 + y^2) \tag{2b}$$

can be shown to have only one equilibrium point at $(0, 0)$. Yet any trajectory starting on the unit circle traverses it clockwise because, when $x^2 + y^2 = 1$, we have $dy/dx = -x/y$ (see Exercise 2). If a trajectory starts inside the unit circle, it spirals outward, asymptotically approaching the circle as $t \to \infty$. If a trajectory starts outside the unit circle, it spirals inward, again asymptotically approaching the circle as $t \to \infty$. The circle $x^2 + y^2 = 1$ is called a **limit cycle** of the system (Figure 9.34). In this system, the values of $x$ and $y$ eventually become periodic.

# Exercises 9.5

1. List three important considerations that are ignored in the competitive-hunter model as presented in the text.

2. For the system (2a) and (2b), show that any trajectory starting on the unit circle $x^2 + y^2 = 1$ will traverse the unit circle in a periodic solution. First introduce polar coordinates and rewrite the system as $dr/dt = r(1 - r^2)$ and $-d\theta/dt = -1$.

3. Develop a model for the growth of trout and bass, assuming that in isolation trout demonstrate exponential decay [so that $a < 0$ in Equations (1a) and (1b)] and that the bass population grows logistically with a population limit $M$. Analyze graphically the motion in the vicinity of the rest points in your model. Is coexistence possible?

4. How might the competitive-hunter model be validated? Include a discussion of how the various constants $a$, $b$, $m$, and $n$ might be estimated. How could state conservation authorities use the model to ensure the survival of both species?

5. Consider another competitive-hunter model defined by

$$\frac{dx}{dt} = a\left(1 - \frac{x}{k_1}\right)x - bxy,$$

$$\frac{dy}{dt} = m\left(1 - \frac{y}{k_2}\right)y - nxy,$$

where $x$ and $y$ represent trout and bass populations, respectively.

   a. What assumptions are implicitly being made about the growth of trout and bass in the absence of competition?

   b. Interpret the constants $a$, $b$, $m$, $n$, $k_1$, and $k_2$ in terms of the physical problem.

   c. Perform a graphical analysis:

   i) Find the possible equilibrium levels.

   ii) Determine whether coexistence is possible.

   iii) Pick several typical starting points and sketch typical trajectories in the phase plane.

   iv) Interpret the outcomes predicted by your graphical analysis in terms of the constants $a$, $b$, $m$, $n$, $k_1$, and $k_2$.

   *Note:* When you get to part (iii), you should realize that five cases exist. You will need to analyze all five cases.

6. **An economic model** Consider the following economic model. Let $P$ be the price of a single item on the market. Let $Q$ be the quantity of the item available on the market. Both $P$ and $Q$ are functions of time. If one considers price and quantity as two interacting species, the following model might be proposed:

$$\frac{dP}{dt} = aP\left(\frac{b}{Q} - P\right),$$

$$\frac{dQ}{dt} = cQ(fP - Q),$$

where $a$, $b$, $c$, and $f$ are positive constants. Justify and discuss the adequacy of the model.

   a. If $a = 1$, $b = 20{,}000$, $c = 1$, and $f = 30$, find the equilibrium points of this system. If possible, classify each equilibrium point with respect to its stability. If a point cannot be readily classified, give some explanation.

   b. Perform a graphical stability analysis to determine what will happen to the levels of $P$ and $Q$ as time increases.

   c. Give an economic interpretation of the curves that determine the equilibrium points.

7. **Two trajectories approach equilibrium** Show that the two trajectories leading to $(m/n, a/b)$ shown in Figure 9.31 are unique by carrying out the following steps.

   a. From system (1a) and (1b) apply the Chain Rule to derive the following equation:

$$\frac{dy}{dx} = \frac{(m - nx)y}{(a - by)x}.$$

   b. Separate the variables, integrate, and exponentiate to obtain

$$y^a e^{-by} = Kx^m e^{-nx},$$

   where $K$ is a constant of integration.

   c. Let $f(y) = y^a/e^{by}$ and $g(x) = x^m/e^{nx}$. Show that $f(y)$ has a unique maximum of $M_y = (a/eb)^a$ when $y = a/b$ as shown in Figure 9.35. Similarly, show that $g(x)$ has a unique maximum $M_x = (m/en)^m$ when $x = m/n$, also shown in Figure 9.35.

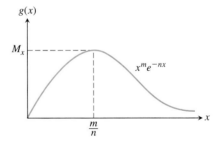

**FIGURE 9.35** Graphs of the functions $f(y) = y^a/e^{by}$ and $g(x) = x^m/e^{nx}$.

   d. Consider what happens as $(x, y)$ approaches $(m/n, a/b)$. Take limits in part (b) as $x \to m/n$ and $y \to a/b$ to show that either

$$\lim_{\substack{x \to m/n \\ y \to a/b}} \left[\left(\frac{y^a}{e^{by}}\right)\left(\frac{e^{nx}}{x^m}\right)\right] = K$$

   or $M_y/M_x = K$. Thus any solution trajectory that approaches $(m/n, a/b)$ must satisfy

$$\frac{y^a}{e^{by}} = \left(\frac{M_y}{M_x}\right)\left(\frac{x^m}{e^{nx}}\right).$$

   e. Show that only one trajectory can approach $(m/n, a/b)$ from below the line $y = a/b$. Pick $y_0 < a/b$. From Figure 9.35 you can see that $f(y_0) < M_y$, which implies that

$$\frac{M_y}{M_x}\left(\frac{x^m}{e^{nx}}\right) = y_0{}^a/e^{by_0} < M_y.$$

This in turn implies that

$$\frac{x^m}{e^{nx}} < M_x.$$

Figure 9.35 tells you that for $g(x)$ there is a unique value $x_0 < m/n$ satisfying this last inequality. That is, for each $y < a/b$ there is a unique value of $x$ satisfying the equation in part (d). Thus there can exist only one trajectory solution approaching $(m/n, a/b)$ from below, as shown in Figure 9.36.

**f.** Use a similar argument to show that the solution trajectory leading to $(m/n, a/b)$ is unique if $y_0 > a/b$.

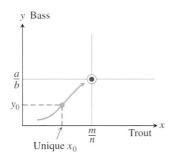

**FIGURE 9.36** For any $y < a/b$ only one solution trajectory leads to the rest point $(m/n, a/b)$.

**8.** Show that the second-order differential equation $y'' = F(x, y, y')$ can be reduced to a system of two first-order differential equations

$$\frac{dy}{dx} = z,$$

$$\frac{dz}{dx} = F(x, y, z).$$

Can something similar be done to the $n$th-order differential equation $y^{(n)} = F\left(x, y, y', y'', \ldots, y^{(n-1)}\right)$?

### Lotka-Volterra Equations for a Predator-Prey Model

In 1925 Lotka and Volterra introduced the *predator-prey* equations, a system of equations that models the populations of two species, one of which preys on the other. Let $x(t)$ represent the number of rabbits living in a region at time $t$, and $y(t)$ the number of foxes in the same region. As time passes, the number of rabbits increases at a rate proportional to their population, and decreases at a rate proportional to the number of encounters between rabbits and foxes. The foxes, which compete for food, increase in number at a rate proportional to the number of encounters with rabbits but decrease at a rate proportional to the number of foxes. The number of encounters between rabbits and foxes is assumed to be proportional to the product of the two populations. These assumptions lead to the autonomous system

$$\frac{dx}{dt} = (a - by)x$$

$$\frac{dy}{dt} = (-c + dx)y$$

where $a, b, c, d$ are positive constants. The values of these constants vary according to the specific situation being modeled. We can study the nature of the population changes without setting these constants to specific values.

**9.** What happens to the rabbit population if there are no foxes present?

**10.** What happens to the fox population if there are no rabbits present?

**11.** Show that $(0, 0)$ and $(c/d, a/b)$ are equilibrium points. Explain the meaning of each of these points.

**12.** Show, by differentiating, that the function

$$C(t) = a \ln y(t) - by(t) - dx(t) + c \ln x(t)$$

is constant when $x(t)$ and $y(t)$ are positive and satisfy the predator-prey equations.

While $x$ and $y$ may change over time, $C(t)$ does not. Thus, $C$ is a *conserved quantity* and its existence gives a *conservation law*. A trajectory that begins at a point $(x, y)$ at time $t = 0$ gives a value of $C$ that remains unchanged at future times. Each value of the constant $C$ gives a trajectory for the autonomous system, and these trajectories close up, rather than spiraling inward or outward. The rabbit and fox populations oscillate through repeated cycles along a fixed trajectory. Figure 9.37 shows several trajectories for the predator-prey system.

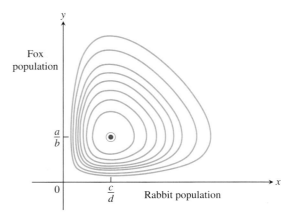

**FIGURE 9.37** Some trajectories along which $C$ is conserved.

**13.** Using a procedure similar to that in the text for the competitive-hunter model, show that each trajectory is traversed in a counterclockwise direction as time $t$ increases.

Along each trajectory, both the rabbit and fox populations fluctuate between their maximum and minimum levels. The maximum and minimum levels for the rabbit population occur where the trajectory intersects the horizontal line $y = a/b$. For the fox population, they occur where the trajectory intersects the vertical line $x = c/d$. When the rabbit population is at its maximum, the fox population is below its maximum value. As the rabbit population declines from this point in time, we move counterclockwise around the trajectory, and the fox population grows until it reaches its maximum value. At this point the rabbit population has declined to $x = c/d$ and is no longer at its peak value. We see that the fox population reaches its maximum value at a later time than the rabbits. The predator population *lags behind* that of the prey in achieving its maximum values. This lag effect is shown in Figure 9.38, which graphs both $x(t)$ and $y(t)$.

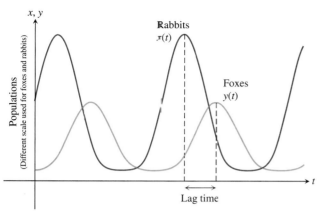

**FIGURE 9.38** The fox and rabbit populations oscillate periodically, with the maximum fox population lagging the maximum rabbit population.

**14.** At some time during a trajectory cycle, a wolf invades the rabbit-fox territory, eats some rabbits, and then leaves. Does this mean that the fox population will from then on have a lower maximum value? Explain your answer.

## Chapter **9**    Questions to Guide Your Review

**1.** What is a first-order differential equation? When is a function a solution of such an equation?

**2.** What is a general solution? A particular solution?

**3.** What is the slope field of a differential equation $y' = f(x, y)$? What can we learn from such fields?

**4.** Describe Euler's method for solving the initial value problem $y' = f(x, y)$, $y(x_0) = y_0$ numerically. Give an example. Comment on the method's accuracy. Why might you want to solve an initial value problem numerically?

**5.** How do you solve linear first-order differential equations?

**6.** What is an orthogonal trajectory of a family of curves? Describe how one is found for a given family of curves.

**7.** What is an autonomous differential equation? What are its equilibrium values? How do they differ from critical points? What is a stable equilibrium value? Unstable?

**8.** How do you construct the phase line for an autonomous differential equation? How does the phase line help you produce a graph which qualitatively depicts a solution to the differential equation?

**9.** Why is the exponential model unrealistic for predicting long-term population growth? How does the logistic model correct for the deficiency in the exponential model for population growth? What is the logistic differential equation? What is the form of its solution? Describe the graph of the logistic solution.

**10.** What is an autonomous system of differential equations? What is a solution to such a system? What is a trajectory of the system?

## Chapter **9**    Practice Exercises

In Exercises 1–16 solve the differential equation.

**1.** $y' = xe^y \sqrt{x - 2}$

**2.** $y' = xye^{x^2}$

**3.** $\sec x \, dy + x \cos^2 y \, dx = 0$

**4.** $2x^2 \, dx - 3\sqrt{y} \csc x \, dy = 0$

**5.** $y' = \dfrac{e^y}{xy}$

**6.** $y' = xe^{x-y} \csc y$

**7.** $x(x - 1) \, dy - y \, dx = 0$

**8.** $y' = (y^2 - 1)x^{-1}$

**9.** $2y' - y = xe^{x/2}$

**10.** $\dfrac{y'}{2} + y = e^{-x} \sin x$

**11.** $xy' + 2y = 1 - x^{-1}$

**12.** $xy' - y = 2x \ln x$

**13.** $(1 + e^x) \, dy + (ye^x + e^{-x}) \, dx = 0$

**14.** $e^{-x} \, dy + (e^{-x}y - 4x) \, dx = 0$

**15.** $(x + 3y^2) \, dy + y \, dx = 0$ (*Hint:* $d(xy) = y \, dx + x \, dy$)

**16.** $x \, dy + (3y - x^{-2} \cos x) \, dx = 0, \quad x > 0$

**Initial Value Problems**

In Exercises 17–22 solve the initial value problem.

**17.** $(x + 1)\dfrac{dy}{dx} + 2y = x, \quad x > -1, \quad y(0) = 1$

**18.** $x\dfrac{dy}{dx} + 2y = x^2 + 1, \quad x > 0, \quad y(1) = 1$

**19.** $\dfrac{dy}{dx} + 3x^2y = x^2, \quad y(0) = -1$

**20.** $x \, dy + (y - \cos x) \, dx = 0, \quad y\left(\dfrac{\pi}{2}\right) = 0$

**21.** $xy' + (x - 2)y = 3x^3e^{-x}, \quad y(1) = 0$

**22.** $y \, dx + (3x - xy + 2) \, dy = 0, \quad y(2) = -1, \quad y < 0$

## Euler's Method

In Exercises 23 and 24, use Euler's method to solve the initial value problem on the given interval starting at $x_0$ with $dx = 0.1$.

☐T **23.** $y' = y + \cos x, \quad y(0) = 0; \quad 0 \le x \le 2; \quad x_0 = 0$

☐T **24.** $y' = (2 - y)(2x + 3), \quad y(-3) = 1; \quad -3 \le x \le -1; \quad x_0 = -3$

In Exercises 25 and 26, use Euler's method with $dx = 0.05$ to estimate $y(c)$ where $y$ is the solution to the given initial value problem.

☐T **25.** $c = 3; \quad \dfrac{dy}{dx} = \dfrac{x - 2y}{x + 1}, \quad y(0) = 1$

☐T **26.** $c = 4; \quad \dfrac{dy}{dx} = \dfrac{x^2 - 2y + 1}{x}, \quad y(1) = 1$

In Exercises 27 and 28, use Euler's method to solve the initial value problem graphically, starting at $x_0 = 0$ with

    **a.** $dx = 0.1$.         **b.** $dx = -0.1$.

☐T **27.** $\dfrac{dy}{dx} = \dfrac{1}{e^{x+y+2}}, \quad y(0) = -2$

☐T **28.** $\dfrac{dy}{dx} = -\dfrac{x^2 + y}{e^y + x}, \quad y(0) = 0$

## Slope Fields

In Exercises 29–32, sketch part of the equation's slope field. Then add to your sketch the solution curve that passes through the point $P(1, -1)$. Use Euler's method with $x_0 = 1$ and $dx = 0.2$ to estimate $y(2)$. Round your answers to four decimal places. Find the exact value of $y(2)$ for comparison.

**29.** $y' = x$                **30.** $y' = 1/x$

**31.** $y' = xy$             **32.** $y' = 1/y$

## Autonomous Differential Equations and Phase Lines

In Exercises 33 and 34:

    **a.** Identify the equilibrium values. Which are stable and which are unstable?

    **b.** Construct a phase line. Identify the signs of $y'$ and $y''$.

    **c.** Sketch a representative selection of solution curves.

**33.** $\dfrac{dy}{dx} = y^2 - 1$          **34.** $\dfrac{dy}{dx} = y - y^2$

## Applications

**35. Escape velocity** The gravitational attraction $F$ exerted by an airless moon on a body of mass $m$ at a distance $s$ from the moon's center is given by the equation $F = -mg\, R^2 s^{-2}$, where $g$ is the acceleration of gravity at the moon's surface and $R$ is the moon's radius (see accompanying figure). The force $F$ is negative because it acts in the direction of decreasing $s$.

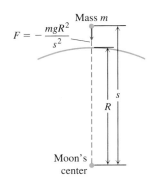

**a.** If the body is projected vertically upward from the moon's surface with an initial velocity $v_0$ at time $t = 0$, use Newton's second law, $F = ma$, to show that the body's velocity at position $s$ is given by the equation

$$v^2 = \frac{2gR^2}{s} + v_0{}^2 - 2gR.$$

Thus, the velocity remains positive as long as $v_0 \ge \sqrt{2gR}$. The velocity $v_0 = \sqrt{2gR}$ is the moon's **escape velocity**. A body projected upward with this velocity or a greater one will escape from the moon's gravitational pull.

**b.** Show that if $v_0 = \sqrt{2gR}$, then

$$s = R\left(1 + \frac{3v_0}{2R}t\right)^{2/3}.$$

**36. Coasting to a stop** Table 9.6 shows the distance $s$ (meters) coasted on inline skates in $t$ (seconds) by Johnathon Krueger. Find a model for his position in the form of Equation (2) of Section 9.3. His initial velocity was $v_0 = 0.86$ m/s, his mass $m = 30.84$ kg, and his total coasting distance 0.97 m.

**TABLE 9.6** Johnathon Krueger skating data

| $t$ (s) | $s$ (m) | $t$ (s) | $s$ (m) | $t$ (s) | $s$ (m) |
|---|---|---|---|---|---|
| 0 | 0 | 0.93 | 0.61 | 1.86 | 0.93 |
| 0.13 | 0.08 | 1.06 | 0.68 | 2.00 | 0.94 |
| 0.27 | 0.19 | 1.20 | 0.74 | 2.13 | 0.95 |
| 0.40 | 0.28 | 1.33 | 0.79 | 2.26 | 0.96 |
| 0.53 | 0.36 | 1.46 | 0.83 | 2.39 | 0.96 |
| 0.67 | 0.45 | 1.60 | 0.87 | 2.53 | 0.97 |
| 0.80 | 0.53 | 1.73 | 0.90 | 2.66 | 0.97 |

# Chapter 9    Additional and Advanced Exercises

## Theory and Applications

**1. Transport through a cell membrane** Under some conditions, the result of the movement of a dissolved substance across a cell's membrane is described by the equation

$$\frac{dy}{dt} = k\frac{A}{V}(c - y).$$

In this equation, $y$ is the concentration of the substance inside the cell and $dy/dt$ is the rate at which $y$ changes over time. The letters

$k$, $A$, $V$, and $c$ stand for constants, $k$ being the *permeability coefficient* (a property of the membrane), $A$ the surface area of the membrane, $V$ the cell's volume, and $c$ the concentration of the substance outside the cell. The equation says that the rate at which the concentration changes within the cell is proportional to the difference between it and the outside concentration.

**a.** Solve the equation for $y(t)$, using $y_0$ to denote $y(0)$.

**b.** Find the steady-state concentration, $\lim_{t\to\infty} y(t)$.

2. **Height of a rocket** If an external force $F$ acts upon a system whose mass varies with time, Newton's law of motion is

$$\frac{d(mv)}{dt} = F + (v + u)\frac{dm}{dt}.$$

In this equation, $m$ is the mass of the system at time $t$, $v$ is its velocity, and $v + u$ is the velocity of the mass that is entering (or leaving) the system at the rate $dm/dt$. Suppose that a rocket of initial mass $m_0$ starts from rest, but is driven upward by firing some of its mass directly backward at the constant rate of $dm/dt = -b$ units per second and at constant speed relative to the rocket $u = -c$. The only external force acting on the rocket is $F = -mg$ due to gravity. Under these assumptions, show that the height of the rocket above the ground at the end of $t$ seconds ($t$ small compared to $m_0/b$) is

$$y = c\left[t + \frac{m_0 - bt}{b}\ln\frac{m_0 - bt}{m_0}\right] - \frac{1}{2}gt^2.$$

3. **a.** Assume that $P(x)$ and $Q(x)$ are continuous over the interval $[a, b]$. Use the Fundamental Theorem of Calculus, Part 1, to show that any function $y$ satisfying the equation

$$v(x)y = \int v(x)Q(x)\,dx + C$$

for $v(x) = e^{\int P(x)\,dx}$ is a solution to the first-order linear equation

$$\frac{dy}{dx} + P(x)y = Q(x).$$

**b.** If $C = y_0 v(x_0) - \int_{x_0}^{x} v(t)Q(t)\,dt$, then show that any solution $y$ in part (a) satisfies the initial condition $y(x_0) = y_0$.

4. (*Continuation of Exercise 3.*) Assume the hypotheses of Exercise 3, and assume that $y_1(x)$ and $y_2(x)$ are both solutions to the first-order linear equation satisfying the initial condition $y(x_0) = y_0$.

**a.** Verify that $y(x) = y_1(x) - y_2(x)$ satisfies the initial value problem

$$y' + P(x)y = 0, \quad y(x_0) = 0.$$

**b.** For the integrating factor $v(x) = e^{\int P(x)\,dx}$, show that

$$\frac{d}{dx}\left(v(x)[y_1(x) - y_2(x)]\right) = 0.$$

Conclude that $v(x)[y_1(x) - y_2(x)] \equiv$ constant.

**c.** From part (a), we have $y_1(x_0) - y_2(x_0) = 0$. Since $v(x) > 0$ for $a < x < b$, use part (b) to establish that $y_1(x) - y_2(x) \equiv 0$ on the interval $(a, b)$. Thus $y_1(x) = y_2(x)$ for all $a < x < b$.

### Homogeneous Equations

A first-order differential equation of the form

$$\frac{dy}{dx} = F\left(\frac{y}{x}\right)$$

is called *homogeneous*. It can be transformed into an equation whose variables are separable by defining the new variable $v = y/x$. Then, $y = vx$ and

$$\frac{dy}{dx} = v + x\frac{dv}{dx}.$$

Substitution into the original differential equation and collecting terms with like variables then gives the separable equation

$$\frac{dx}{x} + \frac{dv}{v - F(v)} = 0.$$

After solving this separable equation, the solution of the original equation is obtained when we replace $v$ by $y/x$.

Solve the homogeneous equations in Exercises 5–10. First put the equation in the form of a homogeneous equation.

5. $(x^2 + y^2)\,dx + xy\,dy = 0$

6. $x^2\,dy + (y^2 - xy)\,dx = 0$

7. $(xe^{y/x} + y)\,dx - x\,dy = 0$

8. $(x + y)\,dy + (x - y)\,dx = 0$

9. $y' = \dfrac{y}{x} + \cos\dfrac{y - x}{x}$

10. $\left(x\sin\dfrac{y}{x} - y\cos\dfrac{y}{x}\right)dx + x\cos\dfrac{y}{x}\,dy = 0$

---

## Chapter **9** Technology Application Projects

**Mathematica/Maple Modules:**

***Drug Dosages: Are They Effective? Are They Safe?***
Formulate and solve an initial value model for the absorption of a drug in the bloodstream.

***First-Order Differential Equations and Slope Fields***
Plot slope fields and solution curves for various initial conditions to selected first-order differential equations.

# 10

# Infinite Sequences and Series

**OVERVIEW** Up to this point in calculus, we have focused on the derivative and integral of a function. Now we introduce a third key topic of interest in the analysis and computation of functions, called *infinite series*. Such series give us precise ways to express many numbers and functions, both familiar and new, as arithmetic sums with infinitely many terms. For example, we will learn that

$$\frac{\pi}{4} = 1 - \frac{1}{3} + \frac{1}{5} - \frac{1}{7} + \frac{1}{9} - \cdots$$

and

$$\cos x = 1 - \frac{x^2}{2} + \frac{x^4}{24} - \frac{x^6}{720} - \frac{x^8}{40,320} - \cdots.$$

Often scientists and engineers simplify a problem by replacing a function with an approximation using the first few terms of a series that expresses it. One method represents a known differentiable function $f(x)$ as an infinite series in powers of $x$, so it looks like a "polynomial with infinitely many terms," as we see with the cosine function given above. Moreover, the method extends our knowledge of how to evaluate, differentiate, and integrate polynomials, so we can work with even more general functions than any encountered before. These new functions are commonly solutions to differential equations arising in important applications of mathematics to science and engineering.

Everyone knows how to add two numbers together, or even several. But how do you add together infinitely many numbers? Or more generally, how do you add infinitely many powers of $x$? In this chapter we answer these questions, which are part of the theory of infinite sequences and series. As with the differential and integral calculus, limits play a major role in the development of infinite series.

## 10.1 Sequences

HISTORICAL ESSAY

Sequences and Series

Sequences are fundamental to the study of infinite series and many applications of mathematics. We have already seen an example of a sequence when we studied Newton's Method in Section 4.6. There we produced a sequence of approximations $x_n$ that became closer and closer to the root of a differentiable function. Now we will explore general sequences of numbers and the conditions under which they converge to a finite number.

### Representing Sequences

A sequence is a list of numbers

$$a_1, a_2, a_3, \ldots, a_n, \ldots$$

in a given order. Each of $a_1$, $a_2$, $a_3$ and so on represents a number. These are the **terms** of the sequence. For example, the sequence

$$2, 4, 6, 8, 10, 12, \ldots, 2n, \ldots$$

has first term $a_1 = 2$, second term $a_2 = 4$, and $n$th term $a_n = 2n$. The integer $n$ is called the **index** of $a_n$, and indicates where $a_n$ occurs in the list. Order is important. The sequence 2, 4, 6, 8 . . . is not the same as the sequence 4, 2, 6, 8 . . . .

We can think of the sequence

$$a_1, a_2, a_3, \ldots, a_n, \ldots$$

as a function that sends 1 to $a_1$, 2 to $a_2$, 3 to $a_3$, and in general sends the positive integer $n$ to the $n$th term $a_n$. More precisely, an **infinite sequence** of numbers is a function whose domain is the set of positive integers.

The function associated with the sequence

$$2, 4, 6, 8, 10, 12, \ldots, 2n, \ldots$$

sends 1 to $a_1 = 2$, 2 to $a_2 = 4$, and so on. The general behavior of this sequence is described by the formula $a_n = 2n$.

We can equally well make the domain the integers larger than a given number $n_0$, and we allow sequences of this type also. For example, the sequence

$$12, 14, 16, 18, 20, 22 \ldots$$

is described by the formula $a_n = 10 + 2n$. It can also be described by the simpler formula $b_n = 2n$, where the index $n$ starts at 6 and increases. To allow such simpler formulas, we let the first index of the sequence be any integer. In the sequence above, $\{a_n\}$ starts with $a_1$ while $\{b_n\}$ starts with $b_6$.

Sequences can be described by writing rules that specify their terms, such as

$$a_n = \sqrt{n}, \quad b_n = (-1)^{n+1}\frac{1}{n}, \quad c_n = \frac{n-1}{n}, \quad d_n = (-1)^{n+1},$$

or by listing terms:

$$\{a_n\} = \left\{ \sqrt{1}, \sqrt{2}, \sqrt{3}, \ldots, \sqrt{n}, \ldots \right\}$$

$$\{b_n\} = \left\{ 1, -\frac{1}{2}, \frac{1}{3}, -\frac{1}{4}, \ldots, (-1)^{n-1}\frac{1}{n}, \ldots \right\}$$

$$\{c_n\} = \left\{ 0, \frac{1}{2}, \frac{2}{3}, \frac{3}{4}, \frac{4}{5}, \ldots, \frac{n-1}{n}, \ldots \right\}$$

$$\{d_n\} = \left\{ 1, -1, 1, -1, 1, -1, \ldots, (-1)^{n+1}, \ldots \right\}.$$

We also sometimes write a sequence using its rule, as with

$$\{a_n\} = \left\{ \sqrt{n} \right\}_{n=1}^{\infty}.$$

Figure 10.1 shows two ways to represent sequences graphically. The first marks the first few points from $a_1, a_2, a_3, \ldots, a_n, \ldots$ on the real axis. The second method shows the graph of the function defining the sequence. The function is defined only on integer inputs, and the graph consists of some points in the $xy$-plane located at $(1, a_1)$, $(2, a_2), \ldots, (n, a_n), \ldots$.

## Convergence and Divergence

Sometimes the numbers in a sequence approach a single value as the index $n$ increases. This happens in the sequence

$$\left\{ 1, \frac{1}{2}, \frac{1}{3}, \frac{1}{4}, \ldots, \frac{1}{n}, \ldots \right\}$$

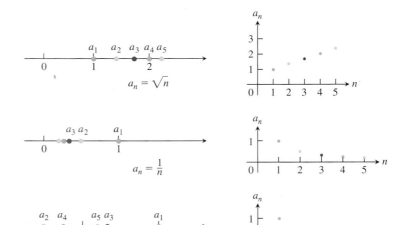

**FIGURE 10.1**   Sequences can be represented as points on the real line or as points in the plane where the horizontal axis $n$ is the index number of the term and the vertical axis $a_n$ is its value.

whose terms approach 0 as $n$ gets large, and in the sequence

$$\left\{0, \frac{1}{2}, \frac{2}{3}, \frac{3}{4}, \frac{4}{5}, \ldots, 1 - \frac{1}{n}, \ldots\right\}$$

whose terms approach 1. On the other hand, sequences like

$$\left\{\sqrt{1}, \sqrt{2}, \sqrt{3}, \ldots, \sqrt{n}, \ldots\right\}$$

have terms that get larger than any number as $n$ increases, and sequences like

$$\{1, -1, 1, -1, 1, -1, \ldots, (-1)^{n+1}, \ldots\}$$

bounce back and forth between 1 and $-1$, never converging to a single value. The following definition captures the meaning of having a sequence converge to a limiting value. It says that if we go far enough out in the sequence, by taking the index $n$ to be larger than some value $N$, the difference between $a_n$ and the limit of the sequence becomes less than any preselected number $\epsilon > 0$.

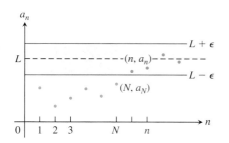

**FIGURE 10.2**   In the representation of a sequence as points in the plane, $a_n \to L$ if $y = L$ is a horizontal asymptote of the sequence of points $\{(n, a_n)\}$. In this figure, all the $a_n$'s after $a_N$ lie within $\epsilon$ of $L$.

---

**DEFINITIONS**   The sequence $\{a_n\}$ **converges** to the number $L$ if for every positive number $\epsilon$ there corresponds an integer $N$ such that for all $n$,

$$n > N \quad \Rightarrow \quad |a_n - L| < \epsilon.$$

If no such number $L$ exists, we say that $\{a_n\}$ **diverges**.

If $\{a_n\}$ converges to $L$, we write $\lim_{n \to \infty} a_n = L$, or simply $a_n \to L$, and call $L$ the **limit** of the sequence (Figure 10.2).

---

The definition is very similar to the definition of the limit of a function $f(x)$ as $x$ tends to $\infty$ ($\lim_{x \to \infty} f(x)$ in Section 2.6). We will exploit this connection to calculate limits of sequences.

**EXAMPLE 1**   Show that

**(a)** $\lim_{n \to \infty} \frac{1}{n} = 0$    **(b)** $\lim_{n \to \infty} k = k$    (any constant $k$)

**Solution**

**(a)** Let $\epsilon > 0$ be given. We must show that there exists an integer $N$ such that for all $n$,

$$n > N \quad \Rightarrow \quad \left| \frac{1}{n} - 0 \right| < \epsilon.$$

This implication will hold if $(1/n) < \epsilon$ or $n > 1/\epsilon$. If $N$ is any integer greater than $1/\epsilon$, the implication will hold for all $n > N$. This proves that $\lim_{n\to\infty}(1/n) = 0$.

**(b)** Let $\epsilon > 0$ be given. We must show that there exists an integer $N$ such that for all $n$,

$$n > N \quad \Rightarrow \quad |k - k| < \epsilon.$$

Since $k - k = 0$, we can use any positive integer for $N$ and the implication will hold. This proves that $\lim_{n\to\infty} k = k$ for any constant $k$. ■

**EXAMPLE 2**    Show that the sequence $\{1, -1, 1, -1, 1, -1, \ldots, (-1)^{n+1}, \ldots\}$ diverges.

**Solution**    Suppose the sequence converges to some number $L$. By choosing $\epsilon = 1/2$ in the definition of the limit, all terms $a_n$ of the sequence with index $n$ larger than some $N$ must lie within $\epsilon = 1/2$ of $L$. Since the number 1 appears repeatedly as every other term of the sequence, we must have that the number 1 lies within the distance $\epsilon = 1/2$ of $L$. It follows that $|L - 1| < 1/2$, or equivalently, $1/2 < L < 3/2$. Likewise, the number $-1$ appears repeatedly in the sequence with arbitrarily high index. So we must also have that $|L - (-1)| < 1/2$, or equivalently, $-3/2 < L < -1/2$. But the number $L$ cannot lie in both of the intervals $(1/2, 3/2)$ and $(-3/2, -1/2)$ because they have no overlap. Therefore, no such limit $L$ exists and so the sequence diverges.

Note that the same argument works for any positive number $\epsilon$ smaller than 1, not just $1/2$. ■

The sequence $\{\sqrt{n}\}$ also diverges, but for a different reason. As $n$ increases, its terms become larger than any fixed number. We describe the behavior of this sequence by writing

$$\lim_{n\to\infty} \sqrt{n} = \infty.$$

In writing infinity as the limit of a sequence, we are not saying that the differences between the terms $a_n$ and $\infty$ become small as $n$ increases. Nor are we asserting that there is some number infinity that the sequence approaches. We are merely using a notation that captures the idea that $a_n$ eventually gets and stays larger than any fixed number as $n$ gets large (see Figure 10.3a). The terms of a sequence might also decrease to negative infinity, as in Figure 10.3b.

---

> **DEFINITION**    The sequence $\{a_n\}$ **diverges to infinity** if for every number $M$ there is an integer $N$ such that for all $n$ larger than $N$, $a_n > M$. If this condition holds we write
>
> $$\lim_{n\to\infty} a_n = \infty \qquad \text{or} \qquad a_n \to \infty.$$
>
> Similarly, if for every number $m$ there is an integer $N$ such that for all $n > N$ we have $a_n < m$, then we say $\{a_n\}$ **diverges to negative infinity** and write
>
> $$\lim_{n\to\infty} a_n = -\infty \qquad \text{or} \qquad a_n \to -\infty.$$

---

A sequence may diverge without diverging to infinity or negative infinity, as we saw in Example 2. The sequences $\{1, -2, 3, -4, 5, -6, 7, -8, \ldots\}$ and $\{1, 0, 2, 0, 3, 0, \ldots\}$ are also examples of such divergence.

The convergence or divergence of a sequence is not affected by the values of any number of its initial terms (whether we omit or change the first 10, 1000, or even the first million terms does not matter). From Figure 10.2, we can see that only the part of the sequence that remains after discarding some initial number of terms determines whether the sequence has a limit and the value of that limit when it does exist.

(a)

(b)

**FIGURE 10.3**    (a) The sequence diverges to $\infty$ because no matter what number $M$ is chosen, the terms of the sequence after some index $N$ all lie in the yellow band above $M$. (b) The sequence diverges to $-\infty$ because all terms after some index $N$ lie below any chosen number $m$.

## Calculating Limits of Sequences

Since sequences are functions with domain restricted to the positive integers, it is not surprising that the theorems on limits of functions given in Chapter 2 have versions for sequences.

---

**THEOREM 1** Let $\{a_n\}$ and $\{b_n\}$ be sequences of real numbers, and let $A$ and $B$ be real numbers. The following rules hold if $\lim_{n \to \infty} a_n = A$ and $\lim_{n \to \infty} b_n = B$.

1. *Sum Rule:* $\qquad\qquad\qquad\quad \lim_{n \to \infty}(a_n + b_n) = A + B$

2. *Difference Rule:* $\qquad\qquad\quad \lim_{n \to \infty}(a_n - b_n) = A - B$

3. *Constant Multiple Rule:* $\qquad \lim_{n \to \infty}(k \cdot b_n) = k \cdot B$    (any number $k$)

4. *Product Rule:* $\qquad\qquad\qquad \lim_{n \to \infty}(a_n \cdot b_n) = A \cdot B$

5. *Quotient Rule:* $\qquad\qquad\qquad \lim_{n \to \infty}\dfrac{a_n}{b_n} = \dfrac{A}{B}$    if $B \neq 0$

---

The proof is similar to that of Theorem 1 of Section 2.2 and is omitted.

**EXAMPLE 3** By combining Theorem 1 with the limits of Example 1, we have:

**(a)** $\lim_{n \to \infty}\left(-\dfrac{1}{n}\right) = -1 \cdot \lim_{n \to \infty}\dfrac{1}{n} = -1 \cdot 0 = 0$    Constant Multiple Rule and Example 1a

**(b)** $\lim_{n \to \infty}\left(\dfrac{n-1}{n}\right) = \lim_{n \to \infty}\left(1 - \dfrac{1}{n}\right) = \lim_{n \to \infty}1 - \lim_{n \to \infty}\dfrac{1}{n} = 1 - 0 = 1$    Difference Rule and Example 1a

**(c)** $\lim_{n \to \infty}\dfrac{5}{n^2} = 5 \cdot \lim_{n \to \infty}\dfrac{1}{n} \cdot \lim_{n \to \infty}\dfrac{1}{n} = 5 \cdot 0 \cdot 0 = 0$    Product Rule

**(d)** $\lim_{n \to \infty}\dfrac{4 - 7n^6}{n^6 + 3} = \lim_{n \to \infty}\dfrac{(4/n^6) - 7}{1 + (3/n^6)} = \dfrac{0 - 7}{1 + 0} = -7.$    Sum and Quotient Rules    ∎

Be cautious in applying Theorem 1. It does not say, for example, that each of the sequences $\{a_n\}$ and $\{b_n\}$ have limits if their sum $\{a_n + b_n\}$ has a limit. For instance, $\{a_n\} = \{1, 2, 3, \dots\}$ and $\{b_n\} = \{-1, -2, -3, \dots\}$ both diverge, but their sum $\{a_n + b_n\} = \{0, 0, 0, \dots\}$ clearly converges to 0.

One consequence of Theorem 1 is that every nonzero multiple of a divergent sequence $\{a_n\}$ diverges. For suppose, to the contrary, that $\{ca_n\}$ converges for some number $c \neq 0$. Then, by taking $k = 1/c$ in the Constant Multiple Rule in Theorem 1, we see that the sequence

$$\left\{\dfrac{1}{c} \cdot ca_n\right\} = \{a_n\}$$

converges. Thus, $\{ca_n\}$ cannot converge unless $\{a_n\}$ also converges. If $\{a_n\}$ does not converge, then $\{ca_n\}$ does not converge.

The next theorem is the sequence version of the Sandwich Theorem in Section 2.2. You are asked to prove the theorem in Exercise 109. (See Figure 10.4.)

---

**THEOREM 2—The Sandwich Theorem for Sequences** Let $\{a_n\}$, $\{b_n\}$, and $\{c_n\}$ be sequences of real numbers. If $a_n \leq b_n \leq c_n$ holds for all $n$ beyond some index $N$, and if $\lim_{n \to \infty} a_n = \lim_{n \to \infty} c_n = L$, then $\lim_{n \to \infty} b_n = L$ also.

---

An immediate consequence of Theorem 2 is that, if $|b_n| \leq c_n$ and $c_n \to 0$, then $b_n \to 0$ because $-c_n \leq b_n \leq c_n$. We use this fact in the next example.

**EXAMPLE 4** Since $1/n \to 0$, we know that

**(a)** $\dfrac{\cos n}{n} \to 0$        because        $-\dfrac{1}{n} \leq \dfrac{\cos n}{n} \leq \dfrac{1}{n};$

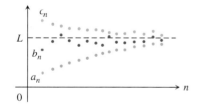

**FIGURE 10.4** The terms of sequence $\{b_n\}$ are sandwiched between those of $\{a_n\}$ and $\{c_n\}$, forcing them to the same common limit $L$.

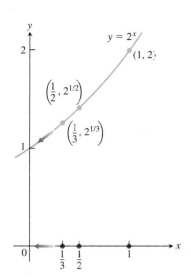

**FIGURE 10.5** As $n \to \infty$, $1/n \to 0$ and $2^{1/n} \to 2^0$ (Example 6). The terms of $\{1/n\}$ are shown on the $x$-axis; the terms of $\{2^{1/n}\}$ are shown as the $y$-values on the graph of $f(x) = 2^x$.

**(b)** $\dfrac{1}{2^n} \to 0$      because      $0 \leq \dfrac{1}{2^n} \leq \dfrac{1}{n};$

**(c)** $(-1)^n \dfrac{1}{n} \to 0$      because      $-\dfrac{1}{n} \leq (-1)^n \dfrac{1}{n} \leq \dfrac{1}{n}.$    ■

The application of Theorems 1 and 2 is broadened by a theorem stating that applying a continuous function to a convergent sequence produces a convergent sequence. We state the theorem, leaving the proof as an exercise (Exercise 110).

> **THEOREM 3—The Continuous Function Theorem for Sequences** Let $\{a_n\}$ be a sequence of real numbers. If $a_n \to L$ and if $f$ is a function that is continuous at $L$ and defined at all $a_n$, then $f(a_n) \to f(L)$.

**EXAMPLE 5**    Show that $\sqrt{(n+1)/n} \to 1$.

**Solution**    We know that $(n+1)/n \to 1$. Taking $f(x) = \sqrt{x}$ and $L = 1$ in Theorem 3 gives $\sqrt{(n+1)/n} \to \sqrt{1} = 1$.    ■

**EXAMPLE 6**    The sequence $\{1/n\}$ converges to 0. By taking $a_n = 1/n$, $f(x) = 2^x$, and $L = 0$ in Theorem 3, we see that $2^{1/n} = f(1/n) \to f(L) = 2^0 = 1$. The sequence $\{2^{1/n}\}$ converges to 1 (Figure 10.5).    ■

### Using L'Hôpital's Rule

The next theorem formalizes the connection between $\lim_{n \to \infty} a_n$ and $\lim_{x \to \infty} f(x)$. It enables us to use l'Hôpital's Rule to find the limits of some sequences.

> **THEOREM 4**    Suppose that $f(x)$ is a function defined for all $x \geq n_0$ and that $\{a_n\}$ is a sequence of real numbers such that $a_n = f(n)$ for $n \geq n_0$. Then
> $$\lim_{x \to \infty} f(x) = L \quad \Rightarrow \quad \lim_{n \to \infty} a_n = L.$$

**Proof**    Suppose that $\lim_{x \to \infty} f(x) = L$. Then for each positive number $\epsilon$ there is a number $M$ such that for all $x$,

$$x > M \quad \Rightarrow \quad |f(x) - L| < \epsilon.$$

Let $N$ be an integer greater than $M$ and greater than or equal to $n_0$. Then

$$n > N \quad \Rightarrow \quad a_n = f(n) \quad \text{and} \quad |a_n - L| = |f(n) - L| < \epsilon. \quad ■$$

**EXAMPLE 7**    Show that

$$\lim_{n \to \infty} \frac{\ln n}{n} = 0.$$

**Solution**    The function $(\ln x)/x$ is defined for all $x \geq 1$ and agrees with the given sequence at positive integers. Therefore, by Theorem 4, $\lim_{n \to \infty} (\ln n)/n$ will equal $\lim_{x \to \infty} (\ln x)/x$ if the latter exists. A single application of l'Hôpital's Rule shows that

$$\lim_{x \to \infty} \frac{\ln x}{x} = \lim_{x \to \infty} \frac{1/x}{1} = \frac{0}{1} = 0.$$

We conclude that $\lim_{n \to \infty} (\ln n)/n = 0$.    ■

When we use l'Hôpital's Rule to find the limit of a sequence, we often treat $n$ as a continuous real variable and differentiate directly with respect to $n$. This saves us from having to rewrite the formula for $a_n$ as we did in Example 7.

**EXAMPLE 8**    Does the sequence whose $n$th term is

$$a_n = \left(\frac{n+1}{n-1}\right)^n$$

converge? If so, find $\lim_{n\to\infty} a_n$.

**Solution**    The limit leads to the indeterminate form $1^\infty$. We can apply l'Hôpital's Rule if we first change the form to $\infty \cdot 0$ by taking the natural logarithm of $a_n$:

$$\ln a_n = \ln\left(\frac{n+1}{n-1}\right)^n$$

$$= n \ln\left(\frac{n+1}{n-1}\right).$$

Then,

$$\lim_{n\to\infty} \ln a_n = \lim_{n\to\infty} n \ln\left(\frac{n+1}{n-1}\right) \qquad \infty \cdot 0 \text{ form}$$

$$= \lim_{n\to\infty} \frac{\ln\left(\dfrac{n+1}{n-1}\right)}{1/n} \qquad \frac{0}{0} \text{ form}$$

$$= \lim_{n\to\infty} \frac{-2/(n^2-1)}{-1/n^2} \qquad \text{L'Hôpital's Rule: differentiate numerator and denominator.}$$

$$= \lim_{n\to\infty} \frac{2n^2}{n^2-1} = 2.$$

Since $\ln a_n \to 2$ and $f(x) = e^x$ is continuous, Theorem 4 tells us that

$$a_n = e^{\ln a_n} \to e^2.$$

The sequence $\{a_n\}$ converges to $e^2$.    ∎

## Commonly Occurring Limits

The next theorem gives some limits that arise frequently.

---

**THEOREM 5**    The following six sequences converge to the limits listed below:

**1.** $\displaystyle\lim_{n\to\infty} \frac{\ln n}{n} = 0$                      **2.** $\displaystyle\lim_{n\to\infty} \sqrt[n]{n} = 1$

**3.** $\displaystyle\lim_{n\to\infty} x^{1/n} = 1$    $(x > 0)$        **4.** $\displaystyle\lim_{n\to\infty} x^n = 0$    $(|x| < 1)$

**5.** $\displaystyle\lim_{n\to\infty} \left(1 + \frac{x}{n}\right)^n = e^x$    (any $x$)     **6.** $\displaystyle\lim_{n\to\infty} \frac{x^n}{n!} = 0$    (any $x$)

In Formulas (3) through (6), $x$ remains fixed as $n \to \infty$.

---

**Proof**    The first limit was computed in Example 7. The next two can be proved by taking logarithms and applying Theorem 4 (Exercises 107 and 108). The remaining proofs are given in Appendix 5.    ∎

**Factorial Notation**

The notation $n!$ ("$n$ factorial") means the product $1 \cdot 2 \cdot 3 \cdots n$ of the integers from 1 to $n$. Notice that $(n + 1)! = (n + 1) \cdot n!$. Thus, $4! = 1 \cdot 2 \cdot 3 \cdot 4 = 24$ and $5! = 1 \cdot 2 \cdot 3 \cdot 4 \cdot 5 = 5 \cdot 4! = 120$. We define $0!$ to be 1. Factorials grow even faster than exponentials, as the table suggests. The values in the table are rounded.

| $n$ | $e^n$ | $n!$ |
|---|---|---|
| 1 | 3 | 1 |
| 5 | 148 | 120 |
| 10 | 22,026 | 3,628,800 |
| 20 | $4.9 \times 10^8$ | $2.4 \times 10^{18}$ |

**EXAMPLE 9**    These are examples of the limits in Theorem 5.

(a) $\dfrac{\ln \left(n^2\right)}{n} = \dfrac{2 \ln n}{n} \to 2 \cdot 0 = 0$    Formula 1

(b) $\sqrt[n]{n^2} = n^{2/n} = \left(n^{1/n}\right)^2 \to (1)^2 = 1$    Formula 2

(c) $\sqrt[n]{3n} = 3^{1/n}\left(n^{1/n}\right) \to 1 \cdot 1 = 1$    Formula 3 with $x = 3$ and Formula 2

(d) $\left(-\dfrac{1}{2}\right)^n \to 0$    Formula 4 with $x = -\dfrac{1}{2}$

(e) $\left(\dfrac{n-2}{n}\right)^n = \left(1 + \dfrac{-2}{n}\right)^n \to e^{-2}$    Formula 5 with $x = -2$

(f) $\dfrac{100^n}{n!} \to 0$    Formula 6 with $x = 100$    ∎

## Recursive Definitions

So far, we have calculated each $a_n$ directly from the value of $n$. But sequences are often defined **recursively** by giving

1. The value(s) of the initial term or terms, and

2. A rule, called a **recursion formula**, for calculating any later term from terms that precede it.

## EXAMPLE 10

(a) The statements $a_1 = 1$ and $a_n = a_{n-1} + 1$ for $n > 1$ define the sequence $1, 2, 3, \ldots,$ $n, \ldots$ of positive integers. With $a_1 = 1$, we have $a_2 = a_1 + 1 = 2, a_3 = a_2 + 1 = 3$, and so on.

(b) The statements $a_1 = 1$ and $a_n = n \cdot a_{n-1}$ for $n > 1$ define the sequence $1, 2, 6, 24, \ldots, n!, \ldots$ of factorials. With $a_1 = 1$, we have $a_2 = 2 \cdot a_1 = 2$, $a_3 = 3 \cdot a_2 = 6, a_4 = 4 \cdot a_3 = 24$, and so on.

(c) The statements $a_1 = 1$, $a_2 = 1$, and $a_{n+1} = a_n + a_{n-1}$ for $n > 2$ define the sequence $1, 1, 2, 3, 5, \ldots$ of **Fibonacci numbers**. With $a_1 = 1$ and $a_2 = 1$, we have $a_3 = 1 + 1 = 2, a_4 = 2 + 1 = 3, a_5 = 3 + 2 = 5$, and so on.

(d) As we can see by applying Newton's method (see Exercise 135), the statements $x_0 = 1$ and $x_{n+1} = x_n - \left[(\sin x_n - x_n^2)/(\cos x_n - 2x_n)\right]$ for $n > 0$ define a sequence that, when it converges, gives a solution to the equation $\sin x - x^2 = 0$.    ∎

## Bounded Monotonic Sequences

Two concepts that play a key role in determining the convergence of a sequence are those of a *bounded* sequence and a *monotonic* sequence.

> **DEFINITIONS**    A sequence $\{a_n\}$ is **bounded from above** if there exists a number $M$ such that $a_n \leq M$ for all $n$. The number $M$ is an **upper bound** for $\{a_n\}$. If $M$ is an upper bound for $\{a_n\}$ but no number less than $M$ is an upper bound for $\{a_n\}$, then $M$ is the **least upper bound** for $\{a_n\}$.
>
> A sequence $\{a_n\}$ is **bounded from below** if there exists a number $m$ such that $a_n \geq m$ for all $n$. The number $m$ is a **lower bound** for $\{a_n\}$. If $m$ is a lower bound for $\{a_n\}$ but no number greater than $m$ is a lower bound for $\{a_n\}$, then $m$ is the **greatest lower bound** for $\{a_n\}$.
>
> If $\{a_n\}$ is bounded from above and below, then $\{a_n\}$ is **bounded**. If $\{a_n\}$ is not bounded, then we say that $\{a_n\}$ is an **unbounded** sequence.

## EXAMPLE 11

**(a)** The sequence $1, 2, 3, \ldots, n, \ldots$ has no upper bound because it eventually surpasses every number $M$. However, it is bounded below by every real number less than or equal to 1. The number $m = 1$ is the greatest lower bound of the sequence.

**(b)** The sequence $\dfrac{1}{2}, \dfrac{2}{3}, \dfrac{3}{4}, \ldots, \dfrac{n}{n+1}, \ldots$ is bounded above by every real number greater than or equal to 1. The upper bound $M = 1$ is the least upper bound (Exercise 127). The sequence is also bounded below by every number less than or equal to $\dfrac{1}{2}$, which is its greatest lower bound. ∎

If a sequence $\{a_n\}$ converges to the number $L$, then by definition there is a number $N$ such that $|a_n - L| < 1$ if $n > N$. That is,

$$L - 1 < a_n < L + 1 \quad \text{for } n > N.$$

If $M$ is a number larger than $L + 1$ and all of the finitely many numbers $a_1, a_2, \ldots, a_N$, then for every index $n$ we have $a_n \le M$ so that $\{a_n\}$ is bounded from above. Similarly, if $m$ is a number smaller than $L - 1$ and all of the numbers $a_1, a_2, \ldots, a_N$, then $m$ is a lower bound of the sequence. Therefore, all convergent sequences are bounded.

Although it is true that every convergent sequence is bounded, there are bounded sequences that fail to converge. One example is the bounded sequence $\{(-1)^{n+1}\}$ discussed in Example 2. The problem here is that some bounded sequences bounce around in the band determined by any lower bound $m$ and any upper bound $M$ (Figure 10.6). An important type of sequence that does not behave that way is one for which each term is at least as large, or at least as small, as its predecessor.

> **DEFINITIONS** A sequence $\{a_n\}$ is **nondecreasing** if $a_n \le a_{n+1}$ for all $n$. That is, $a_1 \le a_2 \le a_3 \le \ldots$. The sequence is **nonincreasing** if $a_n \ge a_{n+1}$ for all $n$. The sequence $\{a_n\}$ is **monotonic** if it is either nondecreasing or nonincreasing.

## EXAMPLE 12

**(a)** The sequence $1, 2, 3, \ldots, n, \ldots$ is nondecreasing.

**(b)** The sequence $\dfrac{1}{2}, \dfrac{2}{3}, \dfrac{3}{4}, \ldots, \dfrac{n}{n+1}, \ldots$ is nondecreasing.

**(c)** The sequence $1, \dfrac{1}{2}, \dfrac{1}{4}, \dfrac{1}{8}, \ldots, \dfrac{1}{2^n}, \ldots$ is nonincreasing.

**(d)** The constant sequence $3, 3, 3, \ldots, 3, \ldots$ is both nondecreasing and nonincreasing.

**(e)** The sequence $1, -1, 1, -1, 1, -1, \ldots$ is not monotonic. ∎

A nondecreasing sequence that is bounded from above always has a least upper bound. Likewise, a nonincreasing sequence bounded from below always has a greatest lower bound. These results are based on the *completeness* property of the real numbers, discussed in Appendix 6. We now prove that if $L$ is the least upper bound of a nondecreasing sequence then the sequence converges to $L$, and that if $L$ is the greatest lower bound of a nonincreasing sequence then the sequence converges to $L$.

> **THEOREM 6—The Monotonic Sequence Theorem** If a sequence $\{a_n\}$ is both bounded and monotonic, then the sequence converges.

**Proof** Suppose $\{a_n\}$ is nondecreasing, $L$ is its least upper bound, and we plot the points $(1, a_1), (2, a_2), \ldots, (n, a_n), \ldots$ in the $xy$-plane. If $M$ is an upper bound of the sequence, all these points will lie on or below the line $y = M$ (Figure 10.7). The line $y = L$ is the

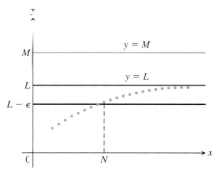

**Convergent sequences are bounded**

**FIGURE 10.6** Some bounded sequences bounce around between their bounds and fail to converge to any limiting value.

**FIGURE 10.7** If the terms of a nondecreasing sequence have an upper bound $M$, they have a limit $L \le M$.

lowest such line. None of the points $(n, a_n)$ lies above $y = L$, but some do lie above any lower line $y = L - \epsilon$, if $\epsilon$ is a positive number. The sequence converges to $L$ because

**a.** $a_n \leq L$ for *all* values of $n$, and

**b.** given any $\epsilon > 0$, there exists at least one integer $N$ for which $a_N > L - \epsilon$.

The fact that $\{a_n\}$ is nondecreasing tells us further that

$$a_n \geq a_N > L - \epsilon \qquad \text{for all } n \geq N.$$

Thus, *all* the numbers $a_n$ beyond the $N$th number lie within $\epsilon$ of $L$. This is precisely the condition for $L$ to be the limit of the sequence $\{a_n\}$.

The proof for nonincreasing sequences bounded from below is similar. ∎

It is important to realize that Theorem 6 does not say that convergent sequences are monotonic. The sequence $\{(-1)^{n+1}/n\}$ converges and is bounded, but it is not monotonic since it alternates between positive and negative values as it tends toward zero. What the theorem does say is that a nondecreasing sequence converges when it is bounded from above, but it diverges to infinity otherwise.

# Exercises 10.1

## Finding Terms of a Sequence

Each of Exercises 1–6 gives a formula for the $n$th term $a_n$ of a sequence $\{a_n\}$. Find the values of $a_1, a_2, a_3,$ and $a_4$.

**1.** $a_n = \dfrac{1 - n}{n^2}$

**2.** $a_n = \dfrac{1}{n!}$

**3.** $a_n = \dfrac{(-1)^{n+1}}{2n - 1}$

**4.** $a_n = 2 + (-1)^n$

**5.** $a_n = \dfrac{2^n}{2^{n+1}}$

**6.** $a_n = \dfrac{2^n - 1}{2^n}$

Each of Exercises 7–12 gives the first term or two of a sequence along with a recursion formula for the remaining terms. Write out the first ten terms of the sequence.

**7.** $a_1 = 1, \quad a_{n+1} = a_n + (1/2^n)$

**8.** $a_1 = 1, \quad a_{n+1} = a_n/(n + 1)$

**9.** $a_1 = 2, \quad a_{n+1} = (-1)^{n+1}a_n/2$

**10.** $a_1 = -2, \quad a_{n+1} = na_n/(n + 1)$

**11.** $a_1 = a_2 = 1, \quad a_{n+2} = a_{n+1} + a_n$

**12.** $a_1 = 2, \quad a_2 = -1, \quad a_{n+2} = a_{n+1}/a_n$

## Finding a Sequence's Formula

In Exercises 13–26, find a formula for the $n$th term of the sequence.

**13.** The sequence $1, -1, 1, -1, 1, \ldots$     1's with alternating signs

**14.** The sequence $-1, 1, -1, 1, -1, \ldots$     1's with alternating signs

**15.** The sequence $1, -4, 9, -16, 25, \ldots$     Squares of the positive integers, with alternating signs

**16.** The sequence $1, -\dfrac{1}{4}, \dfrac{1}{9}, -\dfrac{1}{16}, \dfrac{1}{25}, \ldots$     Reciprocals of squares of the positive integers, with alternating signs

**17.** $\dfrac{1}{9}, \dfrac{2}{12}, \dfrac{2^2}{15}, \dfrac{2^3}{18}, \dfrac{2^4}{21}, \ldots$     Powers of 2 divided by multiples of 3

**18.** $-\dfrac{3}{2}, -\dfrac{1}{6}, \dfrac{1}{12}, \dfrac{3}{20}, \dfrac{5}{30}, \ldots$     Integers differing by 2 divided by products of consecutive integers

**19.** The sequence $0, 3, 8, 15, 24, \ldots$     Squares of the positive integers diminished by 1

**20.** The sequence $-3, -2, -1, 0, 1, \ldots$     Integers, beginning with $-3$

**21.** The sequence $1, 5, 9, 13, 17, \ldots$     Every other odd positive integer

**22.** The sequence $2, 6, 10, 14, 18, \ldots$     Every other even positive integer

**23.** $\dfrac{5}{1}, \dfrac{8}{2}, \dfrac{11}{6}, \dfrac{14}{24}, \dfrac{17}{120}, \ldots$     Integers differing by 3 divided by factorials

**24.** $\dfrac{1}{25}, \dfrac{8}{125}, \dfrac{27}{625}, \dfrac{64}{3125}, \dfrac{125}{15{,}625}, \ldots$     Cubes of positive integers divided by powers of 5

**25.** The sequence $1, 0, 1, 0, 1, \ldots$     Alternating 1's and 0's

**26.** The sequence $0, 1, 1, 2, 2, 3, 3, 4, \ldots$     Each positive integer repeated

## Convergence and Divergence

Which of the sequences $\{a_n\}$ in Exercises 27–90 converge, and which diverge? Find the limit of each convergent sequence.

**27.** $a_n = 2 + (0.1)^n$

**28.** $a_n = \dfrac{n + (-1)^n}{n}$

**29.** $a_n = \dfrac{1 - 2n}{1 + 2n}$

**30.** $a_n = \dfrac{2n + 1}{1 - 3\sqrt{n}}$

**31.** $a_n = \dfrac{1 - 5n^4}{n^4 + 8n^3}$

**32.** $a_n = \dfrac{n + 3}{n^2 + 5n + 6}$

**33.** $a_n = \dfrac{n^2 - 2n + 1}{n - 1}$

**34.** $a_n = \dfrac{1 - n^3}{70 - 4n^2}$

**35.** $a_n = 1 + (-1)^n$

**36.** $a_n = (-1)^n \left(1 - \dfrac{1}{n}\right)$

**37.** $a_n = \left(\dfrac{n+1}{2n}\right)\left(1 - \dfrac{1}{n}\right)$

**38.** $a_n = \left(2 - \dfrac{1}{2^n}\right)\left(3 + \dfrac{1}{2^n}\right)$

**39.** $a_n = \dfrac{(-1)^{n+1}}{2n - 1}$

**40.** $a_n = \left(-\dfrac{1}{2}\right)^n$

**41.** $a_n = \sqrt{\dfrac{2n}{n+1}}$

**42.** $a_n = \dfrac{1}{(0.9)^n}$

**43.** $a_n = \sin\left(\dfrac{\pi}{2} + \dfrac{1}{n}\right)$

**44.** $a_n = n\pi \cos(n\pi)$

**45.** $a_n = \dfrac{\sin n}{n}$

**46.** $a_n = \dfrac{\sin^2 n}{2^n}$

**47.** $a_n = \dfrac{n}{2^n}$

**48.** $a_n = \dfrac{3^n}{n^3}$

**49.** $a_n = \dfrac{\ln(n+1)}{\sqrt{n}}$

**50.** $a_n = \dfrac{\ln n}{\ln 2n}$

**51.** $a_n = 8^{1/n}$

**52.** $a_n = (0.03)^{1/n}$

**53.** $a_n = \left(1 + \dfrac{7}{n}\right)^n$

**54.** $a_n = \left(1 - \dfrac{1}{n}\right)^n$

**55.** $a_n = \sqrt[n]{10n}$

**56.** $a_n = \sqrt[n]{n^2}$

**57.** $a_n = \left(\dfrac{3}{n}\right)^{1/n}$

**58.** $a_n = (n + 4)^{1/(n+4)}$

**59.** $a_n = \dfrac{\ln n}{n^{1/n}}$

**60.** $a_n = \ln n - \ln(n+1)$

**61.** $a_n = \sqrt[n]{4^n n}$

**62.** $a_n = \sqrt[n]{3^{2n+1}}$

**63.** $a_n = \dfrac{n!}{n^n}$ (*Hint:* Compare with $1/n$.)

**64.** $a_n = \dfrac{(-4)^n}{n!}$

**65.** $a_n = \dfrac{n!}{10^{6n}}$

**66.** $a_n = \dfrac{n!}{2^n \cdot 3^n}$

**67.** $a_n = \left(\dfrac{1}{n}\right)^{1/(\ln n)}$

**68.** $a_n = \ln\left(1 + \dfrac{1}{n}\right)^n$

**69.** $a_n = \left(\dfrac{3n+1}{3n-1}\right)^n$

**70.** $a_n = \left(\dfrac{n}{n+1}\right)^n$

**71.** $a_n = \left(\dfrac{x^n}{2n+1}\right)^{1/n}, \quad x > 0$

**72.** $a_n = \left(1 - \dfrac{1}{n^2}\right)^n$

**73.** $a_n = \dfrac{3^n \cdot 6^n}{2^{-n} \cdot n!}$

**74.** $a_n = \dfrac{(10/11)^n}{(9/10)^n + (11/12)^n}$

**75.** $a_n = \tanh n$

**76.** $a_n = \sinh(\ln n)$

**77.** $a_n = \dfrac{n^2}{2n-1} \sin \dfrac{1}{n}$

**78.** $a_n = n\left(1 - \cos\dfrac{1}{n}\right)$

**79.** $a_n = \sqrt{n} \sin \dfrac{1}{\sqrt{n}}$

**80.** $a_n = (3^n + 5^n)^{1/n}$

**81.** $a_n = \tan^{-1} n$

**82.** $a_n = \dfrac{1}{\sqrt{n}} \tan^{-1} n$

**83.** $a_n = \left(\dfrac{1}{3}\right)^n + \dfrac{1}{\sqrt{2^n}}$

**84.** $a_n = \sqrt[n]{n^2 + n}$

**85.** $a_n = \dfrac{(\ln n)^{200}}{n}$

**86.** $a_n = \dfrac{(\ln n)^5}{\sqrt{n}}$

**87.** $a_n = n - \sqrt{n^2 - n}$

**88.** $a_n = \dfrac{1}{\sqrt{n^2 - 1} - \sqrt{n^2 + n}}$

**89.** $a_n = \dfrac{1}{n} \displaystyle\int_1^n \dfrac{1}{x} dx$

**90.** $a_n = \displaystyle\int_1^n \dfrac{1}{x^p} dx, \quad p > 1$

## Recursively Defined Sequences

In Exercises 91–98, assume that each sequence converges and find its limit.

**91.** $a_1 = 2, \quad a_{n+1} = \dfrac{72}{1 + a_n}$

**92.** $a_1 = -1, \quad a_{n+1} = \dfrac{a_n + 6}{a_n + 2}$

**93.** $a_1 = -4, \quad a_{n+1} = \sqrt{8 + 2a_n}$

**94.** $a_1 = 0, \quad a_{n+1} = \sqrt{8 + 2a_n}$

**95.** $a_1 = 5, \quad a_{n+1} = \sqrt{5a_n}$

**96.** $a_1 = 3, \quad a_{n+1} = 12 - \sqrt{a_n}$

**97.** $2, 2 + \dfrac{1}{2}, 2 + \dfrac{1}{2 + \dfrac{1}{2}}, 2 + \dfrac{1}{2 + \dfrac{1}{2 + \dfrac{1}{2}}}, \dots$

**98.** $\sqrt{1}, \sqrt{1 + \sqrt{1}}, \sqrt{1 + \sqrt{1 + \sqrt{1}}},$
$\sqrt{1 + \sqrt{1 + \sqrt{1 + \sqrt{1}}}}, \dots$

## Theory and Examples

**99.** The first term of a sequence is $x_1 = 1$. Each succeeding term is the sum of all those that come before it:
$$x_{n+1} = x_1 + x_2 + \cdots + x_n.$$
Write out enough early terms of the sequence to deduce a general formula for $x_n$ that holds for $n \geq 2$.

**100.** A sequence of rational numbers is described as follows:
$$\dfrac{1}{1}, \dfrac{3}{2}, \dfrac{7}{5}, \dfrac{17}{12}, \dots, \dfrac{a}{b}, \dfrac{a + 2b}{a + b}, \dots.$$
Here the numerators form one sequence, the denominators form a second sequence, and their ratios form a third sequence. Let $x_n$ and $y_n$ be, respectively, the numerator and the denominator of the $n$th fraction $r_n = x_n/y_n$.

**a.** Verify that $x_1^2 - 2y_1^2 = -1$, $x_2^2 - 2y_2^2 = +1$ and, more generally, that if $a^2 - 2b^2 = -1$ or $+1$, then
$$(a + 2b)^2 - 2(a + b)^2 = +1 \quad \text{or} \quad -1,$$
respectively.

**b.** The fractions $r_n = x_n/y_n$ approach a limit as $n$ increases. What is that limit? (*Hint:* Use part (a) to show that $r_n^2 - 2 = \pm(1/y_n)^2$ and that $y_n$ is not less than $n$.)

**101. Newton's method** The following sequences come from the recursion formula for Newton's method,
$$x_{n+1} = x_n - \dfrac{f(x_n)}{f'(x_n)}.$$

Do the sequences converge? If so, to what value? In each case, begin by identifying the function $f$ that generates the sequence.

**a.** $x_0 = 1$, $x_{n+1} = x_n - \dfrac{x_n^2 - 2}{2x_n} = \dfrac{x_n}{2} + \dfrac{1}{x_n}$

**b.** $x_0 = 1$, $x_{n+1} = x_n - \dfrac{\tan x_n - 1}{\sec^2 x_n}$

**c.** $x_0 = 1$, $x_{n+1} = x_n - 1$

**102. a.** Suppose that $f(x)$ is differentiable for all $x$ in $[0, 1]$ and that $f(0) = 0$. Define sequence $\{a_n\}$ by the rule $a_n = nf(1/n)$. Show that $\lim_{n \to \infty} a_n = f'(0)$. Use the result in part (a) to find the limits of the following sequences $\{a_n\}$.

**b.** $a_n = n \tan^{-1} \dfrac{1}{n}$      **c.** $a_n = n(e^{1/n} - 1)$

**d.** $a_n = n \ln\left(1 + \dfrac{2}{n}\right)$

**103. Pythagorean triples** A triple of positive integers $a$, $b$, and $c$ is called a **Pythagorean triple** if $a^2 + b^2 = c^2$. Let $a$ be an odd positive integer and let

$$b = \left\lfloor \dfrac{a^2}{2} \right\rfloor \quad \text{and} \quad c = \left\lceil \dfrac{a^2}{2} \right\rceil$$

be, respectively, the integer floor and ceiling for $a^2/2$.

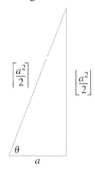

**a.** Show that $a^2 + b^2 = c^2$. (*Hint:* Let $a = 2n + 1$ and express $b$ and $c$ in terms of $n$.)

**b.** By direct calculation, or by appealing to the accompanying figure, find

$$\lim_{a \to \infty} \dfrac{\left\lfloor \dfrac{a^2}{2} \right\rfloor}{\left\lceil \dfrac{a^2}{2} \right\rceil}.$$

**104. The $n$th root of $n!$**

**a.** Show that $\lim_{n \to \infty} (2n\pi)^{1/(2n)} = 1$ and hence, using Stirling's approximation (Chapter 8, Additional Exercise 52a), that

$$\sqrt[n]{n!} \approx \dfrac{n}{e} \quad \text{for large values of } n.$$

**T b.** Test the approximation in part (a) for $n = 40, 50, 60, \ldots$, as far as your calculator will allow.

**105. a.** Assuming that $\lim_{n \to \infty} (1/n^c) = 0$ if $c$ is any positive constant, show that

$$\lim_{n \to \infty} \dfrac{\ln n}{n^c} = 0$$

if $c$ is any positive constant.

**b.** Prove that $\lim_{n \to \infty} (1/n^c) = 0$ if $c$ is any positive constant. (*Hint:* If $\epsilon = 0.001$ and $c = 0.04$, how large should $N$ be to ensure that $|1/n^c - 0| < \epsilon$ if $n > N$?)

**106. The zipper theorem** Prove the "zipper theorem" for sequences: If $\{a_n\}$ and $\{b_n\}$ both converge to $L$, then the sequence

$$a_1, b_1, a_2, b_2, \ldots, a_n, b_n, \ldots$$

converges to $L$.

**107.** Prove that $\lim_{n \to \infty} \sqrt[n]{n} = 1$.

**108.** Prove that $\lim_{n \to \infty} x^{1/n} = 1$, $(x > 0)$.

**109.** Prove Theorem 2.      **110.** Prove Theorem 3.

In Exercises 111–114, determine if the sequence is monotonic and if it is bounded.

**111.** $a_n = \dfrac{3n + 1}{n + 1}$      **112.** $a_n = \dfrac{(2n + 3)!}{(n + 1)!}$

**113.** $a_n = \dfrac{2^n 3^n}{n!}$      **114.** $a_n = 2 - \dfrac{2}{n} - \dfrac{1}{2^n}$

Which of the sequences in Exercises 115–124 converge, and which diverge? Give reasons for your answers.

**115.** $a_n = 1 - \dfrac{1}{n}$      **116.** $a_n = n - \dfrac{1}{n}$

**117.** $a_n = \dfrac{2^n - 1}{2^n}$      **118.** $a_n = \dfrac{2^n - 1}{3^n}$

**119.** $a_n = ((-1)^n + 1)\left(\dfrac{n + 1}{n}\right)$

**120.** The first term of a sequence is $x_1 = \cos(1)$. The next terms are $x_2 = x_1$ or $\cos(2)$, whichever is larger; and $x_3 = x_2$ or $\cos(3)$, whichever is larger (farther to the right). In general,

$$x_{n+1} = \max\{x_n, \cos(n + 1)\}.$$

**121.** $a_n = \dfrac{1 + \sqrt{2n}}{\sqrt{n}}$      **122.** $a_n = \dfrac{n + 1}{n}$

**123.** $a_n = \dfrac{4^{n+1} + 3^n}{4^n}$

**124.** $a_1 = 1$, $a_{n+1} = 2a_n - 3$

In Exercises 125–126, use the definition of convergence to prove the given limit.

**125.** $\lim_{n \to \infty} \dfrac{\sin n}{n} = 0$      **126.** $\lim_{n \to \infty} \left(1 - \dfrac{1}{n^2}\right) = 1$

**127. The sequence $\{n/(n + 1)\}$ has a least upper bound of 1** Show that if $M$ is a number less than 1, then the terms of $\{n/(n + 1)\}$ eventually exceed $M$. That is, if $M < 1$ there is an integer $N$ such that $n/(n + 1) > M$ whenever $n > N$. Since $n/(n + 1) < 1$ for every $n$, this proves that 1 is a least upper bound for $\{n/(n + 1)\}$.

**128. Uniqueness of least upper bounds** Show that if $M_1$ and $M_2$ are least upper bounds for the sequence $\{a_n\}$, then $M_1 = M_2$. That is, a sequence cannot have two different least upper bounds.

**129.** Is it true that a sequence $\{a_n\}$ of positive numbers must converge if it is bounded from above? Give reasons for your answer.

**130.** Prove that if $\{a_n\}$ is a convergent sequence, then to every positive number $\epsilon$ there corresponds an integer $N$ such that for all $m$ and $n$,

$$m > N \quad \text{and} \quad n > N \quad \Rightarrow \quad |a_m - a_n| < \epsilon.$$

**131. Uniqueness of limits** Prove that limits of sequences are unique. That is, show that if $L_1$ and $L_2$ are numbers such that $a_n \to L_1$ and $a_n \to L_2$, then $L_1 = L_2$.

**132. Limits and subsequences** If the terms of one sequence appear in another sequence in their given order, we call the first sequence a **subsequence** of the second. Prove that if two subsequences of a sequence $\{a_n\}$ have different limits $L_1 \neq L_2$, then $\{a_n\}$ diverges.

**133.** For a sequence $\{a_n\}$ the terms of even index are denoted by $a_{2k}$ and the terms of odd index by $a_{2k+1}$. Prove that if $a_{2k} \to L$ and $a_{2k+1} \to L$, then $a_n \to L$.

**134.** Prove that a sequence $\{a_n\}$ converges to 0 if and only if the sequence of absolute values $\{|a_n|\}$ converges to 0.

**135. Sequences generated by Newton's method** Newton's method, applied to a differentiable function $f(x)$, begins with a starting value $x_0$ and constructs from it a sequence of numbers $\{x_n\}$ that under favorable circumstances converges to a zero of $f$. The recursion formula for the sequence is

$$x_{n+1} = x_n - \frac{f(x_n)}{f'(x_n)}.$$

**a.** Show that the recursion formula for $f(x) = x^2 - a, a > 0$, can be written as $x_{n+1} = (x_n + a/x_n)/2$.

**T** **b.** Starting with $x_0 = 1$ and $a = 3$, calculate successive terms of the sequence until the display begins to repeat. What number is being approximated? Explain.

**T** **136. A recursive definition of $\pi/2$** If you start with $x_1 = 1$ and define the subsequent terms of $\{x_n\}$ by the rule $x_n = x_{n-1} + \cos x_{n-1}$, you generate a sequence that converges rapidly to $\pi/2$. **(a)** Try it. **(b)** Use the accompanying figure to explain why the convergence is so rapid.

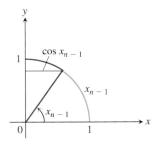

### COMPUTER EXPLORATIONS

Use a CAS to perform the following steps for the sequences in Exercises 137–148.

**a.** Calculate and then plot the first 25 terms of the sequence. Does the sequence appear to be bounded from above or below? Does it appear to converge or diverge? If it does converge, what is the limit $L$?

**b.** If the sequence converges, find an integer $N$ such that $|a_n - L| \leq 0.01$ for $n \geq N$. How far in the sequence do you have to get for the terms to lie within 0.0001 of $L$?

**137.** $a_n = \sqrt[n]{n}$

**138.** $a_n = \left(1 + \frac{0.5}{n}\right)^n$

**139.** $a_1 = 1, \quad a_{n+1} = a_n + \frac{1}{5^n}$

**140.** $a_1 = 1, \quad a_{n+1} = a_n + (-2)^n$

**141.** $a_n = \sin n$

**142.** $a_n = n \sin \frac{1}{n}$

**143.** $a_n = \frac{\sin n}{n}$

**144.** $a_n = \frac{\ln n}{n}$

**145.** $a_n = (0.9999)^n$

**146.** $a_n = (123456)^{1/n}$

**147.** $a_n = \frac{8^n}{n!}$

**148.** $a_n = \frac{n^{41}}{19^n}$

# $10.2$ Infinite Series

An *infinite series* is the sum of an infinite sequence of numbers

$$a_1 + a_2 + a_3 + \cdots + a_n + \cdots$$

The goal of this section is to understand the meaning of such an infinite sum and to develop methods to calculate it. Since there are infinitely many terms to add in an infinite series, we cannot just keep adding to see what comes out. Instead we look at the result of summing just the first $n$ terms of the sequence. The sum of the first $n$ terms

$$s_n = a_1 + a_2 + a_3 + \cdots + a_n$$

is an ordinary finite sum and can be calculated by normal addition. It is called the *nth partial sum*. As $n$ gets larger, we expect the partial sums to get closer and closer to a limiting value in the same sense that the terms of a sequence approach a limit, as discussed in Section 10.1.

For example, to assign meaning to an expression like

$$1 + \frac{1}{2} + \frac{1}{4} + \frac{1}{8} + \frac{1}{16} + \cdots$$

we add the terms one at a time from the beginning and look for a pattern in how these partial sums grow.

| Partial sum | | Value | Suggestive expression for partial sum |
|---|---|---|---|
| First: | $s_1 = 1$ | $1$ | $2 - 1$ |
| Second: | $s_2 = 1 + \dfrac{1}{2}$ | $\dfrac{3}{2}$ | $2 - \dfrac{1}{2}$ |
| Third: | $s_3 = 1 + \dfrac{1}{2} + \dfrac{1}{4}$ | $\dfrac{7}{4}$ | $2 - \dfrac{1}{4}$ |
| $\vdots$ | $\vdots$ | $\vdots$ | $\vdots$ |
| $n$th: | $s_n = 1 + \dfrac{1}{2} + \dfrac{1}{4} + \cdots + \dfrac{1}{2^{n-1}}$ | $\dfrac{2^n - 1}{2^{n-1}}$ | $2 - \dfrac{1}{2^{n-1}}$ |

Indeed there is a pattern. The partial sums form a sequence whose $n$th term is

$$s_n = 2 - \frac{1}{2^{n-1}}.$$

This sequence of partial sums converges to 2 because $\lim_{n\to\infty}(1/2^{n-1}) = 0$. We say

$$\text{``the sum of the infinite series } 1 + \frac{1}{2} + \frac{1}{4} + \cdots + \frac{1}{2^{n-1}} + \cdots \text{ is 2.''}$$

Is the sum of any finite number of terms in this series equal to 2? No. Can we actually add an infinite number of terms one by one? No. But we can still define their sum by defining it to be the limit of the sequence of partial sums as $n \to \infty$, in this case 2 (Figure 10.8). Our knowledge of sequences and limits enables us to break away from the confines of finite sums.

---

**DEFINITIONS**   Given a sequence of numbers $\{a_n\}$, an expression of the form

$$a_1 + a_2 + a_3 + \cdots + a_n + \cdots$$

is an **infinite series**. The number $a_n$ is the **$n$th term** of the series. The sequence $\{s_n\}$ defined by

$$s_1 = a_1$$
$$s_2 = a_1 + a_2$$
$$\vdots$$
$$s_n = a_1 + a_2 + \cdots + a_n = \sum_{k=1}^{n} a_k$$
$$\vdots$$

is the **sequence of partial sums** of the series, the number $s_n$ being the **$n$th partial sum**. If the sequence of partial sums converges to a limit $L$, we say that the series **converges** and that its **sum** is $L$. In this case, we also write

$$a_1 + a_2 + \cdots + a_n + \cdots = \sum_{n=1}^{\infty} a_n = L.$$

If the sequence of partial sums of the series does not converge, we say that the series **diverges**.

---

**FIGURE 10.8** As the lengths $1, \frac{1}{2}, \frac{1}{4}, \frac{1}{8}, \dots$ are added one by one, the sum approaches 2.

When we begin to study a given series $a_1 + a_2 + \cdots + a_n + \cdots$, we might not know whether it converges or diverges. In either case, it is convenient to use sigma notation to write the series as

$$\sum_{n=1}^{\infty} a_n, \qquad \sum_{k=1}^{\infty} a_k, \qquad \text{or} \qquad \sum a_n$$

A useful shorthand when summation from 1 to $\infty$ is understood

## Geometric Series

**Geometric series** are series of the form

$$a + ar + ar^2 + \cdots + ar^{n-1} + \cdots = \sum_{n=1}^{\infty} ar^{n-1}$$

in which $a$ and $r$ are fixed real numbers and $a \neq 0$. The series can also be written as $\sum_{n=0}^{\infty} ar^n$. The **ratio** $r$ can be positive, as in

$$1 + \frac{1}{2} + \frac{1}{4} + \cdots + \left(\frac{1}{2}\right)^{n-1} + \cdots, \qquad r = 1/2, a = 1$$

or negative, as in

$$1 - \frac{1}{3} + \frac{1}{9} - \cdots + \left(-\frac{1}{3}\right)^{n-1} + \cdots. \qquad r = -1/3, a = 1$$

If $r = 1$, the $n$th partial sum of the geometric series is

$$s_n = a + a(1) + a(1)^2 + \cdots + a(1)^{n-1} = na,$$

and the series diverges because $\lim_{n \to \infty} s_n = \pm\infty$, depending on the sign of $a$. If $r = -1$, the series diverges because the $n$th partial sums alternate between $a$ and 0. If $|r| \neq 1$, we can determine the convergence or divergence of the series in the following way:

$$s_n = a + ar + ar^2 + \cdots + ar^{n-1}$$

$$rs_n = ar + ar^2 + \cdots + ar^{n-1} + ar^n \qquad \text{Multiply } s_n \text{ by } r.$$

$$s_n - rs_n = a - ar^n \qquad \text{Subtract } rs_n \text{ from } s_n. \text{ Most of the terms on the right cancel.}$$

$$s_n(1 - r) = a(1 - r^n) \qquad \text{Factor.}$$

$$s_n = \frac{a(1 - r^n)}{1 - r}, \qquad (r \neq 1). \qquad \text{We can solve for } s_n \text{ if } r \neq 1.$$

If $|r| < 1$, then $r^n \to 0$ as $n \to \infty$ (as in Section 10.1) and $s_n \to a/(1 - r)$. If $|r| > 1$, then $|r^n| \to \infty$ and the series diverges.

---

If $|r| < 1$, the geometric series $a + ar + ar^2 + \cdots + ar^{n-1} + \cdots$ converges to $a/(1 - r)$:

$$\sum_{n=1}^{\infty} ar^{n-1} = \frac{a}{1 - r}, \qquad |r| < 1.$$

If $|r| \geq 1$, the series diverges.

---

We have determined when a geometric series converges or diverges, and to what value. Often we can determine that a series converges without knowing the value to which it converges, as we will see in the next several sections. The formula $a/(1 - r)$ for the sum of a geometric series applies *only* when the summation index begins with $n = 1$ in the expression $\sum_{n=1}^{\infty} ar^{n-1}$ (or with the index $n = 0$ if we write the series as $\sum_{n=0}^{\infty} ar^n$).

**EXAMPLE 1**    The geometric series with $a = 1/9$ and $r = 1/3$ is

$$\frac{1}{9} + \frac{1}{27} + \frac{1}{81} + \cdots = \sum_{n=1}^{\infty} \frac{1}{9}\left(\frac{1}{3}\right)^{n-1} = \frac{1/9}{1 - (1/3)} = \frac{1}{6}. \qquad \blacksquare$$

**EXAMPLE 2**    The series

$$\sum_{n=0}^{\infty} \frac{(-1)^n 5}{4^n} = 5 - \frac{5}{4} + \frac{5}{16} - \frac{5}{64} + \cdots$$

is a geometric series with $a = 5$ and $r = -1/4$. It converges to

$$\frac{a}{1 - r} = \frac{5}{1 + (1/4)} = 4. \qquad \blacksquare$$

**EXAMPLE 3**    You drop a ball from $a$ meters above a flat surface. Each time the ball hits the surface after falling a distance $h$, it rebounds a distance $rh$, where $r$ is positive but less than 1. Find the total distance the ball travels up and down (Figure 10.9).

**Solution**    The total distance is

$$s = a + \underbrace{2ar + 2ar^2 + 2ar^3 + \cdots}_{\text{This sum is } 2ar/(1 - r).} = a + \frac{2ar}{1 - r} = a\frac{1 + r}{1 - r}.$$

If $a = 6$ m and $r = 2/3$, for instance, the distance is

$$s = 6 \cdot \frac{1 + (2/3)}{1 - (2/3)} = 6\left(\frac{5/3}{1/3}\right) = 30 \text{ m.} \qquad \blacksquare$$

**EXAMPLE 4**    Express the repeating decimal $5.232323\ldots$ as the ratio of two integers.

**Solution**    From the definition of a decimal number, we get a geometric series

$$5.232323\ldots = 5 + \frac{23}{100} + \frac{23}{(100)^2} + \frac{23}{(100)^3} + \cdots$$

$$= 5 + \frac{23}{100}\underbrace{\left(1 + \frac{1}{100} + \left(\frac{1}{100}\right)^2 + \cdots\right)}_{1/(1 - 0.01)} \qquad \substack{a = 1, \\ r = 1/100}$$

$$= 5 + \frac{23}{100}\left(\frac{1}{0.99}\right) = 5 + \frac{23}{99} = \frac{518}{99} \qquad \blacksquare$$

Unfortunately, formulas like the one for the sum of a convergent geometric series are rare and we usually have to settle for an estimate of a series' sum (more about this later). The next example, however, is another case in which we can find the sum exactly.

**EXAMPLE 5**    Find the sum of the "telescoping" series $\sum_{n=1}^{\infty} \frac{1}{n(n + 1)}$.

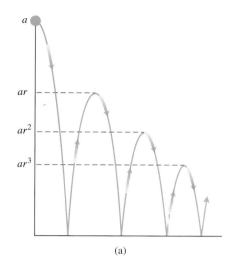

$a$

$ar$

$ar^2$

$ar^3$

(a)

(b)

**FIGURE 10.9**  (a) Example 3 shows how to use a geometric series to calculate the total vertical distance traveled by a bouncing ball if the height of each rebound is reduced by the factor $r$. (b) A stroboscopic photo of a bouncing ball. (*Source: PSSC Physics*, 2nd ed., Reprinted by permission of Educational Development Center, Inc.)

**Solution** We look for a pattern in the sequence of partial sums that might lead to a formula for $s_k$. The key observation is the partial fraction decomposition

$$\frac{1}{n(n+1)} = \frac{1}{n} - \frac{1}{n+1},$$

so

$$\sum_{n=1}^{k} \frac{1}{n(n+1)} = \sum_{n=1}^{k} \left( \frac{1}{n} - \frac{1}{n+1} \right)$$

and

$$s_k = \left( \frac{1}{1} - \frac{1}{2} \right) + \left( \frac{1}{2} - \frac{1}{3} \right) + \left( \frac{1}{3} - \frac{1}{4} \right) + \cdots + \left( \frac{1}{k} - \frac{1}{k+1} \right).$$

Removing parentheses and canceling adjacent terms of opposite sign collapses the sum to

$$s_k = 1 - \frac{1}{k+1}.$$

We now see that $s_k \to 1$ as $k \to \infty$. The series converges, and its sum is 1:

$$\sum_{n=1}^{\infty} \frac{1}{n(n+1)} = 1. \qquad \blacksquare$$

### The *n*th-Term Test for a Divergent Series

One reason that a series may fail to converge is that its terms don't become small.

**EXAMPLE 6**    The series

$$\sum_{n=1}^{\infty} \frac{n+1}{n} = \frac{2}{1} + \frac{3}{2} + \frac{4}{3} + \cdots + \frac{n+1}{n} + \cdots$$

diverges because the partial sums eventually outgrow every preassigned number. Each term is greater than 1, so the sum of $n$ terms is greater than $n$. $\qquad \blacksquare$

We now show that $\lim_{n \to \infty} a_n$ must equal zero if the series $\sum_{n=1}^{\infty} a_n$ converges. To see why, let $S$ represent the series' sum and $s_n = a_1 + a_2 + \cdots + a_n$ the $n$th partial sum. When $n$ is large, both $s_n$ and $s_{n-1}$ are close to $S$, so their difference, $a_n$, is close to zero. More formally,

$$a_n = s_n - s_{n-1} \quad \rightarrow \quad S - S = 0. \qquad \begin{array}{l} \text{Difference Rule} \\ \text{for sequences} \end{array}$$

This establishes the following theorem.

---

**THEOREM 7**    If $\displaystyle\sum_{n=1}^{\infty} a_n$ converges, then $a_n \to 0$.

---

Theorem 7 leads to a test for detecting the kind of divergence that occurred in Example 6.

---

**The *n*th-Term Test for Divergence**

$\displaystyle\sum_{n=1}^{\infty} a_n$ diverges if $\displaystyle\lim_{n \to \infty} a_n$ fails to exist or is different from zero.

---

**EXAMPLE 7**    The following are all examples of divergent series.

**(a)** $\displaystyle\sum_{n=1}^{\infty} n^2$ diverges because $n^2 \to \infty$.

**(b)** $\displaystyle\sum_{n=1}^{\infty} \frac{n+1}{n}$ diverges because $\dfrac{n+1}{n} \to 1.$   $\lim_{n\to\infty} a_n \neq 0$

**(c)** $\displaystyle\sum_{n=1}^{\infty} (-1)^{n+1}$ diverges because $\lim_{n\to\infty} (-1)^{n+1}$ does not exist.

**(d)** $\displaystyle\sum_{n=1}^{\infty} \frac{-n}{2n+5}$ diverges because $\lim_{n\to\infty} \dfrac{-n}{2n+5} = -\dfrac{1}{2} \neq 0.$   ∎

**EXAMPLE 8**    The series

$$1 + \underbrace{\frac{1}{2} + \frac{1}{2}}_{2 \text{ terms}} + \underbrace{\frac{1}{4} + \frac{1}{4} + \frac{1}{4} + \frac{1}{4}}_{4 \text{ terms}} + \cdots + \underbrace{\frac{1}{2^n} + \frac{1}{2^n} + \cdots + \frac{1}{2^n}}_{2^n \text{ terms}} + \cdots$$

diverges because the terms can be grouped into infinitely many clusters each of which adds to 1, so the partial sums increase without bound. However, the terms of the series form a sequence that converges to 0. Example 1 of Section 10.3 shows that the harmonic series $\sum 1/n$ also behaves in this manner.   ∎

## Combining Series

Whenever we have two convergent series, we can add them term by term, subtract them term by term, or multiply them by constants to make new convergent series.

---

**THEOREM 8**    If $\sum a_n = A$ and $\sum b_n = B$ are convergent series, then

1. *Sum Rule:*  $\quad\quad\quad\quad\quad \sum (a_n + b_n) = \sum a_n + \sum b_n = A + B$
2. *Difference Rule:*  $\quad\quad\quad \sum (a_n - b_n) = \sum a_n - \sum b_n = A - B$
3. *Constant Multiple Rule:*  $\quad \sum k a_n = k \sum a_n = kA$   (any number $k$).

---

**Proof**    The three rules for series follow from the analogous rules for sequences in Theorem 1, Section 10.1. To prove the Sum Rule for series, let

$$A_n = a_1 + a_2 + \cdots + a_n, \quad B_n = b_1 + b_2 + \cdots + b_n.$$

Then the partial sums of $\sum (a_n + b_n)$ are

$$s_n = (a_1 + b_1) + (a_2 + b_2) + \cdots + (a_n + b_n)$$
$$= (a_1 + \cdots + a_n) + (b_1 + \cdots + b_n)$$
$$= A_n + B_n.$$

Since $A_n \to A$ and $B_n \to B$, we have $s_n \to A + B$ by the Sum Rule for sequences. The proof of the Difference Rule is similar.

To prove the Constant Multiple Rule for series, observe that the partial sums of $\sum k a_n$ form the sequence

$$s_n = k a_1 + k a_2 + \cdots + k a_n = k(a_1 + a_2 + \cdots + a_n) = k A_n,$$

which converges to $kA$ by the Constant Multiple Rule for sequences.   ∎

As corollaries of Theorem 8, we have the following results. We omit the proofs.

---

1. Every nonzero constant multiple of a divergent series diverges.

2. If $\sum a_n$ converges and $\sum b_n$ diverges, then $\sum (a_n + b_n)$ and $\sum (a_n - b_n)$ both diverge.

---

**Caution** Remember that $\sum(a_n + b_n)$ can converge when $\sum a_n$ and $\sum b_n$ both diverge. For example, $\sum a_n = 1 + 1 + 1 + \cdots$ and $\sum b_n = (-1) + (-1) + (-1) + \cdots$ diverge, whereas $\sum(a_n + b_n) = 0 + 0 + 0 + \cdots$ converges to 0.

**EXAMPLE 9** Find the sums of the following series.

(a) $\displaystyle\sum_{n=1}^{\infty} \frac{3^{n-1} - 1}{6^{n-1}} = \sum_{n=1}^{\infty} \left( \frac{1}{2^{n-1}} - \frac{1}{6^{n-1}} \right)$

$\displaystyle = \sum_{n=1}^{\infty} \frac{1}{2^{n-1}} - \sum_{n=1}^{\infty} \frac{1}{6^{n-1}}$    Difference Rule

$\displaystyle = \frac{1}{1 - (1/2)} - \frac{1}{1 - (1/6)}$    Geometric series with $a = 1$ and $r = 1/2, 1/6$

$\displaystyle = 2 - \frac{6}{5} = \frac{4}{5}$

(b) $\displaystyle\sum_{n=0}^{\infty} \frac{4}{2^n} = 4 \sum_{n=0}^{\infty} \frac{1}{2^n}$    Constant Multiple Rule

$\displaystyle = 4 \left( \frac{1}{1 - (1/2)} \right)$    Geometric series with $a = 1, r = 1/2$

$\displaystyle = 8$    ∎

## Adding or Deleting Terms

We can add a finite number of terms to a series or delete a finite number of terms without altering the series' convergence or divergence, although in the case of convergence this will usually change the sum. If $\sum_{n=1}^{\infty} a_n$ converges, then $\sum_{n=k}^{\infty} a_n$ converges for any $k > 1$ and

$$\sum_{n=1}^{\infty} a_n = a_1 + a_2 + \cdots + a_{k-1} + \sum_{n=k}^{\infty} a_n.$$

Conversely, if $\sum_{n=k}^{\infty} a_n$ converges for any $k > 1$, then $\sum_{n=1}^{\infty} a_n$ converges. Thus,

$$\sum_{n=1}^{\infty} \frac{1}{5^n} = \frac{1}{5} + \frac{1}{25} + \frac{1}{125} + \sum_{n=4}^{\infty} \frac{1}{5^n}$$

and

$$\sum_{n=4}^{\infty} \frac{1}{5^n} = \left( \sum_{n=1}^{\infty} \frac{1}{5^n} \right) - \frac{1}{5} - \frac{1}{25} - \frac{1}{125}.$$

The convergence or divergence of a series is not affected by its first few terms. Only the "tail" of the series, the part that remains when we sum beyond some finite number of initial terms, influences whether it converges or diverges.

## Reindexing

As long as we preserve the order of its terms, we can reindex any series without altering its convergence. To raise the starting value of the index $h$ units, replace the $n$ in the formula for $a_n$ by $n - h$:

$$\sum_{n=1}^{\infty} a_n = \sum_{n=1+h}^{\infty} a_{n-h} = a_1 + a_2 + a_3 + \cdots.$$

To lower the starting value of the index $h$ units, replace the $n$ in the formula for $a_n$ by $n + h$:

$$\sum_{n=1}^{\infty} a_n = \sum_{n=1-h}^{\infty} a_{n+h} = a_1 + a_2 + a_3 + \cdots.$$

We saw this reindexing in starting a geometric series with the index $n = 0$ instead of the index $n = 1$, but we can use any other starting index value as well. We usually give preference to indexings that lead to simple expressions.

**EXAMPLE 10**    We can write the geometric series

$$\sum_{n=1}^{\infty} \frac{1}{2^{n-1}} = 1 + \frac{1}{2} + \frac{1}{4} + \cdots$$

as

$$\sum_{n=0}^{\infty} \frac{1}{2^n}, \qquad \sum_{n=5}^{\infty} \frac{1}{2^{n-5}}, \qquad \text{or even} \qquad \sum_{n=-4}^{\infty} \frac{1}{2^{n+4}}.$$

The partial sums remain the same no matter what indexing we choose to use.    ∎

# Exercises 10.2

### Finding $n$th Partial Sums

In Exercises 1–6, find a formula for the $n$th partial sum of each series and use it to find the series' sum if the series converges.

**1.** $2 + \frac{2}{3} + \frac{2}{9} + \frac{2}{27} + \cdots + \frac{2}{3^{n-1}} + \cdots$

**2.** $\frac{9}{100} + \frac{9}{100^2} + \frac{9}{100^3} + \cdots + \frac{9}{100^n} + \cdots$

**3.** $1 - \frac{1}{2} + \frac{1}{4} - \frac{1}{8} + \cdots + (-1)^{n-1}\frac{1}{2^{n-1}} + \cdots$

**4.** $1 - 2 + 4 - 8 + \cdots + (-1)^{n-1} 2^{n-1} + \cdots$

**5.** $\frac{1}{2 \cdot 3} + \frac{1}{3 \cdot 4} + \frac{1}{4 \cdot 5} + \cdots + \frac{1}{(n + 1)(n + 2)} + \cdots$

**6.** $\frac{5}{1 \cdot 2} + \frac{5}{2 \cdot 3} + \frac{5}{3 \cdot 4} + \cdots + \frac{5}{n(n + 1)} + \cdots$

### Series with Geometric Terms

In Exercises 7–14, write out the first eight terms of each series to show how the series starts. Then find the sum of the series or show that it diverges.

**7.** $\sum_{n=0}^{\infty} \frac{(-1)^n}{4^n}$

**8.** $\sum_{n=2}^{\infty} \frac{1}{4^n}$

**9.** $\sum_{n=1}^{\infty} \left(1 - \frac{7}{4^n}\right)$

**10.** $\sum_{n=0}^{\infty} (-1)^n \frac{5}{4^n}$

**11.** $\sum_{n=0}^{\infty} \left(\frac{5}{2^n} + \frac{1}{3^n}\right)$

**12.** $\sum_{n=0}^{\infty} \left(\frac{5}{2^n} - \frac{1}{3^n}\right)$

**13.** $\sum_{n=0}^{\infty} \left(\frac{1}{2^n} + \frac{(-1)^n}{5^n}\right)$

**14.** $\sum_{n=0}^{\infty} \left(\frac{2^{n+1}}{5^n}\right)$

In Exercises 15–18, determine if the geometric series converges or diverges. If a series converges, find its sum.

**15.** $1 + \left(\frac{2}{5}\right) + \left(\frac{2}{5}\right)^2 + \left(\frac{2}{5}\right)^3 + \left(\frac{2}{5}\right)^4 + \cdots$

**16.** $1 + (-3) + (-3)^2 + (-3)^3 + (-3)^4 - \cdots$

**17.** $\left(\frac{1}{8}\right) + \left(\frac{1}{8}\right)^2 + \left(\frac{1}{8}\right)^3 + \left(\frac{1}{8}\right)^4 + \left(\frac{1}{8}\right)^5 + \cdots$

**18.** $\left(\frac{-2}{3}\right)^2 + \left(\frac{-2}{3}\right)^3 + \left(\frac{-2}{3}\right)^4 + \left(\frac{-2}{3}\right)^5 + \left(\frac{-2}{3}\right)^6 + \cdots$

### Repeating Decimals

Express each of the numbers in Exercises 19–26 as the ratio of two integers.

**19.** $0.\overline{23} = 0.23\ 23\ 23 \ldots$

**20.** $0.\overline{234} = 0.234\ 234\ 234 \ldots$

**21.** $0.\overline{7} = 0.7777 \ldots$

**22.** $0.\overline{d} = 0.dddd \ldots, \quad$ where $d$ is a digit

**23.** $0.0\overline{6} = 0.06666 \ldots$

**24.** $1.\overline{414} = 1.414\ 414\ 414 \ldots$

**25.** $1.24\overline{123} = 1.24\ 123\ 123\ 123 \ldots$

**26.** $3.\overline{142857} = 3.142857\ 142857 \ldots$

### Using the $n$th-Term Test

In Exercises 27–34, use the $n$th-Term Test for divergence to show that the series is divergent, or state that the test is inconclusive.

**27.** $\sum_{n=1}^{\infty} \frac{n}{n + 10}$

**28.** $\sum_{n=1}^{\infty} \frac{n(n + 1)}{(n + 2)(n + 3)}$

**29.** $\sum_{n=0}^{\infty} \frac{1}{n + 4}$

**30.** $\sum_{n=1}^{\infty} \frac{n}{n^2 + 3}$

**31.** $\sum_{n=1}^{\infty} \cos \frac{1}{n}$

**32.** $\sum_{n=0}^{\infty} \frac{e^n}{e^n + n}$

**33.** $\sum_{n=1}^{\infty} \ln \frac{1}{n}$

**34.** $\sum_{n=0}^{\infty} \cos n\pi$

### Telescoping Series

In Exercises 35–40, find a formula for the $n$th partial sum of the series and use it to determine if the series converges or diverges. If a series converges, find its sum.

**35.** $\sum_{n=1}^{\infty} \left(\frac{1}{n} - \frac{1}{n + 1}\right)$

**36.** $\sum_{n=1}^{\infty} \left(\frac{3}{n^2} - \frac{3}{(n + 1)^2}\right)$

**37.** $\sum_{n=1}^{\infty} \left(\ln \sqrt{n + 1} - \ln \sqrt{n}\right)$

**38.** $\displaystyle\sum_{n=1}^{\infty} (\tan(n) - \tan(n-1))$

**39.** $\displaystyle\sum_{n=1}^{\infty} \left( \cos^{-1}\left(\frac{1}{n+1}\right) - \cos^{-1}\left(\frac{1}{n+2}\right) \right)$

**40.** $\displaystyle\sum_{n=1}^{\infty} \left( \sqrt{n+4} - \sqrt{n+3} \right)$

Find the sum of each series in Exercises 41–48.

**41.** $\displaystyle\sum_{n=1}^{\infty} \frac{4}{(4n-3)(4n+1)}$

**42.** $\displaystyle\sum_{n=1}^{\infty} \frac{6}{(2n-1)(2n+1)}$

**43.** $\displaystyle\sum_{n=1}^{\infty} \frac{40n}{(2n-1)^2(2n+1)^2}$

**44.** $\displaystyle\sum_{n=1}^{\infty} \frac{2n+1}{n^2(n+1)^2}$

**45.** $\displaystyle\sum_{n=1}^{\infty} \left( \frac{1}{\sqrt{n}} - \frac{1}{\sqrt{n+1}} \right)$

**46.** $\displaystyle\sum_{n=1}^{\infty} \left( \frac{1}{2^{1/n}} - \frac{1}{2^{1/(n+1)}} \right)$

**47.** $\displaystyle\sum_{n=1}^{\infty} \left( \frac{1}{\ln(n+2)} - \frac{1}{\ln(n+1)} \right)$

**48.** $\displaystyle\sum_{n=1}^{\infty} (\tan^{-1}(n) - \tan^{-1}(n+1))$

## Convergence or Divergence

Which series in Exercises 49–68 converge, and which diverge? Give reasons for your answers. If a series converges, find its sum.

**49.** $\displaystyle\sum_{n=0}^{\infty} \left( \frac{1}{\sqrt{2}} \right)^n$

**50.** $\displaystyle\sum_{n=0}^{\infty} (\sqrt{2})^n$

**51.** $\displaystyle\sum_{n=1}^{\infty} (-1)^{n+1} \frac{3}{2^n}$

**52.** $\displaystyle\sum_{n=1}^{\infty} (-1)^{n+1} n$

**53.** $\displaystyle\sum_{n=0}^{\infty} \cos\left(\frac{n\pi}{2}\right)$

**54.** $\displaystyle\sum_{n=0}^{\infty} \frac{\cos n\pi}{5^n}$

**55.** $\displaystyle\sum_{n=0}^{\infty} e^{-2n}$

**56.** $\displaystyle\sum_{n=1}^{\infty} \ln\frac{1}{3^n}$

**57.** $\displaystyle\sum_{n=1}^{\infty} \frac{2}{10^n}$

**58.** $\displaystyle\sum_{n=0}^{\infty} \frac{1}{x^n}, \quad |x| > 1$

**59.** $\displaystyle\sum_{n=0}^{\infty} \frac{2^n - 1}{3^n}$

**60.** $\displaystyle\sum_{n=1}^{\infty} \left( 1 - \frac{1}{n} \right)^n$

**61.** $\displaystyle\sum_{n=0}^{\infty} \frac{n!}{1000^n}$

**62.** $\displaystyle\sum_{n=1}^{\infty} \frac{n^n}{n!}$

**63.** $\displaystyle\sum_{n=1}^{\infty} \frac{2^n + 3^n}{4^n}$

**64.** $\displaystyle\sum_{n=1}^{\infty} \frac{2^n + 4^n}{3^n + 4^n}$

**65.** $\displaystyle\sum_{n=1}^{\infty} \ln\left(\frac{n}{n+1}\right)$

**66.** $\displaystyle\sum_{n=1}^{\infty} \ln\left(\frac{n}{2n+1}\right)$

**67.** $\displaystyle\sum_{n=0}^{\infty} \left(\frac{e}{\pi}\right)^n$

**68.** $\displaystyle\sum_{n=0}^{\infty} \frac{e^{n\pi}}{\pi^{ne}}$

## Geometric Series with a Variable x

In each of the geometric series in Exercises 69–72, write out the first few terms of the series to find $a$ and $r$, and find the sum of the series. Then express the inequality $|r| < 1$ in terms of $x$ and find the values of $x$ for which the inequality holds and the series converges.

**69.** $\displaystyle\sum_{n=0}^{\infty} (-1)^n x^n$

**70.** $\displaystyle\sum_{n=0}^{\infty} (-1)^n x^{2n}$

**71.** $\displaystyle\sum_{n=0}^{\infty} 3\left(\frac{x-1}{2}\right)^n$

**72.** $\displaystyle\sum_{n=0}^{\infty} \frac{(-1)^n}{2}\left(\frac{1}{3+\sin x}\right)^n$

In Exercises 73–78, find the values of $x$ for which the given geometric series converges. Also, find the sum of the series (as a function of $x$) for those values of $x$.

**73.** $\displaystyle\sum_{n=0}^{\infty} 2^n x^n$

**74.** $\displaystyle\sum_{n=0}^{\infty} (-1)^n x^{-2n}$

**75.** $\displaystyle\sum_{n=0}^{\infty} (-1)^n (x+1)^n$

**76.** $\displaystyle\sum_{n=0}^{\infty} \left(-\frac{1}{2}\right)^n (x-3)^n$

**77.** $\displaystyle\sum_{n=0}^{\infty} \sin^n x$

**78.** $\displaystyle\sum_{n=0}^{\infty} (\ln x)^n$

## Theory and Examples

**79.** The series in Exercise 5 can also be written as

$$\sum_{n=1}^{\infty} \frac{1}{(n+1)(n+2)} \quad \text{and} \quad \sum_{n=-1}^{\infty} \frac{1}{(n+3)(n+4)}.$$

Write it as a sum beginning with **(a)** $n = -2$, **(b)** $n = 0$, **(c)** $n = 5$.

**80.** The series in Exercise 6 can also be written as

$$\sum_{n=1}^{\infty} \frac{5}{n(n+1)} \quad \text{and} \quad \sum_{n=0}^{\infty} \frac{5}{(n+1)(n+2)}.$$

Write it as a sum beginning with **(a)** $n = -1$, **(b)** $n = 3$, **(c)** $n = 20$.

**81.** Make up an infinite series of nonzero terms whose sum is

   **a.** 1    **b.** −3    **c.** 0.

**82.** (*Continuation of Exercise 81.*) Can you make an infinite series of nonzero terms that converges to any number you want? Explain.

**83.** Show by example that $\sum(a_n/b_n)$ may diverge even though $\sum a_n$ and $\sum b_n$ converge and no $b_n$ equals 0.

**84.** Find convergent geometric series $A = \sum a_n$ and $B = \sum b_n$ that illustrate the fact that $\sum a_n b_n$ may converge without being equal to $AB$.

**85.** Show by example that $\sum(a_n/b_n)$ may converge to something other than $A/B$ even when $A = \sum a_n$, $B = \sum b_n \neq 0$, and no $b_n$ equals 0.

**86.** If $\sum a_n$ converges and $a_n > 0$ for all $n$, can anything be said about $\sum(1/a_n)$? Give reasons for your answer.

**87.** What happens if you add a finite number of terms to a divergent series or delete a finite number of terms from a divergent series? Give reasons for your answer.

**88.** If $\sum a_n$ converges and $\sum b_n$ diverges, can anything be said about their term-by-term sum $\sum(a_n + b_n)$? Give reasons for your answer.

**89.** Make up a geometric series $\sum ar^{n-1}$ that converges to the number 5 if

   **a.** $a = 2$    **b.** $a = 13/2$.

**90.** Find the value of $b$ for which

$$1 + e^b + e^{2b} + e^{3b} + \cdots = 9.$$

**91.** For what values of $r$ does the infinite series

$$1 + 2r + r^2 + 2r^3 + r^4 + 2r^5 + r^6 + \cdots$$

converge? Find the sum of the series when it converges.

**92.** The accompanying figure shows the first five of a sequence of squares. The outermost square has an area of 4 m$^2$. Each of the other squares is obtained by joining the midpoints of the sides of the squares before it. Find the sum of the areas of all the squares.

**93. Drug dosage** A patient takes a 300 mg tablet for the control of high blood pressure every morning at the same time. The concentration of the drug in the patient's system decays exponentially at a constant hourly rate of $k = 0.12$.

**a.** How many milligrams of the drug are in the patient's system just before the second tablet is taken? Just before the third tablet is taken?

**b.** In the long run, after taking the medication for at least six months, what quantity of drug is in the patient's body just before taking the next regularly scheduled morning tablet?

**94.** Show that the error $(L - s_n)$ obtained by replacing a convergent geometric series with one of its partial sums $s_n$ is $ar^n/(1 - r)$.

**95. The Cantor set** To construct this set, we begin with the closed interval $[0, 1]$. From that interval, remove the middle open interval $(1/3, 2/3)$, leaving the two closed intervals $[0, 1/3]$ and $[2/3, 1]$. At the second step we remove the open middle third interval from each of those remaining. From $[0, 1/3]$ we remove the open interval $(1/9, 2/9)$, and from $[2/3, 1]$ we remove $(7/9, 8/9)$, leaving behind the four closed intervals $[0, 1/9]$,

$[2/9, 1/3]$, $[2/3, 7/9]$, and $[8/9, 1]$. At the next step, we remove the middle open third interval from each closed interval left behind, so $(1/27, 2/27)$ is removed from $[0, 1/9]$, leaving the closed intervals $[0, 1/27]$ and $[2/27, 1/9]$; $(7/27, 8/27)$ is removed from $[2/9, 1/3]$, leaving behind $[2/9, 7/27]$ and $[8/27, 1/3]$, and so forth. We continue this process repeatedly without stopping, at each step removing the open third interval from every closed interval remaining behind from the preceding step. The numbers remaining in the interval $[0, 1]$, after all open middle third intervals have been removed, are the points in the Cantor set (named after Georg Cantor, 1845–1918). The set has some interesting properties.

**a.** The Cantor set contains infinitely many numbers in $[0, 1]$. List 12 numbers that belong to the Cantor set.

**b.** Show, by summing an appropriate geometric series, that the total length of all the open middle third intervals that have been removed from $[0, 1]$ is equal to 1.

**96. Helga von Koch's snowflake curve** Helga von Koch's snowflake is a curve of infinite length that encloses a region of finite area. To see why this is so, suppose the curve is generated by starting with an equilateral triangle whose sides have length 1.

**a.** Find the length $L_n$ of the $n$th curve $C_n$ and show that $\lim_{n \to \infty} L_n = \infty$.

**b.** Find the area $A_n$ of the region enclosed by $C_n$ and show that $\lim_{n \to \infty} A_n = (8/5) A_1$.

$C_1 \qquad C_2 \qquad C_3 \qquad C_4$

# 10.3 The Integral Test

The most basic question we can ask about a series is whether it converges or not. In this section and the next two, we study this question, starting with series that have nonnegative terms. Such a series converges if its sequence of partial sums is bounded. If we establish that a given series does converge, we generally do not have a formula available for its sum. So for a convergent series, we need to investigate the error involved when using a partial sum to approximate its total sum.

## Nondecreasing Partial Sums

Suppose that $\sum_{n=1}^{\infty} a_n$ is an infinite series with $a_n \geq 0$ for all $n$. Then each partial sum is greater than or equal to its predecessor because $s_{n+1} = s_n + a_n$, so

$$s_1 \leq s_2 \leq s_3 \leq \cdots \leq s_n \leq s_{n+1} \leq \cdots .$$

Since the partial sums form a nondecreasing sequence, the Monotonic Sequence Theorem (Theorem 6, Section 10.1) gives the following result.

> **COROLLARY OF THEOREM 6** A series $\sum_{n=1}^{\infty} a_n$ of nonnegative terms converges if and only if its partial sums are bounded from above.

**EXAMPLE 1** As an application of the above corollary, consider the **harmonic series**

$$\sum_{n=1}^{\infty} \frac{1}{n} = 1 + \frac{1}{2} + \frac{1}{3} + \cdots + \frac{1}{n} + \cdots .$$

Although the $n$th term $1/n$ does go to zero, the series diverges because there is no upper bound for its partial sums. To see why, group the terms of the series in the following way:

$$1 + \frac{1}{2} + \underbrace{\left(\frac{1}{3} + \frac{1}{4}\right)}_{>\frac{2}{4} = \frac{1}{2}} + \underbrace{\left(\frac{1}{5} + \frac{1}{6} + \frac{1}{7} + \frac{1}{8}\right)}_{>\frac{4}{8} = \frac{1}{2}} + \underbrace{\left(\frac{1}{9} + \frac{1}{10} + \cdots + \frac{1}{16}\right)}_{>\frac{8}{16} = \frac{1}{2}} + \cdots .$$

The sum of the first two terms is 1.5. The sum of the next two terms is $1/3 + 1/4$, which is greater than $1/4 + 1/4 = 1/2$. The sum of the next four terms is $1/5 + 1/6 + 1/7 + 1/8$, which is greater than $1/8 + 1/8 + 1/8 + 1/8 = 1/2$. The sum of the next eight terms is $1/9 + 1/10 + 1/11 + 1/12 + 1/13 + 1/14 + 1/15 + 1/16$, which is greater than $8/16 = 1/2$. The sum of the next 16 terms is greater than $16/32 = 1/2$, and so on. In general, the sum of $2^n$ terms ending with $1/2^{n+1}$ is greater than $2^n/2^{n+1} = 1/2$. If $n = 2^k$, the partial sum $s_n$ is greater than $k/2$, so the sequence of partial sums is not bounded from above. The harmonic series diverges. ∎

### The Integral Test

We now introduce the Integral Test with a series that is related to the harmonic series, but whose $n$th term is $1/n^2$ instead of $1/n$.

**EXAMPLE 2** Does the following series converge?

$$\sum_{n=1}^{\infty} \frac{1}{n^2} = 1 + \frac{1}{4} + \frac{1}{9} + \frac{1}{16} + \cdots + \frac{1}{n^2} + \cdots$$

**Solution** We determine the convergence of $\sum_{n=1}^{\infty}(1/n^2)$ by comparing it with $\int_1^{\infty}(1/x^2)\,dx$. To carry out the comparison, we think of the terms of the series as values of the function $f(x) = 1/x^2$ and interpret these values as the areas of rectangles under the curve $y = 1/x^2$.

As Figure 10.10 shows,

$$s_n = \frac{1}{1^2} + \frac{1}{2^2} + \frac{1}{3^2} + \cdots + \frac{1}{n^2}$$

$$= f(1) + f(2) + f(3) + \cdots + f(n)$$

$$< f(1) + \int_1^n \frac{1}{x^2}\,dx \qquad \text{Rectangle areas sum to less than area under graph.}$$

$$< 1 + \int_1^{\infty} \frac{1}{x^2}\,dx \qquad \int_1^n (1/x^2)\,dx < \int_1^{\infty}(1/x^2)\,dx$$

$$< 1 + 1 = 2. \qquad \text{As in Section 8.8, Example 3,} \ \int_1^{\infty}(1/x^2)\,dx = 1.$$

Thus the partial sums of $\sum_{n=1}^{\infty}(1/n^2)$ are bounded from above (by 2) and the series converges. ∎

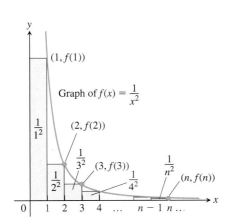

**FIGURE 10.10** The sum of the areas of the rectangles under the graph of $f(x) = 1/x^2$ is less than the area under the graph (Example 2).

**Caution**
The series and integral need not have the same value in the convergent case. You will see in Example 6 that

$$\sum_{n=1}^{\infty}(1/n^2) \neq \int_1^{\infty}(1/x^2)\,dx = 1.$$

**THEOREM 9—The Integral Test** Let $\{a_n\}$ be a sequence of positive terms. Suppose that $a_n = f(n)$, where $f$ is a continuous, positive, decreasing function of $x$ for all $x \geq N$ ($N$ a positive integer). Then the series $\sum_{n=N}^{\infty} a_n$ and the integral $\int_N^{\infty} f(x)\,dx$ both converge or both diverge.

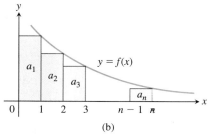

**FIGURE 10.11**   Subject to the conditions of the Integral Test, the series $\sum_{n=1}^{\infty} a_n$ and the integral $\int_1^{\infty} f(x)\, dx$ both converge or both diverge.

**Proof**   We establish the test for the case $N = 1$. The proof for general $N$ is similar.

We start with the assumption that $f$ is a decreasing function with $f(n) = a_n$ for every $n$. This leads us to observe that the rectangles in Figure 10.11a, which have areas $a_1, a_2, \ldots, a_n$, collectively enclose more area than that under the curve $y = f(x)$ from $x = 1$ to $x = n + 1$. That is,

$$\int_1^{n+1} f(x)\, dx \le a_1 + a_2 + \cdots + a_n.$$

In Figure 10.11b the rectangles have been faced to the left instead of to the right. If we momentarily disregard the first rectangle of area $a_1$, we see that

$$a_2 + a_3 + \cdots + a_n \le \int_1^n f(x)\, dx.$$

If we include $a_1$, we have

$$a_1 + a_2 + \cdots + a_n \le a_1 + \int_1^n f(x)\, dx.$$

Combining these results gives

$$\int_1^{n+1} f(x)\, dx \le a_1 + a_2 + \cdots + a_n \le a_1 + \int_1^n f(x)\, dx.$$

These inequalities hold for each $n$, and continue to hold as $n \to \infty$.

If $\int_1^{\infty} f(x)\, dx$ is finite, the right-hand inequality shows that $\sum a_n$ is finite. If $\int_1^{\infty} f(x)\, dx$ is infinite, the left-hand inequality shows that $\sum a_n$ is infinite. Hence the series and the integral are both finite or both infinite. ∎

**EXAMPLE 3**   Show that the **$p$-series**

$$\sum_{n=1}^{\infty} \frac{1}{n^p} = \frac{1}{1^p} + \frac{1}{2^p} + \frac{1}{3^p} + \cdots + \frac{1}{n^p} + \cdots$$

($p$ a real constant) converges if $p > 1$, and diverges if $p \le 1$.

**Solution**   If $p > 1$, then $f(x) = 1/x^p$ is a positive decreasing function of $x$. Since

$$\int_1^{\infty} \frac{1}{x^p}\, dx = \int_1^{\infty} x^{-p}\, dx = \lim_{b \to \infty} \left[ \frac{x^{-p+1}}{-p+1} \right]_1^b$$

$$= \frac{1}{1-p} \lim_{b \to \infty} \left( \frac{1}{b^{p-1}} - 1 \right)$$

$$= \frac{1}{1-p}(0 - 1) = \frac{1}{p-1}, \qquad \begin{array}{l} b^{p-1} \to \infty \text{ as } b \to \infty \\ \text{because } p - 1 > 0. \end{array}$$

the series converges by the Integral Test. We emphasize that the sum of the $p$-series is *not* $1/(p - 1)$. The series converges, but we don't know the value it converges to.

If $p \le 0$, the series diverges by the $n$th-term test. If $0 < p < 1$, then $1 - p > 0$ and

$$\int_1^{\infty} \frac{1}{x^p}\, dx = \frac{1}{1-p} \lim_{b \to \infty} (b^{1-p} - 1) = \infty.$$

The series diverges by the Integral Test.

**The $p$-series** $\displaystyle\sum_{n=1}^{\infty} \frac{1}{n^p}$

converges if $p > 1$, diverges if $p \leq 1$.

If $p = 1$, we have the (divergent) harmonic series

$$1 + \frac{1}{2} + \frac{1}{3} + \cdots + \frac{1}{n} + \cdots .$$

We have convergence for $p > 1$ but divergence for all other values of $p$. ◼

The $p$-series with $p = 1$ is the **harmonic series** (Example 1). The $p$-Series Test shows that the harmonic series is just *barely* divergent; if we increase $p$ to 1.000000001, for instance, the series converges!

The slowness with which the partial sums of the harmonic series approach infinity is impressive. For instance, it takes more than 178 million terms of the harmonic series to move the partial sums beyond 20. (See also Exercise 43b.)

**EXAMPLE 4**    The series $\sum_{n=1}^{\infty}(1/(n^2 + 1))$ is not a $p$-series, but it converges by the Integral Test. The function $f(x) = 1/(x^2 + 1)$ is positive, continuous, and decreasing for $x \geq 1$, and

$$\int_1^{\infty} \frac{1}{x^2 + 1}\,dx = \lim_{b \to \infty} \Big[\arctan x\Big]_1^b$$

$$= \lim_{b \to \infty} \Big[\arctan b - \arctan 1\Big]$$

$$= \frac{\pi}{2} - \frac{\pi}{4} = \frac{\pi}{4}.$$

The Integral Test tells us the series converges, but $\pi/4$ is *not* the sum of the series. ◼

**EXAMPLE 5**    Determine the convergence or divergence of the series.

**(a)** $\displaystyle\sum_{n=1}^{\infty} n e^{-n^2}$      **(b)** $\displaystyle\sum_{n=1}^{\infty} \frac{1}{2^{\ln n}}$

**Solutions**

**(a)** We apply the Integral Test and find that

$$\int_1^{\infty} \frac{x}{e^{x^2}}\,dx = \frac{1}{2}\int_1^{\infty} \frac{du}{e^u} \qquad u = x^2, du = 2x\,dx$$

$$= \lim_{b \to \infty}\left[-\frac{1}{2}e^{-u}\right]_1^b$$

$$= \lim_{b \to \infty}\left(-\frac{1}{2e^b} + \frac{1}{2e}\right) = \frac{1}{2e}.$$

Since the integral converges, the series also converges.

**(b)** Again applying the Integral Test,

$$\int_1^{\infty} \frac{dx}{2^{\ln x}} = \int_0^{\infty} \frac{e^u\,du}{2^u} \qquad u = \ln x, x = e^u, dx = e^u\,du$$

$$= \int_0^{\infty} \left(\frac{e}{2}\right)^u du$$

$$= \lim_{b \to \infty} \frac{1}{\ln\left(\frac{e}{2}\right)}\left(\left(\frac{e}{2}\right)^b - 1\right) = \infty. \qquad (e/2) > 1$$

The improper integral diverges, so the series diverges also. ◼

### Error Estimation

For some convergent series, such as the geometric series or the telescoping series in Example 5 of Section 10.2, we can actually find the total sum of the series. That is, we can find the limiting value $S$ of the sequence of partial sums. For most convergent series, however, we cannot easily find the total sum. Nevertheless, we can *estimate* the sum by adding the first $n$ terms to get $s_n$, but we need to know how far off $s_n$ is from the total sum $S$.

Suppose that a series $\Sigma a_n$ is shown to be convergent by the Integral Test, and we want to estimate the size of the **remainder** $R_n$ measuring the difference between the total sum $S$ of the series and its $n$th partial sum $s_n$. That is, we wish to estimate

$$R_n = S - s_n = a_{n+1} + a_{n+2} + a_{n+3} + \cdots.$$

To get a lower bound for the remainder, we compare the sum of the areas of the rectangles with the area under the curve $y = f(x)$ for $x \geq n$ (see Figure 10.11a). We see that

$$R_n = a_{n+1} + a_{n+2} + a_{n+3} + \cdots \geq \int_{n+1}^{\infty} f(x)\, dx.$$

Similarly, from Figure 10.11b, we find an upper bound with

$$R_n = a_{n+1} + a_{n+2} + a_{n+3} + \cdots \leq \int_{n}^{\infty} f(x)\, dx.$$

These comparisons prove the following result, giving bounds on the size of the remainder.

---

**Bounds for the Remainder in the Integral Test**

Suppose $\{a_k\}$ is a sequence of positive terms with $a_k = f(k)$, where $f$ is a continuous positive decreasing function of $x$ for all $x \geq n$, and that $\Sigma a_n$ converges to $S$. Then the remainder $R_n = S - s_n$ satisfies the inequalities

$$\int_{n+1}^{\infty} f(x)\, dx \leq R_n \leq \int_{n}^{\infty} f(x)\, dx. \tag{1}$$

---

If we add the partial sum $s_n$ to each side of the inequalities in (1), we get

$$s_n + \int_{n+1}^{\infty} f(x)\, dx \leq S \leq s_n + \int_{n}^{\infty} f(x)\, dx \tag{2}$$

since $s_n + R_n = S$. The inequalities in (2) are useful for estimating the error in approximating the sum of a series known to converge by the Integral Test. The error can be no larger than the length of the interval containing $S$, with endpoints given by (2).

**EXAMPLE 6**    Estimate the sum of the series $\Sigma(1/n^2)$ using the inequalities in (2) and $n = 10$.

**Solution**    We have that

$$\int_{n}^{\infty} \frac{1}{x^2}\, dx = \lim_{b \to \infty} \left[ -\frac{1}{x} \right]_{n}^{b} = \lim_{b \to \infty} \left( -\frac{1}{b} + \frac{1}{n} \right) = \frac{1}{n}.$$

Using this result with the inequalities in (2), we get

$$s_{10} + \frac{1}{11} \leq S \leq s_{10} + \frac{1}{10}.$$

Taking $s_{10} = 1 + (1/4) + (1/9) + (1/16) + \cdots + (1/100) \approx 1.54977$, these last inequalities give

$$1.64068 \leq S \leq 1.64977.$$

If we approximate the sum $S$ by the midpoint of this interval, we find that

$$\sum_{n=1}^{\infty} \frac{1}{n^2} \approx 1.6452.$$

The *p*-series for $p = 2$

$$\sum_{n=1}^{\infty} \frac{1}{n^2} = \frac{\pi^2}{6} \approx 1.64493$$

The error in this approximation is then less than half the length of the interval, so the error is less than 0.005. Using a trigonometric *Fourier series* (studied in advanced calculus), it can be shown that $S$ is equal to $\pi^2/6 \approx 1.64493$. ∎

# Exercises 10.3

## Applying the Integral Test

Use the Integral Test to determine if the series in Exercises 1–10 converge or diverge. Be sure to check that the conditions of the Integral Test are satisfied.

**1.** $\sum_{n=1}^{\infty} \frac{1}{n^2}$  **2.** $\sum_{n=1}^{\infty} \frac{1}{n^{0.2}}$  **3.** $\sum_{n=1}^{\infty} \frac{1}{n^2 + 4}$

**4.** $\sum_{n=1}^{\infty} \frac{1}{n + 4}$  **5.** $\sum_{n=1}^{\infty} e^{-2n}$  **6.** $\sum_{n=2}^{\infty} \frac{1}{n(\ln n)^2}$

**7.** $\sum_{n=1}^{\infty} \frac{n}{n^2 + 4}$  **8.** $\sum_{n=2}^{\infty} \frac{\ln (n^2)}{n}$

**9.** $\sum_{n=1}^{\infty} \frac{n^2}{e^{n/3}}$  **10.** $\sum_{n=2}^{\infty} \frac{n - 4}{n^2 - 2n + 1}$

## Determining Convergence or Divergence

Which of the series in Exercises 11–40 converge, and which diverge? Give reasons for your answers. (When you check an answer, remember that there may be more than one way to determine the series' convergence or divergence.)

**11.** $\sum_{n=1}^{\infty} \frac{1}{10^n}$  **12.** $\sum_{n=1}^{\infty} e^{-n}$  **13.** $\sum_{n=1}^{\infty} \frac{n}{n + 1}$

**14.** $\sum_{n=1}^{\infty} \frac{5}{n + 1}$  **15.** $\sum_{n=1}^{\infty} \frac{3}{\sqrt{n}}$  **16.** $\sum_{n=1}^{\infty} \frac{-2}{n\sqrt{n}}$

**17.** $\sum_{n=1}^{\infty} -\frac{1}{8^n}$  **18.** $\sum_{n=1}^{\infty} \frac{-8}{n}$  **19.** $\sum_{n=2}^{\infty} \frac{\ln n}{n}$

**20.** $\sum_{n=2}^{\infty} \frac{\ln n}{\sqrt{n}}$  **21.** $\sum_{n=1}^{\infty} \frac{2^n}{3^n}$  **22.** $\sum_{n=1}^{\infty} \frac{5^n}{4^n + 3}$

**23.** $\sum_{n=0}^{\infty} \frac{-2}{n + 1}$  **24.** $\sum_{n=1}^{\infty} \frac{1}{2n - 1}$  **25.** $\sum_{n=1}^{\infty} \frac{2^n}{n + 1}$

**26.** $\sum_{n=1}^{\infty} \left(1 + \frac{1}{n}\right)^n$  **27.** $\sum_{n=2}^{\infty} \frac{\sqrt{n}}{\ln n}$  **28.** $\sum_{n=1}^{\infty} \frac{1}{\sqrt{n}(\sqrt{n} + 1)}$

**29.** $\sum_{n=1}^{\infty} \frac{1}{(\ln 2)^n}$  **30.** $\sum_{n=1}^{\infty} \frac{1}{(\ln 3)^n}$

**31.** $\sum_{n=3}^{\infty} \frac{(1/n)}{(\ln n)\sqrt{\ln^2 n - 1}}$  **32.** $\sum_{n=1}^{\infty} \frac{1}{n(1 + \ln^2 n)}$

**33.** $\sum_{n=1}^{\infty} n \sin \frac{1}{n}$  **34.** $\sum_{n=1}^{\infty} n \tan \frac{1}{n}$

**35.** $\sum_{n=1}^{\infty} \frac{e^n}{1 + e^{2n}}$  **36.** $\sum_{n=1}^{\infty} \frac{2}{1 + e^n}$

**37.** $\sum_{n=1}^{\infty} \frac{8 \tan^{-1} n}{1 + n^2}$  **38.** $\sum_{n=1}^{\infty} \frac{n}{n^2 + 1}$

**39.** $\sum_{n=1}^{\infty} \text{sech} \, n$  **40.** $\sum_{n=1}^{\infty} \text{sech}^2 n$

## Theory and Examples

For what values of $a$, if any, do the series in Exercises 41 and 42 converge?

**41.** $\sum_{n=1}^{\infty} \left(\frac{a}{n + 2} - \frac{1}{n + 4}\right)$  **42.** $\sum_{n=3}^{\infty} \left(\frac{1}{n - 1} - \frac{2a}{n + 1}\right)$

**43. a.** Draw illustrations like those in Figures 10.11a and 10.11b to show that the partial sums of the harmonic series satisfy the inequalities

$$\ln (n + 1) = \int_1^{n+1} \frac{1}{x} dx \leq 1 + \frac{1}{2} + \cdots + \frac{1}{n}$$

$$\leq 1 + \int_1^n \frac{1}{x} dx = 1 + \ln n.$$

T **b.** There is absolutely no empirical evidence for the divergence of the harmonic series even though we know it diverges. The partial sums just grow too slowly. To see what we mean, suppose you had started with $s_1 = 1$ the day the universe was formed, 13 billion years ago, and added a new term every *second*. About how large would the partial sum $s_n$ be today, assuming a 365-day year?

**44.** Are there any values of $x$ for which $\sum_{n=1}^{\infty}(1/nx)$ converges? Give reasons for your answer.

**45.** Is it true that if $\sum_{n=1}^{\infty} a_n$ is a divergent series of positive numbers, then there is also a divergent series $\sum_{n=1}^{\infty} b_n$ of positive numbers with $b_n < a_n$ for every $n$? Is there a "smallest" divergent series of positive numbers? Give reasons for your answers.

**46.** (*Continuation of Exercise 45.*) Is there a "largest" convergent series of positive numbers? Explain.

**47.** $\sum_{n=1}^{\infty} \left(1/\sqrt{n + 1}\right)$ diverges

**a.** Use the accompanying graph to show that the partial sum

$s_{50} = \sum_{n=1}^{50} \left(1/\sqrt{n + 1}\right)$ satisfies

$$\int_1^{51} \frac{1}{\sqrt{x + 1}} dx < s_{50} < \int_0^{50} \frac{1}{\sqrt{x + 1}} dx.$$

Conclude that $11.5 < s_{50} < 12.3$.

**b.** What should $n$ be in order that the partial sum
$$s_n = \sum_{i=1}^{n} \left(1/\sqrt{i+1}\right) \text{ satisfy } s_n > 1000?$$

**48.** $\sum_{n=1}^{\infty} (1/n^4)$ **converges**

**a.** Use the accompanying graph to find an upper bound for the error if $s_{30} = \sum_{n=1}^{30} (1/n^4)$ is used to estimate the value of $\sum_{n=1}^{\infty} (1/n^4)$.

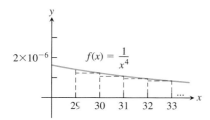

**b.** Find $n$ so that the partial sum $s_n = \sum_{i=1}^{n} (1/i^4)$ estimates the value of $\sum_{n=1}^{\infty} (1/n^4)$ with an error of at most 0.000001.

**49.** Estimate the value of $\sum_{n=1}^{\infty} (1/n^3)$ to within 0.01 of its exact value.

**50.** Estimate the value of $\sum_{n=2}^{\infty} (1/(n^2 + 4))$ to within 0.1 of its exact value.

**51.** How many terms of the convergent series $\sum_{n=1}^{\infty} (1/n^{1.1})$ should be used to estimate its value with error at most 0.00001?

**52.** How many terms of the convergent series $\sum_{n=4}^{\infty} (1/n(\ln n)^3)$ should be used to estimate its value with error at most 0.01?

**53. The Cauchy condensation test** The Cauchy condensation test says: Let $\{a_n\}$ be a nonincreasing sequence ($a_n \geq a_{n+1}$ for all $n$) of positive terms that converges to 0. Then $\sum a_n$ converges if and only if $\sum 2^n a_{2^n}$ converges. For example, $\sum(1/n)$ diverges because $\sum 2^n \cdot (1/2^n) = \sum 1$ diverges. Show why the test works.

**54.** Use the Cauchy condensation test from Exercise 53 to show that

**a.** $\sum_{n=2}^{\infty} \dfrac{1}{n \ln n}$ diverges;

**b.** $\sum_{n=1}^{\infty} \dfrac{1}{n^p}$ converges if $p > 1$ and diverges if $p \leq 1$.

**55. Logarithmic $p$-series**

**a.** Show that the improper integral
$$\int_2^{\infty} \frac{dx}{x(\ln x)^p} \quad (p \text{ a positive constant})$$
converges if and only if $p > 1$.

**b.** What implications does the fact in part (a) have for the convergence of the series
$$\sum_{n=2}^{\infty} \frac{1}{n(\ln n)^p}?$$
Give reasons for your answer.

**56.** (*Continuation of Exercise 55.*) Use the result in Exercise 55 to determine which of the following series converge and which diverge. Support your answer in each case.

**a.** $\sum_{n=2}^{\infty} \dfrac{1}{n(\ln n)}$

**b.** $\sum_{n=2}^{\infty} \dfrac{1}{n(\ln n)^{1.01}}$

**c.** $\sum_{n=2}^{\infty} \dfrac{1}{n \ln (n^3)}$

**d.** $\sum_{n=2}^{\infty} \dfrac{1}{n(\ln n)^3}$

**57. Euler's constant** Graphs like those in Figure 10.11 suggest that as $n$ increases there is little change in the difference between the sum
$$1 + \frac{1}{2} + \cdots + \frac{1}{n}$$
and the integral
$$\ln n = \int_1^n \frac{1}{x} dx.$$
To explore this idea, carry out the following steps.

**a.** By taking $f(x) = 1/x$ in the proof of Theorem 9, show that
$$\ln(n+1) \leq 1 + \frac{1}{2} + \cdots + \frac{1}{n} \leq 1 + \ln n$$
or
$$0 < \ln(n+1) - \ln n \leq 1 + \frac{1}{2} + \cdots + \frac{1}{n} - \ln n \leq 1.$$
Thus, the sequence
$$a_n = 1 + \frac{1}{2} + \cdots + \frac{1}{n} - \ln n$$
is bounded from below and from above.

**b.** Show that
$$\frac{1}{n+1} < \int_n^{n+1} \frac{1}{x} dx = \ln(n+1) - \ln n,$$
and use this result to show that the sequence $\{a_n\}$ in part (a) is decreasing.

Since a decreasing sequence that is bounded from below converges, the numbers $a_n$ defined in part (a) converge:
$$1 + \frac{1}{2} + \cdots + \frac{1}{n} - \ln n \to \gamma.$$
The number $\gamma$, whose value is $0.5772\ldots$, is called *Euler's constant*.

**58.** Use the Integral Test to show that the series
$$\sum_{n=0}^{\infty} e^{-n^2}$$
converges.

**59. a.** For the series $\sum (1/n^3)$, use the inequalities in Equation (2) with $n = 10$ to find an interval containing the sum $S$.

**b.** As in Example 5, use the midpoint of the interval found in part (a) to approximate the sum of the series. What is the maximum error for your approximation?

**60.** Repeat Exercise 59 using the series $\sum (1/n^4)$.

**61. Area** Consider the sequence $\{1/n\}_{n=1}^{\infty}$. On each subinterval $(1/(n+1), 1/n)$ within the interval $[0, 1]$, erect the rectangle with area $a_n$ having height $1/n$ and width equal to the length of the subinterval. Find the total area $\sum a_n$ of all the rectangles. (*Hint:* Use the result of Example 5 in Section 10.2.)

**62. Area** Repeat Exercise 61, using trapezoids instead of rectangles. That is, on the subinterval $(1/(n+1), 1/n)$, let $a_n$ denote the area of the trapezoid having heights $y = 1/(n+1)$ at $x = 1/(n+1)$ and $y = 1/n$ at $x = 1/n$.

# $10.4$ Comparison Tests

We have seen how to determine the convergence of geometric series, $p$-series, and a few others. We can test the convergence of many more series by comparing their terms to those of a series whose convergence is known.

> **THEOREM 10—The Comparison Test** Let $\Sigma a_n$, $\Sigma c_n$, and $\Sigma d_n$ be series with nonnegative terms. Suppose that for some integer $N$
>
> $$d_n \leq a_n \leq c_n \qquad \text{for all} \qquad n > N.$$
>
> (a) If $\Sigma c_n$ converges, then $\Sigma a_n$ also converges.
> (b) If $\Sigma d_n$ diverges, then $\Sigma a_n$ also diverges.

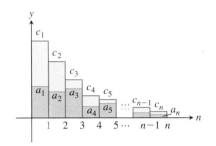

**FIGURE 10.12** If the total area $\Sigma c_n$ of the taller $c_n$ rectangles is finite, then so is the total area $\Sigma a_n$ of the shorter $a_n$ rectangles.

**Proof**    In Part (a), the partial sums of $\Sigma a_n$ are bounded above by

$$M = a_1 + a_2 + \cdots + a_N + \sum_{n=N+1}^{\infty} c_n.$$

They therefore form a nondecreasing sequence with a limit $L \leq M$. That is, if $\Sigma c_n$ converges, then so does $\Sigma a_n$. Figure 10.12 depicts this result, where each term of each series is interpreted as the area of a rectangle (just as we did for the Integral Test in Figure 10.11).

In Part (b), the partial sums of $\Sigma a_n$ are not bounded from above. If they were, the partial sums for $\Sigma d_n$ would be bounded by

$$M^* = d_1 + d_2 + \cdots + d_N + \sum_{n=N+1}^{\infty} a_n$$

and $\Sigma d_n$ would have to converge instead of diverge. ∎

**EXAMPLE 1**    We apply Theorem 10 to several series.

(a) The series

$$\sum_{n=1}^{\infty} \frac{5}{5n - 1}$$

diverges because its $n$th term

$$\frac{5}{5n - 1} = \frac{1}{n - \dfrac{1}{5}} > \frac{1}{n}$$

is greater than the $n$th term of the divergent harmonic series.

(b) The series

$$\sum_{n=0}^{\infty} \frac{1}{n!} = 1 + \frac{1}{1!} + \frac{1}{2!} + \frac{1}{3!} + \cdots$$

converges because its terms are all positive and less than or equal to the corresponding terms of

$$1 + \sum_{n=0}^{\infty} \frac{1}{2^n} = 1 + 1 + \frac{1}{2} + \frac{1}{2^2} + \cdots.$$

The geometric series on the left converges and we have

$$1 + \sum_{n=0}^{\infty} \frac{1}{2^n} = 1 + \frac{1}{1 - (1/2)} = 3.$$

The fact that 3 is an upper bound for the partial sums of $\sum_{n=0}^{\infty}(1/n!)$ does not mean that the series converges to 3. As we will see in Section 10.9, the series converges to $e$.

**(c)** The series

$$5 + \frac{2}{3} + \frac{1}{7} + 1 + \frac{1}{2 + \sqrt{1}} + \frac{1}{4 + \sqrt{2}} + \frac{1}{8 + \sqrt{3}} + \cdots + \frac{1}{2^n + \sqrt{n}} + \cdots$$

converges. To see this, we ignore the first three terms and compare the remaining terms with those of the convergent geometric series $\sum_{n=0}^{\infty}(1/2^n)$. The term $1/(2^n + \sqrt{n})$ of the truncated sequence is less than the corresponding term $1/2^n$ of the geometric series. We see that term by term we have the comparison

$$1 + \frac{1}{2 + \sqrt{1}} + \frac{1}{4 + \sqrt{2}} + \frac{1}{8 + \sqrt{3}} + \cdots \leq 1 + \frac{1}{2} + \frac{1}{4} + \frac{1}{8} + \cdots .$$

So the truncated series and the original series converge by an application of the Comparison Test. ∎

## The Limit Comparison Test

We now introduce a comparison test that is particularly useful for series in which $a_n$ is a rational function of $n$.

> **THEOREM 11—Limit Comparison Test**  Suppose that $a_n > 0$ and $b_n > 0$ for all $n \geq N$ ($N$ an integer).
>
> **1.** If $\lim\limits_{n \to \infty} \dfrac{a_n}{b_n} = c > 0$, then $\sum a_n$ and $\sum b_n$ both converge or both diverge.
>
> **2.** If $\lim\limits_{n \to \infty} \dfrac{a_n}{b_n} = 0$ and $\sum b_n$ converges, then $\sum a_n$ converges.
>
> **3.** If $\lim\limits_{n \to \infty} \dfrac{a_n}{b_n} = \infty$ and $\sum b_n$ diverges, then $\sum a_n$ diverges.

**Proof**  We will prove Part 1. Parts 2 and 3 are left as Exercises 55a and b.

Since $c/2 > 0$, there exists an integer $N$ such that for all $n$

$$n > N \implies \left| \frac{a_n}{b_n} - c \right| < \frac{c}{2}.$$

Limit definition with $\epsilon = c/2, L = c$, and $a_n$ replaced by $a_n/b_n$

Thus, for $n > N$,

$$-\frac{c}{2} < \frac{a_n}{b_n} - c < \frac{c}{2},$$

$$\frac{c}{2} < \frac{a_n}{b_n} < \frac{3c}{2},$$

$$\left(\frac{c}{2}\right)b_n < a_n < \left(\frac{3c}{2}\right)b_n.$$

If $\sum b_n$ converges, then $\sum (3c/2)b_n$ converges and $\sum a_n$ converges by the Direct Comparison Test. If $\sum b_n$ diverges, then $\sum (c/2)b_n$ diverges and $\sum a_n$ diverges by the Direct Comparison Test. ∎

**EXAMPLE 2**    Which of the following series converge, and which diverge?

**(a)** $\dfrac{3}{4} + \dfrac{5}{9} + \dfrac{7}{16} + \dfrac{9}{25} + \cdots = \sum\limits_{n=1}^{\infty} \dfrac{2n + 1}{(n + 1)^2} = \sum\limits_{n=1}^{\infty} \dfrac{2n + 1}{n^2 + 2n + 1}$

**(b)** $\dfrac{1}{1} + \dfrac{1}{3} + \dfrac{1}{7} + \dfrac{1}{15} + \cdots = \displaystyle\sum_{n=1}^{\infty} \dfrac{1}{2^n - 1}$

**(c)** $\dfrac{1 + 2 \ln 2}{9} + \dfrac{1 + 3 \ln 3}{14} + \dfrac{1 + 4 \ln 4}{21} + \cdots = \displaystyle\sum_{n=2}^{\infty} \dfrac{1 + n \ln n}{n^2 + 5}$

**Solution** We apply the Limit Comparison Test to each series.

**(a)** Let $a_n = (2n + 1)/(n^2 + 2n + 1)$. For large $n$, we expect $a_n$ to behave like $2n/n^2 = 2/n$ since the leading terms dominate for large $n$, so we let $b_n = 1/n$. Since

$$\sum_{n=1}^{\infty} b_n = \sum_{n=1}^{\infty} \frac{1}{n} \text{ diverges}$$

and

$$\lim_{n \to \infty} \frac{a_n}{b_n} = \lim_{n \to \infty} \frac{2n^2 + n}{n^2 + 2n + 1} = 2,$$

$\sum a_n$ diverges by Part 1 of the Limit Comparison Test. We could just as well have taken $b_n = 2/n$, but $1/n$ is simpler.

**(b)** Let $a_n = 1/(2^n - 1)$. For large $n$, we expect $a_n$ to behave like $1/2^n$, so we let $b_n = 1/2^n$. Since

$$\sum_{n=1}^{\infty} b_n = \sum_{n=1}^{\infty} \frac{1}{2^n} \text{ converges}$$

and

$$\lim_{n \to \infty} \frac{a_n}{b_n} = \lim_{n \to \infty} \frac{2^n}{2^n - 1}$$

$$= \lim_{n \to \infty} \frac{1}{1 - (1/2^n)}$$

$$= 1,$$

$\sum a_n$ converges by Part 1 of the Limit Comparison Test.

**(c)** Let $a_n = (1 + n \ln n)/(n^2 + 5)$. For large $n$, we expect $a_n$ to behave like $(n \ln n)/n^2 = (\ln n)/n$, which is greater than $1/n$ for $n \geq 3$, so we let $b_n = 1/n$. Since

$$\sum_{n=2}^{\infty} b_n = \sum_{n=2}^{\infty} \frac{1}{n} \text{ diverges}$$

and

$$\lim_{n \to \infty} \frac{a_n}{b_n} = \lim_{n \to \infty} \frac{n + n^2 \ln n}{n^2 + 5}$$

$$= \infty,$$

$\sum a_n$ diverges by Part 3 of the Limit Comparison Test. ∎

**EXAMPLE 3** Does $\displaystyle\sum_{n=1}^{\infty} \dfrac{\ln n}{n^{3/2}}$ converge?

**Solution** Because $\ln n$ grows more slowly than $n^c$ for any positive constant $c$ (Section 10.1, Exercise 105), we can compare the series to a convergent $p$-series. To get the $p$-series, we see that

$$\frac{\ln n}{n^{3/2}} < \frac{n^{1/4}}{n^{3/2}} = \frac{1}{n^{5/4}}$$

for $n$ sufficiently large. Then taking $a_n = (\ln n)/n^{3/2}$ and $b_n = 1/n^{5/4}$, we have

$$\lim_{n\to\infty} \frac{a_n}{b_n} = \lim_{n\to\infty} \frac{\ln n}{n^{1/4}}$$

$$= \lim_{n\to\infty} \frac{1/n}{(1/4)n^{-3/4}} \qquad \text{l'Hôpital's Rule}$$

$$= \lim_{n\to\infty} \frac{4}{n^{1/4}} = 0.$$

Since $\Sigma b_n = \Sigma(1/n^{5/4})$ is a $p$-series with $p > 1$, it converges, so $\Sigma a_n$ converges by Part 2 of the Limit Comparison Test. ∎

# Exercises 10.4

## Comparison Test

In Exercises 1–8, use the Comparison Test to determine if each series converges or diverges.

**1.** $\displaystyle\sum_{n=1}^{\infty} \frac{1}{n^2 + 30}$

**2.** $\displaystyle\sum_{n=1}^{\infty} \frac{n-1}{n^4 + 2}$

**3.** $\displaystyle\sum_{n=2}^{\infty} \frac{1}{\sqrt{n} - 1}$

**4.** $\displaystyle\sum_{n=2}^{\infty} \frac{n+2}{n^2 - n}$

**5.** $\displaystyle\sum_{n=1}^{\infty} \frac{\cos^2 n}{n^{3/2}}$

**6.** $\displaystyle\sum_{n=1}^{\infty} \frac{1}{n3^n}$

**7.** $\displaystyle\sum_{n=1}^{\infty} \sqrt{\frac{n+4}{n^4 + 4}}$

**8.** $\displaystyle\sum_{n=1}^{\infty} \frac{\sqrt{n}+1}{\sqrt{n^2 + 3}}$

## Limit Comparison Test

In Exercises 9–16, use the Limit Comparison Test to determine if each series converges or diverges.

**9.** $\displaystyle\sum_{n=1}^{\infty} \frac{n-2}{n^3 - n^2 + 3}$

(*Hint:* Limit Comparison with $\Sigma_{n=1}^{\infty}(1/n^2)$)

**10.** $\displaystyle\sum_{n=1}^{\infty} \sqrt{\frac{n+1}{n^2 + 2}}$

(*Hint:* Limit Comparison with $\Sigma_{n=1}^{\infty}(1/\sqrt{n})$)

**11.** $\displaystyle\sum_{n=2}^{\infty} \frac{n(n+1)}{(n^2 + 1)(n-1)}$

**12.** $\displaystyle\sum_{n=1}^{\infty} \frac{2^n}{3 + 4^n}$

**13.** $\displaystyle\sum_{n=1}^{\infty} \frac{5^n}{\sqrt{n} \, 4^n}$

**14.** $\displaystyle\sum_{n=1}^{\infty} \left(\frac{2n+3}{5n+4}\right)^n$

**15.** $\displaystyle\sum_{n=2}^{\infty} \frac{1}{\ln n}$

(*Hint:* Limit Comparison with $\Sigma_{n=2}^{\infty}(1/n)$)

**16.** $\displaystyle\sum_{n=1}^{\infty} \ln\left(1 + \frac{1}{n^2}\right)$

(*Hint:* Limit Comparison with $\Sigma_{n=1}^{\infty}(1/n^2)$)

## Determining Convergence or Divergence

Which of the series in Exercises 17–54 converge, and which diverge? Use any method, and give reasons for your answers.

**17.** $\displaystyle\sum_{n=1}^{\infty} \frac{1}{2\sqrt{n} + \sqrt[3]{n}}$

**18.** $\displaystyle\sum_{n=1}^{\infty} \frac{3}{n + \sqrt{n}}$

**19.** $\displaystyle\sum_{n=1}^{\infty} \frac{\sin^2 n}{2^n}$

**20.** $\displaystyle\sum_{n=1}^{\infty} \frac{1 + \cos n}{n^2}$

**21.** $\displaystyle\sum_{n=1}^{\infty} \frac{2n}{3n - 1}$

**22.** $\displaystyle\sum_{n=1}^{\infty} \frac{n+1}{n^2\sqrt{n}}$

**23.** $\displaystyle\sum_{n=1}^{\infty} \frac{10n + 1}{n(n+1)(n+2)}$

**24.** $\displaystyle\sum_{n=3}^{\infty} \frac{5n^3 - 3n}{n^2(n-2)(n^2 + 5)}$

**25.** $\displaystyle\sum_{n=1}^{\infty} \left(\frac{n}{3n+1}\right)^n$

**26.** $\displaystyle\sum_{n=1}^{\infty} \frac{1}{\sqrt{n^3 + 2}}$

**27.** $\displaystyle\sum_{n=3}^{\infty} \frac{1}{\ln(\ln n)}$

**28.** $\displaystyle\sum_{n=1}^{\infty} \frac{(\ln n)^2}{n^3}$

**29.** $\displaystyle\sum_{n=2}^{\infty} \frac{1}{\sqrt{n} \ln n}$

**30.** $\displaystyle\sum_{n=1}^{\infty} \frac{(\ln n)^2}{n^{3/2}}$

**31.** $\displaystyle\sum_{n=1}^{\infty} \frac{1}{1 + \ln n}$

**32.** $\displaystyle\sum_{n=2}^{\infty} \frac{\ln(n+1)}{n+1}$

**33.** $\displaystyle\sum_{n=2}^{\infty} \frac{1}{n\sqrt{n^2 - 1}}$

**34.** $\displaystyle\sum_{n=1}^{\infty} \frac{\sqrt{n}}{n^2 + 1}$

**35.** $\displaystyle\sum_{n=1}^{\infty} \frac{1-n}{n2^n}$

**36.** $\displaystyle\sum_{n=1}^{\infty} \frac{n + 2^n}{n^2 2^n}$

**37.** $\displaystyle\sum_{n=1}^{\infty} \frac{1}{3^{n-1} + 1}$

**38.** $\displaystyle\sum_{n=1}^{\infty} \frac{3^{n-1} + 1}{3^n}$

**39.** $\displaystyle\sum_{n=1}^{\infty} \frac{n+1}{n^2 + 3n} \cdot \frac{1}{5n}$

**40.** $\displaystyle\sum_{n=1}^{\infty} \frac{2^n + 3^n}{3^n + 4^n}$

**41.** $\displaystyle\sum_{n=1}^{\infty} \frac{2^n - n}{n2^n}$

**42.** $\displaystyle\sum_{n=1}^{\infty} \frac{\ln n}{\sqrt{n} \, e^n}$

**43.** $\displaystyle\sum_{n=2}^{\infty} \frac{1}{n!}$

(*Hint:* First show that $(1/n!) \le (1/n(n-1))$ for $n \ge 2$.)

**44.** $\displaystyle\sum_{n=1}^{\infty} \frac{(n-1)!}{(n+2)!}$

**45.** $\displaystyle\sum_{n=1}^{\infty} \sin\frac{1}{n}$

**46.** $\displaystyle\sum_{n=1}^{\infty} \tan\frac{1}{n}$

**47.** $\displaystyle\sum_{n=1}^{\infty} \frac{\tan^{-1} n}{n^{1.1}}$

**48.** $\displaystyle\sum_{n=1}^{\infty} \frac{\sec^{-1} n}{n^{1.3}}$

**49.** $\displaystyle\sum_{n=1}^{\infty} \frac{\coth n}{n^2}$

**50.** $\displaystyle\sum_{n=1}^{\infty} \frac{\tanh n}{n^2}$

**51.** $\displaystyle\sum_{n=1}^{\infty} \frac{1}{n\sqrt[n]{n}}$

**52.** $\displaystyle\sum_{n=1}^{\infty} \frac{\sqrt[n]{n}}{n^2}$

**53.** $\displaystyle\sum_{n=1}^{\infty} \frac{1}{1 + 2 + 3 + \cdots + n}$

**54.** $\displaystyle\sum_{n=1}^{\infty} \frac{1}{1 + 2^2 + 3^2 + \cdots + n^2}$

## Theory and Examples

**55.** Prove **(a)** Part 2 and **(b)** Part 3 of the Limit Comparison Test.

**56.** If $\Sigma_{n=1}^{\infty} a_n$ is a convergent series of nonnegative numbers, can anything be said about $\Sigma_{n=1}^{\infty}(a_n/n)$? Explain.

**57.** Suppose that $a_n > 0$ and $b_n > 0$ for $n \ge N$ ($N$ an integer). If $\lim_{n\to\infty}(a_n/b_n) = \infty$ and $\Sigma a_n$ converges, can anything be said about $\Sigma b_n$? Give reasons for your answer.

**58.** Prove that if $\Sigma a_n$ is a convergent series of nonnegative terms, then $\Sigma a_n^2$ converges.

**59.** Suppose that $a_n > 0$ and $\lim_{n \to \infty} a_n = \infty$. Prove that $\Sigma a_n$ diverges.

**60.** Suppose that $a_n > 0$ and $\lim_{n \to \infty} n^2 a_n = 0$. Prove that $\Sigma a_n$ converges.

**61.** Show that $\sum_{n=2}^{\infty} ((\ln n)^q / n^p)$ converges for $-\infty < q < \infty$ and $p > 1$.

(*Hint:* Limit Comparison with $\sum_{n=2}^{\infty} 1/n^r$ for $1 < r < p$.)

**62.** (*Continuation of Exercise 61.*) Show that $\sum_{n=2}^{\infty} ((\ln n)^q / n^p)$ diverges for $-\infty < q < \infty$ and $0 < p < 1$.

(*Hint:* Limit Comparison with an appropriate *p*-series.)

**63. Decimal numbers** Any real number in the interval $[0, 1]$ can be represented by a decimal (not necessarily unique) as

$$0.d_1 d_2 d_3 d_4 \ldots = \frac{d_1}{10} + \frac{d_2}{10^2} + \frac{d_3}{10^3} + \frac{d_4}{10^4} + \cdots,$$

where $d_i$ is one of the integers 0, 1, 2, 3, ... , 9. Prove that the series on the right-hand side always converges.

**64.** If $\Sigma a_n$ is a convergent series of positive terms, prove that $\Sigma \sin (a_n)$ converges.

In Exercises 65–70, use the results of Exercises 61 and 62 to determine if each series converges or diverges.

**65.** $\sum_{n=2}^{\infty} \frac{(\ln n)^3}{n^4}$

**66.** $\sum_{n=2}^{\infty} \sqrt{\frac{\ln n}{n}}$

**67.** $\sum_{n=2}^{\infty} \frac{(\ln n)^{1000}}{n^{1.001}}$

**68.** $\sum_{n=2}^{\infty} \frac{(\ln n)^{1/5}}{n^{0.99}}$

**69.** $\sum_{n=2}^{\infty} \frac{1}{n^{1.1}(\ln n)^3}$

**70.** $\sum_{n=2}^{\infty} \frac{1}{\sqrt{n} \cdot \ln n}$

**COMPUTER EXPLORATIONS**

**71.** It is not yet known whether the series

$$\sum_{n=1}^{\infty} \frac{1}{n^3 \sin^2 n}$$

converges or diverges. Use a CAS to explore the behavior of the series by performing the following steps.

**a.** Define the sequence of partial sums

$$s_k = \sum_{n=1}^{k} \frac{1}{n^3 \sin^2 n}.$$

What happens when you try to find the limit of $s_k$ as $k \to \infty$? Does your CAS find a closed form answer for this limit?

**b.** Plot the first 100 points $(k, s_k)$ for the sequence of partial sums. Do they appear to converge? What would you estimate the limit to be?

**c.** Next plot the first 200 points $(k, s_k)$. Discuss the behavior in your own words.

**d.** Plot the first 400 points $(k, s_k)$. What happens when $k = 355$? Calculate the number $355/113$. Explain from you calculation what happened at $k = 355$. For what values of $k$ would you guess this behavior might occur again?

**72. a.** Use Theorem 8 to show that

$$S = \sum_{n=1}^{\infty} \frac{1}{n(n + 1)} + \sum_{n=1}^{\infty} \left( \frac{1}{n^2} - \frac{1}{n(n + 1)} \right)$$

where $S = \sum_{n=1}^{\infty} (1/n^2)$, the sum of a convergent *p*-series.

**b.** From Example 5, Section 10.2, show that

$$S = 1 + \sum_{n=1}^{\infty} \frac{1}{n^2(n + 1)}.$$

**c.** Explain why taking the first $M$ terms in the series in part (b) gives a better approximation to $S$ than taking the first $M$ terms in the original series $\sum_{n=1}^{\infty} (1/n^2)$.

**d.** We know the exact value of $S$ is $\pi^2/6$. Which of the sums

$$\sum_{n=1}^{1000000} \frac{1}{n^2} \quad \text{or} \quad 1 + \sum_{n=1}^{1000} \frac{1}{n^2(n + 1)}$$

gives a better approximation to $S$?

# 10.5 Absolute Convergence; The Ratio and Root Tests

When some of the terms of a series are positive and others are negative, the series may or may not converge. For example, the geometric series

$$5 - \frac{5}{4} + \frac{5}{16} - \frac{5}{64} + \cdots = \sum_{n=0}^{\infty} 5 \left( \frac{-1}{4} \right)^n \qquad (1)$$

converges (since $|r| = \frac{1}{4} < 1$), whereas the different geometric series

$$1 - \frac{5}{4} + \frac{25}{16} - \frac{125}{64} + \cdots = \sum_{n=0}^{\infty} \left( \frac{-5}{4} \right)^n \qquad (2)$$

diverges (since $|r| = 5/4 > 1$). In series (1), there is some cancelation in the partial sums, which may be assisting the convergence property of the series. However, if we make all of the terms positive in series (1) to form the new series

$$5 + \frac{5}{4} + \frac{5}{16} + \frac{5}{64} + \cdots = \sum_{n=0}^{\infty} \left| 5 \left( \frac{-1}{4} \right)^n \right| = \sum_{n=0}^{\infty} 5 \left( \frac{1}{4} \right)^n,$$

we see that it still converges. For a general series with both positive and negative terms, we can apply the tests for convergence studied before to the series of absolute values of its terms. In doing so, we are led naturally to the following concept.

---

> **DEFINITION**    A series $\sum a_n$ **converges absolutely** (is **absolutely convergent**) if the corresponding series of absolute values, $\sum |a_n|$, converges.

---

So the geometric series (1) is absolutely convergent. We observed, too, that it is also convergent. This situation is always true: An absolutely convergent series is convergent as well, which we now prove.

---

> **THEOREM 12—The Absolute Convergence Test**    If $\displaystyle\sum_{n=1}^{\infty} |a_n|$ converges, then $\displaystyle\sum_{n=1}^{\infty} a_n$ converges.

---

**Proof**    For each $n$,

$$-|a_n| \le a_n \le |a_n|, \qquad \text{so} \qquad 0 \le a_n + |a_n| \le 2|a_n|.$$

If $\sum_{n=1}^{\infty} |a_n|$ converges, then $\sum_{n=1}^{\infty} 2|a_n|$ converges and, by the Direct Comparison Test, the nonnegative series $\sum_{n=1}^{\infty} (a_n + |a_n|)$ converges. The equality $a_n = (a_n + |a_n|) - |a_n|$ now lets us express $\sum_{n=1}^{\infty} a_n$ as the difference of two convergent series:

$$\sum_{n=1}^{\infty} a_n = \sum_{n=1}^{\infty} (a_n + |a_n| - |a_n|) = \sum_{n=1}^{\infty} (a_n + |a_n|) - \sum_{n=1}^{\infty} |a_n|.$$

Therefore, $\sum_{n=1}^{\infty} a_n$ converges. ∎

**EXAMPLE 1**    This example gives two series that converge absolutely.

**(a)** For $\displaystyle\sum_{n=1}^{\infty} (-1)^{n+1} \frac{1}{n^2} = 1 - \frac{1}{4} + \frac{1}{9} - \frac{1}{16} + \cdots$, the corresponding series of absolute values is the convergent series

$$\sum_{n=1}^{\infty} \frac{1}{n^2} = 1 + \frac{1}{4} + \frac{1}{9} + \frac{1}{16} + \cdots.$$

The original series converges because it converges absolutely.

**(b)** For $\displaystyle\sum_{n=1}^{\infty} \frac{\sin n}{n^2} = \frac{\sin 1}{1} + \frac{\sin 2}{4} + \frac{\sin 3}{9} + \cdots$, which contains both positive and negative terms, the corresponding series of absolute values is

$$\sum_{n=1}^{\infty} \left| \frac{\sin n}{n^2} \right| = \frac{|\sin 1|}{1} + \frac{|\sin 2|}{4} + \cdots,$$

which converges by comparison with $\sum_{n=1}^{\infty} (1/n^2)$ because $|\sin n| \le 1$ for every $n$. The original series converges absolutely; therefore it converges. ∎

Caution    Be careful when using Theorem 12. A convergent series need *not* converge absolutely, as you will see in the next section.

## The Ratio Test

The Ratio Test measures the rate of growth (or decline) of a series by examining the ratio $a_{n+1}/a_n$. For a geometric series $\sum ar^n$, this rate is a constant $((ar^{n+1})/(ar^n) = r)$, and the series converges if and only if its ratio is less than 1 in absolute value. The Ratio Test is a powerful rule extending that result.

---

**THEOREM 13—The Ratio Test**    Let $\sum a_n$ be any series and suppose that

$$\lim_{n \to \infty} \left| \frac{a_{n+1}}{a_n} \right| = \rho.$$

Then **(a)** the series *converges absolutely* if $\rho < 1$, **(b)** the series *diverges* if $\rho > 1$ or $\rho$ is infinite, **(c)** the test is *inconclusive* if $\rho = 1$.

---

**Proof**

**(a)  $\rho < 1$.** Let $r$ be a number between $\rho$ and 1. Then the number $\epsilon = r - \rho$ is positive. Since

$$\left| \frac{a_{n+1}}{a_n} \right| \to \rho,$$

$|a_{n+1}/a_n|$ must lie within $\epsilon$ of $\rho$ when $n$ is large enough, say, for all $n \geq N$. In particular,

$$\left| \frac{a_{n+1}}{a_n} \right| < \rho + \epsilon = r, \qquad \text{when } n \geq N.$$

That is,

$$|a_{N+1}| < r|a_N|,$$
$$|a_{N+2}| < r|a_{N+1}| < r^2|a_N|,$$
$$|a_{N+3}| < r|a_{N+2}| < r^3|a_N|,$$
$$\vdots$$
$$|a_{N+m}| < r|a_{N+m-1}| < r^m|a_N|.$$

Therefore.

$$\sum_{m=N}^{\infty} |a_m| = \sum_{m=0}^{\infty} |a_{N+m}| \leq \sum_{m=0}^{\infty} |a_N| \, r^m = |a_N| \sum_{m=0}^{\infty} r^m.$$

The geometric series on the right-hand side converges because $0 < r < 1$, so the series of absolute values $\sum_{m=N}^{\infty} |a_m|$ converges by the Comparison Test. Because adding or deleting finitely many terms in a series does not affect its convergence or divergence property, the series $\sum_{n=1}^{\infty} |a_n|$ also converges. That is, the series $\sum a_n$ is absolutely convergent.

**(b)  $1 < \rho \leq \infty$.** From some index $M$ on,

$$\left| \frac{a_{n+1}}{a_n} \right| > 1 \qquad \text{and} \qquad |a_M| < |a_{M+1}| < |a_{M+2}| < \cdots.$$

The terms of the series do not approach zero as $n$ becomes infinite, and the series diverges by the $n$th-Term Test.

**(c)  $\rho = 1$.** The two series

$$\sum_{n=1}^{\infty} \frac{1}{n} \qquad \text{and} \qquad \sum_{n=1}^{\infty} \frac{1}{n^2}$$

show that some other test for convergence must be used when $\rho = 1$.

$$\text{For } \sum_{n=1}^{\infty} \frac{1}{n}: \quad \left|\frac{a_{n+1}}{a_n}\right| = \frac{1/(n+1)}{1/n} = \frac{n}{n+1} \to 1.$$

$$\text{For } \sum_{n=1}^{\infty} \frac{1}{n^2}: \quad \left|\frac{a_{n+1}}{a_n}\right| = \frac{1/(n+1)^2}{1/n^2} = \left(\frac{n}{n+1}\right)^2 \to 1^2 = 1.$$

In both cases, $\rho = 1$, yet the first series diverges, whereas the second converges. ∎

The Ratio Test is often effective when the terms of a series contain factorials of expressions involving $n$ or expressions raised to a power involving $n$.

**EXAMPLE 2**  Investigate the convergence of the following series.

**(a)** $\displaystyle\sum_{n=0}^{\infty} \frac{2^n + 5}{3^n}$  **(b)** $\displaystyle\sum_{n=1}^{\infty} \frac{(2n)!}{n!n!}$  **(c)** $\displaystyle\sum_{n=1}^{\infty} \frac{4^n n! n!}{(2n)!}$

**Solution**  We apply the Ratio Test to each series.

**(a)** For the series $\sum_{n=0}^{\infty} (2^n + 5)/3^n$,

$$\left|\frac{a_{n+1}}{a_n}\right| = \frac{(2^{n+1} + 5)/3^{n+1}}{(2^n + 5)/3^n} = \frac{1}{3} \cdot \frac{2^{n+1} + 5}{2^n + 5} = \frac{1}{3} \cdot \left(\frac{2 + 5 \cdot 2^{-n}}{1 + 5 \cdot 2^{-n}}\right) \to \frac{1}{3} \cdot \frac{2}{1} = \frac{2}{3}.$$

The series converges absolutely (and thus converges) because $\rho = 2/3$ is less than 1. This does *not* mean that $2/3$ is the sum of the series. In fact,

$$\sum_{n=0}^{\infty} \frac{2^n + 5}{3^n} = \sum_{n=0}^{\infty} \left(\frac{2}{3}\right)^n + \sum_{n=0}^{\infty} \frac{5}{3^n} = \frac{1}{1 - (2/3)} + \frac{5}{1 - (1/3)} = \frac{21}{2}.$$

**(b)** If $a_n = \dfrac{(2n)!}{n!n!}$, then $a_{n+1} = \dfrac{(2n+2)!}{(n+1)!(n+1)!}$ and

$$\left|\frac{a_{n+1}}{a_n}\right| = \frac{n!n!(2n+2)(2n+1)(2n)!}{(n+1)!(n+1)!(2n)!}$$

$$= \frac{(2n+2)(2n+1)}{(n+1)(n+1)} = \frac{4n+2}{n+1} \to 4.$$

The series diverges because $\rho = 4$ is greater than 1.

**(c)** If $a_n = 4^n n! n!/(2n)!$, then

$$\left|\frac{a_{n+1}}{a_n}\right| = \frac{4^{n+1}(n+1)!(n+1)!}{(2n+2)(2n+1)(2n)!} \cdot \frac{(2n)!}{4^n n! n!}$$

$$= \frac{4(n+1)(n+1)}{(2n+2)(2n+1)} = \frac{2(n+1)}{2n+1} \to 1.$$

Because the limit is $\rho = 1$, we cannot decide from the Ratio Test whether the series converges. When we notice that $a_{n+1}/a_n = (2n+2)/(2n+1)$, we conclude that $a_{n+1}$ is always greater than $a_n$ because $(2n+2)/(2n+1)$ is always greater than 1. Therefore, all terms are greater than or equal to $a_1 = 2$, and the $n$th term does not approach zero as $n \to \infty$. The series diverges. ∎

## The Root Test

The convergence tests we have so far for $\Sigma a_n$ work best when the formula for $a_n$ is relatively simple. However, consider the series with the terms

$$a_n = \begin{cases} n/2^n, & n \text{ odd} \\ 1/2^n, & n \text{ even.} \end{cases}$$

To investigate convergence we write out several terms of the series:

$$\sum_{n=1}^{\infty} a_n = \frac{1}{2^1} + \frac{1}{2^2} + \frac{3}{2^3} + \frac{1}{2^4} + \frac{5}{2^5} + \frac{1}{2^6} + \frac{7}{2^7} + \cdots$$

$$= \frac{1}{2} + \frac{1}{4} + \frac{3}{8} + \frac{1}{16} + \frac{5}{32} + \frac{1}{64} + \frac{7}{128} + \cdots.$$

Clearly, this is not a geometric series. The $n$th term approaches zero as $n \to \infty$, so the $n$th-Term Test does not tell us if the series diverges. The Integral Test does not look promising. The Ratio Test produces

$$\left| \frac{a_{n+1}}{a_n} \right| = \begin{cases} \dfrac{1}{2n}, & n \text{ odd} \\ \dfrac{n+1}{2}, & n \text{ even} \end{cases}$$

As $n \to \infty$, the ratio is alternately small and large and has no limit. However, we will see that the following test establishes that the series converges.

---

**THEOREM 14—The Root Test** Let $\sum a_n$ be any series and suppose that

$$\lim_{n \to \infty} \sqrt[n]{|a_n|} = \rho.$$

Then **(a)** the series *converges absolutely* if $\rho < 1$, **(b)** the series *diverges* if $\rho > 1$ or $\rho$ is infinite, **(c)** the test is *inconclusive* if $\rho = 1$.

---

**Proof**

**(a)** $\rho < 1.$ Choose an $\epsilon > 0$ so small that $\rho + \epsilon < 1$. Since $\sqrt[n]{|a_n|} \to \rho$, the terms $\sqrt[n]{|a_n|}$ eventually get to within $\epsilon$ of $\rho$. So there exists an index $M$ such that

$$\sqrt[n]{|a_n|} < \rho + \epsilon \qquad \text{when } n \geq M.$$

Then it is also true that

$$|a_n| < (\rho + \epsilon)^n \qquad \text{for } n \geq M.$$

Now, $\sum_{n=M}^{\infty} (\rho + \epsilon)^n$, a geometric series with ratio $(\rho + \epsilon) < 1$, converges. By the Comparison Test, $\sum_{n=M}^{\infty} |a_n|$ converges, from which it follows that

$$\sum_{n=1}^{\infty} |a_n| = |a_1| + \cdots + |a_{M-1}| + \sum_{n=M}^{\infty} |a_n|$$

converges. Therefore, $\sum a_n$ converges absolutely.

**(b)** $1 < \rho \leq \infty.$ For all indices beyond some integer $M$, we have $\sqrt[n]{|a_n|} > 1$, so that $|a_n| > 1$ for $n > M$. The terms of the series do not converge to zero. The series diverges by the $n$th-Term Test.

**(c)** $\rho = 1.$ The series $\sum_{n=1}^{\infty} (1/n)$ and $\sum_{n=1}^{\infty} (1/n^2)$ show that the test is not conclusive when $\rho = 1$. The first series diverges and the second converges, but in both cases $\sqrt[n]{|a_n|} \to 1$. ■

**EXAMPLE 3** Consider again the series with terms $a_n = \begin{cases} n/2^n, & n \text{ odd} \\ 1/2^n, & n \text{ even.} \end{cases}$

Does $\sum a_n$ converge?

**Solution**   We apply the Root Test, finding that

$$\sqrt[n]{|a_n|} = \begin{cases} \sqrt[n]{n}/2, & n \text{ odd} \\ 1/2, & n \text{ even}. \end{cases}$$

Therefore,

$$\frac{1}{2} \le \sqrt[n]{|a_n|} \le \frac{\sqrt[n]{n}}{2}.$$

Since $\sqrt[n]{n} \to 1$ (Section 10.1, Theorem 5), we have $\lim_{n\to\infty} \sqrt[n]{|a_n|} = 1/2$ by the Sandwich Theorem. The limit is less than 1, so the series converges absolutely by the Root Test.  ■

**EXAMPLE 4**    Which of the following series converge, and which diverge?

**(a)** $\displaystyle\sum_{n=1}^{\infty} \frac{n^2}{2^n}$    **(b)** $\displaystyle\sum_{n=1}^{\infty} \frac{2^n}{n^3}$    **(c)** $\displaystyle\sum_{n=1}^{\infty} \left(\frac{1}{1+n}\right)^n$

**Solution**   We apply the Root Test to each series, noting that each series has positive terms.

**(a)** $\displaystyle\sum_{n=1}^{\infty} \frac{n^2}{2^n}$ converges because $\sqrt[n]{\dfrac{n^2}{2^n}} = \dfrac{\sqrt[n]{n^2}}{\sqrt[n]{2^n}} = \dfrac{(\sqrt[n]{n})^2}{2} \to \dfrac{1^2}{2} < 1.$

**(b)** $\displaystyle\sum_{n=1}^{\infty} \frac{2^n}{n^3}$ diverges because $\sqrt[n]{\dfrac{2^n}{n^3}} = \dfrac{2}{(\sqrt[n]{n})^3} \to \dfrac{2}{1^3} > 1.$

**(c)** $\displaystyle\sum_{n=1}^{\infty} \left(\frac{1}{1+n}\right)^n$ converges because $\sqrt[n]{\left(\dfrac{1}{1+n}\right)^n} = \dfrac{1}{1+n} \to 0 < 1.$  ■

# Exercises 10.5

**Using the Ratio Test**

In Exercises 1–8, use the Ratio Test to determine if each series converges absolutely or diverges.

**1.** $\displaystyle\sum_{n=1}^{\infty} \frac{2^n}{n!}$

**2.** $\displaystyle\sum_{n=1}^{\infty} (-1)^n \frac{n+2}{3^n}$

**3.** $\displaystyle\sum_{n=1}^{\infty} \frac{(n-1)!}{(n+1)^2}$

**4.** $\displaystyle\sum_{n=1}^{\infty} \frac{2^{n+1}}{n3^{n-1}}$

**5.** $\displaystyle\sum_{n=1}^{\infty} \frac{n^4}{(-4)^n}$

**6.** $\displaystyle\sum_{n=2}^{\infty} \frac{3^{n+2}}{\ln n}$

**7.** $\displaystyle\sum_{n=1}^{\infty} (-1)^n \frac{n^2(n+2)!}{n!\,3^{2n}}$

**8.** $\displaystyle\sum_{n=1}^{\infty} \frac{n5^n}{(2n+3)\ln(n+1)}$

**Using the Root Test**

In Exercises 9–16, use the Root Test to determine if each series converges absolutely or diverges.

**9.** $\displaystyle\sum_{n=1}^{\infty} \frac{7}{(2n+5)^n}$

**10.** $\displaystyle\sum_{n=1}^{\infty} \frac{4^n}{(3n)^n}$

**11.** $\displaystyle\sum_{n=1}^{\infty} \left(\frac{4n+3}{3n-5}\right)^n$

**12.** $\displaystyle\sum_{n=1}^{\infty} \left(-\ln\left(e^2 + \frac{1}{n}\right)\right)^{n+1}$

**13.** $\displaystyle\sum_{n=1}^{\infty} \frac{-8}{(3+(1/n))^{2n}}$

**14.** $\displaystyle\sum_{n=1}^{\infty} \sin^n\left(\frac{1}{\sqrt{n}}\right)$

**15.** $\displaystyle\sum_{n=1}^{\infty} (-1)^n \left(1 - \frac{1}{n}\right)^{n^2}$

(*Hint:* $\displaystyle\lim_{n\to\infty} (1 + x/n)^n = e^x$)

**16.** $\displaystyle\sum_{n=2}^{\infty} \frac{(-1)^n}{n^{1+n}}$

**Determining Convergence or Divergence**

In Exercises 17–44, use any method to determine if the series converges or diverges. Give reasons for your answer.

**17.** $\displaystyle\sum_{n=1}^{\infty} \frac{n^{\sqrt{2}}}{2^n}$

**18.** $\displaystyle\sum_{n=1}^{\infty} (-1)^n n^2 e^{-n}$

**19.** $\displaystyle\sum_{n=1}^{\infty} n!(-e)^{-n}$

**20.** $\displaystyle\sum_{n=1}^{\infty} \frac{n!}{10^n}$

**21.** $\displaystyle\sum_{n=1}^{\infty} \frac{n^{10}}{10^n}$

**22.** $\displaystyle\sum_{n=1}^{\infty} \left(\frac{n-2}{n}\right)^n$

**23.** $\sum_{n=1}^{\infty} \frac{2 + (-1)^n}{1.25^n}$

**24.** $\sum_{n=1}^{\infty} \frac{(-2)^n}{3^n}$

**25.** $\sum_{n=1}^{\infty} (-1)^n \left(1 - \frac{3}{n}\right)^n$

**26.** $\sum_{n=1}^{\infty} \left(1 - \frac{1}{3n}\right)^n$

**27.** $\sum_{n=1}^{\infty} \frac{\ln n}{n^3}$

**28.** $\sum_{n=1}^{\infty} \frac{(-\ln n)^n}{n^n}$

**29.** $\sum_{n=1}^{\infty} \left(\frac{1}{n} - \frac{1}{n^2}\right)$

**30.** $\sum_{n=1}^{\infty} \left(\frac{1}{n} - \frac{1}{n^2}\right)^n$

**31.** $\sum_{n=1}^{\infty} \frac{e^n}{n^e}$

**32.** $\sum_{n=1}^{\infty} \frac{n \ln n}{(-2)^n}$

**33.** $\sum_{n=1}^{\infty} \frac{(n+1)(n+2)}{n!}$

**34.** $\sum_{n=1}^{\infty} e^{-n}(n^3)$

**35.** $\sum_{n=1}^{\infty} \frac{(n+3)!}{3!n!3^n}$

**36.** $\sum_{n=1}^{\infty} \frac{n2^n(n+1)!}{3^n n!}$

**37.** $\sum_{n=1}^{\infty} \frac{n!}{(2n+1)!}$

**38.** $\sum_{n=1}^{\infty} \frac{n!}{(-n)^n}$

**39.** $\sum_{n=2}^{\infty} \frac{-n}{(\ln n)^n}$

**40.** $\sum_{n=2}^{\infty} \frac{n}{(\ln n)^{(n/2)}}$

**41.** $\sum_{n=1}^{\infty} \frac{n! \ln n}{n(n+2)!}$

**42.** $\sum_{n=1}^{\infty} \frac{(-3)^n}{n^3 2^n}$

**43.** $\sum_{n=1}^{\infty} \frac{(n!)^2}{(2n)!}$

**44.** $\sum_{n=1}^{\infty} \frac{(2n+3)(2^n+3)}{3^n+2}$

**Recursively Defined Terms**   Which of the series $\sum_{n=1}^{\infty} a_n$ defined by the formulas in Exercises 45–54 converge, and which diverge? Give reasons for your answers.

**45.** $a_1 = 2, \quad a_{n+1} = \frac{1 + \sin n}{n} a_n$

**46.** $a_1 = 1, \quad a_{n+1} = \frac{1 + \tan^{-1} n}{n} a_n$

**47.** $a_1 = \frac{1}{3}, \quad a_{n+1} = \frac{3n - 1}{2n + 5} a_n$

**48.** $a_1 = 3, \quad a_{n+1} = \frac{n}{n + 1} a_n$

**49.** $a_1 = 2, \quad a_{n+1} = \frac{2}{n} a_n$

**50.** $a_1 = 5, \quad a_{n+1} = \frac{\sqrt[n]{n}}{2} a_n$

**51.** $a_1 = 1, \quad a_{n+1} = \frac{1 + \ln n}{n} a_n$

**52.** $a_1 = \frac{1}{2}, \quad a_{n+1} = \frac{n + \ln n}{n + 10} a_n$

**53.** $a_1 = \frac{1}{3}, \quad a_{n+1} = \sqrt[n]{a_n}$

**54.** $a_1 = \frac{1}{2}, \quad a_{n+1} = (a_n)^{n+1}$

**Convergence or Divergence**
Which of the series in Exercises 55–62 converge, and which diverge? Give reasons for your answers.

**55.** $\sum_{n=1}^{\infty} \frac{2^n n! n!}{(2n)!}$

**56.** $\sum_{n=1}^{\infty} \frac{(-1)^n (3n)!}{n!(n + 1)!(n + 2)!}$

**57.** $\sum_{n=1}^{\infty} \frac{(n!)^n}{(n^n)^2}$

**58.** $\sum_{n=1}^{\infty} (-1)^n \frac{(n!)^n}{n^{(n^2)}}$

**59.** $\sum_{n=1}^{\infty} \frac{n^n}{2^{(n^2)}}$

**60.** $\sum_{n=1}^{\infty} \frac{n^n}{(2^n)^2}$

**61.** $\sum_{n=1}^{\infty} \frac{1 \cdot 3 \cdot \cdots \cdot (2n - 1)}{4^n 2^n n!}$

**62.** $\sum_{n=1}^{\infty} \frac{1 \cdot 3 \cdot \cdots \cdot (2n - 1)}{[2 \cdot 4 \cdot \cdots \cdot (2n)](3^n + 1)}$

**Theory and Examples**

**63.** Neither the Ratio Test nor the Root Test helps with *p*-series. Try them on

$$\sum_{n=1}^{\infty} \frac{1}{n^p}$$

and show that both tests fail to provide information about convergence.

**64.** Show that neither the Ratio Test nor the Root Test provides information about the convergence of

$$\sum_{n=2}^{\infty} \frac{1}{(\ln n)^p} \quad (p \text{ constant}).$$

**65.** Let $a_n = \begin{cases} n/2^n, & \text{if } n \text{ is a prime number} \\ 1/2^n, & \text{otherwise.} \end{cases}$

Does $\sum a_n$ converge? Give reasons for your answer.

**66.** Show that $\sum_{n=1}^{\infty} 2^{(n^2)}/n!$ diverges. Recall from the Laws of Exponents that $2^{(n^2)} = (2^n)^n$.

# $10.6$ Alternating Series and Conditional Convergence

A series in which the terms are alternately positive and negative is an **alternating series**. Here are three examples:

$$1 - \frac{1}{2} + \frac{1}{3} - \frac{1}{4} + \frac{1}{5} - \cdots + \frac{(-1)^{n+1}}{n} + \cdots \tag{1}$$

$$-2 + 1 - \frac{1}{2} + \frac{1}{4} - \frac{1}{8} + \cdots + \frac{(-1)^n 4}{2^n} + \cdots \tag{2}$$

$$1 - 2 + 3 - 4 + 5 - 6 + \cdots + (-1)^{n+1} n + \cdots \tag{3}$$

We see from these examples that the $n$th term of an alternating series is of the form

$$a_n = (-1)^{n+1}u_n \qquad \text{or} \qquad a_n = (-1)^n u_n$$

where $u_n = |a_n|$ is a positive number.

Series (1), called the **alternating harmonic series**, converges, as we will see in a moment. Series (2), a geometric series with ratio $r = -1/2$, converges to $-2/[1 + (1/2)] = -4/3$. Series (3) diverges because the $n$th term does not approach zero.

We prove the convergence of the alternating harmonic series by applying the Alternating Series Test. This test is for *convergence* of an alternating series and cannot be used to conclude that such a series diverges. The test is also valid for the alternating series $-u_1 + u_2 - u_3 + \cdots$, like the one in Series (2) given above.

---

**THEOREM 15—The Alternating Series Test**    The series

$$\sum_{n=1}^{\infty}(-1)^{n+1}u_n = u_1 - u_2 + u_3 - u_4 + \cdots$$

converges if all three of the following conditions are satisfied:

1. The $u_n$'s are all positive.
2. The positive $u_n$'s are (eventually) nonincreasing: $u_n \geq u_{n+1}$ for all $n \geq N$, for some integer $N$.
3. $u_n \to 0$.

---

**Proof**    Assume $N = 1$. If $n$ is an even integer, say $n = 2m$, then the sum of the first $n$ terms is

$$\begin{aligned} s_{2m} &= (u_1 - u_2) + (u_3 - u_4) + \cdots + (u_{2m-1} - u_{2m}) \\ &= u_1 - (u_2 - u_3) - (u_4 - u_5) - \cdots - (u_{2m-2} - u_{2m-1}) - u_{2m}. \end{aligned}$$

The first equality shows that $s_{2m}$ is the sum of $m$ nonnegative terms, since each term in parentheses is positive or zero. Hence $s_{2m+2} \geq s_{2m}$, and the sequence $\{s_{2m}\}$ is nondecreasing. The second equality shows that $s_{2m} \leq u_1$. Since $\{s_{2m}\}$ is nondecreasing and bounded from above, it has a limit, say

$$\lim_{m \to \infty} s_{2m} = L. \tag{4}$$

If $n$ is an odd integer, say $n = 2m + 1$, then the sum of the first $n$ terms is $s_{2m+1} = s_{2m} + u_{2m+1}$. Since $u_n \to 0$,

$$\lim_{m \to \infty} u_{2m+1} = 0$$

and, as $m \to \infty$,

$$s_{2m+1} = s_{2m} + u_{2m+1} \to L + 0 = L. \tag{5}$$

Combining the results of Equations (4) and (5) gives $\lim_{n \to \infty} s_n = L$ (Section 10.1, Exercise 133). ∎

**EXAMPLE 1**    The alternating harmonic series

$$\sum_{n=1}^{\infty}(-1)^{n+1}\frac{1}{n} = 1 - \frac{1}{2} + \frac{1}{3} - \frac{1}{4} + \cdots$$

clearly satisfies the three requirements of Theorem 15 with $N = 1$; it therefore converges. ∎

Rather than directly verifying the definition $u_n \geq u_{n+1}$, a second way to show that the sequence $\{u_n\}$ is nonincreasing is to define a differentiable function $f(x)$ satisfying $f(n) = u_n$. That is, the values of $f$ match the values of the sequence at every positive

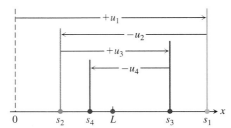

**FIGURE 10.13** The partial sums of an alternating series that satisfies the hypotheses of Theorem 15 for $N = 1$ straddle the limit from the beginning.

integer $n$. If $f'(x) \leq 0$ for all $x$ greater than or equal to some positive integer $N$, then $f(x)$ is nonincreasing for $x \geq N$. It follows that $f(n) \geq f(n + 1)$, or $u_n \geq u_{n+1}$, for $n \geq N$.

**EXAMPLE 2**     Consider the sequence where $u_n = 10n/(n^2 + 16)$. Define $f(x) = 10x/(x^2 + 16)$. Then from the Derivative Quotient Rule,

$$f'(x) = \frac{10(16 - x^2)}{(x^2 + 16)^2} \leq 0 \qquad \text{whenever } x \geq 4.$$

It follows that $u_n \geq u_{n+1}$ for $n \geq 4$. That is, the sequence $\{u_n\}$ is nonincreasing for $n \geq 4$.                                                                                   ∎

A graphical interpretation of the partial sums (Figure 10.13) shows how an alternating series converges to its limit $L$ when the three conditions of Theorem 15 are satisfied with $N = 1$. Starting from the origin of the $x$-axis, we lay off the positive distance $s_1 = u_1$. To find the point corresponding to $s_2 = u_1 - u_2$, we back up a distance equal to $u_2$. Since $u_2 \leq u_1$, we do not back up any farther than the origin. We continue in this seesaw fashion, backing up or going forward as the signs in the series demand. But for $n \geq N$, each forward or backward step is shorter than (or at most the same size as) the preceding step because $u_{n+1} \leq u_n$. And since the $n$th term approaches zero as $n$ increases, the size of step we take forward or backward gets smaller and smaller. We oscillate across the limit $L$, and the amplitude of oscillation approaches zero. The limit $L$ lies between any two successive sums $s_n$ and $s_{n+1}$ and hence differs from $s_n$ by an amount less than $u_{n+1}$.

Because

$$\left| L - s_n \right| < u_{n+1} \qquad \text{for } n \geq N,$$

we can make useful estimates of the sums of convergent alternating series.

---

**THEOREM 16—The Alternating Series Estimation Theorem**     If the alternating series $\sum_{n=1}^{\infty}(-1)^{n+1}u_n$ satisfies the three conditions of Theorem 15, then for $n \geq N$,

$$s_n = u_1 - u_2 + \cdots + (-1)^{n+1}u_n$$

approximates the sum $L$ of the series with an error whose absolute value is less than $u_{n+1}$, the absolute value of the first unused term. Furthermore, the sum $L$ lies between any two successive partial sums $s_n$ and $s_{n+1}$, and the remainder, $L - s_n$, has the same sign as the first unused term.

---

We leave the verification of the sign of the remainder for Exercise 61.

**EXAMPLE 3**     We try Theorem 16 on a series whose sum we know:

$$\sum_{n=0}^{\infty}(-1)^n\frac{1}{2^n} = 1 - \frac{1}{2} + \frac{1}{4} - \frac{1}{8} + \frac{1}{16} - \frac{1}{32} + \frac{1}{64} - \frac{1}{128} + \frac{1}{256} - \cdots.$$

The theorem says that if we truncate the series after the eighth term, we throw away a total that is positive and less than $1/256$. The sum of the first eight terms is $s_8 = 0.6640625$ and the sum of the first nine terms is $s_9 = 0.66796875$. The sum of the geometric series is

$$\frac{1}{1 - (-1/2)} = \frac{1}{3/2} = \frac{2}{3},$$

and we note that $0.6640625 < (2/3) < 0.66796875$. The difference, $(2/3) - 0.6640625 = 0.0026041666\ldots$, is positive and is less than $(1/256) = 0.00390625$.                                                                                   ∎

## Conditional Convergence

If we replace all the negative terms in the alternating series in Example 3, changing them to positive terms instead, we obtain the geometric series $\sum 1/2^n$. The original series and the new series of absolute values both converge (although to different sums). For an absolutely convergent series, changing infinitely many of the negative terms in the series to positive values does not change its property of still being a convergent series. Other convergent series may behave differently. The convergent alternating harmonic series has infinitely many negative terms, but if we change its negative terms to positive values, the resulting series is the divergent harmonic series. So the presence of infinitely many negative terms is essential to the convergence of the alternating harmonic series. The following terminology distinguishes these two types of convergent series.

---

**DEFINITION** A convergent series that is not absolutely convergent is **conditionally convergent**.

---

The alternating harmonic series is conditionally convergent, or **converges conditionally**. The next example extends that result to the alternating $p$-series.

**EXAMPLE 4** If $p$ is a positive constant, the sequence $\{1/n^p\}$ is a decreasing sequence with limit zero. Therefore the alternating $p$-series

$$\sum_{n=1}^{\infty} \frac{(-1)^{n-1}}{n^p} = 1 - \frac{1}{2^p} + \frac{1}{3^p} - \frac{1}{4^p} + \cdots, \quad p > 0$$

converges.

If $p > 1$, the series converges absolutely as an ordinary $p$-series. If $0 < p \leq 1$, the series converges conditionally by the alternating series test. For instance,

$$\text{Absolute convergence } (p = 3/2): \quad 1 - \frac{1}{2^{3/2}} + \frac{1}{3^{3/2}} - \frac{1}{4^{3/2}} + \cdots$$

$$\text{Conditional convergence } (p = 1/2): \quad 1 - \frac{1}{\sqrt{2}} + \frac{1}{\sqrt{3}} - \frac{1}{\sqrt{4}} + \cdots \qquad \blacksquare$$

We need to be careful when using a conditionally convergent series. We have seen with the alternating harmonic series that altering the signs of infinitely many terms of a conditionally convergent series can change its convergence status. Even more, simply changing the order of occurrence of infinitely many of its terms can also have a significant effect, as we now discuss.

## Rearranging Series

We can always rearrange the terms of a *finite* sum. The same result is true for an infinite series that is absolutely convergent (see Exercise 70 for an outline of the proof).

---

**THEOREM 17—The Rearrangement Theorem for Absolutely Convergent Series** If $\sum_{n=1}^{\infty} a_n$ converges absolutely, and $b_1, b_2, \ldots, b_n, \ldots$ is any arrangement of the sequence $\{a_n\}$, then $\sum b_n$ converges absolutely and

$$\sum_{n=1}^{\infty} b_n = \sum_{n=1}^{\infty} a_n.$$

---

On the other hand, if we rearrange the terms of a conditionally convergent series, we can get different results. In fact, for any real number $r$, a given conditionally convergent series can be rearranged so its sum is equal to $r$. (We omit the proof of this fact.) Here's an example of summing the terms of a conditionally convergent series with different orderings, with each ordering giving a different value for the sum.

**EXAMPLE 5**   We know that the alternating harmonic series $\sum_{n=1}^{\infty}(-1)^{n+1}/n$ converges to some number $L$. Moreover, by Theorem 16, $L$ lies between the successive partial sums $s_2 = 1/2$ and $s_3 = 5/6$, so $L \neq 0$. If we multiply the series by 2 we obtain

$$2L = 2\sum_{n=1}^{\infty}\frac{(-1)^{n+1}}{n} = 2\left(1 - \frac{1}{2} + \frac{1}{3} - \frac{1}{4} + \frac{1}{5} - \frac{1}{6} + \frac{1}{7} - \frac{1}{8} + \frac{1}{9} - \frac{1}{10} + \frac{1}{11} - \cdots\right)$$

$$= 2 - 1 + \frac{2}{3} - \frac{1}{2} + \frac{2}{5} - \frac{1}{3} + \frac{2}{7} - \frac{1}{4} + \frac{2}{9} - \frac{1}{5} + \frac{2}{11} - \cdots.$$

Now we change the order of this last sum by grouping each pair of terms with the same odd denominator, but leaving the negative terms with the even denominators as they are placed (so the denominators are the positive integers in their natural order). This rearrangement gives

$$(2 - 1) - \frac{1}{2} + \left(\frac{2}{3} - \frac{1}{3}\right) - \frac{1}{4} + \left(\frac{2}{5} - \frac{1}{5}\right) - \frac{1}{6} + \left(\frac{2}{7} - \frac{1}{7}\right) - \frac{1}{8} + \cdots$$

$$= \left(1 - \frac{1}{2} + \frac{1}{3} - \frac{1}{4} + \frac{1}{5} - \frac{1}{6} + \frac{1}{7} - \frac{1}{8} + \frac{1}{9} - \frac{1}{10} + \frac{1}{11} - \cdots\right)$$

$$= \sum_{n=1}^{\infty}\frac{(-1)^{n+1}}{n} = L.$$

So by rearranging the terms of the conditionally convergent series $\sum_{n=1}^{\infty} 2(-1)^{n+1}/n$, the series becomes $\sum_{n=1}^{\infty}(-1)^{n+1}/n$, which is the alternating harmonic series itself. If the two series are the same, it would imply that $2L = L$, which is clearly false since $L \neq 0$. ∎

Example 5 shows that we cannot rearrange the terms of a conditionally convergent series and expect the new series to be the same as the original one. When we use a conditionally convergent series, the terms must be added together in the order they are given to obtain a correct result. To the contrary, Theorem 17 guarantees that the terms of an absolutely convergent series can be summed in any order without affecting the result.

## Summary of Tests

We have developed a variety of tests to determine convergence or divergence for an infinite series of constants. There are other tests we have not presented which are sometimes given in more advanced courses. Here is a summary of the tests we have considered.

---

1. **The $n$th-Term Test:** If it is not true that $a_n \to 0$, then the series diverges.

2. **Geometric series:** $\sum ar^n$ converges if $|r| < 1$; otherwise it diverges.

3. **$p$-series:** $\sum 1/n^p$ converges if $p > 1$; otherwise it diverges.

4. **Series with nonnegative terms:** Try the Integral Test or try comparing to a known series with the Comparison Test or the Limit Comparison Test. Try the Ratio or Root Test.

5. **Series with some negative terms:** Does $\sum |a_n|$ converge by the Ratio or Root Test, or by another of the tests listed above? Remember, absolute convergence implies convergence.

6. **Alternating series:** $\sum a_n$ converges if the series satisfies the conditions of the Alternating Series Test.

---

# Exercises 10.6

## Determining Convergence or Divergence

In Exercises 1–14, determine if the alternating series converges or diverges. Some of the series do not satisfy the conditions of the Alternating Series Test.

**1.** $\displaystyle\sum_{n=1}^{\infty}(-1)^{n+1}\frac{1}{\sqrt{n}}$

**2.** $\displaystyle\sum_{n=1}^{\infty}(-1)^{n+1}\frac{1}{n^{3/2}}$

**3.** $\displaystyle\sum_{n=1}^{\infty}(-1)^{n+1}\frac{1}{n3^n}$

**4.** $\displaystyle\sum_{n=2}^{\infty}(-1)^{n}\frac{4}{(\ln n)^2}$

**5.** $\displaystyle\sum_{n=1}^{\infty}(-1)^{n}\frac{n}{n^2+1}$

**6.** $\displaystyle\sum_{n=1}^{\infty}(-1)^{n+1}\frac{n^2+5}{n^2+4}$

**7.** $\displaystyle\sum_{n=1}^{\infty}(-1)^{n+1}\frac{2^n}{n^2}$

**8.** $\displaystyle\sum_{n=1}^{\infty}(-1)^{n}\frac{10^n}{(n+1)!}$

**9.** $\displaystyle\sum_{n=1}^{\infty}(-1)^{n+1}\left(\frac{n}{10}\right)^n$

**10.** $\displaystyle\sum_{n=2}^{\infty}(-1)^{n+1}\frac{1}{\ln n}$

**11.** $\displaystyle\sum_{n=1}^{\infty}(-1)^{n+1}\frac{\ln n}{n}$

**12.** $\displaystyle\sum_{n=1}^{\infty}(-1)^{n}\ln\left(1+\frac{1}{n}\right)$

**13.** $\displaystyle\sum_{n=1}^{\infty}(-1)^{n+1}\frac{\sqrt{n}+1}{n+1}$

**14.** $\displaystyle\sum_{n=1}^{\infty}(-1)^{n+1}\frac{3\sqrt{n}+1}{\sqrt{n}+1}$

## Absolute and Conditional Convergence

Which of the series in Exercises 15–48 converge absolutely, which converge, and which diverge? Give reasons for your answers.

**15.** $\displaystyle\sum_{n=1}^{\infty}(-1)^{n+1}(0.1)^n$

**16.** $\displaystyle\sum_{n=1}^{\infty}(-1)^{n+1}\frac{(0.1)^n}{n}$

**17.** $\displaystyle\sum_{n=1}^{\infty}(-1)^{n}\frac{1}{\sqrt{n}}$

**18.** $\displaystyle\sum_{n=1}^{\infty}\frac{(-1)^n}{1+\sqrt{n}}$

**19.** $\displaystyle\sum_{n=1}^{\infty}(-1)^{n+1}\frac{n}{n^3+1}$

**20.** $\displaystyle\sum_{n=1}^{\infty}(-1)^{n+1}\frac{n!}{2^n}$

**21.** $\displaystyle\sum_{n=1}^{\infty}(-1)^{n}\frac{1}{n+3}$

**22.** $\displaystyle\sum_{n=1}^{\infty}(-1)^{n}\frac{\sin n}{n^2}$

**23.** $\displaystyle\sum_{n=1}^{\infty}(-1)^{n+1}\frac{3+n}{5+n}$

**24.** $\displaystyle\sum_{n=1}^{\infty}\frac{(-2)^{n+1}}{n+5^n}$

**25.** $\displaystyle\sum_{n=1}^{\infty}(-1)^{n+1}\frac{1+n}{n^2}$

**26.** $\displaystyle\sum_{n=1}^{\infty}(-1)^{n+1}\left(\sqrt[n]{10}\right)$

**27.** $\displaystyle\sum_{n=1}^{\infty}(-1)^{n}n^2(2/3)^n$

**28.** $\displaystyle\sum_{n=2}^{\infty}(-1)^{n+1}\frac{1}{n\ln n}$

**29.** $\displaystyle\sum_{n=1}^{\infty}(-1)^{n}\frac{\tan^{-1}n}{n^2+1}$

**30.** $\displaystyle\sum_{n=1}^{\infty}(-1)^{n}\frac{\ln n}{n-\ln n}$

**31.** $\displaystyle\sum_{n=1}^{\infty}(-1)^{n}\frac{n}{n+1}$

**32.** $\displaystyle\sum_{n=1}^{\infty}(-5)^{-n}$

**33.** $\displaystyle\sum_{n=1}^{\infty}\frac{(-100)^n}{n!}$

**34.** $\displaystyle\sum_{n=1}^{\infty}\frac{(-1)^{n-1}}{n^2+2n+1}$

**35.** $\displaystyle\sum_{n=1}^{\infty}\frac{\cos n\pi}{n\sqrt{n}}$

**36.** $\displaystyle\sum_{n=1}^{\infty}\frac{\cos n\pi}{n}$

**37.** $\displaystyle\sum_{n=1}^{\infty}\frac{(-1)^n(n+1)^n}{(2n)^n}$

**38.** $\displaystyle\sum_{n=1}^{\infty}\frac{(-1)^{n+1}(n!)^2}{(2n)!}$

**39.** $\displaystyle\sum_{n=1}^{\infty}(-1)^{n}\frac{(2n)!}{2^n n!n}$

**40.** $\displaystyle\sum_{n=1}^{\infty}(-1)^{n}\frac{(n!)^2 3^n}{(2n+1)!}$

**41.** $\displaystyle\sum_{n=1}^{\infty}(-1)^{n}\left(\sqrt{n+1}-\sqrt{n}\right)$

**42.** $\displaystyle\sum_{n=1}^{\infty}(-1)^{n}\left(\sqrt{n^2+n}-n\right)$

**43.** $\displaystyle\sum_{n=1}^{\infty}(-1)^{n}\left(\sqrt{n+\sqrt{n}}-\sqrt{n}\right)$

**44.** $\displaystyle\sum_{n=1}^{\infty}\frac{(-1)^n}{\sqrt{n}+\sqrt{n+1}}$

**45.** $\displaystyle\sum_{n=1}^{\infty}(-1)^{n}\operatorname{sech}n$

**46.** $\displaystyle\sum_{n=1}^{\infty}(-1)^{n}\operatorname{csch}n$

**47.** $\dfrac{1}{4}-\dfrac{1}{6}+\dfrac{1}{8}-\dfrac{1}{10}+\dfrac{1}{12}-\dfrac{1}{14}+\cdots$

**48.** $1+\dfrac{1}{4}-\dfrac{1}{9}-\dfrac{1}{16}+\dfrac{1}{25}+\dfrac{1}{36}-\dfrac{1}{49}-\dfrac{1}{64}+\cdots$

## Error Estimation

In Exercises 49–52, estimate the magnitude of the error involved in using the sum of the first four terms to approximate the sum of the entire series.

**49.** $\displaystyle\sum_{n=1}^{\infty}(-1)^{n+1}\frac{1}{n}$

**50.** $\displaystyle\sum_{n=1}^{\infty}(-1)^{n+1}\frac{1}{10^n}$

**51.** $\displaystyle\sum_{n=1}^{\infty}(-1)^{n+1}\frac{(0.01)^n}{n}$   As you will see in Section 10.7, the sum is ln (1.01).

**52.** $\dfrac{1}{1+t}=\displaystyle\sum_{n=0}^{\infty}(-1)^n t^n,\quad 0<t<1$

In Exercises 53–56, determine how many terms should be used to estimate the sum of the entire series with an error of less than 0.001.

**53.** $\displaystyle\sum_{n=1}^{\infty}(-1)^{n}\frac{1}{n^2+3}$

**54.** $\displaystyle\sum_{n=1}^{\infty}(-1)^{n+1}\frac{n}{n^2+1}$

**55.** $\displaystyle\sum_{n=1}^{\infty}(-1)^{n+1}\frac{1}{(n+3\sqrt{n})^3}$

**56.** $\displaystyle\sum_{n=1}^{\infty}(-1)^{n}\frac{1}{\ln(\ln(n+2))}$

T Approximate the sums in Exercises 57 and 58 with an error of magnitude less than $5\times10^{-6}$.

**57.** $\displaystyle\sum_{n=0}^{\infty}(-1)^{n}\frac{1}{(2n)!}$   As you will see in Section 10.9, the sum is cos 1, the cosine of 1 radian.

**58.** $\displaystyle\sum_{n=0}^{\infty}(-1)^{n}\frac{1}{n!}$   As you will see in Section 10.9, the sum is $e^{-1}$.

## Theory and Examples

**59. a.** The series

$$\frac{1}{3}-\frac{1}{2}+\frac{1}{9}-\frac{1}{4}+\frac{1}{27}-\frac{1}{8}+\cdots+\frac{1}{3^n}-\frac{1}{2^n}+\cdots$$

does not meet one of the conditions of Theorem 14. Which one?

**b.** Use Theorem 17 to find the sum of the series in part (a).

**T 60.** The limit $L$ of an alternating series that satisfies the conditions of Theorem 15 lies between the values of any two consecutive partial sums. This suggests using the average

$$\frac{s_n + s_{n+1}}{2} = s_n + \frac{1}{2}(-1)^{n+2}a_{n+1}$$

to estimate $L$. Compute

$$s_{20} + \frac{1}{2} \cdot \frac{1}{21}$$

as an approximation to the sum of the alternating harmonic series. The exact sum is $\ln 2 = 0.69314718\ldots.$

**61. The sign of the remainder of an alternating series that satisfies the conditions of Theorem 15** Prove the assertion in Theorem 16 that whenever an alternating series satisfying the conditions of Theorem 15 is approximated with one of its partial sums, then the remainder (sum of the unused terms) has the same sign as the first unused term. (*Hint:* Group the remainder's terms in consecutive pairs.)

**62.** Show that the sum of the first $2n$ terms of the series

$$1 - \frac{1}{2} + \frac{1}{2} - \frac{1}{3} + \frac{1}{3} - \frac{1}{4} + \frac{1}{4} - \frac{1}{5} + \frac{1}{5} - \frac{1}{6} + \cdots$$

is the same as the sum of the first $n$ terms of the series

$$\frac{1}{1 \cdot 2} + \frac{1}{2 \cdot 3} + \frac{1}{3 \cdot 4} + \frac{1}{4 \cdot 5} + \frac{1}{5 \cdot 6} + \cdots.$$

Do these series converge? What is the sum of the first $2n + 1$ terms of the first series? If the series converge, what is their sum?

**63.** Show that if $\sum_{n=1}^{\infty} a_n$ diverges, then $\sum_{n=1}^{\infty} |a_n|$ diverges.

**64.** Show that if $\sum_{n=1}^{\infty} a_n$ converges absolutely, then

$$\left| \sum_{n=1}^{\infty} a_n \right| \le \sum_{n=1}^{\infty} |a_n|.$$

**65.** Show that if $\sum_{n=1}^{\infty} a_n$ and $\sum_{n=1}^{\infty} b_n$ both converge absolutely, then so do the following.

**a.** $\sum_{n=1}^{\infty}(a_n + b_n)$      **b.** $\sum_{n=1}^{\infty}(a_n - b_n)$

**c.** $\sum_{n=1}^{\infty} ka_n$ ($k$ any number)

**66.** Show by example that $\sum_{n=1}^{\infty} a_n b_n$ may diverge even if $\sum_{n=1}^{\infty} a_n$ and $\sum_{n=1}^{\infty} b_n$ both converge.

**67.** If $\sum a_n$ converges absolutely, prove that $\sum a_n^2$ converges.

**68.** Does the series

$$\sum_{n=1}^{\infty}\left(\frac{1}{n} - \frac{1}{n^2}\right)$$

converge or diverge? Justify your answer.

**T 69.** In the alternating harmonic series, suppose the goal is to arrange the terms to get a new series that converges to $-1/2$. Start the new arrangement with the first negative term, which is $-1/2$. Whenever you have a sum that is less than or equal to $-1/2$, start introducing positive terms, taken in order, until the new total is greater than $-1/2$. Then add negative terms until the total is less than or equal to $-1/2$ again. Continue this process until your partial sums have been above the target at least three times and finish at or below it. If $s_n$ is the sum of the first $n$ terms of your new series, plot the points $(n, s_n)$ to illustrate how the sums are behaving.

**70. Outline of the proof of the Rearrangement Theorem (Theorem 17)**

**a.** Let $\epsilon$ be a positive real number, let $L = \sum_{n=1}^{\infty} a_n$, and let $s_k = \sum_{n=1}^{k} a_n$. Show that for some index $N_1$ and for some index $N_2 \ge N_1$,

$$\sum_{n=N_1}^{\infty} |a_n| < \frac{\epsilon}{2} \quad \text{and} \quad |s_{N_2} - L| < \frac{\epsilon}{2}.$$

Since all the terms $a_1, a_2, \ldots, a_{N_1}$ appear somewhere in the sequence $\{b_n\}$, there is an index $N_3 \ge N_2$ such that if $n \ge N_3$, then $\left(\sum_{k=1}^{n} b_k\right) - s_{N_2}$ is at most a sum of terms $a_m$ with $m \ge N_1$. Therefore, if $n \ge N_3$,

$$\left| \sum_{k=1}^{n} b_k - L \right| \le \left| \sum_{k=1}^{n} b_k - s_{N_2} \right| + |s_{N_2} - L|$$

$$\le \sum_{k=N_1}^{\infty} |a_k| + |s_{N_2} - L| < \epsilon.$$

**b.** The argument in part (a) shows that if $\sum_{n=1}^{\infty} a_n$ converges absolutely then $\sum_{n=1}^{\infty} b_n$ converges and $\sum_{n=1}^{\infty} b_n = \sum_{n=1}^{\infty} a_n$. Now show that because $\sum_{n=1}^{\infty} |a_n|$ converges, $\sum_{n=1}^{\infty} |b_n|$ converges to $\sum_{n=1}^{\infty} |a_n|$.

# $10.7$ Power Series

Now that we can test many infinite series of numbers for convergence, we can study sums that look like "infinite polynomials." We call these sums *power series* because they are defined as infinite series of powers of some variable, in our case $x$. Like polynomials, power series can be added, subtracted, multiplied, differentiated, and integrated to give new power series. With power series we can extend the methods of calculus we have developed to a vast array of functions, making the techniques of calculus applicable in a much wider setting.

### Power Series and Convergence

We begin with the formal definition, which specifies the notation and terminology used for power series.

---

**DEFINITIONS**  A **power series about $x = 0$** is a series of the form

$$\sum_{n=0}^{\infty} c_n x^n = c_0 + c_1 x + c_2 x^2 + \cdots + c_n x^n + \cdots.$$  (1)

A **power series about $x = a$** is a series of the form

$$\sum_{n=0}^{\infty} c_n (x - a)^n = c_0 + c_1(x - a) + c_2(x - a)^2 + \cdots + c_n(x - a)^n + \cdots$$  (2)

in which the **center** $a$ and the **coefficients** $c_0, c_1, c_2, \ldots, c_n, \ldots$ are constants.

---

Equation (1) is the special case obtained by taking $a = 0$ in Equation (2). We will see that a power series defines a function $f(x)$ on a certain interval where it converges. Moreover, this function will be shown to be continuous and differentiable over the interior of that interval.

**EXAMPLE 1**  Taking all the coefficients to be 1 in Equation (1) gives the geometric power series

$$\sum_{n=0}^{\infty} x^n = 1 + x + x^2 + \cdots + x^n + \cdots.$$

This is the geometric series with first term 1 and ratio $x$. It converges to $1/(1 - x)$ for $|x| < 1$. We express this fact by writing

$$\frac{1}{1 - x} = 1 + x + x^2 + \cdots + x^n + \cdots, \qquad -1 < x < 1.$$  (3)

■

**Power Series for a Reciprocal**

$$\frac{1}{1 - x} = \sum_{n=0}^{\infty} x^n, \quad |x| < 1$$

Up to now, we have used Equation (3) as a formula for the sum of the series on the right. We now change the focus: We think of the partial sums of the series on the right as polynomials $P_n(x)$ that approximate the function on the left. For values of $x$ near zero, we need take only a few terms of the series to get a good approximation. As we move toward $x = 1$, or $-1$, we must take more terms. Figure 10.14 shows the graphs of

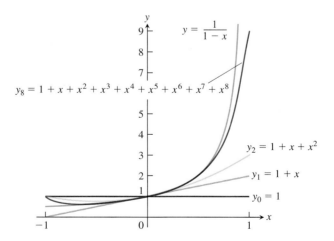

**FIGURE 10.14**  The graphs of $f(x) = 1/(1 - x)$ in Example 1 and four of its polynomial approximations.

$f(x) = 1/(1 - x)$ and the approximating polynomials $y_n = P_n(x)$ for $n = 0, 1, 2,$ and 8. The function $f(x) = 1/(1 - x)$ is not continuous on intervals containing $x = 1$, where it has a vertical asymptote. The approximations do not apply when $x \geq 1$.

**EXAMPLE 2**    The power series

$$1 - \frac{1}{2}(x - 2) + \frac{1}{4}(x - 2)^2 + \cdots + \left(-\frac{1}{2}\right)^n (x - 2)^n + \cdots \qquad (4)$$

matches Equation (2) with $a = 2, c_0 = 1, c_1 = -1/2, c_2 = 1/4, \ldots, c_n = (-1/2)^n$. This is a geometric series with first term 1 and ratio $r = -\dfrac{x - 2}{2}$. The series converges for $\left|\dfrac{x - 2}{2}\right| < 1$ or $0 < x < 4$. The sum is

$$\frac{1}{1 - r} = \frac{1}{1 + \dfrac{x - 2}{2}} = \frac{2}{x},$$

so

$$\frac{2}{x} = 1 - \frac{(x - 2)}{2} + \frac{(x - 2)^2}{4} - \cdots + \left(-\frac{1}{2}\right)^n (x - 2)^n + \cdots, \qquad 0 < x < 4.$$

Series (4) generates useful polynomial approximations of $f(x) = 2/x$ for values of $x$ near 2:

$$P_0(x) = 1$$

$$P_1(x) = 1 - \frac{1}{2}(x - 2) = 2 - \frac{x}{2}$$

$$P_2(x) = 1 - \frac{1}{2}(x - 2) + \frac{1}{4}(x - 2)^2 = 3 - \frac{3x}{2} + \frac{x^2}{4},$$

and so on (Figure 10.15).

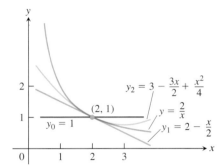

**FIGURE 10.15**    The graphs of $f(x) = 2/x$ and its first three polynomial approximations (Example 2).

The following example illustrates how we test a power series for convergence by using the Ratio Test to see where it converges and diverges.

**EXAMPLE 3**    For what values of $x$ do the following power series converge?

(a) $\displaystyle\sum_{n=1}^{\infty} (-1)^{n-1} \frac{x^n}{n} = x - \frac{x^2}{2} + \frac{x^3}{3} - \cdots$

(b) $\displaystyle\sum_{n=1}^{\infty} (-1)^{n-1} \frac{x^{2n-1}}{2n - 1} = x - \frac{x^3}{3} + \frac{x^5}{5} - \cdots$

(c) $\displaystyle\sum_{n=0}^{\infty} \frac{x^n}{n!} = 1 + x + \frac{x^2}{2!} + \frac{x^3}{3!} + \cdots$

(d) $\displaystyle\sum_{n=0}^{\infty} n! x^n = 1 + x + 2! x^2 + 3! x^3 + \cdots$

**Solution**    Apply the Ratio Test to the series $\sum |u_n|$, where $u_n$ is the $n$th term of the power series in question.

(a) $\left|\dfrac{u_{n+1}}{u_n}\right| = \left|\dfrac{x^{n+1}}{n + 1} \cdot \dfrac{n}{x}\right| = \dfrac{n}{n + 1}|x| \to |x|.$

The series converges absolutely for $|x| < 1$. It diverges if $|x| > 1$ because the $n$th term does not converge to zero. At $x = 1$, we get the alternating harmonic series $1 - 1/2 + 1/3 - 1/4 + \cdots$, which converges. At $x = -1$, we get $-1 - 1/2 -$

$1/3 - 1/4 - \cdots$, the negative of the harmonic series; it diverges. Series (a) converges for $-1 < x \leq 1$ and diverges elsewhere.

**(b)** $\left| \dfrac{u_{n+1}}{u_n} \right| = \left| \dfrac{x^{2n+1}}{2n+1} \cdot \dfrac{2n-1}{x^{2n-1}} \right| = \dfrac{2n-1}{2n+1} x^2 \rightarrow x^2.$  $\quad 2(n+1) - 1 = 2n+1$

The series converges absolutely for $x^2 < 1$. It diverges for $x^2 > 1$ because the $n$th term does not converge to zero. At $x = 1$ the series becomes $1 - 1/3 + 1/5 - 1/7 + \cdots$, which converges by the Alternating Series Theorem. It also converges at $x = -1$ because it is again an alternating series that satisfies the conditions for convergence. The value at $x = -1$ is the negative of the value at $x = 1$. Series (b) converges for $-1 \leq x \leq 1$ and diverges elsewhere.

**(c)** $\left| \dfrac{u_{n+1}}{u_n} \right| = \left| \dfrac{x^{n+1}}{(n+1)!} \cdot \dfrac{n!}{x^n} \right| = \dfrac{|x|}{n+1} \rightarrow 0$ for every $x$.  $\quad \dfrac{n!}{(n+1)!} = \dfrac{1 \cdot 2 \cdot 3 \cdots n}{1 \cdot 2 \cdot 3 \cdots n \cdot (n+1)}$

The series converges absolutely for all $x$.

**(d)** $\left| \dfrac{u_{n+1}}{u_n} \right| = \left| \dfrac{(n+1)! x^{n+1}}{n! x^n} \right| = (n+1)|x| \rightarrow \infty$ unless $x = 0$.

The series diverges for all values of $x$ except $x = 0$.

The previous example illustrated how a power series might converge. The next result shows that if a power series converges at more than one value, then it converges over an entire interval of values. The interval might be finite or infinite and contain one, both, or none of its endpoints. We will see that each endpoint of a finite interval must be tested independently for convergence or divergence.

> **THEOREM 18—The Convergence Theorem for Power Series**  If the power series
>
> $$\sum_{n=0}^{\infty} a_n x^n = a_0 + a_1 x + a_2 x^2 + \cdots \text{ converges at } x = c \neq 0, \text{ then it converges}$$
>
> absolutely for all $x$ with $|x| < |c|$. If the series diverges at $x = d$, then it diverges for all $x$ with $|x| > |d|$.

**Proof**  The proof uses the Comparison Test, with the given series compared to a converging geometric series.

Suppose the series $\sum_{n=0}^{\infty} a_n c^n$ converges. Then $\lim_{n \to \infty} a_n c^n = 0$ by the $n$th-Term Test. Hence, there is an integer $N$ such that $|a_n c^n| < 1$ for all $n > N$, so that

$$|a_n| < \frac{1}{|c|^n} \qquad \text{for } n > N. \tag{5}$$

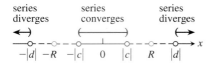

**FIGURE 10.16** Convergence of $\sum a_n x^n$ at $x = c$ implies absolute convergence on the interval $-|c| < x < |c|$; divergence at $x = d$ implies divergence for $|x| > |d|$. The corollary to Theorem 18 asserts the existence of a radius of convergence $R \geq 0$. For $|x| < R$ the series converges absolutely and for $|x| > R$ it diverges.

Now take any $x$ such that $|x| < |c|$, so that $|x|/|c| < 1$. Multiplying both sides of Equation (5) by $|x|^n$ gives

$$|a_n|\,|x|^n < \frac{|x|^n}{|c|^n} \qquad \text{for } n > N.$$

Since $|x/c| < 1$, it follows that the geometric series $\sum_{n=0}^{\infty} |x/c|^n$ converges. By the Comparison Test (Theorem 10), the series $\sum_{n=0}^{\infty} |a_n|\,|x^n|$ converges, so the original power series $\sum_{n=0}^{\infty} a_n x^n$ converges absolutely for $-|c| < x < |c|$ as claimed by the theorem. (See Figure 10.16.)

Now suppose that the series $\sum_{n=0}^{\infty} a_n x^n$ diverges at $x = d$. If $x$ is a number with $|x| > |d|$ and the series converges at $x$, then the first half of the theorem shows that the series also converges at $d$, contrary to our assumption. So the series diverges for all $x$ with $|x| > |d|$. ∎

To simplify the notation, Theorem 18 deals with the convergence of series of the form $\sum a_n x^n$. For series of the form $\sum a_n(x - a)^n$ we can replace $x - a$ by $x'$ and apply the results to the series $\sum a_n(x')^n$.

### The Radius of Convergence of a Power Series

The theorem we have just proved and the examples we have studied lead to the conclusion that a power series $\sum c_n(x - a)^n$ behaves in one of three possible ways. It might converge only at $x = a$, or converge everywhere, or converge on some interval of radius $R$ centered at $x = a$. We prove this as a Corollary to Theorem 18.

---

**COROLLARY TO THEOREM 18** The convergence of the series $\sum c_n(x - a)^n$ is described by one of the following three cases:

1. There is a positive number $R$ such that the series diverges for $x$ with $|x - a| > R$ but converges absolutely for $x$ with $|x - a| < R$. The series may or may not converge at either of the endpoints $x = a - R$ and $x = a + R$.

2. The series converges absolutely for every $x$ ($R = \infty$).

3. The series converges at $x = a$ and diverges elsewhere ($R = 0$).

---

**Proof** We first consider the case where $a = 0$, so that we have a power series $\sum_{n=0}^{\infty} c_n x^n$ centered at 0. If the series converges everywhere we are in Case 2. If it converges only at $x = 0$ then we are in Case 3. Otherwise there is a nonzero number $d$ such that $\sum_{n=0}^{\infty} c_n d^n$ diverges. Let $S$ be the set of values of $x$ for which $\sum_{n=0}^{\infty} c_n x^n$ converges. The set $S$ does not include any $x$ with $|x| > |d|$, since Theorem 18 implies the series diverges at all such values. So the set $S$ is bounded. By the Completeness Property of the Real Numbers (Appendix 6) $S$ has a least upper bound $R$. (This is the smallest number with the property that all elements of $S$ are less than or equal to $R$.) Since we are not in Case 3, the series converges at some number $b \neq 0$ and, by Theorem 18, also on the open interval $(-|b|, |b|)$. Therefore $R > 0$.

If $|x| < R$ then there is a number $c$ in $S$ with $|x| < c < R$, since otherwise $R$ would not be the least upper bound for $S$. The series converges at $c$ since $c \in S$, so by Theorem 18 the series converges absolutely at $x$.

Now suppose $|x| > R$. If the series converges at $x$, then Theorem 18 implies it converges absolutely on the open interval $(-|x|, |x|)$, so that $S$ contains this interval. Since $R$

is an upper bound for $S$, it follows that $|x| \le R$, which is a contradiction. So if $|x| > R$ then the series diverges. This proves the theorem for power series centered at $a = 0$.

For a power series centered at an arbitrary point $x = a$, set $x' = x - a$ and repeat the argument above, replacing $x$ with $x'$. Since $x' = 0$ when $x = a$, convergence of the series $\sum_{n=0}^{\infty} |c_n(x')^n|$ on a radius $R$ open interval centered at $x' = 0$ corresponds to convergence of the series $\sum_{n=0}^{\infty} |c_n(x - a)^n|$ on a radius $R$ open interval centered at $x = a$. ∎

$R$ is called the **radius of convergence** of the power series, and the interval of radius $R$ centered at $x = a$ is called the **interval of convergence**. The interval of convergence may be open, closed, or half-open, depending on the particular series. At points $x$ with $|x - a| < R$, the series converges absolutely. If the series converges for all values of $x$, we say its radius of convergence is infinite. If it converges only at $x = a$, we say its radius of convergence is zero.

---

How to Test a Power Series for Convergence

1. *Use the Ratio Test (or Root Test) to find the interval where the series converges absolutely.* Ordinarily, this is an open interval

$$|x - a| < R \qquad \text{or} \qquad a - R < x < a + R.$$

2. *If the interval of absolute convergence is finite, test for convergence or divergence at each endpoint, as in Examples 3a and b.* Use a Comparison Test, the Integral Test, or the Alternating Series Test.

3. *If the interval of absolute convergence is $a - R < x < a + R$, the series diverges for $|x - a| > R$* (it does not even converge conditionally) because the $n$th term does not approach zero for those values of $x$.

---

## Operations on Power Series

On the intersection of their intervals of convergence, two power series can be added and subtracted term by term just like series of constants (Theorem 8). They can be multiplied just as we multiply polynomials, but we often limit the computation of the product to the first few terms, which are the most important. The following result gives a formula for the coefficients in the product, but we omit the proof. (Power series can also be divided in a way similar to division of polynomials, but we do not give a formula for the general coefficient here.)

---

**THEOREM 19—The Series Multiplication Theorem for Power Series** If $A(x) = \sum_{n=0}^{\infty} a_n x^n$ and $B(x) = \sum_{n=0}^{\infty} b_n x^n$ converge absolutely for $|x| < R$, and

$$c_n = a_0 b_n + a_1 b_{n-1} + a_2 b_{n-2} + \cdots + a_{n-1} b_1 + a_n b_0 = \sum_{k=0}^{n} a_k b_{n-k},$$

then $\sum_{n=0}^{\infty} c_n x^n$ converges absolutely to $A(x)B(x)$ for $|x| < R$:

$$\left( \sum_{n=0}^{\infty} a_n x^n \right) \cdot \left( \sum_{n=0}^{\infty} b_n x^n \right) = \sum_{n=0}^{\infty} c_n x^n.$$

---

Finding the general coefficient $c_n$ in the product of two power series can be very tedious and the term may be unwieldy. The following computation provides an illustration of a product where we find the first few terms by multiplying the terms of the second series by each term of the first series:

$$\left(\sum_{n=0}^{\infty} x^n\right) \cdot \left(\sum_{n=0}^{\infty} (-1)^n \frac{x^{n+1}}{n+1}\right)$$

$$= (1 + x + x^2 + \cdots)\left(x - \frac{x^2}{2} + \frac{x^3}{3} - \cdots\right) \qquad \text{Multiply second series} \ldots$$

$$= \underbrace{\left(x - \frac{x^2}{2} + \frac{x^3}{3} - \cdots\right)}_{\text{by } 1} + \underbrace{\left(x^2 - \frac{x^3}{2} + \frac{x^4}{3} - \cdots\right)}_{\text{by } x} + \underbrace{\left(x^3 - \frac{x^4}{2} + \frac{x^5}{3} - \cdots\right)}_{\text{by } x^2} + \cdots$$

$$= x + \frac{x^2}{2} + \frac{5x^3}{6} - \frac{x^4}{6} \cdots. \qquad \text{and gather the first four powers.}$$

We can also substitute a function $f(x)$ for $x$ in a convergent power series.

---

**THEOREM 20** If $\sum_{n=0}^{\infty} a_n x^n$ converges absolutely for $|x| < R$, then $\sum_{n=0}^{\infty} a_n(f(x))^n$ converges absolutely for any continuous function $f$ on $|f(x)| < R$.

---

Since $1/(1-x) = \sum_{n=0}^{\infty} x^n$ converges absolutely for $|x| < 1$, it follows from Theorem 20 that $1/(1-4x^2) = \sum_{n=0}^{\infty} (4x^2)^n$ converges absolutely for $|4x^2| < 1$ or $|x| < 1/2$.

A theorem from advanced calculus says that a power series can be differentiated term by term at each interior point of its interval of convergence.

---

**THEOREM 21—The Term-by-Term Differentiation Theorem** If $\sum c_n(x-a)^n$ has radius of convergence $R > 0$, it defines a function

$$f(x) = \sum_{n=0}^{\infty} c_n(x-a)^n \qquad \text{on the interval} \qquad a - R < x < a + R.$$

This function $f$ has derivatives of all orders inside the interval, and we obtain the derivatives by differentiating the original series term by term:

$$f'(x) = \sum_{n=1}^{\infty} nc_n(x-a)^{n-1},$$

$$f''(x) = \sum_{n=2}^{\infty} n(n-1)c_n(x-a)^{n-2},$$

and so on. Each of these derived series converges at every point of the interval $a - R < x < a + R$.

---

**EXAMPLE 4** Find series for $f'(x)$ and $f''(x)$ if

$$f(x) = \frac{1}{1-x} = 1 + x + x^2 + x^3 + x^4 + \cdots + x^n + \cdots$$

$$= \sum_{n=0}^{\infty} x^n, \qquad -1 < x < 1.$$

**Solution** We differentiate the power series on the right term by term:

$$f'(x) = \frac{1}{(1-x)^2} = 1 + 2x + 3x^2 + 4x^3 + \cdots + nx^{n-1} + \cdots$$

$$= \sum_{n=1}^{\infty} nx^{n-1}, \qquad -1 < x < 1;$$

$$f''(x) = \frac{2}{(1-x)^3} = 2 + 6x + 12x^2 + \cdots + n(n-1)x^{n-2} + \cdots$$

$$= \sum_{n=2}^{\infty} n(n-1)x^{n-2}, \qquad -1 < x < 1. \qquad \blacksquare$$

Caution   Term-by-term differentiation might not work for other kinds of series. For example, the trigonometric series

$$\sum_{n=1}^{\infty} \frac{\sin{(n!x)}}{n^2}$$

converges for all $x$. But if we differentiate term by term we get the series

$$\sum_{n=1}^{\infty} \frac{n!\cos{(n!x)}}{n^2},$$

which diverges for all $x$. This is not a power series since it is not a sum of positive integer powers of $x$.

It is also true that a power series can be integrated term by term throughout its interval of convergence. This result is proved in a more advanced course.

---

**THEOREM 22—The Term-by-Term Integration Theorem**   Suppose that

$$f(x) = \sum_{n=0}^{\infty} c_n(x-a)^n$$

converges for $a - R < x < a + R\,(R > 0)$. Then

$$\sum_{n=0}^{\infty} c_n \frac{(x-a)^{n+1}}{n+1}$$

converges for $a - R < x < a + R$ and

$$\int f(x)\,dx = \sum_{n=0}^{\infty} c_n \frac{(x-a)^{n+1}}{n+1} + C$$

for $a - R < x < a + R$.

---

**EXAMPLE 5**   Identify the function

$$f(x) = \sum_{n=0}^{\infty} \frac{(-1)^n x^{2n+1}}{2n+1} = x - \frac{x^3}{3} + \frac{x^5}{5} - \cdots, \qquad -1 \le x \le 1.$$

**Solution**   We differentiate the original series term by term and get

$$f'(x) = 1 - x^2 + x^4 - x^6 + \cdots, \qquad -1 < x < 1. \qquad \text{Theorem 21}$$

This is a geometric series with first term 1 and ratio $-x^2$, so

$$f'(x) = \frac{1}{1-(-x^2)} = \frac{1}{1+x^2}.$$

We can now integrate $f'(x) = 1/(1+x^2)$ to get

$$\int f'(x)\,dx = \int \frac{dx}{1+x^2} = \tan^{-1}x + C.$$

**The Number $\pi$ as a Series**

$$\frac{\pi}{4} = \tan^{-1}1 = \sum_{n=0}^{\infty} \frac{(-1)^n}{2n+1}$$

The series for $f(x)$ is zero when $x = 0$, so $C = 0$. Hence

$$f(x) = x - \frac{x^3}{3} + \frac{x^5}{5} - \frac{x^7}{7} + \cdots = \tan^{-1}x, \qquad -1 < x < 1. \qquad (6)$$

It can be shown that the series also converges to $\tan^{-1} x$ at the endpoints $x = \pm 1$, but we omit the proof. ∎

Notice that the original series in Example 5 converges at both endpoints of the original interval of convergence, but Theorem 22 can guarantee the convergence of the differentiated series only inside the interval.

**EXAMPLE 6** The series

$$\frac{1}{1+t} = 1 - t + t^2 - t^3 + \cdots$$

converges on the open interval $-1 < t < 1$. Therefore,

$$\ln(1+x) = \int_0^x \frac{1}{1+t}\,dt = t - \frac{t^2}{2} + \frac{t^3}{3} - \frac{t^4}{4} + \cdots \Big]_0^x \qquad \text{Theorem 22}$$

$$= x - \frac{x^2}{2} + \frac{x^3}{3} - \frac{x^4}{4} + \cdots$$

or

$$\ln(1+x) = \sum_{n=1}^{\infty} \frac{(-1)^{n-1} x^n}{n}, \qquad -1 < x < 1.$$

**Alternating Harmonic Series Sum**

$$\ln 2 = \sum_{n=1}^{\infty} \frac{(-1)^{n-1}}{n}$$

It can also be shown that the series converges at $x = 1$ to the number $\ln 2$, but that was not guaranteed by the theorem. ∎

---

## Exercises 10.7

### Intervals of Convergence

In Exercises 1–36, **(a)** find the series' radius and interval of convergence. For what values of $x$ does the series converge **(b)** absolutely, **(c)** conditionally?

**1.** $\displaystyle\sum_{n=0}^{\infty} x^n$

**2.** $\displaystyle\sum_{n=0}^{\infty} (x+5)^n$

**3.** $\displaystyle\sum_{n=0}^{\infty} (-1)^n (4x+1)^n$

**4.** $\displaystyle\sum_{n=1}^{\infty} \frac{(3x-2)^n}{n}$

**5.** $\displaystyle\sum_{n=0}^{\infty} \frac{(x-2)^n}{10^n}$

**6.** $\displaystyle\sum_{n=0}^{\infty} (2x)^n$

**7.** $\displaystyle\sum_{n=0}^{\infty} \frac{nx^n}{n+2}$

**8.** $\displaystyle\sum_{n=1}^{\infty} \frac{(-1)^n (x+2)^n}{n}$

**9.** $\displaystyle\sum_{n=1}^{\infty} \frac{x^n}{n\sqrt{n}\,3^n}$

**10.** $\displaystyle\sum_{n=1}^{\infty} \frac{(x-1)^n}{\sqrt{n}}$

**11.** $\displaystyle\sum_{n=0}^{\infty} \frac{(-1)^n x^n}{n!}$

**12.** $\displaystyle\sum_{n=0}^{\infty} \frac{3^n x^n}{n!}$

**13.** $\displaystyle\sum_{n=1}^{\infty} \frac{4^n x^{2n}}{n}$

**14.** $\displaystyle\sum_{n=1}^{\infty} \frac{(x-1)^n}{n^3 3^n}$

**15.** $\displaystyle\sum_{n=0}^{\infty} \frac{x^n}{\sqrt{n^2+3}}$

**16.** $\displaystyle\sum_{n=0}^{\infty} \frac{(-1)^n x^{n+1}}{\sqrt{n+3}}$

**17.** $\displaystyle\sum_{n=0}^{\infty} \frac{n(x+3)^n}{5^n}$

**18.** $\displaystyle\sum_{n=0}^{\infty} \frac{nx^n}{4^n(n^2+1)}$

**19.** $\displaystyle\sum_{n=0}^{\infty} \frac{\sqrt{n}\,x^n}{3^n}$

**20.** $\displaystyle\sum_{n=1}^{\infty} \sqrt[n]{n}\,(2x+5)^n$

**21.** $\displaystyle\sum_{n=1}^{\infty} (2+(-1)^n)\cdot(x+1)^{n-1}$

**22.** $\displaystyle\sum_{n=1}^{\infty} \frac{(-1)^n 3^{2n}(x-2)^n}{3n}$

**23.** $\displaystyle\sum_{n=1}^{\infty} \left(1+\frac{1}{n}\right)^n x^n$

**24.** $\displaystyle\sum_{n=1}^{\infty} (\ln n) x^n$

**25.** $\displaystyle\sum_{n=1}^{\infty} n^n x^n$

**26.** $\displaystyle\sum_{n=0}^{\infty} n!(x-4)^n$

**27.** $\displaystyle\sum_{n=1}^{\infty} \frac{(-1)^{n+1}(x+2)^n}{n2^n}$

**28.** $\displaystyle\sum_{n=0}^{\infty} (-2)^n(n+1)(x-1)^n$

**29.** $\displaystyle\sum_{n=2}^{\infty} \frac{x^n}{n(\ln n)^2}$  Get the information you need about $\sum 1/(n(\ln n)^2)$ from Section 10.3, Exercise 55.

**30.** $\displaystyle\sum_{n=2}^{\infty} \frac{x^n}{n\ln n}$  Get the information you need about $\sum 1/(n\ln n)$ from Section 10.3, Exercise 54.

**31.** $\displaystyle\sum_{n=1}^{\infty} \frac{(4x-5)^{2n+1}}{n^{3/2}}$

**32.** $\displaystyle\sum_{n=1}^{\infty} \frac{(3x+1)^{n+1}}{2n+2}$

**33.** $\displaystyle\sum_{n=1}^{\infty} \frac{1}{2\cdot4\cdot6\cdots(2n)}\, x^n$

**34.** $\displaystyle\sum_{n=1}^{\infty} \frac{3\cdot5\cdot7\cdots(2n+1)}{n^2\cdot2^n}\, x^{n+1}$

**35.** $\displaystyle\sum_{n=1}^{\infty} \frac{1+2+3+\cdots+n}{1^2+2^2+3^2+\cdots+n^2}\, x^n$

**36.** $\displaystyle\sum_{n=1}^{\infty} \left(\sqrt{n+1}-\sqrt{n}\right)(x-3)^n$

In Exercises 37–40, find the series' radius of convergence.

**37.** $\displaystyle\sum_{n=1}^{\infty} \frac{n!}{3\cdot6\cdot9\cdots3n}\, x^n$

**38.** $\displaystyle\sum_{n=1}^{\infty} \left(\frac{2\cdot4\cdot6\cdots(2n)}{2\cdot5\cdot8\cdots(3n-1)}\right)^2 x^n$

**39.** $\displaystyle\sum_{n=1}^{\infty} \frac{(n!)^2}{2^n(2n)!}\, x^n$

**40.** $\displaystyle\sum_{n=1}^{\infty} \left(\frac{n}{n+1}\right)^{n^2} x^n$

(*Hint:* Apply the Root Test.)

In Exercises 41–48, use Theorem 20 to find the series' interval of convergence and, within this interval, the sum of the series as a function of $x$.

**41.** $\displaystyle\sum_{n=0}^{\infty} 3^n x^n$

**42.** $\displaystyle\sum_{n=0}^{\infty} (e^x - 4)^n$

**43.** $\displaystyle\sum_{n=0}^{\infty} \frac{(x-1)^{2n}}{4^n}$

**44.** $\displaystyle\sum_{n=0}^{\infty} \frac{(x+1)^{2n}}{9^n}$

**45.** $\displaystyle\sum_{n=0}^{\infty} \left(\frac{\sqrt{x}}{2} - 1\right)^n$

**46.** $\displaystyle\sum_{n=0}^{\infty} (\ln x)^n$

**47.** $\displaystyle\sum_{n=0}^{\infty} \left(\frac{x^2+1}{3}\right)^n$

**48.** $\displaystyle\sum_{n=0}^{\infty} \left(\frac{x^2-1}{2}\right)^n$

### Using the Geometric Series

**49.** In Example 2 we represented the function $f(x) = 2/x$ as a power series about $x = 2$. Use a geometric series to represent $f(x)$ as a power series about $x = 1$, and find its interval of convergence.

**50.** Use a geometric series to represent each of the given functions as a power series about $x = 0$, and find their intervals of convergence.

    **a.** $f(x) = \dfrac{5}{3 - x}$     **b.** $g(x) = \dfrac{3}{x - 2}$

**51.** Represent the function $g(x)$ in Exercise 50 as a power series about $x = 5$, and find the interval of convergence.

**52. a.** Find the interval of convergence of the power series

$$\sum_{n=0}^{\infty} \frac{8}{4^{n+2}} x^n.$$

    **b.** Represent the power series in part (a) as a power series about $x = 3$ and identify the interval of convergence of the new series. (Later in the chapter you will understand why the new interval of convergence does not necessarily include all of the numbers in the original interval of convergence.)

### Theory and Examples

**53.** For what values of $x$ does the series

$$1 - \frac{1}{2}(x-3) + \frac{1}{4}(x-3)^2 + \cdots + \left(-\frac{1}{2}\right)^n(x-3)^n + \cdots$$

converge? What is its sum? What series do you get if you differentiate the given series term by term? For what values of $x$ does the new series converge? What is its sum?

**54.** If you integrate the series in Exercise 53 term by term, what new series do you get? For what values of $x$ does the new series converge, and what is another name for its sum?

**55.** The series

$$\sin x = x - \frac{x^3}{3!} + \frac{x^5}{5!} - \frac{x^7}{7!} + \frac{x^9}{9!} - \frac{x^{11}}{11!} + \cdots$$

converges to $\sin x$ for all $x$.

    **a.** Find the first six terms of a series for $\cos x$. For what values of $x$ should the series converge?

    **b.** By replacing $x$ by $2x$ in the series for $\sin x$, find a series that converges to $\sin 2x$ for all $x$.

    **c.** Using the result in part (a) and series multiplication, calculate the first six terms of a series for $2 \sin x \cos x$. Compare your answer with the answer in part (b).

**56.** The series

$$e^x = 1 + x + \frac{x^2}{2!} + \frac{x^3}{3!} + \frac{x^4}{4!} + \frac{x^5}{5!} + \cdots$$

converges to $e^x$ for all $x$.

    **a.** Find a series for $(d/dx)e^x$. Do you get the series for $e^x$? Explain your answer.

    **b.** Find a series for $\int e^x\,dx$. Do you get the series for $e^x$? Explain your answer.

    **c.** Replace $x$ by $-x$ in the series for $e^x$ to find a series that converges to $e^{-x}$ for all $x$. Then multiply the series for $e^x$ and $e^{-x}$ to find the first six terms of a series for $e^{-x}\cdot e^x$.

**57.** The series

$$\tan x = x + \frac{x^3}{3} + \frac{2x^5}{15} + \frac{17x^7}{315} + \frac{62x^9}{2835} + \cdots$$

converges to $\tan x$ for $-\pi/2 < x < \pi/2$.

    **a.** Find the first five terms of the series for $\ln|\sec x|$. For what values of $x$ should the series converge?

    **b.** Find the first five terms of the series for $\sec^2 x$. For what values of $x$ should this series converge?

    **c.** Check your result in part (b) by squaring the series given for $\sec x$ in Exercise 58.

**58.** The series

$$\sec x = 1 + \frac{x^2}{2} + \frac{5}{24}x^4 + \frac{61}{720}x^6 + \frac{277}{8064}x^8 + \cdots$$

converges to $\sec x$ for $-\pi/2 < x < \pi/2$.

    **a.** Find the first five terms of a power series for the function $\ln|\sec x + \tan x|$. For what values of $x$ should the series converge?

**b.** Find the first four terms of a series for $\sec x \tan x$. For what values of $x$ should the series converge?

**c.** Check your result in part (b) by multiplying the series for $\sec x$ by the series given for $\tan x$ in Exercise 57.

**59. Uniqueness of convergent power series**

**a.** Show that if two power series $\sum_{n=0}^{\infty} a_n x^n$ and $\sum_{n=0}^{\infty} b_n x^n$ are convergent and equal for all values of $x$ in an open interval $(-c, c)$, then $a_n = b_n$ for every $n$. (*Hint:* Let $f(x) = \sum_{n=0}^{\infty} a_n x^n = \sum_{n=0}^{\infty} b_n x^n$. Differentiate term by term to show that $a_n$ and $b_n$ both equal $f^{(n)}(0)/(n!)$.)

**b.** Show that if $\sum_{n=0}^{\infty} a_n x^n = 0$ for all $x$ in an open interval $(-c, c)$, then $a_n = 0$ for every $n$.

**60. The sum of the series $\sum_{n=0}^{\infty} (n^2/2^n)$** To find the sum of this series, express $1/(1 - x)$ as a geometric series, differentiate both sides of the resulting equation with respect to $x$, multiply both sides of the result by $x$, differentiate again, multiply by $x$ again, and set $x$ equal to $1/2$. What do you get?

# 10.8 Taylor and Maclaurin Series

We have seen how geometric series can be used to generate a power series for a few functions having a special form, like $f(x) = 1/(1 - x)$ or $g(x) = 3/(x - 2)$. Now we expand our capability to represent a function with a power series. This section shows how functions that are infinitely differentiable generate power series called *Taylor series*. In many cases, these series provide useful polynomial approximations of the generating functions. Because they are used routinely by mathematicians and scientists, Taylor series are considered one of the most important themes of infinite series.

## Series Representations

We know from Theorem 21 that within its interval of convergence $I$ the sum of a power series is a continuous function with derivatives of all orders. But what about the other way around? If a function $f(x)$ has derivatives of all orders on an interval, can it be expressed as a power series on at least part of that interval? And if it can, what are its coefficients?

We can answer the last question readily if we assume that $f(x)$ is the sum of a power series about $x = a$,

$$f(x) = \sum_{n=0}^{\infty} a_n(x - a)^n$$

$$= a_0 + a_1(x - a) + a_2(x - a)^2 + \cdots + a_n(x - a)^n + \cdots$$

with a positive radius of convergence. By repeated term-by-term differentiation within the interval of convergence $I$, we obtain

$$f'(x) = a_1 + 2a_2(x - a) + 3a_3(x - a)^2 + \cdots + na_n(x - a)^{n-1} + \cdots,$$

$$f''(x) = 1 \cdot 2a_2 + 2 \cdot 3a_3(x - a) + 3 \cdot 4a_4(x - a)^2 + \cdots,$$

$$f'''(x) = 1 \cdot 2 \cdot 3a_3 + 2 \cdot 3 \cdot 4a_4(x - a) + 3 \cdot 4 \cdot 5a_5(x - a)^2 + \cdots,$$

with the $n$th derivative, for all $n$, being

$$f^{(n)}(x) = n!a_n + \text{a sum of terms with } (x - a) \text{ as a factor.}$$

Since these equations all hold at $x = a$, we have

$$f'(a) = a_1, \qquad f''(a) = 1 \cdot 2a_2, \qquad f'''(a) = 1 \cdot 2 \cdot 3a_3,$$

and, in general,

$$f^{(n)}(a) = n!a_n.$$

These formulas reveal a pattern in the coefficients of any power series $\sum_{n=0}^{\infty} a_n(x - a)^n$ that converges to the values of $f$ on $I$ ("represents $f$ on $I$"). If there *is* such a series (still an open question), then there is only one such series, and its $n$th coefficient is

$$a_n = \frac{f^{(n)}(a)}{n!}.$$

If $f$ has a series representation, then the series must be

$$f(x) = f(a) + f'(a)(x - a) + \frac{f''(a)}{2!}(x - a)^2$$

$$+ \cdots + \frac{f^{(n)}(a)}{n!}(x - a)^n + \cdots. \tag{1}$$

But if we start with an arbitrary function $f$ that is infinitely differentiable on an interval containing $x = a$ and use it to generate the series in Equation (1), will the series then converge to $f(x)$ at each $x$ in the interval of convergence? The answer is maybe—for some functions it will but for other functions it will not (as we will see in Example 4).

### Taylor and Maclaurin Series

The series on the right-hand side of Equation (1) is the most important and useful series we will study in this chapter.

> **DEFINITIONS**  Let $f$ be a function with derivatives of all orders throughout some interval containing $a$ as an interior point. Then the **Taylor series generated by $f$ at $x = a$** is
>
> $$\sum_{k=0}^{\infty} \frac{f^{(k)}(a)}{k!}(x - a)^k = f(a) + f'(a)(x - a) + \frac{f''(a)}{2!}(x - a)^2$$
>
> $$+ \cdots + \frac{f^{(n)}(a)}{n!}(x - a)^n + \cdots.$$
>
> The **Maclaurin series of $f$** is the Taylor series generated by $f$ at $x = 0$, or
>
> $$\sum_{k=0}^{\infty} \frac{f^{(k)}(0)}{k!}x^k = f(0) + f'(0)x + \frac{f''(0)}{2!}x^2 + \cdots + \frac{f^{(n)}(0)}{n!}x^n + \cdots.$$

The Maclaurin series generated by $f$ is often just called the Taylor series of $f$.

**EXAMPLE 1**   Find the Taylor series generated by $f(x) = 1/x$ at $a = 2$. Where, if anywhere, does the series converge to $1/x$?

**Solution**   We need to find $f(2), f'(2), f''(2), \ldots$. Taking derivatives we get

$$f(x) = x^{-1}, \quad f'(x) = -x^{-2}, \quad f''(x) = 2!x^{-3}, \quad \cdots, \quad f^{(n)}(x) = (-1)^n n! x^{-(n+1)},$$

so that

$$f(2) = 2^{-1} = \frac{1}{2}, \quad f'(2) = -\frac{1}{2^2}, \quad \frac{f''(2)}{2!} = 2^{-3} = \frac{1}{2^3}, \quad \cdots, \quad \frac{f^{(n)}(2)}{n!} = \frac{(-1)^n}{2^{n+1}}.$$

The Taylor series is

$$f(2) + f'(2)(x - 2) + \frac{f''(2)}{2!}(x - 2)^2 + \cdots + \frac{f^{(n)}(2)}{n!}(x - 2)^n + \cdots$$

$$= \frac{1}{2} - \frac{(x - 2)}{2^2} + \frac{(x - 2)^2}{2^3} - \cdots + (-1)^n \frac{(x - 2)^n}{2^{n+1}} + \cdots.$$

This is a geometric series with first term $1/2$ and ratio $r = -(x - 2)/2$. It converges absolutely for $|x - 2| < 2$ and its sum is

$$\frac{1/2}{1 + (x - 2)/2} = \frac{1}{2 + (x - 2)} = \frac{1}{x}.$$

In this example the Taylor series generated by $f(x) = 1/x$ at $a = 2$ converges to $1/x$ for $|x - 2| < 2$ or $0 < x < 4$.  ∎

## Taylor Polynomials

The linearization of a differentiable function $f$ at a point $a$ is the polynomial of degree one given by

$$P_1(x) = f(a) + f'(a)(x - a).$$

In Section 3.9 we used this linearization to approximate $f(x)$ at values of $x$ near $a$. If $f$ has derivatives of higher order at $a$, then it has higher-order polynomial approximations as well, one for each available derivative. These polynomials are called the Taylor polynomials of $f$.

> **DEFINITION**  Let $f$ be a function with derivatives of order $k$ for $k = 1, 2, \ldots, N$ in some interval containing $a$ as an interior point. Then for any integer $n$ from 0 through $N$, the **Taylor polynomial of order $n$** generated by $f$ at $x = a$ is the polynomial
>
> $$P_n(x) = f(a) + f'(a)(x - a) + \frac{f''(a)}{2!}(x - a)^2 + \cdots$$
>
> $$+ \frac{f^{(k)}(a)}{k!}(x - a)^k + \cdots + \frac{f^{(n)}(a)}{n!}(x - a)^n.$$

We speak of a Taylor polynomial of *order n* rather than *degree n* because $f^{(n)}(a)$ may be zero. The first two Taylor polynomials of $f(x) = \cos x$ at $x = 0$, for example, are $P_0(x) = 1$ and $P_1(x) = 1$. The first-order Taylor polynomial has degree zero, not one.

Just as the linearization of $f$ at $x = a$ provides the best linear approximation of $f$ in the neighborhood of $a$, the higher-order Taylor polynomials provide the "best" polynomial approximations of their respective degrees. (See Exercise 40.)

**EXAMPLE 2**  Find the Taylor series and the Taylor polynomials generated by $f(x) = e^x$ at $x = 0$.

**Solution**  Since $f^{(n)}(x) = e^x$ and $f^{(n)}(0) = 1$ for every $n = 0, 1, 2, \ldots$, the Taylor series generated by $f$ at $x = 0$ (see Figure 10.17) is

$$f(0) + f'(0)x + \frac{f''(0)}{2!}x^2 + \cdots + \frac{f^{(n)}(0)}{n!}x^n + \cdots$$

$$= 1 + x + \frac{x^2}{2} + \cdots + \frac{x^n}{n!} + \cdots$$

$$= \sum_{k=0}^{\infty} \frac{x^k}{k!}.$$

This is also the Maclaurin series for $e^x$. In the next section we will see that the series converges to $e^x$ at every $x$.

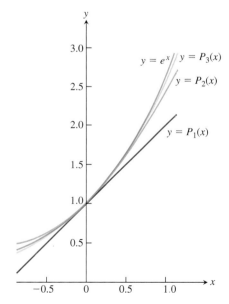

**FIGURE 10.17**  The graph of $f(x) = e^x$ and its Taylor polynomials

$P_1(x) = 1 + x$

$P_2(x) = 1 + x + (x^2/2!)$

$P_3(x) = 1 + x + (x^2/2!) + (x^3/3!).$

Notice the very close agreement near the center $x = 0$ (Example 2).

The Taylor polynomial of order $n$ at $x = 0$ is

$$P_n(x) = 1 + x + \frac{x^2}{2} + \cdots + \frac{x^n}{n!}.$$

**EXAMPLE 3**    Find the Taylor series and Taylor polynomials generated by $f(x) = \cos x$ at $x = 0$.

**Solution**    The cosine and its derivatives are

$$\begin{aligned}
f(x) &= \cos x, & f'(x) &= -\sin x, \\
f''(x) &= -\cos x, & f^{(3)}(x) &= \sin x, \\
&\;\;\vdots & &\;\;\vdots \\
f^{(2n)}(x) &= (-1)^n \cos x, & f^{(2n+1)}(x) &= (-1)^{n+1} \sin x.
\end{aligned}$$

At $x = 0$, the cosines are 1 and the sines are 0, so

$$f^{(2n)}(0) = (-1)^n, \qquad f^{(2n+1)}(0) = 0.$$

The Taylor series generated by $f$ at 0 is

$$f(0) + f'(0)x + \frac{f''(0)}{2!}x^2 + \frac{f'''(0)}{3!}x^3 + \cdots + \frac{f^{(n)}(0)}{n!}x^n + \cdots$$

$$= 1 + 0 \cdot x - \frac{x^2}{2!} + 0 \cdot x^3 + \frac{x^4}{4!} + \cdots + (-1)^n \frac{x^{2n}}{(2n)!} + \cdots$$

$$= \sum_{k=0}^{\infty} \frac{(-1)^k x^{2k}}{(2k)!}.$$

This is also the Maclaurin series for $\cos x$. Notice that only even powers of $x$ occur in the Taylor series generated by the cosine function, which is consistent with the fact that it is an even function. In Section 10.9, we will see that the series converges to $\cos x$ at every $x$.

Because $f^{(2n+1)}(0) = 0$, the Taylor polynomials of orders $2n$ and $2n + 1$ are identical:

$$P_{2n}(x) = P_{2n+1}(x) = 1 - \frac{x^2}{2!} + \frac{x^4}{4!} - \cdots + (-1)^n \frac{x^{2n}}{(2n)!}.$$

Figure 10.18 shows how well these polynomials approximate $f(x) = \cos x$ near $x = 0$. Only the right-hand portions of the graphs are given because the graphs are symmetric about the $y$-axis.

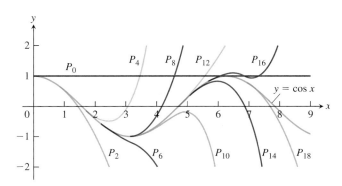

**FIGURE 10.18**    The polynomials

$$P_{2n}(x) = \sum_{k=0}^{n} \frac{(-1)^k x^{2k}}{(2k)!}$$

converge to $\cos x$ as $n \to \infty$. We can deduce the behavior of $\cos x$ arbitrarily far away solely from knowing the values of the cosine and its derivatives at $x = 0$ (Example 3).

**EXAMPLE 4** It can be shown (though not easily) that

$$f(x) = \begin{cases} 0, & x = 0 \\ e^{-1/x^2}, & x \neq 0 \end{cases}$$

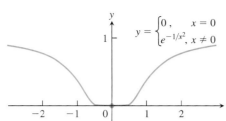

(Figure 10.19) has derivatives of all orders at $x = 0$ and that $f^{(n)}(0) = 0$ for all $n$. This means that the Taylor series generated by $f$ at $x = 0$ is

$$f(0) + f'(0)x + \frac{f''(0)}{2!}x^2 + \cdots + \frac{f^{(n)}(0)}{n!}x^n + \cdots$$

$$= 0 + 0 \cdot x + 0 \cdot x^2 + \cdots + 0 \cdot x^n + \cdots$$

$$= 0 + 0 + \cdots + 0 + \cdots.$$

**FIGURE 10.19** The graph of the continuous extension of $y = e^{-1/x^2}$ is so flat at the origin that all of its derivatives there are zero (Example 4). Therefore its Taylor series, which is zero everywhere, is not the function itself.

The series converges for every $x$ (its sum is 0) but converges to $f(x)$ only at $x = 0$. That is, the Taylor series generated by $f(x)$ in this example is *not* equal to the function $f(x)$ over the entire interval of convergence. ∎

Two questions still remain.

1. For what values of $x$ can we normally expect a Taylor series to converge to its generating function?

2. How accurately do a function's Taylor polynomials approximate the function on a given interval?

The answers are provided by a theorem of Taylor in the next section.

## Exercises 10.8

### Finding Taylor Polynomials

In Exercises 1–10, find the Taylor polynomials of orders 0, 1, 2, and 3 generated by $f$ at $a$.

1. $f(x) = e^{2x}$, $a = 0$
2. $f(x) = \sin x$, $a = 0$
3. $f(x) = \ln x$, $a = 1$
4. $f(x) = \ln(1 + x)$, $a = 0$
5. $f(x) = 1/x$, $a = 2$
6. $f(x) = 1/(x + 2)$, $a = 0$
7. $f(x) = \sin x$, $a = \pi/4$
8. $f(x) = \tan x$, $a = \pi/4$
9. $f(x) = \sqrt{x}$, $a = 4$
10. $f(x) = \sqrt{1 - x}$, $a = 0$

### Finding Taylor Series at $x = 0$ (Maclaurin Series)

Find the Maclaurin series for the functions in Exercises 11–22.

11. $e^{-x}$
12. $xe^x$
13. $\dfrac{1}{1 + x}$
14. $\dfrac{2 + x}{1 - x}$
15. $\sin 3x$
16. $\sin \dfrac{x}{2}$
17. $7 \cos(-x)$
18. $5 \cos \pi x$
19. $\cosh x = \dfrac{e^x + e^{-x}}{2}$
20. $\sinh x = \dfrac{e^x - e^{-x}}{2}$
21. $x^4 - 2x^3 - 5x + 4$
22. $\dfrac{x^2}{x + 1}$

### Finding Taylor and Maclaurin Series

In Exercises 23–32, find the Taylor series generated by $f$ at $x = a$.

23. $f(x) = x^3 - 2x + 4$, $a = 2$
24. $f(x) = 2x^3 + x^2 + 3x - 8$, $a = 1$

25. $f(x) = x^4 + x^2 + 1$, $a = -2$
26. $f(x) = 3x^5 - x^4 + 2x^3 + x^2 - 2$, $a = -1$
27. $f(x) = 1/x^2$, $a = 1$
28. $f(x) = 1/(1 - x)^3$, $a = 0$
29. $f(x) = e^x$, $a = 2$
30. $f(x) = 2^x$, $a = 1$
31. $f(x) = \cos(2x + (\pi/2))$, $a = \pi/4$
32. $f(x) = \sqrt{x + 1}$, $a = 0$

In Exercises 33–36, find the first three nonzero terms of the Maclaurin series for each function and the values of $x$ for which the series converges absolutely.

33. $f(x) = \cos x - (2/(1 - x))$
34. $f(x) = (1 - x + x^2)e^x$
35. $f(x) = (\sin x) \ln(1 + x)$
36. $f(x) = x \sin^2 x$

### Theory and Examples

37. Use the Taylor series generated by $e^x$ at $x = a$ to show that

$$e^x = e^a \left[ 1 + (x - a) + \frac{(x - a)^2}{2!} + \cdots \right].$$

38. (*Continuation of Exercise 37.*) Find the Taylor series generated by $e^x$ at $x = 1$. Compare your answer with the formula in Exercise 37.

39. Let $f(x)$ have derivatives through order $n$ at $x = a$. Show that the Taylor polynomial of order $n$ and its first $n$ derivatives have the same values that $f$ and its first $n$ derivatives have at $x = a$.

**40. Approximation properties of Taylor polynomials** Suppose that $f(x)$ is differentiable on an interval centered at $x = a$ and that $g(x) = b_0 + b_1(x - a) + \cdots + b_n(x - a)^n$ is a polynomial of degree $n$ with constant coefficients $b_0, \ldots, b_n$. Let $E(x) = f(x) - g(x)$. Show that if we impose on $g$ the conditions

**i)** $E(a) = 0$       The approximation error is zero at $x = a$.

**ii)** $\displaystyle\lim_{x \to a} \frac{E(x)}{(x - a)^n} = 0,$    The error is negligible when compared to $(x - a)^n$.

then

$$g(x) = f(a) + f'(a)(x - a) + \frac{f''(a)}{2!}(x - a)^2 + \cdots$$
$$+ \frac{f^{(n)}(a)}{n!}(x - a)^n.$$

Thus, the Taylor polynomial $P_n(x)$ is the only polynomial of degree less than or equal to $n$ whose error is both zero at $x = a$ and negligible when compared with $(x - a)^n$.

**Quadratic Approximations** The Taylor polynomial of order 2 generated by a twice-differentiable function $f(x)$ at $x = a$ is called the *quadratic approximation* of $f$ at $x = a$. In Exercises 41–46, find the **(a)** linearization (Taylor polynomial of order 1) and **(b)** quadratic approximation of $f$ at $x = 0$.

**41.** $f(x) = \ln(\cos x)$      **42.** $f(x) = e^{\sin x}$

**43.** $f(x) = 1/\sqrt{1 - x^2}$      **44.** $f(x) = \cosh x$

**45.** $f(x) = \sin x$      **46.** $f(x) = \tan x$

# 10.9 Convergence of Taylor Series

In the last section we asked when a Taylor series for a function can be expected to converge to that (generating) function. We answer the question in this section with the following theorem.

---

**THEOREM 23—Taylor's Theorem** If $f$ and its first $n$ derivatives $f', f'', \ldots, f^{(n)}$ are continuous on the closed interval between $a$ and $b$, and $f^{(n)}$ is differentiable on the open interval between $a$ and $b$, then there exists a number $c$ between $a$ and $b$ such that

$$f(b) = f(a) + f'(a)(b - a) + \frac{f''(a)}{2!}(b - a)^2 + \cdots$$
$$+ \frac{f^{(n)}(a)}{n!}(b - a)^n + \frac{f^{(n+1)}(c)}{(n + 1)!}(b - a)^{n+1}.$$

---

Taylor's Theorem is a generalization of the Mean Value Theorem (Exercise 45). There is a proof of Taylor's Theorem at the end of this section.

When we apply Taylor's Theorem, we usually want to hold $a$ fixed and treat $b$ as an independent variable. Taylor's formula is easier to use in circumstances like these if we change $b$ to $x$. Here is a version of the theorem with this change.

---

**Taylor's Formula**

If $f$ has derivatives of all orders in an open interval $I$ containing $a$, then for each positive integer $n$ and for each $x$ in $I$,

$$f(x) = f(a) + f'(a)(x - a) + \frac{f''(a)}{2!}(x - a)^2 + \cdots$$
$$+ \frac{f^{(n)}(a)}{n!}(x - a)^n + R_n(x), \tag{1}$$

where

$$R_n(x) = \frac{f^{(n+1)}(c)}{(n + 1)!}(x - a)^{n+1} \qquad \text{for some } c \text{ between } a \text{ and } x. \tag{2}$$

---

When we state Taylor's theorem this way, it says that for each $x \in I$,

$$f(x) = P_n(x) + R_n(x).$$

The function $R_n(x)$ is determined by the value of the $(n + 1)$st derivative $f^{(n+1)}$ at a point $c$ that depends on both $a$ and $x$, and that lies somewhere between them. For any value of $n$ we want, the equation gives both a polynomial approximation of $f$ of that order and a formula for the error involved in using that approximation over the interval $I$.

Equation (1) is called **Taylor's formula**. The function $R_n(x)$ is called the **remainder of order $n$** or the **error term** for the approximation of $f$ by $P_n(x)$ over $I$.

---

If $R_n(x) \to 0$ as $n \to \infty$ for all $x \in I$, we say that the Taylor series generated by $f$ at $x = a$ **converges** to $f$ on $I$, and we write

$$f(x) = \sum_{k=0}^{\infty} \frac{f^{(k)}(a)}{k!} (x - a)^k.$$

---

Often we can estimate $R_n$ without knowing the value of $c$, as the following example illustrates.

**EXAMPLE 1** Show that the Taylor series generated by $f(x) = e^x$ at $x = 0$ converges to $f(x)$ for every real value of $x$.

**Solution** The function has derivatives of all orders throughout the interval $I = (-\infty, \infty)$. Equations (1) and (2) with $f(x) = e^x$ and $a = 0$ give

$$e^x = 1 + x + \frac{x^2}{2!} + \cdots + \frac{x^n}{n!} + R_n(x) \qquad \text{Polynomial from Section 10.8, Example 2}$$

and

$$R_n(x) = \frac{e^c}{(n + 1)!} x^{n+1} \qquad \text{for some } c \text{ between } 0 \text{ and } x.$$

Since $e^x$ is an increasing function of $x$, $e^c$ lies between $e^0 = 1$ and $e^x$. When $x$ is negative, so is $c$, and $e^c < 1$. When $x$ is zero, $e^x = 1$ so that $R_n(x) = 0$. When $x$ is positive, so is $c$, and $e^c < e^x$. Thus, for $R_n(x)$ given as above,

$$|R_n(x)| \leq \frac{|x|^{n+1}}{(n + 1)!} \qquad \text{when } x \leq 0, \qquad e^c < 1$$

and

$$|R_n(x)| < e^x \frac{x^{n+1}}{(n + 1)!} \qquad \text{when } x > 0. \qquad e^c < e^x$$

Finally, because

$$\lim_{n \to \infty} \frac{x^{n+1}}{(n + 1)!} = 0 \qquad \text{for every } x, \qquad \text{Section 10.1, Theorem 5}$$

$\lim_{n \to \infty} R_n(x) = 0$, and the series converges to $e^x$ for every $x$. Thus,

$$e^x = \sum_{k=0}^{\infty} \frac{x^k}{k!} = 1 + x + \frac{x^2}{2!} + \cdots + \frac{x^k}{k!} + \cdots. \qquad (3)$$

∎

**The Number $e$ as a Series**

$$e = \sum_{n=0}^{\infty} \frac{1}{n!}$$

We can use the result of Example 1 with $x = 1$ to write

$$e = 1 + 1 + \frac{1}{2!} + \cdots + \frac{1}{n!} + R_n(1),$$

where for some $c$ between 0 and 1,

$$R_n(1) = e^c \frac{1}{(n+1)!} < \frac{3}{(n+1)!}. \qquad e^c < e^1 < 3$$

## Estimating the Remainder

It is often possible to estimate $R_n(x)$ as we did in Example 1. This method of estimation is so convenient that we state it as a theorem for future reference.

---

**THEOREM 24—The Remainder Estimation Theorem**   If there is a positive constant $M$ such that $\left| f^{(n+1)}(t) \right| \leq M$ for all $t$ between $x$ and $a$, inclusive, then the remainder term $R_n(x)$ in Taylor's Theorem satisfies the inequality

$$\left| R_n(x) \right| \leq M \frac{\left| x - a \right|^{n+1}}{(n+1)!}.$$

If this inequality holds for every $n$ and the other conditions of Taylor's Theorem are satisfied by $f$, then the series converges to $f(x)$.

---

The next two examples use Theorem 24 to show that the Taylor series generated by the sine and cosine functions do in fact converge to the functions themselves.

**EXAMPLE 2**   Show that the Taylor series for $\sin x$ at $x = 0$ converges for all $x$.

**Solution**   The function and its derivatives are

$$f(x) = \sin x, \qquad f'(x) = \cos x,$$

$$f''(x) = -\sin x, \qquad f'''(x) = -\cos x,$$

$$\vdots \qquad\qquad\qquad \vdots$$

$$f^{(2k)}(x) = (-1)^k \sin x, \qquad f^{(2k+1)}(x) = (-1)^k \cos x,$$

so

$$f^{(2k)}(0) = 0 \qquad \text{and} \qquad f^{(2k+1)}(0) = (-1)^k.$$

The series has only odd-powered terms and, for $n = 2k + 1$, Taylor's Theorem gives

$$\sin x = x - \frac{x^3}{3!} + \frac{x^5}{5!} - \cdots + \frac{(-1)^k x^{2k+1}}{(2k+1)!} + R_{2k+1}(x).$$

All the derivatives of $\sin x$ have absolute values less than or equal to 1, so we can apply the Remainder Estimation Theorem with $M = 1$ to obtain

$$\left| R_{2k+1}(x) \right| \leq 1 \cdot \frac{\left| x \right|^{2k+2}}{(2k+2)!}.$$

From Theorem 5, Rule 6, we have $\left( \left| x \right|^{2k+2}/(2k+2)! \right) \to 0$ as $k \to \infty$, whatever the value of $x$, so $R_{2k+1}(x) \to 0$ and the Maclaurin series for $\sin x$ converges to $\sin x$ for every $x$. Thus,

$$\sin x = \sum_{k=0}^{\infty} \frac{(-1)^k x^{2k+1}}{(2k+1)!} = x - \frac{x^3}{3!} + \frac{x^5}{5!} - \frac{x^7}{7!} + \cdots. \qquad (4)$$

**EXAMPLE 3**    Show that the Taylor series for $\cos x$ at $x = 0$ converges to $\cos x$ for every value of $x$.

**Solution**    We add the remainder term to the Taylor polynomial for $\cos x$ (Section 10.8, Example 3) to obtain Taylor's formula for $\cos x$ with $n = 2k$:

$$\cos x = 1 - \frac{x^2}{2!} + \frac{x^4}{4!} - \cdots + (-1)^k \frac{x^{2k}}{(2k)!} + R_{2k}(x).$$

Because the derivatives of the cosine have absolute value less than or equal to 1, the Remainder Estimation Theorem with $M = 1$ gives

$$\left| R_{2k}(x) \right| \le 1 \cdot \frac{|x|^{2k+1}}{(2k+1)!}.$$

For every value of $x$, $R_{2k}(x) \to 0$ as $k \to \infty$. Therefore, the series converges to $\cos x$ for every value of $x$. Thus,

$$\cos x = \sum_{k=0}^{\infty} \frac{(-1)^k x^{2k}}{(2k)!} = 1 - \frac{x^2}{2!} + \frac{x^4}{4!} - \frac{x^6}{6!} + \cdots . \tag{5}$$

∎

## Using Taylor Series

Since every Taylor series is a power series, the operations of adding, subtracting, and multiplying Taylor series are all valid on the intersection of their intervals of convergence.

**EXAMPLE 4**    Using known series, find the first few terms of the Taylor series for the given function using power series operations.

**(a)** $\dfrac{1}{3}(2x + x \cos x)$ \qquad\qquad **(b)** $e^x \cos x$

**Solution**

**(a)** $\dfrac{1}{3}(2x + x \cos x) = \dfrac{2}{3}x + \dfrac{1}{3}x\left(1 - \dfrac{x^2}{2!} + \dfrac{x^4}{4!} - \cdots + (-1)^k \dfrac{x^{2k}}{(2k)!} + \cdots\right)$

$$= \frac{2}{3}x + \frac{1}{3}x - \frac{x^3}{3!} + \frac{x^5}{3 \cdot 4!} - \cdots = x - \frac{x^3}{6} + \frac{x^5}{72} - \cdots$$

**(b)** $e^x \cos x = \left(1 + x + \dfrac{x^2}{2!} + \dfrac{x^3}{3!} + \dfrac{x^4}{4!} + \cdots\right) \cdot \left(1 - \dfrac{x^2}{2!} + \dfrac{x^4}{4!} - \cdots\right)$   Multiply the first series by each term of the second series.

$$= \left(1 + x + \frac{x^2}{2!} + \frac{x^3}{3!} + \frac{x^4}{4!} + \cdots\right) - \left(\frac{x^2}{2!} + \frac{x^3}{2!} + \frac{x^4}{2!2!} + \frac{x^5}{2!3!} + \cdots\right)$$

$$+ \left(\frac{x^4}{4!} + \frac{x^5}{4!} + \frac{x^6}{2!4!} + \cdots\right) + \cdots$$

$$= 1 + x - \frac{x^3}{3} - \frac{x^4}{6} + \cdots$$

∎

By Theorem 20, we can use the Taylor series of the function $f$ to find the Taylor series of $f(u(x))$ where $u(x)$ is any continuous function. The Taylor series resulting from this substitution will converge for all $x$ such that $u(x)$ lies within the interval of convergence of the Taylor series of $f$. For instance, we can find the Taylor series for $\cos 2x$ by substituting $2x$ for $x$ in the Taylor series for $\cos x$:

$$\cos 2x = \sum_{k=0}^{\infty} \frac{(-1)^k (2x)^{2k}}{(2k)!} = 1 - \frac{(2x)^2}{2!} + \frac{(2x)^4}{4!} - \frac{(2x)^6}{6!} + \cdots \qquad \text{Eq. (5) with } 2x \text{ for } x$$

$$= 1 - \frac{2^2 x^2}{2!} + \frac{2^4 x^4}{4!} - \frac{2^6 x^6}{6!} + \cdots$$

$$= \sum_{k=0}^{\infty} (-1)^k \frac{2^{2k} x^{2k}}{(2k)!}.$$

**EXAMPLE 5**  For what values of $x$ can we replace $\sin x$ by $x - (x^3/3!)$ with an error of magnitude no greater than $3 \times 10^{-4}$?

**Solution**  Here we can take advantage of the fact that the Taylor series for $\sin x$ is an alternating series for every nonzero value of $x$. According to the Alternating Series Estimation Theorem (Section 10.6), the error in truncating

$$\sin x = x - \frac{x^3}{3!} + \frac{x^5}{5!} - \cdots$$

after $(x^3/3!)$ is no greater than

$$\left| \frac{x^5}{5!} \right| = \frac{|x|^5}{120}.$$

Therefore the error will be less than or equal to $3 \times 10^{-4}$ if

$$\frac{|x|^5}{120} < 3 \times 10^{-4} \qquad \text{or} \qquad |x| < \sqrt[5]{360 \times 10^{-4}} \approx 0.514. \qquad \text{Rounded down, to be safe}$$

The Alternating Series Estimation Theorem tells us something that the Remainder Estimation Theorem does not: namely, that the estimate $x - (x^3/3!)$ for $\sin x$ is an underestimate when $x$ is positive, because then $x^5/120$ is positive.

Figure 10.20 shows the graph of $\sin x$, along with the graphs of a number of its approximating Taylor polynomials. The graph of $P_3(x) = x - (x^3/3!)$ is almost indistinguishable from the sine curve when $0 \le x \le 1$. ∎

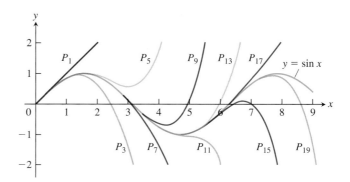

**FIGURE 10.20**  The polynomials

$$P_{2n+1}(x) = \sum_{k=0}^{n} \frac{(-1)^k x^{2k+1}}{(2k+1)!}$$

converge to $\sin x$ as $n \to \infty$. Notice how closely $P_3(x)$ approximates the sine curve for $x \le 1$ (Example 5).

### A Proof of Taylor's Theorem

We prove Taylor's theorem assuming $a < b$. The proof for $a > b$ is nearly the same.

The Taylor polynomial

$$P_n(x) = f(a) + f'(a)(x - a) + \frac{f''(a)}{2!}(x - a)^2 + \cdots + \frac{f^{(n)}(a)}{n!}(x - a)^n$$

and its first $n$ derivatives match the function $f$ and its first $n$ derivatives at $x = a$. We do not disturb that matching if we add another term of the form $K(x - a)^{n+1}$, where $K$ is any constant, because such a term and its first $n$ derivatives are all equal to zero at $x = a$. The new function

$$\phi_n(x) = P_n(x) + K(x - a)^{n+1}$$

and its first $n$ derivatives still agree with $f$ and its first $n$ derivatives at $x = a$.

We now choose the particular value of $K$ that makes the curve $y = \phi_n(x)$ agree with the original curve $y = f(x)$ at $x = b$. In symbols,

$$f(b) = P_n(b) + K(b - a)^{n+1}, \quad \text{or} \quad K = \frac{f(b) - P_n(b)}{(b - a)^{n+1}}. \tag{6}$$

With $K$ defined by Equation (6), the function

$$F(x) = f(x) - \phi_n(x)$$

measures the difference between the original function $f$ and the approximating function $\phi_n$ for each $x$ in $[a, b]$.

We now use Rolle's Theorem (Section 4.2). First, because $F(a) = F(b) = 0$ and both $F$ and $F'$ are continuous on $[a, b]$, we know that

$$F'(c_1) = 0 \qquad \text{for some } c_1 \text{ in } (a, b).$$

Next, because $F'(a) = F'(c_1) = 0$ and both $F'$ and $F''$ are continuous on $[a, c_1]$, we know that

$$F''(c_2) = 0 \qquad \text{for some } c_2 \text{ in } (a, c_1).$$

Rolle's Theorem, applied successively to $F'', F''', \ldots, F^{(n-1)}$, implies the existence of

$$c_3 \quad \text{in } (a, c_2) \qquad \text{such that } F'''(c_3) = 0,$$
$$c_4 \quad \text{in } (a, c_3) \qquad \text{such that } F^{(4)}(c_4) = 0,$$
$$\vdots$$
$$c_n \quad \text{in } (a, c_{n-1}) \qquad \text{such that } F^{(n)}(c_n) = 0.$$

Finally, because $F^{(n)}$ is continuous on $[a, c_n]$ and differentiable on $(a, c_n)$, and $F^{(n)}(a) = F^{(n)}(c_n) = 0$, Rolle's Theorem implies that there is a number $c_{n+1}$ in $(a, c_n)$ such that

$$F^{(n+1)}(c_{n+1}) = 0. \tag{7}$$

If we differentiate $F(x) = f(x) - P_n(x) - K(x - a)^{n+1}$ a total of $n + 1$ times, we get

$$F^{(n+1)}(x) = f^{(n+1)}(x) - 0 - (n + 1)!K. \tag{8}$$

Equations (7) and (8) together give

$$K = \frac{f^{(n+1)}(c)}{(n + 1)!} \qquad \text{for some number } c = c_{n+1} \text{ in } (a, b). \tag{9}$$

Equations (6) and (9) give

$$f(b) = P_n(b) + \frac{f^{(n+1)}(c)}{(n+1)!}(b-a)^{n+1}.$$

This concludes the proof. ∎

## Exercises 10.9

### Finding Taylor Series

Use substitution (as in Example 4) to find the Taylor series at $x = 0$ of the functions in Exercises 1–10.

**1.** $e^{-5x}$     **2.** $e^{-x/2}$     **3.** $5 \sin(-x)$

**4.** $\sin\left(\dfrac{\pi x}{2}\right)$     **5.** $\cos 5x^2$     **6.** $\cos\left(x^{2/3}/\sqrt{2}\right)$

**7.** $\ln(1 + x^2)$     **8.** $\tan^{-1}(3x^4)$     **9.** $\dfrac{1}{1 + \frac{3}{4}x^3}$

**10.** $\dfrac{1}{2 - x}$

Use power series operations to find the Taylor series at $x = 0$ for the functions in Exercises 11–28.

**11.** $xe^x$     **12.** $x^2 \sin x$     **13.** $\dfrac{x^2}{2} - 1 + \cos x$

**14.** $\sin x - x + \dfrac{x^3}{3!}$     **15.** $x \cos \pi x$     **16.** $x^2 \cos(x^2)$

**17.** $\cos^2 x$ (*Hint:* $\cos^2 x = (1 + \cos 2x)/2$.)

**18.** $\sin^2 x$     **19.** $\dfrac{x^2}{1 - 2x}$     **20.** $x \ln(1 + 2x)$

**21.** $\dfrac{1}{(1 - x)^2}$     **22.** $\dfrac{2}{(1 - x)^3}$     **23.** $x \tan^{-1} x^2$

**24.** $\sin x \cdot \cos x$     **25.** $e^x + \dfrac{1}{1 + x}$     **26.** $\cos x - \sin x$

**27.** $\dfrac{x}{3} \ln(1 + x^2)$           **28.** $\ln(1 + x) - \ln(1 - x)$

Find the first four nonzero terms in the Maclaurin series for the functions in Exercises 29–34.

**29.** $e^x \sin x$     **30.** $\dfrac{\ln(1 + x)}{1 - x}$     **31.** $(\tan^{-1} x)^2$

**32.** $\cos^2 x \cdot \sin x$     **33.** $e^{\sin x}$     **34.** $\sin(\tan^{-1} x)$

### Error Estimates

**35.** Estimate the error if $P_3(x) = x - (x^3/6)$ is used to estimate the value of $\sin x$ at $x = 0.1$.

**36.** Estimate the error if $P_4(x) = 1 + x + (x^2/2) + (x^3/6) + (x^4/24)$ is used to estimate the value of $e^x$ at $x = 1/2$.

**37.** For approximately what values of $x$ can you replace $\sin x$ by $x - (x^3/6)$ with an error of magnitude no greater than $5 \times 10^{-4}$? Give reasons for your answer.

**38.** If $\cos x$ is replaced by $1 - (x^2/2)$ and $|x| < 0.5$, what estimate can be made of the error? Does $1 - (x^2/2)$ tend to be too large, or too small? Give reasons for your answer.

**39.** How close is the approximation $\sin x = x$ when $|x| < 10^{-3}$? For which of these values of $x$ is $x < \sin x$?

**40.** The estimate $\sqrt{1 + x} = 1 + (x/2)$ is used when $x$ is small. Estimate the error when $|x| < 0.01$.

**41.** The approximation $e^x = 1 + x + (x^2/2)$ is used when $x$ is small. Use the Remainder Estimation Theorem to estimate the error when $|x| < 0.1$.

**42.** (*Continuation of Exercise 41.*) When $x < 0$, the series for $e^x$ is an alternating series. Use the Alternating Series Estimation Theorem to estimate the error that results from replacing $e^x$ by $1 + x + (x^2/2)$ when $-0.1 < x < 0$. Compare your estimate with the one you obtained in Exercise 41.

### Theory and Examples

**43.** Use the identity $\sin^2 x = (1 - \cos 2x)/2$ to obtain the Maclaurin series for $\sin^2 x$. Then differentiate this series to obtain the Maclaurin series for $2 \sin x \cos x$. Check that this is the series for $\sin 2x$.

**44.** (*Continuation of Exercise 43.*) Use the identity $\cos^2 x = \cos 2x + \sin^2 x$ to obtain a power series for $\cos^2 x$.

**45. Taylor's Theorem and the Mean Value Theorem** Explain how the Mean Value Theorem (Section 4.2, Theorem 4) is a special case of Taylor's Theorem.

**46. Linearizations at inflection points** Show that if the graph of a twice-differentiable function $f(x)$ has an inflection point at $x = a$, then the linearization of $f$ at $x = a$ is also the quadratic approximation of $f$ at $x = a$. This explains why tangent lines fit so well at inflection points.

**47. The (second) second derivative test** Use the equation

$$f(x) = f(a) + f'(a)(x - a) + \frac{f''(c_2)}{2}(x - a)^2$$

to establish the following test.

Let $f$ have continuous first and second derivatives and suppose that $f'(a) = 0$. Then

**a.** $f$ has a local maximum at $a$ if $f'' \le 0$ throughout an interval whose interior contains $a$;

**b.** $f$ has a local minimum at $a$ if $f'' \ge 0$ throughout an interval whose interior contains $a$.

**48. A cubic approximation** Use Taylor's formula with $a = 0$ and $n = 3$ to find the standard cubic approximation of $f(x) = 1/(1 - x)$ at $x = 0$. Give an upper bound for the magnitude of the error in the approximation when $|x| \leq 0.1$.

**49. a.** Use Taylor's formula with $n = 2$ to find the quadratic approximation of $f(x) = (1 + x)^k$ at $x = 0$ ($k$ a constant).

**b.** If $k = 3$, for approximately what values of $x$ in the interval $[0, 1]$ will the error in the quadratic approximation be less than $1/100$?

**50. Improving approximations of $\pi$**

**a.** Let $P$ be an approximation of $\pi$ accurate to $n$ decimals. Show that $P + \sin P$ gives an approximation correct to $3n$ decimals. (*Hint:* Let $P = \pi + x$.)

T **b.** Try it with a calculator.

**51. The Taylor series generated by $f(x) = \sum_{n=0}^{\infty} a_n x^n$ is $\sum_{n=0}^{\infty} a_n x^n$** A function defined by a power series $\sum_{n=0}^{\infty} a_n x^n$ with a radius of convergence $R > 0$ has a Taylor series that converges to the function at every point of $(-R, R)$. Show this by showing that the Taylor series generated by $f(x) = \sum_{n=0}^{\infty} a_n x^n$ is the series $\sum_{n=0}^{\infty} a_n x^n$ itself.

An immediate consequence of this is that series like

$$x \sin x = x^2 - \frac{x^4}{3!} + \frac{x^6}{5!} - \frac{x^8}{7!} + \cdots$$

and

$$x^2 e^x = x^2 + x^3 + \frac{x^4}{2!} + \frac{x^5}{3!} + \cdots,$$

obtained by multiplying Taylor series by powers of $x$, as well as series obtained by integration and differentiation of convergent power series, are themselves the Taylor series generated by the functions they represent.

**52. Taylor series for even functions and odd functions** (*Continuation of Section 10.7, Exercise 59.*) Suppose that $f(x) = \sum_{n=0}^{\infty} a_n x^n$ converges for all $x$ in an open interval $(-R, R)$. Show that

**a.** If $f$ is even, then $a_1 = a_3 = a_5 = \cdots = 0$, i.e., the Taylor series for $f$ at $x = 0$ contains only even powers of $x$.

**b.** If $f$ is odd, then $a_0 = a_2 = a_4 = \cdots = 0$, i.e., the Taylor series for $f$ at $x = 0$ contains only odd powers of $x$.

**COMPUTER EXPLORATIONS**

Taylor's formula with $n = 1$ and $a = 0$ gives the linearization of a function at $x = 0$. With $n = 2$ and $n = 3$ we obtain the standard quadratic and cubic approximations. In these exercises we explore the errors associated with these approximations. We seek answers to two questions:

**a.** For what values of $x$ can the function be replaced by each approximation with an error less than $10^{-2}$?

**b.** What is the maximum error we could expect if we replace the function by each approximation over the specified interval?

Using a CAS, perform the following steps to aid in answering questions (a) and (b) for the functions and intervals in Exercises 53–58.

*Step 1:* Plot the function over the specified interval.

*Step 2:* Find the Taylor polynomials $P_1(x)$, $P_2(x)$, and $P_3(x)$ at $x = 0$.

*Step 3:* Calculate the $(n + 1)$st derivative $f^{(n+1)}(c)$ associated with the remainder term for each Taylor polynomial. Plot the derivative as a function of $c$ over the specified interval and estimate its maximum absolute value, $M$.

*Step 4:* Calculate the remainder $R_n(x)$ for each polynomial. Using the estimate $M$ from Step 3 in place of $f^{(n+1)}(c)$, plot $R_n(x)$ over the specified interval. Then estimate the values of $x$ that answer question (a).

*Step 5:* Compare your estimated error with the actual error $E_n(x) = |f(x) - P_n(x)|$ by plotting $E_n(x)$ over the specified interval. This will help answer question (b).

*Step 6:* Graph the function and its three Taylor approximations together. Discuss the graphs in relation to the information discovered in Steps 4 and 5.

**53.** $f(x) = \dfrac{1}{\sqrt{1 + x}}, \quad |x| \leq \dfrac{3}{4}$

**54.** $f(x) = (1 + x)^{3/2}, \quad -\dfrac{1}{2} \leq x \leq 2$

**55.** $f(x) = \dfrac{x}{x^2 + 1}, \quad |x| \leq 2$

**56.** $f(x) = (\cos x)(\sin 2x), \quad |x| \leq 2$

**57.** $f(x) = e^{-x}\cos 2x, \quad |x| \leq 1$

**58.** $f(x) = e^{x/3}\sin 2x, \quad |x| \leq 2$

# $10.10$ The Binomial Series and Applications of Taylor Series

We can use Taylor series to solve problems that would otherwise be intractable. For example, many functions have antiderivatives that cannot be expressed using familiar functions. In this section we show how to evaluate integrals of such functions by giving them as Taylor series. We also show how to use Taylor series to evaluate limits that lead to indeterminate forms and how Taylor series can be used to extend the exponential function from real to complex numbers. We begin with a discussion of the binomial series, which comes from the Taylor series of the function $f(x) = (1 + x)^m$, and conclude the section with Table 10.1, which lists some commonly used Taylor series.

## The Binomial Series for Powers and Roots

The Taylor series generated by $f(x) = (1 + x)^m$, when $m$ is constant, is

$$1 + mx + \frac{m(m - 1)}{2!}x^2 + \frac{m(m - 1)(m - 2)}{3!}x^3 + \cdots$$

$$+ \frac{m(m - 1)(m - 2) \cdots (m - k + 1)}{k!}x^k + \cdots. \qquad (1)$$

This series, called the **binomial series**, converges absolutely for $|x| < 1$. To derive the series, we first list the function and its derivatives:

$$f(x) = (1 + x)^m$$

$$f'(x) = m(1 + x)^{m-1}$$

$$f''(x) = m(m - 1)(1 + x)^{m-2}$$

$$f'''(x) = m(m - 1)(m - 2)(1 + x)^{m-3}$$

$$\vdots$$

$$f^{(k)}(x) = m(m - 1)(m - 2) \cdots (m - k + 1)(1 + x)^{m-k}.$$

We then evaluate these at $x = 0$ and substitute into the Taylor series formula to obtain Series (1).

If $m$ is an integer greater than or equal to zero, the series stops after $(m + 1)$ terms because the coefficients from $k = m + 1$ on are zero.

If $m$ is not a positive integer or zero, the series is infinite and converges for $|x| < 1$. To see why, let $u_k$ be the term involving $x^k$. Then apply the Ratio Test for absolute convergence to see that

$$\left| \frac{u_{k+1}}{u_k} \right| = \left| \frac{m - k}{k + 1}x \right| \rightarrow |x| \qquad \text{as } k \rightarrow \infty.$$

Our derivation of the binomial series shows only that it is generated by $(1 + x)^m$ and converges for $|x| < 1$. The derivation does not show that the series converges to $(1 + x)^m$. It does, but we leave the proof to Exercise 58. The following formulation gives a succinct way to express the series.

---

**The Binomial Series**

For $-1 < x < 1$,

$$(1 + x)^m = 1 + \sum_{k=1}^{\infty} \binom{m}{k}x^k,$$

where we define

$$\binom{m}{1} = m, \qquad \binom{m}{2} = \frac{m(m - 1)}{2!},$$

and

$$\binom{m}{k} = \frac{m(m - 1)(m - 2) \cdots (m - k + 1)}{k!} \qquad \text{for } k \geq 3.$$

---

**EXAMPLE 1**　　If $m = -1$,

$$\binom{-1}{1} = -1, \qquad \binom{-1}{2} = \frac{-1(-2)}{2!} = 1,$$

and

$$\binom{-1}{k} = \frac{-1(-2)(-3)\cdots(-1-k+1)}{k!} = (-1)^k \left(\frac{k!}{k!}\right) = (-1)^k.$$

With these coefficient values and with $x$ replaced by $-x$, the binomial series formula gives the familiar geometric series

$$(1+x)^{-1} = 1 + \sum_{k=1}^{\infty}(-1)^k x^k = 1 - x + x^2 - x^3 + \cdots + (-1)^k x^k + \cdots. \qquad \blacksquare$$

**EXAMPLE 2** We know from Section 3.9, Example 1, that $\sqrt{1+x} \approx 1 + (x/2)$ for $|x|$ small. With $m = 1/2$, the binomial series gives quadratic and higher-order approximations as well, along with error estimates that come from the Alternating Series Estimation Theorem:

$$(1+x)^{1/2} = 1 + \frac{x}{2} + \frac{\left(\frac{1}{2}\right)\left(-\frac{1}{2}\right)}{2!}x^2 + \frac{\left(\frac{1}{2}\right)\left(-\frac{1}{2}\right)\left(-\frac{3}{2}\right)}{3!}x^3$$

$$+ \frac{\left(\frac{1}{2}\right)\left(-\frac{1}{2}\right)\left(-\frac{3}{2}\right)\left(-\frac{5}{2}\right)}{4!}x^4 + \cdots$$

$$= 1 + \frac{x}{2} - \frac{x^2}{8} + \frac{x^3}{16} - \frac{5x^4}{128} + \cdots.$$

Substitution for $x$ gives still other approximations. For example,

$$\sqrt{1-x^2} \approx 1 - \frac{x^2}{2} - \frac{x^4}{8} \qquad \text{for } |x^2| \text{ small}$$

$$\sqrt{1-\frac{1}{x}} \approx 1 - \frac{1}{2x} - \frac{1}{8x^2} \qquad \text{for } \left|\frac{1}{x}\right| \text{ small, that is, } |x| \text{ large.} \qquad \blacksquare$$

## Evaluating Nonelementary Integrals

Sometimes we can use a familiar Taylor series to find the sum of a given power series in terms of a known function. For example,

$$x^2 - \frac{x^6}{3!} + \frac{x^{10}}{5!} - \frac{x^{14}}{7!} + \cdots = (x^2) - \frac{(x^2)^3}{3!} + \frac{(x^2)^5}{5!} - \frac{(x^2)^7}{7!} + \cdots = \sin x^2.$$

Additional examples are provided in Exercises 59–62.

Taylor series can be used to express nonelementary integrals in terms of series. Integrals like $\int \sin x^2 \, dx$ arise in the study of the diffraction of light.

**EXAMPLE 3** Express $\int \sin x^2 \, dx$ as a power series.

**Solution** From the series for $\sin x$ we substitute $x^2$ for $x$ to obtain

$$\sin x^2 = x^2 - \frac{x^6}{3!} + \frac{x^{10}}{5!} - \frac{x^{14}}{7!} + \frac{x^{18}}{9!} - \cdots.$$

Therefore,

$$\int \sin x^2 \, dx = C + \frac{x^3}{3} - \frac{x^7}{7\cdot 3!} + \frac{x^{11}}{11\cdot 5!} - \frac{x^{15}}{15\cdot 7!} + \frac{x^{19}}{19\cdot 9!} - \cdots. \qquad \blacksquare$$

**EXAMPLE 4**    Estimate $\int_0^1 \sin x^2 \, dx$ with an error of less than 0.001.

**Solution**    From the indefinite integral in Example 3, we easily find that

$$\int_0^1 \sin x^2 \, dx = \frac{1}{3} - \frac{1}{7 \cdot 3!} + \frac{1}{11 \cdot 5!} - \frac{1}{15 \cdot 7!} + \frac{1}{19 \cdot 9!} - \cdots.$$

The series on the right-hand side alternates, and we find by numerical evaluations that

$$\frac{1}{11 \cdot 5!} \approx 0.00076$$

is the first term to be numerically less than 0.001. The sum of the preceding two terms gives

$$\int_0^1 \sin x^2 \, dx \approx \frac{1}{3} - \frac{1}{42} \approx 0.310.$$

With two more terms we could estimate

$$\int_0^1 \sin x^2 \, dx \approx 0.310268$$

with an error of less than $10^{-6}$. With only one term beyond that we have

$$\int_0^1 \sin x^2 \, dx \approx \frac{1}{3} - \frac{1}{42} + \frac{1}{1320} - \frac{1}{75600} + \frac{1}{6894720} \approx 0.310268303,$$

with an error of about $1.08 \times 10^{-9}$. To guarantee this accuracy with the error formula for the Trapezoidal Rule would require using about 8000 subintervals.    ■

## Arctangents

In Section 10.7, Example 5, we found a series for $\tan^{-1} x$ by differentiating to get

$$\frac{d}{dx} \tan^{-1} x = \frac{1}{1 + x^2} = 1 - x^2 + x^4 - x^6 + \cdots$$

and then integrating to get

$$\tan^{-1} x = x - \frac{x^3}{3} + \frac{x^5}{5} - \frac{x^7}{7} + \cdots.$$

However, we did not prove the term-by-term integration theorem on which this conclusion depended. We now derive the series again by integrating both sides of the finite formula

$$\frac{1}{1 + t^2} = 1 - t^2 + t^4 - t^6 + \cdots + (-1)^n t^{2n} + \frac{(-1)^{n+1} t^{2n+2}}{1 + t^2}, \tag{2}$$

in which the last term comes from adding the remaining terms as a geometric series with first term $a = (-1)^{n+1} t^{2n+2}$ and ratio $r = -t^2$. Integrating both sides of Equation (2) from $t = 0$ to $t = x$ gives

$$\tan^{-1} x = x - \frac{x^3}{3} + \frac{x^5}{5} - \frac{x^7}{7} + \cdots + (-1)^n \frac{x^{2n+1}}{2n + 1} + R_n(x),$$

where

$$R_n(x) = \int_0^x \frac{(-1)^{n+1} t^{2n+2}}{1 + t^2} \, dt.$$

The denominator of the integrand is greater than or equal to 1; hence

$$|R_n(x)| \leq \int_0^{|x|} t^{2n+2} \, dt = \frac{|x|^{2n+3}}{2n + 3}.$$

If $|x| \le 1$, the right side of this inequality approaches zero as $n \to \infty$. Therefore $\lim_{n \to \infty} R_n(x) = 0$ if $|x| \le 1$ and

$$\tan^{-1} x = \sum_{n=0}^{\infty} \frac{(-1)^n x^{2n+1}}{2n + 1}, \qquad |x| \le 1.$$

$$\tan^{-1} x = x - \frac{x^3}{3} + \frac{x^5}{5} - \frac{x^7}{7} + \cdots, \qquad |x| \le 1. \tag{3}$$

We take this route instead of finding the Taylor series directly because the formulas for the higher-order derivatives of $\tan^{-1} x$ are unmanageable. When we put $x = 1$ in Equation (3), we get **Leibniz's formula**:

$$\frac{\pi}{4} = 1 - \frac{1}{3} + \frac{1}{5} - \frac{1}{7} + \frac{1}{9} - \cdots + \frac{(-1)^n}{2n + 1} + \cdots.$$

Because this series converges very slowly, it is not used in approximating $\pi$ to many decimal places. The series for $\tan^{-1} x$ converges most rapidly when $x$ is near zero. For that reason, people who use the series for $\tan^{-1} x$ to compute $\pi$ use various trigonometric identities.

For example, if

$$\alpha = \tan^{-1} \frac{1}{2} \qquad \text{and} \qquad \beta = \tan^{-1} \frac{1}{3},$$

then

$$\tan(\alpha + \beta) = \frac{\tan \alpha + \tan \beta}{1 - \tan \alpha \tan \beta} = \frac{\frac{1}{2} + \frac{1}{3}}{1 - \frac{1}{6}} = 1 = \tan \frac{\pi}{4}$$

and

$$\frac{\pi}{4} = \alpha + \beta = \tan^{-1} \frac{1}{2} + \tan^{-1} \frac{1}{3}.$$

Now Equation (3) may be used with $x = 1/2$ to evaluate $\tan^{-1}(1/2)$ and with $x = 1/3$ to give $\tan^{-1}(1/3)$. The sum of these results, multiplied by 4, gives $\pi$.

## Evaluating Indeterminate Forms

We can sometimes evaluate indeterminate forms by expressing the functions involved as Taylor series.

**EXAMPLE 5**    Evaluate

$$\lim_{x \to 1} \frac{\ln x}{x - 1}.$$

**Solution**    We represent $\ln x$ as a Taylor series in powers of $x - 1$. This can be accomplished by calculating the Taylor series generated by $\ln x$ at $x = 1$ directly or by replacing $x$ by $x - 1$ in the series for $\ln(1 + x)$ in Section 10.7, Example 6. Either way, we obtain

$$\ln x = (x - 1) - \frac{1}{2}(x - 1)^2 + \cdots,$$

from which we find that

$$\lim_{x \to 1} \frac{\ln x}{x - 1} = \lim_{x \to 1} \left( 1 - \frac{1}{2}(x - 1) + \cdots \right) = 1. \qquad \blacksquare$$

Of course, this particular limit can be evaluated using l'Hôpital's Rule just as well.

**EXAMPLE 6**    Evaluate

$$\lim_{x \to 0} \frac{\sin x - \tan x}{x^3}.$$

**Solution**    The Taylor series for $\sin x$ and $\tan x$, to terms in $x^5$, are

$$\sin x = x - \frac{x^3}{3!} + \frac{x^5}{5!} - \cdots, \qquad \tan x = x + \frac{x^3}{3} + \frac{2x^5}{15} + \cdots.$$

Subtracting the series term by term, it follows that

$$\sin x - \tan x = -\frac{x^3}{2} - \frac{x^5}{8} - \cdots = x^3 \left( -\frac{1}{2} - \frac{x^2}{8} - \cdots \right).$$

Division of both sides by $x^3$ and taking limits then gives

$$\lim_{x \to 0} \frac{\sin x - \tan x}{x^3} = \lim_{x \to 0} \left( -\frac{1}{2} - \frac{x^2}{8} - \cdots \right)$$

$$= -\frac{1}{2}. \qquad \blacksquare$$

If we apply series to calculate $\lim_{x \to 0}((1/\sin x) - (1/x))$, we not only find the limit successfully but also discover an approximation formula for $\csc x$.

**EXAMPLE 7**    Find $\lim\limits_{x \to 0} \left( \dfrac{1}{\sin x} - \dfrac{1}{x} \right)$.

**Solution**    Using algebra and the Taylor series for $\sin x$, we have

$$\frac{1}{\sin x} - \frac{1}{x} = \frac{x - \sin x}{x \sin x} = \frac{x - \left( x - \dfrac{x^3}{3!} + \dfrac{x^5}{5!} - \cdots \right)}{x \cdot \left( x - \dfrac{x^3}{3!} + \dfrac{x^5}{5!} - \cdots \right)}$$

$$= \frac{x^3 \left( \dfrac{1}{3!} - \dfrac{x^2}{5!} + \cdots \right)}{x^2 \left( 1 - \dfrac{x^2}{3!} + \cdots \right)} = x \cdot \frac{\dfrac{1}{3!} - \dfrac{x^2}{5!} + \cdots}{1 - \dfrac{x^2}{3!} + \cdots}.$$

Therefore,

$$\lim_{x \to 0} \left( \frac{1}{\sin x} - \frac{1}{x} \right) = \lim_{x \to 0} \left( x \cdot \frac{\dfrac{1}{3!} - \dfrac{x^2}{5!} + \cdots}{1 - \dfrac{x^2}{3!} + \cdots} \right) = 0.$$

From the quotient on the right, we can see that if $|x|$ is small, then

$$\frac{1}{\sin x} - \frac{1}{x} \approx x \cdot \frac{1}{3!} = \frac{x}{6} \qquad \text{or} \qquad \csc x \approx \frac{1}{x} + \frac{x}{6}. \qquad \blacksquare$$

### Euler's Identity

A complex number is a number of the form $a + bi$, where $a$ and $b$ are real numbers and $i = \sqrt{-1}$ (see Appendix 7). If we substitute $x = i\theta$ ($\theta$ real) in the Taylor series for $e^x$ and use the relations

$$i^2 = -1, \qquad i^3 = i^2 i = -i, \qquad i^4 = i^2 i^2 = 1, \qquad i^5 = i^4 i = i,$$

and so on, to simplify the result, we obtain

$$e^{i\theta} = 1 + \frac{i\theta}{1!} + \frac{i^2\theta^2}{2!} + \frac{i^3\theta^3}{3!} + \frac{i^4\theta^4}{4!} + \frac{i^5\theta^5}{5!} + \frac{i^6\theta^6}{6!} + \cdots$$

$$= \left(1 - \frac{\theta^2}{2!} + \frac{\theta^4}{4!} - \frac{\theta^6}{6!} + \cdots\right) + i\left(\theta - \frac{\theta^3}{3!} + \frac{\theta^5}{5!} - \cdots\right) = \cos\theta + i\sin\theta.$$

This does not *prove* that $e^{i\theta} = \cos\theta + i\sin\theta$ because we have not yet defined what it means to raise $e$ to an imaginary power. Rather, it says how to define $e^{i\theta}$ to be consistent with other things we know about the exponential function for real numbers.

---

**DEFINITION**

For any real number $\theta$, $e^{i\theta} = \cos\theta + i\sin\theta$. (4)

---

Equation (4), called **Euler's identity**, enables us to define $e^{a+bi}$ to be $e^a \cdot e^{bi}$ for any complex number $a + bi$. One consequence of the identity is the equation

$$e^{i\pi} = -1.$$

When written in the form $e^{i\pi} + 1 = 0$, this equation combines five of the most important constants in mathematics.

---

**TABLE 10.1** Frequently used Taylor series

$$\frac{1}{1-x} = 1 + x + x^2 + \cdots + x^n + \cdots = \sum_{n=0}^{\infty} x^n, \qquad |x| < 1$$

$$\frac{1}{1+x} = 1 - x + x^2 - \cdots + (-x)^n + \cdots = \sum_{n=0}^{\infty} (-1)^n x^n, \qquad |x| < 1$$

$$e^x = 1 - x + \frac{x^2}{2!} + \cdots + \frac{x^n}{n!} + \cdots = \sum_{n=0}^{\infty} \frac{x^n}{n!}, \qquad |x| < \infty$$

$$\sin x = x - \frac{x^3}{3!} + \frac{x^5}{5!} - \cdots + (-1)^n \frac{x^{2n+1}}{(2n+1)!} + \cdots = \sum_{n=0}^{\infty} \frac{(-1)^n x^{2n+1}}{(2n+1)!}, \qquad |x| < \infty$$

$$\cos x = 1 - \frac{x^2}{2!} + \frac{x^4}{4!} - \cdots + (-1)^n \frac{x^{2n}}{(2n)!} + \cdots = \sum_{n=0}^{\infty} \frac{(-1)^n x^{2n}}{(2n)!}, \qquad |x| < \infty$$

$$\ln(1+x) = x - \frac{x^2}{2} + \frac{x^3}{3} - \cdots + (-1)^{n-1} \frac{x^n}{n} + \cdots = \sum_{n=1}^{\infty} \frac{(-1)^{n-1} x^n}{n}, \qquad -1 < x \le 1$$

$$\tan^{-1} x = x - \frac{x^3}{3} + \frac{x^5}{5} - \cdots + (-1)^n \frac{x^{2n+1}}{2n+1} + \cdots = \sum_{n=0}^{\infty} \frac{(-1)^n x^{2n+1}}{2n+1}, \qquad |x| \le 1$$

# Exercises 10.10

## Binomial Series

Find the first four terms of the binomial series for the functions in Exercises 1–10.

**1.** $(1 + x)^{1/2}$ 　　**2.** $(1 + x)^{1/3}$ 　　**3.** $(1 - x)^{-3}$

**4.** $(1 - 2x)^{1/2}$ 　　**5.** $\left(1 + \dfrac{x}{2}\right)^{-2}$ 　　**6.** $\left(1 - \dfrac{x}{3}\right)^{4}$

**7.** $(1 + x^3)^{-1/2}$ 　　**8.** $(1 + x^2)^{-1/3}$

**9.** $\left(1 + \dfrac{1}{x}\right)^{1/2}$ 　　**10.** $\dfrac{x}{\sqrt[3]{1 + x}}$

Find the binomial series for the functions in Exercises 11–14.

**11.** $(1 + x)^4$ 　　**12.** $(1 + x^2)^3$

**13.** $(1 - 2x)^3$ 　　**14.** $\left(1 - \dfrac{x}{2}\right)^{4}$

## Approximations and Nonelementary Integrals

T In Exercises 15–18, use series to estimate the integrals' values with an error of magnitude less than $10^{-5}$. (The answer section gives the integrals' values rounded to seven decimal places.)

**15.** $\displaystyle\int_0^{0.6} \sin x^2 \, dx$ 　　**16.** $\displaystyle\int_0^{0.4} \dfrac{e^{-x} - 1}{x} dx$

**17.** $\displaystyle\int_0^{0.5} \dfrac{1}{\sqrt{1 + x^4}} dx$ 　　**18.** $\displaystyle\int_0^{0.35} \sqrt[3]{1 + x^2} \, dx$

T Use series to approximate the values of the integrals in Exercises 19–22 with an error of magnitude less than $10^{-8}$.

**19.** $\displaystyle\int_0^{0.1} \dfrac{\sin x}{x} dx$ 　　**20.** $\displaystyle\int_0^{0.1} e^{-x^2} \, dx$

**21.** $\displaystyle\int_0^{0.1} \sqrt{1 + x^4} \, dx$ 　　**22.** $\displaystyle\int_0^{1} \dfrac{1 - \cos x}{x^2} dx$

**23.** Estimate the error if $\cos t^2$ is approximated by $1 - \dfrac{t^4}{2} + \dfrac{t^8}{4!}$ in the integral $\int_0^1 \cos t^2 \, dt$.

**24.** Estimate the error if $\cos \sqrt{t}$ is approximated by $1 - \dfrac{t}{2} + \dfrac{t^2}{4!} - \dfrac{t^3}{6!}$ in the integral $\int_0^1 \cos \sqrt{t} \, dt$.

In Exercises 25–28, find a polynomial that will approximate $F(x)$ throughout the given interval with an error of magnitude less than $10^{-3}$.

**25.** $F(x) = \displaystyle\int_0^x \sin t^2 \, dt, \quad [0, 1]$

**26.** $F(x) = \displaystyle\int_0^x t^2 e^{-t^2} \, dt, \quad [0, 1]$

**27.** $F(x) = \displaystyle\int_0^x \tan^{-1} t \, dt, \quad$ **(a)** $[0, 0.5]$ 　**(b)** $[0, 1]$

**28.** $F(x) = \displaystyle\int_0^x \dfrac{\ln(1 + t)}{t} dt, \quad$ **(a)** $[0, 0.5]$ 　**(b)** $[0, 1]$

## Indeterminate Forms

Use series to evaluate the limits in Exercises 29–40.

**29.** $\displaystyle\lim_{x\to 0} \dfrac{e^x - (1 + x)}{x^2}$ 　　**30.** $\displaystyle\lim_{x\to 0} \dfrac{e^x - e^{-x}}{x}$

**31.** $\displaystyle\lim_{t\to 0} \dfrac{1 - \cos t - (t^2/2)}{t^4}$ 　　**32.** $\displaystyle\lim_{\theta\to 0} \dfrac{\sin\theta - \theta + (\theta^3/6)}{\theta^5}$

**33.** $\displaystyle\lim_{y\to 0} \dfrac{y - \tan^{-1} y}{y^3}$ 　　**34.** $\displaystyle\lim_{y\to 0} \dfrac{\tan^{-1} y - \sin y}{y^3 \cos y}$

**35.** $\displaystyle\lim_{x\to\infty} x^2 \left(e^{-1/x^2} - 1\right)$ 　　**36.** $\displaystyle\lim_{x\to\infty} (x + 1) \sin \dfrac{1}{x + 1}$

**37.** $\displaystyle\lim_{x\to 0} \dfrac{\ln(1 + x^2)}{1 - \cos x}$ 　　**38.** $\displaystyle\lim_{x\to 2} \dfrac{x^2 - 4}{\ln(x - 1)}$

**39.** $\displaystyle\lim_{x\to 0} \dfrac{\sin 3x^2}{1 - \cos 2x}$ 　　**40.** $\displaystyle\lim_{x\to 0} \dfrac{\ln(1 + x^3)}{x \cdot \sin x^2}$

## Using Table 10.1

In Exercises 41–52, use Table 10.1 to find the sum of each series.

**41.** $1 + 1 + \dfrac{1}{2!} + \dfrac{1}{3!} + \dfrac{1}{4!} + \cdots$

**42.** $\left(\dfrac{1}{4}\right)^3 + \left(\dfrac{1}{4}\right)^4 + \left(\dfrac{1}{4}\right)^5 + \left(\dfrac{1}{4}\right)^6 + \cdots$

**43.** $1 - \dfrac{3^2}{4^2 \cdot 2!} + \dfrac{3^4}{4^4 \cdot 4!} - \dfrac{3^6}{4^6 \cdot 6!} + \cdots$

**44.** $\dfrac{1}{2} - \dfrac{1}{2 \cdot 2^2} + \dfrac{1}{3 \cdot 2^3} - \dfrac{1}{4 \cdot 2^4} + \cdots$

**45.** $\dfrac{\pi}{3} - \dfrac{\pi^3}{3^3 \cdot 3!} + \dfrac{\pi^5}{3^5 \cdot 5!} - \dfrac{\pi^7}{3^7 \cdot 7!} + \cdots$

**46.** $\dfrac{2}{3} - \dfrac{2^3}{3^3 \cdot 3} + \dfrac{2^5}{3^5 \cdot 5} - \dfrac{2^7}{3^7 \cdot 7} + \cdots$

**47.** $x^3 + x^4 + x^5 + x^6 + \cdots$

**48.** $1 - \dfrac{3^2 x^2}{2!} + \dfrac{3^4 x^4}{4!} - \dfrac{3^6 x^6}{6!} + \cdots$

**49.** $x^3 - x^5 + x^7 - x^9 + x^{11} - \cdots$

**50.** $x^2 - 2x^3 + \dfrac{2^2 x^4}{2!} - \dfrac{2^3 x^5}{3!} + \dfrac{2^4 x^6}{4!} - \cdots$

**51.** $-1 + 2x - 3x^2 + 4x^3 - 5x^4 + \cdots$

**52.** $1 + \dfrac{x}{2} + \dfrac{x^2}{3} + \dfrac{x^3}{4} + \dfrac{x^4}{5} + \cdots$

## Theory and Examples

**53.** Replace $x$ by $-x$ in the Taylor series for $\ln(1 + x)$ to obtain a series for $\ln(1 - x)$. Then subtract this from the Taylor series for $\ln(1 + x)$ to show that for $|x| < 1$,

$$\ln\dfrac{1 + x}{1 - x} = 2\left(x + \dfrac{x^3}{3} + \dfrac{x^5}{5} + \cdots\right).$$

**54.** How many terms of the Taylor series for $\ln(1 + x)$ should you add to be sure of calculating $\ln(1.1)$ with an error of magnitude less than $10^{-8}$? Give reasons for your answer.

**55.** According to the Alternating Series Estimation Theorem, how many terms of the Taylor series for $\tan^{-1} 1$ would you have to add to be sure of finding $\pi/4$ with an error of magnitude less than $10^{-3}$? Give reasons for your answer.

**56.** Show that the Taylor series for $f(x) = \tan^{-1} x$ diverges for $|x| > 1$.

T **57. Estimating Pi**    About how many terms of the Taylor series for $\tan^{-1} x$ would you have to use to evaluate each term on the right-hand side of the equation

$$\pi = 48 \tan^{-1}\frac{1}{18} + 32 \tan^{-1}\frac{1}{57} - 20 \tan^{-1}\frac{1}{239}$$

with an error of magnitude less than $10^{-6}$? In contrast, the convergence of $\sum_{n=1}^{\infty}(1/n^2)$ to $\pi^2/6$ is so slow that even 50 terms will not yield two-place accuracy.

**58.** Use the following steps to prove that the binomial series in Equation (1) converges to $(1 + x)^m$.

  **a.** Differentiate the series

$$f(x) = 1 + \sum_{k=1}^{\infty}\binom{m}{k}x^k$$

  to show that

$$f'(x) = \frac{mf(x)}{1 + x}, \quad -1 < x < 1.$$

  **b.** Define $g(x) = (1 + x)^{-m} f(x)$ and show that $g'(x) = 0$.

  **c.** From part (b), show that

$$f(x) = (1 + x)^m.$$

**59. a.** Use the binomial series and the fact that

$$\frac{d}{dx}\sin^{-1} x = (1 - x^2)^{-1/2}$$

  to generate the first four nonzero terms of the Taylor series for $\sin^{-1} x$. What is the radius of convergence?

  **b. Series for $\cos^{-1} x$**    Use your result in part (a) to find the first five nonzero terms of the Taylor series for $\cos^{-1} x$.

**60. a. Series for $\sinh^{-1} x$**    Find the first four nonzero terms of the Taylor series for

$$\sinh^{-1} x = \int_0^x \frac{dt}{\sqrt{1 + t^2}}.$$

T **b.** Use the first *three* terms of the series in part (a) to estimate $\sinh^{-1} 0.25$. Give an upper bound for the magnitude of the estimation error.

**61.** Obtain the Taylor series for $1/(1 + x)^2$ from the series for $-1/(1 + x)$.

**62.** Use the Taylor series for $1/(1 - x^2)$ to obtain a series for $2x/(1 - x^2)^2$.

T **63. Estimating Pi**    The English mathematician Wallis discovered the formula

$$\frac{\pi}{4} = \frac{2 \cdot 4 \cdot 4 \cdot 6 \cdot 6 \cdot 8 \cdot \cdots}{3 \cdot 3 \cdot 5 \cdot 5 \cdot 7 \cdot 7 \cdot \cdots}.$$

Find $\pi$ to two decimal places with this formula.

**64. The complete elliptic integral of the first kind** is the integral

$$K = \int_0^{\pi/2} \frac{d\theta}{\sqrt{1 - k^2 \sin^2 \theta}},$$

where $0 < k < 1$ is constant.

  **a.** Show that the first four terms of the binomial series for $1/\sqrt{1 - x}$ are

$$(1 - x)^{-1/2} = 1 + \frac{1}{2}x + \frac{1 \cdot 3}{2 \cdot 4}x^2 + \frac{1 \cdot 3 \cdot 5}{2 \cdot 4 \cdot 6}x^3 + \cdots.$$

  **b.** From part (a) and the reduction integral Formula 67 at the back of the book, show that

$$K = \frac{\pi}{2}\left[1 + \left(\frac{1}{2}\right)^2 k^2 + \left(\frac{1 \cdot 3}{2 \cdot 4}\right)^2 k^4 + \left(\frac{1 \cdot 3 \cdot 5}{2 \cdot 4 \cdot 6}\right)^2 k^6 + \cdots\right].$$

**65. Series for $\sin^{-1} x$**    Integrate the binomial series for $(1 - x^2)^{-1/2}$ to show that for $|x| < 1$,

$$\sin^{-1} x = x + \sum_{n=1}^{\infty} \frac{1 \cdot 3 \cdot 5 \cdot \cdots \cdot (2n - 1)}{2 \cdot 4 \cdot 6 \cdot \cdots \cdot (2n)}\frac{x^{2n+1}}{2n + 1}.$$

**66. Series for $\tan^{-1} x$ for $|x| > 1$**    Derive the series

$$\tan^{-1} x = \frac{\pi}{2} - \frac{1}{x} + \frac{1}{3x^3} - \frac{1}{5x^5} + \cdots, \quad x > 1$$

$$\tan^{-1} x = -\frac{\pi}{2} - \frac{1}{x} + \frac{1}{3x^3} - \frac{1}{5x^5} + \cdots, \quad x < -1,$$

by integrating the series

$$\frac{1}{1 + t^2} = \frac{1}{t^2} \cdot \frac{1}{1 + (1/t^2)} = \frac{1}{t^2} - \frac{1}{t^4} + \frac{1}{t^6} - \frac{1}{t^8} + \cdots$$

in the first case from $x$ to $\infty$ and in the second case from $-\infty$ to $x$.

### Euler's Identity

**67.** Use Equation (4) to write the following powers of $e$ in the form $a + bi$.

  **a.** $e^{-i\pi}$      **b.** $e^{i\pi/4}$      **c.** $e^{-i\pi/2}$

**68.** Use Equation (4) to show that

$$\cos \theta = \frac{e^{i\theta} + e^{-i\theta}}{2} \quad \text{and} \quad \sin \theta = \frac{e^{i\theta} - e^{-i\theta}}{2i}.$$

**69.** Establish the equations in Exercise 68 by combining the formal Taylor series for $e^{i\theta}$ and $e^{-i\theta}$.

**70.** Show that

  **a.** $\cosh i\theta = \cos \theta$,      **b.** $\sinh i\theta = i \sin \theta$.

**71.** By multiplying the Taylor series for $e^x$ and $\sin x$, find the terms through $x^5$ of the Taylor series for $e^x \sin x$. This series is the imaginary part of the series for

$$e^x \cdot e^{ix} = e^{(1+i)x}.$$

Use this fact to check your answer. For what values of $x$ should the series for $e^x \sin x$ converge?

**72.** When $a$ and $b$ are real, we define $e^{(a+ib)x}$ with the equation

$$e^{(a+ib)x} = e^{ax} \cdot e^{ibx} = e^{ax}(\cos bx + i \sin bx).$$

Differentiate the right-hand side of this equation to show that

$$\frac{d}{dx}e^{(a+ib)x} = (a + ib)e^{(a+ib)x}.$$

Thus the familiar rule $(d/dx)e^{kx} = ke^{kx}$ holds for $k$ complex as well as real.

**73.** Use the definition of $e^{i\theta}$ to show that for any real numbers $\theta$, $\theta_1$, and $\theta_2$,

**a.** $e^{i\theta_1}e^{i\theta_2} = e^{i(\theta_1+\theta_2)}$,         **b.** $e^{-i\theta} = 1/e^{i\theta}$.

**74.** Two complex numbers $a + ib$ and $c + id$ are equal if and only if $a = c$ and $b = d$. Use this fact to evaluate

$$\int e^{ax}\cos bx\, dx \quad \text{and} \quad \int e^{ax}\sin bx\, dx$$

from

$$\int e^{(a+ib)x}\, dx = \frac{a - ib}{a^2 + b^2}e^{(a+ib)x} + C,$$

where $C = C_1 + iC_2$ is a complex constant of integration.

# Chapter **10** Questions to Guide Your Review

**1.** What is an infinite sequence? What does it mean for such a sequence to converge? To diverge? Give examples.

**2.** What is a monotonic sequence? Under what circumstances does such a sequence have a limit? Give examples.

**3.** What theorems are available for calculating limits of sequences? Give examples.

**4.** What theorem sometimes enables us to use l'Hôpital's Rule to calculate the limit of a sequence? Give an example.

**5.** What are the six commonly occurring limits in Theorem 5 that arise frequently when you work with sequences and series?

**6.** What is an infinite series? What does it mean for such a series to converge? To diverge? Give examples.

**7.** What is a geometric series? When does such a series converge? Diverge? When it does converge, what is its sum? Give examples.

**8.** Besides geometric series, what other convergent and divergent series do you know?

**9.** What is the $n$th-Term Test for Divergence? What is the idea behind the test?

**10.** What can be said about term-by-term sums and differences of convergent series? About constant multiples of convergent and divergent series?

**11.** What happens if you add a finite number of terms to a convergent series? A divergent series? What happens if you delete a finite number of terms from a convergent series? A divergent series?

**12.** How do you reindex a series? Why might you want to do this?

**13.** Under what circumstances will an infinite series of nonnegative terms converge? Diverge? Why study series of nonnegative terms?

**14.** What is the Integral Test? What is the reasoning behind it? Give an example of its use.

**15.** When do $p$-series converge? Diverge? How do you know? Give examples of convergent and divergent $p$-series.

**16.** What are the Direct Comparison Test and the Limit Comparison Test? What is the reasoning behind these tests? Give examples of their use.

**17.** What are the Ratio and Root Tests? Do they always give you the information you need to determine convergence or divergence? Give examples.

**18.** What is absolute convergence? Conditional convergence? How are the two related?

**19.** What is an alternating series? What theorem is available for determining the convergence of such a series?

**20.** How can you estimate the error involved in approximating the sum of an alternating series with one of the series' partial sums? What is the reasoning behind the estimate?

**21.** What do you know about rearranging the terms of an absolutely convergent series? Of a conditionally convergent series?

**22.** What is a power series? How do you test a power series for convergence? What are the possible outcomes?

**23.** What are the basic facts about

    **a.** sums, differences, and products of power series?

    **b.** substitution of a function for $x$ in a power series?

    **c.** term-by-term differentiation of power series?

    **d.** term-by-term integration of power series?

    Give examples.

**24.** What is the Taylor series generated by a function $f(x)$ at a point $x = a$? What information do you need about $f$ to construct the series? Give an example.

**25.** What is a Maclaurin series?

**26.** Does a Taylor series always converge to its generating function? Explain.

**27.** What are Taylor polynomials? Of what use are they?

**28.** What is Taylor's formula? What does it say about the errors involved in using Taylor polynomials to approximate functions? In particular, what does Taylor's formula say about the error in a linearization? A quadratic approximation?

**29.** What is the binomial series? On what interval does it converge? How is it used?

**30.** How can you sometimes use power series to estimate the values of nonelementary definite integrals? To find limits?

**31.** What are the Taylor series for $1/(1 - x)$, $1/(1 + x)$, $e^x$, $\sin x$, $\cos x$, $\ln(1 + x)$, and $\tan^{-1}x$? How do you estimate the errors involved in replacing these series with their partial sums?

## Determining Convergence of Sequences

Which of the sequences whose $n$th terms appear in Exercises 1–18 converge, and which diverge? Find the limit of each convergent sequence.

**1.** $a_n = 1 + \dfrac{(-1)^n}{n}$

**2.** $a_n = \dfrac{1 - (-1)^n}{\sqrt{n}}$

**3.** $a_n = \dfrac{1 - 2^n}{2^n}$

**4.** $a_n = 1 + (0.9)^n$

**5.** $a_n = \sin \dfrac{n\pi}{2}$

**6.** $a_n = \sin n\pi$

**7.** $a_n = \dfrac{\ln(n^2)}{n}$

**8.** $a_n = \dfrac{\ln(2n + 1)}{n}$

**9.** $a_n = \dfrac{n + \ln n}{n}$

**10.** $a_n = \dfrac{\ln(2n^3 + 1)}{n}$

**11.** $a_n = \left(\dfrac{n - 5}{n}\right)^n$

**12.** $a_n = \left(1 + \dfrac{1}{n}\right)^{-n}$

**13.** $a_n = \sqrt[n]{\dfrac{3^n}{n}}$

**14.** $a_n = \left(\dfrac{3}{n}\right)^{1/n}$

**15.** $a_n = n(2^{1/n} - 1)$

**16.** $a_n = \sqrt[n]{2n + 1}$

**17.** $a_n = \dfrac{(n + 1)!}{n!}$

**18.** $a_n = \dfrac{(-4)^n}{n!}$

## Convergent Series

Find the sums of the series in Exercises 19–24.

**19.** $\displaystyle\sum_{n=3}^{\infty} \dfrac{1}{(2n - 3)(2n - 1)}$

**20.** $\displaystyle\sum_{n=2}^{\infty} \dfrac{-2}{n(n + 1)}$

**21.** $\displaystyle\sum_{n=1}^{\infty} \dfrac{9}{(3n - 1)(3n + 2)}$

**22.** $\displaystyle\sum_{n=3}^{\infty} \dfrac{-8}{(4n - 3)(4n + 1)}$

**23.** $\displaystyle\sum_{n=0}^{\infty} e^{-n}$

**24.** $\displaystyle\sum_{n=1}^{\infty} (-1)^n \dfrac{3}{4^n}$

## Determining Convergence of Series

Which of the series in Exercises 25–40 converge absolutely, which converge conditionally, and which diverge? Give reasons for your answers.

**25.** $\displaystyle\sum_{n=1}^{\infty} \dfrac{1}{\sqrt{n}}$

**26.** $\displaystyle\sum_{n=1}^{\infty} \dfrac{-5}{n}$

**27.** $\displaystyle\sum_{n=1}^{\infty} \dfrac{(-1)^n}{\sqrt{n}}$

**28.** $\displaystyle\sum_{n=1}^{\infty} \dfrac{1}{2n^3}$

**29.** $\displaystyle\sum_{n=1}^{\infty} \dfrac{(-1)^n}{\ln(n + 1)}$

**30.** $\displaystyle\sum_{n=2}^{\infty} \dfrac{1}{n(\ln n)^2}$

**31.** $\displaystyle\sum_{n=1}^{\infty} \dfrac{\ln n}{n^3}$

**32.** $\displaystyle\sum_{n=3}^{\infty} \dfrac{\ln n}{\ln(\ln n)}$

**33.** $\displaystyle\sum_{n=1}^{\infty} \dfrac{(-1)^n}{n\sqrt{n^2 + 1}}$

**34.** $\displaystyle\sum_{n=1}^{\infty} \dfrac{(-1)^n 3n^2}{n^3 + 1}$

**35.** $\displaystyle\sum_{n=1}^{\infty} \dfrac{n + 1}{n!}$

**36.** $\displaystyle\sum_{n=1}^{\infty} \dfrac{(-1)^n(n^2 + 1)}{2n^2 + n - 1}$

**37.** $\displaystyle\sum_{n=1}^{\infty} \dfrac{(-3)^n}{n!}$

**38.** $\displaystyle\sum_{n=1}^{\infty} \dfrac{2^n 3^n}{n^n}$

**39.** $\displaystyle\sum_{n=1}^{\infty} \dfrac{1}{\sqrt{n(n + 1)(n + 2)}}$

**40.** $\displaystyle\sum_{n=2}^{\infty} \dfrac{1}{n\sqrt{n^2 - 1}}$

## Power Series

In Exercises 41–50, **(a)** find the series' radius and interval of convergence. Then identify the values of $x$ for which the series converges **(b)** absolutely and **(c)** conditionally.

**41.** $\displaystyle\sum_{n=1}^{\infty} \dfrac{(x + 4)^n}{n3^n}$

**42.** $\displaystyle\sum_{n=1}^{\infty} \dfrac{(x - 1)^{2n-2}}{(2n - 1)!}$

**43.** $\displaystyle\sum_{n=1}^{\infty} \dfrac{(-1)^{n-1}(3x - 1)^n}{n^2}$

**44.** $\displaystyle\sum_{n=0}^{\infty} \dfrac{(n + 1)(2x + 1)^n}{(2n + 1)2^n}$

**45.** $\displaystyle\sum_{n=1}^{\infty} \dfrac{x^n}{n^n}$

**46.** $\displaystyle\sum_{n=1}^{\infty} \dfrac{x^n}{\sqrt{n}}$

**47.** $\displaystyle\sum_{n=0}^{\infty} \dfrac{(n + 1)x^{2n-1}}{3^n}$

**48.** $\displaystyle\sum_{n=0}^{\infty} \dfrac{(-1)^n(x - 1)^{2n+1}}{2n + 1}$

**49.** $\displaystyle\sum_{n=1}^{\infty} (\operatorname{csch} n)x^n$

**50.** $\displaystyle\sum_{n=1}^{\infty} (\coth n)x^n$

## Maclaurin Series

Each of the series in Exercises 51–56 is the value of the Taylor series at $x = 0$ of a function $f(x)$ at a particular point. What function and what point? What is the sum of the series?

**51.** $1 - \dfrac{1}{4} + \dfrac{1}{16} - \cdots + (-1)^n \dfrac{1}{4^n} + \cdots$

**52.** $\dfrac{2}{3} - \dfrac{4}{18} + \dfrac{8}{81} - \cdots + (-1)^{n-1} \dfrac{2^n}{n3^n} + \cdots$

**53.** $\pi - \dfrac{\pi^3}{3!} + \dfrac{\pi^5}{5!} - \cdots + (-1)^n \dfrac{\pi^{2n+1}}{(2n + 1)!} + \cdots$

**54.** $1 - \dfrac{\pi^2}{9 \cdot 2!} + \dfrac{\pi^4}{81 \cdot 4!} - \cdots + (-1)^n \dfrac{\pi^{2n}}{3^{2n}(2n)!} + \cdots$

**55.** $1 + \ln 2 + \dfrac{(\ln 2)^2}{2!} + \cdots + \dfrac{(\ln 2)^n}{n!} + \cdots$

**56.** $\dfrac{1}{\sqrt{3}} - \dfrac{1}{9\sqrt{3}} + \dfrac{1}{45\sqrt{3}} - \cdots$
$+ (-1)^{n-1} \dfrac{1}{(2n - 1)(\sqrt{3})^{2n-1}} + \cdots$

Find Taylor series at $x = 0$ for the functions in Exercises 57–64.

**57.** $\dfrac{1}{1 - 2x}$

**58.** $\dfrac{1}{1 + x^3}$

**59.** $\sin \pi x$

**60.** $\sin \dfrac{2x}{3}$

**61.** $\cos \left(x^{5/3}\right)$

**62.** $\cos \dfrac{x^3}{\sqrt{5}}$

**63.** $e^{(\pi x/2)}$

**64.** $e^{-x^2}$

### Taylor Series

In Exercises 65–68, find the first four nonzero terms of the Taylor series generated by $f$ at $x = a$.

**65.** $f(x) = \sqrt{3 + x^2}$ at $x = -1$

**66.** $f(x) = 1/(1 - x)$ at $x = 2$

**67.** $f(x) = 1/(x + 1)$ at $x = 3$

**68.** $f(x) = 1/x$ at $x = a > 0$

### Nonelementary Integrals

Use series to approximate the values of the integrals in Exercises 69–72 with an error of magnitude less than $10^{-8}$. (The answer section gives the integrals' values rounded to 10 decimal places.)

**69.** $\displaystyle\int_0^{1/2} e^{-x^3}\, dx$

**70.** $\displaystyle\int_0^1 x \sin\left(x^3\right) dx$

**71.** $\displaystyle\int_0^{1/2} \dfrac{\tan^{-1} x}{x}\, dx$

**72.** $\displaystyle\int_0^{1/64} \dfrac{\tan^{-1} x}{\sqrt{x}}\, dx$

### Using Series to Find Limits

In Exercises 73–78:

   **a.** Use power series to evaluate the limit.

   T  **b.** Then use a grapher to support your calculation.

**73.** $\displaystyle\lim_{x\to 0} \dfrac{7 \sin x}{e^{2x} - 1}$

**74.** $\displaystyle\lim_{\theta\to 0} \dfrac{e^{\theta} - e^{-\theta} - 2\theta}{\theta - \sin \theta}$

**75.** $\displaystyle\lim_{t\to 0} \left(\dfrac{1}{2 - 2\cos t} - \dfrac{1}{t^2}\right)$

**76.** $\displaystyle\lim_{h\to 0} \dfrac{(\sin h)/h - \cos h}{h^2}$

**77.** $\displaystyle\lim_{z\to 0} \dfrac{1 - \cos^2 z}{\ln(1 - z) + \sin z}$

**78.** $\displaystyle\lim_{y\to 0} \dfrac{y^2}{\cos y - \cosh y}$

### Theory and Examples

**79.** Use a series representation of $\sin 3x$ to find values of $r$ and $s$ for which

$$\lim_{x\to 0} \left(\dfrac{\sin 3x}{x^3} + \dfrac{r}{x^2} + s\right) = 0.$$

T **80.** Compare the accuracies of the approximations $\sin x \approx x$ and $\sin x \approx 6x/(6 + x^2)$ by comparing the graphs of $f(x) = \sin x - x$ and $g(x) = \sin x - (6x/(6 + x^2))$. Describe what you find.

**81.** Find the radius of convergence of the series

$$\sum_{n=1}^{\infty} \dfrac{2 \cdot 5 \cdot 8 \cdot \cdots \cdot (3n - 1)}{2 \cdot 4 \cdot 6 \cdot \cdots \cdot (2n)} x^n.$$

**82.** Find the radius of convergence of the series

$$\sum_{n=1}^{\infty} \dfrac{3 \cdot 5 \cdot 7 \cdot \cdots \cdot (2n + 1)}{4 \cdot 9 \cdot 14 \cdot \cdots \cdot (5n - 1)} (x - 1)^n.$$

**83.** Find a closed-form formula for the $n$th partial sum of the series $\sum_{n=2}^{\infty} \ln\left(1 - (1/n^2)\right)$ and use it to determine the convergence or divergence of the series.

**84.** Evaluate $\sum_{k=2}^{\infty}(1/(k^2 - 1))$ by finding the limits as $n \to \infty$ of the series' $n$th partial sum.

**85. a.** Find the interval of convergence of the series

$$y = 1 + \dfrac{1}{6}x^3 + \dfrac{1}{180}x^6 + \cdots$$
$$+ \dfrac{1 \cdot 4 \cdot 7 \cdot \cdots \cdot (3n - 2)}{(3n)!} x^{3n} + \cdots.$$

   **b.** Show that the function defined by the series satisfies a differential equation of the form

$$\dfrac{d^2 y}{dx^2} = x^a y + b$$

   and find the values of the constants $a$ and $b$.

**86. a.** Find the Maclaurin series for the function $x^2/(1 + x)$.

   **b.** Does the series converge at $x = 1$? Explain.

**87.** If $\sum_{n=1}^{\infty} a_n$ and $\sum_{n=1}^{\infty} b_n$ are convergent series of nonnegative numbers, can anything be said about $\sum_{n=1}^{\infty} a_n b_n$? Give reasons for your answer.

**88.** If $\sum_{n=1}^{\infty} a_n$ and $\sum_{n=1}^{\infty} b_n$ are divergent series of nonnegative numbers, can anything be said about $\sum_{n=1}^{\infty} a_n b_n$? Give reasons for your answer.

**89.** Prove that the sequence $\{x_n\}$ and the series $\sum_{k=1}^{\infty}(x_{k+1} - x_k)$ both converge or both diverge.

**90.** Prove that $\sum_{n=1}^{\infty}(a_n/(1 + a_n))$ converges if $a_n > 0$ for all $n$ and $\sum_{n=1}^{\infty} a_n$ converges.

**91.** Suppose that $a_1, a_2, a_3, \ldots, a_n$ are positive numbers satisfying the following conditions:

   **i)** $a_1 \geq a_2 \geq a_3 \geq \cdots$;

   **ii)** the series $a_2 + a_4 + a_8 + a_{16} + \cdots$ diverges.

   Show that the series

$$\dfrac{a_1}{1} + \dfrac{a_2}{2} + \dfrac{a_3}{3} + \cdots$$

diverges.

**92.** Use the result in Exercise 91 to show that

$$1 + \sum_{n=2}^{\infty} \dfrac{1}{n \ln n}$$

diverges.

## Chapter **10** Additional and Advanced Exercises

### Determining Convergence of Series

Which of the series $\sum_{n=1}^{\infty} a_n$ defined by the formulas in Exercises 1–4 converge, and which diverge? Give reasons for your answers.

**1.** $\sum_{n=1}^{\infty} \dfrac{1}{(3n-2)^{n+(1/2)}}$

**2.** $\sum_{n=1}^{\infty} \dfrac{(\tan^{-1} n)^2}{n^2+1}$

**3.** $\sum_{n=1}^{\infty} (-1)^n \tanh n$

**4.** $\sum_{n=2}^{\infty} \dfrac{\log_n(n!)}{n^3}$

Which of the series $\sum_{n=1}^{\infty} a_n$ defined by the formulas in Exercises 5–8 converge, and which diverge? Give reasons for your answers.

**5.** $a_1 = 1, \quad a_{n+1} = \dfrac{n(n+1)}{(n+2)(n+3)} a_n$

(*Hint:* Write out several terms, see which factors cancel, and then generalize.)

**6.** $a_1 = a_2 = 7, \quad a_{n+1} = \dfrac{n}{(n-1)(n+1)} a_n \quad$ if $n \geq 2$

**7.** $a_1 = a_2 = 1, \quad a_{n+1} = \dfrac{1}{1 + a_n} \quad$ if $n \geq 2$

**8.** $a_n = 1/3^n \quad$ if $n$ is odd, $\quad a_n = n/3^n \quad$ if $n$ is even

### Choosing Centers for Taylor Series

Taylor's formula

$$f(x) = f(a) + f'(a)(x-a) + \frac{f''(a)}{2!}(x-a)^2 + \cdots$$

$$+ \frac{f^{(n)}(a)}{n!}(x-a)^n + \frac{f^{(n+1)}(c)}{(n+1)!}(x-a)^{n+1}$$

expresses the value of $f$ at $x$ in terms of the values of $f$ and its derivatives at $x = a$. In numerical computations, we therefore need $a$ to be a point where we know the values of $f$ and its derivatives. We also need $a$ to be close enough to the values of $f$ we are interested in to make $(x-a)^{n+1}$ so small we can neglect the remainder.

In Exercises 9–14, what Taylor series would you choose to represent the function near the given value of $x$? (There may be more than one good answer.) Write out the first four nonzero terms of the series you choose.

**9.** $\cos x \quad$ near $\quad x = 1$

**10.** $\sin x \quad$ near $\quad x = 6.3$

**11.** $e^x \quad$ near $\quad x = 0.4$

**12.** $\ln x \quad$ near $\quad x = 1.3$

**13.** $\cos x \quad$ near $\quad x = 69$

**14.** $\tan^{-1} x \quad$ near $\quad x = 2$

### Theory and Examples

**15.** Let $a$ and $b$ be constants with $0 < a < b$. Does the sequence $\{(a^n + b^n)^{1/n}\}$ converge? If it does converge, what is the limit?

**16.** Find the sum of the infinite series

$$1 + \frac{2}{10} + \frac{3}{10^2} + \frac{7}{10^3} + \frac{2}{10^4} + \frac{3}{10^5} + \frac{7}{10^6} + \frac{2}{10^7}$$

$$+ \frac{3}{10^8} + \frac{7}{10^9} + \cdots.$$

**17.** Evaluate

$$\sum_{n=0}^{\infty} \int_n^{n+1} \frac{1}{1+x^2} \, dx.$$

**18.** Find all values of $x$ for which

$$\sum_{n=1}^{\infty} \frac{nx^n}{(n+1)(2x+1)^n}$$

converges absolutely.

T **19. a.** Does the value of

$$\lim_{n \to \infty} \left(1 - \frac{\cos(a/n)}{n}\right)^n, \quad a \text{ constant},$$

appear to depend on the value of $a$? If so, how?

**b.** Does the value of

$$\lim_{n \to \infty} \left(1 - \frac{\cos(a/n)}{bn}\right)^n, \quad a \text{ and } b \text{ constant}, b \neq 0,$$

appear to depend on the value of $b$? If so, how?

**c.** Use calculus to confirm your findings in parts (a) and (b).

**20.** Show that if $\sum_{n=1}^{\infty} a_n$ converges, then

$$\sum_{n=1}^{\infty} \left(\frac{1 + \sin(a_n)}{2}\right)^n$$

converges.

**21.** Find a value for the constant $b$ that will make the radius of convergence of the power series

$$\sum_{n=2}^{\infty} \frac{b^n x^n}{\ln n}$$

equal to 5.

**22.** How do you know that the functions $\sin x$, $\ln x$, and $e^x$ are not polynomials? Give reasons for your answer.

**23.** Find the value of $a$ for which the limit

$$\lim_{x \to 0} \frac{\sin(ax) - \sin x - x}{x^3}$$

is finite and evaluate the limit.

**24.** Find values of $a$ and $b$ for which

$$\lim_{x \to 0} \frac{\cos(ax) - b}{2x^2} = -1.$$

**25. Raabe's (or Gauss's) Test** The following test, which we state without proof, is an extension of the Ratio Test.

*Raabe's Test:* If $\sum_{n=1}^{\infty} u_n$ is a series of positive constants and there exist constants $C$, $K$, and $N$ such that

$$\frac{u_n}{u_{n+1}} = 1 + \frac{C}{n} + \frac{f(n)}{n^2},$$

where $|f(n)| < K$ for $n \geq N$, then $\sum_{n=1}^{\infty} u_n$ converges if $C > 1$ and diverges if $C \leq 1$.

Show that the results of Raabe's Test agree with what you know about the series $\sum_{n=1}^{\infty} (1/n^2)$ and $\sum_{n=1}^{\infty} (1/n)$.

**26.** (*Continuation of Exercise 25.*) Suppose that the terms of $\sum_{n=1}^{\infty} u_n$ are defined recursively by the formulas

$$u_1 = 1, \quad u_{n+1} = \frac{(2n-1)^2}{(2n)(2n+1)} u_n.$$

Apply Raabe's Test to determine whether the series converges.

**27.** If $\sum_{n=1}^{\infty} a_n$ converges, and if $a_n \neq 1$ and $a_n > 0$ for all $n$,

**a.** Show that $\sum_{n=1}^{\infty} a_n^2$ converges.

**b.** Does $\sum_{n=1}^{\infty} a_n/(1 - a_n)$ converge? Explain.

**28.** (*Continuation of Exercise 27.*) If $\sum_{n=1}^{\infty} a_n$ converges, and if $1 > a_n > 0$ for all $n$, show that $\sum_{n=1}^{\infty} \ln(1 - a_n)$ converges.

(*Hint:* First show that $|\ln(1 - a_n)| \leq a_n/(1 - a_n)$.)

**29. Nicole Oresme's Theorem** Prove Nicole Oresme's Theorem that

$$1 + \frac{1}{2} \cdot 2 + \frac{1}{4} \cdot 3 + \cdots + \frac{n}{2^{n-1}} + \cdots = 4.$$

(*Hint:* Differentiate both sides of the equation $1/(1 - x) = 1 + \sum_{n=1}^{\infty} x^n$.)

**30. a.** Show that

$$\sum_{n=1}^{\infty} \frac{n(n+1)}{x^n} = \frac{2x^2}{(x-1)^3}$$

for $|x| > 1$ by differentiating the identity

$$\sum_{n=1}^{\infty} x^{n+1} = \frac{x^2}{1 - x}$$

twice, multiplying the result by $x$, and then replacing $x$ by $1/x$.

**b.** Use part (a) to find the real solution greater than 1 of the equation

$$x = \sum_{n=1}^{\infty} \frac{n(n+1)}{x^n}.$$

**31. Quality control**

**a.** Differentiate the series

$$\frac{1}{1 - x} = 1 + x + x^2 + \cdots + x^n + \cdots$$

to obtain a series for $1/(1 - x)^2$.

**b.** In one throw of two dice, the probability of getting a roll of 7 is $p = 1/6$. If you throw the dice repeatedly, the probability that a 7 will appear for the first time at the $n$th throw is $q^{n-1}p$, where $q = 1 - p = 5/6$. The expected number of throws until a 7 first appears is $\sum_{n=1}^{\infty} nq^{n-1}p$. Find the sum of this series.

**c.** As an engineer applying statistical control to an industrial operation, you inspect items taken at random from the assembly line. You classify each sampled item as either "good" or "bad." If the probability of an item's being good is $p$ and of an item's being bad is $q = 1 - p$, the probability that the first bad item found is the $n$th one inspected is $p^{n-1}q$. The average number inspected up to and including the first bad item found is $\sum_{n=1}^{\infty} np^{n-1}q$. Evaluate this sum, assuming $0 < p < 1$.

**32. Expected value** Suppose that a random variable $X$ may assume the values $1, 2, 3, \ldots$, with probabilities $p_1, p_2, p_3, \ldots$, where $p_k$ is the probability that $X$ equals $k$ ($k = 1, 2, 3, \ldots$). Suppose also that $p_k \geq 0$ and that $\sum_{k=1}^{\infty} p_k = 1$. The **expected value** of $X$, denoted by $E(X)$, is the number $\sum_{k=1}^{\infty} kp_k$, provided the series converges. In each of the following cases, show that $\sum_{k=1}^{\infty} p_k = 1$ and find $E(X)$ if it exists. (*Hint:* See Exercise 31.)

**a.** $p_k = 2^{-k}$ \qquad **b.** $p_k = \dfrac{5^{k-1}}{6^k}$

**c.** $p_k = \dfrac{1}{k(k+1)} = \dfrac{1}{k} - \dfrac{1}{k+1}$

T **33. Safe and effective dosage** The concentration in the blood resulting from a single dose of a drug normally decreases with time as the drug is eliminated from the body. Doses may therefore need to be repeated periodically to keep the concentration from dropping below some particular level. One model for the effect of repeated doses gives the residual concentration just before the $(n+1)$st dose as

$$R_n = C_0 e^{-kt_0} + C_0 e^{-2kt_0} + \cdots + C_0 e^{-nkt_0},$$

where $C_0 =$ the change in concentration achievable by a single dose (mg/mL), $k =$ the *elimination constant* (h$^{-1}$), and $t_0 =$ time between doses (h). See the accompanying figure.

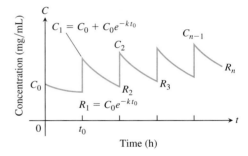

**a.** Write $R_n$ in closed from as a single fraction, and find $R = \lim_{n\to\infty} R_n$.

**b.** Calculate $R_1$ and $R_{10}$ for $C_0 = 1$ mg/mL, $k = 0.1$ h$^{-1}$, and $t_0 = 10$ h. How good an estimate of $R$ is $R_{10}$?

**c.** If $k = 0.01$ h$^{-1}$ and $t_0 = 10$ h, find the smallest $n$ such that $R_n > (1/2)R$. Use $C_0 = 1$ mg/mL.

(*Source: Prescribing Safe and Effective Dosage,* B. Horelick and S. Koont, COMAP, Inc., Lexington, MA.)

**34. Time between drug doses** (*Continuation of Exercise 33.*) If a drug is known to be ineffective below a concentration $C_L$ and harmful above some higher concentration $C_H$, one need to find values of $C_0$ and $t_0$ that will produce a concentration that is safe (not above $C_H$) but effective (not below $C_L$). See the accompanying figure. We therefore want to find values for $C_0$ and $t_0$ for which

$$R = C_L \quad \text{and} \quad C_0 + R = C_H.$$

Thus $C_0 = C_H - C_L$. When these values are substituted in the equation for $R$ obtained in part (a) of Exercise 33, the resulting equation simplifies to

$$t_0 = \frac{1}{k} \ln \frac{C_H}{C_L}.$$

To reach an effective level rapidly, one might administer a "loading" dose that would produce a concentration of $C_H$ mg/mL. This could be followed every $t_0$ hours by a dose that raises the concentration by $C_0 = C_H - C_L$ mg/mL.

**a.** Verify the preceding equation for $t_0$.

**b.** If $k = 0.05\ h^{-1}$ and the highest safe concentration is $e$ times the lowest effective concentration, find the length of time between doses that will ensure safe and effective concentrations.

**c.** Given $C_H = 2$ mg/mL, $C_L = 0.5$ mg/mL, and $k = 0.02\ h^{-1}$, determine a scheme for administering the drug.

**d.** Suppose that $k = 0.2\ h^{-1}$ and that the smallest effective concentration is $0.03$ mg/mL. A single dose that produces a concentration of $0.1$ mg/mL is administered. About how long will the drug remain effective?

---

# Chapter 10    Technology Application Projects

## Mathematica/Maple Modules:

### Bouncing Ball

The model predicts the height of a bouncing ball, and the time until it stops bouncing.

### Taylor Polynomial Approximations of a Function

A graphical animation shows the convergence of the Taylor polynomials to functions having derivatives of all orders over an interval in their domains.

# 11

# Parametric Equations and Polar Coordinates

**OVERVIEW** In this chapter we study new ways to define curves in the plane. Instead of thinking of a curve as the graph of a function or equation, we consider a more general way of thinking of a curve as the path of a moving particle whose position is changing over time. Then each of the $x$- and $y$-coordinates of the particle's position becomes a function of a third variable $t$. We can also change the way in which points in the plane themselves are described by using *polar coordinates* rather than the rectangular or Cartesian system. Both of these new tools are useful for describing motion, like that of planets and satellites, or projectiles moving in the plane or space. In addition, we review the geometric definitions and standard equations of parabolas, ellipses, and hyperbolas. These curves are called *conic sections*, or *conics*, and model the paths traveled by projectiles, planets, or any other object moving under the sole influence of a gravitational or electromagnetic force.

## 11.1 Parametrizations of Plane Curves

In previous chapters, we have studied curves as the graphs of functions or of equations involving the two variables $x$ and $y$. We are now going to introduce another way to describe a curve by expressing both coordinates as functions of a third variable $t$.

### Parametric Equations

Figure 11.1 shows the path of a moving particle in the $xy$-plane. Notice that the path fails the vertical line test, so it cannot be described as the graph of a function of the variable $x$. However, we can sometimes describe the path by a pair of equations, $x = f(t)$ and $y = g(t)$, where $f$ and $g$ are continuous functions. When studying motion, $t$ usually denotes time. Equations like these describe more general curves than those described by a single function, and they provide not only the graph of the path traced out but also the location of the particle $(x, y) = (f(t), g(t))$ at any time $t$.

> **DEFINITION** If $x$ and $y$ are given as functions
>
> $$x = f(t), \qquad y = g(t)$$
>
> over an interval $I$ of $t$-values, then the set of points $(x, y) = (f(t), g(t))$ defined by these equations is a **parametric curve**. The equations are **parametric equations** for the curve.

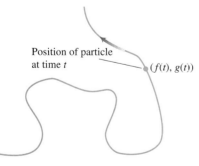

**FIGURE 11.1** The curve or path traced by a particle moving in the $xy$-plane is not always the graph of a function or single equation.

The variable $t$ is a **parameter** for the curve, and its domain $I$ is the **parameter interval**. If $I$ is a closed interval, $a \le t \le b$, the point $(f(a), g(a))$ is the **initial point** of the curve and $(f(b), g(b))$ is the **terminal point**. When we give parametric equations and a parameter

**655**

interval for a curve, we say that we have **parametrized** the curve. The equations and interval together constitute a **parametrization** of the curve. A given curve can be represented by different sets of parametric equations. (See Exercises 19 and 20.)

**EXAMPLE 1**  Sketch the curve defined by the parametric equations

$$x = t^2, \qquad y = t + 1, \qquad -\infty < t < \infty.$$

**Solution**  We make a brief table of values (Table 11.1), plot the points $(x, y)$, and draw a smooth curve through them (Figure 11.2). Each value of $t$ gives a point $(x, y)$ on the curve, such as $t = 1$ giving the point $(1, 2)$ recorded in Table 11.1. If we think of the curve as the path of a moving particle, then the particle moves along the curve in the direction of the arrows shown in Figure 11.2. Although the time intervals in the table are equal, the consecutive points plotted along the curve are not at equal arc length distances. The reason for this is that the particle slows down as it gets nearer to the $y$-axis along the lower branch of the curve as $t$ increases, and then speeds up after reaching the $y$-axis at $(0, 1)$ and moving along the upper branch. Since the interval of values for $t$ is all real numbers, there is no initial point and no terminal point for the curve.  ∎

**TABLE 11.1** Values of $x = t^2$ and $y = t + 1$ for selected values of $t$.

| $t$ | $x$ | $y$ |
|-----|-----|-----|
| $-3$ | 9 | $-2$ |
| $-2$ | 4 | $-1$ |
| $-1$ | 1 | 0 |
| 0 | 0 | 1 |
| 1 | 1 | 2 |
| 2 | 4 | 3 |
| 3 | 9 | 4 |

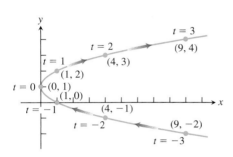

**FIGURE 11.2**  The curve given by the parametric equations $x = t^2$ and $y = t + 1$ (Example 1).

**EXAMPLE 2**  Identify geometrically the curve in Example 1 (Figure 11.2) by eliminating the parameter $t$ and obtaining an algebraic equation in $x$ and $y$.

**Solution**  We solve the equation $y = t + 1$ for the parameter $t$ and substitute the result into the parametric equation for $x$. This procedure gives $t = y - 1$ and

$$x = t^2 = (y - 1)^2 = y^2 - 2y + 1.$$

The equation $x = y^2 - 2y + 1$ represents a parabola, as displayed in Figure 11.2. It is sometimes quite difficult, or even impossible, to eliminate the parameter from a pair of parametric equations, as we did here.  ∎

**EXAMPLE 3**  Graph the parametric curves

**(a)**  $x = \cos t, \qquad y = \sin t, \qquad 0 \le t \le 2\pi.$

**(b)**  $x = a \cos t, \qquad y = a \sin t, \qquad 0 \le t \le 2\pi.$

**Solution**

**(a)**  Since $x^2 + y^2 = \cos^2 t + \sin^2 t = 1$, the parametric curve lies along the unit circle $x^2 + y^2 = 1$. As $t$ increases from 0 to $2\pi$, the point $(x, y) = (\cos t, \sin t)$ starts at $(1, 0)$ and traces the entire circle once counterclockwise (Figure 11.3).

**Figure 11.3**  The equations $x = \cos t$ and $y = \sin t$ describe motion on the circle $x^2 + y^2 = 1$. The arrow shows the direction of increasing $t$ (Example 3).

**(b)** For $x = a \cos t$, $y = a \sin t$, $0 \leq t \leq 2\pi$, we have $x^2 + y^2 = a^2 \cos^2 t + a^2 \sin^2 t = a^2$. The parametrization describes a motion that begins at the point $(a, 0)$ and traverses the circle $x^2 + y^2 = a^2$ once counterclockwise, returning to $(a, 0)$ at $t = 2\pi$. The graph is a circle centered at the origin with radius $r = |a|$ and coordinate points $(a \cos t, a \sin t)$. ∎

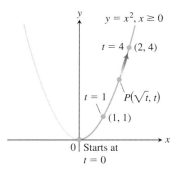

**FIGURE 11.4** The equations $x = \sqrt{t}$ and $y = t$ and the interval $t \geq 0$ describe the path of a particle that traces the right-hand half of the parabola $y = x^2$ (Example 4).

**EXAMPLE 4** The position $P(x, y)$ of a particle moving in the $xy$-plane is given by the equations and parameter interval

$$x = \sqrt{t}, \qquad y = t, \qquad t \geq 0.$$

Identify the path traced by the particle and describe the motion.

**Solution** We try to identify the path by eliminating $t$ between the equations $x = \sqrt{t}$ and $y = t$, which might produce a recognizable algebraic relation between $x$ and $y$. We find that

$$y = t = \left(\sqrt{t}\right)^2 = x^2.$$

Thus, the particle's position coordinates satisfy the equation $y = x^2$, so the particle moves along the parabola $y = x^2$.

It would be a mistake, however, to conclude that the particle's path is the entire parabola $y = x^2$; it is only half the parabola. The particle's $x$-coordinate is never negative. The particle starts at $(0, 0)$ when $t = 0$ and rises into the first quadrant as $t$ increases (Figure 11.4). The parameter interval is $[0, \infty)$ and there is no terminal point. ∎

The graph of any function $y = f(x)$ can always be given a **natural parametrization** $x = t$ and $y = f(t)$. The domain of the parameter in this case is the same as the domain of the function $f$.

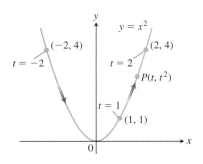

**FIGURE 11.5** The path defined by $x = t, y = t^2, -\infty < t < \infty$ is the entire parabola $y = x^2$ (Example 5).

**EXAMPLE 5** A parametrization of the graph of the function $f(x) = x^2$ is given by

$$x = t, \qquad y = f(t) = t^2, \qquad -\infty < t < \infty.$$

When $t \geq 0$, this parametrization gives the same path in the $xy$-plane as we had in Example 4. However, since the parameter $t$ here can now also be negative, we obtain the left-hand part of the parabola as well; that is, we have the entire parabolic curve. For this parametrization, there is no starting point and no terminal point (Figure 11.5). ∎

Notice that a parametrization also specifies *when* (the value of the parameter) a particle moving along the curve is *located* at a specific point along the curve. In Example 4, the point $(2, 4)$ is reached when $t = 4$; in Example 5, it is reached "earlier" when $t = 2$. You can see the implications of this aspect of parametrizations when considering the possibility of two objects coming into collision: they have to be at the exact same location point $P(x, y)$ for some (possibly different) values of their respective parameters. We will say more about this aspect of parametrizations when we study motion in Chapter 13.

**EXAMPLE 6** Find a parametrization for the line through the point $(a, b)$ having slope $m$.

**Solution** A Cartesian equation of the line is $y - b = m(x - a)$. If we set the parameter $t = x - a$, we find that $x = a + t$ and $y - b = mt$. That is,

$$x = a + t, \qquad y = b + mt, \qquad -\infty < t < \infty$$

parametrizes the line. This parametrization differs from the one we would obtain by the natural parametrization in Example 5 when $t = x$. However, both parametrizations describe the same line. ∎

**TABLE 11.2** Values of
$x = t + (1/t)$ and $y = t - (1/t)$
for selected values of $t$.

| $t$ | $1/t$ | $x$ | $y$ |
|-----|-------|-----|------|
| 0.1 | 10.0 | 10.1 | −9.9 |
| 0.2 | 5.0 | 5.2 | −4.8 |
| 0.4 | 2.5 | 2.9 | −2.1 |
| 1.0 | 1.0 | 2.0 | 0.0 |
| 2.0 | 0.5 | 2.5 | 1.5 |
| 5.0 | 0.2 | 5.2 | 4.8 |
| 10.0 | 0.1 | 10.1 | 9.9 |

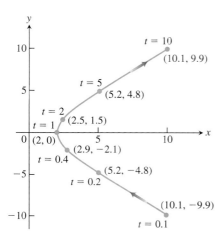

**FIGURE 11.6** The curve for
$x = t + (1/t)$, $y = t - (1/t)$, $t > 0$
in Example 7. (The part shown is for
$0.1 \leq t \leq 10$.)

**HISTORICAL BIOGRAPHY**

Christian Huygens
(1629–1695)

**EXAMPLE 7**    Sketch and identify the path traced by the point $P(x, y)$ if

$$x = t + \frac{1}{t}, \qquad y = t - \frac{1}{t}, \qquad t > 0.$$

**Solution**    We make a brief table of values in Table 11.2, plot the points, and draw a smooth curve through them, as we did in Example 1. Next we eliminate the parameter $t$ from the equations. The procedure is more complicated than in Example 2. Taking the difference between $x$ and $y$ as given by the parametric equations, we find that

$$x - y = \left(t + \frac{1}{t}\right) - \left(t - \frac{1}{t}\right) = \frac{2}{t}.$$

If we add the two parametric equations, we get

$$x + y = \left(t + \frac{1}{t}\right) + \left(t - \frac{1}{t}\right) = 2t.$$

We can then eliminate the parameter $t$ by multiplying these last equations together:

$$(x - y)(x + y) = \left(\frac{2}{t}\right)(2t) = 4,$$

or, expanding the expression on the left-hand side, we obtain a standard equation for a hyperbola (reviewed in Section 11.6):

$$x^2 - y^2 = 4. \tag{1}$$

Thus the coordinates of all the points $P(x, y)$ described by the parametric equations satisfy Equation (1). However, Equation (1) does not require that the $x$-coordinate be positive. So there are points $(x, y)$ on the hyperbola that do not satisfy the parametric equation $x = t + (1/t)$, $t > 0$, for which $x$ is always positive. That is, the parametric equations do not yield any points on the left branch of the hyperbola given by Equation (1), points where the $x$-coordinate would be negative. For small positive values of $t$, the path lies in the fourth quadrant and rises into the first quadrant as $t$ increases, crossing the $x$-axis when $t = 1$ (see Figure 11.6). The parameter domain is $(0, \infty)$ and there is no starting point and no terminal point for the path.    ∎

Examples 4, 5, and 6 illustrate that a given curve, or portion of it, can be represented by different parametrizations. In the case of Example 7, we can also represent the right-hand branch of the hyperbola by the parametrization

$$x = \sqrt{4 + t^2}, \qquad y = t, \qquad -\infty < t < \infty,$$

which is obtained by solving Equation (1) for $x \geq 0$ and letting $y$ be the parameter. Still another parametrization for the right-hand branch of the hyperbola given by Equation (1) is

$$x = 2 \sec t, \qquad y = 2 \tan t, \qquad -\frac{\pi}{2} < t < \frac{\pi}{2}.$$

This parametrization follows from the trigonometric identity $\sec^2 t - \tan^2 t = 1$, so

$$x^2 - y^2 = 4 \sec^2 t - 4 \tan^2 t = 4(\sec^2 t - \tan^2 t) = 4.$$

As $t$ runs between $-\pi/2$ and $\pi/2$, $x = \sec t$ remains positive and $y = \tan t$ runs between $-\infty$ and $\infty$, so $P$ traverses the hyperbola's right-hand branch. It comes in along the branch's lower half as $t \to 0^-$, reaches $(2, 0)$ at $t = 0$, and moves out into the first quadrant as $t$ increases steadily toward $\pi/2$. This is the same hyperbola branch for which a portion is shown in Figure 11.6.

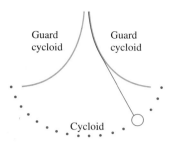

**FIGURE 11.7** In Huygens' pendulum clock, the bob swings in a cycloid, so the frequency is independent of the amplitude.

## Cycloids

The problem with a pendulum clock whose bob swings in a circular arc is that the frequency of the swing depends on the amplitude of the swing. The wider the swing, the longer it takes the bob to return to center (its lowest position).

This does not happen if the bob can be made to swing in a *cycloid*. In 1673, Christian Huygens designed a pendulum clock whose bob would swing in a cycloid, a curve we define in Example 8. He hung the bob from a fine wire constrained by guards that caused it to draw up as it swung away from center (Figure 11.7), and we describe the path parametrically in the next example.

**EXAMPLE 8** A wheel of radius $a$ rolls along a horizontal straight line. Find parametric equations for the path traced by a point $P$ on the wheel's circumference. The path is called a **cycloid**.

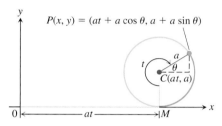

**FIGURE 11.8** The position of $P(x, y)$ on the rolling wheel at angle $t$ (Example 8).

**Solution** We take the line to be the $x$-axis, mark a point $P$ on the wheel, start the wheel with $P$ at the origin, and roll the wheel to the right. As parameter, we use the angle $t$ through which the wheel turns, measured in radians. Figure 11.8 shows the wheel a short while later when its base lies $at$ units from the origin. The wheel's center $C$ lies at $(at, a)$ and the coordinates of $P$ are

$$x = at + a\cos\theta, \qquad y = a + a\sin\theta.$$

To express $\theta$ in terms of $t$, we observe that $t + \theta = 3\pi/2$ in the figure, so that

$$\theta = \frac{3\pi}{2} - t.$$

This makes

$$\cos\theta = \cos\left(\frac{3\pi}{2} - t\right) = -\sin t, \qquad \sin\theta = \sin\left(\frac{3\pi}{2} - t\right) = -\cos t.$$

The equations we seek are

$$x = at - a\sin t, \qquad y = a - a\cos t.$$

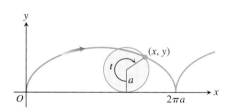

**FIGURE 11.9** The cycloid curve $x = a(t - \sin t)$, $y = a(1 - \cos t)$, for $t \geq 0$.

These are usually written with the $a$ factored out:

$$x = a(t - \sin t), \qquad y = a(1 - \cos t). \tag{2}$$

Figure 11.9 shows the first arch of the cycloid and part of the next. ∎

## Brachistochrones and Tautochrones

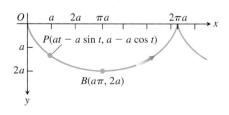

**FIGURE 11.10** Turning Figure 11.9 upside down, the $y$-axis points downward, indicating the direction of the gravitational force. Equations (2) still describe the curve parametrically.

If we turn Figure 11.9 upside down, Equations (2) still apply and the resulting curve (Figure 11.10) has two interesting physical properties. The first relates to the origin $O$ and the point $B$ at the bottom of the first arch. Among all smooth curves joining these points, the cycloid is the curve along which a frictionless bead, subject only to the force of gravity, will slide from $O$ to $B$ the fastest. This makes the cycloid a **brachistochrone** ("brah-*kiss*-toe-krone"), or shortest-time curve for these points. The second property is that even if you start the bead partway down the curve toward $B$, it will still take the bead the same amount of time to reach $B$. This makes the cycloid a **tautochrone** ("*taw*-toe-krone"), or same-time curve for $O$ and $B$.

Are there any other brachistochrones joining $O$ and $B$, or is the cycloid the only one? We can formulate this as a mathematical question in the following way. At the start, the kinetic energy of the bead is zero, since its velocity (speed) is zero. The work done by gravity in moving the bead from $(0, 0)$ to any other point $(x, y)$ in the plane is $mgy$, and this must equal the change in kinetic energy. (See Exercise 23 in Section 6.5.) That is,

$$mgy = \frac{1}{2}mv^2 - \frac{1}{2}m(0)^2.$$

Thus, the speed of the bead when it reaches $(x, y)$ has to be

$$v = \sqrt{2gy}.$$

That is,

$$\frac{ds}{dT} = \sqrt{2gy} \qquad \text{\small{$ds$ is the arc length differential along the bead's path and $T$ represents time.}}$$

or

$$dT = \frac{ds}{\sqrt{2gy}} = \frac{\sqrt{1 + (dy/dx)^2}\, dx}{\sqrt{2gy}}. \qquad (3)$$

The time $T_f$ it takes the bead to slide along a particular path $y = f(x)$ from $O$ to $B(a\pi, 2a)$ is

$$T_f = \int_{x=0}^{x=a\pi} \sqrt{\frac{1 + (dy/dx)^2}{2gy}}\, dx. \qquad (4)$$

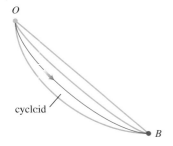

$O$

cyclcid

$B$

**FIGURE 11.11** The cycloid is the unique curve which minimizes the time it takes for a frictionless bead to slide from point $O$ to $B$.

What curves $y = f(x)$, if any, minimize the value of this integral?

At first sight, we might guess that the straight line joining $O$ and $B$ would give the shortest time, but perhaps not. There might be some advantage in having the bead fall vertically at first to build up its speed faster. With a higher speed, the bead could travel a longer path and still reach $B$ first. Indeed, this is the right idea. The solution, from a branch of mathematics known as the *calculus of variations*, is that the original cycloid from $O$ to $B$ is the one and only brachistochrone for $O$ and $B$ (Figure 11.11).

In the next section we show how to find the arc length differential $ds$ for a parametrized curve. Once we know how to find $ds$, we can calculate the time given by the right-hand side of Equation (4) for the cycloid. This calculation gives the amount of time it takes a frictionless bead to slide down the cycloid to $B$ after it is released from rest at $O$. The time turns out to be equal to $\pi\sqrt{a/g}$, where $a$ is the radius of the wheel defining the particular cycloid. Moreover, if we start the bead at some lower point on the cycloid, corresponding to a parameter value $t_0 > 0$, we can integrate the parametric form of $ds/\sqrt{2gy}$ in Equation (3) over the interval $[t_0, \pi]$ to find the time it takes the bead to reach the point $B$. That calculation results in the same time $T = \pi\sqrt{a/g}$. It takes the bead the same amount of time to reach $B$ no matter where it starts, which makes the cycloid a tautochrone. Beads starting simultaneously from $O$, $A$, and $C$ in Figure 11.12, for instance, will all reach $B$ at exactly the same time. This is the reason why Huygens' pendulum clock in Figure 11.7 is independent of the amplitude of the swing.

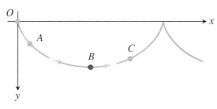

$O$

$x$

$A$

$B$

$C$

$y$

**FIGURE 11.12** Beads released simultaneously on the upside-down cycloid at $O$, $A$, and $C$ will reach $B$ at the same time.

# Exercises 11.1

### Finding Cartesian from Parametric Equations

Exercises 1–18 give parametric equations and parameter intervals for the motion of a particle in the $xy$-plane. Identify the particle's path by finding a Cartesian equation for it. Graph the Cartesian equation. (The graphs will vary with the equation used.) Indicate the portion of the graph traced by the particle and the direction of motion.

1. $x = 3t,\quad y = 9t^2,\quad -\infty < t < \infty$

2. $x = -\sqrt{t},\quad y = t,\quad t \geq 0$

3. $x = 2t - 5,\quad y = 4t - 7,\quad -\infty < t < \infty$

4. $x = 3 - 3t,\quad y = 2t,\quad 0 \leq t \leq 1$

5. $x = \cos 2t,\quad y = \sin 2t,\quad 0 \leq t \leq \pi$

6. $x = \cos(\pi - t),\quad y = \sin(\pi - t),\quad 0 \leq t \leq \pi$

7. $x = 4 \cos t,\quad y = 2 \sin t,\quad 0 \leq t \leq 2\pi$

8. $x = 4 \sin t,\quad y = 5 \cos t,\quad 0 \leq t \leq 2\pi$

9. $x = \sin t,\quad y = \cos 2t,\quad -\dfrac{\pi}{2} \leq t \leq \dfrac{\pi}{2}$

10. $x = 1 + \sin t,\quad y = \cos t - 2,\quad 0 \leq t \leq \pi$

11. $x = t^2,\quad y = t^6 - 2t^4,\quad -\infty < t < \infty$

12. $x = \dfrac{t}{t - 1},\quad y = \dfrac{t - 2}{t + 1},\quad -1 < t < 1$

13. $x = t,\quad y = \sqrt{1 - t^2},\quad -1 \leq t \leq 0$

14. $x = \sqrt{t + 1},\quad y = \sqrt{t},\quad t \geq 0$

15. $x = \sec^2 t - 1,\quad y = \tan t,\quad -\pi/2 < t < \pi/2$

16. $x = -\sec t,\quad y = \tan t,\quad -\pi/2 < t < \pi/2$

17. $x = -\cosh t,\quad y = \sinh t,\quad -\infty < t < \infty$

18. $x = 2 \sinh t,\quad y = 2 \cosh t,\quad -\infty < t < \infty$

### Finding Parametric Equations

19. Find parametric equations and a parameter interval for the motion of a particle that starts at $(a, 0)$ and traces the circle $x^2 + y^2 = a^2$
    a. once clockwise.
    b. once counterclockwise.
    c. twice clockwise.
    d. twice counterclockwise.

    (There are many ways to do these, so your answers may not be the same as the ones in the back of the book.)

20. Find parametric equations and a parameter interval for the motion of a particle that starts at $(a, 0)$ and traces the ellipse $(x^2/a^2) + (y^2/b^2) = 1$
    a. once clockwise.
    b. once counterclockwise.
    c. twice clockwise.
    d. twice counterclockwise.

    (As in Exercise 19, there are many correct answers.)

In Exercises 21–26, find a parametrization for the curve.

21. the line segment with endpoints $(-1, -3)$ and $(4, 1)$

22. the line segment with endpoints $(-1, 3)$ and $(3, -2)$

23. the lower half of the parabola $x - 1 = y^2$

24. the left half of the parabola $y = x^2 + 2x$

25. the ray (half line) with initial point $(2, 3)$ that passes through the point $(-1, -1)$

26. the ray (half line) with initial point $(-1, 2)$ that passes through the point $(0, 0)$

27. Find parametric equations and a parameter interval for the motion of a particle starting at the point $(2, 0)$ and tracing the top half of the circle $x^2 + y^2 = 4$ four times.

28. Find parametric equations and a parameter interval for the motion of a particle that moves along the graph of $y = x^2$ in the following way: Beginning at $(0, 0)$ it moves to $(3, 9)$, and then travels back and forth from $(3, 9)$ to $(-3, 9)$ infinitely many times.

29. Find parametric equations for the semicircle

$$x^2 + y^2 = a^2,\quad y > 0,$$

using as parameter the slope $t = dy/dx$ of the tangent to the curve at $(x, y)$.

30. Find parametric equations for the circle

$$x^2 + y^2 = a^2,$$

using as parameter the arc length $s$ measured counterclockwise from the point $(a, 0)$ to the point $(x, y)$.

31. Find a parametrization for the line segment joining points $(0, 2)$ and $(4, 0)$ using the angle $\theta$ in the accompanying figure as the parameter.

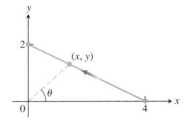

32. Find a parametrization for the curve $y = \sqrt{x}$ with terminal point $(0, 0)$ using the angle $\theta$ in the accompanying figure as the parameter.

33. Find a parametrization for the circle $(x - 2)^2 + y^2 = 1$ starting at $(1, 0)$ and moving clockwise once around the circle, using the central angle $\theta$ in the accompanying figure as the parameter.

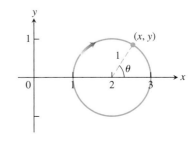

**34.** Find a parametrization for the circle $x^2 + y^2 = 1$ starting at $(1, 0)$ and moving counterclockwise to the terminal point $(0, 1)$, using the angle $\theta$ in the accompanying figure as the parameter.

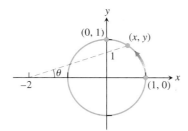

**35. The witch of Maria Agnesi**  The bell-shaped witch of Maria Agnesi can be constructed in the following way. Start with a circle of radius 1, centered at the point $(0, 1)$, as shown in the accompanying figure. Choose a point $A$ on the line $y = 2$ and connect it to the origin with a line segment. Call the point where the segment crosses the circle $B$. Let $P$ be the point where the vertical line through $A$ crosses the horizontal line through $B$. The witch is the curve traced by $P$ as $A$ moves along the line $y = 2$. Find parametric equations and a parameter interval for the witch by expressing the coordinates of $P$ in terms of $t$, the radian measure of the angle that segment $OA$ makes with the positive $x$-axis. The following equalities (which you may assume) will help.

**a.** $x = AQ$  **b.** $y = 2 - AB \sin t$
**c.** $AB \cdot OA = (AQ)^2$

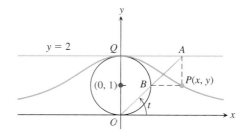

**36. Hypocycloid**  When a circle rolls on the inside of a fixed circle, any point $P$ on the circumference of the rolling circle describes a *hypocycloid*. Let the fixed circle be $x^2 + y^2 = a^2$, let the radius of the rolling circle be $b$, and let the initial position of the tracing point $P$ be $A(a, 0)$. Find parametric equations for the hypocycloid, using as the parameter the angle $\theta$ from the positive $x$-axis to the line joining the circles' centers. In particular, if $b = a/4$, as in the accompanying figure, show that the hypocycloid is the astroid

$$x = a \cos^3 \theta, \quad y = a \sin^3 \theta.$$

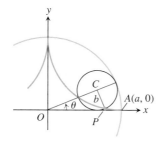

**37.** As the point $N$ moves along the line $y = a$ in the accompanying figure, $P$ moves in such a way that $OP = MN$. Find parametric equations for the coordinates of $P$ as functions of the angle $t$ that the line $ON$ makes with the positive $y$-axis.

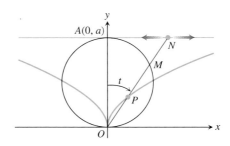

**38. Trochoids**  A wheel of radius $a$ rolls along a horizontal straight line without slipping. Find parametric equations for the curve traced out by a point $P$ on a spoke of the wheel $b$ units from its center. As parameter, use the angle $\theta$ through which the wheel turns. The curve is called a *trochoid*, which is a cycloid when $b = a$.

### Distance Using Parametric Equations

**39.** Find the point on the parabola $x = t, y = t^2, -\infty < t < \infty$, closest to the point $(2, 1/2)$. (*Hint:* Minimize the square of the distance as a function of $t$.)

**40.** Find the point on the ellipse $x = 2 \cos t, y = \sin t, 0 \le t \le 2\pi$ closest to the point $(3/4, 0)$. (*Hint:* Minimize the square of the distance as a function of $t$.)

### T GRAPHER EXPLORATIONS

If you have a parametric equation grapher, graph the equations over the given intervals in Exercises 41–48.

**41. Ellipse**  $x = 4 \cos t, \quad y = 2 \sin t, \quad$ over
  **a.** $0 \le t \le 2\pi$
  **b.** $0 \le t \le \pi$
  **c.** $-\pi/2 \le t \le \pi/2$.

**42. Hyperbola branch**  $x = \sec t$ (enter as $1/\cos(t)$), $y = \tan t$ (enter as $\sin(t)/\cos(t)$), over
  **a.** $-1.5 \le t \le 1.5$
  **b.** $-0.5 \le t \le 0.5$
  **c.** $-0.1 \le t \le 0.1$.

**43. Parabola**  $x = 2t + 3, \quad y = t^2 - 1, \quad -2 \le t \le 2$

**44. Cycloid**  $x = t - \sin t, \quad y = 1 - \cos t, \quad$ over
  **a.** $0 \le t \le 2\pi$
  **b.** $0 \le t \le 4\pi$
  **c.** $\pi \le t \le 3\pi$.

**45. Deltoid**

$$x = 2 \cos t + \cos 2t, \quad y = 2 \sin t - \sin 2t; \quad 0 \le t \le 2\pi$$

What happens if you replace 2 with $-2$ in the equations for $x$ and $y$? Graph the new equations and find out.

**46. A nice curve**

$$x = 3 \cos t + \cos 3t, \quad y = 3 \sin t - \sin 3t; \quad 0 \le t \le 2\pi$$

What happens if you replace 3 with $-3$ in the equations for $x$ and $y$? Graph the new equations and find out.

**47. a. Epicycloid**

$x = 9 \cos t - \cos 9t, \quad y = 9 \sin t - \sin 9t; \quad 0 \le t \le 2\pi$

   **b. Hypocycloid**

$x = 8 \cos t + 2 \cos 4t, \quad y = 8 \sin t - 2 \sin 4t; \quad 0 \le t \le 2\pi$

   **c. Hypotrochoid**

$x = \cos t + 5 \cos 3t, \quad y = 6 \cos t - 5 \sin 3t; \quad 0 \le t \le 2\pi$

**48. a.** $x = 6 \cos t + 5 \cos 3t, \quad y = 6 \sin t - 5 \sin 3t;$ $\quad 0 \le t \le 2\pi$

   **b.** $x = 6 \cos 2t + 5 \cos 6t, \quad y = 6 \sin 2t - 5 \sin 6t;$ $\quad 0 \le t \le \pi$

   **c.** $x = 6 \cos t + 5 \cos 3t, \quad y = 6 \sin 2t - 5 \sin 3t;$ $\quad 0 \le t \le 2\pi$

   **d.** $x = 6 \cos 2t + 5 \cos 6t, \quad y = 6 \sin 4t - 5 \sin 6t;$ $\quad 0 \le t \le \pi$

# 11.2 Calculus with Parametric Curves

In this section we apply calculus to parametric curves. Specifically, we find slopes, lengths, and areas associated with parametrized curves.

## Tangents and Areas

A parametrized curve $x = f(t)$ and $y = g(t)$ is **differentiable** at $t$ if $f$ and $g$ are differentiable at $t$. At a point on a differentiable parametrized curve where $y$ is also a differentiable function of $x$, the derivatives $dy/dt$, $dx/dt$, and $dy/dx$ are related by the Chain Rule:

$$\frac{dy}{dt} = \frac{dy}{dx} \cdot \frac{dx}{dt}.$$

If $dx/dt \ne 0$, we may divide both sides of this equation by $dx/dt$ to solve for $dy/dx$.

---

**Parametric Formula for $dy/dx$**

If all three derivatives exist and $dx/dt \ne 0$,

$$\frac{dy}{dx} = \frac{dy/dt}{dx/dt}. \tag{1}$$

---

If parametric equations define $y$ as a twice-differentiable function of $x$, we can apply Equation (1) to the function $dy/dx = y'$ to calculate $d^2y/dx^2$ as a function of $t$:

$$\frac{d^2y}{dx^2} = \frac{d}{dx}(y') = \frac{dy'/dt}{dx/dt}. \quad \text{Eq. (1) with } y' \text{ in place of } y$$

---

**Parametric Formula for $d^2y/dx^2$**

If the equations $x = f(t), y = g(t)$ define $y$ as a twice-differentiable function of $x$, then at any point where $dx/dt \ne 0$ and $y' = dy/dx$,

$$\frac{d^2y}{dx^2} = \frac{dy'/dt}{dx/dt}. \tag{2}$$

---

**FIGURE 11.13** The curve in Example 1 is the right-hand branch of the hyperbola $x^2 - y^2 = 1$.

**EXAMPLE 1**    Find the tangent to the curve

$$x = \sec t, \qquad y = \tan t, \qquad -\frac{\pi}{2} < t < \frac{\pi}{2},$$

at the point $\left( \sqrt{2}, 1 \right)$, where $t = \pi/4$ (Figure 11.13).

**Solution** The slope of the curve at $t$ is

$$\frac{dy}{dx} = \frac{dy/dt}{dx/dt} = \frac{\sec^2 t}{\sec t \tan t} = \frac{\sec t}{\tan t}. \qquad \text{Eq. (1)}$$

Setting $t$ equal to $\pi/4$ gives

$$\frac{dy}{dx}\bigg|_{t=\pi/4} = \frac{\sec(\pi/4)}{\tan(\pi/4)}$$

$$= \frac{\sqrt{2}}{1} = \sqrt{2}.$$

The tangent line is

$$y - 1 = \sqrt{2}\left(x - \sqrt{2}\right)$$
$$y = \sqrt{2}\,x - 2 + 1$$
$$y = \sqrt{2}\,x - 1.$$　■

**EXAMPLE 2**　Find $d^2y/dx^2$ as a function of $t$ if $x = t - t^2$ and $y = t - t^3$.

**Finding $d^2y/dx^2$ in Terms of $t$**
1. Express $y' = dy/dx$ in terms of $t$.
2. Find $dy'/dt$.
3. Divide $dy'/dt$ by $dx/dt$.

**Solution**

1. Express $y' = dy/dx$ in terms of $t$.

$$y' = \frac{dy}{dx} = \frac{dy/dt}{dx/dt} = \frac{1 - 3t^2}{1 - 2t}$$

2. Differentiate $y'$ with respect to $t$.

$$\frac{dy'}{dt} = \frac{d}{dt}\left(\frac{1 - 3t^2}{1 - 2t}\right) = \frac{2 - 6t + 6t^2}{(1 - 2t)^2} \qquad \text{Derivative Quotient Rule}$$

3. Divide $dy'/dt$ by $dx/dt$.

$$\frac{d^2y}{dx^2} = \frac{dy'/dt}{dx/dt} = \frac{(2 - 6t + 6t^2)/(1 - 2t)^2}{1 - 2t} = \frac{2 - 6t + 6t^2}{(1 - 2t)^3} \qquad \text{Eq. (2)}$$　■

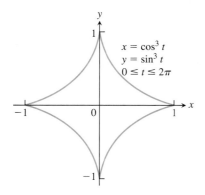

$x = \cos^3 t$
$y = \sin^3 t$
$0 \le t \le 2\pi$

**FIGURE 11.14** The astroid in Example 3.

**EXAMPLE 3**　Find the area enclosed by the astroid (Figure 11.14)

$$x = \cos^3 t, \qquad y = \sin^3 t, \qquad 0 \le t \le 2\pi.$$

**Solution** By symmetry, the enclosed area is 4 times the area beneath the curve in the first quadrant where $0 \le t \le \pi/2$. We can apply the definite integral formula for area studied in Chapter 5, using substitution to express the curve and differential $dx$ in terms of the parameter $t$. So,

$$A = 4 \int_0^1 y \, dx$$

$$= 4 \int_0^{\pi/2} \sin^3 t \cdot 3 \cos^2 t \sin t \, dt \qquad \text{Substitution for } y \text{ and } dx$$

$$= 12 \int_0^{\pi/2} \left( \frac{1 - \cos 2t}{2} \right)^2 \left( \frac{1 + \cos 2t}{2} \right) dt \qquad \sin^4 t = \left( \frac{1 - \cos 2t}{2} \right)^2$$

$$= \frac{3}{2} \int_0^{\pi/2} (1 - 2 \cos 2t + \cos^2 2t)(1 + \cos 2t) \, dt \qquad \text{Expand squared term.}$$

$$= \frac{3}{2} \int_0^{\pi/2} (1 - \cos 2t - \cos^2 2t + \cos^3 2t) \, dt \qquad \text{Multiply terms.}$$

$$= \frac{3}{2} \left[ \int_0^{\pi/2} (1 - \cos 2t) \, dt - \int_0^{\pi/2} \cos^2 2t \, dt + \int_0^{\pi/2} \cos^3 2t \, dt \right]$$

$$= \frac{3}{2} \left[ \left( t - \frac{1}{2} \sin 2t \right) - \frac{1}{2} \left( t + \frac{1}{4} \sin 2t \right) + \frac{1}{2} \left( \sin 2t - \frac{1}{3} \sin^3 2t \right) \right]_0^{\pi/2} \qquad \begin{matrix} \text{Section 8.3,} \\ \text{Example 3} \end{matrix}$$

$$= \frac{3}{2} \left[ \left( \frac{\pi}{2} - 0 - 0 - 0 \right) - \frac{1}{2} \left( \frac{\pi}{2} + 0 - 0 - 0 \right) + \frac{1}{2} (0 - 0 - 0 + 0) \right] \qquad \text{Evaluate.}$$

$$= \frac{3\pi}{8}.$$

■

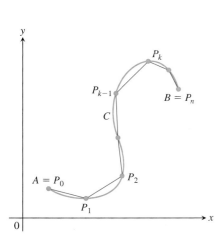

**FIGURE 11.15** The length of the smooth curve C from A to B is approximated by the sum of the lengths of the polygonal path (straight-line segments) starting at $A = P_0$, then to $P_1$, and so on, ending at $B = P_n$.

## Length of a Parametrically Defined Curve

Let $C$ be a curve given parametrically by the equations

$$x = f(t) \qquad \text{and} \qquad y = g(t), \qquad a \le t \le b.$$

We assume the functions $f$ and $g$ are **continuously differentiable** (meaning they have continuous first derivatives) on the interval $[a, b]$. We also assume that the derivatives $f'(t)$ and $g'(t)$ are not simultaneously zero, which prevents the curve $C$ from having any corners or cusps. Such a curve is called a **smooth curve**. We subdivide the path (or arc) $AB$ into $n$ pieces at points $A = P_0, P_1, P_2, \ldots, P_n = B$ (Figure 11.15). These points correspond to a partition of the interval $[a, b]$ by $a = t_0 < t_1 < t_2 < \cdots < t_n = b$, where $P_k = (f(t_k), g(t_k))$. Join successive points of this subdivision by straight-line segments (Figure 11.15). A representative line segment has length

$$L_k = \sqrt{(\Delta x_k)^2 + (\Delta y_k)^2}$$

$$= \sqrt{[f(t_k) - f(t_{k-1})]^2 + [g(t_k) - g(t_{k-1})]^2}$$

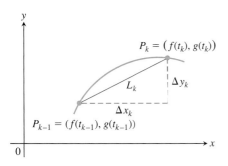

**FIGURE 11.16** The arc $P_{k-1}P_k$ is approximated by the straight-line segment shown here, which has length $L_k = \sqrt{(\Delta x_k)^2 + (\Delta y_k)^2}$.

(see Figure 11.16). If $\Delta t_k$ is small, the length $L_k$ is approximately the length of arc $P_{k-1}P_k$. By the Mean Value Theorem there are numbers $t_k^*$ and $t_k^{**}$ in $[t_{k-1}, t_k]$ such that

$$\Delta x_k = f(t_k) - f(t_{k-1}) = f'(t_k^*) \, \Delta t_k,$$

$$\Delta y_k = g(t_k) - g(t_{k-1}) = g'(t_k^{**}) \, \Delta t_k.$$

Assuming the path from $A$ to $B$ is traversed exactly once as $t$ increases from $t = a$ to $t = b$, with no doubling back or retracing, an approximation to the (yet to be defined) "length" of the curve $AB$ is the sum of all the lengths $L_k$:

$$\sum_{k=1}^{n} L_k = \sum_{k=1}^{n} \sqrt{(\Delta x_k)^2 + (\Delta y_k)^2}$$

$$= \sum_{k=1}^{n} \sqrt{[f'(t_k^*)]^2 + [g'(t_k^{**})]^2} \, \Delta t_k.$$

Although this last sum on the right is not exactly a Riemann sum (because $f'$ and $g'$ are evaluated at different points), it can be shown that its limit, as the norm of the partition tends to zero and the number of segments $n \to \infty$, is the definite integral

$$\lim_{\|P\| \to 0} \sum_{k=1}^{n} \sqrt{[f'(t_k^*)]^2 + [g'(t_k^{**})]^2} \, \Delta t_k = \int_a^b \sqrt{[f'(t)]^2 + [g'(t)]^2} \, dt.$$

Therefore, it is reasonable to define the length of the curve from $A$ to $B$ as this integral.

---

**DEFINITION** If a curve $C$ is defined parametrically by $x = f(t)$ and $y = g(t)$, $a \le t \le b$, where $f'$ and $g'$ are continuous and not simultaneously zero on $[a, b]$, and $C$ is traversed exactly once as $t$ increases from $t = a$ to $t = b$, then **the length of $C$** is the definite integral

$$L = \int_a^b \sqrt{[f'(t)]^2 + [g'(t)]^2} \, dt.$$

---

A smooth curve $C$ does not double back or reverse the direction of motion over the time interval $[a, b]$ since $(f')^2 + (g')^2 > 0$ throughout the interval. At a point where a curve does start to double back on itself, either the curve fails to be differentiable or both derivatives must simultaneously equal zero. We will examine this phenomenon in Chapter 13, where we study tangent vectors to curves.

If $x = f(t)$ and $y = g(t)$, then using the Leibniz notation we have the following result for arc length:

$$L = \int_a^b \sqrt{\left(\frac{dx}{dt}\right)^2 + \left(\frac{dy}{dt}\right)^2} \, dt. \tag{3}$$

If there are two different parametrizations for a curve $C$ whose length we want to find, it does not matter which one we use. However, the parametrization we choose must meet the conditions stated in the definition of the length of $C$ (see Exercise 41 for an example).

**EXAMPLE 4** Using the definition, find the length of the circle of radius $r$ defined parametrically by

$$x = r \cos t \quad \text{and} \quad y = r \sin t, \quad 0 \le t \le 2\pi.$$

**Solution** As $t$ varies from 0 to $2\pi$, the circle is traversed exactly once, so the circumference is

$$L = \int_0^{2\pi} \sqrt{\left(\frac{dx}{dt}\right)^2 + \left(\frac{dy}{dt}\right)^2} \, dt.$$

We find

$$\frac{dx}{dt} = -r \sin t, \qquad \frac{dy}{dt} = r \cos t$$

and

$$\left(\frac{dx}{dt}\right)^2 + \left(\frac{dy}{dt}\right)^2 = r^2(\sin^2 t + \cos^2 t) = r^2.$$

So

$$L = \int_0^{2\pi} \sqrt{r^2}\, dt = r\left[t\right]_0^{2\pi} = 2\pi r.$$    ■

**EXAMPLE 5**    Find the length of the astroid (Figure 11.14)

$$x = \cos^3 t, \qquad y = \sin^3 t, \qquad 0 \le t \le 2\pi.$$

**Solution**    Because of the curve's symmetry with respect to the coordinate axes, its length is four times the length of the first-quadrant portion. We have

$$x = \cos^3 t, \qquad\qquad y = \sin^3 t$$

$$\left(\frac{dx}{dt}\right)^2 = \left[3\cos^2 t(-\sin t)\right]^2 = 9\cos^4 t \sin^2 t$$

$$\left(\frac{dy}{dt}\right)^2 = \left[3\sin^2 t(\cos t)\right]^2 = 9\sin^4 t \cos^2 t$$

$$\sqrt{\left(\frac{dx}{dt}\right)^2 + \left(\frac{dy}{dt}\right)^2} = \sqrt{9\cos^2 t \sin^2 t \underbrace{(\cos^2 t + \sin^2 t)}_{1}}$$

$$= \sqrt{9\cos^2 t \sin^2 t}$$

$$= 3\left|\cos t \sin t\right| \qquad\qquad \text{\small cos } t \text{ sin } t \ge 0 \text{ for}$$
$$\text{\small } 0 \le t \le \pi/2$$

$$= 3\cos t \sin t.$$

Therefore,

$$\text{Length of first-quadrant portion} = \int_0^{\pi/2} 3\cos t \sin t\, dt$$

$$= \frac{3}{2}\int_0^{\pi/2} \sin 2t\, dt \qquad\qquad \text{\small cos } t \text{ sin } t =$$
$$\text{\small } (1/2)\sin 2t$$

$$= -\frac{3}{4}\cos 2t \Bigg]_0^{\pi/2} = \frac{3}{2}.$$

The length of the astroid is four times this: $4(3/2) = 6$.    ■

**EXAMPLE 6**    Find the perimeter of the ellipse $\dfrac{x^2}{a^2} + \dfrac{y^2}{b^2} = 1$.

**Solution**    Parametrically, we represent the ellipse by the equations $x = a \sin t$ and $y = b \cos t$, $a > b$ and $0 \le t \le 2\pi$. Then,

$$\left(\frac{dx}{dt}\right)^2 + \left(\frac{dy}{dt}\right)^2 = a^2 \cos^2 t + b^2 \sin^2 t$$

$$= a^2 - (a^2 - b^2)\sin^2 t$$

$$= a^2[1 - e^2 \sin^2 t] \qquad e = 1 - \frac{b^2}{a^2}, \text{ not the}$$
$$\text{number } 2.71828\dots$$

From Equation (3), the perimeter is given by

$$P = 4a \int_0^{\pi/2} \sqrt{1 - e^2 \sin^2 t}\, dt.$$

(We investigate the meaning of $e$ in Section 11.7.) The integral for $P$ is nonelementary and known as the *complete elliptic integral of the second kind.* We can compute its value to within any degree of accuracy using infinite series in the following way. From the binomial expansion for $\sqrt{1 - x}$ in Section 10.10, we have

$$\sqrt{1 - e^2 \sin^2 t} = 1 - \frac{1}{2}e^2 \sin^2 t - \frac{1}{2\cdot 4}e^4 \sin^4 t - \cdots, \qquad |e\sin t| \le e < 1$$

Then to each term in this last expression we apply the integral Formula 157 (at the back of the book) for $\int_0^{\pi/2} \sin^n t\, dt$ when $n$ is even, giving the perimeter

$$P = 4a \int_0^{\pi/2} \sqrt{1 - e^2 \sin^2 t}\, dt$$

$$= 4a\left[\frac{\pi}{2} - \left(\frac{1}{2}e^2\right)\left(\frac{1}{2}\cdot\frac{\pi}{2}\right) - \left(\frac{1}{2\cdot 4}e^4\right)\left(\frac{1\cdot 3}{2\cdot 4}\cdot\frac{\pi}{2}\right) - \left(\frac{1\cdot 3}{2\cdot 4\cdot 6}e^6\right)\left(\frac{1\cdot 3\cdot 5}{2\cdot 4\cdot 6}\cdot\frac{\pi}{2}\right) - \cdots\right]$$

$$= 2\pi a\left[1 - \left(\frac{1}{2}\right)^2 e^2 - \left(\frac{1\cdot 3}{2\cdot 4}\right)^2 \frac{e^4}{3} - \left(\frac{1\cdot 3\cdot 5}{2\cdot 4\cdot 6}\right)^2 \frac{e^6}{5} - \cdots\right].$$

Since $e < 1$, the series on the right-hand side converges by comparison with the geometric series $\sum_{n=1}^{\infty} (e^2)^n$. ∎

## Length of a Curve $y = f(x)$

The length formula in Section 6.3 is a special case of Equation (3). Given a continuously differentiable function $y = f(x)$, $a \le x \le b$, we can assign $x = t$ as a parameter. The graph of the function $f$ is then the curve $C$ defined parametrically by

$$x = t \qquad \text{and} \qquad y = f(t), \qquad a \le t \le b,$$

a special case of what we considered before. Then,

$$\frac{dx}{dt} = 1 \qquad \text{and} \qquad \frac{dy}{dt} = f'(t).$$

From Equation (1), we have

$$\frac{dy}{dx} = \frac{dy/dt}{dx/dt} = f'(t),$$

giving

$$\left(\frac{dx}{dt}\right)^2 + \left(\frac{dy}{dt}\right)^2 = 1 + [f'(t)]^2$$

$$= 1 + [f'(x)]^2. \qquad t = x$$

Substitution into Equation (3) gives the arc length formula for the graph of $y = f(x)$, in agreement with Equation (3) in Section 6.3.

## The Arc Length Differential

Consistent with our discussion in Section 6.3, we can define the arc length function for a parametrically defined curve $x = f(t)$ and $y = g(t)$, $a \le t \le b$, by

$$s(t) = \int_a^t \sqrt{[f'(z)]^2 + [g'(z)]^2}\, dz.$$

Then, by the Fundamental Theorem of Calculus,

$$\frac{ds}{dt} = \sqrt{[f'(t)]^2 + [g'(t)]^2} = \sqrt{\left(\frac{dx}{dt}\right)^2 + \left(\frac{dy}{dt}\right)^2}.$$

The differential of arc length is

$$ds = \sqrt{\left(\frac{dx}{dt}\right)^2 + \left(\frac{dy}{dt}\right)^2}\, dt. \tag{4}$$

Equation (4) is often abbreviated to

$$ds = \sqrt{dx^2 + dy^2}.$$

Just as in Section 6.3, we can integrate the differential $ds$ between appropriate limits to find the total length of a curve.

Here's an example where we use the arc length formula to find the centroid of an arc.

**EXAMPLE 7**     Find the centroid of the first-quadrant arc of the astroid in Example 5.

**Solution**     We take the curve's density to be $\delta = 1$ and calculate the curve's mass and moments about the coordinate axes as we did in Section 6.6.

The distribution of mass is symmetric about the line $y = x$, so $\bar{x} = \bar{y}$. A typical segment of the curve (Figure 11.17) has mass

$$dm = 1 \cdot ds = \sqrt{\left(\frac{dx}{dt}\right)^2 + \left(\frac{dy}{dt}\right)^2}\, dt = 3 \cos t \sin t\, dt. \quad \text{From Example 5}$$

The curve's mass is

$$M = \int_0^{\pi/2} dm = \int_0^{\pi/2} 3 \cos t \sin t\, dt = \frac{3}{2}. \quad \text{Again from Example 5}$$

The curve's moment about the $x$-axis is

$$M_x = \int \tilde{y}\, dm = \int_0^{\pi/2} \sin^3 t \cdot 3 \cos t \sin t\, dt$$

$$= 3 \int_0^{\pi/2} \sin^4 t \cos t\, dt = 3 \cdot \frac{\sin^5 t}{5}\Big]_0^{\pi/2} = \frac{3}{5}.$$

It follows that

$$\bar{y} = \frac{M_x}{M} = \frac{3/5}{3/2} = \frac{2}{5}.$$

The centroid is the point $(2/5, 2/5)$.

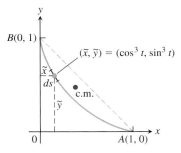

**FIGURE 11.17**   The centroid (c.m.) of the astroid arc in Example 7.

**EXAMPLE 8**  Find the time $T_c$ it takes for a frictionless bead to slide along the cycloid $x = a(t - \sin t)$, $y = a(1 - \cos t)$ from $t = 0$ to $t = \pi$ (see Figure 11.12).

**Solution**  From Equation (3) in Section 11.1, we want to find the time

$$T_c = \int_{t=0}^{t=\pi} \frac{ds}{\sqrt{2gy}}$$

for $ds$ and $y$ expressed parametrically in terms of the parameter $t$. For the cycloid, $dx/dt = a(1 - \cos t)$ and $dy/dt = a \sin t$, so

$$ds = \sqrt{\left(\frac{dx}{dt}\right)^2 + \left(\frac{dy}{dt}\right)^2} \, dt$$

$$= \sqrt{a^2 (1 - 2\cos t + \cos^2 t + \sin^2 t)} \, dt$$

$$= \sqrt{a^2 (2 - 2\cos t)} \, dt.$$

Substituting for $ds$ and $y$ in the integrand, it follows that

$$T_c = \int_0^\pi \sqrt{\frac{a^2(2 - 2\cos t)}{2ga(1 - \cos t)}} \, dt \qquad \scriptstyle y = a(1 - \cos t)$$

$$= \int_0^\pi \sqrt{\frac{a}{g}} \, dt = \pi\sqrt{\frac{a}{g}},$$

which is the amount of time it takes the frictionless bead to slide down the cycloid to $B$ after it is released from rest at $O$ (see Figure 11.12).  ∎

## Areas of Surfaces of Revolution

In Section 6.4 we found integral formulas for the area of a surface when a curve is revolved about a coordinate axis. Specifically, we found that the surface area is $S = \int 2\pi y \, ds$ for revolution about the $x$-axis, and $S = \int 2\pi x \, ds$ for revolution about the $y$-axis. If the curve is parametrized by the equations $x = f(t)$ and $y = g(t)$, $a \le t \le b$, where $f$ and $g$ are continuously differentiable and $(f')^2 + (g')^2 > 0$ on $[a, b]$, then the arc length differential $ds$ is given by Equation (4). This observation leads to the following formulas for area of surfaces of revolution for smooth parametrized curves.

---

**Area of Surface of Revolution for Parametrized Curves**

If a smooth curve $x = f(t)$, $y = g(t)$, $a \le t \le b$, is traversed exactly once as $t$ increases from $a$ to $b$, then the areas of the surfaces generated by revolving the curve about the coordinate axes are as follows.

**1. Revolution about the $x$-axis ($y \ge 0$):**

$$S = \int_a^b 2\pi y \sqrt{\left(\frac{dx}{dt}\right)^2 + \left(\frac{dy}{dt}\right)^2} \, dt \tag{5}$$

**2. Revolution about the $y$-axis ($x \ge 0$):**

$$S = \int_a^b 2\pi x \sqrt{\left(\frac{dx}{dt}\right)^2 + \left(\frac{dy}{dt}\right)^2} \, dt \tag{6}$$

---

As with length, we can calculate surface area from any convenient parametrization that meets the stated criteria.

Circle
$x = \cos t$
$y = 1 + \sin t$
$0 \le t \le 2\pi$

$y$

$\bullet (0, 1)$

$x$

**FIGURE 11.18** In Example 9 we calculate the area of the surface of revolution
swept out by this parametrized curve.

**EXAMPLE 9**   The standard parametrization of the circle of radius 1 centered at the
point $(0, 1)$ in the $xy$-plane is

$$x = \cos t, \qquad y = 1 + \sin t, \qquad 0 \le t \le 2\pi.$$

Use this parametrization to find the area of the surface swept out by revolving the circle
about the $x$-axis (Figure 11.18).

**Solution**   We evaluate the formula

$$S = \int_a^b 2\pi y \sqrt{\left(\frac{dx}{dt}\right)^2 + \left(\frac{dy}{dt}\right)^2}\, dt \qquad \begin{array}{l}\text{Eq. (5) for revolution}\\ \text{about the } x\text{-axis;}\\ y = 1 + \sin t \ge 0\end{array}$$

$$= \int_0^{2\pi} 2\pi(1 + \sin t) \underbrace{\sqrt{(-\sin t)^2 + (\cos t)^2}}_{1}\, dt$$

$$= 2\pi \int_0^{2\pi} (1 + \sin t)\, dt$$

$$= 2\pi\Big[t - \cos t\Big]_0^{2\pi} = 4\pi^2. \qquad \blacksquare$$

## Exercises 11.2

**Tangents to Parametrized Curves**

In Exercises 1–14, find an equation for the line tangent to the curve at
the point defined by the given value of $t$. Also, find the value of $d^2y/dx^2$
at this point.

**1.** $x = 2\cos t, \quad y = 2\sin t, \quad t = \pi/4$

**2.** $x = \sin 2\pi t, \quad y = \cos 2\pi t, \quad t = -1/6$

**3.** $x = 4\sin t, \quad y = 2\cos t, \quad t = \pi/4$

**4.** $x = \cos t, \quad y = \sqrt{3}\cos t, \quad t = 2\pi/3$

**5.** $x = t, \quad y = \sqrt{t}, \quad t = 1/4$

**6.** $x = \sec^2 t - 1, \quad y = \tan t, \quad t = -\pi/4$

**7.** $x = \sec t, \quad y = \tan t, \quad t = \pi/6$

**8.** $x = -\sqrt{t + 1}, \quad y = \sqrt{3t}, \quad t = 3$

**9.** $x = 2t^2 + 3, \quad y = t^4, \quad t = -1$

**10.** $x = 1/t, \quad y = -2 + \ln t, \quad t = 1$

**11.** $x = t - \sin t, \quad y = 1 - \cos t, \quad t = \pi/3$

**12.** $x = \cos t, \quad y = 1 + \sin t, \quad t = \pi/2$

**13.** $x = \dfrac{1}{t + 1}, \quad y = \dfrac{t}{t - 1}, \quad t = 2$

**14.** $x = t + e^t, \quad y = 1 - e^t, \quad t = 0$

**Implicitly Defined Parametrizations**

Assuming that the equations in Exercises 15–20 define $x$ and $y$ implicitly as differentiable functions $x = f(t), \ y = g(t)$, find the slope of
the curve $x = f(t), \ y = g(t)$ at the given value of $t$.

**15.** $x^3 + 2t^2 = 9, \quad 2y^3 - 3t^2 = 4, \quad t = 2$

**16.** $x = \sqrt{5 - \sqrt{t}}, \quad y(t - 1) = \sqrt{t}, \quad t = 4$

**17.** $x + 2x^{3/2} = t^2 + t, \quad y\sqrt{t + 1} + 2t\sqrt{y} = 4, \quad t = 0$

**18.** $x \sin t + 2x = t, \quad t \sin t - 2t = y, \quad t = \pi$

**19.** $x = t^3 + t, \quad y + 2t^3 = 2x + t^2, \quad t = 1$

**20.** $t = \ln(x - t), \quad y = te^t, \quad t = 0$

**Area**

**21.** Find the area under one arch of the cycloid

$$x = a(t - \sin t), \quad y = a(1 - \cos t).$$

**22.** Find the area enclosed by the $y$-axis and the curve

$$x = t - t^2, \quad y = 1 + e^{-t}.$$

**23.** Find the area enclosed by the ellipse

$$x = a\cos t, \quad y = b\sin t, \quad 0 \le t \le 2\pi.$$

**24.** Find the area under $y = x^3$ over $[0, 1]$ using the following
parametrizations.

**a.** $x = t^2, \quad y = t^6$ \qquad **b.** $x = t^3, \quad y = t^9$

**Lengths of Curves**

Find the lengths of the curves in Exercises 25–30.

**25.** $x = \cos t, \quad y = t + \sin t, \quad 0 \le t \le \pi$

**26.** $x = t^3, \quad y = 3t^2/2, \quad 0 \le t \le \sqrt{3}$

**27.** $x = t^2/2, \quad y = (2t + 1)^{3/2}/3, \quad 0 \le t \le 4$

**28.** $x = (2t + 3)^{3/2}/3, \quad y = t + t^2/2, \quad 0 \le t \le 3$

**29.** $x = 8\cos t + 8t\sin t$ \qquad **30.** $x = \ln(\sec t + \tan t) - \sin t$
$\quad\ y = 8\sin t - 8t\cos t,$ \qquad\qquad $y = \cos t, \quad 0 \le t \le \pi/3$
$\quad\ 0 \le t \le \pi/2$

**Surface Area**

Find the areas of the surfaces generated by revolving the curves in
Exercises 31–34 about the indicated axes.

**31.** $x = \cos t, \quad y = 2 + \sin t, \quad 0 \le t \le 2\pi; \quad x$-axis

**32.** $x = (2/3)t^{3/2}, \quad y = 2\sqrt{t}, \quad 0 \le t \le \sqrt{3}; \quad$ *y*-axis

**33.** $x = t + \sqrt{2}, \quad y = (t^2/2) + \sqrt{2}t, \quad -\sqrt{2} \le t \le \sqrt{2}; \quad$ *y*-axis

**34.** $x = \ln(\sec t + \tan t) - \sin t, y = \cos t, 0 \le t \le \pi/3; \ x\text{-axis}$

**35. A cone frustum** The line segment joining the points (0, 1) and (2, 2) is revolved about the *x*-axis to generate a frustum of a cone. Find the surface area of the frustum using the parametrization $x = 2t, y = t + 1, 0 \le t \le 1$. Check your result with the geometry formula: Area $= \pi(r_1 + r_2)(\text{slant height})$.

**36. A cone** The line segment joining the origin to the point $(h, r)$ is revolved about the *x*-axis to generate a cone of height *h* and base radius *r*. Find the cone's surface area with the parametric equations $x = ht, \ y = rt, \ 0 \le t \le 1$. Check your result with the geometry formula: Area $= \pi r(\text{slant height})$.

### Centroids

**37.** Find the coordinates of the centroid of the curve
$$x = \cos t + t \sin t, \quad y = \sin t - t \cos t, \quad 0 \le t \le \pi/2.$$

**38.** Find the coordinates of the centroid of the curve
$$x = e^t \cos t, \quad y = e^t \sin t, \quad 0 \le t \le \pi.$$

**39.** Find the coordinates of the centroid of the curve
$$x = \cos t, \quad y = t + \sin t, \quad 0 \le t \le \pi.$$

**T** **40.** Most centroid calculations for curves are done with a calculator or computer that has an integral evaluation program. As a case in point, find, to the nearest hundredth, the coordinates of the centroid of the curve
$$x = t^3, \quad y = 3t^2/2, \quad 0 \le t \le \sqrt{3}.$$

### Theory and Examples

**41. Length is independent of parametrization** To illustrate the fact that the numbers we get for length do not depend on the way we parametrize our curves (except for the mild restrictions preventing doubling back mentioned earlier), calculate the length of the semicircle $y = \sqrt{1 - x^2}$ with these two different parametrizations:

**a.** $x = \cos 2t, \quad y = \sin 2t, \quad 0 \le t \le \pi/2.$

**b.** $x = \sin \pi t, \quad y = \cos \pi t, \quad -1/2 \le t \le 1/2.$

**42. a.** Show that the Cartesian formula
$$L = \int_c^d \sqrt{1 + \left(\frac{dx}{dy}\right)^2} \, dy$$

for the length of the curve $x = g(y), c \le y \le d$ (Section 6.3, Equation 4), is a special case of the parametric length formula
$$L = \int_a^b \sqrt{\left(\frac{dx}{dt}\right)^2 + \left(\frac{dy}{dt}\right)^2} \, dt.$$

Use this result to find the length of each curve.

**b.** $x = y^{3/2}, \quad 0 \le y \le 4/3$

**c.** $x = \frac{3}{2} y^{2/3}, \quad 0 \le y \le 1$

**43.** The curve with parametric equations
$$x = (1 + 2 \sin \theta) \cos \theta, \quad y = (1 + 2 \sin \theta) \sin \theta$$

is called a *limaçon* and is shown in the accompanying figure. Find the points $(x, y)$ and the slopes of the tangent lines at these points for

**a.** $\theta = 0.$ **b.** $\theta = \pi/2.$ **c.** $\theta = 4\pi/3.$

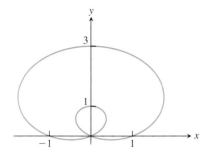

**44.** The curve with parametric equations
$$x = t, \quad y = 1 - \cos t, \quad 0 \le t \le 2\pi$$

is called a *sinusoid* and is shown in the accompanying figure. Find the point $(x, y)$ where the slope of the tangent line is

**a.** largest. **b.** smallest.

**T** The curves in Exercises 45 and 46 are called *Bowditch curves* or *Lissajous figures*. In each case, find the point in the interior of the first quadrant where the tangent to the curve is horizontal, and find the equations of the two tangents at the origin.

**45.** **46.**

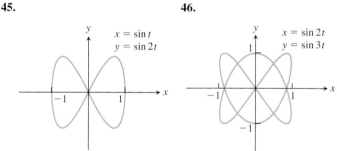

**47. Cycloid**

**a.** Find the length of one arch of the cycloid
$$x = a(t - \sin t), \quad y = a(1 - \cos t).$$

**b.** Find the area of the surface generated by revolving one arch of the cycloid in part (a) about the *x*-axis for $a = 1$.

**48. Volume** Find the volume swept out by revolving the region bounded by the *x*-axis and one arch of the cycloid
$$x = t - \sin t, \quad y = 1 - \cos t$$

about the *x*-axis.

### COMPUTER EXPLORATIONS

In Exercises 49–52, use a CAS to perform the following steps for the given curve over the closed interval.

**a.** Plot the curve together with the polygonal path approximations for $n = 2, 4, 8$ partition points over the interval. (See Figure 11.15.)

**b.** Find the corresponding approximation to the length of the curve by summing the lengths of the line segments.

**c.** Evaluate the length of the curve using an integral. Compare your approximations for $n = 2, 4, 8$ with the actual length given by the integral. How does the actual length compare with the approximations as $n$ increases? Explain your answer.

**49.** $x = \dfrac{1}{3}t^3, \quad y = \dfrac{1}{2}t^2, \quad 0 \le t \le 1$

**50.** $x = 2t^3 - 16t^2 + 25t + 5, \quad y = t^2 + t - 3, \quad 0 \le t \le 6$

**51.** $x = t - \cos t, \quad y = 1 + \sin t, \quad -\pi \le t \le \pi$

**52.** $x = e^t \cos t, \quad y = e^t \sin t, \quad 0 \le t \le \pi$

# 11.3 Polar Coordinates

In this section we study polar coordinates and their relation to Cartesian coordinates. You will see that polar coordinates are very useful for calculating many multiple integrals studied in Chapter 15. They are also useful in describing the paths of planets and satellites.

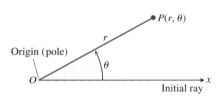

**FIGURE 11.19**  To define polar coordinates for the plane, we start with an origin, called the pole, and an initial ray.

## Definition of Polar Coordinates

To define polar coordinates, we first fix an **origin** $O$ (called the **pole**) and an **initial ray** from $O$ (Figure 11.19). Usually the positive $x$-axis is chosen as the initial ray. Then each point $P$ can be located by assigning to it a **polar coordinate pair** $(r, \theta)$ in which $r$ gives the directed distance from $O$ to $P$ and $\theta$ gives the directed angle from the initial ray to ray $OP$. So we label the point $P$ as

$$P(r, \theta)$$

Directed distance from $O$ to $P$    Directed angle from initial ray to $OP$

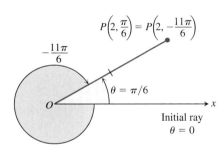

**FIGURE 11.20**  Polar coordinates are not unique.

As in trigonometry, $\theta$ is positive when measured counterclockwise and negative when measured clockwise. The angle associated with a given point is not unique. While a point in the plane has just one pair of Cartesian coordinates, it has infinitely many pairs of polar coordinates. For instance, the point 2 units from the origin along the ray $\theta = \pi/6$ has polar coordinates $r = 2, \ \theta = \pi/6$. It also has coordinates $r = 2, \theta = -11\pi/6$ (Figure 11.20). In some situations we allow $r$ to be negative. That is why we use directed distance in defining $P(r, \theta)$. The point $P(2, 7\pi/6)$ can be reached by turning $7\pi/6$ radians counterclockwise from the initial ray and going forward 2 units (Figure 11.21). It can also be reached by turning $\pi/6$ radians counterclockwise from the initial ray and going *backward* 2 units. So the point also has polar coordinates $r = -2, \theta = \pi/6$.

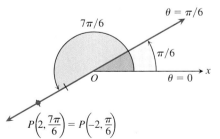

**FIGURE 11.21**  Polar coordinates can have negative $r$-values.

**EXAMPLE 1**    Find all the polar coordinates of the point $P(2, \pi/6)$.

**Solution**   We sketch the initial ray of the coordinate system, draw the ray from the origin that makes an angle of $\pi/6$ radians with the initial ray, and mark the point $(2, \pi/6)$ (Figure 11.22). We then find the angles for the other coordinate pairs of $P$ in which $r = 2$ and $r = -2$.

For $r = 2$, the complete list of angles is

$$\frac{\pi}{6}, \quad \frac{\pi}{6} \pm 2\pi, \quad \frac{\pi}{6} \pm 4\pi, \quad \frac{\pi}{6} \pm 6\pi, \ldots.$$

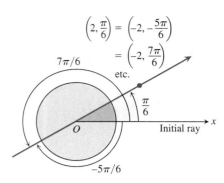

**FIGURE 11.22** The point $P(2, \pi/6)$ has infinitely many polar coordinate pairs (Example 1).

For $r = -2$, the angles are

$$-\frac{5\pi}{6}, \quad -\frac{5\pi}{6} \pm 2\pi, \quad -\frac{5\pi}{6} \pm 4\pi, \quad -\frac{5\pi}{6} \pm 6\pi, \dots.$$

The corresponding coordinate pairs of $P$ are

$$\left(2, \frac{\pi}{6} + 2n\pi\right), \qquad n = 0, \pm 1, \pm 2, \dots$$

and

$$\left(-2, -\frac{5\pi}{6} + 2n\pi\right), \qquad n = 0, \pm 1, \pm 2, \dots.$$

When $n = 0$, the formulas give $(2, \pi/6)$ and $(-2, -5\pi/6)$. When $n = 1$, they give $(2, 13\pi/6)$ and $(-2, 7\pi/6)$, and so on. ∎

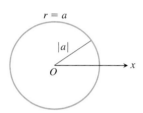

**FIGURE 11.23** The polar equation for a circle is $r = a$.

## Polar Equations and Graphs

If we hold $r$ fixed at a constant value $r = a \neq 0$, the point $P(r, \theta)$ will lie $|a|$ units from the origin $O$. As $\theta$ varies over any interval of length $2\pi$, $P$ then traces a circle of radius $|a|$ centered at $O$ (Figure 11.23).

    If we hold $\theta$ fixed at a constant value $\theta = \theta_0$ and let $r$ vary between $-\infty$ and $\infty$, the point $P(r, \theta)$ traces the line through $O$ that makes an angle of measure $\theta_0$ with the initial ray. (See Figure 11.21 for an example.)

**EXAMPLE 2** A circle or line can have more than one polar equation.

**(a)** $r = 1$ and $r = -1$ are equations for the circle of radius 1 centered at $O$.

**(b)** $\theta = \pi/6, \theta = 7\pi/6$, and $\theta = -5\pi/6$ are equations for the line in Figure 11.22. ∎

    Equations of the form $r = a$ and $\theta = \theta_0$ can be combined to define regions, segments, and rays.

**EXAMPLE 3** Graph the sets of points whose polar coordinates satisfy the following conditions.

**(a)** $1 \leq r \leq 2$     and     $0 \leq \theta \leq \dfrac{\pi}{2}$

**(b)** $-3 \leq r \leq 2$     and     $\theta = \dfrac{\pi}{4}$

**(c)** $\dfrac{2\pi}{3} \leq \theta \leq \dfrac{5\pi}{6}$     (no restriction on $r$)

**Solution** The graphs are shown in Figure 11.24. ∎

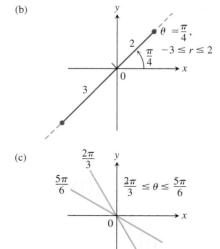

**FIGURE 11.24** The graphs of typical inequalities in $r$ and $\theta$ (Example 3).

## Relating Polar and Cartesian Coordinates

When we use both polar and Cartesian coordinates in a plane, we place the two origins together and take the initial polar ray as the positive $x$-axis. The ray $\theta = \pi/2, r > 0$,

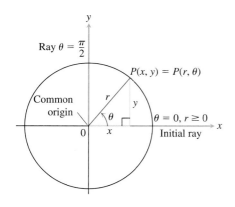

**FIGURE 11.25** The usual way to relate polar and Cartesian coordinates.

becomes the positive $y$-axis (Figure 11.25). The two coordinate systems are then related by the following equations.

---

**Equations Relating Polar and Cartesian Coordinates**

$$x = r \cos \theta, \qquad y = r \sin \theta, \qquad r^2 = x^2 + y^2, \qquad \tan \theta = \frac{y}{x}$$

---

The first two of these equations uniquely determine the Cartesian coordinates $x$ and $y$ given the polar coordinates $r$ and $\theta$. On the other hand, if $x$ and $y$ are given, the third equation gives two possible choices for $r$ (a positive and a negative value). For each $(x, y) \neq (0, 0)$, there is a unique $\theta \in [0, 2\pi)$ satisfying the first two equations, each then giving a polar coordinate representation of the Cartesian point $(x, y)$. The other polar coordinate representations for the point can be determined from these two, as in Example 1.

**EXAMPLE 4**  Here are some plane curves expressed in terms of both polar coordinate and Cartesian coordinate equations.

| Polar equation | Cartesian equivalent |
|---|---|
| $r \cos \theta = 2$ | $x = 2$ |
| $r^2 \cos \theta \sin \theta = 4$ | $xy = 4$ |
| $r^2 \cos^2 \theta - r^2 \sin^2 \theta = 1$ | $x^2 - y^2 = 1$ |
| $r = 1 + 2r \cos \theta$ | $y^2 - 3x^2 - 4x - 1 = 0$ |
| $r = 1 - \cos \theta$ | $x^4 + y^4 + 2x^2 y^2 + 2x^3 + 2xy^2 - y^2 = 0$ |

Some curves are more simply expressed with polar coordinates; others are not. ∎

**EXAMPLE 5**  Find a polar equation for the circle $x^2 + (y - 3)^2 = 9$ (Figure 11.26).

**Solution**  We apply the equations relating polar and Cartesian coordinates:

$$x^2 + (y - 3)^2 = 9$$
$$x^2 + y^2 - 6y + 9 = 9 \qquad \text{Expand } (y - 3)^2.$$
$$x^2 + y^2 - 6y = 0 \qquad \text{Cancelation}$$
$$r^2 - 6r \sin \theta = 0 \qquad x^2 + y^2 = r^2,\ y = r \sin \theta$$
$$r = 0 \quad \text{or} \quad r - 6 \sin \theta = 0$$
$$r = 6 \sin \theta \qquad \text{Includes both possibilities}$$ ∎

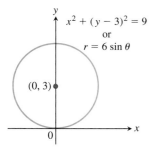

**FIGURE 11.26** The circle in Example 5.

**EXAMPLE 6**  Replace the following polar equations by equivalent Cartesian equations and identify their graphs.

**(a)** $r \cos \theta = -4$

**(b)** $r^2 = 4r \cos \theta$

**(c)** $r = \dfrac{4}{2 \cos \theta - \sin \theta}$

**Solution**  We use the substitutions $r \cos \theta = x$, $r \sin \theta = y$, and $r^2 = x^2 + y^2$.

**(a)** $r \cos \theta = -4$

The Cartesian equation:    $r \cos \theta = -4$
$$x = -4 \qquad \text{Substitution}$$

The graph:    Vertical line through $x = -4$ on the $x$-axis

**(b)** $r^2 = 4r \cos \theta$

| The Cartesian equation: | $r^2 = 4r \cos \theta$ | |
| :-- | :-- | :-- |
| | $x^2 + y^2 = 4x$ | Substitution |
| | $x^2 - 4x + y^2 = 0$ | |
| | $x^2 - 4x + 4 + y^2 = 4$ | Completing the square |
| | $(x - 2)^2 + y^2 = 4$ | Factoring |

The graph:    Circle, radius 2, center $(h, k) = (2, 0)$

**(c)** $r = \dfrac{4}{2 \cos \theta - \sin \theta}$

| The Cartesian equation: | $r(2 \cos \theta - \sin \theta) = 4$ | |
| :-- | :-- | :-- |
| | $2r \cos \theta - r \sin \theta = 4$ | Multiplying by $r$ |
| | $2x - y = 4$ | Substitution |
| | $y = 2x - 4$ | Solve for $y$. |

The graph:    Line, slope $m = 2$, $y$-intercept $b = -4$    ■

# Exercises 11.3

## Polar Coordinates

**1.** Which polar coordinate pairs label the same point?

**a.** $(3, 0)$    **b.** $(-3, 0)$    **c.** $(2, 2\pi/3)$

**d.** $(2, 7\pi/3)$    **e.** $(-3, \pi)$    **f.** $(2, \pi/3)$

**g.** $(-3, 2\pi)$    **h.** $(-2, -\pi/3)$

**2.** Which polar coordinate pairs label the same point?

**a.** $(-2, \pi/3)$    **b.** $(2, -\pi/3)$    **c.** $(r, \theta)$

**d.** $(r, \theta + \pi)$    **e.** $(-r, \theta)$    **f.** $(2, -2\pi/3)$

**g.** $(-r, \theta + \pi)$    **h.** $(-2, 2\pi/3)$

**3.** Plot the following points (given in polar coordinates). Then find all the polar coordinates of each point.

**a.** $(2, \pi/2)$    **b.** $(2, 0)$

**c.** $(-2, \pi/2)$    **d.** $(-2, 0)$

**4.** Plot the following points (given in polar coordinates). Then find all the polar coordinates of each point.

**a.** $(3, \pi/4)$    **b.** $(-3, \pi/4)$

**c.** $(3, -\pi/4)$    **d.** $(-3, -\pi/4)$

## Polar to Cartesian Coordinates

**5.** Find the Cartesian coordinates of the points in Exercise 1.

**6.** Find the Cartesian coordinates of the following points (given in polar coordinates).

**a.** $\left(\sqrt{2}, \pi/4\right)$    **b.** $(1, 0)$

**c.** $(0, \pi/2)$    **d.** $\left(-\sqrt{2}, \pi/4\right)$

**e.** $(-3, 5\pi/6)$    **f.** $(5, \tan^{-1}(4/3))$

**g.** $(-1, 7\pi)$    **h.** $\left(2\sqrt{3}, 2\pi/3\right)$

## Cartesian to Polar Coordinates

**7.** Find the polar coordinates, $0 \le \theta < 2\pi$ and $r \ge 0$, of the following points given in Cartesian coordinates.

**a.** $(1, 1)$    **b.** $(-3, 0)$

**c.** $\left(\sqrt{3}, -1\right)$    **d.** $(-3, 4)$

**8.** Find the polar coordinates, $-\pi \le \theta < \pi$ and $r \ge 0$, of the following points given in Cartesian coordinates.

**a.** $(-2, -2)$    **b.** $(0, 3)$

**c.** $\left(-\sqrt{3}, 1\right)$    **d.** $(5, -12)$

**9.** Find the polar coordinates, $0 \le \theta < 2\pi$ and $r \le 0$, of the following points given in Cartesian coordinates.

**a.** $(3, 3)$    **b.** $(-1, 0)$

**c.** $\left(-1, \sqrt{3}\right)$    **d.** $(4, -3)$

**10.** Find the polar coordinates, $-\pi \le \theta < 2\pi$ and $r \le 0$, of the following points given in Cartesian coordinates.

**a.** $(-2, 0)$    **b.** $(1, 0)$

**c.** $(0, -3)$    **d.** $\left(\dfrac{\sqrt{3}}{2}, \dfrac{1}{2}\right)$

## Graphing Sets of Polar Coordinate Points

Graph the sets of points whose polar coordinates satisfy the equations and inequalities in Exercises 11–26.

**11.** $r = 2$      **12.** $0 \le r \le 2$

**13.** $r \ge 1$      **14.** $1 \le r \le 2$

**15.** $0 \le \theta \le \pi/6, \quad r \ge 0$      **16.** $\theta = 2\pi/3, \quad r \le -2$

**17.** $\theta = \pi/3, \quad -1 \le r \le 3$      **18.** $\theta = 11\pi/4, \quad r \ge -1$

**19.** $\theta = \pi/2, \quad r \ge 0$      **20.** $\theta = \pi/2, \quad r \le 0$

**21.** $0 \le \theta \le \pi, \quad r = 1$      **22.** $0 \le \theta \le \pi, \quad r = -1$

**23.** $\pi/4 \le \theta \le 3\pi/4, \quad 0 \le r \le 1$

**24.** $-\pi/4 \le \theta \le \pi/4, \quad -1 \le r \le 1$

**25.** $-\pi/2 \le \theta \le \pi/2, \quad 1 \le r \le 2$

**26.** $0 \le \theta \le \pi/2, \quad 1 \le |r| \le 2$

**Polar to Cartesian Equations**

Replace the polar equations in Exercises 27–52 with equivalent Cartesian equations. Then describe or identify the graph.

**27.** $r \cos \theta = 2$      **28.** $r \sin \theta = -1$

**29.** $r \sin \theta = 0$      **30.** $r \cos \theta = 0$

**31.** $r = 4 \csc \theta$      **32.** $r = -3 \sec \theta$

**33.** $r \cos \theta + r \sin \theta = 1$      **34.** $r \sin \theta = r \cos \theta$

**35.** $r^2 = 1$      **36.** $r^2 = 4r \sin \theta$

**37.** $r = \dfrac{5}{\sin \theta - 2 \cos \theta}$      **38.** $r^2 \sin 2\theta = 2$

**39.** $r = \cot \theta \csc \theta$      **40.** $r = 4 \tan \theta \sec \theta$

**41.** $r = \csc \theta \, e^{r \cos \theta}$      **42.** $r \sin \theta = \ln r + \ln \cos \theta$

**43.** $r^2 + 2r^2 \cos \theta \sin \theta = 1$      **44.** $\cos^2 \theta = \sin^2 \theta$

**45.** $r^2 = -4r \cos \theta$      **46.** $r^2 = -6r \sin \theta$

**47.** $r = 8 \sin \theta$      **48.** $r = 3 \cos \theta$

**49.** $r = 2 \cos \theta + 2 \sin \theta$      **50.** $r = 2 \cos \theta - \sin \theta$

**51.** $r \sin \left( \theta + \dfrac{\pi}{6} \right) = 2$      **52.** $r \sin \left( \dfrac{2\pi}{3} - \theta \right) = 5$

**Cartesian to Polar Equations**

Replace the Cartesian equations in Exercises 53–66 with equivalent polar equations.

**53.** $x = 7$      **54.** $y = 1$      **55.** $x = y$

**56.** $x - y = 3$      **57.** $x^2 + y^2 = 4$      **58.** $x^2 - y^2 = 1$

**59.** $\dfrac{x^2}{9} + \dfrac{y^2}{4} = 1$      **60.** $xy = 2$

**61.** $y^2 = 4x$      **62.** $x^2 + xy + y^2 = 1$

**63.** $x^2 + (y - 2)^2 = 4$      **64.** $(x - 5)^2 + y^2 = 25$

**65.** $(x - 3)^2 + (y + 1)^2 = 4$      **66.** $(x + 2)^2 + (y - 5)^2 = 16$

**67.** Find all polar coordinates of the origin.

**68. Vertical and horizontal lines**

     **a.** Show that every vertical line in the $xy$-plane has a polar equation of the form $r = a \sec \theta$.

     **b.** Find the analogous polar equation for horizontal lines in the $xy$-plane.

# 11.4 Graphing Polar Coordinate Equations

It is often helpful to graph an equation expressed in polar coordinates in the Cartesian $xy$-plane. This section describes some techniques for graphing these equations using symmetries and tangents to the graph.

## Symmetry

Figure 11.27 illustrates the standard polar coordinate tests for symmetry. The following summary says how the symmetric points are related.

---

**Symmetry Tests for Polar Graphs in the Cartesian $xy$-Plane**

**1.** *Symmetry about the x-axis:* If the point $(r, \theta)$ lies on the graph, then the point $(r, -\theta)$ or $(-r, \pi - \theta)$ lies on the graph (Figure 11.27a).

**2.** *Symmetry about the y-axis:* If the point $(r, \theta)$ lies on the graph, then the point $(r, \pi - \theta)$ or $(-r, -\theta)$ lies on the graph (Figure 11.27b).

**3.** *Symmetry about the origin:* If the point $(r, \theta)$ lies on the graph, then the point $(-r, \theta)$ or $(r, \theta + \pi)$ lies on the graph (Figure 11.27c).

---

(a) About the $x$-axis

(b) About the $y$-axis

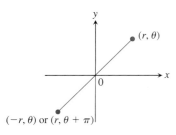

(c) About the origin

**FIGURE 11.27** Three tests for symmetry in polar coordinates.

## Slope

The slope of a polar curve $r = f(\theta)$ in the $xy$-plane is still given by $dy/dx$, which is not $r' = df/d\theta$. To see why, think of the graph of $f$ as the graph of the parametric equations

$$x = r \cos \theta = f(\theta) \cos \theta, \qquad y = r \sin \theta = f(\theta) \sin \theta.$$

If $f$ is a differentiable function of $\theta$, then so are $x$ and $y$ and, when $dx/d\theta \neq 0$, we can calculate $dy/dx$ from the parametric formula

$$\frac{dy}{dx} = \frac{dy/d\theta}{dx/d\theta} \qquad \text{Section 11.2, Eq. (1) with } t = \theta$$

$$= \frac{\dfrac{d}{d\theta}(f(\theta) \cdot \sin \theta)}{\dfrac{d}{d\theta}(f(\theta) \cdot \cos \theta)}$$

$$= \frac{\dfrac{df}{d\theta} \sin \theta + f(\theta) \cos \theta}{\dfrac{df}{d\theta} \cos \theta - f(\theta) \sin \theta} \qquad \text{Product Rule for derivatives}$$

Therefore we see that $dy/dx$ is not the same as $df/d\theta$.

---

**Slope of the Curve $r = f(\theta)$ in the Cartesian $xy$-Plane**

$$\left.\frac{dy}{dx}\right|_{(r,\,\theta)} = \frac{f'(\theta) \sin \theta + f(\theta) \cos \theta}{f'(\theta) \cos \theta - f(\theta) \sin \theta}$$

provided $dx/d\theta \neq 0$ at $(r, \theta)$.

---

If the curve $r = f(\theta)$ passes through the origin at $\theta = \theta_0$, then $f(\theta_0) = 0$, and the slope equation gives

$$\left.\frac{dy}{dx}\right|_{(0,\,\theta_0)} = \frac{f'(\theta_0) \sin \theta_0}{f'(\theta_0) \cos \theta_0} = \tan \theta_0.$$

If the graph of $r = f(\theta)$ passes through the origin at the value $\theta = \theta_0$, the slope of the curve there is $\tan \theta_0$. The reason we say "slope at $(0, \theta_0)$" and not just "slope at the origin" is that a polar curve may pass through the origin (or any point) more than once, with different slopes at different $\theta$-values. This is not the case in our first example, however.

**EXAMPLE 1** Graph the curve $r = 1 - \cos \theta$ in the Cartesian $xy$-plane.

**Solution** The curve is symmetric about the $x$-axis because

$$(r, \theta) \text{ on the graph} \Rightarrow r = 1 - \cos \theta$$
$$\Rightarrow r = 1 - \cos(-\theta) \qquad \cos \theta = \cos(-\theta)$$
$$\Rightarrow (r, -\theta) \text{ on the graph.}$$

| $\theta$ | $r = 1 - \cos\theta$ |
|---|---|
| 0 | 0 |
| $\dfrac{\pi}{3}$ | $\dfrac{1}{2}$ |
| $\dfrac{\pi}{2}$ | 1 |
| $\dfrac{2\pi}{3}$ | $\dfrac{3}{2}$ |
| $\pi$ | 2 |

(a)

(b)

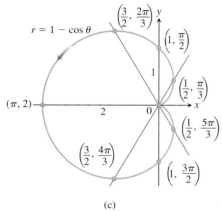

(c)

**FIGURE 11.28** The steps in graphing the cardioid $r = 1 - \cos\theta$ (Example 1). The arrow shows the direction of increasing $\theta$.

As $\theta$ increases from 0 to $\pi$, $\cos\theta$ decreases from 1 to $-1$, and $r = 1 - \cos\theta$ increases from a minimum value of 0 to a maximum value of 2. As $\theta$ continues on from $\pi$ to $2\pi$, $\cos\theta$ increases from $-1$ back to 1 and $r$ decreases from 2 back to 0. The curve starts to repeat when $\theta = 2\pi$ because the cosine has period $2\pi$.

The curve leaves the origin with slope $\tan(0) = 0$ and returns to the origin with slope $\tan(2\pi) = 0$.

We make a table of values from $\theta = 0$ to $\theta = \pi$, plot the points, draw a smooth curve through them with a horizontal tangent at the origin, and reflect the curve across the $x$-axis to complete the graph (Figure 11.28). The curve is called a *cardioid* because of its heart shape.

**EXAMPLE 2**    Graph the curve $r^2 = 4\cos\theta$ in the Cartesian $xy$-plane.

**Solution**    The equation $r^2 = 4\cos\theta$ requires $\cos\theta \geq 0$, so we get the entire graph by running $\theta$ from $-\pi/2$ to $\pi/2$. The curve is symmetric about the $x$-axis because

$$(r, \theta) \text{ on the graph} \Rightarrow r^2 = 4\cos\theta$$
$$\Rightarrow r^2 = 4\cos(-\theta) \qquad \cos\theta = \cos(-\theta)$$
$$\Rightarrow (r, -\theta) \text{ on the graph.}$$

The curve is also symmetric about the origin because

$$(r, \theta) \text{ on the graph} \Rightarrow r^2 = 4\cos\theta$$
$$\Rightarrow (-r)^2 = 4\cos\theta$$
$$\Rightarrow (-r, \theta) \text{ on the graph.}$$

Together, these two symmetries imply symmetry about the $y$-axis.

The curve passes through the origin when $\theta = -\pi/2$ and $\theta = \pi/2$. It has a vertical tangent both times because $\tan\theta$ is infinite.

For each value of $\theta$ in the interval between $-\pi/2$ and $\pi/2$, the formula $r^2 = 4\cos\theta$ gives two values of $r$:

$$r = \pm 2\sqrt{\cos\theta}.$$

We make a short table of values, plot the corresponding points, and use information about symmetry and tangents to guide us in connecting the points with a smooth curve (Figure 11.29).

| $\theta$ | $\cos\theta$ | $r = \pm 2\sqrt{\cos\theta}$ |
|---|---|---|
| 0 | 1 | $\pm 2$ |
| $\pm\dfrac{\pi}{6}$ | $\dfrac{\sqrt{3}}{2}$ | $\approx \pm 1.9$ |
| $\pm\dfrac{\pi}{4}$ | $\dfrac{1}{\sqrt{2}}$ | $\approx \pm 1.7$ |
| $\pm\dfrac{\pi}{3}$ | $\dfrac{1}{2}$ | $\approx \pm 1.4$ |
| $\pm\dfrac{\pi}{2}$ | 0 | 0 |

(a)

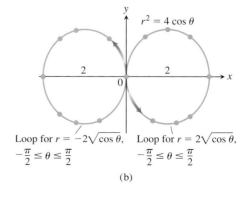

(b)

**FIGURE 11.29** The graph of $r^2 = 4\cos\theta$. The arrows show the direction of increasing $\theta$. The values of $r$ in the table are rounded (Example 2).

(a)

(b)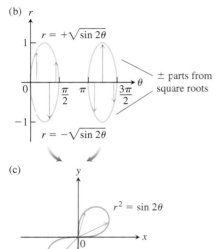

(c)

**FIGURE 11.30** To plot $r = f(\theta)$ in the Cartesian $r\theta$-plane in (b), we first plot $r^2 = \sin 2\theta$ in the $r^2\theta$-plane in (a) and then ignore the values of $\theta$ for which $\sin 2\theta$ is negative. The radii from the sketch in (b) cover the polar graph of the lemniscate in (c) twice (Example 3).

## Converting a Graph from the $r\theta$- to $xy$-Plane

One way to graph a polar equation $r = f(\theta)$ in the $xy$-plane is to make a table of $(r, \theta)$-values, plot the corresponding points there, and connect them in order of increasing $\theta$. This can work well if enough points have been plotted to reveal all the loops and dimples in the graph. Another method of graphing is to

1. first graph the function $r = f(\theta)$ in the *Cartesian $r\theta$-plane*,

2. then use that Cartesian graph as a "table" and guide to sketch the *polar* coordinate graph in the $xy$-plane.

This method is sometimes better than simple point plotting because the first Cartesian graph, even when hastily drawn, shows at a glance where $r$ is positive, negative, and nonexistent, as well as where $r$ is increasing and decreasing. Here's an example.

**EXAMPLE 3** Graph the *lemniscate* curve $r^2 = \sin 2\theta$ in the Cartesian $xy$-plane.

**Solution** Here we begin by plotting $r^2$ (not $r$) as a function of $\theta$ in the Cartesian $r^2\theta$-plane. See Figure 11.30a. We pass from there to the graph of $r = \pm\sqrt{\sin 2\theta}$ in the $r\theta$-plane (Figure 11.30b), and then draw the polar graph (Figure 11.30c). The graph in Figure 11.30b "covers" the final polar graph in Figure 11.30c twice. We could have managed with either loop alone, with the two upper halves, or with the two lower halves. The double covering does no harm, however, and we actually learn a little more about the behavior of the function this way. ∎

### USING TECHNOLOGY   Graphing Polar Curves Parametrically

For complicated polar curves we may need to use a graphing calculator or computer to graph the curve. If the device does not plot polar graphs directly, we can convert $r = f(\theta)$ into parametric form using the equations

$$x = r\cos\theta = f(\theta)\cos\theta, \qquad y = r\sin\theta = f(\theta)\sin\theta.$$

Then we use the device to draw a parametrized curve in the Cartesian $xy$-plane. It may be necessary to use the parameter $t$ rather than $\theta$ for the graphing device.

## Exercises 11.4

**Symmetries and Polar Graphs**

Identify the symmetries of the curves in Exercises 1–12. Then sketch the curves in the $xy$-plane.

**1.** $r = 1 + \cos\theta$        **2.** $r = 2 - 2\cos\theta$

**3.** $r = 1 - \sin\theta$        **4.** $r = 1 + \sin\theta$

**5.** $r = 2 + \sin\theta$        **6.** $r = 1 + 2\sin\theta$

**7.** $r = \sin(\theta/2)$        **8.** $r = \cos(\theta/2)$

**9.** $r^2 = \cos\theta$        **10.** $r^2 = \sin\theta$

**11.** $r^2 = -\sin\theta$        **12.** $r^2 = -\cos\theta$

Graph the lemniscates in Exercises 13–16. What symmetries do these curves have?

**13.** $r^2 = 4\cos 2\theta$        **14.** $r^2 = 4\sin 2\theta$

**15.** $r^2 = -\sin 2\theta$        **16.** $r^2 = -\cos 2\theta$

**Slopes of Polar Curves in the $xy$-Plane**

Find the slopes of the curves in Exercises 17–20 at the given points. Sketch the curves along with their tangents at these points.

**17. Cardioid**   $r = -1 + \cos\theta; \quad \theta = \pm\pi/2$

**18. Cardioid**   $r = -1 + \sin\theta; \quad \theta = 0, \pi$

**19. Four-leaved rose**   $r = \sin 2\theta; \quad \theta = \pm\pi/4, \pm 3\pi/4$

**20. Four-leaved rose**   $r = \cos 2\theta; \quad \theta = 0, \pm\pi/2, \pi$

**Graphing Limaçons**

Graph the limaçons in Exercises 21–24. Limaçon ("*lee*-ma-sahn") is Old French for "snail." You will understand the name when you graph the limaçons in Exercise 21. Equations for limaçons have the form $r = a \pm b \cos \theta$ or $r = a \pm b \sin \theta$. There are four basic shapes.

**21. Limaçons with an inner loop**

    **a.** $r = \dfrac{1}{2} + \cos \theta$         **b.** $r = \dfrac{1}{2} + \sin \theta$

**22. Cardioids**

    **a.** $r = 1 - \cos \theta$         **b.** $r = -1 + \sin \theta$

**23. Dimpled limaçons**

    **a.** $r = \dfrac{3}{2} + \cos \theta$         **b.** $r = \dfrac{3}{2} - \sin \theta$

**24. Oval limaçons**

    **a.** $r = 2 + \cos \theta$         **b.** $r = -2 + \sin \theta$

**Graphing Polar Regions and Curves in the *xy*-Plane**

**25.** Sketch the region defined by the inequalities $-1 \le r \le 2$ and $-\pi/2 \le \theta \le \pi/2$.

**26.** Sketch the region defined by the inequalities $0 \le r \le 2 \sec \theta$ and $-\pi/4 \le \theta \le \pi/4$.

In Exercises 27 and 28, sketch the region defined by the inequality.

**27.** $0 \le r \le 2 - 2 \cos \theta$     **28.** $0 \le r^2 \le \cos \theta$

T **29.** Which of the following has the same graph as $r = 1 - \cos \theta$?

    **a.** $r = -1 - \cos \theta$     **b.** $r = 1 + \cos \theta$

Confirm your answer with algebra.

T **30.** Which of the following has the same graph as $r = \cos 2\theta$?

    **a.** $r = -\sin (2\theta + \pi/2)$     **b.** $r = -\cos (\theta/2)$

Confirm your answer with algebra.

T **31. A rose within a rose** Graph the equation $r = 1 - 2 \sin 3\theta$.

T **32. The nephroid of Freeth** Graph the nephroid of Freeth:

$$r = 1 + 2 \sin \frac{\theta}{2}.$$

T **33. Roses** Graph the roses $r = \cos m\theta$ for $m = 1/3, 2, 3,$ and $7$.

T **34. Spirals** Polar coordinates are just the thing for defining spirals. Graph the following spirals.

    **a.** $r = \theta$

    **b.** $r = -\theta$

    **c.** A logarithmic spiral: $r = e^{\theta/10}$

    **d.** A hyperbolic spiral: $r = 8/\theta$

    **e.** An equilateral hyperbola: $r = \pm 10/\sqrt{\theta}$

    (Use different colors for the two branches.)

T **35.** Graph the equation $r = \sin \left( \frac{8}{7} \theta \right)$ for $0 \le \theta \le 14\pi$.

T **36.** Graph the equation

$$r = \sin^2 (2.3\theta) + \cos^4 (2.3\theta)$$

for $0 \le \theta \le 10\pi$.

# $11.5$ Areas and Lengths in Polar Coordinates

This section shows how to calculate areas of plane regions and lengths of curves in polar coordinates. The defining ideas are the same as before, but the formulas are different in polar versus Cartesian coordinates.

## Area in the Plane

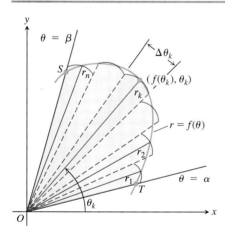

**FIGURE 11.31** To derive a formula for the area of region $OTS$, we approximate the region with fan-shaped circular sectors.

The region $OTS$ in Figure 11.31 is bounded by the rays $\theta = \alpha$ and $\theta = \beta$ and the curve $r = f(\theta)$. We approximate the region with $n$ nonoverlapping fan-shaped circular sectors based on a partition $P$ of angle $TOS$. The typical sector has radius $r_k = f(\theta_k)$ and central angle of radian measure $\Delta \theta_k$. Its area is $\Delta \theta_k / 2\pi$ times the area of a circle of radius $r_k$, or

$$A_k = \frac{1}{2} r_k^2 \, \Delta \theta_k = \frac{1}{2} \left( f(\theta_k) \right)^2 \Delta \theta_k.$$

The area of region $OTS$ is approximately

$$\sum_{k=1}^{n} A_k = \sum_{k=1}^{n} \frac{1}{2} \left( f(\theta_k) \right)^2 \Delta \theta_k.$$

If $f$ is continuous, we expect the approximations to improve as the norm of the partition $P$ goes to zero, where the norm of $P$ is the largest value of $\Delta \theta_k$. We are then led to the following formula defining the region's area:

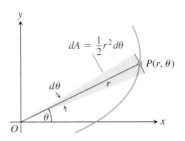

**FIGURE 11.32** The area differential $dA$ for the curve $r = f(\theta)$.

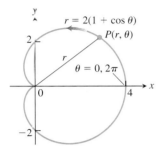

**FIGURE 11.33** The cardioid in Example 1.

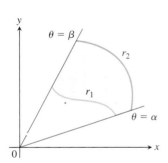

**FIGURE 11.34** The area of the shaded region is calculated by subtracting the area of the region between $r_1$ and the origin from the area of the region between $r_2$ and the origin.

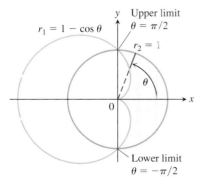

**FIGURE 11.35** The region and limits of integration in Example 2.

$$A = \lim_{\|P\| \to 0} \sum_{k=1}^{n} \frac{1}{2}\left(f(\theta_k)\right)^2 \Delta\theta_k$$

$$= \int_{\alpha}^{\beta} \frac{1}{2}\left(f(\theta)\right)^2 d\theta.$$

---

**Area of the Fan-Shaped Region Between the Origin and the Curve $r = f(\theta)$, $\alpha \le \theta \le \beta$**

$$A = \int_{\alpha}^{\beta} \frac{1}{2}r^2\, d\theta$$

This is the integral of the **area differential** (Figure 11.32)

$$dA = \frac{1}{2}r^2\, d\theta = \frac{1}{2}\left(f(\theta)\right)^2 d\theta.$$

---

**EXAMPLE 1** Find the area of the region in the $xy$-plane enclosed by the cardioid $r = 2(1 + \cos\theta)$.

**Solution** We graph the cardioid (Figure 11.33) and determine that the radius $OP$ sweeps out the region exactly once as $\theta$ runs from 0 to $2\pi$. The area is therefore

$$\int_{\theta=0}^{\theta=2\pi} \frac{1}{2}r^2\, d\theta = \int_{0}^{2\pi} \frac{1}{2}\cdot 4(1 + \cos\theta)^2\, d\theta$$

$$= \int_{0}^{2\pi} 2(1 + 2\cos\theta + \cos^2\theta)\, d\theta$$

$$= \int_{0}^{2\pi} \left(2 + 4\cos\theta + 2\cdot\frac{1 + \cos 2\theta}{2}\right) d\theta$$

$$= \int_{0}^{2\pi} (3 + 4\cos\theta + \cos 2\theta)\, d\theta$$

$$= \left[3\theta + 4\sin\theta + \frac{\sin 2\theta}{2}\right]_{0}^{2\pi} = 6\pi - 0 = 6\pi. \quad \blacksquare$$

To find the area of a region like the one in Figure 11.34, which lies between two polar curves $r_1 = r_1(\theta)$ and $r_2 = r_2(\theta)$ from $\theta = \alpha$ to $\theta = \beta$, we subtract the integral of $(1/2)r_1^2\, d\theta$ from the integral of $(1/2)r_2^2\, d\theta$. This leads to the following formula.

---

**Area of the Region $0 \le r_1(\theta) \le r \le r_2(\theta)$, $\alpha \le \theta \le \beta$**

$$A = \int_{\alpha}^{\beta} \frac{1}{2}r_2^2\, d\theta - \int_{\alpha}^{\beta} \frac{1}{2}r_1^2\, d\theta = \int_{\alpha}^{\beta} \frac{1}{2}\left(r_2^2 - r_1^2\right) d\theta \qquad (1)$$

---

**EXAMPLE 2** Find the area of the region that lies inside the circle $r = 1$ and outside the cardioid $r = 1 - \cos\theta$.

**Solution** We sketch the region to determine its boundaries and find the limits of integration (Figure 11.35). The outer curve is $r_2 = 1$, the inner curve is $r_1 = 1 - \cos\theta$, and $\theta$ runs from $-\pi/2$ to $\pi/2$. The area, from Equation (1), is

$$A = \int_{-\pi/2}^{\pi/2} \frac{1}{2}\left(r_2{}^2 - r_1{}^2\right) d\theta$$

$$= 2\int_{0}^{\pi/2} \frac{1}{2}\left(r_2{}^2 - r_1{}^2\right) d\theta \qquad \text{Symmetry}$$

$$= \int_{0}^{\pi/2} \left(1 - (1 - 2\cos\theta + \cos^2\theta)\right) d\theta \qquad \text{Square } r_1.$$

$$= \int_{0}^{\pi/2} (2\cos\theta - \cos^2\theta)\, d\theta = \int_{0}^{\pi/2}\left(2\cos\theta - \frac{1 + \cos 2\theta}{2}\right) d\theta$$

$$= \left[2\sin\theta - \frac{\theta}{2} - \frac{\sin 2\theta}{4}\right]_{0}^{\pi/2} = 2 - \frac{\pi}{4}. \qquad \blacksquare$$

The fact that we can represent a point in different ways in polar coordinates requires extra care in deciding when a point lies on the graph of a polar equation and in determining the points in which polar graphs intersect. (We needed intersection points in Example 2.) In Cartesian coordinates, we can always find the points where two curves cross by solving their equations simultaneously. In polar coordinates, the story is different. Simultaneous solution may reveal some intersection points without revealing others, so it is sometimes difficult to find all points of intersection of two polar curves. One way to identify all the points of intersection is to graph the equations.

## Length of a Polar Curve

We can obtain a polar coordinate formula for the length of a curve $r = f(\theta)$, $\alpha \le \theta \le \beta$, by parametrizing the curve as

$$x = r\cos\theta = f(\theta)\cos\theta, \qquad y = r\sin\theta = f(\theta)\sin\theta, \qquad \alpha \le \theta \le \beta. \qquad (2)$$

The parametric length formula, Equation (3) from Section 11.2, then gives the length as

$$L = \int_{\alpha}^{\beta} \sqrt{\left(\frac{dx}{d\theta}\right)^2 + \left(\frac{dy}{d\theta}\right)^2}\, d\theta.$$

This equation becomes

$$L = \int_{\alpha}^{\beta} \sqrt{r^2 + \left(\frac{dr}{d\theta}\right)^2}\, d\theta$$

when Equations (2) are substituted for $x$ and $y$ (Exercise 29).

---

**Length of a Polar Curve**

If $r = f(\theta)$ has a continuous first derivative for $\alpha \le \theta \le \beta$ and if the point $P(r, \theta)$ traces the curve $r = f(\theta)$ exactly once as $\theta$ runs from $\alpha$ to $\beta$, then the length of the curve is

$$L = \int_{\alpha}^{\beta} \sqrt{r^2 + \left(\frac{dr}{d\theta}\right)^2}\, d\theta. \qquad (3)$$

---

**EXAMPLE 3**    Find the length of the cardioid $r = 1 - \cos\theta$.

**Solution**    We sketch the cardioid to determine the limits of integration (Figure 11.36). The point $P(r, \theta)$ traces the curve once, counterclockwise as $\theta$ runs from 0 to $2\pi$, so these are the values we take for $\alpha$ and $\beta$.

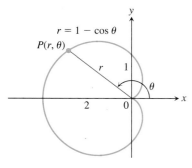

**FIGURE 11.36**    Calculating the length of a cardioid (Example 3).

With

$$r = 1 - \cos\theta, \qquad \frac{dr}{d\theta} = \sin\theta,$$

we have

$$r^2 + \left(\frac{dr}{d\theta}\right)^2 = (1 - \cos\theta)^2 + (\sin\theta)^2$$

$$= 1 - 2\cos\theta + \underbrace{\cos^2\theta + \sin^2\theta}_{1} = 2 - 2\cos\theta$$

and

$$L = \int_{\alpha}^{\beta} \sqrt{r^2 + \left(\frac{dr}{d\theta}\right)^2}\, d\theta = \int_0^{2\pi} \sqrt{2 - 2\cos\theta}\, d\theta$$

$$= \int_0^{2\pi} \sqrt{4\sin^2\frac{\theta}{2}}\, d\theta \qquad 1 - \cos\theta = 2\sin^2(\theta/2)$$

$$= \int_0^{2\pi} 2\left|\sin\frac{\theta}{2}\right| d\theta$$

$$= \int_0^{2\pi} 2\sin\frac{\theta}{2}\, d\theta \qquad \sin(\theta/2) \ge 0 \quad \text{for} \quad 0 \le \theta \le 2\pi$$

$$= \left[-4\cos\frac{\theta}{2}\right]_0^{2\pi} = 4 + 4 = 8. \qquad \blacksquare$$

## Exercises 11.5

### Finding Polar Areas

Find the areas of the regions in Exercises 1–8.

**1.** Bounded by the spiral $r = \theta$ for $0 \le \theta \le \pi$

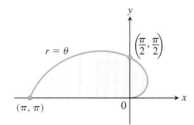

**2.** Bounded by the circle $r = 2\sin\theta$ for $\pi/4 \le \theta \le \pi/2$

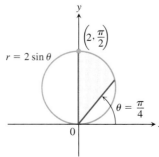

**3.** Inside the oval limaçon $r = 4 + 2\cos\theta$

**4.** Inside the cardioid $r = a(1 + \cos\theta)$, $a > 0$

**5.** Inside one leaf of the four-leaved rose $r = \cos 2\theta$

**6.** Inside one leaf of the three-leaved rose $r = \cos 3\theta$

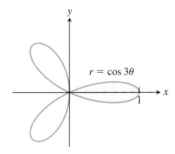

**7.** Inside one loop of the lemniscate $r^2 = 4\sin 2\theta$

**8.** Inside the six-leaved rose $r^2 = 2\sin 3\theta$

Find the areas of the regions in Exercises 9–18.

**9.** Shared by the circles $r = 2\cos\theta$ and $r = 2\sin\theta$

**10.** Shared by the circles $r = 1$ and $r = 2\sin\theta$

**11.** Shared by the circle $r = 2$ and the cardioid $r = 2(1 - \cos\theta)$

**12.** Shared by the cardioids $r = 2(1 + \cos\theta)$ and $r = 2(1 - \cos\theta)$

**13.** Inside the lemniscate $r^2 = 6\cos 2\theta$ and outside the circle $r = \sqrt{3}$

14. Inside the circle $r = 3a \cos \theta$ and outside the cardioid $r = a(1 + \cos \theta)$, $a > 0$

15. Inside the circle $r = -2 \cos \theta$ and outside the circle $r = 1$

16. Inside the circle $r = 6$ above the line $r = 3 \csc \theta$

17. Inside the circle $r = 4 \cos \theta$ and to the right of the vertical line $r = \sec \theta$

18. Inside the circle $r = 4 \sin \theta$ and below the horizontal line $r = 3 \csc \theta$

19. **a.** Find the area of the shaded region in the accompanying figure.

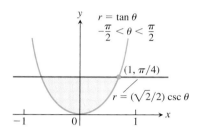

**b.** It looks as if the graph of $r = \tan \theta$, $-\pi/2 < \theta < \pi/2$, could be asymptotic to the lines $x = 1$ and $x = -1$. Is it? Give reasons for your answer.

20. The area of the region that lies inside the cardioid curve $r = \cos \theta + 1$ and outside the circle $r = \cos \theta$ is not

$$\frac{1}{2} \int_0^{2\pi} [(\cos \theta + 1)^2 - \cos^2 \theta] \, d\theta = \pi.$$

Why not? What *is* the area? Give reasons for your answers.

### Finding Lengths of Polar Curves

Find the lengths of the curves in Exercises 21–28.

21. The spiral $r = \theta^2$, $0 \le \theta \le \sqrt{5}$

22. The spiral $r = e^\theta/\sqrt{2}$, $0 \le \theta \le \pi$

23. The cardioid $r = 1 + \cos \theta$

24. The curve $r = a \sin^2(\theta/2)$, $0 \le \theta \le \pi$, $a > 0$

25. The parabolic segment $r = 6/(1 + \cos \theta)$, $0 \le \theta \le \pi/2$

26. The parabolic segment $r = 2/(1 - \cos \theta)$, $\pi/2 \le \theta \le \pi$

27. The curve $r = \cos^3(\theta/3)$, $0 \le \theta \le \pi/4$

28. The curve $r = \sqrt{1 + \sin 2\theta}$, $0 \le \theta \le \pi\sqrt{2}$

29. **The length of the curve $r = f(\theta)$, $\alpha \le \theta \le \beta$** Assuming that the necessary derivatives are continuous, show how the substitutions

$$x = f(\theta) \cos \theta, \quad y = f(\theta) \sin \theta$$

(Equations 2 in the text) transform

$$L = \int_\alpha^\beta \sqrt{\left(\frac{dx}{d\theta}\right)^2 + \left(\frac{dy}{d\theta}\right)^2} \, d\theta$$

into

$$L = \int_\alpha^\beta \sqrt{r^2 + \left(\frac{dr}{d\theta}\right)^2} \, d\theta.$$

30. **Circumferences of circles** As usual, when faced with a new formula, it is a good idea to try it on familiar objects to be sure it gives results consistent with past experience. Use the length formula in Equation (3) to calculate the circumferences of the following circles ($a > 0$).

**a.** $r = a$    **b.** $r = a \cos \theta$    **c.** $r = a \sin \theta$

### Theory and Examples

31. **Average value** If $f$ is continuous, the average value of the polar coordinate $r$ over the curve $r = f(\theta)$, $\alpha \le \theta \le \beta$, with respect to $\theta$ is given by the formula

$$r_{av} = \frac{1}{\beta - \alpha} \int_\alpha^\beta f(\theta) \, d\theta.$$

Use this formula to find the average value of $r$ with respect to $\theta$ over the following curves ($a > 0$).

**a.** The cardioid $r = a(1 - \cos \theta)$

**b.** The circle $r = a$

**c.** The circle $r = a \cos \theta$, $-\pi/2 \le \theta \le \pi/2$

32. **$r = f(\theta)$ vs. $r = 2f(\theta)$** Can anything be said about the relative lengths of the curves $r = f(\theta)$, $\alpha \le \theta \le \beta$, and $r = 2f(\theta)$, $\alpha \le \theta \le \beta$? Give reasons for your answer.

# 11.6 Conic Sections

In this section we define and review parabolas, ellipses, and hyperbolas geometrically and derive their standard Cartesian equations. These curves are called *conic sections* or *conics* because they are formed by cutting a double cone with a plane (Figure 11.37). This geometry method was the only way they could be described by Greek mathematicians who did not have our tools of Cartesian or polar coordinates. In the next section we express the conics in polar coordinates.

## Parabolas

> **DEFINITIONS** A set that consists of all the points in a plane equidistant from a given fixed point and a given fixed line in the plane is a **parabola**. The fixed point is the **focus** of the parabola. The fixed line is the **directrix**.

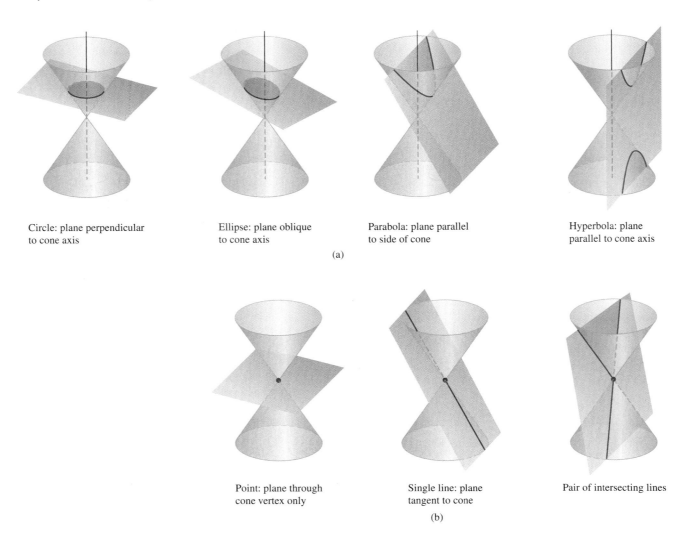

Circle: plane perpendicular
to cone axis

Ellipse: plane oblique
to cone axis

Parabola: plane parallel
to side of cone

Hyperbola: plane
parallel to cone axis

(a)

Point: plane through
cone vertex only

Single line: plane
tangent to cone

Pair of intersecting lines

(b)

**FIGURE 11.37** The standard conic sections (a) are the curves in which a plane cuts a *double* cone. Hyperbolas come in two parts, called *branches*. The point and lines obtained by passing the plane through the cone's vertex (b) are *degenerate* conic sections.

If the focus $F$ lies on the directrix $L$, the parabola is the line through $F$ perpendicular to $L$. We consider this to be a degenerate case and assume henceforth that $F$ does not lie on $L$.

A parabola has its simplest equation when its focus and directrix straddle one of the coordinate axes. For example, suppose that the focus lies at the point $F(0, p)$ on the positive $y$-axis and that the directrix is the line $y = -p$ (Figure 11.38). In the notation of the figure, a point $P(x, y)$ lies on the parabola if and only if $PF = PQ$. From the distance formula,

$$PF = \sqrt{(x - 0)^2 + (y - p)^2} = \sqrt{x^2 + (y - p)^2}$$
$$PQ = \sqrt{(x - x)^2 + (y - (-p))^2} = \sqrt{(y + p)^2}.$$

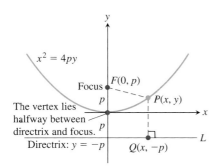

**FIGURE 11.38** The standard form of the parabola $x^2 = 4py$, $p > 0$.

When we equate these expressions, square, and simplify, we get

$$y = \frac{x^2}{4p} \qquad \text{or} \qquad x^2 = 4py. \qquad \text{Standard form} \qquad (1)$$

These equations reveal the parabola's symmetry about the $y$-axis. We call the $y$-axis the **axis** of the parabola (short for "axis of symmetry").

The point where a parabola crosses its axis is the **vertex**. The vertex of the parabola $x^2 = 4py$ lies at the origin (Figure 11.38). The positive number $p$ is the parabola's **focal length**.

If the parabola opens downward, with its focus at $(0, -p)$ and its directrix the line $y = p$, then Equations (1) become

$$y = -\frac{x^2}{4p} \quad \text{and} \quad x^2 = -4py.$$

By interchanging the variables $x$ and $y$, we obtain similar equations for parabolas opening to the right or to the left (Figure 11.39).

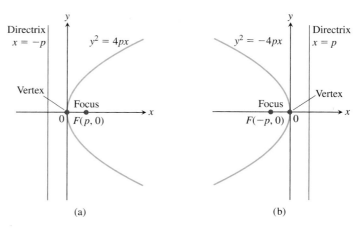

**FIGURE 11.39** (a) The parabola $y^2 = 4px$. (b) The parabola $y^2 = -4px$.

**EXAMPLE 1** Find the focus and directrix of the parabola $y^2 = 10x$.

**Solution** We find the value of $p$ in the standard equation $y^2 = 4px$:

$$4p = 10, \quad \text{so} \quad p = \frac{10}{4} = \frac{5}{2}.$$

Then we find the focus and directrix for this value of $p$:

$$\text{Focus:} \quad (p, 0) = \left(\frac{5}{2}, 0\right)$$

$$\text{Directrix:} \quad x = -p \quad \text{or} \quad x = -\frac{5}{2}. \qquad \blacksquare$$

## Ellipses

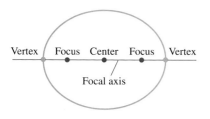

**FIGURE 11.40** Points on the focal axis of an ellipse.

**DEFINITIONS** An **ellipse** is the set of points in a plane whose distances from two fixed points in the plane have a constant sum. The two fixed points are the **foci** of the ellipse.

The line through the foci of an ellipse is the ellipse's **focal axis**. The point on the axis halfway between the foci is the **center**. The points where the focal axis and ellipse cross are the ellipse's **vertices** (Figure 11.40).

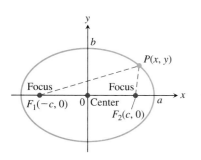

**FIGURE 11.41** The ellipse defined by the equation $PF_1 + PF_2 = 2a$ is the graph of the equation $(x^2/a^2) + (y^2/b^2) = 1$, where $b^2 = a^2 - c^2$.

If the foci are $F_1(-c, 0)$ and $F_2(c, 0)$ (Figure 11.41), and $PF_1 + PF_2$ is denoted by $2a$, then the coordinates of a point $P$ on the ellipse satisfy the equation

$$\sqrt{(x + c)^2 + y^2} + \sqrt{(x - c)^2 + y^2} = 2a.$$

To simplify this equation, we move the second radical to the right-hand side, square, isolate the remaining radical, and square again, obtaining

$$\frac{x^2}{a^2} + \frac{y^2}{a^2 - c^2} = 1. \tag{2}$$

Since $PF_1 + PF_2$ is greater than the length $F_1F_2$ (by the triangle inequality for triangle $PF_1F_2$), the number $2a$ is greater than $2c$. Accordingly, $a > c$ and the number $a^2 - c^2$ in Equation (2) is positive.

The algebraic steps leading to Equation (2) can be reversed to show that every point $P$ whose coordinates satisfy an equation of this form with $0 < c < a$ also satisfies the equation $PF_1 + PF_2 = 2a$. A point therefore lies on the ellipse if and only if its coordinates satisfy Equation (2).

If

$$b = \sqrt{a^2 - c^2}, \tag{3}$$

then $a^2 - c^2 = b^2$ and Equation (2) takes the form

$$\frac{x^2}{a^2} + \frac{y^2}{b^2} = 1. \tag{4}$$

Equation (4) reveals that this ellipse is symmetric with respect to the origin and both coordinate axes. It lies inside the rectangle bounded by the lines $x = \pm a$ and $y = \pm b$. It crosses the axes at the points $(\pm a, 0)$ and $(0, \pm b)$. The tangents at these points are perpendicular to the axes because

$$\frac{dy}{dx} = -\frac{b^2 x}{a^2 y} \qquad \text{Obtained from Eq. (4)}$$
$$\text{by implicit differentiation}$$

is zero if $x = 0$ and infinite if $y = 0$.

The **major axis** of the ellipse in Equation (4) is the line segment of length $2a$ joining the points $(\pm a, 0)$. The **minor axis** is the line segment of length $2b$ joining the points $(0, \pm b)$. The number $a$ itself is the **semimajor axis**, the number $b$ the **semiminor axis**. The number $c$, found from Equation (3) as

$$c = \sqrt{a^2 - b^2},$$

is the **center-to-focus distance** of the ellipse. If $a = b$, the ellipse is a circle.

**EXAMPLE 2**     The ellipse

$$\frac{x^2}{16} + \frac{y^2}{9} = 1 \tag{5}$$

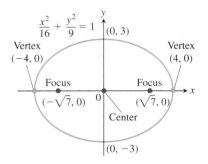

$$\frac{x^2}{16} + \frac{y^2}{9} = 1$$

**FIGURE 11.42** An ellipse with its major axis horizontal (Example 2).

(Figure 11.42) has

Semimajor axis:  $a = \sqrt{16} = 4$,     Semiminor axis:  $b = \sqrt{9} = 3$

Center-to-focus distance:  $c = \sqrt{16 - 9} = \sqrt{7}$

Foci:  $(\pm c, 0) = \left(\pm\sqrt{7}, 0\right)$

Vertices:  $(\pm a, 0) = (\pm 4, 0)$

Center:  $(0, 0)$.                                                              ∎

If we interchange $x$ and $y$ in Equation (5), we have the equation

$$\frac{x^2}{9} + \frac{y^2}{16} = 1. \tag{6}$$

The major axis of this ellipse is now vertical instead of horizontal, with the foci and vertices on the $y$-axis. There is no confusion in analyzing Equations (5) and (6). If we find the intercepts on the coordinate axes, we will know which way the major axis runs because it is the longer of the two axes.

Standard-Form Equations for Ellipses Centered at the Origin

*Foci on the x-axis:*  $\dfrac{x^2}{a^2} + \dfrac{y^2}{b^2} = 1$  $(a > b)$

Center-to-focus distance:  $c = \sqrt{a^2 - b^2}$
Foci:  $(\pm c, 0)$
Vertices:  $(\pm a, 0)$

*Foci on the y-axis:*  $\dfrac{x^2}{b^2} + \dfrac{y^2}{a^2} = 1$  $(a > b)$

Center-to-focus distance:  $c = \sqrt{a^2 - b^2}$
Foci:  $(0, \pm c)$
Vertices:  $(0, \pm a)$

In each case, $a$ is the semimajor axis and $b$ is the semiminor axis.

## Hyperbolas

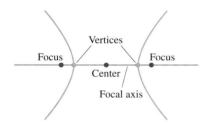

**FIGURE 11.43** Points on the focal axis of a hyperbola.

**DEFINITIONS**  A **hyperbola** is the set of points in a plane whose distances from two fixed points in the plane have a constant difference. The two fixed points are the **foci** of the hyperbola.

The line through the foci of a hyperbola is the **focal axis**. The point on the axis halfway between the foci is the hyperbola's **center**. The points where the focal axis and hyperbola cross are the **vertices** (Figure 11.43).

If the foci are $F_1(-c, 0)$ and $F_2(c, 0)$ (Figure 11.44) and the constant difference is $2a$, then a point $(x, y)$ lies on the hyperbola if and only if

$$\sqrt{(x + c)^2 + y^2} - \sqrt{(x - c)^2 + y^2} = \pm 2a. \tag{7}$$

To simplify this equation, we move the second radical to the right-hand side, square, isolate the remaining radical, and square again, obtaining

$$\frac{x^2}{a^2} + \frac{y^2}{a^2 - c^2} = 1. \tag{8}$$

So far, this looks just like the equation for an ellipse. But now $a^2 - c^2$ is negative because $2a$, being the difference of two sides of triangle $PF_1F_2$, is less than $2c$, the third side.

The algebraic steps leading to Equation (8) can be reversed to show that every point $P$ whose coordinates satisfy an equation of this form with $0 < a < c$ also satisfies Equation (7). A point therefore lies on the hyperbola if and only if its coordinates satisfy Equation (8).

If we let $b$ denote the positive square root of $c^2 - a^2$,

$$b = \sqrt{c^2 - a^2}, \tag{9}$$

then $a^2 - c^2 = -b^2$ and Equation (8) takes the more compact form

$$\frac{x^2}{a^2} - \frac{y^2}{b^2} = 1. \tag{10}$$

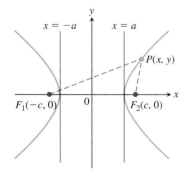

**FIGURE 11.44** Hyperbolas have two branches. For points on the right-hand branch of the hyperbola shown here, $PF_1 - PF_2 = 2a$. For points on the left-hand branch, $PF_2 - PF_1 = 2a$. We then let $b = \sqrt{c^2 - a^2}$.

The differences between Equation (10) and the equation for an ellipse (Equation 4) are the minus sign and the new relation

$$c^2 = a^2 + b^2. \qquad \text{From Eq. (9)}$$

Like the ellipse, the hyperbola is symmetric with respect to the origin and coordinate axes. It crosses the $x$-axis at the points $(\pm a, 0)$. The tangents at these points are vertical because

$$\frac{dy}{dx} = \frac{b^2 x}{a^2 y} \qquad \begin{array}{l}\text{Obtained from Eq. (10) by} \\ \text{implicit differentiation}\end{array}$$

is infinite when $y = 0$. The hyperbola has no $y$-intercepts; in fact, no part of the curve lies between the lines $x = -a$ and $x = a$.

The lines

$$y = \pm \frac{b}{a} x$$

are the two **asymptotes** of the hyperbola defined by Equation (10). The fastest way to find the equations of the asymptotes is to replace the 1 in Equation (10) by 0 and solve the new equation for $y$:

$$\underbrace{\frac{x^2}{a^2} - \frac{y^2}{b^2} = 1}_{\text{hyperbola}} \rightarrow \underbrace{\frac{x^2}{a^2} - \frac{y^2}{b^2} = 0}_{\text{0 for 1}} \rightarrow \underbrace{y = \pm \frac{b}{a} x.}_{\text{asymptotes}}$$

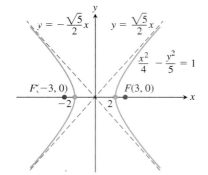

FIGURE 11.45 The hyperbola and its asymptotes in Example 3.

**EXAMPLE 3**    The equation

$$\frac{x^2}{4} - \frac{y^2}{5} = 1 \qquad (11)$$

is Equation (10) with $a^2 = 4$ and $b^2 = 5$ (Figure 11.45). We have

Center-to-focus distance:   $c = \sqrt{a^2 + b^2} = \sqrt{4 + 5} = 3$

Foci:   $(\pm c, 0) = (\pm 3, 0),$     Vertices:   $(\pm a, 0) = (\pm 2, 0)$

Center:   $(0, 0)$

Asymptotes:   $\dfrac{x^2}{4} - \dfrac{y^2}{5} = 0$   or   $y = \pm \dfrac{\sqrt{5}}{2} x.$   ∎

If we interchange $x$ and $y$ in Equation (11), the foci and vertices of the resulting hyperbola will lie along the $y$-axis. We still find the asymptotes in the same way as before, but now their equations will be $y = \pm 2x/\sqrt{5}$.

---

**Standard-Form Equations for Hyperbolas Centered at the Origin**

*Foci on the x-axis:*   $\dfrac{x^2}{a^2} - \dfrac{y^2}{b^2} = 1$     *Foci on the y-axis:*   $\dfrac{y^2}{a^2} - \dfrac{x^2}{b^2} = 1$

Center-to-focus distance:   $c = \sqrt{a^2 + b^2}$         Center-to-focus distance:   $c = \sqrt{a^2 + b^2}$

Foci:   $(\pm c, 0)$                         Foci:   $(0, \pm c)$

Vertices:   $(\pm a, 0)$                     Vertices:   $(0, \pm a)$

Asymptotes:   $\dfrac{x^2}{a^2} - \dfrac{y^2}{b^2} = 0$   or   $y = \pm \dfrac{b}{a} x$     Asymptotes:   $\dfrac{y^2}{a^2} - \dfrac{x^2}{b^2} = 0$   or   $y = \pm \dfrac{a}{b} x$

Notice the difference in the asymptote equations ($b/a$ in the first, $a/b$ in the second).

We shift conics using the principles reviewed in Section 1.2, replacing $x$ by $x + h$ and $y$ by $y + k$.

**EXAMPLE 4**    Show that the equation $x^2 - 4y^2 + 2x + 8y - 7 = 0$ represents a hyperbola. Find its center, asymptotes, and foci.

**Solution**    We reduce the equation to standard form by completing the square in $x$ and $y$ as follows:

$$(x^2 + 2x) - 4(y^2 - 2y) = 7$$

$$(x^2 + 2x + 1) - 4(y^2 - 2y + 1) = 7 + 1 - 4$$

$$\frac{(x + 1)^2}{4} - (y - 1)^2 = 1.$$

This is the standard form Equation (10) of a hyperbola with $x$ replaced by $x + 1$ and $y$ replaced by $y - 1$. The hyperbola is shifted one unit to the left and one unit upward, and it has center $x + 1 = 0$ and $y - 1 = 0$, or $x = -1$ and $y = 1$. Moreover,

$$a^2 = 4, \qquad b^2 = 1, \qquad c^2 = a^2 + b^2 = 5,$$

so the asymptotes are the two lines

$$\frac{x + 1}{2} - (y - 1) = 0 \qquad \text{and} \qquad \frac{x + 1}{2} + (y - 1) = 0,$$

or

$$y - 1 = \pm \frac{1}{2}(x + 1).$$

The shifted foci have coordinates $\left(-1 \pm \sqrt{5}, 1\right)$.    ■

# Exercises 11.6

**Identifying Graphs**

Match the parabolas in Exercises 1–4 with the following equations:

$$x^2 = 2y, \quad x^2 = -6y, \quad y^2 = 8x, \quad y^2 = -4x.$$

Then find each parabola's focus and directrix.

Match each conic section in Exercises 5–8 with one of these equations:

$$\frac{x^2}{4} + \frac{y^2}{9} = 1, \qquad \frac{x^2}{2} + y^2 = 1,$$

$$\frac{y^2}{4} - x^2 = 1, \qquad \frac{x^2}{4} - \frac{y^2}{9} = 1.$$

Then find the conic section's foci and vertices. If the conic section is a hyperbola, find its asymptotes as well.

**1.**

**2.**

**3.**

**4.**

**5.**

**6.**

**7.** **8.**

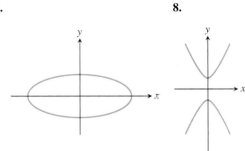

## Parabolas

Exercises 9–16 give equations of parabolas. Find each parabola's focus and directrix. Then sketch the parabola. Include the focus and directrix in your sketch.

**9.** $y^2 = 12x$  **10.** $x^2 = 6y$  **11.** $x^2 = -8y$

**12.** $y^2 = -2x$  **13.** $y = 4x^2$  **14.** $y = -8x^2$

**15.** $x = -3y^2$  **16.** $x = 2y^2$

## Ellipses

Exercises 17–24 give equations for ellipses. Put each equation in standard form. Then sketch the ellipse. Include the foci in your sketch.

**17.** $16x^2 + 25y^2 = 400$  **18.** $7x^2 + 16y^2 = 112$

**19.** $2x^2 + y^2 = 2$  **20.** $2x^2 + y^2 = 4$

**21.** $3x^2 + 2y^2 = 6$  **22.** $9x^2 + 10y^2 = 90$

**23.** $6x^2 + 9y^2 = 54$  **24.** $169x^2 + 25y^2 = 4225$

Exercises 25 and 26 give information about the foci and vertices of ellipses centered at the origin of the $xy$-plane. In each case, find the ellipse's standard-form equation from the given information.

**25.** Foci: $\left(\pm\sqrt{2}, 0\right)$  Vertices: $(\pm 2, 0)$

**26.** Foci: $(0, \pm 4)$  Vertices: $(0, \pm 5)$

## Hyperbolas

Exercises 27–34 give equations for hyperbolas. Put each equation in standard form and find the hyperbola's asymptotes. Then sketch the hyperbola. Include the asymptotes and foci in your sketch.

**27.** $x^2 - y^2 = 1$  **28.** $9x^2 - 16y^2 = 144$

**29.** $y^2 - x^2 = 8$  **30.** $y^2 - x^2 = 4$

**31.** $8x^2 - 2y^2 = 16$  **32.** $y^2 - 3x^2 = 3$

**33.** $8y^2 - 2x^2 = 16$  **34.** $64x^2 - 36y^2 = 2304$

Exercises 35–38 give information about the foci, vertices, and asymptotes of hyperbolas centered at the origin of the $xy$-plane. In each case, find the hyperbola's standard-form equation from the information given.

**35.** Foci: $\left(0, \pm\sqrt{2}\right)$

 Asymptotes: $y = \pm x$

**36.** Foci: $(\pm 2, 0)$

 Asymptotes: $y = \pm\dfrac{1}{\sqrt{3}}x$

**37.** Vertices: $(\pm 3, 0)$

 Asymptotes: $y = \pm\dfrac{4}{3}x$

**38.** Vertices: $(0, \pm 2)$

 Asymptotes: $y = \pm\dfrac{1}{2}x$

## Shifting Conic Sections

You may wish to review Section 1.2 before solving Exercises 39–56.

**39.** The parabola $y^2 = 8x$ is shifted down 2 units and right 1 unit to generate the parabola $(y + 2)^2 = 8(x - 1)$.

 **a.** Find the new parabola's vertex, focus, and directrix.

 **b.** Plot the new vertex, focus, and directrix, and sketch in the parabola.

**40.** The parabola $x^2 = -4y$ is shifted left 1 unit and up 3 units to generate the parabola $(x + 1)^2 = -4(y - 3)$.

 **a.** Find the new parabola's vertex, focus, and directrix.

 **b.** Plot the new vertex, focus, and directrix, and sketch in the parabola.

**41.** The ellipse $(x^2/16) + (y^2/9) = 1$ is shifted 4 units to the right and 3 units up to generate the ellipse

$$\frac{(x - 4)^2}{16} + \frac{(y - 3)^2}{9} = 1.$$

 **a.** Find the foci, vertices, and center of the new ellipse.

 **b.** Plot the new foci, vertices, and center, and sketch in the new ellipse.

**42.** The ellipse $(x^2/9) + (y^2/25) = 1$ is shifted 3 units to the left and 2 units down to generate the ellipse

$$\frac{(x + 3)^2}{9} + \frac{(y + 2)^2}{25} = 1.$$

 **a.** Find the foci, vertices, and center of the new ellipse.

 **b.** Plot the new foci, vertices, and center, and sketch in the new ellipse.

**43.** The hyperbola $(x^2/16) - (y^2/9) = 1$ is shifted 2 units to the right to generate the hyperbola

$$\frac{(x - 2)^2}{16} - \frac{y^2}{9} = 1.$$

 **a.** Find the center, foci, vertices, and asymptotes of the new hyperbola.

 **b.** Plot the new center, foci, vertices, and asymptotes, and sketch in the hyperbola.

**44.** The hyperbola $(y^2/4) - (x^2/5) = 1$ is shifted 2 units down to generate the hyperbola

$$\frac{(y + 2)^2}{4} - \frac{x^2}{5} = 1.$$

 **a.** Find the center, foci, vertices, and asymptotes of the new hyperbola.

 **b.** Plot the new center, foci, vertices, and asymptotes, and sketch in the hyperbola.

Exercises 45–48 give equations for parabolas and tell how many units up or down and to the right or left each parabola is to be shifted. Find an equation for the new parabola, and find the new vertex, focus, and directrix.

**45.** $y^2 = 4x$, left 2, down 3  **46.** $y^2 = -12x$, right 4, up 3

**47.** $x^2 = 8y$, right 1, down 7  **48.** $x^2 = 6y$, left 3, down 2

Exercises 49–52 give equations for ellipses and tell how many units up or down and to the right or left each ellipse is to be shifted. Find an equation for the new ellipse, and find the new foci, vertices, and center.

**49.** $\dfrac{x^2}{6} + \dfrac{y^2}{9} = 1$,  left 2, down 1

**50.** $\dfrac{x^2}{2} + y^2 = 1$,  right 3, up 4

**51.** $\dfrac{x^2}{3} + \dfrac{y^2}{2} = 1$,  right 2, up 3

**52.** $\dfrac{x^2}{16} + \dfrac{y^2}{25} = 1$,  left 4, down 5

Exercises 53–56 give equations for hyperbolas and tell how many units up or down and to the right or left each hyperbola is to be shifted. Find an equation for the new hyperbola, and find the new center, foci, vertices, and asymptotes.

**53.** $\dfrac{x^2}{4} - \dfrac{y^2}{5} = 1$,  right 2, up 2

**54.** $\dfrac{x^2}{16} - \dfrac{y^2}{9} = 1$,  left 2, down 1

**55.** $y^2 - x^2 = 1$,  left 1, down 1

**56.** $\dfrac{y^2}{3} - x^2 = 1$,  right 1, up 3

Find the center, foci, vertices, asymptotes, and radius, as appropriate, of the conic sections in Exercises 57–68.

**57.** $x^2 + 4x + y^2 = 12$

**58.** $2x^2 + 2y^2 - 28x + 12y + 114 = 0$

**59.** $x^2 + 2x + 4y - 3 = 0$  **60.** $y^2 - 4y - 8x - 12 = 0$

**61.** $x^2 + 5y^2 + 4x = 1$  **62.** $9x^2 + 6y^2 + 36y = 0$

**63.** $x^2 + 2y^2 - 2x - 4y = -1$

**64.** $4x^2 + y^2 + 8x - 2y = -1$

**65.** $x^2 - y^2 - 2x + 4y = 4$  **66.** $x^2 - y^2 + 4x - 6y = 6$

**67.** $2x^2 - y^2 + 6y = 3$  **68.** $y^2 - 4x^2 + 16x = 24$

**Theory and Examples**

**69.** If lines are drawn parallel to the coordinate axes through a point $P$ on the parabola $y^2 = kx$, $k > 0$, the parabola partitions the rectangular region bounded by these lines and the coordinate axes into two smaller regions, $A$ and $B$.

  **a.** If the two smaller regions are revolved about the $y$-axis, show that they generate solids whose volumes have the ratio 4:1.

  **b.** What is the ratio of the volumes generated by revolving the regions about the $x$-axis?

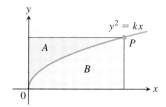

**70. Suspension bridge cables hang in parabolas**  The suspension bridge cable shown in the accompanying figure supports a uniform load of $w$ newtons per horizontal meter. It can be shown that if $H$ is the horizontal tension of the cable at the origin, then the curve of the cable satisfies the equation

$$\frac{dy}{dx} = \frac{w}{H}x.$$

Show that the cable hangs in a parabola by solving this differential equation subject to the initial condition that $y = 0$ when $x = 0$.

**71. The width of a parabola at the focus**  Show that the number $4p$ is the *width* of the parabola $x^2 = 4py$ $(p > 0)$ at the focus by showing that the line $y = p$ cuts the parabola at points that are $4p$ units apart.

**72. The asymptotes of $(x^2/a^2) - (y^2/b^2) = 1$**  Show that the vertical distance between the line $y = (b/a)x$ and the upper half of the right-hand branch $y = (b/a)\sqrt{x^2 - a^2}$ of the hyperbola $(x^2/a^2) - (y^2/b^2) = 1$ approaches 0 by showing that

$$\lim_{x \to \infty}\left(\frac{b}{a}x - \frac{b}{a}\sqrt{x^2 - a^2}\right) = \frac{b}{a}\lim_{x \to \infty}\left(x - \sqrt{x^2 - a^2}\right) = 0.$$

Similar results hold for the remaining portions of the hyperbola and the lines $y = \pm(b/a)x$.

**73. Area**  Find the dimensions of the rectangle of largest area that can be inscribed in the ellipse $x^2 + 4y^2 = 4$ with its sides parallel to the coordinate axes. What is the area of the rectangle?

**74. Volume**  Find the volume of the solid generated by revolving the region enclosed by the ellipse $9x^2 + 4y^2 = 36$ about the **(a)** $x$-axis, **(b)** $y$-axis.

**75. Volume**  The "triangular" region in the first quadrant bounded by the $x$-axis, the line $x = 4$, and the hyperbola $9x^2 - 4y^2 = 36$ is revolved about the $x$-axis to generate a solid. Find the volume of the solid.

**76. Tangents**  Show that the tangents to the curve $y^2 = 4px$ from any point on the line $x = -p$ are perpendicular.

**77. Tangents**  Find equations for the tangents to the circle $(x - 2)^2 + (y - 1)^2 = 5$ at the points where the circle crosses the coordinate axes.

**78. Volume**  The region bounded on the left by the $y$-axis, on the right by the hyperbola $x^2 - y^2 = 1$, and above and below by the lines $y = \pm 3$ is revolved about the $y$-axis to generate a solid. Find the volume of the solid.

**79. Centroid**  Find the centroid of the region that is bounded below by the $x$-axis and above by the ellipse $(x^2/9) + (y^2/16) = 1$.

**80. Surface area**  The curve $y = \sqrt{x^2 + 1}$, $0 \le x \le \sqrt{2}$, which is part of the upper branch of the hyperbola $y^2 - x^2 = 1$, is revolved about the $x$-axis to generate a surface. Find the area of the surface.

81. **The reflective property of parabolas** The accompanying figure shows a typical point $P(x_0, y_0)$ on the parabola $y^2 = 4px$. The line $L$ is tangent to the parabola at $P$. The parabola's focus lies at $F(p, 0)$. The ray $L'$ extending from $P$ to the right is parallel to the $x$-axis. We show that light from $F$ to $P$ will be reflected out along $L'$ by showing that $\beta$ equals $\alpha$. Establish this equality by taking the following steps.

   **a.** Show that $\tan \beta = 2p/y_0$.

   **b.** Show that $\tan \phi = y_0/(x_0 - p)$.

   **c.** Use the identity

   $$\tan \alpha = \frac{\tan \phi - \tan \beta}{1 - \tan \phi \tan \beta}$$

   to show that $\tan \alpha = 2p/y_0$.

   Since $\alpha$ and $\beta$ are both acute, $\tan \beta = \tan \alpha$ implies $\beta = \alpha$.

This reflective property of parabolas is used in applications like car headlights, radio telescopes, and satellite TV dishes.

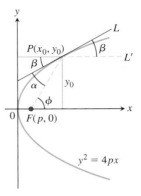

# 11.7 Conics in Polar Coordinates

Polar coordinates are especially important in astronomy and astronautical engineering because satellites, moons, planets, and comets all move approximately along ellipses, parabolas, and hyperbolas that can be described with a single relatively simple polar coordinate equation. We develop that equation here after first introducing the idea of a conic section's *eccentricity*. The eccentricity reveals the conic section's type (circle, ellipse, parabola, or hyperbola) and the degree to which it is "squashed" or flattened.

## Eccentricity

Although the center-to-focus distance $c$ does not appear in the standard Cartesian equation

$$\frac{x^2}{a^2} + \frac{y^2}{b^2} = 1, \quad (a > b)$$

for an ellipse, we can still determine $c$ from the equation $c = \sqrt{a^2 - b^2}$. If we fix $a$ and vary $c$ over the interval $0 \le c \le a$, the resulting ellipses will vary in shape. They are circles if $c = 0$ (so that $a = b$) and flatten, becoming more oblong, as $c$ increases. If $c = a$, the foci and vertices overlap and the ellipse degenerates into a line segment. Thus we are led to consider the ratio $e = c/a$. We use this ratio for hyperbolas as well, except in this case $c$ equals $\sqrt{a^2 + b^2}$ instead of $\sqrt{a^2 - b^2}$. We define these ratios with the term *eccentricity*.

---

**DEFINITION**

The **eccentricity** of the ellipse $(x^2/a^2) + (y^2/b^2) = 1 \ (a > b)$ is

$$e = \frac{c}{a} = \frac{\sqrt{a^2 - b^2}}{a}.$$

The **eccentricity** of the hyperbola $(x^2/a^2) - (y^2/b^2) = 1$ is

$$e = \frac{c}{a} = \frac{\sqrt{a^2 + b^2}}{a}.$$

The **eccentricity** of a parabola is $e = 1$.

---

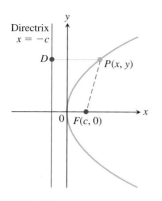

**FIGURE 11.46** The distance from the focus $F$ to any point $P$ on a parabola equals the distance from $P$ to the nearest point $D$ on the directrix, so $PF = PD$.

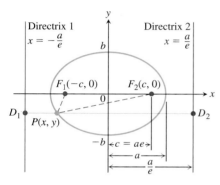

**FIGURE 11.47** The foci and directrices of the ellipse $(x^2/a^2) + (y^2/b^2) = 1$. Directrix 1 corresponds to focus $F_1$ and directrix 2 to focus $F_2$.

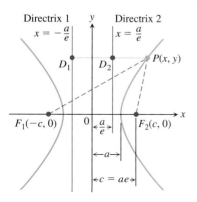

**FIGURE 11.48** The foci and directrices of the hyperbola $(x^2/a^2) - (y^2/b^2) = 1$. No matter where $P$ lies on the hyperbola, $PF_1 = e \cdot PD_1$ and $PF_2 = e \cdot PD_2$.

Whereas a parabola has one focus and one directrix, each **ellipse** has two foci and two **directrices**. These are the lines perpendicular to the major axis at distances $\pm a/e$ from the center. From Figure 11.46 we see that a parabola has the property

$$PF = 1 \cdot PD \tag{1}$$

for any point $P$ on it, where $F$ is the focus and $D$ is the point nearest $P$ on the directrix. For an ellipse, it can be shown that the equations that replace Equation (1) are

$$PF_1 = e \cdot PD_1, \qquad PF_2 = e \cdot PD_2. \tag{2}$$

Here, $e$ is the eccentricity, $P$ is any point on the ellipse, $F_1$ and $F_2$ are the foci, and $D_1$ and $D_2$ are the points on the directrices nearest $P$ (Figure 11.47).

In both Equations (2) the directrix and focus must correspond; that is, if we use the distance from $P$ to $F_1$, we must also use the distance from $P$ to the directrix at the same end of the ellipse. The directrix $x = -a/e$ corresponds to $F_1(-c, 0)$, and the directrix $x = a/e$ corresponds to $F_2(c, 0)$.

As with the ellipse, it can be shown that the lines $x = \pm a/e$ act as **directrices** for the **hyperbola** and that

$$PF_1 = e \cdot PD_1 \qquad \text{and} \qquad PF_2 = e \cdot PD_2. \tag{3}$$

Here $P$ is any point on the hyperbola, $F_1$ and $F_2$ are the foci, and $D_1$ and $D_2$ are the points nearest $P$ on the directrices (Figure 11.48).

In both the ellipse and the hyperbola, the eccentricity is the ratio of the distance between the foci to the distance between the vertices (because $c/a = 2c/2a$).

$$\text{Eccentricity} = \frac{\text{distance between foci}}{\text{distance between vertices}}$$

In an ellipse, the foci are closer together than the vertices and the ratio is less than 1. In a hyperbola, the foci are farther apart than the vertices and the ratio is greater than 1.

The "focus–directrix" equation $PF = e \cdot PD$ unites the parabola, ellipse, and hyperbola in the following way. Suppose that the distance $PF$ of a point $P$ from a fixed point $F$ (the focus) is a constant multiple of its distance from a fixed line (the directrix). That is, suppose

$$PF = e \cdot PD, \tag{4}$$

where $e$ is the constant of proportionality. Then the path traced by $P$ is

**(a)** a *parabola* if $e = 1$,

**(b)** an *ellipse* of eccentricity $e$ if $e < 1$, and

**(c)** a *hyperbola* of eccentricity $e$ if $e > 1$.

As $e$ increases ($e \to 1^-$), ellipses become more oblong, and ($e \to \infty$) hyperbolas flatten toward two lines parallel to the directrix. There are no coordinates in Equation (4), and when we try to translate it into Cartesian coordinate form, it translates in different ways depending on the size of $e$. However, as we are about to see, in polar coordinates the equation $PF = e \cdot PD$ translates into a single equation regardless of the value of $e$.

Given the focus and corresponding directrix of a hyperbola centered at the origin and with foci on the $x$-axis, we can use the dimensions shown in Figure 11.48 to find $e$.

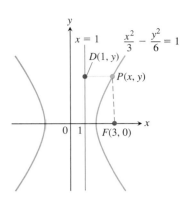

**FIGURE 11.49** The hyperbola and directrix in Example 1.

Knowing $e$, we can derive a Cartesian equation for the hyperbola from the equation $PF = e \cdot PD$, as in the next example. We can find equations for ellipses centered at the origin and with foci on the $x$-axis in a similar way, using the dimensions shown in Figure 11.47.

**EXAMPLE 1** Find a Cartesian equation for the hyperbola centered at the origin that has a focus at $(3, 0)$ and the line $x = 1$ as the corresponding directrix.

**Solution** We first use the dimensions shown in Figure 11.48 to find the hyperbola's eccentricity. The focus is (see Figure 11.49)

$$(c, 0) = (3, 0), \quad \text{so} \quad c = 3.$$

Again from Figure 11.48, the directrix is the line

$$x = \frac{a}{e} = 1, \quad \text{so} \quad a = e.$$

When combined with the equation $e = c/a$ that defines eccentricity, these results give

$$e = \frac{c}{a} = \frac{3}{e}, \quad \text{so} \quad e^2 = 3 \quad \text{and} \quad e = \sqrt{3}.$$

Knowing $e$, we can now derive the equation we want from the equation $PF = e \cdot PD$. In the coordinates of Figure 11.49, we have

$$PF = e \cdot PD \qquad \text{Eq. (4)}$$
$$\sqrt{(x - 3)^2 + (y - 0)^2} = \sqrt{3}\,|x - 1| \qquad e = \sqrt{3}$$
$$x^2 - 6x + 9 + y^2 = 3(x^2 - 2x + 1) \qquad \text{Square both sides.}$$
$$2x^2 - y^2 = 6$$
$$\frac{x^2}{3} - \frac{y^2}{6} = 1. \qquad \blacksquare$$

## Polar Equations

To find a polar equation for an ellipse, parabola, or hyperbola, we place one focus at the origin and the corresponding directrix to the right of the origin along the vertical line $x = k$ (Figure 11.50). In polar coordinates, this makes

$$PF = r$$

and

$$PD = k - FB = k - r\cos\theta.$$

The conic's focus–directrix equation $PF = e \cdot PD$ then becomes

$$r = e(k - r\cos\theta),$$

which can be solved for $r$ to obtain the following expression.

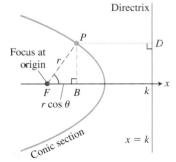

**FIGURE 11.50** If a conic section is put in the position with its focus placed at the origin and a directrix perpendicular to the initial ray and right of the origin, we can find its polar equation from the conic's focus–directrix equation.

---

**Polar Equation for a Conic with Eccentricity $e$**

$$r = \frac{ke}{1 + e\cos\theta}, \qquad (5)$$

where $x = k > 0$ is the vertical directrix.

---

**EXAMPLE 2**    Here are polar equations for three conics. The eccentricity values identifying the conic are the same for both polar and Cartesian coordinates.

$$e = \frac{1}{2}: \quad \text{ellipse} \qquad r = \frac{k}{2 + \cos\theta}$$

$$e = 1: \quad \text{parabola} \qquad r = \frac{k}{1 + \cos\theta}$$

$$e = 2: \quad \text{hyperbola} \qquad r = \frac{2k}{1 + 2\cos\theta} \qquad \blacksquare$$

You may see variations of Equation (5), depending on the location of the directrix. If the directrix is the line $x = -k$ to the left of the origin (the origin is still a focus), we replace Equation (5) with

$$r = \frac{ke}{1 - e\cos\theta}.$$

The denominator now has a $(-)$ instead of a $(+)$. If the directrix is either of the lines $y = k$ or $y = -k$, the equations have sines in them instead of cosines, as shown in Figure 11.51.

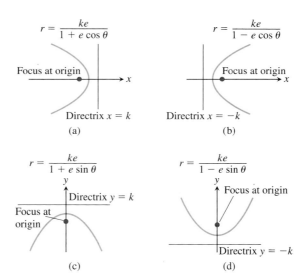

**FIGURE 11.51**  Equations for conic sections with eccentricity $e > 0$ but different locations of the directrix. The graphs here show a parabola, so $e = 1$.

**EXAMPLE 3**    Find an equation for the hyperbola with eccentricity $3/2$ and directrix $x = 2$.

**Solution**    We use Equation (5) with $k = 2$ and $e = 3/2$:

$$r = \frac{2(3/2)}{1 + (3/2)\cos\theta} \qquad \text{or} \qquad r = \frac{6}{2 + 3\cos\theta}. \qquad \blacksquare$$

**EXAMPLE 4**    Find the directrix of the parabola

$$r = \frac{25}{10 + 10\cos\theta}.$$

**Solution** We divide the numerator and denominator by 10 to put the equation in standard polar form:

$$r = \frac{5/2}{1 + \cos \theta}.$$

This is the equation

$$r = \frac{ke}{1 + e \cos \theta}$$

with $k = 5/2$ and $e = 1$. The equation of the directrix is $x = 5/2$. ∎

From the ellipse diagram in Figure 11.52, we see that $k$ is related to the eccentricity $e$ and the semimajor axis $a$ by the equation

$$k = \frac{a}{e} - ea.$$

From this, we find that $ke = a(1 - e^2)$. Replacing $ke$ in Equation (5) by $a(1 - e^2)$ gives the standard polar equation for an ellipse.

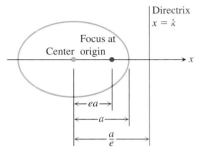

**FIGURE 11.52** In an ellipse with semimajor axis $a$, the focus–directrix distance is $k = (a/e) - ea$, so $ke = a(1 - e^2)$.

---

**Polar Equation for the Ellipse with Eccentricity $e$ and Semimajor Axis $a$**

$$r = \frac{a(1 - e^2)}{1 + e \cos \theta} \tag{6}$$

---

Notice that when $e = 0$, Equation (6) becomes $r = a$, which represents a circle.

## Lines

Suppose the perpendicular from the origin to line $L$ meets $L$ at the point $P_0(r_0, \theta_0)$, with $r_0 \geq 0$ (Figure 11.53). Then, if $P(r, \theta)$ is any other point on $L$, the points $P$, $P_0$, and $O$ are the vertices of a right triangle, from which we can read the relation

$$r_0 = r \cos (\theta - \theta_0).$$

---

**The Standard Polar Equation for Lines**

If the point $P_0(r_0, \theta_0)$ is the foot of the perpendicular from the origin to the line $L$, and $r_0 \geq 0$, then an equation for $L$ is

$$r \cos (\theta - \theta_0) = r_0. \tag{7}$$

---

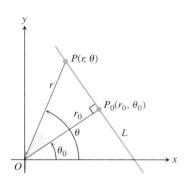

**FIGURE 11.53** We can obtain a polar equation for line $L$ by reading the relation $r_0 = r \cos (\theta - \theta_0)$ from the right triangle $OP_0P$.

For example, if $\theta_0 = \pi/3$ and $r_0 = 2$, we find that

$$r \cos \left( \theta - \frac{\pi}{3} \right) = 2$$

$$r \left( \cos \theta \cos \frac{\pi}{3} + \sin \theta \sin \frac{\pi}{3} \right) = 2$$

$$\frac{1}{2} r \cos \theta + \frac{\sqrt{3}}{2} r \sin \theta = 2, \qquad \text{or} \qquad x + \sqrt{3}\, y = 4.$$

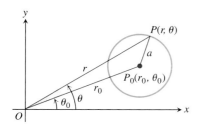

**FIGURE 11.54** We can get a polar equation for this circle by applying the Law of Cosines to triangle $OP_0P$.

## Circles

To find a polar equation for the circle of radius $a$ centered at $P_0(r_0, \theta_0)$, we let $P(r, \theta)$ be a point on the circle and apply the Law of Cosines to triangle $OP_0P$ (Figure 11.54). This gives

$$a^2 = r_0{}^2 + r^2 - 2r_0r \cos(\theta - \theta_0).$$

If the circle passes through the origin, then $r_0 = a$ and this equation simplifies to

$$a^2 = a^2 + r^2 - 2ar \cos(\theta - \theta_0)$$
$$r^2 = 2ar \cos(\theta - \theta_0)$$
$$r = 2a \cos(\theta - \theta_0).$$

If the circle's center lies on the positive $x$-axis, $\theta_0 = 0$ and we get the further simplification

$$r = 2a \cos\theta. \tag{8}$$

If the center lies on the positive $y$-axis, $\theta = \pi/2$, $\cos(\theta - \pi/2) = \sin\theta$, and the equation $r = 2a \cos(\theta - \theta_0)$ becomes

$$r = 2a \sin\theta. \tag{9}$$

Equations for circles through the origin centered on the negative $x$- and $y$-axes can be obtained by replacing $r$ with $-r$ in the above equations.

**EXAMPLE 5** Here are several polar equations given by Equations (8) and (9) for circles through the origin and having centers that lie on the $x$- or $y$-axis.

| Radius | Center (polar coordinates) | Polar equation |
|---|---|---|
| 3 | $(3, 0)$ | $r = 6 \cos\theta$ |
| 2 | $(2, \pi/2)$ | $r = 4 \sin\theta$ |
| 1/2 | $(-1/2, 0)$ | $r = -\cos\theta$ |
| 1 | $(-1, \pi/2)$ | $r = -2 \sin\theta$ |

# Exercises 11.7

## Ellipses and Eccentricity

In Exercises 1–8, find the eccentricity of the ellipse. Then find and graph the ellipse's foci and directrices.

1. $16x^2 + 25y^2 = 400$
2. $7x^2 + 16y^2 = 112$
3. $2x^2 + y^2 = 2$
4. $2x^2 + y^2 = 4$
5. $3x^2 + 2y^2 = 6$
6. $9x^2 + 10y^2 = 90$
7. $6x^2 + 9y^2 = 54$
8. $169x^2 + 25y^2 = 4225$

Exercises 9–12 give the foci or vertices and the eccentricities of ellipses centered at the origin of the $xy$-plane. In each case, find the ellipse's standard-form equation in Cartesian coordinates.

9. Foci: $(0, \pm 3)$
   Eccentricity: 0.5
10. Foci: $(\pm 8, 0)$
    Eccentricity: 0.2
11. Vertices: $(0, \pm 70)$
    Eccentricity: 0.1
12. Vertices: $(\pm 10, 0)$
    Eccentricity: 0.24

Exercises 13–16 give foci and corresponding directrices of ellipses centered at the origin of the $xy$-plane. In each case, use the dimensions in Figure 11.47 to find the eccentricity of the ellipse. Then find the ellipse's standard-form equation in Cartesian coordinates.

13. Focus: $(\sqrt{5}, 0)$
    Directrix: $x = \dfrac{9}{\sqrt{5}}$
14. Focus: $(4, 0)$
    Directrix: $x = \dfrac{16}{3}$
15. Focus: $(-4, 0)$
    Directrix: $x = -16$
16. Focus: $(-\sqrt{2}, 0)$
    Directrix: $x = -2\sqrt{2}$

## Hyperbolas and Eccentricity

In Exercises 17–24, find the eccentricity of the hyperbola. Then find and graph the hyperbola's foci and directrices.

**17.** $x^2 - y^2 = 1$

**18.** $9x^2 - 16y^2 = 144$

**19.** $y^2 - x^2 = 8$

**20.** $y^2 - x^2 = 4$

**21.** $8x^2 - 2y^2 = 16$

**22.** $y^2 - 3x^2 = 3$

**23.** $8y^2 - 2x^2 = 16$

**24.** $64x^2 - 36y^2 = 2304$

Exercises 25–28 give the eccentricities and the vertices or foci of hyperbolas centered at the origin of the $xy$-plane. In each case, find the hyperbola's standard-form equation in Cartesian coordinates.

**25.** Eccentricity: 3

  Vertices: $(0, \pm 1)$

**26.** Eccentricity: 2

  Vertices: $(\pm 2, 0)$

**27.** Eccentricity: 3

  Foci: $(\pm 3, 0)$

**28.** Eccentricity: 1.25

  Foci: $(0, \pm 5)$

## Eccentricities and Directrices

Exercises 29–36 give the eccentricities of conic sections with one focus at the origin along with the directrix corresponding to that focus. Find a polar equation for each conic section.

**29.** $e = 1, \quad x = 2$

**30.** $e = 1, \quad y = 2$

**31.** $e = 5, \quad y = -6$

**32.** $e = 2, \quad x = 4$

**33.** $e = 1/2, \quad x = 1$

**34.** $e = 1/4, \quad x = -2$

**35.** $e = 1/5, \quad y = -10$

**36.** $e = 1/3, \quad y = 6$

## Parabolas and Ellipses

Sketch the parabolas and ellipses in Exercises 37–44. Include the directrix that corresponds to the focus at the origin. Label the vertices with appropriate polar coordinates. Label the centers of the ellipses as well.

**37.** $r = \dfrac{1}{1 + \cos \theta}$

**38.** $r = \dfrac{6}{2 + \cos \theta}$

**39.** $r = \dfrac{25}{10 - 5 \cos \theta}$

**40.** $r = \dfrac{4}{2 - 2 \cos \theta}$

**41.** $r = \dfrac{400}{16 + 8 \sin \theta}$

**42.** $r = \dfrac{12}{3 + 3 \sin \theta}$

**43.** $r = \dfrac{8}{2 - 2 \sin \theta}$

**44.** $r = \dfrac{4}{2 - \sin \theta}$

## Lines

Sketch the lines in Exercises 45–48 and find Cartesian equations for them.

**45.** $r \cos \left( \theta - \dfrac{\pi}{4} \right) = \sqrt{2}$

**46.** $r \cos \left( \theta + \dfrac{3\pi}{4} \right) = 1$

**47.** $r \cos \left( \theta - \dfrac{2\pi}{3} \right) = 3$

**48.** $r \cos \left( \theta + \dfrac{\pi}{3} \right) = 2$

Find a polar equation in the form $r \cos (\theta - \theta_0) = r_0$ for each of the lines in Exercises 49–52.

**49.** $\sqrt{2}x + \sqrt{2}y = 6$

**50.** $\sqrt{3}x - y = 1$

**51.** $y = -5$

**52.** $x = -4$

## Circles

Sketch the circles in Exercises 53–56. Give polar coordinates for their centers and identify their radii.

**53.** $r = 4 \cos \theta$

**54.** $r = 6 \sin \theta$

**55.** $r = -2 \cos \theta$

**56.** $r = -8 \sin \theta$

Find polar equations for the circles in Exercises 57–64. Sketch each circle in the coordinate plane and label it with both its Cartesian and polar equations.

**57.** $(x - 6)^2 + y^2 = 36$

**58.** $(x + 2)^2 + y^2 = 4$

**59.** $x^2 + (y - 5)^2 = 25$

**60.** $x^2 + (y + 7)^2 = 49$

**61.** $x^2 + 2x + y^2 = 0$

**62.** $x^2 - 16x + y^2 = 0$

**63.** $x^2 + y^2 + y = 0$

**64.** $x^2 + y^2 - \dfrac{4}{3}y = 0$

## Examples of Polar Equations

T Graph the lines and conic sections in Exercises 65–74.

**65.** $r = 3 \sec (\theta - \pi/3)$

**66.** $r = 4 \sec (\theta + \pi/6)$

**67.** $r = 4 \sin \theta$

**68.** $r = -2 \cos \theta$

**69.** $r = 8/(4 + \cos \theta)$

**70.** $r = 8/(4 + \sin \theta)$

**71.** $r = 1/(1 - \sin \theta)$

**72.** $r = 1/(1 + \cos \theta)$

**73.** $r = 1/(1 + 2 \sin \theta)$

**74.** $r = 1/(1 + 2 \cos \theta)$

**75. Perihelion and aphelion** A planet travels about its sun in an ellipse whose semimajor axis has length $a$. (See accompanying figure.)

  **a.** Show that $r = a(1 - e)$ when the planet is closest to the sun and that $r = a(1 + e)$ when the planet is farthest from the sun.

  **b.** Use the data in the table in Exercise 76 to find how close each planet in our solar system comes to the sun and how far away each planet gets from the sun.

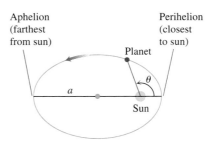

**76. Planetary orbits** Use the data in the table below and Equation (6) to find polar equations for the orbits of the planets.

| Planet | Semimajor axis (astronomical units) | Eccentricity |
|---|---|---|
| Mercury | 0.3871 | 0.2056 |
| Venus | 0.7233 | 0.0068 |
| Earth | 1.000 | 0.0167 |
| Mars | 1.524 | 0.0934 |
| Jupiter | 5.203 | 0.0484 |
| Saturn | 9.539 | 0.0543 |
| Uranus | 19.18 | 0.0460 |
| Neptune | 30.06 | 0.0082 |

# Chapter 11  Questions to Guide Your Review

1. What is a parametrization of a curve in the $xy$-plane? Does a function $y = f(x)$ always have a parametrization? Are parametrizations of a curve unique? Give examples.

2. Give some typical parametrizations for lines, circles, parabolas, ellipses, and hyperbolas. How might the parametrized curve differ from the graph of its Cartesian equation?

3. What is a cycloid? What are typical parametric equations for cycloids? What physical properties account for the importance of cycloids?

4. What is the formula for the slope $dy/dx$ of a parametrized curve $x = f(t), y = g(t)$? When does the formula apply? When can you expect to be able to find $d^2y/dx^2$ as well? Give examples.

5. How can you sometimes find the area bounded by a parametrized curve and one of the coordinate axes?

6. How do you find the length of a smooth parametrized curve $x = f(t), y = g(t), a \le t \le b$? What does smoothness have to do with length? What else do you need to know about the parametrization in order to find the curve's length? Give examples.

7. What is the arc length function for a smooth parametrized curve? What is its arc length differential?

8. Under what conditions can you find the area of the surface generated by revolving a curve $x = f(t), y = g(t), a \le t \le b$, about the $x$-axis? the $y$-axis? Give examples.

9. What are polar coordinates? What equations relate polar coordinates to Cartesian coordinates? Why might you want to change from one coordinate system to the other?

10. What consequence does the lack of uniqueness of polar coordinates have for graphing? Give an example.

11. How do you graph equations in polar coordinates? Include in your discussion symmetry, slope, behavior at the origin, and the use of Cartesian graphs. Give examples.

12. How do you find the area of a region $0 \le r_1(\theta) \le r \le r_2(\theta)$, $\alpha \le \theta \le \beta$, in the polar coordinate plane? Give examples.

13. Under what conditions can you find the length of a curve $r = f(\theta), \alpha \le \theta \le \beta$, in the polar coordinate plane? Give an example of a typical calculation.

14. What is a parabola? What are the Cartesian equations for parabolas whose vertices lie at the origin and whose foci lie on the coordinate axes? How can you find the focus and directrix of such a parabola from its equation?

15. What is an ellipse? What are the Cartesian equations for ellipses centered at the origin with foci on one of the coordinate axes? How can you find the foci, vertices, and directrices of such an ellipse from its equation?

16. What is a hyperbola? What are the Cartesian equations for hyperbolas centered at the origin with foci on one of the coordinate axes? How can you find the foci, vertices, and directrices of such an ellipse from its equation?

17. What is the eccentricity of a conic section? How can you classify conic sections by eccentricity? How does eccentricity change the shape of ellipses and hyperbolas?

18. Explain the equation $PF = e \cdot PD$.

19. What are the standard equations for lines and conic sections in polar coordinates? Give examples.

# Chapter 11  Practice Exercises

## Identifying Parametric Equations in the Plane

Exercises 1–6 give parametric equations and parameter intervals for the motion of a particle in the $xy$-plane. Identify the particle's path by finding a Cartesian equation for it. Graph the Cartesian equation and indicate the direction of motion and the portion traced by the particle.

1. $x = t/2, \quad y = t + 1; \quad -\infty < t < \infty$

2. $x = \sqrt{t}, \quad y = 1 - \sqrt{t}; \quad t \ge 0$

3. $x = (1/2)\tan t, \quad y = (1/2)\sec t; \quad -\pi/2 < t < \pi/2$

4. $x = -2\cos t, \quad y = 2\sin t; \quad 0 \le t \le \pi$

5. $x = -\cos t, \quad y = \cos^2 t; \quad 0 \le t \le \pi$

6. $x = 4\cos t, \quad y = 9\sin t; \quad 0 \le t \le 2\pi$

## Finding Parametric Equations and Tangent Lines

7. Find parametric equations and a parameter interval for the motion of a particle in the $xy$-plane that traces the ellipse $16x^2 + 9y^2 = 144$ once counterclockwise. (There are many ways to do this.)

8. Find parametric equations and a parameter interval for the motion of a particle that starts at the point $(-2, 0)$ in the $xy$-plane and traces the circle $x^2 + y^2 = 4$ three times clockwise. (There are many ways to do this.)

In Exercises 9 and 10, find an equation for the line in the $xy$-plane that is tangent to the curve at the point corresponding to the given value of $t$. Also, find the value of $d^2y/dx^2$ at this point.

9. $x = (1/2)\tan t, \quad y = (1/2)\sec t; \quad t = \pi/3$

10. $x = 1 + 1/t^2, \quad y = 1 - 3/t; \quad t = 2$

11. Eliminate the parameter to express the curve in the form $y = f(x)$.

   a. $x = 4t^2, \quad y = t^3 - 1$

   b. $x = \cos t, \quad y = \tan t$

12. Find parametric equations for the given curve.

   a. Line through $(1, -2)$ with slope 3

   b. $(x - 1)^2 + (y + 2)^2 = 9$

   c. $y = 4x^2 - x$

   d. $9x^2 + 4y^2 = 36$

## Lengths of Curves

Find the lengths of the curves in Exercises 13–19.

**13.** $y = x^{1/2} - (1/3)x^{3/2}, \quad 1 \le x \le 4$

**14.** $x = y^{2/3}, \quad 1 \le y \le 8$

**15.** $y = (5/12)x^{6/5} - (5/8)x^{4/5}, \quad 1 \le x \le 32$

**16.** $x = (y^3/12) + (1/y), \quad 1 \le y \le 2$

**17.** $x = 5 \cos t - \cos 5t, \quad y = 5 \sin t - \sin 5t, \quad 0 \le t \le \pi/2$

**18.** $x = t^3 - 6t^2, \quad y = t^3 + 6t^2, \quad 0 \le t \le 1$

**19.** $x = 3 \cos \theta, \quad y = 3 \sin \theta, \quad 0 \le \theta \le \dfrac{3\pi}{2}$

**20.** Find the length of the enclosed loop $x = t^2, y = (t^3/3) - t$ shown here. The loop starts at $t = -\sqrt{3}$ and ends at $t = \sqrt{3}$.

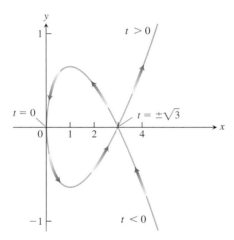

## Surface Areas

Find the areas of the surfaces generated by revolving the curves in Exercises 21 and 22 about the indicated axes.

**21.** $x = t^2/2, \quad y = 2t, \quad 0 \le t \le \sqrt{5}; \quad x\text{-axis}$

**22.** $x = t^2 + 1/(2t), \quad y = 4\sqrt{t}, \quad 1/\sqrt{2} \le t \le 1; \quad y\text{-axis}$

## Polar to Cartesian Equations

Sketch the lines in Exercises 23–28. Also, find a Cartesian equation for each line.

**23.** $r \cos\left(\theta + \dfrac{\pi}{3}\right) = 2\sqrt{3}$   **24.** $r \cos\left(\theta - \dfrac{3\pi}{4}\right) = \dfrac{\sqrt{2}}{2}$

**25.** $r = 2 \sec \theta$   **26.** $r = -\sqrt{2} \sec \theta$

**27.** $r = -(3/2) \csc \theta$   **28.** $r = \left(3\sqrt{3}\right) \csc \theta$

Find Cartesian equations for the circles in Exercises 29–32. Sketch each circle in the coordinate plane and label it with both its Cartesian and polar equations.

**29.** $r = -4 \sin \theta$   **30.** $r = 3\sqrt{3} \sin \theta$

**31.** $r = 2\sqrt{2} \cos \theta$   **32.** $r = -6 \cos \theta$

## Cartesian to Polar Equations

Find polar equations for the circles in Exercises 33–36. Sketch each circle in the coordinate plane and label it with both its Cartesian and polar equations.

**33.** $x^2 + y^2 + 5y = 0$   **34.** $x^2 + y^2 - 2y = 0$

**35.** $x^2 + y^2 - 3x = 0$   **36.** $x^2 + y^2 + 4x = 0$

## Graphs in Polar Coordinates

Sketch the regions defined by the polar coordinate inequalities in Exercises 37 and 38.

**37.** $0 \le r \le 6 \cos \theta$   **38.** $-4 \sin \theta \le r \le 0$

Match each graph in Exercises 39–46 with the appropriate equation (a)–(l). There are more equations than graphs, so some equations will not be matched.

**a.** $r = \cos 2\theta$   **b.** $r \cos \theta = 1$   **c.** $r = \dfrac{6}{1 - 2 \cos \theta}$

**d.** $r = \sin 2\theta$   **e.** $r = \theta$   **f.** $r^2 = \cos 2\theta$

**g.** $r = 1 + \cos \theta$   **h.** $r = 1 - \sin \theta$   **i.** $r = \dfrac{2}{1 - \cos \theta}$

**j.** $r^2 = \sin 2\theta$   **k.** $r = -\sin \theta$   **l.** $r = 2 \cos \theta + 1$

**39.** Four-leaved rose   **40.** Spiral

**41.** Limaçon   **42.** Lemniscate

**43.** Circle   **44.** Cardioid

**45.** Parabola   **46.** Lemniscate

## Area in Polar Coordinates

Find the areas of the regions in the polar coordinate plane described in Exercises 47–50.

**47.** Enclosed by the limaçon $r = 2 - \cos \theta$

**48.** Enclosed by one leaf of the three-leaved rose $r = \sin 3\theta$

**49.** Inside the "figure eight" $r = 1 + \cos 2\theta$ and outside the circle $r = 1$

**50.** Inside the cardioid $r = 2(1 + \sin \theta)$ and outside the circle $r = 2 \sin \theta$

## Length in Polar Coordinates

Find the lengths of the curves given by the polar coordinate equations in Exercises 51–54.

**51.** $r = -1 + \cos\theta$

**52.** $r = 2\sin\theta + 2\cos\theta, \quad 0 \le \theta \le \pi/2$

**53.** $r = 8\sin^3(\theta/3), \quad 0 \le \theta \le \pi/4$

**54.** $r = \sqrt{1 + \cos 2\theta}, \quad -\pi/2 \le \theta \le \pi/2$

## Graphing Conic Sections

Sketch the parabolas in Exercises 55–58. Include the focus and directrix in each sketch.

**55.** $x^2 = -4y$

**56.** $x^2 = 2y$

**57.** $y^2 = 3x$

**58.** $y^2 = -(8/3)x$

Find the eccentricities of the ellipses and hyperbolas in Exercises 59–62. Sketch each conic section. Include the foci, vertices, and asymptotes (as appropriate) in your sketch.

**59.** $16x^2 + 7y^2 = 112$

**60.** $x^2 + 2y^2 = 4$

**61.** $3x^2 - y^2 = 3$

**62.** $5y^2 - 4x^2 = 20$

Exercises 63–68 give equations for conic sections and tell how many units up or down and to the right or left each curve is to be shifted. Find an equation for the new conic section, and find the new foci, vertices, centers, and asymptotes, as appropriate. If the curve is a parabola, find the new directrix as well.

**63.** $x^2 = -12y, \quad$ right 2, up 3

**64.** $y^2 = 10x, \quad$ left $1/2$, down 1

**65.** $\dfrac{x^2}{9} + \dfrac{y^2}{25} = 1, \quad$ left 3, down 5

**66.** $\dfrac{x^2}{169} + \dfrac{y^2}{144} = 1, \quad$ right 5, up 12

**67.** $\dfrac{y^2}{8} - \dfrac{x^2}{2} = 1, \quad$ right 2, up $2\sqrt{2}$

**68.** $\dfrac{x^2}{36} - \dfrac{y^2}{64} = 1, \quad$ left 10, down 3

## Identifying Conic Sections

Complete the squares to identify the conic sections in Exercises 69–76. Find their foci, vertices, centers, and asymptotes (as appropriate). If the curve is a parabola, find its directrix as well.

**69.** $x^2 - 4x - 4y^2 = 0$

**70.** $4x^2 - y^2 + 4y = 8$

**71.** $y^2 - 2y + 16x = -49$

**72.** $x^2 - 2x + 8y = -17$

**73.** $9x^2 + 16y^2 + 54x - 64y = -1$

**74.** $25x^2 + 9y^2 - 100x + 54y = 44$

**75.** $x^2 + y^2 - 2x - 2y = 0$

**76.** $x^2 + y^2 + 4x + 2y = 1$

## Conics in Polar Coordinates

Sketch the conic sections whose polar coordinate equations are given in Exercises 77–80. Give polar coordinates for the vertices and, in the case of ellipses, for the centers as well.

**77.** $r = \dfrac{2}{1 + \cos\theta}$

**78.** $r = \dfrac{8}{2 + \cos\theta}$

**79.** $r = \dfrac{6}{1 - 2\cos\theta}$

**80.** $r = \dfrac{12}{3 + \sin\theta}$

Exercises 81–84 give the eccentricities of conic sections with one focus at the origin of the polar coordinate plane, along with the directrix for that focus. Find a polar equation for each conic section.

**81.** $e = 2, \quad r\cos\theta = 2$

**82.** $e = 1, \quad r\cos\theta = -4$

**83.** $e = 1/2, \quad r\sin\theta = 2$

**84.** $e = 1/3, \quad r\sin\theta = -6$

## Theory and Examples

**85.** Find the volume of the solid generated by revolving the region enclosed by the ellipse $9x^2 + 4y^2 = 36$ about **(a)** the $x$-axis, **(b)** the $y$-axis.

**86.** The "triangular" region in the first quadrant bounded by the $x$-axis, the line $x = 4$, and the hyperbola $9x^2 - 4y^2 = 36$ is revolved about the $x$-axis to generate a solid. Find the volume of the solid.

**87.** Show that the equations $x = r\cos\theta, y = r\sin\theta$ transform the polar equation

$$r = \frac{k}{1 + e\cos\theta}$$

into the Cartesian equation

$$(1 - e^2)x^2 + y^2 + 2kex - k^2 = 0.$$

**88. Archimedes spirals**  The graph of an equation of the form $r = a\theta$, where $a$ is a nonzero constant, is called an *Archimedes spiral*. Is there anything special about the widths between the successive turns of such a spiral?

---

## Chapter **11**  Additional and Advanced Exercises

### Finding Conic Sections

**1.** Find an equation for the parabola with focus $(4, 0)$ and directrix $x = 3$. Sketch the parabola together with its vertex, focus, and directrix.

**2.** Find the vertex, focus, and directrix of the parabola
$$x^2 - 6x - 12y + 9 = 0.$$

**3.** Find an equation for the curve traced by the point $P(x, y)$ if the distance from $P$ to the vertex of the parabola $x^2 = 4y$ is twice the distance from $P$ to the focus. Identify the curve.

**4.** A line segment of length $a + b$ runs from the $x$-axis to the $y$-axis. The point $P$ on the segment lies $a$ units from one end and $b$ units from the other end. Show that $P$ traces an ellipse as the ends of the segment slide along the axes.

**5.** The vertices of an ellipse of eccentricity 0.5 lie at the points $(0, \pm 2)$. Where do the foci lie?

**6.** Find an equation for the ellipse of eccentricity $2/3$ that has the line $x = 2$ as a directrix and the point $(4, 0)$ as the corresponding focus.

**7.** One focus of a hyperbola lies at the point $(0, -7)$ and the corresponding directrix is the line $y = -1$. Find an equation for the hyperbola if its eccentricity is **(a)** 2, **(b)** 5.

**8.** Find an equation for the hyperbola with foci $(0, -2)$ and $(0, 2)$ that passes through the point $(12, 7)$.

**9.** Show that the line
$$b^2xx_1 + a^2yy_1 - a^2b^2 = 0$$
is tangent to the ellipse $b^2x^2 + a^2y^2 - a^2b^2 = 0$ at the point $(x_1, y_1)$ on the ellipse.

**10.** Show that the line
$$b^2xx_1 - a^2yy_1 - a^2b^2 = 0$$
is tangent to the hyperbola $b^2x^2 - a^2y^2 - a^2b^2 = 0$ at the point $(x_1, y_1)$ on the hyperbola.

### Equations and Inequalities

What points in the $xy$-plane satisfy the equations and inequalities in Exercises 11–16? Draw a figure for each exercise.

**11.** $(x^2 - y^2 - 1)(x^2 + y^2 - 25)(x^2 + 4y^2 - 4) = 0$

**12.** $(x + y)(x^2 + y^2 - 1) = 0$

**13.** $(x^2/9) + (y^2/16) \leq 1$

**14.** $(x^2/9) - (y^2/16) \leq 1$

**15.** $(9x^2 + 4y^2 - 36)(4x^2 + 9y^2 - 16) \leq 0$

**16.** $(9x^2 + 4y^2 - 36)(4x^2 + 9y^2 - 16) > 0$

### Polar Coordinates

**17. a.** Find an equation in polar coordinates for the curve
$$x = e^{2t} \cos t, \quad y = e^{2t} \sin t; \quad -\infty < t < \infty.$$

**b.** Find the length of the curve from $t = 0$ to $t = 2\pi$.

**18.** Find the length of the curve $r = 2 \sin^3(\theta/3)$, $0 \leq \theta \leq 3\pi$, in the polar coordinate plane.

Exercises 19–22 give the eccentricities of conic sections with one focus at the origin of the polar coordinate plane, along with the directrix for that focus. Find a polar equation for each conic section.

**19.** $e = 2, \quad r \cos \theta = 2$      **20.** $e = 1, \quad r \cos \theta = -4$

**21.** $e = 1/2, \quad r \sin \theta = 2$      **22.** $e = 1/3, \quad r \sin \theta = -6$

### Theory and Examples

**23. Epicycloids** When a circle rolls externally along the circumference of a second, fixed circle, any point $P$ on the circumference of the rolling circle describes an *epicycloid*, as shown here. Let the fixed circle have its center at the origin $O$ and have radius $a$.

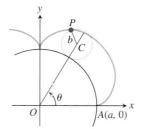

Let the radius of the rolling circle be $b$ and let the initial position of the tracing point $P$ be $A(a, 0)$. Find parametric equations for the epicycloid, using as the parameter the angle $\theta$ from the positive $x$-axis to the line through the circles' centers.

**24.** Find the centroid of the region enclosed by the $x$-axis and the cycloid arch
$$x = a(t - \sin t), \quad y = a(1 - \cos t); \quad 0 \leq t \leq 2\pi.$$

**The Angle Between the Radius Vector and the Tangent Line to a Polar Coordinate Curve** In Cartesian coordinates, when we want to discuss the direction of a curve at a point, we use the angle $\phi$ measured counterclockwise from the positive $x$-axis to the tangent line. In polar coordinates, it is more convenient to calculate the angle $\psi$ from the *radius vector* to the tangent line (see the accompanying figure). The angle $\phi$ can then be calculated from the relation
$$\phi = \theta + \psi, \tag{1}$$
which comes from applying the Exterior Angle Theorem to the triangle in the accompanying figure.

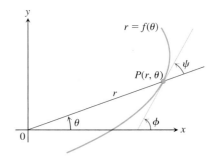

Suppose the equation of the curve is given in the form $r = f(\theta)$, where $f(\theta)$ is a differentiable function of $\theta$. Then
$$x = r \cos \theta \quad \text{and} \quad y = r \sin \theta \tag{2}$$
are differentiable functions of $\theta$ with
$$\frac{dx}{d\theta} = -r \sin \theta + \cos \theta \frac{dr}{d\theta},$$
$$\frac{dy}{d\theta} = r \cos \theta + \sin \theta \frac{dr}{d\theta}. \tag{3}$$

Since $\psi = \phi - \theta$ from (1),
$$\tan \psi = \tan(\phi - \theta) = \frac{\tan \phi - \tan \theta}{1 + \tan \phi \tan \theta}.$$

Furthermore,
$$\tan \phi = \frac{dy}{dx} = \frac{dy/d\theta}{dx/d\theta}$$

because $\tan \phi$ is the slope of the curve at $P$. Also,
$$\tan \theta = \frac{y}{x}.$$

Hence
$$\tan \psi = \frac{\dfrac{dy/d\theta}{dx/d\theta} - \dfrac{y}{x}}{1 + \dfrac{y}{x}\dfrac{dy/d\theta}{dx/d\theta}} = \frac{x\dfrac{dy}{d\theta} - y\dfrac{dx}{d\theta}}{x\dfrac{dx}{d\theta} + y\dfrac{dy}{d\theta}}. \tag{4}$$

The numerator in the last expression in Equation (4) is found from Equations (2) and (3) to be
$$x\frac{dy}{d\theta} - y\frac{dx}{d\theta} = r^2.$$

Similarly, the denominator is

$$x\frac{dx}{d\theta} + y\frac{dy}{d\theta} = r\frac{dr}{d\theta}.$$

When we substitute these into Equation (4), we obtain

$$\tan \psi = \frac{r}{dr/d\theta}. \tag{5}$$

This is the equation we use for finding $\psi$ as a function of $\theta$.

**25.** Show, by reference to a figure, that the angle $\beta$ between the tangents to two curves at a point of intersection may be found from the formula

$$\tan \beta = \frac{\tan \psi_2 - \tan \psi_1}{1 + \tan \psi_2 \tan \psi_1}. \tag{6}$$

When will the two curves intersect at right angles?

**26.** Find the value of $\tan \psi$ for the curve $r = \sin^4(\theta/4)$.

**27.** Find the angle between the radius vector to the curve $r = 2a \sin 3\theta$ and its tangent when $\theta = \pi/6$.

T **28. a.** Graph the hyperbolic spiral $r\theta = 1$. What appears to happen to $\psi$ as the spiral winds in around the origin?

  **b.** Confirm your finding in part (a) analytically.

**29.** The circles $r = \sqrt{3} \cos \theta$ and $r = \sin \theta$ intersect at the point $\left(\sqrt{3}/2, \pi/3\right)$. Show that their tangents are perpendicular there.

**30.** Find the angle at which the cardioid $r = a(1 - \cos \theta)$ crosses the ray $\theta = \pi/2$.

# Chapter 11    Technology Application Projects

**Mathematica / Maple Modules:**

**_Radar Tracking of a Moving Object_**
**Part I:** Convert from polar to Cartesian coordinates.

**_Parametric and Polar Equations with a Figure Skater_**
**Part I:** Visualize position, velocity, and acceleration to analyze motion defined by parametric equations.
**Part II:** Find and analyze the equations of motion for a figure skater tracing a polar plot.

# 12

# Vectors and the Geometry of Space

**OVERVIEW** This chapter is foundational to the study of multivariable calculus. To apply calculus in many real-world situations and in higher mathematics, we need an analytic geometry to describe three-dimensional space. To accomplish this objective, we introduce three-dimensional coordinate systems and vectors. Building on what we already know about coordinates in the $xy$-plane, we establish coordinates in space by adding a third axis that measures distance above and below the $xy$-plane. Then we define vectors and use them to study the analytic geometry of space. Vectors provide simple ways to define equations for lines, planes, curves, and surfaces in space. We use these geometric concepts throughout the remainder of the text to study motion in space and the calculus of functions of several variables and vector fields, with their many important applications in science, engineering, operations research, economics, and higher mathematics.

## 12.1 Three-Dimensional Coordinate Systems

To locate a point in space, we use three mutually perpendicular coordinate axes, arranged as in Figure 12.1. The axes shown there make a *right-handed* coordinate frame. When you hold your right hand so that the fingers curl from the positive $x$-axis toward the positive $y$-axis, your thumb points along the positive $z$-axis. So when you look down on the $xy$-plane from the positive direction of the $z$-axis, positive angles in the plane are measured counterclockwise from the positive $x$-axis and around the positive $z$-axis. (In a *left-handed* coordinate frame, the $z$-axis would point downward in Figure 12.1 and angles in the plane would be positive when measured clockwise from the positive $x$-axis. Right-handed and left-handed coordinate frames are not equivalent.)

The Cartesian coordinates $(x, y, z)$ of a point $P$ in space are the values at which the planes through $P$ perpendicular to the axes cut the axes. Cartesian coordinates for space are also called **rectangular coordinates** because the axes that define them meet at right angles. Points on the $x$-axis have $y$- and $z$-coordinates equal to zero. That is, they have coordinates of the form $(x, 0, 0)$. Similarly, points on the $y$-axis have coordinates of the form $(0, y, 0)$, and points on the $z$-axis have coordinates of the form $(0, 0, z)$.

The planes determined by the coordinates axes are the **$xy$-plane**, whose standard equation is $z = 0$; the **$yz$-plane**, whose standard equation is $x = 0$; and the **$xz$-plane**, whose standard equation is $y = 0$. They meet at the **origin** $(0, 0, 0)$ (Figure 12.2). The origin is also identified by simply 0 or sometimes the letter $O$.

The three **coordinate planes** $x = 0$, $y = 0$, and $z = 0$ divide space into eight cells called **octants**. The octant in which the point coordinates are all positive is called the **first octant**; there is no convention for numbering the other seven octants.

The points in a plane perpendicular to the $x$-axis all have the same $x$-coordinate, this being the number at which that plane cuts the $x$-axis. The $y$- and $z$-coordinates can be any numbers. Similarly, the points in a plane perpendicular to the $y$-axis have a common

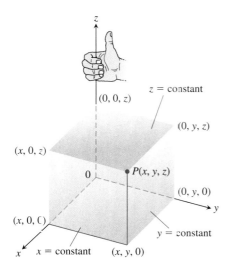

**FIGURE 12.1** The Cartesian coordinate system is right-handed.

**706**

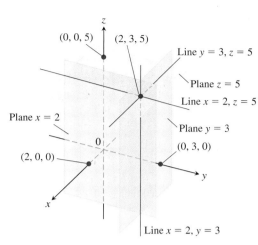

**FIGURE 12.2**   The planes $x = 0$, $y = 0$, and $z = 0$ divide space into eight octants.

**FIGURE 12.3**   The planes $x = 2$, $y = 3$, and $z = 5$ determine three lines through the point $(2, 3, 5)$.

$y$-coordinate and the points in a plane perpendicular to the $z$-axis have a common $z$-coordinate. To write equations for these planes, we name the common coordinate's value. The plane $x = 2$ is the plane perpendicular to the $x$-axis at $x = 2$. The plane $y = 3$ is the plane perpendicular to the $y$-axis at $y = 3$. The plane $z = 5$ is the plane perpendicular to the $z$-axis at $z = 5$. Figure 12.3 shows the planes $x = 2$, $y = 3$, and $z = 5$, together with their intersection point $(2, 3, 5)$.

The planes $x = 2$ and $y = 3$ in Figure 12.3 intersect in a line parallel to the $z$-axis. This line is described by the *pair* of equations $x = 2$, $y = 3$. A point $(x, y, z)$ lies on the line if and only if $x = 2$ and $y = 3$. Similarly, the line of intersection of the planes $y = 3$ and $z = 5$ is described by the equation pair $y = 3$, $z = 5$. This line runs parallel to the $x$-axis. The line of intersection of the planes $x = 2$ and $z = 5$, parallel to the $y$-axis, is described by the equation pair $x = 2$, $z = 5$.

In the following examples, we match coordinate equations and inequalities with the sets of points they define in space.

**EXAMPLE 1**    We interpret these equations and inequalities geometrically.

**(a)** $z \geq 0$    The half-space consisting of the points on and above the $xy$-plane.

**(b)** $x = -3$    The plane perpendicular to the $x$-axis at $x = -3$. This plane lies parallel to the $yz$-plane and 3 units behind it.

**(c)** $z = 0, x \leq 0, y \geq 0$    The second quadrant of the $xy$-plane.

**(d)** $x \geq 0, y \geq 0, z \geq 0$    The first octant.

**(e)** $-1 \leq y \leq 1$    The slab between the planes $y = -1$ and $y = 1$ (planes included).

**(f)** $y = -2, z = 2$    The line in which the planes $y = -2$ and $z = 2$ intersect. Alternatively, the line through the point $(0, -2, 2)$ parallel to the $x$-axis.    ∎

**EXAMPLE 2**    What points $P(x, y, z)$ satisfy the equations

$$x^2 + y^2 = 4 \quad \text{and} \quad z = 3?$$

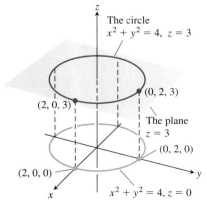

**FIGURE 12.4**   The circle $x^2 + y^2 = 4$ in the plane $z = 3$ (Example 2).

**Solution**   The points lie in the horizontal plane $z = 3$ and, in this plane, make up the circle $x^2 + y^2 = 4$. We call this set of points "the circle $x^2 + y^2 = 4$ in the plane $z = 3$" or, more simply, "the circle $x^2 + y^2 = 4$, $z = 3$" (Figure 12.4).    ∎

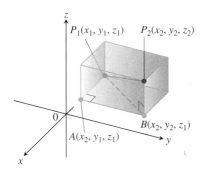

FIGURE 12.5 We find the distance between $P_1$ and $P_2$ by applying the Pythagorean theorem to the right triangles $P_1AB$ and $P_1BP_2$.

## Distance and Spheres in Space

The formula for the distance between two points in the $xy$-plane extends to points in space.

---

**The Distance Between $P_1(x_1, y_1, z_1)$ and $P_2(x_2, y_2, z_2)$**

$$|P_1P_2| = \sqrt{(x_2 - x_1)^2 + (y_2 - y_1)^2 + (z_2 - z_1)^2}$$

---

**Proof** We construct a rectangular box with faces parallel to the coordinate planes and the points $P_1$ and $P_2$ at opposite corners of the box (Figure 12.5). If $A(x_2, y_1, z_1)$ and $B(x_2, y_2, z_1)$ are the vertices of the box indicated in the figure, then the three box edges $P_1A$, $AB$, and $BP_2$ have lengths

$$|P_1A| = |x_2 - x_1|, \qquad |AB| = |y_2 - y_1|, \qquad |BP_2| = |z_2 - z_1|.$$

Because triangles $P_1BP_2$ and $P_1AB$ are both right-angled, two applications of the Pythagorean theorem give

$$|P_1P_2|^2 = |P_1B|^2 + |BP_2|^2 \qquad \text{and} \qquad |P_1B|^2 = |P_1A|^2 + |AB|^2$$

(see Figure 12.5). So

$$\begin{aligned} |P_1P_2|^2 &= |P_1B|^2 + |BP_2|^2 \\ &= |P_1A|^2 + |AB|^2 + |BP_2|^2 && \text{Substitute } |P_1B|^2 = |P_1A|^2 + |AB|^2. \\ &= |x_2 - x_1|^2 + |y_2 - y_1|^2 + |z_2 - z_1|^2 \\ &= (x_2 - x_1)^2 + (y_2 - y_1)^2 + (z_2 - z_1)^2. \end{aligned}$$

Therefore

$$|P_1P_2| = \sqrt{(x_2 - x_1)^2 + (y_2 - y_1)^2 + (z_2 - z_1)^2}. \qquad \blacksquare$$

**EXAMPLE 3** The distance between $P_1(2, 1, 5)$ and $P_2(-2, 3, 0)$ is

$$\begin{aligned} |P_1P_2| &= \sqrt{(-2 - 2)^2 + (3 - 1)^2 + (0 - 5)^2} \\ &= \sqrt{16 + 4 + 25} \\ &= \sqrt{45} \approx 6.708. \end{aligned}$$

We can use the distance formula to write equations for spheres in space (Figure 12.6). A point $P(x, y, z)$ lies on the sphere of radius $a$ centered at $P_0(x_0, y_0, z_0)$ precisely when $|P_0P| = a$ or

$$(x - x_0)^2 + (y - y_0)^2 + (z - z_0)^2 = a^2.$$

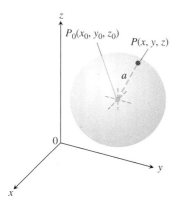

FIGURE 12.6 The sphere of radius $a$ centered at the point $(x_0, y_0, z_0)$.

---

**The Standard Equation for the Sphere of Radius $a$ and Center $(x_0, y_0, z_0)$**

$$(x - x_0)^2 + (y - y_0)^2 + (z - z_0)^2 = a^2$$

---

**EXAMPLE 4** Find the center and radius of the sphere

$$x^2 + y^2 + z^2 + 3x - 4z + 1 = 0.$$

**Solution** We find the center and radius of a sphere the way we find the center and radius of a circle: Complete the squares on the $x$-, $y$-, and $z$-terms as necessary and write each quadratic as a squared linear expression. Then, from the equation in standard form, read off the center and radius. For the sphere here, we have

$$x^2 + y^2 + z^2 + 3x - 4z + 1 = 0$$

$$(x^2 + 3x) + y^2 + (z^2 - 4z) = -1$$

$$\left(x^2 + 3x + \left(\frac{3}{2}\right)^2\right) + y^2 + \left(z^2 - 4z + \left(\frac{-4}{2}\right)^2\right) = -1 + \left(\frac{3}{2}\right)^2 + \left(\frac{-4}{2}\right)^2$$

$$\left(x + \frac{3}{2}\right)^2 + y^2 + (z - 2)^2 = -1 + \frac{9}{4} + 4 = \frac{21}{4}.$$

From this standard form, we read that $x_0 = -3/2$, $y_0 = 0$, $z_0 = 2$, and $a = \sqrt{21}/2$. The center is $(-3/2, 0, 2)$. The radius is $\sqrt{21}/2$. ■

**EXAMPLE 5**    Here are some geometric interpretations of inequalities and equations involving spheres.

(a)   $x^2 + y^2 + z^2 < 4$            The interior of the sphere $x^2 + y^2 + z^2 = 4$.

(b)   $x^2 + y^2 + z^2 \le 4$            The solid ball bounded by the sphere $x^2 + y^2 + z^2 = 4$. Alternatively, the sphere $x^2 + y^2 + z^2 = 4$ together with its interior.

(c)   $x^2 + y^2 + z^2 > 4$            The exterior of the sphere $x^2 + y^2 + z^2 = 4$.

(d)   $x^2 + y^2 + z^2 = 4, z \le 0$      The lower hemisphere cut from the sphere $x^2 + y^2 + z^2 = 4$ by the $xy$-plane (the plane $z = 0$). ■

Just as polar coordinates give another way to locate points in the $xy$-plane (Section 11.3), alternative coordinate systems, different from the Cartesian coordinate system developed here, exist for three-dimensional space. We examine two of these coordinate systems in Section 15.7.

# Exercises 12.1

## Geometric Interpretations of Equations

In Exercises 1–16, give a geometric description of the set of points in space whose coordinates satisfy the given pairs of equations.

**1.** $x = 2$,   $y = 3$
**2.** $x = -1$,   $z = 0$
**3.** $y = 0$,   $z = 0$
**4.** $x = 1$,   $y = 0$
**5.** $x^2 + y^2 = 4$,   $z = 0$
**6.** $x^2 + y^2 = 4$,   $z = -2$
**7.** $x^2 + z^2 = 4$,   $y = 0$
**8.** $y^2 + z^2 = 1$,   $x = 0$
**9.** $x^2 + y^2 + z^2 = 1$,   $x = 0$
**10.** $x^2 + y^2 + z^2 = 25$,   $y = -4$
**11.** $x^2 + y^2 + (z + 3)^2 = 25$,   $z = 0$
**12.** $x^2 + (y - 1)^2 + z^2 = 4$,   $y = 0$
**13.** $x^2 + y^2 = 4$,   $z = y$
**14.** $x^2 + y^2 + z^2 = 4$,   $y = x$
**15.** $y = x^2$,   $z = 0$
**16.** $z = y^2$,   $x = 1$

## Geometric Interpretations of Inequalities and Equations

In Exercises 17–24, describe the sets of points in space whose coordinates satisfy the given inequalities or combinations of equations and inequalities.

**17. a.** $x \ge 0$,   $y \ge 0$,   $z = 0$    **b.** $x \ge 0$,   $y \le 0$,   $z = 0$

**18. a.** $0 \le x \le 1$              **b.** $0 \le x \le 1$,   $0 \le y \le 1$
    **c.** $0 \le x \le 1$,   $0 \le y \le 1$,   $0 \le z \le 1$

**19. a.** $x^2 + y^2 + z^2 \le 1$     **b.** $x^2 + y^2 + z^2 > 1$

**20. a.** $x^2 + y^2 \le 1$,   $z = 0$    **b.** $x^2 + y^2 \le 1$,   $z = 3$
    **c.** $x^2 + y^2 \le 1$,   no restriction on $z$

**21. a.** $1 \le x^2 + y^2 + z^2 \le 4$
    **b.** $x^2 + y^2 + z^2 \le 1$,   $z \ge 0$

**22. a.** $x = y$,   $z = 0$          **b.** $x = y$,   no restriction on $z$

**23. a.** $y \ge x^2$,   $z \ge 0$       **b.** $x \le y^2$,   $0 \le z \le 2$

**24. a.** $z = 1 - y$,   no restriction on $x$
    **b.** $z = y^3$,   $x = 2$

In Exercises 25–34, describe the given set with a single equation or with a pair of equations.

**25.** The plane perpendicular to the

    **a.** $x$-axis at $(3, 0, 0)$         **b.** $y$-axis at $(0, -1, 0)$

    **c.** $z$-axis at $(0, 0, -2)$

**26.** The plane through the point $(3, -1, 2)$ perpendicular to the

    **a.** $x$-axis     **b.** $y$-axis     **c.** $z$-axis

**27.** The plane through the point $(3, -1, 1)$ parallel to the

    **a.** $xy$-plane     **b.** $yz$-plane     **c.** $xz$-plane

**28.** The circle of radius 2 centered at $(0, 0, 0)$ and lying in the

    **a.** $xy$-plane     **b.** $yz$-plane     **c.** $xz$-plane

**29.** The circle of radius 2 centered at $(0, 2, 0)$ and lying in the

    **a.** $xy$-plane     **b.** $yz$-plane     **c.** plane $y = 2$

**30.** The circle of radius 1 centered at $(-3, 4, 1)$ and lying in a plane parallel to the

    **a.** $xy$-plane     **b.** $yz$-plane     **c.** $xz$-plane

**31.** The line through the point $(1, 3, -1)$ parallel to the

    **a.** $x$-axis     **b.** $y$-axis     **c.** $z$-axis

**32.** The set of points in space equidistant from the origin and the point $(0, 2, 0)$

**33.** The circle in which the plane through the point $(1, 1, 3)$ perpendicular to the $z$-axis meets the sphere of radius 5 centered at the origin

**34.** The set of points in space that lie 2 units from the point $(0, 0, 1)$ and, at the same time, 2 units from the point $(0, 0, -1)$

### Inequalities to Describe Sets of Points

Write inequalities to describe the sets in Exercises 35–40.

**35.** The slab bounded by the planes $z = 0$ and $z = 1$ (planes included)

**36.** The solid cube in the first octant bounded by the coordinate planes and the planes $x = 2$, $y = 2$, and $z = 2$

**37.** The half-space consisting of the points on and below the $xy$-plane

**38.** The upper hemisphere of the sphere of radius 1 centered at the origin

**39.** The **(a)** interior and **(b)** exterior of the sphere of radius 1 centered at the point $(1, 1, 1)$

**40.** The closed region bounded by the spheres of radius 1 and radius 2 centered at the origin. (*Closed* means the spheres are to be included. Had we wanted the spheres left out, we would have asked for the *open* region bounded by the spheres. This is analogous to the way we use *closed* and *open* to describe intervals: *closed* means endpoints included, *open* means endpoints left out. Closed sets include boundaries; open sets leave them out.)

### Distance

In Exercises 41–46, find the distance between points $P_1$ and $P_2$.

**41.** $P_1(1, 1, 1)$,     $P_2(3, 3, 0)$

**42.** $P_1(-1, 1, 5)$,     $P_2(2, 5, 0)$

**43.** $P_1(1, 4, 5)$,     $P_2(4, -2, 7)$

**44.** $P_1(3, 4, 5)$,     $P_2(2, 3, 4)$

**45.** $P_1(0, 0, 0)$,     $P_2(2, -2, -2)$

**46.** $P_1(5, 3, -2)$,     $P_2(0, 0, 0)$

### Spheres

Find the centers and radii of the spheres in Exercises 47–50.

**47.** $(x + 2)^2 + y^2 + (z - 2)^2 = 8$

**48.** $(x - 1)^2 + \left(y + \dfrac{1}{2}\right)^2 + (z + 3)^2 = 25$

**49.** $\left(x - \sqrt{2}\right)^2 + \left(y - \sqrt{2}\right)^2 + \left(z + \sqrt{2}\right)^2 = 2$

**50.** $x^2 + \left(y + \dfrac{1}{3}\right)^2 + \left(z - \dfrac{1}{3}\right)^2 = \dfrac{16}{9}$

Find equations for the spheres whose centers and radii are given in Exercises 51–54.

| | Center | Radius |
|---|---|---|
| **51.** | $(1, 2, 3)$ | $\sqrt{14}$ |
| **52.** | $(0, -1, 5)$ | $2$ |
| **53.** | $\left(-1, \dfrac{1}{2}, -\dfrac{2}{3}\right)$ | $\dfrac{4}{9}$ |
| **54.** | $(0, -7, 0)$ | $7$ |

Find the centers and radii of the spheres in Exercises 55–58.

**55.** $x^2 + y^2 + z^2 + 4x - 4z = 0$

**56.** $x^2 + y^2 + z^2 - 6y + 8z = 0$

**57.** $2x^2 + 2y^2 + 2z^2 + x + y + z = 9$

**58.** $3x^2 + 3y^2 + 3z^2 + 2y - 2z = 9$

### Theory and Examples

**59.** Find a formula for the distance from the point $P(x, y, z)$ to the

    **a.** $x$-axis.     **b.** $y$-axis.     **c.** $z$-axis.

**60.** Find a formula for the distance from the point $P(x, y, z)$ to the

    **a.** $xy$-plane.     **b.** $yz$-plane.     **c.** $xz$-plane.

**61.** Find the perimeter of the triangle with vertices $A(-1, 2, 1)$, $B(1, -1, 3)$, and $C(3, 4, 5)$.

**62.** Show that the point $P(3, 1, 2)$ is equidistant from the points $A(2, -1, 3)$ and $B(4, 3, 1)$.

**63.** Find an equation for the set of all points equidistant from the planes $y = 3$ and $y = -1$.

**64.** Find an equation for the set of all points equidistant from the point $(0, 0, 2)$ and the $xy$-plane.

**65.** Find the point on the sphere $x^2 + (y - 3)^2 + (z + 5)^2 = 4$ nearest

    **a.** the $xy$-plane.         **b.** the point $(0, 7, -5)$.

**66.** Find the point equidistant from the points $(0, 0, 0)$, $(0, 4, 0)$, $(3, 0, 0)$, and $(2, 2, -3)$.

# 12.2 Vectors

Some of the things we measure are determined simply by their magnitudes. To record mass, length, or time, for example, we need only write down a number and name an appropriate unit of measure. We need more information to describe a force, displacement, or velocity. To describe a force, we need to record the direction in which it acts as well as how large it is. To describe a body's displacement, we have to say in what direction it moved as well as how far. To describe a body's velocity, we have to know where the body is headed as well as how fast it is going. In this section we show how to represent things that have both magnitude and direction in the plane or in space.

## Component Form

A quantity such as force, displacement, or velocity is called a **vector** and is represented by a **directed line segment** (Figure 12.7). The arrow points in the direction of the action and its length gives the magnitude of the action in terms of a suitably chosen unit. For example, a force vector points in the direction in which the force acts and its length is a measure of the force's strength; a velocity vector points in the direction of motion and its length is the speed of the moving object. Figure 12.8 displays the velocity vector $\mathbf{v}$ at a specific location for a particle moving along a path in the plane or in space. (This application of vectors is studied in Chapter 13.)

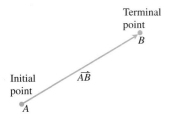

**FIGURE 12.7** The directed line segment $\overrightarrow{AB}$ is called a vector.

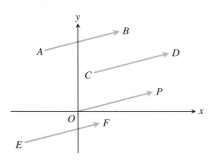

**FIGURE 12.9** The four arrows in the plane (directed line segments) shown here have the same length and direction. They therefore represent the same vector, and we write $\overrightarrow{AB} = \overrightarrow{CD} = \overrightarrow{OP} = \overrightarrow{EF}$.

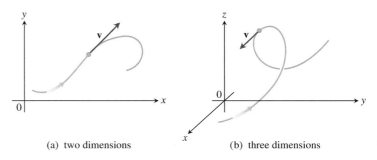

(a) two dimensions    (b) three dimensions

**FIGURE 12.8** The velocity vector of a particle moving along a path (a) in the plane (b) in space. The arrowhead on the path indicates the direction of motion of the particle.

> **DEFINITIONS** The vector represented by the directed line segment $\overrightarrow{AB}$ has **initial point** $A$ and **terminal point** $B$ and its **length** is denoted by $|\overrightarrow{AB}|$. Two vectors are **equal** if they have the same length and direction.

The arrows we use when we draw vectors are understood to represent the same vector if they have the same length, are parallel, and point in the same direction (Figure 12.9) regardless of the initial point.

In textbooks, vectors are usually written in lowercase, boldface letters, for example $\mathbf{u}$, $\mathbf{v}$, and $\mathbf{w}$. Sometimes we use uppercase boldface letters, such as $\mathbf{F}$, to denote a force vector. In handwritten form, it is customary to draw small arrows above the letters, for example $\vec{u}$, $\vec{v}$, $\vec{w}$, and $\vec{F}$.

We need a way to represent vectors algebraically so that we can be more precise about the direction of a vector. Let $\mathbf{v} = \overrightarrow{PQ}$. There is one directed line segment equal to $\overrightarrow{PQ}$ whose initial point is the origin (Figure 12.10). It is the representative of $\mathbf{v}$ in **standard position** and is the vector we normally use to represent $\mathbf{v}$. We can specify $\mathbf{v}$ by writing the

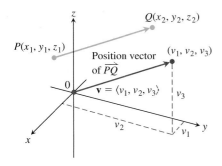

**FIGURE 12.10** A vector $\overrightarrow{PQ}$ in standard position has its initial point at the origin. The directed line segments $\overrightarrow{PQ}$ and $\mathbf{v}$ are parallel and have the same length.

coordinates of its terminal point $(v_1, v_2, v_3)$ when **v** is in standard position. If **v** is a vector in the plane its terminal point $(v_1, v_2)$ has two coordinates.

---

**DEFINITION** If **v** is a **two-dimensional** vector in the plane equal to the vector with initial point at the origin and terminal point $(v_1, v_2)$, then the **component form** of **v** is

$$\mathbf{v} = \langle v_1, v_2 \rangle.$$

If **v** is a **three-dimensional** vector equal to the vector with initial point at the origin and terminal point $(v_1, v_2, v_3)$, then the **component form** of **v** is

$$\mathbf{v} = \langle v_1, v_2, v_3 \rangle.$$

---

So a two-dimensional vector is an ordered pair $\mathbf{v} = \langle v_1, v_2 \rangle$ of real numbers, and a three-dimensional vector is an ordered triple $\mathbf{v} = \langle v_1, v_2, v_3 \rangle$ of real numbers. The numbers $v_1$, $v_2$, and $v_3$ are the **components** of **v**.

If $\mathbf{v} = \langle v_1, v_2, v_3 \rangle$ is represented by the directed line segment $\overrightarrow{PQ}$, where the initial point is $P(x_1, y_1, z_1)$ and the terminal point is $Q(x_2, y_2, z_2)$, then $x_1 + v_1 = x_2$, $y_1 + v_2 = y_2$, and $z_1 + v_3 = z_2$ (see Figure 12.10). Thus, $v_1 = x_2 - x_1$, $v_2 = y_2 - y_1$, and $v_3 = z_2 - z_1$ are the components of $\overrightarrow{PQ}$.

In summary, given the points $P(x_1, y_1, z_1)$ and $Q(x_2, y_2, z_2)$, the standard position vector $\mathbf{v} = \langle v_1, v_2, v_3 \rangle$ equal to $\overrightarrow{PQ}$ is

$$\mathbf{v} = \langle x_2 - x_1, y_2 - y_1, z_2 - z_1 \rangle.$$

If **v** is two-dimensional with $P(x_1, y_1)$ and $Q(x_2, y_2)$ as points in the plane, then $\mathbf{v} = \langle x_2 - x_1, y_2 - y_1 \rangle$. There is no third component for planar vectors. With this understanding, we will develop the algebra of three-dimensional vectors and simply drop the third component when the vector is two-dimensional (a planar vector).

Two vectors are equal if and only if their standard position vectors are identical. Thus $\langle u_1, u_2, u_3 \rangle$ and $\langle v_1, v_2, v_3 \rangle$ are **equal** if and only if $u_1 = v_1$, $u_2 = v_2$, and $u_3 = v_3$.

The **magnitude** or **length** of the vector $\overrightarrow{PQ}$ is the length of any of its equivalent directed line segment representations. In particular, if $\mathbf{v} = \langle x_2 - x_1, y_2 - y_1, z_2 - z_1 \rangle$ is the standard position vector for $\overrightarrow{PQ}$, then the distance formula gives the magnitude or length of **v**, denoted by the symbol $|\mathbf{v}|$ or $\|\mathbf{v}\|$.

---

The **magnitude** or **length** of the vector $\mathbf{v} = \overrightarrow{PQ}$ is the nonnegative number

$$|\mathbf{v}| = \sqrt{v_1{}^2 + v_2{}^2 + v_3{}^2} = \sqrt{(x_2 - x_1)^2 + (y_2 - y_1)^2 + (z_2 - z_1)^2}$$

(see Figure 12.10).

---

The only vector with length 0 is the **zero vector** $\mathbf{0} = \langle 0, 0 \rangle$ or $\mathbf{0} = \langle 0, 0, 0 \rangle$. This vector is also the only vector with no specific direction.

**EXAMPLE 1** Find the **(a)** component form and **(b)** length of the vector with initial point $P(-3, 4, 1)$ and terminal point $Q(-5, 2, 2)$.

**Solution**

**(a)** The standard position vector **v** representing $\overrightarrow{PQ}$ has components

$$v_1 = x_2 - x_1 = -5 - (-3) = -2, \qquad v_2 = y_2 - y_1 = 2 - 4 = -2,$$

and

$$v_3 = z_2 - z_1 = 2 - 1 = 1.$$

The component form of $\overrightarrow{PQ}$ is

$$\mathbf{v} = \langle -2, -2, 1 \rangle.$$

**(b)** The length or magnitude of $\mathbf{v} = \overrightarrow{PQ}$ is

$$|\mathbf{v}| = \sqrt{(-2)^2 + (-2)^2 + (1)^2} = \sqrt{9} = 3.$$ ■

**EXAMPLE 2**　A small cart is being pulled along a smooth horizontal floor with a 20-N force $\mathbf{F}$ making a 45° angle to the floor (Figure 12.11). What is the *effective* force moving the cart forward?

**Solution**　The effective force is the horizontal component of $\mathbf{F} = \langle a, b \rangle$, given by

$$a = |\mathbf{F}| \cos 45° = (20)\left(\frac{\sqrt{2}}{2}\right) \approx 14.14 \text{ N}.$$

Notice that $\mathbf{F}$ is a two-dimensional vector. ■

**FIGURE 12.11**　The force pulling the cart forward is represented by the vector **F** whose horizontal component is the effective force (Example 2).

## Vector Algebra Operations

Two principal operations involving vectors are *vector addition* and *scalar multiplication*. A **scalar** is simply a real number, and is called such when we want to draw attention to its differences from vectors. Scalars can be positive, negative, or zero and are used to "scale" a vector by multiplication.

---

**DEFINITIONS**　Let $\mathbf{u} = \langle u_1, u_2, u_3 \rangle$ and $\mathbf{v} = \langle v_1, v_2, v_3 \rangle$ be vectors with $k$ a scalar.

　　**Addition:**　　　　$\mathbf{u} + \mathbf{v} = \langle u_1 + v_1, u_2 + v_2, u_3 + v_3 \rangle$

　　**Scalar multiplication:**　$k\mathbf{u} = \langle ku_1, ku_2, ku_3 \rangle$

---

We add vectors by adding the corresponding components of the vectors. We multiply a vector by a scalar by multiplying each component by the scalar. The definitions apply to planar vectors except there are only two components, $\langle u_1, u_2 \rangle$ and $\langle v_1, v_2 \rangle$.

The definition of vector addition is illustrated geometrically for planar vectors in Figure 12.12a, where the initial point of one vector is placed at the terminal point of the other. Another interpretation is shown in Figure 12.12b (called the **parallelogram law** of

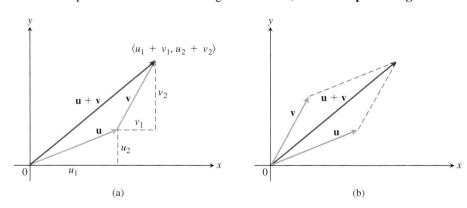

**FIGURE 12.12**　(a) Geometric interpretation of the vector sum. (b) The parallelogram law of vector addition in which both vectors are in standard position.

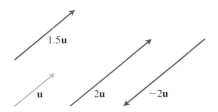

**FIGURE 12.13** Scalar multiples of **u**.

(a)

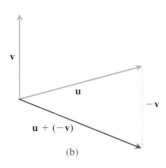

(b)

**FIGURE 12.14** (a) The vector **u** − **v**, when added to **v**, gives **u**. (b) **u** − **v** = **u** + (−**v**).

addition), where the sum, called the **resultant vector**, is the diagonal of the parallelogram. In physics, forces add vectorially as do velocities, accelerations, and so on. So the force acting on a particle subject to two gravitational forces, for example, is obtained by adding the two force vectors.

Figure 12.13 displays a geometric interpretation of the product $k\mathbf{u}$ of the scalar $k$ and vector **u**. If $k > 0$, then $k\mathbf{u}$ has the same direction as **u**; if $k < 0$, then the direction of $k\mathbf{u}$ is opposite to that of **u**. Comparing the lengths of **u** and $k\mathbf{u}$, we see that

$$|k\mathbf{u}| = \sqrt{(ku_1)^2 + (ku_2)^2 + (ku_3)^2} = \sqrt{k^2(u_1{}^2 + u_2{}^2 + u_3{}^2)}$$
$$= \sqrt{k^2}\sqrt{u_1{}^2 + u_2{}^2 + u_3{}^2} = |k||\mathbf{u}|.$$

The length of $k\mathbf{u}$ is the absolute value of the scalar $k$ times the length of **u**. The vector $(-1)\mathbf{u} = -\mathbf{u}$ has the same length as **u** but points in the opposite direction.

The **difference u − v** of two vectors is defined by

$$\mathbf{u} - \mathbf{v} = \mathbf{u} + (-\mathbf{v}).$$

If $\mathbf{u} = \langle u_1, u_2, u_3 \rangle$ and $\mathbf{v} = \langle v_1, v_2, v_3 \rangle$, then

$$\mathbf{u} - \mathbf{v} = \langle u_1 - v_1, u_2 - v_2, u_3 - v_3 \rangle.$$

Note that $(\mathbf{u} - \mathbf{v}) + \mathbf{v} = \mathbf{u}$, so adding the vector $(\mathbf{u} - \mathbf{v})$ to **v** gives **u** (Figure 12.14a). Figure 12.14b shows the difference **u** − **v** as the sum **u** + (−**v**).

**EXAMPLE 3**    Let $\mathbf{u} = \langle -1, 3, 1 \rangle$ and $\mathbf{v} = \langle 4, 7, 0 \rangle$. Find the components of

**(a)** $2\mathbf{u} + 3\mathbf{v}$    **(b)** $\mathbf{u} - \mathbf{v}$    **(c)** $\left|\dfrac{1}{2}\mathbf{u}\right|$.

**Solution**

**(a)** $2\mathbf{u} + 3\mathbf{v} = 2\langle -1, 3, 1 \rangle + 3\langle 4, 7, 0 \rangle = \langle -2, 6, 2 \rangle + \langle 12, 21, 0 \rangle = \langle 10, 27, 2 \rangle$

**(b)** $\mathbf{u} - \mathbf{v} = \langle -1, 3, 1 \rangle - \langle 4, 7, 0 \rangle = \langle -1 - 4, 3 - 7, 1 - 0 \rangle = \langle -5, -4, 1 \rangle$

**(c)** $\left|\dfrac{1}{2}\mathbf{u}\right| = \left|\left\langle -\dfrac{1}{2}, \dfrac{3}{2}, \dfrac{1}{2} \right\rangle\right| = \sqrt{\left(-\dfrac{1}{2}\right)^2 + \left(\dfrac{3}{2}\right)^2 + \left(\dfrac{1}{2}\right)^2} = \dfrac{1}{2}\sqrt{11}.$    ∎

Vector operations have many of the properties of ordinary arithmetic.

---

**Properties of Vector Operations**

Let **u**, **v**, **w** be vectors and $a$, $b$ be scalars.

**1.** $\mathbf{u} + \mathbf{v} = \mathbf{v} + \mathbf{u}$        **2.** $(\mathbf{u} + \mathbf{v}) + \mathbf{w} = \mathbf{u} + (\mathbf{v} + \mathbf{w})$

**3.** $\mathbf{u} + \mathbf{0} = \mathbf{u}$        **4.** $\mathbf{u} + (-\mathbf{u}) = \mathbf{0}$

**5.** $0\mathbf{u} = \mathbf{0}$        **6.** $1\mathbf{u} = \mathbf{u}$

**7.** $a(b\mathbf{u}) = (ab)\mathbf{u}$        **8.** $a(\mathbf{u} + \mathbf{v}) = a\mathbf{u} + a\mathbf{v}$

**9.** $(a + b)\mathbf{u} = a\mathbf{u} + b\mathbf{u}$

---

These properties are readily verified using the definitions of vector addition and multiplication by a scalar. For instance, to establish Property 1, we have

$$\mathbf{u} + \mathbf{v} = \langle u_1, u_2, u_3 \rangle + \langle v_1, v_2, v_3 \rangle$$
$$= \langle u_1 + v_1, u_2 + v_2, u_3 + v_3 \rangle$$
$$= \langle v_1 + u_1, v_2 + u_2, v_3 + u_3 \rangle$$
$$= \langle v_1, v_2, v_3 \rangle + \langle u_1, u_2, u_3 \rangle$$
$$= \mathbf{v} + \mathbf{u}.$$

When three or more space vectors lie in the same plane, we say they are **coplanar** vectors. For example, the vectors **u**, **v**, and **u** + **v** are always coplanar.

### Unit Vectors

A vector **v** of length 1 is called a **unit vector**. The **standard unit vectors** are

$$\mathbf{i} = \langle 1, 0, 0 \rangle, \quad \mathbf{j} = \langle 0, 1, 0 \rangle, \quad \text{and} \quad \mathbf{k} = \langle 0, 0, 1 \rangle.$$

Any vector $\mathbf{v} = \langle v_1, v_2, v_3 \rangle$ can be written as a *linear combination* of the standard unit vectors as follows:

$$\begin{aligned}
\mathbf{v} = \langle v_1, v_2, v_3 \rangle &= \langle v_1, 0, 0 \rangle + \langle 0, v_2, 0 \rangle + \langle 0, 0, v_3 \rangle \\
&= v_1 \langle 1, 0, 0 \rangle + v_2 \langle 0, 1, 0 \rangle + v_3 \langle 0, 0, 1 \rangle \\
&= v_1 \mathbf{i} + v_2 \mathbf{j} + v_3 \mathbf{k}.
\end{aligned}$$

We call the scalar (or number) $v_1$ the **i-component** of the vector **v**, $v_2$ the **j-component**, and $v_3$ the **k-component**. In component form, the vector from $P_1(x_1, y_1, z_1)$ to $P_2(x_2, y_2, z_2)$ is

$$\overrightarrow{P_1P_2} = (x_2 - x_1)\mathbf{i} + (y_2 - y_1)\mathbf{j} + (z_2 - z_1)\mathbf{k}$$

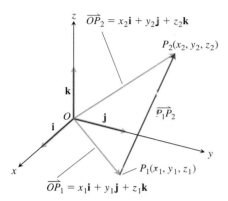

**FIGURE 12.15** The vector from $P_1$ to $P_2$ is $\overrightarrow{P_1P_2} = (x_2 - x_1)\mathbf{i} + (y_2 - y_1)\mathbf{j} + (z_2 - z_1)\mathbf{k}$.

(Figure 12.15).

Whenever $\mathbf{v} \neq \mathbf{0}$, its length $|\mathbf{v}|$ is not zero and

$$\left| \frac{1}{|\mathbf{v}|} \mathbf{v} \right| = \frac{1}{|\mathbf{v}|} |\mathbf{v}| = 1.$$

That is, $\mathbf{v}/|\mathbf{v}|$ is a unit vector in the direction of **v**, called **the direction** of the nonzero vector **v**.

**EXAMPLE 4**  Find a unit vector **u** in the direction of the vector from $P_1(1, 0, 1)$ to $P_2(3, 2, 0)$.

**Solution**  We divide $\overrightarrow{P_1P_2}$ by its length:

$$\overrightarrow{P_1P_2} = (3 - 1)\mathbf{i} + (2 - 0)\mathbf{j} + (0 - 1)\mathbf{k} = 2\mathbf{i} + 2\mathbf{j} - \mathbf{k}$$

$$|\overrightarrow{P_1P_2}| = \sqrt{(2)^2 + (2)^2 + (-1)^2} = \sqrt{4 + 4 + 1} = \sqrt{9} = 3$$

$$\mathbf{u} = \frac{\overrightarrow{P_1P_2}}{|\overrightarrow{P_1P_2}|} = \frac{2\mathbf{i} + 2\mathbf{j} - \mathbf{k}}{3} = \frac{2}{3}\mathbf{i} + \frac{2}{3}\mathbf{j} - \frac{1}{3}\mathbf{k}.$$

The unit vector **u** is the direction of $\overrightarrow{P_1P_2}$. ∎

**EXAMPLE 5**  If $\mathbf{v} = 3\mathbf{i} - 4\mathbf{j}$ is a velocity vector, express **v** as a product of its speed times its direction of motion.

**Solution**  Speed is the magnitude (length) of **v**:

$$|\mathbf{v}| = \sqrt{(3)^2 + (-4)^2} = \sqrt{9 + 16} = 5.$$

The unit vector $\mathbf{v}/|\mathbf{v}|$ is the direction of **v**:

$$\frac{\mathbf{v}}{|\mathbf{v}|} = \frac{3\mathbf{i} - 4\mathbf{j}}{5} = \frac{3}{5}\mathbf{i} - \frac{4}{5}\mathbf{j}.$$

So

$$\mathbf{v} = 3\mathbf{i} - 4\mathbf{j} = 5\left(\frac{3}{5}\mathbf{i} - \frac{4}{5}\mathbf{j}\right).$$

Length   Direction of motion
(speed)

In summary, we can express any nonzero vector $\mathbf{v}$ in terms of its two important features, length and direction, by writing $\mathbf{v} = |\mathbf{v}|\frac{\mathbf{v}}{|\mathbf{v}|}$.

---

If $\mathbf{v} \ne \mathbf{0}$, then

1. $\frac{\mathbf{v}}{|\mathbf{v}|}$ is a unit vector called the **direction** of $\mathbf{v}$;

2. the equation $\mathbf{v} = |\mathbf{v}|\frac{\mathbf{v}}{|\mathbf{v}|}$ expresses $\mathbf{v}$ as its length times its direction.

---

**EXAMPLE 6**   A force of 6 newtons is applied in the direction of the vector $\mathbf{v} = 2\mathbf{i} + 2\mathbf{j} - \mathbf{k}$. Express the force $\mathbf{F}$ as a product of its magnitude and direction.

**Solution**   The force vector has magnitude 6 and direction $\frac{\mathbf{v}}{|\mathbf{v}|}$, so

$$\mathbf{F} = 6\frac{\mathbf{v}}{|\mathbf{v}|} = 6\frac{2\mathbf{i} + 2\mathbf{j} - \mathbf{k}}{\sqrt{2^2 + 2^2 + (-1)^2}} = 6\frac{2\mathbf{i} + 2\mathbf{j} - \mathbf{k}}{3}$$

$$= 6\left(\frac{2}{3}\mathbf{i} + \frac{2}{3}\mathbf{j} - \frac{1}{3}\mathbf{k}\right).$$

### Midpoint of a Line Segment

Vectors are often useful in geometry. For example, the coordinates of the midpoint of a line segment are found by averaging.

---

The **midpoint** $M$ of the line segment joining points $P_1(x_1, y_1, z_1)$ and $P_2(x_2, y_2, z_2)$ is the point

$$\left(\frac{x_1 + x_2}{2}, \frac{y_1 + y_2}{2}, \frac{z_1 + z_2}{2}\right).$$

---

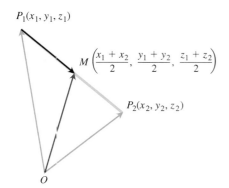

$P_1(x_1, y_1, z_1)$

$M\left(\frac{x_1 + x_2}{2}, \frac{y_1 + y_2}{2}, \frac{z_1 + z_2}{2}\right)$

$P_2(x_2, y_2, z_2)$

$O$

**FIGURE 12.16**   The coordinates of the midpoint are the averages of the coordinates of $P_1$ and $P_2$.

To see why, observe (Figure 12.16) that

$$\overrightarrow{OM} = \overrightarrow{OP_1} + \frac{1}{2}(\overrightarrow{P_1P_2}) = \overrightarrow{OP_1} + \frac{1}{2}(\overrightarrow{OP_2} - \overrightarrow{OP_1})$$

$$= \frac{1}{2}(\overrightarrow{OP_1} + \overrightarrow{OP_2})$$

$$= \frac{x_1 + x_2}{2}\mathbf{i} + \frac{y_1 + y_2}{2}\mathbf{j} + \frac{z_1 + z_2}{2}\mathbf{k}.$$

**EXAMPLE 7**   The midpoint of the segment joining $P_1(3, -2, 0)$ and $P_2(7, 4, 4)$ is

$$\left(\frac{3 + 7}{2}, \frac{-2 + 4}{2}, \frac{0 + 4}{2}\right) = (5, 1, 2).$$

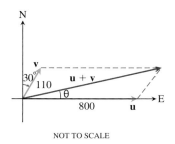

**FIGURE 12.17** Vectors representing the velocities of the airplane **u** and tailwind **v** in Example 8.

## Applications

An important application of vectors occurs in navigation.

**EXAMPLE 8**   A jet airliner, flying due east at 800 km/h in still air, encounters a 110 km/h tailwind blowing in the direction 60° north of east. The airplane holds its compass heading due east but, because of the wind, acquires a new ground speed and direction. What are they?

**Solution**   If **u** is the velocity of the airplane alone and **v** is the velocity of the tailwind, then $|\mathbf{u}| = 800$ and $|\mathbf{v}| = 110$ (Figure 12.17). The velocity of the airplane with respect to the ground is given by the magnitude and direction of the resultant vector **u** + **v**. If we let the positive $x$-axis represent east and the positive $y$-axis represent north, then the component forms of **u** and **v** are

$$\mathbf{u} = \langle 800, 0 \rangle \quad \text{and} \quad \mathbf{v} = \langle 110 \cos 60°, 110 \sin 60° \rangle = \langle 55, 55\sqrt{3} \rangle.$$

Therefore,

$$\mathbf{u} + \mathbf{v} = \langle 855, 55\sqrt{3} \rangle = 855\mathbf{i} + 55\sqrt{3}\,\mathbf{j}$$

$$|\mathbf{u} + \mathbf{v}| = \sqrt{855^2 + (55\sqrt{3})^2} \approx 860.3$$

and

$$\theta = \tan^{-1} \frac{55\sqrt{3}}{855} \approx 6.4°. \qquad \text{Figure 12.17}$$

The new ground speed of the airplane is about 860.3 km/h, and its new direction is about 6.4° north of east. ∎

Another important application occurs in physics and engineering when several forces are acting on a single object.

**EXAMPLE 9**   A 75-N weight is suspended by two wires, as shown in Figure 12.18a. Find the forces $\mathbf{F}_1$ and $\mathbf{F}_2$ acting in both wires.

**Solution**   The force vectors $\mathbf{F}_1$ and $\mathbf{F}_2$ have magnitudes $|\mathbf{F}_1|$ and $|\mathbf{F}_2|$ and components that are measured in newtons. The resultant force is the sum $\mathbf{F}_1 + \mathbf{F}_2$ and must be equal in magnitude and acting in the opposite (or upward) direction to the weight vector **w** (see Figure 12.18b). It follows from the figure that

$$\mathbf{F}_1 = \langle -|\mathbf{F}_1| \cos 55°, |\mathbf{F}_1| \sin 55° \rangle \quad \text{and} \quad \mathbf{F}_2 = \langle |\mathbf{F}_2| \cos 40°, |\mathbf{F}_2| \sin 40° \rangle.$$

Since $\mathbf{F}_1 + \mathbf{F}_2 = \langle 0, 75 \rangle$, the resultant vector leads to the system of equations

$$-|\mathbf{F}_1| \cos 55° + |\mathbf{F}_2| \cos 40° = 0$$

$$|\mathbf{F}_1| \sin 55° + |\mathbf{F}_2| \sin 40° = 75.$$

Solving for $|\mathbf{F}_2|$ in the first equation and substituting the result into the second equation, we get

$$|\mathbf{F}_2| = \frac{|\mathbf{F}_1| \cos 55°}{\cos 40°} \quad \text{and} \quad |\mathbf{F}_1| \sin 55° + \frac{|\mathbf{F}_1| \cos 55°}{\cos 40°} \sin 40° = 75.$$

It follows that

$$|\mathbf{F}_1| = \frac{75}{\sin 55° + \cos 55° \tan 40°} \approx 57.67 \text{ N},$$

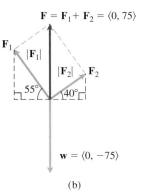

**FIGURE 12.18**   The suspended weight in Example 9.

and

$$|\mathbf{F}_2| = \frac{75 \cos 55°}{\sin 55° \cos 40° + \cos 55° \sin 40°}$$

$$= \frac{75 \cos 55°}{\sin (55° + 40°)} \approx 43.18 \text{ N}.$$

The force vectors are then $\mathbf{F}_1 = \langle -33.08, 47.24 \rangle$ and $\mathbf{F}_2 = \langle 33.08, 27.76 \rangle$.   ■

# Exercises 12.2

## Vectors in the Plane

In Exercises 1–8, let $\mathbf{u} = \langle 3, -2 \rangle$ and $\mathbf{v} = \langle -2, 5 \rangle$. Find the **(a)** component form and **(b)** magnitude (length) of the vector.

1. $3\mathbf{u}$

2. $-2\mathbf{v}$

3. $\mathbf{u} + \mathbf{v}$

4. $\mathbf{u} - \mathbf{v}$

5. $2\mathbf{u} - 3\mathbf{v}$

6. $-2\mathbf{u} + 5\mathbf{v}$

7. $\frac{3}{5}\mathbf{u} + \frac{4}{5}\mathbf{v}$

8. $-\frac{5}{13}\mathbf{u} + \frac{12}{13}\mathbf{v}$

In Exercises 9–16, find the component form of the vector.

9. The vector $\overrightarrow{PQ}$, where $P = (1, 3)$ and $Q = (2, -1)$

10. The vector $\overrightarrow{OP}$ where $O$ is the origin and $P$ is the midpoint of segment $RS$, where $R = (2, -1)$ and $S = (-4, 3)$

11. The vector from the point $A = (2, 3)$ to the origin

12. The sum of $\overrightarrow{AB}$ and $\overrightarrow{CD}$, where $A = (1, -1), B = (2, 0)$, $C = (-1, 3)$, and $D = (-2, 2)$

13. The unit vector that makes an angle $\theta = 2\pi/3$ with the positive $x$-axis

14. The unit vector that makes an angle $\theta = -3\pi/4$ with the positive $x$-axis

15. The unit vector obtained by rotating the vector $\langle 0, 1 \rangle$ 120° counterclockwise about the origin

16. The unit vector obtained by rotating the vector $\langle 1, 0 \rangle$ 135° counterclockwise about the origin

## Vectors in Space

In Exercises 17–22, express each vector in the form $\mathbf{v} = v_1\mathbf{i} + v_2\mathbf{j} + v_3\mathbf{k}$.

17. $\overrightarrow{P_1P_2}$ if $P_1$ is the point $(5, 7, -1)$ and $P_2$ is the point $(2, 9, -2)$

18. $\overrightarrow{P_1P_2}$ if $P_1$ is the point $(1, 2, 0)$ and $P_2$ is the point $(-3, 0, 5)$

19. $\overrightarrow{AB}$ if $A$ is the point $(-7, -8, 1)$ and $B$ is the point $(-10, 8, 1)$

20. $\overrightarrow{AB}$ if $A$ is the point $(1, 0, 3)$ and $B$ is the point $(-1, 4, 5)$

21. $5\mathbf{u} - \mathbf{v}$ if $\mathbf{u} = \langle 1, 1, -1 \rangle$ and $\mathbf{v} = \langle 2, 0, 3 \rangle$

22. $-2\mathbf{u} + 3\mathbf{v}$ if $\mathbf{u} = \langle -1, 0, 2 \rangle$ and $\mathbf{v} = \langle 1, 1, 1 \rangle$

## Geometric Representations

In Exercises 23 and 24, copy vectors $\mathbf{u}, \mathbf{v}$, and $\mathbf{w}$ head to tail as needed to sketch the indicated vector.

23.

  **a.** $\mathbf{u} + \mathbf{v}$          **b.** $\mathbf{u} + \mathbf{v} + \mathbf{w}$

  **c.** $\mathbf{u} - \mathbf{v}$          **d.** $\mathbf{u} - \mathbf{w}$

24.

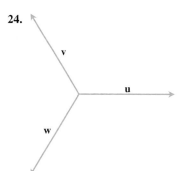

  **a.** $\mathbf{u} - \mathbf{v}$          **b.** $\mathbf{u} - \mathbf{v} + \mathbf{w}$

  **c.** $2\mathbf{u} - \mathbf{v}$         **d.** $\mathbf{u} + \mathbf{v} + \mathbf{w}$

## Length and Direction

In Exercises 25–30, express each vector as a product of its length and direction.

25. $2\mathbf{i} + \mathbf{j} - 2\mathbf{k}$

26. $9\mathbf{i} - 2\mathbf{j} + 6\mathbf{k}$

27. $5\mathbf{k}$

28. $\frac{3}{5}\mathbf{i} + \frac{4}{5}\mathbf{k}$

29. $\frac{1}{\sqrt{6}}\mathbf{i} - \frac{1}{\sqrt{6}}\mathbf{j} - \frac{1}{\sqrt{6}}\mathbf{k}$

30. $\frac{\mathbf{i}}{\sqrt{3}} + \frac{\mathbf{j}}{\sqrt{3}} + \frac{\mathbf{k}}{\sqrt{3}}$

**31.** Find the vectors whose lengths and directions are given. Try to do the calculations without writing.

| Length | Direction |
| --- | --- |
| **a.** 2 | $\mathbf{i}$ |
| **b.** $\sqrt{3}$ | $-\mathbf{k}$ |
| **c.** $\dfrac{1}{2}$ | $\dfrac{3}{5}\mathbf{j} + \dfrac{4}{5}\mathbf{k}$ |
| **d.** 7 | $\dfrac{6}{7}\mathbf{i} - \dfrac{2}{7}\mathbf{j} + \dfrac{3}{7}\mathbf{k}$ |

**32.** Find the vectors whose lengths and directions are given. Try to do the calculations without writing.

| Length | Direction |
| --- | --- |
| **a.** 7 | $-\mathbf{j}$ |
| **b.** $\sqrt{2}$ | $-\dfrac{3}{5}\mathbf{i} - \dfrac{4}{5}\mathbf{k}$ |
| **c.** $\dfrac{13}{12}$ | $\dfrac{3}{13}\mathbf{i} - \dfrac{4}{13}\mathbf{j} - \dfrac{12}{13}\mathbf{k}$ |
| **d.** $a > 0$ | $\dfrac{1}{\sqrt{2}}\mathbf{i} + \dfrac{1}{\sqrt{3}}\mathbf{j} - \dfrac{1}{\sqrt{6}}\mathbf{k}$ |

**33.** Find a vector of magnitude 7 in the direction of $\mathbf{v} = 12\mathbf{i} - 5\mathbf{k}$.

**34.** Find a vector of magnitude 3 in the direction opposite to the direction of $\mathbf{v} = (1/2)\mathbf{i} - (1/2)\mathbf{j} - (1/2)\mathbf{k}$.

**Direction and Midpoints**

In Exercises 35–38, find

   **a.** the direction of $\overrightarrow{P_1 P_2}$ and

   **b.** the midpoint of line segment $P_1 P_2$.

**35.** $P_1(-1, 1, 5)$    $P_2(2, 5, 0)$

**36.** $P_1(1, 4, 5)$    $P_2(4, -2, 7)$

**37.** $P_1(3, 4, 5)$    $P_2(2, 3, 4)$

**38.** $P_1(0, 0, 0)$    $P_2(2, -2, -2)$

**39.** If $\overrightarrow{AB} = \mathbf{i} + 4\mathbf{j} - 2\mathbf{k}$ and $B$ is the point $(5, 1, 3)$, find $A$.

**40.** If $\overrightarrow{AB} = -7\mathbf{i} + 3\mathbf{j} + 8\mathbf{k}$ and $A$ is the point $(-2, -3, 6)$, find $B$.

**Theory and Applications**

**41. Linear combination** Let $\mathbf{u} = 2\mathbf{i} + \mathbf{j}, \mathbf{v} = \mathbf{i} + \mathbf{j}$, and $\mathbf{w} = \mathbf{i} - \mathbf{j}$. Find scalars $a$ and $b$ such that $\mathbf{u} = a\mathbf{v} + b\mathbf{w}$.

**42. Linear combination** Let $\mathbf{u} = \mathbf{i} - 2\mathbf{j}, \mathbf{v} = 2\mathbf{i} + 3\mathbf{j}$, and $\mathbf{w} = \mathbf{i} + \mathbf{j}$. Write $\mathbf{u} = \mathbf{u}_1 + \mathbf{u}_2$, where $\mathbf{u}_1$ is parallel to $\mathbf{v}$ and $\mathbf{u}_2$ is parallel to $\mathbf{w}$. (See Exercise 41.)

**43. Velocity** An airplane is flying in the direction 25° west of north at 800 km/h. Find the component form of the velocity of the airplane, assuming that the positive $x$-axis represents due east and the positive $y$-axis represents due north.

**44.** (*Continuation of Example 8.*) What speed and direction should the jetliner in Example 8 have in order for the resultant vector to be 800 km/h due east?

**45.** Consider a 100-N weight suspended by two wires as shown in the accompanying figure. Find the magnitudes and components of the force vectors $\mathbf{F}_1$ and $\mathbf{F}_2$.

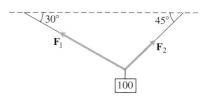

**46.** Consider a 50-N weight suspended by two wires as shown in the accompanying figure. If the magnitude of vector $\mathbf{F}_1$ is 35 N, find angle $\alpha$ and the magnitude of vector $\mathbf{F}_2$.

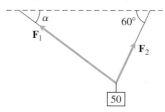

**47.** Consider a $w$-N weight suspended by two wires as shown in the accompanying figure. If the magnitude of vector $\mathbf{F}_2$ is 100 N, find $w$ and the magnitude of vector $\mathbf{F}_1$.

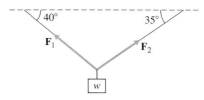

**48.** Consider a 25-N weight suspended by two wires as shown in the accompanying figure. If the magnitudes of vectors $\mathbf{F}_1$ and $\mathbf{F}_2$ are both 75 N, then angles $\alpha$ and $\beta$ are equal. Find $\alpha$.

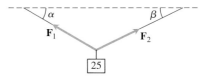

**49. Location** A bird flies from its nest 5 km in the direction 60° north of east, where it stops to rest on a tree. It then flies 10 km in the direction due southeast and lands atop a telephone pole. Place an $xy$-coordinate system so that the origin is the bird's nest, the $x$-axis points east, and the $y$-axis points north.

   **a.** At what point is the tree located?

   **b.** At what point is the telephone pole?

**50.** Use similar triangles to find the coordinates of the point $Q$ that divides the segment from $P_1(x_1, y_1, z_1)$ to $P_2(x_2, y_2, z_2)$ into two lengths whose ratio is $p/q = r$.

**51. Medians of a triangle** Suppose that $A$, $B$, and $C$ are the corner points of the thin triangular plate of constant density shown here.

   **a.** Find the vector from $C$ to the midpoint $M$ of side $AB$.

   **b.** Find the vector from $C$ to the point that lies two-thirds of the way from $C$ to $M$ on the median $CM$.

   **c.** Find the coordinates of the point in which the medians of $\triangle ABC$ intersect. According to Exercise 19, Section 6.6, this point is the plate's center of mass. (See the accompanying figure.)

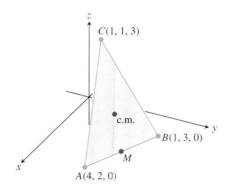

**52.** Find the vector from the origin to the point of intersection of the medians of the triangle whose vertices are

$$A(1, -1, 2), \quad B(2, 1, 3), \quad \text{and} \quad C(-1, 2, -1).$$

**53.** Let *ABCD* be a general, not necessarily planar, quadrilateral in space. Show that the two segments joining the midpoints of opposite sides of *ABCD* bisect each other. (*Hint:* Show that the segments have the same midpoint.)

**54.** Vectors are drawn from the center of a regular *n*-sided polygon in the plane to the vertices of the polygon. Show that the sum of the vectors is zero. (*Hint:* What happens to the sum if you rotate the polygon about its center?)

**55.** Suppose that *A*, *B*, and *C* are vertices of a triangle and that *a*, *b*, and *c* are, respectively, the midpoints of the opposite sides. Show that $\overrightarrow{Aa} + \overrightarrow{Bb} + \overrightarrow{Cc} = 0$.

**56. Unit vectors in the plane** Show that a unit vector in the plane can be expressed as $\mathbf{u} = (\cos\theta)\mathbf{i} + (\sin\theta)\mathbf{j}$, obtained by rotating $\mathbf{i}$ through an angle $\theta$ in the counterclockwise direction. Explain why this form gives *every* unit vector in the plane.

# $12.3$ The Dot Product

**FIGURE 12.19** The magnitude of the force **F** in the direction of vector **v** is the length $|\mathbf{F}| \cos\theta$ of the projection of **F** onto **v**

If a force **F** is applied to a particle moving along a path, we often need to know the magnitude of the force in the direction of motion. If **v** is parallel to the tangent line to the path at the point where **F** is applied, then we want the magnitude of **F** in the direction of **v**. Figure 12.19 shows that the scalar quantity we seek is the length $|\mathbf{F}| \cos\theta$, where $\theta$ is the angle between the two vectors **F** and **v**.

In this section we show how to calculate easily the angle between two vectors directly from their components. A key part of the calculation is an expression called the *dot product*. Dot products are also called *inner* or *scalar* products because the product results in a scalar, not a vector. After investigating the dot product, we apply it to finding the projection of one vector onto another (as displayed in Figure 12.19) and to finding the work done by a constant force acting through a displacement.

## Angle Between Vectors

When two nonzero vectors **u** and **v** are placed so their initial points coincide, they form an angle $\theta$ of measure $0 \le \theta \le \pi$ (Figure 12.20). If the vectors do not lie along the same line, the angle $\theta$ is measured in the plane containing both of them. If they do lie along the same line, the angle between them is 0 if they point in the same direction and $\pi$ if they point in opposite directions. The angle $\theta$ is the **angle between u and v**. Theorem 1 gives a formula to determine this angle.

**FIGURE 12.20** The angle between **u** and **v**.

> **THEOREM 1—Angle Between Two Vectors** The angle $\theta$ between two nonzero vectors $\mathbf{u} = \langle u_1, u_2, u_3 \rangle$ and $\mathbf{v} = \langle v_1, v_2, v_3 \rangle$ is given by
>
> $$\theta = \cos^{-1}\left(\frac{u_1 v_1 + u_2 v_2 + u_3 v_3}{|\mathbf{u}||\mathbf{v}|}\right).$$

We use the law of cosines to prove Theorem 1, but before doing so, we focus attention on the expression $u_1 v_1 + u_2 v_2 + u_3 v_3$ in the calculation for $\theta$. This expression is the sum of the products of the corresponding components for the vectors **u** and **v**.

> **DEFINITION**    The **dot product u · v** ("**u dot v**") of vectors $\mathbf{u} = \langle u_1, u_2, u_3 \rangle$ and $\mathbf{v} = \langle v_1, v_2, v_3 \rangle$ is the scalar
>
> $$\mathbf{u} \cdot \mathbf{v} = u_1 v_1 + u_2 v_2 + u_3 v_3.$$

**EXAMPLE 1**    We illustrate the definition.

**(a)**  $\langle 1, -2, -1 \rangle \cdot \langle -6, 2, -3 \rangle = (1)(-6) + (-2)(2) + (-1)(-3)$

$$= -6 - 4 + 3 = -7$$

**(b)**  $\left( \dfrac{1}{2} \mathbf{i} + 3\mathbf{j} + \mathbf{k} \right) \cdot (4\mathbf{i} - \mathbf{j} + 2\mathbf{k}) = \left( \dfrac{1}{2} \right)(4) + (3)(-1) + (1)(2) = 1$     ∎

The dot product of a pair of two-dimensional vectors is defined in a similar fashion:

$$\langle u_1, u_2 \rangle \cdot \langle v_1, v_2 \rangle = u_1 v_1 + u_2 v_2.$$

We will see throughout the remainder of the book that the dot product is a key tool for many important geometric and physical calculations in space (and the plane), not just for finding the angle between two vectors.

**Proof of Theorem 1**    Applying the law of cosines (Equation (8), Section 1.3) to the triangle in Figure 12.21, we find that

$$|\mathbf{w}|^2 = |\mathbf{u}|^2 + |\mathbf{v}|^2 - 2|\mathbf{u}||\mathbf{v}| \cos \theta \qquad \text{Law of cosines}$$
$$2|\mathbf{u}||\mathbf{v}| \cos \theta = |\mathbf{u}|^2 + |\mathbf{v}|^2 - |\mathbf{w}|^2.$$

Because $\mathbf{w} = \mathbf{u} - \mathbf{v}$, the component form of $\mathbf{w}$ is $\langle u_1 - v_1, u_2 - v_2, u_3 - v_3 \rangle$. So

$$|\mathbf{u}|^2 = \left( \sqrt{u_1{}^2 + u_2{}^2 + u_3{}^2} \right)^2 = u_1{}^2 + u_2{}^2 + u_3{}^2$$
$$|\mathbf{v}|^2 = \left( \sqrt{v_1{}^2 + v_2{}^2 + v_3{}^2} \right)^2 = v_1{}^2 + v_2{}^2 + v_3{}^2$$
$$|\mathbf{w}|^2 = \left( \sqrt{(u_1 - v_1)^2 + (u_2 - v_2)^2 + (u_3 - v_3)^2} \right)^2$$
$$= (u_1 - v_1)^2 + (u_2 - v_2)^2 + (u_3 - v_3)^2$$
$$= u_1{}^2 - 2u_1 v_1 + v_1{}^2 + u_2{}^2 - 2u_2 v_2 + v_2{}^2 + u_3{}^2 - 2u_3 v_3 + v_3{}^2$$

and

$$|\mathbf{u}|^2 + |\mathbf{v}|^2 - |\mathbf{w}|^2 = 2(u_1 v_1 + u_2 v_2 + u_3 v_3).$$

Therefore,

$$2|\mathbf{u}||\mathbf{v}| \cos \theta = |\mathbf{u}|^2 + |\mathbf{v}|^2 - |\mathbf{w}|^2 = 2(u_1 v_1 + u_2 v_2 + u_3 v_3)$$
$$|\mathbf{u}||\mathbf{v}| \cos \theta = u_1 v_1 + u_2 v_2 + u_3 v_3$$
$$\cos \theta = \frac{u_1 v_1 + u_2 v_2 + u_3 v_3}{|\mathbf{u}||\mathbf{v}|}.$$

Since $0 \le \theta < \pi$, we have

$$\theta = \cos^{-1} \left( \frac{u_1 v_1 + u_2 v_2 + u_3 v_3}{|\mathbf{u}||\mathbf{v}|} \right).$$     ∎

**FIGURE 12.21**  The parallelogram law of addition of vectors gives $\mathbf{w} = \mathbf{u} - \mathbf{v}$.

> **The Angle Between Two Nonzero Vectors u and v**
>
> $$\theta = \cos^{-1} \left( \frac{\mathbf{u} \cdot \mathbf{v}}{|\mathbf{u}||\mathbf{v}|} \right)$$

**EXAMPLE 2**    Find the angle between $\mathbf{u} = \mathbf{i} - 2\mathbf{j} - 2\mathbf{k}$ and $\mathbf{v} = 6\mathbf{i} + 3\mathbf{j} + 2\mathbf{k}$.

**Solution**    We use the formula above:

$$\mathbf{u} \cdot \mathbf{v} = (1)(6) + (-2)(3) + (-2)(2) = 6 - 6 - 4 = -4$$
$$|\mathbf{u}| = \sqrt{(1)^2 + (-2)^2 + (-2)^2} = \sqrt{9} = 3$$
$$|\mathbf{v}| = \sqrt{(6)^2 + (3)^2 + (2)^2} = \sqrt{49} = 7$$
$$\theta = \cos^{-1}\left(\frac{\mathbf{u} \cdot \mathbf{v}}{|\mathbf{u}||\mathbf{v}|}\right) = \cos^{-1}\left(\frac{-4}{(3)(7)}\right) \approx 1.76 \text{ radians or } 100.98°. \qquad \blacksquare$$

The angle formula applies to two-dimensional vectors as well. Note that the angle $\theta$ is acute if $\mathbf{u} \cdot \mathbf{v} > 0$ and obtuse if $\mathbf{u} \cdot \mathbf{v} < 0$.

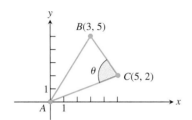

**FIGURE 12.22**    The triangle in Example 3.

**EXAMPLE 3**    Find the angle $\theta$ in the triangle $ABC$ determined by the vertices $A = (0, 0)$, $B = (3, 5)$, and $C = (5, 2)$ (Figure 12.22).

**Solution**    The angle $\theta$ is the angle between the vectors $\overrightarrow{CA}$ and $\overrightarrow{CB}$. The component forms of these two vectors are

$$\overrightarrow{CA} = \langle -5, -2 \rangle \qquad \text{and} \qquad \overrightarrow{CB} = \langle -2, 3 \rangle.$$

First we calculate the dot product and magnitudes of these two vectors.

$$\overrightarrow{CA} \cdot \overrightarrow{CB} = (-5)(-2) + (-2)(3) = 4$$
$$|\overrightarrow{CA}| = \sqrt{(-5)^2 + (-2)^2} = \sqrt{29}$$
$$|\overrightarrow{CB}| = \sqrt{(-2)^2 + (3)^2} = \sqrt{13}$$

Then applying the angle formula, we have

$$\theta = \cos^{-1}\left(\frac{\overrightarrow{CA} \cdot \overrightarrow{CB}}{|\overrightarrow{CA}||\overrightarrow{CB}|}\right)$$
$$= \cos^{-1}\left(\frac{4}{(\sqrt{29})(\sqrt{13})}\right)$$
$$\approx 78.1° \text{ or } 1.36 \text{ radians}. \qquad \blacksquare$$

## Orthogonal Vectors

Two nonzero vectors $\mathbf{u}$ and $\mathbf{v}$ are perpendicular if the angle between them is $\pi/2$. For such vectors, we have $\mathbf{u} \cdot \mathbf{v} = 0$ because $\cos(\pi/2) = 0$. The converse is also true. If $\mathbf{u}$ and $\mathbf{v}$ are nonzero vectors with $\mathbf{u} \cdot \mathbf{v} = |\mathbf{u}||\mathbf{v}|\cos\theta = 0$, then $\cos\theta = 0$ and $\theta = \cos^{-1} 0 = \pi/2$. The following definition also allows for one or both of the vectors to be the zero vector.

---

**DEFINITION**    Vectors $\mathbf{u}$ and $\mathbf{v}$ are orthogonal if $\mathbf{u} \cdot \mathbf{v} = 0$.

---

**EXAMPLE 4**    To determine if two vectors are orthogonal, calculate their dot product.

**(a)** $\mathbf{u} = \langle 3, -2 \rangle$ and $\mathbf{v} = \langle 4, 6 \rangle$ are orthogonal because $\mathbf{u} \cdot \mathbf{v} = (3)(4) + (-2)(6) = 0$.

**(b)** $\mathbf{u} = 3\mathbf{i} - 2\mathbf{j} + \mathbf{k}$ and $\mathbf{v} = 2\mathbf{j} + 4\mathbf{k}$ are orthogonal because $\mathbf{u} \cdot \mathbf{v} = (3)(0) + (-2)(2) + (1)(4) = 0$.

**(c)  0** is orthogonal to every vector **u** since

$$\mathbf{0} \cdot \mathbf{u} = \langle 0, 0, 0 \rangle \cdot \langle u_1, u_2, u_3 \rangle$$
$$= (0)(u_1) + (0)(u_2) + (0)(u_3)$$
$$= 0.$$ ∎

### Dot Product Properties and Vector Projections

The dot product obeys many of the laws that hold for ordinary products of real numbers (scalars).

---

**Properties of the Dot Product**

If **u**, **v**, and **w** are any vectors and $c$ is a scalar, then

1. $\mathbf{u} \cdot \mathbf{v} = \mathbf{v} \cdot \mathbf{u}$
2. $(c\mathbf{u}) \cdot \mathbf{v} = \mathbf{u} \cdot (c\mathbf{v}) = c(\mathbf{u} \cdot \mathbf{v})$
3. $\mathbf{u} \cdot (\mathbf{v} + \mathbf{w}) = \mathbf{u} \cdot \mathbf{v} + \mathbf{u} \cdot \mathbf{w}$
4. $\mathbf{u} \cdot \mathbf{u} = |\mathbf{u}|^2$
5. $\mathbf{0} \cdot \mathbf{u} = 0.$

---

**Proofs of Properties 1 and 3**    The properties are easy to prove using the definition. For instance, here are the proofs of Properties 1 and 3.

1.  $\mathbf{u} \cdot \mathbf{v} = u_1 v_1 + u_2 v_2 + u_3 v_3 = v_1 u_1 + v_2 u_2 + v_3 u_3 = \mathbf{v} \cdot \mathbf{u}$

3.  $\mathbf{u} \cdot (\mathbf{v} + \mathbf{w}) = \langle u_1, u_2, u_3 \rangle \cdot \langle v_1 + w_1, v_2 + w_2, v_3 + w_3 \rangle$
$$= u_1(v_1 + w_1) + u_2(v_2 + w_2) + u_3(v_3 + w_3)$$
$$= u_1 v_1 + u_1 w_1 + u_2 v_2 + u_2 w_2 + u_3 v_3 + u_3 w_3$$
$$= (u_1 v_1 + u_2 v_2 + u_3 v_3) + (u_1 w_1 + u_2 w_2 + u_3 w_3)$$
$$= \mathbf{u} \cdot \mathbf{v} + \mathbf{u} \cdot \mathbf{w}$$ ∎

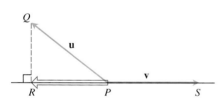

**FIGURE 12.23**   The vector projection of **u** onto **v**.

We now return to the problem of projecting one vector onto another, posed in the opening to this section. The **vector projection** of $\mathbf{u} = \overrightarrow{PQ}$ onto a nonzero vector $\mathbf{v} = \overrightarrow{PS}$ (Figure 12.23) is the vector $\overrightarrow{PR}$ determined by dropping a perpendicular from $Q$ to the line $PS$. The notation for this vector is

$$\text{proj}_{\mathbf{v}} \, \mathbf{u} \qquad \text{(``the vector projection of } \mathbf{u} \text{ onto } \mathbf{v}\text{'').}$$

If **u** represents a force, then $\text{proj}_{\mathbf{v}} \, \mathbf{u}$ represents the effective force in the direction of **v** (Figure 12.24).

If the angle $\theta$ between **u** and **v** is acute, $\text{proj}_{\mathbf{v}} \, \mathbf{u}$ has length $|\mathbf{u}| \cos\theta$ and direction $\mathbf{v}/|\mathbf{v}|$ (Figure 12.25). If $\theta$ is obtuse, $\cos\theta < 0$ and $\text{proj}_{\mathbf{v}} \, \mathbf{u}$ has length $-|\mathbf{u}| \cos\theta$ and direction $-\mathbf{v}/|\mathbf{v}|$. In both cases,

$$\text{proj}_{\mathbf{v}} \, \mathbf{u} = \left( |\mathbf{u}| \cos\theta \right) \frac{\mathbf{v}}{|\mathbf{v}|}$$

$$= \left( \frac{\mathbf{u} \cdot \mathbf{v}}{|\mathbf{v}|} \right) \frac{\mathbf{v}}{|\mathbf{v}|} \qquad \quad |\mathbf{u}| \cos\theta = \frac{|\mathbf{u}||\mathbf{v}|\cos\theta}{|\mathbf{v}|} = \frac{\mathbf{u} \cdot \mathbf{v}}{|\mathbf{v}|}$$

$$= \left( \frac{\mathbf{u} \cdot \mathbf{v}}{|\mathbf{v}|^2} \right) \mathbf{v}.$$

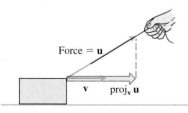

**FIGURE 12.24**   If we pull on the box with force **u**, the effective force moving the box forward in the direction **v** is the projection of **u** onto **v**.

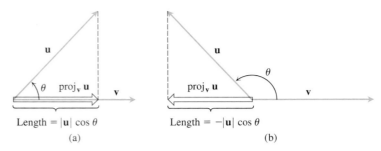

**FIGURE 12.25** The length of $\text{proj}_\mathbf{v}\,\mathbf{u}$ is (a) $|\mathbf{u}|\cos\theta$ if $\cos\theta \geq 0$ and (b) $-|\mathbf{u}|\cos\theta$ if $\cos\theta < 0$.

The number $|\mathbf{u}|\cos\theta$ is called the **scalar component of u in the direction of v** (or of **u onto v**). To summarize,

---

The vector projection of **u** onto **v** is the vector

$$\text{proj}_\mathbf{v}\,\mathbf{u} = \left(\frac{\mathbf{u}\cdot\mathbf{v}}{|\mathbf{v}|^2}\right)\mathbf{v}. \tag{1}$$

The scalar component of **u** in the direction of **v** is the scalar

$$|\mathbf{u}|\cos\theta = \frac{\mathbf{u}\cdot\mathbf{v}}{|\mathbf{v}|} = \mathbf{u}\cdot\frac{\mathbf{v}}{|\mathbf{v}|}. \tag{2}$$

---

Note that both the vector projection of **u** onto **v** and the scalar component of **u** onto **v** depend only on the direction of the vector **v** and not its length (because we dot **u** with $\mathbf{v}/|\mathbf{v}|$, which is the direction of **v**).

**EXAMPLE 5**    Find the vector projection of $\mathbf{u} = 6\mathbf{i} + 3\mathbf{j} + 2\mathbf{k}$ onto $\mathbf{v} = \mathbf{i} - 2\mathbf{j} - 2\mathbf{k}$ and the scalar component of **u** in the direction of **v**.

**Solution**    We find $\text{proj}_\mathbf{v}\,\mathbf{u}$ from Equation (1):

$$\text{proj}_\mathbf{v}\,\mathbf{u} = \frac{\mathbf{u}\cdot\mathbf{v}}{\mathbf{v}\cdot\mathbf{v}}\,\mathbf{v} = \frac{6 - 6 - 4}{1 + 4 + 4}\,(\mathbf{i} - 2\mathbf{j} - 2\mathbf{k})$$

$$= -\frac{4}{9}(\mathbf{i} - 2\mathbf{j} - 2\mathbf{k}) = -\frac{4}{9}\mathbf{i} + \frac{8}{9}\mathbf{j} + \frac{8}{9}\mathbf{k}.$$

We find the scalar component of **u** in the direction of **v** from Equation (2):

$$|\mathbf{u}|\cos\theta = \mathbf{u}\cdot\frac{\mathbf{v}}{|\mathbf{v}|} = (6\mathbf{i} + 3\mathbf{j} + 2\mathbf{k})\cdot\left(\frac{1}{3}\mathbf{i} - \frac{2}{3}\mathbf{j} - \frac{2}{3}\mathbf{k}\right)$$

$$= 2 - 2 - \frac{4}{3} = -\frac{4}{3}.$$  ∎

Equations (1) and (2) also apply to two-dimensional vectors. We demonstrate this in the next example.

**EXAMPLE 6**    Find the vector projection of a force $\mathbf{F} = 5\mathbf{i} + 2\mathbf{j}$ onto $\mathbf{v} = \mathbf{i} - 3\mathbf{j}$ and the scalar component of **F** in the direction of **v**.

**Solution**  The vector projection is

$$\text{proj}_{\mathbf{v}}\,\mathbf{F} = \left(\frac{\mathbf{F}\cdot\mathbf{v}}{|\mathbf{v}|^2}\right)\mathbf{v}$$

$$= \frac{5-6}{1+9}(\mathbf{i}-3\mathbf{j}) = -\frac{1}{10}(\mathbf{i}-3\mathbf{j})$$

$$= -\frac{1}{10}\mathbf{i} + \frac{3}{10}\mathbf{j}.$$

The scalar component of $\mathbf{F}$ in the direction of $\mathbf{v}$ is

$$|\mathbf{F}|\cos\theta = \frac{\mathbf{F}\cdot\mathbf{v}}{|\mathbf{v}|} = \frac{5-6}{\sqrt{1+9}} = -\frac{1}{\sqrt{10}}. \qquad\blacksquare$$

A routine calculation (see Exercise 29) verifies that the vector $\mathbf{u} - \text{proj}_{\mathbf{v}}\,\mathbf{u}$ is orthogonal to the projection vector $\text{proj}_{\mathbf{v}}\,\mathbf{u}$ (which has the same direction as $\mathbf{v}$). So the equation

$$\mathbf{u} = \text{proj}_{\mathbf{v}}\,\mathbf{u} + (\mathbf{u} - \text{proj}_{\mathbf{v}}\,\mathbf{u}) = \underbrace{\left(\frac{\mathbf{u}\cdot\mathbf{v}}{|\mathbf{v}|^2}\right)\mathbf{v}}_{\text{Parallel to v}} + \underbrace{\left(\mathbf{u} - \left(\frac{\mathbf{u}\cdot\mathbf{v}}{|\mathbf{v}|^2}\right)\mathbf{v}\right)}_{\text{Orthogonal to v}}$$

expresses $\mathbf{u}$ as a sum of orthogonal vectors.

## Work

In Chapter 6, we calculated the work done by a constant force of magnitude $F$ in moving an object through a distance $d$ as $W = Fd$. That formula holds only if the force is directed along the line of motion. If a force $\mathbf{F}$ moving an object through a displacement $\mathbf{D} = \overrightarrow{PQ}$ has some other direction, the work is performed by the component of $\mathbf{F}$ in the direction of $\mathbf{D}$. If $\theta$ is the angle between $\mathbf{F}$ and $\mathbf{D}$ (Figure 12.26), then

$$\text{Work} = \left(\begin{array}{c}\text{scalar component of }\mathbf{F}\\ \text{in the direction of }\mathbf{D}\end{array}\right)(\text{length of }\mathbf{D})$$

$$= (|\mathbf{F}|\cos\theta)|\mathbf{D}|$$

$$= \mathbf{F}\cdot\mathbf{D}.$$

**FIGURE 12.26**  The work done by a constant force $\mathbf{F}$ during a displacement $\mathbf{D}$ is $(|\mathbf{F}|\cos\theta)|\mathbf{D}|$, which is the dot product $\mathbf{F}\cdot\mathbf{D}$.

---

**DEFINITION**  The **work** done by a constant force $\mathbf{F}$ acting through a displacement $\mathbf{D} = \overrightarrow{PQ}$ is

$$W = \mathbf{F}\cdot\mathbf{D}.$$

---

**EXAMPLE 7**  If $|\mathbf{F}| = 40$ N, $|\mathbf{D}| = 3$ m, and $\theta = 60°$, the work done by $\mathbf{F}$ in acting from $P$ to $Q$ is

$$\text{Work} = \mathbf{F}\cdot\mathbf{D} \qquad\qquad \text{Definition}$$

$$= |\mathbf{F}||\mathbf{D}|\cos\theta$$

$$= (40)(3)\cos 60° \qquad \text{Given values}$$

$$= (120)(1/2) = 60 \text{ J}. \qquad\blacksquare$$

We encounter more challenging work problems in Chapter 16 when we learn to find the work done by a variable force along a more general *path* in space.

# Exercises 12.3

## Dot Product and Projections

In Exercises 1–8, find

    **a.** $\mathbf{v} \cdot \mathbf{u}$, $|\mathbf{v}|$, $|\mathbf{u}|$

    **b.** the cosine of the angle between $\mathbf{v}$ and $\mathbf{u}$

    **c.** the scalar component of $\mathbf{u}$ in the direction of $\mathbf{v}$

    **d.** the vector $\text{proj}_\mathbf{v}\, \mathbf{u}$.

1. $\mathbf{v} = 2\mathbf{i} - 4\mathbf{j} + \sqrt{5}\mathbf{k}$,    $\mathbf{u} = -2\mathbf{i} + 4\mathbf{j} - \sqrt{5}\mathbf{k}$
2. $\mathbf{v} = (3/5)\mathbf{i} + (4/5)\mathbf{k}$,    $\mathbf{u} = 5\mathbf{i} + 12\mathbf{j}$
3. $\mathbf{v} = 10\mathbf{i} + 11\mathbf{j} - 2\mathbf{k}$,    $\mathbf{u} = 3\mathbf{j} + 4\mathbf{k}$
4. $\mathbf{v} = 2\mathbf{i} + 10\mathbf{j} - 11\mathbf{k}$,    $\mathbf{u} = 2\mathbf{i} + 2\mathbf{j} + \mathbf{k}$
5. $\mathbf{v} = 5\mathbf{j} - 3\mathbf{k}$,    $\mathbf{u} = \mathbf{i} + \mathbf{j} + \mathbf{k}$
6. $\mathbf{v} = -\mathbf{i} + \mathbf{j}$,    $\mathbf{u} = \sqrt{2}\mathbf{i} + \sqrt{3}\mathbf{j} + 2\mathbf{k}$
7. $\mathbf{v} = 5\mathbf{i} + \mathbf{j}$,    $\mathbf{u} = 2\mathbf{i} + \sqrt{17}\mathbf{j}$
8. $\mathbf{v} = \left\langle \dfrac{1}{\sqrt{2}}, \dfrac{1}{\sqrt{3}} \right\rangle$,    $\mathbf{u} = \left\langle \dfrac{1}{\sqrt{2}}, -\dfrac{1}{\sqrt{3}} \right\rangle$

## Angle Between Vectors

$\boxed{\text{T}}$ Find the angles between the vectors in Exercises 9–12 to the nearest hundredth of a radian.

9. $\mathbf{u} = 2\mathbf{i} + \mathbf{j}$,    $\mathbf{v} = \mathbf{i} + 2\mathbf{j} - \mathbf{k}$
10. $\mathbf{u} = 2\mathbf{i} - 2\mathbf{j} + \mathbf{k}$,    $\mathbf{v} = 3\mathbf{i} + 4\mathbf{k}$
11. $\mathbf{u} = \sqrt{3}\mathbf{i} - 7\mathbf{j}$,    $\mathbf{v} = \sqrt{3}\mathbf{i} + \mathbf{j} - 2\mathbf{k}$
12. $\mathbf{u} = \mathbf{i} + \sqrt{2}\mathbf{j} - \sqrt{2}\mathbf{k}$,    $\mathbf{v} = -\mathbf{i} + \mathbf{j} + \mathbf{k}$

13. **Triangle** Find the measures of the angles of the triangle whose vertices are $A = (-1, 0)$, $B = (2, 1)$, and $C = (1, -2)$.

14. **Rectangle** Find the measures of the angles between the diagonals of the rectangle whose vertices are $A = (1, 0)$, $B = (0, 3)$, $C = (3, 4)$, and $D = (4, 1)$.

15. **Direction angles and direction cosines** The *direction angles* $\alpha$, $\beta$, and $\gamma$ of a vector $\mathbf{v} = a\mathbf{i} + b\mathbf{j} + c\mathbf{k}$ are defined as follows:

    $\alpha$ is the angle between $\mathbf{v}$ and the positive $x$-axis $(0 \le \alpha \le \pi)$

    $\beta$ is the angle between $\mathbf{v}$ and the positive $y$-axis $(0 \le \beta \le \pi)$

    $\gamma$ is the angle between $\mathbf{v}$ and the positive $z$-axis $(0 \le \gamma \le \pi)$.

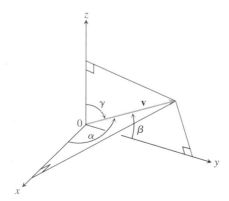

    **a.** Show that

$$\cos\alpha = \frac{a}{|\mathbf{v}|}, \qquad \cos\beta = \frac{b}{|\mathbf{v}|}, \qquad \cos\gamma = \frac{c}{|\mathbf{v}|},$$

    and $\cos^2\alpha + \cos^2\beta + \cos^2\gamma = 1$. These cosines are called the *direction cosines* of $\mathbf{v}$.

    **b. Unit vectors are built from direction cosines** Show that if $\mathbf{v} = a\mathbf{i} + b\mathbf{j} + c\mathbf{k}$ is a unit vector, then $a$, $b$, and $c$ are the direction cosines of $\mathbf{v}$.

16. **Water main construction** A water main is to be constructed with a 20% grade in the north direction and a 10% grade in the east direction. Determine the angle $\theta$ required in the water main for the turn from north to east.

## Theory and Examples

17. **Sums and differences** In the accompanying figure, it looks as if $\mathbf{v}_1 + \mathbf{v}_2$ and $\mathbf{v}_1 - \mathbf{v}_2$ are orthogonal. Is this mere coincidence, or are there circumstances under which we may expect the sum of two vectors to be orthogonal to their difference? Give reasons for your answer.

18. **Orthogonality on a circle** Suppose that $AB$ is the diameter of a circle with center $O$ and that $C$ is a point on one of the two arcs joining $A$ and $B$. Show that $\overrightarrow{CA}$ and $\overrightarrow{CB}$ are orthogonal.

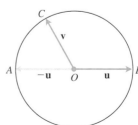

19. **Diagonals of a rhombus** Show that the diagonals of a rhombus (parallelogram with sides of equal length) are perpendicular.

20. **Perpendicular diagonals** Show that squares are the only rectangles with perpendicular diagonals.

21. **When parallelograms are rectangles** Prove that a parallelogram is a rectangle if and only if its diagonals are equal in length. (This fact is often exploited by carpenters.)

22. **Diagonal of parallelogram** Show that the indicated diagonal of the parallelogram determined by vectors **u** and **v** bisects the angle between **u** and **v** if $|\mathbf{u}| = |\mathbf{v}|$.

23. **Projectile motion** A gun with muzzle velocity of 400 m/s is fired at an angle of 8° above the horizontal. Find the horizontal and vertical components of the velocity.

24. **Inclined plane** Suppose that a box is being towed up an inclined plane as shown in the figure. Find the force **w** needed to make the component of the force parallel to the inclined plane equal to 2.5 N.

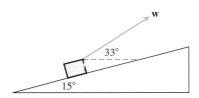

25. **a. Cauchy-Schwartz inequality** Since $\mathbf{u} \cdot \mathbf{v} = |\mathbf{u}||\mathbf{v}|\cos\theta$, show that the inequality $|\mathbf{u} \cdot \mathbf{v}| \le |\mathbf{u}||\mathbf{v}|$ holds for any vectors **u** and **v**.

   **b.** Under what circumstances, if any, does $|\mathbf{u} \cdot \mathbf{v}|$ equal $|\mathbf{u}||\mathbf{v}|$? Give reasons for your answer.

26. **Dot multiplication is positive definite** Show that dot multiplication of vectors is *positive definite*; that is, show that $\mathbf{u} \cdot \mathbf{u} \ge 0$ for every vector **u** and that $\mathbf{u} \cdot \mathbf{u} = 0$ if and only if $\mathbf{u} = \mathbf{0}$.

27. **Orthogonal unit vectors** If $\mathbf{u}_1$ and $\mathbf{u}_2$ are orthogonal unit vectors and $\mathbf{v} = a\mathbf{u}_1 + b\mathbf{u}_2$, find $\mathbf{v} \cdot \mathbf{u}_1$.

28. **Cancellation in dot products** In real-number multiplication, if $uv_1 = uv_2$ and $u \ne 0$, we can cancel the $u$ and conclude that $v_1 = v_2$. Does the same rule hold for the dot product? That is, if $\mathbf{u} \cdot \mathbf{v}_1 = \mathbf{u} \cdot \mathbf{v}_2$ and $\mathbf{u} \ne \mathbf{0}$, can you conclude that $\mathbf{v}_1 = \mathbf{v}_2$? Give reasons for your answer.

29. Using the definition of the projection of **u** onto **v**, show by direct calculation that $(\mathbf{u} - \text{proj}_\mathbf{v}\,\mathbf{u}) \cdot \text{proj}_\mathbf{v}\,\mathbf{u} = 0$.

30. A force $\mathbf{F} = 2\mathbf{i} + \mathbf{j} - 3\mathbf{k}$ is applied to a spacecraft with velocity vector $\mathbf{v} = 3\mathbf{i} - \mathbf{j}$. Express **F** as a sum of a vector parallel to **v** and a vector orthogonal to **v**.

### Equations for Lines in the Plane

31. **Line perpendicular to a vector** Show that $\mathbf{v} = a\mathbf{i} + b\mathbf{j}$ is perpendicular to the line $ax + by = c$ by establishing that the slope of the vector **v** is the negative reciprocal of the slope of the given line.

32. **Line parallel to a vector** Show that the vector $\mathbf{v} = a\mathbf{i} + b\mathbf{j}$ is parallel to the line $bx - ay = c$ by establishing that the slope of the line segment representing **v** is the same as the slope of the given line.

In Exercises 33–36, use the result of Exercise 31 to find an equation for the line through $P$ perpendicular to **v**. Then sketch the line. Include **v** in your sketch *as a vector starting at the origin*.

33. $P(2, 1)$,  $\mathbf{v} = \mathbf{i} + 2\mathbf{j}$        34. $P(-1, 2)$,  $\mathbf{v} = -2\mathbf{i} - \mathbf{j}$

35. $P(-2, -7)$,  $\mathbf{v} = -2\mathbf{i} + \mathbf{j}$    36. $P(11, 10)$,  $\mathbf{v} = 2\mathbf{i} - 3\mathbf{j}$

In Exercises 37–40, use the result of Exercise 32 to find an equation for the line through $P$ parallel to **v**. Then sketch the line. Include **v** in your sketch *as a vector starting at the origin*.

37. $P(-2, 1)$,  $\mathbf{v} = \mathbf{i} - \mathbf{j}$       38. $P(0, -2)$,  $\mathbf{v} = 2\mathbf{i} + 3\mathbf{j}$

39. $P(1, 2)$,  $\mathbf{v} = -\mathbf{i} - 2\mathbf{j}$       40. $P(1, 3)$,  $\mathbf{v} = 3\mathbf{i} - 2\mathbf{j}$

### Work

41. **Work along a line** Find the work done by a force $\mathbf{F} = 5\mathbf{i}$ (magnitude 5 N) in moving an object along the line from the origin to the point $(1, 1)$ (distance in meters).

42. **Locomotive** The Union Pacific's *Big Boy* locomotive could pull 6000-tonne trains with a tractive effort (pull) of 602,148 N. At this level of effort, about how much work did *Big Boy* do on the (approximately straight) 605-km journey from San Francisco to Los Angeles?

43. **Inclined plane** How much work does it take to slide a crate 20 m along a loading dock by pulling on it with a 200-N force at an angle of 30° from the horizontal?

44. **Sailboat** The wind passing over a boat's sail exerted a 1000 N magnitude force **F** as shown here. How much work did the wind perform in moving the boat forward 1 km? Answer in joules.

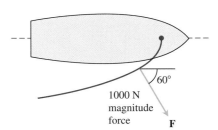

### Angles Between Lines in the Plane

The **acute angle between intersecting lines** that do not cross at right angles is the same as the angle determined by vectors normal to the lines or by the vectors parallel to the lines.

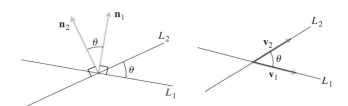

Use this fact and the results of Exercise 31 or 32 to find the acute angles between the lines in Exercises 45–50.

**45.** $3x + y = 5$,   $2x - y = 4$

**46.** $y = \sqrt{3}x - 1$,   $y = -\sqrt{3}x + 2$

**47.** $\sqrt{3}x - y = -2$,   $x - \sqrt{3}y = 1$

**48.** $x + \sqrt{3}y = 1$,   $\left(1 - \sqrt{3}\right)x + \left(1 + \sqrt{3}\right)y = 8$

**49.** $3x - 4y = 3$,   $x - y = 7$

**50.** $12x + 5y = 1$,   $2x - 2y = 3$

# $12.4$ The Cross Product

In studying lines in the plane, when we needed to describe how a line was tilting, we used the notions of slope and angle of inclination. In space, we want a way to describe how a *plane* is tilting. We accomplish this by multiplying two vectors in the plane together to get a third vector perpendicular to the plane. The direction of this third vector tells us the "inclination" of the plane. The product we use to multiply the vectors together is the *vector* or *cross product*, the second of the two vector multiplication methods. We study the cross product in this section.

## The Cross Product of Two Vectors in Space

**FIGURE 12.27** The construction of $\mathbf{u} \times \mathbf{v}$.

We start with two nonzero vectors $\mathbf{u}$ and $\mathbf{v}$ in space. If $\mathbf{u}$ and $\mathbf{v}$ are not parallel, they determine a plane. We select a unit vector $\mathbf{n}$ perpendicular to the plane by the **right-hand rule**. This means that we choose $\mathbf{n}$ to be the unit (normal) vector that points the way your right thumb points when your fingers curl through the angle $\theta$ from $\mathbf{u}$ to $\mathbf{v}$ (Figure 12.27). Then we define a new vector as follows.

---

**DEFINITION**    The **cross product** $\mathbf{u} \times \mathbf{v}$ ("**u cross v**") is the vector

$$\mathbf{u} \times \mathbf{v} = (|\mathbf{u}||\mathbf{v}| \sin \theta) \, \mathbf{n}.$$

---

Unlike the dot product, the cross product is a vector. For this reason it's also called the **vector product** of $\mathbf{u}$ and $\mathbf{v}$, and applies *only* to vectors in space. The vector $\mathbf{u} \times \mathbf{v}$ is orthogonal to both $\mathbf{u}$ and $\mathbf{v}$ because it is a scalar multiple of $\mathbf{n}$.

There is a straightforward way to calculate the cross product of two vectors from their components. The method does not require that we know the angle between them (as suggested by the definition), but we postpone that calculation momentarily so we can focus first on the properties of the cross product.

Since the sines of $0$ and $\pi$ are both zero, it makes sense to define the cross product of two parallel nonzero vectors to be $\mathbf{0}$. If one or both of $\mathbf{u}$ and $\mathbf{v}$ are zero, we also define $\mathbf{u} \times \mathbf{v}$ to be zero. This way, the cross product of two vectors $\mathbf{u}$ and $\mathbf{v}$ is zero if and only if $\mathbf{u}$ and $\mathbf{v}$ are parallel or one or both of them are zero.

---

**Parallel Vectors**

Nonzero vectors $\mathbf{u}$ and $\mathbf{v}$ are parallel if and only if $\mathbf{u} \times \mathbf{v} = \mathbf{0}$.

---

The cross product obeys the following laws.

---

**Properties of the Cross Product**

If $\mathbf{u}$, $\mathbf{v}$, and $\mathbf{w}$ are any vectors and $r$, $s$ are scalars, then

**1.** $(r\mathbf{u}) \times (s\mathbf{v}) = (rs)(\mathbf{u} \times \mathbf{v})$

**2.** $\mathbf{u} \times (\mathbf{v} + \mathbf{w}) = \mathbf{u} \times \mathbf{v} + \mathbf{u} \times \mathbf{w}$

**3.** $\mathbf{v} \times \mathbf{u} = -(\mathbf{u} \times \mathbf{v})$

**4.** $(\mathbf{v} + \mathbf{w}) \times \mathbf{u} = \mathbf{v} \times \mathbf{u} + \mathbf{w} \times \mathbf{u}$

**5.** $\mathbf{0} \times \mathbf{u} = \mathbf{0}$

**6.** $\mathbf{u} \times (\mathbf{v} \times \mathbf{w}) = (\mathbf{u} \cdot \mathbf{w})\mathbf{v} - (\mathbf{u} \cdot \mathbf{v})\mathbf{w}$

---

To visualize Property 3, for example, notice that when the fingers of your right hand curl through the angle $\theta$ from $\mathbf{v}$ to $\mathbf{u}$, your thumb points the opposite way; the unit vector we choose in forming $\mathbf{v} \times \mathbf{u}$ is the negative of the one we choose in forming $\mathbf{u} \times \mathbf{v}$ (Figure 12.28).

Property 1 can be verified by applying the definition of cross product to both sides of the equation and comparing the results. Property 2 is proved in Appendix 8. Property 4 follows by multiplying both sides of the equation in Property 2 by $-1$ and reversing the order of the products using Property 3. Property 5 is a definition. As a rule, cross product multiplication is *not associative* so $(\mathbf{u} \times \mathbf{v}) \times \mathbf{w}$ does not generally equal $\mathbf{u} \times (\mathbf{v} \times \mathbf{w})$. (See Additional Exercise 17.)

When we apply the definition and Property 3 to calculate the pairwise cross products of $\mathbf{i}$, $\mathbf{j}$, and $\mathbf{k}$, we find (Figure 12.29)

**FIGURE 12.28** The construction of $\mathbf{v} \times \mathbf{u}$.

$$\mathbf{i} \times \mathbf{j} = -(\mathbf{j} \times \mathbf{i}) = \mathbf{k}$$
$$\mathbf{j} \times \mathbf{k} = -(\mathbf{k} \times \mathbf{j}) = \mathbf{i}$$
$$\mathbf{k} \times \mathbf{i} = -(\mathbf{i} \times \mathbf{k}) = \mathbf{j}$$

and

$$\mathbf{i} \times \mathbf{i} = \mathbf{j} \times \mathbf{j} = \mathbf{k} \times \mathbf{k} = \mathbf{0}.$$

Diagram for recalling cross products

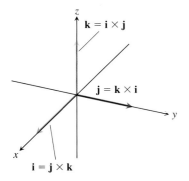

**FIGURE 12.29** The pairwise cross products of $\mathbf{i}$, $\mathbf{j}$, and $\mathbf{k}$.

### $|\mathbf{u} \times \mathbf{v}|$ Is the Area of a Parallelogram

Because $\mathbf{n}$ is a unit vector, the magnitude of $\mathbf{u} \times \mathbf{v}$ is

$$|\mathbf{u} \times \mathbf{v}| = |\mathbf{u}||\mathbf{v}| |\sin\theta||\mathbf{n}| = |\mathbf{u}||\mathbf{v}| \sin\theta.$$

This is the area of the parallelogram determined by $\mathbf{u}$ and $\mathbf{v}$ (Figure 12.30), $|\mathbf{u}|$ being the base of the parallelogram and $|\mathbf{v}||\sin\theta|$ the height.

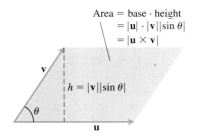

**FIGURE 12.30** The parallelogram determined by $\mathbf{u}$ and $\mathbf{v}$.

### Determinant Formula for $\mathbf{u} \times \mathbf{v}$

Our next objective is to calculate $\mathbf{u} \times \mathbf{v}$ from the components of $\mathbf{u}$ and $\mathbf{v}$ relative to a Cartesian coordinate system.

Suppose that

$$\mathbf{u} = u_1 \mathbf{i} + u_2 \mathbf{j} + u_3 \mathbf{k} \qquad \text{and} \qquad \mathbf{v} = v_1 \mathbf{i} + v_2 \mathbf{j} + v_3 \mathbf{k}.$$

Then the distributive laws and the rules for multiplying $\mathbf{i}$, $\mathbf{j}$, and $\mathbf{k}$ tell us that

$$\begin{aligned}
\mathbf{u} \times \mathbf{v} &= (u_1 \mathbf{i} + u_2 \mathbf{j} + u_3 \mathbf{k}) \times (v_1 \mathbf{i} + v_2 \mathbf{j} + v_3 \mathbf{k}) \\
&= u_1 v_1 \mathbf{i} \times \mathbf{i} + u_1 v_2 \mathbf{i} \times \mathbf{j} + u_1 v_3 \mathbf{i} \times \mathbf{k} \\
&\quad + u_2 v_1 \mathbf{j} \times \mathbf{i} + u_2 v_2 \mathbf{j} \times \mathbf{j} + u_2 v_3 \mathbf{j} \times \mathbf{k} \\
&\quad + u_3 v_1 \mathbf{k} \times \mathbf{i} + u_3 v_2 \mathbf{k} \times \mathbf{j} + u_3 v_3 \mathbf{k} \times \mathbf{k} \\
&= (u_2 v_3 - u_3 v_2)\mathbf{i} - (u_1 v_3 - u_3 v_1)\mathbf{j} + (u_1 v_2 - u_2 v_1)\mathbf{k}.
\end{aligned}$$

The component terms in the last line are hard to remember, but they are the same as the terms in the expansion of the symbolic determinant

$$\begin{vmatrix} \mathbf{i} & \mathbf{j} & \mathbf{k} \\ u_1 & u_2 & u_3 \\ v_1 & v_2 & v_3 \end{vmatrix}.$$

**Determinants**

$2 \times 2$ and $3 \times 3$ determinants are evaluated as follows:

$$\begin{vmatrix} a & b \\ c & d \end{vmatrix} = ad - bc$$

$$\begin{vmatrix} a_1 & a_2 & a_3 \\ b_1 & b_2 & b_3 \\ c_1 & c_2 & c_3 \end{vmatrix} = a_1 \begin{vmatrix} b_2 & b_3 \\ c_2 & c_3 \end{vmatrix}$$

$$- a_2 \begin{vmatrix} b_1 & b_3 \\ c_1 & c_3 \end{vmatrix} + a_3 \begin{vmatrix} b_1 & b_2 \\ c_1 & c_2 \end{vmatrix}$$

So we restate the calculation in this easy-to-remember form.

---

**Calculating the Cross Product as a Determinant**

If $\mathbf{u} = u_1\mathbf{i} + u_2\mathbf{j} + u_3\mathbf{k}$ and $\mathbf{v} = v_1\mathbf{i} + v_2\mathbf{j} + v_3\mathbf{k}$, then

$$\mathbf{u} \times \mathbf{v} = \begin{vmatrix} \mathbf{i} & \mathbf{j} & \mathbf{k} \\ u_1 & u_2 & u_3 \\ v_1 & v_2 & v_3 \end{vmatrix}.$$

---

**EXAMPLE 1**   Find $\mathbf{u} \times \mathbf{v}$ and $\mathbf{v} \times \mathbf{u}$ if $\mathbf{u} = 2\mathbf{i} + \mathbf{j} + \mathbf{k}$ and $\mathbf{v} = -4\mathbf{i} + 3\mathbf{j} + \mathbf{k}$.

**Solution**   We expand the symbolic determinant:

$$\mathbf{u} \times \mathbf{v} = \begin{vmatrix} \mathbf{i} & \mathbf{j} & \mathbf{k} \\ 2 & 1 & 1 \\ -4 & 3 & 1 \end{vmatrix} = \begin{vmatrix} 1 & 1 \\ 3 & 1 \end{vmatrix} \mathbf{i} - \begin{vmatrix} 2 & 1 \\ -4 & 1 \end{vmatrix} \mathbf{j} + \begin{vmatrix} 2 & 1 \\ -4 & 3 \end{vmatrix} \mathbf{k}$$

$$= -2\mathbf{i} - 6\mathbf{j} + 10\mathbf{k}$$

$$\mathbf{v} \times \mathbf{u} = -(\mathbf{u} \times \mathbf{v}) = 2\mathbf{i} + 6\mathbf{j} - 10\mathbf{k} \qquad \text{Property 3} \qquad \blacksquare$$

**EXAMPLE 2**   Find a vector perpendicular to the plane of $P(1, -1, 0)$, $Q(2, 1, -1)$, and $R(-1, 1, 2)$ (Figure 12.31).

**Solution**   The vector $\overrightarrow{PQ} \times \overrightarrow{PR}$ is perpendicular to the plane because it is perpendicular to both vectors. In terms of components,

$$\overrightarrow{PQ} = (2 - 1)\mathbf{i} + (1 + 1)\mathbf{j} + (-1 - 0)\mathbf{k} = \mathbf{i} + 2\mathbf{j} - \mathbf{k}$$

$$\overrightarrow{PR} = (-1 - 1)\mathbf{i} + (1 + 1)\mathbf{j} + (2 - 0)\mathbf{k} = -2\mathbf{i} + 2\mathbf{j} + 2\mathbf{k}$$

$$\overrightarrow{PQ} \times \overrightarrow{PR} = \begin{vmatrix} \mathbf{i} & \mathbf{j} & \mathbf{k} \\ 1 & 2 & -1 \\ -2 & 2 & 2 \end{vmatrix} = \begin{vmatrix} 2 & -1 \\ 2 & 2 \end{vmatrix} \mathbf{i} - \begin{vmatrix} 1 & -1 \\ -2 & 2 \end{vmatrix} \mathbf{j} + \begin{vmatrix} 1 & 2 \\ -2 & 2 \end{vmatrix} \mathbf{k}$$

$$= 6\mathbf{i} + 6\mathbf{k}. \qquad \blacksquare$$

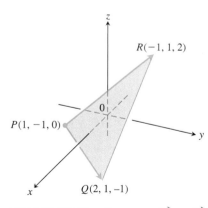

**FIGURE 12.31**   The vector $\overrightarrow{PQ} \times \overrightarrow{PR}$ is perpendicular to the plane of triangle $PQR$ (Example 2). The area of triangle $PQR$ is half of $\left|\overrightarrow{PQ} \times \overrightarrow{PR}\right|$ (Example 3).

**EXAMPLE 3**   Find the area of the triangle with vertices $P(1, -1, 0)$, $Q(2, 1, -1)$, and $R(-1, 1, 2)$ (Figure 12.31).

**Solution**   The area of the parallelogram determined by $P$, $Q$, and $R$ is

$$\left|\overrightarrow{PQ} \times \overrightarrow{PR}\right| = \left|6\mathbf{i} + 6\mathbf{k}\right| \qquad \text{Values from Example 2}$$

$$= \sqrt{(6)^2 + (6)^2} = \sqrt{2 \cdot 36} = 6\sqrt{2}.$$

The triangle's area is half of this, or $3\sqrt{2}$. $\qquad \blacksquare$

**EXAMPLE 4**   Find a unit vector perpendicular to the plane of $P(1, -1, 0)$, $Q(2, 1, -1)$, and $R(-1, 1, 2)$.

**Solution**   Since $\overrightarrow{PQ} \times \overrightarrow{PR}$ is perpendicular to the plane, its direction $\mathbf{n}$ is a unit vector perpendicular to the plane. Taking values from Examples 2 and 3, we have

$$\mathbf{n} = \frac{\overrightarrow{PQ} \times \overrightarrow{PR}}{\left|\overrightarrow{PQ} \times \overrightarrow{PR}\right|} = \frac{6\mathbf{i} + 6\mathbf{k}}{6\sqrt{2}} = \frac{1}{\sqrt{2}}\mathbf{i} + \frac{1}{\sqrt{2}}\mathbf{k}. \qquad \blacksquare$$

For ease in calculating the cross product using determinants, we usually write vectors in the form $\mathbf{v} = v_1\mathbf{i} + v_2\mathbf{j} + v_3\mathbf{k}$ rather than as ordered triples $\mathbf{v} = \langle v_1, v_2, v_3 \rangle$.

### Torque

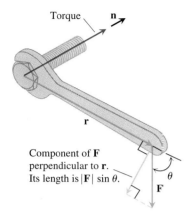

Torque    $\mathbf{n}$

$\mathbf{r}$

Component of $\mathbf{F}$ perpendicular to $\mathbf{r}$. Its length is $|\mathbf{F}| \sin\theta$.

$\theta$

$\mathbf{F}$

**FIGURE 12.32** The torque vector describes the tendency of the force $\mathbf{F}$ to drive the bolt forward.

When we turn a bolt by applying a force $\mathbf{F}$ to a wrench (Figure 12.32), we produce a torque that causes the bolt to rotate. The **torque vector** points in the direction of the axis of the bolt according to the right-hand rule (so the rotation is counterclockwise when viewed from the *tip* of the vector). The magnitude of the torque depends on how far out on the wrench the force is applied and on how much of the force is perpendicular to the wrench at the point of application. The number we use to measure the torque's magnitude is the product of the length of the lever arm $\mathbf{r}$ and the scalar component of $\mathbf{F}$ perpendicular to $\mathbf{r}$. In the notation of Figure 12.32,

$$\text{Magnitude of torque vector} = |\mathbf{r}||\mathbf{F}|\sin\theta,$$

or $|\mathbf{r} \times \mathbf{F}|$. If we let $\mathbf{n}$ be a unit vector along the axis of the bolt in the direction of the torque, then a complete description of the torque vector is $\mathbf{r} \times \mathbf{F}$, or

$$\text{Torque vector} = (|\mathbf{r}||\mathbf{F}|\sin\theta)\,\mathbf{n}.$$

Recall that we defined $\mathbf{u} \times \mathbf{v}$ to be $\mathbf{0}$ when $\mathbf{u}$ and $\mathbf{v}$ are parallel. This is consistent with the torque interpretation as well. If the force $\mathbf{F}$ in Figure 12.32 is parallel to the wrench, meaning that we are trying to turn the bolt by pushing or pulling along the line of the wrench's handle, the torque produced is zero.

3 m bar

$P$    $Q$

70°

20 N magnitude force    $\mathbf{F}$

**FIGURE 12.33** The magnitude of the torque exerted by $\mathbf{F}$ at $P$ is about 56.4 N·m (Example 5). The bar rotates counterclockwise around $P$.

**EXAMPLE 5**    The magnitude of the torque generated by force $\mathbf{F}$ at the pivot point $P$ in Figure 12.33 is

$$|\overrightarrow{PQ} \times \mathbf{F}| = |\overrightarrow{PQ}||\mathbf{F}|\sin 70°$$
$$\approx (3)(20)(0.94)$$
$$\approx 56.4\ \text{N·m}.$$

In this example the torque vector is pointing out of the page toward you.    ∎

### Triple Scalar or Box Product

The product $(\mathbf{u} \times \mathbf{v}) \cdot \mathbf{w}$ is called the **triple scalar product** of $\mathbf{u}$, $\mathbf{v}$, and $\mathbf{w}$ (in that order). As you can see from the formula

$$|(\mathbf{u} \times \mathbf{v}) \cdot \mathbf{w}| = |\mathbf{u} \times \mathbf{v}||\mathbf{w}||\cos\theta|,$$

the absolute value of this product is the volume of the parallelepiped (parallelogram-sided box) determined by $\mathbf{u}$, $\mathbf{v}$, and $\mathbf{w}$ (Figure 12.34). The number $|\mathbf{u} \times \mathbf{v}|$ is the area of the base

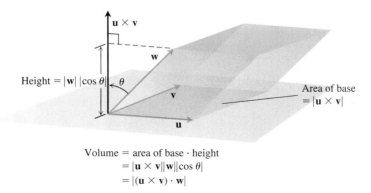

$\mathbf{u} \times \mathbf{v}$

$\mathbf{w}$

Height $= |\mathbf{w}||\cos\theta|$    $\theta$

$\mathbf{v}$

Area of base $= |\mathbf{u} \times \mathbf{v}|$

$\mathbf{u}$

Volume = area of base · height
$= |\mathbf{u} \times \mathbf{v}||\mathbf{w}||\cos\theta|$
$= |(\mathbf{u} \times \mathbf{v}) \cdot \mathbf{w}|$

**FIGURE 12.34** The number $|(\mathbf{u} \times \mathbf{v}) \cdot \mathbf{w}|$ is the volume of a parallelepiped.

parallelogram. The number $|\mathbf{w}||\cos\theta|$ is the parallelepiped's height. Because of this geometry, $(\mathbf{u} \times \mathbf{v}) \cdot \mathbf{w}$ is also called the **box product** of $\mathbf{u}$, $\mathbf{v}$, and $\mathbf{w}$.

By treating the planes of $\mathbf{v}$ and $\mathbf{w}$ and of $\mathbf{w}$ and $\mathbf{u}$ as the base planes of the parallelepiped determined by $\mathbf{u}$, $\mathbf{v}$, and $\mathbf{w}$, we see that

$$(\mathbf{u} \times \mathbf{v}) \cdot \mathbf{w} = (\mathbf{v} \times \mathbf{w}) \cdot \mathbf{u} = (\mathbf{w} \times \mathbf{u}) \cdot \mathbf{v}.$$

> The dot and cross may be interchanged in a triple scalar product without altering is value.

Since the dot product is commutative, we also have

$$(\mathbf{u} \times \mathbf{v}) \cdot \mathbf{w} = \mathbf{u} \cdot (\mathbf{v} \times \mathbf{w}).$$

The triple scalar product can be evaluated as a determinant:

$$(\mathbf{u} \times \mathbf{v}) \cdot \mathbf{w} = \left[ \begin{vmatrix} u_2 & u_3 \\ v_2 & v_3 \end{vmatrix} \mathbf{i} - \begin{vmatrix} u_1 & u_3 \\ v_1 & v_3 \end{vmatrix} \mathbf{j} + \begin{vmatrix} u_1 & u_2 \\ v_1 & v_2 \end{vmatrix} \mathbf{k} \right] \cdot \mathbf{w}$$

$$= w_1 \begin{vmatrix} u_2 & u_3 \\ v_2 & v_3 \end{vmatrix} - w_2 \begin{vmatrix} u_1 & u_3 \\ v_1 & v_3 \end{vmatrix} + w_3 \begin{vmatrix} u_1 & u_2 \\ v_1 & v_2 \end{vmatrix}$$

$$= \begin{vmatrix} u_1 & u_2 & u_3 \\ v_1 & v_2 & v_3 \\ w_1 & w_2 & w_3 \end{vmatrix}.$$

---

**Calculating the Triple Scalar Product as a Determinant**

$$(\mathbf{u} \times \mathbf{v}) \cdot \mathbf{w} = \begin{vmatrix} u_1 & u_2 & u_3 \\ v_1 & v_2 & v_3 \\ w_1 & w_2 & w_3 \end{vmatrix}$$

---

**EXAMPLE 6** Find the volume of the box (parallelepiped) determined by $\mathbf{u} = \mathbf{i} + 2\mathbf{j} - \mathbf{k}$, $\mathbf{v} = -2\mathbf{i} + 3\mathbf{k}$, and $\mathbf{w} = 7\mathbf{j} - 4\mathbf{k}$.

**Solution** Using the rule for calculating a $3 \times 3$ determinant, we find

$$(\mathbf{u} \times \mathbf{v}) \cdot \mathbf{w} = \begin{vmatrix} 1 & 2 & -1 \\ -2 & 0 & 3 \\ 0 & 7 & -4 \end{vmatrix} = (1) \begin{vmatrix} 0 & 3 \\ 7 & -4 \end{vmatrix} - (2) \begin{vmatrix} -2 & 3 \\ 0 & -4 \end{vmatrix} + (-1) \begin{vmatrix} -2 & 0 \\ 0 & 7 \end{vmatrix} = -23.$$

The volume is $|(\mathbf{u} \times \mathbf{v}) \cdot \mathbf{w}| = 23$ units cubed. ∎

# Exercises 12.4

**Cross Product Calculations**

In Exercises 1–8, find the length and direction (when defined) of $\mathbf{u} \times \mathbf{v}$ and $\mathbf{v} \times \mathbf{u}$.

1. $\mathbf{u} = 2\mathbf{i} - 2\mathbf{j} - \mathbf{k}$, $\quad \mathbf{v} = \mathbf{i} - \mathbf{k}$

2. $\mathbf{u} = 2\mathbf{i} + 3\mathbf{j}$, $\quad \mathbf{v} = -\mathbf{i} + \mathbf{j}$

3. $\mathbf{u} = 2\mathbf{i} - 2\mathbf{j} + 4\mathbf{k}$, $\quad \mathbf{v} = -\mathbf{i} + \mathbf{j} - 2\mathbf{k}$

4. $\mathbf{u} = \mathbf{i} + \mathbf{j} - \mathbf{k}$, $\quad \mathbf{v} = 0$

5. $\mathbf{u} = 2\mathbf{i}$, $\quad \mathbf{v} = -3\mathbf{j}$

6. $\mathbf{u} = \mathbf{i} \times \mathbf{j}$, $\quad \mathbf{v} = \mathbf{j} \times \mathbf{k}$

7. $\mathbf{u} = -8\mathbf{i} - 2\mathbf{j} - 4\mathbf{k}$, $\quad \mathbf{v} = 2\mathbf{i} + 2\mathbf{j} + \mathbf{k}$

8. $\mathbf{u} = \frac{3}{2}\mathbf{i} - \frac{1}{2}\mathbf{j} + \mathbf{k}$, $\quad \mathbf{v} = \mathbf{i} + \mathbf{j} + 2\mathbf{k}$

In Exercises 9–14, sketch the coordinate axes and then include the vectors $\mathbf{u}$, $\mathbf{v}$, and $\mathbf{u} \times \mathbf{v}$ as vectors starting at the origin.

9. $\mathbf{u} = \mathbf{i}$, $\quad \mathbf{v} = \mathbf{j}$ $\qquad$ 10. $\mathbf{u} = \mathbf{i} - \mathbf{k}$, $\quad \mathbf{v} = \mathbf{j}$

11. $\mathbf{u} = \mathbf{i} - \mathbf{k}$, $\quad \mathbf{v} = \mathbf{j} + \mathbf{k}$ $\qquad$ 12. $\mathbf{u} = 2\mathbf{i} - \mathbf{j}$, $\quad \mathbf{v} = \mathbf{i} + 2\mathbf{j}$

13. $\mathbf{u} = \mathbf{i} + \mathbf{j}$, $\quad \mathbf{v} = \mathbf{i} - \mathbf{j}$ $\qquad$ 14. $\mathbf{u} = \mathbf{j} + 2\mathbf{k}$, $\quad \mathbf{v} = \mathbf{i}$

**Triangles in Space**
In Exercises 15–18,

a. Find the area of the triangle determined by the points $P$, $Q$, and $R$.

b. Find a unit vector perpendicular to plane $PQR$.

15. $P(1, -1, 2)$,   $Q(2, 0, -1)$,   $R(0, 2, 1)$
16. $P(1, 1, 1)$,   $Q(2, 1, 3)$,   $R(3, -1, 1)$
17. $P(2, -2, 1)$,   $Q(3, -1, 2)$,   $R(3, -1, 1)$
18. $P(-2, 2, 0)$,   $Q(0, 1, -1)$,   $R(-1, 2, -2)$

**Triple Scalar Products**
In Exercises 19–22, verify that $(\mathbf{u} \times \mathbf{v}) \cdot \mathbf{w} = (\mathbf{v} \times \mathbf{w}) \cdot \mathbf{u} = (\mathbf{w} \times \mathbf{u}) \cdot \mathbf{v}$ and find the volume of the parallelepiped (box) determined by $\mathbf{u}$, $\mathbf{v}$, and $\mathbf{w}$.

| $\mathbf{u}$ | $\mathbf{v}$ | $\mathbf{w}$ |
|---|---|---|
| 19. $2\mathbf{i}$ | $2\mathbf{j}$ | $2\mathbf{k}$ |
| 20. $\mathbf{i} - \mathbf{j} + \mathbf{k}$ | $2\mathbf{i} + \mathbf{j} - 2\mathbf{k}$ | $-\mathbf{i} + 2\mathbf{j} - \mathbf{k}$ |
| 21. $2\mathbf{i} + \mathbf{j}$ | $2\mathbf{i} - \mathbf{j} + \mathbf{k}$ | $\mathbf{i} + 2\mathbf{k}$ |
| 22. $\mathbf{i} + \mathbf{j} - 2\mathbf{k}$ | $-\mathbf{i} - \mathbf{k}$ | $2\mathbf{i} + 4\mathbf{j} - 2\mathbf{k}$ |

**Theory and Examples**

23. **Parallel and perpendicular vectors**  Let $\mathbf{u} = 5\mathbf{i} - \mathbf{j} + \mathbf{k}$, $\mathbf{v} = \mathbf{j} - 5\mathbf{k}$, $\mathbf{w} = -15\mathbf{i} + 3\mathbf{j} - 3\mathbf{k}$. Which vectors, if any, are **(a)** perpendicular? **(b)** Parallel? Give reasons for your answers.

24. **Parallel and perpendicular vectors**  Let $\mathbf{u} = \mathbf{i} + 2\mathbf{j} - \mathbf{k}$, $\mathbf{v} = -\mathbf{i} + \mathbf{j} + \mathbf{k}$, $\mathbf{w} = \mathbf{i} + \mathbf{k}$, $\mathbf{r} = -(\pi/2)\mathbf{i} - \pi\mathbf{j} + (\pi/2)\mathbf{k}$. Which vectors, if any, are **(a)** perpendicular? **(b)** Parallel? Give reasons for your answers.

In Exercises 25 and 26, find the magnitude of the torque exerted by $\mathbf{F}$ on the bolt at $P$ if $|\overrightarrow{PQ}| = 20$ cm and $|\mathbf{F}| = 15$ N. Answer in newton-meters.

25.                                26.

27. Which of the following are *always true*, and which are *not always true*? Give reasons for your answers.

a. $|\mathbf{u}| = \sqrt{\mathbf{u} \cdot \mathbf{u}}$

b. $\mathbf{u} \cdot \mathbf{u} = |\mathbf{u}|$

c. $\mathbf{u} \times \mathbf{0} = \mathbf{0} \times \mathbf{u} = \mathbf{0}$

d. $\mathbf{u} \times (-\mathbf{u}) = \mathbf{0}$

e. $\mathbf{u} \times \mathbf{v} = \mathbf{v} \times \mathbf{u}$

f. $\mathbf{u} \times (\mathbf{v} + \mathbf{w}) = \mathbf{u} \times \mathbf{v} + \mathbf{u} \times \mathbf{w}$

g. $(\mathbf{u} \times \mathbf{v}) \cdot \mathbf{v} = 0$

h. $(\mathbf{u} \times \mathbf{v}) \cdot \mathbf{w} = \mathbf{u} \cdot (\mathbf{v} \times \mathbf{w})$

28. Which of the following are *always true*, and which are *not always true*? Give reasons for your answers.

a. $\mathbf{u} \cdot \mathbf{v} = \mathbf{v} \cdot \mathbf{u}$         b. $\mathbf{u} \times \mathbf{v} = -(\mathbf{v} \times \mathbf{u})$

c. $(-\mathbf{u}) \times \mathbf{v} = -(\mathbf{u} \times \mathbf{v})$

d. $(c\mathbf{u}) \cdot \mathbf{v} = \mathbf{u} \cdot (c\mathbf{v}) = c(\mathbf{u} \cdot \mathbf{v})$   (any number $c$)

e. $c(\mathbf{u} \times \mathbf{v}) = (c\mathbf{u}) \times \mathbf{v} = \mathbf{u} \times (c\mathbf{v})$   (any number $c$)

f. $\mathbf{u} \cdot \mathbf{u} = |\mathbf{u}|^2$

g. $(\mathbf{u} \times \mathbf{u}) \cdot \mathbf{u} = 0$

h. $(\mathbf{u} \times \mathbf{v}) \cdot \mathbf{u} = \mathbf{v} \cdot (\mathbf{u} \times \mathbf{v})$

29. Given nonzero vectors $\mathbf{u}$, $\mathbf{v}$, and $\mathbf{w}$, use dot product and cross product notation, as appropriate, to describe the following.

a. The vector projection of $\mathbf{u}$ onto $\mathbf{v}$

b. A vector orthogonal to $\mathbf{u}$ and $\mathbf{v}$

c. A vector orthogonal to $\mathbf{u} \times \mathbf{v}$ and $\mathbf{w}$

d. The volume of the parallelepiped determined by $\mathbf{u}$, $\mathbf{v}$, and $\mathbf{w}$

e. A vector orthogonal to $\mathbf{u} \times \mathbf{v}$ and $\mathbf{u} \times \mathbf{w}$

f. A vector of length $|\mathbf{u}|$ in the direction of $\mathbf{v}$

30. Compute $(\mathbf{i} \times \mathbf{j}) \times \mathbf{j}$ and $\mathbf{i} \times (\mathbf{j} \times \mathbf{j})$. What can you conclude about the associativity of the cross product?

31. Let $\mathbf{u}$, $\mathbf{v}$, and $\mathbf{w}$ be vectors. Which of the following make sense, and which do not? Give reasons for your answers.

a. $(\mathbf{u} \times \mathbf{v}) \cdot \mathbf{w}$

b. $\mathbf{u} \times (\mathbf{v} \cdot \mathbf{w})$

c. $\mathbf{u} \times (\mathbf{v} \times \mathbf{w})$

d. $\mathbf{u} \cdot (\mathbf{v} \cdot \mathbf{w})$

32. **Cross products of three vectors**  Show that except in degenerate cases, $(\mathbf{u} \times \mathbf{v}) \times \mathbf{w}$ lies in the plane of $\mathbf{u}$ and $\mathbf{v}$, whereas $\mathbf{u} \times (\mathbf{v} \times \mathbf{w})$ lies in the plane of $\mathbf{v}$ and $\mathbf{w}$. What *are* the degenerate cases?

33. **Cancelation in cross products**  If $\mathbf{u} \times \mathbf{v} = \mathbf{u} \times \mathbf{w}$ and $\mathbf{u} \neq \mathbf{0}$, then does $\mathbf{v} = \mathbf{w}$? Give reasons for your answer.

34. **Double cancelation**  If $\mathbf{u} \neq \mathbf{0}$ and if $\mathbf{u} \times \mathbf{v} = \mathbf{u} \times \mathbf{w}$ and $\mathbf{u} \cdot \mathbf{v} = \mathbf{u} \cdot \mathbf{w}$, then does $\mathbf{v} = \mathbf{w}$? Give reasons for your answer.

**Area of a Parallelogram**
Find the areas of the parallelograms whose vertices are given in Exercises 35–40.

35. $A(1, 0)$,   $B(0, 1)$,   $C(-1, 0)$,   $D(0, -1)$
36. $A(0, 0)$,   $B(7, 3)$,   $C(9, 8)$,   $D(2, 5)$
37. $A(-1, 2)$,   $B(2, 0)$,   $C(7, 1)$,   $D(4, 3)$
38. $A(-6, 0)$,   $B(1, -4)$,   $C(3, 1)$,   $D(-4, 5)$
39. $A(0, 0, 0)$,   $B(3, 2, 4)$,   $C(5, 1, 4)$,   $D(2, -1, 0)$
40. $A(1, 0, -1)$,   $B(1, 7, 2)$,   $C(2, 4, -1)$,   $D(0, 3, 2)$

**Area of a Triangle**
Find the areas of the triangles whose vertices are given in Exercises 41–47.

41. $A(0, 0)$,   $B(-2, 3)$,   $C(3, 1)$
42. $A(-1, -1)$,   $B(3, 3)$,   $C(2, 1)$
43. $A(-5, 3)$,   $B(1, -2)$,   $C(6, -2)$
44. $A(-6, 0)$,   $B(10, -5)$,   $C(-2, 4)$
45. $A(1, 0, 0)$,   $B(0, 2, 0)$,   $C(0, 0, -1)$
46. $A(0, 0, 0)$,   $B(-1, 1, -1)$,   $C(3, 0, 3)$
47. $A(1, -1, 1)$,   $B(0, 1, 1)$,   $C(1, 0, -1)$

**48.** Find the volume of a parallelepiped if four of its eight vertices are $A(0, 0, 0)$, $B(1, 2, 0)$, $C(0, -3, 2)$, and $D(3, -4, 5)$.

**49. Triangle area** Find a $2 \times 2$ determinant formula for the area of the triangle in the $xy$-plane with vertices at $(0, 0)$, $(a_1, a_2)$, and $(b_1, b_2)$. Explain your work.

**50. Triangle area** Find a concise $3 \times 3$ determinant formula that gives the area of a triangle in the $xy$-plane having vertices $(a_1, a_2)$, $(b_1, b_2)$, and $(c_1, c_2)$.

# $12.5$ Lines and Planes in Space

This section shows how to use scalar and vector products to write equations for lines, line segments, and planes in space. We will use these representations throughout the rest of the book in studying the calculus of curves and surfaces in space.

### Lines and Line Segments in Space

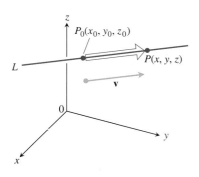

**FIGURE 12.35** A point $P$ lies on $L$ through $P_0$ parallel to $\mathbf{v}$ if and only if $\overrightarrow{P_0P}$ is a scalar multiple of $\mathbf{v}$.

In the plane, a line is determined by a point and a number giving the slope of the line. In space a line is determined by a point and a *vector* giving the direction of the line.

Suppose that $L$ is a line in space passing through a point $P_0(x_0, y_0, z_0)$ parallel to a vector $\mathbf{v} = v_1\mathbf{i} + v_2\mathbf{j} + v_3\mathbf{k}$. Then $L$ is the set of all points $P(x, y, z)$ for which $\overrightarrow{P_0P}$ is parallel to $\mathbf{v}$ (Figure 12.35). Thus, $\overrightarrow{P_0P} = t\mathbf{v}$ for some scalar parameter $t$. The value of $t$ depends on the location of the point $P$ along the line, and the domain of $t$ is $(-\infty, \infty)$. The expanded form of the equation $\overrightarrow{P_0P} = t\mathbf{v}$ is

$$(x - x_0)\mathbf{i} + (y - y_0)\mathbf{j} + (z - z_0)\mathbf{k} = t(v_1\mathbf{i} + v_2\mathbf{j} + v_3\mathbf{k}),$$

which can be rewritten as

$$x\mathbf{i} + y\mathbf{j} + z\mathbf{k} = x_0\mathbf{i} + y_0\mathbf{j} + z_0\mathbf{k} + t(v_1\mathbf{i} + v_2\mathbf{j} + v_3\mathbf{k}). \tag{1}$$

If $\mathbf{r}(t)$ is the position vector of a point $P(x, y, z)$ on the line and $\mathbf{r}_0$ is the position vector of the point $P_0(x_0, y_0, z_0)$, then Equation (1) gives the following vector form for the equation of a line in space.

---

**Vector Equation for a Line**

**A vector equation for the line $L$ through $P_0(x_0, y_0, z_0)$ parallel to $\mathbf{v}$ is**

$$\mathbf{r}(t) = \mathbf{r}_0 + t\mathbf{v}, \qquad -\infty < t < \infty, \tag{2}$$

where $\mathbf{r}$ is the position vector of a point $P(x, y, z)$ on $L$ and $\mathbf{r}_0$ is the position vector of $P_0(x_0, y_0, z_0)$.

---

Equating the corresponding components of the two sides of Equation (1) gives three scalar equations involving the parameter $t$:

$$x = x_0 + tv_1, \qquad y = y_0 + tv_2, \qquad z = z_0 + tv_3.$$

These equations give us the standard parametrization of the line for the parameter interval $-\infty < t < \infty$.

---

**Parametric Equations for a Line**

**The standard parametrization of the line through $P_0(x_0, y_0, z_0)$ parallel to $\mathbf{v} = v_1\mathbf{i} + v_2\mathbf{j} + v_3\mathbf{k}$ is**

$$x = x_0 + tv_1, \quad y = y_0 + tv_2, \quad z = z_0 + tv_3, \quad -\infty < t < \infty \tag{3}$$

---

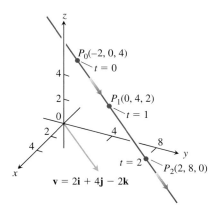

**FIGURE 12.36** Selected points and parameter values on the line in Example 1. The arrows show the direction of increasing $t$.

**EXAMPLE 1**     Find parametric equations for the line through $(-2, 0, 4)$ parallel to $\mathbf{v} = 2\mathbf{i} + 4\mathbf{j} - 2\mathbf{k}$ (Figure 12.36).

**Solution**     With $P_0(x_0, y_0, z_0)$ equal to $(-2, 0, 4)$ and $v_1\mathbf{i} + v_2\mathbf{j} + v_3\mathbf{k}$ equal to $2\mathbf{i} + 4\mathbf{j} - 2\mathbf{k}$, Equations (3) become

$$x = -2 + 2t, \qquad y = 4t, \qquad z = 4 - 2t. \qquad \blacksquare$$

**EXAMPLE 2**     Find parametric equations for the line through $P(-3, 2, -3)$ and $Q(1, -1, 4)$.

**Solution**     The vector

$$\begin{aligned}
\overrightarrow{PQ} &= (1 - (-3))\mathbf{i} + (-1 - 2)\mathbf{j} + (4 - (-3))\mathbf{k} \\
&= 4\mathbf{i} - 3\mathbf{j} + 7\mathbf{k}
\end{aligned}$$

is parallel to the line, and Equations (3) with $(x_0, y_0, z_0) = (-3, 2, -3)$ give

$$x = -3 + 4t, \qquad y = 2 - 3t, \qquad z = -3 + 7t.$$

We could have chosen $Q(1, -1, 4)$ as the "base point" and written

$$x = 1 + 4t, \qquad y = -1 - 3t, \qquad z = 4 + 7t.$$

These equations serve as well as the first; they simply place you at a different point on the line for a given value of $t$. $\qquad \blacksquare$

Notice that parametrizations are not unique. Not only can the "base point" change, but so can the parameter. The equations $x = -3 + 4t^3$, $y = 2 - 3t^3$, and $z = -3 + 7t^3$ also parametrize the line in Example 2.

To parametrize a line segment joining two points, we first parametrize the line through the points. We then find the $t$-values for the endpoints and restrict $t$ to lie in the closed interval bounded by these values. The line equations together with this added restriction parametrize the segment.

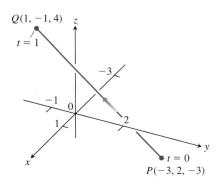

**FIGURE 12.37** Example 3 derives a parametrization of line segment $PQ$. The arrow shows the direction of increasing $t$.

**EXAMPLE 3**     Parametrize the line segment joining the points $P(-3, 2, -3)$ and $Q(1, -1, 4)$ (Figure 12.37).

**Solution**     We begin with equations for the line through $P$ and $Q$, taking them, in this case, from Example 2:

$$x = -3 + 4t, \qquad y = 2 - 3t, \qquad z = -3 + 7t.$$

We observe that the point

$$(x, y, z) = (-3 + 4t, 2 - 3t, -3 + 7t)$$

on the line passes through $P(-3, 2, -3)$ at $t = 0$ and $Q(1, -1, 4)$ at $t = 1$. We add the restriction $0 \le t \le 1$ to parametrize the segment:

$$x = -3 + 4t, \qquad y = 2 - 3t, \qquad z = -3 + 7t, \qquad 0 \le t \le 1. \qquad \blacksquare$$

The vector form (Equation (2)) for a line in space is more revealing if we think of a line as the path of a particle starting at position $P_0(x_0, y_0, z_0)$ and moving in the direction of vector $\mathbf{v}$. Rewriting Equation (2), we have

$$\begin{aligned}
\mathbf{r}(t) &= \mathbf{r}_0 + t\mathbf{v} \\
&= \mathbf{r}_0 + t|\mathbf{v}| \frac{\mathbf{v}}{|\mathbf{v}|}.
\end{aligned} \qquad (4)$$

Initial position        Time        Speed        Direction

In other words, the position of the particle at time $t$ is its initial position plus its distance moved (speed $\times$ time) in the direction $\mathbf{v}/|\mathbf{v}|$ of its straight-line motion.

**EXAMPLE 4**   A helicopter is to fly directly from a helipad at the origin in the direction of the point $(1, 1, 1)$ at a speed of 60 m/s. What is the position of the helicopter after 10 s?

**Solution**   We place the origin at the starting position (helipad) of the helicopter. Then the unit vector

$$\mathbf{u} = \frac{1}{\sqrt{3}}\mathbf{i} + \frac{1}{\sqrt{3}}\mathbf{j} + \frac{1}{\sqrt{3}}\mathbf{k}$$

gives the flight direction of the helicopter. From Equation (4), the position of the helicopter at any time $t$ is

$$\mathbf{r}(t) = \mathbf{r}_0 + t(\text{speed})\mathbf{u}$$

$$= \mathbf{0} + t(60)\left(\frac{1}{\sqrt{3}}\mathbf{i} + \frac{1}{\sqrt{3}}\mathbf{j} + \frac{1}{\sqrt{3}}\mathbf{k}\right)$$

$$= 20\sqrt{3}t(\mathbf{i} + \mathbf{j} + \mathbf{k}).$$

When $t = 10$ s,

$$\mathbf{r}(10) = 200\sqrt{3}(\mathbf{i} + \mathbf{j} + \mathbf{k})$$

$$= \left\langle 200\sqrt{3}, 200\sqrt{3}, 200\sqrt{3}\right\rangle.$$

After 10 s of flight from the origin toward $(1, 1, 1)$, the helicopter is located at the point $\left(200\sqrt{3}, 200\sqrt{3}, 200\sqrt{3}\right)$ in space. It has traveled a distance of $(60 \text{ m/s})(10 \text{ s}) = 600$ m, which is the length of the vector $\mathbf{r}(10)$.   ∎

### The Distance from a Point to a Line in Space

To find the distance from a point $S$ to a line that passes through a point $P$ parallel to a vector $\mathbf{v}$, we find the absolute value of the scalar component of $\overrightarrow{PS}$ in the direction of a vector normal to the line (Figure 12.38). In the notation of the figure, the absolute value of the scalar component is $|\overrightarrow{PS}|\sin\theta$, which is $\dfrac{|\overrightarrow{PS} \times \mathbf{v}|}{|\mathbf{v}|}$.

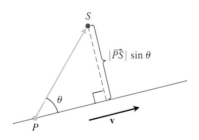

**FIGURE 12.38** The distance from $S$ to the line through $P$ parallel to $\mathbf{v}$ is $|\overrightarrow{PS}|\sin\theta$, where $\theta$ is the angle between $\overrightarrow{PS}$ and $\mathbf{v}$.

---

**Distance from a Point $S$ to a Line Through $P$ Parallel to v**

$$d = \frac{|\overrightarrow{PS} \times \mathbf{v}|}{|\mathbf{v}|} \tag{5}$$

---

**EXAMPLE 5**   Find the distance from the point $S(1, 1, 5)$ to the line

$$L: \qquad x = 1 + t, \qquad y = 3 - t, \qquad z = 2t.$$

**Solution**   We see from the equations for $L$ that $L$ passes through $P(1, 3, 0)$ parallel to $\mathbf{v} = \mathbf{i} - \mathbf{j} + 2\mathbf{k}$. With

$$\overrightarrow{PS} = (1 - 1)\mathbf{i} + (1 - 3)\mathbf{j} + (5 - 0)\mathbf{k} = -2\mathbf{j} + 5\mathbf{k}$$

and

$$\vec{PS} \times \mathbf{v} = \begin{vmatrix} \mathbf{i} & \mathbf{j} & \mathbf{k} \\ 0 & -2 & 5 \\ 1 & -1 & 2 \end{vmatrix} = \mathbf{i} + 5\mathbf{j} + 2\mathbf{k},$$

Equation (5) gives

$$d = \frac{|\vec{PS} \times \mathbf{v}|}{|\mathbf{v}|} = \frac{\sqrt{1 + 25 + 4}}{\sqrt{1 + 1 + 4}} = \frac{\sqrt{30}}{\sqrt{6}} = \sqrt{5}. \qquad \blacksquare$$

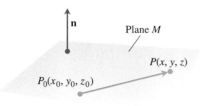

**FIGURE 12.39** The standard equation for a plane in space is defined in terms of a vector normal to the plane: A point $P$ lies in the plane through $P_0$ normal to $\mathbf{n}$ if and only if $\mathbf{n} \cdot \vec{P_0P} = 0$.

## An Equation for a Plane in Space

A plane in space is determined by knowing a point on the plane and its "tilt" or orientation. This "tilt" is defined by specifying a vector that is perpendicular or normal to the plane.

Suppose that plane $M$ passes through a point $P_0(x_0, y_0, z_0)$ and is normal to the nonzero vector $\mathbf{n} = A\mathbf{i} + B\mathbf{j} + C\mathbf{k}$. Then $M$ is the set of all points $P(x, y, z)$ for which $\vec{P_0P}$ is orthogonal to $\mathbf{n}$ (Figure 12.39). Thus, the dot product $\mathbf{n} \cdot \vec{P_0P} = 0$. This equation is equivalent to

$$(A\mathbf{i} + B\mathbf{j} + C\mathbf{k}) \cdot [(x - x_0)\mathbf{i} + (y - y_0)\mathbf{j} + (z - z_0)\mathbf{k}] = 0,$$

so the plane $M$ consists of the points $(x, y, z)$ satisfying

$$A(x - x_0) + B(y - y_0) + C(z - z_0) = 0.$$

---

**Equation for a Plane**

The plane through $P_0(x_0, y_0, z_0)$ normal to $\mathbf{n} = A\mathbf{i} + B\mathbf{j} + C\mathbf{k}$ has

| | |
|---|---|
| **Vector equation:** | $\mathbf{n} \cdot \vec{P_0P} = 0$ |
| **Component equation:** | $A(x - x_0) + B(y - y_0) + C(z - z_0) = 0$ |
| **Component equation simplified:** | $Ax + By + Cz = D,$    where |
| | $D = Ax_0 + By_0 + Cz_0$ |

---

**EXAMPLE 6** Find an equation for the plane through $P_0(-3, 0, 7)$ perpendicular to $\mathbf{n} = 5\mathbf{i} + 2\mathbf{j} - \mathbf{k}$.

**Solution** The component equation is

$$5(x - (-3)) + 2(y - 0) + (-1)(z - 7) = 0.$$

Simplifying, we obtain

$$5x + 15 + 2y - z + 7 = 0$$
$$5x + 2y - z = -22. \qquad \blacksquare$$

Notice in Example 6 how the components of $\mathbf{n} = 5\mathbf{i} + 2\mathbf{j} - \mathbf{k}$ became the coefficients of $x$, $y$, and $z$ in the equation $5x + 2y - z = -22$. The vector $\mathbf{n} = A\mathbf{i} + B\mathbf{j} + C\mathbf{k}$ is normal to the plane $Ax + By + Cz = D$.

**EXAMPLE 7**    Find an equation for the plane through $A(0, 0, 1)$, $B(2, 0, 0)$, and $C(0, 3, 0)$.

**Solution**    We find a vector normal to the plane and use it with one of the points (it does not matter which) to write an equation for the plane.

The cross product

$$\overrightarrow{AB} \times \overrightarrow{AC} = \begin{vmatrix} \mathbf{i} & \mathbf{j} & \mathbf{k} \\ 2 & 0 & -1 \\ 0 & 3 & -1 \end{vmatrix} = 3\mathbf{i} + 2\mathbf{j} + 6\mathbf{k}$$

is normal to the plane. We substitute the components of this vector and the coordinates of $A(0, 0, 1)$ into the component form of the equation to obtain

$$3(x - 0) + 2(y - 0) + 6(z - 1) = 0$$
$$3x + 2y + 6z = 6.$$ ∎

## Lines of Intersection

Just as lines are parallel if and only if they have the same direction, two planes are **parallel** if and only if their normals are parallel, or $\mathbf{n}_1 = k\mathbf{n}_2$ for some scalar $k$. Two planes that are not parallel intersect in a line.

**EXAMPLE 8**    Find a vector parallel to the line of intersection of the planes $3x - 6y - 2z = 15$ and $2x + y - 2z = 5$.

**Solution**    The line of intersection of two planes is perpendicular to both planes' normal vectors $\mathbf{n}_1$ and $\mathbf{n}_2$ (Figure 12.40) and therefore parallel to $\mathbf{n}_1 \times \mathbf{n}_2$. Turning this around, $\mathbf{n}_1 \times \mathbf{n}_2$ is a vector parallel to the planes' line of intersection. In our case,

$$\mathbf{n}_1 \times \mathbf{n}_2 = \begin{vmatrix} \mathbf{i} & \mathbf{j} & \mathbf{k} \\ 3 & -6 & -2 \\ 2 & 1 & -2 \end{vmatrix} = 14\mathbf{i} + 2\mathbf{j} + 15\mathbf{k}.$$

Any nonzero scalar multiple of $\mathbf{n}_1 \times \mathbf{n}_2$ will do as well. ∎

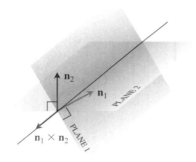

**FIGURE 12.40** How the line of intersection of two planes is related to the planes' normal vectors (Example 8).

**EXAMPLE 9**    Find parametric equations for the line in which the planes $3x - 6y - 2z = 15$ and $2x + y - 2z = 5$ intersect.

**Solution**    We find a vector parallel to the line and a point on the line and use Equations (3).

Example 8 identifies $\mathbf{v} = 14\mathbf{i} + 2\mathbf{j} + 15\mathbf{k}$ as a vector parallel to the line. To find a point on the line, we can take any point common to the two planes. Substituting $z = 0$ in the plane equations and solving for $x$ and $y$ simultaneously identifies one of these points as $(3, -1, 0)$. The line is

$$x = 3 + 14t, \qquad y = -1 + 2t, \qquad z = 15t.$$

The choice $z = 0$ is arbitrary and we could have chosen $z = 1$ or $z = -1$ just as well. Or we could have let $x = 0$ and solved for $y$ and $z$. The different choices would simply give different parametrizations of the same line. ∎

Sometimes we want to know where a line and a plane intersect. For example, if we are looking at a flat plate and a line segment passes through it, we may be interested in knowing what portion of the line segment is hidden from our view by the plate. This application is used in computer graphics (Exercise 74).

**EXAMPLE 10**    Find the point where the line

$$x = \frac{8}{3} + 2t, \qquad y = -2t, \qquad z = 1 + t$$

intersects the plane $3x + 2y + 6z = 6$.

**Solution**    The point

$$\left(\frac{8}{3} + 2t, -2t, 1 + t\right)$$

lies in the plane if its coordinates satisfy the equation of the plane, that is, if

$$3\left(\frac{8}{3} + 2t\right) + 2(-2t) + 6(1 + t) = 6$$

$$8 + 6t - 4t + 6 + 6t = 6$$

$$8t = -8$$

$$t = -1.$$

The point of intersection is

$$(x, y, z)\big|_{t=-1} = \left(\frac{8}{3} - 2, 2, 1 - 1\right) = \left(\frac{2}{3}, 2, 0\right). \qquad \blacksquare$$

### The Distance from a Point to a Plane

If $P$ is a point on a plane with normal $\mathbf{n}$, then the distance from any point $S$ to the plane is the length of the vector projection of $\overrightarrow{PS}$ onto $\mathbf{n}$. That is, the distance from $S$ to the plane is

$$d = \left|\overrightarrow{PS} \cdot \frac{\mathbf{n}}{|\mathbf{n}|}\right| \tag{6}$$

where $\mathbf{n} = A\mathbf{i} + B\mathbf{j} + C\mathbf{k}$ is normal to the plane.

**EXAMPLE 11**    Find the distance from $S(1, 1, 3)$ to the plane $3x + 2y + 6z = 6$.

**Solution**    We find a point $P$ in the plane and calculate the length of the vector projection of $\overrightarrow{PS}$ onto a vector $\mathbf{n}$ normal to the plane (Figure 12.41). The coefficients in the equation $3x + 2y + 6z = 6$ give

$$\mathbf{n} = 3\mathbf{i} + 2\mathbf{j} + 6\mathbf{k}.$$

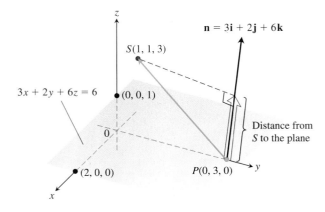

**FIGURE 12.41**    The distance from $S$ to the plane is the length of the vector projection of $\overrightarrow{PS}$ onto $\mathbf{n}$ (Example 11).

The points on the plane easiest to find from the plane's equation are the intercepts. If we take $P$ to be the $y$-intercept $(0, 3, 0)$, then

$$\overrightarrow{PS} = (1 - 0)\mathbf{i} + (1 - 3)\mathbf{j} + (3 - 0)\mathbf{k}$$
$$= \mathbf{i} - 2\mathbf{j} + 3\mathbf{k},$$
$$|\mathbf{n}| = \sqrt{(3)^2 + (2)^2 + (6)^2} = \sqrt{49} = 7.$$

The distance from $S$ to the plane is

$$d = \left| \overrightarrow{PS} \cdot \frac{\mathbf{n}}{|\mathbf{n}|} \right| \qquad \text{Length of } \text{proj}_{\mathbf{n}}\,\overrightarrow{PS}$$

$$= \left| (\mathbf{i} - 2\mathbf{j} + 3\mathbf{k}) \cdot \left( \frac{3}{7}\mathbf{i} + \frac{2}{7}\mathbf{j} + \frac{6}{7}\mathbf{k} \right) \right|$$

$$= \left| \frac{3}{7} - \frac{4}{7} + \frac{18}{7} \right| = \frac{17}{7}. \qquad \blacksquare$$

## Angles Between Planes

The angle between two intersecting planes is defined to be the acute angle between their normal vectors (Figure 12.42).

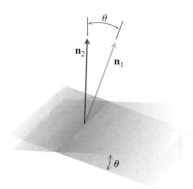

**EXAMPLE 12**     Find the angle between the planes $3x - 6y - 2z = 15$ and $2x + y - 2z = 5$.

**Solution**   The vectors

$$\mathbf{n}_1 = 3\mathbf{i} - 6\mathbf{j} - 2\mathbf{k}, \qquad \mathbf{n}_2 = 2\mathbf{i} + \mathbf{j} - 2\mathbf{k}$$

are normals to the planes. The angle between them is

**FIGURE 12.42**   The angle between two planes is obtained from the angle between their normals.

$$\theta = \cos^{-1}\left( \frac{\mathbf{n}_1 \cdot \mathbf{n}_2}{|\mathbf{n}_1||\mathbf{n}_2|} \right)$$

$$= \cos^{-1}\left( \frac{4}{21} \right)$$

$$\approx 1.38 \text{ radians.} \qquad \text{About 79 degrees} \qquad \blacksquare$$

## Exercises 12.5

### Lines and Line Segments

Find parametric equations for the lines in Exercises 1–12.

1. The line through the point $P(3, -4, -1)$ parallel to the vector $\mathbf{i} + \mathbf{j} + \mathbf{k}$

2. The line through $P(1, 2, -1)$ and $Q(-1, 0, 1)$

3. The line through $P(-2, 0, 3)$ and $Q(3, 5, -2)$

4. The line through $P(1, 2, 0)$ and $Q(1, 1, -1)$

5. The line through the origin parallel to the vector $2\mathbf{j} + \mathbf{k}$

6. The line through the point $(3, -2, 1)$ parallel to the line $x = 1 + 2t, y = 2 - t, z = 3t$

7. The line through $(1, 1, 1)$ parallel to the $z$-axis

8. The line through $(2, 4, 5)$ perpendicular to the plane $3x + 7y - 5z = 21$

9. The line through $(0, -7, 0)$ perpendicular to the plane $x + 2y + 2z = 13$

10. The line through $(2, 3, 0)$ perpendicular to the vectors $\mathbf{u} = \mathbf{i} + 2\mathbf{j} + 3\mathbf{k}$ and $\mathbf{v} = 3\mathbf{i} + 4\mathbf{j} + 5\mathbf{k}$

11. The $x$-axis          12. The $z$-axis

Find parametrizations for the line segments joining the points in Exercises 13–20. Draw coordinate axes and sketch each segment, indicating the direction of increasing $t$ for your parametrization.

13. $(0, 0, 0), \quad (1, 1, 3/2)$       14. $(0, 0, 0), \quad (1, 0, 0)$

15. $(1, 0, 0), \quad (1, 1, 0)$       16. $(1, 1, 0), \quad (1, 1, 1)$

17. $(0, 1, 1), \quad (0, -1, 1)$       18. $(0, 2, 0), \quad (3, 0, 0)$

19. $(2, 0, 2), \quad (0, 2, 0)$       20. $(1, 0, -1), \quad (0, 3, 0)$

**Planes**

Find equations for the planes in Exercises 21–26.

21. The plane through $P_0(0, 2, -1)$ normal to $\mathbf{n} = 3\mathbf{i} - 2\mathbf{j} - \mathbf{k}$

22. The plane through $(1, -1, 3)$ parallel to the plane

$$3x + y + z = 7$$

23. The plane through $(1, 1, -1)$, $(2, 0, 2)$, and $(0, -2, 1)$

24. The plane through $(2, 4, 5)$, $(1, 5, 7)$, and $(-1, 6, 8)$

25. The plane through $P_0(2, 4, 5)$ perpendicular to the line

$$x = 5 + t, \quad y = 1 + 3t, \quad z = 4t$$

26. The plane through $A(1, -2, 1)$ perpendicular to the vector from the origin to $A$

27. Find the point of intersection of the lines $x = 2t + 1$, $y = 3t + 2$, $z = 4t + 3$, and $x = s + 2, y = 2s + 4, z = -4s - 1$, and then find the plane determined by these lines.

28. Find the point of intersection of the lines $x = t, y = -t + 2$, $z = t + 1$, and $x = 2s + 2, y = s + 3, z = 5s + 6$, and then find the plane determined by these lines.

In Exercises 29 and 30, find the plane containing the intersecting lines.

29. $L1: x = -1 + t, \quad y = 2 + t, \quad z = 1 - t; \quad -\infty < t < \infty$

$\quad L2: x = 1 - 4s, \quad y = 1 + 2s, \quad z = 2 - 2s; \quad -\infty < s < \infty$

30. $L1: x = t, \quad y = 3 - 3t, \quad z = -2 - t; \quad -\infty < t < \infty$

$\quad L2: x = 1 + s, \quad y = 4 + s, \quad z = -1 + s; \quad -\infty < s < \infty$

31. Find a plane through $P_0(2, 1, -1)$ and perpendicular to the line of intersection of the planes $2x + y - z = 3, x + 2y + z = 2$.

32. Find a plane through the points $P_1(1, 2, 3)$, $P_2(3, 2, 1)$ and perpendicular to the plane $4x - y + 2z = 7$.

**Distances**

In Exercises 33–38, find the distance from the point to the line.

33. $(0, 0, 12); \quad x = 4t, \quad y = -2t, \quad z = 2t$

34. $(0, 0, 0); \quad x = 5 + 3t, \quad y = 5 + 4t, \quad z = -3 - 5t$

35. $(2, 1, 3); \quad x = 2 + 2t, \quad y = 1 + 6t, \quad z = 3$

36. $(2, 1, -1); \quad x = 2t, \quad y = 1 + 2t, \quad z = 2t$

37. $(3, -1, 4); \quad x = 4 - t, \quad y = 3 + 2t, \quad z = -5 + 3t$

38. $(-1, 4, 3); \quad x = 10 + 4t, \quad y = -3, \quad z = 4t$

In Exercises 39–44, find the distance from the point to the plane.

39. $(2, -3, 4), \quad x + 2y + 2z = 13$

40. $(0, 0, 0), \quad 3x + 2y + 6z = 6$

41. $(0, 1, 1), \quad 4y + 3z = -12$

42. $(2, 2, 3), \quad 2x + y + 2z = 4$

43. $(0, -1, 0), \quad 2x + y + 2z = 4$

44. $(1, 0, -1), \quad -4x + y + z = 4$

45. Find the distance from the plane $x + 2y + 6z = 1$ to the plane $x + 2y + 6z = 10$.

46. Find the distance from the line $x = 2 + t, y = 1 + t$, $z = -(1/2) - (1/2)t$ to the plane $x + 2y + 6z = 10$.

**Angles**

Find the angles between the planes in Exercises 47 and 48.

47. $x + y = 1, \quad 2x + y - 2z = 2$

48. $5x + y - z = 10, \quad x - 2y + 3z = -1$

T  Use a calculator to find the acute angles between the planes in Exercises 49–52 to the nearest hundredth of a radian.

49. $2x + 2y + 2z = 3, \quad 2x - 2y - z = 5$

50. $x + y + z = 1, \quad z = 0$  (the $xy$-plane)

51. $2x + 2y - z = 3, \quad x + 2y + z = 2$

52. $4y + 3z = -12, \quad 3x + 2y + 6z = 6$

**Intersecting Lines and Planes**

In Exercises 53–56, find the point in which the line meets the plane.

53. $x = 1 - t, \quad y = 3t, \quad z = 1 + t; \quad 2x - y + 3z = 6$

54. $x = 2, \quad y = 3 + 2t, \quad z = -2 - 2t; \quad 6x + 3y - 4z = -12$

55. $x = 1 + 2t, \quad y = 1 + 5t, \quad z = 3t; \quad x + y + z = 2$

56. $x = -1 + 3t, \quad y = -2, \quad z = 5t; \quad 2x - 3z = 7$

Find parametrizations for the lines in which the planes in Exercises 57–60 intersect.

57. $x + y + z = 1, \quad x + y = 2$

58. $3x - 6y - 2z = 3, \quad 2x + y - 2z = 2$

59. $x - 2y + 4z = 2, \quad x + y - 2z = 5$

60. $5x - 2y = 11, \quad 4y - 5z = -17$

Given two lines in space, either they are parallel, they intersect, or they are skew (lie in parallel planes). In Exercises 61 and 62, determine whether the lines, taken two at a time, are parallel, intersect, or are skew. If they intersect, find the point of intersection. Otherwise, find the distance between the two lines.

61. $L1: x = 3 + 2t, \quad y = -1 + 4t, \quad z = 2 - t; \quad -\infty < t < \infty$

$\quad L2: x = 1 + 4s, y = 1 + 2s, z = -3 + 4s; \quad -\infty < s < \infty$

$\quad L3: x = 3 + 2r, \quad y = 2 + r, \quad z = -2 + 2r; \quad -\infty < r < \infty$

62. $L1: x = 1 + 2t, \quad y = -1 - t, \quad z = 3t; \quad -\infty < t < \infty$

$\quad L2: x = 2 - s, \quad y = 3s, \quad z = 1 + s; \quad -\infty < s < \infty$

$\quad L3: x = 5 + 2r, \quad y = 1 - r, \quad z = 8 + 3r; \quad -\infty < r < \infty$

**Theory and Examples**

63. Use Equations (3) to generate a parametrization of the line through $P(2, -4, 7)$ parallel to $\mathbf{v}_1 = 2\mathbf{i} - \mathbf{j} + 3\mathbf{k}$. Then generate another parametrization of the line using the point $P_2(-2, -2, 1)$ and the vector $\mathbf{v}_2 = -\mathbf{i} + (1/2)\mathbf{j} - (3/2)\mathbf{k}$.

64. Use the component form to generate an equation for the plane through $P_1(4, 1, 5)$ normal to $\mathbf{n}_1 = \mathbf{i} - 2\mathbf{j} + \mathbf{k}$. Then generate another equation for the same plane using the point $P_2(3, -2, 0)$ and the normal vector $\mathbf{n}_2 = -\sqrt{2}\mathbf{i} + 2\sqrt{2}\mathbf{j} - \sqrt{2}\mathbf{k}$.

65. Find the points in which the line $x = 1 + 2t, y = -1 - t$, $z = 3t$ meets the coordinate planes. Describe the reasoning behind your answer.

66. Find equations for the line in the plane $z = 3$ that makes an angle of $\pi/6$ rad with $\mathbf{i}$ and an angle of $\pi/3$ rad with $\mathbf{j}$. Describe the reasoning behind your answer.

67. Is the line $x = 1 - 2t, y = 2 + 5t, z = -3t$ parallel to the plane $2x + y - z = 8$? Give reasons for your answer.

**68.** How can you tell when two planes $A_1x + B_1y + C_1z = D_1$ and $A_2x + B_2y + C_2z = D_2$ are parallel? Perpendicular? Give reasons for your answer.

**69.** Find two different planes whose intersection is the line $x = 1 + t, y = 2 - t, z = 3 + 2t$. Write equations for each plane in the form $Ax + By + Cz = D$.

**70.** Find a plane through the origin that is perpendicular to the plane $M: 2x + 3y + z = 12$ in a right angle. How do you know that your plane is perpendicular to $M$?

**71.** The graph of $(x/a) + (y/b) + (z/c) = 1$ is a plane for any nonzero numbers $a$, $b$, and $c$. Which planes have an equation of this form?

**72.** Suppose $L_1$ and $L_2$ are disjoint (nonintersecting) nonparallel lines. Is it possible for a nonzero vector to be perpendicular to both $L_1$ and $L_2$? Give reasons for your answer.

**73. Perspective in computer graphics** In computer graphics and perspective drawing, we need to represent objects seen by the eye in space as images on a two-dimensional plane. Suppose that the eye is at $E(x_0, 0, 0)$ as shown here and that we want to represent a point $P_1(x_1, y_1, z_1)$ as a point on the $yz$-plane. We do this by projecting $P_1$ onto the plane with a ray from $E$. The point $P_1$ will be portrayed as the point $P(0, y, z)$. The problem for us as graphics designers is to find $y$ and $z$ given $E$ and $P_1$.

**a.** Write a vector equation that holds between $\overrightarrow{EP}$ and $\overrightarrow{EP_1}$. Use the equation to express $y$ and $z$ in terms of $x_0$, $x_1$, $y_1$, and $z_1$.

**b.** Test the formulas obtained for $y$ and $z$ in part (a) by investigating their behavior at $x_1 = 0$ and $x_1 = x_0$ and by seeing what happens as $x_0 \to \infty$. What do you find?

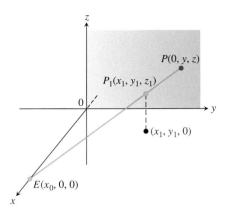

**74. Hidden lines in computer graphics** Here is another typical problem in computer graphics. Your eye is at $(4, 0, 0)$. You are looking at a triangular plate whose vertices are at $(1, 0, 1)$, $(1, 1, 0)$, and $(-2, 2, 2)$. The line segment from $(1, 0, 0)$ to $(0, 2, 2)$ passes through the plate. What portion of the line segment is hidden from your view by the plate? (This is an exercise in finding intersections of lines and planes.)

# $12.6$ Cylinders and Quadric Surfaces

Up to now, we have studied two special types of surfaces: spheres and planes. In this section, we extend our inventory to include a variety of cylinders and quadric surfaces. Quadric surfaces are surfaces defined by second-degree equations in $x$, $y$, and $z$. Spheres are quadric surfaces, but there are others of equal interest which will be needed in Chapters 14–16.

## Cylinders

A **cylinder** is a surface that is generated by moving a straight line along a given planar curve while holding the line parallel to a given fixed line. The curve is called a **generating curve** for the cylinder (Figure 12.43). In solid geometry, where *cylinder* means *circular cylinder*, the generating curves are circles, but now we allow generating curves of any kind. The cylinder in our first example is generated by a parabola.

**EXAMPLE 1** Find an equation for the cylinder made by the lines parallel to the $z$-axis that pass through the parabola $y = x^2, z = 0$ (Figure 12.44).

**Solution** The point $P_0(x_0, x_0^2, 0)$ lies on the parabola $y = x^2$ in the $xy$-plane. Then, for any value of $z$, the point $Q(x_0, x_0^2, z)$ lies on the cylinder because it lies on the line $x = x_0, y = x_0^2$ through $P_0$ parallel to the $z$-axis. Conversely, any point $Q(x_0, x_0^2, z)$ whose $y$-coordinate is the square of its $x$-coordinate lies on the cylinder because it lies on the line $x = x_0, y = x_0^2$ through $P_0$ parallel to the $z$-axis (Figure 12.44).

Regardless of the value of $z$, therefore, the points on the surface are the points whose coordinates satisfy the equation $y = x^2$. This makes $y = x^2$ an equation for the cylinder. Because of this, we call the cylinder "the cylinder $y = x^2$." ∎

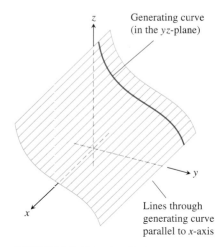

**FIGURE 12.43** A cylinder and generating curve.

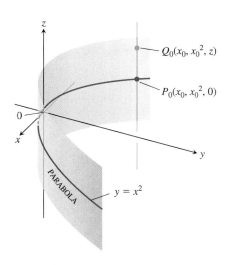

**FIGURE 12.44** Every point of the cylinder in Example 1 has coordinates of the form $(x_0, x_0^2, z)$. We call it "the cylinder $y = x^2$."

As Example 1 suggests, any curve $f(x, y) = c$ in the $xy$-plane defines a cylinder parallel to the $z$-axis whose equation is also $f(x, y) = c$. For instance, the equation $x^2 + y^2 = 1$ defines the circular cylinder made by the lines parallel to the $z$-axis that pass through the circle $x^2 + y^2 = 1$ in the $xy$-plane.

In a similar way, any curve $g(x, z) = c$ in the $xz$-plane defines a cylinder parallel to the $y$-axis whose space equation is also $g(x, z) = c$. Any curve $h(y, z) = c$ defines a cylinder parallel to the $x$-axis whose space equation is also $h(y, z) = c$. The axis of a cylinder need not be parallel to a coordinate axis, however.

## Quadric Surfaces

A **quadric surface** is the graph in space of a second-degree equation in $x$, $y$, and $z$. In this section we study quadric surfaces given by the equation

$$Ax^2 + By^2 + Cz^2 + Dz = E,$$

where $A$, $B$, $C$, $D$, and $E$ are constants. The basic quadric surfaces are **ellipsoids**, **paraboloids**, **elliptical cones**, and **hyperboloids**. Spheres are special cases of ellipsoids. We present a few examples illustrating how to sketch a quadric surface, and then give a summary table of graphs of the basic types.

**EXAMPLE 2**    The **ellipsoid**

$$\frac{x^2}{a^2} + \frac{y^2}{b^2} + \frac{z^2}{c^2} = 1$$

(Figure 12.45) cuts the coordinate axes at $(\pm a, 0, 0)$, $(0, \pm b, 0)$, and $(0, 0, \pm c)$. It lies within the rectangular box defined by the inequalities $|x| \leq a$, $|y| \leq b$, and $|z| \leq c$. The surface is symmetric with respect to each of the coordinate planes because each variable in the defining equation is squared.

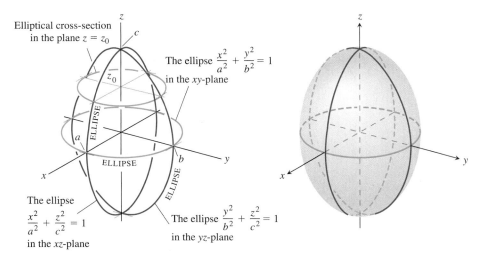

**FIGURE 12.45**    The ellipsoid

$$\frac{x^2}{a^2} + \frac{y^2}{b^2} + \frac{z^2}{c^2} = 1$$

in Example 2 has elliptical cross-sections in each of the three coordinate planes.

The curves in which the three coordinate planes cut the surface are ellipses. For example,

$$\frac{x^2}{a^2} + \frac{y^2}{b^2} = 1 \qquad \text{when} \qquad z = 0.$$

The curve cut from the surface by the plane $z = z_0$, $|z_0| < c$, is the ellipse

$$\frac{x^2}{a^2(1 - (z_0/c)^2)} + \frac{y^2}{b^2(1 - (z_0/c)^2)} = 1.$$

If any two of the semiaxes $a$, $b$, and $c$ are equal, the surface is an **ellipsoid of revolution**. If all three are equal, the surface is a sphere.  ■

**EXAMPLE 3**   The **hyperbolic paraboloid**

$$\frac{y^2}{b^2} - \frac{x^2}{a^2} = \frac{z}{c}, \qquad c > 0$$

has symmetry with respect to the planes $x = 0$ and $y = 0$ (Figure 12.46). The cross-sections in these planes are

$$x = 0: \quad \text{the parabola } z = \frac{c}{b^2}y^2. \tag{1}$$

$$y = 0: \quad \text{the parabola } z = -\frac{c}{a^2}x^2. \tag{2}$$

In the plane $x = 0$, the parabola opens upward from the origin. The parabola in the plane $y = 0$ opens downward.

If we cut the surface by a plane $z = z_0 > 0$, the cross-section is a hyperbola,

$$\frac{y^2}{b^2} - \frac{x^2}{a^2} = \frac{z_0}{c},$$

with its focal axis parallel to the $y$-axis and its vertices on the parabola in Equation (1). If $z_0$ is negative, the focal axis is parallel to the $x$-axis and the vertices lie on the parabola in Equation (2).

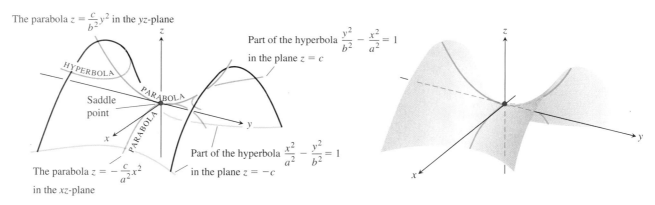

**FIGURE 12.46**   The hyperbolic paraboloid $(y^2/b^2) - (x^2/a^2) = z/c$, $c > 0$. The cross-sections in planes perpendicular to the $z$-axis above and below the $xy$-plane are hyperbolas. The cross-sections in planes perpendicular to the other axes are parabolas.

Near the origin, the surface is shaped like a saddle or mountain pass. To a person traveling along the surface in the $yz$-plane the origin looks like a minimum. To a person traveling the $xz$-plane the origin looks like a maximum. Such a point is called a **saddle point** of a surface. We will say more about saddle points in Section 14.7.  ■

Table 12.1 shows graphs of the six basic types of quadric surfaces. Each surface shown is symmetric with respect to the $z$-axis, but other coordinate axes can serve as well (with appropriate changes to the equation).

**TABLE 12.1** Graphs of Quadric Surfaces

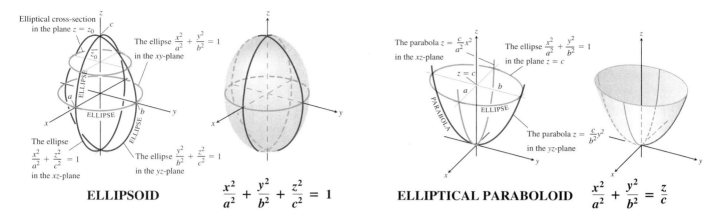

| ELLIPSOID | $\dfrac{x^2}{a^2} + \dfrac{y^2}{b^2} + \dfrac{z^2}{c^2} = 1$ | ELLIPTICAL PARABOLOID | $\dfrac{x^2}{a^2} + \dfrac{y^2}{b^2} = \dfrac{z}{c}$ |

| ELLIPTICAL CONE | $\dfrac{x^2}{a^2} + \dfrac{y^2}{b^2} = \dfrac{z^2}{c^2}$ | HYPERBOLOID OF ONE SHEET | $\dfrac{x^2}{a^2} + \dfrac{y^2}{b^2} - \dfrac{z^2}{c^2} = 1$ |

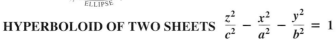

| HYPERBOLOID OF TWO SHEETS | $\dfrac{z^2}{c^2} - \dfrac{x^2}{a^2} - \dfrac{y^2}{b^2} = 1$ | HYPERBOLIC PARABOLOID | $\dfrac{y^2}{b^2} - \dfrac{x^2}{a^2} = \dfrac{z}{c}, c > 0$ |

## Exercises 12.6

### Matching Equations with Surfaces

In Exercises 1–12, match the equation with the surface it defines. Also, identify each surface by type (paraboloid, ellipsoid, etc.). The surfaces are labeled (a)–(1).

**1.** $x^2 + y^2 + 4z^2 = 10$       **2.** $z^2 + 4y^2 - 4x^2 = 4$

**3.** $9y^2 + z^2 = 16$           **4.** $y^2 + z^2 = x^2$

**5.** $x = y^2 - z^2$            **6.** $x = -y^2 - z^2$

**7.** $x^2 + 2z^2 = 8$           **8.** $z^2 + x^2 - y^2 = 1$

**9.** $x = z^2 - y^2$            **10.** $z = -4x^2 - y^2$

**11.** $x^2 + 4z^2 = y^2$         **12.** $9x^2 + 4y^2 + 2z^2 = 36$

**a.**

**b.**

**c.**

**d.**

**e.**

**f.**

**g.**

**h.**

**i.**

**j.**

**k.**

**l.**

### Drawing

Sketch the surfaces in Exercises 13–44.

#### CYLINDERS

**13.** $x^2 + y^2 = 4$           **14.** $z = y^2 - 1$

**15.** $x^2 + 4z^2 = 16$         **16.** $4x^2 + y^2 = 36$

#### ELLIPSOIDS

**17.** $9x^2 + y^2 + z^2 = 9$      **18.** $4x^2 + 4y^2 + z^2 = 16$

**19.** $4x^2 + 9y^2 + 4z^2 = 36$    **20.** $9x^2 + 4y^2 + 36z^2 = 36$

#### PARABOLOIDS AND CONES

**21.** $z = x^2 + 4y^2$          **22.** $z = 8 - x^2 - y^2$

**23.** $x = 4 - 4y^2 - z^2$       **24.** $y = 1 - x^2 - z^2$

**25.** $x^2 + y^2 = z^2$         **26.** $4x^2 + 9z^2 = 9y^2$

#### HYPERBOLOIDS

**27.** $x^2 + y^2 - z^2 = 1$      **28.** $y^2 + z^2 - x^2 = 1$

**29.** $z^2 - x^2 - y^2 = 1$      **30.** $(y^2/4) - (x^2/4) - z^2 = 1$

#### HYPERBOLIC PARABOLOIDS

**31.** $y^2 - x^2 = z$          **32.** $x^2 - y^2 = z$

#### ASSORTED

**33.** $z = 1 + y^2 - x^2$        **34.** $4x^2 + 4y^2 = z^2$

**35.** $y = -(x^2 + z^2)$        **36.** $16x^2 + 4y^2 = 1$

**37.** $x^2 + y^2 - z^2 = 4$      **38.** $x^2 + z^2 = y$

**39.** $x^2 + z^2 = 1$          **40.** $16y^2 + 9z^2 = 4x^2$

**41.** $z = -(x^2 + y^2)$        **42.** $y^2 - x^2 - z^2 = 1$

**43.** $4y^2 + z^2 - 4x^2 = 4$     **44.** $x^2 + y^2 = z$

### Theory and Examples

**45. a.** Express the area $A$ of the cross-section cut from the ellipsoid

$$x^2 + \frac{y^2}{4} + \frac{z^2}{9} = 1$$

by the plane $z = c$ as a function of $c$. (The area of an ellipse with semiaxes $a$ and $b$ is $\pi ab$.)

**b.** Use slices perpendicular to the $z$-axis to find the volume of the ellipsoid in part (a).

**c.** Now find the volume of the ellipsoid

$$\frac{x^2}{a^2} + \frac{y^2}{b^2} + \frac{z^2}{c^2} = 1.$$

Does your formula give the volume of a sphere of radius $a$ if $a = b = c$?

**46.** The barrel shown here is shaped like an ellipsoid with equal pieces cut from the ends by planes perpendicular to the $z$-axis. The cross-sections perpendicular to the $z$-axis are circular. The barrel is $2h$ units high, its midsection radius is $R$, and its end radii are both $r$. Find a formula for the barrel's volume. Then check two things. First, suppose the sides of the barrel are straightened to turn the barrel into a cylinder of radius $R$ and height $2h$. Does your formula give the cylinder's volume? Second, suppose $r = 0$ and $h = R$ so the barrel is a sphere. Does your formula give the sphere's volume?

**47.** Show that the volume of the segment cut from the paraboloid

$$\frac{x^2}{a^2} + \frac{y^2}{b^2} = \frac{z}{c}$$

by the plane $z = h$ equals half the segment's base times its altitude.

**48. a.** Find the volume of the solid bounded by the hyperboloid

$$\frac{x^2}{a^2} + \frac{y^2}{b^2} - \frac{z^2}{c^2} = 1$$

and the planes $z = 0$ and $z = h, h > 0$.

**b.** Express your answer in part (a) in terms of $h$ and the areas $A_0$ and $A_h$ of the regions cut by the hyperboloid from the planes $z = 0$ and $z = h$.

**c.** Show that the volume in part (a) is also given by the formula

$$V = \frac{h}{6}(A_0 + 4A_m + A_h),$$

where $A_m$ is the area of the region cut by the hyperboloid from the plane $z = h/2$.

**Viewing Surfaces**

T Plot the surfaces in Exercises 49–52 over the indicated domains. If you can, rotate the surface into different viewing positions.

**49.** $z = y^2, \quad -2 \le x \le 2, \quad -0.5 \le y \le 2$

**50.** $z = 1 - y^2, \quad -2 \le x \le 2, \quad -2 \le y \le 2$

**51.** $z = x^2 + y^2, \quad -3 \le x \le 3, \quad -3 \le y \le 3$

**52.** $z = x^2 + 2y^2$   over

    **a.** $-3 \le x \le 3, \quad -3 \le y \le 3$

    **b.** $-1 \le x \le 1, \quad -2 \le y \le 3$

    **c.** $-2 \le x \le 2, \quad -2 \le y \le 2$

    **d.** $-2 \le x \le 2, \quad -1 \le y \le 1$

**COMPUTER EXPLORATIONS**

Use a CAS to plot the surfaces in Exercises 53–58. Identify the type of quadric surface from your graph.

**53.** $\dfrac{x^2}{9} + \dfrac{y^2}{36} = 1 - \dfrac{z^2}{25}$     **54.** $\dfrac{x^2}{9} - \dfrac{z^2}{9} = 1 - \dfrac{y^2}{16}$

**55.** $5x^2 = z^2 - 3y^2$     **56.** $\dfrac{y^2}{16} = 1 - \dfrac{x^2}{9} + z$

**57.** $\dfrac{x^2}{9} - 1 = \dfrac{y^2}{16} + \dfrac{z^2}{2}$     **58.** $y - \sqrt{4 - z^2} = 0$

---

# Chapter 12   Questions to Guide Your Review

**1.** When do directed line segments in the plane represent the same vector?

**2.** How are vectors added and subtracted geometrically? Algebraically?

**3.** How do you find a vector's magnitude and direction?

**4.** If a vector is multiplied by a positive scalar, how is the result related to the original vector? What if the scalar is zero? Negative?

**5.** Define the *dot product* (*scalar product*) of two vectors. Which algebraic laws are satisfied by dot products? Give examples. When is the dot product of two vectors equal to zero?

**6.** What geometric interpretation does the dot product have? Give examples.

**7.** What is the vector projection of a vector **u** onto a vector **v**? Give an example of a useful application of a vector projection.

**8.** Define the *cross product* (*vector product*) of two vectors. Which algebraic laws are satisfied by cross products, and which are not? Give examples. When is the cross product of two vectors equal to zero?

**9.** What geometric or physical interpretations do cross products have? Give examples.

**10.** What is the determinant formula for calculating the cross product of two vectors relative to the Cartesian **i**, **j**, **k**-coordinate system? Use it in an example.

**11.** How do you find equations for lines, line segments, and planes in space? Give examples. Can you express a line in space by a single equation? A plane?

**12.** How do you find the distance from a point to a line in space? From a point to a plane? Give examples.

**13.** What are box products? What significance do they have? How are they evaluated? Give an example.

**14.** How do you find equations for spheres in space? Give examples.

**15.** How do you find the intersection of two lines in space? A line and a plane? Two planes? Give examples.

**16.** What is a cylinder? Give examples of equations that define cylinders in Cartesian coordinates.

**17.** What are quadric surfaces? Give examples of different kinds of ellipsoids, paraboloids, cones, and hyperboloids (equations and sketches).

# Chapter **12** Practice Exercises

## Vector Calculations in Two Dimensions

In Exercises 1–4, let $\mathbf{u} = \langle -3, 4 \rangle$ and $\mathbf{v} = \langle 2, -5 \rangle$. Find **(a)** the component form of the vector and **(b)** its magnitude.

1. $5\mathbf{u} + 9\mathbf{v}$
2. $\mathbf{u} - \mathbf{v}$
3. $-7\mathbf{u}$
4. $9\mathbf{v}$

In Exercises 5–8, find the component form of the vector.

5. The vector obtained by rotating $\langle 0, 1 \rangle$ through an angle of $2\pi/3$ radians
6. The unit vector that makes an angle of $\pi/6$ radian with the positive $x$-axis
7. The vector 2 units long in the direction $4\mathbf{i} - \mathbf{j}$
8. The vector 5 units long in the direction opposite to the direction of $(3/5)\mathbf{i} + (4/5)\mathbf{j}$

Express the vectors in Exercises 9–12 in terms of their lengths and directions.

9. $\sqrt{3}\mathbf{i} + \sqrt{3}\mathbf{j}$
10. $-3\mathbf{i} + 2\mathbf{j}$
11. Velocity vector $\mathbf{v} = (-2\sin t)\mathbf{i} + (2\cos t)\mathbf{j}$ when $t = \pi/2$.
12. Velocity vector $\mathbf{v} = (e^t\cos t - e^t\sin t)\mathbf{i} + (e^t\sin t + e^t\cos t)\mathbf{j}$ when $t = \ln 2$.

## Vector Calculations in Three Dimensions

Express the vectors in Exercises 13 and 14 in terms of their lengths and directions.

13. $4\mathbf{i} + 5\mathbf{j} - 3\mathbf{k}$
14. $\mathbf{i} - 5\mathbf{j} + \mathbf{k}$
15. Find a vector 8 units long in the direction of $\mathbf{v} = 3\mathbf{i} + 2\mathbf{j} - \mathbf{k}$.
16. Find a vector 6 units long in the direction along the direction of $\mathbf{v} = (2/5)\mathbf{i} + (3/5)\mathbf{j} + (4/5)\mathbf{k}$.

In Exercises 17 and 18, find $|\mathbf{v}|$, $|\mathbf{u}|$, $\mathbf{v} \cdot \mathbf{u}$, $\mathbf{u} \cdot \mathbf{v}$, $\mathbf{v} \times \mathbf{u}$, $\mathbf{u} \times \mathbf{v}$, $|\mathbf{v} \times \mathbf{u}|$, the angle between $\mathbf{v}$ and $\mathbf{u}$, the scalar component of $\mathbf{u}$ in the direction of $\mathbf{v}$, and the vector projection of $\mathbf{u}$ onto $\mathbf{v}$.

17. $\mathbf{v} = \mathbf{i} - \mathbf{j}$
    $\mathbf{u} = 3\mathbf{i} - \mathbf{j} + 5\mathbf{k}$
18. $\mathbf{v} = \mathbf{i} - \mathbf{j} - 3\mathbf{k}$
    $\mathbf{u} = -\mathbf{j} - \mathbf{k}$

In Exercises 19 and 20, find $\text{proj}_{\mathbf{v}}\,\mathbf{u}$.

19. $\mathbf{v} = 4\mathbf{i} - \mathbf{j} + \mathbf{k}$
    $\mathbf{u} = \mathbf{i} - \mathbf{j} + 7\mathbf{k}$
20. $\mathbf{u} = \mathbf{j} + 5\mathbf{k}$
    $\mathbf{v} = \mathbf{i} - \mathbf{j} - \mathbf{k}$

In Exercises 21 and 22, draw coordinate axes and then sketch $\mathbf{u}$, $\mathbf{v}$, and $\mathbf{u} \times \mathbf{v}$ as vectors at the origin.

21. $\mathbf{u} = \mathbf{i}$, $\mathbf{v} = \mathbf{i} + \mathbf{j}$
22. $\mathbf{u} = \mathbf{i} - \mathbf{j}$, $\mathbf{v} = \mathbf{i} + \mathbf{j}$
23. If $|\mathbf{v}| = 2$, $|\mathbf{w}| = 3$, and the angle between $\mathbf{v}$ and $\mathbf{w}$ is $\pi/3$, find $|\mathbf{v} - 2\mathbf{w}|$.
24. For what value or values of $a$ will the vectors $\mathbf{u} = 2\mathbf{i} + 4\mathbf{j} - 5\mathbf{k}$ and $\mathbf{v} = -4\mathbf{i} - 8\mathbf{j} + a\mathbf{k}$ be parallel?

In Exercises 25 and 26, find **(a)** the area of the parallelogram determined by vectors $\mathbf{u}$ and $\mathbf{v}$ and **(b)** the volume of the parallelepiped determined by the vectors $\mathbf{u}$, $\mathbf{v}$, and $\mathbf{w}$.

25. $\mathbf{u} = \mathbf{i} - \mathbf{j} - \mathbf{k}$; $\mathbf{v} = 3\mathbf{i} + 4\mathbf{j} - 5\mathbf{k}$; $\mathbf{w} = 2\mathbf{i} + 2\mathbf{j} - \mathbf{k}$
26. $\mathbf{u} = \mathbf{j} - \mathbf{k}$; $\mathbf{v} = \mathbf{k}$; $\mathbf{w} = \mathbf{i} - \mathbf{j} - \mathbf{k}$

## Lines, Planes, and Distances

27. Suppose that $\mathbf{n}$ is normal to a plane and that $\mathbf{v}$ is parallel to the plane. Describe how you would find a vector $\mathbf{n}$ that is both perpendicular to $\mathbf{v}$ and parallel to the plane.
28. Find a vector in the plane parallel to the line $ax + by = c$.

In Exercises 29 and 30, find the distance from the point to the line.

29. $(2, 2, 0)$; $x = -t$, $y = t$, $z = -1 + t$
30. $(0, 4, 1)$; $x = 2 + t$, $y = 2 + t$, $z = t$
31. Parametrize the line that passes through the point $(1, 2, 3)$ parallel to the vector $\mathbf{v} = -3\mathbf{i} + 7\mathbf{k}$.
32. Parametrize the line segment joining the points $P(1, 2, 0)$ and $Q(1, 3, -1)$.

In Exercises 33 and 34, find the distance from the point to the plane.

33. $(6, 0, -6)$, $x - y = 4$
34. $(3, 0, 10)$, $2x + 3y + z = 2$
35. Find an equation for the plane that passes through the point $(3, -2, 1)$ normal to the vector $\mathbf{n} = 2\mathbf{i} + \mathbf{j} + \mathbf{k}$.
36. Find an equation for the plane that passes through the point $(-1, 6, 0)$ perpendicular to the line $x = -1 + t, y = 6 - 2t$, $z = 3t$.

In Exercises 37 and 38, find an equation for the plane through points $P$, $Q$, and $R$.

37. $P(1, -1, 2)$, $Q(2, 1, 3)$, $R(-1, 2, -1)$
38. $P(1, 0, 0)$, $Q(0, 1, 0)$, $R(0, 0, 1)$
39. Find the points in which the line $x = 1 + 2t, y = -1 - t$, $z = 3t$ meets the three coordinate planes.
40. Find the point in which the line through the origin perpendicular to the plane $2x - y - z = 4$ meets the plane $3x - 5y + 2z = 6$.
41. Find the acute angle between the planes $x = 7$ and $x + y + \sqrt{2}z = -3$.
42. Find the acute angle between the planes $x + y = 1$ and $y + z = 1$.
43. Find parametric equations for the line in which the planes $x + 2y + z = 1$ and $x - y + 2z = -8$ intersect.
44. Show that the line in which the planes

$$x + 2y - 2z = 5 \quad \text{and} \quad 5x - 2y - z = 0$$

intersect is parallel to the line

$$x = -3 + 2t, \quad y = 3t, \quad z = 1 + 4t.$$

**45.** The planes $3x + 6z = 1$ and $2x + 2y - z = 3$ intersect in a line.

  **a.** Show that the planes are orthogonal.

  **b.** Find equations for the line of intersection.

**46.** Find an equation for the plane that passes through the point $(1, 2, 3)$ parallel to $\mathbf{u} = 2\mathbf{i} + 3\mathbf{j} + \mathbf{k}$ and $\mathbf{v} = \mathbf{i} - \mathbf{j} + 2\mathbf{k}$.

**47.** Is $\mathbf{v} = 2\mathbf{i} - 4\mathbf{j} + \mathbf{k}$ related in any special way to the plane $2x + y = 5$? Give reasons for your answer.

**48.** The equation $\mathbf{n} \cdot \overrightarrow{P_0P} = 0$ represents the plane through $P_0$ normal to $\mathbf{n}$. What set does the inequality $\mathbf{n} \cdot \overrightarrow{P_0P} > 0$ represent?

**49.** Find the distance from the point $P(1, 4, 0)$ to the plane through $A(0, 0, 0)$, $B(2, 0, -1)$, and $C(2, -1, 0)$.

**50.** Find the distance from the point $(2, 2, 3)$ to the plane $2x + 3y + 5z = 0$.

**51.** Find a vector parallel to the plane $2x - y - z = 4$ and orthogonal to $\mathbf{i} + \mathbf{j} + \mathbf{k}$.

**52.** Find a unit vector orthogonal to $\mathbf{A}$ in the plane of $\mathbf{B}$ and $\mathbf{C}$ if $\mathbf{A} = 2\mathbf{i} - \mathbf{j} + \mathbf{k}$, $\mathbf{B} = \mathbf{i} + 2\mathbf{j} + \mathbf{k}$, and $\mathbf{C} = \mathbf{i} + \mathbf{j} - 2\mathbf{k}$.

**53.** Find a vector of magnitude 2 parallel to the line of intersection of the planes $x + 2y + z - 1 = 0$ and $x - y + 2z + 7 = 0$.

**54.** Find the point in which the line through the origin perpendicular to the plane $2x - y - z = 4$ meets the plane $3x - 5y + 2z = 6$.

**55.** Find the point in which the line through $P(3, 2, 1)$ normal to the plane $2x - y + 2z = -2$ meets the plane.

**56.** What angle does the line of intersection of the planes $2x + y - z = 0$ and $x + y + 2z = 0$ make with the positive $x$-axis?

**57.** The line

$$L: \quad x = 3 + 2t, \quad y = 2t, \quad z = t$$

intersects the plane $x + 3y - z = -4$ in a point $P$. Find the coordinates of $P$ and find equations for the line in the plane through $P$ perpendicular to $L$.

**58.** Show that for every real number $k$ the plane

$$x - 2y + z + 3 + k(2x - y - z + 1) = 0$$

contains the line of intersection of the planes

$$x - 2y + z + 3 = 0 \quad \text{and} \quad 2x - y - z + 1 = 0.$$

**59.** Find an equation for the plane through $A(-2, 0, -3)$ and $B(1, -2, 1)$ that lies parallel to the line through $C(-2, -13/5, 26/5)$ and $D(16/5, -13/5, 0)$.

**60.** Is the line $x = 1 + 2t, y = -2 + 3t, z = -5t$ related in any way to the plane $-4x - 6y + 10z = 9$? Give reasons for your answer.

**61.** Which of the following are equations for the plane through the points $P(1, 1, -1)$, $Q(3, 0, 2)$, and $R(-2, 1, 0)$?

  **a.** $(2\mathbf{i} - 3\mathbf{j} + 3\mathbf{k}) \cdot ((x + 2)\mathbf{i} + (y - 1)\mathbf{j} + z\mathbf{k}) = 0$

  **b.** $x = 3 - t, \quad y = -11t, \quad z = 2 - 3t$

  **c.** $(x + 2) + 11(y - 1) = 3z$

  **d.** $(2\mathbf{i} - 3\mathbf{j} + 3\mathbf{k}) \times ((x + 2)\mathbf{i} + (y - 1)\mathbf{j} + z\mathbf{k}) = \mathbf{0}$

  **e.** $(2\mathbf{i} - \mathbf{j} + 3\mathbf{k}) \times (-3\mathbf{i} + \mathbf{k}) \cdot ((x + 2)\mathbf{i} + (y - 1)\mathbf{j} + z\mathbf{k})$
$= 0$

**62.** The parallelogram shown here has vertices at $A(2, -1, 4)$, $B(1, 0, -1)$, $C(1, 2, 3)$, and $D$. Find

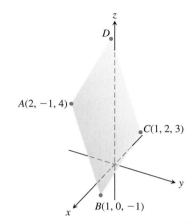

  **a.** the coordinates of $D$,

  **b.** the cosine of the interior angle at $B$,

  **c.** the vector projection of $\overrightarrow{BA}$ onto $\overrightarrow{BC}$,

  **d.** the area of the parallelogram,

  **e.** an equation for the plane of the parallelogram,

  **f.** the areas of the orthogonal projections of the parallelogram on the three coordinate planes.

**63. Distance between skew lines** Find the distance between the line $L_1$ through the points $A(1, 0, -1)$ and $B(-1, 1, 0)$ and the line $L_2$ through the points $C(3, 1, -1)$ and $D(4, 5, -2)$. The distance is to be measured along the line perpendicular to the two lines. First find a vector $\mathbf{n}$ perpendicular to both lines. Then project $\overrightarrow{AC}$ onto $\mathbf{n}$.

**64.** (*Continuation of Exercise 63.*) Find the distance between the line through $A(4, 0, 2)$ and $B(2, 4, 1)$ and the line through $C(1, 3, 2)$ and $D(2, 2, 4)$.

Quadric Surfaces

Identify and sketch the surfaces in Exercises 65–76.

**65.** $x^2 + y^2 + z^2 = 4$      **66.** $x^2 + (y - 1)^2 + z^2 = 1$

**67.** $4x^2 + 4y^2 + z^2 = 4$      **68.** $36x^2 + 9y^2 + 4z^2 = 36$

**69.** $z = -(x^2 + y^2)$      **70.** $y = -(x^2 + z^2)$

**71.** $x^2 + y^2 = z^2$      **72.** $x^2 + z^2 = y^2$

**73.** $x^2 + y^2 - z^2 = 4$      **74.** $4y^2 + z^2 - 4x^2 = 4$

**75.** $y^2 - x^2 - z^2 = 1$      **76.** $z^2 - x^2 - y^2 = 1$

# Chapter **12** Additional and Advanced Exercises

1. **Submarine hunting** Two surface ships on maneuvers are trying to determine a submarine's course and speed to prepare for an aircraft intercept. As shown here, ship $A$ is located at $(4, 0, 0)$, whereas ship $B$ is located at $(0, 5, 0)$. All coordinates are given in thousands of meters. Ship $A$ locates the submarine in the direction of the vector $2\mathbf{i} + 3\mathbf{j} - (1/3)\mathbf{k}$, and ship $B$ locates it in the direction of the vector $18\mathbf{i} - 6\mathbf{j} - \mathbf{k}$. Four minutes ago, the submarine was located at $(2, -1, -1/3)$. The aircraft is due in 20 min. Assuming that the submarine moves in a straight line at a constant speed, to what position should the surface ships direct the aircraft?

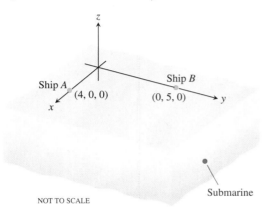

2. **A helicopter rescue** Two helicopters, $H_1$ and $H_2$, are traveling together. At time $t = 0$, they separate and follow different straight-line paths given by

$$H_1: \quad x = 6 + 40t, \quad y = -3 + 10t, \quad z = -3 + 2t$$
$$H_2: \quad x = 6 + 110t, \quad y = -3 + 4t, \quad z = -3 + t.$$

Time $t$ is measured in hours, and all coordinates are measured in kilometers. Due to system malfunctions, $H_2$ stops its flight at $(446, 13, 1)$ and, in a negligible amount of time, lands at $(446, 13, 0)$. Two hours later, $H_1$ is advised of this fact and heads toward $H_2$ at 150 km/h. How long will it take $H_1$ to reach $H_2$?

3. **Torque** The operator's manual for the Toro® 53-cm lawnmower says "tighten the spark plug to 20.4 N·m." If you are installing the plug with a 26.5-cm socket wrench that places the center of your hand 23 cm from the axis of the spark plug, about how hard should you pull? Answer in newtons.

4. **Rotating body** The line through the origin and the point $A(1, 1, 1)$ is the axis of rotation of a rigid body rotating with a constant angular speed of $3/2$ rad/s. The rotation appears to be clockwise

when we look toward the origin from $A$. Find the velocity $\mathbf{v}$ of the point of the body that is at the position $B(1, 3, 2)$.

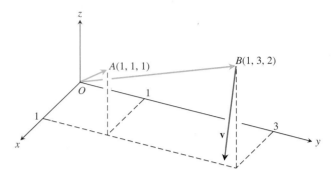

5. Consider the weight suspended by two wires in each diagram. Find the magnitudes and components of vectors $\mathbf{F}_1$ and $\mathbf{F}_2$, and angles $\alpha$ and $\beta$.

   a.

   b.

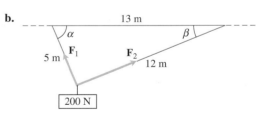

   (*Hint:* This triangle is a right triangle.)

6. Consider a weight of $w$ N suspended by two wires in the diagram, where $\mathbf{T}_1$ and $\mathbf{T}_2$ are force vectors directed along the wires.

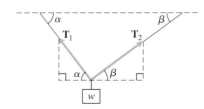

   a. Find the vectors $\mathbf{T}_1$ and $\mathbf{T}_2$ and show that their magnitudes are

   $$|\mathbf{T}_1| = \frac{w \cos \beta}{\sin (\alpha + \beta)}$$

   and

   $$|\mathbf{T}_2| = \frac{w \cos \alpha}{\sin (\alpha + \beta)}.$$

**b.** For a fixed $\beta$ determine the value of $\alpha$ which minimizes the magnitude $|\mathbf{T}_1|$.

**c.** For a fixed $\alpha$ determine the value of $\beta$ which minimizes the magnitude $|\mathbf{T}_2|$.

**7. Determinants and planes**

**a.** Show that

$$\begin{vmatrix} x_1 - x & y_1 - y & z_1 - z \\ x_2 - x & y_2 - y & z_2 - z \\ x_3 - x & y_3 - y & z_3 - z \end{vmatrix} = 0$$

is an equation for the plane through the three noncollinear points $P_1(x_1, y_1, z_1)$, $P_2(x_2, y_2, z_2)$, and $P_3(x_3, y_3, z_3)$.

**b.** What set of points in space is described by the equation

$$\begin{vmatrix} x & y & z & 1 \\ x_1 & y_1 & z_1 & 1 \\ x_2 & y_2 & z_2 & 1 \\ x_3 & y_3 & z_3 & 1 \end{vmatrix} = 0?$$

**8. Determinants and lines** Show that the lines

$$x = a_1 s + b_1, \quad y = a_2 s + b_2, \quad z = a_3 s + b_3, \quad -\infty < s < \infty$$

and

$$x = c_1 t + d_1, \quad y = c_2 t + d_2, \quad z = c_3 t + d_3, \quad -\infty < t < \infty,$$

intersect or are parallel if and only if

$$\begin{vmatrix} a_1 & c_1 & b_1 - d_1 \\ a_2 & c_2 & b_2 - d_2 \\ a_3 & c_3 & b_3 - d_3 \end{vmatrix} = 0.$$

**9.** Consider a regular tetrahedron of side length 2.

**a.** Use vectors to find the angle $\theta$ formed by the base of the tetrahedron and any one of its other edges.

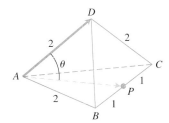

**b.** Use vectors to find the angle $\theta$ formed by any two adjacent faces of the tetrahedron. This angle is commonly referred to as a dihedral angle.

**10.** In the figure here, $D$ is the midpoint of side $AB$ of triangle $ABC$, and $E$ is one-third of the way between $C$ and $B$. Use vectors to prove that $F$ is the midpoint of line segment $CD$.

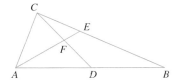

**11.** Use vectors to show that the distance from $P_1(x_1, y_1)$ to the line $ax + by = c$ is

$$d = \frac{|ax_1 + by_1 - c|}{\sqrt{a^2 + b^2}}.$$

**12. a.** Use vectors to show that the distance from $P_1(x_1, y_1, z_1)$ to the plane $Ax + By + Cz = D$ is

$$d = \frac{|Ax_1 + By_1 + Cz_1 - D|}{\sqrt{A^2 + B^2 + C^2}}.$$

**b.** Find an equation for the sphere that is tangent to the planes $x + y + z = 3$ and $x + y + z = 9$ if the planes $2x - y = 0$ and $3x - z = 0$ pass through the center of the sphere.

**13. a. Distance between parallel planes** Show that the distance between the parallel planes $Ax + By + Cz = D_1$ and $Ax + By + Cz = D_2$ is

$$d = \frac{|D_1 - D_2|}{|A\mathbf{i} + B\mathbf{j} + C\mathbf{k}|}.$$

**b.** Find the distance between the planes $2x + 3y - z = 6$ and $2x + 3y - z = 12$.

**c.** Find an equation for the plane parallel to the plane $2x - y + 2z = -4$ if the point $(3, 2, -1)$ is equidistant from the two planes.

**d.** Write equations for the planes that lie parallel to and 5 units away from the plane $x - 2y + z = 3$.

**14.** Prove that four points $A$, $B$, $C$, and $D$ are coplanar (lie in a common plane) if and only if $\overrightarrow{AD} \cdot (\overrightarrow{AB} \times \overrightarrow{BC}) = 0$.

**15. The projection of a vector on a plane** Let $P$ be a plane in space and let $\mathbf{v}$ be a vector. The vector projection of $\mathbf{v}$ onto the plane $P$, $\text{proj}_P \mathbf{v}$, can be defined informally as follows. Suppose the sun is shining so that its rays are normal to the plane $P$. Then $\text{proj}_P \mathbf{v}$ is the "shadow" of $\mathbf{v}$ onto $P$. If $P$ is the plane $x + 2y + 6z = 6$ and $\mathbf{v} = \mathbf{i} + \mathbf{j} + \mathbf{k}$, find $\text{proj}_P \mathbf{v}$.

**16.** The accompanying figure shows nonzero vectors $\mathbf{v}$, $\mathbf{w}$, and $\mathbf{z}$, with $\mathbf{z}$ orthogonal to the line $L$, and $\mathbf{v}$ and $\mathbf{w}$ making equal angles $\beta$ with $L$. Assuming $|\mathbf{v}| = |\mathbf{w}|$, find $\mathbf{w}$ in terms of $\mathbf{v}$ and $\mathbf{z}$.

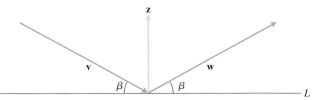

**17. Triple vector products** The *triple vector products* $(\mathbf{u} \times \mathbf{v}) \times \mathbf{w}$ and $\mathbf{u} \times (\mathbf{v} \times \mathbf{w})$ are usually not equal, although the formulas for evaluating them from components are similar:

$$(\mathbf{u} \times \mathbf{v}) \times \mathbf{w} = (\mathbf{u} \cdot \mathbf{w})\mathbf{v} - (\mathbf{v} \cdot \mathbf{w})\mathbf{u}.$$

$$\mathbf{u} \times (\mathbf{v} \times \mathbf{w}) = (\mathbf{u} \cdot \mathbf{w})\mathbf{v} - (\mathbf{u} \cdot \mathbf{v})\mathbf{w}.$$

Verify each formula for the following vectors by evaluating its two sides and comparing the results.

| | $\mathbf{u}$ | $\mathbf{v}$ | $\mathbf{w}$ |
|---|---|---|---|
| **a.** | $2\mathbf{i}$ | $2\mathbf{j}$ | $2\mathbf{k}$ |
| **b.** | $\mathbf{i} - \mathbf{j} + \mathbf{k}$ | $2\mathbf{i} + \mathbf{j} - 2\mathbf{k}$ | $-\mathbf{i} + 2\mathbf{j} - \mathbf{k}$ |
| **c.** | $2\mathbf{i} + \mathbf{j}$ | $2\mathbf{i} - \mathbf{j} + \mathbf{k}$ | $\mathbf{i} + 2\mathbf{k}$ |
| **d.** | $\mathbf{i} + \mathbf{j} - 2\mathbf{k}$ | $-\mathbf{i} - \mathbf{k}$ | $2\mathbf{i} + 4\mathbf{j} - 2\mathbf{k}$ |

**18. Cross and dot products** Show that if **u**, **v**, **w**, and **r** are any vectors, then

a. $\mathbf{u} \times (\mathbf{v} \times \mathbf{w}) + \mathbf{v} \times (\mathbf{w} \times \mathbf{u}) + \mathbf{w} \times (\mathbf{u} \times \mathbf{v}) = \mathbf{0}$

b. $\mathbf{u} \times \mathbf{v} = (\mathbf{u} \cdot \mathbf{v} \times \mathbf{i})\mathbf{i} + (\mathbf{u} \cdot \mathbf{v} \times \mathbf{j})\mathbf{j} + (\mathbf{u} \cdot \mathbf{v} \times \mathbf{k})\mathbf{k}$

c. $(\mathbf{u} \times \mathbf{v}) \cdot (\mathbf{w} \times \mathbf{r}) = \begin{vmatrix} \mathbf{u} \cdot \mathbf{w} & \mathbf{v} \cdot \mathbf{w} \\ \mathbf{u} \cdot \mathbf{r} & \mathbf{v} \cdot \mathbf{r} \end{vmatrix}.$

**19. Cross and dot products** Prove or disprove the formula

$$\mathbf{u} \times (\mathbf{u} \times (\mathbf{u} \times \mathbf{v})) \cdot \mathbf{w} = -|\mathbf{u}|^2 \mathbf{u} \cdot \mathbf{v} \times \mathbf{w}.$$

**20.** By forming the cross product of two appropriate vectors, derive the trigonometric identity

$$\sin(A - B) = \sin A \cos B - \cos A \sin B.$$

**21.** Use vectors to prove that

$$(a^2 + b^2)(c^2 + d^2) \geq (ac + bd)^2$$

for any four numbers $a$, $b$, $c$, and $d$. (*Hint:* Let $\mathbf{u} = a\mathbf{i} + b\mathbf{j}$ and $\mathbf{v} = c\mathbf{i} + d\mathbf{j}$.)

**22. Dot multiplication is positive definite** Show that dot multiplication of vectors is *positive definite*; that is, show $\mathbf{u} \cdot \mathbf{u} \geq 0$ for every vector $\mathbf{u}$ and that $\mathbf{u} \cdot \mathbf{u} = 0$ if and only if $\mathbf{u} = \mathbf{0}$.

**23.** Show that $|\mathbf{u} + \mathbf{v}| \leq |\mathbf{u}| + |\mathbf{v}|$ for any vectors **u** and **v**.

**24.** Show that $\mathbf{w} = |\mathbf{v}|\mathbf{u} + |\mathbf{u}|\mathbf{v}$ bisects the angle between **u** and **v**.

**25.** Show that $|\mathbf{v}|\mathbf{u} + |\mathbf{u}|\mathbf{v}$ and $|\mathbf{v}|\mathbf{u} - |\mathbf{u}|\mathbf{v}$ are orthogonal.

## Chapter **12** Technology Application Projects

**Mathematica/Maple Modules:**

***Using Vectors to Represent Lines and Find Distances***
**Parts I and II**: Learn the advantages of interpreting lines as vectors.
**Part III**: Use vectors to find the distance from a point to a line.

***Putting a Scene in Three Dimensions onto a Two-Dimensional Canvas***
Use the concept of planes in space to obtain a two-dimensional image.

***Getting Started in Plotting in 3D***
**Part I**: Use the vector definition of lines and planes to generate graphs and equations, and to compare different forms for the equations of a single line.
**Part II**: Plot functions that are defined implicitly.

# 13

# Vector-Valued Functions and Motion in Space

**OVERVIEW** Now that we have learned about vectors and the geometry of space, we can combine these ideas with our earlier study of functions. In this chapter we introduce the calculus of vector-valued functions. The domains of these functions are sets of real numbers, as before, but their ranges consist of vectors, not scalars. When a vector-valued function changes, the change can occur in both magnitude and direction, so the derivative is itself a vector. The integral of a vector-valued function is also a vector. We use the calculus of these functions to describe the paths and motions of objects moving in a plane or in space, so their velocities and accelerations are given by vectors. We also introduce new scalars that quantify the turning and twisting in the path of an object moving in space.

## 13.1 Curves in Space and Their Tangents

When a particle moves through space during a time interval $I$, we think of the particle's coordinates as functions defined on $I$:

$$x = f(t), \qquad y = g(t), \qquad z = h(t), \qquad t \in I. \tag{1}$$

The points $(x, y, z) = (f(t), g(t), h(t))$, $t \in I$, make up the **curve** in space that we call the particle's **path**. The equations and interval in Equation (1) parametrize the curve. 

A curve in space can also be represented in vector form. The vector

$$\mathbf{r}(t) = \overrightarrow{OP} = f(t)\mathbf{i} + g(t)\mathbf{j} + h(t)\mathbf{k} \tag{2}$$

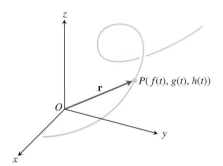

**FIGURE 13.1** The position vector $\mathbf{r} = \overrightarrow{OP}$ of a particle moving through space is a function of time.

from the origin to the particle's position $P(f(t), g(t), h(t))$ at time $t$ is the particle's position vector (Figure 13.1). The functions $f$, $g$, and $h$ are the **component functions** (components) of the position vector. We think of the particle's path as the curve traced by $\mathbf{r}$ during the time interval $I$. Figure 13.2 displays several space curves generated by a computer graphing program. It would not be easy to plot these curves by hand.

Equation (2) defines $\mathbf{r}$ as a vector function of the real variable $t$ on the interval $I$. More generally, a **vector-valued function** or **vector function** on a domain set $D$ is a rule that assigns a vector in space to each element in $D$. For now, the domains will be intervals of real numbers resulting in a space curve. Later, in Chapter 16, the domains will be regions in the plane. Vector functions will then represent surfaces in space. Vector functions on a domain in the plane or space also give rise to "vector fields," which are important to the study of the flow of a fluid, gravitational fields, and electromagnetic phenomena. We investigate vector fields and their applications in Chapter 16.

Real-valued functions are called **scalar functions** to distinguish them from vector functions. The components of $\mathbf{r}$ in Equation (2) are scalar functions of $t$. The domain of a vector-valued function is the common domain of its components.

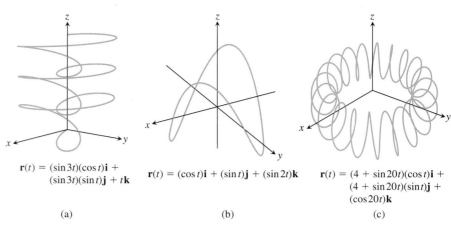

$$\mathbf{r}(t) = (\sin 3t)(\cos t)\mathbf{i} + (\sin 3t)(\sin t)\mathbf{j} + t\mathbf{k}$$

(a)

$$\mathbf{r}(t) = (\cos t)\mathbf{i} + (\sin t)\mathbf{j} + (\sin 2t)\mathbf{k}$$

(b)

$$\mathbf{r}(t) = (4 + \sin 20t)(\cos t)\mathbf{i} + (4 + \sin 20t)(\sin t)\mathbf{j} + (\cos 20t)\mathbf{k}$$

(c)

**FIGURE 13.2** Space curves are defined by the position vectors $\mathbf{r}(t)$.

**EXAMPLE 1** Graph the vector function

$$\mathbf{r}(t) = (\cos t)\mathbf{i} + (\sin t)\mathbf{j} + t\mathbf{k}.$$

**Solution** The vector function

$$\mathbf{r}(t) = (\cos t)\mathbf{i} + (\sin t)\mathbf{j} + t\mathbf{k}$$

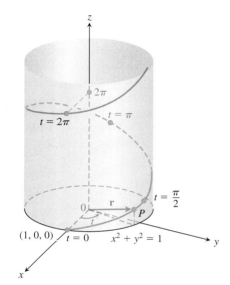

**FIGURE 13.3** The upper half of the helix $\mathbf{r}(t) = (\cos t)\mathbf{i} + (\sin t)\mathbf{j} + t\mathbf{k}$ (Example 1).

is defined for all real values of $t$. The curve traced by $\mathbf{r}$ winds around the circular cylinder $x^2 + y^2 = 1$ (Figure 13.3). The curve lies on the cylinder because the $\mathbf{i}$- and $\mathbf{j}$-components of $\mathbf{r}$, being the $x$- and $y$-coordinates of the tip of $\mathbf{r}$, satisfy the cylinder's equation:

$$x^2 + y^2 = (\cos t)^2 + (\sin t)^2 = 1.$$

The curve rises as the $\mathbf{k}$-component $z = t$ increases. Each time $t$ increases by $2\pi$, the curve completes one turn around the cylinder. The curve is called a **helix** (from an old Greek word for "spiral"). The equations

$$x = \cos t, \qquad y = \sin t, \qquad z = t$$

parametrize the helix, the interval $-\infty < t < \infty$ being understood. Figure 13.4 shows more helices. Note how constant multiples of the parameter $t$ can change the number of turns per unit of time.   ∎

$$\mathbf{r}(t) = (\cos t)\mathbf{i} + (\sin t)\mathbf{j} + t\mathbf{k}$$

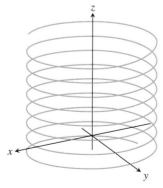

$$\mathbf{r}(t) = (\cos t)\mathbf{i} + (\sin t)\mathbf{j} + 0.3t\mathbf{k}$$

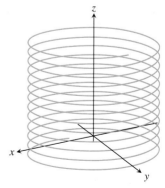

$$\mathbf{r}(t) = (\cos 5t)\mathbf{i} + (\sin 5t)\mathbf{j} + t\mathbf{k}$$

**FIGURE 13.4** Helices spiral upward around a cylinder, like coiled springs.

## Limits and Continuity

The way we define limits of vector-valued functions is similar to the way we define limits of real-valued functions.

> **DEFINITION**    Let $\mathbf{r}(t) = f(t)\mathbf{i} + g(t)\mathbf{j} + h(t)\mathbf{k}$ be a vector function with domain $D$, and $\mathbf{L}$ a vector. We say that $\mathbf{r}$ has **limit L** as $t$ approaches $t_0$ and write
>
> $$\lim_{t \to t_0} \mathbf{r}(t) = \mathbf{L}$$
>
> if, for every number $\epsilon > 0$, there exists a corresponding number $\delta > 0$ such that for all $t \in D$
>
> $$|\mathbf{r}(t) - \mathbf{L}| < \epsilon \quad \text{whenever} \quad 0 < |t - t_0| < \delta.$$

If $\mathbf{L} = L_1\mathbf{i} + L_2\mathbf{j} + L_3\mathbf{k}$, then it can be shown that $\lim_{t \to t_0} \mathbf{r}(t) = \mathbf{L}$ precisely when

$$\lim_{t \to t_0} f(t) = L_1, \qquad \lim_{t \to t_0} g(t) = L_2, \qquad \text{and} \qquad \lim_{t \to t_0} h(t) = L_3.$$

We omit the proof. The equation

To calculate the limit of a vector function, we find the limit of each component scalar function.

$$\lim_{t \to t_0} \mathbf{r}(t) = \left( \lim_{t \to t_0} f(t) \right)\mathbf{i} + \left( \lim_{t \to t_0} g(t) \right)\mathbf{j} + \left( \lim_{t \to t_0} h(t) \right)\mathbf{k} \qquad (3)$$

provides a practical way to calculate limits of vector functions.

**EXAMPLE 2**    If $\mathbf{r}(t) = (\cos t)\mathbf{i} + (\sin t)\mathbf{j} + t\mathbf{k}$, then

$$\lim_{t \to \pi/4} \mathbf{r}(t) = \left( \lim_{t \to \pi/4} \cos t \right)\mathbf{i} + \left( \lim_{t \to \pi/4} \sin t \right)\mathbf{j} + \left( \lim_{t \to \pi/4} t \right)\mathbf{k}$$

$$= \frac{\sqrt{2}}{2}\mathbf{i} + \frac{\sqrt{2}}{2}\mathbf{j} + \frac{\pi}{4}\mathbf{k}. \qquad \blacksquare$$

We define continuity for vector functions the same way we define continuity for scalar functions defined over an interval.

> **DEFINITION**    A vector function $\mathbf{r}(t)$ is **continuous at a point** $t = t_0$ in its domain if $\lim_{t \to t_0} \mathbf{r}(t) = \mathbf{r}(t_0)$. The function is **continuous** if it is continuous over its interval domain.

From Equation (3), we see that $\mathbf{r}(t)$ is continuous at $t = t_0$ if and only if each component function is continuous there (Exercise 31).

**EXAMPLE 3**

**(a)** All the space curves shown in Figures 13.2 and 13.4 are continuous because their component functions are continuous at every value of $t$ in $(-\infty, \infty)$.

**(b)** The function

$$\mathbf{g}(t) = (\cos t)\mathbf{i} + (\sin t)\mathbf{j} + \lfloor t \rfloor\mathbf{k}$$

is discontinuous at every integer, where the greatest integer function $\lfloor t \rfloor$ is discontinuous. $\blacksquare$

### Derivatives and Motion

Suppose that $\mathbf{r}(t) = f(t)\mathbf{i} + g(t)\mathbf{j} + h(t)\mathbf{k}$ is the position vector of a particle moving along a curve in space and that $f$, $g$, and $h$ are differentiable functions of $t$. Then the difference between the particle's positions at time $t$ and time $t + \Delta t$ is

$$\Delta\mathbf{r} = \mathbf{r}(t + \Delta t) - \mathbf{r}(t)$$

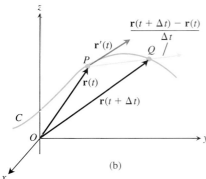

**FIGURE 13.5** As $\Delta t \to 0$, the point $Q$ approaches the point $P$ along the curve $C$. In the limit, the vector $\overrightarrow{PQ} / \Delta t$ becomes the tangent vector $\mathbf{r}'(t)$.

(Figure 13.5a). In terms of components, we find

$$
\begin{aligned}
\Delta \mathbf{r} &= \mathbf{r}(t + \Delta t) - \mathbf{r}(t) \\
&= [ f(t + \Delta t)\mathbf{i} + g(t + \Delta t)\mathbf{j} + h(t + \Delta t)\mathbf{k} ] \\
&\quad - [ f(t)\mathbf{i} + g(t)\mathbf{j} + h(t)\mathbf{k} ] \\
&= [ f(t + \Delta t) - f(t) ]\mathbf{i} + [ g(t + \Delta t) - g(t) ]\mathbf{j} + [ h(t + \Delta t) - h(t) ]\mathbf{k}.
\end{aligned}
$$

As $\Delta t$ approaches zero, three things seem to happen simultaneously. First, $Q$ approaches $P$ along the curve. Second, the secant line $PQ$ seems to approach a limiting position tangent to the curve at $P$. Third, the quotient $\Delta \mathbf{r}/\Delta t$ (Figure 13.5b) approaches the limit

$$
\begin{aligned}
\lim_{\Delta t \to 0} \frac{\Delta \mathbf{r}}{\Delta t} &= \left[ \lim_{\Delta t \to 0} \frac{f(t + \Delta t) - f(t)}{\Delta t} \right]\mathbf{i} + \left[ \lim_{\Delta t \to 0} \frac{g(t + \Delta t) - g(t)}{\Delta t} \right]\mathbf{j} \\
&\quad + \left[ \lim_{\Delta t \to 0} \frac{h(t + \Delta t) - h(t)}{\Delta t} \right]\mathbf{k} \\
&= \left[ \frac{df}{dt} \right]\mathbf{i} + \left[ \frac{dg}{dt} \right]\mathbf{j} + \left[ \frac{dh}{dt} \right]\mathbf{k}.
\end{aligned}
$$

These observations lead us to the following definition.

---

**DEFINITION** The vector function $\mathbf{r}(t) = f(t)\mathbf{i} + g(t)\mathbf{j} + h(t)\mathbf{k}$ has a **derivative (is differentiable) at $t$** if $f$, $g$, and $h$ have derivatives at $t$. The derivative is the vector function

$$
\mathbf{r}'(t) = \frac{d\mathbf{r}}{dt} = \lim_{\Delta t \to 0} \frac{\mathbf{r}(t + \Delta t) - \mathbf{r}(t)}{\Delta t} = \frac{df}{dt}\mathbf{i} + \frac{dg}{dt}\mathbf{j} + \frac{dh}{dt}\mathbf{k}.
$$

---

A vector function $\mathbf{r}$ is **differentiable** if it is differentiable at every point of its domain. The curve traced by $\mathbf{r}$ is **smooth** if $d\mathbf{r}/dt$ is continuous and never $\mathbf{0}$, that is, if $f$, $g$, and $h$ have continuous first derivatives that are not simultaneously 0.

The geometric significance of the definition of derivative is shown in Figure 13.5. The points $P$ and $Q$ have position vectors $\mathbf{r}(t)$ and $\mathbf{r}(t + \Delta t)$, and the vector $\overrightarrow{PQ}$ is represented by $\mathbf{r}(t + \Delta t) - \mathbf{r}(t)$. For $\Delta t > 0$, the scalar multiple $(1/\Delta t)(\mathbf{r}(t + \Delta t) - \mathbf{r}(t))$ points in the same direction as the vector $\overrightarrow{PQ}$. As $\Delta t \to 0$, this vector approaches a vector that is tangent to the curve at $P$ (Figure 13.5b). The vector $\mathbf{r}'(t)$, when different from $\mathbf{0}$, is defined to be the vector **tangent** to the curve at $P$. The **tangent line** to the curve at a point $(f(t_0), g(t_0), h(t_0))$ is defined to be the line through the point parallel to $\mathbf{r}'(t_0)$. We require $d\mathbf{r}/dt \neq \mathbf{0}$ for a smooth curve to make sure the curve has a continuously turning tangent at each point. On a smooth curve, there are no sharp corners or cusps.

A curve that is made up of a finite number of smooth curves pieced together in a continuous fashion is called **piecewise smooth** (Figure 13.6).

Look once again at Figure 13.5. We drew the figure for $\Delta t$ positive, so $\Delta \mathbf{r}$ points forward, in the direction of the motion. The vector $\Delta \mathbf{r}/\Delta t$, having the same direction as $\Delta \mathbf{r}$, points forward too. Had $\Delta t$ been negative, $\Delta \mathbf{r}$ would have pointed backward, against the direction of motion. The quotient $\Delta \mathbf{r}/\Delta t$, however, being a negative scalar multiple of $\Delta \mathbf{r}$, would once again have pointed forward. No matter how $\Delta \mathbf{r}$ points, $\Delta \mathbf{r}/\Delta t$ points forward and we expect the vector $d\mathbf{r}/dt = \lim_{\Delta t \to 0} \Delta \mathbf{r}/\Delta t$, when different from $\mathbf{0}$, to do the same. This means that the derivative $d\mathbf{r}/dt$, which is the rate of change of position with respect to time, always points in the direction of motion. For a smooth curve, $d\mathbf{r}/dt$ is never zero; the particle does not stop or reverse direction.

**FIGURE 13.6** A piecewise smooth curve made up of five smooth curves connected end to end in a continuous fashion. The curve here is not smooth at the points joining the five smooth curves.

> **DEFINITIONS** If **r** is the position vector of a particle moving along a smooth curve in space, then
>
> $$\mathbf{v}(t) = \frac{d\mathbf{r}}{dt}$$
>
> is the particle's **velocity vector**, tangent to the curve. At any time $t$, the direction of **v** is the **direction of motion**, the magnitude of **v** is the particle's **speed**, and the derivative $\mathbf{a} = d\mathbf{v}/dt$, when it exists, is the particle's **acceleration vector**. In summary,
>
> 1. Velocity is the derivative of position:    $\mathbf{v} = \dfrac{d\mathbf{r}}{dt}$.
>
> 2. Speed is the magnitude of velocity:    Speed $= |\mathbf{v}|$.
>
> 3. Acceleration is the derivative of velocity:    $\mathbf{a} = \dfrac{d\mathbf{v}}{dt} = \dfrac{d^2\mathbf{r}}{dt^2}$.
>
> 4. The unit vector $\mathbf{v}/|\mathbf{v}|$ is the direction of motion at time $t$.

**EXAMPLE 4**    Find the velocity, speed, and acceleration of a particle whose motion in space is given by the position vector $\mathbf{r}(t) = 2 \cos t\, \mathbf{i} + 2 \sin t\, \mathbf{j} + 5 \cos^2 t\, \mathbf{k}$. Sketch the velocity vector $\mathbf{v}(7\pi/4)$.

**Solution**    The velocity and acceleration vectors at time $t$ are

$$\mathbf{v}(t) = \mathbf{r}'(t) = -2 \sin t\, \mathbf{i} + 2 \cos t\, \mathbf{j} - 10 \cos t \sin t\, \mathbf{k}$$
$$= -2 \sin t\, \mathbf{i} + 2 \cos t\, \mathbf{j} - 5 \sin 2t\, \mathbf{k},$$

$$\mathbf{a}(t) = \mathbf{r}''(t) = -2 \cos t\, \mathbf{i} - 2 \sin t\, \mathbf{j} - 10 \cos 2t\, \mathbf{k},$$

and the speed is

$$|\mathbf{v}(t)| = \sqrt{(-2 \sin t)^2 + (2 \cos t)^2 + (-5 \sin 2t)^2} = \sqrt{4 + 25 \sin^2 2t}.$$

When $t = 7\pi/4$, we have

$$\mathbf{v}\left(\frac{7\pi}{4}\right) = \sqrt{2}\,\mathbf{i} + \sqrt{2}\,\mathbf{j} + 5\mathbf{k}, \qquad \mathbf{a}\left(\frac{7\pi}{4}\right) = -\sqrt{2}\,\mathbf{i} + \sqrt{2}\,\mathbf{j}, \qquad \left|\mathbf{v}\left(\frac{7\pi}{4}\right)\right| = \sqrt{29}.$$

A sketch of the curve of motion, and the velocity vector when $t = 7\pi/4$, can be seen in Figure 13.7.    ∎

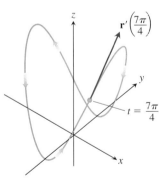

**FIGURE 13.7** The curve and the velocity vector when $t = 7\pi/4$ for the motion given in Example 4.

We can express the velocity of a moving particle as the product of its speed and direction:

$$\text{Velocity} = |\mathbf{v}|\left(\frac{\mathbf{v}}{|\mathbf{v}|}\right) = (\text{speed})(\text{direction}).$$

## Differentiation Rules

Because the derivatives of vector functions may be computed component by component, the rules for differentiating vector functions have the same form as the rules for differentiating scalar functions.

> **Differentiation Rules for Vector Functions**
>
> Let **u** and **v** be differentiable vector functions of $t$, **C** a constant vector, $c$ any scalar, and $f$ any differentiable scalar function.
>
> 1. *Constant Function Rule:* $\qquad \dfrac{d}{dt}\mathbf{C} = \mathbf{0}$
>
> 2. *Scalar Multiple Rules:* $\qquad \dfrac{d}{dt}\big[\,c\mathbf{u}(t)\,\big] = c\mathbf{u}'(t)$
>
> $\qquad\qquad\qquad\qquad\qquad \dfrac{d}{dt}\big[\,f(t)\mathbf{u}(t)\,\big] = f'(t)\mathbf{u}(t) + f(t)\mathbf{u}'(t)$
>
> 3. *Sum Rule:* $\qquad\qquad \dfrac{d}{dt}\big[\,\mathbf{u}(t) + \mathbf{v}(t)\,\big] = \mathbf{u}'(t) + \mathbf{v}'(t)$
>
> 4. *Difference Rule:* $\qquad \dfrac{d}{dt}\big[\,\mathbf{u}(t) - \mathbf{v}(t)\,\big] = \mathbf{u}'(t) - \mathbf{v}'(t)$
>
> 5. *Dot Product Rule:* $\qquad \dfrac{d}{dt}\big[\,\mathbf{u}(t) \cdot \mathbf{v}(t)\,\big] = \mathbf{u}'(t) \cdot \mathbf{v}(t) + \mathbf{u}(t) \cdot \mathbf{v}'(t)$
>
> 6. *Cross Product Rule:* $\qquad \dfrac{d}{dt}\big[\,\mathbf{u}(t) \times \mathbf{v}(t)\,\big] = \mathbf{u}'(t) \times \mathbf{v}(t) + \mathbf{u}(t) \times \mathbf{v}'(t)$
>
> 7. *Chain Rule:* $\qquad\qquad \dfrac{d}{dt}\big[\,\mathbf{u}(f(t))\,\big] = f'(t)\mathbf{u}'(f(t))$

When you use the Cross Product Rule, remember to preserve the order of the factors. If **u** comes first on the left side of the equation, it must also come first on the right or the signs will be wrong.

We will prove the product rules and Chain Rule but leave the rules for constants, scalar multiples, sums, and differences as exercises.

**Proof of the Dot Product Rule**     Suppose that

$$\mathbf{u} = u_1(t)\mathbf{i} + u_2(t)\mathbf{j} + u_3(t)\mathbf{k}$$

and

$$\mathbf{v} = v_1(t)\mathbf{i} + v_2(t)\mathbf{j} + v_3(t)\mathbf{k}.$$

Then

$$\frac{d}{dt}(\mathbf{u} \cdot \mathbf{v}) = \frac{d}{dt}(u_1 v_1 + u_2 v_2 + u_3 v_3)$$

$$= \underbrace{u_1' v_1 + u_2' v_2 + u_3' v_3}_{\mathbf{u}' \cdot \mathbf{v}} + \underbrace{u_1 v_1' + u_2 v_2' + u_3 v_3'}_{\mathbf{u} \cdot \mathbf{v}'}. \qquad\blacksquare$$

**Proof of the Cross Product Rule**     We model the proof after the proof of the Product Rule for scalar functions. According to the definition of derivative,

$$\frac{d}{dt}(\mathbf{u} \times \mathbf{v}) = \lim_{h \to 0} \frac{\mathbf{u}(t + h) \times \mathbf{v}(t + h) - \mathbf{u}(t) \times \mathbf{v}(t)}{h}.$$

To change this fraction into an equivalent one that contains the difference quotients for the derivatives of **u** and **v**, we subtract and add $\mathbf{u}(t) \times \mathbf{v}(t + h)$ in the numerator. Then

$$\frac{d}{dt}(\mathbf{u} \times \mathbf{v})$$

$$= \lim_{h \to 0} \frac{\mathbf{u}(t + h) \times \mathbf{v}(t + h) - \mathbf{u}(t) \times \mathbf{v}(t + h) + \mathbf{u}(t) \times \mathbf{v}(t + h) - \mathbf{u}(t) \times \mathbf{v}(t)}{h}$$

$$= \lim_{h \to 0} \left[ \frac{\mathbf{u}(t + h) - \mathbf{u}(t)}{h} \times \mathbf{v}(t + h) + \mathbf{u}(t) \times \frac{\mathbf{v}(t + h) - \mathbf{v}(t)}{h} \right]$$

$$= \lim_{h \to 0} \frac{\mathbf{u}(t + h) - \mathbf{u}(t)}{h} \times \lim_{h \to 0} \mathbf{v}(t + h) + \lim_{h \to 0} \mathbf{u}(t) \times \lim_{h \to 0} \frac{\mathbf{v}(t + h) - \mathbf{v}(t)}{h}.$$

The last of these equalities holds because the limit of the cross product of two vector functions is the cross product of their limits if the latter exist (Exercise 32). As $h$ approaches zero, $\mathbf{v}(t + h)$ approaches $\mathbf{v}(t)$ because $\mathbf{v}$, being differentiable at $t$, is continuous at $t$ (Exercise 33). The two fractions approach the values of $d\mathbf{u}/dt$ and $d\mathbf{v}/dt$ at $t$. In short,

$$\frac{d}{dt}(\mathbf{u} \times \mathbf{v}) = \frac{d\mathbf{u}}{dt} \times \mathbf{v} + \mathbf{u} \times \frac{d\mathbf{v}}{dt}. \qquad\blacksquare$$

**Proof of the Chain Rule**    Suppose that $\mathbf{u}(s) = a(s)\mathbf{i} + b(s)\mathbf{j} + c(s)\mathbf{k}$ is a differentiable vector function of $s$ and that $s = f(t)$ is a differentiable scalar function of $t$. Then $a$, $b$, and $c$ are differentiable functions of $t$, and the Chain Rule for differentiable real-valued functions gives

$$\begin{aligned}
\frac{d}{dt}\big[\mathbf{u}(s)\big] &= \frac{da}{dt}\mathbf{i} + \frac{db}{dt}\mathbf{j} + \frac{dc}{dt}\mathbf{k} \\
&= \frac{da}{ds}\frac{ds}{dt}\mathbf{i} + \frac{db}{ds}\frac{ds}{dt}\mathbf{j} + \frac{dc}{ds}\frac{ds}{dt}\mathbf{k} \\
&= \frac{ds}{dt}\left(\frac{da}{ds}\mathbf{i} + \frac{db}{ds}\mathbf{j} + \frac{dc}{ds}\mathbf{k}\right) \\
&= \frac{ds}{dt}\frac{d\mathbf{u}}{ds} \\
&= f'(t)\mathbf{u}'(f(t)). \qquad\qquad s = f(t) \qquad\blacksquare
\end{aligned}$$

As an algebraic convenience, we sometimes write the product of a scalar $c$ and a vector $\mathbf{v}$ as $\mathbf{v}c$ instead of $c\mathbf{v}$. This permits us, for instance, to write the Chain Rule in a familiar form:

$$\frac{d\mathbf{u}}{dt} = \frac{d\mathbf{u}}{ds}\frac{ds}{dt},$$

where $s = f(t)$.

### Vector Functions of Constant Length

When we track a particle moving on a sphere centered at the origin (Figure 13.8), the position vector has a constant length equal to the radius of the sphere. The velocity vector $d\mathbf{r}/dt$, tangent to the path of motion, is tangent to the sphere and hence perpendicular to $\mathbf{r}$. This is always the case for a differentiable vector function of constant length: The vector and its first derivative are orthogonal. By direct calculation,

$$\begin{aligned}
\mathbf{r}(t) \cdot \mathbf{r}(t) &= c^2 & |\mathbf{r}(t)| = c \text{ is constant.} \\
\frac{d}{dt}\big[\mathbf{r}(t) \cdot \mathbf{r}(t)\big] &= 0 & \text{Differentiate both sides.} \\
\mathbf{r}'(t) \cdot \mathbf{r}(t) + \mathbf{r}(t) \cdot \mathbf{r}'(t) &= 0 & \text{Rule 5 with } \mathbf{r}(t) = \mathbf{u}(t) = \mathbf{v}(t) \\
2\mathbf{r}'(t) \cdot \mathbf{r}(t) &= 0.
\end{aligned}$$

The vectors $\mathbf{r}'(t)$ and $\mathbf{r}(t)$ are orthogonal because their dot product is 0. In summary,

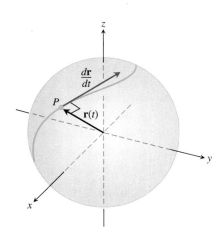

**FIGURE 13.8** If a particle moves on a sphere in such a way that its position $\mathbf{r}$ is a differentiable function of time, then $\mathbf{r} \cdot (d\mathbf{r}/dt) = 0$.

> If $\mathbf{r}$ is a differentiable vector function of $t$ of constant length, then
> $$\mathbf{r} \cdot \frac{d\mathbf{r}}{dt} = 0. \qquad (4)$$

We will use this observation repeatedly in Section 13.4. The converse is also true (see Exercise 27).

# Exercises 13.1

### Motion in the Plane

In Exercises 1–4, $\mathbf{r}(t)$ is the position of a particle in the $xy$-plane at time $t$. Find an equation in $x$ and $y$ whose graph is the path of the particle. Then find the particle's velocity and acceleration vectors at the given value of $t$.

**1.** $\mathbf{r}(t) = (t + 1)\mathbf{i} + (t^2 - 1)\mathbf{j}, \quad t = 1$

**2.** $\mathbf{r}(t) = \dfrac{t}{t + 1}\mathbf{i} + \dfrac{1}{t}\mathbf{j}, \quad t = -\dfrac{1}{2}$

**3.** $\mathbf{r}(t) = e^t\mathbf{i} + \dfrac{2}{9}e^{2t}\mathbf{j}, \quad t = \ln 3$

**4.** $\mathbf{r}(t) = (\cos 2t)\mathbf{i} + (3 \sin 2t)\mathbf{j}, \quad t = 0$

Exercises 5–8 give the position vectors of particles moving along various curves in the $xy$-plane. In each case, find the particle's velocity and acceleration vectors at the stated times and sketch them as vectors on the curve.

**5. Motion on the circle $x^2 + y^2 = 1$**

$$\mathbf{r}(t) = (\sin t)\mathbf{i} + (\cos t)\mathbf{j}; \quad t = \pi/4 \text{ and } \pi/2$$

**6. Motion on the circle $x^2 + y^2 = 16$**

$$\mathbf{r}(t) = \left(4\cos\frac{t}{2}\right)\mathbf{i} + \left(4\sin\frac{t}{2}\right)\mathbf{j}; \quad t = \pi \text{ and } 3\pi/2$$

**7. Motion on the cycloid $x = t - \sin t, \quad y = 1 - \cos t$**

$$\mathbf{r}(t) = (t - \sin t)\mathbf{i} + (1 - \cos t)\mathbf{j}; \quad t = \pi \text{ and } 3\pi/2$$

**8. Motion on the parabola $y = x^2 + 1$**

$$\mathbf{r}(t) = t\mathbf{i} + (t^2 + 1)\mathbf{j}; \quad t = -1, 0, \text{ and } 1$$

**Motion in Space**

In Exercises 9–14, $\mathbf{r}(t)$ is the position of a particle in space at time $t$. Find the particle's velocity and acceleration vectors. Then find the particle's speed and direction of motion at the given value of $t$. Write the particle's velocity at that time as the product of its speed and direction.

**9.** $\mathbf{r}(t) = (t + 1)\mathbf{i} + (t^2 - 1)\mathbf{j} + 2t\mathbf{k}, \quad t = 1$

**10.** $\mathbf{r}(t) = (1 + t)\mathbf{i} + \dfrac{t^2}{\sqrt{2}}\mathbf{j} + \dfrac{t^3}{3}\mathbf{k}, \quad t = 1$

**11.** $\mathbf{r}(t) = (2\cos t)\mathbf{i} + (3\sin t)\mathbf{j} + 4t\mathbf{k}, \quad t = \pi/2$

**12.** $\mathbf{r}(t) = (\sec t)\mathbf{i} + (\tan t)\mathbf{j} + \dfrac{4}{3}t\mathbf{k}, \quad t = \pi/6$

**13.** $\mathbf{r}(t) = (2\ln(t + 1))\mathbf{i} + t^2\mathbf{j} + \dfrac{t^2}{2}\mathbf{k}, \quad t = 1$

**14.** $\mathbf{r}(t) = (e^{-t})\mathbf{i} + (2\cos 3t)\mathbf{j} + (2\sin 3t)\mathbf{k}, \quad t = 0$

In Exercises 15–18, $\mathbf{r}(t)$ is the position of a particle in space at time $t$. Find the angle between the velocity and acceleration vectors at time $t = 0$.

**15.** $\mathbf{r}(t) = (3t + 1)\mathbf{i} + \sqrt{3}t\mathbf{j} + t^2\mathbf{k}$

**16.** $\mathbf{r}(t) = \left(\dfrac{\sqrt{2}}{2}t\right)\mathbf{i} + \left(\dfrac{\sqrt{2}}{2}t - 16t^2\right)\mathbf{j}$

**17.** $\mathbf{r}(t) = (\ln(t^2 + 1))\mathbf{i} + (\tan^{-1} t)\mathbf{j} + \sqrt{t^2 + 1}\,\mathbf{k}$

**18.** $\mathbf{r}(t) = \dfrac{4}{9}(1 + t)^{3/2}\mathbf{i} + \dfrac{4}{9}(1 - t)^{3/2}\mathbf{j} + \dfrac{1}{3}t\mathbf{k}$

**Tangents to Curves**

As mentioned in the text, the **tangent line** to a smooth curve $\mathbf{r}(t) = f(t)\mathbf{i} + g(t)\mathbf{j} + h(t)\mathbf{k}$ at $t = t_0$ is the line that passes through the point $(f(t_0), g(t_0), h(t_0))$ parallel to $\mathbf{v}(t_0)$, the curve's velocity vector at $t_0$. In Exercises 19–22, find parametric equations for the line that is tangent to the given curve at the given parameter value $t = t_0$.

**19.** $\mathbf{r}(t) = (\sin t)\mathbf{i} + (t^2 - \cos t)\mathbf{j} + e^t\mathbf{k}, \quad t_0 = 0$

**20.** $\mathbf{r}(t) = t^2\mathbf{i} + (2t - 1)\mathbf{j} + t^3\mathbf{k}, \quad t_0 = 2$

**21.** $\mathbf{r}(t) = \ln t\,\mathbf{i} + \dfrac{t - 1}{t + 2}\mathbf{j} + t\ln t\,\mathbf{k}, \quad t_0 = 1$

**22.** $\mathbf{r}(t) = (\cos t)\mathbf{i} + (\sin t)\mathbf{j} + (\sin 2t)\mathbf{k}, \quad t_0 = \dfrac{\pi}{2}$

**Theory and Examples**

**23. Motion along a circle** Each of the following equations in parts (a)–(e) describes the motion of a particle having the same path, namely the unit circle $x^2 + y^2 = 1$. Although the path of each particle in parts (a)–(e) is the same, the behavior, or "dynamics," of each particle is different. For each particle, answer the following questions.

   i) Does the particle have constant speed? If so, what is its constant speed?

   ii) Is the particle's acceleration vector always orthogonal to its velocity vector?

   iii) Does the particle move clockwise or counterclockwise around the circle?

   iv) Does the particle begin at the point $(1, 0)$?

   **a.** $\mathbf{r}(t) = (\cos t)\mathbf{i} + (\sin t)\mathbf{j}, \quad t \geq 0$

   **b.** $\mathbf{r}(t) = \cos(2t)\mathbf{i} + \sin(2t)\mathbf{j}, \quad t \geq 0$

   **c.** $\mathbf{r}(t) = \cos(t - \pi/2)\mathbf{i} + \sin(t - \pi/2)\mathbf{j}, \quad t \geq 0$

   **d.** $\mathbf{r}(t) = (\cos t)\mathbf{i} - (\sin t)\mathbf{j}, \quad t \geq 0$

   **e.** $\mathbf{r}(t) = \cos(t^2)\mathbf{i} + \sin(t^2)\mathbf{j}, \quad t \geq 0$

**24. Motion along a circle** Show that the vector-valued function

$$\mathbf{r}(t) = (2\mathbf{i} + 2\mathbf{j} + \mathbf{k})$$

$$+ \cos t\left(\frac{1}{\sqrt{2}}\mathbf{i} - \frac{1}{\sqrt{2}}\mathbf{j}\right) + \sin t\left(\frac{1}{\sqrt{3}}\mathbf{i} + \frac{1}{\sqrt{3}}\mathbf{j} + \frac{1}{\sqrt{3}}\mathbf{k}\right)$$

describes the motion of a particle moving in the circle of radius 1 centered at the point $(2, 2, 1)$ and lying in the plane $x + y - 2z = 2$.

**25. Motion along a parabola** A particle moves along the top of the parabola $y^2 = 2x$ from left to right at a constant speed of 5 units per second. Find the velocity of the particle as it moves through the point $(2, 2)$.

**26. Motion along a cycloid** A particle moves in the $xy$-plane in such a way that its position at time $t$ is

$$\mathbf{r}(t) = (t - \sin t)\mathbf{i} + (1 - \cos t)\mathbf{j}.$$

   **T**   **a.** Graph $\mathbf{r}(t)$. The resulting curve is a cycloid.

   **b.** Find the maximum and minimum values of $|\mathbf{v}|$ and $|\mathbf{a}|$. (*Hint:* Find the extreme values of $|\mathbf{v}|^2$ and $|\mathbf{a}|^2$ first and take square roots later.)

**27.** Let $\mathbf{r}$ be a differentiable vector function of $t$. Show that if $\mathbf{r} \cdot (d\mathbf{r}/dt) = 0$ for all $t$, then $|\mathbf{r}|$ is constant.

**28. Derivatives of triple scalar products**

   **a.** Show that if $\mathbf{u}$, $\mathbf{v}$, and $\mathbf{w}$ are differentiable vector functions of $t$, then

$$\frac{d}{dt}(\mathbf{u} \cdot \mathbf{v} \times \mathbf{w}) = \frac{d\mathbf{u}}{dt} \cdot \mathbf{v} \times \mathbf{w} + \mathbf{u} \cdot \frac{d\mathbf{v}}{dt} \times \mathbf{w} + \mathbf{u} \cdot \mathbf{v} \times \frac{d\mathbf{w}}{dt}.$$

   **b.** Show that

$$\frac{d}{dt}\left(\mathbf{r} \cdot \frac{d\mathbf{r}}{dt} \times \frac{d^2\mathbf{r}}{dt^2}\right) = \mathbf{r} \cdot \left(\frac{d\mathbf{r}}{dt} \times \frac{d^3\mathbf{r}}{dt^3}\right).$$

(*Hint:* Differentiate on the left and look for vectors whose products are zero.)

**29.** Prove the two Scalar Multiple Rules for vector functions.

**30.** Prove the Sum and Difference Rules for vector functions.

**31. Component test for continuity at a point**  Show that the vector function $\mathbf{r}$ defined by $\mathbf{r}(t) = f(t)\mathbf{i} + g(t)\mathbf{j} + h(t)\mathbf{k}$ is continuous at $t = t_0$ if and only if $f$, $g$, and $h$ are continuous at $t_0$.

**32. Limits of cross products of vector functions**  Suppose that $\mathbf{r}_1(t) = f_1(t)\mathbf{i} + f_2(t)\mathbf{j} + f_3(t)\mathbf{k}$, $\mathbf{r}_2(t) = g_1(t)\mathbf{i} + g_2(t)\mathbf{j} + g_3(t)\mathbf{k}$, $\lim_{t \to t_0} \mathbf{r}_1(t) = \mathbf{A}$, and $\lim_{t \to t_0} \mathbf{r}_2(t) = \mathbf{B}$. Use the determinant formula for cross products and the Limit Product Rule for scalar functions to show that

$$\lim_{t \to t_0} (\mathbf{r}_1(t) \times \mathbf{r}_2(t)) = \mathbf{A} \times \mathbf{B}.$$

**33. Differentiable vector functions are continuous**  Show that if $\mathbf{r}(t) = f(t)\mathbf{i} + g(t)\mathbf{j} + h(t)\mathbf{k}$ is differentiable at $t = t_0$, then it is continuous at $t_0$ as well.

**34. Constant Function Rule**  Prove that if $\mathbf{u}$ is the vector function with the constant value $\mathbf{C}$, then $d\mathbf{u}/dt = \mathbf{0}$.

**COMPUTER EXPLORATIONS**

Use a CAS to perform the following steps in Exercises 35–38.

**a.** Plot the space curve traced out by the position vector $\mathbf{r}$.

**b.** Find the components of the velocity vector $d\mathbf{r}/dt$.

**c.** Evaluate $d\mathbf{r}/dt$ at the given point $t_0$ and determine the equation of the tangent line to the curve at $\mathbf{r}(t_0)$.

**d.** Plot the tangent line together with the curve over the given interval.

**35.** $\mathbf{r}(t) = (\sin t - t \cos t)\mathbf{i} + (\cos t + t \sin t)\mathbf{j} + t^2\mathbf{k}$, $0 \le t \le 6\pi$, $t_0 = 3\pi/2$

**36.** $\mathbf{r}(t) = \sqrt{2}t\mathbf{i} + e^t\mathbf{j} + e^{-t}\mathbf{k}$, $-2 \le t \le 3$, $t_0 = 1$

**37.** $\mathbf{r}(t) = (\sin 2t)\mathbf{i} + (\ln(1 + t))\mathbf{j} + t\mathbf{k}$, $0 \le t \le 4\pi$, $t_0 = \pi/4$

**38.** $\mathbf{r}(t) = (\ln(t^2 + 2))\mathbf{i} + (\tan^{-1} 3t)\mathbf{j} + \sqrt{t^2 + 1}\ \mathbf{k}$, $-3 \le t \le 5$, $t_0 = 3$

In Exercises 39 and 40, you will explore graphically the behavior of the helix

$$\mathbf{r}(t) = (\cos at)\mathbf{i} + (\sin at)\mathbf{j} + bt\mathbf{k}$$

as you change the values of the constants $a$ and $b$. Use a CAS to perform the steps in each exercise.

**39.** Set $b = 1$. Plot the helix $\mathbf{r}(t)$ together with the tangent line to the curve at $t = 3\pi/2$ for $a = 1$, 2, 4, and 6 over the interval $0 \le t \le 4\pi$. Describe in your own words what happens to the graph of the helix and the position of the tangent line as $a$ increases through these positive values.

**40.** Set $a = 1$. Plot the helix $\mathbf{r}(t)$ together with the tangent line to the curve at $t = 3\pi/2$ for $b = 1/4$, 1/2, 2, and 4 over the interval $0 \le t \le 4\pi$. Describe in your own words what happens to the graph of the helix and the position of the tangent line as $b$ increases through these positive values.

# 13.2 Integrals of Vector Functions; Projectile Motion

In this section we investigate integrals of vector functions and their application to motion along a path in space or in the plane.

## Integrals of Vector Functions

A differentiable vector function $\mathbf{R}(t)$ is an **antiderivative** of a vector function $\mathbf{r}(t)$ on an interval $I$ if $d\mathbf{R}/dt = \mathbf{r}$ at each point of $I$. If $\mathbf{R}$ is an antiderivative of $\mathbf{r}$ on $I$, it can be shown, working one component at a time, that every antiderivative of $\mathbf{r}$ on $I$ has the form $\mathbf{R} + \mathbf{C}$ for some constant vector $\mathbf{C}$ (Exercise 41). The set of all antiderivatives of $\mathbf{r}$ on $I$ is the **indefinite integral** of $\mathbf{r}$ on $I$.

---

**DEFINITION**  The **indefinite integral** of $\mathbf{r}$ with respect to $t$ is the set of all antiderivatives of $\mathbf{r}$, denoted by $\int \mathbf{r}(t)\, dt$. If $\mathbf{R}$ is any antiderivative of $\mathbf{r}$, then

$$\int \mathbf{r}(t)\, dt = \mathbf{R}(t) + \mathbf{C}.$$

---

The usual arithmetic rules for indefinite integrals apply.

**EXAMPLE 1**    To integrate a vector function, we integrate each of its components.

$$\int ((\cos t)\mathbf{i} + \mathbf{j} - 2t\mathbf{k})\, dt = \left(\int \cos t\, dt\right)\mathbf{i} + \left(\int dt\right)\mathbf{j} - \left(\int 2t\, dt\right)\mathbf{k} \qquad (1)$$

$$= (\sin t + C_1)\mathbf{i} + (t + C_2)\mathbf{j} - (t^2 + C_3)\mathbf{k} \qquad (2)$$

$$= (\sin t)\mathbf{i} + t\mathbf{j} - t^2\mathbf{k} + \mathbf{C} \qquad {\scriptstyle C = C_1\mathbf{i} + C_2\mathbf{j} - C_3\mathbf{k}}$$

As in the integration of scalar functions, we recommend that you skip the steps in Equations (1) and (2) and go directly to the final form. Find an antiderivative for each component and add a *constant vector* at the end. ∎

Definite integrals of vector functions are best defined in terms of components. The definition is consistent with how we compute limits and derivatives of vector functions.

---

**DEFINITION**    If the components of $\mathbf{r}(t) = f(t)\mathbf{i} + g(t)\mathbf{j} + h(t)\mathbf{k}$ are integrable over $[a, b]$, then so is $\mathbf{r}$, and the **definite integral** of $\mathbf{r}$ from $a$ to $b$ is

$$\int_a^b \mathbf{r}(t)\, dt = \left(\int_a^b f(t)\, dt\right)\mathbf{i} + \left(\int_a^b g(t)\, dt\right)\mathbf{j} + \left(\int_a^b h(t)\, dt\right)\mathbf{k}.$$

---

**EXAMPLE 2**    As in Example 1, we integrate each component.

$$\int_0^\pi ((\cos t)\mathbf{i} + \mathbf{j} - 2t\mathbf{k})\, dt = \left(\int_0^\pi \cos t\, dt\right)\mathbf{i} + \left(\int_0^\pi dt\right)\mathbf{j} - \left(\int_0^\pi 2t\, dt\right)\mathbf{k}$$

$$= \left[\sin t\right]_0^\pi \mathbf{i} + \left[t\right]_0^\pi \mathbf{j} - \left[t^2\right]_0^\pi \mathbf{k}$$

$$= [0 - 0]\mathbf{i} + [\pi - 0]\mathbf{j} - [\pi^2 - 0^2]\mathbf{k}$$

$$= \pi\mathbf{j} - \pi^2\mathbf{k} \qquad ∎$$

The Fundamental Theorem of Calculus for continuous vector functions says that

$$\int_a^b \mathbf{r}(t)\, dt = \mathbf{R}(t)\Big]_a^b = \mathbf{R}(b) - \mathbf{R}(a)$$

where $\mathbf{R}$ is any antiderivative of $\mathbf{r}$, so that $\mathbf{R}'(t) = \mathbf{r}(t)$ (Exercise 42). Notice that an antiderivative of a vector function is also a vector function, whereas a definite integral of a vector function is a single constant vector.

**EXAMPLE 3**    Suppose we do not know the path of a hang glider, but only its acceleration vector $\mathbf{a}(t) = -(3\cos t)\mathbf{i} - (3\sin t)\mathbf{j} + 2\mathbf{k}$. We also know that initially (at time $t = 0$) the glider departed from the point $(4, 0, 0)$ with velocity $\mathbf{v}(0) = 3\mathbf{j}$. Find the glider's position as a function of $t$.

**Solution**    Our goal is to find $\mathbf{r}(t)$ knowing

The differential equation:    $\mathbf{a} = \dfrac{d^2\mathbf{r}}{dt^2} = -(3\cos t)\mathbf{i} - (3\sin t)\mathbf{j} + 2\mathbf{k}$

The initial conditions:    $\mathbf{v}(0) = 3\mathbf{j}$   and   $\mathbf{r}(0) = 4\mathbf{i} + 0\mathbf{j} + 0\mathbf{k}$.

Integrating both sides of the differential equation with respect to $t$ gives

$$\mathbf{v}(t) = -(3\sin t)\mathbf{i} + (3\cos t)\mathbf{j} + 2t\mathbf{k} + \mathbf{C}_1.$$

We use $\mathbf{v}(0) = 3\mathbf{j}$ to find $\mathbf{C}_1$:

$$3\mathbf{j} = -(3\sin 0)\mathbf{i} + (3\cos 0)\mathbf{j} + (0)\mathbf{k} + \mathbf{C}_1$$

$$3\mathbf{j} = 3\mathbf{j} + \mathbf{C}_1$$

$$\mathbf{C}_1 = \mathbf{0}.$$

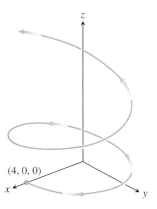

FIGURE 13.9 The path of the hang glider in Example 3. Although the path spirals around the *z*-axis, it is not a helix.

The glider's velocity as a function of time is

$$\frac{d\mathbf{r}}{dt} = \mathbf{v}(t) = -(3 \sin t)\mathbf{i} + (3 \cos t)\mathbf{j} + 2t\mathbf{k}.$$

Integrating both sides of this last differential equation gives

$$\mathbf{r}(t) = (3 \cos t)\mathbf{i} + (3 \sin t)\mathbf{j} + t^2\mathbf{k} + \mathbf{C}_2.$$

We then use the initial condition $\mathbf{r}(0) = 4\mathbf{i}$ to find $\mathbf{C}_2$:

$$4\mathbf{i} = (3 \cos 0)\mathbf{i} + (3 \sin 0)\mathbf{j} + (0^2)\mathbf{k} + \mathbf{C}_2$$
$$4\mathbf{i} = 3\mathbf{i} + (0)\mathbf{j} + (0)\mathbf{k} + \mathbf{C}_2$$
$$\mathbf{C}_2 = \mathbf{i}.$$

The glider's position as a function of $t$ is

$$\mathbf{r}(t) = (1 + 3 \cos t)\mathbf{i} + (3 \sin t)\mathbf{j} + t^2\mathbf{k}.$$

This is the path of the glider shown in Figure 13.9. Although the path resembles that of a helix due to its spiraling nature around the *z*-axis, it is not a helix because of the way it is rising. (We say more about this in Section 13.5.) ∎

## The Vector and Parametric Equations for Ideal Projectile Motion

A classic example of integrating vector functions is the derivation of the equations for the motion of a projectile. In physics, projectile motion describes how an object fired at some angle from an initial position, and acted upon by only the force of gravity, moves in a vertical coordinate plane. In the classic example, we ignore the effects of any frictional drag on the object, which may vary with its speed and altitude, and also the fact that the force of gravity changes slightly with the projectile's changing height. In addition, we ignore the long-distance effects of Earth turning beneath the projectile, such as in a rocket launch or the firing of a projectile from a cannon. Ignoring these effects gives us a reasonable approximation of the motion in most cases.

To derive equations for projectile motion, we assume that the projectile behaves like a particle moving in a vertical coordinate plane and that the only force acting on the projectile during its flight is the constant force of gravity, which always points straight down. We assume that the projectile is launched from the origin at time $t = 0$ into the first quadrant with an initial velocity $\mathbf{v}_0$ (Figure 13.10). If $\mathbf{v}_0$ makes an angle $\alpha$ with the horizontal, then

$$\mathbf{v}_0 = (|\mathbf{v}_0| \cos \alpha)\mathbf{i} + (|\mathbf{v}_0| \sin \alpha)\mathbf{j}.$$

If we use the simpler notation $v_0$ for the initial speed $|\mathbf{v}_0|$, then

$$\mathbf{v}_0 = (v_0 \cos \alpha)\mathbf{i} + (v_0 \sin \alpha)\mathbf{j}. \tag{3}$$

The projectile's initial position is

$$\mathbf{r}_0 = 0\mathbf{i} + 0\mathbf{j} = \mathbf{0}. \tag{4}$$

Newton's second law of motion says that the force acting on the projectile is equal to the projectile's mass $m$ times its acceleration, or $m(d^2\mathbf{r}/dt^2)$ if $\mathbf{r}$ is the projectile's position vector and $t$ is time. If the force is solely the gravitational force $-mg\mathbf{j}$, then

$$m\frac{d^2\mathbf{r}}{dt^2} = -mg\mathbf{j} \quad \text{and} \quad \frac{d^2\mathbf{r}}{dt^2} = -g\mathbf{j},$$

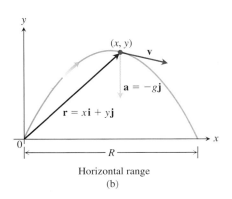

FIGURE 13.10 (a) Position, velocity, acceleration, and launch angle at $t = 0$. (b) Position, velocity, and acceleration at a later time $t$.

where $g$ is the acceleration due to gravity. We find $\mathbf{r}$ as a function of $t$ by solving the following initial value problem.

Differential equation: $\quad \dfrac{d^2\mathbf{r}}{dt^2} = -g\mathbf{j}$

Initial conditions: $\quad \mathbf{r} = \mathbf{r}_0 \quad$ and $\quad \dfrac{d\mathbf{r}}{dt} = \mathbf{v}_0 \quad$ when $t = 0$

The first integration gives

$$\dfrac{d\mathbf{r}}{dt} = -(gt)\mathbf{j} + \mathbf{v}_0.$$

A second integration gives

$$\mathbf{r} = -\dfrac{1}{2}gt^2\mathbf{j} + \mathbf{v}_0 t + \mathbf{r}_0.$$

Substituting the values of $\mathbf{v}_0$ and $\mathbf{r}_0$ from Equations (3) and (4) gives

$$\mathbf{r} = -\dfrac{1}{2}gt^2\mathbf{j} + \underbrace{(v_0\cos\alpha)t\mathbf{i} + (v_0\sin\alpha)t\mathbf{j}}_{\mathbf{v}_0 t} + \mathbf{0}.$$

Collecting terms, we have

---

**Ideal Projectile Motion Equation**

$$\mathbf{r} = (v_0\cos\alpha)t\mathbf{i} + \left((v_0\sin\alpha)t - \dfrac{1}{2}gt^2\right)\mathbf{j}. \qquad (5)$$

---

Equation (5) is the *vector equation* of the path for ideal projectile motion. The angle $\alpha$ is the projectile's **launch angle** (**firing angle, angle of elevation**), and $v_0$, as we said before, is the projectile's **initial speed**. The components of $\mathbf{r}$ give the parametric equations

$$x = (v_0\cos\alpha)t \qquad \text{and} \qquad y = (v_0\sin\alpha)t - \dfrac{1}{2}gt^2, \qquad (6)$$

where $x$ is the distance downrange and $y$ is the height of the projectile at time $t \geq 0$.

**EXAMPLE 4** A projectile is fired from the origin over horizontal ground at an initial speed of 500 m/s and a launch angle of 60°. Where will the projectile be 10 s later?

**Solution** We use Equation (5) with $v_0 = 500$, $\alpha = 60°$, $g = 9.8$, and $t = 10$ to find the projectile's components 10 s after firing.

$$\mathbf{r} = (v_0\cos\alpha)t\mathbf{i} + \left((v_0\sin\alpha)t - \dfrac{1}{2}gt^2\right)\mathbf{j}$$

$$= (500)\left(\dfrac{1}{2}\right)(10)\mathbf{i} + \left((500)\left(\dfrac{\sqrt{3}}{2}\right)10 - \left(\dfrac{1}{2}\right)(9.8)(100)\right)\mathbf{j}$$

$$\approx 2500\mathbf{i} + 3840\mathbf{j}$$

Ten seconds after firing, the projectile is about 3840 m above ground and 2500 m downrange from the origin.

Ideal projectiles move along parabolas, as we now deduce from Equations (6). If we substitute $t = x/(v_0 \cos \alpha)$ from the first equation into the second, we obtain the Cartesian coordinate equation

$$y = -\left(\frac{g}{2v_0{}^2 \cos^2 \alpha}\right)x^2 + (\tan \alpha)x.$$

This equation has the form $y = ax^2 + bx$, so its graph is a parabola.

A projectile reaches its highest point when its vertical velocity component is zero. When fired over horizontal ground, the projectile lands when its vertical component equals zero in Equation (5), and the **range** $R$ is the distance from the origin to the point of impact. We summarize the results here, which you are asked to verify in Exercise 27.

---

**Height, Flight Time, and Range for Ideal Projectile Motion**

For ideal projectile motion when an object is launched from the origin over a horizontal surface with initial speed $v_0$ and launch angle $\alpha$:

$$\text{Maximum height:} \qquad y_{\max} = \frac{(v_0 \sin \alpha)^2}{2g}$$

$$\text{Flight time:} \qquad t = \frac{2v_0 \sin \alpha}{g}$$

$$\text{Range:} \qquad R = \frac{v_0{}^2}{g} \sin 2\alpha.$$

---

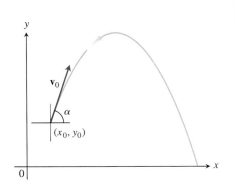

**FIGURE 13.11** The path of a projectile fired from $(x_0, y_0)$ with an initial velocity $\mathbf{v}_0$ at an angle of $\alpha$ degrees with the horizontal.

If we fire our ideal projectile from the point $(x_0, y_0)$ instead of the origin (Figure 13.11), the position vector for the path of motion is

$$\mathbf{r} = (x_0 + (v_0 \cos \alpha)t)\mathbf{i} + \left(y_0 + (v_0 \sin \alpha)t - \frac{1}{2}gt^2\right)\mathbf{j}, \qquad (7)$$

as you are asked to show in Exercise 29.

## Projectile Motion with Wind Gusts

The next example shows how to account for another force acting on a projectile, due to a gust of wind. We also assume that the path of the baseball in Example 5 lies in a vertical plane.

**EXAMPLE 5**     A baseball is hit when it is 1 m above the ground. It leaves the bat with initial speed of 50 m/s, making an angle of 20° with the horizontal. At the instant the ball is hit, an instantaneous gust of wind blows in the horizontal direction directly opposite the direction the ball is taking toward the outfield, adding a component of $-2.5\mathbf{i}$ (m/s) to the ball's initial velocity (2.5 m/s = 9 km/h).

**(a)** Find a vector equation (position vector) for the path of the baseball.

**(b)** How high does the baseball go, and when does it reach maximum height?

**(c)** Assuming that the ball is not caught, find its range and flight time.

**Solution**

**(a)** Using Equation (3) and accounting for the gust of wind, the initial velocity of the baseball is

$$\mathbf{v}_0 = (v_0 \cos \alpha)\mathbf{i} + (v_0 \sin \alpha)\mathbf{j} - 2.5\mathbf{i}$$
$$= (50 \cos 20°)\mathbf{i} + (50 \sin 20°)\mathbf{j} - (2.5)\mathbf{i}$$
$$= (50 \cos 20° - 2.5)\mathbf{i} + (50 \sin 20°)\mathbf{j}.$$

The initial position is $\mathbf{r}_0 = 0\mathbf{i} + 1\mathbf{j}$. Integration of $d^2\mathbf{r}/dt^2 = -g\mathbf{j}$ gives

$$\frac{d\mathbf{r}}{dt} = -(gt)\mathbf{j} + \mathbf{v}_0.$$

A second integration gives

$$\mathbf{r} = -\frac{1}{2}gt^2\mathbf{j} + \mathbf{v}_0 t + \mathbf{r}_0.$$

Substituting the values of $\mathbf{v}_0$ and $\mathbf{r}_0$ into the last equation gives the position vector of the baseball.

$$\mathbf{r} = -\frac{1}{2}gt^2\mathbf{j} + \mathbf{v}_0 t + \mathbf{r}_0$$
$$= -4.9t^2\mathbf{j} + (50 \cos 20° - 2.5)t\mathbf{i} + (50 \sin 20°)t\mathbf{j} + 1\mathbf{j}$$
$$= (50 \cos 20° - 2.5)t\mathbf{i} + \left(1 + (50 \sin 20°)t - 4.9t^2\right)\mathbf{j}.$$

**(b)** The baseball reaches its highest point when the vertical component of velocity is zero, or

$$\frac{dy}{dt} = 50 \sin 20° - 9.8t = 0.$$

Solving for $t$ we find

$$t = \frac{50 \sin 20°}{9.8} \approx 1.75 \text{ s}.$$

Substituting this time into the vertical component for $\mathbf{r}$ gives the maximum height

$$y_{max} = 1 + (50 \sin 20°)(1.75) - 4.9(1.75)^2$$
$$\approx 15.9 \text{ m}.$$

That is, the maximum height of the baseball is about 15.9 m, reached about 1.75 s after leaving the bat.

**(c)** To find when the baseball lands, we set the vertical component for $\mathbf{r}$ equal to 0 and solve for $t$:

$$1 + (50 \sin 20°)t - 4.9t^2 = 0$$
$$1 + (17.1)t - 4.9t^2 = 0.$$

The solution values are about $t = 3.55$ s and $t = -0.06$ s. Substituting the positive time into the horizontal component for $\mathbf{r}$, we find the range

$$R = (50 \cos 20° - 2.5)(3.55)$$
$$\approx 157.8 \text{ m}.$$

Thus, the horizontal range is about 157.8 m, and the flight time is about 3.55 s.    ■

In Exercises 37 and 38, we consider projectile motion when there is air resistance slowing down the flight.

# Exercises 13.2

## Integrating Vector-Valued Functions

Evaluate the integrals in Exercises 1–10.

**1.** $\int_0^1 \left[ t^3 \mathbf{i} + 7\mathbf{j} + (t + 1)\mathbf{k} \right] dt$

**2.** $\int_1^2 \left[ (6 - 6t)\mathbf{i} + 3\sqrt{t}\mathbf{j} + \left( \frac{4}{t^2} \right)\mathbf{k} \right] dt$

**3.** $\int_{-\pi/4}^{\pi/4} \left[ (\sin t)\mathbf{i} + (1 + \cos t)\mathbf{j} + (\sec^2 t)\mathbf{k} \right] dt$

**4.** $\int_0^{\pi/3} \left[ (\sec t \tan t)\mathbf{i} + (\tan t)\mathbf{j} + (2 \sin t \cos t)\mathbf{k} \right] dt$

**5.** $\int_1^4 \left[ \frac{1}{t}\mathbf{i} + \frac{1}{5 - t}\mathbf{j} + \frac{1}{2t}\mathbf{k} \right] dt$

**6.** $\int_0^1 \left[ \frac{2}{\sqrt{1 - t^2}}\mathbf{i} + \frac{\sqrt{3}}{1 + t^2}\mathbf{k} \right] dt$

**7.** $\int_0^1 \left[ te^{t^2} \mathbf{i} + e^{-t}\mathbf{j} + \mathbf{k} \right] dt$

**8.** $\int_1^{\ln 3} \left[ te^t \mathbf{i} + e^t \mathbf{j} + \ln t \, \mathbf{k} \right] dt$

**9.** $\int_0^{\pi/2} \left[ \cos t \, \mathbf{i} - \sin 2t \, \mathbf{j} + \sin^2 t \, \mathbf{k} \right] dt$

**10.** $\int_0^{\pi/4} \left[ \sec t \, \mathbf{i} + \tan^2 t \, \mathbf{j} - t \sin t \, \mathbf{k} \right] dt$

## Initial Value Problems

Solve the initial value problems in Exercises 11–16 for **r** as a vector function of $t$.

**11.** Differential equation: $\quad \dfrac{d\mathbf{r}}{dt} = -t\mathbf{i} - t\mathbf{j} - t\mathbf{k}$

Initial condition: $\quad \mathbf{r}(0) = \mathbf{i} + 2\mathbf{j} + 3\mathbf{k}$

**12.** Differential equation: $\quad \dfrac{d\mathbf{r}}{dt} = (180t)\mathbf{i} + (180t - 16t^2)\mathbf{j}$

Initial condition: $\quad \mathbf{r}(0) = 100\mathbf{j}$

**13.** Differential equation: $\quad \dfrac{d\mathbf{r}}{dt} = \dfrac{3}{2}(t + 1)^{1/2}\mathbf{i} + e^{-t}\mathbf{j} + \dfrac{1}{t + 1}\mathbf{k}$

Initial condition: $\quad \mathbf{r}(0) = \mathbf{k}$

**14.** Differential equation: $\quad \dfrac{d\mathbf{r}}{dt} = (t^3 + 4t)\mathbf{i} + t\mathbf{j} + 2t^2\mathbf{k}$

Initial condition: $\quad \mathbf{r}(0) = \mathbf{i} + \mathbf{j}$

**15.** Differential equation: $\quad \dfrac{d^2\mathbf{r}}{dt^2} = -32\mathbf{k}$

Initial conditions: $\quad \mathbf{r}(0) = 100\mathbf{k}$ and

$\left. \dfrac{d\mathbf{r}}{dt} \right|_{t=0} = 8\mathbf{i} + 8\mathbf{j}$

**16.** Differential equation: $\quad \dfrac{d^2\mathbf{r}}{dt^2} = -(\mathbf{i} + \mathbf{j} + \mathbf{k})$

Initial conditions: $\quad \mathbf{r}(0) = 10\mathbf{i} + 10\mathbf{j} + 10\mathbf{k}$ and

$\left. \dfrac{d\mathbf{r}}{dt} \right|_{t=0} = \mathbf{0}$

## Motion Along a Straight Line

**17.** At time $t = 0$, a particle is located at the point $(1, 2, 3)$. It travels in a straight line to the point $(4, 1, 4)$, has speed 2 at $(1, 2, 3)$ and constant acceleration $3\mathbf{i} - \mathbf{j} + \mathbf{k}$. Find an equation for the position vector $\mathbf{r}(t)$ of the particle at time $t$.

**18.** A particle traveling in a straight line is located at the point $(1, -1, 2)$ and has speed 2 at time $t = 0$. The particle moves toward the point $(3, 0, 3)$ with constant acceleration $2\mathbf{i} + \mathbf{j} + \mathbf{k}$. Find its position vector $\mathbf{r}(t)$ at time $t$.

## Projectile Motion

Projectile flights in the following exercises are to be treated as ideal unless stated otherwise. All launch angles are assumed to be measured from the horizontal. All projectiles are assumed to be launched from the origin over a horizontal surface unless stated otherwise.

**19. Travel time**   A projectile is fired at a speed of 840 m/s at an angle of 60°. How long will it take to get 21 km downrange?

**20. Range and height versus speed**

  **a.** Show that doubling a projectile's initial speed at a given launch angle multiplies its range by 4.

  **b.** By about what percentage should you increase the initial speed to double the height and range?

**21. Flight time and height**   A projectile is fired with an initial speed of 500 m/s at an angle of elevation of 45°.

  **a.** When and how far away will the projectile strike?

  **b.** How high overhead will the projectile be when it is 5 km downrange?

  **c.** What is the greatest height reached by the projectile?

**22. Throwing a baseball**   A baseball is thrown from the stands 9.8 m above the field at an angle of 30° up from the horizontal. When and how far away will the ball strike the ground if its initial speed is 9.8 m/s?

**23. Firing golf balls**   A spring gun at ground level fires a golf ball at an angle of 45°. The ball lands 10 m away.

  **a.** What was the ball's initial speed?

  **b.** For the same initial speed, find the two firing angles that make the range 6 m.

**24. Beaming electrons**   An electron in a TV tube is beamed horizontally at a speed of $5 \times 10^6$ m/s toward the face of the tube 40 cm away. About how far will the electron drop before it hits?

**25. Equal-range firing angles**   What two angles of elevation will enable a projectile to reach a target 16 km downrange on the same level as the gun if the projectile's initial speed is 400 m/s?

**26. Finding muzzle speed**   Find the muzzle speed of a gun whose maximum range is 24.5 km.

**27.** Verify the results given in the text (following Example 4) for the maximum height, flight time, and range for ideal projectile motion.

**28. Colliding marbles**    The accompanying figure shows an experiment with two marbles. Marble $A$ was launched toward marble $B$ with launch angle $\alpha$ and initial speed $v_0$. At the same instant, marble $B$ was released to fall from rest at $R \tan \alpha$ units directly above a spot $R$ units downrange from $A$. The marbles were found to collide regardless of the value of $v_0$. Was this mere coincidence, or must this happen? Give reasons for your answer.

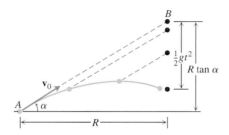

**29. Firing from $(x_0, y_0)$**    Derive the equations

$$x = x_0 + (v_0 \cos \alpha)t,$$

$$y = y_0 + (v_0 \sin \alpha)t - \frac{1}{2}gt^2$$

(see Equation (7) in the text) by solving the following initial value problem for a vector $\mathbf{r}$ in the plane.

Differential equation:    $\dfrac{d^2\mathbf{r}}{dt^2} = -g\mathbf{j}$

Initial conditions:    $\mathbf{r}(0) = x_0\mathbf{i} + y_0\mathbf{j}$

$$\frac{d\mathbf{r}}{dt}(0) = (v_0 \cos \alpha)\mathbf{i} + (v_0 \sin \alpha)\mathbf{j}$$

**30. Where trajectories crest**    For a projectile fired from the ground at launch angle $\alpha$ with initial speed $v_0$, consider $\alpha$ as a variable and $v_0$ as a fixed constant. For each $\alpha$, $0 < \alpha < \pi/2$, we obtain a parabolic trajectory as shown in the accompanying figure. Show that the points in the plane that give the maximum heights of these parabolic trajectories all lie on the ellipse

$$x^2 + 4\left(y - \frac{v_0{}^2}{4g}\right)^2 = \frac{v_0{}^4}{4g^2},$$

where $x \geq 0$.

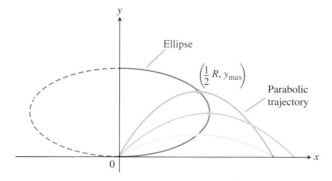

**31. Launching downhill**    An ideal projectile is launched straight down an inclined plane as shown in the accompanying figure.

  **a.** Show that the greatest downhill range is achieved when the initial velocity vector bisects angle $AOR$.

  **b.** If the projectile were fired uphill instead of down, what launch angle would maximize its range? Give reasons for your answer.

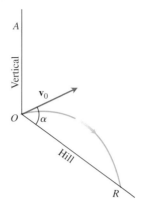

**32. Elevated green**    A golf ball is hit with an initial speed of 35.5 m/s at an angle of elevation of 45° from the tee to a green that is elevated 14 m above the tee as shown in the diagram. Assuming that the pin, 112 m downrange, does not get in the way, where will the ball land in relation to the pin?

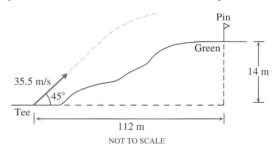

NOT TO SCALE

**33. Volleyball**    A volleyball is hit when it is 1.3 m above the ground and 4 m from a 2-m-high net. It leaves the point of impact with an initial velocity of 12 m/s at an angle of 27° and slips by the opposing team untouched.

  **a.** Find a vector equation for the path of the volleyball.

  **b.** How high does the volleyball go, and when does it reach maximum height?

  **c.** Find its range and flight time.

  **d.** When is the volleyball 2.3 m above the ground? How far (ground distance) is the volleyball from where it will land?

  **e.** Suppose that the net is raised to 2.5 m. Does this change things? Explain.

**34. Shot put**    In Moscow in 1987, Natalya Lisouskaya set a women's world record by putting an 4 kg shot 22.63 m. Assuming that she launched the shot at a 40° angle to the horizontal from 2 m above the ground, what was the shot's initial speed?

**35. Model train**    The accompanying multiflash photograph shows a model train engine moving at a constant speed on a straight horizontal track. As the engine moved along, a marble was fired into the air by a spring in the engine's smokestack. The marble, which continued to move with the same forward speed as the engine, rejoined the engine 1 s after it was fired. Measure the angle the marble's path made with the horizontal and use the information to find how high the marble went and how fast the engine was moving.

*Source: PSSC Physics*, 2nd ed., Reprinted by permission of Educational Development Center, Inc.

**36. Hitting a baseball under a wind gust**   A baseball is hit when it is 0.8 m above the ground. It leaves the bat with an initial velocity of 40 m/s at a launch angle of 23°. At the instant the ball is hit, an instantaneous gust of wind blows against the ball, adding a component of $-4\mathbf{i}$ (m/s) to the ball's initial velocity. A 5-m-high fence lies 90 m from home plate in the direction of the flight.

   **a.** Find a vector equation for the path of the baseball.

   **b.** How high does the baseball go, and when does it reach maximum height?

   **c.** Find the range and flight time of the baseball, assuming that the ball is not caught.

   **d.** When is the baseball 6 m high? How far (ground distance) is the baseball from home plate at that height?

   **e.** Has the batter hit a home run? Explain.

### Projectile Motion with Linear Drag

The main force affecting the motion of a projectile, other than gravity, is air resistance. This slowing down force is **drag force**, and it acts in a direction *opposite* to the velocity of the projectile (see accompanying figure). For projectiles moving through the air at relatively low speeds, however, the drag force is (very nearly) proportional to the speed (to the first power) and so is called **linear**.

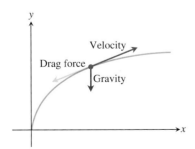

**37. Linear drag**   Derive the equations

$$x = \frac{v_0}{k}(1 - e^{-kt}) \cos \alpha$$

$$y = \frac{v_0}{k}(1 - e^{-kt})(\sin \alpha) + \frac{g}{k^2}(1 - kt - e^{-kt})$$

by solving the following initial value problem for a vector $\mathbf{r}$ in the plane.

Differential equation:    $\dfrac{d^2\mathbf{r}}{dt^2} = -g\mathbf{j} - k\mathbf{v} = -g\mathbf{j} - k\dfrac{d\mathbf{r}}{dt}$

Initial conditions:     $\mathbf{r}(0) = \mathbf{0}$

$$\left.\frac{d\mathbf{r}}{dt}\right|_{t=0} = \mathbf{v}_0 = (v_0 \cos \alpha)\mathbf{i} + (v_0 \sin \alpha)\mathbf{j}$$

The **drag coefficient** $k$ is a positive constant representing resistance due to air density, $v_0$ and $\alpha$ are the projectile's initial speed and launch angle, and $g$ is the acceleration of gravity.

**38. Hitting a baseball with linear drag**   Consider the baseball problem in Example 5 when there is linear drag (see Exercise 37). Assume a drag coefficient $k = 0.12$, but no gust of wind.

   **a.** From Exercise 37, find a vector form for the path of the baseball.

   **b.** How high does the baseball go, and when does it reach maximum height?

   **c.** Find the range and flight time of the baseball.

   **d.** When is the baseball 9 m high? How far (ground distance) is the baseball from home plate at that height?

   **e.** A 3-m-high outfield fence is 115 m from home plate in the direction of the flight of the baseball. The outfielder can jump and catch any ball up to 3.3 m off the ground to stop it from going over the fence. Has the batter hit a home run?

### Theory and Examples

**39.** Establish the following properties of integrable vector functions.

   **a.** The *Constant Scalar Multiple Rule:*

$$\int_a^b k\mathbf{r}(t)\, dt = k \int_a^b \mathbf{r}(t)\, dt \quad \text{(any scalar } k)$$

   The *Rule for Negatives,*

$$\int_a^b (-\mathbf{r}(t))\, dt = -\int_a^b \mathbf{r}(t)\, dt,$$

   is obtained by taking $k = -1$.

   **b.** The *Sum and Difference Rules:*

$$\int_a^b (\mathbf{r}_1(t) \pm \mathbf{r}_2(t))\, dt = \int_a^b \mathbf{r}_1(t)\, dt \pm \int_a^b \mathbf{r}_2(t)\, dt$$

   **c.** The *Constant Vector Multiple Rules:*

$$\int_a^b \mathbf{C} \cdot \mathbf{r}(t)\, dt = \mathbf{C} \cdot \int_a^b \mathbf{r}(t)\, dt \quad \text{(any constant vector } \mathbf{C})$$

   and

$$\int_a^b \mathbf{C} \times \mathbf{r}(t)\, dt = \mathbf{C} \times \int_a^b \mathbf{r}(t)\, dt \quad \text{(any constant vector } \mathbf{C})$$

**40. Products of scalar and vector functions** Suppose that the scalar function $u(t)$ and the vector function $\mathbf{r}(t)$ are both defined for $a \leq t \leq b$.

  **a.** Show that $u\mathbf{r}$ is continuous on $[a, b]$ if $u$ and $\mathbf{r}$ are continuous on $[a, b]$.

  **b.** If $u$ and $\mathbf{r}$ are both differentiable on $[a, b]$, show that $u\mathbf{r}$ is differentiable on $[a, b]$ and that

$$\frac{d}{dt}(u\mathbf{r}) = u\frac{d\mathbf{r}}{dt} + \mathbf{r}\frac{du}{dt}.$$

**41. Antiderivatives of vector functions**

  **a.** Use Corollary 2 of the Mean Value Theorem for scalar functions to show that if two vector functions $\mathbf{R}_1(t)$ and $\mathbf{R}_2(t)$ have identical derivatives on an interval $I$, then the functions differ by a constant vector value throughout $I$.

  **b.** Use the result in part (a) to show that if $\mathbf{R}(t)$ is any antiderivative of $\mathbf{r}(t)$ on $I$, then any other antiderivative of $\mathbf{r}$ on $I$ equals $\mathbf{R}(t) + \mathbf{C}$ for some constant vector $\mathbf{C}$.

**42. The Fundamental Theorem of Calculus** The Fundamental Theorem of Calculus for scalar functions of a real variable holds for vector functions of a real variable as well. Prove this by using the theorem for scalar functions to show first that if a vector function $\mathbf{r}(t)$ is continuous for $a \leq t \leq b$, then

$$\frac{d}{dt}\int_a^t \mathbf{r}(\tau)\, d\tau = \mathbf{r}(t)$$

at every point $t$ of $(a, b)$. Then use the conclusion in part (b) of Exercise 41 to show that if $\mathbf{R}$ is any antiderivative of $\mathbf{r}$ on $[a, b]$ then

$$\int_a^b \mathbf{r}(t)\, dt = \mathbf{R}(b) - \mathbf{R}(a).$$

**43. Hitting a baseball with linear drag under a wind gust** Consider again the baseball problem in Example 5. This time assume a drag coefficient of 0.08 *and* an instantaneous gust of wind that adds a component of $-5\mathbf{i}$ (m/s) to the initial velocity at the instant the baseball is hit.

  **a.** Find a vector equation for the path of the baseball.

  **b.** How high does the baseball go, and when does it reach maximum height?

  **c.** Find the range and flight time of the baseball.

  **d.** When is the baseball 10 m high? How far (ground distance) is the baseball from home plate at that height?

  **e.** A 6-m-high outfield fence is 120 m from home plate in the direction of the flight of the baseball. Has the batter hit a home run? If "yes," what change in the horizontal component of the ball's initial velocity would have kept the ball in the park? If "no," what change would have allowed it to be a home run?

**44. Height versus time** Show that a projectile attains three-quarters of its maximum height in half the time it takes to reach the maximum height.

# $13.3$ Arc Length in Space

In this and the next two sections, we study the mathematical features of a curve's shape that describe the sharpness of its turning and its twisting.

## Arc Length Along a Space Curve

**FIGURE 13.12** Smooth curves can be scaled like number lines, the coordinate of each point being its directed distance along the curve from a preselected base point.

One of the features of smooth space and plane curves is that they have a measurable length. This enables us to locate points along these curves by giving their directed distance $s$ along the curve from some base point, the way we locate points on coordinate axes by giving their directed distance from the origin (Figure 13.12). This is what we did for plane curves in Section 13.2.

To measure distance along a smooth curve in space, we add a $z$-term to the formula we use for curves in the plane.

---

> **DEFINITION** The **length** of a smooth curve $\mathbf{r}(t) = x(t)\mathbf{i} + y(t)\mathbf{j} + z(t)\mathbf{k}$, $a \leq t \leq b$, that is traced exactly once as $t$ increases from $t = a$ to $t = b$, is
>
> $$L = \int_a^b \sqrt{\left(\frac{dx}{dt}\right)^2 + \left(\frac{dy}{dt}\right)^2 + \left(\frac{dz}{dt}\right)^2}\, dt. \tag{1}$$

---

Just as for plane curves, we can calculate the length of a curve in space from any convenient parametrization that meets the stated conditions. We omit the proof.

The square root in Equation (1) is $|\mathbf{v}|$, the length of a velocity vector $d\mathbf{r}/dt$. This enables us to write the formula for length a shorter way.

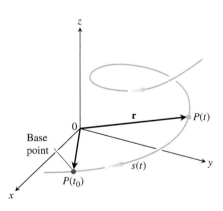

**FIGURE 13.13**　The helix in Example 1, $\mathbf{r}(t) = (\cos t)\mathbf{i} + (\sin t)\mathbf{j} + t\mathbf{k}$.

---

**Arc Length Formula**

$$L = \int_a^b |\mathbf{v}|\, dt \qquad (2)$$

---

**EXAMPLE 1**　A glider is soaring upward along the helix $\mathbf{r}(t) = (\cos t)\mathbf{i} + (\sin t)\mathbf{j} + t\mathbf{k}$. How long is the glider's path from $t = 0$ to $t = 2\pi$?

**Solution**　The path segment during this time corresponds to one full turn of the helix (Figure 13.13). The length of this portion of the curve is

$$L = \int_a^b |\mathbf{v}|\, dt = \int_0^{2\pi} \sqrt{(-\sin t)^2 + (\cos t)^2 + (1)^2}\, dt$$

$$= \int_0^{2\pi} \sqrt{2}\, dt = 2\pi\sqrt{2} \text{ units of length.}$$

This is $\sqrt{2}$ times the circumference of the circle in the $xy$-plane over which the helix stands.　■

If we choose a base point $P(t_0)$ on a smooth curve $C$ parametrized by $t$, each value of $t$ determines a point $P(t) = (x(t), y(t), z(t))$ on $C$ and a "directed distance"

$$s(t) = \int_{t_0}^t |\mathbf{v}(\tau)|\, d\tau,$$

measured along $C$ from the base point (Figure 13.14). This is the arc length function we defined in Section 11.2 for plane curves that have no $z$-component. If $t > t_0$, $s(t)$ is the distance along the curve from $P(t_0)$ to $P(t)$. If $t < t_0$, $s(t)$ is the negative of the distance. Each value of $s$ determines a point on $C$, and this parametrizes $C$ with respect to $s$. We call $s$ an **arc length parameter** for the curve. The parameter's value increases in the direction of increasing $t$. We will see that the arc length parameter is particularly effective for investigating the turning and twisting nature of a space curve.

---

**Arc Length Parameter with Base Point** $P(t_0)$

$$s(t) = \int_{t_0}^t \sqrt{[x'(\tau)]^2 + [y'(\tau)]^2 + [z'(\tau)]^2}\, d\tau = \int_{t_0}^t |\mathbf{v}(\tau)|\, d\tau \qquad (3)$$

---

**FIGURE 13.14**　The directed distance along the curve from $P(t_0)$ to any point $P(t)$ is

$$s(t) = \int_{t_0}^t |\mathbf{v}(\tau)|\, d\tau.$$

We use the Greek letter $\tau$ ("tau") as the variable of integration in Equation (3) because the letter $t$ is already in use as the upper limit.

If a curve $\mathbf{r}(t)$ is already given in terms of some parameter $t$ and $s(t)$ is the arc length function given by Equation (3), then we may be able to solve for $t$ as a function of $s$: $t = t(s)$. Then the curve can be reparametrized in terms of $s$ by substituting for $t$: $\mathbf{r} = \mathbf{r}(t(s))$. The new parametrization identifies a point on the curve with its directed distance along the curve from the base point.

**EXAMPLE 2**　This is an example for which we can actually find the arc length parametrization of a curve. If $t_0 = 0$, the arc length parameter along the helix

$$\mathbf{r}(t) = (\cos t)\mathbf{i} + (\sin t)\mathbf{j} + t\mathbf{k}$$

from $t_0$ to $t$ is

$$s(t) = \int_{t_0}^{t} |\mathbf{v}(\tau)| \, d\tau \qquad \text{Eq. (3)}$$

$$= \int_{0}^{t} \sqrt{2} \, d\tau \qquad \text{Value from Example 1}$$

$$= \sqrt{2} \, t.$$

Solving this equation for $t$ gives $t = s/\sqrt{2}$. Substituting into the position vector $\mathbf{r}$ gives the following arc length parametrization for the helix:

$$\mathbf{r}(t(s)) = \left(\cos \frac{s}{\sqrt{2}}\right)\mathbf{i} + \left(\sin \frac{s}{\sqrt{2}}\right)\mathbf{j} + \frac{s}{\sqrt{2}}\mathbf{k}. \qquad \blacksquare$$

Unlike Example 2, the arc length parametrization is generally difficult to find analytically for a curve already given in terms of some other parameter $t$. Fortunately, however, we rarely need an exact formula for $s(t)$ or its inverse $t(s)$.

## Speed on a Smooth Curve

Since the derivatives beneath the radical in Equation (3) are continuous (the curve is smooth), the Fundamental Theorem of Calculus tells us that $s$ is a differentiable function of $t$ with derivative

$$\frac{ds}{dt} = |\mathbf{v}(t)|. \qquad (4)$$

Equation (4) says that the speed with which a particle moves along its path is the magnitude of $\mathbf{v}$, consistent with what we know.

Although the base point $P(t_0)$ plays a role in defining $s$ in Equation (3), it plays no role in Equation (4). The rate at which a moving particle covers distance along its path is independent of how far away it is from the base point.

Notice that $ds/dt > 0$ since, by definition, $|\mathbf{v}|$ is never zero for a smooth curve. We see once again that $s$ is an increasing function of $t$.

## Unit Tangent Vector

We already know the velocity vector $\mathbf{v} = d\mathbf{r}/dt$ is tangent to the curve $\mathbf{r}(t)$ and that the vector

$$\mathbf{T} = \frac{\mathbf{v}}{|\mathbf{v}|}$$

is therefore a unit vector tangent to the (smooth) curve, called the **unit tangent vector** (Figure 13.15). The unit tangent vector $\mathbf{T}$ is a differentiable function of $t$ whenever $\mathbf{v}$ is a differentiable function of $t$. As we will see in Section 13.5, $\mathbf{T}$ is one of three unit vectors in a traveling reference frame that is used to describe the motion of objects traveling in three dimensions.

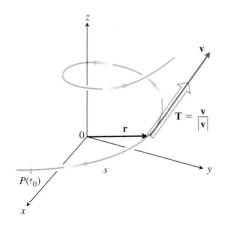

**FIGURE 13.15** We find the unit tangent vector $\mathbf{T}$ by dividing $\mathbf{v}$ by $|\mathbf{v}|$.

**EXAMPLE 3** Find the unit tangent vector of the curve

$$\mathbf{r}(t) = (1 + 3 \cos t)\mathbf{i} + (3 \sin t)\mathbf{j} + t^2\mathbf{k}$$

representing the path of the glider in Example 3, Section 13.2.

**Solution** In that example, we found

$$\mathbf{v} = \frac{d\mathbf{r}}{dt} = -(3 \sin t)\mathbf{i} + (3 \cos t)\mathbf{j} + 2t\mathbf{k}$$

and

$$|\mathbf{v}| = \sqrt{9 + 4t^2}.$$

Thus,

$$\mathbf{T} = \frac{\mathbf{v}}{|\mathbf{v}|} = -\frac{3 \sin t}{\sqrt{9 + 4t^2}}\mathbf{i} + \frac{3 \cos t}{\sqrt{9 + 4t^2}}\mathbf{j} + \frac{2t}{\sqrt{9 + 4t^2}}\mathbf{k}. \qquad ■$$

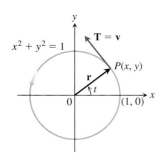

**FIGURE 13.16** Counterclockwise motion around the unit circle.

For the counterclockwise motion

$$\mathbf{r}(t) = (\cos t)\mathbf{i} + (\sin t)\mathbf{j}$$

around the unit circle, we see that

$$\mathbf{v} = (-\sin t)\mathbf{i} + (\cos t)\mathbf{j}$$

is already a unit vector, so $\mathbf{T} = \mathbf{v}$ and $\mathbf{T}$ is orthogonal to $\mathbf{r}$ (Figure 13.16).

The velocity vector is the change in the position vector $\mathbf{r}$ with respect to time $t$, but how does the position vector change with respect to arc length? More precisely, what is the derivative $d\mathbf{r}/ds$? Since $ds/dt > 0$ for the curves we are considering, $s$ is one-to-one and has an inverse that gives $t$ as a differentiable function of $s$ (Section 7.1). The derivative of the inverse is

$$\frac{dt}{ds} = \frac{1}{ds/dt} = \frac{1}{|\mathbf{v}|}.$$

This makes $\mathbf{r}$ a differentiable function of $s$ whose derivative can be calculated with the Chain Rule to be

$$\frac{d\mathbf{r}}{ds} = \frac{d\mathbf{r}}{dt}\frac{dt}{ds} = \mathbf{v}\frac{1}{|\mathbf{v}|} = \frac{\mathbf{v}}{|\mathbf{v}|} = \mathbf{T}. \qquad (5)$$

This equation says that $d\mathbf{r}/ds$ is the unit tangent vector in the direction of the velocity vector $\mathbf{v}$ (Figure 13.15).

# Exercises 13.3

### Finding Tangent Vectors and Lengths

In Exercises 1–8, find the curve's unit tangent vector. Also, find the length of the indicated portion of the curve.

1. $\mathbf{r}(t) = (2 \cos t)\mathbf{i} + (2 \sin t)\mathbf{j} + \sqrt{5}t\mathbf{k}, \quad 0 \le t \le \pi$

2. $\mathbf{r}(t) = (6 \sin 2t)\mathbf{i} + (6 \cos 2t)\mathbf{j} + 5t\mathbf{k}, \quad 0 \le t \le \pi$

3. $\mathbf{r}(t) = t\mathbf{i} + (2/3)t^{3/2}\mathbf{k}, \quad 0 \le t \le 8$

4. $\mathbf{r}(t) = (2 + t)\mathbf{i} - (t + 1)\mathbf{j} + t\mathbf{k}, \quad 0 \le t \le 3$

5. $\mathbf{r}(t) = (\cos^3 t)\mathbf{j} + (\sin^3 t)\mathbf{k}, \quad 0 \le t \le \pi/2$

6. $\mathbf{r}(t) = 6t^3\mathbf{i} - 2t^3\mathbf{j} - 3t^3\mathbf{k}, \quad 1 \le t \le 2$

7. $\mathbf{r}(t) = (t \cos t)\mathbf{i} + (t \sin t)\mathbf{j} + (2\sqrt{2}/3)t^{3/2}\mathbf{k}, \quad 0 \le t \le \pi$

8. $\mathbf{r}(t) = (t \sin t + \cos t)\mathbf{i} + (t \cos t - \sin t)\mathbf{j}, \quad \sqrt{2} \le t \le 2$

9. Find the point on the curve

$$\mathbf{r}(t) = (5 \sin t)\mathbf{i} + (5 \cos t)\mathbf{j} + 12t\mathbf{k}$$

at a distance $26\pi$ units along the curve from the point $(0, 5, 0)$ in the direction of increasing arc length.

10. Find the point on the curve

$$\mathbf{r}(t) = (12 \sin t)\mathbf{i} - (12 \cos t)\mathbf{j} + 5t\mathbf{k}$$

at a distance $13\pi$ units along the curve from the point $(0, -12, 0)$ in the direction opposite to the direction of increasing arc length.

### Arc Length Parameter

In Exercises 11–14, find the arc length parameter along the curve from the point where $t = 0$ by evaluating the integral

$$s = \int_0^t |\mathbf{v}(\tau)| \, d\tau$$

from Equation (3). Then find the length of the indicated portion of the curve.

11. $\mathbf{r}(t) = (4 \cos t)\mathbf{i} + (4 \sin t)\mathbf{j} + 3t\mathbf{k}, \quad 0 \le t \le \pi/2$

12. $\mathbf{r}(t) = (\cos t + t \sin t)\mathbf{i} + (\sin t - t \cos t)\mathbf{j}, \quad \pi/2 \le t \le \pi$

13. $\mathbf{r}(t) = (e^t \cos t)\mathbf{i} + (e^t \sin t)\mathbf{j} + e^t\mathbf{k}, \quad -\ln 4 \le t \le 0$

14. $\mathbf{r}(t) = (1 + 2t)\mathbf{i} + (1 + 3t)\mathbf{j} + (6 - 6t)\mathbf{k}, \quad -1 \le t \le 0$

**Theory and Examples**

**15. Arc length**    Find the length of the curve

$$\mathbf{r}(t) = \left(\sqrt{2}\,t\right)\mathbf{i} + \left(\sqrt{2}\,t\right)\mathbf{j} + (1 - t^2)\mathbf{k}$$

from $(0, 0, 1)$ to $\left(\sqrt{2}, \sqrt{2}, 0\right)$.

**16. Length of helix**    The length $2\pi\sqrt{2}$ of the turn of the helix in Example 1 is also the length of the diagonal of a square $2\pi$ units on a side. Show how to obtain this square by cutting away and flattening a portion of the cylinder around which the helix winds.

**17. Ellipse**

a. Show that the curve $\mathbf{r}(t) = (\cos t)\mathbf{i} + (\sin t)\mathbf{j} + (1 - \cos t)\mathbf{k}$, $0 \le t \le 2\pi$, is an ellipse by showing that it is the intersection of a right circular cylinder and a plane. Find equations for the cylinder and plane.

b. Sketch the ellipse on the cylinder. Add to your sketch the unit tangent vectors at $t = 0, \pi/2, \pi$, and $3\pi/2$.

c. Show that the acceleration vector always lies parallel to the plane (orthogonal to a vector normal to the plane). Thus, if you draw the acceleration as a vector attached to the ellipse, it will lie in the plane of the ellipse. Add the acceleration vectors for $t = 0, \pi/2, \pi$, and $3\pi/2$ to your sketch.

d. Write an integral for the length of the ellipse. Do not try to evaluate the integral; it is nonelementary.

T e. **Numerical integrator**    Estimate the length of the ellipse to two decimal places.

**18. Length is independent of parametrization**    To illustrate that the length of a smooth space curve does not depend on the parametrization you use to compute it, calculate the length of one turn of the helix in Example 1 with the following parametrizations.

a. $\mathbf{r}(t) = (\cos 4t)\mathbf{i} + (\sin 4t)\mathbf{j} + 4t\mathbf{k}$,    $0 \le t \le \pi/2$

b. $\mathbf{r}(t) = \left[\cos (t/2)\right]\mathbf{i} + \left[\sin (t/2)\right]\mathbf{j} + (t/2)\mathbf{k}$,    $0 \le t \le 4\pi$

c. $\mathbf{r}(t) = (\cos t)\mathbf{i} - (\sin t)\mathbf{j} - t\mathbf{k}$,    $-2\pi \le t \le 0$

**19. The involute of a circle**    If a string wound around a fixed circle is unwound while held taut in the plane of the circle, its end $P$ traces an *involute* of the circle. In the accompanying figure, the circle in question is the circle $x^2 + y^2 = 1$ and the tracing point starts at $(1, 0)$. The unwound portion of the string is tangent to the circle at $Q$, and $t$ is the radian measure of the angle from the positive $x$-axis to segment $OQ$. Derive the parametric equations

$$x = \cos t + t \sin t, \quad y = \sin t - t \cos t, \quad t > 0$$

of the point $P(x, y)$ for the involute.

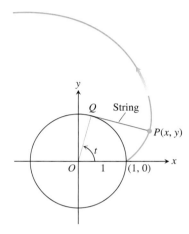

**20.** (*Continuation of Exercise 19.*) Find the unit tangent vector to the involute of the circle at the point $P(x, y)$.

**21. Distance along a line**    Show that if $\mathbf{u}$ is a unit vector, then the arc length parameter along the line $\mathbf{r}(t) = P_0 + t\mathbf{u}$ from the point $P_0(x_0, y_0, z_0)$ where $t = 0$, is $t$ itself.

**22.** Use Simpson's Rule with $n = 10$ to approximate the length of arc of $\mathbf{r}(t) = t\mathbf{i} + t^2\mathbf{j} + t^3\mathbf{k}$ from the origin to the point $(2, 4, 8)$.

# $13.4$  Curvature and Normal Vectors of a Curve

In this section we study how a curve turns or bends. To gain perspective, we look first at curves in the coordinate plane. Then we consider curves in space.

## Curvature of a Plane Curve

As a particle moves along a smooth curve in the plane, $\mathbf{T} = d\mathbf{r}/ds$ turns as the curve bends. Since $\mathbf{T}$ is a unit vector, its length remains constant and only its direction changes as the particle moves along the curve. The rate at which $\mathbf{T}$ turns per unit of length along the curve is called the *curvature* (Figure 13.17). The traditional symbol for the curvature function is the Greek letter $\kappa$ ("kappa").

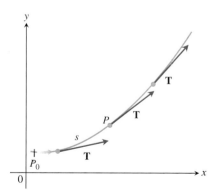

**FIGURE 13.17**    As $P$ moves along the curve in the direction of increasing arc length, the unit tangent vector turns. The value of $|d\mathbf{T}/ds|$ at $P$ is called the *curvature* of the curve at $P$.

> **DEFINITION**    If $\mathbf{T}$ is the unit vector of a smooth curve, the **curvature** function of the curve is
>
> $$\kappa = \left|\frac{d\mathbf{T}}{ds}\right|.$$

If $|d\mathbf{T}/ds|$ is large, $\mathbf{T}$ turns sharply as the particle passes through $P$, and the curvature at $P$ is large. If $|d\mathbf{T}/ds|$ is close to zero, $\mathbf{T}$ turns more slowly and the curvature at $P$ is smaller.

If a smooth curve $\mathbf{r}(t)$ is already given in terms of some parameter $t$ other than the arc length parameter $s$, we can calculate the curvature as

$$\kappa = \left|\frac{d\mathbf{T}}{ds}\right| = \left|\frac{d\mathbf{T}}{dt}\frac{dt}{ds}\right| \qquad \text{Chain Rule}$$

$$= \frac{1}{|ds/dt|}\left|\frac{d\mathbf{T}}{dt}\right|$$

$$= \frac{1}{|\mathbf{v}|}\left|\frac{d\mathbf{T}}{dt}\right|. \qquad \frac{ds}{dt} = |\mathbf{v}|$$

---

**Formula for Calculating Curvature**

If $\mathbf{r}(t)$ is a smooth curve, then the curvature is the scalar function

$$\kappa = \frac{1}{|\mathbf{v}|}\left|\frac{d\mathbf{T}}{dt}\right|, \qquad (1)$$

where $\mathbf{T} = \mathbf{v}/|\mathbf{v}|$ is the unit tangent vector.

---

Testing the definition, we see in Examples 1 and 2 below that the curvature is constant for straight lines and circles.

**EXAMPLE 1**   A straight line is parametrized by $\mathbf{r}(t) = \mathbf{C} + t\mathbf{v}$ for constant vectors $\mathbf{C}$ and $\mathbf{v}$. Thus, $\mathbf{r}'(t) = \mathbf{v}$, and the unit tangent vector $\mathbf{T} = \mathbf{v}/|\mathbf{v}|$ is a constant vector that always points in the same direction and has derivative $\mathbf{0}$ (Figure 13.18). It follows that, for any value of the parameter $t$, the curvature of the straight line is zero:

$$\kappa = \frac{1}{|\mathbf{v}|}\left|\frac{d\mathbf{T}}{dt}\right| = \frac{1}{|\mathbf{v}|}|\mathbf{0}| = 0. \qquad \blacksquare$$

**EXAMPLE 2**   Here we find the curvature of a circle. We begin with the parametrization

$$\mathbf{r}(t) = (a\cos t)\mathbf{i} + (a\sin t)\mathbf{j}$$

of a circle of radius $a$. Then,

$$\mathbf{v} = \frac{d\mathbf{r}}{dt} = -(a\sin t)\mathbf{i} + (a\cos t)\mathbf{j}$$

$$|\mathbf{v}| = \sqrt{(-a\sin t)^2 + (a\cos t)^2} = \sqrt{a^2} = |a| = a. \qquad \begin{array}{l}\text{Since } a > 0, \\ |a| = a.\end{array}$$

From this we find

$$\mathbf{T} = \frac{\mathbf{v}}{|\mathbf{v}|} = -(\sin t)\mathbf{i} + (\cos t)\mathbf{j}$$

$$\frac{d\mathbf{T}}{dt} = -(\cos t)\mathbf{i} - (\sin t)\mathbf{j}$$

$$\left|\frac{d\mathbf{T}}{dt}\right| = \sqrt{\cos^2 t + \sin^2 t} = 1.$$

Hence, for any value of the parameter $t$, the curvature of the circle is

$$\kappa = \frac{1}{|\mathbf{v}|}\left|\frac{d\mathbf{T}}{dt}\right| = \frac{1}{a}(1) = \frac{1}{a} = \frac{1}{\text{radius}}. \qquad \blacksquare$$

Although the formula for calculating $\kappa$ in Equation (1) is also valid for space curves, in the next section we find a computational formula that is usually more convenient to apply.

**FIGURE 13.18**   Along a straight line, $\mathbf{T}$ always points in the same direction. The curvature, $|d\mathbf{T}/ds|$, is zero (Example 1).

Among the vectors orthogonal to the unit tangent vector **T** is one of particular significance because it points in the direction in which the curve is turning. Since **T** has constant length (namely, 1), the derivative $d\mathbf{T}/ds$ is orthogonal to **T** (Equation 4, Section 13.1). Therefore, if we divide $d\mathbf{T}/ds$ by its length $\kappa$, we obtain a *unit* vector **N** orthogonal to **T** (Figure 13.19).

---

**DEFINITION**  At a point where $\kappa \neq 0$, the **principal unit normal** vector for a smooth curve in the plane is

$$\mathbf{N} = \frac{1}{\kappa}\frac{d\mathbf{T}}{ds}.$$

---

The vector $d\mathbf{T}/ds$ points in the direction in which **T** turns as the curve bends. Therefore, if we face in the direction of increasing arc length, the vector $d\mathbf{T}/ds$ points toward the right if **T** turns clockwise and toward the left if **T** turns counterclockwise. In other words, the principal normal vector **N** will point toward the concave side of the curve (Figure 13.19).

If a smooth curve $\mathbf{r}(t)$ is already given in terms of some parameter $t$ other than the arc length parameter $s$, we can use the Chain Rule to calculate **N** directly:

$$\mathbf{N} = \frac{d\mathbf{T}/ds}{|d\mathbf{T}/ds|}$$

$$= \frac{(d\mathbf{T}/dt)(dt/ds)}{|d\mathbf{T}/dt||dt/ds|}$$

$$= \frac{d\mathbf{T}/dt}{|d\mathbf{T}/dt|}. \qquad \frac{dt}{ds} = \frac{1}{ds/dt} > 0 \text{ cancels.}$$

This formula enables us to find **N** without having to find $\kappa$ and $s$ first.

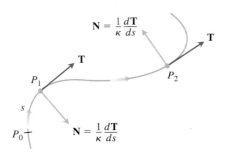

**FIGURE 13.19**  The vector $d\mathbf{T}/ds$, normal to the curve, always points in the direction in which **T** is turning. The unit normal vector **N** is the direction of $d\mathbf{T}/ds$.

---

**Formula for Calculating N**
If $\mathbf{r}(t)$ is a smooth curve, then the principal unit normal is

$$\mathbf{N} = \frac{d\mathbf{T}/dt}{|d\mathbf{T}/dt|}, \tag{2}$$

where $\mathbf{T} = \mathbf{v}/|\mathbf{v}|$ is the unit tangent vector.

---

**EXAMPLE 3**  Find **T** and **N** for the circular motion

$$\mathbf{r}(t) = (\cos 2t)\mathbf{i} + (\sin 2t)\mathbf{j}.$$

**Solution**  We first find **T**:

$$\mathbf{v} = -(2\sin 2t)\mathbf{i} + (2\cos 2t)\mathbf{j}$$

$$|\mathbf{v}| = \sqrt{4\sin^2 2t + 4\cos^2 2t} = 2$$

$$\mathbf{T} = \frac{\mathbf{v}}{|\mathbf{v}|} = -(\sin 2t)\mathbf{i} + (\cos 2t)\mathbf{j}.$$

From this we find

$$\frac{d\mathbf{T}}{dt} = -(2\cos 2t)\mathbf{i} - (2\sin 2t)\mathbf{j}$$

$$\left|\frac{d\mathbf{T}}{dt}\right| = \sqrt{4\cos^2 2t + 4\sin^2 2t} = 2$$

and

$$\mathbf{N} = \frac{d\mathbf{T}/dt}{|d\mathbf{T}/dt|}$$

$$= -(\cos 2t)\mathbf{i} - (\sin 2t)\mathbf{j}. \qquad \text{Eq. (2)}$$

Notice that $\mathbf{T} \cdot \mathbf{N} = 0$, verifying that $\mathbf{N}$ is orthogonal to $\mathbf{T}$. Notice too, that for the circular motion here, $\mathbf{N}$ points from $\mathbf{r}(t)$ toward the circle's center at the origin. ∎

### Circle of Curvature for Plane Curves

The **circle of curvature** or **osculating circle** at a point $P$ on a plane curve where $\kappa \neq 0$ is the circle in the plane of the curve that

1. is tangent to the curve at $P$ (has the same tangent line the curve has)
2. has the same curvature the curve has at $P$
3. has center that lies toward the concave or inner side of the curve (as in Figure 13.20).

The **radius of curvature** of the curve at $P$ is the radius of the circle of curvature, which, according to Example 2, is

$$\text{Radius of curvature} = \rho = \frac{1}{\kappa}.$$

To find $\rho$, we find $\kappa$ and take the reciprocal. The **center of curvature** of the curve at $P$ is the center of the circle of curvature.

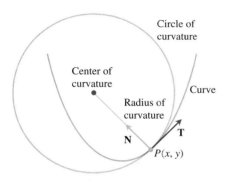

**FIGURE 13.20** The center of the osculating circle at $P(x, y)$ lies toward the inner side of the curve.

**EXAMPLE 4** Find and graph the osculating circle of the parabola $y = x^2$ at the origin.

**Solution** We parametrize the parabola using the parameter $t = x$ (Section 11.1, Example 5):

$$\mathbf{r}(t) = t\mathbf{i} + t^2\mathbf{j}.$$

First we find the curvature of the parabola at the origin, using Equation (1):

$$\mathbf{v} = \frac{d\mathbf{r}}{dt} = \mathbf{i} + 2t\mathbf{j}$$

$$|\mathbf{v}| = \sqrt{1 + 4t^2}$$

so that

$$\mathbf{T} = \frac{\mathbf{v}}{|\mathbf{v}|} = (1 + 4t^2)^{-1/2}\mathbf{i} + 2t(1 + 4t^2)^{-1/2}\mathbf{j}.$$

From this we find

$$\frac{d\mathbf{T}}{dt} = -4t(1 + 4t^2)^{-3/2}\mathbf{i} + \left[2(1 + 4t^2)^{-1/2} - 8t^2(1 + 4t^2)^{-3/2}\right]\mathbf{j}.$$

At the origin, $t = 0$, so the curvature is

$$\kappa(0) = \frac{1}{|\mathbf{v}(0)|}\left|\frac{d\mathbf{T}}{dt}(0)\right| \qquad \text{Eq. (1)}$$

$$= \frac{1}{\sqrt{1}}|0\mathbf{i} + 2\mathbf{j}|$$

$$= (1)\sqrt{0^2 + 2^2} = 2.$$

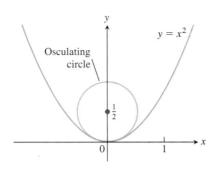

**FIGURE 13.21** The osculating circle for the parabola $y = x^2$ at the origin (Example 4).

Therefore, the radius of curvature is $1/\kappa = 1/2$. At the origin we have $t = 0$ and $\mathbf{T} = \mathbf{i}$, so $\mathbf{N} = \mathbf{j}$. Thus the center of the circle is $(0, 1/2)$. The equation of the osculating circle is, then,

$$(x - 0)^2 + \left(y - \frac{1}{2}\right)^2 = \left(\frac{1}{2}\right)^2.$$

You can see from Figure 13.21 that the osculating circle is a better approximation to the parabola at the origin than is the tangent line approximation $y = 0$.                    ◼

## Curvature and Normal Vectors for Space Curves

If a smooth curve in space is specified by the position vector $\mathbf{r}(t)$ as a function of some parameter $t$, and if $s$ is the arc length parameter of the curve, then the unit tangent vector $\mathbf{T}$ is $d\mathbf{r}/ds = \mathbf{v}/|\mathbf{v}|$. The **curvature** in space is then defined to be

$$\kappa = \left|\frac{d\mathbf{T}}{ds}\right| = \frac{1}{|\mathbf{v}|}\left|\frac{d\mathbf{T}}{dt}\right| \tag{3}$$

just as for plane curves. The vector $d\mathbf{T}/ds$ is orthogonal to $\mathbf{T}$, and we define the **principal unit normal** to be

$$\mathbf{N} = \frac{1}{\kappa}\frac{d\mathbf{T}}{ds} = \frac{d\mathbf{T}/dt}{|d\mathbf{T}/dt|}. \tag{4}$$

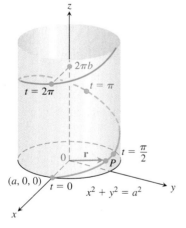

**FIGURE 13.22** The helix

$$\mathbf{r}(t) = (a \cos t)\mathbf{i} + (a \sin t)\mathbf{j} + bt\mathbf{k},$$

drawn with $a$ and $b$ positive and $t \geq 0$ (Example 5).

**EXAMPLE 5**    Find the curvature for the helix (Figure 13.22)

$$\mathbf{r}(t) = (a \cos t)\mathbf{i} + (a \sin t)\mathbf{j} + bt\mathbf{k}, \qquad a, b \geq 0, \qquad a^2 + b^2 \neq 0.$$

**Solution**    We calculate $\mathbf{T}$ from the velocity vector $\mathbf{v}$:

$$\mathbf{v} = -(a \sin t)\mathbf{i} + (a \cos t)\mathbf{j} + b\mathbf{k}$$

$$|\mathbf{v}| = \sqrt{a^2 \sin^2 t + a^2 \cos^2 t + b^2} = \sqrt{a^2 + b^2}$$

$$\mathbf{T} = \frac{\mathbf{v}}{|\mathbf{v}|} = \frac{1}{\sqrt{a^2 + b^2}}[-(a \sin t)\mathbf{i} + (a \cos t)\mathbf{j} + b\mathbf{k}].$$

Then using Equation (3),

$$\kappa = \frac{1}{|\mathbf{v}|}\left|\frac{d\mathbf{T}}{dt}\right|$$

$$= \frac{1}{\sqrt{a^2 + b^2}}\left|\frac{1}{\sqrt{a^2 + b^2}}[-(a \cos t)\mathbf{i} - (a \sin t)\mathbf{j}]\right|$$

$$= \frac{a}{a^2 + b^2}|-(\cos t)\mathbf{i} - (\sin t)\mathbf{j}|$$

$$= \frac{a}{a^2 + b^2}\sqrt{(\cos t)^2 + (\sin t)^2} = \frac{a}{a^2 + b^2}.$$

From this equation, we see that increasing $b$ for a fixed $a$ decreases the curvature. Decreasing $a$ for a fixed $b$ eventually decreases the curvature as well.

   If $b = 0$, the helix reduces to a circle of radius $a$ and its curvature reduces to $1/a$, as it should. If $a = 0$, the helix becomes the $z$-axis, and its curvature reduces to 0, again as it should.                    ◼

**EXAMPLE 6**　　Find **N** for the helix in Example 5 and describe how the vector is pointing.

**Solution**　We have

$$\frac{d\mathbf{T}}{dt} = -\frac{1}{\sqrt{a^2 + b^2}}\left[(a \cos t)\mathbf{i} + (a \sin t)\mathbf{j}\right] \qquad \text{Example 5}$$

$$\left|\frac{d\mathbf{T}}{dt}\right| = \frac{1}{\sqrt{a^2 + b^2}}\sqrt{a^2 \cos^2 t + a^2 \sin^2 t} = \frac{a}{\sqrt{a^2 + b^2}}$$

$$\mathbf{N} = \frac{d\mathbf{T}/dt}{|d\mathbf{T}/dt|} \qquad \text{Eq. (4)}$$

$$= -\frac{\sqrt{a^2 + b^2}}{a} \cdot \frac{1}{\sqrt{a^2 + b^2}}\left[(a \cos t)\mathbf{i} + (a \sin t)\mathbf{j}\right]$$

$$= -(\cos t)\mathbf{i} - (\sin t)\mathbf{j}.$$

Thus, **N** is parallel to the $xy$-plane and always points toward the $z$-axis.　∎

# Exercises **13.4**

**Plane Curves**

Find **T**, **N**, and $\kappa$ for the plane curves in Exercises 1–4.

**1.** $\mathbf{r}(t) = t\mathbf{i} + (\ln \cos t)\mathbf{j}, \quad -\pi/2 < t < \pi/2$

**2.** $\mathbf{r}(t) = (\ln \sec t)\mathbf{i} + t\mathbf{j}, \quad -\pi/2 < t < \pi/2$

**3.** $\mathbf{r}(t) = (2t + 3)\mathbf{i} + (5 - t^2)\mathbf{j}$

**4.** $\mathbf{r}(t) = (\cos t + t \sin t)\mathbf{i} + (\sin t - t \cos t)\mathbf{j}, \quad t > 0$

**5. A formula for the curvature of the graph of a function in the $xy$-plane**

   **a.** The graph $y = f(x)$ in the $xy$-plane automatically has the parametrization $x = x, y = f(x)$, and the vector formula $\mathbf{r}(x) = x\mathbf{i} + f(x)\mathbf{j}$. Use this formula to show that if $f$ is a twice-differentiable function of $x$, then

   $$\kappa(x) = \frac{|f''(x)|}{\left[1 + (f'(x))^2\right]^{3/2}}.$$

   **b.** Use the formula for $\kappa$ in part (a) to find the curvature of $y = \ln(\cos x), -\pi/2 < x < \pi/2$. Compare your answer with the answer in Exercise 1.

   **c.** Show that the curvature is zero at a point of inflection.

**6. A formula for the curvature of a parametrized plane curve**

   **a.** Show that the curvature of a smooth curve $\mathbf{r}(t) = f(t)\mathbf{i} + g(t)\mathbf{j}$ defined by twice-differentiable functions $x = f(t)$ and $y = g(t)$ is given by the formula

   $$\kappa = \frac{|\dot{x}\ddot{y} - \dot{y}\ddot{x}|}{(\dot{x}^2 + \dot{y}^2)^{3/2}}.$$

   The dots in the formula denote differentiation with respect to $t$, one derivative for each dot. Apply the formula to find the curvatures of the following curves.

   **b.** $\mathbf{r}(t) = t\mathbf{i} + (\ln \sin t)\mathbf{j}, \quad 0 < t < \pi$

   **c.** $\mathbf{r}(t) = \left[\tan^{-1}(\sinh t)\right]\mathbf{i} + (\ln \cosh t)\mathbf{j}$

**7. Normals to plane curves**

   **a.** Show that $\mathbf{n}(t) = -g'(t)\mathbf{i} + f'(t)\mathbf{j}$ and $-\mathbf{n}(t) = g'(t)\mathbf{i} - f'(t)\mathbf{j}$ are both normal to the curve $\mathbf{r}(t) = f(t)\mathbf{i} + g(t)\mathbf{j}$ at the point $(f(t), g(t))$.

   To obtain **N** for a particular plane curve, we can choose the one of **n** or $-\mathbf{n}$ from part (a) that points toward the concave side of the curve, and make it into a unit vector. (See Figure 13.19.) Apply this method to find **N** for the following curves.

   **b.** $\mathbf{r}(t) = t\mathbf{i} + e^{2t}\mathbf{j}$

   **c.** $\mathbf{r}(t) = \sqrt{4 - t^2}\,\mathbf{i} + t\mathbf{j}, \quad -2 \le t \le 2$

**8.** (*Continuation of Exercise 7.*)

   **a.** Use the method of Exercise 7 to find **N** for the curve $\mathbf{r}(t) = t\mathbf{i} + (1/3)t^3\mathbf{j}$ when $t < 0$; when $t > 0$.

   **b.** Calculate **N** for $t \ne 0$ directly from **T** using Equation (4) for the curve in part (a). Does **N** exist at $t = 0$? Graph the curve and explain what is happening to **N** as $t$ passes from negative to positive values.

**Space Curves**

Find **T**, **N**, and $\kappa$ for the space curves in Exercises 9–16.

**9.** $\mathbf{r}(t) = (3 \sin t)\mathbf{i} + (3 \cos t)\mathbf{j} + 4t\mathbf{k}$

**10.** $\mathbf{r}(t) = (\cos t + t \sin t)\mathbf{i} + (\sin t - t \cos t)\mathbf{j} + 3\mathbf{k}$

**11.** $\mathbf{r}(t) = (e^t \cos t)\mathbf{i} + (e^t \sin t)\mathbf{j} + 2\mathbf{k}$

**12.** $\mathbf{r}(t) = (6 \sin 2t)\mathbf{i} + (6 \cos 2t)\mathbf{j} + 5t\mathbf{k}$

**13.** $\mathbf{r}(t) = (t^3/3)\mathbf{i} + (t^2/2)\mathbf{j}, \quad t > 0$

**14.** $\mathbf{r}(t) = (\cos^3 t)\mathbf{i} + (\sin^3 t)\mathbf{j}, \quad 0 < t < \pi/2$

**15.** $\mathbf{r}(t) = t\mathbf{i} + (a \cosh(t/a))\mathbf{j}, \quad a > 0$

**16.** $\mathbf{r}(t) = (\cosh t)\mathbf{i} - (\sinh t)\mathbf{j} + t\mathbf{k}$

**More on Curvature**

**17.** Show that the parabola $y = ax^2, a \ne 0$, has its largest curvature at its vertex and has no minimum curvature. (*Note:* Since the curvature of a curve remains the same if the curve is translated or rotated, this result is true for any parabola.)

18. Show that the ellipse $x = a \cos t$, $y = b \sin t$, $a > b > 0$, has its largest curvature on its major axis and its smallest curvature on its minor axis. (As in Exercise 17, the same is true for any ellipse.)

19. **Maximizing the curvature of a helix** In Example 5, we found the curvature of the helix $\mathbf{r}(t) = (a \cos t)\mathbf{i} + (a \sin t)\mathbf{j} + bt\mathbf{k}$ $(a, b \geq 0)$ to be $\kappa = a/(a^2 + b^2)$. What is the largest value $\kappa$ can have for a given value of $b$? Give reasons for your answer.

20. **Total curvature** We find the **total curvature** of the portion of a smooth curve that runs from $s = s_0$ to $s = s_1 > s_0$ by integrating $\kappa$ from $s_0$ to $s_1$. If the curve has some other parameter, say $t$, then the total curvature is

$$K = \int_{s_0}^{s_1} \kappa \, ds = \int_{t_0}^{t_1} \kappa \frac{ds}{dt} dt = \int_{t_0}^{t_1} \kappa |\mathbf{v}| \, dt,$$

where $t_0$ and $t_1$ correspond to $s_0$ and $s_1$. Find the total curvatures of

a. The portion of the helix $\mathbf{r}(t) = (3 \cos t)\mathbf{i} + (3 \sin t)\mathbf{j} + t\mathbf{k}$, $0 \leq t \leq 4\pi$.

b. The parabola $y = x^2$, $-\infty < x < \infty$.

21. Find an equation for the circle of curvature of the curve $\mathbf{r}(t) = t\mathbf{i} + (\sin t)\mathbf{j}$ at the point $(\pi/2, 1)$. (The curve parametrizes the graph of $y = \sin x$ in the $xy$-plane.)

22. Find an equation for the circle of curvature of the curve $\mathbf{r}(t) = (2\ln t)\mathbf{i} - [t + (1/t)]\mathbf{j}$, $e^{-2} \leq t \leq e^2$, at the point $(0, -2)$, where $t = 1$.

[T] The formula

$$\kappa(x) = \frac{|f''(x)|}{\left[1 + (f'(x))^2\right]^{3/2}},$$

derived in Exercise 5, expresses the curvature $\kappa(x)$ of a twice-differentiable plane curve $y = f(x)$ as a function of $x$. Find the curvature function of each of the curves in Exercises 23–26. Then graph $f(x)$ together with $\kappa(x)$ over the given interval. You will find some surprises.

23. $y = x^2$, $-2 \leq x \leq 2$      24. $y = x^4/4$, $-2 \leq x \leq 2$

25. $y = \sin x$, $0 \leq x \leq 2\pi$      26. $y = e^x$, $-1 \leq x \leq 2$

27. **Osculating circle** Show that the center of the osculating circle for the parabola $y = x^2$ at the point $(a, a^2)$ is located at $\left(-4a^3, 3a^2 + \dfrac{1}{2}\right)$.

28. **Osculating circle** Find a parametrization of the osculating circle for the parabola $y = x^2$ when $x = 1$.

**COMPUTER EXPLORATIONS**

In Exercises 29–36 you will use a CAS to explore the osculating circle at a point $P$ on a plane curve where $\kappa \neq 0$. Use a CAS to perform the following steps:

a. Plot the plane curve given in parametric or function form over the specified interval to see what it looks like.

b. Calculate the curvature $\kappa$ of the curve at the given value $t_0$ using the appropriate formula from Exercise 5 or 6. Use the parametrization $x = t$ and $y = f(t)$ if the curve is given as a function $y = f(x)$.

c. Find the unit normal vector $\mathbf{N}$ at $t_0$. Notice that the signs of the components of $\mathbf{N}$ depend on whether the unit tangent vector $\mathbf{T}$ is turning clockwise or counterclockwise at $t = t_0$. (See Exercise 7.)

d. If $\mathbf{C} = a\mathbf{i} + b\mathbf{j}$ is the vector from the origin to the center $(a, b)$ of the osculating circle, find the center $\mathbf{C}$ from the vector equation

$$\mathbf{C} = \mathbf{r}(t_0) + \frac{1}{\kappa(t_0)}\mathbf{N}(t_0).$$

The point $P(x_0, y_0)$ on the curve is given by the position vector $\mathbf{r}(t_0)$.

e. Plot implicitly the equation $(x - a)^2 + (y - b)^2 = 1/\kappa^2$ of the osculating circle. Then plot the curve and osculating circle together. You may need to experiment with the size of the viewing window, but be sure the axes are equally scaled.

29. $\mathbf{r}(t) = (3 \cos t)\mathbf{i} + (5 \sin t)\mathbf{j}$, $0 \leq t \leq 2\pi$, $t_0 = \pi/4$

30. $\mathbf{r}(t) = (\cos^3 t)\mathbf{i} + (\sin^3 t)\mathbf{j}$, $0 \leq t \leq 2\pi$, $t_0 = \pi/4$

31. $\mathbf{r}(t) = t^2\mathbf{i} + (t^3 - 3t)\mathbf{j}$, $-4 \leq t \leq 4$, $t_0 = 3/5$

32. $\mathbf{r}(t) = (t^3 - 2t^2 - t)\mathbf{i} + \dfrac{3t}{\sqrt{1 + t^2}}\mathbf{j}$, $-2 \leq t \leq 5$, $t_0 = 1$

33. $\mathbf{r}(t) = (2t - \sin t)\mathbf{i} + (2 - 2\cos t)\mathbf{j}$, $0 \leq t \leq 3\pi$, $t_0 = 3\pi/2$

34. $\mathbf{r}(t) = (e^{-t} \cos t)\mathbf{i} + (e^{-t} \sin t)\mathbf{j}$, $0 \leq t \leq 6\pi$, $t_0 = \pi/4$

35. $y = x^2 - x$, $-2 \leq x \leq 5$, $x_0 = 1$

36. $y = x(1 - x)^{2/5}$, $-1 \leq x \leq 2$, $x_0 = 1/2$

# 13.5 Tangential and Normal Components of Acceleration

If you are traveling along a space curve, the Cartesian $\mathbf{i}$, $\mathbf{j}$, and $\mathbf{k}$ coordinate system for representing the vectors describing your motion is not truly relevant to you. What is meaningful instead are the vectors representative of your forward direction (the unit tangent vector $\mathbf{T}$), the direction in which your path is turning (the unit normal vector $\mathbf{N}$), and the tendency of your motion to "twist" out of the plane created by these vectors in the direction perpendicular to this plane (defined by the *unit binormal vector* $\mathbf{B} = \mathbf{T} \times \mathbf{N}$). Expressing the acceleration vector along the curve as a linear combination of this **TNB** frame of mutually orthogonal unit vectors traveling with the motion (Figure 13.23) is particularly revealing of the nature of the path and motion along it.

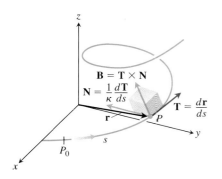

**FIGURE 13.23** The **TNB** frame of mutually orthogonal unit vectors traveling along a curve in space.

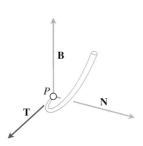

**FIGURE 13.24** The vectors **T**, **N**, and **B** (in that order) make a right-handed frame of mutually orthogonal unit vectors in space.

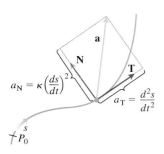

**FIGURE 13.25** The tangential and normal components of acceleration. The acceleration **a** always lies in the plane of **T** and **N**, orthogonal to **B**.

## The TNB Frame

The **binormal vector** of a curve in space is $\mathbf{B} = \mathbf{T} \times \mathbf{N}$, a unit vector orthogonal to both **T** and **N** (Figure 13.24). Together **T**, **N**, and **B** define a moving right-handed vector frame that plays a significant role in calculating the paths of particles moving through space. It is called the **Frenet** ("fre-*nay*") **frame** (after Jean-Frédéric Frenet, 1816–1900), or the **TNB frame**.

## Tangential and Normal Components of Acceleration

When an object is accelerated by gravity, brakes, or a combination of rocket motors, we usually want to know how much of the acceleration acts in the direction of motion, in the tangential direction **T**. We can calculate this using the Chain Rule to rewrite **v** as

$$\mathbf{v} = \frac{d\mathbf{r}}{dt} = \frac{d\mathbf{r}}{ds}\frac{ds}{dt} = \mathbf{T}\frac{ds}{dt}.$$

Then we differentiate both ends of this string of equalities to get

$$\mathbf{a} = \frac{d\mathbf{v}}{dt} = \frac{d}{dt}\left(\mathbf{T}\frac{ds}{dt}\right) = \frac{d^2s}{dt^2}\mathbf{T} + \frac{ds}{dt}\frac{d\mathbf{T}}{dt}$$

$$= \frac{d^2s}{dt^2}\mathbf{T} + \frac{ds}{dt}\left(\frac{d\mathbf{T}}{ds}\frac{ds}{dt}\right) = \frac{d^2s}{dt^2}\mathbf{T} + \frac{ds}{dt}\left(\kappa\mathbf{N}\frac{ds}{dt}\right) \qquad \frac{d\mathbf{T}}{ds} = \kappa\mathbf{N}$$

$$= \frac{d^2s}{dt^2}\mathbf{T} + \kappa\left(\frac{ds}{dt}\right)^2\mathbf{N}.$$

---

**DEFINITION** If the acceleration vector is written as

$$\mathbf{a} = a_{\mathrm{T}}\mathbf{T} + a_{\mathrm{N}}\mathbf{N}, \tag{1}$$

then

$$a_{\mathrm{T}} = \frac{d^2s}{dt^2} = \frac{d}{dt}|\mathbf{v}| \qquad \text{and} \qquad a_{\mathrm{N}} = \kappa\left(\frac{ds}{dt}\right)^2 = \kappa|\mathbf{v}|^2 \tag{2}$$

are the **tangential** and **normal** scalar components of acceleration.

---

Notice that the binormal vector **B** does not appear in Equation (1). No matter how the path of the moving object we are watching may appear to twist and turn in space, the acceleration **a** *always lies in the plane of* **T** *and* **N** orthogonal to **B**. The equation also tells us exactly how much of the acceleration takes place tangent to the motion $(d^2s/dt^2)$ and how much takes place normal to the motion $[\kappa(ds/dt)^2]$ (Figure 13.25).

What information can we discover from Equations (2)? By definition, acceleration **a** is the rate of change of velocity **v**, and in general, both the length and direction of **v** change as an object moves along its path. The tangential component of acceleration $a_{\mathrm{T}}$ measures the rate of change of the *length* of **v** (that is, the change in the speed). The normal component of acceleration $a_{\mathrm{N}}$ measures the rate of change of the *direction* of **v**.

Notice that the normal scalar component of the acceleration is the curvature times the *square* of the speed. This explains why you have to hold on when your car makes a sharp (large $\kappa$), high-speed (large $|\mathbf{v}|$) turn. If you double the speed of your car, you will experience four times the normal component of acceleration for the same curvature.

If an object moves in a circle at a constant speed, $d^2s/dt^2$ is zero and all the acceleration points along **N** toward the circle's center. If the object is speeding up or slowing down, **a** has a nonzero tangential component (Figure 13.26).

To calculate $a_{\mathrm{N}}$, we usually use the formula $a_{\mathrm{N}} = \sqrt{|\mathbf{a}|^2 - a_{\mathrm{T}}^2}$, which comes from solving the equation $|\mathbf{a}|^2 = \mathbf{a}\cdot\mathbf{a} = a_{\mathrm{T}}^2 + a_{\mathrm{N}}^2$ for $a_{\mathrm{N}}$. With this formula, we can find $a_{\mathrm{N}}$ without having to calculate $\kappa$ first.

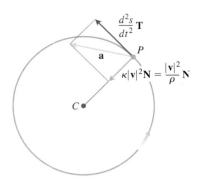

**FIGURE 13.26** The tangential and normal components of the acceleration of an object that is speeding up as it moves counterclockwise around a circle of radius $\rho$.

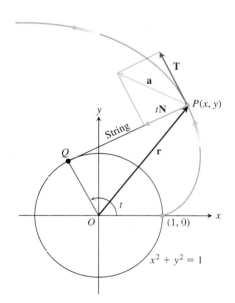

**FIGURE 13.27** The tangential and normal components of the acceleration of the motion $\mathbf{r}(t) = (\cos t + t \sin t)\mathbf{i} + (\sin t - t \cos t)\mathbf{j}$, for $t > 0$. If a string wound around a fixed circle is unwound while held taut in the plane of the circle, its end $P$ traces an involute of the circle (Example 1).

---

> **Formula for Calculating the Normal Component of Acceleration**
>
> $$a_\mathrm{N} = \sqrt{|\mathbf{a}|^2 - a_\mathrm{T}{}^2} \qquad (3)$$

**EXAMPLE 1** Without finding $\mathbf{T}$ and $\mathbf{N}$, write the acceleration of the motion

$$\mathbf{r}(t) = (\cos t + t \sin t)\mathbf{i} + (\sin t - t \cos t)\mathbf{j}, \qquad t > 0$$

in the form $\mathbf{a} = a_\mathrm{T}\mathbf{T} + a_\mathrm{N}\mathbf{N}$. (The path of the motion is the involute of the circle in Figure 13.27. See also Section 13.3, Exercise 19.)

**Solution** We use the first of Equations (2) to find $a_\mathrm{T}$:

$$\mathbf{v} = \frac{d\mathbf{r}}{dt} = (-\sin t + \sin t + t \cos t)\mathbf{i} + (\cos t - \cos t + t \sin t)\mathbf{j}$$

$$= (t \cos t)\mathbf{i} + (t \sin t)\mathbf{j}$$

$$|\mathbf{v}| = \sqrt{t^2 \cos^2 t + t^2 \sin^2 t} = \sqrt{t^2} = |t| = t \qquad t > 0$$

$$a_\mathrm{T} = \frac{d}{dt}|\mathbf{v}| = \frac{d}{dt}(t) = 1. \qquad \text{Eq. (2)}$$

Knowing $a_\mathrm{T}$, we use Equation (3) to find $a_\mathrm{N}$:

$$\mathbf{a} = (\cos t - t \sin t)\mathbf{i} + (\sin t + t \cos t)\mathbf{j}$$

$$|\mathbf{a}|^2 = t^2 + 1 \qquad \text{After some algebra}$$

$$a_\mathrm{N} = \sqrt{|\mathbf{a}|^2 - a_\mathrm{T}{}^2}$$

$$= \sqrt{(t^2 + 1) - (1)} = \sqrt{t^2} = t.$$

We then use Equation (1) to find $\mathbf{a}$:

$$\mathbf{a} = a_\mathrm{T}\mathbf{T} + a_\mathrm{N}\mathbf{N} = (1)\mathbf{T} + (t)\mathbf{N} = \mathbf{T} + t\mathbf{N}. \qquad \blacksquare$$

## Torsion

How does $d\mathbf{B}/ds$ behave in relation to $\mathbf{T}$, $\mathbf{N}$, and $\mathbf{B}$? From the rule for differentiating a cross product in Section 13.1, we have

$$\frac{d\mathbf{B}}{ds} = \frac{d(\mathbf{T} \times \mathbf{N})}{ds} = \frac{d\mathbf{T}}{ds} \times \mathbf{N} + \mathbf{T} \times \frac{d\mathbf{N}}{ds}.$$

Since $\mathbf{N}$ is the direction of $d\mathbf{T}/ds$, $(d\mathbf{T}/ds) \times \mathbf{N} = \mathbf{0}$ and

$$\frac{d\mathbf{B}}{ds} = \mathbf{0} + \mathbf{T} \times \frac{d\mathbf{N}}{ds} = \mathbf{T} \times \frac{d\mathbf{N}}{ds}.$$

From this we see that $d\mathbf{B}/ds$ is orthogonal to $\mathbf{T}$, since a cross product is orthogonal to its factors.

Since $d\mathbf{B}/ds$ is also orthogonal to $\mathbf{B}$ (the latter has constant length), it follows that $d\mathbf{B}/ds$ is orthogonal to the plane of $\mathbf{B}$ and $\mathbf{T}$. In other words, $d\mathbf{B}/ds$ is parallel to $\mathbf{N}$, so $d\mathbf{B}/ds$ is a scalar multiple of $\mathbf{N}$. In symbols,

$$\frac{d\mathbf{B}}{ds} = -\tau\mathbf{N}.$$

The negative sign in this equation is traditional. The scalar $\tau$ is called the *torsion* along the curve. Notice that

$$\frac{d\mathbf{B}}{ds} \cdot \mathbf{N} = -\tau\mathbf{N} \cdot \mathbf{N} = -\tau(1) = -\tau.$$

We use this equation for our next definition.

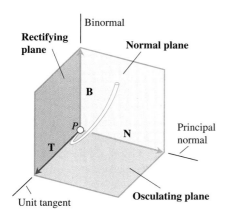

**FIGURE 13.28** The names of the three planes determined by **T**, **N**, and **B**.

> **DEFINITION**    Let $\mathbf{B} = \mathbf{T} \times \mathbf{N}$. The **torsion** function of a smooth curve is
>
> $$\tau = -\frac{d\mathbf{B}}{ds} \cdot \mathbf{N}. \qquad (4)$$

Unlike the curvature $\kappa$, which is never negative, the torsion $\tau$ may be positive, negative, or zero.

The three planes determined by **T**, **N**, and **B** are named and shown in Figure 13.28. The curvature $\kappa = |d\mathbf{T}/ds|$ can be thought of as the rate at which the normal plane turns as the point $P$ moves along its path. Similarly, the torsion $\tau = -(d\mathbf{B}/ds) \cdot \mathbf{N}$ is the rate at which the osculating plane turns about **T** as $P$ moves along the curve. Torsion measures how the curve twists.

Look at Figure 13.29. If $P$ is a train climbing up a curved track, the rate at which the headlight turns from side to side per unit distance is the curvature of the track. The rate at which the engine tends to twist out of the plane formed by **T** and **N** is the torsion. It can be shown that a space curve is a helix if and only if it has constant nonzero curvature and constant nonzero torsion.

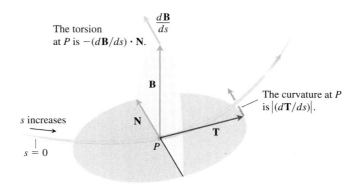

**FIGURE 13.29** Every moving body travels with a **TNB** frame that characterizes the geometry of its path of motion.

## Formulas for Computing Curvature and Torsion

We now give easy-to-use formulas for computing the curvature and torsion of a smooth curve. From Equations (1) and (2), we have

$$\mathbf{v} \times \mathbf{a} = \left(\frac{ds}{dt}\mathbf{T}\right) \times \left[\frac{d^2s}{dt^2}\mathbf{T} + \kappa\left(\frac{ds}{dt}\right)^2\mathbf{N}\right] \qquad \begin{array}{l} v = dr/dt = \\ (ds/dt)\mathbf{T} \end{array}$$

$$= \left(\frac{ds}{dt}\frac{d^2s}{dt^2}\right)(\mathbf{T} \times \mathbf{T}) + \kappa\left(\frac{ds}{dt}\right)^3(\mathbf{T} \times \mathbf{N})$$

$$= \kappa\left(\frac{ds}{dt}\right)^3\mathbf{B}. \qquad \begin{array}{l} \mathbf{T} \times \mathbf{T} = \mathbf{0} \text{ and} \\ \mathbf{T} \times \mathbf{N} = \mathbf{B} \end{array}$$

It follows that

$$|\mathbf{v} \times \mathbf{a}| = \kappa\left|\frac{ds}{dt}\right|^3|\mathbf{B}| = \kappa|\mathbf{v}|^3. \qquad \frac{ds}{dt} = |\mathbf{v}| \quad \text{and} \quad |\mathbf{B}| = 1$$

Solving for $\kappa$ gives the following formula.

---

**Vector Formula for Curvature**

$$\kappa = \frac{|\mathbf{v} \times \mathbf{a}|}{|\mathbf{v}|^3} \tag{5}$$

---

Equation (5) calculates the curvature, a geometric property of the curve, from the velocity and acceleration of any vector representation of the curve in which $|\mathbf{v}|$ is different from zero. From any formula for motion along a curve, no matter how variable the motion may be (as long as $\mathbf{v}$ is never zero), we can calculate a geometric property of the curve that seems to have nothing to do with the way the curve is parametrically defined.

The most widely used formula for torsion, derived in more advanced texts, is given in a determinant form.

---

**Formula for Torsion**

$$\tau = \frac{\begin{vmatrix} \dot{x} & \dot{y} & \dot{z} \\ \ddot{x} & \ddot{y} & \ddot{z} \\ \dddot{x} & \dddot{y} & \dddot{z} \end{vmatrix}}{|\mathbf{v} \times \mathbf{a}|^2} \qquad (\text{if } \mathbf{v} \times \mathbf{a} \neq \mathbf{0}) \tag{6}$$

---

**Newton's Dot Notation for Derivatives**
The dots in Equation (6) denote differentiation with respect to $t$, one derivative for each dot. Thus, $\dot{x}$ ("x dot") means $dx/dt$, $\ddot{x}$ ("x double dot") means $d^2x/dt^2$, and $\dddot{x}$ ("x triple dot") means $d^3x/dt^3$. Similarly, $\dot{y} = dy/dt$, and so on.

This formula calculates the torsion directly from the derivatives of the component functions $x = f(t), y = g(t), z = h(t)$ that make up $\mathbf{r}$. The determinant's first row comes from $\mathbf{v}$, the second row comes from $\mathbf{a}$, and the third row comes from $\dot{\mathbf{a}} = d\mathbf{a}/dt$. Newton's dot notation for derivatives is traditional.

**EXAMPLE 2** Use Equations (5) and (6) to find the curvature $\kappa$ and torsion $\tau$ for the helix

$$\mathbf{r}(t) = (a \cos t)\mathbf{i} + (a \sin t)\mathbf{j} + bt\,\mathbf{k}, \qquad a, b \geq 0, \qquad a^2 + b^2 \neq 0.$$

**Solution** We calculate the curvature with Equation (5):

$$\mathbf{v} = -(a \sin t)\mathbf{i} + (a \cos t)\mathbf{j} + b\mathbf{k}$$

$$\mathbf{a} = -(a \cos t)\mathbf{i} - (a \sin t)\mathbf{j}$$

$$\mathbf{v} \times \mathbf{a} = \begin{vmatrix} \mathbf{i} & \mathbf{j} & \mathbf{k} \\ -a \sin t & a \cos t & b \\ -a \cos t & -a \sin t & 0 \end{vmatrix}$$

$$= (ab \sin t)\mathbf{i} - (ab \cos t)\mathbf{j} + a^2\mathbf{k}$$

$$\kappa = \frac{|\mathbf{v} \times \mathbf{a}|}{|\mathbf{v}|^3} = \frac{\sqrt{a^2b^2 + a^4}}{(a^2 + b^2)^{3/2}} = \frac{a\sqrt{a^2 + b^2}}{(a^2 + b^2)^{3/2}} = \frac{a}{a^2 + b^2}. \tag{7}$$

Notice that Equation (7) agrees with the result in Example 5 in Section 13.4, where we calculated the curvature directly from its definition.

To evaluate Equation (6) for the torsion, we find the entries in the determinant by differentiating $\mathbf{r}$ with respect to $t$. We already have $\mathbf{v}$ and $\mathbf{a}$, and

$$\dot{\mathbf{a}} = \frac{d\mathbf{a}}{dt} = (a \sin t)\mathbf{i} - (a \cos t)\mathbf{j}.$$

Hence,

$$\tau = \frac{\begin{vmatrix} \dot{x} & \dot{y} & \dot{z} \\ \ddot{x} & \ddot{y} & \ddot{z} \\ \dddot{x} & \dddot{y} & \dddot{z} \end{vmatrix}}{|\mathbf{v} \times \mathbf{a}|^2} = \frac{\begin{vmatrix} -a \sin t & a \cos t & b \\ -a \cos t & -a \sin t & 0 \\ a \sin t & -a \cos t & 0 \end{vmatrix}}{\left(a\sqrt{a^2 + b^2}\right)^2} \quad \text{Value of } |\mathbf{v} \times \mathbf{a}| \text{ from Eq. (7)}$$

$$= \frac{b(a^2 \cos^2 t + a^2 \sin^2 t)}{a^2(a^2 + b^2)}$$

$$= \frac{b}{a^2 + b^2}.$$

From this last equation we see that the torsion of a helix about a circular cylinder is constant. In fact, constant curvature and constant torsion characterize the helix among all curves in space. ∎

---

**Computation Formulas for Curves in Space**

*Unit tangent vector:*

$$\mathbf{T} = \frac{\mathbf{v}}{|\mathbf{v}|}$$

*Principal unit normal vector:*

$$\mathbf{N} = \frac{d\mathbf{T}/dt}{|d\mathbf{T}/dt|}$$

*Binormal vector:*

$$\mathbf{B} = \mathbf{T} \times \mathbf{N}$$

*Curvature:*

$$\kappa = \left|\frac{d\mathbf{T}}{ds}\right| = \frac{|\mathbf{v} \times \mathbf{a}|}{|\mathbf{v}|^3}$$

*Torsion:*

$$\tau = -\frac{d\mathbf{B}}{ds} \cdot \mathbf{N} = \frac{\begin{vmatrix} \dot{x} & \dot{y} & \dot{z} \\ \ddot{x} & \ddot{y} & \ddot{z} \\ \dddot{x} & \dddot{y} & \dddot{z} \end{vmatrix}}{|\mathbf{v} \times \mathbf{a}|^2}$$

*Tangential and normal scalar components of acceleration:*

$$\mathbf{a} = a_{\mathrm{T}}\mathbf{T} + a_{\mathrm{N}}\mathbf{N}$$

$$a_{\mathrm{T}} = \frac{d}{dt}|\mathbf{v}|$$

$$a_{\mathrm{N}} = \kappa|\mathbf{v}|^2 = \sqrt{|\mathbf{a}|^2 - a_{\mathrm{T}}^2}$$

---

# Exercises 13.5

**Finding Tangential and Normal Components**

In Exercises 1 and 2, write **a** in the form $\mathbf{a} = a_{\mathrm{T}}\mathbf{T} + a_{\mathrm{N}}\mathbf{N}$ without finding **T** and **N**.

**1.** $\mathbf{r}(t) = (a \cos t)\mathbf{i} + (a \sin t)\mathbf{j} + bt\mathbf{k}$

**2.** $\mathbf{r}(t) = (1 + 3t)\mathbf{i} + (t - 2)\mathbf{j} - 3t\mathbf{k}$

In Exercises 3–6, write **a** in the form $\mathbf{a} = a_{\mathrm{T}}\mathbf{T} + a_{\mathrm{N}}\mathbf{N}$ at the given value of $t$ without finding **T** and **N**.

**3.** $\mathbf{r}(t) = (t + 1)\mathbf{i} + 2t\mathbf{j} + t^2\mathbf{k}, \quad t = 1$

**4.** $\mathbf{r}(t) = (t \cos t)\mathbf{i} + (t \sin t)\mathbf{j} + t^2\mathbf{k}, \quad t = 0$

**5.** $\mathbf{r}(t) = t^2\mathbf{i} + (t + (1/3)t^3)\mathbf{j} + (t - (1/3)t^3)\mathbf{k}, \quad t = 0$

**6.** $\mathbf{r}(t) = (e^t \cos t)\mathbf{i} + (e^t \sin t)\mathbf{j} + \sqrt{2}e^t\mathbf{k}, \quad t = 0$

**Finding the TNB Frame**

In Exercises 7 and 8, find **r**, **T**, **N**, and **B** at the given value of $t$. Then find equations for the osculating, normal, and rectifying planes at that value of $t$.

**7.** $\mathbf{r}(t) = (\cos t)\mathbf{i} + (\sin t)\mathbf{j} - \mathbf{k}, \quad t = \pi/4$

**8.** $\mathbf{r}(t) = (\cos t)\mathbf{i} + (\sin t)\mathbf{j} + t\mathbf{k}, \quad t = 0$

In Exercises 9–16 of Section 13.4, you found $\mathbf{T}$, $\mathbf{N}$, and $\kappa$. Now, in the following Exercises 9–16, find $\mathbf{B}$ and $\tau$ for these space curves.

9. $\mathbf{r}(t) = (3 \sin t)\mathbf{i} + (3 \cos t)\mathbf{j} + 4t\mathbf{k}$

10. $\mathbf{r}(t) = (\cos t + t \sin t)\mathbf{i} + (\sin t - t \cos t)\mathbf{j} + 3\mathbf{k}$

11. $\mathbf{r}(t) = (e^t \cos t)\mathbf{i} + (e^t \sin t)\mathbf{j} + 2\mathbf{k}$

12. $\mathbf{r}(t) = (6 \sin 2t)\mathbf{i} + (6 \cos 2t)\mathbf{j} + 5t\mathbf{k}$

13. $\mathbf{r}(t) = (t^3/3)\mathbf{i} + (t^2/2)\mathbf{j}, \quad t > 0$

14. $\mathbf{r}(t) = (\cos^3 t)\mathbf{i} + (\sin^3 t)\mathbf{j}, \quad 0 < t < \pi/2$

15. $\mathbf{r}(t) = t\mathbf{i} + (a \cosh(t/a))\mathbf{j}, \quad a > 0$

16. $\mathbf{r}(t) = (\cosh t)\mathbf{i} - (\sinh t)\mathbf{j} + t\mathbf{k}$

### Physical Applications

17. The speedometer on your car reads a steady 35 km/h. Could you be accelerating? Explain.

18. Can anything be said about the acceleration of a particle that is moving at a constant speed? Give reasons for your answer.

19. Can anything be said about the speed of a particle whose acceleration is always orthogonal to its velocity? Give reasons for your answer.

20. An object of mass $m$ travels along the parabola $y = x^2$ with a constant speed of 10 units/s. What is the force on the object due to its acceleration at $(0, 0)$? at $(2^{1/2}, 2)$? Write your answers in terms of $\mathbf{i}$ and $\mathbf{j}$. (Remember Newton's law, $\mathbf{F} = m\mathbf{a}$.)

### Theory and Examples

21. Show that $\kappa$ and $\tau$ are both zero for the line

$$\mathbf{r}(t) = (x_0 + At)\mathbf{i} + (y_0 + Bt)\mathbf{j} + (z_0 + Ct)\mathbf{k}.$$

22. Show that a moving particle will move in a straight line if the normal component of its acceleration is zero.

23. **A sometime shortcut to curvature** If you already know $|a_N|$ and $|\mathbf{v}|$, then the formula $a_N = \kappa|\mathbf{v}|^2$ gives a convenient way to find the curvature. Use it to find the curvature and radius of curvature of the curve

$$\mathbf{r}(t) = (\cos t + t \sin t)\mathbf{i} + (\sin t - t \cos t)\mathbf{j}, \quad t > 0.$$

(Take $a_N$ and $|\mathbf{v}|$ from Example 1.)

24. What can be said about the torsion of a smooth plane curve $\mathbf{r}(t) = f(t)\mathbf{i} + g(t)\mathbf{j}$? Give reasons for your answer.

25. **Differentiable curves with zero torsion lie in planes** That a sufficiently differentiable curve with zero torsion lies in a plane is a special case of the fact that a particle whose velocity remains perpendicular to a fixed vector $\mathbf{C}$ moves in a plane perpendicular to $\mathbf{C}$. This, in turn, can be viewed as the following result.

Suppose $\mathbf{r}(t) = f(t)\mathbf{i} + g(t)\mathbf{j} + h(t)\mathbf{k}$ is twice differentiable for all $t$ in an interval $[a, b]$, that $\mathbf{r} = 0$ when $t = a$, and that $\mathbf{v} \cdot \mathbf{k} = 0$ for all $t$ in $[a, b]$. Show that $h(t) = 0$ for all $t$ in $[a, b]$. (*Hint:* Start with $\mathbf{a} = d^2\mathbf{r}/dt^2$ and apply the initial conditions in reverse order.)

26. **A formula that calculates $\tau$ from $\mathbf{B}$ and $\mathbf{v}$** If we start with the definition $\tau = -(d\mathbf{B}/ds) \cdot \mathbf{N}$ and apply the Chain Rule to rewrite $d\mathbf{B}/ds$ as

$$\frac{d\mathbf{B}}{ds} = \frac{d\mathbf{B}}{dt}\frac{dt}{ds} = \frac{d\mathbf{B}}{dt}\frac{1}{|\mathbf{v}|},$$

we arrive at the formula

$$\tau = -\frac{1}{|\mathbf{v}|}\left(\frac{d\mathbf{B}}{dt} \cdot \mathbf{N}\right).$$

Use the formula to find the torsion of the helix in Example 2.

### COMPUTER EXPLORATIONS

Rounding the answers to four decimal places, use a CAS to find $\mathbf{v}$, $\mathbf{a}$, speed, $\mathbf{T}$, $\mathbf{N}$, $\mathbf{B}$, $\kappa$, $\tau$, and the tangential and normal components of acceleration for the curves in Exercises 27–30 at the given values of $t$.

27. $\mathbf{r}(t) = (t \cos t)\mathbf{i} + (t \sin t)\mathbf{j} + t\mathbf{k}, \quad t = \sqrt{3}$

28. $\mathbf{r}(t) = (e^t \cos t)\mathbf{i} + (e^t \sin t)\mathbf{j} + e^t\mathbf{k}, \quad t = \ln 2$

29. $\mathbf{r}(t) = (t - \sin t)\mathbf{i} + (1 - \cos t)\mathbf{j} + \sqrt{-t}\mathbf{k}, \quad t = -3\pi$

30. $\mathbf{r}(t) = (3t - t^2)\mathbf{i} + (3t^2)\mathbf{j} + (3t + t^3)\mathbf{k}, \quad t = 1$

# $13.6$ Velocity and Acceleration in Polar Coordinates

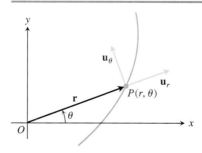

**FIGURE 13.30** The length of $\mathbf{r}$ is the positive polar coordinate $r$ of the point $P$. Thus, $\mathbf{u}_r$, which is $\mathbf{r}/|\mathbf{r}|$, is also $\mathbf{r}/r$. Equations (1) express $\mathbf{u}_r$ and $\mathbf{u}_\theta$ in terms of $\mathbf{i}$ and $\mathbf{j}$.

In this section we derive equations for velocity and acceleration in polar coordinates. These equations are useful for calculating the paths of planets and satellites in space, and we use them to examine Kepler's three laws of planetary motion.

## Motion in Polar and Cylindrical Coordinates

When a particle at $P(r, \theta)$ moves along a curve in the polar coordinate plane, we express its position, velocity, and acceleration in terms of the moving unit vectors

$$\mathbf{u}_r = (\cos \theta)\mathbf{i} + (\sin \theta)\mathbf{j}, \qquad \mathbf{u}_\theta = -(\sin \theta)\mathbf{i} + (\cos \theta)\mathbf{j}, \tag{1}$$

shown in Figure 13.30. The vector $\mathbf{u}_r$ points along the position vector $\overrightarrow{OP}$, so $\mathbf{r} = r\mathbf{u}_r$. The vector $\mathbf{u}_\theta$, orthogonal to $\mathbf{u}_r$, points in the direction of increasing $\theta$.

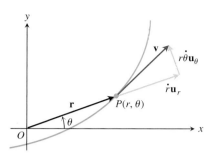

**FIGURE 13.31** In polar coordinates, the velocity vector is

$$\mathbf{v} = \dot{r}\mathbf{u}_r + r\dot{\theta}\mathbf{u}_\theta.$$

We find from Equations (1) that

$$\frac{d\mathbf{u}_r}{d\theta} = -(\sin\theta)\mathbf{i} + (\cos\theta)\mathbf{j} = \mathbf{u}_\theta$$

$$\frac{d\mathbf{u}_\theta}{d\theta} = -(\cos\theta)\mathbf{i} - (\sin\theta)\mathbf{j} = -\mathbf{u}_r.$$

When we differentiate $\mathbf{u}_r$ and $\mathbf{u}_\theta$ with respect to $t$ to find how they change with time, the Chain Rule gives

$$\dot{\mathbf{u}}_r = \frac{d\mathbf{u}_r}{d\theta}\dot{\theta} = \dot{\theta}\mathbf{u}_\theta, \qquad \dot{\mathbf{u}}_\theta = \frac{d\mathbf{u}_\theta}{d\theta}\dot{\theta} = -\dot{\theta}\mathbf{u}_r. \tag{2}$$

Hence, we can express the velocity vector in terms of $\mathbf{u}_r$ and $\mathbf{u}_\theta$ as

$$\mathbf{v} = \dot{\mathbf{r}} = \frac{d}{dt}(r\mathbf{u}_r) = \dot{r}\mathbf{u}_r + r\dot{\mathbf{u}}_r = \dot{r}\mathbf{u}_r + r\dot{\theta}\mathbf{u}_\theta.$$

See Figure 13.31. As in the previous section, we use Newton's dot notation for time derivatives to keep the formulas as simple as we can: $\dot{\mathbf{u}}_r$ means $d\mathbf{u}_r/dt$, $\dot{\theta}$ means $d\theta/dt$, and so on.

The acceleration is

$$\mathbf{a} = \dot{\mathbf{v}} = (\ddot{r}\mathbf{u}_r + \dot{r}\dot{\mathbf{u}}_r) + (\dot{r}\dot{\theta}\mathbf{u}_\theta + r\ddot{\theta}\mathbf{u}_\theta + r\dot{\theta}\dot{\mathbf{u}}_\theta).$$

When Equations (2) are used to evaluate $\dot{\mathbf{u}}_r$ and $\dot{\mathbf{u}}_\theta$ and the components are separated, the equation for acceleration in terms of $\mathbf{u}_r$ and $\mathbf{u}_\theta$ becomes

$$\mathbf{a} = (\ddot{r} - r\dot{\theta}^2)\mathbf{u}_r + (r\ddot{\theta} + 2\dot{r}\dot{\theta})\mathbf{u}_\theta.$$

To extend these equations of motion to space, we add $z\mathbf{k}$ to the right-hand side of the equation $\mathbf{r} = r\mathbf{u}_r$. Then, in these *cylindrical coordinates*, we have

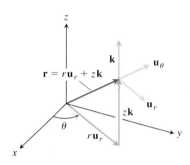

**FIGURE 13.32** Position vector and basic unit vectors in cylindrical coordinates. Notice that $|\mathbf{r}| \neq r$ if $z \neq 0$ since $|\mathbf{r}| = \sqrt{r^2 + z^2}$.

$$
\begin{array}{lll}
\textit{Position:} & \mathbf{r} = r\mathbf{u}_r + z\mathbf{k} & \\
\textit{Velocity:} & \mathbf{v} = \dot{r}\mathbf{u}_r + r\dot{\theta}\mathbf{u}_\theta + \dot{z}\mathbf{k} & \tag{3} \\
\textit{Acceleration:} & \mathbf{a} = (\ddot{r} - r\dot{\theta}^2)\mathbf{u}_r + (r\ddot{\theta} + 2\dot{r}\dot{\theta})\mathbf{u}_\theta + \ddot{z}\mathbf{k} &
\end{array}
$$

The vectors $\mathbf{u}_r$, $\mathbf{u}_\theta$, and $\mathbf{k}$ make a right-handed frame (Figure 13.32) in which

$$\mathbf{u}_r \times \mathbf{u}_\theta = \mathbf{k}, \qquad \mathbf{u}_\theta \times \mathbf{k} = \mathbf{u}_r, \qquad \mathbf{k} \times \mathbf{u}_r = \mathbf{u}_\theta.$$

### Planets Move in Planes

Newton's law of gravitation says that if $\mathbf{r}$ is the radius vector from the center of a sun of mass $M$ to the center of a planet of mass $m$, then the force $\mathbf{F}$ of the gravitational attraction between the planet and sun is

$$\mathbf{F} = -\frac{GmM}{|\mathbf{r}|^2}\frac{\mathbf{r}}{|\mathbf{r}|}$$

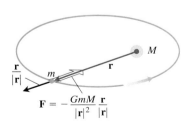

**FIGURE 13.33** The force of gravity is directed along the line joining the centers of mass.

(Figure 13.33). The number $G$ is the **universal gravitational constant**. If we measure mass in kilograms, force in newtons, and distance in meters, $G$ is about $6.6738 \times 10^{-11}$ $\text{Nm}^2\,\text{kg}^{-2}$.

Combining the gravitation law with Newton's second law, $\mathbf{F} = m\ddot{\mathbf{r}}$, for the force acting on the planet gives

$$m\ddot{\mathbf{r}} = -\frac{GmM}{|\mathbf{r}|^2}\frac{\mathbf{r}}{|\mathbf{r}|},$$

$$\ddot{\mathbf{r}} = -\frac{GM}{|\mathbf{r}|^2}\frac{\mathbf{r}}{|\mathbf{r}|}.$$

The planet is therefore accelerated toward the sun's center of mass at all times.

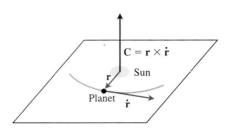

**FIGURE 13.34** A planet that obeys Newton's laws of gravitation and motion travels in the plane through the sun's center of mass perpendicular to $\mathbf{C} = \mathbf{r} \times \dot{\mathbf{r}}$.

Since $\ddot{\mathbf{r}}$ is a scalar multiple of $\mathbf{r}$, we have

$$\mathbf{r} \times \ddot{\mathbf{r}} = \mathbf{0}.$$

From this last equation,

$$\frac{d}{dt}(\mathbf{r} \times \dot{\mathbf{r}}) = \dot{\mathbf{r}} \times \dot{\mathbf{r}} + \mathbf{r} \times \ddot{\mathbf{r}} = \underbrace{\mathbf{r} \times \ddot{\mathbf{r}}}_{0} = \mathbf{0}.$$

It follows that

$$\mathbf{r} \times \dot{\mathbf{r}} = \mathbf{C} \tag{4}$$

for some constant vector $\mathbf{C}$.

Equation (4) tells us that $\mathbf{r}$ and $\dot{\mathbf{r}}$ always lie in a plane perpendicular to $\mathbf{C}$. Hence, the planet moves in a fixed plane through the center of its sun (Figure 13.34). We next see how Kepler's laws describe the motion in a precise way.

### Kepler's First Law (Ellipse Law)

*Kepler's first law* says that a planet's path is an ellipse with the sun at one focus. The eccentricity of the ellipse is

$$e = \frac{r_0 v_0{}^2}{GM} - 1 \tag{5}$$

and the polar equation (see Section 11.7, Equation (5)) is

$$r = \frac{(1 + e)r_0}{1 + e \cos \theta}. \tag{6}$$

Here $v_0$ is the speed when the planet is positioned at its minimum distance $r_0$ from the sun. We omit the lengthy proof. The sun's mass $M$ is $1.99 \times 10^{30}$ kg.

### Kepler's Second Law (Equal Area Law)

*Kepler's second law* says that the radius vector from the sun to a planet (the vector $\mathbf{r}$ in our model) sweeps out equal areas in equal times, as displayed in Figure 13.35. In that figure, we assume the plane of the planet is the $xy$-plane, so the unit vector in the direction of $\mathbf{C}$ is $\mathbf{k}$. We introduce polar coordinates in the plane, choosing as initial line $\theta = 0$, the direction $\mathbf{r}$ when $|\mathbf{r}| = r$ is a minimum value. Then at $t = 0$, we have $r(0) = r_0$ being a minimum so

$$\dot{r}\big|_{t=0} = \frac{dr}{dt}\bigg|_{t=0} = 0 \quad \text{and} \quad v_0 = |\mathbf{v}|_{t=0} = \big[r\dot\theta\big]_{t=0}. \qquad \text{Eq. (3), } \dot{z} = 0$$

To derive Kepler's second law, we use Equation (3) to evaluate the cross product $\mathbf{C} = \mathbf{r} \times \dot{\mathbf{r}}$ from Equation (4):

$$\mathbf{C} = \mathbf{r} \times \dot{\mathbf{r}} = \mathbf{r} \times \mathbf{v}$$

$$= r\mathbf{u}_r \times (\dot{r}\mathbf{u}_r + r\dot\theta\mathbf{u}_\theta) \qquad \text{Eq. (3), } \dot{z} = 0$$

$$= r\dot{r}\underbrace{(\mathbf{u}_r \times \mathbf{u}_r)}_{0} + r(r\dot\theta)\underbrace{(\mathbf{u}_r \times \mathbf{u}_\theta)}_{\mathbf{k}}$$

$$= r(r\dot\theta)\mathbf{k}. \tag{7}$$

Setting $t$ equal to zero shows that

$$\mathbf{C} = \big[r(r\dot\theta)\big]_{t=0}\mathbf{k} = r_0 v_0 \mathbf{k}.$$

Substituting this value for $\mathbf{C}$ in Equation (7) gives

$$r_0 v_0 \mathbf{k} = r^2\dot\theta\mathbf{k}, \quad \text{or} \quad r^2\dot\theta = r_0 v_0.$$

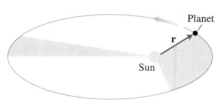

**FIGURE 13.35** The line joining a planet to its sun sweeps over equal areas in equal times.

This is where the area comes in. The area differential in polar coordinates is

$$dA = \frac{1}{2}r^2\, d\theta$$

(Section 11.5). Accordingly, $dA/dt$ has the constant value

$$\frac{dA}{dt} = \frac{1}{2}r^2\dot{\theta} = \frac{1}{2}r_0v_0. \tag{8}$$

So $dA/dt$ is constant, giving Kepler's second law.

### Kepler's Third Law (Time–Distance Law)

The time $T$ it takes a planet to go around its sun once is the planet's **orbital period**. *Kepler's third law* says that $T$ and the orbit's semimajor axis $a$ are related by the equation

$$\frac{T^2}{a^3} = \frac{4\pi^2}{GM}.$$

Since the right-hand side of this equation is constant within a given solar system, the ratio of $T^2$ to $a^3$ *is the same for every planet in the system.*

Here is a partial derivation of Kepler's third law. The area enclosed by the planet's elliptical orbit is calculated as follows:

$$\text{Area} = \int_0^T dA$$

$$= \int_0^T \frac{1}{2}r_0v_0\, dt \qquad \text{Eq. (8)}$$

$$= \frac{1}{2}Tr_0v_0.$$

If $b$ is the semiminor axis, the area of the ellipse is $\pi ab$, so

$$T = \frac{2\pi ab}{r_0v_0} = \frac{2\pi a^2}{r_0v_0}\sqrt{1 - e^2}. \qquad \substack{\text{For any ellipse,} \\ b = a\sqrt{1-e^2}.} \tag{9}$$

It remains only to express $a$ and $e$ in terms of $r_0$, $v_0$, $G$, and $M$. Equation (5) does this for $e$. For $a$, we observe that setting $\theta$ equal to $\pi$ in Equation (6) gives

$$r_{\max} = r_0\frac{1 + e}{1 - e}.$$

Hence, from Figure 13.36,

$$2a = r_0 + r_{\max} = \frac{2r_0}{1 - e} = \frac{2r_0GM}{2GM - r_0v_0{}^2}. \tag{10}$$

Squaring both sides of Equation (9) and substituting the results of Equations (5) and (10) produces Kepler's third law (Exercise 9).

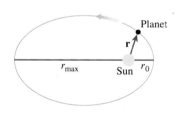

**FIGURE 13.36** The length of the major axis of the ellipse is $2a = r_0 + r_{\max}$.

---

## Exercises 13.6

In Exercises 1–5, find the velocity and acceleration vectors in terms of $\mathbf{u}_r$ and $\mathbf{u}_\theta$.

1. $r = a(1 - \cos\theta)$ and $\dfrac{d\theta}{dt} = 3$

2. $r = a\sin 2\theta$ and $\dfrac{d\theta}{dt} = 2t$

3. $r = e^{a\theta}$ and $\dfrac{d\theta}{dt} = 2$

4. $r = a(1 + \sin t)$ and $\theta = 1 - e^{-t}$

5. $r = 2\cos 4t$ and $\theta = 2t$

6. **Type of orbit** For what values of $v_0$ in Equation (5) is the orbit in Equation (6) a circle? An ellipse? A parabola? A hyperbola?

7. **Circular orbits** Show that a planet in a circular orbit moves with a constant speed. (*Hint:* This is a consequence of one of Kepler's laws.)

8. Suppose that $\mathbf{r}$ is the position vector of a particle moving along a plane curve and $dA/dt$ is the rate at which the vector sweeps

out area. Without introducing coordinates, and assuming the necessary derivatives exist, give a geometric argument based on increments and limits for the validity of the equation

$$\frac{dA}{dt} = \frac{1}{2}|\mathbf{r} \times \dot{\mathbf{r}}|.$$

**9. Kepler's third law** Complete the derivation of Kepler's third law (the part following Equation (10)).

**10.** Do the data in the accompanying table support Kepler's third law? Give reasons for your answer.

| | Semimajor axis | |
| Planet | $a$ ($10^{10}$ m) | Period $T$ (years) |
| --- | --- | --- |
| Mercury | 5.79 | 0.241 |
| Venus | 10.81 | 0.615 |
| Mars | 22.78 | 1.881 |
| Saturn | 142.70 | 29.457 |

**11. Earth's major axis** Estimate the length of the major axis of Earth's orbit if its orbital period is 365.256 days.

**12.** Estimate the length of the major axis of the orbit of Uranus if its orbital period is 84 years.

**13.** The eccentricity of Earth's orbit is $e = 0.0167$, so the orbit is nearly circular, with radius approximately $150 \times 10^6$ km. Find the rate $dA/dt$ in units of $km^2/s$ satisfying Kepler's second law.

**14. Jupiter's orbital period** Estimate the orbital period of Jupiter, assuming that $a = 77.8 \times 10^{10}$ m.

**15. Mass of Jupiter** Io is one of the moons of Jupiter. It has a semimajor axis of $0.042 \times 10^{10}$ m and an orbital period of 1.769 days. Use these data to estimate the mass of Jupiter.

**16. Distance from Earth to the moon** The period of the moon's rotation around Earth is $2.36055 \times 10^6$ s. Estimate the distance to the moon.

---

## Chapter **13** Questions to Guide Your Review

**1.** State the rules for differentiating and integrating vector functions. Give examples.

**2.** How do you define and calculate the velocity, speed, direction of motion, and acceleration of a body moving along a sufficiently differentiable space curve? Give an example.

**3.** What is special about the derivatives of vector functions of constant length? Give an example.

**4.** What are the vector and parametric equations for ideal projectile motion? How do you find a projectile's maximum height, flight time, and range? Give examples.

**5.** How do you define and calculate the length of a segment of a smooth space curve? Give an example. What mathematical assumptions are involved in the definition?

**6.** How do you measure distance along a smooth curve in space from a preselected base point? Give an example.

**7.** What is a differentiable curve's unit tangent vector? Give an example.

**8.** Define curvature, circle of curvature (osculating circle), center of curvature, and radius of curvature for twice-differentiable curves in the plane. Give examples. What curves have zero curvature? Constant curvature?

**9.** What is a plane curve's principal normal vector? When is it defined? Which way does it point? Give an example.

**10.** How do you define $\mathbf{N}$ and $\kappa$ for curves in space? How are these quantities related? Give examples.

**11.** What is a curve's binormal vector? Give an example. How is this vector related to the curve's torsion? Give an example.

**12.** What formulas are available for writing a moving object's acceleration as a sum of its tangential and normal components? Give an example. Why might one want to write the acceleration this way? What if the object moves at a constant speed? At a constant speed around a circle?

**13.** State Kepler's laws.

---

## Chapter **13** Practice Exercises

**Motion in the Plane**

In Exercises 1 and 2, graph the curves and sketch their velocity and acceleration vectors at the given values of $t$. Then write $\mathbf{a}$ in the form $\mathbf{a} = a_T\mathbf{T} + a_N\mathbf{N}$ without finding $\mathbf{T}$ and $\mathbf{N}$, and find the value of $\kappa$ at the given values of $t$.

**1.** $\mathbf{r}(t) = (4 \cos t)\mathbf{i} + (\sqrt{2} \sin t)\mathbf{j}, \quad t = 0$ and $\pi/4$

**2.** $\mathbf{r}(t) = (\sqrt{3} \sec t)\mathbf{i} + (\sqrt{3} \tan t)\mathbf{j}, \quad t = 0$

**3.** The position of a particle in the plane at time $t$ is

$$\mathbf{r} = \frac{1}{\sqrt{1 + t^2}}\mathbf{i} + \frac{t}{\sqrt{1 + t^2}}\mathbf{j}.$$

Find the particle's highest speed.

**4.** Suppose $\mathbf{r}(t) = (e^t \cos t)\mathbf{i} + (e^t \sin t)\mathbf{j}$. Show that the angle between $\mathbf{r}$ and $\mathbf{a}$ never changes. What *is* the angle?

5. **Finding curvature** At point $P$, the velocity and acceleration of a particle moving in the plane are $\mathbf{v} = 3\mathbf{i} + 4\mathbf{j}$ and $\mathbf{a} = 5\mathbf{i} + 15\mathbf{j}$. Find the curvature of the particle's path at $P$.

6. Find the point on the curve $y = e^x$ where the curvature is greatest.

7. A particle moves around the unit circle in the $xy$-plane. Its position at time $t$ is $\mathbf{r} = x\mathbf{i} + y\mathbf{j}$, where $x$ and $y$ are differentiable functions of $t$. Find $dy/dt$ if $\mathbf{v} \cdot \mathbf{i} = y$. Is the motion clockwise or counterclockwise?

8. You send a message through a pneumatic tube that follows the curve $9y = x^3$ (distance in meters). At the point $(3, 3)$, $\mathbf{v} \cdot \mathbf{i} = 4$ and $\mathbf{a} \cdot \mathbf{i} = -2$. Find the values of $\mathbf{v} \cdot \mathbf{j}$ and $\mathbf{a} \cdot \mathbf{j}$ at $(3, 3)$.

9. **Characterizing circular motion** A particle moves in the plane so that its velocity and position vectors are always orthogonal. Show that the particle moves in a circle centered at the origin.

10. **Speed along a cycloid** A circular wheel with radius 1 m and center $C$ rolls to the right along the $x$-axis at a half-turn per second. (See the accompanying figure.) At time $t$ seconds, the position vector of the point $P$ on the wheel's circumference is

$$\mathbf{r} = (\pi t - \sin \pi t)\mathbf{i} + (1 - \cos \pi t)\mathbf{j}.$$

a. Sketch the curve traced by $P$ during the interval $0 \le t \le 3$.

b. Find $\mathbf{v}$ and $\mathbf{a}$ at $t = 0, 1, 2,$ and 3 and add these vectors to your sketch.

c. At any given time, what is the forward speed of the topmost point of the wheel? Of $C$?

**Projectile Motion**

11. **Shot put** A shot leaves the thrower's hand 2 m above the ground at a $45°$ angle at $14$ m/s. Where is it 3 s later?

12. **Javelin** A javelin leaves the thrower's hand 2.5 m above the ground at a $45°$ angle at $24$ m/s. How high does it go?

13. A golf ball is hit with an initial speed $v_0$ at an angle $\alpha$ to the horizontal from a point that lies at the foot of a straight-sided hill that is inclined at an angle $\phi$ to the horizontal, where

$$0 < \phi < \alpha < \frac{\pi}{2}.$$

Show that the ball lands at a distance

$$\frac{2v_0{}^2 \cos \alpha}{g \cos^2 \phi} \sin (\alpha - \phi),$$

measured up the face of the hill. Hence, show that the greatest range that can be achieved for a given $v_0$ occurs when

$\alpha = (\phi/2) + (\pi/4)$, i.e., when the initial velocity vector bisects the angle between the vertical and the hill.

T 14. **Javelin** In Potsdam in 1988, Petra Felke of (then) East Germany set a women's world record by throwing a javelin 80 m.

a. Assuming that Felke launched the javelin at a $40°$ angle to the horizontal 2 m above the ground, what was the javelin's initial speed?

b. How high did the javelin go?

**Motion in Space**

Find the lengths of the curves in Exercises 15 and 16.

15. $\mathbf{r}(t) = (2 \cos t)\mathbf{i} + (2 \sin t)\mathbf{j} + t^2\mathbf{k}, \quad 0 \le t \le \pi/4$

16. $\mathbf{r}(t) = (3 \cos t)\mathbf{i} + (3 \sin t)\mathbf{j} + 2t^{3/2}\mathbf{k}, \quad 0 \le t \le 3$

In Exercises 17–20, find $\mathbf{T}, \mathbf{N}, \mathbf{B}, \kappa,$ and $\tau$ at the given value of $t$.

17. $\mathbf{r}(t) = \frac{4}{9}(1 + t)^{3/2}\mathbf{i} + \frac{4}{9}(1 - t)^{3/2}\mathbf{j} + \frac{1}{3}t\mathbf{k}, \quad t = 0$

18. $\mathbf{r}(t) = (e^t \sin 2t)\mathbf{i} + (e^t \cos 2t)\mathbf{j} + 2e^t\mathbf{k}, \quad t = 0$

19. $\mathbf{r}(t) = t\mathbf{i} + \frac{1}{2}e^{2t}\mathbf{j}, \quad t = \ln 2$

20. $\mathbf{r}(t) = (3 \cosh 2t)\mathbf{i} + (3 \sinh 2t)\mathbf{j} + 6t\mathbf{k}, \quad t = \ln 2$

In Exercises 21 and 22, write $\mathbf{a}$ in the form $\mathbf{a} = a_{\mathrm{T}}\mathbf{T} + a_{\mathrm{N}}\mathbf{N}$ at $t = 0$ without finding $\mathbf{T}$ and $\mathbf{N}$.

21. $\mathbf{r}(t) = (2 + 3t + 3t^2)\mathbf{i} + (4t + 4t^2)\mathbf{j} - (6 \cos t)\mathbf{k}$

22. $\mathbf{r}(t) = (2 + t)\mathbf{i} + (t + 2t^2)\mathbf{j} + (1 + t^2)\mathbf{k}$

23. Find $\mathbf{T}, \mathbf{N}, \mathbf{B}, \kappa,$ and $\tau$ as functions of $t$ if

$$\mathbf{r}(t) = (\sin t)\mathbf{i} + \left(\sqrt{2} \cos t\right)\mathbf{j} + (\sin t)\mathbf{k}.$$

24. At what times in the interval $0 \le t \le \pi$ are the velocity and acceleration vectors of the motion $\mathbf{r}(t) = \mathbf{i} + (5 \cos t)\mathbf{j} + (3 \sin t)\mathbf{k}$ orthogonal?

25. The position of a particle moving in space at time $t \ge 0$ is

$$\mathbf{r}(t) = 2\mathbf{i} + \left(4 \sin \frac{t}{2}\right)\mathbf{j} + \left(3 - \frac{t}{\pi}\right)\mathbf{k}.$$

Find the first time $\mathbf{r}$ is orthogonal to the vector $\mathbf{i} - \mathbf{j}$.

26. Find equations for the osculating, normal, and rectifying planes of the curve $\mathbf{r}(t) = t\mathbf{i} + t^2\mathbf{j} + t^3\mathbf{k}$ at the point $(1, 1, 1)$.

27. Find parametric equations for the line that is tangent to the curve $\mathbf{r}(t) = e^t\mathbf{i} + (\sin t)\mathbf{j} + \ln(1 - t)\mathbf{k}$ at $t = 0$.

28. Find parametric equations for the line tangent to the helix $\mathbf{r}(t) = \left(\sqrt{2} \cos t\right)\mathbf{i} + \left(\sqrt{2} \sin t\right)\mathbf{j} + t\mathbf{k}$ at the point where $t = \pi/4$.

**Theory and Examples**

29. **Synchronous curves** By eliminating $\alpha$ from the ideal projectile equations

$$x = (v_0 \cos \alpha)t, \quad y = (v_0 \sin \alpha)t - \frac{1}{2}gt^2,$$

show that $x^2 + (y + gt^2/2)^2 = v_0{}^2t^2$. This shows that projectiles launched simultaneously from the origin at the same initial speed will, at any given instant, all lie on the circle of radius $v_0t$ centered at $(0, -gt^2/2)$, regardless of their launch angle. These circles are the *synchronous curves* of the launching.

**30. Radius of curvature** Show that the radius of curvature of a twice-differentiable plane curve $\mathbf{r}(t) = f(t)\mathbf{i} + g(t)\mathbf{j}$ is given by the formula

$$\rho = \frac{\sqrt{\dot{x}^2 + \dot{y}^2}}{\sqrt{\ddot{x}^2 + \ddot{y}^2 - \ddot{s}^2}}, \quad \text{where} \quad \ddot{s} = \frac{d}{dt}\sqrt{\dot{x}^2 + \dot{y}^2}.$$

**31. An alternative definition of curvature in the plane** An alternative definition gives the curvature of a sufficiently differentiable plane curve to be $|d\phi/ds|$, where $\phi$ is the angle between $\mathbf{T}$ and $\mathbf{i}$ (Figure 13.37a). Figure 13.37b shows the distance $s$ measured

(a)

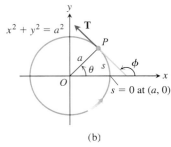

(b)

**FIGURE 13.37** Figures for Exercise 31.

counterclockwise around the circle $x^2 + y^2 = a^2$ from the point $(a, 0)$ to a point $P$, along with the angle $\phi$ at $P$. Calculate the circle's curvature using the alternative definition. (*Hint:* $\phi = \theta + \pi/2$.)

**32. The view from *Skylab 4*** What percentage of Earth's surface area could the astronauts see when *Skylab 4* was at its apogee height, 437 km above the surface? To find out, model the visible surface as the surface generated by revolving the circular arc $GT$, shown here, about the $y$-axis. Then carry out these steps:

**1.** Use similar triangles in the figure to show that $y_0/6380 = 6380/(6380 + 437)$. Solve for $y_0$.

**2.** To four significant digits, calculate the visible area as

$$VA = \int_{y_0}^{6380} 2\pi x \sqrt{1 + \left(\frac{dx}{dy}\right)^2}\, dy.$$

**3.** Express the result as a percentage of Earth's surface area.

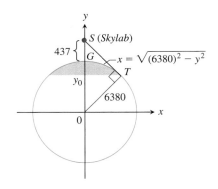

---

## Chapter **13** Additional and Advanced Exercises

### Applications

**1.** A frictionless particle $P$, starting from rest at time $t = 0$ at the point $(a, 0, 0)$, slides down the helix

$$\mathbf{r}(\theta) = (a \cos \theta)\mathbf{i} + (a \sin \theta)\mathbf{j} + b\theta\mathbf{k} \quad (a, b > 0)$$

under the influence of gravity, as in the accompanying figure. The $\theta$ in this equation is the cylindrical coordinate $\theta$ and the helix is the curve $r = a$, $z = b\theta$, $\theta \geq 0$, in cylindrical coordinates. We assume $\theta$ to be a differentiable function of $t$ for the motion. The law of conservation of energy tells us that the particle's speed after it has fallen straight down a distance $z$ is $\sqrt{2gz}$, where $g$ is the constant acceleration of gravity.

**a.** Find the angular velocity $d\theta/dt$ when $\theta = 2\pi$.

**b.** Express the particle's $\theta$- and $z$-coordinates as functions of $t$.

**c.** Express the tangential and normal components of the velocity $d\mathbf{r}/dt$ and acceleration $d^2\mathbf{r}/dt^2$ as functions of $t$. Does the acceleration have any nonzero component in the direction of the binormal vector $\mathbf{B}$?

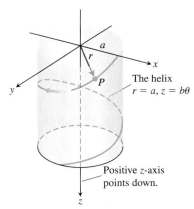

**2.** Suppose the curve in Exercise 1 is replaced by the conical helix $r = a\theta$, $z = b\theta$ shown in the accompanying figure.

**a.** Express the angular velocity $d\theta/dt$ as a function of $\theta$.

**b.** Express the distance the particle travels along the helix as a function of $\theta$.

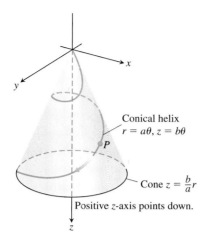

Conical helix
$r = a\theta, z = b\theta$

$P$

Cone $z = \dfrac{b}{a} r$

Positive $z$-axis points down.

## Motion in Polar and Cylindrical Coordinates

**3.** Deduce from the orbit equation

$$r = \frac{(1 + e)r_0}{1 + e \cos \theta}$$

that a planet is closest to its sun when $\theta = 0$ and show that $r = r_0$ at that time.

**T** **4. A Kepler equation** The problem of locating a planet in its orbit at a given time and date eventually leads to solving "Kepler" equations of the form

$$f(x) = x - 1 - \frac{1}{2} \sin x = 0.$$

   **a.** Show that this particular equation has a solution between $x = 0$ and $x = 2$.

   **b.** With your computer or calculator in radian mode, use Newton's method to find the solution to as many places as you can.

**5.** In Section 13.6, we found the velocity of a particle moving in the plane to be

$$\mathbf{v} = \dot{x}\mathbf{i} + \dot{y}\mathbf{j} = \dot{r}\mathbf{u}_r + r\dot{\theta}\mathbf{u}_\theta.$$

   **a.** Express $\dot{x}$ and $\dot{y}$ in terms of $\dot{r}$ and $r\dot{\theta}$ by evaluating the dot products $\mathbf{v} \cdot \mathbf{i}$ and $\mathbf{v} \cdot \mathbf{j}$.

   **b.** Express $\dot{r}$ and $r\dot{\theta}$ in terms of $\dot{x}$ and $\dot{y}$ by evaluating the dot products $\mathbf{v} \cdot \mathbf{u}_r$ and $\mathbf{v} \cdot \mathbf{u}_\theta$.

**6.** Express the curvature of a twice-differentiable curve $r = f(\theta)$ in the polar coordinate plane in terms of $f$ and its derivatives.

**7.** A slender rod through the origin of the polar coordinate plane rotates (in the plane) about the origin at the rate of 3 rad/min. A beetle starting from the point (2, 0) crawls along the rod toward the origin at the rate of 1 cm/min.

   **a.** Find the beetle's acceleration and velocity in polar form when it is halfway to (1 cm from) the origin.

**T** **b.** To the nearest millimeter, what will be the length of the path the beetle has traveled by the time it reaches the origin?

**8. Arc length in cylindrical coordinates**

   **a.** Show that when you express $ds^2 = dx^2 + dy^2 + dz^2$ in terms of cylindrical coordinates, you get $ds^2 = dr^2 + r^2 \, d\theta^2 + dz^2$.

   **b.** Interpret this result geometrically in terms of the edges and a diagonal of a box. Sketch the box.

   **c.** Use the result in part (a) to find the length of the curve $r = e^\theta, z = e^\theta, 0 \le \theta \le \ln 8$.

**9. Unit vectors for position and motion in cylindrical coordinates** When the position of a particle moving in space is given in cylindrical coordinates, the unit vectors we use to describe its position and motion are

$$\mathbf{u}_r = (\cos \theta)\mathbf{i} + (\sin \theta)\mathbf{j}, \qquad \mathbf{u}_\theta = -(\sin \theta)\mathbf{i} + (\cos \theta)\mathbf{j},$$

and $\mathbf{k}$ (see accompanying figure). The particle's position vector is then $\mathbf{r} = r\mathbf{u}_r + z\mathbf{k}$, where $r$ is the positive polar distance coordinate of the particle's position.

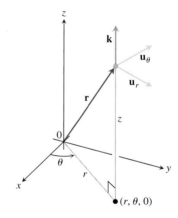

   **a.** Show that $\mathbf{u}_r$, $\mathbf{u}_\theta$, and $\mathbf{k}$, in this order, form a right-handed frame of unit vectors.

   **b.** Show that

$$\frac{d\mathbf{u}_r}{d\theta} = \mathbf{u}_\theta \quad \text{and} \quad \frac{d\mathbf{u}_\theta}{d\theta} = -\mathbf{u}_r.$$

   **c.** Assuming that the necessary derivatives with respect to $t$ exist, express $\mathbf{v} = \dot{\mathbf{r}}$ and $\mathbf{a} = \ddot{\mathbf{r}}$ in terms of $\mathbf{u}_r$, $\mathbf{u}_\theta$, $\mathbf{k}$, $\dot{r}$, and $\dot{\theta}$.

   **d. Conservation of angular momentum** Let $\mathbf{r}(t)$ denote the position in space of a moving object at time $t$. Suppose the force acting on the object at time $t$ is

$$\mathbf{F}(t) = -\frac{c}{|\mathbf{r}(t)|^3} \mathbf{r}(t),$$

   where $c$ is a constant. In physics the **angular momentum** of an object at time $t$ is defined to be $\mathbf{L}(t) = \mathbf{r}(t) \times m\mathbf{v}(t)$, where $m$ is the mass of the object and $\mathbf{v}(t)$ is the velocity. Prove that angular momentum is a conserved quantity; i.e., prove that $\mathbf{L}(t)$ is a constant vector, independent of time. Remember Newton's law $\mathbf{F} = m\mathbf{a}$. (This is a calculus problem, not a physics problem.)

## Chapter 13 Technology Application Projects

Mathematica / Maple Modules:

### *Radar Tracking of a Moving Object*

Visualize position, velocity, and acceleration vectors to analyze motion.

### *Parametric and Polar Equations with a Figure Skater*

Visualize position, velocity, and acceleration vectors to analyze motion.

### *Moving in Three Dimensions*

Compute distance traveled, speed, curvature, and torsion for motion along a space curve. Visualize and compute the tangential, normal, and binormal vectors associated with motion along a space curve.

# 14

# Partial Derivatives

**OVERVIEW** Many functions depend on more than one independent variable. For instance, the volume of a right circular cylinder is a function $V = \pi r^2 h$ of its radius and its height, so it is a function $V(r, h)$ of two variables $r$ and $h$. The speed of sound through seawater is primarily a function of salinity $S$ and temperature $T$. The surface area of the human body is a function of its height $h$ and weight $w$. The monthly payment on a home mortgage is a function of the principal borrowed $P$, the interest rate $i$, and the term $t$ of the loan.

In this chapter we extend the basic ideas of single-variable differential calculus to functions of several variables. Their derivatives are more varied and interesting because of the different ways the variables can interact. The applications of these derivatives are also more varied than for single-variable calculus, and in the next chapter we will see that the same is true for integrals involving several variables.

## 14.1 Functions of Several Variables

Real-valued functions of several independent real variables are defined analogously to functions in the single-variable case. Points in the domain are ordered pairs (triples, quadruples, $n$-tuples) of real numbers, and values in the range are real numbers as we have worked with all along.

> **DEFINITIONS** Suppose $D$ is a set of $n$-tuples of real numbers $(x_1, x_2, \ldots, x_n)$. A **real-valued function** $f$ on $D$ is a rule that assigns a unique (single) real number
>
> $$w = f(x_1, x_2, \ldots, x_n)$$
>
> to each element in $D$. The set $D$ is the function's **domain**. The set of $w$-values taken on by $f$ is the function's **range**. The symbol $w$ is the **dependent variable** of $f$, and $f$ is said to be a function of the $n$ **independent variables** $x_1$ to $x_n$. We also call the $x_j$'s the function's **input variables** and call $w$ the function's **output variable**.

If $f$ is a function of two independent variables, we usually call the independent variables $x$ and $y$ and the dependent variable $z$, and we picture the domain of $f$ as a region in the $xy$-plane (Figure 14.1). If $f$ is a function of three independent variables, we call the independent variables $x$, $y$, and $z$ and the dependent variable $w$, and we picture the domain as a region in space.

In applications, we tend to use letters that remind us of what the variables stand for. To say that the volume of a right circular cylinder is a function of its radius and height, we might write $V = f(r, h)$. To be more specific, we might replace the notation $f(r, h)$ by the formula

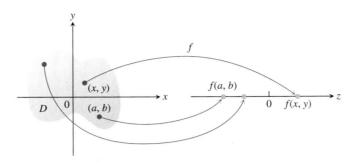

**FIGURE 14.1**  An arrow diagram for the function $z = f(x, y)$.

that calculates the value of $V$ from the values of $r$ and $h$, and write $V = \pi r^2 h$. In either case, $r$ and $h$ would be the independent variables and $V$ the dependent variable of the function.

As usual, we evaluate functions defined by formulas by substituting the values of the independent variables in the formula and calculating the corresponding value of the dependent variable. For example, the value of $f(x, y, z) = \sqrt{x^2 + y^2 + z^2}$ at the point $(3, 0, 4)$ is

$$f(3, 0, 4) = \sqrt{(3)^2 + (0)^2 + (4)^2} = \sqrt{25} = 5.$$

## Domains and Ranges

In defining a function of more than one variable, we follow the usual practice of excluding inputs that lead to complex numbers or division by zero. If $f(x, y) = \sqrt{y - x^2}$, $y$ cannot be less than $x^2$. If $f(x, y) = 1/(xy)$, $xy$ cannot be zero. The domain of a function is assumed to be the largest set for which the defining rule generates real numbers, unless the domain is otherwise specified explicitly. The range consists of the set of output values for the dependent variable.

## EXAMPLE 1

**(a)** These are functions of two variables. Note the restrictions that may apply to their domains in order to obtain a real value for the dependent variable $z$.

| Function | Domain | Range |
|---|---|---|
| $z = \sqrt{y - x^2}$ | $y \geq x^2$ | $[0, \infty)$ |
| $z = \dfrac{1}{xy}$ | $xy \neq 0$ | $(-\infty, 0) \cup (0, \infty)$ |
| $z = \sin xy$ | Entire plane | $[-1, 1]$ |

**(b)** These are functions of three variables with restrictions on some of their domains.

| Function | Domain | Range |
|---|---|---|
| $w = \sqrt{x^2 + y^2 + z^2}$ | Entire space | $[0, \infty)$ |
| $w = \dfrac{1}{x^2 + y^2 + z^2}$ | $(x, y, z) \neq (0, 0, 0)$ | $(0, \infty)$ |
| $w = xy \ln z$ | Half-space $z > 0$ | $(-\infty, \infty)$ |

## Functions of Two Variables

Regions in the plane can have interior points and boundary points just like intervals on the real line. Closed intervals $[a, b]$ include their boundary points, open intervals $(a, b)$ don't include their boundary points, and intervals such as $[a, b)$ are neither open nor closed.

(a) Interior point

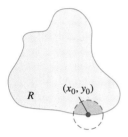

(b) Boundary point

**FIGURE 14.2** Interior points and boundary points of a plane region $R$. An interior point is necessarily a point of $R$. A boundary point of $R$ need not belong to $R$.

**DEFINITIONS** A point $(x_0, y_0)$ in a region (set) $R$ in the $xy$-plane is an **interior point** of $R$ if it is the center of a disk of positive radius that lies entirely in $R$ (Figure 14.2). A point $(x_0, y_0)$ is a **boundary point** of $R$ if every disk centered at $(x_0, y_0)$ contains points that lie outside of $R$ as well as points that lie in $R$. (The boundary point itself need not belong to $R$.)

The interior points of a region, as a set, make up the **interior** of the region. The region's boundary points make up its **boundary**. A region is **open** if it consists entirely of interior points. A region is **closed** if it contains all its boundary points (Figure 14.3).

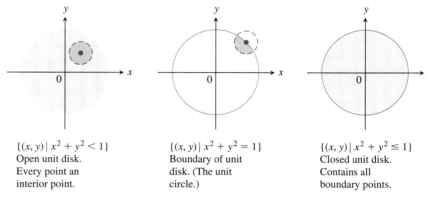

$\{(x, y) \mid x^2 + y^2 < 1\}$
Open unit disk.
Every point an
interior point.

$\{(x, y) \mid x^2 + y^2 = 1\}$
Boundary of unit
disk. (The unit
circle.)

$\{(x, y) \mid x^2 + y^2 \leq 1\}$
Closed unit disk.
Contains all
boundary points.

**FIGURE 14.3** Interior points and boundary points of the unit disk in the plane.

As with a half-open interval of real numbers $[a, b)$, some regions in the plane are neither open nor closed. If you start with the open disk in Figure 14.3 and add to it some, but not all, of its boundary points, the resulting set is neither open nor closed. The boundary points that *are* there keep the set from being open. The absence of the remaining boundary points keeps the set from being closed.

**DEFINITIONS** A region in the plane is **bounded** if it lies inside a disk of finite radius. A region is **unbounded** if it is not bounded.

Examples of *bounded* sets in the plane include line segments, triangles, interiors of triangles, rectangles, circles, and disks. Examples of *unbounded* sets in the plane include lines, coordinate axes, the graphs of functions defined on infinite intervals, quadrants, half-planes, and the plane itself.

**EXAMPLE 2** Describe the domain of the function $f(x, y) = \sqrt{y - x^2}$.

**Solution** Since $f$ is defined only where $y - x^2 \geq 0$, the domain is the closed, unbounded region shown in Figure 14.4. The parabola $y = x^2$ is the boundary of the domain. The points above the parabola make up the domain's interior. ∎

### Graphs, Level Curves, and Contours of Functions of Two Variables

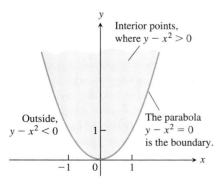

**FIGURE 14.4** The domain of $f(x, y)$ in Example 2 consists of the shaded region and its bounding parabola.

There are two standard ways to picture the values of a function $f(x, y)$. One is to draw and label curves in the domain on which $f$ has a constant value. The other is to sketch the surface $z = f(x, y)$ in space.

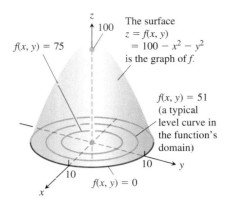

**FIGURE 14.5** The graph and selected level curves of the function $f(x, y)$ in Example 3.

**DEFINITIONS** The set of points in the plane where a function $f(x, y)$ has a constant value $f(x, y) = c$ is called a **level curve** of $f$. The set of all points $(x, y, f(x, y))$ in space, for $(x, y)$ in the domain of $f$, is called the **graph** of $f$. The graph of $f$ is also called the **surface $z = f(x, y)$.**

**EXAMPLE 3** Graph $f(x, y) = 100 - x^2 - y^2$ and plot the level curves $f(x, y) = 0$, $f(x, y) = 51$, and $f(x, y) = 75$ in the domain of $f$ in the plane.

**Solution** The domain of $f$ is the entire $xy$-plane, and the range of $f$ is the set of real numbers less than or equal to 100. The graph is the paraboloid $z = 100 - x^2 - y^2$, the positive portion of which is shown in Figure 14.5.

The level curve $f(x, y) = 0$ is the set of points in the $xy$-plane at which

$$f(x, y) = 100 - x^2 - y^2 = 0, \qquad \text{or} \qquad x^2 + y^2 = 100,$$

which is the circle of radius 10 centered at the origin. Similarly, the level curves $f(x, y) = 51$ and $f(x, y) = 75$ (Figure 14.5) are the circles

$$f(x, y) = 100 - x^2 - y^2 = 51, \qquad \text{or} \qquad x^2 + y^2 = 49$$
$$f(x, y) = 100 - x^2 - y^2 = 75, \qquad \text{or} \qquad x^2 + y^2 = 25.$$

The level curve $f(x, y) = 100$ consists of the origin alone. (It is still a level curve.)

If $x^2 + y^2 > 100$, then the values of $f(x, y)$ are negative. For example, the circle $x^2 + y^2 = 144$, which is the circle centered at the origin with radius 12, gives the constant value $f(x, y) = -44$ and is a level curve of $f$. ∎

The curve in space in which the plane $z = c$ cuts a surface $z = f(x, y)$ is made up of the points that represent the function value $f(x, y) = c$. It is called the **contour curve** $f(x, y) = c$ to distinguish it from the level curve $f(x, y) = c$ in the domain of $f$. Figure 14.6 shows the contour curve $f(x, y) = 75$ on the surface $z = 100 - x^2 - y^2$ defined by the function $f(x, y) = 100 - x^2 - y^2$. The contour curve lies directly above the circle $x^2 + y^2 = 25$, which is the level curve $f(x, y) = 75$ in the function's domain.

Not everyone makes this distinction, however, and you may wish to call both kinds of curves by a single name and rely on context to convey which one you have in mind. On most maps, for example, the curves that represent constant elevation (height above sea level) are called contours, not level curves (Figure 14.7).

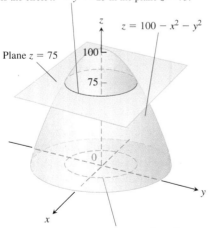

The contour curve $f(x, y) = 100 - x^2 - y^2 = 75$ is the circle $x^2 + y^2 = 25$ in the plane $z = 75$.

The level curve $f(x, y) = 100 - x^2 - y^2 = 75$ is the circle $x^2 + y^2 = 25$ in the $xy$-plane.

**FIGURE 14.6** A plane $z = c$ parallel to the $xy$-plane intersecting a surface $z = f(x, y)$ produces a contour curve.

### Functions of Three Variables

In the plane, the points where a function of two independent variables has a constant value $f(x, y) = c$ make a curve in the function's domain. In space, the points where a function of three independent variables has a constant value $f(x, y, z) = c$ make a surface in the function's domain.

**DEFINITION** The set of points $(x, y, z)$ in space where a function of three independent variables has a constant value $f(x, y, z) = c$ is called a **level surface** of $f$.

Since the graphs of functions of three variables consist of points $(x, y, z, f(x, y, z))$ lying in a four-dimensional space, we cannot sketch them effectively in our three-dimensional frame of reference. We can see how the function behaves, however, by looking at its three-dimensional level surfaces.

**EXAMPLE 4** Describe the level surfaces of the function

$$f(x, y, z) = \sqrt{x^2 + y^2 + z^2}.$$

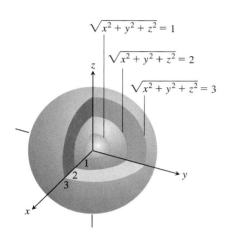

$$\sqrt{x^2 + y^2 + z^2} = 1$$

$$\sqrt{x^2 + y^2 + z^2} = 2$$

$$\sqrt{x^2 + y^2 + z^2} = 3$$

**FIGURE 14.8** The level surfaces of $f(x, y, z) = \sqrt{x^2 + y^2 + z^2}$ are concentric spheres (Example 4).

**FIGURE 14.7** Contours on Mt. Washington in New Hampshire. (Reprinted by permission of the Appalachian Mountain Club.)

**Solution** The value of $f$ is the distance from the origin to the point $(x, y, z)$. Each level surface $\sqrt{x^2 + y^2 + z^2} = c, c > 0$, is a sphere of radius $c$ centered at the origin. Figure 14.8 shows a cutaway view of three of these spheres. The level surface $\sqrt{x^2 + y^2 + z^2} = 0$ consists of the origin alone.

We are not graphing the function here; we are looking at level surfaces in the function's domain. The level surfaces show how the function's values change as we move through its domain. If we remain on a sphere of radius $c$ centered at the origin, the function maintains a constant value, namely $c$. If we move from a point on one sphere to a point on another, the function's value changes. It increases if we move away from the origin and decreases if we move toward the origin. The way the values change depends on the direction we take. The dependence of change on direction is important. We return to it in Section 14.5. ∎

The definitions of interior, boundary, open, closed, bounded, and unbounded for regions in space are similar to those for regions in the plane. To accommodate the extra dimension, we use solid balls of positive radius instead of disks.

(a) Interior point

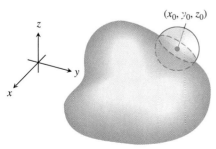

(b) Boundary point

**FIGURE 14.9** Interior points and boundary points of a region in space. As with regions in the plane, a boundary point need not belong to the space region $R$.

> **DEFINITIONS** A point $(x_0, y_0, z_0)$ in a region $R$ in space is an **interior point** of $R$ if it is the center of a solid ball that lies entirely in $R$ (Figure 14.9a). A point $(x_0, y_0, z_0)$ is a **boundary point** of $R$ if every solid ball centered at $(x_0, y_0, z_0)$ contains points that lie outside of $R$ as well as points that lie inside $R$ (Figure 14.9b). The **interior** of $R$ is the set of interior points of $R$. The **boundary** of $R$ is the set of boundary points of $R$.
>
> A region is **open** if it consists entirely of interior points. A region is **closed** if it contains its entire boundary.

Examples of *open* sets in space include the interior of a sphere, the open half-space $z > 0$, the first octant (where $x$, $y$, and $z$ are all positive), and space itself. Examples of *closed* sets in space include lines, planes, and the closed half-space $z \geq 0$. A solid sphere with part of its boundary removed or a solid cube with a missing face, edge, or corner point is *neither open nor closed*.

Functions of more than three independent variables are also important. For example, the temperature on a surface in space may depend not only on the location of the point $P(x, y, z)$ on the surface but also on the time $t$ when it is visited, so we would write $T = f(x, y, z, t)$.

## Computer Graphing

Three-dimensional graphing software makes it possible to graph functions of two variables. We can often get information more quickly from a graph than from a formula, since the surfaces reveal increasing and decreasing behavior, and high points or low points.

**EXAMPLE 5** The temperature $w$ beneath the Earth's surface is a function of the depth $x$ beneath the surface and the time $t$ of the year. If we measure $x$ in meters and $t$ as the number of days elapsed from the expected date of the yearly highest surface temperature, we can model the variation in temperature with the function

$$w = \cos(1.7 \times 10^{-2}t - 0.6x)e^{-0.6x}.$$

(The temperature at 0 m is scaled to vary from $+1$ to $-1$, so that the variation at $x$ meters can be interpreted as a fraction of the variation at the surface.)

Figure 14.10 shows a graph of the function. At a depth of 5 m, the variation (change in vertical amplitude in the figure) is about 5% of the surface variation. At 8 m, there is almost no variation during the year.

The graph also shows that the temperature 5 m below the surface is about half a year out of phase with the surface temperature. When the temperature is lowest on the surface (late January, say), it is at its highest 5 m below. Five meters below the ground, the seasons are reversed. ∎

Figure 14.11 shows computer-generated graphs of a number of functions of two variables together with their level curves.

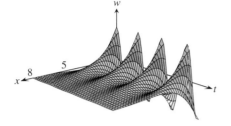

**FIGURE 14.10** This graph shows the seasonal variation of the temperature below ground as a fraction of surface temperature (Example 5).

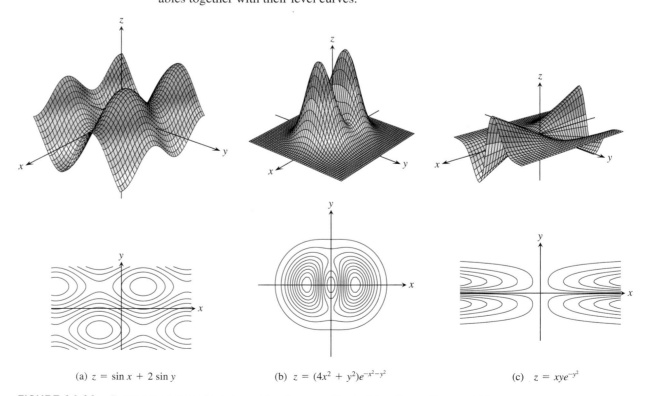

(a) $z = \sin x + 2 \sin y$    (b) $z = (4x^2 + y^2)e^{-x^2-y^2}$    (c) $z = xye^{-y^2}$

**FIGURE 14.11** Computer-generated graphs and level curves of typical functions of two variables.

# Exercises 14.1

### Domain, Range, and Level Curves

In Exercises 1–4, find the specific function values.

**1.** $f(x, y) = x^2 + xy^3$

    **a.** $f(0, 0)$               **b.** $f(-1, 1)$

    **c.** $f(2, 3)$               **d.** $f(-3, -2)$

**2.** $f(x, y) = \sin(xy)$

    **a.** $f\left(2, \dfrac{\pi}{6}\right)$          **b.** $f\left(-3, \dfrac{\pi}{12}\right)$

    **c.** $f\left(\pi, \dfrac{1}{4}\right)$           **d.** $f\left(-\dfrac{\pi}{2}, -7\right)$

**3.** $f(x, y, z) = \dfrac{x - y}{y^2 + z^2}$

    **a.** $f(3, -1, 2)$       **b.** $f\left(1, \dfrac{1}{2}, -\dfrac{1}{4}\right)$

    **c.** $f\left(0, -\dfrac{1}{3}, 0\right)$     **d.** $f(2, 2, 100)$

**4.** $f(x, y, z) = \sqrt{49 - x^2 - y^2 - z^2}$

    **a.** $f(0, 0, 0)$         **b.** $f(2, -3, 6)$

    **c.** $f(-1, 2, 3)$      **d.** $f\left(\dfrac{4}{\sqrt{2}}, \dfrac{5}{\sqrt{2}}, \dfrac{6}{\sqrt{2}}\right)$

In Exercises 5–12, find and sketch the domain for each function.

**5.** $f(x, y) = \sqrt{y - x - 2}$

**6.** $f(x, y) = \ln(x^2 + y^2 - 4)$

**7.** $f(x, y) = \dfrac{(x - 1)(y + 2)}{(y - x)(y - x^3)}$

**8.** $f(x, y) = \dfrac{\sin(xy)}{x^2 + y^2 - 25}$

**9.** $f(x, y) = \cos^{-1}(y - x^2)$

**10.** $f(x, y) = \ln(xy + x - y - 1)$

**11.** $f(x, y) = \sqrt{(x^2 - 4)(y^2 - 9)}$

**12.** $f(x, y) = \dfrac{1}{\ln(4 - x^2 - y^2)}$

In Exercises 13–16, find and sketch the level curves $f(x, y) = c$ on the same set of coordinate axes for the given values of $c$. We refer to these level curves as a contour map.

**13.** $f(x, y) = x + y - 1$,   $c = -3, -2, -1, 0, 1, 2, 3$

**14.** $f(x, y) = x^2 + y^2$,   $c = 0, 1, 4, 9, 16, 25$

**15.** $f(x, y) = xy$,   $c = -9, -4, -1, 0, 1, 4, 9$

**16.** $f(x, y) = \sqrt{25 - x^2 - y^2}$,   $c = 0, 1, 2, 3, 4$

In Exercises 17–30, **(a)** find the function's domain, **(b)** find the function's range, **(c)** describe the function's level curves, **(d)** find the boundary of the function's domain, **(e)** determine if the domain is an open region, a closed region, or neither, and **(f)** decide if the domain is bounded or unbounded.

**17.** $f(x, y) = y - x$       **18.** $f(x, y) = \sqrt{y - x}$

**19.** $f(x, y) = 4x^2 + 9y^2$    **20.** $f(x, y) = x^2 - y^2$

**21.** $f(x, y) = xy$              **22.** $f(x, y) = y/x^2$

**23.** $f(x, y) = \dfrac{1}{\sqrt{16 - x^2 - y^2}}$    **24.** $f(x, y) = \sqrt{9 - x^2 - y^2}$

**25.** $f(x, y) = \ln(x^2 + y^2)$     **26.** $f(x, y) = e^{-(x^2 + y^2)}$

**27.** $f(x, y) = \sin^{-1}(y - x)$    **28.** $f(x, y) = \tan^{-1}\left(\dfrac{y}{x}\right)$

**29.** $f(x, y) = \ln(x^2 + y^2 - 1)$   **30.** $f(x, y) = \ln(9 - x^2 - y^2)$

### Matching Surfaces with Level Curves

Exercises 31–36 show level curves for the functions graphed in (a)–(f) on the following page. Match each set of curves with the appropriate function.

**31.**                     **32.**

**33.**                     **34.**

**35.**                     **36.**

**a.**

$z = (\cos x)(\cos y)\, e^{-\sqrt{x^2 - y^2}/4}$

**b.**

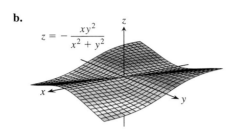

$z = -\dfrac{xy^2}{x^2 + y^2}$

**c.**

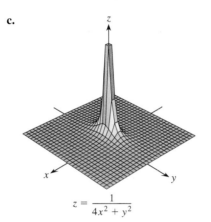

$z = \dfrac{1}{4x^2 + y^2}$

**d.**

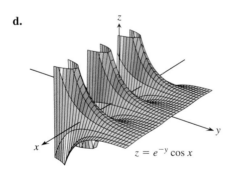

$z = e^{-y} \cos x$

**e.**

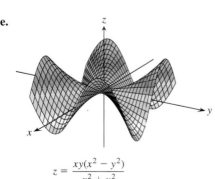

$z = \dfrac{xy(x^2 - y^2)}{x^2 + y^2}$

**f.**

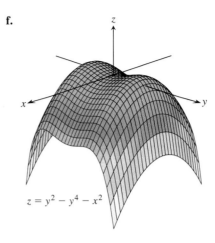

$z = y^2 - y^4 - x^2$

### Functions of Two Variables

Display the values of the functions in Exercises 37–48 in two ways:
**(a)** by sketching the surface $z = f(x, y)$ and **(b)** by drawing an assortment of level curves in the function's domain. Label each level curve with its function value.

**37.** $f(x, y) = y^2$      **38.** $f(x, y) = \sqrt{x}$

**39.** $f(x, y) = x^2 + y^2$      **40.** $f(x, y) = \sqrt{x^2 + y^2}$

**41.** $f(x, y) = x^2 - y$      **42.** $f(x, y) = 4 - x^2 - y^2$

**43.** $f(x, y) = 4x^2 + y^2$      **44.** $f(x, y) = 6 - 2x - 3y$

**45.** $f(x, y) = 1 - |y|$      **46.** $f(x, y) = 1 - |x| - |y|$

**47.** $f(x, y) = \sqrt{x^2 + y^2 + 4}$      **48.** $f(x, y) = \sqrt{x^2 + y^2 - 4}$

### Finding Level Curves

In Exercises 49–52, find an equation for and sketch the graph of the level curve of the function $f(x, y)$ that passes through the given point.

**49.** $f(x, y) = 16 - x^2 - y^2, \quad \left(2\sqrt{2}, \sqrt{2}\right)$

**50.** $f(x, y) = \sqrt{x^2 - 1}, \quad (1, 0)$

**51.** $f(x, y) = \sqrt{x + y^2 - 3}, \quad (3, -1)$

**52.** $f(x, y) = \dfrac{2y - x}{x + y + 1}, \quad (-1, 1)$

### Sketching Level Surfaces

In Exercises 53–60, sketch a typical level surface for the function.

**53.** $f(x, y, z) = x^2 + y^2 + z^2$      **54.** $f(x, y, z) = \ln\left(x^2 + y^2 + z^2\right)$

**55.** $f(x, y, z) = x + z$      **56.** $f(x, y, z) = z$

**57.** $f(x, y, z) = x^2 + y^2$      **58.** $f(x, y, z) = y^2 + z^2$

**59.** $f(x, y, z) = z - x^2 - y^2$

**60.** $f(x, y, z) = (x^2/25) + (y^2/16) + (z^2/9)$

### Finding Level Surfaces

In Exercises 61–64, find an equation for the level surface of the function through the given point.

**61.** $f(x, y, z) = \sqrt{x - y} - \ln z, \quad (3, -1, 1)$

**62.** $f(x, y, z) = \ln\left(x^2 + y + z^2\right), \quad (-1, 2, 1)$

**63.** $g(x, y, z) = \sqrt{x^2 + y^2 + z^2}$,  $\left(1, -1, \sqrt{2}\right)$

**64.** $g(x, y, z) = \dfrac{x - y + z}{2x + y - z}$,  $(1, 0, -2)$

In Exercises 65–68, find and sketch the domain of $f$. Then find an equation for the level curve or surface of the function passing through the given point.

**65.** $f(x, y) = \displaystyle\sum_{n=0}^{\infty} \left(\dfrac{x}{y}\right)^n$,  $(1, 2)$

**66.** $g(x, y, z) = \displaystyle\sum_{n=0}^{\infty} \dfrac{(x + y)^n}{n!z^n}$,  $(\ln 4, \ln 9, 2)$

**67.** $f(x, y) = \displaystyle\int_x^y \dfrac{d\theta}{\sqrt{1 - \theta^2}}$,  $(0, 1)$

**68.** $g(x, y, z) = \displaystyle\int_x^y \dfrac{dt}{1 + t^2} + \int_0^z \dfrac{d\theta}{\sqrt{4 - \theta^2}}$,  $\left(0, 1, \sqrt{3}\right)$

**COMPUTER EXPLORATIONS**

Use a CAS to perform the following steps for each of the functions in Exercises 69–72.

**a.** Plot the surface over the given rectangle.

**b.** Plot several level curves in the rectangle.

**c.** Plot the level curve of $f$ through the given point.

**69.** $f(x, y) = x \sin \dfrac{y}{2} + y \sin 2x$,  $0 \le x \le 5\pi$,  $0 \le y \le 5\pi$,  $P(3\pi, 3\pi)$

**70.** $f(x, y) = (\sin x)(\cos y)e^{\sqrt{x^2 + y^2}/8}$,  $0 \le x \le 5\pi$,  $0 \le y \le 5\pi$,  $P(4\pi, 4\pi)$

**71.** $f(x, y) = \sin(x + 2 \cos y)$,  $-2\pi \le x \le 2\pi$,  $-2\pi \le y \le 2\pi$,  $P(\pi, \pi)$

**72.** $f(x, y) = e^{(x^{0.1} - y)} \sin(x^2 + y^2)$,  $0 \le x \le 2\pi$,  $-2\pi \le y \le \pi$,  $P(\pi, -\pi)$

Use a CAS to plot the implicitly defined level surfaces in Exercises 73–76.

**73.** $4 \ln(x^2 + y^2 + z^2) = 1$     **74.** $x^2 + z^2 = 1$

**75.** $x + y^2 - 3z^2 = 1$

**76.** $\sin\left(\dfrac{x}{2}\right) - (\cos y)\sqrt{x^2 + z^2} = 2$

**Parametrized Surfaces**  Just as you describe curves in the plane parametrically with a pair of equations $x = f(t), y = g(t)$ defined on some parameter interval $I$, you can sometimes describe surfaces in space with a triple of equations $x = f(u, v), y = g(u, v), z = h(u, v)$ defined on some parameter rectangle $a \le u \le b, c \le v \le d$. Many computer algebra systems permit you to plot such surfaces in *parametric mode*. (Parametrized surfaces are discussed in detail in Section 16.5.) Use a CAS to plot the surfaces in Exercises 77–80. Also plot several level curves in the $xy$-plane.

**77.** $x = u \cos v$,  $y = u \sin v$,  $z = u$,  $0 \le u \le 2$,  $0 \le v \le 2\pi$

**78.** $x = u \cos v$,  $y = u \sin v$,  $z = v$,  $0 \le u \le 2$,  $0 \le v \le 2\pi$

**79.** $x = (2 + \cos u) \cos v$,  $y = (2 + \cos u) \sin v$,  $z = \sin u$,  $0 \le u \le 2\pi$,  $0 \le v \le 2\pi$

**80.** $x = 2 \cos u \cos v$,  $y = 2 \cos u \sin v$,  $z = 2 \sin u$,  $0 \le u \le 2\pi$,  $0 \le v \le \pi$

# 14.2 Limits and Continuity in Higher Dimensions

This section treats limits and continuity for multivariable functions. These ideas are analogous to limits and continuity for single-variable functions, but including more independent variables leads to additional complexity and important differences requiring some new ideas.

## Limits for Functions of Two Variables

If the values of $f(x, y)$ lie arbitrarily close to a fixed real number $L$ for all points $(x, y)$ sufficiently close to a point $(x_0, y_0)$, we say that $f$ approaches the limit $L$ as $(x, y)$ approaches $(x_0, y_0)$. This is similar to the informal definition for the limit of a function of a single variable. Notice, however, that if $(x_0, y_0)$ lies in the interior of $f$'s domain, $(x, y)$ can approach $(x_0, y_0)$ from any direction. For the limit to exist, the same limiting value must be obtained whatever direction of approach is taken. We illustrate this issue in several examples following the definition.

> **DEFINITION** We say that a function $f(x, y)$ approaches the **limit** $L$ as $(x, y)$ approaches $(x_0, y_0)$, and write
>
> $$\lim_{(x, y) \to (x_0, y_0)} f(x, y) = L$$
>
> if, for every number $\epsilon > 0$, there exists a corresponding number $\delta > 0$ such that for all $(x, y)$ in the domain of $f$,
>
> $$\left| f(x, y) - L \right| < \epsilon \quad \text{whenever} \quad 0 < \sqrt{(x - x_0)^2 + (y - y_0)^2} < \delta.$$

The definition of limit says that the distance between $f(x, y)$ and $L$ becomes arbitrarily small whenever the distance from $(x, y)$ to $(x_0, y_0)$ is made sufficiently small (but not 0). The definition applies to interior points $(x_0, y_0)$ as well as boundary points of the domain of $f$, although a boundary point need not lie within the domain. The points $(x, y)$ that approach $(x_0, y_0)$ are always taken to be in the domain of $f$. See Figure 14.12.

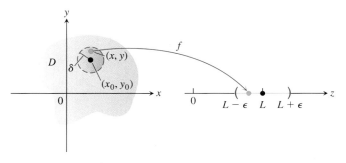

**FIGURE 14.12** In the limit definition, $\delta$ is the radius of a disk centered at $(x_0, y_0)$. For all points $(x, y)$ within this disk, the function values $f(x, y)$ lie inside the corresponding interval $(L - \epsilon, L + \epsilon)$.

As for functions of a single variable, it can be shown that

$$\lim_{(x, y) \to (x_0, y_0)} x = x_0$$

$$\lim_{(x, y) \to (x_0, y_0)} y = y_0$$

$$\lim_{(x, y) \to (x_0, y_0)} k = k \quad \text{(any number } k\text{)}.$$

For example, in the first limit statement above, $f(x, y) = x$ and $L = x_0$. Using the definition of limit, suppose that $\epsilon > 0$ is chosen. If we let $\delta$ equal this $\epsilon$, we see that

$$0 < \sqrt{(x - x_0)^2 + (y - y_0)^2} < \delta = \epsilon$$

implies

$$\sqrt{(x - x_0)^2} < \epsilon \qquad \scriptstyle (x - x_0)^2 \leq (x - x_0)^2 + (y - y_0)^2$$
$$\left| x - x_0 \right| < \epsilon \qquad \scriptstyle \sqrt{a^2} = |a|$$
$$\left| f(x, y) - x_0 \right| < \epsilon. \qquad \scriptstyle x = f(x, y)$$

That is,

$$\left| f(x, y) - x_0 \right| < \epsilon \quad \text{whenever} \quad 0 < \sqrt{(x - x_0)^2 + (y - y_0)^2} < \delta.$$

So a $\delta$ has been found satisfying the requirement of the definition, and

$$\lim_{(x, y) \to (x_0, y_0)} f(x, y) = \lim_{(x, y) \to (x_0, y_0)} x = x_0.$$

As with single-variable functions, the limit of the sum of two functions is the sum of their limits (when they both exist), with similar results for the limits of the differences, constant multiples, products, quotients, powers, and roots.

---

**THEOREM 1—Properties of Limits of Functions of Two Variables**    The following rules hold if $L$, $M$, and $k$ are real numbers and

$$\lim_{(x, y) \to (x_0, y_0)} f(x, y) = L \quad \text{and} \quad \lim_{(x, y) \to (x_0, y_0)} g(x, y) = M.$$

1. *Sum Rule:*                   $\lim\limits_{(x, y) \to (x_0, y_0)} (f(x, y) + g(x, y)) = L + M$

2. *Difference Rule:*              $\lim\limits_{(x, y) \to (x_0, y_0)} (f(x, y) - g(x, y)) = L - M$

3. *Constant Multiple Rule:*       $\lim\limits_{(x, y) \to (x_0, y_0)} kf(x, y) = kL$     (any number $k$)

4. *Product Rule:*                $\lim\limits_{(x, y) \to (x_0, y_0)} (f(x, y) \cdot g(x, y)) = L \cdot M$

5. *Quotient Rule:*               $\lim\limits_{(x, y) \to (x_0, y_0)} \dfrac{f(x, y)}{g(x, y)} = \dfrac{L}{M}, \quad M \neq 0$

6. *Power Rule:*                  $\lim\limits_{(x, y) \to (x_0, y_0)} [f(x, y)]^n = L^n$, $n$ a positive integer

7. *Root Rule:*                   $\lim\limits_{(x, y) \to (x_0, y_0)} \sqrt[n]{f(x, y)} = \sqrt[n]{L} = L^{1/n}$,

$n$ a positive integer, and if $n$ is even,
we assume that $L > 0$.

---

While we won't prove Theorem 1 here, we give an informal discussion of why it's true. If $(x, y)$ is sufficiently close to $(x_0, y_0)$, then $f(x, y)$ is close to $L$ and $g(x, y)$ is close to $M$ (from the informal interpretation of limits). It is then reasonable that $f(x, y) + g(x, y)$ is close to $L + M$; $f(x, y) - g(x, y)$ is close to $L - M$; $kf(x, y)$ is close to $kL$; $f(x, y)g(x, y)$ is close to $LM$; and $f(x, y)/g(x, y)$ is close to $L/M$ if $M \neq 0$.

When we apply Theorem 1 to polynomials and rational functions, we obtain the useful result that the limits of these functions as $(x, y) \to (x_0, y_0)$ can be calculated by evaluating the functions at $(x_0, y_0)$. The only requirement is that the rational functions be defined at $(x_0, y_0)$.

**EXAMPLE 1**    In this example, we can combine the three simple results following the limit definition with the results in Theorem 1 to calculate the limits. We simply substitute the $x$- and $y$-values of the point being approached into the functional expression to find the limiting value.

**(a)**   $\lim\limits_{(x, y) \to (0, 1)} \dfrac{x - xy + 3}{x^2 y + 5xy - y^3} = \dfrac{0 - (0)(1) + 3}{(0)^2(1) + 5(0)(1) - (1)^3} = -3$

**(b)**   $\lim\limits_{(x, y) \to (3, -4)} \sqrt{x^2 + y^2} = \sqrt{(3)^2 + (-4)^2} = \sqrt{25} = 5$    ∎

**EXAMPLE 2**    Find $\lim\limits_{(x, y) \to (0, 0)} \dfrac{x^2 - xy}{\sqrt{x} - \sqrt{y}}$.

**Solution** Since the denominator $\sqrt{x} - \sqrt{y}$ approaches 0 as $(x, y) \to (0, 0)$, we cannot use the Quotient Rule from Theorem 1. If we multiply numerator and denominator by $\sqrt{x} + \sqrt{y}$, however, we produce an equivalent fraction whose limit we *can* find:

$$\lim_{(x, y) \to (0,0)} \frac{x^2 - xy}{\sqrt{x} - \sqrt{y}} = \lim_{(x, y) \to (0,0)} \frac{(x^2 - xy)(\sqrt{x} + \sqrt{y})}{(\sqrt{x} - \sqrt{y})(\sqrt{x} + \sqrt{y})} \qquad \text{Multiply by a form equal to 1.}$$

$$= \lim_{(x, y) \to (0,0)} \frac{x(x - y)(\sqrt{x} + \sqrt{y})}{x - y} \qquad \text{Algebra}$$

$$= \lim_{(x, y) \to (0,0)} x(\sqrt{x} + \sqrt{y}) \qquad \text{Cancel the nonzero factor } (x - y).$$

$$= 0(\sqrt{0} + \sqrt{0}) = 0 \qquad \text{Known limit values}$$

We can cancel the factor $(x - y)$ because the path $y = x$ (along which $x - y = 0$) is *not* in the domain of the function

$$f(x, y) = \frac{x^2 - xy}{\sqrt{x} - \sqrt{y}}.$$

∎

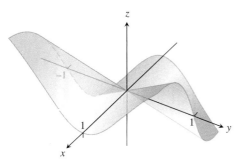

**FIGURE 14.13** The surface graph shows the limit of the function in Example 3 must be 0, if it exists.

**EXAMPLE 3** Find $\displaystyle\lim_{(x, y) \to (0,0)} \frac{4xy^2}{x^2 + y^2}$ if it exists.

**Solution** We first observe that along the line $x = 0$, the function always has value 0 when $y \neq 0$. Likewise, along the line $y = 0$, the function has value 0 provided $x \neq 0$. So if the limit does exist as $(x, y)$ approaches $(0, 0)$, the value of the limit must be 0 (see Figure 14.13). To see if this is true, we apply the definition of limit.

Let $\epsilon > 0$ be given, but arbitrary. We want to find a $\delta > 0$ such that

$$\left| \frac{4xy^2}{x^2 + y^2} - 0 \right| < \epsilon \qquad \text{whenever} \qquad 0 < \sqrt{x^2 + y^2} < \delta$$

or

$$\frac{4|x|y^2}{x^2 + y^2} < \epsilon \qquad \text{whenever} \qquad 0 < \sqrt{x^2 + y^2} < \delta.$$

Since $y^2 \leq x^2 + y^2$ we have that

$$\frac{4|x|y^2}{x^2 + y^2} \leq 4|x| = 4\sqrt{x^2} \leq 4\sqrt{x^2 + y^2}. \qquad \frac{y^2}{x^2 + y^2} \leq 1$$

So if we choose $\delta = \epsilon/4$ and let $0 < \sqrt{x^2 + y^2} < \delta$, we get

$$\left| \frac{4xy^2}{x^2 + y^2} - 0 \right| \leq 4\sqrt{x^2 + y^2} < 4\delta = 4\left(\frac{\epsilon}{4}\right) = \epsilon.$$

It follows from the definition that

$$\lim_{(x, y) \to (0,0)} \frac{4xy^2}{x^2 + y^2} = 0.$$

∎

**EXAMPLE 4**   If $f(x, y) = \dfrac{y}{x}$, does $\displaystyle\lim_{(x, y)\to(0, 0)} f(x, y)$ exist?

**Solution**   The domain of $f$ does not include the $y$-axis, so we do not consider any points $(x, y)$ where $x = 0$ in the approach toward the origin $(0, 0)$. Along the $x$-axis, the value of the function is $f(x, 0) = 0$ for all $x \neq 0$. So if the limit does exist as $(x, y) \to (0, 0)$, the value of the limit must be $L = 0$. On the other hand, along the line $y = x$, the value of the function is $f(x, x) = x/x = 1$ for all $x \neq 0$. That is, the function $f$ approaches the value 1 along the line $y = x$. This means that for every disk of radius $\delta$ centered at $(0, 0)$, the disk will contain points $(x, 0)$ on the $x$-axis where the value of the function is 0, and also points $(x, x)$ along the line $y = x$ where the value of the function is 1. So no matter how small we choose $\delta$ as the radius of the disk in Figure 14.12, there will be points within the disk for which the function values differ by 1. Therefore, the limit cannot exist because we can take $\epsilon$ to be any number less than 1 in the limit definition and deny that $L = 0$ or 1, or any other real number. The limit does not exist because we have different limiting values along different paths approaching the point $(0, 0)$. ∎

## Continuity

As with functions of a single variable, continuity is defined in terms of limits.

---

**DEFINITION**   A function $f(x, y)$ is **continuous at the point $(x_0, y_0)$** if

1. $f$ is defined at $(x_0, y_0)$,
2. $\displaystyle\lim_{(x, y)\to(x_0, y_0)} f(x, y)$ exists,
3. $\displaystyle\lim_{(x, y)\to(x_0, y_0)} f(x, y) = f(x_0, y_0)$.

A function is **continuous** if it is continuous at every point of its domain.

---

As with the definition of limit, the definition of continuity applies at boundary points as well as interior points of the domain of $f$. The only requirement is that each point $(x, y)$ near $(x_0, y_0)$ be in the domain of $f$.

A consequence of Theorem 1 is that algebraic combinations of continuous functions are continuous at every point at which all the functions involved are defined. This means that sums, differences, constant multiples, products, quotients, and powers of continuous functions are continuous where defined. In particular, polynomials and rational functions of two variables are continuous at every point at which they are defined.

**EXAMPLE 5**   Show that

$$f(x, y) = \begin{cases} \dfrac{2xy}{x^2 + y^2}, & (x, y) \neq (0, 0) \\ 0, & (x, y) = (0, 0) \end{cases}$$

is continuous at every point except the origin (Figure 14.14).

**Solution**   The function $f$ is continuous at any point $(x, y) \neq (0, 0)$ because its values are then given by a rational function of $x$ and $y$ and the limiting value is obtained by substituting the values of $x$ and $y$ into the functional expression.

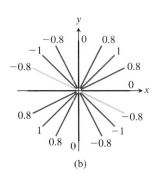

**FIGURE 14.14**   (a) The graph of
$$f(x, y) = \begin{cases} \dfrac{2xy}{x^2 + y^2}, & (x, y) \neq (0, 0) \\ 0, & (x, y) = (0, 0). \end{cases}$$
The function is continuous at every point except the origin. (b) The values of $f$ are different constants along each line $y = mx, x \neq 0$ (Example 5).

At $(0, 0)$, the value of $f$ is defined, but $f$, we claim, has no limit as $(x, y) \to (0, 0)$. The reason is that different paths of approach to the origin can lead to different results, as we now see.

For every value of $m$, the function $f$ has a constant value on the "punctured" line $y = mx, x \neq 0$, because

$$f(x, y)\Big|_{y=mx} = \frac{2xy}{x^2 + y^2}\Big|_{y=mx} = \frac{2x(mx)}{x^2 + (mx)^2} = \frac{2mx^2}{x^2 + m^2x^2} = \frac{2m}{1 + m^2}.$$

Therefore, $f$ has this number as its limit as $(x, y)$ approaches $(0, 0)$ along the line:

$$\lim_{\substack{(x, y) \to (0,0) \\ \text{along } y=mx}} f(x, y) = \lim_{(x, y) \to (0,0)} \left[ f(x, y)\Big|_{y=mx} \right] = \frac{2m}{1 + m^2}.$$

This limit changes with each value of the slope $m$. There is therefore no single number we may call the limit of $f$ as $(x, y)$ approaches the origin. The limit fails to exist, and the function is not continuous. ■

Examples 4 and 5 illustrate an important point about limits of functions of two or more variables. For a limit to exist at a point, the limit must be the same along every approach path. This result is analogous to the single-variable case where both the left- and right-sided limits had to have the same value. For functions of two or more variables, if we ever find paths with different limits, we know the function has no limit at the point they approach.

> **Two-Path Test for Nonexistence of a Limit**
>
> If a function $f(x, y)$ has different limits along two different paths in the domain of $f$ as $(x, y)$ approaches $(x_0, y_0)$, then $\lim_{(x, y) \to (x_0, y_0)} f(x, y)$ does not exist.

**EXAMPLE 6**  Show that the function

$$f(x, y) = \frac{2x^2y}{x^4 + y^2}$$

(Figure 14.15) has no limit as $(x, y)$ approaches $(0, 0)$.

**Solution**  The limit cannot be found by direct substitution, which gives the indeterminate form $0/0$. We examine the values of $f$ along parabolic curves that end at $(0, 0)$. Along the curve $y = kx^2, x \neq 0$, the function has the constant value

$$f(x, y)\Big|_{y=kx^2} = \frac{2x^2y}{x^4 + y^2}\Big|_{y=kx^2} = \frac{2x^2(kx^2)}{x^4 + (kx^2)^2} = \frac{2kx^4}{x^4 + k^2x^4} = \frac{2k}{1 + k^2}.$$

Therefore,

$$\lim_{\substack{(x, y) \to (0,0) \\ \text{along } y=kx^2}} f(x, y) = \lim_{(x, y) \to (0,0)} \left[ f(x, y)\Big|_{y=kx^2} \right] = \frac{2k}{1 + k^2}.$$

This limit varies with the path of approach. If $(x, y)$ approaches $(0, 0)$ along the parabola $y = x^2$, for instance, $k = 1$ and the limit is 1. If $(x, y)$ approaches $(0, 0)$ along the $x$-axis, $k = 0$ and the limit is 0. By the two-path test, $f$ has no limit as $(x, y)$ approaches $(0, 0)$. ■

It can be shown that the function in Example 6 has limit 0 along every path $y = mx$ (Exercise 53). We conclude that

> Having the same limit along all straight lines approaching $(x_0, y_0)$ does not imply a limit exists at $(x_0, y_0)$.

(a)

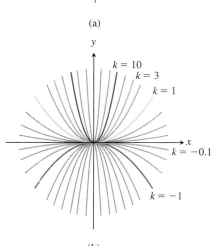

(b)

**FIGURE 14.15** (a) The graph of $f(x, y) = 2x^2y/(x^4 + y^2)$. (b) Along each path $y = kx^2$ the value of $f$ is constant, but varies with $k$ (Example 6).

Whenever it is correctly defined, the composite of continuous functions is also continuous. The only requirement is that each function be continuous where it is applied. The proof, omitted here, is similar to that for functions of a single variable (Theorem 9 in Section 2.5).

---

**Continuity of Composites**

If $f$ is continuous at $(x_0, y_0)$ and $g$ is a single-variable function continuous at $f(x_0, y_0)$, then the composite function $h = g \circ f$ defined by $h(x, y) = g(f(x, y))$ is continuous at $(x_0, y_0)$.

---

For example, the composite functions

$$e^{x-y}, \qquad \cos \frac{xy}{x^2 + 1}, \qquad \ln(1 + x^2 y^2)$$

are continuous at every point $(x, y)$.

### Functions of More Than Two Variables

The definitions of limit and continuity for functions of two variables and the conclusions about limits and continuity for sums, products, quotients, powers, and composites all extend to functions of three or more variables. Functions like

$$\ln(x + y + z) \qquad \text{and} \qquad \frac{y \sin z}{x - 1}$$

are continuous throughout their domains, and limits like

$$\lim_{P \to (1,0,-1)} \frac{e^{x+z}}{z^2 + \cos \sqrt{xy}} = \frac{e^{1-1}}{(-1)^2 + \cos 0} = \frac{1}{2},$$

where $P$ denotes the point $(x, y, z)$, may be found by direct substitution.

### Extreme Values of Continuous Functions on Closed, Bounded Sets

The Extreme Value Theorem (Theorem 1, Section 4.1) states that a function of a single variable that is continuous throughout a closed, bounded interval $[a, b]$ takes on an absolute maximum value and an absolute minimum value at least once in $[a, b]$. The same holds true of a function $z = f(x, y)$ that is continuous on a closed, bounded set $R$ in the plane (like a line segment, a disk, or a filled-in triangle). The function takes on an absolute maximum value at some point in $R$ and an absolute minimum value at some point in $R$. The function may take on a maximum or minimum value more than once over $R$.

Similar results hold for functions of three or more variables. A continuous function $w = f(x, y, z)$, for example, must take on absolute maximum and minimum values on any closed, bounded set (solid ball or cube, spherical shell, rectangular solid) on which it is defined. We will learn how to find these extreme values in Section 14.7.

## Exercises 14.2

### Limits with Two Variables

Find the limits in Exercises 1–12.

**1.** $\displaystyle\lim_{(x, y) \to (0,0)} \frac{3x^2 - y^2 + 5}{x^2 + y^2 + 2}$

**2.** $\displaystyle\lim_{(x, y) \to (0,4)} \frac{x}{\sqrt{y}}$

**3.** $\displaystyle\lim_{(x, y) \to (3,4)} \sqrt{x^2 + y^2 - 1}$

**4.** $\displaystyle\lim_{(x, y) \to (2, -3)} \left( \frac{1}{x} + \frac{1}{y} \right)^2$

**5.** $\displaystyle\lim_{(x, y) \to (0, \pi/4)} \sec x \tan y$

**6.** $\displaystyle\lim_{(x, y) \to (0,0)} \cos \frac{x^2 + y^3}{x + y + 1}$

**7.** $\lim\limits_{(x,\,y)\to(0,\ln 2)} e^{x-y}$

**8.** $\lim\limits_{(x,\,y)\to(1,1)} \ln|1 + x^2 y^2|$

**9.** $\lim\limits_{(x,\,y)\to(0,0)} \dfrac{e^y \sin x}{x}$

**10.** $\lim\limits_{(x,\,y)\to(1/27,\,\pi^3)} \cos\sqrt[3]{xy}$

**11.** $\lim\limits_{(x,\,y)\to(1,\,\pi/6)} \dfrac{x \sin y}{x^2 + 1}$

**12.** $\lim\limits_{(x,\,y)\to(\pi/2,0)} \dfrac{\cos y + 1}{y - \sin x}$

### Limits of Quotients
Find the limits in Exercises 13–24 by rewriting the fractions first.

**13.** $\lim\limits_{\substack{(x,\,y)\to(1,1) \\ x\neq y}} \dfrac{x^2 - 2xy + y^2}{x - y}$

**14.** $\lim\limits_{\substack{(x,\,y)\to(1,1) \\ x\neq y}} \dfrac{x^2 - y^2}{x - y}$

**15.** $\lim\limits_{\substack{(x,\,y)\to(1,1) \\ x\neq 1}} \dfrac{xy - y - 2x + 2}{x - 1}$

**16.** $\lim\limits_{\substack{(x,\,y)\to(2,\,-4) \\ x\neq -4,\, x\neq x^2}} \dfrac{y + 4}{x^2 y - xy + 4x^2 - 4x}$

**17.** $\lim\limits_{\substack{(x,\,y)\to(0,0) \\ x\neq y}} \dfrac{x - y + 2\sqrt{x} - 2\sqrt{y}}{\sqrt{x} - \sqrt{y}}$

**18.** $\lim\limits_{\substack{(x,\,y)\to(2,2) \\ x+y\neq 4}} \dfrac{x + y - 4}{\sqrt{x + y} - 2}$

**19.** $\lim\limits_{\substack{(x,\,y)\to(2,0) \\ 2x-y\neq 4}} \dfrac{\sqrt{2x - y} - 2}{2x - y - 4}$

**20.** $\lim\limits_{\substack{(x,\,y)\to(4,3) \\ x\neq y+1}} \dfrac{\sqrt{x} - \sqrt{y + 1}}{x - y - 1}$

**21.** $\lim\limits_{(x,\,y)\to(0,0)} \dfrac{\sin(x^2 + y^2)}{x^2 + y^2}$

**22.** $\lim\limits_{(x,\,y)\to(0,0)} \dfrac{1 - \cos(xy)}{xy}$

**23.** $\lim\limits_{(x,\,y)\to(1,-1)} \dfrac{x^3 + y^3}{x + y}$

**24.** $\lim\limits_{(x,\,y)\to(2,2)} \dfrac{x - y}{x^4 - y^4}$

### Limits with Three Variables
Find the limits in Exercises 25–30.

**25.** $\lim\limits_{P\to(1,3,4)} \left( \dfrac{1}{x} + \dfrac{1}{y} + \dfrac{1}{z} \right)$

**26.** $\lim\limits_{P\to(1,-1,-1)} \dfrac{2xy + yz}{x^2 + z^2}$

**27.** $\lim\limits_{P\to(\pi,\pi,0)} (\sin^2 x + \cos^2 y + \sec^2 z)$

**28.** $\lim\limits_{P\to(-1/4,\pi,\pi/2,2)} \tan^{-1} xyz$

**29.** $\lim\limits_{P\to(\pi,0,3)} ze^{-2y} \cos 2x$

**30.** $\lim\limits_{P\to(2,-3,6)} \ln\sqrt{x^2 + y^2 + z^2}$

### Continuity for Two Variables
At what points $(x, y)$ in the plane are the functions in Exercises 31–34 continuous?

**31. a.** $f(x, y) = \sin(x + y)$ **b.** $f(x, y) = \ln(x^2 + y^2)$

**32. a.** $f(x, y) = \dfrac{x + y}{x - y}$ **b.** $f(x, y) = \dfrac{y}{x^2 + 1}$

**33. a.** $g(x, y) = \sin\dfrac{1}{xy}$ **b.** $g(x, y) = \dfrac{x + y}{2 + \cos x}$

**34. a.** $g(x, y) = \dfrac{x^2 + y^2}{x^2 - 3x + 2}$ **b.** $g(x, y) = \dfrac{1}{x^2 - y}$

### Continuity for Three Variables
At what points $(x, y, z)$ in space are the functions in Exercises 35–40 continuous?

**35. a.** $f(x, y, z) = x^2 + y^2 - 2z^2$
   **b.** $f(x, y, z) = \sqrt{x^2 + y^2 - 1}$

**36. a.** $f(x, y, z) = \ln xyz$ **b.** $f(x, y, z) = e^{x+y} \cos z$

**37. a.** $h(x, y, z) = xy \sin\dfrac{1}{z}$ **b.** $h(x, y, z) = \dfrac{1}{x^2 + z^2 - 1}$

**38. a.** $h(x, y, z) = \dfrac{1}{|y| + |z|}$ **b.** $h(x, y, z) = \dfrac{1}{|xy| + |z|}$

**39. a.** $h(x, y, z) = \ln(z - x^2 - y^2 - 1)$

   **b.** $h(x, y, z) = \dfrac{1}{z - \sqrt{x^2 + y^2}}$

**40. a.** $h(x, y, z) = \sqrt{4 - x^2 - y^2 - z^2}$

   **b.** $h(x, y, z) = \dfrac{1}{4 - \sqrt{x^2 + y^2 + z^2 - 9}}$

### No Limit Exists at the Origin
By considering different paths of approach, show that the functions in Exercises 41–48 have no limit as $(x, y) \to (0, 0)$.

**41.** $f(x, y) = -\dfrac{x}{\sqrt{x^2 + y^2}}$

**42.** $f(x, y) = \dfrac{x^4}{x^4 + y^2}$

**43.** $f(x, y) = \dfrac{x^4 - y^2}{x^4 + y^2}$

**44.** $f(x, y) = \dfrac{xy}{|xy|}$

**45.** $g(x, y) = \dfrac{x - y}{x + y}$

**46.** $g(x, y) = \dfrac{x^2 - y}{x - y}$

**47.** $h(x, y) = \dfrac{x^2 + y}{y}$

**48.** $h(x, y) = \dfrac{x^2 y}{x^4 + y^2}$

### Theory and Examples
In Exercises 49 and 50, show that the limits do not exist.

**49.** $\lim\limits_{(x,\,y)\to(1,1)} \dfrac{xy^2 - 1}{y - 1}$

**50.** $\lim\limits_{(x,\,y)\to(1,-1)} \dfrac{xy + 1}{x^2 - y^2}$

**51.** Let $f(x, y) = \begin{cases} 1, & y \geq x^4 \\ 1, & y \leq 0 \\ 0, & \text{otherwise.} \end{cases}$

Find each of the following limits, or explain that the limit does not exist.

 **a.** $\lim\limits_{(x,\,y)\to(0,1)} f(x, y)$

 **b.** $\lim\limits_{(x,\,y)\to(2,3)} f(x, y)$

 **c.** $\lim\limits_{(x,\,y)\to(0,0)} f(x, y)$

**52.** Let $f(x, y) = \begin{cases} x^2, & x \geq 0 \\ x^3, & x < 0 \end{cases}$.

Find the following limits.

**a.** $\lim\limits_{(x, y)\to(3, -2)} f(x, y)$

**b.** $\lim\limits_{(x, y)\to(-2, 1)} f(x, y)$

**c.** $\lim\limits_{(x, y)\to(0,0)} f(x, y)$

**53.** Show that the function in Example 6 has limit 0 along every straight line approaching $(0, 0)$.

**54.** If $f(x_0, y_0) = 3$, what can you say about

$$\lim\limits_{(x, y)\to(x_0, y_0)} f(x, y)$$

if $f$ is continuous at $(x_0, y_0)$? If $f$ is not continuous at $(x_0, y_0)$? Give reasons for your answers.

**The Sandwich Theorem** for functions of two variables states that if $g(x, y) \leq f(x, y) \leq h(x, y)$ for all $(x, y) \neq (x_0, y_0)$ in a disk centered at $(x_0, y_0)$ and if $g$ and $h$ have the same finite limit $L$ as $(x, y) \to (x_0, y_0)$, then

$$\lim\limits_{(x, y)\to(x_0, y_0)} f(x, y) = L.$$

Use this result to support your answers to the questions in Exercises 55–58.

**55.** Does knowing that

$$1 - \frac{x^2 y^2}{3} < \frac{\tan^{-1} xy}{xy} < 1$$

tell you anything about

$$\lim\limits_{(x, y)\to(0,0)} \frac{\tan^{-1} xy}{xy}?$$

Give reasons for your answer.

**56.** Does knowing that

$$2|xy| - \frac{x^2 y^2}{6} < 4 - 4\cos\sqrt{|xy|} < 2|xy|$$

tell you anything about

$$\lim\limits_{(x, y)\to(0,0)} \frac{4 - 4\cos\sqrt{|xy|}}{|xy|}?$$

Give reasons for your answer.

**57.** Does knowing that $\left|\sin(1/x)\right| \leq 1$ tell you anything about

$$\lim\limits_{(x, y)\to(0,0)} y\sin\frac{1}{x}?$$

Give reasons for your answer.

**58.** Does knowing that $\left|\cos(1/y)\right| \leq 1$ tell you anything about

$$\lim\limits_{(x, y)\to(0,0)} x\cos\frac{1}{y}?$$

Give reasons for your answer.

**59.** (*Continuation of Example 5.*)

**a.** Reread Example 5. Then substitute $m = \tan\theta$ into the formula

$$f(x, y)\Big|_{y=mx} = \frac{2m}{1 + m^2}$$

and simplify the result to show how the value of $f$ varies with the line's angle of inclination.

**b.** Use the formula you obtained in part (a) to show that the limit of $f$ as $(x, y) \to (0, 0)$ along the line $y = mx$ varies from $-1$ to 1 depending on the angle of approach.

**60. Continuous extension**  Define $f(0, 0)$ in a way that extends

$$f(x, y) = xy\frac{x^2 - y^2}{x^2 + y^2}$$

to be continuous at the origin.

### Changing Variables to Polar Coordinates

If you cannot make any headway with $\lim_{(x, y)\to(0,0)} f(x, y)$ in rectangular coordinates, try changing to polar coordinates. Substitute $x = r\cos\theta$, $y = r\sin\theta$, and investigate the limit of the resulting expression as $r \to 0$. In other words, try to decide whether there exists a number $L$ satisfying the following criterion:

Given $\epsilon > 0$, there exists a $\delta > 0$ such that for all $r$ and $\theta$,

$$|r| < \delta \quad \Rightarrow \quad |f(r, \theta) - L| < \epsilon. \tag{1}$$

If such an $L$ exists, then

$$\lim\limits_{(x, y)\to(0,0)} f(x, y) = \lim\limits_{r\to0} f(r\cos\theta, r\sin\theta) = L.$$

For instance,

$$\lim\limits_{(x, y)\to(0,0)} \frac{x^3}{x^2 + y^2} = \lim\limits_{r\to0} \frac{r^3\cos^3\theta}{r^2} = \lim\limits_{r\to0} r\cos^3\theta = 0.$$

To verify the last of these equalities, we need to show that Equation (1) is satisfied with $f(r, \theta) = r\cos^3\theta$ and $L = 0$. That is, we need to show that given any $\epsilon > 0$, there exists a $\delta > 0$ such that for all $r$ and $\theta$,

$$|r| < \delta \quad \Rightarrow \quad |r\cos^3\theta - 0| < \epsilon.$$

Since

$$|r\cos^3\theta| = |r||\cos^3\theta| \leq |r| \cdot 1 = |r|,$$

the implication holds for all $r$ and $\theta$ if we take $\delta = \epsilon$.

In contrast,

$$\frac{x^2}{x^2 + y^2} = \frac{r^2\cos^2\theta}{r^2} = \cos^2\theta$$

takes on all values from 0 to 1 regardless of how small $|r|$ is, so that $\lim_{(x, y)\to(0,0)} x^2/(x^2 + y^2)$ does not exist.

In each of these instances, the existence or nonexistence of the limit as $r \to 0$ is fairly clear. Shifting to polar coordinates does not always help, however, and may even tempt us to false conclusions. For example, the limit may exist along every straight line (or ray) $\theta = $ constant and yet fail to exist in the broader sense. Example 5 illustrates this point. In polar coordinates, $f(x, y) = (2x^2 y)/(x^4 + y^2)$ becomes

$$f(r\cos\theta, r\sin\theta) = \frac{r\cos\theta\sin 2\theta}{r^2\cos^4\theta + \sin^2\theta}$$

for $r \neq 0$. If we hold $\theta$ constant and let $r \to 0$, the limit is 0. On the path $y = x^2$, however, we have $r \sin \theta = r^2 \cos^2 \theta$ and

$$f(r \cos \theta, r \sin \theta) = \frac{r \cos \theta \sin 2\theta}{r^2 \cos^4 \theta + (r \cos^2 \theta)^2}$$

$$= \frac{2r \cos^2 \theta \sin \theta}{2r^2 \cos^4 \theta} = \frac{r \sin \theta}{r^2 \cos^2 \theta} = 1.$$

In Exercises 61–66, find the limit of $f$ as $(x, y) \to (0, 0)$ or show that the limit does not exist.

**61.** $f(x, y) = \dfrac{x^3 - xy^2}{x^2 + y^2}$

**62.** $f(x, y) = \cos\left(\dfrac{x^3 - y^3}{x^2 + y^2}\right)$

**63.** $f(x, y) = \dfrac{y^2}{x^2 + y^2}$

**64.** $f(x, y) = \dfrac{2x}{x^2 + x + y^2}$

**65.** $f(x, y) = \tan^{-1}\left(\dfrac{|x| + |y|}{x^2 + y^2}\right)$

**66.** $f(x, y) = \dfrac{x^2 - y^2}{x^2 + y^2}$

In Exercises 67 and 68, define $f(0, 0)$ in a way that extends $f$ to be continuous at the origin.

**67.** $f(x, y) = \ln\left(\dfrac{3x^2 - x^2y^2 + 3y^2}{x^2 + y^2}\right)$

**68.** $f(x, y) = \dfrac{3x^2y}{x^2 + y^2}$

**Using the Limit Definition**

Each of Exercises 69–74 gives a function $f(x, y)$ and a positive number $\epsilon$. In each exercise, show that there exists a $\delta > 0$ such that for all $(x, y)$,

$$\sqrt{x^2 + y^2} < \delta \quad \Rightarrow \quad |f(x, y) - f(0, 0)| < \epsilon.$$

**69.** $f(x, y) = x^2 + y^2, \quad \epsilon = 0.01$

**70.** $f(x, y) = y/(x^2 + 1), \quad \epsilon = 0.05$

**71.** $f(x, y) = (x + y)/(x^2 + 1), \quad \epsilon = 0.01$

**72.** $f(x, y) = (x + y)/(2 + \cos x), \quad \epsilon = 0.02$

**73.** $f(x, y) = \dfrac{xy^2}{x^2 + y^2}$ and $f(0, 0) = 0, \quad \epsilon = 0.04$

**74.** $f(x, y) = \dfrac{x^3 + y^4}{x^2 + y^2}$ and $f(0, 0) = 0, \quad \epsilon = 0.02$

Each of Exercises 75–78 gives a function $f(x, y, z)$ and a positive number $\epsilon$. In each exercise, show that there exists a $\delta > 0$ such that for all $(x, y, z)$,

$$\sqrt{x^2 + y^2 + z^2} < \delta \quad \Rightarrow \quad |f(x, y, z) - f(0, 0, 0)| < \epsilon.$$

**75.** $f(x, y, z) = x^2 + y^2 + z^2, \quad \epsilon = 0.015$

**76.** $f(x, y, z) = xyz, \quad \epsilon = 0.008$

**77.** $f(x, y, z) = \dfrac{x + y + z}{x^2 + y^2 + z^2 + 1}, \quad \epsilon = 0.015$

**78.** $f(x, y, z) = \tan^2 x + \tan^2 y + \tan^2 z, \quad \epsilon = 0.03$

**79.** Show that $f(x, y, z) = x + y - z$ is continuous at every point $(x_0, y_0, z_0)$.

**80.** Show that $f(x, y, z) = x^2 + y^2 + z^2$ is continuous at the origin.

# 14.3 Partial Derivatives

The calculus of several variables is similar to single-variable calculus applied to several variables one at a time. When we hold all but one of the independent variables of a function constant and differentiate with respect to that one variable, we get a "partial" derivative. This section shows how partial derivatives are defined and interpreted geometrically, and how to calculate them by applying the rules for differentiating functions of a single variable. The idea of *differentiability* for functions of several variables requires more than the existence of the partial derivatives because a point can be approached from so many different directions. However, we will see that differentiable functions of several variables behave in the same way as differentiable single-variable functions, so they are continuous and can be well approximated by linear functions.

## Partial Derivatives of a Function of Two Variables

If $(x_0, y_0)$ is a point in the domain of a function $f(x, y)$, the vertical plane $y = y_0$ will cut the surface $z = f(x, y)$ in the curve $z = f(x, y_0)$ (Figure 14.16). This curve is the graph of the function $z = f(x, y_0)$ in the plane $y = y_0$. The horizontal coordinate in this plane is $x$; the vertical coordinate is $z$. The $y$-value is held constant at $y_0$, so $y$ is not a variable.

We define the partial derivative of $f$ with respect to $x$ at the point $(x_0, y_0)$ as the ordinary derivative of $f(x, y_0)$ with respect to $x$ at the point $x = x_0$. To distinguish partial derivatives from ordinary derivatives we use the symbol $\partial$ rather than the $d$ previously used. In the definition, $h$ represents a real number, positive or negative.

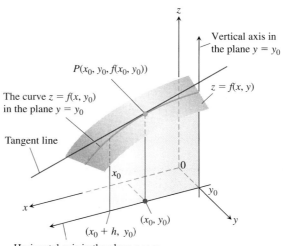

**FIGURE 14.16** The intersection of the plane $y = y_0$ with the surface $z = f(x, y)$, viewed from above the first quadrant of the $xy$-plane.

---

**DEFINITION** The **partial derivative of $f(x, y)$ with respect to $x$ at the point $(x_0, y_0)$** is

$$\frac{\partial f}{\partial x}\bigg|_{(x_0, y_0)} = \lim_{h \to 0} \frac{f(x_0 + h, y_0) - f(x_0, y_0)}{h},$$

provided the limit exists.

---

An equivalent expression for the partial derivative is

$$\frac{d}{dx} f(x, y_0)\bigg|_{x = x_0}.$$

The slope of the curve $z = f(x, y_0)$ at the point $P(x_0, y_0, f(x_0, y_0))$ in the plane $y = y_0$ is the value of the partial derivative of $f$ with respect to $x$ at $(x_0, y_0)$. (In Figure 14.16 this slope is negative.) The tangent line to the curve at $P$ is the line in the plane $y = y_0$ that passes through $P$ with this slope. The partial derivative $\partial f / \partial x$ at $(x_0, y_0)$ gives the rate of change of $f$ with respect to $x$ when $y$ is held fixed at the value $y_0$.

We use several notations for the partial derivative:

$$\frac{\partial f}{\partial x}(x_0, y_0) \text{ or } f_x(x_0, y_0), \qquad \frac{\partial z}{\partial x}\bigg|_{(x_0, y_0)}, \qquad \text{and} \qquad f_x, \frac{\partial f}{\partial x}, z_x, \text{ or } \frac{\partial z}{\partial x}.$$

The definition of the partial derivative of $f(x, y)$ with respect to $y$ at a point $(x_0, y_0)$ is similar to the definition of the partial derivative of $f$ with respect to $x$. We hold $x$ fixed at the value $x_0$ and take the ordinary derivative of $f(x_0, y)$ with respect to $y$ at $y_0$.

---

**DEFINITION** The **partial derivative of $f(x, y)$ with respect to $y$ at the point $(x_0, y_0)$** is

$$\frac{\partial f}{\partial y}\bigg|_{(x_0, y_0)} = \frac{d}{dy} f(x_0, y)\bigg|_{y = y_0} = \lim_{h \to 0} \frac{f(x_0, y_0 + h) - f(x_0, y_0)}{h},$$

provided the limit exists.

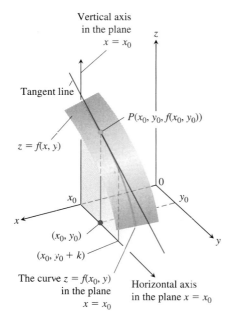

**FIGURE 14.17** The intersection of the plane $x = x_0$ with the surface $z = f(x, y)$, viewed from above the first quadrant of the $xy$-plane.

The slope of the curve $z = f(x_0, y)$ at the point $P(x_0, y_0, f(x_0, y_0))$ in the vertical plane $x = x_0$ (Figure 14.17) is the partial derivative of $f$ with respect to $y$ at $(x_0, y_0)$. The tangent line to the curve at $P$ is the line in the plane $x = x_0$ that passes through $P$ with this slope. The partial derivative gives the rate of change of $f$ with respect to $y$ at $(x_0, y_0)$ when $x$ is held fixed at the value $x_0$.

The partial derivative with respect to $y$ is denoted the same way as the partial derivative with respect to $x$:

$$\frac{\partial f}{\partial y}(x_0, y_0), \qquad f_y(x_0, y_0), \qquad \frac{\partial f}{\partial y}, \qquad f_y.$$

Notice that we now have two tangent lines associated with the surface $z = f(x, y)$ at the point $P(x_0, y_0, f(x_0, y_0))$ (Figure 14.18). Is the plane they determine tangent to the surface at $P$? We will see that it is for the *differentiable* functions defined at the end of this section, and we will learn how to find the tangent plane in Section 14.6. First we have to learn more about partial derivatives themselves.

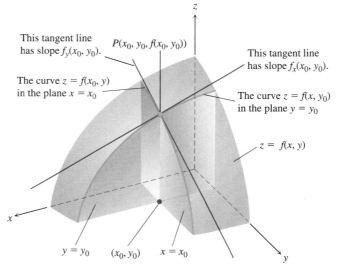

**FIGURE 14.18** Figures 14.16 and 14.17 combined. The tangent lines at the point $(x_0, y_0, f(x_0, y_0))$ determine a plane that, in this picture at least, appears to be tangent to the surface.

## Calculations

The definitions of $\partial f/\partial x$ and $\partial f/\partial y$ give us two different ways of differentiating $f$ at a point: with respect to $x$ in the usual way while treating $y$ as a constant and with respect to $y$ in the usual way while treating $x$ as a constant. As the following examples show, the values of these partial derivatives are usually different at a given point $(x_0, y_0)$.

**EXAMPLE 1** Find the values of $\partial f/\partial x$ and $\partial f/\partial y$ at the point $(4, -5)$ if

$$f(x, y) = x^2 + 3xy + y - 1.$$

**Solution** To find $\partial f/\partial x$, we treat $y$ as a constant and differentiate with respect to $x$:

$$\frac{\partial f}{\partial x} = \frac{\partial}{\partial x}(x^2 + 3xy + y - 1) = 2x + 3 \cdot 1 \cdot y + 0 - 0 = 2x + 3y.$$

The value of $\partial f/\partial x$ at $(4, -5)$ is $2(4) + 3(-5) = -7$.

To find $\partial f/\partial y$, we treat $x$ as a constant and differentiate with respect to $y$:

$$\frac{\partial f}{\partial y} = \frac{\partial}{\partial y}\,(x^2 + 3xy + y - 1) = 0 + 3 \cdot x \cdot 1 + 1 - 0 = 3x + 1.$$

The value of $\partial f/\partial y$ at $(4, -5)$ is $3(4) + 1 = 13$. ∎

**EXAMPLE 2**    Find $\partial f/\partial y$ as a function if $f(x, y) = y \sin xy$.

**Solution**   We treat $x$ as a constant and $f$ as a product of $y$ and $\sin xy$:

$$\frac{\partial f}{\partial y} = \frac{\partial}{\partial y}\,(y \sin xy) = y \frac{\partial}{\partial y} \sin xy + (\sin xy) \frac{\partial}{\partial y}\,(y)$$

$$= (y \cos xy) \frac{\partial}{\partial y}\,(xy) + \sin xy = xy \cos xy + \sin xy. \quad ∎$$

**EXAMPLE 3**    Find $f_x$ and $f_y$ as functions if

$$f(x, y) = \frac{2y}{y + \cos x}.$$

**Solution**   We treat $f$ as a quotient. With $y$ held constant, we get

$$f_x = \frac{\partial}{\partial x}\left(\frac{2y}{y + \cos x}\right) = \frac{(y + \cos x) \frac{\partial}{\partial x}\,(2y) - 2y \frac{\partial}{\partial x}\,(y + \cos x)}{(y + \cos x)^2}$$

$$= \frac{(y + \cos x)(0) - 2y(-\sin x)}{(y + \cos x)^2} = \frac{2y \sin x}{(y + \cos x)^2}.$$

With $x$ held constant, we get

$$f_y = \frac{\partial}{\partial y}\left(\frac{2y}{y + \cos x}\right) = \frac{(y + \cos x) \frac{\partial}{\partial y}\,(2y) - 2y \frac{\partial}{\partial y}\,(y + \cos x)}{(y + \cos x)^2}$$

$$= \frac{(y + \cos x)(2) - 2y(1)}{(y + \cos x)^2} = \frac{2 \cos x}{(y + \cos x)^2}. \quad ∎$$

Implicit differentiation works for partial derivatives the way it works for ordinary derivatives, as the next example illustrates.

**EXAMPLE 4**    Find $\partial z/\partial x$ if the equation

$$yz - \ln z = x + y$$

defines $z$ as a function of the two independent variables $x$ and $y$ and the partial derivative exists.

**Solution**   We differentiate both sides of the equation with respect to $x$, holding $y$ constant and treating $z$ as a differentiable function of $x$:

$$\frac{\partial}{\partial x}\,(yz) - \frac{\partial}{\partial x} \ln z = \frac{\partial x}{\partial x} + \frac{\partial y}{\partial x}$$

$$y \frac{\partial z}{\partial x} - \frac{1}{z} \frac{\partial z}{\partial x} = 1 + 0 \qquad \text{With } y \text{ constant,}$$
$$\qquad\qquad\qquad\qquad\qquad\qquad \frac{\partial}{\partial x}\,(yz) = y \frac{\partial z}{\partial x}.$$

$$\left(y - \frac{1}{z}\right) \frac{\partial z}{\partial x} = 1$$

$$\frac{\partial z}{\partial x} = \frac{z}{yz - 1}. \quad ∎$$

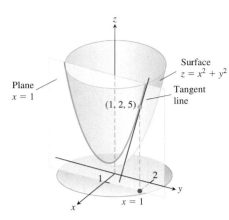

**FIGURE 14.19** The tangent to the curve of intersection of the plane $x = 1$ and surface $z = x^2 + y^2$ at the point $(1, 2, 5)$ (Example 5).

**EXAMPLE 5** The plane $x = 1$ intersects the paraboloid $z = x^2 + y^2$ in a parabola. Find the slope of the tangent to the parabola at $(1, 2, 5)$ (Figure 14.19).

**Solution** The slope is the value of the partial derivative $\partial z / \partial y$ at $(1, 2)$:

$$\left. \frac{\partial z}{\partial y} \right|_{(1,2)} = \left. \frac{\partial}{\partial y} (x^2 + y^2) \right|_{(1,2)} = \left. 2y \right|_{(1,2)} = 2(2) = 4.$$

As a check, we can treat the parabola as the graph of the single-variable function $z = (1)^2 + y^2 = 1 + y^2$ in the plane $x = 1$ and ask for the slope at $y = 2$. The slope, calculated now as an ordinary derivative, is

$$\left. \frac{dz}{dy} \right|_{y=2} = \left. \frac{d}{dy} (1 + y^2) \right|_{y=2} = \left. 2y \right|_{y=2} = 4. \qquad \blacksquare$$

### Functions of More Than Two Variables

The definitions of the partial derivatives of functions of more than two independent variables are like the definitions for functions of two variables. They are ordinary derivatives with respect to one variable, taken while the other independent variables are held constant.

**EXAMPLE 6** If $x$, $y$, and $z$ are independent variables and

$$f(x, y, z) = x \sin (y + 3z),$$

then

$$\frac{\partial f}{\partial z} = \frac{\partial}{\partial z} [x \sin (y + 3z)] = x \frac{\partial}{\partial z} \sin (y + 3z)$$

$$= x \cos (y + 3z) \frac{\partial}{\partial z} (y + 3z) = 3x \cos (y + 3z). \qquad \blacksquare$$

**EXAMPLE 7** If resistors of $R_1$, $R_2$, and $R_3$ ohms are connected in parallel to make an $R$-ohm resistor, the value of $R$ can be found from the equation

$$\frac{1}{R} = \frac{1}{R_1} + \frac{1}{R_2} + \frac{1}{R_3}$$

(Figure 14.20). Find the value of $\partial R / \partial R_2$ when $R_1 = 30$, $R_2 = 45$, and $R_3 = 90$ ohms.

**Solution** To find $\partial R / \partial R_2$, we treat $R_1$ and $R_3$ as constants and, using implicit differentiation, differentiate both sides of the equation with respect to $R_2$:

$$\frac{\partial}{\partial R_2} \left( \frac{1}{R} \right) = \frac{\partial}{\partial R_2} \left( \frac{1}{R_1} + \frac{1}{R_2} + \frac{1}{R_3} \right)$$

$$- \frac{1}{R^2} \frac{\partial R}{\partial R_2} = 0 - \frac{1}{R_2{}^2} + 0$$

$$\frac{\partial R}{\partial R_2} = \frac{R^2}{R_2{}^2} = \left( \frac{R}{R_2} \right)^2 .$$

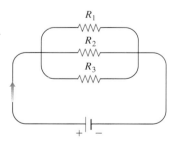

**FIGURE 14.20** Resistors arranged this way are said to be connected in parallel (Example 7). Each resistor lets a portion of the current through. Their equivalent resistance $R$ is calculated with the formula

$$\frac{1}{R} = \frac{1}{R_1} + \frac{1}{R_2} + \frac{1}{R_3} .$$

When $R_1 = 30$, $R_2 = 45$, and $R_3 = 90$,

$$\frac{1}{R} = \frac{1}{30} + \frac{1}{45} + \frac{1}{90} = \frac{3 + 2 + 1}{90} = \frac{6}{90} = \frac{1}{15},$$

so $R = 15$ and

$$\frac{\partial R}{\partial R_2} = \left(\frac{15}{45}\right)^2 = \left(\frac{1}{3}\right)^2 = \frac{1}{9}.$$

Thus at the given values, a small change in the resistance $R_2$ leads to a change in $R$ about $1/9$th as large. ◾

### Partial Derivatives and Continuity

A function $f(x, y)$ can have partial derivatives with respect to both $x$ and $y$ at a point without the function being continuous there. This is different from functions of a single variable, where the existence of a derivative implies continuity. If the partial derivatives of $f(x, y)$ exist and are continuous throughout a disk centered at $(x_0, y_0)$, however, then $f$ *is* continuous at $(x_0, y_0)$, as we see at the end of this section.

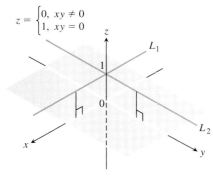

$z = \begin{cases} 0, & xy \neq 0 \\ 1, & xy = 0 \end{cases}$

**FIGURE 14.21** The graph of

$$f(x, y) = \begin{cases} 0, & xy \neq 0 \\ 1, & xy = 0 \end{cases}$$

consists of the lines $L_1$ and $L_2$ and the four open quadrants of the $xy$-plane. The function has partial derivatives at the origin but is not continuous there (Example 8).

**EXAMPLE 8**  Let

$$f(x, y) = \begin{cases} 0, & xy \neq 0 \\ 1, & xy = 0 \end{cases}$$

(Figure 14.21).

**(a)** Find the limit of $f$ as $(x, y)$ approaches $(0, 0)$ along the line $y = x$.

**(b)** Prove that $f$ is not continuous at the origin.

**(c)** Show that both partial derivatives $\partial f/\partial x$ and $\partial f/\partial y$ exist at the origin.

**Solution**

**(a)** Since $f(x, y)$ is constantly zero along the line $y = x$ (except at the origin), we have

$$\lim_{(x, y) \to (0,0)} f(x, y) \bigg|_{y=x} = \lim_{(x, y) \to (0,0)} 0 = 0.$$

**(b)** Since $f(0, 0) = 1$, the limit in part (a) proves that $f$ is not continuous at $(0, 0)$.

**(c)** To find $\partial f/\partial x$ at $(0, 0)$, we hold $y$ fixed at $y = 0$. Then $f(x, y) = 1$ for all $x$, and the graph of $f$ is the line $L_1$ in Figure 14.21. The slope of this line at any $x$ is $\partial f/\partial x = 0$. In particular, $\partial f/\partial x = 0$ at $(0, 0)$. Similarly, $\partial f/\partial y$ is the slope of line $L_2$ at any $y$, so $\partial f/\partial y = 0$ at $(0, 0)$. ◾

Example 8 notwithstanding, it is still true in higher dimensions that *differentiability* at a point implies continuity. What Example 8 suggests is that we need a stronger requirement for differentiability in higher dimensions than the mere existence of the partial derivatives. We define differentiability for functions of two variables (which is slightly more complicated than for single-variable functions) at the end of this section and then revisit the connection to continuity.

### Second-Order Partial Derivatives

When we differentiate a function $f(x, y)$ twice, we produce its second-order derivatives. These derivatives are usually denoted by

$$\frac{\partial^2 f}{\partial x^2} \text{ or } f_{xx}, \qquad \frac{\partial^2 f}{\partial y^2} \text{ or } f_{yy},$$

$$\frac{\partial^2 f}{\partial x \partial y} \text{ or } f_{yx}, \quad \text{and} \quad \frac{\partial^2 f}{\partial y \partial x} \text{ or } f_{xy}.$$

The defining equations are

$$\frac{\partial^2 f}{\partial x^2} = \frac{\partial}{\partial x}\left(\frac{\partial f}{\partial x}\right), \qquad \frac{\partial^2 f}{\partial x \partial y} = \frac{\partial}{\partial x}\left(\frac{\partial f}{\partial y}\right),$$

and so on. Notice the order in which the mixed partial derivatives are taken:

$$\frac{\partial^2 f}{\partial x \partial y} \qquad \text{Differentiate first with respect to } y, \text{ then with respect to } x.$$

$$f_{yx} = (f_y)_x \qquad \text{Means the same thing.}$$

**EXAMPLE 9**    If $f(x, y) = x \cos y + y e^x$, find the second-order derivatives

$$\frac{\partial^2 f}{\partial x^2}, \qquad \frac{\partial^2 f}{\partial y \partial x}, \qquad \frac{\partial^2 f}{\partial y^2}, \qquad \text{and} \qquad \frac{\partial^2 f}{\partial x \partial y}.$$

**Solution**    The first step is to calculate both first partial derivatives.

$$\frac{\partial f}{\partial x} = \frac{\partial}{\partial x}(x \cos y + y e^x) \qquad\qquad \frac{\partial f}{\partial y} = \frac{\partial}{\partial y}(x \cos y + y e^x)$$

$$= \cos y + y e^x \qquad\qquad\qquad\qquad = -x \sin y + e^x$$

Now we find both partial derivatives of each first partial:

$$\frac{\partial^2 f}{\partial y \partial x} = \frac{\partial}{\partial y}\left(\frac{\partial f}{\partial x}\right) = -\sin y + e^x \qquad \frac{\partial^2 f}{\partial x \partial y} = \frac{\partial}{\partial x}\left(\frac{\partial f}{\partial y}\right) = -\sin y + e^x$$

$$\frac{\partial^2 f}{\partial x^2} = \frac{\partial}{\partial x}\left(\frac{\partial f}{\partial x}\right) = y e^x. \qquad \frac{\partial^2 f}{\partial y^2} = \frac{\partial}{\partial y}\left(\frac{\partial f}{\partial y}\right) = -x \cos y. \qquad ∎$$

## The Mixed Derivative Theorem

You may have noticed that the "mixed" second-order partial derivatives

$$\frac{\partial^2 f}{\partial y \partial x} \qquad \text{and} \qquad \frac{\partial^2 f}{\partial x \partial y}$$

in Example 9 are equal. This is not a coincidence. They must be equal whenever $f, f_x, f_y, f_{xy}$, and $f_{yx}$ are continuous, as stated in the following theorem. However, the mixed derivatives can be different when the continuity conditions are not satisfied (see Exercise 72).

> **THEOREM 2—The Mixed Derivative Theorem**    If $f(x, y)$ and its partial derivatives $f_x, f_y, f_{xy}$, and $f_{yx}$ are defined throughout an open region containing a point $(a, b)$ and are all continuous at $(a, b)$, then
>
> $$f_{xy}(a, b) = f_{yx}(a, b).$$

Theorem 2 is also known as Clairaut's Theorem, named after the French mathematician Alexis Clairaut, who discovered it. A proof is given in Appendix 9. Theorem 2 says that to calculate a mixed second-order derivative, we may differentiate in either order, provided the continuity conditions are satisfied. This ability to proceed in different order sometimes simplifies our calculations.

**EXAMPLE 10**    Find $\partial^2 w / \partial x \partial y$ if

$$w = xy + \frac{e^y}{y^2 + 1}.$$

**Solution**   The symbol $\partial^2 w/\partial x \partial y$ tells us to differentiate first with respect to $y$ and then with respect to $x$. However, if we interchange the order of differentiation and differentiate first with respect to $x$ we get the answer more quickly. In two steps,

$$\frac{\partial w}{\partial x} = y \qquad \text{and} \qquad \frac{\partial^2 w}{\partial y \partial x} = 1.$$

If we differentiate first with respect to $y$, we obtain $\partial^2 w/\partial x \partial y = 1$ as well. We can differentiate in either order because the conditions of Theorem 2 hold for $w$ at all points $(x_0, y_0)$.  ∎

## Partial Derivatives of Still Higher Order

Although we will deal mostly with first- and second-order partial derivatives, because these appear the most frequently in applications, there is no theoretical limit to how many times we can differentiate a function as long as the derivatives involved exist. Thus, we get third- and fourth-order derivatives denoted by symbols like

$$\frac{\partial^3 f}{\partial x \partial y^2} = f_{yyx},$$

$$\frac{\partial^4 f}{\partial x^2 \partial y^2} = f_{yyxx},$$

and so on. As with second-order derivatives, the order of differentiation is immaterial as long as all the derivatives through the order in question are continuous.

**EXAMPLE 11**     Find $f_{yxyz}$ if $f(x, y, z) = 1 - 2xy^2z + x^2y$.

**Solution**   We first differentiate with respect to the variable $y$, then $x$, then $y$ again, and finally with respect to $z$:

$$f_y = -4xyz + x^2$$
$$f_{yx} = -4yz + 2x$$
$$f_{yxy} = -4z$$
$$f_{yxyz} = -4.$$  ∎

## Differentiability

The concept of *differentiability* for functions of several variables is more complicated than for single-variable functions because a point in the domain can be approached along more than one path. In defining the partial derivatives for a function of two variables, we intersected the surface of the graph with vertical planes parallel to the $xz$- and $yz$-planes, creating a curve on each plane, called a *trace*. The partial derivatives were seen as the slopes of the two tangent lines to these trace curves at the point on the surface corresponding to the point $(x_0, y_0)$ being approached in the domain. (See Figure 14.18.) For a differentiable function, it would seem reasonable to assume that if we were to rotate slightly one of these vertical planes, keeping it vertical but no longer parallel to its coordinate plane, then a smooth trace curve would appear on that plane that would have a tangent line at the point on the surface having a slope differing just slightly from what it was before (when the plane was parallel to its coordinate plane). However, the mere existence of the original partial derivative does not guarantee that result. For example, the surface might have a "fissure" in the direction of the new plane, so the trace curve is not even continuous at $(x_0, y_0)$, let alone having a tangent line at the corresponding point on the curve. Just as having a limit in the $x$- and $y$-coordinate directions does not imply the function itself has a limit at $(x_0, y_0)$, as we see in Figure 14.21, so is it the case that the existence of both partial derivatives is not enough by itself to ensure derivatives exist for trace curves in other vertical planes. For the existence of differentiability, a property is needed to ensure that no abrupt change occurs in the function resulting from small changes in the independent variables along any path approaching $(x_0, y_0)$, paths along which *both* variables $x$ and $y$ are allowed to change, rather than just one of them at a time. We saw a way of thinking about the change in a function in Section 3.9.

In our study of functions of a single variable, we found that if a function $y = f(x)$ is differentiable at $x = x_0$, then the change $\Delta y$ resulting in the change of $x$ from $x_0$ to $x_0 + \Delta x$ is close to the change $\Delta L$ along the tangent line (or linear approximation $L$ of the function $f$ at $x_0$). That is, from Equation (1) in Section 3.9,

$$\Delta y = f'(x_0)\Delta x + \epsilon\Delta x$$

in which $\epsilon \to 0$ as $\Delta x \to 0$. The extension of this result is what we use to *define* differentiability for functions of two variables.

---

**DEFINITION**     A function $z = f(x, y)$ is **differentiable at** $(x_0, y_0)$ if $f_x(x_0, y_0)$ and $f_y(x_0, y_0)$ exist and $\Delta z$ satisfies an equation of the form

$$\Delta z = f_x(x_0, y_0)\Delta x + f_y(x_0, y_0)\Delta y + \epsilon_1\Delta x + \epsilon_2\Delta y$$

in which each of $\epsilon_1, \epsilon_2 \to 0$ as both $\Delta x, \Delta y \to 0$. We call $f$ **differentiable** if it is differentiable at every point in its domain, and say that its graph is a **smooth surface**.

---

The following theorem (proved in Appendix 9) and its accompanying corollary tell us that functions with *continuous* first partial derivatives at $(x_0, y_0)$ are differentiable there, and they are closely approximated locally by a linear function. We study this approximation in Section 14.6.

---

**THEOREM 3—The Increment Theorem for Functions of Two Variables**     Suppose that the first partial derivatives of $f(x, y)$ are defined throughout an open region $R$ containing the point $(x_0, y_0)$ and that $f_x$ and $f_y$ are continuous at $(x_0, y_0)$. Then the change

$$\Delta z = f(x_0 + \Delta x, y_0 + \Delta y) - f(x_0, y_0)$$

in the value of $f$ that results from moving from $(x_0, y_0)$ to another point $(x_0 + \Delta x, y_0 + \Delta y)$ in $R$ satisfies an equation of the form

$$\Delta z = f_x(x_0, y_0)\Delta x + f_y(x_0, y_0)\Delta y + \epsilon_1\Delta x + \epsilon_2\Delta y$$

in which each of $\epsilon_1, \epsilon_2 \to 0$ as both $\Delta x, \Delta y \to 0$.

---

**COROLLARY OF THEOREM 3**     If the partial derivatives $f_x$ and $f_y$ of a function $f(x, y)$ are continuous throughout an open region $R$, then $f$ is differentiable at every point of $R$.

---

If $z = f(x, y)$ is differentiable, then the definition of differentiability ensures that $\Delta z = f(x_0 + \Delta x, y_0 + \Delta y) - f(x_0, y_0)$ approaches 0 as $\Delta x$ and $\Delta y$ approach 0. This tells us that a function of two variables is continuous at every point where it is differentiable.

---

**THEOREM 4—Differentiability Implies Continuity**     If a function $f(x, y)$ is differentiable at $(x_0, y_0)$, then $f$ is continuous at $(x_0, y_0)$.

As we can see from Corollary 3 and Theorem 4, a function $f(x, y)$ must be continuous at a point $(x_0, y_0)$ if $f_x$ and $f_y$ are continuous throughout an open region containing $(x_0, y_0)$. Remember, however, that it is still possible for a function of two variables to be discontinuous at a point where its first partial derivatives exist, as we saw in Example 8. Existence alone of the partial derivatives at that point is not enough, but continuity of the partial derivatives guarantees differentiability.

# Exercises 14.3

## Calculating First-Order Partial Derivatives

In Exercises 1–22, find $\partial f / \partial x$ and $\partial f / \partial y$.

**1.** $f(x, y) = 2x^2 - 3y - 4$      **2.** $f(x, y) = x^2 - xy + y^2$

**3.** $f(x, y) = (x^2 - 1)(y + 2)$

**4.** $f(x, y) = 5xy - 7x^2 - y^2 + 3x - 6y + 2$

**5.** $f(x, y) = (xy - 1)^2$      **6.** $f(x, y) = (2x - 3y)^3$

**7.** $f(x, y) = \sqrt{x^2 + y^2}$      **8.** $f(x, y) = (x^3 + (y/2))^{2/3}$

**9.** $f(x, y) = 1/(x + y)$      **10.** $f(x, y) = x/(x^2 + y^2)$

**11.** $f(x, y) = (x + y)/(xy - 1)$   **12.** $f(x, y) = \tan^{-1}(y/x)$

**13.** $f(x, y) = e^{(x+y+1)}$      **14.** $f(x, y) = e^{-x} \sin(x + y)$

**15.** $f(x, y) = \ln(x + y)$      **16.** $f(x, y) = e^{xy} \ln y$

**17.** $f(x, y) = \sin^2(x - 3y)$      **18.** $f(x, y) = \cos^2(3x - y^2)$

**19.** $f(x, y) = x^y$      **20.** $f(x, y) = \log_y x$

**21.** $f(x, y) = \displaystyle\int_x^y g(t)\, dt$    ($g$ continuous for all $t$)

**22.** $f(x, y) = \displaystyle\sum_{n=0}^{\infty} (xy)^n$    ($|xy| < 1$)

In Exercises 23–34, find $f_x$, $f_y$, and $f_z$.

**23.** $f(x, y, z) = 1 + xy^2 - 2z^2$

**24.** $f(x, y, z) = xy + yz + xz$

**25.** $f(x, y, z) = x - \sqrt{y^2 + z^2}$

**26.** $f(x, y, z) = (x^2 + y^2 + z^2)^{-1/2}$

**27.** $f(x, y, z) = \sin^{-1}(xyz)$

**28.** $f(x, y, z) = \sec^{-1}(x + yz)$

**29.** $f(x, y, z) = \ln(x + 2y + 3z)$

**30.** $f(x, y, z) = yz \ln(xy)$

**31.** $f(x, y, z) = e^{-(x^2+y^2+z^2)}$

**32.** $f(x, y, z) = e^{-xyz}$

**33.** $f(x, y, z) = \tanh(x + 2y + 3z)$

**34.** $f(x, y, z) = \sinh(xy - z^2)$

In Exercises 35–40, find the partial derivative of the function with respect to each variable.

**35.** $f(t, \alpha) = \cos(2\pi t - \alpha)$

**36.** $g(u, v) = v^2 e^{(2u/v)}$

**37.** $h(\rho, \phi, \theta) = \rho \sin \phi \cos \theta$

**38.** $g(r, \theta, z) = r(1 - \cos \theta) - z$

**39. Work done by the heart**   (Section 3.9, Exercise 61)

$$W(P, V, \delta, v, g) = PV + \frac{V \delta v^2}{2g}$$

**40. Wilson lot size formula**   (Section 4.5, Exercise 53)

$$A(c, h, k, m, q) = \frac{km}{q} + cm + \frac{hq}{2}$$

## Calculating Second-Order Partial Derivatives

Find all the second-order partial derivatives of the functions in Exercises 41–50.

**41.** $f(x, y) = x + y + xy$      **42.** $f(x, y) = \sin xy$

**43.** $g(x, y) = x^2 y + \cos y + y \sin x$

**44.** $h(x, y) = xe^y + y + 1$      **45.** $r(x, y) = \ln(x + y)$

**46.** $s(x, y) = \tan^{-1}(y/x)$      **47.** $w = x^2 \tan(xy)$

**48.** $w = ye^{x^2 - y}$      **49.** $w = x \sin(x^2 y)$

**50.** $w = \dfrac{x - y}{x^2 + y}$

## Mixed Partial Derivatives

In Exercises 51–54, verify that $w_{xy} = w_{yx}$.

**51.** $w = \ln(2x + 3y)$      **52.** $w = e^x + x \ln y + y \ln x$

**53.** $w = xy^2 + x^2 y^3 + x^3 y^4$    **54.** $w = x \sin y + y \sin x + xy$

**55.** Which order of differentiation will calculate $f_{xy}$ faster: $x$ first or $y$ first? Try to answer without writing anything down.

   **a.** $f(x, y) = x \sin y + e^y$

   **b.** $f(x, y) = 1/x$

   **c.** $f(x, y) = y + (x/y)$

   **d.** $f(x, y) = y + x^2 y + 4y^3 - \ln(y^2 + 1)$

   **e.** $f(x, y) = x^2 + 5xy + \sin x + 7e^x$

   **f.** $f(x, y) = x \ln xy$

**56.** The fifth-order partial derivative $\partial^5 f / \partial x^2 \partial y^3$ is zero for each of the following functions. To show this as quickly as possible, which variable would you differentiate with respect to first: $x$ or $y$? Try to answer without writing anything down.

   **a.** $f(x, y) = y^2 x^4 e^x + 2$

   **b.** $f(x, y) = y^2 + y(\sin x - x^4)$

   **c.** $f(x, y) = x^2 + 5xy + \sin x + 7e^x$

   **d.** $f(x, y) = xe^{y^2/2}$

## Using the Partial Derivative Definition

In Exercises 57–60, use the limit definition of partial derivative to compute the partial derivatives of the functions at the specified points.

**57.** $f(x, y) = 1 - x + y - 3x^2y$, $\dfrac{\partial f}{\partial x}$ and $\dfrac{\partial f}{\partial y}$ at $(1, 2)$

**58.** $f(x, y) = 4 + 2x - 3y - xy^2$, $\dfrac{\partial f}{\partial x}$ and $\dfrac{\partial f}{\partial y}$ at $(-2, 1)$

**59.** $f(x, y) = \sqrt{2x + 3y - 1}$, $\dfrac{\partial f}{\partial x}$ and $\dfrac{\partial f}{\partial y}$ at $(-2, 3)$

**60.** $f(x, y) = \begin{cases} \dfrac{\sin(x^3 + y^4)}{x^2 + y^2}, & (x, y) \neq (0, 0) \\ 0, & (x, y) = (0, 0), \end{cases}$

$\dfrac{\partial f}{\partial x}$ and $\dfrac{\partial f}{\partial y}$ at $(0, 0)$

**61.** Let $f(x, y) = 2x + 3y - 4$. Find the slope of the line tangent to this surface at the point $(2, -1)$ and lying in the **a.** plane $x = 2$ **b.** plane $y = -1$.

**62.** Let $f(x, y) = x^2 + y^3$. Find the slope of the line tangent to this surface at the point $(-1, 1)$ and lying in the **a.** plane $x = -1$ **b.** plane $y = 1$.

**63. Three variables** Let $w = f(x, y, z)$ be a function of three independent variables and write the formal definition of the partial derivative $\partial f / \partial z$ at $(x_0, y_0, z_0)$. Use this definition to find $\partial f / \partial z$ at $(1, 2, 3)$ for $f(x, y, z) = x^2yz^2$.

**64. Three variables** Let $w = f(x, y, z)$ be a function of three independent variables and write the formal definition of the partial derivative $\partial f / \partial y$ at $(x_0, y_0, z_0)$. Use this definition to find $\partial f / \partial y$ at $(-1, 0, 3)$ for $f(x, y, z) = -2xy^2 + yz^2$.

## Differentiating Implicitly

**65.** Find the value of $\partial z / \partial x$ at the point $(1, 1, 1)$ if the equation

$$xy + z^3x - 2yz = 0$$

defines $z$ as a function of the two independent variables $x$ and $y$ and the partial derivative exists.

**66.** Find the value of $\partial x / \partial z$ at the point $(1, -1, -3)$ if the equation

$$xz + y \ln x - x^2 + 4 = 0$$

defines $x$ as a function of the two independent variables $y$ and $z$ and the partial derivative exists.

Exercises 67 and 68 are about the triangle shown here.

**67.** Express $A$ implicitly as a function of $a$, $b$, and $c$ and calculate $\partial A / \partial a$ and $\partial A / \partial b$.

**68.** Express $a$ implicitly as a function of $A$, $b$, and $B$ and calculate $\partial a / \partial A$ and $\partial a / \partial B$.

**69. Two dependent variables** Express $v_x$ in terms of $u$ and $y$ if the equations $x = v \ln u$ and $y = u \ln v$ define $u$ and $v$ as functions of the independent variables $x$ and $y$, and if $v_x$ exists. (*Hint:* Differentiate both equations with respect to $x$ and solve for $v_x$ by eliminating $u_x$.)

**70. Two dependent variables** Find $\partial x / \partial u$ and $\partial y / \partial u$ if the equations $u = x^2 - y^2$ and $v = x^2 - y$ define $x$ and $y$ as functions of the independent variables $u$ and $v$, and the partial derivatives exist. (See the hint in Exercise 69.) Then let $s = x^2 + y^2$ and find $\partial s / \partial u$.

## Theory and Examples

**71.** Let $f(x, y) = \begin{cases} y^3, & y \geq 0 \\ -y^2, & y < 0. \end{cases}$

Find $f_x, f_y, f_{xy}$, and $f_{yx}$, and state the domain for each partial derivative.

**72.** Let $f(x, y) = \begin{cases} xy \dfrac{x^2 - y^2}{x^2 + y^2}, & \text{if } (x, y) \neq 0, \\ 0, & \text{if } (x, y) = 0. \end{cases}$

**a.** Show that $\dfrac{\partial f}{\partial y}(x, 0) = x$ for all $x$, and $\dfrac{\partial f}{\partial x}(0, y) = -y$ for all $y$.

**b.** Show that $\dfrac{\partial^2 f}{\partial y \partial x}(0, 0) \neq \dfrac{\partial^2 f}{\partial x \partial y}(0, 0)$.

The graph of $f$ is shown on page 788.

The **three-dimensional Laplace equation**

$$\frac{\partial^2 f}{\partial x^2} + \frac{\partial^2 f}{\partial y^2} + \frac{\partial^2 f}{\partial z^2} = 0$$

is satisfied by steady-state temperature distributions $T = f(x, y, z)$ in space, by gravitational potentials, and by electrostatic potentials. The **two-dimensional Laplace equation**

$$\frac{\partial^2 f}{\partial x^2} + \frac{\partial^2 f}{\partial y^2} = 0,$$

obtained by dropping the $\partial^2 f / \partial z^2$ term from the previous equation, describes potentials and steady-state temperature distributions in a plane (see the accompanying figure). The plane (a) may be treated as a thin slice of the solid (b) perpendicular to the $z$-axis.

Boundary temperatures controlled

Show that each function in Exercises 73–80 satisfies a Laplace equation.

**73.** $f(x, y, z) = x^2 + y^2 - 2z^2$

**74.** $f(x, y, z) = 2z^3 - 3(x^2 + y^2)z$

**75.** $f(x, y) = e^{-2y} \cos 2x$

**76.** $f(x, y) = \ln \sqrt{x^2 + y^2}$

**77.** $f(x, y) = 3x + 2y - 4$

**78.** $f(x, y) = \tan^{-1} \dfrac{x}{y}$

**79.** $f(x, y, z) = (x^2 + y^2 + z^2)^{-1/2}$

**80.** $f(x, y, z) = e^{3x+4y} \cos 5z$

**The Wave Equation**    If we stand on an ocean shore and take a snapshot of the waves, the picture shows a regular pattern of peaks and valleys in an instant of time. We see periodic vertical motion in space, with respect to distance. If we stand in the water, we can feel the rise and fall of the water as the waves go by. We see periodic vertical motion in time. In physics, this beautiful symmetry is expressed by the **one-dimensional wave equation**

$$\frac{\partial^2 w}{\partial t^2} = c^2 \frac{\partial^2 w}{\partial x^2},$$

where $w$ is the wave height, $x$ is the distance variable, $t$ is the time variable, and $c$ is the velocity with which the waves are propagated.

In our example, $x$ is the distance across the ocean's surface, but in other applications, $x$ might be the distance along a vibrating string, distance through air (sound waves), or distance through space (light waves). The number $c$ varies with the medium and type of wave.

Show that the functions in Exercises 81–87 are all solutions of the wave equation.

**81.** $w = \sin(x + ct)$

**82.** $w = \cos(2x + 2ct)$

**83.** $w = \sin(x + ct) + \cos(2x + 2ct)$

**84.** $w = \ln(2x + 2ct)$

**85.** $w = \tan(2x - 2ct)$

**86.** $w = 5 \cos(3x + 3ct) + e^{x+ct}$

**87.** $w = f(u)$, where $f$ is a differentiable function of $u$, and $u = a(x + ct)$, where $a$ is a constant

**88.** Does a function $f(x, y)$ with continuous first partial derivatives throughout an open region $R$ have to be continuous on $R$? Give reasons for your answer.

**89.** If a function $f(x, y)$ has continuous second partial derivatives throughout an open region $R$, must the first-order partial derivatives of $f$ be continuous on $R$? Give reasons for your answer.

**90. The heat equation**    An important partial differential equation that describes the distribution of heat in a region at time $t$ can be represented by the *one-dimensional heat equation*

$$\frac{\partial f}{\partial t} = \frac{\partial^2 f}{\partial x^2}.$$

Show that $u(x, t) = \sin(\alpha x) \cdot e^{-\beta t}$ satisfies the heat equation for constants $\alpha$ and $\beta$. What is the relationship between $\alpha$ and $\beta$ for this function to be a solution?

**91.** Let $f(x, y) = \begin{cases} \dfrac{xy^2}{x^2 + y^4}, & (x, y) \neq (0, 0) \\ 0, & (x, y) = (0, 0). \end{cases}$

Show that $f_x(0, 0)$ and $f_y(0, 0)$ exist, but $f$ is not differentiable at $(0, 0)$. (*Hint:* Use Theorem 4 and show that $f$ is not continuous at $(0, 0)$.)

**92.** Let $f(x, y) = \begin{cases} 0, & x^2 < y < 2x^2 \\ 1, & \text{otherwise.} \end{cases}$

Show that $f_x(0, 0)$ and $f_y(0, 0)$ exist, but $f$ is not differentiable at $(0, 0)$.

# 14.4 The Chain Rule

The Chain Rule for functions of a single variable studied in Section 3.6 says that when $w = f(x)$ is a differentiable function of $x$ and $x = g(t)$ is a differentiable function of $t$, $w$ is a differentiable function of $t$ and $dw/dt$ can be calculated by the formula

$$\frac{dw}{dt} = \frac{dw}{dx}\frac{dx}{dt}.$$

For this composite function $w(t) = f(g(t))$, we can think of $t$ as the independent variable and $x = g(t)$ as the "intermediate variable," because $t$ determines the value of $x$ which in turn gives the value of $w$ from the function $f$. We display the Chain Rule in a "branch diagram" in the margin on the next page.

For functions of several variables the Chain Rule has more than one form, which depends on how many independent and intermediate variables are involved. However, once the variables are taken into account, the Chain Rule works in the same way we just discussed.

To find $dw/dt$, we read down the route from $w$ to $t$, multiplying derivatives along the way.

**Chain Rule**

$w = f(x)$ — Dependent variable

$\dfrac{dw}{dx}$

$x$ — Intermediate variable

$\dfrac{dx}{dt}$

$t$ — Independent variable

$\dfrac{dw}{dt} = \dfrac{dw}{dx}\dfrac{dx}{dt}$

Each of $\dfrac{\partial f}{\partial x}, \dfrac{\partial w}{\partial x}, f_x$ indicates the partial derivative of $f$ with respect to $x$.

## Functions of Two Variables

The Chain Rule formula for a differentiable function $w = f(x, y)$ when $x = x(t)$ and $y = y(t)$ are both differentiable functions of $t$ is given in the following theorem.

---

**THEOREM 5—Chain Rule For Functions of One Independent Variable and Two Intermediate Variables**    If $w = f(x, y)$ is differentiable and if $x = x(t), y = y(t)$ are differentiable functions of $t$, then the composite $w = f(x(t), y(t))$ is a differentiable function of $t$ and

$$\frac{dw}{dt} = f_x(x(t), y(t)) \cdot x'(t) + f_y(x(t), y(t)) \cdot y'(t),$$

or

$$\frac{dw}{dt} = \frac{\partial f}{\partial x}\frac{dx}{dt} + \frac{\partial f}{\partial y}\frac{dy}{dt}.$$

---

**Proof**    The proof consists of showing that if $x$ and $y$ are differentiable at $t = t_0$, then $w$ is differentiable at $t_0$ and

$$\left(\frac{dw}{dt}\right)_{t_0} = \left(\frac{\partial w}{\partial x}\right)_{P_0}\left(\frac{dx}{dt}\right)_{t_0} + \left(\frac{\partial w}{\partial y}\right)_{P_0}\left(\frac{dy}{dt}\right)_{t_0},$$

where $P_0 = (x(t_0), y(t_0))$. The subscripts indicate where each of the derivatives is to be evaluated.

Let $\Delta x, \Delta y,$ and $\Delta w$ be the increments that result from changing $t$ from $t_0$ to $t_0 + \Delta t$. Since $f$ is differentiable (see the definition in Section 14.3),

$$\Delta w = \left(\frac{\partial w}{\partial x}\right)_{P_0}\Delta x + \left(\frac{\partial w}{\partial y}\right)_{P_0}\Delta y + \epsilon_1\Delta x + \epsilon_2\Delta y,$$

where $\epsilon_1, \epsilon_2 \to 0$ as $\Delta x, \Delta y \to 0$. To find $dw/dt$, we divide this equation through by $\Delta t$ and let $\Delta t$ approach zero. The division gives

$$\frac{\Delta w}{\Delta t} = \left(\frac{\partial w}{\partial x}\right)_{P_0}\frac{\Delta x}{\Delta t} + \left(\frac{\partial w}{\partial y}\right)_{P_0}\frac{\Delta y}{\Delta t} + \epsilon_1\frac{\Delta x}{\Delta t} + \epsilon_2\frac{\Delta y}{\Delta t}.$$

Letting $\Delta t$ approach zero gives

$$\left(\frac{dw}{dt}\right)_{t_0} = \lim_{\Delta t \to 0}\frac{\Delta w}{\Delta t}$$

$$= \left(\frac{\partial w}{\partial x}\right)_{P_0}\left(\frac{dx}{dt}\right)_{t_0} + \left(\frac{\partial w}{\partial y}\right)_{P_0}\left(\frac{dy}{dt}\right)_{t_0} + 0\cdot\left(\frac{dx}{dt}\right)_{t_0} + 0\cdot\left(\frac{dy}{dt}\right)_{t_0}. \qquad \blacksquare$$

Often we write $\partial w/\partial x$ for the partial derivative $\partial f/\partial x$, so we can rewrite the Chain Rule in Theorem 5 in the form

$$\frac{dw}{dt} = \frac{\partial w}{\partial x}\frac{dx}{dt} + \frac{\partial w}{\partial y}\frac{dy}{dt}.$$

To remember the Chain Rule, picture the diagram below. To find $dw/dt$, start at $w$ and read down each route to $t$, multiplying derivatives along the way. Then add the products.

**Chain Rule**

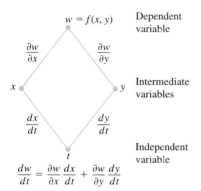

$$\frac{dw}{dt} = \frac{\partial w}{\partial x}\frac{dx}{dt} + \frac{\partial w}{\partial y}\frac{dy}{dt}$$

However, the meaning of the dependent variable $w$ is different on each side of the preceding equation. On the left-hand side, it refers to the composite function $w = f(x(t), y(t))$ as a function of the single variable $t$. On the right-hand side, it refers to the function $w = f(x, y)$ as a function of the two variables $x$ and $y$. Moreover, the single derivatives $dw/dt$, $dx/dt$, and $dy/dt$ are being evaluated at a point $t_0$, whereas the partial derivatives $\partial w/\partial x$ and $\partial w/\partial y$ are being evaluated at the point $(x_0, y_0)$, with $x_0 = x(t_0)$ and $y_0 = y(t_0)$. With that understanding, we will use both of these forms interchangeably throughout the text whenever no confusion will arise.

The **branch diagram** in the margin provides a convenient way to remember the Chain Rule. The "true" independent variable in the composite function is $t$, whereas $x$ and $y$ are *intermediate variables* (controlled by $t$) and $w$ is the dependent variable.

A more precise notation for the Chain Rule shows where the various derivatives in Theorem 5 are evaluated:

$$\frac{dw}{dt}(t_0) = \frac{\partial f}{\partial x}(x_0, y_0) \cdot \frac{dx}{dt}(t_0) + \frac{\partial f}{\partial y}(x_0, y_0) \cdot \frac{dy}{dt}(t_0).$$

**EXAMPLE 1** Use the Chain Rule to find the derivative of

$$w = xy$$

with respect to $t$ along the path $x = \cos t$, $y = \sin t$. What is the derivative's value at $t = \pi/2$?

**Solution** We apply the Chain Rule to find $dw/dt$ as follows:

$$\frac{dw}{dt} = \frac{\partial w}{\partial x}\frac{dx}{dt} + \frac{\partial w}{\partial y}\frac{dy}{dt}$$

$$= \frac{\partial(xy)}{\partial x} \cdot \frac{d}{dt}(\cos t) + \frac{\partial(xy)}{\partial y} \cdot \frac{d}{dt}(\sin t)$$

$$= (y)(-\sin t) + (x)(\cos t)$$

$$= (\sin t)(-\sin t) + (\cos t)(\cos t)$$

$$= -\sin^2 t + \cos^2 t$$

$$= \cos 2t.$$

In this example, we can check the result with a more direct calculation. As a function of $t$,

$$w = xy = \cos t \sin t = \frac{1}{2}\sin 2t,$$

so

$$\frac{dw}{dt} = \frac{d}{dt}\left(\frac{1}{2}\sin 2t\right) = \frac{1}{2} \cdot 2 \cos 2t = \cos 2t.$$

In either case, at the given value of $t$,

$$\left(\frac{dw}{dt}\right)_{t=\pi/2} = \cos\left(2 \cdot \frac{\pi}{2}\right) = \cos \pi = -1. \qquad \blacksquare$$

## Functions of Three Variables

You can probably predict the Chain Rule for functions of three intermediate variables, as it only involves adding the expected third term to the two-variable formula.

---

**THEOREM 6—Chain Rule for Functions of One Independent Variable and Three Intermediate Variables** If $w = f(x, y, z)$ is differentiable and $x$, $y$, and $z$ are differentiable functions of $t$, then $w$ is a differentiable function of $t$ and

$$\frac{dw}{dt} = \frac{\partial w}{\partial x}\frac{dx}{dt} + \frac{\partial w}{\partial y}\frac{dy}{dt} + \frac{\partial w}{\partial z}\frac{dz}{dt}.$$

---

Here we have three routes from $w$ to $t$ instead of two, but finding $dw/dt$ is still the same. Read down each route, multiplying derivatives along the way; then add.

**Chain Rule**

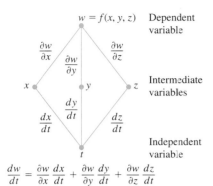

$$\frac{dw}{dt} = \frac{\partial w}{\partial x}\frac{dx}{dt} + \frac{\partial w}{\partial y}\frac{dy}{dt} + \frac{\partial w}{\partial z}\frac{dz}{dt}$$

The proof is identical with the proof of Theorem 5 except that there are now three intermediate variables instead of two. The branch diagram we use for remembering the new equation is similar as well, with three routes from $w$ to $t$.

**EXAMPLE 2**  Find $dw/dt$ if

$$w = xy + z, \qquad x = \cos t, \qquad y = \sin t, \qquad z = t.$$

In this example the values of $w(t)$ are changing along the path of a helix (Section 13.1) as $t$ changes. What is the derivative's value at $t = 0$?

**Solution**  Using the Chain Rule for three intermediate variables, we have

$$\frac{dw}{dt} = \frac{\partial w}{\partial x}\frac{dx}{dt} + \frac{\partial w}{\partial y}\frac{dy}{dt} + \frac{\partial w}{\partial z}\frac{dz}{dt}$$
$$= (y)(-\sin t) + (x)(\cos t) + (1)(1) \qquad \text{Substitute for}$$
$$= (\sin t)(-\sin t) + (\cos t)(\cos t) + 1 \qquad \text{the intermediate}$$
$$= -\sin^2 t + \cos^2 t + 1 = 1 + \cos 2t, \qquad \text{variables.}$$

so

$$\left(\frac{dw}{dt}\right)_{t=0} = 1 + \cos(0) = 2. \qquad \blacksquare$$

For a physical interpretation of change along a curve, think of an object whose position is changing with time $t$. If $w = T(x, y, z)$ is the temperature at each point $(x, y, z)$ along a curve $C$ with parametric equations $x = x(t)$, $y = y(t)$, and $z = z(t)$, then the composite function $w = T(x(t), y(t), z(t))$ represents the temperature relative to $t$ along the curve. The derivative $dw/dt$ is then the instantaneous rate of change of temperature due to the motion along the curve, as calculated in Theorem 6.

### Functions Defined on Surfaces

If we are interested in the temperature $w = f(x, y, z)$ at points $(x, y, z)$ on the earth's surface, we might prefer to think of $x$, $y$, and $z$ as functions of the variables $r$ and $s$ that give the points' longitudes and latitudes. If $x = g(r, s)$, $y = h(r, s)$, and $z = k(r, s)$, we could then express the temperature as a function of $r$ and $s$ with the composite function

$$w = f(g(r, s), h(r, s), k(r, s)).$$

Under the conditions stated below, $w$ has partial derivatives with respect to both $r$ and $s$ that can be calculated in the following way.

---

**THEOREM 7—Chain Rule for Two Independent Variables and Three Intermediate Variables**  Suppose that $w = f(x, y, z)$, $x = g(r, s)$, $y = h(r, s)$, and $z = k(r, s)$. If all four functions are differentiable, then $w$ has partial derivatives with respect to $r$ and $s$, given by the formulas

$$\frac{\partial w}{\partial r} = \frac{\partial w}{\partial x}\frac{\partial x}{\partial r} + \frac{\partial w}{\partial y}\frac{\partial y}{\partial r} + \frac{\partial w}{\partial z}\frac{\partial z}{\partial r}$$

$$\frac{\partial w}{\partial s} = \frac{\partial w}{\partial x}\frac{\partial x}{\partial s} + \frac{\partial w}{\partial y}\frac{\partial y}{\partial s} + \frac{\partial w}{\partial z}\frac{\partial z}{\partial s}.$$

---

The first of these equations can be derived from the Chain Rule in Theorem 6 by holding $s$ fixed and treating $r$ as $t$. The second can be derived in the same way, holding $r$ fixed and treating $s$ as $t$. The branch diagrams for both equations are shown in Figure 14.22.

**FIGURE 14.22** Composite function and branch diagrams for Theorem 7.

**EXAMPLE 3**    Express $\partial w / \partial r$ and $\partial w / \partial s$ in terms of $r$ and $s$ if

$$w = x + 2y + z^2, \qquad x = \frac{r}{s}, \qquad y = r^2 + \ln s, \qquad z = 2r.$$

**Solution**    Using the formulas in Theorem 7, we find

$$\frac{\partial w}{\partial r} = \frac{\partial w}{\partial x}\frac{\partial x}{\partial r} + \frac{\partial w}{\partial y}\frac{\partial y}{\partial r} + \frac{\partial w}{\partial z}\frac{\partial z}{\partial r}$$

$$= (1)\left(\frac{1}{s}\right) + (2)(2r) + (2z)(2) \qquad \text{Substitute for intermediate variable } z.$$

$$= \frac{1}{s} + 4r + (4r)(2) = \frac{1}{s} + 12r$$

$$\frac{\partial w}{\partial s} = \frac{\partial w}{\partial x}\frac{\partial x}{\partial s} + \frac{\partial w}{\partial y}\frac{\partial y}{\partial s} + \frac{\partial w}{\partial z}\frac{\partial z}{\partial s}$$

$$= (1)\left(-\frac{r}{s^2}\right) + (2)\left(\frac{1}{s}\right) + (2z)(0) = \frac{2}{s} - \frac{r}{s^2}. \qquad \blacksquare$$

If $f$ is a function of two intermediate variables instead of three, each equation in Theorem 7 becomes correspondingly one term shorter.

**Chain Rule**

**FIGURE 14.23**    Branch diagram for the equation

$$\frac{\partial w}{\partial r} = \frac{\partial w}{\partial x}\frac{\partial x}{\partial r} + \frac{\partial w}{\partial y}\frac{\partial y}{\partial r}.$$

---

If $w = f(x, y)$, $x = g(r, s)$, and $y = h(r, s)$, then

$$\frac{\partial w}{\partial r} = \frac{\partial w}{\partial x}\frac{\partial x}{\partial r} + \frac{\partial w}{\partial y}\frac{\partial y}{\partial r} \qquad \text{and} \qquad \frac{\partial w}{\partial s} = \frac{\partial w}{\partial x}\frac{\partial x}{\partial s} + \frac{\partial w}{\partial y}\frac{\partial y}{\partial s}.$$

---

Figure 14.23 shows the branch diagram for the first of these equations. The diagram for the second equation is similar; just replace $r$ with $s$.

**EXAMPLE 4**    Express $\partial w / \partial r$ and $\partial w / \partial s$ in terms of $r$ and $s$ if

$$w = x^2 + y^2, \qquad x = r - s, \qquad y = r + s.$$

**Solution**  The preceding discussion gives the following.

$$\frac{\partial w}{\partial r} = \frac{\partial w}{\partial x}\frac{\partial x}{\partial r} + \frac{\partial w}{\partial y}\frac{\partial y}{\partial r} \qquad \frac{\partial w}{\partial s} = \frac{\partial w}{\partial x}\frac{\partial x}{\partial s} + \frac{\partial w}{\partial y}\frac{\partial y}{\partial s}$$

$$= (2x)(1) + (2y)(1) \qquad\qquad = (2x)(-1) + (2y)(1)$$

$$= 2(r - s) + 2(r + s) \qquad\quad = -2(r - s) + 2(r + s)$$

$$= 4r \qquad\qquad\qquad\qquad\qquad = 4s$$

Substitute for the intermediate variables.

If $f$ is a function of a single intermediate variable $x$, our equations are even simpler.

**Chain Rule**

FIGURE 14.24 Branch diagram for differentiating $f$ as a composite function of $r$ and $s$ with one intermediate variable.

If $w = f(x)$ and $x = g(r, s)$, then

$$\frac{\partial w}{\partial r} = \frac{dw}{dx}\frac{\partial x}{\partial r} \qquad \text{and} \qquad \frac{\partial w}{\partial s} = \frac{dw}{dx}\frac{\partial x}{\partial s}.$$

In this case, we use the ordinary (single-variable) derivative, $dw/dx$. The branch diagram is shown in Figure 14.24.

## Implicit Differentiation Revisited

The two-variable Chain Rule in Theorem 5 leads to a formula that takes some of the algebra out of implicit differentiation. Suppose that

1. The function $F(x, y)$ is differentiable and
2. The equation $F(x, y) = 0$ defines $y$ implicitly as a differentiable function of $x$, say $y = h(x)$.

Since $w = F(x, y) = 0$, the derivative $dw/dx$ must be zero. Computing the derivative from the Chain Rule (branch diagram in Figure 14.25), we find

$$0 = \frac{dw}{dx} = F_x\frac{dx}{dx} + F_y\frac{dy}{dx} \qquad \text{Theorem 5 with } t = x \text{ and } f = F$$

$$= F_x \cdot 1 + F_y \cdot \frac{dy}{dx}.$$

If $F_y = \partial w/\partial y \neq 0$, we can solve this equation for $dy/dx$ to get

$$\frac{dy}{dx} = -\frac{F_x}{F_y}.$$

We state this result formally.

FIGURE 14.25 Branch diagram for differentiating $w = F(x, y)$ with respect to $x$. Setting $dw/dx = 0$ leads to a simple computational formula for implicit differentiation (Theorem 8).

**THEOREM 8—A Formula for Implicit Differentiation**  Suppose that $F(x, y)$ is differentiable and that the equation $F(x, y) = 0$ defines $y$ as a differentiable function of $x$. Then at any point where $F_y \neq 0$,

$$\frac{dy}{dx} = -\frac{F_x}{F_y}. \tag{1}$$

**EXAMPLE 5**     Use Theorem 8 to find $dy/dx$ if $y^2 - x^2 - \sin xy = 0$.

**Solution**     Take $F(x, y) = y^2 - x^2 - \sin xy$. Then

$$\frac{dy}{dx} = -\frac{F_x}{F_y} = -\frac{-2x - y \cos xy}{2y - x \cos xy}$$

$$= \frac{2x + y \cos xy}{2y - x \cos xy}.$$

This calculation is significantly shorter than a single-variable calculation using implicit differentiation.                                                                        ∎

The result in Theorem 8 is easily extended to three variables. Suppose that the equation $F(x, y, z) = 0$ defines the variable $z$ implicitly as a function $z = f(x, y)$. Then for all $(x, y)$ in the domain of $f$, we have $F(x, y, f(x, y)) = 0$. Assuming that $F$ and $f$ are differentiable functions, we can use the Chain Rule to differentiate the equation $F(x, y, z) = 0$ with respect to the independent variable $x$:

$$0 = \frac{\partial F}{\partial x}\frac{\partial x}{\partial x} + \frac{\partial F}{\partial y}\frac{\partial y}{\partial x} + \frac{\partial F}{\partial z}\frac{\partial z}{\partial x}$$

$$= F_x \cdot 1 + F_y \cdot 0 + F_z \cdot \frac{\partial z}{\partial x},$$   $y$ is constant when differentiating with respect to $x$.

so

$$F_x + F_z \frac{\partial z}{\partial x} = 0.$$

A similar calculation for differentiating with respect to the independent variable $y$ gives

$$F_y + F_z \frac{\partial z}{\partial y} = 0.$$

Whenever $F_z \neq 0$, we can solve these last two equations for the partial derivatives of $z = f(x, y)$ to obtain

$$\frac{\partial z}{\partial x} = -\frac{F_x}{F_z} \quad \text{and} \quad \frac{\partial z}{\partial y} = -\frac{F_y}{F_z}. \tag{2}$$

An important result from advanced calculus, called the **Implicit Function Theorem**, states the conditions for which our results in Equations (2) are valid. If the partial derivatives $F_x$, $F_y$, and $F_z$ are continuous throughout an open region $R$ in space containing the point $(x_0, y_0, z_0)$, and if for some constant $c$, $F(x_0, y_0, z_0) = c$ and $F_z(x_0, y_0, z_0) \neq 0$, then the equation $F(x, y, z) = c$ defines $z$ implicitly as a differentiable function of $x$ and $y$ near $(x_0, y_0, z_0)$, and the partial derivatives of $z$ are given by Equations (2).

**EXAMPLE 6**     Find $\frac{\partial z}{\partial x}$ and $\frac{\partial z}{\partial y}$ at $(0, 0, 0)$ if $x^3 + z^2 + ye^{xz} + z \cos y = 0$.

**Solution**     Let $F(x, y, z) = x^3 + z^2 + ye^{xz} + z \cos y$. Then

$$F_x = 3x^2 + zye^{xz}, \qquad F_y = e^{xz} - z \sin y, \qquad \text{and} \qquad F_z = 2z + xye^{xz} + \cos y.$$

Since $F(0, 0, 0) = 0$, $F_z(0, 0, 0) = 1 \neq 0$, and all first partial derivatives are continuous, the Implicit Function Theorem says that $F(x, y, z) = 0$ defines $z$ as a differentiable function of $x$ and $y$ near the point $(0, 0, 0)$. From Equations (2),

$$\frac{\partial z}{\partial x} = -\frac{F_x}{F_z} = -\frac{3x^2 + zye^{xz}}{2z + xye^{xz} + \cos y} \quad \text{and} \quad \frac{\partial z}{\partial y} = -\frac{F_y}{F_z} = -\frac{e^{xz} - z \sin y}{2z + xye^{xz} + \cos y}.$$

At $(0, 0, 0)$ we find

$$\frac{\partial z}{\partial x} = -\frac{0}{1} = 0 \quad \text{and} \quad \frac{\partial z}{\partial y} = -\frac{1}{1} = -1. \qquad \blacksquare$$

### Functions of Many Variables

We have seen several different forms of the Chain Rule in this section, but each one is just a special case of one general formula. When solving particular problems, it may help to draw the appropriate branch diagram by placing the dependent variable on top, the intermediate variables in the middle, and the selected independent variable at the bottom. To find the derivative of the dependent variable with respect to the selected independent variable, start at the dependent variable and read down each route of the branch diagram to the independent variable, calculating and multiplying the derivatives along each route. Then add the products found for the different routes.

In general, suppose that $w = f(x, y, \ldots, v)$ is a differentiable function of the intermediate variables $x, y, \ldots, v$ (a finite set) and the $x, y, \ldots, v$ are differentiable functions of the independent variables $p, q, \ldots, t$ (another finite set). Then $w$ is a differentiable function of the variables $p$ through $t$, and the partial derivatives of $w$ with respect to these variables are given by equations of the form

$$\frac{\partial w}{\partial p} = \frac{\partial w}{\partial x}\frac{\partial x}{\partial p} + \frac{\partial w}{\partial y}\frac{\partial y}{\partial p} + \cdots + \frac{\partial w}{\partial v}\frac{\partial v}{\partial p}.$$

The other equations are obtained by replacing $p$ by $q, \ldots, t$, one at a time.

One way to remember this equation is to think of the right-hand side as the dot product of two vectors with components

$$\underbrace{\left(\frac{\partial w}{\partial x}, \frac{\partial w}{\partial y}, \ldots, \frac{\partial w}{\partial v}\right)}_{\substack{\text{Derivatives of } w \text{ with} \\ \text{respect to the} \\ \text{intermediate variables}}} \quad \text{and} \quad \underbrace{\left(\frac{\partial x}{\partial p}, \frac{\partial y}{\partial p}, \ldots, \frac{\partial v}{\partial p}\right)}_{\substack{\text{Derivatives of the intermediate} \\ \text{variables with respect to the} \\ \text{selected independent variable}}}.$$

## Exercises 14.4

### Chain Rule: One Independent Variable

In Exercises 1–6, (a) express $dw/dt$ as a function of $t$, both by using the Chain Rule and by expressing $w$ in terms of $t$ and differentiating directly with respect to $t$. Then (b) evaluate $dw/dt$ at the given value of $t$.

1. $w = x^2 + y^2, \quad x = \cos t, \quad y = \sin t; \quad t = \pi$

2. $w = x^2 + y^2, \quad x = \cos t + \sin t, \quad y = \cos t - \sin t; \quad t = 0$

3. $w = \frac{x}{z} + \frac{y}{z}, \quad x = \cos^2 t, \quad y = \sin^2 t, \quad z = 1/t; \quad t = 3$

4. $w = \ln(x^2 + y^2 + z^2), \quad x = \cos t, \quad y = \sin t, \quad z = 4\sqrt{t}; \quad t = 3$

5. $w = 2ye^x - \ln z, \quad x = \ln(t^2 + 1), \quad y = \tan^{-1} t, \quad z = e^t; \quad t = 1$

6. $w = z - \sin xy, \quad x = t, \quad y = \ln t, \quad z = e^{t-1}; \quad t = 1$

### Chain Rule: Two and Three Independent Variables

In Exercises 7 and 8, (a) express $\partial z/\partial u$ and $\partial z/\partial v$ as functions of $u$ and $v$ both by using the Chain Rule and by expressing $z$ directly in terms of $u$ and $v$ before differentiating. Then (b) evaluate $\partial z/\partial u$ and $\partial z/\partial v$ at the given point $(u, v)$.

7. $z = 4e^x \ln y, \quad x = \ln(u \cos v), \quad y = u \sin v;$
$(u, v) = (2, \pi/4)$

**8.** $z = \tan^{-1}(x/y), \quad x = u \cos v, \quad y = u \sin v;$
$(u, v) = (1.3, \pi/6)$

In Exercises 9 and 10, **(a)** express $\partial w/\partial u$ and $\partial w/\partial v$ as functions of $u$ and $v$ both by using the Chain Rule and by expressing $w$ directly in terms of $u$ and $v$ before differentiating. Then **(b)** evaluate $\partial w/\partial u$ and $\partial w/\partial v$ at the given point $(u, v)$.

**9.** $w = xy + yz + xz, \quad x = u + v, \quad y = u - v, \quad z = uv;$
$(u, v) = (1/2, 1)$

**10.** $w = \ln(x^2 + y^2 + z^2), \quad x = ue^v \sin u, \quad y = ue^v \cos u,$
$z = ue^v; \quad (u, v) = (-2, 0)$

In Exercises 11 and 12, **(a)** express $\partial u/\partial x$, $\partial u/\partial y$, and $\partial u/\partial z$ as functions of $x$, $y$, and $z$ both by using the Chain Rule and by expressing $u$ directly in terms of $x$, $y$, and $z$ before differentiating. Then **(b)** evaluate $\partial u/\partial x$, $\partial u/\partial y$, and $\partial u/\partial z$ at the given point $(x, y, z)$.

**11.** $u = \dfrac{p - q}{q - r}, \quad p = x + y + z, \quad q = x - y + z,$
$r = x + y - z; \quad (x, y, z) = \left(\sqrt{3}, 2, 1\right)$

**12.** $u = e^{qr} \sin^{-1} p, \quad p = \sin x, \quad q = z^2 \ln y, \quad r = 1/z;$
$(x, y, z) = (\pi/4, 1/2, -1/2)$

**Using a Branch Diagram**

In Exercises 13–24, draw a branch diagram and write a Chain Rule formula for each derivative.

**13.** $\dfrac{dz}{dt}$ for $z = f(x, y), \quad x = g(t), \quad y = h(t)$

**14.** $\dfrac{dz}{dt}$ for $z = f(u, v, w), \quad u = g(t), \quad v = h(t), \quad w = k(t)$

**15.** $\dfrac{\partial w}{\partial u}$ and $\dfrac{\partial w}{\partial v}$ for $w = h(x, y, z), \quad x = f(u, v), \quad y = g(u, v),$
$z = k(u, v)$

**16.** $\dfrac{\partial w}{\partial x}$ and $\dfrac{\partial w}{\partial y}$ for $w = f(r, s, t), \quad r = g(x, y), \quad s = h(x, y),$
$t = k(x, y)$

**17.** $\dfrac{\partial w}{\partial u}$ and $\dfrac{\partial w}{\partial v}$ for $w = g(x, y), \quad x = h(u, v), \quad y = k(u, v)$

**18.** $\dfrac{\partial w}{\partial x}$ and $\dfrac{\partial w}{\partial y}$ for $w = g(u, v), \quad u = h(x, y), \quad v = k(x, y)$

**19.** $\dfrac{\partial z}{\partial t}$ and $\dfrac{\partial z}{\partial s}$ for $z = f(x, y), \quad x = g(t, s), \quad y = h(t, s)$

**20.** $\dfrac{\partial y}{\partial r}$ for $y = f(u), \quad u = g(r, s)$

**21.** $\dfrac{\partial w}{\partial s}$ and $\dfrac{\partial w}{\partial t}$ for $w = g(u), \quad u = h(s, t)$

**22.** $\dfrac{\partial w}{\partial p}$ for $w = f(x, y, z, v), \quad x = g(p, q), \quad y = h(p, q),$
$z = j(p, q), \quad v = k(p, q)$

**23.** $\dfrac{\partial w}{\partial r}$ and $\dfrac{\partial w}{\partial s}$ for $w = f(x, y), \quad x = g(r), \quad y = h(s)$

**24.** $\dfrac{\partial w}{\partial s}$ for $w = g(x, y), \quad x = h(r, s, t), \quad y = k(r, s, t)$

**Implicit Differentiation**

Assuming that the equations in Exercises 25–28 define $y$ as a differentiable function of $x$, use Theorem 8 to find the value of $dy/dx$ at the given point.

**25.** $x^3 - 2y^2 + xy = 0, \quad (1, 1)$

**26.** $xy + y^2 - 3x - 3 = 0, \quad (-1, 1)$

**27.** $x^2 + xy + y^2 - 7 = 0, \quad (1, 2)$

**28.** $xe^y + \sin xy + y - \ln 2 = 0, \quad (0, \ln 2)$

Find the values of $\partial z/\partial x$ and $\partial z/\partial y$ at the points in Exercises 29–32.

**29.** $z^3 - xy + yz + y^3 - 2 = 0, \quad (1, 1, 1)$

**30.** $\dfrac{1}{x} + \dfrac{1}{y} + \dfrac{1}{z} - 1 = 0, \quad (2, 3, 6)$

**31.** $\sin(x + y) + \sin(y + z) + \sin(x + z) = 0, \quad (\pi, \pi, \pi)$

**32.** $xe^y + ye^z + 2 \ln x - 2 - 3 \ln 2 = 0, \quad (1, \ln 2, \ln 3)$

**Finding Partial Derivatives at Specified Points**

**33.** Find $\partial w/\partial r$ when $r = 1, s = -1$ if $w = (x + y + z)^2$, $x = r - s, y = \cos(r + s), z = \sin(r + s)$.

**34.** Find $\partial w/\partial v$ when $u = -1, v = 2$ if $w = xy + \ln z$, $x = v^2/u, y = u + v, z = \cos u$.

**35.** Find $\partial w/\partial v$ when $u = 0, v = 0$ if $w = x^2 + (y/x)$, $x = u - 2v + 1, y = 2u + v - 2$.

**36.** Find $\partial z/\partial u$ when $u = 0, v = 1$ if $z = \sin xy + x \sin y$, $x = u^2 + v^2, y = uv$.

**37.** Find $\partial z/\partial u$ and $\partial z/\partial v$ when $u = \ln 2, v = 1$ if $z = 5 \tan^{-1} x$ and $x = e^u + \ln v$.

**38.** Find $\partial z/\partial u$ and $\partial z/\partial v$ when $u = 1, v = -2$ if $z = \ln q$ and $q = \sqrt{v + 3} \tan^{-1} u$.

**Theory and Examples**

**39.** Assume that $w = f(s^3 + t^2)$ and $f'(x) = e^x$. Find $\dfrac{\partial w}{\partial t}$ and $\dfrac{\partial w}{\partial s}$.

**40.** Assume that $w = f\left(ts^2, \dfrac{s}{t}\right), \dfrac{\partial f}{\partial x}(x, y) = xy$, and $\dfrac{\partial f}{\partial y}(x, y) = \dfrac{x^2}{2}$.
Find $\dfrac{\partial w}{\partial t}$ and $\dfrac{\partial w}{\partial s}$.

**41. Changing voltage in a circuit** The voltage $V$ in a circuit that satisfies the law $V = IR$ is slowly dropping as the battery wears out. At the same time, the resistance $R$ is increasing as the resistor heats up. Use the equation

$$\frac{dV}{dt} = \frac{\partial V}{\partial I}\frac{dI}{dt} + \frac{\partial V}{\partial R}\frac{dR}{dt}$$

to find how the current is changing at the instant when $R = 600$ ohms, $I = 0.04$ amp, $dR/dt = 0.5$ ohm/sec, and $dV/dt = -0.01$ volt/sec.

**42. Changing dimensions in a box** The lengths $a$, $b$, and $c$ of the edges of a rectangular box are changing with time. At the instant in question, $a = 1$ m, $b = 2$ m, $c = 3$ m, $da/dt = db/dt = 1$ m/sec, and $dc/dt = -3$ m/sec. At what rates are the box's volume $V$ and surface area $S$ changing at that instant? Are the box's interior diagonals increasing in length or decreasing?

**43.** If $f(u, v, w)$ is differentiable and $u = x - y, v = y - z$, and $w = z - x$, show that

$$\frac{\partial f}{\partial x} + \frac{\partial f}{\partial y} + \frac{\partial f}{\partial z} = 0.$$

**44. Polar coordinates** Suppose that we substitute polar coordinates $x = r \cos \theta$ and $y = r \sin \theta$ in a differentiable function $w = f(x, y)$.

a. Show that

$$\frac{\partial w}{\partial r} = f_x \cos \theta + f_y \sin \theta$$

and

$$\frac{1}{r} \frac{\partial w}{\partial \theta} = -f_x \sin \theta + f_y \cos \theta.$$

b. Solve the equations in part (a) to express $f_x$ and $f_y$ in terms of $\partial w/\partial r$ and $\partial w/\partial \theta$.

c. Show that

$$(f_x)^2 + (f_y)^2 = \left(\frac{\partial w}{\partial r}\right)^2 + \frac{1}{r^2}\left(\frac{\partial w}{\partial \theta}\right)^2.$$

**45. Laplace equations** Show that if $w = f(u, v)$ satisfies the Laplace equation $f_{uu} + f_{vv} = 0$ and if $u = (x^2 - y^2)/2$ and $v = xy$, then $w$ satisfies the Laplace equation $w_{xx} + w_{yy} = 0$.

**46. Laplace equations** Let $w = f(u) + g(v)$, where $u = x + iy$, $v = x - iy$, and $i = \sqrt{-1}$. Show that $w$ satisfies the Laplace equation $w_{xx} + w_{yy} = 0$ if all the necessary functions are differentiable.

**47. Extreme values on a helix** Suppose that the partial derivatives of a function $f(x, y, z)$ at points on the helix $x = \cos t, y = \sin t$, $z = t$ are

$$f_x = \cos t, \qquad f_y = \sin t, \qquad f_z = t^2 + t - 2.$$

At what points on the curve, if any, can $f$ take on extreme values?

**48. A space curve** Let $w = x^2 e^{2y} \cos 3z$. Find the value of $dw/dt$ at the point $(1, \ln 2, 0)$ on the curve $x = \cos t, y = \ln(t + 2)$, $z = t$.

**49. Temperature on a circle** Let $T = f(x, y)$ be the temperature at the point $(x, y)$ on the circle $x = \cos t, y = \sin t, 0 \le t \le 2\pi$ and suppose that

$$\frac{\partial T}{\partial x} = 8x - 4y, \qquad \frac{\partial T}{\partial y} = 8y - 4x.$$

a. Find where the maximum and minimum temperatures on the circle occur by examining the derivatives $dT/dt$ and $d^2T/dt^2$.

b. Suppose that $T = 4x^2 - 4xy + 4y^2$. Find the maximum and minimum values of $T$ on the circle.

**50. Temperature on an ellipse** Let $T = g(x, y)$ be the temperature at the point $(x, y)$ on the ellipse

$$x = 2\sqrt{2} \cos t, \qquad y = \sqrt{2} \sin t, \qquad 0 \le t \le 2\pi,$$

and suppose that

$$\frac{\partial T}{\partial x} = y, \qquad \frac{\partial T}{\partial y} = x.$$

a. Locate the maximum and minimum temperatures on the ellipse by examining $dT/dt$ and $d^2T/dt^2$.

b. Suppose that $T = xy - 2$. Find the maximum and minimum values of $T$ on the ellipse.

**Differentiating Integrals** Under mild continuity restrictions, it is true that if

$$F(x) = \int_a^b g(t, x)\, dt,$$

then $F'(x) = \int_a^b g_x(t, x)\, dt$. Using this fact and the Chain Rule, we can find the derivative of

$$F(x) = \int_a^{f(x)} g(t, x)\, dt$$

by letting

$$G(u, x) = \int_a^u g(t, x)\, dt,$$

where $u = f(x)$. Find the derivatives of the functions in Exercises 51 and 52.

**51.** $F(x) = \int_0^{x^2} \sqrt{t^4 + x^3}\, dt$     **52.** $F(x) = \int_{x^2}^1 \sqrt{t^3 + x^2}\, dt$

# 14.5 Directional Derivatives and Gradient Vectors

If you look at the map (Figure 14.26) showing contours within Yosemite National Park in California, you will notice that the streams flow perpendicular to the contours. The streams are following paths of steepest descent so the waters reach lower elevations as quickly as possible. Therefore, the fastest instantaneous rate of change in a stream's elevation above sea level has a particular direction. In this section, you will see why this direction, called the "downhill" direction, is perpendicular to the contours.

## Directional Derivatives in the Plane

We know from Section 14.4 that if $f(x, y)$ is differentiable, then the rate at which $f$ changes with respect to $t$ along a differentiable curve $x = g(t), y = h(t)$ is

$$\frac{df}{dt} = \frac{\partial f}{\partial x} \frac{dx}{dt} + \frac{\partial f}{\partial y} \frac{dy}{dt}.$$

At any point $P_0(x_0, y_0) = P_0(g(t_0), h(t_0))$, this equation gives the rate of change of $f$ with respect to increasing $t$ and therefore depends, among other things, on the direction of

**FIGURE 14.26** Contours within Yosemite National Park in California show streams, which follow paths of steepest descent, running perpendicular to the contours. (*Source:* Yosemite National Park Map from U.S. Geological Survey, http://www.usgs.gov)

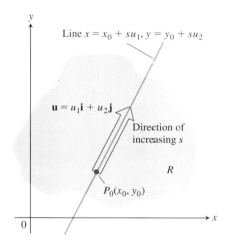

**FIGURE 14.27** The rate of change of $f$ in the direction of **u** at a point $P_0$ is the rate at which $f$ changes along this line at $P_0$.

motion along the curve. If the curve is a straight line and $t$ is the arc length parameter along the line measured from $P_0$ in the direction of a given unit vector **u**, then $df/dt$ is the rate of change of $f$ with respect to distance in its domain in the direction of **u**. By varying **u**, we find the rates at which $f$ changes with respect to distance as we move through $P_0$ in different directions. We now define this idea more precisely.

Suppose that the function $f(x, y)$ is defined throughout a region $R$ in the $xy$-plane, that $P_0(x_0, y_0)$ is a point in $R$, and that $\mathbf{u} = u_1\mathbf{i} + u_2\mathbf{j}$ is a unit vector. Then the equations

$$x = x_0 + su_1, \qquad y = y_0 + su_2$$

parametrize the line through $P_0$ parallel to **u**. If the parameter $s$ measures arc length from $P_0$ in the direction of **u**, we find the rate of change of $f$ at $P_0$ in the direction of **u** by calculating $df/ds$ at $P_0$ (Figure 14.27).

---

**DEFINITION** The **derivative of $f$ at $P_0(x_0, y_0)$ in the direction of the unit vector $\mathbf{u} = u_1\mathbf{i} + u_2\mathbf{j}$** is the number

$$\left(\frac{df}{ds}\right)_{\mathbf{u}, P_0} = \lim_{s \to 0} \frac{f(x_0 + su_1, y_0 + su_2) - f(x_0, y_0)}{s}, \qquad (1)$$

provided the limit exists.

---

The **directional derivative** defined by Equation (1) is also denoted by

$$(D_\mathbf{u}f)_{P_0}. \qquad \text{"The derivative of } f \text{ at } P_0 \text{ in the direction of } \mathbf{u}\text{"}$$

The partial derivatives $f_x(x_0, y_0)$ and $f_y(x_0, y_0)$ are the directional derivatives of $f$ at $P_0$ in the **i** and **j** directions. This observation can be seen by comparing Equation (1) to the definitions of the two partial derivatives given in Section 14.3.

**EXAMPLE 1** Using the definition, find the derivative of

$$f(x, y) = x^2 + xy$$

at $P_0(1, 2)$ in the direction of the unit vector $\mathbf{u} = \left(1/\sqrt{2}\right)\mathbf{i} + \left(1/\sqrt{2}\right)\mathbf{j}$.

**Solution** Applying the definition in Equation (1), we obtain

$$\left(\frac{df}{ds}\right)_{\mathbf{u}, P_0} = \lim_{s \to 0} \frac{f(x_0 + su_1, y_0 + su_2) - f(x_0, y_0)}{s} \qquad \text{Eq. (1)}$$

$$= \lim_{s \to 0} \frac{f\left(1 + s \cdot \dfrac{1}{\sqrt{2}}, 2 + s \cdot \dfrac{1}{\sqrt{2}}\right) - f(1, 2)}{s}$$

$$= \lim_{s \to 0} \frac{\left(1 + \dfrac{s}{\sqrt{2}}\right)^2 + \left(1 + \dfrac{s}{\sqrt{2}}\right)\left(2 + \dfrac{s}{\sqrt{2}}\right) - (1^2 + 1 \cdot 2)}{s}$$

$$= \lim_{s \to 0} \frac{\left(1 + \dfrac{2s}{\sqrt{2}} + \dfrac{s^2}{2}\right) + \left(2 + \dfrac{3s}{\sqrt{2}} + \dfrac{s^2}{2}\right) - 3}{s}$$

$$= \lim_{s \to 0} \frac{\dfrac{5s}{\sqrt{2}} + s^2}{s} = \lim_{s \to 0} \left(\dfrac{5}{\sqrt{2}} + s\right) = \dfrac{5}{\sqrt{2}}.$$

The rate of change of $f(x, y) = x^2 + xy$ at $P_0(1, 2)$ in the direction $\mathbf{u}$ is $5/\sqrt{2}$. ∎

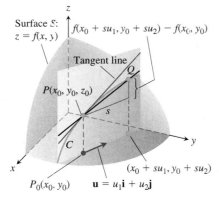

**FIGURE 14.28** The slope of the trace curve $C$ at $P_0$ is $\lim_{Q \to P}$ slope $(PQ)$; this is the directional derivative

$$\left(\frac{df}{ds}\right)_{\mathbf{u}, P_0} = (D_{\mathbf{u}}f)_{P_0}.$$

## Interpretation of the Directional Derivative

The equation $z = f(x, y)$ represents a surface $S$ in space. If $z_0 = f(x_0, y_0)$, then the point $P(x_0, y_0, z_0)$ lies on $S$. The vertical plane that passes through $P$ and $P_0(x_0, y_0)$ parallel to $\mathbf{u}$ intersects $S$ in a curve $C$ (Figure 14.28). The rate of change of $f$ in the direction of $\mathbf{u}$ is the slope of the tangent to $C$ at $P$ in the right-handed system formed by the vectors $\mathbf{u}$ and $\mathbf{k}$.

When $\mathbf{u} = \mathbf{i}$, the directional derivative at $P_0$ is $\partial f/\partial x$ evaluated at $(x_0, y_0)$. When $\mathbf{u} = \mathbf{j}$, the directional derivative at $P_0$ is $\partial f/\partial y$ evaluated at $(x_0, y_0)$. The directional derivative generalizes the two partial derivatives. We can now ask for the rate of change of $f$ in any direction $\mathbf{u}$, not just the directions $\mathbf{i}$ and $\mathbf{j}$.

For a physical interpretation of the directional derivative, suppose that $T = f(x, y)$ is the temperature at each point $(x, y)$ over a region in the plane. Then $f(x_0, y_0)$ is the temperature at the point $P_0(x_0, y_0)$ and $(D_{\mathbf{u}}f)_{P_0}$ is the instantaneous rate of change of the temperature at $P_0$ stepping off in the direction $\mathbf{u}$.

## Calculation and Gradients

We now develop an efficient formula to calculate the directional derivative for a differentiable function $f$. We begin with the line

$$x = x_0 + su_1, \qquad y = y_0 + su_2, \qquad (2)$$

through $P_0(x_0, y_0)$, parametrized with the arc length parameter $s$ increasing in the direction of the unit vector $\mathbf{u} = u_1\mathbf{i} + u_2\mathbf{j}$. Then by the Chain Rule we find

$$\left(\frac{df}{ds}\right)_{\mathbf{u}, P_0} = \left(\frac{\partial f}{\partial x}\right)_{P_0}\frac{dx}{ds} + \left(\frac{\partial f}{\partial y}\right)_{P_0}\frac{dy}{ds} \qquad \text{Chain Rule for differentiable } f$$

$$= \left(\frac{\partial f}{\partial x}\right)_{P_0} u_1 + \left(\frac{\partial f}{\partial y}\right)_{P_0} u_2 \qquad \begin{array}{l}\text{From Eqs. (2), } dx/ds = u_1 \\ \text{and } dy/ds = u_2\end{array}$$

$$= \underbrace{\left[\left(\frac{\partial f}{\partial x}\right)_{P_0}\mathbf{i} + \left(\frac{\partial f}{\partial y}\right)_{P_0}\mathbf{j}\right]}_{\text{Gradient of } f \text{ at } P_0} \cdot \underbrace{\left[u_1\mathbf{i} + u_2\mathbf{j}\right]}_{\text{Direction } \mathbf{u}}. \tag{3}$$

Equation (3) says that the derivative of a differentiable function $f$ in the direction of $\mathbf{u}$ at $P_0$ is the dot product of $\mathbf{u}$ with the special vector, which we now define.

---

**DEFINITION** The **gradient vector (gradient)** of $f(x, y)$ at a point $P_0(x_0, y_0)$ is the vector

$$\nabla f = \frac{\partial f}{\partial x}\mathbf{i} + \frac{\partial f}{\partial y}\mathbf{j}$$

obtained by evaluating the partial derivatives of $f$ at $P_0$.

---

The notation $\nabla f$ is read "grad $f$" as well as "gradient of $f$" and "del $f$." The symbol $\nabla$ by itself is read "del." Another notation for the gradient is grad $f$. Using the gradient notation, we restate Equation (3) as a theorem.

---

**THEOREM 9—The Directional Derivative Is a Dot Product** If $f(x, y)$ is differentiable in an open region containing $P_0(x_0, y_0)$, then

$$\left(\frac{df}{ds}\right)_{\mathbf{u}, P_0} = (\nabla f)_{P_0} \cdot \mathbf{u}, \tag{4}$$

the dot product of the gradient $\nabla f$ at $P_0$ and $\mathbf{u}$. In brief, $D_{\mathbf{u}}f = \nabla f \cdot \mathbf{u}$.

---

**EXAMPLE 2** Find the derivative of $f(x, y) = xe^y + \cos(xy)$ at the point $(2, 0)$ in the direction of $\mathbf{v} = 3\mathbf{i} - 4\mathbf{j}$.

**Solution** Recall that the direction of a vector $\mathbf{v}$ is the unit vector obtained by dividing $\mathbf{v}$ by its length:

$$\mathbf{u} = \frac{\mathbf{v}}{|\mathbf{v}|} = \frac{\mathbf{v}}{5} = \frac{3}{5}\mathbf{i} - \frac{4}{5}\mathbf{j}.$$

The partial derivatives of $f$ are everywhere continuous and at $(2, 0)$ are given by

$$f_x(2, 0) = (e^y - y\sin(xy))_{(2,0)} = e^0 - 0 = 1$$
$$f_y(2, 0) = (xe^y - x\sin(xy))_{(2,0)} = 2e^0 - 2\cdot 0 = 2.$$

The gradient of $f$ at $(2, 0)$ is

$$\nabla f|_{(2,0)} = f_x(2, 0)\mathbf{i} + f_y(2, 0)\mathbf{j} = \mathbf{i} + 2\mathbf{j}$$

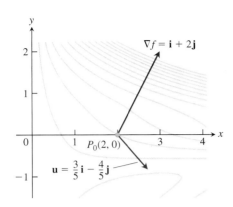

**FIGURE 14.29** Picture $\nabla f$ as a vector in the domain of $f$. The figure shows a number of level curves of $f$. The rate at which $f$ changes at $(2, 0)$ in the direction $\mathbf{u}$ is $\nabla f \cdot \mathbf{u} = -1$, which is the component of $\nabla f$ in the direction of unit vector $\mathbf{u}$ (Example 2).

(Figure 14.29). The derivative of $f$ at $(2, 0)$ in the direction of $\mathbf{v}$ is therefore

$$(D_{\mathbf{u}}f)_{(2,0)} = \nabla f|_{(2,0)} \cdot \mathbf{u} \qquad \text{Eq. (4) with the } (D_{\mathbf{u}}f)_{P_0} \text{ notation}$$

$$= (\mathbf{i} + 2\mathbf{j}) \cdot \left(\frac{3}{5}\mathbf{i} - \frac{4}{5}\mathbf{j}\right) = \frac{3}{5} - \frac{8}{5} = -1. \qquad \blacksquare$$

Evaluating the dot product in the brief version of Equation (4) gives

$$D_{\mathbf{u}}f = \nabla f \cdot \mathbf{u} = |\nabla f||\mathbf{u}| \cos \theta = |\nabla f| \cos \theta,$$

where $\theta$ is the angle between the vectors $\mathbf{u}$ and $\nabla f$, and reveals the following properties.

---

**Properties of the Directional Derivative** $D_{\mathbf{u}}f = \nabla f \cdot \mathbf{u} = |\nabla f| \cos \theta$

1. The function $f$ increases most rapidly when $\cos \theta = 1$ or when $\theta = 0$ and $\mathbf{u}$ is the direction of $\nabla f$. That is, at each point $P$ in its domain, $f$ increases most rapidly in the direction of the gradient vector $\nabla f$ at $P$. The derivative in this direction is

$$D_{\mathbf{u}}f = |\nabla f| \cos (0) = |\nabla f|.$$

2. Similarly, $f$ decreases most rapidly in the direction of $-\nabla f$. The derivative in this direction is $D_{\mathbf{u}}f = |\nabla f|\cos (\pi) = -|\nabla f|$.

3. Any direction $\mathbf{u}$ orthogonal to a gradient $\nabla f \neq 0$ is a direction of zero change in $f$ because $\theta$ then equals $\pi/2$ and

$$D_{\mathbf{u}}f = |\nabla f|\cos (\pi/2) = |\nabla f| \cdot 0 = 0.$$

---

As we discuss later, these properties hold in three dimensions as well as two.

**EXAMPLE 3** Find the directions in which $f(x, y) = (x^2/2) + (y^2/2)$

**(a)** increases most rapidly at the point $(1, 1)$, and

**(b)** decreases most rapidly at $(1, 1)$.

**(c)** What are the directions of zero change in $f$ at $(1, 1)$?

**Solution**

**(a)** The function increases most rapidly in the direction of $\nabla f$ at $(1, 1)$. The gradient there is

$$(\nabla f)_{(1,1)} = (x\mathbf{i} + y\mathbf{j})_{(1,1)} = \mathbf{i} + \mathbf{j}.$$

Its direction is

$$\mathbf{u} = \frac{\mathbf{i} + \mathbf{j}}{|\mathbf{i} + \mathbf{j}|} = \frac{\mathbf{i} + \mathbf{j}}{\sqrt{(1)^2 + (1)^2}} = \frac{1}{\sqrt{2}}\mathbf{i} + \frac{1}{\sqrt{2}}\mathbf{j}.$$

**(b)** The function decreases most rapidly in the direction of $-\nabla f$ at $(1, 1)$, which is

$$-\mathbf{u} = -\frac{1}{\sqrt{2}}\mathbf{i} - \frac{1}{\sqrt{2}}\mathbf{j}.$$

**(c)** The directions of zero change at $(1, 1)$ are the directions orthogonal to $\nabla f$:

$$\mathbf{n} = -\frac{1}{\sqrt{2}}\mathbf{i} + \frac{1}{\sqrt{2}}\mathbf{j} \qquad \text{and} \qquad -\mathbf{n} = \frac{1}{\sqrt{2}}\mathbf{i} - \frac{1}{\sqrt{2}}\mathbf{j}.$$

See Figure 14.30. $\blacksquare$

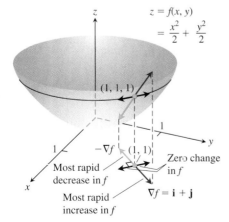

**FIGURE 14.30** The direction in which $f(x, y)$ increases most rapidly at $(1, 1)$ is the direction of $\nabla f|_{(1,1)} = \mathbf{i} + \mathbf{j}$. It corresponds to the direction of steepest ascent on the surface at $(1, 1, 1)$ (Example 3).

## Gradients and Tangents to Level Curves

If a differentiable function $f(x, y)$ has a constant value $c$ along a smooth curve $\mathbf{r} = g(t)\mathbf{i} + h(t)\mathbf{j}$ (making the curve part of a level curve of $f$), then $f(g(t), h(t)) = c$. Differentiating both sides of this equation with respect to $t$ leads to the equations

$$\frac{d}{dt} f(g(t), h(t)) = \frac{d}{dt} (c)$$

$$\frac{\partial f}{\partial x} \frac{dg}{dt} + \frac{\partial f}{\partial y} \frac{dh}{dt} = 0 \qquad \text{Chain Rule}$$

$$\underbrace{\left( \frac{\partial f}{\partial x} \mathbf{i} + \frac{\partial f}{\partial y} \mathbf{j} \right)}_{\nabla f} \cdot \underbrace{\left( \frac{dg}{dt} \mathbf{i} + \frac{dh}{dt} \mathbf{j} \right)}_{\frac{d\mathbf{r}}{dt}} = 0. \qquad (5)$$

Equation (5) says that $\nabla f$ is normal to the tangent vector $d\mathbf{r}/dt$, so it is normal to the curve.

The level curve $f(x, y) = f(x_0, y_0)$

$(x_0, y_0)$

$\nabla f(x_0, y_0)$

---

> At every point $(x_0, y_0)$ in the domain of a differentiable function $f(x, y)$, the gradient of $f$ is normal to the level curve through $(x_0, y_0)$ (Figure 14.31).

**FIGURE 14.31** The gradient of a differentiable function of two variables at a point is always normal to the function's level curve through that point.

Equation (5) validates our observation that streams flow perpendicular to the contours in topographical maps (see Figure 14.26). Since the downflowing stream will reach its destination in the fastest way, it must flow in the direction of the negative gradient vectors from Property 2 for the directional derivative. Equation (5) tells us these directions are perpendicular to the level curves.

This observation also enables us to find equations for tangent lines to level curves. They are the lines normal to the gradients. The line through a point $P_0(x_0, y_0)$ normal to a vector $\mathbf{N} = A\mathbf{i} + B\mathbf{j}$ has the equation

$$A(x - x_0) + B(y - y_0) = 0$$

(Exercise 39). If $\mathbf{N}$ is the gradient $(\nabla f)_{(x_0, y_0)} = f_x(x_0, y_0)\mathbf{i} + f_y(x_0, y_0)\mathbf{j}$, the equation gives the following formula.

---

> **Tangent Line to a Level Curve**
>
> $$f_x(x_0, y_0)(x - x_0) + f_y(x_0, y_0)(y - y_0) = 0 \qquad (6)$$

---

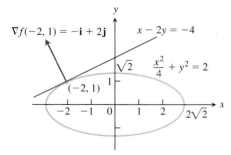

$\nabla f(-2, 1) = -\mathbf{i} + 2\mathbf{j}$

$x - 2y = -4$

$\frac{x^2}{4} + y^2 = 2$

$(-2, 1)$

**FIGURE 14.32** We can find the tangent to the ellipse $(x^2/4) + y^2 = 2$ by treating the ellipse as a level curve of the function $f(x, y) = (x^2/4) + y^2$ (Example 4).

**EXAMPLE 4**    Find an equation for the tangent to the ellipse

$$\frac{x^2}{4} + y^2 = 2$$

(Figure 14.32) at the point $(-2, 1)$.

**Solution**    The ellipse is a level curve of the function

$$f(x, y) = \frac{x^2}{4} + y^2.$$

The gradient of $f$ at $(-2, 1)$ is

$$\nabla f|_{(-2,1)} = \left(\frac{x}{2}\mathbf{i} + 2y\mathbf{j}\right)_{(-2,1)} = -\mathbf{i} + 2\mathbf{j}.$$

The tangent to the ellipse at $(-2, 1)$ is the line

$$(-1)(x + 2) + (2)(y - 1) = 0 \qquad \text{Eq. (6)}$$
$$x - 2y = -4. \qquad\qquad\qquad ∎$$

If we know the gradients of two functions $f$ and $g$, we automatically know the gradients of their sum, difference, constant multiples, product, and quotient. You are asked to establish the following rules in Exercise 40. Notice that these rules have the same form as the corresponding rules for derivatives of single-variable functions.

---

**Algebra Rules for Gradients**

1. *Sum Rule:* $\qquad\qquad\qquad \nabla(f + g) = \nabla f + \nabla g$

2. *Difference Rule:* $\qquad\quad\ \nabla(f - g) = \nabla f - \nabla g$

3. *Constant Multiple Rule:* $\quad \nabla(kf) = k\nabla f \qquad$ (any number $k$)

4. *Product Rule:* $\qquad\qquad \nabla(fg) = f\nabla g + g\nabla f \qquad$ Scalar multipliers on left

5. *Quotient Rule:* $\qquad\qquad \nabla\left(\dfrac{f}{g}\right) = \dfrac{g\nabla f - f\nabla g}{g^2}$ $\qquad$ of gradients

---

**EXAMPLE 5** We illustrate two of the rules with

$$f(x, y) = x - y \qquad g(x, y) = 3y$$
$$\nabla f = \mathbf{i} - \mathbf{j} \qquad\quad \nabla g = 3\mathbf{j}.$$

We have

1. $\nabla(f - g) = \nabla(x - 4y) = \mathbf{i} - 4\mathbf{j} = \nabla f - \nabla g \qquad$ Rule 2

2. $\nabla(fg) = \nabla(3xy - 3y^2) = 3y\mathbf{i} + (3x - 6y)\mathbf{j}$
$\qquad\quad = 3y(\mathbf{i} - \mathbf{j}) + 3y\mathbf{j} + (3x - 6y)\mathbf{j} \qquad g\nabla f$ plus terms . . .
$\qquad\quad = 3y(\mathbf{i} - \mathbf{j}) + (3x - 3y)\mathbf{j} \qquad\qquad$ simplified.
$\qquad\quad = 3y(\mathbf{i} - \mathbf{j}) + (x - y)3\mathbf{j} = g\nabla f + f\nabla g \qquad$ Rule 4 $\qquad ∎$

## Functions of Three Variables

For a differentiable function $f(x, y, z)$ and a unit vector $\mathbf{u} = u_1\mathbf{i} + u_2\mathbf{j} + u_3\mathbf{k}$ in space, we have

$$\nabla f = \frac{\partial f}{\partial x}\mathbf{i} + \frac{\partial f}{\partial y}\mathbf{j} + \frac{\partial f}{\partial z}\mathbf{k}$$

and

$$D_{\mathbf{u}}f = \nabla f \cdot \mathbf{u} = \frac{\partial f}{\partial x}u_1 + \frac{\partial f}{\partial y}u_2 + \frac{\partial f}{\partial z}u_3.$$

The directional derivative can once again be written in the form

$$D_{\mathbf{u}}f = \nabla f \cdot \mathbf{u} = |\nabla f||u|\cos\theta = |\nabla f|\cos\theta,$$

so the properties listed earlier for functions of two variables extend to three variables. At any given point, $f$ increases most rapidly in the direction of $\nabla f$ and decreases most rapidly in the direction of $-\nabla f$. In any direction orthogonal to $\nabla f$, the derivative is zero.

## EXAMPLE 6

**(a)** Find the derivative of $f(x, y, z) = x^3 - xy^2 - z$ at $P_0(1, 1, 0)$ in the direction of $\mathbf{v} = 2\mathbf{i} - 3\mathbf{j} + 6\mathbf{k}$.

**(b)** In what directions does $f$ change most rapidly at $P_0$, and what are the rates of change in these directions?

**Solution**

**(a)** The direction of $\mathbf{v}$ is obtained by dividing $\mathbf{v}$ by its length:

$$|\mathbf{v}| = \sqrt{(2)^2 + (-3)^2 + (6)^2} = \sqrt{49} = 7$$

$$\mathbf{u} = \frac{\mathbf{v}}{|\mathbf{v}|} = \frac{2}{7}\mathbf{i} - \frac{3}{7}\mathbf{j} + \frac{6}{7}\mathbf{k}.$$

The partial derivatives of $f$ at $P_0$ are

$$f_x = (3x^2 - y^2)_{(1,1,0)} = 2, \qquad f_y = -2xy|_{(1,1,0)} = -2, \qquad f_z = -1|_{(1,1,0)} = -1.$$

The gradient of $f$ at $P_0$ is

$$\nabla f|_{(1,1,0)} = 2\mathbf{i} - 2\mathbf{j} - \mathbf{k}.$$

The derivative of $f$ at $P_0$ in the direction of $\mathbf{v}$ is therefore

$$(D_{\mathbf{u}}f)_{(1,1,0)} = \nabla f|_{(1,1,0)} \cdot \mathbf{u} = (2\mathbf{i} - 2\mathbf{j} - \mathbf{k}) \cdot \left(\frac{2}{7}\mathbf{i} - \frac{3}{7}\mathbf{j} + \frac{6}{7}\mathbf{k}\right)$$

$$= \frac{4}{7} + \frac{6}{7} - \frac{6}{7} = \frac{4}{7}.$$

**(b)** The function increases most rapidly in the direction of $\nabla f = 2\mathbf{i} - 2\mathbf{j} - \mathbf{k}$ and decreases most rapidly in the direction of $-\nabla f$. The rates of change in the directions are, respectively,

$$|\nabla f| = \sqrt{(2)^2 + (-2)^2 + (-1)^2} = \sqrt{9} = 3 \qquad \text{and} \qquad -|\nabla f| = -3. \qquad \blacksquare$$

### The Chain Rule for Paths

If $\mathbf{r}(t) = x(t)\mathbf{i} + y(t)\mathbf{j} + z(t)\mathbf{k}$ is a smooth path $C$, and $w = f(\mathbf{r}(t))$ is a scalar function evaluated along $C$, then according to the Chain Rule, Theorem 6 in Section 14.4,

$$\frac{dw}{dt} = \frac{\partial w}{\partial x}\frac{dx}{dt} + \frac{\partial w}{\partial y}\frac{dy}{dt} + \frac{\partial w}{\partial z}\frac{dz}{dt}.$$

The partial derivatives on the right-hand side of the above equation are evaluated along the curve $\mathbf{r}(t)$, and the derivatives of the intermediate variables are evaluated at $t$. If we express this equation using vector notation, we have

> **The Derivative Along a Path**
>
> $$\frac{d}{dt}f(\mathbf{r}(t)) = \nabla f(\mathbf{r}(t)) \cdot \mathbf{r}'(t). \tag{7}$$

What Equation (7) says is that the derivative of the composite function $f(\mathbf{r}(t))$ is the "derivative" (gradient) of the outside function $f$ "times" (dot product) the derivative of the inside function $\mathbf{r}$. This is analogous to the "Outside-Inside" Rule for derivatives of composite functions studied in Section 3.6. That is, the multivariable Chain Rule for paths has exactly *the same form* as the rule for single-variable differential calculus when appropriate interpretations are given to the meanings of the terms and operations involved.

## Exercises 14.5

### Calculating Gradients

In Exercises 1–6, find the gradient of the function at the given point. Then sketch the gradient together with the level curve that passes through the point.

**1.** $f(x, y) = y - x$, $(2, 1)$ **2.** $f(x, y) = \ln(x^2 + y^2)$, $(1, 1)$

**3.** $g(x, y) = xy^2$, $(2, -1)$ **4.** $g(x, y) = \dfrac{x^2}{2} - \dfrac{y^2}{2}$, $(\sqrt{2}, 1)$

**5.** $f(x, y) = \sqrt{2x + 3y}$, $(-1, 2)$

**6.** $f(x, y) = \tan^{-1}\dfrac{\sqrt{x}}{y}$, $(4, -2)$

In Exercises 7–10, find $\nabla f$ at the given point.

**7.** $f(x, y, z) = x^2 + y^2 - 2z^2 + z \ln x$, $(1, 1, 1)$

**8.** $f(x, y, z) = 2z^3 - 3(x^2 + y^2)z + \tan^{-1} xz$, $(1, 1, 1)$

**9.** $f(x, y, z) = (x^2 + y^2 + z^2)^{-1/2} + \ln(xyz)$, $(-1, 2, -2)$

**10.** $f(x, y, z) = e^{x+y} \cos z + (y + 1) \sin^{-1} x$, $(0, 0, \pi/6)$

### Finding Directional Derivatives

In Exercises 11–18, find the derivative of the function at $P_0$ in the direction of $\mathbf{u}$.

**11.** $f(x, y) = 2xy - 3y^2$, $P_0(5, 5)$, $\mathbf{u} = 4\mathbf{i} + 3\mathbf{j}$

**12.** $f(x, y) = 2x^2 + y^2$, $P_0(-1, 1)$, $\mathbf{u} = 3\mathbf{i} - 4\mathbf{j}$

**13.** $g(x, y) = \dfrac{x - y}{xy + 2}$, $P_0(1, -1)$, $\mathbf{u} = 12\mathbf{i} + 5\mathbf{j}$

**14.** $h(x, y) = \tan^{-1}(y/x) + \sqrt{3}\sin^{-1}(xy/2)$, $P_0(1, 1)$, $\mathbf{u} = 3\mathbf{i} - 2\mathbf{j}$

**15.** $f(x, y, z) = xy + yz + zx$, $P_0(1, -1, 2)$, $\mathbf{u} = 3\mathbf{i} + 6\mathbf{j} - 2\mathbf{k}$

**16.** $f(x, y, z) = x^2 + 2y^2 - 3z^2$, $P_0(1, 1, 1)$, $\mathbf{u} = \mathbf{i} + \mathbf{j} + \mathbf{k}$

**17.** $g(x, y, z) = 3e^x \cos yz$, $P_0(0, 0, 0)$, $\mathbf{u} = 2\mathbf{i} + \mathbf{j} - 2\mathbf{k}$

**18.** $h(x, y, z) = \cos xy + e^{yz} + \ln zx$, $P_0(1, 0, 1/2)$, $\mathbf{u} = \mathbf{i} + 2\mathbf{j} + 2\mathbf{k}$

In Exercises 19–24, find the directions in which the functions increase and decrease most rapidly at $P_0$. Then find the derivatives of the functions in these directions.

**19.** $f(x, y) = x^2 + xy + y^2$, $P_0(-1, 1)$

**20.** $f(x, y) = x^2y + e^{xy} \sin y$, $P_0(1, 0)$

**21.** $f(x, y, z) = (x/y) - yz$, $P_0(4, 1, 1)$

**22.** $g(x, y, z) = xe^y + z^2$, $P_0(1, \ln 2, 1/2)$

**23.** $f(x, y, z) = \ln xy + \ln yz + \ln xz$, $P_0(1, 1, 1)$

**24.** $h(x, y, z) = \ln(x^2 + y^2 - 1) + y + 6z$, $P_0(1, 1, 0)$

### Tangent Lines to Level Curves

In Exercises 25–28, sketch the curve $f(x, y) = c$ together with $\nabla f$ and the tangent line at the given point. Then write an equation for the tangent line.

**25.** $x^2 + y^2 = 4$, $(\sqrt{2}, \sqrt{2})$

**26.** $x^2 - y = 1$, $(\sqrt{2}, 1)$

**27.** $xy = -4$, $(2, -2)$

**28.** $x^2 - xy + y^2 = 7$, $(-1, 2)$

### Theory and Examples

**29.** Let $f(x, y) = x^2 - xy + y^2 - y$. Find the directions $\mathbf{u}$ and the values of $D_{\mathbf{u}} f(1, -1)$ for which

**a.** $D_{\mathbf{u}} f(1, -1)$ is largest **b.** $D_{\mathbf{u}} f(1, -1)$ is smallest

**c.** $D_{\mathbf{u}} f(1, -1) = 0$ **d.** $D_{\mathbf{u}} f(1, -1) = 4$

**e.** $D_{\mathbf{u}} f(1, -1) = -3$

**30.** Let $f(x, y) = \dfrac{(x - y)}{(x + y)}$. Find the directions $\mathbf{u}$ and the values of $D_{\mathbf{u}} f\left(-\dfrac{1}{2}, \dfrac{3}{2}\right)$ for which

**a.** $D_{\mathbf{u}} f\left(-\dfrac{1}{2}, \dfrac{3}{2}\right)$ is largest **b.** $D_{\mathbf{u}} f\left(-\dfrac{1}{2}, \dfrac{3}{2}\right)$ is smallest

**c.** $D_{\mathbf{u}} f\left(-\dfrac{1}{2}, \dfrac{3}{2}\right) = 0$ **d.** $D_{\mathbf{u}} f\left(-\dfrac{1}{2}, \dfrac{3}{2}\right) = -2$

**e.** $D_{\mathbf{u}} f\left(-\dfrac{1}{2}, \dfrac{3}{2}\right) = 1$

**31. Zero directional derivative** In what direction is the derivative of $f(x, y) = xy + y^2$ at $P(3, 2)$ equal to zero?

**32. Zero directional derivative** In what directions is the derivative of $f(x, y) = (x^2 - y^2)/(x^2 + y^2)$ at $P(1, 1)$ equal to zero?

**33.** Is there a direction $\mathbf{u}$ in which the rate of change of $f(x, y) = x^2 - 3xy + 4y^2$ at $P(1, 2)$ equals 14? Give reasons for your answer.

**34. Changing temperature along a circle** Is there a direction $\mathbf{u}$ in which the rate of change of the temperature function $T(x, y, z) = 2xy - yz$ (temperature in degrees Celsius, distance in meters) at $P(1, -1, 1)$ is $-3°C/m$? Give reasons for your answer.

**35.** The derivative of $f(x, y)$ at $P_0(1, 2)$ in the direction of $\mathbf{i} + \mathbf{j}$ is $2\sqrt{2}$ and in the direction of $-2\mathbf{j}$ is $-3$. What is the derivative of $f$ in the direction of $-\mathbf{i} - 2\mathbf{j}$? Give reasons for your answer.

**36.** The derivative of $f(x, y, z)$ at a point $P$ is greatest in the direction of $\mathbf{v} = \mathbf{i} + \mathbf{j} - \mathbf{k}$. In this direction, the value of the derivative is $2\sqrt{3}$.

**a.** What is $\nabla f$ at $P$? Give reasons for your answer.

**b.** What is the derivative of $f$ at $P$ in the direction of $\mathbf{i} + \mathbf{j}$?

**37. Directional derivatives and scalar components** How is the derivative of a differentiable function $f(x, y, z)$ at a point $P_0$ in the direction of a unit vector $\mathbf{u}$ related to the scalar component of $(\nabla f)_{P_0}$ in the direction of $\mathbf{u}$? Give reasons for your answer.

**38. Directional derivatives and partial derivatives** Assuming that the necessary derivatives of $f(x, y, z)$ are defined, how are $D_{\mathbf{i}} f$, $D_{\mathbf{j}} f$, and $D_{\mathbf{k}} f$ related to $f_x$, $f_y$, and $f_z$? Give reasons for your answer.

**39. Lines in the xy-plane** Show that $A(x - x_0) + B(y - y_0) = 0$ is an equation for the line in the $xy$-plane through the point $(x_0, y_0)$ normal to the vector $\mathbf{N} = A\mathbf{i} + B\mathbf{j}$.

**40. The algebra rules for gradients** Given a constant $k$ and the gradients

$$\nabla f = \frac{\partial f}{\partial x}\mathbf{i} + \frac{\partial f}{\partial y}\mathbf{j} + \frac{\partial f}{\partial z}\mathbf{k}, \qquad \nabla g = \frac{\partial g}{\partial x}\mathbf{i} + \frac{\partial g}{\partial y}\mathbf{j} + \frac{\partial g}{\partial z}\mathbf{k},$$

establish the algebra rules for gradients.

# 14.6 Tangent Planes and Differentials

In single-variable differential calculus we saw how the derivative defined the tangent line to the graph of a differentiable function at a point on the graph. The tangent line then provided for a linearization of the function at the point. In this section, we will see analogously how the gradient defines the *tangent plane* to the level surface of a function $w = f(x, y, z)$ at a point on the surface. In the same way as before, the tangent plane then provides for a linearization of $f$ at the point and defines the total differential of the function.

## Tangent Planes and Normal Lines

If $\mathbf{r}(t) = x(t)\mathbf{i} + y(t)\mathbf{j} + z(t)\mathbf{k}$ is a smooth curve on the level surface $f(x, y, z) = c$ of a differentiable function $f$, we found in Equation (7) of the last section that

$$\frac{d}{dt} f(\mathbf{r}(t)) = \nabla f(\mathbf{r}(t)) \cdot \mathbf{r}'(t).$$

Since $f$ is constant along the curve $\mathbf{r}$, the derivative on the left-hand side of the equation is 0, so the gradient $\nabla f$ is orthogonal to the curve's velocity vector $\mathbf{r}'$.

Now let us restrict our attention to the curves that pass through $P_0$ (Figure 14.33). All the velocity vectors at $P_0$ are orthogonal to $\nabla f$ at $P_0$, so the curves' tangent lines all lie in the plane through $P_0$ normal to $\nabla f$. We now define this plane.

**FIGURE 14.33** The gradient $\nabla f$ is orthogonal to the velocity vector of every smooth curve in the surface through $P_0$. The velocity vectors at $P_0$ therefore lie in a common plane, which we call the tangent plane at $P_0$.

> **DEFINITIONS** The **tangent plane** at the point $P_0(x_0, y_0, z_0)$ on the level surface $f(x, y, z) = c$ of a differentiable function $f$ is the plane through $P_0$ normal to $\nabla f|_{P_0}$.
>
> The **normal line** of the surface at $P_0$ is the line through $P_0$ parallel to $\nabla f|_{P_0}$.

From Section 12.5, the tangent plane and normal line have the following equations:

> **Tangent Plane to** $f(x, y, z) = c$ **at** $P_0(x_0, y_0, z_0)$
>
> $$f_x(P_0)(x - x_0) + f_y(P_0)(y - y_0) + f_z(P_0)(z - z_0) = 0 \qquad (1)$$
>
> **Normal Line to** $f(x, y, z) = c$ **at** $P_0(x_0, y_0, z_0)$
>
> $$x = x_0 + f_x(P_0)t, \qquad y = y_0 + f_y(P_0)t, \qquad z = z_0 + f_z(P_0)t \qquad (2)$$

**EXAMPLE 1**     Find the tangent plane and normal line of the level surface

$$f(x, y, z) = x^2 + y^2 + z - 9 = 0 \qquad \text{A circular paraboloid}$$

at the point $P_0(1, 2, 4)$.

**Solution**     The surface is shown in Figure 14.34.

The tangent plane is the plane through $P_0$ perpendicular to the gradient of $f$ at $P_0$. The gradient is

$$\nabla f|_{P_0} = (2x\mathbf{i} + 2y\mathbf{j} + \mathbf{k})_{(1,2,4)} = 2\mathbf{i} + 4\mathbf{j} + \mathbf{k}.$$

The tangent plane is therefore the plane

$$2(x - 1) + 4(y - 2) + (z - 4) = 0, \qquad \text{or} \qquad 2x + 4y + z = 14.$$

The line normal to the surface at $P_0$ is

$$x = 1 + 2t, \qquad y = 2 + 4t, \qquad z = 4 + t.$$

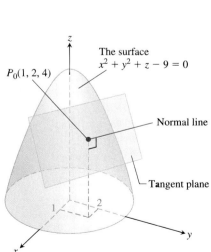

**FIGURE 14.34** The tangent plane and normal line to this level surface at $P_0$ (Example 1).

To find an equation for the plane tangent to a smooth surface $z = f(x, y)$ at a point $P_0(x_0, y_0, z_0)$ where $z_0 = f(x_0, y_0)$, we first observe that the equation $z = f(x, y)$ is equivalent to $f(x, y) - z = 0$. The surface $z = f(x, y)$ is therefore the zero level surface of the function $F(x, y, z) = f(x, y) - z$. The partial derivatives of $F$ are

$$F_x = \frac{\partial}{\partial x} (f(x, y) - z) = f_x - 0 = f_x$$

$$F_y = \frac{\partial}{\partial y} (f(x, y) - z) = f_y - 0 = f_y$$

$$F_z = \frac{\partial}{\partial z} (f(x, y) - z) = 0 - 1 = -1.$$

The formula

$$F_x(P_0)(x - x_0) + F_y(P_0)(y - y_0) + F_z(P_0)(z - z_0) = 0$$

for the plane tangent to the level surface at $P_0$ therefore reduces to

$$f_x(x_0, y_0)(x - x_0) + f_y(x_0, y_0)(y - y_0) - (z - z_0) = 0.$$

---

**Plane Tangent to a Surface $z = f(x, y)$ at $(x_0, y_0, f(x_0, y_0))$**

The plane tangent to the surface $z = f(x, y)$ of a differentiable function $f$ at the point $P_0(x_0, y_0, z_0) = (x_0, y_0, f(x_0, y_0))$ is

$$f_x(x_0, y_0)(x - x_0) + f_y(x_0, y_0)(y - y_0) - (z - z_0) = 0. \qquad (3)$$

---

**EXAMPLE 2**    Find the plane tangent to the surface $z = x \cos y - ye^x$ at $(0, 0, 0)$.

**Solution**    We calculate the partial derivatives of $f(x, y) = x \cos y - ye^x$ and use Equation (3):

$$f_x(0, 0) = (\cos y - ye^x)_{(0,0)} = 1 - 0 \cdot 1 = 1$$
$$f_y(0, 0) = (-x \sin y - e^x)_{(0,0)} = 0 - 1 = -1.$$

The tangent plane is therefore

$$1 \cdot (x - 0) - 1 \cdot (y - 0) - (z - 0) = 0, \qquad \text{Eq. (3)}$$

or

$$x - y - z = 0. \qquad \blacksquare$$

**EXAMPLE 3**    The surfaces

$$f(x, y, z) = x^2 + y^2 - 2 = 0 \qquad \text{A cylinder}$$

and

$$g(x, y, z) = x + z - 4 = 0 \qquad \text{A plane}$$

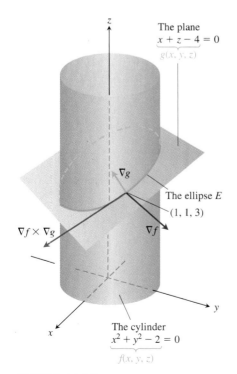

The plane
$x + z - 4 = 0$
$g(x, y, z)$

$\nabla g$

The ellipse $E$
$(1, 1, 3)$

$\nabla f \times \nabla g$    $\nabla f$

The cylinder
$x^2 + y^2 - 2 = 0$
$f(x, y, z)$

**FIGURE 14.35** This cylinder and plane intersect in an ellipse $E$ (Example 3).

meet in an ellipse $E$ (Figure 14.35). Find parametric equations for the line tangent to $E$ at the point $P_0(1, 1, 3)$.

**Solution**   The tangent line is orthogonal to both $\nabla f$ and $\nabla g$ at $P_0$, and therefore parallel to $\mathbf{v} = \nabla f \times \nabla g$. The components of $\mathbf{v}$ and the coordinates of $P_0$ give us equations for the line. We have

$$\nabla f|_{(1,1,3)} = (2x\mathbf{i} + 2y\mathbf{j})_{(1,1,3)} = 2\mathbf{i} + 2\mathbf{j}$$

$$\nabla g|_{(1,1,3)} = (\mathbf{i} + \mathbf{k})_{(1,1,3)} = \mathbf{i} + \mathbf{k}$$

$$\mathbf{v} = (2\mathbf{i} + 2\mathbf{j}) \times (\mathbf{i} + \mathbf{k}) = \begin{vmatrix} \mathbf{i} & \mathbf{j} & \mathbf{k} \\ 2 & 2 & 0 \\ 1 & 0 & 1 \end{vmatrix} = 2\mathbf{i} - 2\mathbf{j} - 2\mathbf{k}.$$

The tangent line to the ellipse of intersection is

$$x = 1 + 2t, \qquad y = 1 - 2t, \qquad z = 3 - 2t. \qquad \blacksquare$$

## Estimating Change in a Specific Direction

The directional derivative plays the role of an ordinary derivative when we want to estimate how much the value of a function $f$ changes if we move a small distance $ds$ from a point $P_0$ to another point nearby. If $f$ were a function of a single variable, we would have

$$df = f'(P_0)\, ds. \qquad \text{Ordinary derivative} \times \text{increment}$$

For a function of two or more variables, we use the formula

$$df = (\nabla f|_{P_0} \cdot \mathbf{u})\, ds, \qquad \text{Directional derivative} \times \text{increment}$$

where $\mathbf{u}$ is the direction of the motion away from $P_0$.

---

**Estimating the Change in $f$ in a Direction u**

To estimate the change in the value of a differentiable function $f$ when we move a small distance $ds$ from a point $P_0$ in a particular direction $\mathbf{u}$, use the formula

$$df = \underbrace{(\nabla f|_{P_0} \cdot \mathbf{u})}_{\substack{\text{Directional} \\ \text{derivative}}}\ \underbrace{ds}_{\substack{\text{Distance} \\ \text{increment}}}$$

---

**EXAMPLE 4**   Estimate how much the value of

$$f(x, y, z) = y \sin x + 2yz$$

will change if the point $P(x, y, z)$ moves 0.1 unit from $P_0(0, 1, 0)$ straight toward $P_1(2, 2, -2)$.

**Solution**   We first find the derivative of $f$ at $P_0$ in the direction of the vector $\overrightarrow{P_0P_1} = 2\mathbf{i} + \mathbf{j} - 2\mathbf{k}$. The direction of this vector is

$$\mathbf{u} = \frac{\overrightarrow{P_0P_1}}{|\overrightarrow{P_0P_1}|} = \frac{\overrightarrow{P_0P_1}}{3} = \frac{2}{3}\mathbf{i} + \frac{1}{3}\mathbf{j} - \frac{2}{3}\mathbf{k}.$$

The gradient of $f$ at $P_0$ is

$$\nabla f|_{(0,1,0)} = ((y \cos x)\mathbf{i} + (\sin x + 2z)\mathbf{j} + 2y\mathbf{k})_{(0,1,0)} = \mathbf{i} + 2\mathbf{k}.$$

Therefore,

$$\nabla f|_{P_0} \cdot \mathbf{u} = (\mathbf{i} + 2\mathbf{k}) \cdot \left( \frac{2}{3}\mathbf{i} + \frac{1}{3}\mathbf{j} - \frac{2}{3}\mathbf{k} \right) = \frac{2}{3} - \frac{4}{3} = -\frac{2}{3}.$$

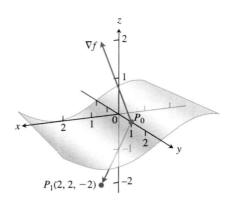

**FIGURE 14.36** As $P(x, y, z)$ moves off the level surface at $P_0$ by 0.1 unit directly toward $F_1$, the function $f$ changes value by approximately $-0.067$ unit (Example 4).

The change $df$ in $f$ that results from moving $ds = 0.1$ unit away from $P_0$ in the direction of $\mathbf{u}$ is approximately

$$df = (\nabla f|_{P_0} \cdot \mathbf{u})(ds) = \left(-\frac{2}{3}\right)(0.1) \approx -0.067 \text{ unit}.$$

See Figure 14.36. ∎

### How to Linearize a Function of Two Variables

Functions of two variables can be complicated, and we sometimes need to approximate them with simpler ones that give the accuracy required for specific applications without being so difficult to work with. We do this in a way that is similar to the way we find linear replacements for functions of a single variable (Section 3.9).

Suppose the function we wish to approximate is $z = f(x, y)$ near a point $(x_0, y_0)$ at which we know the values of $f$, $f_x$, and $f_y$ and at which $f$ is differentiable. If we move from $(x_0, y_0)$ to any nearby point $(x, y)$ by increments $\Delta x = x - x_0$ and $\Delta y = y - y_0$ (see Figure 14.37), then the definition of differentiability from Section 14.3 gives the change

$$f(x, y) - f(x_0, y_0) = f_x(x_0, y_0)\Delta x + f_y(x_0, y_0)\Delta y + \epsilon_1\Delta x + \epsilon_2\Delta y,$$

where $\epsilon_1, \epsilon_2 \to 0$ as $\Delta x, \Delta y \to 0$. If the increments $\Delta x$ and $\Delta y$ are small, the products $\epsilon_1\Delta x$ and $\epsilon_2\Delta y$ will eventually be smaller still and we have the approximation

$$f(x, y) \approx \underbrace{f(x_0, y_0) + f_x(x_0, y_0)(x - x_0) + f_y(x_0, y_0)(y - y_0)}_{L(x, y)}.$$

In other words, as long as $\Delta x$ and $\Delta y$ are small, $f$ will have approximately the same value as the linear function $L$.

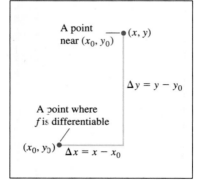

**FIGURE 14.37** If $f$ is differentiable at $(x_0, y_0)$, then the value of $f$ at any point $(x, y)$ nearby is approximately $f(x_0, y_0) + f_x(x_0, y_0)\Delta x + f_y(x_0, y_0)\Delta y$.

---

**DEFINITIONS** The **linearization** of a function $f(x, y)$ at a point $(x_0, y_0)$ where $f$ is differentiable is the function

$$L(x, y) = f(x_0, y_0) + f_x(x_0, y_0)(x - x_0) + f_y(x_0, y_0)(y - y_0).$$

The approximation

$$f(x, y) \approx L(x, y)$$

is the **standard linear approximation** of $f$ at $(x_0, y_0)$.

---

From Equation (3), we find that the plane $z = L(x, y)$ is tangent to the surface $z = f(x, y)$ at the point $(x_0, y_0)$. Thus, the linearization of a function of two variables is a tangent-*plane* approximation in the same way that the linearization of a function of a single variable is a tangent-*line* approximation. (See Exercise 55.)

**EXAMPLE 5** Find the linearization of

$$f(x, y) = x^2 - xy + \frac{1}{2}y^2 + 3$$

at the point $(3, 2)$.

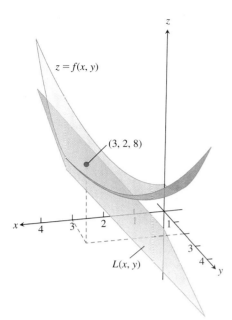

**FIGURE 14.38** The tangent plane $L(x, y)$ represents the linearization of $f(x, y)$ in Example 5.

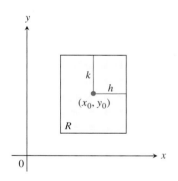

**FIGURE 14.39** The rectangular region $R$: $|x - x_0| \leq h$, $|y - y_0| \leq k$ in the $xy$-plane.

**Solution** We first evaluate $f$, $f_x$, and $f_y$ at the point $(x_0, y_0) = (3, 2)$:

$$f(3, 2) = \left(x^2 - xy + \frac{1}{2}y^2 + 3\right)_{(3,2)} = 8$$

$$f_x(3, 2) = \frac{\partial}{\partial x}\left(x^2 - xy + \frac{1}{2}y^2 + 3\right)_{(3,2)} = (2x - y)_{(3,2)} = 4$$

$$f_y(3, 2) = \frac{\partial}{\partial y}\left(x^2 - xy + \frac{1}{2}y^2 + 3\right)_{(3,2)} = (-x + y)_{(3,2)} = -1,$$

giving

$$L(x, y) = f(x_0, y_0) + f_x(x_0, y_0)(x - x_0) + f_y(x_0, y_0)(y - y_0)$$
$$= 8 + (4)(x - 3) + (-1)(y - 2) = 4x - y - 2.$$

The linearization of $f$ at $(3, 2)$ is $L(x, y) = 4x - y - 2$ (see Figure 14.38).

When approximating a differentiable function $f(x, y)$ by its linearization $L(x, y)$ at $(x_0, y_0)$, an important question is how accurate the approximation might be.

If we can find a common upper bound $M$ for $|f_{xx}|$, $|f_{yy}|$, and $|f_{xy}|$ on a rectangle $R$ centered at $(x_0, y_0)$ (Figure 14.39), then we can bound the error $E$ throughout $R$ by using a simple formula (derived in Section 14.9). The **error** is defined by $E(x, y) = f(x, y) - L(x, y)$.

---

**The Error in the Standard Linear Approximation**

If $f$ has continuous first and second partial derivatives throughout an open set containing a rectangle $R$ centered at $(x_0, y_0)$ and if $M$ is any upper bound for the values of $|f_{xx}|$, $|f_{yy}|$, and $|f_{xy}|$ on $R$, then the error $E(x, y)$ incurred in replacing $f(x, y)$ on $R$ by its linearization

$$L(x, y) = f(x_0, y_0) + f_x(x_0, y_0)(x - x_0) + f_y(x_0, y_0)(y - y_0)$$

satisfies the inequality

$$|E(x, y)| \leq \frac{1}{2}M(|x - x_0| + |y - y_0|)^2.$$

---

To make $|E(x, y)|$ small for a given $M$, we just make $|x - x_0|$ and $|y - y_0|$ small.

### Differentials

Recall from Section 3.9 that for a function of a single variable, $y = f(x)$, we defined the change in $f$ as $x$ changes from $a$ to $a + \Delta x$ by

$$\Delta f = f(a + \Delta x) - f(a)$$

and the differential of $f$ as

$$df = f'(a)\Delta x.$$

We now consider the differential of a function of two variables.

Suppose a differentiable function $f(x, y)$ and its partial derivatives exist at a point $(x_0, y_0)$. If we move to a nearby point $(x_0 + \Delta x, y_0 + \Delta y)$, the change in $f$ is

$$\Delta f = f(x_0 + \Delta x, y_0 + \Delta y) - f(x_0, y_0).$$

A straightforward calculation from the definition of $L(x, y)$, using the notation $x - x_0 = \Delta x$ and $y - y_0 = \Delta y$, shows that the corresponding change in $L$ is

$$\Delta L = L(x_0 + \Delta x, y_0 + \Delta y) - L(x_0, y_0)$$
$$= f_x(x_0, y_0)\Delta x + f_y(x_0, y_0)\Delta y.$$

The **differentials** $dx$ and $dy$ are independent variables, so they can be assigned any values. Often we take $dx = \Delta x = x - x_0$, and $dy = \Delta y = y - y_0$. We then have the following definition of the differential or *total* differential of $f$.

---

**DEFINITION** If we move from $(x_0, y_0)$ to a point $(x_0 + dx, y_0 + dy)$ nearby, the resulting change

$$df = f_x(x_0, y_0)\, dx + f_y(x_0, y_0)\, dy$$

in the linearization of $f$ is called the **total differential of** $f$.

---

**EXAMPLE 6** Suppose that a cylindrical can is designed to have a radius of 1 cm and a height of 5 cm, but that the radius and height are off by the amounts $dr = +0.03$ and $dh = -0.1$. Estimate the resulting absolute change in the volume of the can.

**Solution** To estimate the absolute change in $V = \pi r^2 h$, we use

$$\Delta V \approx dV = V_r(r_0, h_0)\, dr + V_h(r_0, h_0)\, dh.$$

With $V_r = 2\pi rh$ and $V_h = \pi r^2$, we get

$$dV = 2\pi r_0 h_0\, dr + \pi r_0{}^2\, dh = 2\pi(1)(5)(0.03) + \pi(1)^2(-0.1)$$
$$= 0.3\pi - 0.1\pi = 0.2\pi \approx 0.63 \text{ cm}^3 \qquad \blacksquare$$

**EXAMPLE 7** Your company manufactures stainless steel right circular cylindrical storage tanks that are 2.5 m high with a radius of 0.5 m. How sensitive are the tanks' volumes to small variations in height and radius?

**Solution** With $V = \pi r^2 h$, the total differential gives the approximation for the change in volume as

$$dV = V_r(0.5, 2.5)\, dr + V_h(0.5, 2.5)\, dh$$
$$= (2\pi rh)_{(0.5,\, 2.5)}\, dr + (\pi r^2)_{(0.5,\, 2.5)}\, dh$$
$$= 2.5\pi\, dr + 0.25\pi\, dh.$$

Thus, a 1-unit change in $r$ will change $V$ by about $2.5\pi$ units. A 1-unit change in $h$ will change $V$ by about $0.25\pi$ units. The tank's volume is 10 times more sensitive to a small change in $r$ than it is to a small change of equal size in $h$. As a quality control engineer concerned with being sure the tanks have the correct volume, you would want to pay special attention to their radii.

In contrast, if the values of $r$ and $h$ are reversed to make $r = 2.5$ and $h = 0.5$, then the total differential in $V$ becomes

$$dV = (2\pi rh)_{(2.5,\, 0.5)}\, dr + (\pi r^2)_{(2.5,\, 0.5)}\, dh = 2.5\pi\, dr + 6.25\pi\, dh.$$

Now the volume is more sensitive to changes in $h$ than to changes in $r$ (Figure 14.40).

The general rule is that functions are most sensitive to small changes in the variables that generate the largest partial derivatives. $\qquad \blacksquare$

**FIGURE 14.40** The volume of cylinder (a) is more sensitive to a small change in $r$ than it is to an equally small change in $h$. The volume of cylinder (b) is more sensitive to small changes in $h$ than it is to small changes in $r$ (Example 7).

### Functions of More Than Two Variables

Analogous results hold for differentiable functions of more than two variables.

1. The **linearization** of $f(x, y, z)$ at a point $P_0(x_0, y_0, z_0)$ is

$$L(x, y, z) = f(P_0) + f_x(P_0)(x - x_0) + f_y(P_0)(y - y_0) + f_z(P_0)(z - z_0).$$

2. Suppose that $R$ is a closed rectangular solid centered at $P_0$ and lying in an open region on which the second partial derivatives of $f$ are continuous. Suppose also that $|f_{xx}|, |f_{yy}|, |f_{zz}|, |f_{xy}|, |f_{xz}|,$ and $|f_{yz}|$ are all less than or equal to $M$ throughout $R$. Then the **error** $E(x, y, z) = f(x, y, z) - L(x, y, z)$ in the approximation of $f$ by $L$ is bounded throughout $R$ by the inequality

$$|E| \leq \frac{1}{2}M(|x - x_0| + |y - y_0| + |z - z_0|)^2.$$

3. If the second partial derivatives of $f$ are continuous and if $x$, $y$, and $z$ change from $x_0, y_0,$ and $z_0$ by small amounts $dx$, $dy$, and $dz$, the **total differential**

$$df = f_x(P_0)\,dx + f_y(P_0)\,dy + f_z(P_0)\,dz$$

gives a good approximation of the resulting change in $f$.

**EXAMPLE 8**   Find the linearization $L(x, y, z)$ of

$$f(x, y, z) = x^2 - xy + 3 \sin z$$

at the point $(x_0, y_0, z_0) = (2, 1, 0)$. Find an upper bound for the error incurred in replacing $f$ by $L$ on the rectangular region

$$R: \quad |x - 2| \leq 0.01, \qquad |y - 1| \leq 0.02, \qquad |z| \leq 0.01.$$

**Solution**   Routine calculations give

$$f(2, 1, 0) = 2, \qquad f_x(2, 1, 0) = 3, \qquad f_y(2, 1, 0) = -2, \qquad f_z(2, 1, 0) = 3.$$

Thus,

$$L(x, y, z) = 2 + 3(x - 2) + (-2)(y - 1) + 3(z - 0) = 3x - 2y + 3z - 2.$$

Since

$$f_{xx} = 2, \qquad f_{yy} = 0, \qquad f_{zz} = -3 \sin z, \qquad f_{xy} = -1, \qquad f_{xz} = 0, \qquad f_{yz} = 0,$$

and $|-3 \sin z| \leq 3 \sin 0.01 \approx 0.03$, we may take $M = 2$ as a bound on the second partials. Hence, the error incurred by replacing $f$ by $L$ on $R$ satisfies

$$|E| \leq \frac{1}{2}(2)(0.01 + 0.02 + 0.01)^2 = 0.0016. \qquad \blacksquare$$

## Exercises 14.6

### Tangent Planes and Normal Lines to Surfaces

In Exercises 1–8, find equations for the

**(a)** tangent plane and

**(b)** normal line at the point $P_0$ on the given surface.

1. $x^2 + y^2 + z^2 = 3, \quad P_0(1, 1, 1)$

2. $x^2 + y^2 - z^2 = 18, \quad P_0(3, 5, -4)$

3. $2z - x^2 = 0, \quad P_0(2, 0, 2)$

4. $x^2 + 2xy - y^2 + z^2 = 7 \quad P_0(1, -1, 3)$

5. $\cos \pi x - x^2y + e^{xz} + yz = 4, \quad P_0(0, 1, 2)$

6. $x^2 - xy - y^2 - z = 0, \quad P_0(1, 1, -1)$

7. $x + y + z = 1, \quad P_0(0, 1, 0)$

8. $x^2 + y^2 - 2xy - x + 3y - z = -4, \quad P_0(2, -3, 18)$

In Exercises 9–12, find an equation for the plane that is tangent to the given surface at the given point.

9. $z = \ln(x^2 + y^2), \quad (1, 0, 0)$

10. $z = e^{-(x^2+y^2)}, \quad (0, 0, 1)$

11. $z = \sqrt{y - x}, \quad (1, 2, 1)$

12. $z = 4x^2 + y^2, \quad (1, 1, 5)$

## Tangent Lines to Intersecting Surfaces

In Exercises 13–18, find parametric equations for the line tangent to the curve of intersection of the surfaces at the given point.

**13.** Surfaces: $x + y^2 + 2z = 4$, $x = 1$

   Point: $(1, 1, 1)$

**14.** Surfaces: $xyz = 1$, $x^2 + 2y^2 + 3z^2 = 6$

   Point: $(1, 1, 1)$

**15.** Surfaces: $x^2 + 2y + 2z = 4$, $y = 1$

   Point: $(1, 1, 1/2)$

**16.** Surfaces: $x + y^2 + z = 2$, $y = 1$

   Point: $(1/2, 1, 1/2)$

**17.** Surfaces: $x^3 + 3x^2y^2 + y^3 + 4xy - z^2 = 0$,

   $x^2 + y^2 + z^2 = 11$

   Point: $(1, 1, 3)$

**18.** Surfaces: $x^2 + y^2 = 4$, $x^2 + y^2 - z = 0$

   Point: $\left(\sqrt{2}, \sqrt{2}, 4\right)$

## Estimating Change

**19.** By about how much will

$$f(x, y, z) = \ln\sqrt{x^2 + y^2 + z^2}$$

change if the point $P(x, y, z)$ moves from $P_0(3, 4, 12)$ a distance of $ds = 0.1$ unit in the direction of $3\mathbf{i} + 6\mathbf{j} - 2\mathbf{k}$?

**20.** By about how much will

$$f(x, y, z) = e^x \cos yz$$

change as the point $P(x, y, z)$ moves from the origin a distance of $ds = 0.1$ unit in the direction of $2\mathbf{i} + 2\mathbf{j} - 2\mathbf{k}$?

**21.** By about how much will

$$g(x, y, z) = x + x \cos z - y \sin z + y$$

change if the point $P(x, y, z)$ moves from $P_0(2, -1, 0)$ a distance of $ds = 0.2$ unit toward the point $P_1(0, 1, 2)$?

**22.** By about how much will

$$h(x, y, z) = \cos(\pi xy) + xz^2$$

change if the point $P(x, y, z)$ moves from $P_0(-1, -1, -1)$ a distance of $ds = 0.1$ unit toward the origin?

**23. Temperature change along a circle** Suppose that the Celsius temperature at the point $(x, y)$ in the $xy$-plane is $T(x, y) = x \sin 2y$ and that distance in the $xy$-plane is measured in meters. A particle is moving *clockwise* around the circle of radius 1 m centered at the origin at the constant rate of 2 m/sec.

  **a.** How fast is the temperature experienced by the particle changing in degrees Celsius per meter at the point $P\left(1/2, \sqrt{3}/2\right)$?

  **b.** How fast is the temperature experienced by the particle changing in degrees Celsius per second at $P$?

**24. Changing temperature along a space curve** The Celsius temperature in a region in space is given by $T(x, y, z) = 2x^2 - xyz$. A particle is moving in this region and its position at time $t$ is given by $x = 2t^2$, $y = 3t$, $z = -t^2$, where time is measured in seconds and distance in meters.

  **a.** How fast is the temperature experienced by the particle changing in degrees Celsius per meter when the particle is at the point $P(8, 6, -4)$?

  **b.** How fast is the temperature experienced by the particle changing in degrees Celsius per second at $P$?

## Finding Linearizations

In Exercises 25–30, find the linearization $L(x, y)$ of the function at each point.

**25.** $f(x, y) = x^2 + y^2 + 1$ at   **a.** $(0, 0)$,   **b.** $(1, 1)$

**26.** $f(x, y) = (x + y + 2)^2$ at   **a.** $(0, 0)$,   **b.** $(1, 2)$

**27.** $f(x, y) = 3x - 4y + 5$ at   **a.** $(0, 0)$,   **b.** $(1, 1)$

**28.** $f(x, y) = x^3y^4$ at   **a.** $(1, 1)$,   **b.** $(0, 0)$

**29.** $f(x, y) = e^x \cos y$ at   **a.** $(0, 0)$,   **b.** $(0, \pi/2)$

**30.** $f(x, y) = e^{2y-x}$ at   **a.** $(0, 0)$,   **b.** $(1, 2)$

**31. Wind chill factor** Wind chill, a measure of the apparent temperature felt on exposed skin, is a function of air temperature and wind speed. The precise formula, updated by the National Weather Service in 2001 and based on modern heat transfer theory, a human face model, and skin tissue resistance, is (after unit conversion)

$$W = W(v, T) = 13.13 + 0.6215\,T - 11.36\,v^{0.16}$$
$$+ 0.396\,T \cdot v^{0.16},$$

where $T$ is air temperature in °C and $v$ is wind speed in km/h. A partial wind chill chart is given.

| | | | | $T(°C)$ | | | |
|---|---|---|---|---|---|---|---|
| | **5** | **0** | **−5** | **−10** | **−15** | **−20** | **−25** |
| **10** | 2.7 | −3.3 | −9.3 | −15.2 | −21.2 | −27.2 | −33.1 |
| **20** | 1.1 | −5.2 | −11.5 | −17.8 | −24.1 | −30.4 | −36.7 |
| **30** | 0.1 | −6.4 | −13.0 | −19.5 | −26.0 | −32.5 | −39.0 |
| **40** | −0.7 | −7.4 | −14.0 | −20.7 | −27.4 | −34.1 | −40.8 |
| **50** | −1.3 | −8.1 | −14.9 | −21.7 | −28.5 | −35.4 | −42.2 |
| **60** | −1.8 | −8.7 | −15.7 | −22.6 | −29.5 | −36.4 | −43.3 |

$v$ (km/h) labels the rows.

  **a.** Use the table to find $W(30, -5)$, $W(50, -25)$, and $W(30, -10)$.

  **b.** Use the formula to find $W(15, -40)$, $W(80, -40)$, and $W(90, 0)$.

  **c.** Find the linearization $L(v, T)$ of the function $W(v, T)$ at the point $(40, -10)$.

  **d.** Use $L(v, T)$ in part (c) to estimate the following wind chill values.

   **i)** $W(39, -9)$   **ii)** $W(42, -12)$

   **iii)** $W(10, -25)$ (Explain why this value is much different from the value found in the table.)

**32.** Find the linearization $L(v, T)$ of the function $W(v, T)$ in Exercise 31 at the point $(50, -20)$. Use it to estimate the following wind chill values.

  **a.** $W(49, -22)$

  **b.** $W(53, -19)$

  **c.** $W(60, -30)$

### Bounding the Error in Linear Approximations

In Exercises 33–38, find the linearization $L(x, y)$ of the function $f(x, y)$ at $P_0$. Then find an upper bound for the magnitude $|E|$ of the error in the approximation $f(x, y) \approx L(x, y)$ over the rectangle $R$.

**33.** $f(x, y) = x^2 - 3xy + 5$ at $P_0(2, 1)$,

$R$: $|x - 2| \leq 0.1$, $|y - 1| \leq 0.1$

**34.** $f(x, y) = (1/2)x^2 + xy + (1/4)y^2 + 3x - 3y + 4$ at $P_0(2, 2)$,

$R$: $|x - 2| \leq 0.1$, $|y - 2| \leq 0.1$

**35.** $f(x, y) = 1 + y + x \cos y$ at $P_0(0, 0)$,

$R$: $|x| \leq 0.2$, $|y| \leq 0.2$

(Use $|\cos y| \leq 1$ and $|\sin y| \leq 1$ in estimating $E$.)

**36.** $f(x, y) = xy^2 + y \cos (x - 1)$ at $P_0(1, 2)$,

$R$: $|x - 1| \leq 0.1$, $|y - 2| \leq 0.1$

**37.** $f(x, y) = e^x \cos y$ at $P_0(0, 0)$,

$R$: $|x| \leq 0.1$, $|y| \leq 0.1$

(Use $e^x \leq 1.11$ and $|\cos y| \leq 1$ in estimating $E$.)

**38.** $f(x, y) = \ln x + \ln y$ at $P_0(1, 1)$,

$R$: $|x - 1| \leq 0.2$, $|y - 1| \leq 0.2$

### Linearizations for Three Variables

Find the linearizations $L(x, y, z)$ of the functions in Exercises 39–44 at the given points.

**39.** $f(x, y, z) = xy + yz + xz$ at

  **a.** $(1, 1, 1)$      **b.** $(1, 0, 0)$      **c.** $(0, 0, 0)$

**40.** $f(x, y, z) = x^2 + y^2 + z^2$ at

  **a.** $(1, 1, 1)$      **b.** $(0, 1, 0)$      **c.** $(1, 0, 0)$

**41.** $f(x, y, z) = \sqrt{x^2 + y^2 + z^2}$ at

  **a.** $(1, 0, 0)$      **b.** $(1, 1, 0)$      **c.** $(1, 2, 2)$

**42.** $f(x, y, z) = (\sin xy)/z$ at

  **a.** $(\pi/2, 1, 1)$      **b.** $(2, 0, 1)$

**43.** $f(x, y, z) = e^x + \cos (y + z)$ at

  **a.** $(0, 0, 0)$      **b.** $\left(0, \frac{\pi}{2}, 0\right)$      **c.** $\left(0, \frac{\pi}{4}, \frac{\pi}{4}\right)$

**44.** $f(x, y, z) = \tan^{-1} (xyz)$ at

  **a.** $(1, 0, 0)$      **b.** $(1, 1, 0)$      **c.** $(1, 1, 1)$

In Exercises 45–48, find the linearization $L(x, y, z)$ of the function $f(x, y, z)$ at $P_0$. Then find an upper bound for the magnitude of the error $E$ in the approximation $f(x, y, z) \approx L(x, y, z)$ over the region $R$.

**45.** $f(x, y, z) = xz - 3yz + 2$ at $P_0(1, 1, 2)$,

$R$: $|x - 1| \leq 0.01$, $|y - 1| \leq 0.01$, $|z - 2| \leq 0.02$

**46.** $f(x, y, z) = x^2 + xy + yz + (1/4)z^2$ at $P_0(1, 1, 2)$,

$R$: $|x - 1| \leq 0.01$, $|y - 1| \leq 0.01$, $|z - 2| \leq 0.08$

**47.** $f(x, y, z) = xy + 2yz - 3xz$ at $P_0(1, 1, 0)$,

$R$: $|x - 1| \leq 0.01$, $|y - 1| \leq 0.01$, $|z| \leq 0.01$

**48.** $f(x, y, z) = \sqrt{2} \cos x \sin (y + z)$ at $P_0(0, 0, \pi/4)$,

$R$: $|x| \leq 0.01$, $|y| \leq 0.01$, $|z - \pi/4| \leq 0.01$

### Estimating Error; Sensitivity to Change

**49. Estimating maximum error** Suppose that $T$ is to be found from the formula $T = x (e^y + e^{-y})$, where $x$ and $y$ are found to be 2 and $\ln 2$ with maximum possible errors of $|dx| = 0.1$ and $|dy| = 0.02$. Estimate the maximum possible error in the computed value of $T$.

**50. Variation in electrical resistance** The resistance $R$ produced by wiring resistors of $R_1$ and $R_2$ ohms in parallel (see accompanying figure) can be calculated from the formula

$$\frac{1}{R} = \frac{1}{R_1} + \frac{1}{R_2}.$$

  **a.** Show that

$$dR = \left(\frac{R}{R_1}\right)^2 dR_1 + \left(\frac{R}{R_2}\right)^2 dR_2.$$

  **b.** You have designed a two-resistor circuit, like the one shown, to have resistances of $R_1 = 100$ ohms and $R_2 = 400$ ohms, but there is always some variation in manufacturing and the resistors received by your firm will probably not have these exact values. Will the value of $R$ be more sensitive to variation in $R_1$ or to variation in $R_2$? Give reasons for your answer.

  **c.** In another circuit like the one shown, you plan to change $R_1$ from 20 to 20.1 ohms and $R_2$ from 25 to 24.9 ohms. By about what percentage will this change $R$?

**51.** You plan to calculate the area of a long, thin rectangle from measurements of its length and width. Which dimension should you measure more carefully? Give reasons for your answer.

**52. a.** Around the point $(1, 0)$, is $f(x, y) = x^2(y + 1)$ more sensitive to changes in $x$ or to changes in $y$? Give reasons for your answer.

  **b.** What ratio of $dx$ to $dy$ will make $df$ equal zero at $(1, 0)$?

**53. Value of a $2 \times 2$ determinant** If $|a|$ is much greater than $|b|$, $|c|$, and $|d|$, to which of $a$, $b$, $c$, and $d$ is the value of the determinant

$$f(a, b, c, d) = \begin{vmatrix} a & b \\ c & d \end{vmatrix}$$

most sensitive? Give reasons for your answer.

**54. The Wilson lot size formula** The Wilson lot size formula in economics says that the most economical quantity $Q$ of goods (radios, shoes, brooms, whatever) for a store to order is given by the formula $Q = \sqrt{2KM/h}$, where $K$ is the cost of placing the order, $M$ is the number of items sold per week, and $h$ is the weekly holding cost for each item (cost of space, utilities, security, and so on). To which of the variables $K$, $M$, and $h$ is $Q$ most sensitive near the point $(K_0, M_0, h_0) = (2, 20, 0.05)$? Give reasons for your answer.

### Theory and Examples

**55. The linearization of $f(x, y)$ is a tangent-plane approximation** Show that the tangent plane at the point $P_0(x_0, y_0, f(x_0, y_0))$ on the surface $z = f(x, y)$ defined by a differentiable function $f$ is the plane

$$f_x(x_0, y_0)(x - x_0) + f_y(x_0, y_0)(y - y_0) - (z - f(x_0, y_0)) = 0$$

or

$$z = f(x_0, y_0) + f_x(x_0, y_0)(x - x_0) + f_y(x_0, y_0)(y - y_0).$$

Thus, the tangent plane at $P_0$ is the graph of the linearization of $f$ at $P_0$ (see accompanying figure).

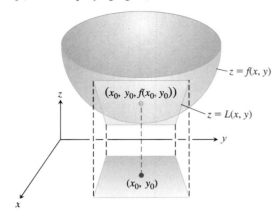

**56. Change along the involute of a circle** Find the derivative of $f(x, y) = x^2 + y^2$ in the direction of the unit tangent vector of the curve

$$\mathbf{r}(t) = (\cos t + t \sin t)\mathbf{i} + (\sin t - t \cos t)\mathbf{j}, \qquad t > 0.$$

**57. Tangent curves** A smooth curve is *tangent* to the surface at a point of intersection if its velocity vector is orthogonal to $\nabla f$ there.
   Show that the curve

$$\mathbf{r}(t) = \sqrt{t}\,\mathbf{i} + \sqrt{t}\,\mathbf{j} + (2t - 1)\mathbf{k}$$

is tangent to the surface $x^2 + y^2 - z = 1$ when $t = 1$.

**58. Normal curves** A smooth curve is *normal* to a surface $f(x, y, z) = c$ at a point of intersection if the curve's velocity vector is a nonzero scalar multiple of $\nabla f$ at the point.
   Show that the curve

$$\mathbf{r}(t) = \sqrt{t}\,\mathbf{i} + \sqrt{t}\,\mathbf{j} - \frac{1}{4}(t + 3)\mathbf{k}$$

is normal to the surface $x^2 + y^2 - z = 3$ when $t = 1$.

# 14.7 Extreme Values and Saddle Points

Continuous functions of two variables assume extreme values on closed, bounded domains (see Figures 14.41 and 14.42). We see in this section that we can narrow the search for these extreme values by examining the functions' first partial derivatives. A function of two variables can assume extreme values only at domain boundary points or at interior domain points where both first partial derivatives are zero or where one or both of the first partial derivatives fail to exist. However, the vanishing of derivatives at an interior point $(a, b)$ does not always signal the presence of an extreme value. The surface that is the graph of the function might be shaped like a saddle right above $(a, b)$ and cross its tangent plane there.

## Derivative Tests for Local Extreme Values

To find the local extreme values of a function of a single variable, we look for points where the graph has a horizontal tangent line. At such points, we then look for local maxima, local minima, and points of inflection. For a function $f(x, y)$ of two variables, we look for points where the surface $z = f(x, y)$ has a horizontal tangent *plane*. At such points, we then look for local maxima, local minima, and saddle points. We begin by defining maxima and minima.

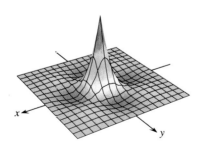

**FIGURE 14.41**   The function

$$z = (\cos x)(\cos y)e^{-\sqrt{x^2 + y^2}}$$

has a maximum value of 1 and a minimum value of about $-0.067$ on the square region $|x| \leq 3\pi/2$, $|y| \leq 3\pi/2$.

---

**DEFINITIONS**   Let $f(x, y)$ be defined on a region $R$ containing the point $(a, b)$. Then ·

1. $f(a, b)$ is a **local maximum** value of $f$ if $f(a, b) \geq f(x, y)$ for all domain points $(x, y)$ in an open disk centered at $(a, b)$.
2. $f(a, b)$ is a **local minimum** value of $f$ if $f(a, b) \leq f(x, y)$ for all domain points $(x, y)$ in an open disk centered at $(a, b)$.

---

Local maxima correspond to mountain peaks on the surface $z = f(x, y)$ and local minima correspond to valley bottoms (Figure 14.43). At such points the tangent planes, when they exist, are horizontal. Local extrema are also called **relative extrema**.

As with functions of a single variable, the key to identifying the local extrema is the First Derivative Test, which we next state and prove.

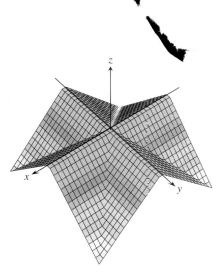

FIGURE 14.42   The "roof surface"
$$z = \frac{1}{2}\left(\big|\,|x| - |y|\,\big| - |x| - |y|\right)$$
has a maximum value of 0 and a minimum value of $-a$ on the square region $|x| \le a$, $|y| \le a$.

Local maxima
(no greater value of $f$ nearby)

Local minimum
(no smaller value
of $f$ nearby)

FIGURE 14.43   A local maximum occurs at a mountain peak and a local minimum occurs at a valley low point.

> **THEOREM 10—First Derivative Test for Local Extreme Values**   If $f(x, y)$ has a local maximum or minimum value at an interior point $(a, b)$ of its domain and if the first partial derivatives exist there, then $f_x(a, b) = 0$ and $f_y(a, b) = 0$.

**Proof**    If $f$ has a local extremum at $(a, b)$, then the function $g(x) = f(x, b)$ has a local extremum at $x = a$ (Figure 14.44). Therefore, $g'(a) = 0$ (Chapter 4, Theorem 2). Now $g'(a) = f_x(a, b)$, so $f_x(a, b) = 0$. A similar argument with the function $h(y) = f(a, y)$ shows that $f_y(a, b) = 0$.  ∎

If we substitute the values $f_x(a, b) = 0$ and $f_y(a, b) = 0$ into the equation

$$f_x(a, b)(x - a) + f_y(a, b)(y - b) - (z - f(a, b)) = 0$$

for the tangent plane to the surface $z = f(x, y)$ at $(a, b)$, the equation reduces to

$$0 \cdot (x - a) + 0 \cdot (y - b) - z + f(a, b) = 0$$

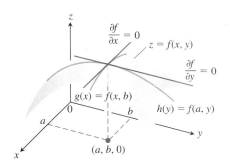

FIGURE 14.44   If a local maximum of $f$ occurs at $x = a$, $y = b$, then the first partial derivatives $f_x(a, b)$ and $f_y(a, b)$ are both zero.

or

$$z = f(a, b).$$

Thus, Theorem 10 says that the surface does indeed have a horizontal tangent plane at a local extremum, provided there is a tangent plane there.

> **DEFINITION**   An interior point of the domain of a function $f(x, y)$ where both $f_x$ and $f_y$ are zero or where one or both of $f_x$ and $f_y$ do not exist is a **critical point** of $f$.

Theorem 10 says that the only points where a function $f(x, y)$ can assume extreme values are critical points and boundary points. As with differentiable functions of a single variable, not every critical point gives rise to a local extremum. A differentiable function of a single variable might have a point of inflection. A differentiable function of two variables might have a *saddle point*, with the graph of $f$ crossing the tangent plane defined there.

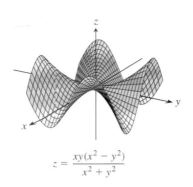

$$z = \frac{xy(x^2 - y^2)}{x^2 + y^2}$$

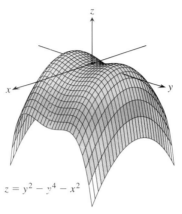

$$z = y^2 - y^4 - x^2$$

**FIGURE 14.45** Saddle points at the origin.

**DEFINITION** A differentiable function $f(x, y)$ has a **saddle point** at a critical point $(a, b)$ if in every open disk centered at $(a, b)$ there are domain points $(x, y)$ where $f(x, y) > f(a, b)$ and domain points $(x, y)$ where $f(x, y) < f(a, b)$. The corresponding point $(a, b, f(a, b))$ on the surface $z = f(x, y)$ is called a saddle point of the surface (Figure 14.45).

**EXAMPLE 1** Find the local extreme values of $f(x, y) = x^2 + y^2 - 4y + 9$.

**Solution** The domain of $f$ is the entire plane (so there are no boundary points) and the partial derivatives $f_x = 2x$ and $f_y = 2y - 4$ exist everywhere. Therefore, local extreme values can occur only where

$$f_x = 2x = 0 \quad \text{and} \quad f_y = 2y - 4 = 0.$$

The only possibility is the point $(0, 2)$, where the value of $f$ is 5. Since $f(x, y) = x^2 + (y - 2)^2 + 5$ is never less than 5, we see that the critical point $(0, 2)$ gives a local minimum (Figure 14.46). ∎

**EXAMPLE 2** Find the local extreme values (if any) of $f(x, y) = y^2 - x^2$.

**Solution** The domain of $f$ is the entire plane (so there are no boundary points) and the partial derivatives $f_x = -2x$ and $f_y = 2y$ exist everywhere. Therefore, local extrema can occur only at the origin $(0, 0)$ where $f_x = 0$ and $f_y = 0$. Along the positive $x$-axis, however, $f$ has the value $f(x, 0) = -x^2 < 0$; along the positive $y$-axis, $f$ has the value $f(0, y) = y^2 > 0$. Therefore, every open disk in the $xy$-plane centered at $(0, 0)$ contains points where the function is positive and points where it is negative. The function has a saddle point at the origin and no local extreme values (Figure 14.47a). Figure 14.47b displays the level curves (they are hyperbolas) of $f$, and shows the function decreasing and increasing in an alternating fashion among the four groupings of hyperbolas. ∎

That $f_x = f_y = 0$ at an interior point $(a, b)$ of $R$ does not guarantee $f$ has a local extreme value there. If $f$ and its first and second partial derivatives are continuous on $R$, however, we may be able to learn more from the following theorem, proved in Section 14.9.

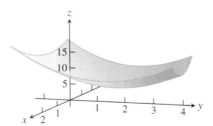

**FIGURE 14.46** The graph of the function $f(x, y) = x^2 + y^2 - 4y + 9$ is a paraboloid which has a local minimum value of 5 at the point $(0, 2)$ (Example 1).

**THEOREM 11—Second Derivative Test for Local Extreme Values** Suppose that $f(x, y)$ and its first and second partial derivatives are continuous throughout a disk centered at $(a, b)$ and that $f_x(a, b) = f_y(a, b) = 0$. Then

i) $f$ has a **local maximum** at $(a, b)$ if $f_{xx} < 0$ and $f_{xx}f_{yy} - f_{xy}^2 > 0$ at $(a, b)$.

ii) $f$ has a **local minimum** at $(a, b)$ if $f_{xx} > 0$ and $f_{xx}f_{yy} - f_{xy}^2 > 0$ at $(a, b)$.

iii) $f$ has a **saddle point** at $(a, b)$ if $f_{xx}f_{yy} - f_{xy}^2 < 0$ at $(a, b)$.

iv) **the test is inconclusive** at $(a, b)$ if $f_{xx}f_{yy} - f_{xy}^2 = 0$ at $(a, b)$. In this case, we must find some other way to determine the behavior of $f$ at $(a, b)$.

The expression $f_{xx}f_{yy} - f_{xy}^2$ is called the **discriminant** or **Hessian** of $f$. It is sometimes easier to remember it in determinant form,

$$f_{xx}f_{yy} - f_{xy}^2 = \begin{vmatrix} f_{xx} & f_{xy} \\ f_{xy} & f_{yy} \end{vmatrix}.$$

Theorem 11 says that if the discriminant is positive at the point $(a, b)$, then the surface curves the same way in all directions: downward if $f_{xx} < 0$, giving rise to a local maximum, and

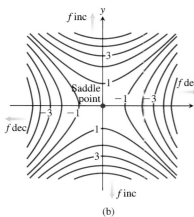

**FIGURE 14.47** (a) The origin is a saddle point of the function $f(x, y) = y^2 - x^2$. There are no local extreme values (Example 2). (b) Level curves for the function $f$ in Example 2.

upward if $f_{xx} > 0$, giving a local minimum. On the other hand, if the discriminant is negative at $(a, b)$, then the surface curves up in some directions and down in others, so we have a saddle point.

**EXAMPLE 3** Find the local extreme values of the function

$$f(x, y) = xy - x^2 - y^2 - 2x - 2y + 4.$$

**Solution** The function is defined and differentiable for all $x$ and $y$, and its domain has no boundary points. The function therefore has extreme values only at the points where $f_x$ and $f_y$ are simultaneously zero. This leads to

$$f_x = y - 2x - 2 = 0, \qquad f_y = x - 2y - 2 = 0,$$

or

$$x = y = -2.$$

Therefore, the point $(-2, -2)$ is the only point where $f$ may take on an extreme value. To see if it does so, we calculate

$$f_{xx} = -2, \qquad f_{yy} = -2, \qquad f_{xy} = 1.$$

The discriminant of $f$ at $(a, b) = (-2, -2)$ is

$$f_{xx}f_{yy} - f_{xy}{}^2 = (-2)(-2) - (1)^2 = 4 - 1 = 3.$$

The combination

$$f_{xx} < 0 \qquad \text{and} \qquad f_{xx}f_{yy} - f_{xy}{}^2 > 0$$

tells us that $f$ has a local maximum at $(-2, -2)$. The value of $f$ at this point is $f(-2, -2) = 8$. ∎

**EXAMPLE 4** Find the local extreme values of $f(x, y) = 3y^2 - 2y^3 - 3x^2 + 6xy$.

**Solution** Since $f$ is differentiable everywhere, it can assume extreme values only where

$$f_x = 6y - 6x = 0 \qquad \text{and} \qquad f_y = 6y - 6y^2 + 6x = 0.$$

From the first of these equations we find $x = y$, and substitution for $y$ into the second equation then gives

$$6x - 6x^2 + 6x = 0 \qquad \text{or} \qquad 6x (2 - x) = 0.$$

The two critical points are therefore $(0, 0)$ and $(2, 2)$.

To classify the critical points, we calculate the second derivatives:

$$f_{xx} = -6, \qquad f_{yy} = 6 - 12y, \qquad f_{xy} = 6.$$

The discriminant is given by

$$f_{xx}f_{yy} - f_{xy}{}^2 = (-36 + 72y) - 36 = 72(y - 1).$$

At the critical point $(0, 0)$ we see that the value of the discriminant is the negative number $-72$, so the function has a saddle point at the origin. At the critical point $(2, 2)$ we see that the discriminant has the positive value 72. Combining this result with the negative value of the second partial $f_{xx} = -6$, Theorem 11 says that the critical point $(2, 2)$ gives a local maximum value of $f(2, 2) = 12 - 16 - 12 + 24 = 8$. A graph of the surface is shown in Figure 14.48. ∎

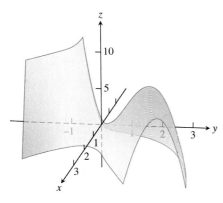

**FIGURE 14.48** The surface $z = 3y^2 - 2y^3 - 3x^2 + 6xy$ has a saddle point at the origin and a local maximum at the point $(2, 2)$ (Example 4).

**EXAMPLE 5** Find the critical points of the function $f(x, y) = 10xye^{-(x^2+y^2)}$ and use the Second Derivative Test to classify each point as one where a saddle, local minimum, or local maximum occurs.

**Solution** First we find the partial derivatives $f_x$ and $f_y$ and set them simultaneously to zero in seeking the critical points:

$$f_x = 10ye^{-(x^2+y^2)} - 20x^2ye^{-(x^2+y^2)} = 10y(1 - 2x^2)e^{-(x^2+y^2)} = 0 \Rightarrow y = 0 \text{ or } 1 - 2x^2 = 0,$$

$$f_y = 10xe^{-(x^2+y^2)} - 20xy^2e^{-(x^2+y^2)} = 10x(1 - 2y^2)e^{-(x^2+y^2)} = 0 \Rightarrow x = 0 \text{ or } 1 - 2y^2 = 0.$$

Since both partial derivatives are continuous everywhere, the only critical points are

$$(0, 0), \left(\frac{1}{\sqrt{2}}, \frac{1}{\sqrt{2}}\right), \left(-\frac{1}{\sqrt{2}}, \frac{1}{\sqrt{2}}\right), \left(\frac{1}{\sqrt{2}}, -\frac{1}{\sqrt{2}}\right), \text{ and } \left(-\frac{1}{\sqrt{2}}, -\frac{1}{\sqrt{2}}\right).$$

Next we calculate the second partial derivatives in order to evaluate the discriminant at each critical point:

$$f_{xx} = -20xy(1 - 2x^2)e^{-(x^2+y^2)} - 40xye^{-(x^2+y^2)} = -20xy(3 - 2x^2)e^{-(x^2+y^2)},$$

$$f_{xy} = f_{yx} = 10(1 - 2x^2)e^{-(x^2+y^2)} - 20y^2(1 - 2x^2)e^{-(x^2+y^2)} = 10(1 - 2x^2)(1 - 2y^2)e^{-(x^2+y^2)},$$

$$f_{yy} = -20xy(1 - 2y^2)e^{-(x^2+y^2)} - 40xye^{-(x^2+y^2)} = -20xy(3 - 2y^2)e^{-(x^2+y^2)}.$$

The following table summarizes the values needed by the Second Derivative Test.

| Critical Point | $f_{xx}$ | $f_{xy}$ | $f_{yy}$ | Discriminant $D$ |
|---|---|---|---|---|
| $(0, 0)$ | $0$ | $10$ | $0$ | $-100$ |
| $\left(\dfrac{1}{\sqrt{2}}, \dfrac{1}{\sqrt{2}}\right)$ | $-\dfrac{20}{e}$ | $0$ | $-\dfrac{20}{e}$ | $\dfrac{400}{e^2}$ |
| $\left(-\dfrac{1}{\sqrt{2}}, \dfrac{1}{\sqrt{2}}\right)$ | $\dfrac{20}{e}$ | $0$ | $\dfrac{20}{e}$ | $\dfrac{400}{e^2}$ |
| $\left(\dfrac{1}{\sqrt{2}}, -\dfrac{1}{\sqrt{2}}\right)$ | $\dfrac{20}{e}$ | $0$ | $\dfrac{20}{e}$ | $\dfrac{400}{e^2}$ |
| $\left(-\dfrac{1}{\sqrt{2}}, -\dfrac{1}{\sqrt{2}}\right)$ | $-\dfrac{20}{e}$ | $0$ | $-\dfrac{20}{e}$ | $\dfrac{400}{e^2}$ |

$z = 10xye^{-(x^2 + y^2)}$

**FIGURE 14.49** A graph of the function in Example 5.

From the table we find that $D < 0$ at the critical point $(0, 0)$, giving a saddle; $D > 0$ and $f_{xx} < 0$ at the critical points $\left(1/\sqrt{2}, 1/\sqrt{2}\right)$ and $\left(-1/\sqrt{2}, -1/\sqrt{2}\right)$, giving local maximum values there; and $D > 0$ and $f_{xx} > 0$ at the critical points $\left(-1/\sqrt{2}, 1/\sqrt{2}\right)$ and $\left(1/\sqrt{2}, -1/\sqrt{2}\right)$, each giving local minimum values. A graph of the surface is shown in Figure 14.49.

### Absolute Maxima and Minima on Closed Bounded Regions

We organize the search for the absolute extrema of a continuous function $f(x, y)$ on a closed and bounded region $R$ into three steps.

1. *List the interior points of $R$* where $f$ may have local maxima and minima and evaluate $f$ at these points. These are the critical points of $f$.

2. *List the boundary points of $R$* where $f$ has local maxima and minima and evaluate $f$ at these points. We show how to do this in the next example.

3. *Look through the lists* for the maximum and minimum values of $f$. These will be the absolute maximum and minimum values of $f$ on $R$. Since absolute maxima and minima are also local maxima and minima, the absolute maximum and minimum values of $f$ appear somewhere in the lists made in Steps 1 and 2.

**EXAMPLE 6**    Find the absolute maximum and minimum values of

$$f(x, y) = 2 + 2x + 4y - x^2 - y^2$$

on the triangular region in the first quadrant bounded by the lines $x = 0$, $y = 0$, and $y = 9 - x$.

**Solution**    Since $f$ is differentiable, the only places where $f$ can assume these values are points inside the triangle (Figure 14.50a) where $f_x = f_y = 0$ and points on the boundary.

**(a) Interior points.** For these we have

$$f_x = 2 - 2x = 0, \qquad f_y = 4 - 2y = 0,$$

yielding the single point $(x, y) = (1, 2)$. The value of $f$ there is

$$f(1, 2) = 7.$$

**(b) Boundary points.** We take the triangle one side at a time:

**i)** On the segment $OA$, $y = 0$. The function

$$f(x, y) = f(x, 0) = 2 + 2x - x^2$$

may now be regarded as a function of $x$ defined on the closed interval $0 \le x \le 9$. Its extreme values (we know from Chapter 4) may occur at the endpoints

$$\begin{array}{lll} x = 0 & \text{where} & f(0, 0) = 2 \\ x = 9 & \text{where} & f(9, 0) = 2 + 18 - 81 = -61 \end{array}$$

and at the interior points where $f'(x, 0) = 2 - 2x = 0$. The only interior point where $f'(x, 0) = 0$ is $x = 1$, where

$$f(x, 0) = f(1, 0) = 3.$$

**ii)** On the segment $OB$, $x = 0$ and

$$f(x, y) = f(0, y) = 2 + 4y - y^2.$$

As in part i), we consider $f(0, y)$ as a function of $y$ defined on the closed interval $[0, 9]$. Its extreme values can occur at the endpoints or at interior points where $f'(0, y) = 0$. Since $f'(0, y) = 4 - 2y$, the only interior point where $f'(0, y) = 0$ occurs at $(0, 2)$, with $f(0, 2) = 6$. So the candidates for this segment are

$$f(0, 0) = 2, \qquad f(0, 9) = -43, \qquad f(0, 2) = 6.$$

**iii)** We have already accounted for the values of $f$ at the endpoints of $AB$, so we need only look at the interior points of the line segment $AB$. With $y = 9 - x$, we have

$$f(x, y) = 2 + 2x + 4(9 - x) - x^2 - (9 - x)^2 = -43 + 16x - 2x^2.$$

Setting $f'(x, 9 - x) = 16 - 4x = 0$ gives

$$x = 4.$$

At this value of $x$,

$$y = 9 - 4 = 5 \qquad \text{and} \qquad f(x, y) = f(4, 5) = -11.$$

**Summary**    We list all the function value candidates: $7, 2, -61, 3, -43, 6, -11$. The maximum is 7, which $f$ assumes at $(1, 2)$. The minimum is $-61$, which $f$ assumes at $(9, 0)$. See Figure 14.50b. ∎

Solving extreme value problems with algebraic constraints on the variables usually requires the method of Lagrange multipliers introduced in the next section. But sometimes we can solve such problems directly, as in the next example.

(a)

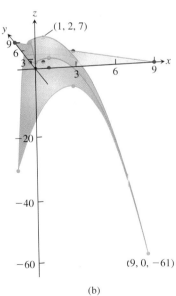

(b)

**FIGURE 14.50**    (a) This triangular region is the domain of the function in Example 6. (b) The graph of the function in Example 6. The blue points are the candidates for maxima or minima.

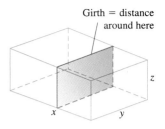

Girth = distance around here

**FIGURE 14.51** The box in Example 7.

**EXAMPLE 7** A delivery company accepts only rectangular boxes the sum of whose length and girth (perimeter of a cross-section) does not exceed 270 cm. Find the dimensions of an acceptable box of largest volume.

**Solution** Let $x$, $y$, and $z$ represent the length, width, and height of the rectangular box, respectively. Then the girth is $2y + 2z$. We want to maximize the volume $V = xyz$ of the box (Figure 14.51) satisfying $x + 2y + 2z = 270$ (the largest box accepted by the delivery company). Thus, we can write the volume of the box as a function of two variables:

$$V(y, z) = (270 - 2y - 2z)yz \qquad \begin{matrix} V = xyz \text{ and} \\ x = 270 - 2y - 2z \end{matrix}$$

$$= 270yz - 2y^2z - 2yz^2.$$

Setting the first partial derivatives equal to zero,

$$V_y(y, z) = 270z - 4yz - 2z^2 = (270 - 4y - 2z)z = 0$$
$$V_z(y, z) = 270y - 2y^2 - 4yz = (270 - 2y - 4z)y = 0,$$

gives the critical points $(0, 0)$, $(0, 135)$, $(135, 0)$, and $(45, 45)$. The volume is zero at $(0, 0)$, $(0, 135)$, and $(135, 0)$, which are not maximum values. At the point $(45, 45)$, we apply the Second Derivative Test (Theorem 11):

$$V_{yy} = -4z, \qquad V_{zz} = -4y, \qquad V_{yz} = 270 - 4y - 4z.$$

Then

$$V_{yy}V_{zz} - V_{yz}^2 = 16yz - 4(135 - 2y - 2z)^2.$$

Thus,

$$V_{yy}(45, 45) = -4(45) < 0$$

and

$$\left[V_{yy}V_{zz} - V_{yz}^2\right]_{(45, 45)} = 16(45)(45) - 4(-45)^2 > 0$$

imply that $(45, 45)$ gives a maximum volume. The dimensions of the package are $x = 270 - 2(45) - 2(45) = 90$ cm, $y = 45$ cm, and $z = 45$ cm. The maximum volume is $V = (90)(45)(45) = 182{,}250$ cm$^3$, or 182.25 liters. ∎

Despite the power of Theorem 11, we urge you to remember its limitations. It does not apply to boundary points of a function's domain, where it is possible for a function to have extreme values along with nonzero derivatives. Also, it does not apply to points where either $f_x$ or $f_y$ fails to exist.

---

**Summary of Max-Min Tests**

The extreme values of $f(x, y)$ can occur only at

i) **boundary points** of the domain of $f$

ii) **critical points** (interior points where $f_x = f_y = 0$ or points where $f_x$ or $f_y$ fails to exist).

If the first- and second-order partial derivatives of $f$ are continuous throughout a disk centered at a point $(a, b)$ and $f_x(a, b) = f_y(a, b) = 0$, the nature of $f(a, b)$ can be tested with the **Second Derivative Test**:

i) $f_{xx} < 0$ and $f_{xx}f_{yy} - f_{xy}^2 > 0$ at $(a, b) \Rightarrow$ **local maximum**

ii) $f_{xx} > 0$ and $f_{xx}f_{yy} - f_{xy}^2 > 0$ at $(a, b) \Rightarrow$ **local minimum**

iii) $f_{xx}f_{yy} - f_{xy}^2 < 0$ at $(a, b) \Rightarrow$ **saddle point**

iv) $f_{xx}f_{yy} - f_{xy}^2 = 0$ at $(a, b) \Rightarrow$ **test is inconclusive**

# Exercises 14.7

### Finding Local Extrema

Find all the local maxima, local minima, and saddle points of the functions in Exercises 1–30.

**1.** $f(x, y) = x^2 + xy + y^2 + 3x - 3y + 4$

**2.** $f(x, y) = 2xy - 5x^2 - 2y^2 + 4x + 4y - 4$

**3.** $f(x, y) = x^2 + xy + 3x + 2y + 5$

**4.** $f(x, y) = 5xy - 7x^2 + 3x - 6y + 2$

**5.** $f(x, y) = 2xy - x^2 - 2y^2 + 3x + 4$

**6.** $f(x, y) = x^2 - 4xy + y^2 + 6y + 2$

**7.** $f(x, y) = 2x^2 + 3xy + 4y^2 - 5x + 2y$

**8.** $f(x, y) = x^2 - 2xy + 2y^2 - 2x + 2y + 1$

**9.** $f(x, y) = x^2 - y^2 - 2x + 4y + 6$

**10.** $f(x, y) = x^2 + 2xy$

**11.** $f(x, y) = \sqrt{56x^2 - 8y^2 - 16x - 31} + 1 - 8x$

**12.** $f(x, y) = 1 - \sqrt[3]{x^2 + y^2}$

**13.** $f(x, y) = x^3 - y^3 - 2xy + 6$

**14.** $f(x, y) = x^3 + 3xy + y^3$

**15.** $f(x, y) = 6x^2 - 2x^3 + 3y^2 + 6xy$

**16.** $f(x, y) = x^3 + y^3 + 3x^2 - 3y^2 - 8$

**17.** $f(x, y) = x^3 + 3xy^2 - 15x + y^3 - 15y$

**18.** $f(x, y) = 2x^3 + 2y^3 - 9x^2 + 3y^2 - 12y$

**19.** $f(x, y) = 4xy - x^4 - y^4$

**20.** $f(x, y) = x^4 + y^4 + 4xy$

**21.** $f(x, y) = \dfrac{1}{x^2 + y^2 - 1}$

**22.** $f(x, y) = \dfrac{1}{x} + xy + \dfrac{1}{y}$

**23.** $f(x, y) = y \sin x$

**24.** $f(x, y) = e^{2x} \cos y$

**25.** $f(x, y) = e^{x^2 + y^2 - 4x}$

**26.** $f(x, y) = e^y - ye^x$

**27.** $f(x, y) = e^{-y}(x^2 + y^2)$

**28.** $f(x, y) = e^x(x^2 - y^2)$

**29.** $f(x, y) = 2 \ln x + \ln y - 4x - y$

**30.** $f(x, y) = \ln (x + y) + x^2 - y$

### Finding Absolute Extrema

In Exercises 31–38, find the absolute maxima and minima of the functions on the given domains.

**31.** $f(x, y) = 2x^2 - 4x + y^2 - 4y + 1$ on the closed triangular plate bounded by the lines $x = 0, y = 2, y = 2x$ in the first quadrant

**32.** $D(x, y) = x^2 - xy + y^2 + 1$ on the closed triangular plate in the first quadrant bounded by the lines $x = 0, y = 4, y = x$

**33.** $f(x, y) = x^2 + y^2$ on the closed triangular plate bounded by the lines $x = 0, y = 0, y + 2x = 2$ in the first quadrant

**34.** $T(x, y) = x^2 + xy + y^2 - 6x$ on the rectangular plate $0 \le x \le 5, -3 \le y \le 3$

**35.** $T(x, y) = x^2 + xy + y^2 - 6x + 2$ on the rectangular plate $0 \le x \le 5, -3 \le y \le 0$

**36.** $f(x, y) = 48xy - 32x^3 - 24y^2$ on the rectangular plate $0 \le x \le 1, 0 \le y \le 1$

**37.** $f(x, y) = (4x - x^2) \cos y$ on the rectangular plate $1 \le x \le 3$, $-\pi/4 \le y \le \pi/4$ (see accompanying figure)

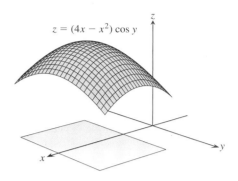

$z = (4x - x^2) \cos y$

**38.** $f(x, y) = 4x - 8xy + 2y + 1$ on the triangular plate bounded by the lines $x = 0, y = 0, x + y = 1$ in the first quadrant

**39.** Find two numbers $a$ and $b$ with $a \le b$ such that

$$\int_a^b (6 - x - x^2) \, dx$$

has its largest value.

**40.** Find two numbers $a$ and $b$ with $a \le b$ such that

$$\int_a^b (24 - 2x - x^2)^{1/3} \, dx$$

has its largest value.

**41. Temperatures** A flat circular plate has the shape of the region $x^2 + y^2 \le 1$. The plate, including the boundary where $x^2 + y^2 = 1$, is heated so that the temperature at the point $(x, y)$ is

$$T(x, y) = x^2 + 2y^2 - x.$$

Find the temperatures at the hottest and coldest points on the plate.

**42.** Find the critical point of

$$f(x, y) = xy + 2x - \ln x^2 y$$

in the open first quadrant ($x > 0, y > 0$) and show that $f$ takes on a minimum there.

### Theory and Examples

**43.** Find the maxima, minima, and saddle points of $f(x, y)$, if any, given that

**a.** $f_x = 2x - 4y$ and $f_y = 2y - 4x$

**b.** $f_x = 2x - 2$ and $f_y = 2y - 4$

**c.** $f_x = 9x^2 - 9$ and $f_y = 2y + 4$

Describe your reasoning in each case.

**44.** The discriminant $f_{xx}f_{yy} - f_{xy}^2$ is zero at the origin for each of the following functions, so the Second Derivative Test fails there. Determine whether the function has a maximum, a minimum, or neither at the origin by imagining what the surface $z = f(x, y)$ looks like. Describe your reasoning in each case.

**a.** $f(x, y) = x^2 y^2$

**b.** $f(x, y) = 1 - x^2 y^2$

**c.** $f(x, y) = xy^2$

**d.** $f(x, y) = x^3 y^2$

**e.** $f(x, y) = x^3 y^3$

**f.** $f(x, y) = x^4 y^4$

**45.** Show that $(0, 0)$ is a critical point of $f(x, y) = x^2 + kxy + y^2$ no matter what value the constant $k$ has. (*Hint:* Consider two cases: $k = 0$ and $k \neq 0$.)

**46.** For what values of the constant $k$ does the Second Derivative Test guarantee that $f(x, y) = x^2 + kxy + y^2$ will have a saddle point at $(0, 0)$? A local minimum at $(0, 0)$? For what values of $k$ is the Second Derivative Test inconclusive? Give reasons for your answers.

**47.** If $f_x(a, b) = f_y(a, b) = 0$, must $f$ have a local maximum or minimum value at $(a, b)$? Give reasons for your answer.

**48.** Can you conclude anything about $f(a, b)$ if $f$ and its first and second partial derivatives are continuous throughout a disk centered at the critical point $(a, b)$ and $f_{xx}(a, b)$ and $f_{yy}(a, b)$ differ in sign? Give reasons for your answer.

**49.** Among all the points on the graph of $z = 10 - x^2 - y^2$ that lie above the plane $x + 2y + 3z = 0$, find the point farthest from the plane.

**50.** Find the point on the graph of $z = x^2 + y^2 + 10$ nearest the plane $x + 2y - z = 0$.

**51.** Find the point on the plane $3x + 2y + z = 6$ that is nearest the origin.

**52.** Find the minimum distance from the point $(2, -1, 1)$ to the plane $x + y - z = 2$.

**53.** Find three numbers whose sum is 9 and whose sum of squares is a minimum.

**54.** Find three positive numbers whose sum is 3 and whose product is a maximum.

**55.** Find the maximum value of $s = xy + yz + xz$ where $x + y + z = 6$.

**56.** Find the minimum distance from the cone $z = \sqrt{x^2 + y^2}$ to the point $(-6, 4, 0)$.

**57.** Find the dimensions of the rectangular box of maximum volume that can be inscribed inside the sphere $x^2 + y^2 + z^2 = 4$.

**58.** Among all closed rectangular boxes of volume $27 \text{ cm}^3$, what is the smallest surface area?

**59.** You are to construct an open rectangular box from $12 \text{ m}^2$ of material. What dimensions will result in a box of maximum volume?

**60.** Consider the function $f(x, y) = x^2 + y^2 + 2xy - x - y + 1$ over the square $0 \leq x \leq 1$ and $0 \leq y \leq 1$.

   **a.** Show that $f$ has an absolute minimum along the line segment $2x + 2y = 1$ in this square. What *is* the absolute minimum value?

   **b.** Find the absolute maximum value of $f$ over the square.

**Extreme Values on Parametrized Curves** To find the extreme values of a function $f(x, y)$ on a curve $x = x(t)$, $y = y(t)$, we treat $f$ as a function of the single variable $t$ and use the Chain Rule to find where $df/dt$ is zero. As in any other single-variable case, the extreme values of $f$ are then found among the values at the

   **a.** critical points (points where $df/dt$ is zero or fails to exist), and

   **b.** endpoints of the parameter domain.

Find the absolute maximum and minimum values of the following functions on the given curves.

**61.** Functions:

   **a.** $f(x, y) = x + y$   **b.** $g(x, y) = xy$   **c.** $h(x, y) = 2x^2 + y^2$

Curves:

   **i)** The semicircle $x^2 + y^2 = 4$,   $y \geq 0$

   **ii)** The quarter circle $x^2 + y^2 = 4$,   $x \geq 0$,   $y \geq 0$

Use the parametric equations $x = 2 \cos t, y = 2 \sin t$.

**62.** Functions:

   **a.** $f(x, y) = 2x + 3y$

   **b.** $g(x, y) = xy$

   **c.** $h(x, y) = x^2 + 3y^2$

Curves:

   **i)** The semiellipse $(x^2/9) + (y^2/4) = 1$,   $y \geq 0$

   **ii)** The quarter ellipse $(x^2/9) + (y^2/4) = 1$,   $x \geq 0$,   $y \geq 0$

Use the parametric equations $x = 3 \cos t, y = 2 \sin t$.

**63.** Function: $f(x, y) = xy$

Curves:

   **i)** The line $x = 2t$,   $y = t + 1$

   **ii)** The line segment $x = 2t$,   $y = t + 1$,   $-1 \leq t \leq 0$

   **iii)** The line segment $x = 2t$,   $y = t + 1$,   $0 \leq t \leq 1$

**64.** Functions:

   **a.** $f(x, y) = x^2 + y^2$

   **b.** $g(x, y) = 1/(x^2 + y^2)$

Curves:

   **i)** The line $x = t$,   $y = 2 - 2t$

   **ii)** The line segment $x = t$,   $y = 2 - 2t$,   $0 \leq t \leq 1$

**65. Least squares and regression lines** When we try to fit a line $y = mx + b$ to a set of numerical data points $(x_1, y_1)$, $(x_2, y_2), \ldots, (x_n, y_n)$, we usually choose the line that minimizes the sum of the squares of the vertical distances from the points to the line. In theory, this means finding the values of $m$ and $b$ that minimize the value of the function

$$w = (mx_1 + b - y_1)^2 + \cdots + (mx_n + b - y_n)^2. \quad (1)$$

(See the accompanying figure.) Show that the values of $m$ and $b$ that do this are

$$m = \frac{\left(\sum x_k\right)\left(\sum y_k\right) - n\sum x_k y_k}{\left(\sum x_k\right)^2 - n\sum x_k^2}, \quad (2)$$

$$b = \frac{1}{n}\left(\sum y_k - m\sum x_k\right), \quad (3)$$

with all sums running from $k = 1$ to $k = n$. Many scientific calculators have these formulas built in, enabling you to find $m$ and $b$ with only a few keystrokes after you have entered the data.

The line $y = mx + b$ determined by these values of $m$ and $b$ is called the **least squares line**, **regression line**, or **trend line** for the data under study. Finding a least squares line lets you

**1.** summarize data with a simple expression,

**2.** predict values of $y$ for other, experimentally untried values of $x$,

**3.** handle data analytically.

We demonstrated these ideas with a variety of applications in Section 1.4.

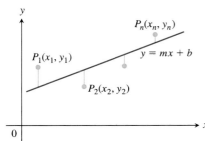

In Exercises 66–68, use Equations (2) and (3) to find the least squares line for each set of data points. Then use the linear equation you obtain to predict the value of $y$ that would correspond to $x = 4$.

**66.** $(-2, 0)$, $(0, 2)$, $(2, 3)$     **67.** $(-1, 2)$, $(0, 1)$, $(3, -4)$

**68.** $(0, 0)$, $(1, 2)$, $(2, 3)$

**COMPUTER EXPLORATIONS**

In Exercises 69–74, you will explore functions to identify their local extrema. Use a CAS to perform the following steps:

**a.** Plot the function over the given rectangle.

**b.** Plot some level curves in the rectangle.

**c.** Calculate the function's first partial derivatives and use the CAS equation solver to find the critical points. How do the critical points relate to the level curves plotted in part (b)? Which critical points, if any, appear to give a saddle point? Give reasons for your answer.

**d.** Calculate the function's second partial derivatives and find the discriminant $f_{xx}f_{yy} - f_{xy}^2$.

**e.** Using the max-min tests, classify the critical points found in part (c). Are your findings consistent with your discussion in part (c)?

**69.** $f(x, y) = x^2 + y^3 - 3xy$,  $-5 \le x \le 5$,  $-5 \le y \le 5$

**70.** $f(x, y) = x^3 - 3xy^2 + y^2$,  $-2 \le x \le 2$,  $-2 \le y \le 2$

**71.** $f(x, y) = x^4 + y^2 - 8x^2 - 6y + 16$,  $-3 \le x \le 3$,  $-6 \le y \le 6$

**72.** $f(x, y) = 2x^4 + y^4 - 2x^2 - 2y^2 + 3$,  $-3/2 \le x \le 3/2$,  $-3/2 \le y \le 3/2$

**73.** $f(x, y) = 5x^6 + 18x^5 - 30x^4 + 30xy^2 - 120x^3$,  $-4 \le x \le 3$,  $-2 \le y \le 2$

**74.** $f(x, y) = \begin{cases} x^5 \ln (x^2 + y^2), & (x, y) \ne (0, 0) \\ 0, & (x, y) = (0, 0) \end{cases}$,  $-2 \le x \le 2$,  $-2 \le y \le 2$

# $14.8$ Lagrange Multipliers

**HISTORICAL BIOGRAPHY**

Joseph Louis Lagrange
(1736–1813)

Sometimes we need to find the extreme values of a function whose domain is constrained to lie within some particular subset of the plane—for example, a disk, a closed triangular region, or along a curve. We saw an instance of this situation in Example 6 of the previous section. Here we explore a powerful method for finding extreme values of constrained functions: the method of *Lagrange multipliers*.

## Constrained Maxima and Minima

To gain some insight, we first consider a problem where a constrained minimum can be found by eliminating a variable.

**EXAMPLE 1**    Find the point $p(x, y, z)$ on the plane $2x + y - z - 5 = 0$ that is closest to the origin.

**Solution**    The problem asks us to find the minimum value of the function

$$|\overrightarrow{OP}| = \sqrt{(x - 0)^2 + (y - 0)^2 + (z - 0)^2}$$
$$= \sqrt{x^2 + y^2 + z^2}$$

subject to the constraint that

$$2x + y - z - 5 = 0.$$

Since $|\overrightarrow{OP}|$ has a minimum value wherever the function

$$f(x, y, z) = x^2 + y^2 + z^2$$

has a minimum value, we may solve the problem by finding the minimum value of $f(x, y, z)$ subject to the constraint $2x + y - z - 5 = 0$ (thus avoiding square roots). If we regard $x$ and $y$ as the independent variables in this equation and write $z$ as

$$z = 2x + y - 5,$$

our problem reduces to one of finding the points $(x, y)$ at which the function

$$h(x, y) = f(x, y, 2x + y - 5) = x^2 + y^2 + (2x + y - 5)^2$$

has its minimum value or values. Since the domain of $h$ is the entire $xy$-plane, the First Derivative Test of Section 14.7 tells us that any minima that $h$ might have must occur at points where

$$h_x = 2x + 2(2x + y - 5)(2) = 0, \qquad h_y = 2y + 2(2x + y - 5) = 0.$$

This leads to

$$10x + 4y = 20, \qquad 4x + 4y = 10,$$

and the solution

$$x = \frac{5}{3}, \qquad y = \frac{5}{6}.$$

We may apply a geometric argument together with the Second Derivative Test to show that these values minimize $h$. The $z$-coordinate of the corresponding point on the plane $z = 2x + y - 5$ is

$$z = 2\left(\frac{5}{3}\right) + \frac{5}{6} - 5 = -\frac{5}{6}.$$

Therefore, the point we seek is

Closest point: $\qquad P\left(\frac{5}{3}, \frac{5}{6}, -\frac{5}{6}\right).$

The distance from $P$ to the origin is $5/\sqrt{6} \approx 2.04$. ∎

Attempts to solve a constrained maximum or minimum problem by substitution, as we might call the method of Example 1, do not always go smoothly. This is one of the reasons for learning the new method of this section.

**EXAMPLE 2**    Find the points on the hyperbolic cylinder $x^2 - z^2 - 1 = 0$ that are closest to the origin.

**Solution 1**    The cylinder is shown in Figure 14.52. We seek the points on the cylinder closest to the origin. These are the points whose coordinates minimize the value of the function

$$f(x, y, z) = x^2 + y^2 + z^2 \qquad \text{Square of the distance}$$

subject to the constraint that $x^2 - z^2 - 1 = 0$. If we regard $x$ and $y$ as independent variables in the constraint equation, then

$$z^2 = x^2 - 1$$

and the values of $f(x, y, z) = x^2 + y^2 + z^2$ on the cylinder are given by the function

$$h(x, y) = x^2 + y^2 + (x^2 - 1) = 2x^2 + y^2 - 1.$$

To find the points on the cylinder whose coordinates minimize $f$, we look for the points in the $xy$-plane whose coordinates minimize $h$. The only extreme value of $h$ occurs where

$$h_x = 4x = 0 \qquad \text{and} \qquad h_y = 2y = 0,$$

that is, at the point $(0, 0)$. But there are no points on the cylinder where both $x$ and $y$ are zero. What went wrong?

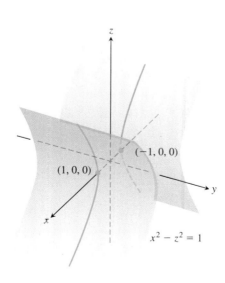

**FIGURE 14.52**    The hyperbolic cylinder $x^2 - z^2 - 1 = 0$ in Example 2.

$(-1, 0, 0)$

$(1, 0, 0)$

$x^2 - z^2 = 1$

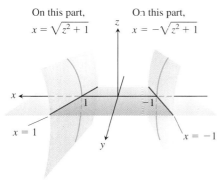

**The hyperbolic cylinder** $x^2 - z^2 = 1$

On this part,
$x = \sqrt{z^2 + 1}$

On this part,
$x = -\sqrt{z^2 + 1}$

$x = 1$

$x = -1$

**FIGURE 14.53** The region in the $xy$-plane from which the first two coordinates of the points $(x, y, z)$ on the hyperbolic cylinder $x^2 - z^2 = 1$ are selected excludes the band $-1 < x < 1$ in the $xy$-plane (Example 2).

What happened was that the First Derivative Test found (as it should have) the point *in the domain of h* where $h$ has a minimum value. We, on the other hand, want the points *on the cylinder* where $h$ has a minimum value. Although the domain of $h$ is the entire $xy$-plane, the domain from which we can select the first two coordinates of the points $(x, y, z)$ on the cylinder is restricted to the "shadow" of the cylinder on the $xy$-plane; it does not include the band between the lines $x = -1$ and $x = 1$ (Figure 14.53).

We can avoid this problem if we treat $y$ and $z$ as independent variables (instead of $x$ and $y$) and express $x$ in terms of $y$ and $z$ as

$$x^2 = z^2 + 1.$$

With this substitution, $f(x, y, z) = x^2 + y^2 + z^2$ becomes

$$k(y, z) = (z^2 + 1) + y^2 + z^2 = 1 + y^2 + 2z^2$$

and we look for the points where $k$ takes on its smallest value. The domain of $k$ in the $yz$-plane now matches the domain from which we select the $y$- and $z$-coordinates of the points $(x, y, z)$ on the cylinder. Hence, the points that minimize $k$ in the plane will have corresponding points on the cylinder. The smallest values of $k$ occur where

$$k_y = 2y = 0 \qquad \text{and} \qquad k_z = 4z = 0,$$

or where $y = z = 0$. This leads to

$$x^2 = z^2 + 1 = 1, \qquad x = \pm 1.$$

The corresponding points on the cylinder are $(\pm 1, 0, 0)$. We can see from the inequality

$$k(y, z) = 1 + y^2 + 2z^2 \geq 1$$

that the points $(\pm 1, 0, 0)$ give a minimum value for $k$. We can also see that the minimum distance from the origin to a point on the cylinder is 1 unit.

**Solution 2** Another way to find the points on the cylinder closest to the origin is to imagine a small sphere centered at the origin expanding like a soap bubble until it just touches the cylinder (Figure 14.54). At each point of contact, the cylinder and sphere have the same tangent plane and normal line. Therefore, if the sphere and cylinder are represented as the level surfaces obtained by setting

$$f(x, y, z) = x^2 + y^2 + z^2 - a^2 \qquad \text{and} \qquad g(x, y, z) = x^2 - z^2 - 1$$

equal to 0, then the gradients $\nabla f$ and $\nabla g$ will be parallel where the surfaces touch. At any point of contact, we should therefore be able to find a scalar $\lambda$ ("lambda") such that

$$\nabla f = \lambda \nabla g,$$

or

$$2x\mathbf{i} + 2y\mathbf{j} + 2z\mathbf{k} = \lambda(2x\mathbf{i} - 2z\mathbf{k}).$$

Thus, the coordinates $x$, $y$, and $z$ of any point of tangency will have to satisfy the three scalar equations

$$2x = 2\lambda x, \qquad 2y = 0, \qquad 2z = -2\lambda z.$$

For what values of $\lambda$ will a point $(x, y, z)$ whose coordinates satisfy these scalar equations also lie on the surface $x^2 - z^2 - 1 = 0$? To answer this question, we use our knowledge that no point on the surface has a zero $x$-coordinate to conclude that $x \neq 0$. Hence, $2x = 2\lambda x$ only if

$$2 = 2\lambda, \qquad \text{or} \qquad \lambda = 1.$$

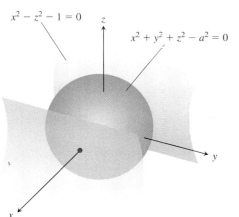

$x^2 - z^2 - 1 = 0$

$x^2 + y^2 + z^2 - a^2 = 0$

**FIGURE 14.54** A sphere expanding like a soap bubble centered at the origin until it just touches the hyperbolic cylinder $x^2 - z^2 - 1 = 0$ (Example 2).

For $\lambda = 1$, the equation $2z = -2\lambda z$ becomes $2z = -2z$. If this equation is to be satisfied as well, $z$ must be zero. Since $y = 0$ also (from the equation $2y = 0$), we conclude that the points we seek all have coordinates of the form

$$(x, 0, 0).$$

What points on the surface $x^2 - z^2 = 1$ have coordinates of this form? The answer is the points $(x, 0, 0)$ for which

$$x^2 - (0)^2 = 1, \qquad x^2 = 1, \qquad \text{or} \qquad x = \pm 1.$$

The points on the cylinder closest to the origin are the points $(\pm 1, 0, 0)$. ∎

## The Method of Lagrange Multipliers

In Solution 2 of Example 2, we used the **method of Lagrange multipliers**. The method says that the local extreme values of a function $f(x, y, z)$ whose variables are subject to a constraint $g(x, y, z) = 0$ are to be found on the surface $g = 0$ among the points where

$$\nabla f = \lambda \nabla g$$

for some scalar $\lambda$ (called a **Lagrange multiplier**).

To explore the method further and see why it works, we first make the following observation, which we state as a theorem.

---

**THEOREM 12—The Orthogonal Gradient Theorem** Suppose that $f(x, y, z)$ is differentiable in a region whose interior contains a smooth curve

$$C: \quad \mathbf{r}(t) = x(t)\mathbf{i} + y(t)\mathbf{j} + z(t)\mathbf{k}.$$

If $P_0$ is a point on $C$ where $f$ has a local maximum or minimum relative to its values on $C$, then $\nabla f$ is orthogonal to $C$ at $P_0$.

---

**Proof** We show that $\nabla f$ is orthogonal to the curve's tangent vector $\mathbf{r}'$ at $P_0$. The values of $f$ on $C$ are given by the composite $f(x(t), y(t), z(t))$, whose derivative with respect to $t$ is

$$\frac{df}{dt} = \frac{\partial f}{\partial x}\frac{dx}{dt} + \frac{\partial f}{\partial y}\frac{dy}{dt} + \frac{\partial f}{\partial z}\frac{dz}{dt} = \nabla f \cdot \mathbf{r}'.$$

At any point $P_0$ where $f$ has a local maximum or minimum relative to its values on the curve, $df/dt = 0$, so

$$\nabla f \cdot \mathbf{r}' = 0. \qquad ∎$$

By dropping the $z$-terms in Theorem 12, we obtain a similar result for functions of two variables.

---

**COROLLARY** At the points on a smooth curve $\mathbf{r}(t) = x(t)\mathbf{i} + y(t)\mathbf{j}$ where a differentiable function $f(x, y)$ takes on its local maxima and minima relative to its values on the curve, $\nabla f \cdot \mathbf{r}' = 0$.

---

Theorem 12 is the key to the method of Lagrange multipliers. Suppose that $f(x, y, z)$ and $g(x, y, z)$ are differentiable and that $P_0$ is a point on the surface $g(x, y, z) = 0$ where $f$ has a local maximum or minimum value relative to its other values on the surface. We assume also that $\nabla g \neq \mathbf{0}$ at points on the surface $g(x, y, z) = 0$. Then $f$ takes on a local maximum or minimum at $P_0$ relative to its values on every differentiable curve through $P_0$

on the surface $g(x, y, z) = 0$. Therefore, $\nabla f$ is orthogonal to the tangent vector of every such differentiable curve through $P_0$. So is $\nabla g$, moreover (because $\nabla g$ is orthogonal to the level surface $g = 0$, as we saw in Section 14.5). Therefore, at $P_0$, $\nabla f$ is some scalar multiple $\lambda$ of $\nabla g$.

---

**The Method of Lagrange Multipliers**

Suppose that $f(x, y, z)$ and $g(x, y, z)$ are differentiable and $\nabla g \neq \mathbf{0}$ when $g(x, y, z) = 0$. To find the local maximum and minimum values of $f$ subject to the constraint $g(x, y, z) = 0$ (if these exist), find the values of $x, y, z$, and $\lambda$ that simultaneously satisfy the equations

$$\nabla f = \lambda \nabla g \qquad \text{and} \qquad g(x, y, z) = 0. \qquad (1)$$

For functions of two independent variables, the condition is similar, but without the variable $z$.

---

Some care must be used in applying this method. An extreme value may not actually exist (Exercise 45).

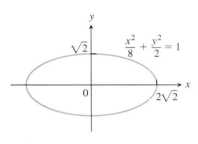

**FIGURE 14.55** Example 3 shows how to find the largest and smallest values of the product $xy$ on this ellipse.

**EXAMPLE 3**   Find the greatest and smallest values that the function

$$f(x, y) = xy$$

takes on the ellipse (Figure 14.55)

$$\frac{x^2}{8} + \frac{y^2}{2} = 1.$$

**Solution**   We want to find the extreme values of $f(x, y) = xy$ subject to the constraint

$$g(x, y) = \frac{x^2}{8} + \frac{y^2}{2} - 1 = 0.$$

To do so, we first find the values of $x, y$, and $\lambda$ for which

$$\nabla f = \lambda \nabla g \qquad \text{and} \qquad g(x, y) = 0.$$

The gradient equation in Equations (1) gives

$$y\mathbf{i} + x\mathbf{j} = \frac{\lambda}{4}x\mathbf{i} + \lambda y\mathbf{j},$$

from which we find

$$y = \frac{\lambda}{4}x, \qquad x = \lambda y, \qquad \text{and} \qquad y = \frac{\lambda}{4}(\lambda y) = \frac{\lambda^2}{4}y,$$

so that $y = 0$ or $\lambda = \pm 2$. We now consider these two cases.

**Case 1:**   If $y = 0$, then $x = y = 0$. But $(0, 0)$ is not on the ellipse. Hence, $y \neq 0$.
**Case 2:**   If $y \neq 0$, then $\lambda = \pm 2$ and $x = \pm 2y$. Substituting this in the equation $g(x, y) = 0$ gives

$$\frac{(\pm 2y)^2}{8} + \frac{y^2}{2} = 1, \qquad 4y^2 + 4y^2 = 8 \qquad \text{and} \qquad y = \pm 1.$$

The function $f(x, y) = xy$ therefore takes on its extreme values on the ellipse at the four points $(\pm 2, 1)$, $(\pm 2, -1)$. The extreme values are $xy = 2$ and $xy = -2$.

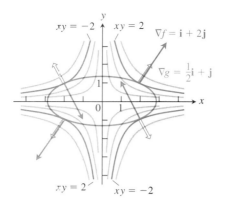

**FIGURE 14.56** When subjected to the constraint $g(x, y) = x^2/8 + y^2/2 - 1 = 0$, the function $f(x, y) = xy$ takes on extreme values at the four points $(\pm 2, \pm 1)$. These are the points on the ellipse when $\nabla f$ (red) is a scalar multiple of $\nabla g$ (blue) (Example 3).

**The Geometry of the Solution** The level curves of the function $f(x, y) = xy$ are the hyperbolas $xy = c$ (Figure 14.56). The farther the hyperbolas lie from the origin, the larger the absolute value of $f$. We want to find the extreme values of $f(x, y)$, given that the point $(x, y)$ also lies on the ellipse $x^2 + 4y^2 = 8$. Which hyperbolas intersecting the ellipse lie farthest from the origin? The hyperbolas that just graze the ellipse, the ones that are tangent to it, are farthest. At these points, any vector normal to the hyperbola is normal to the ellipse, so $\nabla f = y\mathbf{i} + x\mathbf{j}$ is a multiple $(\lambda = \pm 2)$ of $\nabla g = (x/4)\mathbf{i} + y\mathbf{j}$. At the point $(2, 1)$, for example,

$$\nabla f = \mathbf{i} + 2\mathbf{j}, \qquad \nabla g = \frac{1}{2}\mathbf{i} + \mathbf{j}, \qquad \text{and} \qquad \nabla f = 2\nabla g.$$

At the point $(-2, 1)$,

$$\nabla f = \mathbf{i} - 2\mathbf{j}, \qquad \nabla g = -\frac{1}{2}\mathbf{i} + \mathbf{j}, \qquad \text{and} \qquad \nabla f = -2\nabla g. \qquad \blacksquare$$

**EXAMPLE 4** Find the maximum and minimum values of the function $f(x, y) = 3x + 4y$ on the circle $x^2 + y^2 = 1$.

**Solution** We model this as a Lagrange multiplier problem with

$$f(x, y) = 3x + 4y, \qquad g(x, y) = x^2 + y^2 - 1$$

and look for the values of $x$, $y$, and $\lambda$ that satisfy the equations

$$\nabla f = \lambda \nabla g: \quad 3\mathbf{i} + 4\mathbf{j} = 2x\lambda\mathbf{i} + 2y\lambda\mathbf{j}$$
$$g(x, y) = 0: \quad x^2 + y^2 - 1 = 0.$$

The gradient equation in Equations (1) implies that $\lambda \neq 0$ and gives

$$x = \frac{3}{2\lambda}, \qquad y = \frac{2}{\lambda}.$$

These equations tell us, among other things, that $x$ and $y$ have the same sign. With these values for $x$ and $y$, the equation $g(x, y) = 0$ gives

$$\left(\frac{3}{2\lambda}\right)^2 + \left(\frac{2}{\lambda}\right)^2 - 1 = 0,$$

so

$$\frac{9}{4\lambda^2} + \frac{4}{\lambda^2} = 1, \qquad 9 + 16 = 4\lambda^2, \qquad 4\lambda^2 = 25, \qquad \text{and} \qquad \lambda = \pm\frac{5}{2}.$$

Thus,

$$x = \frac{3}{2\lambda} = \pm\frac{3}{5}, \qquad y = \frac{2}{\lambda} = \pm\frac{4}{5},$$

and $f(x, y) = 3x + 4y$ has extreme values at $(x, y) = \pm(3/5, 4/5)$.

By calculating the value of $3x + 4y$ at the points $\pm(3/5, 4/5)$, we see that its maximum and minimum values on the circle $x^2 + y^2 = 1$ are

$$3\left(\frac{3}{5}\right) + 4\left(\frac{4}{5}\right) = \frac{25}{5} = 5 \qquad \text{and} \qquad 3\left(-\frac{3}{5}\right) + 4\left(-\frac{4}{5}\right) = -\frac{25}{5} = -5.$$

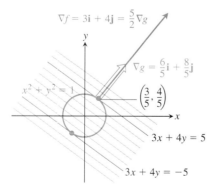

**FIGURE 14.57** The function $f(x, y) = 3x + 4y$ takes on its largest value on the unit circle $g(x, y) = x^2 + y^2 - 1 = 0$ at the point $(3/5, 4/5)$ and its smallest value at the point $(-3/5, -4/5)$ (Example 4). At each of these points, $\nabla f$ is a scalar multiple of $\nabla g$. The figure shows the gradients at the first point but not the second.

**The Geometry of the Solution** The level curves of $f(x, y) = 3x + 4y$ are the lines $3x + 4y = c$ (Figure 14.57). The farther the lines lie from the origin, the larger the absolute value of $f$. We want to find the extreme values of $f(x, y)$ given that the point $(x, y)$ also lies on the circle $x^2 + y^2 = 1$. Which lines intersecting the circle lie farthest from the origin? The lines tangent to the circle are farthest. At the points of tangency, any vector normal to the line is normal to the circle, so the gradient $\nabla f = 3\mathbf{i} + 4\mathbf{j}$ is a multiple $(\lambda = \pm 5/2)$ of the gradient $\nabla g = 2x\mathbf{i} + 2y\mathbf{j}$. At the point $(3/5, 4/5)$, for example,

$$\nabla f = 3\mathbf{i} + 4\mathbf{j}, \qquad \nabla g = \frac{6}{5}\mathbf{i} + \frac{8}{5}\mathbf{j}, \qquad \text{and} \qquad \nabla f = \frac{5}{2}\nabla g. \qquad \blacksquare$$

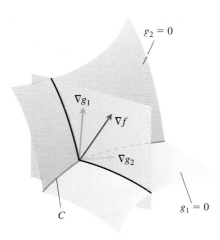

**FIGURE 14.58** The vectors $\nabla g_1$ and $\nabla g_2$ lie in a plane perpendicular to the curve $C$ because $\nabla g_1$ is normal to the surface $g_1 = 0$ and $\nabla g_2$ is normal to the surface $g_2 = 0$.

## Lagrange Multipliers with Two Constraints

Many problems require us to find the extreme values of a differentiable function $f(x, y, z)$ whose variables are subject to two constraints. If the constraints are

$$g_1(x, y, z) = 0 \qquad \text{and} \qquad g_2(x, y, z) = 0$$

and $g_1$ and $g_2$ are differentiable, with $\nabla g_1$ not parallel to $\nabla g_2$, we find the constrained local maxima and minima of $f$ by introducing two Lagrange multipliers $\lambda$ and $\mu$ (mu, pronounced "mew"). That is, we locate the points $P(x, y, z)$ where $f$ takes on its constrained extreme values by finding the values of $x, y, z, \lambda$, and $\mu$ that simultaneously satisfy the three equations

$$\nabla f = \lambda \nabla g_1 + \mu \nabla g_2, \qquad g_1(x, y, z) = 0, \qquad g_2(x, y, z) = 0 \qquad (2)$$

Equations (2) have a nice geometric interpretation. The surfaces $g_1 = 0$ and $g_2 = 0$ (usually) intersect in a smooth curve, say $C$ (Figure 14.58). Along this curve we seek the points where $f$ has local maximum and minimum values relative to its other values on the curve. These are the points where $\nabla f$ is normal to $C$, as we saw in Theorem 12. But $\nabla g_1$ and $\nabla g_2$ are also normal to $C$ at these points because $C$ lies in the surfaces $g_1 = 0$ and $g_2 = 0$. Therefore, $\nabla f$ lies in the plane determined by $\nabla g_1$ and $\nabla g_2$, which means that $\nabla f = \lambda \nabla g_1 + \mu \nabla g_2$ for some $\lambda$ and $\mu$. Since the points we seek also lie in both surfaces, their coordinates must satisfy the equations $g_1(x, y, z) = 0$ and $g_2(x, y, z) = 0$, which are the remaining requirements in Equations (2).

**EXAMPLE 5**     The plane $x + y + z = 1$ cuts the cylinder $x^2 + y^2 = 1$ in an ellipse (Figure 14.59). Find the points on the ellipse that lie closest to and farthest from the origin.

**Solution**     We find the extreme values of

$$f(x, y, z) = x^2 + y^2 + z^2$$

(the square of the distance from $(x, y, z)$ to the origin) subject to the constraints

$$g_1(x, y, z) = x^2 + y^2 - 1 = 0 \qquad (3)$$
$$g_2(x, y, z) = x + y + z - 1 = 0. \qquad (4)$$

The gradient equation in Equations (2) then gives

$$\nabla f = \lambda \nabla g_1 + \mu \nabla g_2$$
$$2x\mathbf{i} + 2y\mathbf{j} + 2z\mathbf{k} = \lambda(2x\mathbf{i} + 2y\mathbf{j}) + \mu(\mathbf{i} + \mathbf{j} + \mathbf{k})$$
$$2x\mathbf{i} + 2y\mathbf{j} + 2z\mathbf{k} = (2\lambda x + \mu)\mathbf{i} + (2\lambda y + \mu)\mathbf{j} + \mu\mathbf{k}$$

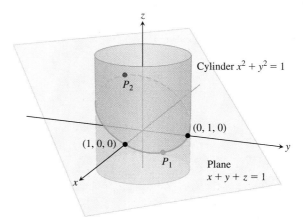

**FIGURE 14.59** On the ellipse where the plane and cylinder meet, we find the points closest to and farthest from the origin (Example 5).

or

$$2x = 2\lambda x + \mu, \qquad 2y = 2\lambda y + \mu, \qquad 2z = \mu. \qquad (5)$$

The scalar equations in Equations (5) yield

$$2x = 2\lambda x + 2z \Longrightarrow (1 - \lambda)x = z,$$
$$2y = 2\lambda y + 2z \Longrightarrow (1 - \lambda)y = z. \qquad (6)$$

Equations (6) are satisfied simultaneously if either $\lambda = 1$ and $z = 0$ or $\lambda \neq 1$ and $x = y = z/(1 - \lambda)$.

If $z = 0$, then solving Equations (3) and (4) simultaneously to find the corresponding points on the ellipse gives the two points $(1, 0, 0)$ and $(0, 1, 0)$. This makes sense when you look at Figure 14.59.

If $x = y$, then Equations (3) and (4) give

$$x^2 + x^2 - 1 = 0 \qquad\qquad x + x + z - 1 = 0$$
$$2x^2 = 1 \qquad\qquad z = 1 - 2x$$
$$x = \pm\frac{\sqrt{2}}{2} \qquad\qquad z = 1 \mp \sqrt{2}.$$

The corresponding points on the ellipse are

$$P_1 = \left(\frac{\sqrt{2}}{2}, \frac{\sqrt{2}}{2}, 1 - \sqrt{2}\right) \qquad \text{and} \qquad P_2 = \left(-\frac{\sqrt{2}}{2}, -\frac{\sqrt{2}}{2}, 1 + \sqrt{2}\right).$$

Here we need to be careful, however. Although $P_1$ and $P_2$ both give local maxima of $f$ on the ellipse, $P_2$ is farther from the origin than $P_1$.

The points on the ellipse closest to the origin are $(1, 0, 0)$ and $(0, 1, 0)$. The point on the ellipse farthest from the origin is $P_2$. (See Figure 14.59.) ∎

# Exercises 14.8

## Two Independent Variables with One Constraint

1. **Extrema on an ellipse** Find the points on the ellipse $x^2 + 2y^2 = 1$ where $f(x, y) = xy$ has its extreme values.

2. **Extrema on a circle** Find the extreme values of $f(x, y) = xy$ subject to the constraint $g(x, y) = x^2 + y^2 - 10 = 0$.

3. **Maximum on a line** Find the maximum value of $f(x, y) = 49 - x^2 - y^2$ on the line $x + 3y = 10$.

4. **Extrema on a line** Find the local extreme values of $f(x, y) = x^2y$ on the line $x + y = 3$.

5. **Constrained minimum** Find the points on the curve $xy^2 = 54$ nearest the origin.

6. **Constrained minimum** Find the points on the curve $x^2y = 2$ nearest the origin.

7. Use the method of Lagrange multipliers to find

   a. **Minimum on a hyperbola** The minimum value of $x + y$, subject to the constraints $xy = 16, x > 0, y > 0$

   b. **Maximum on a line** The maximum value of $xy$, subject to the constraint $x + y = 16$.

   Comment on the geometry of each solution.

8. **Extrema on a curve** Find the points on the curve $x^2 + xy + y^2 = 1$ in the $xy$-plane that are nearest to and farthest from the origin.

9. **Minimum surface area with fixed volume** Find the dimensions of the closed right circular cylindrical can of smallest surface area whose volume is $16\pi$ cm$^3$.

10. **Cylinder in a sphere** Find the radius and height of the open right circular cylinder of largest surface area that can be inscribed in a sphere of radius $a$. What *is* the largest surface area?

11. **Rectangle of greatest area in an ellipse** Use the method of Lagrange multipliers to find the dimensions of the rectangle of greatest area that can be inscribed in the ellipse $x^2/16 + y^2/9 = 1$ with sides parallel to the coordinate axes.

12. **Rectangle of longest perimeter in an ellipse** Find the dimensions of the rectangle of largest perimeter that can be inscribed in the ellipse $x^2/a^2 + y^2/b^2 = 1$ with sides parallel to the coordinate axes. What *is* the largest perimeter?

13. **Extrema on a circle** Find the maximum and minimum values of $x^2 + y^2$ subject to the constraint $x^2 - 2x + y^2 - 4y = 0$.

14. **Extrema on a circle** Find the maximum and minimum values of $3x - y + 6$ subject to the constraint $x^2 + y^2 = 4$.

15. **Ant on a metal plate** The temperature at a point $(x, y)$ on a metal plate is $T(x, y) = 4x^2 - 4xy + y^2$. An ant on the plate walks around the circle of radius 5 centered at the origin. What are the highest and lowest temperatures encountered by the ant?

16. **Cheapest storage tank** Your firm has been asked to design a storage tank for liquid petroleum gas. The customer's specifications call for a cylindrical tank with hemispherical ends, and the tank is to hold 8000 m$^3$ of gas. The customer also wants to use the smallest amount of material possible in building the tank. What radius and height do you recommend for the cylindrical portion of the tank?

### Three Independent Variables with One Constraint

17. **Minimum distance to a point** Find the point on the plane $x + 2y + 3z = 13$ closest to the point $(1, 1, 1)$.

18. **Maximum distance to a point** Find the point on the sphere $x^2 + y^2 + z^2 = 4$ farthest from the point $(1, -1, 1)$.

19. **Minimum distance to the origin** Find the minimum distance from the surface $x^2 - y^2 - z^2 = 1$ to the origin.

20. **Minimum distance to the origin** Find the point on the surface $z = xy + 1$ nearest the origin.

21. **Minimum distance to the origin** Find the points on the surface $z^2 = xy + 4$ closest to the origin.

22. **Minimum distance to the origin** Find the point(s) on the surface $xyz = 1$ closest to the origin.

23. **Extrema on a sphere** Find the maximum and minimum values of

$$f(x, y, z) = x - 2y + 5z$$

on the sphere $x^2 + y^2 + z^2 = 30$.

24. **Extrema on a sphere** Find the points on the sphere $x^2 + y^2 + z^2 = 25$ where $f(x, y, z) = x + 2y + 3z$ has its maximum and minimum values.

25. **Minimizing a sum of squares** Find three real numbers whose sum is 9 and the sum of whose squares is as small as possible.

26. **Maximizing a product** Find the largest product the positive numbers $x$, $y$, and $z$ can have if $x + y + z^2 = 16$.

27. **Rectangular box of largest volume in a sphere** Find the dimensions of the closed rectangular box with maximum volume that can be inscribed in the unit sphere.

28. **Box with vertex on a plane** Find the volume of the largest closed rectangular box in the first octant having three faces in the coordinate planes and a vertex on the plane $x/a + y/b + z/c = 1$, where $a > 0$, $b > 0$, and $c > 0$.

29. **Hottest point on a space probe** A space probe in the shape of the ellipsoid

$$4x^2 + y^2 + 4z^2 = 16$$

enters Earth's atmosphere and its surface begins to heat. After 1 hour, the temperature at the point $(x, y, z)$ on the probe's surface is

$$T(x, y, z) = 3x^2 + 4yz - 16z + 600.$$

Find the hottest point on the probe's surface.

30. **Extreme temperatures on a sphere** Suppose that the Celsius temperature at the point $(x, y, z)$ on the sphere $x^2 + y^2 + z^2 = 1$ is $T = 400xyz^2$. Locate the highest and lowest temperatures on the sphere.

31. **Cobb-Douglas production function** During the 1920s, Charles Cobb and Paul Douglas modeled total production output $P$ (of a firm, industry, or entire economy) as a function of labor hours involved $x$ and capital invested $y$ (which includes the monetary worth of all buildings and equipment). The Cobb-Douglas production function is given by

$$P(x, y) = kx^\alpha y^{1-\alpha},$$

where $k$ and $\alpha$ are constants representative of a particular firm or economy.

a. Show that a doubling of both labor and capital results in a doubling of production $P$.

b. Suppose a particular firm has the production function for $k = 120$ and $\alpha = 3/4$. Assume that each unit of labor costs \$250 and each unit of capital costs \$400, and that the total expenses for all costs cannot exceed \$100,000. Find the maximum production level for the firm.

32. (*Continuation of Exercise 31.*) If the cost of a unit of labor is $c_1$ and the cost of a unit of capital is $c_2$, and if the firm can spend only $B$ dollars as its total budget, then production $P$ is constrained by $c_1 x + c_2 y = B$. Show that the maximum production level subject to the constraint occurs at the point

$$x = \frac{\alpha B}{c_1} \quad \text{and} \quad y = \frac{(1 - \alpha)B}{c_2}.$$

33. **Maximizing a utility function: an example from economics** In economics, the usefulness or *utility* of amounts $x$ and $y$ of two capital goods $G_1$ and $G_2$ is sometimes measured by a function $U(x, y)$. For example, $G_1$ and $G_2$ might be two chemicals a pharmaceutical company needs to have on hand and $U(x, y)$ the gain from manufacturing a product whose synthesis requires different amounts of the chemicals depending on the process used. If $G_1$ costs $a$ dollars per kilogram, $G_2$ costs $b$ dollars per kilogram, and the total amount allocated for the purchase of $G_1$ and $G_2$ together is $c$ dollars, then the company's managers want to maximize $U(x, y)$ given that $ax + by = c$. Thus, they need to solve a typical Lagrange multiplier problem.

   Suppose that

$$U(x, y) = xy + 2x$$

and that the equation $ax + by = c$ simplifies to

$$2x + y = 30.$$

Find the maximum value of $U$ and the corresponding values of $x$ and $y$ subject to this latter constraint.

34. **Blood types** Human blood types are classified by three gene forms A, B, and O. Blood types AA, BB, and OO are *homozygous*, and blood types AB, AO, and BO are *heterozygous*. If $p$, $q$, and $r$ represent the proportions of the three gene forms to the population, respectively, then the *Hardy-Weinberg Law* asserts that the proportion $Q$ of heterozygous persons in any specific population is modeled by

$$Q(p, q, r) = 2(pq + pr + qr),$$

subject to $p + q + r = 1$. Find the maximum value of $Q$.

35. **Length of a beam** In Section 4.5, Exercise 39, we posed a problem of finding the length $L$ of the shortest beam that can reach over a wall of height $h$ to a tall building located $k$ units from the wall. Use Lagrange multipliers to show that

$$L = \left( h^{2/3} + k^{2/3} \right)^{3/2}.$$

36. **Locating a radio telescope** You are in charge of erecting a radio telescope on a newly discovered planet. To minimize interference, you want to place it where the magnetic field of the planet is weakest. The planet is spherical, with a radius of 6 units. Based on a coordinate system whose origin is at the center of the planet, the strength of the magnetic field is given by $M(x, y, z) = 6x - y^2 + xz + 60$. Where should you locate the radio telescope?

**Extreme Values Subject to Two Constraints**

37. Maximize the function $f(x, y, z) = x^2 + 2y - z^2$ subject to the constraints $2x - y = 0$ and $y + z = 0$.

38. Minimize the function $f(x, y, z) = x^2 + y^2 + z^2$ subject to the constraints $x + 2y + 3z = 6$ and $x + 3y + 9z = 9$.

39. **Minimum distance to the origin** Find the point closest to the origin on the line of intersection of the planes $y + 2z = 12$ and $x + y = 6$.

40. **Maximum value on line of intersection** Find the maximum value that $f(x, y, z) = x^2 + 2y - z^2$ can have on the line of intersection of the planes $2x - y = 0$ and $y + z = 0$.

41. **Extrema on a curve of intersection** Find the extreme values of $f(x, y, z) = x^2yz + 1$ on the intersection of the plane $z = 1$ with the sphere $x^2 + y^2 + z^2 = 10$.

42. a. **Maximum on line of intersection** Find the maximum value of $w = xyz$ on the line of intersection of the two planes $x + y + z = 40$ and $x + y - z = 0$.

    b. Give a geometric argument to support your claim that you have found a maximum, and not a minimum, value of $w$.

43. **Extrema on a circle of intersection** Find the extreme values of the function $f(x, y, z) = xy + z^2$ on the circle in which the plane $y - x = 0$ intersects the sphere $x^2 + y^2 + z^2 = 4$.

44. **Minimum distance to the origin** Find the point closest to the origin on the curve of intersection of the plane $2y + 4z = 5$ and the cone $z^2 = 4x^2 + 4y^2$.

**Theory and Examples**

45. **The condition $\nabla f = \lambda \nabla g$ is not sufficient** Although $\nabla f = \lambda \nabla g$ is a necessary condition for the occurrence of an extreme value of $f(x, y)$ subject to the conditions $g(x, y) = 0$ and $\nabla g \neq \mathbf{0}$, it does not in itself guarantee that one exists. As a case in point, try using the method of Lagrange multipliers to find a maximum value of $f(x, y) = x + y$ subject to the constraint that $xy = 16$. The method will identify the two points $(4, 4)$ and $(-4, -4)$ as candidates for the location of extreme values. Yet the sum $(x + y)$ has no maximum value on the hyperbola $xy = 16$. The farther you go from the origin on this hyperbola in the first quadrant, the larger the sum $f(x, y) = x + y$ becomes.

46. **A least squares plane** The plane $z = Ax + By + C$ is to be "fitted" to the following points $(x_k, y_k, z_k)$:

    $$(0, 0, 0), \quad (0, 1, 1), \quad (1, 1, 1), \quad (1, 0, -1).$$

Find the values of $A$, $B$, and $C$ that minimize

$$\sum_{k=1}^{4} (Ax_k + By_k + C - z_k)^2,$$

the sum of the squares of the deviations.

47. a. **Maximum on a sphere** Show that the maximum value of $a^2b^2c^2$ on a sphere of radius $r$ centered at the origin of a Cartesian $abc$-coordinate system is $(r^2/3)^3$.

    b. **Geometric and arithmetic means** Using part (a), show that for nonnegative numbers $a$, $b$, and $c$,

    $$(abc)^{1/3} \leq \frac{a + b + c}{3};$$

    that is, the *geometric mean* of three nonnegative numbers is less than or equal to their *arithmetic mean*.

48. **Sum of products** Let $a_1, a_2, \ldots, a_n$ be $n$ positive numbers. Find the maximum of $\sum_{i=1}^{n} a_i x_i$ subject to the constraint $\sum_{i=1}^{n} x_i^2 = 1$.

**COMPUTER EXPLORATIONS**

In Exercises 49–54, use a CAS to perform the following steps implementing the method of Lagrange multipliers for finding constrained extrema:

a. Form the function $h = f - \lambda_1 g_1 - \lambda_2 g_2$, where $f$ is the function to optimize subject to the constraints $g_1 = 0$ and $g_2 = 0$.

b. Determine all the first partial derivatives of $h$, including the partials with respect to $\lambda_1$ and $\lambda_2$, and set them equal to 0.

c. Solve the system of equations found in part (b) for all the unknowns, including $\lambda_1$ and $\lambda_2$.

d. Evaluate $f$ at each of the solution points found in part (c) and select the extreme value subject to the constraints asked for in the exercise.

49. Minimize $f(x, y, z) = xy + yz$ subject to the constraints $x^2 + y^2 - 2 = 0$ and $x^2 + z^2 - 2 = 0$.

50. Minimize $f(x, y, z) = xyz$ subject to the constraints $x^2 + y^2 - 1 = 0$ and $x - z = 0$.

51. Maximize $f(x, y, z) = x^2 + y^2 + z^2$ subject to the constraints $2y + 4z - 5 = 0$ and $4x^2 + 4y^2 - z^2 = 0$.

52. Minimize $f(x, y, z) = x^2 + y^2 + z^2$ subject to the constraints $x^2 - xy + y^2 - z^2 - 1 = 0$ and $x^2 + y^2 - 1 = 0$.

53. Minimize $f(x, y, z, w) = x^2 + y^2 + z^2 + w^2$ subject to the constraints $2x - y + z - w - 1 = 0$ and $x + y - z + w - 1 = 0$.

54. Determine the distance from the line $y = x + 1$ to the parabola $y^2 = x$. (*Hint:* Let $(x, y)$ be a point on the line and $(w, z)$ a point on the parabola. You want to minimize $(x - w)^2 + (y - z)^2$.)

# $14.9$ Taylor's Formula for Two Variables

In this section we use Taylor's formula to derive the Second Derivative Test for local extreme values (Section 14.7) and the error formula for linearizations of functions of two independent variables (Section 14.6). The use of Taylor's formula in these derivations leads to an extension of the formula that provides polynomial approximations of all orders for functions of two independent variables.

## Derivation of the Second Derivative Test

Let $f(x, y)$ have continuous partial derivatives in an open region $R$ containing a point $P(a, b)$ where $f_x = f_y = 0$ (Figure 14.60). Let $h$ and $k$ be increments small enough to put the

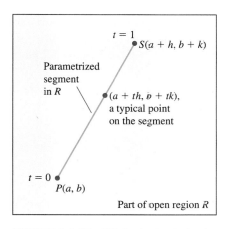

**FIGURE 14.60** We begin the derivation of the Second Derivative Test at $P(a, b)$ by parametrizing a typical line segment from $P$ to a point $S$ nearby.

point $S(a + h, b + k)$ and the line segment joining it to $P$ inside $R$. We parametrize the segment $PS$ as

$$x = a + th, \qquad y = b + tk, \qquad 0 \le t \le 1.$$

If $F(t) = f(a + th, b + tk)$, the Chain Rule gives

$$F'(t) = f_x \frac{dx}{dt} + f_y \frac{dy}{dt} = hf_x + kf_y.$$

Since $f_x$ and $f_y$ are differentiable (they have continuous partial derivatives), $F'$ is a differentiable function of $t$ and

$$F'' = \frac{\partial F'}{\partial x} \frac{dx}{dt} + \frac{\partial F'}{\partial y} \frac{dy}{dt} = \frac{\partial}{\partial x}(hf_x + kf_y) \cdot h + \frac{\partial}{\partial y}(hf_x + kf_y) \cdot k$$

$$= h^2 f_{xx} + 2hkf_{xy} + k^2 f_{yy}. \qquad {\scriptstyle f_{xy} = f_{yx}}$$

Since $F$ and $F'$ are continuous on $[0, 1]$ and $F'$ is differentiable on $(0, 1)$, we can apply Taylor's formula with $n = 2$ and $a = 0$ to obtain

$$F(1) = F(0) + F'(0)(1 - 0) + F''(c) \frac{(1 - 0)^2}{2}$$

$$= F(0) + F'(0) + \frac{1}{2} F''(c) \tag{1}$$

for some $c$ between 0 and 1. Writing Equation (1) in terms of $f$ gives

$$f(a + h, b + k) = f(a, b) + hf_x(a, b) + kf_y(a, b)$$

$$+ \frac{1}{2}\left(h^2 f_{xx} + 2hkf_{xy} + k^2 f_{yy}\right)\bigg|_{(a+ch, b+ck)}. \tag{2}$$

Since $f_x(a, b) = f_y(a, b) = 0$, this reduces to

$$f(a + h, b + k) - f(a, b) = \frac{1}{2}\left(h^2 f_{xx} + 2hkf_{xy} + k^2 f_{yy}\right)\bigg|_{(a+ch, b+ck)}. \tag{3}$$

The presence of an extremum of $f$ at $(a, b)$ is determined by the sign of $f(a + h, b + k) - f(a, b)$. By Equation (3), this is the same as the sign of

$$Q(c) = \left(h^2 f_{xx} + 2hkf_{xy} + k^2 f_{yy}\right)\big|_{(a+ch, b+ck)}.$$

Now, if $Q(0) \ne 0$, the sign of $Q(c)$ will be the same as the sign of $Q(0)$ for sufficiently small values of $h$ and $k$. We can predict the sign of

$$Q(0) = h^2 f_{xx}(a, b) + 2hkf_{xy}(a, b) + k^2 f_{yy}(a, b) \tag{4}$$

from the signs of $f_{xx}$ and $f_{xx}f_{yy} - f_{xy}^2$ at $(a, b)$. Multiply both sides of Equation (4) by $f_{xx}$ and rearrange the right-hand side to get

$$f_{xx}Q(0) = (hf_{xx} + kf_{xy})^2 + (f_{xx}f_{yy} - f_{xy}^2)k^2. \tag{5}$$

From Equation (5) we see that

1. If $f_{xx} < 0$ and $f_{xx}f_{yy} - f_{xy}^2 > 0$ at $(a, b)$, then $Q(0) < 0$ for all sufficiently small nonzero values of $h$ and $k$, and $f$ has a *local maximum* value at $(a, b)$.

2. If $f_{xx} > 0$ and $f_{xx}f_{yy} - f_{xy}^2 > 0$ at $(a, b)$, then $Q(0) > 0$ for all sufficiently small nonzero values of $h$ and $k$, and $f$ has a *local minimum* value at $(a, b)$.

3. If $f_{xx}f_{yy} - f_{xy}^2 < 0$ at $(a, b)$, there are combinations of arbitrarily small nonzero values of $h$ and $k$ for which $Q(0) > 0$, and other values for which $Q(0) < 0$. Arbitrarily close to the point $P_0(a, b, f(a, b))$ on the surface $z = f(x, y)$ there are points above $P_0$ and points below $P_0$, so $f$ has a *saddle point* at $(a, b)$.

4. If $f_{xx}f_{yy} - f_{xy}^2 = 0$, another test is needed. The possibility that $Q(0)$ equals zero prevents us from drawing conclusions about the sign of $Q(c)$.

## The Error Formula for Linear Approximations

We want to show that the difference $E(x, y)$ between the values of a function $f(x, y)$ and its linearization $L(x, y)$ at $(x_0, y_0)$ satisfies the inequality

$$|E(x, y)| \leq \frac{1}{2} M(|x - x_0| + |y - y_0|)^2.$$

The function $f$ is assumed to have continuous second partial derivatives throughout an open set containing a closed rectangular region $R$ centered at $(x_0, y_0)$. The number $M$ is an upper bound for $|f_{xx}|$, $|f_{yy}|$, and $|f_{xy}|$ on $R$.

The inequality we want comes from Equation (2). We substitute $x_0$ and $y_0$ for $a$ and $b$, and $x - x_0$ and $y - y_0$ for $h$ and $k$, respectively, and rearrange the result as

$$f(x, y) = \underbrace{f(x_0, y_0) + f_x(x_0, y_0)(x - x_0) + f_y(x_0, y_0)(y - y_0)}_{\text{linearization } L(x, y)}$$

$$\underbrace{+ \frac{1}{2}\left((x - x_0)^2 f_{xx} + 2(x - x_0)(y - y_0)f_{xy} + (y - y_0)^2 f_{yy}\right)\Big|_{(x_0 + c(x - x_0),\, y_0 + c(y - y_0))}}_{\text{error } E(x, y)}.$$

This equation reveals that

$$|E| \leq \frac{1}{2}\left(|x - x_0|^2 |f_{xx}| + 2|x - x_0||y - y_0||f_{xy}| + |y - y_0|^2 |f_{yy}|\right).$$

Hence, if $M$ is an upper bound for the values of $|f_{xx}|$, $|f_{xy}|$, and $|f_{yy}|$ on $R$,

$$|E| \leq \frac{1}{2}\left(|x - x_0|^2 M + 2|x - x_0||y - y_0|M + |y - y_0|^2 M\right)$$

$$= \frac{1}{2}M(|x - x_0| + |y - y_0|)^2.$$

## Taylor's Formula for Functions of Two Variables

The formulas derived earlier for $F'$ and $F''$ can be obtained by applying to $f(x, y)$ the operators

$$\left(h\frac{\partial}{\partial x} + k\frac{\partial}{\partial y}\right) \quad \text{and} \quad \left(h\frac{\partial}{\partial x} + k\frac{\partial}{\partial y}\right)^2 = h^2\frac{\partial^2}{\partial x^2} + 2hk\frac{\partial^2}{\partial x\,\partial y} + k^2\frac{\partial^2}{\partial y^2}.$$

These are the first two instances of a more general formula,

$$F^{(n)}(t) = \frac{d^n}{dt^n}F(t) = \left(h\frac{\partial}{\partial x} + k\frac{\partial}{\partial y}\right)^n f(x, y), \tag{6}$$

which says that applying $d^n/dt^n$ to $F(t)$ gives the same result as applying the operator

$$\left(h\frac{\partial}{\partial x} + k\frac{\partial}{\partial y}\right)^n$$

to $f(x, y)$ after expanding it by the Binomial Theorem.

If partial derivatives of $f$ through order $n + 1$ are continuous throughout a rectangular region centered at $(a, b)$, we may extend the Taylor formula for $F(t)$ to

$$F(t) = F(0) + F'(0)t + \frac{F''(0)}{2!}t^2 + \cdots + \frac{F^{(n)}(0)}{n!}t^{(n)} + \text{remainder},$$

and take $t = 1$ to obtain

$$F(1) = F(0) + F'(0) + \frac{F''(0)}{2!} + \cdots + \frac{F^{(n)}(0)}{n!} + \text{remainder}.$$

When we replace the first $n$ derivatives on the right of this last series by their equivalent expressions from Equation (6) evaluated at $t = 0$ and add the appropriate remainder term, we arrive at the following formula.

---

**Taylor's Formula for $f(x, y)$ at the Point $(a, b)$**

Suppose $f(x, y)$ and its partial derivatives through order $n + 1$ are continuous throughout an open rectangular region $R$ centered at a point $(a, b)$. Then, throughout $R$,

$$f(a + h, b + k) = f(a, b) + (hf_x + kf_y)\big|_{(a, b)} + \frac{1}{2!}\left(h^2 f_{xx} + 2hk f_{xy} + k^2 f_{yy}\right)\big|_{(a, b)}$$

$$+ \frac{1}{3!}\left(h^3 f_{xxx} + 3h^2 k f_{xxy} + 3hk^2 f_{xyy} + k^3 f_{yyy}\right)\big|_{(a, b)} + \cdots + \frac{1}{n!}\left(h\frac{\partial}{\partial x} + k\frac{\partial}{\partial y}\right)^n f\bigg|_{(a, b)}$$

$$+ \frac{1}{(n + 1)!}\left(h\frac{\partial}{\partial x} + k\frac{\partial}{\partial y}\right)^{n+1} f\bigg|_{(a+ch, b+ck)}. \tag{7}$$

---

The first $n$ derivative terms are evaluated at $(a, b)$. The last term is evaluated at some point $(a + ch, b + ck)$ on the line segment joining $(a, b)$ and $(a + h, b + k)$.

If $(a, b) = (0, 0)$ and we treat $h$ and $k$ as independent variables (denoting them now by $x$ and $y$), then Equation (7) assumes the following form.

---

**Taylor's Formula for $f(x, y)$ at the Origin**

$$f(x, y) = f(0, 0) + xf_x + yf_y + \frac{1}{2!}\left(x^2 f_{xx} + 2xy f_{xy} + y^2 f_{yy}\right)$$

$$+ \frac{1}{3!}\left(x^3 f_{xxx} + 3x^2 y f_{xxy} + 3xy^2 f_{xyy} + y^3 f_{yyy}\right) + \cdots + \frac{1}{n!}\left(x^n \frac{\partial^n f}{\partial x^n} + nx^{n-1}y\frac{\partial^n f}{\partial x^{n-1}\partial y} + \cdots + y^n\frac{\partial^n f}{\partial y^n}\right)$$

$$+ \frac{1}{(n + 1)!}\left(x^{n+1}\frac{\partial^{n+1} f}{\partial x^{n+1}} + (n + 1)x^n y\frac{\partial^{n+1} f}{\partial x^n \partial y} + \cdots + y^{n+1}\frac{\partial^{n+1} f}{\partial y^{n+1}}\right)\bigg|_{(cx, cy)} \tag{8}$$

---

The first $n$ derivative terms are evaluated at $(0, 0)$. The last term is evaluated at a point on the line segment joining the origin and $(x, y)$.

Taylor's formula provides polynomial approximations of two-variable functions. The first $n$ derivative terms give the polynomial; the last term gives the approximation error. The first three terms of Taylor's formula give the function's linearization. To improve on the linearization, we add higher-power terms.

**EXAMPLE 1**    Find a quadratic approximation to $f(x, y) = \sin x \sin y$ near the origin. How accurate is the approximation if $|x| \le 0.1$ and $|y| \le 0.1$?

**Solution**    We take $n = 2$ in Equation (8):

$$f(x, y) = f(0, 0) + (xf_x + yf_y) + \frac{1}{2}(x^2 f_{xx} + 2xy f_{xy} + y^2 f_{yy})$$

$$+ \frac{1}{6}(x^3 f_{xxx} + 3x^2 y f_{xxy} + 3xy^2 f_{xyy} + y^3 f_{yyy})_{(cx, cy)}.$$

Calculating the values of the partial derivatives,

$$f(0, 0) = \sin x \sin y\big|_{(0,0)} = 0, \qquad f_{xx}(0, 0) = -\sin x \sin y\big|_{(0,0)} = 0,$$

$$f_x(0, 0) = \cos x \sin y\big|_{(0,0)} = 0, \qquad f_{xy}(0, 0) = \cos x \cos y\big|_{(0,0)} = 1,$$

$$f_y(0, 0) = \sin x \cos y\big|_{(0,0)} = 0, \qquad f_{yy}(0, 0) = -\sin x \sin y\big|_{(0,0)} = 0,$$

we have the result

$$\sin x \sin y \approx 0 + 0 + 0 + \frac{1}{2}(x^2(0) + 2xy(1) + y^2(0)), \quad \text{or} \quad \sin x \sin y \approx xy.$$

The error in the approximation is

$$E(x, y) = \frac{1}{6}(x^3 f_{xxx} + 3x^2 y f_{xxy} + 3xy^2 f_{xyy} + y^3 f_{yyy})\bigg|_{(cx, cy)}.$$

The third derivatives never exceed 1 in absolute value because they are products of sines and cosines. Also, $|x| \le 0.1$ and $|y| \le 0.1$. Hence

$$|E(x, y)| \le \frac{1}{6}((0.1)^3 + 3(0.1)^3 + 3(0.1)^3 + (0.1)^3) = \frac{8}{6}(0.1)^3 \le 0.00134$$

(rounded up). The error will not exceed 0.00134 if $|x| \le 0.1$ and $|y| \le 0.1$.    ■

## Exercises 14.9

**Finding Quadratic and Cubic Approximations**
In Exercises 1–10, use Taylor's formula for $f(x, y)$ at the origin to find quadratic and cubic approximations of $f$ near the origin.

**1.** $f(x, y) = xe^y$

**2.** $f(x, y) = e^x \cos y$

**3.** $f(x, y) = y \sin x$

**4.** $f(x, y) = \sin x \cos y$

**5.** $f(x, y) = e^x \ln(1 + y)$

**6.** $f(x, y) = \ln(2x + y + 1)$

**7.** $f(x, y) = \sin(x^2 + y^2)$

**8.** $f(x, y) = \cos(x^2 + y^2)$

**9.** $f(x, y) = \dfrac{1}{1 - x - y}$

**10.** $f(x, y) = \dfrac{1}{1 - x - y + xy}$

**11.** Use Taylor's formula to find a quadratic approximation of $f(x, y) = \cos x \cos y$ at the origin. Estimate the error in the approximation if $|x| \le 0.1$ and $|y| \le 0.1$.

**12.** Use Taylor's formula to find a quadratic approximation of $e^x \sin y$ at the origin. Estimate the error in the approximation if $|x| \le 0.1$ and $|y| \le 0.1$.

# 14.10 Partial Derivatives with Constrained Variables

In finding partial derivatives of functions like $w = f(x, y)$, we have assumed $x$ and $y$ to be independent. In many applications, however, this is not the case. For example, the internal energy $U$ of a gas may be expressed as a function $U = f(P, V, T)$ of pressure $P$, volume $V$, and temperature $T$. If the individual molecules of the gas do not interact, however, $P$, $V$, and $T$ obey (and are constrained by) the ideal gas law

$$PV = nRT \qquad (n \text{ and } R \text{ constant}),$$

and fail to be independent. In this section we learn how to find partial derivatives in situations like this, which occur in economics, engineering, and physics.

### Decide Which Variables Are Dependent and Which Are Independent

If the variables in a function $w = f(x, y, z)$ are constrained by a relation like the one imposed on $x$, $y$, and $z$ by the equation $z = x^2 + y^2$, the geometric meanings and the numerical values of the partial derivatives of $f$ will depend on which variables are chosen to be dependent and which are chosen to be independent. To see how this choice can affect the outcome, we consider the calculation of $\partial w / \partial x$ when $w = x^2 + y^2 + z^2$ and $z = x^2 + y^2$.

**EXAMPLE 1**    Find $\partial w / \partial x$ if $w = x^2 + y^2 + z^2$ and $z = x^2 + y^2$.

**Solution**    We are given two equations in the four unknowns $x$, $y$, $z$, and $w$. Like many such systems, this one can be solved for two of the unknowns (the dependent variables) in terms of the others (the independent variables). In being asked for $\partial w / \partial x$, we are told that $w$ is to be a dependent variable and $x$ an independent variable. The possible choices for the other variables come down to

| Dependent | Independent |
|-----------|-------------|
| $w, z$ | $x, y$ |
| $w, y$ | $x, z$ |

In either case, we can express $w$ explicitly in terms of the selected independent variables. We do this by using the second equation $z = x^2 + y^2$ to eliminate the remaining dependent variable in the first equation.

In the first case, the remaining dependent variable is $z$. We eliminate it from the first equation by replacing it by $x^2 + y^2$. The resulting expression for $w$ is

$$w = x^2 + y^2 + z^2 = x^2 + y^2 + \left(x^2 + y^2\right)^2$$
$$= x^2 + y^2 + x^4 + 2x^2y^2 + y^4$$

and

$$\frac{\partial w}{\partial x} = 2x + 4x^3 + 4xy^2. \tag{1}$$

This is the formula for $\partial w / \partial x$ when $x$ and $y$ are the independent variables.

In the second case, where the independent variables are $x$ and $z$ and the remaining dependent variable is $y$, we eliminate the dependent variable $y$ in the expression for $w$ by replacing $y^2$ in the second equation by $z - x^2$. This gives

$$w = x^2 + y^2 + z^2 = x^2 + \left(z - x^2\right) + z^2 = z + z^2$$

and

$$\frac{\partial w}{\partial x} = 0. \tag{2}$$

This is the formula for $\partial w / \partial x$ when $x$ and $z$ are the independent variables.

The formulas for $\partial w / \partial x$ in Equations (1) and (2) are genuinely different. We cannot change either formula into the other by using the relation $z = x^2 + y^2$. There is not just one $\partial w / \partial x$, there are two, and we see that the original instruction to find $\partial w / \partial x$ was incomplete. *Which* $\partial w / \partial x$? we ask.

The geometric interpretations of Equations (1) and (2) help to explain why the equations differ. The function $w = x^2 + y^2 + z^2$ measures the square of the distance from the point $(x, y, z)$ to the origin. The condition $z = x^2 + y^2$ says that the point $(x, y, z)$ lies on the paraboloid of revolution shown in Figure 14.61. What does it mean to calculate $\partial w / \partial x$ at a point $P(x, y, z)$ that can move only on this surface? What is the value of $\partial w / \partial x$ when the coordinates of $P$ are, say, $(1, 0, 1)$?

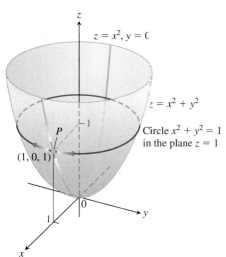

**FIGURE 14.61**    If $P$ is constrained to lie on the paraboloid $z = x^2 + y^2$, the value of the partial derivative of $w = x^2 + y^2 + z^2$ with respect to $x$ at $P$ depends on the direction of motion (Example 1). (1) As $x$ changes, with $y = 0$, $P$ moves up or down the surface on the parabola $z = x^2$ in the $xz$-plane with $\partial w / \partial x = 2x + 4x^3$. (2) As $x$ changes, with $z = 1$, $P$ moves on the circle $x^2 + y^2 = 1$, $z = 1$, and $\partial w / \partial x = 0$.

If we take $x$ and $y$ to be independent, then we find $\partial w / \partial x$ by holding $y$ fixed (at $y = 0$ in this case) and letting $x$ vary. Hence, $P$ moves along the parabola $z = x^2$ in the $xz$-plane. As $P$ moves on this parabola, $w$, which is the square of the distance from $P$ to the origin, changes. We calculate $\partial w / \partial x$ in this case (our first solution above) to be

$$\frac{\partial w}{\partial x} = 2x + 4x^3 + 4xy^2.$$

At the point $P(1, 0, 1)$, the value of this derivative is

$$\frac{\partial w}{\partial x} = 2 + 4 + 0 = 6.$$

If we take $x$ and $z$ to be independent, then we find $\partial w / \partial x$ by holding $z$ fixed while $x$ varies. Since the $z$-coordinate of $P$ is 1, varying $x$ moves $P$ along a circle in the plane $z = 1$. As $P$ moves along this circle, its distance from the origin remains constant, and $w$, being the square of this distance, does not change. That is,

$$\frac{\partial w}{\partial x} = 0,$$

as we found in our second solution. ■

### How to Find $\partial w / \partial x$ When the Variables in $w = f(x, y, z)$ Are Constrained by Another Equation

As we saw in Example 1, a typical routine for finding $\partial w / \partial x$ when the variables in the function $w = f(x, y, z)$ are related by another equation has three steps. These steps apply to finding $\partial w / \partial y$ and $\partial w / \partial z$ as well.

---

1. *Decide* which variables are to be dependent and which are to be independent. (In practice, the decision is based on the physical or theoretical context of our work. In the exercises at the end of this section, we say which variables are which.)
2. *Eliminate* the other dependent variable(s) in the expression for $w$.
3. *Differentiate* as usual.

---

If we cannot carry out Step 2 after deciding which variables are dependent, we differentiate the equations as they are and try to solve for $\partial w / \partial x$ afterward. The next example shows how this is done.

**EXAMPLE 2**    Find $\partial w / \partial x$ at the point $(x, y, z) = (2, -1, 1)$ if

$$w = x^2 + y^2 + z^2, \qquad z^3 - xy + yz + y^3 = 1,$$

and $x$ and $y$ are the independent variables.

**Solution**    It is not convenient to eliminate $z$ in the expression for $w$. We therefore differentiate both equations implicitly with respect to $x$, treating $x$ and $y$ as independent variables and $w$ and $z$ as dependent variables. This gives

$$\frac{\partial w}{\partial x} = 2x + 2z \frac{\partial z}{\partial x} \tag{3}$$

and

$$3z^2 \frac{\partial z}{\partial x} - y + y \frac{\partial z}{\partial x} + 0 = 0. \tag{4}$$

These equations may now be combined to express $\partial w / \partial x$ in terms of $x$, $y$, and $z$. We solve Equation (4) for $\partial z / \partial x$ to get

$$\frac{\partial z}{\partial x} = \frac{y}{y + 3z^2}$$

and substitute into Equation (3) to get

$$\frac{\partial w}{\partial x} = 2x + \frac{2yz}{y + 3z^2}.$$

The value of this derivative at $(x, y, z) = (2, -1, 1)$ is

$$\left(\frac{\partial w}{\partial x}\right)_{(2,-1,1)} = 2(2) + \frac{2(-1)(1)}{-1 + 3(1)^2} = 4 + \frac{-2}{2} = 3. \qquad \blacksquare$$

## Notation

To show what variables are assumed to be independent in calculating a derivative, we can use the following notation:

$$\left(\frac{\partial w}{\partial x}\right)_y \qquad \partial w / \partial x \text{ with } x \text{ and } y \text{ independent}$$

$$\left(\frac{\partial f}{\partial y}\right)_{x, t} \qquad \partial f / \partial y \text{ with } y, x, \text{ and } t \text{ independent}$$

**EXAMPLE 3**   Find $(\partial w / \partial x)_{y, z}$ if $w = x^2 + y - z + \sin t$ and $x + y = t$.

**Solution**   With $x$, $y$, $z$ independent, we have

$$t = x + y, \qquad w = x^2 + y - z + \sin (x + y)$$

$$\left(\frac{\partial w}{\partial x}\right)_{y, z} = 2x + 0 - 0 + \cos (x + y) \frac{\partial}{\partial x} (x + y)$$

$$= 2x + \cos (x + y). \qquad \blacksquare$$

## Arrow Diagrams

In solving problems like the one in Example 3, it often helps to start with an arrow diagram that shows how the variables and functions are related. If

$$w = x^2 + y - z + \sin t \qquad \text{and} \qquad x + y = t$$

and we are asked to find $\partial w / \partial x$ when $x$, $y$, and $z$ are independent, the appropriate diagram is one like this:

$$\begin{pmatrix} x \\ y \\ z \end{pmatrix} \rightarrow \begin{pmatrix} x \\ y \\ z \\ t \end{pmatrix} \rightarrow w \qquad \qquad (5)$$

Independent        Intermediate        Dependent
variables            variables            variable

To avoid confusion between the independent and intermediate variables with the same symbolic names in the diagram, it is helpful to rename the intermediate variables (so they are seen as *functions* of the independent variables). Thus, let $u = x$, $v = y$, and $s = z$ denote the renamed intermediate variables. With this notation, the arrow diagram becomes

$$\begin{pmatrix} x \\ y \\ z \end{pmatrix} \longrightarrow \begin{pmatrix} u \\ v \\ s \\ t \end{pmatrix} \longrightarrow w \tag{6}$$

Independent    Intermediate    Dependent
variables    variables and    variable
    relations

$u = x$
$v = y$
$s = z$
$t = x + y$

The diagram shows the independent variables on the left, the intermediate variables and their relation to the independent variables in the middle, and the dependent variable on the right. The function $w$ now becomes

$$w = u^2 + v - s + \sin t,$$

where

$$u = x, \qquad v = y, \qquad s = z, \qquad \text{and} \qquad t = x + y.$$

To find $\partial w / \partial x$, we apply the four-variable form of the Chain Rule to $w$, guided by the arrow diagram in Equation (6):

$$\begin{aligned}
\frac{\partial w}{\partial x} &= \frac{\partial w}{\partial u}\frac{\partial u}{\partial x} + \frac{\partial w}{\partial v}\frac{\partial v}{\partial x} + \frac{\partial w}{\partial s}\frac{\partial s}{\partial x} + \frac{\partial w}{\partial t}\frac{\partial t}{\partial x} \\
&= (2u)(1) + (1)(0) + (-1)(0) + (\cos t)(1) \\
&= 2u + \cos t \\
&= 2x + \cos (x + y). \qquad \text{Substituting the original independent}\\
&\qquad\qquad\qquad\qquad\qquad\quad \text{variables } u = x \text{ and } t = x + y
\end{aligned}$$

## Exercises 14.10

### Finding Partial Derivatives with Constrained Variables

In Exercises 1–3, begin by drawing a diagram that shows the relations among the variables.

**1.** If $w = x^2 + y^2 + z^2$ and $z = x^2 + y^2$, find

**a.** $\left(\dfrac{\partial w}{\partial y}\right)_z$     **b.** $\left(\dfrac{\partial w}{\partial z}\right)_x$     **c.** $\left(\dfrac{\partial w}{\partial z}\right)_y$.

**2.** If $w = x^2 + y - z + \sin t$ and $x + y = t$, find

**a.** $\left(\dfrac{\partial w}{\partial y}\right)_{x,z}$     **b.** $\left(\dfrac{\partial w}{\partial y}\right)_{z,t}$     **c.** $\left(\dfrac{\partial w}{\partial z}\right)_{x,y}$

**d.** $\left(\dfrac{\partial w}{\partial z}\right)_{y,t}$     **e.** $\left(\dfrac{\partial w}{\partial t}\right)_{x,z}$     **f.** $\left(\dfrac{\partial w}{\partial t}\right)_{y,z}$.

**3.** Let $U = f(P, V, T)$ be the internal energy of a gas that obeys the ideal gas law $PV = nRT$ ($n$ and $R$ constant). Find

**a.** $\left(\dfrac{\partial U}{\partial P}\right)_V$     **b.** $\left(\dfrac{\partial U}{\partial T}\right)_V$.

**4.** Find

**a.** $\left(\dfrac{\partial w}{\partial x}\right)_y$     **b.** $\left(\dfrac{\partial w}{\partial z}\right)_y$

at the point $(x, y, z) = (0, 1, \pi)$ if

$$w = x^2 + y^2 + z^2 \qquad \text{and} \qquad y \sin z + z \sin x = 0.$$

**5.** Find

**a.** $\left(\dfrac{\partial w}{\partial y}\right)_x$     **b.** $\left(\dfrac{\partial w}{\partial y}\right)_z$

at the point $(w, x, y, z) = (4, 2, 1, -1)$ if

$$w = x^2 y^2 + yz - z^3 \qquad \text{and} \qquad x^2 + y^2 + z^2 = 6.$$

**6.** Find $(\partial u / \partial y)_x$ at the point $(u, v) = \left(\sqrt{2}, 1\right)$ if $x = u^2 + v^2$ and $y = uv$.

**7.** Suppose that $x^2 + y^2 = r^2$ and $x = r \cos \theta$, as in polar coordinates. Find

$$\left(\frac{\partial x}{\partial r}\right)_{\theta} \quad \text{and} \quad \left(\frac{\partial r}{\partial x}\right)_{y}.$$

**8.** Suppose that

$$w = x^2 - y^2 + 4z + t \quad \text{and} \quad x + 2z + t = 25.$$

Show that the equations

$$\frac{\partial w}{\partial x} = 2x - 1 \quad \text{and} \quad \frac{\partial w}{\partial x} = 2x - 2$$

each give $\partial w / \partial x$, depending on which variables are chosen to be dependent and which variables are chosen to be independent. Identify the independent variables in each case.

### Theory and Examples

**9.** Establish the fact, widely used in hydrodynamics, that if $f(x, y, z) = 0$, then

$$\left(\frac{\partial x}{\partial y}\right)_{z} \left(\frac{\partial y}{\partial z}\right)_{x} \left(\frac{\partial z}{\partial x}\right)_{y} = -1.$$

(*Hint:* Express all the derivatives in terms of the formal partial derivatives $\partial f / \partial x$, $\partial f / \partial y$, and $\partial f / \partial z$.)

**10.** If $z = x + f(u)$, where $u = xy$, show that

$$x\frac{\partial z}{\partial x} - y\frac{\partial z}{\partial y} = x.$$

**11.** Suppose that the equation $g(x, y, z) = 0$ determines $z$ as a differentiable function of the independent variables $x$ and $y$ and that $g_z \neq 0$. Show that

$$\left(\frac{\partial z}{\partial y}\right)_{x} = -\frac{\partial g / \partial y}{\partial g / \partial z}.$$

**12.** Suppose that $f(x, y, z, w) = 0$ and $g(x, y, z, w) = 0$ determine $z$ and $w$ as differentiable functions of the independent variables $x$ and $y$, and suppose that

$$\frac{\partial f}{\partial z}\frac{\partial g}{\partial w} - \frac{\partial f}{\partial w}\frac{\partial g}{\partial z} \neq 0.$$

Show that

$$\left(\frac{\partial z}{\partial x}\right)_{y} = -\frac{\dfrac{\partial f}{\partial x}\dfrac{\partial g}{\partial w} - \dfrac{\partial f}{\partial w}\dfrac{\partial g}{\partial x}}{\dfrac{\partial f}{\partial z}\dfrac{\partial g}{\partial w} - \dfrac{\partial f}{\partial w}\dfrac{\partial g}{\partial z}}$$

and

$$\left(\frac{\partial w}{\partial y}\right)_{x} = -\frac{\dfrac{\partial f}{\partial z}\dfrac{\partial g}{\partial y} - \dfrac{\partial f}{\partial y}\dfrac{\partial g}{\partial z}}{\dfrac{\partial f}{\partial z}\dfrac{\partial g}{\partial w} - \dfrac{\partial f}{\partial w}\dfrac{\partial g}{\partial z}}.$$

---

## Chapter 14 Questions to Guide Your Review

**1.** What is a real-valued function of two independent variables? Three independent variables? Give examples.

**2.** What does it mean for sets in the plane or in space to be open? Closed? Give examples. Give examples of sets that are neither open nor closed.

**3.** How can you display the values of a function $f(x, y)$ of two independent variables graphically? How do you do the same for a function $f(x, y, z)$ of three independent variables?

**4.** What does it mean for a function $f(x, y)$ to have limit $L$ as $(x, y) \to (x_0, y_0)$? What are the basic properties of limits of functions of two independent variables?

**5.** When is a function of two (three) independent variables continuous at a point in its domain? Give examples of functions that are continuous at some points but not others.

**6.** What can be said about algebraic combinations and composites of continuous functions?

**7.** Explain the two-path test for nonexistence of limits.

**8.** How are the partial derivatives $\partial f / \partial x$ and $\partial f / \partial y$ of a function $f(x, y)$ defined? How are they interpreted and calculated?

**9.** How does the relation between first partial derivatives and continuity of functions of two independent variables differ from the relation between first derivatives and continuity for real-valued functions of a single independent variable? Give an example.

**10.** What is the Mixed Derivative Theorem for mixed second-order partial derivatives? How can it help in calculating partial derivatives of second and higher orders? Give examples.

**11.** What does it mean for a function $f(x, y)$ to be differentiable? What does the Increment Theorem say about differentiability?

**12.** How can you sometimes decide from examining $f_x$ and $f_y$ that a function $f(x, y)$ is differentiable? What is the relation between the differentiability of $f$ and the continuity of $f$ at a point?

**13.** What is the general Chain Rule? What form does it take for functions of two independent variables? Three independent variables? Functions defined on surfaces? How do you diagram these different forms? Give examples. What pattern enables one to remember all the different forms?

**14.** What is the derivative of a function $f(x, y)$ at a point $P_0$ in the direction of a unit vector $\mathbf{u}$? What rate does it describe? What geometric interpretation does it have? Give examples.

**15.** What is the gradient vector of a differentiable function $f(x, y)$? How is it related to the function's directional derivatives? State the analogous results for functions of three independent variables.

**16.** How do you find the tangent line at a point on a level curve of a differentiable function $f(x, y)$? How do you find the tangent plane and normal line at a point on a level surface of a differentiable function $f(x, y, z)$? Give examples.

**17.** How can you use directional derivatives to estimate change?

**18.** How do you linearize a function $f(x, y)$ of two independent variables at a point $(x_0, y_0)$? Why might you want to do this? How do you linearize a function of three independent variables?

**19.** What can you say about the accuracy of linear approximations of functions of two (three) independent variables?

**20.** If $(x, y)$ moves from $(x_0, y_0)$ to a point $(x_0 + dx, y_0 + dy)$ nearby, how can you estimate the resulting change in the value of a differentiable function $f(x, y)$? Give an example.

**21.** How do you define local maxima, local minima, and saddle points for a differentiable function $f(x, y)$? Give examples.

**22.** What derivative tests are available for determining the local extreme values of a function $f(x, y)$? How do they enable you to narrow your search for these values? Give examples.

**23.** How do you find the extrema of a continuous function $f(x, y)$ on a closed bounded region of the $xy$-plane? Give an example.

**24.** Describe the method of Lagrange multipliers and give examples.

**25.** How does Taylor's formula for a function $f(x, y)$ generate polynomial approximations and error estimates?

**26.** If $w = f(x, y, z)$, where the variables $x$, $y$, and $z$ are constrained by an equation $g(x, y, z) = 0$, what is the meaning of the notation $(\partial w/\partial x)_y$? How can an arrow diagram help you calculate this partial derivative with constrained variables? Give examples.

## Chapter **14** Practice Exercises

### Domain, Range, and Level Curves

In Exercises 1–4, find the domain and range of the given function and identify its level curves. Sketch a typical level curve.

**1.** $f(x, y) = 9x^2 + y^2$

**2.** $f(x, y) = e^{x+y}$

**3.** $g(x, y) = 1/xy$

**4.** $g(x, y) = \sqrt{x^2 - y}$

In Exercises 5–8, find the domain and range of the given function and identify its level surfaces. Sketch a typical level surface.

**5.** $f(x, y, z) = x^2 + y^2 - z$

**6.** $g(x, y, z) = x^2 + 4y^2 + 9z^2$

**7.** $h(x, y, z) = \dfrac{1}{x^2 + y^2 + z^2}$

**8.** $k(x, y, z) = \dfrac{1}{x^2 + y^2 + z^2 + 1}$

### Evaluating Limits

Find the limits in Exercises 9–14.

**9.** $\displaystyle\lim_{(x,y)\to(\pi, \ln 2)} e^y \cos x$

**10.** $\displaystyle\lim_{(x,y)\to(0,0)} \dfrac{2 + y}{x + \cos y}$

**11.** $\displaystyle\lim_{(x,y)\to(1,1)} \dfrac{x - y}{x^2 - y^2}$

**12.** $\displaystyle\lim_{(x,y)\to(1,1)} \dfrac{x^3 y^3 - 1}{xy - 1}$

**13.** $\displaystyle\lim_{P\to(1, -1, e)} \ln|x + y + z|$

**14.** $\displaystyle\lim_{P\to(1,-1,-1)} \tan^{-1}(x + y + z)$

By considering different paths of approach, show that the limits in Exercises 15 and 16 do not exist.

**15.** $\displaystyle\lim_{\substack{(x,y)\to(0,0)\\ y\neq x^2}} \dfrac{y}{x^2 - y}$

**16.** $\displaystyle\lim_{\substack{(x,y)\to(0,0)\\ xy\neq 0}} \dfrac{x^2 + y^2}{xy}$

**17. Continuous extension** Let $f(x, y) = (x^2 - y^2)/(x^2 + y^2)$ for $(x, y) \neq (0, 0)$. Is it possible to define $f(0, 0)$ in a way that makes $f$ continuous at the origin? Why?

**18. Continuous extension** Let

$$f(x, y) = \begin{cases} \dfrac{\sin(x - y)}{|x| + |y|}, & |x| + |y| \neq 0 \\ 0, & (x, y) = (0, 0). \end{cases}$$

Is $f$ continuous at the origin? Why?

### Partial Derivatives

In Exercises 19–24, find the partial derivative of the function with respect to each variable.

**19.** $f(r, \theta) = r \sin \theta - r \cos \theta$

**20.** $f(x, y) = \dfrac{1}{2}\ln(x^2 - y^2) + \sin^{-1}\dfrac{y}{x}$

**21.** $f(R_1, R_2, R_3) = \dfrac{1}{R_1} + \dfrac{1}{R_2} + \dfrac{1}{R_3}$

**22.** $f(x, y, z) = \cos(4\pi x - y + 5z)$

**23.** $P(n, R, T, V) = \dfrac{nRT}{V}$ (the ideal gas law)

**24.** $f(r, l, T, w) = \dfrac{1}{2rl}\sqrt{\dfrac{T}{\pi w}}$

### Second-Order Partials

Find the second-order partial derivatives of the functions in Exercises 25–28.

**25.** $g(x, y) = y + \dfrac{x}{y}$

**26.** $f(x, y) = e^y - x \cos y$

**27.** $g(x, y) = y - xy - 8y^3 + \ln(y^2 - 1)$

**28.** $g(x, y) = x^2 - 9xy - \sin y + 5e^x$

### Chain Rule Calculations

**29.** Find $dw/dt$ at $t = 0$ if $w = \sin(xy + \pi)$, $x = e^t$, and $y = \ln(t + 1)$.

**30.** Find $dw/dt$ at $t = 1$ if $w = xe^y + y \sin z - \cos z$, $x = 2\sqrt{t}$, $y = t - 1 + \ln t$, and $z = \pi t$.

**31.** Find $\partial w/\partial r$ and $\partial w/\partial s$ when $r = \pi$ and $s = 0$ if $w = \sin(2x - y)$, $x = r + \sin s$, $y = rs$.

**32.** Find $\partial w/\partial u$ and $\partial w/\partial v$ when $u = v = 0$ if $w = \ln\sqrt{1 + x^2} - \tan^{-1} x$ and $x = 2e^u \cos v$.

**33.** Find the value of the derivative of $f(x, y, z) = xy + yz + xz$ with respect to $t$ on the curve $x = \cos t$, $y = \sin t$, $z = \cos 2t$ at $t = 1$.

**34.** Show that if $w = f(s)$ is any differentiable function of $s$ and if $s = y + 5x$, then

$$\frac{\partial w}{\partial x} - 5\frac{\partial w}{\partial y} = 0.$$

## Implicit Differentiation

Assuming that the equations in Exercises 35 and 36 define $y$ as a differentiable function of $x$, find the value of $dy/dx$ at point $P$.

**35.** $1 - x - y^2 - \sin xy = 0$, $P(0, 1)$

**36.** $2xy + e^{x+y} - 2 = 0$, $P(0, \ln 2)$

## Directional Derivatives

In Exercises 37–40, find the directions in which $f$ increases and decreases most rapidly at $P_0$ and find the derivative of $f$ in each direction. Also, find the derivative of $f$ at $P_0$ in the direction of the vector $\mathbf{v}$.

**37.** $f(x, y) = \cos x \cos y$, $P_0(\pi/4, \pi/4)$, $\mathbf{v} = 3\mathbf{i} + 4\mathbf{j}$

**38.** $f(x, y) = x^2 e^{-2y}$, $P_0(1, 0)$, $\mathbf{v} = \mathbf{i} + \mathbf{j}$

**39.** $f(x, y, z) = \ln(2x + 3y + 6z)$, $P_0(-1, -1, 1)$, $\mathbf{v} = 2\mathbf{i} + 3\mathbf{j} + 6\mathbf{k}$

**40.** $f(x, y, z) = x^2 + 3xy - z^2 + 2y + z + 4$, $P_0(0, 0, 0)$, $\mathbf{v} = \mathbf{i} + \mathbf{j} + \mathbf{k}$

**41. Derivative in velocity direction** Find the derivative of $f(x, y, z) = xyz$ in the direction of the velocity vector of the helix

$$\mathbf{r}(t) = (\cos 3t)\mathbf{i} + (\sin 3t)\mathbf{j} + 3t\mathbf{k}$$

at $t = \pi/3$.

**42. Maximum directional derivative** What is the largest value that the directional derivative of $f(x, y, z) = xyz$ can have at the point $(1, 1, 1)$?

**43. Directional derivatives with given values** At the point $(1, 2)$, the function $f(x, y)$ has a derivative of 2 in the direction toward $(2, 2)$ and a derivative of $-2$ in the direction toward $(1, 1)$.

  **a.** Find $f_x(1, 2)$ and $f_y(1, 2)$.

  **b.** Find the derivative of $f$ at $(1, 2)$ in the direction toward the point $(4, 6)$.

**44.** Which of the following statements are true if $f(x, y)$ is differentiable at $(x_0, y_0)$? Give reasons for your answers.

  **a.** If $\mathbf{u}$ is a unit vector, the derivative of $f$ at $(x_0, y_0)$ in the direction of $\mathbf{u}$ is $(f_x(x_0, y_0)\mathbf{i} + f_y(x_0, y_0)\mathbf{j}) \cdot \mathbf{u}$.

  **b.** The derivative of $f$ at $(x_0, y_0)$ in the direction of $\mathbf{u}$ is a vector.

  **c.** The directional derivative of $f$ at $(x_0, y_0)$ has its greatest value in the direction of $\nabla f$.

  **d.** At $(x_0, y_0)$, vector $\nabla f$ is normal to the curve $f(x, y) = f(x_0, y_0)$.

## Gradients, Tangent Planes, and Normal Lines

In Exercises 45 and 46, sketch the surface $f(x, y, z) = c$ together with $\nabla f$ at the given points.

**45.** $x^2 + y + z^2 = 0$; $(0, -1, \pm 1)$, $(0, 0, 0)$

**46.** $y^2 + z^2 = 4$; $(2, \pm 2, 0)$, $(2, 0, \pm 2)$

In Exercises 47 and 48, find an equation for the plane tangent to the level surface $f(x, y, z) = c$ at the point $P_0$. Also, find parametric equations for the line that is normal to the surface at $P_0$.

**47.** $x^2 - y - 5z = 0$, $P_0(2, -1, 1)$

**48.** $x^2 + y^2 + z = 4$, $P_0(1, 1, 2)$

In Exercises 49 and 50, find an equation for the plane tangent to the surface $z = f(x, y)$ at the given point.

**49.** $z = \ln(x^2 + y^2)$, $(0, 1, 0)$

**50.** $z = 1/(x^2 + y^2)$, $(1, 1, 1/2)$

In Exercises 51 and 52, find equations for the lines that are tangent and normal to the level curve $f(x, y) = c$ at the point $P_0$. Then sketch the lines and level curve together with $\nabla f$ at $P_0$.

**51.** $y - \sin x = 1$, $P_0(\pi, 1)$  **52.** $\dfrac{y^2}{2} - \dfrac{x^2}{2} = \dfrac{3}{2}$, $P_0(1, 2)$

## Tangent Lines to Curves

In Exercises 53 and 54, find parametric equations for the line that is tangent to the curve of intersection of the surfaces at the given point.

**53.** Surfaces: $x^2 + 2y + 2z = 4$, $y = 1$

  Point: $(1, 1, 1/2)$

**54.** Surfaces: $x + y^2 + z = 2$, $y = 1$

  Point: $(1/2, 1, 1/2)$

## Linearizations

In Exercises 55 and 56, find the linearization $L(x, y)$ of the function $f(x, y)$ at the point $P_0$. Then find an upper bound for the magnitude of the error $E$ in the approximation $f(x, y) \approx L(x, y)$ over the rectangle $R$.

**55.** $f(x, y) = \sin x \cos y$, $P_0(\pi/4, \pi/4)$

  $R: \left|x - \dfrac{\pi}{4}\right| \le 0.1$, $\left|y - \dfrac{\pi}{4}\right| \le 0.1$

**56.** $f(x, y) = xy - 3y^2 + 2$, $P_0(1, 1)$

  $R: |x - 1| \le 0.1$, $|y - 1| \le 0.2$

Find the linearizations of the functions in Exercises 57 and 58 at the given points.

**57.** $f(x, y, z) = xy + 2yz - 3xz$ at $(1, 0, 0)$ and $(1, 1, 0)$

**58.** $f(x, y, z) = \sqrt{2} \cos x \sin(y + z)$ at $(0, 0, \pi/4)$ and $(\pi/4, \pi/4, 0)$

## Estimates and Sensitivity to Change

**59. Measuring the volume of a pipeline** You plan to calculate the volume inside a stretch of pipeline that is about 36 cm in diameter and 1 km long. With which measurement should you be more careful, the length or the diameter? Why?

**60. Sensitivity to change** Is $f(x, y) = x^2 - xy + y^2 - 3$ more sensitive to changes in $x$ or to changes in $y$ when it is near the point $(1, 2)$? How do you know?

**61. Change in an electrical circuit** Suppose that the current $I$ (amperes) in an electrical circuit is related to the voltage $V$ (volts) and the resistance $R$ (ohms) by the equation $I = V/R$. If the voltage drops from 24 to 23 volts and the resistance drops from 100 to 80 ohms, will $I$ increase or decrease? By about how much? Is the change in $I$ more sensitive to change in the voltage or to change in the resistance? How do you know?

**62. Maximum error in estimating the area of an ellipse** If $a = 10$ cm and $b = 16$ cm to the nearest millimeter, what should you expect the maximum percentage error to be in the calculated area $A = \pi ab$ of the ellipse $x^2/a^2 + y^2/b^2 = 1$?

**63. Error in estimating a product** Let $y = uv$ and $z = u + v$, where $u$ and $v$ are positive independent variables.

**a.** If $u$ is measured with an error of 2% and $v$ with an error of 3%, about what is the percentage error in the calculated value of $y$?

**b.** Show that the percentage error in the calculated value of $z$ is less than the percentage error in the value of $y$.

**64. Cardiac index** To make different people comparable in studies of cardiac output, researchers divide the measured cardiac output by the body surface area to find the *cardiac index C*:

$$C = \frac{\text{cardiac output}}{\text{body surface area}}.$$

The body surface area $B$ of a person with weight $w$ and height $h$ is approximated by the formula

$$B = 71.84 w^{0.425} h^{0.725},$$

which gives $B$ in square centimeters when $w$ is measured in kilograms and $h$ in centimeters. You are about to calculate the cardiac index of a person 180 cm tall, weighing 70 kg, with cardiac output of 7 L/min. Which will have a greater effect on the calculation, a 1-kg error in measuring the weight or a 1-cm error in measuring the height?

## Local Extrema

Test the functions in Exercises 65–70 for local maxima and minima and saddle points. Find each function's value at these points.

**65.** $f(x, y) = x^2 - xy + y^2 + 2x + 2y - 4$

**66.** $f(x, y) = 5x^2 + 4xy - 2y^2 + 4x - 4y$

**67.** $f(x, y) = 2x^3 + 3xy + 2y^3$

**68.** $f(x, y) = x^3 + y^3 - 3xy + 15$

**69.** $f(x, y) = x^3 + y^3 + 3x^2 - 3y^2$

**70.** $f(x, y) = x^4 - 8x^2 + 3y^2 - 6y$

## Absolute Extrema

In Exercises 71–78, find the absolute maximum and minimum values of $f$ on the region $R$.

**71.** $f(x, y) = x^2 + xy + y^2 - 3x + 3y$

    $R$: The triangular region cut from the first quadrant by the line $x + y = 4$

**72.** $f(x, y) = x^2 - y^2 - 2x + 4y + 1$

    $R$: The rectangular region in the first quadrant bounded by the coordinate axes and the lines $x = 4$ and $y = 2$

**73.** $f(x, y) = y^2 - xy - 3y + 2x$

    $R$: The square region enclosed by the lines $x = \pm 2$ and $y = \pm 2$

**74.** $f(x, y) = 2x + 2y - x^2 - y^2$

    $R$: The square region bounded by the coordinate axes and the lines $x = 2$, $y = 2$ in the first quadrant

**75.** $f(x, y) = x^2 - y^2 - 2x + 4y$

    $R$: The triangular region bounded below by the $x$-axis, above by the line $y = x + 2$, and on the right by the line $x = 2$

**76.** $f(x, y) = 4xy - x^4 - y^4 + 16$

    $R$: The triangular region bounded below by the line $y = -2$, above by the line $y = x$, and on the right by the line $x = 2$

**77.** $f(x, y) = x^3 + y^3 + 3x^2 - 3y^2$

    $R$: The square region enclosed by the lines $x = \pm 1$ and $y = \pm 1$

**78.** $f(x, y) = x^3 + 3xy + y^3 + 1$

    $R$: The square region enclosed by the lines $x = \pm 1$ and $y = \pm 1$

## Lagrange Multipliers

**79. Extrema on a circle** Find the extreme values of $f(x, y) = x^3 + y^2$ on the circle $x^2 + y^2 = 1$.

**80. Extrema on a circle** Find the extreme values of $f(x, y) = xy$ on the circle $x^2 + y^2 = 1$.

**81. Extrema in a disk** Find the extreme values of $f(x, y) = x^2 + 3y^2 + 2y$ on the unit disk $x^2 + y^2 \le 1$.

**82. Extrema in a disk** Find the extreme values of $f(x, y) = x^2 + y^2 - 3x - xy$ on the disk $x^2 + y^2 \le 9$.

**83. Extrema on a sphere** Find the extreme values of $f(x, y, z) = x - y + z$ on the unit sphere $x^2 + y^2 + z^2 = 1$.

**84. Minimum distance to origin** Find the points on the surface $x^2 - zy = 4$ closest to the origin.

**85. Minimizing cost of a box** A closed rectangular box is to have volume $V$ cm$^3$. The cost of the material used in the box is $a$ cents/cm$^2$ for top and bottom, $b$ cents/cm$^2$ for front and back, and $c$ cents/cm$^2$ for the remaining sides. What dimensions minimize the total cost of materials?

**86. Least volume** Find the plane $x/a + y/b + z/c = 1$ that passes through the point $(2, 1, 2)$ and cuts off the least volume from the first octant.

**87. Extrema on curve of intersecting surfaces** Find the extreme values of $f(x, y, z) = x(y + z)$ on the curve of intersection of the right circular cylinder $x^2 + y^2 = 1$ and the hyperbolic cylinder $xz = 1$.

**88. Minimum distance to origin on curve of intersecting plane and cone** Find the point closest to the origin on the curve of intersection of the plane $x + y + z = 1$ and the cone $z^2 = 2x^2 + 2y^2$.

## Theory and Examples

**89.** Let $w = f(r, \theta)$, $r = \sqrt{x^2 + y^2}$, and $\theta = \tan^{-1}(y/x)$. Find $\partial w/\partial x$ and $\partial w/\partial y$ and express your answers in terms of $r$ and $\theta$.

**90.** Let $z = f(u, v)$, $u = ax + by$, and $v = ax - by$. Express $z_x$ and $z_y$ in terms of $f_u, f_v$, and the constants $a$ and $b$.

**91.** If $a$ and $b$ are constants, $w = u^3 + \tanh u + \cos u$, and $u = ax + by$, show that

$$a \frac{\partial w}{\partial y} = b \frac{\partial w}{\partial x}.$$

**92. Using the Chain Rule** If $w = \ln(x^2 + y^2 + 2z)$, $x = r + s$, $y = r - s$, and $z = 2rs$, find $w_r$ and $w_s$ by the Chain Rule. Then check your answer another way.

**93. Angle between vectors** The equations $e^u \cos v - x = 0$ and $e^u \sin v - y = 0$ define $u$ and $v$ as differentiable functions of $x$ and $y$. Show that the angle between the vectors

$$\frac{\partial u}{\partial x} \mathbf{i} + \frac{\partial u}{\partial y} \mathbf{j} \quad \text{and} \quad \frac{\partial v}{\partial x} \mathbf{i} + \frac{\partial v}{\partial y} \mathbf{j}$$

is constant.

**94. Polar coordinates and second derivatives** Introducing polar coordinates $x = r \cos \theta$ and $y = r \sin \theta$ changes $f(x, y)$ to $g(r, \theta)$. Find the value of $\partial^2 g / \partial \theta^2$ at the point $(r, \theta) = (2, \pi/2)$, given that

$$\frac{\partial f}{\partial x} = \frac{\partial f}{\partial y} = \frac{\partial^2 f}{\partial x^2} = \frac{\partial^2 f}{\partial y^2} = 1$$

at that point.

**95. Normal line parallel to a plane** Find the points on the surface

$$(y + z)^2 + (z - x)^2 = 16$$

where the normal line is parallel to the $yz$-plane.

**96. Tangent plane parallel to $xy$-plane** Find the points on the surface

$$xy + yz + zx - x - z^2 = 0$$

where the tangent plane is parallel to the $xy$-plane.

**97. When gradient is parallel to position vector** Suppose that $\nabla f(x, y, z)$ is always parallel to the position vector $x\mathbf{i} + y\mathbf{j} + z\mathbf{k}$. Show that $f(0, 0, a) = f(0, 0, -a)$ for any $a$.

**98. One-sided directional derivative in all directions, but no gradient** The *one-sided directional derivative of $f$ at $P(x_0, y_0, z_0)$ in the direction $\mathbf{u} = u_1\mathbf{i} + u_2\mathbf{j} + u_3\mathbf{k}$* is the number

$$\lim_{s \to 0^+} \frac{f(x_0 + su_1, y_0 + su_2, z_0 + su_3) - f(x_0, y_0, z_0)}{s}.$$

Show that the one-sided directional derivative of

$$f(x, y, z) = \sqrt{x^2 + y^2 + z^2}$$

at the origin equals 1 in any direction but that $f$ has no gradient vector at the origin.

**99. Normal line through origin** Show that the line normal to the surface $xy + z = 2$ at the point $(1, 1, 1)$ passes through the origin.

**100. Tangent plane and normal line**
   **a.** Sketch the surface $x^2 - y^2 + z^2 = 4$.
   **b.** Find a vector normal to the surface at $(2, -3, 3)$. Add the vector to your sketch.
   **c.** Find equations for the tangent plane and normal line at $(2, -3, 3)$.

**Partial Derivatives with Constrained Variables**
In Exercises 101 and 102, begin by drawing a diagram that shows the relations among the variables.

**101.** If $w = x^2 e^{yz}$ and $z = x^2 - y^2$ find
   **a.** $\left(\dfrac{\partial w}{\partial y}\right)_z$     **b.** $\left(\dfrac{\partial w}{\partial z}\right)_x$     **c.** $\left(\dfrac{\partial w}{\partial z}\right)_y.$

**102.** Let $U = f(P, V, T)$ be the internal energy of a gas that obeys the ideal gas law $PV = nRT$ ($n$ and $R$ constant). Find
   **a.** $\left(\dfrac{\partial U}{\partial T}\right)_P$     **b.** $\left(\dfrac{\partial U}{\partial V}\right)_T.$

---

# Chapter **14**  Additional and Advanced Exercises

**Partial Derivatives**

**1. Function with saddle at the origin** If you did Exercise 60 in Section 14.2, you know that the function

$$f(x, y) = \begin{cases} xy \dfrac{x^2 - y^2}{x^2 + y^2}, & (x, y) \neq (0, 0) \\ 0, & (x, y) = (0, 0) \end{cases}$$

(see the accompanying figure) is continuous at $(0, 0)$. Find $f_{xy}(0, 0)$ and $f_{yx}(0, 0)$.

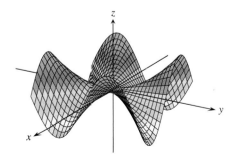

**2. Finding a function from second partials** Find a function $w = f(x, y)$ whose first partial derivatives are $\partial w / \partial x = 1 + e^x \cos y$ and $\partial w / \partial y = 2y - e^x \sin y$ and whose value at the point $(\ln 2, 0)$ is $\ln 2$.

**3. A proof of Leibniz's Rule** Leibniz's Rule says that if $f$ is continuous on $[a, b]$ and if $u(x)$ and $v(x)$ are differentiable functions of $x$ whose values lie in $[a, b]$, then

$$\frac{d}{dx} \int_{u(x)}^{v(x)} f(t) \, dt = f(v(x)) \frac{dv}{dx} - f(u(x)) \frac{du}{dx}.$$

Prove the rule by setting

$$g(u, v) = \int_u^v f(t) \, dt, \quad u = u(x), \quad v = v(x)$$

and calculating $dg / dx$ with the Chain Rule.

**4. Finding a function with constrained second partials** Suppose that $f$ is a twice-differentiable function of $r$, that $r = \sqrt{x^2 + y^2 + z^2}$, and that

$$f_{xx} + f_{yy} + f_{zz} = 0.$$

Show that for some constants $a$ and $b$,

$$f(r) = \frac{a}{r} + b.$$

**5. Homogeneous functions** A function $f(x, y)$ is *homogeneous of degree n* ($n$ a nonnegative integer) if $f(tx, ty) = t^n f(x, y)$ for all $t$, $x$, and $y$. For such a function (sufficiently differentiable), prove that

**a.** $x\dfrac{\partial f}{\partial x} + y\dfrac{\partial f}{\partial y} = nf(x, y)$

**b.** $x^2 \left(\dfrac{\partial^2 f}{\partial x^2}\right) + 2xy\left(\dfrac{\partial^2 f}{\partial x \partial y}\right) + y^2 \left(\dfrac{\partial^2 f}{\partial y^2}\right) = n(n - 1)f.$

**6. Surface in polar coordinates** Let

$$f(r, \theta) = \begin{cases} \dfrac{\sin 6r}{6r}, & r \neq 0 \\ 1, & r = 0, \end{cases}$$

where $r$ and $\theta$ are polar coordinates. Find

**a.** $\lim\limits_{r \to 0} f(r, \theta)$  **b.** $f_r(0, 0)$  **c.** $f_\theta(r, \theta), \quad r \neq 0.$

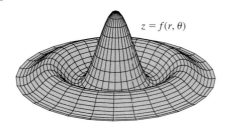

$z = f(r, \theta)$

### Gradients and Tangents

**7. Properties of position vectors** Let $\mathbf{r} = x\mathbf{i} + y\mathbf{j} + z\mathbf{k}$ and let $r = |\mathbf{r}|$.

**a.** Show that $\nabla r = \mathbf{r}/r.$

**b.** Show that $\nabla(r^n) = nr^{n-2}\mathbf{r}.$

**c.** Find a function whose gradient equals $\mathbf{r}$.

**d.** Show that $\mathbf{r} \cdot d\mathbf{r} = r \, dr.$

**e.** Show that $\nabla(\mathbf{A} \cdot \mathbf{r}) = \mathbf{A}$ for any constant vector $\mathbf{A}$.

**8. Gradient orthogonal to tangent** Suppose that a differentiable function $f(x, y)$ has the constant value $c$ along the differentiable curve $x = g(t), y = h(t)$; that is,

$$f(g(t), h(t)) = c$$

for all values of $t$. Differentiate both sides of this equation with respect to $t$ to show that $\nabla f$ is orthogonal to the curve's tangent vector at every point on the curve.

**9. Curve tangent to a surface** Show that the curve

$$\mathbf{r}(t) = (\ln t)\mathbf{i} + (t \ln t)\mathbf{j} + t\mathbf{k}$$

is tangent to the surface

$$xz^2 - yz + \cos xy = 1$$

at $(0, 0, 1)$.

**10. Curve tangent to a surface** Show that the curve

$$\mathbf{r}(t) = \left(\dfrac{t^3}{4} - 2\right)\mathbf{i} + \left(\dfrac{4}{t} - 3\right)\mathbf{j} + \cos(t - 2)\mathbf{k}$$

is tangent to the surface

$$x^3 + y^3 + z^3 - xyz = 0$$

at $(0, -1, 1)$.

### Extreme Values

**11. Extrema on a surface** Show that the only possible maxima and minima of $z$ on the surface $z = x^3 + y^3 - 9xy + 27$ occur at $(0, 0)$ and $(3, 3)$. Show that neither a maximum nor a minimum

occurs at $(0, 0)$. Determine whether $z$ has a maximum or a minimum at $(3, 3)$.

**12. Maximum in closed first quadrant** Find the maximum value of $f(x, y) = 6xye^{-(2x+3y)}$ in the closed first quadrant (includes the nonnegative axes).

**13. Minimum volume cut from first octant** Find the minimum volume for a region bounded by the planes $x = 0, y = 0, z = 0$ and a plane tangent to the ellipsoid

$$\dfrac{x^2}{a^2} + \dfrac{y^2}{b^2} + \dfrac{z^2}{c^2} = 1$$

at a point in the first octant.

**14. Minimum distance from a line to a parabola in $xy$-plane** By minimizing the function $f(x, y, u, v) = (x - u)^2 + (y - v)^2$ subject to the constraints $y = x + 1$ and $u = v^2$, find the minimum distance in the $xy$-plane from the line $y = x + 1$ to the parabola $y^2 = x$.

### Theory and Examples

**15. Boundedness of first partials implies continuity** Prove the following theorem: If $f(x, y)$ is defined in an open region $R$ of the $xy$-plane and if $f_x$ and $f_y$ are bounded on $R$, then $f(x, y)$ is continuous on $R$. (The assumption of boundedness is essential.)

**16.** Suppose that $\mathbf{r}(t) = g(t)\mathbf{i} + h(t)\mathbf{j} + k(t)\mathbf{k}$ is a smooth curve in the domain of a differentiable function $f(x, y, z)$. Describe the relation between $df/dt$, $\nabla f$, and $\mathbf{v} = d\mathbf{r}/dt$. What can be said about $\nabla f$ and $\mathbf{v}$ at interior points of the curve where $f$ has extreme values relative to its other values on the curve? Give reasons for your answer.

**17. Finding functions from partial derivatives** Suppose that $f$ and $g$ are functions of $x$ and $y$ such that

$$\dfrac{\partial f}{\partial y} = \dfrac{\partial g}{\partial x} \quad \text{and} \quad \dfrac{\partial f}{\partial x} = \dfrac{\partial g}{\partial y},$$

and suppose that

$$\dfrac{\partial f}{\partial x} = 0, \quad f(1, 2) = g(1, 2) = 5, \quad \text{and} \quad f(0, 0) = 4.$$

Find $f(x, y)$ and $g(x, y)$.

**18. Rate of change of the rate of change** We know that if $f(x, y)$ is a function of two variables and if $\mathbf{u} = a\mathbf{i} + b\mathbf{j}$ is a unit vector, then $D_{\mathbf{u}}f(x, y) = f_x(x, y)a + f_y(x, y)b$ is the rate of change of $f(x, y)$ at $(x, y)$ in the direction of $\mathbf{u}$. Give a similar formula for the rate of change *of the rate of change of* $f(x, y)$ at $(x, y)$ in the direction $\mathbf{u}$.

**19. Path of a heat-seeking particle** A heat-seeking particle has the property that at any point $(x, y)$ in the plane it moves in the direction of maximum temperature increase. If the temperature at $(x, y)$ is $T(x, y) = -e^{-2y}\cos x$, find an equation $y = f(x)$ for the path of a heat-seeking particle at the point $(\pi/4, 0)$.

**20. Velocity after a ricochet** A particle traveling in a straight line with constant velocity $\mathbf{i} + \mathbf{j} - 5\mathbf{k}$ passes through the point $(0, 0, 30)$ and hits the surface $z = 2x^2 + 3y^2$. The particle ricochets off the surface, the angle of reflection being equal to the angle of incidence. Assuming no loss of speed, what is the velocity of the particle after the ricochet? Simplify your answer.

**21. Directional derivatives tangent to a surface** Let $S$ be the surface that is the graph of $f(x, y) = 10 - x^2 - y^2$. Suppose that the temperature in space at each point $(x, y, z)$ is $T(x, y, z) = x^2y + y^2z + 4x + 14y + z$.

  **a.** Among all the possible directions tangential to the surface $S$ at the point $(0, 0, 10)$, which direction will make the rate of change of temperature at $(0, 0, 10)$ a maximum?

  **b.** Which direction tangential to $S$ at the point $(1, 1, 8)$ will make the rate of change of temperature a maximum?

**22. Drilling another borehole** On a flat surface of land, geologists drilled a borehole straight down and hit a mineral deposit at 300 m. They drilled a second borehole 30 m to the north of the first and hit the mineral deposit at 285 m. A third borehole 30 m east of the first borehole struck the mineral deposit at 307.5 m. The geologists have reasons to believe that the mineral deposit is in the shape of a dome, and for the sake of economy, they would like to find where the deposit is closest to the surface. Assuming the surface to be the $xy$-plane, in what direction from the first borehole would you suggest the geologists drill their fourth borehole?

**The one-dimensional heat equation** If $w(x, t)$ represents the temperature at position $x$ at time $t$ in a uniform wire with perfectly insulated sides, then the partial derivatives $w_{xx}$ and $w_t$ satisfy a differential equation of the form

$$w_{xx} = \frac{1}{c^2} w_t.$$

This equation is called the *one-dimensional heat equation*. The value of the positive constant $c^2$ is determined by the material from which the wire is made.

**23.** Find all solutions of the one-dimensional heat equation of the form $w = e^{rt} \sin \pi x$, where $r$ is a constant.

**24.** Find all solutions of the one-dimensional heat equation that have the form $w = e^{rt} \sin kx$ and satisfy the conditions that $w(0, t) = 0$ and $w(L, t) = 0$. What happens to these solutions as $t \to \infty$?

---

## Chapter **14** Technology Application Projects

Mathematica/Maple Modules:

*Plotting Surfaces*
Efficiently generate plots of surfaces, contours, and level curves.

*Exploring the Mathematics Behind Skateboarding: Analysis of the Directional Derivative*
The path of a skateboarder is introduced, first on a level plane, then on a ramp, and finally on a paraboloid. Compute, plot, and analyze the directional derivative in terms of the skateboarder.

*Looking for Patterns and Applying the Method of Least Squares to Real Data*
Fit a line to a set of numerical data points by choosing the line that minimizes the sum of the squares of the vertical distances from the points to the line.

*Lagrange Goes Skateboarding: How High Does He Go?*
Revisit and analyze the skateboarders' adventures for maximum and minimum heights from both a graphical and analytic perspective using Lagrange multipliers.

# 15

# Multiple Integrals

**OVERVIEW** In this chapter we define the *double integral* of a function of two variables $f(x, y)$ over a region in the plane as the limit of approximating Riemann sums. Just as a single integral represents signed area, so does a double integral represent signed volume. Double integrals can be evaluated using the Fundamental Theorem of Calculus studied in Section 5.4, but now the evaluations are done twice by integrating with respect to each of the variables $x$ and $y$ in turn. Double integrals can be used to find areas of more general regions in the plane than those encountered in Chapter 5. Moreover, just as the Substitution Rule could simplify finding single integrals, we can sometimes use polar coordinates to simplify computing a double integral. We study more general substitutions for evaluating double integrals as well.

We also define *triple integrals* for a function of three variables $f(x, y, z)$ over a region in space. Triple integrals can be used to find volumes of still more general regions in space, and their evaluation is like that of double integrals with yet a third evaluation. *Cylindrical* or *spherical coordinates* can sometimes be used to simplify the calculation of a triple integral, and we investigate those techniques. Double and triple integrals have a number of additional applications, such as calculating the average value of a multivariable function, and finding moments and centers of mass for more general regions than those encountered before.

## 15.1 Double and Iterated Integrals over Rectangles

In Chapter 5 we defined the definite integral of a continuous function $f(x)$ over an interval $[a, b]$ as a limit of Riemann sums. In this section we extend this idea to define the *double integral* of a continuous function of two variables $f(x, y)$ over a bounded rectangle $R$ in the plane. The Riemann sums for the integral of a single-variable function $f(x)$ are obtained by partitioning a finite interval into thin subintervals, multiplying the width of each subinterval by the value of $f$ at a point $c_k$ inside that subinterval, and then adding together all the products. A similar method of partitioning, multiplying, and summing is used to construct double integrals as limits of approximating Riemann sums.

### Double Integrals

We begin our investigation of double integrals by considering the simplest type of planar region, a rectangle. We consider a function $f(x, y)$ defined on a rectangular region $R$,

$$R: \quad a \leq x \leq b, \quad c \leq y \leq d.$$

We subdivide $R$ into small rectangles using a network of lines parallel to the $x$- and $y$-axes (Figure 15.1). The lines divide $R$ into $n$ rectangular pieces, where the number of such pieces $n$ gets large as the width and height of each piece gets small. These rectangles form

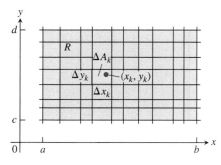

**FIGURE 15.1** Rectangular grid partitioning the region $R$ into small rectangles of area $\Delta A_k = \Delta x_k \, \Delta y_k$.

a **partition** of $R$. A small rectangular piece of width $\Delta x$ and height $\Delta y$ has area $\Delta A = \Delta x \Delta y$. If we number the small pieces partitioning $R$ in some order, then their areas are given by numbers $\Delta A_1, \Delta A_2, \ldots, \Delta A_n$, where $\Delta A_k$ is the area of the $k$th small rectangle.

To form a Riemann sum over $R$, we choose a point $(x_k, y_k)$ in the $k$th small rectangle, multiply the value of $f$ at that point by the area $\Delta A_k$, and add together the products:

$$S_n = \sum_{k=1}^{n} f(x_k, y_k)\,\Delta A_k.$$

Depending on how we pick $(x_k, y_k)$ in the $k$th small rectangle, we may get different values for $S_n$.

We are interested in what happens to these Riemann sums as the widths and heights of all the small rectangles in the partition of $R$ approach zero. The **norm** of a partition $P$, written $\|P\|$, is the largest width or height of any rectangle in the partition. If $\|P\| = 0.1$ then all the rectangles in the partition of $R$ have width at most 0.1 and height at most 0.1. Sometimes the Riemann sums converge as the norm of $P$ goes to zero, written $\|P\| \to 0$. The resulting limit is then written as

$$\lim_{\|P\| \to 0} \sum_{k=1}^{n} f(x_k, y_k)\,\Delta A_k.$$

As $\|P\| \to 0$ and the rectangles get narrow and short, their number $n$ increases, so we can also write this limit as

$$\lim_{n \to \infty} \sum_{k=1}^{n} f(x_k, y_k)\,\Delta A_k,$$

with the understanding that $\|P\| \to 0$, and hence $\Delta A_k \to 0$, as $n \to \infty$.

Many choices are involved in a limit of this kind. The collection of small rectangles is determined by the grid of vertical and horizontal lines that determine a rectangular partition of $R$. In each of the resulting small rectangles there is a choice of an arbitrary point $(x_k, y_k)$ at which $f$ is evaluated. These choices together determine a single Riemann sum. To form a limit, we repeat the whole process again and again, choosing partitions whose rectangle widths and heights both go to zero and whose number goes to infinity.

When a limit of the sums $S_n$ exists, giving the same limiting value no matter what choices are made, then the function $f$ is said to be **integrable** and the limit is called the **double integral** of $f$ over $R$, written as

$$\iint_R f(x, y)\,dA \qquad \text{or} \qquad \iint_R f(x, y)\,dx\,dy.$$

It can be shown that if $f(x, y)$ is a continuous function throughout $R$, then $f$ is integrable, as in the single-variable case discussed in Chapter 5. Many discontinuous functions are also integrable, including functions that are discontinuous only on a finite number of points or smooth curves. We leave the proof of these facts to a more advanced text.

## Double Integrals as Volumes

When $f(x, y)$ is a positive function over a rectangular region $R$ in the $xy$-plane, we may interpret the double integral of $f$ over $R$ as the volume of the 3-dimensional solid region over the $xy$-plane bounded below by $R$ and above by the surface $z = f(x, y)$ (Figure 15.2). Each term $f(x_k, y_k)\Delta A_k$ in the sum $S_n = \sum f(x_k, y_k)\Delta A_k$ is the volume of a vertical rectangular box that approximates the volume of the portion of the solid that stands directly above the base $\Delta A_k$. The sum $S_n$ thus approximates what we want to call the total volume of the solid. We *define* this volume to be

$$\text{Volume} = \lim_{n \to \infty} S_n = \iint_R f(x, y)\,dA,$$

where $\Delta A_k \to 0$ as $n \to \infty$.

**FIGURE 15.2** Approximating solids with rectangular boxes leads us to define the volumes of more general solids as double integrals. The volume of the solid shown here is the double integral of $f(x, y)$ over the base region $R$.

As you might expect, this more general method of calculating volume agrees with the methods in Chapter 6, but we do not prove this here. Figure 15.3 shows Riemann sum approximations to the volume becoming more accurate as the number $n$ of boxes increases.

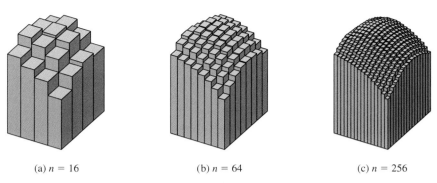

(a) $n = 16$        (b) $n = 64$        (c) $n = 256$

**FIGURE 15.3** As $n$ increases, the Riemann sum approximations approach the total volume of the solid shown in Figure 15.2.

## Fubini's Theorem for Calculating Double Integrals

Suppose that we wish to calculate the volume under the plane $z = 4 - x - y$ over the rectangular region $R$: $0 \le x \le 2, 0 \le y \le 1$ in the $xy$-plane. If we apply the method of slicing from Section 6.1, with slices perpendicular to the $x$-axis (Figure 15.4), then the volume is

$$\int_{x=0}^{x=2} A(x) \, dx, \tag{1}$$

where $A(x)$ is the cross-sectional area at $x$. For each value of $x$, we may calculate $A(x)$ as the integral

$$A(x) = \int_{y=0}^{y=1} (4 - x - y) \, dy, \tag{2}$$

which is the area under the curve $z = 4 - x - y$ in the plane of the cross-section at $x$. In calculating $A(x)$, $x$ is held fixed and the integration takes place with respect to $y$. Combining Equations (1) and (2), we see that the volume of the entire solid is

$$\text{Volume} = \int_{x=0}^{x=2} A(x) \, dx = \int_{x=0}^{x=2} \left( \int_{y=0}^{y=1} (4 - x - y) \, dy \right) dx$$

$$= \int_{x=0}^{x=2} \left[ 4y - xy - \frac{y^2}{2} \right]_{y=0}^{y=1} dx = \int_{x=0}^{x=2} \left( \frac{7}{2} - x \right) dx$$

$$= \left[ \frac{7}{2}x - \frac{x^2}{2} \right]_0^2 = 5.$$

If we just wanted to write a formula for the volume, without carrying out any of the integrations, we could write

$$\text{Volume} = \int_0^2 \int_0^1 (4 - x - y) \, dy \, dx. \tag{3}$$

The expression on the right, called an **iterated** or **repeated integral**, says that the volume is obtained by integrating $4 - x - y$ with respect to $y$ from $y = 0$ to $y = 1$, holding $x$ fixed, and then integrating the resulting expression in $x$ with respect to $x$ from $x = 0$ to $x = 2$. The limits of integration 0 and 1 are associated with $y$, so they are placed on the integral closest to $dy$. The other limits of integration, 0 and 2, are associated with the variable $x$, so they are placed on the outside integral symbol that is paired with $dx$.

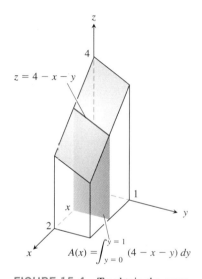

**FIGURE 15.4** To obtain the cross-sectional area $A(x)$, we hold $x$ fixed and integrate with respect to $y$.

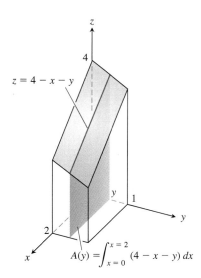

$$z = 4 - x - y$$

$$A(y) = \int_{x=0}^{x=2} (4 - x - y)\, dx$$

**FIGURE 15.5** To obtain the cross-sectional area $A(y)$, we hold $y$ fixed and integrate with respect to $x$.

What would have happened if we had calculated the volume by slicing with planes perpendicular to the $y$-axis (Figure 15.5)? As a function of $y$, the typical cross-sectional area is

$$A(y) = \int_{x=0}^{x=2} (4 - x - y)\, dx = \left[ 4x - \frac{x^2}{2} - xy \right]_{x=0}^{x=2} = 6 - 2y. \qquad (4)$$

The volume of the entire solid is therefore

$$\text{Volume} = \int_{y=0}^{y=1} A(y)\, dy = \int_{y=0}^{y=1} (6 - 2y)\, dy = \left[ 6y - y^2 \right]_0^1 = 5,$$

in agreement with our earlier calculation.

Again, we may give a formula for the volume as an iterated integral by writing

$$\text{Volume} = \int_0^1 \int_0^2 (4 - x - y)\, dx\, dy.$$

The expression on the right says we can find the volume by integrating $4 - x - y$ with respect to $x$ from $x = 0$ to $x = 2$ as in Equation (4) and integrating the result with respect to $y$ from $y = 0$ to $y = 1$. In this iterated integral, the order of integration is first $x$ and then $y$, the reverse of the order in Equation (3).

What do these two volume calculations with iterated integrals have to do with the double integral

$$\iint_R (4 - x - y)\, dA$$

over the rectangle $R: 0 \le x \le 2, 0 \le y \le 1$? The answer is that both iterated integrals give the value of the double integral. This is what we would reasonably expect, since the double integral measures the volume of the same region as the two iterated integrals. A theorem published in 1907 by Guido Fubini says that the double integral of any continuous function over a rectangle can be calculated as an iterated integral in either order of integration. (Fubini proved his theorem in greater generality, but this is what it says in our setting.)

---

**THEOREM 1—Fubini's Theorem (First Form)**   If $f(x, y)$ is continuous throughout the rectangular region $R: a \le x \le b, c \le y \le d$, then

$$\iint_R f(x, y)\, dA = \int_c^d \int_a^b f(x, y)\, dx\, dy = \int_a^b \int_c^d f(x, y)\, dy\, dx.$$

---

Fubini's Theorem says that double integrals over rectangles can be calculated as iterated integrals. Thus, we can evaluate a double integral by integrating with respect to one variable at a time using the Fundamental Theorem of Calculus.

Fubini's Theorem also says that we may calculate the double integral by integrating in *either* order, a genuine convenience. When we calculate a volume by slicing, we may use either planes perpendicular to the $x$-axis or planes perpendicular to the $y$-axis.

**EXAMPLE 1**   Calculate $\iint_R f(x, y)\, dA$ for

$$f(x, y) = 100 - 6x^2 y \qquad \text{and} \qquad R: \ 0 \le x \le 2, \ -1 \le y \le 1.$$

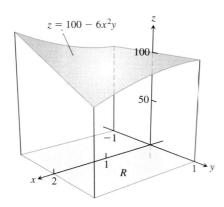

**FIGURE 15.6** The double integral $\iint_R f(x, y)\, dA$ gives the volume under this surface over the rectangular region $R$ (Example 1).

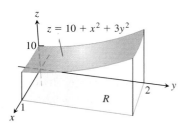

**FIGURE 15.7** The double integral $\iint_R f(x, y)\, dA$ gives the volume under this surface over the rectangular region $R$ (Example 2).

**Solution** Figure 15.6 displays the volume beneath the surface. By Fubini's Theorem,

$$\iint_R f(x, y)\, dA = \int_{-1}^{1}\int_{0}^{2} (100 - 6x^2 y)\, dx\, dy = \int_{-1}^{1}\left[ 100x - 2x^3 y \right]_{x=0}^{x=2} dy$$

$$= \int_{-1}^{1} (200 - 16y)\, dy = \left[ 200y - 8y^2 \right]_{-1}^{1} = 400.$$

Reversing the order of integration gives the same answer:

$$\int_{0}^{2}\int_{-1}^{1} (100 - 6x^2 y)\, dy\, dx = \int_{0}^{2}\left[ 100y - 3x^2 y^2 \right]_{y=-1}^{y=1} dx$$

$$= \int_{0}^{2} \left[ (100 - 3x^2) - (-100 - 3x^2) \right] dx$$

$$= \int_{0}^{2} 200\, dx = 400. \qquad \blacksquare$$

**EXAMPLE 2** Find the volume of the region bounded above by the elliptical paraboloid $z = 10 + x^2 + 3y^2$ and below by the rectangle $R: 0 \le x \le 1, 0 \le y \le 2$.

**Solution** The surface and volume are shown in Figure 15.7. The volume is given by the double integral

$$V = \iint_R (10 + x^2 + 3y^2)\, dA = \int_{0}^{1}\int_{0}^{2} (10 + x^2 + 3y^2)\, dy\, dx$$

$$= \int_{0}^{1}\left[ 10y + x^2 y + y^3 \right]_{y=0}^{y=2} dx$$

$$= \int_{0}^{1} (20 + 2x^2 + 8)\, dx = \left[ 20x + \frac{2}{3}x^3 + 8x \right]_{0}^{1} = \frac{86}{3}. \qquad \blacksquare$$

# Exercises 15.1

### Evaluating Iterated Integrals
In Exercises 1–14, evaluate the iterated integral.

**1.** $\displaystyle\int_{1}^{2}\int_{0}^{4} 2xy\, dy\, dx$

**2.** $\displaystyle\int_{0}^{2}\int_{-1}^{1} (x - y)\, dy\, dx$

**3.** $\displaystyle\int_{-1}^{0}\int_{-1}^{1} (x + y + 1)\, dx\, dy$

**4.** $\displaystyle\int_{0}^{1}\int_{0}^{1}\left(1 - \frac{x^2 + y^2}{2}\right) dx\, dy$

**5.** $\displaystyle\int_{0}^{3}\int_{0}^{2} (4 - y^2)\, dy\, dx$

**6.** $\displaystyle\int_{0}^{3}\int_{-2}^{0} (x^2 y - 2xy)\, dy\, dx$

**7.** $\displaystyle\int_{0}^{1}\int_{0}^{1} \frac{y}{1 + xy}\, dx\, dy$

**8.** $\displaystyle\int_{1}^{4}\int_{0}^{4}\left(\frac{x}{2} + \sqrt{y}\right) dx\, dy$

**9.** $\displaystyle\int_{0}^{\ln 2}\int_{1}^{\ln 5} e^{2x+y}\, dy\, dx$

**10.** $\displaystyle\int_{0}^{1}\int_{1}^{2} xye^x\, dy\, dx$

**11.** $\displaystyle\int_{-1}^{2}\int_{0}^{\pi/2} y \sin x\, dx\, dy$

**12.** $\displaystyle\int_{\pi}^{2\pi}\int_{0}^{\pi} (\sin x + \cos y)\, dx\, dy$

**13.** $\displaystyle\int_{1}^{4}\int_{1}^{e} \frac{\ln x}{xy}\, dx\, dy$

**14.** $\displaystyle\int_{-1}^{2}\int_{1}^{2} x \ln y\, dy\, dx$

### Evaluating Double Integrals over Rectangles
In Exercises 15–22, evaluate the double integral over the given region $R$.

**15.** $\displaystyle\iint_R (6y^2 - 2x)\, dA, \qquad R: 0 \le x \le 1, \quad 0 \le y \le 2$

**16.** $\displaystyle\iint_R \left(\frac{\sqrt{x}}{y^2}\right) dA, \qquad R: 0 \le x \le 4, \quad 1 \le y \le 2$

**17.** $\displaystyle\iint_R xy \cos y\, dA, \qquad R: -1 \le x \le 1, \quad 0 \le y \le \pi$

**18.** $\displaystyle\iint_R y \sin (x + y)\, dA, \qquad R: -\pi \le x \le 0, \quad 0 \le y \le \pi$

**19.** $\displaystyle\iint_R e^{x-y}\, dA,$     $R:$   $0 \le x \le \ln 2, \;\; 0 \le y \le \ln 2$

**20.** $\displaystyle\iint_R xye^{xy^2}\, dA,$     $R:$   $0 \le x \le 2, \;\; 0 \le y \le 1$

**21.** $\displaystyle\iint_R \frac{xy^3}{x^2 + 1}\, dA,$     $R:$   $0 \le x \le 1, \;\; 0 \le y \le 2$

**22.** $\displaystyle\iint_R \frac{y}{x^2 y^2 + 1}\, dA,$     $R:$   $0 \le x \le 1, \;\; 0 \le y \le 1$

In Exercises 23 and 24, integrate $f$ over the given region.

**23. Square**   $f(x, y) = 1/(xy)$   over   the   square   $1 \le x \le 2,$   $1 \le y \le 2$

**24. Rectangle**   $f(x, y) = y \cos xy$ over the rectangle $0 \le x \le \pi,$   $0 \le y \le 1$

**25.** Find the volume of the region bounded above by the paraboloid $z = x^2 + y^2$ and below by the square $R: -1 \le x \le 1,$ $-1 \le y \le 1.$

**26.** Find the volume of the region bounded above by the elliptical paraboloid $z = 16 - x^2 - y^2$ and below by the square $R: 0 \le x \le 2, 0 \le y \le 2.$

**27.** Find the volume of the region bounded above by the plane $z = 2 - x - y$ and below by the square $R: 0 \le x \le 1,$ $0 \le y \le 1.$

**28.** Find the volume of the region bounded above by the plane $z = y/2$ and below by the rectangle $R: 0 \le x \le 4, 0 \le y \le 2.$

**29.** Find the volume of the region bounded above by the surface $z = 2 \sin x \cos y$ and below by the rectangle $R: 0 \le x \le \pi/2,$ $0 \le y \le \pi/4.$

**30.** Find the volume of the region bounded above by the surface $z = 4 - y^2$ and below by the rectangle $R: 0 \le x \le 1,$ $0 \le y \le 2.$

**31.** Find a value of the constant $k$ so that $\displaystyle\int_1^2 \int_0^3 kx^2 y \, dx\, dy = 1.$

**32.** Evaluate $\displaystyle\int_{-1}^1 \int_0^{\pi/2} x \sin \sqrt{y} \, dy\, dx.$

**33.** Use Fubini's Theorem to evaluate

$$\int_0^2 \int_0^1 \frac{x}{1 + xy} \, dx\, dy.$$

**34.** Use Fubini's Theorem to evaluate

$$\int_0^1 \int_0^3 xe^{xy} \, dx\, dy.$$

**T** **35.** Use a software application to compute the integrals

**a.** $\displaystyle\int_0^1 \int_0^2 \frac{y - x}{(x + y)^3} \, dx\, dy$

**b.** $\displaystyle\int_0^2 \int_0^1 \frac{y - x}{(x + y)^3} \, dy\, dx$

Explain why your results do not contradict Fubini's Theorem.

**36.** If $f(x, y)$ is continuous over $R: a \le x \le b, c \le y \le d$ and

$$F(x, y) = \int_a^x \int_c^y f(u, v) \, dv\, du$$

on the interior of $R$, find the second partial derivatives $F_{xy}$ and $F_{yx}.$

# $15.2$ Double Integrals over General Regions

In this section we define and evaluate double integrals over bounded regions in the plane which are more general than rectangles. These double integrals are also evaluated as iterated integrals, with the main practical problem being that of determining the limits of integration. Since the region of integration may have boundaries other than line segments parallel to the coordinate axes, the limits of integration often involve variables, not just constants.

## Double Integrals over Bounded, Nonrectangular Regions

To define the double integral of a function $f(x, y)$ over a bounded, nonrectangular region $R$, such as the one in Figure 15.8, we again begin by covering $R$ with a grid of small rectangular cells whose union contains all points of $R$. This time, however, we cannot exactly fill $R$ with a finite number of rectangles lying inside $R$, since its boundary is curved, and some of the small rectangles in the grid lie partly outside $R$. A partition of $R$ is formed by taking the rectangles that lie completely inside it, not using any that are either partly or completely outside. For commonly arising regions, more and more of $R$ is included as the norm of a partition (the largest width or height of any rectangle used) approaches zero.

Once we have a partition of $R$, we number the rectangles in some order from 1 to $n$ and let $\Delta A_k$ be the area of the $k$th rectangle. We then choose a point $(x_k, y_k)$ in the $k$th rectangle and form the Riemann sum

$$S_n = \sum_{k=1}^n f(x_k, y_k) \, \Delta A_k.$$

**FIGURE 15.8** A rectangular grid partitioning a bounded, nonrectangular region into rectangular cells.

As the norm of the partition forming $S_n$ goes to zero, $\|P\| \to 0$, the width and height of each enclosed rectangle goes to zero and their number goes to infinity. If $f(x, y)$ is a continuous function, then these Riemann sums converge to a limiting value, not dependent on any of the choices we made. This limit is called the **double integral** of $f(x, y)$ over $R$:

$$\lim_{\|P\| \to 0} \sum_{k=1}^{n} f(x_k, y_k) \, \Delta A_k = \iint_R f(x, y) \, dA.$$

The nature of the boundary of $R$ introduces issues not found in integrals over an interval. When $R$ has a curved boundary, the $n$ rectangles of a partition lie inside $R$ but do not cover all of $R$. In order for a partition to approximate $R$ well, the parts of $R$ covered by small rectangles lying partly outside $R$ must become negligible as the norm of the partition approaches zero. This property of being nearly filled in by a partition of small norm is satisfied by all the regions that we will encounter. There is no problem with boundaries made from polygons, circles, ellipses, and from continuous graphs over an interval, joined end to end. A curve with a "fractal" type of shape would be problematic, but such curves arise rarely in most applications. A careful discussion of which type of regions $R$ can be used for computing double integrals is left to a more advanced text.

## Volumes

If $f(x, y)$ is positive and continuous over $R$, we define the volume of the solid region between $R$ and the surface $z = f(x, y)$ to be $\iint_R f(x, y) \, dA$, as before (Figure 15.9).

If $R$ is a region like the one shown in the $xy$-plane in Figure 15.10, bounded "above" and "below" by the curves $y = g_2(x)$ and $y = g_1(x)$ and on the sides by the lines $x = a, x = b$, we may again calculate the volume by the method of slicing. We first calculate the cross-sectional area

$$A(x) = \int_{y=g_1(x)}^{y=g_2(x)} f(x, y) \, dy$$

and then integrate $A(x)$ from $x = a$ to $x = b$ to get the volume as an iterated integral:

$$V = \int_a^b A(x) \, dx = \int_a^b \int_{g_1(x)}^{g_2(x)} f(x, y) \, dy \, dx. \tag{1}$$

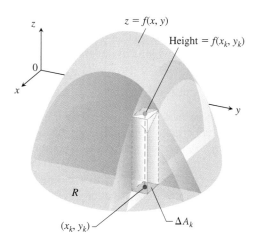

$$\text{Volume} = \lim \sum f(x_k, y_k) \, \Delta A_k = \iint_R f(x, y) \, dA$$

**FIGURE 15.9** We define the volumes of solids with curved bases as a limit of approximating rectangular boxes.

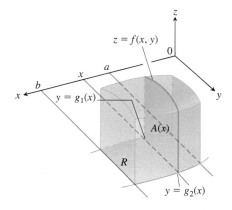

**FIGURE 15.10** The area of the vertical slice shown here is $A(x)$. To calculate the volume of the solid, we integrate this area from $x = a$ to $x = b$:

$$\int_a^b A(x) \, dx = \int_a^b \int_{g_1(x)}^{g_2(x)} f(x, y) \, dy \, dx.$$

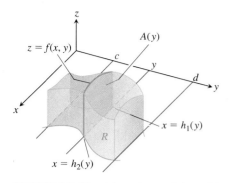

**FIGURE 15.11** The volume of the solid shown here is

$$\int_c^d A(y)\, dy = \int_c^d \int_{h_1(y)}^{h_2(y)} f(x, y)\, dx\, dy.$$

For a given solid, Theorem 2 says we can calculate the volume as in Figure 15.10, or in the way shown here. Both calculations have the same result.

Similarly, if $R$ is a region like the one shown in Figure 15.11, bounded by the curves $x = h_2(y)$ and $x = h_1(y)$ and the lines $y = c$ and $y = d$, then the volume calculated by slicing is given by the iterated integral

$$\text{Volume} = \int_c^d \int_{h_1(y)}^{h_2(y)} f(x, y)\, dx\, dy. \tag{2}$$

That the iterated integrals in Equations (1) and (2) both give the volume that we defined to be the double integral of $f$ over $R$ is a consequence of the following stronger form of Fubini's Theorem.

---

**THEOREM 2—Fubini's Theorem (Stronger Form)** Let $f(x, y)$ be continuous on a region $R$.

1. If $R$ is defined by $a \le x \le b$, $g_1(x) \le y \le g_2(x)$, with $g_1$ and $g_2$ continuous on $[a, b]$, then

$$\iint_R f(x, y)\, dA = \int_a^b \int_{g_1(x)}^{g_2(x)} f(x, y)\, dy\, dx.$$

2. If $R$ is defined by $c \le y \le d$, $h_1(y) \le x \le h_2(y)$, with $h_1$ and $h_2$ continuous on $[c, d]$, then

$$\iint_R f(x, y)\, dA = \int_c^d \int_{h_1(y)}^{h_2(y)} f(x, y)\, dx\, dy.$$

---

**EXAMPLE 1** Find the volume of the prism whose base is the triangle in the $xy$-plane bounded by the $x$-axis and the lines $y = x$ and $x = 1$ and whose top lies in the plane

$$z = f(x, y) = 3 - x - y.$$

**Solution** See Figure 15.12. For any $x$ between 0 and 1, $y$ may vary from $y = 0$ to $y = x$ (Figure 15.12b). Hence,

$$V = \int_0^1 \int_0^x (3 - x - y)\, dy\, dx = \int_0^1 \left[ 3y - xy - \frac{y^2}{2} \right]_{y=0}^{y=x} dx$$

$$= \int_0^1 \left( 3x - \frac{3x^2}{2} \right) dx = \left[ \frac{3x^2}{2} - \frac{x^3}{2} \right]_{x=0}^{x=1} = 1.$$

When the order of integration is reversed (Figure 15.12c), the integral for the volume is

$$V = \int_0^1 \int_y^1 (3 - x - y)\, dx\, dy = \int_0^1 \left[ 3x - \frac{x^2}{2} - xy \right]_{x=y}^{x=1} dy$$

$$= \int_0^1 \left( 3 - \frac{1}{2} - y - 3y + \frac{y^2}{2} + y^2 \right) dy$$

$$= \int_0^1 \left( \frac{5}{2} - 4y + \frac{3}{2}y^2 \right) dy = \left[ \frac{5}{2}y - 2y^2 + \frac{y^3}{2} \right]_{y=0}^{y=1} = 1.$$

The two integrals are equal, as they should be. ∎

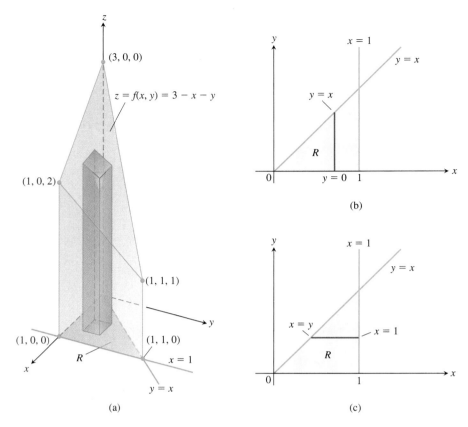

**FIGURE 15.12** (a) Prism with a triangular base in the $xy$-plane. The volume of this prism is defined as a double integral over $R$. To evaluate it as an iterated integral, we may integrate first with respect to $y$ and then with respect to $x$, or the other way around (Example 1). (b) Integration limits of

$$\int_{x=0}^{x=1} \int_{y=0}^{y=x} f(x, y)\, dy\, dx.$$

If we integrate first with respect to $y$, we integrate along a vertical line through $R$ and then integrate from left to right to include all the vertical lines in $R$. (c) Integration limits of

$$\int_{y=0}^{y=1} \int_{x=y}^{x=1} f(x, y)\, dx\, dy.$$

If we integrate first with respect to $x$, we integrate along a horizontal line through $R$ and then integrate from bottom to top to include all the horizontal lines in $R$.

Although Fubini's Theorem assures us that a double integral may be calculated as an iterated integral in either order of integration, the value of one integral may be easier to find than the value of the other. The next example shows how this can happen.

**EXAMPLE 2** Calculate

$$\iint_R \frac{\sin x}{x}\, dA,$$

where $R$ is the triangle in the $xy$-plane bounded by the $x$-axis, the line $y = x$, and the line $x = 1$.

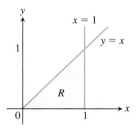

**FIGURE 15.13**   The region of integration in Example 2.

(a)

(b)

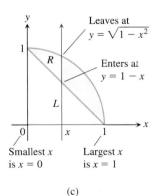

(c)

**FIGURE 15.14**   Finding the limits of integration when integrating first with respect to $y$ and then with respect to $x$.

**Solution**   The region of integration is shown in Figure 15.13. If we integrate first with respect to $y$ and then with respect to $x$, we find

$$\int_0^1 \left( \int_0^x \frac{\sin x}{x} \, dy \right) dx = \int_0^1 \left( y \frac{\sin x}{x} \right]_{y=0}^{y=x} \right) dx = \int_0^1 \sin x \, dx$$

$$= -\cos(1) + 1 \approx 0.46.$$

If we reverse the order of integration and attempt to calculate

$$\int_0^1 \int_y^1 \frac{\sin x}{x} \, dx \, dy,$$

we run into a problem because $\int ((\sin x)/x) \, dx$ cannot be expressed in terms of elementary functions (there is no simple antiderivative).

There is no general rule for predicting which order of integration will be the good one in circumstances like these. If the order you first choose doesn't work, try the other. Sometimes neither order will work, and then we need to use numerical approximations.    ∎

## Finding Limits of Integration

We now give a procedure for finding limits of integration that applies for many regions in the plane. Regions that are more complicated, and for which this procedure fails, can often be split up into pieces on which the procedure works.

**Using Vertical Cross-Sections**   When faced with evaluating $\iint_R f(x, y) \, dA$, integrating first with respect to $y$ and then with respect to $x$, do the following three steps:

1. *Sketch.* Sketch the region of integration and label the bounding curves (Figure 15.14a).

2. *Find the y-limits of integration.* Imagine a vertical line $L$ cutting through $R$ in the direction of increasing $y$. Mark the $y$-values where $L$ enters and leaves. These are the $y$-limits of integration and are usually functions of $x$ (instead of constants) (Figure 15.14b).

3. *Find the x-limits of integration.* Choose $x$-limits that include all the vertical lines through $R$. The integral shown here (see Figure 15.14c) is

$$\iint_R f(x, y) \, dA = \int_{x=0}^{x=1} \int_{y=1-x}^{y=\sqrt{1-x^2}} f(x, y) \, dy \, dx.$$

**Using Horizontal Cross-Sections**   To evaluate the same double integral as an iterated integral with the order of integration reversed, use horizontal lines instead of vertical lines in Steps 2 and 3 (see Figure 15.15). The integral is

$$\iint_R f(x, y) \, dA = \int_0^1 \int_{1-y}^{\sqrt{1-y^2}} f(x, y) \, dx \, dy.$$

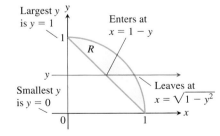

**FIGURE 15.15**   Finding the limits of integration when integrating first with respect to $x$ and then with respect to $y$.

(a)

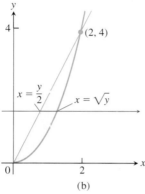

(b)

**FIGURE 15.16** Region of integration for Example 3.

**EXAMPLE 3**    Sketch the region of integration for the integral

$$\int_0^2 \int_{x^2}^{2x} (4x + 2) \, dy \, dx$$

and write an equivalent integral with the order of integration reversed.

**Solution**    The region of integration is given by the inequalities $x^2 \le y \le 2x$ and $0 \le x \le 2$. It is therefore the region bounded by the curves $y = x^2$ and $y = 2x$ between $x = 0$ and $x = 2$ (Figure 15.16a).

To find limits for integrating in the reverse order, we imagine a horizontal line passing from left to right through the region. It enters at $x = y/2$ and leaves at $x = \sqrt{y}$. To include all such lines, we let $y$ run from $y = 0$ to $y = 4$ (Figure 15.16b). The integral is

$$\int_0^4 \int_{y/2}^{\sqrt{y}} (4x + 2) \, dx \, dy.$$

The common value of these integrals is 8.    ■

### Properties of Double Integrals

Like single integrals, double integrals of continuous functions have algebraic properties that are useful in computations and applications.

---

If $f(x, y)$ and $g(x, y)$ are continuous on the bounded region $R$, then the following properties hold.

1. *Constant Multiple:* $\displaystyle\iint\limits_R cf(x, y) \, dA = c \iint\limits_R f(x, y) \, dA$    (any number $c$)

2. *Sum and Difference:*

$$\iint\limits_R (f(x, y) \pm g(x, y)) \, dA = \iint\limits_R f(x, y) \, dA \pm \iint\limits_R g(x, y) \, dA$$

3. *Domination:*

   (a) $\displaystyle\iint\limits_R f(x, y) \, dA \ge 0$    if    $f(x, y) \ge 0$ on $R$

   (b) $\displaystyle\iint\limits_R f(x, y) \, dA \ge \iint\limits_R g(x, y) \, dA$    if    $f(x, y) \ge g(x, y)$ on $R$

4. *Additivity:* $\displaystyle\iint\limits_R f(x, y) \, dA = \iint\limits_{R_1} f(x, y) \, dA + \iint\limits_{R_2} f(x, y) \, dA$

   if $R$ is the union of two nonoverlapping regions $R_1$ and $R_2$

---

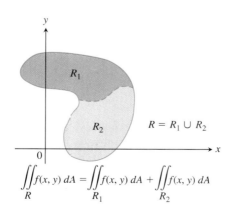

**FIGURE 15.17** The Additivity Property for rectangular regions holds for regions bounded by smooth curves.

Property 4 assumes that the region of integration $R$ is decomposed into nonoverlapping regions $R_1$ and $R_2$ with boundaries consisting of a finite number of line segments or smooth curves. Figure 15.17 illustrates an example of this property.

The idea behind these properties is that integrals behave like sums. If the function $f(x, y)$ is replaced by its constant multiple $cf(x, y)$, then a Riemann sum for $f$

$$S_n = \sum_{k=1}^{n} f(x_k, y_k) \, \Delta A_k$$

is replaced by a Riemann sum for $cf$

$$\sum_{k=1}^{n} cf(x_k, y_k) \, \Delta A_k = c \sum_{k=1}^{n} f(x_k, y_k) \, \Delta A_k = cS_n.$$

Taking limits as $n \to \infty$ shows that $c \lim_{n \to \infty} S_n = c \iint_R f \, dA$ and $\lim_{n \to \infty} cS_n = \iint_R cf \, dA$ are equal. It follows that the Constant Multiple Property carries over from sums to double integrals.

The other properties are also easy to verify for Riemann sums, and carry over to double integrals for the same reason. While this discussion gives the idea, an actual proof that these properties hold requires a more careful analysis of how Riemann sums converge.

(a)

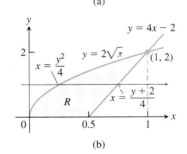

(b)

**FIGURE 15.18** (a) The solid "wedge-like" region whose volume is found in Example 4. (b) The region of integration $R$ showing the order $dx\,dy$.

**EXAMPLE 4**     Find the volume of the wedgelike solid that lies beneath the surface $z = 16 - x^2 - y^2$ and above the region $R$ bounded by the curve $y = 2\sqrt{x}$, the line $y = 4x - 2$, and the $x$-axis.

**Solution**     Figure 15.18a shows the surface and the "wedgelike" solid whose volume we want to calculate. Figure 15.18b shows the region of integration in the $xy$-plane. If we integrate in the order $dy\,dx$ (first with respect to $y$ and then with respect to $x$), two integrations will be required because $y$ varies from $y = 0$ to $y = 2\sqrt{x}$ for $0 \le x \le 0.5$, and then varies from $y = 4x - 2$ to $y = 2\sqrt{x}$ for $0.5 \le x \le 1$. So we choose to integrate in the order $dx\,dy$, which requires only one double integral whose limits of integration are indicated in Figure 15.18b. The volume is then calculated as the iterated integral:

$$\iint_R \left( 16 - x^2 - y^2 \right) dA$$

$$= \int_0^2 \int_{y^2/4}^{(y+2)/4} \left( 16 - x^2 - y^2 \right) dx\,dy$$

$$= \int_0^2 \left[ 16x - \frac{x^3}{3} - xy^2 \right]_{x=y^2/4}^{x=(y+2)/4} dx$$

$$= \int_0^2 \left[ 4(y + 2) - \frac{(y + 2)^3}{3 \cdot 64} - \frac{(y + 2)y^2}{4} - 4y^2 + \frac{y^6}{3 \cdot 64} + \frac{y^4}{4} \right] dy$$

$$= \left[ \frac{191y}{24} + \frac{63y^2}{32} - \frac{145y^3}{96} - \frac{49y^4}{768} + \frac{y^5}{20} + \frac{y^7}{1344} \right]_0^2 = \frac{20803}{1680} \approx 12.4. \qquad \blacksquare$$

Our development of the double integral has focused on its representation of the volume of the solid region between $R$ and the surface $z = f(x, y)$ of a positive continuous function. Just as we saw with signed area in the case of single integrals, when $f(x_k, y_k)$ is negative, then the product $f(x_k, y_k) \Delta A_k$ is the negative of the volume of the rectangular box shown in Figure 15.9 that was used to form the approximating Riemann sum. So for an arbitrary continuous function $f$ defined over $R$, the limit of any Riemann sum represents the *signed* volume (not the total volume) of the solid region between $R$ and the surface. The double integral has other interpretations as well, and in the next section we will see how it is used to calculate the area of a general region in the plane.

## Exercises 15.2

### Sketching Regions of Integration

In Exercises 1–8, sketch the described regions of integration.

1. $0 \le x \le 3, \quad 0 \le y \le 2x$
2. $-1 \le x \le 2, \quad x - 1 \le y \le x^2$
3. $-2 \le y \le 2, \quad y^2 \le x \le 4$
4. $0 \le y \le 1, \quad y \le x \le 2y$
5. $0 \le x \le 1, \quad e^x \le y \le e$
6. $1 \le x \le e^2, \quad 0 \le y \le \ln x$
7. $0 \le y \le 1, \quad 0 \le x \le \sin^{-1} y$
8. $0 \le y \le 8, \quad \dfrac{1}{4}y \le x \le y^{1/3}$

### Finding Limits of Integration

In Exercises 9–18, write an iterated integral for $\iint_R dA$ over the described region $R$ using (a) vertical cross-sections, (b) horizontal cross-sections.

9.

10.

11.

12.

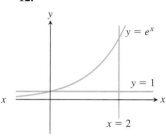

13. Bounded by $y = \sqrt{x}, y = 0$, and $x = 9$
14. Bounded by $y = \tan x, x = 0$, and $y = 1$
15. Bounded by $y = e^{-x}, y = 1$, and $x = \ln 3$
16. Bounded by $y = 0, x = 0, y = 1$, and $y = \ln x$
17. Bounded by $y = 3 - 2x, y = x$, and $x = 0$
18. Bounded by $y = x^2$ and $y = x + 2$

### Finding Regions of Integration and Double Integrals

In Exercises 19–24, sketch the region of integration and evaluate the integral.

19. $\displaystyle\int_0^\pi \int_0^x x \sin y \, dy \, dx$

20. $\displaystyle\int_0^\pi \int_0^{\sin x} y \, dy \, dx$

21. $\displaystyle\int_1^{\ln 8} \int_0^{\ln y} e^{x+y} \, dx \, dy$

22. $\displaystyle\int_1^2 \int_y^{y^2} dx \, dy$

23. $\displaystyle\int_0^1 \int_0^{y^2} 3y^3 e^{xy} \, dx \, dy$

24. $\displaystyle\int_1^4 \int_0^{\sqrt{x}} \frac{3}{2} e^{y/\sqrt{x}} \, dy \, dx$

In Exercises 25–28, integrate $f$ over the given region.

25. **Quadrilateral**  $f(x, y) = x/y$ over the region in the first quadrant bounded by the lines $y = x, y = 2x, x = 1$, and $x = 2$

26. **Triangle**  $f(x, y) = x^2 + y^2$ over the triangular region with vertices $(0, 0), (1, 0)$, and $(0, 1)$

27. **Triangle**  $f(u, v) = v - \sqrt{u}$ over the triangular region cut from the first quadrant of the $uv$-plane by the line $u + v = 1$

28. **Curved region**  $f(s, t) = e^s \ln t$ over the region in the first quadrant of the $st$-plane that lies above the curve $s = \ln t$ from $t = 1$ to $t = 2$

Each of Exercises 29–32 gives an integral over a region in a Cartesian coordinate plane. Sketch the region and evaluate the integral.

29. $\displaystyle\int_{-2}^0 \int_v^{-v} 2 \, dp \, dv$  (the $pv$-plane)

30. $\displaystyle\int_0^1 \int_0^{\sqrt{1-s^2}} 8t \, dt \, ds$  (the $st$-plane)

31. $\displaystyle\int_{-\pi/3}^{\pi/3} \int_0^{\sec t} 3 \cos t \, du \, dt$  (the $tu$-plane)

32. $\displaystyle\int_0^{3/2} \int_1^{4-2u} \frac{4 - 2u}{v^2} \, dv \, du$  (the $uv$-plane)

### Reversing the Order of Integration

In Exercises 33–46, sketch the region of integration and write an equivalent double integral with the order of integration reversed.

33. $\displaystyle\int_0^1 \int_2^{4-2x} dy \, dx$

34. $\displaystyle\int_0^2 \int_{y-2}^0 dx \, dy$

35. $\displaystyle\int_0^1 \int_y^{\sqrt{y}} dx \, dy$

36. $\displaystyle\int_0^1 \int_{1-x}^{1-x^2} dy \, dx$

37. $\displaystyle\int_0^1 \int_1^{e^x} dy \, dx$

38. $\displaystyle\int_0^{\ln 2} \int_{e^y}^2 dx \, dy$

39. $\displaystyle\int_0^{3/2} \int_0^{9-4x^2} 16x \, dy \, dx$

40. $\displaystyle\int_0^2 \int_0^{4-y^2} y \, dx \, dy$

41. $\displaystyle\int_0^1 \int_{-\sqrt{1-y^2}}^{\sqrt{1-y^2}} 3y \, dx \, dy$

42. $\displaystyle\int_0^2 \int_{-\sqrt{4-x^2}}^{\sqrt{4-x^2}} 6x \, dy \, dx$

43. $\displaystyle\int_1^e \int_0^{\ln x} xy \, dy \, dx$

44. $\displaystyle\int_0^{\pi/6} \int_{\sin x}^{1/2} xy^2 \, dy \, dx$

45. $\displaystyle\int_0^3 \int_1^{e^y} (x + y) \, dx \, dy$

46. $\displaystyle\int_0^{\sqrt{3}} \int_0^{\tan^{-1} y} \sqrt{xy} \, dx \, dy$

In Exercises 47–56, sketch the region of integration, reverse the order of integration, and evaluate the integral.

**47.** $\displaystyle\int_0^\pi \int_x^\pi \frac{\sin y}{y}\, dy\, dx$

**48.** $\displaystyle\int_0^2 \int_x^2 2y^2 \sin xy\, dy\, dx$

**49.** $\displaystyle\int_0^1 \int_y^1 x^2 e^{xy}\, dx\, dy$

**50.** $\displaystyle\int_0^2 \int_0^{4-x^2} \frac{xe^{2y}}{4-y}\, dy\, dx$

**51.** $\displaystyle\int_0^{2\sqrt{\ln 3}} \int_{y/2}^{\sqrt{\ln 3}} e^{x^2}\, dx\, dy$

**52.** $\displaystyle\int_0^3 \int_{\sqrt{x/3}}^1 e^{y^3}\, dy\, dx$

**53.** $\displaystyle\int_0^{1/16} \int_{y^{1/4}}^{1/2} \cos(16\pi x^5)\, dx\, dy$

**54.** $\displaystyle\int_0^8 \int_{\sqrt[3]{x}}^2 \frac{dy\, dx}{y^4+1}$

**55. Square region** $\iint_R (y - 2x^2)\, dA$ where $R$ is the region bounded by the square $|x| + |y| = 1$

**56. Triangular region** $\iint_R xy\, dA$ where $R$ is the region bounded by the lines $y = x$, $y = 2x$, and $x + y = 2$

**Volume Beneath a Surface** $z = f(x, y)$

**57.** Find the volume of the region bounded above by the paraboloid $z = x^2 + y^2$ and below by the triangle enclosed by the lines $y = x$, $x = 0$, and $x + y = 2$ in the $xy$-plane.

**58.** Find the volume of the solid that is bounded above by the cylinder $z = x^2$ and below by the region enclosed by the parabola $y = 2 - x^2$ and the line $y = x$ in the $xy$-plane.

**59.** Find the volume of the solid whose base is the region in the $xy$-plane that is bounded by the parabola $y = 4 - x^2$ and the line $y = 3x$, while the top of the solid is bounded by the plane $z = x + 4$.

**60.** Find the volume of the solid in the first octant bounded by the coordinate planes, the cylinder $x^2 + y^2 = 4$, and the plane $z + y = 3$.

**61.** Find the volume of the solid in the first octant bounded by the coordinate planes, the plane $x = 3$, and the parabolic cylinder $z = 4 - y^2$.

**62.** Find the volume of the solid cut from the first octant by the surface $z = 4 - x^2 - y$.

**63.** Find the volume of the wedge cut from the first octant by the cylinder $z = 12 - 3y^2$ and the plane $x + y = 2$.

**64.** Find the volume of the solid cut from the square column $|x| + |y| \le 1$ by the planes $z = 0$ and $3x + z = 3$.

**65.** Find the volume of the solid that is bounded on the front and back by the planes $x = 2$ and $x = 1$, on the sides by the cylinders $y = \pm 1/x$, and above and below by the planes $z = x + 1$ and $z = 0$.

**66.** Find the volume of the solid bounded on the front and back by the planes $x = \pm\pi/3$, on the sides by the cylinders $y = \pm\sec x$, above by the cylinder $z = 1 + y^2$, and below by the $xy$-plane.

In Exercises 67 and 68, sketch the region of integration and the solid whose volume is given by the double integral.

**67.** $\displaystyle\int_0^3 \int_0^{2-2x/3} \left(1 - \frac{1}{3}x - \frac{1}{2}y\right) dy\, dx$

**68.** $\displaystyle\int_0^4 \int_{-\sqrt{16-y^2}}^{\sqrt{16-y^2}} \sqrt{25 - x^2 - y^2}\, dx\, dy$

**Integrals over Unbounded Regions**

Improper double integrals can often be computed similarly to improper integrals of one variable. The first iteration of the following improper integrals is conducted just as if they were proper integrals. One then evaluates an improper integral of a single variable by taking appropriate limits, as in Section 8.8. Evaluate the improper integrals in Exercises 69–72 as iterated integrals.

**69.** $\displaystyle\int_1^\infty \int_{e^{-x}}^1 \frac{1}{x^3 y}\, dy\, dx$

**70.** $\displaystyle\int_{-1}^1 \int_{-1/\sqrt{1-x^2}}^{1/\sqrt{1-x^2}} (2y + 1)\, dy\, dx$

**71.** $\displaystyle\int_{-\infty}^\infty \int_{-\infty}^\infty \frac{1}{(x^2 + 1)(y^2 + 1)}\, dx\, dy$

**72.** $\displaystyle\int_0^\infty \int_0^\infty xe^{-(x+2y)}\, dx\, dy$

**Approximating Integrals with Finite Sums**

In Exercises 73 and 74, approximate the double integral of $f(x, y)$ over the region $R$ partitioned by the given vertical lines $x = a$ and horizontal lines $y = c$. In each subrectangle, use $(x_k, y_k)$ as indicated for your approximation.

$$\iint_R f(x, y)\, dA \approx \sum_{k=1}^n f(x_k, y_k)\, \Delta A_k$$

**73.** $f(x, y) = x + y$ over the region $R$ bounded above by the semicircle $y = \sqrt{1 - x^2}$ and below by the $x$-axis, using the partition $x = -1, -1/2, 0, 1/4, 1/2, 1$ and $y = 0, 1/2, 1$ with $(x_k, y_k)$ the lower left corner in the $k$th subrectangle (provided the subrectangle lies within $R$)

**74.** $f(x, y) = x + 2y$ over the region $R$ inside the circle $(x - 2)^2 + (y - 3)^2 = 1$ using the partition $x = 1, 3/2, 2, 5/2, 3$ and $y = 2, 5/2, 3, 7/2, 4$ with $(x_k, y_k)$ the center (centroid) in the $k$th subrectangle (provided the subrectangle lies within $R$)

**Theory and Examples**

**75. Circular sector** Integrate $f(x, y) = \sqrt{4 - x^2}$ over the smaller sector cut from the disk $x^2 + y^2 \le 4$ by the rays $\theta = \pi/6$ and $\theta = \pi/2$.

**76. Unbounded region** Integrate $f(x, y) = 1/[(x^2 - x)(y - 1)^{2/3}]$ over the infinite rectangle $2 \le x < \infty$, $0 \le y \le 2$.

**77. Noncircular cylinder** A solid right (noncircular) cylinder has its base $R$ in the $xy$-plane and is bounded above by the paraboloid $z = x^2 + y^2$. The cylinder's volume is

$$V = \int_0^1 \int_0^y (x^2 + y^2)\, dx\, dy + \int_1^2 \int_0^{2-y} (x^2 + y^2)\, dx\, dy.$$

Sketch the base region $R$ and express the cylinder's volume as a single iterated integral with the order of integration reversed. Then evaluate the integral to find the volume.

**78. Converting to a double integral** Evaluate the integral

$$\int_0^2 (\tan^{-1} \pi x - \tan^{-1} x) \, dx.$$

(*Hint·* Write the integrand as an integral.)

**79. Maximizing a double integral** What region $R$ in the $xy$-plane maximizes the value of

$$\iint_R (4 - x^2 - 2y^2) \, dA?$$

Give reasons for your answer.

**80. Minimizing a double integral** What region $R$ in the $xy$-plane minimizes the value of

$$\iint_R (x^2 + y^2 - 9) \, dA?$$

Give reasons for your answer.

**81.** Is it possible to evaluate the integral of a continuous function $f(x, y)$ over a rectangular region in the $xy$-plane and get different answers depending on the order of integration? Give reasons for your answer.

**82.** How would you evaluate the double integral of a continuous function $f(x, y)$ over the region $R$ in the $xy$-plane enclosed by the triangle with vertices $(0, 1)$, $(2, 0)$, and $(1, 2)$? Give reasons for your answer.

**83. Unbounded region** Prove that

$$\int_{-\infty}^{\infty} \int_{-\infty}^{\infty} e^{-x^2 - y^2} \, dx \, dy = \lim_{b \to \infty} \int_{-b}^{b} \int_{-b}^{b} e^{-x^2 - y^2} \, dx \, dy$$

$$= 4 \left( \int_0^{\infty} e^{-x^2} \, dx \right)^2.$$

**84. Improper double integral** Evaluate the improper integral

$$\int_0^1 \int_0^3 \frac{x^2}{(y - 1)^{2/3}} \, dy \, dx.$$

**COMPUTER EXPLORATIONS**

Use a CAS double-integral evaluator to estimate the values of the integrals in Exercises 85–88.

**85.** $\int_1^3 \int_1^x \frac{1}{xy} \, dy \, dx$    **86.** $\int_0^1 \int_0^1 e^{-(x^2 + y^2)} \, dy \, dx$

**87.** $\int_0^1 \int_0^1 \tan^{-1} xy \, dy \, dx$

**88.** $\int_{-1}^1 \int_0^{\sqrt{1-x^2}} 3\sqrt{1 - x^2 - y^2} \, dy \, dx$

Use a CAS double-integral evaluator to find the integrals in Exercises 89–94. Then reverse the order of integration and evaluate, again with a CAS.

**89.** $\int_0^1 \int_{2y}^4 e^{x^2} \, dx \, dy$

**90.** $\int_0^3 \int_{x^2}^9 x \cos(y^2) \, dy \, dx$

**91.** $\int_0^2 \int_{y^3}^{4\sqrt{2y}} (x^2 y - xy^2) \, dx \, dy$

**92.** $\int_0^2 \int_0^{4-y^2} e^{xy} \, dx \, dy$

**93.** $\int_1^2 \int_0^{x^2} \frac{1}{x + y} \, dy \, dx$    **94.** $\int_1^2 \int_{y^3}^8 \frac{1}{\sqrt{x^2 + y^2}} \, dx \, dy$

# 15.3 Area by Double Integration

In this section we show how to use double integrals to calculate the areas of bounded regions in the plane, and to find the average value of a function of two variables.

### Areas of Bounded Regions in the Plane

If we take $f(x, y) = 1$ in the definition of the double integral over a region $R$ in the preceding section, the Riemann sums reduce to

$$S_n = \sum_{k=1}^n f(x_k, y_k) \, \Delta A_k = \sum_{k=1}^n \Delta A_k. \tag{1}$$

This is simply the sum of the areas of the small rectangles in the partition of $R$, and approximates what we would like to call the area of $R$. As the norm of a partition of $R$ approaches zero, the height and width of all rectangles in the partition approach zero, and the coverage of $R$ becomes increasingly complete (Figure 15.8). We define the area of $R$ to be the limit

$$\lim_{\|P\| \to 0} \sum_{k=1}^n \Delta A_k = \iint_R dA. \tag{2}$$

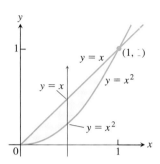

**FIGURE 15.19** The region in Example 1.

(a)

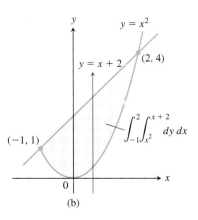

(b)

**FIGURE 15.20** Calculating this area takes (a) two double integrals if the first integration is with respect to $x$, but (b) only one if the first integration is with respect to $y$ (Example 2).

**DEFINITION** The **area** of a closed, bounded plane region $R$ is

$$A = \iint_R dA.$$

As with the other definitions in this chapter, the definition here applies to a greater variety of regions than does the earlier single-variable definition of area, but it agrees with the earlier definition on regions to which they both apply. To evaluate the integral in the definition of area, we integrate the constant function $f(x, y) = 1$ over $R$.

**EXAMPLE 1** Find the area of the region $R$ bounded by $y = x$ and $y = x^2$ in the first quadrant.

**Solution** We sketch the region (Figure 15.19), noting where the two curves intersect at the origin and $(1, 1)$, and calculate the area as

$$A = \int_0^1 \int_{x^2}^x dy\, dx = \int_0^1 \left[ y \right]_{x^2}^x dx$$
$$= \int_0^1 (x - x^2)\, dx = \left[ \frac{x^2}{2} - \frac{x^3}{3} \right]_0^1 = \frac{1}{6}.$$

Notice that the single-variable integral $\int_0^1 (x - x^2)\, dx$, obtained from evaluating the inside iterated integral, is the integral for the area between these two curves using the method of Section 5.6. ∎

**EXAMPLE 2** Find the area of the region $R$ enclosed by the parabola $y = x^2$ and the line $y = x + 2$.

**Solution** If we divide $R$ into the regions $R_1$ and $R_2$ shown in Figure 15.20a, we may calculate the area as

$$A = \iint_{R_1} dA + \iint_{R_2} dA = \int_0^1 \int_{-\sqrt{y}}^{\sqrt{y}} dx\, dy + \int_1^4 \int_{y-2}^{\sqrt{y}} dx\, dy.$$

On the other hand, reversing the order of integration (Figure 15.20b) gives

$$A = \int_{-1}^2 \int_{x^2}^{x+2} dy\, dx.$$

This second result, which requires only one integral, is simpler and is the only one we would bother to write down in practice. The area is

$$A = \int_{-1}^2 \left[ y \right]_{x^2}^{x+2} dx = \int_{-1}^2 (x + 2 - x^2)\, dx = \left[ \frac{x^2}{2} + 2x - \frac{x^3}{3} \right]_{-1}^2 = \frac{9}{2}.$$ ∎

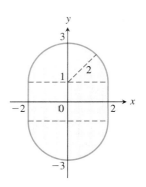

**FIGURE 15.21** The playing field described by the region $R$ in Example 3

**EXAMPLE 3** Find the area of the playing field described by
$R: -2 \leq x \leq 2, \; -1 - \sqrt{4 - x^2} \leq y \leq 1 + \sqrt{4 - x^2}$, using

**(a)** Fubini's Theorem **(b)** Simple geometry.

**Solution** The region $R$ is shown in Figure 15.21.

**(a)** From the symmetries observed in the figure, we see that the area of $R$ is 4 times its area in the first quadrant. Using Fubini's Theorem, we have

$$A = \iint_R dA = 4 \int_0^2 \int_0^{1 + \sqrt{4 - x^2}} dy \; dx$$

$$= 4 \int_0^2 \left(1 + \sqrt{4 - x^2}\right) dx$$

$$= 4\left[ x + \frac{x}{2}\sqrt{4 - x^2} + \frac{4}{2} \sin^{-1} \frac{x}{2} \right]_0^2 \quad \text{Integral Table Formula 45}$$

$$= 4\left( 2 + 0 + 2 \cdot \frac{\pi}{2} - 0 \right) = 8 + 4\pi.$$

**(b)** The region $R$ consists of a rectangle mounted on two sides by half disks of radius 2. The area can be computed by summing the area of the $4 \times 2$ rectangle and the area of a circle of radius 2, so

$$A = 8 + \pi 2^2 = 8 + 4\pi. \qquad \blacksquare$$

## Average Value

The average value of an integrable function of one variable on a closed interval is the integral of the function over the interval divided by the length of the interval. For an integrable function of two variables defined on a bounded region in the plane, the average value is the integral over the region divided by the area of the region. This can be visualized by thinking of the function as giving the height at one instant of some water sloshing around in a tank whose vertical walls lie over the boundary of the region. The average height of the water in the tank can be found by letting the water settle down to a constant height. The height is then equal to the volume of water in the tank divided by the area of $R$. We are led to define the average value of an integrable function $f$ over a region $R$ as follows:

$$\textbf{Average value of } f \text{ over } R = \frac{1}{\text{area of } R} \iint_R f \; dA. \tag{3}$$

If $f$ is the temperature of a thin plate covering $R$, then the double integral of $f$ over $R$ divided by the area of $R$ is the plate's average temperature. If $f(x, y)$ is the distance from the point $(x, y)$ to a fixed point $P$, then the average value of $f$ over $R$ is the average distance of points in $R$ from $P$.

**EXAMPLE 4** Find the average value of $f(x, y) = x \cos xy$ over the rectangle $R: 0 \leq x \leq \pi, 0 \leq y \leq 1$.

**Solution**    The value of the integral of $f$ over $R$ is

$$\int_0^\pi \int_0^1 x \cos xy \, dy \, dx = \int_0^\pi \left[ \sin xy \right]_{y=0}^{y=1} dx \qquad \left| \int x \cos xy \, dy = \sin xy + C \right.$$

$$= \int_0^\pi (\sin x - 0) \, dx = -\cos x \Big]_0^\pi = 1 + 1 = 2.$$

The area of $R$ is $\pi$. The average value of $f$ over $R$ is $2/\pi$.    ■

## Exercises 15.3

### Area by Double Integrals

In Exercises 1–12, sketch the region bounded by the given lines and curves. Then express the region's area as an iterated double integral and evaluate the integral.

1. The coordinate axes and the line $x + y = 2$
2. The lines $x = 0$, $y = 2x$, and $y = 4$
3. The parabola $x = -y^2$ and the line $y = x + 2$
4. The parabola $x = y - y^2$ and the line $y = -x$
5. The curve $y = e^x$ and the lines $y = 0$, $x = 0$, and $x = \ln 2$
6. The curves $y = \ln x$ and $y = 2 \ln x$ and the line $x = e$, in the first quadrant
7. The parabolas $x = y^2$ and $x = 2y - y^2$
8. The parabolas $x = y^2 - 1$ and $x = 2y^2 - 2$
9. The lines $y = x$, $y = x/3$, and $y = 2$
10. The lines $y = 1 - x$ and $y = 2$ and the curve $y = e^x$
11. The lines $y = 2x$, $y = x/2$, and $y = 3 - x$
12. The lines $y = x - 2$ and $y = -x$ and the curve $y = \sqrt{x}$

### Identifying the Region of Integration

The integrals and sums of integrals in Exercises 13–18 give the areas of regions in the $xy$-plane. Sketch each region, label each bounding curve with its equation, and give the coordinates of the points where the curves intersect. Then find the area of the region.

13. $\int_0^6 \int_{y^2/3}^{2y} dx \, dy$

14. $\int_0^3 \int_{-x}^{x(2-x)} dy \, dx$

15. $\int_0^{\pi/4} \int_{\sin x}^{\cos x} dy \, dx$

16. $\int_{-1}^2 \int_{y^2}^{y+2} dx \, dy$

17. $\int_{-1}^0 \int_{-2x}^{1-x} dy \, dx + \int_0^2 \int_{-x/2}^{1-x} dy \, dx$

18. $\int_0^2 \int_{x^2-4}^0 dy \, dx + \int_0^4 \int_0^{\sqrt{x}} dy \, dx$

### Finding Average Values

19. Find the average value of $f(x, y) = \sin(x + y)$ over

    **a.** the rectangle $0 \le x \le \pi$, $0 \le y \le \pi$.

    **b.** the rectangle $0 \le x \le \pi$, $0 \le y \le \pi/2$.

20. Which do you think will be larger, the average value of $f(x, y) = xy$ over the square $0 \le x \le 1$, $0 \le y \le 1$, or the

average value of $f$ over the quarter circle $x^2 + y^2 \le 1$ in the first quadrant? Calculate them to find out.

21. Find the average height of the paraboloid $z = x^2 + y^2$ over the square $0 \le x \le 2$, $0 \le y \le 2$.

22. Find the average value of $f(x, y) = 1/(xy)$ over the square $\ln 2 \le x \le 2 \ln 2$, $\ln 2 \le y \le 2 \ln 2$.

### Theory and Examples

23. **Geometric area**    Find the area of the region

    $$R: 0 \le x \le 2, \ 2 - x \le y \le \sqrt{4 - x^2},$$

    using **(a)** Fubini's Theorem, **(b)** simple geometry.

24. **Geometric area**    Find the area of the circular washer with outer radius 2 and inner radius 1, using **(a)** Fubini's Theorem, **(b)** simple geometry.

25. **Bacterium population**    If $f(x, y) = (10{,}000e^y)/(1 + |x|/2)$ represents the "population density" of a certain bacterium on the $xy$-plane, where $x$ and $y$ are measured in centimeters, find the total population of bacteria within the rectangle $-5 \le x \le 5$ and $-2 \le y \le 0$.

26. **Regional population**    If $f(x, y) = 100(y + 1)$ represents the population density of a planar region on Earth, where $x$ and $y$ are measured in kilometers, find the number of people in the region bounded by the curves $x = y^2$ and $x = 2y - y^2$.

27. **Average temperature in Texas**    According to the *Texas Almanac*, Texas has 254 counties and a National Weather Service station in each county. Assume that at time $t_0$, each of the 254 weather stations recorded the local temperature. Find a formula that would give a reasonable approximation of the average temperature in Texas at time $t_0$. Your answer should involve information that you would expect to be readily available in the *Texas Almanac*.

28. If $y = f(x)$ is a nonnegative continuous function over the closed interval $a \le x \le b$, show that the double integral definition of area for the closed plane region bounded by the graph of $f$, the vertical lines $x = a$ and $x = b$, and the $x$-axis agrees with the definition for area beneath the curve in Section 5.3.

29. Suppose $f(x, y)$ is continuous over a region $R$ in the plane and that the area $A(R)$ of the region is defined. If there are constants $m$ and $M$ such that $m \le f(x, y) \le M$ for all $(x, y) \in R$, prove that

    $$mA(R) \le \iint_R f(x, y) \, dA \le MA(R).$$

30. Suppose $f(x, y)$ is continuous and nonnegative over a region $R$ in the plane with a defined area $A(R)$. If $\iint_R f(x, y) \, dA = 0$, prove that $f(x, y) = 0$ at every point $(x, y) \in R$.

# 15.4 Double Integrals in Polar Form

Double integrals are sometimes easier to evaluate if we change to polar coordinates. This section shows how to accomplish the change and how to evaluate double integrals over regions whose boundaries are given by polar equations.

### Integrals in Polar Coordinates

When we defined the double integral of a function over a region $R$ in the $xy$-plane, we began by cutting $R$ into rectangles whose sides were parallel to the coordinate axes. These were the natural shapes to use because their sides have either constant $x$-values or constant $y$-values. In polar coordinates, the natural shape is a "polar rectangle" whose sides have constant $r$- and $\theta$-values. To avoid ambiguities when describing the region of integration with polar coordinates, we use polar coordinate points $(r, \theta)$ where $r \geq 0$.

Suppose that a function $f(r, \theta)$ is defined over a region $R$ that is bounded by the rays $\theta = \alpha$ and $\theta = \beta$ and by the continuous curves $r = g_1(\theta)$ and $r = g_2(\theta)$. Suppose also that $0 \leq g_1(\theta) \leq g_2(\theta) \leq a$ for every value of $\theta$ between $\alpha$ and $\beta$. Then $R$ lies in a fan-shaped region $Q$ defined by the inequalities $0 \leq r \leq a$ and $\alpha \leq \theta \leq \beta$, where $0 \leq \beta - \alpha \leq 2\pi$. See Figure 15.22.

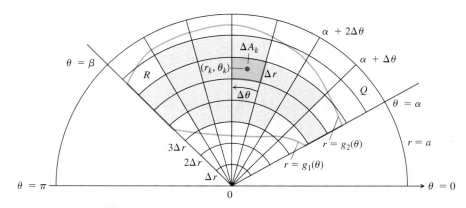

**FIGURE 15.22** The region $R$: $g_1(\theta) \leq r \leq g_2(\theta)$, $\alpha \leq \theta \leq \beta$, is contained in the fan-shaped region $Q$: $0 \leq r \leq a$, $\alpha \leq \theta \leq \beta$, where $0 \leq \beta - \alpha \leq 2\pi$. The partition of $Q$ by circular arcs and rays induces a partition of $R$.

We cover $Q$ by a grid of circular arcs and rays. The arcs are cut from circles centered at the origin, with radii $\Delta r, 2\Delta r, \ldots, m\Delta r$, where $\Delta r = a/m$. The rays are given by

$$\theta = \alpha, \qquad \theta = \alpha + \Delta\theta, \qquad \theta = \alpha + 2\Delta\theta, \qquad \ldots, \qquad \theta = \alpha + m'\Delta\theta = \beta,$$

where $\Delta\theta = (\beta - \alpha)/m'$. The arcs and rays partition $Q$ into small patches called "polar rectangles."

We number the polar rectangles that lie inside $R$ (the order does not matter), calling their areas $\Delta A_1, \Delta A_2, \ldots, \Delta A_n$. We let $(r_k, \theta_k)$ be any point in the polar rectangle whose area is $\Delta A_k$. We then form the sum

$$S_n = \sum_{k=1}^{n} f(r_k, \theta_k)\,\Delta A_k.$$

If $f$ is continuous throughout $R$, this sum will approach a limit as we refine the grid to make $\Delta r$ and $\Delta\theta$ go to zero. The limit is called the double integral of $f$ over $R$. In symbols,

$$\lim_{n \to \infty} S_n = \iint\limits_{R} f(r, \theta)\, dA.$$

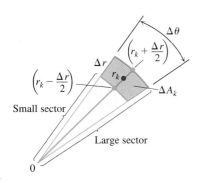

**FIGURE 15.23** The observation that

$$\Delta A_k = \begin{pmatrix} \text{area of} \\ \text{large sector} \end{pmatrix} - \begin{pmatrix} \text{area of} \\ \text{small sector} \end{pmatrix}$$

leads to the formula $\Delta A_k = r_k \, \Delta r \, \Delta \theta$.

To evaluate this limit, we first have to write the sum $S_n$ in a way that expresses $\Delta A_k$ in terms of $\Delta r$ and $\Delta \theta$. For convenience we choose $r_k$ to be the average of the radii of the inner and outer arcs bounding the $k$th polar rectangle $\Delta A_k$. The radius of the inner arc bounding $\Delta A_k$ is then $r_k - (\Delta r/2)$ (Figure 15.23). The radius of the outer arc is $r_k + (\Delta r/2)$.

The area of a wedge-shaped sector of a circle having radius $r$ and angle $\theta$ is

$$A = \frac{1}{2} \theta \cdot r^2,$$

as can be seen by multiplying $\pi r^2$, the area of the circle, by $\theta/2\pi$, the fraction of the circle's area contained in the wedge. So the areas of the circular sectors subtended by these arcs at the origin are

$$\text{Inner radius:} \quad \frac{1}{2} \left( r_k - \frac{\Delta r}{2} \right)^2 \Delta \theta$$

$$\text{Outer radius:} \quad \frac{1}{2} \left( r_k + \frac{\Delta r}{2} \right)^2 \Delta \theta.$$

Therefore,

$$\Delta A_k = \text{area of large sector} - \text{area of small sector}$$

$$= \frac{\Delta \theta}{2} \left[ \left( r_k + \frac{\Delta r}{2} \right)^2 - \left( r_k - \frac{\Delta r}{2} \right)^2 \right] = \frac{\Delta \theta}{2} (2 r_k \, \Delta r) = r_k \, \Delta r \, \Delta \theta.$$

Combining this result with the sum defining $S_n$ gives

$$S_n = \sum_{k=1}^{n} f(r_k, \theta_k) \, r_k \, \Delta r \, \Delta \theta.$$

As $n \to \infty$ and the values of $\Delta r$ and $\Delta \theta$ approach zero, these sums converge to the double integral

$$\lim_{n \to \infty} S_n = \iint_R f(r, \theta) \, r \, dr \, d\theta.$$

A version of Fubini's Theorem says that the limit approached by these sums can be evaluated by repeated single integrations with respect to $r$ and $\theta$ as

$$\iint_R f(r, \theta) \, dA = \int_{\theta = \alpha}^{\theta = \beta} \int_{r = g_1(\theta)}^{r = g_2(\theta)} f(r, \theta) \, r \, dr \, d\theta.$$

### Finding Limits of Integration

The procedure for finding limits of integration in rectangular coordinates also works for polar coordinates. To evaluate $\iint_R f(r, \theta) \, dA$ over a region $R$ in polar coordinates, integrating first with respect to $r$ and then with respect to $\theta$, take the following steps.

1. *Sketch.* Sketch the region and label the bounding curves (Figure 15.24a).

2. *Find the r-limits of integration.* Imagine a ray $L$ from the origin cutting through $R$ in the direction of increasing $r$. Mark the $r$-values where $L$ enters and leaves $R$. These are the $r$-limits of integration. They usually depend on the angle $\theta$ that $L$ makes with the positive $x$-axis (Figure 15.24b).

3. *Find the θ-limits of integration.* Find the smallest and largest $\theta$-values that bound $R$. These are the $\theta$-limits of integration (Figure 15.24c). The polar iterated integral is

$$\iint_R f(r, \theta) \, dA = \int_{\theta = \pi/4}^{\theta = \pi/2} \int_{r = \sqrt{2} \csc \theta}^{r = 2} f(r, \theta) \, r \, dr \, d\theta.$$

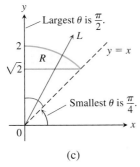

**FIGURE 15.24** Finding the limits of integration in polar coordinates.

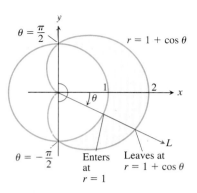

**FIGURE 15.25** Finding the limits of integration in polar coordinates for the region in Example 1.

**EXAMPLE 1** Find the limits of integration for integrating $f(r, \theta)$ over the region $R$ that lies inside the cardioid $r = 1 + \cos \theta$ and outside the circle $r = 1$.

**Solution**

1. We first sketch the region and label the bounding curves (Figure 15.25).

2. Next we find the *r-limits of integration*. A typical ray from the origin enters $R$ where $r = 1$ and leaves where $r = 1 + \cos \theta$.

3. Finally we find the $\theta$-*limits of integration*. The rays from the origin that intersect $R$ run from $\theta = -\pi/2$ to $\theta = \pi/2$. The integral is

$$\int_{-\pi/2}^{\pi/2} \int_{1}^{1+\cos\theta} f(r, \theta)\, r\, dr\, d\theta. \qquad \blacksquare$$

If $f(r, \theta)$ is the constant function whose value is 1, then the integral of $f$ over $R$ is the area of $R$.

**Area Differential in Polar Coordinates**

$$dA = r\, dr\, d\theta$$

---

**Area in Polar Coordinates**

The area of a closed and bounded region $R$ in the polar coordinate plane is

$$A = \iint_R r\, dr\, d\theta.$$

---

This formula for area is consistent with all earlier formulas, although we do not prove this fact.

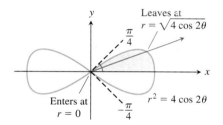

**FIGURE 15.26** To integrate over the shaded region, we run $r$ from 0 to $\sqrt{4 \cos 2\theta}$ and $\theta$ from 0 to $\pi/4$ (Example 2).

**EXAMPLE 2** Find the area enclosed by the lemniscate $r^2 = 4 \cos 2\theta$.

**Solution** We graph the lemniscate to determine the limits of integration (Figure 15.26) and see from the symmetry of the region that the total area is 4 times the first-quadrant portion.

$$A = 4 \int_0^{\pi/4} \int_0^{\sqrt{4\cos 2\theta}} r\, dr\, d\theta = 4 \int_0^{\pi/4} \left[\frac{r^2}{2}\right]_{r=0}^{r=\sqrt{4\cos 2\theta}} d\theta$$

$$= 4 \int_0^{\pi/4} 2 \cos 2\theta\, d\theta = 4 \sin 2\theta \Big]_0^{\pi/4} = 4. \qquad \blacksquare$$

### Changing Cartesian Integrals into Polar Integrals

The procedure for changing a Cartesian integral $\iint_R f(x, y)\, dx\, dy$ into a polar integral has two steps. First substitute $x = r \cos \theta$ and $y = r \sin \theta$, and replace $dx\, dy$ by $r\, dr\, d\theta$ in the Cartesian integral. Then supply polar limits of integration for the boundary of $R$. The Cartesian integral then becomes

$$\iint_R f(x, y)\, dx\, dy = \iint_G f(r \cos \theta, r \sin \theta)\, r\, dr\, d\theta,$$

where $G$ denotes the same region of integration now described in polar coordinates. This is like the substitution method in Chapter 5 except that there are now two variables to substitute for instead of one. Notice that the area differential $dx\, dy$ is not replaced by $dr\, d\theta$ but by $r\, dr\, d\theta$. A more general discussion of changes of variables (substitutions) in multiple integrals is given in Section 15.8.

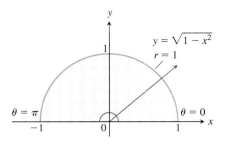

**FIGURE 15.27** The semicircular region in Example 3 is the region

$$0 \le r \le 1, \qquad 0 \le \theta \le \pi.$$

**EXAMPLE 3**     Evaluate

$$\iint\limits_{R} e^{x^2+y^2} \, dy \, dx,$$

where $R$ is the semicircular region bounded by the $x$-axis and the curve $y = \sqrt{1 - x^2}$ (Figure 15.27).

**Solution**   In Cartesian coordinates, the integral in question is a nonelementary integral and there is no direct way to integrate $e^{x^2+y^2}$ with respect to either $x$ or $y$. Yet this integral and others like it are important in mathematics—in statistics, for example—and we need to find a way to evaluate it. Polar coordinates save the day. Substituting $x = r \cos \theta$, $y = r \sin \theta$ and replacing $dy \, dx$ by $r \, dr \, d\theta$ enables us to evaluate the integral as

$$\iint\limits_{R} e^{x^2+y^2} \, dy \, dx = \int_0^\pi \int_0^1 e^{r^2} r \, dr \, d\theta = \int_0^\pi \left[ \frac{1}{2} e^{r^2} \right]_0^1 d\theta$$

$$= \int_0^\pi \frac{1}{2} (e - 1) \, d\theta = \frac{\pi}{2} (e - 1).$$

The $r$ in the $r \, dr \, d\theta$ was just what we needed to integrate $e^{r^2}$. Without it, we would have been unable to find an antiderivative for the first (innermost) iterated integral. ∎

**EXAMPLE 4**     Evaluate the integral

$$\int_0^1 \int_0^{\sqrt{1-x^2}} (x^2 + y^2) \, dy \, dx.$$

**Solution**   Integration with respect to $y$ gives

$$\int_0^1 \left( x^2 \sqrt{1 - x^2} + \frac{(1 - x^2)^{3/2}}{3} \right) dx,$$

an integral difficult to evaluate without tables.

Things go better if we change the original integral to polar coordinates. The region of integration in Cartesian coordinates is given by the inequalities $0 \le y \le \sqrt{1 - x^2}$ and $0 \le x \le 1$, which correspond to the interior of the unit quarter circle $x^2 + y^2 = 1$ in the first quadrant. (See Figure 15.27, first quadrant.) Substituting the polar coordinates $x = r \cos \theta$, $y = r \sin \theta$, $0 \le \theta \le \pi/2$, and $0 \le r \le 1$, and replacing $dx \, dy$ by $r \, dr \, d\theta$ in the double integral, we get

$$\int_0^1 \int_0^{\sqrt{1-x^2}} (x^2 + y^2) \, dy \, dx = \int_0^{\pi/2} \int_0^1 (r^2) \, r \, dr \, d\theta$$

$$= \int_0^{\pi/2} \left[ \frac{r^4}{4} \right]_{r=0}^{r=1} d\theta = \int_0^{\pi/2} \frac{1}{4} \, d\theta = \frac{\pi}{8}.$$

Why is the polar coordinate transformation so effective here? One reason is that $x^2 + y^2$ simplifies to $r^2$. Another is that the limits of integration become constants. ∎

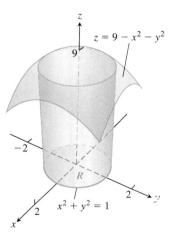

**FIGURE 15.28** The solid region in Example 5.

**EXAMPLE 5**     Find the volume of the solid region bounded above by the paraboloid $z = 9 - x^2 - y^2$ and below by the unit circle in the $xy$-plane.

**Solution**   The region of integration $R$ is the unit circle $x^2 + y^2 = 1$, which is described in polar coordinates by $r = 1$, $0 \le \theta \le 2\pi$. The solid region is shown in Figure 15.28. The volume is given by the double integral

$$\iint\limits_{R} (9 - x^2 - y^2)\, dA = \int_{0}^{2\pi} \int_{0}^{1} (9 - r^2)\, r\, dr\, d\theta$$

$$= \int_{0}^{2\pi} \int_{0}^{1} (9r - r^3)\, dr\, d\theta$$

$$= \int_{0}^{2\pi} \left[ \frac{9}{2} r^2 - \frac{1}{4} r^4 \right]_{r=0}^{r=1} d\theta$$

$$= \frac{17}{4} \int_{0}^{2\pi} d\theta = \frac{17\pi}{2}. \qquad \blacksquare$$

**EXAMPLE 6**    Using polar integration, find the area of the region $R$ in the $xy$-plane enclosed by the circle $x^2 + y^2 + 4$, above the line $y = 1$, and below the line $y = \sqrt{3}x$.

**Solution**    A sketch of the region $R$ is shown in Figure 15.29. First we note that the line $y = \sqrt{3}x$ has slope $\sqrt{3} = \tan\theta$, so $\theta = \pi/3$. Next we observe that the line $y = 1$ intersects the circle $x^2 + y^2 = 4$ when $x^2 + 1 = 4$, or $x = \sqrt{3}$. Moreover, the radial line from the origin through the point $(\sqrt{3}, 1)$ has slope $1/\sqrt{3} = \tan\theta$, giving its angle of inclination as $\theta = \pi/6$. This information is shown in Figure 15.29.

Now, for the region $R$, as $\theta$ varies from $\pi/6$ to $\pi/3$, the polar coordinate $r$ varies from the horizontal line $y = 1$ to the circle $x^2 + y^2 = 4$. Substituting $r\sin\theta$ for $y$ in the equation for the horizontal line, we have $r\sin\theta = 1$, or $r = \csc\theta$, which is the polar equation of the line. The polar equation for the circle is $r = 2$. So in polar coordinates, for $\pi/6 \le \theta \le \pi/3$, $r$ varies from $r = \csc\theta$ to $r = 2$. It follows that the iterated integral for the area then gives

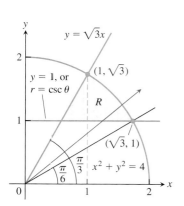

$y = \sqrt{3}x$

$y = 1$, or $r = \csc\theta$

$(1, \sqrt{3})$

$R$

$(\sqrt{3}, 1)$

$x^2 + y^2 = 4$

**FIGURE 15.29**    The region $R$ in Example 6.

$$\iint\limits_{R} dA = \int_{\pi/6}^{\pi/3} \int_{\csc\theta}^{2} r\, dr\, d\theta$$

$$= \int_{\pi/6}^{\pi/3} \left[ \frac{1}{2} r^2 \right]_{r=\csc\theta}^{r=2} d\theta$$

$$= \int_{\pi/6}^{\pi/3} \frac{1}{2} \left[ 4 - \csc^2\theta \right] d\theta$$

$$= \frac{1}{2} \left[ 4\theta + \cot\theta \right]_{\pi/6}^{\pi/3}$$

$$= \frac{1}{2} \left( \frac{4\pi}{3} + \frac{1}{\sqrt{3}} \right) - \frac{1}{2} \left( \frac{4\pi}{6} + \sqrt{3} \right) = \frac{\pi - \sqrt{3}}{3}. \qquad \blacksquare$$

# Exercises 15.4

## Regions in Polar Coordinates

In Exercises 1–8, describe the given region in polar coordinates.

**1.**

**2.**

**3.**

**4.**

**5.**

**6.**

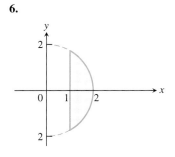

7. The region enclosed by the circle $x^2 + y^2 = 2x$

8. The region enclosed by the semicircle $x^2 + y^2 = 2y, y \geq 0$

**Evaluating Polar Integrals**

In Exercises 9–22, change the Cartesian integral into an equivalent polar integral. Then evaluate the polar integral.

**9.** $\displaystyle\int_{-1}^{1}\int_{0}^{\sqrt{1-x^2}} dy\, dx$

**10.** $\displaystyle\int_{0}^{1}\int_{0}^{\sqrt{1-y^2}} (x^2 + y^2)\, dx\, dy$

**11.** $\displaystyle\int_{0}^{2}\int_{0}^{\sqrt{4-y^2}} (x^2 + y^2)\, dx\, dy$

**12.** $\displaystyle\int_{-a}^{a}\int_{-\sqrt{a^2-x^2}}^{\sqrt{a^2-x^2}} dy\, dx$

**13.** $\displaystyle\int_{0}^{6}\int_{0}^{y} x\, dx\, dy$

**14.** $\displaystyle\int_{0}^{2}\int_{0}^{x} y\, dy\, dx$

**15.** $\displaystyle\int_{1}^{\sqrt{3}}\int_{1}^{x} dy\, dx$

**16.** $\displaystyle\int_{\sqrt{2}}^{2}\int_{\sqrt{4-y^2}}^{y} dx\, dy$

**17.** $\displaystyle\int_{-1}^{0}\int_{-\sqrt{1-x^2}}^{0} \frac{2}{1 + \sqrt{x^2 + y^2}}\, dy\, dx$

**18.** $\displaystyle\int_{-1}^{1}\int_{-\sqrt{1-x^2}}^{\sqrt{1-x^2}} \frac{2}{(1 + x^2 + y^2)^2}\, dy\, dx$

**19.** $\displaystyle\int_{0}^{\ln 2}\int_{0}^{\sqrt{(\ln 2)^2 - y^2}} e^{\sqrt{x^2 + y^2}}\, dx\, dy$

**20.** $\displaystyle\int_{-1}^{1}\int_{-\sqrt{1-y^2}}^{\sqrt{1-y^2}} \ln(x^2 + y^2 + 1)\, dx\, dy$

**21.** $\displaystyle\int_{0}^{1}\int_{x}^{\sqrt{2-x^2}} (x + 2y)\, dy\, dx$

**22.** $\displaystyle\int_{1}^{2}\int_{0}^{\sqrt{2x-x^2}} \frac{1}{(x^2 + y^2)^2}\, dy\, dx$

In Exercises 23–26, sketch the region of integration and convert each polar integral or sum of integrals to a Cartesian integral or sum of integrals. Do not evaluate the integrals.

**23.** $\displaystyle\int_{0}^{\pi/2}\int_{0}^{1} r^3 \sin\theta \cos\theta\, dr\, d\theta$

**24.** $\displaystyle\int_{\pi/6}^{\pi/2}\int_{1}^{\csc\theta} r^2 \cos\theta\, dr\, d\theta$

**25.** $\displaystyle\int_{0}^{\pi/4}\int_{0}^{2\sec\theta} r^5 \sin^2\theta\, dr\, d\theta$

**26.** $\displaystyle\int_{0}^{\tan^{-1}\frac{4}{3}}\int_{0}^{3\sec\theta} r^7\, dr\, d\theta + \int_{\tan^{-1}\frac{4}{3}}^{\pi/2}\int_{0}^{4\csc\theta} r^7\, dr\, d\theta$

**Area in Polar Coordinates**

27. Find the area of the region cut from the first quadrant by the curve $r = 2(2 - \sin 2\theta)^{1/2}$.

28. **Cardioid overlapping a circle**  Find the area of the region that lies inside the cardioid $r = 1 + \cos\theta$ and outside the circle $r = 1$.

29. **One leaf of a rose**  Find the area enclosed by one leaf of the rose $r = 12\cos 3\theta$.

30. **Snail shell**  Find the area of the region enclosed by the positive $x$-axis and spiral $r = 4\theta/3, 0 \leq \theta \leq 2\pi$. The region looks like a snail shell.

31. **Cardioid in the first quadrant**  Find the area of the region cut from the first quadrant by the cardioid $r = 1 + \sin\theta$.

32. **Overlapping cardioids**  Find the area of the region common to the interiors of the cardioids $r = 1 + \cos\theta$ and $r = 1 - \cos\theta$.

**Average Values**

In polar coordinates, the **average value** of a function over a region $R$ (Section 15.3) is given by

$$\frac{1}{\text{Area}(R)} \iint_R f(r, \theta)\, r\, dr\, d\theta.$$

33. **Average height of a hemisphere**  Find the average height of the hemispherical surface $z = \sqrt{a^2 - x^2 - y^2}$ above the disk $x^2 + y^2 \leq a^2$ in the $xy$-plane.

34. **Average height of a cone**  Find the average height of the (single) cone $z = \sqrt{x^2 + y^2}$ above the disk $x^2 + y^2 \leq a^2$ in the $xy$-plane.

35. **Average distance from interior of disk to center**  Find the average distance from a point $P(x, y)$ in the disk $x^2 + y^2 \leq a^2$ to the origin.

36. **Average distance squared from a point in a disk to a point in its boundary**  Find the average value of the *square* of the distance from the point $P(x, y)$ in the disk $x^2 + y^2 \leq 1$ to the boundary point $A(1, 0)$.

**Theory and Examples**

37. **Converting to a polar integral**  Integrate $f(x, y) = [\ln(x^2 + y^2)]/\sqrt{x^2 + y^2}$ over the region $1 \leq x^2 + y^2 \leq e$.

38. **Converting to a polar integral**  Integrate $f(x, y) = [\ln(x^2 + y^2)]/(x^2 + y^2)$ over the region $1 \leq x^2 + y^2 \leq e^2$.

39. **Volume of noncircular right cylinder**  The region that lies inside the cardioid $r = 1 + \cos\theta$ and outside the circle $r = 1$ is the base of a solid right cylinder. The top of the cylinder lies in the plane $z = x$. Find the cylinder's volume.

40. **Volume of noncircular right cylinder**  The region enclosed by the lemniscate $r^2 = 2\cos 2\theta$ is the base of a solid right cylinder whose top is bounded by the sphere $z = \sqrt{2 - r^2}$. Find the cylinder's volume.

**41. Converting to polar integrals**

a. The usual way to evaluate the improper integral
$I = \int_0^\infty e^{-x^2} dx$ is first to calculate its square:

$$I^2 = \left( \int_0^\infty e^{-x^2} dx \right) \left( \int_0^\infty e^{-y^2} dy \right) = \int_0^\infty \int_0^\infty e^{-(x^2+y^2)} dx\, dy.$$

Evaluate the last integral using polar coordinates and solve the resulting equation for $I$.

b. Evaluate

$$\lim_{x \to \infty} \operatorname{erf}(x) = \lim_{x \to \infty} \int_0^x \frac{2 e^{-t^2}}{\sqrt{\pi}} dt.$$

**42. Converting to a polar integral**   Evaluate the integral

$$\int_0^\infty \int_0^\infty \frac{1}{(1 + x^2 + y^2)^2} dx\, dy.$$

**43. Existence**   Integrate the function $f(x, y) = 1/(1 - x^2 - y^2)$ over the disk $x^2 + y^2 \le 3/4$. Does the integral of $f(x, y)$ over the disk $x^2 + y^2 \le 1$ exist? Give reasons for your answer.

**44. Area formula in polar coordinates**   Use the double integral in polar coordinates to derive the formula

$$A = \int_\alpha^\beta \frac{1}{2} r^2 \, d\theta$$

for the area of the fan-shaped region between the origin and polar curve $r = f(\theta)$, $\alpha \le \theta \le \beta$.

**45. Average distance to a given point inside a disk**   Let $P_0$ be a point inside a circle of radius $a$ and let $h$ denote the distance from $P_0$ to the center of the circle. Let $d$ denote the distance from an arbitrary point $P$ to $P_0$. Find the average value of $d^2$ over the region enclosed by the circle. (*Hint:* Simplify your work by placing the center of the circle at the origin and $P_0$ on the $x$-axis.)

**46. Area**   Suppose that the area of a region in the polar coordinate plane is

$$A = \int_{\pi/4}^{3\pi/4} \int_{\csc\theta}^{2\sin\theta} r \, dr \, d\theta.$$

Sketch the region and find its area.

**47.** Evaluate the integral $\iint_R \sqrt{x^2 + y^2}\, dA$, where $R$ is the region inside the upper semicircle of radius 2 centered at the origin, but outside the circle $x^2 + (y - 1)^2 = 1$.

**48.** Evaluate the integral $\iint_R (x^2 + y^2)^{-2}\, dA$, where $R$ is the region inside the circle $x^2 + y^2 = 2$ for $x \le -1$.

**COMPUTER EXPLORATIONS**

In Exercises 49–52, use a CAS to change the Cartesian integrals into an equivalent polar integral and evaluate the polar integral. Perform the following steps in each exercise.

a. Plot the Cartesian region of integration in the $xy$-plane.

b. Change each boundary curve of the Cartesian region in part (a) to its polar representation by solving its Cartesian equation for $r$ and $\theta$.

c. Using the results in part (b), plot the polar region of integration in the $r\theta$-plane.

d. Change the integrand from Cartesian to polar coordinates. Determine the limits of integration from your plot in part (c) and evaluate the polar integral using the CAS integration utility.

**49.** $\displaystyle \int_0^1 \int_x^1 \frac{y}{x^2 + y^2} \, dy \, dx$    **50.** $\displaystyle \int_0^1 \int_0^{x/2} \frac{x}{x^2 + y^2} \, dy \, dx$

**51.** $\displaystyle \int_0^1 \int_{-y/3}^{y/3} \frac{y}{\sqrt{x^2 + y^2}} \, dx \, dy$    **52.** $\displaystyle \int_0^1 \int_y^{2-y} \sqrt{x + y} \, dx \, dy$

# 15.5 Triple Integrals in Rectangular Coordinates

Just as double integrals allow us to deal with more general situations than could be handled by single integrals, triple integrals enable us to solve still more general problems. We use triple integrals to calculate the volumes of three-dimensional shapes and the average value of a function over a three-dimensional region. Triple integrals also arise in the study of vector fields and fluid flow in three dimensions, as we will see in Chapter 16.

## Triple Integrals

If $F(x, y, z)$ is a function defined on a closed bounded region $D$ in space, such as the region occupied by a solid ball or a lump of clay, then the integral of $F$ over $D$ may be defined in the following way. We partition a rectangular boxlike region containing $D$ into rectangular cells by planes parallel to the coordinate axes (Figure 15.30). We number the cells that lie completely inside $D$ from 1 to $n$ in some order, the $k$th cell having dimensions $\Delta x_k$ by $\Delta y_k$ by $\Delta z_k$ and volume $\Delta V_k = \Delta x_k \Delta y_k \Delta z_k$. We choose a point $(x_k, y_k, z_k)$ in each cell and form the sum

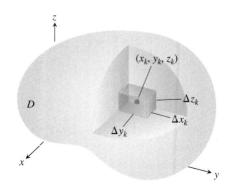

**FIGURE 15.30**   Partitioning a solid with rectangular cells of volume $\Delta V_k$.

$$S_n = \sum_{k=1}^n F(x_k, y_k, z_k)\, \Delta V_k. \qquad (1)$$

We are interested in what happens as $D$ is partitioned by smaller and smaller cells, so that $\Delta x_k$, $\Delta y_k$, $\Delta z_k$, and the norm of the partition $\|P\|$, the largest value among $\Delta x_k$, $\Delta y_k$, $\Delta z_k$, all approach zero. When a single limiting value is attained, no matter how the partitions and points $(x_k, y_k, z_k)$ are chosen, we say that $F$ is **integrable** over $D$. As before, it can be shown that when $F$ is continuous and the bounding surface of $D$ is formed from finitely many smooth surfaces joined together along finitely many smooth curves, then $F$ is integrable. As $\|P\| \to 0$ and the number of cells $n$ goes to $\infty$, the sums $S_n$ approach a limit. We call this limit the **triple integral of $F$ over $D$** and write

$$\lim_{n \to \infty} S_n = \iiint\limits_D F(x, y, z)\, dV \quad \text{or} \quad \lim_{\|P\| \to 0} S_n = \iiint\limits_D F(x, y, z)\, dx\, dy\, dz.$$

The regions $D$ over which continuous functions are integrable are those having "reasonably smooth" boundaries.

## Volume of a Region in Space

If $F$ is the constant function whose value is 1, then the sums in Equation (1) reduce to

$$S_n = \sum F(x_k, y_k, z_k)\, \Delta V_k = \sum 1 \cdot \Delta V_k = \sum \Delta V_k .$$

As $\Delta x_k$, $\Delta y_k$, and $\Delta z_k$ approach zero, the cells $\Delta V_k$ become smaller and more numerous and fill up more and more of $D$. We therefore define the volume of $D$ to be the triple integral

$$\lim_{n \to \infty} \sum_{k=1}^{n} \Delta V_k = \iiint\limits_D dV.$$

---

> **DEFINITION**    The **volume** of a closed, bounded region $D$ in space is
>
> $$V = \iiint\limits_D dV.$$

---

This definition is in agreement with our previous definitions of volume, although we omit the verification of this fact. As we see in a moment, this integral enables us to calculate the volumes of solids enclosed by curved surfaces. These are more general solids than the ones encountered before (Chapter 6 and Section 15.2).

## Finding Limits of Integration in the Order *dz dy dx*

We evaluate a triple integral by applying a three-dimensional version of Fubini's Theorem (Section 15.2) to evaluate it by three repeated single integrations. As with double integrals, there is a geometric procedure for finding the limits of integration for these iterated integrals.

To evaluate

$$\iiint\limits_D F(x, y, z)\, dV$$

over a region $D$, integrate first with respect to $z$, then with respect to $y$, and finally with respect to $x$. (You might choose a different order of integration, but the procedure is similar, as we illustrate in Example 2.)

1. *Sketch.* Sketch the region $D$ along with its "shadow" $R$ (vertical projection) in the $xy$-plane. Label the upper and lower bounding surfaces of $D$ and the upper and lower bounding curves of $R$.

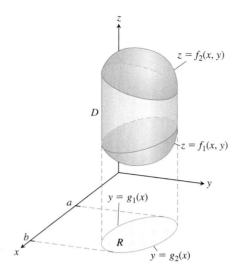

**2.** *Find the z-limits of integration.* Draw a line $M$ passing through a typical point $(x, y)$ in $R$ parallel to the $z$-axis. As $z$ increases, $M$ enters $D$ at $z = f_1(x, y)$ and leaves at $z = f_2(x, y)$. These are the $z$-limits of integration.

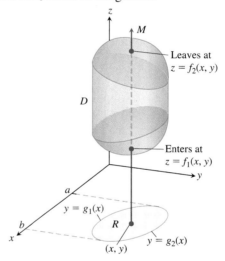

**3.** *Find the y-limits of integration.* Draw a line $L$ through $(x, y)$ parallel to the $y$-axis. As $y$ increases, $L$ enters $R$ at $y = g_1(x)$ and leaves at $y = g_2(x)$. These are the $y$-limits of integration.

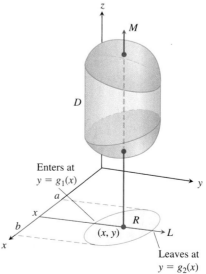

4. *Find the x-limits of integration.* Choose x-limits that include all lines through $R$ parallel to the y-axis ($x = a$ and $x = b$ in the preceding figure). These are the x-limits of integration. The integral is

$$\int_{x=a}^{x=b} \int_{y=g_1(x)}^{y=g_2(x)} \int_{z=f_1(x,y)}^{z=f_2(x,y)} F(x, y, z) \, dz \, dy \, dx.$$

Follow similar procedures if you change the order of integration. The "shadow" of region $D$ lies in the plane of the last two variables with respect to which the iterated integration takes place.

The preceding procedure applies whenever a solid region $D$ is bounded above and below by a surface, and when the "shadow" region $R$ is bounded by a lower and upper curve. It does not apply to regions with complicated holes through them, although sometimes such regions can be subdivided into simpler regions for which the procedure does apply.

**EXAMPLE 1**    Find the volume of the region $D$ enclosed by the surfaces $z = x^2 + 3y^2$ and $z = 8 - x^2 - y^2$.

**Solution**    The volume is

$$V = \iiint_D dz \, dy \, dx,$$

the integral of $F(x, y, z) = 1$ over $D$. To find the limits of integration for evaluating the integral, we first sketch the region. The surfaces (Figure 15.31) intersect on the elliptical cylinder $x^2 + 3y^2 = 8 - x^2 - y^2$ or $x^2 + 2y^2 = 4$, $z > 0$. The boundary of the region $R$, the projection of $D$ onto the xy-plane, is an ellipse with the same equation: $x^2 + 2y^2 = 4$. The "upper" boundary of $R$ is the curve $y = \sqrt{(4 - x^2)/2}$. The lower boundary is the curve $y = -\sqrt{(4 - x^2)/2}$.

Now we find the z-limits of integration. The line $M$ passing through a typical point $(x, y)$ in $R$ parallel to the z-axis enters $D$ at $z = x^2 + 3y^2$ and leaves at $z = 8 - x^2 - y^2$.

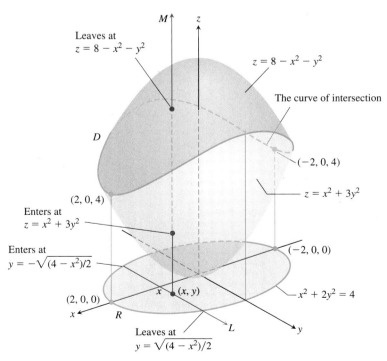

**FIGURE 15.31**    The volume of the region enclosed by two paraboloids, calculated in Example 1.

Next we find the *y*-limits of integration. The line *L* through $(x, y)$ parallel to the *y*-axis enters *R* at $y = -\sqrt{(4 - x^2)/2}$ and leaves at $y = \sqrt{(4 - x^2)/2}$.

Finally we find the *x*-limits of integration. As *L* sweeps across *R*, the value of *x* varies from $x = -2$ at $(-2, 0, 0)$ to $x = 2$ at $(2, 0, 0)$. The volume of *D* is

$$
V = \iiint_D dz\, dy\, dx
$$

$$
= \int_{-2}^{2} \int_{-\sqrt{(4-x^2)/2}}^{\sqrt{(4-x^2)/2}} \int_{x^2+3y^2}^{8-x^2-y^2} dz\, dy\, dx
$$

$$
= \int_{-2}^{2} \int_{-\sqrt{(4-x^2)/2}}^{\sqrt{(4-x^2)/2}} (8 - 2x^2 - 4y^2)\, dy\, dx
$$

$$
= \int_{-2}^{2} \left[ (8 - 2x^2)y - \frac{4}{3}y^3 \right]_{y=-\sqrt{(4-x^2)/2}}^{y=\sqrt{(4-x^2)/2}} dx
$$

$$
= \int_{-2}^{2} \left( 2(8 - 2x^2)\sqrt{\frac{4 - x^2}{2}} - \frac{8}{3}\left(\frac{4 - x^2}{2}\right)^{3/2} \right) dx
$$

$$
= \int_{-2}^{2} \left[ 8\left(\frac{4 - x^2}{2}\right)^{3/2} - \frac{8}{3}\left(\frac{4 - x^2}{2}\right)^{3/2} \right] dx = \frac{4\sqrt{2}}{3} \int_{-2}^{2} (4 - x^2)^{3/2}\, dx
$$

$$
= 8\pi\sqrt{2}. \qquad \text{\small After integration with the substitution } x = 2\sin u \qquad ■
$$

In the next example, we project *D* onto the *xz*-plane instead of the *xy*-plane, to show how to use a different order of integration.

**EXAMPLE 2** Set up the limits of integration for evaluating the triple integral of a function $F(x, y, z)$ over the tetrahedron *D* with vertices $(0, 0, 0)$, $(1, 1, 0)$, $(0, 1, 0)$, and $(0, 1, 1)$. Use the order of integration *dy dz dx*.

**Solution** We sketch *D* along with its "shadow" *R* in the *xz*-plane (Figure 15.32). The upper (right-hand) bounding surface of *D* lies in the plane $y = 1$. The lower (left-hand) bounding surface lies in the plane $y = x + z$. The upper boundary of *R* is the line $z = 1 - x$. The lower boundary is the line $z = 0$.

First we find the *y*-limits of integration. The line through a typical point $(x, z)$ in *R* parallel to the *y*-axis enters *D* at $y = x + z$ and leaves at $y = 1$.

Next we find the *z*-limits of integration. The line *L* through $(x, z)$ parallel to the *z*-axis enters *R* at $z = 0$ and leaves at $z = 1 - x$.

Finally we find the *x*-limits of integration. As *L* sweeps across *R*, the value of *x* varies from $x = 0$ to $x = 1$. The integral is

$$
\int_{0}^{1} \int_{0}^{1-x} \int_{x+z}^{1} F(x, y, z)\, dy\, dz\, dx. \qquad ■
$$

**EXAMPLE 3** Integrate $F(x, y, z) = 1$ over the tetrahedron *D* in Example 2 in the order *dz dy dx*, and then integrate in the order *dy dz dx*.

**Solution** First we find the *z*-limits of integration. A line *M* parallel to the *z*-axis through a typical point $(x, y)$ in the *xy*-plane "shadow" enters the tetrahedron at $z = 0$ and exits through the upper plane where $z = y - x$ (Figure 15.33).

Next we find the *y*-limits of integration. On the *xy*-plane, where $z = 0$, the sloped side of the tetrahedron crosses the plane along the line $y = x$. A line *L* through $(x, y)$ parallel to the *y*-axis enters the shadow in the *xy*-plane at $y = x$ and exits at $y = 1$ (Figure 15.33).

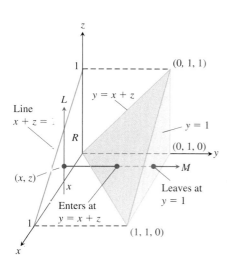

**FIGURE 15.32** Finding the limits of integration for evaluating the triple integral of a function defined over the tetrahedron *D* (Examples 2 and 3).

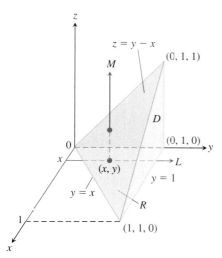

**FIGURE 15.33** The tetrahedron in Example 3 showing how the limits of integration are found for the order *dz dy dx*.

Finally we find the $x$-limits of integration. As the line $L$ parallel to the $y$-axis in the previous step sweeps out the shadow, the value of $x$ varies from $x = 0$ to $x = 1$ at the point $(1, 1, 0)$ (see Figure 15.33). The integral is

$$\int_0^1 \int_x^1 \int_0^{y-x} F(x, y, z) \, dz \, dy \, dx.$$

For example, if $F(x, y, z) = 1$, we would find the volume of the tetrahedron to be

$$V = \int_0^1 \int_x^1 \int_0^{y-x} dz \, dy \, dx$$

$$= \int_0^1 \int_x^1 (y - x) \, dy \, dx$$

$$= \int_0^1 \left[ \frac{1}{2} y^2 - xy \right]_{y=x}^{y=1} dx$$

$$= \int_0^1 \left( \frac{1}{2} - x + \frac{1}{2} x^2 \right) dx$$

$$= \left[ \frac{1}{2} x - \frac{1}{2} x^2 + \frac{1}{6} x^3 \right]_0^1$$

$$= \frac{1}{6}.$$

We get the same result by integrating with the order $dy \, dz \, dx$. From Example 2,

$$V = \int_0^1 \int_0^{1-x} \int_{x+z}^1 dy \, dz \, dx$$

$$= \int_0^1 \int_0^{1-x} (1 - x - z) \, dz \, dx$$

$$= \int_0^1 \left[ (1 - x)z - \frac{1}{2} z^2 \right]_{z=0}^{z=1-x} dx$$

$$= \int_0^1 \left[ (1 - x)^2 - \frac{1}{2} (1 - x)^2 \right] dx$$

$$= \frac{1}{2} \int_0^1 (1 - x)^2 \, dx$$

$$= -\frac{1}{6} (1 - x)^3 \Big]_0^1 = \frac{1}{6}. \qquad \blacksquare$$

## Average Value of a Function in Space

The average value of a function $F$ over a region $D$ in space is defined by the formula

$$\textbf{Average value of } F \text{ over } D = \frac{1}{\text{volume of } D} \iiint_D F \, dV. \qquad (2)$$

For example, if $F(x, y, z) = \sqrt{x^2 + y^2 + z^2}$, then the average value of $F$ over $D$ is the average distance of points in $D$ from the origin. If $F(x, y, z)$ is the temperature at $(x, y, z)$ on a solid that occupies a region $D$ in space, then the average value of $F$ over $D$ is the average temperature of the solid.

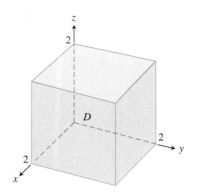

**FIGURE 15.34** The region of integration in Example 4.

**EXAMPLE 4**   Find the average value of $F(x, y, z) = xyz$ throughout the cubical region $D$ bounded by the coordinate planes and the planes $x = 2$, $y = 2$, and $z = 2$ in the first octant.

**Solution**   We sketch the cube with enough detail to show the limits of integration (Figure 15.34). We then use Equation (2) to calculate the average value of $F$ over the cube.

The volume of the region $D$ is $(2)(2)(2) = 8$. The value of the integral of $F$ over the cube is

$$\int_0^2 \int_0^2 \int_0^2 xyz \, dx \, dy \, dz = \int_0^2 \int_0^2 \left[ \frac{x^2}{2} yz \right]_{x=0}^{x=2} dy \, dz = \int_0^2 \int_0^2 2yz \, dy \, dz$$

$$= \int_0^2 \left[ y^2 z \right]_{y=0}^{y=2} dz = \int_0^2 4z \, dz = \left[ 2z^2 \right]_0^2 = 8.$$

With these values, Equation (2) gives

$$\begin{array}{c} \text{Average value of} \\ xyz \text{ over the cube} \end{array} = \frac{1}{\text{volume}} \iiint_{\text{cube}} xyz \, dV = \left( \frac{1}{8} \right)(8) = 1.$$

In evaluating the integral, we chose the order $dx \, dy \, dz$, but any of the other five possible orders would have done as well.   ■

## Properties of Triple Integrals

Triple integrals have the same algebraic properties as double and single integrals. Simply replace the double integrals in the four properties given in Section 15.2, page 880, with triple integrals.

## Exercises 15.5

### Triple Integrals in Different Iteration Orders

1. Evaluate the integral in Example 2 taking $F(x, y, z) = 1$ to find the volume of the tetrahedron in the order $dz \, dx \, dy$.

2. **Volume of rectangular solid**   Write six different iterated triple integrals for the volume of the rectangular solid in the first octant bounded by the coordinate planes and the planes $x = 1$, $y = 2$, and $z = 3$. Evaluate one of the integrals.

3. **Volume of tetrahedron**   Write six different iterated triple integrals for the volume of the tetrahedron cut from the first octant by the plane $6x + 3y + 2z = 6$. Evaluate one of the integrals.

4. **Volume of solid**   Write six different iterated triple integrals for the volume of the region in the first octant enclosed by the cylinder $x^2 + z^2 = 4$ and the plane $y = 3$. Evaluate one of the integrals.

5. **Volume enclosed by paraboloids**   Let $D$ be the region bounded by the paraboloids $z = 8 - x^2 - y^2$ and $z = x^2 + y^2$. Write six different triple iterated integrals for the volume of $D$. Evaluate one of the integrals.

6. **Volume inside paraboloid beneath a plane**   Let $D$ be the region bounded by the paraboloid $z = x^2 + y^2$ and the plane $z = 2y$. Write triple iterated integrals in the order $dz \, dx \, dy$ and $dz \, dy \, dx$ that give the volume of $D$. Do not evaluate either integral.

### Evaluating Triple Iterated Integrals

Evaluate the integrals in Exercises 7–20.

7. $\displaystyle\int_0^1 \int_0^1 \int_0^1 (x^2 + y^2 + z^2) \, dz \, dy \, dx$

8. $\displaystyle\int_0^{\sqrt{2}} \int_0^{3y} \int_{x^2+3y^2}^{8-x^2-y^2} dz \, dx \, dy$

9. $\displaystyle\int_1^e \int_1^{e^2} \int_1^{e^3} \frac{1}{xyz} dx \, dy \, dz$

10. $\displaystyle\int_0^1 \int_0^{3-3x} \int_0^{3-3x-y} dz \, dy \, dx$

11. $\displaystyle\int_0^{\pi/6} \int_0^1 \int_{-2}^3 y \sin z \, dx \, dy \, dz$

12. $\displaystyle\int_{-1}^1 \int_0^1 \int_0^2 (x + y + z) \, dy \, dx \, dz$

13. $\displaystyle\int_0^3 \int_0^{\sqrt{9-x^2}} \int_0^{\sqrt{9-x^2}} dz \, dy \, dx$

14. $\displaystyle\int_0^2 \int_{-\sqrt{4-y^2}}^{\sqrt{4-y^2}} \int_0^{2x+y} dz \, dx \, dy$

15. $\displaystyle\int_0^1 \int_0^{2-x} \int_0^{2-x-y} dz \, dy \, dx$

16. $\displaystyle\int_0^1 \int_0^{1-x^2} \int_3^{4-x^2-y} x \, dz \, dy \, dx$

17. $\displaystyle\int_0^\pi \int_0^\pi \int_0^\pi \cos(u + v + w) \, du \, dv \, dw$   ($uvw$-space)

18. $\displaystyle\int_0^1 \int_1^{\sqrt{e}} \int_1^e se^s \ln r \frac{(\ln t)^2}{t} dt \, dr \, ds$   ($rst$-space)

**19.** $\int_0^{\pi/4} \int_0^{\ln \sec v} \int_{-\infty}^{2t} e^x \, dx \, dt \, dv$   (*tvx*-space)

**20.** $\int_0^7 \int_0^2 \int_0^{\sqrt{4-q^2}} \frac{q}{r+1} \, dp \, dq \, dr$   (*pqr*-space)

**Finding Equivalent Iterated Integrals**

**21.** Here is the region of integration of the integral

$$\int_{-1}^1 \int_{x^2}^1 \int_0^{1-y} dz \, dy \, dx.$$

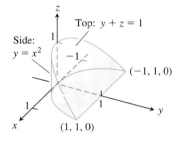

Rewrite the integral as an equivalent iterated integral in the order

**a.** *dy dz dx*      **b.** *dy dx dz*

**c.** *dx dy dz*      **d.** *dx dz dy*

**e.** *dz dx dy.*

**22.** Here is the region of integration of the integral

$$\int_0^1 \int_{-1}^0 \int_0^{y^2} dz \, dy \, dx.$$

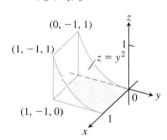

Rewrite the integral as an equivalent iterated integral in the order

**a.** *dy dz dx*      **b.** *dy dx dz*

**c.** *dx dy dz*      **d.** *dx dz dy*

**e.** *dz dx dy.*

**Finding Volumes Using Triple Integrals**

Find the volumes of the regions in Exercises 23–36.

**23.** The region between the cylinder $z = y^2$ and the *xy*-plane that is bounded by the planes $x = 0, x = 1, y = -1, y = 1$

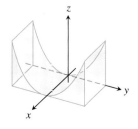

**24.** The region in the first octant bounded by the coordinate planes and the planes $x + z = 1, y + 2z = 2$

**25.** The region in the first octant bounded by the coordinate planes, the plane $y + z = 2$, and the cylinder $x = 4 - y^2$

**26.** The wedge cut from the cylinder $x^2 + y^2 = 1$ by the planes $z = -y$ and $z = 0$

**27.** The tetrahedron in the first octant bounded by the coordinate planes and the plane passing through $(1, 0, 0), (0, 2, 0)$, and $(0, 0, 3)$

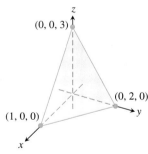

**28.** The region in the first octant bounded by the coordinate planes, the plane $y = 1 - x$, and the surface $z = \cos(\pi x/2), 0 \le x \le 1$

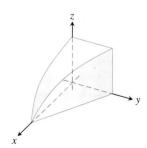

**29.** The region common to the interiors of the cylinders $x^2 + y^2 = 1$ and $x^2 + z^2 = 1$, one-eighth of which is shown in the accompanying figure

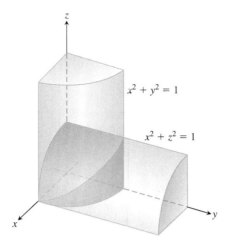

**30.** The region in the first octant bounded by the coordinate planes and the surface $z = 4 - x^2 - y$

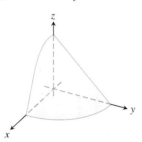

**31.** The region in the first octant bounded by the coordinate planes, the plane $x + y = 4$, and the cylinder $y^2 + 4z^2 = 16$

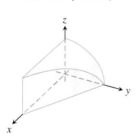

**32.** The region cut from the cylinder $x^2 + y^2 = 4$ by the plane $z = 0$ and the plane $x + z = 3$

**33.** The region between the planes $x + y + 2z = 2$ and $2x + 2y + z = 4$ in the first octant

**34.** The finite region bounded by the planes $z = x$, $x + z = 8$, $z = y$, $y = 8$, and $z = 0$

**35.** The region cut from the solid elliptical cylinder $x^2 + 4y^2 \leq 4$ by the $xy$-plane and the plane $z = x + 2$

**36.** The region bounded in back by the plane $x = 0$, on the front and sides by the parabolic cylinder $x = 1 - y^2$, on the top by the paraboloid $z = x^2 + y^2$, and on the bottom by the $xy$-plane

**Average Values**

In Exercises 37–40, find the average value of $F(x, y, z)$ over the given region.

**37.** $F(x, y, z) = x^2 + 9$ over the cube in the first octant bounded by the coordinate planes and the planes $x = 2$, $y = 2$, and $z = 2$

**38.** $F(x, y, z) = x + y - z$ over the rectangular solid in the first octant bounded by the coordinate planes and the planes $x = 1$, $y = 1$, and $z = 2$

**39.** $F(x, y, z) = x^2 + y^2 + z^2$ over the cube in the first octant bounded by the coordinate planes and the planes $x = 1$, $y = 1$, and $z = 1$

**40.** $F(x, y, z) = xyz$ over the cube in the first octant bounded by the coordinate planes and the planes $x = 2$, $y = 2$, and $z = 2$

**Changing the Order of Integration**

Evaluate the integrals in Exercises 41–44 by changing the order of integration in an appropriate way.

**41.** $\displaystyle\int_0^4 \int_0^1 \int_{2y}^2 \frac{4 \cos (x^2)}{2\sqrt{z}} \, dx \, dy \, dz$

**42.** $\displaystyle\int_0^1 \int_0^1 \int_{x^2}^1 12xze^{zy^2} \, dy \, dx \, dz$

**43.** $\displaystyle\int_0^1 \int_{\sqrt[3]{z}}^1 \int_0^{\ln 3} \frac{\pi e^{2x} \sin \pi y^2}{y^2} \, dx \, dy \, dz$

**44.** $\displaystyle\int_0^2 \int_0^{4-x^2} \int_0^x \frac{\sin 2z}{4 - z} \, dy \, dz \, dx$

**Theory and Examples**

**45. Finding an upper limit of an iterated integral** Solve for $a$:

$$\int_0^1 \int_0^{4-a-x^2} \int_a^{4-x^2-y} dz \, dy \, dx = \frac{4}{15}.$$

**46. Ellipsoid** For what value of $c$ is the volume of the ellipsoid $x^2 + (y/2)^2 + (z/c)^2 = 1$ equal to $8\pi$?

**47. Minimizing a triple integral** What domain $D$ in space minimizes the value of the integral

$$\iiint_D (4x^2 + 4y^2 + z^2 - 4) \, dV \, ?$$

Give reasons for your answer.

**48. Maximizing a triple integral** What domain $D$ in space maximizes the value of the integral

$$\iiint_D (1 - x^2 - y^2 - z^2) \, dV \, ?$$

Give reasons for your answer.

**COMPUTER EXPLORATIONS**

In Exercises 49–52, use a CAS integration utility to evaluate the triple integral of the given function over the specified solid region.

**49.** $F(x, y, z) = x^2y^2z$ over the solid cylinder bounded by $x^2 + y^2 = 1$ and the planes $z = 0$ and $z = 1$

**50.** $F(x, y, z) = |xyz|$ over the solid bounded below by the paraboloid $z = x^2 + y^2$ and above by the plane $z = 1$

**51.** $F(x, y, z) = \dfrac{z}{(x^2 + y^2 + z^2)^{3/2}}$ over the solid bounded below by the cone $z = \sqrt{x^2 + y^2}$ and above by the plane $z = 1$

**52.** $F(x, y, z) = x^4 + y^2 + z^2$ over the solid sphere $x^2 + y^2 + z^2 \le 1$

# 15.6 Moments and Centers of Mass

This section shows how to calculate the masses and moments of two- and three-dimensional objects in Cartesian coordinates. Section 15.7 gives the calculations for cylindrical and spherical coordinates. The definitions and ideas are similar to the single-variable case we studied in Section 6.6, but now we can consider more realistic situations.

## Masses and First Moments

If $\delta(x, y, z)$ is the density (mass per unit volume) of an object occupying a region $D$ in space, the integral of $\delta$ over $D$ gives the **mass** of the object. To see why, imagine partitioning the object into $n$ mass elements like the one in Figure 15.35. The object's mass is the limit

$$M = \lim_{n \to \infty} \sum_{k=1}^{n} \Delta m_k = \lim_{n \to \infty} \sum_{k=1}^{n} \delta(x_k, y_k, z_k) \, \Delta V_k = \iiint_D \delta(x, y, z) \, dV.$$

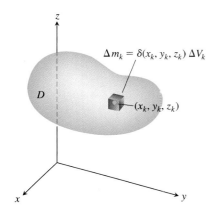

FIGURE 15.35  To define an object's mass, we first imagine it to be partitioned into a finite number of mass elements $\Delta m_k$.

The *first moment* of a solid region $D$ about a coordinate plane is defined as the triple integral over $D$ of the distance from a point $(x, y, z)$ in $D$ to the plane multiplied by the density of the solid at that point. For instance, the first moment about the $yz$-plane is the integral

$$M_{yz} = \iiint_D x\delta(x, y, z) \, dV.$$

The *center of mass* is found from the first moments. For instance, the $x$-coordinate of the center of mass is $\bar{x} = M_{yz}/M$.

For a two-dimensional object, such as a thin, flat plate, we calculate first moments about the coordinate axes by simply dropping the $z$-coordinate. So the first moment about the $y$-axis is the double integral over the region $R$ forming the plate of the distance from the axis multiplied by the density, or

$$M_y = \iint_R x\delta(x, y) \, dA.$$

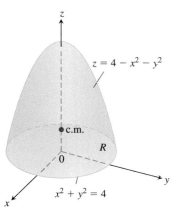

FIGURE 15.36  Finding the center of mass of a solid (Example 1).

Table 15.1 summarizes the formulas.

**EXAMPLE 1**    Find the center of mass of a solid of constant density $\delta$ bounded below by the disk $R: x^2 + y^2 \le 4$ in the plane $z = 0$ and above by the paraboloid $z = 4 - x^2 - y^2$ (Figure 15.36).

**TABLE 15.1** Mass and first moment formulas

**THREE-DIMENSIONAL SOLID**

**Mass:** $\quad M = \iiint\limits_{D} \delta \, dV \qquad \delta = \delta(x, y, z)$ is the density at $(x, y, z)$.

**First moments about the coordinate planes:**

$$M_{yz} = \iiint\limits_{D} x \, \delta \, dV, \qquad M_{xz} = \iiint\limits_{D} y \, \delta \, dV, \qquad M_{xy} = \iiint\limits_{D} z \, \delta \, dV$$

**Center of mass:**

$$\bar{x} = \frac{M_{yz}}{M}, \qquad \bar{y} = \frac{M_{xz}}{M}, \qquad \bar{z} = \frac{M_{xy}}{M}$$

**TWO-DIMENSIONAL PLATE**

**Mass:** $\quad M = \iint\limits_{R} \delta \, dA \qquad \delta = \delta(x, y)$ is the density at $(x, y)$.

**First moments:** $\quad M_y = \iint\limits_{R} x \, \delta \, dA, \qquad M_x = \iint\limits_{R} y \, \delta \, dA$

**Center of mass:** $\quad \bar{x} = \frac{M_y}{M}, \qquad \bar{y} = \frac{M_x}{M}$

**Solution** By symmetry $\bar{x} = \bar{y} = 0$. To find $\bar{z}$, we first calculate

$$M_{xy} = \iiint\limits_{R} \int_{z=0}^{z=4-x^2-y^2} z \, \delta \, dz \, dy \, dx = \iint\limits_{R} \left[ \frac{z^2}{2} \right]_{z=0}^{z=4-x^2-y^2} \delta \, dy \, dx$$

$$= \frac{\delta}{2} \iint\limits_{R} (4 - x^2 - y^2)^2 \, dy \, dx$$

$$= \frac{\delta}{2} \int_0^{2\pi} \int_0^2 (4 - r^2)^2 r \, dr \, d\theta \qquad \text{Polar coordinates simplify the integration.}$$

$$= \frac{\delta}{2} \int_0^{2\pi} \left[ -\frac{1}{6}(4 - r^2)^3 \right]_{r=0}^{r=2} d\theta = \frac{16\delta}{3} \int_0^{2\pi} d\theta = \frac{32\pi\delta}{3}.$$

A similar calculation gives the mass

$$M = \iiint\limits_{R} \int_0^{4-x^2-y^2} \delta \, dz \, dy \, dx = 8\pi\delta.$$

Therefore $\bar{z} = (M_{xy}/M) = 4/3$ and the center of mass is $(\bar{x}, \bar{y}, \bar{z}) = (0, 0, 4/3)$. ∎

When the density of a solid object or plate is constant (as in Example 1), the center of mass is called the **centroid** of the object. To find a centroid, we set $\delta$ equal to 1 and proceed to find $\bar{x}$, $\bar{y}$, and $\bar{z}$ as before, by dividing first moments by masses. These calculations are also valid for two-dimensional objects.

**EXAMPLE 2** Find the centroid of the region in the first quadrant that is bounded above by the line $y = x$ and below by the parabola $y = x^2$.

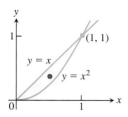

**FIGURE 15.37** The centroid of this region is found in Example 2.

**Solution** We sketch the region and include enough detail to determine the limits of integration (Figure 15.37). We then set $\delta$ equal to 1 and evaluate the appropriate formulas from Table 15.1:

$$M = \int_0^1 \int_{x^2}^x 1 \, dy \, dx = \int_0^1 \Big[ y \Big]_{y=x^2}^{y=x} dx = \int_0^1 (x - x^2) \, dx = \left[ \frac{x^2}{2} - \frac{x^3}{3} \right]_0^1 = \frac{1}{6}$$

$$M_x = \int_0^1 \int_{x^2}^x y \, dy \, dx = \int_0^1 \left[ \frac{y^2}{2} \right]_{y=x^2}^{y=x} dx$$

$$= \int_0^1 \left( \frac{x^2}{2} - \frac{x^4}{2} \right) dx = \left[ \frac{x^3}{6} - \frac{x^5}{10} \right]_0^1 = \frac{1}{15}$$

$$M_y = \int_0^1 \int_{x^2}^x x \, dy \, dx = \int_0^1 \Big[ xy \Big]_{y=x^2}^{y=x} dx = \int_0^1 (x^2 - x^3) \, dx = \left[ \frac{x^3}{3} - \frac{x^4}{4} \right]_0^1 = \frac{1}{12}.$$

From these values of $M$, $M_x$, and $M_y$, we find

$$\bar{x} = \frac{M_y}{M} = \frac{1/12}{1/6} = \frac{1}{2} \qquad \text{and} \qquad \bar{y} = \frac{M_x}{M} = \frac{1/15}{1/6} = \frac{2}{5}.$$

The centroid is the point $(1/2, 2/5)$.  ∎

### Moments of Inertia

An object's first moments (Table 15.1) tell us about balance and about the torque the object experiences about different axes in a gravitational field. If the object is a rotating shaft, however, we are more likely to be interested in how much energy is stored in the shaft or about how much energy is generated by a shaft rotating at a particular angular velocity. This is where the second moment or moment of inertia comes in.

Think of partitioning the shaft into small blocks of mass $\Delta m_k$ and let $r_k$ denote the distance from the $k$th block's center of mass to the axis of rotation (Figure 15.38). If the shaft rotates at a constant angular velocity of $\omega = d\theta/dt$ radians per second, the block's center of mass will trace its orbit at a linear speed of

$$v_k = \frac{d}{dt}(r_k\theta) = r_k \frac{d\theta}{dt} = r_k\omega.$$

The block's kinetic energy will be approximately

$$\frac{1}{2}\Delta m_k v_k^2 = \frac{1}{2}\Delta m_k (r_k\omega)^2 = \frac{1}{2}\omega^2 r_k^2 \, \Delta m_k.$$

The kinetic energy of the shaft will be approximately

$$\sum \frac{1}{2}\omega^2 r_k^2 \, \Delta m_k.$$

The integral approached by these sums as the shaft is partitioned into smaller and smaller blocks gives the shaft's kinetic energy:

$$\text{KE}_{\text{shaft}} = \int \frac{1}{2}\omega^2 r^2 \, dm = \frac{1}{2}\omega^2 \int r^2 \, dm. \tag{1}$$

The factor

$$I = \int r^2 \, dm$$

is the *moment of inertia* of the shaft about its axis of rotation, and we see from Equation (1) that the shaft's kinetic energy is

$$\text{KE}_{\text{shaft}} = \frac{1}{2}I\omega^2.$$

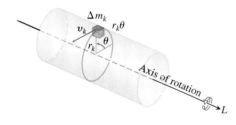

**FIGURE 15.38** To find an integral for the amount of energy stored in a rotating shaft, we first imagine the shaft to be partitioned into small blocks. Each block has its own kinetic energy. We add the contributions of the individual blocks to find the kinetic energy of the shaft.

The moment of inertia of a shaft resembles in some ways the inertial mass of a locomotive. To start a locomotive with mass $m$ moving at a linear velocity $v$, we need to provide a kinetic energy of $KE = (1/2)mv^2$. To stop the locomotive we have to remove this amount of energy. To start a shaft with moment of inertia $I$ rotating at an angular velocity $\omega$, we need to provide a kinetic energy of $KE = (1/2)I\omega^2$. To stop the shaft we have to take this amount of energy back out. The shaft's moment of inertia is analogous to the locomotive's mass. What makes the locomotive hard to start or stop is its mass. What makes the shaft hard to start or stop is its moment of inertia. The moment of inertia depends not only on the mass of the shaft but also on its distribution. Mass that is farther away from the axis of rotation contributes more to the moment of inertia.

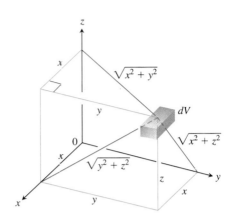

**FIGURE 15.39** Distances from $dV$ to the coordinate planes and axes.

We now derive a formula for the moment of inertia for a solid in space. If $r(x, y, z)$ is the distance from the point $(x, y, z)$ in $D$ to a line $L$, then the moment of inertia of the mass $\Delta m_k = \delta(x_k, y_k, z_k)\Delta V_k$ about the line $L$ (as in Figure 15.38) is approximately $\Delta I_k = r^2(x_k, y_k, z_k)\Delta m_k$. **The moment of inertia about $L$** of the entire object is

$$I_L = \lim_{n\to\infty} \sum_{k=1}^{n} \Delta I_k = \lim_{n\to\infty} \sum_{k=1}^{n} r^2(x_k, y_k, z_k)\delta(x_k, y_k, z_k)\,\Delta V_k = \iiint_D r^2\delta\,dV.$$

If $L$ is the $x$-axis, then $r^2 = y^2 + z^2$ (Figure 15.39) and

$$I_x = \iiint_D (y^2 + z^2)\,\delta(x, y, z)\,dV.$$

Similarly, if $L$ is the $y$-axis or $z$-axis we have

$$I_y = \iiint_D (x^2 + z^2)\,\delta(x, y, z)\,dV \qquad \text{and} \qquad I_z = \iiint_D (x^2 + y^2)\,\delta(x, y, z)\,dV.$$

Table 15.2 summarizes the formulas for these moments of inertia (second moments because they invoke the *squares* of the distances). It shows the definition of the *polar moment* about the origin as well.

**EXAMPLE 3**   Find $I_x, I_y, I_z$ for the rectangular solid of constant density $\delta$ shown in Figure 15.40.

**Solution**   The formula for $I_x$ gives

$$I_x = \int_{-c/2}^{c/2}\int_{-b/2}^{b/2}\int_{-a/2}^{a/2} (y^2 + z^2)\,\delta\,dx\,dy\,dz.$$

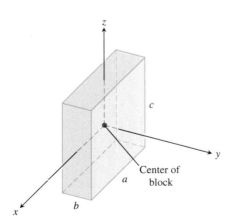

**FIGURE 15.40** Finding $I_x, I_y,$ and $I_z$ for the block shown here. The origin lies at the center of the block (Example 3).

We can avoid some of the work of integration by observing that $(y^2 + z^2)\delta$ is an even function of $x$, $y$, and $z$ since $\delta$ is constant. The rectangular solid consists of eight symmetric pieces, one in each octant. We can evaluate the integral on one of these pieces and then multiply by 8 to get the total value.

$$I_x = 8\int_0^{c/2}\int_0^{b/2}\int_0^{a/2} (y^2 + z^2)\,\delta\,dx\,dy\,dz = 4a\delta\int_0^{c/2}\int_0^{b/2} (y^2 + z^2)\,dy\,dz$$

$$= 4a\delta\int_0^{c/2}\left[\frac{y^3}{3} + z^2y\right]_{y=0}^{y=b/2} dz$$

$$= 4a\delta\int_0^{c/2}\left(\frac{b^3}{24} + \frac{z^2b}{2}\right) dz$$

$$= 4a\delta\left(\frac{b^3c}{48} + \frac{c^3b}{48}\right) = \frac{abc\delta}{12}(b^2 + c^2) = \frac{M}{12}(b^2 + c^2). \qquad M = abc\delta$$

---

**TABLE 15.2** Moments of inertia (second moments) formulas

**THREE-DIMENSIONAL SOLID**

About the $x$-axis: $\quad I_x = \iiint (y^2 + z^2)\, \delta\, dV \qquad \delta = \delta(x, y, z)$

About the $y$-axis: $\quad I_y = \iiint (x^2 + z^2)\, \delta\, dV$

About the $z$-axis: $\quad I_z = \iiint (x^2 + y^2)\, \delta\, dV$

About a line $L$: $\quad I_L = \iiint r^2(x, y, z)\, \delta\, dV \qquad r(x, y, z) = $ distance from the point $(x, y, z)$ to line $L$

**TWO-DIMENSIONAL PLATE**

About the $x$-axis: $\quad I_x = \iint y^2\, \delta\, dA \qquad \delta = \delta(x, y)$

About the $y$-axis: $\quad I_y = \iint x^2\, \delta\, dA$

About a line $L$: $\quad I_L = \iint r^2(x, y)\, \delta\, dA \qquad r(x, y) = $ distance from $(x, y)$ to $L$

About the origin (polar moment): $\quad I_0 = \iint (x^2 + y^2)\, \delta\, dA = I_x + I_y$

---

Similarly,

$$I_y = \frac{M}{12}(a^2 + c^2) \qquad \text{and} \qquad I_z = \frac{M}{12}(a^2 + b^2). \qquad \blacksquare$$

**EXAMPLE 4**    A thin plate covers the triangular region bounded by the $x$-axis and the lines $x = 1$ and $y = 2x$ in the first quadrant. The plate's density at the point $(x, y)$ is $\delta(x, y) = 6x + 6y + 6$. Find the plate's moments of inertia about the coordinate axes and the origin.

**Solution**    We sketch the plate and put in enough detail to determine the limits of integration for the integrals we have to evaluate (Figure 15.41). The moment of inertia about the $x$-axis is

$$I_x = \int_0^1 \int_0^{2x} y^2 \delta(x, y)\, dy\, dx = \int_0^1 \int_0^{2x} (6xy^2 + 6y^3 + 6y^2)\, dy\, dx$$

$$= \int_0^1 \left[ 2xy^3 + \frac{3}{2}y^4 + 2y^3 \right]_{y=0}^{y=2x} dx = \int_0^1 (40x^4 + 16x^3)\, dx$$

$$= \left[ 8x^5 + 4x^4 \right]_0^1 = 12.$$

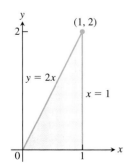

**FIGURE 15.41**    The triangular region covered by the plate in Example 4.

Beam A

Axis

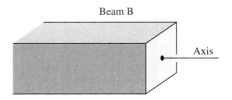

Beam B

Axis

**FIGURE 15.42** The greater the polar moment of inertia of the cross-section of a beam about the beam's longitudinal axis, the stiffer the beam. Beams A and B have the same cross-sectional area, but A is stiffer.

Similarly, the moment of inertia about the $y$-axis is

$$I_y = \int_0^1 \int_0^{2x} x^2 \delta(x, y) \, dy \, dx = \frac{39}{5}.$$

Notice that we integrate $y^2$ times density in calculating $I_x$ and $x^2$ times density to find $I_y$. Since we know $I_x$ and $I_y$, we do not need to evaluate an integral to find $I_0$; we can use the equation $I_0 = I_x + I_y$ from Table 15.2 instead:

$$I_0 = 12 + \frac{39}{5} = \frac{60 + 39}{5} = \frac{99}{5}.$$

The moment of inertia also plays a role in determining how much a horizontal metal beam will bend under a load. The stiffness of the beam is a constant times $I$, the moment of inertia of a typical cross-section of the beam about the beam's longitudinal axis. The greater the value of $I$, the stiffer the beam and the less it will bend under a given load. That is why we use I-beams instead of beams whose cross-sections are square. The flanges at the top and bottom of the beam hold most of the beam's mass away from the longitudinal axis to increase the value of $I$ (Figure 15.42).

# Exercises **15.6**

### Plates of Constant Density

1. **Finding a center of mass** Find the center of mass of a thin plate of density $\delta = 3$ bounded by the lines $x = 0$, $y = x$, and the parabola $y = 2 - x^2$ in the first quadrant.

2. **Finding moments of inertia** Find the moments of inertia about the coordinate axes of a thin rectangular plate of constant density $\delta$ bounded by the lines $x = 3$ and $y = 3$ in the first quadrant.

3. **Finding a centroid** Find the centroid of the region in the first quadrant bounded by the $x$-axis, the parabola $y^2 = 2x$, and the line $x + y = 4$.

4. **Finding a centroid** Find the centroid of the triangular region cut from the first quadrant by the line $x + y = 3$.

5. **Finding a centroid** Find the centroid of the region cut from the first quadrant by the circle $x^2 + y^2 = a^2$.

6. **Finding a centroid** Find the centroid of the region between the $x$-axis and the arch $y = \sin x$, $0 \le x \le \pi$.

7. **Finding moments of inertia** Find the moment of inertia about the $x$-axis of a thin plate of density $\delta = 1$ bounded by the circle $x^2 + y^2 = 4$. Then use your result to find $I_y$ and $I_0$ for the plate.

8. **Finding a moment of inertia** Find the moment of inertia with respect to the $y$-axis of a thin sheet of constant density $\delta = 1$ bounded by the curve $y = (\sin^2 x)/x^2$ and the interval $\pi \le x \le 2\pi$ of the $x$-axis.

9. **The centroid of an infinite region** Find the centroid of the infinite region in the second quadrant enclosed by the coordinate axes and the curve $y = e^x$. (Use improper integrals in the mass-moment formulas.)

10. **The first moment of an infinite plate** Find the first moment about the $y$-axis of a thin plate of density $\delta(x, y) = 1$ covering the infinite region under the curve $y = e^{-x^2/2}$ in the first quadrant.

### Plates with Varying Density

11. **Finding a moment of inertia** Find the moment of inertia about the $x$-axis of a thin plate bounded by the parabola $x = y - y^2$ and the line $x + y = 0$ if $\delta(x, y) = x + y$.

12. **Finding mass** Find the mass of a thin plate occupying the smaller region cut from the ellipse $x^2 + 4y^2 = 12$ by the parabola $x = 4y^2$ if $\delta(x, y) = 5x$.

13. **Finding a center of mass** Find the center of mass of a thin triangular plate bounded by the $y$-axis and the lines $y = x$ and $y = 2 - x$ if $\delta(x, y) = 6x + 3y + 3$.

14. **Finding a center of mass and moment of inertia** Find the center of mass and moment of inertia about the $x$-axis of a thin plate bounded by the curves $x = y^2$ and $x = 2y - y^2$ if the density at the point $(x, y)$ is $\delta(x, y) = y + 1$.

15. **Center of mass, moment of inertia** Find the center of mass and the moment of inertia about the $y$-axis of a thin rectangular plate cut from the first quadrant by the lines $x = 6$ and $y = 1$ if $\delta(x, y) = x + y + 1$.

16. **Center of mass, moment of inertia** Find the center of mass and the moment of inertia about the $y$-axis of a thin plate bounded by the line $y = 1$ and the parabola $y = x^2$ if the density is $\delta(x, y) = y + 1$.

17. **Center of mass, moment of inertia** Find the center of mass and the moment of inertia about the $y$-axis of a thin plate bounded by the $x$-axis, the lines $x = \pm 1$, and the parabola $y = x^2$ if $\delta(x, y) = 7y + 1$.

**18. Center of mass, moment of inertia** Find the center of mass and the moment of inertia about the $x$-axis of a thin rectangular plate bounded by the lines $x = 0, x = 20, y = -1$, and $y = 1$ if $\delta(x, y) = 1 + (x/20)$.

**19. Center of mass, moments of inertia** Find the center of mass, the moment of inertia about the coordinate axes, and the polar moment of inertia of a thin triangular plate bounded by the lines $y = x, y = -x$, and $y = 1$ if $\delta(x, y) = y + 1$.

**20. Center of mass, moments of inertia** Repeat Exercise 19 for $\delta(x, y) = 3x^2 + 1$.

Solids with Constant Density

**21. Moments of inertia** Find the moments of inertia of the rectangular solid shown here with respect to its edges by calculating $I_x, I_y$, and $I_z$.

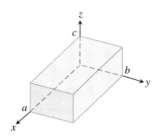

**22. Moments of inertia** The coordinate axes in the figure run through the centroid of a solid wedge parallel to the labeled edges. Find $I_x, I_y$, and $I_z$ if $a = b = 6$ and $c = 4$.

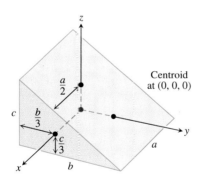

Centroid at $(0, 0, 0)$

**23. Center of mass and moments of inertia** A solid "trough" of constant density is bounded below by the surface $z = 4y^2$, above by the plane $z = 4$, and on the ends by the planes $x = 1$ and $x = -1$. Find the center of mass and the moments of inertia with respect to the three axes.

**24. Center of mass** A solid of constant density is bounded below by the plane $z = 0$, on the sides by the elliptical cylinder $x^2 + 4y^2 = 4$, and above by the plane $z = 2 - x$ (see the accompanying figure).

**a.** Find $\bar{x}$ and $\bar{y}$.

**b.** Evaluate the integral

$$M_{xy} = \int_{-2}^{2} \int_{-(1/2)\sqrt{4-x^2}}^{(1/2)\sqrt{4-x^2}} \int_{0}^{2-x} z \, dz \, dy \, dx$$

using integral tables to carry out the final integration with respect to $x$. Then divide $M_{xy}$ by $M$ to verify that $\bar{z} = 5/4$.

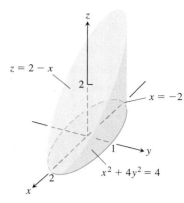

**25. a. Center of mass** Find the center of mass of a solid of constant density bounded below by the paraboloid $z = x^2 + y^2$ and above by the plane $z = 4$.

**b.** Find the plane $z = c$ that divides the solid into two parts of equal volume. This plane does not pass through the center of mass.

**26. Moments** A solid cube, 2 units on a side, is bounded by the planes $x = \pm 1, z = \pm 1, y = 3$, and $y = 5$. Find the center of mass and the moments of inertia about the coordinate axes.

**27. Moment of inertia about a line** A wedge like the one in Exercise 22 has $a = 4, b = 6$, and $c = 3$. Make a quick sketch to check for yourself that the square of the distance from a typical point $(x, y, z)$ of the wedge to the line $L: z = 0, y = 6$ is $r^2 = (y - 6)^2 + z^2$. Then calculate the moment of inertia of the wedge about $L$.

**28. Moment of inertia about a line** A wedge like the one in Exercise 22 has $a = 4, b = 6$, and $c = 3$. Make a quick sketch to check for yourself that the square of the distance from a typical point $(x, y, z)$ of the wedge to the line $L: x = 4, y = 0$ is $r^2 = (x - 4)^2 + y^2$. Then calculate the moment of inertia of the wedge about $L$.

Solids with Varying Density

In Exercises 29 and 30, find

    **a.** the mass of the solid.     **b.** the center of mass.

**29.** A solid region in the first octant is bounded by the coordinate planes and the plane $x + y + z = 2$. The density of the solid is $\delta(x, y, z) = 2x$.

**30.** A solid in the first octant is bounded by the planes $y = 0$ and $z = 0$ and by the surfaces $z = 4 - x^2$ and $x = y^2$ (see the accompanying figure). Its density function is $\delta(x, y, z) = kxy, k$ a constant.

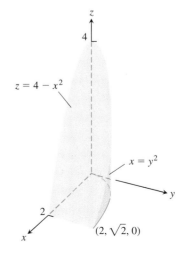

In Exercises 31 and 32, find

   **a.** the mass of the solid.

   **b.** the center of mass.

   **c.** the moments of inertia about the coordinate axes.

**31.** A solid cube in the first octant is bounded by the coordinate planes and by the planes $x = 1$, $y = 1$, and $z = 1$. The density of the cube is $\delta(x, y, z) = x + y + z + 1$.

**32.** A wedge like the one in Exercise 22 has dimensions $a = 2$, $b = 5$, and $c = 3$. The density is $\delta(x, y, z) = x + 1$. Notice that if the density is constant, the center of mass will be $(0, 0, 0)$.

**33. Mass** Find the mass of the solid bounded by the planes $x + z = 1$, $x - z = -1$, $y = 0$, and the surface $y = \sqrt{z}$. The density of the solid is $\delta(x, y, z) = 2y + 5$.

**34. Mass** Find the mass of the solid region bounded by the parabolic surfaces $z = 16 - 2x^2 - 2y^2$ and $z = 2x^2 + 2y^2$ if the density of the solid is $\delta(x, y, z) = \sqrt{x^2 + y^2}$.

### Theory and Examples

**The Parallel Axis Theorem** Let $L_{c.m.}$ be a line through the center of mass of a body of mass $m$ and let $L$ be a parallel line $h$ units away from $L_{c.m.}$. The *Parallel Axis Theorem* says that the moments of inertia $I_{c.m.}$ and $I_L$ of the body about $L_{c.m.}$ and $L$ satisfy the equation

$$I_L = I_{c.m.} + mh^2. \qquad (2)$$

As in the two-dimensional case, the theorem gives a quick way to calculate one moment when the other moment and the mass are known.

**35. Proof of the Parallel Axis Theorem**

   **a.** Show that the first moment of a body in space about any plane through the body's center of mass is zero. (*Hint:* Place the body's center of mass at the origin and let the plane be the $yz$-plane. What does the formula $\bar{x} = M_{yz}/M$ then tell you?)

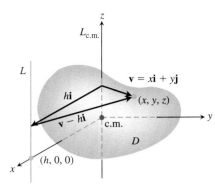

   **b.** To prove the Parallel Axis Theorem, place the body with its center of mass at the origin, with the line $L_{c.m.}$ along the $z$-axis and the line $L$ perpendicular to the $xy$-plane at the point $(h, 0, 0)$. Let $D$ be the region of space occupied by the body. Then, in the notation of the figure,

$$I_L = \iiint_D |\mathbf{v} - h\mathbf{i}|^2 \, dm.$$

Expand the integrand in this integral and complete the proof.

**36.** The moment of inertia about a diameter of a solid sphere of constant density and radius $a$ is $(2/5)ma^2$, where $m$ is the mass of the sphere. Find the moment of inertia about a line tangent to the sphere.

**37.** The moment of inertia of the solid in Exercise 21 about the $z$-axis is $I_z = abc(a^2 + b^2)/3$.

   **a.** Use Equation (2) to find the moment of inertia of the solid about the line parallel to the $z$-axis through the solid's center of mass.

   **b.** Use Equation (2) and the result in part (a) to find the moment of inertia of the solid about the line $x = 0$, $y = 2b$.

**38.** If $a = b = 6$ and $c = 4$, the moment of inertia of the solid wedge in Exercise 22 about the $x$-axis is $I_x = 208$. Find the moment of inertia of the wedge about the line $y = 4$, $z = -4/3$ (the edge of the wedge's narrow end).

# 15.7 Triple Integrals in Cylindrical and Spherical Coordinates

When a calculation in physics, engineering, or geometry involves a cylinder, cone, or sphere, we can often simplify our work by using cylindrical or spherical coordinates, which are introduced in this section. The procedure for transforming to these coordinates and evaluating the resulting triple integrals is similar to the transformation to polar coordinates in the plane studied in Section 15.4.

### Integration in Cylindrical Coordinates

We obtain cylindrical coordinates for space by combining polar coordinates in the $xy$-plane with the usual $z$-axis. This assigns to every point in space one or more coordinate triples of the form $(r, \theta, z)$, as shown in Figure 15.43. Here we require $r \geq 0$.

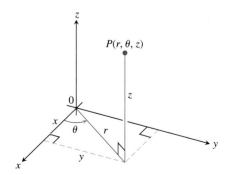

**FIGURE 15.43** The cylindrical coordinates of a point in space are $r$, $\theta$, and $z$.

> **DEFINITION** **Cylindrical coordinates** represent a point $P$ in space by ordered triples $(r, \theta, z)$ in which $r \geq 0$,
>
> **1.** $r$ and $\theta$ are polar coordinates for the vertical projection of $P$ on the $xy$-plane
>
> **2.** $z$ is the rectangular vertical coordinate.

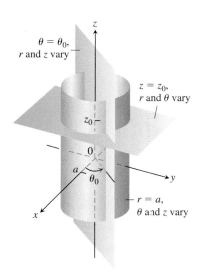

**FIGURE 15.44** Constant-coordinate equations in cylindrical coordinates yield cylinders and planes.

The values of $x$, $y$, $r$, and $\theta$ in rectangular and cylindrical coordinates are related by the usual equations.

---

**Equations Relating Rectangular $(x, y, z)$ and Cylindrical $(r, \theta, z)$ Coordinates**

$$x = r \cos \theta, \qquad y = r \sin \theta, \qquad z = z,$$
$$r^2 = x^2 + y^2, \qquad \tan \theta = y/x$$

---

In cylindrical coordinates, the equation $r = a$ describes not just a circle in the $xy$-plane but an entire cylinder about the $z$-axis (Figure 15.44). The $z$-axis is given by $r = 0$. The equation $\theta = \theta_0$ describes the plane that contains the $z$-axis and makes an angle $\theta_0$ with the positive $x$-axis. And, just as in rectangular coordinates, the equation $z = z_0$ describes a plane perpendicular to the $z$-axis.

Cylindrical coordinates are good for describing cylinders whose axes run along the $z$-axis and planes that either contain the $z$-axis or lie perpendicular to the $z$-axis. Surfaces like these have equations of constant coordinate value:

$$r = 4 \qquad \text{Cylinder, radius 4, axis the } z\text{-axis}$$
$$\theta = \frac{\pi}{3} \qquad \text{Plane containing the } z\text{-axis}$$
$$z = 2. \qquad \text{Plane perpendicular to the } z\text{-axis}$$

When computing triple integrals over a region $D$ in cylindrical coordinates, we partition the region into $n$ small cylindrical wedges, rather than into rectangular boxes. In the $k$th cylindrical wedge, $r$, $\theta$ and $z$ change by $\Delta r_k$, $\Delta \theta_k$, and $\Delta z_k$, and the largest of these numbers among all the cylindrical wedges is called the **norm** of the partition. We define the triple integral as a limit of Riemann sums using these wedges. The volume of such a cylindrical wedge $\Delta V_k$ is obtained by taking the area $\Delta A_k$ of its base in the $r\theta$-plane and multiplying by the height $\Delta z$ (Figure 15.45).

**FIGURE 15.45** In cylindrical coordinates the volume of the wedge is approximated by the product $\Delta V = \Delta z \, r \, \Delta r \, \Delta \theta$.

For a point $(r_k, \theta_k, z_k)$ in the center of the $k$th wedge, we calculated in polar coordinates that $\Delta A_k = r_k \, \Delta r_k \, \Delta \theta_k$. So $\Delta V_k = \Delta z_k \, r_k \, \Delta r_k \, \Delta \theta_k$ and a Riemann sum for $f$ over $D$ has the form

$$S_n = \sum_{k=1}^{n} f(r_k, \theta_k, z_k) \, \Delta z_k \, r_k \, \Delta r_k \, \Delta \theta_k.$$

The triple integral of a function $f$ over $D$ is obtained by taking a limit of such Riemann sums with partitions whose norms approach zero:

---

$$\lim_{n \to \infty} S_n = \iiint_D f \, dV = \iiint_D f \, dz \, r \, dr \, d\theta.$$

---

**Volume Differential in Cylindrical Coordinates**

$$dV = dz \, r \, dr \, d\theta$$

Triple integrals in cylindrical coordinates are then evaluated as iterated integrals, as in the following example. Although the definition of cylindrical coordinates makes sense without any restrictions on $\theta$, in most situations when integrating, we will need to restrict $\theta$ to an interval of length $2\pi$. So we impose the requirement that $\alpha \le \theta \le \beta$, where $0 \le \beta - \alpha \le 2\pi$.

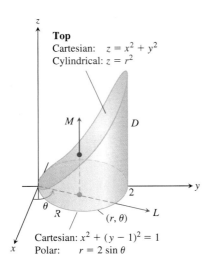

**FIGURE 15.46** Finding the limits of integration for evaluating an integral in cylindrical coordinates (Example 1).

**EXAMPLE 1** Find the limits of integration in cylindrical coordinates for integrating a function $f(r, \theta, z)$ over the region $D$ bounded below by the plane $z = 0$, laterally by the circular cylinder $x^2 + (y - 1)^2 = 1$, and above by the paraboloid $z = x^2 + y^2$.

**Solution** The base of $D$ is also the region's projection $R$ on the $xy$-plane. The boundary of $R$ is the circle $x^2 + (y - 1)^2 = 1$. Its polar coordinate equation is

$$x^2 + (y - 1)^2 = 1$$
$$x^2 + y^2 - 2y + 1 = 1$$
$$r^2 - 2r \sin \theta = 0$$
$$r = 2 \sin \theta.$$

The region is sketched in Figure 15.46.

We find the limits of integration, starting with the $z$-limits. A line $M$ through a typical point $(r, \theta)$ in $R$ parallel to the $z$-axis enters $D$ at $z = 0$ and leaves at $z = x^2 + y^2 = r^2$.

Next we find the $r$-limits of integration. A ray $L$ through $(r, \theta)$ from the origin enters $R$ at $r = 0$ and leaves at $r = 2 \sin \theta$.

Finally we find the $\theta$-limits of integration. As $L$ sweeps across $R$, the angle $\theta$ it makes with the positive $x$-axis runs from $\theta = 0$ to $\theta = \pi$. The integral is

$$\iiint\limits_{D} f(r, \theta, z) \, dV = \int_0^\pi \int_0^{2 \sin \theta} \int_0^{r^2} f(r, \theta, z) \, dz \, r \, dr \, d\theta. \qquad \blacksquare$$

Example 1 illustrates a good procedure for finding limits of integration in cylindrical coordinates. The procedure is summarized as follows.

### How to Integrate in Cylindrical Coordinates

To evaluate

$$\iiint\limits_{D} f(r, \theta, z) \, dV$$

over a region $D$ in space in cylindrical coordinates, integrating first with respect to $z$, then with respect to $r$, and finally with respect to $\theta$, take the following steps.

1. *Sketch.* Sketch the region $D$ along with its projection $R$ on the $xy$-plane. Label the surfaces and curves that bound $D$ and $R$.

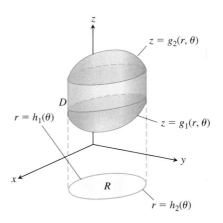

2. *Find the z-limits of integration.* Draw a line $M$ through a typical point $(r, \theta)$ of $R$ parallel to the z-axis. As $z$ increases, $M$ enters $D$ at $z = g_1(r, \theta)$ and leaves at $z = g_2(r, \theta)$. These are the z-limits of integration.

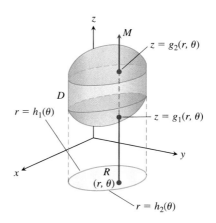

3. *Find the r-limits of integration.* Draw a ray $L$ through $(r, \theta)$ from the origin. The ray enters $R$ at $r = h_1(\theta)$ and leaves at $r = h_2(\theta)$. These are the r-limits of integration.

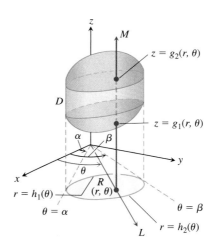

4. *Find the θ-limits of integration.* As $L$ sweeps across $R$, the angle $\theta$ it makes with the positive x-axis runs from $\theta = \alpha$ to $\theta = \beta$. These are the θ-limits of integration. The integral is

$$\iiint_D f(r, \theta, z) \, dV = \int_{\theta=\alpha}^{\theta=\beta} \int_{r=h_1(\theta)}^{r=h_2(\theta)} \int_{z=g_1(r, \theta)}^{z=g_2(r, \theta)} f(r, \theta, z) \, dz \, r \, dr \, d\theta.$$

**EXAMPLE 2**     Find the centroid $(\delta = 1)$ of the solid enclosed by the cylinder $x^2 + y^2 = 4$, bounded above by the paraboloid $z = x^2 + y^2$, and bounded below by the xy-plane.

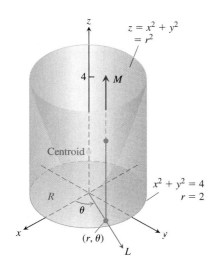

**FIGURE 15.47** Example 2 shows how to find the centroid of this solid.

**Solution**   We sketch the solid, bounded above by the paraboloid $z = r^2$ and below by the plane $z = 0$ (Figure 15.47). Its base $R$ is the disk $0 \leq r \leq 2$ in the xy-plane.

The solid's centroid $(\bar{x}, \bar{y}, \bar{z})$ lies on its axis of symmetry, here the $z$-axis. This makes $\bar{x} = \bar{y} = 0$. To find $\bar{z}$, we divide the first moment $M_{xy}$ by the mass $M$.

To find the limits of integration for the mass and moment integrals, we continue with the four basic steps. We completed our initial sketch. The remaining steps give the limits of integration.

*The z-limits.* A line $M$ through a typical point $(r, \theta)$ in the base parallel to the $z$-axis enters the solid at $z = 0$ and leaves at $z = r^2$.

*The r-limits.* A ray $L$ through $(r, \theta)$ from the origin enters $R$ at $r = 0$ and leaves at $r = 2$.

*The θ-limits.* As $L$ sweeps over the base like a clock hand, the angle $\theta$ it makes with the positive $x$-axis runs from $\theta = 0$ to $\theta = 2\pi$. The value of $M_{xy}$ is

$$M_{xy} = \int_0^{2\pi} \int_0^2 \int_0^{r^2} z \, dz \, r \, dr \, d\theta = \int_0^{2\pi} \int_0^2 \left[\frac{z^2}{2}\right]_0^{r^2} r \, dr \, d\theta$$

$$= \int_0^{2\pi} \int_0^2 \frac{r^5}{2} \, dr \, d\theta = \int_0^{2\pi} \left[\frac{r^6}{12}\right]_0^2 d\theta = \int_0^{2\pi} \frac{16}{3} \, d\theta = \frac{32\pi}{3}.$$

The value of $M$ is

$$M = \int_0^{2\pi} \int_0^2 \int_0^{r^2} dz \, r \, dr \, d\theta = \int_0^{2\pi} \int_0^2 \left[z\right]_0^{r^2} r \, dr \, d\theta$$

$$= \int_0^{2\pi} \int_0^2 r^3 \, dr \, d\theta = \int_0^{2\pi} \left[\frac{r^4}{4}\right]_0^2 d\theta = \int_0^{2\pi} 4 \, d\theta = 8\pi.$$

Therefore,

$$\bar{z} = \frac{M_{xy}}{M} = \frac{32\pi}{3} \frac{1}{8\pi} = \frac{4}{3},$$

and the centroid is $(0, 0, 4/3)$. Notice that the centroid lies on the $z$-axis, outside the solid. ∎

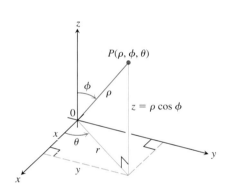

**FIGURE 15.48** The spherical coordinates $\rho$, $\phi$, and $\theta$ and their relation to $x$, $y$, $z$, and $r$.

## Spherical Coordinates and Integration

Spherical coordinates locate points in space with two angles and one distance, as shown in Figure 15.48. The first coordinate, $\rho = |\overrightarrow{OP}|$, is the point's distance from the origin and is never negative. The second coordinate, $\phi$, is the angle $\overrightarrow{OP}$ makes with the positive $z$-axis. It is required to lie in the interval $[0, \pi]$. The third coordinate is the angle $\theta$ as measured in cylindrical coordinates.

---

**DEFINITION** **Spherical coordinates** represent a point $P$ in space by ordered triples $(\rho, \phi, \theta)$ in which

1. $\rho$ is the distance from $P$ to the origin ($\rho \geq 0$).
2. $\phi$ is the angle $\overrightarrow{OP}$ makes with the positive $z$-axis ($0 \leq \phi \leq \pi$).
3. $\theta$ is the angle from cylindrical coordinates.

---

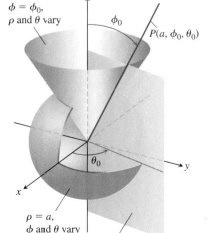

**FIGURE 15.49** Constant-coordinate equations in spherical coordinates yield spheres, single cones, and half-planes.

On maps of the earth, $\theta$ is related to the meridian of a point on the earth and $\phi$ to its latitude, while $\rho$ is related to elevation above the earth's surface.

The equation $\rho = a$ describes the sphere of radius $a$ centered at the origin (Figure 15.49). The equation $\phi = \phi_0$ describes a single cone whose vertex lies at the origin and whose axis lies along the $z$-axis. (We broaden our interpretation to include the $xy$-plane as the cone $\phi = \pi/2$.) If $\phi_0$ is greater than $\pi/2$, the cone $\phi = \phi_0$ opens downward. The equation $\theta = \theta_0$ describes the half-plane that contains the $z$-axis and makes an angle $\theta_0$ with the positive $x$-axis.

> **Equations Relating Spherical Coordinates to Cartesian and Cylindrical Coordinates**
>
> $$r = \rho \sin \phi, \qquad x = r \cos \theta = \rho \sin \phi \cos \theta,$$
> $$z = \rho \cos \phi, \qquad y = r \sin \theta = \rho \sin \phi \sin \theta, \tag{1}$$
> $$\rho = \sqrt{x^2 + y^2 + z^2} = \sqrt{r^2 + z^2}.$$

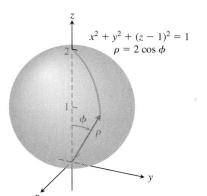

$$x^2 + y^2 + (z-1)^2 = 1$$
$$\rho = 2 \cos \phi$$

**FIGURE 15.50** The sphere in Example 3.

**EXAMPLE 3**    Find a spherical coordinate equation for the sphere $x^2 + y^2 + (z - 1)^2 = 1$.

**Solution**    We use Equations (1) to substitute for $x$, $y$, and $z$:

$$x^2 + y^2 + (z - 1)^2 = 1$$
$$\rho^2 \sin^2 \phi \cos^2 \theta + \rho^2 \sin^2 \phi \sin^2 \theta + (\rho \cos \phi - 1)^2 = 1 \qquad \text{Eqs. (1)}$$
$$\rho^2 \sin^2 \phi \underbrace{(\cos^2 \theta + \sin^2 \theta)}_{1} + \rho^2 \cos^2 \phi - 2\rho \cos \phi + 1 = 1$$
$$\rho^2 \underbrace{(\sin^2 \phi + \cos^2 \phi)}_{1} = 2\rho \cos \phi$$
$$\rho^2 = 2\rho \cos \phi$$
$$\rho = 2 \cos \phi. \qquad \rho > 0$$

The angle $\phi$ varies from 0 at the north pole of the sphere to $\pi/2$ at the south pole; the angle $\theta$ does not appear in the expression for $\rho$, reflecting the symmetry about the $z$-axis (see Figure 15.50). ∎

**EXAMPLE 4**    Find a spherical coordinate equation for the cone $z = \sqrt{x^2 + y^2}$.

**Solution 1**    *Use geometry.* The cone is symmetric with respect to the $z$-axis and cuts the first quadrant of the $yz$-plane along the line $z = y$. The angle between the cone and the positive $z$-axis is therefore $\pi/4$ radians. The cone consists of the points whose spherical coordinates have $\phi$ equal to $\pi/4$, so its equation is $\phi = \pi/4$. (See Figure 15.51.)

**Solution 2**    *Use algebra.* If we use Equations (1) to substitute for $x$, $y$, and $z$ we obtain the same result:

$$z = \sqrt{x^2 + y^2}$$
$$\rho \cos \phi = \sqrt{\rho^2 \sin^2 \phi} \qquad \text{Example 3}$$
$$\rho \cos \phi = \rho \sin \phi \qquad \rho > 0, \sin \phi \geq 0$$
$$\cos \phi = \sin \phi$$
$$\phi = \frac{\pi}{4}. \qquad 0 \leq \phi \leq \pi$$
∎

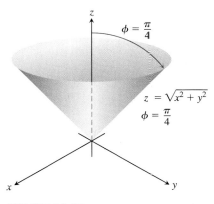

$$\phi = \frac{\pi}{4}$$
$$z = \sqrt{x^2 + y^2}$$
$$\phi = \frac{\pi}{4}$$

**FIGURE 15.51** The cone in Example 4.

Spherical coordinates are useful for describing spheres centered at the origin, half-planes hinged along the $z$-axis, and cones whose vertices lie at the origin and whose axes lie along the $z$-axis. Surfaces like these have equations of constant coordinate value:

$$\rho = 4 \qquad \text{Sphere, radius 4, center at origin}$$

$$\phi = \frac{\pi}{3} \qquad \text{Cone opening up from the origin, making an angle of } \pi/3 \text{ radians with the positive } z\text{-axis}$$

$$\theta = \frac{\pi}{3}. \qquad \text{Half-plane, hinged along the } z\text{-axis, making an angle of } \pi/3 \text{ radians with the positive } x\text{-axis}$$

When computing triple integrals over a region $D$ in spherical coordinates, we partition the region into $n$ spherical wedges. The size of the $k$th spherical wedge, which contains a

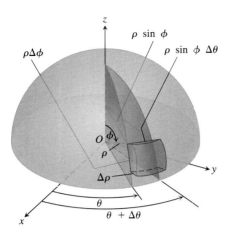

**FIGURE 15.52** In spherical coordinates we use the volume of a spherical wedge, which closely approximates that of a cube.

point $(\rho_k, \phi_k, \theta_k)$, is given by the changes $\Delta\rho_k$, $\Delta\phi_k$, and $\Delta\theta_k$ in $\rho$, $\phi$, and $\theta$. Such a spherical wedge has one edge a circular arc of length $\rho_k \Delta\phi_k$, another edge a circular arc of length $\rho_k \sin\phi_k \Delta\theta_k$, and thickness $\Delta\rho_k$. The spherical wedge closely approximates a cube of these dimensions when $\Delta\rho_k$, $\Delta\phi_k$, and $\Delta\theta_k$ are all small (Figure 15.52). It can be shown that the volume of this spherical wedge $\Delta V_k$ is $\Delta V_k = \rho_k{}^2 \sin\phi_k \, \Delta\rho_k \, \Delta\phi_k \, \Delta\theta_k$ for $(\rho_k, \phi_k, \theta_k)$, a point chosen inside the wedge.

The corresponding Riemann sum for a function $f(\rho, \phi, \theta)$ is

$$S_n = \sum_{k=1}^{n} f(\rho_k, \phi_k, \theta_k) \, \rho_k{}^2 \sin\phi_k \, \Delta\rho_k \, \Delta\phi_k \, \Delta\theta_k.$$

As the norm of a partition approaches zero, and the spherical wedges get smaller, the Riemann sums have a limit when $f$ is continuous:

$$\lim_{n\to\infty} S_n = \iiint_D f(\rho, \phi, \theta) \, dV = \iiint_D f(\rho, \phi, \theta) \, \rho^2 \sin\phi \, d\rho \, d\phi \, d\theta.$$

**Volume Differential in Spherical Coordinates**

$$dV = \rho^2 \sin\phi \, d\rho \, d\phi \, d\theta$$

To evaluate integrals in spherical coordinates, we usually integrate first with respect to $\rho$. The procedure for finding the limits of integration is as follows. We restrict our attention to integrating over domains that are solids of revolution about the $z$-axis (or portions thereof) and for which the limits for $\theta$ and $\phi$ are constant. As with cylindrical coordinates, we restrict $\theta$ in the form $\alpha \le \theta \le \beta$ and $0 \le \beta - \alpha \le 2\pi$.

### How to Integrate in Spherical Coordinates

To evaluate

$$\iiint_D f(\rho, \phi, \theta) \, dV$$

over a region $D$ in space in spherical coordinates, integrating first with respect to $\rho$, then with respect to $\phi$, and finally with respect to $\theta$, take the following steps.

1.  *Sketch.* Sketch the region $D$ along with its projection $R$ on the $xy$-plane. Label the surfaces that bound $D$.

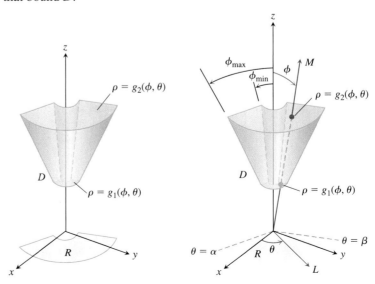

2. *Find the $\rho$-limits of integration.* Draw a ray $M$ from the origin through $D$, making an angle $\phi$ with the positive $z$-axis. Also draw the projection of $M$ on the $xy$-plane (call the projection $L$). The ray $L$ makes an angle $\theta$ with the positive $x$-axis. As $\rho$ increases, $M$ enters $D$ at $\rho = g_1(\phi, \theta)$ and leaves at $\rho = g_2(\phi, \theta)$. These are the $\rho$-limits of integration shown in the above figure.

3. *Find the $\phi$-limits of integration.* For any given $\theta$, the angle $\phi$ that $M$ makes with the $z$-axis runs from $\phi = \phi_{\min}$ to $\phi = \phi_{\max}$. These are the $\phi$-limits of integration.

4. *Find the $\theta$-limits of integration.* The ray $L$ sweeps over $R$ as $\theta$ runs from $\alpha$ to $\beta$. These are the $\theta$-limits of integration. The integral is

$$\iiint_D f(\rho, \phi, \theta)\, dV = \int_{\theta=\alpha}^{\theta=\beta} \int_{\phi=\phi_{\min}}^{\phi=\phi_{\max}} \int_{\rho=g_1(\phi, \theta)}^{\rho=g_2(\phi, \theta)} f(\rho, \phi, \theta)\, \rho^2 \sin \phi\, d\rho\, d\phi\, d\theta.$$

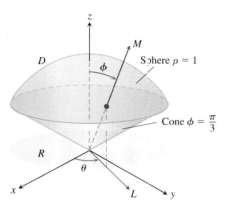

**FIGURE 15.53** The ice cream cone in Example 5.

**EXAMPLE 5**    Find the volume of the "ice cream cone" $D$ cut from the solid sphere $\rho \le 1$ by the cone $\phi = \pi/3$.

**Solution**    The volume is $V = \iiint_D \rho^2 \sin \phi\, d\rho\, d\phi\, d\theta$, the integral of $f(\rho, \phi, \theta) = 1$ over $D$.

To find the limits of integration for evaluating the integral, we begin by sketching $D$ and its projection $R$ on the $xy$-plane (Figure 15.53).

*The $\rho$-limits of integration.* We draw a ray $M$ from the origin through $D$, making an angle $\phi$ with the positive $z$-axis. We also draw $L$, the projection of $M$ on the $xy$-plane, along with the angle $\theta$ that $L$ makes with the positive $x$-axis. Ray $M$ enters $D$ at $\rho = 0$ and leaves at $\rho = 1$.

*The $\phi$-limits of integration.* The cone $\phi = \pi/3$ makes an angle of $\pi/3$ with the positive $z$-axis. For any given $\theta$, the angle $\phi$ can run from $\phi = 0$ to $\phi = \pi/3$.

*The $\theta$-limits of integration.* The ray $L$ sweeps over $R$ as $\theta$ runs from 0 to $2\pi$. The volume is

$$V = \iiint_D \rho^2 \sin \phi\, d\rho\, d\phi\, d\theta = \int_0^{2\pi} \int_0^{\pi/3} \int_0^1 \rho^2 \sin \phi\, d\rho\, d\phi\, d\theta$$

$$= \int_0^{2\pi} \int_0^{\pi/3} \left[ \frac{\rho^3}{3} \right]_0^1 \sin \phi\, d\phi\, d\theta = \int_0^{2\pi} \int_0^{\pi/3} \frac{1}{3} \sin \phi\, d\phi\, d\theta$$

$$= \int_0^{2\pi} \left[ -\frac{1}{3} \cos \phi \right]_0^{\pi/3} d\theta = \int_0^{2\pi} \left( -\frac{1}{6} + \frac{1}{3} \right) d\theta = \frac{1}{6}(2\pi) = \frac{\pi}{3}. \qquad \blacksquare$$

**EXAMPLE 6**    A solid of constant density $\delta = 1$ occupies the region $D$ in Example 5. Find the solid's moment of inertia about the $z$-axis.

**Solution**    In rectangular coordinates, the moment is

$$I_z = \iiint_D (x^2 + y^2)\, dV.$$

In spherical coordinates, $x^2 + y^2 = (\rho \sin \phi \cos \theta)^2 + (\rho \sin \phi \sin \theta)^2 = \rho^2 \sin^2 \phi$. Hence,

$$I_z = \iiint_D (\rho^2 \sin^2 \phi)\, \rho^2 \sin \phi\, d\rho\, d\phi\, d\theta = \iiint_D \rho^4 \sin^3 \phi\, d\rho\, d\phi\, d\theta.$$

For the region $D$ in Example 5, this becomes

$$I_z = \int_0^{2\pi} \int_0^{\pi/3} \int_0^1 \rho^4 \sin^3 \phi \, d\rho \, d\phi \, d\theta = \int_0^{2\pi} \int_0^{\pi/3} \left[ \frac{\rho^5}{5} \right]_0^1 \sin^3 \phi \, d\phi \, d\theta$$

$$= \frac{1}{5} \int_0^{2\pi} \int_0^{\pi/3} (1 - \cos^2 \phi) \sin \phi \, d\phi \, d\theta = \frac{1}{5} \int_0^{2\pi} \left[ -\cos \phi + \frac{\cos^3 \phi}{3} \right]_0^{\pi/3} d\theta$$

$$= \frac{1}{5} \int_0^{2\pi} \left( -\frac{1}{2} + 1 + \frac{1}{24} - \frac{1}{3} \right) d\theta = \frac{1}{5} \int_0^{2\pi} \frac{5}{24} d\theta = \frac{1}{24} (2\pi) = \frac{\pi}{12}. \quad \blacksquare$$

---

### Coordinate Conversion Formulas

| CYLINDRICAL TO RECTANGULAR | SPHERICAL TO RECTANGULAR | SPHERICAL TO CYLINDRICAL |
|---|---|---|
| $x = r \cos \theta$ | $x = \rho \sin \phi \cos \theta$ | $r = \rho \sin \phi$ |
| $y = r \sin \theta$ | $y = \rho \sin \phi \sin \theta$ | $z = \rho \cos \phi$ |
| $z = z$ | $z = \rho \cos \phi$ | $\theta = \theta$ |

Corresponding formulas for $dV$ in triple integrals:

$$dV = dx \, dy \, dz$$
$$= dz \, r \, dr \, d\theta$$
$$= \rho^2 \sin \phi \, d\rho \, d\phi \, d\theta$$

---

In the next section we offer a more general procedure for determining $dV$ in cylindrical and spherical coordinates. The results, of course, will be the same.

## Exercises 15.7

### Evaluating Integrals in Cylindrical Coordinates

Evaluate the cylindrical coordinate integrals in Exercises 1–6.

**1.** $\int_0^{2\pi} \int_0^1 \int_r^{\sqrt{2-r^2}} dz \, r \, dr \, d\theta$

**2.** $\int_0^{2\pi} \int_0^3 \int_{r^2/3}^{\sqrt{18-r^2}} dz \, r \, dr \, d\theta$

**3.** $\int_0^{2\pi} \int_0^{\theta/2\pi} \int_0^{3+24r^2} dz \, r \, dr \, d\theta$

**4.** $\int_0^{\pi} \int_0^{\theta/\pi} \int_{-\sqrt{4-r^2}}^{3\sqrt{4-r^2}} z \, dz \, r \, dr \, d\theta$

**5.** $\int_0^{2\pi} \int_0^1 \int_r^{1/\sqrt{2-r^2}} 3 \, dz \, r \, dr \, d\theta$

**6.** $\int_0^{2\pi} \int_0^1 \int_{-1/2}^{1/2} (r^2 \sin^2 \theta + z^2) \, dz \, r \, dr \, d\theta$

### Changing the Order of Integration in Cylindrical Coordinates

The integrals we have seen so far suggest that there are preferred orders of integration for cylindrical coordinates, but other orders usually work well and are occasionally easier to evaluate. Evaluate the integrals in Exercises 7–10.

**7.** $\int_0^{2\pi} \int_0^3 \int_0^{z/3} r^3 \, dr \, dz \, d\theta$

**8.** $\int_{-1}^1 \int_0^{2\pi} \int_0^{1+\cos\theta} 4r \, dr \, d\theta \, dz$

**9.** $\int_0^1 \int_0^{\sqrt{z}} \int_0^{2\pi} (r^2 \cos^2 \theta + z^2) r \, d\theta \, dr \, dz$

**10.** $\int_0^2 \int_{r-2}^{\sqrt{4-r^2}} \int_0^{2\pi} (r \sin \theta + 1) r \, d\theta \, dz \, dr$

**11.** Let $D$ be the region bounded below by the plane $z = 0$, above by the sphere $x^2 + y^2 + z^2 = 4$, and on the sides by the cylinder $x^2 + y^2 = 1$. Set up the triple integrals in cylindrical coordinates that give the volume of $D$ using the following orders of integration.

    **a.** $dz \, dr \, d\theta$     **b.** $dr \, dz \, d\theta$     **c.** $d\theta \, dz \, dr$

**12.** Let $D$ be the region bounded below by the cone $z = \sqrt{x^2 + y^2}$ and above by the paraboloid $z = 2 - x^2 - y^2$. Set up the triple integrals in cylindrical coordinates that give the volume of $D$ using the following orders of integration.

    **a.** $dz \, dr \, d\theta$     **b.** $dr \, dz \, d\theta$     **c.** $d\theta \, dz \, dr$

### Finding Iterated Integrals in Cylindrical Coordinates

**13.** Give the limits of integration for evaluating the integral

$$\iiint f(r, \theta, z) \, dz \, r \, dr \, d\theta$$

as an iterated integral over the region that is bounded below by the plane $z = 0$, on the side by the cylinder $r = \cos\theta$, and on top by the paraboloid $z = 3r^2$.

**14.** Convert the integral

$$\int_{-1}^{1}\int_{0}^{\sqrt{1-y^2}}\int_{0}^{x}(x^2 + y^2)\,dz\,dx\,dy$$

to an equivalent integral in cylindrical coordinates and evaluate the result.

In Exercises 15–20, set up the iterated integral for evaluating $\iiint_D f(r, \theta, z)\,dz\,r\,dr\,d\theta$ over the given region $D$.

**15.** $D$ is the right circular cylinder whose base is the circle $r = 2\sin\theta$ in the $xy$-plane and whose top lies in the plane $z = 4 - y$.

**16.** $D$ is the right circular cylinder whose base is the circle $r = 3\cos\theta$ and whose top lies in the plane $z = 5 - x$.

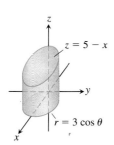

**17.** $D$ is the solid right cylinder whose base is the region in the $xy$-plane that lies inside the cardioid $r = 1 + \cos\theta$ and outside the circle $r = 1$ and whose top lies in the plane $z = 4$.

**18.** $D$ is the solid right cylinder whose base is the region between the circles $r = \cos\theta$ and $r = 2\cos\theta$ and whose top lies in the plane $z = 3 - y$.

**19.** $D$ is the prism whose base is the triangle in the $xy$-plane bounded by the $x$-axis and the lines $y = x$ and $x = 1$ and whose top lies in the plane $z = 2 - y$.

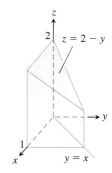

**20.** $D$ is the prism whose base is the triangle in the $xy$-plane bounded by the $y$-axis and the lines $y = x$ and $y = 1$ and whose top lies in the plane $z = 2 - x$.

**Evaluating Integrals in Spherical Coordinates**

Evaluate the spherical coordinate integrals in Exercises 21–26.

**21.** $\displaystyle\int_{0}^{\pi}\int_{0}^{\pi}\int_{0}^{2\sin\phi}\rho^2\sin\phi\,d\rho\,d\phi\,d\theta$

**22.** $\displaystyle\int_{0}^{2\pi}\int_{0}^{\pi/4}\int_{0}^{2}(\rho\cos\phi)\,\rho^2\sin\phi\,d\rho\,d\phi\,d\theta$

**23.** $\displaystyle\int_{0}^{2\pi}\int_{0}^{\pi}\int_{0}^{(1-\cos\phi)/2}\rho^2\sin\phi\,d\rho\,d\phi\,d\theta$

**24.** $\displaystyle\int_{0}^{3\pi/2}\int_{0}^{\pi}\int_{0}^{1}5\rho^3\sin^3\phi\,d\rho\,d\phi\,d\theta$

**25.** $\displaystyle\int_{0}^{2\pi}\int_{0}^{\pi/3}\int_{\sec\phi}^{2}3\rho^2\sin\phi\,d\rho\,d\phi\,d\theta$

**26.** $\int_0^{2\pi} \int_0^{\pi/4} \int_0^{\sec\phi} (\rho \cos\phi) \rho^2 \sin\phi \, d\rho \, d\phi \, d\theta$

**Changing the Order of Integration in Spherical Coordinates**
The previous integrals suggest there are preferred orders of integration for spherical coordinates, but other orders give the same value and are occasionally easier to evaluate. Evaluate the integrals in Exercises 27–30.

**27.** $\int_0^2 \int_{-\pi}^0 \int_{\pi/4}^{\pi/2} \rho^3 \sin 2\phi \, d\phi \, d\theta \, d\rho$

**28.** $\int_{\pi/6}^{\pi/3} \int_{\csc\phi}^{2\csc\phi} \int_0^{2\pi} \rho^2 \sin\phi \, d\theta \, d\rho \, d\phi$

**29.** $\int_0^1 \int_0^{\pi} \int_0^{\pi/4} 12\rho \sin^3\phi \, d\phi \, d\theta \, d\rho$

**30.** $\int_{\pi/6}^{\pi/2} \int_{-\pi/2}^{\pi/2} \int_{\csc\phi}^{2} 5\rho^4 \sin^3\phi \, d\rho \, d\theta \, d\phi$

**31.** Let $D$ be the region in Exercise 11. Set up the triple integrals in spherical coordinates that give the volume of $D$ using the following orders of integration.

    **a.** $d\rho \, d\phi \, d\theta$            **b.** $d\phi \, d\rho \, d\theta$

**32.** Let $D$ be the region bounded below by the cone $z = \sqrt{x^2 + y^2}$ and above by the plane $z = 1$. Set up the triple integrals in spherical coordinates that give the volume of $D$ using the following orders of integration.

    **a.** $d\rho \, d\phi \, d\theta$            **b.** $d\phi \, d\rho \, d\theta$

**Finding Iterated Integrals in Spherical Coordinates**
In Exercises 33–38, (a) find the spherical coordinate limits for the integral that calculates the volume of the given solid and then (b) evaluate the integral.

**33.** The solid between the sphere $\rho = \cos\phi$ and the hemisphere $\rho = 2, z \geq 0$

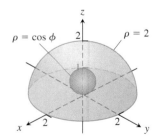

**34.** The solid bounded below by the hemisphere $\rho = 1, z \geq 0$, and above by the cardioid of revolution $\rho = 1 + \cos\phi$

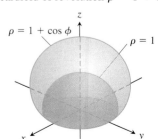

**35.** The solid enclosed by the cardioid of revolution $\rho = 1 - \cos\phi$

**36.** The upper portion cut from the solid in Exercise 35 by the $xy$-plane

**37.** The solid bounded below by the sphere $\rho = 2 \cos\phi$ and above by the cone $z = \sqrt{x^2 + y^2}$

**38.** The solid bounded below by the $xy$-plane, on the sides by the sphere $\rho = 2$, and above by the cone $\phi = \pi/3$

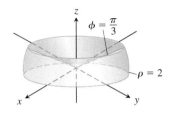

**Finding Triple Integrals**
**39.** Set up triple integrals for the volume of the sphere $\rho = 2$ in (a) spherical, (b) cylindrical, and (c) rectangular coordinates.

**40.** Let $D$ be the region in the first octant that is bounded below by the cone $\phi = \pi/4$ and above by the sphere $\rho = 3$. Express the volume of $D$ as an iterated triple integral in (a) cylindrical and (b) spherical coordinates. Then (c) find $V$.

**41.** Let $D$ be the smaller cap cut from a solid ball of radius 2 units by a plane 1 unit from the center of the sphere. Express the volume of $D$ as an iterated triple integral in (a) spherical, (b) cylindrical, and (c) rectangular coordinates. Then (d) find the volume by evaluating one of the three triple integrals.

**42.** Express the moment of inertia $I_z$ of the solid hemisphere $x^2 + y^2 + z^2 \leq 1, z \geq 0$, as an iterated integral in (a) cylindrical and (b) spherical coordinates. Then (c) find $I_z$.

**Volumes**
Find the volumes of the solids in Exercises 43–48.

**43.**

**44.**

**45.**

**46.**

**47.**

$z = \sqrt{1 - x^2 - y^2}$

$r = \sin \theta$

**48.**

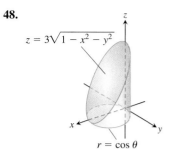

$z = 3\sqrt{1 - x^2 - y^2}$

$r = \cos \theta$

**49. Sphere and cones** Find the volume of the portion of the solid sphere $\rho \leq a$ that lies between the cones $\phi = \pi/3$ and $\phi = 2\pi/3$.

**50. Sphere and half-planes** Find the volume of the region cut from the solid sphere $\rho \leq a$ by the half-planes $\theta = 0$ and $\theta = \pi/6$ in the first octant.

**51. Sphere and plane** Find the volume of the smaller region cut from the solid sphere $\rho \leq 2$ by the plane $z = 1$.

**52. Cone and planes** Find the volume of the solid enclosed by the cone $z = \sqrt{x^2 + y^2}$ between the planes $z = 1$ and $z = 2$.

**53. Cylinder and paraboloid** Find the volume of the region bounded below by the plane $z = 0$, laterally by the cylinder $x^2 + y^2 = 1$, and above by the paraboloid $z = x^2 + y^2$.

**54. Cylinder and paraboloids** Find the volume of the region bounded below by the paraboloid $z = x^2 + y^2$, laterally by the cylinder $x^2 + y^2 = 1$, and above by the paraboloid $z = x^2 + y^2 + 1$.

**55. Cylinder and cones** Find the volume of the solid cut from the thick-walled cylinder $1 \leq x^2 + y^2 \leq 2$ by the cones $z = \pm\sqrt{x^2 + y^2}$.

**56. Sphere and cylinder** Find the volume of the region that lies inside the sphere $x^2 + y^2 + z^2 = 2$ and outside the cylinder $x^2 + y^2 = 1$.

**57. Cylinder and planes** Find the volume of the region enclosed by the cylinder $x^2 + y^2 = 4$ and the planes $z = 0$ and $y + z = 4$.

**58. Cylinder and planes** Find the volume of the region enclosed by the cylinder $x^2 + y^2 = 4$ and the planes $z = 0$ and $x + y + z = 4$.

**59. Region trapped by paraboloids** Find the volume of the region bounded above by the paraboloid $z = 5 - x^2 - y^2$ and below by the paraboloid $z = 4x^2 + 4y^2$.

**60. Paraboloid and cylinder** Find the volume of the region bounded above by the paraboloid $z = 9 - x^2 - y^2$, below by the $xy$-plane, and lying *outside* the cylinder $x^2 + y^2 = 1$.

**61. Cylinder and sphere** Find the volume of the region cut from the solid cylinder $x^2 + y^2 \leq 1$ by the sphere $x^2 + y^2 + z^2 = 4$.

**62. Sphere and paraboloid** Find the volume of the region bounded above by the sphere $x^2 + y^2 + z^2 = 2$ and below by the paraboloid $z = x^2 + y^2$.

**Average Values**

**63.** Find the average value of the function $f(r, \theta, z) = r$ over the region bounded by the cylinder $r = 1$ between the planes $z = -1$ and $z = 1$.

**64.** Find the average value of the function $f(r, \theta, z) = r$ over the solid ball bounded by the sphere $r^2 + z^2 = 1$. (This is the sphere $x^2 + y^2 + z^2 = 1$.)

**65.** Find the average value of the function $f(\rho, \phi, \theta) = \rho$ over the solid ball $\rho \leq 1$.

**66.** Find the average value of the function $f(\rho, \phi, \theta) = \rho \cos \phi$ over the solid upper ball $\rho \leq 1, 0 \leq \phi \leq \pi/2$.

**Masses, Moments, and Centroids**

**67. Center of mass** A solid of constant density is bounded below by the plane $z = 0$, above by the cone $z = r, r \geq 0$, and on the sides by the cylinder $r = 1$. Find the center of mass.

**68. Centroid** Find the centroid of the region in the first octant that is bounded above by the cone $z = \sqrt{x^2 + y^2}$, below by the plane $z = 0$, and on the sides by the cylinder $x^2 + y^2 = 4$ and the planes $x = 0$ and $y = 0$.

**69. Centroid** Find the centroid of the solid in Exercise 38.

**70. Centroid** Find the centroid of the solid bounded above by the sphere $\rho = a$ and below by the cone $\phi = \pi/4$.

**71. Centroid** Find the centroid of the region that is bounded above by the surface $z = \sqrt{r}$, on the sides by the cylinder $r = 4$, and below by the $xy$-plane.

**72. Centroid** Find the centroid of the region cut from the solid ball $r^2 + z^2 \leq 1$ by the half-planes $\theta = -\pi/3, r \geq 0$, and $\theta = \pi/3$, $r \geq 0$.

**73. Moment of inertia of solid cone** Find the moment of inertia of a right circular cone of base radius 1 and height 1 about an axis through the vertex parallel to the base. (Take $\delta = 1$.)

**74. Moment of inertia of solid sphere** Find the moment of inertia of a solid sphere of radius $a$ about a diameter. (Take $\delta = 1$.)

**75. Moment of inertia of solid cone** Find the moment of inertia of a right circular cone of base radius $a$ and height $h$ about its axis. (*Hint:* Place the cone with its vertex at the origin and its axis along the $z$-axis.)

**76. Variable density** A solid is bounded on the top by the paraboloid $z = r^2$, on the bottom by the plane $z = 0$, and on the sides by the cylinder $r = 1$. Find the center of mass and the moment of inertia about the $z$-axis if the density is

    **a.** $\delta(r, \theta, z) = z$           **b.** $\delta(r, \theta, z) = r$.

**77. Variable density** A solid is bounded below by the cone $z = \sqrt{x^2 + y^2}$ and above by the plane $z = 1$. Find the center of mass and the moment of inertia about the $z$-axis if the density is

    **a.** $\delta(r, \theta, z) = z$           **b.** $\delta(r, \theta, z) = z^2$.

**78. Variable density** A solid ball is bounded by the sphere $\rho = a$. Find the moment of inertia about the $z$-axis if the density is

    **a.** $\delta(\rho, \phi, \theta) = \rho^2$      **b.** $\delta(\rho, \phi, \theta) = r = \rho \sin \phi$.

**79. Centroid of solid semiellipsoid** Show that the centroid of the solid semiellipsoid of revolution $(r^2/a^2) + (z^2/h^2) \leq 1, z \geq 0$, lies on the $z$-axis three-eighths of the way from the base to the top. The special case $h = a$ gives a solid hemisphere. Thus, the centroid of a solid hemisphere lies on the axis of symmetry three-eighths of the way from the base to the top.

**80. Centroid of solid cone** Show that the centroid of a solid right circular cone is one-fourth of the way from the base to the vertex. (In general, the centroid of a solid cone or pyramid is one-fourth of the way from the centroid of the base to the vertex.)

**81. Density of center of a planet** A planet is in the shape of a sphere of radius $R$ and total mass $M$ with spherically symmetric density distribution that increases linearly as one approaches its center.

What is the density at the center of this planet if the density at its edge (surface) is taken to be zero?

82. **Mass of planet's atmosphere** A spherical planet of radius $R$ has an atmosphere whose density is $\mu = \mu_0 e^{-ch}$, where $h$ is the altitude above the surface of the planet, $\mu_0$ is the density at sea level, and $c$ is a positive constant. Find the mass of the planet's atmosphere.

**Theory and Examples**

83. **Vertical planes in cylindrical coordinates**

    a. Show that planes perpendicular to the $x$-axis have equations of the form $r = a \sec \theta$ in cylindrical coordinates.

    b. Show that planes perpendicular to the $y$-axis have equations of the form $r = b \csc \theta$.

84. (*Continuation of Exercise 83.*) Find an equation of the form $r = f(\theta)$ in cylindrical coordinates for the plane $ax + by = c$, $c \neq 0$.

85. **Symmetry** What symmetry will you find in a surface that has an equation of the form $r = f(z)$ in cylindrical coordinates? Give reasons for your answer.

86. **Symmetry** What symmetry will you find in a surface that has an equation of the form $\rho = f(\phi)$ in spherical coordinates? Give reasons for your answer.

# $15.8$ Substitutions in Multiple Integrals

This section introduces the ideas involved in coordinate transformations to evaluate multiple integrals by substitution. The method replaces complicated integrals by ones that are easier to evaluate. Substitutions accomplish this by simplifying the integrand, the limits of integration, or both. A thorough discussion of multivariable transformations and substitutions is best left to a more advanced course, but our introduction here shows how the substitutions just studied reflect the general idea derived for single integral calculus.

## Substitutions in Double Integrals

The polar coordinate substitution of Section 15.4 is a special case of a more general substitution method for double integrals, a method that pictures changes in variables as transformations of regions.

Suppose that a region $G$ in the $uv$-plane is transformed into the region $R$ in the $xy$-plane by equations of the form

$$x = g(u, v), \qquad y = h(u, v),$$

as suggested in Figure 15.54. We assume the transformation is one-to-one on the interior of $G$. We call $R$ the **image** of $G$ under the transformation, and $G$ the **preimage** of $R$. Any function $f(x, y)$ defined on $R$ can be thought of as a function $f(g(u, v), h(u, v))$ defined on $G$ as well. How is the integral of $f(x, y)$ over $R$ related to the integral of $f(g(u, v), h(u, v))$ over $G$?

To gain some insight into the question, let's look again at the single variable case. To be consistent with how we are using them now, we interchange the variables $x$ and $u$ used in the substitution method for single integrals in Chapter 5, so the equation is stated here as

$$\int_{g(a)}^{g(b)} f(x)\, dx = \int_a^b f(g(u))\, g'(u)\, du. \qquad {\scriptstyle x = g(u),\ dx = g'(u)\, du}$$

To propose an analogue for substitution in a double integral $\iint_R f(x, y)\, dx\, dy$, we need a derivative factor like $g'(u)$ as a multiplier that transforms the area element $du\, dv$ in the region $G$ to its corresponding area element $dx\, dy$ in the region $R$. Let's denote this factor by $J$. In continuing with our analogy, it is reasonable to assume that $J$ is a function of both variables $u$ and $v$, just as $g'$ is a function of the single variable $u$. Moreover, $J$ should register instantaneous change, so partial derivatives are going to be involved in its expression. Since four partial derivatives are associated with the transforming equations $x = g(u, v)$ and $y = h(u, v)$, it is also reasonable to assume that the factor $J(u, v)$ we seek includes them all. These features are captured in the following definition constructed from the partial derivatives, and named after the German mathematician Carl Jacobi.

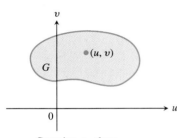

$x = g(u, v)$
$y = h(u, v)$

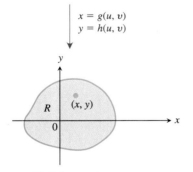

**FIGURE 15.54** The equations $x = g(u, v)$ and $y = h(u, v)$ allow us to change an integral over a region $R$ in the $xy$-plane into an integral over a region $G$ in the $uv$-plane.

**HISTORICAL BIOGRAPHY**

Carl Gustav Jacob Jacobi
(1804–1851)

> **DEFINITION** The **Jacobian determinant** or **Jacobian** of the coordinate transformation $x = g(u, v), y = h(u, v)$ is
>
> $$J(u, v) = \begin{vmatrix} \dfrac{\partial x}{\partial u} & \dfrac{\partial x}{\partial v} \\[2mm] \dfrac{\partial y}{\partial u} & \dfrac{\partial y}{\partial v} \end{vmatrix} = \dfrac{\partial x}{\partial u}\dfrac{\partial y}{\partial v} - \dfrac{\partial y}{\partial u}\dfrac{\partial x}{\partial v}. \qquad (1)$$

The Jacobian can also be denoted by

$$J(u, v) = \frac{\partial(x, y)}{\partial(u, v)}$$

**Differential Area Change Substituting**
$x = g(u, v), y = h(u, v)$

$$dx\, dy = \left| \frac{\partial(x, y)}{\partial(u, v)} \right| du\, dv$$

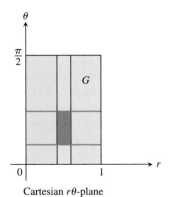

Cartesian $r\theta$-plane

$$\begin{aligned} x &= r\cos\theta \\ y &= r\sin\theta \end{aligned}$$

Cartesian $xy$-plane

**FIGURE 15.55** The equations $x = r\cos\theta, y = r\sin\theta$ transform $G$ into $R$. The Jacobian factor $r$, calculated in Example 1, scales the differential rectangle $dr\, d\theta$ in $G$ to match with the differential area element $dx\, dy$ in $R$.

to help us remember how the determinant in Equation (1) is constructed from the partial derivatives of $x$ and $y$. The array of partial derivatives in Equation (1) behaves just like the derivative $g'$ in the single variable situation. The Jacobian measures how much the transformation is expanding or contracting the area around the point $(u, v)$. Effectively, the factor $|J|$ converts the area of the differential rectangle $du\, dv$ in $G$ to match with its corresponding differential area $dx\, dy$ in $R$. We note that, in general, the value of the scaling factor $|J|$ depends on the point $(u, v)$ in $G$. Our examples to follow will show how it scales the differential area $du\, dv$ for specific transformations.

With the definition of the Jacobian determinant, we can now answer our original question concerning the relationship of the integral of $f(x, y)$ over the region $R$ to the integral of $f(g(u, v), h(u, v))$ over $G$.

> **THEOREM 3—Substitution for Double Integrals** Suppose that $f(x, y)$ is continuous over the region $R$. Let $G$ be the preimage of $R$ under the transformation $x = g(u, v), y = h(u, v)$, assumed to be one-to-one on the interior of $G$. If the functions $g$ and $h$ have continuous first partial derivatives within the interior of $G$, then
>
> $$\iint\limits_R f(x, y)\, dx\, dy = \iint\limits_G f(g(u, v), h(u, v)) \left| \frac{\partial(x, y)}{\partial(u, v)} \right| du\, dv. \qquad (2)$$

The derivation of Equation (2) is intricate and properly belongs to a course in advanced calculus, so we do not derive it here. We now present examples illustrating the substitution method defined by the equation.

**EXAMPLE 1** Find the Jacobian for the polar coordinate transformation $x = r\cos\theta$, $y = r\sin\theta$, and use Equation (2) to write the Cartesian integral $\iint_R f(x, y)\, dx\, dy$ as a polar integral.

**Solution** Figure 15.55 shows how the equations $x = r\cos\theta, y = r\sin\theta$ transform the rectangle $G: 0 \leq r \leq 1, 0 \leq \theta \leq \pi/2$, into the quarter circle $R$ bounded by $x^2 + y^2 = 1$ in the first quadrant of the $xy$-plane.

For polar coordinates, we have $r$ and $\theta$ in place of $u$ and $v$. With $x = r \cos \theta$ and $y = r \sin \theta$, the Jacobian is

$$J(r, \theta) = \begin{vmatrix} \dfrac{\partial x}{\partial r} & \dfrac{\partial x}{\partial \theta} \\[2mm] \dfrac{\partial y}{\partial r} & \dfrac{\partial y}{\partial \theta} \end{vmatrix} = \begin{vmatrix} \cos \theta & -r \sin \theta \\ \sin \theta & r \cos \theta \end{vmatrix} = r(\cos^2 \theta + \sin^2 \theta) = r.$$

Since we assume $r \geq 0$ when integrating in polar coordinates, $|J(r, \theta)| = |r| = r$, so that Equation (2) gives

$$\iint_R f(x, y) \, dx \, dy = \iint_G f(r \cos \theta, r \sin \theta) \, r \, dr \, d\theta. \tag{3}$$

This is the same formula we derived independently using a geometric argument for polar area in Section 15.4.

Notice that the integral on the right-hand side of Equation (3) is not the integral of $f(r \cos \theta, r \sin \theta)$ over a region in the polar coordinate plane. It is the integral of the product of $f(r \cos \theta, r \sin \theta)$ and $r$ over a region $G$ in the *Cartesian $r\theta$-plane.* ∎

Here is an example of a substitution in which the image of a rectangle under the coordinate transformation is a trapezoid. Transformations like this one are called **linear transformations** and their Jacobians are constant throughout $G$.

**EXAMPLE 2** Evaluate

$$\int_0^4 \int_{x=y/2}^{x=(y/2)+1} \frac{2x - y}{2} dx \, dy$$

by applying the transformation

$$u = \frac{2x - y}{2}, \qquad v = \frac{y}{2} \tag{4}$$

and integrating over an appropriate region in the $uv$-plane.

**Solution** We sketch the region $R$ of integration in the $xy$-plane and identify its boundaries (Figure 15.56).

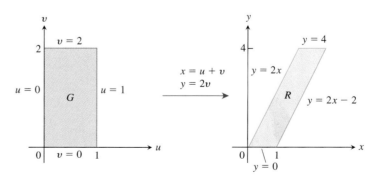

**FIGURE 15.56** The equations $x = u + v$ and $y = 2v$ transform $G$ into $R$. Reversing the transformation by the equations $u = (2x - y)/2$ and $v = y/2$ transforms $R$ into $G$ (Example 2).

To apply Equation (2), we need to find the corresponding $uv$-region $G$ and the Jacobian of the transformation. To find them, we first solve Equations (4) for $x$ and $y$ in terms of $u$ and $v$. From those equations it is easy to find algebraically that

$$x = u + v, \qquad y = 2v. \tag{5}$$

We then find the boundaries of $G$ by substituting these expressions into the equations for the boundaries of $R$ (Figure 15.56).

| xy-equations for the boundary of R | Corresponding uv-equations for the boundary of G | Simplified uv-equations |
|---|---|---|
| $x = y/2$ | $u + v = 2v/2 = v$ | $u = 0$ |
| $x = (y/2) + 1$ | $u + v = (2v/2) + 1 = v + 1$ | $u = 1$ |
| $y = 0$ | $2v = 0$ | $v = 0$ |
| $y = 4$ | $2v = 4$ | $v = 2$ |

From Equations (5) the Jacobian of the transformation is

$$J(u, v) = \begin{vmatrix} \dfrac{\partial x}{\partial u} & \dfrac{\partial x}{\partial v} \\ \dfrac{\partial y}{\partial u} & \dfrac{\partial y}{\partial v} \end{vmatrix} = \begin{vmatrix} \dfrac{\partial}{\partial u}(u + v) & \dfrac{\partial}{\partial v}(u + v) \\ \dfrac{\partial}{\partial u}(2v) & \dfrac{\partial}{\partial v}(2v) \end{vmatrix} = \begin{vmatrix} 1 & 1 \\ 0 & 2 \end{vmatrix} = 2.$$

We now have everything we need to apply Equation (2):

$$\int_0^4 \int_{x=y/2}^{x=(y/2)+1} \frac{2x - y}{2} \, dx \, dy = \int_{v=0}^{v=2} \int_{u=0}^{u=1} u \, |J(u, v)| \, du \, dv$$

$$= \int_0^2 \int_0^1 (u)(2) \, du \, dv = \int_0^2 \left[ u^2 \right]_0^1 dv = \int_0^2 dv = 2. \quad \blacksquare$$

**EXAMPLE 3**  Evaluate

$$\int_0^1 \int_0^{1-x} \sqrt{x + y} \, (y - 2x)^2 \, dy \, dx.$$

**Solution**  We sketch the region $R$ of integration in the $xy$-plane and identify its boundaries (Figure 15.57). The integrand suggests the transformation $u = x + y$ and $v = y - 2x$. Routine algebra produces $x$ and $y$ as functions of $u$ and $v$:

$$x = \frac{u}{3} - \frac{v}{3}, \qquad y = \frac{2u}{3} + \frac{v}{3}. \qquad (6)$$

From Equations (6), we can find the boundaries of the $uv$-region $G$ (Figure 15.57).

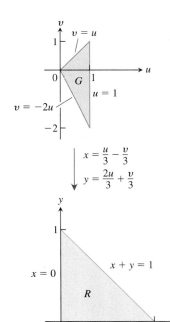

**FIGURE 15.57**  The equations $x = (u/3) - (v/3)$ and $y = (2u/3) + (v/3)$ transform $G$ into $R$. Reversing the transformation by the equations $u = x + y$ and $v = y - 2x$ transforms $R$ into $G$ (Example 3).

| xy-equations for the boundary of R | Corresponding uv-equations for the boundary of G | Simplified uv-equations |
|---|---|---|
| $x + y = 1$ | $\left(\dfrac{u}{3} - \dfrac{v}{3}\right) + \left(\dfrac{2u}{3} + \dfrac{v}{3}\right) = 1$ | $u = 1$ |
| $x = 0$ | $\dfrac{u}{3} - \dfrac{v}{3} = 0$ | $v = u$ |
| $y = 0$ | $\dfrac{2u}{3} + \dfrac{v}{3} = 0$ | $v = -2u$ |

The Jacobian of the transformation in Equations (6) is

$$J(u, v) = \begin{vmatrix} \dfrac{\partial x}{\partial u} & \dfrac{\partial x}{\partial v} \\[2mm] \dfrac{\partial y}{\partial u} & \dfrac{\partial y}{\partial v} \end{vmatrix} = \begin{vmatrix} \dfrac{1}{3} & -\dfrac{1}{3} \\[2mm] \dfrac{2}{3} & \dfrac{1}{3} \end{vmatrix} = \dfrac{1}{3}.$$

Applying Equation (2), we evaluate the integral:

$$\int_0^1 \int_0^{1-x} \sqrt{x + y}\,(y - 2x)^2 \, dy \, dx = \int_{u=0}^{u=1} \int_{v=-2u}^{v=u} u^{1/2} v^2 \, \left| J(u, v) \right| \, dv \, du$$

$$= \int_0^1 \int_{-2u}^{u} u^{1/2} v^2 \left(\frac{1}{3}\right) dv \, du = \frac{1}{3} \int_0^1 u^{1/2} \left[\frac{1}{3} v^3 \right]_{v=-2u}^{v=u} du$$

$$= \frac{1}{9} \int_0^1 u^{1/2} (u^3 + 8u^3) \, du = \int_0^1 u^{7/2} \, du = \frac{2}{9} u^{9/2} \Big]_0^1 = \frac{2}{9}. \qquad \blacksquare$$

In the next example we illustrate a nonlinear transformation of coordinates resulting from simplifying the form of the integrand. Like the polar coordinates' transformation, nonlinear transformations can map a straight-line boundary of a region into a curved boundary (or vice versa with the inverse transformation). In general, nonlinear transformations are more complex to analyze than linear ones, and a complete treatment is left to a more advanced course.

**EXAMPLE 4** Evaluate the integral

$$\int_1^2 \int_{1/y}^{y} \sqrt{\frac{y}{x}}\, e^{\sqrt{xy}} \, dx \, dy.$$

**Solution** The square root terms in the integrand suggest that we might simplify the integration by substituting $u = \sqrt{xy}$ and $v = \sqrt{y/x}$. Squaring these equations, we readily have $u^2 = xy$ and $v^2 = y/x$, which imply that $u^2 v^2 = y^2$ and $u^2/v^2 = x^2$. So we obtain the transformation (in the same ordering of the variables as discussed before)

$$x = \frac{u}{v} \qquad \text{and} \qquad y = uv,$$

with $u > 0$ and $v > 0$. Let's first see what happens to the integrand itself under this transformation. The Jacobian of the transformation is not constant and we find

$$J(u, v) = \begin{vmatrix} \dfrac{\partial x}{\partial u} & \dfrac{\partial x}{\partial v} \\[2mm] \dfrac{\partial y}{\partial u} & \dfrac{\partial y}{\partial v} \end{vmatrix} = \begin{vmatrix} \dfrac{1}{v} & -\dfrac{u}{v^2} \\[2mm] v & u \end{vmatrix} = \dfrac{2u}{v}.$$

If $G$ is the region of integration in the $uv$-plane, then by Equation (2) the transformed double integral under the substitution is

$$\iint_R \sqrt{\frac{y}{x}}\, e^{\sqrt{xy}} \, dx \, dy = \iint_G v e^u \, \frac{2u}{v} \, du \, dv = \iint_G 2u e^u \, du \, dv.$$

The transformed integrand function is easier to integrate than the original one, so we proceed to determine the limits of integration for the transformed integral.

The region of integration $R$ of the original integral in the $xy$-plane is shown in Figure 15.58. From the substitution equations $u = \sqrt{xy}$ and $v = \sqrt{y/x}$, we see that the image of the left-hand boundary $xy = 1$ for $R$ is the vertical line segment $u = 1, 2 \ge v \ge 1$, in $G$ (see Figure 15.59). Likewise, the right-hand boundary $y = x$ of $R$ maps to the horizontal line segment $v = 1, 1 \le u \le 2$, in $G$. Finally, the horizontal top boundary $y = 2$ of $R$

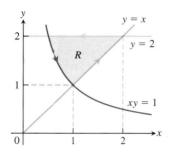

**FIGURE 15.58** The region of integration $R$ in Example 4.

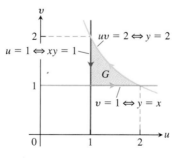

**FIGURE 15.59** The boundaries of the region $G$ correspond to those of region $R$ in Figure 15.58. Notice as we move counterclockwise around the region $R$, we also move counterclockwise around the region $G$. The inverse transformation equations $u = \sqrt{xy}$, $v = \sqrt{y/x}$ produce the region $G$ from the region $R$.

maps to $uv = 2, 1 \leq v \leq 2$, in $G$. As we move counterclockwise around the boundary of the region $R$, we also move counterclockwise around the boundary of $G$, as shown in Figure 15.59. Knowing the region of integration $G$ in the $uv$-plane, we can now write equivalent iterated integrals:

$$\int_1^2 \int_{1/y}^y \sqrt{\frac{y}{x}}\, e^{\sqrt{xy}}\, dx\, dy = \int_1^2 \int_1^{2/u} 2ue^u\, dv\, du. \qquad \text{Note the order of integration.}$$

We now evaluate the transformed integral on the right-hand side,

$$\int_1^2 \int_1^{2/u} 2ue^u\, dv\, du = 2 \int_1^2 \left[ vue^u \right]_{v=1}^{v=2/u} du$$

$$= 2 \int_1^2 (2e^u - ue^u)\, du$$

$$= 2 \int_1^2 (2 - u)e^u\, du$$

$$= 2 \left[ (2 - u)e^u + e^u \right]_{u=1}^{u=2} \qquad \text{Integrate by parts.}$$

$$= 2(e^2 - (e + e)) = 2e(e - 2). \qquad \blacksquare$$

### Substitutions in Triple Integrals

The cylindrical and spherical coordinate substitutions in Section 15.7 are special cases of a substitution method that pictures changes of variables in triple integrals as transformations of three-dimensional regions. The method is like the method for double integrals given by Equation (2) except that now we work in three dimensions instead of two.

Suppose that a region $G$ in $uvw$-space is transformed one-to-one into the region $D$ in $xyz$-space by differentiable equations of the form

$$x = g(u, v, w), \qquad y = h(u, v, w), \qquad z = k(u, v, w),$$

as suggested in Figure 15.60. Then any function $F(x, y, z)$ defined on $D$ can be thought of as a function

$$F(g(u, v, w), h(u, v, w), k(u, v, w)) = H(u, v, w)$$

defined on $G$. If $g$, $h$, and $k$ have continuous first partial derivatives, then the integral of $F(x, y, z)$ over $D$ is related to the integral of $H(u, v, w)$ over $G$ by the equation

$$\iiint_D F(x, y, z)\, dx\, dy\, dz = \iiint_G H(u, v, w) |J(u, v, w)|\, du\, dv\, dw. \qquad (7)$$

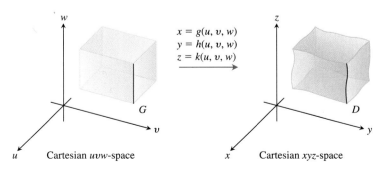

**FIGURE 15.60** The equations $x = g(u, v, w)$, $y = h(u, v, w)$, and $z = k(u, v, w)$ allow us to change an integral over a region $D$ in Cartesian $xyz$-space into an integral over a region $G$ in Cartesian $uvw$-space using Equation (7).

Cube with sides
parallel to the
coordinate axes

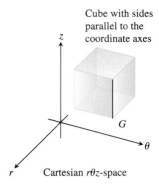

Cartesian $r\theta z$-space

$x = r \cos \theta$
$y = r \sin \theta$
$z = z$

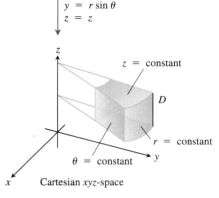

Cartesian $xyz$-space

**FIGURE 15.61** The equations
$x = r \cos \theta, y = r \sin \theta,$ and $z = z$
transform the cube $G$ into a cylindrical
wedge $D$.

The factor $J(u, v, w)$, whose absolute value appears in this equation, is the **Jacobian determinant**

$$J(u, v, w) = \begin{vmatrix} \dfrac{\partial x}{\partial u} & \dfrac{\partial x}{\partial v} & \dfrac{\partial x}{\partial w} \\ \dfrac{\partial y}{\partial u} & \dfrac{\partial y}{\partial v} & \dfrac{\partial y}{\partial w} \\ \dfrac{\partial z}{\partial u} & \dfrac{\partial z}{\partial v} & \dfrac{\partial z}{\partial w} \end{vmatrix} = \frac{\partial(x, y, z)}{\partial(u, v, w)}.$$

This determinant measures how much the volume near a point in $G$ is being expanded or contracted by the transformation from $(u, v, w)$ to $(x, y, z)$ coordinates. As in the two-dimensional case, the derivation of the change-of-variable formula in Equation (7) is omitted.

For cylindrical coordinates, $r$, $\theta$, and $z$ take the place of $u$, $v$, and $w$. The transformation from Cartesian $r\theta z$-space to Cartesian $xyz$-space is given by the equations

$$x = r \cos \theta, \qquad y = r \sin \theta, \qquad z = z$$

(Figure 15.61). The Jacobian of the transformation is

$$J(r, \theta, z) = \begin{vmatrix} \dfrac{\partial x}{\partial r} & \dfrac{\partial x}{\partial \theta} & \dfrac{\partial x}{\partial z} \\ \dfrac{\partial y}{\partial r} & \dfrac{\partial y}{\partial \theta} & \dfrac{\partial y}{\partial z} \\ \dfrac{\partial z}{\partial r} & \dfrac{\partial z}{\partial \theta} & \dfrac{\partial z}{\partial z} \end{vmatrix} = \begin{vmatrix} \cos \theta & -r \sin \theta & 0 \\ \sin \theta & r \cos \theta & 0 \\ 0 & 0 & 1 \end{vmatrix}$$

$$= r \cos^2 \theta + r \sin^2 \theta = r.$$

The corresponding version of Equation (7) is

$$\iiint_D F(x, y, z) \, dx \, dy \, dz = \iiint_G H(r, \theta, z) |r| \, dr \, d\theta \, dz.$$

We can drop the absolute value signs because $r \geq 0$.

For spherical coordinates, $\rho$, $\phi$, and $\theta$ take the place of $u$, $v$, and $w$. The transformation from Cartesian $\rho\phi\theta$-space to Cartesian $xyz$-space is given by

$$x = \rho \sin \phi \cos \theta, \qquad y = \rho \sin \phi \sin \theta, \qquad z = \rho \cos \phi$$

(Figure 15.62). The Jacobian of the transformation (see Exercise 23) is

$$J(\rho, \phi, \theta) = \begin{vmatrix} \dfrac{\partial x}{\partial \rho} & \dfrac{\partial x}{\partial \phi} & \dfrac{\partial x}{\partial \theta} \\ \dfrac{\partial y}{\partial \rho} & \dfrac{\partial y}{\partial \phi} & \dfrac{\partial y}{\partial \theta} \\ \dfrac{\partial z}{\partial \rho} & \dfrac{\partial z}{\partial \phi} & \dfrac{\partial z}{\partial \theta} \end{vmatrix} = \rho^2 \sin \phi.$$

The corresponding version of Equation (7) is

$$\iiint_D F(x, y, z) \, dx \, dy \, dz = \iiint_G H(\rho, \phi, \theta) \, |\rho^2 \sin \phi| \, d\rho \, d\phi \, d\theta.$$

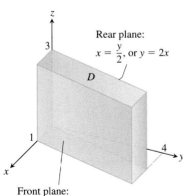

Cube with sides parallel to the coordinate axes

$$x = \rho \sin \phi \cos \theta$$
$$y = \rho \sin \phi \sin \theta$$
$$z = \rho \cos \phi$$

Cartesian $\rho\phi\theta$-space

Cartesian $xyz$-space

**FIGURE 15.62** The equations $x = \rho \sin \phi \cos \theta$, $y = \rho \sin \phi \sin \theta$, and $z = \rho \cos \phi$ transform the cube $G$ into the spherical wedge $D$.

We can drop the absolute value signs because $\sin \phi$ is never negative for $0 \le \phi \le \pi$. Note that this is the same result we obtained in Section 15.7.

Here is an example of another substitution. Although we could evaluate the integral in this example directly, we have chosen it to illustrate the substitution method in a simple (and fairly intuitive) setting.

**EXAMPLE 5** Evaluate

$$\int_0^3 \int_0^4 \int_{x=y/2}^{x=(y/2)+1} \left( \frac{2x - y}{2} + \frac{z}{3} \right) dx \, dy \, dz$$

by applying the transformation

$$u = (2x - y)/2, \qquad v = y/2, \qquad w = z/3 \tag{8}$$

and integrating over an appropriate region in $uvw$-space.

**Solution** We sketch the region $D$ of integration in $xyz$-space and identify its boundaries (Figure 15.63). In this case, the bounding surfaces are planes.

To apply Equation (7), we need to find the corresponding $uvw$-region $G$ and the Jacobian of the transformation. To find them, we first solve Equations (8) for $x$, $y$, and $z$ in terms of $u$, $v$, and $w$. Routine algebra gives

$$x = u + v, \qquad y = 2v, \qquad z = 3w. \tag{9}$$

We then find the boundaries of $G$ by substituting these expressions into the equations for the boundaries of $D$:

**FIGURE 15.63** The equations $x = u + v$, $y = 2v$, and $z = 3w$ transform $G$ into $D$. Reversing the transformation by the equations $u = (2x - y)/2$, $v = y/2$, and $w = z/3$ transforms $D$ into $G$ (Example 5).

| $xyz$-equations for the boundary of $D$ | Corresponding $uvw$-equations for the boundary of $G$ | Simplified $uvw$-equations |
|---|---|---|
| $x = y/2$ | $u + v = 2v/2 = v$ | $u = 0$ |
| $x = (y/2) + 1$ | $u + v = (2v/2) + 1 = v + 1$ | $u = 1$ |
| $y = 0$ | $2v = 0$ | $v = 0$ |
| $y = 4$ | $2v = 4$ | $v = 2$ |
| $z = 0$ | $3w = 0$ | $w = 0$ |
| $z = 3$ | $3w = 3$ | $w = 1$ |

The Jacobian of the transformation, again from Equations (9), is

$$J(u, v, w) = \begin{vmatrix} \dfrac{\partial x}{\partial u} & \dfrac{\partial x}{\partial v} & \dfrac{\partial x}{\partial w} \\[6pt] \dfrac{\partial y}{\partial u} & \dfrac{\partial y}{\partial v} & \dfrac{\partial y}{\partial w} \\[6pt] \dfrac{\partial z}{\partial u} & \dfrac{\partial z}{\partial v} & \dfrac{\partial z}{\partial w} \end{vmatrix} = \begin{vmatrix} 1 & 1 & 0 \\ 0 & 2 & 0 \\ 0 & 0 & 3 \end{vmatrix} = 6.$$

We now have everything we need to apply Equation (7):

$$\int_0^3 \int_0^4 \int_{x=y/2}^{x=(y/2)+1} \left( \frac{2x - y}{2} + \frac{z}{3} \right) dx\, dy\, dz$$

$$= \int_0^1 \int_0^2 \int_0^1 (u + w)\, |J(u, v, w)|\, du\, dv\, dw$$

$$= \int_0^1 \int_0^2 \int_0^1 (u + w)(6)\, du\, dv\, dw = 6 \int_0^1 \int_0^2 \left[ \frac{u^2}{2} + uw \right]_0^1 dv\, dw$$

$$= 6 \int_0^1 \int_0^2 \left( \frac{1}{2} + w \right) dv\, dw = 6 \int_0^1 \left[ \frac{v}{2} + vw \right]_0^2 dw = 6 \int_0^1 (1 + 2w)\, dw$$

$$= 6 \left[ w + w^2 \right]_0^1 = 6(2) = 12. \qquad \blacksquare$$

# Exercises 15.8

### Jacobians and Transformed Regions in the Plane

**1. a.** Solve the system

$$u = x - y, \qquad v = 2x + y$$

for $x$ and $y$ in terms of $u$ and $v$. Then find the value of the Jacobian $\partial(x, y)/\partial(u, v)$.

**b.** Find the image under the transformation $u = x - y$, $v = 2x + y$ of the triangular region with vertices $(0, 0)$, $(1, 1)$, and $(1, -2)$ in the $xy$-plane. Sketch the transformed region in the $uv$-plane.

**2. a.** Solve the system

$$u = x + 2y, \qquad v = x - y$$

for $x$ and $y$ in terms of $u$ and $v$. Then find the value of the Jacobian $\partial(x, y)/\partial(u, v)$.

**b.** Find the image under the transformation $u = x + 2y$, $v = x - y$ of the triangular region in the $xy$-plane bounded by the lines $y = 0$, $y = x$, and $x + 2y = 2$. Sketch the transformed region in the $uv$-plane.

**3. a.** Solve the system

$$u = 3x + 2y, \qquad v = x + 4y$$

for $x$ and $y$ in terms of $u$ and $v$. Then find the value of the Jacobian $\partial(x, y)/\partial(u, v)$.

**b.** Find the image under the transformation $u = 3x + 2y$, $v = x + 4y$ of the triangular region in the $xy$-plane bounded

by the $x$-axis, the $y$-axis, and the line $x + y = 1$. Sketch the transformed region in the $uv$-plane.

**4. a.** Solve the system

$$u = 2x - 3y, \qquad v = -x + y$$

for $x$ and $y$ in terms of $u$ and $v$. Then find the value of the Jacobian $\partial(x, y)/\partial(u, v)$.

**b.** Find the image under the transformation $u = 2x - 3y$, $v = -x + y$ of the parallelogram $R$ in the $xy$-plane with boundaries $x = -3$, $x = 0$, $y = x$, and $y = x + 1$. Sketch the transformed region in the $uv$-plane.

### Substitutions in Double Integrals

**5.** Evaluate the integral

$$\int_0^4 \int_{x=y/2}^{x=(y/2)+1} \frac{2x - y}{2}\, dx\, dy$$

from Example 1 directly by integration with respect to $x$ and $y$ to confirm that its value is 2.

**6.** Use the transformation in Exercise 1 to evaluate the integral

$$\iint_R (2x^2 - xy - y^2)\, dx\, dy$$

for the region $R$ in the first quadrant bounded by the lines $y = -2x + 4$, $y = -2x + 7$, $y = x - 2$, and $y = x + 1$.

**7.** Use the transformation in Exercise 3 to evaluate the integral

$$\iint_R (3x^2 + 14xy + 8y^2)\, dx\, dy$$

for the region $R$ in the first quadrant bounded by the lines $y = -(3/2)x + 1$, $y = -(3/2)x + 3$, $y = -(1/4)x$, and $y = -(1/4)x + 1$.

**8.** Use the transformation and parallelogram $R$ in Exercise 4 to evaluate the integral

$$\iint_R 2(x - y)\, dx\, dy.$$

**9.** Let $R$ be the region in the first quadrant of the $xy$-plane bounded by the hyperbolas $xy = 1$, $xy = 9$ and the lines $y = x$, $y = 4x$. Use the transformation $x = u/v$, $y = uv$ with $u > 0$ and $v > 0$ to rewrite

$$\iint_R \left(\sqrt{\frac{y}{x}} + \sqrt{xy}\right) dx\, dy$$

as an integral over an appropriate region $G$ in the $uv$-plane. Then evaluate the $uv$-integral over $G$.

**10. a.** Find the Jacobian of the transformation $x = u$, $y = uv$ and sketch the region $G$: $1 \le u \le 2$, $1 \le uv \le 2$, in the $uv$-plane.

   **b.** Then use Equation (2) to transform the integral

$$\int_1^2 \int_1^2 \frac{y}{x}\, dy\, dx$$

into an integral over $G$, and evaluate both integrals.

**11. Polar moment of inertia of an elliptical plate**   A thin plate of constant density covers the region bounded by the ellipse $x^2/a^2 + y^2/b^2 = 1$, $a > 0$, $b > 0$, in the $xy$-plane. Find the first moment of the plate about the origin. (*Hint:* Use the transformation $x = ar\cos\theta$, $y = br\sin\theta$.)

**12. The area of an ellipse**   The area $\pi ab$ of the ellipse $x^2/a^2 + y^2/b^2 = 1$ can be found by integrating the function $f(x, y) = 1$ over the region bounded by the ellipse in the $xy$-plane. Evaluating the integral directly requires a trigonometric substitution. An easier way to evaluate the integral is to use the transformation $x = au$, $y = bv$ and evaluate the transformed integral over the disk $G$: $u^2 + v^2 \le 1$ in the $uv$-plane. Find the area this way.

**13.** Use the transformation in Exercise 2 to evaluate the integral

$$\int_0^{2/3} \int_y^{2-2y} (x + 2y)e^{(y-x)}\, dx\, dy$$

by first writing it as an integral over a region $G$ in the $uv$-plane.

**14.** Use the transformation $x = u + (1/2)v$, $y = v$ to evaluate the integral

$$\int_0^2 \int_{y/2}^{(y+4)/2} y^3(2x - y)e^{(2x-y)^2}\, dx\, dy$$

by first writing it as an integral over a region $G$ in the $uv$-plane.

**15.** Use the transformation $x = u/v$, $y = uv$ to evaluate the integral sum

$$\int_1^2 \int_{1/y}^y (x^2 + y^2)\, dx\, dy + \int_2^4 \int_{y/4}^{4/y} (x^2 + y^2)\, dx\, dy.$$

**16.** Use the transformation $x = u^2 - v^2$, $y = 2uv$ to evaluate the integral

$$\int_0^1 \int_0^{2\sqrt{1-x}} \sqrt{x^2 + y^2}\, dy\, dx.$$

(*Hint:* Show that the image of the triangular region $G$ with vertices $(0, 0)$, $(1, 0)$, $(1, 1)$ in the $uv$-plane is the region of integration $R$ in the $xy$-plane defined by the limits of integration.)

### Substitutions in Triple Integrals

**17.** Evaluate the integral in Example 5 by integrating with respect to $x$, $y$, and $z$.

**18. Volume of an ellipsoid**   Find the volume of the ellipsoid

$$\frac{x^2}{a^2} + \frac{y^2}{b^2} + \frac{z^2}{c^2} = 1.$$

(*Hint:* Let $x = au$, $y = bv$, and $z = cw$. Then find the volume of an appropriate region in $uvw$-space.)

**19.** Evaluate

$$\iiint |xyz|\, dx\, dy\, dz$$

over the solid ellipsoid

$$\frac{x^2}{a^2} + \frac{y^2}{b^2} + \frac{z^2}{c^2} \le 1.$$

(*Hint:* Let $x = au$, $y = bv$, and $z = cw$. Then integrate over an appropriate region in $uvw$-space.)

**20.** Let $D$ be the region in $xyz$-space defined by the inequalities

$$1 \le x \le 2, \quad 0 \le xy \le 2, \quad 0 \le z \le 1.$$

Evaluate

$$\iiint_D (x^2y + 3xyz)\, dx\, dy\, dz$$

by applying the transformation

$$u = x, \quad v = xy, \quad w = 3z$$

and integrating over an appropriate region $G$ in $uvw$-space.

### Theory and Examples

**21.** Find the Jacobian $\partial(x, y)/\partial(u, v)$ of the transformation

   **a.** $x = u\cos v$, $\quad y = u\sin v$

   **b.** $x = u\sin v$, $\quad y = u\cos v$.

**22.** Find the Jacobian $\partial(x, y, z)/\partial(u, v, w)$ of the transformation

   **a.** $x = u\cos v$, $\quad y = u\sin v$, $\quad z = w$

   **b.** $x = 2u - 1$, $\quad y = 3v - 4$, $\quad z = (1/2)(w - 4)$.

23. Evaluate the appropriate determinant to show that the Jacobian of the transformation from Cartesian $\rho\phi\theta$-space to Cartesian $xyz$-space is $\rho^2 \sin\phi$.

24. **Substitutions in single integrals** How can substitutions in single definite integrals be viewed as transformations of regions? What is the Jacobian in such a case? Illustrate with an example.

25. **Centroid of a solid semiellipsoid** Assuming the result that the centroid of a solid hemisphere lies on the axis of symmetry three-eighths of the way from the base toward the top, show, by transforming the appropriate integrals, that the center of mass of a solid semiellipsoid $(x^2/a^2) + (y^2/b^2) + (z^2/c^2) \leq 1$, $z \geq 0$, lies on the $z$-axis three-eighths of the way from the base toward the top. (You can do this without evaluating any of the integrals.)

26. **Cylindrical shells** In Section 6.2, we learned how to find the volume of a solid of revolution using the shell method; namely, if the region between the curve $y = f(x)$ and the $x$-axis from $a$ to $b$ $(0 < a < b)$ is revolved about the $y$-axis, the volume of the resulting solid is $\int_a^b 2\pi x f(x)\, dx$. Prove that finding volumes by

using triple integrals gives the same result. (*Hint:* Use cylindrical coordinates with the roles of $y$ and $z$ changed.)

27. **Inverse transform** The equations $x = g(u, v)$, $y = h(u, v)$ in Figure 15.54 transform the region $G$ in the $uv$-plane into the region $R$ in the $xy$-plane. Since the substitution transformation is one-to-one with continuous first partial derivatives, it has an inverse transformation and there are equations $u = \alpha(x, y)$, $v = \beta(x, y)$ with continuous first partial derivatives transforming $R$ back into $G$. Moreover, the Jacobian determinants of the transformations are related reciprocally by

$$\frac{\partial(x, y)}{\partial(u, v)} = \left(\frac{\partial(u, v)}{\partial(x, y)}\right)^{-1}. \tag{10}$$

Equation (10) is proved in advanced calculus. Use it to find the area of the region $R$ in the first quadrant of the $xy$-plane bounded by the lines $y = 2x$, $2y = x$, and the curves $xy = 2$, $2xy = 1$ for $u = xy$ and $v = y/x$.

28. (*Continuation of Exercise 27.*) For the region $R$ described in Exercise 27, evaluate the integral $\iint_R y^2\, dA$.

---

## Chapter **15** Questions to Guide Your Review

1. Define the double integral of a function of two variables over a bounded region in the coordinate plane.

2. How are double integrals evaluated as iterated integrals? Does the order of integration matter? How are the limits of integration determined? Give examples.

3. How are double integrals used to calculate areas and average values. Give examples.

4. How can you change a double integral in rectangular coordinates into a double integral in polar coordinates? Why might it be worthwhile to do so? Give an example.

5. Define the triple integral of a function $f(x, y, z)$ over a bounded region in space.

6. How are triple integrals in rectangular coordinates evaluated? How are the limits of integration determined? Give an example.

7. How are double and triple integrals in rectangular coordinates used to calculate volumes, average values, masses, moments, and centers of mass? Give examples.

8. How are triple integrals defined in cylindrical and spherical coordinates? Why might one prefer working in one of these coordinate systems to working in rectangular coordinates?

9. How are triple integrals in cylindrical and spherical coordinates evaluated? How are the limits of integration found? Give examples.

10. How are substitutions in double integrals pictured as transformations of two-dimensional regions? Give a sample calculation.

11. How are substitutions in triple integrals pictured as transformations of three-dimensional regions? Give a sample calculation.

---

## Chapter **15** Practice Exercises

**Evaluating Double Iterated Integrals**
In Exercises 1–4, sketch the region of integration and evaluate the double integral.

1. $\displaystyle\int_1^{10}\int_0^{1/y} y e^{xy}\, dx\, dy$

2. $\displaystyle\int_0^1\int_0^{x^3} e^{y/x}\, dy\, dx$

3. $\displaystyle\int_0^{3/2}\int_{-\sqrt{9-4t^2}}^{\sqrt{9-4t^2}} t\, ds\, dt$

4. $\displaystyle\int_0^1\int_{\sqrt{y}}^{2-\sqrt{y}} xy\, dx\, dy$

In Exercises 5–8, sketch the region of integration and write an equivalent integral with the order of integration reversed. Then evaluate both integrals.

5. $\displaystyle\int_0^4\int_{-\sqrt{4-y}}^{(y-4)/2} dx\, dy$

6. $\displaystyle\int_0^1\int_{x^2}^{x} \sqrt{x}\, dy\, dx$

7. $\displaystyle\int_0^{3/2}\int_{-\sqrt{9-4y^2}}^{\sqrt{9-4y^2}} y\, dx\, dy$

8. $\displaystyle\int_0^2\int_0^{4-x^2} 2x\, dy\, dx$

Evaluate the integrals in Exercises 9–12.

**9.** $\int_0^1 \int_{2y}^2 4 \cos\left(x^2\right) dx\, dy$      **10.** $\int_0^2 \int_{y/2}^1 e^{x^2}\, dx\, dy$

**11.** $\int_0^8 \int_{\sqrt[3]{x}}^2 \frac{dy\, dx}{y^4 + 1}$      **12.** $\int_0^1 \int_{\sqrt[3]{y}}^1 \frac{2\pi \sin \pi x^2}{x^2}\, dx\, dy$

### Areas and Volumes Using Double Integrals

**13. Area between line and parabola** Find the area of the region enclosed by the line $y = 2x + 4$ and the parabola $y = 4 - x^2$ in the $xy$-plane.

**14. Area bounded by lines and parabola** Find the area of the "triangular" region in the $xy$-plane that is bounded on the right by the parabola $y = x^2$, on the left by the line $x + y = 2$, and above by the line $y = 4$.

**15. Volume of the region under a paraboloid** Find the volume under the paraboloid $z = x^2 + y^2$ above the triangle enclosed by the lines $y = x, x = 0$, and $x + y = 2$ in the $xy$-plane.

**16. Volume of the region under a parabolic cylinder** Find the volume under the parabolic cylinder $z = x^2$ above the region enclosed by the parabola $y = 6 - x^2$ and the line $y = x$ in the $xy$-plane.

### Average Values

Find the average value of $f(x, y) = xy$ over the regions in Exercises 17 and 18.

**17.** The square bounded by the lines $x = 1, y = 1$ in the first quadrant

**18.** The quarter circle $x^2 + y^2 \le 1$ in the first quadrant

### Polar Coordinates

Evaluate the integrals in Exercises 19 and 20 by changing to polar coordinates.

**19.** $\int_{-1}^1 \int_{-\sqrt{1-x^2}}^{\sqrt{1-x^2}} \frac{2\, dy\, dx}{(1 + x^2 + y^2)^2}$

**20.** $\int_{-1}^1 \int_{-\sqrt{1-y^2}}^{\sqrt{1-y^2}} \ln\left(x^2 + y^2 + 1\right) dx\, dy$

**21. Integrating over a lemniscate** Integrate the function $f(x, y) = 1/(1 + x^2 + y^2)^2$ over the region enclosed by one loop of the lemniscate $(x^2 + y^2)^2 - (x^2 - y^2) = 0$.

**22.** Integrate $f(x, y) = 1/(1 + x^2 + y^2)^2$ over

    **a. Triangular region** The triangle with vertices $(0, 0), (1, 0)$, and $\left(1, \sqrt{3}\right)$.

    **b. First quadrant** The first quadrant of the $xy$-plane.

### Evaluating Triple Iterated Integrals

Evaluate the integrals in Exercises 23–26.

**23.** $\int_0^\pi \int_0^\pi \int_0^\pi \cos(x + y + z)\, dx\, dy\, dz$

**24.** $\int_{\ln 6}^{\ln 7} \int_0^{\ln 2} \int_{\ln 4}^{\ln 5} e^{(x+y+z)}\, dz\, dy\, dx$

**25.** $\int_0^1 \int_0^{x^2} \int_0^{x+y} (2x - y - z)\, dz\, dy\, dx$

**26.** $\int_1^e \int_1^x \int_0^z \frac{2y}{z^3}\, dy\, dz\, dx$

### Volumes and Average Values Using Triple Integrals

**27. Volume** Find the volume of the wedge-shaped region enclosed on the side by the cylinder $x = -\cos y, -\pi/2 \le y \le \pi/2$, on the top by the plane $z = -2x$, and below by the $xy$-plane.

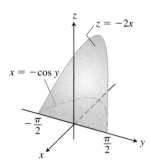

**28. Volume** Find the volume of the solid that is bounded above by the cylinder $z = 4 - x^2$, on the sides by the cylinder $x^2 + y^2 = 4$, and below by the $xy$-plane.

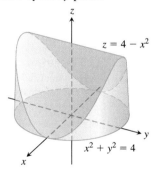

**29. Average value** Find the average value of $f(x, y, z) = 30xz\sqrt{x^2 + y}$ over the rectangular solid in the first octant bounded by the coordinate planes and the planes $x = 1, y = 3, z = 1$.

**30. Average value** Find the average value of $\rho$ over the solid sphere $\rho \le a$ (spherical coordinates).

### Cylindrical and Spherical Coordinates

**31. Cylindrical to rectangular coordinates** Convert

$$\int_0^{2\pi} \int_0^{\sqrt 2} \int_r^{\sqrt{4-r^2}} 3\, dz\, r\, dr\, d\theta, \qquad r \ge 0$$

to **(a)** rectangular coordinates with the order of integration $dz\, dx\, dy$ and **(b)** spherical coordinates. Then **(c)** evaluate one of the integrals.

**32. Rectangular to cylindrical coordinates** **(a)** Convert to cylindrical coordinates. Then **(b)** evaluate the new integral.

$$\int_0^1 \int_{-\sqrt{1-x^2}}^{\sqrt{1-x^2}} \int_{-(x^2+y^2)}^{(x^2+y^2)} 21xy^2\, dz\, dy\, dx$$

**33. Rectangular to spherical coordinates** **(a)** Convert to spherical coordinates. Then **(b)** evaluate the new integral.

$$\int_{-1}^1 \int_{-\sqrt{1-x^2}}^{\sqrt{1-x^2}} \int_{\sqrt{x^2+y^2}}^1 dz\, dy\, dx$$

**34. Rectangular, cylindrical, and spherical coordinates** Write an iterated triple integral for the integral of $f(x, y, z) = 6 + 4y$ over the region in the first octant bounded by the cone $z = \sqrt{x^2 + y^2}$, the cylinder $x^2 + y^2 = 1$, and the coordinate planes in **(a)** rectangular coordinates, **(b)** cylindrical coordinates, and **(c)** spherical coordinates. Then **(d)** find the integral of $f$ by evaluating one of the triple integrals.

**35. Cylindrical to rectangular coordinates**   Set up an integral in rectangular coordinates equivalent to the integral

$$\int_0^{\pi/2} \int_1^{\sqrt{3}} \int_1^{\sqrt{4-r^2}} r^3 (\sin\theta \cos\theta) z^2 \, dz \, dr \, d\theta.$$

Arrange the order of integration to be $z$ first, then $y$, then $x$.

**36. Rectangular to cylindrical coordinates**   The volume of a solid is

$$\int_0^2 \int_0^{\sqrt{2x-x^2}} \int_{-\sqrt{4-x^2-y^2}}^{\sqrt{4-x^2-y^2}} dz \, dy \, dx.$$

**a.** Describe the solid by giving equations for the surfaces that form its boundary.

**b.** Convert the integral to cylindrical coordinates but do not evaluate the integral.

**37. Spherical versus cylindrical coordinates**   Triple integrals involving spherical shapes do not always require spherical coordinates for convenient evaluation. Some calculations may be accomplished more easily with cylindrical coordinates. As a case in point, find the volume of the region bounded above by the sphere $x^2 + y^2 + z^2 = 8$ and below by the plane $z = 2$ by using **(a)** cylindrical coordinates and **(b)** spherical coordinates.

### Masses and Moments

**38. Finding $I_z$ in spherical coordinates**   Find the moment of inertia about the $z$-axis of a solid of constant density $\delta = 1$ that is bounded above by the sphere $\rho = 2$ and below by the cone $\phi = \pi/3$ (spherical coordinates).

**39. Moment of inertia of a "thick" sphere**   Find the moment of inertia of a solid of constant density $\delta$ bounded by two concentric spheres of radii $a$ and $b$ ($a < b$) about a diameter.

**40. Moment of inertia of an apple**   Find the moment of inertia about the $z$-axis of a solid of density $\delta = 1$ enclosed by the spherical coordinate surface $\rho = 1 - \cos\phi$. The solid is the red curve rotated about the $z$-axis in the accompanying figure.

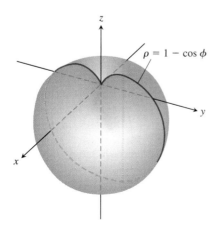

**41. Centroid**   Find the centroid of the "triangular" region bounded by the lines $x = 2, y = 2$ and the hyperbola $xy = 2$ in the $xy$-plane.

**42. Centroid**   Find the centroid of the region between the parabola $x + y^2 - 2y = 0$ and the line $x + 2y = 0$ in the $xy$-plane.

**43. Polar moment**   Find the polar moment of inertia about the origin of a thin triangular plate of constant density $\delta = 3$ bounded by the $y$-axis and the lines $y = 2x$ and $y = 4$ in the $xy$-plane.

**44. Polar moment**   Find the polar moment of inertia about the center of a thin rectangular sheet of constant density $\delta = 1$ bounded by the lines

    **a.** $x = \pm 2$, $\quad y = \pm 1$ in the $xy$-plane

    **b.** $x = \pm a$, $\quad y = \pm b$ in the $xy$-plane.

    (*Hint:* Find $I_x$. Then use the formula for $I_x$ to find $I_y$, and add the two to find $I_0$.)

**45. Inertial moment**   Find the moment of inertia about the $x$-axis of a thin plate of constant density $\delta$ covering the triangle with vertices $(0, 0)$, $(3, 0)$, and $(3, 2)$ in the $xy$-plane.

**46. Plate with variable density**   Find the center of mass and the moments of inertia about the coordinate axes of a thin plate bounded by the line $y = x$ and the parabola $y = x^2$ in the $xy$-plane if the density is $\delta(x, y) = x + 1$.

**47. Plate with variable density**   Find the mass and first moments about the coordinate axes of a thin square plate bounded by the lines $x = \pm 1, y = \pm 1$ in the $xy$-plane if the density is $\delta(x, y) = x^2 + y^2 + 1/3$.

**48. Triangles with same inertial moment**   Find the moment of inertia about the $x$-axis of a thin triangular plate of constant density $\delta$ whose base lies along the interval $[0, b]$ on the $x$-axis and whose vertex lies on the line $y = h$ above the $x$-axis. As you will see, it does not matter where on the line this vertex lies. All such triangles have the same moment of inertia about the $x$-axis.

**49. Centroid**   Find the centroid of the region in the polar coordinate plane defined by the inequalities $0 \le r \le 3, -\pi/3 \le \theta \le \pi/3$.

**50. Centroid**   Find the centroid of the region in the first quadrant bounded by the rays $\theta = 0$ and $\theta = \pi/2$ and the circles $r = 1$ and $r = 3$.

**51. a. Centroid**   Find the centroid of the region in the polar coordinate plane that lies inside the cardioid $r = 1 + \cos\theta$ and outside the circle $r = 1$.

    **b.** Sketch the region and show the centroid in your sketch.

**52. a. Centroid**   Find the centroid of the plane region defined by the polar coordinate inequalities $0 \le r \le a, -\alpha \le \theta \le \alpha$ $(0 < \alpha \le \pi)$. How does the centroid move as $\alpha \to \pi^-$?

    **b.** Sketch the region for $\alpha = 5\pi/6$ and show the centroid in your sketch.

### Substitutions

**53.** Show that if $u = x - y$ and $v = y$, then for any continuous $f$

$$\int_0^\infty \int_0^x e^{-sx} f(x - y, y) \, dy \, dx = \int_0^\infty \int_0^\infty e^{-s(u+v)} f(u, v) \, du \, dv.$$

**54.** What relationship must hold between the constants $a$, $b$, and $c$ to make

$$\int_{-\infty}^\infty \int_{-\infty}^\infty e^{-(ax^2 + 2bxy + cy^2)} \, dx \, dy = 1?$$

(*Hint:* Let $s = \alpha x + \beta y$ and $t = \gamma x + \delta y$, where $(\alpha\delta - \beta\gamma)^2 = ac - b^2$. Then $ax^2 + 2bxy + cy^2 = s^2 + t^2$.)

# Chapter **15**   Additional and Advanced Exercises

## Volumes

1. **Sand pile: double and triple integrals**   The base of a sand pile covers the region in the *xy*-plane that is bounded by the parabola $x^2 + y = 6$ and the line $y = x$. The height of the sand above the point $(x, y)$ is $x^2$. Express the volume of sand as **(a)** a double integral, **(b)** a triple integral. Then **(c)** find the volume.

2. **Water in a hemispherical bowl**   A hemispherical bowl of radius 5 cm is filled with water to within 3 cm of the top. Find the volume of water in the bowl.

3. **Solid cylindrical region between two planes**   Find the volume of the portion of the solid cylinder $x^2 + y^2 \le 1$ that lies between the planes $z = 0$ and $x + y + z = 2$.

4. **Sphere and paraboloid**   Find the volume of the region bounded above by the sphere $x^2 + y^2 + z^2 = 2$ and below by the paraboloid $z = x^2 + y^2$.

5. **Two paraboloids**   Find the volume of the region bounded above by the paraboloid $z = 3 - x^2 - y^2$ and below by the paraboloid $z = 2x^2 + 2y^2$.

6. **Spherical coordinates**   Find the volume of the region enclosed by the spherical coordinate surface $\rho = 2 \sin \phi$ (see accompanying figure).

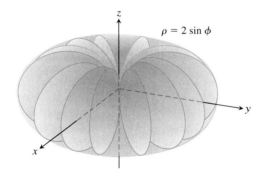

$\rho = 2 \sin \phi$

7. **Hole in sphere**   A circular cylindrical hole is bored through a solid sphere, the axis of the hole being a diameter of the sphere. The volume of the remaining solid is

$$V = 2 \int_0^{2\pi} \int_0^{\sqrt{3}} \int_1^{\sqrt{4-z^2}} r \, dr \, dz \, d\theta.$$

   **a.** Find the radius of the hole and the radius of the sphere.

   **b.** Evaluate the integral.

8. **Sphere and cylinder**   Find the volume of material cut from the solid sphere $r^2 + z^2 \le 9$ by the cylinder $r = 3 \sin \theta$.

9. **Two paraboloids**   Find the volume of the region enclosed by the surfaces $z = x^2 + y^2$ and $z = (x^2 + y^2 + 1)/2$.

10. **Cylinder and surface $z = xy$**   Find the volume of the region in the first octant that lies between the cylinders $r = 1$ and $r = 2$ and that is bounded below by the *xy*-plane and above by the surface $z = xy$.

## Changing the Order of Integration

11. Evaluate the integral

$$\int_0^{\infty} \frac{e^{-ax} - e^{-bx}}{x} \, dx.$$

   (*Hint:* Use the relation

$$\frac{e^{-ax} - e^{-bx}}{x} = \int_a^b e^{-xy} \, dy$$

   to form a double integral and evaluate the integral by changing the order of integration.)

12. **a. Polar coordinates**   Show, by changing to polar coordinates, that

$$\int_0^{a \sin \beta} \int_{y \cot \beta}^{\sqrt{a^2 - y^2}} \ln (x^2 + y^2) \, dx \, dy = a^2 \beta \left( \ln a - \frac{1}{2} \right),$$

   where $a > 0$ and $0 < \beta < \pi/2$.

   **b.** Rewrite the Cartesian integral with the order of integration reversed.

13. **Reducing a double to a single integral**   By changing the order of integration, show that the following double integral can be reduced to a single integral:

$$\int_0^x \int_0^u e^{m(x-t)} f(t) \, dt \, du = \int_0^x (x - t) e^{m(x-t)} f(t) \, dt.$$

   Similarly, it can be shown that

$$\int_0^x \int_0^v \int_0^u e^{m(x-t)} f(t) \, dt \, du \, dv = \int_0^x \frac{(x - t)^2}{2} e^{m(x-t)} f(t) \, dt.$$

14. **Transforming a double integral to obtain constant limits**   Sometimes a multiple integral with variable limits can be changed into one with constant limits. By changing the order of integration, show that

$$\int_0^1 f(x) \left( \int_0^x g(x - y) f(y) \, dy \right) dx$$

$$= \int_0^1 f(y) \left( \int_y^1 g(x - y) f(x) \, dx \right) dy$$

$$= \frac{1}{2} \int_0^1 \int_0^1 g( |x - y| ) f(x) f(y) \, dx \, dy.$$

## Masses and Moments

15. **Minimizing polar inertia**   A thin plate of constant density is to occupy the triangular region in the first quadrant of the *xy*-plane having vertices $(0, 0)$, $(a, 0)$, and $(a, 1/a)$. What value of $a$ will minimize the plate's polar moment of inertia about the origin?

16. **Polar inertia of triangular plate**   Find the polar moment of inertia about the origin of a thin triangular plate of constant

density $\delta = 3$ bounded by the $y$-axis and the lines $y = 2x$ and $y = 4$ in the $xy$-plane.

**17. Mass and polar inertia of a counterweight** The counterweight of a flywheel of constant density 1 has the form of the smaller segment cut from a circle of radius $a$ by a chord at a distance $b$ from the center ($b < a$). Find the mass of the counterweight and its polar moment of inertia about the center of the wheel.

**18. Centroid of a boomerang** Find the centroid of the boomerang-shaped region between the parabolas $y^2 = -4(x - 1)$ and $y^2 = -2(x - 2)$ in the $xy$-plane.

**Theory and Examples**

**19.** Evaluate

$$\int_0^a \int_0^b e^{\max(b^2x^2,\, a^2y^2)} \, dy \, dx,$$

where $a$ and $b$ are positive numbers and

$$\max(b^2x^2, a^2y^2) = \begin{cases} b^2x^2 & \text{if } b^2x^2 \geq a^2y^2 \\ a^2y^2 & \text{if } b^2x^2 < a^2y^2. \end{cases}$$

**20.** Show that

$$\iint \frac{\partial^2 F(x, y)}{\partial x \, \partial y} \, dx \, dy$$

over the rectangle $x_0 \leq x \leq x_1$, $y_0 \leq y \leq y_1$, is

$$F(x_1, y_1) - F(x_0, y_1) - F(x_1, y_0) + F(x_0, y_0).$$

**21.** Suppose that $f(x, y)$ can be written as a product $f(x, y) = F(x)G(y)$ of a function of $x$ and a function of $y$. Then the integral of $f$ over the rectangle $R: a \leq x \leq b, c \leq y \leq d$ can be evaluated as a product as well, by the formula

$$\iint_R f(x, y) \, dA = \left( \int_a^b F(x) \, dx \right)\left( \int_c^d G(y) \, dy \right). \tag{1}$$

The argument is that

$$\iint_R f(x, y) \, dA = \int_c^d \left( \int_a^b F(x)G(y) \, dx \right) dy \tag{i}$$

$$= \int_c^d \left( G(y) \int_a^b F(x) \, dx \right) dy \tag{ii}$$

$$= \int_c^d \left( \int_a^b F(x) \, dx \right) G(y) \, dy \tag{iii}$$

$$= \left( \int_a^b F(x) \, dx \right) \int_c^d G(y) \, dy. \tag{iv}$$

**a.** Give reasons for steps (i) through (iv).

When it applies, Equation (1) can be a time-saver. Use it to evaluate the following integrals.

**b.** $\int_0^{\ln 2} \int_0^{\pi/2} e^x \cos y \, dy \, dx$   **c.** $\int_1^2 \int_{-1}^1 \frac{x}{y^2} dx \, dy$

**22.** Let $D_{\mathbf{u}}f$ denote the derivative of $f(x, y) = (x^2 + y^2)/2$ in the direction of the unit vector $\mathbf{u} = u_1\mathbf{i} + u_2\mathbf{j}$.

**a. Finding average value** Find the average value of $D_{\mathbf{u}}f$ over the triangular region cut from the first quadrant by the line $x + y = 1$.

**b. Average value and centroid** Show in general that the average value of $D_{\mathbf{u}}f$ over a region in the $xy$-plane is the value of $D_{\mathbf{u}}f$ at the centroid of the region.

**23. The value of $\Gamma(1/2)$** The gamma function,

$$\Gamma(x) = \int_0^\infty t^{x-1} e^{-t} \, dt,$$

extends the factorial function from the nonnegative integers to other real values. Of particular interest in the theory of differential equations is the number

$$\Gamma\left(\frac{1}{2}\right) = \int_0^\infty t^{(1/2)-1} e^{-t} \, dt = \int_0^\infty \frac{e^{-t}}{\sqrt{t}} dt. \tag{2}$$

**a.** If you have not yet done Exercise 41 in Section 15.4, do it now to show that

$$I = \int_0^\infty e^{-y^2} \, dy = \frac{\sqrt{\pi}}{2}.$$

**b.** Substitute $y = \sqrt{t}$ in Equation (2) to show that $\Gamma(1/2) = 2I = \sqrt{\pi}$.

**24. Total electrical charge over circular plate** The electrical charge distribution on a circular plate of radius $R$ meters is $\sigma(r, \theta) = kr(1 - \sin\theta)$ coulomb/m² ($k$ a constant). Integrate $\sigma$ over the plate to find the total charge $Q$.

**25. A parabolic rain gauge** A bowl is in the shape of the graph of $z = x^2 + y^2$ from $z = 0$ to $z = 30$ cm. You plan to calibrate the bowl to make it into a rain gauge. What height in the bowl would correspond to 3 cm of rain? 9 cm of rain?

**26. Water in a satellite dish** A parabolic satellite dish is 2 m wide and 1/2 m deep. Its axis of symmetry is tilted 30 degrees from the vertical.

**a.** Set up, but do not evaluate, a triple integral in rectangular coordinates that gives the amount of water the satellite dish will hold. (*Hint:* Put your coordinate system so that the satellite dish is in "standard position" and the plane of the water level is slanted.) (*Caution:* The limits of integration are not "nice.")

**b.** What would be the smallest tilt of the satellite dish so that it holds no water?

**27. An infinite half-cylinder** Let $D$ be the interior of the infinite right circular half-cylinder of radius 1 with its single-end face suspended 1 unit above the origin and its axis the ray from $(0, 0, 1)$ to $\infty$. Use cylindrical coordinates to evaluate

$$\iiint_D z(r^2 + z^2)^{-5/2} \, dV.$$

**28. Hypervolume** We have learned that $\int_a^b 1 \, dx$ is the length of the interval $[a, b]$ on the number line (one-dimensional space), $\iint_R 1 \, dA$ is the area of region $R$ in the $xy$-plane (two-dimensional space), and $\iiint_D 1 \, dV$ is the volume of the region $D$ in three-dimensional space ($xyz$-space). We could continue: If $Q$ is a region in 4-space ($xyzw$-space), then $\iiiint_Q 1 \, dV$ is the "hypervolume" of $Q$. Use your generalizing abilities and a Cartesian coordinate system of 4-space to find the hypervolume inside the unit 4-dimensional sphere $x^2 + y^2 + z^2 + w^2 = 1$.

# Chapter **15**   Technology Application Projects

Mathematica/Maple Modules:

***Take Your Chances: Try the Monte Carlo Technique for Numerical Integration in Three Dimensions***
Use the Monte Carlo technique to integrate numerically in three dimensions.

***Means and Moments and Exploring New Plotting Techniques, Part II***
Use the method of moments in a form that makes use of geometric symmetry as well as multiple integration.

# 16

# Integrals and Vector Fields

**OVERVIEW**  In this chapter we extend the theory of integration over coordinate lines and planes to general curves and surfaces in space. The resulting theory of line and surface integrals gives powerful mathematical tools for science and engineering. Line integrals are used to find the work done by a force in moving an object along a path, and to find the mass of a curved wire with variable density. Surface integrals are used to find the rate of flow of a fluid across a surface. We present the fundamental theorems of vector integral calculus, and discuss their mathematical consequences and physical applications. In the final analysis, the key theorems are shown as generalized interpretations of the Fundamental Theorem of Calculus.

## 16.1 Line Integrals

To calculate the total mass of a wire lying along a curve in space, or to find the work done by a variable force acting along such a curve, we need a more general notion of integral than was defined in Chapter 5. We need to integrate over a curve $C$ rather than over an interval $[a, b]$. These more general integrals are called *line integrals* (although *path* integrals might be more descriptive). We make our definitions for space curves, with curves in the $xy$-plane being the special case with $z$-coordinate identically zero.

Suppose that $f(x, y, z)$ is a real-valued function we wish to integrate over the curve $C$ lying within the domain of $f$ and parametrized by $\mathbf{r}(t) = g(t)\mathbf{i} + h(t)\mathbf{j} + k(t)\mathbf{k}, a \le t \le b$. The values of $f$ along the curve are given by the composite function $f(g(t), h(t), k(t))$. We are going to integrate this composite with respect to arc length from $t = a$ to $t = b$. To begin, we first partition the curve $C$ into a finite number $n$ of subarcs (Figure 16.1). The typical subarc has length $\Delta s_k$. In each subarc we choose a point $(x_k, y_k, z_k)$ and form the sum

$$S_n = \sum_{k=1}^{n} f(x_k, y_k, z_k)\, \Delta s_k,$$

**FIGURE 16.1**  The curve $\mathbf{r}(t)$ partitioned into small arcs from $t = a$ to $t = b$. The length of a typical subarc is $\Delta s_k$.

which is similar to a Riemann sum. Depending on how we partition the curve $C$ and pick $(x_k, y_k, z_k)$ in the $k$th subarc, we may get different values for $S_n$. If $f$ is continuous and the functions $g$, $h$, and $k$ have continuous first derivatives, then these sums approach a limit as $n$ increases and the lengths $\Delta s_k$ approach zero. This limit gives the following definition, similar to that for a single integral. In the definition, we assume that the partition satisfies $\Delta s_k \to 0$ as $n \to \infty$.

**DEFINITION**    If $f$ is defined on a curve C given parametrically by $\mathbf{r}(t) = g(t)\mathbf{i} + h(t)\mathbf{j} + k(t)\mathbf{k}$, $a \le t \le b$, then the **line integral of $f$ over $C$** is

$$\int_C f(x, y, z)\, ds = \lim_{n \to \infty} \sum_{k=1}^{n} f(x_k, y_k, z_k)\, \Delta s_k, \tag{1}$$

provided this limit exists.

If the curve $C$ is smooth for $a \le t \le b$ (so $\mathbf{v} = d\mathbf{r}/dt$ is continuous and never $\mathbf{0}$) and the function $f$ is continuous on $C$, then the limit in Equation (1) can be shown to exist. We can then apply the Fundamental Theorem of Calculus to differentiate the arc length equation,

$$s(t) = \int_a^t |\mathbf{v}(\tau)|\, d\tau, \qquad \text{Eq. (3) of Section 13.3 with } t_0 = a$$

to express $ds$ in Equation (1) as $ds = |\mathbf{v}(t)|\, dt$ and evaluate the integral of $f$ over $C$ as

$$\frac{ds}{dt} = |\mathbf{v}| = \sqrt{\left(\frac{dx}{dt}\right)^2 + \left(\frac{dy}{dt}\right)^2 + \left(\frac{dz}{dt}\right)^2}$$

$$\int_C f(x, y, z)\, ds = \int_a^b f(g(t), h(t), k(t)) |\mathbf{v}(t)|\, dt. \tag{2}$$

Notice that the integral on the right side of Equation (2) is just an ordinary (single) definite integral, as defined in Chapter 5, where we are integrating with respect to the parameter $t$. The formula evaluates the line integral on the left side correctly no matter what parametrization is used, as long as the parametrization is smooth. Note that the parameter $t$ defines a direction along the path. The starting point on $C$ is the position $\mathbf{r}(a)$ and movement along the path is in the direction of increasing $t$ (see Figure 16.1).

$f(\mathbf{r}(t)) = f(g(t), h(t), k(t))$

> **How to Evaluate a Line Integral**
>
> To integrate a continuous function $f(x, y, z)$ over a curve $C$:
>
> **1.** Find a smooth parametrization of $C$,
>
> $$\mathbf{r}(t) = g(t)\mathbf{i} + h(t)\mathbf{j} + k(t)\mathbf{k}, \qquad a \le t \le b.$$
>
> **2.** Evaluate the integral as
>
> $$\int_C f(x, y, z)\, ds = \int_a^b f(g(t), h(t), k(t)) |\mathbf{v}(t)|\, dt.$$

If $f$ has the constant value 1, then the integral of $f$ over $C$ gives the length of $C$ from $t = a$ to $t = b$ in Figure 16.1. We also write $f(\mathbf{r}(t))$ for the evaluation $f(g(t), h(t), k(t))$ along the curve $\mathbf{r}$.

**EXAMPLE 1**    Integrate $f(x, y, z) = x - 3y^2 + z$ over the line segment $C$ joining the origin to the point $(1, 1, 1)$ (Figure 16.2).

**Solution**    We choose the simplest parametrization we can think of:

$$\mathbf{r}(t) = t\mathbf{i} + t\mathbf{j} + t\mathbf{k}, \qquad 0 \le t \le 1.$$

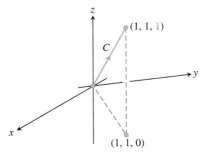

**FIGURE 16.2**    The integration path in Example 1.

The components have continuous first derivatives and $|\mathbf{v}(t)| = |\mathbf{i} + \mathbf{j} + \mathbf{k}| = \sqrt{1^2 + 1^2 + 1^2} = \sqrt{3}$ is never 0, so the parametrization is smooth. The integral of $f$ over $C$ is

$$\int_C f(x, y, z)\, ds = \int_0^1 f(t, t, t)\left(\sqrt{3}\right) dt \qquad \text{Eq. (2), } ds = |\mathbf{v}(t)|\, dt = \sqrt{3}\, dt$$

$$= \int_0^1 (t - 3t^2 + t)\sqrt{3}\, dt$$

$$= \sqrt{3}\int_0^1 (2t - 3t^2)\, dt = \sqrt{3}\left[t^2 - t^3\right]_0^1 = 0. \qquad \blacksquare$$

## Additivity

Line integrals have the useful property that if a piecewise smooth curve $C$ is made by joining a finite number of smooth curves $C_1, C_2, \ldots, C_n$ end to end (Section 13.1), then the integral of a function over $C$ is the sum of the integrals over the curves that make it up:

$$\int_C f\, ds = \int_{C_1} f\, ds + \int_{C_2} f\, ds + \cdots + \int_{C_n} f\, ds. \qquad (3)$$

**EXAMPLE 2**    Figure 16.3 shows another path from the origin to $(1, 1, 1)$, the union of line segments $C_1$ and $C_2$. Integrate $f(x, y, z) = x - 3y^2 + z$ over $C_1 \cup C_2$.

**Solution**    We choose the simplest parametrizations for $C_1$ and $C_2$ we can find, calculating the lengths of the velocity vectors as we go along:

$$C_1\colon\quad \mathbf{r}(t) = t\mathbf{i} + t\mathbf{j}, \quad 0 \le t \le 1; \quad |\mathbf{v}| = \sqrt{1^2 + 1^2} = \sqrt{2}$$

$$C_2\colon\quad \mathbf{r}(t) = \mathbf{i} + \mathbf{j} + t\mathbf{k}, \quad 0 \le t \le 1; \quad |\mathbf{v}| = \sqrt{0^2 + 0^2 + 1^2} = 1.$$

With these parametrizations we find that

$$\int_{C_1 \cup C_2} f(x, y, z)\, ds = \int_{C_1} f(x, y, z)\, ds + \int_{C_2} f(x, y, z)\, ds \qquad \text{Eq. (3)}$$

$$= \int_0^1 f(t, t, 0)\sqrt{2}\, dt + \int_0^1 f(1, 1, t)(1)\, dt \qquad \text{Eq. (2)}$$

$$= \int_0^1 (t - 3t^2 + 0)\sqrt{2}\, dt + \int_0^1 (1 - 3 + t)(1)\, dt$$

$$= \sqrt{2}\left[\frac{t^2}{2} - t^3\right]_0^1 + \left[\frac{t^2}{2} - 2t\right]_0^1 = -\frac{\sqrt{2}}{2} - \frac{3}{2}. \qquad \blacksquare$$

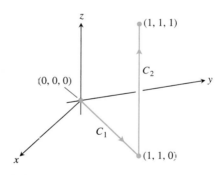

**FIGURE 16.3**    The path of integration in Example 2.

Notice three things about the integrations in Examples 1 and 2. First, as soon as the components of the appropriate curve were substituted into the formula for $f$, the integration became a standard integration with respect to $t$. Second, the integral of $f$ over $C_1 \cup C_2$ was obtained by integrating $f$ over each section of the path and adding the results. Third, the integrals of $f$ over $C$ and $C_1 \cup C_2$ had different values. We investigate this third observation in Section 16.3.

> The value of the line integral along a path joining two points can change if you change the path between them.

**EXAMPLE 3**    Find the line integral of $f(x, y, z) = 2xy + \sqrt{z}$ over the helix $\mathbf{r}(t) = \cos t\mathbf{i} + \sin t\mathbf{j} + t\mathbf{k}, 0 \le t \le \pi$.

**Solution**    For the helix we find, $\mathbf{v}(t) = \mathbf{r}'(t) = -\sin t\mathbf{i} + \cos t\mathbf{j} + \mathbf{k}$ and $|\mathbf{v}(t)| = \sqrt{(-\sin t)^2 + (\cos t)^2 + 1} = \sqrt{2}$. Evaluating the function $f$ along the path, we obtain

$$f(\mathbf{r}(t)) = f(\cos t, \sin t, t) = 2 \cos t \sin t + \sqrt{t} = \sin 2t + \sqrt{t}.$$

The line integral is given by

$$\int_C f(x, y, z)\, ds = \int_0^\pi \left( \sin 2t + \sqrt{t} \right)\sqrt{2}\, dt$$

$$= \sqrt{2}\left[ -\frac{1}{2}\cos 2t + \frac{2}{3}t^{3/2} \right]_0^\pi$$

$$= \frac{2\sqrt{2}}{3}\pi^{3/2} \approx 5.25. \qquad \blacksquare$$

## Mass and Moment Calculations

We treat coil springs and wires as masses distributed along smooth curves in space. The distribution is described by a continuous density function $\delta(x, y, z)$ representing mass per unit length. When a curve $C$ is parametrized by $\mathbf{r}(t) = x(t)\mathbf{i} + y(t)\mathbf{j} + z(t)\mathbf{k}, a \le t \le b$, then $x, y$, and $z$ are functions of the parameter $t$, the density is the function $\delta(x(t), y(t), z(t))$, and the arc length differential is given by

$$ds = \sqrt{\left( \frac{dx}{dt} \right)^2 + \left( \frac{dy}{dt} \right)^2 + \left( \frac{dz}{dt} \right)^2}\, dt.$$

(See Section 13.3.) The spring's or wire's mass, center of mass, and moments are then calculated with the formulas in Table 16.1, with the integrations in terms of the parameter $t$ over the interval $[a, b]$. For example, the formula for mass becomes

$$M = \int_a^b \delta(x(t), y(t), z(t))\sqrt{\left( \frac{dx}{dt} \right)^2 + \left( \frac{dy}{dt} \right)^2 + \left( \frac{dz}{dt} \right)^2}\, dt.$$

These formulas also apply to thin rods, and their derivations are similar to those in Section 6.6. Notice how alike the formulas are to those in Tables 15.1 and 15.2 for double and triple integrals. The double integrals for planar regions, and the triple integrals for solids, become line integrals for coil springs, wires, and thin rods.

Notice that the element of mass $dm$ is equal to $\delta\, ds$ in the table rather than $\delta\, dV$ as in Table 15.1, and that the integrals are taken over the curve $C$.

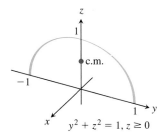

**FIGURE 16.4** Example 4 shows how to find the center of mass of a circular arch of variable density.

**EXAMPLE 4**    A slender metal arch, denser at the bottom than top, lies along the semicircle $y^2 + z^2 = 1, z \ge 0$, in the $yz$-plane (Figure 16.4). Find the center of the arch's mass if the density at the point $(x, y, z)$ on the arch is $\delta(x, y, z) = 2 - z$.

**Solution**    We know that $\bar{x} = 0$ and $\bar{y} = 0$ because the arch lies in the $yz$-plane with its mass distributed symmetrically about the $z$-axis. To find $\bar{z}$, we parametrize the circle as

$$\mathbf{r}(t) = (\cos t)\mathbf{j} + (\sin t)\mathbf{k}, \qquad 0 \le t \le \pi.$$

---

**TABLE 16.1** Mass and moment formulas for coil springs, wires, and thin rods lying along a smooth curve $C$ in space

---

**Mass:** $\quad M = \int_C \delta \, ds \qquad$ $\delta = \delta(x, y, z)$ is the density at $(x, y, z)$

**First moments about the coordinate planes:**

$$M_{yz} = \int_C x \, \delta \, ds, \qquad M_{xz} = \int_C y \, \delta \, ds, \qquad M_{xy} = \int_C z \, \delta \, ds$$

**Coordinates of the center of mass:**

$$\bar{x} = M_{yz}/M, \qquad \bar{y} = M_{xz}/M, \qquad \bar{z} = M_{xy}/M$$

**Moments of inertia about axes and other lines:**

$$I_x = \int_C (y^2 + z^2) \delta \, ds, \qquad I_y = \int_C (x^2 + z^2) \delta \, ds, \qquad I_z = \int_C (x^2 + y^2) \delta \, ds,$$

$$I_L = \int_C r^2 \delta \, ds \qquad r(x, y, z) = \text{distance from the point } (x, y, z) \text{ to line } L$$

---

For this parametrization,

$$|\mathbf{v}(t)| = \sqrt{\left(\frac{dx}{dt}\right)^2 + \left(\frac{dy}{dt}\right)^2 + \left(\frac{dz}{dt}\right)^2} = \sqrt{(0)^2 + (-\sin t)^2 + (\cos t)^2} = 1,$$

so $ds = |\mathbf{v}| \, dt = dt$.

The formulas in Table 16.1 then give

$$M = \int_C \delta \, ds = \int_C (2 - z) \, ds = \int_0^\pi (2 - \sin t) \, dt = 2\pi - 2$$

$$M_{xy} = \int_C z \delta \, ds = \int_C z(2 - z) \, ds = \int_0^\pi (\sin t)(2 - \sin t) \, dt$$

$$= \int_0^\pi (2 \sin t - \sin^2 t) \, dt = \frac{8 - \pi}{2} \qquad \text{Routine integration}$$

$$\bar{z} = \frac{M_{xy}}{M} = \frac{8 - \pi}{2} \cdot \frac{1}{2\pi - 2} = \frac{8 - \pi}{4\pi - 4} \approx 0.57.$$

With $\bar{z}$ to the nearest hundredth, the center of mass is $(0, 0, 0.57)$. ∎

## Line Integrals in the Plane

There is an interesting geometric interpretation for line integrals in the plane. If $C$ is a smooth curve in the $xy$-plane parametrized by $\mathbf{r}(t) = x(t)\mathbf{i} + y(t)\mathbf{j}$, $a \le t \le b$, we generate a cylindrical surface by moving a straight line along $C$ orthogonal to the plane, holding the line parallel to the $z$-axis, as in Section 12.6. If $z = f(x, y)$ is a nonnegative continuous function over a region in the plane containing the curve $C$, then the graph of $f$ is a surface that lies above the plane. The cylinder cuts through this surface, forming a curve on it that lies above the curve $C$ and follows its winding nature. The part of the cylindrical surface that lies beneath the surface curve and above the $xy$-plane is like a "winding wall" or "fence" standing on the curve $C$ and orthogonal to the plane. At any point $(x, y)$ along the curve, the height of the wall is $f(x, y)$. We show the wall in Figure 16.5, where the "top" of

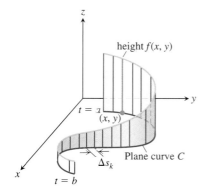

height $f(x, y)$

Plane curve $C$

**FIGURE 16.5** The line integral $\int_C f \, ds$ gives the area of the portion of the cylindrical surface or "wall" beneath $z = f(x, y) \ge 0$.

the wall is the curve lying on the surface $z = f(x, y)$. (We do not display the surface formed by the graph of $f$ in the figure, only the curve on it that is cut out by the cylinder.) From the definition

$$\int_C f \, ds = \lim_{n \to \infty} \sum_{k=1}^{n} f(x_k, y_k) \, \Delta s_k,$$

where $\Delta s_k \to 0$ as $n \to \infty$, we see that the line integral $\int_C f \, ds$ is the area of the wall shown in the figure.

## Exercises 16.1

### Graphs of Vector Equations

Match the vector equations in Exercises 1–8 with the graphs (a)–(h) given here.

**a.**

**b.**

**c.**

**d.**

**e.**

**f.**

**g.**

**h.**

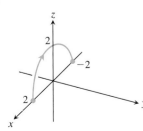

1. $\mathbf{r}(t) = t\mathbf{i} + (1 - t)\mathbf{j}, \quad 0 \le t \le 1$
2. $\mathbf{r}(t) = \mathbf{i} + \mathbf{j} + t\mathbf{k}, \quad -1 \le t \le 1$
3. $\mathbf{r}(t) = (2 \cos t)\mathbf{i} + (2 \sin t)\mathbf{j}, \quad 0 \le t \le 2\pi$
4. $\mathbf{r}(t) = t\mathbf{i}, \quad -1 \le t \le 1$
5. $\mathbf{r}(t) = t\mathbf{i} + t\mathbf{j} + t\mathbf{k}, \quad 0 \le t \le 2$
6. $\mathbf{r}(t) = t\mathbf{j} + (2 - 2t)\mathbf{k}, \quad 0 \le t \le 1$
7. $\mathbf{r}(t) = (t^2 - 1)\mathbf{j} + 2t\mathbf{k}, \quad -1 \le t \le 1$
8. $\mathbf{r}(t) = (2 \cos t)\mathbf{i} + (2 \sin t)\mathbf{k}, \quad 0 \le t \le \pi$

### Evaluating Line Integrals over Space Curves

9. Evaluate $\int_C (x + y) \, ds$ where $C$ is the straight-line segment $x = t, y = (1 - t), z = 0$, from $(0, 1, 0)$ to $(1, 0, 0)$.

10. Evaluate $\int_C (x - y + z - 2) \, ds$ where $C$ is the straight-line segment $x = t, y = (1 - t), z = 1$, from $(0, 1, 1)$ to $(1, 0, 1)$.

11. Evaluate $\int_C (xy + y + z) \, ds$ along the curve $\mathbf{r}(t) = 2t\mathbf{i} + t\mathbf{j} + (2 - 2t)\mathbf{k}, 0 \le t \le 1$.

12. Evaluate $\int_C \sqrt{x^2 + y^2} \, ds$ along the curve $\mathbf{r}(t) = (4 \cos t)\mathbf{i} + (4 \sin t)\mathbf{j} + 3t\mathbf{k}, -2\pi \le t \le 2\pi$.

13. Find the line integral of $f(x, y, z) = x + y + z$ over the straight-line segment from $(1, 2, 3)$ to $(0, -1, 1)$.

14. Find the line integral of $f(x, y, z) = \sqrt{3}/(x^2 + y^2 + z^2)$ over the curve $\mathbf{r}(t) = t\mathbf{i} + t\mathbf{j} + t\mathbf{k}, 1 \le t \le \infty$.

15. Integrate $f(x, y, z) = x + \sqrt{y} - z^2$ over the path from $(0, 0, 0)$ to $(1, 1, 1)$ (see accompanying figure) given by

$$C_1: \quad \mathbf{r}(t) = t\mathbf{i} + t^2\mathbf{j}, \quad 0 \le t \le 1$$
$$C_2: \quad \mathbf{r}(t) = \mathbf{i} + \mathbf{j} + t\mathbf{k}, \quad 0 \le t \le 1$$

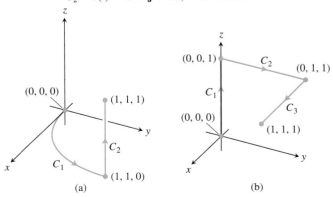

The paths of integration for Exercises 15 and 16.

**16.** Integrate $f(x, y, z) = x + \sqrt{y} - z^2$ over the path from $(0, 0, 0)$ to $(1, 1, 1)$ (see accompanying figure) given by

$$C_1: \quad \mathbf{r}(t) = t\mathbf{k}, \quad 0 \le t \le 1$$
$$C_2: \quad \mathbf{r}(t) = t\mathbf{j} + \mathbf{k}, \quad 0 \le t \le 1$$
$$C_3: \quad \mathbf{r}(t) = t\mathbf{i} + \mathbf{j} + \mathbf{k}, \quad 0 \le t \le 1$$

**17.** Integrate $f(x, y, z) = (x + y + z)/(x^2 + y^2 + z^2)$ over the path $\mathbf{r}(t) = t\mathbf{i} + t\mathbf{j} + t\mathbf{k}, 0 < a \le t \le b$.

**18.** Integrate $f(x, y, z) = -\sqrt{x^2 + z^2}$ over the circle

$$\mathbf{r}(t) = (a \cos t)\mathbf{j} + (a \sin t)\mathbf{k}, \qquad 0 \le t \le 2\pi.$$

### Line Integrals over Plane Curves

**19.** Evaluate $\int_C x \, ds$, where $C$ is

**a.** the straight-line segment $x = t, y = t/2$, from $(0, 0)$ to $(4, 2)$.

**b.** the parabolic curve $x = t, y = t^2$, from $(0, 0)$ to $(2, 4)$.

**20.** Evaluate $\int_C \sqrt{x + 2y} \, ds$, where $C$ is

**a.** the straight-line segment $x = t, y = 4t$, from $(0, 0)$ to $(1, 4)$.

**b.** $C_1 \cup C_2$; $C_1$ is the line segment from $(0, 0)$ to $(1, 0)$ and $C_2$ is the line segment from $(1, 0)$ to $(1, 2)$.

**21.** Find the line integral of $f(x, y) = ye^{x^2}$ along the curve $\mathbf{r}(t) = 4t\mathbf{i} - 3t\mathbf{j}, -1 \le t \le 2$.

**22.** Find the line integral of $f(x, y) = x - y + 3$ along the curve $\mathbf{r}(t) = (\cos t)\mathbf{i} + (\sin t)\mathbf{j}, 0 \le t \le 2\pi$.

**23.** Evaluate $\displaystyle\int_C \frac{x^2}{y^{4/3}} \, ds$, where $C$ is the curve $x = t^2, y = t^3$, for $1 \le t \le 2$.

**24.** Find the line integral of $f(x, y) = \sqrt{y}/x$ along the curve $\mathbf{r}(t) = t^3\mathbf{i} + t^4\mathbf{j}, 1/2 \le t \le 1$.

**25.** Evaluate $\int_C (x + \sqrt{y}) \, ds$ where $C$ is given in the accompanying figure.

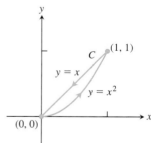

**26.** Evaluate $\displaystyle\int_C \frac{1}{x^2 + y^2 + 1} \, ds$ where $C$ is given in the accompanying figure.

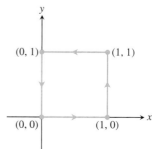

In Exercises 27–30, integrate $f$ over the given curve.

**27.** $f(x, y) = x^3/y, \quad C: \quad y = x^2/2, \quad 0 \le x \le 2$

**28.** $f(x, y) = (x + y^2)/\sqrt{1 + x^2}, \quad C: \quad y = x^2/2$ from $(1, 1/2)$ to $(0, 0)$

**29.** $f(x, y) = x + y, \quad C: \quad x^2 + y^2 = 4$ in the first quadrant from $(2, 0)$ to $(0, 2)$

**30.** $f(x, y) = x^2 - y, \quad C: \quad x^2 + y^2 = 4$ in the first quadrant from $(0, 2)$ to $(\sqrt{2}, \sqrt{2})$

**31.** Find the area of one side of the "winding wall" standing orthogonally on the curve $y = x^2, 0 \le x \le 2$, and beneath the curve on the surface $f(x, y) = x + \sqrt{y}$.

**32.** Find the area of one side of the "wall" standing orthogonally on the curve $2x + 3y = 6, 0 \le x \le 6$, and beneath the curve on the surface $f(x, y) = 4 + 3x + 2y$.

### Masses and Moments

**33. Mass of a wire** Find the mass of a wire that lies along the curve $\mathbf{r}(t) = (t^2 - 1)\mathbf{j} + 2t\mathbf{k}, 0 \le t \le 1$, if the density is $\delta = (3/2)t$.

**34. Center of mass of a curved wire** A wire of density $\delta(x, y, z) = 15\sqrt{y + 2}$ lies along the curve $\mathbf{r}(t) = (t^2 - 1)\mathbf{j} + 2t\mathbf{k}, -1 \le t \le 1$. Find its center of mass. Then sketch the curve and center of mass together.

**35. Mass of wire with variable density** Find the mass of a thin wire lying along the curve $\mathbf{r}(t) = \sqrt{2}t\mathbf{i} + \sqrt{2}t\mathbf{j} + (4 - t^2)\mathbf{k}, 0 \le t \le 1$, if the density is **(a)** $\delta = 3t$ and **(b)** $\delta = 1$.

**36. Center of mass of wire with variable density** Find the center of mass of a thin wire lying along the curve $\mathbf{r}(t) = t\mathbf{i} + 2t\mathbf{j} + (2/3)t^{3/2}\mathbf{k}, 0 \le t \le 2$, if the density is $\delta = 3\sqrt{5 + t}$.

**37. Moment of inertia of wire hoop** A circular wire hoop of constant density $\delta$ lies along the circle $x^2 + y^2 = a^2$ in the $xy$-plane. Find the hoop's moment of inertia about the $z$-axis.

**38. Inertia of a slender rod** A slender rod of constant density lies along the line segment $\mathbf{r}(t) = t\mathbf{j} + (2 - 2t)\mathbf{k}, 0 \le t \le 1$, in the $yz$-plane. Find the moments of inertia of the rod about the three coordinate axes.

**39. Two springs of constant density** A spring of constant density $\delta$ lies along the helix

$$\mathbf{r}(t) = (\cos t)\mathbf{i} + (\sin t)\mathbf{j} + t\mathbf{k}, \qquad 0 \le t \le 2\pi.$$

**a.** Find $I_z$.

**b.** Suppose that you have another spring of constant density $\delta$ that is twice as long as the spring in part (a) and lies along the helix for $0 \le t \le 4\pi$. Do you expect $I_z$ for the longer spring to be the same as that for the shorter one, or should it be different? Check your prediction by calculating $I_z$ for the longer spring.

**40. Wire of constant density** A wire of constant density $\delta = 1$ lies along the curve

$$\mathbf{r}(t) = (t \cos t)\mathbf{i} + (t \sin t)\mathbf{j} + (2\sqrt{2}/3)t^{3/2}\mathbf{k}, \qquad 0 \le t \le 1.$$

Find $\bar{z}$ and $I_z$.

**41. The arch in Example 4** Find $I_x$ for the arch in Example 4.

**42. Center of mass and moments of inertia for wire with variable density** Find the center of mass and the moments of inertia about the coordinate axes of a thin wire lying along the curve

$$\mathbf{r}(t) = t\mathbf{i} + \frac{2\sqrt{2}}{3}t^{3/2}\mathbf{j} + \frac{t^2}{2}\mathbf{k}, \qquad 0 \le t \le 2,$$

if the density is $\delta = 1/(t + 1)$.

**COMPUTER EXPLORATIONS**

In Exercises 43–46, use a CAS to perform the following steps to evaluate the line integrals.

  **a.** Find $ds = |\mathbf{v}(t)|\, dt$ for the path $\mathbf{r}(t) = g(t)\mathbf{i} + h(t)\mathbf{j} + k(t)\mathbf{k}$.

  **b.** Express the integrand $f(g(t), h(t), k(t))|\mathbf{v}(t)|$ as a function of the parameter $t$.

  **c.** Evaluate $\int_C f\, ds$ using Equation (2) in the text.

**43.** $f(x, y, z) = \sqrt{1 + 30x^2 + 10y}; \quad \mathbf{r}(t) = t\mathbf{i} + t^2\mathbf{j} + 3t^2\mathbf{k},$
  $0 \le t \le 2$

**44.** $f(x, y, z) = \sqrt{1 + x^3 + 5y^3}; \quad \mathbf{r}(t) = t\mathbf{i} + \frac{1}{3}t^2\mathbf{j} + \sqrt{t}\mathbf{k},$
  $0 \le t \le 2$

**45.** $f(x, y, z) = x\sqrt{y} - 3z^2; \quad \mathbf{r}(t) = (\cos 2t)\mathbf{i} + (\sin 2t)\mathbf{j} + 5t\mathbf{k},$
  $0 \le t \le 2\pi$

**46.** $f(x, y, z) = \left(1 + \frac{9}{4}z^{1/3}\right)^{1/4}; \quad \mathbf{r}(t) = (\cos 2t)\mathbf{i} + (\sin 2t)\mathbf{j} +$
  $t^{5/2}\mathbf{k}, \quad 0 \le t \le 2\pi$

# 16.2 Vector Fields and Line Integrals: Work, Circulation, and Flux

Gravitational and electric forces have both a direction and a magnitude. They are represented by a vector at each point in their domain, producing a *vector field*. In this section we show how to compute the work done in moving an object through such a field by using a line integral involving the vector field. We also discuss velocity fields, such as the vector field representing the velocity of a flowing fluid in its domain. A line integral can be used to find the rate at which the fluid flows along or across a curve within the domain.

## Vector Fields

**FIGURE 16.6** Velocity vectors of a flow around an airfoil in a wind tunnel.

Suppose a region in the plane or in space is occupied by a moving fluid, such as air or water. The fluid is made up of a large number of particles, and at any instant of time, a particle has a velocity $\mathbf{v}$. At different points of the region at a given (same) time, these velocities can vary. We can think of a velocity vector being attached to each point of the fluid representing the velocity of a particle at that point. Such a fluid flow is an example of a *vector field*. Figure 16.6 shows a velocity vector field obtained from air flowing around an airfoil in a wind tunnel. Figure 16.7 shows a vector field of velocity vectors along the streamlines of water moving through a contracting channel. Vector fields are also associated with forces such as gravitational attraction (Figure 16.8), and with magnetic fields, electric fields, and there are also purely mathematical fields.

Generally, a **vector field** is a function that assigns a vector to each point in its domain. A vector field on a three-dimensional domain in space might have a formula like

$$\mathbf{F}(x, y, z) = M(x, y, z)\mathbf{i} + N(x, y, z)\mathbf{j} + P(x, y, z)\mathbf{k}.$$

The field is **continuous** if the **component functions** $M$, $N$, and $P$ are continuous; it is **differentiable** if each of the component functions is differentiable. The formula for a field of two-dimensional vectors could look like

$$\mathbf{F}(x, y) = M(x, y)\mathbf{i} + N(x, y)\mathbf{j}.$$

**FIGURE 16.7** Streamlines in a contracting channel. The water speeds up as the channel narrows and the velocity vectors increase in length.

We encountered another type of vector field in Chapter 13. The tangent vectors $\mathbf{T}$ and normal vectors $\mathbf{N}$ for a curve in space both form vector fields along the curve. Along a curve $\mathbf{r}(t)$ they might have a component formula similar to the velocity field expression

$$\mathbf{v}(t) = f(t)\mathbf{i} + g(t)\mathbf{j} + h(t)\mathbf{k}.$$

If we attach the gradient vector $\nabla f$ of a scalar function $f(x, y, z)$ to each point of a level surface of the function, we obtain a three-dimensional field on the surface. If we attach the velocity vector to each point of a flowing fluid, we have a three-dimensional

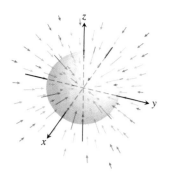

**FIGURE 16.8** Vectors in a gravitational field point toward the center of mass that gives the source of the field.

field defined on a region in space. These and other fields are illustrated in Figures 16.6–16.15. To sketch the fields, we picked a representative selection of domain points and drew the vectors attached to them. The arrows are drawn with their tails, not their heads, attached to the points where the vector functions are evaluated.

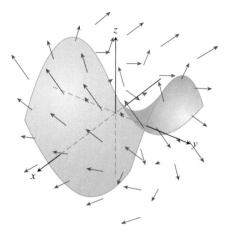

**FIGURE 16.9** A surface, like a mesh net or parachute, in a vector field representing water or wind flow velocity vectors. The arrows show the direction and their lengths indicate speed.

$f(x, y, z) = c$

**FIGURE 16.10** The field of gradient vectors $\nabla f$ on a surface $f(x, y, z) = c$.

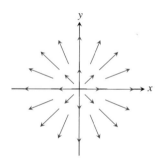

**FIGURE 16.11** The radial field $\mathbf{F} = x\mathbf{i} + y\mathbf{j}$ of position vectors of points in the plane. Notice the convention that an arrow is drawn with its tail, not its head, at the point where $\mathbf{F}$ is evaluated.

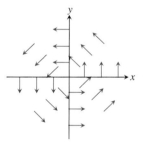

**FIGURE 16.12** A "spin" field of rotating unit vectors

$$\mathbf{F} = (-y\mathbf{i} + x\mathbf{j})/(x^2 + y^2)^{1/2}$$

in the plane. The field is not defined at the origin.

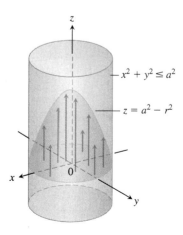

**FIGURE 16.13** The flow of fluid in a long cylindrical pipe. The vectors $\mathbf{v} = (a^2 - r^2)\mathbf{k}$ inside the cylinder that have their bases in the $xy$-plane have their tips on the paraboloid $z = a^2 - r^2$.

### Gradient Fields

The gradient vector of a differentiable scalar-valued function at a point gives the direction of greatest increase of the function. An important type of vector field is formed by all the gradient vectors of the function (see Section 14.5). We define the **gradient field** of a differentiable function $f(x, y, z)$ to be the field of gradient vectors

$$\nabla f = \frac{\partial f}{\partial x}\mathbf{i} + \frac{\partial f}{\partial y}\mathbf{j} + \frac{\partial f}{\partial z}\mathbf{k}.$$

At each point $(x, y, z)$, the gradient field gives a vector pointing in the direction of greatest increase of $f$, with magnitude being the value of the directional derivative in that direction. The gradient field is not always a force field or a velocity field.

WIND SPEED, M/S

0  2  4  6  8  10 12 14 16+

**FIGURE 16.15**   NASA's *Seasat* used radar to take 350,000 wind measurements over the world's oceans. The arrows show wind direction; their length and the color contouring indicate speed. Notice the heavy storm south of Greenland.

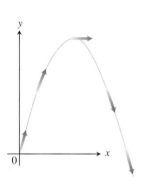

**FIGURE 16.14**   The velocity vectors $\mathbf{v}(t)$ of a projectile's motion make a vector field along the trajectory.

**EXAMPLE 1**    Suppose that the temperature $T$ at each point $(x, y, z)$ in a region of space is given by

$$T = 100 - x^2 - y^2 - z^2,$$

and that $\mathbf{F}(x, y, z)$ is defined to be the gradient of $T$. Find the vector field $\mathbf{F}$.

**Solution**    The gradient field $\mathbf{F}$ is the field $\mathbf{F} = \nabla T = -2x\mathbf{i} - 2y\mathbf{j} - 2z\mathbf{k}$. At each point in space, the vector field $\mathbf{F}$ gives the direction for which the increase in temperature is greatest. ∎

### Line Integrals of Vector Fields

In Section 16.1 we defined the line integral of a scalar function $f(x, y, z)$ over a path $C$. We turn our attention now to the idea of a line integral of a vector field $\mathbf{F}$ along the curve $C$. Such line integrals have important applications in studying fluid flows, and electrical or gravitational fields.

Assume that the vector field $\mathbf{F} = M(x, y, z)\mathbf{i} + N(x, y, z)\mathbf{j} + P(x, y, z)\mathbf{k}$ has continuous components, and that the curve $C$ has a smooth parametrization $\mathbf{r}(t) = g(t)\mathbf{i} + h(t)\mathbf{j} + k(t)\mathbf{k}$, $a \le t \le b$. As discussed in Section 16.1, the parametrization $\mathbf{r}(t)$ defines a direction (or orientation) along $C$ which we call the **forward direction**. At each point along the path $C$, the tangent vector $\mathbf{T} = d\mathbf{r}/ds = \mathbf{v}/|\mathbf{v}|$ is a unit vector tangent to the path and pointing in this forward direction. (The vector $\mathbf{v} = d\mathbf{r}/dt$ is the velocity vector tangent to $C$ at the point, as discussed in Sections 13.1 and 13.3.) Intuitively, the line integral of the vector field is the line integral of the scalar tangential component of **F** along $C$. This tangential component is given by the dot product

$$\mathbf{F} \cdot \mathbf{T} = \mathbf{F} \cdot \frac{d\mathbf{r}}{ds},$$

so we have the following formal definition, where $f = \mathbf{F} \cdot \mathbf{T}$ in Equation (1) of Section 16.1.

---

**DEFINITION** Let **F** be a vector field with continuous components defined along a smooth curve $C$ parametrized by $\mathbf{r}(t)$, $a \le t \le b$. Then the **line integral of F along $C$** is

$$\int_C \mathbf{F} \cdot \mathbf{T} \, ds = \int_C \left( \mathbf{F} \cdot \frac{d\mathbf{r}}{ds} \right) ds = \int_C \mathbf{F} \cdot d\mathbf{r}.$$

---

We evaluate line integrals of vector fields in a way similar to how we evaluate line integrals of scalar functions (Section 16.1).

---

Evaluating the Line Integral of $\mathbf{F} = M\mathbf{i} + N\mathbf{j} + P\mathbf{k}$ Along
$C: \mathbf{r}(t) = g(t)\mathbf{i} + h(t)\mathbf{j} + k(t)\mathbf{k}$

1. Express the vector field **F** in terms of the parametrized curve $C$ as $\mathbf{F}(\mathbf{r}(t))$ by substituting the components $x = g(t)$, $y = h(t)$, $z = k(t)$ of **r** into the scalar components $M(x, y, z)$, $N(x, y, z)$, $P(x, y, z)$ of **F**.
2. Find the derivative (velocity) vector $d\mathbf{r}/dt$.
3. Evaluate the line integral with respect to the parameter $t$, $a \le t \le b$, to obtain

$$\int_C \mathbf{F} \cdot d\mathbf{r} = \int_a^b \mathbf{F}(\mathbf{r}(t)) \cdot \frac{d\mathbf{r}}{dt} \, dt.$$

---

**EXAMPLE 2** Evaluate $\int_C \mathbf{F} \cdot d\mathbf{r}$, where $\mathbf{F}(x, y, z) = z\mathbf{i} + xy\mathbf{j} - y^2\mathbf{k}$ along the curve $C$ given by $\mathbf{r}(t) = t^2\mathbf{i} + t\mathbf{j} + \sqrt{t}\,\mathbf{k}$, $0 \le t \le 1$.

**Solution** We have

$$\mathbf{F}(\mathbf{r}(t)) = \sqrt{t}\,\mathbf{i} + t^3\mathbf{j} - t^2\mathbf{k} \qquad z = \sqrt{t}, xy = t^3, -y^2 = -t^2$$

and

$$\frac{d\mathbf{r}}{dt} = 2t\mathbf{i} + \mathbf{j} + \frac{1}{2\sqrt{t}}\mathbf{k}.$$

Thus,

$$\int_C \mathbf{F} \cdot d\mathbf{r} = \int_0^1 \mathbf{F}(\mathbf{r}(t)) \cdot \frac{d\mathbf{r}}{dt} \, dt$$

$$= \int_0^1 \left( 2t^{3/2} + t^3 - \frac{1}{2}t^{3/2} \right) dt$$

$$= \left[ \left( \frac{3}{2} \right)\left( \frac{2}{5}t^{5/2} \right) + \frac{1}{4}t^4 \right]_0^1 = \frac{17}{20}. \qquad \blacksquare$$

## Line Integrals with Respect to *dx*, *dy*, or *dz*

When analyzing forces or flows, it is often useful to consider each component direction separately. In such situations we want a line integral of a scalar function with respect to one of the coordinates, such as $\int_C M\,dx$. This integral is not the same as the arc length line integral $\int_C M\,ds$ we defined in Section 16.1. To define the integral $\int_C M\,dx$ for the scalar function $M(x, y, z)$, we specify a vector field $\mathbf{F} = M(x, y, z)\mathbf{i}$ having a component only in the $x$-direction, and none in the $y$- or $z$-direction. Then, over the curve $C$ parametrized by $\mathbf{r}(t) = g(t)\mathbf{i} + h(t)\mathbf{j} + k(t)\mathbf{k}$ for $a \le t \le b$, we have $x = g(t)$, $dx = g'(t)\,dt$, and

$$\mathbf{F} \cdot d\mathbf{r} = \mathbf{F} \cdot \frac{d\mathbf{r}}{dt}\,dt = M(x, y, z)g'(t)\,dt = M(x, y, z)\,dx.$$

From the definition of the line integral of $\mathbf{F}$ along $C$, we define

$$\int_C M(x, y, z)\,dx = \int_C \mathbf{F} \cdot d\mathbf{r}, \quad \text{where} \quad \mathbf{F} = M(x, y, z)\mathbf{i}.$$

In the same way, by defining $\mathbf{F} = N(x, y, z)\mathbf{j}$ with a component only in the $y$-direction, or as $\mathbf{F} = P(x, y, z)\mathbf{k}$ with a component only in the $z$-direction, we can obtain the line integrals $\int_C N\,dy$ and $\int_C P\,dz$. Expressing everything in terms of the parameter $t$ along the curve $C$, we have the following formulas for these three integrals:

$$\int_C M(x, y, z)\,dx = \int_a^b M(g(t), h(t), k(t))\,g'(t)\,dt \qquad (1)$$

$$\int_C N(x, y, z)\,dy = \int_a^b N(g(t), h(t), k(t))\,h'(t)\,dt \qquad (2)$$

$$\int_C P(x, y, z)\,dz = \int_a^b P(g(t), h(t), k(t))\,k'(t)\,dt \qquad (3)$$

It often happens that these line integrals occur in combination, and we abbreviate the notation by writing

$$\int_C M(x, y, z)\,dx + \int_C N(x, y, z)\,dy + \int_C P(x, y, z)\,dz = \int_C M\,dx + N\,dy + P\,dz.$$

**EXAMPLE 3** Evaluate the line integral $\int_C -y\,dx + z\,dy + 2x\,dz$, where $C$ is the helix $\mathbf{r}(t) = (\cos t)\mathbf{i} + (\sin t)\mathbf{j} + t\mathbf{k}$, $0 \le t \le 2\pi$.

**Solution** We express everything in terms of the parameter $t$, so $x = \cos t$, $y = \sin t$, $z = t$, and $dx = -\sin t\,dt$, $dy = \cos t\,dt$, $dz = dt$. Then,

$$\int_C -y\,dx + z\,dy + 2x\,dz = \int_0^{2\pi} [(-\sin t)(-\sin t) + t\cos t + 2\cos t]\,dt$$

$$= \int_0^{2\pi} [2\cos t + t\cos t + \sin^2 t]\,dt$$

$$= \left[2\sin t + (t\sin t + \cos t) + \left(\frac{t}{2} - \frac{\sin 2t}{4}\right)\right]_0^{2\pi}$$

$$= [0 + (0 + 1) + (\pi - 0)] - [0 + (0 + 1) + (0 - 0)]$$

$$= \pi. \qquad \blacksquare$$

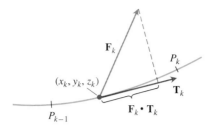

**FIGURE 16.16** The work done along the subarc shown here is approximately $\mathbf{F}_k \cdot \mathbf{T}_k \, \Delta s_k$, where $\mathbf{F}_k = \mathbf{F}(x_k, y_k, z_k)$ and $\mathbf{T}_k = \mathbf{T}(x_k, y_k, z_k)$.

**FIGURE 16.17** The work done by a force $\mathbf{F}$ is the line integral of the scalar component $\mathbf{F} \cdot \mathbf{T}$ over the smooth curve from $A$ to $B$.

## Work Done by a Force over a Curve in Space

Suppose that the vector field $\mathbf{F} = M(x, y, z)\mathbf{i} + N(x, y, z)\mathbf{j} + P(x, y, z)\mathbf{k}$ represents a force throughout a region in space (it might be the force of gravity or an electromagnetic force of some kind) and that

$$\mathbf{r}(t) = g(t)\mathbf{i} + h(t)\mathbf{j} + k(t)\mathbf{k}, \qquad a \le t \le b,$$

is a smooth curve in the region. The formula for the work done by the force in moving an object along the curve is motivated by the same kind of reasoning we used in Chapter 6 to derive the ordinary single integral for the work done by a continuous force of magnitude $F(x)$ directed along an interval of the $x$-axis. For a curve $C$ in space, we define the work done by a continuous force field $\mathbf{F}$ to move an object along $C$ from a point $A$ to another point $B$ as follows.

We divide $C$ into $n$ subarcs $P_{k-1}P_k$ with lengths $\Delta s_k$, starting at $A$ and ending at $B$. We choose any point $(x_k, y_k, z_k)$ in the subarc $P_{k-1}P_k$ and let $\mathbf{T}(x_k, y_k, z_k)$ be the unit tangent vector at the chosen point. The work $W_k$ done to move the object along the subarc $P_{k-1}P_k$ is approximated by the tangential component of the force $\mathbf{F}(x_k, y_k, z_k)$ times the arclength $\Delta s_k$ approximating the distance the object moves along the subarc (see Figure 16.16). The total work done in moving the object from point $A$ to point $B$ is then approximated by summing the work done along each of the subarcs, so

$$W \approx \sum_{k=1}^{n} W_k \approx \sum_{k=1}^{n} \mathbf{F}(x_k, y_k, z_k) \cdot \mathbf{T}(x_k, y_k, z_k) \, \Delta s_k.$$

For any subdivision of $C$ into $n$ subarcs, and for any choice of the points $(x_k, y_k, z_k)$ within each subarc, as $n \to \infty$ and $\Delta s_k \to 0$, these sums approach the line integral

$$\int_C \mathbf{F} \cdot \mathbf{T} \, ds.$$

This is just the line integral of $\mathbf{F}$ along $C$, which defines the total work done.

---

**DEFINITION** Let $C$ be a smooth curve parametrized by $\mathbf{r}(t)$, $a \le t \le b$, and $\mathbf{F}$ be a continuous force field over a region containing $C$. Then the **work** done in moving an object from the point $A = \mathbf{r}(a)$ to the point $B = \mathbf{r}(b)$ along $C$ is

$$W = \int_C \mathbf{F} \cdot \mathbf{T} \, ds = \int_a^b \mathbf{F}(\mathbf{r}(t)) \cdot \frac{d\mathbf{r}}{dt} \, dt. \tag{4}$$

---

The sign of the number we calculate with this integral depends on the direction in which the curve is traversed. If we reverse the direction of motion, then we reverse the direction of $\mathbf{T}$ in Figure 16.17 and change the sign of $\mathbf{F} \cdot \mathbf{T}$ and its integral.

Using the notations we have presented, we can express the work integral in a variety of ways, depending upon what seems most suitable or convenient for a particular discussion. Table 16.2 shows five ways we can write the work integral in Equation (4). In the table, the field components $M$, $N$, and $P$ are functions of the intermediate variables $x$, $y$, and $z$, which in turn are functions of the independent variable $t$ along the curve $C$ in the vector field. So along the curve, $x = g(t)$, $y = h(t)$, and $z = k(t)$ with $dx = g'(t)\,dt$, $dy = h'(t)\,dt$, and $dz = k'(t)\,dt$.

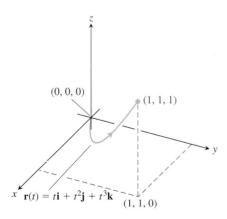

**FIGURE 16.18** The curve in Example 4.

**EXAMPLE 4** Find the work done by the force field $\mathbf{F} = (y - x^2)\mathbf{i} + (z - y^2)\mathbf{j} + (x - z^2)\mathbf{k}$ along the curve $\mathbf{r}(t) = t\mathbf{i} + t^2\mathbf{j} + t^3\mathbf{k}$, $0 \le t \le 1$, from $(0, 0, 0)$ to $(1, 1, 1)$ (Figure 16.18).

---

**TABLE 16.2** Different ways to write the work integral for $\mathbf{F} = M\mathbf{i} + N\mathbf{j} + P\mathbf{k}$ over the curve $C: \mathbf{r}(t) = g(t)\mathbf{i} + h(t)\mathbf{j} + k(t)\mathbf{k}$, $a \le t \le b$

$$W = \int_C \mathbf{F} \cdot \mathbf{T} \, ds \qquad\qquad\qquad \text{The definition}$$

$$= \int_C \mathbf{F} \cdot d\mathbf{r} \qquad\qquad\qquad \text{Vector differential form}$$

$$= \int_a^b \mathbf{F} \cdot \frac{d\mathbf{r}}{dt} \, dt \qquad\qquad\qquad \text{Parametric vector evaluation}$$

$$= \int_a^b \left( Mg'(t) + Nh'(t) + Pk'(t) \right) dt \qquad \text{Parametric scalar evaluation}$$

$$= \int_C M \, dx + N \, dy + P \, dz \qquad\qquad \text{Scalar differential form}$$

---

**Solution** First we evaluate $\mathbf{F}$ on the curve $\mathbf{r}(t)$:

$$\mathbf{F} = (y - x^2)\mathbf{i} + (z - y^2)\mathbf{j} + (x - z^2)\mathbf{k}$$
$$= \underset{0}{\underbrace{(t^2 - t^2)}}\mathbf{i} + (t^3 - t^4)\mathbf{j} + (t - t^6)\mathbf{k}. \qquad \text{Substitute } x = t,$$
$$\text{\small } y = t^2, z = t^3.$$

Then we find $d\mathbf{r}/dt$,

$$\frac{d\mathbf{r}}{dt} = \frac{d}{dt}(t\mathbf{i} + t^2\mathbf{j} + t^3\mathbf{k}) = \mathbf{i} + 2t\mathbf{j} + 3t^2\mathbf{k}.$$

Finally, we find $\mathbf{F} \cdot d\mathbf{r}/dt$ and integrate from $t = 0$ to $t = 1$:

$$\mathbf{F} \cdot \frac{d\mathbf{r}}{dt} = \left[ (t^3 - t^4)\mathbf{j} + (t - t^6)\mathbf{k} \right] \cdot (\mathbf{i} + 2t\mathbf{j} + 3t^2\mathbf{k})$$
$$= (t^3 - t^4)(2t) + (t - t^6)(3t^2) = 2t^4 - 2t^5 + 3t^3 - 3t^8.$$

So,

$$\text{Work} = \int_0^1 (2t^4 - 2t^5 + 3t^3 - 3t^8) \, dt$$

$$= \left[ \frac{2}{5}t^5 - \frac{2}{6}t^6 + \frac{3}{4}t^4 - \frac{3}{9}t^9 \right]_0^1 = \frac{29}{60}. \qquad \blacksquare$$

**EXAMPLE 5** Find the work done by the force field $\mathbf{F} = x\mathbf{i} + y\mathbf{j} + z\mathbf{k}$ in moving an object along the curve $C$ parametrized by $\mathbf{r}(t) = \cos(\pi t)\mathbf{i} + t^2\mathbf{j} + \sin(\pi t)\mathbf{k}$, $0 \le t \le 1$.

**Solution** We begin by writing $\mathbf{F}$ along $C$ as a function of $t$,

$$\mathbf{F}(\mathbf{r}(t)) = \cos(\pi t)\mathbf{i} + t^2\mathbf{j} + \sin(\pi t)\mathbf{k}.$$

Next we compute $d\mathbf{r}/dt$,

$$\frac{d\mathbf{r}}{dt} = -\pi \sin(\pi t)\mathbf{i} + 2t\mathbf{j} + \pi \cos(\pi t)\mathbf{k}.$$

We then calculate the dot product,

$$\mathbf{F}(\mathbf{r}(t)) \cdot \frac{d\mathbf{r}}{dt} = -\pi \sin(\pi t) \cos(\pi t) + 2t^3 + \pi \sin(\pi t) \cos(\pi t) = 2t^3.$$

The work done is the line integral

$$\int_a^b \mathbf{F}(\mathbf{r}(t)) \cdot \frac{d\mathbf{r}}{dt} \, dt = \int_0^1 2t^3 \, dt = \frac{t^4}{2} \Big]_0^1 = \frac{1}{2}. \qquad \blacksquare$$

## Flow Integrals and Circulation for Velocity Fields

Suppose that $\mathbf{F}$ represents the velocity field of a fluid flowing through a region in space (a tidal basin or the turbine chamber of a hydroelectric generator, for example). Under these circumstances, the integral of $\mathbf{F} \cdot \mathbf{T}$ along a curve in the region gives the fluid's flow along, or *circulation* around, the curve. For instance, the vector field in Figure 16.11 gives zero circulation around the unit circle in the plane. By contrast, the vector field in Figure 16.12 gives a nonzero circulation around the unit circle.

---

**DEFINITIONS**   If $\mathbf{r}(t)$ parametrizes a smooth curve $C$ in the domain of a continuous velocity field $\mathbf{F}$, the **flow** along the curve from $A = \mathbf{r}(a)$ to $B = \mathbf{r}(b)$ is

$$\text{Flow} = \int_C \mathbf{F} \cdot \mathbf{T} \, ds. \qquad (5)$$

The integral is called a **flow integral**. If the curve starts and ends at the same point, so that $A = B$, the flow is called the **circulation** around the curve.

---

The direction we travel along $C$ matters. If we reverse the direction, then $\mathbf{T}$ is replaced by $-\mathbf{T}$ and the sign of the integral changes. We evaluate flow integrals the same way we evaluate work integrals.

**EXAMPLE 6**   A fluid's velocity field is $\mathbf{F} = x\mathbf{i} + z\mathbf{j} + y\mathbf{k}$. Find the flow along the helix $\mathbf{r}(t) = (\cos t)\mathbf{i} + (\sin t)\mathbf{j} + t\mathbf{k}, 0 \le t \le \pi/2$.

**Solution**   We evaluate $\mathbf{F}$ on the curve,

$$\mathbf{F} = x\mathbf{i} + z\mathbf{j} + y\mathbf{k} = (\cos t)\mathbf{i} + t\mathbf{j} + (\sin t)\mathbf{k} \qquad \text{Substitute } x = \cos t, z = t, y = \sin t.$$

and then find $d\mathbf{r}/dt$:

$$\frac{d\mathbf{r}}{dt} = (-\sin t)\mathbf{i} + (\cos t)\mathbf{j} + \mathbf{k}.$$

Then we integrate $\mathbf{F} \cdot (d\mathbf{r}/dt)$ from $t = 0$ to $t = \frac{\pi}{2}$:

$$\mathbf{F} \cdot \frac{d\mathbf{r}}{dt} = (\cos t)(-\sin t) + (t)(\cos t) + (\sin t)(1)$$

$$= -\sin t \cos t + t \cos t + \sin t.$$

So,

$$\text{Flow} = \int_{t=a}^{t=b} \mathbf{F} \cdot \frac{d\mathbf{r}}{dt} dt = \int_0^{\pi/2} (-\sin t \cos t + t \cos t + \sin t) \, dt$$

$$= \left[ \frac{\cos^2 t}{2} + t \sin t \right]_0^{\pi/2} = \left( 0 + \frac{\pi}{2} \right) - \left( \frac{1}{2} + 0 \right) = \frac{\pi}{2} - \frac{1}{2}. \qquad \blacksquare$$

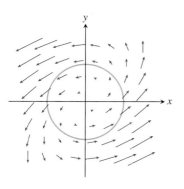

**FIGURE 16.19** The vector field **F** and curve **r**(t) in Example 7.

**EXAMPLE 7**    Find the circulation of the field $\mathbf{F} = (x - y)\mathbf{i} + x\mathbf{j}$ around the circle $\mathbf{r}(t) = (\cos t)\mathbf{i} + (\sin t)\mathbf{j}, 0 \le t \le 2\pi$ (Figure 16.19).

**Solution**    On the circle, $\mathbf{F} = (x - y)\mathbf{i} + x\mathbf{j} = (\cos t - \sin t)\mathbf{i} + (\cos t)\mathbf{j}$, and

$$\frac{d\mathbf{r}}{dt} = (-\sin t)\mathbf{i} + (\cos t)\mathbf{j}.$$

Then

$$\mathbf{F} \cdot \frac{d\mathbf{r}}{dt} = -\sin t \cos t + \underbrace{\sin^2 t + \cos^2 t}_{1}$$

gives

$$\text{Circulation} = \int_0^{2\pi} \mathbf{F} \cdot \frac{d\mathbf{r}}{dt} dt = \int_0^{2\pi} (1 - \sin t \cos t) \, dt$$

$$= \left[ t - \frac{\sin^2 t}{2} \right]_0^{2\pi} = 2\pi.$$

As Figure 16.19 suggests, a fluid with this velocity field is circulating *counterclockwise* around the circle, so the circulation is positive.    ∎

### Flux Across a Simple Closed Plane Curve

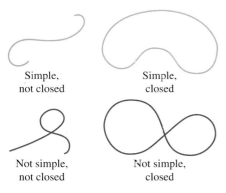

Simple,
not closed

Simple,
closed

Not simple,
not closed

Not simple,
closed

**FIGURE 16.20** Distinguishing curves that are simple or closed. Closed curves are also called loops.

A curve in the $xy$-plane is **simple** if it does not cross itself (Figure 16.20). When a curve starts and ends at the same point, it is a **closed curve** or **loop**. To find the rate at which a fluid is entering or leaving a region enclosed by a smooth simple closed curve $C$ in the $xy$-plane, we calculate the line integral over $C$ of $\mathbf{F} \cdot \mathbf{n}$, the scalar component of the fluid's velocity field in the direction of the curve's outward-pointing normal vector. We use only the normal component of $\mathbf{F}$, while ignoring the tangential component, because the normal component leads to the flow across $C$. The value of this integral is the *flux* of $\mathbf{F}$ across $C$. *Flux* is Latin for *flow*, but many flux calculations involve no motion at all. If $\mathbf{F}$ were an electric field or a magnetic field, for instance, the integral of $\mathbf{F} \cdot \mathbf{n}$ is still called the flux of the field across $C$.

> **DEFINITION**    If $C$ is a smooth simple closed curve in the domain of a continuous vector field $\mathbf{F} = M(x, y)\mathbf{i} + N(x, y)\mathbf{j}$ in the plane, and if $\mathbf{n}$ is the outward-pointing unit normal vector on $C$, the **flux** of $\mathbf{F}$ across $C$ is
>
> $$\text{Flux of } \mathbf{F} \text{ across } C = \int_C \mathbf{F} \cdot \mathbf{n} \, ds. \tag{6}$$

Notice the difference between flux and circulation. The flux of $\mathbf{F}$ across $C$ is the line integral with respect to arc length of $\mathbf{F} \cdot \mathbf{n}$, the scalar component of $\mathbf{F}$ in the direction of the outward normal. The circulation of $\mathbf{F}$ around $C$ is the line integral with respect to arc length of $\mathbf{F} \cdot \mathbf{T}$, the scalar component of $\mathbf{F}$ in the direction of the unit tangent vector. Flux is the integral of the normal component of $\mathbf{F}$; circulation is the integral of the tangential component of $\mathbf{F}$. In Section 16.6 we define flux across a surface.

To evaluate the integral for flux in Equation (6), we begin with a smooth parametrization

$$x = g(t), \qquad y = h(t), \qquad a \le t \le b,$$

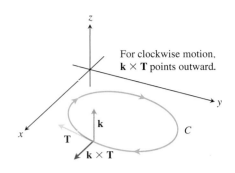

For clockwise motion, $\mathbf{k} \times \mathbf{T}$ points outward.

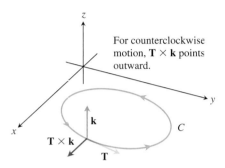

For counterclockwise motion, $\mathbf{T} \times \mathbf{k}$ points outward.

**FIGURE 16.21** To find an outward unit normal vector for a smooth simple curve $C$ in the $xy$-plane that is traversed counterclockwise as $t$ increases, we take $\mathbf{n} = \mathbf{T} \times \mathbf{k}$. For clockwise motion, we take $\mathbf{n} = \mathbf{k} \times \mathbf{T}$.

that traces the curve $C$ exactly once as $t$ increases from $a$ to $b$. We can find the outward unit normal vector $\mathbf{n}$ by crossing the curve's unit tangent vector $\mathbf{T}$ with the vector $\mathbf{k}$. But which order do we choose, $\mathbf{T} \times \mathbf{k}$ or $\mathbf{k} \times \mathbf{T}$? Which one points outward? It depends on which way $C$ is traversed as $t$ increases. If the motion is clockwise, $\mathbf{k} \times \mathbf{T}$ points outward; if the motion is counterclockwise, $\mathbf{T} \times \mathbf{k}$ points outward (Figure 16.21). The usual choice is $\mathbf{n} = \mathbf{T} \times \mathbf{k}$, the choice that assumes counterclockwise motion. Thus, although the value of the integral in Equation (6) does not depend on which way $C$ is traversed, the formulas we are about to derive for computing $\mathbf{n}$ and evaluating the integral assume counterclockwise motion.

In terms of components,

$$\mathbf{n} = \mathbf{T} \times \mathbf{k} = \left(\frac{dx}{ds}\mathbf{i} + \frac{dy}{ds}\mathbf{j}\right) \times \mathbf{k} = \frac{dy}{ds}\mathbf{i} - \frac{dx}{ds}\mathbf{j}.$$

If $\mathbf{F} = M(x, y)\mathbf{i} + N(x, y)\mathbf{j}$, then

$$\mathbf{F} \cdot \mathbf{n} = M(x, y)\frac{dy}{ds} - N(x, y)\frac{dx}{ds}.$$

Hence,

$$\int_C \mathbf{F} \cdot \mathbf{n} \, ds = \int_C \left(M\frac{dy}{ds} - N\frac{dx}{ds}\right) ds = \oint_C M \, dy - N \, dx.$$

We put a directed circle $\circlearrowleft$ on the last integral as a reminder that the integration around the closed curve $C$ is to be in the counterclockwise direction. To evaluate this integral, we express $M$, $dy$, $N$, and $dx$ in terms of the parameter $t$ and integrate from $t = a$ to $t = b$. We do not need to know $\mathbf{n}$ or $ds$ explicitly to find the flux.

---

**Calculating Flux Across a Smooth Closed Plane Curve**

$$(\text{Flux of } \mathbf{F} = M\mathbf{i} + N\mathbf{j} \text{ across } C) = \oint_C M \, dy - N \, dx \qquad (7)$$

The integral can be evaluated from any smooth parametrization $x = g(t)$, $y = h(t)$, $a \leq t \leq b$, that traces $C$ counterclockwise exactly once.

---

**EXAMPLE 8** Find the flux of $\mathbf{F} = (x - y)\mathbf{i} + x\mathbf{j}$ across the circle $x^2 + y^2 = 1$ in the $xy$-plane. (The vector field and curve were shown previously in Figure 16.19.)

**Solution** The parametrization $\mathbf{r}(t) = (\cos t)\mathbf{i} + (\sin t)\mathbf{j}$, $0 \leq t \leq 2\pi$, traces the circle counterclockwise exactly once. We can therefore use this parametrization in Equation (7). With

$$M = x - y = \cos t - \sin t, \qquad dy = d(\sin t) = \cos t \, dt$$
$$N = x = \cos t, \qquad dx = d(\cos t) = -\sin t \, dt,$$

we find

$$\text{Flux} = \oint_C M \, dy - N \, dx = \int_0^{2\pi} (\cos^2 t - \sin t \cos t + \cos t \sin t) \, dt \qquad \text{Eq. (7)}$$

$$= \int_0^{2\pi} \cos^2 t \, dt = \int_0^{2\pi} \frac{1 + \cos 2t}{2} \, dt = \left[\frac{t}{2} + \frac{\sin 2t}{4}\right]_0^{2\pi} = \pi.$$

The flux of $\mathbf{F}$ across the circle is $\pi$. Since the answer is positive, the net flow across the curve is outward. A net inward flow would have given a negative flux. ∎

# Exercises 16.2

### Vector Fields

Find the gradient fields of the functions in Exercises 1–4.

**1.** $f(x, y, z) = (x^2 + y^2 + z^2)^{-1/2}$

**2.** $f(x, y, z) = \ln\sqrt{x^2 + y^2 + z^2}$

**3.** $g(x, y, z) = e^z - \ln(x^2 + y^2)$

**4.** $g(x, y, z) = xy + yz + xz$

**5.** Give a formula $\mathbf{F} = M(x, y)\mathbf{i} + N(x, y)\mathbf{j}$ for the vector field in the plane that has the property that $\mathbf{F}$ points toward the origin with magnitude inversely proportional to the square of the distance from $(x, y)$ to the origin. (The field is not defined at $(0, 0)$.)

**6.** Give a formula $\mathbf{F} = M(x, y)\mathbf{i} + N(x, y)\mathbf{j}$ for the vector field in the plane that has the properties that $\mathbf{F} = \mathbf{0}$ at $(0, 0)$ and that at any other point $(a, b)$, $\mathbf{F}$ is tangent to the circle $x^2 + y^2 = a^2 + b^2$ and points in the clockwise direction with magnitude $|\mathbf{F}| = \sqrt{a^2 + b^2}$.

### Line Integrals of Vector Fields

In Exercises 7–12, find the line integrals of $\mathbf{F}$ from $(0, 0, 0)$ to $(1, 1, 1)$ over each of the following paths in the accompanying figure.

    **a.** The straight-line path $C_1$: $\mathbf{r}(t) = t\mathbf{i} + t\mathbf{j} + t\mathbf{k}$, $\quad 0 \le t \le 1$

    **b.** The curved path $C_2$: $\mathbf{r}(t) = t\mathbf{i} + t^2\mathbf{j} + t^4\mathbf{k}$, $\quad 0 \le t \le 1$

    **c.** The path $C_3 \cup C_4$ consisting of the line segment from $(0, 0, 0)$ to $(1, 1, 0)$ followed by the segment from $(1, 1, 0)$ to $(1, 1, 1)$

**7.** $\mathbf{F} = 3y\mathbf{i} + 2x\mathbf{j} + 4z\mathbf{k}$      **8.** $\mathbf{F} = [1/(x^2 + 1)]\mathbf{j}$

**9.** $\mathbf{F} = \sqrt{z}\,\mathbf{i} - 2x\mathbf{j} + \sqrt{y}\,\mathbf{k}$    **10.** $\mathbf{F} = xy\mathbf{i} + yz\mathbf{j} + xz\mathbf{k}$

**11.** $\mathbf{F} = (3x^2 - 3x)\mathbf{i} + 3z\mathbf{j} + \mathbf{k}$

**12.** $\mathbf{F} = (y + z)\mathbf{i} + (z + x)\mathbf{j} + (x + y)\mathbf{k}$

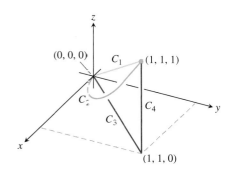

### Line Integrals with Respect to $x$, $y$, and $z$

In Exercises 13–16, find the line integrals along the given path $C$.

**13.** $\displaystyle\int_C (x - y)\,dx$, where $C$: $x = t, y = 2t + 1$, for $0 \le t \le 3$

**14.** $\displaystyle\int_C \frac{x}{y}\,dy$, where $C$: $x = t, y = t^2$, for $1 \le t \le 2$

**15.** $\displaystyle\int_C (x^2 + y^2)\,dy$, where $C$ is given in the accompanying figure

**16.** $\displaystyle\int_C \sqrt{x + y}\,dx$, where $C$ is given in the accompanying figure

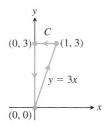

**17.** Along the curve $\mathbf{r}(t) = t\mathbf{i} - \mathbf{j} + t^2\mathbf{k}, 0 \le t \le 1$, evaluate each of the following integrals.

    **a.** $\displaystyle\int_C (x + y - z)\,dx$      **b.** $\displaystyle\int_C (x + y - z)\,dy$

    **c.** $\displaystyle\int_C (x + y - z)\,dz$

**18.** Along the curve $\mathbf{r}(t) = (\cos t)\mathbf{i} + (\sin t)\mathbf{j} - (\cos t)\mathbf{k}, 0 \le t \le \pi$, evaluate each of the following integrals.

    **a.** $\displaystyle\int_C xz\,dx$    **b.** $\displaystyle\int_C xz\,dy$    **c.** $\displaystyle\int_C xyz\,dz$

### Work

In Exercises 19–22, find the work done by $\mathbf{F}$ over the curve in the direction of increasing $t$.

**19.** $\mathbf{F} = xy\mathbf{i} + y\mathbf{j} - yz\mathbf{k}$
    $\mathbf{r}(t) = t\mathbf{i} + t^2\mathbf{j} + t\mathbf{k}, \quad 0 \le t \le 1$

**20.** $\mathbf{F} = 2y\mathbf{i} + 3x\mathbf{j} + (x + y)\mathbf{k}$
    $\mathbf{r}(t) = (\cos t)\mathbf{i} + (\sin t)\mathbf{j} + (t/6)\mathbf{k}, \quad 0 \le t \le 2\pi$

**21.** $\mathbf{F} = z\mathbf{i} + x\mathbf{j} + y\mathbf{k}$
    $\mathbf{r}(t) = (\sin t)\mathbf{i} + (\cos t)\mathbf{j} + t\mathbf{k}, \quad 0 \le t \le 2\pi$

**22.** $\mathbf{F} = 6z\mathbf{i} + y^2\mathbf{j} + 12x\mathbf{k}$
    $\mathbf{r}(t) = (\sin t)\mathbf{i} + (\cos t)\mathbf{j} + (t/6)\mathbf{k}, \quad 0 \le t \le 2\pi$

### Line Integrals in the Plane

**23.** Evaluate $\int_C xy\,dx + (x + y)\,dy$ along the curve $y = x^2$ from $(-1, 1)$ to $(2, 4)$.

**24.** Evaluate $\int_C (x - y)\,dx + (x + y)\,dy$ counterclockwise around the triangle with vertices $(0, 0)$, $(1, 0)$, and $(0, 1)$.

**25.** Evaluate $\int_C \mathbf{F} \cdot \mathbf{T}\,ds$ for the vector field $\mathbf{F} = x^2\mathbf{i} - y\mathbf{j}$ along the curve $x = y^2$ from $(4, 2)$ to $(1, -1)$.

**26.** Evaluate $\int_C \mathbf{F} \cdot d\mathbf{r}$ for the vector field $\mathbf{F} = y\mathbf{i} - x\mathbf{j}$ counterclockwise along the unit circle $x^2 + y^2 = 1$ from $(1, 0)$ to $(0, 1)$.

## Work, Circulation, and Flux in the Plane

**27. Work** Find the work done by the force $\mathbf{F} = xy\mathbf{i} + (y - x)\mathbf{j}$ over the straight line from $(1, 1)$ to $(2, 3)$.

**28. Work** Find the work done by the gradient of $f(x, y) = (x + y)^2$ counterclockwise around the circle $x^2 + y^2 = 4$ from $(2, 0)$ to itself.

**29. Circulation and flux** Find the circulation and flux of the fields

$$\mathbf{F}_1 = x\mathbf{i} + y\mathbf{j} \quad \text{and} \quad \mathbf{F}_2 = -y\mathbf{i} + x\mathbf{j}$$

around and across each of the following curves.

**a.** The circle $\mathbf{r}(t) = (\cos t)\mathbf{i} + (\sin t)\mathbf{j}, \quad 0 \le t \le 2\pi$

**b.** The ellipse $\mathbf{r}(t) = (\cos t)\mathbf{i} + (4 \sin t)\mathbf{j}, \quad 0 \le t \le 2\pi$

**30. Flux across a circle** Find the flux of the fields

$$\mathbf{F}_1 = 2x\mathbf{i} - 3y\mathbf{j} \quad \text{and} \quad \mathbf{F}_2 = 2x\mathbf{i} + (x - y)\mathbf{j}$$

across the circle

$$\mathbf{r}(t) = (a \cos t)\mathbf{i} + (a \sin t)\mathbf{j}, \quad 0 \le t \le 2\pi.$$

In Exercises 31–34, find the circulation and flux of the field $\mathbf{F}$ around and across the closed semicircular path that consists of the semicircular arch $\mathbf{r}_1(t) = (a \cos t)\mathbf{i} + (a \sin t)\mathbf{j}, 0 \le t \le \pi$, followed by the line segment $\mathbf{r}_2(t) = t\mathbf{i}, -a \le t \le a$.

**31.** $\mathbf{F} = x\mathbf{i} + y\mathbf{j}$      **32.** $\mathbf{F} = x^2\mathbf{i} + y^2\mathbf{j}$

**33.** $\mathbf{F} = -y\mathbf{i} + x\mathbf{j}$      **34.** $\mathbf{F} = -y^2\mathbf{i} + x^2\mathbf{j}$

**35. Flow integrals** Find the flow of the velocity field $\mathbf{F} = (x + y)\mathbf{i} - (x^2 + y^2)\mathbf{j}$ along each of the following paths from $(1, 0)$ to $(-1, 0)$ in the $xy$-plane.

**a.** The upper half of the circle $x^2 + y^2 = 1$

**b.** The line segment from $(1, 0)$ to $(-1, 0)$

**c.** The line segment from $(1, 0)$ to $(0, -1)$ followed by the line segment from $(0, -1)$ to $(-1, 0)$

**36. Flux across a triangle** Find the flux of the field $\mathbf{F}$ in Exercise 35 outward across the triangle with vertices $(1, 0), (0, 1), (-1, 0)$.

**37.** Find the flow of the velocity field $\mathbf{F} = y^2\mathbf{i} + 2xy\mathbf{j}$ along each of the following paths from $(0, 0)$ to $(2, 4)$.

**a.**

**b.**

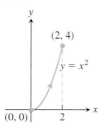

**c.** Use any path from $(0, 0)$ to $(2, 4)$ different from parts (a) and (b).

**38.** Find the circulation of the field $\mathbf{F} = y\mathbf{i} + (x + 2y)\mathbf{j}$ around each of the following closed paths.

**a.**

**b.**

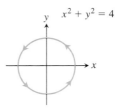

**c.** Use any closed path different from parts (a) and (b).

## Vector Fields in the Plane

**39. Spin field** Draw the spin field

$$\mathbf{F} = -\frac{y}{\sqrt{x^2 + y^2}}\mathbf{i} + \frac{x}{\sqrt{x^2 + y^2}}\mathbf{j}$$

(see Figure 16.12) along with its horizontal and vertical components at a representative assortment of points on the circle $x^2 + y^2 = 4$.

**40. Radial field** Draw the radial field

$$\mathbf{F} = x\mathbf{i} + y\mathbf{j}$$

(see Figure 16.11) along with its horizontal and vertical components at a representative assortment of points on the circle $x^2 + y^2 = 1$.

**41. A field of tangent vectors**

**a.** Find a field $\mathbf{G} = P(x, y)\mathbf{i} + Q(x, y)\mathbf{j}$ in the $xy$-plane with the property that at any point $(a, b) \ne (0, 0)$, $\mathbf{G}$ is a vector of magnitude $\sqrt{a^2 + b^2}$ tangent to the circle $x^2 + y^2 = a^2 + b^2$ and pointing in the counterclockwise direction. (The field is undefined at $(0, 0)$.)

**b.** How is $\mathbf{G}$ related to the spin field $\mathbf{F}$ in Figure 16.12?

**42. A field of tangent vectors**

**a.** Find a field $\mathbf{G} = P(x, y)\mathbf{i} + Q(x, y)\mathbf{j}$ in the $xy$-plane with the property that at any point $(a, b) \ne (0, 0)$, $\mathbf{G}$ is a unit vector tangent to the circle $x^2 + y^2 = a^2 + b^2$ and pointing in the clockwise direction.

**b.** How is $\mathbf{G}$ related to the spin field $\mathbf{F}$ in Figure 16.12?

**43. Unit vectors pointing toward the origin** Find a field $\mathbf{F} = M(x, y)\mathbf{i} + N(x, y)\mathbf{j}$ in the $xy$-plane with the property that at each point $(x, y) \ne (0, 0)$, $\mathbf{F}$ is a unit vector pointing toward the origin. (The field is undefined at $(0, 0)$.)

**44. Two "central" fields** Find a field $\mathbf{F} = M(x, y)\mathbf{i} + N(x, y)\mathbf{j}$ in the $xy$-plane with the property that at each point $(x, y) \ne (0, 0)$, $\mathbf{F}$ points toward the origin and $|\mathbf{F}|$ is **(a)** the distance from $(x, y)$ to the origin, **(b)** inversely proportional to the distance from $(x, y)$ to the origin. (The field is undefined at $(0, 0)$.)

**45. Work and area** Suppose that $f(t)$ is differentiable and positive for $a \le t \le b$. Let $C$ be the path $\mathbf{r}(t) = t\mathbf{i} + f(t)\mathbf{j}, a \le t \le b$, and $\mathbf{F} = y\mathbf{i}$. Is there any relation between the value of the work integral

$$\int_C \mathbf{F} \cdot d\mathbf{r}$$

and the area of the region bounded by the $t$-axis, the graph of $f$, and the lines $t = a$ and $t = b$? Give reasons for your answer.

**46. Work done by a radial force with constant magnitude** A particle moves along the smooth curve $y = f(x)$ from $(a, f(a))$ to

$(b, f(b))$. The force moving the particle has constant magnitude $k$ and always points away from the origin. Show that the work done by the force is

$$\int_C \mathbf{F} \cdot \mathbf{T} \, ds = k\left[ (b^2 + (f(b))^2)^{1/2} - (a^2 + (f(a))^2)^{1/2} \right].$$

**Flow Integrals in Space**

In Exercises 47–50, $\mathbf{F}$ is the velocity field of a fluid flowing through a region in space. Find the flow along the given curve in the direction of increasing $t$.

**47.** $\mathbf{F} = -4xy\mathbf{i} + 8y\mathbf{j} + 2\mathbf{k}$
$\mathbf{r}(t) = t\mathbf{i} + t^2\mathbf{j} + \mathbf{k}, \quad 0 \le t \le 2$

**48.** $\mathbf{F} = x^2\mathbf{i} + yz\mathbf{j} + y^2\mathbf{k}$
$\mathbf{r}(t) = 3t\mathbf{j} + 4t\mathbf{k}, \quad 0 \le t \le 1$

**49.** $\mathbf{F} = (x - z)\mathbf{i} + x\mathbf{k}$
$\mathbf{r}(t) = (\cos t)\mathbf{i} + (\sin t)\mathbf{k}, \quad 0 \le t \le \pi$

**50.** $\mathbf{F} = -y\mathbf{i} + x\mathbf{j} + 2\mathbf{k}$
$\mathbf{r}(t) = (-2\cos t)\mathbf{i} + (2\sin t)\mathbf{j} + 2t\mathbf{k}, \quad 0 \le t \le 2\pi$

**51. Circulation** Find the circulation of $\mathbf{F} = 2x\mathbf{i} + 2z\mathbf{j} + 2y\mathbf{k}$ around the closed path consisting of the following three curves traversed in the direction of increasing $t$.

$C_1: \mathbf{r}(t) = (\cos t)\mathbf{i} + (\sin t)\mathbf{j} + t\mathbf{k}, \quad 0 \le t \le \pi/2$
$C_2: \mathbf{r}(t) = \mathbf{j} + (\pi/2)(1 - t)\mathbf{k}, \quad 0 \le t \le 1$
$C_3: \mathbf{r}(t) = t\mathbf{i} + (1 - t)\mathbf{j}, \quad 0 \le t \le 1$

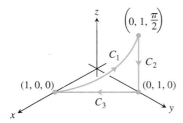

**52. Zero circulation** Let $C$ be the ellipse in which the plane $2x + 3y - z = 0$ meets the cylinder $x^2 + y^2 = 12$. Show, without evaluating either line integral directly, that the circulation of the field $\mathbf{F} = x\mathbf{i} + y\mathbf{j} + z\mathbf{k}$ around $C$ in either direction is zero.

**53. Flow along a curve** The field $\mathbf{F} = xy\mathbf{i} + y\mathbf{j} - yz\mathbf{k}$ is the velocity field of a flow in space. Find the flow from $(0, 0, 0)$ to

$(1, 1, 1)$ along the curve of intersection of the cylinder $y = x^2$ and the plane $z = x$. (*Hint:* Use $t = x$ as the parameter.)

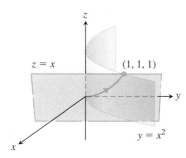

**54. Flow of a gradient field** Find the flow of the field $\mathbf{F} = \nabla(xy^2z^3)$:
a. Once around the curve $C$ in Exercise 52, clockwise as viewed from above
b. Along the line segment from $(1, 1, 1)$ to $(2, 1, -1)$.

**COMPUTER EXPLORATIONS**

In Exercises 55–60, use a CAS to perform the following steps for finding the work done by force $\mathbf{F}$ over the given path:
a. Find $d\mathbf{r}$ for the path $\mathbf{r}(t) = g(t)\mathbf{i} + h(t)\mathbf{j} + k(t)\mathbf{k}$.
b. Evaluate the force $\mathbf{F}$ along the path.
c. Evaluate $\displaystyle\int_C \mathbf{F} \cdot d\mathbf{r}$.

**55.** $\mathbf{F} = xy^6\mathbf{i} + 3x(xy^5 + 2)\mathbf{j}; \quad \mathbf{r}(t) = (2\cos t)\mathbf{i} + (\sin t)\mathbf{j},$
$0 \le t \le 2\pi$

**56.** $\mathbf{F} = \dfrac{3}{1 + x^2}\mathbf{i} + \dfrac{2}{1 + y^2}\mathbf{j}; \quad \mathbf{r}(t) = (\cos t)\mathbf{i} + (\sin t)\mathbf{j},$
$0 \le t \le \pi$

**57.** $\mathbf{F} = (y + yz\cos xyz)\mathbf{i} + (x^2 + xz\cos xyz)\mathbf{j} + (z + xy\cos xyz)\mathbf{k}; \quad \mathbf{r}(t) = (2\cos t)\mathbf{i} + (3\sin t)\mathbf{j} + \mathbf{k},$
$0 \le t \le 2\pi$

**58.** $\mathbf{F} = 2xy\mathbf{i} - y^2\mathbf{j} + ze^x\mathbf{k}; \quad \mathbf{r}(t) = -t\mathbf{i} + \sqrt{t}\mathbf{j} + 3t\mathbf{k},$
$1 \le t \le 4$

**59.** $\mathbf{F} = (2y + \sin x)\mathbf{i} + (z^2 + (1/3)\cos y)\mathbf{j} + x^4\mathbf{k}; \quad \mathbf{r}(t) = (\sin t)\mathbf{i} + (\cos t)\mathbf{j} + (\sin 2t)\mathbf{k}, \quad -\pi/2 \le t \le \pi/2$

**60.** $\mathbf{F} = (x^2y)\mathbf{i} + \dfrac{1}{3}x^3\mathbf{j} + xy\mathbf{k}; \quad \mathbf{r}(t) = (\cos t)\mathbf{i} + (\sin t)\mathbf{j} + (2\sin^2 t - 1)\mathbf{k}, \quad 0 \le t \le 2\pi$

# 16.3 Path Independence, Conservative Fields, and Potential Functions

A **gravitational field G** is a vector field that represents the effect of gravity at a point in space due to the presence of a massive object. The gravitational force on a body of mass $m$ placed in the field is given by $\mathbf{F} = m\mathbf{G}$. Similarly, an **electric field E** is a vector field in space that represents the effect of electric forces on a charged particle placed within it. The force on a body of charge $q$ placed in the field is given by $\mathbf{F} = q\mathbf{E}$. In gravitational and electric fields, the amount of work it takes to move a mass or charge from one point to another depends on the initial and final positions of the object—not on which path is taken between these positions. In this section we study vector fields with this property and the calculation of work integrals associated with them.

### Path Independence

If $A$ and $B$ are two points in an open region $D$ in space, the line integral of $\mathbf{F}$ along $C$ from $A$ to $B$ for a field $\mathbf{F}$ defined on $D$ usually depends on the path $C$ taken, as we saw in Section 16.1. For some special fields, however, the integral's value is the same for all paths from $A$ to $B$.

> **DEFINITIONS** Let $\mathbf{F}$ be a vector field defined on an open region $D$ in space, and suppose that for any two points $A$ and $B$ in $D$ the line integral $\int_C \mathbf{F} \cdot d\mathbf{r}$ along a path $C$ from $A$ to $B$ in $D$ is the same over all paths from $A$ to $B$. Then the integral $\int_C \mathbf{F} \cdot d\mathbf{r}$ is **path independent in $D$** and the field $\mathbf{F}$ is **conservative on $D$**.

The word *conservative* comes from physics, where it refers to fields in which the principle of conservation of energy holds. When a line integral is independent of the path $C$ from point $A$ to point $B$, we sometimes represent the integral by the symbol $\int_A^B$ rather than the usual line integral symbol $\int_C$. This substitution helps us remember the path-independence property.

Under differentiability conditions normally met in practice, we will show that a field $\mathbf{F}$ is conservative if and only if it is the gradient field of a scalar function $f$—that is, if and only if $\mathbf{F} = \nabla f$ for some $f$. The function $f$ then has a special name.

> **DEFINITION** If $\mathbf{F}$ is a vector field defined on $D$ and $\mathbf{F} = \nabla f$ for some scalar function $f$ on $D$, then $f$ is called a **potential function for $\mathbf{F}$**.

A gravitational potential is a scalar function whose gradient field is a gravitational field, an electric potential is a scalar function whose gradient field is an electric field, and so on. As we will see, once we have found a potential function $f$ for a field $\mathbf{F}$, we can evaluate all the line integrals in the domain of $\mathbf{F}$ over any path between $A$ and $B$ by

$$\int_A^B \mathbf{F} \cdot d\mathbf{r} = \int_A^B \nabla f \cdot d\mathbf{r} = f(B) - f(A). \tag{1}$$

If you think of $\nabla f$ for functions of several variables as analogous to the derivative $f'$ for functions of a single variable, then you see that Equation (1) is the vector calculus rendition of the Fundamental Theorem of Calculus formula (also called the Net Change Theorem)

$$\int_a^b f'(x)\, dx = f(b) - f(a).$$

Conservative fields have other important properties. For example, saying that $\mathbf{F}$ is conservative on $D$ is equivalent to saying that the integral of $\mathbf{F}$ around every closed path in $D$ is zero. Certain conditions on the curves, fields, and domains must be satisfied for Equation (1) to be valid. We discuss these conditions next.

### Assumptions on Curves, Vector Fields, and Domains

In order for the computations and results we derive below to be valid, we must assume certain properties for the curves, surfaces, domains, and vector fields we consider. We give these assumptions in the statements of theorems, and they also apply to the examples and exercises unless otherwise stated.

The curves we consider are **piecewise smooth**. Such curves are made up of finitely many smooth pieces connected end to end, as discussed in Section 13.1. We will treat vector fields $\mathbf{F}$ whose components have continuous first partial derivatives.

Simply connected

(a)

Simply connected

(b)

Not simply connected

(c)

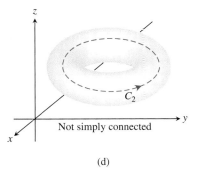

Not simply connected

(d)

**FIGURE 16.22** Four connected regions. In (a) and (b), the regions are simply connected. In (c) and (d), the regions are not simply connected because the curves $C_1$ and $C_2$ cannot be contracted to a point inside the regions containing them.

The domains $D$ we consider are **connected**. For an open region, this means that any two points in $D$ can be joined by a smooth curve that lies in the region. Some results also require $D$ to be **simply connected**, which means that every loop in $D$ can be contracted to a point in $D$ without ever leaving $D$. The plane with a disk removed is a two-dimensional region that is *not* simply connected; a loop in the plane that goes around the disk cannot be contracted to a point without going into the "hole" left by the removed disk (see Figure 16.22c). Similarly, if we remove a line from space, the remaining region $D$ is *not* simply connected. A curve encircling the line cannot be shrunk to a point while remaining inside $D$.

Connectivity and simple connectivity are not the same, and neither property implies the other. Think of connected regions as being in "one piece" and simply connected regions as not having any "loop-catching holes." All of space itself is both connected and simply connected. Figure 16.22 illustrates some of these properties.

**Caution** Some of the results in this chapter can fail to hold if applied to situations where the conditions we've imposed do not hold. In particular, the component test for conservative fields, given later in this section, is not valid on domains that are not simply connected (see Example 5). We do not always require that a domain be simply connected, so the condition will be stated when needed.

### Line Integrals in Conservative Fields

Gradient fields $\mathbf{F}$ are obtained by differentiating a scalar function $f$. A theorem analogous to the Fundamental Theorem of Calculus gives a way to evaluate the line integrals of gradient fields.

> **THEOREM 1—Fundamental Theorem of Line Integrals**  Let $C$ be a smooth curve joining the point $A$ to the point $B$ in the plane or in space and parametrized by $\mathbf{r}(t)$. Let $f$ be a differentiable function with a continuous gradient vector $\mathbf{F} = \nabla f$ on a domain $D$ containing $C$. Then
>
> $$\int_C \mathbf{F} \cdot d\mathbf{r} = f(B) - f(A).$$

Like the Fundamental Theorem, Theorem 1 gives a way to evaluate line integrals without having to take limits of Riemann sums or finding the line integral by the procedure used in Section 16.2. Before proving Theorem 1, we give an example.

**EXAMPLE 1**  Suppose the force field $\mathbf{F} = \nabla f$ is the gradient of the function

$$f(x, y, z) = -\frac{1}{x^2 + y^2 + z^2}.$$

Find the work done by $\mathbf{F}$ in moving an object along a smooth curve $C$ joining $(1, 0, 0)$ to $(0, 0, 2)$ that does not pass through the origin.

**Solution**  An application of Theorem 1 shows that the work done by $\mathbf{F}$ along any smooth curve $C$ joining the two points and not passing through the origin is

$$\int_C \mathbf{F} \cdot d\mathbf{r} = f(0, 0, 2) - f(1, 0, 0) = -\frac{1}{4} - (-1) = \frac{3}{4}.\qquad\blacksquare$$

The gravitational force due to a planet, and the electric force associated with a charged particle, can both be modeled by the field $\mathbf{F}$ given in Example 1 up to a constant that depends on the units of measurement.

**Proof of Theorem 1**    Suppose that $A$ and $B$ are two points in region $D$ and that $C: \mathbf{r}(t) = g(t)\mathbf{i} + h(t)\mathbf{j} + k(t)\mathbf{k}$, $a \le t \le b$, is a smooth curve in $D$ joining $A$ to $B$. In Section 14.5 we found that the derivative of a scalar function $f$ along a path $C$ is the dot product $\nabla f(\mathbf{r}(t)) \cdot \mathbf{r}'(t)$, so we have

$$
\begin{aligned}
\int_C \mathbf{F} \cdot d\mathbf{r} &= \int_A^B \nabla f \cdot d\mathbf{r} \qquad &\mathbf{F} = \nabla f \\
&= \int_{t=a}^{t=b} \nabla f(\mathbf{r}(t)) \cdot \mathbf{r}'(t)\, dt \\
&= \int_a^b \frac{d}{dt} f(\mathbf{r}(t))\, dt \qquad &\text{Eq. (7) of Section 14.5} \\
&= f(\mathbf{r}(b)) - f(\mathbf{r}(a)) \qquad &\text{Net Change Theorem} \\
&= f(B) - f(A). \qquad &\mathbf{r}(a) = A, \mathbf{r}(b) = B \qquad \blacksquare
\end{aligned}
$$

So we see from Theorem 1 that the line integral of a gradient field $\mathbf{F} = \nabla f$ is straightforward to compute once we know the function $f$. Many important vector fields arising in applications are indeed gradient fields. The next result, which follows from Theorem 1, shows that any conservative field is of this type.

---

**THEOREM 2—Conservative Fields are Gradient Fields**    Let $\mathbf{F} = M\mathbf{i} + N\mathbf{j} + P\mathbf{k}$ be a vector field whose components are continuous throughout an open connected region $D$ in space. Then $\mathbf{F}$ is conservative if and only if $\mathbf{F}$ is a gradient field $\nabla f$ for a differentiable function $f$.

---

Theorem 2 says that $\mathbf{F} = \nabla f$ if and only if for any two points $A$ and $B$ in the region $D$, the value of line integral $\int_C \mathbf{F} \cdot d\mathbf{r}$ is independent of the path $C$ joining $A$ to $B$ in $D$.

**Proof of Theorem 2**    If $\mathbf{F}$ is a gradient field, then $\mathbf{F} = \nabla f$ for a differentiable function $f$, and Theorem 1 shows that $\int_C \mathbf{F} \cdot d\mathbf{r} = f(B) - f(A)$. The value of the line integral does not depend on $C$, but only on its endpoints $A$ and $B$. So the line integral is path independent and $\mathbf{F}$ satisfies the definition of a conservative field.

On the other hand, suppose that $\mathbf{F}$ is a conservative vector field. We want to find a function $f$ on $D$ satisfying $\nabla f = \mathbf{F}$. First, pick a point $A$ in $D$ and set $f(A) = 0$. For any other point $B$ in $D$ define $f(B)$ to equal $\int_C \mathbf{F} \cdot d\mathbf{r}$, where $C$ is *any* smooth path in $D$ from $A$ to $B$. The value of $f(B)$ does not depend on the choice of $C$, since $\mathbf{F}$ is conservative. To show that $\nabla f = \mathbf{F}$ we need to demonstrate that $\partial f / \partial x = M$, $\partial f / \partial y = N$, and $\partial f / \partial z = P$.

Suppose that $B$ has coordinates $(x, y, z)$. By definition, the value of the function $f$ at a nearby point $B_0$ located at $(x_0, y, z)$ is $\int_{C_0} \mathbf{F} \cdot d\mathbf{r}$, where $C_0$ is any path from $A$ to $B_0$. We take a path $C = C_0 \cup L$ from $A$ to $B$ formed by first traveling along $C_0$ to arrive at $B_0$ and then traveling along the line segment $L$ from $B_0$ to $B$ (Figure 16.23). When $B_0$ is close to $B$, the segment $L$ lies in $D$ and, since the value $f(B)$ is independent of the path from $A$ to $B$,

$$
f(x, y, z) = \int_{C_0} \mathbf{F} \cdot d\mathbf{r} + \int_L \mathbf{F} \cdot d\mathbf{r}.
$$

Differentiating, we have

$$
\frac{\partial}{\partial x} f(x, y, z) = \frac{\partial}{\partial x} \left( \int_{C_0} \mathbf{F} \cdot d\mathbf{r} + \int_L \mathbf{F} \cdot d\mathbf{r} \right).
$$

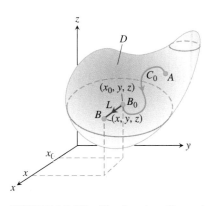

**FIGURE 16.23**    The function $f(x, y, z)$ in the proof of Theorem 2 is computed by a line integral $\int_{C_0} \mathbf{F} \cdot d\mathbf{r} = f(B_0)$ from $A$ to $B_0$, plus a line integral $\int_L \mathbf{F} \cdot d\mathbf{r}$ along a line segment $L$ parallel to the $x$-axis and joining $B_0$ to $B$ located at $(x, y, z)$. The value of $f$ at $A$ is $f(A) = 0$.

Only the last term on the right depends on $x$, so

$$\frac{\partial}{\partial x} f(x, y, z) = \frac{\partial}{\partial x} \int_{L} \mathbf{F} \cdot d\mathbf{r}.$$

Now parametrize $L$ as $\mathbf{r}(t) = t\mathbf{i} + y\mathbf{j} + z\mathbf{k}$, $x_0 \leq t \leq x$. Then $d\mathbf{r}/dt = \mathbf{i}$, $\mathbf{F} \cdot d\mathbf{r}/dt = M$, and $\int_{L} \mathbf{F} \cdot d\mathbf{r} = \int_{x_0}^{x} M(t, y, z) \, dt$. Differentiating then gives

$$\frac{\partial}{\partial x} f(x, y, z) = \frac{\partial}{\partial x} \int_{x_0}^{x} M(t, y, z) \, dt = M(x, y, z)$$

by the Fundamental Theorem of Calculus. The partial derivatives $\partial f / \partial y = N$ and $\partial f / \partial z = P$ follow similarly, showing that $\mathbf{F} = \nabla f$. ∎

**EXAMPLE 2**    Find the work done by the conservative field

$$\mathbf{F} = yz\mathbf{i} + xz\mathbf{j} + xy\mathbf{k} = \nabla f, \quad \text{where} \quad f(x, y, z) = xyz,$$

along any smooth curve $C$ joining the point $A(-1, 3, 9)$ to $B(1, 6, -4)$.

**Solution**    With $f(x, y, z) = xyz$, we have

$$
\begin{aligned}
\int_{C} \mathbf{F} \cdot d\mathbf{r} &= \int_{A}^{B} \nabla f \cdot d\mathbf{r} && \text{\scriptsize $\mathbf{F} = \nabla f$ and path} \\
&&& \text{\scriptsize independence} \\
&= f(B) - f(A) && \text{\scriptsize Theorem 1} \\
&= xyz\big|_{(1,6,-4)} - xyz\big|_{(-1,3,9)} \\
&= (1)(6)(-4) - (-1)(3)(9) \\
&= -24 + 27 = 3.
\end{aligned}
$$ ∎

A very useful property of line integrals in conservative fields comes into play when the path of integration is a closed curve, or loop. We often use the notation $\oint_{C}$ for integration around a closed path (discussed with more detail in the next section).

---

**THEOREM 3—Loop Property of Conservative Fields**    The following statements are equivalent.

**1.** $\oint_{C} \mathbf{F} \cdot d\mathbf{r} = 0$ around every loop (that is, closed curve $C$) in $D$.

**2.** The field $\mathbf{F}$ is conservative on $D$.

---

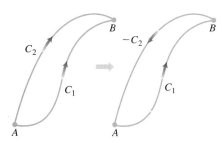

**FIGURE 16.24**  If we have two paths from $A$ to $B$, one of them can be reversed to make a loop.

**Proof that Part 1 ⇒ Part 2**    We want to show that for any two points $A$ and $B$ in $D$, the integral of $\mathbf{F} \cdot d\mathbf{r}$ has the same value over any two paths $C_1$ and $C_2$ from $A$ to $B$. We reverse the direction on $C_2$ to make a path $-C_2$ from $B$ to $A$ (Figure 16.24). Together, $C_1$ and $-C_2$ make a closed loop $C$, and by assumption,

$$\int_{C_1} \mathbf{F} \cdot d\mathbf{r} - \int_{C_2} \mathbf{F} \cdot d\mathbf{r} = \int_{C_1} \mathbf{F} \cdot d\mathbf{r} + \int_{-C_2} \mathbf{F} \cdot d\mathbf{r} = \int_{C} \mathbf{F} \cdot d\mathbf{r} = 0.$$

Thus, the integrals over $C_1$ and $C_2$ give the same value. Note that the definition of $\mathbf{F} \cdot d\mathbf{r}$ shows that changing the direction along a curve reverses the sign of the line integral.

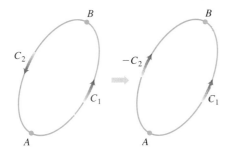

**FIGURE 16.25** If $A$ and $B$ lie on a loop, we can reverse part of the loop to make two paths from $A$ to $B$.

**Proof that Part 2 ⟹ Part 1**    We want to show that the integral of $\mathbf{F} \cdot d\mathbf{r}$ is zero over any closed loop $C$. We pick two points $A$ and $B$ on $C$ and use them to break $C$ into two pieces: $C_1$ from $A$ to $B$ followed by $C_2$ from $B$ back to $A$ (Figure 16.25). Then

$$\oint_C \mathbf{F} \cdot d\mathbf{r} = \int_{C_1} \mathbf{F} \cdot d\mathbf{r} + \int_{C_2} \mathbf{F} \cdot d\mathbf{r} = \int_A^B \mathbf{F} \cdot d\mathbf{r} - \int_A^B \mathbf{F} \cdot d\mathbf{r} = 0. \qquad ■$$

The following diagram summarizes the results of Theorems 2 and 3.

| | Theorem 2 | | Theorem 3 | |
|---|---|---|---|---|
| $\mathbf{F} = \nabla f$ on D | ⟺ | $\mathbf{F}$ conservative on D | ⟺ | $\oint_C \mathbf{F} \cdot d\mathbf{r} = 0$ over any loop in D |

Two questions arise:

1. How do we know whether a given vector field $\mathbf{F}$ is conservative?
2. If $\mathbf{F}$ is in fact conservative, how do we find a potential function $f$ (so that $\mathbf{F} = \nabla f$)?

### Finding Potentials for Conservative Fields

The test for a vector field being conservative involves the equivalence of certain first partial derivatives of the field components.

---

**Component Test for Conservative Fields**

Let $\mathbf{F} = M(x, y, z)\mathbf{i} + N(x, y, z)\mathbf{j} + P(x, y, z)\mathbf{k}$ be a field on an open simply connected domain whose component functions have continuous first partial derivatives. Then, $\mathbf{F}$ is conservative if and only if

$$\frac{\partial P}{\partial y} = \frac{\partial N}{\partial z}, \qquad \frac{\partial M}{\partial z} = \frac{\partial P}{\partial x}, \qquad \text{and} \qquad \frac{\partial N}{\partial x} = \frac{\partial M}{\partial y}. \qquad (2)$$

---

**Proof that Equations (2) hold if F is conservative**    There is a potential function $f$ such that

$$\mathbf{F} = M\mathbf{i} + N\mathbf{j} + P\mathbf{k} = \frac{\partial f}{\partial x}\mathbf{i} + \frac{\partial f}{\partial y}\mathbf{j} + \frac{\partial f}{\partial z}\mathbf{k}.$$

Hence,

$$\frac{\partial P}{\partial y} = \frac{\partial}{\partial y}\left(\frac{\partial f}{\partial z}\right) = \frac{\partial^2 f}{\partial y\, \partial z}$$

$$= \frac{\partial^2 f}{\partial z\, \partial y} \qquad \text{Mixed Derivative Theorem, Section 14.3}$$

$$= \frac{\partial}{\partial z}\left(\frac{\partial f}{\partial y}\right) = \frac{\partial N}{\partial z}.$$

The others in Equations (2) are proved similarly.    ■

The second half of the proof, that Equations (2) imply that $\mathbf{F}$ is conservative, is a consequence of Stokes' Theorem, taken up in Section 16.7, and requires our assumption that the domain of $\mathbf{F}$ be simply connected.

Once we know that $\mathbf{F}$ is conservative, we usually want to find a potential function for $\mathbf{F}$. This requires solving the equation $\nabla f = \mathbf{F}$ or

$$\frac{\partial f}{\partial x}\mathbf{i} + \frac{\partial f}{\partial y}\mathbf{j} + \frac{\partial f}{\partial z}\mathbf{k} = M\mathbf{i} + N\mathbf{j} + P\mathbf{k}$$

for $f$. We accomplish this by integrating the three equations

$$\frac{\partial f}{\partial x} = M, \qquad \frac{\partial f}{\partial y} = N, \qquad \frac{\partial f}{\partial z} = P,$$

as illustrated in the next example.

**EXAMPLE 3**   Show that $\mathbf{F} = (e^x \cos y + yz)\mathbf{i} + (xz - e^x \sin y)\mathbf{j} + (xy + z)\mathbf{k}$ is conservative over its natural domain and find a potential function for it.

**Solution**   The natural domain of $\mathbf{F}$ is all of space, which is open and simply connected. We apply the test in Equations (2) to

$$M = e^x \cos y + yz, \qquad N = xz - e^x \sin y, \qquad P = xy + z$$

and calculate

$$\frac{\partial P}{\partial y} = x = \frac{\partial N}{\partial z}, \qquad \frac{\partial M}{\partial z} = y = \frac{\partial P}{\partial x}, \qquad \frac{\partial N}{\partial x} = -e^x \sin y + z = \frac{\partial M}{\partial y}.$$

The partial derivatives are continuous, so these equalities tell us that $\mathbf{F}$ is conservative, so there is a function $f$ with $\nabla f = \mathbf{F}$ (Theorem 2).

We find $f$ by integrating the equations

$$\frac{\partial f}{\partial x} = e^x \cos y + yz, \qquad \frac{\partial f}{\partial y} = xz - e^x \sin y, \qquad \frac{\partial f}{\partial z} = xy + z. \tag{3}$$

We integrate the first equation with respect to $x$, holding $y$ and $z$ fixed, to get

$$f(x, y, z) = e^x \cos y + xyz + g(y, z).$$

We write the constant of integration as a function of $y$ and $z$ because its value may depend on $y$ and $z$, though not on $x$. We then calculate $\partial f / \partial y$ from this equation and match it with the expression for $\partial f / \partial y$ in Equations (3). This gives

$$-e^x \sin y + xz + \frac{\partial g}{\partial y} = xz - e^x \sin y,$$

so $\partial g / \partial y = 0$. Therefore, $g$ is a function of $z$ alone, and

$$f(x, y, z) = e^x \cos y + xyz + h(z).$$

We now calculate $\partial f / \partial z$ from this equation and match it to the formula for $\partial f / \partial z$ in Equations (3). This gives

$$xy + \frac{dh}{dz} = xy + z, \qquad \text{or} \qquad \frac{dh}{dz} = z,$$

so

$$h(z) = \frac{z^2}{2} + C.$$

Hence,

$$f(x, y, z) = e^x \cos y + xyz + \frac{z^2}{2} + C.$$

We found infinitely many potential functions of **F**, one for each value of $C$. ∎

**EXAMPLE 4**  Show that $\mathbf{F} = (2x - 3)\mathbf{i} - z\mathbf{j} + (\cos z)\mathbf{k}$ is not conservative.

**Solution**  We apply the Component Test in Equations (2) and find immediately that

$$\frac{\partial P}{\partial y} = \frac{\partial}{\partial y}(\cos z) = 0, \qquad \frac{\partial N}{\partial z} = \frac{\partial}{\partial z}(-z) = -1.$$

The two are unequal, so **F** is not conservative. No further testing is required. ∎

**EXAMPLE 5**  Show that the vector field

$$\mathbf{F} = \frac{-y}{x^2 + y^2}\mathbf{i} + \frac{x}{x^2 + y^2}\mathbf{j} + 0\mathbf{k}$$

satisfies the equations in the Component Test, but is not conservative over its natural domain. Explain why this is possible.

**Solution**  We have $M = -y/(x^2 + y^2)$, $N = x/(x^2 + y^2)$, and $P = 0$. If we apply the Component Test, we find

$$\frac{\partial P}{\partial y} = 0 = \frac{\partial N}{\partial z}, \qquad \frac{\partial P}{\partial x} = 0 = \frac{\partial M}{\partial z}, \qquad \text{and} \qquad \frac{\partial M}{\partial y} = \frac{y^2 - x^2}{(x^2 + y^2)^2} = \frac{\partial N}{\partial x}.$$

So it may appear that the field **F** passes the Component Test. However, the test assumes that the domain of **F** is simply connected, which is not the case here. Since $x^2 + y^2$ cannot equal zero, the natural domain is the complement of the $z$-axis and contains loops that cannot be contracted to a point. One such loop is the unit circle $C$ in the $xy$-plane. The circle is parametrized by $\mathbf{r}(t) = (\cos t)\mathbf{i} + (\sin t)\mathbf{j}, 0 \le t \le 2\pi$. This loop wraps around the $z$-axis and cannot be contracted to a point while staying within the complement of the $z$-axis.

To show that **F** is not conservative, we compute the line integral $\oint_C \mathbf{F} \cdot d\mathbf{r}$ around the loop $C$. First we write the field in terms of the parameter $t$:

$$\mathbf{F} = \frac{-y}{x^2 + y^2}\mathbf{i} + \frac{x}{x^2 + y^2}\mathbf{j} = \frac{-\sin t}{\sin^2 t + \cos^2 t}\mathbf{i} + \frac{\cos t}{\sin^2 t + \cos^2 t}\mathbf{j} = (-\sin t)\mathbf{i} + (\cos t)\mathbf{j}.$$

Next we find $d\mathbf{r}/dt = (-\sin t)\mathbf{i} + (\cos t)\mathbf{j}$, and then calculate the line integral as

$$\oint_C \mathbf{F} \cdot d\mathbf{r} = \oint_C \mathbf{F} \cdot \frac{d\mathbf{r}}{dt} dt = \int_0^{2\pi} (\sin^2 t + \cos^2 t)\, dt = 2\pi.$$

Since the line integral of **F** around the loop $C$ is not zero, the field **F** is not conservative, by Theorem 3. The field **F** is displayed in Figure 16.28d in the next section. ∎

Example 5 shows that the Component Test does not apply when the domain of the field is not simply connected. However, if we change the domain in the example so that it is restricted to the ball of radius 1 centered at the point (2, 2, 2), or to any similar ball-shaped region which does not contain a piece of the $z$-axis, then this new domain $D$ *is* simply connected. Now the partial derivative Equations (2), as well as all the assumptions of the Component Test, are satisfied. In this new situation, the field **F** in Example 5 is conservative on $D$.

Just as we must be careful with a function when determining if it satisfies a property throughout its domain (like continuity or the Intermediate Value Property), so must we also be careful with a vector field in determining the properties it may or may not have over its assigned domain.

## Exact Differential Forms

It is often convenient to express work and circulation integrals in the differential form

$$\int_C M\,dx + N\,dy + P\,dz$$

discussed in Section 16.2. Such line integrals are relatively easy to evaluate if $M\,dx + N\,dy + P\,dz$ is the total differential of a function $f$ and $C$ is any path joining the two points from $A$ to $B$. For then

$$\int_C M\,dx + N\,dy + P\,dz = \int_C \frac{\partial f}{\partial x}dx + \frac{\partial f}{\partial y}dy + \frac{\partial f}{\partial z}dz$$

$$= \int_A^B \nabla f \cdot d\mathbf{r} \qquad \nabla f \text{ is conservative.}$$

$$= f(B) - f(A). \qquad \text{Theorem 1}$$

Thus,

$$\int_A^B df = f(B) - f(A),$$

just as with differentiable functions of a single variable.

---

**DEFINITIONS**   Any expression $M(x, y, z)\,dx + N(x, y, z)\,dy + P(x, y, z)\,dz$ is a **differential form**. A differential form is **exact** on a domain $D$ in space if

$$M\,dx + N\,dy + P\,dz = \frac{\partial f}{\partial x}dx + \frac{\partial f}{\partial y}dy + \frac{\partial f}{\partial z}dz = df$$

for some scalar function $f$ throughout $D$.

---

Notice that if $M\,dx + N\,dy + P\,dz = df$ on $D$, then $\mathbf{F} = M\mathbf{i} + N\mathbf{j} + P\mathbf{k}$ is the gradient field of $f$ on $D$. Conversely, if $\mathbf{F} = \nabla f$, then the form $M\,dx + N\,dy + P\,dz$ is exact. The test for the form's being exact is therefore the same as the test for $\mathbf{F}$ being conservative.

---

**Component Test for Exactness of** $M\,dx + N\,dy + P\,dz$

The differential form $M\,dx + N\,dy + P\,dz$ is exact on an open simply connected domain if and only if

$$\frac{\partial P}{\partial y} = \frac{\partial N}{\partial z}, \qquad \frac{\partial M}{\partial z} = \frac{\partial P}{\partial x}, \qquad \text{and} \qquad \frac{\partial N}{\partial x} = \frac{\partial M}{\partial y}.$$

This is equivalent to saying that the field $\mathbf{F} = M\mathbf{i} + N\mathbf{j} + P\mathbf{k}$ is conservative.

---

**EXAMPLE 6**   Show that $y\,dx + x\,dy + 4\,dz$ is exact and evaluate the integral

$$\int_{(1,1,1)}^{(2,3,-1)} y\,dx + x\,dy + 4\,dz$$

over any path from $(1, 1, 1)$ to $(2, 3, -1)$.

**Solution**   We let $M = y, N = x, P = 4$ and apply the Test for Exactness:

$$\frac{\partial P}{\partial y} = 0 = \frac{\partial N}{\partial z}, \qquad \frac{\partial M}{\partial z} = 0 = \frac{\partial P}{\partial x}, \qquad \frac{\partial N}{\partial x} = 1 = \frac{\partial M}{\partial y}.$$

These equalities tell us that $y\,dx + x\,dy + 4\,dz$ is exact, so

$$y\,dx + x\,dy + 4\,dz = df$$

for some function $f$, and the integral's value is $f(2, 3, -1) - f(1, 1, 1)$.
We find $f$ up to a constant by integrating the equations

$$\frac{\partial f}{\partial x} = y, \qquad \frac{\partial f}{\partial y} = x, \qquad \frac{\partial f}{\partial z} = 4. \tag{4}$$

From the first equation we get

$$f(x, y, z) = xy + g(y, z).$$

The second equation tells us that

$$\frac{\partial f}{\partial y} = x + \frac{\partial g}{\partial y} = x, \qquad \text{or} \qquad \frac{\partial g}{\partial y} = 0.$$

Hence, $g$ is a function of $z$ alone, and

$$f(x, y, z) = xy + h(z).$$

The third of Equations (4) tells us that

$$\frac{\partial f}{\partial z} = 0 + \frac{dh}{dz} = 4, \qquad \text{or} \qquad h(z) = 4z + C.$$

Therefore,

$$f(x, y, z) = xy + 4z + C.$$

The value of the line integral is independent of the path taken from $(1, 1, 1)$ to $(2, 3, -1)$, and equals

$$f(2, 3, -1) - f(1, 1, 1) = 2 + C - (5 + C) = -3. \qquad \blacksquare$$

# Exercises 16.3

## Testing for Conservative Fields

Which fields in Exercises 1–6 are conservative, and which are not?

1. $\mathbf{F} = yz\mathbf{i} + xz\mathbf{j} + xy\mathbf{k}$

2. $\mathbf{F} = (y \sin z)\mathbf{i} + (x \sin z)\mathbf{j} + (xy\cos z)\mathbf{k}$

3. $\mathbf{F} = y\mathbf{i} + (x + z)\mathbf{j} - y\mathbf{k}$

4. $\mathbf{F} = -y\mathbf{i} + x\mathbf{j}$

5. $\mathbf{F} = (z + y)\mathbf{i} + z\mathbf{j} + (y + x)\mathbf{k}$

6. $\mathbf{F} = (e^x \cos y)\mathbf{i} - (e^x \sin y)\mathbf{j} + z\mathbf{k}$

## Finding Potential Functions

In Exercises 7–12, find a potential function $f$ for the field $\mathbf{F}$.

7. $\mathbf{F} = 2x\mathbf{i} + 3y\mathbf{j} + 4z\mathbf{k}$

8. $\mathbf{F} = (y + z)\mathbf{i} + (x + z)\mathbf{j} + (x + y)\mathbf{k}$

9. $\mathbf{F} = e^{y+2z}(\mathbf{i} + x\mathbf{j} + 2x\mathbf{k})$

10. $\mathbf{F} = (y \sin z)\mathbf{i} + (x \sin z)\mathbf{j} + (xy\cos z)\mathbf{k}$

11. $\mathbf{F} = (\ln x + \sec^2(x + y))\mathbf{i} +$

$$\left(\sec^2(x + y) + \frac{y}{y^2 + z^2}\right)\mathbf{j} + \frac{z}{y^2 + z^2}\mathbf{k}$$

12. $\mathbf{F} = \dfrac{y}{1 + x^2 y^2}\mathbf{i} + \left(\dfrac{x}{1 + x^2 y^2} + \dfrac{z}{\sqrt{1 - y^2 z^2}}\right)\mathbf{j} +$

$$\left(\dfrac{y}{\sqrt{1 - y^2 z^2}} + \dfrac{1}{z}\right)\mathbf{k}$$

### Exact Differential Forms

In Exercises 13–17, show that the differential forms in the integrals are exact. Then evaluate the integrals.

**13.** $\displaystyle\int_{(0,0,0)}^{(2,3,-6)} 2x\,dx + 2y\,dy + 2z\,dz$

**14.** $\displaystyle\int_{(1,1,2)}^{(3,5,0)} yz\,dx + xz\,dy + xy\,dz$

**15.** $\displaystyle\int_{(0,0,0)}^{(1,2,3)} 2xy\,dx + (x^2 - z^2)\,dy - 2yz\,dz$

**16.** $\displaystyle\int_{(0,0,0)}^{(3,3,1)} 2x\,dx - y^2\,dy - \frac{4}{1+z^2}\,dz$

**17.** $\displaystyle\int_{(1,0,0)}^{(0,1,1)} \sin y\cos x\,dx + \cos y\sin x\,dy + dz$

### Finding Potential Functions to Evaluate Line Integrals

Although they are not defined on all of space $R^3$, the fields associated with Exercises 18–22 are conservative. Find a potential function for each field and evaluate the integrals as in Example 6.

**18.** $\displaystyle\int_{(0,2,1)}^{(1,\pi/2,2)} 2\cos y\,dx + \left(\frac{1}{y} - 2x\sin y\right)dy + \frac{1}{z}\,dz$

**19.** $\displaystyle\int_{(1,1,1)}^{(1,2,3)} 3x^2\,dx + \frac{z^2}{y}\,dy + 2z\ln y\,dz$

**20.** $\displaystyle\int_{(1,2,1)}^{(2,1,1)} (2x\ln y - yz)\,dx + \left(\frac{x^2}{y} - xz\right)dy - xy\,dz$

**21.** $\displaystyle\int_{(1,1,1)}^{(2,2,2)} \frac{1}{y}\,dx + \left(\frac{1}{z} - \frac{x}{y^2}\right)dy - \frac{y}{z^2}\,dz$

**22.** $\displaystyle\int_{(-1,-1,-1)}^{(2,2,2)} \frac{2x\,dx + 2y\,dy + 2z\,dz}{x^2 + y^2 + z^2}$

### Applications and Examples

**23. Revisiting Example 6** Evaluate the integral

$$\int_{(1,1,1)}^{(2,3,-1)} y\,dx + x\,dy + 4\,dz$$

from Example 6 by finding parametric equations for the line segment from $(1, 1, 1)$ to $(2, 3, -1)$ and evaluating the line integral of $\mathbf{F} = y\mathbf{i} + x\mathbf{j} + 4\mathbf{k}$ along the segment. Since $\mathbf{F}$ is conservative, the integral is independent of the path.

**24.** Evaluate

$$\int_C x^2\,dx + yz\,dy + (y^2/2)\,dz$$

along the line segment $C$ joining $(0, 0, 0)$ to $(0, 3, 4)$.

**Independence of path** Show that the values of the integrals in Exercises 25 and 26 do not depend on the path taken from $A$ to $B$.

**25.** $\displaystyle\int_A^B z^2\,dx + 2y\,dy + 2xz\,dz$

**26.** $\displaystyle\int_A^B \frac{x\,dx + y\,dy + z\,dz}{\sqrt{x^2 + y^2 + z^2}}$

In Exercises 27 and 28, find a potential function for $\mathbf{F}$.

**27.** $\mathbf{F} = \dfrac{2x}{y}\mathbf{i} + \left(\dfrac{1-x^2}{y^2}\right)\mathbf{j}$,  $\{(x, y): y > 0\}$

**28.** $\mathbf{F} = (e^x \ln y)\mathbf{i} + \left(\dfrac{e^x}{y} + \sin z\right)\mathbf{j} + (y\cos z)\mathbf{k}$

**29. Work along different paths** Find the work done by $\mathbf{F} = (x^2 + y)\mathbf{i} + (y^2 + x)\mathbf{j} + ze^z\mathbf{k}$ over the following paths from $(1, 0, 0)$ to $(1, 0, 1)$.

   **a.** The line segment $x = 1, y = 0, 0 \le z \le 1$

   **b.** The helix $\mathbf{r}(t) = (\cos t)\mathbf{i} + (\sin t)\mathbf{j} + (t/2\pi)\mathbf{k}, 0 \le t \le 2\pi$

   **c.** The $x$-axis from $(1, 0, 0)$ to $(0, 0, 0)$ followed by the parabola $z = x^2, y = 0$ from $(0, 0, 0)$ to $(1, 0, 1)$

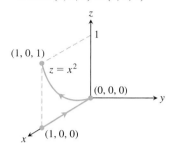

**30. Work along different paths** Find the work done by $\mathbf{F} = e^{yz}\mathbf{i} + (xze^{yz} + z\cos y)\mathbf{j} + (xye^{yz} + \sin y)\mathbf{k}$ over the following paths from $(1, 0, 1)$ to $(1, \pi/2, 0)$.

   **a.** The line segment $x = 1, y = \pi t/2, z = 1 - t, 0 \le t \le 1$

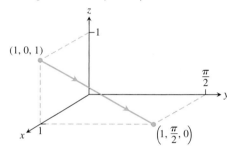

   **b.** The line segment from $(1, 0, 1)$ to the origin followed by the line segment from the origin to $(1, \pi/2, 0)$

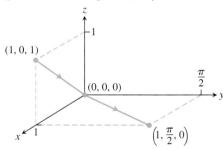

   **c.** The line segment from $(1, 0, 1)$ to $(1, 0, 0)$, followed by the $x$-axis from $(1, 0, 0)$ to the origin, followed by the parabola $y = \pi x^2/2, z = 0$ from there to $(1, \pi/2, 0)$

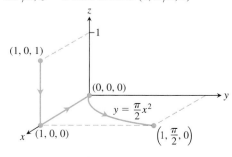

**31. Evaluating a work integral two ways** Let $\mathbf{F} = \nabla(x^3 y^2)$ and let $C$ be the path in the $xy$-plane from $(-1, 1)$ to $(1, 1)$ that consists of the line segment from $(-1, 1)$ to $(0, 0)$ followed by the line segment from $(0, 0)$ to $(1, 1)$. Evaluate $\int_C \mathbf{F} \cdot d\mathbf{r}$ in two ways.

   **a.** Find parametrizations for the segments that make up $C$ and evaluate the integral.

   **b.** Use $f(x, y) = x^3 y^2$ as a potential function for $\mathbf{F}$.

**32. Integral along different paths** Evaluate the line integral $\int_C 2x \cos y \, dx - x^2 \sin y \, dy$ along the following paths $C$ in the $xy$-plane.

   **a.** The parabola $y = (x - 1)^2$ from $(1, 0)$ to $(0, 1)$

   **b.** The line segment from $(-1, \pi)$ to $(1, 0)$

   **c.** The $x$-axis from $(-1, 0)$ to $(1, 0)$

   **d.** The astroid $\mathbf{r}(t) = (\cos^3 t)\mathbf{i} + (\sin^3 t)\mathbf{j}, 0 \le t \le 2\pi$, counterclockwise from $(1, 0)$ back to $(1, 0)$

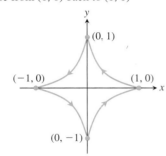

**33. a. Exact differential form** How are the constants $a$, $b$, and $c$ related if the following differential form is exact?

$$(ay^2 + 2czx) \, dx + y(bx + cz) \, dy + (ay^2 + cx^2) \, dz$$

   **b. Gradient field** For what values of $b$ and $c$ will

$$\mathbf{F} = (y^2 + 2czx)\mathbf{i} + y(bx + cz)\mathbf{j} + (y^2 + cx^2)\mathbf{k}$$

   be a gradient field?

**34. Gradient of a line integral** Suppose that $\mathbf{F} = \nabla f$ is a conservative vector field and

$$g(x, y, z) = \int_{(0,0,0)}^{(x,y,z)} \mathbf{F} \cdot d\mathbf{r}.$$

Show that $\nabla g = \mathbf{F}$.

**35. Path of least work** You have been asked to find the path along which a force field $\mathbf{F}$ will perform the least work in moving a particle between two locations. A quick calculation on your part shows $\mathbf{F}$ to be conservative. How should you respond? Give reasons for your answer.

**36. A revealing experiment** By experiment, you find that a force field $\mathbf{F}$ performs only half as much work in moving an object along path $C_1$ from $A$ to $B$ as it does in moving the object along path $C_2$ from $A$ to $B$. What can you conclude about $\mathbf{F}$? Give reasons for your answer.

**37. Work by a constant force** Show that the work done by a constant force field $\mathbf{F} = a\mathbf{i} + b\mathbf{j} + c\mathbf{k}$ in moving a particle along any path from $A$ to $B$ is $W = \mathbf{F} \cdot \overrightarrow{AB}$.

**38. Gravitational field**

   **a.** Find a potential function for the gravitational field

$$\mathbf{F} = -GmM \frac{x\mathbf{i} + y\mathbf{j} + z\mathbf{k}}{(x^2 + y^2 + z^2)^{3/2}}$$

   ($G$, $m$, and $M$ are constants).

   **b.** Let $P_1$ and $P_2$ be points at distance $s_1$ and $s_2$ from the origin. Show that the work done by the gravitational field in part (a) in moving a particle from $P_1$ to $P_2$ is

$$GmM\left(\frac{1}{s_2} - \frac{1}{s_1}\right).$$

# 16.4 Green's Theorem in the Plane

If $\mathbf{F}$ is a conservative field, then we know $\mathbf{F} = \nabla f$ for a differentiable function $f$, and we can calculate the line integral of $\mathbf{F}$ over any path $C$ joining point $A$ to $B$ as $\int_C \mathbf{F} \cdot d\mathbf{r} = f(B) - f(A)$. In this section we derive a method for computing a work or flux integral over a *closed* curve $C$ in the plane when the field $\mathbf{F}$ is *not* conservative. This method comes from Green's Theorem, which allows us to convert the line integral into a double integral over the region enclosed by $C$.

   The discussion is given in terms of velocity fields of fluid flows (a fluid is a liquid or a gas) because they are easy to visualize. However, Green's Theorem applies to any vector field, independent of any particular interpretation of the field, provided the assumptions of the theorem are satisfied. We introduce two new ideas for Green's Theorem: *circulation density* around an axis perpendicular to the plane and *divergence* (or *flux density*).

### Spin Around an Axis: The k-Component of Curl

Suppose that $\mathbf{F}(x, y) = M(x, y)\mathbf{i} + N(x, y)\mathbf{j}$ is the velocity field of a fluid flowing in the plane and that the first partial derivatives of $M$ and $N$ are continuous at each point of a region $R$. Let $(x, y)$ be a point in $R$ and let $A$ be a small rectangle with one corner at $(x, y)$ that, along with its interior, lies entirely in $R$. The sides of the rectangle, parallel to the coordinate axes, have lengths of $\Delta x$ and $\Delta y$. Assume that the components $M$ and $N$ do not

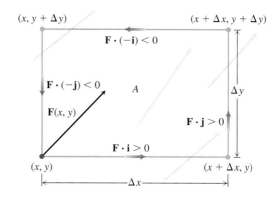

**FIGURE 16.26** The rate at which a fluid flows along the bottom edge of a rectangular region $A$ in the direction $\mathbf{i}$ is approximately $\mathbf{F}(x, y) \cdot \mathbf{i} \, \Delta x$, which is positive for the vector field $\mathbf{F}$ shown here. To approximate the rate of circulation at the point $(x, y)$, we calculate the (approximate) flow rates along each edge in the directions of the red arrows, sum these rates, and then divide the sum by the area of $A$. Taking the limit as $\Delta x \to 0$ and $\Delta y \to 0$ gives the rate of the circulation per unit area.

change sign throughout a small region containing the rectangle $A$. The first idea we use to convey Green's Theorem quantifies the rate at which a floating paddle wheel, with axis perpendicular to the plane, spins at a point in a fluid flowing in a plane region. This idea gives some sense of how the fluid is circulating around axes located at different points and perpendicular to the region. Physicists sometimes refer to this as the *circulation density* of a vector field $\mathbf{F}$ at a point. To obtain it, we consider the velocity field

$$\mathbf{F}(x, y) = M(x, y)\mathbf{i} + N(x, y)\mathbf{j}$$

and the rectangle $A$ in Figure 16.26 (where we assume both components of $\mathbf{F}$ are positive).

The circulation rate of $\mathbf{F}$ around the boundary of $A$ is the sum of flow rates along the sides in the tangential direction. For the bottom edge, the flow rate is approximately

$$\mathbf{F}(x, y) \cdot \mathbf{i} \, \Delta x = M(x, y)\Delta x.$$

This is the scalar component of the velocity $\mathbf{F}(x, y)$ in the tangent direction $\mathbf{i}$ times the length of the segment. The flow rates may be positive or negative depending on the components of $\mathbf{F}$. We approximate the net circulation rate around the rectangular boundary of $A$ by summing the flow rates along the four edges as defined by the following dot products.

|        |   |
|--------|---|
| Top:    | $\mathbf{F}(x, y + \Delta y) \cdot (-\mathbf{i}) \, \Delta x = -M(x, y + \Delta y)\Delta x$ |
| Bottom: | $\mathbf{F}(x, y) \cdot \mathbf{i} \, \Delta x = M(x, y)\Delta x$ |
| Right:  | $\mathbf{F}(x + \Delta x, y) \cdot \mathbf{j} \, \Delta y = N(x + \Delta x, y)\Delta y$ |
| Left:   | $\mathbf{F}(x, y) \cdot (-\mathbf{j}) \, \Delta y = -N(x, y)\Delta y$ |

We sum opposite pairs to get

Top and bottom:
$$-(M(x, y + \Delta y) - M(x, y))\Delta x \approx -\left(\frac{\partial M}{\partial y}\Delta y\right)\Delta x$$

Right and left:
$$(N(x + \Delta x, y) - N(x, y))\Delta y \approx \left(\frac{\partial N}{\partial x}\Delta x\right)\Delta y.$$

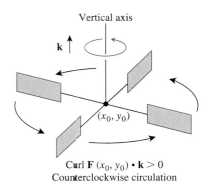

Curl $\mathbf{F}\,(x_0, y_0) \cdot \mathbf{k} > 0$
Counterclockwise circulation

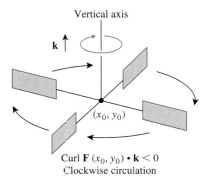

Curl $\mathbf{F}\,(x_0, y_0) \cdot \mathbf{k} < 0$
Clockwise circulation

**FIGURE 16.27** In the flow of an incompressible fluid over a plane region, the **k**-component of the curl measures the rate of the fluid's rotation at a point. The **k**-component of the curl is positive at points where the rotation is counterclockwise and negative where the rotation is clockwise.

Adding these last two equations gives the net circulation rate relative to the counterclockwise orientation,

$$\text{Circulation rate around rectangle} \approx \left( \frac{\partial N}{\partial x} - \frac{\partial M}{\partial y} \right) \Delta x\, \Delta y.$$

We now divide by $\Delta x\, \Delta y$ to estimate the circulation rate per unit area or *circulation density* for the rectangle:

$$\frac{\text{Circulation around rectangle}}{\text{rectangle area}} \approx \frac{\partial N}{\partial x} - \frac{\partial M}{\partial y}.$$

We let $\Delta x$ and $\Delta y$ approach zero to define the *circulation density* of $\mathbf{F}$ at the point $(x, y)$.

If we see a counterclockwise rotation looking downward onto the $xy$-plane from the tip of the unit $\mathbf{k}$ vector, then the circulation density is positive (Figure 16.27). The value of the circulation density is the **k**-component of a more general circulation vector field we define in Section 16.7, called the *curl* of the vector field $\mathbf{F}$. For Green's Theorem, we need only this **k**-component, obtained by taking the dot product of curl $\mathbf{F}$ with $\mathbf{k}$.

---

**DEFINITION** The **circulation density** of a vector field $\mathbf{F} = M\mathbf{i} + N\mathbf{j}$ at the point $(x, y)$ is the scalar expression

$$\frac{\partial N}{\partial x} - \frac{\partial M}{\partial y}. \qquad (1)$$

This expression is also called **the k-component of the curl**, denoted by $(\text{curl } \mathbf{F}) \cdot \mathbf{k}$.

---

If water is moving about a region in the $xy$-plane in a thin layer, then the **k**-component of the curl at a point $(x_0, y_0)$ gives a way to measure how fast and in what direction a small paddle wheel spins if it is put into the water at $(x_0, y_0)$ with its axis perpendicular to the plane, parallel to $\mathbf{k}$ (Figure 16.27). Looking downward onto the $xy$-plane, it spins counterclockwise when $(\text{curl } \mathbf{F}) \cdot \mathbf{k}$ is positive and clockwise when the **k**-component is negative.

**EXAMPLE 1** The following vector fields represent the velocity of a gas flowing in the $xy$-plane. Find the circulation density of each vector field and interpret its physical meaning Figure 16.28 displays the vector fields.

**(a)** *Uniform expansion or compression:* $\mathbf{F}(x, y) = cx\mathbf{i} + cy\mathbf{j}$

**(b)** *Uniform rotation:* $\mathbf{F}(x, y) = -cy\mathbf{i} + cx\mathbf{j}$

**(c)** *Shearing flow:* $\mathbf{F}(x, y) = y\mathbf{i}$

**(d)** *Whirlpool effect:* $\mathbf{F}(x, y) = \dfrac{-y}{x^2 + y^2}\mathbf{i} + \dfrac{x}{x^2 + y^2}\mathbf{j}$

**Solution**

**(a)** *Uniform expansion:* $(\text{curl } \mathbf{F}) \cdot \mathbf{k} = \dfrac{\partial}{\partial k}(cy) - \dfrac{\partial}{\partial y}(cx) = 0$. The gas is not circulating at very small scales.

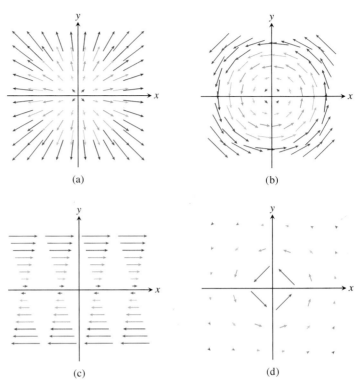

**FIGURE 16.28**  Velocity fields of a gas flowing in the plane (Example 1).

**(b)** *Rotation:* $(\text{curl } \mathbf{F}) \cdot \mathbf{k} = \dfrac{\partial}{\partial x}(cx) - \dfrac{\partial}{\partial y}(-cy) = 2c$. The constant circulation density indicates rotation at every point. If $c > 0$, the rotation is counterclockwise; if $c < 0$, the rotation is clockwise.

**(c)** *Shear:* $(\text{curl } \mathbf{F}) \cdot \mathbf{k} = -\dfrac{\partial}{\partial y}(y) = -1$. The circulation density is constant and negative, so a paddle wheel floating in water undergoing such a shearing flow spins clockwise. The rate of rotation is the same at each point. The average effect of the fluid flow is to push fluid clockwise around each of small circles shown in Figure 16.29.

**(d)** *Whirlpool:*

$$(\text{curl } \mathbf{F}) \cdot \mathbf{k} = \frac{\partial}{\partial x}\left(\frac{x}{x^2 + y^2}\right) - \frac{\partial}{\partial y}\left(\frac{-y}{x^2 + y^2}\right) = \frac{y^2 - x^2}{(x^2 + y^2)^2} - \frac{y^2 - x^2}{(x^2 + y^2)^2} = 0.$$

The circulation density is 0 at every point away from the origin (where the vector field is undefined and the whirlpool effect is taking place), and the gas is not circulating at any point for which the vector field is defined.  ∎

One form of Green's Theorem tells us how circulation density can be used to calculate the line integral for flow in the *xy*-plane. (The flow integral was defined in Section 16.2.) A second form of the theorem tells us how we can calculate the flux integral from *flux density*. We define this idea next, and then we present both versions of the theorem.

### Divergence

Consider again the velocity field $\mathbf{F}(x, y) = M(x, y)\mathbf{i} + N(x, y)\mathbf{j}$ in a domain containing the rectangle $A$, as shown in Figure 16.30. As before, we assume the field components do not change sign throughout a small region containing the rectangle $A$. Our interest now is to determine the rate at which the fluid leaves $A$.

**FIGURE 16.29**  A shearing flow pushes the fluid clockwise around each point (Example 1c).

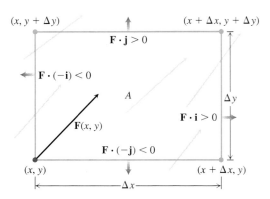

**FIGURE 16.30** The rate at which the fluid leaves the rectangular region $A$ across the bottom edge in the direction of the outward normal $-\mathbf{j}$ is approximately $\mathbf{F}(x, y) \cdot (-\mathbf{j})\, \Delta x$, which is negative for the vector field $\mathbf{F}$ shown here. To approximate the flow rate at the point $(x, y)$, we calculate the (approximate) flow rates across each edge in the directions of the red arrows, sum these rates, and then divide the sum by the area of $A$. Taking the limit as $\Delta x \to 0$ and $\Delta y \to 0$ gives the flow rate per unit area.

The rate at which fluid leaves the rectangle across the bottom edge is approximately (Figure 16.30)

$$\mathbf{F}(x, y) \cdot (-\mathbf{j})\, \Delta x = -N(x, y)\Delta x.$$

This is the scalar component of the velocity at $(x, y)$ in the direction of the outward normal times the length of the segment. If the velocity is in meters per second, for example, the flow rate will be in meters per second times meters or square meters per second. The rates at which the fluid crosses the other three sides in the directions of their outward normals can be estimated in a similar way. The flow rates may be positive or negative depending on the signs of the components of $\mathbf{F}$. We approximate the net flow rate across the rectangular boundary of $A$ by summing the flow rates across the four edges as defined by the following dot products.

**Fluid Flow Rates:**    Top:    $\mathbf{F}(x, y + \Delta y) \cdot \mathbf{j}\, \Delta x = N(x, y + \Delta y)\Delta x$

Bottom:    $\mathbf{F}(x, y) \cdot (-\mathbf{j})\, \Delta x = -N(x, y)\Delta x$

Right:    $\mathbf{F}(x + \Delta x, y) \cdot \mathbf{i}\, \Delta y = M(x + \Delta x, y)\Delta y$

Left:    $\mathbf{F}(x, y) \cdot (-\mathbf{i})\, \Delta y = -M(x, y)\Delta y$

Summing opposite pairs gives

Top and bottom:    $(N(x, y + \Delta y) - N(x, y))\Delta x \approx \left(\dfrac{\partial N}{\partial y}\Delta y\right)\Delta x$

Right and left:    $(M(x + \Delta x, y) - M(x, y))\Delta y \approx \left(\dfrac{\partial M}{\partial x}\Delta x\right)\Delta y.$

Adding these last two equations gives the net effect of the flow rates, or the

$$\text{Flux across rectangle boundary} \approx \left(\frac{\partial M}{\partial x} + \frac{\partial N}{\partial y}\right)\Delta x \Delta y.$$

We now divide by $\Delta x \Delta y$ to estimate the total flux per unit area or *flux density* for the rectangle:

$$\frac{\text{Flux across rectangle boundary}}{\text{rectangle area}} \approx \left( \frac{\partial M}{\partial x} + \frac{\partial N}{\partial y} \right).$$

Finally, we let $\Delta x$ and $\Delta y$ approach zero to define the flux density of $\mathbf{F}$ at the point $(x, y)$. In mathematics, we call the flux density the *divergence* of $\mathbf{F}$. The symbol for it is div $\mathbf{F}$, pronounced "divergence of $\mathbf{F}$" or "div $\mathbf{F}$."

**Source:** div $\mathbf{F}(x_0, y_0) > 0$

A gas expanding at the point $(x_0, y_0)$

**Sink:** div $\mathbf{F}(x_0, y_0) < 0$

A gas compressing at the point $(x_0, y_0)$

**FIGURE 16.31** If a gas is expanding at a point $(x_0, y_0)$, the lines of flow have positive divergence; if the gas is compressing, the divergence is negative.

---

> **DEFINITION** The **divergence (flux density)** of a vector field $\mathbf{F} = M\mathbf{i} + N\mathbf{j}$ at the point $(x, y)$ is
>
> $$\text{div } \mathbf{F} = \frac{\partial M}{\partial x} + \frac{\partial N}{\partial y}. \tag{2}$$

---

A gas is compressible, unlike a liquid, and the divergence of its velocity field measures to what extent it is expanding or compressing at each point. Intuitively, if a gas is expanding at the point $(x_0, y_0)$, the lines of flow would diverge there (hence the name) and, since the gas would be flowing out of a small rectangle about $(x_0, y_0)$, the divergence of $\mathbf{F}$ at $(x_0, y_0)$ would be positive. If the gas were compressing instead of expanding, the divergence would be negative (Figure 16.31).

**EXAMPLE 2** Find the divergence, and interpret what it means, for each vector field in Example 1 representing the velocity of a gas flowing in the $xy$-plane.

**Solution**

**(a)** div $\mathbf{F} = \frac{\partial}{\partial x}(cx) + \frac{\partial}{\partial y}(cy) = 2c$: If $c > 0$, the gas is undergoing uniform expansion; if $c < 0$, it is undergoing uniform compression.

**(b)** div $\mathbf{F} = \frac{\partial}{\partial x}(-cy) + \frac{\partial}{\partial y}(cx) = 0$: The gas is neither expanding nor compressing.

**(c)** div $\mathbf{F} = \frac{\partial}{\partial x}(y) = 0$: The gas is neither expanding nor compressing.

**(d)** div $\mathbf{F} = \frac{\partial}{\partial x}\left(\frac{-y}{x^2 + y^2}\right) + \frac{\partial}{\partial y}\left(\frac{x}{x^2 + y^2}\right) = \frac{2xy}{(x^2 + y^2)^2} - \frac{2xy}{(x^2 + y^2)^2} = 0$: Again, the divergence is zero at all points in the domain of the velocity field. ∎

Cases (b), (c), and (d) of Figure 16.28 are plausible models for the two-dimensional flow of a liquid. In fluid dynamics, when the velocity field of a flowing liquid always has divergence equal to zero, as in those cases, the liquid is said to be **incompressible**.

### Two Forms for Green's Theorem

We introduced the notation $\oint_C$ in Section 16.3 for integration around a closed curve. We elaborate further on the notation here. A simple closed curve $C$ can be traversed in two possible directions. (Recall that a curve is simple if it does not cross itself.) The curve is traversed counterclockwise, and said to be *positively oriented*, if the region it encloses is always to the left of an object as it moves along the path. Otherwise it is traversed clockwise and *negatively oriented*. The line integral of a vector field **F** along *C* reverses sign if we change the orientation. We use the notation

$$\oint_C \mathbf{F}(x, y) \cdot d\mathbf{r}$$

for the line integral when the simple closed curve *C* is traversed counterclockwise, with its positive orientation.

In one form, Green's Theorem says that the counterclockwise circulation of a vector field around a simple closed curve is the double integral of the **k**-component of the curl of the field over the region enclosed by the curve. Recall the defining Equation (5) for circulation in Section 16.2.

---

**THEOREM 4—Green's Theorem (Circulation-Curl or Tangential Form)** Let *C* be a piecewise smooth, simple closed curve enclosing a region *R* in the plane. Let **F** = *M***i** + *N***j** be a vector field with *M* and *N* having continuous first partial derivatives in an open region containing *R*. Then the counterclockwise circulation of **F** around *C* equals the double integral of (curl **F**) · **k** over *R*.

$$\underbrace{\oint_C \mathbf{F} \cdot \mathbf{T} \, ds = \oint_C M \, dx + N \, dy}_{\text{Counterclockwise circulation}} = \underbrace{\iint_R \left( \frac{\partial N}{\partial x} - \frac{\partial M}{\partial y} \right) dx \, dy}_{\text{Curl integral}} \qquad (3)$$

---

A second form of Green's Theorem says that the outward flux of a vector field across a simple closed curve in the plane equals the double integral of the divergence of the field over the region enclosed by the curve. Recall the formulas for flux in Equations (6) and (7) in Section 16.2.

---

**THEOREM 5—Green's Theorem (Flux-Divergence or Normal Form)** Let *C* be a piecewise smooth, simple closed curve enclosing a region *R* in the plane. Let **F** = *M***i** + *N***j** be a vector field with *M* and *N* having continuous first partial derivatives in an open region containing *R*. Then the outward flux of **F** across *C* equals the double integral of div **F** over the region *R* enclosed by *C*.

$$\underbrace{\oint_C \mathbf{F} \cdot \mathbf{n} \, ds = \oint_C M \, dy - N \, dx}_{\text{Outward flux}} = \underbrace{\iint_R \left( \frac{\partial M}{\partial x} + \frac{\partial N}{\partial y} \right) dx \, dy}_{\text{Divergence integral}} \qquad (4)$$

---

The two forms of Green's Theorem are equivalent. Applying Equation (3) to the field **G**₁ = −*N***i** + *M***j** gives Equation (4), and applying Equation (4) to **G**₂ = *N***i** − *M***j** gives Equation (3).

Both forms of Green's Theorem can be viewed as two-dimensional generalizations of the Net Change Theorem in Section 5.4. The counterclockwise circulation of **F** around *C*, defined by the line integral on the left-hand side of Equation (3), is the integral of its rate of change (circulation density) over the region $R$ enclosed by $C$, which is the double integral on the right-hand side of Equation (3). Likewise, the outward flux of **F** across $C$, defined by the line integral on the left-hand side of Equation (4), is the integral of its rate of change (flux density) over the region $R$ enclosed by $C$, which is the double integral on the right-hand side of Equation (4).

**EXAMPLE 3**    Verify both forms of Green's Theorem for the vector field

$$\mathbf{F}(x, y) = (x - y)\mathbf{i} + x\mathbf{j}$$

and the region $R$ bounded by the unit circle

$$C: \quad \mathbf{r}(t) = (\cos t)\mathbf{i} + (\sin t)\mathbf{j}, \quad 0 \le t \le 2\pi.$$

**Solution**    Evaluating $\mathbf{F}(\mathbf{r}(t))$ and computing the partial derivatives of the components of $\mathbf{F}$, we have

$$M = \cos t - \sin t, \qquad dx = d(\cos t) = -\sin t \, dt,$$
$$N = \cos t, \qquad dy = d(\sin t) = \cos t \, dt,$$
$$\frac{\partial M}{\partial x} = 1, \qquad \frac{\partial M}{\partial y} = -1, \qquad \frac{\partial N}{\partial x} = 1, \qquad \frac{\partial N}{\partial y} = 0.$$

The two sides of Equation (3) are

$$\oint_C M \, dx + N \, dy = \int_{t=0}^{t=2\pi} (\cos t - \sin t)(-\sin t \, dt) + (\cos t)(\cos t \, dt)$$

$$= \int_0^{2\pi} (-\sin t \cos t + 1) \, dt = 2\pi$$

$$\iint_R \left( \frac{\partial N}{\partial x} - \frac{\partial M}{\partial y} \right) dx \, dy = \iint_R (1 - (-1)) \, dx \, dy$$

$$= 2 \iint_R dx \, dy = 2(\text{area inside the unit circle}) = 2\pi.$$

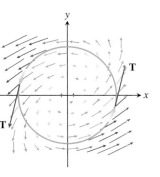

**FIGURE 16.32**    The vector field in Example 3 has a counterclockwise circulation of $2\pi$ around the unit circle.

Figure 16.32 displays the vector field and circulation around $C$.

The two sides of Equation (4) are

$$\oint_C M \, dy - N \, dx = \int_{t=0}^{t=2\pi} (\cos t - \sin t)(\cos t \, dt) - (\cos t)(-\sin t \, dt)$$

$$= \int_0^{2\pi} \cos^2 t \, dt = \pi$$

$$\iint_R \left( \frac{\partial M}{\partial x} + \frac{\partial N}{\partial y} \right) dx \, dy = \iint_R (1 + 0) \, dx \, dy$$

$$= \iint_R dx \, dy = \pi. \qquad \blacksquare$$

### Using Green's Theorem to Evaluate Line Integrals

If we construct a closed curve $C$ by piecing together a number of different curves end to end, the process of evaluating a line integral over $C$ can be lengthy because there are so many different integrals to evaluate. If $C$ bounds a region $R$ to which Green's Theorem applies, however, we can use Green's Theorem to change the line integral around $C$ into one double integral over $R$.

**EXAMPLE 4**     Evaluate the line integral

$$\oint_C xy \, dy - y^2 \, dx,$$

where $C$ is the square cut from the first quadrant by the lines $x = 1$ and $y = 1$.

**Solution**   We can use either form of Green's Theorem to change the line integral into a double integral over the square, where $C$ is the square's boundary and $R$ is its interior.

1. *With the Tangential Form* Equation (3): Taking $M = -y^2$ and $N = xy$ gives the result:

$$\oint_C -y^2 \, dx + xy \, dy = \iint_R (y - (-2y)) \, dx \, dy = \int_0^1 \int_0^1 3y \, dx \, dy$$

$$= \int_0^1 \left[ 3xy \right]_{x=0}^{x=1} dy = \int_0^1 3y \, dy = \frac{3}{2}y^2 \Big]_0^1 = \frac{3}{2}.$$

2. *With the Normal Form* Equation (4): Taking $M = xy, N = y^2$, gives the same result:

$$\oint_C xy \, dy - y^2 \, dx = \iint_R (y + 2y) \, dx \, dy = \frac{3}{2}.$$ ∎

**EXAMPLE 5**     Calculate the outward flux of the vector field $\mathbf{F}(x, y) = 2e^{xy}\mathbf{i} + y^3\mathbf{j}$ across the square bounded by the lines $x = \pm 1$ and $y = \pm 1$.

**Solution**   Calculating the flux with a line integral would take four integrations, one for each side of the square. With Green's Theorem, we can change the line integral to one double integral. With $M = 2e^{xy}, N = y^3$, $C$ the square, and $R$ the square's interior, we have

$$\text{Flux} = \oint_C \mathbf{F} \cdot \mathbf{n} \, ds = \oint_C M \, dy - N \, dx$$

$$= \iint_R \left( \frac{\partial M}{\partial x} + \frac{\partial N}{\partial y} \right) dx \, dy \qquad \text{Green's Theorem, Eq. (4)}$$

$$= \int_{-1}^1 \int_{-1}^1 (2ye^{xy} + 3y^2) \, dx \, dy = \int_{-1}^1 \left[ 2e^{xy} + 3xy^2 \right]_{x=-1}^{x=1} dy$$

$$= \int_{-1}^1 (2e^y + 6y^2 - 2e^{-y}) \, dy = \left[ 2e^y + 2y^3 + 2e^{-y} \right]_{-1}^1 = 4.$$ ∎

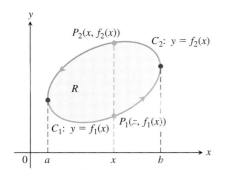

**FIGURE 16.33** The boundary curve $C$ is made up of $C_1$, the graph of $y = f_1(x)$, and $C_2$, the graph of $y = f_2(x)$.

### Proof of Green's Theorem for Special Regions

Let $C$ be a smooth simple closed curve in the $xy$-plane with the property that lines parallel to the axes cut it at no more than two points. Let $R$ be the region enclosed by $C$ and suppose that $M$, $N$, and their first partial derivatives are continuous at every point of some open region containing $C$ and $R$. We want to prove the circulation-curl form of Green's Theorem,

$$\oint_C M\,dx + N\,dy = \iint_R \left( \frac{\partial N}{\partial x} - \frac{\partial M}{\partial y} \right) dx\,dy. \tag{5}$$

Figure 16.33 shows $C$ made up of two directed parts:

$$C_1: \quad y = f_1(x), \quad a \le x \le b, \qquad C_2: \quad y = f_2(x), \quad b \ge x \ge a.$$

For any $x$ between $a$ and $b$, we can integrate $\partial M/\partial y$ with respect to $y$ from $y = f_1(x)$ to $y = f_2(x)$ and obtain

$$\int_{f_1(x)}^{f_2(x)} \frac{\partial M}{\partial y}\,dy = M(x, y)\Big]_{y=f_1(x)}^{y=f_2(x)} = M(x, f_2(x)) - M(x, f_1(x)).$$

We can then integrate this with respect to $x$ from $a$ to $b$:

$$\int_a^b \int_{f_1(x)}^{f_2(x)} \frac{\partial M}{\partial y}\,dy\,dx = \int_a^b \left[ M(x, f_2(x)) - M(x, f_1(x)) \right] dx$$

$$= -\int_b^a M(x, f_2(x))\,dx - \int_a^b M(x, f_1(x))\,dx$$

$$= -\int_{C_2} M\,dx - \int_{C_1} M\,dx$$

$$= -\oint_C M\,dx.$$

Therefore, reversing the order of the equations, we have

$$\oint_C M\,dx = \iint_R \left( -\frac{\partial M}{\partial y} \right) dx\,dy. \tag{6}$$

Equation (6) is half the result we need for Equation (5). We derive the other half by integrating $\partial N/\partial x$ first with respect to $x$ and then with respect to $y$, as suggested by Figure 16.34. This shows the curve $C$ of Figure 16.33 decomposed into the two directed parts $C_1': x = g_1(y), d \ge y \ge c$ and $C_2': x = g_2(y), c \le y \le d$. The result of this double integration is

$$\oint_C N\,dy = \iint_R \frac{\partial N}{\partial x}\,dx\,dy. \tag{7}$$

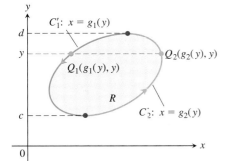

**FIGURE 16.34** The boundary curve $C$ is made up of $C_1'$, the graph of $x = g_1(y)$, and $C_2'$, the graph of $x = g_2(y)$.

Summing Equations (6) and (7) gives Equation (5). This concludes the proof. ∎

Green's Theorem also holds for more general regions, such as those shown in Figure 16.35, but we will not prove this result here. Notice that the region in Figure 16.35(c) is not simply connected. The curves $C_1$ and $C_h$ on its boundary are oriented so that the region $R$ is always on the left-hand side as the curves are traversed in the directions shown, and cancelation occurs over common boundary arcs traversed in opposite directions. With this convention, Green's Theorem is valid for regions that are not simply connected.

(a)

(b)

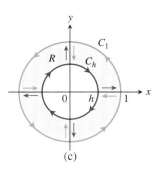
(c)

**FIGURE 16.35**   Other regions to which Green's Theorem applies. In (c) the axes convert the region into four simply connected regions, and we sum the line integrals along the oriented boundaries.

# Exercises 16.4

### Verifying Green's Theorem

In Exercises 1–4, verify the conclusion of Green's Theorem by evaluating both sides of Equations (3) and (4) for the field $\mathbf{F} = M\mathbf{i} + N\mathbf{j}$. Take the domains of integration in each case to be the disk $R: x^2 + y^2 \le a^2$ and its bounding circle $C: \mathbf{r} = (a\cos t)\mathbf{i} + (a\sin t)\mathbf{j}, 0 \le t \le 2\pi$.

**1.** $\mathbf{F} = -y\mathbf{i} + x\mathbf{j}$      **2.** $\mathbf{F} = y\mathbf{i}$

**3.** $\mathbf{F} = 2x\mathbf{i} - 3y\mathbf{j}$      **4.** $\mathbf{F} = -x^2y\mathbf{i} + xy^2\mathbf{j}$

### Circulation and Flux

In Exercises 5–14, use Green's Theorem to find the counterclockwise circulation and outward flux for the field $\mathbf{F}$ and curve $C$.

**5.** $\mathbf{F} = (x - y)\mathbf{i} + (y - x)\mathbf{j}$

   $C$: The square bounded by $x = 0, x = 1, y = 0, y = 1$

**6.** $\mathbf{F} = (x^2 + 4y)\mathbf{i} + (x + y^2)\mathbf{j}$

   $C$: The square bounded by $x = 0, x = 1, y = 0, y = 1$

**7.** $\mathbf{F} = (y^2 - x^2)\mathbf{i} + (x^2 + y^2)\mathbf{j}$

   $C$: The triangle bounded by $y = 0, x = 3$, and $y = x$

**8.** $\mathbf{F} = (x + y)\mathbf{i} - (x^2 + y^2)\mathbf{j}$

   $C$: The triangle bounded by $y = 0, x = 1$, and $y = x$

**9.** $\mathbf{F} = (xy + y^2)\mathbf{i} + (x - y)\mathbf{j}$    **10.** $\mathbf{F} = (x + 3y)\mathbf{i} + (2x - y)\mathbf{j}$

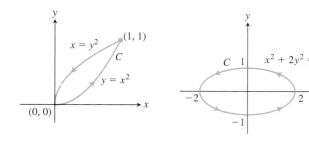

**11.** $\mathbf{F} = x^3y^2\mathbf{i} + \dfrac{1}{2}x^4y\,\mathbf{j}$      **12.** $\mathbf{F} = \dfrac{x}{1 + y^2}\mathbf{i} + \left(\tan^{-1} y\right)\mathbf{j}$

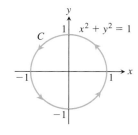

**13.** $\mathbf{F} = (x + e^x \sin y)\mathbf{i} + (x + e^x \cos y)\mathbf{j}$

   $C$: The right-hand loop of the lemniscate $r^2 = \cos 2\theta$

**14.** $\mathbf{F} = \left(\tan^{-1}\dfrac{y}{x}\right)\mathbf{i} + \ln(x^2 + y^2)\mathbf{j}$

   $C$: The boundary of the region defined by the polar coordinate inequalities $1 \le r \le 2, 0 \le \theta \le \pi$

**15.** Find the counterclockwise circulation and outward flux of the field $\mathbf{F} = xy\mathbf{i} + y^2\mathbf{j}$ around and over the boundary of the region enclosed by the curves $y = x^2$ and $y = x$ in the first quadrant.

**16.** Find the counterclockwise circulation and the outward flux of the field $\mathbf{F} = (-\sin y)\mathbf{i} + (x\cos y)\mathbf{j}$ around and over the square cut from the first quadrant by the lines $x = \pi/2$ and $y = \pi/2$.

**17.** Find the outward flux of the field

$$\mathbf{F} = \left(3xy - \frac{x}{1 + y^2}\right)\mathbf{i} + (e^x + \tan^{-1} y)\mathbf{j}$$

across the cardioid $r = a(1 + \cos\theta), a > 0$.

**18.** Find the counterclockwise circulation of $\mathbf{F} = (y + e^x \ln y)\mathbf{i} + (e^x/y)\mathbf{j}$ around the boundary of the region that is bounded above by the curve $y = 3 - x^2$ and below by the curve $y = x^4 + 1$.

**Work**

In Exercises 19 and 20, find the work done by $\mathbf{F}$ in moving a particle once counterclockwise around the given curve.

**19.** $\mathbf{F} = 2xy^3\mathbf{i} + 4x^2y^2\mathbf{j}$

$C$: The boundary of the "triangular" region in the first quadrant enclosed by the $x$-axis, the line $x = 1$, and the curve $y = x^3$

**20.** $\mathbf{F} = (4x - 2y)\mathbf{i} + (2x - 4y)\mathbf{j}$

$C$: The circle $(x - 2)^2 + (y - 2)^2 = 4$

**Using Green's Theorem**

Apply Green's Theorem to evaluate the integrals in Exercises 21–24.

**21.** $\oint_C (y^2 \, dx + x^2 \, dy)$

$C$: The triangle bounded by $x = 0, x + y = 1, y = 0$

**22.** $\oint_C (3y \, dx + 2x \, dy)$

$C$: The boundary of $0 \le x \le \pi, 0 \le y \le \sin x$

**23.** $\oint_C (6y + x) \, dx + (y + 2x) \, dy$

$C$: The circle $(x - 2)^2 + (y - 3)^2 = 4$

**24.** $\oint_C (2x + y^2) \, dx + (2xy + 3y) \, dy$

$C$: Any simple closed curve in the plane for which Green's Theorem holds

**Calculating Area with Green's Theorem** If a simple closed curve $C$ in the plane and the region $R$ it encloses satisfy the hypotheses of Green's Theorem, the area of $R$ is given by

> **Green's Theorem Area Formula**
>
> $$\text{Area of } R = \frac{1}{2} \oint_C x \, dy - y \, dx$$

The reason is that by Equation (4), run backward,

$$\text{Area of } R = \iint_R dy \, dx = \iint_R \left( \frac{1}{2} + \frac{1}{2} \right) dy \, dx$$

$$= \oint_C \frac{1}{2} x \, dy - \frac{1}{2} y \, dx.$$

Use the Green's Theorem area formula given above to find the areas of the regions enclosed by the curves in Exercises 25–28.

**25.** The circle $\mathbf{r}(t) = (a \cos t)\mathbf{i} + (a \sin t)\mathbf{j}, \quad 0 \le t \le 2\pi$

**26.** The ellipse $\mathbf{r}(t) = (a \cos t)\mathbf{i} + (b \sin t)\mathbf{j}, \quad 0 \le t \le 2\pi$

**27.** The astroid $\mathbf{r}(t) = (\cos^3 t)\mathbf{i} + (\sin^3 t)\mathbf{j}, \quad 0 \le t \le 2\pi$

**28.** One arch of the cycloid $x = t - \sin t, \quad y = 1 - \cos t$

**29.** Let $C$ be the boundary of a region on which Green's Theorem holds. Use Green's Theorem to calculate

**a.** $\oint_C f(x) \, dx + g(y) \, dy$

**b.** $\oint_C ky \, dx + hx \, dy \quad (k \text{ and } h \text{ constants}).$

**30. Integral dependent only on area** Show that the value of

$$\oint_C xy^2 \, dx + (x^2y + 2x) \, dy$$

around any square depends only on the area of the square and not on its location in the plane.

**31.** Evaluate the integral

$$\oint_C 4x^3y \, dx + x^4 \, dy$$

for any closed path $C$.

**32.** Evaluate the integral

$$\oint_C -y^3 \, dy + x^3 \, dx$$

for any closed path $C$.

**33. Area as a line integral** Show that if $R$ is a region in the plane bounded by a piecewise smooth, simple closed curve $C$, then

$$\text{Area of } R = \oint_C x \, dy = -\oint_C y \, dx.$$

**34. Definite integral as a line integral** Suppose that a nonnegative function $y = f(x)$ has a continuous first derivative on $[a, b]$. Let $C$ be the boundary of the region in the $xy$-plane that is bounded below by the $x$-axis, above by the graph of $f$, and on the sides by the lines $x = a$ and $x = b$. Show that

$$\int_a^b f(x) \, dx = -\oint_C y \, dx.$$

**35. Area and the centroid** Let $A$ be the area and $\bar{x}$ the $x$-coordinate of the centroid of a region $R$ that is bounded by a piecewise smooth, simple closed curve $C$ in the $xy$-plane. Show that

$$\frac{1}{2} \oint_C x^2 \, dy = -\oint_C xy \, dx = \frac{1}{3} \oint_C x^2 \, dy - xy \, dx = A\bar{x}.$$

**36. Moment of inertia** Let $I_y$ be the moment of inertia about the $y$-axis of the region in Exercise 35. Show that

$$\frac{1}{3} \oint_C x^3 \, dy = -\oint_C x^2y \, dx = \frac{1}{4} \oint_C x^3 \, dy - x^2y \, dx = I_y.$$

**37. Green's Theorem and Laplace's equation** Assuming that all the necessary derivatives exist and are continuous, show that if $f(x, y)$ satisfies the Laplace equation

$$\frac{\partial^2 f}{\partial x^2} + \frac{\partial^2 f}{\partial y^2} = 0,$$

then

$$\oint_C \frac{\partial f}{\partial y}\,dx - \frac{\partial f}{\partial x}\,dy = 0$$

for all closed curves $C$ to which Green's Theorem applies. (The converse is also true: If the line integral is always zero, then $f$ satisfies the Laplace equation.)

**38. Maximizing work**  Among all smooth, simple closed curves in the plane, oriented counterclockwise, find the one along which the work done by

$$\mathbf{F} = \left(\frac{1}{4}x^2y + \frac{1}{3}y^3\right)\mathbf{i} + x\mathbf{j}$$

is greatest. (*Hint:* Where is (curl **F**) · **k** positive?)

**39. Regions with many holes**  Green's Theorem holds for a region $R$ with any finite number of holes as long as the bounding curves are smooth, simple, and closed and we integrate over each component of the boundary in the direction that keeps $R$ on our immediate left as we go along (see accompanying figure).

**a.** Let $f(x, y) = \ln(x^2 + y^2)$ and let $C$ be the circle $x^2 + y^2 = a^2$. Evaluate the flux integral

$$\oint_C \nabla f \cdot \mathbf{n}\,ds.$$

**b.** Let $K$ be an arbitrary smooth, simple closed curve in the plane that does not pass through $(0, 0)$. Use Green's Theorem to show that

$$\oint_K \nabla f \cdot \mathbf{n}\,ds$$

has two possible values, depending on whether $(0, 0)$ lies inside $K$ or outside $K$.

**40. Bendixson's criterion**  The *streamlines* of a planar fluid flow are the smooth curves traced by the fluid's individual particles. The vectors $\mathbf{F} = M(x, y)\mathbf{i} + N(x, y)\mathbf{j}$ of the flow's velocity field are the tangent vectors of the streamlines. Show that if the flow takes place over a simply connected region $R$ (no holes or missing points) and that if $M_x + N_y \neq 0$ throughout $R$, then none of the streamlines in $R$ is closed. In other words, no particle of fluid ever has a closed trajectory in $R$. The criterion $M_x + N_y \neq 0$ is called **Bendixson's criterion** for the nonexistence of closed trajectories.

**41.** Establish Equation (7) to finish the proof of the special case of Green's Theorem.

**42. Curl component of conservative fields**  Can anything be said about the curl component of a conservative two-dimensional vector field? Give reasons for your answer.

**COMPUTER EXPLORATIONS**

In Exercises 43–46, use a CAS and Green's Theorem to find the counterclockwise circulation of the field **F** around the simple closed curve $C$. Perform the following CAS steps.

**a.** Plot $C$ in the $xy$-plane.

**b.** Determine the integrand $(\partial N/\partial x) - (\partial M/\partial y)$ for the tangential form of Green's Theorem.

**c.** Determine the (double integral) limits of integration from your plot in part (a) and evaluate the curl integral for the circulation.

**43.** $\mathbf{F} = (2x - y)\mathbf{i} + (x + 3y)\mathbf{j},$   $C$: The ellipse $x^2 + 4y^2 = 4$

**44.** $\mathbf{F} = (2x^3 - y^3)\mathbf{i} + (x^3 + y^3)\mathbf{j},$   $C$: The ellipse $\dfrac{x^2}{4} + \dfrac{y^2}{9} = 1$

**45.** $\mathbf{F} = x^{-1}e^y\mathbf{i} + (e^y \ln x + 2x)\mathbf{j},$
$C$: The boundary of the region defined by $y = 1 + x^4$ (below) and $y = 2$ (above)

**46.** $\mathbf{F} = xe^y\mathbf{i} + (4x^2 \ln y)\mathbf{j},$
$C$: The triangle with vertices $(0, 0)$, $(2, 0)$, and $(0, 4)$

# 16.5  Surfaces and Area

We have defined curves in the plane in three different ways:

Explicit form:                 $y = f(x)$

Implicit form:                 $F(x, y) = 0$

Parametric vector form:    $\mathbf{r}(t) = f(t)\mathbf{i} + g(t)\mathbf{j},$   $a \le t \le b.$

We have analogous definitions of surfaces in space:

Explicit form:   $z = f(x, y)$

Implicit form:   $F(x, y, z) = 0.$

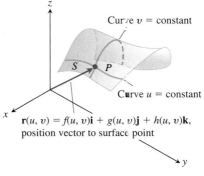

**FIGURE 16.36** A parametrized surface $S$ expressed as a vector function of two variables defined on a region $R$.

There is also a parametric form for surfaces that gives the position of a point on the surface as a vector function of two variables. We discuss this new form in this section and apply the form to obtain the area of a surface as a double integral. Double integral formulas for areas of surfaces given in implicit and explicit forms are then obtained as special cases of the more general parametric formula.

## Parametrizations of Surfaces

Suppose

$$\mathbf{r}(u, v) = f(u, v)\mathbf{i} + g(u, v)\mathbf{j} + h(u, v)\mathbf{k} \qquad (1)$$

is a continuous vector function that is defined on a region $R$ in the $uv$-plane and one-to-one on the interior of $R$ (Figure 16.36). We call the range of $\mathbf{r}$ the **surface** $S$ defined or traced by $\mathbf{r}$. Equation (1) together with the domain $R$ constitutes a **parametrization** of the surface. The variables $u$ and $v$ are the **parameters**, and $R$ is the **parameter domain**. To simplify our discussion, we take $R$ to be a rectangle defined by inequalities of the form $a \le u \le b$, $c \le v \le d$. The requirement that $\mathbf{r}$ be one-to-one on the interior of $R$ ensures that $S$ does not cross itself. Notice that Equation (1) is the vector equivalent of *three* parametric equations:

$$x = f(u, v), \qquad y = g(u, v), \qquad z = h(u, v).$$

**EXAMPLE 1**     Find a parametrization of the cone

$$z = \sqrt{x^2 + y^2}, \qquad 0 \le z \le 1.$$

**Solution**    Here, cylindrical coordinates provide a parametrization. A typical point $(x, y, z)$ on the cone (Figure 16.37) has $x = r\cos\theta$, $y = r\sin\theta$, and $z = \sqrt{x^2 + y^2} = r$, with $0 \le r \le 1$ and $0 \le \theta \le 2\pi$. Taking $u = r$ and $v = \theta$ in Equation (1) gives the parametrization

$$\mathbf{r}(r, \theta) = (r\cos\theta)\mathbf{i} + (r\sin\theta)\mathbf{j} + r\mathbf{k}, \qquad 0 \le r \le 1, \quad 0 \le \theta \le 2\pi.$$

The parametrization is one-to-one on the interior of the domain $R$, though not on the boundary tip of its cone where $r = 0$. ∎

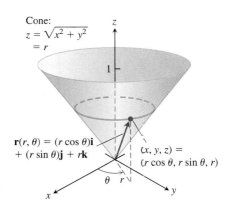

**FIGURE 16.37** The cone in Example 1 can be parametrized using cylindrical coordinates.

**EXAMPLE 2**     Find a parametrization of the sphere $x^2 + y^2 + z^2 = a^2$.

**Solution**    Spherical coordinates provide what we need. A typical point $(x, y, z)$ on the sphere (Figure 16.38) has $x = a\sin\phi\cos\theta$, $y = a\sin\phi\sin\theta$, and $z = a\cos\phi$, $0 \le \phi \le \pi$, $0 \le \theta \le 2\pi$. Taking $u = \phi$ and $v = \theta$ in Equation (1) gives the parametrization

$$\mathbf{r}(\phi, \theta) = (a\sin\phi\cos\theta)\mathbf{i} + (a\sin\phi\sin\theta)\mathbf{j} + (a\cos\phi)\mathbf{k},$$
$$0 \le \phi \le \pi, \quad 0 \le \theta \le 2\pi.$$

Again, the parametrization is one-to-one on the interior of the domain $R$, though not on its boundary "poles" where $\phi = 0$ or $\phi = \pi$. ∎

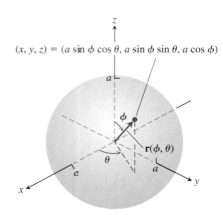

$(x, y, z) = (a \sin \phi \cos \theta, a \sin \phi \sin \theta, a \cos \phi)$

**FIGURE 16.38** The sphere in Example 2 can be parametrized using spherical coordinates.

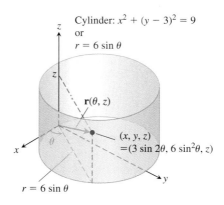

Cylinder: $x^2 + (y - 3)^2 = 9$
or
$r = 6 \sin \theta$

$r = 6 \sin \theta$

**FIGURE 15.39** The cylinder in Example 3 can be parametrized using cylindrical coordinates.

**EXAMPLE 3** Find a parametrization of the cylinder

$$x^2 + (y - 3)^2 = 9, \qquad 0 \le z \le 5.$$

**Solution** In cylindrical coordinates, a point $(x, y, z)$ has $x = r \cos \theta$, $y = r \sin \theta$, and $z = z$. For points on the cylinder $x^2 + (y - 3)^2 = 9$ (Figure 16.39), the equation is the same as the polar equation for the cylinder's base in the $xy$-plane:

$$x^2 + (y^2 - 6y + 9) = 9$$
$$r^2 - 6r \sin \theta = 0 \qquad \substack{x^2 + y^2 = r^2, \\ y = r \sin \theta}$$

or

$$r = 6 \sin \theta, \qquad 0 \le \theta \le \pi.$$

A typical point on the cylinder therefore has

$$x = r \cos \theta = 6 \sin \theta \cos \theta = 3 \sin 2\theta$$
$$y = r \sin \theta = 6 \sin^2 \theta$$
$$z = z.$$

Taking $u = \theta$ and $v = z$ in Equation (1) gives the one-to-one parametrization

$$\mathbf{r}(\theta, z) = (3 \sin 2\theta)\mathbf{i} + (6 \sin^2 \theta)\mathbf{j} + z\mathbf{k}, \qquad 0 \le \theta \le \pi, \quad 0 \le z \le 5. \qquad \blacksquare$$

## Surface Area

Our goal is to find a double integral for calculating the area of a curved surface $S$ based on the parametrization

$$\mathbf{r}(u, v) = f(u, v)\mathbf{i} + g(u, v)\mathbf{j} + h(u, v)\mathbf{k}, \qquad a \le u \le b, \quad c \le v \le d.$$

We need $S$ to be smooth for the construction we are about to carry out. The definition of smoothness involves the partial derivatives of $\mathbf{r}$ with respect to $u$ and $v$:

$$\mathbf{r}_u = \frac{\partial \mathbf{r}}{\partial u} = \frac{\partial f}{\partial u}\mathbf{i} + \frac{\partial g}{\partial u}\mathbf{j} + \frac{\partial h}{\partial u}\mathbf{k}$$

$$\mathbf{r}_v = \frac{\partial \mathbf{r}}{\partial v} = \frac{\partial f}{\partial v}\mathbf{i} + \frac{\partial g}{\partial v}\mathbf{j} + \frac{\partial h}{\partial v}\mathbf{k}.$$

---

**DEFINITION** A parametrized surface $\mathbf{r}(u, v) = f(u, v)\mathbf{i} + g(u, v)\mathbf{j} + h(u, v)\mathbf{k}$ is **smooth** if $\mathbf{r}_u$ and $\mathbf{r}_v$ are continuous and $\mathbf{r}_u \times \mathbf{r}_v$ is never zero on the interior of the parameter domain.

---

The condition that $\mathbf{r}_u \times \mathbf{r}_v$ is never the zero vector in the definition of smoothness means that the two vectors $\mathbf{r}_u$ and $\mathbf{r}_v$ are nonzero and never lie along the same line, so they always determine a plane tangent to the surface. We relax this condition on the boundary of the domain, but this does not affect the area computations.

Now consider a small rectangle $\Delta A_{uv}$ in $R$ with sides on the lines $u = u_0$, $u = u_0 + \Delta u$, $v = v_0$, and $v = v_0 + \Delta v$ (Figure 16.40). Each side of $\Delta A_{uv}$ maps to a curve on the surface $S$, and together these four curves bound a "curved patch element" $\Delta \sigma_{uv}$. In the notation of the figure, the side $v = v_0$ maps to curve $C_1$, the side $u = u_0$ maps to $C_2$, and their common vertex $(u_0, v_0)$ maps to $P_0$.

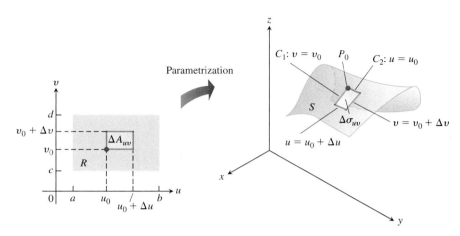

**FIGURE 16.40** A rectangular area element $\Delta A_{uv}$ in the $uv$-plane maps onto a curved patch element $\Delta \sigma_{uv}$ on $S$.

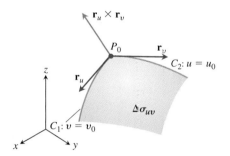

**FIGURE 16.41** A magnified view of a surface patch element $\Delta \sigma_{uv}$.

Figure 16.41 shows an enlarged view of $\Delta \sigma_{uv}$. The partial derivative vector $\mathbf{r}_u(u_0, v_0)$ is tangent to $C_1$ at $P_0$. Likewise, $\mathbf{r}_v(u_0, v_0)$ is tangent to $C_2$ at $P_0$. The cross product $\mathbf{r}_u \times \mathbf{r}_v$ is normal to the surface at $P_0$. (Here is where we begin to use the assumption that $S$ is smooth. We want to be sure that $\mathbf{r}_u \times \mathbf{r}_v \neq \mathbf{0}$.)

We next approximate the surface patch element $\Delta \sigma_{uv}$ by the parallelogram on the tangent plane whose sides are determined by the vectors $\Delta u \mathbf{r}_u$ and $\Delta v \mathbf{r}_v$ (Figure 16.42). The area of this parallelogram is

$$|\Delta u \mathbf{r}_u \times \Delta v \mathbf{r}_v| = |\mathbf{r}_u \times \mathbf{r}_v| \, \Delta u \, \Delta v. \tag{2}$$

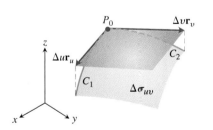

**FIGURE 16.42** The area of the parallelogram determined by the vectors $\Delta u \mathbf{r}_u$ and $\Delta v \mathbf{r}_v$ approximates the area of the surface patch element $\Delta \sigma_{uv}$.

A partition of the region $R$ in the $uv$-plane by rectangular regions $\Delta A_{uv}$ induces a partition of the surface $S$ into surface patch elements $\Delta \sigma_{uv}$. We approximate the area of each surface patch element $\Delta \sigma_{uv}$ by the parallelogram area in Equation (2) and sum these areas together to obtain an approximation of the surface area of $S$:

$$\sum_n |\mathbf{r}_u \times \mathbf{r}_v| \, \Delta u \, \Delta v. \tag{3}$$

As $\Delta u$ and $\Delta v$ approach zero independently, the number of area elements $n$ tends to $\infty$ and the continuity of $\mathbf{r}_u$ and $\mathbf{r}_v$ guarantees that the sum in Equation (3) approaches the double integral $\int_c^d \int_a^b |\mathbf{r}_u \times \mathbf{r}_v| \, du \, dv$. This double integral over the region $R$ defines the area of the surface $S$.

---

**DEFINITION** The **area** of the smooth surface

$$\mathbf{r}(u, v) = f(u, v)\mathbf{i} + g(u, v)\mathbf{j} + h(u, v)\mathbf{k}, \qquad a \leq u \leq b, \quad c \leq v \leq d$$

is

$$A = \iint_R |\mathbf{r}_u \times \mathbf{r}_v| \, dA = \int_c^d \int_a^b |\mathbf{r}_u \times \mathbf{r}_v| \, du \, dv. \tag{4}$$

We can abbreviate the integral in Equation (4) by writing $d\sigma$ for $|\mathbf{r}_u \times \mathbf{r}_v| \, du \, dv$. The surface area differential $d\sigma$ is analogous to the arc length differential $ds$ in Section 13.3.

---

**Surface Area Differential for a Parametrized Surface**

$$d\sigma = |\mathbf{r}_u \times \mathbf{r}_v| \, du \, dv \qquad \iint_S d\sigma \qquad (5)$$

Surface area
differential

Differential formula
for surface area

---

**EXAMPLE 4**    Find the surface area of the cone in Example 1 (Figure 16.37).

**Solution**    In Example 1, we found the parametrization

$$\mathbf{r}(r, \theta) = (r \cos \theta)\mathbf{i} + (r \sin \theta)\mathbf{j} + r\mathbf{k}, \qquad 0 \le r \le 1, \quad 0 \le \theta \le 2\pi.$$

To apply Equation (4), we first find $\mathbf{r}_r \times \mathbf{r}_\theta$:

$$\mathbf{r}_r \times \mathbf{r}_\theta = \begin{vmatrix} \mathbf{i} & \mathbf{j} & \mathbf{k} \\ \cos \theta & \sin \theta & 1 \\ -r \sin \theta & r \cos \theta & 0 \end{vmatrix}$$

$$= -(r \cos \theta)\mathbf{i} - (r \sin \theta)\mathbf{j} + \underbrace{(r \cos^2 \theta + r \sin^2 \theta)}_{r}\mathbf{k}.$$

Thus, $|\mathbf{r}_r \times \mathbf{r}_\theta| = \sqrt{r^2 \cos^2 \theta + r^2 \sin^2 \theta + r^2} = \sqrt{2r^2} = \sqrt{2}r$. The area of the cone is

$$A = \int_0^{2\pi} \int_0^1 |\mathbf{r}_r \times \mathbf{r}_\theta| \, dr \, d\theta \qquad \text{Eq. (4) with } u = r, v = \theta$$

$$= \int_0^{2\pi} \int_0^1 \sqrt{2}r \, dr \, d\theta = \int_0^{2\pi} \frac{\sqrt{2}}{2} d\theta = \frac{\sqrt{2}}{2}(2\pi) = \pi\sqrt{2} \text{ units squared.} \quad \blacksquare$$

**EXAMPLE 5**    Find the surface area of a sphere of radius $a$.

**Solution**    We use the parametrization from Example 2:

$$\mathbf{r}(\phi, \theta) = (a \sin \phi \cos \theta)\mathbf{i} + (a \sin \phi \sin \theta)\mathbf{j} + (a \cos \phi)\mathbf{k},$$

$$0 \le \phi \le \pi, \quad 0 \le \theta \le 2\pi.$$

For $\mathbf{r}_\phi \times \mathbf{r}_\theta$, we get

$$\mathbf{r}_\phi \times \mathbf{r}_\theta = \begin{vmatrix} \mathbf{i} & \mathbf{j} & \mathbf{k} \\ a \cos \phi \cos \theta & a \cos \phi \sin \theta & -a \sin \phi \\ -a \sin \phi \sin \theta & a \sin \phi \cos \theta & 0 \end{vmatrix}$$

$$= (a^2 \sin^2 \phi \cos \theta)\mathbf{i} + (a^2 \sin^2 \phi \sin \theta)\mathbf{j} + (a^2 \sin \phi \cos \phi)\mathbf{k}.$$

Thus,

$$|\mathbf{r}_\phi \times \mathbf{r}_\theta| = \sqrt{a^4 \sin^4 \phi \cos^2 \theta + a^4 \sin^4 \phi \sin^2 \theta + a^4 \sin^2 \phi \cos^2 \phi}$$

$$= \sqrt{a^4 \sin^4 \phi + a^4 \sin^2 \phi \cos^2 \phi} = \sqrt{a^4 \sin^2 \phi (\sin^2 \phi + \cos^2 \phi)}$$

$$= a^2 \sqrt{\sin^2 \phi} = a^2 \sin \phi,$$

since $\sin \phi \geq 0$ for $0 \leq \phi \leq \pi$. Therefore, the area of the sphere is

$$A = \int_0^{2\pi} \int_0^{\pi} a^2 \sin \phi \, d\phi \, d\theta$$

$$= \int_0^{2\pi} \left[ -a^2 \cos \phi \right]_0^{\pi} d\theta = \int_0^{2\pi} 2a^2 \, d\theta = 4\pi a^2 \quad \text{units squared.}$$

This agrees with the well-known formula for the surface area of a sphere. ■

**EXAMPLE 6**   Let $S$ be the "football" surface formed by rotating the curve $x = \cos z$, $y = 0, -\pi/2 \leq z \leq \pi/2$ around the $z$-axis (see Figure 16.43). Find a parametrization for $S$ and compute its surface area.

**Solution**   Example 2 suggests finding a parametrization of $S$ based on its rotation around the $z$-axis. If we rotate a point $(x, 0, z)$ on the curve $x = \cos z$, $y = 0$ about the $z$-axis, we obtain a circle at height $z$ above the $xy$-plane that is centered on the $z$-axis and has radius $r = \cos z$ (see Figure 16.43). The point sweeps out the circle through an angle of rotation $\theta, 0 \leq \theta \leq 2\pi$. We let $(x, y, z)$ be an arbitrary point on this circle, and define the parameters $u = z$ and $v = \theta$. Then we have $x = r \cos \theta = \cos u \cos v$, $y = r \sin \theta = \cos u \sin v$, and $z = u$ giving a parametrization for $S$ as

$$\mathbf{r}(u, v) = \cos u \cos v \, \mathbf{i} + \cos u \sin v \, \mathbf{j} + u \mathbf{k}, \quad -\frac{\pi}{2} \leq u \leq \frac{\pi}{2}, \quad 0 \leq v \leq 2\pi.$$

Next we use Equation (5) to find the surface area of $S$. Differentiation of the parametrization gives

$$\mathbf{r}_u = -\sin u \cos v \, \mathbf{i} - \sin u \sin v \, \mathbf{j} + \mathbf{k}$$

and

$$\mathbf{r}_v = -\cos u \sin v \, \mathbf{i} + \cos u \cos v \, \mathbf{j}.$$

Computing the cross product we have

$$\mathbf{r}_u \times \mathbf{r}_v = \begin{vmatrix} \mathbf{i} & \mathbf{j} & \mathbf{k} \\ -\sin u \cos v & -\sin u \sin v & 1 \\ -\cos u \sin v & \cos u \cos v & 0 \end{vmatrix}$$

$$= -\cos u \cos v \, \mathbf{i} - \cos u \sin v \, \mathbf{j} - (\sin u \cos u \cos^2 v + \cos u \sin u \sin^2 v)\mathbf{k}.$$

Taking the magnitude of the cross product gives

$$|\mathbf{r}_u \times \mathbf{r}_v| = \sqrt{\cos^2 u \, (\cos^2 v + \sin^2 v) + \sin^2 u \cos^2 u}$$

$$= \sqrt{\cos^2 u \, (1 + \sin^2 u)}$$

$$= \cos u \sqrt{1 + \sin^2 u}. \qquad \cos u \geq 0 \text{ for } -\frac{\pi}{2} \leq u \leq \frac{\pi}{2}$$

From Equation (4) the surface area is given by the integral

$$A = \int_0^{2\pi} \int_{-\pi/2}^{\pi/2} \cos u \sqrt{1 + \sin^2 u} \, du \, dv.$$

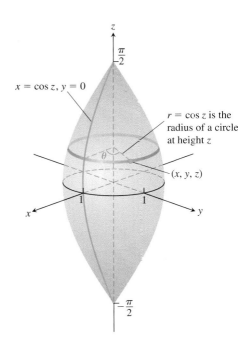

$x = \cos z, y = 0$

$r = \cos z$ is the radius of a circle at height $z$

$(x, y, z)$

**FIGURE 16.43**   The "football" surface in Example 6 obtained by rotating the curve $x = \cos z$ about the $z$-axis.

To evaluate the integral, we substitute $w = \sin u$ and $dw = \cos u\, du$, $-1 \le w \le 1$. Since the surface $S$ is symmetric across the $xy$-plane, we need only integrate with respect to $w$ from 0 to 1, and multiply the result by 2. In summary, we have

$$
\begin{aligned}
A &= 2 \int_0^{2\pi} \int_0^1 \sqrt{1 + w^2} \; dw \; dv \\
&= 2 \int_0^{2\pi} \left[ \frac{w}{2} \sqrt{1 + w^2} + \frac{1}{2} \ln \left( w + \sqrt{1 + w^2} \right) \right]_0^1 dv \qquad \text{Integral Table Formula 35} \\
&= \int_0^{2\pi} 2 \left[ \frac{1}{2} \sqrt{2} + \frac{1}{2} \ln \left( 1 + \sqrt{2} \right) \right] dv \\
&= 2\pi \left[ \sqrt{2} + \ln \left( 1 + \sqrt{2} \right) \right].
\end{aligned}
$$
∎

### Implicit Surfaces

Surfaces are often presented as level sets of a function, described by an equation such as

$$
F(x, y, z) = c,
$$

for some constant $c$. Such a level surface does not come with an explicit parametrization, and is called an *implicitly defined surface*. Implicit surfaces arise, for example, as equipotential surfaces in electric or gravitational fields. Figure 16.44 shows a piece of such a surface. It may be difficult to find explicit formulas for the functions $f$, $g$, and $h$ that describe the surface in the form $\mathbf{r}(u, v) = f(u, v)\mathbf{i} + g(u, v)\mathbf{j} + h(u, v)\mathbf{k}$. We now show how to compute the surface area differential $d\sigma$ for implicit surfaces.

Figure 16.44 shows a piece of an implicit surface $S$ that lies above its "shadow" region $R$ in the plane beneath it. The surface is defined by the equation $F(x, y, z) = c$ and $\mathbf{p}$ is a unit vector normal to the plane region $R$. We assume that the surface is **smooth** ($F$ is differentiable and $\nabla F$ is nonzero and continuous on $S$) and that $\nabla F \cdot \mathbf{p} \ne 0$, so the surface never folds back over itself.

Assume that the normal vector $\mathbf{p}$ is the unit vector $\mathbf{k}$, so the region $R$ in Figure 16.44 lies in the $xy$-plane. By assumption, we then have $\nabla F \cdot \mathbf{p} = \nabla F \cdot \mathbf{k} = F_z \ne 0$ on $S$. The Implicit Function Theorem (see Section 14.4) implies that $S$ is then the graph of a differentiable function $z = h(x, y)$, although the function $h(x, y)$ is not explicitly known. Define the parameters $u$ and $v$ by $u = x$ and $v = y$. Then $z = h(u, v)$ and

$$
\mathbf{r}(u, v) = u\mathbf{i} + v\mathbf{j} + h(u, v)\mathbf{k} \tag{6}
$$

gives a parametrization of the surface $S$. We use Equation (4) to find the area of $S$.

Calculating the partial derivatives of $\mathbf{r}$, we find

$$
\mathbf{r}_u = \mathbf{i} + \frac{\partial h}{\partial u} \mathbf{k} \qquad \text{and} \qquad \mathbf{r}_v = \mathbf{j} + \frac{\partial h}{\partial v} \mathbf{k}.
$$

Applying the Chain Rule for implicit differentiation (see Equation (2) in Section 14.4) to $F(x, y, z) = c$, where $x = u$, $y = v$, and $z = h(u, v)$, we obtain the partial derivatives

$$
\frac{\partial h}{\partial u} = -\frac{F_x}{F_z} \qquad \text{and} \qquad \frac{\partial h}{\partial v} = -\frac{F_y}{F_z}.
$$

Substitution of these derivatives into the derivatives of $\mathbf{r}$ gives

$$
\mathbf{r}_u = \mathbf{i} - \frac{F_x}{F_z} \mathbf{k} \qquad \text{and} \qquad \mathbf{r}_v = \mathbf{j} - \frac{F_y}{F_z} \mathbf{k}.
$$

Surface $F(x, y, z) = c$

$S$

$\mathbf{p}$

$R$

The vertical projection or "shadow" of $S$ on a coordinate plane

**FIGURE 16.44** As we soon see, the area of a surface $S$ in space can be calculated by evaluating a related double integral over the vertical projection or "shadow" of $S$ on a coordinate plane. The unit vector $\mathbf{p}$ is normal to the plane.

From a routine calculation of the cross product we find

$$\mathbf{r}_u \times \mathbf{r}_v = \frac{F_x}{F_z}\mathbf{i} + \frac{F_y}{F_z}\mathbf{j} + \mathbf{k} \qquad F_z \neq 0$$

$$= \frac{1}{F_z}(F_x\mathbf{i} + F_y\mathbf{j} + F_z\mathbf{k})$$

$$= \frac{\nabla F}{F_z} = \frac{\nabla F}{\nabla F \cdot \mathbf{k}}$$

$$= \frac{\nabla F}{\nabla F \cdot \mathbf{p}}. \qquad \mathbf{p} = \mathbf{k}$$

Therefore, the surface area differential is given by

$$d\sigma = |\mathbf{r}_u \times \mathbf{r}_v|\, du\, dv = \frac{|\nabla F|}{|\nabla F \cdot \mathbf{p}|}\, dx\, dy. \qquad u = x \text{ and } v = y$$

We obtain similar calculations if instead the vector $\mathbf{p} = \mathbf{j}$ is normal to the $xz$-plane when $F_y \neq 0$ on $S$, or if $\mathbf{p} = \mathbf{i}$ is normal to the $yz$-plane when $F_x \neq 0$ on $S$. Combining these results with Equation (4) then gives the following general formula.

---

**Formula for the Surface Area of an Implicit Surface**

The area of the surface $F(x, y, z) = c$ over a closed and bounded plane region $R$ is

$$\text{Surface area} = \iint_R \frac{|\nabla F|}{|\nabla F \cdot \mathbf{p}|}\, dA, \qquad (7)$$

where $\mathbf{p} = \mathbf{i}, \mathbf{j},$ or $\mathbf{k}$ is normal to $R$ and $\nabla F \cdot \mathbf{p} \neq 0$.

---

Thus, the area is the double integral over $R$ of the magnitude of $\nabla F$ divided by the magnitude of the scalar component of $\nabla F$ normal to $R$.

We reached Equation (7) under the assumption that $\nabla F \cdot \mathbf{p} \neq 0$ throughout $R$ and that $\nabla F$ is continuous. Whenever the integral exists, however, we define its value to be the area of the portion of the surface $F(x, y, z) = c$ that lies over $R$. (Recall that the projection is assumed to be one-to-one.)

**EXAMPLE 7**    Find the area of the surface cut from the bottom of the paraboloid $x^2 + y^2 - z = 0$ by the plane $z = 4$.

**Solution**   We sketch the surface $S$ and the region $R$ below it in the $xy$-plane (Figure 16.45). The surface $S$ is part of the level surface $F(x, y, z) = x^2 + y^2 - z = 0$, and $R$ is the disk $x^2 + y^2 \leq 4$ in the $xy$-plane. To get a unit vector normal to the plane of $R$, we can take $\mathbf{p} = \mathbf{k}$.

At any point $(x, y, z)$ on the surface, we have

$$F(x, y, z) = x^2 + y^2 - z$$

$$\nabla F = 2x\mathbf{i} + 2y\mathbf{j} - \mathbf{k}$$

$$|\nabla F| = \sqrt{(2x)^2 + (2y)^2 + (-1)^2}$$

$$= \sqrt{4x^2 + 4y^2 + 1}$$

$$|\nabla F \cdot \mathbf{p}| = |\nabla F \cdot \mathbf{k}| = |-1| = 1.$$

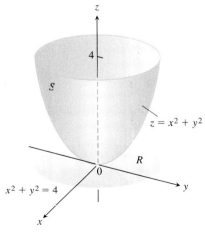

**FIGURE 16.45**   The area of this parabolic surface is calculated in Example 7.

In the region $R$, $dA = dx\, dy$. Therefore,

$$\text{Surface area} = \iint_R \frac{|\nabla F|}{|\nabla F \cdot \mathbf{p}|}\, dA \qquad \text{Eq. (7)}$$

$$= \iint_{x^2+y^2\le 4} \sqrt{4x^2 + 4y^2 + 1}\, dx\, dy$$

$$= \int_0^{2\pi} \int_0^2 \sqrt{4r^2 + 1}\; r\, dr\, d\theta \qquad \text{Polar coordinates}$$

$$= \int_0^{2\pi} \left[ \frac{1}{12}(4r^2 + 1)^{3/2} \right]_0^2 d\theta$$

$$= \int_0^{2\pi} \frac{1}{12}(17^{3/2} - 1)\, d\theta = \frac{\pi}{6}\left(17\sqrt{17} - 1\right). \qquad \blacksquare$$

Example 7 illustrates how to find the surface area for a function $z = f(x, y)$ over a region $R$ in the $xy$-plane. Actually, the surface area differential can be obtained in two ways, and we show this in the next example.

**EXAMPLE 8** Derive the surface area differential $d\sigma$ of the surface $z = f(x, y)$ over a region $R$ in the $xy$-plane **(a)** parametrically using Equation (5), and **(b)** implicitly, as in Equation (7).

**Solution**

**(a)** We parametrize the surface by taking $x = u$, $y = v$, and $z = f(x, y)$ over $R$. This gives the parametrization

$$\mathbf{r}(u, v) = u\mathbf{i} + v\mathbf{j} + f(u, v)\mathbf{k}.$$

Computing the partial derivatives gives $\mathbf{r}_u = \mathbf{i} + f_u\mathbf{k}$, $\mathbf{r}_v = \mathbf{j} + f_v\mathbf{k}$ and

$$\mathbf{r}_u \times \mathbf{r}_v = -f_u\mathbf{i} - f_v\mathbf{j} + \mathbf{k}. \qquad \begin{vmatrix} \mathbf{i} & \mathbf{j} & \mathbf{k} \\ 1 & 0 & f_u \\ 0 & 1 & f_v \end{vmatrix}$$

Then $|\mathbf{r}_u \times \mathbf{r}_v|\, du\, dv = \sqrt{f_u^2 + f_v^2 + 1}\, du\, dv$. Substituting for $u$ and $v$ then gives the surface area differential

$$d\sigma = \sqrt{f_x^2 + f_y^2 + 1}\, dx\, dy.$$

**(b)** We define the implicit function $F(x, y, z) = f(x, y) - z$. Since $(x, y)$ belongs to the region $R$, the unit normal to the plane of $R$ is $\mathbf{p} = \mathbf{k}$. Then $\nabla F = f_x\mathbf{i} + f_y\mathbf{j} - \mathbf{k}$ so that $|\nabla F \cdot \mathbf{p}| = |-1| = 1$, $|\nabla F| = \sqrt{f_x^2 + f_y^2 + 1}$, and $|\nabla F| / |\nabla F \cdot \mathbf{p}| = |\nabla F|$. The surface area differential is again given by

$$d\sigma = \sqrt{f_x^2 + f_y^2 + 1}\, dx\, dy. \qquad \blacksquare$$

The surface area differential derived in Example 8 gives the following formula for calculating the surface area of the graph of a function defined explicitly as $z = f(x, y)$.

---

**Formula for the Surface Area of a Graph $z = f(x, y)$**

For a graph $z = f(x, y)$ over a region $R$ in the $xy$-plane, the surface area formula is

$$A = \iint_R \sqrt{f_x^2 + f_y^2 + 1}\, dx\, dy. \qquad (8)$$

# Exercises 16.5

### Finding Parametrizations

In Exercises 1–16, find a parametrization of the surface. (There are many correct ways to do these, so your answers may not be the same as those in the back of the book.)

1. The paraboloid $z = x^2 + y^2$, $z \le 4$

2. The paraboloid $z = 9 - x^2 - y^2$, $z \ge 0$

3. **Cone frustum** The first-octant portion of the cone $z = \sqrt{x^2 + y^2}/2$ between the planes $z = 0$ and $z = 3$

4. **Cone frustum** The portion of the cone $z = 2\sqrt{x^2 + y^2}$ between the planes $z = 2$ and $z = 4$

5. **Spherical cap** The cap cut from the sphere $x^2 + y^2 + z^2 = 9$ by the cone $z = \sqrt{x^2 + y^2}$

6. **Spherical cap** The portion of the sphere $x^2 + y^2 + z^2 = 4$ in the first octant between the $xy$-plane and the cone $z = \sqrt{x^2 + y^2}$

7. **Spherical band** The portion of the sphere $x^2 + y^2 + z^2 = 3$ between the planes $z = \sqrt{3}/2$ and $z = -\sqrt{3}/2$

8. **Spherical cap** The upper portion cut from the sphere $x^2 + y^2 + z^2 = 8$ by the plane $z = -2$

9. **Parabolic cylinder between planes** The surface cut from the parabolic cylinder $z = 4 - y^2$ by the planes $x = 0$, $x = 2$, and $z = 0$

10. **Parabolic cylinder between planes** The surface cut from the parabolic cylinder $y = x^2$ by the planes $z = 0$, $z = 3$, and $y = 2$

11. **Circular cylinder band** The portion of the cylinder $y^2 + z^2 = 9$ between the planes $x = 0$ and $x = 3$

12. **Circular cylinder band** The portion of the cylinder $x^2 + z^2 = 4$ above the $xy$-plane between the planes $y = -2$ and $y = 2$

13. **Tilted plane inside cylinder** The portion of the plane $x + y + z = 1$

    a. Inside the cylinder $x^2 + y^2 = 9$

    b. Inside the cylinder $y^2 + z^2 = 9$

14. **Tilted plane inside cylinder** The portion of the plane $x - y + 2z = 2$

    a. Inside the cylinder $x^2 + z^2 = 3$

    b. Inside the cylinder $y^2 + z^2 = 2$

15. **Circular cylinder band** The portion of the cylinder $(x - 2)^2 + z^2 = 4$ between the planes $y = 0$ and $y = 3$

16. **Circular cylinder band** The portion of the cylinder $y^2 + (z - 5)^2 = 25$ between the planes $x = 0$ and $x = 10$

### Surface Area of Parametrized Surfaces

In Exercises 17–26, use a parametrization to express the area of the surface as a double integral. Then evaluate the integral. (There are many correct ways to set up the integrals, so your integrals may not be the same as those in the back of the book. They should have the same values, however.)

17. **Tilted plane inside cylinder** The portion of the plane $y + 2z = 2$ inside the cylinder $x^2 + y^2 = 1$

18. **Plane inside cylinder** The portion of the plane $z = -x$ inside the cylinder $x^2 + y^2 = 4$

19. **Cone frustum** The portion of the cone $z = 2\sqrt{x^2 + y^2}$ between the planes $z = 2$ and $z = 6$

20. **Cone frustum** The portion of the cone $z = \sqrt{x^2 + y^2}/3$ between the planes $z = 1$ and $z = 4/3$

21. **Circular cylinder band** The portion of the cylinder $x^2 + y^2 = 1$ between the planes $z = 1$ and $z = 4$

22. **Circular cylinder band** The portion of the cylinder $x^2 + z^2 = 10$ between the planes $y = -1$ and $y = 1$

23. **Parabolic cap** The cap cut from the paraboloid $z = 2 - x^2 - y^2$ by the cone $z = \sqrt{x^2 + y^2}$

24. **Parabolic band** The portion of the paraboloid $z = x^2 + y^2$ between the planes $z = 1$ and $z = 4$

25. **Sawed-off sphere** The lower portion cut from the sphere $x^2 + y^2 + z^2 = 2$ by the cone $z = \sqrt{x^2 + y^2}$

26. **Spherical band** The portion of the sphere $x^2 + y^2 + z^2 = 4$ between the planes $z = -1$ and $z = \sqrt{3}$

### Planes Tangent to Parametrized Surfaces

The tangent plane at a point $P_0(f(u_0, v_0), g(u_0, v_0), h(u_0, v_0))$ on a parametrized surface $\mathbf{r}(u, v) = f(u, v)\mathbf{i} + g(u, v)\mathbf{j} + h(u, v)\mathbf{k}$ is the plane through $P_0$ normal to the vector $\mathbf{r}_u(u_0, v_0) \times \mathbf{r}_v(u_0, v_0)$, the cross product of the tangent vectors $\mathbf{r}_u(u_0, v_0)$ and $\mathbf{r}_v(u_0, v_0)$ at $P_0$. In Exercises 27–30, find an equation for the plane tangent to the surface at $P_0$. Then find a Cartesian equation for the surface and sketch the surface and tangent plane together.

27. **Cone** The cone $\mathbf{r}(r, \theta) = (r \cos \theta)\mathbf{i} + (r \sin \theta)\mathbf{j} + r\mathbf{k}$, $r \ge 0$, $0 \le \theta \le 2\pi$ at the point $P_0(\sqrt{2}, \sqrt{2}, 2)$ corresponding to $(r, \theta) = (2, \pi/4)$

28. **Hemisphere** The hemisphere surface $\mathbf{r}(\phi, \theta) = (4 \sin \phi \cos \theta)\mathbf{i} + (4 \sin \phi \sin \theta)\mathbf{j} + (4 \cos \phi)\mathbf{k}$, $0 \le \phi \le \pi/2$, $0 \le \theta \le 2\pi$, at the point $P_0(\sqrt{2}, \sqrt{2}, 2\sqrt{3})$ corresponding to $(\phi, \theta) = (\pi/6, \pi/4)$

29. **Circular cylinder** The circular cylinder $\mathbf{r}(\theta, z) = (3 \sin 2\theta)\mathbf{i} + (6 \sin^2 \theta)\mathbf{j} + z\mathbf{k}$, $0 \le \theta \le \pi$, at the point $P_0(3\sqrt{3}/2, 9/2, 0)$ corresponding to $(\theta, z) = (\pi/3, 0)$ (See Example 3.)

30. **Parabolic cylinder** The parabolic cylinder surface $\mathbf{r}(x, y) = x\mathbf{i} + y\mathbf{j} - x^2\mathbf{k}$, $-\infty < x < \infty$, $-\infty < y < \infty$, at the point $P_0(1, 2, -1)$ corresponding to $(x, y) = (1, 2)$

### More Parametrizations of Surfaces

31. a. A *torus of revolution* (doughnut) is obtained by rotating a circle $C$ in the $xz$-plane about the $z$-axis in space. (See the accompanying figure.) If $C$ has radius $r > 0$ and center $(R, 0, 0)$, show that a parametrization of the torus is

$$\mathbf{r}(u, v) = ((R + r \cos u)\cos v)\mathbf{i}$$
$$+ ((R + r \cos u)\sin v)\mathbf{j} + (r \sin u)\mathbf{k},$$

where $0 \le u \le 2\pi$ and $0 \le v \le 2\pi$ are the angles in the figure.

**b.** Show that the surface area of the torus is $A = 4\pi^2 Rr$.

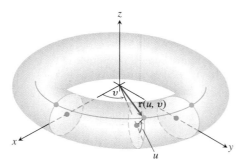

**32. Parametrization of a surface of revolution** Suppose that the parametrized curve $C$: $(f(u), g(u))$ is revolved about the $x$-axis, where $g(u) > 0$ for $a \le u \le b$.

**a.** Show that

$$\mathbf{r}(u, v) = f(u)\mathbf{i} + (g(u)\cos v)\mathbf{j} + (g(u)\sin v)\mathbf{k}$$

is a parametrization of the resulting surface of revolution, where $0 \le v \le 2\pi$ is the angle from the $xy$-plane to the point $\mathbf{r}(u, v)$ on the surface. (See the accompanying figure.) Notice that $f(u)$ measures distance *along* the axis of revolution and $g(u)$ measures distance *from* the axis of revolution.

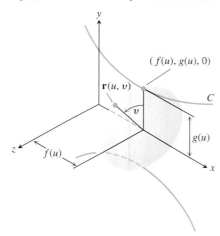

**b.** Find a parametrization for the surface obtained by revolving the curve $x = y^2$, $y \ge 0$, about the $x$-axis.

**33. a. Parametrization of an ellipsoid** The parametrization $x = a\cos\theta$, $y = b\sin\theta$, $0 \le \theta \le 2\pi$ gives the ellipse $(x^2/a^2) + (y^2/b^2) = 1$. Using the angles $\theta$ and $\phi$ in spherical coordinates, show that

$$\mathbf{r}(\theta, \phi) = (a\cos\theta\cos\phi)\mathbf{i} + (b\sin\theta\cos\phi)\mathbf{j} + (c\sin\phi)\mathbf{k}$$

is a parametrization of the ellipsoid $(x^2/a^2) + (y^2/b^2) + (z^2/c^2) = 1$.

**b.** Write an integral for the surface area of the ellipsoid, but do not evaluate the integral.

**34. Hyperboloid of one sheet**

**a.** Find a parametrization for the hyperboloid of one sheet $x^2 + y^2 - z^2 = 1$ in terms of the angle $\theta$ associated with the circle $x^2 + y^2 = r^2$ and the hyperbolic parameter $u$ associated with the hyperbolic function $r^2 - z^2 = 1$. (*Hint:* $\cosh^2 u - \sinh^2 u = 1$.)

**b.** Generalize the result in part (a) to the hyperboloid $(x^2/a^2) + (y^2/b^2) - (z^2/c^2) = 1$.

**35.** (*Continuation of Exercise 34.*) Find a Cartesian equation for the plane tangent to the hyperboloid $x^2 + y^2 - z^2 = 25$ at the point $(x_0, y_0, 0)$, where $x_0^2 + y_0^2 = 25$.

**36. Hyperboloid of two sheets** Find a parametrization of the hyperboloid of two sheets $(z^2/c^2) - (x^2/a^2) - (y^2/b^2) = 1$.

Surface Area for Implicit and Explicit Forms

**37.** Find the area of the surface cut from the paraboloid $x^2 + y^2 - z = 0$ by the plane $z = 2$.

**38.** Find the area of the band cut from the paraboloid $x^2 + y^2 - z = 0$ by the planes $z = 2$ and $z = 6$.

**39.** Find the area of the region cut from the plane $x + 2y + 2z = 5$ by the cylinder whose walls are $x = y^2$ and $x = 2 - y^2$.

**40.** Find the area of the portion of the surface $x^2 - 2z = 0$ that lies above the triangle bounded by the lines $x = \sqrt{3}, y = 0$, and $y = x$ in the $xy$-plane.

**41.** Find the area of the surface $x^2 - 2y - 2z = 0$ that lies above the triangle bounded by the lines $x = 2, y = 0$, and $y = 3x$ in the $xy$-plane.

**42.** Find the area of the cap cut from the sphere $x^2 + y^2 + z^2 = 2$ by the cone $z = \sqrt{x^2 + y^2}$.

**43.** Find the area of the ellipse cut from the plane $z = cx$ ($c$ a constant) by the cylinder $x^2 + y^2 = 1$.

**44.** Find the area of the upper portion of the cylinder $x^2 + z^2 = 1$ that lies between the planes $x = \pm 1/2$ and $y = \pm 1/2$.

**45.** Find the area of the portion of the paraboloid $x = 4 - y^2 - z^2$ that lies above the ring $1 \le y^2 + z^2 \le 4$ in the $yz$-plane.

**46.** Find the area of the surface cut from the paraboloid $x^2 + y + z^2 = 2$ by the plane $y = 0$.

**47.** Find the area of the surface $x^2 - 2\ln x + \sqrt{15}y - z = 0$ above the square $R$: $1 \le x \le 2, 0 \le y \le 1$, in the $xy$-plane.

**48.** Find the area of the surface $2x^{3/2} + 2y^{3/2} - 3z = 0$ above the square $R$: $0 \le x \le 1, 0 \le y \le 1$, in the $xy$-plane.

Find the area of the surfaces in Exercises 49–54.

**49.** The surface cut from the bottom of the paraboloid $z = x^2 + y^2$ by the plane $z = 3$

**50.** The surface cut from the "nose" of the paraboloid $x = 1 - y^2 - z^2$ by the $yz$-plane

**51.** The portion of the cone $z = \sqrt{x^2 + y^2}$ that lies over the region between the circle $x^2 + y^2 = 1$ and the ellipse $9x^2 + 4y^2 = 36$ in the $xy$-plane. (*Hint:* Use formulas from geometry to find the area of the region.)

**52.** The triangle cut from the plane $2x + 6y + 3z = 6$ by the bounding planes of the first octant. Calculate the area three ways, using different explicit forms.

**53.** The surface in the first octant cut from the cylinder $y = (2/3)z^{3/2}$ by the planes $x = 1$ and $y = 16/3$

**54.** The portion of the plane $y + z = 4$ that lies above the region cut from the first quadrant of the $xz$-plane by the parabola $x = 4 - z^2$

**55.** Use the parametrization

$$\mathbf{r}(x, z) = x\mathbf{i} + f(x, z)\mathbf{j} + z\mathbf{k}$$

and Equation (5) to derive a formula for $d\sigma$ associated with the explicit form $y = f(x, z)$.

**56.** Let $S$ be the surface obtained by rotating the smooth curve $y = f(x), a \leq x \leq b$, about the $x$-axis, where $f(x) \geq 0$.

**a.** Show that the vector function

$$\mathbf{r}(x, \theta) = x\mathbf{i} + f(x) \cos \theta\mathbf{j} + f(x) \sin \theta\mathbf{k}$$

is a parametrization of $S$, where $\theta$ is the angle of rotation around the $x$-axis (see the accompanying figure).

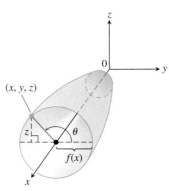

**b.** Use Equation (4) to show that the surface area of this surface of revolution is given by

$$A = \int_a^b 2\pi f(x) \sqrt{1 + [f'(x)]^2}\, dx.$$

# 16.6 Surface Integrals

To compute the mass of a surface, the flow of a liquid across a curved membrane, or the total electrical charge on a surface, we need to integrate a function over a curved surface in space. Such a *surface integral* is the two-dimensional extension of the line integral concept used to integrate over a one-dimensional curve. Like line integrals, surface integrals arise in two forms. One form occurs when we integrate a scalar function over a surface, such as integrating a mass density function defined on a surface to find its total mass. This form corresponds to line integrals of scalar functions defined in Section 16.1, and we used it to find the mass of a thin wire. The second form is for surface integrals of vector fields, analogous to the line integrals for vector fields defined in Section 16.2. An example of this form occurs when we want to measure the net flow of a fluid across a surface submerged in the fluid (just as we previously defined the flux of $\mathbf{F}$ across a curve). In this section we investigate these ideas and some of their applications.

## Surface Integrals

Suppose that the function $G(x, y, z)$ gives the *mass density* (mass per unit area) at each point on a surface $S$. Then we can calculate the total mass of $S$ as an integral in the following way.

Assume, as in Section 16.5, that the surface $S$ is defined parametrically on a region $R$ in the $uv$-plane,

$$\mathbf{r}(u, v) = f(u, v)\mathbf{i} + g(u, v)\mathbf{j} + h(u, v)\mathbf{k}, \qquad (u, v) \in R.$$

In Figure 16.46, we see how a subdivision of $R$ (considered as a rectangle for simplicity) divides the surface $S$ into corresponding curved surface elements, or patches, of area

$$\Delta\sigma_{uv} \approx |\mathbf{r}_u \times \mathbf{r}_v|\, du\, dv.$$

As we did for the subdivisions when defining double integrals in Section 15.2, we number the surface element patches in some order with their areas given by $\Delta\sigma_1, \Delta\sigma_2, \ldots, \Delta\sigma_n$. To form a Riemann sum over $S$, we choose a point $(x_k, y_k, z_k)$ in the $k$th patch, multiply the value of the function $G$ at that point by the area $\Delta\sigma_k$, and add together the products:

$$\sum_{k=1}^{n} G(x_k, y_k, z_k)\, \Delta\sigma_k.$$

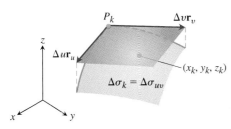

**FIGURE 16.46** The area of the patch $\Delta\sigma_k$ is approximated by the area of the tangent parallelogram determined by the vectors $\Delta u\, \mathbf{r}_u$ and $\Delta v\, \mathbf{r}_v$. The point $(x_k, y_k, z_k)$ lies on the surface patch, beneath the parallelogram shown here.

Depending on how we pick $(x_k, y_k, z_k)$ in the $k$th patch, we may get different values for this Riemann sum. Then we take the limit as the number of surface patches increases, their areas shrink to zero, and both $\Delta u \rightarrow 0$ and $\Delta v \rightarrow 0$. This limit, whenever it exists independent of all choices made, defines the **surface integral of $G$ over the surface $S$** as

$$\iint_S G(x, y, z)\,d\sigma = \lim_{n \to \infty} \sum_{k=1}^{n} G(x_k, y_k, z_k)\,\Delta\sigma_k. \qquad (1)$$

Notice the analogy with the definition of the double integral (Section 15.2) and with the line integral (Section 16.1). If $S$ is a piecewise smooth surface, and $G$ is continuous over $S$, then the surface integral defined by Equation (1) can be shown to exist.

The formula for evaluating the surface integral depends on the manner in which $S$ is described, parametrically, implicitly or explicitly, as discussed in Section 16.5.

---

**Formulas for a Surface Integral of a Scalar Function**

1. For a smooth surface $S$ defined **parametrically** as $\mathbf{r}(u, v) = f(u, v)\mathbf{i} + g(u, v)\mathbf{j} + h(u, v)\mathbf{k}$, $(u, v) \in R$, and a continuous function $G(x, y, z)$ defined on $S$, the surface integral of $G$ over $S$ is given by the double integral over $R$,

$$\iint_S G(x, y, z)\,d\sigma = \iint_R G(f(u, v), g(u, v), h(u, v)) |\mathbf{r}_u \times \mathbf{r}_v|\,du\,dv. \qquad (2)$$

2. For a surface $S$ given **implicitly** by $F(x, y, z) = c$, where $F$ is a continuously differentiable function, with $S$ lying above its closed and bounded shadow region $R$ in the coordinate plane beneath it, the surface integral of the continuous function $G$ over $S$ is given by the double integral over $R$,

$$\iint_S G(x, y, z)\,d\sigma = \iint_R G(x, y, z)\,\frac{|\nabla F|}{|\nabla F \cdot \mathbf{p}|}\,dA, \qquad (3)$$

   where $\mathbf{p}$ is a unit vector normal to $R$ and $\nabla F \cdot \mathbf{p} \neq 0$.

3. For a surface $S$ given **explicitly** as the graph of $z = f(x, y)$, where $f$ is a continuously differentiable function over a region $R$ in the $xy$-plane, the surface integral of the continuous function $G$ over $S$ is given by the double integral over $R$,

$$\iint_S G(x, y, z)\,d\sigma = \iint_R G(x, y, f(x, y)) \sqrt{f_x^2 + f_y^2 + 1}\,dx\,dy. \qquad (4)$$

---

The surface integral in Equation (1) takes on different meanings in different applications. If $G$ has the constant value 1, the integral gives the area of $S$. If $G$ gives the mass density of a thin shell of material modeled by $S$, the integral gives the mass of the shell. If $G$ gives the charge density of a thin shell, then the integral gives the total charge.

**EXAMPLE 1**   Integrate $G(x, y, z) = x^2$ over the cone $z = \sqrt{x^2 + y^2}$, $0 \le z \le 1$.

**Solution**   Using Equation (2) and the calculations from Example 4 in Section 16.5, we have $|\mathbf{r}_r \times \mathbf{r}_\theta| = \sqrt{2}r$ and

$$\iint_S x^2 \, d\sigma = \int_0^{2\pi} \int_0^1 (r^2 \cos^2 \theta)(\sqrt{2}r) \, dr \, d\theta \qquad x = r \cos \theta$$

$$= \sqrt{2} \int_0^{2\pi} \int_0^1 r^3 \cos^2 \theta \, dr \, d\theta$$

$$= \frac{\sqrt{2}}{4} \int_0^{2\pi} \cos^2 \theta \, d\theta = \frac{\sqrt{2}}{4} \left[ \frac{\theta}{2} + \frac{1}{4} \sin 2\theta \right]_0^{2\pi} = \frac{\pi \sqrt{2}}{4}. \qquad \blacksquare$$

Surface integrals behave like other double integrals, the integral of the sum of two functions being the sum of their integrals and so on. The domain Additivity Property takes the form

$$\iint_S G \, d\sigma = \iint_{S_1} G \, d\sigma + \iint_{S_2} G \, d\sigma + \cdots + \iint_{S_n} G \, d\sigma.$$

When $S$ is partitioned by smooth curves into a finite number of smooth patches with non-overlapping interiors (i.e., if $S$ is piecewise smooth), then the integral over $S$ is the sum of the integrals over the patches. Thus, the integral of a function over the surface of a cube is the sum of the integrals over the faces of the cube. We integrate over a turtle shell of welded plates by integrating over one plate at a time and adding the results.

**EXAMPLE 2**   Integrate $G(x, y, z) = xyz$ over the surface of the cube cut from the first octant by the planes $x = 1$, $y = 1$, and $z = 1$ (Figure 16.47).

**Solution**   We integrate $xyz$ over each of the six sides and add the results. Since $xyz = 0$ on the sides that lie in the coordinate planes, the integral over the surface of the cube reduces to

$$\iint_{\substack{\text{Cube} \\ \text{surface}}} xyz \, d\sigma = \iint_{\text{Side } A} xyz \, d\sigma + \iint_{\text{Side } B} xyz \, d\sigma + \iint_{\text{Side } C} xyz \, d\sigma.$$

Side $A$ is the surface $f(x, y, z) = z = 1$ over the square region $R_{xy}$: $0 \le x \le 1$, $0 \le y \le 1$, in the $xy$-plane. For this surface and region,

$$\mathbf{p} = \mathbf{k}, \qquad \nabla f = \mathbf{k}, \qquad |\nabla f| = 1, \qquad |\nabla f \cdot \mathbf{p}| = |\mathbf{k} \cdot \mathbf{k}| = 1$$

$$d\sigma = \frac{|\nabla f|}{|\nabla f \cdot \mathbf{p}|} \, dA = \frac{1}{1} \, dx \, dy = dx \, dy$$

$$xyz = xy(1) = xy$$

and

$$\iint_{\text{Side } A} xyz \, d\sigma = \iint_{R_{xy}} xy \, dx \, dy = \int_0^1 \int_0^1 xy \, dx \, dy = \int_0^1 \frac{y}{2} \, dy = \frac{1}{4}.$$

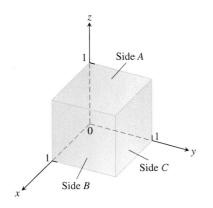

**FIGURE 16.47**   The cube in Example 2.

Symmetry tells us that the integrals of $xyz$ over sides $B$ and $C$ are also $1/4$. Hence,

$$\iint_{\substack{Cube \\ surface}} xyz \, d\sigma = \frac{1}{4} + \frac{1}{4} + \frac{1}{4} = \frac{3}{4}.$$ ∎

**EXAMPLE 3**　Integrate $G(x, y, z) = \sqrt{1 - x^2 - y^2}$ over the "football" surface $S$ formed by rotating the curve $x = \cos z$, $y = 0$, $-\pi/2 \leq z \leq \pi/2$, around the $z$-axis.

**Solution**　The surface is displayed in Figure 16.43, and in Example 6 of Section 16.5 we found the parametrization

$$x = \cos u \cos v, \quad y = \cos u \sin v, \quad z = u, \quad -\frac{\pi}{2} \leq u \leq \frac{\pi}{2} \quad \text{and} \quad 0 \leq v \leq 2\pi,$$

where $v$ represents the angle of rotation from the $xz$-plane about the $z$-axis. Substituting this parametrization into the expression for $G$ gives

$$\sqrt{1 - x^2 - y^2} = \sqrt{1 - (\cos^2 u)(\cos^2 v + \sin^2 v)} = \sqrt{1 - \cos^2 u} = |\sin u|.$$

The surface area differential for the parametrization was found to be (Example 6, Section 16.5)

$$d\sigma = \cos u \sqrt{1 + \sin^2 u} \, du \, dv.$$

These calculations give the surface integral

$$\iint_S \sqrt{1 - x^2 - y^2} \, d\sigma = \int_0^{2\pi} \int_{-\pi/2}^{\pi/2} |\sin u| \cos u \sqrt{1 + \sin^2 u} \, du \, dv$$

$$= 2 \int_0^{2\pi} \int_0^{\pi/2} \sin u \cos u \sqrt{1 + \sin^2 u} \, du \, dv$$

$$= \int_0^{2\pi} \int_1^2 \sqrt{w} \, dw \, dv \qquad \begin{array}{l} w = 1 + \sin^2 u, \\ dw = 2 \sin u \cos u \, du \\ \text{When } u = 0, w = 1. \\ \text{When } u = \pi/2, w = 2. \end{array}$$

$$= 2\pi \cdot \frac{2}{3} w^{3/2} \Big]_1^2 = \frac{4\pi}{3} \left( 2\sqrt{2} - 1 \right).$$ ∎

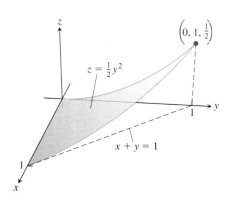

$z = \frac{1}{2}y^2$

$\left(0, 1, \frac{1}{2}\right)$

$x + y = 1$

**FIGURE 16.48**　The surface $S$ in Example 4.

**EXAMPLE 4**　Evaluate $\iint_S \sqrt{x(1 + 2z)} \, d\sigma$　on the portion of the cylinder $z = y^2/2$ over the triangular region $R: x \geq 0, y \geq 0, x + y \leq 1$ in the $xy$-plane (Figure 16.48).

**Solution**　The function $G$ on the surface $S$ is given by

$$G(x, y, z) = \sqrt{x(1 + 2z)} = \sqrt{x}\sqrt{1 + y^2}.$$

With $z = f(x, y) = y^2/2$, we use Equation (4) to evaluate the surface integral:

$$d\sigma = \sqrt{f_x{}^2 + f_y{}^2 + 1} \, dx \, dy = \sqrt{0 + y^2 + 1} \, dx \, dy$$

and

$$\iint\limits_S G(x, y, z)\, d\sigma = \iint\limits_R \left(\sqrt{x}\sqrt{1 + y^2}\right)\sqrt{1 + y^2}\, dx\, dy$$

$$= \int_0^1 \int_0^{1-x} \sqrt{x}\,(1 + y^2)\, dy\, dx$$

$$= \int_0^1 \sqrt{x}\left[(1 - x) + \frac{1}{3}(1 - x)^3\right] dx \qquad \text{Integrate and evaluate.}$$

$$= \int_0^1 \left(\frac{4}{3}x^{1/2} - 2x^{3/2} + x^{5/2} - \frac{1}{3}x^{7/2}\right) dx \qquad \text{Routine algebra}$$

$$= \left[\frac{8}{9}x^{3/2} - \frac{4}{5}x^{5/2} + \frac{2}{7}x^{7/2} - \frac{2}{27}x^{9/2}\right]_0^1$$

$$= \frac{8}{9} - \frac{4}{5} + \frac{2}{7} - \frac{2}{27} = \frac{284}{945} \approx 0.30. \qquad \blacksquare$$

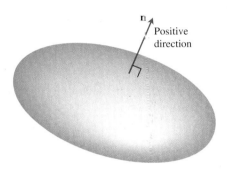

**FIGURE 16.49** Smooth closed surfaces in space are orientable. The outward unit normal vector defines the positive direction at each point.

**FIGURE 16.50** To make a Möbius band, take a rectangular strip of paper *abcd*, give the end *bc* a single twist, and paste the ends of the strip together to match *a* with *c* and *b* with *d*. The Möbius band is a nonorientable or one-sided surface.

## Orientation of a Surface

The curve $C$ in a line integral inherits a natural orientation from its parametrization $\mathbf{r}(t)$ because the parameter belongs to an interval $a \le t \le b$ directed by the real line. The unit tangent vector $\mathbf{T}$ along $C$ points in this forward direction. For a surface $S$, the parametrization $\mathbf{r}(u, v)$ gives a vector $\mathbf{r}_u \times \mathbf{r}_v$ that is normal to the surface, but if $S$ has two "sides," then at each point the negative $-(\mathbf{r}_u \times \mathbf{r}_v)$ is also normal to the surface, so we need to choose which direction to use. For example, if you look at the sphere in Figure 16.38, at any point on the sphere there is a normal vector pointing inward toward the center of the sphere and another opposite normal pointing outward. When we specify which of these normals we are going to use consistently across the entire surface, the surface is given an *orientation*. A smooth surface $S$ is **orientable** (or **two-sided**) if it is possible to define a field of unit normal vectors $\mathbf{n}$ on $S$ which varies continuously with position. Any patch or subportion of an orientable surface is orientable. Spheres and other smooth closed surfaces in space (smooth surfaces that enclose solids) are orientable. By convention, we usually choose $\mathbf{n}$ on a closed surface to point outward.

Once $\mathbf{n}$ has been chosen, we say that we have **oriented** the surface, and we call the surface together with its normal field an **oriented surface**. The vector $\mathbf{n}$ at any point is called the **positive direction** at that point (Figure 16.49).

The Möbius band in Figure 16.50 is not orientable. No matter where you start to construct a continuous unit normal field (shown as the shaft of a thumbtack in the figure), moving the vector continuously around the surface in the manner shown will return it to the starting point with a direction opposite to the one it had when it started out. The vector at that point cannot point both ways and yet it must if the field is to be continuous. We conclude that no such field exists.

## Surface Integrals of Vector Fields

In Section 16.2 we defined the line integral of a vector field along a path $C$ as $\int_C \mathbf{F} \cdot \mathbf{T}\, ds$, where $\mathbf{T}$ is the unit tangent vector to the path pointing in the forward oriented direction. By analogy we now have the following corresponding definition for surface integrals.

---

**DEFINITION** Let **F** be a vector field in three-dimensional space with continuous components defined over a smooth surface $S$ having a chosen field of normal unit vectors **n** orienting $S$. Then the **surface integral of F over S** is

$$\iint_S \mathbf{F} \cdot \mathbf{n} \, d\sigma. \tag{5}$$

---

The surface integral of **F** is also called the **flux** of the vector field across the oriented surface $S$ (analogous to the definition of flux of a vector field in the $xy$-plane across a closed curve in the plane, as defined in Section 16.2). The expression $\mathbf{F} \cdot \mathbf{n} \, d\sigma$ in the integral (5) is also written as $\mathbf{F} \cdot d\boldsymbol{\sigma}$, which corresponds to the notation $\mathbf{F} \cdot d\mathbf{r}$ used for $\mathbf{F} \cdot \mathbf{T} \, ds$ in line integrals for vector fields. If **F** is the velocity field of a three-dimensional fluid flow, then the flux of **F** across $S$ is the net rate at which fluid is crossing $S$ per unit time in the chosen positive direction **n** defined by the orientation of $S$. Fluid flows are discussed in more detail in Section 16.7, so here we focus on several examples calculating surface integrals of vector fields.

**EXAMPLE 5** Find the flux of $\mathbf{F} = yz\mathbf{i} + x\mathbf{j} - z^2\mathbf{k}$ through the parabolic cylinder $y = x^2$, $0 \le x \le 1$, $0 \le z \le 4$, in the direction **n** indicated in Figure 16.51.

**Solution** On the surface we have $x = x$, $y = x^2$, and $z = z$, so we automatically have the parametrization $\mathbf{r}(x, z) = x\mathbf{i} + x^2\mathbf{j} + z\mathbf{k}$, $0 \le x \le 1$, $0 \le z \le 4$. The cross product of tangent vectors is

$$\mathbf{r}_x \times \mathbf{r}_z = \begin{vmatrix} \mathbf{i} & \mathbf{j} & \mathbf{k} \\ 1 & 2x & 0 \\ 0 & 0 & 1 \end{vmatrix} = 2x\mathbf{i} - \mathbf{j}.$$

**FIGURE 16.51** Finding the flux through the surface of a parabolic cylinder (Example 5).

The unit normal vectors pointing outward from the surface as indicated in Figure 16.51 are

$$\mathbf{n} = \frac{\mathbf{r}_x \times \mathbf{r}_z}{|\mathbf{r}_x \times \mathbf{r}_z|} = \frac{2x\mathbf{i} - \mathbf{j}}{\sqrt{4x^2 + 1}}.$$

On the surface, $y = x^2$, so the vector field there is

$$\mathbf{F} = yz\mathbf{i} + x\mathbf{j} - z^2\mathbf{k} = x^2z\mathbf{i} + x\mathbf{j} - z^2\mathbf{k}.$$

Thus,

$$\mathbf{F} \cdot \mathbf{n} = \frac{1}{\sqrt{4x^2 + 1}} \big( (x^2z)(2x) + (x)(-1) + (-z^2)(0) \big)$$

$$= \frac{2x^3z - x}{\sqrt{4x^2 + 1}}.$$

The flux of **F** outward through the surface is

$$\iint\limits_{S} \mathbf{F} \cdot \mathbf{n}\, d\sigma = \int_{0}^{4}\int_{0}^{1} \frac{2x^3z - x}{\sqrt{4x^2 + 1}} \left| \mathbf{r}_x \times \mathbf{r}_z \right| dx\, dz$$

$$= \int_{0}^{4}\int_{0}^{1} \frac{2x^3z - x}{\sqrt{4x^2 + 1}} \sqrt{4x^2 + 1}\, dx\, dz$$

$$= \int_{0}^{4}\int_{0}^{1} (2x^3z - x)\, dx\, dz = \int_{0}^{4} \left[\frac{1}{2}x^4z - \frac{1}{2}x^2\right]_{x=0}^{x=1} dz$$

$$= \int_{0}^{4} \frac{1}{2}(z - 1)\, dz = \frac{1}{4}(z - 1)^2 \Big]_{0}^{4}$$

$$= \frac{1}{4}(9) - \frac{1}{4}(1) = 2. \qquad\blacksquare$$

There is a simple formula for the flux of **F** across a parametrized surface $\mathbf{r}(u, v)$. Since

$$d\sigma = \left| \mathbf{r}_u \times \mathbf{r}_v \right| du\, dv,$$

with the orientation

$$\mathbf{n} = \frac{\mathbf{r}_u \times \mathbf{r}_v}{\left| \mathbf{r}_u \times \mathbf{r}_v \right|}$$

**Flux Across a Parametrized Surface**

$$\text{Flux} = \iint\limits_{R} \mathbf{F} \cdot (\mathbf{r}_u \times \mathbf{r}_v)\, du\, dv$$

it follows that

$$\iint\limits_{S} \mathbf{F} \cdot \mathbf{n}\, d\sigma = \iint\limits_{R} \mathbf{F} \cdot \frac{\mathbf{r}_u \times \mathbf{r}_v}{\left| \mathbf{r}_u \times \mathbf{r}_v \right|} \left| \mathbf{r}_u \times \mathbf{r}_v \right| du\, dv = \iint\limits_{R} \mathbf{F} \cdot (\mathbf{r}_u \times \mathbf{r}_v)\, du\, dv.$$

This integral for flux simplifies the computation in Example 5. Since

$$\mathbf{F} \cdot (\mathbf{r}_x \times \mathbf{r}_z) = (x^2z)(2x) + (x)(-1) = 2x^3z - x,$$

we obtain directly

$$\text{Flux} = \iint\limits_{S} \mathbf{F} \cdot \mathbf{n}\, d\sigma = \int_{0}^{4}\int_{0}^{1} (2x^3z - x)\, dx\, dz = 2$$

in Example 5.

If $S$ is part of a level surface $g(x, y, z) = c$, then **n** may be taken to be one of the two fields

$$\mathbf{n} = \pm \frac{\nabla g}{\left| \nabla g \right|}, \tag{6}$$

depending on which one gives the preferred direction. The corresponding flux is

$$\text{Flux} = \iint\limits_{S} \mathbf{F} \cdot \mathbf{n}\, d\sigma$$

$$= \iint\limits_{R} \left( \mathbf{F} \cdot \frac{\pm \nabla g}{\left| \nabla g \right|} \right) \frac{\left| \nabla g \right|}{\left| \nabla g \cdot \mathbf{p} \right|}\, dA \qquad \text{Eqs. (6) and (3)}$$

$$= \iint\limits_{R} \mathbf{F} \cdot \frac{\pm \nabla g}{\left| \nabla g \cdot \mathbf{p} \right|}\, dA. \tag{7}$$

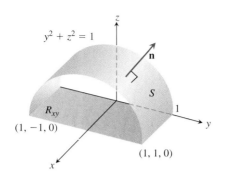

**FIGURE 16.52** Calculating the flux of a vector field outward through the surface S. The area of the shadow region $R_{xy}$ is 2 (Example 6).

**EXAMPLE 6**  Find the flux of $\mathbf{F} = yz\mathbf{j} + z^2\mathbf{k}$ outward through the surface S cut from the cylinder $y^2 + z^2 = 1$, $z \geq 0$, by the planes $x = 0$ and $x = 1$.

**Solution**  The outward normal field on S (Figure 16.52) may be calculated from the gradient of $g(x, y, z) = y^2 + z^2$ to be

$$\mathbf{n} = +\frac{\nabla g}{|\nabla g|} = \frac{2y\mathbf{j} + 2z\mathbf{k}}{\sqrt{4y^2 + 4z^2}} = \frac{2y\mathbf{j} + 2z\mathbf{k}}{2\sqrt{1}} = y\mathbf{j} + z\mathbf{k}.$$

With $\mathbf{p} = \mathbf{k}$, we also have

$$d\sigma = \frac{|\nabla g|}{|\nabla g \cdot \mathbf{k}|} dA = \frac{2}{|2z|} dA = \frac{1}{z} dA.$$

We can drop the absolute value bars because $z \geq 0$ on S.

The value of $\mathbf{F} \cdot \mathbf{n}$ on the surface is

$$\mathbf{F} \cdot \mathbf{n} = (yz\mathbf{j} + z^2\mathbf{k}) \cdot (y\mathbf{j} + z\mathbf{k})$$
$$= y^2 z + z^3 = z(y^2 + z^2) \qquad$$
$$= z. \qquad\qquad y^2 + z^2 = 1 \text{ on } S$$

The surface projects onto the shadow region $R_{xy}$, which is the rectangle in the $xy$-plane shown in Figure 16.52. Therefore, the flux of $\mathbf{F}$ outward through S is

$$\iint_S \mathbf{F} \cdot \mathbf{n} \, d\sigma = \iint_{R_{xy}} (z)\left(\frac{1}{z} dA\right) = \iint_{R_{xy}} dA = \text{area}(R_{xy}) = 2. \qquad \blacksquare$$

### Moments and Masses of Thin Shells

Thin shells of material like bowls, metal drums, and domes are modeled with surfaces. Their moments and masses are calculated with the formulas in Table 16.3. The derivations are similar to those in Section 6.6. The formulas are like those for line integrals in Table 16.1, Section 16.1.

---

**TABLE 16.3** Mass and moment formulas for very thin shells

**Mass:**  $M = \iint_S \delta \, d\sigma$    $\delta = \delta(x, y, z) = $ density at $(x, y, z)$ as mass per unit area

**First moments about the coordinate planes:**

$$M_{yz} = \iint_S x \, \delta \, d\sigma, \qquad M_{xz} = \iint_S y \, \delta \, d\sigma, \qquad M_{xy} = \iint_S z \, \delta \, d\sigma$$

**Coordinates of center of mass:**

$$\bar{x} = M_{yz}/M, \qquad \bar{y} = M_{xz}/M, \qquad \bar{z} = M_{xy}/M$$

**Moments of inertia about coordinate axes:**

$$I_x = \iint_S (y^2 + z^2) \, \delta \, d\sigma, \quad I_y = \iint_S (x^2 + z^2) \, \delta \, d\sigma, \quad I_z = \iint_S (x^2 + y^2) \, \delta \, d\sigma,$$

$$I_L = \iint_S r^2 \delta \, d\sigma \qquad r(x, y, z) = \text{distance from point } (x, y, z) \text{ to line}$$

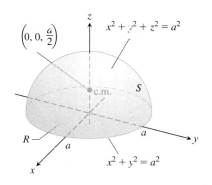

**FIGURE 16.53** The center of mass of a thin hemispherical shell of constant density lies on the axis of symmetry halfway from the base to the top (Example 7).

**EXAMPLE 7**    Find the center of mass of a thin hemispherical shell of radius $a$ and constant density $\delta$.

**Solution**    We model the shell with the hemisphere

$$f(x, y, z) = x^2 + y^2 + z^2 = a^2, \qquad z \ge 0$$

(Figure 16.53). The symmetry of the surface about the $z$-axis tells us that $\bar{x} = \bar{y} = 0$. It remains only to find $\bar{z}$ from the formula $\bar{z} = M_{xy}/M$.

The mass of the shell is

$$M = \iint_S \delta \, d\sigma = \delta \iint_S d\sigma = (\delta)(\text{area of } S) = 2\pi a^2 \delta. \qquad \text{$\delta$ = constant}$$

To evaluate the integral for $M_{xy}$, we take $\mathbf{p} = \mathbf{k}$ and calculate

$$|\nabla f| = |2x\mathbf{i} + 2y\mathbf{j} + 2z\mathbf{k}| = 2\sqrt{x^2 + y^2 + z^2} = 2a$$

$$|\nabla f \cdot \mathbf{p}| = |\nabla f \cdot \mathbf{k}| = |2z| = 2z$$

$$d\sigma = \frac{|\nabla f|}{|\nabla f \cdot \mathbf{p}|} dA = \frac{a}{z} dA.$$

Then

$$M_{xy} = \iint_S z\delta \, d\sigma = \delta \iint_R z\frac{a}{z} dA = \delta a \iint_R dA = \delta a(\pi a^2) = \delta\pi a^3$$

$$\bar{z} = \frac{M_{xy}}{M} = \frac{\pi a^3 \delta}{2\pi a^2 \delta} = \frac{a}{2}.$$

The shell's center of mass is the point $(0, 0, a/2)$.

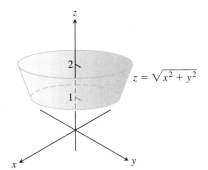

**FIGURE 16.54** The cone frustum formed when the cone $z = \sqrt{x^2 + y^2}$ is cut by the planes $z = 1$ and $z = 2$ (Example 8).

**EXAMPLE 8**    Find the center of mass of a thin shell of density $\delta = 1/z^2$ cut from the cone $z = \sqrt{x^2 + y^2}$ by the planes $z = 1$ and $z = 2$ (Figure 16.54).

**Solution**    The symmetry of the surface about the $z$-axis tells us that $\bar{x} = \bar{y} = 0$. We find $\bar{z} = M_{xy}/M$. Working as in Example 4 of Section 16.5, we have

$$\mathbf{r}(r, \theta) = (r\cos\theta)\mathbf{i} + (r\sin\theta)\mathbf{j} + r\mathbf{k}, \qquad 1 \le r \le 2, \quad 0 \le \theta \le 2\pi,$$

and

$$|\mathbf{r}_r \times \mathbf{r}_\theta| = \sqrt{2}\, r.$$

Therefore,

$$M = \iint_S \delta \, d\sigma = \int_0^{2\pi} \int_1^2 \frac{1}{r^2}\sqrt{2}\, r \, dr \, d\theta$$

$$= \sqrt{2} \int_0^{2\pi} \Big[\ln r\Big]_1^2 \, d\theta = \sqrt{2} \int_0^{2\pi} \ln 2 \, d\theta$$

$$= 2\pi\sqrt{2} \ln 2,$$

$$M_{xy} = \iint_S \delta z \, d\sigma = \int_0^{2\pi} \int_1^2 \frac{1}{r^2} r\sqrt{2}r \, dr \, d\theta$$

$$= \sqrt{2} \int_0^{2\pi} \int_1^2 dr \, d\theta$$

$$= \sqrt{2} \int_0^{2\pi} d\theta = 2\pi\sqrt{2},$$

$$\bar{z} = \frac{M_{xy}}{M} = \frac{2\pi\sqrt{2}}{2\pi\sqrt{2}\ln 2} = \frac{1}{\ln 2}.$$

The shell's center of mass is the point $(0, 0, 1/\ln 2)$. ∎

## Exercises 16.6

### Surface Integrals of Scalar Functions

In Exercises 1–8, integrate the given function over the given surface.

1. **Parabolic cylinder** $G(x, y, z) = x$, over the parabolic cylinder $y = x^2, 0 \le x \le 2, 0 \le z \le 3$

2. **Circular cylinder** $G(x, y, z) = z$, over the cylindrical surface $y^2 + z^2 = 4, z \ge 0, 1 \le x \le 4$

3. **Sphere** $G(x, y, z) = x^2$, over the unit sphere $x^2 + y^2 + z^2 = 1$

4. **Hemisphere** $G(x, y, z) = z^2$, over the hemisphere $x^2 + y^2 + z^2 = a^2, z \ge 0$

5. **Portion of plane** $F(x, y, z) = z$, over the portion of the plane $x + y + z = 4$ that lies above the square $0 \le x \le 1$, $0 \le y \le 1$, in the $xy$-plane

6. **Cone** $F(x, y, z) = z - x$, over the cone $z = \sqrt{x^2 + y^2}$, $0 \le z \le 1$

7. **Parabolic dome** $H(x, y, z) = x^2\sqrt{5 - 4z}$, over the parabolic dome $z = 1 - x^2 - y^2, z \ge 0$

8. **Spherical cap** $H(x, y, z) = yz$, over the part of the sphere $x^2 + y^2 + z^2 = 4$ that lies above the cone $z = \sqrt{x^2 + y^2}$

9. Integrate $G(x, y, z) = x + y + z$ over the surface of the cube cut from the first octant by the planes $x = a, y = a, z = a$.

10. Integrate $G(x, y, z) = y + z$ over the surface of the wedge in the first octant bounded by the coordinate planes and the planes $x = 2$ and $y + z = 1$.

11. Integrate $G(x, y, z) = xyz$ over the surface of the rectangular solid cut from the first octant by the planes $x = a, y = b$, and $z = c$.

12. Integrate $G(x, y, z) = xyz$ over the surface of the rectangular solid bounded by the planes $x = \pm a, y = \pm b$, and $z = \pm c$.

13. Integrate $G(x, y, z) = x + y + z$ over the portion of the plane $2x + 2y + z = 2$ that lies in the first octant.

14. Integrate $G(x, y, z) = x\sqrt{y^2 + 4}$ over the surface cut from the parabolic cylinder $y^2 + 4z = 16$ by the planes $x = 0$, $x = 1$, and $z = 0$.

15. Integrate $G(x, y, z) = z - x$ over the portion of the graph of $z = x + y^2$ above the triangle in the $xy$-plane having vertices (0, 0, 0), (1, 1, 0), and (0, 1, 0). (See accompanying figure.)

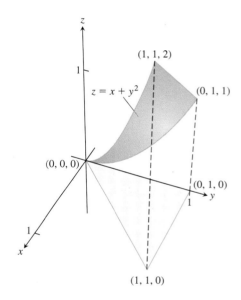

16. Integrate $G(x, y, z) = x$ over the surface given by

$$z = x^2 + y \quad \text{for} \quad 0 \le x \le 1, \quad -1 \le y \le 1.$$

17. Integrate $G(x, y, z) = xyz$ over the triangular surface with vertices (1, 0, 0), (0, 2, 0), and (0, 1, 1).

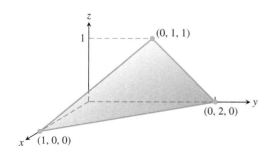

18. Integrate $G(x, y, z) = x - y - z$ over the portion of the plane $x + y = 1$ in the first octant between $z = 0$ and $z = 1$ (see the accompanying figure on the next page).

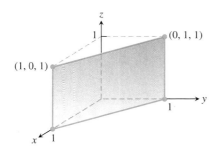

### Finding Flux or Surface Integrals of Vector Fields

In Exercises 19–28, use a parametrization to find the flux $\iint_S \mathbf{F} \cdot \mathbf{n} \, d\sigma$ across the surface in the specified direction.

**19. Parabolic cylinder** $\mathbf{F} = z^2\mathbf{i} + x\mathbf{j} - 3z\mathbf{k}$ outward (normal away from the x-axis) through the surface cut from the parabolic cylinder $z = 4 - y^2$ by the planes $x = 0, x = 1$, and $z = 0$

**20. Parabolic cylinder** $\mathbf{F} = x^2\mathbf{j} - xz\mathbf{k}$ outward (normal away from the yz-plane) through the surface cut from the parabolic cylinder $y = x^2, -1 \le x \le 1$, by the planes $z = 0$ and $z = 2$

**21. Sphere** $\mathbf{F} = z\mathbf{k}$ across the portion of the sphere $x^2 + y^2 + z^2 = a^2$ in the first octant in the direction away from the origin

**22. Sphere** $\mathbf{F} = x\mathbf{i} + y\mathbf{j} + z\mathbf{k}$ across the sphere $x^2 + y^2 + z^2 = a^2$ in the direction away from the origin

**23. Plane** $\mathbf{F} = 2xy\mathbf{i} + 2yz\mathbf{j} + 2xz\mathbf{k}$ upward across the portion of the plane $x + y + z = 2a$ that lies above the square $0 \le x \le a$, $0 \le y \le a$, in the xy-plane

**24. Cylinder** $\mathbf{F} = x\mathbf{i} + y\mathbf{j} + z\mathbf{k}$ outward through the portion of the cylinder $x^2 + y^2 = 1$ cut by the planes $z = 0$ and $z = a$

**25. Cone** $\mathbf{F} = xy\mathbf{i} - z\mathbf{k}$ outward (normal away from the z-axis) through the cone $z = \sqrt{x^2 + y^2}, 0 \le z \le 1$

**26. Cone** $\mathbf{F} = y^2\mathbf{i} + xz\mathbf{j} - \mathbf{k}$ outward (normal away from the z-axis) through the cone $z = 2\sqrt{x^2 + y^2}, 0 \le z \le 2$

**27. Cone frustum** $\mathbf{F} = -x\mathbf{i} - y\mathbf{j} + z^2\mathbf{k}$ outward (normal away from the z-axis) through the portion of the cone $z = \sqrt{x^2 + y^2}$ between the planes $z = 1$ and $z = 2$

**28. Paraboloid** $\mathbf{F} = 4x\mathbf{i} + 4y\mathbf{j} + 2\mathbf{k}$ outward (normal away from the z-axis) through the surface cut from the bottom of the paraboloid $z = x^2 + y^2$ by the plane $z = 1$

In Exercises 29 and 30, find the surface integral of the field $\mathbf{F}$ over the portion of the given surface in the specified direction.

**29.** $\mathbf{F}(x, y, z) = -\mathbf{i} + 2\mathbf{j} - 3\mathbf{k}$

$S$: rectangular surface $z = 0$, $\quad 0 \le x \le 2$, $\quad 0 \le y \le 3$, direction $\mathbf{k}$

**30.** $\mathbf{F}(x, y, z) = yx^2\mathbf{i} - 2\mathbf{j} + xz\mathbf{k}$

$S$: rectangular surface $y = 0$, $\quad -1 \le x \le 2$, $\quad 2 \le z \le 7$, direction $-\mathbf{j}$

In Exercises 31–36, use Equation (7) to find the surface integral of the field $\mathbf{F}$ over the portion of the sphere $x^2 + y^2 + z^2 = a^2$ in the first octant in the direction away from the origin.

**31.** $\mathbf{F}(x, y, z) = z\mathbf{k}$

**32.** $\mathbf{F}(x, y, z) = -y\mathbf{i} + x\mathbf{j}$

**33.** $\mathbf{F}(x, y, z) = y\mathbf{i} - x\mathbf{j} + \mathbf{k}$

**34.** $\mathbf{F}(x, y, z) = zx\mathbf{i} + zy\mathbf{j} + z^2\mathbf{k}$

**35.** $\mathbf{F}(x, y, z) = x\mathbf{i} + y\mathbf{j} + z\mathbf{k}$

**36.** $\mathbf{F}(x, y, z) = \dfrac{x\mathbf{i} + y\mathbf{j} + z\mathbf{k}}{\sqrt{x^2 + y^2 + z^2}}$

**37.** Find the flux of the field $\mathbf{F}(x, y, z) = z^2\mathbf{i} + x\mathbf{j} - 3z\mathbf{k}$ outward through the surface cut from the parabolic cylinder $z = 4 - y^2$ by the planes $x = 0, x = 1$, and $z = 0$.

**38.** Find the flux of the field $\mathbf{F}(x, y, z) = 4x\mathbf{i} + 4y\mathbf{j} + 2\mathbf{k}$ outward (away from the z-axis) through the surface cut from the bottom of the paraboloid $z = x^2 + y^2$ by the plane $z = 1$.

**39.** Let $S$ be the portion of the cylinder $y = e^x$ in the first octant that projects parallel to the x-axis onto the rectangle $R_{yz}$: $1 \le y \le 2$, $0 \le z \le 1$ in the yz-plane (see the accompanying figure). Let $\mathbf{n}$ be the unit vector normal to $S$ that points away from the yz-plane. Find the flux of the field $\mathbf{F}(x, y, z) = -2\mathbf{i} + 2y\mathbf{j} + z\mathbf{k}$ across $S$ in the direction of $\mathbf{n}$.

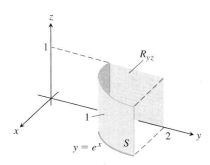

**40.** Let $S$ be the portion of the cylinder $y = \ln x$ in the first octant whose projection parallel to the y-axis onto the xz-plane is the rectangle $R_{xz}$: $1 \le x \le e, 0 \le z \le 1$. Let $\mathbf{n}$ be the unit vector normal to $S$ that points away from the xz-plane. Find the flux of $\mathbf{F} = 2y\mathbf{j} + z\mathbf{k}$ through $S$ in the direction of $\mathbf{n}$.

**41.** Find the outward flux of the field $\mathbf{F} = 2xy\mathbf{i} + 2yz\mathbf{j} + 2xz\mathbf{k}$ across the surface of the cube cut from the first octant by the planes $x = a, y = a, z = a$.

**42.** Find the outward flux of the field $\mathbf{F} = xz\mathbf{i} + yz\mathbf{j} + \mathbf{k}$ across the surface of the upper cap cut from the solid sphere $x^2 + y^2 + z^2 \le 25$ by the plane $z = 3$.

### Moments and Masses

**43. Centroid** Find the centroid of the portion of the sphere $x^2 + y^2 + z^2 = a^2$ that lies in the first octant.

**44. Centroid** Find the centroid of the surface cut from the cylinder $y^2 + z^2 = 9, z \ge 0$, by the planes $x = 0$ and $x = 3$ (resembles the surface in Example 6).

**45. Thin shell of constant density** Find the center of mass and the moment of inertia about the z-axis of a thin shell of constant density $\delta$ cut from the cone $x^2 + y^2 - z^2 = 0$ by the planes $z = 1$ and $z = 2$.

**46. Conical surface of constant density** Find the moment of inertia about the z-axis of a thin shell of constant density $\delta$ cut from the cone $4x^2 + 4y^2 - z^2 = 0, z \ge 0$, by the circular cylinder $x^2 + y^2 = 2x$ (see the accompanying figure).

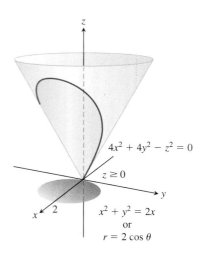

$4x^2 + 4y^2 - z^2 = 0$

$z \geq 0$

$x^2 + y^2 = 2x$

or

$r = 2 \cos \theta$

**47. Spherical shells**

    **a.** Find the moment of inertia about a diameter of a thin spherical shell of radius $a$ and constant density $\delta$. (Work with a hemispherical shell and double the result.)

    **b.** Use the Parallel Axis Theorem (Exercises 15.6) and the result in part (a) to find the moment of inertia about a line tangent to the shell.

**48. Conical Surface** Find the centroid of the lateral surface of a solid cone of base radius $a$ and height $h$ (cone surface minus the base).

# $16.7$ Stokes' Theorem

To calculate the counterclockwise circulation of a two-dimensional vector field $\mathbf{F} = M\mathbf{i} + N\mathbf{j}$ around a simple closed curve in the plane, Green's Theorem says we can compute the double integral over the region enclosed by the curve of the scalar quantity $(\partial N/\partial x - \partial M/\partial y)$. This expression is the $\mathbf{k}$-component of a *curl vector* field, which we define in this section, and it measures the rate of rotation of $\mathbf{F}$ at each point in the region around an axis parallel to $\mathbf{k}$. For a vector field on three-dimensional space, the rotation at each point is around an axis that is parallel to the curl vector at that point. When a closed curve $C$ in space is the boundary of an oriented surface, we will see that the circulation of $\mathbf{F}$ around $C$ is equal to the surface integral of the curl vector field. This result extends Green's Theorem from regions in the plane to general surfaces in space having a smooth boundary curve.

## The Curl Vector Field

Suppose that $\mathbf{F}$ is the velocity field of a fluid flowing in space. Particles near the point $(x, y, z)$ in the fluid tend to rotate around an axis through $(x, y, z)$ that is parallel to a certain vector we are about to define. This vector points in the direction for which the rotation is counterclockwise when viewed looking down onto the plane of the circulation from the tip of the arrow representing the vector. This is the direction your right-hand thumb points when your fingers curl around the axis of rotation in the way consistent with the rotating motion of the particles in the fluid (see Figure 16.55). The length of the vector measures the rate of rotation. The vector is called the **curl vector**, and for the vector field $\mathbf{F} = M\mathbf{i} + N\mathbf{j} + P\mathbf{k}$ it is defined to be

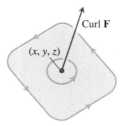

**FIGURE 16.55** The circulation vector at a point $(x, y, z)$ in a plane in a three-dimensional fluid flow. Notice its right-hand relation to the rotating particles in the fluid.

$$\text{curl } \mathbf{F} = \left(\frac{\partial P}{\partial y} - \frac{\partial N}{\partial z}\right)\mathbf{i} + \left(\frac{\partial M}{\partial z} - \frac{\partial P}{\partial x}\right)\mathbf{j} + \left(\frac{\partial N}{\partial x} - \frac{\partial M}{\partial y}\right)\mathbf{k}. \qquad (1)$$

This information is a consequence of Stokes' Theorem, the generalization to space of the circulation-curl form of Green's Theorem and the subject of this section.

Notice that $(\text{curl } \mathbf{F}) \cdot \mathbf{k} = (\partial N/\partial x - \partial M/\partial y)$ is consistent with our definition in Section 16.4 when $\mathbf{F} = M(x, y)\mathbf{i} + N(x, y)\mathbf{j}$. The formula for curl $\mathbf{F}$ in Equation (1) is often expressed with the symbolic operator

$$\nabla = \mathbf{i}\frac{\partial}{\partial x} + \mathbf{j}\frac{\partial}{\partial y} + \mathbf{k}\frac{\partial}{\partial z} \qquad (2)$$

to compute the curl of **F** as

$$\nabla \times \mathbf{F} = \begin{vmatrix} \mathbf{i} & \mathbf{j} & \mathbf{k} \\ \dfrac{\partial}{\partial x} & \dfrac{\partial}{\partial y} & \dfrac{\partial}{\partial z} \\ M & N & P \end{vmatrix}$$

$$= \left( \frac{\partial P}{\partial y} - \frac{\partial N}{\partial z} \right)\mathbf{i} + \left( \frac{\partial M}{\partial z} - \frac{\partial P}{\partial x} \right)\mathbf{j} + \left( \frac{\partial N}{\partial x} - \frac{\partial M}{\partial y} \right)\mathbf{k}.$$

The symbol $\nabla$ is pronounced "del," and we often use this cross product notation to write the curl symbolically as "del cross **F**."

$$\text{curl } \mathbf{F} = \nabla \times \mathbf{F} \qquad (3)$$

**EXAMPLE 1**     Find the curl of $\mathbf{F} = (x^2 - z)\mathbf{i} + xe^z\mathbf{j} + xy\mathbf{k}$.

**Solution**   We use Equation (3) and the determinant form, so

$$\text{curl } \mathbf{F} = \nabla \times \mathbf{F}$$

$$= \begin{vmatrix} \mathbf{i} & \mathbf{j} & \mathbf{k} \\ \dfrac{\partial}{\partial x} & \dfrac{\partial}{\partial y} & \dfrac{\partial}{\partial z} \\ x^2 - z & xe^z & xy \end{vmatrix}$$

$$= \left( \frac{\partial}{\partial y}(xy) - \frac{\partial}{\partial z}(xe^z) \right)\mathbf{i} - \left( \frac{\partial}{\partial x}(xy) - \frac{\partial}{\partial z}(x^2 - z) \right)\mathbf{j}$$

$$+ \left( \frac{\partial}{\partial x}(xe^z) - \frac{\partial}{\partial y}(x^2 - z) \right)\mathbf{k}$$

$$= (x - xe^z)\mathbf{i} - (y + 1)\mathbf{j} + (e^z - 0)\mathbf{k}$$

$$= x(1 - e^z)\mathbf{i} - (y + 1)\mathbf{j} + e^z\mathbf{k}. \qquad \blacksquare$$

As we will see, the operator $\nabla$ has a number of other applications. For instance, when applied to a scalar function $f(x, y, z)$, it gives the gradient of $f$:

$$\nabla f = \frac{\partial f}{\partial x}\mathbf{i} + \frac{\partial f}{\partial y}\mathbf{j} + \frac{\partial f}{\partial z}\mathbf{k}.$$

It is sometimes read as "del $f$" as well as "grad $f$."

### Stokes' Theorem

Stokes' Theorem generalizes Green's Theorem to three dimensions. The circulation-curl form of Green's Theorem relates the counterclockwise circulation of a vector field around a simple closed curve $C$ in the $xy$-plane to a double integral over the plane region $R$ enclosed by $C$. Stokes' Theorem relates the circulation of a vector field around the boundary $C$ of an oriented surface $S$ in space (Figure 16.56) to a surface integral over the surface $S$. We require that the surface be **piecewise smooth**, which means that it is a finite union of smooth surfaces joining along smooth curves.

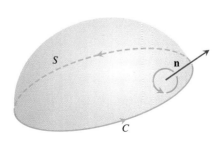

**FIGURE 16.56**   The orientation of the bounding curve $C$ gives it a right-handed relation to the normal field **n**. If the thumb of a right hand points along **n**, the fingers curl in the direction of $C$.

**THEOREM 6—Stokes' Theorem** Let $S$ be a piecewise smooth oriented surface having a piecewise smooth boundary curve $C$. Let $\mathbf{F} = M\mathbf{i} + N\mathbf{j} + P\mathbf{k}$ be a vector field whose components have continuous first partial derivatives on an open region containing $S$. Then the circulation of $\mathbf{F}$ around $C$ in the direction counterclockwise with respect to the surface's unit normal vector $\mathbf{n}$ equals the integral of the curl vector field $\nabla \times \mathbf{F}$ over $S$:

$$\oint_C \mathbf{F} \cdot d\mathbf{r} = \iint_S \nabla \times \mathbf{F} \cdot \mathbf{n}\, d\sigma \qquad (4)$$

Counterclockwise    Curl integral
circulation

Notice from Equation (4) that if two different oriented surfaces $S_1$ and $S_2$ have the same boundary $C$, their curl integrals are equal:

$$\iint_{S_1} \nabla \times \mathbf{F} \cdot \mathbf{n}_1\, d\sigma = \iint_{S_2} \nabla \times \mathbf{F} \cdot \mathbf{n}_2\, d\sigma.$$

Both curl integrals equal the counterclockwise circulation integral on the left side of Equation (4) as long as the unit normal vectors $\mathbf{n}_1$ and $\mathbf{n}_2$ correctly orient the surfaces. So the curl integral is independent of the surface and depends only on circulation along the boundary curve. This independence of surface resembles the path independence for the flow integral of a conservative velocity field along a curve, where the value of the flow integral depends only on the endpoints (that is, the boundary points) of the path. The curl field $\nabla \times \mathbf{F}$ is analogous to the gradient field $\nabla f$ of a scalar function $f$.

If $C$ is a curve in the $xy$-plane, oriented counterclockwise, and $R$ is the region in the $xy$-plane bounded by $C$, then $d\sigma = dx\,dy$ and

$$(\nabla \times \mathbf{F}) \cdot \mathbf{n} = (\nabla \times \mathbf{F}) \cdot \mathbf{k} = \left( \frac{\partial N}{\partial x} - \frac{\partial M}{\partial y} \right).$$

Under these conditions, Stokes' equation becomes

$$\oint_C \mathbf{F} \cdot d\mathbf{r} = \iint_R \left( \frac{\partial N}{\partial x} - \frac{\partial M}{\partial y} \right) dx\,dy,$$

which is the circulation-curl form of the equation in Green's Theorem. Conversely, by reversing these steps we can rewrite the circulation-curl form of Green's Theorem for two-dimensional fields in del notation as

$$\oint_C \mathbf{F} \cdot d\mathbf{r} = \iint_R \nabla \times \mathbf{F} \cdot \mathbf{k}\, dA. \qquad (5)$$

See Figure 16.57.

Green:

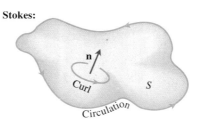

Stokes:

**FIGURE 16.57** Comparison of Green's Theorem and Stokes' Theorem.

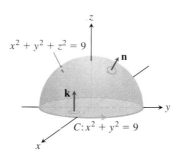

**FIGURE 16.58** A hemisphere and a disk, each with boundary $C$ (Examples 2 and 3).

**EXAMPLE 2** Evaluate Equation (4) for the hemisphere $S: x^2 + y^2 + z^2 = 9,\ z \geq 0$, its bounding circle $C: x^2 + y^2 = 9,\ z = 0$, and the field $\mathbf{F} = y\mathbf{i} - x\mathbf{j}$.

**Solution** The hemisphere looks much like the surface in Figure 16.56 with the bounding circle $C$ in the $xy$-plane (see Figure 16.58). We calculate the counterclockwise circulation around $C$ (as viewed from above) using the parametrization $\mathbf{r}(\theta) = (3 \cos \theta)\mathbf{i} + (3 \sin \theta)\mathbf{j}, 0 \le \theta \le 2\pi$:

$$d\mathbf{r} = (-3 \sin \theta \, d\theta)\mathbf{i} + (3 \cos \theta \, d\theta)\mathbf{j}$$

$$\mathbf{F} = y\mathbf{i} - x\mathbf{j} = (3 \sin \theta)\mathbf{i} - (3 \cos \theta)\mathbf{j}$$

$$\mathbf{F} \cdot d\mathbf{r} = -9 \sin^2 \theta \, d\theta - 9 \cos^2 \theta \, d\theta = -9 \, d\theta$$

$$\oint_C \mathbf{F} \cdot d\mathbf{r} = \int_0^{2\pi} -9 \, d\theta = -18\pi.$$

For the curl integral of $\mathbf{F}$, we have

$$\nabla \times \mathbf{F} = \left( \frac{\partial P}{\partial y} - \frac{\partial N}{\partial z} \right)\mathbf{i} + \left( \frac{\partial M}{\partial z} - \frac{\partial P}{\partial x} \right)\mathbf{j} + \left( \frac{\partial N}{\partial x} - \frac{\partial M}{\partial y} \right)\mathbf{k}$$

$$= (0 - 0)\mathbf{i} + (0 - 0)\mathbf{j} + (-1 - 1)\mathbf{k} = -2\mathbf{k}$$

$$\mathbf{n} = \frac{x\mathbf{i} + y\mathbf{j} + z\mathbf{k}}{\sqrt{x^2 + y^2 + z^2}} = \frac{x\mathbf{i} + y\mathbf{j} + z\mathbf{k}}{3} \qquad \text{Outer unit normal}$$

$$d\sigma = \frac{3}{z} dA \qquad \qquad \text{Section 16.6, Example 7,}$$
$$\text{with } a = 3$$

$$\nabla \times \mathbf{F} \cdot \mathbf{n} \, d\sigma = -\frac{2z}{3} \frac{3}{z} dA = -2 \, dA$$

and

$$\iint_S \nabla \times \mathbf{F} \cdot \mathbf{n} \, d\sigma = \iint_{x^2+y^2 \le 9} -2 \, dA = -18\pi.$$

The circulation around the circle equals the integral of the curl over the hemisphere, as it should from Stokes' Theorem. ∎

The surface integral in Stokes' Theorem can be computed using any surface having boundary curve $C$, provided the surface is properly oriented and lies within the domain of the field $\mathbf{F}$. The next example illustrates this fact for the circulation around the curve $C$ in Example 2.

**EXAMPLE 3** Calculate the circulation around the bounding circle $C$ in Example 2 using the disk of radius 3 centered at the origin in the $xy$-plane as the surface $S$ (instead of the hemisphere). See Figure 16.58.

**Solution** As in Example 2, $\nabla \times \mathbf{F} = -2\mathbf{k}$. For the surface being the described disk in the $xy$-plane, we have the normal vector $\mathbf{n} = \mathbf{k}$ so that

$$\nabla \times \mathbf{F} \cdot \mathbf{n} \, d\sigma = -2\mathbf{k} \cdot \mathbf{k} \, dA = -2 \, dA$$

and

$$\iint_S \nabla \times \mathbf{F} \cdot \mathbf{n} \, d\sigma = \iint_{x^2+y^2 \le 9} -2 \, dA = -18\pi,$$

a simpler calculation than before. ∎

$C: x^2 + y^2 = 4, z = 2$

$S: \mathbf{r}(r, \theta) = (r \cos \theta)\mathbf{i} + (r \sin \theta)\mathbf{j} + r\mathbf{k}$

**FIGURE 16.59** The curve $C$ and cone $S$ in Example 4.

**EXAMPLE 4** Find the circulation of the field $\mathbf{F} = (x^2 - y)\mathbf{i} + 4z\mathbf{j} + x^2\mathbf{k}$ around the curve $C$ in which the plane $z = 2$ meets the cone $z = \sqrt{x^2 + y^2}$, counterclockwise as viewed from above (Figure 16.59).

**Solution**   Stokes' Theorem enables us to find the circulation by integrating over the surface of the cone. Traversing $C$ in the counterclockwise direction viewed from above corresponds to taking the *inner* normal $\mathbf{n}$ to the cone, the normal with a positive $\mathbf{k}$-component.
   We parametrize the cone as

$$\mathbf{r}(r, \theta) = (r\cos\theta)\mathbf{i} + (r\sin\theta)\mathbf{j} + r\mathbf{k}, \qquad 0 \le r \le 2, \quad 0 \le \theta \le 2\pi.$$

We then have

$$\mathbf{n} = \frac{\mathbf{r}_r \times \mathbf{r}_\theta}{|\mathbf{r}_r \times \mathbf{r}_\theta|} = \frac{-(r\cos\theta)\mathbf{i} - (r\sin\theta)\mathbf{j} + r\mathbf{k}}{r\sqrt{2}} \qquad \text{Section 16.5, Example 4}$$

$$= \frac{1}{\sqrt{2}}\big(-(\cos\theta)\mathbf{i} - (\sin\theta)\mathbf{j} + \mathbf{k}\big)$$

$$d\sigma = r\sqrt{2}\,dr\,d\theta \qquad \text{Section 16.5, Example 4}$$

$$\nabla \times \mathbf{F} = -4\mathbf{i} - 2x\mathbf{j} + \mathbf{k} \qquad \text{Routine calculation}$$

$$= -4\mathbf{i} - 2r\cos\theta\mathbf{j} + \mathbf{k}. \qquad x = r\cos\theta$$

Accordingly,

$$\nabla \times \mathbf{F} \cdot \mathbf{n} = \frac{1}{\sqrt{2}}\Big(4\cos\theta + 2r\cos\theta\sin\theta + 1\Big)$$

$$= \frac{1}{\sqrt{2}}\Big(4\cos\theta + r\sin 2\theta + 1\Big)$$

and the circulation is

$$\oint_C \mathbf{F}\cdot d\mathbf{r} = \iint_S \nabla \times \mathbf{F}\cdot\mathbf{n}\,d\sigma \qquad \text{Stokes' Theorem, Eq. (4)}$$

$$= \int_0^{2\pi}\int_0^2 \frac{1}{\sqrt{2}}\Big(4\cos\theta + r\sin 2\theta + 1\Big)\big(r\sqrt{2}\,dr\,d\theta\big) = 4\pi. \qquad \blacksquare$$

**EXAMPLE 5**    The cone used in Example 4 is not the easiest surface to use for calculating the circulation around the bounding circle $C$ lying in the plane $z = 3$. If instead we use the flat disk of radius 3 centered on the $z$-axis and lying in the plane $z = 3$, then the normal vector to the surface $S$ is $\mathbf{n} = \mathbf{k}$. Just as in the computation for Example 4, we still have $\nabla \times \mathbf{F} = -4\mathbf{i} - 2x\mathbf{j} + \mathbf{k}$. However, now we get $\nabla \times \mathbf{F} \cdot \mathbf{n} = 1$, so that

$$\iint_S \nabla \times \mathbf{F}\cdot\mathbf{n}\,d\sigma = \iint_{x^2+y^2 \le 4} 1\,dA = 4\pi. \qquad \text{The shadow is the disk of radius 2 in the } xy\text{-plane.}$$

This result agrees with the circulation value found in Example 4.    $\blacksquare$

**EXAMPLE 6**    Find a parametrization for the surface $S$ formed by the part of the hyperbolic paraboloid $z = y^2 - x^2$ lying inside the cylinder of radius one around the $z$-axis and for the boundary curve $C$ of $S$. (See Figure 16.60.) Then verify Stokes' Theorem for $S$ using the normal having positive $\mathbf{k}$-component and the vector field $\mathbf{F} = y\mathbf{i} - x\mathbf{j} + x^2\mathbf{k}$.

**Solution**    As the unit circle is traversed counterclockwise in the $xy$-plane, the $z$-coordinate of the surface with the curve $C$ as boundary is given by $y^2 - x^2$. A parametrization of $C$ is given by

$$\mathbf{r}(t) = (\cos t)\mathbf{i} + (\sin t)\mathbf{j} + (\sin^2 t - \cos^2 t)\mathbf{k}, \quad 0 \le t \le 2\pi$$

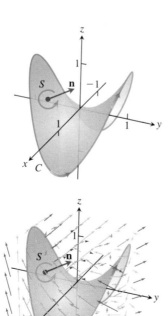

**FIGURE 16.60**   The surface and vector field for Example 6.

with

$$\frac{d\mathbf{r}}{dt} = (-\sin t)\mathbf{i} + (\cos t)\mathbf{j} + (4 \sin t \cos t)\mathbf{k}, \quad 0 \le t \le 2\pi.$$

Along the curve $\mathbf{r}(t)$ the formula for the vector field $\mathbf{F}$ is

$$\mathbf{F} = (\sin t)\mathbf{i} - (\cos t)\mathbf{j} + (\cos^2 t)\mathbf{k}.$$

The counterclockwise circulation along $C$ is the value of the line integral

$$\int_0^{2\pi} \mathbf{F} \cdot \frac{d\mathbf{r}}{dt} dt = \int_0^{2\pi} \left( -\sin^2 t - \cos^2 t + 4 \sin t \cos^3 t \right) dt$$

$$= \int_0^{2\pi} \left( 4 \sin t \cos^3 t - 1 \right) dt$$

$$= \left[ -\cos^4 t - t \right]_0^{2\pi} = -2\pi.$$

We now compute the same quantity by integrating $\nabla \times \mathbf{F} \cdot \mathbf{n}$ over the surface $S$. We use polar coordinates and parametrize $S$ by noting that above the point $(r, \theta)$ in the plane, the $z$–coordinate of $S$ is $y^2 - x^2 = r^2 \sin^2 \theta - r^2 \cos^2 \theta$. A parametrization of $S$ is

$$\mathbf{r}(r, \theta) = (r \cos \theta)\mathbf{i} + (r \sin \theta)\mathbf{j} + r^2(\sin^2 \theta - \cos^2 \theta)\mathbf{k}, \quad 0 \le r \le 1, \quad 0 \le \theta \le 2\pi.$$

We next compute $\nabla \times \mathbf{F} \cdot \mathbf{n} \, d\sigma$. We have

$$\nabla \times \mathbf{F} = \begin{vmatrix} \mathbf{i} & \mathbf{j} & \mathbf{k} \\ \dfrac{\partial}{\partial x} & \dfrac{\partial}{\partial y} & \dfrac{\partial}{\partial z} \\ y & -x & x^2 \end{vmatrix} = -2x\mathbf{j} - 2\mathbf{k} = -(2r \cos \theta)\mathbf{j} - 2\mathbf{k}$$

and

$$\mathbf{r}_r = (\cos \theta)\mathbf{i} + (\sin \theta)\mathbf{j} + 2r(\sin^2 \theta - \cos^2 \theta)\mathbf{k}$$

$$\mathbf{r}_\theta = (-r \sin \theta)\mathbf{i} + (r \cos \theta)\mathbf{j} + 4r^2(\sin \theta \cos \theta)\mathbf{k}$$

$$\mathbf{r}_r \times \mathbf{r}_\theta = \begin{vmatrix} \mathbf{i} & \mathbf{j} & \mathbf{k} \\ \cos \theta & \sin \theta & 2r(\sin^2 \theta - \cos^2 \theta) \\ -r \sin \theta & r \cos \theta & 4r^2(\sin \theta \cos \theta) \end{vmatrix}$$

$$= 2r^2(2 \sin^2 \theta \cos \theta - \sin^2 \theta \cos \theta + \cos^3 \theta)\mathbf{i}$$
$$\quad -2r^2(2 \sin \theta \cos^2 \theta + \sin^3 \theta + \sin \theta \cos^2 \theta)\mathbf{j} + r\mathbf{k}.$$

We now obtain

$$\iint_S \nabla \times \mathbf{F} \cdot \mathbf{n} \, d\sigma = \int_0^{2\pi} \int_0^1 \nabla \times \mathbf{F} \cdot \frac{\mathbf{r}_r \times \mathbf{r}_\theta}{|\mathbf{r}_r \times \mathbf{r}_\theta|} |\mathbf{r}_r \times \mathbf{r}_\theta| \, dr \, d\theta$$

$$= \int_0^{2\pi} \int_0^1 \nabla \times \mathbf{F} \cdot (\mathbf{r}_r \times \mathbf{r}_\theta) \, dr \, d\theta$$

$$= \int_0^{2\pi} \int_0^1 \left[ 4r^3 \left( 2 \sin \theta \cos^3 \theta + \sin^3 \theta \cos \theta + \sin \theta \cos^3 \theta \right) - 2r \right] dr \, d\theta$$

$$= \int_0^{2\pi} \left[ r^4 \left( 3 \sin \theta \cos^3 \theta + \sin^3 \theta \cos \theta \right) - r^2 \right]_{r=0}^{r=1} d\theta \qquad \text{Integrate.}$$

$$= \int_0^{2\pi} \left( 3 \sin \theta \cos^3 \theta + \sin^3 \theta \cos \theta - 1 \right) d\theta \qquad \text{Evaluate.}$$

$$= \left[ -\frac{3}{4} \cos^4 \theta + \frac{1}{4} \sin^4 \theta - \theta \right]_0^{2\pi}$$

$$= \left( -\frac{3}{4} + 0 - 2\pi + \frac{3}{4} - 0 + 0 \right) = -2\pi.$$

So the surface integral of $\nabla \times \mathbf{F} \cdot \mathbf{n}$ over $S$ equals the counterclockwise circulation of $\mathbf{F}$ along $C$, as asserted by Stokes' Theorem. ∎

**EXAMPLE 7** Calculate the circulation of the vector field

$$\mathbf{F} = (x^2 + z)\mathbf{i} + (y^2 + 2x)\mathbf{j} + (z^2 - y)\mathbf{k}$$

along the curve of intersection of the sphere $x^2 + y^2 + z^2 = 1$ with the cone $z = \sqrt{x^2 + y^2}$ traversed in the counterclockwise direction around the $z$-axis when viewed from above.

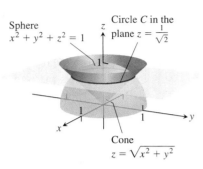

Sphere
$x^2 + y^2 + z^2 = 1$

Circle $C$ in the plane $z = \frac{1}{\sqrt{2}}$

Cone
$z = \sqrt{x^2 + y^2}$

**FIGURE 16.61** Circulation curve $C$ in Example 7.

**Solution** The sphere and cone intersect when $1 = (x^2 + y^2) + z^2 = z^2 + z^2 = 2z^2$, or $z = 1/\sqrt{2}$ (see Figure 16.61). We apply Stokes' Theorem to the curve of intersection $x^2 + y^2 = 1/2$ considered as the boundary of the enclosed disk in the plane $z = 1/\sqrt{2}$. The normal vector to the surface is then $\mathbf{n} = \mathbf{k}$. We calculate the curl vector as

$$\nabla \times \mathbf{F} = \begin{vmatrix} \mathbf{i} & \mathbf{j} & \mathbf{k} \\ \dfrac{\partial}{\partial x} & \dfrac{\partial}{\partial y} & \dfrac{\partial}{\partial z} \\ x^2 + z & y^2 + 2x & z^2 - y \end{vmatrix} = -\mathbf{i} + \mathbf{j} + 2\mathbf{k}, \qquad \text{Routine calculation}$$

so that $\nabla \times \mathbf{F} \cdot \mathbf{k} = 2$. The circulation around the disk is

$$\oint_C \mathbf{F} \cdot d\mathbf{r} = \iint_S \nabla \times \mathbf{F} \cdot \mathbf{k} \, d\sigma$$

$$= \iint_S 2 \, d\sigma = 2 \cdot \text{area of disk} = 2 \cdot \pi \left(\frac{1}{\sqrt{2}}\right)^2 = \pi. \qquad ∎$$

**Paddle Wheel Interpretation of $\nabla \times \mathbf{F}$**

Suppose that $\mathbf{F}$ is the velocity field of a fluid moving in a region $R$ in space containing the closed curve $C$. Then

$$\oint_C \mathbf{F} \cdot d\mathbf{r}$$

is the circulation of the fluid around $C$. By Stokes' Theorem, the circulation is equal to the flux of $\nabla \times \mathbf{F}$ through any suitably oriented surface $S$ with boundary $C$:

$$\oint_C \mathbf{F} \cdot d\mathbf{r} = \iint_S \nabla \times \mathbf{F} \cdot \mathbf{n} \, d\sigma.$$

Suppose we fix a point $Q$ in the region $R$ and a direction $\mathbf{u}$ at $Q$. Take $C$ to be a circle of radius $\rho$, with center at $Q$, whose plane is normal to $\mathbf{u}$. If $\nabla \times \mathbf{F}$ is continuous at $Q$, the average value of the $\mathbf{u}$-component of $\nabla \times \mathbf{F}$ over the circular disk $S$ bounded by $C$ approaches the $\mathbf{u}$-component of $\nabla \times \mathbf{F}$ at $Q$ as the radius $\rho \to 0$:

$$(\nabla \times \mathbf{F} \cdot \mathbf{u})_Q = \lim_{\rho \to 0} \frac{1}{\pi \rho^2} \iint_S \nabla \times \mathbf{F} \cdot \mathbf{u} \, d\sigma.$$

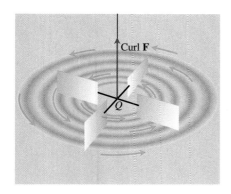

**FIGURE 16.62** A small paddle wheel in a fluid spins fastest at point $Q$ when its axle points in the direction of curl **F**.

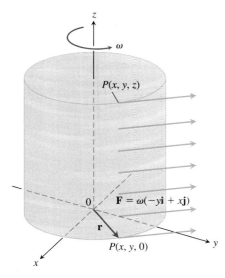

**FIGURE 16.63** A steady rotational flow parallel to the $xy$-plane, with constant angular velocity $\omega$ in the positive (counter-clockwise) direction (Example 8).

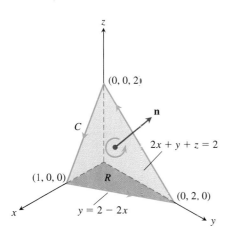

**FIGURE 16.64** The planar surface in Example 9.

If we apply Stokes' Theorem and replace the surface integral by a line integral over $C$, we get

$$(\nabla \times \mathbf{F} \cdot \mathbf{u})_Q = \lim_{\rho \to 0} \frac{1}{\pi \rho^2} \oint_C \mathbf{F} \cdot d\mathbf{r}. \qquad (6)$$

The left-hand side of Equation (6) has its maximum value when **u** is the direction of $\nabla \times \mathbf{F}$. When $\rho$ is small, the limit on the right-hand side of Equation (6) is approximately

$$\frac{1}{\pi \rho^2} \oint_C \mathbf{F} \cdot d\mathbf{r},$$

which is the circulation around $C$ divided by the area of the disk (circulation density). Suppose that a small paddle wheel of radius $\rho$ is introduced into the fluid at $Q$, with its axle directed along **u** (Figure 16.62). The circulation of the fluid around $C$ affects the rate of spin of the paddle wheel. The wheel spins fastest when the circulation integral is maximized; therefore it spins fastest when the axle of the paddle wheel points in the direction of $\nabla \times \mathbf{F}$.

**EXAMPLE 8** A fluid of constant density rotates around the $z$-axis with velocity $\mathbf{F} = \omega(-y\mathbf{i} + x\mathbf{j})$, where $\omega$ is a positive constant called the *angular velocity* of the rotation (Figure 16.63). Find $\nabla \times \mathbf{F}$ and relate it to the circulation density.

**Solution** With $\mathbf{F} = -\omega y\mathbf{i} + \omega x\mathbf{j}$, we find the curl

$$\nabla \times \mathbf{F} = \left(\frac{\partial P}{\partial y} - \frac{\partial N}{\partial z}\right)\mathbf{i} + \left(\frac{\partial M}{\partial z} - \frac{\partial P}{\partial x}\right)\mathbf{j} + \left(\frac{\partial N}{\partial x} - \frac{\partial M}{\partial y}\right)\mathbf{k}$$

$$= (0 - 0)\mathbf{i} + (0 - 0)\mathbf{j} + (\omega - (-\omega))\mathbf{k} = 2\omega\mathbf{k}.$$

By Stokes' Theorem, the circulation of **F** around a circle $C$ of radius $\rho$ bounding a disk $S$ in a plane normal to $\nabla \times \mathbf{F}$, say the $xy$-plane, is

$$\oint_C \mathbf{F} \cdot d\mathbf{r} = \iint_S \nabla \times \mathbf{F} \cdot \mathbf{n} \, d\sigma = \iint_S 2\omega\mathbf{k} \cdot \mathbf{k} \, dx \, dy = (2\omega)(\pi\rho^2).$$

Thus solving this last equation for $2\omega$, we have

$$(\nabla \times \mathbf{F}) \cdot \mathbf{k} = 2\omega = \frac{1}{\pi\rho^2} \oint_C \mathbf{F} \cdot d\mathbf{r},$$

consistent with Equation (6) when $\mathbf{u} = \mathbf{k}$. ∎

**EXAMPLE 9** Use Stokes' Theorem to evaluate $\int_C \mathbf{F} \cdot d\mathbf{r}$, if $\mathbf{F} = xz\mathbf{i} + xy\mathbf{j} + 3xz\mathbf{k}$ and $C$ is the boundary of the portion of the plane $2x + y + z = 2$ in the first octant, traversed counterclockwise as viewed from above (Figure 16.64).

**Solution** The plane is the level surface $f(x, y, z) = 2$ of the function $f(x, y, z) = 2x + y + z$. The unit normal vector

$$\mathbf{n} = \frac{\nabla f}{|\nabla f|} = \frac{(2\mathbf{i} + \mathbf{j} + \mathbf{k})}{|2\mathbf{i} + \mathbf{j} + \mathbf{k}|} = \frac{1}{\sqrt{6}}\left(2\mathbf{i} + \mathbf{j} + \mathbf{k}\right)$$

is consistent with the counterclockwise motion around $C$. To apply Stokes' Theorem, we find

$$\text{curl } \mathbf{F} = \nabla \times \mathbf{F} = \begin{vmatrix} \mathbf{i} & \mathbf{j} & \mathbf{k} \\ \dfrac{\partial}{\partial x} & \dfrac{\partial}{\partial y} & \dfrac{\partial}{\partial z} \\ xz & xy & 3xz \end{vmatrix} = (x - 3z)\mathbf{j} + y\mathbf{k}.$$

On the plane, $z$ equals $2 - 2x - y$, so

$$\nabla \times \mathbf{F} = (x - 3(2 - 2x - y))\mathbf{j} + y\mathbf{k} = (7x + 3y - 6)\mathbf{j} + y\mathbf{k}$$

and

$$\nabla \times \mathbf{F} \cdot \mathbf{n} = \frac{1}{\sqrt{6}}\left( 7x + 3y - 6 + y \right) = \frac{1}{\sqrt{6}}\left( 7x + 4y - 6 \right).$$

The surface area element is

$$d\sigma = \frac{|\nabla f|}{|\nabla f \cdot \mathbf{k}|} dA = \frac{\sqrt{6}}{1} dx\, dy.$$

The circulation is

$$\oint_C \mathbf{F} \cdot d\mathbf{r} = \iint_S \nabla \times \mathbf{F} \cdot \mathbf{n}\, d\sigma \qquad \text{Stokes' Theorem, Eq. (4)}$$

$$= \int_0^1 \int_0^{2-2x} \frac{1}{\sqrt{6}}\left( 7x + 4y - 6 \right)\sqrt{6}\, dy\, dx$$

$$= \int_0^1 \int_0^{2-2x} (7x + 4y - 6)\, dy\, dx = -1. \qquad \blacksquare$$

**EXAMPLE 10**  Let the surface $S$ be the elliptical paraboloid $z = x^2 + 4y^2$ lying beneath the plane $z = 1$ (Figure 16.65). We define the orientation of $S$ by taking the *inner* normal vector $\mathbf{n}$ to the surface, which is the normal having a positive $\mathbf{k}$-component. Find the flux of $\nabla \times \mathbf{F}$ across $S$ in the direction $\mathbf{n}$ for the vector field $\mathbf{F} = y\mathbf{i} - xz\mathbf{j} + xz^2\mathbf{k}$.

**Solution**  We use Stokes' Theorem to calculate the curl integral by finding the equivalent counterclockwise circulation of $\mathbf{F}$ around the curve of intersection $C$ of the paraboloid $z = x^2 + 4y^2$ and the plane $z = 1$, as shown in Figure 16.65. Note that the orientation of $S$ is consistent with traversing $C$ in a counterclockwise direction around the $z$-axis. The curve $C$ is the ellipse $x^2 + 4y^2 = 1$ in the plane $z = 1$. We can parametrize the ellipse by $x = \cos t$, $y = \frac{1}{2}\sin t$, $z = 1$ for $0 \le t \le 2\pi$, so $C$ is given by

$$\mathbf{r}(t) = (\cos t)\mathbf{i} + \frac{1}{2}(\sin t)\mathbf{j} + \mathbf{k}, \qquad 0 \le t \le 2\pi.$$

To compute the circulation integral $\oint_C \mathbf{F} \cdot d\mathbf{r}$, we evaluate $\mathbf{F}$ along $C$ and find the velocity vector $d\mathbf{r}/dt$:

$$\mathbf{F}(\mathbf{r}(t)) = \frac{1}{2}(\sin t)\mathbf{i} - (\cos t)\mathbf{j} + (\cos t)\mathbf{k}$$

and

$$\frac{d\mathbf{r}}{dt} = -(\sin t)\mathbf{i} + \frac{1}{2}(\cos t)\mathbf{j}.$$

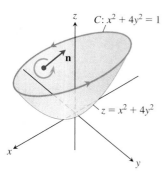

**FIGURE 16.65** The portion of the elliptical paraboloid in Example 10, showing its curve of intersection $C$ with the plane $z = 1$ and its inner normal orientation by $\mathbf{n}$.

Then,

$$\oint_C \mathbf{F} \cdot d\mathbf{r} = \int_0^{2\pi} \mathbf{F}(\mathbf{r}(t)) \cdot \frac{d\mathbf{r}}{dt} \, dt$$

$$= \int_0^{2\pi} \left( -\frac{1}{2} \sin^2 t - \frac{1}{2} \cos^2 t \right) dt$$

$$= -\frac{1}{2} \int_0^{2\pi} dt = -\pi.$$

Therefore, by Stokes' Theorem the flux of the curl across $S$ in the direction $\mathbf{n}$ for the field $\mathbf{F}$ is

$$\iint_S \nabla \times \mathbf{F} \cdot \mathbf{n} \, d\sigma = -\pi. \qquad \blacksquare$$

### Proof Outline of Stokes' Theorem for Polyhedral Surfaces

Let $S$ be a polyhedral surface consisting of a finite number of plane regions or faces. (See Figure 16.66 for examples.) We apply Green's Theorem to each separate face of $S$. There are two types of faces:

**1.** Those that are surrounded on all sides by other faces.

**2.** Those that have one or more edges that are not adjacent to other faces.

The boundary $\Delta$ of $S$ consists of those edges of the type 2 faces that are not adjacent to other faces. In Figure 16.66a, the triangles $EAB$, $BCE$, and $CDE$ represent a part of $S$, with $ABCD$ part of the boundary $\Delta$. Although Green's Theorem was stated for curves in the $xy$-plane, a generalized form applies to plane curves in space, where $\mathbf{n}$ is normal to the plane (instead of $\mathbf{k}$). In the generalized tangential form, the theorem asserts that the line integral of $\mathbf{F}$ around the curve enclosing the plane region $R$ normal to $\mathbf{n}$ equals the double integral of (curl $\mathbf{F}$) $\cdot \mathbf{n}$ over $R$. Applying this generalized form to the three triangles of Figure 16.66a in turn, and adding the results, gives

$$\left( \oint_{EAB} + \oint_{BCE} + \oint_{CDE} \right) \mathbf{F} \cdot d\mathbf{r} = \left( \iint_{EAB} + \iint_{BCE} + \iint_{CDE} \right) \nabla \times \mathbf{F} \cdot \mathbf{n} \, d\sigma. \qquad (7)$$

The three line integrals on the left-hand side of Equation (7) combine into a single line integral taken around the periphery $ABCDE$ because the integrals along interior segments cancel in pairs. For example, the integral along segment $BE$ in triangle $ABE$ is opposite in sign to the integral along the same segment in triangle $EBC$. The same holds for segment $CE$. Hence, Equation (7) reduces to

$$\oint_{ABCDE} \mathbf{F} \cdot d\mathbf{r} = \iint_{ABCDE} \nabla \times \mathbf{F} \cdot \mathbf{n} \, d\sigma.$$

When we apply the generalized form of Green's Theorem to all the faces and add the results, we get

$$\oint_\Delta \mathbf{F} \cdot d\mathbf{r} = \iint_S \nabla \times \mathbf{F} \cdot \mathbf{n} \, d\sigma.$$

This is Stokes' Theorem for the polyhedral surface $S$ in Figure 16.66a. More general polyhedral surfaces are shown in Figure 16.66b and the proof can be extended to them. General smooth surfaces can be obtained as limits of polyhedral surfaces and a complete proof can be found in more advanced texts.

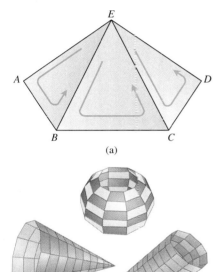

**FIGURE 16.66** (a) Part of a polyhedral surface. (b) Other polyhedral surfaces.

## Stokes' Theorem for Surfaces with Holes

Stokes' Theorem holds for an oriented surface $S$ that has one or more holes (Figure 16.67). The surface integral over $S$ of the normal component of $\nabla \times \mathbf{F}$ equals the sum of the line integrals around all the boundary curves of the tangential component of $\mathbf{F}$, where the curves are to be traced in the direction induced by the orientation of $S$. For such surfaces the theorem is unchanged, but $C$ is considered as a union of simple closed curves.

## An Important Identity

The following identity arises frequently in mathematics and the physical sciences.

**FIGURE 16.67** Stokes' Theorem also holds for oriented surfaces with holes. Consistent with the orientation of $S$, the outer curve is traversed counterclockwise around $\mathbf{n}$ and the inner curves surrounding the holes are traversed clockwise.

$$\text{curl grad } f = \mathbf{0} \qquad \text{or} \qquad \nabla \times \nabla f = \mathbf{0} \qquad (8)$$

Forces arising in the study of electromagnetism and gravity are often associated with a potential function $f$. The identity (8) says that these forces have curl equal to zero. The identity (8) holds for any function $f(x, y, z)$ whose second partial derivatives are continuous. The proof goes like this:

$$\nabla \times \nabla f = \begin{vmatrix} \mathbf{i} & \mathbf{j} & \mathbf{k} \\ \dfrac{\partial}{\partial x} & \dfrac{\partial}{\partial y} & \dfrac{\partial}{\partial z} \\ \dfrac{\partial f}{\partial x} & \dfrac{\partial f}{\partial y} & \dfrac{\partial f}{\partial z} \end{vmatrix} = (f_{zy} - f_{yz})\mathbf{i} - (f_{zx} - f_{xz})\mathbf{j} + (f_{yx} - f_{xy})\mathbf{k}.$$

If the second partial derivatives are continuous, the mixed second derivatives in parentheses are equal (Theorem 2, Section 14.3) and the vector is zero.

## Conservative Fields and Stokes' Theorem

In Section 16.3, we found that a field $\mathbf{F}$ being conservative in an open region $D$ in space is equivalent to the integral of $\mathbf{F}$ around every closed loop in $D$ being zero. This, in turn, is equivalent in *simply connected* open regions to saying that $\nabla \times \mathbf{F} = \mathbf{0}$ (which gives a test for determining if $\mathbf{F}$ is conservative for such regions).

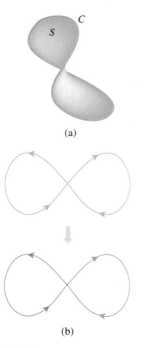

**(a)**

**(b)**

**FIGURE 16.68** (a) In a simply connected open region in space, a simple closed curve $C$ is the boundary of a smooth surface $S$. (b) Smooth curves that cross themselves can be divided into loops to which Stokes' Theorem applies.

> **THEOREM 7—Curl F = 0 Related to the Closed-Loop Property**  If $\nabla \times \mathbf{F} = \mathbf{0}$ at every point of a simply connected open region $D$ in space, then on any piecewise-smooth closed path $C$ in $D$,
>
> $$\oint_C \mathbf{F} \cdot d\mathbf{r} = 0.$$

**Sketch of a Proof**    Theorem 7 can be proved in two steps. The first step is for simple closed curves (loops that do not cross themselves), like the one in Figure 16.68a. A theorem from topology, a branch of advanced mathematics, states that every smooth simple closed curve $C$ in a simply connected open region $D$ is the boundary of a smooth two-sided surface $S$ that also lies in $D$. Hence, by Stokes' Theorem,

$$\oint_C \mathbf{F} \cdot d\mathbf{r} = \iint_S \nabla \times \mathbf{F} \cdot \mathbf{n} \, d\sigma = 0.$$

The second step is for curves that cross themselves, like the one in Figure 16.68b. The idea is to break these into simple loops spanned by orientable surfaces, apply Stokes' Theorem one loop at a time, and add the results. ∎

The following diagram summarizes the results for conservative fields defined on connected, simply connected open regions. For such regions, the four statements are equivalent to each other.

Theorem 2,
Section 16.3

$\mathbf{F}$ conservative on $D$ $\Longleftrightarrow$ $\mathbf{F} = \nabla f$ on $D$

Theorem 3,
Section 16.3 $\updownarrow$

Vector identity (Eq. 8)
(continuous second
partial derivatives)

$\oint_C \mathbf{F} \cdot d\mathbf{r} = 0$ $\Longleftarrow$ $\nabla \times \mathbf{F} = \mathbf{0}$ throughout $D$
over any closed
path in $D$

Theorem 7
Domain's simple
connectivity and
Stokes' Theorem

# Exercises 16.7

### Using Stokes' Theorem to Find Line Integrals

In Exercises 1–6, use the surface integral in Stokes' Theorem to calculate the circulation of the field $\mathbf{F}$ around the curve $C$ in the indicated direction.

1. $\mathbf{F} = x^2\mathbf{i} + 2x\mathbf{j} + z^2\mathbf{k}$

   $C$: The ellipse $4x^2 + y^2 = 4$ in the $xy$-plane, counterclockwise when viewed from above

2. $\mathbf{F} = 2y\mathbf{i} + 3x\mathbf{j} - z^2\mathbf{k}$

   $C$: The circle $x^2 + y^2 = 9$ in the $xy$-plane, counterclockwise when viewed from above

3. $\mathbf{F} = y\mathbf{i} + xz\mathbf{j} + x^2\mathbf{k}$

   $C$: The boundary of the triangle cut from the plane $x + y + z = 1$ by the first octant, counterclockwise when viewed from above

4. $\mathbf{F} = (y^2 + z^2)\mathbf{i} + (x^2 + z^2)\mathbf{j} + (x^2 + y^2)\mathbf{k}$

   $C$: The boundary of the triangle cut from the plane $x + y + z = 1$ by the first octant, counterclockwise when viewed from above

5. $\mathbf{F} = (y^2 + z^2)\mathbf{i} + (x^2 + y^2)\mathbf{j} + (x^2 + y^2)\mathbf{k}$

   $C$: The square bounded by the lines $x = \pm 1$ and $y = \pm 1$ in the $xy$-plane, counterclockwise when viewed from above

6. $\mathbf{F} = x^2y^3\mathbf{i} + \mathbf{j} + z\mathbf{k}$

   $C$: The intersection of the cylinder $x^2 + y^2 = 4$ and the hemisphere $x^2 + y^2 + z^2 = 16, z \geq 0$, counterclockwise when viewed from above

### Integral of the Curl Vector Field

7. Let $\mathbf{n}$ be the outer unit normal of the elliptical shell

   $$S: \quad 4x^2 + 9y^2 + 36z^2 = 36, \qquad z \geq 0,$$

and let

$$\mathbf{F} = y\mathbf{i} + x^2\mathbf{j} + (x^2 + y^4)^{3/2} \sin e^{\sqrt{xyz}}\,\mathbf{k}.$$

Find the value of

$$\iint_S \nabla \times \mathbf{F} \cdot \mathbf{n}\, d\sigma.$$

(*Hint:* One parametrization of the ellipse at the base of the shell is $x = 3 \cos t, y = 2 \sin t, 0 \leq t \leq 2\pi$.)

8. Let $\mathbf{n}$ be the outer unit normal (normal away from the origin) of the parabolic shell

   $$S: \quad 4x^2 + y + z^2 = 4, \qquad y \geq 0,$$

and let

$$\mathbf{F} = \left(-z + \frac{1}{2 + x}\right)\mathbf{i} + (\tan^{-1} y)\mathbf{j} + \left(x + \frac{1}{4 + z}\right)\mathbf{k}.$$

Find the value of

$$\iint_S \nabla \times \mathbf{F} \cdot \mathbf{n}\, d\sigma.$$

9. Let $S$ be the cylinder $x^2 + y^2 = a^2, 0 \leq z \leq h$, together with its top, $x^2 + y^2 \leq a^2, z = h$. Let $\mathbf{F} = -y\mathbf{i} + x\mathbf{j} + x^2\mathbf{k}$. Use Stokes' Theorem to find the flux of $\nabla \times \mathbf{F}$ outward through $S$.

10. Evaluate

$$\iint_S \nabla \times (y\mathbf{i}) \cdot \mathbf{n}\, d\sigma,$$

where $S$ is the hemisphere $x^2 + y^2 + z^2 = 1, z \geq 0$.

11. Suppose $\mathbf{F} = \nabla \times \mathbf{A}$, where

$$\mathbf{A} = \left(y + \sqrt{z}\right)\mathbf{i} + e^{xyz}\mathbf{j} + \cos{(xz)}\mathbf{k}.$$

Determine the flux of $\mathbf{F}$ outward through the hemisphere $x^2 + y^2 + z^2 = 1, z \geq 0$.

12. Repeat Exercise 11 for the flux of $\mathbf{F}$ across the entire unit sphere.

### Stokes' Theorem for Parametrized Surfaces

In Exercises 13–18, use the surface integral in Stokes' Theorem to calculate the flux of the curl of the field $\mathbf{F}$ across the surface $S$ in the direction of the outward unit normal $\mathbf{n}$.

13. $\mathbf{F} = 2z\mathbf{i} + 3x\mathbf{j} + 5y\mathbf{k}$

  $S$:  $\mathbf{r}(r, \theta) = (r \cos \theta)\mathbf{i} + (r \sin \theta)\mathbf{j} + (4 - r^2)\mathbf{k}$,
  $0 \leq r \leq 2, \quad 0 \leq \theta \leq 2\pi$

14. $\mathbf{F} = (y - z)\mathbf{i} + (z - x)\mathbf{j} + (x + z)\mathbf{k}$

  $S$:  $\mathbf{r}(r, \theta) = (r \cos \theta)\mathbf{i} + (r \sin \theta)\mathbf{j} + (9 - r^2)\mathbf{k}$,
  $0 \leq r \leq 3, \quad 0 \leq \theta \leq 2\pi$

15. $\mathbf{F} = x^2y\mathbf{i} + 2y^3z\mathbf{j} + 3z\mathbf{k}$

  $S$:  $\mathbf{r}(r, \theta) = (r \cos \theta)\mathbf{i} + (r \sin \theta)\mathbf{j} + r\mathbf{k}$,
  $0 \leq r \leq 1, \quad 0 \leq \theta \leq 2\pi$

16. $\mathbf{F} = (x - y)\mathbf{i} + (y - z)\mathbf{j} + (z - x)\mathbf{k}$

  $S$:  $\mathbf{r}(r, \theta) = (r \cos \theta)\mathbf{i} + (r \sin \theta)\mathbf{j} + (5 - r)\mathbf{k}$,
  $0 \leq r \leq 5, \quad 0 \leq \theta \leq 2\pi$

17. $\mathbf{F} = 3y\mathbf{i} + (5 - 2x)\mathbf{j} + (z^2 - 2)\mathbf{k}$

  $S$:  $\mathbf{r}(\phi, \theta) = \left(\sqrt{3} \sin \phi \cos \theta\right)\mathbf{i} + \left(\sqrt{3} \sin \phi \sin \theta\right)\mathbf{j} + \left(\sqrt{3} \cos \phi\right)\mathbf{k}, \quad 0 \leq \phi \leq \pi/2, \quad 0 \leq \theta \leq 2\pi$

18. $\mathbf{F} = y^2\mathbf{i} + z^2\mathbf{j} + x\mathbf{k}$

  $S$:  $\mathbf{r}(\phi, \theta) = (2 \sin \phi \cos \theta)\mathbf{i} + (2 \sin \phi \sin \theta)\mathbf{j} + (2 \cos \phi)\mathbf{k}, \quad 0 \leq \phi \leq \pi/2, \quad 0 \leq \theta \leq 2\pi$

### Theory and Examples

19. Let $C$ be the smooth curve $\mathbf{r}(t) = (2 \cos t)\mathbf{i} + (2 \sin t)\mathbf{j} + (3 - 2 \cos^3 t)\mathbf{k}$, oriented to be traversed counterclockwise around the $z$-axis when viewed from above. Let $S$ be the piecewise smooth cylindrical surface $x^2 + y^2 = 4$, below the curve for $z \geq 0$, together with the base disk in the $xy$-plane. Note that $C$ lies on the cylinder $S$ and above the $xy$-plane (see the accompanying figure). Verify Equation (4) in Stokes' Theorem for the vector field $\mathbf{F} = y\mathbf{i} - x\mathbf{j} + x^2\mathbf{k}$.

20. Verify Stokes' Theorem for the vector field $\mathbf{F} = 2xy\mathbf{i} + x\mathbf{j} + (y + z)\mathbf{k}$ and surface $z = 4 - x^2 - y^2, z \geq 0$, oriented with unit normal $\mathbf{n}$ pointing upward.

21. **Zero circulation** Use Equation (8) and Stokes' Theorem to show that the circulations of the following fields around the boundary of any smooth orientable surface in space are zero.

  a. $\mathbf{F} = 2x\mathbf{i} + 2y\mathbf{j} + 2z\mathbf{k}$    b. $\mathbf{F} = \nabla(xy^2z^3)$

  c. $\mathbf{F} = \nabla \times (x\mathbf{i} + y\mathbf{j} + z\mathbf{k})$    d. $\mathbf{F} = \nabla f$

22. **Zero circulation** Let $f(x, y, z) = (x^2 + y^2 + z^2)^{-1/2}$. Show that the clockwise circulation of the field $\mathbf{F} = \nabla f$ around the circle $x^2 + y^2 = a^2$ in the $xy$-plane is zero

  a. by taking $\mathbf{r} = (a \cos t)\mathbf{i} + (a \sin t)\mathbf{j}, 0 \leq t \leq 2\pi$, and integrating $\mathbf{F} \cdot d\mathbf{r}$ over the circle.

  b. by applying Stokes' Theorem.

23. Let $C$ be a simple closed smooth curve in the plane $2x + 2y + z = 2$, oriented as shown here. Show that

$$\oint_C 2y \, dx + 3z \, dy - x \, dz$$

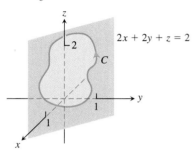

depends only on the area of the region enclosed by $C$ and not on the position or shape of $C$.

24. Show that if $\mathbf{F} = x\mathbf{i} + y\mathbf{j} + z\mathbf{k}$, then $\nabla \times \mathbf{F} = \mathbf{0}$.

25. Find a vector field with twice-differentiable components whose curl is $x\mathbf{i} + y\mathbf{j} + z\mathbf{k}$ or prove that no such field exists.

26. Does Stokes' Theorem say anything special about circulation in a field whose curl is zero? Give reasons for your answer.

27. Let $R$ be a region in the $xy$-plane that is bounded by a piecewise smooth simple closed curve $C$ and suppose that the moments of inertia of $R$ about the $x$- and $y$-axes are known to be $I_x$ and $I_y$. Evaluate the integral

$$\oint_C \nabla(r^4) \cdot \mathbf{n} \, ds,$$

where $r = \sqrt{x^2 + y^2}$, in terms of $I_x$ and $I_y$.

28. **Zero curl, yet the field is not conservative** Show that the curl of

$$\mathbf{F} = \frac{-y}{x^2 + y^2}\mathbf{i} + \frac{x}{x^2 + y^2}\mathbf{j} + z\mathbf{k}$$

is zero but that

$$\oint_C \mathbf{F} \cdot d\mathbf{r}$$

is not zero if $C$ is the circle $x^2 + y^2 = 1$ in the $xy$-plane. (Theorem 7 does not apply here because the domain of $\mathbf{F}$ is not simply connected. The field $\mathbf{F}$ is not defined along the $z$-axis so there is no way to contract $C$ to a point without leaving the domain of $\mathbf{F}$.)

# 16.8 The Divergence Theorem and a Unified Theory

The divergence form of Green's Theorem in the plane states that the net outward flux of a vector field across a simple closed curve can be calculated by integrating the divergence of the field over the region enclosed by the curve. The corresponding theorem in three dimensions, called the *Divergence Theorem*, states that the net outward flux of a vector field across a closed surface in space can be calculated by integrating the divergence of the field over the region enclosed by the surface. In this section we prove the Divergence Theorem and show how it simplifies the calculation of flux, which is the integral of the field over the closed oriented surface. We also derive Gauss's law for flux in an electric field and the continuity equation of hydrodynamics. Finally, we summarize the chapter's vector integral theorems in a single unifying principle generalizing the Fundamental Theorem of Calculus.

## Divergence in Three Dimensions

The **divergence** of a vector field $\mathbf{F} = M(x, y, z)\mathbf{i} + N(x, y, z)\mathbf{j} + P(x, y, z)\mathbf{k}$ is the scalar function

$$\operatorname{div} \mathbf{F} = \nabla \cdot \mathbf{F} = \frac{\partial M}{\partial x} + \frac{\partial N}{\partial y} + \frac{\partial P}{\partial z}. \tag{1}$$

The symbol "div $\mathbf{F}$" is read as "divergence of $\mathbf{F}$" or "div $\mathbf{F}$." The notation $\nabla \cdot \mathbf{F}$ is read "del dot $\mathbf{F}$."

Div $\mathbf{F}$ has the same physical interpretation in three dimensions that it does in two. If $\mathbf{F}$ is the velocity field of a flowing gas, the value of div $\mathbf{F}$ at a point $(x, y, z)$ is the rate at which the gas is compressing or expanding at $(x, y, z)$. The divergence is the flux per unit volume or *flux density* at the point.

**EXAMPLE 1**    The following vector fields represent the velocity of a gas flowing in space. Find the divergence of each vector field and interpret its physical meaning. Figure 16.69 displays the vector fields.

(a)  Expansion: $\mathbf{F}(x, y, z) = x\mathbf{i} + y\mathbf{j} + z\mathbf{k}$

(b)  Compression: $\mathbf{F}(x, y, z) = -x\mathbf{i} - y\mathbf{j} - z\mathbf{k}$

(c)  Rotation about the $z$-axis: $\mathbf{F}(x, y, z) = -y\mathbf{i} + x\mathbf{j}$

(d)  Shearing along parallel horizontal planes: $\mathbf{F}(x, y, z) = z\mathbf{j}$

**Solution**

(a)  $\operatorname{div} \mathbf{F} = \frac{\partial}{\partial x}(x) + \frac{\partial}{\partial y}(y) + \frac{\partial}{\partial z}(z) = 3$: The gas is undergoing constant uniform expansion at all points.

(b)  $\operatorname{div} \mathbf{F} = \frac{\partial}{\partial x}(-x) + \frac{\partial}{\partial y}(-y) + \frac{\partial}{\partial z}(-z) = -3$: The gas is undergoing constant uniform compression at all points.

(c)  $\operatorname{div} \mathbf{F} = \frac{\partial}{\partial x}(-y) + \frac{\partial}{\partial y}(x) = 0$: The gas is neither expanding nor compressing at any point.

(d)  $\operatorname{div} \mathbf{F} = \frac{\partial}{\partial y}(z) = 0$: Again, the divergence is zero at all points in the domain of the velocity field, so the gas is neither expanding nor compressing at any point.    ∎

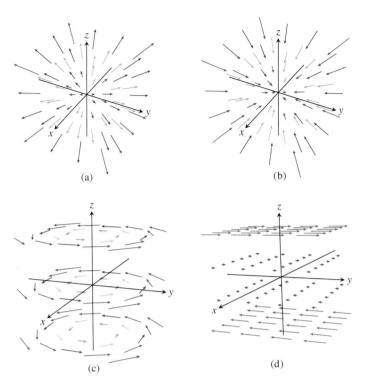

**FIGURE 16.69** Velocity fields of a gas flowing in space (Example 1).

## Divergence Theorem

The Divergence Theorem says that under suitable conditions, the outward flux of a vector field across a closed surface equals the triple integral of the divergence of the field over the three-dimensional region enclosed by the surface.

---

**THEOREM 8—Divergence Theorem**   Let $\mathbf{F}$ be a vector field whose components have continuous first partial derivatives, and let $S$ be a piecewise smooth oriented closed surface. The flux of $\mathbf{F}$ across $S$ in the direction of the surface's outward unit normal field $\mathbf{n}$ equals the triple integral of the divergence $\nabla \cdot \mathbf{F}$ over the region $D$ enclosed by the surface:

$$\iint_S \mathbf{F} \cdot \mathbf{n} \, d\sigma = \iiint_D \nabla \cdot \mathbf{F} \, dV. \tag{2}$$

Outward     Divergence
flux         integral

---

**EXAMPLE 2**   Evaluate both sides of Equation (2) for the expanding vector field $\mathbf{F} = x\mathbf{i} + y\mathbf{j} + z\mathbf{k}$ over the sphere $x^2 + y^2 + z^2 = a^2$ (Figure 16.70).

**Solution**   The outer unit normal to $S$, calculated from the gradient of $f(x, y, z) = x^2 + y^2 + z^2 - a^2$, is

$$\mathbf{n} = \frac{2(x\mathbf{i} + y\mathbf{j} + z\mathbf{k})}{\sqrt{4(x^2 + y^2 + z^2)}} = \frac{x\mathbf{i} + y\mathbf{j} + z\mathbf{k}}{a}. \qquad {\scriptstyle x^2 + y^2 + z^2 = a^2 \text{ on } S}$$

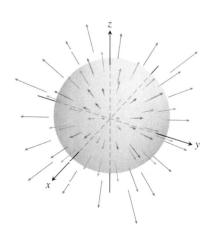

**FIGURE 16.70** A uniformly expanding vector field and a sphere (Example 2).

It follows that

$$\mathbf{F} \cdot \mathbf{n} \, d\sigma = \frac{x^2 + y^2 + z^2}{a} \, d\sigma = \frac{a^2}{a} \, d\sigma = a \, d\sigma.$$

Therefore, the outward flux is

$$\iint_S \mathbf{F} \cdot \mathbf{n} \, d\sigma = \iint_S a \, d\sigma = a \iint_S d\sigma = a(4\pi a^2) = 4\pi a^3. \qquad \text{Area of } S \text{ is } 4\pi a^2.$$

For the right-hand side of Equation (2), the divergence of $\mathbf{F}$ is

$$\nabla \cdot \mathbf{F} = \frac{\partial}{\partial x}(x) + \frac{\partial}{\partial y}(y) + \frac{\partial}{\partial z}(z) = 3,$$

so we obtain the divergence integral,

$$\iiint_D \nabla \cdot \mathbf{F} \, dV = \iiint_D 3 \, dV = 3\left(\frac{4}{3}\pi a^3\right) = 4\pi a^3. \qquad ■$$

Many vector fields of interest in applied science have zero divergence at each point. A common example is the velocity field of a circulating incompressible liquid, since it is neither expanding nor contracting. Other examples include constant vector fields $\mathbf{F} = a\mathbf{i} + b\mathbf{j} + c\mathbf{k}$, and velocity fields for shearing action along a fixed plane (see Example 1d). If $\mathbf{F}$ is a vector field whose divergence is zero at each point in the region $D$, then the integral on the right-hand side of Equation (2) equals 0. So if $S$ is any closed surface for which the Divergence Theorem applies, then the outward flux of $\mathbf{F}$ across $S$ is zero. We state this important application of the Divergence Theorem.

---

**COROLLARY** The outward flux across a piecewise smooth oriented closed surface $S$ is zero for any vector field $\mathbf{F}$ having zero divergence at every point of the region enclosed by the surface.

---

**EXAMPLE 3**   Find the flux of $\mathbf{F} = xy\mathbf{i} + yz\mathbf{j} + xz\mathbf{k}$ outward through the surface of the cube cut from the first octant by the planes $x = 1$, $y = 1$, and $z = 1$.

**Solution**   Instead of calculating the flux as a sum of six separate integrals, one for each face of the cube, we can calculate the flux by integrating the divergence

$$\nabla \cdot \mathbf{F} = \frac{\partial}{\partial x}(xy) + \frac{\partial}{\partial y}(yz) + \frac{\partial}{\partial z}(xz) = y + z + x$$

over the cube's interior:

$$\text{Flux} = \iint_{\substack{\text{Cube} \\ \text{surface}}} \mathbf{F} \cdot \mathbf{n} \, d\sigma = \iiint_{\substack{\text{Cube} \\ \text{interior}}} \nabla \cdot \mathbf{F} \, dV \qquad \text{The Divergence Theorem}$$

$$= \int_0^1 \int_0^1 \int_0^1 (x + y + z) \, dx \, dy \, dz = \frac{3}{2}. \qquad \text{Routine integration} \qquad ■$$

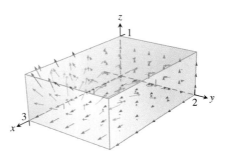

**FIGURE 16.71** The integral of div **F** over this region equals the total flux across the six sides (Example 4).

## EXAMPLE 4

**(a)** Calculate the flux of the vector field

$$\mathbf{F} = x^2\mathbf{i} + 4xyz\mathbf{j} + ze^x\mathbf{k}$$

out of the box-shaped region $D$: $0 \le x \le 3$, $0 \le y \le 2$, $0 \le z \le 1$. (See Figure 16.71.)

**(b)** Integrate div **F** over this region and show that the result is the same value as in part (a), as asserted by the Divergence Theorem.

### Solution

**(a)** The region $D$ has six sides. We calculate the flux across each side in turn. Consider the top side in the plane $z = 1$, having outward normal $\mathbf{n} = \mathbf{k}$. The flux across this side is given by $\mathbf{F} \cdot \mathbf{n} = ze^x$. Since $z = 1$ on this side, the flux at a point $(x, y, z)$ on the top is $e^x$. The total outward flux across this side is given by the surface integral

$$\int_0^2 \int_0^3 e^x \, dx \, dy = 2e^3 - 2. \qquad \text{\small Routine integration}$$

The outward flux across the other sides is computed similarly, and the results are summarized in the following table.

| Side | Unit normal n | F · n | Flux across side |
|------|---------------|-------|------------------|
| $x = 0$ | $-\mathbf{i}$ | $-x^2 = 0$ | $0$ |
| $x = 3$ | $\mathbf{i}$ | $x^2 = 9$ | $18$ |
| $y = 0$ | $-\mathbf{j}$ | $-4xyz = 0$ | $0$ |
| $y = 2$ | $\mathbf{j}$ | $4xyz = 8xz$ | $18$ |
| $z = 0$ | $-\mathbf{k}$ | $-ze^x = 0$ | $0$ |
| $z = 1$ | $\mathbf{k}$ | $ze^x = e^x$ | $2e^3 - 2$ |

The total outward flux is obtained by adding the terms for each of the six sides, giving

$$18 + 18 + 2e^3 - 2 = 34 + 2e^3.$$

**(b)** We first compute the divergence of **F**, obtaining

$$\text{div } \mathbf{F} = \nabla \cdot \mathbf{F} = 2x + 4xz + e^x.$$

The integral of the divergence of **F** over $D$ is

$$\iiint_D \text{div } \mathbf{F} \, dV = \int_0^1 \int_0^2 \int_0^3 (2x + 4xz + e^x) \, dx \, dy \, dz$$

$$= \int_0^1 \int_0^2 (8 + 18z + e^3) \, dy \, dz$$

$$= \int_0^1 (16 + 36z + 2e^3) \, dz$$

$$= 34 + 2e^3.$$

As asserted by the Divergence Theorem, the integral of the divergence over $D$ equals the outward flux across the boundary surface of $D$. ■

### Divergence and the Curl

If **F** is a vector field on three-dimensional space, then the curl $\nabla \times \mathbf{F}$ is also a vector field on three-dimensional space. So we can calculate the divergence of $\nabla \times \mathbf{F}$ using Equation (1). The result of this calculation is always 0.

---

**THEOREM 9**  If $\mathbf{F} = M\mathbf{i} + N\mathbf{j} + P\mathbf{k}$ is a vector field with continuous second partial derivatives, then

$$\text{div}\,(\text{curl}\,\mathbf{F}) = \nabla \cdot (\nabla \times \mathbf{F}) = 0.$$

---

**Proof**  From the definitions of the divergence and curl, we have

$$\text{div}\,(\text{curl}\,\mathbf{F}) = \nabla \cdot (\nabla \times \mathbf{F})$$

$$= \frac{\partial}{\partial x}\left(\frac{\partial P}{\partial y} - \frac{\partial N}{\partial z}\right) + \frac{\partial}{\partial y}\left(\frac{\partial M}{\partial z} - \frac{\partial P}{\partial x}\right) + \frac{\partial}{\partial z}\left(\frac{\partial N}{\partial x} - \frac{\partial M}{\partial y}\right)$$

$$= \frac{\partial^2 P}{\partial x \partial y} - \frac{\partial^2 N}{\partial x \partial z} + \frac{\partial^2 M}{\partial y \partial z} - \frac{\partial^2 P}{\partial y \partial x} + \frac{\partial^2 N}{\partial z \partial x} - \frac{\partial^2 M}{\partial z \partial y}$$

$$= 0,$$

because the mixed second partial derivatives cancel by the Mixed Derivative Theorem in Section 14.3. ∎

Theorem 9 has some interesting applications. If a vector field $\mathbf{G} = \text{curl}\,\mathbf{F}$, then the field **G** must have divergence 0. Saying this another way, if div $\mathbf{G} \neq 0$, then **G** cannot be the curl of any vector field **F** having continuous second partial derivatives. Moreover, if $\mathbf{G} = \text{curl}\,\mathbf{F}$, then the outward flux of **G** across any closed surface $S$ is zero by the corollary to the Divergence Theorem, provided the conditions of the theorem are satisfied. So if there is a closed surface for which the surface integral of the vector field **G** is nonzero, we can conclude that **G** is *not* the curl of some vector field **F**.

### Proof of the Divergence Theorem for Special Regions

To prove the Divergence Theorem, we take the components of **F** to have continuous first partial derivatives. We first assume that $D$ is a convex region with no holes or bubbles, such as a solid ball, cube, or ellipsoid, and that $S$ is a piecewise smooth surface. In addition, we assume that any line perpendicular to the $xy$-plane at an interior point of the region $R_{xy}$ that is the projection of $D$ on the $xy$-plane intersects the surface $S$ in exactly two points, producing surfaces

$$S_1: \quad z = f_1(x, y), \qquad (x, y) \text{ in } R_{xy}$$
$$S_2: \quad z = f_2(x, y), \qquad (x, y) \text{ in } R_{xy},$$

with $f_1 \leq f_2$. We make similar assumptions about the projection of $D$ onto the other coordinate planes. See Figure 16.72, which illustrates these assumptions.

The components of the unit normal vector $\mathbf{n} = n_1\mathbf{i} + n_2\mathbf{j} + n_3\mathbf{k}$ are the cosines of the angles $\alpha$, $\beta$, and $\gamma$ that **n** makes with **i**, **j**, and **k** (Figure 16.73). This is true because all the vectors involved are unit vectors, giving the *direction cosines*

$$n_1 = \mathbf{n} \cdot \mathbf{i} = |\mathbf{n}||\mathbf{i}| \cos \alpha = \cos \alpha$$
$$n_2 = \mathbf{n} \cdot \mathbf{j} = |\mathbf{n}||\mathbf{j}| \cos \beta = \cos \beta$$
$$n_3 = \mathbf{n} \cdot \mathbf{k} = |\mathbf{n}||\mathbf{k}| \cos \gamma = \cos \gamma.$$

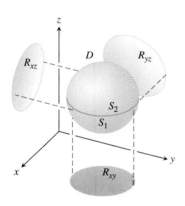

**FIGURE 16.72**  We prove the Divergence Theorem for the kind of three-dimensional region shown here.

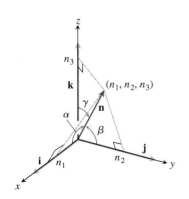

**FIGURE 16.73**  The components of **n** are the cosines of the angles $\alpha$, $\beta$, and $\gamma$ that it makes with **i**, **j**, and **k**.

Thus, the unit normal vector is given by

$$\mathbf{n} = (\cos \alpha)\mathbf{i} + (\cos \beta)\mathbf{j} + (\cos \gamma)\mathbf{k}$$

and

$$\mathbf{F} \cdot \mathbf{n} = M \cos \alpha + N \cos \beta + P \cos \gamma.$$

In component form, the Divergence Theorem states that

$$\underset{S}{\iint} \underbrace{(M \cos \alpha + N \cos \beta + P \cos \gamma) \, d\sigma}_{\mathbf{F} \cdot \mathbf{n}} = \underset{D}{\iiint} \underbrace{\left( \frac{\partial M}{\partial x} + \frac{\partial N}{\partial y} + \frac{\partial P}{\partial z} \right)}_{\text{div } \mathbf{F}} \, dx \, dy \, dz.$$

We prove the theorem by establishing the following three equations:

$$\underset{S}{\iint} M \cos \alpha \, d\sigma = \underset{D}{\iiint} \frac{\partial M}{\partial x} \, dx \, dy \, dz \qquad (3)$$

$$\underset{S}{\iint} N \cos \beta \, d\sigma = \underset{D}{\iiint} \frac{\partial N}{\partial y} \, dx \, dy \, dz \qquad (4)$$

$$\underset{S}{\iint} P \cos \gamma \, d\sigma = \underset{D}{\iiint} \frac{\partial P}{\partial z} \, dx \, dy \, dz \qquad (5)$$

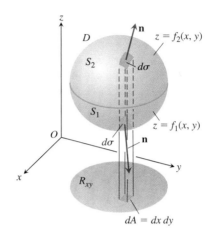

**FIGURE 16.74** The region $D$ enclosed by the surfaces $S_1$ and $S_2$ projects vertically onto $R_{xy}$ in the $xy$-plane.

**Proof of Equation (5)** We prove Equation (5) by converting the surface integral on the left to a double integral over the projection $R_{xy}$ of $D$ on the $xy$-plane (Figure 16.74). The surface $S$ consists of an upper part $S_2$ whose equation is $z = f_2(x, y)$ and a lower part $S_1$ whose equation is $z = f_1(x, y)$. On $S_2$, the outer normal $\mathbf{n}$ has a positive $\mathbf{k}$-component and

$$\cos \gamma \, d\sigma = dx \, dy \qquad \text{because} \qquad d\sigma = \frac{dA}{|\cos \gamma|} = \frac{dx \, dy}{\cos \gamma}.$$

See Figure 16.75. On $S_1$, the outer normal $\mathbf{n}$ has a negative $\mathbf{k}$-component and

$$\cos \gamma \, d\sigma = -dx \, dy.$$

Therefore,

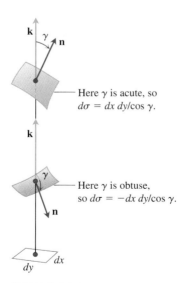

**FIGURE 16.75** An enlarged view of the area patches in Figure 16.74. The relations $d\sigma = \pm dx \, dy / \cos \gamma$ come from Eq. (7) in Section 16.5 with $F = \mathbf{F} \cdot \mathbf{n}$.

$$\underset{S}{\iint} P \cos \gamma \, d\sigma = \underset{S_2}{\iint} P \cos \gamma \, d\sigma + \underset{S_1}{\iint} P \cos \gamma \, d\sigma$$

$$= \underset{R_{xy}}{\iint} P(x, y, f_2(x, y)) \, dx \, dy - \underset{R_{xy}}{\iint} P(x, y, f_1(x, y)) \, dx \, dy$$

$$= \underset{R_{xy}}{\iint} \left[ P(x, y, f_2(x, y)) - P(x, y, f_1(x, y)) \right] dx \, dy$$

$$= \underset{R_{xy}}{\iint} \left[ \int_{f_1(x, y)}^{f_2(x, y)} \frac{\partial P}{\partial z} \, dz \right] dx \, dy = \underset{D}{\iiint} \frac{\partial P}{\partial z} \, dz \, dx \, dy.$$

This proves Equation (5). The proofs for Equations (3) and (4) follow the same pattern; or just permute $x, y, z$; $M, N, P$; $\alpha, \beta, \gamma$, in order, and get those results from Equation (5). This proves the Divergence Theorem for these special regions. ∎

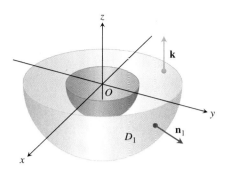

**FIGURE 16.76** The lower half of the solid region between two concentric spheres.

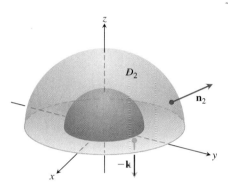

**FIGURE 16.77** The upper half of the solid region between two concentric spheres.

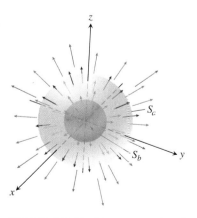

**FIGURE 16.78** Two concentric spheres in an expanding vector field. The outer sphere is $S_a$ and surrounds the inner sphere $S_b$.

### Divergence Theorem for Other Regions

The Divergence Theorem can be extended to regions that can be partitioned into a finite number of simple regions of the type just discussed and to regions that can be defined as limits of simpler regions in certain ways. For an example of one step in such a splitting process, suppose that $D$ is the region between two concentric spheres and that $\mathbf{F}$ has continuously differentiable components throughout $D$ and on the bounding surfaces. Split $D$ by an equatorial plane and apply the Divergence Theorem to each half separately. The bottom half, $D_1$, is shown in Figure 16.76. The surface $S_1$ that bounds $D_1$ consists of an outer hemisphere, a plane washer-shaped base, and an inner hemisphere. The Divergence Theorem says that

$$\iint_{S_1} \mathbf{F} \cdot \mathbf{n}_1 \, d\sigma_1 = \iiint_{D_1} \nabla \cdot \mathbf{F} \, dV_1. \tag{6}$$

The unit normal $\mathbf{n}_1$ that points outward from $D_1$ points away from the origin along the outer surface, equals $\mathbf{k}$ along the flat base, and points toward the origin along the inner surface. Next apply the Divergence Theorem to $D_2$, and its surface $S_2$ (Figure 16.77):

$$\iint_{S_2} \mathbf{F} \cdot \mathbf{n}_2 \, d\sigma_2 = \iiint_{D_2} \nabla \cdot \mathbf{F} \, dV_2. \tag{7}$$

As we follow $\mathbf{n}_2$ over $S_2$, pointing outward from $D_2$, we see that $\mathbf{n}_2$ equals $-\mathbf{k}$ along the washer-shaped base in the $xy$-plane, points away from the origin on the outer sphere, and points toward the origin on the inner sphere. When we add Equations (6) and (7), the integrals over the flat base cancel because of the opposite signs of $\mathbf{n}_1$ and $\mathbf{n}_2$. We thus arrive at the result

$$\iint_S \mathbf{F} \cdot \mathbf{n} \, d\sigma = \iiint_D \nabla \cdot \mathbf{F} \, dV,$$

with $D$ the region between the spheres, $S$ the boundary of $D$ consisting of two spheres, and $\mathbf{n}$ the unit normal to $S$ directed outward from $D$.

**EXAMPLE 5**　Find the net outward flux of the field

$$\mathbf{F} = \frac{x\mathbf{i} + y\mathbf{j} + z\mathbf{k}}{\rho^3}, \qquad \rho = \sqrt{x^2 + y^2 + z^2} \tag{8}$$

across the boundary of the region $D$: $0 < b^2 \le x^2 + y^2 + z^2 \le a^2$ (Figure 16.78).

**Solution**　The flux can be calculated by integrating $\nabla \cdot \mathbf{F}$ over $D$. Note that $\rho \ne 0$ in $D$. We have

$$\frac{\partial \rho}{\partial x} = \frac{1}{2}(x^2 + y^2 + z^2)^{-1/2}(2x) = \frac{x}{\rho}$$

and

$$\frac{\partial M}{\partial x} = \frac{\partial}{\partial x}(x\rho^{-3}) = \rho^{-3} - 3x\rho^{-4}\frac{\partial \rho}{\partial x} = \frac{1}{\rho^3} - \frac{3x^2}{\rho^5}.$$

Similarly,

$$\frac{\partial N}{\partial y} = \frac{1}{\rho^3} - \frac{3y^2}{\rho^5} \quad \text{and} \quad \frac{\partial P}{\partial z} = \frac{1}{\rho^3} - \frac{3z^2}{\rho^5}.$$

Hence,

$$\text{div } \mathbf{F} = \frac{3}{\rho^3} - \frac{3}{\rho^5}(x^2 + y^2 + z^2) = \frac{3}{\rho^3} - \frac{3\rho^2}{\rho^5} = 0.$$

So the net outward flux of **F** across the boundary of $D$ is zero by the corollary to the Divergence Theorem. There is more to learn about this vector field **F**, though. The flux leaving $D$ across the inner sphere $S_b$ is the negative of the flux leaving $D$ across the outer sphere $S_a$ (because the sum of these fluxes is zero). Hence, the flux of **F** across $S_b$ in the direction away from the origin equals the flux of **F** across $S_a$ in the direction away from the origin. Thus, the flux of **F** across a sphere centered at the origin is independent of the radius of the sphere. What is this flux?

To find it, we evaluate the flux integral directly for an arbitrary sphere $S_a$. The outward unit normal on the sphere of radius $a$ is

$$\mathbf{n} = \frac{x\mathbf{i} + y\mathbf{j} + z\mathbf{k}}{\sqrt{x^2 + y^2 + z^2}} = \frac{x\mathbf{i} + y\mathbf{j} + z\mathbf{k}}{a}.$$

Hence, on the sphere,

$$\mathbf{F} \cdot \mathbf{n} = \frac{x\mathbf{i} + y\mathbf{j} + z\mathbf{k}}{a^3} \cdot \frac{x\mathbf{i} + y\mathbf{j} + z\mathbf{k}}{a} = \frac{x^2 + y^2 + z^2}{a^4} = \frac{a^2}{a^4} = \frac{1}{a^2}$$

and

$$\iint_{S_a} \mathbf{F} \cdot \mathbf{n}\, d\sigma = \frac{1}{a^2} \iint_{S_a} d\sigma = \frac{1}{a^2}(4\pi a^2) = 4\pi.$$

The outward flux of **F** in Equation (8) across any sphere centered at the origin is $4\pi$. This result does not contradict the Divergence Theorem because **F** is not continuous at the origin. ∎

### Gauss's Law: One of the Four Great Laws of Electromagnetic Theory

There is still more to be learned from Example 5. In electromagnetic theory, the electric field created by a point charge $q$ located at the origin is

$$\mathbf{E}(x, y, z) = \frac{1}{4\pi\epsilon_0} \frac{q}{|\mathbf{r}|^2}\left(\frac{\mathbf{r}}{|\mathbf{r}|}\right) = \frac{q}{4\pi\epsilon_0} \frac{\mathbf{r}}{|\mathbf{r}|^3} = \frac{q}{4\pi\epsilon_0} \frac{x\mathbf{i} + y\mathbf{j} + z\mathbf{k}}{\rho^3},$$

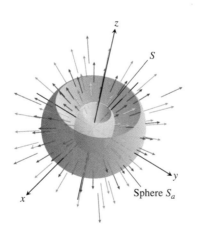

where $\epsilon_0$ is a physical constant, **r** is the position vector of the point $(x, y, z)$, and $\rho = |\mathbf{r}| = \sqrt{x^2 + y^2 + z^2}$. From Equation (8),

$$\mathbf{E} = \frac{q}{4\pi\epsilon_0}\mathbf{F}.$$

The calculations in Example 5 show that the outward flux of **E** across any sphere centered at the origin is $q/\epsilon_0$, but this result is not confined to spheres. The outward flux of **E** across any closed surface $S$ that encloses the origin (and to which the Divergence Theorem applies) is also $q/\epsilon_0$. To see why, we have only to imagine a large sphere $S_a$ centered at the origin and enclosing the surface $S$ (see Figure 16.79). Since

$$\nabla \cdot \mathbf{E} = \nabla \cdot \frac{q}{4\pi\epsilon_0}\mathbf{F} = \frac{q}{4\pi\epsilon_0}\nabla \cdot \mathbf{F} = 0$$

when $\rho > 0$, the triple integral of $\nabla \cdot \mathbf{E}$ over the region $D$ between $S$ and $S_a$ is zero. Hence, by the Divergence Theorem,

$$\iint_{\substack{\text{Boundary} \\ \text{of } D}} \mathbf{E} \cdot \mathbf{n}\, d\sigma = 0.$$

**FIGURE 16.79** A sphere $S_a$ surrounding another surface $S$. The tops of the surfaces are removed for visualization.

So the flux of **E** across $S$ in the direction away from the origin must be the same as the flux of **E** across $S_a$ in the direction away from the origin, which is $q/\epsilon_0$. This statement, called *Gauss's law*, also applies to charge distributions that are more general than the one assumed here, as shown in nearly any physics text. For any closed surface that encloses the origin, we have

$$\text{Gauss's law:} \qquad \iint_S \mathbf{E} \cdot \mathbf{n}\, d\sigma = \frac{q}{\epsilon_0}.$$

### Continuity Equation of Hydrodynamics

Let $D$ be a region in space bounded by a closed oriented surface $S$. If $\mathbf{v}(x, y, z)$ is the velocity field of a fluid flowing smoothly through $D$, $\delta = \delta(t, x, y, z)$ is the fluid's density at $(x, y, z)$ at time $t$, and $\mathbf{F} = \delta\mathbf{v}$, then the **continuity equation** of hydrodynamics states that

$$\nabla \cdot \mathbf{F} + \frac{\partial \delta}{\partial t} = 0.$$

If the functions involved have continuous first partial derivatives, the equation evolves naturally from the Divergence Theorem, as we now demonstrate.

First, the integral

$$\iint_S \mathbf{F} \cdot \mathbf{n} \, d\sigma$$

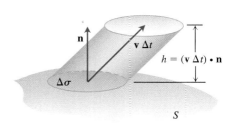

**FIGURE 16.80** The fluid that flows upward through the patch $\Delta\sigma$ in a short time $\Delta t$ fills a "cylinder" whose volume is approximately base $\times$ height $=$ $\mathbf{v} \cdot \mathbf{n} \, \Delta\sigma \, \Delta t$.

is the rate at which mass leaves $D$ across $S$ (leaves because $\mathbf{n}$ is the outer normal). To see why, consider a patch of area $\Delta\sigma$ on the surface (Figure 16.80). In a short time interval $\Delta t$, the volume $\Delta V$ of fluid that flows across the patch is approximately equal to the volume of a cylinder with base area $\Delta\sigma$ and height $(\mathbf{v}\Delta t) \cdot \mathbf{n}$, where $\mathbf{v}$ is a velocity vector rooted at a point of the patch:

$$\Delta V \approx \mathbf{v} \cdot \mathbf{n} \, \Delta\sigma \, \Delta t.$$

The mass of this volume of fluid is about

$$\Delta m \approx \delta \mathbf{v} \cdot \mathbf{n} \, \Delta\sigma \, \Delta t,$$

so the rate at which mass is flowing out of $D$ across the patch is about

$$\frac{\Delta m}{\Delta t} \approx \delta \mathbf{v} \cdot \mathbf{n} \, \Delta\sigma.$$

This leads to the approximation

$$\sum \frac{\Delta m}{\Delta t} \approx \sum \delta \mathbf{v} \cdot \mathbf{n} \, \Delta\sigma$$

as an estimate of the average rate at which mass flows across $S$. Finally, letting $\Delta\sigma \to 0$ and $\Delta t \to 0$ gives the instantaneous rate at which mass leaves $D$ across $S$ as

$$\frac{dm}{dt} = \iint_S \delta \mathbf{v} \cdot \mathbf{n} \, d\sigma,$$

which for our particular flow is

$$\frac{dm}{dt} = \iint_S \mathbf{F} \cdot \mathbf{n} \, d\sigma.$$

Now let $B$ be a solid sphere centered at a point $Q$ in the flow. The average value of $\nabla \cdot \mathbf{F}$ over $B$ is

$$\frac{1}{\text{volume of } B} \iiint_B \nabla \cdot \mathbf{F} \, dV.$$

It is a consequence of the continuity of the divergence that $\nabla \cdot \mathbf{F}$ actually takes on this value at some point $P$ in $B$. Thus, by the Divergence Theorem Equation (2),

$$(\nabla \cdot \mathbf{F})_P = \frac{1}{\text{volume of } B} \iiint_B \nabla \cdot \mathbf{F} \, dV = \frac{\displaystyle\iint_S \mathbf{F} \cdot \mathbf{n} \, d\sigma}{\text{volume of } B}$$

$$= \frac{\text{rate at which mass leaves } B \text{ across its surface } S}{\text{volume of } B}. \tag{9}$$

The last term of the equation describes decrease in mass per unit volume.

Now let the radius of $B$ approach zero while the center $Q$ stays fixed. The left side of Equation (9) converges to $(\nabla \cdot \mathbf{F})_Q$, and the right side converges to $(-\partial \delta / \partial t)_Q$, since $\delta = m/V$. The equality of these two limits is the continuity equation

$$\nabla \cdot \mathbf{F} = -\frac{\partial \delta}{\partial t}.$$

The continuity equation "explains" $\nabla \cdot \mathbf{F}$: The divergence of $\mathbf{F}$ at a point is the rate at which the density of the fluid is decreasing there. The Divergence Theorem

$$\iint\limits_{S} \mathbf{F} \cdot \mathbf{n} \, d\sigma = \iiint\limits_{D} \nabla \cdot \mathbf{F} \, dV$$

now says that the net decrease in density of the fluid in region $D$ (divergence integral) is accounted for by the mass transported across the surface $S$ (outward flux integral). So, the theorem is a statement about conservation of mass (Exercise 31).

## Unifying the Integral Theorems

If we think of a two-dimensional field $\mathbf{F} = M(x, y)\mathbf{i} + N(x, y)\mathbf{j}$ as a three-dimensional field whose $\mathbf{k}$-component is zero, then $\nabla \cdot \mathbf{F} = (\partial M / \partial x) + (\partial N / \partial y)$ and the normal form of Green's Theorem can be written as

$$\oint\limits_{C} \mathbf{F} \cdot \mathbf{n} \, ds = \iint\limits_{R} \left( \frac{\partial M}{\partial x} + \frac{\partial N}{\partial y} \right) dx \, dy = \iint\limits_{R} \nabla \cdot \mathbf{F} \, dA.$$

Similarly, $\nabla \times \mathbf{F} \cdot \mathbf{k} = (\partial N / \partial x) - (\partial M / \partial y)$, so the tangential form of Green's Theorem can be written as

$$\oint\limits_{C} \mathbf{F} \cdot \mathbf{T} \, ds = \iint\limits_{R} \left( \frac{\partial N}{\partial x} - \frac{\partial M}{\partial y} \right) dx \, dy = \iint\limits_{R} \nabla \times \mathbf{F} \cdot \mathbf{k} \, dA.$$

With the equations of Green's Theorem now in del notation, we can see their relationships to the equations in Stokes' Theorem and the Divergence Theorem, all summarized here.

---

### Green's Theorem and Its Generalization to Three Dimensions

**Tangential form of Green's Theorem:** $\displaystyle \oint\limits_{C} \mathbf{F} \cdot \mathbf{T} \, ds = \iint\limits_{R} \nabla \times \mathbf{F} \cdot \mathbf{k} \, dA$

**Stokes' Theorem:** $\displaystyle \oint\limits_{C} \mathbf{F} \cdot \mathbf{T} \, ds = \iint\limits_{S} \nabla \times \mathbf{F} \cdot \mathbf{n} \, d\sigma$

**Normal form of Green's Theorem:** $\displaystyle \oint\limits_{C} \mathbf{F} \cdot \mathbf{n} \, ds = \iint\limits_{R} \nabla \cdot \mathbf{F} \, dA$

**Divergence Theorem:** $\displaystyle \iint\limits_{S} \mathbf{F} \cdot \mathbf{n} \, d\sigma = \iiint\limits_{D} \nabla \cdot \mathbf{F} \, dV$

---

Notice how Stokes' Theorem generalizes the tangential (curl) form of Green's Theorem from a flat surface in the plane to a surface in three-dimensional space. In each case, the surface integral of curl $\mathbf{F}$ over the interior of the oriented surface equals the circulation of $\mathbf{F}$ around the boundary.

Likewise, the Divergence Theorem generalizes the normal (flux) form of Green's Theorem from a two-dimensional region in the plane to a three-dimensional region in space. In each case, the integral of $\nabla \cdot \mathbf{F}$ over the interior of the region equals the total flux of the field across the boundary enclosing the region.

**FIGURE 16.81** The outward unit normals at the boundary of $[a, b]$ in one-dimensional space.

There is still more to be learned here. All these results can be thought of as forms of a *single fundamental theorem*. Think back to the Fundamental Theorem of Calculus in Section 5.4. It says that if $f(x)$ is differentiable on $(a, b)$ and continuous on $[a, b]$, then

$$\int_a^b \frac{df}{dx}\,dx = f(b) - f(a).$$

If we let $\mathbf{F} = f(x)\mathbf{i}$ throughout $[a, b]$, then $(df/dx) = \nabla \cdot \mathbf{F}$. If we define the unit vector field $\mathbf{n}$ normal to the boundary of $[a, b]$ to be $\mathbf{i}$ at $b$ and $-\mathbf{i}$ at $a$ (Figure 16.81), then

$$
\begin{aligned}
f(b) - f(a) &= f(b)\mathbf{i} \cdot (\mathbf{i}) + f(a)\mathbf{i} \cdot (-\mathbf{i}) \\
&= \mathbf{F}(b) \cdot \mathbf{n} + \mathbf{F}(a) \cdot \mathbf{n} \\
&= \text{total outward flux of } \mathbf{F} \text{ across the boundary of } [a, b].
\end{aligned}
$$

The Fundamental Theorem now says that

$$\mathbf{F}(b) \cdot \mathbf{n} + \mathbf{F}(a) \cdot \mathbf{n} = \int_{[a, b]} \nabla \cdot \mathbf{F}\,dx.$$

The Fundamental Theorem of Calculus, the normal form of Green's Theorem, and the Divergence Theorem all say that the integral of the differential operator $\nabla \cdot$ operating on a field $\mathbf{F}$ over a region equals the sum of the normal field components over the boundary enclosing the region. (Here we are interpreting the line integral in Green's Theorem and the surface integral in the Divergence Theorem as "sums" over the boundary.)

Stokes' Theorem and the tangential form of Green's Theorem say that, when things are properly oriented, the surface integral of the differential operator $\nabla \times$ operating on a field equals the sum of the tangential field components over the boundary of the surface.

The beauty of these interpretations is the observance of a single unifying principle, which we might state as follows.

---

**A Unifying Fundamental Theorem of Vector Integral Calculus**

The integral of a differential operator acting on a field over a region equals the sum of the field components appropriate to the operator over the boundary of the region.

---

## Exercises 16.8

**Calculating Divergence**

In Exercises 1–4, find the divergence of the field.

1. The spin field in Figure 16.12
2. The radial field in Figure 16.11
3. The gravitational field in Figure 16.8 and Exercise 38a in Section 16.3
4. The velocity field in Figure 16.13

**Calculating Flux Using the Divergence Theorem**

In Exercises 5–16, use the Divergence Theorem to find the outward flux of **F** across the boundary of the region $D$.

5. **Cube**   $\mathbf{F} = (y - x)\mathbf{i} + (z - y)\mathbf{j} + (y - x)\mathbf{k}$
   $D$:   The cube bounded by the planes $x = \pm 1$, $y = \pm 1$, and $z = \pm 1$

6. $\mathbf{F} = x^2\mathbf{i} + y^2\mathbf{j} + z^2\mathbf{k}$
   a. **Cube**   $D$:   The cube cut from the first octant by the planes $x = 1$, $y = 1$, and $z = 1$

   b. **Cube**   $D$:   The cube bounded by the planes $x = \pm 1$, $y = \pm 1$, and $z = \pm 1$

   c. **Cylindrical can**   $D$:   The region cut from the solid cylinder $x^2 + y^2 \le 4$ by the planes $z = 0$ and $z = 1$

7. **Cylinder and paraboloid**   $\mathbf{F} = y\mathbf{i} + xy\mathbf{j} - z\mathbf{k}$
   $D$:   The region inside the solid cylinder $x^2 + y^2 \le 4$ between the plane $z = 0$ and the paraboloid $z = x^2 + y^2$

8. **Sphere**   $\mathbf{F} = x^2\mathbf{i} + xz\mathbf{j} + 3z\mathbf{k}$
   $D$:   The solid sphere $x^2 + y^2 + z^2 \le 4$

9. **Portion of sphere**   $\mathbf{F} = x^2\mathbf{i} - 2xy\mathbf{j} + 3xz\mathbf{k}$
   $D$:   The region cut from the first octant by the sphere $x^2 + y^2 + z^2 = 4$

10. **Cylindrical can**   $\mathbf{F} = (6x^2 + 2xy)\mathbf{i} + (2y + x^2z)\mathbf{j} + 4x^2y^3\mathbf{k}$
    $D$:   The region cut from the first octant by the cylinder $x^2 + y^2 = 4$ and the plane $z = 3$

**11. Wedge** $\mathbf{F} = 2xz\mathbf{i} - xy\mathbf{j} - z^2\mathbf{k}$

$D$: The wedge cut from the first octant by the plane $y + z = 4$ and the elliptical cylinder $4x^2 + y^2 = 16$

**12. Sphere** $\mathbf{F} = x^3\mathbf{i} + y^3\mathbf{j} + z^3\mathbf{k}$

$D$: The solid sphere $x^2 + y^2 + z^2 \le a^2$

**13. Thick sphere** $\mathbf{F} = \sqrt{x^2 + y^2 + z^2}\,(x\mathbf{i} + y\mathbf{j} + z\mathbf{k})$

$D$: The region $1 \le x^2 + y^2 + z^2 \le 2$

**14. Thick sphere** $\mathbf{F} = (x\mathbf{i} + y\mathbf{j} + z\mathbf{k})/\sqrt{x^2 + y^2 + z^2}$

$D$: The region $1 \le x^2 + y^2 + z^2 \le 4$

**15. Thick sphere** $\mathbf{F} = (5x^3 + 12xy^2)\mathbf{i} + (y^3 + e^y \sin z)\mathbf{j} + (5z^3 + e^y \cos z)\mathbf{k}$

$D$: The solid region between the spheres $x^2 + y^2 + z^2 = 1$ and $x^2 + y^2 + z^2 = 2$

**16. Thick cylinder** $\mathbf{F} = \ln(x^2 + y^2)\mathbf{i} - \left(\dfrac{2z}{x}\tan^{-1}\dfrac{y}{x}\right)\mathbf{j} + z\sqrt{x^2 + y^2}\,\mathbf{k}$

$D$: The thick-walled cylinder $1 \le x^2 + y^2 \le 2, -1 \le z \le 2$

### Theory and Examples

**17. a.** Show that the outward flux of the position vector field $\mathbf{F} = x\mathbf{i} + y\mathbf{j} + z\mathbf{k}$ through a smooth closed surface $S$ is three times the volume of the region enclosed by the surface.

**b.** Let $\mathbf{n}$ be the outward unit normal vector field on $S$. Show that it is not possible for $\mathbf{F}$ to be orthogonal to $\mathbf{n}$ at every point of $S$.

**18.** The base of the closed cubelike surface shown here is the unit square in the $xy$-plane. The four sides lie in the planes $x = 0$, $x = 1$, $y = 0$, and $y = 1$. The top is an arbitrary smooth surface whose identity is unknown. Let $\mathbf{F} = x\mathbf{i} - 2y\mathbf{j} + (z + 3)\mathbf{k}$ and suppose the outward flux of $\mathbf{F}$ through Side $A$ is 1 and through Side $B$ is $-3$. Can you conclude anything about the outward flux through the top? Give reasons for your answer.

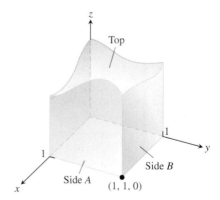

**19.** Let $\mathbf{F} = (y \cos 2x)\mathbf{i} + (y^2 \sin 2x)\mathbf{j} + (x^2y + z)\mathbf{k}$. Is there a vector field $\mathbf{A}$ such that $\mathbf{F} = \nabla \times \mathbf{A}$? Explain your answer.

**20. Outward flux of a gradient field** Let $S$ be the surface of the portion of the solid sphere $x^2 + y^2 + z^2 \le a^2$ that lies in the first octant and let $f(x, y, z) = \ln\sqrt{x^2 + y^2 + z^2}$. Calculate

$$\iint_S \nabla f \cdot \mathbf{n}\, d\sigma.$$

($\nabla f \cdot \mathbf{n}$ is the derivative of $f$ in the direction of outward normal $\mathbf{n}$.)

**21.** Let $\mathbf{F}$ be a field whose components have continuous first partial derivatives throughout a portion of space containing a region $D$ bounded by a smooth closed surface $S$. If $|\mathbf{F}| \le 1$, can any bound be placed on the size of

$$\iiint_D \nabla \cdot \mathbf{F}\, dV?$$

Give reasons for your answer.

**22. Maximum flux** Among all rectangular solids defined by the inequalities $0 \le x \le a, 0 \le y \le b, 0 \le z \le 1$, find the one for which the total flux of $\mathbf{F} = (-x^2 - 4xy)\mathbf{i} - 6yz\mathbf{j} + 12z\mathbf{k}$ outward through the six sides is greatest. What *is* the greatest flux?

**23.** Calculate the net outward flux of the vector field

$$\mathbf{F} = xy\mathbf{i} + (\sin xz + y^2)\mathbf{j} + (e^{xy^2} + x)\mathbf{k}$$

over the surface $S$ surrounding the region $D$ bounded by the planes $y = 0, z = 0, z = 2 - y$ and the parabolic cylinder $z = 1 - x^2$.

**24.** Compute the net outward flux of the vector field $\mathbf{F} = (x\mathbf{i} + y\mathbf{j} + z\mathbf{k})/(x^2 + y^2 + z^2)^{3/2}$ across the ellipsoid $9x^2 + 4y^2 + 6z^2 = 36$.

**25.** Let $\mathbf{F}$ be a differentiable vector field and let $g(x, y, z)$ be a differentiable scalar function. Verify the following identities.

**a.** $\nabla \cdot (g\mathbf{F}) = g\nabla \cdot \mathbf{F} + \nabla g \cdot \mathbf{F}$

**b.** $\nabla \times (g\mathbf{F}) = g\nabla \times \mathbf{F} + \nabla g \times \mathbf{F}$

**26.** Let $\mathbf{F}_1$ and $\mathbf{F}_2$ be differentiable vector fields and let $a$ and $b$ be arbitrary real constants. Verify the following identities.

**a.** $\nabla \cdot (a\mathbf{F}_1 + b\mathbf{F}_2) = a\nabla \cdot \mathbf{F}_1 + b\nabla \cdot \mathbf{F}_2$

**b.** $\nabla \times (a\mathbf{F}_1 + b\mathbf{F}_2) = a\nabla \times \mathbf{F}_1 + b\nabla \times \mathbf{F}_2$

**c.** $\nabla \cdot (\mathbf{F}_1 \times \mathbf{F}_2) = \mathbf{F}_2 \cdot \nabla \times \mathbf{F}_1 - \mathbf{F}_1 \cdot \nabla \times \mathbf{F}_2$

**27.** If $\mathbf{F} = M\mathbf{i} + N\mathbf{j} + P\mathbf{k}$ is a differentiable vector field, we define the notation $\mathbf{F} \cdot \nabla$ to mean

$$M\frac{\partial}{\partial x} + N\frac{\partial}{\partial y} + P\frac{\partial}{\partial z}.$$

For differentiable vector fields $\mathbf{F}_1$ and $\mathbf{F}_2$, verify the following identities.

**a.** $\nabla \times (\mathbf{F}_1 \times \mathbf{F}_2) = (\mathbf{F}_2 \cdot \nabla)\mathbf{F}_1 - (\mathbf{F}_1 \cdot \nabla)\mathbf{F}_2 + (\nabla \cdot \mathbf{F}_2)\mathbf{F}_1 - (\nabla \cdot \mathbf{F}_1)\mathbf{F}_2$

**b.** $\nabla(\mathbf{F}_1 \cdot \mathbf{F}_2) = (\mathbf{F}_1 \cdot \nabla)\mathbf{F}_2 + (\mathbf{F}_2 \cdot \nabla)\mathbf{F}_1 + \mathbf{F}_1 \times (\nabla \times \mathbf{F}_2) + \mathbf{F}_2 \times (\nabla \times \mathbf{F}_1)$

**28. Harmonic functions** A function $f(x, y, z)$ is said to be *harmonic* in a region $D$ in space if it satisfies the Laplace equation

$$\nabla^2 f = \nabla \cdot \nabla f = \frac{\partial^2 f}{\partial x^2} + \frac{\partial^2 f}{\partial y^2} + \frac{\partial^2 f}{\partial z^2} = 0$$

throughout $D$.

**a.** Suppose that $f$ is harmonic throughout a bounded region $D$ enclosed by a smooth surface $S$ and that $\mathbf{n}$ is the chosen unit normal vector on $S$. Show that the integral over $S$ of $\nabla f \cdot \mathbf{n}$, the derivative of $f$ in the direction of $\mathbf{n}$, is zero.

**b.** Show that if $f$ is harmonic on $D$, then

$$\iint_S f \nabla f \cdot \mathbf{n}\, d\sigma = \iiint_D |\nabla f|^2\, dV.$$

**29. Green's first formula** Suppose that $f$ and $g$ are scalar functions with continuous first- and second-order partial derivatives throughout a region $D$ that is bounded by a closed piecewise smooth surface $S$. Show that

$$\iint_S f\, \nabla g \cdot \mathbf{n}\, d\sigma = \iiint_D (f\, \nabla^2 g + \nabla f \cdot \nabla g)\, dV. \qquad (10)$$

Equation (10) is **Green's first formula**. (*Hint:* Apply the Divergence Theorem to the field $\mathbf{F} = f\, \nabla g$.)

**30. Green's second formula** (*Continuation of Exercise 29.*) Interchange $f$ and $g$ in Equation (10) to obtain a similar formula. Then subtract this formula from Equation (10) to show that

$$\iint_S (f\, \nabla g - g\, \nabla f) \cdot \mathbf{n}\, d\sigma = \iiint_D (f\, \nabla^2 g - g\, \nabla^2 f)\, dV. \qquad (11)$$

This equation is **Green's second formula**.

**31. Conservation of mass** Let $\mathbf{v}(t, x, y, z)$ be a continuously differentiable vector field over the region $D$ in space and let $p(t, x, y, z)$ be a continuously differentiable scalar function. The variable $t$ represents the time domain. The Law of Conservation of Mass asserts that

$$\frac{d}{dt} \iiint_D p(t, x, y, z)\, dV = - \iint_S p\mathbf{v} \cdot \mathbf{n}\, d\sigma,$$

where $S$ is the surface enclosing $D$.

**a.** Give a physical interpretation of the conservation of mass law if $\mathbf{v}$ is a velocity flow field and $p$ represents the density of the fluid at point $(x, y, z)$ at time $t$.

**b.** Use the Divergence Theorem and Leibniz's Rule,

$$\frac{d}{dt} \iiint_D p(t, x, y, z)\, dV = \iiint_D \frac{\partial p}{\partial t}\, dV,$$

to show that the Law of Conservation of Mass is equivalent to the continuity equation,

$$\nabla \cdot p\mathbf{v} + \frac{\partial p}{\partial t} = 0.$$

(In the first term $\nabla \cdot p\mathbf{v}$, the variable $t$ is held fixed, and in the second term $\partial p / \partial t$, it is assumed that the point $(x, y, z)$ in $D$ is held fixed.)

**32. The heat diffusion equation** Let $T(t, x, y, z)$ be a function with continuous second derivatives giving the temperature at time $t$ at the point $(x, y, z)$ of a solid occupying a region $D$ in space. If the solid's heat capacity and mass density are denoted by the constants $c$ and $\rho$, respectively, the quantity $c\rho T$ is called the solid's **heat energy per unit volume**.

**a.** Explain why $-\nabla T$ points in the direction of heat flow.

**b.** Let $-k\nabla T$ denote the **energy flux vector**. (Here the constant $k$ is called the **conductivity**.) Assuming the Law of Conservation of Mass with $-k\nabla T = \mathbf{v}$ and $c\rho T = p$ in Exercise 31, derive the diffusion (heat) equation

$$\frac{\partial T}{\partial t} = K\nabla^2 T,$$

where $K = k/(c\rho) > 0$ is the *diffusivity* constant. (Notice that if $T(t, x)$ represents the temperature at time $t$ at position $x$ in a uniform conducting rod with perfectly insulated sides, then $\nabla^2 T = \partial^2 T / \partial x^2$ and the diffusion equation reduces to the one-dimensional heat equation in Chapter 14's Additional Exercises.)

---

## Chapter **16** Questions to Guide Your Review

1. What are line integrals of scalar functions? How are they evaluated? Give examples.

2. How can you use line integrals to find the centers of mass of springs or wires? Explain.

3. What is a vector field? What is the line integral of a vector field? What is a gradient field? Give examples.

4. What is the flow of a vector field along a curve? What is the work done by vector field moving an object along a curve? How do you calculate the work done? Give examples.

5. What is the Fundamental Theorem of line integrals? Explain how it relates to the Fundamental Theorem of Calculus.

6. Specify three properties that are special about conservative fields. How can you tell when a field is conservative?

7. What is special about path independent fields?

8. What is a potential function? Show by example how to find a potential function for a conservative field.

9. What is a differential form? What does it mean for such a form to be exact? How do you test for exactness? Give examples.

10. What is Green's Theorem? Discuss how the two forms of Green's Theorem extend the Net Change Theorem in Chapter 5.

11. How do you calculate the area of a parametrized surface in space? Of an implicitly defined surface $F(x, y, z) = 0$? Of the surface which is the graph of $z = f(x, y)$? Give examples.

12. How do you integrate a scalar function over a parametrized surface? Of surfaces that are defined implicitly or in explicit form? Give examples.

13. What is an oriented surface? What is the surface integral of a vector field in three-dimensional space over an oriented surface? How is it related to the net outward flux of the field? Give examples.

14. What is the curl of a vector field? How can you interpret it?

15. What is Stokes' Theorem? Explain how it generalizes Green's Theorem to three dimensions.

16. What is the divergence of a vector field? How can you interpret it?

17. What is the Divergence Theorem? Explain how it generalizes Green's Theorem to three dimensions.

18. How do Green's Theorem, Stokes' Theorem, and the Divergence Theorem relate to the Fundamental Theorem of Calculus for ordinary single integrals?

# Chapter **16** Practice Exercises

## Evaluating Line Integrals

1. The accompanying figure shows two polygonal paths in space joining the origin to the point $(1, 1, 1)$. Integrate $f(x, y, z) = 2x - 3y^2 - 2z + 3$ over each path.

Path 1

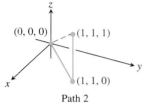
Path 2

2. The accompanying figure shows three polygonal paths joining the origin to the point $(1, 1, 1)$. Integrate $f(x, y, z) = x^2 + y - z$ over each path.

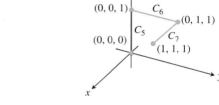

3. Integrate $f(x, y, z) = \sqrt{x^2 + z^2}$ over the circle

$$\mathbf{r}(t) = (a \cos t)\mathbf{j} + (a \sin t)\mathbf{k}, \qquad 0 \le t \le 2\pi.$$

4. Integrate $f(x, y, z) = \sqrt{x^2 + y^2}$ over the involute curve

$$\mathbf{r}(t) = (\cos t + t \sin t)\mathbf{i} + (\sin t - t \cos t)\mathbf{j}, \qquad 0 \le t \le \sqrt{3}.$$

Evaluate the integrals in Exercises 5 and 6.

5. $\displaystyle\int_{(-1,1,1)}^{(4,-3,0)} \frac{dx + dy + dz}{\sqrt{x + y + z}}$
6. $\displaystyle\int_{(1,1,1)}^{(10,3,3)} dx - \sqrt{\frac{z}{y}}dy - \sqrt{\frac{y}{z}}dz$

7. Integrate $\mathbf{F} = -(y \sin z)\mathbf{i} + (x \sin z)\mathbf{j} + (xy \cos z)\mathbf{k}$ around the circle cut from the sphere $x^2 + y^2 + z^2 = 5$ by the plane $z = -1$, clockwise as viewed from above.

8. Integrate $\mathbf{F} = 3x^2y\mathbf{i} + (x^3 + 1)\mathbf{j} + 9z^2\mathbf{k}$ around the circle cut from the sphere $x^2 + y^2 + z^2 = 9$ by the plane $x = 2$.

Evaluate the integrals in Exercises 9 and 10.

9. $\displaystyle\int_C 8x \sin y \, dx - 8y \cos x \, dy$

$C$ is the square cut from the first quadrant by the lines $x = \pi/2$ and $y = \pi/2$.

10. $\displaystyle\int_C y^2 \, dx + x^2 \, dy$

$C$ is the circle $x^2 + y^2 = 4$.

## Finding and Evaluating Surface Integrals

11. **Area of an elliptical region**  Find the area of the elliptical region cut from the plane $x + y + z = 1$ by the cylinder $x^2 + y^2 = 1$.

12. **Area of a parabolic cap**  Find the area of the cap cut from the paraboloid $y^2 + z^2 = 3x$ by the plane $x = 1$.

13. **Area of a spherical cap**  Find the area of the cap cut from the top of the sphere $x^2 + y^2 + z^2 = 1$ by the plane $z = \sqrt{2}/2$.

14. **a. Hemisphere cut by cylinder**  Find the area of the surface cut from the hemisphere $x^2 + y^2 + z^2 = 4, z \ge 0$, by the cylinder $x^2 + y^2 = 2x$.

   **b.** Find the area of the portion of the cylinder that lies inside the hemisphere. (*Hint:* Project onto the $xz$-plane. Or evaluate the integral $\int h \, ds$, where $h$ is the altitude of the cylinder and $ds$ is the element of arc length on the circle $x^2 + y^2 = 2x$ in the $xy$-plane.)

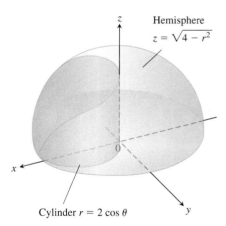

Hemisphere
$z = \sqrt{4 - r^2}$

Cylinder $r = 2 \cos \theta$

15. **Area of a triangle**  Find the area of the triangle in which the plane $(x/a) + (y/b) + (z/c) = 1$ ($a, b, c > 0$) intersects the first octant. Check your answer with an appropriate vector calculation.

16. **Parabolic cylinder cut by planes**  Integrate

   **a.** $g(x, y, z) = \dfrac{yz}{\sqrt{4y^2 + 1}}$     **b.** $g(x, y, z) = \dfrac{z}{\sqrt{4y^2 + 1}}$

   over the surface cut from the parabolic cylinder $y^2 - z = 1$ by the planes $x = 0, x = 3$, and $z = 0$.

17. **Circular cylinder cut by planes**  Integrate $g(x, y, z) = x^4y(y^2 + z^2)$ over the portion of the cylinder $y^2 + z^2 = 25$ that lies in the first octant between the planes $x = 0$ and $x = 1$ and above the plane $z = 3$.

18. **Area of Wyoming**  The state of Wyoming is bounded by the meridians $111°3'$ and $104°3'$ west longitude and by the circles $41°$ and $45°$ north latitude. Assuming that Earth is a sphere of radius $R = 3959$ mi, find the area of Wyoming.

## Parametrized Surfaces

Find parametrizations for the surfaces in Exercises 19–24. (There are many ways to do these, so your answers may not be the same as those in the back of the book.)

**19. Spherical band** The portion of the sphere $x^2 + y^2 + z^2 = 36$ between the planes $z = -3$ and $z = 3\sqrt{3}$

**20. Parabolic cap** The portion of the paraboloid $z = -(x^2 + y^2)/2$ above the plane $z = -2$

**21. Cone** The cone $z = 1 + \sqrt{x^2 + y^2}, z \leq 3$

**22. Plane above square** The portion of the plane $4x + 2y + 4z = 12$ that lies above the square $0 \leq x \leq 2, 0 \leq y \leq 2$ in the first quadrant

**23. Portion of paraboloid** The portion of the paraboloid $y = 2(x^2 + z^2), y \leq 2$ that lies above the $xy$-plane

**24. Portion of hemisphere** The portion of the hemisphere $x^2 + y^2 + z^2 = 10, y \geq 0$, in the first octant

**25. Surface area** Find the area of the surface

$$\mathbf{r}(u, v) = (u + v)\mathbf{i} + (u - v)\mathbf{j} + v\mathbf{k},$$

$$0 \leq u \leq 1, 0 \leq v \leq 1.$$

**26. Surface integral** Integrate $f(x, y, z) = xy - z^2$ over the surface in Exercise 25.

**27. Area of a helicoid** Find the surface area of the helicoid $\mathbf{r}(r, \theta) = (r \cos \theta)\mathbf{i} + (r \sin \theta)\mathbf{j} + \theta\mathbf{k}, \quad 0 \leq \theta \leq 2\pi, \quad 0 \leq r \leq 1$, in the accompanying figure.

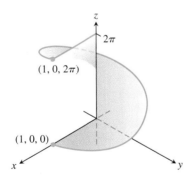

**28. Surface integral** Evaluate the integral $\iint_S \sqrt{x^2 + y^2 + 1}\, d\sigma$, where $S$ is the helicoid in Exercise 27.

## Conservative Fields

Which of the fields in Exercises 29–32 are conservative, and which are not?

**29.** $\mathbf{F} = x\mathbf{i} + y\mathbf{j} + z\mathbf{k}$

**30.** $\mathbf{F} = (x\mathbf{i} + y\mathbf{j} + z\mathbf{k})/(x^2 + y^2 + z^2)^{3/2}$

**31.** $\mathbf{F} = xe^y\mathbf{i} + ye^z\mathbf{j} + ze^x\mathbf{k}$

**32.** $\mathbf{F} = (\mathbf{i} + z\mathbf{j} + y\mathbf{k})/(x + yz)$

Find potential functions for the fields in Exercises 33 and 34.

**33.** $\mathbf{F} = 2\mathbf{i} + (2y + z)\mathbf{j} + (y + 1)\mathbf{k}$

**34.** $\mathbf{F} = (z \cos xz)\mathbf{i} + e^y\mathbf{j} + (x \cos xz)\mathbf{k}$

## Work and Circulation

In Exercises 35 and 36, find the work done by each field along the paths from $(0, 0, 0)$ to $(1, 1, 1)$ in Exercise 1.

**35.** $\mathbf{F} = 2xy\mathbf{i} + \mathbf{j} + x^2\mathbf{k}$     **36.** $\mathbf{F} = 2xy\mathbf{i} + x^2\mathbf{j} + \mathbf{k}$

**37. Finding work in two ways** Find the work done by

$$\mathbf{F} = \frac{x\mathbf{i} + y\mathbf{j}}{(x^2 + y^2)^{3/2}}$$

over the plane curve $\mathbf{r}(t) = (e^t \cos t)\mathbf{i} + (e^t \sin t)\mathbf{j}$ from the point $(1, 0)$ to the point $(e^{2\pi}, 0)$ in two ways:

**a.** By using the parametrization of the curve to evaluate the work integral.

**b.** By evaluating a potential function for $\mathbf{F}$.

**38. Flow along different paths** Find the flow of the field $\mathbf{F} = \nabla(x^2ze^y)$

**a.** once around the ellipse $C$ in which the plane $x + y + z = 1$ intersects the cylinder $x^2 + z^2 = 25$, clockwise as viewed from the positive $y$-axis.

**b.** along the curved boundary of the helicoid in Exercise 27 from $(1, 0, 0)$ to $(1, 0, 2\pi)$.

In Exercises 39 and 40, use the curl integral in Stokes' Theorem to find the circulation of the field $\mathbf{F}$ around the curve $C$ in the indicated direction.

**39. Circulation around an ellipse** $\mathbf{F} = y^2\mathbf{i} - y\mathbf{j} + 3z^2\mathbf{k}$

$C$: The ellipse in which the plane $2x + 6y - 3z = 6$ meets the cylinder $x^2 + y^2 = 1$, counterclockwise as viewed from above

**40. Circulation around a circle** $\mathbf{F} = (x^2 + y)\mathbf{i} + (x + y)\mathbf{j} + (4y^2 - z)\mathbf{k}$

$C$: The circle in which the plane $z = -y$ meets the sphere $x^2 + y^2 + z^2 = 4$, counterclockwise as viewed from above

## Masses and Moments

**41. Wire with different densities** Find the mass of a thin wire lying along the curve $\mathbf{r}(t) = \sqrt{2}t\mathbf{i} + \sqrt{2}t\mathbf{j} + (4 - t^2)\mathbf{k}, 0 \leq t \leq 1$, if the density at $t$ is **(a)** $\delta = 3t$ and **(b)** $\delta = 1$.

**42. Wire with variable density** Find the center of mass of a thin wire lying along the curve $\mathbf{r}(t) = t\mathbf{i} + 2t\mathbf{j} + (2/3)t^{3/2}\mathbf{k}$, $0 \leq t \leq 2$, if the density at $t$ is $\delta = 3\sqrt{5 + t}$.

**43. Wire with variable density** Find the center of mass and the moments of inertia about the coordinate axes of a thin wire lying along the curve

$$\mathbf{r}(t) = t\mathbf{i} + \frac{2\sqrt{2}}{3}t^{3/2}\mathbf{j} + \frac{t^2}{2}\mathbf{k}, \qquad 0 \leq t \leq 2,$$

if the density at $t$ is $\delta = 1/(t + 1)$.

**44. Center of mass of an arch** A slender metal arch lies along the semicircle $y = \sqrt{a^2 - x^2}$ in the $xy$-plane. The density at the point $(x, y)$ on the arch is $\delta(x, y) = 2a - y$. Find the center of mass.

**45. Wire with constant density** A wire of constant density $\delta = 1$ lies along the curve $\mathbf{r}(t) = (e^t \cos t)\mathbf{i} + (e^t \sin t)\mathbf{j} + e^t\mathbf{k}, 0 \leq t \leq \ln 2$. Find $\bar{z}$ and $I_z$.

**46. Helical wire with constant density** Find the mass and center of mass of a wire of constant density $\delta$ that lies along the helix $\mathbf{r}(t) = (2 \sin t)\mathbf{i} + (2 \cos t)\mathbf{j} + 3t\mathbf{k}, 0 \leq t \leq 2\pi$.

**47. Inertia and center of mass of a shell** Find $I_z$ and the center of mass of a thin shell of density $\delta(x, y, z) = z$ cut from the upper portion of the sphere $x^2 + y^2 + z^2 = 25$ by the plane $z = 3$.

**48. Moment of inertia of a cube** Find the moment of inertia about the $z$-axis of the surface of the cube cut from the first octant by the planes $x = 1, y = 1$, and $z = 1$ if the density is $\delta = 1$.

### Flux Across a Plane Curve or Surface

Use Green's Theorem to find the counterclockwise circulation and outward flux for the fields and curves in Exercises 49 and 50.

**49. Square** $\mathbf{F} = (2xy + x)\mathbf{i} + (xy - y)\mathbf{j}$

$C$: The square bounded by $x = 0$, $x = 1$, $y = 0$, $y = 1$

**50. Triangle** $\mathbf{F} = (y - 6x^2)\mathbf{i} + (x + y^2)\mathbf{j}$

$C$: The triangle made by the lines $y = 0$, $y = x$, and $x = 1$

**51. Zero line integral** Show that

$$\oint_C \ln x \sin y \, dy - \frac{\cos y}{x} dx = 0$$

for any closed curve $C$ to which Green's Theorem applies.

**52. a. Outward flux and area** Show that the outward flux of the position vector field $\mathbf{F} = x\mathbf{i} + y\mathbf{j}$ across any closed curve to which Green's Theorem applies is twice the area of the region enclosed by the curve.

   **b.** Let $\mathbf{n}$ be the outward unit normal vector to a closed curve to which Green's Theorem applies. Show that it is not possible for $\mathbf{F} = x\mathbf{i} + y\mathbf{j}$ to be orthogonal to $\mathbf{n}$ at every point of $C$.

In Exercises 53–56, find the outward flux of $\mathbf{F}$ across the boundary of $D$.

**53. Cube** $\mathbf{F} = 2xy\mathbf{i} + 2yz\mathbf{j} + 2xz\mathbf{k}$

$D$: The cube cut from the first octant by the planes $x = 1$, $y = 1$, $z = 1$

**54. Spherical cap** $\mathbf{F} = xz\mathbf{i} + yz\mathbf{j} + \mathbf{k}$

$D$: The entire surface of the upper cap cut from the solid sphere $x^2 + y^2 + z^2 \leq 25$ by the plane $z = 3$

**55. Spherical cap** $\mathbf{F} = -2x\mathbf{i} - 3y\mathbf{j} + z\mathbf{k}$

$D$: The upper region cut from the solid sphere $x^2 + y^2 + z^2 \leq 2$ by the paraboloid $z = x^2 + y^2$

**56. Cone and cylinder** $\mathbf{F} = (6x + y)\mathbf{i} - (x + z)\mathbf{j} + 4yz\mathbf{k}$

$D$: The region in the first octant bounded by the cone $z = \sqrt{x^2 + y^2}$, the cylinder $x^2 + y^2 = 1$, and the coordinate planes

**57. Hemisphere, cylinder, and plane** Let $S$ be the surface that is bounded on the left by the hemisphere $x^2 + y^2 + z^2 = a^2$, $y \leq 0$, in the middle by the cylinder $x^2 + z^2 = a^2$, $0 \leq y \leq a$, and on the right by the plane $y = a$. Find the flux of $\mathbf{F} = y\mathbf{i} + z\mathbf{j} + x\mathbf{k}$ outward across $S$.

**58. Cylinder and planes** Find the outward flux of the field $\mathbf{F} = 3xz^2\mathbf{i} + y\mathbf{j} - z^3\mathbf{k}$ across the surface of the solid in the first octant that is bounded by the cylinder $x^2 + 4y^2 = 16$ and the planes $y = 2z$, $x = 0$, and $z = 0$.

**59. Cylindrical can** Use the Divergence Theorem to find the flux of $\mathbf{F} = xy^2\mathbf{i} + x^2y\mathbf{j} + y\mathbf{k}$ outward through the surface of the region enclosed by the cylinder $x^2 + y^2 = 1$ and the planes $z = 1$ and $z = -1$.

**60. Hemisphere** Find the flux of $\mathbf{F} = (3z + 1)\mathbf{k}$ upward across the hemisphere $x^2 + y^2 + z^2 = a^2$, $z \geq 0$ **(a)** with the Divergence Theorem and **(b)** by evaluating the flux integral directly.

---

## Chapter 16   Additional and Advanced Exercises

### Finding Areas with Green's Theorem

Use the Green's Theorem area formula in Exercises 16.4 to find the areas of the regions enclosed by the curves in Exercises 1–4.

**1.** The limaçon $x = 2\cos t - \cos 2t$, $y = 2\sin t - \sin 2t$, $0 \leq t \leq 2\pi$

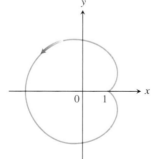

**2.** The deltoid $x = 2\cos t + \cos 2t$, $y = 2\sin t - \sin 2t$, $0 \leq t \leq 2\pi$

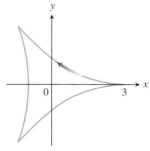

**3.** The eight curve $x = (1/2)\sin 2t$, $y = \sin t$, $0 \leq t \leq \pi$ (one loop)

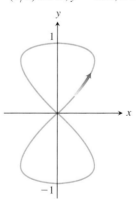

**4.** The teardrop $x = 2a\cos t - a\sin 2t$, $y = b\sin t$, $0 \leq t \leq 2\pi$

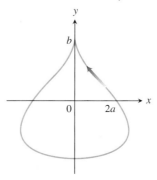

**Theory and Applications**

**5. a.** Give an example of a vector field **F** $(x, y, z)$ that has value **0** at only one point and such that curl **F** is nonzero everywhere. Be sure to identify the point and compute the curl.

**b.** Give an example of a vector field **F** $(x, y, z)$ that has value **0** on precisely one line and such that curl **F** is nonzero everywhere. Be sure to identify the line and compute the curl.

**c.** Give an example of a vector field **F** $(x, y, z)$ that has value **0** on a surface and such that curl **F** is nonzero everywhere. Be sure to identify the surface and compute the curl.

**6.** Find all points $(a, b, c)$ on the sphere $x^2 + y^2 + z^2 = R^2$ where the vector field $\mathbf{F} = yz^2\mathbf{i} + xz^2\mathbf{j} + 2xyz\mathbf{k}$ is normal to the surface and $\mathbf{F}(a, b, c) \neq \mathbf{0}$.

**7.** Find the mass of a spherical shell of radius $R$ such that at each point $(x, y, z)$ on the surface the mass density $\delta(x, y, z)$ is its distance to some fixed point $(a, b, c)$ of the surface.

**8.** Find the mass of a helicoid

$$\mathbf{r}(r, \theta) = (r \cos \theta)\mathbf{i} + (r \sin \theta)\mathbf{j} + \theta\mathbf{k},$$

$0 \leq r \leq 1, 0 \leq \theta \leq 2\pi$, if the density function is $\delta(x, y, z) = 2\sqrt{x^2 + y^2}$. See Practice Exercise 27 for a figure.

**9.** Among all rectangular regions $0 \leq x \leq a, 0 \leq y \leq b$, find the one for which the total outward flux of $\mathbf{F} = (x^2 + 4xy)\mathbf{i} - 6y\mathbf{j}$ across the four sides is least. What *is* the least flux?

**10.** Find an equation for the plane through the origin such that the circulation of the flow field $\mathbf{F} = z\mathbf{i} + x\mathbf{j} + y\mathbf{k}$ around the circle of intersection of the plane with the sphere $x^2 + y^2 + z^2 = 4$ is a maximum.

**11.** A string lies along the circle $x^2 + y^2 = 4$ from $(2, 0)$ to $(0, 2)$ in the first quadrant. The density of the string is $\rho(x, y) = xy$.

**a.** Partition the string into a finite number of subarcs to show that the work done by gravity to move the string straight down to the $x$-axis is given by

$$\text{Work} = \lim_{n\to\infty} \sum_{k=1}^{n} g\, x_k y_k{}^2 \Delta s_k = \int_C g\, xy^2\, ds,$$

where $g$ is the gravitational constant.

**b.** Find the total work done by evaluating the line integral in part (a).

**c.** Show that the total work done equals the work required to move the string's center of mass $(\bar{x}, \bar{y})$ straight down to the $x$-axis.

**12.** A thin sheet lies along the portion of the plane $x + y + z = 1$ in the first octant. The density of the sheet is $\delta(x, y, z) = xy$.

**a.** Partition the sheet into a finite number of subpieces to show that the work done by gravity to move the sheet straight down to the $xy$-plane is given by

$$\text{Work} = \lim_{n\to\infty} \sum_{k=1}^{n} g\, x_k y_k z_k\, \Delta\sigma_k = \iint_S g\, xyz\, d\sigma,$$

where $g$ is the gravitational constant.

**b.** Find the total work done by evaluating the surface integral in part (a).

**c.** Show that the total work done equals the work required to move the sheet's center of mass $(\bar{x}, \bar{y}, \bar{z})$ straight down to the $xy$-plane.

**13. Archimedes' principle** If an object such as a ball is placed in a liquid, it will either sink to the bottom, float, or sink a certain distance and remain suspended in the liquid. Suppose a fluid has constant weight density $w$ and that the fluid's surface coincides with the plane $z = 4$. A spherical ball remains suspended in the fluid and occupies the region $x^2 + y^2 + (z - 2)^2 \leq 1$.

**a.** Show that the surface integral giving the magnitude of the total force on the ball due to the fluid's pressure is

$$\text{Force} = \lim_{n\to\infty} \sum_{k=1}^{n} w(4 - z_k)\, \Delta\sigma_k = \iint_S w(4 - z)\, d\sigma.$$

**b.** Since the ball is not moving, it is being held up by the buoyant force of the liquid. Show that the magnitude of the buoyant force on the sphere is

$$\text{Buoyant force} = \iint_S w(z - 4)\mathbf{k} \cdot \mathbf{n}\, d\sigma,$$

where **n** is the outer unit normal at $(x, y, z)$. This illustrates Archimedes' principle that the magnitude of the buoyant force on a submerged solid equals the weight of the displaced fluid.

**c.** Use the Divergence Theorem to find the magnitude of the buoyant force in part (b).

**14. Fluid force on a curved surface** A cone in the shape of the surface $z = \sqrt{x^2 + y^2}, 0 \leq z \leq 2$ is filled with a liquid of constant weight density $w$. Assuming the $xy$-plane is "ground level," show that the total force on the portion of the cone from $z = 1$ to $z = 2$ due to liquid pressure is the surface integral

$$F = \iint_S w(2 - z)\, d\sigma.$$

Evaluate the integral.

**15. Faraday's law** If $\mathbf{E}(t, x, y, z)$ and $\mathbf{B}(t, x, y, z)$ represent the electric and magnetic fields at point $(x, y, z)$ at time $t$, a basic principle of electromagnetic theory says that $\nabla \times \mathbf{E} = -\partial\mathbf{B}/\partial t$. In this expression $\nabla \times \mathbf{E}$ is computed with $t$ held fixed and $\partial\mathbf{B}/\partial t$ is calculated with $(x, y, z)$ fixed. Use Stokes' Theorem to derive Faraday's law,

$$\oint_C \mathbf{E} \cdot d\mathbf{r} = -\frac{\partial}{\partial t} \iint_S \mathbf{B} \cdot \mathbf{n}\, d\sigma,$$

where $C$ represents a wire loop through which current flows counterclockwise with respect to the surface's unit normal **n**, giving rise to the voltage

$$\oint_C \mathbf{E} \cdot d\mathbf{r}$$

around $C$. The surface integral on the right side of the equation is called the *magnetic flux*, and $S$ is any oriented surface with boundary $C$.

**16.** Let

$$\mathbf{F} = -\frac{GmM}{|\mathbf{r}|^3}\mathbf{r}$$

be the gravitational force field defined for $\mathbf{r} \neq \mathbf{0}$. Use Gauss's law in Section 16.8 to show that there is no continuously differentiable vector field **H** satisfying $\mathbf{F} = \nabla \times \mathbf{H}$.

17. If $f(x, y, z)$ and $g(x, y, z)$ are continuously differentiable scalar functions defined over the oriented surface $S$ with boundary curve $C$, prove that

$$\iint_S (\nabla f \times \nabla g) \cdot \mathbf{n} \, d\sigma = \oint_C f \, \nabla g \cdot d\mathbf{r}.$$

18. Suppose that $\nabla \cdot \mathbf{F}_1 = \nabla \cdot \mathbf{F}_2$ and $\nabla \times \mathbf{F}_1 = \nabla \times \mathbf{F}_2$ over a region $D$ enclosed by the oriented surface $S$ with outward unit normal $\mathbf{n}$ and that $\mathbf{F}_1 \cdot \mathbf{n} = \mathbf{F}_2 \cdot \mathbf{n}$ on $S$. Prove that $\mathbf{F}_1 = \mathbf{F}_2$ throughout $D$.

19. Prove or disprove that if $\nabla \cdot \mathbf{F} = 0$ and $\nabla \times \mathbf{F} = \mathbf{0}$, then $\mathbf{F} = \mathbf{0}$.

20. Let $S$ be an oriented surface parametrized by $\mathbf{r}(u, v)$. Define the notation $d\boldsymbol{\sigma} = \mathbf{r}_u \, du \times \mathbf{r}_v \, dv$ so that $d\boldsymbol{\sigma}$ is a vector normal to the surface. Also, the magnitude $d\sigma = |d\boldsymbol{\sigma}|$ is the element of surface area (by Equation 5 in Section 16.5). Derive the identity

$$d\sigma = (EG - F^2)^{1/2} \, du \, dv$$

where

$$E = |\mathbf{r}_u|^2, \quad F = \mathbf{r}_u \cdot \mathbf{r}_v, \quad \text{and} \quad G = |\mathbf{r}_v|^2.$$

21. Show that the volume $V$ of a region $D$ in space enclosed by the oriented surface $S$ with outward normal $\mathbf{n}$ satisfies the identity

$$V = \frac{1}{3} \iint_S \mathbf{r} \cdot \mathbf{n} \, d\sigma,$$

where $\mathbf{r}$ is the position vector of the point $(x, y, z)$ in $D$.

## Chapter 16 Technology Application Projects

### Mathematica/Maple Modules:

**Work in Conservative and Nonconservative Force Fields**
Explore integration over vector fields and experiment with conservative and nonconservative force functions along different paths in the field.

**How Can You Visualize Green's Theorem?**
Explore integration over vector fields and use parametrizations to compute line integrals. Both forms of Green's Theorem are explored.

**Visualizing and Interpreting the Divergence Theorem**
Verify the Divergence Theorem by formulating and evaluating certain divergence and surface integrals.

# Appendices

## A.1 Real Numbers and the Real Line

This section reviews real numbers, inequalities, intervals, and absolute values.

### Real Numbers

Much of calculus is based on properties of the real number system. **Real numbers** are numbers that can be expressed as decimals, such as

$$-\frac{3}{4} = -0.75000\ldots$$

$$\frac{1}{3} = 0.33333\ldots$$

$$\sqrt{2} = 1.4142\ldots$$

The dots . . . in each case indicate that the sequence of decimal digits goes on forever. Every conceivable decimal expansion represents a real number, although some numbers have two representations. For instance, the infinite decimals .999 . . . and 1.000 . . . represent the same real number 1. A similar statement holds for any number with an infinite tail of 9's.

The real numbers can be represented geometrically as points on a number line called the **real line**.

### RULES FOR INEQUALITIES

If $a$, $b$, and $c$ are real numbers, then:
1. $a < b \Rightarrow a + c < b + c$
2. $a < b \Rightarrow a - c < b - c$
3. $a < b$ and $c > 0 \Rightarrow cc < bc$
4. $a < b$ and $c < 0 \Rightarrow bc < ac$
   Special case: $a < b \Rightarrow -b < -a$
5. $a > 0 \Rightarrow \frac{1}{a} > 0$
6. If $a$ and $b$ are both positive or both negative, then $a < b \Rightarrow \frac{1}{b} < \frac{1}{a}$.

The symbol $\mathbb{R}$ denotes either the real number system or, equivalently, the real line.

The properties of the real number system fall into three categories: algebraic properties, order properties, and completeness. The **algebraic properties** say that the real numbers can be added, subtracted, multiplied, and divided (except by 0) to produce more real numbers under the usual rules of arithmetic. *You can never divide by* 0.

The **order properties** of real numbers are given in Appendix 6. The useful rules at the left can be derived from them, where the symbol $\Rightarrow$ means "implies."

Notice the rules for multiplying an inequality by a number. Multiplying by a positive number preserves the inequality; multiplying by a negative number reverses the inequality. Also, reciprocation reverses the inequality for numbers of the same sign. For example, $2 < 5$ but $-2 > -5$ and $1/2 > 1/5$.

The **completeness property** of the real number system is deeper and harder to define precisely. However, the property is essential to the idea of a limit (Chapter 2). Roughly speaking, it says that there are enough real numbers to "complete" the real number line, in the sense that there are no "holes" or "gaps" in it. Many theorems of calculus would fail if the real number system were not complete. The topic is best saved for a more advanced course, but Appendix 6 hints about what is involved and how the real numbers are constructed.

We distinguish three special subsets of real numbers.

1.  The **natural numbers**, namely 1, 2, 3, 4, . . .
2.  The **integers**, namely 0, $\pm 1$, $\pm 2$, $\pm 3$, . . .
3.  The **rational numbers**, namely the numbers that can be expressed in the form of a fraction $m/n$, where $m$ and $n$ are integers and $n \neq 0$. Examples are

$$\frac{1}{3}, \quad -\frac{4}{9} = \frac{-4}{9} = \frac{4}{-9}, \quad \frac{200}{13}, \quad \text{and} \quad 57 = \frac{57}{1}.$$

The rational numbers are precisely the real numbers with decimal expansions that are either

**(a)**  terminating (ending in an infinite string of zeros), for example,

$$\frac{3}{4} = 0.75000 \ldots = 0.75 \quad \text{or}$$

**(b)**  eventually repeating (ending with a block of digits that repeats over and over), for example,

$$\frac{23}{11} = 2.090909 \ldots = 2.\overline{09} \qquad \text{The bar indicates the block of repeating digits.}$$

A terminating decimal expansion is a special type of repeating decimal, since the ending zeros repeat.

The set of rational numbers has all the algebraic and order properties of the real numbers but lacks the completeness property. For example, there is no rational number whose square is 2; there is a "hole" in the rational line where $\sqrt{2}$ should be.

Real numbers that are not rational are called **irrational numbers**. They are characterized by having nonterminating and nonrepeating decimal expansions. Examples are $\pi$, $\sqrt{2}$, $\sqrt[3]{5}$, and $\log_{10} 3$. Since every decimal expansion represents a real number, it should be clear that there are infinitely many irrational numbers. Both rational and irrational numbers are found arbitrarily close to any point on the real line.

Set notation is very useful for specifying a particular subset of real numbers. A **set** is a collection of objects, and these objects are the **elements** of the set. If $S$ is a set, the notation $a \in S$ means that $a$ is an element of $S$, and $a \notin S$ means that $a$ is not an element of $S$. If $S$ and $T$ are sets, then $S \cup T$ is their **union** and consists of all elements belonging to either $S$ or $T$ (or to both $S$ and $T$). The **intersection** $S \cap T$ consists of all elements belonging to both $S$ and $T$. The **empty set** $\varnothing$ is the set that contains no elements. For example, the intersection of the rational numbers and the irrational numbers is the empty set.

Some sets can be described by *listing* their elements in braces. For instance, the set $A$ consisting of the natural numbers (or positive integers) less than 6 can be expressed as

$$A = \{1, 2, 3, 4, 5\}.$$

The entire set of integers is written as

$$\{0, \pm 1, \pm 2, \pm 3, \ldots\}.$$

Another way to describe a set is to enclose in braces a rule that generates all the elements of the set. For instance, the set

$$A = \{x \,|\, x \text{ is an integer and } 0 < x < 6\}$$

is the set of positive integers less than 6.

### Intervals

A subset of the real line is called an **interval** if it contains at least two numbers and contains all the real numbers lying between any two of its elements. For example, the set of all real numbers $x$ such that $x > 6$ is an interval, as is the set of all $x$ such that $-2 \le x \le 5$. The set of all nonzero real numbers is not an interval; since 0 is absent, the set fails to contain every real number between $-1$ and 1 (for example).

Geometrically, intervals correspond to rays and line segments on the real line, along with the real line itself. Intervals of numbers corresponding to line segments are **finite intervals**; intervals corresponding to rays and the real line are **infinite intervals**.

A finite interval is said to be **closed** if it contains both of its endpoints, **half-open** if it contains one endpoint but not the other, and **open** if it contains neither endpoint. The endpoints are also called **boundary points**; they make up the interval's **boundary**. The remaining points of the interval are **interior points** and together compose the interval's **interior**. Infinite intervals are closed if they contain a finite endpoint, and open otherwise. The entire real line $\mathbb{R}$ is an infinite interval that is both open and closed. Table A.1 summarizes the various types of intervals.

**TABLE A.1** Types of intervals

| Notation | Set description | Type | Picture |
|---|---|---|---|
| $(a, b)$ | $\{x \mid a < x < b\}$ | Open | |
| $[a, b]$ | $\{x \mid a \le x \le b\}$ | Closed | |
| $[a, b)$ | $\{x \mid a \le x < b\}$ | Half-open | |
| $(a, b]$ | $\{x \mid a < x \le b\}$ | Half-open | |
| $(a, \infty)$ | $\{x \mid x > a\}$ | Open | |
| $[a, \infty)$ | $\{x \mid x \ge a\}$ | Closed | |
| $(-\infty, b)$ | $\{x \mid x < b\}$ | Open | |
| $(-\infty, b]$ | $\{x \mid x \le b\}$ | Closed | |
| $(-\infty, \infty)$ | $\mathbb{R}$ (set of all real numbers) | Both open and closed | |

### Solving Inequalities

The process of finding the interval or intervals of numbers that satisfy an inequality in $x$ is called **solving** the inequality.

**EXAMPLE 1**   Solve the following inequalities and show their solution sets on the real line.

**(a)** $2x - 1 < x + 3$   **(b)** $-\dfrac{x}{3} < 2x + 1$   **(c)** $\dfrac{6}{x - 1} \ge 5$

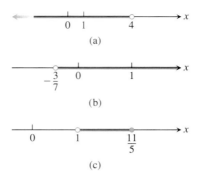

(a)

(b)

(c)

**FIGURE A.1** Solution sets for the in-equalities in Example 1.

**Solution**

**(a)**
$$2x - 1 < x + 3$$
$$2x < x + 4 \qquad \text{Add 1 to both sides.}$$
$$x < 4 \qquad \text{Subtract } x \text{ from both sides.}$$

The solution set is the open interval $(-\infty, 4)$ (Figure A.1a).

**(b)**
$$-\frac{x}{3} < 2x + 1$$
$$-x < 6x + 3 \qquad \text{Multiply both sides by 3.}$$
$$0 < 7x + 3 \qquad \text{Add } x \text{ to both sides.}$$
$$-3 < 7x \qquad \text{Subtract 3 from both sides.}$$
$$-\frac{3}{7} < x \qquad \text{Divide by 7.}$$

The solution set is the open interval $(-3/7, \infty)$ (Figure A.1b).

**(c)**  The inequality $6/(x - 1) \geq 5$ can hold only if $x > 1$, because otherwise $6/(x - 1)$ is undefined or negative. Therefore, $(x - 1)$ is positive and the inequality will be pre-served if we multiply both sides by $(x - 1)$, and we have

$$\frac{6}{x - 1} \geq 5$$
$$6 \geq 5x - 5 \qquad \text{Multiply both sides by } (x - 1).$$
$$11 \geq 5x \qquad \text{Add 5 to both sides.}$$
$$\frac{11}{5} \geq x. \qquad \text{Or } x \leq \frac{11}{5}.$$

The solution set is the half-open interval $(1, 11/5\,]$ (Figure A.1c).   ■

## Absolute Value

The **absolute value** of a number $x$, denoted by $|x|$, is defined by the formula

$$|x| = \begin{cases} x, & x \geq 0 \\ -x, & x < 0. \end{cases}$$

**EXAMPLE 2**   $|3| = 3, \quad |0| = 0, \quad |-5| = -(-5) = 5, \quad \left||-a|\right| = |a|$   ■

Geometrically, the absolute value of $x$ is the distance from $x$ to 0 on the real number line. Since distances are always positive or 0, we see that $|x| \geq 0$ for every real number $x$, and $|x| = 0$ if and only if $x = 0$. Also,

$$|x - y| = \text{the distance between } x \text{ and } y$$

on the real line (Figure A.2).

Since the symbol $\sqrt{a}$ always denotes the *nonnegative* square root of $a$, an alternate definition of $|x|$ is

$$|x| = \sqrt{x^2}.$$

It is important to remember that $\sqrt{a^2} = |a|$. Do not write $\sqrt{a^2} = a$ unless you already know that $a \geq 0$.

The absolute value has the following properties. (You are asked to prove these proper-ties in the exercises.)

**FIGURE A.2** Absolute values give distances between points on the number line.

---

**Absolute Value Properties**

1. $|-a| = |a|$    A number and its additive inverse or negative have the same absolute value.

2. $|ab| = |a||b|$    The absolute value of a product is the product of the absolute values.

3. $\left|\dfrac{a}{b}\right| = \dfrac{|a|}{|b|}$    The absolute value of a quotient is the quotient of the absolute values.

4. $|a + b| \le |a| + |b|$    The **triangle inequality**. The absolute value of the sum of two numbers is less than or equal to the sum of their absolute values.

---

**FIGURE A.3** $|x| < a$ means $x$ lies between $-a$ and $a$.

**ABSOLUTE VALUES AND INTERVALS**

If $a$ is any positive number, then

5. $|x| = a \iff x = \pm a$
6. $|x| < a \iff -a < x < a$
7. $|x| > a \iff x > a \text{ or } x < -a$
8. $|x| \le a \iff -a \le x \le a$
9. $|x| \ge a \iff x \ge a \text{ or } x \le -a$

Note that $|-a| \ne -|a|$. For example, $|-3| = 3$, whereas $-|3| = -3$. If $a$ and $b$ differ in sign, then $|a + b|$ is less than $|a| + |b|$. In all other cases, $|a + b|$ equals $|a| + |b|$. Absolute value bars in expressions like $|-3 + 5|$ work like parentheses: We do the arithmetic inside *before* taking the absolute value.

**EXAMPLE 3**

$$|-3 + 5| = |2| = 2 < |-3| + |5| = 8$$
$$|3 + 5| = |8| = |3| + |5|$$
$$|-3 - 5| = |-8| = 8 = |-3| + |-5|$$    ■

The inequality $|x| < a$ says that the distance from $x$ to 0 is less than the positive number $a$. This means that $x$ must lie between $-a$ and $a$, as we can see from Figure A.3.

The statements in the table are all consequences of the definition of absolute value and are often helpful when solving equations or inequalities involving absolute values.

The symbol $\iff$ is often used by mathematicians to denote the "if and only if" logical relationship. It also means "implies and is implied by."

**EXAMPLE 4**    Solve the equation $|2x - 3| = 7$.

**Solution**    By Property 5, $2x - 3 = \pm 7$, so there are two possibilities:

$$2x - 3 = 7 \qquad 2x - 3 = -7 \qquad \text{Equivalent equations without absolute values}$$
$$2x = 10 \qquad\quad 2x = -4 \qquad \text{Solve as usual.}$$
$$x = 5 \qquad\quad\; x = -2$$

The solutions of $|2x - 3| = 7$ are $x = 5$ and $x = -2$.    ■

**EXAMPLE 5**    Solve the inequality $\left|5 - \dfrac{2}{x}\right| < 1$.

**Solution**    We have

$$\left|5 - \frac{2}{x}\right| < 1 \iff -1 < 5 - \frac{2}{x} < 1 \qquad \text{Property 6}$$
$$\iff -6 < -\frac{2}{x} < -4 \qquad \text{Subtract 5.}$$
$$\iff 3 > \frac{1}{x} > 2 \qquad \text{Multiply by } -\frac{1}{2}.$$
$$\iff \frac{1}{3} < x < \frac{1}{2}. \qquad \text{Take reciprocals.}$$

Notice how the various rules for inequalities were used here. Multiplying by a negative number reverses the inequality. So does taking reciprocals in an inequality in which both sides are positive. The original inequality holds if and only if $(1/3) < x < (1/2)$. The solution set is the open interval $(1/3, 1/2)$. ∎

## Exercises A.1

1. Express $1/9$ as a repeating decimal, using a bar to indicate the repeating digits. What are the decimal representations of $2/9$? $3/9$? $8/9$? $9/9$?

2. If $2 < x < 6$, which of the following statements about $x$ are necessarily true, and which are not necessarily true?
   a. $0 < x < 4$
   b. $0 < x - 2 < 4$
   c. $1 < \frac{x}{2} < 3$
   d. $\frac{1}{6} < \frac{1}{x} < \frac{1}{2}$
   e. $1 < \frac{6}{x} < 3$
   f. $|x - 4| < 2$
   g. $-6 < -x < 2$
   h. $-6 < -x < -2$

In Exercises 3–6, solve the inequalities and show the solution sets on the real line.

3. $-2x > 4$
4. $5x - 3 \le 7 - 3x$
5. $2x - \frac{1}{2} \ge 7x + \frac{7}{6}$
6. $\frac{4}{5}(x - 2) < \frac{1}{3}(x - 6)$

Solve the equations in Exercises 7–9.

7. $|y| = 3$
8. $|2t + 5| = 4$
9. $|8 - 3s| = \frac{9}{2}$

Solve the inequalities in Exercises 10–17, expressing the solution sets as intervals or unions of intervals. Also, show each solution set on the real line.

10. $|x| < 2$
11. $|t - 1| \le 3$
12. $|3y - 7| < 4$
13. $\left|\frac{z}{5} - 1\right| \le 1$
14. $\left|3 - \frac{1}{x}\right| < \frac{1}{2}$
15. $|2s| \ge 4$
16. $|1 - x| > 1$
17. $\left|\frac{r + 1}{2}\right| \ge 1$

Solve the inequalities in Exercises 18–21. Express the solution sets as intervals or unions of intervals and show them on the real line. Use the result $\sqrt{a^2} = |a|$ as appropriate.

18. $x^2 < 2$
19. $4 < x^2 < 9$
20. $(x - 1)^2 < 4$
21. $x^2 - x < 0$

22. Do not fall into the trap of thinking $|-a| = a$. For what real numbers $a$ is this equation true? For what real numbers is it false?

23. Solve the equation $|x - 1| = 1 - x$.

24. **A proof of the triangle inequality** Give the reason justifying each of the numbered steps in the following proof of the triangle inequality.

$$|a + b|^2 = (a + b)^2 \qquad (1)$$
$$= a^2 + 2ab + b^2$$
$$\le a^2 + 2|a||b| + b^2 \qquad (2)$$
$$= |a|^2 + 2|a||b| + |b|^2 \qquad (3)$$
$$= (|a| + |b|)^2$$
$$|a + b| \le |a| + |b| \qquad (4)$$

25. Prove that $|ab| = |a||b|$ for any numbers $a$ and $b$.

26. If $|x| \le 3$ and $x > -1/2$, what can you say about $x$?

27. Graph the inequality $|x| + |y| \le 1$.

28. For any number $a$, prove that $|-a| = |a|$.

29. Let $a$ be any positive number. Prove that $|x| > a$ if and only if $x > a$ or $x < -a$.

30. a. If $b$ is any nonzero real number, prove that $|1/b| = 1/|b|$.
    b. Prove that $\left|\frac{a}{b}\right| = \frac{|a|}{|b|}$ for any numbers $a$ and $b \ne 0$.

# A.2 Mathematical Induction

Many formulas, like

$$1 + 2 + \cdots + n = \frac{n(n + 1)}{2},$$

can be shown to hold for every positive integer $n$ by applying an axiom called the *mathematical induction principle*. A proof that uses this axiom is called a *proof by mathematical induction* or a *proof by induction*.

The steps in proving a formula by induction are the following:

1. Check that the formula holds for $n = 1$.
2. Prove that if the formula holds for any positive integer $n = k$, then it also holds for the next integer, $n = k + 1$.

The induction axiom says that once these steps are completed, the formula holds for all positive integers $n$. By Step 1 it holds for $n = 1$. By Step 2 it holds for $n = 2$, and therefore by Step 2 also for $n = 3$, and by Step 2 again for $n = 4$, and so on. If the first domino falls, and the $k$th domino always knocks over the $(k + 1)$st when it falls, all the dominoes fall.

From another point of view, suppose we have a sequence of statements $S_1, S_2, \ldots, S_n, \ldots$, one for each positive integer. Suppose we can show that assuming any one of the statements to be true implies that the next statement in line is true. Suppose that we can also show that $S_1$ is true. Then we may conclude that the statements are true from $S_1$ on.

**EXAMPLE 1**    Use mathematical induction to prove that for every positive integer $n$,

$$1 + 2 + \cdots + n = \frac{n(n + 1)}{2}.$$

**Solution**    We accomplish the proof by carrying out the two steps above.

1.    The formula holds for $n = 1$ because

$$1 = \frac{1(1 + 1)}{2}.$$

2.    If the formula holds for $n = k$, does it also hold for $n = k + 1$? The answer is yes, as we now show. If

$$1 + 2 + \cdots + k = \frac{k(k + 1)}{2},$$

then

$$1 + 2 + \cdots + k + (k + 1) = \frac{k(k + 1)}{2} + (k + 1) = \frac{k^2 + k + 2k + 2}{2}$$

$$= \frac{(k + 1)(k + 2)}{2} = \frac{(k + 1)((k + 1) + 1)}{2}.$$

The last expression in this string of equalities is the expression $n(n + 1)/2$ for $n = (k + 1)$.

The mathematical induction principle now guarantees the original formula for all positive integers $n$.    ■

In Example 4 of Section 5.2 we gave another proof for the formula giving the sum of the first $n$ integers. However, proof by mathematical induction is more general. It can be used to find the sums of the squares and cubes of the first $n$ integers (Exercises 9 and 10). Here is another example.

**EXAMPLE 2**    Show by mathematical induction that for all positive integers $n$,

$$\frac{1}{2^1} + \frac{1}{2^2} + \cdots + \frac{1}{2^n} = 1 - \frac{1}{2^n}.$$

**Solution**    We accomplish the proof by carrying out the two steps of mathematical induction.

1.    The formula holds for $n = 1$ because

$$\frac{1}{2^1} = 1 - \frac{1}{2^1}.$$

**2.** If

$$\frac{1}{2^1} + \frac{1}{2^2} + \cdots + \frac{1}{2^k} = 1 - \frac{1}{2^k},$$

then

$$\frac{1}{2^1} + \frac{1}{2^2} + \cdots + \frac{1}{2^k} + \frac{1}{2^{k+1}} = 1 - \frac{1}{2^k} + \frac{1}{2^{k+1}} = 1 - \frac{1 \cdot 2}{2^k \cdot 2} + \frac{1}{2^{k+1}}$$

$$= 1 - \frac{2}{2^{k+1}} + \frac{1}{2^{k+1}} = 1 - \frac{1}{2^{k+1}}.$$

Thus, the original formula holds for $n = (k + 1)$ whenever it holds for $n = k$.

With these steps verified, the mathematical induction principle now guarantees the formula for every positive integer $n$. ◼

### Other Starting Integers

Instead of starting at $n = 1$ some induction arguments start at another integer. The steps for such an argument are as follows.

**1.** Check that the formula holds for $n = n_1$ (the first appropriate integer).

**2.** Prove that if the formula holds for any integer $n = k \geq n_1$, then it also holds for $n = (k + 1)$.

Once these steps are completed, the mathematical induction principle guarantees the formula for all $n \geq n_1$.

**EXAMPLE 3**    Show that $n! > 3^n$ if $n$ is large enough.

**Solution**    How large is large enough? We experiment:

| $n$ | 1 | 2 | 3 | 4 | 5 | 6 | 7 |
|-----|---|---|---|----|-----|-----|------|
| $n!$ | 1 | 2 | 6 | 24 | 120 | 720 | 5040 |
| $3^n$ | 3 | 9 | 27 | 81 | 243 | 729 | 2187 |

It looks as if $n! > 3^n$ for $n \geq 7$. To be sure, we apply mathematical induction. We take $n_1 = 7$ in Step 1 and complete Step 2.

Suppose $k! > 3^k$ for some $k \geq 7$. Then

$$(k + 1)! = (k + 1)(k!) > (k + 1)3^k > 7 \cdot 3^k > 3^{k+1}.$$

Thus, for $k \geq 7$,

$$k! > 3^k \quad \text{implies} \quad (k + 1)! > 3^{k+1}.$$

The mathematical induction principle now guarantees $n! \geq 3^n$ for all $n \geq 7$. ◼

### Proof of the Derivative Sum Rule for Sums of Finitely Many Functions

We prove the statement

$$\frac{d}{dx}(u_1 + u_2 + \cdots + u_n) = \frac{du_1}{dx} + \frac{du_2}{dx} + \cdots + \frac{du_n}{dx}$$

by mathematical induction. The statement is true for $n = 2$, as was proved in Section 3.3. This is Step 1 of the induction proof.

Step 2 is to show that if the statement is true for any positive integer $n = k$, where $k \geq n_0 = 2$, then it is also true for $n = k + 1$. So suppose that

$$\frac{d}{dx}(u_1 + u_2 + \cdots + u_k) = \frac{du_1}{dx} + \frac{du_2}{dx} + \cdots + \frac{du_k}{dx}. \tag{1}$$

Then

$$\frac{d}{dx}\underbrace{(u_1 + u_2 + \cdots + u_k}_{\text{Call the function defined by this sum } u.} + \underbrace{u_{k+1})}_{\substack{\text{Call this} \\ \text{function } v.}}$$

$$= \frac{d}{dx}(u_1 + u_2 + \cdots + u_k) + \frac{du_{k+1}}{dx} \qquad \text{Sum Rule for } \frac{d}{dx}(u + v)$$

$$= \frac{du_1}{dx} + \frac{du_2}{dx} + \cdots + \frac{du_k}{dx} + \frac{du_{k+1}}{dx}. \qquad \text{Eq. (1)}$$

With these steps verified, the mathematical induction principle now guarantees the Sum Rule for every integer $n \geq 2$.

# Exercises A.2

1. Assuming that the triangle inequality $|a + b| \leq |a| + |b|$ holds for any two numbers $a$ and $b$, show that

$$|x_1 + x_2 + \cdots + x_n| \leq |x_1| + |x_2| + \cdots + |x_n|$$

for any $n$ numbers.

2. Show that if $r \neq 1$, then

$$1 + r + r^2 + \cdots + r^n = \frac{1 - r^{n+1}}{1 - r}$$

for every positive integer $n$.

3. Use the Product Rule, $\frac{d}{dx}(uv) = u\frac{dv}{dx} + v\frac{du}{dx}$, and the fact that $\frac{d}{dx}(x) = 1$ to show that $\frac{d}{dx}(x^n) = nx^{n-1}$ for every positive integer $n$.

4. Suppose that a function $f(x)$ has the property that $f(x_1x_2) = f(x_1) + f(x_2)$ for any two positive numbers $x_1$ and $x_2$. Show that

$$f(x_1x_2 \cdots x_n) = f(x_1) + f(x_2) + \cdots + f(x_n)$$

for the product of any $n$ positive numbers $x_1, x_2, \ldots, x_n$.

5. Show that

$$\frac{2}{3^1} + \frac{2}{3^2} + \cdots + \frac{2}{3^n} = 1 - \frac{1}{3^n}$$

for all positive integers $n$.

6. Show that $n! > n^3$ if $n$ is large enough.

7. Show that $2^n > n^2$ if $n$ is large enough.

8. Show that $2^n \geq 1/8$ for $n \geq -3$.

9. **Sums of squares**  Show that the sum of the squares of the first $n$ positive integers is

$$\frac{n\left(n + \frac{1}{2}\right)(n + 1)}{3}.$$

10. **Sums of cubes**  Show that the sum of the cubes of the first $n$ positive integers is $(n(n + 1)/2)^2$.

11. **Rules for finite sums**  Show that the following finite sum rules hold for every positive integer $n$. (See Section 5.2.)

    a. $\displaystyle\sum_{k=1}^{n}(a_k + b_k) = \sum_{k=1}^{n} a_k + \sum_{k=1}^{n} b_k$

    b. $\displaystyle\sum_{k=1}^{n}(a_k - b_k) = \sum_{k=1}^{n} a_k - \sum_{k=1}^{n} b_k$

    c. $\displaystyle\sum_{k=1}^{n} ca_k = c \cdot \sum_{k=1}^{n} a_k$  (any number $c$)

    d. $\displaystyle\sum_{k=1}^{n} a_k = n \cdot c$  (if $a_k$ has the constant value $c$)

12. Show that $|x^n| = |x|^n$ for every positive integer $n$ and every real number $x$.

# A.3 Lines, Circles, and Parabolas

This section reviews coordinates, lines, distance, circles, and parabolas in the plane. The notion of increment is also discussed.

## Cartesian Coordinates in the Plane

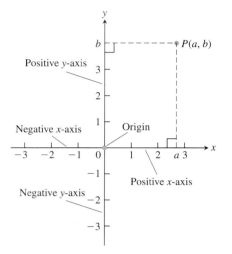

**FIGURE A.4** Cartesian coordinates in the plane are based on two perpendicular axes intersecting at the origin.

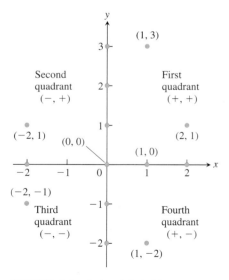

**FIGURE A.5** Points labeled in the xy-coordinate or Cartesian plane. The points on the axes all have coordinate pairs but are usually labeled with single real numbers, (so (1, 0) on the x-axis is labeled as 1). Notice the coordinate sign patterns of the quadrants.

In Appendix 1 we identified the points on the line with real numbers by assigning them coordinates. Points in the plane can be identified with ordered pairs of real numbers. To begin, we draw two perpendicular coordinate lines that intersect at the 0-point of each line. These lines are called **coordinate axes** in the plane. On the horizontal x-axis, numbers are denoted by $x$ and increase to the right. On the vertical y-axis, numbers are denoted by $y$ and increase upward (Figure A.4). Thus "upward" and "to the right" are positive directions, whereas "downward" and "to the left" are considered negative. The **origin** $O$, also labeled 0, of the coordinate system is the point in the plane where $x$ and $y$ are both zero.

If $P$ is any point in the plane, it can be located by exactly one ordered pair of real numbers in the following way. Draw lines through $P$ perpendicular to the two coordinate axes. These lines intersect the axes at points with coordinates $a$ and $b$ (Figure A.4). The ordered pair $(a, b)$ is assigned to the point $P$ and is called its **coordinate pair**. The first number $a$ is the **x-coordinate** (or **abscissa**) of $P$; the second number $b$ is the **y-coordinate** (or **ordinate**) of $P$. The x-coordinate of every point on the y-axis is 0. The y-coordinate of every point on the x-axis is 0. The origin is the point (0, 0).

Starting with an ordered pair $(a, b)$, we can reverse the process and arrive at a corresponding point $P$ in the plane. Often we identify $P$ with the ordered pair and write $P(a, b)$. We sometimes also refer to "the point $(a, b)$" and it will be clear from the context when $(a, b)$ refers to a point in the plane and not to an open interval on the real line. Several points labeled by their coordinates are shown in Figure A.5.

This coordinate system is called the **rectangular coordinate system** or **Cartesian coordinate system** (after the sixteenth-century French mathematician René Descartes). The coordinate axes of this coordinate or Cartesian plane divide the plane into four regions called **quadrants**, numbered counterclockwise as shown in Figure A.5.

The **graph** of an equation or inequality in the variables $x$ and $y$ is the set of all points $P(x, y)$ in the plane whose coordinates satisfy the equation or inequality. When we plot data in the coordinate plane or graph formulas whose variables have different units of measure, we do not need to use the same scale on the two axes. If we plot time vs. thrust for a rocket motor, for example, there is no reason to place the mark that shows 1 s on the time axis the same distance from the origin as the mark that shows 1 N on the thrust axis.

Usually when we graph functions whose variables do not represent physical measurements and when we draw figures in the coordinate plane to study their geometry and trigonometry, we try to make the scales on the axes identical. A vertical unit of distance then looks the same as a horizontal unit. As on a surveyor's map or a scale drawing, line segments that are supposed to have the same length will look as if they do and angles that are supposed to be congruent will look congruent.

Computer displays and calculator displays are another matter. The vertical and horizontal scales on machine-generated graphs usually differ, and there are corresponding distortions in distances, slopes, and angles. Circles may look like ellipses, rectangles may look like squares, right angles may appear to be acute or obtuse, and so on. We discuss these displays and distortions in greater detail in Section 1.4.

## Increments and Straight Lines

When a particle moves from one point in the plane to another, the net changes in its coordinates are called *increments*. They are calculated by subtracting the coordinates of the

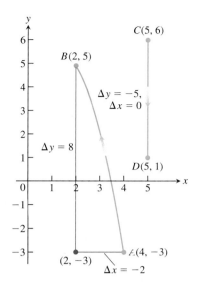

**FIGURE A.6** Coordinate increments may be positive, negative, or zero (Example 1).

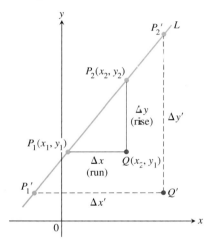

**FIGURE A.7** Triangles $P_1QP_2$ and $P_1'Q'P_2'$ are similar, so the ratio of their sides has the same value for any two points on the line. This common value is the line's slope.

starting point from the coordinates of the ending point. If $x$ changes from $x_1$ to $x_2$, the **increment** in $x$ is

$$\Delta x = x_2 - x_1.$$

**EXAMPLE 1** In going from the point $A(4, -3)$ to the point $B(2, 5)$ the increments in the $x$- and $y$-coordinates are

$$\Delta x = 2 - 4 = -2, \qquad \Delta y = 5 - (-3) = 8.$$

From $C(5, 6)$ to $D(5, 1)$ the coordinate increments are

$$\Delta x = 5 - 5 = 0, \qquad \Delta y = 1 - 6 = -5.$$

See Figure A.6.   ∎

Given two points $P_1(x_1, y_1)$ and $P_2(x_2, y_2)$ in the plane, we call the increments $\Delta x = x_2 - x_1$ and $\Delta y = y_2 - y_1$ the **run** and the **rise**, respectively, between $P_1$ and $P_2$. Two such points always determine a unique straight line (usually called simply a line) passing through them both. We call the line $P_1P_2$.

Any nonvertical line in the plane has the property that the ratio

$$m = \frac{\text{rise}}{\text{run}} = \frac{\Delta y}{\Delta x} = \frac{y_2 - y_1}{x_2 - x_1}$$

has the same value for every choice of the two points $P_1(x_1, y_1)$ and $P_2(x_2, y_2)$ on the line (Figure A.7). This is because the ratios of corresponding sides for similar triangles are equal.

---

**DEFINITION** The constant ratio

$$m = \frac{\text{rise}}{\text{run}} = \frac{\Delta y}{\Delta x} = \frac{y_2 - y_1}{x_2 - x_1}$$

is the **slope** of the nonvertical line $P_1P_2$.

---

The slope tells us the direction (uphill, downhill) and steepness of a line. A line with positive slope rises uphill to the right; one with negative slope falls downhill to the right (Figure A.8). The greater the absolute value of the slope, the more rapid the rise or fall. The slope of a vertical line is *undefined*. Since the run $\Delta x$ is zero for a vertical line, we cannot form the slope ratio $m$.

The direction and steepness of a line can also be measured with an angle. The **angle of inclination** of a line that crosses the $x$-axis is the smallest counterclockwise angle from the $x$-axis to the line (Figure A.9). The inclination of a horizontal line is 0°. The inclination of a vertical line is 90°. If $\phi$ (the Greek letter phi) is the inclination of a line, then $0 \leq \phi < 180°$.

The relationship between the slope $m$ of a nonvertical line and the line's angle of inclination $\phi$ is shown in Figure A.10:

$$m = \tan\phi.$$

Straight lines have relatively simple equations. All points on the *vertical line* through the point $a$ on the $x$-axis have $x$-coordinates equal to $a$. Thus, $x = a$ is an equation for the vertical line. Similarly, $y = b$ is an equation for the *horizontal line* meeting the $y$-axis at $b$. (See Figure A.11.)

We can write an equation for a nonvertical straight line $L$ if we know its slope $m$ and the coordinates of one point $P_1(x_1, y_1)$ on it. If $P(x, y)$ is *any* other point on $L$, then we can

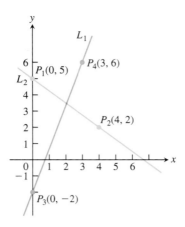

**FIGURE A.8** The slope of $L_1$ is

$$m = \frac{\Delta y}{\Delta x} = \frac{6 - (-2)}{3 - 0} = \frac{8}{3}.$$

That is, $y$ increases 8 units every time $x$ increases 3 units. The slope of $L_2$ is

$$m = \frac{\Delta y}{\Delta x} = \frac{2 - 5}{4 - 0} = \frac{-3}{4}.$$

That is, $y$ decreases 3 units every time $x$ increases 4 units.

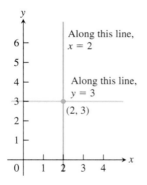

**FIGURE A.11** The standard equations for the vertical and horizontal lines through $(2, 3)$ are $x = 2$ and $y = 3$.

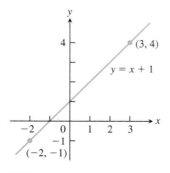

**FIGURE A.12** The line in Example 3.

use the two points $P_1$ and $P$ to compute the slope,

$$m = \frac{y - y_1}{x - x_1}$$

so that

$$y - y_1 = m(x - x_1), \quad \text{or} \quad y = y_1 + m(x - x_1).$$

---

The equation

$$y = y_1 + m(x - x_1)$$

is the **point-slope equation** of the line that passes through the point $(x_1, y_1)$ and has slope $m$.

---

**FIGURE A.9** Angles of inclination are measured counterclockwise from the $x$-axis.

**FIGURE A.10** The slope of a nonvertical line is the tangent of its angle of inclination.

$$m = \frac{\Delta y}{\Delta x} = \tan \phi$$

**EXAMPLE 2** Write an equation for the line through the point $(2, 3)$ with slope $-3/2$.

**Solution** We substitute $x_1 = 2$, $y_1 = 3$, and $m = -3/2$ into the point-slope equation and obtain

$$y = 3 - \frac{3}{2}(x - 2), \quad \text{or} \quad y = -\frac{3}{2}x + 6.$$

When $x = 0$, $y = 6$ so the line intersects the $y$-axis at $y = 6$. ∎

**EXAMPLE 3** Write an equation for the line through $(-2, -1)$ and $(3, 4)$.

**Solution** The line's slope is

$$m = \frac{-1 - 4}{-2 - 3} = \frac{-5}{-5} = 1.$$

We can use this slope with either of the two given points in the point-slope equation:

**With $(x_1, y_1) = (-2, -1)$**
$y = -1 + 1 \cdot (x - (-2))$
$y = -1 + x + 2$
$y = x + 1$

**With $(x_1, y_1) = (3, 4)$**
$y = 4 + 1 \cdot (x - 3)$
$y = 4 + x - 3$
$y = x + 1$

Same result

Either way, $y = x + 1$ is an equation for the line (Figure A.12). ∎

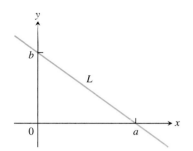

**FIGURE A.13** Line $L$ has $x$-intercept $a$ and $y$-intercept $b$.

The $y$-coordinate of the point where a nonvertical line intersects the $y$-axis is called the **$y$-intercept** of the line. Similarly, the **$x$-intercept** of a nonhorizontal line is the $x$-coordinate of the point where it crosses the $x$-axis (Figure A.13). A line with slope $m$ and $y$-intercept $b$ passes through the point $(0, b)$, so it has equation

$$y = b + m(x - 0), \qquad \text{or, more simply,} \qquad y = mx + b.$$

---

The equation

$$y = mx + b$$

is called the **slope-intercept equation** of the line with slope $m$ and $y$-intercept $b$.

---

Lines with equations of the form $y = mx$ have $y$-intercept 0 and so pass through the origin. Equations of lines are called **linear** equations.

The equation

$$Ax + By = C \qquad (A \text{ and } B \text{ not both } 0)$$

is called the **general linear equation** in $x$ and $y$ because its graph always represents a line and every line has an equation in this form (including lines with undefined slope).

## Parallel and Perpendicular Lines

Lines that are parallel have equal angles of inclination, so they have the same slope (if they are not vertical). Conversely, lines with equal slopes have equal angles of inclination and so are parallel.

If two nonvertical lines $L_1$ and $L_2$ are perpendicular, their slopes $m_1$ and $m_2$ satisfy $m_1 m_2 = -1$, so each slope is the *negative reciprocal* of the other:

$$m_1 = -\frac{1}{m_2}, \qquad m_2 = -\frac{1}{m_1}.$$

To see this, notice by inspecting similar triangles in Figure A.14 that $m_1 = a/h$, and $m_2 = -h/a$. Hence, $m_1 m_2 = (a/h)(-h/a) = -1$.

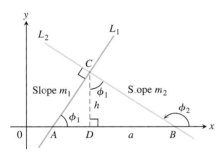

**FIGURE A.14** $\triangle ADC$ is similar to $\triangle CDB$. Hence $\phi_1$ is also the upper angle in $\triangle CDB$. From the sides of $\triangle CDB$, we read $\tan \phi_1 = a/h$.

## Distance and Circles in the Plane

The distance between points in the plane is calculated with a formula that comes from the Pythagorean theorem (Figure A.15).

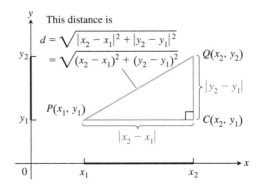

**FIGURE A.15** To calculate the distance between $P(x_1, y_1)$ and $Q(x_2, y_2)$, apply the Pythagorean theorem to triangle $PCQ$.

Distance Formula for Points in the Plane

The distance between $P(x_1, y_1)$ and $Q(x_2, y_2)$ is

$$d = \sqrt{(\Delta x)^2 + (\Delta y)^2} = \sqrt{(x_2 - x_1)^2 + (y_2 - y_1)^2}.$$

### EXAMPLE 4

(a)  The distance between $P(-1, 2)$ and $Q(3, 4)$ is

$$\sqrt{(3 - (-1))^2 + (4 - 2)^2} = \sqrt{(4)^2 + (2)^2} = \sqrt{20} = \sqrt{4 \cdot 5} = 2\sqrt{5}.$$

(b)  The distance from the origin to $P(x, y)$ is

$$\sqrt{(x - 0)^2 + (y - 0)^2} = \sqrt{x^2 + y^2}.$$

By definition, a **circle** of radius $a$ is the set of all points $P(x, y)$ whose distance from some center $C(h, k)$ equals $a$ (Figure A.16). From the distance formula, $P$ lies on the circle if and only if

$$\sqrt{(x - h)^2 + (y - k)^2} = a,$$

so

$$(x - h)^2 + (y - k)^2 = a^2. \qquad (1)$$

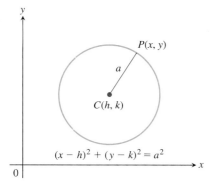

FIGURE A.16   A circle of radius $a$ in the $xy$-plane, with center at $(h, k)$.

Equation (1) is the **standard equation** of a circle with center $(h, k)$ and radius $a$. The circle of radius $a = 1$ and centered at the origin is the **unit circle** with equation

$$x^2 + y^2 = 1.$$

### EXAMPLE 5

(a)  The standard equation for the circle of radius 2 centered at $(3, 4)$ is

$$(x - 3)^2 + (y - 4)^2 = 2^2 = 4.$$

(b)  The circle

$$(x - 1)^2 + (y + 5)^2 = 3$$

has $h = 1, k = -5$, and $a = \sqrt{3}$. The center is the point $(h, k) = (1, -5)$ and the radius is $a = \sqrt{3}$.  ∎

If an equation for a circle is not in standard form, we can find the circle's center and radius by first converting the equation to standard form. The algebraic technique for doing so is *completing the square*.

### EXAMPLE 6   Find the center and radius of the circle

$$x^2 + y^2 + 4x - 6y - 3 = 0.$$

**Solution**     We convert the equation to standard form by completing the squares in $x$ and $y$:

$$x^2 + y^2 + 4x - 6y - 3 = 0$$

Start with the given equation. Gather terms. Move the constant to the right-hand side.

$$(x^2 + 4x) + (y^2 - 6y) = 3$$

$$\left(x^2 + 4x + \left(\frac{4}{2}\right)^2\right) + \left(y^2 - 6y + \left(\frac{-6}{2}\right)^2\right) =$$

Add the square of half the coefficient of $x$ to each side of the equation. Do the same for $y$. The parenthetical expressions on the left-hand side are now perfect squares.

$$3 + \left(\frac{4}{2}\right)^2 + \left(\frac{-6}{2}\right)^2$$

$$(x^2 + 4x + 4) + (y^2 - 6y + 9) = 3 + 4 + 9$$

Write each quadratic as a squared linear expression.

$$(x + 2)^2 + (y - 3)^2 = 16$$

The center is $(-2, 3)$ and the radius is $a = 4$.     ■

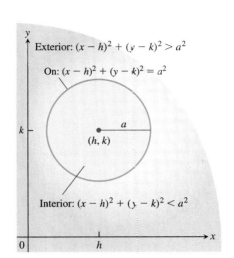

FIGURE A.17   The interior and exterior of the circle $(x - h)^2 + (y - k)^2 = a^2$.

The points $(x, y)$ satisfying the inequality

$$(x - h)^2 + (y - k)^2 < a^2$$

make up the **interior** region of the circle with center $(h, k)$ and radius $a$ (Figure A.17). The circle's **exterior** consists of the points $(x, y)$ satisfying

$$(x - h)^2 + (y - k)^2 > a^2.$$

## Parabolas

The geometric definition and properties of general parabolas are reviewed in Chapter 11. Here we look at parabolas arising as the graphs of equations of the form $y = ax^2 + bx + c$.

**EXAMPLE 7**     Consider the equation $y = x^2$. Some points whose coordinates satisfy this equation are $(0, 0)$, $(1, 1)$, $\left(\frac{3}{2}, \frac{9}{4}\right)$, $(-1, 1)$, $(2, 4)$, and $(-2, 4)$. These points (and all others satisfying the equation) make up a smooth curve called a parabola (Figure A.18).     ■

FIGURE A.18   The parabola $y = x^2$ (Example 7).

The graph of an equation of the form

$$y = ax^2$$

is a **parabola** whose **axis** (axis of symmetry) is the $y$-axis. The parabola's **vertex** (point where the parabola and axis cross) lies at the origin. The parabola opens upward if $a > 0$ and downward if $a < 0$. The larger the value of $|a|$, the narrower the parabola (Figure A.19).

Generally, the graph of $y = ax^2 + bx + c$ is a shifted and scaled version of the parabola $y = x^2$. We discuss shifting and scaling of graphs in more detail in Section 1.2.

---

**The Graph of $y = ax^2 + bx + c,\quad a \neq 0$**

The graph of the equation $y = ax^2 + bx + c, a \neq 0$, is a parabola. The parabola opens upward if $a > 0$ and downward if $a < 0$. The **axis** is the line

$$x = -\frac{b}{2a}. \tag{2}$$

The **vertex** of the parabola is the point where the axis and parabola intersect. Its $x$-coordinate is $x = -b/2a$; its $y$-coordinate is found by substituting $x = -b/2a$ in the parabola's equation.

---

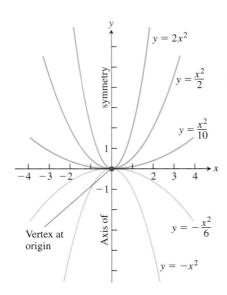

**FIGURE A.19** Besides determining the direction in which the parabola $y = ax^2$ opens, the number $a$ is a scaling factor. The parabola widens as $a$ approaches zero and narrows as $|a|$ becomes large.

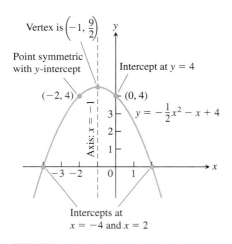

**FIGURE A.20** The parabola in Example 8.

Notice that if $a = 0$, then we have $y = bx + c$, which is an equation for a line. The axis, given by Equation (2), can be found by completing the square.

**EXAMPLE 8** Graph the equation $y = -\frac{1}{2}x^2 - x + 4$.

**Solution** Comparing the equation with $y = ax^2 + bx + c$ we see that

$$a = -\frac{1}{2}, \qquad b = -1, \qquad c = 4.$$

Since $a < 0$, the parabola opens downward. From Equation (2) the axis is the vertical line

$$x = -\frac{b}{2a} = -\frac{(-1)}{2(-1/2)} = -1.$$

When $x = -1$, we have

$$y = -\frac{1}{2}(-1)^2 - (-1) + 4 = \frac{9}{2}.$$

The vertex is $(-1, 9/2)$.

The $x$-intercepts are where $y = 0$:

$$-\frac{1}{2}x^2 - x + 4 = 0$$
$$x^2 + 2x - 8 = 0$$
$$(x - 2)(x + 4) = 0$$
$$x = 2, \qquad x = -4$$

We plot some points, sketch the axis, and use the direction of opening to complete the graph in Figure A.20. ∎

## Ellipses

The geometric definition and properties of general ellipses are reviewed in Chapter 11. Here we relate them to circles. Although they are not the graphs of functions, circles can be stretched horizontally or vertically in the same way as the graphs of functions. The standard equation for a circle of radius $r$ centered at the origin is

$$x^2 + y^2 = r^2.$$

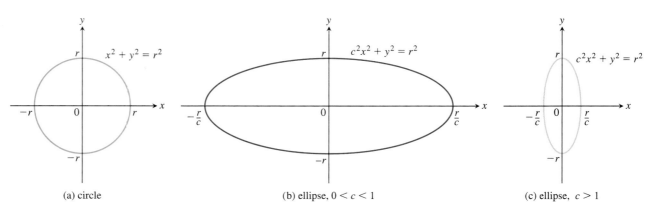

(a) circle

(b) ellipse, $0 < c < 1$

(c) ellipse, $c > 1$

**FIGURE A.21** Horizontal stretching or compression of a circle produces graphs of ellipses.

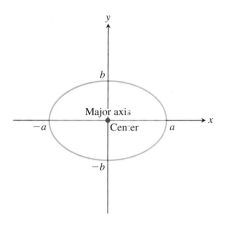

**FIGURE A.22** Graph of the ellipse $\frac{x^2}{a^2} + \frac{y^2}{b^2} = 1$, $a > b$, where the major axis is horizontal.

Substituting $cx$ for $x$ in the standard equation for a circle (Figure A.21) gives

$$c^2x^2 + y^2 = r^2. \tag{3}$$

If $0 < c < 1$, the graph of Equation (3) horizontally stretches the circle; if $c > 1$ the circle is compressed horizontally. In either case, the graph of Equation (3) is an ellipse (Figure A.21). Notice in Figure A.21 that the $y$-intercepts of all three graphs are always $-r$ and $r$. In Figure A.21b, the line segment joining the points $(\pm r/c, 0)$ is called the **major axis** of the ellipse; the **minor axis** is the line segment joining $(0, \pm r)$. The axes of the ellipse are reversed in Figure A.21c: The major axis is the line segment joining the points $(0, \pm r)$, and the minor axis is the line segment joining the points $(\pm r/c, 0)$. In both cases, the major axis is the longer line segment.

If we divide both sides of Equation (3) by $r^2$, we obtain

$$\frac{x^2}{a^2} + \frac{y^2}{b^2} = 1 \tag{4}$$

where $a = r/c$ and $b = r$. If $a > b$, the major axis is horizontal; if $a < b$, the major axis is vertical. The **center** of the ellipse given by Equation (4) is the origin (Figure A.22).

Substituting $x - h$ for $x$, and $y - k$ for $y$, in Equation (4) results in

$$\frac{(x-h)^2}{a^2} + \frac{(y-k)^2}{b^2} = 1. \tag{5}$$

Equation (5) is the **standard equation of an ellipse** with center at $(h, k)$.

# Exercises A.3

### Distance, Slopes, and Lines

In Exercises 1 and 2, a particle moves from $A$ to $B$ in the coordinate plane. Find the increments $\Delta x$ and $\Delta y$ in the particle's coordinates. Also find the distance from $A$ to $B$.

**1.** $A(-3, 2)$,  $B(-1, -2)$
**2.** $A(-3.2, -2)$,  $B(-8.1, -2)$

Describe the graphs of the equations in Exercises 3 and 4.

**3.** $x^2 + y^2 = 1$
**4.** $x^2 + y^2 \leq 3$

Plot the points in Exercises 5 and 6 and find the slope (if any) of the line they determine. Also find the common slope (if any) of the lines perpendicular to line $AB$.

**5.** $A(-1, 2)$,  $B(-2, -1)$
**6.** $A(2, 3)$,  $B(-1, 3)$

In Exercises 7 and 8, find an equation for **(a)** the vertical line and **(b)** the horizontal line through the given point.

**7.** $(-1, 4/3)$
**8.** $\left(0, -\sqrt{2}\right)$

In Exercises 9–15, write an equation for each line described.

**9.** Passes through $(-1, 1)$ with slope $-1$

**10.** Passes through $(3, 4)$ and $(-2, 5)$

**11.** Has slope $-5/4$ and $y$-intercept $6$

**12.** Passes through $(-12, -9)$ and has slope $0$

**13.** Has $y$-intercept $4$ and $x$-intercept $-1$

**14.** Passes through $(5, -1)$ and is parallel to the line $2x + 5y = 15$

**15.** Passes through $(4, 10)$ and is perpendicular to the line $6x - 3y = 5$

In Exercises 16 and 17, find the line's $x$- and $y$-intercepts and use this information to graph the line.

**16.** $3x + 4y = 12$
**17.** $\sqrt{2}x - \sqrt{3}y = \sqrt{6}$

**18.** Is there anything special about the relationship between the lines $Ax + By = C_1$ and $Bx - Ay = C_2$ $(A \neq 0, B \neq 0)$? Give reasons for your answer.

**19.** A particle starts at $A(-2, 3)$ and its coordinates change by increments $\Delta x = 5$, $\Delta y = -6$. Find its new position.

**20.** The coordinates of a particle change by $\Delta x = 5$ and $\Delta y = 6$ as it moves from $A(x, y)$ to $B(3, -3)$. Find $x$ and $y$.

### Circles

In Exercises 21–23, find an equation for the circle with the given center $C(h, k)$ and radius $a$. Then sketch the circle in the $xy$-plane. Include the circle's center in your sketch. Also, label the circle's $x$- and $y$-intercepts, if any, with their coordinate pairs.

**21.** $C(0, 2)$,  $a = 2$
**22.** $C(-1, 5)$,  $a = \sqrt{10}$
**23.** $C\left(-\sqrt{3}, -2\right)$,  $a = 2$

Graph the circles whose equations are given in Exercises 24–26. Label each circle's center and intercepts (if any) with their coordinate pairs.

**24.** $x^2 + y^2 + 4x - 4y + 4 = 0$

**25.** $x^2 + y^2 - 3y - 4 = 0$     **26.** $x^2 + y^2 - 4x + 4y = 0$

## Parabolas

Graph the parabolas in Exercises 27–30. Label the vertex, axis, and intercepts in each case.

**27.** $y = x^2 - 2x - 3$      **28.** $y = -x^2 + 4x$

**29.** $y = -x^2 - 6x - 5$      **30.** $y = \frac{1}{2}x^2 + x + 4$

## Inequalities

Describe the regions defined by the inequalities and pairs of inequalities in Exercises 31–34.

**31.** $x^2 + y^2 > 7$      **32.** $(x - 1)^2 + y^2 \le 4$

**33.** $x^2 + y^2 > 1$,   $x^2 + y^2 < 4$

**34.** $x^2 + y^2 + 6y < 0$,   $y > -3$

**35.** Write an inequality that describes the points that lie inside the circle with center $(-2, 1)$ and radius $\sqrt{6}$.

**36.** Write a pair of inequalities that describe the points that lie inside or on the circle with center $(0, 0)$ and radius $\sqrt{2}$, and on or to the right of the vertical line through $(1, 0)$.

## Theory and Examples

In Exercises 37–40, graph the two equations and find the points at which the graphs intersect.

**37.** $y = 2x$,   $x^2 + y^2 = 1$      **38.** $y - x = 1$,   $y = x^2$

**39.** $y = -x^2$,   $y = 2x^2 - 1$

**40.** $x^2 + y^2 = 1$,   $(x - 1)^2 + y^2 = 1$

**41. Insulation** By measuring slopes in the figure, estimate the temperature change in degrees per centimeter for **(a)** the gypsum wallboard; **(b)** the fiberglass insulation; **(c)** the wood sheathing.

The temperature changes in the wall in Exercises 41 and 42.

**42. Insulation** According to the figure in Exercise 41, which of the materials is the best insulator? The poorest? Explain.

**43. Pressure under water** The pressure $p$ experienced by a diver under water is related to the diver's depth $d$ by an equation of the form $p = kd + 101.3$ ($k$ a constant). At the surface, the pressure is 101.3 kPa. The pressure at 100 meters is about 1106.5 kPa. Find the pressure at 50 meters.

**44. Reflected light** A ray of light comes in along the line $x + y = 1$ from the second quadrant and reflects off the $x$-axis (see the accompanying figure). The angle of incidence is equal to the angle of reflection. Write an equation for the line along which the departing light travels.

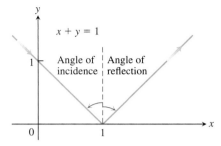

The path of the light ray in Exercise 44. Angles of incidence and reflection are measured from the perpendicular.

**45. Fahrenheit vs. Celsius** In the $FC$-plane, sketch the graph of the equation

$$C = \frac{5}{9}(F - 32)$$

linking Fahrenheit and Celsius temperatures. On the same graph sketch the line $C = F$. Is there a temperature at which a Celsius thermometer gives the same numerical reading as a Fahrenheit thermometer? If so, find it.

**46. The Mt. Washington Cog Railway** Civil engineers calculate the slope of roadbed as the ratio of the distance it rises or falls to the distance it runs horizontally. They call this ratio the **grade** of the roadbed, usually written as a percentage. Along the coast, commercial railroad grades are usually less than 2%. In the mountains, they may go as high as 4%. Highway grades are usually less than 5%.

    The steepest part of the Mt. Washington Cog Railway in New Hampshire has an exceptional 37.1% grade. Along this part of the track, the seats in the front of the car are 4 m above those in the rear. About how far apart are the front and rear rows of seats?

**47.** By calculating the lengths of its sides, show that the triangle with vertices at the points $A(1, 2)$, $B(5, 5)$, and $C(4, -2)$ is isosceles but not equilateral.

**48.** Show that the triangle with vertices $A(0, 0)$, $B\left(1, \sqrt{3}\right)$, and $C(2, 0)$ is equilateral.

**49.** Show that the points $A(2, -1)$, $B(1, 3)$, and $C(-3, 2)$ are vertices of a square, and find the fourth vertex.

**50.** Three different parallelograms have vertices at $(-1, 1)$, $(2, 0)$, and $(2, 3)$. Sketch them and find the coordinates of the fourth vertex of each.

**51.** For what value of $k$ is the line $2x + ky = 3$ perpendicular to the line $4x + y = 1$? For what value of $k$ are the lines parallel?

**52. Midpoint of a line segment** Show that the point with coordinates

$$\left(\frac{x_1 + x_2}{2}, \frac{y_1 + y_2}{2}\right)$$

is the midpoint of the line segment joining $P(x_1, y_1)$ to $Q(x_2, y_2)$.

# A.4 Proofs of Limit Theorems

This appendix proves Theorem 1, Parts 2–5, and Theorem 4 from Section 2.2.

---

**THEOREM 1—Limit Laws**   If $L$, $M$, $c$, and $k$ are real numbers and

$$\lim_{x \to c} f(x) = L \quad \text{and} \quad \lim_{x \to c} g(x) = M, \quad \text{then}$$

1. *Sum Rule:* $\qquad\qquad\qquad\qquad \lim_{x \to c} (f(x) + g(x)) = L + M$

2. *Difference Rule:* $\qquad\qquad\quad \lim_{x \to c} (f(x) - g(x)) = L - M$

3. *Constant Multiple Rule:* $\qquad \lim_{x \to c} (k \cdot f(x)) = k \cdot L$

4. *Product Rule:* $\qquad\qquad\qquad \lim_{x \to c} (f(x) \cdot g(x)) = L \cdot M$

5. *Quotient Rule:* $\qquad\qquad\quad \lim_{x \to c} \dfrac{f(x)}{g(x)} = \dfrac{L}{M}, \qquad M \neq 0$

6. *Power Rule:* $\qquad\qquad\qquad \lim_{x \to c} [f(x)]^n = L^n$, $n$ a positive integer

7. *Root Rule:* $\qquad\qquad\qquad\quad \lim_{x \to c} \sqrt[n]{f(x)} = \sqrt[n]{L} = L^{1/n}$, $n$ a positive integer

(If $n$ is even, we assume that $\lim_{x \to c} f(x) = L > 0$.)

---

We proved the Sum Rule in Section 2.3, and the Power and Root Rules are proved in more advanced texts. We obtain the Difference Rule by replacing $g(x)$ by $-g(x)$ and $M$ by $-M$ in the Sum Rule. The Constant Multiple Rule is the special case $g(x) = k$ of the Product Rule. This leaves only the Product and Quotient Rules.

**Proof of the Limit Product Rule**   We show that for any $\epsilon > 0$ there exists a $\delta > 0$ such that for all $x$ in the intersection $D$ of the domains of $f$ and $g$,

$$0 < |x - c| < \delta \quad \Rightarrow \quad |f(x)g(x) - LM| < \epsilon.$$

Suppose then that $\epsilon$ is a positive number, and write $f(x)$ and $g(x)$ as

$$f(x) = L + (f(x) - L), \qquad g(x) = M + (g(x) - M).$$

Multiply these expressions together and subtract $LM$:

$$
\begin{aligned}
f(x) \cdot g(x) - LM &= (L + (f(x) - L))(M + (g(x) - M)) - LM \\
&= LM + L(g(x) - M) + M(f(x) - L) \\
&\quad + (f(x) - L)(g(x) - M) - LM \\
&= L(g(x) - M) + M(f(x) - L) + (f(x) - L)(g(x) - M). \quad (1)
\end{aligned}
$$

Since $f$ and $g$ have limits $L$ and $M$ as $x \to c$, there exist positive numbers $\delta_1$, $\delta_2$, $\delta_3$, and $\delta_4$ such that for all $x$ in $D$

$$
\begin{aligned}
0 < |x - c| < \delta_1 &\quad \Rightarrow \quad |f(x) - L| < \sqrt{\epsilon/3} \\
0 < |x - c| < \delta_2 &\quad \Rightarrow \quad |g(x) - M| < \sqrt{\epsilon/3} \\
0 < |x - c| < \delta_3 &\quad \Rightarrow \quad |f(x) - L| < \epsilon/(3(1 + |M|)) \\
0 < |x - c| < \delta_4 &\quad \Rightarrow \quad |g(x) - M| < \epsilon/(3(1 + |L|)).
\end{aligned}
\qquad (2)
$$

If we take $\delta$ to be the smallest numbers $\delta_1$ through $\delta_4$, the inequalities on the right-hand side of the Implications (2) will hold simultaneously for $0 < |x - c| < \delta$. Therefore, for all $x$ in $D$, $0 < |x - c| < \delta$ implies

$$|f(x) \cdot g(x) - LM| \qquad \text{\small Triangle inequality applied to Eq. (1)}$$

$$\leq |L||g(x) - M| + |M||f(x) - L| + |f(x) - L||g(x) - M|$$

$$\leq (1 + |L|)|g(x) - M| + (1 + |M|)|f(x) - L| + |f(x) - L||g(x) - M|$$

$$< \frac{\epsilon}{3} + \frac{\epsilon}{3} + \sqrt{\frac{\epsilon}{3}}\sqrt{\frac{\epsilon}{3}} = \epsilon. \qquad \text{\small Values from (2)}$$

This completes the proof of the Limit Product Rule. ∎

**Proof of the Limit Quotient Rule** We show that $\lim_{x \to c}(1/g(x)) = 1/M$. We can then conclude that

$$\lim_{x \to c}\frac{f(x)}{g(x)} = \lim_{x \to c}\left(f(x) \cdot \frac{1}{g(x)}\right) = \lim_{x \to c} f(x) \cdot \lim_{x \to c}\frac{1}{g(x)} = L \cdot \frac{1}{M} = \frac{L}{M}$$

by the Limit Product Rule.

Let $\epsilon > 0$ be given. To show that $\lim_{x \to c}(1/g(x)) = 1/M$, we need to show that there exists a $\delta > 0$ such that for all $x$

$$0 < |x - c| < \delta \quad \Longrightarrow \quad \left|\frac{1}{g(x)} - \frac{1}{M}\right| < \epsilon.$$

Since $|M| > 0$, there exists a positive number $\delta_1$ such that for all $x$

$$0 < |x - c| < \delta_1 \quad \Longrightarrow \quad |g(x) - M| < \frac{M}{2}. \tag{3}$$

For any numbers $A$ and $B$ it can be shown that $|A| - |B| \leq |A - B|$ and $|B| - |A| \leq |A - B|$, from which it follows that $||A| - |B|| \leq |A - B|$. With $A = g(x)$ and $B = M$, this becomes

$$||g(x)| - |M|| \leq |g(x) - M|,$$

which can be combined with the inequality on the right in Implication (3) to get, in turn,

$$|| g(x) | - | M || < \frac{| M |}{2}$$

$$-\frac{|M|}{2} < |g(x)| - |M| < \frac{|M|}{2}$$

$$\frac{|M|}{2} < |g(x)| < \frac{3|M|}{2}$$

$$|M| < 2|g(x)| < 3|M|$$

$$\frac{1}{|g(x)|} < \frac{2}{|M|} < \frac{3}{|g(x)|}. \tag{4}$$

Therefore, $0 < |x - c| < \delta_1$ implies that

$$\left|\frac{1}{g(x)} - \frac{1}{M}\right| = \left|\frac{M - g(x)}{Mg(x)}\right| \leq \frac{1}{|M|} \cdot \frac{1}{|g(x)|} \cdot |M - g(x)|$$

$$< \frac{1}{|M|} \cdot \frac{2}{|M|} \cdot |M - g(x)|. \qquad \text{\small Inequality (4)} \tag{5}$$

Since $(1/2)|M|^2\epsilon > 0$, there exists a number $\delta_2 > 0$ such that for all $x$

$$0 < |x - c| < \delta_2 \quad \Rightarrow \quad |M - g(x)| < \frac{\epsilon}{2}|M|^2. \tag{6}$$

If we take $\delta$ to be the smaller of $\delta_1$ and $\delta_2$, the conclusions in (5) and (6) both hold for all $x$ such that $0 < |x - c| < \delta$. Combining these conclusions gives

$$0 < |x - c| < \delta \quad \Rightarrow \quad \left|\frac{1}{g(x)} - \frac{1}{M}\right| < \epsilon.$$

This concludes the proof of the Limit Quotient Rule.  ∎

---

**THEOREM 4—The Sandwich Theorem**  Suppose that $g(x) \le f(x) \le h(x)$ for all $x$ in some open interval $I$ containing $c$, except possibly at $x = c$ itself. Suppose also that $\lim_{x \to c} g(x) = \lim_{x \to c} h(x) = L$. Then $\lim_{x \to c} f(x) = L$.

---

**Proof for Right-Hand Limits**  Suppose $\lim_{x \to c^+} g(x) = \lim_{x \to c^+} h(x) = L$. Then for any $\epsilon > 0$ there exists a $\delta > 0$ such that for all $x$ the interval $c < x < c + \delta$ is contained in $I$ and the inequality implies

$$L - \epsilon < g(x) < L + \epsilon \quad \text{and} \quad L - \epsilon < h(x) < L + \epsilon.$$

These inequalities combine with the inequality $g(x) \le f(x) \le h(x)$ to give

$$L - \epsilon < g(x) \le f(x) \le h(x) < L + \epsilon,$$
$$L - \epsilon < f(x) < L + \epsilon,$$
$$-\epsilon < f(x) - L < \epsilon.$$

Therefore, for all $x$, the inequality $c < x < c + \delta$ implies $|f(x) - L| < \epsilon$.

**Proof for Left-Hand Limits**  Suppose $\lim_{x \to c^-} g(x) = \lim_{x \to c^-} h(x) = L$. Then for any $\epsilon > 0$ there exists a $\delta > 0$ such that for all $x$ the interval $c - \delta < x < c$ is contained in $I$ and the inequality implies

$$L - \epsilon < g(x) < L + \epsilon \quad \text{and} \quad L - \epsilon < h(x) < L + \epsilon.$$

We conclude as before that for all $x$, $c - \delta < x < c$ implies $|f(x) - L| < \epsilon$.

**Proof for Two-Sided Limits**  If $\lim_{x \to c} g(x) = \lim_{x \to c} h(x) = L$, then $g(x)$ and $h(x)$ both approach $L$ as $x \to c^+$ and as $x \to c^-$; so $\lim_{x \to c^+} f(x) = L$ and $\lim_{x \to c^-} f(x) = L$. Hence $\lim_{x \to c} f(x)$ exists and equals $L$.  ∎

---

# Exercises A.4

1. Suppose that functions $f_1(x)$, $f_2(x)$, and $f_3(x)$ have limits $L_1$, $L_2$, and $L_3$, respectively, as $x \to c$. Show that their sum has limit $L_1 + L_2 + L_3$. Use mathematical induction (Appendix 2) to generalize this result to the sum of any finite number of functions.

2. Use mathematical induction and the Limit Product Rule in Theorem 1 to show that if functions $f_1(x)$, $f_2(x)$, ..., $f_n(x)$ have limits $L_1, L_2, \ldots, L_n$ as $x \to c$, then

$$\lim_{x \to c} f_1(x) \cdot f_2(x) \cdot \cdots \cdot f_n(x) = L_1 \cdot L_2 \cdot \cdots \cdot L_n.$$

3. Use the fact that $\lim_{x \to c} x = c$ and the result of Exercise 2 to show that $\lim_{x \to c} x^n = c^n$ for any integer $n > 1$.

4. **Limits of polynomials**  Use the fact that $\lim_{x \to c}(k) = k$ for any number $k$ together with the results of Exercises 1 and 3 to show that $\lim_{x \to c} f(x) = f(c)$ for any polynomial function

$$f(x) = a_n x^n + a_{n-1} x^{n-1} + \cdots + a_1 x + a_0.$$

5. **Limits of rational functions** Use Theorem 1 and the result of Exercise 4 to show that if $f(x)$ and $g(x)$ are polynomial functions and $g(c) \neq 0$, then

$$\lim_{x \to c} \frac{f(x)}{g(x)} = \frac{f(c)}{g(c)}.$$

6. **Composites of continuous functions** Figure A.23 gives the diagram for a proof that the composite of two continuous func-

tions is continuous. Reconstruct the proof from the diagram. The statement to be proved is this: If $f$ is continuous at $x = c$ and $g$ is continuous at $f(c)$, then $g \circ f$ is continuous at $c$.

Assume that $c$ is an interior point of the domain of $f$ and that $f(c)$ is an interior point of the domain of $g$. This will make the limits involved two-sided. (The arguments for the cases that involve one-sided limits are similar.)

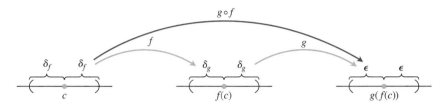

**FIGURE A.23** The diagram for a proof that the composite of two continuous functions is continuous.

# A.5 Commonly Occurring Limits

This appendix verifies limits (4)–(6) in Theorem 5 of Section 10.1.

**Limit 4: If $|x| < 1$, $\lim_{n \to \infty} x^n = 0$** We need to show that to each $\epsilon > 0$ there corresponds an integer $N$ so large that $|x^n| < \epsilon$ for all $n$ greater than $N$. Since $\epsilon^{1/n} \to 1$, while $|x| < 1$, there exists an integer $N$ for which $\epsilon^{1/N} > |x|$. In other words,

$$|x^N| = |x|^N < \epsilon. \tag{1}$$

This is the integer we seek because, if $|x| < 1$, then

$$|x^n| < |x^N| \quad \text{for all } n > N. \tag{2}$$

Combining (1) and (2) produces $|x^n| < \epsilon$ for all $n > N$, concluding the proof. ∎

**Limit 5: For any number $x$, $\lim_{n \to \infty} \left(1 + \dfrac{x}{n}\right)^n = e^x$** Let

$$a_n = \left(1 + \frac{x}{n}\right)^n.$$

Then

$$\ln a_n = \ln\left(1 + \frac{x}{n}\right)^n = n \ln\left(1 + \frac{x}{n}\right) \to x,$$

as we can see by the following application of l'Hôpital's Rule, in which we differentiate with respect to $n$:

$$\lim_{n \to \infty} n \ln\left(1 + \frac{x}{n}\right) = \lim_{n \to \infty} \frac{\ln(1 + x/n)}{1/n}$$

$$= \lim_{n \to \infty} \frac{\left(\dfrac{1}{1 + x/n}\right) \cdot \left(-\dfrac{x}{n^2}\right)}{-1/n^2} = \lim_{n \to \infty} \frac{x}{1 + x/n} = x.$$

Apply Theorem 3, Section 10.1, with $f(x) = e^x$ to conclude that

$$\left(1 + \frac{x}{n}\right)^n = a_n = e^{\ln a_n} \to e^x. \quad ∎$$

**Limit 6: For any number $x$, $\displaystyle\lim_{n\to\infty} \frac{x^n}{n!} = 0$**  Since

$$-\frac{|x|^n}{n!} \leq \frac{x^n}{n!} \leq \frac{|x|^n}{n!},$$

all we need to show is that $|x|^n/n! \to 0$. We can then apply the Sandwich Theorem for Sequences (Section 10.1, Theorem 2) to conclude that $x^n/n! \to 0$.

The first step in showing that $|x|^n/n! \to 0$ is to choose an integer $M > |x|$, so that $(|x|/M) < 1$. By Limit 4, just proved, we then have $(|x|/M)^n \to 0$. We then restrict our attention to values of $n > M$. For these values of $n$, we can write

$$\frac{|x|^n}{n!} = \frac{|x|^n}{\underbrace{1 \cdot 2 \cdot \cdots \cdot M \cdot (M+1) \cdot (M+2) \cdot \cdots \cdot n}_{(n-M)\ \text{factors}}}$$

$$\leq \frac{|x|^n}{M! M^{n-M}} = \frac{|x|^n M^M}{M! M^n} = \frac{M^M}{M!}\left(\frac{|x|}{M}\right)^n.$$

Thus,

$$0 \leq \frac{|x|^n}{n!} \leq \frac{M^M}{M!}\left(\frac{|x|}{M}\right)^n.$$

Now, the constant $M^M/M!$ does not change as $n$ increases. Thus the Sandwich Theorem tells us that $|x|^n/n! \to 0$ because $(|x|/M)^n \to 0$. ∎

# A.6 Theory of the Real Numbers

A rigorous development of calculus is based on properties of the real numbers. Many results about functions, derivatives, and integrals would be false if stated for functions defined only on the rational numbers. In this appendix we briefly examine some basic concepts of the theory of the reals that hint at what might be learned in a deeper, more theoretical study of calculus.

Three types of properties make the real numbers what they are. These are the **algebraic**, **order**, and **completeness** properties. The algebraic properties involve addition and multiplication, subtraction and division. They apply to rational or complex numbers as well as to the reals.

The structure of numbers is built around a set with addition and multiplication operations. The following properties are required of addition and multiplication.

**A1**   $a + (b + c) = (a + b) + c$ for all $a, b, c$.

**A2**   $a + b = b + a$ for all $a, b$.

**A3**   There is a number called "0" such that $a + 0 = a$ for all $a$.

**A4**   For each number $a$, there is a $b$ such that $a + b = 0$.

**M1**   $a(bc) = (ab)c$ for all $a, b, c$.

**M2**   $ab = ba$ for all $a, b$.

**M3**   There is a number called "1" such that $a \cdot 1 = a$ for all $a$.

**M4**   For each nonzero $a$, there is a $b$ such that $ab = 1$.

**D**   $a(b + c) = ab + ac$ for all $a, b, c$.

A1 and M1 are *associative laws*, A2 and M2 are *commutativity laws*, A3 and M3 are *identity laws*, and D is the *distributive law*. Sets that have these algebraic properties are examples of **fields**, and are studied in depth in the area of theoretical mathematics called abstract algebra.

The **order** properties allow us to compare the size of any two numbers. The order properties are

**O1**   For any $a$ and $b$, either $a \leq b$ or $b \leq a$ or both.

**O2**   If $a \leq b$ and $b \leq a$ then $a = b$.

**O3**   If $a \leq b$ and $b \leq c$ then $a \leq c$.

**O4**   If $a \leq b$ then $a + c \leq b + c$.

**O5**   If $a \leq b$ and $0 \leq c$ then $ac \leq bc$.

O3 is the *transitivity law*, and O4 and O5 relate ordering to addition and multiplication.

We can order the reals, the integers, and the rational numbers, but we cannot order the complex numbers. There is no reasonable way to decide whether a number like $i = \sqrt{-1}$ is bigger or smaller than zero. A field in which the size of any two elements can be compared as above is called an **ordered field**. Both the rational numbers and the real numbers are ordered fields, and there are many others.

We can think of real numbers geometrically, lining them up as points on a line. The **completeness property** says that the real numbers correspond to all points on the line, with no "holes" or "gaps." The rationals, in contrast, omit points such as $\sqrt{2}$ and $\pi$, and the integers even leave out fractions like $1/2$. The reals, having the completeness property, omit no points.

What exactly do we mean by this vague idea of missing holes? To answer this we must give a more precise description of completeness. A number $M$ is an **upper bound** for a set of numbers if all numbers in the set are smaller than or equal to $M$. $M$ is a **least upper bound** if it is the smallest upper bound. For example, $M = 2$ is an upper bound for the negative numbers. So is $M = 1$, showing that 2 is not a least upper bound. The least upper bound for the set of negative numbers is $M = 0$. We define a **complete** ordered field to be one in which every nonempty set bounded above has a least upper bound.

If we work with just the rational numbers, the set of numbers less than $\sqrt{2}$ is bounded, but it does not have a rational least upper bound, since any rational upper bound $M$ can be replaced by a slightly smaller rational number that is still larger than $\sqrt{2}$. So the rationals are not complete. In the real numbers, a set that is bounded above always has a least upper bound. The reals are a complete ordered field.

The completeness property is at the heart of many results in calculus. One example occurs when searching for a maximum value for a function on a closed interval $[a, b]$, as in Section 4.1. The function $y = x - x^3$ has a maximum value on $[0, 1]$ at the point $x$ satisfying $1 - 3x^2 = 0$, or $x = \sqrt{1/3}$. If we limited our consideration to functions defined only on rational numbers, we would have to conclude that the function has no maximum, since $\sqrt{1/3}$ is irrational (Figure A.24). The Extreme Value Theorem (Section 4.1), which implies that continuous functions on closed intervals $[a, b]$ have a maximum value, is not true for functions defined only on the rationals.

The Intermediate Value Theorem implies that a continuous function $f$ on an interval $[a, b]$ with $f(a) < 0$ and $f(b) > 0$ must be zero somewhere in $[a, b]$. The function values cannot jump from negative to positive without there being some point $x$ in $[a, b]$ where $f(x) = 0$. The Intermediate Value Theorem also relies on the completeness of the real numbers and is false for continuous functions defined only on the rationals. The function $f(x) = 3x^2 - 1$ has $f(0) = -1$ and $f(1) = 2$, but if we consider $f$ only on the rational numbers, it never equals zero. The only value of $x$ for which $f(x) = 0$ is $x = \sqrt{1/3}$, an irrational number.

We have captured the desired properties of the reals by saying that the real numbers are a complete ordered field. But we're not quite finished. Greek mathematicians in the school of Pythagoras tried to impose another property on the numbers of the real line, the

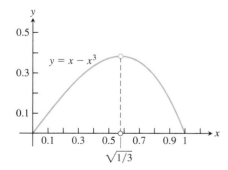

**FIGURE A.24** The maximum value of $y = x - x^3$ on $[0, 1]$ occurs at the irrational number $x = \sqrt{1/3}$.

condition that all numbers are ratios of integers. They learned that their effort was doomed when they discovered irrational numbers such as $\sqrt{2}$. How do we know that our efforts to specify the real numbers are not also flawed, for some unseen reason? The artist Escher drew optical illusions of spiral staircases that went up and up until they rejoined themselves at the bottom. An engineer trying to build such a staircase would find that no structure realized the plans the architect had drawn. Could it be that our design for the reals contains some subtle contradiction, and that no construction of such a number system can be made?

We resolve this issue by giving a specific description of the real numbers and verifying that the algebraic, order, and completeness properties are satisfied in this model. This is called a **construction** of the reals, and just as stairs can be built with wood, stone, or steel, there are several approaches to constructing the reals. One construction treats the reals as all the infinite decimals,

$$a.d_1 d_2 d_3 d_4 \ldots$$

In this approach a real number is an integer $a$ followed by a sequence of decimal digits $d_1, d_2, d_3, \ldots$, each between 0 and 9. This sequence may stop, or repeat in a periodic pattern, or keep going forever with no pattern. In this form, $2.00, 0.3333333 \ldots$ and $3.1415926535898 \ldots$ represent three familiar real numbers. The real meaning of the dots "$\ldots$" following these digits requires development of the theory of sequences and series, as in Chapter 10. Each real number is constructed as the limit of a sequence of rational numbers given by its finite decimal approximations. An infinite decimal is then the same as a series

$$a + \frac{d_1}{10} + \frac{d_2}{100} + \cdots.$$

This decimal construction of the real numbers is not entirely straightforward. It's easy enough to check that it gives numbers that satisfy the completeness and order properties, but verifying the algebraic properties is rather involved. Even adding or multiplying two numbers requires an infinite number of operations. Making sense of division requires a careful argument involving limits of rational approximations to infinite decimals.

A different approach was taken by Richard Dedekind (1831–1916), a German mathematician, who gave the first rigorous construction of the real numbers in 1872. Given any real number $x$, we can divide the rational numbers into two sets: those less than or equal to $x$ and those greater. Dedekind cleverly reversed this reasoning and defined a real number to be a division of the rational numbers into two such sets. This seems like a strange approach, but such indirect methods of constructing new structures from old are common in theoretical mathematics.

These and other approaches can be used to construct a system of numbers having the desired algebraic, order, and completeness properties. A final issue that arises is whether all the constructions give the same thing. Is it possible that different constructions result in different number systems satisfying all the required properties? If yes, which of these is the real numbers? Fortunately, the answer turns out to be no. The reals are the only number system satisfying the algebraic, order, and completeness properties.

Confusion about the nature of the numbers and about limits caused considerable controversy in the early development of calculus. Calculus pioneers such as Newton, Leibniz, and their successors, when looking at what happens to the difference quotient

$$\frac{\Delta y}{\Delta x} = \frac{f(x + \Delta x) - f(x)}{\Delta x}$$

as each of $\Delta y$ and $\Delta x$ approach zero, talked about the resulting derivative being a quotient of two infinitely small quantities. These "infinitesimals," written $dx$ and $dy$, were thought to be some new kind of number, smaller than any fixed number but not zero. Similarly, a definite integral was thought of as a sum of an infinite number of infinitesimals

$$f(x) \cdot dx$$

as $x$ varied over a closed interval. While the approximating difference quotients $\Delta y/\Delta x$ were understood much as today, it was the quotient of infinitesimal quantities, rather than a limit, that was thought to encapsulate the meaning of the derivative. This way of thinking led to logical difficulties, as attempted definitions and manipulations of infinitesimals ran into contradictions and inconsistencies. The more concrete and computable difference quotients did not cause such trouble, but they were thought of merely as useful calculation tools. Difference quotients were used to work out the numerical value of the derivative and to derive general formulas for calculation, but were not considered to be at the heart of the question of what the derivative actually was. Today we realize that the logical problems associated with infinitesimals can be avoided by *defining* the derivative to be the limit of its approximating difference quotients. The ambiguities of the old approach are no longer present, and in the standard theory of calculus, infinitesimals are neither needed nor used.

# A.7 Complex Numbers

Complex numbers are expressed in the form $a + ib$, or $a + bi$, where $a$ and $b$ are real numbers and $i$ is a symbol for $\sqrt{-1}$. Unfortunately, the words "real" and "imaginary" have connotations that somehow place $\sqrt{-1}$ in a less favorable position in our minds than $\sqrt{2}$. As a matter of fact, a good deal of imagination, in the sense of *inventiveness*, has been required to construct the *real* number system, which forms the basis of the calculus (see Appendix 6). In this appendix we review the various stages of this invention. The further invention of a complex number system is then presented.

## The Development of the Real Numbers

The earliest stage of number development was the recognition of the **counting numbers** $1, 2, 3, \ldots$, which we now call the **natural numbers** or the **positive integers**. Certain simple arithmetical operations, such as addition and multiplication, can be performed with these numbers without getting outside the system. By this we mean that if $m$ and $n$ are any positive integers, then

$$m + n = p \qquad \text{and} \qquad mn = q \qquad (1)$$

are also positive integers. Given the two positive integers on the left side of either equation in (1), we can find the corresponding positive integer on the right side. More than this, we can sometimes specify the positive integers $m$ and $p$ and find a positive integer $n$ such that $m + n = p$. For instance, $3 + n = 7$ can be solved when the only numbers we know are the positive integers. But the equation $7 + n = 3$ cannot be solved unless the number system is enlarged.

The number zero and the negative integers were invented to solve equations like $7 + n = 3$. In a civilization that recognizes all the **integers**

$$\ldots, -3, -2, -1, 0, 1, 2, 3, \ldots, \qquad (2)$$

an educated person can always find the missing integer that solves the equation $m + n = p$ when given the other two integers in the equation.

Suppose our educated people also know how to multiply any two of the integers in the list (2). If, in Equations (1), they are given $m$ and $q$, they discover that sometimes they can find $n$ and sometimes they cannot. Using their imagination, they may be inspired to invent still more numbers and introduce fractions, which are just ordered pairs $m/n$ of integers $m$ and $n$. The number zero has special properties that may bother them for a while, but they ultimately discover that it is handy to have all ratios of integers $m/n$, excluding only those having zero in the denominator. This system, called the set of **rational numbers**, is now rich enough for them to perform the **rational operations** of arithmetic:

1.  **(a)** addition
    **(b)** subtraction

2.  **(a)** multiplication
    **(b)** division

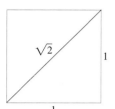

**FIGURE A.25** With a straightedge and compass, it is possible to construct a segment of irrational length.

on any two numbers in the system, *except that they cannot divide by zero* because it is meaningless.

The geometry of the unit square (Figure A.25) and the Pythagorean theorem showed that they could construct a geometric line segment that, in terms of some basic unit of length, has length equal to $\sqrt{2}$. Thus they could solve the equation

$$x^2 = 2$$

by a geometric construction. But then they discovered that the line segment representing $\sqrt{2}$ is an incommensurable quantity. This means that $\sqrt{2}$ cannot be expressed as the ratio of two *integer* multiples of some unit of length. That is, our educated people could not find a rational number solution of the equation $x^2 = 2$.

There *is* no rational number whose square is 2. To see why, suppose that there were such a rational number. Then we could find integers $p$ and $q$ with no common factor other than 1, and such that

$$p^2 = 2q^2. \tag{3}$$

Since $p$ and $q$ are integers, $p$ must be even; otherwise its product with itself would be odd. In symbols, $p = 2p_1$, where $p_1$ is an integer. This leads to $2p_1{}^2 = q^2$, which says $q$ must be even, say $q = 2q_1$, where $q_1$ is an integer. This makes 2 a factor of both $p$ and $q$, contrary to our choice of $p$ and $q$ as integers with no common factor other than 1. Hence there is no rational number whose square is 2.

Although our educated people could not find a rational solution of the equation $x^2 = 2$, they could get a sequence of rational numbers

$$\frac{1}{1}, \frac{7}{5}, \frac{41}{29}, \frac{239}{169}, \cdots, \tag{4}$$

whose squares form a sequence

$$\frac{1}{1}, \frac{49}{25}, \frac{1681}{841}, \frac{57{,}121}{28{,}561}, \cdots, \tag{5}$$

that converges to 2 as its limit. This time their imagination suggested that they needed the concept of a limit of a sequence of rational numbers. If we accept the fact that an increasing sequence that is bounded from above always approaches a limit (Theorem 6, Section 10.1) and observe that the sequence in (4) has these properties, then we want it to have a limit $L$. This would also mean, from (5), that $L^2 = 2$, and hence $L$ is *not* one of our rational numbers. If to the rational numbers we further add the limits of all bounded increasing sequences of rational numbers, we arrive at the system of all "real" numbers. The word *real* is placed in quotes because there is nothing that is either "more real" or "less real" about this system than there is about any other mathematical system.

## The Complex Numbers

Imagination was called upon at many stages during the development of the real number system. In fact, the art of invention was needed at least three times in constructing the systems we have discussed so far:

1.  The *first invented* system: the set of *all integers* as constructed from the counting numbers.

2.  The *second invented* system: the set of *rational numbers $m/n$* as constructed from the integers.

3.  The *third invented* system: the set of all *real numbers $x$* as constructed from the rational numbers.

These invented systems form a hierarchy in which each system contains the previous system. Each system is also richer than its predecessor in that it permits additional operations to be performed without going outside the system:

1. In the system of all integers, we can solve all equations of the form

$$x + a = 0, \tag{6}$$

where $a$ can be any integer.

2. In the system of all rational numbers, we can solve all equations of the form

$$ax + b = 0, \tag{7}$$

provided $a$ and $b$ are rational numbers and $a \neq 0$.

3. In the system of all real numbers, we can solve all of Equations (6) and (7) and, in addition, all quadratic equations

$$ax^2 + bx + c = 0 \quad \text{having} \quad a \neq 0 \quad \text{and} \quad b^2 - 4ac \geq 0. \tag{8}$$

You are probably familiar with the formula that gives the solutions of Equation (8), namely,

$$x = \frac{-b \pm \sqrt{b^2 - 4ac}}{2a}, \tag{9}$$

and are familiar with the further fact that when the discriminant, $b^2 - 4ac$, is negative, the solutions in Equation (9) do *not* belong to any of the systems discussed above. In fact, the very simple quadratic equation

$$x^2 + 1 = 0$$

is impossible to solve if the only number systems that can be used are the three invented systems mentioned so far.

Thus we come to the *fourth invented* system, the set of *all complex numbers $a + ib$*. We could dispense entirely with the symbol $i$ and use the ordered pair notation $(a, b)$. Since, under algebraic operations, the numbers $a$ and $b$ are treated somewhat differently, it is essential to keep the *order* straight. We therefore might say that the **complex number system** consists of the set of all ordered pairs of real numbers $(a, b)$, together with the rules by which they are to be equated, added, multiplied, and so on, listed below. We will use both the $(a, b)$ notation and the notation $a + ib$ in the discussion that follows. We call $a$ the **real part** and $b$ the **imaginary part** of the complex number $(a, b)$.

We make the following definitions.

*Equality*

$a + ib = c + id$      Two complex numbers $(a, b)$
if and only if      and $(c, d)$ are *equal* if and only
$a = c$ and $b = d$.      if $a = c$ and $b = d$.

*Addition*

$(a + ib) + (c + id)$      The *sum* of the two complex
$= (a + c) + i(b + d)$      numbers $(a, b)$ and $(c, d)$ is the
     complex number $(a + c, b + d)$.

*Multiplication*

$(a + ib)(c + id)$      The *product* of two complex
$= (ac - bd) + i(ad + bc)$      numbers $(a, b)$ and $(c, d)$ is the complex
     number $(ac - bd, ad + bc)$.

$c(a + ib) = ac + i(bc)$      The product of a real number $c$
     and the complex number $(a, b)$ is
     the complex number $(ac, bc)$.

The set of all complex numbers $(a, b)$ in which the second number $b$ is zero has all the properties of the set of real numbers $a$. For example, addition and multiplication of $(a, 0)$ and $(c, 0)$ give

$$(a, 0) + (c, 0) = (a + c, 0),$$
$$(a, 0) \cdot (c, 0) = (ac, 0),$$

which are numbers of the same type with imaginary part equal to zero. Also, if we multiply a "real number" $(a, 0)$ and the complex number $(c, d)$, we get

$$(a, 0) \cdot (c, d) = (ac, ad) = a(c, d).$$

In particular, the complex number $(0, 0)$ plays the role of *zero* in the complex number system, and the complex number $(1, 0)$ plays the role of *unity* or *one*.

The number pair $(0, 1)$, which has real part equal to zero and imaginary part equal to one, has the property that its square,

$$(0, 1)(0, 1) = (-1, 0),$$

has real part equal to minus one and imaginary part equal to zero. Therefore, in the system of complex numbers $(a, b)$ there is a number $x = (0, 1)$ whose square can be added to unity $= (1, 0)$ to produce zero $= (0, 0)$, that is,

$$(0, 1)^2 + (1, 0) = (0, 0).$$

The equation

$$x^2 + 1 = 0$$

therefore has a solution $x = (0, 1)$ in this new number system.

You are probably more familiar with the $a + ib$ notation than you are with the notation $(a, b)$. And since the laws of algebra for the ordered pairs enable us to write

$$(a, b) = (a, 0) + (0, b) = a(1, 0) + b(0, 1),$$

while $(1, 0)$ behaves like unity and $(0, 1)$ behaves like a square root of minus one, we need not hesitate to write $a + ib$ in place of $(a, b)$. The $i$ associated with $b$ is like a tracer element that tags the imaginary part of $a + ib$. We can pass at will from the realm of ordered pairs $(a, b)$ to the realm of expressions $a + ib$, and conversely. But there is nothing less "real" about the symbol $(0, 1) = i$ than there is about the symbol $(1, 0) = 1$, once we have learned the laws of algebra in the complex number system of ordered pairs $(a, b)$.

To reduce any rational combination of complex numbers to a single complex number, we apply the laws of elementary algebra, replacing $i^2$ wherever it appears by $-1$. Of course, we cannot divide by the complex number $(0, 0) = 0 + i0$. But if $a + ib \neq 0$, then we may carry out a division as follows:

$$\frac{c + id}{a + ib} = \frac{(c + id)(a - ib)}{(a + ib)(a - ib)} = \frac{(ac + bd) + i(ad - bc)}{a^2 + b^2}.$$

The result is a complex number $x + iy$ with

$$x = \frac{ac + bd}{a^2 + b^2}, \qquad y = \frac{ad - bc}{a^2 + b^2},$$

and $a^2 + b^2 \neq 0$, since $a + ib = (a, b) \neq (0, 0)$.

The number $a - ib$ that is used as the multiplier to clear the $i$ from the denominator is called the **complex conjugate** of $a + ib$. It is customary to use $\bar{z}$ (read "$z$ bar") to denote the complex conjugate of $z$; thus

$$z = a + ib, \qquad \bar{z} = a - ib.$$

Multiplying the numerator and denominator of the fraction $(c + id)/(a + ib)$ by the complex conjugate of the denominator will always replace the denominator by a real number.

**EXAMPLE 1** We give some illustrations of the arithmetic operations with complex numbers.

**(a)** $(2 + 3i) + (6 - 2i) = (2 + 6) + (3 - 2)i = 8 + i$

**(b)** $(2 + 3i) - (6 - 2i) = (2 - 6) + (3 - (-2))i = -4 + 5i$

**(c)** $(2 + 3i)(6 - 2i) = (2)(6) + (2)(-2i) + (3i)(6) + (3i)(-2i)$
$$= 12 - 4i + 18i - 6i^2 = 12 + 14i + 6 = 18 + 14i$$

**(d)** $\dfrac{2 + 3i}{6 - 2i} = \dfrac{2 + 3i}{6 - 2i}\dfrac{6 + 2i}{6 + 2i}$

$$= \frac{12 + 4i + 18i + 6i^2}{36 + 12i - 12i - 4i^2}$$

$$= \frac{6 + 22i}{40} = \frac{3}{20} + \frac{11}{20}i \qquad\blacksquare$$

## Argand Diagrams

There are two geometric representations of the complex number $z = x + iy$:

**1.** as the point $P(x, y)$ in the $xy$-plane

**2.** as the vector $\overrightarrow{OP}$ from the origin to $P$.

In each representation, the $x$-axis is called the **real axis** and the $y$-axis is the **imaginary axis**. Both representations are **Argand diagrams** for $x + iy$ (Figure A.26).

In terms of the polar coordinates of $x$ and $y$, we have

$$x = r \cos \theta, \qquad y = r \sin \theta,$$

and

$$z = x + iy = r(\cos \theta + i \sin \theta). \tag{10}$$

We define the **absolute value** of a complex number $x + iy$ to be the length $r$ of a vector $\overrightarrow{OP}$ from the origin to $P(x, y)$. We denote the absolute value by vertical bars; thus,

$$|x + iy| = \sqrt{x^2 + y^2}.$$

If we always choose the polar coordinates $r$ and $\theta$ so that $r$ is nonnegative, then

$$r = |x + iy|.$$

The polar angle $\theta$ is called the **argument** of $z$ and is written $\theta = \arg z$. Of course, any integer multiple of $2\pi$ may be added to $\theta$ to produce another appropriate angle.

The following equation gives a useful formula connecting a complex number $z$, its conjugate $\bar{z}$, and its absolute value $|z|$, namely,

$$z \cdot \bar{z} = |z|^2.$$

## Euler's Formula

The identity

$$e^{i\theta} = \cos \theta + i \sin \theta,$$

called **Euler's formula**, enables us to rewrite Equation (10) as

$$z = re^{i\theta}.$$

This formula, in turn, leads to the following rules for calculating products, quotients, powers, and roots of complex numbers. It also leads to Argand diagrams for $e^{i\theta}$. Since $\cos \theta + i \sin \theta$ is what we get from Equation (10) by taking $r = 1$, we can say that $e^{i\theta}$ is represented by a unit vector that makes an angle $\theta$ with the positive $x$-axis, as shown in Figure A.27.

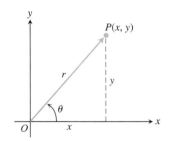

**FIGURE A.26** This Argand diagram represents $z = x + iy$ both as a point $P(x, y)$ and as a vector $\overrightarrow{OP}$.

The notation $\exp(A)$ also stands for $e^A$.

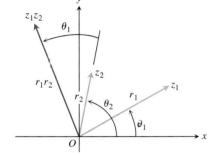

**FIGURE A.27** Argand diagrams for $e^{i\theta} = \cos\theta + i\sin\theta$ (a) as a vector and (b) as a point.

## Products

To multiply two complex numbers, we multiply their absolute values and add their angles. Let

$$z_1 = r_1 e^{i\theta_1}, \qquad z_2 = r_2 e^{i\theta_2}, \tag{11}$$

so that

$$|z_1| = r_1, \qquad \arg z_1 = \theta_1; \qquad |z_2| = r_2, \qquad \arg z_2 = \theta_2.$$

Then

$$z_1 z_2 = r_1 e^{i\theta_1} \cdot r_2 e^{i\theta_2} = r_1 r_2 e^{i(\theta_1 + \theta_2)}$$

and hence

$$\begin{aligned} |z_1 z_2| &= r_1 r_2 = |z_1| \cdot |z_2| \\ \arg(z_1 z_2) &= \theta_1 + \theta_2 = \arg z_1 + \arg z_2. \end{aligned} \tag{12}$$

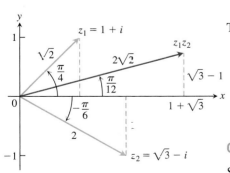

**FIGURE A.28** When $z_1$ and $z_2$ are multiplied, $|z_1 z_2| = r_1 \cdot r_2$ and $\arg(z_1 z_2) = \theta_1 + \theta_2$.

Thus, the product of two complex numbers is represented by a vector whose length is the product of the lengths of the two factors and whose argument is the sum of their arguments (Figure A.28). In particular, from Equation (12) a vector may be rotated counterclockwise through an angle $\theta$ by multiplying it by $e^{i\theta}$. Multiplication by $i$ rotates 90°, by $-1$ rotates 180°, by $-i$ rotates 270°, and so on.

**EXAMPLE 2**    Let $z_1 = 1 + i, z_2 = \sqrt{3} - i$. We plot these complex numbers in an Argand diagram (Figure A.29) from which we read off the polar representations

$$z_1 = \sqrt{2}e^{i\pi/4}, \qquad z_2 = 2e^{-i\pi/6}.$$

Then

$$z_1 z_2 = 2\sqrt{2}\exp\left(\frac{i\pi}{4} - \frac{i\pi}{6}\right) = 2\sqrt{2}\exp\left(\frac{i\pi}{12}\right)$$

$$= 2\sqrt{2}\left(\cos\frac{\pi}{12} + i\sin\frac{\pi}{12}\right) \approx 2.73 + 0.73i. \quad\blacksquare$$

## Quotients

Suppose $r_2 \neq 0$ in Equation (11). Then

$$\frac{z_1}{z_2} = \frac{r_1 e^{i\theta_1}}{r_2 e^{i\theta_2}} = \frac{r_1}{r_2}e^{i(\theta_1 - \theta_2)}.$$

**FIGURE A.29** To multiply two complex numbers, multiply their absolute values and add their arguments.

Hence

$$\left|\frac{z_1}{z_2}\right| = \frac{r_1}{r_2} = \frac{|z_1|}{|z_2|} \quad \text{and} \quad \arg\left(\frac{z_1}{z_2}\right) = \theta_1 - \theta_2 = \arg z_1 - \arg z_2.$$

That is, we divide lengths and subtract angles for the quotient of complex numbers.

**EXAMPLE 3**    Let $z_1 = 1 + i$ and $z_2 = \sqrt{3} - i$, as in Example 2. Then

$$\frac{1 + i}{\sqrt{3} - i} = \frac{\sqrt{2}e^{i\pi/4}}{2e^{-i\pi/6}} = \frac{\sqrt{2}}{2}e^{5\pi i/12} \approx 0.707\left(\cos\frac{5\pi}{12} + i\sin\frac{5\pi}{12}\right)$$

$$\approx 0.183 + 0.683i. \qquad \blacksquare$$

### Powers

If $n$ is a positive integer, we may apply the product formulas in Equation (12) to find

$$z^n = z \cdot z \cdot \cdots \cdot z. \qquad n \text{ factors}$$

With $z = re^{i\theta}$, we obtain

$$z^n = (re^{i\theta})^n = r^n e^{i(\theta + \theta + \cdots + \theta)} \qquad n \text{ summands}$$

$$= r^n e^{in\theta}. \qquad (13)$$

The length $r = |z|$ is raised to the $n$th power and the angle $\theta = \arg z$ is multiplied by $n$.

If we take $r = 1$ in Equation (13), we obtain De Moivre's Theorem.

---

**De Moivre's Theorem**

$$(\cos\theta + i\sin\theta)^n = \cos n\theta + i\sin n\theta. \qquad (14)$$

---

If we expand the left side of De Moivre's equation above by the Binomial Theorem and reduce it to the form $a + ib$, we obtain formulas for $\cos n\theta$ and $\sin n\theta$ as polynomials of degree $n$ in $\cos\theta$ and $\sin\theta$.

**EXAMPLE 4**    If $n = 3$ in Equation (14), we have

$$(\cos\theta + i\sin\theta)^3 = \cos 3\theta + i\sin 3\theta.$$

The left side of this equation expands to

$$\cos^3\theta + 3i\cos^2\theta\sin\theta - 3\cos\theta\sin^2\theta - i\sin^3\theta.$$

The real part of this must equal $\cos 3\theta$ and the imaginary part must equal $\sin 3\theta$. Therefore,

$$\cos 3\theta = \cos^3\theta - 3\cos\theta\sin^2\theta,$$
$$\sin 3\theta = 3\cos^2\theta\sin\theta - \sin^3\theta. \qquad \blacksquare$$

### Roots

If $z = re^{i\theta}$ is a complex number different from zero and $n$ is a positive integer, then there are precisely $n$ different complex numbers $w_0, w_1, \ldots, w_{n-1}$, that are $n$th roots of $z$. To see why, let $w = \rho e^{i\alpha}$ be an $n$th root of $z = re^{i\theta}$, so that

$$w^n = z$$

or

$$\rho^n e^{in\alpha} = re^{i\theta}.$$

Then

$$\rho = \sqrt[n]{r}$$

is the real, positive $n$th root of $r$. For the argument, although we cannot say that $n\alpha$ and $\theta$ must be equal, we can say that they may differ only by an integer multiple of $2\pi$. That is,

$$n\alpha = \theta + 2k\pi, \qquad k = 0, \pm 1, \pm 2, \dots.$$

Therefore,

$$\alpha = \frac{\theta}{n} + k\frac{2\pi}{n}.$$

Hence, all the $n$th roots of $z = re^{i\theta}$ are given by

$$\sqrt[n]{re^{i\theta}} = \sqrt[n]{r} \exp i\left(\frac{\theta}{n} + k\frac{2\pi}{n}\right), \qquad k = 0, \pm 1, \pm 2, \dots. \qquad (15)$$

There might appear to be infinitely many different answers corresponding to the infinitely many possible values of $k$, but $k = n + m$ gives the same answer as $k = m$ in Equation (15). Thus, we need only take $n$ consecutive values for $k$ to obtain all the different $n$th roots of $z$. For convenience, we take

$$k = 0, 1, 2, \dots, n - 1.$$

All the $n$th roots of $re^{i\theta}$ lie on a circle centered at the origin and having radius equal to the real, positive $n$th root of $r$. One of them has argument $\alpha = \theta/n$. The others are uniformly spaced around the circle, each being separated from its neighbors by an angle equal to $2\pi/n$. Figure A.30 illustrates the placement of the three cube roots, $w_0, w_1, w_2$, of the complex number $z = re^{i\theta}$.

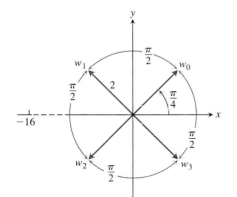

**FIGURE A.30** The three cube roots of $z = re^{i\theta}$.

**EXAMPLE 5** Find the four fourth roots of $-16$.

**Solution** As our first step, we plot the number $-16$ in an Argand diagram (Figure A.31) and determine its polar representation $re^{i\theta}$. Here, $z = -16$, $r = +16$, and $\theta = \pi$. One of the fourth roots of $16e^{i\pi}$ is $2e^{i\pi/4}$. We obtain others by successive additions of $2\pi/4 = \pi/2$ to the argument of this first one. Hence,

$$\sqrt[4]{16 \exp i\pi} = 2 \exp i\left(\frac{\pi}{4}, \frac{3\pi}{4}, \frac{5\pi}{4}, \frac{7\pi}{4}\right),$$

and the four roots are

$$w_0 = 2\left[\cos\frac{\pi}{4} + i\sin\frac{\pi}{4}\right] = \sqrt{2}(1 + i)$$

$$w_1 = 2\left[\cos\frac{3\pi}{4} + i\sin\frac{3\pi}{4}\right] = \sqrt{2}(-1 + i)$$

$$w_2 = 2\left[\cos\frac{5\pi}{4} + i\sin\frac{5\pi}{4}\right] = \sqrt{2}(-1 - i)$$

**FIGURE A.31** The four fourth roots of $-16$.

$$w_3 = 2\left[\cos\frac{7\pi}{4} + i\sin\frac{7\pi}{4}\right] = \sqrt{2}(1 - i).$$

### The Fundamental Theorem of Algebra

One might say that the invention of $\sqrt{-1}$ is all well and good and leads to a number system that is richer than the real number system alone; but where will this process end? Are we also going to invent still more systems so as to obtain $\sqrt[4]{-1}$, $\sqrt[6]{-1}$, and so on? But it turns out this is not necessary. These numbers are already expressible in terms of the complex number system $a + ib$. In fact, the Fundamental Theorem of Algebra says that with the introduction of the complex numbers we now have enough numbers to factor every polynomial into a product of linear factors and so enough numbers to solve every possible polynomial equation.

---

**The Fundamental Theorem of Algebra**

Every polynomial equation of the form

$$a_n z^n + a_{n-1} z^{n-1} + \cdots + a_1 z + a_0 = 0,$$

in which the coefficients $a_0, a_1, \ldots, a_n$ are any complex numbers, whose degree $n$ is greater than or equal to one, and whose leading coefficient $a_n$ is not zero, has exactly $n$ roots in the complex number system, provided each multiple root of multiplicity $m$ is counted as $m$ roots.

---

A proof of this theorem can be found in almost any text on the theory of functions of a complex variable.

## Exercises A.7

### Operations with Complex Numbers

1. **How computers multiply complex numbers** Find $(a, b) \cdot (c, d) = (ac - bd, ad + bc)$.

   **a.** $(2, 3) \cdot (4, -2)$   **b.** $(2, -1) \cdot (-2, 3)$

   **c.** $(-1, -2) \cdot (2, 1)$

   (This is how complex numbers are multiplied by computers.)

2. Solve the following equations for the real numbers, $x$ and $y$.

   **a.** $(3 + 4i)^2 - 2(x - iy) = x + iy$

   **b.** $\left(\dfrac{1 - i}{1 - i}\right)^2 + \dfrac{1}{x + iy} = 1 + i$

   **c.** $(3 - 2i)(x + iy) = 2(x - 2iy) + 2i - 1$

### Graphing and Geometry

3. How may the following complex numbers be obtained from $z = x - iy$ geometrically? Sketch.

   **a.** $\bar{z}$   **b.** $\overline{(-z)}$

   **c.** $-z$   **d.** $1/z$

4. Show that the distance between the two points $z_1$ and $z_2$ in an Argand diagram is $|z_1 - z_2|$.

In Exercises 5–10, graph the points $z = x + iy$ that satisfy the given conditions.

5. **a.** $|z| = 2$   **b.** $|z| < 2$   **c.** $|z| > 2$

6. $|z - 1| = 2$   7. $|z + 1| = 1$

8. $|z + 1| = |z - 1|$   9. $|z + i| = |z - 1|$

10. $|z + 1| \geq |z|$

Express the complex numbers in Exercises 11–14 in the form $re^{i\theta}$, with $r \geq 0$ and $-\pi < \theta \leq \pi$. Draw an Argand diagram for each calculation.

11. $\left(1 + \sqrt{-3}\right)^2$   12. $\dfrac{1 + i}{1 - i}$

13. $\dfrac{1 + i\sqrt{3}}{1 - i\sqrt{3}}$   14. $(2 + 3i)(1 - 2i)$

### Powers and Roots

Use De Moivre's Theorem to express the trigonometric functions in Exercises 15 and 16 in terms of $\cos \theta$ and $\sin \theta$.

15. $\cos 4\theta$   16. $\sin 4\theta$

17. Find the three cube roots of 1.

18. Find the two square roots of $i$.

19. Find the three cube roots of $-8i$.

20. Find the six sixth roots of 64.

21. Find the four solutions of the equation $z^4 - 2z^2 + 4 = 0$.

22. Find the six solutions of the equation $z^6 + 2z^3 + 2 = 0$.

23. Find all solutions of the equation $x^4 + 4x^2 + 16 = 0$.

24. Solve the equation $x^4 + 1 = 0$.

### Theory and Examples

25. **Complex numbers and vectors in the plane** Show with an Argand diagram that the law for adding complex numbers is the same as the parallelogram law for adding vectors.

**26. Complex arithmetic with conjugates** Show that the conjugate of the sum (product, or quotient) of two complex numbers, $z_1$ and $z_2$, is the same as the sum (product, or quotient) of their conjugates.

**27. Complex roots of polynomials with real coefficients come in complex-conjugate pairs**

a. Extend the results of Exercise 26 to show that $f(\bar{z}) = \overline{f(z)}$ if

$$f(z) = a_n z^n + a_{n-1} z^{n-1} + \cdots + a_1 z + a_0$$

is a polynomial with real coefficients $a_0, \ldots, a_n$.

b. If $z$ is a root of the equation $f(z) = 0$, where $f(z)$ is a polynomial with real coefficients as in part (a), show that the conjugate $\bar{z}$ is also a root of the equation. (*Hint:* Let $f(z) = u + iv = 0$; then both $u$ and $v$ are zero. Use the fact that $f(\bar{z}) = \overline{f(z)} = u - iv$.)

**28. Absolute value of a conjugate** Show that $|\bar{z}| = |z|$.

**29. When $z = \bar{z}$** If $z$ and $\bar{z}$ are equal, what can you say about the location of the point $z$ in the complex plane?

**30. Real and imaginary parts** Let $\text{Re}(z)$ denote the real part of $z$ and $\text{Im}(z)$ the imaginary part. Show that the following relations hold for any complex numbers $z$, $z_1$, and $z_2$.

a. $z + \bar{z} = 2\text{Re}(z)$

b. $z - \bar{z} = 2i\text{Im}(z)$

c. $|\text{Re}(z)| \leq |z|$

d. $|z_1 + z_2|^2 = |z_1|^2 + |z_2|^2 + 2\text{Re}(z_1 \bar{z}_2)$

e. $|z_1 + z_2| \leq |z_1| + |z_2|$

# A.8 The Distributive Law for Vector Cross Products

In this appendix we prove the Distributive Law

$$\mathbf{u} \times (\mathbf{v} + \mathbf{w}) = \mathbf{u} \times \mathbf{v} + \mathbf{u} \times \mathbf{w},$$

which is Property 2 in Section 12.4.

**Proof** To derive the Distributive Law, we construct $\mathbf{u} \times \mathbf{v}$ a new way. We draw $\mathbf{u}$ and $\mathbf{v}$ from the common point $O$ and construct a plane $M$ perpendicular to $\mathbf{u}$ at $O$ (Figure A.32). We then project $\mathbf{v}$ orthogonally onto $M$, yielding a vector $\mathbf{v}'$ with length $|\mathbf{v}|\sin\theta$. We rotate $\mathbf{v}'$ 90° about $\mathbf{u}$ in the positive sense to produce a vector $\mathbf{v}''$. Finally, we multiply $\mathbf{v}''$ by the length of $\mathbf{u}$. The resulting vector $|\mathbf{u}|\mathbf{v}''$ is equal to $\mathbf{u} \times \mathbf{v}$ since $\mathbf{v}''$ has the same direction as $\mathbf{u} \times \mathbf{v}$ by its construction (Figure A.32) and

$$|\mathbf{u}||\mathbf{v}''| = |\mathbf{u}||\mathbf{v}'| = |\mathbf{u}||\mathbf{v}|\sin\theta = |\mathbf{u} \times \mathbf{v}|.$$

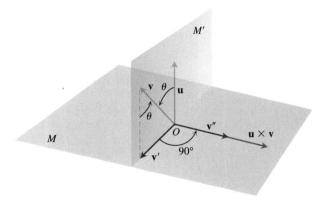

**FIGURE A.32** As explained in the text, $\mathbf{u} \times \mathbf{v} = |\mathbf{u}|\mathbf{v}''$. (The primes used here are purely notational and do not denote derivatives.)

Now each of these three operations, namely,

1. projection onto $M$

2. rotation about $\mathbf{u}$ through 90°

3. multiplication by the scalar $|\mathbf{u}|$

when applied to a triangle whose plane is not parallel to **u**, will produce another triangle. If we start with the triangle whose sides are **v**, **w**, and **v** + **w** (Figure A.33) and apply these three steps, we successively obtain the following:

1.  A triangle whose sides are **v′**, **w′**, and (**v** + **w**)′ satisfying the vector equation

$$\mathbf{v'} + \mathbf{w'} = (\mathbf{v} + \mathbf{w})'$$

2.  A triangle whose sides are **v″**, **w″**, and (**v** + **w**)″ satisfying the vector equation

$$\mathbf{v''} + \mathbf{w''} = (\mathbf{v} + \mathbf{w})''$$

(The double prime on each vector has the same meaning as in Figure A.32.)

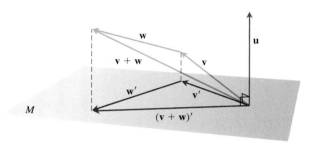

**FIGURE A.33**  The vectors, **v**, **w**, **v** + **w**, and their projections onto a plane perpendicular to **u**.

3.  A triangle whose sides are $|\mathbf{u}|\mathbf{v''}$, $|\mathbf{u}|\mathbf{w''}$, and $|\mathbf{u}|(\mathbf{v} + \mathbf{w})''$ satisfying the vector equation

$$|\mathbf{u}|\mathbf{v''} + |\mathbf{u}|\mathbf{w''} = |\mathbf{u}|(\mathbf{v} + \mathbf{w})''.$$

Substituting $|\mathbf{u}|\mathbf{v''} = \mathbf{u} \times \mathbf{v}$, $|\mathbf{u}|\mathbf{w''} = \mathbf{u} \times \mathbf{w}$, and $|\mathbf{u}|(\mathbf{v} + \mathbf{w})'' = \mathbf{u} \times (\mathbf{v} + \mathbf{w})$ from our discussion above into this last equation gives

$$\mathbf{u} \times \mathbf{v} + \mathbf{u} \times \mathbf{w} = \mathbf{u} \times (\mathbf{v} + \mathbf{w}),$$

which is the law we wanted to establish. ∎

# A.9 The Mixed Derivative Theorem and the Increment Theorem

This appendix derives the Mixed Derivative Theorem (Theorem 2, Section 14.3) and the Increment Theorem for Functions of Two Variables (Theorem 3, Section 14.3). Euler first published the Mixed Derivative Theorem in 1734, in a series of papers he wrote on hydrodynamics.

> **THEOREM 2—The Mixed Derivative Theorem**  If $f(x, y)$ and its partial derivatives $f_x$, $f_y$, $f_{xy}$, and $f_{yx}$ are defined throughout an open region containing a point $(a, b)$ and are all continuous at $(a, b)$, then
>
> $$f_{xy}(a, b) = f_{yx}(a, b).$$

**Proof**   The equality of $f_{xy}(a, b)$ and $f_{yx}(a, b)$ can be established by four applications of the Mean Value Theorem (Theorem 4, Section 4.2). By hypothesis, the point $(a, b)$ lies in the interior of a rectangle $R$ in the $xy$-plane on which $f$, $f_x$, $f_y$, $f_{xy}$, and $f_{yx}$ are all defined.

We let $h$ and $k$ be the numbers such that the point $(a + h, b + k)$ also lies in $R$, and we consider the difference

$$\Delta = F(a + h) - F(a), \tag{1}$$

where

$$F(x) = f(x, b + k) - f(x, b). \tag{2}$$

We apply the Mean Value Theorem to $F$, which is continuous because it is differentiable. Then Equation (1) becomes

$$\Delta = hF'(c_1), \tag{3}$$

where $c_1$ lies between $a$ and $a + h$. From Equation (2),

$$F'(x) = f_x(x, b + k) - f_x(x, b),$$

so Equation (3) becomes

$$\Delta = h[f_x(c_1, b + k) - f_x(c_1, b)]. \tag{4}$$

Now we apply the Mean Value Theorem to the function $g(y) = f_x(c_1, y)$ and have

$$g(b + k) - g(b) = kg'(d_1),$$

or

$$f_x(c_1, b + k) - f_x(c_1, b) = kf_{xy}(c_1, d_1)$$

for some $d_1$ between $b$ and $b + k$. By substituting this into Equation (4), we get

$$\Delta = hkf_{xy}(c_1, d_1) \tag{5}$$

for some point $(c_1, d_1)$ in the rectangle $R'$ whose vertices are the four points $(a, b)$, $(a + h, b)$, $(a + h, b + k)$, and $(a, b + k)$. (See Figure A.34.)

By substituting from Equation (2) into Equation (1), we may also write

$$\begin{aligned}
\Delta &= f(a + h, b + k) - f(a + h, b) - f(a, b + k) + f(a, b) \\
&= [f(a + h, b + k) - f(a, b + k)] - [f(a + h, b) - f(a, b)] \\
&= \phi(b + k) - \phi(b), \tag{6}
\end{aligned}$$

where

$$\phi(y) = f(a + h, y) - f(a, y). \tag{7}$$

The Mean Value Theorem applied to Equation (6) now gives

$$\Delta = k\phi'(d_2) \tag{8}$$

for some $d_2$ between $b$ and $b + k$. By Equation (7),

$$\phi'(y) = f_y(a + h, y) - f_y(a, y). \tag{9}$$

Substituting from Equation (9) into Equation (8) gives

$$\Delta = k[f_y(a + h, d_2) - f_y(a, d_2)].$$

Finally, we apply the Mean Value Theorem to the expression in brackets and get

$$\Delta = khf_{yx}(c_2, d_2) \tag{10}$$

for some $c_2$ between $a$ and $a + h$.

Together, Equations (5) and (10) show that

$$f_{xy}(c_1, d_1) = f_{yx}(c_2, d_2), \tag{11}$$

where $(c_1, d_1)$ and $(c_2, d_2)$ both lie in the rectangle $R'$ (Figure A.34). Equation (11) is not quite the result we want, since it says only that $f_{xy}$ has the same value at $(c_1, d_1)$ that $f_{yx}$ has at $(c_2, d_2)$. The numbers $h$ and $k$ in our discussion, however, may be made as small as

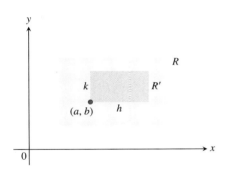

**FIGURE A.34** The key to proving $f_{xy}(a, b) = f_{yx}(a, b)$ is that no matter how small $R'$ is, $f_{xy}$ and $f_{yx}$ take on equal values somewhere inside $R'$ (although not necessarily at the same point).

we wish. The hypothesis that $f_{xy}$ and $f_{yx}$ are both continuous at $(a, b)$ means that $f_{xy}(c_1, d_1) = f_{xy}(a, b) + \epsilon_1$ and $f_{yx}(c_2, d_2) = f_{yx}(a, b) + \epsilon_2$, where each of $\epsilon_1, \epsilon_2 \to 0$ as both $h, k \to 0$. Hence, if we let $h$ and $k \to 0$, we have $f_{xy}(a, b) = f_{yx}(a, b)$. ∎

The equality of $f_{xy}(a, b)$ and $f_{yx}(a, b)$ can be proved with hypotheses weaker than the ones we assumed. For example, it is enough for $f$, $f_x$, and $f_y$ to exist in $R$ and for $f_{xy}$ to be continuous at $(a, b)$. Then $f_{yx}$ will exist at $(a, b)$ and equal $f_{xy}$ at that point.

---

**THEOREM 3—The Increment Theorem for Functions of Two Variables**
Suppose that the first partial derivatives of $f(x, y)$ are defined throughout an open region $R$ containing the point $(x_0, y_0)$ and that $f_x$ and $f_y$ are continuous at $(x_0, y_0)$. Then the change

$$\Delta z = f(x_0 + \Delta x, y_0 + \Delta y) - f(x_0, y_0)$$

in the value of $f$ that results from moving from $(x_0, y_0)$ to another point $(x_0 + \Delta x, y_0 + \Delta y)$ in $R$ satisfies an equation of the form

$$\Delta z = f_x(x_0, y_0) \Delta x + f_y(x_0, y_0) \Delta y + \epsilon_1 \Delta x + \epsilon_2 \Delta y$$

in which each of $\epsilon_1, \epsilon_2 \to 0$ as both $\Delta x, \Delta y \to 0$.

---

**FIGURE A.35** The rectangular region $T$ in the proof of the Increment Theorem. The figure is drawn for $\Delta x$ and $\Delta y$ positive, but either increment might be zero or negative.

**Proof**    We work within a rectangle $T$ centered at $A(x_0, y_0)$ and lying within $R$, and we assume that $\Delta x$ and $\Delta y$ are already so small that the line segment joining $A$ to $B(x_0 + \Delta x, y_0)$ and the line segment joining $B$ to $C(x_0 + \Delta x, y_0 + \Delta y)$ lie in the interior of $T$ (Figure A.35).

We may think of $\Delta z$ as the sum $\Delta z = \Delta z_1 + \Delta z_2$ of two increments, where

$$\Delta z_1 = f(x_0 + \Delta x, y_0) - f(x_0, y_0)$$

is the change in the value of $f$ from $A$ to $B$ and

$$\Delta z_2 = f(x_0 + \Delta x, y_0 + \Delta y) - f(x_0 + \Delta x, y_0)$$

is the change in the value of $f$ from $B$ to $C$ (Figure A.36).

On the closed interval of $x$-values joining $x_0$ to $x_0 + \Delta x$, the function $F(x) = f(x, y_0)$ is a differentiable (and hence continuous) function of $x$, with derivative

$$F'(x) = f_x(x, y_0).$$

By the Mean Value Theorem (Theorem 4, Section 4.2), there is an $x$-value $c$ between $x_0$ and $x_0 + \Delta x$ at which

$$F(x_0 + \Delta x) - F(x_0) = F'(c) \Delta x$$

or

$$f(x_0 + \Delta x, y_0) - f(x_0, y_0) = f_x(c, y_0) \Delta x$$

or

$$\Delta z_1 = f_x(c, y_0) \Delta x. \tag{12}$$

Similarly, $G(y) = f(x_0 + \Delta x, y)$ is a differentiable (and hence continuous) function of $y$ on the closed $y$-interval joining $y_0$ and $y_0 + \Delta y$, with derivative

$$G'(y) = f_y(x_0 + \Delta x, y).$$

Hence, there is a $y$-value $d$ between $y_0$ and $y_0 + \Delta y$ at which

$$G(y_0 + \Delta y) - G(y_0) = G'(d) \Delta y$$

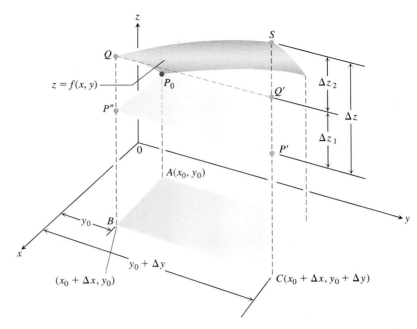

**FIGURE A.36** Part of the surface $z = f(x, y)$ near $P_0(x_0, y_0, f(x_0, y_0))$. The points $P_0$, $P'$, and $P''$ have the same height $z_0 = f(x_0, y_0)$ above the $xy$-plane. The change in $z$ is $\Delta z = P'S$. The change

$$\Delta z_1 = f(x_0 + \Delta x, y_0) - f(x_0, y_0),$$

shown as $P''Q = P'Q'$, is caused by changing $x$ from $x_0$ to $x_0 + \Delta x$ while holding $y$ equal to $y_0$. Then, with $x$ held equal to $x_0 + \Delta x$,

$$\Delta z_2 = f(x_0 + \Delta x, y_0 + \Delta y) - f(x_0 + \Delta x, y_0)$$

is the change in $z$ caused by changing $y_0$ from $y_0 + \Delta y$, which is represented by $Q'S$. The total change in $z$ is the sum of $\Delta z_1$ and $\Delta z_2$.

or

$$f(x_0 + \Delta x, y_0 + \Delta y) - f(x_0 + \Delta x, y) = f_y(x_0 + \Delta x, d)\, \Delta y$$

or

$$\Delta z_2 = f_y(x_0 + \Delta x, d)\, \Delta y. \tag{13}$$

Now, as both $\Delta x$ and $\Delta y \to 0$, we know that $c \to x_0$ and $d \to y_0$. Therefore, since $f_x$ and $f_y$ are continuous at $(x_0, y_0)$, the quantities

$$
\begin{aligned}
\epsilon_1 &= f_x(c, y_0) - f_x(x_0, y_0), \\
\epsilon_2 &= f_y(x_0 + \Delta x, d) - f_y(x_0, y_0)
\end{aligned}
\tag{14}
$$

both approach zero as both $\Delta x$ and $\Delta y \to 0$.

Finally,

$$
\begin{aligned}
\Delta z &= \Delta z_1 + \Delta z_2 \\
&= f_x(c, y_0)\Delta x + f_y(x_0 + \Delta x, d)\Delta y && \text{From Eqs.} \\
& && \text{(12) and (13)} \\
&= \left[ f_x(x_0, y_0) + \epsilon_1 \right]\Delta x + \left[ f_y(x_0, y_0) + \epsilon_2 \right]\Delta y && \text{From Eq. (14)} \\
&= f_x(x_0, y_0)\Delta x + f_y(x_0, y_0)\Delta y + \epsilon_1\Delta x + \epsilon_2\Delta y,
\end{aligned}
$$

where both $\epsilon_1$ and $\epsilon_2 \to 0$ as both $\Delta x$ and $\Delta y \to 0$, which is what we set out to prove. ∎

Analogous results hold for functions of any finite number of independent variables. Suppose that the first partial derivatives of $w = f(x, y, z)$ are defined throughout an open region containing the point $(x_0, y_0, z_0)$ and that $f_x$, $f_y$, and $f_z$ are continuous at $(x_0, y_0, z_0)$. Then

$$\Delta w = f(x_0 + \Delta x, y_0 + \Delta y, z_0 + \Delta z) - f(x_0, y_0, z_0)$$
$$= f_x \Delta x + f_y \Delta y + f_z \Delta z + \epsilon_1 \Delta x + \epsilon_2 \Delta y + \epsilon_3 \Delta z, \tag{15}$$

where $\epsilon_1, \epsilon_2, \epsilon_3 \to 0$ as $\Delta x$, $\Delta y$, and $\Delta z \to 0$.

The partial derivatives $f_x, f_y, f_z$ in Equation (15) are to be evaluated at the point $(x_0, y_0, z_0)$.

Equation (15) can be proved by treating $\Delta w$ as the sum of three increments,

$$\Delta w_1 = f(x_0 + \Delta x, y_0, z_0) - f(x_0, y_0, z_0) \tag{16}$$
$$\Delta w_2 = f(x_0 + \Delta x, y_0 + \Delta y, z_0) - f(x_0 + \Delta x, y_0, z_0) \tag{17}$$
$$\Delta w_3 = f(x_0 + \Delta x, y_0 + \Delta y, z_0 + \Delta z) - f(x_0 + \Delta x, y_0 + \Delta y, z_0), \tag{18}$$

and applying the Mean Value Theorem to each of these separately. Two coordinates remain constant and only one varies in each of these partial increments $\Delta w_1$, $\Delta w_2$, $\Delta w_3$. In Equation (17), for example, only $y$ varies, since $x$ is held equal to $x_0 + \Delta x$ and $z$ is held equal to $z_0$. Since $f(x_0 + \Delta x, y, z_0)$ is a continuous function of $y$ with a derivative $f_y$, it is subject to the Mean Value Theorem, and we have

$$\Delta w_2 = f_y(x_0 + \Delta x, y_1, z_0) \Delta y$$

for some $y_1$ between $y_0$ and $y_0 + \Delta y$.

## Chapter 1

### Section 1.1, pp. 25–27

**1.** $D: (-\infty, \infty)$, $R: [1, \infty)$    **3.** $D: [-2, \infty)$, $R: [0, \infty)$

**5.** $D: (-\infty, 3) \cup (3, \infty)$, $R: (-\infty, 0) \cup (0, \infty)$

**7. (a)** Not a function of $x$ because some values of $x$ have two values of $y$

   **(b)** A function of $x$ because for every $x$ there is only one possible $y$

**9.** $A = \dfrac{\sqrt{3}}{4}x^2$, $p = 3x$   **11.** $x = \dfrac{d}{\sqrt{3}}$, $A = 2d^2$, $V = \dfrac{d^3}{3\sqrt{3}}$

**13.** $L = \dfrac{\sqrt{20x^2 - 20x + 25}}{4}$

**15.** $(-\infty, \infty)$      **17.** $(-\infty, \infty)$

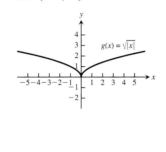

**19.** $(-\infty, 0) \cup (0, \infty)$

**21.** $(-\infty, -5) \cup (-5, -3] \cup [3, 5) \cup (5, \infty)$

**23. (a)** For each positive value of $x$, there are two values of $y$.    **(b)** For each value of $x \neq 0$, there are two values of $y$.

**25.**

**27.**

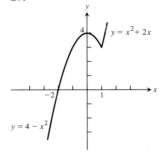

**29. (a)** $f(x) = \begin{cases} x, & 0 \le x \le 1 \\ -x + 2, & 1 < x \le 2 \end{cases}$

   **(b)** $f(x) = \begin{cases} 2, & 0 \le x < 1 \\ 0, & 1 \le x < 2 \\ 2, & 2 \le x < 3 \\ 0, & 3 \le x \le 4 \end{cases}$

**31. (a)** $f(x) = \begin{cases} -x, & -1 \le x < 0 \\ 1, & 0 < x \le 1 \\ -\frac{1}{2}x + \frac{3}{2}, & 1 < x < 3 \end{cases}$

   **(b)** $f(x) = \begin{cases} \frac{1}{2}x, & -2 \le x \le 0 \\ -2x + 2, & 0 < x \le 1 \\ -1, & 1 < x \le 3 \end{cases}$

**33. (a)** $0 \le x < 1$   **(b)** $-1 < x \le 0$     **35.** Yes

**37.** Symmetric about the origin     **39.** Symmetric about the origin

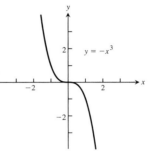

Dec. $-\infty < x < \infty$

Inc. $-\infty < x < 0$ and $0 < x < \infty$

**41.** Symmetric about the $y$-axis     **43.** Symmetric about the origin

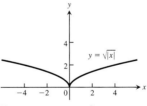

Dec. $-\infty < x \le 0$;
Inc. $0 \le x < \infty$

Inc. $-\infty < x < \infty$

**45.** No symmetry

Dec. $0 \le x < \infty$

**47.** Even    **49.** Even    **51.** Odd    **53.** Even

**55.** Neither    **57.** Neither    **59.** $t = 180$    **61.** $s = 2.4$

**63.** $V = x(14 - 2x)(22 - 2x)$

**65. (a)** $h$   **(b)** $f$   **(c)** $g$     **67. (a)** $(-2, 0) \cup (4, \infty)$

**71.** $C = 5(2 + \sqrt{2})h$

**Section 1.2, pp. 32–35**

**1.** $D_f : -\infty < x < \infty, D_g : x \geq 1, R_f : -\infty < y < \infty,$
$R_g : y \geq 0, D_{f+g} = D_{f \cdot g} = D_g, R_{f+g} : y \geq 1, R_{f \cdot g} : y \geq 0$

**3.** $D_f : -\infty < x < \infty, D_g : -\infty < x < \infty, R_f : y = 2,$
$R_g : y \geq 1, D_{f/g} : -\infty < x < \infty, R_{f/g} : 0 < y \leq 2,$
$D_{g/f} : -\infty < x < \infty, R_{g/f} : y \geq 1/2$

**5.** (a) 2   (b) 22   (c) $x^2 + 2$   (d) $x^2 + 10x + 22$   (e) 5
(f) $-2$   (g) $x + 10$   (h) $x^4 - 6x^2 + 6$

**7.** $13 - 3x$    **9.** $\sqrt{\dfrac{5x + 1}{4x + 1}}$

**11.** (a) $f(g(x))$   (b) $j(g(x))$   (c) $g(g(x))$   (d) $j(j(x))$
(e) $g(h(f(x)))$   (f) $h(j(f(x)))$

**13.**

| | $g(x)$ | $f(x)$ | $(f \circ g)(x)$ |
|---|---|---|---|
| (a) | $x - 7$ | $\sqrt{x}$ | $\sqrt{x - 7}$ |
| (b) | $x + 2$ | $3x$ | $3x + 6$ |
| (c) | $x^2$ | $\sqrt{x - 5}$ | $\sqrt{x^2 - 5}$ |
| (d) | $\dfrac{x}{x - 1}$ | $\dfrac{x}{x - 1}$ | $x$ |
| (e) | $\dfrac{1}{x - 1}$ | $1 + \dfrac{1}{x}$ | $x$ |
| (f) | $\dfrac{1}{x}$ | $\dfrac{1}{x}$ | $x$ |

**15.** (a) 1   (b) 2   (c) $-2$   (d) 0   (e) $-1$   (f) 0

**17.** (a) $f(g(x)) = \sqrt{\dfrac{1}{x} + 1}, \ g(f(x)) = \dfrac{1}{\sqrt{x + 1}}$
(b) $D_{f \circ g} = (-\infty, -1] \cup (0, \infty), D_{g \circ f} = (-1, \infty)$
(c) $R_{f \circ g} = [0, 1) \cup (1, \infty), R_{g \circ f} = (0, \infty)$

**19.** $g(x) = \dfrac{2x}{x - 1}$

**21.** (a) $y = -(x + 7)^2$   (b) $y = -(x - 4)^2$

**23.** (a) Position 4   (b) Position 1   (c) Position 2
(d) Position 3

**25.** $(x + 2)^2 + (y + 3)^2 = 49$    **27.** $y + 1 = (x + 1)^3$

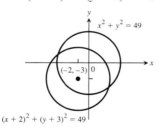

**29.** $y = \sqrt{x + 0.81}$    **31.** $y = 2x$

**33.** $y - 1 = \dfrac{1}{x - 1}$    **35.**

**37.**    **39.**

**41.**    **43.**

**45.**    **47.**

**49.**    **51.**

**53.**

**55.** (a) $D : [0, 2], \quad R : [2, 3]$   (b) $D : [0, 2], \quad R : [-1, 0]$

**(c)** $D: [0, 2], \quad R: [0, 2]$ **(d)** $D: [0, 2], \quad R: [-1, 0]$

$y = 2f(x)$

$y = -f(x)$

**(e)** $D: [-2, 0], \quad R: [0, 1]$ **(f)** $D: [1, 3], R: [0, 1]$

$y = f(x + 2)$

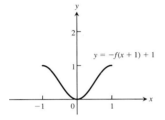

$y = f(x - 1)$

**(g)** $D: [-2, 0], \quad R: [0, 1]$ **(h)** $D: [-1, 1], \quad R: [0, 1]$

$y = f(-x)$

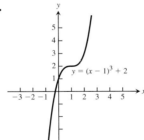

$y = -f(x + 1) + 1$

**57.** $y = 3x^2 - 3$ **59.** $y = \dfrac{1}{2} + \dfrac{1}{2x^2}$ **61.** $y = \sqrt{4x + 1}$

**63.** $y = \sqrt{4 - \dfrac{x^2}{4}}$ **65.** $y = 1 - 27x^3$

**67.**

$y = -\sqrt{2x + 1}$

**69.**

$y = (x - 1)^3 + 2$

**71.**

$y = \dfrac{1}{2^x} - 1$

**73.**

$y = -\sqrt[3]{x}$

**75.**

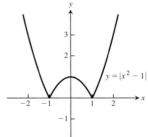

$y = |x^2 - 1|$

**77. (a)** Odd **(b)** Odd **(c)** Odd **(d)** Even **(e)** Even
**(f)** Even **(g)** Even **(h)** Even **(i)** Odd

Section 1.3, pp. 41–43

**1. (a)** $8\pi$ m **(b)** $\dfrac{55\pi}{9}$ m **3.** 8.4 cm (or 84 mm)

**5.**

| $\theta$ | $-\pi$ | $-2\pi/3$ | $0$ | $\pi/2$ | $3\pi/4$ |
|---|---|---|---|---|---|
| $\sin \theta$ | $0$ | $-\dfrac{\sqrt{3}}{2}$ | $0$ | $1$ | $\dfrac{1}{\sqrt{2}}$ |
| $\cos \theta$ | $-1$ | $-\dfrac{1}{2}$ | $1$ | $0$ | $-\dfrac{1}{\sqrt{2}}$ |
| $\tan \theta$ | $0$ | $\sqrt{3}$ | $0$ | UND | $-1$ |
| $\cot \theta$ | UND | $\dfrac{1}{\sqrt{3}}$ | UND | $0$ | $-1$ |
| $\sec \theta$ | $-1$ | $-2$ | $1$ | UND | $-\sqrt{2}$ |
| $\csc \theta$ | UND | $-\dfrac{2}{\sqrt{3}}$ | UND | $1$ | $\sqrt{2}$ |

**7.** $\cos x = -4/5, \tan x = -3/4$

**9.** $\sin x = -\dfrac{\sqrt{8}}{3}, \tan x = -\sqrt{8}$

**11.** $\sin x = -\dfrac{1}{\sqrt{5}}, \cos x = -\dfrac{2}{\sqrt{5}}$

**13.** Period $\pi$ **15.** Period 2

$y = \sin 2x$

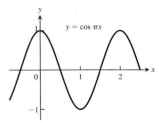

$y = \cos \pi x$

**17.** Period 6 **19.** Period $2\pi$

$y = -\sin \dfrac{\pi x}{3}$

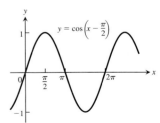

$y = \cos\left(x - \dfrac{\pi}{2}\right)$

**21.** Period $2\pi$

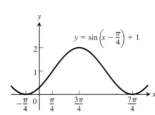

**23.** Period $\pi/2$, symmetric about the origin

**9.** $[-5, 5]$ by $[-6, 6]$

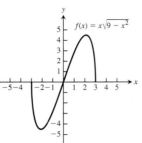

**11.** $[-2, 6]$ by $[-5, 4]$

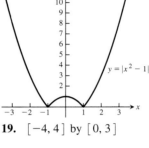

**25.** Period 4, symmetric about the $y$-axis

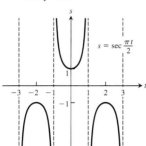

**29.** $D : (-\infty, \infty)$, $R : y = -1, 0, 1$

**13.** $[-2, 8]$ by $[-5, 10]$

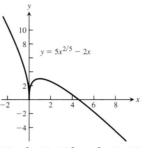

**15.** $[-3, 3]$ by $[0, 10]$

**17.** $[-10, 10]$ by $[-10, 10]$

**19.** $[-4, 4]$ by $[0, 3]$

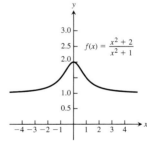

**39.** $-\cos x$    **41.** $-\cos x$    **43.** $\dfrac{\sqrt{6} + \sqrt{2}}{4}$    **45.** $\dfrac{\sqrt{2} + \sqrt{6}}{4}$

**47.** $\dfrac{2 + \sqrt{2}}{4}$    **49.** $\dfrac{2 - \sqrt{3}}{4}$    **51.** $\dfrac{\pi}{3}, \dfrac{2\pi}{3}, \dfrac{4\pi}{3}, \dfrac{5\pi}{3}$

**53.** $\dfrac{\pi}{6}, \dfrac{\pi}{2}, \dfrac{5\pi}{6}, \dfrac{3\pi}{2}$    **59.** $\sqrt{7} \approx 2.65$    **63.** $a = 1.464$

**65.** $A = 2, B = 2\pi$, $C = -\pi, D = -1$

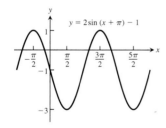

**67.** $A = -\dfrac{2}{\pi}, B = 4$, $C = 0, D = \dfrac{1}{\pi}$

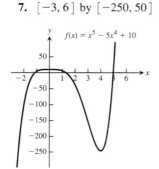

**21.** $[-10, 10]$ by $[-6, 6]$

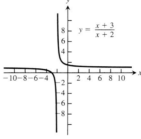

**23.** $[-6, 10]$ by $[-6, 6]$

**Section 1.4, pp. 48–50**

**1.** d    **3.** d

**5.** $[-3, 5]$ by $[-15, 40]$

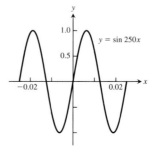

**7.** $[-3, 6]$ by $[-250, 50]$

**25.** $\left[-\dfrac{\pi}{125}, \dfrac{\pi}{125}\right]$ by $[-1.25, 1.25]$

**27.** $[-100\pi, 100\pi]$ by $[-1.25, 1.25]$

**29.** $\left[-\dfrac{\pi}{15}, \dfrac{\pi}{15}\right]$ by $\left[-0.25, 0.25\right]$   **31.**

**33.**                                         **35.**

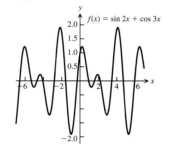

**37.** (a) & (b) $y = 0.55529x - 26.977$

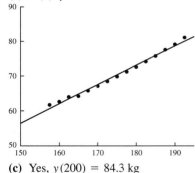

(c) Yes, $y(200) = 84.3$ kg

**39.** (a) & (b) $y = 3814x - 7.4988 \cdot 10^6$

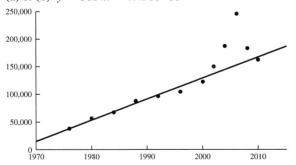

(c) The price of a home within the "bubble" was inflated, in the sense that it exceeded the historical trend.

**41.** (a) & (b) $y = 9.7571 \cdot 10^{-3}x - 19.118$

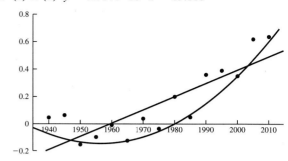

(c) $y = 2.6076 \cdot 10^{-4}x^2 - 1.0203x + 997.90$

Practice Exercises, pp. 50–52

**1.** $A = \pi r^2,\ C = 2\pi r,\ A = \dfrac{C^2}{4\pi}$   **3.** $x = \tan\theta,\ y = \tan^2\theta$

**5.** Origin   **7.** Neither   **9.** Even   **11.** Even
**13.** Odd   **15.** Neither
**17.** (a) Even   (b) Odd   (c) Odd   (d) Even   (e) Even
**19.** (a) Domain: all reals   (b) Range: $[-2, \infty)$
**21.** (a) Domain: $[-4, 4]$   (b) Range: $[0, 4]$
**23.** (a) Domain: all reals   (b) Range: $(-3, \infty)$
**25.** (a) Domain: all reals   (b) Range: $[-3, 1]$
**27.** (a) Domain: $(3, \infty)$   (b) Range: all reals
**29.** (a) Increasing   (b) Neither   (c) Decreasing   (d) Increasing
**31.** (a) Domain: $[-4, 4]$   (b) Range: $[0, 2]$

**33.** $f(x) = \begin{cases} 1 - x, & 0 \le x < 1 \\ 2 - x, & 1 \le x \le 2 \end{cases}$

**35.** (a) 1   (b) $\dfrac{1}{\sqrt{2.5}} = \sqrt{\dfrac{2}{5}}$   (c) $x, x \ne 0$

(d) $\dfrac{1}{\sqrt{1/\sqrt{x + 2} + 2}}$

**37.** (a) $(f \circ g)(x) = -x, x \ge -2, (g \circ f)(x) = \sqrt{4 - x^2}$
(b) Domain $(f \circ g)$: $[-2, \infty)$, domain $(g \circ f)$: $[-2, 2]$
(c) Range $(f \circ g)$: $(-\infty, 2]$, range $(g \circ f)$: $[0, 2]$

**39.**

**41.** Replace the portion for $x < 0$ with the mirror image of the portion for $x > 0$ to make the new graph symmetric with respect to the $y$-axis.

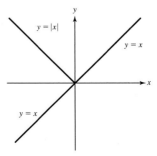

**43.** Reflects the portion for $y < 0$ across the $x$-axis
**45.** Reflects the portion for $y < 0$ across the $x$-axis
**47.** Adds the mirror image of the portion for $x > 0$ to make the new graph symmetric with respect to the $y$-axis

**49.** (a) $y = g(x - 3) + \frac{1}{2}$   (b) $y = g\left(x + \frac{2}{3}\right) - 2$

(c) $y = g(-x)$   (d) $y = -g(x)$   (e) $y = 5g(x)$
(f) $y = g(5x)$

**51.**

**53.**

**55.** Period $\pi$

**57.** Period 2

**59.**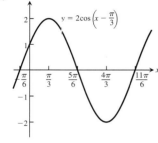

**61.** (a) $a = 1$   $b = \sqrt{3}$   (b) $a = 2\sqrt{3}/3$   $c = 4\sqrt{3}/3$

**63.** (a) $a = \dfrac{b}{\tan B}$   (b) $c = \dfrac{a}{\sin A}$

**65.** $\approx 16.98$ m     **67.** (b) $4\pi$

**1.** Yes. For instance: $f(x) = 1/x$ and $g(x) = 1/x$, or $f(x) = 2x$ and $g(x) = x/2$, or $f(x) = e^x$ and $g(x) = \ln x$.
**3.** If $f(x)$ is odd, then $g(x) = f(x) - 2$ is not odd. Nor is $g(x)$ even, unless $f(x) = 0$ for all $x$. If $f$ is even, then $g(x) = f(x) - 2$ is also even.

**5.**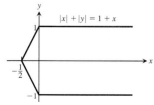

## Chapter 2

**1.** (a) 19     (b) 1

**3.** (a) $-\dfrac{4}{\pi}$   (b) $-\dfrac{3\sqrt{3}}{\pi}$   **5.** 1

**7.** (a) 4     (b) $y = 4x - 9$
**9.** (a) 2     (b) $y = 2x - 7$
**11.** (a) 12     (b) $y = 12x - 16$
**13.** (a) $-9$     (b) $y = -9x - 2$
**15.** Your estimates may not completely agree with these.

(a)

| $PQ_1$ | $PQ_2$ | $PQ_3$ | $PQ_4$ |
|---|---|---|---|
| 43 | 46 | 49 | 50 |

The appropriate units are m/s.

(b) $\approx 50$ m/s or 180 km/h

**17.** (a)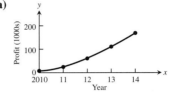

(b) $\approx$ \$56,000/year
(c) $\approx$ \$42,000/year

**19.** (a) $0.414213, 0.449489, \left(\sqrt{1 + h} - 1\right)/h$   (b) $g(x) = \sqrt{x}$

| $1 + h$ | 1.1 | 1.01 | 1.001 | 1.0001 |
|---|---|---|---|---|
| $\sqrt{1 + h}$ | 1.04880 | 1.004987 | 1.0004998 | 1.0000499 |
| $\left(\sqrt{1 + h} - 1\right)/h$ | 0.4880 | 0.4987 | 0.4998 | 0.499 |

| 1.00001 | 1.000001 |
|---|---|
| 1.000005 | 1.0000005 |
| 0.5 | 0.5 |

(c) 0.5     (d) 0.5
**21.** (a) 15 km/h, 3.3 km/h, 10 km/h   (b) 10 km/h, 0 km/h, 4 km/h
(c) 20 km/h when $t = 3.5$ h

**1.** (a) Does not exist. As $x$ approaches 1 from the right, $g(x)$ approaches 0. As $x$ approaches 1 from the left, $g(x)$ approaches 1. There is no single number $L$ that all the values $g(x)$ get arbitrarily close to as $x \to 1$.
(b) 1   (c) 0   (d) $1/2$
**3.** (a) True   (b) True   (c) False   (d) False
(e) False   (f) True   (g) True
**5.** As $x$ approaches 0 from the left, $x/|x|$ approaches $-1$. As $x$ approaches 0 from the right, $x/|x|$ approaches 1. There is no single number $L$ that the function values all get arbitrarily close to as $x \to 0$.
**7.** Nothing can be said.     **9.** No; no; no     **11.** $-4$     **13.** $-8$
**15.** 3     **17.** $-25/2$     **19.** 16     **21.** $3/2$     **23.** $1/10$
**25.** $-7$     **27.** $3/2$     **29.** $-1/2$     **31.** $-1$     **33.** $4/3$
**35.** $1/6$     **37.** 4     **39.** $1/2$     **41.** $3/2$     **43.** $-1$     **45.** 1
**47.** $1/3$     **49.** $\sqrt{4 - \pi}$
**51.** (a) Quotient Rule   (b) Difference and Power Rules
(c) Sum and Constant Multiple Rules
**53.** (a) $-10$   (b) $-20$   (c) $-1$   (d) $5/7$
**55.** (a) 4   (b) $-21$   (c) $-12$   (d) $-7/3$

**57.** 2    **59.** 3    **61.** $1/(2\sqrt{7})$    **63.** $\sqrt{5}$

**65.** **(a)** The limit is 1.

**67.** **(a)** $f(x) = (x^2 - 9)/(x + 3)$

| $x$ | $-3.1$ | $-3.01$ | $-3.001$ | $-3.0001$ | $-3.00001$ | $-3.000001$ |
|-----|--------|---------|----------|-----------|------------|-------------|
| $f(x)$ | $-6.1$ | $-6.01$ | $-6.001$ | $-6.0001$ | $-6.00001$ | $-6.000001$ |

| $x$ | $-2.9$ | $-2.99$ | $-2.999$ | $-2.9999$ | $-2.99999$ | $-2.999999$ |
|-----|--------|---------|----------|-----------|------------|-------------|
| $f(x)$ | $-5.9$ | $-5.99$ | $-5.999$ | $-5.9999$ | $-5.99999$ | $-5.999999$ |

  **(c)** $\lim\limits_{x \to -3} f(x) = -6$

**69.** **(a)** $G(x) = (x + 6)/(x^2 + 4x - 12)$

| $x$ | $-5.9$ | $-5.99$ | $-5.999$ | $-5.9999$ |
|-----|--------|---------|----------|-----------|
| $G(x)$ | $-.126582$ | $-.1251564$ | $-.1250156$ | $-.1250015$ |

| $-5.99999$ | $-5.999999$ |
|------------|-------------|
| $-.1250001$ | $-.1250000$ |

| $x$ | $-6.1$ | $-6.01$ | $-6.001$ | $-6.0001$ |
|-----|--------|---------|----------|-----------|
| $G(x)$ | $-.123456$ | $-.124843$ | $-.124984$ | $-.124998$ |

| $-6.00001$ | $-6.000001$ |
|------------|-------------|
| $-.124999$ | $-.124999$ |

  **(c)** $\lim\limits_{x \to -6} G(x) = -1/8 = -0.125$

**71.** **(a)** $f(x) = (x^2 - 1)/(|x| - 1)$

| $x$ | $-1.1$ | $-1.01$ | $-1.001$ | $-1.0001$ | $-1.00001$ | $-1.000001$ |
|-----|--------|---------|----------|-----------|------------|-------------|
| $f(x)$ | $2.1$ | $2.01$ | $2.001$ | $2.0001$ | $2.00001$ | $2.000001$ |

| $x$ | $-.9$ | $-.99$ | $-.999$ | $-.9999$ | $-.99999$ | $-.999999$ |
|-----|-------|--------|---------|----------|-----------|------------|
| $f(x)$ | $1.9$ | $1.99$ | $1.999$ | $1.9999$ | $1.99999$ | $1.999999$ |

  **(c)** $\lim\limits_{x \to -1} f(x) = 2$

**73.** **(a)** $g(\theta) = (\sin \theta)/\theta$

| $\theta$ | $.1$ | $.01$ | $.001$ | $.0001$ | $.00001$ | $.000001$ |
|----------|------|-------|--------|---------|----------|-----------|
| $g(\theta)$ | $.998334$ | $.999983$ | $.999999$ | $.999999$ | $.999999$ | $.999999$ |

| $\theta$ | $-.1$ | $-.01$ | $-.001$ | $-.0001$ | $-.00001$ | $-.000001$ |
|----------|-------|--------|---------|----------|-----------|------------|
| $g(\theta)$ | $.998334$ | $.999983$ | $.999999$ | $.999999$ | $.999999$ | $.999999$ |

  $\lim\limits_{\theta \to 0} g(\theta) = 1$

**75.** $c = 0, 1, -1$; the limit is 0 at $c = 0$, and 1 at $c = 1, -1$.

**77.** 7    **79.** **(a)** 5    **(b)** 5

**81.** **(a)** 0    **(b)** 0

### Section 2.3, pp. 79–82

**1.** $\delta = 2$

**3.** $\delta = 1/2$

**5.** $\delta = 1/18$

**7.** $\delta = 0.1$    **9.** $\delta = 7/16$    **11.** $\delta = \sqrt{5} - 2$

**13.** $\delta = 0.36$    **15.** $(3.99, 4.01)$,    $\delta = 0.01$

**17.** $(-0.19, 0.21)$,    $\delta = 0.19$    **19.** $(3, 15)$,    $\delta = 5$

**21.** $(10/3, 5)$,    $\delta = 2/3$

**23.** $\left(-\sqrt{4.5}, -\sqrt{3.5}\right)$,    $\delta = \sqrt{4.5} - 2 \approx 0.12$

**25.** $\left(\sqrt{15}, \sqrt{17}\right)$,    $\delta = \sqrt{17} - 4 \approx 0.12$

**27.** $\left(2 - \dfrac{0.03}{m}, 2 + \dfrac{0.03}{m}\right)$,    $\delta = \dfrac{0.03}{m}$

**29.** $\left(\dfrac{1}{2} - \dfrac{c}{m}, \dfrac{c}{m} + \dfrac{1}{2}\right)$,    $\delta = \dfrac{c}{m}$    **31.** $L = -3$,    $\delta = 0.01$

**33.** $L = 4$,    $\delta = 0.05$    **35.** $L = 4$,    $\delta = 0.75$

**55.** $[8.7332, 8.7476]$. To be safe, the left endpoint was rounded up and the right endpoint rounded down.

**59.** The limit does not exist as $x$ approaches 3.

### Section 2.4, pp. 87–89

**1.** **(a)** True    **(b)** True    **(c)** False    **(d)** True
    **(e)** True    **(f)** True    **(g)** False    **(h)** False
    **(i)** False    **(j)** False    **(k)** True    **(l)** False

**3.** **(a)** 2, 1    **(b)** No, $\lim\limits_{x \to 2^-} f(x) \neq \lim\limits_{x \to 2^-} f(x)$
    **(c)** 3, 3    **(d)** Yes, 3

**5.** **(a)** No    **(b)** Yes, 0    **(c)** No

**7.** **(a)**    **(b)** 1, 1    **(c)** Yes, 1

**9.** **(a)** $D : 0 \le x \le 2, R : 0 < y \le 1$ and $y = 2$
    **(b)** $(0, 1) \cup (1, 2)$    **(c)** $x = 2$    **(d)** $x = 0$

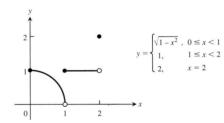

**11.** $\sqrt{3}$    **13.** 1    **15.** $2/\sqrt{5}$    **17.** **(a)** 1    **(b)** $-1$

**19.** **(a)** 1    **(b)** $2/3$    **21.** 1    **23.** $3/4$    **25.** 2    **27.** $1/2$

**29.** 2    **31.** 0    **33.** 1    **35.** $1/2$    **37.** 0    **39.** $3/8$

**41.** 3    **47.** $\delta = \epsilon^2$, $\lim\limits_{x \to 5^+} \sqrt{x - 5} = 0$

**51.** **(a)** 400    **(b)** 399    **(c)** The limit does not exist.

### Section 2.5, pp. 98–100

**1.** No; discontinuous at $x = 2$; not defined at $x = 2$

**3.** Continuous    **5.** **(a)** Yes    **(b)** Yes    **(c)** Yes    **(d)** Yes

**7.** **(a)** No    **(b)** No    **9.** 0    **11.** 1, nonremovable; 0, removable

**13.** All $x$ except $x = 2$    **15.** All $x$ except $x = 3, x = 1$

**17.** All $x$    **19.** All $x$ except $x = 0$

**21.** All $x$ except $n\pi/2$,    $n$ any integer

**23.** All $x$ except $n\pi/2$,    $n$ an odd integer

**25.** All $x \ge -3/2$    **27.** All $x$    **29.** All $x$

**31.** 0; continuous at $x = \pi$    **33.** 1; continuous at $y = 1$

**35.** $\sqrt{2}/2$; continuous at $t = 0$    **37.** 0; continuous at $x = 0$

**39.** $g(3) = 6$    **41.** $f(1) = 3/2$    **43.** $a = 4/3$    **45.** $a = -2, 3$

**47.** $a = 5/2, b = -1/2$    **71.** $x \approx 1.8794, -1.5321, -0.3473$

**73.** $x \approx 1.7549$    **75.** $x \approx 3.5156$    **77.** $x \approx 0.7391$

**1.** (a) 0   (b) $-2$   (c) 2   (d) Does not exist   (e) $-1$
    (f) $\infty$   (g) Does not exist   (h) 1   (i) 0
**3.** (a) $-3$   (b) $-3$    **5.** (a) $1/2$   (b) $1/2$    **7.** (a) $-5/3$
    (b) $-5/3$    **9.** 0    **11.** $-1$    **13.** (a) $2/5$   (b) $2/5$
**15.** (a) 0   (b) 0   **17.** (a) 7   (b) 7    **19.** (a) 0   (b) 0
**21.** (a) $\infty$   (b) $\infty$    **23.** 2    **25.** $\infty$    **27.** 0
**29.** 1    **31.** $\infty$    **33.** 1    **35.** $1/2$    **37.** $\infty$    **39.** $-\infty$
**41.** $-\infty$    **43.** $\infty$    **45.** (a) $\infty$   (b) $-\infty$    **47.** $\infty$
**49.** $\infty$    **51.** $-\infty$    **53.** (a) $\infty$   (b) $-\infty$   (c) $-\infty$    (d) $\infty$
**55.** (a) $-\infty$   (b) $\infty$   (c) 0   (d) $3/2$
**57.** (a) $-\infty$   (b) $1/4$   (c) $1/4$   (d) $1/4$   (e) It will be $-\infty$.
**59.** (a) $-\infty$    (b) $\infty$    **61.** (a) $\infty$   (b) $\infty$   (c) $\infty$   (d) $\infty$

**63.**

**65.**

**67.**

**69.** Here is one possibility.

**71.** Here is one possibility.

**73.** Here is one possibility.

**75.** Here is one possibility.

**79.** At most one

**81.** 0    **83.** $-3/4$    **85.** $5/2$

**93.** (a) For every positive real number $B$ there exists a corresponding number $\delta > 0$ such that for all $x$

$$c - \delta < x < c \implies f(x) > B.$$

(b) For every negative real number $-B$ there exists a corresponding number $\delta > 0$ such that for all $x$

$$c < x < c + \delta \implies f(x) < -B.$$

(c) For every negative real number $-B$ there exists a corresponding number $\delta > 0$ such that for all $x$

$$c - \delta < x < c \implies f(x) < -B.$$

**99.**

**101.**

**103.**

**105.**

**107.**

**109.** At $\infty$: $\infty$, at $-\infty$: 0

**1.** At $x = -1$:   $\lim\limits_{x \to -1^-} f(x) = \lim\limits_{x \to -1^+} f(x) = 1$, so
                $\lim\limits_{x \to -1} f(x) = 1 = f(-1)$; continuous at $x = -1$
    At $x = 0$:    $\lim\limits_{x \to 0^-} f(x) = \lim\limits_{x \to 0^+} f(x) = 0$, so $\lim\limits_{x \to 0} f(x) = 0$.
              However, $f(0) \ne 0$, so $f$ is discontinuous at $x = 0$. The discontinuity can be removed by redefining $f(0)$ to be 0.
    At $x = 1$:    $\lim\limits_{x \to 1^-} f(x) = -1$ and $\lim\limits_{x \to 1^+} f(x) = 1$, so $\lim\limits_{x \to 1} f(x)$
              does not exist. The function is discontinuous at $x = 1$, and the discontinuity is not removable.

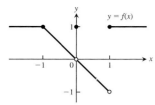

3. (a) $-21$ (b) $49$ (c) $0$ (d) $1$ (e) $1$ (f) $7$
(g) $-7$ (h) $-\dfrac{1}{7}$  5. $4$

7. (a) $(-\infty, +\infty)$ (b) $[0, \infty)$ (c) $(-\infty, 0)$ and $(0, \infty)$
(d) $(0, \infty)$

9. (a) Does not exist (b) $0$  11. $\dfrac{1}{2}$  13. $2x$  15. $-\dfrac{1}{4}$

17. $2/3$  19. $2/\pi$  21. $1$  23. $4$

25. $2$  27. $0$

31. No in both cases, because $\displaystyle\lim_{x\to 1} f(x)$ does not exist, and $\displaystyle\lim_{x\to -1} f(x)$ does not exist.

33. Yes, $f$ does have a continuous extension, to $a = 1$ with $f(1) = 4/3$.

35. No  37. $2/5$  39. $0$  41. $-\infty$  43. $0$  45. $1$

47. (a) $x = 3$ (b) $x = 1$ (c) $x = -4$

### Additional and Advanced Exercises, pp. 116–118

3. $0$; the left-hand limit was needed because the function is undefined for $v > c$.  5. $17.5 < t < 22.5$; within $2.5°C$

13. (a) $B$ (b) $A$ (c) $A$ (d) $A$

21. (a) $\displaystyle\lim_{a\to 0} r_+(a) = 0.5,\ \lim_{a\to -1^+} r_+(a) = 1$
(b) $\displaystyle\lim_{a\to 0} r_-(a)$ does not exist, $\displaystyle\lim_{a\to -1^+} r_-(a) = 1$

25. $0$  27. $1$  29. $4$  31. $y = 2x$  33. $y = x,\ y = -x$

# Chapter 3

## Section 3.1, pp. 122–123

1. $P_1: m_1 = 1,\ P_2: m_2 = 5$  3. $P_1: m_1 = 5/2,\ P_2: m_2 = -1/2$

5. $y = 2x + 5$  7. $y = x + 1$

9. $y = 12x + 16$

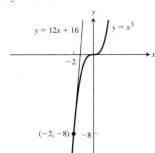

11. $m = 4,\ y - 5 = 4(x - 2)$
13. $m = -2,\ y - 3 = -2(x - 3)$
15. $m = 12,\ y - 8 = 12(t - 2)$
17. $m = \dfrac{1}{4},\ y - 2 = \dfrac{1}{4}(x - 4)$
19. $m = -1$  21. $m = -1/4$

23. (a) It is the rate of change of the number of cells when $t = 5$. The units are the number of cells per hour.
(b) $P'(3)$ because the slope of the curve is greater there.
(c) $51.72 \approx 52$ cells/h

25. $(-2, -5)$  27. $y = -(x + 1),\ y = -(x - 3)$
29. $19.6\ \text{m/s}$  31. $6\pi$  35. Yes  37. Yes
39. (a) Nowhere  41. (a) At $x = 0$  43. (a) Nowhere
45. (a) At $x = 1$  47. (a) At $x = 0$

## Section 3.2, pp. 129–132

1. $-2x, 6, 0, -2$  3. $-\dfrac{2}{t^3}, 2, -\dfrac{1}{4}, -\dfrac{2}{3\sqrt{3}}$

5. $\dfrac{3}{2\sqrt{3\theta}}, \dfrac{3}{2\sqrt{3}}, \dfrac{1}{2}, \dfrac{3}{2\sqrt{2}}$  7. $6x^2$  9. $\dfrac{1}{(2t + 1)^2}$

11. $\dfrac{3}{2} q^{1/2}$  13. $1 - \dfrac{9}{x^2}, 0$  15. $3t^2 - 2t, 5$

17. $\dfrac{-4}{(x - 2)\sqrt{x - 2}},\ y - 4 = -\dfrac{1}{2}(x - 6)$  19. $6$

21. $1/8$  23. $\dfrac{-1}{(x + 2)^2}$  25. $\dfrac{-1}{(x - 1)^2}$  27. (b)  29. (d)

31. (a) $x = 0, 1, 4$
(b)

33.

35. (a) i) $0.77\,°C/h$  ii) $1.43\,°C/h$
iii) $0\,°C/h$  iv) $-1.88\,°C/h$
(b) $3.63\,°C/h$ at 12 P.M., $-4\,°C/h$ at 6 P.M.
(c)

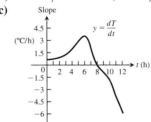

37. Since $\displaystyle\lim_{h\to 0^+} \dfrac{f(0 + h) - f(0)}{h} = 1$
while $\displaystyle\lim_{h\to 0^-} \dfrac{f(0 + h) - f(0)}{h} = 0$,
$f'(0) = \displaystyle\lim_{h\to 0} \dfrac{f(0 + h) - f(0)}{h}$ does not exist and $f(x)$ is not differentiable at $x = 0$.

39. Since $\displaystyle\lim_{h\to 0^+} \dfrac{f(1 + h) - f(1)}{h} = 2$ while
$\displaystyle\lim_{h\to 0^-} \dfrac{f(1 + h) - f(1)}{h} = \dfrac{1}{2},\ f'(1) = \displaystyle\lim_{h\to 0} \dfrac{f(1 + h) - f(1)}{h}$
does not exist and $f(x)$ is not differentiable at $x = 1$.

**41.** Since $f(x)$ is not continuous at $x = 0$, $f(x)$ is not differentiable at $x = 0$.

**43. (a)** $-3 \leq x \leq 2$ **(b)** None **(c)** None

**45. (a)** $-3 \leq x < 0, 0 < x \leq 3$ **(b)** None **(c)** $x = 0$

**47. (a)** $-1 \leq x < 0, 0 < x \leq 2$ **(b)** $x = 0$ **(c)** None

## Section 3.3, pp. 139–141

**1.** $\dfrac{dy}{dx} = -2x, \dfrac{d^2y}{dx^2} = -2$

**3.** $\dfrac{ds}{dt} = 15t^2 - 15t^4, \dfrac{d^2s}{dt^2} = 30t - 60t^3$

**5.** $\dfrac{dy}{dx} = 4x^2 - 1, \dfrac{d^2y}{dx^2} = 8x$

**7.** $\dfrac{dw}{dz} = -\dfrac{6}{z^3} + \dfrac{1}{z^2}, \dfrac{d^2w}{dz^2} = \dfrac{18}{z^4} - \dfrac{2}{z^3}$

**9.** $\dfrac{dy}{dx} = 12x - 10 + 10x^{-3}, \dfrac{d^2y}{dx^2} = 12 - 30x^{-4}$

**11.** $\dfrac{dr}{ds} = \dfrac{-2}{3s^3} + \dfrac{5}{2s^2}, \dfrac{d^2r}{ds^2} = \dfrac{2}{s^4} - \dfrac{5}{s^3}$

**13.** $y' = -5x^4 + 12x^2 - 2x - 3$

**15.** $y' = 3x^2 + 10x + 2 - \dfrac{1}{x^2}$ **17.** $y' = \dfrac{-19}{(3x - 2)^2}$

**19.** $g'(x) = \dfrac{x^2 + x + 4}{(x + 0.5)^2}$ **21.** $\dfrac{dv}{dt} = \dfrac{t^2 - 2t - 1}{(1 + t^2)^2}$

**23.** $f'(s) = \dfrac{1}{\sqrt{s}(\sqrt{s} + 1)^2}$ **25.** $v' = -\dfrac{1}{x^2} + 2x^{-3/2}$

**27.** $y' = \dfrac{-4x^3 - 3x^2 + 1}{(x^2 - 1)^2(x^2 + x + 1)^2}$

**29.** $y' = 2x^3 - 3x - 1, y'' = 6x^2 - 3, y''' = 12x, y^{(4)} = 12,$ $y^{(n)} = 0$ for $n \geq 5$

**31.** $y' = 3x^2 + 8x + 1, y'' = 6x + 8, y''' = 6, y^{(n)} = 0$ for $n \geq 4$

**33.** $y' = 2x - 7x^{-2}, y'' = 2 + 14x^{-3}$

**35.** $\dfrac{dr}{d\theta} = 3\theta^{-4}, \dfrac{d^2r}{d\theta^2} = -12\theta^{-5}$ **37.** $\dfrac{dw}{dz} = -z^{-2} - 1, \dfrac{d^2w}{dz^2} = 2z^{-3}$

**39.** $\dfrac{dp}{dq} = \dfrac{1}{6}q + \dfrac{1}{6}q^{-3} + q^{-5}, \dfrac{d^2p}{dq^2} = \dfrac{1}{6} - \dfrac{1}{2}q^{-4} - 5q^{-6}$

**41. (a)** 13 **(b)** $-7$ **(c)** $7/25$ **(d)** 20

**43. (a)** $y = -\dfrac{x}{8} + \dfrac{5}{4}$ **(b)** $m = -4$ at $(0, 1)$

**(c)** $y = 8x - 15, y = 8x + 17$

**45.** $y = 4x, y = 2$ **47.** $a = 1, b = 1, c = 0$

**49.** $(2, 4)$ **51.** $(0, 0), (4, 2)$ **53. (a)** $y = 2x + 2$ **(c)** $(2, 6)$

**55.** 50 **57.** $a = -3$

**59.** $P'(x) = na_n x^{n-1} + (n - 1)a_{n-1}x^{n-2} + \cdots + 2a_2 x + a_1$

**61.** The Product Rule is then the Constant Multiple Rule, so the latter is a special case of the Product Rule.

**63. (a)** $\dfrac{d}{dx}(uvw) = uvw' + uv'w + u'vw$

**(b)** $\dfrac{d}{dx}(u_1 u_2 u_3 u_4) = u_1 u_2 u_3 u_4' + u_1 u_2 u_3' u_4 + u_1 u_2' u_3 u_4 + u_1' u_2 u_3 u_4$

**(c)** $\dfrac{d}{dx}(u_1 \cdots u_n) = u_1 u_2 \cdots u_{n-1}u_n' + u_1 u_2 \cdots u_{n-2}u_{n-1}'u_n + \cdots + u_1' u_2 \cdots u_n$

**65.** $\dfrac{dP}{dV} = -\dfrac{nRT}{(V - nb)^2} + \dfrac{2an^2}{V^3}$

## Section 3.4, pp. 147–151

**1. (a)** $-2$ m, $-1$ m/s

**(b)** 3 m/s, 1 m/s; 2 m/s², 2 m/s²

**(c)** Changes direction at $t = 3/2$ s

**3. (a)** $-9$ m, $-3$ m/s

**(b)** 3 m/s, 12 m/s; 6 m/s², $-12$ m/s²

**(c)** No change in direction

**5. (a)** $-20$ m, $-5$ m/s

**(b)** 45 m/s, (1/5) m/s; 140 m/s², (4/25) m/s²

**(c)** No change in direction

**7. (a)** $a(1) = -6$ m/s², $a(3) = 6$ m/s²

**(b)** $v(2) = 3$ m/s **(c)** 6 m

**9.** Mars: $\approx 7.5$ s, Jupiter: $\approx 1.2$ s

**11.** $g_s = 0.75$ m/s²

**13. (a)** $v = -9.8t, |v| = 9.8t$ m/s, $a = -9.8$ m/s²

**(b)** $t \approx 3.4$ s

**(c)** $v \approx -33.1$ m/s

**15. (a)** $t = 2, t = 7$ **(b)** $3 \leq t \leq 6$

**(c)** **(d)**

**17. (a)** 57 m/s **(b)** 2 s **(c)** 8 s, 0 m/s

**(d)** 10.8 s, 27 m/s **(e)** 2.8 s

**(f)** Greatest acceleration happens 2 s after launch

**(g)** Constant acceleration between 2 and 10.8 s, $-10$ m/s²

**19. (a)** $\dfrac{4}{7}$ s, 280 cm/s **(b)** 560 cm/s, 980 cm/s²

**(c)** 29.75 flashes/s

**21.** $C$ = position, $A$ = velocity, $B$ = acceleration

**23. (a)** \$110/machine **(b)** \$80 **(c)** \$79.90

**25. (a)** $b'(0) = 10^4$ bacteria/h **(b)** $b'(5) = 0$ bacteria/h

**(c)** $b'(10) = -10^4$ bacteria/h

**27. (a)** $\dfrac{dy}{dt} = \dfrac{t}{12} - 1$

**(b)** The largest value of $\dfrac{dy}{dt}$ is 0 m/h when $t = 12$ and the smallest value of $\dfrac{dy}{dt}$ is $-1$ m/h when $t = 0$.

**(c)**

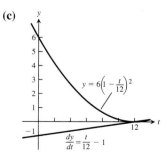

**29.** 0.846 m, 1.482 m, the additional number of meters required to stop the car if speed is increased by 1 km/h

**31.** $t = 25$ s, $D = \dfrac{6250}{9}$ m

**33.**

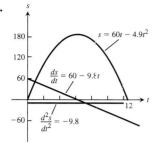

**(a)** $v = 0$ when $t = 6.12$ s
**(b)** $v > 0$ when $0 \le t < 6.12 \Rightarrow$ the object moves up; $v < 0$ when $6.12 < t \le 12.24 \Rightarrow$ the object moves down.
**(c)** The object changes direction at $t = 6.12$ s.
**(d)** The object speeds up on $(6.12, 12.24]$ and slows down on $[0, 6.12)$.
**(e)** The object is moving fastest at the endpoints $t = 0$ and $t = 12.24$ when it is traveling $60$ m/s. It's moving slowest at $t = 6.12$ when the speed is 0.
**(f)** When $t = 6.12$ the object is $s = 183.6$ m from the origin and farthest away.

**35.**

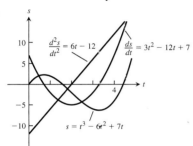

**(a)** $v = 0$ when $t = \dfrac{6 \pm \sqrt{15}}{3}$ s

**(b)** $v < 0$ when $\dfrac{6 - \sqrt{15}}{3} < t < \dfrac{6 + \sqrt{15}}{3} \Rightarrow$ the object moves left; $v > 0$ when $0 \le t < \dfrac{6 - \sqrt{15}}{3}$ or $\dfrac{6 + \sqrt{15}}{3} < t \le 4 \Rightarrow$ the object moves right.

**(c)** The object changes direction at $t = \dfrac{6 \pm \sqrt{15}}{3}$ s.

**(d)** The object speeds up on $\left(\dfrac{6 - \sqrt{15}}{3}, 2\right) \cup \left(\dfrac{6 + \sqrt{15}}{3}, 4\right]$ and slows down on $\left[0, \dfrac{6 - \sqrt{15}}{3}\right) \cup \left(2, \dfrac{6 + \sqrt{15}}{3}\right)$.

**(e)** The object is moving fastest at $t = 0$ and $t = 4$ when it is moving $7$ units/s and slowest at $t = \dfrac{6 \pm \sqrt{15}}{3}$ s.

**(f)** When $t = \dfrac{6 + \sqrt{15}}{3}$ the object is at position $s \approx -6.303$ units and farthest from the origin.

**Section 3.5, pp. 155–157**

**1.** $-10 - 3\sin x$    **3.** $2x \cos x - x^2 \sin x$

**5.** $-\csc x \cot x - \dfrac{2}{\sqrt{x}}$    **7.** $\sin x \sec^2 x + \sin x$

**9.** $x \tan^2 x + \sec x - \dfrac{1}{x^2}$    **11.** $\dfrac{-\csc^2 x}{(1 + \cot x)^2}$

**13.** $4 \tan x \sec x - \csc^2 x$    **15.** $0$

**17.** $3x^2 \sin x \cos x + x^3 \cos^2 x - x^3 \sin^2 x$

**19.** $\sec^2 t - 1$    **21.** $\dfrac{-2 \csc t \cot t}{(1 - \csc t)^2}$    **23.** $-\theta(\theta \cos \theta + 2 \sin \theta)$

**25.** $\sec \theta \csc \theta (\tan \theta - \cot \theta) = \sec^2 \theta - \csc^2 \theta$    **27.** $\sec^2 q$

**29.** $\sec^2 q$    **31.** $\dfrac{q^3 \cos q - q^2 \sin q - q \cos q - \sin q}{(q^2 - 1)^2}$

**33. (a)** $2 \csc^3 x - \csc x$    **(b)** $2 \sec^3 x - \sec x$
**35.**

**37.**

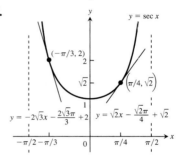

**39.** Yes, at $x = \pi$    **41.** No

**43.** $\left(-\dfrac{\pi}{4}, -1\right); \left(\dfrac{\pi}{4}, 1\right)$

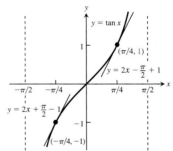

**45. (a)** $y = -x + \pi/2 + 2$    **(b)** $y = 4 - \sqrt{3}$
**47.** $0$    **49.** $\sqrt{3}/2$    **51.** $-1$    **53.** $0$
**55.** $-\sqrt{2}$ m/s, $\sqrt{2}$ m/s, $\sqrt{2}$ m/s², $\sqrt{2}$ m/s³
**57.** $c = 9$    **59.** $\sin x$
**61. (a) i)** $10$ cm    **ii)** $5$ cm    **iii)** $-5\sqrt{2} \approx -7.1$ cm
    **(b) i)** $0$ cm/s    **ii)** $-5\sqrt{3} \approx -8.7$ cm/s
    **iii)** $-5\sqrt{2} \approx -7.1$ cm/s

**Section 3.6, pp. 162–165**

**1.** $12x^3$    **3.** $3 \cos(3x + 1)$    **5.** $\dfrac{\cos x}{2\sqrt{\sin x}}$

**7.** $2\pi x \sec^2(\pi x^2)$

**9.** With $u = (2x + 1)$, $y = u^5$: $\dfrac{dy}{dx} = \dfrac{dy}{du}\dfrac{du}{dx} = 5u^4 \cdot 2 = 10(2x + 1)^4$

**11.** With $u = (1 - (x/7))$, $y = u^{-7}$: $\dfrac{dy}{dx} = \dfrac{dy}{du}\dfrac{du}{dx} =$

$-7u^{-8} \cdot \left(-\dfrac{1}{7}\right) = \left(1 - \dfrac{x}{7}\right)^{-8}$

**13.** With $u = ((x^2/8) + x - (1/x))$, $y = u^4$: $\dfrac{dy}{dx} = \dfrac{dy}{du}\dfrac{du}{dx} =$

$4u^3 \cdot \left(\dfrac{x}{4} + 1 + \dfrac{1}{x^2}\right) = 4\left(\dfrac{x^2}{8} + x - \dfrac{1}{x}\right)^3\left(\dfrac{x}{4} + 1 + \dfrac{1}{x^2}\right)$

**15.** With $u = \tan x$, $y = \sec u$: $\dfrac{dy}{dx} = \dfrac{dy}{du}\dfrac{du}{dx} =$

$(\sec u \tan u)(\sec^2 x) = \sec(\tan x)\tan(\tan x)\sec^2 x$

**17.** With $u = \tan x$, $y = u^3$: $\dfrac{dy}{dx} = \dfrac{dy}{du}\dfrac{du}{dx} = 3u^2 \sec^2 x =$

$3\tan^2 x (\sec^2 x)$

**19.** $-\dfrac{1}{2\sqrt{3-t}}$ **21.** $\dfrac{4}{\pi}(\cos 3t - \sin 5t)$ **23.** $\dfrac{\csc\theta}{\cot\theta + \csc\theta}$

**25.** $2x\sin^4 x + 4x^2\sin^3 x \cos x + \cos^{-2}x + 2x\cos^{-3}x\sin x$

**27.** $(3x - 2)^5 - \dfrac{1}{x^3\left(4 - \dfrac{1}{2x^2}\right)^2}$ **29.** $\dfrac{(4x+3)^3(4x+7)}{(x+1)^4}$

**31.** $\sqrt{x}\sec^2\left(2\sqrt{x}\right) + \tan\left(2\sqrt{x}\right)$ **33.** $\dfrac{x\sec x \tan x + \sec x}{2\sqrt{7 + x\sec x}}$

**35.** $\dfrac{2\sin\theta}{(1 + \cos\theta)^2}$ **37.** $-2\sin(\theta^2)\sin 2\theta + 2\theta\cos(2\theta)\cos(\theta^2)$

**39.** $\left(\dfrac{t+2}{2(t+1)^{3/2}}\right)\cos\left(\dfrac{t}{\sqrt{t+1}}\right)$

**41.** $2\pi\sin(\pi t - 2)\cos(\pi t - 2)$ **43.** $\dfrac{8\sin(2t)}{(1 + \cos 2t)^5}$

**45.** $10t^{10}\tan^9 t \sec^2 t + 10t^9\tan^{10}t$

**47.** $\dfrac{-3t^6(t^2+4)}{(t^3 - 4t)^4}$ **49.** $-2\cos(\cos(2t-5))(\sin(2t-5))$

**51.** $\left(1 + \tan^4\left(\dfrac{t}{12}\right)\right)^2\left(\tan^3\left(\dfrac{t}{12}\right)\sec^2\left(\dfrac{t}{12}\right)\right)$

**53.** $-\dfrac{t\sin(t^2)}{\sqrt{1 + \cos(t^2)}}$ **55.** $6\tan(\sin^3 t)\sec^2(\sin^3 t)\sin^2 t \cos t$

**57.** $3(2t^2 - 5)^3(18t^2 - 5)$ **59.** $\dfrac{6}{x^3}\left(1 + \dfrac{1}{x}\right)\left(1 + \dfrac{2}{x}\right)$

**61.** $2\csc^2(3x - 1)\cot(3x - 1)$ **63.** $16(2x+1)^2(5x+1)$

**65.** $5/2$ **67.** $-\pi/4$ **69.** $0$ **71.** $-5$

**73.** (a) $2/3$ (b) $2\pi + 5$ (c) $15 - 8\pi$ (d) $37/6$ (e) $-1$
(f) $\sqrt{2}/24$ (g) $5/32$ (h) $-5/(3\sqrt{17})$ **75.** $5$

**77.** (a) $1$ (b) $1$ **79.** $y = 1 - 4x$

**81.** (a) $y = \pi x + 2 - \pi$ (b) $\pi/2$

**83.** It multiplies the velocity, acceleration, and jerk by 2, 4, and 8, respectively.

**85.** $v(6) = \dfrac{2}{5}$m/s, $a(6) = -\dfrac{4}{125}$m/s$^2$

### Section 3.7, pp. 169–170

**1.** $\dfrac{-2xy - y^2}{x^2 + 2xy}$ **3.** $\dfrac{1 - 2y}{2x + 2y - 1}$

**5.** $\dfrac{-2x^3 + 3x^2y - xy^2 + x}{x^2y - x^3 + y}$ **7.** $\dfrac{1}{y(x+1)^2}$ **9.** $\cos y \cot y$

**11.** $\dfrac{-\cos^2(xy) - y}{x}$ **13.** $\dfrac{-y^2}{y\sin\left(\dfrac{1}{y}\right) - \cos\left(\dfrac{1}{y}\right) + xy}$

**15.** $-\dfrac{\sqrt{r}}{\sqrt{\theta}}$ **17.** $\dfrac{-r}{\theta}$

**19.** $y' = -\dfrac{x}{y}$, $y'' = \dfrac{-y^2 - x^2}{y^3}$

**21.** $y' = \dfrac{x+1}{y}$, $y'' = \dfrac{y^2 - (x+1)^2}{y^3}$

**23.** $y' = \dfrac{\sqrt{y}}{\sqrt{y} + 1}$, $y'' = \dfrac{1}{2\left(\sqrt{y} + 1\right)^3}$

**25.** $-2$ **27.** $(-2, 1): m = -1, (-2, -1): m = 1$

**29.** (a) $y = \dfrac{7}{4}x - \dfrac{1}{2}$ (b) $y = -\dfrac{4}{7}x + \dfrac{29}{7}$

**31.** (a) $y = 3x + 6$ (b) $y = -\dfrac{1}{3}x + \dfrac{8}{3}$

**33.** (a) $y = \dfrac{6}{7}x + \dfrac{6}{7}$ (b) $y = -\dfrac{7}{6}x - \dfrac{7}{6}$

**35.** (a) $y = -\dfrac{\pi}{2}x + \pi$ (b) $y = \dfrac{2}{\pi}x - \dfrac{2}{\pi} + \dfrac{\pi}{2}$

**37.** (a) $y = 2\pi x - 2\pi$ (b) $y = -\dfrac{x}{2\pi} + \dfrac{1}{2\pi}$

**39.** Points: $\left(-\sqrt{7}, 0\right)$ and $\left(\sqrt{7}, 0\right)$, Slope: $-2$

**41.** $m = -1$ at $\left(\dfrac{\sqrt{3}}{4}, \dfrac{\sqrt{3}}{2}\right)$, $m = \sqrt{3}$ at $\left(\dfrac{\sqrt{3}}{4}, \dfrac{1}{2}\right)$

**43.** $(-3, 2): m = -\dfrac{27}{8}; (-3, -2): m = \dfrac{27}{8}; (3, 2): m = \dfrac{27}{8};$
$(3, -2): m = -\dfrac{27}{8}$

**45.** $(3, -1)$

**51.** $\dfrac{dy}{dx} = -\dfrac{y^3 + 2xy}{x^2 + 3xy^2}$, $\dfrac{dx}{dy} = -\dfrac{x^2 + 3xy^2}{y^3 + 2xy}$, $\dfrac{dx}{dy} = \dfrac{1}{dy/dx}$

### Section 3.8, pp. 175–179

**1.** $\dfrac{dA}{dt} = 2\pi r\dfrac{dr}{dt}$ **3.** $10$ **5.** $-6$ **7.** $-3/2$

**9.** $31/13$ **11.** (a) $-180$ m$^2$/min (b) $-135$ m$^3$/min

**13.** (a) $\dfrac{dV}{dt} = \pi r^2\dfrac{dh}{dt}$ (b) $\dfrac{dV}{dt} = 2\pi hr\dfrac{dr}{dt}$
(c) $\dfrac{dV}{dt} = \pi r^2\dfrac{dh}{dt} + 2\pi hr\dfrac{dr}{dt}$

**15.** (a) $1$ volt/s (b) $-\dfrac{1}{3}$ amp/s
(c) $\dfrac{dR}{dt} = \dfrac{1}{I}\left(\dfrac{dV}{dt} - \dfrac{V}{I}\dfrac{dI}{dt}\right)$
(d) $3/2$ ohms/s, $R$ is increasing.

**17.** (a) $\dfrac{ds}{dt} = \dfrac{x}{\sqrt{x^2 + y^2}}\dfrac{dx}{dt}$
(b) $\dfrac{ds}{dt} = \dfrac{x}{\sqrt{x^2 + y^2}}\dfrac{dx}{dt} + \dfrac{y}{\sqrt{x^2 + y^2}}\dfrac{dy}{dt}$ (c) $\dfrac{dx}{dt} = -\dfrac{y}{x}\dfrac{dy}{dt}$

**19.** (a) $\dfrac{dA}{dt} = \dfrac{1}{2}ab\cos\theta\dfrac{d\theta}{dt}$
(b) $\dfrac{dA}{dt} = \dfrac{1}{2}ab\cos\theta\dfrac{d\theta}{dt} + \dfrac{1}{2}b\sin\theta\dfrac{da}{dt}$
(c) $\dfrac{dA}{dt} = \dfrac{1}{2}ab\cos\theta\dfrac{d\theta}{dt} + \dfrac{1}{2}b\sin\theta\dfrac{da}{dt} + \dfrac{1}{2}a\sin\theta\dfrac{db}{dt}$

**21. (a)** $14 \text{ cm}^2/\text{s}$, increasing   **(b)** $0 \text{ cm}/\text{s}$, constant
**(c)** $-14/13 \text{ cm}/\text{s}$, decreasing
**23. (a)** $-3.6 \text{ m}/\text{s}$   **(b)** $-5.355 \text{ m}^2/\text{s}$   **(c)** $-1 \text{ rad}/\text{s}$
**25.** $6 \text{ m}/\text{s}$

**27. (a)** $\dfrac{dh}{dt} = 11.19 \text{ cm}/\text{min}$   **(b)** $\dfrac{dr}{dt} = 14.92 \text{ cm}/\text{min}$

**29. (a)** $\dfrac{-1}{24\pi} \text{ m}/\text{min}$   **(b)** $r = \sqrt{26y - y^2} \text{ m}$

**(c)** $\dfrac{dr}{dt} = -\dfrac{5}{288\pi} \text{ m}/\text{min}$

**31.** $1 \text{ m}/\text{min}, 40\pi \text{ m}^2/\text{min}$   **33.** $3.16 \text{ m}/\text{s}$
**35.** Increasing at $466/1681 \text{ } L/\text{min}^2$
**37.** $-5 \text{ m}/\text{s}$   **39.** $-441 \text{ m}/\text{s}$
**41.** $\dfrac{5}{72\pi} \text{ cm}/\text{min}, \dfrac{10}{3} \text{ cm}^2/\text{min}$
**43. (a)** $-10/\sqrt{13} \approx -2.774 \text{ m}/\text{s}$
**(b)** $d\theta_1/dt = 5/39 \text{ rad}/\text{s}, d\theta_2/dt = -5/39 \text{ rad}/\text{s}$
**(c)** $d\theta_1/dt = 1/6 \text{ rad}/\text{s}, d\theta_2/dt = -1/6 \text{ rad}/\text{s}$
**45.** $-5.5 \text{ deg}/\text{min}$

Section 3.9, pp. 188–190
**1.** $L(x) = 10x - 13$   **3.** $L(x) = 2$   **5.** $L(x) = x - \pi$
**7.** $2x$   **9.** $-x - 5$   **11.** $\dfrac{1}{12}x + \dfrac{4}{3}$
**13.** $f(0) = 1$. Also, $f'(x) = k(1 + x)^{k-1}$, so $f'(0) = k$. This means the linearization at $x = 0$ is $L(x) = 1 + kx$.
**15. (a)** $1.01$   **(b)** $1.003$
**17.** $\left(3x^2 - \dfrac{3}{2\sqrt{x}}\right) dx$   **19.** $\dfrac{2 - 2x^2}{(1 + x^2)^2} dx$
**21.** $\dfrac{1 - y}{3\sqrt{y} + x} dx$   **23.** $\dfrac{5}{2\sqrt{x}} \cos\left(5\sqrt{x}\right) dx$
**25.** $(4x^2) \sec^2\left(\dfrac{x^3}{3}\right) dx$
**27.** $\dfrac{3}{\sqrt{x}}\left(\csc\left(1 - 2\sqrt{x}\right) \cot\left(1 - 2\sqrt{x}\right)\right) dx$
**29. (a)** $0.41$   **(b)** $0.4$   **(c)** $0.01$
**31. (a)** $0.231$   **(b)** $0.2$   **(c)** $0.031$
**33. (a)** $-1/3$   **(b)** $-2/5$   **(c)** $1/15$
**35.** $dV = 4\pi r_0^2 \, dr$   **37.** $dS = 12x_0 \, dx$   **39.** $dV = 2\pi r_0 h \, dr$
**41. (a)** $0.08\pi \text{ m}^2$   **(b)** $2\%$   **43.** $dV \approx 565.5 \text{ cm}^3$
**45. (a)** $2\%$   **(b)** $4\%$   **47.** $\dfrac{1}{3}\%$   **49.** $3\%$
**51.** The ratio equals $37.52$, so a change in the acceleration of gravity on the moon has about $38$ times the effect that a change of the same magnitude has on Earth.
**53.** Increase $V \approx 40\%$
**55. (a) i)** $b_0 = f(a)$   **ii)** $b_1 = f'(a)$   **iii)** $b_2 = \dfrac{f''(a)}{2}$
**(b)** $Q(x) = 1 + x + x^2$   **(d)** $Q(x) = 1 - (x - 1) + (x - 1)^2$
**(e)** $Q(x) = 1 + \dfrac{x}{2} - \dfrac{x^2}{8}$
**(f)** The linearization of any differentiable function $u(x)$ at $x = a$ is $L(x) = u(a) + u'(a)(x - a) = b_0 + b_1(x - a)$, where $b_0$ and $b_1$ are the coefficients of the constant and linear terms of the quadratic approximation. Thus, the linearization for $f(x)$ at $x = 0$ is $1 + x$; the linearization for $g(x)$ at $x = 1$ is $1 - (x - 1)$ or $2 - x$; and the linearization for $h(x)$ at $x = 0$ is $1 + \dfrac{x}{2}$.

Practice Exercises, pp. 191–196
**1.** $5x^4 - 0.25x + 0.25$   **3.** $3x(x - 2)$
**5.** $2(x + 1)(2x^2 + 4x + 1)$
**7.** $3(\theta^2 + \sec\theta + 1)^2(2\theta + \sec\theta\tan\theta)$
**9.** $\dfrac{1}{2\sqrt{t}\left(1 + \sqrt{t}\right)^2}$   **11.** $2\sec^2 x \tan x$
**13.** $8\cos^3(1 - 2t)\sin(1 - 2t)$   **15.** $5(\sec t)(\sec t + \tan t)^5$
**17.** $\dfrac{\theta\cos\theta + \sin\theta}{\sqrt{2\theta}\sin\theta}$   **19.** $\dfrac{\cos\sqrt{2\theta}}{\sqrt{2\theta}}$
**21.** $x\csc\left(\dfrac{2}{x}\right) + \csc\left(\dfrac{2}{x}\right)\cot\left(\dfrac{2}{x}\right)$
**23.** $\dfrac{1}{2}x^{1/2}\sec(2x)^2\left[16\tan(2x)^2 - x^{-2}\right]$
**25.** $-10x\csc^2(x^2)$   **27.** $8x^3\sin(2x^2)\cos(2x^2) + 2x\sin^2(2x^2)$
**29.** $\dfrac{-(t + 1)}{8t^3}$   **31.** $\dfrac{1 - x}{(x + 1)^3}$   **33.** $\dfrac{-1}{2x^2\left(1 + \dfrac{1}{x}\right)^{1/2}}$
**35.** $\dfrac{-2\sin\theta}{(\cos\theta - 1)^2}$   **37.** $3\sqrt{2x + 1}$   **39.** $-9\left[\dfrac{5x + \cos 2x}{(5x^2 + \sin 2x)^{5/2}}\right]$
**41.** $-\dfrac{y + 2}{x + 3}$   **43.** $\dfrac{-3x^2 - 4y + 2}{4x - 4y^{1/3}}$   **45.** $-\dfrac{y}{x}$
**47.** $\dfrac{1}{2y(x + 1)^2}$   **49.** $\dfrac{dp}{dq} = \dfrac{6q - 4p}{3p^2 + 4q}$
**51.** $\dfrac{dr}{ds} = (2r - 1)(\tan 2s)$
**53. (a)** $\dfrac{d^2y}{dx^2} = \dfrac{-2xy^3 - 2x^4}{y^5}$   **(b)** $\dfrac{d^2y}{dx^2} = \dfrac{-2xy^2 - 1}{x^4y^3}$
**55. (a)** $7$   **(b)** $-2$   **(c)** $5/12$   **(d)** $1/4$   **(e)** $12$   **(f)** $9/2$
**(g)** $3/4$
**57.** $0$   **59.** $\sqrt{3}$   **61.** $-\dfrac{1}{2}$   **63.** $\dfrac{-2}{(2t + 1)^2}$

**65. (a)**

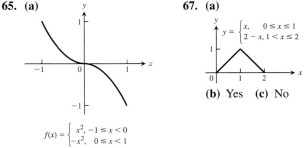

$f(x) = \begin{cases} x^2, & -1 \le x < 0 \\ -x^2, & 0 \le x < 1 \end{cases}$

**(b)** Yes   **(c)** Yes

**67. (a)**

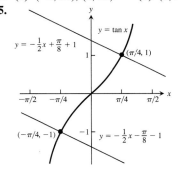

$y = \begin{cases} x, & 0 \le x \le 1 \\ 2 - x, & 1 < x \le 2 \end{cases}$

**(b)** Yes   **(c)** No

**69.** $\left(\dfrac{5}{2}, \dfrac{9}{4}\right)$ and $\left(\dfrac{3}{2}, -\dfrac{1}{4}\right)$   **71.** $(-1, 27)$ and $(2, 0)$
**73. (a)** $(-2, 16), (3, 11)$   **(b)** $(0, 20), (1, 7)$
**75.**
$y = \tan x$
$y = -\dfrac{1}{2}x + \dfrac{\pi}{8} + 1$   $(\pi/4, 1)$
$(-\pi/4, -1)$   $y = -\dfrac{1}{2}x - \dfrac{\pi}{8} - 1$

**77.** $\dfrac{1}{4}$    **79.** 4

**81.** Tangent: $y = -\dfrac{1}{4}x + \dfrac{9}{4}$, normal: $y = 4x - 2$

**83.** Tangent: $y = 2x - 4$, normal: $y = -\dfrac{1}{2}x + \dfrac{7}{2}$

**85.** Tangent: $y = -\dfrac{5}{4}x + 6$, normal: $y = \dfrac{4}{5}x - \dfrac{11}{5}$

**87.** $(1, 1)$: $m = -\dfrac{1}{2}$; $(1, -1)$: $m$ not defined

**89.** $B = $ graph of $f$, $A = $ graph of $f'$

**91.**

**93.** (a) $0, 0$    (b) 1700 rabbits, $\approx 1400$ rabbits
**95.** $-1$    **97.** $1/2$    **99.** 4    **101.** 1
**103.** To make $g$ continuous at the origin, define $g(0) = 1$.
**105.** (a) $\dfrac{dS}{dt} = (4\pi r + 2\pi h)\dfrac{dr}{dt}$    (b) $\dfrac{dS}{dt} = 2\pi r\dfrac{dh}{dt}$

(c) $\dfrac{dS}{dt} = (4\pi r + 2\pi h)\dfrac{dr}{dt} + 2\pi r\dfrac{dh}{dt}$

(d) $\dfrac{dr}{dt} = -\dfrac{r}{2r + h}\dfrac{dh}{dt}$

**107.** $-40 \text{ m}^2/\text{s}$    **109.** $0.02 \text{ ohm}/\text{s}$    **111.** $2 \text{ m}/\text{s}$

**113.** (a) $r = \dfrac{2}{5}h$    (b) $-\dfrac{5}{16\pi} \text{ m}/\text{min}$

**115.** (a) $\dfrac{3}{5} \text{ km}/\text{s}$ or $600 \text{ m}/\text{s}$    (b) $\dfrac{18}{\pi}$ rpm

**117.** (a) $L(x) = 2x + \dfrac{\pi - 2}{2}$

(b) $L(x) = -\sqrt{2}x + \dfrac{\sqrt{2}(4 - \pi)}{4}$

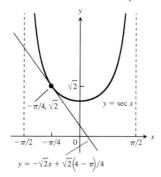

**119.** $L(x) = 1.5x + 0.5$    **121.** $dS = \dfrac{\pi r h_0}{\sqrt{r^2 + h_0^2}}\, dh$

**123.** (a) $4\%$    (b) $8\%$    (c) $12\%$

Additional and Advanced Exercises, pp. 196–198
**1.** (a) $\sin 2\theta = 2\sin\theta\cos\theta$; $2\cos 2\theta = 2\sin\theta(-\sin\theta) + \cos\theta(2\cos\theta)$; $2\cos 2\theta = -2\sin^2\theta + 2\cos^2\theta$; $\cos 2\theta = \cos^2\theta - \sin^2\theta$
(b) $\cos 2\theta = \cos^2\theta - \sin^2\theta$; $-2\sin 2\theta = 2\cos\theta(-\sin\theta) - 2\sin\theta(\cos\theta)$; $\sin 2\theta = \cos\theta\sin\theta + \sin\theta\cos\theta$; $\sin 2\theta = 2\sin\theta\cos\theta$

**3.** (a) $a = 1, b = 0, c = -\dfrac{1}{2}$    (b) $b = \cos a, c = \sin a$

**5.** $h = -4, k = \dfrac{9}{2}, a = \dfrac{5\sqrt{5}}{2}$

**7.** (a) $0.09y$    (b) Increasing at $1\%$ per year
**9.** Answers will vary. Here is one possibility.

**11.** (a) $2 \text{ s}, 19.6 \text{ m}/\text{s}$    (b) $12.25 \text{ s}, 120 \text{ m}$

**15.** (a) $m = -\dfrac{b}{\pi}$    (b) $m = -1, b = \pi$

**17.** (a) $a = \dfrac{3}{4}, b = \dfrac{9}{4}$    **19.** $f$ odd $\Rightarrow f'$ is even

**23.** $h'$ is defined but not continuous at $x = 0$; $k'$ is defined *and* continuous at $x = 0$.
**27.** (a) $0.248 \text{ m}$    (b) $0.02015 \text{ s}$
(c) It will lose about $28.3 \text{ min}/\text{day}$.

# Chapter 4

Section 4.1, pp. 204–207
**1.** Absolute minimum at $x = c_2$; absolute maximum at $x = b$
**3.** Absolute maximum at $x = c$; no absolute minimum
**5.** Absolute minimum at $x = a$; absolute maximum at $x = c$
**7.** No absolute minimum; no absolute maximum
**9.** Absolute maximum at $(0, 5)$    **11.** (c)    **13.** (d)
**15.** Absolute minimum at $x = 0$; no absolute maximum    **17.** Absolute maximum at $x = 2$; no absolute minimum

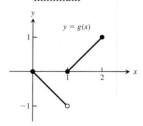

**19.** Absolute maximum at $x = \pi/2$; absolute minimum at $x = 3\pi/2$

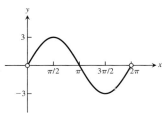

**21.** Absolute maximum: $-3$; absolute minimum: $-19/3$

**23.** Absolute maximum: 3; absolute minimum: $-1$

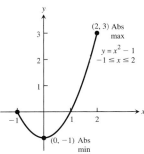

**25.** Absolute maximum: $-0.25$; absolute minimum $-4$

**27.** Absolute maximum: 2; absolute minimum: $-1$

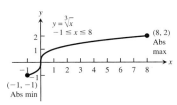

**29.** Absolute maximum: 2; absolute minimum: 0

**31.** Absolute maximum: 1; absolute minimum: $-1$

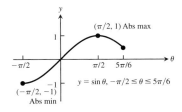

**33.** Absolute maximum $2/\sqrt{3}$; absolute minimum: 1

**35.** Absolute maximum: 2; absolute minimum: $-1$

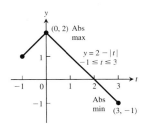

**37.** Increasing on $(0, 8)$, decreasing on $(-1, 0)$; absolute maximum: 16 at $x = 8$; absolute minimum: 0 at $x = 0$

**39.** Increasing on $(-32, 1)$; absolute maximum: 1 at $\theta = 1$; absolute minimum: $-8$ at $\theta = -32$

**41.** $x = 3$

**43.** $x = 1, x = 4$

**45.** $x = 1$

**47.** $x = 0$ and $x = 4$

**49.** Minimum value is 1 at $x = 2$.

**51.** Local maximum at $(-2, 17)$; local minimum at $\left(\dfrac{4}{3}, -\dfrac{41}{27}\right)$

**53.** Minimum value is 0 at $x = -1$ and $x = 1$.

**55.** There is a local minimum at $(0, 1)$.

**57.** Maximum value is $\dfrac{1}{2}$ at $x = 1$; minimum value is $-\dfrac{1}{2}$ at $x = -1$.

**59.**

| Critical point or endpoint | Derivative | Extremum | Value |
|---|---|---|---|
| $x = -\dfrac{4}{5}$ | 0 | Local max | $\dfrac{12}{25}10^{1/3} \approx 1.034$ |
| $x = 0$ | Undefined | Local min | 0 |

**61.**

| Critical point or endpoint | Derivative | Extremum | Value |
|---|---|---|---|
| $x = -2$ | Undefined | Local max | 0 |
| $x = -\sqrt{2}$ | 0 | Minimum | $-2$ |
| $x = \sqrt{2}$ | 0 | Maximum | 2 |
| $x = 2$ | Undefined | Local min | 0 |

**63.**

| Critical point or endpoint | Derivative | Extremum | Value |
|---|---|---|---|
| $x = 1$ | Undefined | Minimum | 2 |

**65.**

| Critical point or endpoint | Derivative | Extremum | Value |
|---|---|---|---|
| $x = -1$ | 0 | Maximum | 5 |
| $x = 1$ | Undefined | Local min | 1 |
| $x = 3$ | 0 | Maximum | 5 |

**67.** (a) No
   (b) The derivative is defined and nonzero for $x \ne 2$. Also, $f(2) = 0$ and $f(x) > 0$ for all $x \ne 2$.
   (c) No, because $(-\infty, \infty)$ is not a closed interval.
   (d) The answers are the same as parts (a) and (b), with 2 replaced by $a$.

**69.** Yes

**71.** $g$ assumes a local maximum at $-c$.

**73.** (a) Maximum value is 144 at $x = 2$.
   (b) The largest volume of the box is 144 cubic units, and it occurs when $x = 2$.

**75.** $\dfrac{v_0^2}{2g} + s_0$

**77.** Maximum value is 11 at $x = 5$; minimum value is 5 on the interval $[-3, 2]$; local maximum at $(-5, 9)$.

**79.** Maximum value is 5 on the interval $[3, \infty)$; minimum value is $-5$ on the interval $(-\infty, -2]$.

**Section 4.2, pp. 211–213**

**1.** $1/2$    **3.** $1$

**5.** $\frac{1}{3}\left(1 + \sqrt{7}\right) \approx 1.22, \frac{1}{3}\left(1 - \sqrt{7}\right) \approx -0.549$

**7.** Does not; $f$ is not differentiable at the interior domain point $x = 0$.

**9.** Does    **11.** Does not; $f$ is not differentiable at $x = -1$.

**15. (a)**

**27.** Yes    **29. (a)** 4    **(b)** 3    **(c)** 3

**31. (a)** $\frac{x^2}{2} + C$    **(b)** $\frac{x^3}{3} + C$    **(c)** $\frac{x^4}{4} + C$

**33. (a)** $\frac{1}{x} + C$    **(b)** $x + \frac{1}{x} + C$    **(c)** $5x - \frac{1}{x} + C$

**35. (a)** $-\frac{1}{2}\cos 2t + C$    **(b)** $2\sin\frac{t}{2} + C$

    **(c)** $-\frac{1}{2}\cos 2t + 2\sin\frac{t}{2} + C$

**37.** $f(x) = x^2 - x$    **39.** $r(\theta) = 8\theta + \cot\theta - 2\pi - 1$

**41.** $s = 4.9t^2 + 5t + 10$    **43.** $s = \dfrac{1 - \cos(\pi t)}{\pi}$

**45.** $16t^2 + 20t + 5$    **47.** $s = \sin(2t) - 3$

**49.** If $T(t)$ is the temperature of the thermometer at time $t$, then $T(0) = -19\,°C$ and $T(14) = 100\,°C$. From the Mean Value Theorem, there exists a $0 < t_0 < 14$ such that $\dfrac{T(14) - T(0)}{14 - 0} = 8.5\,°C/s = T'(t_0)$, the rate at which the temperature was changing at $t = t_0$ as measured by the rising mercury on the thermometer.

**51.** Because its average speed was approximately 7.667 knots, and by the Mean Value Theorem, it must have been going that speed at least once during the trip.

**55.** The conclusion of the Mean Value Theorem yields

$$\frac{\frac{1}{b} - \frac{1}{a}}{b - a} = -\frac{1}{c^2} \Rightarrow c^2\left(\frac{a - b}{ab}\right) = a - b \Rightarrow c = \sqrt{ab}.$$

**59.** $f(x)$ must be zero at least once between $a$ and $b$ by the Intermediate Value Theorem. Now suppose that $f(x)$ is zero twice between $a$ and $b$. Then, by the Mean Value Theorem, $f'(x)$ would have to be zero at least once between the two zeros of $f(x)$, but this can't be true since we are given that $f'(x) \neq 0$ on this interval. Therefore, $f(x)$ is zero once and only once between $a$ and $b$.

**69.** $1.09999 \leq f(0.1) \leq 1.1$

**Section 4.3, pp. 217–218**

**1. (a)** 0, 1

   **(b)** Increasing on $(-\infty, 0)$ and $(1, \infty)$; decreasing on $(0, 1)$

   **(c)** Local maximum at $x = 0$; local minimum at $x = 1$

**3. (a)** $-2, 1$

   **(b)** Increasing on $(-2, 1)$ and $(1, \infty)$; decreasing on $(-\infty, -2)$

   **(c)** No local maximum; local minimum at $x = -2$

**5. (a)** $-2, 1, 3$    **(b)** increasing on $(-2, 1)$ and $(3, \infty)$, decreasing on $(-\infty, -2)$ and $(1, 3)$    **(c)** local maximum at $x = 1$, local minimum at $x = -2, 3$

**7. (a)** 0, 1

   **(b)** Increasing on $(-\infty, -2)$ and $(1, \infty)$; decreasing on $(-2, 0)$ and $(0, 1)$

   **(c)** Local minimum at $x = 1$

**9. (a)** $-2, 2$

   **(b)** Increasing on $(-\infty, -2)$ and $(2, \infty)$; decreasing on $(-2, 0)$ and $(0, 2)$

   **(c)** Local maximum at $x = -2$; local minimum at $x = 2$

**11. (a)** $-2, 0$

   **(b)** Increasing on $(-\infty, -2)$ and $(0, \infty)$; decreasing on $(-2, 0)$

   **(c)** Local maximum at $x = -2$; local minimum at $x = 0$

**13. (a)** $\dfrac{\pi}{2}, \dfrac{2\pi}{3}, \dfrac{4\pi}{3}$

   **(b)** Increasing on $\left(\dfrac{2\pi}{3}, \dfrac{4\pi}{3}\right)$; decreasing on $\left(0, \dfrac{\pi}{2}\right), \left(\dfrac{\pi}{2}, \dfrac{2\pi}{3}\right)$, and $\left(\dfrac{4\pi}{3}, 2\pi\right)$

   **(c)** Local maximum at $x = 0$ and $x = \dfrac{4\pi}{3}$; local minimum at $x = \dfrac{2\pi}{3}$ and $x = 2\pi$

**15. (a)** Increasing on $(-2, 0)$ and $(2, 4)$; decreasing on $(-4, -2)$ and $(0, 2)$

   **(b)** Absolute maximum at $(-4, 2)$; local maximum at $(0, 1)$ and $(4, -1)$; absolute minimum at $(2, -3)$; local minimum at $(-2, 0)$

**17. (a)** Increasing on $(-4, -1), (1/2, 2)$, and $(2, 4)$; decreasing on $(-1, 1/2)$

   **(b)** Absolute maximum at $(4, 3)$; local maximum at $(-1, 2)$ and $(2, 1)$; no absolute minimum; local minimum at $(-4, -1)$ and $(1/2, -1)$

**19. (a)** Increasing on $(-\infty, -1.5)$; decreasing on $(-1.5, \infty)$

   **(b)** Local maximum: 5.25 at $t = -1.5$; absolute maximum: 5.25 at $t = -1.5$

**21. (a)** Decreasing on $(-\infty, 0)$; increasing on $(0, 4/3)$; decreasing on $(4/3, \infty)$

   **(b)** Local minimum at $x = 0$ $(0, 0)$; local maximum at $x = 4/3$ $(4/3, 32/27)$; no absolute extrema

**23. (a)** Decreasing on $(-\infty, 0)$; increasing on $(0, 1/2)$; decreasing on $(1/2, \infty)$

   **(b)** Local minimum at $\theta = 0$ $(0, 0)$; local maximum at $\theta = 1/2$ $(1/2, 1/4)$; no absolute extrema

**25. (a)** Increasing on $(-\infty, \infty)$; never decreasing

   **(b)** No local extrema; no absolute extrema

**27. (a)** Increasing on $(-2, 0)$ and $(2, \infty)$; decreasing on $(-\infty, -2)$ and $(0, 2)$

   **(b)** Local maximum: 16 at $x = 0$; local minimum: 0 at $x = \pm 2$; no absolute maximum; absolute minimum: 0 at $x = \pm 2$

**29. (a)** Increasing on $(-\infty, -1)$; decreasing on $(-1, 0)$; increasing on $(0, 1)$; decreasing on $(1, \infty)$

   **(b)** Local maximum: 0.5 at $x = \pm 1$; local minimum: 0 at $x = 0$; absolute maximum: $1/2$ at $x = \pm 1$; no absolute minimum

**31. (a)** Increasing on $(10, \infty)$; decreasing on $(1, 10)$

   **(b)** Local maximum: 1 at $x = 1$; local minimum: $-8$ at $x = 10$; absolute minimum: $-8$ at $x = 10$

**33. (a)** Decreasing on $\left(-2\sqrt{2}, -2\right)$; increasing on $(-2, 2)$; decreasing on $\left(2, 2\sqrt{2}\right)$

   **(b)** Local minima: $g(-2) = -4, g\left(2\sqrt{2}\right) = 0$; local maxima: $g\left(-2\sqrt{2}\right) = 0, g(2) = 4$; absolute maximum: 4 at $x = 2$; absolute minimum: $-4$ at $x = -2$

**35. (a)** Increasing on $(-\infty, 1)$; decreasing when $1 < x < 2$, decreasing when $2 < x < 3$; discontinuous at $x = 2$; increasing on $(3, \infty)$

**(b)** Local minimum at $x = 3$ $(3, 6)$; local maximum at $x = 1$ $(1, 2)$; no absolute extrema

**37. (a)** Increasing on $(-2, 0)$ and $(0, \infty)$; decreasing on $(-\infty, -2)$

**(b)** Local minimum: $-6\sqrt[3]{2}$ at $x = -2$; no absolute maximum; absolute minimum: $-6\sqrt[3]{2}$ at $x = -2$

**39. (a)** Increasing on $\left(-\infty, -2/\sqrt{7}\right)$ and $\left(2/\sqrt{7}, \infty\right)$; decreasing on $\left(-2/\sqrt{7}, 0\right)$ and $\left(0, 2/\sqrt{7}\right)$

**(b)** Local maximum: $24\sqrt[3]{2}/7^{7/6} \approx 3.12$ at $x = -2/\sqrt{7}$; local minimum: $-24\sqrt[3]{2}/7^{7/6} \approx -3.12$ at $x = 2/\sqrt{7}$; no absolute extrema

**41. (a)** Local maximum: 1 at $x = 1$; local minimum: 0 at $x = 2$

**(b)** Absolute maximum: 1 at $x = 1$; no absolute minimum

**43. (a)** Local maximum: 1 at $x = 1$; local minimum: 0 at $x = 2$

**(b)** No absolute maximum; absolute minimum: 0 at $x = 2$

**45. (a)** Local maxima: $-9$ at $t = -3$ and 16 at $t = 2$; local minimum: $-16$ at $t = -2$

**(b)** Absolute maximum: 16 at $t = 2$; no absolute minimum

**47. (a)** Local minimum: 0 at $x = 0$

**(b)** No absolute maximum; absolute minimum: 0 at $x = 0$

**49. (a)** Local maximum: 5 at $x = 0$; local minimum: 0 at $x = -5$ and $x = 5$

**(b)** Absolute maximum: 5 at $x = 0$; absolute minimum: 0 at $x = -5$ and $x = 5$

**51. (a)** Local maximum: 2 at $x = 0$;

local minimum: $\dfrac{\sqrt{3}}{4\sqrt{3} - 6}$ at $x = 2 - \sqrt{3}$

**(b)** No absolute maximum; an absolute minimum at $x = 2 - \sqrt{3}$

**53. (a)** Local maximum: 1 at $x = \pi/4$;

local maximum: 0 at $x = \pi$;

local minimum: 0 at $x = 0$;

local minimum: $-1$ at $x = 3\pi/4$

**55.** Local maximum: 2 at $x = \pi/6$;

local maximum: $\sqrt{3}$ at $x = 2\pi$;

local minimum: $-2$ at $x = 7\pi/6$;

local minimum: $\sqrt{3}$ at $x = 0$

**57. (a)** Local minimum: $(\pi/3) - \sqrt{3}$ at $x = 2\pi/3$;

local maximum: 0 at $x = 0$;

local maximum: $\pi$ at $x = 2\pi$

**59. (a)** Local minimum: 0 at $x = \pi/4$

**61.** Local maximum: 3 at $\theta = 0$;

local minimum: $-3$ at $\theta = 2\pi$

**63.**

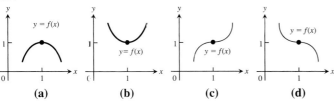

**(a)**     **(b)**     **(c)**     **(d)**

**65. (a)**         **(b)**

 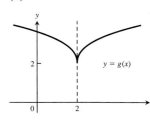

**69.** $a = -2, b = 4$

**Section 4.4, pp. 226–229**

**1.** Local maximum: $3/2$ at $x = -1$; local minimum: $-3$ at $x = 2$; point of inflection at $(1/2, -3/4)$; rising on $(-\infty, -1)$ and $(2, \infty)$; falling on $(-1, 2)$; concave up on $(1/2, \infty)$; concave down on $(-\infty, 1/2)$

**3.** Local maximum: $3/4$ at $x = 0$; local minimum: 0 at $x = \pm 1$; points of inflection at $\left(-\sqrt{3}, \dfrac{3\sqrt[3]{4}}{4}\right)$ and $\left(\sqrt{3}, \dfrac{3\sqrt[3]{4}}{4}\right)$; rising on $(-1, 0)$ and $(1, \infty)$; falling on $(-\infty, -1)$ and $(0, 1)$; concave up on $\left(-\infty, -\sqrt{3}\right)$ and $\left(\sqrt{3}, \infty\right)$; concave down on $\left(-\sqrt{3}, \sqrt{3}\right)$

**5.** Local maxima: $\dfrac{-2\pi}{3} + \dfrac{\sqrt{3}}{2}$ at $x = -2\pi/3$, $\dfrac{\pi}{3} + \dfrac{\sqrt{3}}{2}$ at $x = \pi/3$; local minima: $-\dfrac{\pi}{3} - \dfrac{\sqrt{3}}{2}$ at $x = -\pi/3$, $\dfrac{2\pi}{3} - \dfrac{\sqrt{3}}{2}$ at $x = 2\pi/3$; points of inflection at $(-\pi/2, -\pi/2)$, $(0, 0)$, and $(\pi/2, \pi/2)$; rising on $(-\pi/3, \pi/3)$; falling on $(-2\pi/3, -\pi/3)$ and $(\pi/3, 2\pi/3)$; concave up on $(-\pi/2, 0)$ and $(\pi/2, 2\pi/3)$; concave down on $(-2\pi/3, -\pi/2)$ and $(0, \pi/2)$

**7.** Local maxima: 1 at $x = -\pi/2$ and $x = \pi/2$, 0 at $x = -2\pi$ and $x = 2\pi$; local minima: $-1$ at $x = -3\pi/2$ and $x = 3\pi/2$, 0 at $x = 0$; points of inflection at $(-\pi, 0)$ and $(\pi, 0)$; rising on $(-3\pi/2, -\pi/2)$, $(0, \pi/2)$, and $(3\pi/2, 2\pi)$; falling on $(-2\pi, -3\pi/2)$, $(-\pi/2, 0)$, and $(\pi/2, 3\pi/2)$; concave up on $(-2\pi, -\pi)$ and $(\pi, 2\pi)$; concave down on $(-\pi, 0)$ and $(0, \pi)$

**9.**         **11.**

**13.**         **15.**

**17.**

**19.**

**37.**

**39.**

**21.**

**23.**

**41.**

**43.**

**25.**

**45.**

**47.**

**27.**

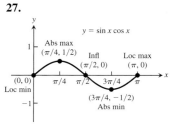

**49.** $y'' = 1 - 2x$

**29.**

**31.**

**51.** $y'' = 3(x - 3)(x - 1)$   **53.** $y'' = 3(x - 2)(x + 2)$

**33.**

**35.**

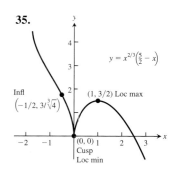

**55.** $y'' = 4(4 - x)(5x^2 - 16x + 8)$

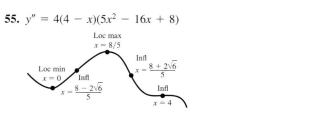

**57.** $y'' = 2 \sec^2 x \tan x$

**59.** $y'' = -\frac{1}{2}\csc^2\frac{\theta}{2}$, $0 < \theta < 2\pi$

**61.** $y'' = 2\tan\theta\sec^2\theta$, $-\frac{\pi}{2} < \theta < \frac{\pi}{2}$

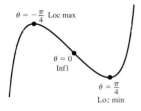

**63.** $y'' = -\sin t$, $0 \le t \le 2\pi$

**65.** $y'' = -\frac{2}{3}(x+1)^{-5/3}$

**67.** $y'' = \frac{1}{3}x^{-2/3} + \frac{2}{3}x^{-5/3}$

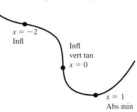

**69.** $y'' = \begin{cases} -2, & x < 0 \\ 2, & x > 0 \end{cases}$

**71.**

**73.**

**75.**

**77.**

**79.**

**81.**

**83.**

**85.**

**87.**

**89.**

**91.**

**93.**

| Point | $y'$ | $y''$ |
|-------|------|-------|
| $P$ | $-$ | $+$ |
| $Q$ | $+$ | $0$ |
| $R$ | $+$ | $-$ |
| $S$ | $0$ | $-$ |
| $T$ | $-$ | $-$ |

**95.**

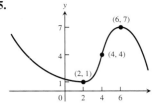

**97. (a)** Towards origin: $0 \le t < 2$ and $6 \le t \le 10$; away from origin: $2 \le t \le 6$ and $10 \le t \le 15$
  **(b)** $t = 2, t = 6, t = 10$
  **(c)** $t = 5, t = 7, t = 13$
  **(d)** Positive: $5 \le t \le 7$, $13 \le t \le 15$; negative: $0 \le t \le 5, 7 \le t \le 13$

**99.** $\approx 60$ thousand units

**101.** Local minimum at $x = 2$; inflection points at $x = 1$ and $x = 5/3$

**105.** $b = -3$      **109.** $-1, 2$

**111.** $a = 1, b = 3, c = 9$

**113.** The zeros of $y' = 0$ and $y'' = 0$ are extrema and points of inflection, respectively. Inflection at $x = 3$, local maximum at $x = 0$, local minimum at $x = 4$.

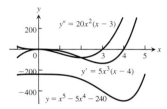

**115.** The zeros of $y' = 0$ and $y'' = 0$ are extrema and points of inflection, respectively. Inflection at $x = -\sqrt[3]{2}$; local maximum at $x = -2$; local minimum at $x = 0$.

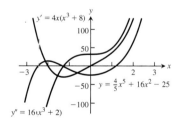

**Section 4.5, pp. 235–241**

**1.** 16 cm, 4 cm by 4 cm

**3. (a)** $(x, 1 - x)$      **(b)** $A(x) = 2x(1 - x)$
  **(c)** $\dfrac{1}{2}$ square units, 1 by $\dfrac{1}{2}$

**5.** $14 \times 35 \times 5$ cm, 2450 cm³

**7.** 80,000 m²; 400 m by 200 m

**9. (a)** The optimum dimensions of the tank are 2 m on the base edges and 1 m deep.
  **(b)** Minimizing the surface area of the tank minimizes its weight for a given wall thickness. The thickness of the steel walls would likely be determined by other considerations such as structural requirements.

**11.** $45 \times 22.5$ cm      **13.** $\dfrac{\pi}{2}$      **15.** $h : r = 8 : \pi$

**17. (a)** $V(x) = 2x(60 - 2x)(45 - 2x)$      **(b)** Domain: $(0, 22.5)$

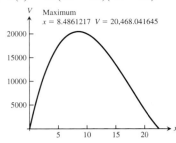

  **(c)** Maximum volume $\approx 20{,}468$ cm³ when $x \approx 8.49$ cm
  **(d)** $V'(x) = 24x^2 - 840x + 5400$, so the critical point is at $x = \dfrac{35 - 5\sqrt{13}}{2}$, which confirms the result in part (c).
  **(e)** $x = 12.5$ cm or $x = 5$ cm

**19.** $\approx 2418.40$ cm³

**21. (a)** $h = 184/3, w = 46$
  **(b)**

**23.** If $r$ is the radius of the hemisphere, $h$ the height of the cylinder, and $V$ the volume, then $r = \left(\dfrac{3V}{8\pi}\right)^{1/3}$ and $h = \left(\dfrac{3V}{\pi}\right)^{1/3}$.

**25. (b)** $x = 16.2$      **(c)** $L \approx 28$ cm

**27.** Radius $= \sqrt{2}$ m, height $= 1$ m, volume $= \dfrac{2\pi}{3}$ m³

**29.** 1      **31.** $\dfrac{9b}{9 + \sqrt{3}\pi}$ m, triangle; $\dfrac{b\sqrt{3}\pi}{9 + \sqrt{3}\pi}$ m, circle

**33.** $\dfrac{3}{2} \times 2$

**35. (a)** 16      **(b)** $-1$

**37. (a)** $v(0) = 29.4$ m/s
  **(b)** 78.4 m at $t = 3$ s
  **(c)** Velocity when $s = 0$ is $v(7) = -39.2$ m/s.

**39.** $\approx 9.58$ m      **41.** 4 m

**43. (a)** $10\sqrt{3} \times 10\sqrt{6}$ cm

**45. (a)** $10\pi \approx 31.42$ cm/s; when $t = 0.5$ s, 1.5 s, 2.5 s, 3.5 s; $s = 0$, acceleration is 0.
  **(b)** 10 cm from rest position; speed is 0.

**47. (a)** $s = ((12 - 12t)^2 + 64t^2)^{1/2}$
  **(b)** $-12$ knots, 8 knots
  **(c)** No
  **(d)** $4\sqrt{13}$. This limit is the square root of the sums of the squares of the individual speeds.

**49.** $x = \dfrac{a}{2}, v = \dfrac{ka^2}{4}$

**51.** $\dfrac{c}{2} + 50$

**53. (a)** $\sqrt{\dfrac{2km}{h}}$      **(b)** $\sqrt{\dfrac{2km}{h}}$

**57.** $2 \times 2 \times \dfrac{3}{2}$ m, \$720      **59.** $M = \dfrac{C}{2}$      **65. (a)** $y = -1$

**67. (a)** The minimum distance is $\dfrac{\sqrt{5}}{2}$.

**(b)** The minimum distance is from the point $(3/2, 0)$ to the point $(1, 1)$ on the graph of $y = \sqrt{x}$, and this occurs at the value $x = 1$, where $D(x)$, the distance squared, has its minimum value.

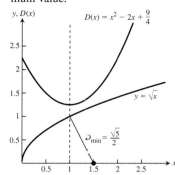

**Section 4.6, pp. 244–246**

**1.** $x_2 = -\dfrac{5}{3}, \dfrac{13}{21}$  **3.** $x_2 = -\dfrac{51}{31}, \dfrac{5763}{4945}$  **5.** $x_2 = \dfrac{2387}{2000}$

**7.** $x_1$, and all later approximations will equal $x_0$.

**9.**

$$y = \begin{cases} \sqrt{x}, & x \geq 0 \\ \sqrt{-x}, & x < 0 \end{cases}$$

**11.** The points of intersection of $y = x^3$ and $y = 3x + 1$ or $y = x^3 - 3x$ and $y = 1$ have the same $x$-values as the roots of part (i) or the solutions of part (iv).  **13.** 1.165561185

**15. (a)** Two  **(b)** 0.35003501505249 and $-1.0261731615301$

**17.** $\pm 1.3065629648764, \pm 0.5411961001462$  **19.** $x \approx 0.45$

**21.** 0.8192  **23.** The root is 1.17951.

**25. (a)** For $x_0 = -2$ or $x_0 = -0.8$, $x_i \to -1$ as $i$ gets large.
**(b)** For $x_0 = -0.5$ or $x_0 = 0.25$, $x_i \to 0$ as $i$ gets large.
**(c)** For $x_0 = 0.8$ or $x_0 = 2$, $x_i \to 1$ as $i$ gets large.
**(d)** For $x_0 = -\sqrt{21}/7$ or $x_0 = \sqrt{21}/7$, Newton's method does not converge. The values of $x_i$ alternate between $-\sqrt{21}/7$ and $\sqrt{21}/7$ as $i$ increases.

**27.** Answers will vary with machine speed.

**Section 4.7, pp. 252–256**

**1. (a)** $x^2$  **(b)** $\dfrac{x^3}{3}$  **(c)** $\dfrac{x^3}{3} - x^2 + x$

**3. (a)** $x^{-3}$  **(b)** $-\dfrac{1}{3}x^{-3}$  **(c)** $-\dfrac{1}{3}x^{-3} + x^2 + 3x$

**5. (a)** $-\dfrac{1}{x}$  **(b)** $-\dfrac{5}{x}$  **(c)** $2x + \dfrac{5}{x}$

**7. (a)** $\sqrt{x^3}$  **(b)** $\sqrt{x}$  **(c)** $\dfrac{2\sqrt{x^3}}{3} + 2\sqrt{x}$

**9. (a)** $x^{2/3}$  **(b)** $x^{1/3}$  **(c)** $x^{-1/3}$

**11. (a)** $\cos(\pi x)$  **(b)** $-3\cos x$  **(c)** $-\dfrac{1}{\pi}\cos(\pi x) + \cos(3x)$

**13. (a)** $\dfrac{1}{2}\tan x$  **(b)** $2\tan\left(\dfrac{x}{3}\right)$  **(c)** $-\dfrac{2}{3}\tan\left(\dfrac{3x}{2}\right)$

**15. (a)** $-\csc x$  **(b)** $\dfrac{1}{5}\csc(5x)$  **(c)** $2\csc\left(\dfrac{\pi x}{2}\right)$

**17.** $\dfrac{x^2}{2} + x + C$  **19.** $t^3 + \dfrac{t^2}{4} + C$  **21.** $\dfrac{x^4}{2} - \dfrac{5x^2}{2} + 7x + C$

**23.** $-\dfrac{1}{x} - \dfrac{x^3}{3} - \dfrac{x}{3} + C$  **25.** $\dfrac{3}{2}x^{2/3} + C$

**27.** $\dfrac{2}{3}x^{3/2} + \dfrac{3}{4}x^{4/3} + C$  **29.** $4y^2 - \dfrac{8}{3}y^{3/4} + C$

**31.** $x^2 + \dfrac{2}{x} + C$  **33.** $2\sqrt{t} - \dfrac{2}{\sqrt{t}} + C$  **35.** $-2\sin t + C$

**37.** $-21\cos\dfrac{\theta}{3} + C$  **39.** $3\cot x + C$  **41.** $-\dfrac{1}{2}\csc\theta + C$

**43.** $4\sec x - 2\tan x + C$  **45.** $-\dfrac{1}{2}\cos 2x + \cot x + C$

**47.** $\dfrac{t}{2} + \dfrac{\sin 4t}{8} + C$  **49.** $\dfrac{3x^{(\sqrt{3}+1)}}{\sqrt{3}+1} + C$  **51.** $\tan\theta + C$

**53.** $-\cot x - x + C$  **55.** $-\cos\theta + \theta + C$

**63. (a)** Wrong: $\dfrac{d}{dx}\left(\dfrac{x^2}{2}\sin x + C\right) = \dfrac{2x}{2}\sin x + \dfrac{x^2}{2}\cos x = x\sin x + \dfrac{x^2}{2}\cos x$

**(b)** Wrong: $\dfrac{d}{dx}(-x\cos x + C) = -\cos x + x\sin x$

**(c)** Right: $\dfrac{d}{dx}(-x\cos x + \sin x + C) = -\cos x + x\sin x + \cos x = x\sin x$

**65. (a)** Wrong: $\dfrac{d}{dx}\left(\dfrac{(2x+1)^3}{3} + C\right) = \dfrac{3(2x+1)^2(2)}{3} = 2(2x+1)^2$

**(b)** Wrong: $\dfrac{d}{dx}((2x+1)^3 + C) = 3(2x+1)^2(2) = 6(2x+1)^2$

**(c)** Right: $\dfrac{d}{dx}((2x+1)^3 + C) = 6(2x+1)^2$

**67.** Right  **69. (b)**  **71.** $y = x^2 - 7x + 10$

**73.** $y = -\dfrac{1}{x} + \dfrac{x^2}{2} - \dfrac{1}{2}$  **75.** $y = 9x^{1/3} + 4$

**77.** $s = t + \sin t + 4$  **79.** $r = \cos(\pi\theta) - 1$

**81.** $v = \dfrac{1}{2}\sec t + \dfrac{1}{2}$  **83.** $y = x^2 - x^3 + 4x + 1$

**85.** $r = \dfrac{1}{t} + 2t - 2$  **87.** $y = x^3 - 4x^2 + 5$

**89.** $y = -\sin t + \cos t + t^3 - 1$  **91.** $y = 2x^{3/2} - 50$

**93.** $y = x - x^{4/3} + \dfrac{1}{2}$  **95.** $y = -\sin x - \cos x - 2$

**97. (a) (i)** 33.2 units, **(ii)** 33.2 units, **(iii)** 33.2 units  **(b)** True

**99.** $t = 30/k, k = 6$

**101. (a)** $v = 10t^{3/2} - 6t^{1/2}$  **(b)** $s = 4t^{5/2} - 4t^{3/2}$

**105. (a)** $-\sqrt{x} + C$  **(b)** $x + C$  **(c)** $\sqrt{x} + C$
**(d)** $-x + C$  **(e)** $x - \sqrt{x} + C$  **(f)** $-x - \sqrt{x} + C$

**Practice Exercises, pp. 257–259**

**1.** No  **3.** No minimum; absolute maximum: $f(1) = 16$; critical points: $x = 1$ and $11/3$

**5.** Yes, except at $x = 0$  **7.** No  **11. (b)** one

**13. (b)** 0.8555 99677 2

**19.** Global minimum value of $\dfrac{1}{2}$ at $x = 2$

**21. (a)** $t = 0, 6, 12$ **(b)** $t = 3, 9$ **(c)** $6 < t < 12$
**(d)** $0 < t < 6, 12 < t < 14$

**23.**

**25.**

**27.**

**29.**

**31.**

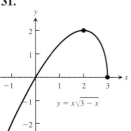

**33. (a)** Local maximum at $x = 4$, local minimum at $x = -4$,
inflection point at $x = 0$
**(b)**

**35. (a)** Local maximum at $x = 0$, local minima at $x = -1$ and
$x = 2$, inflection points at $x = (1 \pm \sqrt{7})/3$
**(b)**

**37. (a)** Local maximum at $x = -\sqrt{2}$, local minimum at $x = \sqrt{2}$,
inflection points at $x = \pm 1$ and $0$
**(b)**

**43.**

**45.**

**47.**

**49.**

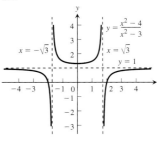

**51. (a)** $0, 36$ **(b)** $18, 18$ **53.** 54 square units
**55.** height $= 2$, radius $= \sqrt{2}$
**57.** $x = 5 - \sqrt{5}$ hundred $\approx 276$ tires,
$y = 2(5 - \sqrt{5})$ hundred $\approx 553$ tires
**59.** Dimensions: base is 15 cm by 30 cm, height $= 5$ cm;
maximum volume $= 2250$ cm$^3$
**61.** $x_5 = 2.1958\,23345$ **63.** $\dfrac{x^6}{6} + 4\dfrac{x^3}{3} + 9x + C$
**65.** $\dfrac{2}{5}t^{5/2} - \dfrac{5}{2t^2} + C$ **67.** $-\dfrac{1}{2(r - 3)^2} + C$ **69.** $2(\theta^3 - 1)^{3/2} + C$
**71.** $-\dfrac{1}{4}(1 - x^5)^{4/5} + C$ **73.** $10 \tan \dfrac{s}{10} + C$
**75.** $-\dfrac{1}{\sqrt{3}} \csc \sqrt{3}\theta + C$ **77.** $\dfrac{1}{2}x + 3 \sin \left(\dfrac{x}{3}\right) + C$
**79.** $y = x - \dfrac{1}{x} - 1$ **81.** $r = 4t^{5/2} + 4t^{3/2} - 8t$

**Additional and Advanced Exercises, pp. 259–261**
**1.** The function is constant on the interval.
**3.** The extreme points will not be at the end of an open interval.
**5. (a)** A local minimum at $x = -1$, points of inflection at $x = 0$
and $x = 2$
**(b)** A local maximum at $x = 0$ and local minima at $x = -1$
and $x = 2$, points of inflection at $x = \dfrac{1 \pm \sqrt{7}}{3}$
**9.** No **11.** $a = 1, b = 0, c = 1$ **13.** Yes
**15.** Drill the hole at $y = h/2$.
**17.** $r = \dfrac{RH}{2(H - R)}$ for $H > 2R, r = R$ if $H \leq 2R$
**19. (a)** $\dfrac{c - b}{2e}$ **(b)** $\dfrac{c + b}{2}$ **(c)** $\dfrac{b^2 - 2bc + c^2 + 4ae}{4e}$
**(d)** $\dfrac{c + b + t}{2}$
**21.** $m_0 = 1 - \dfrac{1}{q}, m_1 = \dfrac{1}{q}$
**23. (a)** $k = 15$ **(b)** 7.5 m
**25.** Yes, $y = x + C$ **27.** $v_0 = \dfrac{2\sqrt{2}}{3}b^{3/4}$

# Chapter 5

## Section 5.1, pp. 271–273

1. (a) 0.125   (b) 0 21875   (c) 0.625   (d) 0.46875
3. (a) 1.066667   (b) 1.283333   (c) 2.666667   (d) 2.083333
5. 0.3125, 0.328125       7. 1.5, 1.574603
9. (a) 87 cm   (b) 87 cm       11. (a) 1180 m   (b) 1300 m
13. (a) 22.862 m/s   (b) 13.867 m/s   (c) 44.893 m
15. $\dfrac{31}{16}$       17. 1
19. (a) Upper = 758 L, lower = 543 L
    (b) Upper = 2363 L, lower = 1693 L
    (c) ≈ 31.4 h, ≈ 32.4 h
21. (a) 2   (b) $2\sqrt{2} \approx 2.828$

   (c) $8\sin\left(\dfrac{\pi}{8}\right) \approx 3 061$

   (d) Each area is less than the area of the circle, $\pi$. As $n$
       increases, the polygon area approaches $\pi$.

## Section 5.2, pp. 279–280

1. $\dfrac{6(1)}{1+1} + \dfrac{6(2)}{2+1} = 7$

3. $\cos(1)\pi + \cos(2)\pi + \cos(3)\pi + \cos(4)\pi = 0$

5. $\sin\pi - \sin\dfrac{\pi}{2} + \sin\dfrac{\pi}{3} = \dfrac{\sqrt{3}-2}{2}$       7. All of them       9. b

11. $\sum\limits_{k=1}^{6} k$   13. $\sum\limits_{k=1}^{4} \dfrac{1}{2^k}$   15. $\sum\limits_{k=1}^{5} (-1)^{k+1}\dfrac{1}{k}$
17. (a) −15   (b) 1   (c) 1   (d) −11   (e) 16
19. (a) 55   (b) 385   (c) 3025
21. −56     23. −73     25. 240     27. 3376
29. (a) 21   (b) 3500   (c) 2620
31. (a) 4n   (b) cn   (c) $(n^2 - n)/2$
33. (a)                                    (b)

(c)

35. (a)                         (b)

(c)                              37. 1.2

39. $\dfrac{2}{3} - \dfrac{1}{2n} - \dfrac{1}{6n^2}, \quad \dfrac{2}{3}$

41. $12 + \dfrac{27n+9}{2n^2}, \quad 12$

43. $\dfrac{5}{6} + \dfrac{6n+1}{6n^2}, \quad \dfrac{5}{6}$

45. $\dfrac{1}{2} + \dfrac{1}{n} + \dfrac{1}{2n^2}, \quad \dfrac{1}{2}$

## Section 5.3, pp. 288–292

1. $\displaystyle\int_0^2 x^2\,dx$   3. $\displaystyle\int_{-7}^5 (x^2 - 3x)\,dx$   5. $\displaystyle\int_2^3 \dfrac{1}{1-x}\,dx$

7. $\displaystyle\int_{-\pi/4}^0 \sec x\,dx$

9. (a) 0   (b) −8   (c) −12   (d) 10   (e) −2   (f) 16
11. (a) 5   (b) $5\sqrt{3}$   (c) −5   (d) −5
13. (a) 4   (b) −4       15. Area = 21 square units
17. Area = $9\pi/2$ square units       19. Area = 2.5 square units
21. Area = 3 square units       23. $b^2/4$       25. $b^2 - a^2$
27. (a) $2\pi$   (b) $\pi$       29. 1/2       31. $3\pi^2/2$       33. 7/3
35. 1/24       37. $3a^2/2$       39. $b/3$       41. −14
43. −2       45. −7/4       47. 7       49. 0
51. Using $n$ subintervals of length $\Delta x = b/n$ and right-endpoint
    values:

$$\text{Area} = \int_0^b 3x^2\,dx = b^3$$

53. Using $n$ subintervals of length $\Delta x = b/n$ and right-endpoint
    values:

$$\text{Area} = \int_0^b 2x\,dx = b^2$$

55. av(f) = 0       57. av(f) = −2       59. av(f) = 1
61. (a) av(g) = −1/2   (b) av(g) = 1   (c) av(g) = 1/4
63. $c(b - a)$       65. $b^3/3 - a^3/3$       67. 9
69. $b^4/4 - a^4/4$       71. $a = 0$ and $b = 1$ maximize the integral.
73. Upper bound = 1, lower bound = 1/2

75. For example, $\displaystyle\int_0^1 \sin(x^2)\,dx \le \int_0^1 dx = 1$

77. $\displaystyle\int_a^b f(x)\,dx \ge \int_a^b 0\,dx = 0$       79. Upper bound = 1/2

## Section 5.4, pp. 300–303

1. −10/3       3. 124/125       5. 753/16       7. 1       9. $2\sqrt{3}$

11. 0       13. −π/4       15. $1 - \dfrac{\pi}{4}$       17. $\dfrac{2-\sqrt{2}}{4}$       19. −8/3

**21.** $-3/4$   **23.** $\sqrt{2} - \sqrt[4]{8} + 1$   **25.** $-1$   **27.** $16$

**29.** $\dfrac{1}{2}$   **31.** $\sqrt{26} - \sqrt{5}$   **33.** $\left(\cos\sqrt{x}\right)\left(\dfrac{1}{2\sqrt{x}}\right)$   **35.** $4t^5$

**37.** $3$   **39.** $\sqrt{1 + x^2}$   **41.** $-\dfrac{1}{2}x^{-1/2}\sin x$   **43.** $0$

**45.** $1$   **47.** $28/3$   **49.** $1/2$   **51.** $\pi$   **53.** $\dfrac{\sqrt{2}\pi}{2}$

**55.** d, since $y' = \dfrac{1}{x}$ and $y(\pi) = \displaystyle\int_\pi^\pi \dfrac{1}{t}\,dt - 3 = -3$

**57.** b, since $y' = \sec x$ and $y(0) = \displaystyle\int_0^0 \sec t\,dt + 4 = 4$

**59.** $y = \displaystyle\int_2^x \sec t\,dt + 3$   **61.** $\dfrac{2}{3}bh$   **63.** \$9.00

**65.** (a) $T(0) = 20°C$, $T(16) = 24°C$,
$\quad T(25) = 30°C$
  (b) $av(T) = 23.33°C$

**67.** $2x - 2$   **69.** $-3x + 5$

**71.** (a) True. Since $f$ is continuous, $g$ is differentiable by Part 1 of the Fundamental Theorem of Calculus.
  (b) True: $g$ is continuous because it is differentiable.
  (c) True, since $g'(1) = f(1) = 0$.
  (d) False, since $g''(1) = f'(1) > 0$.
  (e) True, since $g'(1) = 0$ and $g''(1) = f'(1) > 0$.
  (f) False: $g''(x) = f'(x) > 0$, so $g''$ never changes sign.
  (g) True, since $g'(1) = f(1) = 0$ and $g'(x) = f(x)$ is an increasing function of $x$ (because $f'(x) > 0$).

**73.** (a) $v = \dfrac{ds}{dt} = \dfrac{d}{dt}\displaystyle\int_0^t f(x)\,dx = f(t) \Rightarrow v(5) = f(5) = 2\text{ m/s}$

  (b) $a = df/dt$ is negative, since the slope of the tangent line at $t = 5$ is negative.

  (c) $s = \displaystyle\int_0^3 f(x)\,dx = \dfrac{1}{2}(3)(3) = \dfrac{9}{2}\text{m}$, since the integral is the area of the triangle formed by $y = f(x)$, the $x$-axis, and $x = 3$.

  (d) $t = 6$, since after $t = 6$ to $t = 9$, the region lies below the $x$-axis.

  (e) At $t = 4$ and $t = 7$, since there are horizontal tangents there.

  (f) Toward the origin between $t = 6$ and $t = 9$, since the velocity is negative on this interval. Away from the origin between $t = 0$ and $t = 6$, since the velocity is positive there.

  (g) Right or positive side, because the integral of $f$ from 0 to 9 is positive, there being more area above the $x$-axis than below.

Section 5.5, pp. 308–310

**1.** $\dfrac{1}{6}(2x + 4)^6 + C$   **3.** $-\dfrac{1}{3}(x^2 + 5)^{-3} + C$

**5.** $\dfrac{1}{10}(3x^2 + 4x)^5 + C$   **7.** $-\dfrac{1}{3}\cos 3x + C$

**9.** $\dfrac{1}{2}\sec 2t + C$   **11.** $-6(1 - r^3)^{1/2} + C$

**13.** $\dfrac{1}{3}(x^{3/2} - 1) - \dfrac{1}{6}\sin(2x^{3/2} - 2) + C$

**15.** (a) $-\dfrac{1}{4}(\cot^2 2\theta) + C$   (b) $-\dfrac{1}{4}(\csc^2 2\theta) + C$

**17.** $-\dfrac{1}{3}(3 - 2s)^{3/2} + C$   **19.** $-\dfrac{2}{5}(1 - \theta^2)^{5/4} + C$

**21.** $\left(-2/(1 + \sqrt{x})\right) + C$   **23.** $\dfrac{1}{3}\tan(3x + 2) + C$

**25.** $\dfrac{1}{2}\sin^6\left(\dfrac{x}{3}\right) + C$   **27.** $\left(\dfrac{r^3}{18} - 1\right)^6 + C$

**29.** $-\dfrac{2}{3}\cos(x^{3/2} + 1) + C$   **31.** $\dfrac{1}{2\cos(2t + 1)} + C$

**33.** $-\sin\left(\dfrac{1}{t} - 1\right) + C$   **35.** $-\dfrac{\sin^2(1/\theta)}{2} + C$

**37.** $\dfrac{2}{3}(1 + x)^{3/2} - 2(1 + x)^{1/2} + C$   **39.** $\dfrac{2}{3}\left(2 - \dfrac{1}{x}\right)^{3/2} + C$

**41.** $\dfrac{2}{27}\left(1 - \dfrac{3}{x^3}\right)^{3/2} + C$   **43.** $\dfrac{1}{12}(x - 1)^{12} + \dfrac{1}{11}(x - 1)^{11} + C$

**45.** $-\dfrac{1}{8}(1 - x)^8 + \dfrac{4}{7}(1 - x)^7 - \dfrac{2}{3}(1 - x)^6 + C$

**47.** $\dfrac{1}{5}(x^2 + 1)^{5/2} - \dfrac{1}{3}(x^2 + 1)^{3/2} + C$   **49.** $\dfrac{-1}{4(x^2 - 4)^2} + C$

**51.** (a) $-\dfrac{6}{2 + \tan^3 x} + C$   (b) $-\dfrac{6}{2 + \tan^3 x} + C$

  (c) $-\dfrac{6}{2 + \tan^3 x} + C$

**53.** $\dfrac{1}{6}\sin\sqrt{3(2r - 1)^2 + 6} + C$   **55.** $s = \dfrac{1}{2}(3t^2 - 1)^4 - 5$

**57.** $s = 4t - 2\sin\left(2t + \dfrac{\pi}{6}\right) + 9$

**59.** $s = \sin\left(2t - \dfrac{\pi}{2}\right) + 100t + 1$   **61.** 6 m

Section 5.6, pp. 317–320

**1.** (a) $14/3$   (b) $2/3$   **3.** (a) $1/2$   (b) $-1/2$
**5.** (a) $15/16$   (b) $0$   **7.** (a) $0$   (b) $1/8$   **9.** (a) $4$   (b) $0$
**11.** (a) $506/375$   (b) $86{,}744/375$   **13.** (a) $0$   (b) $0$
**15.** $2\sqrt{3}$   **17.** $3/4$   **19.** $3^{5/2} - 1$   **21.** $3$   **23.** $\pi/3$
**25.** $16/3$   **27.** $2^{5/2}$   **29.** $\pi/2$   **31.** $128/15$
**33.** $4/3$   **35.** $5/6$   **37.** $38/3$   **39.** $49/6$
**41.** $32/3$   **43.** $48/5$   **45.** $8/3$   **47.** $8$
**49.** $5/3$ (There are three intersection points.)   **51.** $18$
**53.** $243/8$   **55.** $8/3$   **57.** $2$   **59.** $104/15$   **61.** $56/15$
**63.** $4$   **65.** $\dfrac{4}{3} - \dfrac{4}{\pi}$   **67.** $\pi/2$   **69.** $2$   **71.** $1/2$
**73.** $1$   **75.** (a) $\left(\pm\sqrt{c}, c\right)$   (b) $c = 4^{2/3}$   (c) $c = 4^{2/3}$
**77.** $11/3$   **79.** $3/4$   **81.** Neither   **83.** $F(6) - F(2)$
**85.** (a) $-3$   (b) $3$   **87.** $I = a/2$

Practice Exercises, pp. 321–323
**1.** (a) About 680 m   (b) $h$ (meters)

**3.** (a) $-1/2$   (b) $31$   (c) $13$   (d) $0$

**5.** $\int_{1}^{5} (2x - 1)^{-1/2} dx = 2$    **7.** $\int_{-\pi}^{0} \cos\frac{x}{2} dx = 2$

**9.** (a) 4   (b) 2   (c) $-2$   (d) $-2\pi$   (e) $8/5$

**11.** $8/3$    **13.** 62    **15.** 1    **17.** $1/6$    **19.** 18    **21.** $9/8$

**23.** $\dfrac{\pi^2}{32} + \dfrac{\sqrt{2}}{2} - 1$    **25.** 4    **27.** $\dfrac{8\sqrt{2} - 7}{6}$

**29.** Min: $-4$, max: 0, area: $27/4$    **31.** $6/5$

**35.** $y = \int_{5}^{x} \left(\dfrac{\sin t}{t}\right) dt - 3$    **37.** $16(\sin x)^{-1/2} + C$

**39.** $3\theta^2 - \theta + \cos(4\theta + 1) + C$    **41.** $\dfrac{x^3}{3} + \dfrac{25}{x} + C$

**43.** $\dfrac{1}{9}\sin(6t^{3/2}) + C$    **45.** 36    **47.** 4    **49.** $1/3$    **51.** $40/49$

**53.** $\left(1 - \dfrac{1^{3/4}}{2}\right)^{4/3}$    **55.** $\pi/2$    **57.** $\pi/4$    **59.** $4\sqrt{3} - \pi$

**61.** $(-1)/8$    **63.** 2    **65.** 1    **67.** (a) $b$   (b) $b$

**71.** $-4°$ C    **73.** $\sqrt{2 + \cos^3 x}$    **75.** $\dfrac{-6}{3 + x^4}$

**77.** Yes    **79.** $-\sqrt{1 + x^2}$

**81.** Cost $\approx 13{,}897.50$ using a lower sum estimate

### Additional and Advanced Exercises, pp. 324–326

**1.** (a) Yes   (b) No    **5.** (a) $1/4$   (b) $\sqrt[3]{12}$

**7.** $f(x) = \dfrac{x}{\sqrt{x^2 + 1}}$    **9.** $y = x^3 + 2x - 4$

**11.** $36/5$    **13.** $\dfrac{1}{2} - \dfrac{2}{\pi}$

**15.** $13/3$

**17.** $1/2$    **19.** $1/6$    **21.** $\int_{0}^{1} f(x)\, dx$    **23.** (b) $\pi r^2$

**25.** (a) 0   (b) $-1$   (c) $-\pi$   (d) $x = 1$
  (e) $y = 2x + 2 - \pi$   (f) $x = -1, x = 2$
  (g) $[-2\pi, 0]$

**27.** $2/x$    **29.** $\dfrac{\sin 4y}{\sqrt{y}} - \dfrac{\sin y}{2\sqrt{y}}$

# Chapter 6

### Section 6.1, pp. 335–338

**1.** 16    **3.** $16/3$    **5.** (a) $2\sqrt{3}$   (b) 8    **7.** (a) 60   (b) 36

**9.** $8\pi$    **11.** 10    **13.** (a) $s^2h$   (b) $s^2h$    **15.** $\dfrac{2\pi}{3}$

**17.** $4 - \pi$    **19.** $\dfrac{32\pi}{5}$    **21.** $36\pi$    **23.** $\pi$

**25.** $\pi\left(\dfrac{\pi}{2} + 2\sqrt{2} - \dfrac{11}{3}\right)$    **27.** $2\pi$    **29.** $2\pi$    **31.** $\dfrac{16\pi}{35}$

**33.** $\pi^2 - 2\pi$    **35.** $\dfrac{2\pi}{3}$    **37.** $\dfrac{117\pi}{5}$    **39.** $\pi(\pi - 2)$

**41.** $\dfrac{4\pi}{3}$    **43.** $8\pi$    **45.** $\dfrac{7\pi}{6}$

**47.** (a) $8\pi$   (b) $\dfrac{32\pi}{5}$   (c) $\dfrac{8\pi}{3}$   (d) $\dfrac{224\pi}{15}$

**49.** (a) $\dfrac{16\pi}{15}$   (b) $\dfrac{56\pi}{15}$   (c) $\dfrac{64\pi}{15}$    **51.** $V = 2a^2b\pi^2$

**53.** (a) $V = \dfrac{\pi h^2(3a - h)}{3}$   (b) $\dfrac{1}{120\pi}$ m/sec

**57.** $V = 3308 \text{ cm}^3$    **59.** $\dfrac{4 - b + a}{2}$

### Section 6.2, pp. 343–345

**1.** $6\pi$    **3.** $2\pi$    **5.** $14\pi/3$    **7.** $8\pi$    **9.** $5\pi/6$

**11.** $\dfrac{7\pi}{15}$    **13.** (b) $4\pi$    **15.** $\dfrac{16\pi}{15}(3\sqrt{2} + 5)$

**17.** $\dfrac{8\pi}{3}$    **19.** $\dfrac{4\pi}{3}$    **21.** $\dfrac{16\pi}{3}$

**23.** (a) $16\pi$   (b) $32\pi$   (c) $28\pi$
  (d) $24\pi$   (e) $60\pi$   (f) $48\pi$

**25.** (a) $\dfrac{27\pi}{2}$   (b) $\dfrac{27\pi}{2}$   (c) $\dfrac{72\pi}{5}$   (d) $\dfrac{108\pi}{5}$

**27.** (a) $\dfrac{6\pi}{5}$   (b) $\dfrac{4\pi}{5}$   (c) $2\pi$   (d) $2\pi$

**29.** (a) About the $x$-axis: $V = \dfrac{2\pi}{15}$; about the $y$-axis: $V = \dfrac{\pi}{6}$

  (b) About the $x$-axis: $V = \dfrac{2\pi}{15}$; about the $y$-axis: $V = \dfrac{\pi}{6}$

**31.** (a) $\dfrac{5\pi}{3}$   (b) $\dfrac{4\pi}{3}$   (c) $2\pi$   (d) $\dfrac{2\pi}{3}$

**33.** (a) $\dfrac{4\pi}{15}$   (b) $\dfrac{7\pi}{30}$

**35.** (a) $\dfrac{24\pi}{5}$   (b) $\dfrac{48\pi}{5}$

**37.** (a) $\dfrac{9\pi}{16}$   (b) $\dfrac{9\pi}{16}$

**39.** Disk: 2 integrals; washer: 2 integrals; shell: 1 integral

**41.** (a) $\dfrac{256\pi}{3}$   (b) $\dfrac{244\pi}{3}$

### Section 6.3, pp. 349–351

**1.** 12    **3.** $\dfrac{53}{6}$    **5.** $\dfrac{123}{32}$    **7.** $\dfrac{99}{8}$    **9.** $\dfrac{53}{6}$    **11.** 2

**13.** (a) $\int_{-1}^{2} \sqrt{1 + 4x^2}\, dx$   (c) $\approx 6.13$

**15.** (a) $\int_{0}^{\pi} \sqrt{1 + \cos^2 y}\, dy$   (c) $\approx 3.82$

**17.** (a) $\int_{-1}^{3} \sqrt{1 + (y + 1)^2}\, dy$   (c) $\approx 9.29$

**19.** (a) $\int_{0}^{\pi/6} \sec x\, dx$   (c) $\approx 0.55$

**21.** (a) $y = \sqrt{x}$ from $(1, 1)$ to $(4, 2)$
  (b) Only one. We know the derivative of the function and the value of the function at one value of $x$.

**23.** 1    **25.** Yes, $f(x) = \pm x + C$ where $C$ is any real number.

**33.** $\displaystyle\int_0^x \sqrt{1 + 9t}\, dt,\ \frac{2}{27}(10^{3/2} - 1)$

### Section 6.4, pp. 354–356

**1.** (a) $\displaystyle 2\pi \int_0^{\pi/4} (\tan x)\sqrt{1 + \sec^4 x}\, dx$    (c) $S \approx 3.84$

(b)

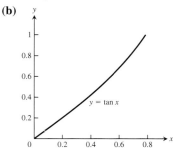

**3.** (a) $\displaystyle 2\pi \int_1^2 \frac{1}{y}\sqrt{1 + y^{-4}}\, dy$    (c) $S \approx 5.02$

(b)

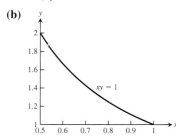

**5.** (a) $\displaystyle 2\pi \int_1^4 (3 - x^{1/2})^2\sqrt{1 + (1 - 3x^{-1/2})^2}\, dx$    (c) $S \approx 63.37$

(b)

**7.** (a) $\displaystyle 2\pi \int_0^{\pi/3}\left(\int_0^y \tan t\, dt\right)\sec y\, dy$    (c) $S \approx 2.08$

(b)

**9.** $4\pi\sqrt{5}$    **11.** $3\pi\sqrt{5}$    **13.** $98\pi/81$    **15.** $2\pi$
**17.** $\pi(\sqrt{8} - 1)/9$    **19.** $35\pi\sqrt{5}/3$

**21.** $(2\pi/3)(2\sqrt{2} - 1)$    **23.** $253\pi/20$
**27.** Order 226.2 liters of each color.

### Section 6.5, pp. 361–365

**1.** 400 N/m    **3.** 4 cm, 0.08 J
**5.** (a) 12,000 N/cm    (b) 6000 N · cm, 18,000 N · cm
**7.** 780 J    **9.** 108,000 J    **11.** 234 J
**13.** (a) 235,200,000 J    (b) 17 h, 46 min    (c) 266 min
    (d) At 9780 N/m³: a) 234,720,000 J  b) 17 h, 44 min
        At 9820 N/m³: a) 235,680,000 J  b) 17 h, 48 min
**15.** 5,772,676.5 J    **17.** 8,977,666 J
**19.** 385,369 J    **21.** 15,073,099.75 J
**25.** 120 J    **27.** 138.72 J    **29.** 2.175 J
**31.** $5.144 \times 10^{10}$ J    **33.** 9146.7 N
**35.** (a) 26,989.2 N    (b) 25,401.6 N    **37.** (a) 182,933 N    (b) 187,600 N
**39.** 5808 N    **41.** (a) 19,600 N    (b) 14,700 N    (c) 16,135 N
**43.** (a) 14,933 N    (b) 2.94 m    **45.** $\dfrac{wb}{2}$
**47.** No. The tank will overflow because the movable end will have
    moved only 2.18 m by the time the tank is full.

### Section 6.6, pp. 374–376

**1.** $\bar{x} = 0, \bar{y} = 12/5$    **3.** $\bar{x} = 1, \bar{y} = -3/5$
**5.** $\bar{x} = 16/105, \bar{y} = 8/15$    **7.** $\bar{x} = 0, \bar{y} = \pi/8$
**9.** $\bar{x} = 1, \bar{y} = -2.5$
**11.** $\bar{x} = \bar{y} = 2/(4 - \pi)$    **13.** $\bar{x} = 3/2, \bar{y} = 1/2$

**15.** (a) $\dfrac{224\pi}{3}$    (b) $\bar{x} = 2, \bar{y} = 0$

(c)

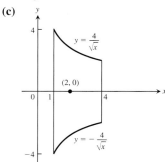

**17.** $\bar{x} = \bar{y} = 1/3$    **19.** $\bar{x} = a/3, \bar{y} = b/3$    **21.** $13\delta/6$
**23.** $\bar{x} = 0, \bar{y} = \dfrac{a\pi}{4}$    **25.** $\bar{x} = 1/2, \bar{y} = 4$
**27.** $\bar{x} = 6/5, \bar{y} = 8/7$    **33.** $V = 32\pi, S = 32\sqrt{2}\pi$    **35.** $4\pi^2$
**37.** $\bar{x} = 0, \bar{y} = \dfrac{2a}{\pi}$    **39.** $\bar{x} = 0, \bar{y} = \dfrac{4b}{3\pi}$
**41.** $\sqrt{2}\pi a^3(4 + 3\pi)/6$    **43.** $\bar{x} = \dfrac{a}{3}, \bar{y} = \dfrac{b}{3}$

### Practice Exercises, pp. 376–378

**1.** $\dfrac{9\pi}{280}$    **3.** $\pi^2$    **5.** $\dfrac{72\pi}{35}$
**7.** (a) $2\pi$    (b) $\pi$    (c) $12\pi/5$    (d) $26\pi/5$
**9.** (a) $8\pi$    (b) $1088\pi/15$    (c) $512\pi/15$
**11.** $\pi(3\sqrt{3} - \pi)/3$
**13.** (a) $16\pi/15$    (b) $8\pi/5$    (c) $8\pi/3$    (d) $32\pi/5$
**15.** $\dfrac{28\pi}{3}$ m³    **17.** $\dfrac{10}{3}$    **19.** $\dfrac{285}{8}$    **21.** $28\pi\sqrt{2}/3$
**23.** $4\pi$    **25.** 4640 J    **27.** $\dfrac{w}{2}(2ar - a^2)$

**29.** 65,680,230 J   **31.** 3,375,000$\pi$ J, 257 s

**33.** $\bar{x} = 0, \bar{y} = 8/5$   **35.** $\bar{x} = 3/2, \bar{y} = 12/5$

**37.** $\bar{x} = 9/5, \bar{y} = 11/10$   **39.** 52,267 J   **41.** 344,960 N

*Additional and Advanced Exercises, pp. 378–379*

**1.** $f(x) = \sqrt{\dfrac{2x - a}{\pi}}$   **3.** $f(x) = \sqrt{C^2 - 1}\, x + a$, where $C \geq 1$

**5.** $\dfrac{\pi}{30\sqrt{2}}$   **7.** 28/3   **9.** $\dfrac{4h\sqrt{3mh}}{3}$

**11.** $\bar{x} = 0, \bar{y} = \dfrac{n}{2n + 1}, (0, 1/2)$

**15. (a)** $\bar{x} = \bar{y} = 4(a^2 + ab + b^2)/(3\pi(a + b))$
    **(b)** $(2a/\pi, 2a/\pi)$

**17.** $\approx$ 365,867 N

# Chapter 7

## Section 7.1, pp. 386–388

**1.** One-to-one   **3.** Not one-to-one   **5.** One-to-one

**7.** Not one-to-one   **9.** One-to-one

**11.** $D: (0, 1]$  $R: [0, \infty)$   **13.** $D: [-1, 1]$  $R: [-\pi/2, \pi/2]$

**15.** $D: [0, 6]$  $R: [0, 3]$   **17. (a)** Symmetric about the line $y = x$

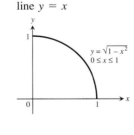

**19.** $f^{-1}(x) = \sqrt{x - 1}$   **21.** $f^{-1}(x) = \sqrt[3]{x + 1}$

**23.** $f^{-1}(x) = \sqrt{x} - 1$

**25.** $f^{-1}(x) = \sqrt[5]{x}$; domain: $-\infty < x < \infty$; range: $-\infty < y < \infty$

**27.** $f^{-1}(x) = 5\sqrt{x - 1}$; domain: $-\infty < x < \infty$; range: $-\infty < y < \infty$

**29.** $f^{-1}(x) = \dfrac{1}{\sqrt{x}}$; domain: $x > 0$; range: $y > 0$

**31.** $f^{-1}(x) = \dfrac{2x + 3}{x - 1}$; domain: $-\infty < x < \infty, x \neq 1$; range: $-\infty < y < \infty, y \neq 2$

**33.** $f^{-1}(x) = 1 - \sqrt{x + 1}$; domain: $-1 \leq x < \infty$; range: $-\infty < y \leq 1$

**35. (a)** $f^{-1}(x) = \dfrac{x}{2} - \dfrac{3}{2}$   **37. (a)** $f^{-1}(x) = -\dfrac{x}{4} + \dfrac{5}{4}$

**(b)**

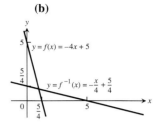

**(c)** $-4, -1/4$

**(c)** 2, 1/2

**39. (b)**

**(c)** Slope of $f$ at $(1, 1) : 3$; slope of $g$ at $(1, 1): 1/3$; slope of $f$ at $(-1, -1): 3$; slope of $g$ at $(-1, -1): 1/3$

**(d)** $y = 0$ is tangent to $y = x^3$ at $x = 0$; $x = 0$ is tangent to $y = \sqrt[3]{x}$ at $x = 0$.

**41.** 1/9   **43.** 3

**45. (a)** $f^{-1}(x) = \dfrac{1}{m}x$
    **(b)** The graph of $f^{-1}$ is the line through the origin with slope $1/m$.

**47. (a)** $f^{-1}(x) = x - 1$

**(b)** $f^{-1}(x) = x - b$. The graph of $f^{-1}$ is a line parallel to the graph of $f$. The graphs of $f$ and $f^{-1}$ lie on opposite sides of the line $y = x$ and are equidistant from that line.

**(c)** Their graphs will be parallel to one another and lie on opposite sides of the line $y = x$ equidistant from that line.

**51.** Increasing, therefore one-to-one; $df^{-1}/dx = \dfrac{1}{9}x^{-2/3}$

**53.** Decreasing, therefore one-to-one; $df^{-1}/dx = -\dfrac{1}{3}x^{-2/3}$

## Section 7.2, pp. 395–396

**1. (a)** $\ln 3 - 2\ln 2$   **(b)** $2(\ln 2 - \ln 3)$   **(c)** $-\ln 2$
    **(d)** $\dfrac{2}{3}\ln 3$   **(e)** $\ln 3 + \dfrac{1}{2}\ln 2$   **(f)** $\dfrac{1}{2}(3\ln 3 - \ln 2)$

**3. (a)** $\ln 5$   **(b)** $\ln (x - 3)$   **(c)** $\ln (t^2)$

**5.** $1/x$   **7.** $2/t$   **9.** $-1/x$   **11.** $\dfrac{1}{\theta + 1}$   **13.** $3/x$

**15.** $2(\ln t) + (\ln t)^2$    **17.** $x^3\ln x$    **19.** $\dfrac{1 - \ln t}{t^2}$

**21.** $\dfrac{1}{x(1 + \ln x)^2}$    **23.** $\dfrac{1}{x\ln x}$    **25.** $2\cos(\ln\theta)$

**27.** $-\dfrac{3x + 2}{2x(x + 1)}$    **29.** $\dfrac{2}{t(1 - \ln t)^2}$    **31.** $\dfrac{\tan(\ln\theta)}{\theta}$

**33.** $\dfrac{10x}{x^2 + 1} + \dfrac{1}{2(1 - x)}$    **35.** $2x\ln|x| - x\ln\dfrac{|x|}{\sqrt{2}}$

**37.** $\ln\left(\dfrac{2}{3}\right)$    **39.** $\ln|y^2 - 25| + C$    **41.** $\ln 3$

**43.** $(\ln 2)^2$    **45.** $\dfrac{1}{\ln 4}$    **47.** $\ln|6 + 3\tan t| + C$

**49.** $\ln 2$    **51.** $\ln 27$    **53.** $\ln(1 + \sqrt{x}) + C$

**55.** $\left(\dfrac{1}{2}\right)\sqrt{x(x + 1)}\left(\dfrac{1}{x} + \dfrac{1}{x + 1}\right) = \dfrac{2x + 1}{2\sqrt{x(x + 1)}}$

**57.** $\left(\dfrac{1}{2}\right)\sqrt{\dfrac{t}{t + 1}}\left(\dfrac{1}{t} - \dfrac{1}{t + 1}\right) = \dfrac{1}{2\sqrt{t}(t + 1)^{3/2}}$

**59.** $\sqrt{\theta + 3}(\sin\theta)\left(\dfrac{1}{2(\theta + 3)} + \cot\theta\right)$

**61.** $t(t + 1)(t + 2)\left[\dfrac{1}{t} + \dfrac{1}{t + 1} + \dfrac{1}{t + 2}\right] = 3t^2 + 6t + 2$

**63.** $\dfrac{\theta + 5}{\theta\cos\theta}\left[\dfrac{1}{\theta + 5} - \dfrac{1}{\theta} + \tan\theta\right]$

**65.** $\dfrac{x\sqrt{x^2 + 1}}{(x + 1)^{2/3}}\left[\dfrac{1}{x} + \dfrac{x}{x^2 + 1} - \dfrac{2}{3(x + 1)}\right]$

**67.** $\dfrac{1}{3}\sqrt[3]{\dfrac{x(x - 2)}{x^2 + 1}}\left(\dfrac{1}{x} + \dfrac{1}{x - 2} - \dfrac{2x}{x^2 + 1}\right)$

**69.** (a) Max = 0 at $x = 0$, min = $-\ln 2$ at $x = \pi/3$
     (b) Max = 1 at $x = 1$, min = $\cos(\ln 2)$ at $x = 1/2$ and $x = 2$

**71.** $\ln 16$    **73.** $4\pi\ln 4$    **75.** $\pi\ln 16$

**77.** (a) $6 + \ln 2$    (b) $8 + \ln 9$

**79.** (a) $\bar{x} \approx 1.44, \bar{y} \approx 0.36$

     (b)

**83.** $y = x + \ln|x| + 2$    **85.** (b) 0.00469

### Section 7.3, pp. 404–407

**1.** (a) $t = -10\ln 3$    (b) $t = -\dfrac{\ln 2}{k}$    (c) $t = \dfrac{\ln 4}{\ln 2}$

**3.** $4(\ln x)^2$    **5.** $-5e^{-5x}$    **7.** $-7e^{(5-7x)}$    **9.** $xe^x$

**11.** $x^2 e^x$    **13.** $2e^\theta\cos\theta$    **15.** $2\theta e^{-\theta^2}\sin(e^{-\theta^2})$

**17.** $\dfrac{1 - t}{t}$    **19.** $1/(1 + e^\theta)$    **21.** $e^{\cos t}(1 - t\sin t)$

**23.** $(\sin x)/x$    **25.** $\dfrac{ye^y\cos x}{1 - ye^y\sin x}$    **27.** $\dfrac{2e^{2x} - \cos(x + 3y)}{3\cos(x + 3y)}$

**29.** $\dfrac{1}{3}e^{3x} - 5e^{-x} + C$    **31.** 1    **33.** $8e^{(x+1)} + C$    **35.** 2

**37.** $2e^{\sqrt{r}} + C$    **39.** $-e^{-t^2} + C$    **41.** $-e^{1/x} + C$    **43.** $e$

**45.** $\dfrac{1}{\pi}e^{\sec\pi t} + C$    **47.** 1    **49.** $\ln(1 + e^r) + C$

**51.** $y = 1 - \cos(e^t - 2)$    **53.** $y = 2(e^{-x} + x) - 1$    **55.** $2^x\ln 2$

**57.** $\left(\dfrac{\ln 5}{2\sqrt{s}}\right)5^{\sqrt{s}}$    **59.** $\pi x^{(\pi - 1)}$    **61.** $-\sqrt{2}\cos\theta^{(\sqrt{2}-1)}\sin\theta$

**63.** $7^{\sec\theta}(\ln 7)^2(\sec\theta\tan\theta)$    **65.** $(3\cos 3t)(2^{\sin 3t})\ln 2$

**67.** $\dfrac{1}{\theta\ln 2}$    **69.** $\dfrac{3}{x\ln 4}$    **71.** $\dfrac{x^2}{\ln 10} + 3x^2\log_{10}x$

**73.** $\dfrac{-2}{(x + 1)(x - 1)}$    **75.** $\sin(\log_7\theta) + \dfrac{1}{\ln 7}\cos(\log_7\theta)$

**77.** $\dfrac{1}{\ln 10}$    **79.** $\dfrac{1}{t}(\log_2 3)3^{\log_2 t}$    **81.** $\dfrac{1}{t}$

**83.** $\dfrac{5^x}{\ln 5} + C$    **85.** $\dfrac{1}{2\ln 2}$    **87.** $\dfrac{1}{\ln 2}$    **89.** $\dfrac{6}{\ln 7}$

**91.** 32760    **93.** $\dfrac{3x^{(\sqrt{3}+1)}}{\sqrt{3} + 1} + C$    **95.** $3^{\sqrt{2}+1}$

**97.** $\dfrac{1}{\ln 10}\left(\dfrac{(\ln x)^2}{2}\right) + C$    **99.** $2(\ln 2)^2$    **101.** $\dfrac{3\ln 2}{2}$

**103.** $\ln 10$    **105.** $(\ln 10)\ln|\ln x| + C$

**107.** $\ln(\ln x), x > 1$    **109.** $-\ln x$

**111.** $(x + 1)^x\left(\dfrac{x}{x + 1} + \ln(x + 1)\right)$    **113.** $(\sqrt{t})^t\left(\dfrac{\ln t}{2} + \dfrac{1}{2}\right)$

**115.** $(\sin x)^x(\ln\sin x + x\cot x)$    **117.** $\cos x^x \cdot x^x(1 + \ln x)$

**119.** Maximum: 1 at $x = 0$, minimum: $2 - 2\ln 2$ at $x = \ln 2$

**121.** (a) Abs max: $\dfrac{1}{e}$ at $x = 1$    (b) $\left(2, \dfrac{2}{e^2}\right)$

**123.** Abs max of $1/(2e)$ assumed at $x = 1/\sqrt{e}$    **125.** 2

**127.** $y = e^{x/2} - 1$    **129.** $\dfrac{e^2 - 1}{2e}$    **131.** $\ln(\sqrt{2} + 1)$

**133.** (a) $\dfrac{d}{dx}(x\ln x - x + C) = x\cdot\dfrac{1}{x} + \ln x - 1 + 0 = \ln x$

     (b) $\dfrac{1}{e - 1}$

**135.** (b) $|\text{error}| \approx 0.02140$
     (c) $L(x) = x + 1$ never overestimates $e^x$.

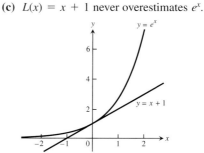

**137.** $2\ln 5$    **139.** $x \approx -0.76666$

**141.** (a) $L(x) = 1 + (\ln 2)x \approx 0.69x + 1$

### Section 7.4, pp. 414–416

**9.** $\dfrac{2}{3}y^{3/2} - x^{1/2} = C$    **11.** $e^y - e^x = C$

**13.** $-x + 2\tan\sqrt{y} = C$    **15.** $e^{-y} + 2e^{\sqrt{x}} = C$

**17.** $y = \sin(x^2 + C)$    **19.** $\dfrac{1}{3}\ln|y^3 - 2| = x^3 + C$

**21.** $4\ln\left(\sqrt{y} + 2\right) = e^{x^2} + C$

**23.** (a) $-0.00001$    (b) 10,536 years    (c) 82%

**25.** 54.88 g    **27.** 19.9 m    **29.** $2.8147498 \times 10^{14}$

**31.** (a) 8 years   (b) 32.02 years   **33.** Yes, $y(20) < 1$
**35.** 15.28 years   **37.** 56,562 years
**41.** (a) 17.5 min   (b) 13.26 min
**43.** $-3°C$   **45.** About 6693 years   **47.** 54.62%
**49.** $\approx 15{,}683$ years

Section 7.5, pp. 423–425
**1.** $-1/4$   **3.** $5/7$   **5.** $1/2$   **7.** $1/4$   **9.** $-23/7$
**11.** $5/7$   **13.** 0   **15.** $-16$   **17.** $-2$   **19.** $1/4$
**21.** 2   **23.** 3   **25.** $-1$   **27.** $\ln 3$   **29.** $\dfrac{1}{\ln 2}$   **31.** $\ln 2$
**33.** 1   **35.** $1/2$   **37.** $\ln 2$   **39.** $-\infty$   **41.** $-1/2$
**43.** $-1$   **45.** 1   **47.** 0   **49.** 2   **51.** $1/e$   **53.** 1
**55.** $1/e$   **57.** $e^{1/2}$   **59.** 1   **61.** $e^3$   **63.** 0   **65.** $+1$
**67.** 3   **69.** 1   **71.** 0   **73.** $\infty$   **75.** (b) is correct.
**77.** (d) is correct.   **79.** $c = \dfrac{27}{10}$   **81.** (b) $\dfrac{-1}{2}$   **83.** $-1$
**87.** (a) $y = 1$   (b) $y = 0, y = \dfrac{3}{2}$
**89.** (a) We should assign the value 1 to $f(x) = (\sin x)^x$ to make it continuous at $x = 0$.

(c) The maximum value of $f(x)$ is close to 1 near the point $x \approx 1.55$ (see the graph in part (a)).

Section 7.6, pp. 434–437
**1.** (a) $\pi/4$   (b) $-\pi/3$   (c) $\pi/6$
**3.** (a) $-\pi/6$   (b) $\pi/4$   (c) $-\pi/3$
**5.** (a) $\pi/3$   (b) $3\pi/4$   (c) $\pi/6$
**7.** (a) $3\pi/4$   (b) $\pi/6$   (c) $2\pi/3$
**9.** $1/\sqrt{2}$   **11.** $-1/\sqrt{3}$   **13.** $\pi/2$   **15.** $\pi/2$   **17.** $\pi/2$
**19.** 0   **21.** $\dfrac{-2x}{\sqrt{1 - x^4}}$   **23.** $\dfrac{\sqrt{2}}{\sqrt{1 - 2t^2}}$
**25.** $\dfrac{1}{|2s + 1|\sqrt{s^2 + s}}$   **27.** $\dfrac{-2x}{(x^2 + 1)\sqrt{x^4 + 2x^2}}$
**29.** $\dfrac{-1}{\sqrt{1 - t^2}}$   **31.** $\dfrac{-1}{2\sqrt{t}(1 + t)}$   **33.** $\dfrac{1}{(\tan^{-1}x)(1 + x^2)}$
**35.** $\dfrac{-e^t}{|e^t|\sqrt{(e^t)^2 - 1}} = \dfrac{-1}{\sqrt{e^{2t} - 1}}$   **37.** $\dfrac{-2s^n}{\sqrt{1 - s^2}}$   **39.** 0
**41.** $\sin^{-1}x$   **43.** $\sin^{-1}\dfrac{x}{3} + C$   **45.** $\dfrac{1}{\sqrt{17}}\tan^{-1}\dfrac{x}{\sqrt{17}} + C$
**47.** $\dfrac{1}{\sqrt{2}}\sec^{-1}\left|\dfrac{5x}{\sqrt{2}}\right| + C$   **49.** $2\pi/3$   **51.** $\pi/16$
**53.** $-\pi/12$   **55.** $\dfrac{3}{2}\sin^{-1}2(r - 1) + C$
**57.** $\dfrac{\sqrt{2}}{2}\tan^{-1}\left(\dfrac{x - 1}{\sqrt{2}}\right) + C$   **59.** $\dfrac{1}{4}\sec^{-1}\left|\dfrac{2x - 1}{2}\right| + C$
**61.** $\pi$   **63.** $\pi/12$   **65.** $\dfrac{1}{2}\sin^{-1} y^2 + C$
**67.** $\sin^{-1}(x - 2) + C$   **69.** $\pi$   **71.** $\dfrac{1}{2}\tan^{-1}\left(\dfrac{y - 1}{2}\right) + C$
**73.** $2\pi$   **75.** $\dfrac{1}{2}\ln(x^2 + 4) + 2\tan^{-1}\dfrac{x}{2} + C$

**77.** $x + \ln(x^2 + 9) - \dfrac{10}{3}\tan^{-1}\dfrac{x}{3} + C$
**79.** $\sec^{-1}|x + 1| + C$   **81.** $e^{\sin^{-1}x} + C$
**83.** $\dfrac{1}{3}(\sin^{-1}x)^3 + C$   **85.** $\ln|\tan^{-1}y| + C$   **87.** $\sqrt{3} - 1$
**89.** $\dfrac{2}{3}\tan^{-1}\left(\dfrac{\tan^{-1}\sqrt{x}}{3}\right) + C$   **91.** 5   **93.** 2   **95.** 1
**97.** 1   **103.** $y = \sin^{-1}x$   **105.** $y = \sec^{-1}x + \dfrac{2\pi}{3}, x > 1$
**107.** (b) $x = \sqrt{5}$   **109.** $\theta = \cos^{-1}\left(\dfrac{1}{\sqrt{3}}\right) \approx 54.7°$
**121.** $\pi^2/2$   **123.** (a) $\pi^2/2$   (b) $2\pi$
**125.** (a) 0.84107   (b) $-0.72973$   (c) 0.46365
**127.** (a) Domain: all real numbers except those having the form $\dfrac{\pi}{2} + k\pi$, where $k$ is an integer

Range: $-\dfrac{\pi}{2} < y < \dfrac{\pi}{2}$

(b) Domain: $-\infty < x < \infty$; Range: $-\infty < y < \infty$

**129.** (a) Domain: $-\infty < x < \infty$;   (b) Domain: $-1 \leq x \leq 1$;
   Range: $0 \leq y \leq \pi$   Range: $-1 \leq y \leq 1$

**131.** The graphs are identical.   **133.**

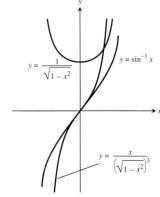

### Section 7.7, pp. 444–446

**1.** $\cosh x = 5/4,\ \tanh x = -3/5,\ \coth x = -5/3,$
$\operatorname{sech} x = 4/5,\ \operatorname{csch} x = -4/3$

**3.** $\sinh x = 8/15,\ \tanh x = 8/17,\ \coth x = 17/8,\ \operatorname{sech} x = 15/17,$
$\operatorname{csch} x = 15/8$

**5.** $x + \dfrac{1}{x}$    **7.** $e^{5x}$    **9.** $e^{4x}$    **13.** $2\cosh\dfrac{x}{3}$

**15.** $\operatorname{sech}^2\sqrt{t} + \dfrac{\tanh\sqrt{t}}{\sqrt{t}}$    **17.** $\coth z$

**19.** $(\ln\operatorname{sech}\theta)(\operatorname{sech}\theta\tanh\theta)$    **21.** $\tanh^3 v$    **23.** $2$

**25.** $\dfrac{1}{2\sqrt{x}(1+x)}$    **27.** $\dfrac{1}{1+\theta} - \tanh^{-1}\theta$

**29.** $\dfrac{1}{2\sqrt{t}} - \coth^{-1}\sqrt{t}$    **31.** $-\operatorname{sech}^{-1}x$    **33.** $\dfrac{\ln 2}{\sqrt{1+\left(\frac{1}{2}\right)^{2\theta}}}$

**35.** $|\sec x|$    **41.** $\dfrac{\cosh 2x}{2} + C$

**43.** $12\sinh\left(\dfrac{x}{2} - \ln 3\right) + C$    **45.** $7\ln\left|e^{x/7} + e^{-x/7}\right| + C$

**47.** $\tanh\left(x - \dfrac{1}{2}\right) + C$    **49.** $-2\operatorname{sech}\sqrt{t} + C$    **51.** $\ln\dfrac{5}{2}$

**53.** $\dfrac{3}{32} + \ln 2$    **55.** $e - e^{-1}$    **57.** $3/4$    **59.** $\dfrac{3}{8} + \ln\sqrt{2}$

**61.** $\ln(2/3)$    **63.** $\dfrac{-\ln 3}{2}$    **65.** $\ln 3$

**67.** (a) $\sinh^{-1}(\sqrt{3})$   (b) $\ln(\sqrt{3}+2)$

**69.** (a) $\coth^{-1}(2) - \coth^{-1}(5/4)$   (b) $\left(\dfrac{1}{2}\right)\ln\left(\dfrac{1}{3}\right)$

**71.** (a) $-\operatorname{sech}^{-1}\left(\dfrac{12}{13}\right) + \operatorname{sech}^{-1}\left(\dfrac{4}{5}\right)$

  (b) $-\ln\left(\dfrac{1+\sqrt{1-(12/13)^2}}{(12/13)}\right) + \ln\left(\dfrac{1+\sqrt{1-(4/5)^2}}{(4/5)}\right)$

  $= -\ln\left(\dfrac{3}{2}\right) + \ln(2) = \ln(4/3)$

**73.** (a) $0$   (b) $0$

**77.** (b) $\sqrt{\dfrac{mg}{k}}$   (c) $70\sqrt{30/47} \approx 55.93$ m/s   **79.** $2\pi$   **81.** $\dfrac{6}{5}$

### Section 7.8, pp. 451–452

**1.** (a) Slower   (b) Slower   (c) Slower   (d) Faster
  (e) Slower   (f) Slower   (g) Same   (h) Slower

**3.** (a) Same   (b) Faster   (c) Same   (d) Same   (e) Slower
  (f) Faster   (g) Slower   (h) Same

**5.** (a) Same   (b) Same   (c) Same   (d) Faster   (e) Faster
  (f) Same   (g) Slower   (h) Faster

**7.** d, a, c, b

**9.** (a) False   (b) False   (c) True   (d) True   (e) True
  (f) True   (g) False   (h) True

**13.** When the degree of $f$ is less than or equal to the degree of $g$.

**15.** $1, 1$

**21.** (b) $\ln(e^{17000000}) = 17{,}000{,}000 < (e^{17\times 10^6})^{1/10^6}$
  $= e^{17} \approx 24{,}154{,}952.75$
  (c) $x \approx 3.4306311 \times 10^{15}$
  (d) They cross at $x \approx 3.4306311 \times 10^{15}$.

**23.** (a) The algorithm that takes $O(n\log_2 n)$ steps
  (b)

**25.** It could take one million for a sequential search; at most 20 steps for a binary search.

### Practice Exercises, pp. 453–455

**1.** $-2e^{-x/5}$    **3.** $xe^{4x}$    **5.** $\dfrac{2\sin\theta\cos\theta}{\sin^2\theta} = 2\cot\theta$    **7.** $\dfrac{2}{(\ln 2)x}$

**9.** $-8^{-t}(\ln 8)$    **11.** $18x^{2.6}$

**13.** $(x+2)^{x+2}(\ln(x+2)+1)$    **15.** $-\dfrac{1}{\sqrt{1-u^2}}$

**17.** $\dfrac{-1}{\sqrt{1-x^2}\cos^{-1}x}$    **19.** $\tan^{-1}(t) + \dfrac{t}{1+t^2} - \dfrac{1}{2t}$

**21.** $\dfrac{1-z}{\sqrt{z^2-1}} + \sec^{-1}z$    **23.** $-1$

**25.** $\dfrac{2(x^2+1)}{\sqrt{\cos 2x}}\left[\dfrac{2x}{x^2+1} + \tan 2x\right]$

**27.** $5\left[\dfrac{(t+1)(t-1)}{(t-2)(t+3)}\right]^5\left[\dfrac{1}{t+1} + \dfrac{1}{t-1} - \dfrac{1}{t-2} - \dfrac{1}{t+3}\right]$

**29.** $\dfrac{1}{\sqrt{\theta}}(\sin\theta)^{\sqrt{\theta}}\left(\dfrac{\ln\sqrt{\sin\theta}}{2} + \theta\cot\theta\right)$    **31.** $\sin(e^x) + C$

**33.** $\tan(e^x - 7) + C$    **35.** $e^{\tan x} + C$    **37.** $\dfrac{-\ln 7}{3}$

**39.** $\ln 8$    **41.** $\ln(9/25)$    **43.** $-[\ln|\cos(\ln v)|] + C$

**45.** $-\dfrac{1}{2}(\ln x)^{-2} + C$    **47.** $-\cot(1+\ln r) + C$

**49.** $\dfrac{1}{2\ln 3}(3^{x^2}) + C$    **51.** $5\ln 9$    **53.** $15/16 + \ln 2$

**55.** $e - 1$    **57.** $1/6$    **59.** $9/14$

**61.** $\dfrac{1}{3}\left[(\ln 4)^3 - (\ln 2)^3\right]$ or $\dfrac{7}{3}(\ln 2)^3$    **63.** $\dfrac{9\ln 2}{4}$    **65.** $\pi$

**67.** $\pi/\sqrt{3}$    **69.** $\sec^{-1}|2y| + C$    **71.** $\pi/12$

**73.** $\sin^{-1}(x+1) + C$    **75.** $\pi/2$    **77.** $\dfrac{1}{3}\sec^{-1}\left(\dfrac{t+1}{3}\right) + C$

**79.** $y = \dfrac{\ln 2}{\ln(3/2)}$    **81.** $y = \ln x - \ln 3$    **83.** $y = \dfrac{1}{1-e^x}$

**85.** $5$    **87.** $0$    **89.** $1$    **91.** $3/7$    **93.** $0$    **95.** $1$

**97.** $\ln 10$    **99.** $\ln 2$    **101.** $5$    **103.** $-\infty$    **105.** $1$

**107.** $1$    **109.** (a) Same rate   (b) Same rate   (c) Faster   (d) Faster
  (e) Same rate   (f) Same rate

**111.** (a) True   (b) False   (c) False   (d) True   (e) True
  (f) True

**113.** $1/3$

**115.** Absolute maximum $= 0$ at $x = e/2$,
  absolute minimum $= -0.5$ at $x = 0.5$

**117.** $1$    **119.** $1/e$ m/s

**121.** $1/\sqrt{2}$ units long by $1/\sqrt{e}$ units high,
  $A = 1/\sqrt{2e} \approx 0.43$ units$^2$

**123. (a)** Absolute maximum of $2/e$ at $x = e^2$; inflection point $(e^{8/3}, (8/3)e^{-4/3})$; concave up on $(e^{8/3}, \infty)$; concave down on $(0, e^{8/3})$

**(b)** Absolute maximum of 1 at $x = 0$; inflection points $(\pm 1/\sqrt{2}, 1/\sqrt{e})$; concave up on $(-\infty, -1/\sqrt{2}) \cup (1/\sqrt{2}, \infty)$; concave down on $(-1/\sqrt{2}, 1/\sqrt{2})$

**(c)** Absolute maximum of 1 at $x = 0$; inflection point $(1, 2/e)$; concave up on $(1, \infty)$; concave down on $(-\infty, 1)$

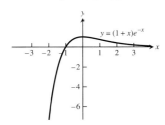

**125.** $y = \left(\tan^{-1}\left(\dfrac{x + C}{2}\right)\right)^2$   **127.** $y^2 = \sin^{-1}(2 \tan x + C)$

**129.** $y = -2 + \ln(2 - e^{-x})$   **131.** $y = 4x - 4\sqrt{x} + 1$

**133.** 18,935 years   **135.** $20(5 - \sqrt{17})$ m

### Additional and Advanced Exercises, pp. 456–457

**1.** $\pi/2$   **3.** $1/\sqrt{e}$   **5.** $\ln 2$   **7. (a)** 1   **(b)** $\pi/2$   **(c)** $\pi$

**9.** $\dfrac{1}{\ln 2}, \dfrac{1}{2\ln 2}, 2 : 1$   **11.** $x = 2$   **13.** $2/17$

**17.** $\bar{x} = \dfrac{\ln 4}{\pi}, \bar{y} = 0$   **19. (b)** $61°$

## Chapter 8

### Section 8.1, pp. 462–463

**1.** $\ln 5$   **3.** $2 \tan x - 2 \sec x - x + C$

**5.** $\sin^{-1} x + \sqrt{1 - x^2} + C$   **7.** $e^{-\cot z} + C$

**9.** $\tan^{-1}(e^z) + C$   **11.** $\pi$   **13.** $t + \cot t + \csc t + C$

**15.** $\sqrt{2}$   **17.** $\dfrac{1}{8}\ln(1 + 4\ln^2 y) + C$

**19.** $\ln|1 + \sin \theta| + C$   **21.** $2t^2 - t + 2\tan^{-1}\left(\dfrac{t}{2}\right) + C$

**23.** $2(\sqrt{2} - 1) \approx 0.82843$   **25.** $\sec^{-1}(e^y) + C$

**27.** $\sin^{-1}(2\ln x) + C$   **29.** $\ln|\sin x| + \ln|\cos x| + C$

**31.** $7 + \ln 8$   **33.** $\left(\sin^{-1} y - \sqrt{1 - y^2}\right]_{-1}^{0} = \dfrac{\pi}{2} - 1$

**35.** $\sec^{-1}\left|\dfrac{x - 1}{7}\right| + C$   **37.** $\dfrac{\theta^3}{3} - \dfrac{\theta^2}{2} + \theta + \dfrac{5}{2}\ln|2\theta - 5| + C$

**39.** $x - \ln(1 + e^x) + C$   **41.** $2\sqrt{2} - \ln\left(3 + 2\sqrt{2}\right)$

**43.** $\ln\left(2 + \sqrt{3}\right)$   **45.** $\bar{x} = 0, \quad \bar{y} = \dfrac{1}{\ln\left(3 + 2\sqrt{2}\right)}$

**47.** $xe^{x^3} + C$   **49.** $\dfrac{1}{30}(x^4 + 1)^{3/2}(3x^4 - 2) + C$

### Section 8.2, pp. 469–471

**1.** $-2x \cos(x/2) + 4 \sin(x/2) + C$

**3.** $t^2 \sin t + 2t \cos t - 2 \sin t + C$

**5.** $\ln 4 - \dfrac{3}{4}$   **7.** $xe^x - e^x + C$

**9.** $-(x^2 + 2x + 2)e^{-x} + C$

**11.** $y \tan^{-1}(y) - \ln\sqrt{1 + y^2} + C$

**13.** $x \tan x + \ln|\cos x| + C$

**15.** $(x^3 - 3x^2 + 6x - 6)e^x + C$   **17.** $(x^2 - 7x + 7)e^x + C$

**19.** $(x^5 - 5x^4 + 20x^3 - 60x^2 + 120x - 120)e^x + C$

**21.** $\dfrac{1}{2}(-e^\theta \cos \theta + e^\theta \sin \theta) + C$

**23.** $\dfrac{e^{2x}}{13}(3 \sin 3x + 2 \cos 3x) + C$

**25.** $\dfrac{2}{3}\left(\sqrt{3s + 9}\,e^{\sqrt{3s+9}} - e^{\sqrt{3s+9}}\right) + C$

**27.** $\dfrac{\pi\sqrt{3}}{3} - \ln(2) - \dfrac{\pi^2}{18}$

**29.** $\dfrac{1}{2}\left[-x \cos(\ln x) + x \sin(\ln x)\right] + C$

**31.** $\dfrac{1}{2}\ln|\sec x^2 + \tan x^2| + C$

**33.** $\dfrac{1}{2}x^2(\ln x)^2 - \dfrac{1}{2}x^2 \ln x + \dfrac{1}{4}x^2 + C$

**35.** $-\dfrac{1}{x}\ln x - \dfrac{1}{x} + C$   **37.** $\dfrac{1}{4}e^{x^4} + C$

**39.** $\dfrac{1}{3}x^2(x^2 + 1)^{3/2} - \dfrac{2}{15}(x^2 + 1)^{5/2} + C$

**41.** $-\dfrac{2}{5}\sin 3x \sin 2x - \dfrac{3}{5}\cos 3x \cos 2x + C$

**43.** $\dfrac{2}{9}x^{3/2}(3 \ln x - 2) + C$

**45.** $2\sqrt{x} \sin\sqrt{x} + 2 \cos\sqrt{x} + C$

**47.** $\dfrac{\pi^2 - 4}{8}$   **49.** $\dfrac{5\pi - 3\sqrt{3}}{9}$

**51.** $\dfrac{1}{2}(x^2 + 1)\tan^{-1} x - \dfrac{x}{2} + C$

**53. (a)** $\pi$   **(b)** $3\pi$   **(c)** $5\pi$   **(d)** $(2n + 1)\pi$

**55.** $2\pi(1 - \ln 2)$   **57. (a)** $\pi(\pi - 2)$   **(b)** $2\pi$

**59. (a)** 1   **(b)** $(e - 2)\pi$   **(c)** $\dfrac{\pi}{2}(e^2 + 9)$

   **(d)** $\bar{x} = \dfrac{1}{4}(e^2 + 1), \bar{y} = \dfrac{1}{2}(e - 2)$

**61.** $\dfrac{1}{2\pi}(1 - e^{-2\pi})$   **63.** $u = x^n, dv = \cos x\, dx$

**65.** $u = x^n, dv = e^{ax}\, dx$   **71.** $x \sin^{-1} x + \cos(\sin^{-1} x) + C$

**73.** $x \sec^{-1} x - \ln|x + \sqrt{x^2 - 1}| + C$   **75.** Yes

**77. (a)** $x \sinh^{-1} x - \cosh(\sinh^{-1} x) + C$
   **(b)** $x \sinh^{-1} x - (1 + x^2)^{1/2} + C$

### Section 8.3, pp. 476–477

**1.** $\dfrac{1}{2}\sin 2x + C$   **3.** $-\dfrac{1}{4}\cos^4 x + C$

**5.** $\dfrac{1}{3}\cos^3 x - \cos x + C$

**7.** $-\cos x + \dfrac{2}{3}\cos^3 x - \dfrac{1}{5}\cos^5 x + C$

**9.** $\sin x - \dfrac{1}{3}\sin^3 x + C$ **11.** $\dfrac{1}{4}\sin^4 x - \dfrac{1}{6}\sin^6 x + C$

**13.** $\dfrac{1}{2}x + \dfrac{1}{4}\sin 2x + C$ **15.** $16/35$ **17.** $3\pi$

**19.** $-4\sin x \cos^3 x + 2\cos x \sin x + 2x + C$

**21.** $-\cos^4 2\theta + C$ **23.** 4 **25.** 2

**27.** $\sqrt{\dfrac{3}{2}} - \dfrac{2}{3}$ **29.** $\dfrac{4}{5}\left(\dfrac{3}{2}\right)^{5/2} - \dfrac{18}{35} - \dfrac{2}{7}\left(\dfrac{3}{2}\right)^{7/2}$ **31.** $\sqrt{2}$

**33.** $\dfrac{1}{2}\tan^2 x + C$ **35.** $\dfrac{1}{3}\sec^3 x + C$ **37.** $\dfrac{1}{3}\tan^3 x + C$

**39.** $2\sqrt{3} + \ln\left(2 + \sqrt{3}\right)$ **41.** $\dfrac{2}{3}\tan\theta + \dfrac{1}{3}\sec^2\theta \tan\theta + C$

**43.** $4/3$ **45.** $2\tan^2 x - 2\ln\left(1 + \tan^2 x\right) + C$

**47.** $\dfrac{1}{4}\tan^4 x - \dfrac{1}{2}\tan^2 x + \ln|\sec x| + C$ **49.** $\dfrac{4}{3} - \ln\sqrt{3}$

**51.** $-\dfrac{1}{10}\cos 5x - \dfrac{1}{2}\cos x + C$ **53.** $\pi$

**55.** $\dfrac{1}{2}\sin x + \dfrac{1}{14}\sin 7x + C$

**57.** $\dfrac{1}{6}\sin 3\theta - \dfrac{1}{4}\sin\theta - \dfrac{1}{20}\sin 5\theta + C$

**59.** $-\dfrac{2}{5}\cos^5\theta + C$ **61.** $\dfrac{1}{4}\cos\theta - \dfrac{1}{20}\cos 5\theta + C$

**63.** $\sec x - \ln|\csc x + \cot x| + C$ **65.** $\cos x + \sec x + C$

**67.** $\dfrac{1}{4}x^2 - \dfrac{1}{4}x \sin 2x - \dfrac{1}{8}\cos 2x + C$ **69.** $\ln\left(1 + \sqrt{2}\right)$

**71.** $\pi^2/2$ **73.** $\bar{x} = \dfrac{4\pi}{3}, \bar{y} = \dfrac{8\pi^2 + 3}{12\pi}$

Section 8.4, pp. 481–482

**1.** $\ln\left|\sqrt{9 + x^2} + x\right| + C$ **3.** $\pi/4$ **5.** $\pi/6$

**7.** $\dfrac{25}{2}\sin^{-1}\left(\dfrac{t}{5}\right) + \dfrac{t\sqrt{25 - t^2}}{2} + C$

**9.** $\dfrac{1}{2}\ln\left|\dfrac{2x}{7} - \dfrac{\sqrt{4x^2 - 49}}{7}\right| + C$

**11.** $7\left[\dfrac{\sqrt{y^2 - 49}}{7} - \sec^{-1}\left(\dfrac{y}{7}\right)\right] + C$ **13.** $\dfrac{\sqrt{x^2 - 1}}{x} + C$

**15.** $-\sqrt{9 - x^2} + C$ **17.** $\dfrac{1}{3}(x^2 + 4)^{3/2} - 4\sqrt{x^2 + 4} + C$

**19.** $\dfrac{-2\sqrt{4 - w^2}}{w} + C$ **21.** $\sin^{-1} x - \sqrt{1 - x^2} + C$

**23.** $4\sqrt{3} - \dfrac{4\pi}{3}$ **25.** $-\dfrac{x}{\sqrt{x^2 - 1}} + C$

**27.** $-\dfrac{1}{5}\left(\dfrac{\sqrt{1 - x^2}}{x}\right)^5 + C$ **29.** $2\tan^{-1} 2x + \dfrac{4x}{(4x^2 + 1)} + C$

**31.** $\dfrac{1}{2}x^2 + \dfrac{1}{2}\ln|x^2 - 1| + C$ **33.** $\dfrac{1}{3}\left(\dfrac{v}{\sqrt{1 - v^2}}\right)^3 + C$

**35.** $\ln 9 - \ln\left(1 + \sqrt{10}\right)$ **37.** $\pi/6$ **39.** $\sec^{-1}|x| + C$

**41.** $\sqrt{x^2 - 1} + C$ **43.** $\dfrac{1}{2}\ln\left|\sqrt{1 + x^4} + x^2\right| + C$

**45.** $4\sin^{-1}\dfrac{\sqrt{x}}{2} + \sqrt{x}\sqrt{4 - x} + C$

**47.** $\dfrac{1}{4}\sin^{-1}\sqrt{x} - \dfrac{1}{4}\sqrt{x}\sqrt{1 - x}\,(1 - 2x) + C$

**49.** $y = 2\left[\dfrac{\sqrt{x^2 - 4}}{2} - \sec^{-1}\left(\dfrac{x}{2}\right)\right]$

**51.** $y = \dfrac{3}{2}\tan^{-1}\left(\dfrac{x}{2}\right) - \dfrac{3\pi}{8}$ **53.** $3\pi/4$

**55. (a)** $\dfrac{1}{12}\left(\pi + 6\sqrt{3} - 12\right)$

**(b)** $\bar{x} = \dfrac{3\sqrt{3} - \pi}{4\left(\pi + 6\sqrt{3} - 12\right)}, \bar{y} = \dfrac{\pi^2 + 12\sqrt{3}\pi - 72}{12\left(\pi + 6\sqrt{3} - 12\right)}$

**57. (a)** $-\dfrac{1}{3}x^2(1 - x^2)^{3/2} - \dfrac{2}{15}(1 - x^2)^{5/2} + C$

**(b)** $-\dfrac{1}{3}(1 - x^2)^{3/2} + \dfrac{1}{5}(1 - x^2)^{5/2} + C$

**(c)** $\dfrac{1}{5}(1 - x^2)^{5/2} - \dfrac{1}{3}(1 - x^2)^{3/2} + C$

Section 8.5, pp. 489–490

**1.** $\dfrac{2}{x - 3} + \dfrac{3}{x - 2}$ **3.** $\dfrac{1}{x + 1} + \dfrac{3}{(x + 1)^2}$

**5.** $\dfrac{-2}{z} + \dfrac{-1}{z^2} + \dfrac{2}{z - 1}$ **7.** $1 + \dfrac{17}{t - 3} + \dfrac{-12}{t - 2}$

**9.** $\dfrac{1}{2}\left[\ln|1 + x| - \ln|1 - x|\right] + C$

**11.** $\dfrac{1}{7}\ln\left|(x + 6)^2(x - 1)^5\right| + C$ **13.** $(\ln 15)/2$

**15.** $-\dfrac{1}{2}\ln|t| + \dfrac{1}{6}\ln|t + 2| + \dfrac{1}{3}\ln|t - 1| + C$ **17.** $3\ln 2 - 2$

**19.** $\dfrac{1}{4}\ln\left|\dfrac{x + 1}{x - 1}\right| - \dfrac{x}{2(x^2 - 1)} + C$ **21.** $(\pi + 2\ln 2)/8$

**23.** $\tan^{-1} y - \dfrac{1}{y^2 + 1} + C$

**25.** $-(s - 1)^{-2} + (s - 1)^{-1} + \tan^{-1} s + C$

**27.** $\dfrac{2}{3}\ln|x - 1| + \dfrac{1}{6}\ln|x^2 + x + 1| - \sqrt{3}\tan^{-1}\left(\dfrac{2x + 1}{\sqrt{3}}\right) + C$

**29.** $\dfrac{1}{4}\ln\left|\dfrac{x - 1}{x + 1}\right| + \dfrac{1}{2}\tan^{-1} x + C$

**31.** $\dfrac{-1}{\theta^2 + 2\theta + 2} + \ln\left(\theta^2 + 2\theta + 2\right) - \tan^{-1}(\theta + 1) + C$

**33.** $x^2 + \ln\left|\dfrac{x - 1}{x}\right| + C$

**35.** $9x + 2\ln|x| + \dfrac{1}{x} + 7\ln|x - 1| + C$

**37.** $\dfrac{y^2}{2} - \ln|y| + \dfrac{1}{2}\ln(1 + y^2) + C$ **39.** $\ln\left(\dfrac{e^t + 1}{e^t + 2}\right) + C$

**41.** $\dfrac{1}{5}\ln\left|\dfrac{\sin y - 2}{\sin y + 3}\right| + C$

**43.** $\dfrac{(\tan^{-1} 2x)^2}{4} - 3\ln|x - 2| + \dfrac{6}{x - 2} + C$

**45.** $\ln\left|\dfrac{\sqrt{x} - 1}{\sqrt{x} + 1}\right| + C$

**47.** $2\sqrt{1 + x} + \ln\left|\dfrac{\sqrt{x + 1} - 1}{\sqrt{x + 1} + 1}\right| + C$

**49.** $\dfrac{1}{4}\ln\left|\dfrac{x^4}{x^4 + 1}\right| + C$ **51.** $x = \ln|t - 2| - \ln|t - 1| + \ln 2$

**53.** $x = \dfrac{6t}{t+2} - 1$    **55.** $3\pi \ln 25$    **57.** 1.10

**59. (a)** $x = \dfrac{1000e^{4t}}{499 + e^{ct}}$    **(b)** 1.55 days

Section 8.6, pp. 495–496

**1.** $\dfrac{2}{\sqrt{3}}\left(\tan^{-1}\sqrt{\dfrac{x-3}{3}}\right) + C$

**3.** $\sqrt{x-2}\left(\dfrac{2(x-2)}{3} + 4\right) + C$    **5.** $\dfrac{(2x-3)^{3/2}(x+1)}{5} + C$

**7.** $\dfrac{-\sqrt{9-4x}}{x} - \dfrac{2}{3}\ln\left|\dfrac{\sqrt{9-4x}-3}{\sqrt{9-4x}+3}\right| + C$

**9.** $\dfrac{(x+2)(2x-6)\sqrt{4x-x^2}}{6} + 4\sin^{-1}\left(\dfrac{x-2}{2}\right) + C$

**11.** $-\dfrac{1}{\sqrt{7}}\ln\left|\dfrac{\sqrt{7}+\sqrt{7+x^2}}{x}\right| + C$

**13.** $\sqrt{4-x^2} - 2\ln\left|\dfrac{2+\sqrt{4-x^2}}{x}\right| + C$

**15.** $\dfrac{e^{2t}}{13}(2\cos 3t + 3\sin 3t) + C$

**17.** $\dfrac{x^2}{2}\cos^{-1}x + \dfrac{1}{4}\sin^{-1}x - \dfrac{1}{4}x\sqrt{1-x^2} + C$

**19.** $\dfrac{x^3}{3}\tan^{-1}x - \dfrac{x^2}{6} + \dfrac{1}{6}\ln(1+x^2) + C$

**21.** $-\dfrac{\cos 5x}{10} - \dfrac{\cos x}{2} + C$

**23.** $8\left[\dfrac{\sin(7t/2)}{7} - \dfrac{\sin(9t/2)}{9}\right] + C$

**25.** $6\sin(\theta/12) + \dfrac{6}{7}\sin(7\theta/12) + C$

**27.** $\dfrac{1}{2}\ln(x^2+1) + \dfrac{x}{2(1+x^2)} + \dfrac{1}{2}\tan^{-1}x + C$

**29.** $\left(x - \dfrac{1}{2}\right)\sin^{-1}\sqrt{x} + \dfrac{1}{2}\sqrt{x-x^2} + C$

**31.** $\sin^{-1}\sqrt{x} - \sqrt{x-x^2} + C$

**33.** $\sqrt{1-\sin^2 t} - \ln\left|\dfrac{1+\sqrt{1-\sin^2 t}}{\sin t}\right| + C$

**35.** $\ln\left|\ln y + \sqrt{3+(\ln y)^2}\right| + C$

**37.** $\ln\left|x+1+\sqrt{x^2+2x+5}\right| + C$

**39.** $\dfrac{x+2}{2}\sqrt{5-4x-x^2} + \dfrac{9}{2}\sin^{-1}\left(\dfrac{x+2}{3}\right) + C$

**41.** $-\dfrac{\sin^4 2x \cos 2x}{10} - \dfrac{2\sin^2 2x \cos 2x}{15} - \dfrac{4\cos 2x}{15} + C$

**43.** $\dfrac{\sin^3 2\theta \cos^2 2\theta}{10} + \dfrac{\sin^3 2\theta}{15} + C$

**45.** $\tan^2 2x - 2\ln|\sec 2x| + C$

**47.** $\dfrac{(\sec \pi x)(\tan \pi x)}{\pi} - \dfrac{1}{\pi}\ln|\sec \pi x + \tan \pi x| + C$

**49.** $\dfrac{-\csc^3 x \cot x}{4} - \dfrac{3\csc x \cot x}{8} - \dfrac{3}{8}\ln|\csc x + \cot x| + C$

**51.** $\dfrac{1}{2}\big[\sec(e^t - 1)\tan(e^t - 1) +$
     $\ln|\sec(e^t - 1) + \tan(e^t - 1)|\big] + C$

**53.** $\sqrt{2} + \ln\left(\sqrt{2}+1\right)$    **55.** $\pi/3$

**57.** $2\pi\sqrt{3} + \pi\sqrt{2}\ln\left(\sqrt{2}+\sqrt{3}\right)$    **59.** $\bar{x} = 4/3, \bar{y} = \ln\sqrt{2}$

**61.** 7.62    **63.** $\pi/8$    **67.** $\pi/4$

Section 8.7, pp. 503–506

**1. I: (a)** 1.5, 0  **(b)** 1.5, 0  **(c)** 0%
    **II: (a)** 1.5, 0  **(b)** 1.5, 0  **(c)** 0%
**3. I: (a)** 2.75, 0.08  **(b)** 2.67, 0.08  **(c)** 0.0312 ≈ 3%
    **II: (a)** 2.67, 0  **(b)** 2.67, 0  **(c)** 0%
**5. I: (a)** 6.25, 0.5  **(b)** 6, 0.25  **(c)** 0.0417 ≈ 4%
    **II: (a)** 6, 0  **(b)** 6, 0  **(c)** 0%
**7. I: (a)** 0.509, 0.03125  **(b)** 0.5, 0.009  **(c)** 0.018 ≈ 2%
    **II: (a)** 0.5, 0.002604  **(b)** 0.5, 0.4794  **(c)** 0%
**9. I: (a)** 1.8961, 0.161  **(b)** 2, 0.1039  **(c)** 0.052 ≈ 5%
    **II: (a)** 2.0045, 0.0066  **(b)** 2, 0.00454  **(c)** 0.2%
**11. (a)** 1  **(b)** 2    **13. (a)** 116  **(b)** 2
**15. (a)** 283  **(b)** 2    **17. (a)** 71  **(b)** 10
**19. (a)** 76  **(b)** 12    **21. (a)** 82  **(b)** 8
**23.** 106.6 m³    **25.** ≈2.55 m
**27. (a)** ≈0.00021  **(b)** ≈1.37079  **(c)** ≈0.015%
**31. (a)** ≈5.870  **(b)** $|E_T| \leq 0.0032$    **33.** 21.07 cm    **35.** 14.4
**39.** ≈28.7 mg

Section 8.8, pp. 515–517

**1.** $\pi/2$    **3.** 2    **5.** 6    **7.** $\pi/2$    **9.** ln 3    **11.** ln 4

**13.** 0    **15.** $\sqrt{3}$    **17.** $\pi$    **19.** $\ln\left(1 + \dfrac{\pi}{2}\right)$

**21.** −1    **23.** 1    **25.** −1/4    **27.** $\pi/2$    **29.** $\pi/3$
**31.** 6    **33.** ln 2    **35.** Diverges    **37.** Diverges
**39.** Converges    **41.** Converges    **43.** Diverges
**45.** Converges    **47.** Converges    **49.** Diverges
**51.** Converges    **53.** Converges    **55.** Diverges
**57.** Converges    **59.** Diverges    **61.** Converges
**63.** Converges
**65. (a)** Converges when $p < 1$    **(b)** Converges when $p > 1$
**67.** 1    **69.** $2\pi$    **71.** ln 2
**73. (a)** $\pi/2$    **(b)** $\pi$    **75. (b)** ≈0.88621
**77. (a)**

**(b)** $\pi/2$

**79. (a)**

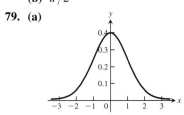

**(b)** ≈0.683, ≈0.954, ≈0.997

**85.** ≈0.16462

## Section 8.9. pp. 528–530

**1.** No **3.** Yes **5.** Yes **7.** Yes **11.** $\approx 0.537$

**13.** $\approx 0.688$ **15.** $\approx 0.0502$ **17.** $\sqrt{21}$ **19.** $\dfrac{1}{2}\ln 2$

**21.** $\dfrac{1}{\pi}, \dfrac{1}{\pi}\left(\tan^{-1}2 - \dfrac{\pi}{4}\right) \approx 0.10242$

**25.** mean $= \dfrac{8}{3} \approx 2.67$, median $= \sqrt{8} \approx 2.83$

**27.** mean $= 2$, median $= \sqrt{2} \approx 1.41$

**29.** $P\left(X < \dfrac{1}{2}\right) \approx 0.3935$

**31.** (a) $\approx 0.57$, so about 57 in every 100 bulbs will fail.
(b) $\approx 832$ h

**33.** $\approx 60$ hydra **35.** (a) $\approx 0.393$ (b) $\approx 0.135$ (c) 0
(d) The probability that any customer waits longer than 3 minutes is $1 - (0.997521)^{200} \approx 0.391 < 1/2$. So the most likely outcome is that all 200 would be served within 3 minutes.

**37.** \$10, 256 **39.** $\approx 323, \approx 262$ **41.** $\approx 0.89435$

**43.** (a) $\approx 16\%$ (b) $\approx 0.23832$ **45.** $\approx 618$ females

**47.** $\approx 61$ adults **49.** $\approx 289$ shafts

**51.** (a) $\approx 0.977$ (b) $\approx 0.159$ (c) $\approx 0.838$

**55.** (a) { LLL, LLD, LDL, DLL, LLU, LUL, ULL, LDD, LDU, LUD, LUU, DLD, DLU, ULD, ULU, DDL, DUL, UDL, UUL, DDD, DDU, DUD, UDD, DUU, UDU, UUD, UUU }
(c) $7/27 \approx 0.26$ (d) $20/27 \approx 0.74$

## Practice Exercises, pp. 531–533

**1.** $(x + 1)(\ln (x + 1)) - (x + 1) + C$

**3.** $\dfrac{1}{2}[2x \sin^{-1} 2x + \cos(\sin^{-1} 2x)] + C$

**5.** $(1 + x^2)e^x - 2x e^x + 2e^x + C$

**7.** $\dfrac{2e^x \sin 2x}{5} + \dfrac{e^x \cos 2x}{5} + C$

**9.** $-2 \ln (x - 2) + 3 \ln (x - 3) + C$

**11.** $\dfrac{1}{4}\ln x - \dfrac{1}{4}\ln\left|x - 2\right| - \dfrac{1}{2(x - 1)} + C$

**13.** $-\ln |1 - \sin x| + \ln |2 - \sin x| + C$

**15.** $-\dfrac{9}{x} + \dfrac{21}{2}(x - 1) + \dfrac{7}{2}(x + 1)$

**17.** $\dfrac{1}{3}\left[\dfrac{5}{4}\ln |x| - \dfrac{5}{4}\ln |x^2 - 4| + \ln\left|\dfrac{x - 2}{x + 2}\right|\right] + C$

**19.** $\dfrac{-1}{3}\dfrac{1}{2}\tan^{-}\left(\dfrac{\theta}{2}\right) + \dfrac{1}{3}\tan^{-1}\theta + C$

**21.** $\dfrac{x^2}{2} + \dfrac{4}{3}\ln |x + 2| + \dfrac{2}{3}\ln |x - 1| + C$

**23.** $\dfrac{x^2}{2} - \dfrac{9}{2}\ln |x + 3| + \dfrac{3}{2}\ln |x + 1| + C$

**25.** $\dfrac{1}{3}\ln\left|\dfrac{\sqrt{x + 1} - 1}{\sqrt{x + 1} + 1}\right| + C$ **27.** $\ln |1 - e^{-s}| + C$

**29.** $-\sqrt{16 - y^2} + C$ **31.** $-\dfrac{1}{2}\ln |4 - x^2| + C$

**33.** $\ln \dfrac{1}{\sqrt{9 - z^2}} + C$ **35.** $\dfrac{1}{6}\ln\left|\dfrac{x + 3}{x - 3}\right| + C$

**37.** $\dfrac{\sin^4 x \cos^7 x}{11} + \dfrac{4}{99}\sin^2 x \cos^7 x + \dfrac{8}{99}\left(\dfrac{-\cos^7 x}{7}\right) + C$

**39.** $\dfrac{\tan^7 x}{7} + C$

**41.** $\dfrac{1}{2}\left[\dfrac{\cos 12\theta}{12} - \dfrac{\cos 2\theta}{2}\right] + C$ **43.** $-4\sqrt{1 + \cos (t/2)} + C$

**45.** At least 16 **47.** $T = \pi, S = \pi$ **49.** $-4°C$

**51.** (a) $\approx 2.42$ L (b) $\approx 24.83$ km/L

**53.** $\pi/2$ **55.** 6 **57.** $\ln 3$ **59.** 2 **61.** $\pi/6$

**63.** Diverges **65.** Diverges **67.** Converges

**69.** $\dfrac{2x^{3/2}}{3} - x + 2\sqrt{x} - 2 \ln\left(\sqrt{x} + 1\right) + C$

**71.** $\dfrac{1}{2}\sin^{-1} (x - 1) + \dfrac{1}{2}(x - 1)\sqrt{2x - x^2} + C$

**73.** $-2 \cot x - \ln |\csc x + \cot x| + \csc x + C$

**75.** $\dfrac{1}{12}\ln\left|\dfrac{3 + v}{3 - v}\right| + \dfrac{1}{6}\tan^{-1}\dfrac{v}{3} + C$

**77.** $\dfrac{\theta \sin (2\theta + 1)}{2} + \dfrac{\cos (2\theta + 1)}{4} + C$ **79.** $\dfrac{1}{4}\sec^2 \theta + C$

**81.** $2\left(\dfrac{(\sqrt{2 - x})^3}{3} - 2\sqrt{2 - x}\right) + C$

**83.** $\tan^{-1}(y - 1) + C$

**85.** $\dfrac{1}{4}\ln |z| - \dfrac{1}{4z} - \dfrac{1}{4}\left[\dfrac{1}{2}\ln (z^2 + 4) + \dfrac{1}{2}\tan^{-1}\left(\dfrac{z}{2}\right)\right] + C$

**87.** $-\dfrac{1}{4}\sqrt{9 - 4t^2} + C$ **89.** $\ln\left(\dfrac{e^t + 1}{e^t + 2}\right) + C$ **91.** $1/4$

**93.** $\dfrac{2}{3}x^{3/2} + C$ **95.** $-\dfrac{1}{5}\tan^{-1}(\cos 5t) + C$

**97.** $2\sqrt{r} - 2 \ln\left(1 + \sqrt{r}\right) + C$

**99.** $\dfrac{1}{2}x^2 - \dfrac{1}{2}\ln (x^2 + 1) + C$

**101.** $\dfrac{2}{3}\ln |x + 1| + \dfrac{1}{6}\ln |x^2 - x + 1| + \dfrac{1}{\sqrt{3}}\tan^{-1}\left(\dfrac{2x - 1}{\sqrt{3}}\right) + C$

**103.** $\dfrac{4}{7}\left(1 + \sqrt{x}\right)^{7/2} - \dfrac{8}{5}\left(1 + \sqrt{x}\right)^{5/2} + \dfrac{4}{3}\left(1 + \sqrt{x}\right)^{3/2} + C$

**105.** $2 \ln |\sqrt{x} + \sqrt{1 + x}| + C$

**107.** $\ln x - \ln |1 + \ln x| + C$

**109.** $\dfrac{1}{2}x^{\ln x} + C$ **111.** $\dfrac{1}{2}\ln\left|\dfrac{1 - \sqrt{1 - x^4}}{x^2}\right| + C$

**113.** (b) $\dfrac{\pi}{4}$ **115.** $x - \dfrac{1}{\sqrt{2}}\tan^{-1}\left(\sqrt{2}\tan x\right) + C$

## Additional and Advanced Exercises, pp. 533–537

**1.** $x(\sin^{-1} x)^2 + 2 (\sin^{-1} x)\sqrt{1 - x^2} - 2x + C$

**3.** $\dfrac{x^2 \sin^{-1} x}{2} + \dfrac{x\sqrt{1 - x^2} - \sin^{-1} x}{4} + C$

**5.** $\dfrac{1}{2}\left(\ln \left(t - \sqrt{1 - t^2}\right) - \sin^{-1} t\right) + C$ **7.** 0

**9.** $\ln (4) - 1$ **11.** 1 **13.** $32\pi/35$ **15.** $2\pi$

**17.** (a) $\pi$ (b) $\pi(2e - 5)$

**19.** (b) $\pi\left(\dfrac{8 (\ln 2)^2}{3} - \dfrac{16 (\ln 2)}{9} + \dfrac{16}{27}\right)$

**21.** $\left(\dfrac{e^2 + 1}{4}, \dfrac{e - 2}{2}\right)$

**23.** $\sqrt{1 + e^2} - \ln\left(\dfrac{\sqrt{1 + e^2}}{e} + \dfrac{1}{e}\right) - \sqrt{2} + \ln\left(1 + \sqrt{2}\right)$

**25.** $\dfrac{12\pi}{5}$ **27.** $a = \dfrac{1}{2}, -\dfrac{\ln 2}{4}$ **29.** $\dfrac{1}{2} < p \le 1$

**33.** $\dfrac{e^{2x}}{13}(3 \sin 3x + 2 \cos 3x) + C$

**35.** $\dfrac{\cos x \sin 3x - 3 \sin x \cos 3x}{8} + C$

**37.** $\dfrac{e^{ax}}{a^2 + b^2}(a \sin bx - b \cos bx) + C$

**39.** $x \ln (ax) - x + C$ **41.** $\dfrac{2}{1 - \tan (x/2)} + C$ **43.** 1

**45.** $\dfrac{\sqrt{3}\pi}{9}$ **47.** $\dfrac{1}{\sqrt{2}} \ln \left| \dfrac{\tan (t/2) + 1 - \sqrt{2}}{\tan (t/2) + 1 + \sqrt{2}} \right| + C$

**49.** $\ln \left| \dfrac{1 + \tan (\theta/2)}{1 - \tan (\theta/2)} \right| + C$

# Chapter 9

## Section 9.1, pp. 544–546

**1.** (d) **3.** (a)

**5.**

**7.** $y' = x - y; y(1) = -1$ **9.** $y' = -(1 + y) \sin x; y(0) = 2$

**11.** $y(\text{exact}) = \dfrac{x}{2} - \dfrac{4}{x}, y_1 = -0.25, y_2 = 0.3, y_3 = 0.75$

**13.** $y(\text{exact}) = 3e^{x(x+2)}, y_1 = 4.2, y_2 = 6.216, y_3 = 9.697$

**15.** $y(\text{exact}) = e^{x^2} + 1, y_1 = 2.0, y_2 = 2.0202, y_3 = 2.0618$

**17.** $y \approx 2.48832$, exact value is $e$.

**19.** $y \approx -0.2272$, exact value is $1/(1 - 2\sqrt{5}) \approx -0.2880$.

**23.**

**25.** **27.**

**35.** Euler's method gives $y \approx 3.45835$; the exact solution is $y = 1 + e \approx 3.71828$.

**37.** $y \approx 1.5000$; exact value is 1.5275.

## Section 9.2, pp. 550–552

**1.** $y = \dfrac{e^x + C}{x}, \quad x > 0$ **3.** $y = \dfrac{C - \cos x}{x^3}, \quad x > 0$

**5.** $y = \dfrac{1}{2} - \dfrac{1}{x} + \dfrac{C}{x^2}, \quad x > 0$ **7.** $y = \dfrac{1}{2}xe^{x/2} + Ce^{x/2}$

**9.** $y = x(\ln x)^2 + Cx$

**11.** $s = \dfrac{t^3}{3(t-1)^4} - \dfrac{t}{(t-1)^4} + \dfrac{C}{(t-1)^4}$

**13.** $r = (\csc \theta)(\ln |\sec \theta| + C), \quad 0 < \theta < \pi/2$

**15.** $y = \dfrac{3}{2} - \dfrac{1}{2}e^{-2t}$ **17.** $y = -\dfrac{1}{\theta}\cos \theta + \dfrac{\pi}{2\theta}$

**19.** $y = 6e^{x^2} - \dfrac{e^{x^2}}{x + 1}$ **21.** $y = y_0 e^{kt}$

**23.** (b) is correct, but (a) is not. **25.** $t = \dfrac{L}{R} \ln 2$ s

**27.** (a) $i = \dfrac{V}{R} - \dfrac{V}{R}e^{-3} = \dfrac{V}{R}(1 - e^{-3}) \approx 0.95\dfrac{V}{R}$ amp **(b)** 86%

**29.** $y = \dfrac{1}{1 + Ce^{-x}}$ **31.** $y^3 = 1 + Cx^{-3}$

## Section 9.3, pp. 557–558

**1.** (a) 168.5 m **(b)** 41.13 s

**3.** $s(t) = 4.91\left(1 - e^{-(22.36/39.92)t}\right)$

**5.** $x^2 + y^2 = C$ **7.** $\ln |y| - \dfrac{1}{2}y^2 = \dfrac{1}{2}x^2 + C$

$kx^2 + y^2 = 1$

**9.** $y = \pm \sqrt{2x + C}$

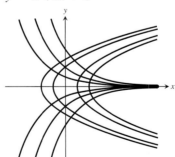

**13.** (a) 4 kg/min **(b)** $(400 + 4t)$ L **(c)** $4\left(\dfrac{y}{100 + t}\right)$ kg/min

**(d)** $\dfrac{dy}{dt} = 4 - \dfrac{4y}{100 + t}, \quad y(0) = 20,$

$y = 0.8(100 + t) - \dfrac{60}{\left(1 + \dfrac{t}{100}\right)^4}$

**(e)** Concentration $= \dfrac{y(25)}{\text{amt. brine in tank}} = \dfrac{75.424}{500} \approx 0.151$ kg/L

**15.** $y(27.8) \approx 5.93$ kg, $t \approx 27.8$ min

**Section 9.4, pp. 564–565**

**1.** $y' = (y + 2)(y - 3)$

(a) $y = -2$ is a stable equilibrium value and $y = 3$ is an unstable equilibrium.

(b) $y'' = 2(y + 2)\left(y - \dfrac{1}{2}\right)(y - 3)$

(c)

**3.** $y' = y^3 - y = (y + 1)y(y - 1)$

(a) $y = -1$ and $y = 1$ are unstable equilibria and $y = 0$ is a stable equilibrium.

(b) $y'' = (3y^2 - 1)y'$
$= 3(y + 1)\left(y + 1/\sqrt{3}\right)y\left(y - 1/\sqrt{3}\right)(y - 1)$

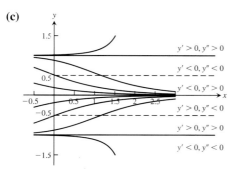

(c)

**5.** $y' = \sqrt{y}, y > 0$

(a) There are no equilibrium values.

(b) $y'' = \dfrac{1}{2}$

(c)

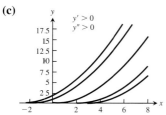

**7.** $y' = (y - 1)(y - 2)(y - 3)$

(a) $y = 1$ and $y = 3$ are unstable equilibria and $y = 2$ is a stable equilibrium.

(b) $y'' = (3y^2 - 12y + 11)(y - 1)(y - 2)(y - 3) =$

$3(y - 1)\left(y - \dfrac{6 - \sqrt{3}}{3}\right)(y - 2)\left(y - \dfrac{6 + \sqrt{3}}{3}\right)(y - 3)$

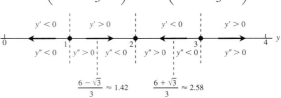

$\dfrac{6 - \sqrt{3}}{3} \approx 1.42 \qquad \dfrac{6 + \sqrt{3}}{3} \approx 2.58$

(c)

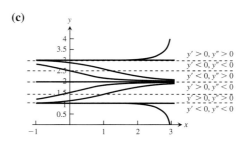

**9.** $\dfrac{dP}{dt} = 1 - 2P$ has a stable equilibrium at $P = \dfrac{1}{2}$;

$\dfrac{d^2P}{dt^2} = -2\dfrac{dP}{dt} = -2(1 - 2P)$.

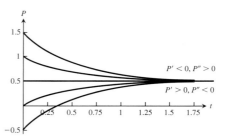

**11.** $\dfrac{dP}{dt} = 2P(P - 3)$ has a stable equilibrium at $P = 0$ and an unstable equilibrium at $P = 3$; $\dfrac{d^2P}{dt^2} = 2(2P - 3)\dfrac{dP}{dt} = 4P(2P - 3)(P - 3)$.

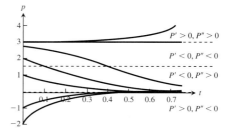

**13.** Before the catastrophe, the population exhibits logistic growth and $P(t)$ increases toward $M_0$, the stable equilibrium. After the catastrophe, the population declines logistically and $P(t)$ decreases toward $M_1$, the new stable equilibrium.

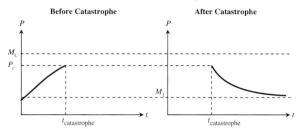

**15.** $\dfrac{dv}{dt} = g - \dfrac{k}{m}v^2$, $g, k, m > 0$ and $v(t) \geq 0$

Equilibrium: $\dfrac{dv}{dt} = g - \dfrac{k}{m}v^2 = 0 \Rightarrow v = \sqrt{\dfrac{mg}{k}}$

Concavity: $\dfrac{d^2v}{dt^2} = -2\left(\dfrac{k}{m}v\right)\dfrac{dv}{dt} = -2\left(\dfrac{k}{m}v\right)\left(g - \dfrac{k}{m}v^2\right)$

**(a)**                    **(b)**

**(c)** $v_{\text{terminal}} = \sqrt{\dfrac{441}{0.2}} = 46.96 \text{ m/s} = 169 \text{ km/h}$ for the 45-kg skydiver

$v_{\text{terminal}} = \sqrt{\dfrac{7.84}{0.2}} = 62.6 \text{ m/s} = 225 \text{ km/h}$ for the 80-kg skydiver

**17.** $F = F_p - F_r; ma = 200 - 50|v|; \dfrac{dv}{dt} = \dfrac{1}{m}(200 - 50|v|)$.

The maximum velocity occurs when $\dfrac{dv}{dt} = 0$ or $v = 4 \text{ m/s}$.

**19.** Phase line:

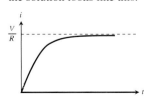

If the switch is closed at $t = 0$, then $i(0) = 0$, and the graph of the solution looks like this:

As $t \to \infty$, $i(t) \to i_{\text{steady state}} = \dfrac{V}{R}$.

Section 9.5, pp. 569–571

**1.** Seasonal variations, nonconformity of the environments, effects of other interactions, unexpected disasters, etc.

**3.** This model assumes that the number of interactions is proportional to the product of $x$ and $y$:

$\dfrac{dx}{dt} = (a - by)x, \quad a < 0,$

$\dfrac{dy}{dt} = m\left(1 - \dfrac{y}{M}\right)y - nxy = y\left(m - \dfrac{m}{M}y - nx\right).$

Rest points are $(0, 0)$, unstable, and $(0, M)$, stable.

**5. (a)** Logistic growth occurs in the absence of the competitor, and involves a simple interaction between the species: Growth dominates the competition when either population is small, so it is difficult to drive either species to extinction.

**(b)** $a$: per capita growth rate for trout
$m$: per capita growth rate for bass
$b$: intensity of competition to the trout
$n$: intensity of competition to the bass
$k_1$: environmental carrying capacity for the trout
$k_2$: environmental carrying capacity for the bass
$\dfrac{a}{b}$: growth versus competition or net growth of trout
$\dfrac{m}{n}$: relative survival of bass

**(c)** $\dfrac{dx}{dt} = 0$ when $x = 0$ or $y = \dfrac{a}{b} - \dfrac{a}{bk_1}x,$

$\dfrac{dy}{dt} = 0$ when $y = 0$ or $y = k_2 - \dfrac{k_2n}{m}x.$

By picking $a/b > k_2$ and $m/n > k_1$, we ensure that an equilibrium point exists inside the first quadrant.

Practice Exercises, pp. 571–572

**1.** $y = -\ln\left(C - \dfrac{2}{5}(x - 2)^{5/2} - \dfrac{4}{3}(x - 2)^{3/2}\right)$

**3.** $\tan y = -x\sin x - \cos x + C$  **5.** $(y + 1)e^{-y} = -\ln|x| + C$

**7.** $y = C\dfrac{x - 1}{x}$  **9.** $y = \dfrac{x^2}{4}e^{x/2} + Ce^{x/2}$

**11.** $y = \dfrac{x^2 - 2x + C}{2x^2}$  **13.** $y = \dfrac{e^{-x} + C}{1 + e^x}$  **15.** $xy + y^3 = C$

**17.** $y = \dfrac{2x^3 + 3x^2 + 6}{6(x + 1)^2}$  **19.** $y = \dfrac{1}{3}(1 - 4e^{-x^3})$

**21.** $y = e^{-x}(3x^3 - 3x^2)$

**23.**

| $x$ | $y$ | $x$ | $y$ |
|-----|-----|-----|-----|
| 0 | 0 | 1.1 | 1.6241 |
| 0.1 | 0.1000 | 1.2 | 1.8319 |
| 0.2 | 0.2095 | 1.3 | 2.0513 |
| 0.3 | 0.3285 | 1.4 | 2.2832 |
| 0.4 | 0.4568 | 1.5 | 2.5285 |
| 0.5 | 0.5946 | 1.6 | 2.7884 |
| 0.6 | 0.7418 | 1.7 | 3.0643 |
| 0.7 | 0.8986 | 1.8 | 3.3579 |
| 0.8 | 1.0649 | 1.9 | 3.6709 |
| 0.9 | 1.2411 | 2.0 | 4.0057 |
| 1.0 | 1.4273 | | |

**25.** $y(3) \approx 0.9131$

**27. (a)**

[−0.2, 4.5] by [−2.5, 0.5]

**(b)** Note that we choose a small interval of $x$-values because the $y$-values decrease very rapidly and our calculator cannot handle the calculations for $x \leq -1$. (This occurs because the analytic solution is $y = -2 + \ln(2 - e^{-x})$, which has an asymptote at $x = -\ln 2 \approx -0.69$. Obviously, the Euler approximations are misleading for $x \leq -0.7$.)

[−1, 0.2] by [−10, 2]

**29.** $y(\text{exact}) = \frac{1}{2}x^2 - \frac{3}{2}$; $y(2) \approx 0.4$; exact value is $\frac{1}{2}$.

**31.** $y(\text{exact}) = -e^{(x^2 - 1)/2}$; $y(2) \approx -3.4192$; exact value is $-e^{3/2} \approx -4.4817$.

**33. (a)** $y = -1$ is stable and $y = 1$ is unstable.

**(b)** $\dfrac{d^2 y}{dx^2} = 2y\dfrac{dy}{dx} = 2y(y^2 - 1)$

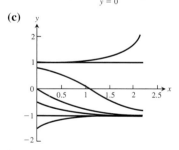

**(c)**

**Additional and Advanced Exercises, pp. 572–573**

**1. (a)** $y = c + (y_0 - c)e^{-k(A/V)t}$
   **(b)** Steady-state solution: $y_\infty = c$
**5.** $x^2(x^2 + 2y^2) = C$
**7.** $\ln|x| + e^{-y/x} = C$
**9.** $\ln|x| - \ln|\sec(y/x - 1) + \tan(y/x - 1)| = C$

# Chapter 10

**Section 10.1, pp. 583–586**

**1.** $a_1 = 0, a_2 = -1/4, a_3 = -2/9, a_4 = -3/16$
**3.** $a_1 = 1, a_2 = -1/3, a_3 = 1/5, a_4 = -1/7$
**5.** $a_1 = 1/2, a_2 = 1/2, a_3 = 1/2, a_4 = 1/2$
**7.** $1, \dfrac{3}{2}, \dfrac{7}{4}, \dfrac{15}{8}, \dfrac{31}{16}, \dfrac{63}{32}, \dfrac{127}{64}, \dfrac{255}{128}, \dfrac{511}{256}, \dfrac{1023}{512}$
**9.** $2, 1, -\dfrac{1}{2}, -\dfrac{1}{4}, \dfrac{1}{8}, \dfrac{1}{16}, -\dfrac{1}{32}, -\dfrac{1}{64}, \dfrac{1}{128}, \dfrac{1}{256}$
**11.** $1, 1, 2, 3, 5, 8, 13, 21, 34, 55$   **13.** $a_n = (-1)^{n+1}, n \geq 1$

**15.** $a_n = (-1)^{n+1}(n)^2, n \geq 1$   **17.** $a_n = \dfrac{2^{n-1}}{3(n+2)}, n \geq 1$
**19.** $a_n = n^2 - 1, n \geq 1$   **21.** $a_n = 4n - 3, n \geq 1$
**23.** $a_n = \dfrac{3n+2}{n!}, n \geq 1$   **25.** $a_n = \dfrac{1 + (-1)^{n+1}}{2}, n \geq 1$
**27.** Converges, 2   **29.** Converges, $-1$   **31.** Converges, $-5$
**33.** Diverges   **35.** Diverges   **37.** Converges, $1/2$
**39.** Converges, 0   **41.** Converges, $\sqrt{2}$   **43.** Converges, 1
**45.** Converges, 0   **47.** Converges, 0   **49.** Converges, 0
**51.** Converges, 1   **53.** Converges, $e^7$   **55.** Converges, 1
**57.** Converges, 1   **59.** Diverges   **61.** Converges, 4
**63.** Converges, 0   **65.** Diverges   **67.** Converges, $e^{-1}$
**69.** Converges, $e^{2/3}$   **71.** Converges, $x$ $(x > 0)$
**73.** Converges, 0   **75.** Converges, 1   **77.** Converges, $1/2$
**79.** Converges, 1   **81.** Converges, $\pi/2$   **83.** Converges, 0
**85.** Converges, 0   **87.** Converges, $1/2$   **89.** Converges, 0
**91.** 8   **93.** 4   **95.** 5   **97.** $1 + \sqrt{2}$   **99.** $x_n = 2^{n-2}$
**101. (a)** $f(x) = x^2 - 2, 1.414213562 \approx \sqrt{2}$
   **(b)** $f(x) = \tan(x) - 1, 0.7853981635 \approx \pi/4$
   **(c)** $f(x) = e^x$, diverges
**103. (b)** 1   **111.** Nondecreasing, bounded
**113.** Not nondecreasing, bounded
**115.** Converges, nondecreasing sequence theorem
**117.** Converges, nondecreasing sequence theorem
**119.** Diverges, definition of divergence   **121.** Converges
**123.** Converges   **135. (b)** $\sqrt{3}$

**Section 10.2, pp. 593–595**

**1.** $s_n = \dfrac{2(1 - (1/3)^n)}{1 - (1/3)}, 3$   **3.** $s_n = \dfrac{1 - (-1/2)^n}{1 - (-1/2)}, 2/3$
**5.** $s_n = \dfrac{1}{2} - \dfrac{1}{n+2}, \dfrac{1}{2}$   **7.** $1 - \dfrac{1}{4} + \dfrac{1}{16} - \dfrac{1}{64} + \cdots, \dfrac{4}{5}$
**9.** $-\dfrac{3}{4} + \dfrac{9}{16} + \dfrac{57}{64} + \dfrac{249}{256} + \cdots$, diverges.
**11.** $(5 + 1) + \left(\dfrac{5}{2} + \dfrac{1}{3}\right) + \left(\dfrac{5}{4} + \dfrac{1}{9}\right) + \left(\dfrac{5}{8} + \dfrac{1}{27}\right) + \cdots, \dfrac{23}{2}$
**13.** $(1 + 1) + \left(\dfrac{1}{2} - \dfrac{1}{5}\right) + \left(\dfrac{1}{4} + \dfrac{1}{25}\right) + \left(\dfrac{1}{8} - \dfrac{1}{125}\right) + \cdots, \dfrac{17}{6}$
**15.** Converges, $5/3$   **17.** Converges, $1/7$   **19.** $23/99$
**21.** $7/9$   **23.** $1/15$   **25.** $41333/33300$   **27.** Diverges
**29.** Inconclusive   **31.** Diverges   **33.** Diverges
**35.** $s_n = 1 - \dfrac{1}{n+1}$; converges, 1   **37.** $s_n = \ln\sqrt{n+1}$; diverges
**39.** $s_n = \dfrac{\pi}{3} - \cos^{-1}\left(\dfrac{1}{n+2}\right)$; converges, $-\dfrac{\pi}{6}$
**41.** 1   **43.** 5   **45.** 1   **47.** $-\dfrac{1}{\ln 2}$   **49.** Converges, $2 + \sqrt{2}$
**51.** Converges, 1   **53.** Diverges   **55.** Converges, $\dfrac{e^2}{e^2 - 1}$
**57.** Converges, $2/9$   **59.** Converges, $3/2$   **61.** Diverges
**63.** Converges, 4   **65.** Diverges   **67.** Converges, $\dfrac{\pi}{\pi - e}$
**69.** $a = 1, r = -x$; converges to $1/(1 + x)$ for $|x| < 1$
**71.** $a = 3, r = (x - 1)/2$; converges to $6/(3 - x)$ for $x$ in $(-1, 3)$
**73.** $|x| < \dfrac{1}{2}, \dfrac{1}{1 - 2x}$   **75.** $-2 < x < 0, \dfrac{1}{2 + x}$
**77.** $x \neq (2k + 1)\dfrac{\pi}{2}, k$ an integer; $\dfrac{1}{1 - \sin x}$
**79. (a)** $\displaystyle\sum_{n=-2}^{\infty} \dfrac{1}{(n+4)(n+5)}$   **(b)** $\displaystyle\sum_{n=0}^{\infty} \dfrac{1}{(n+2)(n+3)}$
   **(c)** $\displaystyle\sum_{n=5}^{\infty} \dfrac{1}{(n-3)(n-2)}$

**89. (a)** $r = 3/5$ **(b)** $r = -3/10$ **91.** $|r| < 1, \dfrac{1 + 2r}{1 - r^2}$

**93. (a)** 16.84 mg, 17.79 mg **(b)** 17.84 mg

**95. (a)** $0, \dfrac{1}{27}, \dfrac{2}{27}, \dfrac{1}{9}, \dfrac{2}{9}, \dfrac{7}{27}, \dfrac{8}{27}, \dfrac{1}{3}, \dfrac{2}{3}, \dfrac{7}{9}, \dfrac{8}{9}, 1$

**(b)** $\displaystyle\sum_{n=1}^{\infty} \dfrac{1}{2}\left(\dfrac{2}{3}\right)^{n-1} = 1$

### Section 10.3, pp. 600–601

**1.** Converges **3.** Converges **5.** Converges **7.** Diverges

**9.** Converges **11.** Converges; geometric series, $r = \dfrac{1}{10} < 1$

**13.** Diverges; $\lim\limits_{n\to\infty} \dfrac{n}{n+1} = 1 \neq 0$

**15.** Diverges; $p$-series, $p < 1$

**17.** Converges; geometric series, $r = \dfrac{1}{8} < 1$

**19.** Diverges; Integral Test

**21.** Converges; geometric series, $r = 2/3 < 1$

**23.** Diverges; Integral Test

**25.** Diverges; $\lim\limits_{n\to\infty} \dfrac{2^n}{n+1} \neq 0$

**27.** Diverges; $\lim_{n\to\infty} \left(\sqrt{n}/\ln n\right) \neq 0$

**29.** Diverges; geometric series, $r = \dfrac{1}{\ln 2} > 1$

**31.** Converges; Integral Test **33.** Diverges; $n$th-Term Test
**35.** Converges; Integral Test **37.** Converges; Integral Test
**39.** Converges; Integral Test **41.** $a = 1$

**43. (a)**

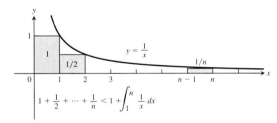

**(b)** $\approx 41.55$

**45.** True **47. (b)** $n \geq 251{,}415$

**49.** $s_8 = \displaystyle\sum_{n=1}^{8} \dfrac{1}{n^3} \approx 1.195$ **51.** $10^{60}$

**59. (a)** $1.20166 \leq S \leq 1.20253$ **(b)** $S \approx 1.2021$, error $< 0.0005$

**61.** $\left(\dfrac{\pi^2}{6} - 1\right) \approx 0.64493$

### Section 10.4, pp. 605–606

**1.** Converges; compare with $\Sigma(1/n^2)$

**3.** Diverges; compare with $\Sigma\left(1/\sqrt{n}\right)$

**5.** Converges; compare with $\Sigma(1/n^{3/2})$

**7.** Converges; compare with $\displaystyle\Sigma\sqrt{\dfrac{n+4n}{n^4 + 0}} = \sqrt{5}\,\Sigma\dfrac{1}{n^{3/2}}$

**9.** Converges **11.** Diverges; limit comparison with $\Sigma(1/n)$

**13.** Diverges; limit comparison with $\Sigma\left(1/\sqrt{n}\right)$
**15.** Diverges **17.** Diverges; limit comparison with $\Sigma\left(1/\sqrt{n}\right)$
**19.** Converges; compare with $\Sigma(1/2^n)$
**21.** Diverges; $n$th-Term Test
**23.** Converges; compare with $\Sigma(1/n^2)$

**25.** Converges; $\left(\dfrac{n}{3n+1}\right)^n < \left(\dfrac{n}{3n}\right)^n = \left(\dfrac{1}{3}\right)^n$

**27.** Diverges; direct comparison with $\Sigma(1/n)$
**29.** Diverges; limit comparison with $\Sigma(1/n)$
**31.** Diverges; limit comparison with $\Sigma(1/n)$
**33.** Converges; compare with $\Sigma(1/n^{3/2})$

**35.** Converges; $\dfrac{1}{n2^n} \leq \dfrac{1}{2^n}$

**37.** Converges; $\dfrac{1}{3^{n-1} + 1} < \dfrac{1}{3^{n-1}}$

**39.** Converges; comparison with $\Sigma(1/5n^2)$
**41.** Diverges; comparison with $\Sigma(1/n)$

**43.** Converges; comparison with $\displaystyle\Sigma\dfrac{1}{n(n-1)}$ or limit comparison with $\Sigma(1/n^2)$

**45.** Diverges; limit comparison with $\Sigma(1/n)$

**47.** Converges; $\dfrac{\tan^{-1} n}{n^{1.1}} < \dfrac{\pi/2}{n^{1.1}}$

**49.** Converges; compare with $\Sigma(1/n^2)$
**51.** Diverges; limit comparison with $\Sigma(1/n)$
**53.** Converges; limit comparison with $\Sigma(1/n^2)$
**65.** Converges **67.** Converges **69.** Converges

### Section 10.5, pp. 611–612

**1.** Converges **3.** Diverges **5.** Converges **7.** Converges
**9.** Converges **11.** Diverges **13.** Converges **15.** Converges
**17.** Converges; Ratio Test **19.** Diverges; Ratio Test
**21.** Converges; Ratio Test
**23.** Converges; compare with $\Sigma(3/(1.25)^n)$

**25.** Diverges; $\lim\limits_{n\to\infty}\left(1 - \dfrac{3}{n}\right)^n = e^{-3} \neq 0$

**27.** Converges; compare with $\Sigma(1/n^2)$
**29.** Diverges; compare with $\Sigma(1/(2n))$
**31.** Diverges; $a_n \nrightarrow 0$ **33.** Converges; Ratio Test
**35.** Converges; Ratio Test **37.** Converges; Ratio Test
**39.** Converges; Root Test **41.** Converges; compare with $\Sigma(1/n^2)$
**43.** Converges; Ratio Test **45.** Converges; Ratio Test
**47.** Diverges; Ratio Test **49.** Converges; Ratio Test

**51.** Converges; Ratio Test **53.** Diverges; $a_n = \left(\dfrac{1}{3}\right)^{(1/n!)} \to 1$

**55.** Converges; Ratio Test **57.** Diverges; Root Test
**59.** Converges; Root Test **61.** Converges; Ratio Test **65.** Yes

### Section 10.6, pp. 617–618

**1.** Converges by Alternating Series Test
**3.** Converges; Alternating Series Test
**5.** Converges; Alternating Series Test
**7.** Diverges; $a_n \nrightarrow 0$
**9.** Diverges; $a_n \nrightarrow 0$
**11.** Converges; Alternating Series Test
**13.** Converges by Alternating Series Test
**15.** Converges absolutely. Series of absolute values is a convergent geometric series.
**17.** Converges conditionally; $1/\sqrt{n} \to 0$ but $\displaystyle\sum_{n=1}^{\infty}\dfrac{1}{\sqrt{n}}$ diverges.

**19.** Converges absolutely; compare with $\sum_{n=1}^{\infty}(1/n^2)$.

**21.** Converges conditionally; $1/(n+3) \to 0$ but $\sum_{n=1}^{\infty}\dfrac{1}{n+3}$

diverges (compare with $\sum_{n=1}^{\infty}(1/n)$).

**23.** Diverges; $\dfrac{3+n}{5+n} \to 1$

**25.** Converges conditionally; $\left(\dfrac{1}{n^2}+\dfrac{1}{n}\right) \to 0$ but $(1+n)/n^2 > 1/n$

**27.** Converges absolutely; Ratio Test

**29.** Converges absolutely by Integral Test

**31.** Diverges; $a_n \nrightarrow 0$ **33.** Converges absolutely by Ratio Test

**35.** Converges absolutely, since $\left|\dfrac{\cos n\pi}{n\sqrt{n}}\right| = \left|\dfrac{(-1)^{n+1}}{n^{3/2}}\right| = \dfrac{1}{n^{3/2}}$

(convergent $p$-series)

**37.** Converges absolutely by Root Test **39.** Diverges; $a_n \to \infty$

**41.** Converges conditionally; $\sqrt{n+1} - \sqrt{n} = 1/(\sqrt{n}+\sqrt{n+1}) \to 0$, but series of absolute values diverges $\left(\text{compare with } \sum\left(1/\sqrt{n}\right)\right)$.

**43.** Diverges, $a_n \to 1/2 \neq 0$

**45.** Converges absolutely; $\operatorname{sech} n = \dfrac{2}{e^n + e^{-n}} = \dfrac{2e^n}{e^{2n}+1} <$

$\dfrac{2e^n}{e^{2n}} = \dfrac{2}{e^n}$, a term from a convergent geometric series.

**47.** Converges conditionally; $\sum(-1)^{n+1}\dfrac{1}{2(n+1)}$ converges by

Alternating Series Test; $\sum\dfrac{1}{2(n+1)}$ diverges by limit comparison with $\sum(1/n)$.

**49.** $|\text{Error}| < 0.2$ **51.** $|\text{Error}| < 2 \times 10^{-11}$

**53.** $n \geq 31$ **55.** $n \geq 4$ **57.** $0.54030$

**59.** (a) $a_n \geq a_{n+1}$ (b) $-1/2$

Section 10.7, pp. 626–628

**1.** (a) $1, -1 < x < 1$ (b) $-1 < x < 1$ (c) none

**3.** (a) $1/4, -1/2 < x < 0$ (b) $-1/2 < x < 0$ (c) none

**5.** (a) $10, -8 < x < 12$ (b) $-8 < x < 12$ (c) none

**7.** (a) $1, -1 < x < 1$ (b) $-1 < x < 1$ (c) none

**9.** (a) $3, -3 \leq x \leq 3$ (b) $-3 \leq x \leq 3$ (c) none

**11.** (a) $\infty$, for all $x$ (b) for all $x$ (c) none

**13.** (a) $1/2, -1/2 < x < 1/2$ (b) $-1/2 < x < 1/2$ (c) none

**15.** (a) $1, -1 \leq x < 1$ (b) $-1 < x < 1$ (c) $x = -1$

**17.** (a) $5, -8 < x < 2$ (b) $-8 < x < 2$ (c) none

**19.** (a) $3, -3 < x < 3$ (b) $-3 < x < 3$ (c) none

**21.** (a) $1, -2 < x < 0$ (b) $-2 < x < 0$ (c) none

**23.** (a) $1, -1 < x < 1$ (b) $-1 < x < 1$ (c) none

**25.** (a) $0, x = 0$ (b) $x = 0$ (c) none

**27.** (a) $2, -4 < x \leq 0$ (b) $-4 < x < 0$ (c) $x = 0$

**29.** (a) $1, -1 \leq x \leq 1$ (b) $-1 \leq x \leq 1$ (c) none

**31.** (a) $1/4, 1 \leq x \leq 3/2$ (b) $1 \leq x \leq 3/2$ (c) none

**33.** (a) $\infty$, for all $x$ (b) for all $x$ (c) none

**35.** (a) $1, -1 \leq x < 1$ (b) $-1 < x < 1$ (c) $-1$

**37.** $3$ **39.** $8$ **41.** $-1/3 < x < 1/3, 1/(1-3x)$

**43.** $-1 < x < 3, 4/(3 + 2x - x^2)$

**45.** $0 < x < 16, 2/(4 - \sqrt{x})$

**47.** $-\sqrt{2} < x < \sqrt{2}, 3/(2 - x^2)$

**49.** $\dfrac{2}{x} = \sum_{n=0}^{\infty} 2(-1)^n(x-1)^n, \ 0 < x < 2$

**51.** $\sum_{n=0}^{\infty}\left(-\tfrac{1}{3}\right)^r(x-5)^n, \ 2 < x < 8$

**53.** $1 < x < 5, 2/(x-1), \sum_{n=1}^{\infty}\left(-\tfrac{1}{2}\right)^n n(x-3)^{n-1}$,

$1 < x < 5, -2/(x-1)^2$

**55.** (a) $\cos x = 1 - \dfrac{x^2}{2!} + \dfrac{x^4}{4!} - \dfrac{x^6}{6!} + \dfrac{x^8}{8!} - \dfrac{x^{10}}{10!} + \cdots;$ converges

for all $x$

(b) Same answer as part (c)

(c) $2x - \dfrac{2^3 x^3}{3!} + \dfrac{2^5 x^5}{5!} - \dfrac{2^7 x^7}{7!} + \dfrac{2^9 x^9}{9!} - \dfrac{2^{11} x^{11}}{11!} + \cdots$

**57.** (a) $\dfrac{x^2}{2} + \dfrac{x^4}{12} + \dfrac{x^6}{45} + \dfrac{17x^8}{2520} + \dfrac{31x^{10}}{14175}, -\dfrac{\pi}{2} < x < \dfrac{\pi}{2}$

(b) $1 + x^2 + \dfrac{2x^4}{3} + \dfrac{17x^6}{45} + \dfrac{62x^8}{315} + \cdots, -\dfrac{\pi}{2} < x < \dfrac{\pi}{2}$

Section 10.8, pp. 632–633

**1.** $P_0(x) = 1, P_1(x) = 1 + 2x, P_2(x) = 1 + 2x + 2x^2,$

$P_3(x) = 1 + 2x + 2x^2 + \dfrac{4}{3}x^3$

**3.** $P_0(x) = 0, P_1(x) = x - 1, P_2(x) = (x-1) - \dfrac{1}{2}(x-1)^2,$

$P_3(x) = (x-1) - \dfrac{1}{2}(x-1)^2 + \dfrac{1}{3}(x-1)^3$

**5.** $P_0(x) = \dfrac{1}{2}, P_1(x) = \dfrac{1}{2} - \dfrac{1}{4}(x-2),$

$P_2(x) = \dfrac{1}{2} - \dfrac{1}{4}(x-2) + \dfrac{1}{8}(x-2)^2,$

$P_3(x) = \dfrac{1}{2} - \dfrac{1}{4}(x-2) + \dfrac{1}{8}(x-2)^2 - \dfrac{1}{16}(x-2)^3$

**7.** $P_0(x) = \dfrac{\sqrt{2}}{2}, P_1(x) = \dfrac{\sqrt{2}}{2} + \dfrac{\sqrt{2}}{2}\left(x - \dfrac{\pi}{4}\right),$

$P_2(x) = \dfrac{\sqrt{2}}{2} + \dfrac{\sqrt{2}}{2}\left(x - \dfrac{\pi}{4}\right) - \dfrac{\sqrt{2}}{4}\left(x - \dfrac{\pi}{4}\right)^2,$

$P_3(x) = \dfrac{\sqrt{2}}{2} + \dfrac{\sqrt{2}}{2}\left(x - \dfrac{\pi}{4}\right) - \dfrac{\sqrt{2}}{4}\left(x - \dfrac{\pi}{4}\right)^2$

$\qquad - \dfrac{\sqrt{2}}{12}\left(x - \dfrac{\pi}{4}\right)^3$

**9.** $P_0(x) = 2, P_1(x) = 2 + \dfrac{1}{4}(x-4),$

$P_2(x) = 2 + \dfrac{1}{4}(x-4) - \dfrac{1}{64}(x-4)^2,$

$P_3(x) = 2 + \dfrac{1}{4}(x-4) - \dfrac{1}{64}(x-4)^2 + \dfrac{1}{512}(x-4)^3$

**11.** $\sum_{n=0}^{\infty}\dfrac{(-x)^n}{n!} = 1 - x + \dfrac{x^2}{2!} - \dfrac{x^3}{3!} + \dfrac{x^4}{4!} - \cdots$

**13.** $\sum_{n=0}^{\infty}(-1)^n x^n = 1 - x + x^2 - x^3 + \cdots$

**15.** $\sum_{n=0}^{\infty}\dfrac{(-1)^n 3^{2n+1} x^{2n+1}}{(2n+1)!}$ **17.** $7\sum_{n=0}^{\infty}\dfrac{(-1)^n x^{2n}}{(2n)!}$ **19.** $\sum_{n=0}^{\infty}\dfrac{x^{2n}}{(2n)!}$

**21.** $x^4 - 2x^3 - 5x + 4$

**23.** $8 + 10(x-2) + 6(x-2)^2 + (x-2)^3$

**25.** $21 - 36(x+2) + 25(x+2)^2 - 8(x+2)^3 + (x+2)^4$

**27.** $\sum_{n=0}^{\infty}(-1)^n(n+1)(x-1)^n$ **29.** $\sum_{n=0}^{\infty}\dfrac{e^2}{n!}(x-2)^n$

**31.** $\sum_{n=0}^{\infty}(-1)^{n+1}\dfrac{2^{2n}}{(2n)!}\left(x - \dfrac{\pi}{4}\right)^{2n}$

**33.** $-1 - 2x - \dfrac{5}{2}x^2 - \cdots, -1 < x < 1$

**35.** $x^2 - \dfrac{1}{2}x^3 + \dfrac{1}{6}x^4 + \cdots, -1 < x < 1$

**41.** $L(x) = 0$, $Q(x) = -x^2/2$     **43.** $L(x) = 1$, $Q(x) = 1 + x^2/2$
**45.** $L(x) = x$, $Q(x) = x$

Section 10.9, pp. 639–640

**1.** $\displaystyle\sum_{n=0}^{\infty} \frac{(-5x)^n}{n!} = 1 - 5x + \frac{5^2 x^2}{2!} - \frac{5^3 x^3}{3!} + \cdots$

**3.** $\displaystyle\sum_{n=0}^{\infty} \frac{5(-1)^n(-x)^{2n+1}}{(2n+1)!} = \sum_{n=0}^{\infty} \frac{5(-1)^{n+1} x^{2n+1}}{(2n+1)!}$
$= -5x + \frac{5x^3}{3!} - \frac{5x^5}{5!} + \frac{5x^7}{7!} - \cdots$

**5.** $\displaystyle\sum_{n=0}^{\infty} \frac{(-1)^n(5x^2)^{2n}}{(2n)!} = 1 - \frac{25x^4}{2!} + \frac{625x^8}{4!} - \cdots$

**7.** $\displaystyle\sum_{n=1}^{\infty} (-1)^{n+1}\frac{x^{2n}}{n} = x^2 - \frac{x^4}{2} + \frac{x^6}{3} - \frac{x^8}{4} + \cdots$

**9.** $\displaystyle\sum_{n=0}^{\infty} (-1)^n \left(\frac{3}{4}\right)^n x^{3n} = 1 - \frac{3}{4}x^3 + \frac{3^2}{4^2}x^6 - \frac{3^3}{4^3}x^9 + \cdots$

**11.** $\displaystyle\sum_{n=0}^{\infty} \frac{x^{n+1}}{n!} = x + x^2 + \frac{x^3}{2!} + \frac{x^4}{3!} + \frac{x^5}{4!} + \cdots$

**13.** $\displaystyle\sum_{n=2}^{\infty} \frac{(-1)^n x^{2n}}{(2n)!} = \frac{x^4}{4!} - \frac{x^6}{6!} + \frac{x^8}{8!} - \frac{x^{10}}{10!} + \cdots$

**15.** $x - \frac{\pi^2 x^3}{2!} + \frac{\pi^4 x^5}{4!} - \frac{\pi^6 x^7}{6!} + \cdots = \displaystyle\sum_{n=0}^{\infty} \frac{(-1)^n \pi^{2n} x^{2n+1}}{(2n)!}$

**17.** $1 + \displaystyle\sum_{n=1}^{\infty} \frac{(-1)^n (2x)^{2n}}{2 \cdot (2n)!} =$
$1 - \frac{(2x)^2}{2 \cdot 2!} + \frac{(2x)^4}{2 \cdot 4!} - \frac{(2x)^6}{2 \cdot 6!} + \frac{(2x)^8}{2 \cdot 8!} - \cdots$

**19.** $x^2 \displaystyle\sum_{n=0}^{\infty} (2x)^n = x^2 + 2x^3 + 4x^4 + \cdots$

**21.** $\displaystyle\sum_{n=1}^{\infty} nx^{n-1} = 1 + 2x + 3x^2 + 4x^3 + \cdots$

**23.** $\displaystyle\sum_{n=1}^{\infty} (-1)^{n+1} \frac{x^{4n-1}}{2n-1} = x^3 - \frac{x^7}{3} + \frac{x^{11}}{5} - \frac{x^{15}}{7} + \cdots$

**25.** $\displaystyle\sum_{n=0}^{\infty} \left(\frac{1}{n!} + (-1)^n\right) x^n = 2 + \frac{3}{2}x^2 - \frac{5}{6}x^3 + \frac{25}{24}x^4 - \cdots$

**27.** $\displaystyle\sum_{n=1}^{\infty} \frac{(-1)^{n-1} x^{2n+1}}{3n} = \frac{x^3}{3} - \frac{x^5}{6} + \frac{x^7}{9} - \cdots$

**29.** $x + x^2 + \frac{x^3}{3} - \frac{x^5}{30} + \cdots$

**31.** $x^2 - \frac{2}{3}x^4 + \frac{23}{45}x^6 - \frac{44}{105}x^8 + \cdots$

**33.** $1 + x + \frac{1}{2}x^2 - \frac{1}{8}x^4 + \cdots$

**35.** $|\text{Error}| \le \frac{1}{10^4 \cdot 4!} < 4.2 \times 10^{-6}$

**37.** $|x| < (0.06)^{1/5} < 0.56968$

**39.** $|\text{Error}| < (10^{-3})^2/6 < 1.67 \times 10^{-10}$,  $-10^{-3} < x < 0$

**41.** $|\text{Error}| < (30.1)(0.1)^3/6 < 1.87 \times 10^{-4}$

**49.** (a) $Q(x) = 1 + kx + \frac{k(k-1)}{2}x^2$  (b) $0 \le x < 100^{-1/3}$

Section 10.10, pp. 647–649

**1.** $1 + \frac{x}{2} - \frac{x^2}{8} + \frac{x^3}{16}$     **3.** $1 + 3x + 6x^2 + 10x^3$

**5.** $1 - x + \frac{3x^2}{4} - \frac{x^3}{2}$     **7.** $1 - \frac{x^3}{2} + \frac{3x^6}{8} - \frac{5x^9}{16}$

**9.** $1 + \frac{1}{2x} - \frac{1}{8x^2} + \frac{1}{16x^3}$

**11.** $(1+x)^4 = 1 + 4x + 6x^2 + 4x^3 + x^4$
**13.** $(1-2x)^3 = 1 - 6x + 12x^2 - 8x^3$
**15.** $0.0713362$     **17.** $0.4969536$     **19.** $0.0999445$     **21.** $0.10000$

**23.** $\frac{1}{13 \cdot 6!} \approx 0.00011$     **25.** $\frac{x^3}{3} - \frac{x^7}{7 \cdot 3!} + \frac{x^{11}}{11 \cdot 5!}$

**27.** (a) $\frac{x^2}{2} - \frac{x^4}{12}$
(b) $\frac{x^2}{2} - \frac{x^4}{3 \cdot 4} + \frac{x^6}{5 \cdot 6} - \frac{x^8}{7 \cdot 8} + \cdots + (-1)^{15} \frac{x^{32}}{31 \cdot 32}$

**29.** $1/2$     **31.** $-1/24$     **33.** $1/3$     **35.** $-1$     **37.** $2$

**39.** $3/2$     **41.** $e$     **43.** $\cos \frac{3}{4}$     **45.** $\frac{\sqrt{3}}{2}$     **47.** $\frac{x^3}{1-x}$

**49.** $\frac{x^3}{1+x^2}$     **51.** $\frac{-1}{(1+x)^2}$     **55.** $500$ terms     **57.** $4$ terms

**59.** (a) $x + \frac{x^3}{6} + \frac{3x^5}{40} + \frac{5x^7}{112}$, radius of convergence $= 1$
(b) $\frac{\pi}{2} - x - \frac{x^3}{6} - \frac{3x^5}{40} - \frac{5x^7}{112}$

**61.** $1 - 2x + 3x^2 - 4x^3 + \cdots$
**67.** (a) $-1$  (b) $\left(1/\sqrt{2}\right)(1+i)$  (c) $-i$
**71.** $x + x^2 + \frac{1}{3}x^3 - \frac{1}{30}x^5 + \cdots$, for all $x$

Practice Exercises, pp. 650–651

**1.** Converges to 1     **3.** Converges to $-1$     **5.** Diverges
**7.** Converges to 0     **9.** Converges to 1     **11.** Converges to $e^{-5}$
**13.** Converges to 3     **15.** Converges to $\ln 2$     **17.** Diverges
**19.** $1/6$     **21.** $3/2$     **23.** $e/(e-1)$     **25.** Diverges
**27.** Converges conditionally     **29.** Converges conditionally
**31.** Converges absolutely     **33.** Converges absolutely
**35.** Converges absolutely     **37.** Converges absolutely
**39.** Converges absolutely
**41.** (a) $3, -7 \le x < -1$  (b) $-7 < x < -1$  (c) $x = -7$
**43.** (a) $1/3, 0 \le x \le 2/3$  (b) $0 \le x \le 2/3$  (c) None
**45.** (a) $\infty$, for all $x$  (b) For all $x$  (c) None
**47.** (a) $\sqrt{3}, -\sqrt{3} < x < \sqrt{3}$  (b) $-\sqrt{3} < x < \sqrt{3}$  (c) None
**49.** (a) $e, -e < x < e$  (b) $-e < x < e$  (c) Empty set

**51.** $\frac{1}{1+x}, \frac{1}{4}, \frac{4}{5}$     **53.** $\sin x, \pi, 0$     **55.** $e^x, \ln 2, 2$     **57.** $\displaystyle\sum_{n=0}^{\infty} 2^n x^n$

**59.** $\displaystyle\sum_{n=0}^{\infty} \frac{(-1)^n \pi^{2n+1} x^{2n+1}}{(2n+1)!}$     **61.** $\displaystyle\sum_{n=0}^{\infty} \frac{(-1)^n x^{10n/3}}{(2n)!}$     **63.** $\displaystyle\sum_{n=0}^{\infty} \frac{((\pi x)/2)^n}{n!}$

**65.** $2 - \frac{(x+1)}{2 \cdot 1!} + \frac{3(x+1)^2}{2^3 \cdot 2!} + \frac{9(x+1)^3}{2^5 \cdot 3!} + \cdots$

**67.** $\frac{1}{4} - \frac{1}{4^2}(x-3) + \frac{1}{4^3}(x-3)^2 - \frac{1}{4^4}(x-3)^3$

**69.** $0.4849171431$     **71.** $0.4872223583$     **73.** $7/2$     **75.** $1/12$
**77.** $-2$     **79.** $r = -3, s = 9/2$     **81.** $2/3$

**83.** $\ln\left(\frac{n+1}{2n}\right)$; the series converges to $\ln\left(\frac{1}{2}\right)$.

**85.** (a) $\infty$  (b) $a = 1, b = 0$     **87.** It converges.

Additional and Advanced Exercises, pp. 652–654

**1.** Converges; Comparison Test     **3.** Diverges; $n$th-Term Test
**5.** Converges; Comparison Test     **7.** Diverges; $n$th-Term Test

**9.** With $a = \pi/3$, $\cos x = \dfrac{1}{2} - \dfrac{\sqrt{3}}{2}(x - \pi/3) - \dfrac{1}{4}(x - \pi/3)^2$

$\qquad + \dfrac{\sqrt{3}}{12}(x - \pi/3)^3 + \cdots$

**11.** With $a = 0$, $e^x = 1 + x + \dfrac{x^2}{2!} + \dfrac{x^3}{3!} + \cdots$

**13.** With $a = 22\pi$, $\cos x = 1 - \dfrac{1}{2}(x - 22\pi)^2 + \dfrac{1}{4!}(x - 22\pi)^4$

$\qquad - \dfrac{1}{6!}(x - 22\pi)^6 + \cdots$

**15.** Converges, limit $= b$    **17.** $\pi/2$    **21.** $b = \pm\dfrac{1}{5}$

**23.** $a = 2$, $L = -7/6$    **27.** **(b)** Yes

**31.** **(a)** $\displaystyle\sum_{n=1}^{\infty} nx^{n-1}$    **(b)** 6    **(c)** $\dfrac{1}{q}$

**33.** **(a)** $R_n = C_0 e^{-kt_0}(1 - e^{-nkt_0})/(1 - e^{-kt_0})$,
$\qquad R = C_0(e^{-kt_0})/(1 - e^{-kt_0}) = C_0/(e^{kt_0} - 1)$
**(b)** $R_1 = 1/e \approx 0.368$,
$\qquad R_{10} = R(1 - e^{-10}) \approx R(0.9999546) \approx 0.58195$;
$\qquad R \approx 0.58198$; $0 < (R - R_{10})/R < 0.0001$
**(c)** 7

# Chapter 11

## Section 11.1, pp. 661–663

**1.**

**3.**

**5.**

**7.**

**9.**

**11.**

**13.**

**15.**

**17.**

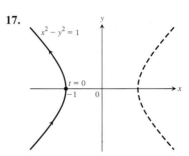

**19.** **(a)** $x = a \cos t$, $y = -a \sin t$, $0 \le t \le 2\pi$
**(b)** $x = a \cos t$, $y = a \sin t$, $0 \le t \le 2\pi$
**(c)** $x = a \cos t$, $y = -a \sin t$, $0 \le t \le 4\pi$
**(d)** $x = a \cos t$, $y = a \sin t$, $0 \le t \le 4\pi$
**21.** Possible answer: $x = -1 + 5t$, $y = -3 + 4t$, $0 \le t \le 1$
**23.** Possible answer: $x = t^2 + 1$, $y = t$, $t \le 0$
**25.** Possible answer: $x = 2 - 3t$, $y = 3 - 4t$, $t \ge 0$
**27.** Possible answer: $x = 2 \cos t$, $y = 2|\sin t|$, $0 \le t \le 4\pi$
**29.** Possible answer: $x = \dfrac{-at}{\sqrt{1 + t^2}}$, $y = \dfrac{a}{\sqrt{1 + t^2}}$, $-\infty < t < \infty$
**31.** Possible answer: $x = \dfrac{4}{1 + 2\tan\theta}$, $y = \dfrac{4\tan\theta}{1 + 2\tan\theta}$,
$\qquad 0 \le \theta < \pi/2$ and $x = 0$, $y = 2$ if $\theta = \pi/2$
**33.** Possible answer: $x = 2 - \cos t$, $y = \sin t$, $0 \le t \le 2\pi$
**35.** $x = 2 \cot t$, $y = 2 \sin^2 t$, $0 < t < \pi$
**37.** $x = a \sin^2 t \tan t$, $y = a \sin^2 t$, $0 \le t < \pi/2$    **39.** $(1, 1)$

## Section 11.2, pp. 671–673

**1.** $y = -x + 2\sqrt{2}$, $\dfrac{d^2y}{dx^2} = -\sqrt{2}$

**3.** $y = -\dfrac{1}{2}x + 2\sqrt{2}$, $\dfrac{d^2y}{dx^2} = -\dfrac{\sqrt{2}}{4}$

**5.** $y = x + \dfrac{1}{4}$, $\dfrac{d^2y}{dx^2} = -2$    **7.** $y = 2x - \sqrt{3}$, $\dfrac{d^2y}{dx^2} = -3\sqrt{3}$

**9.** $y = x - 4$, $\dfrac{d^2y}{dx^2} = \dfrac{1}{2}$

**11.** $y = \sqrt{3}x - \dfrac{\pi\sqrt{3}}{3} + 2$, $\dfrac{d^2y}{dx^2} = -4$

**13.** $y = 9x - 1$, $\dfrac{d^2y}{dx^2} = 108$    **15.** $-\dfrac{3}{16}$    **17.** $-6$

**19.** 1    **21.** $3a^2\pi$    **23.** $|ab|\pi$    **25.** 4    **27.** 12

**29.** $\pi^2$    **31.** $8\pi^2$    **33.** $\dfrac{52\pi}{3}$    **35.** $3\pi\sqrt{5}$

**37.** $(\bar{x}, \bar{y}) = \left(\dfrac{12}{\pi} - \dfrac{24}{\pi^2}, \dfrac{24}{\pi^2} - 2\right)$

**39.** $(\bar{x}, \bar{y}) = \left(\dfrac{1}{3}, \pi - \dfrac{4}{3}\right)$    **41. (a)** $\pi$   **(b)** $\pi$

**43. (a)** $x = 1$,   $y = 0$,   $\dfrac{dy}{dx} = \dfrac{1}{2}$    **(b)** $x = 0$,   $y = 3$,   $\dfrac{dy}{dx} = 0$

   **(c)** $x = \dfrac{\sqrt{3} - 1}{2}$,   $y = \dfrac{3 - \sqrt{3}}{2}$,   $\dfrac{dy}{dx} = \dfrac{2\sqrt{3} - 1}{\sqrt{3} - 2}$

**45.** $\left(\dfrac{\sqrt{2}}{2}, 1\right)$,   $y = 2x$ at $t = 0$, $y = -2x$ at $t = \pi$

**47. (a)** $8a$   **(b)** $\dfrac{64\pi}{3}$

**Section 11.3, pp. 676–677**

**1.** a, e; b, g; c, h; d, f      **3.**

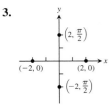

   **(a)** $\left(2, \dfrac{\pi}{2} + 2n\pi\right)$ and $\left(-2, \dfrac{\pi}{2} + (2n + 1)\pi\right)$, $n$ an integer

   **(b)** $(2, 2n\pi)$ and $(-2, (2n + 1)\pi)$, $n$ an integer

   **(c)** $\left(2, \dfrac{3\pi}{2} + 2n\pi\right)$ and $\left(-2, \dfrac{3\pi}{2} + (2n + 1)\pi\right)$, $n$ an integer

   **(d)** $(2, (2n + 1)\pi)$ and $(-2, 2n\pi)$, $n$ an integer

**5. (a)** $(3, 0)$   **(b)** $(-3, 0)$   **(c)** $\left(-1, \sqrt{3}\right)$   **(d)** $\left(1, \sqrt{3}\right)$
   **(e)** $(3, 0)$   **(f)** $\left(1, \sqrt{3}\right)$   **(g)** $(-3, 0)$   **(h)** $\left(-1, \sqrt{3}\right)$

**7. (a)** $\left(\sqrt{2}, \dfrac{\pi}{4}\right)$   **(b)** $(3, \pi)$   **(c)** $\left(2, \dfrac{11\pi}{6}\right)$

   **(d)** $\left(5, \pi - \tan^{-1}\dfrac{4}{3}\right)$

**9. (a)** $\left(-3\sqrt{2}, \dfrac{5\pi}{4}\right)$   **(b)** $(-1, 0)$   **(c)** $\left(-2, \dfrac{5\pi}{3}\right)$

   **(d)** $\left(-5, \pi - \tan^{-1}\dfrac{3}{4}\right)$

**11.**

**13.**

**15.**

**17.**

**19.**

**21.**

**23.**

**25.**

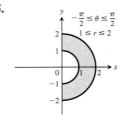

**27.** $x = 2$, vertical line through $(2, 0)$    **29.** $y = 0$, the $x$-axis
**31.** $y = 4$, horizontal line through $(0, 4)$
**33.** $x + y = 1$, line, $m = -1$, $b = 1$
**35.** $x^2 + y^2 = 1$, circle, $C(0, 0)$, radius 1
**37.** $y - 2x = 5$, line, $m = 2$, $b = 5$
**39.** $y^2 = x$, parabola, vertex $(0, 0)$, opens right
**41.** $y = e^x$, graph of natural exponential function
**43.** $x + y = \pm 1$, two straight lines of slope $-1$, $y$-intercepts
   $b = \pm 1$
**45.** $(x + 2)^2 + y^2 = 4$, circle, $C(-2, 0)$, radius 2
**47.** $x^2 + (y - 4)^2 = 16$, circle, $C(0, 4)$, radius 4
**49.** $(x - 1)^2 + (y - 1)^2 = 2$, circle, $C(1, 1)$, radius $\sqrt{2}$
**51.** $\sqrt{3}y + x = 4$    **53.** $r \cos \theta = 7$    **55.** $\theta = \pi/4$
**57.** $r = 2$ or $r = -2$    **59.** $4r^2 \cos^2 \theta + 9r^2 \sin^2 \theta = 36$
**61.** $r \sin^2 \theta = 4 \cos \theta$    **63.** $r = 4 \sin \theta$
**65.** $r^2 = 6r \cos \theta - 2r \sin \theta - 6$
**67.** $(0, \theta)$, where $\theta$ is any angle

**Section 11.4, pp. 680–681**

**1.** $x$-axis               **3.** $y$-axis

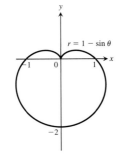

**5.** $y$-axis           **7.** $x$-axis, $y$-axis, origin

**9.** *x*-axis, *y*-axis, origin

**11.** *y*-axis, *x*-axis, origin

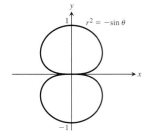

**13.** *x*-axis, *y*-axis, origin    **15.** Origin
**17.** The slope at $(-1, \pi/2)$ is $-1$, at $(-1, -\pi/2)$ is $1$.

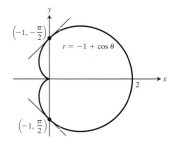

**19.** The slope at $(1, \pi/4)$ is $-1$, at $(-1, -\pi/4)$ is $1$, at $(-1, 3\pi/4)$ is $1$, at $(1, -3\pi/4)$ is $-1$.

**21. (a)**

**(b)**

**23. (a)**

**(b)**

**25.**

**27.**

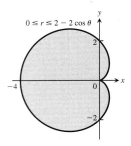

**29.** Equation (a)

Section 11.5, pp. 684–685

**1.** $\dfrac{1}{6}\pi^3$    **3.** $18\pi$    **5.** $\dfrac{\pi}{8}$    **7.** $2$    **9.** $\dfrac{\pi}{2} - 1$

**11.** $5\pi - 8$    **13.** $3\sqrt{3} - \pi$    **15.** $\dfrac{\pi}{3} + \dfrac{\sqrt{3}}{2}$

**17.** $\dfrac{8\pi}{3} + \sqrt{3}$    **19. (a)** $\dfrac{3}{2} - \dfrac{\pi}{4}$    **21.** $19/3$    **23.** $8$

**25.** $3\left(\sqrt{2} + \ln\left(1 + \sqrt{2}\right)\right)$    **27.** $\dfrac{\pi}{8} + \dfrac{3}{8}$

**31. (a)** $a$    **(b)** $a$    **(c)** $2a/\pi$

Section 11.6, pp. 691–694

**1.** $y^2 = 8x$,    $F(2, 0)$, directrix: $x = -2$
**3.** $x^2 = -6y$,    $F(0, -3/2)$, directrix: $y = 3/2$

**5.** $\dfrac{x^2}{4} - \dfrac{y^2}{9} = 1$,    $F\left(\pm\sqrt{13}, 0\right)$,    $V(\pm 2, 0)$,

   asymptotes: $y = \pm\dfrac{3}{2}x$

**7.** $\dfrac{x^2}{2} + y^2 = 1$,    $F(\pm 1, 0)$,    $V\left(\pm\sqrt{2}, 0\right)$

**9.**

**11.**

**13.**

**15.**

**17.**

**19.**

**21.**

**23.**

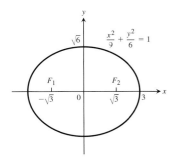

**25.** $\dfrac{x^2}{4} + \dfrac{y^2}{2} = 1$

**27.** Asymptotes: $y = \pm x$

**29.** Asymptotes: $y = \pm x$

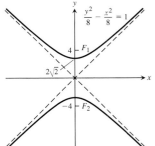

**31.** Asymptotes: $y = \pm 2x$

**33.** Asymptotes: $y = \pm x/2$

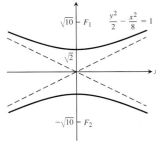

**35.** $y^2 - x^2 = 1$   **37.** $\dfrac{x^2}{9} - \dfrac{y^2}{16} = 1$

**39. (a)** Vertex: $(1, -2)$; focus: $(3, -2)$; directrix: $x = -1$

**(b)**

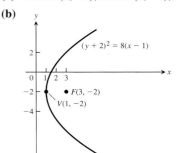

**41. (a)** Foci: $\left(4 \pm \sqrt{7}, 3\right)$; vertices: $(8, 3)$ and $(0, 3)$; center: $(4, 3)$

**(b)**

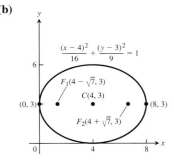

**43. (a)** Center: $(2, 0)$; foci: $(7, 0)$ and $(-3, 0)$; vertices: $(6, 0)$ and $(-2, 0)$; asymptotes: $y = \pm \dfrac{3}{4}(x - 2)$

**(b)**

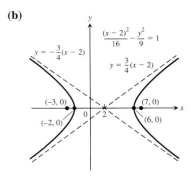

**45.** $(y + 3)^2 = 4(x + 2)$,   $V(-2, -3)$,   $F(-1, -3)$, directrix: $x = -3$

**47.** $(x - 1)^2 = 8(y + 7)$,   $V(1, -7)$,   $F(1, -5)$, directrix: $y = -9$

**49.** $\dfrac{(x + 2)^2}{6} + \dfrac{(y + 1)^2}{9} = 1$,   $F\left(-2, \pm\sqrt{3} - 1\right)$, $V(-2, \pm 3 - 1)$, $C(-2, -1)$

**51.** $\dfrac{(x - 2)^2}{3} + \dfrac{(y - 3)^2}{2} = 1$,   $F(3, 3)$ and $F(1, 3)$, $V\left(\pm\sqrt{3} + 2, 3\right)$, $C(2, 3)$

**53.** $\dfrac{(x - 2)^2}{4} - \dfrac{(y - 2)^2}{5} = 1$,   $C(2, 2)$,   $F(5, 2)$ and $F(-1, 2)$, $V(4, 2)$ and $V(0, 2)$; asymptotes: $(y - 2) = \pm\dfrac{\sqrt{5}}{2}(x - 2)$

**55.** $(y + 1)^2 - (x + 1)^2 = 1$, $C(-1, -1)$, $F(-1, \sqrt{2} - 1)$
and $F(-1, -\sqrt{2} - 1)$, $V(-1, 0)$ and $V(-1, -2)$; asymptotes
$(y + 1) = \pm(x + 1)$

**57.** $C(-2, 0)$, $a = 4$    **59.** $V(-1, 1)$, $F(-1, 0)$

**61.** Ellipse: $\dfrac{(x + 2)^2}{5} + y^2 = 1$, $C(-2, 0)$, $F(0, 0)$ and
$F(-4, 0)$, $V(\sqrt{5} - 2, 0)$ and $V(-\sqrt{5} - 2, 0)$

**63.** Ellipse: $\dfrac{(x - 1)^2}{2} + (y - 1)^2 = 1$, $C(1, 1)$, $F(2, 1)$ and
$F(0, 1)$, $V(\sqrt{2} + 1, 1)$ and $V(-\sqrt{2} + 1, 1)$

**65.** Hyperbola: $(x - 1)^2 - (y - 2)^2 = 1$, $C(1, 2)$,
$F(1 + \sqrt{2}, 2)$ and $F(1 - \sqrt{2}, 2)$, $V(2, 2)$ and
$V(0, 2)$; asymptotes: $(y - 2) = \pm(x - 1)$

**67.** Hyperbola: $\dfrac{(y - 3)^2}{6} - \dfrac{x^2}{3} = 1$, $C(0, 3)$, $F(0, 6)$
and $F(0, 0)$, $V(0, \sqrt{6} + 3)$ and $V(0, -\sqrt{6} + 3)$;
asymptotes: $y = \sqrt{2}x + 3$ or $y = -\sqrt{2}x + 3$

**69. (b)**$1 : 1$    **73.** Length $= 2\sqrt{2}$, width $= \sqrt{2}$, area $= 4$

**75.** $24\pi$

**77.** $x = 0, y = 0$: $y = -2x$; $x = 0, y = 2$: $y = 2x + 2$;
$x = 4, y = 0$: $y = 2x - 8$

**79.** $\bar{x} = 0$, $\bar{y} = \dfrac{16}{3\pi}$

## Section 11.7, pp. 699–700

**1.** $e = \dfrac{3}{5}$, $F(\pm 3, 0)$;
directrices are $x = \pm \dfrac{25}{3}$.

**3.** $e = \dfrac{1}{\sqrt{2}}$; $F(0, \pm 1)$;
directrices are $y = \pm 2$.

**5.** $e = \dfrac{1}{\sqrt{3}}$; $F(0, \pm 1)$;
directrices are $y = \pm 3$.

**7.** $e = \dfrac{\sqrt{3}}{3}$; $F(\pm\sqrt{3}, 0)$;
directrices are $x = \pm 3\sqrt{3}$.

**9.** $\dfrac{x^2}{27} + \dfrac{y^2}{36} = 1$    **11.** $\dfrac{x^2}{4851} + \dfrac{y^2}{4900} = 1$

**13.** $\dfrac{x^2}{9} + \dfrac{y^2}{4} = 1$    **15.** $\dfrac{x^2}{64} + \dfrac{y^2}{48} = 1$

**17.** $e = \sqrt{2}$; $F(\pm\sqrt{2}, 0)$;
directrices are $x = \pm\dfrac{1}{\sqrt{2}}$.

**19.** $e = \sqrt{2}$; $F(0, \pm 4)$;
directrices are $y = \pm 2$.

**21.** $e = \sqrt{5}$; $F(\pm\sqrt{10}, 0)$;
directrices are $x = \pm\dfrac{2}{\sqrt{10}}$.

**23.** $e = \sqrt{5}$; $F(0, \pm\sqrt{10})$;
directrices are $y = \pm\dfrac{2}{\sqrt{10}}$.

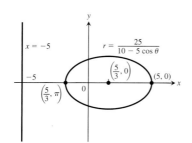

**25.** $y^2 - \dfrac{x^2}{8} = 1$    **27.** $x^2 - \dfrac{y^2}{8} = 1$    **29.** $r = \dfrac{2}{1 + \cos\theta}$

**31.** $r = \dfrac{30}{1 - 5\sin\theta}$    **33.** $r = \dfrac{1}{2 + \cos\theta}$    **35.** $r = \dfrac{10}{5 - \sin\theta}$

**37.**

**39.**

**41.**

**43.**

**45.** $y = 2 - x$

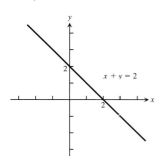

**47.** $y = \dfrac{\sqrt{3}}{3}x + 2\sqrt{3}$

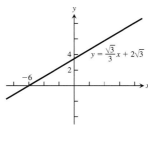

**49.** $r\cos\left(\theta - \dfrac{\pi}{4}\right) = 3$    **51.** $r\cos\left(\theta + \dfrac{\pi}{2}\right) = 5$

**53.**

**55.**

**57.** $r = 12\cos\theta$

**59.** $r = 10\sin\theta$

**61.** $r = -2\cos\theta$

**63.** $r = -\sin\theta$

**65.**

**67.**

**69.**

**71.**

**73.**

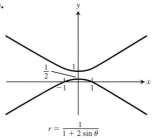

**75. (b)**

| Planet | Perihelion | Aphelion |
| --- | --- | --- |
| Mercury | 0.3075 AU | 0.4667 AU |
| Venus | 0.7184 AU | 0.7282 AU |
| Earth | 0.9833 AU | 1.0167 AU |
| Mars | 1.3817 AU | 1.6663 AU |
| Jupiter | 4.9512 AU | 5.4548 AU |
| Saturn | 9.0210 AU | 10.0570 AU |
| Uranus | 18.2977 AU | 20.0623 AU |
| Neptune | 29.8135 AU | 30.3065 AU |

**Practice Exercises, pp. 701–703**

**1.**

**3.**

**5.**

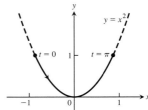

**7.** $x = 3\cos t, \quad y = 4\sin t, \quad 0 \le t \le 2\pi$

**9.** $y = \dfrac{\sqrt{3}}{2}x + \dfrac{1}{4}, \dfrac{1}{4}$

**11. (a)** $y = \dfrac{\pm |x|^{3/2}}{8} - 1$   **(b)** $y = \dfrac{\pm\sqrt{1-x^2}}{x}$

**13.** $\dfrac{10}{3}$   **15.** $\dfrac{285}{8}$   **17.** 10   **19.** $\dfrac{9\pi}{2}$   **21.** $\dfrac{76\pi}{3}$

**23.** $y = \dfrac{\sqrt{3}}{3}x - 4$   **25.** $x = 2$

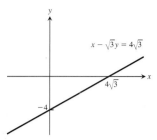

**27.** $y = -\dfrac{3}{2}$   **29.** $x^2 + (y+2)^2 = 4$

   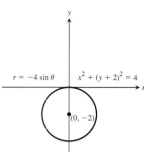

**31.** $\left(x - \sqrt{2}\right)^2 + y^2 = 2$   **33.** $r = -5\sin\theta$

   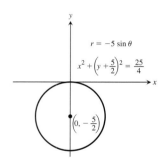

**35.** $r = 3\cos\theta$   **37.**

   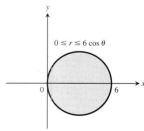

**39.** d   **41.** l   **43.** k   **45.** i   **47.** $\dfrac{9}{2}\pi$   **49.** $2 + \dfrac{\pi}{4}$

**51.** 8   **53.** $\pi - 3$

**55.** Focus is $(0, -1)$, directrix is $y = 1$.

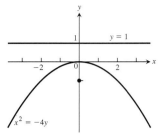

**57.** Focus is $\left(\dfrac{3}{4}, 0\right)$, directrix is $x = -\dfrac{3}{4}$.

**59.** $e = \dfrac{3}{4}$

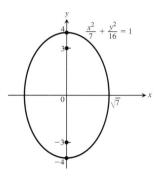

**61.** $e = 2$; the asymptotes are $y = \pm\sqrt{3}\,x$.

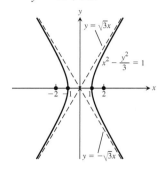

**63.** $(x - 2)^2 = -12(y - 3)$, $V(2, 3)$, $F(2, 0)$, directrix is $y = 6$.

**65.** $\dfrac{(x+3)^2}{9} + \dfrac{(y+5)^2}{25} = 1$, $C(-3, -5)$, $F(-3, -1)$ and $F(-3, -9)$, $V(-3, -10)$ and $V(-3, 0)$.

**67.** $\dfrac{\left(y - 2\sqrt{2}\right)^2}{8} - \dfrac{(x - 2)^2}{2} = 1$, $C\left(2, 2\sqrt{2}\right)$, $F\left(2, 2\sqrt{2} \pm \sqrt{10}\right)$, $V\left(2, 4\sqrt{2}\right)$ and $V(2, 0)$, the asymptotes are $y = 2x - 4 + 2\sqrt{2}$ and $y = -2x + 4 + 2\sqrt{2}$.

**69.** Hyperbola: $C(2, 0)$, $V(0, 0)$ and $V(4, 0)$, the foci are $F\left(2 \pm \sqrt{5}, 0\right)$, and the asymptotes are $y = \pm\dfrac{x - 2}{2}$.

**71.** Parabola: $V(-3, 1)$, $F(-7, 1)$, and the directrix is $x = 1$.

**73.** Ellipse: $C(-3, 2)$, $F\left(-3 \pm \sqrt{7}, 2\right)$, $V(1, 2)$ and $V(-7, 2)$

**75.** Circle: $C(1, 1)$ and radius $= \sqrt{2}$

**77.** $V(1, 0)$   **79.** $V(2, \pi)$ and $V(6, \pi)$

   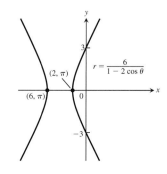

**81.** $r = \dfrac{4}{1 + 2\cos\theta}$   **83.** $r = \dfrac{2}{2 + \sin\theta}$

**85. (a)** $24\pi$   **(b)** $16\pi$

**Additional and Advanced Exercises, pp. 703–705**

**1.** $x - \dfrac{7}{2} = \dfrac{y^2}{2}$

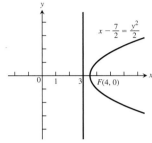

**3.** $3x^2 + 3y^2 - 8y - 4 = 0$   **5.** $F(0, \pm 1)$

**7. (a)** $\dfrac{(y-1)^2}{16} - \dfrac{x^2}{48} = 1$   **(b)** $\dfrac{\left(y + \dfrac{3}{4}\right)^2}{\left(\dfrac{25}{16}\right)} - \dfrac{x^2}{\left(\dfrac{75}{2}\right)} = 1$

**11.**                                    **13.**

**15.**

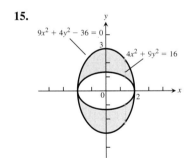

**17. (a)** $r = e^{2\theta}$   **(b)** $\dfrac{\sqrt{5}}{2}(e^{4\pi} - 1)$

**19.** $r = \dfrac{4}{1 + 2\cos\theta}$   **21.** $r = \dfrac{2}{2 + \sin\theta}$

**23.** $x = (a + b)\cos\theta - b\cos\left(\dfrac{a+b}{b}\theta\right),$

$y = (a + b)\sin\theta - b\sin\left(\dfrac{a+b}{b}\theta\right)$

**27.** $\dfrac{\pi}{2}$

# Chapter 12

Section 12.1, pp. 709–710
   **1.** The line through the point $(2, 3, 0)$ parallel to the $z$-axis
   **3.** The $x$-axis
   **5.** The circle $x^2 + y^2 = 4$ in the $xy$-plane
   **7.** The circle $x^2 + z^2 = 4$ in the $xz$-plane
   **9.** The circle $y^2 + z^2 = 1$ in the $yz$-plane
   **11.** The circle $x^2 + y^2 = 16$ in the $xy$-plane
   **13.** The ellipse formed by the intersection of the cylinder
      $x^2 + y^2 = 4$ and the plane $z = y$
   **15.** The parabola $y = x^2$ in the $xy$-plane
   **17. (a)** The first quadrant of the $xy$-plane
      **(b)** The fourth quadrant of the $xy$-plane
   **19. (a)** The ball of radius 1 centered at the origin
      **(b)** All points more than 1 unit from the origin
   **21. (a)** The ball of radius 2 centered at the origin with the interior of
      the ball of radius 1 centered at the origin removed
      **(b)** The solid upper hemisphere of radius 1 centered at the ori-
      gin
   **23. (a)** The region on or inside the parabola $y = x^2$ in the $xy$-plane
      and all points above this region
      **(b)** The region on or to the left of the parabola $x = y^2$ in the
      $xy$-plane and all points above it that are 2 units or less away
      from the $xy$-plane
   **25. (a)** $x = 3$   **(b)** $y = -1$   **(c)** $z = -2$
   **27. (a)** $z = 1$   **(b)** $x = 3$   **(c)** $y = -1$
   **29. (a)** $x^2 + (y - 2)^2 = 4, z = 0$
      **(b)** $(y - 2)^2 + z^2 = 4, x = 0$   **(c)** $x^2 + z^2 = 4, y = 2$
   **31. (a)** $y = 3, z = -1$   **(b)** $x = 1, z = -1$
      **(c)** $x = 1, y = 3$
   **33.** $x^2 + y^2 + z^2 = 25, z = 3$   **35.** $0 \le z \le 1$   **37.** $z \le 0$
   **39. (a)** $(x - 1)^2 + (y - 1)^2 + (z - 1)^2 < 1$
      **(b)** $(x - 1)^2 + (y - 1)^2 + (z - 1)^2 > 1$
   **41.** $3$   **43.** $7$   **45.** $2\sqrt{3}$   **47.** $C(-2, 0, 2), a = 2\sqrt{2}$
   **49.** $C(\sqrt{2}, \sqrt{2}, -\sqrt{2}), a = \sqrt{2}$
   **51.** $(x - 1)^2 + (y - 2)^2 + (z - 3)^2 = 14$

   **53.** $(x + 1)^2 + \left(y - \dfrac{1}{2}\right)^2 + \left(z + \dfrac{2}{3}\right)^2 = \dfrac{16}{81}$

   **55.** $C(-2, 0, 2), a = \sqrt{8}$   **57.** $C\left(-\dfrac{1}{4}, -\dfrac{1}{4}, -\dfrac{1}{4}\right), a = \dfrac{5\sqrt{3}}{4}$

   **59. (a)** $\sqrt{y^2 + z^2}$   **(b)** $\sqrt{x^2 + z^2}$   **(c)** $\sqrt{x^2 + y^2}$
   **61.** $\sqrt{17} + \sqrt{33} + 6$   **63.** $y = 1$
   **65. (a)** $(0, 3, -3)$   **(b)** $(0, 5, -5)$

Section 12.2, pp. 718–720
   **1. (a)** $\langle 9, -6 \rangle$   **(b)** $3\sqrt{13}$   **3. (a)** $\langle 1, 3 \rangle$   **(b)** $\sqrt{10}$
   **5. (a)** $\langle 12, -19 \rangle$   **(b)** $\sqrt{505}$
   **7. (a)** $\left\langle \dfrac{1}{5}, \dfrac{14}{5} \right\rangle$   **(b)** $\dfrac{\sqrt{197}}{5}$   **9.** $\langle 1, -4 \rangle$
   **11.** $\langle -2, -3 \rangle$   **13.** $\left\langle -\dfrac{1}{2}, \dfrac{\sqrt{3}}{2} \right\rangle$   **15.** $\left\langle -\dfrac{\sqrt{3}}{2}, -\dfrac{1}{2} \right\rangle$
   **17.** $-3\mathbf{i} + 2\mathbf{j} - \mathbf{k}$   **19.** $-3\mathbf{i} + 16\mathbf{j}$
   **21.** $3\mathbf{i} + 5\mathbf{j} - 8\mathbf{k}$

**23.** The vector **v** is horizontal and 25 mm long. The vectors **u** and **w** are 17 mm long. **w** is vertical and **u** makes a 45° angle with the horizontal. All vectors must be drawn to scale.

**(a)**    **(b)**

**(c)**    **(d)**

**25.** $3\left(\dfrac{2}{3}\mathbf{i} + \dfrac{1}{3}\mathbf{j} - \dfrac{2}{3}\mathbf{k}\right)$   **27.** $5(\mathbf{k})$

**29.** $\sqrt{\dfrac{1}{2}}\left(\dfrac{1}{\sqrt{3}}\mathbf{i} - \dfrac{1}{\sqrt{3}}\mathbf{j} - \dfrac{1}{\sqrt{3}}\mathbf{k}\right)$

**31.** (a) $2\mathbf{i}$   (b) $-\sqrt{3}\mathbf{k}$   (c) $\dfrac{3}{10}\mathbf{j} + \dfrac{2}{5}\mathbf{k}$   (d) $6\mathbf{i} - 2\mathbf{j} + 3\mathbf{k}$

**33.** $\dfrac{7}{13}(12\mathbf{i} - 5\mathbf{k})$

**35.** (a) $\dfrac{3}{5\sqrt{2}}\mathbf{i} + \dfrac{4}{5\sqrt{2}}\mathbf{j} - \dfrac{1}{\sqrt{2}}\mathbf{k}$   (b) $(1/2, 3, 5/2)$

**37.** (a) $-\dfrac{1}{\sqrt{3}}\mathbf{i} - \dfrac{1}{\sqrt{3}}\mathbf{j} - \dfrac{1}{\sqrt{3}}\mathbf{k}$   (b) $\left(\dfrac{5}{2}, \dfrac{7}{2}, \dfrac{9}{2}\right)$

**39.** $A(4, -3, 5)$   **41.** $a = \dfrac{3}{2}, b = \dfrac{1}{2}$

**43.** $\approx \langle -338\,095, 725.046 \rangle$

**45.** $|\mathbf{F}_1| = \dfrac{100\cos 45°}{\sin 75°} \approx 73.205 \text{ N}$,

$|\mathbf{F}_2| = \dfrac{100\cos 30°}{\sin 75°} \approx 89.658 \text{ N}$,

$\mathbf{F}_1 = \langle -|\mathbf{F}_1|\cos 30°, |\mathbf{F}_1|\sin 30° \rangle \approx \langle -63.397, 36.603 \rangle$,

$\mathbf{F}_2 = \langle |\mathbf{F}_2|\cos 45°, |\mathbf{F}_2|\sin 45° \rangle \approx \langle 63.397, 63.397 \rangle$

**47.** $w = \dfrac{100\sin 75°}{\cos 40°} \approx 126.093 \text{ N}$,

$|\mathbf{F}_1| = \dfrac{w\cos 35°}{\sin 75°} \approx 106.933 \text{ N}$

**49.** (a) $(5\cos 60°, 5\sin 60°) = \left(\dfrac{5}{2}, \dfrac{5\sqrt{3}}{2}\right)$

(b) $(5\cos 60° + 10\cos 315°, 5\sin 60° + 10\sin 315°) =$

$\left(\dfrac{5 + 10\sqrt{2}}{2}, \dfrac{5\sqrt{3} - 10\sqrt{2}}{2}\right)$

**51.** (a) $\dfrac{3}{2}\mathbf{i} + \dfrac{3}{2}\mathbf{j} - 3\mathbf{k}$   (b) $\mathbf{i} + \mathbf{j} - 2\mathbf{k}$   (c) $(2, 2, 1)$

**Section 12.3, pp. 726–728**

**1.** (a) $-25, 5, 5$   (b) $-1$   (c) $-5$   (d) $-2\mathbf{i} + 4\mathbf{j} - \sqrt{5}\mathbf{k}$

**3.** (a) $25, 15, 5$   (b) $\dfrac{1}{3}$   (c) $\dfrac{5}{3}$   (d) $\dfrac{1}{9}(10\mathbf{i} + 11\mathbf{j} - 2\mathbf{k})$

**5.** (a) $2, \sqrt{34}, \sqrt{3}$   (b) $\dfrac{2}{\sqrt{3}\sqrt{34}}$   (c) $\dfrac{2}{\sqrt{34}}$

(d) $\dfrac{1}{17}(5\mathbf{j} - 3\mathbf{k})$

**7.** (a) $10 + \sqrt{17}, \sqrt{26}, \sqrt{21}$   (b) $\dfrac{10 + \sqrt{17}}{\sqrt{546}}$

(c) $\dfrac{10 + \sqrt{17}}{\sqrt{26}}$   (d) $\dfrac{10 + \sqrt{17}}{26}(5\mathbf{i} + \mathbf{j})$

**9.** 0.75 rad   **11.** 1.77 rad

**13.** Angle at $A = \cos^{-1}\left(\dfrac{1}{\sqrt{5}}\right) \approx 63.435$ degrees, angle at

$B = \cos^{-1}\left(\dfrac{3}{5}\right) \approx 53.130$ degrees, angle at

$C = \cos^{-1}\left(\dfrac{1}{\sqrt{5}}\right) \approx 63.435$ degrees.

**23.** Horizontal component: $\approx 396$ m/s, vertical component: $\approx 55.7$ m/s

**25.** (a) Since $|\cos\theta| \leq 1$, we have $|\mathbf{u}\cdot\mathbf{v}| = |\mathbf{u}||\mathbf{v}||\cos\theta| \leq |\mathbf{u}||\mathbf{v}|(1) = |\mathbf{u}||\mathbf{v}|$.

(b) We have equality precisely when $|\cos\theta| = 1$ or when one or both of **u** and **v** are **0**. In the case of nonzero vectors, we have equality when $\theta = 0$ or $\pi$, that is, when the vectors are parallel.

**27.** $a$

**33.** $x + 2y = 4$   **35.** $-2x + y = -3$

   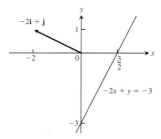

**37.** $x + y = -1$   **39.** $2x - y = 0$

**41.** 5 J   **43.** 3464 J   **45.** $\dfrac{\pi}{4}$   **47.** $\dfrac{\pi}{6}$   **49.** 0.14

**Section 12.4, pp. 732–734**

**1.** $|\mathbf{u} \times \mathbf{v}| = 3$, direction is $\dfrac{2}{3}\mathbf{i} + \dfrac{1}{3}\mathbf{j} + \dfrac{2}{3}\mathbf{k}$; $|\mathbf{v} \times \mathbf{u}| = 3$,

direction is $-\dfrac{2}{3}\mathbf{i} - \dfrac{1}{3}\mathbf{j} - \dfrac{2}{3}\mathbf{k}$

**3.** $|\mathbf{u} \times \mathbf{v}| = 0$, no direction; $|\mathbf{v} \times \mathbf{u}| = 0$, no direction

**5.** $|\mathbf{u} \times \mathbf{v}| = 6$, direction is $-\mathbf{k}$; $|\mathbf{v} \times \mathbf{u}| = 6$, direction is $\mathbf{k}$

**7.** $|\mathbf{u} \times \mathbf{v}| = 6\sqrt{5}$, direction is $\dfrac{1}{\sqrt{5}}\mathbf{i} - \dfrac{2}{\sqrt{5}}\mathbf{k}$; $|\mathbf{v} \times \mathbf{u}| = 6\sqrt{5}$,

direction is $-\dfrac{1}{\sqrt{5}}\mathbf{i} + \dfrac{2}{\sqrt{5}}\mathbf{k}$

**9.**

**11.**

**13.**

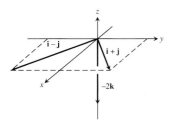

**15.** (a) $2\sqrt{6}$    (b) $\pm\dfrac{1}{\sqrt{6}}(2\mathbf{i}+\mathbf{j}+\mathbf{k})$

**17.** (a) $\dfrac{\sqrt{2}}{2}$    (b) $\pm\dfrac{1}{\sqrt{2}}(\mathbf{i}-\mathbf{j})$

**19.** 8    **21.** 7    **23.** (a) None    (b) $\mathbf{u}$ and $\mathbf{w}$

**25.** $1.5\sqrt{3}$ N·m

**27.** (a) True    (b) Not always true    (c) True    (d) True
  (e) Not always true    (f) True    (g) True    (h) True

**29.** (a) $\text{proj}_{\mathbf{v}}\,\mathbf{u}=\dfrac{\mathbf{u}\cdot\mathbf{v}}{\mathbf{v}\cdot\mathbf{v}}\mathbf{v}$    (b) $\pm\,\mathbf{u}\times\mathbf{v}$    (c) $\pm(\mathbf{u}\times\mathbf{v})\times\mathbf{w}$
  (d) $\left|(\mathbf{u}\times\mathbf{v})\cdot\mathbf{w}\right|$    (e) $(\mathbf{u}\times\mathbf{v})\times(\mathbf{u}\times\mathbf{w})$    (f) $|\mathbf{u}|\dfrac{\mathbf{v}}{|\mathbf{v}|}$

**31.** (a) Yes    (b) No    (c) Yes    (d) No

**33.** No, $\mathbf{v}$ need not equal $\mathbf{w}$. For example, $\mathbf{i}+\mathbf{j}\neq-\mathbf{i}+\mathbf{j}$, but
  $\mathbf{i}\times(\mathbf{i}+\mathbf{j})=\mathbf{i}\times\mathbf{i}+\mathbf{i}\times\mathbf{j}=\mathbf{0}+\mathbf{k}=\mathbf{k}$ and
  $\mathbf{i}\times(-\mathbf{i}+\mathbf{j})=-\mathbf{i}\times\mathbf{i}+\mathbf{i}\times\mathbf{j}=\mathbf{0}+\mathbf{k}=\mathbf{k}$.

**35.** 2    **37.** 13    **39.** $\sqrt{129}$    **41.** $\dfrac{11}{2}$    **43.** $\dfrac{25}{2}$

**45.** $\dfrac{3}{2}$    **47.** $\dfrac{\sqrt{21}}{2}$

**49.** If $\mathbf{A}=a_1\mathbf{i}+a_2\mathbf{j}$ and $\mathbf{B}=b_1\mathbf{i}+b_2\mathbf{j}$, then

$$\mathbf{A}\times\mathbf{B}=\begin{vmatrix}\mathbf{i}&\mathbf{j}&\mathbf{k}\\a_1&a_2&0\\b_1&b_2&0\end{vmatrix}=\begin{vmatrix}a_1&a_2\\b_1&b_2\end{vmatrix}\mathbf{k}$$

and the triangle's area is

$$\frac{1}{2}\left|\mathbf{A}\times\mathbf{B}\right|=\pm\frac{1}{2}\begin{vmatrix}a_1&a_2\\b_1&b_2\end{vmatrix}.$$

The applicable sign is $(+)$ if the acute angle from $\mathbf{A}$ to $\mathbf{B}$ runs counterclockwise in the $xy$-plane, and $(-)$ if it runs clockwise.

**Section 12.5, pp. 740–742**

  **1.** $x=3+t,\;\;y=-4+t,\;\;z=-1+t$
  **3.** $x=-2+5t,\;\;y=5t,\;\;z=3-5t$
  **5.** $x=0,\;\;y=2t,\;\;z=t$
  **7.** $x=1,\;\;y=1,\;\;z=1+t$
  **9.** $x=t,\;\;y=-7+2t,\;\;z=2t$
  **11.** $x=t,\;\;y=0,\;\;z=0$

**13.** $x=t,\;\;y=t,\;\;z=\dfrac{3}{2}t,$
  $0\le t\le 1$

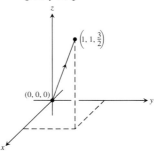

**15.** $x=1,\;\;y=1+t,$
  $z=0,\;\;-1\le t\le 0$

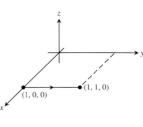

**17.** $x=0,\;\;y=1-2t,$
  $z=1,\;\;0\le t\le 1$

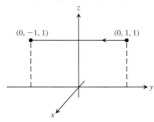

**19.** $x=2-2t,\;\;y=2t,$
  $z=2-2t,\;\;0\le t\le 1$

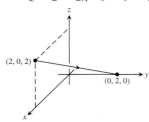

**21.** $3x-2y-z=-3$    **23.** $7x-5y-4z=6$
**25.** $x+3y+4z=34$    **27.** $(1,2,3),\,-20x+12y+z=7$
**29.** $y+z=3$    **31.** $x-y+z=0$    **33.** $2\sqrt{30}$    **35.** 0
**37.** $\dfrac{9\sqrt{42}}{7}$    **39.** 3    **41.** $19/5$    **43.** $5/3$    **45.** $9/\sqrt{41}$
**47.** $\pi/4$    **49.** 1.38 rad    **51.** 0.82 rad    **53.** $\left(\dfrac{3}{2},-\dfrac{3}{2},\dfrac{1}{2}\right)$
**55.** $(1,1,0)$    **57.** $x=1-t,\;\;y=1+t,\;\;z=-1$
**59.** $x=4,\;\;y=3+6t,\;\;z=1+3t$
**61.** $L1$ intersects $L2$; $L2$ is parallel to $L3$, $\sqrt{5}/3$; $L1$ and $L3$ are skew, $10\sqrt{2}/3$
**63.** $x=2+2t,\;\;y=-4-t,\;\;z=7+3t;\;\;x=-2-t,$
  $y=-2+(1/2)t,\;\;z=1-(3/2)t$
**65.** $\left(0,-\dfrac{1}{2},-\dfrac{3}{2}\right),\,(-1,0,-3),\,(1,-1,0)$
**69.** Many possible answers. One possibility: $x+y=3$ and
  $2y+z=7$.
**71.** $(x/a)+(y/b)+(z/c)=1$ describes all planes *except* those through the origin or parallel to a coordinate axis.

**Section 12.6, pp. 746–747**

  **1.** (d), ellipsoid    **3.** (a), cylinder    **5.** (l), hyperbolic paraboloid
  **7.** (b), cylinder    **9.** (k), hyperbolic paraboloid    **11.** (h), cone

**13.**

**15.**

**17.**

$9x^2 + y^2 + z^2 = 9$

**19.**

$4x^2 + 9y^2 + 4z^2 = 36$

**21.**

$z = x^2 + 4y^2$

**23.**

$x = 4 - 4y^2 - z^2$

**25.**

$x^2 + y^2 = z^2$

**27.**

$x^2 + y^2 - z^2 = 1$

**29.**

$z^2 - x^2 - y^2 = 1$

**31.**

$y^2 - x^2 = z$

**33.**

$z^2 = 1 + y^2 - x^2$

**35.**

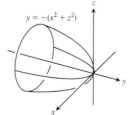

$y = -(x^2 + z^2)$

**37.**

$x^2 + y^2 - z^2 = 4$

**39.**

$x^2 + z^2 = 1$

**41.**

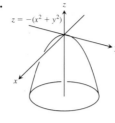

$z = -(x^2 + y^2)$

**43.**

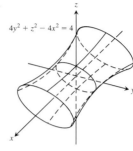

$4y^2 + z^2 - 4x^2 = 4$

**45.** (a) $\dfrac{2\pi(9 - c^2)}{9}$  (b) $8\pi$  (c) $\dfrac{4\pi abc}{3}$

Practice Exercises, pp. 748–749

**1.** (a) $(-3, -28)$  (b) $\sqrt{793}$

**3.** (a) $(42, -49)$  (b) $\sqrt{4165}$

**5.** $\left(-\dfrac{\sqrt{3}}{2}, -\dfrac{1}{2}\right)$ [assuming counterclockwise]

**7.** $\left(\dfrac{8}{\sqrt{17}}, -\dfrac{2}{\sqrt{17}}\right)$

**9.** Length $= \sqrt{6}$, direction is $\dfrac{\mathbf{i}}{\sqrt{2}} + \dfrac{\mathbf{j}}{\sqrt{2}}$.

**11.** $\mathbf{v}\,(\pi/2) = 2(-\mathbf{i})$

**13.** Length $= 5\sqrt{2}$, direction is $\dfrac{4}{5\sqrt{2}}\mathbf{i} + \dfrac{1}{\sqrt{2}}\mathbf{j} + \dfrac{3}{5\sqrt{2}}\mathbf{k}$.

**15.** $8\,\dfrac{3\mathbf{i} + 2\mathbf{j} - \mathbf{k}}{\sqrt{14}}$

**17.** $|\mathbf{v}| = \sqrt{2}$, $|\mathbf{u}| = \sqrt{35}$, $\mathbf{v} \cdot \mathbf{u} = \mathbf{u} \cdot \mathbf{v} = 4$, $\mathbf{v} \times \mathbf{u} = -5\mathbf{i} + 5\mathbf{j} - 2\mathbf{k}$,

$\mathbf{u} \times \mathbf{v} = 5\mathbf{i} - 5\mathbf{j} - 2\mathbf{k}$, $|\mathbf{v} \times \mathbf{u}| = \sqrt{54}$,

$\theta = \cos^{-1}\left(\dfrac{\mathbf{v} \cdot \mathbf{u}}{|\mathbf{v}||\mathbf{u}|}\right) = \cos^{-1}\left(\dfrac{4}{\sqrt{2}\,\sqrt{35}}\right) = \cos^{-1}\left(\dfrac{2\sqrt{2}}{\sqrt{35}}\right)$,

$|\mathbf{u}| \cos\theta = 2\sqrt{2}$, $\text{proj}_{\mathbf{v}}\mathbf{u} = 2(\mathbf{i} - \mathbf{j})$

**19.** $\dfrac{2}{3}\,(4\mathbf{i} - \mathbf{j} + \mathbf{k})$

**21.** $\mathbf{u} \times \mathbf{v} = \mathbf{k}$

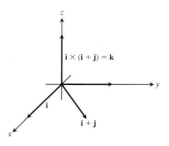

**23.** $2\sqrt{7}$    **25.** (a) $\sqrt{134}$   (b) 15    **29.** $\sqrt{78}/3$

**31.** $x = 1 - 3t$,  $y = 2$,  $z = 3 + 7t$    **33.** $\sqrt{2}$

**35.** $2x + y + z = 5$    **37.** $-9x + y + 7z = 4$

**39.** $\left(0, -\dfrac{1}{2}, -\dfrac{3}{2}\right), (-1, 0, -3), (1, -1, 0)$    **41.** $\pi/3$

**43.** $x = -5 + 5t$,  $y = 3 - t$,  $z = -3t$

**45.** (b) $x = -12t$,  $y = 19/12 + 15t$,  $z = 1/6 + 6t$

**47.** Yes; **v** is parallel to the plane.

**49.** 3    **51.** $-3\mathbf{j} + 3\mathbf{k}$

**53.** $\dfrac{2}{\sqrt{35}} (5\mathbf{i} - \mathbf{j} - 3\mathbf{k})$    **55.** $\left(\dfrac{11}{9}, \dfrac{26}{9}, -\dfrac{7}{9}\right)$

**57.** $(1, -2, -1); x = 1 - 5t$,  $y = -2 + 3t$,  $z = -1 + 4t$

**59.** $2x + 7y + 2z + 10 = 0$

**61.** (a) No  (b) No  (c) No  (d) No  (e) Yes

**63.** $11/\sqrt{107}$

**65.**

$x^2 + y^2 + z^2 = 4$

**67.**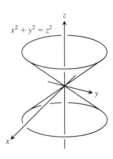

$4x^2 + 4y^2 + z^2 = 4$

**69.**

$z = -(x^2 + y^2)$

**71.** 

$x^2 + y^2 = z^2$

**73.** 

$x^2 + y^2 - z^2 = 4$

**75.** 

$y^2 - x^2 - z^2 = 1$

**1.** $(26, 23, -1/3)$    **3.** $|\mathbf{F}| = 88.7$ N

**5.** (a) $|\mathbf{F}_1| = 80$ N,   $|\mathbf{F}_2| = 60$ N,   $\mathbf{F}_1 = \langle -48, 64 \rangle$,

$\mathbf{F}_2 = \langle 48, 36 \rangle$,   $\alpha = \tan^{-1}\dfrac{4}{3}$,   $\beta = \tan^{-1}\dfrac{3}{4}$

(b) $|\mathbf{F}_1| = \dfrac{2400}{13} \approx 184.615$ N, $|\mathbf{F}_2| = \dfrac{1000}{13} \approx 76.923$ N,

$\mathbf{F}_1 = \left\langle \dfrac{-12{,}000}{169}, \dfrac{28{,}800}{169} \right\rangle \approx \langle -71.006, 170.414 \rangle$,

$\mathbf{F}_2 = \left\langle \dfrac{12{,}000}{169}, \dfrac{5000}{169} \right\rangle \approx \langle 71.006, 29.586 \rangle$,

$\alpha = \tan^{-1}\dfrac{12}{5}$, $\beta = \tan^{-1}\dfrac{5}{12}$

**9.** (a) $\theta = \tan^{-1}\sqrt{2} \approx 54.74°$  (b) $\theta = \tan^{-1}2\sqrt{2} \approx 70.53°$

**13.** (b) $\dfrac{6}{\sqrt{14}}$  (c) $2x - y + 2z = 8$

(d) $x - 2y + z = 3 + 5\sqrt{6}$ and $x - 2y + z = 3 - 5\sqrt{6}$

**15.** $\dfrac{32}{41}\mathbf{i} + \dfrac{23}{41}\mathbf{j} - \dfrac{13}{41}\mathbf{k}$

**17.** (a) $0, 0$  (b) $-10\mathbf{i} - 2\mathbf{j} + 6\mathbf{k}, -9\mathbf{i} - 2\mathbf{j} + 7\mathbf{k}$

(c) $-4\mathbf{i} - 6\mathbf{j} + 2\mathbf{k}, \mathbf{i} - 2\mathbf{j} - 4\mathbf{k}$

(d) $-10\mathbf{i} - 10\mathbf{k}, -12\mathbf{i} - 4\mathbf{j} - 8\mathbf{k}$

**19.** The formula is always true.

# Chapter 13

**1.** $y = x^2 - 2x$,  $\mathbf{v} = \mathbf{i} + 2\mathbf{j}$,  $\mathbf{a} = 2\mathbf{j}$

**3.** $y = \dfrac{2}{9}x^2$,  $\mathbf{v} = 3\mathbf{i} + 4\mathbf{j}$,  $\mathbf{a} = 3\mathbf{i} + 8\mathbf{j}$

**5.** $t = \dfrac{\pi}{4}: \mathbf{v} = \dfrac{\sqrt{2}}{2}\mathbf{i} - \dfrac{\sqrt{2}}{2}\mathbf{j}$,  $\mathbf{a} = \dfrac{-\sqrt{2}}{2}\mathbf{i} - \dfrac{\sqrt{2}}{2}\mathbf{j}$;

$t = \pi/2: \mathbf{v} = -\mathbf{j}$,  $\mathbf{a} = -\mathbf{i}$

**7.** $t = \pi: \mathbf{v} = 2\mathbf{i}$,  $\mathbf{a} = -\mathbf{j}$;   $t = \dfrac{3\pi}{2}: \mathbf{v} = \mathbf{i} - \mathbf{j}$,  $\mathbf{a} = -\mathbf{i}$

**9.** $\mathbf{v} = \mathbf{i} + 2t\mathbf{j} + 2\mathbf{k}; \mathbf{a} = 2\mathbf{j}$; speed: 3; direction: $\dfrac{1}{3}\mathbf{i} + \dfrac{2}{3}\mathbf{j} + \dfrac{2}{3}\mathbf{k}; \mathbf{v}(1) = 3\left(\dfrac{1}{3}\mathbf{i} + \dfrac{2}{3}\mathbf{j} + \dfrac{2}{3}\mathbf{k}\right)$

**11.** $\mathbf{v} = (-2\sin t)\mathbf{i} + (3\cos t)\mathbf{j} + 4\mathbf{k}$;

$\mathbf{a} = (-2\cos t)\mathbf{i} - (3\sin t)\mathbf{j}$; speed: $2\sqrt{5}$;

direction: $\left(-1/\sqrt{5}\right)\mathbf{i} + \left(2/\sqrt{5}\right)\mathbf{k}$;

$\mathbf{v}(\pi/2) = 2\sqrt{5}\left[\left(-1/\sqrt{5}\right)\mathbf{i} + \left(2/\sqrt{5}\right)\mathbf{k}\right]$

**13.** $\mathbf{v} = \left(\dfrac{2}{t+1}\right)\mathbf{i} + 2t\mathbf{j} + t\mathbf{k}$; $\mathbf{a} = \left(\dfrac{-2}{(t+1)^2}\right)\mathbf{i} + 2\mathbf{j} + \mathbf{k}$;

speed: $\sqrt{6}$; direction: $\dfrac{1}{\sqrt{6}}\mathbf{i} + \dfrac{2}{\sqrt{6}}\mathbf{j} + \dfrac{1}{\sqrt{6}}\mathbf{k}$;

$\mathbf{v}(1) = \sqrt{6}\left(\dfrac{1}{\sqrt{6}}\mathbf{i} + \dfrac{2}{\sqrt{6}}\mathbf{j} + \dfrac{1}{\sqrt{6}}\mathbf{k}\right)$

**15.** $\pi/2$    **17.** $\pi/2$

**19.** $x = t$, $y = -1$, $z = 1 + t$    **21.** $x = t$, $y = \dfrac{1}{3}t$, $z = t$

**23.** **(a)** **(i):** It has constant speed 1.   **(ii):** Yes
 **(iii):** Counterclockwise   **(iv):** Yes
 **(b)** **(i):** It has constant speed 2.   **(ii):** Yes
 **(iii):** Counterclockwise   **(iv):** Yes
 **(c)** **(i):** It has constant speed 1.   **(ii):** Yes
 **(iii):** Counterclockwise
 **(iv):** It starts at $(0, -1)$ instead of $(1, 0)$.
 **(d)** **(i):** It has constant speed 1.   **(ii):** Yes
 **(iii):** Clockwise   **(iv):** Yes
 **(e)** **(i):** It has variable speed.   **(ii):** No
 **(iii):** Counterclockwise   **(iv):** Yes

**25.** $\mathbf{v} = 2\sqrt{5}\mathbf{i} + \sqrt{5}\mathbf{j}$

**Section 13.2, pp. 767–770**

**1.** $(1/4)\mathbf{i} + 7\mathbf{j} + (3/2)\mathbf{k}$    **3.** $\left(\dfrac{\pi + 2\sqrt{2}}{2}\right)\mathbf{j} + 2\mathbf{k}$

**5.** $(\ln 4)\mathbf{i} + (\ln 4)\mathbf{j} + (\ln 2)\mathbf{k}$

**7.** $\dfrac{e-1}{2}\mathbf{i} + \dfrac{e-1}{e}\mathbf{j} + \mathbf{k}$    **9.** $\mathbf{i} - \mathbf{j} + \dfrac{\pi}{4}\mathbf{k}$

**11.** $\mathbf{r}(t) = \left(\dfrac{-t^2}{2} + 1\right)\mathbf{i} + \left(\dfrac{-t^2}{2} + 2\right)\mathbf{j} + \left(\dfrac{-t^2}{2} + 3\right)\mathbf{k}$

**13.** $\mathbf{r}(t) = ((t+1)^{3/2} - 1)\mathbf{i} + (-e^{-t} + 1)\mathbf{j} + (\ln(t+1) + 1)\mathbf{k}$

**15.** $\mathbf{r}(t) = 8t\mathbf{i} + 8t\mathbf{j} + (-16t^2 + 100)\mathbf{k}$

**17.** $\mathbf{r}(t) = \left(\dfrac{3}{2}t^2 + \dfrac{6}{\sqrt{11}}t + 1\right)\mathbf{i} - \left(\dfrac{1}{2}t^2 + \dfrac{2}{\sqrt{11}}t - 2\right)\mathbf{j}$

$+\left(\dfrac{1}{2}t^2 + \dfrac{2}{\sqrt{11}}t + 3\right)\mathbf{k} = \left(\dfrac{1}{2}t^2 + \dfrac{2t}{\sqrt{11}}\right)(3\mathbf{i} - \mathbf{j} + \mathbf{k})$

$+(\mathbf{i} + 2\mathbf{j} + 3\mathbf{k})$

**19.** 50 s

**21.** **(a)** 72.2 s; 25,510 m    **(b)** 4020 m    **(c)** 6378 m

**23.** **(a)** $v_0 \approx 9.9$ m/s    **(b)** $\alpha \approx 18.4°$ or $71.6°$

**25.** $39.3°$ or $50.7°$    **31.** **(b)** $\mathbf{v}_0$ would bisect $\angle AOR$.

**33.** **(a)** (Assuming that "$x$" is zero at the point of impact)
 $\mathbf{r}(t) = (x(t))\mathbf{i} + (y(t))\mathbf{j}$, where $x(t) = (12 \cos 27°)t$ and
 $y(t) = 1.3 + (12 \sin 27°)t - 4.9t^2$.
 **(b)** At $t \approx 0.556$ s, it reaches its maximum height of about
 2.814 m.
 **(c)** Range $\approx 14$ m; flight time $\approx 1.31$ s
 **(d)** At $t \approx 0.232$ and $t \approx 0.880$ s, when it is $\approx 11.52$ and
 $\approx 4.59$ m from where it will land
 **(e)** Yes. It changes things because the ball won't clear the net.

**35.** 1.225 m, 5.45 m/s

**43.** **(a)** $\mathbf{r}(t) = (x(t))\mathbf{i} + (y(t))\mathbf{j}$; where

$x(t) = \left(\dfrac{1}{0.08}\right)(1 - e^{-0.08t})(50 \cos 20° - 5)$ and

$y(t) = 1 + \left(\dfrac{50}{0.08}\right)(1 - e^{-0.08t})(\sin 20°)$

$+\left(\dfrac{9.8}{0.08^2}\right)(1 - 0.08t - e^{-0.08t})$

 **(b)** At $t \approx 1.633$ s it reaches a maximum height of about
 14.66 m.
 **(c)** Range $\approx 125.11$ m; flight time $\approx 3.404$ s
 **(d)** At $t \approx 0.670$ and 2.622 s, when it is about 27.39 and 99.30
 m from home plate
 **(e)** No

**Section 13.3, pp. 773–774**

**1.** $\mathbf{T} = \left(-\dfrac{2}{3}\sin t\right)\mathbf{i} + \left(\dfrac{2}{3}\cos t\right)\mathbf{j} + \dfrac{\sqrt{5}}{3}\mathbf{k}, 3\pi$

**3.** $\mathbf{T} = \dfrac{1}{\sqrt{1+t}}\mathbf{i} + \dfrac{\sqrt{t}}{\sqrt{1+t}}\mathbf{k}, \dfrac{52}{3}$    **5.** $\mathbf{T} = -\cos t\mathbf{j} + \sin t\mathbf{k}, \dfrac{3}{2}$

**7.** $\mathbf{T} = \left(\dfrac{\cos t - t\sin t}{t+1}\right)\mathbf{i} + \left(\dfrac{\sin t + t\cos t}{t+1}\right)\mathbf{j}$

$+\left(\dfrac{\sqrt{2}t^{1/2}}{t+1}\right)\mathbf{k}, \dfrac{\pi^2}{2} + \pi$

**9.** $(0, 5, 24\pi)$

**11.** $s(t) = 5t$, $L = \dfrac{5\pi}{2}$

**13.** $s(t) = \sqrt{3}e^t - \sqrt{3}$, $L = \dfrac{3\sqrt{3}}{4}$    **15.** $\sqrt{2} + \ln(1 + \sqrt{2})$

**17.** **(a)** Cylinder is $x^2 + y^2 = 1$; plane is $x + z = 1$.
 **(b)** and **(c)**

 **(d)** $L = \displaystyle\int_0^{2\pi} \sqrt{1 + \sin^2 t}\, dt$   **(e)** $L \approx 7.64$

**Section 13.4, pp. 779–780**

**1.** $\mathbf{T} = (\cos t)\mathbf{i} - (\sin t)\mathbf{j}$,  $\mathbf{N} = (-\sin t)\mathbf{i} - (\cos t)\mathbf{j}$,  $\kappa = \cos t$

**3.** $\mathbf{T} = \dfrac{1}{\sqrt{1+t^2}}\mathbf{i} - \dfrac{t}{\sqrt{1+t^2}}\mathbf{j}$,  $\mathbf{N} = \dfrac{-t}{\sqrt{1+t^2}}\mathbf{i} -$

$\kappa = \dfrac{1}{2(\sqrt{1+t^2})^3}$

**5.** **(b)** $\cos x$

**7. (b)** $\mathbf{N} = \dfrac{-2e^{2t}}{\sqrt{1 + 4e^{4t}}}\,\mathbf{i} + \dfrac{1}{\sqrt{1 + 4e^{4t}}}\,\mathbf{j}$

   **(c)** $\mathbf{N} = -\dfrac{1}{2}\left(\sqrt{4 - t^2}\,\mathbf{i} + t\mathbf{j}\right)$

**9.** $\mathbf{T} = \dfrac{3\cos t}{5}\,\mathbf{i} - \dfrac{3\sin t}{5}\,\mathbf{j} + \dfrac{4}{5}\,\mathbf{k}$, $\mathbf{N} = (-\sin t)\mathbf{i} - (\cos t)\mathbf{j}$,

   $\kappa = \dfrac{3}{25}$

**11.** $\mathbf{T} = \left(\dfrac{\cos t - \sin t}{\sqrt{2}}\right)\mathbf{i} + \left(\dfrac{\cos t + \sin t}{\sqrt{2}}\right)\mathbf{j}$,

   $\mathbf{N} = \left(\dfrac{-\cos t - \sin t}{\sqrt{2}}\right)\mathbf{i} + \left(\dfrac{-\sin t + \cos t}{\sqrt{2}}\right)\mathbf{j}$,  $\kappa = \dfrac{1}{e^t\sqrt{2}}$

**13.** $\mathbf{T} = \dfrac{t}{\sqrt{t^2 + 1}}\,\mathbf{i} - \dfrac{1}{\sqrt{t^2 + 1}}\,\mathbf{j}$,  $\mathbf{N} = \dfrac{\mathbf{i}}{\sqrt{t^2 + 1}} - \dfrac{t\mathbf{j}}{\sqrt{t^2 + 1}}$,

   $\kappa = \dfrac{1}{t(t^2 + 1)^{3/2}}$

**15.** $\mathbf{T} = \left(\text{sech}\,\dfrac{t}{a}\right)\mathbf{i} + \left(\tanh\dfrac{t}{a}\right)\mathbf{j}$,

   $\mathbf{N} = \left(-\tanh\dfrac{t}{a}\right)\mathbf{i} + \left(\text{sech}\,\dfrac{t}{a}\right)\mathbf{j}$,

   $\kappa = \dfrac{1}{a}\,\text{sech}^2\,\dfrac{t}{a}$

**19.** $1/(2b)$

**21.** $\left(x - \dfrac{\pi}{2}\right)^2 + y^2 = 1$

**23.** $\kappa(x) = 2/(1 + 4x^2)^{3/2}$

**25.** $\kappa(x) = |\sin x|/(1 + \cos^2 x)^{3/2}$

Section 13.5, pp. 785–786

**1.** $\mathbf{a} = |a|\mathbf{N}$   **3.** $\mathbf{a}(1) = \dfrac{4}{3}\,\mathbf{T} + \dfrac{2\sqrt{5}}{3}\,\mathbf{N}$   **5.** $\mathbf{a}(0) = 2\mathbf{N}$

**7.** $\mathbf{r}\left(\dfrac{\pi}{4}\right) = \dfrac{\sqrt{2}}{2}\,\mathbf{i} + \dfrac{\sqrt{2}}{2}\,\mathbf{j} - \mathbf{k}$, $\mathbf{T}\left(\dfrac{\pi}{4}\right) = -\dfrac{\sqrt{2}}{2}\,\mathbf{i} + \dfrac{\sqrt{2}}{2}\,\mathbf{j}$,

   $\mathbf{N}\left(\dfrac{\pi}{4}\right) = -\dfrac{\sqrt{2}}{2}\,\mathbf{i} - \dfrac{\sqrt{2}}{2}\,\mathbf{j}$, $\mathbf{B}\left(\dfrac{\pi}{4}\right) = \mathbf{k}$; osculating plane:

   $z = -1$; normal plane: $-x + y = 0$; rectifying plane:
   $x + y = \sqrt{2}$

**9.** $\mathbf{B} = \left(\dfrac{4}{5}\cos t\right)\mathbf{i} - \left(\dfrac{4}{5}\sin t\right)\mathbf{j} - \dfrac{3}{5}\,\mathbf{k}$, $\tau = -\dfrac{4}{25}$

**11.** $\mathbf{B} = \mathbf{k}$, $\tau = 0$   **13.** $\mathbf{B} = -\mathbf{k}$, $\tau = 0$   **15.** $\mathbf{B} = \mathbf{k}$, $\tau = 0$

**17.** Yes. If the car is moving on a curved path ($\kappa \neq 0$), then
   $a_N = \kappa|\mathbf{v}|^2 \neq 0$ and $\mathbf{a} \neq \mathbf{0}$.

**23.** $\kappa = \dfrac{1}{t}$, $\rho = t$

**27.** Components of $\mathbf{v}$: $-1.8701, 0.7089, 1.0000$
   Components of $\mathbf{a}$: $-1.6960, -2.0307, 0$
   Speed: 2.2361; Components of $\mathbf{T}$: $-0.8364, 0.3170, 0.4472$
   Components of $\mathbf{N}$: $-0.4143, -0.8998, -0.1369$
   Components of $\mathbf{B}$: $0.3590, -0.2998, 0.8839$; Curvature: 0.5060
   Torsion: 0.2813; Tangential component of acceleration: 0.7746
   Normal component of acceleration: 2.5298

**29.** Components of $\mathbf{v}$: $2.0000, 0, -0.1629$
   Components of $\mathbf{a}$: $0, -1.0000, -0.0086$; Speed: 2.0066
   Components of $\mathbf{T}$: $0.9967, 0, -0.0812$
   Components of $\mathbf{N}$: $-0.0007, -1.0000, -0.0086$
   Components of $\mathbf{B}$: $-0.0812, 0.0086, 0.9967$;
   Curvature: 0.2484
   Torsion: 0.0411; Tangential component of acceleration: 0.0007
   Normal component of acceleration: 1.0000

Section 13.6, pp. 789–790

**1.** $\mathbf{v} = (3a\sin\theta)\mathbf{u}_r + 3a(1 - \cos\theta)\mathbf{u}_\theta$
   $\mathbf{a} = 9a(2\cos\theta - 1)\mathbf{u}_r + (18a\sin\theta)\mathbf{u}_\theta$

**3.** $\mathbf{v} = 2ae^{a\theta}\mathbf{u}_r + 2e^{a\theta}\mathbf{u}_\theta$
   $\mathbf{a} = 4e^{a\theta}(a^2 - 1)\mathbf{u}_r + 8ae^{a\theta}\mathbf{u}_\theta$

**5.** $\mathbf{v} = (-8\sin 4t)\mathbf{u}_r + (4\cos 4t)\mathbf{u}_\theta$
   $\mathbf{a} = (-40\cos 4t)\mathbf{u}_r - (32\sin 4t)\mathbf{u}_\theta$

**11.** $\approx 29.93 \times 10^{10}$ m   **13.** $\approx 2.25 \times 10^9\,\text{km}^2/\text{s}$

**15.** $\approx 1.876 \times 10^{27}$ kg

Practice Exercises, pp. 790–792

**1.** $\dfrac{x^2}{16} + \dfrac{y^2}{2} = 1$

At $t = 0$: $a_T = 0$, $a_N = 4$, $\kappa = 2$;

At $t = \dfrac{\pi}{4}$: $a_T = \dfrac{7}{3}$, $a_N = \dfrac{4\sqrt{2}}{3}$, $\kappa = \dfrac{4\sqrt{2}}{27}$

**3.** $|\mathbf{v}|_{\max} = 1$   **5.** $\kappa = 1/5$   **7.** $dy/dt = -x$; clockwise

**11.** Shot put is on the ground, about 21.88 m from the stopboard.

**15.** Length $= \dfrac{\pi}{4}\sqrt{1 + \dfrac{\pi^2}{16}} + \ln\left(\dfrac{\pi}{4} + \sqrt{1 + \dfrac{\pi^2}{16}}\right)$

**17.** $\mathbf{T}(0) = \dfrac{2}{3}\,\mathbf{i} - \dfrac{2}{3}\,\mathbf{j} + \dfrac{1}{3}\,\mathbf{k}$;   $\mathbf{N}(0) = \dfrac{1}{\sqrt{2}}\,\mathbf{i} + \dfrac{1}{\sqrt{2}}\,\mathbf{j}$;

   $\mathbf{B}(0) = -\dfrac{1}{3\sqrt{2}}\,\mathbf{i} + \dfrac{1}{3\sqrt{2}}\,\mathbf{j} + \dfrac{4}{3\sqrt{2}}\,\mathbf{k}$;   $\kappa = \dfrac{\sqrt{2}}{3}$; $\tau = \dfrac{1}{6}$

**19.** $\mathbf{T}(\ln 2) = \dfrac{1}{\sqrt{17}}\,\mathbf{i} + \dfrac{4}{\sqrt{17}}\,\mathbf{j}$;   $\mathbf{N}(\ln 2) = -\dfrac{4}{\sqrt{17}}\,\mathbf{i} + \dfrac{1}{\sqrt{17}}\,\mathbf{j}$;

   $\mathbf{B}(\ln 2) = \mathbf{k}$; $\kappa = \dfrac{8}{17\sqrt{17}}$; $\tau = 0$

**21.** $\mathbf{a}(0) = 10\mathbf{T} + 6\mathbf{N}$

**23.** $\mathbf{T} = \left(\dfrac{1}{\sqrt{2}}\cos t\right)\mathbf{i} - (\sin t)\mathbf{j} + \left(\dfrac{1}{\sqrt{2}}\cos t\right)\mathbf{k}$;

   $\mathbf{N} = \left(-\dfrac{1}{\sqrt{2}}\sin t\right)\mathbf{i} - (\cos t)\mathbf{j} - \left(\dfrac{1}{\sqrt{2}}\sin t\right)\mathbf{k}$;

   $\mathbf{B} = \dfrac{1}{\sqrt{2}}\,\mathbf{i} - \dfrac{1}{\sqrt{2}}\,\mathbf{k}$; $\kappa = \dfrac{1}{\sqrt{2}}$; $\tau = 0$

**25.** $\dfrac{\pi}{3}$   **27.** $x = 1 + t$,  $y = t$,  $z = -t$   **31.** $\kappa = \dfrac{1}{a}$

**Additional and Advanced Exercises, pp. 792–793**

**1.** **(a)** $\left.\dfrac{d\theta}{dt}\right|_{\theta=2\pi} = 2\sqrt{\dfrac{\pi gb}{a^2 + b^2}}$

**(b)** $\theta = \dfrac{gbt^2}{2(a^2 + b^2)}, \quad z = \dfrac{gb^2t^2}{2(a^2 + b^2)}$

**(c)** $\mathbf{v}(t) = \dfrac{gbt}{\sqrt{a^2 + b^2}}\mathbf{T};$

$\dfrac{d^2\mathbf{r}}{dt^2} = \dfrac{bg}{\sqrt{a^2 + b^2}}\mathbf{T} + a\left(\dfrac{bgt}{a^2 + b^2}\right)^2\mathbf{N}$

There is no component in the direction of **B**.

**5.** **(a)** $\dfrac{dx}{dt} = \dot{r}\cos\theta - r\dot{\theta}\sin\theta, \dfrac{dy}{dt} = \dot{r}\sin\theta + r\dot{\theta}\cos\theta$

**(b)** $\dfrac{dr}{dt} = \dot{x}\cos\theta + \dot{y}\sin\theta, r\dfrac{d\theta}{dt} = -\dot{x}\sin\theta + \dot{y}\cos\theta$

**7.** **(a)** $\mathbf{a}(1) = -9\mathbf{u}_r - 6\mathbf{u}_\theta, \mathbf{v}(1) = -\mathbf{u}_r + 3\mathbf{u}_\theta$ **(b)** 6.5 cm

**9.** **(c)** $\mathbf{v} = \dot{r}\mathbf{u}_r + r\dot{\theta}\mathbf{u}_\theta + \dot{z}\mathbf{k}, \mathbf{a} = (\ddot{r} - r\dot{\theta}^2)\mathbf{u}_r +$
$(r\ddot{\theta} + 2\dot{r}\dot{\theta})\mathbf{u}_\theta + \ddot{z}\mathbf{k}$

# Chapter 14

**Section 14.1, pp. 801–803**

**1.** **(a)** 0 **(b)** 0 **(c)** 58 **(d)** 33
**3.** **(a)** 4/5 **(b)** 8/5 **(c)** 3 **(d)** 0
**5.** Domain: all points $(x, y)$ on or above line $y = x + 2$
**7.** Domain: all points $(x, y)$ not lying on the graph of $y = x$ or $y = x^3$

**9.** Domain: all points $(x, y)$ satisfying $x^2 - 1 \leq y \leq x^2 + 1$

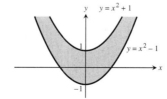

**11.** Domain: all points $(x, y)$ for which
$(x - 2)(x + 2)(y - 3)(y + 3) \geq 0$

**13.**

**15.**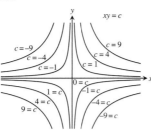

**17.** **(a)** All points in the $xy$-plane **(b)** All reals
**(c)** The lines $y - x = c$ **(d)** No boundary points
**(e)** Both open and closed **(f)** Unbounded
**19.** **(a)** All points in the $xy$-plane **(b)** $z \geq 0$
**(c)** For $f(x, y) = 0$, the origin; for $f(x, y) \neq 0$, ellipses with the center $(0, 0)$, and major and minor axes along the $x$- and $y$-axes, respectively
**(d)** No boundary points **(e)** Both open and closed
**(f)** Unbounded
**21.** **(a)** All points in the $xy$-plane **(b)** All reals
**(c)** For $f(x, y) = 0$, the $x$- and $y$-axes; for $f(x, y) \neq 0$, hyperbolas with the $x$- and $y$-axes as asymptotes
**(d)** No boundary points **(e)** Both open and closed
**(f)** Unbounded
**23.** **(a)** All $(x, y)$ satisfying $x^2 + y^2 < 16$ **(b)** $z \geq 1/4$
**(c)** Circles centered at the origin with radii $r < 4$
**(d)** Boundary is the circle $x^2 + y^2 = 16$
**(e)** Open **(f)** Bounded
**25.** **(a)** $(x, y) \neq (0, 0)$ **(b)** All reals
**(c)** The circles with center $(0, 0)$ and radii $r > 0$
**(d)** Boundary is the single point $(0, 0)$
**(e)** Open **(f)** Unbounded
**27.** **(a)** All $(x, y)$ satisfying $-1 \leq y - x \leq 1$
**(b)** $-\pi/2 \leq z \leq \pi/2$
**(c)** Straight lines of the form $y - x = c$ where $-1 \leq c \leq 1$
**(d)** Boundary is two straight lines $y = 1 + x$ and $y = -1 + x$
**(e)** Closed **(f)** Unbounded
**29.** **(a)** Domain: all points $(x, y)$ outside the circle $x^2 + y^2 = 1$
**(b)** Range: all reals
**(c)** Circles centered at the origin with radii $r > 1$
**(d)** Boundary: $x^2 + y^2 = 1$
**(e)** Open **(f)** Unbounded
**31.** **(f)** **33.** **(a)** **35.** **(d)**
**37.** **(a)**  **(b)**

**39. (a)**

**(b)**

**41. (a)**

**(b)**

**43. (a)**

**(b)**

**45. (a)**

**(b)**

**47. (a)**

**(b)**

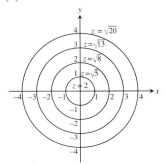

**49.** $x^2 + y^2 = 10$

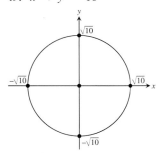

**51.** $x + y^2 = 4$

**53.**

$f(x, y, z) = x^2 + y^2 + z^2 = 1$

**55.**

$f(x, y, z) = x + z = 1$

**57.**

$f(x, y, z) = x^2 + y^2 = 1$

**59.**

$f(x, y, z) = z - x^2 - y^2 = 1$
or $z = x^2 + y^2 + 1$

**61.** $\sqrt{x - y} - \ln z = 2$   **63.** $x^2 + y^2 + z^2 = 4$

**65.** Domain: all points $(x, y)$ satisfying $|x| < |y|$

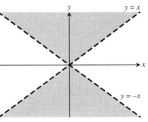

level curve: $y = 2x$

**67.** Domain: all points $(x, y)$ satisfying $-1 \le x \le 1$ and $-1 \le y \le 1$

level curve:
$\sin^{-1} y - \sin^{-1} x = \dfrac{\pi}{2}$

Section 14.2, pp. 809–812

**1.** $5/2$   **3.** $2\sqrt{6}$   **5.** 1   **7.** $1/2$   **9.** 1
**11.** $1/4$   **13.** 0   **15.** $-1$   **17.** 2   **19.** $1/4$
**21.** 1   **23.** 3   **25.** $19/12$   **27.** 2   **29.** 3
**31. (a)** All $(x, y)$   **(b)** All $(x, y)$ except $(0, 0)$
**33. (a)** All $(x, y)$ except where $x = 0$ or $y = 0$   **(b)** All $(x, y)$
**35. (a)** All $(x, y, z)$
  **(b)** All $(x, y, z)$ except the interior of the cylinder $x^2 + y^2 = 1$

**37. (a)** All $(x, y, z)$ with $z \neq 0$    **(b)** All $(x, y, z)$ with $x^2 + z^2 \neq 1$

**39. (a)** All points $(x, y, z)$ satisfying $z > x^2 + y^2 + 1$

     **(b)** All points $(x, y, z)$ satisfying $z \neq \sqrt{x^2 + y^2}$

**41.** Consider paths along $y = x$, $x > 0$, and along $y = x$, $x < 0$.

**43.** Consider the paths $y = kx^2$, $k$ a constant.

**45.** Consider the paths $y = mx$, $m$ a constant, $m \neq -1$.

**47.** Consider the paths $y = kx^2$, $k$ a constant, $k \neq 0$.

**49.** Consider the paths $x = 1$ and $y = x$.

**51. (a)** 1    **(b)** 0    **(c)** Does not exist

**55.** The limit is 1.      **57.** The limit is 0.

**59. (a)** $f(x, y)\big|_{y=mx} = \sin 2\theta$ where $\tan \theta = m$    **61.** 0

**63.** Does not exist    **65.** $\pi/2$    **67.** $f(0, 0) = \ln 3$

**69.** $\delta = 0.1$    **71.** $\delta = 0.005$    **73.** $\delta = 0.04$

**75.** $\delta = \sqrt{0.015}$    **77.** $\delta = 0.005$

### Section 14.3, pp. 821–823

**1.** $\dfrac{\partial f}{\partial x} = 4x, \dfrac{\partial f}{\partial y} = -3$    **3.** $\dfrac{\partial f}{\partial x} = 2x(y + 2), \dfrac{\partial f}{\partial y} = x^2 - 1$

**5.** $\dfrac{\partial f}{\partial x} = 2y(xy - 1), \dfrac{\partial f}{\partial y} = 2x(xy - 1)$

**7.** $\dfrac{\partial f}{\partial x} = \dfrac{x}{\sqrt{x^2 + y^2}}, \dfrac{\partial f}{\partial y} = \dfrac{y}{\sqrt{x^2 + y^2}}$

**9.** $\dfrac{\partial f}{\partial x} = \dfrac{-1}{(x + y)^2}, \dfrac{\partial f}{\partial y} = \dfrac{-1}{(x + y)^2}$

**11.** $\dfrac{\partial f}{\partial x} = \dfrac{-y^2 - 1}{(xy - 1)^2}, \dfrac{\partial f}{\partial y} = \dfrac{-x^2 - 1}{(xy - 1)^2}$

**13.** $\dfrac{\partial f}{\partial x} = e^{x+y+1}, \dfrac{\partial f}{\partial y} = e^{x+y+1}$    **15.** $\dfrac{\partial f}{\partial x} = \dfrac{1}{x + y}, \dfrac{\partial f}{\partial y} = \dfrac{1}{x + y}$

**17.** $\dfrac{\partial f}{\partial x} = 2\sin(x - 3y)\cos(x - 3y),$

   $\dfrac{\partial f}{\partial y} = -6\sin(x - 3y)\cos(x - 3y)$

**19.** $\dfrac{\partial f}{\partial x} = yx^{y-1}, \dfrac{\partial f}{\partial y} = x^y \ln x$    **21.** $\dfrac{\partial f}{\partial x} = -g(x), \dfrac{\partial f}{\partial y} = g(y)$

**23.** $f_x = y^2, f_y = 2xy, f_z = -4z$

**25.** $f_x = 1, f_y = -y(y^2 + z^2)^{-1/2}, f_z = -z(y^2 + z^2)^{-1/2}$

**27.** $f_x = \dfrac{yz}{\sqrt{1 - x^2y^2z^2}}, f_y = \dfrac{xz}{\sqrt{1 - x^2y^2z^2}}, f_z = \dfrac{xy}{\sqrt{1 - x^2y^2z^2}}$

**29.** $f_x = \dfrac{1}{x + 2y + 3z}, f_y = \dfrac{2}{x + 2y + 3z}, f_z = \dfrac{3}{x + 2y + 3z}$

**31.** $f_x = -2xe^{-(x^2+y^2+z^2)}, f_y = -2ye^{-(x^2+y^2+z^2)}, f_z = -2ze^{-(x^2+y^2+z^2)}$

**33.** $f_x = \mathrm{sech}^2(x + 2y + 3z), f_y = 2\,\mathrm{sech}^2(x + 2y + 3z),$

   $f_z = 3\,\mathrm{sech}^2(x + 2y + 3z)$

**35.** $\dfrac{\partial f}{\partial t} = -2\pi \sin(2\pi t - \alpha), \dfrac{\partial f}{\partial \alpha} = \sin(2\pi t - \alpha)$

**37.** $\dfrac{\partial h}{\partial \rho} = \sin\phi \cos\theta, \dfrac{\partial h}{\partial \phi} = \rho\cos\phi\cos\theta, \dfrac{\partial h}{\partial \theta} = -\rho\sin\phi\sin\theta$

**39.** $W_P(P, V, \delta, v, g) = V, W_V(P, V, \delta, v, g) = P + \dfrac{\delta v^2}{2g},$

   $W_\delta(P, V, \delta, v, g) = \dfrac{Vv^2}{2g}, W_v(P, V, \delta, v, g) = \dfrac{V\delta v}{g},$

   $W_g(P, V, \delta, v, g) = -\dfrac{V\delta v^2}{2g^2}$

**41.** $\dfrac{\partial f}{\partial x} = 1 + y, \dfrac{\partial f}{\partial y} = 1 + x, \dfrac{\partial^2 f}{\partial x^2} = 0, \dfrac{\partial^2 f}{\partial y^2} = 0, \dfrac{\partial^2 f}{\partial y\,\partial x} = \dfrac{\partial^2 f}{\partial x\,\partial y} = 1$

**43.** $\dfrac{\partial g}{\partial x} = 2xy + y\cos x, \dfrac{\partial g}{\partial y} = x^2 - \sin y + \sin x,$

   $\dfrac{\partial^2 g}{\partial x^2} = 2y - y\sin x, \dfrac{\partial^2 g}{\partial y^2} = -\cos y,$

   $\dfrac{\partial^2 g}{\partial y\,\partial x} = \dfrac{\partial^2 g}{\partial x\,\partial y} = 2x + \cos x$

**45.** $\dfrac{\partial r}{\partial x} = \dfrac{1}{x + y}, \dfrac{\partial r}{\partial y} = \dfrac{1}{x + y}, \dfrac{\partial^2 r}{\partial x^2} = \dfrac{-1}{(x + y)^2}, \dfrac{\partial^2 r}{\partial y^2} = \dfrac{-1}{(x + y)^2},$

   $\dfrac{\partial^2 r}{\partial y\,\partial x} = \dfrac{\partial^2 r}{\partial x\,\partial y} = \dfrac{-1}{(x + y)^2}$

**47.** $\dfrac{\partial w}{\partial x} = x^2y\sec^2(xy) + 2x\tan(xy), \dfrac{\partial w}{\partial y} = x^3\sec^2(xy),$

   $\dfrac{\partial^2 w}{\partial y\,\partial x} = \dfrac{\partial^2 w}{\partial x\,\partial y} = 2x^3y\sec^2(xy)\tan(xy) + 3x^2\sec^2(xy)$

   $\dfrac{\partial^2 w}{\partial x^2} = 4xy\sec^2(xy) + 2x^2y^2\sec^2(xy)\tan(xy) + 2\tan(xy)$

   $\dfrac{\partial^2 w}{\partial y^2} = 2x^4\sec^2(xy)\tan(xy)$

**49.** $\dfrac{\partial w}{\partial x} = \sin(x^2y) + 2x^2y\cos(x^2y), \dfrac{\partial w}{\partial y} = x^3\cos(x^2y),$

   $\dfrac{\partial^2 w}{\partial y\,\partial x} = \dfrac{\partial^2 w}{\partial x\,\partial y} = 3x^2\cos(x^2y) - 2x^4y\sin(x^2y)$

   $\dfrac{\partial^2 w}{\partial x^2} = 6xy\cos(x^2y) - 4x^3y^2\sin(x^2y)$

   $\dfrac{\partial^2 w}{\partial y^2} = -x^5\sin(x^2y)$

**51.** $\dfrac{\partial w}{\partial x} = \dfrac{2}{2x + 3y}, \dfrac{\partial w}{\partial y} = \dfrac{3}{2x + 3y}, \dfrac{\partial^2 w}{\partial y\,\partial x} = \dfrac{\partial^2 w}{\partial x\,\partial y} = \dfrac{-6}{(2x + 3y)^2}$

**53.** $\dfrac{\partial w}{\partial x} = y^2 + 2xy^3 + 3x^2y^4, \dfrac{\partial w}{\partial y} = 2xy + 3x^2y^2 + 4x^3y^3,$

   $\dfrac{\partial^2 w}{\partial y\,\partial x} = \dfrac{\partial^2 w}{\partial x\,\partial y} = 2y + 6xy^2 + 12x^2y^3$

**55. (a)** $x$ first    **(b)** $y$ first    **(c)** $x$ first

   **(d)** $x$ first    **(e)** $y$ first    **(f)** $y$ first

**57.** $f_x(1, 2) = -13, f_y(1, 2) = -2$

**59.** $f_x(-2, 3) = 1/2, f_y(-2, 3) = 3/4$    **61. (a)** 3    **(b)** 2

**63.** 12    **65.** $-2$    **67.** $\dfrac{\partial A}{\partial a} = \dfrac{a}{bc\sin A}, \dfrac{\partial A}{\partial b} = \dfrac{c\cos A - b}{bc\sin A}$

**69.** $v_x = \dfrac{\ln v}{(\ln u)(\ln v) - 1}$

**71.** $f_x(x, y) = 0$ for all points $(x, y)$,

   $f_y(x, y) = \begin{cases} 3y^2, & y \geq 0 \\ -2y, & y < 0 \end{cases},$

   $f_{xy}(x, y) = f_{yx}(x, y) = 0$ for all points $(x, y)$

**89.** Yes

### Section 14.4, pp. 830–832

**1. (a)** $\dfrac{dw}{dt} = 0,$    **(b)** $\dfrac{dw}{dt}(\pi) = 0$

**3. (a)** $\dfrac{dw}{dt} = 1,$    **(b)** $\dfrac{dw}{dt}(3) = 1$

**5. (a)** $\dfrac{dw}{dt} = 4t\tan^{-1}t + 1$, **(b)** $\dfrac{dw}{dt}(1) = \pi + 1$

**7. (a)** $\dfrac{\partial z}{\partial u} = 4\cos v\ln(u\sin v) + 4\cos v$,

$\dfrac{\partial z}{\partial v} = -4u\sin v\ln(u\sin v) + \dfrac{4u\cos^2 v}{\sin v}$

**(b)** $\dfrac{\partial z}{\partial u} = \sqrt{2}(\ln 2 + 2)$, $\dfrac{\partial z}{\partial v} = -2\sqrt{2}(\ln 2 - 2)$

**9. (a)** $\dfrac{\partial w}{\partial u} = 2u + 4uv$, $\dfrac{\partial w}{\partial v} = -2v + 2u^2$

**(b)** $\dfrac{\partial w}{\partial u} = 3$, $\dfrac{\partial w}{\partial v} = -\dfrac{3}{2}$

**11. (a)** $\dfrac{\partial u}{\partial x} = 0$, $\dfrac{\partial u}{\partial y} = \dfrac{z}{(z-y)^2}$, $\dfrac{\partial u}{\partial z} = \dfrac{-y}{(z-y)^2}$

**(b)** $\dfrac{\partial u}{\partial x} = 0$, $\dfrac{\partial u}{\partial y} = 1$, $\dfrac{\partial u}{\partial z} = -2$

**13.** $\dfrac{dz}{dt} = \dfrac{\partial z}{\partial x}\dfrac{dx}{dt} + \dfrac{\partial z}{\partial y}\dfrac{dy}{dt}$

**15.** $\dfrac{\partial w}{\partial u} = \dfrac{\partial w}{\partial x}\dfrac{\partial x}{\partial u} + \dfrac{\partial w}{\partial y}\dfrac{\partial y}{\partial u} + \dfrac{\partial w}{\partial z}\dfrac{\partial z}{\partial u}$,

$\dfrac{\partial w}{\partial v} = \dfrac{\partial w}{\partial x}\dfrac{\partial x}{\partial v} + \dfrac{\partial w}{\partial y}\dfrac{\partial y}{\partial v} + \dfrac{\partial w}{\partial z}\dfrac{\partial z}{\partial v}$

 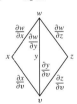

**17.** $\dfrac{\partial w}{\partial u} = \dfrac{\partial w}{\partial x}\dfrac{\partial x}{\partial u} + \dfrac{\partial w}{\partial y}\dfrac{\partial y}{\partial u}$, $\dfrac{\partial w}{\partial v} = \dfrac{\partial w}{\partial x}\dfrac{\partial x}{\partial v} + \dfrac{\partial w}{\partial y}\dfrac{\partial y}{\partial v}$.

**19.** $\dfrac{\partial z}{\partial t} = \dfrac{\partial z}{\partial x}\dfrac{\partial x}{\partial t} + \dfrac{\partial z}{\partial y}\dfrac{\partial y}{\partial t}$, $\dfrac{\partial z}{\partial s} = \dfrac{\partial z}{\partial x}\dfrac{\partial x}{\partial s} + \dfrac{\partial z}{\partial y}\dfrac{\partial y}{\partial s}$

**21.** $\dfrac{\partial w}{\partial s} = \dfrac{dw}{du}\dfrac{\partial u}{\partial s}$, $\dfrac{\partial w}{\partial t} = \dfrac{dw}{du}\dfrac{\partial u}{\partial t}$

**23.** $\dfrac{\partial w}{\partial r} = \dfrac{\partial w}{\partial x}\dfrac{\partial x}{\partial r} + \dfrac{\partial w}{\partial y}\dfrac{\partial y}{\partial r} = \dfrac{\partial w}{\partial x}\dfrac{\partial x}{\partial r}$ since $\dfrac{\partial y}{\partial r} = 0$,

$\dfrac{\partial w}{\partial s} = \dfrac{\partial w}{\partial x}\dfrac{\partial x}{\partial s} + \dfrac{\partial w}{\partial y}\dfrac{\partial y}{\partial s} = \dfrac{\partial w}{\partial y}\dfrac{\partial y}{\partial s}$ since $\dfrac{\partial x}{\partial s} = 0$

 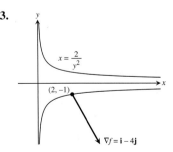

**25.** $4/3$ **27.** $-4/5$ **29.** $\dfrac{\partial z}{\partial x} = \dfrac{1}{4}$, $\dfrac{\partial z}{\partial y} = -\dfrac{3}{4}$

**31.** $\dfrac{\partial z}{\partial x} = -1$, $\dfrac{\partial z}{\partial y} = -1$ **33.** $12$ **35.** $-7$

**37.** $\dfrac{\partial z}{\partial u} = 2$, $\dfrac{\partial z}{\partial v} = 1$ **39.** $\dfrac{\partial w}{\partial t} = 2t\,e^{s^3+t^2}$, $\dfrac{\partial w}{\partial s} = 3s^2\,e^{s^3+t^2}$

**41.** $-0.00005$ amp/s

**47.** $(\cos 1, \sin 1, 1)$ and $(\cos(-2), \sin(-2), -2)$

**49. (a)** Maximum at $\left(-\dfrac{\sqrt{2}}{2}, \dfrac{\sqrt{2}}{2}\right)$ and $\left(\dfrac{\sqrt{2}}{2}, -\dfrac{\sqrt{2}}{2}\right)$; minimum

at $\left(\dfrac{\sqrt{2}}{2}, \dfrac{\sqrt{2}}{2}\right)$ and $\left(-\dfrac{\sqrt{2}}{2}, -\dfrac{\sqrt{2}}{2}\right)$

**(b)** Max $= 6$, min $= 2$

**51.** $2x\sqrt{x^8 + x^3} + \displaystyle\int_0^{x^2} \dfrac{3x^2}{2\sqrt{t^4 + x^3}}\,dt$

**Section 14.5, p. 840**

**1.**

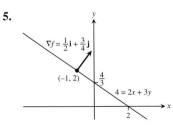

**3.**

**5.**

**7.** $\nabla f = 3\mathbf{i} + 2\mathbf{j} - 4\mathbf{k}$    **9.** $\nabla f = -\dfrac{26}{27}\mathbf{i} + \dfrac{23}{54}\mathbf{j} - \dfrac{23}{54}\mathbf{k}$

**11.** $-4$    **13.** $21/13$    **15.** $3$    **17.** $2$

**19.** $\mathbf{u} = -\dfrac{1}{\sqrt{2}}\mathbf{i} + \dfrac{1}{\sqrt{2}}\mathbf{j}, (D_{\mathbf{u}}f)_{P_0} = \sqrt{2}; -\mathbf{u} = \dfrac{1}{\sqrt{2}}\mathbf{i} - \dfrac{1}{\sqrt{2}}\mathbf{j},$

     $(D_{-\mathbf{u}}f)_{P_0} = -\sqrt{2}$

**21.** $\mathbf{u} = \dfrac{1}{3\sqrt{3}}\mathbf{i} - \dfrac{5}{3\sqrt{3}}\mathbf{j} - \dfrac{1}{3\sqrt{3}}\mathbf{k}, (D_{\mathbf{u}}f)_{P_0} = 3\sqrt{3};$

     $-\mathbf{u} = -\dfrac{1}{3\sqrt{3}}\mathbf{i} + \dfrac{5}{3\sqrt{3}}\mathbf{j} + \dfrac{1}{3\sqrt{3}}\mathbf{k}, (D_{-\mathbf{u}}f)_{P_0} = -3\sqrt{3}$

**23.** $\mathbf{u} = \dfrac{1}{\sqrt{3}}(\mathbf{i} + \mathbf{j} + \mathbf{k}), (D_{\mathbf{u}}f)_{P_0} = 2\sqrt{3};$

     $-\mathbf{u} = -\dfrac{1}{\sqrt{3}}(\mathbf{i} + \mathbf{j} + \mathbf{k}), (D_{-\mathbf{u}}f)_{P_0} = -2\sqrt{3}$

**25.**       **27.**

**29. (a)** $\mathbf{u} = \dfrac{3}{5}\mathbf{i} - \dfrac{4}{5}\mathbf{j}, D_{\mathbf{u}}f(1, -1) = 5$

   **(b)** $\mathbf{u} = -\dfrac{3}{5}\mathbf{i} + \dfrac{4}{5}\mathbf{j}, D_{\mathbf{u}}f(1, -1) = -5$

   **(c)** $\mathbf{u} = \dfrac{4}{5}\mathbf{i} + \dfrac{3}{5}\mathbf{j}, \mathbf{u} = -\dfrac{4}{5}\mathbf{i} - \dfrac{3}{5}\mathbf{j}$

   **(d)** $\mathbf{u} = -\mathbf{j}, \mathbf{u} = \dfrac{24}{25}\mathbf{i} - \dfrac{7}{25}\mathbf{j}$

   **(e)** $\mathbf{u} = -\mathbf{i}, \mathbf{u} = \dfrac{7}{25}\mathbf{i} + \dfrac{24}{25}\mathbf{j}$

**31.** $\mathbf{u} = \dfrac{7}{\sqrt{53}}\mathbf{i} - \dfrac{2}{\sqrt{53}}\mathbf{j}, -\mathbf{u} = -\dfrac{7}{\sqrt{53}}\mathbf{i} + \dfrac{2}{\sqrt{53}}\mathbf{j}$

**33.** No, the maximum rate of change is $\sqrt{185} < 14$.

**35.** $-7/\sqrt{5}$

### Section 14.6, pp. 847–850

**1. (a)** $x + y + z = 3$
   **(b)** $x = 1 + 2t, y = 1 + 2t, z = 1 + 2t$

**3. (a)** $2x - z - 2 = 0$
   **(b)** $x = 2 - 4t, y = 0, z = 2 + 2t$

**5. (a)** $2x + 2y + z - 4 = 0$
   **(b)** $x = 2t, y = 1 + 2t, z = 2 + t$

**7. (a)** $x + y + z - 1 = 0$    **(b)** $x = t, y = 1 + t, z = t$

**9.** $2x - z - 2 = 0$    **11.** $x - y + 2z - 1 = 0$

**13.** $x = 1, y = 1 + 2t, z = 1 - 2t$

**15.** $x = 1 - 2t, y = 1, z = \dfrac{1}{2} + 2t$

**17.** $x = 1 + 90t, y = 1 - 90t, z = 3$

**19.** $df = \dfrac{9}{11,830} \approx 0.0008$    **21.** $dg = 0$

**23. (a)** $\dfrac{\sqrt{3}}{2}\sin\sqrt{3} - \dfrac{1}{2}\cos\sqrt{3} \approx 0.935°\text{C/m}$

   **(b)** $\sqrt{3}\sin\sqrt{3} - \cos\sqrt{3} \approx 1.87°\text{C/s}$

**25. (a)** $L(x, y) = 1$    **(b)** $L(x, y) = 2x + 2y - 1$

**27. (a)** $L(x, y) = 3x - 4y + 5$
   **(b)** $L(x, y) = 3x - 4y + 5$

**29. (a)** $L(x, y) = 1 + x$    **(b)** $L(x, y) = -y + \dfrac{\pi}{2}$

**31. (a)** $W(30, -5) = -13°\text{C}, W(50, -25) = -42.2°\text{C},$
     $W(30, -10) = -19.5°\text{C}$
   **(b)** $W(15, -40) \approx -53.7°\text{C}, W(80, -40) \approx -66.6°\text{C},$
     $W(90, 0) \approx -10.2°\text{C}$
   **(c)** $L(v, T) \approx 1.336\,T - 0.11v - 2.952$
   **(d)**   **i)** $L(39, -9) \approx -19.3°\text{C}$
      **ii)** $L(42, -12) \approx -23.6°\text{C}$
      **iii)** $L(10, -25) \approx -37.5°\text{C}$

**33.** $L(x, y) = 7 + x - 6y; 0.06$    **35.** $L(x, y) = x + y + 1; 0.08$

**37.** $L(x, y) = 1 + x; 0.0222$

**39. (a)** $L(x, y, z) = 2x + 2y + 2z - 3$    **(b)** $L(x, y, z) = y + z$
   **(c)** $L(x, y, z) = 0$

**41. (a)** $L(x, y, z) = x$    **(b)** $L(x, y, z) = \dfrac{1}{\sqrt{2}}x + \dfrac{1}{\sqrt{2}}y$

   **(c)** $L(x, y, z) = \dfrac{1}{3}x + \dfrac{2}{3}y + \dfrac{2}{3}z$

**43. (a)** $L(x, y, z) = 2 + x$

   **(b)** $L(x, y, z) = x - y - z + \dfrac{\pi}{2} + 1$

   **(c)** $L(x, y, z) = x - y - z + \dfrac{\pi}{2} + 1$

**45.** $L(x, y, z) = 2x - 6y - 2z + 6, 0.0024$

**47.** $L(x, y, z) = x + y - z - 1, 0.00135$

**49.** Maximum error (estimate) $\leq 0.31$ in magnitude

**51.** Pay more attention to the smaller of the two dimensions. It will generate the larger partial derivative.

**53.** $f$ is most sensitive to a change in $d$.

### Section 14.7, pp. 857–859

**1.** $f(-3, 3) = -5$, local minimum    **3.** $f(-2, 1)$, saddle point

**5.** $f\left(3, \dfrac{3}{2}\right) = \dfrac{17}{2}$, local maximum

**7.** $f(2, -1) = -6$, local minimum    **9.** $f(1, 2)$, saddle point

**11.** $f\left(\dfrac{16}{7}, 0\right) = -\dfrac{16}{7}$, local maximum

**13.** $f(0, 0)$, saddle point; $f\left(-\dfrac{2}{3}, \dfrac{2}{3}\right) = \dfrac{170}{27}$, local maximum

**15.** $f(0, 0) = 0$, local minimum; $f(1, -1)$, saddle point

**17.** $f(0, \pm\sqrt{5})$, saddle points; $f(-2, -1) = 30$, local maximum; $f(2, 1) = -30$, local minimum

**19.** $f(0, 0)$, saddle point; $f(1, 1) = 2, f(-1, -1) = 2$, local maxima

**21.** $f(0, 0) = -1$, local maximum

**23.** $f(n\pi, 0)$, saddle points, for every integer $n$

**25.** $f(2, 0) = e^{-4}$, local minimum

**27.** $f(0, 0) = 0$, local minimum; $f(0, 2)$, saddle point

**29.** $f\left(\dfrac{1}{2}, 1\right) = \ln\left(\dfrac{1}{4}\right) - 3$, local maximum

**31.** Absolute maximum: 1 at $(0, 0)$; absolute minimum: $-5$ at $(1, 2)$

**33.** Absolute maximum: 4 at $(0, 2)$; absolute minimum: 0 at $(0, 0)$

**35.** Absolute maximum: 11 at $(0, -3)$; absolute minimum: $-10$ at $(4, -2)$

**37.** Absolute maximum: 4 at $(2, 0)$; absolute minimum: $\dfrac{3\sqrt{2}}{2}$ at

$\left(3, -\dfrac{\pi}{4}\right), \left(3, \dfrac{\pi}{4}\right), \left(1, -\dfrac{\pi}{4}\right)$, and $\left(1, \dfrac{\pi}{4}\right)$

**39.** $a = -3, b = 2$

**41.** Hottest is $2\dfrac{1}{4}^{\circ}$ at $\left(-\dfrac{1}{2}, \dfrac{\sqrt{3}}{2}\right)$ and $\left(-\dfrac{1}{2}, -\dfrac{\sqrt{3}}{2}\right)$; coldest is

$-\dfrac{1}{4}^{\circ}$ at $\left(\dfrac{1}{2}, 0\right)$.

**43.** (a) $f(0, 0)$, saddle point    (b) $f(1, 2)$, local minimum
(c) $f(1, -2)$, local minimum; $f(-1, -2)$, saddle point

**49.** $\left(\dfrac{1}{6}, \dfrac{1}{3}, \dfrac{355}{36}\right)$    **51.** $\left(\dfrac{9}{7}, \dfrac{6}{7}, \dfrac{3}{7}\right)$    **53.** 3, 3, 3    **55.** 12

**57.** $\dfrac{4}{\sqrt{3}} \times \dfrac{4}{\sqrt{3}} \times \dfrac{4}{\sqrt{3}}$    **59.** 2 m $\times$ 2 m $\times$ 1 m

**61.** (a) On the semicircle, max $f = 2\sqrt{2}$ at $t = \pi/4$, min $f = -2$
at $t = \pi$. On the quarter circle, max $f = 2\sqrt{2}$ at $t = \pi/4$,
min $f = 2$ at $t = 0, \pi/2$.
(b) On the semicircle, max $g = 2$ at $t = \pi/4$, min $g = -2$ at
$t = 3\pi/4$. On the quarter circle, max $g = 2$ at $t = \pi/4$,
min $g = 0$ at $t = 0, \pi/2$.
(c) On the semicircle, max $h = 8$ at $t = 0, \pi$; min $h = 4$
at $t = \pi/2$. On the quarter circle, max $h = 8$ at $t = 0$,
min $h = 4$ at $t = \pi/2$.

**63.** i) min $f = -1/2$ at $t = -1/2$; no max
ii) max $f = 0$ at $t = -1, 0$; min $f = -1/2$ at $t = -1/2$
iii) max $f = 4$ at $t = 1$; min $f = 0$ at $t = 0$

**67.** $y = -\dfrac{20}{13}x + \dfrac{9}{13}, \; y|_{x=4} = -\dfrac{71}{13}$

### Section 14.8, pp. 866–868

**1.** $\left(\pm\dfrac{1}{\sqrt{2}}, \dfrac{1}{2}\right), \left(\pm\dfrac{1}{\sqrt{2}}, -\dfrac{1}{2}\right)$    **3.** 39    **5.** $\left(3, \pm 3\sqrt{2}\right)$

**7.** (a) 8    (b) 64
**9.** $r = 2$ cm, $h = 4$ cm    **11.** Length $= 4\sqrt{2}$, width $= 3\sqrt{2}$
**13.** $f(0, 0) = 0$ is minimum; $f(2, 4) = 20$ is maximum.
**15.** Lowest $= 0°$, highest $= 125°$

**17.** $\left(\dfrac{3}{2}, 2, \dfrac{5}{2}\right)$    **19.** 1    **21.** $(0, 0, 2), (0, 0, -2)$

**23.** $f(1, -2, 5) = 30$ is maximum; $f(-1, 2, -5) = -30$ is minimum.

**25.** 3, 3, 3    **27.** $\dfrac{2}{\sqrt{3}}$ by $\dfrac{2}{\sqrt{3}}$ by $\dfrac{2}{\sqrt{3}}$ units

**29.** $(\pm 4/3, -4/3, -4/3)$    **31.** $\approx 24{,}322$ units

**33.** $U(8, 14) = \$128$    **37.** $f(2/3, 4/3, -4/3) = \dfrac{4}{3}$

**39.** $(2, 4, 4)$    **41.** Maximum is $1 + 6\sqrt{3}$ at $\left(\pm\sqrt{6}, \sqrt{3}, 1\right)$;
minimum is $1 - 6\sqrt{3}$ at $\left(\pm\sqrt{6}, -\sqrt{3}, 1\right)$.

**43.** Maximum is 4 at $(0, 0, \pm 2)$; minimum is 2 at
$\left(\pm\sqrt{2}, \pm\sqrt{2}, 0\right)$.

**1.** Quadratic: $x + xy$; cubic: $x + xy + \dfrac{1}{2}xy^2$

**3.** Quadratic: $xy$; cubic: $xy$

**5.** Quadratic: $y + \dfrac{1}{2}(2xy - y^2)$;

cubic: $y + \dfrac{1}{2}(2xy - y^2) + \dfrac{1}{6}(3x^2y - 3xy^2 + 2y^3)$

**7.** Quadratic: $\dfrac{1}{2}(2x^2 + 2y^2) = x^2 + y^2$; cubic: $x^2 + y^2$

**9.** Quadratic: $1 + (x + y) + (x + y)^2$;
cubic: $1 + (x + y) + (x + y)^2 + (x + y)^3$

**11.** Quadratic: $1 - \dfrac{1}{2}x^2 - \dfrac{1}{2}y^2$; $E(x, y) \le 0.00134$

**1.** (a) 0    (b) $1 + 2z$    (c) $1 + 2z$

**3.** (a) $\dfrac{\partial U}{\partial P} + \dfrac{\partial U}{\partial T}\left(\dfrac{V}{nR}\right)$    (b) $\dfrac{\partial U}{\partial P}\left(\dfrac{nR}{V}\right) + \dfrac{\partial U}{\partial T}$

**5.** (a) 5    (b) 5    **7.** $\left(\dfrac{\partial x}{\partial r}\right)_\theta = \cos\theta$    $\left(\dfrac{\partial r}{\partial x}\right)_y = \dfrac{x}{\sqrt{x^2 + y^2}}$

**1.** Domain: all points in the $xy$-plane; range: $z \ge 0$. Level curves
are ellipses with major axis along the $y$-axis and minor axis
along the $x$-axis.

**3.** Domain: all $(x, y)$ such that $x \ne 0$ and $y \ne 0$; range: $z \ne 0$.
Level curves are hyperbolas with the $x$- and $y$-axes as asymptotes.

**5.** Domain: all points in $xyz$-space; range: all real numbers. Level
surfaces are paraboloids of revolution with the $z$-axis as axis.

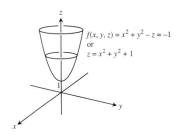

**7.** Domain: all $(x, y, z)$ such that $(x, y, z) \neq (0, 0, 0)$; range: positive real numbers. Level surfaces are spheres with center $(0, 0, 0)$ and radius $r > 0$.

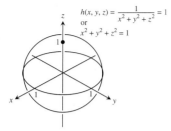

**9.** $-2$    **11.** $1/2$    **13.** $1$    **15.** Let $y = kx^2, k \neq 1$

**17.** No; $\lim_{(x,y) \to (0,0)} f(x, y)$ does not exist.

**19.** $\dfrac{\partial f}{\partial r} = \sin \theta - \cos \theta, \dfrac{\partial f}{\partial \theta} = r \cos \theta + r \sin \theta$

**21.** $\dfrac{\partial f}{\partial R_1} = -\dfrac{1}{R_1^{\,2}}, \dfrac{\partial f}{\partial R_2} = -\dfrac{1}{R_2^{\,2}}, \dfrac{\partial f}{\partial R_3} = -\dfrac{1}{R_3^{\,2}}$

**23.** $\dfrac{\partial P}{\partial n} = \dfrac{RT}{V}, \dfrac{\partial P}{\partial R} = \dfrac{nT}{V}, \dfrac{\partial P}{\partial T} = \dfrac{nR}{V}, \dfrac{\partial P}{\partial V} = -\dfrac{nRT}{V^2}$

**25.** $\dfrac{\partial^2 g}{\partial x^2} = 0, \dfrac{\partial^2 g}{\partial y^2} = \dfrac{2x}{y^3}, \dfrac{\partial^2 g}{\partial y \, \partial x} = \dfrac{\partial^2 g}{\partial x \, \partial y} = -\dfrac{1}{y^2}$

**27.** $\dfrac{\partial^2 g}{\partial x^2} = 0, \dfrac{\partial^2 g}{\partial y^2} = -48y + 2 \left[ \dfrac{(y^2 - 1)(1) - y(2y)}{(y^2 + 1)^2} \right]$

$= -48y - 2 \left[ \dfrac{y^2 + 1}{(y^2 - 1)^2} \right]$

**29.** $\dfrac{dw}{dt} \Big|_{t=0} = -1$

**31.** $\dfrac{\partial w}{\partial r} \Big|_{(r, s)=(\pi, 0)} = 2, \dfrac{\partial w}{\partial s} \Big|_{(r, s)=(\pi, 0)} = 2 - \pi$

**33.** $\dfrac{df}{dt} \Big|_{t=1} = -(\sin 1 + \cos 2)(\sin 1) + (\cos 1 + \cos 2)(\cos 1)$

$- 2(\sin 1 + \cos 1)(\sin 2)$

**35.** $\dfrac{dy}{dx} \Big|_{(x, y)=(0,1)} = -1$

**37.** Increases most rapidly in the direction $\mathbf{u} = -\dfrac{\sqrt{2}}{2} \mathbf{i} - \dfrac{\sqrt{2}}{2} \mathbf{j}$;

decreases most rapidly in the direction $-\mathbf{u} = \dfrac{\sqrt{2}}{2} \mathbf{i} + \dfrac{\sqrt{2}}{2} \mathbf{j}$;

$D_{\mathbf{u}} f = \dfrac{\sqrt{2}}{2}; D_{-\mathbf{u}} f = -\dfrac{\sqrt{2}}{2}; D_{\mathbf{u_1}} f = -\dfrac{7}{10}$ where $\mathbf{u_1} = \dfrac{\mathbf{v}}{|\mathbf{v}|}$

**39.** Increases most rapidly in the direction $\mathbf{u} = \dfrac{2}{7} \mathbf{i} + \dfrac{3}{7} \mathbf{j} + \dfrac{6}{7} \mathbf{k}$;

decreases most rapidly in the direction $-\mathbf{u} = -\dfrac{2}{7} \mathbf{i} - \dfrac{3}{7} \mathbf{j} - \dfrac{6}{7} \mathbf{k}$;

$D_{\mathbf{u}} f = 7; D_{-\mathbf{u}} f = -7; D_{\mathbf{u_1}} f = 7$ where $\mathbf{u_1} = \dfrac{\mathbf{v}}{|\mathbf{v}|}$

**41.** $\pi / \sqrt{2}$

**43.** (a) $f_x(1, 2) = f_y(1, 2) = 2$    (b) $14/5$

**45.**

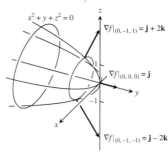

**47.** Tangent: $4x - y - 5z = 4$; normal line:
$x = 2 + 4t, y = -1 - t, z = 1 - 5t$

**49.** $2y - z - 2 = 0$

**51.** Tangent: $x + y = \pi + 1$; normal line: $y = x - \pi + 1$

**53.** $x = 1 - 2t, y = 1, z = 1/2 + 2t$

**55.** Answers will depend on the upper bound used for
$|f_{xx}|, |f_{xy}|, |f_{yy}|$. With $M = \sqrt{2}/2, |E| \leq 0.0142$. With
$M = 1, |E| \leq 0.02$.

**57.** $L(x, y, z) = y - 3z, L(x, y, z) = x + y - z - 1$

**59.** Be more careful with the diameter.

**61.** $dI = 0.038$, % change in $I = 15.83\%$, more sensitive to voltage change

**63.** (a) $5\%$    **65.** Local minimum of $-8$ at $(-2, -2)$

**67.** Saddle point at $(0, 0), f(0, 0) = 0$; local maximum of $1/4$ at $(-1/2, -1/2)$

**69.** Saddle point at $(0, 0), f(0, 0) = 0$; local minimum of $-4$ at $(0, 2)$; local maximum of $4$ at $(-2, 0)$; saddle point at $(-2, 2)$, $f(-2, 2) = 0$

**71.** Absolute maximum: $28$ at $(0, 4)$; absolute minimum: $-9/4$ at $(3/2, 0)$

**73.** Absolute maximum: $18$ at $(2, -2)$; absolute minimum: $-17/4$ at $(-2, 1/2)$

**75.** Absolute maximum: $8$ at $(-2, 0)$; absolute minimum: $-1$ at $(1, 0)$

**77.** Absolute maximum: $4$ at $(1, 0)$; absolute minimum: $-4$ at $(0, -1)$

**79.** Absolute maximum: $1$ at $(0, \pm 1)$ and $(1, 0)$; absolute minimum: $-1$ at $(-1, 0)$

**81.** Maximum: $5$ at $(0, 1)$; minimum: $-1/3$ at $(0, -1/3)$

**83.** Maximum: $\sqrt{3}$ at $\left( \dfrac{1}{\sqrt{3}}, -\dfrac{1}{\sqrt{3}}, \dfrac{1}{\sqrt{3}} \right)$; minimum: $-\sqrt{3}$ at
$\left( -\dfrac{1}{\sqrt{3}}, \dfrac{1}{\sqrt{3}}, -\dfrac{1}{\sqrt{3}} \right)$

**85.** Width $= \left( \dfrac{c^2 V}{ab} \right)^{1/3}$, depth $= \left( \dfrac{b^2 V}{ac} \right)^{1/3}$, height $= \left( \dfrac{a^2 V}{bc} \right)^{1/3}$

**87.** Maximum: $\dfrac{3}{2}$ at $\left(\dfrac{1}{\sqrt{2}}, \dfrac{1}{\sqrt{2}}, \sqrt{2}\right)$ and $\left(-\dfrac{1}{\sqrt{2}}, -\dfrac{1}{\sqrt{2}}, -\sqrt{2}\right)$;

minimum: $\dfrac{1}{2}$ at $\left(-\dfrac{1}{\sqrt{2}}, \dfrac{1}{\sqrt{2}}, -\sqrt{2}\right)$ and $\left(\dfrac{1}{\sqrt{2}}, -\dfrac{1}{\sqrt{2}}, \sqrt{2}\right)$

**89.** $\dfrac{\partial w}{\partial x} = \cos\theta \dfrac{\partial w}{\partial r} - \dfrac{\sin\theta}{r}\dfrac{\partial w}{\partial \theta}, \dfrac{\partial w}{\partial y} = \sin\theta \dfrac{\partial w}{\partial r} + \dfrac{\cos\theta}{r}\dfrac{\partial w}{\partial \theta}$

**95.** $(t, -t \pm 4, t), t$ a real number

**101.** **(a)** $(2y + x^2 z)e^{yz}$   **(b)** $x^2 e^{yz}\left(y - \dfrac{z}{2y}\right)$   **(c)** $(1 + x^2 y)e^{yz}$

**Additional and Advanced Exercises, pp. 881–883**

**1.** $f_{xy}(0, 0) = -1, f_{yx}(0, 0) = 1$

**7.** **(c)** $\dfrac{r^2}{2} = \dfrac{1}{2}(x^2 + y^2 + z^2)$   **13.** $V = \dfrac{\sqrt{3}abc}{2}$

**17.** $f(x, y) = \dfrac{y}{2} + 4, g(x, y) = \dfrac{x}{2} + \dfrac{9}{2}$

**19.** $y = 2\ln|\sin x| + \ln 2$

**21.** **(a)** $\dfrac{1}{\sqrt{53}}(2\mathbf{i} + 7\mathbf{j})$   **(b)** $\dfrac{-1}{\sqrt{29{,}097}}(98\mathbf{i} - 127\mathbf{j} + 58\mathbf{k})$

**23.** $w = e^{-c^2\pi^2 t}\sin\pi x$

# Chapter 15

**Section 15.1, pp. 888–889**

**1.** 24   **3.** 1   **5.** 16   **7.** $2\ln 2 - 1$   **9.** $(3/2)(5 - e)$
**11.** $3/2$   **13.** $\ln 2$   **15.** 14   **17.** 0   **19.** $1/2$
**21.** $2\ln 2$   **23.** $(\ln 2)^2$   **25.** $8/3$   **27.** 1   **29.** $\sqrt{2}$

**31.** $2/27$   **33.** $\dfrac{3}{2}\ln 3 - 1$   **35.** **(a)** $1/3$   **(b)** $2/3$

**Section 15.2, pp. 896–898**

**1.**

**3.**

**5.**

**7.**

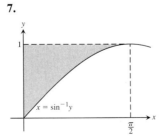

**9.** **(a)** $0 \le x \le 2, x^3 \le y \le 8$
**(b)** $0 \le y \le 8, 0 \le x \le y^{1/3}$

**11.** **(a)** $0 \le x \le 3, x^2 \le y \le 3x$
**(b)** $0 \le y \le 9, \dfrac{y}{3} \le x \le \sqrt{y}$

**13.** **(a)** $0 \le x \le 9, 0 \le y \le \sqrt{x}$
**(b)** $0 \le y \le 3, y^2 \le x \le 9$

**15.** **(a)** $0 \le x \le \ln 3, e^{-x} \le y \le 1$
**(b)** $\dfrac{1}{3} \le y \le 1, -\ln y \le x \le \ln 3$

**17.** **(a)** $0 \le x \le 1, x \le y \le 3 - 2x$
**(b)** $0 \le y \le 1, 0 \le x \le y \cup 1 \le y \le 3, 0 \le x \le \dfrac{3 - y}{2}$

**19.** $\dfrac{\pi^2}{2} + 2$

**21.** $8\ln 8 - 16 + e$

**23.** $e - 2$

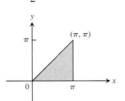

**25.** $\dfrac{3}{2}\ln 2$   **27.** $-1/10$

**29.** 8

**31.** $2\pi$

**33.** $\displaystyle\int_2^4 \int_0^{(4-y)/2} dx\, dy$

**35.** $\displaystyle\int_0^1 \int_{x^2}^x dy\, dx$

**37.** $\int_1^e \int_{\ln y}^1 dx\, dy$

**39.** $\int_0^9 \int_0^{(\sqrt{9-y})/2} 16x\, dx\, dy$

**41.** $\int_{-1}^1 \int_0^{\sqrt{1-x^2}} 3y\, dy\, dx$

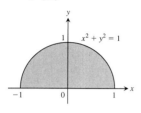

**43.** $\int_0^1 \int_{e^y}^e xy\, dx\, dy$

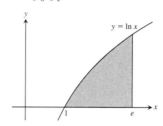

**45.** $\int_1^{e^3} \int_{\ln x}^3 (x+y)\, dy\, dx$

**47.** 2

**49.** $\dfrac{e-2}{2}$

**51.** 2

**53.** $1/(80\pi)$

**55.** $-2/3$

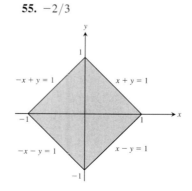

**57.** $4/3$    **59.** $625/12$    **61.** 16    **63.** 20    **65.** $2(1 + \ln 2)$

**67.**

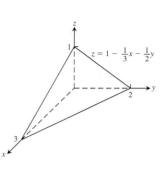

**69.** 1    **71.** $\pi^2$    **73.** $-\dfrac{3}{32}$    **75.** $\dfrac{20\sqrt{3}}{9}$

**77.** $\int_0^1 \int_x^{2-x} (x^2 + y^2)\, dy\, dx = \dfrac{4}{3}$

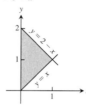

**79.** $R$ is the set of points $(x, y)$ such that $x^2 + 2y^2 < 4$.

**81.** No, by Fubini's Theorem, the two orders of integration must give the same result.

**85.** 0.603    **87.** 0.233

**Section 15.3, p. 901**

**1.** $\int_0^2 \int_0^{2-x} dy\, dx = 2$   or

$\int_0^2 \int_0^{2-y} dx\, dy = 2$

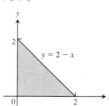

**3.** $\int_{-2}^1 \int_{y-2}^{-y^2} dx\, dy = \dfrac{9}{2}$

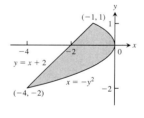

**5.** $\int_0^{\ln 2} \int_0^{e^x} dy\, dx = 1$

**7.** $\int_0^1 \int_{y^2}^{2y-y^2} dx\, dy = \dfrac{1}{3}$

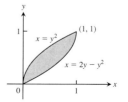

**9.** $\int_0^2 \int_y^{3y} 1 \, dx \, dy = 4$  or

$\int_0^2 \int_{x/3}^x 1 \, dy \, dx + \int_2^6 \int_{x/3}^2 1 \, dy \, dx = 4$

**11.** $\int_0^1 \int_{x/2}^{2x} 1 \, dy \, dx + \int_1^2 \int_{x/2}^{3-x} 1 \, dy \, dx = \dfrac{3}{2}$  or

$\int_0^1 \int_{y/2}^{2y} 1 \, dx \, dy + \int_1^2 \int_{y/2}^{3-y} 1 \, dx \, dy = \dfrac{3}{2}$

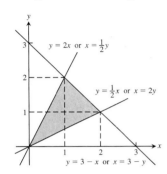

**13.** 12      **15.** $\sqrt{2} - 1$

**17.** $\dfrac{3}{2}$

**19. (a)** 0   **(b)** $4/\pi^2$   **21.** 8/3   **23.** $\pi - 2$
**25.** $40{,}000(1 - e^{-2}) \ln(7/2) \approx 43{,}329$

---

Section 15.4, pp. 906–908

**1.** $\dfrac{\pi}{2} \le \theta \le 2\pi, 0 \le r \le 9$   **3.** $\dfrac{\pi}{4} \le \theta \le \dfrac{3\pi}{4}, 0 \le r \le \csc\theta$

**5.** $0 \le \theta \le \dfrac{\pi}{6}, 1 \le r \le 2\sqrt{3}\sec\theta$;

$\quad \dfrac{\pi}{6} \le \theta \le \dfrac{\pi}{2}, 1 \le r \le 2\csc\theta$

**7.** $-\dfrac{\pi}{2} \le \theta \le \dfrac{\pi}{2}, 0 \le r \le 2\cos\theta$   **9.** $\dfrac{\pi}{2}$

**11.** $2\pi$    **13.** 36    **15.** $2 - \sqrt{3}$    **17.** $(1 - \ln 2)\,\pi$

**19.** $(2\ln 2 - 1)(\pi/2)$    **21.** $\dfrac{2(1 + \sqrt{2})}{3}$

**23.**

$\int_0^1 \int_0^{\sqrt{1-x^2}} xy \, dy \, dx$  or  $\int_0^1 \int_0^{\sqrt{1-y^2}} xy \, dx \, dy$

**25.**

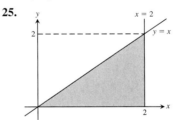

$\int_0^2 \int_0^x y^2 (x^2 + y^2) \, dy \, dx$  or  $\int_0^2 \int_y^2 y^2 (x^2 + y^2) \, dx \, dy$

**27.** $2(\pi - 2)$    **29.** $12\pi$    **31.** $(3\pi/8) + 1$    **33.** $\dfrac{2a}{3}$

**35.** $\dfrac{2a}{3}$    **37.** $2\pi(2 - \sqrt{e})$    **39.** $\dfrac{4}{3} + \dfrac{5\pi}{8}$

**41. (a)** $\dfrac{\sqrt{\pi}}{2}$   **(b)** 1    **43.** $\pi \ln 4$, no    **45.** $\dfrac{1}{2}(a^2 + 2h^2)$

**47.** $\dfrac{8}{9}(3\pi - 4)$

---

Section 15.5, pp. 914–917

**1.** $1/6$

**3.** $\displaystyle\int_0^1 \int_0^{2-2x} \int_0^{3-3x-3y/2} dz \, dy \, dx, \ \int_0^2 \int_0^{1-y/2} \int_0^{3-3x-3y/2} dz \, dx \, dy,$

$\displaystyle\int_0^1 \int_0^{3-3x} \int_0^{2-2x-2z/3} dy \, dz \, dx, \ \int_0^3 \int_0^{1-z/3} \int_0^{2-2x-2z/3} dy \, dx \, dz,$

$\displaystyle\int_0^2 \int_0^{3-3y/2} \int_0^{1-y/2-z/3} dx \, dz \, dy, \ \int_0^3 \int_0^{2-2z/3} \int_0^{1-y/2-z/3} dx \, dy \, dz.$

The value of all six integrals is 1.

**5.** $\displaystyle\int_{-2}^{2}\int_{-\sqrt{4-x^2}}^{\sqrt{4-x^2}}\int_{x^2+y^2}^{8-x^2-y^2} 1\,dz\,dx\,dy,\quad \int_{-2}^{2}\int_{-\sqrt{4-y^2}}^{\sqrt{4-y^2}}\int_{x^2+y^2}^{8-x^2-y^2} 1\,dz\,dx\,dy,$

$\displaystyle\int_{-2}^{2}\int_{4}^{8-y^2}\int_{-\sqrt{8-z-y^2}}^{\sqrt{8-z-y^2}} 1\,dx\,dz\,dy + \int_{-2}^{2}\int_{y^2}^{4}\int_{-\sqrt{z-y^2}}^{\sqrt{z-y^2}} 1\,dx\,dz\,dy,$

$\displaystyle\int_{4}^{8}\int_{-\sqrt{8-z}}^{\sqrt{8-z}}\int_{-\sqrt{8-z-y^2}}^{\sqrt{8-z-y^2}} 1\,dx\,dy\,dz + \int_{0}^{4}\int_{-\sqrt{z}}^{\sqrt{z}}\int_{-\sqrt{z-y^2}}^{\sqrt{z-y^2}} 1\,dx\,dy\,dz,$

$\displaystyle\int_{-2}^{2}\int_{4}^{8-x^2}\int_{-\sqrt{8-z-x^2}}^{\sqrt{8-z-x^2}} 1\,dy\,dz\,dx + \int_{-2}^{2}\int_{x^2}^{4}\int_{-\sqrt{z-x^2}}^{\sqrt{z-x^2}} 1\,dy\,dz\,dx,$

$\displaystyle\int_{4}^{8}\int_{-\sqrt{8-z}}^{\sqrt{8-z}}\int_{-\sqrt{8-z-x^2}}^{\sqrt{8-z-x^2}} 1\,dy\,dx\,dz + \int_{0}^{4}\int_{-\sqrt{z}}^{\sqrt{z}}\int_{-\sqrt{z-x^2}}^{\sqrt{z-x^2}} 1\,dy\,dx\,dz.$

The value of all six integrals is $16\pi$.

**7.** 1    **9.** 6    **11.** $\dfrac{5(2-\sqrt{3})}{4}$    **13.** 18

**15.** $7/6$    **17.** 0    **19.** $\dfrac{1}{2}-\dfrac{\pi}{8}$

**21. (a)** $\displaystyle\int_{-1}^{1}\int_{0}^{1-x^2}\int_{x^2}^{1-z} dy\,dz\,dx$    **(b)** $\displaystyle\int_{0}^{1}\int_{-\sqrt{1-z}}^{\sqrt{1-z}}\int_{x^2}^{1-z} dy\,dx\,dz$

**(c)** $\displaystyle\int_{0}^{1}\int_{0}^{1-z}\int_{-\sqrt{y}}^{\sqrt{y}} dx\,dy\,dz$    **(d)** $\displaystyle\int_{0}^{1}\int_{0}^{1-y}\int_{-\sqrt{y}}^{\sqrt{y}} dx\,dz\,dy$

**(e)** $\displaystyle\int_{0}^{1}\int_{-\sqrt{y}}^{\sqrt{y}}\int_{0}^{1-y} dz\,dx\,dy$

**23.** $2/3$    **25.** $20/3$    **27.** 1    **29.** $16/3$    **31.** $8\pi - \dfrac{32}{3}$

**33.** 2    **35.** $4\pi$    **37.** $31/3$    **39.** 1    **41.** $2\sin 4$

**43.** 4    **45.** $a = 3$ or $a = 13/3$

**47.** The domain is the set of all points $(x, y, z)$ such that $4x^2 + 4y^2 + z^2 \le 4$.

### Section 15.6, pp. 922–924
**1.** $\bar{x} = 5/14, \bar{y} = 38/35$    **3.** $\bar{x} = 64/35, \bar{y} = 5/7$
**5.** $\bar{x} = \bar{y} = 4a/(3\pi)$    **7.** $I_x = I_y = 4\pi, I_0 = 8\pi$
**9.** $\bar{x} = -1, \bar{y} = 1/4$    **11.** $I_x = 64/105$
**13.** $\bar{x} = 3/8, \bar{y} = 17/16$    **15.** $\bar{x} = 11/3, \bar{y} = 14/27, I_y = 432$
**17.** $\bar{x} = 0, \bar{y} = 13/31, I_y = 7/5$
**19.** $\bar{x} = 0, \bar{y} = 7/10; I_x = 9/10, I_y = 3/10, I_0 = 6/5$
**21.** $I_x = \dfrac{M}{3}(b^2 + c^2), I_y = \dfrac{M}{3}(a^2 + c^2), I_z = \dfrac{M}{3}(a^2 + b^2)$
**23.** $\bar{x} = \bar{y} = 0, \bar{z} = 12/5, I_x = 7904/105 \approx 75.28,$
$I_y = 4832/63 \approx 76.70, I_z = 256/45 \approx 5.69$
**25. (a)** $\bar{x} = \bar{y} = 0, \bar{z} = 8/3$    **(b)** $c = 2\sqrt{2}$
**27.** $I_L = 1386$
**29. (a)** $4/3$    **(b)** $\bar{x} = 4/5, \bar{y} = \bar{z} = 2/5$
**31. (a)** $5/2$    **(b)** $\bar{x} = \bar{y} = \bar{z} = 8/15$    **(c)** $I_x = I_y = I_z = 11/6$
**33.** 3

**37. (a)** $I_{\text{c.m.}} = \dfrac{abc(a^2 + b^2)}{12}, R_{\text{c.m.}} = \sqrt{\dfrac{a^2 + b^2}{12}}$

**(b)** $I_L = \dfrac{abc(a^2 + 7b^2)}{3}, R_L = \sqrt{\dfrac{a^2 + 7b^2}{3}}$

### Section 15.7, pp. 932–936
**1.** $\dfrac{4\pi(\sqrt{2}-1)}{3}$    **3.** $\dfrac{17\pi}{5}$    **5.** $\pi(6\sqrt{2}-8)$    **7.** $\dfrac{3\pi}{10}$

**9.** $\pi/3$

**11. (a)** $\displaystyle\int_{0}^{2\pi}\int_{0}^{1}\int_{0}^{\sqrt{4-r^2}} r\,dz\,dr\,d\theta$

**(b)** $\displaystyle\int_{0}^{2\pi}\int_{0}^{\sqrt{3}}\int_{0}^{1} r\,dr\,dz\,d\theta + \int_{0}^{2\pi}\int_{\sqrt{3}}^{2}\int_{0}^{\sqrt{4-z^2}} r\,dr\,dz\,d\theta$

**(c)** $\displaystyle\int_{0}^{1}\int_{0}^{\sqrt{4-r^2}}\int_{0}^{2\pi} r\,d\theta\,dz\,dr$

**13.** $\displaystyle\int_{-\pi/2}^{\pi/2}\int_{0}^{\cos\theta}\int_{0}^{3r^2} f(r, \theta, z)\,dz\,r\,dr\,d\theta$

**15.** $\displaystyle\int_{0}^{\pi}\int_{0}^{2\sin\theta}\int_{0}^{4-r\sin\theta} f(r, \theta, z)\,dz\,r\,dr\,d\theta$

**17.** $\displaystyle\int_{-\pi/2}^{\pi/2}\int_{1}^{1+\cos\theta}\int_{0}^{4} f(r, \theta, z)\,dz\,r\,dr\,d\theta$

**19.** $\displaystyle\int_{0}^{\pi/4}\int_{0}^{\sec\theta}\int_{0}^{2-r\sin\theta} f(r, \theta, z)\,dz\,r\,dr\,d\theta$    **21.** $\pi^2$

**23.** $\pi/3$    **25.** $5\pi$    **27.** $2\pi$    **29.** $\left(\dfrac{8-5\sqrt{2}}{2}\right)\pi$

**31. (a)** $\displaystyle\int_{0}^{2\pi}\int_{0}^{\pi/6}\int_{0}^{2} \rho^2 \sin\phi\,d\rho\,d\phi\,d\theta +$

$\displaystyle\int_{0}^{2\pi}\int_{\pi/6}^{\pi/2}\int_{0}^{\csc\phi} \rho^2 \sin\phi\,d\rho\,d\phi\,d\theta$

**(b)** $\displaystyle\int_{0}^{2\pi}\int_{1}^{2}\int_{\pi/6}^{\sin^{-1}(1/\rho)} \rho^2 \sin\phi\,d\phi\,d\rho\,d\theta +$

$\displaystyle\int_{0}^{2\pi}\int_{0}^{2}\int_{0}^{\pi/6} \rho^2 \sin\phi\,d\phi\,d\rho\,d\theta +$

$\displaystyle\int_{0}^{2\pi}\int_{0}^{1}\int_{\pi/6}^{\pi/2} \rho^2 \sin\phi\,d\phi\,d\rho\,d\theta$

**33.** $\displaystyle\int_{0}^{2\pi}\int_{0}^{\pi/2}\int_{\cos\phi}^{2} \rho^2 \sin\phi\,d\rho\,d\phi\,d\theta = \dfrac{31\pi}{6}$

**35.** $\displaystyle\int_{0}^{2\pi}\int_{0}^{\pi}\int_{0}^{1-\cos\phi} \rho^2 \sin\phi\,d\rho\,d\phi\,d\theta = \dfrac{8\pi}{3}$

**37.** $\displaystyle\int_{0}^{2\pi}\int_{\pi/4}^{\pi/2}\int_{0}^{2\cos\phi} \rho^2 \sin\phi\,d\rho\,d\phi\,d\theta = \dfrac{\pi}{3}$

**39. (a)** $\displaystyle 8\int_{0}^{\pi/2}\int_{0}^{\pi/2}\int_{0}^{2} \rho^2 \sin\phi\,d\rho\,d\phi\,d\theta$

**(b)** $\displaystyle 8\int_{0}^{\pi/2}\int_{0}^{2}\int_{0}^{\sqrt{4-r^2}} r\,dz\,dr\,d\theta$

**(c)** $\displaystyle 8\int_{0}^{2}\int_{0}^{\sqrt{4-x^2}}\int_{0}^{\sqrt{4-x^2-y^2}} dz\,dy\,dx$

**41. (a)** $\displaystyle\int_0^{2\pi}\int_0^{\pi/3}\int_{\sec\phi}^{2}\rho^2\sin\phi\,d\rho\,d\phi\,d\theta$

**(b)** $\displaystyle\int_0^{2\pi}\int_0^{\sqrt{3}}\int_1^{\sqrt{4-r^2}}r\,dz\,dr\,d\theta$

**(c)** $\displaystyle\int_{-\sqrt{3}}^{\sqrt{3}}\int_{-\sqrt{3-x^2}}^{\sqrt{3-x^2}}\int_1^{\sqrt{4-x^2-y^2}}dz\,dy\,dx$    **(d)** $5\pi/3$

**43.** $8\pi/3$    **45.** $9/4$    **47.** $\dfrac{3\pi-4}{18}$    **49.** $\dfrac{2\pi a^3}{3}$

**51.** $5\pi/3$    **53.** $\pi/2$    **55.** $\dfrac{4(2\sqrt{2}-1)\pi}{3}$    **57.** $16\pi$

**59.** $5\pi/2$    **61.** $\dfrac{4\pi(8-3\sqrt{3})}{3}$    **63.** $2/3$    **65.** $3/4$

**67.** $\bar{x}=\bar{y}=0, \bar{z}=3/8$    **69.** $(\bar{x},\bar{y},\bar{z})=(0,0,3/8)$

**71.** $\bar{x}=\bar{y}=0, \bar{z}=5/6$    **73.** $I_x=\pi/4$    **75.** $\dfrac{a^4h\pi}{10}$

**77. (a)** $(\bar{x},\bar{y},\bar{z})=\left(0,0,\dfrac{4}{5}\right), I_z=\dfrac{\pi}{12}$

**(b)** $(\bar{x},\bar{y},\bar{z})=\left(0,0,\dfrac{5}{6}\right), I_z=\dfrac{\pi}{14}$

**81.** $\dfrac{3M}{\pi R^3}$

**85.** The surface's equation $r=f(z)$ tells us that the point $(r,\theta,z)=(f(z),\theta,z)$ will lie on the surface for all $\theta$. In particular, $(f(z),\theta+\pi,z)$ lies on the surface whenever $(f(z),\theta,z)$ lies on the surface, so the surface is symmetric with respect to the $z$-axis.

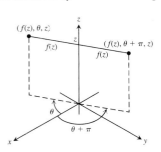

Section 15.8, pp. 944–946

**1. (a)** $x=\dfrac{u+v}{3}, y=\dfrac{v-2u}{3}; \dfrac{1}{3}$

**(b)** Triangular region with boundaries $u=0$, $v=0$, and $u+v=3$

**3. (a)** $x=\dfrac{1}{5}(2u-v), y=\dfrac{1}{10}(3v-u); \dfrac{1}{10}$

**(b)** Triangular region with boundaries $3v=u$, $v=2u$, and $3u+v=10$

**7.** $64/5$    **9.** $\displaystyle\int_1^2\int_1^3(u+v)\dfrac{2u}{v}\,du\,dv=8+\dfrac{52}{3}\ln 2$

**11.** $\dfrac{\pi ab(a^2+b^2)}{4}$    **13.** $\dfrac{1}{3}\left(1+\dfrac{3}{e^2}\right)\approx 0.4687$

**15.** $\dfrac{225}{16}$    **17.** $12$    **19.** $\dfrac{a^2b^2c^2}{6}$

**21. (a)** $\begin{vmatrix}\cos v & -u\sin v\\ \sin v & u\cos v\end{vmatrix}=u\cos^2 v+u\sin^2 v=u$

**(b)** $\begin{vmatrix}\sin v & u\cos v\\ \cos v & -u\sin v\end{vmatrix}=-u\sin^2 v-u\cos^2 v=-u$

**27.** $\dfrac{3}{2}\ln 2$

Practice Exercises, pp. 946–948

**1.** $9e-9$    **3.** $9/2$

**5.** $\displaystyle\int_{-2}^{0}\int_{2x+4}^{4-x^2}dy\,dx=\dfrac{4}{3}$    **7.** $\displaystyle\int_{-3}^{3}\int_0^{(1/2)\sqrt{9-x^2}}y\,dy\,dx=\dfrac{9}{2}$

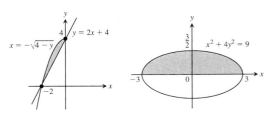

**9.** $\sin 4$    **11.** $\dfrac{\ln 17}{4}$    **13.** $4/3$    **15.** $4/3$    **17.** $1/4$

**19.** $\pi$    **21.** $\dfrac{\pi-2}{4}$    **23.** $0$    **25.** $8/35$    **27.** $\pi/2$

**29.** $\dfrac{2(31-3^{5/2})}{3}$

**31. (a)** $\displaystyle\int_{-\sqrt{2}}^{\sqrt{2}}\int_{-\sqrt{2-y^2}}^{\sqrt{2-y^2}}\int_{\sqrt{x^2+y^2}}^{\sqrt{4-x^2-y^2}}3\,dz\,dx\,dy$

**(b)** $\displaystyle\int_0^{2\pi}\int_0^{\pi/4}\int_0^{2}3\rho^2\sin\phi\,d\rho\,d\phi\,d\theta$    **(c)** $2\pi(8-4\sqrt{2})$

**33.** $\displaystyle\int_0^{2\pi}\int_0^{\pi/4}\int_0^{\sec\phi}\rho^2\sin\phi\,d\rho\,d\phi\,d\theta=\dfrac{\pi}{3}$

**35.** $\displaystyle\int_0^{1}\int_{\sqrt{1-x^2}}^{\sqrt{3-x^2}}\int_1^{\sqrt{4-x^2-y^2}}z^2xy\,dz\,dy\,dx$

$+\displaystyle\int_1^{\sqrt{3}}\int_0^{\sqrt{3-x^2}}\int_1^{\sqrt{4-x^2-y^2}}z^2xy\,dz\,dy\,dx$

**37. (a)** $\dfrac{8\pi(4\sqrt{2}-5)}{3}$    **(b)** $\dfrac{8\pi(4\sqrt{2}-5)}{3}$

**39.** $I_z=\dfrac{8\pi\delta(b^5-a^5)}{15}$

**41.** $\bar{x}=\bar{y}=\dfrac{1}{2-\ln 4}$    **43.** $I_0=104$    **45.** $I_x=2\delta$

**47.** $M=4, M_x=0, M_y=0$    **49.** $\bar{x}=\dfrac{3\sqrt{3}}{\pi}, \bar{y}=0$

**51. (a)** $\bar{x} = \dfrac{15\pi + 32}{6\pi + 48}, \bar{y} = 0$

**(b)**

Additional and Advanced Exercises, pp. 949–950

**1. (a)** $\displaystyle\int_{-3}^{2}\int_{x}^{6-x^2} x^2\, dy\, dx$   **(b)** $\displaystyle\int_{-3}^{2}\int_{x}^{6-x^2}\int_{0}^{x^2} dz\, dy\, dx$

   **(c)** $125/4$

**3.** $2\pi$    **5.** $3\pi/2$

**7. (a)** Hole radius $= 1$, sphere radius $= 2$   **(b)** $4\sqrt{3}\pi$

**9.** $\pi/4$    **11.** $\ln\left(\dfrac{b}{a}\right)$    **15.** $1/\sqrt[4]{3}$

**17.** Mass $= a^2\cos^{-1}\left(\dfrac{b}{a}\right) - b\sqrt{a^2 - b^2},$

$I_0 = \dfrac{a^4}{2}\cos^{-1}\left(\dfrac{b}{a}\right) - \dfrac{b^3}{2}\sqrt{a^2 - b^2} - \dfrac{b^3}{6}(a^2 - b^2)^{3/2}$

**19.** $\dfrac{1}{ab}(e^{a^2b^2} - 1)$    **21. (b)** 1   **(c)** 0

**25.** $h = \sqrt{180}$ cm, $h = \sqrt{540}$ cm    **27.** $2\pi\left[\dfrac{1}{3} - \left(\dfrac{1}{3}\right)\dfrac{\sqrt{2}}{2}\right]$

# Chapter 16

Section 16.1, pp. 957–959

**1.** Graph (c)    **3.** Graph (g)    **5.** Graph (d)    **7.** Graph (f)

**9.** $\sqrt{2}$    **11.** $\dfrac{13}{2}$    **13.** $3\sqrt{14}$    **15.** $\dfrac{1}{6}(5\sqrt{5} + 9)$

**17.** $\sqrt{3}\ln\left(\dfrac{b}{a}\right)$    **19. (a)** $4\sqrt{5}$   **(b)** $\dfrac{1}{12}(17^{3/2} - 1)$

**21.** $\dfrac{15}{32}(e^{16} - e^{64})$    **23.** $\dfrac{1}{27}(40^{3/2} - 13^{3/2})$

**25.** $\dfrac{1}{6}(5^{3/2} + 7\sqrt{2} - 1)$    **27.** $\dfrac{10\sqrt{5} - 2}{3}$    **29.** 8

**31.** $\dfrac{1}{6}(17^{3/2} - 1)$    **33.** $2\sqrt{2} - 1$

**35. (a)** $4\sqrt{2} - 2$   **(b)** $\sqrt{2} + \ln(1 + \sqrt{2})$    **37.** $I_z = 2\pi\delta a^3$

**39. (a)** $I_z = 2\pi\sqrt{2}\delta$   **(b)** $I_z = 4\pi\sqrt{2}\delta$    **41.** $I_x = 2\pi - 2$

Section 16.2, pp. 969–971

**1.** $\nabla f = -(x\mathbf{i} + y\mathbf{j} + z\mathbf{k})(x^2 + y^2 + z^2)^{-3/2}$

**3.** $\nabla g = -\left(\dfrac{2x}{x^2 + y^2}\right)\mathbf{i} - \left(\dfrac{2y}{x^2 + y^2}\right)\mathbf{j} + e^z\mathbf{k}$

**5.** $\mathbf{F} = -\dfrac{kx}{(x^2 + y^2)^{3/2}}\mathbf{i} - \dfrac{ky}{(x^2 + y^2)^{3/2}}\mathbf{j}$, any $k > 0$

**7. (a)** $9/2$   **(b)** $13/3$   **(c)** $9/2$

**9. (a)** $1/3$   **(b)** $-1/5$   **(c)** 0

**11. (a)** 2   **(b)** $3/2$   **(c)** $1/2$

**13.** $-15/2$    **15.** 36    **17. (a)** $-5/6$   **(b)** 0   **(c)** $-7/12$

**19.** $1/2$    **21.** $-\pi$    **23.** $69/4$    **25.** $-39/2$    **27.** $25/6$

**29. (a)** $\text{Circ}_1 = 0$, $\text{circ}_2 = 2\pi$, $\text{flux}_1 = 2\pi$, $\text{flux}_2 = 0$

   **(b)** $\text{Circ}_1 = 0$, $\text{circ}_2 = 8\pi$, $\text{flux}_1 = 8\pi$, $\text{flux}_2 = 0$

**31.** Circ $= 0$, flux $= a^2\pi$    **33.** Circ $= a^2\pi$, flux $= 0$

**35. (a)** $-\dfrac{\pi}{2}$   **(b)** 0   **(c)** 1    **37. (a)** 32   **(b)** 32   **(c)** 32

**39.**

**41. (a)** $\mathbf{G} = -y\mathbf{i} + x\mathbf{j}$   **(b)** $\mathbf{G} = \sqrt{x^2 + y^2}\,\mathbf{F}$

**43.** $\mathbf{F} = -\dfrac{x\mathbf{i} + y\mathbf{j}}{\sqrt{x^2 + y^2}}$    **47.** 48    **49.** $\pi$    **51.** 0    **53.** $\dfrac{1}{2}$

Section 16.3, pp. 980–982

**1.** Conservative    **3.** Not conservative    **5.** Not conservative

**7.** $f(x, y, z) = x^2 + \dfrac{3y^2}{2} + 2z^2 + C$    **9.** $f(x, y, z) = xe^{y+2z} + C$

**11.** $f(x, y, z) = x\ln x - x + \tan(x + y) + \dfrac{1}{2}\ln(y^2 + z^2) + C$

**13.** 49    **15.** $-16$    **17.** 1    **19.** $9\ln 2$    **21.** 0    **23.** $-3$

**27.** $\mathbf{F} = \nabla\left(\dfrac{x^2 - 1}{y}\right)$    **29. (a)** 1   **(b)** 1   **(c)** 1

**31. (a)** 2   **(b)** 2    **33. (a)** $c = b = 2a$   **(b)** $c = b = 2$

**35.** It does not matter what path you use. The work will be the same on any path because the field is conservative.

**37.** The force $\mathbf{F}$ is conservative because all partial derivatives of $M$, $N$, and $P$ are zero. $f(x, y, z) = ax + by + cz + C; A = (xa, ya, za)$ and $B = (xb, yb, zb)$. Therefore, $\int \mathbf{F}\cdot d\mathbf{r} = f(B) - f(A) = a(xb - xa) + b(yb - ya) + c(zb - za) = \mathbf{F}\cdot\overrightarrow{AB}$.

Section 16.4, pp. 992–994

**1.** Flux $= 0$, circ $= 2\pi a^2$    **3.** Flux $= -\pi a^2$, circ $= 0$

**5.** Flux $= 2$, circ $= 0$    **7.** Flux $= -9$, circ $= 9$

**9.** Flux $= -11/60$, circ $= -7/60$

**11.** Flux $= 64/9$, circ $= 0$    **13.** Flux $= 1/2$, circ $= 1/2$

**15.** Flux $= 1/5$, circ $= -1/12$    **17.** 0    **19.** $2/33$    **21.** 0

**23.** $-16\pi$    **25.** $\pi a^2$    **27.** $3\pi/8$

**29. (a)** 0 if $C$ is traversed counterclockwise

   **(b)** $(h - k)(\text{area of the region})$    **39. (a)** 0

Section 16.5, pp. 1003–1005

**1.** $\mathbf{r}(r, \theta) = (r\cos\theta)\mathbf{i} + (r\sin\theta)\mathbf{j} + r^2\mathbf{k}, 0 \le r \le 2$,
$0 \le \theta \le 2\pi$

**3.** $\mathbf{r}(r, \theta) = (r\cos\theta)\mathbf{i} + (r\sin\theta)\mathbf{j} + (r/2)\mathbf{k}, 0 \le r \le 6$,
$0 \le \theta \le \pi/2$

**5.** $\mathbf{r}(r, \theta) = (r\cos\theta)\mathbf{i} + (r\sin\theta)\mathbf{j} + \sqrt{9 - r^2}\,\mathbf{k}$,
$0 \le r \le 3\sqrt{2}/2, 0 \le \theta \le 2\pi$; Also:
$\mathbf{r}(\phi, \theta) = (3\sin\phi\cos\theta)\mathbf{i} + (3\sin\phi\sin\theta)\mathbf{j} + (3\cos\phi)\mathbf{k}, 0 \le \phi \le \pi/4, 0 \le \theta \le 2\pi$

**7.** $\mathbf{r}(\phi, \theta) = \left(\sqrt{3}\sin\phi\cos\theta\right)\mathbf{i} + \left(\sqrt{3}\sin\phi\sin\theta\right)\mathbf{j} +$
$\left(\sqrt{3}\cos\phi\right)\mathbf{k}, \ \pi/3 \le \phi \le 2\pi/3, \ 0 \le \theta \le 2\pi$

**9.** $\mathbf{r}(x, y) = x\mathbf{i} + y\mathbf{j} + (4 - y^2)\mathbf{k}, \ 0 \le x \le 2, -2 \le y \le 2$

**11.** $\mathbf{r}(u, v) = u\mathbf{i} + (3\cos v)\mathbf{j} + (3\sin v)\mathbf{k}, \ 0 \le u \le 3,$
$0 \le v \le 2\pi$

**13. (a)** $\mathbf{r}(r, \theta) = (r\cos\theta)\mathbf{i} + (r\sin\theta)\mathbf{j} + (1 - r\cos\theta - r\sin\theta)\mathbf{k},$
$0 \le r \le 3, \ 0 \le \theta \le 2\pi$
**(b)** $\mathbf{r}(u, v) = (1 - u\cos v - u\sin v)\mathbf{i} + (u\cos v)\mathbf{j} +$
$(u\sin v)\mathbf{k}, \ 0 \le u \le 3, \ 0 \le v \le 2\pi$

**15.** $\mathbf{r}(u, v) = (4\cos^2 v)\mathbf{i} + u\mathbf{j} + (4\cos v\sin v)\mathbf{k}, \ 0 \le u \le 3,$
$-(\pi/2) \le v \le (\pi/2);$ Another way: $\mathbf{r}(u, v) = (2 + 2\cos v)\mathbf{i}$
$+ u\mathbf{j} + (2\sin v)\mathbf{k}, \ 0 \le u \le 3, \ 0 \le v \le 2\pi$

**17.** $\displaystyle\int_0^{2\pi}\int_0^1 \frac{\sqrt{5}}{2} r\, dr\, d\theta = \frac{\pi\sqrt{5}}{2}$

**19.** $\displaystyle\int_0^{2\pi}\int_1^3 r\sqrt{5}\, dr\, d\theta = 8\pi\sqrt{5}$
**21.** $\displaystyle\int_0^{2\pi}\int_1^4 1\, du\, dv = 6\pi$

**23.** $\displaystyle\int_0^{2\pi}\int_0^1 u\sqrt{4u^2 + 1}\, du\, dv = \frac{\left(5\sqrt{5} - 1\right)}{6}\pi$

**25.** $\displaystyle\int_0^{2\pi}\int_{\pi/4}^{\pi} 2\sin\phi\, d\phi\, d\theta = \left(4 + 2\sqrt{2}\right)\pi$

**27.**

**29.**

**33. (b)** $A = \displaystyle\int_0^{2\pi}\int_0^{\pi} \left[a^2 b^2 \sin^2\phi\cos^2\phi + b^2 c^2\cos^4\phi\cos^2\theta + \right.$
$\left. a^2 c^2\cos^4\phi\sin^2\theta\right]^{1/2} d\phi\, d\theta$

**35.** $x_0 x + y_0 y = 25$   **37.** $13\pi/3$   **39.** 4

**41.** $6\sqrt{6} - 2\sqrt{2}$   **43.** $\pi\sqrt{c^2 + 1}$

**45.** $\frac{\pi}{6}\left(17\sqrt{17} - 5\sqrt{5}\right)$   **47.** $3 + 2\ln 2$

**49.** $\frac{\pi}{6}\left(13\sqrt{13} - 1\right)$   **51.** $5\pi\sqrt{2}$   **53.** $\frac{2}{3}\left(5\sqrt{5} - 1\right)$

## Section 16.6, pp. 1014–1016

**1.** $\displaystyle\iint_S x\, d\sigma = \int_0^3\int_0^2 u\sqrt{4u^2 + 1}\, du\, dv = \frac{17\sqrt{17} - 1}{4}$

**3.** $\displaystyle\iint_S x^2\, d\sigma = \int_0^{2\pi}\int_0^{\pi} \sin^3\phi\cos^2\theta\, d\phi\, d\theta = \frac{4\pi}{3}$

**5.** $\displaystyle\iint_S z\, d\sigma = \int_0^1\int_0^1 (4 - u - v)\sqrt{3}\, dv\, du = 3\sqrt{3}$
(for $x = u, y = v$)

**7.** $\displaystyle\iint_S x^2\sqrt{5 - 4z}\, d\sigma = \int_0^1\int_0^{2\pi} u^2\cos^2 v \cdot \sqrt{4u^2 + 1} \cdot$
$u\sqrt{4u^2 + 1}\, dv\, du = \int_0^1\int_0^{2\pi} u^3(4u^2 + 1)\cos^2 v\, dv\, du = \frac{11\pi}{12}$

**9.** $9a^3$   **11.** $\frac{abc}{4}(ab + ac + bc)$   **13.** 2

**15.** $\frac{1}{30}\left(\sqrt{2} + 6\sqrt{6}\right)$   **17.** $\sqrt{6}/30$   **19.** $-32$   **21.** $\frac{\pi a^3}{6}$

**23.** $13a^4/6$   **25.** $2\pi/3$   **27.** $-73\pi/6$   **29.** 18

**31.** $\frac{\pi a^3}{6}$   **33.** $\frac{\pi a^2}{4}$   **35.** $\frac{\pi a^3}{2}$   **37.** $-32$   **39.** $-4$

**41.** $3a^4$   **43.** $\left(\frac{a}{2}, \frac{a}{2}, \frac{a}{2}\right)$

**45.** $(\bar{x}, \bar{y}, \bar{z}) = \left(0, 0, \frac{14}{9}\right), I_z = \frac{15\pi\sqrt{2}}{2}\delta$

**47. (a)** $\frac{8\pi}{3}a^4\delta$   **(b)** $\frac{20\pi}{3}a^4\delta$

## Section 16.7, pp. 1027–1028

**1.** $4\pi$   **3.** $-5/6$   **5.** 0   **7.** $-6\pi$   **9.** $2\pi a^2$

**11.** $-\pi$   **13.** $12\pi$   **15.** $-\pi/4$   **17.** $-15\pi$   **19.** $-8\pi$

**27.** $16I_y + 16I_x$

## Section 16.8, pp. 1039–1041

**1.** 0   **3.** 0   **5.** $-16$   **7.** $-8\pi$   **9.** $3\pi$   **11.** $-40/3$

**13.** $12\pi$   **15.** $12\pi\left(4\sqrt{2} - 1\right)$   **19.** No

**21.** The integral's value never exceeds the surface area of $S$.

**23.** $184/35$

## Practice Exercises, pp. 1042–1044

**1.** Path 1: $2\sqrt{3}$; path 2: $1 + 3\sqrt{2}$   **3.** $4a^2$   **5.** 0

**7.** $8\pi\sin(1)$   **9.** 0   **11.** $\pi\sqrt{3}$

**13.** $2\pi\left(1 - \frac{1}{\sqrt{2}}\right)$   **15.** $\frac{abc}{2}\sqrt{\frac{1}{a^2} + \frac{1}{b^2} + \frac{1}{c^2}}$   **17.** 50

**19.** $\mathbf{r}(\phi, \theta) = (6\sin\phi\cos\theta)\mathbf{i} + (6\sin\phi\sin\theta)\mathbf{j} + (6\cos\phi)\mathbf{k},$
$\frac{\pi}{6} \le \phi \le \frac{2\pi}{3}, \ 0 \le \theta \le 2\pi$

**21.** $\mathbf{r}(r, \theta) = (r\cos\theta)\mathbf{i} + (r\sin\theta)\mathbf{j} + (1 + r)\mathbf{k}, \ 0 \le r \le 2,$
$0 \le \theta \le 2\pi$

**23.** $\mathbf{r}(u, v) = (u\cos v)\mathbf{i} + 2u^2\mathbf{j} + (u\sin v)\mathbf{k}, \ 0 \le u \le 1,$
$0 \le v \le \pi$

**25.** $\sqrt{6}$   **27.** $\pi\left[\sqrt{2} + \ln\left(1 + \sqrt{2}\right)\right]$   **29.** Conservative

**31.** Not conservative   **33.** $f(x, y, z) = y^2 + yz + 2x + z$

**35.** Path 1: 2; path 2: 8/3   **37. (a)** $1 - e^{-2\pi}$   **(b)** $1 - e^{-2\pi}$

**39.** 0   **41. (a)** $4\sqrt{2} - 2$   **(b)** $\sqrt{2} + \ln\left(1 + \sqrt{2}\right)$

**43.** $(\bar{x}, \bar{y}, \bar{z}) = \left(1, \frac{16}{15}, \frac{2}{3}\right); I_x = \frac{232}{45}, I_y = \frac{64}{15}, I_z = \frac{56}{9}$

**45.** $\bar{z} = \frac{3}{2}, I_z = \frac{7\sqrt{3}}{3}$   **47.** $(\bar{x}, \bar{y}, \bar{z}) = (0, 0, 49/12), I_z = 640\pi$

**49.** Flux: $3/2$; circ: $-1/2$   **53.** 3   **55.** $\frac{2\pi}{3}\left(7 - 8\sqrt{2}\right)$

**57.** 0   **59.** $\pi$

**Additional and Advanced Exercises, pp. 1044–1046**

**1.** $6\pi$    **3.** $2/3$

**5. (a)** $\mathbf{F}(x, y, z) = z\mathbf{i} + x\mathbf{j} + y\mathbf{k}$
**(b)** $\mathbf{F}(x, y, z) = z\mathbf{i} + y\mathbf{k}$    **(c)** $\mathbf{F}(x, y, z) = z\mathbf{i}$

**7.** $\dfrac{16\pi R^3}{3}$    **9.** $a = 2, b = 1.$ The minimum flux is $-4.$

**11. (b)** $\dfrac{16}{3}g$

**(c)** Work $= \left(\displaystyle\int_C gxy\, ds\right) \bar{y} = g\displaystyle\int_C xy^2\, ds = \dfrac{16}{3}g$

**13. (c)** $\dfrac{4}{3}\pi w$    **19.** False if $\mathbf{F} = y\mathbf{i} + x\mathbf{j}$

# Appendices

**1.** $0.\overline{1}, 0.\overline{2}, 0.\overline{3}, 0.\overline{8}, 0.\overline{9}$ or $1$

**3.** $x < -2$

**5.** $x \le -\dfrac{1}{3}$

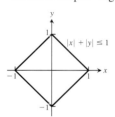

**7.** $3, -3$    **9.** $7/6, 25/6$

**11.** $-2 \le t \le 4$    **13.** $0 \le z \le 10$

**15.** $(-\infty, -2] \cup [2, \infty)$    **17.** $(-\infty, -3] \cup [1, \infty)$

**19.** $(-3, -2) \cup (2, 3)$    **21.** $(0, 1)$    **23.** $(-\infty, 1]$

**27.** The graph of $|x| + |y| \le 1$ is the interior and boundary of the "diamond-shaped" region.

**1.** $2, -4; 2\sqrt{5}$    **3.** Unit circle

**5.** $m_\perp = -\dfrac{1}{3}$

**7. (a)** $x = -1$    **(b)** $y = 4/3$    **9.** $y = -x$

**11.** $y = -\dfrac{5}{4}x + 6$    **13.** $y = 4x + 4$    **15.** $y = -\dfrac{x}{2} + 12$

**17.** $x$-intercept $= \sqrt{3},$ $y$-intercept $= -\sqrt{2}$

**19.** $(3, -3)$

**21.** $x^2 + (y - 2)^2 = 4$    **23.** $\left(x + \sqrt{3}\right)^2 + (y + 2)^2 = 4$

**25.** $x^2 + (y - 3/2)^2 = 25/4$    **27.**

**29.**

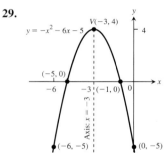

**31.** Exterior points of a circle of radius $\sqrt{7},$ centered at the origin

**33.** The washer between the circles $x^2 + y^2 = 1$ and $x^2 + y^2 = 4$ (points with distance from the origin between 1 and 2)

**35.** $(x + 2)^2 + (y - 1)^2 < 6$

**37.** $\left(\dfrac{1}{\sqrt{5}}, \dfrac{2}{\sqrt{5}}\right),$ $\left(-\dfrac{1}{\sqrt{5}}, -\dfrac{2}{\sqrt{5}}\right)$

**39.** $\left(-\dfrac{1}{\sqrt{3}}, -\dfrac{1}{3}\right),$ $\left(\dfrac{1}{\sqrt{3}}, -\dfrac{1}{3}\right)$

**41. (a)** $\approx -0.1$ °C/mm    **(b)** $\approx -0.31$ °C/mm
**(c)** $\approx -0.15$ °C/mm    **43.** 603.9 kPa

**45.** Yes: $C = F = -40°$

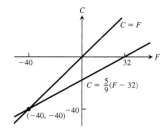

**51.** $k = -8, \quad k = 1/2$

Appendix 7, pp. AP-34–AP-35

**1. (a)** $(14, 8)$ **(b)** $(-1, 8)$ **(c)** $(0, -5)$

**3. (a)** By reflecting $z$ across the real axis
   **(b)** By reflecting $z$ across the imaginary axis
   **(c)** By reflecting $z$ across the real axis and then multiplying the length of the vector by $1/|z|^2$

**5. (a)** Points on the circle $x^2 + y^2 = 4$
   **(b)** Points inside the circle $x^2 + y^2 = 4$
   **(c)** Points outside the circle $x^2 + y^2 = 4$

**7.** Points on a circle of radius 1, center $(-1, 0)$

**9.** Points on the line $y = -x$     **11.** $4e^{2\pi i/3}$     **13.** $1e^{2\pi i/3}$

**15.** $\cos^4\theta - 6\cos^2\theta\sin^2\theta + \sin^4\theta$

**17.** $1, -\dfrac{1}{2} \pm \dfrac{\sqrt{3}}{2}i$     **19.** $2i, -\sqrt{3} - i, \sqrt{3} - i$

**21.** $\dfrac{\sqrt{6}}{2} \pm \dfrac{\sqrt{2}}{2}i, -\dfrac{\sqrt{6}}{2} \pm \dfrac{\sqrt{2}}{2}i$     **23.** $1 \pm \sqrt{3}i, -1 \pm \sqrt{3}i$

# Credits

# Index

# A Brief Table of Integrals

**1.** $\int k\,dx = kx + C$ (any number $k$)

**2.** $\int x^n\,dx = \dfrac{x^{n+1}}{n+1} + C$ $(n \neq -1)$

**3.** $\int \dfrac{dx}{x} = \ln|x| + C$

**4.** $\int e^x\,dx = e^x + C$

**5.** $\int a^x\,dx = \dfrac{a^x}{\ln a} + C$ $(a > 0, a \neq 1)$

**6.** $\int \sin x\,dx = -\cos x + C$

**7.** $\int \cos x\,dx = \sin x + C$

**8.** $\int \sec^2 x\,dx = \tan x + C$

**9.** $\int \csc^2 x\,dx = -\cot x + C$

**10.** $\int \sec x \tan x\,dx = \sec x + C$

**11.** $\int \csc x \cot x\,dx = -\csc x + C$

**12.** $\int \tan x\,dx = \ln|\sec x| + C$

**13.** $\int \cot x\,dx = \ln|\sin x| - C$

**14.** $\int \sinh x\,dx = \cosh x + C$

**15.** $\int \cosh x\,dx = \sinh x + C$

**16.** $\int \dfrac{dx}{\sqrt{a^2 - x^2}} = \sin^{-1}\dfrac{x}{a} + C$

**17.** $\int \dfrac{dx}{a^2 + x^2} = \dfrac{1}{a}\tan^{-1}\dfrac{x}{a} + C$

**18.** $\int \dfrac{dx}{x\sqrt{x^2 - a^2}} = \dfrac{1}{a}\sec^{-1}\left|\dfrac{x}{a}\right| + C$

**19.** $\int \dfrac{dx}{\sqrt{a^2 + x^2}} = \sinh^{-1}\dfrac{x}{a} + C$ $(a > 0)$

**20.** $\int \dfrac{dx}{\sqrt{x^2 - a^2}} = \cosh^{-1}\dfrac{x}{a} + C$ $(x > a > 0)$

**21.** $\int (ax + b)^n\,dx = \dfrac{(ax + b)^{n+1}}{a(n+1)} + C,\quad n \neq -1$

**22.** $\int x(ax + b)^n\,dx = \dfrac{(ax + b)^{n+1}}{a^2}\left[\dfrac{ax + b}{n + 2} - \dfrac{b}{n + 1}\right] + C,\quad n \neq -1, -2$

**23.** $\int (ax + b)^{-1}\,dx = \dfrac{1}{a}\ln|ax + b| + C$

**24.** $\int x(ax + b)^{-1}\,dx = \dfrac{x}{a} - \dfrac{b}{a^2}\ln|ax + b| + C$

**25.** $\int x(ax + b)^{-2}\,dx = \dfrac{1}{a^2}\left[\ln|ax + b| + \dfrac{b}{ax + b}\right] + C$

**26.** $\int \dfrac{dx}{x(ax + b)} = \dfrac{1}{b}\ln\left|\dfrac{x}{ax + b}\right| + C$

**27.** $\int (\sqrt{ax + b})^n\,dx = \dfrac{2}{a}\dfrac{(\sqrt{ax + b})^{n+2}}{n + 2} + C,\quad n \neq -2$

**28.** $\int \dfrac{\sqrt{ax + b}}{x}\,dx = 2\sqrt{ax + b} + b\int \dfrac{dx}{x\sqrt{ax + b}}$

**29. (a)** $\displaystyle\int \frac{dx}{x\sqrt{ax+b}} = \frac{1}{\sqrt{b}}\ln\left|\frac{\sqrt{ax+b}-\sqrt{b}}{\sqrt{ax+b}+\sqrt{b}}\right| + C$    **(b)** $\displaystyle\int \frac{dx}{x\sqrt{ax-b}} = \frac{2}{\sqrt{b}}\tan^{-1}\sqrt{\frac{ax-b}{b}} + C$

**30.** $\displaystyle\int \frac{\sqrt{ax+b}}{x^2}\,dx = -\frac{\sqrt{ax+b}}{x} + \frac{a}{2}\int \frac{dx}{x\sqrt{ax+b}} + C$    **31.** $\displaystyle\int \frac{dx}{x^2\sqrt{ax+b}} = -\frac{\sqrt{ax+b}}{bx} - \frac{a}{2b}\int \frac{dx}{x\sqrt{ax+b}} + C$

**Forms Involving $a^2 + x^2$**

**32.** $\displaystyle\int \frac{dx}{a^2+x^2} = \frac{1}{a}\tan^{-1}\frac{x}{a} + C$    **33.** $\displaystyle\int \frac{dx}{(a^2+x^2)^2} = \frac{x}{2a^2(a^2+x^2)} + \frac{1}{2a^3}\tan^{-1}\frac{x}{a} + C$

**34.** $\displaystyle\int \frac{dx}{\sqrt{a^2+x^2}} = \sinh^{-1}\frac{x}{a} + C = \ln\left(x + \sqrt{a^2+x^2}\right) + C$

**35.** $\displaystyle\int \sqrt{a^2+x^2}\,dx = \frac{x}{2}\sqrt{a^2+x^2} + \frac{a^2}{2}\ln\left(x + \sqrt{a^2+x^2}\right) + C$

**36.** $\displaystyle\int x^2\sqrt{a^2+x^2}\,dx = \frac{x}{8}(a^2+2x^2)\sqrt{a^2+x^2} - \frac{a^4}{8}\ln\left(x + \sqrt{a^2+x^2}\right) + C$

**37.** $\displaystyle\int \frac{\sqrt{a^2+x^2}}{x}\,dx = \sqrt{a^2+x^2} - a\ln\left|\frac{a+\sqrt{a^2+x^2}}{x}\right| + C$

**38.** $\displaystyle\int \frac{\sqrt{a^2+x^2}}{x^2}\,dx = \ln\left(x+\sqrt{a^2+x^2}\right) - \frac{\sqrt{a^2+x^2}}{x} + C$

**39.** $\displaystyle\int \frac{x^2}{\sqrt{a^2+x^2}}\,dx = -\frac{a^2}{2}\ln\left(x+\sqrt{a^2+x^2}\right) + \frac{x\sqrt{a^2+x^2}}{2} + C$

**40.** $\displaystyle\int \frac{dx}{x\sqrt{a^2+x^2}} = -\frac{1}{a}\ln\left|\frac{a+\sqrt{a^2+x^2}}{x}\right| + C$    **41.** $\displaystyle\int \frac{dx}{x^2\sqrt{a^2+x^2}} = -\frac{\sqrt{a^2+x^2}}{a^2x} + C$

**Forms Involving $a^2 - x^2$**

**42.** $\displaystyle\int \frac{dx}{a^2-x^2} = \frac{1}{2a}\ln\left|\frac{x+a}{x-a}\right| + C$    **43.** $\displaystyle\int \frac{dx}{(a^2-x^2)^2} = \frac{x}{2a^2(a^2-x^2)} + \frac{1}{4a^3}\ln\left|\frac{x+a}{x-a}\right| + C$

**44.** $\displaystyle\int \frac{dx}{\sqrt{a^2-x^2}} = \sin^{-1}\frac{x}{a} + C$    **45.** $\displaystyle\int \sqrt{a^2-x^2}\,dx = \frac{x}{2}\sqrt{a^2-x^2} + \frac{a^2}{2}\sin^{-1}\frac{x}{a} + C$

**46.** $\displaystyle\int x^2\sqrt{a^2-x^2}\,dx = \frac{a^4}{8}\sin^{-1}\frac{x}{a} - \frac{1}{8}x\sqrt{a^2-x^2}\,(a^2-2x^2) + C$

**47.** $\displaystyle\int \frac{\sqrt{a^2-x^2}}{x}\,dx = \sqrt{a^2-x^2} - a\ln\left|\frac{a+\sqrt{a^2-x^2}}{x}\right| + C$    **48.** $\displaystyle\int \frac{\sqrt{a^2-x^2}}{x^2}\,dx = -\sin^{-1}\frac{x}{a} - \frac{\sqrt{a^2-x^2}}{x} + C$

**49.** $\displaystyle\int \frac{x^2}{\sqrt{a^2-x^2}}\,dx = \frac{a^2}{2}\sin^{-1}\frac{x}{a} - \frac{1}{2}x\sqrt{a^2-x^2} + C$    **50.** $\displaystyle\int \frac{dx}{x\sqrt{a^2-x^2}} = -\frac{1}{a}\ln\left|\frac{a+\sqrt{a^2-x^2}}{x}\right| + C$

**51.** $\displaystyle\int \frac{dx}{x^2\sqrt{a^2-x^2}} = -\frac{\sqrt{a^2-x^2}}{a^2x} + C$

**Forms Involving $x^2 - a^2$**

**52.** $\displaystyle\int \frac{dx}{\sqrt{x^2-a^2}} = \ln\left|x+\sqrt{x^2-a^2}\right| + C$

**53.** $\displaystyle\int \sqrt{x^2-a^2}\,dx = \frac{x}{2}\sqrt{x^2-a^2} - \frac{a^2}{2}\ln\left|x+\sqrt{x^2-a^2}\right| + C$

**54.** $\int \left( \sqrt{x^2 - a^2} \right)^n dx = \dfrac{x\left( \sqrt{x^2 - a^2} \right)^n}{n + 1} - \dfrac{na^2}{n + 1} \int \left( \sqrt{x^2 - a^2} \right)^{n-2} dx, \quad n \neq -1$

**55.** $\int \dfrac{dx}{\left( \sqrt{x^2 - a^2} \right)^n} = \dfrac{x\left( \sqrt{x^2 - a^2} \right)^{2-n}}{(2 - n)a^2} - \dfrac{n - 3}{(n - 2)a^2} \int \dfrac{dx}{\left( \sqrt{x^2 - a^2} \right)^{n-2}}, \quad n \neq 2$

**56.** $\int x\left( \sqrt{x^2 - a^2} \right)^n dx = \dfrac{\left( \sqrt{x^2 - a^2} \right)^{n+2}}{n + 2} + C, \quad n \neq -2$

**57.** $\int x^2 \sqrt{x^2 - a^2}\, dx = \dfrac{x}{8} (2x^2 - a^2)\sqrt{x^2 - a^2} - \dfrac{a^4}{8} \ln \left| x + \sqrt{x^2 - a^2} \right| + C$

**58.** $\int \dfrac{\sqrt{x^2 - a^2}}{x}\, dx = \sqrt{x^2 - a^2} - a \sec^{-1} \left| \dfrac{x}{a} \right| + C$

**59.** $\int \dfrac{\sqrt{x^2 - a^2}}{x^2}\, dx = \ln \left| x + \sqrt{x^2 - a^2} \right| - \dfrac{\sqrt{x^2 - a^2}}{x} + C$

**60.** $\int \dfrac{x^2}{\sqrt{x^2 - a^2}}\, dx = \dfrac{a^2}{2} \ln \left| x + \sqrt{x^2 - a^2} \right| + \dfrac{x}{2}\sqrt{x^2 - a^2} + C$

**61.** $\int \dfrac{dx}{x\sqrt{x^2 - a^2}} = \dfrac{1}{a} \sec^{-1} \left| \dfrac{x}{a} \right| + C = \dfrac{1}{a} \cos^{-1} \left| \dfrac{a}{x} \right| + C$     **62.** $\int \dfrac{dx}{x^2\sqrt{x^2 - a^2}} = \dfrac{\sqrt{x^2 - a^2}}{a^2 x} + C$

Trigonometric Forms

**63.** $\int \sin ax\, dx = -\dfrac{1}{a} \cos ax + C$

**64.** $\int \cos ax\, dx = \dfrac{1}{a} \sin ax + C$

**65.** $\int \sin^2 ax\, dx = \dfrac{x}{2} - \dfrac{\sin 2ax}{4a} + C$

**66.** $\int \cos^2 ax\, dx = \dfrac{x}{2} + \dfrac{\sin 2ax}{4a} + C$

**67.** $\int \sin^n ax\, dx = -\dfrac{\sin^{n-1} ax \cos ax}{na} + \dfrac{n - 1}{n} \int \sin^{n-2} ax\, dx$

**68.** $\int \cos^n ax\, dx = \dfrac{\cos^{n-1} ax \sin ax}{na} + \dfrac{n - 1}{n} \int \cos^{n-2} ax\, dx$

**69. (a)** $\int \sin ax \cos bx\, dx = -\dfrac{\cos(a + b)x}{2(a + b)} - \dfrac{\cos(a - b)x}{2(a - b)} + C, \quad a^2 \neq b^2$

**(b)** $\int \sin ax \sin bx\, dx = \dfrac{\sin(a - b)x}{2(a - b)} - \dfrac{\sin(a + b)x}{2(a + b)} + C, \quad a^2 \neq b^2$

**(c)** $\int \cos ax \cos bx\, dx = \dfrac{\sin(a - b)x}{2(a - b)} + \dfrac{\sin(a + b)x}{2(a + b)} + C, \quad a^2 \neq b^2$

**70.** $\int \sin ax \cos ax\, dx = -\dfrac{\cos 2ax}{4a} + C$

**71.** $\int \sin^n ax \cos ax\, dx = \dfrac{\sin^{n+1} ax}{(n + 1)a} + C, \quad n \neq -1$

**72.** $\int \dfrac{\cos ax}{\sin ax}\, dx = \dfrac{1}{a} \ln |\sin ax| + C$

**73.** $\int \cos^n ax \sin ax\, dx = -\dfrac{\cos^{n+1} ax}{(n + 1)a} + C, \quad n \neq -1$

**74.** $\int \dfrac{\sin ax}{\cos ax}\, dx = -\dfrac{1}{a} \ln |\cos ax| + C$

**75.** $\int \sin^n ax \cos^m ax\, dx = -\dfrac{\sin^{n-1} ax \cos^{m+1} ax}{a(m + n)} + \dfrac{n - 1}{m + n} \int \sin^{n-2} ax \cos^m ax\, dx, \quad n \neq -m \quad \text{(reduces } \sin^n ax\text{)}$

**76.** $\int \sin^n ax \cos^m ax\, dx = \dfrac{\sin^{n+1} ax \cos^{m-1} ax}{a(m + n)} + \dfrac{m - 1}{m + n} \int \sin^n ax \cos^{m-2} ax\, dx, \quad m \neq -n \quad \text{(reduces } \cos^m ax\text{)}$

**77.** $\int \dfrac{dx}{b - c \sin ax} = \dfrac{-2}{a\sqrt{b^2 - c^2}} \tan^{-1}\left[\sqrt{\dfrac{b-c}{b+c}} \tan\left(\dfrac{\pi}{4} - \dfrac{ax}{2}\right)\right] + C, \quad b^2 > c^2$

**78.** $\int \dfrac{dx}{b - c \sin ax} = \dfrac{-1}{a\sqrt{c^2 - b^2}} \ln\left|\dfrac{c + b\sin ax + \sqrt{c^2 - b^2}\cos ax}{b + c\sin ax}\right| + C, \quad b^2 < c^2$

**79.** $\int \dfrac{dx}{1 - \sin ax} = -\dfrac{1}{a}\tan\left(\dfrac{\pi}{4} - \dfrac{ax}{2}\right) + C$   **80.** $\int \dfrac{dx}{1 - \sin ax} = \dfrac{1}{a}\tan\left(\dfrac{\pi}{4} + \dfrac{ax}{2}\right) + C$

**81.** $\int \dfrac{dx}{b - c \cos ax} = \dfrac{2}{a\sqrt{b^2 - c^2}} \tan^{-1}\left[\sqrt{\dfrac{b-c}{b+c}} \tan\dfrac{ax}{2}\right] + C, \quad b^2 > c^2$

**82.** $\int \dfrac{dx}{b - c \cos ax} = \dfrac{1}{a\sqrt{c^2 - b^2}} \ln\left|\dfrac{c + b\cos ax + \sqrt{c^2 - b^2}\sin ax}{b + c\cos ax}\right| + C, \quad b^2 < c^2$

**83.** $\int \dfrac{dx}{1 - \cos ax} = \dfrac{1}{a}\tan\dfrac{ax}{2} + C$   **84.** $\int \dfrac{dx}{1 - \cos ax} = -\dfrac{1}{a}\cot\dfrac{ax}{2} + C$

**85.** $\int x \sin ax \, dx = \dfrac{1}{a^2}\sin ax - \dfrac{x}{a}\cos ax + C$   **86.** $\int x \cos ax \, dx = \dfrac{1}{a^2}\cos ax + \dfrac{x}{a}\sin ax + C$

**87.** $\int x^n \sin ax \, dx = -\dfrac{x^n}{a}\cos ax + \dfrac{n}{a}\int x^{n-1}\cos ax \, dx$   **88.** $\int x^n \cos ax \, dx = \dfrac{x^n}{a}\sin ax - \dfrac{n}{a}\int x^{n-1}\sin ax \, dx$

**89.** $\int \tan ax \, dx = \dfrac{1}{a}\ln|\sec ax| + C$   **90.** $\int \cot ax \, dx = \dfrac{1}{a}\ln|\sin ax| + C$

**91.** $\int \tan^2 ax \, dx = \dfrac{1}{a}\tan ax - x + C$   **92.** $\int \cot^2 ax \, dx = -\dfrac{1}{a}\cot ax - x + C$

**93.** $\int \tan^n ax \, dx = \dfrac{\tan^{n-1} ax}{a(n-1)} - \int \tan^{n-2} ax \, dx, \quad n \neq 1$   **94.** $\int \cot^n ax \, dx = -\dfrac{\cot^{n-1} ax}{a(n-1)} - \int \cot^{n-2} ax \, dx, \quad n \neq 1$

**95.** $\int \sec ax \, dx = \dfrac{1}{a}\ln|\sec ax + \tan ax| + C$   **96.** $\int \csc ax \, dx = -\dfrac{1}{a}\ln|\csc ax + \cot ax| + C$

**97.** $\int \sec^2 ax \, dx = \dfrac{1}{a}\tan ax + C$   **98.** $\int \csc^2 ax \, dx = -\dfrac{1}{a}\cot ax + C$

**99.** $\int \sec^n ax \, dx = \dfrac{\sec^{n-2} ax \tan ax}{a(n-1)} + \dfrac{n-2}{n-1}\int \sec^{n-2} ax \, dx, \quad n \neq 1$

**100.** $\int \csc^n ax \, dx = -\dfrac{\csc^{n-2} ax \cot ax}{a(n-1)} + \dfrac{n-2}{n-1}\int \csc^{n-2} ax \, dx, \quad n \neq 1$

**101.** $\int \sec^n ax \tan ax \, dx = \dfrac{\sec^n ax}{na} + C, \quad n \neq 0$   **102.** $\int \csc^n ax \cot ax \, dx = -\dfrac{\csc^n ax}{na} + C, \quad n \neq 0$

Inverse Trigonometric Forms

**103.** $\int \sin^{-1} ax \, dx = x \sin^{-1} ax + \dfrac{1}{a}\sqrt{1 - a^2x^2} + C$   **104.** $\int \cos^{-1} ax \, dx = x \cos^{-1} ax - \dfrac{1}{a}\sqrt{1 - a^2x^2} + C$

**105.** $\int \tan^{-1} ax \, dx = x \tan^{-1} ax - \dfrac{1}{2a}\ln(1 + a^2x^2) + C$

**106.** $\int x^n \sin^{-1} ax \, dx = \dfrac{x^{n+1}}{n+1}\sin^{-1} ax - \dfrac{a}{n+1}\int \dfrac{x^{n+1}\, dx}{\sqrt{1 - a^2x^2}}, \quad n \neq -1$

**107.** $\int x^n \cos^{-1} ax \, dx = \dfrac{x^{n+1}}{n+1}\cos^{-1} ax + \dfrac{a}{n+1}\int \dfrac{x^{n+1}\, dx}{\sqrt{1 - a^2x^2}}, \quad n \neq -1$

**108.** $\int x^n \tan^{-1} ax \, dx = \dfrac{x^{n+1}}{n+1}\tan^{-1} ax - \dfrac{a}{n+1}\int \dfrac{x^{n+1}\, dx}{1 + a^2x^2}, \quad n \neq -1$

### Exponential and Logarithmic Forms

**109.** $\displaystyle\int e^{ax}\,dx = \frac{1}{a}e^{ax} + C$

**110.** $\displaystyle\int b^{ax}\,dx = \frac{1}{a\ln b}b^{ax} + C, \quad b > 0, b \neq 1$

**111.** $\displaystyle\int xe^{ax}\,dx = \frac{e^{ax}}{a^2}(ax - 1) + C$

**112.** $\displaystyle\int x^n e^{ax}\,dx = \frac{1}{a}x^n e^{ax} - \frac{n}{a}\int x^{n-1}e^{ax}\,dx$

**113.** $\displaystyle\int x^n b^{ax}\,dx = \frac{x^n b^{ax}}{a\ln b} - \frac{n}{a\ln b}\int x^{n-1}b^{ax}\,dx, \quad b > 0, b \neq 1$

**114.** $\displaystyle\int e^{ax}\sin bx\,dx = \frac{e^{ax}}{a^2 + b^2}(a\sin bx - b\cos bx) + C$

**115.** $\displaystyle\int e^{ax}\cos bx\,dx = \frac{e^{ax}}{a^2 + b^2}(a\cos bx + b\sin bx) + C$

**116.** $\displaystyle\int \ln ax\,dx = x\ln ax - x + C$

**117.** $\displaystyle\int x^n(\ln ax)^m\,dx = \frac{x^{n+1}(\ln ax)^m}{n + 1} - \frac{m}{n + 1}\int x^n(\ln ax)^{m-1}\,dx, \quad n \neq -1$

**118.** $\displaystyle\int x^{-1}(\ln ax)^m\,dx = \frac{(\ln ax)^{m+1}}{m + 1} + C, \quad m \neq -1$

**119.** $\displaystyle\int \frac{dx}{x\ln ax} = \ln|\ln ax| + C$

### Forms Involving $\sqrt{2ax - x^2}$, $a > 0$

**120.** $\displaystyle\int \frac{dx}{\sqrt{2ax - x^2}} = \sin^{-1}\left(\frac{x - a}{a}\right) + C$

**121.** $\displaystyle\int \sqrt{2ax - x^2}\,dx = \frac{x - a}{2}\sqrt{2ax - x^2} + \frac{a^2}{2}\sin^{-1}\left(\frac{x - a}{a}\right) + C$

**122.** $\displaystyle\int \left(\sqrt{2ax - x^2}\right)^n\,dx = \frac{(x - a)\left(\sqrt{2ax - x^2}\right)^n}{n + 1} + \frac{na^2}{n + 1}\int \left(\sqrt{2ax - x^2}\right)^{n-2}\,dx$

**123.** $\displaystyle\int \frac{dx}{\left(\sqrt{2ax - x^2}\right)^n} = \frac{(x - a)\left(\sqrt{2ax - x^2}\right)^{2-n}}{(n - 2)a^2} + \frac{n - 3}{(n - 2)a^2}\int \frac{dx}{\left(\sqrt{2ax - x^2}\right)^{n-2}}$

**124.** $\displaystyle\int x\sqrt{2ax - x^2}\,dx = \frac{(x + a)(2x - 3a)\sqrt{2ax - x^2}}{6} + \frac{a^3}{2}\sin^{-1}\left(\frac{x - a}{a}\right) + C$

**125.** $\displaystyle\int \frac{\sqrt{2ax - x^2}}{x}\,dx = \sqrt{2ax - x^2} + a\sin^{-1}\left(\frac{x - a}{a}\right) + C$

**126.** $\displaystyle\int \frac{\sqrt{2ax - x^2}}{x^2}\,dx = -2\sqrt{\frac{2a - x}{x}} - \sin^{-1}\left(\frac{x - a}{a}\right) + C$

**127.** $\displaystyle\int \frac{x\,dx}{\sqrt{2ax - x^2}} = a\sin^{-1}\left(\frac{x - a}{a}\right) - \sqrt{2ax - x^2} + C$

**128.** $\displaystyle\int \frac{dx}{x\sqrt{2ax - x^2}} = -\frac{1}{a}\sqrt{\frac{2a - x}{x}} + C$

### Hyperbolic Forms

**129.** $\displaystyle\int \sinh ax\,dx = \frac{1}{a}\cosh ax + C$

**130.** $\displaystyle\int \cosh ax\,dx = \frac{1}{a}\sinh ax + C$

**131.** $\displaystyle\int \sinh^2 ax\,dx = \frac{\sinh 2ax}{4a} - \frac{x}{2} + C$

**132.** $\displaystyle\int \cosh^2 ax\,dx = \frac{\sinh 2ax}{4a} + \frac{x}{2} + C$

**133.** $\displaystyle\int \sinh^n ax\,dx = \frac{\sinh^{n-1} ax\cosh ax}{na} - \frac{n - 1}{n}\int \sinh^{n-2} ax\,dx, \quad n \neq 0$

**134.** $\displaystyle\int \cosh^n ax \, dx = \frac{\cosh^{n-1} ax \sinh ax}{na} + \frac{n-1}{n}\int \cosh^{n-2} ax \, dx, \quad n \neq 0$

**135.** $\displaystyle\int x \sinh ax \, dx = \frac{x}{a}\cosh ax - \frac{1}{a^2}\sinh ax + C$      **136.** $\displaystyle\int x \cosh ax \, dx = \frac{x}{a}\sinh ax - \frac{1}{a^2}\cosh ax + C$

**137.** $\displaystyle\int x^n \sinh ax \, dx = \frac{x^n}{a}\cosh ax - \frac{n}{a}\int x^{n-1}\cosh ax \, dx$      **138.** $\displaystyle\int x^n \cosh ax \, dx = \frac{x^n}{a}\sinh ax - \frac{n}{a}\int x^{n-1}\sinh ax \, dx$

**139.** $\displaystyle\int \tanh ax \, dx = \frac{1}{a}\ln(\cosh ax) + C$      **140.** $\displaystyle\int \coth ax \, dx = \frac{1}{a}\ln|\sinh ax| + C$

**141.** $\displaystyle\int \tanh^2 ax \, dx = x - \frac{1}{a}\tanh ax + C$      **142.** $\displaystyle\int \coth^2 ax \, dx = x - \frac{1}{a}\coth ax + C$

**143.** $\displaystyle\int \tanh^n ax \, dx = -\frac{\tanh^{n-1} ax}{(n-1)a} + \int \tanh^{n-2} ax \, dx, \quad n \neq 1$

**144.** $\displaystyle\int \coth^n ax \, dx = -\frac{\coth^{n-1} ax}{(n-1)a} + \int \coth^{n-2} ax \, dx, \quad n \neq 1$

**145.** $\displaystyle\int \operatorname{sech} ax \, dx = \frac{1}{a}\sin^{-1}(\tanh ax) + C$      **146.** $\displaystyle\int \operatorname{csch} ax \, dx = \frac{1}{a}\ln\left|\tanh \frac{ax}{2}\right| + C$

**147.** $\displaystyle\int \operatorname{sech}^2 ax \, dx = \frac{1}{a}\tanh ax + C$      **148.** $\displaystyle\int \operatorname{csch}^2 ax \, dx = -\frac{1}{a}\coth ax + C$

**149.** $\displaystyle\int \operatorname{sech}^n ax \, dx = \frac{\operatorname{sech}^{n-2} ax \tanh ax}{(n-1)a} + \frac{n-2}{n-1}\int \operatorname{sech}^{n-2} ax \, dx, \quad n \neq 1$

**150.** $\displaystyle\int \operatorname{csch}^n ax \, dx = -\frac{\operatorname{csch}^{n-2} ax \coth ax}{(n-1)a} - \frac{n-2}{n-1}\int \operatorname{csch}^{n-2} ax \, dx, \quad n \neq 1$

**151.** $\displaystyle\int \operatorname{sech}^n ax \tanh ax \, dx = -\frac{\operatorname{sech}^n ax}{na} + C, \quad n \neq 0$      **152.** $\displaystyle\int \operatorname{csch}^n ax \coth ax \, dx = -\frac{\operatorname{csch}^n ax}{na} + C, \quad n \neq 0$

**153.** $\displaystyle\int e^{ax}\sinh bx \, dx = \frac{e^{ax}}{2}\left[\frac{e^{bx}}{a+b} - \frac{e^{-bx}}{a-b}\right] + C, \quad a^2 \neq b^2$

**154.** $\displaystyle\int e^{ax}\cosh bx \, dx = \frac{e^{ax}}{2}\left[\frac{e^{bx}}{a+b} + \frac{e^{-bx}}{a-b}\right] + C, \quad a^2 \neq b^2$

**Some Definite Integrals**

**155.** $\displaystyle\int_0^\infty x^{n-1}e^{-x} \, dx = \Gamma(n) = (n-1)!, \quad n > 0$      **156.** $\displaystyle\int_0^\infty e^{-ax^2} \, dx = \frac{1}{2}\sqrt{\frac{\pi}{a}}, \quad a > 0$

**157.** $\displaystyle\int_0^{\pi/2}\sin^n x \, dx = \int_0^{\pi/2}\cos^n x \, dx = \begin{cases} \dfrac{1\cdot3\cdot5\cdot\,\cdots\,\cdot(n-1)}{2\cdot4\cdot6\cdot\,\cdots\,\cdot n}\cdot\dfrac{\pi}{2}, & \text{if } n \text{ is an even integer} \geq 2 \\[2mm] \dfrac{2\cdot4\cdot6\cdot\,\cdots\,\cdot(n-1)}{3\cdot5\cdot7\cdot\,\cdots\,\cdot n}, & \text{if } n \text{ is an odd integer} \geq 3 \end{cases}$

# Basic Formulas and Rules

## Trigonometry Formulas

### Definitions and Fundamental Identities

Sine: $\quad \sin\theta = \dfrac{y}{r} = \dfrac{1}{\csc\theta}$

Cosine: $\quad \cos\theta = \dfrac{x}{r} = \dfrac{1}{\sec\theta}$

Tangent: $\quad \tan\theta = \dfrac{y}{x} = \dfrac{1}{\cot\theta}$

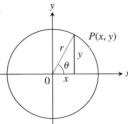

### Identities

$\sin(-\theta) = -\sin\theta, \quad \cos(-\theta) = \cos\theta$

$\sin^2\theta + \cos^2\theta = 1, \quad \sec^2\theta = 1 + \tan^2\theta, \quad \csc^2\theta = 1 + \cot^2\theta$

$\sin 2\theta = 2\sin\theta\cos\theta, \quad \cos 2\theta = \cos^2\theta - \sin^2\theta$

$\cos^2\theta = \dfrac{1 + \cos 2\theta}{2}, \quad \sin^2\theta = \dfrac{1 - \cos 2\theta}{2}$

$\sin(A + B) = \sin A \cos B + \cos A \sin B$

$\sin(A - B) = \sin A \cos B - \cos A \sin B$

$\cos(A + B) = \cos A \cos B - \sin A \sin B$

$\cos(A - B) = \cos A \cos B + \sin A \sin B$

$\tan(A + B) = \dfrac{\tan A + \tan B}{1 - \tan A \tan B}$

$\tan(A - B) = \dfrac{\tan A - \tan B}{1 + \tan A \tan B}$

$\sin\left(A - \dfrac{\pi}{2}\right) = -\cos A, \qquad \cos\left(A - \dfrac{\pi}{2}\right) = \sin A$

$\sin\left(A + \dfrac{\pi}{2}\right) = \cos A, \qquad \cos\left(A + \dfrac{\pi}{2}\right) = -\sin A$

$\sin A \sin B = \dfrac{1}{2}\cos(A - B) - \dfrac{1}{2}\cos(A + B)$

$\cos A \cos B = \dfrac{1}{2}\cos(A - B) + \dfrac{1}{2}\cos(A + B)$

$\sin A \cos B = \dfrac{1}{2}\sin(A - B) + \dfrac{1}{2}\sin(A + B)$

$\sin A + \sin B = 2\sin\dfrac{1}{2}(A + B)\cos\dfrac{1}{2}(A - B)$

$\sin A - \sin B = 2\cos\dfrac{1}{2}(A + B)\sin\dfrac{1}{2}(A - B)$

$\cos A + \cos B = 2\cos\dfrac{1}{2}(A + B)\cos\dfrac{1}{2}(A - B)$

$\cos A - \cos B = -2\sin\dfrac{1}{2}(A + B)\sin\dfrac{1}{2}(A - B)$

## Trigonometric Functions

### Radian Measure

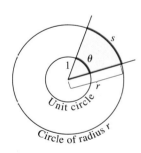

$\dfrac{s}{r} = \dfrac{\theta}{1} = \theta \quad$ or $\quad \theta = \dfrac{s}{r},$

$180° = \pi$ radians.

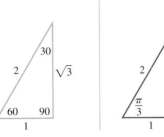

The angles of two common triangles, in degrees and radians.

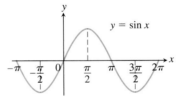

Domain: $(-\infty, \infty)$
Range: $[-1, 1]$

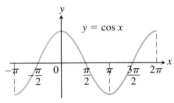

Domain: $(-\infty, \infty)$
Range: $[-1, 1]$

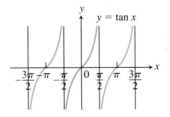

Domain: All real numbers except odd integer multiples of $\pi/2$
Range: $(-\infty, \infty)$

Domain: All real numbers except odd integer multiples of $\pi/2$
Range: $(-\infty, -1] \cup [1, \infty)$

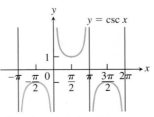

Domain: $x \neq 0, \pm\pi, \pm 2\pi, \ldots$
Range: $(-\infty, -1] \cup [1, \infty)$

Domain: $x \neq 0, \pm\pi, \pm 2\pi, \ldots$
Range: $(-\infty, \infty)$

## SERIES

### Tests for Convergence of Infinite Series

1. **The $n$th-Term Test:** Unless $a_n \to 0$, the series diverges.
2. **Geometric series:** $\sum ar^n$ converges if $|r| < 1$; otherwise it diverges.
3. **$p$-series:** $\sum 1/n^p$ converges if $p > 1$; otherwise it diverges.
4. **Series with nonnegative terms:** Try the Integral Test, Ratio Test, or Root Test. Try comparing to a known series with the Comparison Test or the Limit Comparison Test.
5. **Series with some negative terms:** Does $\sum |a_n|$ converge? If yes, so does $\sum a_n$ since absolute convergence implies convergence.
6. **Alternating series:** $\sum a_n$ converges if the series satisfies the conditions of the Alternating Series Test.

### Taylor Series

$$\frac{1}{1-x} = 1 + x + x^2 + \cdots + x^n + \cdots = \sum_{n=0}^{\infty} x^n, \qquad |x| < 1$$

$$\frac{1}{1+x} = 1 - x + x^2 - \cdots + (-x)^n + \cdots = \sum_{n=0}^{\infty} (-1)^n x^n, \qquad |x| < 1$$

$$e^x = 1 + x + \frac{x^2}{2!} + \cdots + \frac{x^n}{n!} + \cdots = \sum_{n=0}^{\infty} \frac{x^n}{n!}, \qquad |x| < \infty$$

$$\sin x = x - \frac{x^3}{3!} + \frac{x^5}{5!} - \cdots + (-1)^n \frac{x^{2n+1}}{(2n+1)!} + \cdots = \sum_{n=0}^{\infty} \frac{(-1)^n x^{2n+1}}{(2n+1)!}, \qquad |x| < \infty$$

$$\cos x = 1 - \frac{x^2}{2!} + \frac{x^4}{4!} - \cdots + (-1)^n \frac{x^{2n}}{(2n)!} + \cdots = \sum_{n=0}^{\infty} \frac{(-1)^n x^{2n}}{(2n)!}, \qquad |x| < \infty$$

$$\ln(1+x) = x - \frac{x^2}{2} + \frac{x^3}{3} - \cdots + (-1)^{n-1} \frac{x^n}{n} + \cdots = \sum_{n=1}^{\infty} \frac{(-1)^{n-1} x^n}{n}, \qquad -1 < x \le 1$$

$$\ln \frac{1+x}{1-x} = 2 \tanh^{-1} x = 2\left( x + \frac{x^3}{3} + \frac{x^5}{5} + \cdots + \frac{x^{2n+1}}{2n+1} + \cdots \right) = 2 \sum_{n=0}^{\infty} \frac{x^{2n+1}}{2n+1}, \qquad |x| < 1$$

$$\tan^{-1} x = x - \frac{x^3}{3} + \frac{x^5}{5} - \cdots + (-1)^n \frac{x^{2n+1}}{2n+1} + \cdots = \sum_{n=0}^{\infty} \frac{(-1)^n x^{2n+1}}{2n+1}, \qquad |x| \le 1$$

### Binomial Series

$$(1+x)^m = 1 + mx + \frac{m(m-1)x^2}{2!} + \frac{m(m-1)(m-2)x^3}{3!} + \cdots + \frac{m(m-1)(m-2)\cdots(m-k+1)x^k}{k!} + \cdots$$

$$= 1 + \sum_{k=1}^{\infty} \binom{m}{k} x^k, \qquad |x| < 1,$$

where

$$\binom{m}{1} = m, \qquad \binom{m}{2} = \frac{m(m-1)}{2!}, \qquad \binom{m}{k} = \frac{m(m-1)\cdots(m-k+1)}{k!} \qquad \text{for } k \ge 3.$$

## VECTOR OPERATOR FORMULAS (CARTESIAN FORM)

### Formulas for Grad, Div, Curl, and the Laplacian

| | **Cartesian** $(x, y, z)$ <br> **i, j,** and **k** are unit vectors in the directions of increasing $x$, $y$, and $z$. $M$, $N$, and $P$ are the scalar components of $\mathbf{F}(x, y, z)$ in these directions. |
|---|---|
| **Gradient** | $\nabla f = \dfrac{\partial f}{\partial x}\mathbf{i} + \dfrac{\partial f}{\partial y}\mathbf{j} + \dfrac{\partial f}{\partial z}\mathbf{k}$ |
| **Divergence** | $\nabla \cdot \mathbf{F} = \dfrac{\partial M}{\partial x} + \dfrac{\partial N}{\partial y} + \dfrac{\partial P}{\partial z}$ |
| **Curl** | $\nabla \times \mathbf{F} = \begin{vmatrix} \mathbf{i} & \mathbf{j} & \mathbf{k} \\ \dfrac{\partial}{\partial x} & \dfrac{\partial}{\partial y} & \dfrac{\partial}{\partial z} \\ M & N & P \end{vmatrix}$ |
| **Laplacian** | $\nabla^2 f = \dfrac{\partial^2 f}{\partial x^2} + \dfrac{\partial^2 f}{\partial y^2} + \dfrac{\partial^2 f}{\partial z^2}$ |

### Vector Triple Products

$(\mathbf{u} \times \mathbf{v}) \cdot \mathbf{w} = (\mathbf{v} \times \mathbf{w}) \cdot \mathbf{u} = (\mathbf{w} \times \mathbf{u}) \cdot \mathbf{v}$

$\mathbf{u} \times (\mathbf{v} \times \mathbf{w}) = (\mathbf{u} \cdot \mathbf{w})\mathbf{v} - (\mathbf{u} \cdot \mathbf{v})\mathbf{w}$

### The Fundamental Theorem of Line Integrals

Part 1  Let $\mathbf{F} = M\mathbf{i} + N\mathbf{j} + P\mathbf{k}$ be a vector field whose components are continuous throughout an open connected region $D$ in space. Then there exists a differentiable function $f$ such that

$$\mathbf{F} = \nabla f = \frac{\partial f}{\partial x}\mathbf{i} + \frac{\partial f}{\partial y}\mathbf{j} + \frac{\partial f}{\partial z}\mathbf{k}$$

if and only if for all points $A$ and $B$ in $D$ the value of $\int_A^B \mathbf{F} \cdot d\mathbf{r}$ is independent of the path joining $A$ to $B$ in $D$.

Part 2  If the integral is independent of the path from $A$ to $B$, its value is

$$\int_A^B \mathbf{F} \cdot d\mathbf{r} = f(B) - f(A).$$

### Green's Theorem and Its Generalization to Three Dimensions

Tangential form of Green's Theorem:  $\displaystyle\oint_C \mathbf{F} \cdot \mathbf{T}\, ds = \iint_R \nabla \times \mathbf{F} \cdot \mathbf{k}\, dA$

Stokes' Theorem:  $\displaystyle\oint_C \mathbf{F} \cdot \mathbf{T}\, ds = \iint_S \nabla \times \mathbf{F} \cdot \mathbf{n}\, d\sigma$

Normal form of Green's Theorem:  $\displaystyle\oint_C \mathbf{F} \cdot \mathbf{n}\, ds = \iint_R \nabla \cdot \mathbf{F}\, dA$

Divergence Theorem:  $\displaystyle\iint_S \mathbf{F} \cdot \mathbf{n}\, d\sigma = \iiint_D \nabla \cdot \mathbf{F}\, dV$

### Vector Identities

In the identities here, $f$ and $g$ are differentiable scalar functions, $\mathbf{F}$, $\mathbf{F}_1$, and $\mathbf{F}_2$ are differentiable vector fields, and $a$ and $b$ are real constants.

$\nabla \times (\nabla f) = \mathbf{0}$

$\nabla(fg) = f\nabla g + g\nabla f$

$\nabla \cdot (g\mathbf{F}) = g\nabla \cdot \mathbf{F} + \nabla g \cdot \mathbf{F}$

$\nabla \times (g\mathbf{F}) = g\nabla \times \mathbf{F} + \nabla g \times \mathbf{F}$

$\nabla \cdot (a\mathbf{F}_1 + b\mathbf{F}_2) = a\nabla \cdot \mathbf{F}_1 + b\nabla \cdot \mathbf{F}_2$

$\nabla \times (a\mathbf{F}_1 + b\mathbf{F}_2) = a\nabla \times \mathbf{F}_1 + b\nabla \times \mathbf{F}_2$

$\nabla(\mathbf{F}_1 \cdot \mathbf{F}_2) = (\mathbf{F}_1 \cdot \nabla)\mathbf{F}_2 + (\mathbf{F}_2 \cdot \nabla)\mathbf{F}_1 +$
$\mathbf{F}_1 \times (\nabla \times \mathbf{F}_2) + \mathbf{F}_2 \times (\nabla \times \mathbf{F}_1)$

$\nabla \cdot (\mathbf{F}_1 \times \mathbf{F}_2) = \mathbf{F}_2 \cdot \nabla \times \mathbf{F}_1 - \mathbf{F}_1 \cdot \nabla \times \mathbf{F}_2$

$\nabla \times (\mathbf{F}_1 \times \mathbf{F}_2) = (\mathbf{F}_2 \cdot \nabla)\mathbf{F}_1 - (\mathbf{F}_1 \cdot \nabla)\mathbf{F}_2 +$
$(\nabla \cdot \mathbf{F}_2)\mathbf{F}_1 - (\nabla \cdot \mathbf{F}_1)\mathbf{F}_2$

$\nabla \times (\nabla \times \mathbf{F}) = \nabla(\nabla \cdot \mathbf{F}) - (\nabla \cdot \nabla)\mathbf{F} = \nabla(\nabla \cdot \mathbf{F}) - \nabla^2\mathbf{F}$

$(\nabla \times \mathbf{F}) \times \mathbf{F} = (\mathbf{F} \cdot \nabla)\mathbf{F} - \dfrac{1}{2}\nabla(\mathbf{F} \cdot \mathbf{F})$

## BASIC ALGEBRA FORMULAS

### Arithmetic Operations

$$a(b + c) = ab + ac, \qquad \frac{a}{b} \cdot \frac{c}{d} = \frac{ac}{bd}$$

$$\frac{a}{b} + \frac{c}{d} = \frac{ad + bc}{bd}, \qquad \frac{a/b}{c/d} = \frac{a}{b} \cdot \frac{d}{c}$$

### Laws of Signs

$$-(-a) = a, \qquad \frac{-a}{b} = -\frac{a}{b} = \frac{a}{-b}$$

### Zero   Division by zero is not defined.

$$\text{If } a \neq 0: \quad \frac{0}{a} = 0, \quad a^0 = 1, \quad 0^a = 0$$

$$\text{For any number } a: a \cdot 0 = 0 \cdot a = 0$$

### Laws of Exponents

$$a^m a^n = a^{m+n}, \qquad (ab)^m = a^m b^m, \qquad (a^m)^n = a^{mn}, \qquad a^{m/n} = \sqrt[n]{a^m} = \left(\sqrt[n]{a}\right)^m$$

If $a \neq 0$,

$$\frac{a^m}{a^n} = a^{m-n}, \qquad a^0 = 1, \qquad a^{-m} = \frac{1}{a^m}.$$

### The Binomial Theorem   For any positive integer $n$,

$$(a + b)^n = a^n + na^{n-1}b + \frac{n(n-1)}{1 \cdot 2} a^{n-2} b^2$$

$$+ \frac{n(n-1)(n-2)}{1 \cdot 2 \cdot 3} a^{n-3} b^3 + \cdots + nab^{n-1} + b^n.$$

For instance,

$$(a + b)^2 = a^2 + 2ab + b^2, \qquad\qquad (a - b)^2 = a^2 - 2ab + b^2$$
$$(a + b)^3 = a^3 + 3a^2b + 3ab^2 + b^3, \qquad (a - b)^3 = a^3 - 3a^2b + 3ab^2 - b^3.$$

### Factoring the Difference of Like Integer Powers, $n > 1$

$$a^n - b^n = (a - b)(a^{n-1} + a^{n-2}b + a^{n-3}b^2 + \cdots + ab^{n-2} + b^{n-1})$$

For instance,

$$a^2 - b^2 = (a - b)(a + b),$$
$$a^3 - b^3 = (a - b)(a^2 + ab + b^2),$$
$$a^4 - b^4 = (a - b)(a^3 + a^2b + ab^2 + b^3).$$

### Completing the Square   If $a \neq 0$,

$$ax^2 + bx + c = au^2 + C \qquad \left(u = x + (b/2a), C = c - \frac{b^2}{4a}\right)$$

### The Quadratic Formula   If $a \neq 0$ and $ax^2 + bx + c = 0$, then

$$x = \frac{-b \pm \sqrt{b^2 - 4ac}}{2a}.$$

## GEOMETRY FORMULAS

$A$ = area, $B$ = area of base, $C$ = circumference, $S$ = lateral area or surface area, $V$ = volume

### Triangle

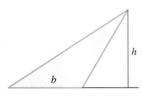

$$A = \frac{1}{2}bh$$

### Similar Triangles

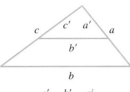

$$\frac{a'}{a} = \frac{b'}{b} = \frac{c'}{c}$$

### Pythagorean Theorem

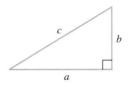

$$a^2 + b^2 = c^2$$

### Parallelogram

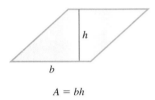

$$A = bh$$

### Trapezoid

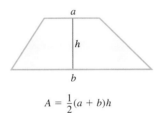

$$A = \frac{1}{2}(a + b)h$$

### Circle

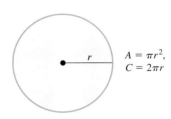

$$A = \pi r^2, \quad C = 2\pi r$$

### Any Cylinder or Prism with Parallel Bases

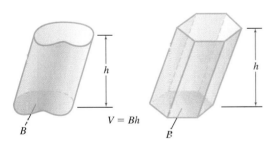

$$V = Bh$$

### Right Circular Cylinder

$$V = \pi r^2 h$$
$$S = 2\pi rh = \text{Area of side}$$

### Any Cone or Pyramid

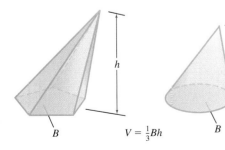

$$V = \frac{1}{3}Bh$$

### Right Circular Cone

$$V = \frac{1}{3}\pi r^2 h$$
$$S = \pi rs = \text{Area of side}$$

### Sphere

$$V = \frac{4}{3}\pi r^3, \quad S = 4\pi r^2$$

## LIMITS

### General Laws

If $L$, $M$, $c$, and $k$ are real numbers and

$$\lim_{x \to c} f(x) = L \quad \text{and} \quad \lim_{x \to c} g(x) = M, \quad \text{then}$$

*Sum Rule:* $\quad\quad\quad\quad\quad \lim_{x \to c} (f(x) + g(x)) = L + M$

*Difference Rule:* $\quad\quad\quad \lim_{x \to c} (f(x) - g(x)) = L - M$

*Product Rule:* $\quad\quad\quad\, \lim_{x \to c} (f(x) \cdot g(x)) = L \cdot M$

*Constant Multiple Rule:* $\quad \lim_{x \to c} (k \cdot f(x)) = k \cdot L$

*Quotient Rule:* $\quad\quad\quad \lim_{x \to c} \dfrac{f(x)}{g(x)} = \dfrac{L}{M}, \quad M \neq 0$

### The Sandwich Theorem

If $g(x) \leq f(x) \leq h(x)$ in an open interval containing $c$, except possibly at $x = c$, and if

$$\lim_{x \to c} g(x) = \lim_{x \to c} h(x) = L,$$

then $\lim_{x \to c} f(x) = L$.

### Inequalities

If $f(x) \leq g(x)$ in an open interval containing $c$, except possibly at $x = c$, and both limits exist, then

$$\lim_{x \to c} f(x) \leq \lim_{x \to c} g(x).$$

### Continuity

If $g$ is continuous at $L$ and $\lim_{x \to c} f(x) = L$, then

$$\lim_{x \to c} g(f(x)) = g(L).$$

### Specific Formulas

If $P(x) = a_n x^n + a_{n-1} x^{n-1} + \cdots + a_0$, then

$$\lim_{x \to c} P(x) = P(c) = a_n c^n + a_{n-1} c^{n-1} + \cdots + a_0.$$

If $P(x)$ and $Q(x)$ are polynomials and $Q(c) \neq 0$, then

$$\lim_{x \to c} \frac{P(x)}{Q(x)} = \frac{P(c)}{Q(c)}.$$

If $f(x)$ is continuous at $x = c$, then

$$\lim_{x \to c} f(x) = f(c).$$

$$\lim_{x \to 0} \frac{\sin x}{x} = 1 \quad \text{and} \quad \lim_{x \to 0} \frac{1 - \cos x}{x} = 0$$

### L'Hôpital's Rule

If $f(a) = g(a) = 0$, both $f'$ and $g'$ exist in an open interval $I$ containing $a$, and $g'(x) \neq 0$ on $I$ if $x \neq a$, then

$$\lim_{x \to a} \frac{f(x)}{g(x)} = \lim_{x \to a} \frac{f'(x)}{g'(x)},$$

assuming the limit on the right side exists.

## DIFFERENTIATION RULES

### General Formulas

Assume $u$ and $v$ are differentiable functions of $x$.

Constant: $\dfrac{d}{dx}(c) = 0$

Sum: $\dfrac{d}{dx}(u + v) = \dfrac{du}{dx} + \dfrac{dv}{dx}$

Difference: $\dfrac{d}{dx}(u - v) = \dfrac{du}{dx} - \dfrac{dv}{dx}$

Constant Multiple: $\dfrac{d}{dx}(cu) = c\dfrac{du}{dx}$

Product: $\dfrac{d}{dx}(uv) = u\dfrac{dv}{dx} + v\dfrac{du}{dx}$

Quotient: $\dfrac{d}{dx}\left(\dfrac{u}{v}\right) = \dfrac{v\dfrac{du}{dx} - u\dfrac{dv}{dx}}{v^2}$

Power: $\dfrac{d}{dx}x^n = nx^{n-1}$

Chain Rule: $\dfrac{d}{dx}(f(g(x)) = f'(g(x)) \cdot g'(x)$

### Trigonometric Functions

$\dfrac{d}{dx}(\sin x) = \cos x$ $\qquad \dfrac{d}{dx}(\cos x) = -\sin x$

$\dfrac{d}{dx}(\tan x) = \sec^2 x$ $\qquad \dfrac{d}{dx}(\sec x) = \sec x \tan x$

$\dfrac{d}{dx}(\cot x) = -\csc^2 x$ $\qquad \dfrac{d}{dx}(\csc x) = -\csc x \cot x$

### Exponential and Logarithmic Functions

$\dfrac{d}{dx}e^x = e^x$ $\qquad \dfrac{d}{dx}\ln x = \dfrac{1}{x}$

$\dfrac{d}{dx}a^x = a^x \ln a$ $\qquad \dfrac{d}{dx}(\log_a x) = \dfrac{1}{x \ln a}$

### Inverse Trigonometric Functions

$\dfrac{d}{dx}(\sin^{-1} x) = \dfrac{1}{\sqrt{1 - x^2}}$ $\qquad \dfrac{d}{dx}(\cos^{-1} x) = -\dfrac{1}{\sqrt{1 - x^2}}$

$\dfrac{d}{dx}(\tan^{-1} x) = \dfrac{1}{1 + x^2}$ $\qquad \dfrac{d}{dx}(\sec^{-1} x) = \dfrac{1}{|x|\sqrt{x^2 - 1}}$

$\dfrac{d}{dx}(\cot^{-1} x) = -\dfrac{1}{1 + x^2}$ $\qquad \dfrac{d}{dx}(\csc^{-1} x) = -\dfrac{1}{|x|\sqrt{x^2 - 1}}$

### Hyperbolic Functions

$\dfrac{d}{dx}(\sinh x) = \cosh x$ $\qquad \dfrac{d}{dx}(\cosh x) = \sinh x$

$\dfrac{d}{dx}(\tanh x) = \operatorname{sech}^2 x$ $\qquad \dfrac{d}{dx}(\operatorname{sech} x) = -\operatorname{sech} x \tanh x$

$\dfrac{d}{dx}(\coth x) = -\operatorname{csch}^2 x$ $\qquad \dfrac{d}{dx}(\operatorname{csch} x) = -\operatorname{csch} x \coth x$

### Inverse Hyperbolic Functions

$\dfrac{d}{dx}(\sinh^{-1} x) = \dfrac{1}{\sqrt{1 + x^2}}$ $\quad \dfrac{d}{dx}(\cosh^{-1} x) = \dfrac{1}{\sqrt{x^2 - 1}}$

$\dfrac{d}{dx}(\tanh^{-1} x) = \dfrac{1}{1 - x^2}$ $\quad \dfrac{d}{dx}(\operatorname{sech}^{-1} x) = -\dfrac{1}{x\sqrt{1 - x^2}}$

$\dfrac{d}{dx}(\coth^{-1} x) = \dfrac{1}{1 - x^2}$ $\quad \dfrac{d}{dx}(\operatorname{csch}^{-1} x) = -\dfrac{1}{|x|\sqrt{1 + x^2}}$

### Parametric Equations

If $x = f(t)$ and $y = g(t)$ are differentiable, then

$$y' = \frac{dy}{dx} = \frac{dy/dt}{dx/dt} \quad \text{and} \quad \frac{d^2y}{dx^2} = \frac{dy'/dt}{dx/dt}.$$

## INTEGRATION RULES

### General Formulas

*Zero:*
$$\int_a^a f(x)\,dx = 0$$

*Order of Integration:*
$$\int_b^a f(x)\,dx = -\int_a^b f(x)\,dx$$

*Constant Multiples:*
$$\int_a^b kf(x)\,dx = k\int_a^b f(x)\,dx \qquad \text{(Any number } k\text{)}$$

$$\int_a^b -f(x)\,dx = -\int_a^b f(x)\,dx \qquad (k = -1)$$

*Sums and Differences:*
$$\int_a^b (f(x) \pm g(x))\,dx = \int_a^b f(x)\,dx \pm \int_a^b g(x)\,dx$$

*Additivity:*
$$\int_a^b f(x)\,dx + \int_b^c f(x)\,dx = \int_a^c f(x)\,dx$$

*Max-Min Inequality:* If max $f$ and min $f$ are the maximum and minimum values of $f$ on $[a, b]$, then
$$\min f \cdot (b - a) \le \int_a^b f(x)\,dx \le \max f \cdot (b - a).$$

*Domination:* $\quad f(x) \ge g(x) \quad$ on $\quad [a, b] \quad$ implies $\quad \displaystyle\int_a^b f(x)\,dx \ge \int_a^b g(x)\,dx$

$$f(x) \ge 0 \quad \text{on} \quad [a, b] \quad \text{implies} \quad \int_a^b f(x)\,dx \ge 0$$

---

### The Fundamental Theorem of Calculus

Part 1   If $f$ is continuous on $[a, b]$, then $F(x) = \int_a^x f(t)\,dt$ is continuous on $[a, b]$ and differentiable on $(a, b)$ and its derivative is $f(x)$:

$$F'(x) = \frac{d}{dx}\int_a^x f(t)\,dt = f(x).$$

Part 2   If $f$ is continuous at every point of $[a, b]$ and $F$ is any antiderivative of $f$ on $[a, b]$, then

$$\int_a^b f(x)\,dx = F(b) - F(a).$$

---

### Substitution in Definite Integrals

$$\int_a^b f(g(x)) \cdot g'(x)\,dx = \int_{g(a)}^{g(b)} f(u)\,du$$

### Integration by Parts

$$\int_a^b f(x)g'(x)\,dx = f(x)g(x)\Big]_a^b - \int_a^b f'(x)g(x)\,dx$$